Optical Science and Engineering

Optical Science and Engineering explores the fundamental principles and advanced topics within the field of optical science and engineering. Designed to bridge the gap between foundational theory and modern techniques, this book provides a comprehensive introduction to essential concepts, alongside practical applications. It integrates extensive numerical examples—primarily using MATLAB—to illustrate key theoretical principles, ensuring a hands-on approach to learning.

Features:

- Covers core topics such as geometrical optics, wave interference, electromagnetic optics, radiometry and photometry/colorimetry fundamentals, scalar diffraction in the Rayleigh-Sommerfeld, Fresnel and Fraunhofer regimes, Fourier Optics, vector diffraction and introduction of photonic crystals
- Introduces advanced numerical electromagnetic methods, including Finite-Difference Frequency-Domain (FDFD), Boundary Element Method (BEM), and Rigorous Coupled-Wave Analysis (RCWA)
- Provides step-by-step MATLAB examples to enhance understanding and application of theoretical concepts
- Includes appendices for each chapter with supplementary material for deeper exploration and clarification of key topics

This textbook is an essential resource for undergraduate and graduate students studying optics, as well as engineers and researchers seeking a deeper understanding of optical science and engineering. With its blend of theoretical insights and computational examples, it serves as both a foundational learning tool and a practical reference for those working in academia and industry.

Series in Optics and Optoelectronics

Handbook of Optoelectronic Device Modeling and Simulation (Two-Volume Set)
Joachim Piprek

Handbook of Optoelectronics, Second Edition (Three-Volume Set)
John P. Dakin, Robert G. W. Brown

Optical MEMS, Nanophotonics, and Their Applications
Guangya Zhou, Chengkuo Lee

Thin-Film Optical Filters, Fifth Edition
H. Angus Macleod

Laser Spectroscopy and Laser Imaging
An Introduction
Helmut H. Telle, Ángel González Ureña

Fourier Optics in Image Processing
Neil Collings

Holography
Principles and Applications
Raymond K. Kostuk

An Introduction to Quantum Optics, Second Edition
Photon and Biphoton Physics
Yanhua Shih

Polarized Light and the Mueller Matrix Approach, Second Edition
José J. Gil, Razvigor Ossikovski

Introduction to Holography, Second Edition
Vincent Toal

Optical Coherence Tomography in Dentistry
Scientific Developments to Clinical Applications
Anderson S. L. Gomes, Denise M. Zezell, Cláudia C. B. O. Mota, John M. Girkin

Holography: Principles and Applications, Second Edition
Raymond K. Kostuk

CAD-Based Optical Design with Quadoa
Rafael G. González-Acuña

Anisotropy of Metamaterials: Beyond Conventional Paradigms
Vladimir Mityushev, Tatjana Gric, Radoslaw Kycia and Natalia Rylko

Optical Science and Engineering
Elias Glytsis

Advanced Optical Techniques in Biosciences
Nirmal Mazumder, Yury Kistenev and Igor Lednev

For more information about this series, please visit:
https://www.crcpress.com/Series-in-Optics-and-Optoelectronics/book-series/TFOPTICSOPT

Optical Science and Engineering

Elias N. Glytsis

CRC Press
Taylor & Francis Group
Boca Raton London New York

CRC Press is an imprint of the
Taylor & Francis Group, an **informa** business

Designed cover image: Elias N. Glytsis

First edition published 2026
by CRC Press
2385 NW Executive Center Drive, Suite 320, Boca Raton FL 33431

and by CRC Press
4 Park Square, Milton Park, Abingdon, Oxon, OX14 4RN

CRC Press is an imprint of Taylor & Francis Group, LLC

ISBN: 978-1-032-79062-6 (hbk)
ISBN: 978-1-032-79064-0 (pbk)
ISBN: 978-1-003-49033-3 (ebk)

DOI: 10.1201/9781003490333

Typeset in CMR10 font
by KnowledgeWorks Global Ltd.

Publisher's note: This book has been prepared from camera-ready copy provided by the authors.

To my children Eleanna and Nikolas, and my wife Dimitra.

Contents

Author

Elias N. Glytsis received his undergraduate diploma from the National Technical University of Athens (Greece) in Electrical Engineering in 1982, and his MSEE and PhD degrees from Georgia Institute of Technology in 1984 and 1987, respectively. From January 1988 till June 2003 (with leave of absence since 2007) he served as a Professor at the School of Electrical and Computer Engineering of Georgia Institute of Technology (starting from the rank of Assistant Professor and getting promoted to the rank of Professor with tenure in July 2000). In July 2003 he joined the faculty at the School of Electrical and Computer Engineering of National Technical University of Athens in Greece at the rank of a tenured Professor. Professor Glytsis' research interests are in the electromagnetic theory of holographic and diffractive and in general photonic structures and devices, optical interconnections, fiber grating devices, integrated optics, generalized anisotropic materials and applications, numerical methods in electromagnetic and optics, stochastic analysis of photonic structures, electromagnetic structures in electric power transmission, quantum heterostructures based on optical analogies, and design-optimization-integration software. He has published more than 130 journal publications, 3 book chapters, and more than 110 conference papers and he coholds 17 U.S. patents. According to Stanford's research (`https://elsevier.digitalcommonsdata.com/datasets/btchxktzyw/7`), Prof. Glytsis has been in the top 2% of highly cited scientists in the world (National Technical University of Athens announcement URL: `https://www.ntua.gr/el/component/k2/item/4272-159-erevnites-tou-ethnikoy-metsoviou-polytexneiou-anamesa-stous-koryfaious-erevnites-pagkosmios-gia-to-etos-2023`). He has been Co-guest editor of two special issues of the Optical Society of America A on grating diffraction. He has been a Topical editor of the Journal of Optical Society of America A on Scattering and Grating Diffraction 1992–1997. He has also been a guest editor of the Microelectronics Journal of a special issue on quasibound states in quantum heterostructure devices. Dr. Glytsis is a senior member of IEEE, a Fellow of the Optical Society of America, a member of EURATOM-Hellenic Republic, and a member of the Greek Society of Professional Engineers.

Preface

During my early years as a young Assistant Professor at the School of Electrical and Computer Engineering (ECE) at Georgia Tech (GT), I had the privilege of teaching the "Optical Engineering" course for the first time. This senior-level elective, originally developed by my former advisor and colleague, Professor Thomas K. Gaylord, covered nearly all aspects of optics, with the exception of quantum optics. The course spanned topics ranging from optical sources and measurements, coherence theory, radiometry and photometry, geometrical optics, and aberrations, to physical optics, interference, scalar diffraction, Fourier optics, and even an introduction to light-matter interactions, including electro-optic and acousto-optic effects, as well as holography. I taught this course every year from 1989 to 1994.

When I joined the School of ECE at the National Technical University of Athens (NTUA) in 2003, I introduced a similar course entitled "Optical Science and Engineering" as an elective for students pursuing the Telecommunications direction of study. This course is offered in the third year of a five-year curriculum leading to an integrated Master's degree in ECE. While the new course was based on the GT "Optical Engineering" curriculum, I adapted the material to fit the constraints of a single semester and expanded certain topics, such as geometrical optics, interference, and scalar diffraction. Despite these modifications, I often felt that I did not have sufficient time in class to delve deeply into many of the theoretical details or fully present the more intricate concepts I was teaching.

Throughout the years, I utilized several well-established textbooks, including "Introduction to Optics" by Pedrotti and "Optics" by Hecht. However, I consistently felt the need for more comprehensive explanations and theoretical clarity, particularly in areas that were often abbreviated or simplified in classroom lectures. About ten years ago, I began writing detailed notes, initially focusing on blackbody radiation and geometrical optics. Over time, these notes expanded to cover a wide range of topics, including wave and electromagnetic optics, interference, anisotropy, scalar and vector diffraction, Fourier optics, and metamaterials. My aim in these notes was to provide the detailed explanations and theoretical rigor that were often missing from traditional classroom presentations.

The result is a book that targets both undergraduate and graduate students in Optics, designed to fill the gap between seminal works such as "Principles of Optics" by Born and Wolf and more conventional textbooks like Pedrotti's "Introduction to Optics," Hecht's "Optics," and Jenkins and White's "Fundamentals of Optics." I believe this book will be valuable not only to students but also to young engineers and researchers in the field of Optics who seek deeper insights into the subject matter.

What sets this book apart is its inclusion of numerous numerical examples, primarily using MATLAB, to illustrate how to apply the relevant theory. These examples cover a wide range of topics, from basic concepts such as paraxial optics, wave interference, and scalar diffraction in both the Fresnel and Fraunhofer regimes, to more advanced topics like numerical vectorial electromagnetic methods for diffraction, including Finite-Difference Frequency-Domain (FDFD), Boundary Element Method (BEM), and Rigorous Coupled-Wave Analysis (RCWA). The book also addresses multilayer anisotropic thin films, combined coherent and incoherent layer interference, and the emerging field of metamaterials. Another distinctive feature of the book is the inclusion of Appendices following each chapter.

These appendices provide supplementary material and supporting information relevant to the topics covered, allowing for deeper exploration and clarification of key concepts. This additional content serves as a valuable resource for readers seeking a more comprehensive understanding of the subject matter. Except for the first chapter, the remaining chapters in this book are relatively long, a departure from the structure found in similar textbooks where material is often subdivided into smaller chapters. However, I have deliberately chosen to organize the content into major chapters, each containing a substantial number of sections and subsections. This structure allows for a more cohesive presentation of complex topics, ensuring that related concepts are grouped together and explored in-depth without unnecessary fragmentation.

Ultimately, this book aspires to be a comprehensive resource, bridging the gap between foundational knowledge and cutting-edge techniques in modern optics.

Specifically, the material in the book is organized into the following chapters:

- **Chapter 1—Brief History of Optics:** This chapter presents a brief history of optics, tracing its development from ancient times to the modern era. It highlights key milestones that have been critical in shaping the field of Optical Science and Technology and, more broadly, advancing human knowledge. The chapter also includes a comprehensive list of Nobel Laureates whose discoveries have significantly contributed to the progress of optics.
- **Chapter 2—Geometrical Optics:** This chapter covers all the fundamental topics of geometrical optics. It begins by introducing Huygen's and Fermat's principles, laying the foundation for understanding refractive and reflective prisms, as well as imaging concepts such as Cartesian reflective and refractive surfaces. The chapter delves into paraxial optics theory, discussing both thin and thick lenses and incorporating the *ABCD* matrix theory. Additionally, it explores essential optical systems, such as the photographic camera (with a brief history of photography), simple magnifiers, the compound microscope, telescopes, and the optics of the human eye. A significant portion is dedicated to the introduction of aberrations, with a strong emphasis on spherical aberrations for various types of spherical lenses. Lastly, the chapter provides a concise introduction to exact ray tracing techniques.
- **Chapter 3—Physical and Electromagnetic Optics:** This chapter provides a comprehensive exploration of physical and electromagnetic optics. It begins with a concise review of electromagnetism, covering plane wave solutions in both isotropic and anisotropic media (including general incident polarization), the index ellipsoid, as well as reflection and transmission at a planar boundary, Poynting's theorem, and various plane wave polarizations. The chapter then introduces wave propagation in lossy media and presents the classical oscillator model for dielectrics and the Drude model for metals. A substantial portion is devoted to light interference, addressing collinear and non-collinear waves, including cases with the same or differing frequencies. Detailed discussions are provided on interferometers such as the Fabry-Perot, Michelson, and Mach-Zehnder, along with an introduction to Fourier Transform Spectroscopy. The chapter also covers optical interference filters, exploring both the transfer matrix and scattering matrix techniques, alongside numerical optimization schemes for anti-reflective filters. A significant section extends the analysis to optical multilayers for anisotropic materials, with a detailed treatment of the 4×4 Berreman's method. Coherence theory is introduced, covering the basics of temporal and spatial coherence, as well as correlation functions in the time and frequency domains. A unique section (not commonly found in other textbooks) includes the analysis of incoherent reflection and transmission, along with the combination of coherent and incoherent layers in optical structures, which holds significant practical

importance. Finally, a brief but informative section on Gaussian beams is presented at the end of the chapter.

- **Chapter 4—Light Measurement and Emission: Radiometry, Photometry, and Blackbody Radiation:** This chapter provides an overview of emission and absorption radiative processes, laying the groundwork for the introduction to radiometry and photometry. The fundamental radiometric and photometric quantities are defined, followed by a discussion on irradiation from both point and extended sources, along with the definition and significance of a Lambertian source. The conservation laws of radiance are also thoroughly examined. The chapter proceeds with a review of the human eye's response to light, emphasizing luminous efficiency and the eye's sensitivity to different wavelengths. Key concepts of colorimetry are introduced, including RGB and XYZ color spaces, the mixture of colors, and color differences. Additionally, the chapter covers the basics of blackbody radiation and presents thermal radiation laws such as Kirchhoff's law and the concept of color temperature. The chapter concludes with a brief discussion on the cosmic microwave background radiation, offering insights into its significance in cosmology.
- **Chapter 5—Scalar Diffraction** In this chapter, the fundamentals of scalar diffraction are explored in detail, beginning with Helmholtz's and Kirchhoff's integral theorems. The Rayleigh-Sommerfeld formulations are then introduced in both three and two dimensions. From these more accurate formulations, the Fresnel and Fraunhofer approximations are derived. The scalar plane wave spectrum method is also illustrated, along with Babinet's principle for scalar diffraction. A substantial section is dedicated to the diffraction by circular or elliptical apertures, obstacles, and annuli. Both on-axis and off-axis cases are presented, considering incident plane waves or converging/diverging spherical waves. This is followed by another extensive section covering diffraction by rectangular apertures, obstacles, and annuli, with Rayleigh-Sommerfeld as well as Fresnel and Fraunhofer formulations. Edge diffraction and diffraction by multiple slits are also discussed. Further, a section is devoted to diffraction by triangular and polygonal apertures and obstacles, including a subsection dedicated to the analytical Fraunhofer approximation for a triangular aperture using boundary waves. Serrated circular apertures and obstacles are analyzed as well. Another key section examines the analysis of Fresnel zone plates, considering both plane wave and spherical wave incidence. T he chapter then addresses scalar diffraction by one-dimensional and two-dimensional (crossed) gratings. Lastly, the chapter concludes with a section on diffractive lenses, offering a comprehensive analysis of their diffraction behavior.
- **Chapter 6—Fourier Optics:** This chapter presents the fundamentals of Fourier Optics, introducing the spatial frequency domain and its significance in understanding optical systems. The linear systems approach to an optical system's response is discussed in detail, emphasizing the role of spatial frequencies. The Fourier transform property of a thin lens is derived, and simple optical systems composed of converging thin lenses are examined to illustrate this concept. The impulse response and point spread functions (PSFs) of a converging lens are defined, providing insight into the behavior of optical imaging systems. Classical imaging setups, including two- or four-lens systems, are also illustrated, offering practical examples. The chapter then summarizes diffraction-limited imaging for both monochromatic and polychromatic light, focusing on an optical system's spatial frequency response. Both coherent and incoherent systems are analyzed, with a brief review of the effects of aberrations on imaging quality. A comparative analysis of the performance of coherent versus incoherent optical systems is also provided. The chapter concludes with several key applications of Fourier Optics, such as Zernike phase contrast microscopy, the basics of holography, spatial filtering techniques, and

the VanderLugt correlator, demonstrating the broad utility of Fourier Optics in modern optical science and engineering.

- **Chapter 7—Vector Diffraction Basics and Selected Methods:** This chapter addresses the broad and complex subject of vector diffraction, which spans a wide range of research topics. Given the depth of this field, a full treatment would require a specialized book. As such, this chapter offers a concise overview, focusing on several key methods with which the author has direct experience. The chapter begins with an introduction to vector diffraction integrals for apertures, covering important formulations such as Kirchhoff's, Stratton-Chu's, Kottler's, and Franz's methods. A straightforward example involving circular apertures is included to illustrate these principles. Babinet's principle is then extended to vector electromagnetic fields, providing a bridge between scalar and vector diffraction theories. Following this, the vectorial plane wave spectrum method is introduced, and a hierarchy of the most prominent integral and differential methods is presented. Two techniques are discussed in more depth: the Boundary Element Method (BEM) for integral approaches and the Finite-Difference Frequency-Domain (FDFD) method for differential solutions, both exemplified through two-dimensional problems. A significant portion of the chapter is dedicated to vector diffraction by gratings. The grating equation and Bragg condition are defined, followed by a detailed presentation of Rigorous Coupled-Wave Analysis (RCWA) for surface-relief gratings. This method is extended to cover cascaded and holographic gratings, including cases involving three-dimensional incidence (conical diffraction) and two-dimensional (crossed) gratings. Special attention is given to high-spatial-frequency (subwavelength) gratings and their homogenization models, particularly the effective medium theory. Additionally, Kogelnik's two-wave coupled-wave theory is briefly discussed, providing analytical insights into the characterization of gratings as either "thick" or "thin." The final section of this chapter explores the theory of photonic crystals in one and two dimensions. A detailed explanation of the plane wave expansion method is provided, offering a foundation for understanding the bandgap properties in these structures.

In all chapters (except the first, which is a brief history of optics), a dedicated problem section is included. Ideally, more problems should have been provided, given the extensive range of topics covered in the book. However, developing new problems is a time-consuming task, and adding more at this stage would have significantly delayed the publication. As a result, each chapter typically contains 15-20 problems that represent the core concepts and most critical parts of the theory discussed. Some of these problems require extensive programming efforts, particularly in cases involving vectorial methods.

All of the figures in this book were created from scratch by me, using PowerPoint or MATLAB for diagramming and visualizations. A small number of figures (six in total), mostly related to the human eye, were borrowed from publicly available resources, including Wikipedia, the European Space Agency, and a specific organization dedicated to the human eye. Proper attribution and acknowledgment are given for these figures wherever they appear in the text.

I would also like to acknowledge the assistance provided by Artificial Intelligence, specifically the online platform *chatGPT*, for enhancing the English language in various sections of this book, particularly in the introductory materials.

When I first conceived this book, I had planned to include the software used to generate all the examples presented. Most of these codes, written primarily in MATLAB and Fortran, were initially intended for research purposes and were not designed for general users. Reworking the software into a user-friendly format would be a separate and monumental undertaking. Therefore, I decided to reserve the idea of releasing the necessary software alongside an extended set of problems for a potential future edition of the book, should such an opportunity arise. In this current edition, the problem sets aim to engage readers with the most important theoretical aspects, providing practical and computational challenges that deepen understanding without overwhelming the reader with excessive programming demands.

Finally, I would like to express my deepest gratitude to my former advisor, mentor, and dear friend, Professor Tom Gaylord of Georgia Tech, who originally developed the undergraduate course that inspired the creation of this book. His guidance and vision served as the foundation for this ambitious undertaking, and his unwavering support has been a constant source of motivation.

I would also like to acknowledge the contributions of the students who enrolled in the elective course associated with this book. Their insightful comments and thoughtful suggestions, particularly regarding the undergraduate component, have been invaluable in shaping some of the topics presented here.

Lastly, I am profoundly thankful to my family, especially my children, for their patience, understanding, and unwavering support throughout the years it took to complete this project. Their encouragement has been the cornerstone of this journey, and without them, this book would not have been possible.

ELIAS N. GLYTSIS
Professor
School of Electrical and Computer Engineering
National Technical University of Athens
February 2025

1

Brief History of Optics

1.1 Introduction

The history of optics, the study of light and vision, spans millennia and has been pivotal in the development of science. Early ideas about light were often intertwined with religious beliefs and can be traced back thousands of years [1]. A comprehensive review of the history of Optics can be found in Ref. [2–5]. This brief chapter summarizes and reviews the key milestones in the field of optics. The review is organized into the following chronological periods:

- Ancient Optics
- Geometric Optics, Euclid, and Ptolemy
- Islamic Era
- The Middle Ages and Early Renaissance
- Early Modern Period: Kepler's Optics
- Newton's Ray Optics and Corpuscular Hypothesis
- 19th-Century Optics: Wave Theory of Optics
- Modern Developments and Quantum Optics

The history of optics traces a journey from ancient philosophical speculations to modern quantum technologies. Its study has been crucial not only for understanding light but also for driving technological advancements that shape the modern world. For instance, optics is central to a wide range of contemporary technologies, including fiber optic communications, laser technology, modern astronomy, sensing and spectroscopy, monitoring, biomedical imaging and diagnostics, and lighting, among many others. Finally, a dedicated section focuses on the speed of light and its measurement, as it has played a pivotal role in advancing the field of optics.

1.2 Ancient Optics

Many ancient civilizations were fascinated by light, and this interest was often reflected in their religious beliefs. For instance, in ancient Egypt, Ra, the Sun God, was believed to emanate light from his eye to illuminate the day [1]. In the Bible, light also holds a divine nature, as it is mentioned on the first day of creation in Genesis. Light was deemed good and separated from darkness, which was associated with negativity. It is remarkable that in the biblical account of creation, light appears first (possibly alluding to an event like the "Big Bang"), and everything else follows.

DOI: 10.1201/9781003490333-1

Early evidence of polished lenses includes the Nimrud lens, discovered in Assyrian ruins. This 8th-century BC piece of rock crystal was unearthed in 1850 by Austen Henry Layard at the Assyrian palace of Nimrud, in present-day Iraq. The lens suggests that ancient cultures may have utilized magnification or sunlight focusing for practical purposes. However, their understanding of optics was likely more practical than theoretical.

Greek philosophers were among the first to explore the nature of light, motivated by their desire to understand vision. They viewed the process of seeing as the result of an interaction between the eyes of the observer and the object being observed. Philosophers like Pythagoras (6th and 5th centuries BCE) proposed that sight involved rays emanating from the eyes, while Democritus (5th and 4th centuries BCE) and others believed that light consisted of streams of atoms reaching the eye from the object. Empedocles (5th century BCE) and his followers suggested that vision was a combined effect, with something emanating both from the eye and the object. Plato's (4th century BCE) extramission theory held that vision occurred because the eyes emitted beams of light, which fused with ambient light to form a sentient "body of vision." Aristotle (4th century BCE), on the other hand, believed that light traveled in straight lines but did not attribute this to any particular substance or medium.

1.3 Geometric Optics, Euclid, and Ptolemy

Euclid was the first to articulate several important properties of light. *Euclid's Optics* is a treatise on the geometry of vision, written by the Greek mathematician around 300 BCE. The earliest surviving manuscript of Optics is in Greek and dates from the 10th century CE. The work primarily addresses the geometry of vision, with minimal reference to the physical or psychological aspects of sight. It is largely a treatise on angular perspective, regarded as an application of Euclidean geometry. The typical problems discussed involve the apparent ratios of lengths positioned at various distances from the eye or at different inclinations relative to the visual axis [2].

Later developments in Greek optical geometry encompassed a much wider range of problems and approaches. By the 3rd century BCE, Archimedes is reported to have developed a geometrical theory of mirrors, along with some observations on refraction. This included a theory of burning mirrors and the challenge of determining the shape of a mirror that would cause all sun rays to converge at a single burning point [2, 4].

Hero of Alexandria (circa 100 CE) was an ancient Greek engineer and mathematician credited with formulating the law of reflection, which states that the angle of incidence equals the angle of reflection. Hero derived this law using the principle of least distance, which asserts that light when reflected from a surface, follows the shortest possible path between two points. This concept predates the more formalized wave and ray optics theories developed much later, yet it represents a remarkable insight into the behavior of light [1].

Ptolemy, the Greek-Egyptian mathematician, astronomer, and philosopher, made significant contributions to the field of optics through his work titled *Optics*, written around the 2nd century CE. In this treatise, he systematically studied the behavior of light, focusing on several key phenomena, including reflection, refraction, and the properties of vision. Ptolemy's work is of great importance in the early history of optics. The Greek text has been lost completely. However, fragments of the work survive only in the form of Latin and Arabic translations, which were presumably based on the Greek original.

The ancient Greeks established foundational concepts of optics through their philosophical inquiries and geometric investigations, even though they never fully understood the mechanisms of vision. Nevertheless, their work significantly influenced later scholars during the Islamic Golden Age and the Renaissance, shaping the evolution of optical science and enhancing our understanding of light and vision.

1.4 Islamic Era

The Islamic Golden Age (roughly the 8th to the 14th century CE) saw remarkable advancements in the field of optics, driven by scholars who built upon and expanded the knowledge inherited from the Greeks. This period is characterized by empirical investigations, theoretical advancements, and the establishment of optics as a distinct scientific discipline.

Alhazen (Ibn al-Haytham) (965–1040 CE), often regarded as the "father of optics," made groundbreaking contributions to the understanding of light and vision. His most famous work, the *Book of Optics*, laid the foundation for modern optics. Alhazen rejected the earlier extramission theory (which proposed that light emitted from the eye allows vision) in favor of the intromission theory, which posited that light enters the eye from external sources. He emphasized the importance of experimentation and observation, establishing a scientific method for studying optical phenomena. He introduced the method of testing hypotheses with experiments, laying the groundwork for the modern scientific method. Alhazen described the anatomy of the eye based on dissections, although he erroneously concluded that vision was produced on the surface of the crystalline lens. Alhazen also studied the refraction of light and its interaction with lenses, explored the principles of camera obscura, and developed theories about how the eye perceives images. It is noteworthy that Alhazen's results were primarily based on experimental observations, in contrast to the works of the ancient Greeks. In summary, his rigorous approaches significantly influenced future generations of scientists such as Jabir ibn Hayyan (Geber) (circa 721–815 CE), who made contributions regarding the interaction of light and materials, and Nasir al-Din al-Tusi (1201–1274 CE), who advanced the understanding of light's behavior in various media.

During the Islamic Golden Age, significant advancements were made in the development of optical instruments. The production and refinement of lenses for magnification and light focusing became increasingly sophisticated, leading to improved optical devices. Furthermore, the concept of the camera obscura was developed, demonstrating the principles of light and projection. This device allowed images from the outside world to be projected onto a surface inside a darkened room.

In summary, the Islamic era of optics represented a crucial period in the history of science, characterized by rigorous experimentation, theoretical advancements, and the development of optical instruments. The contributions made during this time significantly shaped the field of optics and influenced later developments in both the Islamic world and Europe.

1.5 The Middle Ages and Early Renaissance

Despite the difficulties and the reluctance of their transmission, Greek sources fueled the discussion of optical matters until the Renaissance. The diversity of Greek theories of vision was preserved, likely because varying motivations favored different theories. Astronomers and geometers naturally focused on the theories of Euclid and Ptolemy, while philosophers and theologians leaned toward the ideas of Plato or Aristotle. In the Christian world, the general decline of natural philosophy long hindered any creative departure from established ideas. However, the 12th and 13th centuries in Europe experienced a profound cultural transformation, driven by the numerous translations of Greek and Arabic sources. Whereas earlier Christian theology had compromised the dissemination and development of Greek philosophy, the new scholars of the 12th century were eager to learn from nature and from ancient and foreign texts [2, 4].

Roger Bacon (1214–1292) was an English philosopher who emphasized empirical observation in his study of optics. He proposed that the speed of light is finite and that light is propagated through a medium in a manner analogous to the propagation of sound. In addition, Bacon studied the magnification of small objects using convex lenses and suggested that they could find application in the correction of defective eyesight. He attributed the phenomenon of the rainbow to the reflection of sunlight from individual raindrops.

The Polish natural scientist and philosopher Witelo (1220–1278) completed his *Perspectiva*, which would remain a standard text on optics for several centuries. Among other contributions, Witelo described a method for machining parabolic mirrors from iron and conducted careful observations on refraction. He recognized that the angle of refraction is not proportional to the angle of incidence, although he was unaware of the phenomenon of total internal reflection.

By the 16th century, advances in lens-making led to the invention of devices like the telescope and microscope, revolutionizing science. Zacharias Janssen (Netherlands, circa 1590) constructed a compound microscope featuring a converging objective lens and a diverging eye lens. Hans Lipperhey (1570–1619), a German-Dutch spectacle maker, is commonly associated with the invention of the telescope (circa 1608), as he was the first to attempt to obtain a patent for it; however, it remains unclear whether he was the first to build one.

Galileo Galilei (1564–1642), an Italian astronomer, physicist, and engineer, is often credited with significantly advancing the fields of optics and astronomy through his improvements to the telescope and groundbreaking astronomical observations, including the moons of Jupiter, the phases of Venus, the surface of the Moon, and sunspots. He developed a refracting telescope that utilized a convex lens as the objective and a concave lens as the eyepiece. His early telescopes had a magnification power of about 8x, which he later improved to 20x or more. Galileo's work with the telescope revolutionized astronomy and transformed humanity's understanding of the universe. His observations provided critical support for the heliocentric model and challenged long-standing beliefs about the nature of celestial bodies. Despite facing opposition from the Catholic Church for his views, Galileo's contributions to optics and astronomy have had a lasting impact, influencing generations of scientists and shaping the course of scientific thought. His innovations in telescope design and observational astronomy marked the beginning of a new era in the study of the cosmos.

1.6 Early Modern Period: Kepler's Optics

Johannes Kepler (1571–1630) was a German astronomer, mathematician, astrologer, natural philosopher, and writer of music. He made significant contributions to the field of optics, laying the groundwork for modern optical science. His work, particularly in understanding the behavior of light and image formation, was pivotal [1, 4, 6]. Building upon the earlier research of Alhazen and others, Kepler developed a more systematic understanding of how light bends when it passes through different media, contributing to the formulation of the law of refraction. In his work *Dioptrice* (1604), Kepler analyzed the human eye as an optical system, describing how the eye focuses light onto the retina, likening it to a camera obscura. He noted that the image formed on the retina is inverted, a realization that helped clarify the mechanisms of vision and contributed to the understanding of image formation in optical systems. Kepler designed a new type of telescope known as the Keplerian telescope, which utilized two convex lenses—one as the objective lens and the other as the eyepiece. This design provided a wider field of view and greater magnification than earlier telescopes. Applying his understanding of optics to astronomy, Kepler conducted studies on planetary motion, resulting in his three laws of planetary motion, which revolutionized the understanding of celestial mechanics. He was among the first to apply a mathematical approach to both optics and astronomy. His integration of geometry and optics enabled more accurate predictions and descriptions of optical phenomena, paving the way for future scientific inquiry. Kepler's work in optics significantly influenced later scientists, including Isaac Newton and Christiaan Huygens. His mathematical formulations and empirical methods established a foundation that shaped the future study of light, vision, and optics.

Willebrord Snellius (often referred to simply as Snell) (1580–1626) was a Dutch mathematician and astronomer renowned for his groundbreaking contributions to the field of optics, particularly in the study of refraction (1621). Snell's Law quantitatively describes how light bends when transitioning from one medium to another. This law serves as the foundation for understanding how lenses operate and is fundamental to the design of optical instruments, including glasses, cameras, and telescopes. Although Snell is credited with the law of refraction, the principles had been observed earlier by ancient Greek scholars such as Ptolemy and by Islamic scholars. Snell's significant contribution lay in providing a more precise mathematical formulation of these principles. He conducted experiments to measure the angles of refraction for various substances, including air, water, and glass. His empirical approach established a reliable method for determining the refractive index of different materials. In summary, his work was foundational for later studies in optics and helped explain natural phenomena like rainbows.

1.7 Newton's Ray Optics and Corpuscular Hypothesis

Sir Isaac Newton (1643–1727) was an English polymath active in various fields, including mathematics, physics, and astronomy. His work, *Opticks* (1704) [7], is considered a cornerstone in the study of light and color. In contrast to *Principia*, which focused on mechanics and gravitation, *Opticks* explored the nature and behavior of light. Structured around a series of propositions, experiments, and observations, this work covered a broad spectrum of topics related to optics. Newton proposed that light consists of particles, or "corpuscles," a view that stood in opposition to Christiaan Huygens' wave theory of light. Although the wave theory later gained widespread acceptance, Newton's particle theory remained influential for centuries. Through his experiments with prisms, Newton demonstrated that white light is a mixture of different colors, each with distinct refractive properties. He argued that color is not an intrinsic property of objects but rather a characteristic of light itself; objects appear colored due to their ability to selectively reflect certain wavelengths of light while absorbing others. Newton also conducted detailed studies on how light interacts with reflective and refractive surfaces, such as mirrors and lenses. He formulated mathematical laws governing these interactions based on the assumption of light's corpuscular nature. Despite his focus on particles, Newton also explored phenomena such as interference and diffraction, interpreting them within the framework of his particle theory. A notable example is "Newton's rings," a pattern observed when light passes through a thin, varying gap between a spherical and flat transparent surface. This observation suggests that Newton's understanding of light was, in some respects, closer to the wave theory than is commonly assumed [1].

In summary, *Opticks* was pivotal not only in advancing the understanding of light but also in its methodological approach. Newton's emphasis on experimentation laid the groundwork for the modern scientific approach to the study of optics.

1.8 19th-Century Optics: Wave Theory of Optics

Even before Newton's contributions to optics, Francesco Maria Grimaldi (1618–1663), an Italian Jesuit priest, mathematician, and physicist, was among the earliest to observe the diffraction of light in 1665. By passing white light through small apertures, Grimaldi noted that light appeared in areas where the shadow was expected, based on the rectilinear propagation of light particles. He referred to this phenomenon as "diffraction." This observation led him to conclude that light behaves like a fluid, exhibiting wave-like motion, although he remained somewhat unconvinced of this conclusion. Around the same time, Robert Hooke, a contemporary of Newton, also supported a wave theory for the propagation of light. The analogy with sound waves was frequently employed to challenge the hypothesis of light's wave nature, as light waves do not exhibit the same ability to bend around corners as sound waves do. However, Grimaldi's experiments demonstrated that light propagation was not strictly rectilinear, as suggested by Newton's corpuscular approach [1].

Christiaan Huygens (1629–1695) was a Dutch mathematician, physicist, engineer, and astronomer who proposed that light behaves as a wave rather than as a particle [8]. This groundbreaking idea challenged the particle theory of light advocated by Isaac Newton. Huygens' wave theory effectively explained various optical phenomena, including reflection and refraction. He formulated Huygens' Principle, which states that every point on a wavefront can be considered a source of secondary wavelets. The new wavefront is then formed by the envelope of these wavelets. This principle is crucial for understanding wave propagation and phenomena such as diffraction. Huygens also derived the laws of refraction, which describe how light bends when transitioning between different media. He explained this bending in terms of the interactions between light wavefronts and the boundaries of these media. In addition to his theoretical contributions, Huygens made significant advancements in the design of telescopes and microscopes, improving the optical instruments of his time. His innovations in lens design enhanced the quality of images produced by these devices. In 1678, he published his book *Traité de la Lumiére* [9]. This work systematically laid out his theories on light, wave propagation, and optics. It remains a foundational text in the history of optical science.

The downfall of Newton's corpuscular theory of light was accomplished by the interference experiments conducted by Thomas Young. Thomas Young (1773–1829) was a British polymath who made notable contributions to various fields including vision and light. Young is best known for proposing the wave theory of light, challenging the then-prevailing particle theory. His experiments and theories laid the groundwork for understanding light as a wave phenomenon. In 1801, Young conducted his famous double-slit experiment, which demonstrated the interference of light waves. When light passed through two closely spaced slits, it created a pattern of alternating bright and dark fringes on a screen, providing clear evidence of the wave nature of light [10]. By analyzing the interference patterns produced in his double-slit experiment, Young was able to estimate the wavelengths of various colors of light. For the outmost red, he found a wavelength of 0.71 μm and for the blue 0.42 μm which are indeed close to the wavelengths that are currently accepted. Young made important contributions to the understanding of color vision. He proposed a trichromatic theory of color perception, suggesting that the human eye has three types of color receptors, corresponding to red, green, and blue light. Young's work on interference not only advanced optics but also laid the foundation for later developments in wave theory, including the understanding of light diffraction and the principles of holography.

Augustin-Jean Fresnel (1788–1827) was a French civil engineer and physicist whose pioneering research in optics led to the widespread acceptance of the wave theory of light. Fresnel conducted experiments similar to those of Grimaldi and Newton, using small lenses to focus sunlight as an intense source for his observations. He embraced the idea that light propagates according to Huygens' principle, leading to interference and diffraction effects. Fresnel submitted his mathematical model to the French Academy of Sciences in a competition aimed at resolving the ongoing debate between the particle (corpuscle) and wave theories of light. One of the evaluators, Siméon Denis Poisson, applied Fresnel's theory and deduced that a bright spot should appear at the center of a shadow cast by a circular opaque disk, a result he believed to be impossible, thus rejecting Fresnel's wave theory. However, Francois Arago, head of the evaluation committee, performed the experiment suggested by Poisson and successfully observed the bright spot, thereby confirming Fresnel's prediction. This result became known as the "spot of Arago" or the "spot of Poisson" and marked a significant milestone in the confirmation of the wave theory of light. As a result, Fresnel won the competition. Fresnel's contributions extended beyond this experiment. In collaboration with Arago, he incorporated polarization phenomena into his wave theory using birefringent crystals. They attempted to cause interference between the two refracted beams from the birefringent crystal but observed no interference. This led them to conclude that the two beams exhibited waves with orthogonal transverse polarizations, contrary to the previously held belief that light waves were longitudinal. This discovery further solidified the understanding of light's wave nature and its transverse polarization properties [1].

In his *Traité de la Lumiére* [9], Christiaan Huygens proposed that light is a wave propagating through a medium known as the luminiferous aether (or ether). This aether was postulated to explain how light, which appeared to behave as a wave, could travel through the vacuum of space. At the time, it was widely believed that all waves required a medium for propagation, and the aether was proposed to serve this role for light. The hypothesis of the aether became the subject of considerable debate, as it required the existence of an invisible, all-pervasive material that had no observable interaction with physical objects. As the study of light advanced, particularly in the 19th century, the properties required of the aether grew increasingly contradictory. Despite these challenges, there was no alternative theory to replace it. However, by the late 19th century, the aether hypothesis came under serious scrutiny. The turning point came with the negative results of the Michelson–Morley experiment in 1887, which failed to detect any evidence of the aether's existence. This finding was confirmed by further experiments conducted into the 1920s, ultimately leading to the abandonment of the aether theory and paving the way for new models of light propagation, such as Einstein's theory of relativity.

During this period, James Clerk Maxwell (1831–1879), a Scottish physicist, developed the classical theory of electromagnetic radiation, becoming the first to describe electricity, magnetism, and light as different manifestations of the same phenomenon [11,12]. Maxwell frequently attended lectures at the Royal Institution, where he regularly interacted with Michael Faraday and was influenced by Faraday's theories on the lines of force in electricity and magnetism. Motivated by these ideas, Maxwell sought to unify electricity and magnetism within a single theoretical framework. In 1861–1862, Maxwell accomplished this by formulating his theory, which initially consisted of twenty equations. These equations, when expressed in contemporary terms, are condensed into what are now known as Maxwell's four famous equations, central to the field of electromagnetics. This simplification of Maxwell's original work was largely credited to Oliver Heaviside, who streamlined the theory into the four partial differential equations that are still used today. Using his equations, Maxwell derived a wave equation for electromagnetic fields and discovered that the combination of vacuum permittivity and vacuum permeability yielded the speed of light in a vacuum with remarkable accuracy. From this, he concluded that light waves must be electromagnetic in nature, establishing a profound connection between electromagnetism and optics. This insight revolutionized the understanding of light, solidifying its place as a form of electromagnetic radiation.

Later, Heinrich Rudolf Hertz (1857–1894), a German physicist, became the first to conclusively demonstrate the existence of the electromagnetic waves predicted by James Clerk Maxwell's equations. In 1888, Hertz conducted experiments on the emission and reception of electromagnetic waves at MHz frequencies, which are significantly lower than the frequencies of visible light. His experiments confirmed that these waves obeyed the same laws of reflection and refraction as light and propagated at the speed of light. This experimental verification provided strong support for Maxwell's theory and solidified the understanding that light and electromagnetic waves are part of the same phenomenon [1].

In 1895, Wilhelm Roentgen was investigating the movement of high-speed electrons in an evacuated glass vessel when he discovered a new form of radiation. The impact of the electrons on the glass envelope emitted radiation that could penetrate solid materials and cast shadows of internal structures, such as human bones, onto photographic film. Due to the unknown nature of this radiation, it was named X-rays. X-rays were later identified as a form of electromagnetic radiation when, in 1912, Max von Laue demonstrated their interference effects and successfully measured their wavelength [4].

1.9 Modern Developments and Quantum Optics

In a statement attributed to Lord Kelvin during his 1900 address to the British Association for the Advancement of Science, he remarked, "There is nothing new to be discovered in physics now. All that remains is more and more precise measurement." At the time, the classical theories of mechanics, electromagnetism, thermodynamics, and light were well-established, leading many to believe that the fundamental laws of nature had been fully uncovered. This sentiment reflected the prevailing confidence in the completeness of classical physics at the dawn of the 20th century [5]. However, some significant challenges loomed over the horizon of classical physics, which involved light. The first was the Rayleigh–Jeans ultraviolet catastrophe and the nature of blackbody radiation. Others were the photoelectric effect, the Compton scattering, the failed Michelson-Morley experiment to detect aether and the atomic spectra. Scientists were thinking at the time that this issue would be resolved with minor changes in Physics. However, these seemingly minor problems required an entire revolution for the laws of physics. And the result was the introduction of the quantum theory and the relativity which have become cornerstones of modern science and provided the intellectual bedrock for our 21st century world.

Specifically, the Rayleigh–Jeans formula for blackbody radiation accurately matched experimental observations at low frequencies but failed dramatically at high frequencies. According to Rayleigh and Jeans, the emitted energy per unit volume and per unit frequency for a blackbody radiator was proportional to the fourth power of the frequency. This implied that at higher frequencies, the energy would approach infinity, leading to unphysical predictions in the ultraviolet region of the spectrum. This discrepancy became known as the Rayleigh–Jeans ultraviolet catastrophe. By 1900, this failure had raised significant doubts about the validity of classical physics and thermodynamics. The solution came from Max Planck (1858–1947), who introduced a groundbreaking quantum theory of radiation. Planck's radiation formula successfully matched the experimental blackbody radiation spectra across the entire frequency spectrum, revolutionizing physics and laying the foundation for quantum mechanics.

The revival of the particle theory of light, and the birth of the modern concept of the photon, can be credited to Albert Einstein (1879–1955). Einstein's contributions to science are monumental. He is a foundational figure in both quantum mechanics and the theory of relativity. His insights fundamentally changed how we understand the nature of light. In 1905, Einstein proposed the idea of light quanta, now known as photons, in his explanation of the photoelectric effect. This work provided crucial evidence that light could behave as discrete packets of energy, reviving the particle theory of light and laying the groundwork for quantum theory. Einstein's impact on our understanding of light is immeasurable, as his work bridged the wave-particle duality of light and expanded the boundaries of classical physics [5].

Therefore, light interacts with matter as though it consists of particles, each carrying a small quantum of energy. These light quanta were later named "photons" by Gilbert Lewis (1875–1946) in 1926. However, this did not signal a return to the classical corpuscular theory of light. Instead, light was now understood to possess a dual nature, behaving either as a wave or a particle depending on the specific experiment being conducted. This wave-particle duality, a cornerstone of quantum theory, fundamentally redefined the nature of light and continues to influence modern physics [1].

Louis de Broglie (1892–1987), a French aristocrat and physicist, made groundbreaking contributions to quantum theory. In his 1924 PhD thesis, de Broglie postulated the wave nature of electrons, suggesting that all matter possesses wave properties. This idea, known as the de Broglie hypothesis, introduced the concept that particles, such as electrons, can exhibit both wave-like and particle-like behavior, depending on the context. Louis de Broglie assumed that for particles the same relations are valid as for the photon, specifically:

$$E = h\nu = \frac{hc}{\lambda_0} = pc, \qquad \text{with,} \quad \lambda_0 \nu = c, \tag{1.1}$$

where $h = 6.62607015 \times 10^{-34}$ Jsec is the Planck's constant, E is the photon energy, λ_0 is the freespace wavelength, c is the light speed in freespace, and p is the photon momentum. Based on this the de Broglie wavelength λ_B can be defined as:

$$\lambda_B = \frac{h}{p}, \tag{1.2}$$

where p is the relativistic momentum of the particle. The energy and momentum relations for a particle of non-zero rest mass are given by:

$$E = \sqrt{p^2c^2 + m_0^2c^4} = mc^2 = (KE) + m_0c^2 \tag{1.3}$$

$$pc = \sqrt{(KE)^2 + 2(KE)m_0c^2}, \tag{1.4}$$

$$m = \frac{m_0}{\sqrt{1 - u^2/c^2}}, \tag{1.5}$$

where m_0 is the rest mass of the particle and (KE) is its kinetic energy. Equation (1.2) suggests that just as photons (light particles) exhibit both wave and particle characteristics, all forms of matter, including electrons, can similarly exhibit wave-like behavior under appropriate conditions. This hypothesis was experimentally confirmed by the electron diffraction experiments of Davisson and Germer in 1927.

Albert Michelson (1852–1931) and Edward Morley (1838–1923) conducted a famous experiment in 1887 to definitively test the existence of the luminiferous ether, a medium once thought necessary for the propagation of light. They used an interferometer, now known as the Michelson interferometer, to achieve this. In the experiment, white light was sent through a half-silvered mirror, splitting it into two beams. One beam traveled straight to a mirror in one arm, while the other propagated at right angles to another mirror. Both beams then recombined at the beam splitter after traveling equal distances. This recombination produced an interference pattern of constructive and destructive interference. If the Earth moved through the ether, it was expected that the beam traveling along the direction of motion through the ether would take longer to complete its path than the beam traveling perpendicularly. This time difference would cause a shift in the interference pattern. However, the Michelson-Morley experiment produced a null result, i.e., no significant shift in the interference pattern was observed. This finding contradicted the ether theory and suggested that light's speed is constant in all directions, regardless of Earth's motion through space. This result was pivotal in leading to the development of Einstein's special theory of relativity [5].

In 1905, Albert Einstein formulated the theory of special relativity, introducing two key postulates that revolutionized classical physics. The first postulate states that the laws of physics are the same for all observers in inertial (constant speed) frames of reference. The second postulate asserts that the speed of light, c, is constant for all observers, regardless of their motion relative to the light source. From these postulates, Einstein derived the Lorentz transformation, which describes how measurements of space and time change for observers moving relative to one another. This led to the concepts of length contraction and time dilation, which represent a departure from Isaac Newton's idea of absolute space and time. The advent of special relativity, combined with the null result of the Michelson-Morley experiment, led to the abolition of the concept of the luminiferous aether, as it became unnecessary in explaining the propagation of light. This marked a profound shift in the understanding of the fundamental nature of the universe [5].

Einstein also developed the general theory of relativity (1915), providing a geometric theory of gravitation that serves as the modern framework for understanding gravity in contemporary physics. This theory extends the principles of special relativity and refines Newton's law of universal gravitation, offering a unified description of gravity as a property of space and time, referred to collectively as four-dimensional spacetime. In contrast to classical mechanics, where gravity is treated as a force, general relativity describes gravity as the curvature of spacetime caused by mass and energy. Massive objects bend spacetime, and this curvature influences the motion of other objects, which is perceived as gravitational attraction. The Einstein field equations mathematically describe how mass and energy determine spacetime curvature, forming the theoretical foundation of general relativity [5]. General relativity has led to groundbreaking predictions, such as the bending of light by gravity, time dilation near massive objects, and the existence of black holes and gravitational waves—many of which have been confirmed by experiments and observations. One of the most famous confirmations came after World War I when Sir Arthur Eddington organized an expedition to Principe Island near Africa to observe the solar eclipse on May 29, 1919. Eddington's team measured the bending of light from distant stars by the Sun's gravity, confirming Einstein's predicted values. Einstein became an overnight international celebrity and an iconic figure. More recently, the first direct detection of gravitational waves occurred on September 14, 2015, and the discovery was announced by the LIGO and Virgo collaborations on February 11, 2016 [13]. This event confirmed a major prediction of general relativity and opened a new era in the study of the universe.

In 1802, William Wollaston closely analyzed sunlight's spectrum using a prism, discovering that it contained several dark lines. These lines, later rediscovered by Joseph Fraunhofer in 1814, came to be known as Fraunhofer lines. This discovery sparked further interest in the study of spectral lines and led to the development of the field of spectroscopy. Notable scientists such as Gustav Kirchhoff and Robert Bunsen (1860, 1861), Anders Jonas Angstrom (1868), Johann Balmer (1885), and Johannes Rydberg (1888) contributed to the discovery and study of spectral lines emitted by various sources. However, the physical mechanisms behind these lines remained a mystery. With the advent of quantum theory in 1900, Max Planck revolutionized our understanding of radiation. Building on Planck's work, Niels Bohr proposed a model of the atom in 1913, in which electrons orbited the nucleus in discrete energy levels, much like planets orbiting a star. Transitions between these orbits corresponded to the emission or absorption of light quanta, or photons, explaining the observed spectral lines.

The Bohr model was further refined by Werner Heisenberg, Max Born, Erwin Schrödinger, and others with the development of quantum mechanics, introducing the concepts of wavefunctions and quantum probabilities. These advancements provided a deeper understanding of the interaction between electromagnetic radiation and matter, explaining phenomena like atomic spectra with greater precision. This theoretical progress laid the foundation for the invention of the laser by Theodore Maiman in 1960. Maiman's work was based on the theoretical predictions made in 1955 by Charles H. Townes and Arthur Leonard Schawlow, who proposed the concept of stimulated emission for amplifying light. The laser became a transformative technology in optics, leading to innovations across various fields, including telecommunications, medicine, and manufacturing.

One of the most recent major contributions to the understanding of light came through quantum electrodynamics (QED), the quantum theory of electromagnetic interactions (late 1940s). QED was developed by several prominent scientists, including Paul Dirac, Wolfgang Pauli, Eugene Wigner, Pascual Jordan, Werner Heisenberg, Enrico Fermi, Hans Bethe, Shinichiro Tomonaga, Julian Schwinger, Richard Feynman, and Freeman Dyson. This theory, based on the quantization of electromagnetic fields, provides a comprehensive framework for understanding the nature of light at the quantum level.

1.10 Measurement of Speed of Light

Measuring the speed of light has played a pivotal role in advancing optics and deepening our understanding of how light behaves in different media [4]. Initially, there was a debate among scientists about whether light traveled at a finite speed or instantaneously. The Greek philosopher Empedocles argued for a finite speed, while Aristotle and others believed light's propagation was instantaneous. This debate continued throughout the Middle Ages.

Galileo was the first to propose an experiment to measure the speed of light, which he believed to be finite. His experiment required two individuals, A and B, positioned on hilltops several miles apart. Person A would uncover a lantern, and as soon as person B saw the light, they would uncover their own lantern in response. The speed of light would then be calculated as twice the distance between the two people, divided by the time it took for person A to see person B's light after uncovering their own lantern. However, attempts to implement Galileo's proposal were unsuccessful, as the speed of light is far too fast to be accurately measured by human reaction times.

The first successful attempt to measure the speed of light came in 1676 when Danish astronomer Ole Rømer (1644-1710) made a groundbreaking observation. While studying the motion of Jupiter's moon Io and its eclipses, Rømer noticed variations in the timing of these eclipses based on Earth's position relative to Jupiter. He deduced that these discrepancies occurred because light took time to travel over the changing distances between the planets, leading to the first rough estimate of the speed of light [14]. Rømer estimated the speed of light to be around 214,000 km/s, an error of around 30%, which is understandable as planetary distances were not precisely known at that time. Rømer's merit resides in demonstrating that the speed of light was finite.

In 1727, the British astronomer James Bradley calculated the speed of light by carefully measuring the apparent change in a star's position based on its alignment with Earth's orbital velocity. This phenomenon, known as the angle of aberration of starlight, is analogous to the apparent shift in the direction of raindrops observed from a moving convertible compared to a stationary one. Bradley's measured value for the speed of light was remarkably close to the accepted value of approximately 300,000 km/s today.

In the 19th century, the French physicist Hippolyte Fizeau made significant strides in measuring the speed of light using a toothed wheel and a distant mirror. When the wheel rotated at the appropriate speed, it would return to a position that allowed the reflected light to pass through the next opening in the wheel. By measuring the distance between the wheel and the mirror, as well as the speed of the wheel's rotation, Fizeau was able to determine a more accurate value for the speed of light in air (1849). Léon Foucault refined this technique in 1862 by employing rotating mirrors instead of a toothed wheel. His method provided even more precise measurements and demonstrated that light traveled slower in water than in air, thereby supporting the wave theory of light.

In 1958, Froome [15] measured the speed of light at $299,792.50 \pm 0.10\,\text{km/s}$ using a microwave interferometer and a Kerr cell shutter. Later, Ken Evenson and his team at NIST Boulder developed techniques to produce laser light with a very stable frequency, enabling them to measure the speed of light with remarkable precision, arriving at a value of $299,792,456.2 \pm 1.1$ meters per second [16]. Evenson also employed interferometric techniques at visible frequencies. In this method, a laser source is directed at a half-silvered mirror, which splits the light beam into two perpendicular rays that travel to two stationary mirrors and back. These mirrors are positioned equidistant from the half-silvered mirror, ensuring that both light rays take the same amount of time to return. When the beams meet at the half-silvered mirror, they undergo constructive interference, resulting in a new wave that possesses twice the amplitude. The experiment is then repeated, but this time one of the stationary mirrors is moved back, increasing the distance to the half-silvered mirror. When this mirror is shifted back by exactly half a wavelength of the light wave, it leads to destructive interference, causing the waves to cancel each other out and resulting in no detectable signal. In laser interferometry, a laser source with a known frequency is utilized, and the wavelength of the laser light is determined by measuring the distance the stationary mirror must be moved to achieve destructive interference. The speed of light is subsequently calculated by multiplying the frequency by the measured wavelength. Finally, Evenson refined this approach to measure the speed of light with unprecedented accuracy, achieving a value of $299,792,458 \pm 4 \times 10^{-9}\,\text{m/s}$ [17].

Since 1983, the speed of light, denoted as c, has been defined in the International System of Units (SI) as exactly 299,792,458 m/s. This definition establishes the meter as the precise distance that light travels in a vacuum in 1/299792458 of a second.

1.11 Nobel Laureates in Optics-Related Topics

Numerous Nobel laureates have made significant contributions to the fields of optics and photonics. Their work has advanced these areas, ranging from foundational quantum theory to cutting-edge laser technologies and imaging systems. While the majority of these contributions are within the field of Physics, there have also been notable advancements in Chemistry and Medicine. Table 1.1 provides a summary of these Nobel laureates [18].

TABLE 1.1: Nobel Laureates Related to the Field of Optics

Laureate Names	Contribution	Field	Year
W. Rötgen	X-Rays	Physics	1901
H. Lorentz, P. Zeeman	Connection Optical and Electromagnetic Phenomena	Physics	1902
J. W. Strut (Lord Rayleigh)	Gases Densities - Argon	Physics	1904
A. Michelson	Optical Precision Instruments and Metrology/Spectroscopy	Physics	1907
G. Lippman	Method of Reproducing Colors Photographically based on the Phenomenon of Interference	Physics	1908
W. H. Bragg, W. L. Bragg	Analysis of Crystal Structure by means of X-rays	Physics	1915
A. Einstein	Discovery of the Law of the Photoelectric Effect	Physics	1921
C. V. Raman	Scattering of Light and Discovery of the Effect Named After Him	Physics	1930
F. Zernicke	Phase-Constrast Microscopy	Physics	1953
C. H. Townes, N. G. Basov, A. M. Prokhorov	Fundamental work in the field of Quantum Electronics, Maser Principle	Physics	1964
A. Kaslter	Discovery and Development of Optical Methods for Studying Hertzian Resonances in Atoms	Physics	1966
D. Gabor	Invention and Development of the Holographic Method	Physics	1971
M. Ryle, A. Hewish	Pioneering Research in Radio Astrophysics	Physics	1974
A. A. Penzias, R. W. Wilson	Discovery of Cosmic Microwave Background Radiation	Physics	1978
A. Cormack, G. Housefield	Development of Computer Assisted Tomography	Medicine	1979
N. Bloembergen, A. Schawlow, K. Siegbahn	Contribution to the Development of Laser Spectroscopy	Physics	1981
E. Ruska	Electron optics, Design of the First Electron Microscope	Physics	1986

Nobel Laureates Related to the Field of Optics (Continued)

Laureate Names	Contribution	Fields	Year
S. Chu, C. Cohen-Tannoudji, W. Phillips	Development of Methods to Cool and Trap Atoms with Laser Light	Physics	1997
A. H. Zewail	Studies of the Transition States of Chemical Reactions using Femtosecond Spectroscopy	Chemistry	1999
Z. Alferov, H. Kroemer	Development of Semiconductor Heterostructures for High-Speed- and Opto-Electronics	Physics	2000
P. C. Lauterbur, P. Mansfield	Discoveries Concerning Magnetic Resonance Imaging	Medicine	2003
R. J. Glauber, J. L. Hall, T. W. Hänsch	Quantum Theory of Optical Coherence, Laser Precision Spectroscopy	Physics	2005
C. K. Kao, W. S. Boyle, G. E. Smith	Groundbreaking Achievements Concerning the Transmission of Light in Fibers for Optical Communication	Physics	2009
I. Akasaki, H. Amano, S. Nakamura	Invention of Efficient Blue Light-Emitting Diodes which has Enabled Bright and Energy-Saving White Light Sources	Physics	2014
E. Betzig, S. W. Hell, W. E. Moerner	Development of Super-Resolved Fluorescence Microscopy	Chemistry	2014
R. Weiss, B. C. Barish, K. S. Thorne	Decisive Contributions to the LIGO Detector and the Observation of Gravitational Waves	Physics	2017
A. Ashkin, G. Mourou, D. Strickland	Ground Breaking Inventions in the Field of Laser Physics (Optical Tweezers, High-Intensity Ultra-Short Laser Pulses)	Physics	2018
A. Aspect, J. F. Clauser, A. Zeilinger	Experiments with Entangled Photons, Establishing the Violation of Bell Inequalities and Pioneering Quantum Information Science	Physics	2022
F. Krausz, P. Agostini, A. L'Huiller	Experimental Methods that Generate Attosecond Pulses of Light for the Study of Electron Dynamics in Matter	Physics	2023

2

Geometrical Optics

2.1 Introduction

Historically, light was initially believed to be composed of particles that propagate in straight lines. This view was dominant from ancient times until the early 19th century when Young's famous two-slit experiment revealed the wave nature of light. A short description of the historical development of optical science was presented in the chapter *Brief History of Optics* (Chapter 1) based on several references such as Refs. [1, 2] and [4]. Today, it is well-established that light exhibits both wave-like and particle-like behavior, depending on the specific application or experimental conditions. This is known as the principle of wave-particle duality. The most rigorous and complete description of light is provided by Quantum Electrodynamics (QED), the quantum field theory of electromagnetic radiation, also known as *Quantum Optics*. However, in most optical applications, where quantum effects are negligible (as in cases involving large numbers of photons), the classical theory of *Electromagnetic Optics* is sufficiently accurate. When polarization effects can be neglected, the electromagnetic vector field can be approximated as a scalar field, leading to the simpler but highly effective *Physical or Wave Optics*.

Moreover, when light (particularly in the visible region of the electromagnetic spectrum, with wavelengths around one micron) interacts with objects whose dimensions are much larger than the wavelength, its behavior can be accurately described by rays that follow geometrical rules. This domain is known as *Ray Optics* or *Geometrical Optics*. A visual representation of the different theoretical frameworks for describing light is presented in Fig. 2.1 (adapted from [19]). Geometrical Optics can be considered a special case of physical (wave) optics when the wavelength of light tends to zero [20]:

$$\lim_{\lambda_0 \to 0} \{\text{Physical Optics}\} = \{\text{Geometrical Optics}\}\,. \tag{2.1}$$

Within the framework of Geometrical Optics, light is understood to propagate from its source along paths known as rays. When the refractive index of the medium remains constant, these rays follow straight-line trajectories. Although the concept of a ray is highly effective within geometrical optics, it remains somewhat abstract, as light cannot be confined to a thickness-less line [20]. One of the closest real-world approximations of a ray is the narrow, pencil-like beam of a laser. As rays pass through an optical system, their paths are influenced by variations in the refractive index, as well as by reflections and refractions at material boundaries. This is particularly evident in optical systems composed of multiple lenses, where the trajectory of light is shaped by these interactions. The fundamental law governing the propagation of rays in geometrical optics is based on Fermat's Principle. This principle asserts that light travels between two points along the path that requires the least optical time, or equivalently, the path that minimizes the optical path length. Before delving into Fermat's Principle, it is beneficial to first introduce Huygen's Principle. Huygen's Principle provides a foundational framework for understanding wave propagation, which plays a crucial role in both geometrical and wave optics.

DOI: 10.1201/9781003490333-2

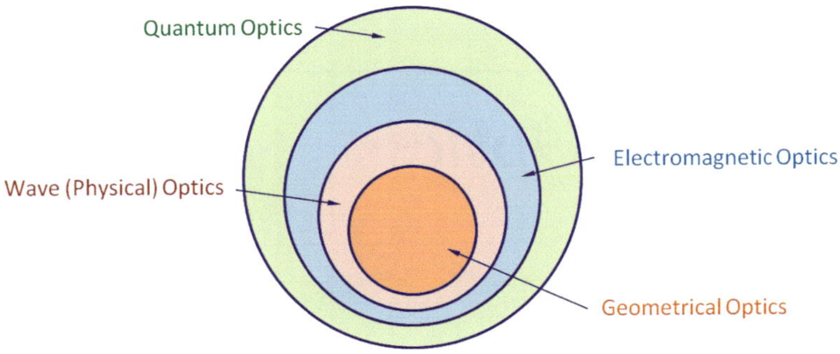

FIGURE 2.1: A diagram of the various theories of optics starting with Quantum Optics, which is the most rigorous, up to Geometrical (ray) Optics which is the simplest. Every circle inside a larger circle is a subset of the larger circle (adapted from Ref. [19]).

2.2 Huygens' Principle

Christiaan Huygens developed the geometrical theory of wave propagation in optics in his seminal work, *Traité de la Lumière* [9] which was published in 1690. In his approach, Huygens assumed that each point on a given wavefront of light could be considered as a secondary point source that emits spherical waves. The envelope of all these spherical secondary waves forms the new wavefront, providing a clear geometrical method to trace the propagation of light waves. This approach is illustrated in Fig. 2.2, where the wavefronts are depicted evolving as the envelope of the secondary wavelets. This principle not only explains how light propagates but also provides insights into the phenomena of reflection, refraction, and diffraction within the framework of wave optics. In Fig. 2.2 a wavefront of

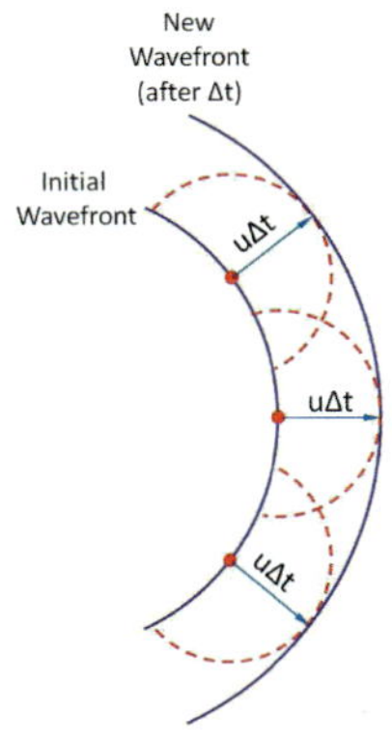

FIGURE 2.2: Each point in the initial wavefront acts like a secondary point source emitting wavelets. The new wavefront is formed by the superposition of all these secondary wavelets (toward the propagation direction only).

the light wave is shown at an initial time t. Each point on this initial wavefront acts as a source of secondary spherical waves, propagating in the direction of light travel with the same frequency as the original wavefront. Three such points are shown, along with their corresponding spherical waves, which propagate with the speed of light in the medium u over a time interval Δt. The envelope of these secondary waves forms the new wavefront at time $t + \Delta t$. While Huygens' principle provides a useful geometric framework for light propagation, it does not account for the interference between secondary waves. As a result, Huygens was unable to predict diffraction effects, which later refinements of his theory would address. Interestingly, Huygens' principle holds in spaces with an odd number of spatial dimensions, as described in [21–23]. In simple terms, this principle implies that an instantaneous signal remains instantaneous for every observer at later times at any point in 3D space, meaning there are no residual disturbances (or "after-ripples"). An example in 3D space is light, or more generally, electromagnetic radiation. In contrast, in even-dimensional spaces, an instantaneous signal does not remain instantaneous for all observers at later times. Residual waves, or after-ripples, follow the main wavefront. A familiar example of this in 2D space is the ripple pattern on a water surface after a pebble is dropped, where waves continue to propagate after the initial disturbance.

Huygens' principle was enhanced by Augustin-Jean Fresnel [24], who introduced the concept of interference for the amplitudes of the secondary waves. This improvement allowed Fresnel to successfully explain the phenomenon of diffraction, which had been unaccounted for in Huygens' original formulation. Later, Gustav Kirchhoff [25] further refined the Huygens-Fresnel theory, addressing several difficulties that Fresnel's formulation encountered. Kirchhoff's contribution provided a more mathematically rigorous foundation for the theory, making it a cornerstone in the study of wave propagation and diffraction. In a more recent development, Miller [26] proposed a correction to the Huygens principle, particularly addressing the assumption of forward-only propagation of secondary wavelets, as initially conceived by Huygens.

Huygens' principle can be utilized to prove both the law of reflection and the law of refraction, which are fundamental in the theory of Geometrical Optics. These two laws describe how light behaves when encountering a boundary between different media, and they play a critical role in understanding optical systems. In the next subsections, simple proofs of these laws, using Huygens' principle, will be presented.

2.2.1 Law of Reflection based on Huygens' Principle

Assume two rays that are reflected at a planar interface as shown in Fig. 2.3. Let AB be an initial wavefront where the point A is assumed to be infinitesimally after the ray is reflected by the planar interface. The points A and B serve as secondary point sources emanating spherical wavelets. Let CD a reflected wavefront after a time Δt. The distances $AD = u\Delta t$ and $BC = u\Delta t$ are traveled in the upper medium at the same time Δt with the same velocity u. The triangles ABC and ACD are orthogonal (wavefronts are perpendicular to

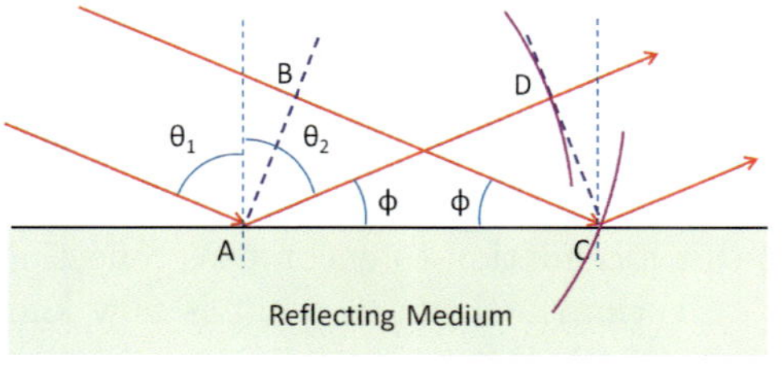

FIGURE 2.3: The law of reflection can be explained based on the Huygens' principle.

the ray direction) and have a common side, AC, and two equal sides $AD = BC = u\Delta t$. Therefore, these triangles are equal and the angles $D\hat{A}C = A\hat{C}B = \phi$. Then $\theta_1 + \phi = \pi/2 = \theta_2 + \phi \Longrightarrow \theta_1 = \theta_2$. Therefore, the angle of incidence is equal to the angle of reflection.

2.2.2 Law of Refraction based on Huygens' Principle

Assume two rays that are refracted at a planar interface between two homogeneous media with refractive indices n_1 and n_2, respectively, as shown in Fig. 2.4. Let AB be an initial wavefront where the point A is assumed to be infinitesimally before the ray is refracted by the planar interface while the same holds for point C. The points A and B serve as secondary point sources emanating spherical wavelets. Let CD a refracted wavefront after

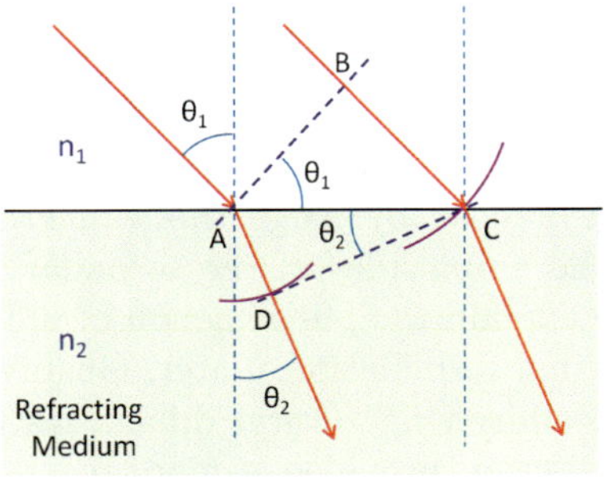

FIGURE 2.4: The law of refraction can be explained based on the Huygens' principle.

a time Δt. The distance $AD = u_2 \Delta t$ and the distance $BC = u_1 \Delta t$ are traveled in the lower and upper medium at the same time Δt with velocities u_1 and u_2, respectively. The angles θ_1 and $B\hat{A}C$ are equal since they have perpendicular sides. The same argument holds for the angles θ_2 and $A\hat{C}D$. Therefore the following equations can be written:

$$\sin\theta_1 = \frac{BC}{AC} = \frac{u_1 \Delta t}{AC} = \frac{c\Delta t}{n_1 AC}, \tag{2.2}$$

$$\sin\theta_2 = \frac{AD}{AC} = \frac{u_2 \Delta t}{AC} = \frac{c\Delta t}{n_2 AC}. \tag{2.3}$$

From the previous equations its straightforward to show that $n_1 \sin\theta_1 = n_2 \sin\theta_2$ which is the Snell's law or the law of refraction (in the previous equations the velocity of light in a homogeneous isotropic medium of refractive index n is expressed as $u = c/n$).

2.3 Fermat's Principle

Heron of Alexandria was among the first to formulate the idea that light propagates between two points in a homogeneous medium by following the shortest path. About a thousand years later, Alhazen expanded this formulation to encompass reflection and refraction. However, around 1662 AD, Pierre de Fermat articulated what is now known as Fermat's Principle, or the Least Time Principle, which states that light travels along the path that minimizes the travel time. The most modern interpretation of Fermat's Principle asserts that the light path is at an extremum, meaning that the optical length between two points taken by a light ray is stationary with respect to first-order variations of nearby paths. Let's consider an inhomogeneous isotropic medium of refractive index $n(\vec{r}) = n(x, y, z)$ as shown in Fig. 2.5. The optical path length, OPL, of a ray from point A to point B can be written as:

$$OPL = \int_A^B n\left(\vec{r}(s)\right) ds, \tag{2.4}$$

where ds is the elementary ray length and $\vec{r}(s)$ the vector specifying the path of the ray. The ray path can be parameterized through the variable t, i.e., $\vec{r} = x(t)\hat{\imath}_x + y(t)\hat{\imath}_y + z(t)\hat{\imath}_z$,

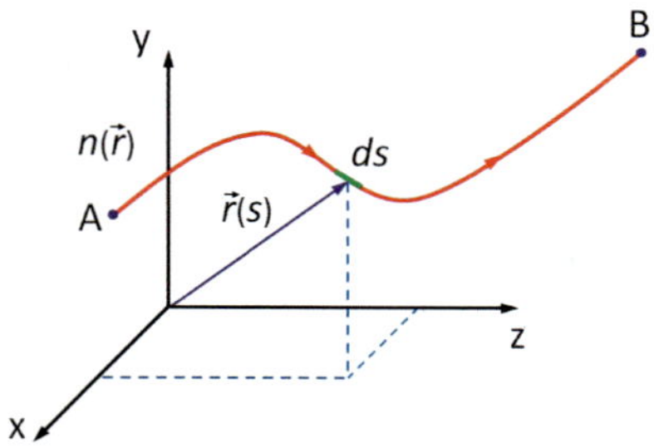

FIGURE 2.5: A ray of light propagates inside an inhomogeneous isotropic medium of refractive index $n(\vec{r})$ from point A to point B. The position of the ray is defined parametrically using the position vector $\vec{r}(s)$ that depends on the parameter s that expresses the length of the ray. The ray follows Fermat's principle.

and $ds = (dx^2 + dy^2 + dz^2)^{1/2} = (\dot{x}^2 + \dot{y}^2 + \dot{z}^2)^{1/2}dt$, where $\dot{w} = dw/dt$ (with $w = x, y, z$), and $\hat{\imath}_x, \hat{\imath}_y, \hat{\imath}_z$ are the unit vectors along the x, y, and z-axis, respectively. Then, Eq. (2.4) can be written in the form:

$$OPL = \int_A^B \underbrace{n(x,y,z)\ \sqrt{\dot{x}^2 + \dot{y}^2 + \dot{z}^2}}_{= F(x,y,z,\dot{x},\dot{y},\dot{z})}\ dt = \int_A^B F(x,y,z,\dot{x},\dot{y},\dot{z})\ dt. \tag{2.5}$$

The above OPL should be stationary according to Fermat's principle. This means that its first order variation $\delta[OPL] = 0$. Using the calculus of variations [27] it is straightforward to show that the following Euler-Lagrange equations should be satisfied:

$$\frac{\partial F}{\partial x} - \frac{d}{dt}\left(\frac{\partial F}{\partial \dot{x}}\right) = 0,$$

$$\frac{\partial F}{\partial y} - \frac{d}{dt}\left(\frac{\partial F}{\partial \dot{y}}\right) = 0,$$

$$\frac{\partial F}{\partial z} - \frac{d}{dt}\left(\frac{\partial F}{\partial \dot{z}}\right) = 0.$$

Using the definition for F shown in Eq. (2.5) the Euler-Lagrange equations become:

$$\begin{aligned} \frac{\partial n}{\partial x} - \frac{d}{ds}\left(n\frac{dx}{ds}\right) &= 0, \\ \frac{\partial n}{\partial y} - \frac{d}{ds}\left(n\frac{dy}{ds}\right) &= 0, \\ \frac{\partial n}{\partial z} - \frac{d}{ds}\left(n\frac{dz}{ds}\right) &= 0. \end{aligned} \tag{2.6}$$

The last equations can be written in the following vector ray equation:

$$\frac{d}{ds}\left(n\frac{d\vec{r}}{ds}\right) = \vec{\nabla}n. \tag{2.7}$$

Equation (2.7) can be re-written in the following computationally more convenient form using the transformation $t = \int ds/n$ [28],

$$\frac{d^2\vec{r}}{dt^2} = n\vec{\nabla}n = \frac{1}{2}\vec{\nabla}n^2. \tag{2.8}$$

For example, assume propagation between points A and B (Fig. 2.5) in a homogeneous isotropic medium. In this case $\vec{\nabla}n = 0$ and the solution of Eq. (2.8) becomes $\vec{r}(t) = \vec{a}t + \vec{b}$ where $\vec{a}$ and $\vec{b}$ two constant vectors. The points A and B can be specified by their position vectors $\vec{r}_A = x_A\hat{\imath}_x + y_A\hat{\imath}_y + z_A\hat{\imath}_z$ and $\vec{r}_B = x_B\hat{\imath}_x + y_B\hat{\imath}_y + z_B\hat{\imath}_z$ [where w_A and w_B, $(w = x, y, z)$ are the coordinates of points A and B in the xyz coordinate system]. Then assuming that $\vec{r}(t = 0) = \vec{r}_A$ and $\vec{r}(t = 1) = \vec{r}_B$ it is straightforward to show that $\vec{a} = \vec{r}_B - \vec{r}_A$ and $\vec{b} = \vec{r}_A$, and the solution for the ray is of the form $\vec{r}(t) = (\vec{r}_B - \vec{r}_A)t + \vec{r}_A$ which is obviously a straight line between points A and B.

2.3.1 Law of Reflection based on Fermat's Principle

According to Fermat's principle, the path taken by the ray from point A to point B via point O (as shown in Fig. 2.6(a)) on the planar mirror must make the optical path length stationary. In this case, the medium is homogeneous and isotropic with a constant refractive index n, so the optical path length is proportional to the geometric path length. Thus, Fermat's principle reduces to the requirement that the total geometric distance traveled by the light is at a stationary point (i.e., minimized). The distances of points A and B from the flat reflective surface are known and are denoted as $a = AA'$ and $b = BB'$ as indicated in Fig. 2.6(a). In addition, the horizontal distance between points A and B is d and it is also known since the A and B are considered constant. The ray AOB should satisfy Fermat's principle, i.e., the following conditions should be satisfied:

$$OPL = \int_A^B n\left(\vec{r}(s)\right) ds = n\int_A^O ds + n\int_O^B ds = n(AO) + n(BO) \Longrightarrow$$

$$\delta[OPL] = 0 \Longrightarrow n\frac{d(AO)}{dx} + n\frac{d(OB)}{dx} = 0 \Longrightarrow$$

$$\frac{d}{dx}\left(\sqrt{x^2 + a^2}\right) = -\frac{d}{dx}\left(\sqrt{(d - x)^2 + b^2}\right) \Longrightarrow \frac{x}{\sqrt{x^2 + a^2}} = \frac{d - x}{\sqrt{(d - x)^2 + b^2}} \Longrightarrow$$

$$\sin\alpha_1 = \sin\alpha_2 \Longrightarrow \alpha_1 = \alpha_2,$$

therefore the angle of incidence α_1 should be equal to the angle of reflection α_2. The optical path of the ray in this case is minimized and the Fermat's principle is equivalent to the minimization of the time that it takes for the ray to travel from A to B via reflection on the mirror. Summarizing, Fermat's principle provides a simple and elegant derivation of the law of reflection.

2.3.2 Law of Refraction based on Fermat's Principle

In the case of refraction, Fermat's principle can also be used to derive Snell's law. In this case let's consider the refraction of a ray from point A to point B via a planar interface between two homogeneous isotropic media with refractive indices n_1 and n_2 as shown in Fig. 2.6(b). Assume that the refraction occurs at point O as shown in Fig. 2.6(b). Again, the vertical distances of points A and B are known and are $a = AA'$ and $b = BB'$ as shown.

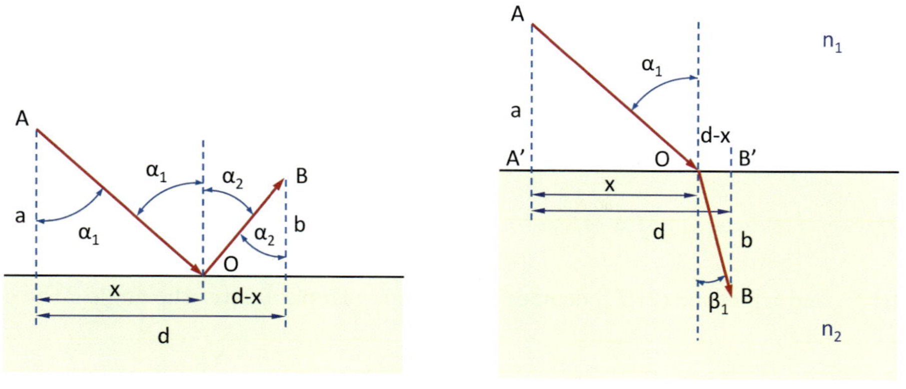

(a) (b)

FIGURE 2.6: (a) Applying Fermat's principle upon reflection at a planar boundary. (b) Applying Fermat's principle upon refraction at a planar boundary.

In addition, the horizontal distance between points A and B is d and it is also known since the A and B are considered constant. To apply Fermat's principle, the optical path length (OPL) must be stationary. The ray AOB should satisfy Fermat's principle, i.e., the following equations should be satisfied:

$$OPL = \int_A^B n\left(\vec{r}(s)\right) ds = n_1 \int_A^O ds + n_2 \int_O^B ds = n_1(AO) + n_2(BO) \Longrightarrow$$

$$\delta[OPL] = 0 \Longrightarrow n_1 \frac{d(AO)}{dx} + n_2 \frac{d(OB)}{dx} = 0 \Longrightarrow$$

$$n_1 \frac{d}{dx}\left(\sqrt{x^2+a^2}\right) = -n_2 \frac{d}{dx}\left(\sqrt{(d-x)^2+b^2}\right) \Longrightarrow n_1 \frac{x}{\sqrt{x^2+a^2}} = n_2 \frac{d-x}{\sqrt{(d-x)^2+b^2}} \Longrightarrow$$

$$n_1 \sin\alpha_1 = n_2 \sin\beta_1.$$

Therefore the angle of incidence α_1 and the angle of refraction β_1 are related via Snell's law. As in the case of reflection, the optical path of the ray is minimized and the Fermat's principle is equivalent to the minimization of the time that it takes for the ray to travel from A to B via refraction at the planar interface between the two media.

2.3.3 Special Cases of Fermat's Principle

The initial formulation of Fermat's Principle stated that light travels along the path of least time. However, as mentioned earlier, the modern interpretation of Fermat's Principle asserts that the ray's path should be stationary. This means that the optical path of a ray could either be a minimum (the least time case, which is the most common), a maximum (the greatest time case, which is less common), or even constant for all paths. In this subsection, a case of maximum time is also presented [29]. The geometry under consideration is shown in Fig. 2.7(a). A ray of light reflects off a spherical mirror with radius b and passes through points A and B. According to the law of reflection, the path ACB will be the preferable path, as any other path, such as the AQB does not satisfy the reflection law (or Fermat's principle). However, a path like AQB is smaller than ACB. To demonstrate this, one can construct an ellipse with its foci at points A and B. The length of the major semi-axis of this ellipse is $a = (d^2 + b^2)^{1/2}$ as shown in Fig. 2.7(a). If the extension of the segment AQ intersects the ellipse at point P then $AP + PB = AC + CB$ since both points P and Q belong to the ellipse. Therefore $AC + CB = AP + PB > AQ + QB$ for every point Q that would be an alternative for the reflection to occur. As a result the actual ray path $AC+CB$ represents a maximum in this case. The same can be shown if Eq. (2.4) is applied,

$$OPL = n\int_A^B ds = n(AQ + QB) = n\left[\sqrt{b^2 + d^2 - 2bd\cos\theta} + \sqrt{b^2 + d^2 + 2bd\cos\theta}\right].$$

In order to find when the OPL becomes stationary with parameter the angle θ the equation

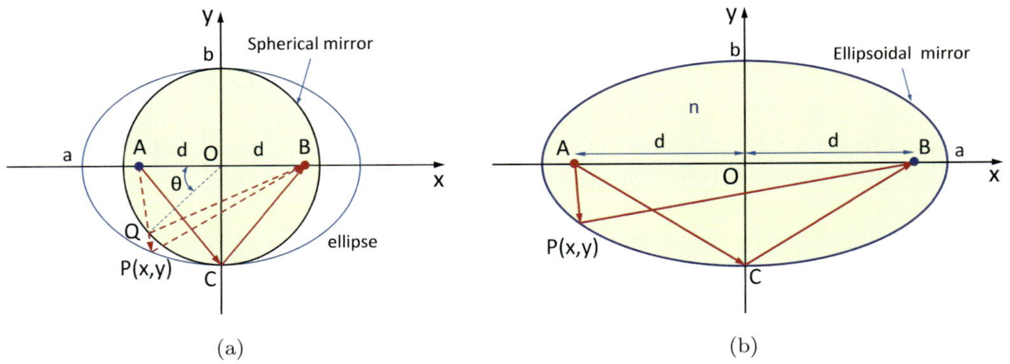

FIGURE 2.7: (a) Application of Fermat's principle upon reflection on a spherical mirror where the optical path and therefore the time of travel of light is maximized. (b) Application of Fermat's principle upon reflection on an elliptical mirror where the optical path and therefore the time of travel of light is stationary.

$d(OPL)/d\theta = 0$ should be solved. The solutions of this equation are $\theta = 0$ (in which case light goes from A to B through the point of intersection of the diameter AB with the spherical mirror) and $\theta = \pi/2$ (through the point C shown. For the first solution, the optical path becomes minimum while for the second solution, the optical path becomes maximum. A special case is when points A and B coincide with the center of the spherical mirror. In this case every possible point Q on the spherical mirror is satisfying Fermat's principle and the optical path is constant i.e., is neither minimum nor maximum. Another example of an optical path that is stationary (constant) is illustrated in Fig. 2.7(b). In this case, the reflection from one focal point of an ellipsoidal mirror to the other focal point is considered. All optical paths for rays emanating from point A and arriving at point B via reflection are equal, a characteristic property of the ellipse. Therefore, this serves as another example of a constant optical path.

2.4 Prisms

Optical prisms are versatile elements made from transparent materials like various types of glass, designed to refract or reflect light. The simplest form, a triangular prism, typically consists of three intersecting surfaces, two of which are responsible for bending or reflecting the light. Prisms can be employed in various ways, such as dispersing an incident light beam into its spectral components, reflecting light, or even splitting light into beams of different polarizations. Two primary categories of prisms exist: (a) Dispersive Prisms: These are used to separate incident light into its constituent wavelengths, or spectral components, and (b) Reflecting Prisms: These are designed to alter the direction or orientation of a light beam, whether by flipping, inverting, rotating, deviating, or displacing it. Prisms play a critical role across a wide array of optical devices, including microscopes, cameras, telescopes, binoculars, periscopes, and ophthalmologic equipment, and in numerous scientific experiments involving light and optics.

2.4.1 Refractive Prisms

In this section, the refractive properties of the simplest optical prism, a prism with a triangular cross-section, will be analyzed. A triangular cross-section prism with isosceles geometry is shown in Fig. 2.8(a). Without any loss of generality, the prism is assumed to be in an environment where the surrounding refractive index is n_0 usually air, while the prism's material has a refractive index n_p. A light ray enters the prism from the left at an angle α_1 as shown on the figure. Upon entering the prism, the ray is refracted at points A and B before it exits through the right surface. The goal is to calculate the angular deviation angle δ between the direction of the incident and the exiting rays, as shown in Fig. 2.8(a). According to Fermat's principle, which, as previously discussed, can be applied here using Snell's law at each refractive interface, the following relationships can be established with reference to Fig. 2.8(a):

$$\begin{aligned}
\delta &= \delta_1 + \delta_2 = (\alpha_1 - \beta_1) + (\beta_2 - \alpha_2) = \alpha_1 + \beta_2 - (\beta_1 + \alpha_2) = \\
&= \alpha_1 + \beta_2 - \sigma, \qquad \text{where,} \qquad (2.9)\\
\beta_2 &= \sin^{-1}\left(\frac{n_p}{n_0}\sin\alpha_2\right), \\
\alpha_2 &= \sigma - \beta_1 = \sigma - \sin^{-1}\left(\frac{n_0}{n_p}\sin\alpha_1\right),
\end{aligned}$$

where finally $\delta = \delta(\alpha_1)$. An example angle of deviation δ as a function of the angle of incidence α_1 is shown in Fig. 2.8(b) for the case of a borosilicate crown glass (BK7) prism with an apex angle of $\sigma = 30°$. From this figure it can be observed that there is a minimum angle of deviation δ_{min}. This minimum deviation angle can be calculated from the minimum of the function $\delta(\alpha_1)$:

$$\begin{aligned}
&\frac{d\delta(\alpha_1^*)}{d\alpha_1} = 0 \Longrightarrow 1 + \frac{d\beta_2}{d\alpha_1} = 0 \Longrightarrow \\
&\frac{d\beta_2}{d\alpha_1} = -1 \Longrightarrow \frac{d\beta_2}{d\alpha_2}\frac{d\alpha_2}{d\alpha_1} = -1 \qquad \text{where,} \\
&\frac{d\beta_2}{d\alpha_2} = \frac{1}{\left[1 - \left(\frac{n_p}{n_0}\sin\alpha_2^*\right)^2\right]^{1/2}}\frac{n_p}{n_0}\cos\alpha_2^*, \\
&\frac{d\alpha_2}{d\alpha_1} = \frac{-1}{\left[1 - \left(\frac{n_0}{n_p}\sin\alpha_1^*\right)^2\right]^{1/2}}\frac{n_0}{n_p}\cos\alpha_1^*.
\end{aligned}$$

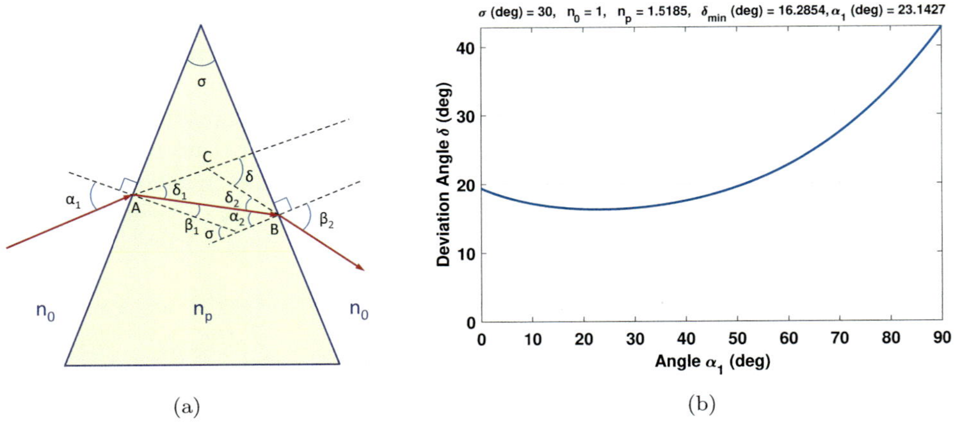

FIGURE 2.8: (a) A generic diagram of a prism of triangular cross-section. The apex angle of the prism is σ. The refractive index of the prism material is n_p while the surrounding prism medium has a refractive index of n_0. A ray is incident from the left at an angle α_1 as shown, and after refracting from both sides of the prisms exits at an angle β_2 and it is deviated from its original direction by an angle δ. (b) The deviation angle δ of a BK7 prism of apex angle $\sigma = 30°$ in the air as a function of the angle of incidence α_1. The refractive index of the prism $n_p = 1.5185$ and corresponds to a freespace wavelength of $\lambda_0 = 550\,\text{nm}$. The minimum angle of deviation is $\delta_{min} = 16.2854°$ and it occurs for an angle of incidence $\alpha_1 = 23.1427°$.

where α_1^* and α_2^* are the incident angle at the left interface and at the right interface at the minimum deviation. Then the condition for the minimum becomes:

$$\frac{d\beta_2}{d\alpha_2}\frac{d\alpha_2}{d\alpha_1} = -1 \Longrightarrow$$

$$\left[1-\left(\frac{n_p}{n_0}\sin\alpha_2^*\right)^2\right]^{1/2}\left[1-\left(\frac{n_0}{n_p}\sin\alpha_1^*\right)^2\right]^{1/2} = \cos\alpha_1^*\cos\alpha_2^* \Longrightarrow$$

$$\left(\frac{n_p^2}{n_0^2}-1\right)\sin^2\alpha_2^* = \left(1-\frac{n_0^2}{n_p^2}\right)\sin^2\alpha_1^* \Longrightarrow$$

$$\left(1-\frac{n_0^2}{n_p^2}\right)\sin^2\beta_2^* = \left(1-\frac{n_0^2}{n_p^2}\right)\sin^2\alpha_1^* \Longrightarrow$$

$$|\sin\beta_2^*| = |\sin\alpha_1^*| \Longrightarrow \beta_2^* = \alpha_1^*,$$

where the Snell's law $n_p\sin\alpha_2^* = n_0\sin\beta_2^*$ was used. The term β_2^* is the angle of the exiting ray from the right interface at the minimum deviation. It is obvious that since $\beta_2^* = \alpha_1^*$ then $\beta_1^* = \alpha_2^* = \sigma/2$ at the optimum. Consequently, the minimum deviation angle occurs at the condition of symmetry and the δ_{min} is given by:

$$\delta_{min} = 2\sin^{-1}\left[\frac{n_p}{n_0}\sin\frac{\sigma}{2}\right] - \sigma, \tag{2.10}$$

$$\alpha_1^* = \sin^{-1}\left[\frac{n_p}{n_0}\sin\frac{\sigma}{2}\right] = \beta_2^*. \tag{2.11}$$

When the apex angle σ of the prism is small, the equation for the minimum deviation angle δ_{min} can be simplified using the small-angle approximation, $\sin\theta \simeq \theta$ for small angles in radians. Under this approximation, the equation for the minimum deviation angle can be expressed in the commonly referenced form:

$$\delta_{min} \simeq \left(\frac{n_p}{n_0} - 1\right)\sigma. \tag{2.12}$$

This approximation is often used in practical applications where the deviation is slight and the angles involved are small. It is interesting to mention that in order to avoid total internal reflection at the right interface the following conditions much hold:

$$\alpha_2 \leq \theta_{cr} = \sin^{-1}\left(\frac{n_0}{n_p}\right) \Longrightarrow$$

$$\alpha_1 \geq \sin^{-1}\left[\frac{n_p}{n_0}\sin(\sigma - \theta_{cr})\right] = \alpha_{1,min} \qquad \text{and,}$$

$$\sigma \leq 2\theta_{cr}.$$

For example, for the case shown in Fig. 2.8(b), $\theta_{cr} = 41.1890°$ and $\alpha_{1,min} = -17.1370°$ while $\sigma = 30° < 2\theta_{cr} = 82.3780°$.

Due to the dispersion in prism materials, typically made of glass or plastic, polychromatic light (such as white light) will separate into its spectral components. This separation is a result of the wavelength-dependent refractive indices that cause different refraction angles for each spectral component. The dispersive power, Δ_w ($w = e, a$, "exact" and "approximate," respectively), of a prism, is defined as:

$$\Delta_e = \frac{\delta_{min,F} - \delta_{min,C}}{\delta_{min,D}}, \tag{2.13}$$

$$\Delta_a \simeq \frac{n_F - n_C}{n_D - n_0}, \tag{2.14}$$

where n_F, n_C, and n_C are the prism material refractive index at the characteristic wavelengths $\lambda_F = 486.1\,\text{nm}$, $\lambda_C = 656.3\,\text{nm}$, and $\lambda_D = 589.2\,\text{nm}$. The wavelengths λ_F and λ_C correspond to *dark lines* due to the absorption by hydrogen atoms, and the λ_D corresponds to the dark line due to the absorption of sodium (in the sun's outer atmosphere) [20]. This measure quantifies the prism's ability to separate different wavelengths, with a higher Δ_w indicating greater dispersion. The inverse of the dispersive power for an optical material, including those used in prisms, is defined as the *Abbe number*, V. Specifically, V is given by the formula:

$$V = \frac{1}{\Delta_w}, \tag{2.15}$$

where Δ_w can be either Δ_e or Δ_a. In practical applications, the refractive index n_0 is typically assumed to be 1, corresponding to air. The Abbe number is used to characterize the dispersive properties of an optical material. Specifically, high dispersion flint glasses have $V < 55$ while lower dispersion crown glasses have $V > 55$. Simple examples of the dispersive properties of an equilateral dispersive prism are shown in Fig. 2.9(a) for BK7 prism material, and in Fig. 2.9(b) for a high dispersion N-SF11 glass. The dispersive properties of the above optical materials can be described using the Sellmeier equation (see Appendix 2A and Table 2A.1). For both cases the angle of incidence is $\alpha_1 = 70°$. The corresponding dispersive powers for the BK7 prism are $\Delta_e = 0.0184$ (Abbe number $V_e = 1/\Delta_e = 54.48$) and $\Delta_a = 0.0156$ (Abbe number $V_a = 1/\Delta_a = 64.13$). The corresponding dispersive powers for the N-SF11 prism are $\Delta_e = 0.0594$ (Abbe number $V_e = 1/\Delta_e = 16.83$) and $\Delta_a = 0.0390$ (Abbe number $V_a = 1/\Delta_a = 25.66$). In broader applications for optical materials, the Abbe number is customarily represented by V_a, as deviation angles are generally specific to prism configurations.

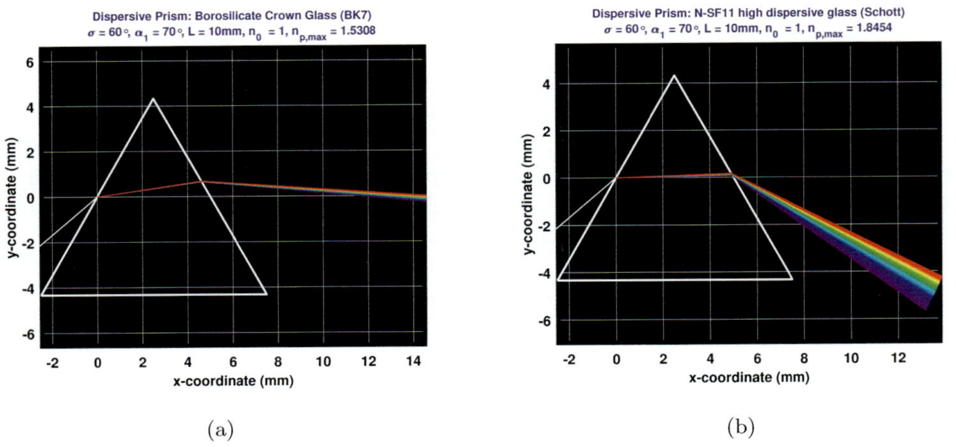

FIGURE 2.9: (a) The dispersive power of an equilateral prism of borosilicate crown glass (BK7) in the visible wavelength range. The angle of incidence is $\alpha_1 = 70^\circ$ for all wavelengths. This prism has low dispersive power (its Abbe number is $V_a = 64.13$). (b) The dispersive power of an equilateral prism of high dispersive glass (N-SF11) in the visible wavelength range. The angle of incidence is $\alpha_1 = 70^\circ$ for all wavelengths. This prism has much larger dispersive power than its BK7 counterpart (its Abbe number is $V_a = 25.66$). For both (a) and (b) the input rays are represented with white color (visible sunlight). However, the refracted parts of the rays are represented with a color analogous to their wavelength. It is evident that the purple wavelengths are deviated more due to the normal dispersion of the glasses of the prism.

2.4.2 Reflective Prisms

Reflective prisms are widely used in optical systems due to their ability to modify light paths effectively. Typically fabricated from transparent materials with highly polished surfaces, reflective prisms are employed for several purposes, including [30]: (a) bending light and changing its direction around obstacles; (b) Folding the light path for more compact system designs; (c) inverting a formed image in order to have the proper orientation; (d) combining or splitting optical beams; and (e) applying anamorphic transformations to shape the beam differently along various axes. These tasks rely on carefully designed geometrical configurations that utilize reflections and refractions. Reflections are achieved either through thin metal layers applied to some surfaces or by utilizing total internal reflection within the prism. In the following section some simple rules are presented in order to obtain a rudimentary set of tools for analyzing and designing optical systems involving reflective mostly prisms.

2.4.2.1 Matrix Method for Rays Reflected by Multiple Mirror-Surfaces

Initially, a matrix method is presented for following a ray's path through many reflecting surfaces (mirror-like). The method is based on the vectorial reflection and refraction approach presented in Appendix 2B. Assume a planar mirror at any position in space and xyz a coordinate system as shown in Fig. 2.10. A ray, with its direction represented by the unit vector $\hat{\imath}$, is incident on the reflective side of the planar mirror. The direction of the reflected ray is represented by the unit vector $\hat{r}$. The direction normal to the mirror surface is denoted by the unit vector $\hat{n}$. According to the Appendix 2B and Eq. (2B.1) the unit vectors shown are related by the expression $\hat{r} = \hat{\imath} - 2(\hat{n} \cdot \hat{\imath})\hat{n}$. Now let's introduce a matrix representation of the last equation. Assume that the incident and the reflected unit vectors $\hat{\imath}$ and $\hat{r}$ can be represented by the column matrices $\tilde{\imath}^T = [i_x, i_y, i_z]^T$ and $\tilde{r}^T = [r_x, r_y, r_z]^T$,

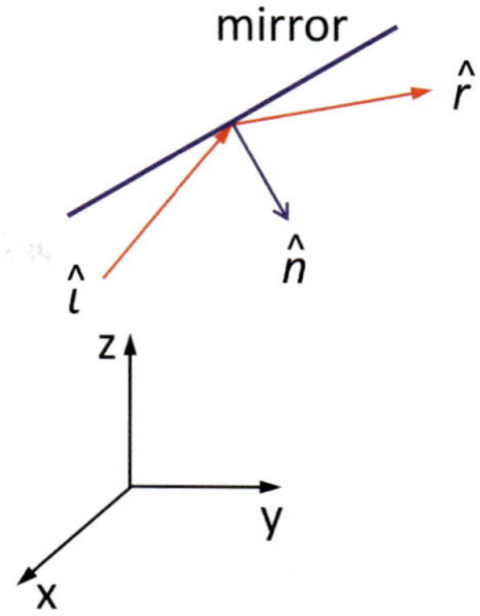

FIGURE 2.10: Assume a planar mirror and a Cartesian coordinate system. A ray, with its direction represented by the unit vector $\hat{\imath}$, is incident on the reflective side of the planar mirror. The direction of the reflected ray is represented by the unit vector $\hat{r}$. The direction normal to the mirror surface is denoted by the unit vector $\hat{n}$.

respectively (T represents the transpose matrix). Similarly, the normal to the mirror unit vector $\hat{n}$ can also be represented as $\tilde{n}^T = [n_x, n_y, n_z]^T$. Then Eq. (2B.1) can be written in the following matrix form [30–33]:

$$\begin{bmatrix} r_x \\ r_y \\ r_z \end{bmatrix} = \begin{bmatrix} i_x \\ i_y \\ i_z \end{bmatrix} - 2(n_x i_x + n_y i_y + n_z i_z) \begin{bmatrix} n_x \\ n_y \\ n_z \end{bmatrix}$$

$$= \underbrace{\begin{bmatrix} 1-2n_x^2 & -2n_x n_y & -2n_x n_z \\ -2n_x n_y & 1-2n_y^2 & -2n_y n_z \\ -2n_x n_z & -2n_y n_z & 1-2n_z^2 \end{bmatrix}}_{=\tilde{M}} \begin{bmatrix} i_x \\ i_y \\ i_z \end{bmatrix}$$

$$\tilde{r} = \tilde{M}\ \tilde{\imath} = \left[\tilde{I} - 2\tilde{n}^T\tilde{n}\right]\ \tilde{\imath}. \tag{2.16}$$

In the previous equation $\tilde{I}$ represents the unit matrix. It can be shown that the determinant of the matrix $\tilde{M}$ is -1. The sign of -1 represents that an image via a single mirror is converted from right-hand to left-hand image and vice-versa. This result can be easily generalized for any number of mirrors by sequential application of the result of Eq. (2.16). Let $\tilde{\imath}$ represent the incident ray to a series of m mirrors (counted from 1 to m sequentially). Furthermore, let $\tilde{r}_i$ represent the ray reflected by mirror i. Then the following equation can be written [30, 31]:

$$\tilde{r}_m = \tilde{M}_m\ \tilde{r}_{m-1} = \tilde{M}_m \tilde{M}_{m-1}\ \tilde{r}_{m-2} = \cdots = \tilde{M}_m \tilde{M}_{m-1} \cdots \tilde{M}_1\ \tilde{\imath} \implies$$

$$\tilde{r}_m = \left[\prod_{i=1}^{m} \tilde{M}_{m-i+1}\right]\ \tilde{\imath}\ =\ \tilde{M}_{total}\ \tilde{\imath}. \tag{2.17}$$

Since the $\tilde{M}_{total}$ matrix is a product of individual reflection matrices of the form of Eq. (2.16), its determinant follows the pattern $det[\tilde{M}_{total}] = (-1)^m$, where m is the number of reflections. Therefore if an image of an object is obtained via an even number of reflections the determinant is $+1$ and the object-to-image parity is conserved. This means that a right-handed object will remain right-handed in the image, and a left-handed object will remain left-handed. In the case that the number of reflections is odd, i.e., the determinant is -1, then the object-to-image parity is reversed. In this case, a right-handed object will appear left-handed in the image, and a left-handed object will appear right-handed.

The conversion of a right-handed object to a left-handed image via a single reflection is demonstrated in Fig. 2.11(a) [30, 31]. The linear object A_oB_o from being right-handed (B_o is to the right of A_o as it is seen by the observer) after reflection from the mirror, its image A_iB_i is left-handed (B_i is to the left of A_i as it is seen by the same observer). This illustrates the fundamental rule that both the object and its image are seen from the same side. For objects containing readable characters, like the letter **R** shown in Fig. 2.11(b), a single reflection (or any odd number of reflections) causes the image to become left-handed, rendering it non-readable. In this case, the image of **R** is mirrored and cannot be correctly read from any orientation.

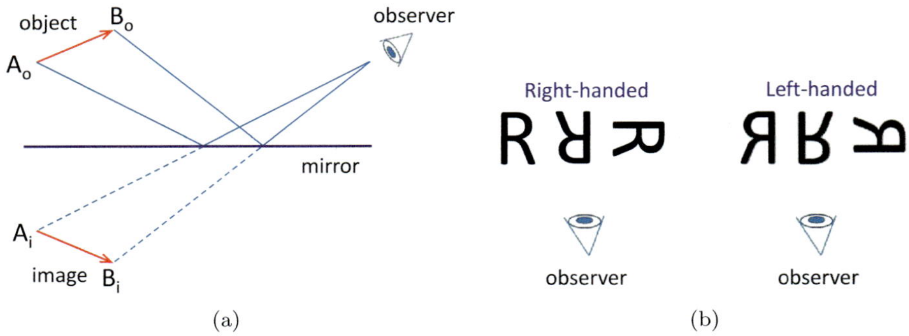

FIGURE 2.11: (a) The image of a simple linear object A_oB_o via a planar mirror changes its parity and becomes left-handed. (b)The right-handed letter **R** becomes left-handed after a single reflection independent of its original orientation.

2.4.2.2 Ray Tracing Through Multiple Mirror-Surfaces

In some cases, determining the orientation of an image, in addition to its parity (whether it is right- or left-handed), is necessary. The basic setup for this analysis is shown in Fig. 2.12. Let point $P(x_o, y_o, z_o)$ an object point and PM the incident ray extending from P to the mirror. The ray is reflected at point $M(x_m, x_m, z_m)$ located at the mirror surface. Here, the

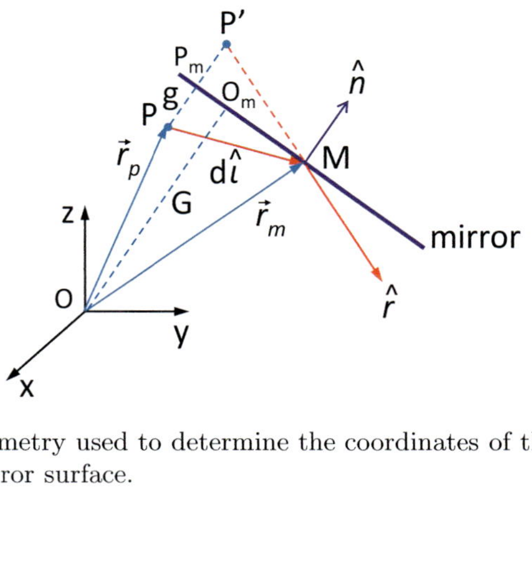

FIGURE 2.12: The geometry used to determine the coordinates of the intersection point M of the incident ray to the mirror surface.

distance from the origin to the mirror is denoted by $OO_m = G$ and the distance of the point P from the mirror is denoted by $PP_m = g$. The point $P'(x_i, y_i, z_i)$ represents the image of P after reflection in the mirror. Initially, the relation between the coordinates of points M and P will be determined. From Fig. 2.12 the following equations can be written [30,31,33]:

$$\vec{r}_m = \vec{r}_p + d\hat{\imath} \tag{2.18}$$

$$d\,(\hat{\imath}\cdot\hat{n}) = g \;=\; G - \vec{r}_p\cdot\hat{n},$$

Now assume that $\hat{\imath} = i_x\hat{\imath}_x + i_y\hat{\imath}_y + i_z\hat{\imath}_z$ and $\hat{n} = n_x\hat{\imath}_x + n_y\hat{\imath}_y + n_z\hat{\imath}_z$. Then the previous equation can be written in a matrix form as:

$$\begin{bmatrix} 1 \\ x_m \\ y_m \\ z_m \end{bmatrix} = \begin{bmatrix} 1 & 0 & 0 & 0 \\ \dfrac{Gi_x}{(\hat{\imath}\cdot\hat{n})} & 1-\dfrac{n_x i_x}{(\hat{\imath}\cdot\hat{n})} & -\dfrac{n_y i_x}{(\hat{\imath}\cdot\hat{n})} & -\dfrac{n_z i_x}{(\hat{\imath}\cdot\hat{n})} \\ \dfrac{Gi_y}{(\hat{\imath}\cdot\hat{n})} & -\dfrac{n_x i_y}{(\hat{\imath}\cdot\hat{n})} & 1-\dfrac{n_y i_y}{(\hat{\imath}\cdot\hat{n})} & -\dfrac{n_z i_y}{(\hat{\imath}\cdot\hat{n})} \\ \dfrac{Gi_z}{(\hat{\imath}\cdot\hat{n})} & -\dfrac{n_x i_z}{(\hat{\imath}\cdot\hat{n})} & \dfrac{n_y i_z}{(\hat{\imath}\cdot\hat{n})} & 1-\dfrac{n_z i_z}{(\hat{\imath}\cdot\hat{n})} \end{bmatrix} \begin{bmatrix} 1 \\ x_o \\ y_o \\ z_o \end{bmatrix} \tag{2.19}$$

In a similar manner the relation between the coordinates of the image point P' and the initial point P can be established. From Fig. 2.12 the following equation can be written:

$$\vec{r}_p^{\,\prime} = \vec{r}_p + 2g\hat{n} = \vec{r}_p + 2\,(G - \vec{r}_p\cdot\vec{n})\,\hat{n}. \tag{2.20}$$

The last equation can be written in the following matrix form:

$$\begin{bmatrix} 1 \\ x_i \\ y_i \\ z_i \end{bmatrix} = \begin{bmatrix} 1 & 0 & 0 & 0 \\ 2Gn_x & 1-2n_x^2 & -2n_x n_y & -2n_x n_z \\ 2Gn_y & -2n_x n_y & 1-2n_y^2 & -2n_y n_z \\ 2Gn_z & -2n_x n_z & -2n_y n_z & 1-2n_z^2 \end{bmatrix} \begin{bmatrix} 1 \\ x_o \\ y_o \\ z_o \end{bmatrix}. \tag{2.21}$$

Since the image orientation is the primary concern, rather than its exact position, it is convenient to set $G = 0$, meaning the mirror is positioned at the origin of the coordinate system. In this case the last equation can be simplified as follows:

$$\begin{bmatrix} x_i \\ y_i \\ z_i \end{bmatrix} = \begin{bmatrix} 1-2n_x^2 & -2n_x n_y & -2n_x n_z \\ -2n_x n_y & 1-2n_y^2 & -2n_y n_z \\ -2n_x n_z & -2n_y n_z & 1-2n_z^2 \end{bmatrix} \begin{bmatrix} x_o \\ y_o \\ z_o \end{bmatrix} = \tilde{M} \begin{bmatrix} x_o \\ y_o \\ z_o \end{bmatrix}, \tag{2.22}$$

where Eq. (2.16) was used. The last result can be generalized even if a series of m reflections surfaces are used, and the role of $\tilde{M}$ is played by $\tilde{M}_{total}$ as shown in Eq. (2.17). Then, the relation between the P and P' coordinates becomes:

$$\begin{bmatrix} x_i \\ y_i \\ z_i \end{bmatrix} = \begin{bmatrix} a_{11} & a_{12} & a_{13} \\ a_{21} & a_{22} & a_{23} \\ a_{31} & a_{32} & a_{33} \end{bmatrix} \begin{bmatrix} x_o \\ y_o \\ z_o \end{bmatrix} = \tilde{M}_{total} \begin{bmatrix} x_o \\ y_o \\ z_o \end{bmatrix}. \tag{2.23}$$

In order to determine the orientation of the actual viewed image it is necessary to find how the object coordinate system $x_o y_o z_o = xyz$ is transformed into the image coordinate system $x_i y_i z_i$ [30, 31, 33]. Therefore the transformation between the unit vectors of $x_o y_o z_o$ $(\hat{\imath}_{ox}, \hat{\imath}_{oy}, \hat{\imath}_{oz})$ and $x_i y_i z_i$ $(\hat{\imath}_{ix}, \hat{\imath}_{iy}, \hat{\imath}_{iz})$ is sought. This is accomplished easily as follows: Now consider the point $P(x_0, y_0, z_0) = (1, 0, 0)$ in object space corresponding to $\vec{r}_P = \hat{\imath}_{ox} = \hat{\imath}_x$ (unit vector along x direction. The resulting image point is $P'(x_i, y_i, z_i) = (a_{11}, a_{21}, a_{31})$. Repeating the process for points $P(x_0, y_0, z_0) = (0, 1, 0)$ and $P(x_0, y_0, z_0) = (0, 0, 1)$ in object space the following equation can be written:

$$\begin{bmatrix} \hat{\imath}_{ix} \\ \hat{\imath}_{iy} \\ \hat{\imath}_{iz} \end{bmatrix} = \begin{bmatrix} a_{11} & a_{21} & a_{31} \\ a_{12} & a_{22} & a_{32} \\ a_{13} & a_{23} & a_{33} \end{bmatrix} \begin{bmatrix} \hat{\imath}_{ox} \\ \hat{\imath}_{oy} \\ \hat{\imath}_{oz} \end{bmatrix} = \tilde{M}_{total}^T \begin{bmatrix} \hat{\imath}_{ox} \\ \hat{\imath}_{oy} \\ \hat{\imath}_{oz} \end{bmatrix}. \tag{2.24}$$

Sometimes the observer (viewer) could be located in a different coordinate system from the one convenient for the object. The observer (viewer) $x_v y_v z_v$ $(\hat{\imath}_{vx}, \hat{\imath}_{vy}, \hat{\imath}_{vz})$ and the object coordinate $x_o y_o z_o$ $(\hat{\imath}_{ox}, \hat{\imath}_{oy}, \hat{\imath}_{oz})$ systems can be related via a linear orthonormal transformation $\tilde{A}$ and the following relations can be written:

$$\begin{bmatrix} \hat{\imath}_{ix} \\ \hat{\imath}_{iy} \\ \hat{\imath}_{iz} \end{bmatrix} = \tilde{M}_{total}^T \tilde{A}^T \begin{bmatrix} \hat{\imath}_{vx} \\ \hat{\imath}_{vy} \\ \hat{\imath}_{vz} \end{bmatrix}. \tag{2.25}$$

Example 2.4.1:

In order to apply the previous analysis assume a rectangular 45°-90°-45° reflective prism in a two-reflections configuration as shown in Fig. 2.13(b). Verify the prism operation using the previously presented analysis.

Solution: The viewer coordinate system $x_v y_v z_v$ coincides with the object coordinate system implying that the transformation matrix $\tilde{A}$ in Eq. (2.25) is a unity matrix. The coordinates of the object point A_o are $(0, y_o, 0)$. In order to find the coordinates of the A_{i1} point Eq. (2.21) is utilized. In this case $G = 0$ since the M_1 mirror passes from the origin O of the $x_v y_v z_v$ coordinate system. The unit vector $\hat{n}_1 = 0\hat{\imath}_x - (1/\sqrt{2})\hat{\imath}_y + (1/\sqrt{2})\hat{\imath}_z$. Then the coordinates of A_{i1} are:

$$\begin{bmatrix} 1 \\ x_{i1} \\ y_{i1} \\ z_{i1} \end{bmatrix} = \begin{bmatrix} 1 & 0 & 0 & 0 \\ 0 & 1 & 0 & 0 \\ 0 & 0 & 0 & 1 \\ 0 & 0 & 1 & 0 \end{bmatrix} \begin{bmatrix} 1 \\ 0 \\ y_o \\ 0 \end{bmatrix} = \begin{bmatrix} 1 \\ 0 \\ 0 \\ y_o \end{bmatrix}. \tag{2.26}$$

Now the coordinates of the final image A_i (after the two reflections on mirrors M_1 and M_2) can be determined. In this case point A_{i1} plays the role of the object point. The distance $G = a + b$ (see Fig. 2.13(a)) and $\hat{n}_2 = 0\hat{\imath}_x + (1/\sqrt{2})\hat{\imath}_y + (1/\sqrt{2})\hat{\imath}_z$. Then the coordinates of A_i are:

$$\begin{bmatrix} 1 \\ x_i \\ y_i \\ z_i \end{bmatrix} = \begin{bmatrix} 1 & 0 & 0 & 0 \\ 0 & 1 & 0 & 0 \\ 2(a+b)/\sqrt{2} & 0 & 0 & -1 \\ 2(a+b)/\sqrt{2} & 0 & -1 & 0 \end{bmatrix} \begin{bmatrix} 1 \\ 0 \\ 0 \\ y_o \end{bmatrix} = \begin{bmatrix} 1 \\ 0 \\ 2(a+b)/\sqrt{2} - y_o \\ 2(a+b)/\sqrt{2} \end{bmatrix}. \tag{2.27}$$

Obviously, the same procedure can be followed for point B_o which has coordinates $(0, y_o + L, 0)$ (assuming that the vector object has a length of L. Then it is straightforward to verify

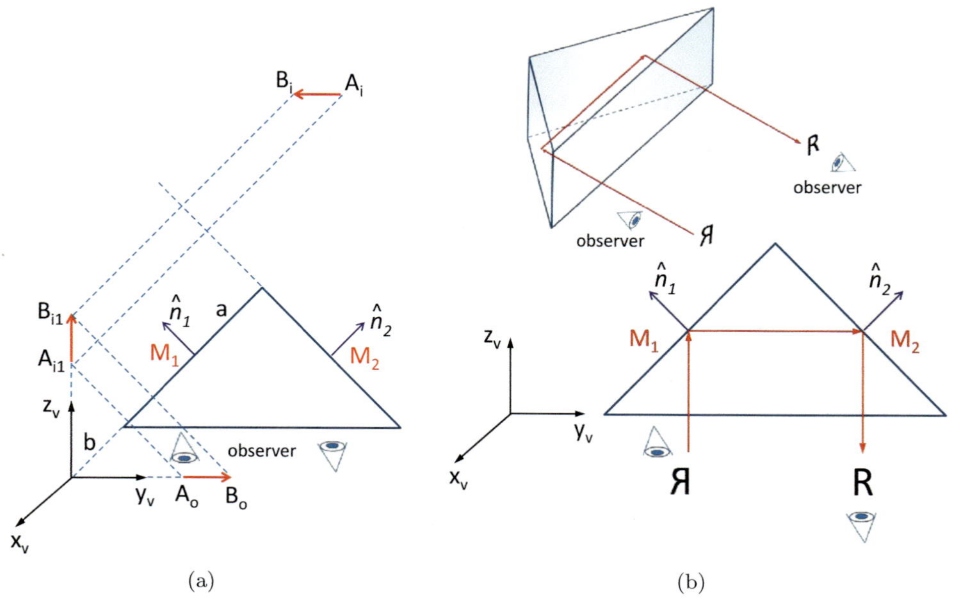

FIGURE 2.13: A rectangular 45°-90°-45° reflective prism in two-reflection configuration. The object and the image should be viewed with the light coming towards the observer from either the object or the image. The equal sides of the prism have length a and its lower left corner is at a distance b from the origin. (a) When coordinates of the image are sought. (b) When orientation of the image is sought.

the positions of the various points shown in Fig. 2.13(a). Of course, from the point-by-point approach, the final image remains right-handed as the initial object is (see the observers positions shown).

Alternatively, if only the orientation of the image is required direct application of Eq. (2.25) can be utilized. The mirror matrices $\tilde{M}_1$, $\tilde{M}_2$ and the overall matrix $\tilde{M}_{total}$ can be straightforwardly determined as follows:

$$\tilde{M}_1 = \left[\tilde{I} - 2\tilde{n}_1^T\tilde{n}_1\right] = \begin{bmatrix} 1 & 0 & 0 \\ 0 & 0 & 1 \\ 0 & 1 & 0 \end{bmatrix}$$

$$\tilde{M}_2 = \left[\tilde{I} - 2\tilde{n}_2^T\tilde{n}_2\right] = \begin{bmatrix} 1 & 0 & 0 \\ 0 & 0 & -1 \\ 0 & -1 & 0 \end{bmatrix}$$

$$\tilde{M}_{total} = \tilde{M}_2\tilde{M}_1 = \begin{bmatrix} 1 & 0 & 0 \\ 0 & -1 & 0 \\ 0 & 0 & -1 \end{bmatrix} = \tilde{M}_{total}^T,$$

where $\tilde{I}$ is the unit matrix. Then from Eq. (2.25) the unit vectors of the final image coordinate system are related to the unit vectors of the viewer's coordinate system as follows: $\hat{\imath}_{xi} = \hat{\imath}_{xv}$, $\hat{\imath}_{yi} = -\hat{\imath}_{yv}$, and $\hat{\imath}_{zi} = -\hat{\imath}_{zv}$, which verify the image orientation of the letter R

shown in Fig. 2.13(b). Furthermore, $det(\tilde{M}_{total}) = 1$ which retains the parity of the object to the image.

2.4.2.3 Some Common Reflective-Refractive Prisms

- **Rectangular Prism—60/45 degrees angle**: Right-angle reflecting prisms are designed to deflect an incident ray of light by 90 degrees or 180 degrees. Two examples of such prisms are illustrated in Figs. 2.14a and 2.14b. The 45-degree rectangular prism, commonly referred to as the Porro prism, is named after its inventor, Ignazio Porro (1801-1875). These prisms are frequently utilized in various optical systems, including telescopes and periscopes, due to their ability to invert and reflect images while maintaining a compact design. In applications like telescopes, the Porro prism enables a more straightforward and compact arrangement of optics, improving the overall viewing experience. In periscopes, these prisms allow light to be redirected around obstacles, making them valuable in both military and civilian contexts.

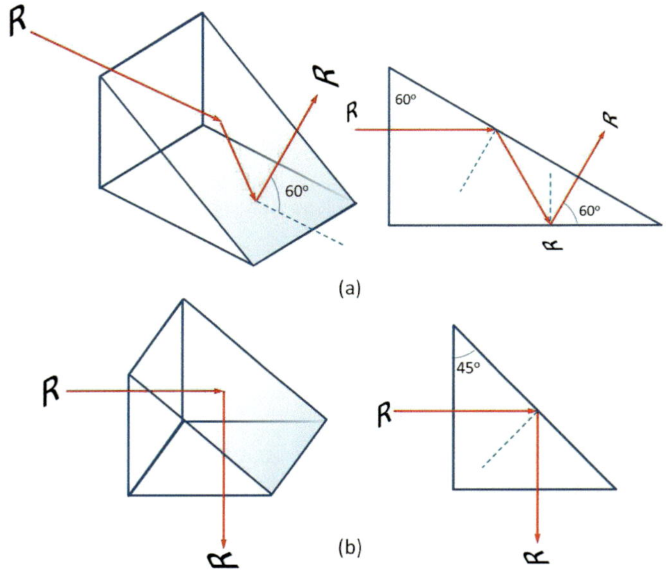

FIGURE 2.14: Rectangular reflective prisms: (a) with a 60-degrees angle; (b) with a 45-degrees angle.

- **Amici-Roof Prism**: Amici prisms, named after their inventor, the Italian astronomer Giovanni Battista Amici (1786-1863), are also referred to as "roof prisms" or "right-angle roof prisms." These prisms are capable of reverting and inverting the image of an object while bending the line of sight through a 90-degree angle. Due to their unique optical properties, Amici prisms are commonly used in various applications, including terrestrial telescopes, viewing systems, and rangefinders. They provide a clear and correctly oriented image, making them particularly useful for observing distant objects. Example prisms can be found in Figs. 2.15a and 2.15b.
- **Penta Prism**: A Penta prism is a five-sided reflecting prism designed to deviate a beam of light by a constant 90 degrees, regardless of whether the incident beam approaches the prism at a right angle. The reflections in a Penta prism are facilitated by thin reflective metallic layers deposited on the reflecting surfaces, rather than relying on total internal

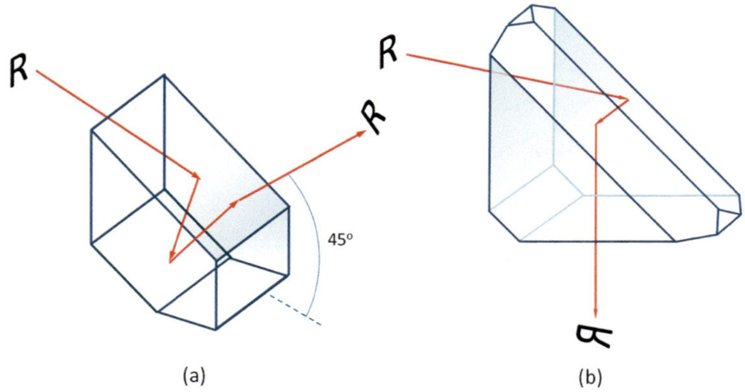

FIGURE 2.15: Amici-Roof reflective prisms with various orientations: (a) deflection of 45 degrees; (b) deflection of 90 degrees.

reflection. To minimize unwanted reflections, antireflection coatings can be applied to the prism's surfaces. One of the notable features of the Penta prism is that the 90-degree deviation angle remains constant, even if the prism is rotated around an axis parallel to the line of intersection of its two reflecting faces. Penta prisms are commonly utilized in various applications, including surveying, alignment, range finding, and other optical systems. An example Penta prism is shown in Fig. 2.16.

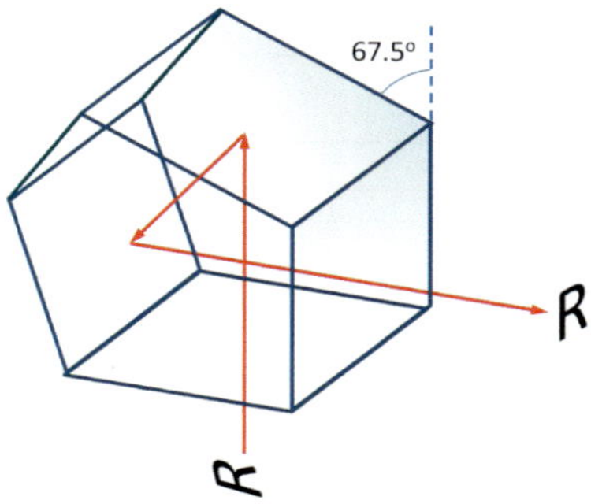

FIGURE 2.16: A Penta reflective prism.

- **Dove (reflecting-refracting) Prism**: Dove prisms serve primarily two applications in optical systems: (a) Image Rotation: Dove prisms are often used as rotators to rotate an image without deviating the beam. When the prism is rotated about the axis of the incident parallel (collimated) ray, the image undergoes a rotation that is twice the angle of the prism's rotation. This property makes Dove prisms particularly useful in applications requiring precise control of image orientation. It is crucial that these prisms are employed with parallel or collimated beams, and their large reflective surfaces must be kept very clean to ensure optimal performance. (b) Retroreflector: In this role, the Dove prism functions similarly to a right-angle prism. It reflects light back toward its source, regardless of the angle of incidence, making it useful in applications requiring feedback of the light signal. Example Dove prisms are shown in Figs. 2.17.
- **Rhomboid Prism**: The Rhomboid prism is designed to displace the line of sight without introducing any angular deviation or altering the orientation of the image. Its unique

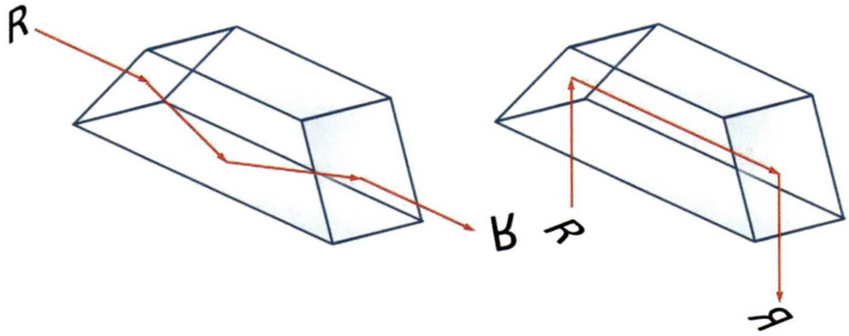

FIGURE 2.17: Dove reflective prisms.

shape and reflective properties enable it to redirect light while maintaining the original path and image characteristics. This capability makes the Rhomboid prism useful in various optical applications, such as: (a) Beam Displacement: It can shift the position of a beam in an optical system without affecting its angle of incidence or the characteristics of the image. (b) Optical Systems: Rhomboid prisms are often utilized in applications where precise alignment of optical paths is essential, such as in laser systems, binoculars, and surveying instruments. An example rhomboid prism is shown in Figs. 2.18.

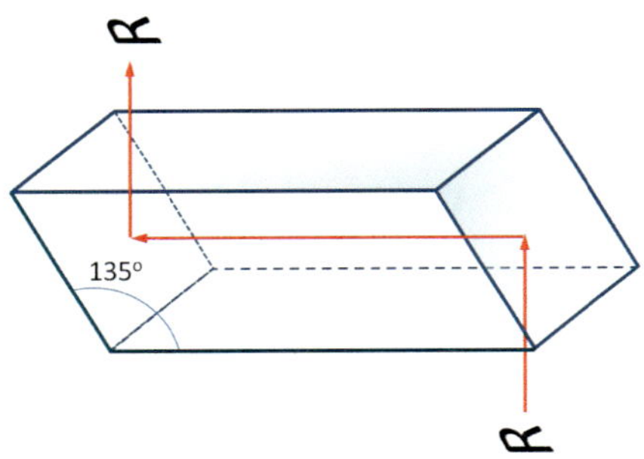

FIGURE 2.18: A Rhomboid reflective prism.

- **Corner-Cube Prism**: The corner-cube prism, also known as a retroreflector, features three mutually perpendicular surfaces that facilitate a unique light reflection process. When light enters through the hypotenuse face, it is reflected on each of the three surfaces sequentially, ultimately emerging through the hypotenuse face parallel to the incident beam. This remarkable property allows it to maintain the same orientation of the incoming beam, regardless of its angle of incidence. For its special performance, it is often used to distance measurement, optical signal processing, and laser interferometer. Corner-cube prisms are used in spectroscopy, satellites, surveying, laser resonator cavities, electronic distance measurement, as well as space applications. An example corner-cube prism is shown in Figs. 2.19.
- **Anamorphic Set of Prisms**: Anamorphic prisms consist of two prisms that work together to modify the shape of an incident light beam. By altering the angles between the incident beam and the two prisms, it is possible to expand or contract the beam in one direction while leaving the other direction unchanged. This property makes anamorphic prisms particularly useful in various optical applications such as beam shaping,

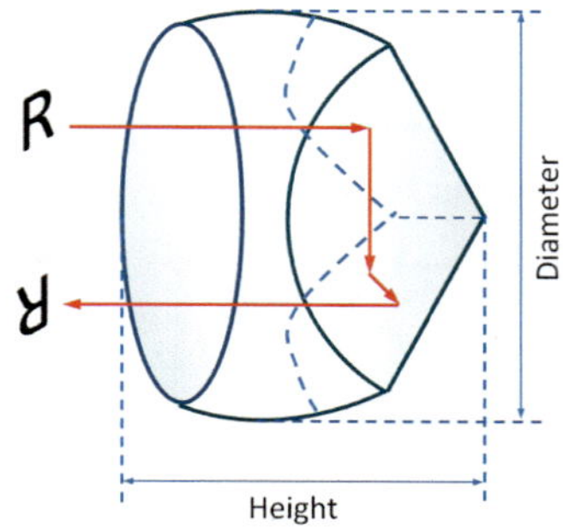

FIGURE 2.19: A Corner-Cube reflective prism.

laser beam expansion, and specilized imaging systems in microscopy, and endoscopy. It is very easy to turn elliptical bean into circular beam. An example of an anamorphic prism is shown in Fig. 2.20(a).

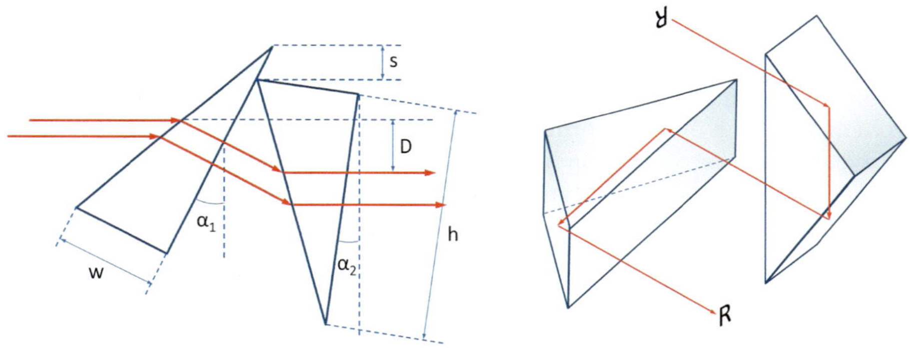

(a) (b)

FIGURE 2.20: (a) An Anamorphic reflective prism. (b) An Erecting reflective prism.

- **Erecting Prism**: Erecting prisms are optical elements designed to invert an image that has been flipped or rotated during its passage through an optical system, returning it to an upright position. A common example of an erecting prism is the Porro prism, which is frequently utilized in binoculars and telescopes. By incorporating erecting prisms into optical designs, manufacturers can create systems that are more intuitive and functional for users across various fields, including astronomy, nature observation, and photography. An example of a two-Porro prism erecting configuration is shown in Fig. 2.20(b). An application in binoculars is shown in Fig. 2.21.

2.5 Imaging Fundamentals

An optical imaging system collects light from an object which scatters the incident light or emits light by itself. The collected light from the object via the optical system is concentrated on a detector (could be a screen, a film, or a digital light detector such as a CCD or CMOS array). This light concentration via the optical system forms what is called *an image* of the

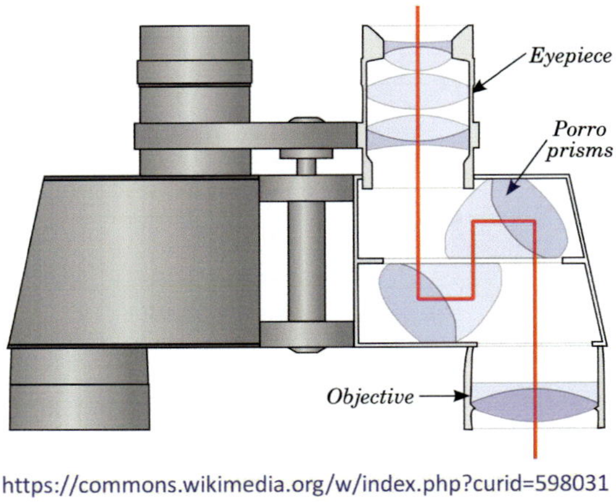

FIGURE 2.21: An erecting prism configuration in binoculars (from https://commons.wikimedia.org/w/index.php?curid=598031).

object. An imaging optical system could be comprised of a variety of refracting surfaces (lenses), reflecting surfaces (mirrors), various media (homogeneous or inhomogeneous), or other components (such as filters, aperture stops, etc.) and its purpose is to concentrate the light from an object in order to form its image. Examples of imaging systems are abundant in nature and technology, with the human eye and the camera being two common ones. A generic diagram of an imaging system is shown in Fig. 2.22. A point O, of an object (at the object plane), emanates electromagnetic radiation (light which is either scattered/reflected or even generated by the object) and part of this light is concentrated by the optical system at the image point I. In this figure the image point I is a real image since actual rays are passing through it and therefore electromagnetic energy arrives at point I coming from point O. However, there are cases in which the point I is formed by ray extensions since the rays after passing from the optical system could diverge. In the latter case point I is called a *virtual image* of point O and of course there is not any electromagnetic energy concentrated at that point. An example, of a virtual image is the image of an object in front of a flat mirror. The image appears to be behind the flat mirror however, no electromagnetic energy is behind the mirror! In an imaging system, the points O and I located in the object and image planes, respectively, are termed *conjugate points* [20]. Each point on the object plane has a corresponding conjugate point on the image plane, ensuring that the spatial relationships of the object are preserved in the image. For precise image formation, it is crucial that all rays emanating from an object point O at a specific time reach the image

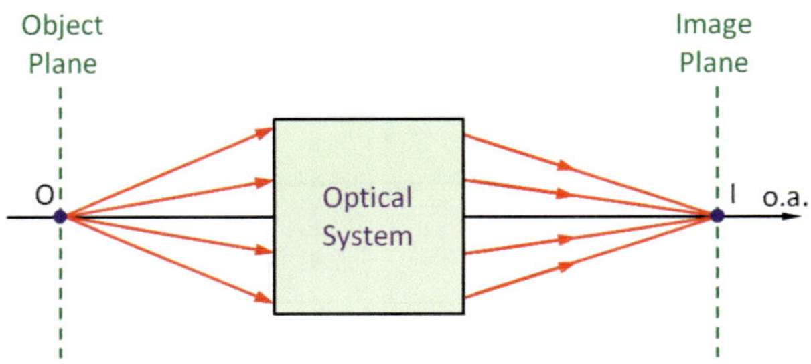

FIGURE 2.22: Imaging through an optical system. The point O is a point of the object plane and the point I corresponds to its image in the image plane.

point I simultaneously. If the rays arrive at varying times, random phase shifts can occur, resulting in destructive interference that causes blurring or even prevents image formation. Thus, to achieve sharp imaging, these rays must arrive at I isochronously, i.e., at the same time [20], ensuring their paths satisfy Fermat's principle (as it was presented in Sec. 2.3) and that their optical path lengths are stationary. However, perfect image formation cannot be achieved in practice for three main reasons: (a) Scattering, partial absorption, and reflection at optical surfaces can prevent some rays from reaching the image point. Additionally, only a portion of the light from the object is captured by the optical system. (b) All optical systems inherently suffer from *optical aberrations*, which reduce image quality. Chromatic aberration, for instance, arises from wavelength-dependent variations in optical properties due to material dispersion. Even in monochromatic conditions, geometrical considerations alone introduce inherent aberrations. These will be discussed further in Sec. 2.9. (c) Since light is electromagnetic radiation is prone to diffraction effects which limit the resolution of the optical system and contribute to the imperfection of the image formation. No optical system is free of these imperfections, and so designs are optimized to provide the best possible image quality within given operational constraints. In this chapter, only geometrical optics considerations are addressed, excluding scattering, absorption, reflection losses, and diffraction effects. The following subsections will introduce basic reflective and refractive systems that, under specific conditions, can be free of monochromatic aberrations, offering an idealized framework for elementary image formation.

2.5.1 Cartesian Reflective Surfaces

For point objects, within the framework of geometrical optics, specialized surfaces can theoretically produce perfect imaging. These surfaces are known as *Cartesian surfaces*. Examples include mirrors with ellipsoidal, hyperboloidal, and paraboloidal shapes, as illustrated in Fig. 2.23.

For the ellipsoidal mirror, the foci A and B are conjugate points (object-image relation). Assume that A is the object point and B is the image point. Every ray emanating from A passes through B. According to Fermat's principle, the optical path length OPL of any ray satisfies the following equation:

$$OPL = n(AP) + n(PB) = n\left[\sqrt{(x+d)^2+y^2} + \sqrt{(x-d)^2+y^2}\right] =$$

$$n(AC) + n(CB) = n2\left[\sqrt{b^2+d^2}\right] = n(AD) + n(DB) = n(AE) + n(EB) = n\,[2a] \Longrightarrow$$

$$\frac{x^2}{a^2} + \frac{y^2}{b^2} = 1, \qquad \text{with } b^2 + d^2 = a^2,$$

where n is the refractive index inside the ellipsoidal mirror and a and b the ellipsoidal semi-axes lengths. This is an equation for an ellipse in the xy plane. In this case it is assumed to be a two-dimensional geometry. However, the extension to three dimensions can be easily done if one considers the three-dimensional ellipsoid that is produced by the revolution of the two-dimensional one around the x or y-axis. Another example is the hyperboloid mirror shown in Fig. 2.23(b). In this case the point B is a virtual image of point A which is assumed to be the object point. Since the extensions of the rays that are reflected by the mirror pass through B a spherical wavefront of radius R is also shown (looks like the reflected rays are coming from B, which is the virtual image of A). Let D and D' the intersection of the spherical wavefront with the reflected rays [34,35]. According to Fermat's principle, the optical paths APD and $AP'D'$ should be equal. It is assumed that the optical paths are equal to K, where K depends on the choice of R. The optical paths can be written for any

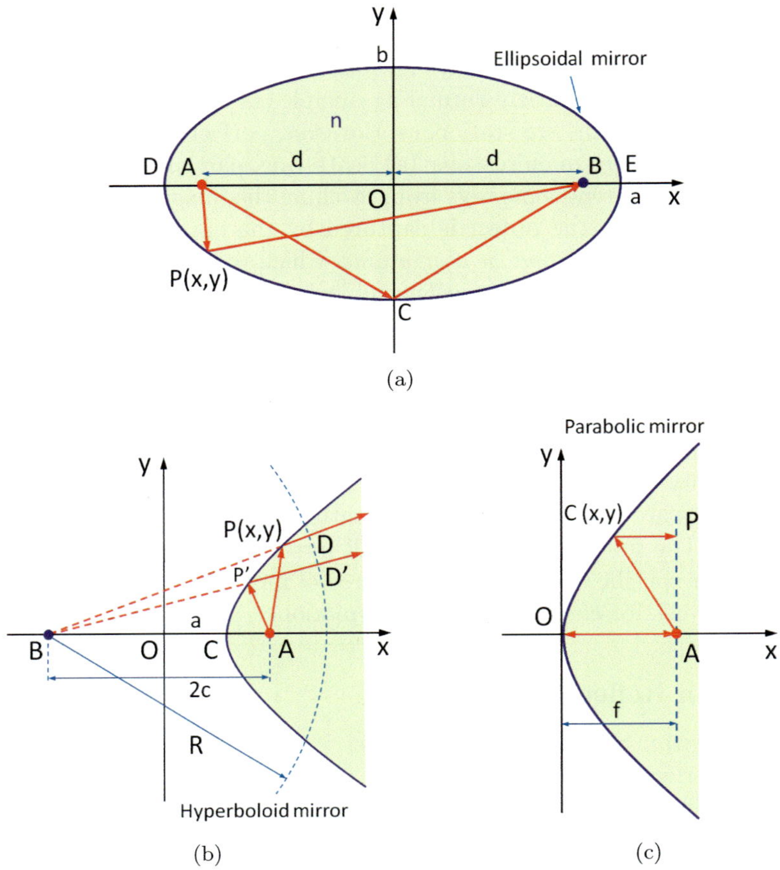

FIGURE 2.23: (a) An ellipsoid mirror. (b) A hyperboloid mirror. (c) A parabolic mirror.

two rays as follows:

$$AP + PD = AP' + P'D' = K \Longrightarrow$$
$$AP + (R - BP) = AP' + (R - BP') = K \Longrightarrow$$
$$BP - AP = BP' - AP' = R - K = K' \Longrightarrow$$
$$BP - AP = K' \Longrightarrow$$
$$\sqrt{(x+c)^2 + y^2} - \sqrt{(x-c)^2 + y^2} = K' = CB - CA = (c+a) - (c-a) = 2a \Longrightarrow$$
$$\frac{x^2}{a^2} - \frac{y^2}{b^2} = 1, \qquad \text{with} \quad c^2 = a^2 + b^2,$$

where K' is a constant equal to $2a$ and a is the distance of the vertex C of the hyperboloid from the origin $[a = (OC)]$. This is an equation of a hyperbola in the xy plane. As with the case of the ellipsoid it is assumed to be a two-dimensional geometry. However, the extension to three dimensions can be easily done if one considers the three-dimensional hyperboloid that is produced by the revolution of the two-dimensional hyperboloid around the x-axis.

Another special case is the parabolic mirror as shown in Fig. 2.23(c). In this case point A represents again the point object and the image (which is assumed as previously to be point B) is at infinity. An example ray, $(AC) + (CP)$, is also shown. According to Fermat's principle, for perfect imaging, all rays reflected by the mirror should intersect the axis at point A with equal optical path lengths. This condition ensures that rays reflected from different points on the parabolic mirror surface will travel the same total distance from A to infinity, forming a collimated beam. Mathematically, this requirement can be expressed by setting up an equation for the optical path length of any ray, $(AC) + (CP)$, such that it remains constant for all points C on the mirror. Then, the following should be true:

$$\begin{aligned} AC + CP = 2OA = 2f &\Longrightarrow \\ \sqrt{(x-f)^2 + y^2} + (f - x) = 2f &\Longrightarrow \\ y^2 - 4fx = 0, & \end{aligned}$$

which of course satisfies an equation of a parabola in the xy plane. Again it can be extended into the three dimensions by revolution of the parabola around the x-axis.

Under geometrical optics considerations, the points B represent perfect images of the corresponding object points A for all the reflective mirror types discussed (ellipsoidal, hyperboloidal, and paraboloidal). However, this perfect imaging condition holds only for point objects precisely located at the focal point of the mirror. Real objects, having finite size, consist of many points that may not coincide exactly with the mirror's focal point, resulting in imperfect imaging for points outside this ideal position. Consequently, such specialized mirrors are limited to applications where precise imaging of a point source is required. In general optical systems, the more easily manufactured spherical mirrors are favored, especially under paraxial approximation conditions. In this regime, spherical mirrors can closely approximate the ideal image formation characteristics of their specialized counterparts, making them practical for a wider range of imaging applications.

According to the paraxial approximation (also referred as "Gaussian Optics") all rays propagate, reflect, or refract within an optical system at small angles with respect to the optical axis (o.a., usually is the axis of symmetry of the optical system). Therefore, all trigonometric functions are approximated by the first terms of their corresponding Taylor series expansions. Specifically, $\sin\theta \simeq \theta$, $\tan\theta \simeq \theta$, and $\cos\theta \simeq 1$. The paraxial approximation significantly simplifies the analysis of optical systems by reducing non-linearities in the trigonometric functions. While the paraxial approximation streamlines the mathematical treatment of optical systems, practical design requires consideration of deviations from paraxial conditions. Such deviations introduce aberrations that affect image quality, as will be discussed in Sec. 2.9. Thus, except where aberrations are explicitly considered, the paraxial approximation is used throughout most of this chapter.

2.5.2 Spherical Reflective Surface - Paraxial Approximation

A spherical convex mirror of radius R is shown in Fig. 2.24. Let point O be an object point from which a ray OA is emitted and subsequently reflects off the spherical mirror at point A. The extension of the reflected ray, denoted by AA', intersects the optical axis at the point I which corresponds to the virtual image of O. The intersection of the spherical mirror with the optical axis (o.a.) corresponds to the vertex V of the spherical mirror. The relation between the distance of the object s and the distance of the image s' from the vertex is sought. Using the law of reflection the following equations can be written:

$$\begin{aligned} \theta_1 = \theta_2 \Longrightarrow (\alpha_1 + \phi) &= (\alpha_2 - \phi) \Longrightarrow \\ \alpha_1 - \alpha_2 = -2\phi \Longrightarrow \frac{h}{s} - \frac{h}{s'} &= -2\frac{h}{R} \Longrightarrow \\ \frac{1}{s} + \frac{1}{s'} = -\frac{2}{R}, & \end{aligned} \tag{2.28}$$

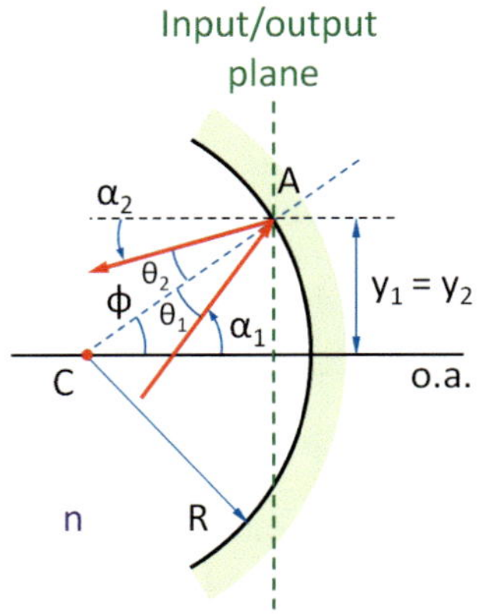

FIGURE 2.24: A spherical convex mirror of radius R. It is assumed that all angles are small with respect to the optical axis in order for the paraxial approximation to hold.

where the paraxial approximation (for all angles $\sin\theta \simeq \tan\theta \simeq \theta$) was used. In addition, due to the paraxial approximation the distance VQ was considered negligible (which is true for rays near the optical axis of the mirror). Furthermore, a sign convention was adopted. I.e., the object distance s is positive if the object is to the left of V (real object). Similarly, the image distance s' is positive when I is also to the left of V (i.e., a real image is formed). Otherwise, the distances s and s' become negative (virtual object and virtual image). In the example case shown in Fig. 2.24 $s > 0$ and $s' < 0$. Furthermore, a sign convention for the radius of curvature R is also adopted. Specifically, $R > 0$ when the center C is to the right of V (convex mirror as the one shown), while $R < 0$ if C is to the left of V (concave mirror). The focal distance of a spherical mirror, under the paraxial approximation, can be defined as $f = -R/2$, using the sign convention. A summary of the sign convention is shown in Table 2.1. The formation of an image of a transverse object is shown in Fig. 2.25(a) (for a convex mirror) and Fig. 2.25(b) (for a concave mirror). Usually, the intersection of at least two rays is needed for the graphical formation of the image. In this case a parallel ray (to the optical axis) and a chief ray (incident at the vertex V of the mirror) are selected. For the convex mirror a virtual image is formed while for the concave one, a real image is formed. From Eq. (2.28) for a real object, the convex mirror will always form a virtual image since $R > 0$ and $s > 0$ according to the sign convention and therefore $s' < 0$ for any distance of the object. In the case of the concave mirror, for the selected object distance a real image is shown. However, if $s < R$ then $s' < 0$ and a virtual image will be formed even for the concave mirror. This is straightforward to determine from Eq. (2.28) and to draw a ray diagram. From the ray diagrams, the transverse magnification m is defined as:

$$m = \frac{h'}{h} = -\frac{s'}{s}, \tag{2.29}$$

where the convention of negative magnification means that the image is inverted. For positive magnification the image is upright (i.e., in the same direction with the object). It can be observed from Fig. 2.25 that a real image is inverted while a virtual image is upright.

2.5.3 Cartesian Refractive Surfaces

Cartesian refractive surfaces can indeed be more complex than their reflective counterparts. Consider initially the refractive surfaces shown in Fig. 2.26. These surfaces for parallel

TABLE 2.1: Convention Summary for Spherical Reflective and Refractive Surfaces

Spherical Mirrors (Light from Left-to-Right)		
R	**s**	**s′**
$R > 0$	$s > 0$	$s' > 0$
(convex)	real (left of V)	real (left of V)
$R < 0$	$s < 0$	$s' < 0$
(concave)	virtual (right of V)	virtual (right of V)
Magnification	$m > 0$ (upright)	$m < 0$ (inverted)
	$m = \frac{h'}{h} = -\frac{s'}{s}$	
Paraxial Equation	$\frac{1}{s} + \frac{1}{s'} = -\frac{2}{R} = \frac{1}{f}$	
Spherical Refractive Surface (Light from Left-to-Right)		
R	**s**	**s′**
$R > 0$	$s > 0$	$s' > 0$
(convex)	real (left of V)	real (right of V)
$R < 0$	$s < 0$	$s' < 0$
(concave)	virtual (right of V)	virtual (left of V)
Magnification	$m > 0$ (upright)	$m < 0$ (inverted)
	$m = \frac{h'}{h} = -\frac{n_1}{n_2}\frac{s'}{s}$	
Paraxial Equation	$\frac{n_1}{s} + \frac{n_2}{s'} = \frac{n_2 - n_1}{R}$	

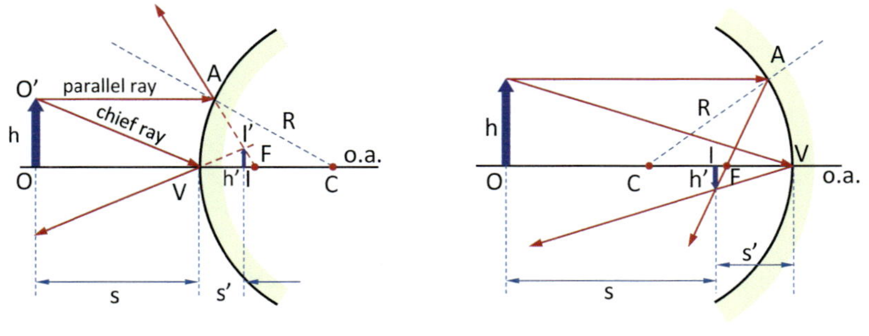

(a) (b)

FIGURE 2.25: (a) Image formation in the case of a convex spherical mirror. (b) Image formation in the case of a concave spherical mirror.

incident rays focus on the second focal point F_2. For example, for the case of $n_1 < n_2$ shown in Fig. 2.26(a) using Fermat's principle the shape of the refractive surface $y(x)$ can be determined as follows:

$$n_1(BA) + n_2(AF_2) = n_1(B'A') + n_2(A'F_2) = n_2 f \Longrightarrow$$
$$n_1 x + n_2\sqrt{(f-x)^2 + y^2} = n_2 f \Longrightarrow$$
$$\left(x - \frac{n_2}{n_1+n_2} f\right)^2 + \frac{n_2^2}{n_2^2 - n_1^2} y^2 = \left(\frac{n_2}{n_1+n_2}\right)^2 f^2, \tag{2.30}$$

where the last equation represents an ellipsoid of revolution around the x-axis which in this example is assumed to be the optical axis. Similarly, for the case of $n_1 > n_2$ shown in Fig.

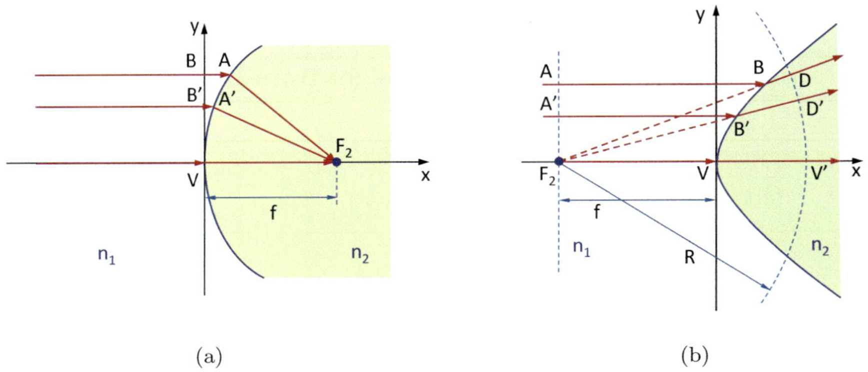

FIGURE 2.26: (a) An ellipsoidal refractive surface for the second focal point F_2 (real) in the case that $n_1 < n_2$. (b) A hyperboloidal refractive surface for the second focal point F_2 (virtual) in the case that $n_1 > n_2$.

2.26(b) using Fermat's principle the shape of the refractive surface $y(x)$ can be determined as follows:

$$n_1(AB) + n_2(BD) = n_1(A'B') + n_2(B'D') = n_1 f + n_2(VV') \Longrightarrow$$
$$n_1(AB) + n_2[R - (F_2B)] = n_1(A'B') + n_2[R - (F_2B')] = n_1 f + n_2(R - f) \Longrightarrow$$
$$n_1(AB) - n_2(F_2B) = n_1(A'B') - n_2(F_2B') = n_1 f - n_2 f \Longrightarrow$$
$$n_1(f + x) - n_2\sqrt{(f+x)^2 + y^2} = (n_1 - n_2) f \Longrightarrow$$
$$\left(x + \frac{n_2}{n_1+n_2} f\right)^2 - \frac{n_2^2}{n_1^2 - n_2^2} y^2 = \left(\frac{n_2}{n_1+n_2}\right)^2 f^2, \tag{2.31}$$

where the last equation represents a hyperboloid of revolution around the x-axis which is also the optical axis. Example surfaces along with the incident and the refracted rays are shown in Figs. 2.27(a) for $n_1 = 1$, $n_2 = 1.5$ and $f = 10\,\text{cm}$ (real F_2 case - ellipsoid) and 2.27(b) for $n_1 = 1.5$, $n_2 = 1$ and $f = 10\,\text{cm}$ (virtual F_2 case - hyperboloid). Similar refractive surfaces can be found for the F_1 focal point case (the focal points F_1 and F_2 are defined at the spherical refraction and the thin lens definitions later in this chapter). For $n_1 > n_2$ and a real F_1 point (to the left of the refractive surface), the refractive surface equation is (ellipsoid):

$$\left(x + \frac{n_1}{n_1+n_2} f\right)^2 + \frac{n_1^2}{n_1^2 - n_2^2} y^2 = \left(\frac{n_1}{n_1+n_2}\right)^2 f^2, \tag{2.32}$$

while for $n_1 < n_2$ and a virtual F_1 point (to the right of the refractive surface) the refractive surface equation is (hyperboloid):

$$\left(x - \frac{n_1}{n_1 + n_2} f\right)^2 - \frac{n_1^2}{n_2^2 - n_1^2} y^2 = \left(\frac{n_1}{n_1 + n_2}\right)^2 f^2. \tag{2.33}$$

Examples of such Cartesian refractive surfaces are shown in Fig. 2.28(a) (real F_1 case - ellipsoid) and Fig. 2.28(b) (virtual F_1 case - hyperboloid), respectively. Of course, these surfaces could be combined to form aspheric lenses.

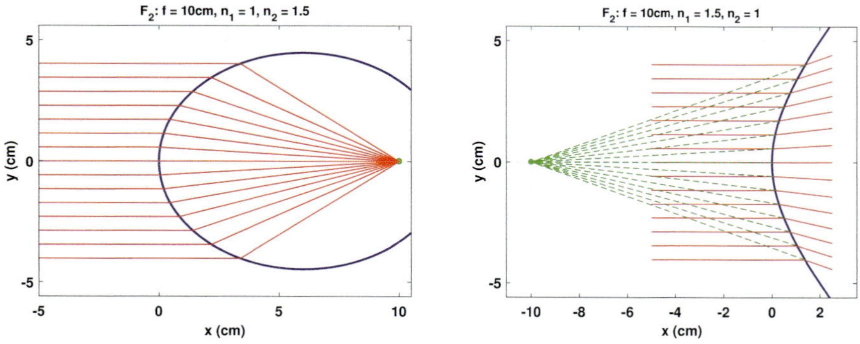

(a) (b)

FIGURE 2.27: (a) An example of an ellipsoidal refractive surface for the second focal point F_2 (real) in the case that $n_1 = 1$, $n_2 = 1.5$, and $f = 10\,cm$. (b) A hyperboloidal refractive surface for the second focal point F_2 (virtual) in the case that $n_1 = 1.5$, $n_2 = 1$ and $f = 10\,cm$.

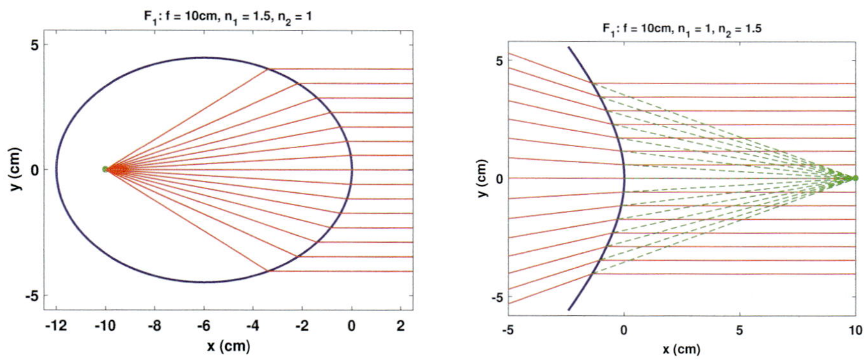

(a) (b)

FIGURE 2.28: (a) An example of an ellipsoidal refractive surface for the first focal point F_1 (real) in the case that $n_1 = 1.5$, $n_2 = 1$, and $f = 10\,cm$. (b) A hyperboloidal refractive surface for the first focal point F_1 (virtual) in the case that $n_1 = 1$, $n_2 = 1.5$ and $f = 10\,cm$.

2.5.4 Cartesian Ovoids

Cartesian refractive surfaces, similar to the previously defined Cartesian refractors for the focal points F_1 and F_2, can be defined when the position of an object distance s_o and an image distance s_i are desired. The resulting refractive surfaces are called *Cartesian ovoids* and have been described initially by R. Descartes. A general case is shown in Fig. 2.29(a) for

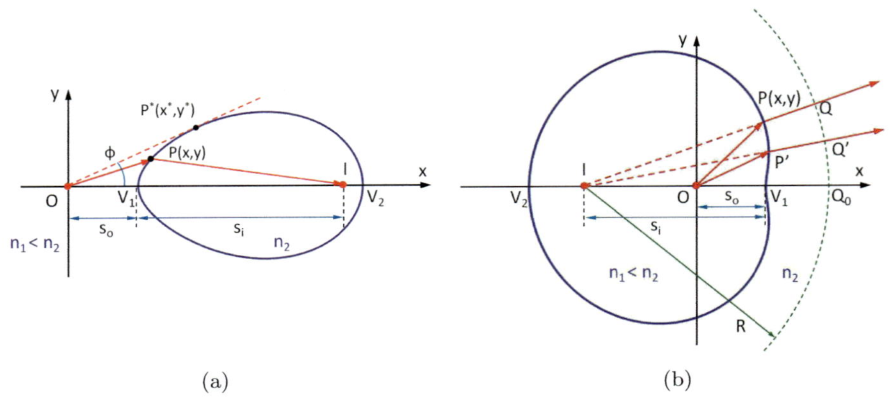

FIGURE 2.29: (a) A general Cartesian ovoid for a real image and $n_1 < n_2$. (b) A general Cartesian ovoid for a virtual image and $n_1 < n_2$.

a real image, and Fig. 2.29(b), for a virtual image, respectively. Both figures correspond to the case of $n_1 < n_2$. The equations specifying the refractive surface can be easily determined using Fermat's principle as it was done in previous sections. For example, for the case of a real image, the following equations can be written:

$$n_1(OP) + n_2(PI) = n_1 s_o + n_2 s_i \Longrightarrow$$

$$n_1\sqrt{x^2 + y^2} + n_2\sqrt{(s_o + s_i - x)^2 + y^2} = n_1 s_o + n_2 s_i, \tag{2.34}$$

where it is assumed that the object is located at the beginning of the xy coordinate system. The vertex points V_1 and V_2 are the points of the ovoid for $y = 0$. These points can be determined by setting $y = 0$ in Eq. (2.34) which results in the following:

$$n_1|x| + n_2|s_o + s_i - x| = n_1 s_o + n_2 s_i,$$

which has the solutions

$$x = s_o, \qquad \text{for} \quad 0 < x < s_o + s_i \tag{2.35}$$

$$x = s_o + \frac{2n_2}{n_1 + n_2} s_i, \qquad \text{for} \quad s_o + s_i < x < \infty. \tag{2.36}$$

Equation (2.35) shows the position of V_1, and Eq. (2.36) shows the position of V_2. An interesting question for this Cartesian ovoid is to find the maximum angle at that a ray from object O can be incident on the refractive surface. This angle can be defined from the tangent from O to the ovoid's surface. Let the tangent have an angle ϕ with respect to the x-axis. If the tangent touches the ovoid at point (x^*, y^*) then $\tan\phi = \alpha = y^*/x^*$. Inserting this relation into Eq. (2.34) the following equation is derived (setting $s_0 + s_i = D$):

$$\sqrt{x^2 + \alpha^2 x^2} + \frac{n_i}{n_0}\sqrt{(D - x)^2 + \alpha^2 x^2} = \frac{n_1 s_0 + n_2 s_i}{n_1} = A. \tag{2.37}$$

Re-arranging the above equation and squaring both sides it is straightforward to derive the following second-order polynomial equation with respect to x (setting $\kappa = n_2/n_1$):

$$(1 + \alpha^2)(\kappa^2 - 1)x^2 + \left[2A(1 + \alpha^2)^{1/2} - 2D\kappa^2\right] x + D^2\kappa^2 - A^2 = 0. \tag{2.38}$$

The last equation should have a single solution with respect to x in order to correspond to the tangent to the ovoid surface. Therefore, the discriminant of the polynomial of Eq. (2.38) should be zero. This can be expressed from the following equation:

$$\mathscr{D} = 4\left[A(1+\alpha^2)^{1/2} - D\kappa^2\right]^2 - 4(D^2\kappa^2 - A^2)(1+\alpha^2)(\kappa^2-1) = 0.$$

Setting $p = (1+\alpha^2)^{1/2}$ the last equation is written as

$$\left[A^2 - (D^2\kappa^2 - A^2)(\kappa^2-1)\right]p^2 - \left[2AD\kappa^2\right]p + D^2\kappa^4 = 0 \Longrightarrow$$
$$a_1p^2 + a_2p + a_3 = 0 \Longrightarrow$$

$$p = (1+\alpha^2)^{1/2} = \frac{-a_2 \pm \sqrt{a_2^2 - 4a_1a_3}}{2a_1} \Longrightarrow \alpha = \pm\sqrt{p^2-1},$$

where the root for $|\alpha| < 1$ is the one that corresponds to the solution of interest. Then the intersection point is given by

$$x^* = -\frac{2A(1+\alpha^2)^{1/2} - 2D\kappa^2}{(1+\alpha^2)(\kappa^2-1)}, \qquad y^* = \pm\alpha x^*.$$

At the point (x^*, y^*) the ray entering the medium of refractive index n_2 has an angle with respect to the normal of the ovoid surface [at (x^*, y^*)], equal to the critical angle $\theta_{cr} = \sin^{-1}(n_1/n_2)$. An example case of a Cartesian ovoid for a real image at $s_i = 10\,\text{cm}$, and object distance of $s_o = 5\,\text{cm}$ and $n_1 = 1$, $n_2 = 1.5$ is shown in Fig. 2.30(a). Some sample rays emanating from the object are illustrated as they refract through the ovoid surface and converge at the image point. These rays, upon encountering the ovoid surface, are refracted in accordance with Snell's law and guided such that they intersect precisely at the designated image point, showcasing the effectiveness of the ovoid surface in controlling light paths for specified object and image distances. Next, let's examine the case of the virtual image. The corresponding geometry is shown in Fig. 2.29(b). Assume a spherical

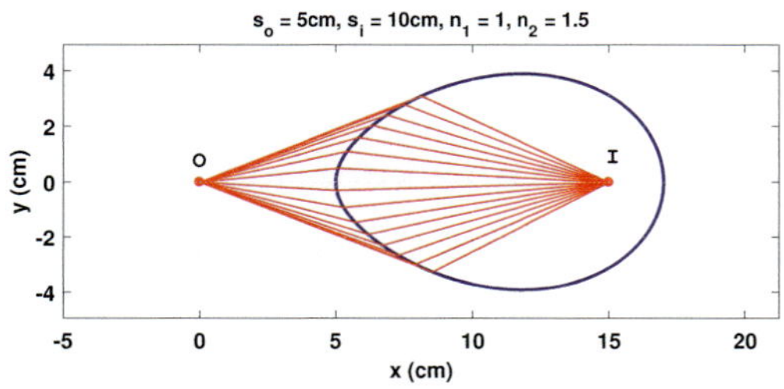

(a)

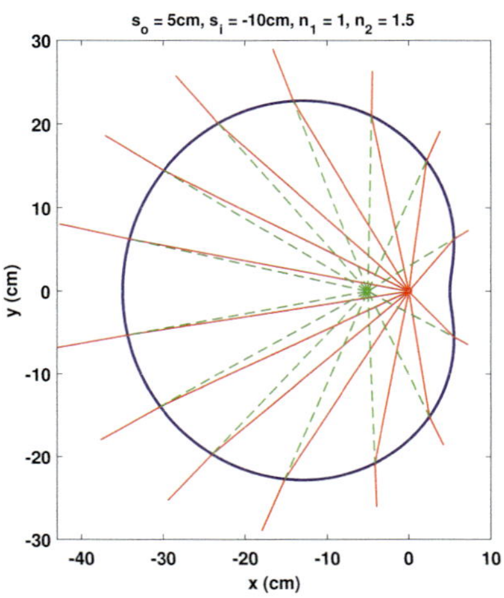

(b)

FIGURE 2.30: (a) An example of a Cartesian ovoid for a real image at $s_i = 10\,\text{cm}$, and object distance of $s_o = 5\,\text{cm}$ and $n_1 = 1$, $n_2 = 1.5$. (b) An example of a Cartesian ovoid for a virtual image at $s_i = -10\,\text{cm}$, and object distance of $s_o = 5\,\text{cm}$ and $n_1 = 1$, $n_2 = 1.5$.

wavefront of radius R emanating from the virtual image point I. Then applying Fermat's principle, the following equations can be written:

$$n_1(OP) + n_2(PQ) = n_1(OP') + n_2(P'Q') = n_1 s_o + n_2(V_1 Q_0) \Longrightarrow$$

$$n_1(OP) + n_2[R - (IP)] = n_1(OP') + n_2[R - (IP')] = n_1 s_o + n_2(R - |s_i|) \Longrightarrow$$

$$n_1\sqrt{x^2 + y^2} - n_2\sqrt{(x + |s_i| - s_o)^2 + y^2} = n_1 s_o - n_2|s_i|, \tag{2.39}$$

where $s_i < 0$ for the virtual image and the object is again considered to be at the origin of the coordinate system. The vertex points V_1 and V_2 are the points of the ovoid for $y = 0$. These points are given by:

$$x = s_o, \qquad \text{for} \quad x > \max\{0, s_o - |s_i|\} \tag{2.40}$$

$$x = \frac{n_1 + n_2}{n_2 - n_1} s_o - \frac{2n_2}{n_2 - n_1}|s_i|, \qquad \text{for} \quad x < \min\{0, s_o - |s_i|\}. \tag{2.41}$$

An example case of a Cartesian ovoid for a virtual image at $s_i = -10\,\text{cm}$, an object distance of $s_o = 5\,\text{cm}$, and $n_1 = 1$, $n_2 = 1.5$ is shown in Fig. 2.30(b). Some sample rays emanating from the object are also shown as they get refracted by the ovoid surface and their extensions pass through the virtual image point. Equation (2.41) is valid when $|s_i| \geq s_o$. In the case that $|s_i| < s_o$ the resulting equation becomes

$$x = s_o - \frac{2n_2}{n_2 + n_1}|s_i|, \qquad \text{for} \quad 0 < x < s_o - |s_i|.$$

The previous equations hold also for the case that $n_1 > n_2$. However, the resulting Cartesian ovoid surfaces look different. Example cases for real and virtual images are shown in Fig. 2.31 for $s_o = 5\,\text{cm}$, $s_i = \pm 10\,\text{cm}$, $n_1 = 1.5$, and $n_2 = 1$.

Despite the precise imaging capabilities of Cartesian ovoids for individual points, their complex geometries present significant challenges in manufacturing. Additionally, while they ensure aberration-free imaging for a single point (i.e., no aberrations but scattering and diffraction are unavoidable), real objects possess finite dimensions, which means that aberrations cannot be entirely eliminated, even with these surfaces. As a result, simpler spherical refractive surfaces are predominantly used in most imaging optical systems, particularly under the paraxial approximation, which allows for a more manageable design while achieving acceptable image quality. The following section delves into these spherical surfaces and their application within the paraxial framework.

2.5.5 Spherical Refractive Surface - Paraxial Approximation

Let's consider the spherical interface or radius R, between two media with refractive indices n_1 and n_2 as shown in Fig. 2.32. Assume the point O corresponds to an object point from which rays are emanating. A ray from O is incident on the spherical boundary at point A as shown. The distance of O from the vertex V is s. The ray will refract according to the Snell's law (or equivalently according to Fermat's principle) and its extension intersects the optical axis at the point I (virtual image in this case) which is at a distance s' from vertex V. Assuming that all angles are small with respect to the optical axis so that the paraxial approximation holds ($\sin\theta \simeq \theta$ and $\tan\theta \simeq \theta$), the following equations can be written:

$$n_1 \sin\theta_1 = n_2 \sin\theta_2 \Longrightarrow n_1\theta_1 \simeq n_2\theta_2 \Longrightarrow$$

$$n_1(\alpha_1 - \phi) = n_2(\alpha_2 - \phi) \Longrightarrow$$

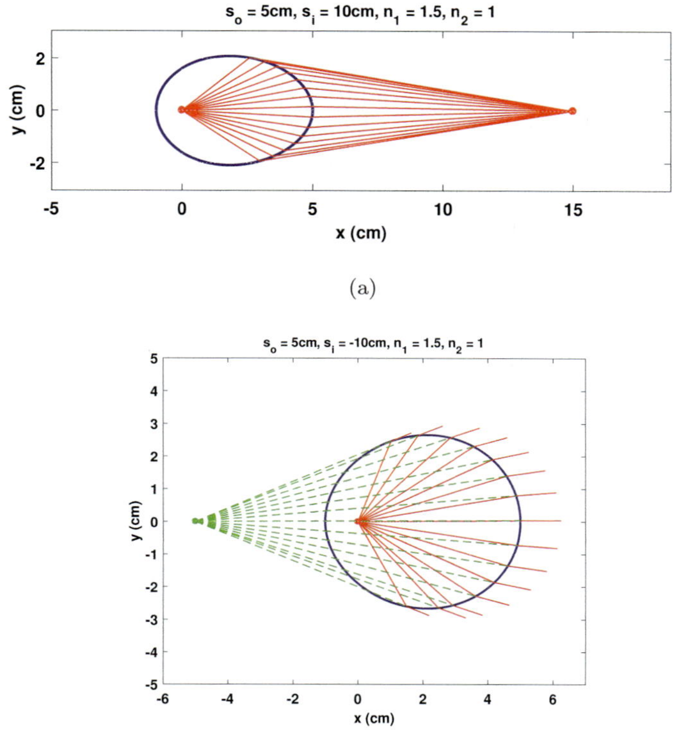

(a)

(b)

FIGURE 2.31: (a) An example of a Cartesian ovoid for a real image at $s_i = 10\,\text{cm}$, and object distance of $s_o = 5\,\text{cm}$ and $n_1 = 1.5$, $n_2 = 1$. (b) An example of a Cartesian ovoid for a virtual image at $s_i = -10\,\text{cm}$, and object distance of $s_o = 5\,\text{cm}$ and $n_1 = 1.5$, $n_2 = 1$.

$$n_1\left(\frac{h}{s} - \frac{h}{R}\right) = n_2\left(\frac{h}{s'} - \frac{h}{R}\right) \Longrightarrow$$

$$\frac{n_1}{s} - \frac{n_2}{s'} = \frac{n_1 - n_2}{R} \Longrightarrow$$

$$\frac{n_1}{s} + \frac{n_2}{s'} = \frac{n_2 - n_1}{R}, \qquad \text{(with sign convention)} \tag{2.42}$$

where the sign convention shown in Table 2.1 is used. Furthermore, simple geometric relations in triangles CAO and CAI were utilized. For example, for the picture shown in Fig. 2.32, $s > 0$ (real object), $s' < 0$ (virtual image), and $R < 0$ (concave surface for light propagation from left to right). If $R \to \infty$ (planar interface) then $s' = -(n_2/n_1)s$.

Some special cases of Eq. (2.42) are of interest for the image formation via such interfaces. Specifically, if $s = \infty$ then the incident ray is parallel to the optical axis and after refraction at the spherical interface passes through the *second focal point* F_2 (at distance f_2 from the vertex V). On the other hand, if $s' = \infty$, i.e., when the refracted rays emerge from the interface parallel to the optical axis, then the incident ray passes through the *first focal point* F_1 (at distance f_1 from the vertex V). The distances f_1 and f_2 are given by:

$$f_1 = \frac{n_1}{n_2 - n_1} R, \tag{2.43}$$

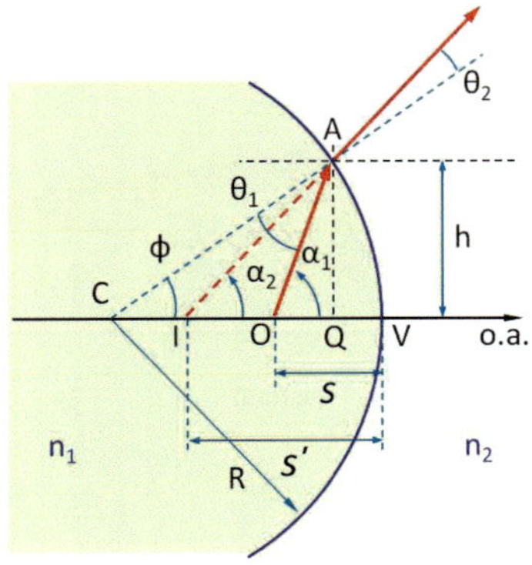

FIGURE 2.32: A spherical refractive surface of radius R between two media of refractive indices n_1 and n_2. It is assumed that all the angles with respect to the optical axis are small.

$$f_2 = \frac{n_2}{n_2 - n_1} R, \tag{2.44}$$

where f_1 and f_2 can be positive or negative. Specifically, for $n_1 < n_2$, a convex spherical interface ($R > 0$) gives $f_2 > 0$ and $f_1 > 0$, and the points F_2 and F_1 are located to the right and to the left of the vertex V as shown in Fig. 2.33(a). On the other hand, for $n_1 < n_2$, a concave spherical interface ($R < 0$) gives $f_2 < 0$ and $f_1 < 0$, and the points F_2 and F_1 are located to the left and to the right of the vertex V as shown in Fig. 2.33(b). When $n_1 > n_2$ the opposite, from the previous results, situation occurs. Similarly to the case of spherical

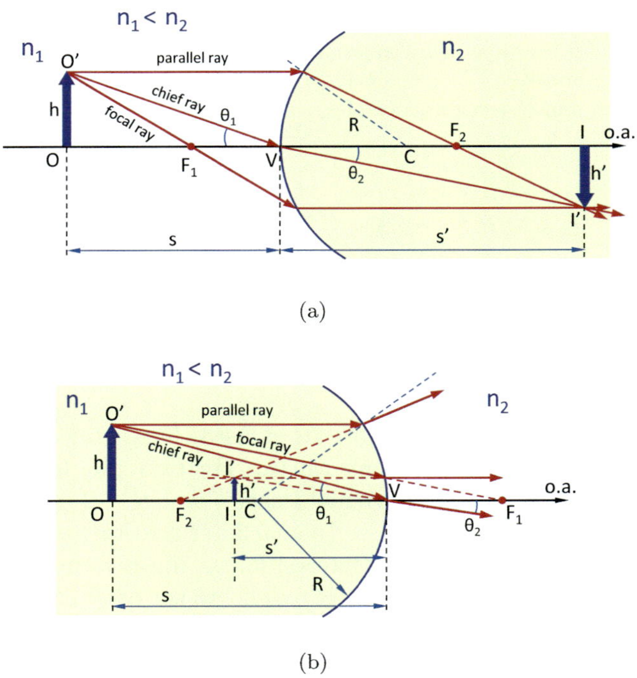

FIGURE 2.33: (a) Example of imaging via a spherical convex refractive surface of radius R (real image formation). (b) Example of imaging via a spherical concave refractive surface of radius R (virtual image formation). For both cases $n_1 < n_2$.

mirrors, a transverse magnification m can be defined. From the figures 2.33(a) and 2.33(b) the following equations can be written:

$$n_1 \sin\theta_1 = n_2 \sin\theta_2 \Longrightarrow n_1\theta_1 \simeq n_2\theta_2 \Longrightarrow$$

$$n_1 \frac{h}{s} = n_2 \frac{h'}{s'} \Longrightarrow$$

$$m = \frac{h'}{h} = -\frac{n_1}{n_2}\frac{s'}{s}, \tag{2.45}$$

where the transverse magnification, m, is positive when the image is upright (in the same direction with the object, virtual image in Fig. 2.33(b)), and negative when the image is inverted (in the opposite direction from the object, real image in Fig. 2.33(a)).

2.6 Thin Lenses

A *thin lens* is comprised of two spherical refractive surfaces of radii R_1 and R_2, separated by a distance d which is considered negligible as compared to its radii (i.e., $d \ll R_1, R_2$). An example of a biconvex spherical thin lens is shown in Fig. 2.34. Assume a point object O located at distance s from the left vertex V_1 of the thin lens. The position of the image point I, at a distance s' from the right vertex V_2, is to be determined. In this case the paraxial equation for the image formation by a spherical interface, Eq. (2.42), can be applied sequentially for the two refractive surfaces as follows:

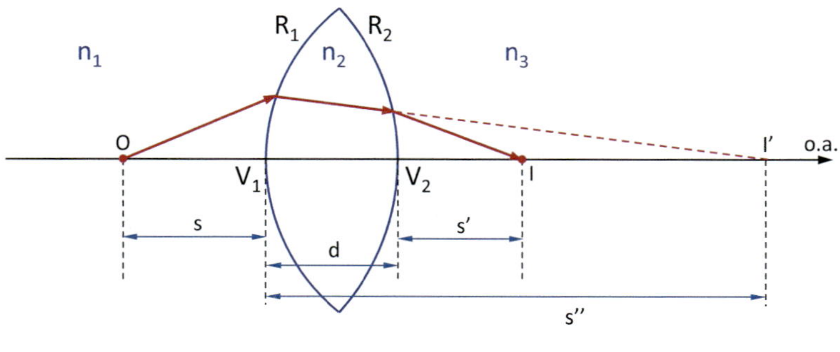

FIGURE 2.34: A thin lens comprised of two spherical surfaces of radii of curvature R_1 and R_2 respectively. The thickness of the lens is $d \ll R_1, R_2$ and can be neglected. All rays are at small angles with respect to the optical axis so that the paraxial approximation holds. The refractive indices are n_1, n_2, and n_3 for the left region, the lens, and the right region respectively.

$$\frac{n_1}{s} + \frac{n_2}{s''} = \frac{n_2 - n_1}{R_1},$$

$$\frac{n_2}{-(s'' - d)} + \frac{n_3}{s'} = \frac{n_3 - n_2}{R_2} \Longrightarrow$$

$$\frac{n_1}{s} + \frac{n_3}{s'} = \frac{n_2 - n_1}{R_1} + \frac{n_3 - n_2}{R_2}, \tag{2.46}$$

where the image from the first refractive surface was considered as the virtual object for the second refractive surface [this is the reason for the $-(s'' - d)$ term] and the distance d

was assumed negligible. In the common case that $n_1 = n_3$, the last equation becomes,

$$\frac{1}{s} + \frac{1}{s'} = \frac{n_2 - n_1}{n_1}\left(\frac{1}{R_1} - \frac{1}{R_2}\right) = \frac{1}{f}, \tag{2.47}$$

where f is the focal distance of the thin lens. The focal lengths f_1 and f_2 of the thin lens can be specified for image and object distance at infinity, respectively. In the case of $n_1 = n_3$ (equal refractive indices in the regions to the left and the right of the lens) $f = f_1 = f_2$ with f as specified in Eq. (2.47). If $n_1 \neq n_3$ then the focal distances f_1 and f_2 are given by

$$\frac{1}{f_1} = \frac{1}{n_1}\left(\frac{n_2 - n_1}{R_1} - \frac{n_2 - n_3}{R_2}\right), \tag{2.48}$$

$$\frac{1}{f_2} = \frac{1}{n_3}\left(\frac{n_2 - n_1}{R_1} - \frac{n_2 - n_3}{R_2}\right). \tag{2.49}$$

For positive lenses $f_1, f_2 > 0$, and the F_1 is to the left of the vertex V while the F_2 point is to the right of V. On the other hand, for negative lenses $f_1, f_2 < 0$, and the F_1 is to the right of the vertex V while the F_2 point is to the left of V. In all the previous thin lens equations the sign convention similar to the spherical mirrors and refractive surfaces is also used. For example, real objects are to the left of the lens ($s > 0$) while real images are to the right of the lens ($s' > 0$). Alternatively, virtual objects are to the right of the lens ($s < 0$) and virtual images are to the left of the lens ($s' < 0$). The transverse magnification, m, is defined for the spherical mirrors and refractive surfaces as

$$m = \frac{h'}{h} = -\frac{n_1}{n_3}\frac{s'}{s}, \tag{2.50}$$

where h and h' are the transverse heights of the object and of the image, respectively. When the image is inverted $m < 0$ while when the image is upright $m > 0$. A summary of the thin lens conventions is given in Table 2.2 for light propagating from left to right and for $n_1 = n_3 < n_2$.

Example image formation is shown in Fig. 2.35 for positive lens and in Fig. 2.36 for negative lens. For a real object ($s > 0$) the negative lens produces always a virtual upright image smaller in size than the object ($0 < m < 1$). On the other hand, a positive lens can produce either real or virtual images depending on the distance of the object from the lens. Specifically, if $2f < s < \infty$ (see Fig. 2.35(a)) a real inverted image is formed of a smaller size than the object ($-1 < m < 0$). In the case that $f < s < 2f$ (see Fig. 2.35(b)) a real inverted image is formed with a larger size than the object ($m < -1$). In the case that $0 < s < f$ (see Fig. 2.35(c)) a virtual image is formed larger than the object ($m > 1$). This last case is the case of the simple magnifier.

Newtonian Thin Lens Equation: The object and the image distances can be measured from the focal points F_1 and F_2 respectively as shown in Fig. 2.37. In this case an alternative to the thin-lens equation, Eq. (2.47), can be derived. Using the similar triangles shown in the figure the following equations can easily be written:

$$\frac{h'}{h} = \frac{f}{x},$$

$$\frac{h'}{h} = \frac{x'}{f},$$

$$xx' = f^2 \quad \text{and} \quad m = -\frac{f}{x} = -\frac{x'}{f}, \tag{2.51}$$

where a sign convention is assumed for the distances x and x'. Specifically, x, x' are considered positive if the object is to the left of F_1 and the image is to the right of F_2. Otherwise, these distances are considered negative.

TABLE 2.2: Convention Summary for Thin Lenses

Thin Lens Sign Conventions (Light from Left-to-Right)						
$i = 1, 2$	$R_i > 0$ (convex)			$R_i < 0$ (concave)		
$s > 0$ (real object - left of V), $s < 0$ (virtual object - right of V)						
$s' > 0$ (real image - right of V), $s' < 0$ (virtual image - left of V)						
$f > 0$ (converging/positive lens), $f < 0$ (diverging/negative lens)						
Thin-Lens Equation	$\frac{1}{s} + \frac{1}{s'} = \frac{1}{f} = \frac{n_2 - n_1}{n_1}\left(\frac{1}{R_1} - \frac{1}{R_2}\right)$					
Magnification	$m = \frac{h'}{h} = -\frac{s'}{s}$					
Positive Lenses						
Lens Type	**R**$_1$	**R**$_2$	**s**	**s**$'$	**m**	**f**
Biconvex	$+$	$-$	$+$	$\pm$	$\mp$	$+$
Plano-Convex	∞	$-$	$+$	$\pm$	$\mp$	$+$
Convex-Plano	$+$	∞	$+$	$\pm$	$\mp$	$+$
Positive Meniscus ($R_1 < R_2$)	$+$	$+$	$+$	$\pm$	$\mp$	$+$
Comment				**s**$'$	**m**	
if $s > 2f$				$+$	$-$	
if $f < s < 2f$				$+$	$-$	
if $0 < s < f$				$-$	$+$	
Negative Lenses						
Lens Type	**R**$_1$	**R**$_2$	**s**	**s**$'$	**m**	**f**
Biconcave	$-$	$+$	$+$	$-$	$+$	$-$
Plano-Concave	∞	$+$	$+$	$-$	$+$	$-$
Concave-Plano	$-$	∞	$+$	$-$	$+$	$-$
Negative Meniscus ($\lvert R_1\rvert < \lvert R_2\rvert$)	$-$	$-$	$+$	$-$	$+$	$-$

Longitudinal Magnification: Since objects typically possess width in addition to height, a longitudinal magnification can also be defined. By differentiating both sides of the thin lens equation, Eq. (2.47), the expression for longitudinal magnification is obtained as follows:

$$-\frac{\Delta s}{s^2} - \frac{\Delta s'}{s'^2} = 0 \Longrightarrow m_L = \frac{\Delta s'}{\Delta s} = -\frac{s'^2}{s^2} = -m^2, \tag{2.52}$$

where m_L is the longitudinal magnification. It can be observed that the "$-$" sign means that a real image is opposite directed except being inverted. A similar relationship for

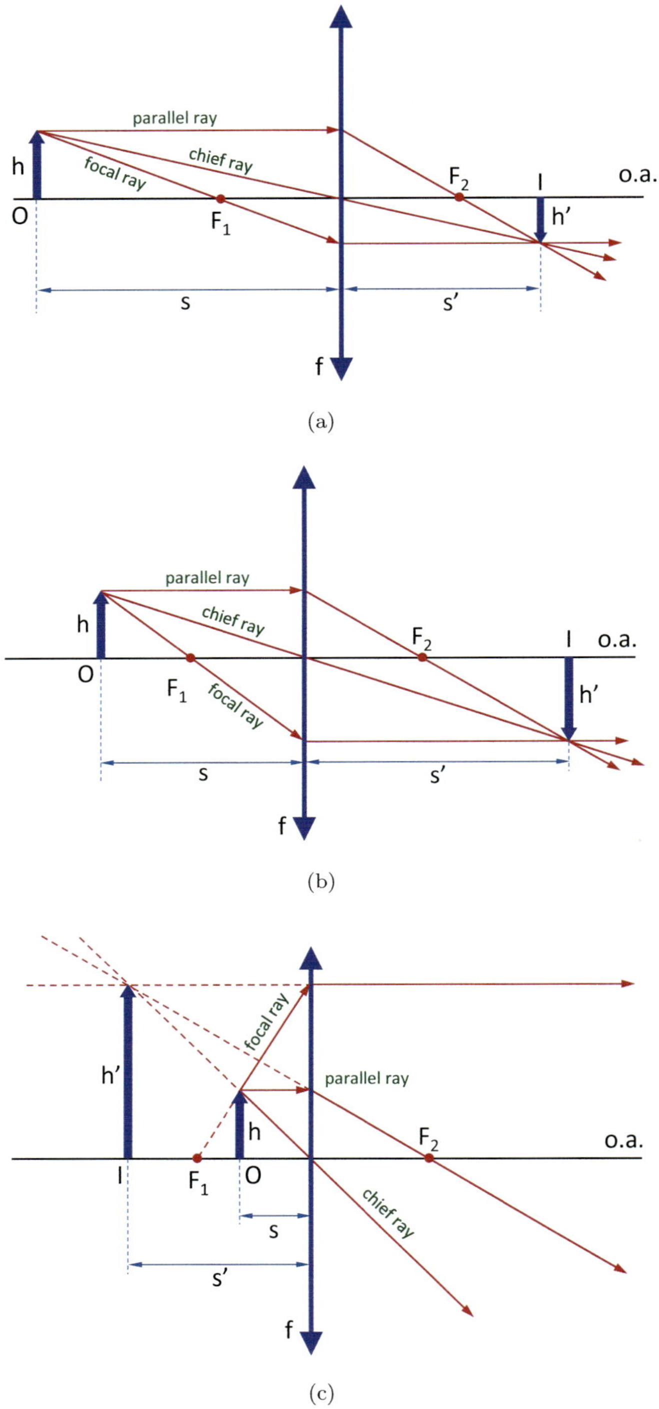

FIGURE 2.35: (a) Example of imaging via a positive thin lens of focal distance $f > 0$ and object distance $s > 2f$ (real image). (b) Example of imaging via a positive thin lens of focal distance $f > 0$ and object distance $f < s < 2f$ (real image). (c) Example of imaging via a positive thin lens of focal distance $f > 0$ and object distance $0 < s < f$ (virtual image).

longitudinal magnification can also be derived using the Newtonian formula for a thin lens:

$$x' = \frac{f^2}{x} \Longrightarrow m_L = \frac{\Delta x'}{\Delta x} = -\frac{f^2}{x^2} = -m^2, \tag{2.53}$$

which is similar to the one derived from the common thin lens equation.

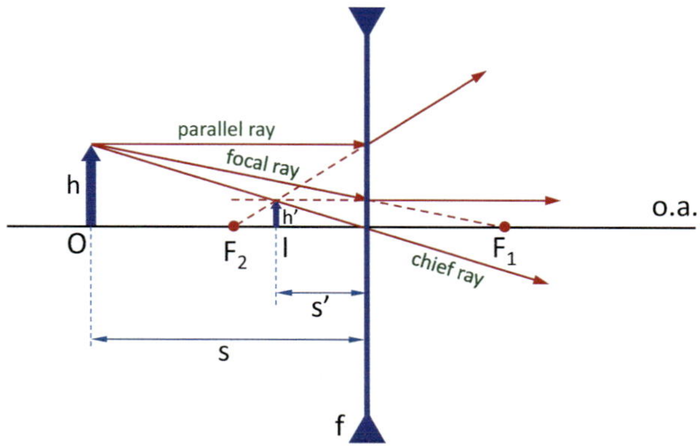

FIGURE 2.36: An example of imaging with a diverging (negative) thin lens of focal length $f < 0$ (always a virtual image).

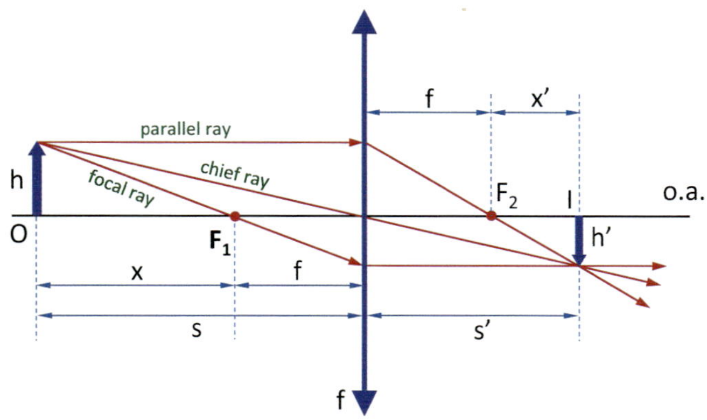

FIGURE 2.37: Thin lens imaging with the Newtonian distances x, x', and f shown ($n_1 = n_3$).

2.7 ABCD Matrices

In order to describe the path of a ray through an optical system under the paraxial approximation, the *Matrix Method* is a very simple and convenient approach. In this method the position of a ray in an optical system can be described by its distance from the optical axis (typically the z-axis) of the system, denoted as y, and its slope $\alpha = dy/dz$ with respect to the optical axis (see Fig. 2.38). Therefore, the position of a ray at each coordinate z of the optical axis inside an optical system can be described by the vector $[y, \alpha]^T$. When $y > 0$

the ray is above the optical axis while when $y < 0$ the ray is below the optical axis. When the slope α of the ray is positive its position y is increasing as it propagates an infinitesimal distance dz. Conversely, if the slope α of the ray is negative its position y is decreasing as it propagates an infinitesimal distance dz. Since the ray's description only involves the distance from the optical axis and the slope, it is assumed that all rays are confined to the same plane as the optical axis, referred to as "meridional" rays. If a ray is not coplanar with the optical axis, it is termed a "skew" ray, which requires more complex analysis. For a deeper understanding, readers are encouraged to consult more advanced texts on optical design, such as those in Refs. [36–38]. In an optical system, as illustrated in Fig. 2.39, the output ray described by the vector $[y_1, \alpha_1]^T$ can be related to the input ray (described by the vector $[y_0, \alpha_0]^T$) via the following equation:

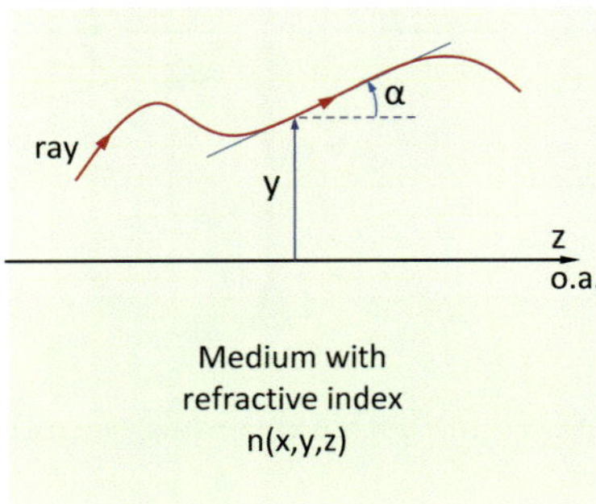

FIGURE 2.38: A point of a ray can be described with its distance y from the optical axis and its slope $\alpha = dy/dz$.

$$\begin{bmatrix} y_1 \\ \alpha_1 \end{bmatrix} = \begin{bmatrix} A & B \\ C & D \end{bmatrix} \begin{bmatrix} y_0 \\ \alpha_0 \end{bmatrix} = \tilde{M} \begin{bmatrix} y_0 \\ \alpha_0 \end{bmatrix}, \tag{2.54}$$

where the 2×2 matrix $\tilde{M}$ is called the $ABCD$ matrix of the optical system for the input and output planes defined. It should be emphasized that the $ABCD$ matrix depends on the elements of the optical system and the definition of the input and output planes. Of course, the presented $ABCD$ matrix is valid under the paraxial approximation and for meridional rays. Any optical system can be broken down into simpler optical components, or different

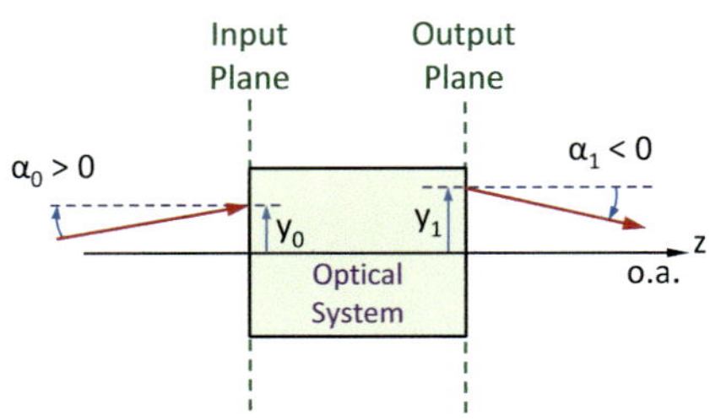

FIGURE 2.39: An optical system that is defined between an input and an output plane can be described under the paraxial approximation with an $ABCD$ matrix.

optical systems can be combined in sequence to create more complex systems. For instance, in advanced camera lenses, multiple groups of lenses are arranged in a cascade to form the complete lens system. A general representation of such a cascade combination of optical systems is illustrated in Fig. 2.40. In this diagram, N optical systems are cascaded to form the optical system which is included between the input and the output planes. Assuming that the $ABCD$ matrix of each subsystem is known, the overall $ABCD$ matrix of the cascaded combination can be determined using the following formula:

$$\begin{aligned}\begin{bmatrix} y_N \\ \alpha_N \end{bmatrix} &= \begin{bmatrix} A_N & B_N \\ C_N & D_N \end{bmatrix}\begin{bmatrix} y_{N-1} \\ \alpha_{N-1} \end{bmatrix} = \begin{bmatrix} A_N & B_N \\ C_N & D_N \end{bmatrix}\begin{bmatrix} A_{N-1} & B_{N-1} \\ C_{N-1} & D_{N-1} \end{bmatrix}\begin{bmatrix} y_{N-2} \\ \alpha_{N-2} \end{bmatrix} = \cdots \\ &= \underbrace{\begin{bmatrix} A_N & B_N \\ C_N & D_N \end{bmatrix}}_{=\tilde{M}_N}\underbrace{\begin{bmatrix} A_{N-1} & B_{N-1} \\ C_{N-1} & D_{N-1} \end{bmatrix}}_{=\tilde{M}_{N-1}}\cdots\underbrace{\begin{bmatrix} A_1 & B_1 \\ C_1 & D_1 \end{bmatrix}}_{=\tilde{M}_1}\begin{bmatrix} y_0 \\ \alpha_0 \end{bmatrix} = \\ &= \left[\prod_{i=1}^{N} \tilde{M}_{N+1-i}\right]\begin{bmatrix} y_0 \\ \alpha_0 \end{bmatrix} = \tilde{M}_{total}\begin{bmatrix} y_0 \\ \alpha_0 \end{bmatrix}, \end{aligned} \tag{2.55}$$

where $\tilde{M}_{total}$ is the overall 2×2 $ABCD$ matrix of the cascaded combination of the N optical subsystems. It is worth mentioning that the multiplication of the matrices is done in the reverse direction from the direction of propagation of the light rays.

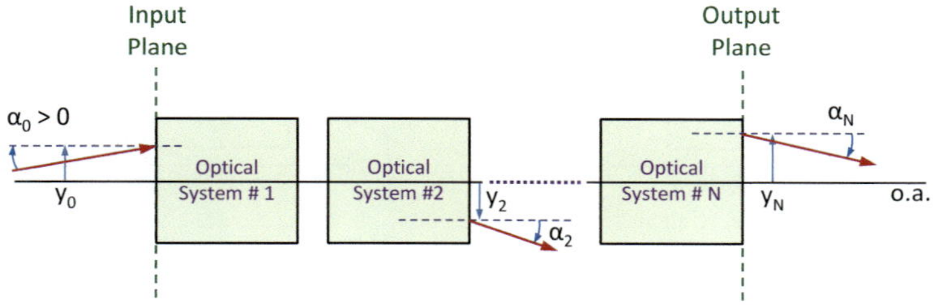

FIGURE 2.40: A cascaded combination of N optical systems each one described by its own $ABCD$ matrix.

The significance of each of the 4 elements of an $ABCD$ matrix of an optical system is outlined in Fig. 2.41. Specifically, for the defined input and output planes, the input, $[y_i, \alpha_i]^T$, and output, $[y_f, \alpha_f]^T$, ray characteristics are related by:

$$\begin{bmatrix} y_f \\ \alpha_f \end{bmatrix} = \begin{bmatrix} A & B \\ C & D \end{bmatrix}\begin{bmatrix} y_i \\ \alpha_i \end{bmatrix}. \tag{2.56}$$

Now the following special cases can be identified:

- $\boldsymbol{D = 0}$: In this case $\alpha_f = Cy_i$, and if y_i is fixed, then all output rays have the same slope with respect to the optical axis. i.e., all rays leaving the object from y_i, at any slope, exit parallel (with slope α_f) from the output plane of the optical system. In this case the input plane coincides with the *First Focal Plane* of the optical system.
- $\boldsymbol{A = 0}$: In this case $y_f = B\alpha_i$, and if α_i is fixed, then all output rays pass from the same point y_f in the output plane independently of their height at the input plane. In this case the output plane coincides with the *Second Focal Plane* of the optical system.
- $\boldsymbol{B = 0}$: In this case $y_f = Ay_i$, and if y_i is fixed, then all output rays pass from the same point y_f in the output plane independently of their slope at the input plane. These points are called object and image points and the optical system is an *Imaging System*. In addition, $A = y_f/y_i$ is the transverse magnification of the optical system.

- $\boldsymbol{C = 0}$: In this case $\alpha_f = D\alpha_i$, and if α_i is fixed, then all output rays exit with the same slope α_f at the output plane independently of their heights at the input plane. The input and output planes form what is called as a *Telescopic System*. In addition, $D = \alpha_f/\alpha_i$ is the angular magnification of the optical system.

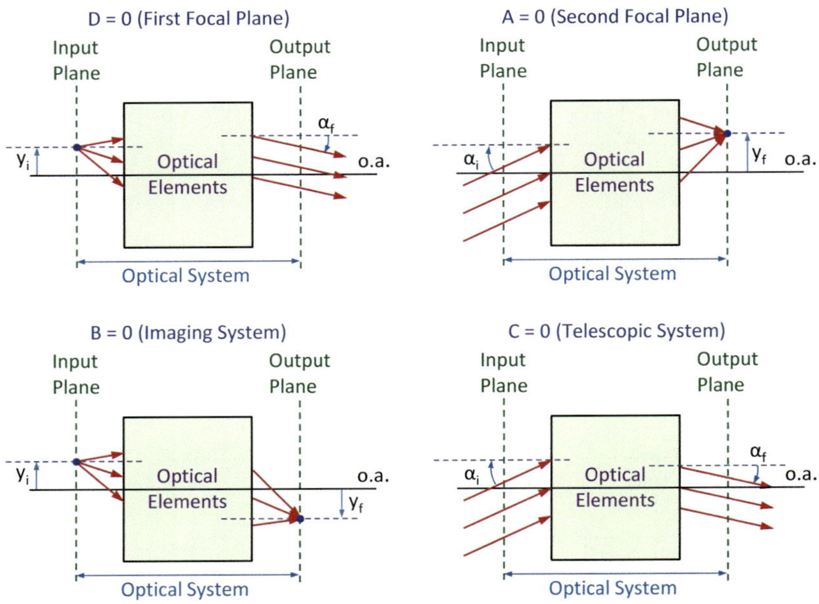

FIGURE 2.41: The significance of an element of the $ABCD$ matrix being zero.

In the next subsections some elementary $ABCD$ matrices are defined. These are (a) the $ABCD$ matrix for the propagation of a ray in a homogeneous medium, (b) the $ABCD$ matrix for the refraction of a ray on a spherical interface between two media, (c) the $ABCD$ matrix for the refraction of a ray on a planar interface between two media, (d) the $ABCD$ matrix for the refraction of a ray through a thin lens, (e) the $ABCD$ matrix for the reflection of a ray at a spherical mirror, and (f) propagation through a "lens-like" medium. Based on these elementary $ABCD$ matrices sophisticated optical systems (such as camera lenses) can be easily analyzed under the paraxial approximation.

2.7.1 Propagation in a Homogeneous Medium

Consider a homogeneous isotropic medium of refractive index n where a ray of light propagates from point A to point B as illustrated in Fig. 2.42. At point A, the ray is at a distance y_1 from the optical axis with a positive slope α_1. After propagating a distance L parallel to the optical axis, the ray reaches point B, located at a distance y_2 from the optical axis and with a slope α_2. To determine the $ABCD$ matrix for this propagation, note that due to the homogeneity of the medium, the ray follows a straight path, as guaranteed by Fermat's principle (see Sec. 2.3). This rectilinear propagation implies that the slope remains constant throughout the distance L, allowing the following relationships to be established:

$$y_2 = y_1 + L \tan \alpha_1 \simeq y_1 + L\alpha_1,$$

$$\alpha_2 = \alpha_1, \qquad \text{and the resulting } ABCD \text{ matrix is}$$

$$\tilde{M} = \begin{bmatrix} 1 & L \\ 0 & 1 \end{bmatrix}. \tag{2.57}$$

In the above derivation the paraxial approximation was used since $\tan \alpha_1 \simeq \alpha_1$ was utilized. Observe that the determinant of the $ABCD$ matrix is $\det(\tilde{M}) = AD - BC = 1$.

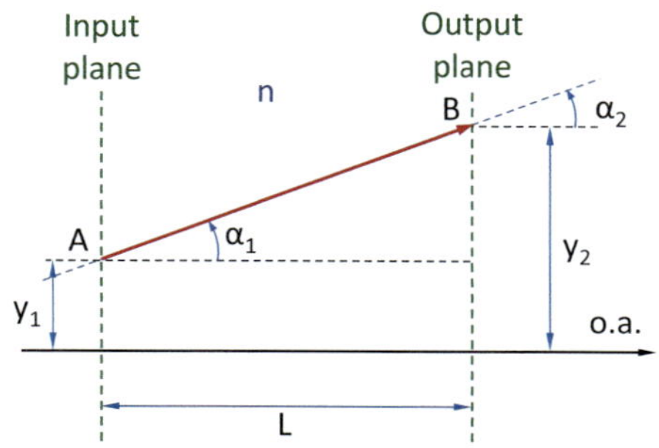

FIGURE 2.42: Propagation of a light ray in a homogeneous medium of refractive index n for a distance L. The ray at the input plane has a distance from the optical axis of y_1 and a slope of α_1.

2.7.2 Refraction at a Spherical/Planar Interface

Consider a spherical interface of radius R separating two homogeneous isotropic media of refractive indices n_1 and n_2, as shown in Fig. 2.43. A ray originates in the medium with index n_1 and is incident on the spherical interface at point A, located at a distance y_1 from the optical axis with an initial slope of α_1. Upon refraction into the medium of index n_2, the ray emerges from the same point A, maintaining the same vertical distance $y_2 = y_1$, but with a modified slope α_2 due to refraction. Assuming the paraxial approximations the following steps can be written in order to determine the corresponding $ABCD$ matrix:

$$\begin{aligned} y_2 &= y_1 \\ n_1 \sin\theta_1 &= n_2 \sin\theta_2 \Longrightarrow n_1\theta_1 \simeq n_2\theta_2 \Longrightarrow \\ n_1(\alpha_1 + \phi) &= n_2(\alpha_2 + \phi) \Longrightarrow \\ \alpha_2 &= \left(\frac{n_1}{n_2} - 1\right)\phi + \frac{n_1}{n_2}\alpha_1 \Longrightarrow \\ \alpha_2 &= \left(\frac{n_1}{n_2} - 1\right)\frac{y_1}{R} + \frac{n_1}{n_2}\alpha_1, \qquad \text{resulting in} \end{aligned}$$

$$\tilde{M} = \begin{bmatrix} 1 & 0 \\ \left(\frac{n_1}{n_2} - 1\right)\frac{1}{R} & \frac{n_1}{n_2} \end{bmatrix}. \tag{2.58}$$

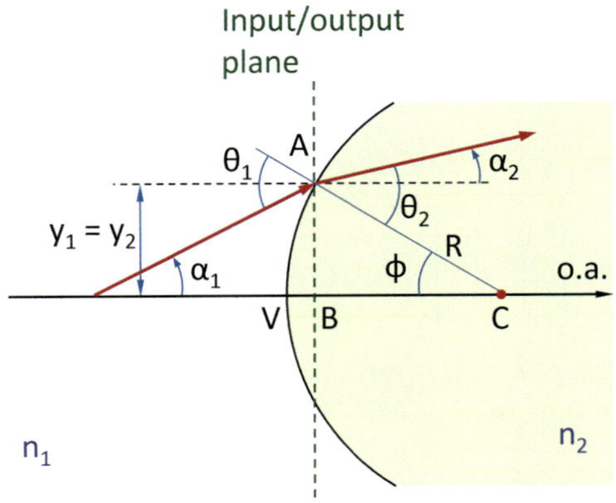

FIGURE 2.43: Refraction of a ray incident on a spherical interface of radius R between two media of refractive indices n_1 and n_2 respectively.

A special case of the above $ABCD$ matrix is when the refracting interface is planar. In this case $R \to \infty$ and the resulting $ABCD$ matrix become:

$$\tilde{M} = \begin{bmatrix} 1 & 0 \\ 0 & \dfrac{n_1}{n_2} \end{bmatrix}. \tag{2.59}$$

The determinant of the $ABCD$ matrices for both spherical or planar interfaces is $\det(\tilde{M}) = AD - BC = n_1/n_2$. This result is significant, as it implies that the determinant depends only on the refractive indices of the media on either side of the interface, regardless of the specific shape of the interface (spherical or planar). For the given matrices, it is also essential to follow the established sign convention, particularly when dealing with the radius of curvature R. Therefore, $R > 0$ for convex and $R < 0$ for concave spherical surface, respectively.

2.7.3 Propagation through a Thin Lens

Now consider a thin lens of focal length f. Without loss of generality a positive thin lens is assumed as illustrated in Fig. 2.44. A ray is incident on the thin lens from the left at a height from the optical axis y_1 and a positive slope of α_1. Since the lens is thin the ray emerges from the right surface of the thin lens at the same height (i.e., $y_2 = y_1$), but with a slope α_2. Using the thin lens equation the following expressions can be written:

$$\begin{aligned} y_2 &= y_1 \\ |\alpha_2| &\simeq \frac{|y_1|}{s_2} = |y_1|\left(\frac{1}{f} - \frac{1}{s_1}\right) = \\ &\quad \frac{|y_1|}{f} - \frac{|y_1|}{s_1} = \frac{|y_1|}{f} - \alpha_1 \Longrightarrow \\ \alpha_2 &= -\frac{y_1}{f} + \alpha_1, \qquad \text{resulting in} \\ \tilde{M} &= \begin{bmatrix} 1 & 0 \\ -\dfrac{1}{f} & 1 \end{bmatrix}. \end{aligned} \tag{2.60}$$

In the above derivation it was taken into account that $\alpha_2 < 0$ and the final equation assumes the sign convention for the thin lens, i.e., $f > 0$ for the positive lens (converging) and $f < 0$ for the negative lens (diverging). Again observe that the determinant of the $ABCD$ matrix for the thin lens is $\det(\tilde{M}) = AD - BC = 1$.

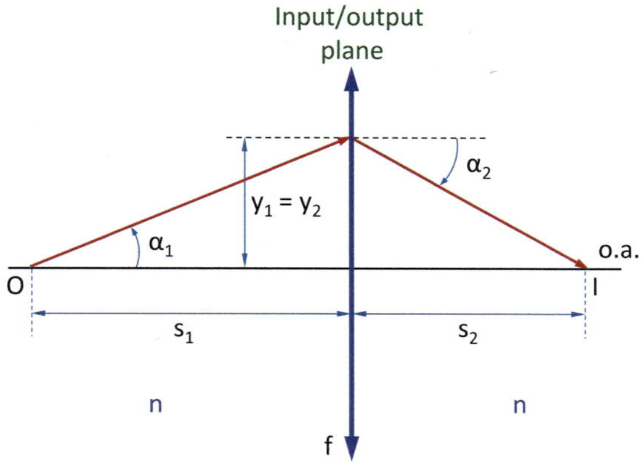

FIGURE 2.44: Propagation of a ray through a positive thin lens of focal distance f.

2.7.4 Reflection at a Spherical/Planar Mirror

For a concave spherical mirror of radius R, a ray incident from the left at point A at a height y_1 from the optical axis is reflected according to the law of reflection as shown in Fig. 2.45. Since the reflection occurs at point A, the distance of the reflected ray from the optical axis remains $y_2 = y_1$. However, the slope of the reflected ray α_2 will change due to the curvature of the mirror. Using the paraxial approximation and the law of reflection the following equations can be written:

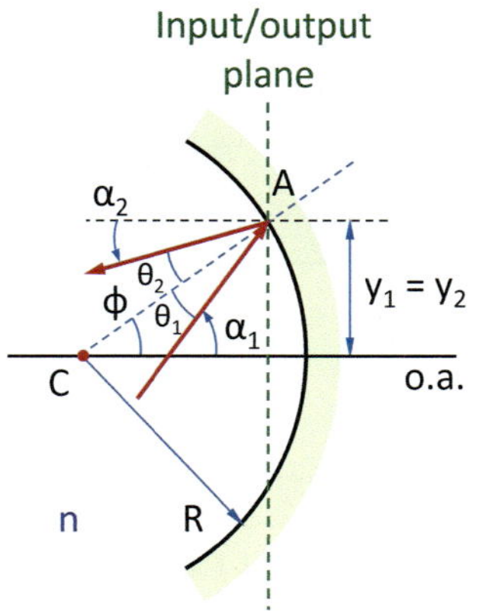

FIGURE 2.45: Reflection of a ray incident on a spherical mirror of radius R.

$$y_2 = y_1$$
$$\theta_2 = \theta_1 \Longrightarrow \phi - |\alpha_2| = \alpha_1 - \phi \Longrightarrow$$
$$|\alpha_2| = 2\phi - \alpha_1 = 2\frac{y_1}{|R|} - \alpha_1 = -2\frac{y_1}{R} - \alpha_1 \Longrightarrow$$

$$\alpha_2 = \frac{2}{R} y_1 + \alpha_1, \qquad \text{resulting in}$$

$$\tilde{M} = \begin{bmatrix} 1 & 0 \\ \dfrac{2}{R} & 1 \end{bmatrix}. \tag{2.61}$$

In the last matrix the convention for R defined previously was used ($R < 0$ in the case shown in Fig. 2.45). Again observe that the determinant of the $ABCD$ matrix for the thin lens is $\det(\tilde{M}) = AD - BC = 1$. As a special case the reflection matrix of a planar mirror can be determined if $R \to \infty$ and the resulting $\tilde{M}$ is the 2×2 unity matrix.

2.7.5 Propagation through a Lens-like-Medium

A lens-like medium is characterized by a refractive index $n(r)$ that varies radially, depending only on the distance r from the optical axis and remaining constant along the optical axis. This radial dependency creates a gradient in refractive index that can focus or defocus rays, similarly to a traditional lens but without a physical surface. The refractive index of a lens-like medium can typically be described by the following equation [39, 40]:

$$n(r) = n_0 \left(1 - \frac{k_2}{2k_0} r^2\right), \tag{2.62}$$

where n_0 is the refractive index along the optical axis, $k_0 = 2\pi/\lambda_0$ (λ_0 is the freespace wavelength) and k_2 a constant, positive or negative, for decreasing or increasing the refractive index along the radial direction, respectively. In Fig. 2.46 a generic lens-like medium with a radially varying refractive index is depicted. The purpose is to determine the corresponding $ABCD$ matrix representation, which allows for the treatment of such a medium in the matrix formalism of ray optics. Assume that the input and the output planes are just to the

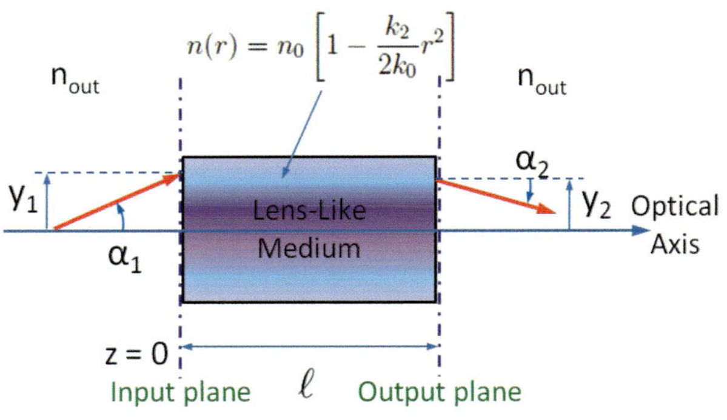

FIGURE 2.46: Propagation of a ray through a lens-like medium.

right of the left boundary and to the left of the right boundary in order to avoid refraction at the planar boundaries. Therefore, the $ABCD$ matrix representation is restricted for the propagation inside the lens-like medium. The behavior of a ray inside an inhomogeneous medium is described by Fermat's principle expressed by Eq. (2.7). Assuming paraxial ray propagation the d/ds derivative in Eq. (2.7) can be approximated by d/dz where z the coordinate representing the optical axis (propagation axis). Then the radial distance of the ray from the optical axis (y distance in the $ABCD$ terminology used) satisfies the equation:

$$\frac{d^2 r}{dz^2} + \frac{k_2}{k_0} r = 0. \tag{2.63}$$

For the initial conditions of the left ray being $r(z=0)=r_0=y_1$ and $r'(z=0)=r'_0=\alpha_1$, Eq. (2.63) has the following solution:

$$\begin{aligned} r(z) &= r_0\cos(gz)+r'_0\frac{1}{g}\sin(gz), \\ r'(z) &= -r_0 g\sin(gz)+r'_0\cos(gz), \end{aligned}$$

where $g=\sqrt{k_2/k_0}$, $r(z=\ell)=y_2$ and $r'(z=\ell)=\alpha_2$ (see Fig. 2.46). The *ABCD* matrix describing the paraxial propagation through the lens-like medium can be determined by writing the last equations in the following matrix form:

$$\begin{bmatrix} r(z) \\ r'(z) \end{bmatrix} = \begin{bmatrix} \cos(gz) & \frac{1}{g}\sin(gz) \\ -g\sin(gz) & \cos(gz) \end{bmatrix} \begin{bmatrix} r_0 \\ r'_0 \end{bmatrix} = \begin{bmatrix} A & B \\ C & D \end{bmatrix} \begin{bmatrix} r_0 \\ r'_0 \end{bmatrix}, \tag{2.64}$$

where z represents the propagation distance within the medium. In case that $z=\ell$ the matrix *ABCD* is shown in Table 2.3. It is interesting to notice that if $k_2>0$ (refractive index decreases away from the optical axis; always $k_0>0$ since it represents the wavenumber of freespace) the solutions of Eq. (2.64) are oscillatory which means that the ray remains always at finite distance from the optical axis. On the other hand, if $k_2<0$ (refractive index increases away from the optical axis) the parameter g becomes imaginary and the cosine and sine solutions convert into hyperbolic cosines and sines. In this case the ray distance from the optical axis increases exponentially with the propagation distance.

Again it can be observed that the determinant of the *ABCD* matrix for a lens-like medium is unity. More generally, this principle applies to any optical element where there is no refractive index change between entry and exit media, as shown by the general form $\det(\tilde{M})=AD-BC=n_i/n_f$ where n_i and n_f are the initial (left region of the optical system) and the final (right region of the optical system) refractive indices. Table 2.3 provides a comprehensive summary of *ABCD* matrices for all basic optical elements, offering a useful reference for constructing and analyzing complex optical systems by combining these elementary matrices.

2.7.6 Cardinal Points of an Optical System

Assume an optical system as the generic one shown in Fig. 2.47. A parallel ray incident from the left intersects the optical axis at point F_2 upon exiting the system. It can be assumed that all the refractive power of the optical system is concentrated at a fictitious plane perpendicular to the optical axis, passing through point H_2. This plane is defined by the intersection points of the extensions of the incident parallel rays and the extensions of the exiting rays that pass through F_2. The point F_2 and the corresponding plane perpendicular to the optical axis that passes through F_2 are referred to as the *second focal point* and the *second focal plane*, respectively. Similarly, the point H_2 and the plane perpendicular to the optical axis that passes through H_2 are designated as the *second principal point* and the *second principal plane*, respectively.

Conversely, for an input ray that passes through point F_1 (the focal ray) and exits parallel to the optical axis, it can be assumed that all the refractive power of the optical system is concentrated at a fictitious plane perpendicular to the optical axis, passing through point H_1. This plane is defined by the intersection points of the extensions of the incident rays passing through F_1 and the extensions of the corresponding parallel output rays. The point F_1 and the plane perpendicular to the optical axis that passes through F_1 are referred to as the *first focal point* and the *first focal plane*, respectively. Similarly, the point H_1 and the

TABLE 2.3: Some elementary $ABCD$ matrices ($g = \sqrt{\frac{k_2}{k_0}}$)

Configuration	Physical Geometry	ABCD Matrix
Homogeneous Layer	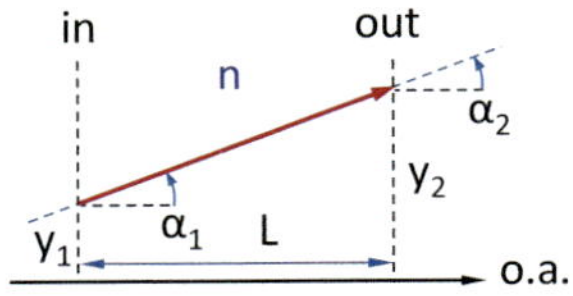	$\tilde{M} = \begin{bmatrix} 1 & L \\ 0 & 1 \end{bmatrix}$
Spherical Refractive Surface	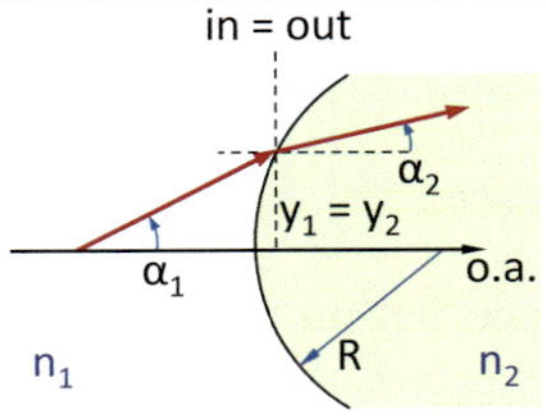	$\tilde{M} = \begin{bmatrix} 1 & 0 \\ \left(\frac{n_1}{n_2} - 1\right)\frac{1}{R} & \frac{n_1}{n_2} \end{bmatrix}$
Planar Refractive Surface	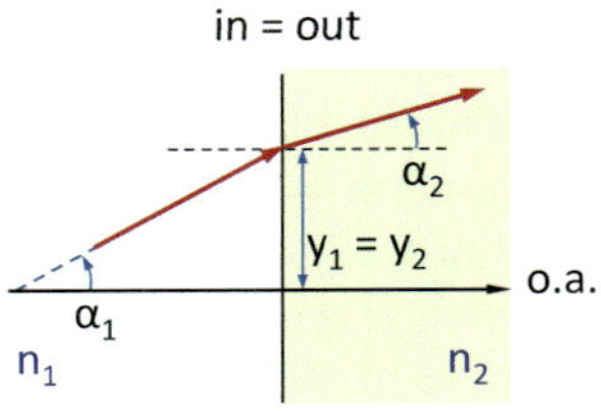	$\tilde{M} = \begin{bmatrix} 1 & 0 \\ 0 & \frac{n_1}{n_2} \end{bmatrix}$
Thin Lens	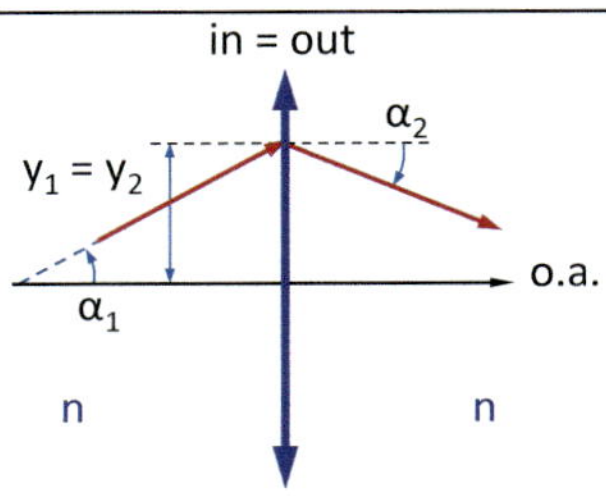	$\tilde{M} = \begin{bmatrix} 1 & 0 \\ -\frac{1}{f} & 1 \end{bmatrix}$
Spherical Mirror	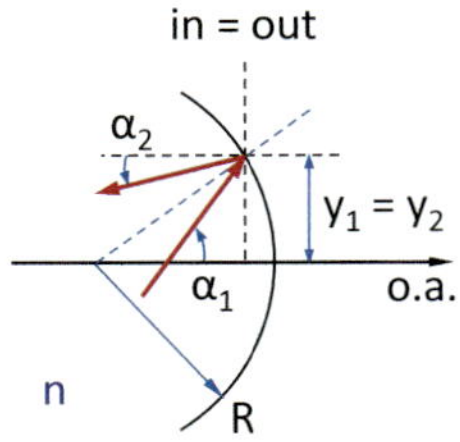	$\tilde{M} = \begin{bmatrix} 1 & 0 \\ \frac{2}{R} & 1 \end{bmatrix}$
Lenslike Medium	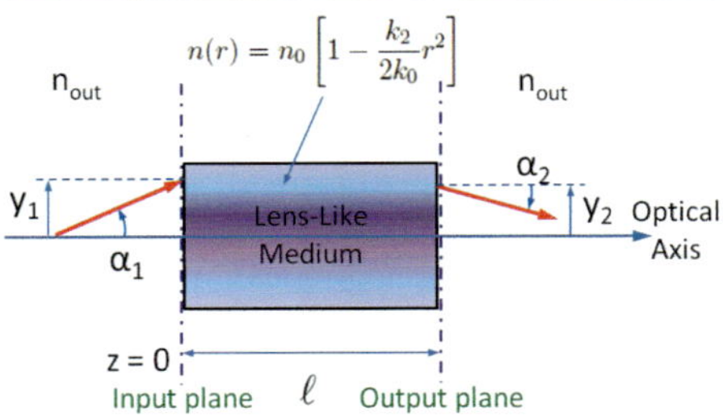	$\tilde{M} = \begin{bmatrix} \cos(g\ell) & \frac{1}{g}\sin(g\ell) \\ -g\sin(g\ell) & \cos(g\ell) \end{bmatrix}$

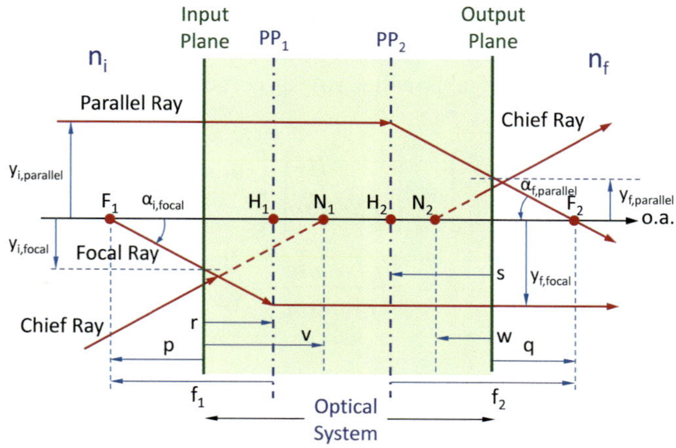

FIGURE 2.47: The definition of the cardinal points of an optical system. The cardinal points are F_1, F_2 (focal points), H_1, H_2 (principal points), and N_1, N_2 (nodal points).

plane perpendicular to the optical axis that passes through H_1 are designated as the *first principal point* and the *first principal plane*, respectively.

Finally, let us consider a ray incident from the left, with its extension passing through point N_1. When the exiting ray from the optical system is parallel to the incident ray, its extension passes through point N_2 (see Fig. 2.47). This situation resembles the case of the chief ray passing through the center of a thin lens, exiting the lens unaffected due to its infinitesimally small thickness. The points N_1 and N_2 are called the *first and the second nodal points*, respectively. The corresponding planes, which are perpendicular to the optical axis and pass through N_1 and N_2, are referred to as the *first and second nodal planes.* The six points $\{F_1, F_2, H_1, H_2, N_1, N_2\}$ are collectively known as the *cardinal points* of the optical system, and their respective perpendicular planes to the optical axis are termed *cardinal planes.*

Given the elements of the $ABCD$ matrix of an optical system, it is straightforward to determine its cardinal points and planes. For example, let's consider the distance p shown in Fig. 2.47 which specifies the position of the F_1 point of the optical system. Considering, the focal ray shown the following equations can be written:

$$\begin{bmatrix} y_{f,focal} \\ 0 \end{bmatrix} = \begin{bmatrix} A & B \\ C & D \end{bmatrix} \begin{bmatrix} y_{i,focal} \\ \alpha_{i,focal} \end{bmatrix} \Longrightarrow$$

$$0 = Cy_{i,focal} + D\alpha_{i,focal} \Longrightarrow \frac{y_{i,focal}}{\alpha_{i,focal}} = -\frac{D}{C},$$

$$|p| = \frac{y_{i,focal}}{\alpha_{i,focal}} \Longrightarrow \quad p = -\frac{y_{i,focal}}{\alpha_{i,focal}} = \frac{D}{C}, \tag{2.65}$$

$$r = \frac{y_{f,focal} - y_{i,focal}}{\alpha_{i,focal}} = \frac{Ay_{i,focal} + B\alpha_{i,focal} - y_{i,focal}}{\alpha_{i,focal}} \Longrightarrow$$

$$r = \frac{-AD + BC}{C} + \frac{D}{C} = \frac{D - (n_i/n_f)}{C}, \tag{2.66}$$

$$f_1 = p - r = \frac{n_i/n_f}{C}, \tag{2.67}$$

where the determinant $AD-BC = n_i/n_f$ was used. The convention here is that all distances that point to the left are considered negative, and all distances that point to the right are considered positive. Similarly, if a parallel ray is considered (see Fig. 2.47) the following equations can be written:

$$\begin{bmatrix} y_{f,parallel} \\ \alpha_{f,parallel} \end{bmatrix} = \begin{bmatrix} A & B \\ C & D \end{bmatrix} \begin{bmatrix} y_{i,parallel} \\ 0 \end{bmatrix} \Longrightarrow$$

$$|q| = \frac{y_{f,parallel}}{|\alpha_{f,parallel}} = \frac{A}{C} \Longrightarrow$$

$$q = -\frac{A}{C}, \tag{2.68}$$

$$s = \frac{y_{i,parallel} - y_{f,parallel}}{\alpha_{f,parallel}} = \frac{1-A}{C}, \tag{2.69}$$

$$f_2 = q - s = -\frac{A}{C} - \frac{1-A}{C} = -\frac{1}{C}. \tag{2.70}$$

Using a similar approach the distances v and w can be determined as

$$v = \frac{D-1}{C}, \tag{2.71}$$

$$w = \frac{(n_i/n_f) - A}{C}. \tag{2.72}$$

A summary of the position of the cardinal points of an optical system from its $ABCD$ matrix elements is shown in Table 2.4.

TABLE 2.4: Cardinal Points from $ABCD$ Elements

Cardinal Points Location	
$p = \frac{D}{C}$	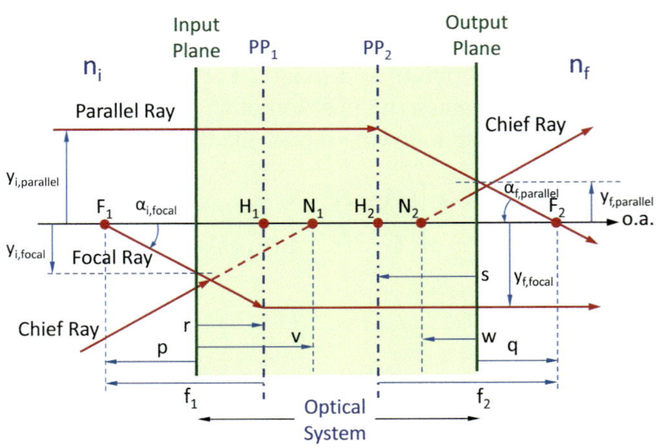
$r = \frac{D-(n_i/n_f)}{C}$	
$q = -\frac{A}{C}$	
$s = \frac{1-A}{C}$	
$v = \frac{D-1}{C}$	
$w = \frac{(n_i/n_f)-A}{C}$	
$f_1 = \frac{n_i/n_f}{C}$	
$f_2 = -\frac{1}{C}$	

The cardinal points can be used to analyze an optical system under the paraxial approximation. To specify the object-image relationship, consider the general optical system illustrated in Fig. 2.48. The two-dimensional object OO' of height h is on the left side of the optical system. Assume that the image II' of height h' is a real image on the right side of the optical system. Next, consider a parallel ray, a focal ray, and a chief ray emanating from the object and converging onto the image, all under the paraxial approximation. From

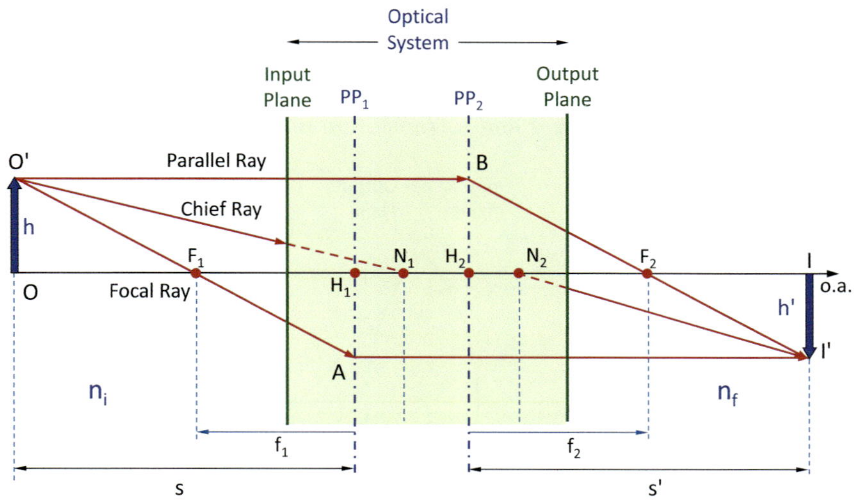

FIGURE 2.48: Equivalent thin lens equation for an optical system based on the cardinal points and the paraxial approximation.

the similar orthogonal triangles $O'OF_1$ and F_1H_1A and the H_2BF_2 and F_2II' the following equations can be written (consider all distances as absolute distances without any sign convention at this point):

$$\frac{h}{s-f_1} = \frac{h'}{f_1},$$

$$\frac{h}{f_2} = \frac{h'}{s'-f_2}, \qquad \text{resulting in}$$

$$\frac{f_1}{s} + \frac{f_2}{s'} = 1. \tag{2.73}$$

In the case that $n_i = n_f$ it is straightforward to show that the nodal points coincide with the principal points, i.e., $N_1 \equiv H_1$ and $N_2 \equiv H_2$. In addition $f_2 = -f_1 = f$ where f is the focal distance of the optical system. Then the equivalent thin lens equation [Eq. (2.73)] becomes:

$$\frac{1}{s} + \frac{1}{s'} = \frac{1}{f}, \tag{2.74}$$

implying that the simple thin lens equation can also be applied to a complicated optical system where the distances of the object and of the image are measured with respect to principal points H_1 and H_2, respectively. In addition, Eq. (2.74) can be generalized utilizing the sign convention used for the thin lens equation. I.e., an object distance is positive when the object is to the left of H_1 (real object) and an image distance is positive when the image is to the right of H_2 (real image). In the case of negative s or s' the object and/or the image are virtual. Furthermore, if $f > 0$ corresponds to F_2 being to the right of H_2 (positive distance) and F_1 to the left of H_1. In the opposite case, where $f < 0$, the system is a diverging one.

2.7.7 Thick Lens

As an example of the $ABCD$ matrix approach, consider the case of a thick lens, defined as a lens with significant width. A biconvex thick lens is depicted in Fig. 2.49, featuring a left radius of curvature R_1, a right radius of curvature R_2, and a thickness d. The refractive index of the lens is n_L, and it is surrounded by homogeneous media with refractive indices n_1 and n_2. Assume that the input plane of the lens is tangential to its left vertex, while the output plane is tangential to its right vertex. The $ABCD$ matrix of the optical system composed of the thick lens is represented as the product of three elementary matrices: two spherical refractive surface matrices and a translation matrix:

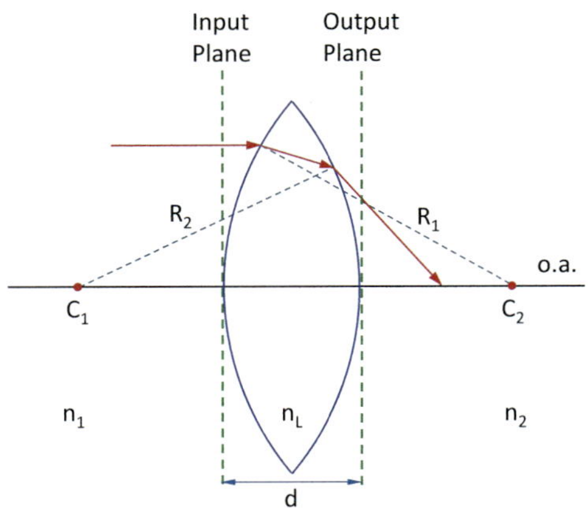

FIGURE 2.49: The optical system consisting of a thick lens with radii of curvature R_1 and R_2 and thickness d. The refractive index of the lens is n_L and the left and right regions refractive indices are n_1 and n_2, respectively.

$$\tilde{M} = \begin{bmatrix} A & B \\ C & D \end{bmatrix} = \begin{bmatrix} 1 & 0 \\ \left(\frac{n_L}{n_2} - 1\right)\frac{1}{R_2} & \frac{n_L}{n_2} \end{bmatrix} \begin{bmatrix} 1 & d \\ 0 & 1 \end{bmatrix} \begin{bmatrix} 1 & 0 \\ \left(\frac{n_1}{n_L} - 1\right)\frac{1}{R_1} & \frac{n_1}{n_L} \end{bmatrix}. \tag{2.75}$$

After the multiplication the resulting A, B, C, and D elements of the thick lens are given by:

$$A = 1 + \frac{d}{R_1}\left(\frac{n_1}{n_L} - 1\right) \tag{2.76}$$

$$B = d\,\frac{n_1}{n_L} \tag{2.77}$$

$$C = \left(\frac{n_L}{n_2} - 1\right)\frac{1}{R_2} + \frac{n_L}{n_2}\left(\frac{n_1}{n_L} - 1\right)\frac{1}{R_1} + \frac{d}{R_1 R_2}\left(\frac{n_L}{n_2} - 1\right)\left(\frac{n_1}{n_L} - 1\right) \tag{2.78}$$

$$D = d\,\frac{n_1}{n_L}\left(\frac{n_1}{n_L} - 1\right)\frac{1}{R_2} + \frac{n_1}{n_2}. \tag{2.79}$$

Then, the effective focal length of the thick lens is expressed by:

$$\frac{1}{f_2} = -C = \frac{n_L - n_1}{n_2}\frac{1}{R_1} - \frac{n_L - n_2}{n_2}\frac{1}{R_2} + \frac{d}{R_1 R_2}\frac{(n_L - n_1)(n_L - n_2)}{n_2 n_L}, \tag{2.80}$$

where $f_1 = (n_1/n_2)(1/C)$ is the first focal distance. The second focal distance of the thick lens as measured from its right vertex point is $q = -A/C = Af_2$. When the refractive indices of the surrounding indices are equal, $n_1 = n_2$, then $f_2 = -f_1 = f$. In the previous expressions the sign conventions for R_1 and R_2 hold as they were described previously and therefore all the expressions are valid for any type of lens and not only for the biconvex one considered in Fig. 2.49. Of course, in the case that $n_1 = n_2 = 1$ (the thick lens is surrounded by air) and $d \to 0$ the expression for the focal distance reduces to that of a thin lens focal length.

Example 2.7.1:

A system of two positive thin lenses is given as shown in Fig. 2.50. The left thin lens has a focal distance of $f_1 = 50\,\text{mm}$ (converging) and the right thin lens has a focal distance of $f_2 = 25\,\text{mm}$ (converging also). The two thin lenses are separated by 40 mm. An object is placed at a distance of 75 mm to the left of the left thin lens. Find the position and magnification of the final image using (a) the method of *ABCD* matrices, (b) the sequential use of the thin lens equation, and (c) the method of cardinal points.

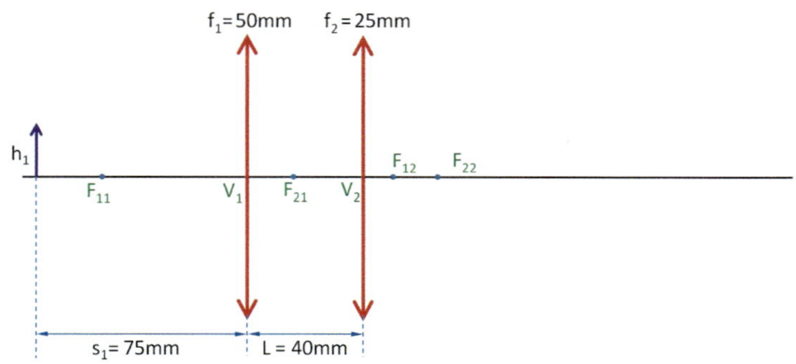

FIGURE 2.50: The optical system consisting of two thin lenses of focal distances of 50mm and 25mm respectively, separated by a distance of 40mm.

Solution: **(a) Method of ABCD Matrices:** The matrix *ABCD* of the two thin lenses separated by a distance $L = 40\,\text{mm}$ can be determined by the multiplication of three basic *ABCD* matrices, the right thin lens matrix, the propagation by distance L matrix, and the left thin lens matrix. Multiplying the three matrices the *ABCD* of the optical system comprising the two thin lenses can be calculated as follows:

$$\begin{bmatrix} A_{TL} & B_{TL} \\ C_{TL} & D_{TL} \end{bmatrix} = \begin{bmatrix} 1 & 0 \\ -\dfrac{1}{25} & 1 \end{bmatrix} \begin{bmatrix} 1 & 40 \\ 0 & 1 \end{bmatrix} \begin{bmatrix} 1 & 0 \\ -\dfrac{1}{50} & 1 \end{bmatrix} = \begin{bmatrix} 0.20 & 40\,mm \\ -0.028\,mm^{-1} & -0.6 \end{bmatrix}.$$

Now it is assumed that the image is formed at a distance x to the right of the right thin lens. Then if the object plane is considered as the input plane and the image plane as the output plane of the overall optical system, the overall *ABCD* matrix can be determined as follows (see Figure 2.51):

$$\begin{bmatrix} A & B \\ C & D \end{bmatrix} = \begin{bmatrix} 1 & x \\ 0 & 1 \end{bmatrix} \begin{bmatrix} A_{TL} & B_{TL} \\ C_{TL} & D_{TL} \end{bmatrix} \begin{bmatrix} 1 & 75 \\ 0 & 1 \end{bmatrix} = \begin{bmatrix} 0.20 - 0.028x & 55 - 2.7x(mm) \\ -0.028\,mm^{-1} & 2.7 \end{bmatrix}.$$

To achieve image formation, it is necessary for the overall *ABCD* matrix element B to equal zero. When $B = 0$, the value of A provides the overall magnification of the image.

Thus,

$$B = 0 \Longrightarrow 55 - 2.7x = 0 \Longrightarrow x = 20.37\,mm,$$

while the magnification, m, is given by:

$$m = A(x = 20.37) = 0.2 - 0.028x = -0.37.$$

Therefore, the image is real and inverted.

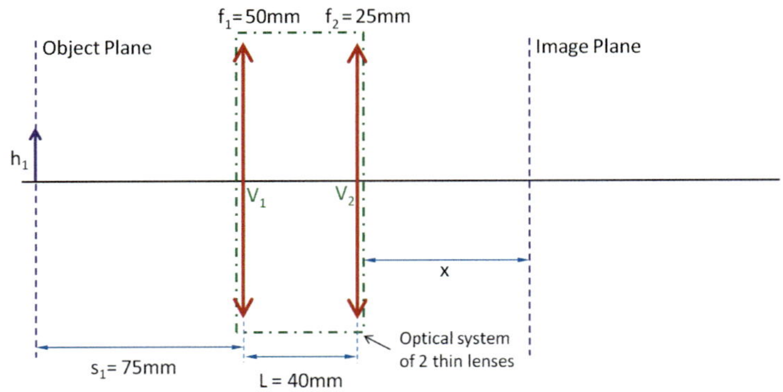

FIGURE 2.51: The optical system consisting of two positive thin lenses of focal distances of 50mm and 25mm respectively, separated by a distance of 40mm is represented by the dashed rectangle and an equivalent $ABCD$ matrix. The image plane is considered to be at the unknown distance x away from the right thin lens.

(b) Method of Cascaded Thin Lenses: In this approach, the thin lens equation is applied sequentially for each thin lens. First, the equation is applied to the left lens, where the object distance is $s_1 = 75$ mm and the focal length is $f_1 = +50$ mm. Applying the thin lens equation in this case the image distance s_1' due to the first lens (if the second lens is temporarily neglected) can be determined from the equation:

$$\frac{1}{s_1} + \frac{1}{s_1'} = \frac{1}{f_1} \Longrightarrow \frac{1}{75} + \frac{1}{s_1'} = \frac{1}{50} \Longrightarrow s_1' = 150\,mm.$$

The image formed by the first lens serves as the object for the second (right) lens. Since the image is to the right of the right lens it will act as a virtual object for the second lens. Therefore, $s_2 = -(150 - 40) = -110$ mm. Applying the thin lens equation for the second lens the final image location can be determined:

$$\frac{1}{s_2} + \frac{1}{s_2'} = \frac{1}{f_2} \Longrightarrow \frac{1}{-110} + \frac{1}{s_2'} = \frac{1}{25} \Longrightarrow s_2' = 20.37\,mm.$$

Therefore, the final image is formed 20.37 mm to the right of the right lens and is a real image. The magnification can be found as follows:

$$m = \frac{h_2'}{h_1} = \frac{h_1'}{h_1}\frac{h_2'}{h_2 = h_1'} = \left(-\frac{s_1'}{s_1}\right)\left(-\frac{s_2'}{s_2}\right) = -\frac{150}{75}\left(-\frac{20.37}{-110}\right) = -0.37,$$

implying the final image is inverted (and real as it was determined through s_2'). A ray diagram of the image formation by the two lenses is shown in Fig. 2.52.

(c) Method Using Cardinal Points: In order to determine the cardinal points ($H_1 \equiv N_1$, $H_2 \equiv N_2$, F_1, and F_2), the matrix $ABCD$ of the optical system comprised of the two thin

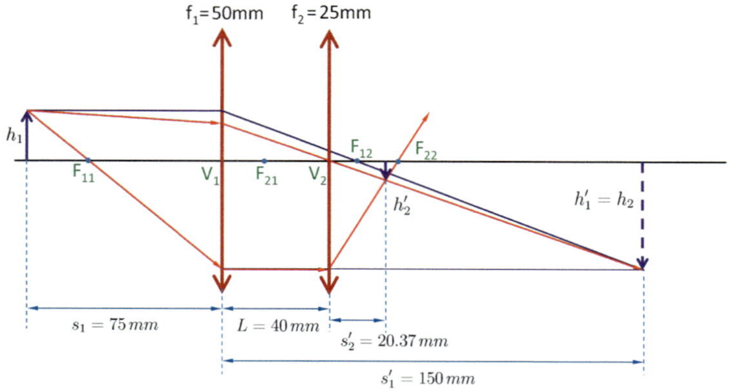

FIGURE 2.52: A ray diagram for the two-lenses system.

lenses must be used. This matrix was computed earlier and is equal to:

$$\begin{bmatrix} A_{TL} & B_{TL} \\ C_{TL} & D_{TL} \end{bmatrix} = \begin{bmatrix} 0.20 & 40\,mm \\ -0.028\,mm^{-1} & -0.6 \end{bmatrix}.$$

Using the elements of the matrix the signed distances, f_1, f_2, r, and s (see Fig. 2.53) can be determined as follows:

$$f_1 = \frac{n_0/n_f}{C_{TL}} = \frac{1}{-0.028} = -35.714\,mm,$$

$$f_2 = -\frac{1}{C_{TL}} = -\frac{1}{-0.028} = +35.714\,mm,$$

$$r = \frac{D_{TL} - (n_i/n_f)}{C_{TL}} = \frac{-0.6 - 1}{-0.028} = +57.1428\,mm,$$

$$s = \frac{1 - A_{TL}}{C_{TL}} = \frac{1 - 0.2}{-0.028} = -28.5714\,mm,$$

Now the thin lens equation can be used by suitably selecting the object and image distances s_o and s_i by measuring them from the first and the second principal planes respectively. In this case $s_o = 75 + r = 132.1428$ mm. The focal distance of the system is $f = 35.714$ mm. Then the thin lens equation can be written as follows:

$$\frac{1}{s_o} + \frac{1}{s_i} = \frac{1}{f} \Longrightarrow \frac{1}{132.1428} + \frac{1}{s_i} = \frac{1}{35.714} \Longrightarrow s_i = 48.9412\,mm.$$

Therefore the position of the real image ($s_i > 0$) from the right thin lens is $s_i - |s| = 48.9412 - 28.5714 = 20.37$ mm. The magnification is $m = -s_i/s_o = -48.9412/132.1428 = -0.37$ as it was computed with the other methods. A ray diagram using the cardinal points is shown in Fig. 2.54.

Example 2.7.2:

Example Imaging with a Nikkor 50mm Lens. In this example a real optical lens is used for imaging. The lens is illustrated in Fig. 2.55, where the cardinal points are also indicated. An object, O, is placed at a distance $L_O = 100$ mm in front of the front (left) lens surface

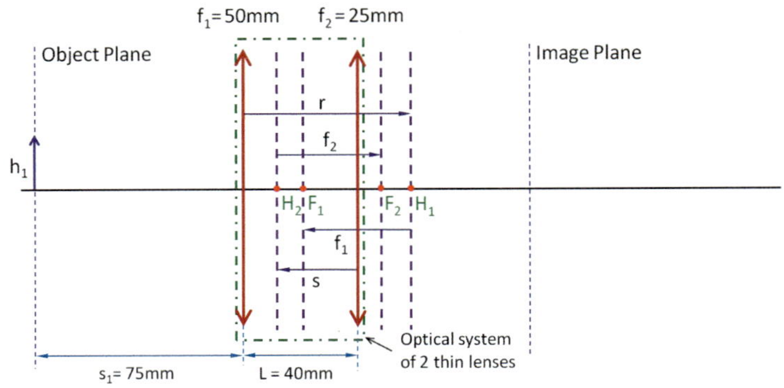

FIGURE 2.53: Important distances for the determination of the cardinal points of the two-lenses optical system. The nodal points N_1 and N_2 coincide with the principal points H_1 and H_2 respectively.

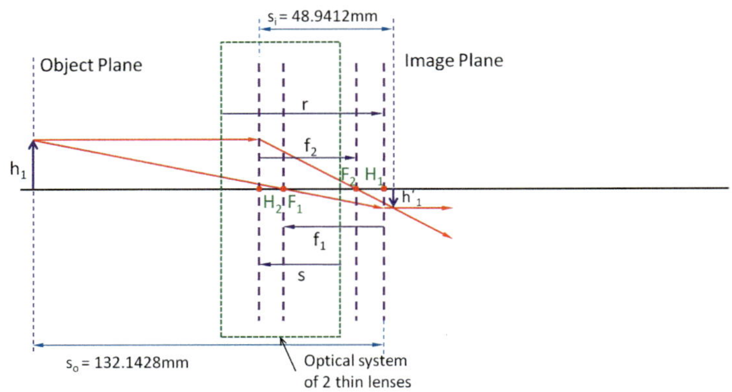

FIGURE 2.54: A ray diagram that shows the formation of the image through the use of the cardinal points of the two-lenses system.

(from vertex V_1) of the Nikkor-H auto 50mm f/2 photographic lens. The goal is to calculate the location of the image, I, with respect to the rear surface of the rear lens (distance L_i), as well as its magnification.
Solution: In order to apply the equivalent thin lens equation, Eq. (2.74), using the cardinal points, it is necessary to specify the distance of the object from the first principal, s_O, as shown in the figure. The distances s_o and s_i are determined as follows (all distances are in millimeters):

$$s_o = L_O + [34.1 - (51.6 - 7.3 - 38.1)] = L_O + 27.9 = 127.9\,mm \Longrightarrow$$

$$\frac{1}{s_o} + \frac{1}{s_i} = \frac{1}{f} \Longrightarrow \frac{1}{127.9} + \frac{1}{s_i} = \frac{1}{51.6} \Longrightarrow s_i = 86.5\,mm.$$

The distance of the image from the right vertex V_2, L_I is easily determined from s_i as follows: $L_I = s_i - (51.6 - 38.1) = 86.5 - 13.5 = 73\,\text{mm}$. Since $s_i > 0$ the image is real and inverted. The transverse magnification $m = -s_i/s_o = -86.5/127.9 = -0.676$. This example demonstrates how, with an understanding of the cardinal points, paraxial imaging can be easily achieved by appropriately applying the thin lens equation.

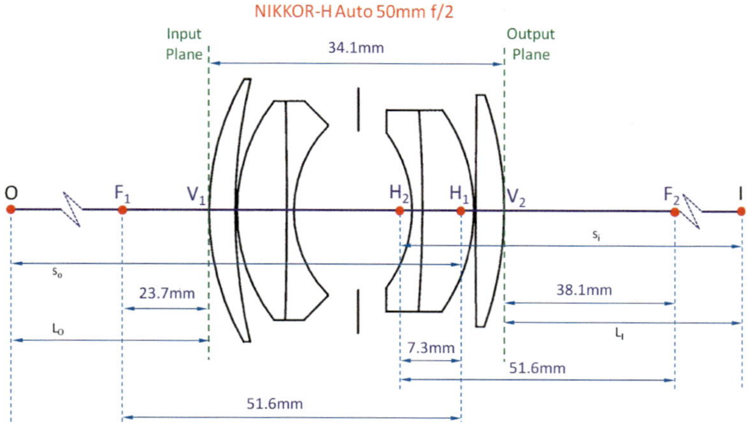

FIGURE 2.55: A Nikkor H-auto 50mm $f/2$ lens imaging [41]. The cardinal points H_1, H_2, F_1, and F_2 are shown in the figure (air is assumed that surrounds the lens). An object O is at a distance $L_O = 100\,\text{mm}$ from the left vertex V_1 of the lens. The image is formed at a distance L_i from the right vertex V_2 of the lens.

2.8 Basic Optical Systems

Optical systems are typically composed of various fundamental components, such as lenses, prisms, mirrors, filters, apertures, and pinholes, each serving a distinct function (for instance, forming an image in a camera). This section provides an overview of some of the most fundamental optical systems. Because most optical systems incorporate apertures, or stops, it is essential first to understand the role and significance of aperture stops, along with the related concepts of entrance and exit pupils. Similarly, understanding field stops, along with entrance and exit windows, is important for all optical instruments. After introducing these essential definitions, this section presents several foundational optical systems: (a) the camera, (b) the magnifier, (c) the compound microscope, (d) the optical telescope, and (e) the human eye.

2.8.1 Stops, Pupils, and Windows

Analyzing optical systems through geometrical optics involves tracing rays that originate from an object and converge at the image. However, depending on the object point's location and the angle of each ray relative to the optical axis, many rays may be blocked by the system's limiting apertures and thus never reach the image. In addition to the physical boundaries of the optical elements themselves, apertures (also referred to as stops, which are geometric boundaries) are often introduced into optical systems. These apertures serve various purposes, such as controlling the light entering the system, setting the field of view, managing aberrations, and defining image boundaries. The presence of apertures in an optical system has two primary effects [20, 42, 43]: (a) Field of View Limitation: Apertures limit the light cone from the object that can be seen through the optical system, effectively restricting the field of view. (b) Image Brightness Control: By selectively limiting the rays that pass through the system's stops, apertures control the irradiance at the image plane, affecting the image's brightness. The following definitions are essential for understanding the role of apertures in optical systems.

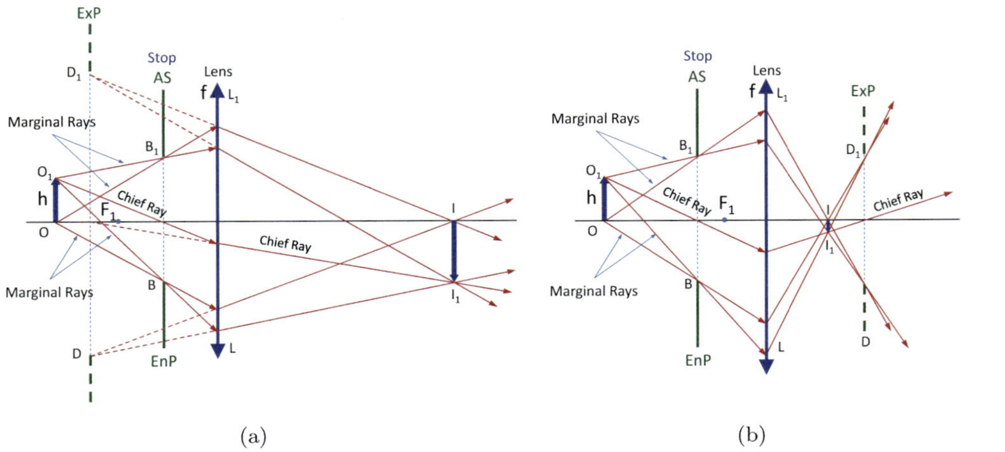

FIGURE 2.56: (a) Stop in front of lens with virtual exit pupil. (b) Stop in front of lens with real exit pupil.

Aperture stop (AS): The aperture stop in an optical system is either a dedicated aperture (stop) or the physical size limitation of one of the optical elements within the system. The aperture stop defines the maximum cone of rays from an axial point on the object that can reach its conjugate image point. Examples of aperture stops include the diaphragm in a camera and the iris in the human eye.

Entrance Pupil (EnP): The entrance pupil of an optical system is the apparent limiting aperture as "seen" by rays coming from the object space. The EnP of an optical system represents the image of the aperture stop (AS) as viewed from an axial point on the object, through any optical elements that precede the AS in the system.

Exit Pupil (ExP): The exit pupil of an optical system is the effective limiting aperture as "seen" by rays arriving from the image space. The ExP of an optical system represents the image of the aperture stop (AS), as formed by the optical elements positioned after the AS in the direction toward the image space.

To demonstrate the previous definitions a simple optical system is considered. This system is comprised of a single lens of focal length $f > 0$, with a stop in front of it, as shown in Fig. 2.56(a). In this setup, the stop is positioned to the left of the lens at a distance less than the focal length. Consider an object OO_1 and its real image II_1 (the object distance is between f and $2f$ in this ray diagram). The stop is the actual limiting aperture for the object rays and therefore consist of the AS. Since no optical elements precede the AS, the entrance pupil (EnP) coincides with the AS itself. The exit pupil (ExP), on the other hand, is the image of the AS as viewed through the lens from the image plane. In this configuration, the ExP appears as a virtual image, as shown in Fig. 2.56(a). The *chief ray* is defined as the ray originating from any off-axis object point that passes through the axial point of the EnP plane (the intersection of the EnP plane with the optical axis). For example, the chief ray from object point O_1 passes through the midpoint of the BB_1 distance. When extended backward from the lens, this chief ray intersects the ExP plane at its axial point, passing through the midpoint of the DD_1 distance, due to the object-image relationship between the EnP and ExP. A *marginal ray* is a ray from an object point that passes through the edge of the EnP. Some marginal rays are also illustrated in Fig. 2.56(a).

A similar case involves positioning the stop again to the left of the lens, but at a distance greater than the lens's focal length, as shown in Fig. 2.56(b). Here, the stop continues to serve as the aperture stop (AS), and the entrance pupil (EnP) coincides with it. However, in this case, the exit pupil (ExP) is a real image, forming on the right side of the lens, as depicted in Fig. 2.56(b). The ExP defines the solid angle of rays that converge on the on-axis image point [43]. In this second configuration, a larger cone of light contributes to image formation, meaning that rays at wider angles relative to the optical axis pass through the system. This increase in angular range tends to amplify aberration effects. Notably, as the stop moves closer to the object, the limiting light cone and the AS may eventually be defined by the lens's edge.

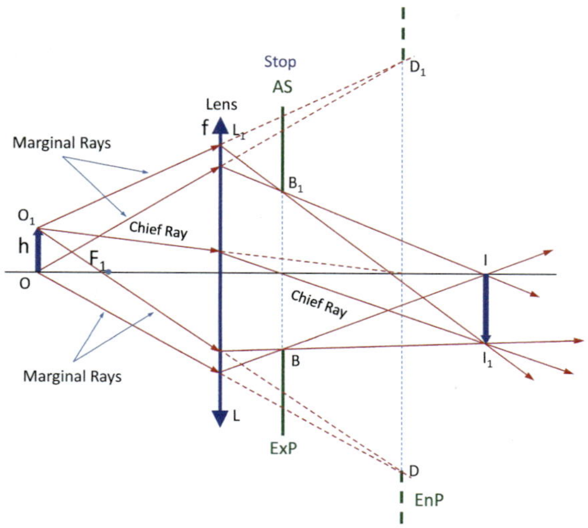

(a)

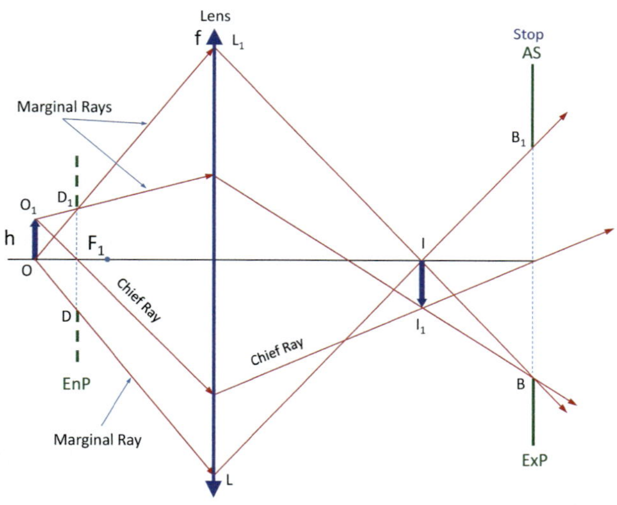

(b)

FIGURE 2.57: (a) Stop in the back of lens with virtual entrance pupil. (b) Stop in the back of lens with real entrance pupil.

Another example considers the stop positioned behind the lens (to the right). The two possible configurations for this case are shown in Figs. 2.57(a) and 2.57(b). Here, the stop serves as the AS, as the lens is assumed to be significantly larger in size. Since no optical elements are present after the AS, the ExP coincides with the AS itself. In contrast, the

EnP is the image of the AS as formed by the lens. Depending on the distance of the AS from the lens, this EnP can be either virtual (as in Fig. 2.57(a)) or real (as in Fig. 2.57(b)). In both cases, the EnP represents the axial cone of light originating from an on-axis object point that, passing through the lens and the AS, reaches the image plane.

In the previous examples, the AS either coincided with the EnP when positioned to the left of the lens or with the ExP when placed to the right. However, in complex optical systems containing multiple elements and stops, identifying the AS requires a systematic approach. To locate the AS, each stop and the rim of each optical element should be imaged to the left (opposite to the direction of ray propagation from the object, assumed here from left to right) using all optical elements preceding the stop or rim of interest. For each image, the solid angle subtended by an axial object point O should be determined. The image with the smallest solid angle defines the actual entrance pupil, and the corresponding stop or rim represents the AS. Similarly, the ExP can be found by imaging all stops and element rims to the right (in the same direction as the rays propagate from the object) using all optical elements following the stop or rim of interest. For each image, the solid angle subtended by an axial image point I should be evaluated. The image with the smallest solid angle represents the actual exit pupil, with the corresponding stop or rim identified as the AS. Both methods should ideally identify the same optical element or stop as the AS, as the EnP and ExP maintain an object-image relationship.

At this point the following additional definitions pertinent to optical systems are introduced:

Field Stop (FS): The field stop defines the solid angle formed by the chief rays emanating from off-axis objects, effectively limiting the field of view (FoV) of an optical instrument. An example of a field stop is the view an individual has from inside a building looking through a window; in this case, the window serves as the FS, restricting the field of view that the observer can perceive.

Entrance Window (EnW): The entrance window of an optical system is the image of the field stop as observed through all the optical elements located before the field stop in the object space.

Exit Window (ExW): The exit window is the image of the field stop as viewed through all the optical elements positioned after the field stop in the image space.

As an example, consider the single-lens optical system depicted in Fig. 2.58. In this configuration, a stop is placed in front of the lens, functioning as both the AS and the EnP. The image size is constrained by the size of the detector, which is a reasonable consideration. The cone of light, defined by the angle of the field of view (FoV), is formed by the chief rays that pass through the edges of the detector. This cone specifies the field of view, while the detector size itself acts as the FS. The FS also serves as the ExW at the image plane, while the image of the FS in the object space, as formed by the lens, corresponds to the EnW at the object plane.

In more complex optical systems, determining the FS, EnW, and ExW may not be straightforward. In such cases, a procedure analogous to that used for identifying the AS can be employed. Specifically, all optical element rims and stops are imaged into the object space using all elements that precede them. The angle subtended by each image at the on-axis position of the EnP is then evaluated. The image corresponding to the smallest angle is identified as the EnW. The physical object (stop or rim) that corresponds to this EnW serves as the FS. Furthermore, the image of the FS, as formed by all subsequent optical elements in the image space, corresponds to the ExW. In essence, the FS acts as a new AS when the object is positioned at the EnP [38].

Another effect to consider is that a cone of light rays originating from an off-axis point of the object may not always converge at the image point, as it can be obstructed by the stops or the rims of the optical elements in the system. This obstruction results in a

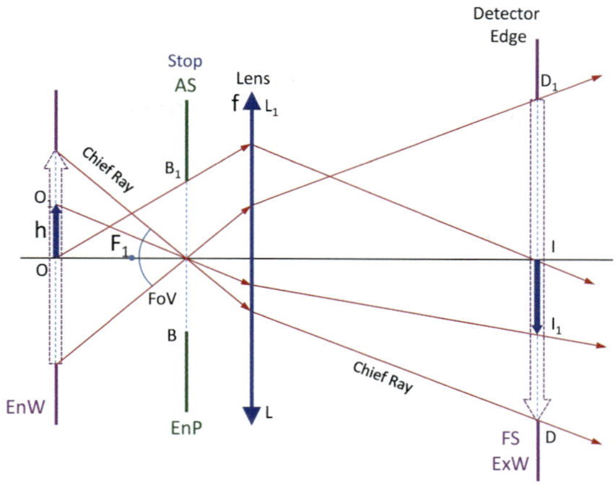

FIGURE 2.58: The definition of the field stop (FS) is due to the detector size. The AS is in front of a single lens. The entrance and exit windows are also shown. The entrance pupil coincides with the AS but the exit pupil is not shown. The field of view (FoV) is also shown.

reduction of light intensity at the edge points of the image, leading to a phenomenon known as *Vignetting*.

Example 2.8.1:

An optical system comprising of two positive lenses $L1$ and $L2$ of focal lengths f_1 and f_2, respectively, is shown in Fig. 2.59. The object plane coincides with the first focal plane of the left lens and the image plane coincides with the second focal plane of the right lens. The separation between the lenses is $f_1 + f_2$. A stop S_1 of diameter D_1 is placed centered at distance $L = f_1/2$ from the right of the $L1$ lens. Another stop S_2 of diameter D_2 is placed centered on the image plane. The diameters of the lenses are equal to $D = 2D_1$ with $D_1 = 1.5D_2$. Furthermore, $f_2 < f_1 < 4f_2$. Determine the AS, entrance, and exit pupils as well as the FS and entrance and exit windows of this optical system.

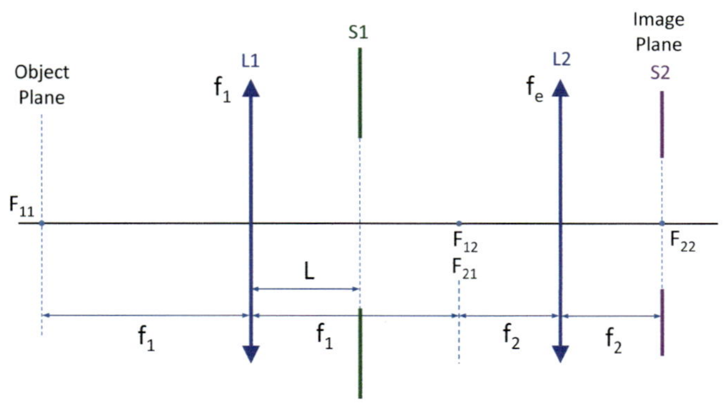

FIGURE 2.59: An example optical system of two-positive lenses $L1$ and $L2$ of focal lengths f_1 and f_2, respectively. The object plane and image plane are shown as well as the stops $S1$ and $S2$.

Solution: Initially, the AS should be determined. For this reason every element should be imaged in the object space and the angle subtended from axial object point with every image should be determined. The first element is the left lens. Since there are no other optics preceding it in object space its image EnP_{L_1} is itself. Therefore the angle $\theta_{L_1} = \tan^{-1}(D/2f_1)$. The second element is the stop $S1$. Its image in object space is given by:

$$\frac{1}{L} + \frac{1}{s'_1} = \frac{1}{f_1} \implies s'_1 = -f_1,$$

and then the corresponding image is EnP_{S_1} at f_1 distance to the right of $L1$ lens with magnification $m_{S_1} = -s'_1/(L = f_1/2) = 2$. Then the corresponding angle is $\theta_{S_1} = \tan^{-1}(D_1 m_{S_1}/2)/2f_1 = \tan^{-1}(D_1/2f_1) < \theta_{L_1}$. Similarly, for the right lens:

$$\frac{1}{f_1+f_2} + \frac{1}{s'_1} = \frac{1}{f_1} \implies s'_1 = \frac{f_1(f_1+f_2)}{f_2}.$$

The corresponding image $\text{EnP}_{L_2} = s'_1$ to the left of $L1$ and the magnification is $m_{L_2} = -s'_1/(f_1+f_2) = -f_1/f_2$. Then the subtended angle is $\theta_{L_2} = \tan^{-1}\{(D(f_1/f_2)/2)/(s'_1 - f_1)\} = \tan^{-1}\{D/2f_1\} = \theta_{L_1}$. The last check is for the stop $S2$. The image of this in the object space is

$$\frac{1}{f_2} + \frac{1}{s'_2} = \frac{1}{f_2} \implies s'_2 = \infty.$$

$$\frac{1}{\infty} + \frac{1}{s'_1} = \frac{1}{f_1} \implies s'_1 = f_1.$$

The corresponding image $\text{EnP}_{S_2} = f_1$ to the left of $L1$ and the magnification is $m_{S_2} = f_1/f_2$. I.e., the image of $S2$ coincides with the object plane. Then the subtended angle is $\theta_{S_2} = \pi$. Therefore the smallest subtended angle is the θ_{S_1} and the $S1$ stop is the AS. Then $\text{EnP}_{S_1} =$ EnP which is located at distance f_1 to the right of $L1$ lens with diameter $D_{EnP} = 2D_1$. The exit pupil is the image of AS with the optics succeeding it and therefore,

$$\frac{1}{f_1/2+f_2} + \frac{1}{s'_2} = \frac{1}{f_2} \implies s'_2 = f_2 + 2\frac{f_2^2}{f_1}.$$

Then the resulting ExP diameter is $D_{ExP} = D_1(2f_2/f_1)$.

A similar procedure is followed for the determination of the FS. For the $L1$ lens its image to object space is itself, and the angle subtended from the axial point of the EnP is $\theta_{L_1} = \tan^{-1}\{D/2f_1\}$. For the $S1$ which is the AS the EnP is its image and the angle $\theta_{S_1} = \pi$. For the lens $L2$ the image is at distance $s'_1 = f_1(f_1+f_2)/f_2$ to the left of $L1$ lens and the corresponding angle from the axial point of EnP is $\theta_{L_2} = \tan^{-1}\{(D/2)(f_1/f_2)/(s'_1+f_1)\} = \tan^{-1}\{D/2(f_1+2f_2)\}$. Finally, for the $S2$ stop the image is the object plane and the corresponding angle from the axial EnP point is $\theta_{S_2} = \tan^{-1}\{(D_2/2)(f_1/f_2)/2f_1\} = \tan^{-1}\{D/12f_2\} < \theta_{L_2}$ (due to the $f_2 < f_1 < 4f_2$ constraint). Consequently the $S2$ aperture is the FS. Then the object plane is the EnW and the image plane is the ExW. The field of view angle is FoV$= 2\tan^{-1}\{D_2/4f_2\}$. On the other hand, the numerical aperture corresponds to the marginal rays cone of light that is entering the system and is $NA = \sin\theta = \sin\{\tan^{-1}(D_1/2f_1)\}$. The NA aperture is defined in Sec. 2.8.4 and in Appendix 2C.

2.8.2 Camera

Humans have long been captivated by the nature of light and the processes involved in image formation and recording. The precursor to the modern camera is the "camera obscura" (Latin for "dark chamber"). The fundamental operation of the camera obscura is illustrated

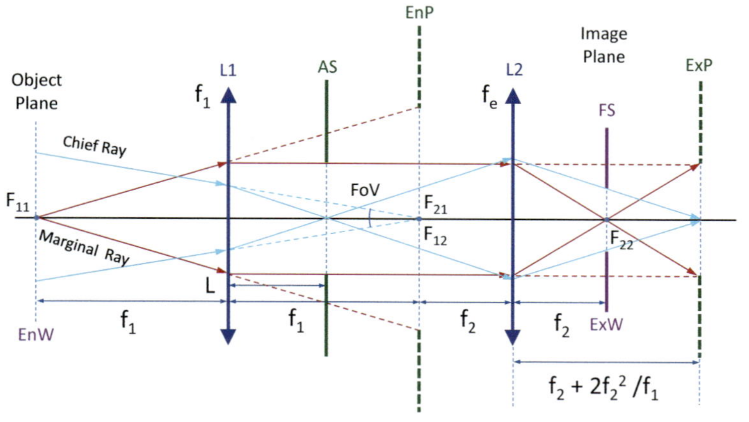

FIGURE 2.60: The chief (blue) and marginal (red) ray diagrams of the optical system of Example 2.8.1. The AS, EnP, ExP, FS, EnW, and ExW are also shown in the diagram. The field of view is the angle FoV.

in Fig. 2.61. This device consists of a closed box with a small hole on one side. Light rays from an object (such as the tree depicted in Fig. 2.61) pass through the hole and project onto the opposite wall of the box, thereby creating an inverted real image. Before delving into the basic functions of contemporary cameras, it is worthwhile to summarize the key dates and events that mark the evolution of the camera and photography, the art of recording images [44].

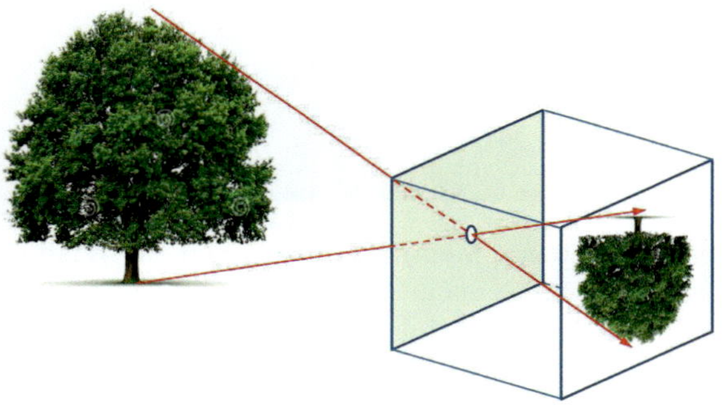

FIGURE 2.61: The basic diagram of the camera obscura. Rays from the object (oak tree) pass through a small hole and form an inverted real image on the back plane of the camera obscura (the tree in the picture was selected from url-link: `https://www.dreamstime.com/royalty-free-stock-photos-oak-tree-image7507678`).

2.8.2.1 Brief History of Photography

The natural occurrences of the camera obscura are believed to have inspired some of the cave paintings from the Paleolithic era. The earliest documented mention of the camera obscura effect comes from Mozi, a Chinese philosopher of the 5th century BCE. A century later, Aristotle (384–322 BCE) observed the crescent-shaped images of the sun projected onto the ground beneath a tree during a solar eclipse, attributing this phenomenon to the small openings between the leaves, which acted as pinholes in the context of the camera obscura [45]. In 1038, the Arab scholar Alhazen described a working model of the camera obscura in his work, *Perspectiva*. For several centuries, the concept of the camera obscura was employed for viewing solar eclipses, and by the 16th century, it served as a valuable aid for drawing. Drawings and paintings, produced using the camera obscura, can be considered the ancestors of modern photographs, although the camera obscura itself could not store images. In 1490, Leonardo da Vinci (1452–1519) was the first to propose that painters could utilize the camera obscura in their work. The first documented mention of a convex lens used in conjunction with the camera obscura came from the Italian mathematician Girolamo Cardano (1501-1576). The addition of a mirror to the camera obscura was first suggested by Ettore Ausonio (1520-1570). In 1604, German astronomer Johannes Kepler (1571-1630) developed the first portable camera obscura, which took the form of a tent with a sheet of paper inside onto which the projected image could be rendered [45]. In 1685, Johann Zahn (1641-1707) likely designed the first camera obscura that could be manually focused by adjusting the lens instead of moving the screen [45].

Although the camera obscura evolved over the years, it was unable to permanently record images. The observation of various light-sensitive substances was known as early as the 17th century. However, photography is primarily a 19th-century invention. The first permanent recorded image was created by the Frenchman Joseph Nicéphore Niépce in 1826. Niépce used a thin asphalt coating on polished pewter to produce a permanent positive image. This photograph, titled as "View from the Window at Le Gras" (shown in Fig. 2.62) is notable for its extremely long exposure time, which possibly lasted up to two days [44]!

FIGURE 2.62: The first photograph entitled "View from the Window at Le Gras," produced by Joseph Nicéphore Niépce in 1826 (manually enhanced version, Rebecca A. Moss, Coordinator of Visual Resources and Digital Content Library, via email. College of Liberal Arts Office of Information Technology, University of Minnesota. `http://www.dcl.umn.edu`) (the picture was taken from url-link: `https://commons.wikimedia.org/w/index.php?curid=107219`).

Around the same time, Louis Jacques Mandé Daguerre contacted Niépce and they formed a partnership in 1829. By the early 1830s, Daguerre and Niépce discovered that light would darken polished silver that had been exposed to iodine fumes. After Niépce's death, Daguerre made significant advancements in 1835 by realizing that a silver iodide plate required only a fraction of the exposure time compared to previous methods. He also found that the recorded image could be stabilized by immersing it in a bath of sodium chloride [44]. This process resulted in the creation of the *Daguerrotype*, a recorded photograph that revolutionized the field of photography. A sample Daguerrotype is shown in Fig. 2.63 which is recognized as the first photograph to include a person. This was possible due to the dramatic reduction in exposure time - Daguerrotypes required exposure times on the order of 10 to 20 minutes, compared to the several hours needed in the in the Niépce process. This Daguerrotype is known as the "View of the Boulevard du Temple." By 1841, William Henry Fox Talbot had developed a printing process that utilized sensitized paper, significantly reducing exposure times to just seconds. This innovation led to the creation of the

FIGURE 2.63: The first photograph including a person (on the pavement at lower left), by Daguerre, in 1838, known as "View of the Boulevard du Temple." By Louis Daguerre—Scanned from The Photography Book, Phaidon Press, London, 1997 (Public Domain) (the picture was taken from url-link: `https://commons.wikimedia.org/w/index.php?curid=5005681`).

"calotype" [44]. By 1850, the Englishman Frederic Scott Archer advanced the field further by coating plates with a sticky fluid, resulting in the "collodion" process, which employed silver salts and reduced exposure times even further. By the mid-1850s, several variants of the collodion process had been invented, accommodating both positive and negative image formation. In 1871, Richard L. Maddox, also from England, discovered the gelatin dry-plate process for photography, which would serve as the precursor to the analog films widely used throughout the major part of the 20th century.

The photographic processes of the 19th century were primarily sensitive to ultraviolet, violet, and blue wavelengths. Enhancing spectral sensitivity was crucial for the advancement of color photography. In 1861, James Clerk Maxwell demonstrated additive color synthesis by using red, blue, and green filters to create the first color image of a ribbon (shown in Fig. 2.64(a)). This pioneering work laid the groundwork for future developments in color photography.

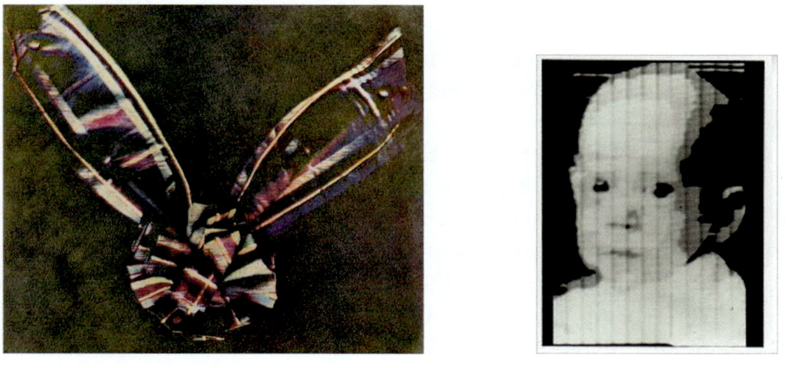

(a) (b)

FIGURE 2.64: (a) The first color photograph of a ribbon, produced according to James-Clerk Maxwell additive color process (1861) (the picture was taken from the url-link: `https://upload.wikimedia.org/wikipedia/commons/thumb/7/7f/Tartan_Ribbon.jpg/220px-Tartan_Ribbon.jpg`). (b) The first digital photograph produced of scanned photograph via a digital computer by Russell Kirch at the U.S. National Bureau of Standards in 1957 (the picture was taken from the url-link: `https://upload.wikimedia.org/wikipedia/commons/thumb/1/1a/NBSFirstScanImage.jpg/170px-NBSFirstScanImage.jpg`).

By 1877, American Eadweard Muybridge developed a fast shutter process that enabled him to capture photographs of moving objects, paving the way for modern cinematography. In 1900, George Eastman, of the Eastman Kodak Company, introduced the first widely available camera, known as the "Brownie." This innovation made photography accessible to amateurs, rather than being limited to professional photographers. Throughout the 20th century, many additional cameras and photographic lenses were introduced, accompanied by advancements in analog films of various speeds.

The advent of the space age significantly accelerated the development of digital photography. The first digital photograph (illustrated in Fig. 2.64(b)) was created in 1957 by Russell Kirsch at the U.S. National Bureau of Standards using a digital computer to scan photographs. A major breakthrough in digital imaging came with the invention of the charge-coupled device (CCD) at Bell Labs by Willard Boyle and George E. Smith in 1969. This technology was further refined by Fairchild Semiconductor in 1973. In 1975, Steven Sasson, an engineer at Eastman Kodak, built the first digital camera using the CCD sensor from Fairchild Semiconductor. Remarkably, this early digital camera weighed 4 kg and could capture an image of 100×100 pixels in 23 seconds, recording it on tape! The first commercial digital camera emerged in 1988. These early digital cameras utilized JPEG and MPEG standards to compress images and video for efficient storage. As technology progressed, CCD sensors were gradually replaced by Complementary Metal-Oxide-Semiconductor (CMOS) sensors, which reduced camera costs and enabled their incorporation into mobile phones. Further information about both CCD and CMOS sensors can be found in Ref. [46]. Today, digital cameras have largely prevailed over analog film cameras in most everyday applications.

2.8.2.2 Camera Basic Operation

The concept of the camera obscura can also be described as a lensless pinhole camera, allowing the formation of an image of an object, such as a tree, through rays passing through an aperture. An example schematic of the lensless pinhole camera is shown in Fig. 2.65. This configuration creates an image without the assistance of a lens. However, since the aperture has a finite diameter, multiple rays from the same object point converge at different points on the image plane. This phenomenon leads to a blurred image on the right side of the configuration. A common solution is to decrease the size of the aperture diameter, which initially helps by restricting many rays from a single point to pass through the aperture. Nonetheless, this approach has its drawbacks: reducing the aperture diameter also diminishes the amount of light entering the system, resulting in a faint image. When using analog film or a digital sensor for image capture, longer exposure times become necessary. Further reduction of the aperture diameter typically worsens image quality due to diffraction effects originating from the aperture. In practice, depending on the object's size and distance from the aperture (or pinhole) plane, there exists an optimal aperture size. For instance, an aperture diameter of approximately 0.5 mm is near optimal for an object distance of 25 cm [42]. To enhance image quality and reduce exposure times, modern cameras utilize lenses, typically in the form of a lens system. An example camera schematic is depicted in Fig. 2.66. This illustration features a typical lens system positioned in front of the sensor plane (for digital cameras) or the film location (for older analog cameras). Additionally, it includes other essential components such as the viewfinder and shutter mechanism. The incorporation of lenses allows for better control over light convergence, resulting in sharper images with less distortion compared to a lensless pinhole camera. By adjusting the focal length and aperture settings of the lens system, photographers can manipulate depth of field, exposure time, and other critical aspects of image formation, significantly improving the overall photographic experience.

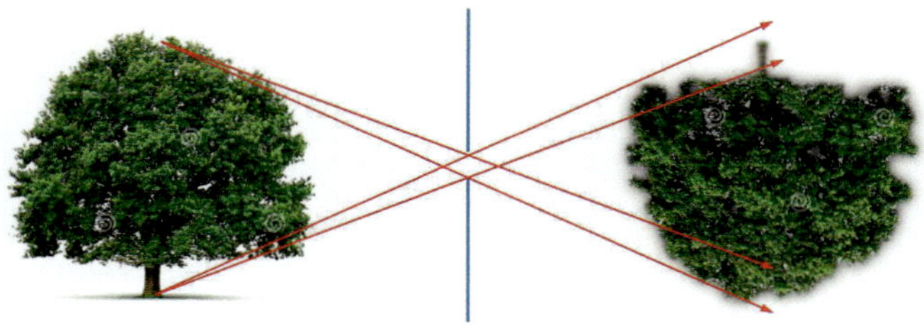

FIGURE 2.65: A schematic of the formation of the image of a tree through an aperture. Since various rays from the same point of the tree passing through the aperture end at different points, the image is blurred. The amount of blurring depends on the diameter of the aperture.

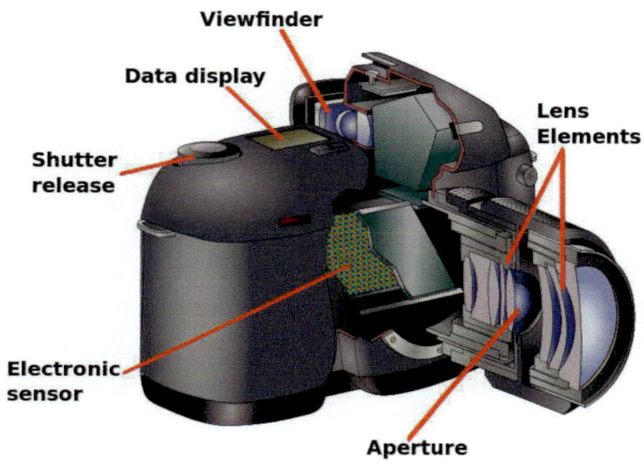

FIGURE 2.66: A schematic of the basic components of a modern camera (by Astrocog) (the picture was taken from url-link: `https://commons.wikimedia.org/w/index.php?curid=29334470`).

At this point, the f-number (or f-stop) for a camera can be defined using Fig. 2.67. In this illustration, a planar object is imaged through a thin lens (or a system of lenses, utilizing the paraxial approximation as discussed in Sec. 2.7.6) onto an image plane. The object and image distances from the thin lens are denoted as s and s', respectively, with the understanding that in the case of a real lens or lens system, these distances are measured from the first and second principal points. From the thin lens equation, the following relations can be established:

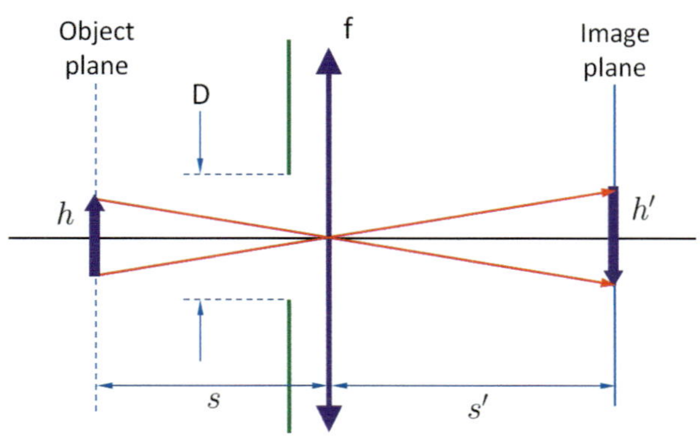

FIGURE 2.67: A basic configuration of forming an image of an object with the help of a lens.

$$\frac{1}{s} + \frac{1}{s'} = \frac{1}{f} \Longrightarrow s' = \frac{sf}{s-f}$$

$$m = \frac{h'}{h} = -\frac{s'}{s} \Longrightarrow h' = h\left|-\frac{f}{s-f}\right| \simeq h\frac{f}{s},$$

since usually $f \ll s$, and m is the transverse magnification. The irradiance, E_e, that arrives at the image plane is approximately proportional to the light power of the image divided by its area:

$$E_e \simeq \frac{k\pi D^2/4}{\pi h'^2/4} \sim \left(\frac{D}{h'}\right)^2 \simeq \left(\frac{D}{hf}\right)^2 s^2$$

$$E_e \sim \left(\frac{D}{f}\right)^2 = \frac{1}{A^2},$$

$$A = \frac{f}{D} \quad \text{(f-number).} \tag{2.81}$$

In the last equation the *f-number* or *f/stop* is defined as it is commonly used for cameras (more general definition of the f-number is presented in Appendix 2C). The exposure that reaches the image plane represents the total light energy (or the total number of photons) required to form the image. Exposure can be mathematically expressed as the product of irradiance and exposure time (i.e., the duration for which the camera aperture remains open). Consequently, exposure is influenced by the inverse square of the f-number, which can be derived from the previously established equations. This relationship implies that a smaller f-number allows for a shorter exposure time to achieve the same total exposure. The total exposure is contingent upon the type of sensor utilized for image recording, which can either be analog (such as the film used predominantly throughout the 20th century) or digital (like the CCD or CMOS sensors employed in contemporary cameras). In commercial cameras, standard f-numbers are typically used, which commonly differ by a factor of $\sqrt{2}$, leading to exposures that vary by a factor of 2. A list of the most standard f-numbers and their corresponding irradiances is presented in Table 2.5. For instance, a photograph taken with an f-number of $A = 1.4$ necessitates twice the exposure time compared to a photograph captured at $A = 1$. It is important to note that the choice of f-number also affects the depth of field, a concept that will be explored further in this section.

2.8.2.3 Circle of Confusion

The human eye has a limited resolution, with the closest comfortable viewing distance for normal vision being approximately 25 cm. At this distance, a person with normal eyesight can resolve about 5 line pairs per millimeter (lp/mm). This finite resolution gives rise to the concept of the "*circle of confusion*" (*CoC*), defined as the largest blurred spot that the human eye still perceives as a single point (i.e., appearing in focus) when viewed from the closest comfortable distance.

The circle of confusion, $CoC = c$, depends on both the size of the sensor (analog or digital) used to capture the image and the viewing conditions, including image size and viewing distance. For instance, when an image is viewed from a distance of 25 cm a CoC of 0.2 mm may be sufficient. Assuming a viewing angle of 60 deg, the diameter of the viewed image can be calculated as $2 \times 25\tan(60°/2) = 28.9$ cm. This dimension aligns with the diagonal of an 8 in × 10 in image, which measures approximately $(8^2 + 10^2)^{1/2} \times 2.54$ cm/in $= 32.52$ cm (for reference $A4$ paper is 8.25 in × 11.75 in = 21 cm × 29.7 cm). An 8 in × 10 in image can be produced from a 35 mm (film or digital sensor) with a diagonal of $(24^2 + 36^2)^{1/2} = 43.26$ mm.

TABLE 2.5: Common f-numbers

$\mathbf{f}/D = A$	$\mathbf{A}^2$	**Irradiance, E** $= k/A^2$
1	1	E_0
1.4	2	$E_0/2$
2	4	$E_0/4$
2.8	8	$E_0/8$
4	16	$E_0/16$
5.6	32	$E_0/32$
8	64	$E_0/64$
11	128	$E_0/128$
16	256	$E_0/256$
22	512	$E_0/512$
32	1024	$E_0/1024$

Considering the magnification $m = 32.52/4.326 = 7.52$, if the 8 in × 10 in image has a $CoC = 0.2\,\text{mm}$, the corresponding CoC on the sensor should be $CoC = 0.2\,\text{mm}/m \simeq 0.026\,\text{mm}$. For larger images, it is reasonable to assume that the viewing distance would increase proportionally, allowing for a larger CoC. However, if a larger image is viewed at the original 25 cm distance, a smaller CoC on the sensor would be necessary to maintain resolution. In photography the CoC can be approximately given by $CoC = c = d/1500$ (or $d/1730$) where d is the diagonal of the film or the digital sensor used. Some common CoC for various sensor sizes are shown in Table 2.6 [47]. The following equation can also be written for the CoC based on the visual acuity of the human eye and the standard distance of 25 cm:

$$CoC = \frac{\text{viewing distance (cm)}}{25\,\text{cm}} \frac{1}{\#\text{lp/mm}} \frac{1}{m}, \tag{2.82}$$

where m is the magnification, and #lp/mm is the resolution required (for example previously 5 lp/mm was used). It should be noted that the CoC is also influenced by diffraction effects, as light behaves like a wave and thus cannot be focused on an infinitely small point. Additionally, the CoC is affected by the optical system's aberrations and, naturally, by the pixel size of a digital detector (or grain size in analog film). These factors can impact the practical values of the CoC. For instance, the CoC value calculated from Eq. (2.82) cannot be smaller than the diffraction limit. The diffraction-limited resolution, detailed in the *Scalar Diffraction* chapter (Chapter 5), is based on the Rayleigh criterion for a circular aperture. According to this limit, the CoC is given by $CoC = 2 \times 1.22\lambda_0(f/D)$, where λ_0 is the wavelength of the light (in the case of white light the average wavelength of about 0.55 μm can be used), and f/D is the f-number.

2.8.2.4 Depth of Field and Depth of Focus

The *Depth of Field* (DoF) and the *Depth of focus* (Dof) are distinct concepts related to the image quality of an imaging system, such as a camera. The DoF refers to the range of axial depth in object space within which an object can be moved without introducing detectable blurring in the image, as long as the image plane remains fixed. In contrast, the Dof is the range of axial depth in image space where the image remains sharp as the position of the analog film or digital sensor is adjusted, with the object plane held constant. Both the DoF and the Dof are influenced by the acceptable CoC and other system parameters, such as

TABLE 2.6: Common Circles of Confusion (from Ref. [47])

Smaller Format		
Image Format	Frame Size	**CoC** (mm)
Nikon 1 Series	8.8 mm × 13.2 mm	0.011
Four Thirds System	13.5 mm × 18 mm	0.015
APS-C	15.0 mm × 22.5 mm	0.018
APS-C Canon	14.8 mm × 22.2 mm	0.018
APS-C Nikon/Pentax/Sony	15.7 mm × 23.6 mm	0.019
APS-H Canon	19.0 mm × 28.7 mm	0.023
35 mm	24 mm × 36 mm	0.029
Medium Format		
Image Format	Frame Size	**CoC** (mm)
645 (6 × 4.5)	56 mm × 42 mm	0.047
6 × 6	56 mm × 56 mm	0.053
6 × 7	56 mm × 69 mm	0.059
6 × 9	56 mm × 84 mm	0.067
6 × 12	56 mm × 112 mm	0.083
6 × 12	56 mm × 168 mm	0.120
Large Format		
Image Format	Frame Size	**CoC** (mm)
4 × 5	102 mm × 127 mm	0.110
5 × 7	127 mm × 178 mm	0.150
8 × 10	203 mm × 254 mm	0.220

the f-number, focal length, and object distance. Given a specified CoC, the DoF and Dof can be determined using the thin lens equation for an optical system. The derivation of the DoF and Dof can be visualized in Fig. 2.68(a). Based on the definitions of distances shown in Fig. 2.68(a), the following equations can be written:

$$\frac{1}{s} + \frac{1}{s'} = \frac{1}{f},$$

$$\frac{1}{z_f} + \frac{1}{z'_f} = \frac{1}{f},$$

$$\frac{1}{z_n} + \frac{1}{z'_n} = \frac{1}{f},$$

where s and s' are the initial object and image distances for the thin lens of focal length f (it is reminded that this thin lens can in general represent an optical imaging system under the paraxial approximation using the cardinal points). The distance z_f is the object distance in the far side of the DoF and the z_n is the object distance in the near side of the DoF. Similarly, distances z'_f and z'_n are the image distances corresponding to z_f and z_n object distances, respectively. The $DoF = z_f - z_n$ as illustrated in Fig. 2.68(a). These distances must be related to the $CoC = c$ and the aperture diameter D using similar triangles. Consequently, additional equations can be readily established:

$$\frac{c}{D} = \frac{s' - z'_f}{z'_f} \Longrightarrow \frac{1}{z'_f} = \frac{1}{s'}\left(1 + \frac{c}{D}\right),$$

$$\frac{c}{D} = \frac{z'_n - s'}{z'_n} \Longrightarrow \frac{1}{z'_n} = \frac{1}{s'}\left(1 - \frac{c}{D}\right).$$

Based on the previous equations the distances z_f and z_n can be calculated as follows:

$$\frac{1}{z_f} = \frac{1}{f} - \frac{1}{s'}\left(1 + \frac{c}{D}\right) = \frac{1}{s}\left(1 + \frac{c}{D}\right) - \frac{1}{f}\frac{c}{D},$$

$$\frac{1}{z_n} = \frac{1}{f} - \frac{1}{s'}\left(1 - \frac{c}{D}\right) = \frac{1}{s}\left(1 - \frac{c}{D}\right) + \frac{1}{f}\frac{c}{D}.$$

From the last equations it can be observed that there is an object distance $s = H$ for which $z_f \to \infty$. This special object distance is known as *hyperfocal distance* and is given by:

$$H = f + \frac{f^2}{c}\frac{1}{(f/D)} = f + \frac{f^2}{c}\frac{1}{A}. \tag{2.83}$$

Since usually $c = CoC \ll f$, the hyperfocal distance can be approximated as $H_a = f^2 c/A$ [i.e., neglecting f of Eq. (2.83)]. When the object distance is set to the hyperfocal distance H, the DoF becomes infinite! This can be depicted in Fig. 2.68(b). In this case the $z_n = H/2$. For object distances $s > H$ the DoF remains infinite (the equations yield $z_f < 0$) as illustrated in Fig. 2.68(c). The actual DoF can be readily calculated and expressed in various forms, as listed below:

$$DoF = z_f - z_n = \frac{2f^2cs(s-f)A}{f^4 - c^2A^2(s-f)^2}, \tag{2.84}$$

$$= \frac{2Acf^2(1-m)}{f^2m^2 - c^2A^2}, \tag{2.85}$$

$$= \frac{2f\left[Hs^2f + cAsH(s-H)\right]}{(f^2 - c^2A^2)\,H(H-s) + (f + cA)^2\,s(H-s)}, \tag{2.86}$$

where it is reminded that $m = s'/s$ is the transverse magnification. Of course, all the above expressions become infinite when $s = H$. In addition, negative values of DoF imply infinite DoF as it was stated previously.

The depth of focus, Dof, can be easily determined with the help of Fig. 2.69 which is compatible with the definition given earlier. Using the thin-lens equations (under the paraxial approximation for optical systems) the following equations can be written ($m = -s'/s$, and similar triangles are utilized):

$$\frac{1}{s} + \frac{1}{s'} = \frac{1}{f} \Longrightarrow s' = (1-m)f,$$

$$\frac{Dof/2}{s'} = \frac{c}{D} \Longrightarrow Dof = 2c(1-m)\left(\frac{f}{D}\right) = 2c(1-m)A. \tag{2.87}$$

The concept of longitudinal (or axial) magnification [see Eq. (2.52)] can be utilized to relate Dof and DoF. Since $|\Delta s'| = m^2|\Delta s|$, using Eq. (2.87) the following relation between DoF and Dof can be written:

$$DoF = \frac{1}{m^2}Dof = 2c\,\frac{1-m}{m^2}\left(\frac{f}{D}\right) = 2c\,\frac{1-m}{m^2}\,A. \tag{2.88}$$

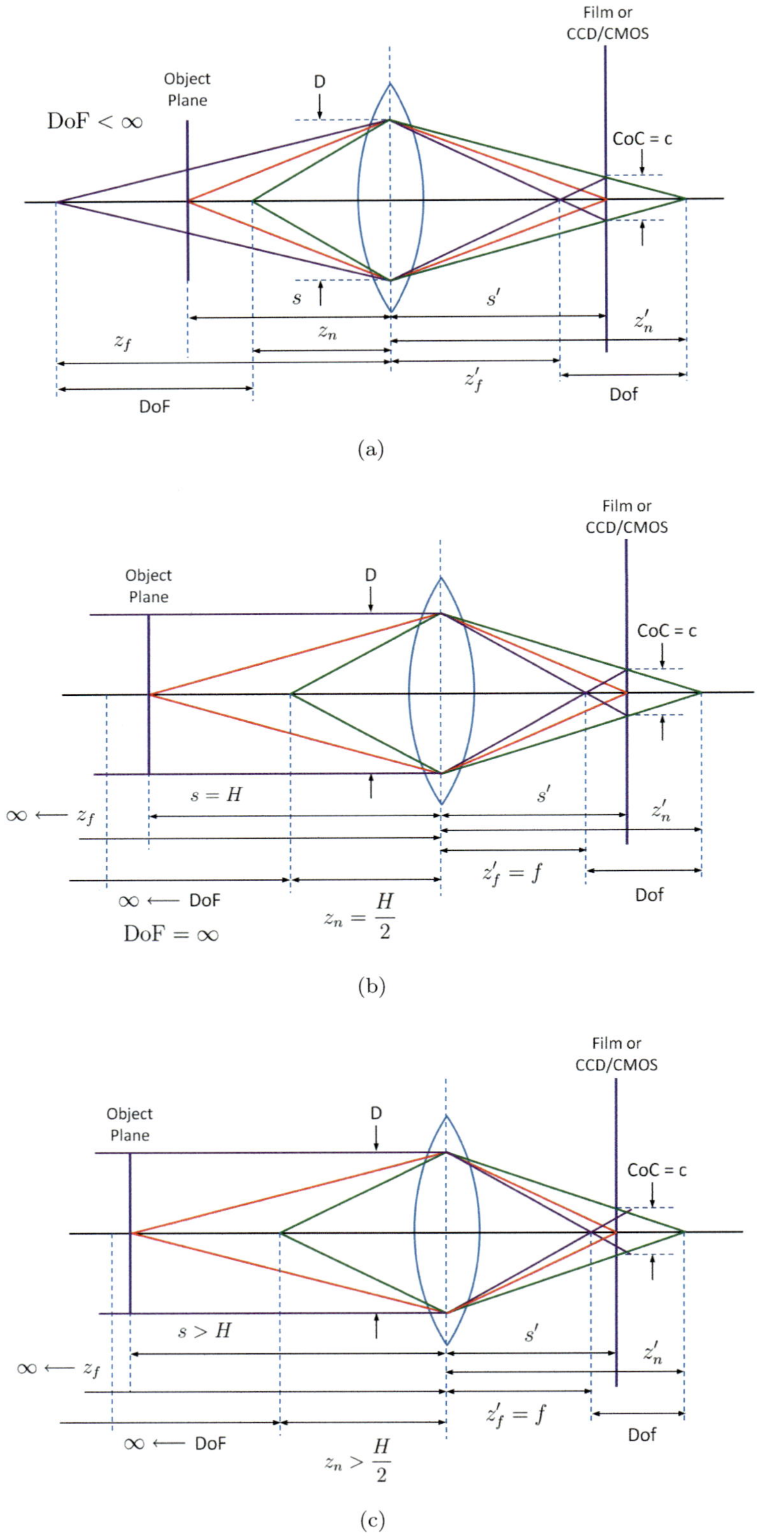

FIGURE 2.68: (a) The definition of the depth of field DoF and the depth of focus, Dof, for a given $CoC = c$. (b) Definition of hyperfocal distance, H for infinite DoF. (c) Focal distance greater than H.

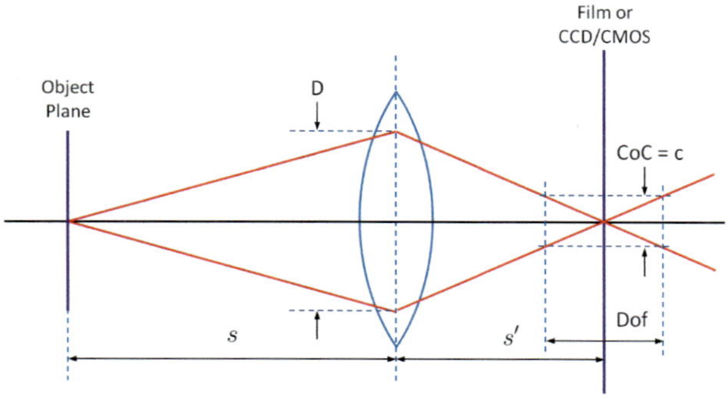

FIGURE 2.69: A schematic for the determination of the depth of focus using its definition.

Naturally, the expression for DoF of Eq. (2.88) is an approximation compared to that in Eq. (2.84). The two expressions converge when both the DoF and Dof are small. At this stage, it is useful to consider some examples to illustrate the dependence of both DoF and Dof of various parameters of the optical system.

Example 2.8.2:

Assume an imaging lens of focal length $f = 50\,\text{mm}$ and an f-number of $f/16$. Using a 35 mm standard film or digital sensor size, the circle of confusion is given by $CoC = c \simeq 1500/\sqrt{24^2 + 36^2} = 0.029\,\text{mm}$. Find the DoF, the Dof, and the transverse magnification as a function of the object distance. Then, assuming a constant object distance of $s = 2\,\text{m}$, find the DoF and the Dof as a function of the camera's f-number.

Solution: The DoF is shown in Fig. 2.70 as a function of the object distance $s < H$ where $H = 5.4379\,\text{m}$ (hyperfocal distance). Of course, as $s \to H$ the $DoF \to \infty$ as expected from the definition of the hyperfocal distance. Observe, the exact value [Eq. (2.84)] and the approximate value [Eq. (2.88)] of DoF differ significantly as the object distance increases. Similarly, the Dof is shown in Fig. 2.71(a) and the resulting magnification in Fig. 2.71(a),

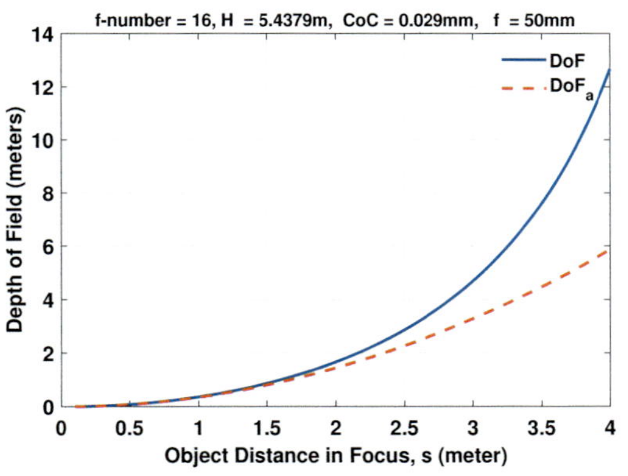

FIGURE 2.70: The depth of field DoF as a function of the object distance for a lens system with $f = 50\,\text{mm}$ and $f/16$. The object distance is retained smaller than the hyperfocal distance.

respectively. Observe that the two equations [Eq. (2.87) and $Dof = z'_n - z'_f$] for the determination of Dof give almost identical results in Fig. 2.71(a). To examine the effect of f-number variation for a fixed imaging lens, the object distance is set to a constant $s = 2\,\text{m}$ while the f-number is varied. The results showing the impact of f-number on DoF and Dof are

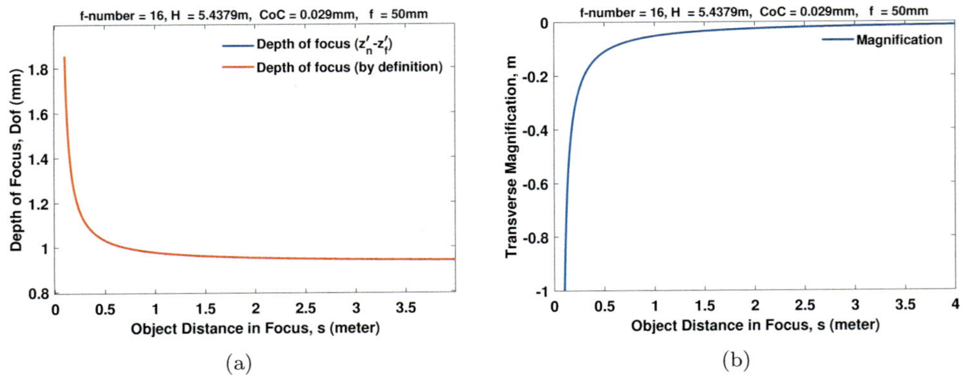

FIGURE 2.71: (a) The depth of focus Dof as a function of the object distance for a lens system with $f = 50\,\text{mm}$ and $f/16$. The object distance is retained smaller than the hyperfocal distance (both curves practically coincide). (b) The magnification of the system as a function of the object distance.

presented in Figs. 2.72(a) and 2.72(b), respectively. Generally, it is observed that the DoF increases with an increasing f-number. Since Dof is small, the two expressions practically coincide, as in the previous case with varying object distance. Example photographs taken with a DSLR camera using a 50 mm focal length at various f-number settings are shown in Fig. 2.73. These images demonstrate the variation in depth of field.

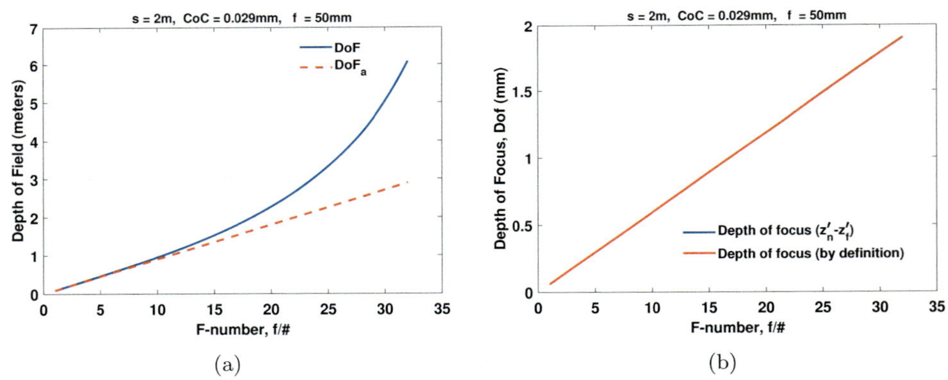

FIGURE 2.72: (a) The depth of field, DoF, as a function of the f-number of the lens system with $f = 50\,\text{mm}$ and $s = 2\,\text{m}$. (b) The depth of focus, Dof, as a function of the f-number of the lens system with $f = 50\,\text{mm}$ and $s = 2\,\text{m}$.

Example 2.8.3:

In this example let's consider a microscope objective of 40X with a numerical aperture of $NA = 0.65$, and a focal length of $f = 4.5\,\text{mm}$. Find the DoF, the Dof, and the transverse magnification as a function of the object distance.

Solution: Using Eq. (2C.2) (of Appendix 2C), which gives the f-number in terms of the numerical aperture, the corresponding f-number is $1/2(NA) = 1/1.30 = 0.76923$. For the CoC the diffraction limit can be utilized. In this case $c = 2 \times 1.22(f/D)\lambda_0 = 2 \times 1.22 \times 0.76923 \times 0.55\mu\text{m} = 0.0010323\,\text{mm}$ (where for the white light, as an approximation, the middle wavelength of $0.55\,\mu\text{m}$ is used). The DoF, Dof and the transverse magnification,

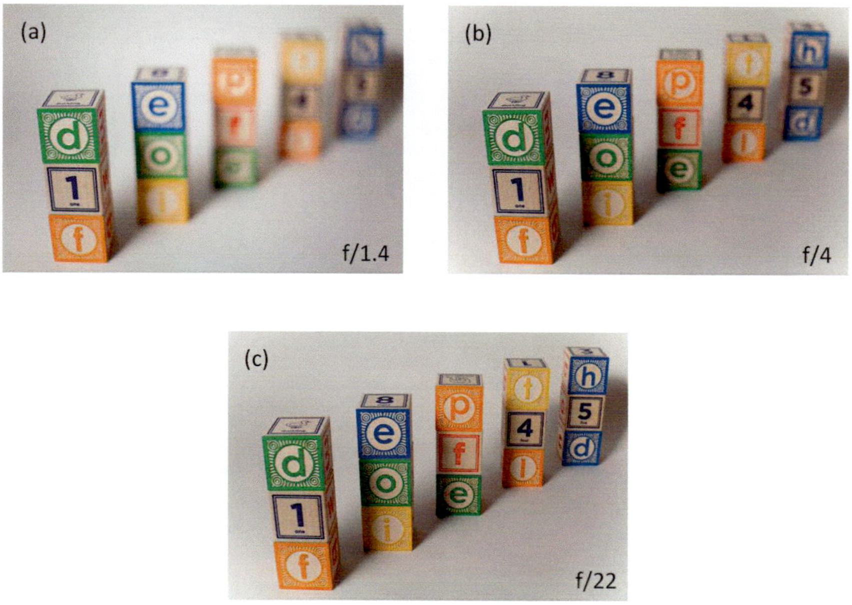

FIGURE 2.73: Demonstration of various DoF in photographs taken with a 50 mm DSLR camera with f-numbers of (a) $f/1.4$, (b) $f/4$, and (c) $f/22$ (by Alex1ruff-Wikipedia) (the pictures were taken from url-link: `https://commons.wikimedia.org/w/index.php?curid=48367959`, `https://commons.wikimedia.org/w/index.php?curid=48367974`, `https://commons.wikimedia.org/w/index.php?curid=48367982`).

m, are shown in Figs. 2.74(a), 2.74(b), and 2.74(c), respectively, as the object distance is varied. In this case the object distance should be measured from the first principle plane and therefore is much greater than the working distance of the objective. From the plots it can be seen that the depth of field is really very small (of the order of a μm) which is expected for microscope objectives. On the other hand, the depth of focus is greater in the order of mm. In the DoF and Dof plots, the exact and approximate expressions practically coincide since both the DoF and the Dof are small.

2.8.2.5 Angle of View

The *angle of view*, AOV, is an important parameter in a camera, as it corresponds to the visual perception of an object's size as it is projected onto the image plane [48]. The AOV represents the angular extent of a scene captured by a camera lens on film or a digital sensor. The AOV can be specified in the horizontal, vertical, or diagonal direction of the detector. In Fig. 2.75, the dimension d denotes either the horizontal, vertical, or diagonal size of the film or sensor. Using this figure, it is straightforward to calculate the AOV by determining the angle ϕ:

$$\tan\frac{\phi}{2} = \frac{d/2}{f} \Longrightarrow \phi = 2\tan^{-1}\left(\frac{d}{2f}\right), \tag{2.89}$$

where it is assumed in the final equation that the object is at infinity. If the object is at a finite distance, the focal length f in Eq. (2.89) should be replaced by the image distance $s' = (1-m)f$ where m is the transverse magnification (noting that $m < 0$ for real images).

Numerous excellent example photographs illustrating the angle of view, depth of field, depth of focus, and other topics related to cameras and photography can be found in the literature, such as in Refs. [48], [49].

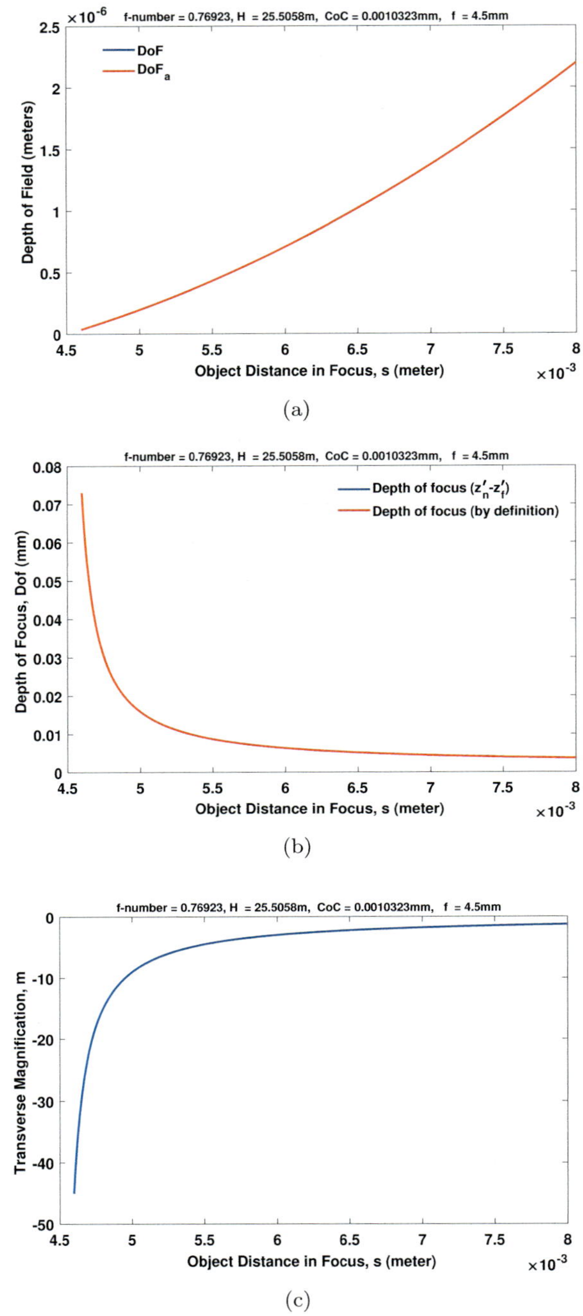

FIGURE 2.74: (a) The depth of field, DoF, as a function of the object distance for a 40X microscope objective with $f = 4.50\,\text{mm}$ and $f/D = 0.79623$. (b) Similarly, the depth of focus, Dof, as a function of the object distance for the same microscope objective. (c) The magnification as a function of the object distance. Notice that for a magnification of $m = -40$ the object distance is $s \simeq 4.61265\text{mm}$ which should correspond to a working distance of $\sim 0.4 - 0.7\text{mm}$. However, in order to find the exact working distance, the position of the first principal plane is necessary, and such information depends on the exact configuration of the lenses and the materials used for the microscope objective (and it was not known in this example, which was used only for demonstration purposes).

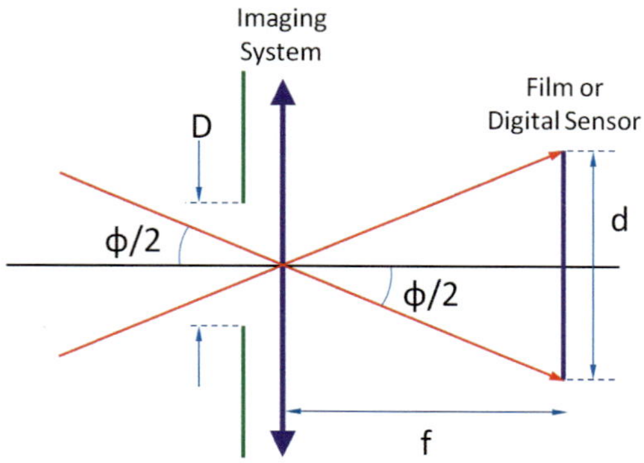

FIGURE 2.75: The angle ϕ that covers the diagonal, d, of the detector size (either an analog film or a digital sensor) specifies the *angle of view.*

Example 2.8.4:

In this example for demonstration purposes the AOV for a standard 35 mm (36mm × 24mm) sensor size is calculated for various focal lengths lenses (adapted from Ref. [49]).
Solution: In this example, the diagonal AOV is determined. Therefore, $d = \sqrt{36^2 + 24^2} = 43.267\,\text{mm}$. The parameters of this example were taken from Ref. [49]. Using Eq. (2.89) the diagonal AOV can be computed and it is tabulated for various focal length lenses in Table 2.7. It can be observed that the large focal length lenses (telephoto lenses) have small AOV. Example photographs for the various lenses of a scene are shown in Ref. [49].

TABLE 2.7: Angle of View for various focal length lenses [49]

f (mm)	AOV (deg)
17	103.68
28	75.38
50	46.79
85	28.56
135	18.21
300	8.25
500	4.95
1000	2.48

2.8.3 Simple Magnifier

When a human observer examines an object, it can be brought closer and closer, causing the retinal image to enlarge as the object approaches. However, the eye's lens can accommodate focus only up to a certain limit, typically around the standard close viewing distance of $d = 25\,\text{cm}$ (for a healthy human eye). Bringing the object closer than this distance results in a blurred retinal image. If the object is small, a positive lens, commonly referred to as *a magnifier* or *a magnifying glass* [20, 42], can be used to help. This type of lens, also known as a reading glass, creates an image larger than what can be achieved with the unaided eye.

In order to illustrate the viewing situation let's consider Fig. 2.76a. The angular extent of the object as observed by the eye is approximately $\theta \simeq h/d$ where h is the object's height. Figure 2.76b represents positive lens can enhance the apparent size of the object. In this case, the observed angle is $\theta' \simeq h/f$. The ratio between the angles θ' and θ is defined as the angular magnification, m_θ, and is given by:

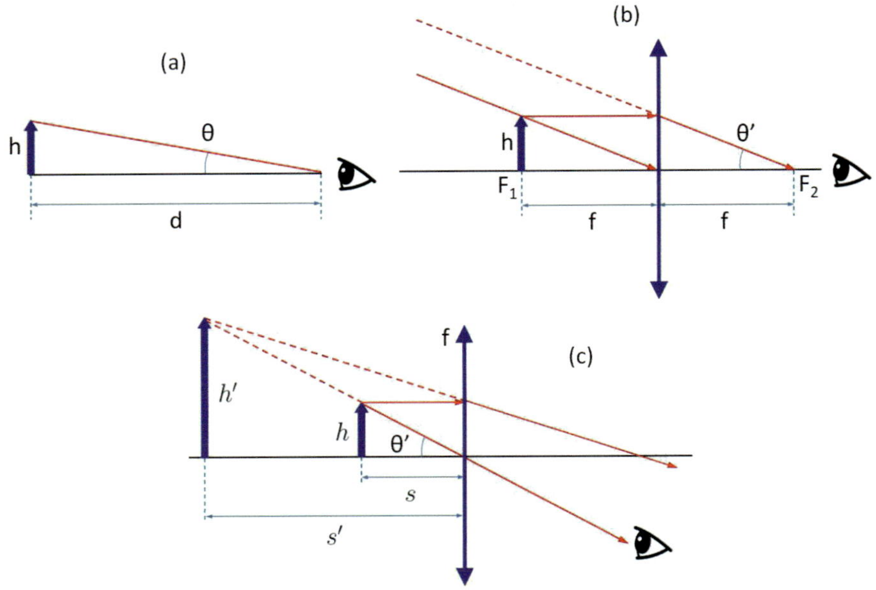

FIGURE 2.76: (a) Viewing an object of height h from the standard viewing distance (for a healthy unaided eye) of $d = 25\,\text{cm}$. The viewing angle is θ. (b) Viewing the same object via a positive lens of focal length f. The object is positioned at the first focal point of the lens. The effective viewing angle is now θ'. (c) Viewing the same object via a positive lens with the object positioned at a distance $s < f$ from the left side of the lens. Again the effective viewing angle is θ'.

$$m_\theta = \frac{\theta'}{\theta} = \frac{h/f}{h/d} = \frac{d}{f}. \tag{2.90}$$

The object can indeed be positioned at a distance from the lens that is less than its focal length, as shown in Fig. 2.76c. In this case the distance $s' = -d = -25\,\text{cm}$ (due to the virtual image formed by the lens; it is important to note that the final image perceived by the observer's retina is real and inverted). Based on this setup, the angular magnification in this more general case can be determined as follows:

$$\frac{1}{s} + \frac{1}{s'} = \frac{1}{f} \Longrightarrow s = \frac{df}{d+f},$$

$$m_\theta = \frac{\theta'}{\theta} = \frac{h/s}{h/d} = 1 + \frac{d}{f}, \tag{2.91}$$

where it is reminded that $d = 25\,\text{cm}$. For instance, common magnifiers have focal lengths of $f = 10\,\text{cm}$ which corresponds to an angular magnification of 2.5 to 3.5, depending on the actual accommodation distance used by the observer while examining the object. These magnifiers are commonly referred to as 2.5X to 3.5X. More complex magnifiers may utilize multiple lenses [42].

2.8.4 Basic Compound Microscope

The compound microscope can achieve significantly higher magnifications than the previously discussed magnifiers. Its invention is generally credited to the Dutch spectacle maker Zacharias Janssen, around 1590 [42]. The basic configuration of a compound microscope is illustrated in Fig. 2.77. A compound microscope consists of two lenses (or, in practice, lens systems): the objective lens, with focal length f_o, and the eyepiece lens, with focal length f_e. Both focal lengths are positive. The separation between these two lenses is $\mathcal{L}$, and the distance between the second focal point of the objective, F_o, and the first focal point of the eyepiece, F_e, is L (in many older microscope models, this length is set to $L = 160\,\text{mm}$). The angular magnification of the compound microscope can be calculated as follows:

$$m_\theta = \frac{\theta'}{\theta} = \frac{h'/s_e}{h/d} = \frac{h'}{h}\frac{d}{s_e} = -\frac{s'_o}{s_o}\frac{d}{s_e}, \quad (\text{where } \; d = 25\,\text{cm})$$

$$\frac{1}{s_o} + \frac{1}{s'_o} = \frac{1}{f_o} \Longrightarrow \frac{s'_o}{s_o} = \frac{\mathcal{L} - s_e}{f_o} - 1,$$

$$\frac{1}{s_e} + \frac{1}{s'_e = -d} = \frac{1}{f_e} \Longrightarrow \frac{d}{s_e} = 1 + \frac{d}{f_e},$$

$$m_\theta = -\left(\frac{\mathcal{L} - s_e}{f_o} - 1\right)\left(1 + \frac{d}{f_e}\right). \tag{2.92}$$

In the case that $\mathcal{L} = L + f_o + f_e$, as it is shown in Fig. 2.77, then the angular magnification becomes:

$$m_\theta = -\left(\frac{L + f_e - s_e}{f_o}\right)\left(1 + \frac{d}{f_e}\right). \tag{2.93}$$

A special case is when $s_e = f_e$ (the image from the objective falls on the first focal plane of the eyepiece). In this case Eq. (2.93) becomes:

$$m_\theta = -\frac{L}{f_o}\frac{d}{f_e}, \tag{2.94}$$

which is listed in many textbooks (for example Refs. [20] and [42]).

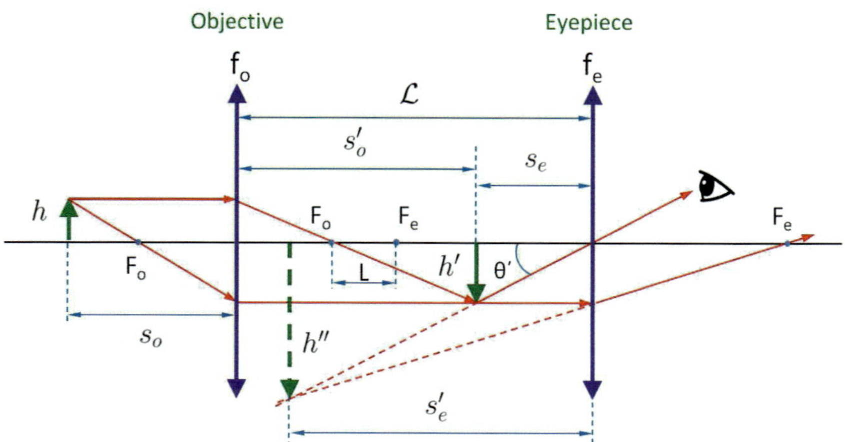

FIGURE 2.77: The basic configuration of the compound microscope. The left lens is the objective and has a focal length f_o. The right lens is the eyepiece and has a focal length f_e. The separation between the second focal point F_o and the first focal point F_e is L. The total tube length is $\mathcal{L}$ (separation of the two lenses). The focal lengths f_o and f_e are both positive.

Many modern microscopes include an *infinity-corrected* system [50]. In this configuration, an additional lens, commonly referred to as a "tube" lens, is placed between the objective and the eyepiece lens. A simplified layout of an infinity-corrected microscope is shown in Fig. 2.78. In the infinity space, the light rays emerging from the objective are parallel, allowing manufacturers to integrate additional components such as vertical illuminators, differential interference contrast (DIC) prisms, polarizers, and retardation plates. In this case, the magnification can be calculated as follows (with the help of Fig. 2.78):

$$\begin{aligned}
\frac{1}{s_o = f_o} + \frac{1}{s_o'} &= \frac{1}{f_o} \Longrightarrow s_o' = \infty \\
\frac{1}{s_t = -\infty} + \frac{1}{s_t'} &= \frac{1}{f_t} \Longrightarrow s_t' = f_t \\
\frac{1}{s_e = L_2 - f_t} + \frac{1}{s_e'} &= \frac{1}{f_e} \Longrightarrow s_e' = \frac{f_e(L_2 - f_t)}{L_2 - f_t - f_e} \\
m &= \frac{h'}{h}\frac{h''}{h'} = -\frac{f_t}{f_o}\left(-\frac{s_e'}{L_2 - f_t}\right), \qquad (2.95)
\end{aligned}$$

where the relation $\theta = h/f_o = h'/f_t$ was used. It is reminded that $s_e' < 0$ since the final image is virtual (as shown in Fig. 2.78).

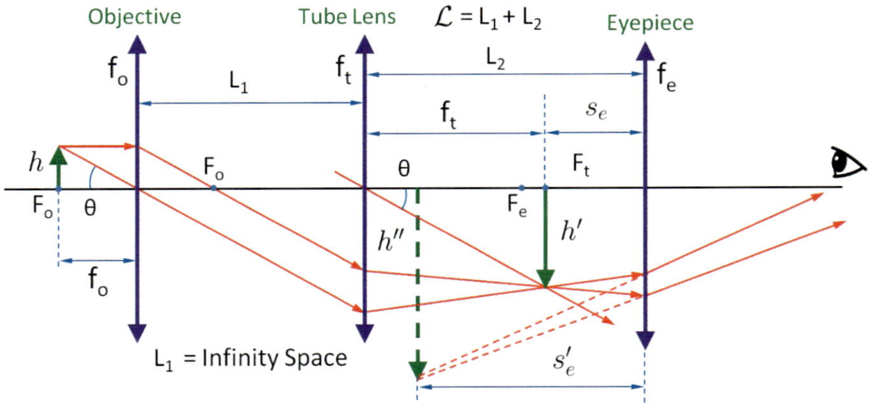

FIGURE 2.78: The basic configuration of the compound microscope with infinity correction with the help of the tube lens of focal length f_t. The left lens is the objective and has a focal length f_o. The right lens is the eyepiece and has a focal length f_e. The separation between the objective and the tube lens is L_1 (infinity space). The total tube length is $\mathcal{L} = L_1 + L_2$. The focal lengths f_o, f_t, and f_e are all positive.

Example 2.8.5:

Assume a compound microscope with objective and eyepiece lenses of focal lengths $f_o = 6\,\text{mm}$ and $f_e = 50\,\text{mm}$, respectively. The separation between the lenses is $\mathcal{L} = 230\,\text{mm}$ and the object is at a distance $s_o = 6.20\,\text{mm}$ away from the objective lens. What is the magnification of this compound microscope?

<u>Solution:</u> Starting from the thin lens equations for the objective and the eyepiece, the following expressions can be easily obtained:

$$\frac{1}{s_o = 6.20} + \frac{1}{s_o'} = \frac{1}{f_o = 6} \Longrightarrow s_o' = 186\,\text{mm}$$

$$\frac{1}{s_e = \mathcal{L} - s'_o = 230 - 186} + \frac{1}{s'_e} = \frac{1}{f_e = 50} \Longrightarrow s'_e = 366.67\,\text{mm}$$

$$m = \frac{h'}{h}\frac{h''}{h'} = -\left(\frac{\mathcal{L} - s_e}{f_o} - 1\right)\left(1 + \frac{|s'_e|}{f_e}\right) = -250.$$

Therefore, the final magnification of the compound microscope is $m = -250$.

Example 2.8.6:

In this example, a compound microscope with an infinity-corrected objective, achieved via a tube lens, is considered. As in Example 2.8.5, the objective and eyepiece lenses have focal lengths of $f_o = 6\,\text{mm}$ and $f_e = 50\,\text{mm}$, respectively. The tube lens has a focal length of $f_t = 200\,\text{mm}$. The separation between the objective and the tube lens is $L_1 = 250\,\text{mm}$, while the separation between the tube lens and the eyepiece is $L_2 = 240\,\text{mm}$. The goal is to determine the overall magnification of this microscope.

Solution: In the infinity-corrected objective, the object is at a distance $s_o = f_o$ from the objective which results in parallel rays in the infinity space between the objective and the tube lens (assume all distances in mm).

$$\frac{1}{s_o = f_o = 6} + \frac{1}{s'_o} = \frac{1}{f_o = 6} \Longrightarrow s'_o = \infty$$

$$\frac{1}{s_t = -\infty} + \frac{1}{s'_t} = \frac{1}{f_t} \Longrightarrow s'_t = f_t = 200\,\text{mm}$$

$$\frac{1}{s_e = L_2 - f_t} + \frac{1}{s'_e} = \frac{1}{f_e} \Longrightarrow s'_e = \frac{f_e(L_2 - f_t)}{L_2 - f_t - f_e} = -200\,\text{mm}$$

$$m = \frac{h'}{h}\frac{h''}{h'} = -\frac{f_t}{f_o}\left(-\frac{s'_e}{L_2 - f_t}\right) = -166.667.$$

Therefore, the final magnification of the compound microscope is $m = -166.667$. In case that it is desired the final image to be formed at the standard viewing distance ($d = 250\,\text{mm}$) from the eyepiece (virtual image) then the distance L_2 should be determined as follows:

$$\frac{1}{s_e = L_2 - f_t} + \frac{1}{s'_e = -d} = \frac{1}{f_e} \Longrightarrow L_2 = f_t + \frac{f_e d}{f_e + d} = 241.667\,\text{mm}$$

$$m = \frac{h'}{h}\frac{h''}{h'} = -\frac{f_t}{f_o}\left(-\frac{-d}{L_2 - f_t}\right) = -200.$$

Microscope objectives typically have focal lengths ranging from 2 mm to 40 mm and produce a real image at a fixed plane behind the objective lens. Specimens are positioned very close to the objective's first focal point. The medium surrounding the specimen may be air or a specialized immersion liquid, which enhances the light-gathering ability of the objective and thereby increases its numerical aperture (NA). The numerical aperture defines the maximum cone of light that a lens can capture (as illustrated in Fig. 2.79). The NA is defined as:

$$NA = n \sin\theta, \tag{2.96}$$

where n is the refractive index of the surrounding lens medium and θ the half-angle of the maximum cone of light that can be captured by the lens.

In order to achieve diffraction-limited resolution of the specimen's image the Rayleigh criterion is commonly used where the minimum resolvable distance of the object is $\rho =$

$1.22\lambda/(2NA)$ (refer to the *Scalar Diffraction* Chapter (Chapter 5), and Appendix 2C). Therefore, to decrease ρ (and thereby increase resolution), a larger NA is required. Since $NA < 1$ in air, immersion liquids with refractive indices $n > 1$ are used to achieve $NA > 1$ for higher resolution.

Several types of objective lenses are available, each designed with different levels of aberration correction. For instance, *achromat* objectives correct chromatic aberration for two wavelengths (red and blue) [50]. Other objectives correct additional aberrations (and are typically more costly), including *Plan achromat*, *Fluorite*, *Plan Fluorite*, and *Plan Apochromat* lenses [50].

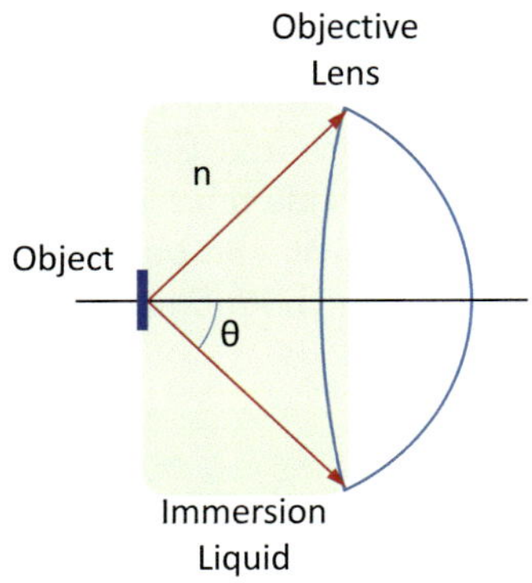

FIGURE 2.79: The numerical aperture (NA) of a microscope objective corresponds to the cone of light of half-angle θ that the lens can capture. To increase NA the lens is immersed in a higher index liquid (usually specialized immersion oils).

Eyepiece lenses (or ocular lenses) generally have focal lengths ranging from 10 mm to 25 mm. Both objective and eyepiece lenses are typically specified by their individual magnification factors. For an objective lens, the magnification is approximately $m_o \simeq L/f_o$, and for the eyepiece lens, it is $m_e \simeq d/f_e$ [see Eq. (2.94)]. For example, if $L = 160\,\text{mm}$ and $f_o = 4\,\text{mm}$, the objective magnification is $160/4 = 40 \equiv 40X$. Similarly, for an eyepiece with $d = 250\,\text{mm}$ and $f_e = 25\,\text{mm}$, the resulting magnification is $250/25 = 10 \equiv 10X$. The overall microscope magnification in this example case is $400X$. Typically, useful overall microscope magnifications range from $500X - 1500X$. While higher magnifications are technically possible, they do not enhance resolution due to diffraction limits. Instead, they simply enlarge the image without improving clarity, an effect known as *empty magnification*. High-quality optical microscopes achieve resolutions on the order of $0.20\,\mu\text{m}$. Table 2.8 provides some common, effective magnification values [50], where the overall magnification is given by $m \simeq m_o m_e NA$.

Optical microscopes use visible light to view a specimen. Depending on the illumination process of the specimen there are different types of microscopy (and correspondingly microscopes). The most common optical microscopy types are:

- **Bright Field Microscopy:** Bright field microscopy is the most common type of microscopy Light from the microscope's illumination source, typically a halogen lamp or LED, passes through the specimen and is collected by the objective lens. In this technique, contrast is primarily achieved through the varying absorption levels across different parts of the specimen. However, bright field microscopy often provides limited contrast, especially for biological specimens, which can make detailed observation challenging.

TABLE 2.8: Range of useful magnification (500–1000× NA) [50].

		Eyepiece				
Objective	NA	10X	12.5X	15X	20X	25X
2.5X	0.08	–	–	–	•	•
4X	0.12	–	–	•	•	•
10X	0.35	•	•	•	•	•
25X	0.55	•	•	•	•	–
40X	0.70	•	•	•	–	–
60X	0.95	•	•	•	–	–
100X	1.42	•	•	–	–	–

The "•" corresponds to a useful combination while "–" corresponds to a non-practical combination.

- **Dark Field Microscopy:** Dark field microscopy is designed to prevent unscattered light from reaching the objective lens, allowing only light scattered by the specimen to be collected. This effect is typically achieved by placing a specialized aperture in front of the microscope's illumination source, which blocks the central light beam. As a result, the background appears dark, highlighting the illuminated specimen, hence the name "dark field microscopy." This technique is particularly suited for biological specimens, though the resulting image generally contains low light levels.
- **Reflected Light Microscopy:** For opaque specimens where light cannot pass through, illumination is applied from above, and the light reflected off the specimen is collected by the objective lens. This technique is often combined with fluorescence or polarized light microscopy to enhance imaging quality and reveal additional specimen details.
- **Fluorescence Microscopy:** Fluorescence microscopy utilizes the property of fluorescence, where organic or inorganic specimens absorb light and then re-emit it at a longer wavelength. In this technique, the emitted fluorescent light from the specimen, whether transmitted or reflected, is collected by the objective lens to form a detailed and high-contrast image. This method is particularly useful for highlighting specific structures or molecules within a specimen.
- **Phase Contrast Microscopy:** Transparent or colorless biological specimens do not significantly alter the intensity of transmitted or reflected light, resulting in low-contrast images. Phase contrast microscopy addresses this limitation by converting phase changes of the light waves into intensity variations, thereby enhancing image contrast. This is achieved by treating the scattered light from the specimen differently from the background light through the use of phase plates, which introduce additional phase shifts. As a result, light waves from the specimen undergo constructive interference, while those from the background experience destructive interference, leading to improved visibility of the specimen's details.
- **Differential Interference Contrast (DIC) Microscopy:** Also known as Nomarski microscopy, DIC microscopy is used to produce images of transparent or colorless specimens. This technique is similar to phase contrast microscopy but has some key differences. In DIC microscopy, a polarized light source is split into two orthogonally polarized, mutually coherent beams, which are then spatially displaced at the specimen's plane. These beams are recombined before being collected by the microscope's objective. Unlike

phase contrast microscopy, which generates images based on the optical path lengths of the specimen, DIC microscopy creates images that reflect the gradient of the optical path lengths. This results in images that exhibit a pseudo three-dimensional relief shading, enhancing the visual depth of the specimen.

- **Confocal Microscopy:** Confocal microscopy, also known as confocal laser scanning microscopy, utilizes a spatial pinhole to block out-of-focus light at the image plane, allowing for precise control of the depth of field. By capturing multiple two-dimensional images at various depths within the specimen, three-dimensional reconstructions can be generated using specialized techniques. In this method, specimens are illuminated with laser light, which induces fluorescence, resulting in scattered light that is collected by the microscope's objective. Confocal microscopy finds applications across various fields, including life sciences, semiconductor inspection, and materials science.

2.8.5 Basic Telescope

A telescope is an optical instrument designed for observing distant objects. There are three main types of telescopes: refractive telescopes, which utilize lenses; reflecting telescopes, which employ mirrors; and catadioptric telescopes, which combine both reflective and refractive elements. The invention of the first telescopes is credited to Hans Lippershey in the Netherlands in 1608. Shortly after Lippershey's invention, Galileo constructed his own telescopes using refractive lenses, leading to the first significant astronomical observations.

2.8.5.1 Refractive Telescopes

The basic refractive telescope, commonly known as the *Keplerian* or astronomical telescope, is depicted in Fig. 2.80. This telescope consists of two positive lenses: the objective lens with a focal length of f_o, and the eyepiece lens of focal length f_e. The second focal point of the objective lens, F_o, coincides with the first focal point, F_e, of the eyepiece lens. The angles θ and θ' represent the input and output angles of the bunch of parallel rays relative to the optical axis, respectively. The angular magnification of this configuration can be determined as follows:

$$m = \frac{\theta'}{\theta} = -\frac{h'/f_o}{h'/f_e} = -\frac{f_o}{f_e}, \tag{2.97}$$

where the negative sign is used since the final image is inverted. At this point, for educational main purposes, the $ABCD$ matrix approach (see Sec. 2.7) is employed to analyze a refractive telescope and derive the same result as in Eq. (2.97). Assuming that the distance between the object and the eyepiece lens is L, the following equations can be formulated using the left surface of the objective lens as the input plane and the right surface of the eyepiece lens as the output plane:

$$\begin{bmatrix} A & B \\ C & D \end{bmatrix} = \begin{bmatrix} 1 & 0 \\ -1/f_e & 1 \end{bmatrix} \begin{bmatrix} 1 & L \\ 0 & 1 \end{bmatrix} \begin{bmatrix} 1 & 0 \\ -1/f_o & 1 \end{bmatrix} = \begin{bmatrix} 1 - \dfrac{L}{f_o} & L \\ -\dfrac{1}{f_o} - \dfrac{1}{f_e} + \dfrac{L}{f_o f_e} & 1 - \dfrac{L}{f_e} \end{bmatrix}.$$

In order for the system to be telescopic $C = 0$ (see Sec. 2.7). Using this condition the following results are obtained:

$$C = 0 \Longrightarrow L = f_o + f_e,$$

$$m = D \Longrightarrow D = 1 - \frac{L}{f_e} = -\frac{f_o}{f_e},$$

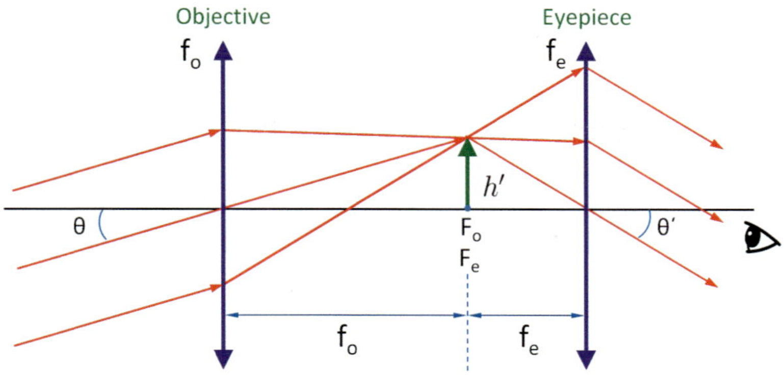

FIGURE 2.80: The basic configuration of a refractive astronomical telescope. The object is assumed to be at infinite distance to the left of the objective lens. The left lens is the objective and has a focal length f_o. The right lens is the eyepiece and has a focal length f_e. The second focal point, F_o, of the objective coincides with the first focal point, F_e, of the eyepiece. Both lenses are positive.

which are the same results as previously with the addition of the constrain $L = f_o + f_e$ which is derived from the telescopic property of the optical system. In the previous equation it was assumed that both lenses are positive. However, if the eyepiece is selected to be a diverging lens, then $f_e < 0$, and the overall telescope length becomes smaller than f_o ($L < f_o$) with the addition of the upright image. This refractive telescope is known as *Galilean* telescope and has the advantage that the images are upright which makes it more suitable for terrestrial observations. The basic Galilean refractive telescope is shown in Fig. 2.81.

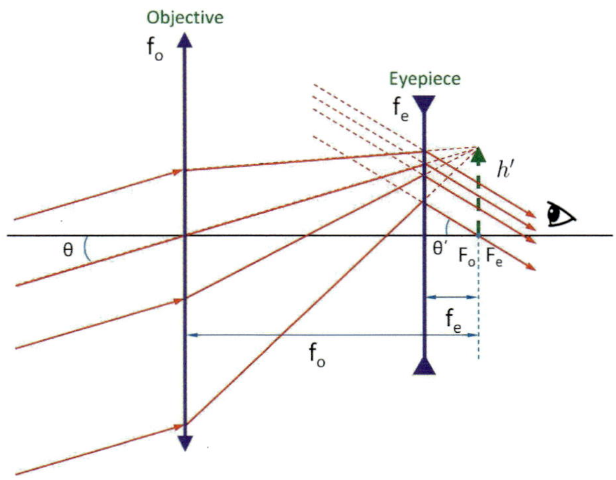

FIGURE 2.81: The basic configuration of a Galilean refractive telescope. The object is assumed to be at infinite distance to the left of the objective lens. The left lens is the objective and has a focal length f_o. The right lens is the eyepiece and has a focal length $f_e < 0$. The second focal point, F_o, of the objective coincides with the first focal point, F_e, of the eyepiece.

Example 2.8.7:

The purpose of this example is to determine the Field Stop (FS) and the Field of View (FoV) for both a Keplerian and a Galilean refractive telescope, which utilize the same objective lens and provide identical magnification. The geometric configuration is illustrated in Fig. 2.82(a). In the case of a Galilean telescope, the eyepiece lens is negative, resulting in $f_e < 0$. The diameters of the objective lens, eyepiece lens, and iris stop are denoted as D_o, D_e, and D_s, respectively.

Solution: In order to determine the FS and the FoV the angles subtended from the axial point of the EnP to the image in object space of the eyepiece rim, θ_e, and the iris stop, θ_s, must be calculated. Then the FS is the one that corresponds to the minimum angle which becomes the FoV. The AS and the EnP coincide with the rim of the objective lens which is

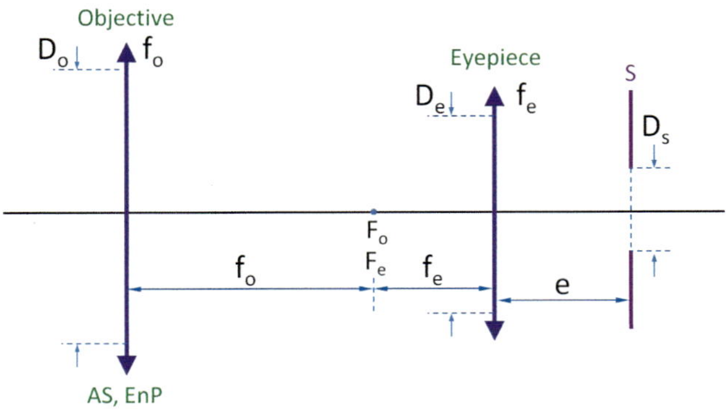

(a)

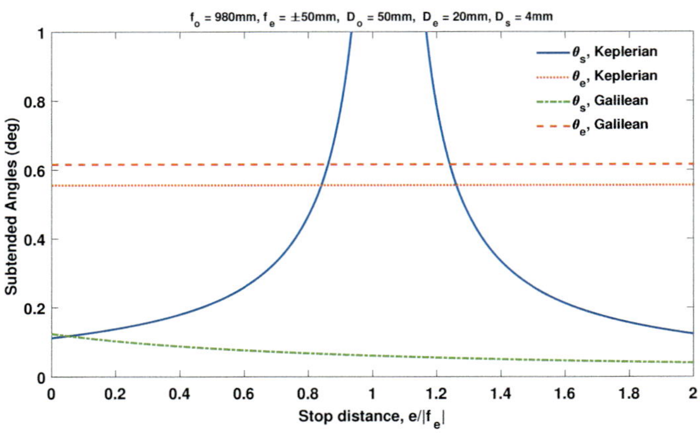

(b)

FIGURE 2.82: (a) The basic geometry of a Keplerian refracting telescope with an iris stop at a distance e from the eyepiece lens of focal length $f_e > 0$. The objective lens has a focal length of f_o. In the Galilean version the eyepiece lens is negative, and its corresponding focal length is $f_e < 0$. (b) The half-angle subtended from the axial point of the EnP (center of objective lens) for the object space image of eyepiece rim, θ_e, and iris stop θ_s, respectively, for a Keplerian and a Galilean refractive telescope with the same objective and the same magnification. The FoV is $2\min\{\theta_s, \theta_e\}$ for each telescope type. The angle θ_e for the Galilean telescope is always greater than θ_s and therefore θ_s is the FoV in this case. The parameters of the example are $D_o = 50\,\text{mm}$, $f_o = 980\,\text{mm}$, $D_e = 20\,\text{mm}$, $f_e = \pm 50\,\text{mm}$, and $D_s = 4\,\text{mm}$ (human eye iris diameter under good illumination). It is evident that the Galilean telescope has a smaller FoV for the parameters selected for all distances e except near the left edge of the figure, where it can be slightly larger.

common for the parameters of refractive Keplerian and Galilean telescopes. The subtended angle θ_e from the axial point of the EnP to the image in object space of the eyepiece lens can be determined as follows:

$$\frac{1}{f_o + f_e} + \frac{1}{s'_o} = \frac{1}{f_o} \implies s'_o = \frac{f_o}{f_e}(f_o + f_e),$$

$$m = -\frac{s'_o}{f_o + f_e} = -\frac{f_o}{f_e},$$

$$\theta_e = \tan^{-1}\left\{\frac{(D_e/2)(f_o/f_e)}{(f_o/f_e)(f_o + f_e)}\right\} = \tan^{-1}\left\{\frac{D_e}{2(f_o + f_e)}\right\}.$$

The above angle is the half-angle of the total angle subtended by the axial EnP point. Similarly, the half-angle θ_s can be calculated as follows:

$$\frac{1}{e} + \frac{1}{s'_e} = \frac{1}{f_e} \implies s'_e = \frac{ef_e}{e - f_e},$$

$$m_e = -\frac{s'_e}{e} = -\frac{f_e}{e - f_e},$$

$$\frac{1}{f_o + f_e - s'_e} + \frac{1}{s'_o} = \frac{1}{f_o} \implies s'_o = \frac{f_o(f_o + f_e - s'_e)}{f_e - s'_e} = -m(\mathcal{L} + me),$$

$$\text{where} \quad \mathcal{L} = f_o + f_e, \qquad m = -\frac{f_o}{f_e},$$

$$\theta_s = \tan^{-1}\left\{\frac{D_s/2}{|\mathcal{L} + me|}\right\}.$$

The FS and the FoV correspond to the $\min\{\theta_s, \theta_e\}$. It is interesting to notice that for the Keplerian telescope configuration FoV$= \theta_e$ when $-(D_s/D_e)\mathcal{L} < me + \mathcal{L} < (D_s/D_e)\mathcal{L}$ as it is evident in Fig. 2.82(b). The parameters of the example are $D_o = 50$ mm, $f_o = 980$ mm, $D_e = 20$ mm, $f_e = \pm 50$ mm, and $D_s = 4$ mm (human eye iris diameter under good illumination). For these parameters and for the Keplerian telescope, the FoV becomes the rim of the eyepiece lens when $0.8408 < e/|f_e| < 1.2612$. The FoV of the Keplerian telescope is always greater than the corresponding FoV of the Galilean counter one for $e/|f_e| < 1/|m| = 0.051$ as is shown in Fig. 2.82(b). It is reminded that the FoV is twice the angles shown in Fig. 2.82(b).

2.8.5.2 Reflecting Telescopes

To effectively observe faint and distant celestial objects, telescopes, particularly astronomical ones, must be designed to maximize light collection. The light-gathering capacity of a telescope is directly proportional to the size of its primary objective lens or mirror. Refractive telescopes, known as refractors, face several challenges when it comes to constructing large-diameter objectives. In practice, refractors typically do not exceed diameters of a few tens of centimeters due to these constraints. One notable exception is the refractor at the *Yerkes* Observatory in Williams Bay, Wisconsin, USA, which has an objective lens with a diameter of 1.02 meters (40 inches), making it the largest refractor in the world. Such large lenses are difficult to manufacture, are significantly heavy, and can be prone to cracking due to temperature variations and their own weight. Consequently, the practicalities of construction and maintenance make large-diameter refractors impractical for most astronomical applications. In response to these limitations, Isaac Newton pioneered the use of reflective telescopes in 1668, utilizing a spherical mirror along with a smaller diagonal mirror for directing the light to an eyepiece (see Fig. 2.83(a)). These telescopes, now known as *Newtonian* telescopes, take advantage of mirrors, which can be constructed much larger than lenses without the same weight and structural issues. This innovation opened new possibilities for astronomical observation, allowing for the construction of larger telescopes capable of gathering more light from distant stars and galaxies.

The most common reflecting telescope configurations are:

- **Newtonian Telescope Configuration:** As previously mentioned, Newtonian telescopes consist of a concave primary mirror and a flat diagonal secondary mirror, as illustrated in Fig. 2.83(a). One significant advantage of Newtonian telescopes is their immunity to chromatic aberrations, which can occur in refractive telescopes due to the dispersion of light through lenses. Because Newtonian telescopes utilize only reflective elements, they eliminate this type of distortion. Additionally, the construction of Newtonian telescopes is generally more straightforward. They feature only one complex surface, which is the primary mirror, in contrast to other types of telescopes such as Gregorian and Cassegrain, which incorporate multiple refractive or reflective surfaces. This simplicity in design typically results in lower manufacturing costs, making Newtonian telescopes a more economical choice for astronomers seeking to build larger, more efficient observing instruments.
- **Gregorian Telescope Configuration:** The Gregorian telescope represents another significant configuration of a reflecting telescope, as depicted in Fig. 2.83(b). This design was conceived by the Scottish mathematician and astronomer James Gregory in the 17th century and was first constructed by Robert Hooke in 1673. The Gregorian telescope is characterized by a primary mirror, which is a concave paraboloidal mirror that collects and focuses incoming light. The light is then directed to a secondary mirror, which is a concave ellipsoidal mirror. This secondary mirror reflects the light back through an aperture located at the center of the primary mirror, allowing it to be viewed through an eyepiece lens. One of the notable advantages of the Gregorian telescope is that it produces an upright image, making it well-suited for terrestrial observations where orientation is important. Additionally, this configuration allows for a much shorter telescope tube length relative to the effective focal length of the system, which contributes to a more compact design.
- **Cassegrain Telescope Configuration:** The Cassegrain telescope is a widely used optical design that combines a primary concave mirror with a secondary convex mirror, as illustrated in Fig. 2.83(c). This configuration was initially developed by Laurent Cassegrain in 1672. In the classic Cassegrain design, a parabolic reflector serves as the primary mirror, while the secondary mirror is typically hyperbolic. This arrangement has the effect of placing the focal point behind the primary mirror, which is advantageous for accessibility and observation. Moreover, the use of a convex secondary mirror allows for a significantly longer effective focal length relative to the physical length of the telescope system. The Cassegrain design is favored for its compactness and versatility, making it suitable for various astronomical applications, including both amateur and professional observations. Its ability to achieve a long focal length in a shorter overall length contributes to its popularity among telescope manufacturers and users alike.

The theoretical analysis of reflecting telescopes incorporating aspheric surfaces for aberration correction is indeed complex and typically requires advanced optical design techniques. For those interested in further reading, Ref. [51] provides valuable insights. For educational purposes, a simple paraxial analysis of a Cassegrain reflecting telescope can be undertaken using the $ABCD$ matrix method. As depicted in Fig. 2.84, the system consists of a concave primary mirror with a radius of curvature $R_p < 0$ and diameter D_p, paired with a convex secondary mirror with a radius of curvature $R_s > 0$ and diameter D_s. The separation of

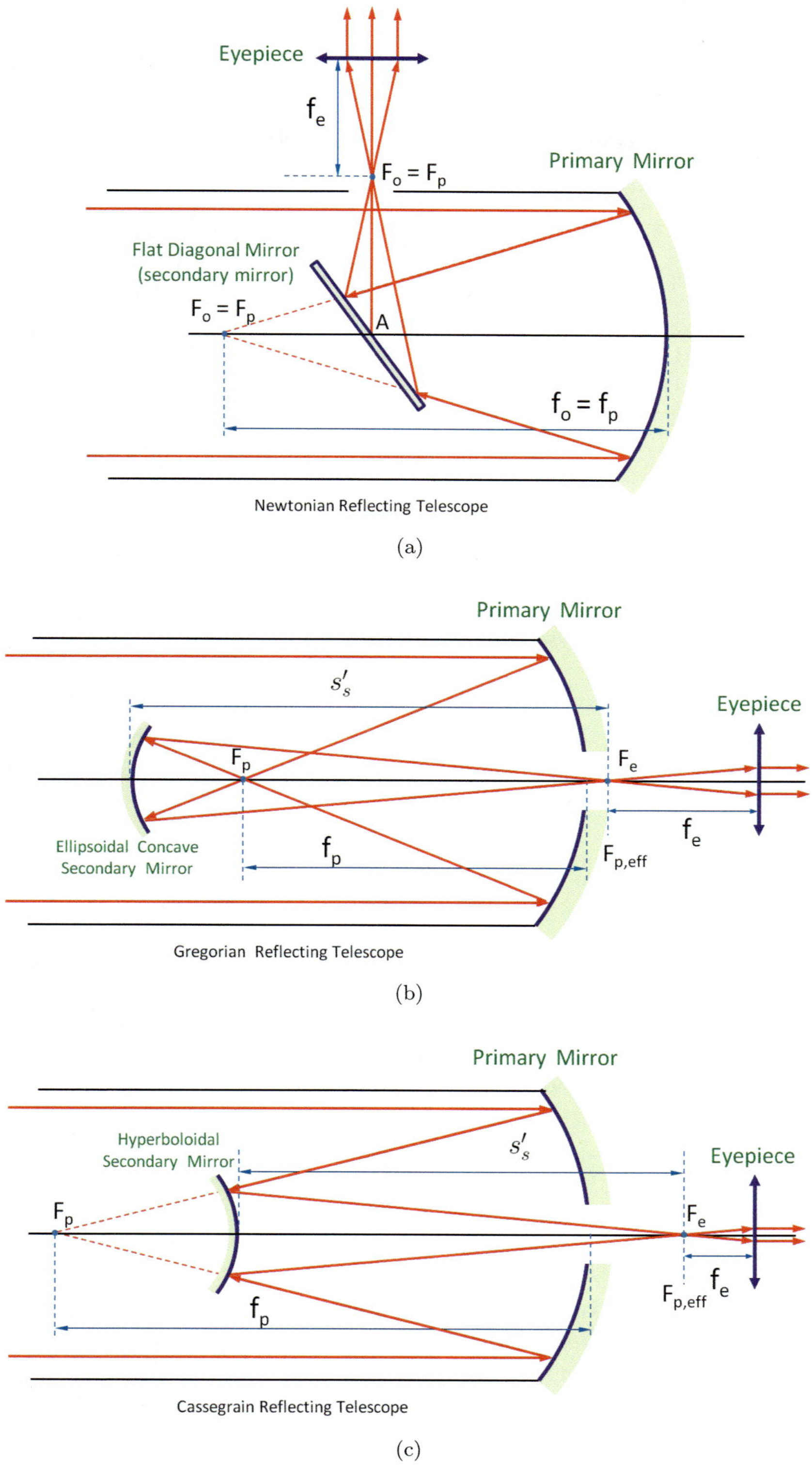

FIGURE 2.83: Basic configurations of reflecting telescopes. (a) Newtonian type. (b) Gregorian type. (c) Cassegrain type.

the two mirrors is denoted as L. To determine the effective focal length f_{eff} of the combined system, the $ABCD$ matrix approach can be employed. Let $F_{p,eff}$ the second focal point of the system of the two mirrors as shown in Fig. 2.84. For the application of the

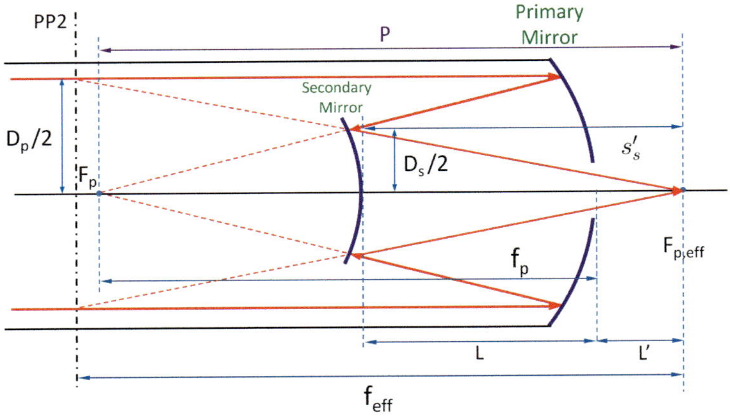

FIGURE 2.84: The basic configuration of a Cassegrain reflecting telescope. The f_{eff} is the effective focal length of the combination of the primary and secondary mirror, and PP_2 is the position of the second principal plane with the input plane the plane just to the left of the primary mirror and the output plane the plane passing through the effective focal point $F_{p,eff}$.

$ABCD$ method, it is assumed that the input plane is located infinitesimally to the left of the primary mirror, while the output plane passes through $F_{p,eff}$ and is perpendicular to the optical axis (the second focal plane). The overall $ABCD$ matrix representing the system is determined as follows:

$$\begin{bmatrix} A & B \\ C & D \end{bmatrix} = \begin{bmatrix} 1 & L+L' \\ 0 & 1 \end{bmatrix} \begin{bmatrix} 1 & 0 \\ -1/f_s & 1 \end{bmatrix} \begin{bmatrix} 1 & L \\ 0 & 1 \end{bmatrix} \begin{bmatrix} 1 & 0 \\ -1/f_p & 1 \end{bmatrix}.$$

In the above equation $f_p = -R_p/2 > 0$ (concave primary mirror) and $f_s = -R_s/2 < 0$ (convex secondary mirror). According to the theory of the $ABCD$ elements (see Sec. 2.7) in order to determine the $F_{p,eff}$ focal point it is necessary to set $A = 0$. This condition can be used to determine L' and then to calculate the effective focal length as $f_{eff} = -1/C$. The resulting expressions are summarized below:

$$L' = \frac{L^2 - (f_p + 2f_s)L + f_p f_s}{f_p + f_s - L}, \tag{2.98}$$

$$f_{eff} = -\frac{1}{C} = \frac{f_p f_s}{f_p + f_s - L}. \tag{2.99}$$

At this point, some simple paraxial optics are employed to determine the radii of curvature of the primary and secondary mirrors, given the effective focal length of the system, as well as the secondary magnification and the distance P shown in Fig. 2.84. Assuming that the light rays are incident on the primary mirror parallel to the optical axis, they will converge at point F_p. This point serves as the virtual object for the secondary mirror, which must satisfy the equation $1/(-s_s) + 1/s'_s = 1/f_s$ with $s_s = f_p - L$. The second principal plane PP_2, is to the left of the output plane a distance $s = (1 - A)/C = 1/C = -f_{eff}$ (since $A = 0$, in the expression of Table 2.4). Using similar orthogonal triangles, the following expressions can be written [51]:

$$\frac{D_s/2}{D_p/2} = \frac{s'_s}{f_{eff}} = \frac{s_s}{f_p} \implies$$

$$f_p = -\frac{-s_s}{s'_s} f_{eff} = \frac{f_{eff}}{m_s} \implies$$

$$R_p = -\frac{2f_{eff}}{m_s}, \tag{2.100}$$

where $m_s = -s'_s/(-s_s)$ is the magnification of the secondary mirror. Next, the radius of the secondary mirror is determined. Again the following equations can be easily written:

$$s_s + s'_s = P \implies s_s + m_s s_s = P \implies s_s = \frac{P}{m_s + 1}$$

$$\frac{1}{-s_s} + \frac{1}{s'_s} = \frac{1}{f_s} \implies f_s = -\frac{m_s}{m_s - 1} s_s \implies$$

$$f_s = -\frac{m_s}{m_s^2 - 1} P \implies R_s = +\frac{2m_s}{m_s^2 - 1} P. \tag{2.101}$$

Usually, $m_s > 1$ since $s'_s \gg s_s$, and this results in $R_s > 0$, as it is expected for a convex mirror. At the same time $R_p < 0$, as it is expected for a concave mirror. If the f_s and P are given from the previous equation the necessary magnification m_s can be determined as follows:

$$m_s = \frac{1 + \left[1 + 4(f_s/P)^2\right]^{1/2}}{2(-f_s/P)}. \tag{2.102}$$

Similar expressions can be easily derived for the Gregorian configuration too. The overall magnification of a reflecting telescope can also be expressed by Eq. (2.97), where the role of f_o is played by the effective focal length f_{eff}.

Example 2.8.8:

To illustrate the previous results, let us consider the parameters of the Hubble Space Telescope [52]. The Hubble telescope is a modified Cassegrain configuration that is known as Ritchey-Chrétien configuration since the mirrors are not spherical but hyperboloidal to reduce aberrations. For the purpose of this example, spherical mirrors are assumed for simplicity. The diameter of the primary concave mirror is $D_p = 2.4\,\text{m}$ and its radius of curvature is $R_p = 11.04\,\text{m}$. In contrast, the secondary convex mirror has a diameter of $D_s = 0.31\,\text{m}$ and a radius of curvature $R_s = 1.36\,\text{m}$. Following the sign convention used throughout this book, $R_p < 0$ and $R_s > 0$. The distance between the primary and the secondary mirror is reported in Ref. [52] as $L = 4.91\,\text{m}$. However, since this analysis involves paraxial optics with spherical mirrors, this distance has been slightly adjusted to obtain the effective focal point $F_{p,eff}$ (see Fig. 2.84) approximately $L' \simeq 1.5\,\text{m}$ behind the primary mirror. Thus, the distance L was selected as $L = 4.9052\,\text{m}$ which is marginally smaller than the actual distance due to the approximations made in this example.

<u>Solution:</u> Using $f_p = -R_p/2 = +5.52\,\text{m}$ and $f_s = -R_s/2 = -0.68\,\text{m}$ and Eq. (2.99) the effective focal length becomes $f_{eff} = 57.57\,\text{m}$ (in Ref. [52] the effective focal length is listed as 57.6 m). Furthermore, from Eq. (2.98) $L' = 1.5068\,\text{m}$ (about 1.5 m is cited in Ref. [52]). Using $P = f_p + L' = 5.52 + 1.5068 = 7.0268\,m$ and $f_s = -0.68\,\text{m}$, Eq. (2.102) predicts $m_s = 10.4294$ (a magnification of 10.43 is listed in Ref. [52]). Of course, using the magnification m_s the radii R_p and R_s are compatible with the results of Eqs. (2.100) and (2.101),

respectively. The magnification of the Hubble space telescope for an eyepiece of focal length $f_e = 25\,\text{mm}$ is given by $m = -f_{eff}/f_e = -57.57/0.025 = -2302.8$.

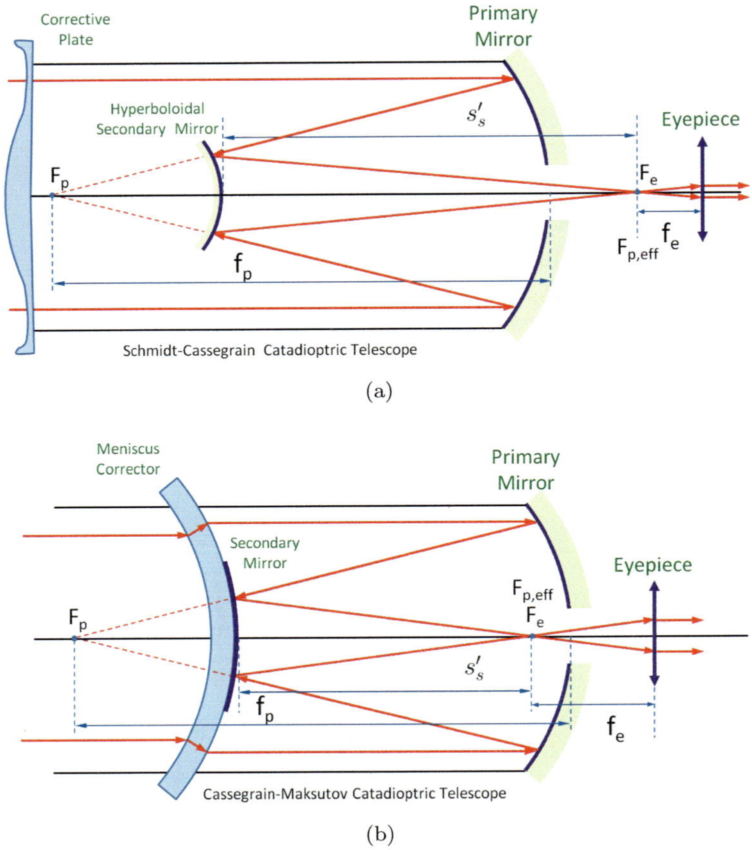

FIGURE 2.85: Basic types of catadioptric telescopes. (a) Schmidt-Cassegrain type. (b) Maksutov-Cassegrain type.

2.8.5.3 Catadioptric Telescopes

Another class of telescopes is characterized as catadioptric. A catadioptric optical system combines both refractive elements (such as lenses and phase plates) and reflective elements (such as mirrors). Some representative popular catadioptric configurations (among many) are:

- **Schmidt-Cassegrain Configuration:** This configuration is illustrated in Fig. 2.85(a). These telescopes represent popular commercial designs for amateur astronomical observations. The configuration corrects the spherical aberration of the primary mirror by utilizing a thin corrective glass plate. This corrective plate appropriately deviates light rays at various distances from the optical axis, effectively eliminating the spherical aberration associated with the primary mirror.
- **Maksutov-Cassegrain Configuration:** This configuration is shown in Fig. 2.85(b). In this design, the telescope combines a spherical mirror with a weakly negative meniscus lens. The negative lens, which typically has a large diameter, is positioned at the entrance

pupil of the telescope. This configuration effectively corrects off-axis aberrations, such as coma, commonly encountered in reflecting telescopes, while also addressing chromatic aberration. Patented in 1941 by D. D. Maksutov, a Russian optician, telescopes of this configuration have become popular choices for amateur astronomical observations.

2.8.5.4 True Field of View

A distant object is undoubtedly magnified when observed through a telescope. However, as the object is magnified further, the view through the telescope may appear smaller. A similar effect is observed in cameras equipped with telephoto lenses, where the angle of view is defined (see Sec. 2.8.2.5). In telescopes, a useful parameter is the true field of view (TFOV). The TFOV indicates the extent of the object (typically an area of the sky) that an observer can see when looking through the telescope with a specific eyepiece in place. Each combination of telescope and eyepiece yields its own TFOV, usually expressed in angular terms. The apparent field of view (AFOV) is the angular measurement that describes how much of the object (area of the sky) can be seen when observing through the eyepiece alone, without it being attached to the telescope. The relationship between the AFOV and TFOV is expressed by:

$$\mathrm{TFOV} = \frac{\mathrm{AFOV}}{m}, \tag{2.103}$$

where m is the telescope magnification [see Eq. (2.97)]. Common eyepieces range in AFOV between 45 deg to 70 deg. In many cases where a camera is used to record the image from a telescope, the apparent field of view (AFOV) coincides with the angle of view defined in Sec. 2.8.2.5. In this case the angle of view depends on the detector size d [see Eq. (2.89)]. Of course, the TFOV can be expressed in the horizontal, the vertical, or the diagonal direction. This is the reason that sometimes the TFOV is defined as $\mathrm{TFOV}_h \times \mathrm{TFOV}_v$ where TFOV_h and TFOV_v are the TFOV in the horizontal and the vertical directions, respectively. For example, for the Hubble Space Telescope the listed TFOV in the visible wavelength (up to $1.1\,\mu$m) is 202 *arcsec* × 202 *arcsec* which is equivalent to 0.056 deg × 0.056 deg for its wide field camera.

Example 2.8.9:

Assume that an amateur astronomer wants to observe the full size of the moon with his telescope. If the full size of the moon as seen from Earth subtends an angle of 0.5 deg what magnification of the telescope is needed assuming that the eyepiece used has an AFOV of 50 deg?

Solution: Using Eq. (2.103) the magnification should be $m = \mathrm{AFOV}/TFOV = 50/0.5 = 100$.

2.8.5.5 Telescope Mountings

In astronomical observations of various celestial objects, a telescope must gather as much light as possible. Distant astronomical objects, such as stars, nebulae, and galaxies, are typically very faint, necessitating long observation periods. However, due to the Earth's rotation, the telescope must track this movement to remain focused on the same area of the sky, allowing for the formation of sharp images. Telescope mountings facilitate this process. A telescope mounting refers to the mechanical structure that supports and guides a telescope. These mountings are designed to bear the telescope's weight while enabling precise pointing as the Earth rotates. Over the years, numerous types of telescope mountings have been developed. The study of telescope mountings is a specialized topic and is beyond the scope of this book. Here, only some basic principles and mountings are presented. Interested readers are encouraged to consult available literature, such as Refs. [53, 54]. Astronomical telescopes are usually mounted in one of two ways:

- **Equatorial:** In the equatorial mounting configuration, one axis, known as the polar axis, is aligned parallel to the Earth's rotation axis. Perpendicular to the polar axis is the declination axis. The telescope can move around both the polar and declination axes via the mounting (see Fig. 2.86(a)). The Earth's rotation is countered by adjusting the telescope tube around the polar axis in the opposite direction to the Earth's rotation. This adjustment occurs at a constant rate, matching the constant speed of the Earth's rotation. Equatorial mountings tend to be inherently heavy. The Hale telescope, located on Mount Palomar and measuring 5 meters, is considered the practical upper limit for this type of mounting [53].
- **Altazimuth or altitude-azimuth:** In the altitude-azimuth mounting configuration, commonly referred to as "altazimuth," the telescope tube is oriented by rotating around two axes: a vertical axis (the azimuth axis) and a horizontal axis (the altitude axis), as illustrated in Fig. 2.86(b). In contrast to the equatorial mounting configuration, the altazimuth design is sturdier and simpler. The reduction in both mass and cost is substantial, making it the standard mounting choice even for mid-class telescopes [53]. When tracking an object in the sky, the altazimuth configuration requires rotation in both axes at varying rates, which facilitates computer-controlled adjustments.

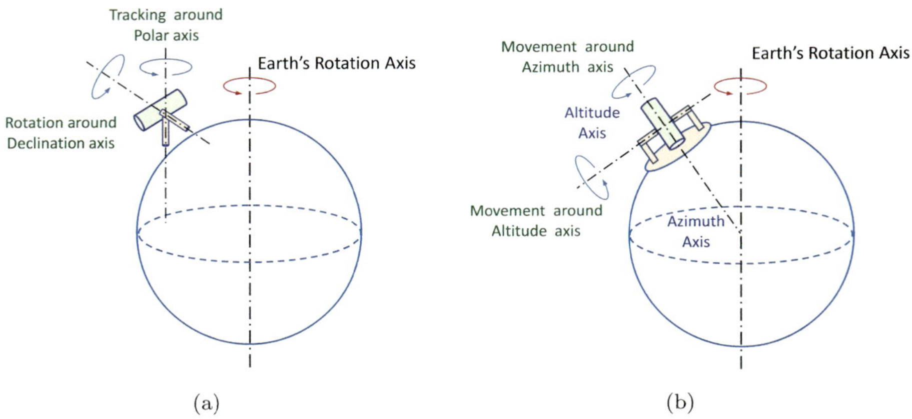

FIGURE 2.86: Basic types of telescope mountings: (a) Equatorial type. (b) Altitude-Azimuth type.

2.8.6 Optics of the Human Eye

The human eye can also be regarded as an optical instrument. From the perspective of optical systems, the human eye is a relatively simple optical system with limited image-forming capabilities due to aberrations. However, the optical processing performed by the human brain significantly enhances the images perceived through the retina, resulting in exceptional vision quality. The purpose of this section is to analyze the average human eye as an optical system using paraxial optics. The parameters of the human eye utilized in this section are taken from Ref. [55].

The human eye consists of two positive lenses: the cornea (with radii of curvature R_1 and R_2 and thickness d_1) and the crystalline lens (of radii of curvature R_3 and R_3 and

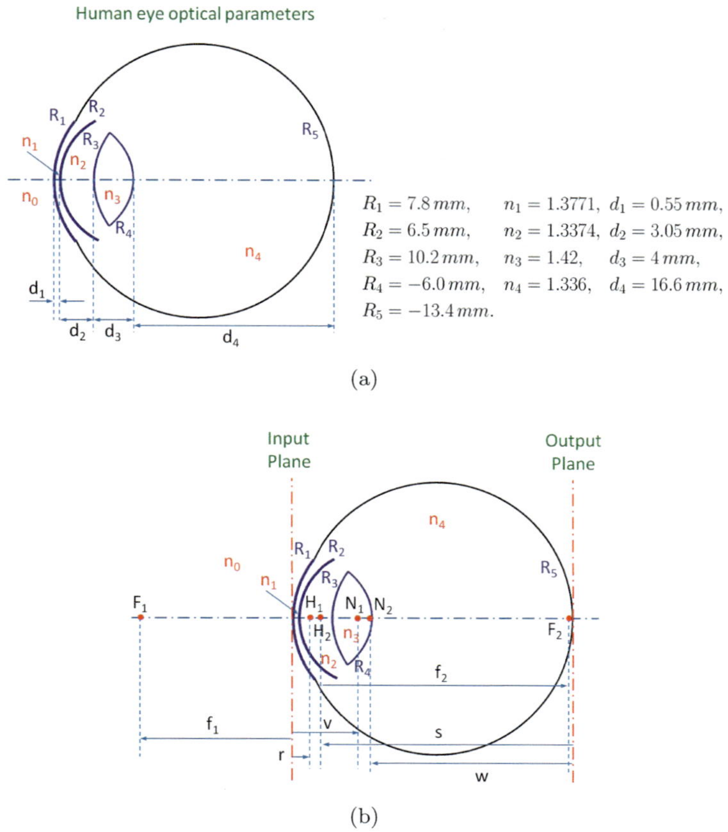

FIGURE 2.87: (a) Human eye optical parameters (average eye). (b) Human eye cardinal points.

thickness d_3) as shown in Fig. 2.87(a). These two lenses form a real image on the retina (the back surface of the eye), where the human photoreceptors (cones and rods, discussed in the chapter on *Light Measurement and Emission: Radiometry, Photometry, and Blackbody Radiation*, Chapter 4) are located. After refracting through the cornea, light passes through a chamber of length d_2, filled with a watery liquid called the *aqueous humor*, which has a refractive index n_2 (close to that of water). Inside the aqueous humor is the *iris* (not shown in Fig. 2.87(a)), which serves as an aperture stop, controlling the amount of light entering the crystalline lens through the opening known as the *pupil.* The crystalline lens itself is composed of concentric layers (resembling an onion) with varying refractive indices. For this simplified analysis, an average refractive index n_3 is assumed. The crystalline lens is controlled by small ligaments and muscles that enable it to adjust its optical power through a process called *accommodation.* This adjustment is achieved by varying the radii of curvature R_3 and R_4 as well as the thickness d_3. Behind the crystalline lens lies the main chamber of the eye, known as the *vitreous humor*, which contains a gel-like liquid with a refractive index n_4. The back (right) boundary of the vitreous humor is the retina, containing the photoreceptors (cones and rods) that detect the image formed. All cones and rods connect to the optic nerve, which transmits signals to the back of the brain, where optical signal processing occurs to produce the sense of vision. The area where the optic nerve exits the eye lacks receptors, creating a *blind spot* where any incident light is undetected. The optical axis of the eye, shown in Fig. 2.87(a), intersects the retina in an area densely populated with receptors called the *macula.*

To determine the effective focal length of the human eye, the cardinal points of the eye are calculated using the $ABCD$ matrix approach (see Sec. 2.7). The parameters, including radii of curvature, refractive indices, and distances, are shown in Fig. 2.87(a) [55]. For completeness, are also listed in the text: $R_1 = 7.8\,\text{mm}$, $R_2 = 6.5\,\text{mm}$, $R_3 = 10.2\,\text{mm}$, $R_4 = -6\,\text{mm}$, $R_5 = -13.4\,\text{mm}$, $n_0 = 1$ (air), $n_1 = 1.3771$, $n_2 = 1.3374$, $n_3 = 1.42$, $n_4 = 1.336$, $d_1 = 0.55\,\text{mm}$, $d_2 = 3.05\,\text{mm}$, $d_3 = 4\,\text{mm}$, and $d_4 = 16.6\,\text{mm}$. The crystalline lens parameters R_3, R_4, and d_3 correspond to a relaxed eye (viewing distant objects). By multiplying the 8 individual $ABCD$ matrices, the overall system matrix $\tilde{M}_T$, is obtained, with the input plane at the left surface of the cornea and the output plane at the right surface of the retina. This matrix enables the calculation of the effective focal length:

$$\tilde{M}_T = \begin{bmatrix} A & B \\ C & D \end{bmatrix} = \begin{bmatrix} 1 & d_4 \\ 0 & 1 \end{bmatrix} \begin{bmatrix} 1 & 0 \\ \left(\dfrac{n_3}{n_4} - 1\right)\dfrac{1}{R_4} & \dfrac{n_3}{n_4} \end{bmatrix} \begin{bmatrix} 1 & d_3 \\ 0 & 1 \end{bmatrix} \begin{bmatrix} 1 & 0 \\ \left(\dfrac{n_2}{n_3} - 1\right)\dfrac{1}{R_3} & \dfrac{n_2}{n_3} \end{bmatrix}$$

$$\begin{bmatrix} 1 & d_2 \\ 0 & 1 \end{bmatrix} \begin{bmatrix} 1 & 0 \\ \left(\dfrac{n_1}{n_2} - 1\right)\dfrac{1}{R_2} & \dfrac{n_1}{n_2} \end{bmatrix} \begin{bmatrix} 1 & d_1 \\ 0 & 1 \end{bmatrix} \begin{bmatrix} 1 & 0 \\ \left(\dfrac{n_0}{n_1} - 1\right)\dfrac{1}{R_1} & \dfrac{n_0}{n_1} \end{bmatrix}$$

$$= \begin{bmatrix} -1.5468 \times 10^{-4} & 16.6856\,\text{mm} \\ -0.0449\,(1/\text{mm}) & 0.6770 \end{bmatrix}.$$

Using the equations of Table 2.4 the position of all cardinal points can be specified. Specifically, the following distances (with the convention defined in Sec. 2.7.6) are computed $p = -15.0886\,\text{mm}$, $r = 1.5946\,\text{mm}$, $q = -0.0034\,\text{mm}$, $s = -22.2922\,\text{mm}$, $v = 7.2001\,\text{mm}$, $w = -16.6867\,\text{mm}$, $f_1 = -16.6832\,\text{mm}$ and $f_2 = 22.2888\,\text{mm}$. The approximate location of all cardinal points, F_1, H_1, N_1, F_2, H_2, and N_2 is shown in Fig. 2.87(b). It is observed that since the input and output refractive indices are different the front focal length $|f_1|$ and the back focal length f_2 are not equal. With these parameters the back focal length of the human eye is 22.2888 mm which corresponds to an optical power of $1/0.0222888 = 44.87 \simeq 45$ Diopters. Of course, through accommodation, the crystalline lens can increase the optical power of the eye.

To illustrate image formation in the human eye, let us assume an object located at a distance of $d_0 = 10\,\text{m}$ away from the eye. The image location can be determined using the previously listed parameters. By applying the matrix approach, the following equations are formulated:

$$\tilde{M}'_T = \begin{bmatrix} 1 & \Delta d_4 \\ 0 & 1 \end{bmatrix} \begin{bmatrix} A & B \\ C & D \end{bmatrix} \begin{bmatrix} 1 & d_0 \\ 0 & 1 \end{bmatrix} = \begin{bmatrix} A + \Delta d_4 C & A d_0 + B + \Delta d_4 (C d_0 + D) \\ C & C d_0 + D \end{bmatrix}.$$

Setting the overall B-element of the system matrix to zero to achieve imaging, we find $(\tilde{M}'_T)_{12} = 0 \rightarrow \Delta d_4 = 0.0338\,\text{mm}$. This result indicates that the final image is ideally formed 0.0338 mm behind the retina (where ideally Δd_4 should be zero). However, this is merely an example calculation, as the crystalline lens can adjust its optical power to bring the image precisely onto the retina ($\Delta d_4 = 0$). To verify this, it is insightful to use Eq. (2.73) to reach the same conclusion. Here, the object distance is given by $s_o = d_0 + r = 10001.5946\,\text{mm}$ with $f_1 = 16.6832\,\text{mm}$ [using the absolute value of f_1 is used in Eq. (2.73)] and s_i (image position) is measured from H_2. Applying the equation yields:

$$\frac{f_1}{s_o} + \frac{f_2}{s_i} = 1 \Longrightarrow s_i = 22.3260\,\text{mm}.$$

Since s_i is measured from H_2, the $\Delta d_4 = s_i - |s| = 22.3260 - 22.2922 = 0.0338\,\text{mm}$ as it was determined using the matrix method.

Another noteworthy calculation is to determine the refractive power of only the crystalline lens. Using Eq. (2.80) for a thick lens, with the previously listed parameters, an optical power of 16.3 Diopters is obtained for the crystalline lens. This value represents approximately one-third of the total refractive power of the human eye ($16.3/45 \simeq 36\%$). However, the crystalline lens has the unique ability to adjust, or "accommodate," by altering its radii of curvature and thickness, thereby varying its refractive power. This accommodation enables the eye to focus images onto the retina for objects at different distances. The accommodative range typically varies between 2 to 14 diopters, depending on age. Unfortunately, with increasing age, the lens becomes more rigid, reducing its accommodative ability, a condition that results in farsightedness (or hyperopia). A comparison between a normal and a farsighted eye is shown in Fig. 2.88(a). Farsightedness can be corrected by placing positive-power lenses in front of the cornea to increase the overall optical power, thereby allowing light to focus on the retina.

Conversely, there are cases where the cornea and crystalline lens possess more optical power than necessary. In such instances, light from an object at infinity is focused in front of the retina, in the vitreous humor region, causing a blurred image at the retina. This condition is known as shortsightedness (or myopia). A comparison between a normal and a shortsighted eye is illustrated in Fig. 2.88(b). Shortsightedness can be corrected by placing negative-power lenses in front of the cornea to reduce the overall optical power, allowing proper focus on the retina. Other refractive errors, such as astigmatism, where variations in optical power occur in two orthogonal directions due to abnormalities in the cornea and/or crystalline lens, also cause blurred retinal images. However, these topics lie beyond the scope of this book.

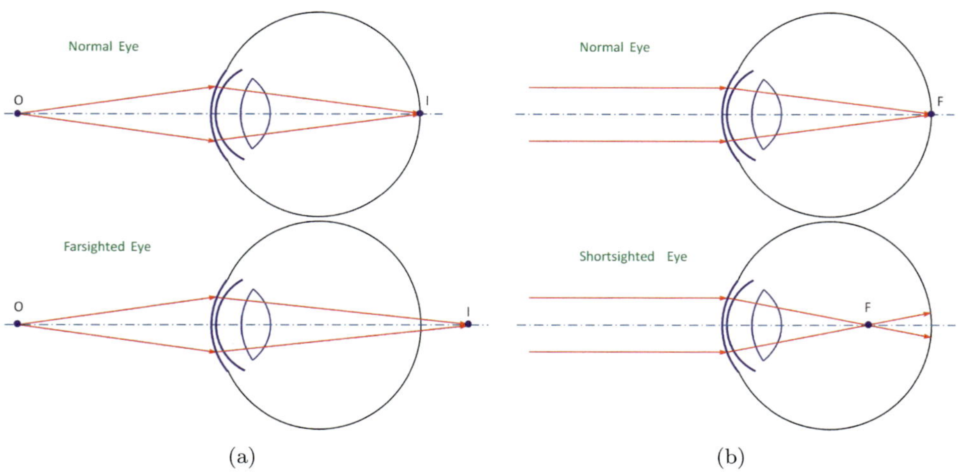

FIGURE 2.88: (a) Comparison of the normal eye and the farsighted eye for an object O. In the normal eye through crystalline lens accommodation, the image is formed on the retina. However, due to lack of sufficient accommodation, in the case of the farsighted eye, the image is formed behind the retina and therefore the retina detects a blurred image. For simplicity, the rays in the eye are shown without accounting for the various refractions. (b) Comparison of the normal eye and the shortsighted eye for an object at infinity. In the normal eye with a relaxed crystalline lens, the image is formed on the retina. However, due to larger than necessary optical power of the crystalline lens (or the cornea or both), in the case of the shortsighted eye, the image is formed before the retina and therefore the retina detects a blurred image. Again for simplicity, the rays in the eye are shown without accounting for the various refractions.

2.9 Aberrations Basics

No optical imaging system is perfect. A single point on an object may not be reproduced as a single point in the image due to inherent imperfections. These imperfections can be broadly classified into three main categories:

- **Diffraction Effects:** These arise from the wave nature of light, limiting the resolution of optical systems. Diffraction causes light to spread as it passes through an aperture, preventing perfect point-to-point imaging. This effect is especially significant in systems with small apertures, where diffraction can limit the system's resolving power and impact image sharpness.
- **Lens Aberrations:** These arise in the realm of geometrical optics. In paraxial optics, ray-tracing equations are accurate to the first order with respect to the inclination angles of rays and the normals to refracting or reflecting surfaces. However, when higher-order terms are included in the trigonometric functions ($\sin\theta = \theta - \theta^3/3! + \theta^5/5! - \cdots$), substantial deviations from paraxial predictions emerge. These deviations imply that rays originating from a single point on an object do not precisely converge to form a single point in the image, and the transverse magnification is no longer consistent across planes. Additionally, the system's optical properties may vary with wavelength. These departures from ideal paraxial behavior are known as *lens aberrations*.
- **Irregularities:** These arise from issues in quality control, such as inhomogeneities in the glass or surface defects. Such irregularities can cause scattering losses and degrade image quality, compounding the impact of other effects like diffraction and aberrations.

In the following sections, both wave and ray aberrations are defined, and a brief analysis is provided. For a more comprehensive and detailed treatment of the various aberrations, readers are referred to specialized texts such as Refs. [56–60].

2.9.1 Wave and Ray Aberration

Initially, the concepts of wave and ray aberrations are introduced, and their interrelationship is established [57, 59]. The wave aberration of a given image point represents the deviation of the actual optical wavefront at the exit pupil from the ideal spherical wavefront. In contrast, ray aberrations describe the deviation between the actual rays and the corresponding paraxial rays. The ray aberrations of an optical system can be determined through exact ray tracing techniques, tracing the rays through the optical system up to the image plane.

However, ray aberrations are related to wave aberrations and, as a result, can be expressed as functions of the latter. Consider an optical system and an object point P as shown in Fig. 2.89 [57]. The image point is denoted as P'. In an ideal, aberration-free optical system, the diverging spherical wavefronts from point P are converted by the system into converging spherical wavefronts at point P'. However, in real optical systems, the exiting wavefronts are generally not spherical, which represents the wave aberrations.

In this section, a brief description of the possible aberrations in a rotationally symmetric optical system is presented. Consider a realistic optical system where aberrations occur. In Fig. 2.90(a) a point $P_w(x_p, y_p, z_p)$ on the actual wavefront (W) that passes from the center O of the exit pupil of the system is shown. The coordinate system is Cartesian with z being the optical axis of the system. A reference spherical wavefront R is also shown, passing through the point O. Point $P(0, 0, R)$ (where R is the radius of the reference spherical wavefront) is the image point produced by the spherical wavefront and corresponds to the paraxial imaging point. The plane normal to the optical axis at point P is the paraxial

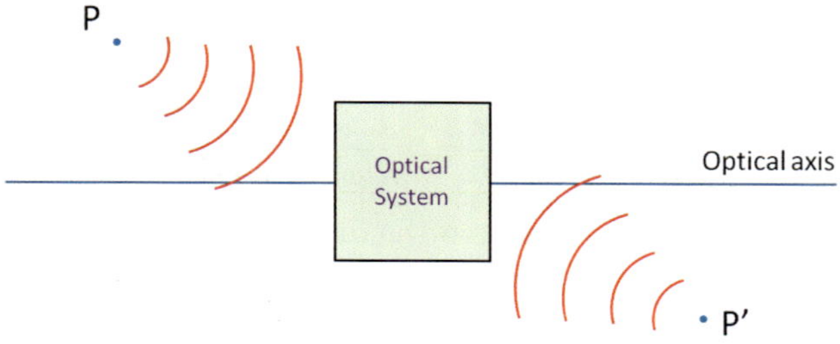

FIGURE 2.89: An ideal optical system without wave aberrations. All diverging spherical wavefronts from object point P are converted via the optical system into converging spherical wavefronts to the image point P'.

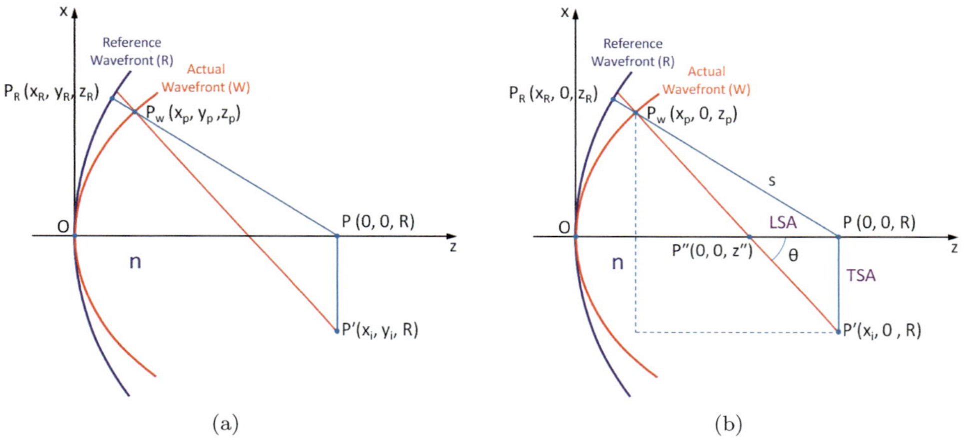

FIGURE 2.90: (a) Diagram for the determination of the wavefront aberration and the corresponding transverse ray aberration ($TSA = PP'$). The reference wavefront is spherical with radius $R = PP_R$. (b) Diagram for the determination of the longitudinal ray aberration ($LSA = PP''$). All rays in this case are in a meridional plane, which is assumed to be the xz plane. The actual ray P_wP' is normal to the actual wavefront W.

image plane. The ray from the actual wavefront passing through $P_w(x_p, y_p, z_p)$ intersects the paraxial image plane at point $P'(x_i, y_i, R)$. The wave aberration is defined as the optical path difference between the actual ray and the reference ray, which can be expressed as [57]:

$$W = n(P_RP - P_wP) = n(R - s) \implies s = R - \frac{W}{n}, \tag{2.104}$$

where $P_RP = R$ and $P_wP = s$ as it is also shown in Fig. 2.90(a). Assume that the function $F(x_p, y_p, z_p) = 0$ represents the surface of the actual wavefront W. In addition, the distance s satisfies the equation $x_p^2 + y_p^2 + (z_p - R)^2 - s^2 = 0$. It is assumed that the coordinates x_p and y_p are independent coordinates and z_p is the dependent one. Then the following equations can be written:

$$\frac{\partial F}{\partial x_p}dx_p + \frac{\partial F}{\partial y_p}dy_p + \frac{\partial F}{\partial z_p}\underbrace{\left(\frac{\partial z_p}{\partial x_p}dx_p + \frac{\partial z_p}{\partial y_p}dy_p\right)}_{= dz_p} = 0 \implies$$

$$\left(\frac{\partial F}{\partial x_p}+\frac{\partial F}{\partial z_p}\frac{\partial z_p}{\partial x_p}\right)dx_p+\left(\frac{\partial F}{\partial y_p}+\frac{\partial F}{\partial z_p}\frac{\partial z_p}{\partial y_p}\right)dy_p=0 \implies$$

$$\frac{\partial z_p}{\partial x_p}=-\frac{\partial F/\partial x_p}{\partial F/\partial z_p},\quad \frac{\partial z_p}{\partial y_p}=-\frac{\partial F/\partial y_p}{\partial F/\partial z_p}. \tag{2.105}$$

Similarly, the following equations related to the distance s can be written:

$$x_p dx_p + y_p dy_p + (z_p - R)dz_p - s ds = 0 \implies$$

$$x_p dx_p + y_p dy_p + (z_p - R)\left(\frac{\partial z_p}{\partial x_p}dx_p+\frac{\partial z_p}{\partial y_p}dy_p\right) - s\left(\frac{\partial s}{\partial x_p}dx_p+\frac{\partial s}{\partial y_p}dy_p\right)=0 \implies$$

$$x_p+(z_p-R)\frac{\partial z_p}{\partial x_p}-s\frac{\partial s}{\partial x_p}=0,\qquad y_p+(z_p-R)\frac{\partial z_p}{\partial y_p}-s\frac{\partial s}{\partial y_p}=0.$$

Using Eqs. (2.105) the latter equations are written as follows:

$$s\frac{\partial s}{\partial x_p}=x_p+(z_p-R)\left(\frac{\partial F/\partial x_p}{\partial F/\partial z_p}\right), \tag{2.106}$$

$$s\frac{\partial s}{\partial y_p}=y_p+(z_p-R)\left(\frac{\partial F/\partial y_p}{\partial F/\partial z_p}\right). \tag{2.107}$$

At this point the fact that the ray P_wP is perpendicular to the actual wavefront is utilized. The direction of P_wP is parallel to the gradient $\vec{\nabla}F = (\partial F/\partial x_p)\hat{\imath}_x + (\partial F/\partial y_p)\hat{\imath}_y + (\partial F/\partial z_p)\hat{\imath}_z$. Then the unit vector in the direction of the ray P_wP is $\hat{\imath}_n = \vec{\nabla}F/|\vec{\nabla}F| = [(x_i - x_p)/s]\hat{\imath}_x + [(y_i - y_p)/s]\hat{\imath}_y + [(R - z_p)/s]\hat{\imath}_z$. Then the following equations can easily be obtained:

$$\frac{\partial F/\partial x_p}{\partial F/\partial z_p}=-\frac{x_i-x_p}{R-z_p},$$

$$\frac{\partial F/\partial y_p}{\partial F/\partial z_p}=-\frac{y_i-y_p}{R-z_p}.$$

The "$-$" sign is inserted since the gradient $\vec{\nabla}F$ points towards increasing R while the ray direction is opposite. Using the last equations in Eqs. (2.106) and (2.107), the following final equations for x_i and y_i are obtained:

$$x_i = s\frac{\partial s}{\partial x_p}=\left(R-\frac{W}{n}\right)\left(-\frac{1}{n}\frac{\partial W}{\partial x_p}\right)\simeq -\frac{R}{n}\frac{\partial W}{\partial x_p}, \tag{2.108}$$

$$y_i = s\frac{\partial s}{\partial y_p}=\left(R-\frac{W}{n}\right)\left(-\frac{1}{n}\frac{\partial W}{\partial y_p}\right)\simeq -\frac{R}{n}\frac{\partial W}{\partial y_p}, \tag{2.109}$$

where the last approximation is usually true since $R \gg W/n$. Therefore, the transverse ray aberration (TSA) is defined as:

$$(x_i, y_i) = -\frac{R}{n}\left(\frac{\partial W}{\partial x_p},\frac{\partial W}{\partial y_p}\right),\qquad TSA = (x_i^2 + y_i^2)^{1/2}. \tag{2.110}$$

The "$-$" sign is the result of the selection in Fig. 2.90(a) of the actual wavefront to be leading the reference wavefront. In other textbooks the opposite is selected, resulting in a "+" sign in Eq. (2.110). In the case of Fig. 2.90(a), $x_i < 0$ since it is shown below the z

axis, and therefore $\partial W/\partial x_p > 0$ must hold. For the longitudinal spherical aberration Fig. 2.90(b) is used. In this figure, all rays are assumed to be meridional, meaning they belong to a plane that includes the optical axis. In Fig. 2.90(b) this plane is the xz plane. The longitudinal spherical aberration LSA is the difference $z'' - R$ and according to Fig. 2.90(b) should be negative. From the figure the following equations can be written:

$$LSA = z'' - R = \frac{-x_i}{\tan\theta} = -x_i \frac{R - z_p}{x_i - x_p} \simeq -R \frac{x_i}{x_i - x_p}, \tag{2.111}$$

since usually $z_p \ll R$. For rotationally symmetrical optical systems, the function $W(x_p, y_p)$ can be written in the form $W(r_{T_p}, \phi_p)$ where r_{T_p} and ϕ_p are the polar coordinates measured in $x_p y_p$ planes (see Fig. 2.91), i.e., $x_p = r_{T_p} \cos\phi_p$, $y_p = r_{T_p} \sin\phi_p$ or $r_{T_p} = (x_p^2 + y_p^2)^{1/2}$ and $\phi_p = \tan^{-1}(y_p/x_p)$. Then, the transverse spherical aberration can be expressed in the form:

$$(x_i, y_i) = -\frac{R}{n}\left(\cos\phi_p \frac{\partial W}{\partial r_{T_p}} - \frac{\sin\phi_p}{r_{T_p}} \frac{\partial W}{\partial \phi_p},\ \sin\phi_p \frac{\partial W}{\partial r_{T_p}} + \frac{\cos\phi_p}{r_{T_p}} \frac{\partial W}{\partial \phi_p}\right), \tag{2.112}$$

and $TSA = (x_i^2 + y_i^2)^{1/2}$. Even if the relation between the wave aberration and the ray aberration was established from the previous analysis the truth is that neither can actually be analytically determined for a general optical system. Even for very simple optical systems, such as a thick lens, the determination of the wave aberration and the ray aberrations is very cumbersome. The usage of exact ray tracing methods is necessary for the actual determination of the ray aberrations. However, some simplifications can be made in order to classify the various aberration types. These simplifications are based on the fact that usually the optical systems have rotational symmetry around their optical axis and this can restrict the form of the function $W(x_o, y_o; x_p, y_p)$, or in polar coordinates $W(r_{T_o}, \theta_o; r_{T_p}, \theta_p)$, where the (x_o, y_o) are the emanating ray coordinates from the object point and (x_p, y_p) the corresponding ray point in the exit pupil of the optical system (see Fig. 2.91). Since the wave aberration function W has rotational symmetry, it can be represented by a power series expansion of the form [56–61]:

$$\begin{aligned} W(\vec{r}_{T_o}, \vec{r}_{T_p}) &= \sum_{\ell=0}^{\infty}\sum_{p=0}^{\infty}\sum_{m=0}^{\infty} C_{\ell pm} \left(\vec{r}_{T_o}\cdot\vec{r}_{T_o}\right)^{\ell} \left(\vec{r}_{T_p}\cdot\vec{r}_{T_p}\right)^{p} \left(\vec{r}_{T_o}\cdot\vec{r}_{T_p}\right)^{m} \\ &= \sum_{\ell=0}^{\infty}\sum_{p=0}^{\infty}\sum_{m=0}^{\infty} C_{\ell pm} r_{T_o}^{2\ell+m} r_{T_p}^{2p+m} \cos^m\phi_p \\ &= \sum_{n=0}^{\infty}\sum_{k=0}^{\infty}\sum_{m=0}^{\infty} C'_{nkm} r_{T_o}^{k} r_{T_p}^{n} \cos^m\phi_p, \end{aligned}$$

where $2p + m = n$, $2\ell + m = k$, and $\vec{r}_{T_o}$, $\vec{r}_{T_p}$ are vectors on the $x_o y_o$ object plane and $x_p y_p$ exit pupil plane, respectively, as shown in Fig. 2.91. Without loss of generality it can be assumed that $\vec{r}_{T_o}$ has $\phi_o = 0$ as shown in Fig. 2.91. Therefore, $r_{T_o} = |\vec{r}_{T_o}| = h$ and represents the height of the object point from the optical axis. However, it can be shown [57] that h is proportional to the image point height $h' = r_{T_i} = |\vec{r}_{T_i}|$ (h'/h equals to the transverse magnification) and it can be replaced in the power series expansion. I.e., the aberration function W can be expressed as:

$$W(h', r_{T_p}, \phi_p) = \sum_{k=0}^{\infty}\sum_{n=0}^{\infty}\sum_{m=0}^{\infty} W_{knm} (h')^k r_{T_p}^{n} \cos^m\phi_p. \tag{2.113}$$

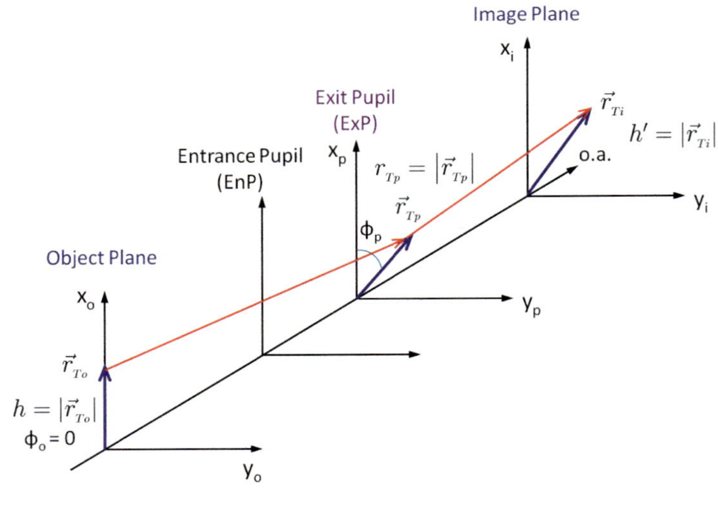

FIGURE 2.91: The Cartesian coordinate systems are related to the object plane, entrance pupil (EnP) plane, exit pupil (ExP) plane, and image plane. An object point is represented by the vector $\vec{r}_{To}$. A ray from the object point passes from the exit pupil from a point that is represented by the vector $\vec{r}_{Tp}$ and crosses the image plane at a point that is represented by the vector $\vec{r}_{Ti}$.

In the above series the degree of the terms that are nonzero is $4, 6, 8, \cdots$ [57]. This means that the power of the product $r_{Tp}h'$ is of degree $4, 6, 8, \cdots$, and the coefficients W_{knm} depend on the optical system under study and are usually difficult to be analytically determined. The coefficients W_{k00} are all zero since they do not depend on r_{Tp} and for $r_{Tp} = 0$ the wave aberration is zero since both the reference spherical wavefront and the actual aberrated wavefront pass from the center of the exit pupil where of course $r_{Tp} = 0$. These aberrations are known as primary or *Seidel* (when the degree of the terms retained is 4), secondary (when the degree of the terms retained is 6), tertiary (when the degree of the terms retained is 8), etc. The primary or Seidel aberrations can be represented by the wave aberration function W:

$$W = W_{040}r_{Tp}^4 + W_{131}h'r_{Tp}^3\cos\phi_p + W_{222}(h')^2r_{Tp}^2\cos^2\phi_p + W_{220}(h')^2r_{Tp}^2 + W_{311}(h')^3r_{Tp}\cos\phi_p. \tag{2.114}$$

The 5 terms appearing in Eq. (2.114) correspond to the five primary aberrations, i.e., spherical, coma, astigmatism, field curvature, and distortion, and are summarized in Table 2.9.

TABLE 2.9: Summary of Seidel Aberrations

$k = 2\ell + m$	$n = 2p + m$	m	ℓ	p	Term	Name
0	4	0	0	2	$W_{040}r_{Tp}^4$	Spherical
1	3	1	0	1	$W_{131}h'r_{Tp}^3\cos\phi_p$	Coma
2	2	2	0	0	$W_{222}(h')^2r_{Tp}^2\cos^2\phi_p$	Astigmatism
2	2	0	1	1	$W_{220}(h')^2r_{Tp}^2$	Field Curvature
3	1	1	1	0	$W_{311}(h')^3r_{Tp}\cos\phi_p$	Distortion

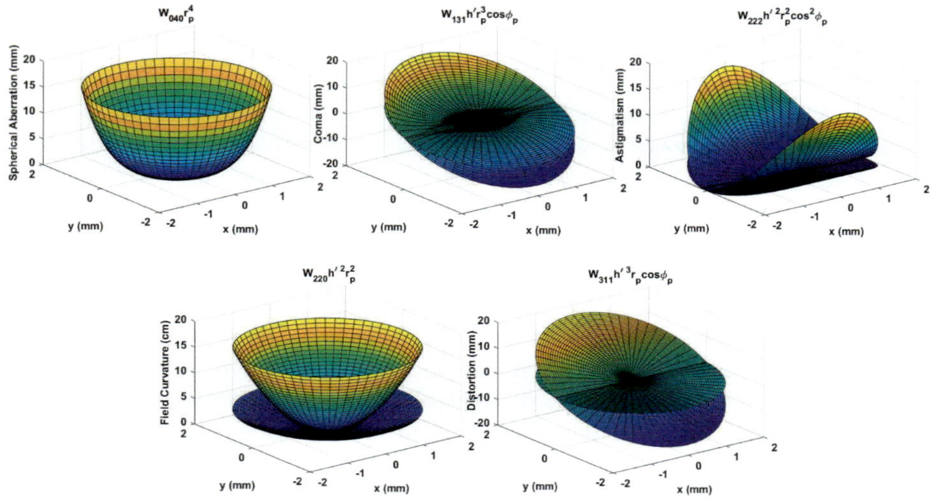

FIGURE 2.92: The five primary or Seidel wave aberrations that are represented in Table 2.9. All except the spherical aberration term are shown for two values of h' as an example.

2.9.1.1 Spherical Aberration via Wave Primary Aberrations

In the case of spherical aberration, the wave aberration is $W_{sph} = W_{040} r_{Tp}^4$ and the corresponding values of x_i and y_i are given by [using Eqs. (2.112)]:

$$x_i = -\frac{R}{n}\left(\frac{\partial W_{sph}}{\partial r_{Tp}} \cos\phi_p\right) = -\frac{4R}{n} W_{040}\ r_{Tp}^3 \cos\phi_p,$$

$$y_i = -\frac{R}{n}\left(\frac{\partial W_{sph}}{\partial r_{Tp}} \sin\phi_p\right) = -\frac{4R}{n} W_{040}\ r_{Tp}^3 \sin\phi_p,$$

$$TSA = (x_i^2 + y_i^2)^{1/2} = \frac{4R}{n}\ |W_{040}|\ r_{Tp}^3, \tag{2.115}$$

$$LSA = -R\frac{x_i}{x_i - x_p} \simeq R\frac{x_i}{x_p} = -\frac{4R^2}{n} W_{040}\ r_{Tp}^2. \tag{2.116}$$

Therefore, the transverse spherical ray aberration depends on r_{Tp}^3 while the longitudinal spherical wave aberration depends on r_{Tp}^2. It is important to note that for cases as shown in Fig. 2.90(b) when $LSA < 0$, special care must be taken when $W_{040} < 0$ to retain the value of LSA negative. This ensures that the actual rays reach the optical axis before the paraxial rays.

2.9.1.2 Coma via Wave Primary Aberrations

In the case of comma aberration, the wave aberration is $W_{coma} = W_{131} h' r_{Tp}^3 \cos\phi_p$ and the corresponding values of x_i and y_i are given by [using Eqs. (2.112)]:

$$x_i = -\frac{R}{n}\left(W_{131} h' r_{Tp}^2\right)(2 + \cos 2\phi_p) = \rho\,(2 + \cos 2\phi_p), \tag{2.117}$$

$$y_i = -\frac{R}{n}\left(W_{131} h' r_{Tp}^2\right)\sin 2\phi_p = \rho \sin 2\phi_p, \tag{2.118}$$

where $\rho = -(R/n)W_{131}h'r_{_{Tp}}^2$. The above equations satisfy the following equation:

$$(x_i - \rho)^2 + y_i^2 = \rho^2. \tag{2.119}$$

In the case where $\rho < 0$, implying $W_{131} > 0$, various points are shown on the exit pupil plane and the corresponding points on the image plane. Specifically, the points A, B, C, D, E, F, G, H are shown on the exit pupil plane, and the corresponding points (with the same notation) on the image plane. For example, the polar coordinates on the exit pupil plane of point A are $(r_{_{Tp}}, 0)$ and the corresponding point in the image plane has rectangular coordinates $(x_i, y_i) = (3\rho, 0)$. Similarly, the positions of all the points on the image plane can be easily verified. In Figs. 2.93(b) and 2.93(c) various coma circles are shown on the image plane for $\rho < 0$ and $\rho > 0$, respectively. It is straightforward to verify that the tangent from the origin of the image plane to all coma circles forms an angle of $\pi/6$ from the x_i axis as is shown. This is a distinctive feature of the coma aberration, indicating the angular deviation of the image points due to the aberration.

2.9.1.3 Astigmatism and Field Curvature via Wave Primary Aberrations

The astigmatism wave aberration is closely related to the field curvature wave aberration, and they are often studied together due to their interdependence in optical systems. [43]. The combined wave aberration for astigmatism and field curvature can be expressed as $W_{ast+fc} = W_{222}(h')^2 r_{_{Tp}}^2 \cos^2\phi_p + W_{220}(h')^2 r_{_{Tp}}^2$, and the corresponding values of x_i and y_i are given by [using Eqs. (2.112)]:

$$x_i = -\frac{2R}{n}(h')^2 r_{_{Tp}} (W_{222} + W_{220}) \cos\phi_p = a_x r_{_{Tp}} \cos\phi_p, \tag{2.120}$$

$$y_i = -\frac{2R}{n}(h')^2 r_{_{Tp}}\, W_{220} \sin\phi_p = a_y r_{_{Tp}} \sin\phi_p, \tag{2.121}$$

The previous equations can be combined to reveal that the locus of points that have the same radial distance $r_{_{Tp}}$ in the exit pupil plane is an ellipse:

$$\frac{x_i^2}{a_x^2 r_{_{Tp}}^2} + \frac{y_i^2}{a_y^2 r_{_{Tp}}^2} = 1. \tag{2.122}$$

The terms $|a_x| r_{_{Tp}}$ and $|a_y| r_{_{Tp}}$ are the semi-axes lengths of the ellipse on the image plane as shown in Fig. 2.94(b). The terms a_x and a_y are defined as $a_x = -(2R/n)(h')^2(W_{222} + W_{220})$ and $a_y = -(2R/n)(h')^2 W_{220}$, respectively. To better understand the astigmatism and field curvature aberrations, one approach is to examine the intersections of the actual rays with planes that are shifted and parallel to the image plane. These shifts help visualize how the rays deviate from the ideal image plane due to the aberrations. The relationship between the shifted planes and the ray positions is illustrated in Figs. 2.94(a) and 2.94(b). The shifts of the actual ray positions, x_i' and y_i', can be related to the plane shift ΔS. This plane shift describes the difference in distance between the ideal image plane (in the absence of aberrations) and the shifted planes where the actual rays intersect, as shown in Fig. 2.94(a). The x_i' and y_i' can be expressed as follows:

$$x_i' \simeq x_i - \Delta S\,\frac{x_p - x_i}{R} \simeq x_i - \frac{\Delta S}{R} x_p = \underbrace{\left(a_x - \frac{\Delta S}{R}\right)}_{= b_x} x_p = b_x x_p, \tag{2.123}$$

$$y_i' \simeq y_i - \Delta S\,\frac{y_p - x_i}{R} \simeq y_i - \frac{\Delta S}{R} y_p = \underbrace{\left(a_y - \frac{\Delta S}{R}\right)}_{= b_y} y_p = b_y y_p. \tag{2.124}$$

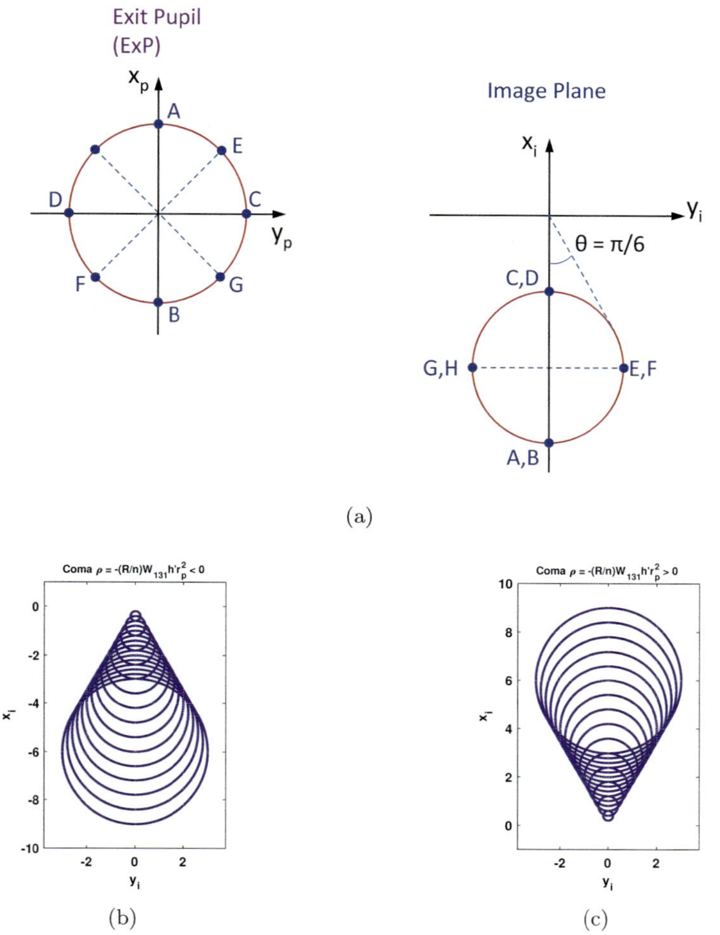

FIGURE 2.93: (a) Example points A, B, C, D, E, F, G, H are shown on the exit pupil plane at a constant distance r_{T_p}. The corresponding points are also shown in the image plane. The relation between the points in the exit pupil and the image planes are related via Eqs. (2.117) and (2.118). (b) Coma circles for various r_{T_p} values (or equivalently ρ values) for $\rho < 0$. (c) Coma circles for various r_{T_p} values (or equivalently ρ values) for $\rho > 0$.

In the last equations the distance along the optical axis between x_p and x_i is approximately R (same for the y points). In addition, $|x_i| \ll |x_p|$ and $|y_i| \ll |y_p|$ which usually holds. Obviously the new points x_i' and y_i' on the shifted plane from an ellipse for constant r_{T_p}, i.e., $(x_i'^2/b_x^2 r_{T_p}^2) + (y_i'^2/b_y^2 r_{T_p}^2) = 1$. However, there is the parameter ΔS that can be varied to make either b_x or b_y equal to zero. In these cases if $b_x = 0$ for $\Delta S = \Delta S_t$ then $x_i' = 0$ and y_i' varies along the y_i direction between $\pm b_y(\text{for } \Delta S = \Delta S_t) r_{T_p}$. This is called the *tangential image plane.* On the other hand, if $b_y = 0$ for $\Delta S = \Delta S_s$ then $y_i' = 0$ and x_i' varies along the x_i direction between $\pm b_x(\text{for } \Delta S = \Delta S_s) r_{T_p}$. This is called the *saggital image plane.* The special shifts ΔS_s and ΔS_t are given by:

$$\Delta S_t = -\frac{2R^2}{n}(h')^2\left(W_{222} + W_{220}\right), \tag{2.125}$$

$$\Delta S_s = -\frac{2R^2}{n}(h')^2 W_{220}. \tag{2.126}$$

Of course, the absolute values of the previous equations should be used to find the actual distances, otherwise, the sign indicates the shift to the left or to the right of the image plane. The tangential and saggital image planes are shown in Fig. 2.94(b). A special shift occurs when $\Delta S = \Delta S_c = (1/2)(\Delta S_t + \Delta S_s) = -[R^2(h')^2/n](W_{222} + 2W_{220})$. In this case $b_x = a_x - \Delta S_c/R = -[R(h')^2/n]W_{222}$ and $b_y = a_y - \Delta S_c/R = +[R(h')^2/n]W_{222}$ which means $|b_x| = |b_y|$ and the ellipse becomes a circle which is called the *Circle of Least Confusion* (CoC) with a radius of $r_c = [R(h')^2/n]|W_{222}|r_{T_p}$. The CoC is also shown in Fig. 2.94(b). It should be mentioned that because ΔS_t [Eq. (2.125)] and ΔS_s [Eq.

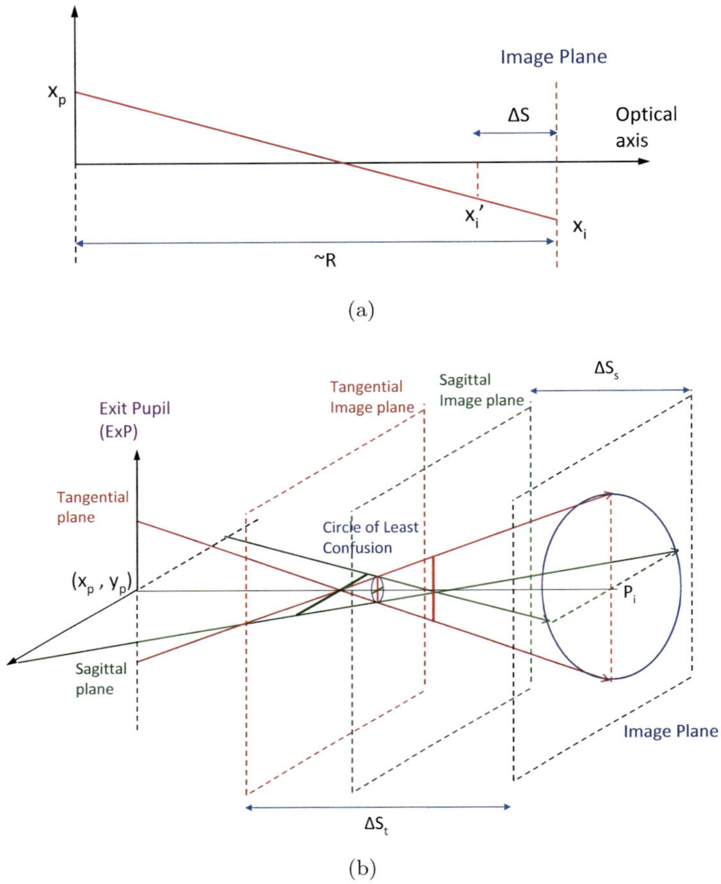

FIGURE 2.94: (a) Intersection of the actual ray, x_i' with a plane parallel to the image (paraxial) plane that is displaced by ΔS. The horizontal distance between the x_p and x_i points is approximately R (the radius of curvature of the ideal spherical wave). (b) The saggital and the tangential images along with the plane of the circle of least confusion.

(2.126)] depend on h' the resulting tangential and saggital focal surfaces are not planar as it is schematically shown in Fig. 2.95. When there is no astigmatism, i.e., $W_{222} = 0$, the tangential and saggital surfaces coincide since $\Delta S_s = \Delta S_t = \Delta S_c = -(2R^2/n)(h')^2 W_{220}$. However, since these shifts depend on h' the surfaces remain non-planar. In the absence of astigmatism case, the common tangential and saggital focal surfaces are called *Petzval surface* and the image remains focused at this surface.

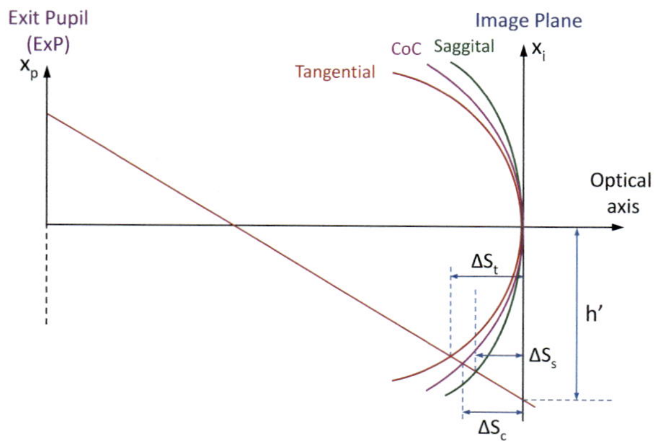

FIGURE 2.95: The saggital (at distance ΔS_s from image plane), the tangential (at distance ΔS_t from image plane), and the CoC (at distance ΔS_c from image plane), surfaces of an optical system with astigmatism and field curvature. When there is no astigmatism the saggital and tangential surfaces coincide, i.e., $\Delta S_s = \Delta S_t = \Delta S_c$, and form the *Petzval surface.*

2.9.1.4 Distortion via Wave Primary Aberrations

In this case of distortion wave aberration, the aberration function is $W_{dis} = W_{311}(h')^3 r_{T_p} \cos\phi_p$ and the corresponding values of x_i and y_i are given by [using Eqs. (2.112)]:

$$x_i = -\frac{R}{n} W_{311}(h')^3, \tag{2.127}$$

$$y_i = 0, \tag{2.128}$$

The above results imply that the distortion aberration depends only on the paraxial image height h'. To understand the effect of the distortion aberration consider a square of size a that is formed in the paraxial image plane as shown in Fig. 2.96(a). Let's focus on the upper horizontal section of the square. The effect of the distortion aberration depends on the paraxial image height $h' = h'(\phi_i) = (a/2)/\cos\phi_i$. Then using Eq. (2.127) the distortion aberration is $D = -(R/n)W_{311}[h'(\phi_i)]^3$ (it is assumed positive in the figure). The corresponding x_i and y_i are given by:

$$x_i(\phi_i) = \frac{a}{2} + D\cos\phi_i = \frac{a}{2} - \frac{R}{n} W_{311}\left(\frac{a/2}{\cos\phi_i}\right)^3 \cos\phi_i,$$

$$y_i(\phi_i) = h'(\phi_i)\sin\phi_i + D\sin\phi_i = \frac{a}{2}\tan\phi_i - \frac{R}{n} W_{311}\left(\frac{a/2}{\cos\phi_i}\right)^3 \sin\phi_i.$$

Applying the last equations, the distorted square is shown in Fig. 2.96(b) along with the unchanged paraxial square. In this figure two cases are shown: (a) When $D < 0$ ($W_{311} > 0$), the barrel distortion occurs, and (b) when $D > 0$ ($W_{311} < 0$), the pincushion distortion occurs. Summarizing at this point, the five basic aberrations in an optical system (Spherical Aberration Coma, Astigmatism, Curvature of field, and Distortion) were defined and a basic analysis was presented. These first five aberration types occur even for monochromatic light and are called *monochromatic aberrations.* However, when white light is used

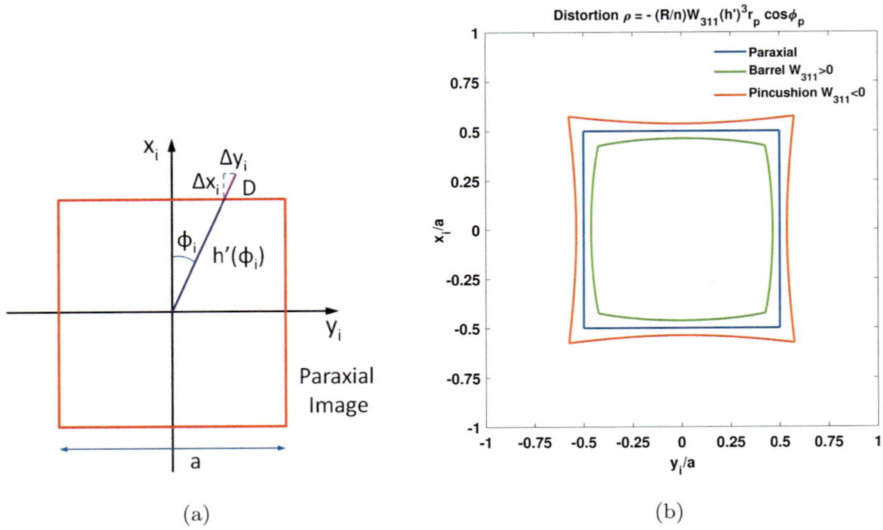

FIGURE 2.96: (a) A square of size a is formed in the paraxial image plane. The distortion depends on $h'(\phi_i)$. The distortion aberration is $D = -(R/n)W_{311}(h')^3$ that leads to aberrations Δx_i and Δy_i. (b) The resulting distortion of the square when $D < 0$ (barrel distortion) and $D > 0$ (pincushion distortion).

(which is common in imaging systems) additional aberrations occur due to the dispersion of the materials that the various optical elements are comprised of. These aberrations occur only for multi-chromatic light and are called *chromatic aberrations*. Chromatic aberrations cause issues such as color fringing, where different wavelengths of light focus at different points. The chromatic aberration is discussed in Sec. 2.9.4 of this chapter. These first three aberrations (spherical aberration, coma, and astigmatism) cause image blur, while the last two (curvature of field and distortion) cause dislocations (distortion) but not blur.

In the upcoming sections, the focus will be on discussing wave and ray aberrations for simpler optical systems, such as a spherical interface or a single thick lens. More attention will be given to spherical aberration, as it is relatively straightforward to analyze using meridional rays. For off-axis rays, an exact ray tracing approach will be more appropriate for studying the various aberrations in the system. An algorithm for exact ray tracing for spherical refractive interfaces will be presented in Sec. 2.10. This method will allow for a more precise calculation of the aberrations for simple systems, such as a spherical refractive surface. In the next section the aberrations of a single spherical refractive surface between two optical media is presented. This discussion will tie into the general principles of wave and ray aberrations introduced earlier, offering further insight into how these aberrations manifest in simple systems.

2.9.2 Aberration of a Spherical Refractive Interface

The simplest problem for calculating the aberration of a spherical refractive interface involves considering a spherical interface of radius R, between two isotropic, homogeneous media with refractive indices n_1 and n_2 respectively. Two cases are examined. (a) The case of meridional rays: Meridional rays are those rays that lie in the plane containing the optical axis of the system. These rays are typically used to analyze spherical aberration, which is the deviation of rays that strike different parts of the spherical interface. (b) The case of skew rays (off-axis): Skew rays are rays that do not lie in the plane containing the optical axis. These off-axis rays result in various monochromatic aberrations. In addition to spherical aberration, off-axis rays introduce all other monochromatic aberrations such as coma, astigmatism, curvature of field, and distortion. These aberrations can be more complex to analyze because the rays do not follow simple paths that can be described by meridional geometry.

2.9.2.1 Meridional Rays (On-Axis)

In this section it is assumed that rays emanate from a point object O and lie within a plane that includes the optical axis of the system, as depicted in Fig. 2.97. These rays are referred to as *meridional* rays. In this configuration, the AS, EnP, and ExP are all positioned at the location of the spherical interface, as shown in Fig. 2.97. This setup simplifies the analysis by concentrating on how meridional rays interact with a single refractive surface.

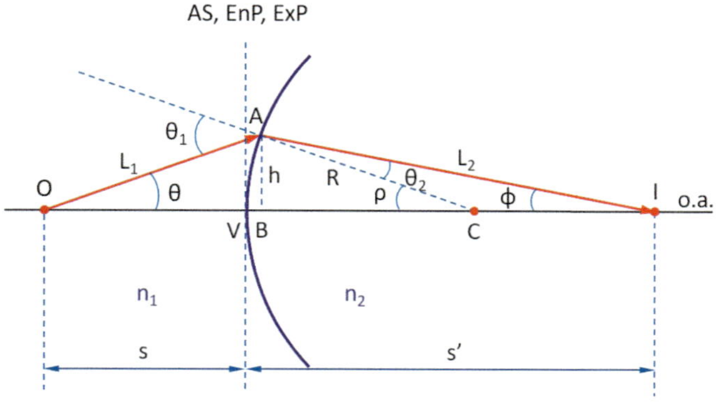

FIGURE 2.97: A spherical refractive interface of radius R between two media of refractive indices of n_1 and n_2, respectively. A ray emanating from point object O is incident on the interface at point A, is refracted, and intersects the optical axis at point I. The AS, EnP, and ExP coincide with the position of the spherical interface.

Consider a meridional ray originating from a point object O located at a distance s from the vertex V of a spherical refractive interface. This ray strikes the interface at point A and, upon refraction, continues into the right-hand region, intersecting the optical axis at point I which is the image of O. The objective is to determine the image distance s' as a function of the height h or of the angle θ (see Fig. 2.97). This dependency represents the spherical aberration for this simple spherical refractive interface. From a paraxial perspective, the image distance s' is given by $(n_1/s) + (n_2/s') = (n_2 - n_1)/R$ [Eq. (2.42)]. However, due to spherical aberration, the exact image position varies with h or θ. This variation in image position with the incident ray parameters reflects the deviation from paraxial behavior, which is characteristic of spherical aberration. The exact solution for s' can be derived using Fermat's Principle. Let the optical path length from point O to A and to I be denoted as OPL. It is straightforward to write the following expressions:

$$OPL = n_1(OA) + n_2(AI) = n_1 L_1 + n_2 L_2, \qquad \text{where}$$

$$L_1 = \left[R^2 + (R+s)^2 - 2R(R+s)\cos\rho\right]^{1/2}, \qquad \text{from OAC}$$

$$L_2 = \left[R^2 + (s'-R)^2 + 2R(s'-R)\cos\rho\right]^{1/2}, \qquad \text{from IAC}$$

$$\frac{d(OPL)}{dh} = 0 \Longrightarrow n_1\frac{dL_1}{dh} + n_2\frac{dL_2}{dh} = 0 \Longrightarrow \left[n_1\frac{dL_1}{d\rho} + n_2\frac{dL_2}{d\rho}\right]\frac{d\rho}{dh} = 0 \Longrightarrow$$

$$n_1\frac{R(R+s)\sin\rho}{L_1} = n_2\frac{R(s'-R)\sin\rho}{L_2}, \tag{2.129}$$

where the last equation needs to be solved for s'. This can be easily accomplished as follows (squaring the last equation):

$$n_1^2\frac{R^2(R+s)^2\sin^2\rho}{L_1^2} = n_2^2\frac{R^2(s'-R)^2\sin^2\rho}{L_2^2} \Longrightarrow$$

$$\frac{1}{n_1^2}\frac{L_1^2}{(R+s)^2\sin^2\rho} = \frac{1}{n_2^2}\frac{L_2^2}{(s'-R)^2\sin^2\rho} \Longrightarrow$$

$$\frac{n_2^2}{n_1^2}\frac{L_1^2}{(R+s)^2\sin^2\rho} = \frac{R^2+(s'-R)^2+2R(s'-R)\cos\rho}{(s'-R)^2\sin^2\rho} \Longrightarrow$$

$$\frac{R^2}{\sin^2\rho}\frac{1}{(s'-R)^2} + 2R\frac{\cos\rho}{\sin^2\rho}\frac{1}{(s'-R)} + \frac{1}{\sin^2\rho} = \frac{n_2^2}{n_1^2}\frac{L_1^2}{(R+s)^2\sin^2\rho}.$$

Setting $x = 1/(s'-R)$, and $\sin\rho = h/R$ (see Fig. 2.97) the last equation can be transformed into a second-degree polynomial equation:

$$x^2 + \beta x + \gamma = 0, \qquad \text{where}$$

$$\beta = 2\frac{\sqrt{R^2-h^2}}{R^2},$$

$$\gamma = \frac{1}{R^2}\left[1 - \frac{n_2^2}{n_1^2}\frac{L_1^2}{(R+s)^2}\right],$$

$$\Delta = \beta^2 - 4\gamma = \frac{4}{R^2}\left[-\frac{h^2}{R^2} + \frac{n_2^2}{n_1^2}\frac{L_1^2}{(R+s)^2}\right],$$

$$x_{1,2} = \left.\frac{1}{s'-R}\right|_{1,2} = \frac{-\beta \pm \sqrt{\Delta}}{2},$$

where the last equation gives real solutions when the discriminant Δ is positive. Of course, for every ray, it should be only one solution accepted. In the case of a real image as shown in Fig. 2.97 there must be only the solution

$$x_1 = (1/2)(-\beta + \sqrt{\Delta}) \to s' = (1/x_1) + R. \tag{2.130}$$

The image distance s' can indeed be determined more easily using basic trigonometry, as demonstrated below. First, consider the triangle OAC in Fig. 2.97. Using the law of sines, we can determine the angles θ_1, θ_2, ρ, and ϕ, which are necessary for calculating s'. Snell's law provides the relationship between the incident angle θ_1 and the refracted angle θ_2. With these angles determined, the image distance s' can be calculated, taking into account the variation of these angles with the incident ray's height h or the angle θ. This derivation is shown in the following expressions:

$$\frac{\sin\theta}{R = AC} = \frac{\sin(\pi-\theta_1)}{R+s = OC} = \frac{\sin\rho}{OA} \Longrightarrow \sin\theta_1 = \frac{R+s}{R}\sin\theta,$$

$$n_1\sin\theta_1 = n_2\sin\theta_2 \Longrightarrow \sin\theta_2 = \frac{n_1\sin\theta_1}{n_2} = \frac{n_1}{n_2}\frac{R+s}{R}\sin\theta,$$

$$OAC \Longrightarrow \rho = \pi - \theta - (\pi - \theta_1) = \theta_1 - \theta,$$

$$ACI \Longrightarrow \phi = \rho - \theta_2 = \theta_1 - \theta_2 - \theta.$$

Now using the law of sines in the OAI triangle and the previous relations it is straightforward to write the following expressions:

$$\frac{\sin(\pi - \theta_1 + \theta_2)}{OI = s + s'} = \frac{\sin\phi}{OA} = \frac{\sin\theta}{AI},$$

$$OA = R\,\frac{\sin\rho}{\sin\theta} = R\,\frac{\sin(\theta_1 - \theta)}{\sin\theta}.$$

From the last two equations, it is simple to solve with respect to the image position s' as follows:

$$s' = R\,\frac{\sin(\theta_1 - \theta)}{\sin\theta}\,\frac{\sin(\theta_1 - \theta_2)}{\sin\phi} - s. \tag{2.131}$$

The last expression for s' is exact since no approximation was involved in determining it. Of course, it was assumed that $n_1 < n_2$ in order for the Fig. 2.97 to be valid.

To achieve a real image with point I positioned to the right of V as shown, specific conditions on the object distance s, the angle of incidence θ, and the refractive indices n_1 and n_2 are required. A particular case arises when the refracted ray becomes parallel to the optical axis, meaning the image position goes to infinity ($s' = \pm\infty$). For a given object distance s, a ray emerging after refraction in the region with refractive index n_2 will be parallel to the optical axis if the following condition is met:

$$n_1 \sin\theta_1 = n_2 \sin\theta_2 = n_2 \sin\rho = n_2 \sin(\theta_1 - \theta) \Longrightarrow$$

$$n_1 \sin\theta_1 = n_2(\sin\theta_1 \cos\theta - \cos\theta_1 \sin\theta) \Longrightarrow$$

$$n_1 \sin\theta_1 = n_2 \left[\sin\theta_1 (1 - \sin^2\theta)^{1/2} - (1 - \sin^2\theta_1)^{1/2} \sin\theta\right], \quad \text{where} \tag{2.132}$$

$$\sin\theta_1 = \frac{R + s}{R}\sin\theta = \alpha \sin\theta,$$

where $\alpha = (R + s)/R$. Using the previous expressions it is straightforward to find the angle $\theta = \theta_{\|}$ for which the refracted ray would be parallel to the optical axis. From Eq. (2.132) setting $x = \sin\theta$ it can be transformed into the following form (for $n_1 < n_2$):

$$2n_1 n_2 \alpha (1 - \alpha^2 x^2)^{1/2} = (n_2^2 - n_1^2)\alpha^2 - n_2^2 \qquad \text{which requires}$$

$$(n_2^2 - n_1^2)\alpha^2 - n_2^2 \geq 0 \Longrightarrow \alpha \geq \frac{n_2}{\left(n_2^2 - n_1^2\right)^{1/2}} \Longrightarrow \frac{s}{R} \geq \frac{n_2}{\left(n_2^2 - n_1^2\right)^{1/2}} - 1 \qquad \text{and}$$

$$\frac{(n_2^2 - n_1^2)\alpha^2 - n_2^2}{2n_1 n_2 \alpha} \leq 1 \Longrightarrow -\frac{n_2}{n_1 + n_2} \leq \alpha \leq \frac{n_2}{n_2 - n_1} \Longrightarrow \frac{s}{R} \leq \frac{n_1}{n_2 - n_1},$$

and summarizing,

$$\frac{n_2}{\left(n_2^2 - n_1^2\right)^{1/2}} - 1 \leq \frac{s}{R} \leq \frac{n_1}{n_2 - n_1}, \qquad \text{and} \tag{2.133}$$

$$x^2 = \sin^2\theta_{\|} = \frac{4n_1^2 n_2^2 \alpha^2 - [(n_2^2 - n_1^2)\alpha^2 - n_2^2]^2}{4n_1^2 n_2^2 \alpha^4}. \tag{2.134}$$

Therefore, the angle $\theta_{\|}$ exists when the constraint described in Eq. (2.133) is satisfied (for $n_1 < n_2$). The maximum angle $\theta = \theta_{max}$ for a ray originating from a given object position corresponds to the ray that is tangent to the spherical interface between the two materials. This maximum angle is defined by the geometry of the system nd is given by:

$$\theta_{max} = \sin^{-1}\left(\frac{R}{R+s}\right).$$

Approach using the third-order (Seidel) approximation: This configuration of the spherical interface is also approached in the literature using the "third-order" Seidel approximation [20,42,57,58,62,63]. In this case the wave aberration $W(A)$ (corresponding to the spherical aberration as in Sec. 2.9.1) is defined as:

$$W(A) = (n_1 L_1 + n_2 L_2) - (n_1 s + n_2 s') \qquad \text{where}$$

$$L_1 = \left[R^2 + (R+s)^2 - 2R(R+s)\cos\rho\right]^{1/2}, \qquad \text{from OAC}$$

$$L_2 = \left[R^2 + (s'-R)^2 + 2R(s'-R)\cos\rho\right]^{1/2}, \qquad \text{from IAC}$$

$$\cos\rho = (1-\sin^2\rho)^{1/2} = \left(1 - \frac{h^2}{R^2}\right)^{1/2} \implies$$

$$\cos\rho \simeq 1 - \frac{h^2}{2R^2} - \frac{h^4}{8R^4} \qquad \text{resulting in}$$

$$L_1 \simeq s\left[1 + \frac{h^2(R+s)}{Rs^2} + \frac{h^4(R+s)}{4R^3s^2}\right]^{1/2}$$

$$L_2 \simeq s'\left[1 + \frac{h^2(R-s')}{Rs'^2} + \frac{h^4(R-s')}{4R^3s'^2}\right]^{1/2}.$$

Using the Taylor series $(1+x)^{1/2} \simeq 1 + (x/2) - (x^2/8)$ and retaining terms up to h^4, the following equations can be derived:

$$L_1 \simeq s\left[1 + \frac{h^2(R+s)}{2Rs^2} + \frac{h^4(R+s)}{8R^3s^2} - \frac{h^4(R+s)^2}{8R^2s^4}\right],$$

$$L_2 \simeq s'\left[1 + \frac{h^2(R-s')}{2Rs'^2} + \frac{h^4(R-s')}{8R^3s'^2} - \frac{h^4(R-s')^2}{8R^2s'^4}\right], \qquad \text{resulting in}$$

$$W(A) \simeq \frac{h^2}{2}\left[\frac{n_1}{s} + \frac{n_2}{s'} - \frac{n_2-n_1}{R}\right]$$
$$- \frac{h^4}{8}\left[\frac{n_1}{s}\left(\frac{1}{s} + \frac{1}{R}\right)^2 + \frac{n_2}{s'}\left(\frac{1}{s'} - \frac{1}{R}\right)^2 - \frac{1}{R^2}\left(\frac{n_1}{s} + \frac{n_2}{s'} - \frac{n_2-n_1}{R}\right)\right]. \quad (2.135)$$

From the last equation, it is obvious that the spherical aberration is of the order h^4 when the paraxial approximation $s' = s'_p$, is used, i.e., when $(n_1/s) + (n_2/s'_p) = (n_2 - n_1)/R$. In order to relate this result with the general theory presented in Sec. 2.9.1 it should be noticed that $h \simeq r_{T_p}$, since h is not measured in the $x_p y_p$ plane (or $h \simeq x_p$ since $y_p = 0$ for meridional rays). The corresponding wave aberration is given by:

$$W(r_{T_p}) = r_{T_p}^4 \underbrace{\left\{-\frac{1}{8}\left[\frac{n_1}{s}\left(\frac{1}{s} + \frac{1}{R}\right)^2 + \frac{n_2}{s'_p}\left(\frac{1}{s'_p} - \frac{1}{R}\right)^2\right]\right\}}_{= W_{040}} = W_{040} r_{T_p}^4 < 0. \quad (2.136)$$

In the above equation s and s'_p satisfy the paraxial equation for the spherical interface. It is observed that $W(r_{T_p}) < 0$ which implies that the actual wavefront is leading the paraxial reference spherical wavefront. This situation is similar to the one shown in Figs. 2.90(a) and 2.90(b). Therefore, TSA and LSA should be negative which means that the actual rays intersect the optical axis before the paraxial ones. The corresponding longitudinal ray aberration (LSA) can be estimated with the help of Eq. (2.116) and the position of the image $s'(h)$ (due to spherical aberration) is given by:

$$s'(h) \simeq s'_p + \underbrace{\frac{4R_{eff}^2}{n_2} W_{040} h^2}_{LSA<0} - \left(R - \frac{h}{\tan\rho}\right) \text{sign}(s'_p), \tag{2.137}$$

where $R_{eff} \simeq s'_p$. The last term in Eq. (2.137) was added since in the derivation of Eq. (2.116) the $z_p = 0$. The function sign(u) represents the sign of u, and s'_p represents the paraxial image distance. Furthermore, the sign that was utilized in Eq. (2.116) was such to make $LSA < 0$ since the actual wavefront is leading the reference wavefront. The above LSA corresponds to the Seidel 3rd-order approximation.

Alternatively, a third-order approximation of the image location s' can be determined with the utilization of Fermat's principle for the optical path length $OPL = n_1L_1 + n_2L_2$, with third-order approximations used in the expressions of L_1 and L_2 distances. Therefore,

$$\begin{aligned}
&OPL = n_1L_1 + n_2L_2 \quad \text{with} \\
&L_1 \simeq s\left[1 + \frac{h^2(R+s)}{2Rs^2} + \frac{h^4(R+s)}{8R^3s^2} - \frac{h^4(R+s)^2}{8R^2s^4}\right], \\
&L_2 \simeq s'\left[1 + \frac{h^2(R-s')}{2Rs'^2} + \frac{h^4(R-s')}{8R^3s'^2} - \frac{h^4(R-s')^2}{8R^2s'^4}\right], \\
&\frac{d(OPL)}{dh} = 0 \Longrightarrow \\
&n_1\frac{R+s}{Rs} + n_1\frac{h^2}{2}\frac{R+s}{R^3s} - n_1\frac{h^2}{2}\frac{(R+s)^2}{R^2s^3} \\
&+ n_2\frac{R-s'}{Rs'} + n_2\frac{h^2}{2}\frac{R-s'}{R^3s'} - n_2\frac{h^2}{2}\frac{(R-s')^2}{R^2s'^3} = 0 \Longrightarrow \\
&X\left(1 + \frac{h^2}{2R^2}\right) = \frac{h^2}{2}\left[\frac{n_1}{s}\left(\frac{1}{s} + \frac{1}{R}\right)^2 + \frac{n_2}{s'}\left(\frac{1}{s'} - \frac{1}{R}\right)^2\right] \quad \text{where} \\
&X = \frac{n_1}{s} + \frac{n_2}{s'} - \frac{(n_2 - n_1)}{R}.
\end{aligned} \tag{2.138}$$

To solve for s' more approximations are necessary. Initially, the coefficient of X is assumed to be unity neglecting the $h^2/2R^2$ term. Then, the following steps are made:

$$X = \frac{h^2}{2}\left[\frac{n_1}{s}\left(\frac{1}{s} + \frac{1}{R}\right)^2 + \frac{n_2}{s'}\left(\frac{1}{s'} - \frac{1}{R}\right)^2\right],$$

$$\frac{n_2}{f_p} = \frac{n_2 - n_1}{R},$$

where f_p corresponds to the second focal distance f_2 under the paraxial approximation. Then the following steps can be taken:

$$\frac{n_1}{s} + \frac{n_2}{s'} \simeq \frac{n_2 - n_1}{R} = \frac{n_2}{f_p} \Longrightarrow$$

$$\frac{1}{s'} = \frac{1}{f_p} - \frac{n_1/n_2}{s}$$

$$\frac{n_2}{s'}\left(\frac{1}{s'} - \frac{1}{R}\right)^2 = \left(\frac{n_2}{f_p} - \frac{n_1}{s}\right)\left(\frac{1}{f_p} - \frac{n_1/n_2}{s} - \frac{1}{R}\right)^2 \Longrightarrow$$

$$\frac{n_2}{s'}\left(\frac{1}{s'} - \frac{1}{R}\right)^2 = \left(\frac{n_2}{f_p} - \frac{n_1}{s}\right)\frac{n_1^2}{n_2^2}\left(\frac{1}{R} + \frac{1}{s}\right)^2 \Longrightarrow$$

$$X = \frac{h^2}{2}\left(\frac{1}{R} + \frac{1}{s}\right)^2 \underbrace{\left[\frac{n_1}{s} + \frac{n_1^2}{n_2^2}\left(\frac{n_2}{f_p} - \frac{n_1}{s}\right)\right]}_{=B}.$$

At this point the quantity B can be manipulated as follows:

$$B = \frac{R}{f_p}\left[\frac{n_1 f_p}{Rs} + \frac{n_1^2}{n_2^2}\left(\frac{n_2}{R} - \frac{n_1 f_p}{Rs}\right)\right] = \frac{R}{f_p}\frac{n_1^2}{n_2}\left[\frac{n_2}{n_1}\frac{f_p}{Rs} + \frac{1}{n_2}\left(\frac{n_2}{R} - \frac{n_1 f_p}{Rs}\right)\right] =$$

$$= \frac{R}{f_p}\frac{n_1^2}{n_2}\left[\frac{n_2^2}{n_1(n_2 - n_1)}\frac{1}{s} + \frac{1}{R} - \frac{n_1}{n_2 - n_1}\frac{1}{s}\right] = \frac{R}{f_p}\frac{n_1^2}{n_2}\left[\frac{1}{R} + \frac{n_1 + n_2}{n_1}\frac{1}{s}\right].$$

Combining the last expression with the expression for X the final form of the third-order approximation for the image location becomes:

$$\frac{n_1}{s} + \frac{n_2}{s'(h)} = \frac{n_2 - n_1}{R} + \frac{h^2}{2}\left[\frac{R}{f_p}\frac{n_1^2}{n_2}\left(\frac{1}{R} + \frac{1}{s}\right)^2\left(\frac{1}{R} + \frac{n_1 + n_2}{n_1}\frac{1}{s}\right)\right]. \tag{2.139}$$

It is worth mentioning that the same equation in Ref. [62] is shown with $n_2 - n_1$ in the last term instead of the $n_2 + n_1$ appearing in Eq. (2.139). From Eq. (2.139) the existence of $\theta_{||}$ can be predicted if there is a value of h that can make $|s'(h)| = \infty$. The value for h exists only if $s < n_1 R/(n_2 - n_1) = f_1$ (where f_1 the paraxial left focal distance). This $h = h_{||}$ value is given by:

$$h_{||} = \left\{\frac{2\left(\frac{n_1}{s} - \frac{n_2 - n_1}{R}\right)}{\frac{R}{f_p}\frac{n_1^2}{n_2}\left(\frac{1}{R} + \frac{1}{s}\right)^2\left(\frac{1}{R} + \frac{n_1 + n_2}{n_1}\frac{1}{s}\right)}\right\}^{1/2}. \tag{2.140}$$

It is straightforward to calculate the approximate third-order $\theta_{||}$ from the equation:

$$\theta_{||} = \tan^{-1}\left\{\frac{\frac{h_{||}}{R}}{\frac{R+s}{R} - \left\{1 - \left(\frac{h_{||}}{R}\right)^2\right\}^{1/2}}\right\}. \tag{2.141}$$

An interesting point is to seek the exact solution of Eq. (2.138) instead of making the approximations leading to Eq. (2.139). This solution could be referred as the exact third-order approximation and it is discussed in Appendix 2D.

Example 2.9.1:

A spherical refractive interface of radius $R = 100\,\text{mm}$ is considered between two media with refractive indices $n_1 = 1$, and $n_2 = 1.5$ in accordance to Fig. 2.97. Two point object O positions are examined: (a) when $s = 300\,\text{mm}$ and (b) when $s = 100\,\text{mm}$. The goal is to determine the image position as a function of the incident ray's angle θ.

Solution: The following image distances are easily calculated: (1) the paraxial approximation image distance, s'_p [given by Eq. (2.42)]; (2) the exact distance $s'(h)$ [as given by Eqs. (2.130) or (2.131)—both are equivalent]; (3) the third-order approximation [given by Eq. (2.139)] and the exact solution of third-order approximation [Eq. (2.138)]; and (4) the image distance $s'(h)$ based on the third-order Seidel approximation [given by Eq. (2.137)]. The results for $s = 300\,\text{mm}$ are shown in Fig. 2.98(a). In this case $\alpha = R+s/R = 4 > n_1/(n_2-n_1) = 2$ and therefore $\theta_{\|}$ does not exist for any angle $|\theta| \leq \theta_{max} = \sin^{-1}(R/(R+s)) = 14.48°$. In this case the 3rd-order Seidel approximation predicts a decreasing $s'(h)$ (or $s'(\theta)$) which also as the angle θ increases, becomes negative! In Fig. 2.98(a) the 3rd-order Seidel approximation is shown up to the angle θ that remains positive since no transition to negative $s'(h)$ is expected from the exact ray analysis. However, in the (b) case where the object distance becomes $s = 100\,\text{mm}$, the $\alpha = R+s/R = 1 < n_1/(n_2-n_1) = 2$ and $\alpha = 1 > n_2/(n_2^2-n_1^2)^{1/2}-1 = 0.34$ and therefore there is a transition at $\theta_{\|} = 26.38°$ [Eq. (2.134)] from a virtual image ($s' < 0$) to a real image ($s' > 0$). This is shown in Fig. 2.98(b). Of course, the paraxial and the 3rd-order Seidel approximation fail to predict this change of the image position. Observe that the third-order approximation as well as the exact third order solution have different transitions due to the various approximations. The predicted $\theta_{\|}$ values from Eqs. (2.141) and from Appendix 2D Eq. (2D.4), are $\theta_{\|} = 25.73°$ and $\theta_{\|} = 24.84°$ for the third-order approximation and the third-order exact solution, respectively, which agree with the results shown in Fig. 2.98(b). The maximum permissible angle θ is $\theta_{max} = 30°$. In these plots it is obvious that all approximate solutions have considerable differences with the exact solution as the angle θ or θ_2 becomes larger. In order to visualize better the comparison between the paraxial and the third-order Seidel approximation, Fig. 2.98(c) is shown, which is the same as Fig. 2.98(b) but including only the paraxial and the third-order Seidel approximations.

2.9.2.2 Skew Rays (Off-Axis)

Now, consider a skew ray that originates from the object point O_1 on object OO_1, as illustrated in Figs. 2.99 (which provides a 3D representation of the geometry) and 2.100 (which shows a simplified geometric view of the skew ray). The wave aberration function W can then be defined as the difference in optical path length between the skew ray O_1QI_1 and the meridional ray O_1VI_1. This wave aberration function W quantifies the deviation of the wavefront associated with the skew ray from the reference wavefront defined by the meridional ray. Therefore, W can be expressed as:

$$W = (O_1QI_1) - (O_1VI_1) = [(O_1QI_1) - (O_1V_1I)] - [(O_1VI_1) - (O_1V_1I)]$$
$$\simeq W_{040}(QV_1)^4 - W_{040}(VV_1)^4 = W_{040}\underbrace{\left[(QV_1)^4 - (VV_1)^4\right]}_{=\Delta},$$

where W_{040} is defined in Eq. (2.136) under the third-order approximation. This expression for W provides insight into the difference in phase or optical distance between the skew and meridional rays, which is essential for understanding off-axis aberrations in the optical system. The distances (QV_1) and (VV_1) can be related using the cosine law (with the help

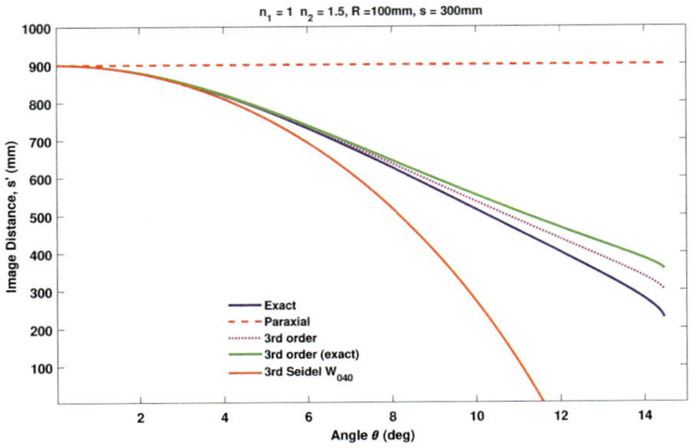

(a)

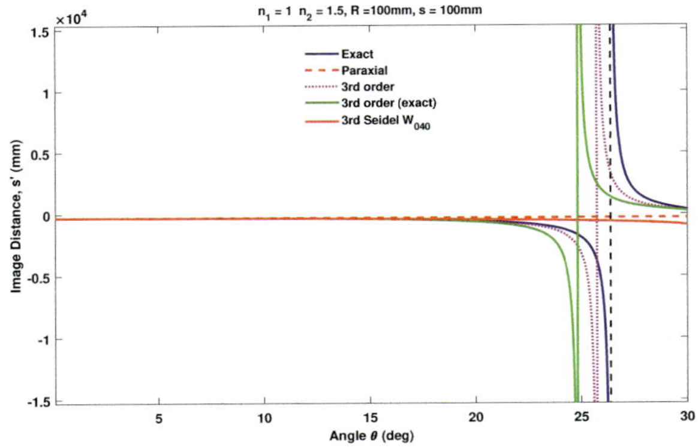

(b)

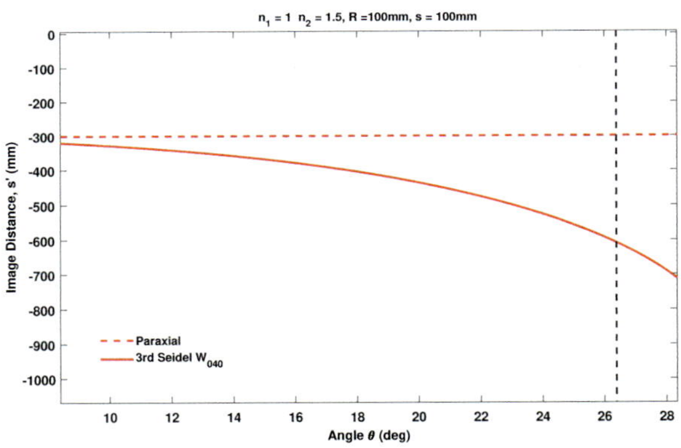

(c)

FIGURE 2.98: (a) The spherical aberration for a spherical refractive interface of radius $R = 100\,\text{mm}$ with $n_1 = 1$, $n_2 = 1.5$, and object distance $s = 300\,\text{mm}$. The paraxial, the exact, the third-order approximation, the exact third-order, and the 3rd-order Seidel solutions are all shown as functions of the angle θ that the ray emanates from the object. Of course, negative angles, i.e., $-\theta$ have the same spherical aberration. (b) Same as (a) but for object distance $s = 100\,\text{mm}$. In this case a transition from a virtual to a real image position occurs at the position of the vertical dashed line ($\theta = \theta_{||}$). (c) Same as Fig. 2.98(b) but including only the paraxial and the 3rd-order Seidel terms. Both paraxial approximation and Seidel fail to predict the change of the image position from virtual to real when the angle of incidence increases.

of Fig. 2.99(b)) as follows:

$$(QV_1)^2 = (VV_1)^2 + r_{_{Tp}}^2 + 2r_{_{Tp}}(VV_1)\cos\phi_p \Longrightarrow$$
$$(QV_1)^4 = (VV_1)^4 + r_{_{Tp}}^4 + 4r_{_{Tp}}^2(VV_1)^2\cos^2\phi_p + 2(VV_1)^2 r_{_{Tp}}^2 + 4(VV_1)^3 r_{_{Tp}}\cos\phi_p + 4(VV_1)r_{_{Tp}}^3\cos\phi_p.$$

At this stage, the distance (VV_1) can be approximated and expressed as a function of

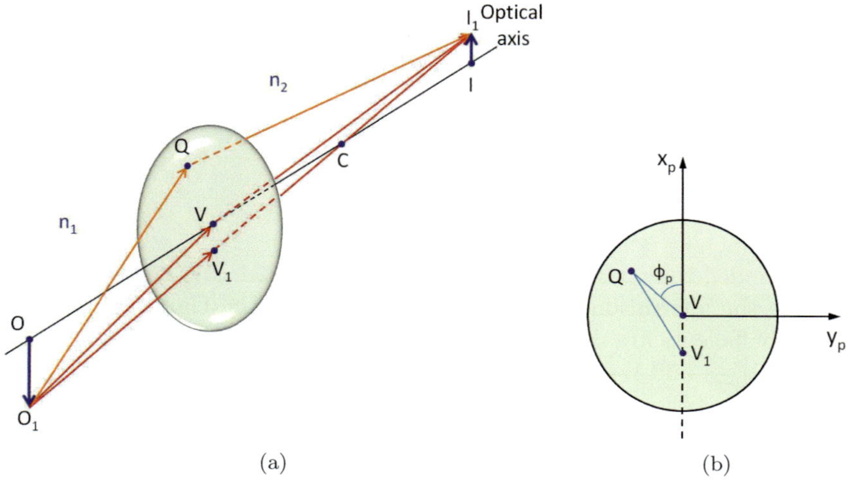

FIGURE 2.99: (a) A skew ray from an object point is incident at point Q on the spherical refractive interface. (b) A detail on the ExP plane (x_p, y_p) of the incident rays.

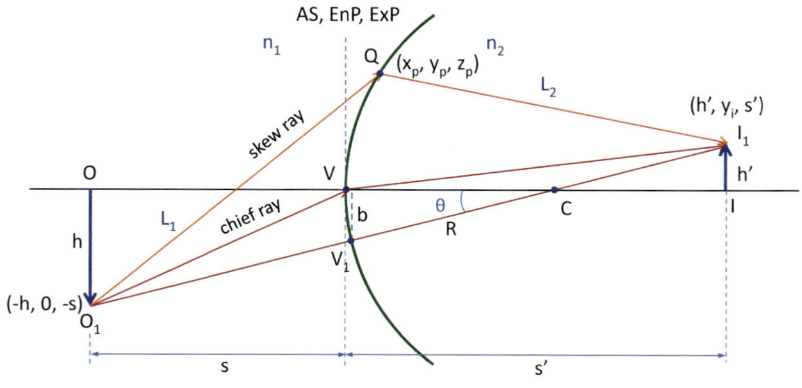

FIGURE 2.100: A simpler picture of the incident rays in the case of skew ray incidence.

h'. Specifically, from triangle VV_1C the distance $(VV_1) = 2R\sin(\theta/2)$. Since the angle θ is considered small, the distance (VV_1) can be approximated by the distance of point V_1 from the optical axis b (see Fig. 2.100). By examining the similar orthogonal triangles that share the angle θ, it can be shown that $b = h'R/(s' - R)$ [57]. This expression simplifies to $(VV_1) \simeq b = [R/(s' - R)]h' = \beta h'$ (where $s' = s'_p$). Then the wave aberration W is expressed as:

$$W(Q) = W_{040}\left[r_{T_p}^4 + 4\beta^2 r_{T_p}^2 (h')^2 \cos^2\phi_p + 2\beta^2 r_{T_p}^2 (h')^2 + 4\beta^3 r_{T_p} (h')^3 \cos\phi_p + 4\beta r_{T_p}^3 h' \cos\phi_p\right]. \tag{2.142}$$

From Eq. (2.142) the 3rd-order Seidel aberration coefficients are:

$$W_{040} = -\frac{1}{8}\left[\frac{n_1}{s}\left(\frac{1}{s}+\frac{1}{R}\right)^2 + \frac{n_2}{s'_p}\left(\frac{1}{s'_p}-\frac{1}{R}\right)^2\right] \tag{2.143}$$

$$W_{131} = 4\beta W_{040} \tag{2.144}$$

$$W_{222} = 4\beta^2 W_{040} \tag{2.145}$$

$$W_{220} = 2\beta^2 W_{040} \tag{2.146}$$

$$W_{311} = 4\beta^3 W_{040}, \tag{2.147}$$

where it is reminded that $\beta = R/(s'_p - R)$. According to Ref. [57] a correction is needed in order to define the aberration coefficients with respect to the Gaussian image point since the previous coefficients are defined with respect to the Petzval image point. In this case, the coefficients W_{220} and W_{311} are slightly modified and are given by:

$$W_{220} = 2\beta^2 W_{040} - \frac{n_2(n_2 - n_1)}{2n_1 R s_p'^2} \tag{2.148}$$

$$W_{311} = 4\beta^3 W_{040} - \frac{n_2(n_2 - n_1)\beta}{2n_1 R s_p'^2}. \tag{2.149}$$

At this point, it is useful to approach the skew ray problem using Fermat's principle, similar to the method applied in Sec. 2.9.2.1 for meridional rays. In this case Fig. 2.100 is used for the analysis. In this case, coordinates are assigned to key points as follows: the object point O_1 is located at $(-h, 0, -s)$, the point of refraction Q at (x_p, y_p, z_p), and the image pointI_1 at (h', y_i, s'), as indicated in the figure. The coordinate system $x_p y_p z_p$ is centered at the vertex point V of the spherical interface. Using these coordinates, the optical path length OPL from O_1 to Q and then to I_1 can be expressed as:

$$OPL = n_1 L_1 + n_2 L_2 =$$

$$= n_1 \underbrace{\left[(x_p + h)^2 + y_p^2 + (s + z_p)^2\right]^{1/2}}_{= L_1} + n_2 \underbrace{\left[(x_p - h')^2 + (y_i - y_p)^2 + (s' - z_p)^2\right]^{1/2}}_{= L_2}.$$

According to Fermat's principle the optical path length OPL should be stationary to the ray path starting from point O_1 and through Q ends at I_1. Therefore, the following conditions should hold:

$$\frac{\partial(OPL)}{\partial x_p} = 0 \Longrightarrow \frac{n_1}{L_1}\left[(x_p + h) + (s + z_p)\frac{\partial z_p}{\partial x_p}\right] + \frac{n_2}{L_2}\left[(x_p - h') + (s' - z_p)\left(-\frac{\partial z_p}{\partial x_p}\right)\right] = 0, \tag{2.150}$$

$$\frac{\partial(OPL)}{\partial y_p} = 0 \Longrightarrow \frac{n_1}{L_1}\left[y_p + (s + z_p)\frac{\partial z_p}{\partial y_p}\right] + \frac{n_2}{L_2}\left[(y_p - y_i) + (s' - z_p)\left(-\frac{\partial z_p}{\partial y_p}\right)\right] = 0. \tag{2.151}$$

The point $Q(x_p, y_p, z_p)$ is on the spherical refractive surface defined by the equation $x_p^2 + y_p^2 + (z_p - R)^2 = R^2$ and the partial derivatives needed are $\partial z_p/\partial x_p = -x_p/(z_p - R)$ and $\partial z_p/\partial y_p = -y_p/(z_p - R)$. Then Eqs. (2.150) and (2.151) form a nonlinear system of two equations with unknowns the values of y_i and s' given the x_p, y_p, h, s, and h'. It is reminded that unknowns are also in the expressions of L_1 and L_2. The system of these equations can be solved easily numerically. In the special case where $x_p, y_p, z_p \simeq 0$ (paraxial rays) Eq. (2.150) gives $h'/h = (n_1/n_2)(s'/s)$ as expected from the paraxial theory (the sign is already included in the coordinate $-h$ of point O_1). Equation (2.151) in this limit is satisfied without providing additional information. To better understand these equations, let's apply the theory to the example 2.9.1 from before, now considering skew rays. This will involve calculating the skew ray aberrations and comparing them to the expected paraxial behavior.

Example 2.9.2:

A spherical refractive interface of radius $R = 100\,\text{mm}$ is considered between two media with refractive indices $n_1 = 1$ and $n_2 = 1.5$ in accordance with Fig. 2.100. In this case the position of point O is specified by the distance $s = 300\,\text{mm}$. The image position at height $h' = h(s_p'/s)(n_1/n_2)$ (the paraxial image height) is sought through the specification of variables s' and y_i when the rays emanate from point O_1 and pass from a varying point $Q(x_p, y_p, z_p)$.

Solution: The following image distances are easily calculated: (1) the paraxial approximation image distance, $s_p' = 900\,mm$ [given by Eq. (2.42)]; (2) the exact distance $s' = s'(x_p, y_p)$ and displacement y_i [as given by the solution of Eqs. (2.150) and (2.151)]. For the 3rd-order Seidel approximation Eqs. (2.143)–(2.149) are utilized for the determination of the y_i based on the aberration coefficients. It is assumed that $x_p = r_{T_p} \cos\phi_p$ and $y_p = r_{T_p} \sin\phi_p$ and the 3rd-order Seidel s' and y_i are given by

$$s' \simeq s_p' + \frac{4R_{eff}^2}{n_2} W_{040} r_{T_p}^2 - \left[R - \left(R^2 - r_{T_p}^2 \right)^{1/2} \right] \text{sign}(s_p') \tag{2.152}$$

$$y_i \simeq -\frac{4R_{eff}}{n_2} W_{040} r_{T_p}^3 \sin\phi_p - \frac{R_{eff}}{n_2} W_{131} r_{T_p}^2 \sin(2\phi_p) - \frac{2R_{eff}}{n_2} W_{220} h' r_{T_p} \sin\phi_p. \tag{2.153}$$

In this setup, the effective radius $R_{eff} \simeq s_p' = 900\,\text{mm}$. Initially, it is assumed that $\phi_p = 0$ or π and r_{T_p} varies from 0 to 5 mm. These rays have $y_p = 0$, meaning that the rays only vary with x_p along the plane, resulting in meridional rays (those lying in the optical axis plane). For such rays, since $y_p = 0$, the transverse position y_i of the image point will also be zero, regardless of whether we use the approximate Seidel method or the exact solution based on Fermat's principle. The image position s' along z-axis is shown in Fig. 2.101. Next, the position of Q is set such that $x_p = 0$ but y_p is varying from 0 to ± 5 mm (therefore $\phi_p = \pm\pi/2$ and r_{T_p} varying from 0 to 5 mm). The resulting image position s' and the transverse image position y_i as functions of y_p are shown in Figs. 2.102(a) and 2.102(b), respectively. These figures display the exact results based on Fermat's principle compared with the third-order Seidel approximation for spherical aberration. In this case the rays are truly skew rays. As expected, there is a significant difference between the exact results and the third-order Seidel approximation when the rays are skewed, with increasing disparity as y_p grows larger. This discrepancy highlights the limitations of the third-order Seidel approximation for handling skew rays, especially when the rays move further from the optical axis. The Seidel approximation, being based on the paraxial approximation and the expansion to third order, becomes less accurate as the angle of incidence increases, leading to noticeable errors in both the image position s' and the transverse position y_i. This demonstrates the importance of using exact methods, like Fermat's principle, for accurate

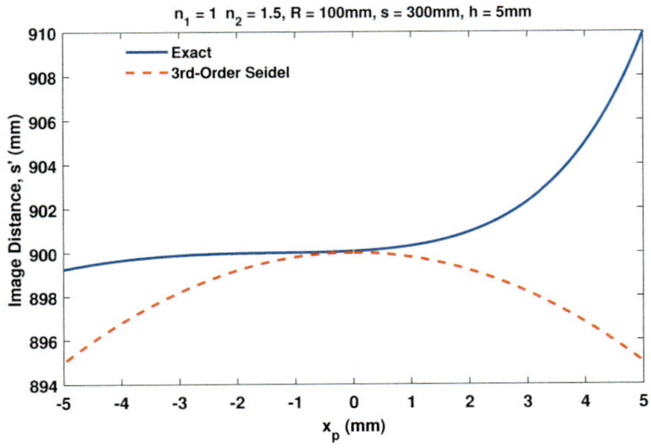

FIGURE 2.101: The exact solution of the s' as a function of x_p (for $y_p = 0$) along with the 3rd-order Seidel approximation. Parameters are defined in example 2.9.2.

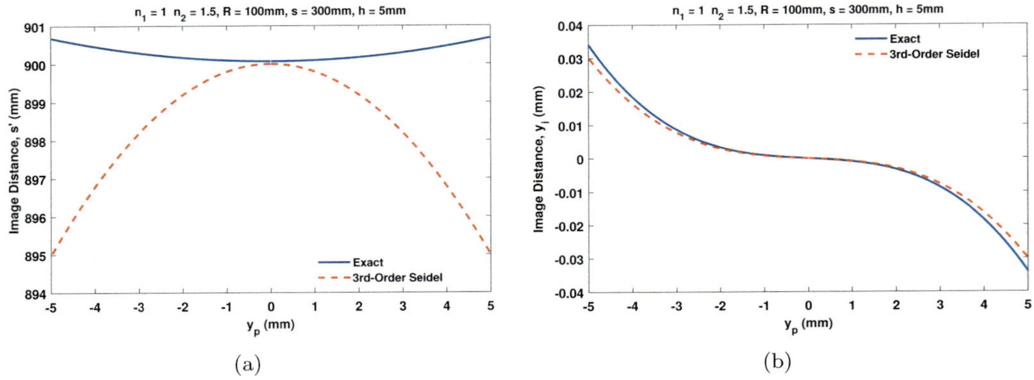

FIGURE 2.102: (a) The exact solution of the s' as a function of y_p (for $x_p = 0$) along with the 3rd-order Seidel approximation. (b) The exact solution of the y_i as a function of y_p (for $x_p = 0$) along with the 3rd-order Seidel approximation.

results when dealing with skew rays. Finally, the points Q were selected such as $r_{T_p} = 5\,\text{mm}$ and ϕ_p can vary from 0 to 2π. This setup allows the examination of the behavior of the image position s' and the transverse image position y_i as functions of the azimuthal angle ϕ_p. The exact and third-order Seidel approximation results for both s' and y_i are shown as functions of the angle ϕ_p in Figs. 2.103(a) and 2.103(b), respectively. Since r_{T_p} is constant in this analysis, the third-order Seidel approximation for s' does not vary as a function of ϕ_p as seen in Fig. 2.103(a).

2.9.3 Spherical Aberration of Thick Lenses

Spherical aberration in an optical system occurs when rays that are parallel to the optical axis do not converge to a single focal point. Instead, rays closer to the axis focus at a point farther from the lens, while rays farther from the axis focus at a point closer to the lens. This results in a blur and is known as positive spherical aberration. The opposite, where the rays focus in a way that the focal points appear closer to the lens, is referred to as negative spherical aberration. However, even though the rays do not all meet at a single point, they do reach a minimum convergence diameter, which is considered the "best" focus. This point of minimum diameter convergence is known as *circle of least confusion.*

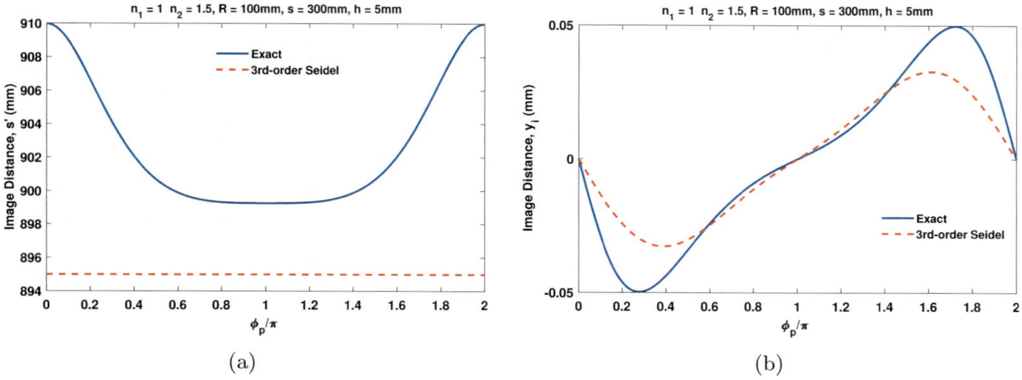

FIGURE 2.103: (a) The exact solution of the s' as a function of ϕ_p (for $r_{T_p} = 5\,\text{mm}$) along with the 3rd-order Seidel approximation. (b) The exact solution of the y_i as a function of ϕ_p (for $r_{T_p} = 5\,\text{mm}$) along with the 3rd-order Seidel approximation.

In the following subsections, the spherical aberration of simple thick lenses is analyzed. In general, spherical aberration in a lens system can be minimized by carefully selecting the shape of the lens surfaces. The aberration can be completely eliminated by using aspheric surfaces, which are specially designed to counteract the effects of spherical aberration. Another method for minimizing spherical aberration is to use a contact doublet lens configuration, where two lenses with different curvatures are used together. Additionally, spherical aberration can be reduced by "stopping down" the aperture of the optical system, which limits the range of rays passing through the lens and reduces the effects of aberration.

2.9.3.1 Plano-Convex Lens

Consider a plano-convex lens with a radius of curvature R_2 on the curved side (with $R_1 = \infty$ for the flat side) and placed between two homogeneous media with refractive indices n_1 and n_2. The lens material has a refractive index n_L, where $n_L > n_1, n_2$. The lens has a thickness d. When a monochromatic ray enters the flat surface parallel to the optical axis at a height h, it refracts through the lens and exits on the curved side, intersecting the optical axis at a distance f_2, as illustrated in Fig. 2.104. This distance f_2 is defined as the focal distance for this incident ray and depends on the height h at which the ray enters the lens. The value of f_2 can be determined without relying on any approximations and may then be compared with the focal distance predicted by paraxial optics. This analysis will yield f_2 as a function of the ray height h, revealing the variation in focal distance due to spherical aberration in this lens geometry. Since the incident ray is parallel to the optical axis, its direction remains unchanged after refraction at point A on the flat surface of the lens (refer to Fig. 2.104). However, upon reaching the second boundary at the curved surface, the refraction alters the ray's direction, as seen in the figure. Specifically, the angles ϕ_1 and ϕ_2 can be easily determined as follows:

$$\sin\phi_1 = \frac{h}{R_2} \Longrightarrow \phi_1 = \sin^{-1}\left(\frac{h}{R_2}\right), \tag{2.154}$$

$$n_L \sin\phi_1 = n_2 \sin\phi_2 \Longrightarrow \phi_2 = \sin^{-1}\left(\frac{n_L \sin\phi_1}{n_2}\right) = \sin^{-1}\left(\frac{n_L}{n_2}\frac{h}{R_2}\right). \tag{2.155}$$

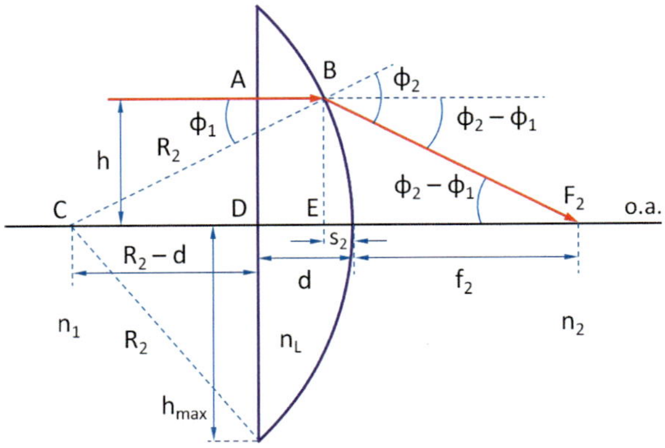

FIGURE 2.104: A plano-convex lens of refractive index n_L, radius R_2, and thickness d between two homogeneous media of refractive indices n_1 and n_2, respectively. A ray parallel to the optical axis is incident on the flat surface of the lens at height h and after refracting though the lens intersects the optical axis at distance f_2. At the flat surface there is no angular change in refraction due to normal incidence. However, there is an angular change of the ray in the refraction that occurs at the spherical surface and the ray's new direction is denoted by the angle ϕ_2.

From triangle BEF_2 the distance $F_2E = h/\tan(\phi_2 - \phi_1)$ while $s_2 = R_2 - \left(R_2^2 - h^2\right)^{1/2}$. The focal distance f_2 is easily found by:

$$f_2 = F_2E - s_2 = \frac{h}{\tan(\phi_2 - \phi_1)} - R_2 + \left(R_2^2 - h^2\right)^{1/2} \Longrightarrow$$

$$f_2 = \frac{h}{\tan\left[\sin^{-1}\left(\dfrac{n_L}{n_2}\dfrac{h}{R_2}\right) - \sin^{-1}\left(\dfrac{h}{R_2}\right)\right]} - R_2 + \left(R_2^2 - h^2\right)^{1/2}. \tag{2.156}$$

Equation (2.156) is valid for angles ϕ_1 that ϕ_2 exists, i.e., when no total internal reflection takes place at point B. From Fig. 2.104 the maximum allowable height $h_{max}(d)$ is given by:

$$h_{max}(d) = \left[R_2^2 - (R_2 - d)^2\right]^{1/2} = \left[d(2R_2 - d)\right]^{1/2}. \tag{2.157}$$

In order to avoid total internal reflection at point B the following inequality should hold:

$$\phi_1 \le \phi_{cr} = \sin^{-1}\left(\frac{n_2}{n_L}\right) \Longrightarrow \quad \sin(\phi_1) \le \sin(\phi_{cr}) = \frac{n_2}{n_L} \Longrightarrow$$

$$\frac{h}{R_2} \le \frac{n_2}{n_L} \Longrightarrow \quad h \le R_2\frac{n_2}{n_L} \Longrightarrow \quad h_{max}(d) \le R_2\frac{n_2}{n_L} = h_{max,TIR} \Longrightarrow \tag{2.158}$$

$$\left[d(2R_2 - d)\right]^{1/2} \le R_2\frac{n_2}{n_L} \quad \Longrightarrow \quad d^2 - 2R_2 d + R_2^2\,\frac{n_2^2}{n_L^2} \ge 0 \Longrightarrow$$

$$0 < d \le d_1 \quad \text{and} \quad d \ge d_2 \quad \text{where,} \tag{2.159}$$

$$d_1 = \frac{R_2}{n_L}\left[n_L - \left(n_L^2 - n_2^2\right)^{1/2}\right], \quad \text{and } d_2 = \frac{R_2}{n_L}\left[n_L + \left(n_L^2 - n_2^2\right)^{1/2}\right].$$

From a practical standpoint, the first inequality in Eq. (2.159), which specifies $0 < d \le d_1$, is essential. The second inequality, $d \ge d_2$, in contrast, corresponds to the complementary plano-convex lens. This alternate configuration is not typically preferred in practical applications because it resembles a lens in the form of a sphere with a small portion removed to create the flat left surface of the lens.

Example 2.9.3:

Consider a plano-convex thick lens with the following parameters: $R_2 = 25\,\text{mm}$, $n_1 = n_2 = 1.0$ (air), $n_L = 1.5$ (glass), and thickness d. The spherical aberration $f_2(h)$ is sought for $d = 5\,\text{mm}$ along with ray diagrams for $d = 5\,\text{mm}$ and $d = 10\,\text{mm}$.
Solution: The spherical aberration $f_2 = f_2(h)$ is calculated from Eq. (2.156) and is plotted in Fig. 2.105. The critical thickness, in this case, is $d_1 = 6.3661\,\text{mm}$, and since $d < d_1$ no total internal reflection occurs for all allowable values of h, i.e., for $-h_{max} \leq h \leq h_{max}$ where $h_{max} = 15\,\text{mm}$ [from Eq. (2.157)] and $h_{max,TIR} = 16.67\,\text{mm}$ [from Eq. (2.158)]. The paraxial approximation values can be found from the theory of $ABCD$ for a thick lens. Using the matrix approach for a thick lens of left radius R_1 and right radius R_2, refractive index n_L, thickness d, surrounded by homogeneous materials of refractive indices n_1 and n_2, respectively, the elements A and C are:

$$A = 1 + \left(\frac{n_1}{n_L} - 1\right)\frac{d}{R_1} \xrightarrow{R_1 \to \infty} A = 1, \tag{2.160}$$

$$C = \left(\frac{n_L}{n_2} - 1\right)\frac{1}{R_2} + \frac{n_L}{n_2}\left(\frac{n_1}{n_L} - 1\right)\frac{1}{R_1} + \frac{d}{R_1 R_2}\frac{(n_L - n_2)(n_1 - n_L)}{n_2 n_L}$$

$$\xrightarrow{R_1 \to \infty} C = \left(\frac{n_L}{n_2} - 1\right)\frac{1}{R_2}. \tag{2.161}$$

The resulting $f_{2,par}$ for the paraxial approximation (measured from the right vertex of the lens) is given by:

$$f_{2,par} = -\frac{1}{C} + \frac{1-A}{C} = -\frac{A}{C} = -R_2\frac{n_2}{n_L - n_2}, \tag{2.162}$$

which results in $f_{2,par} = 50\,\text{mm}$ for the numerical values given (in this case the sign convention should be used, i.e., $R_2 = -25\,\text{mm}$). It is evident from Fig. 2.105 that for small values of h, $f_2 \simeq f_{2,par}$ as it is expected. The ray diagram for the case of $d = 5\,\text{mm}$ is shown in Fig. 2.106(a). Since $d < d_1$ all allowable values of h correspond to parallel rays that are

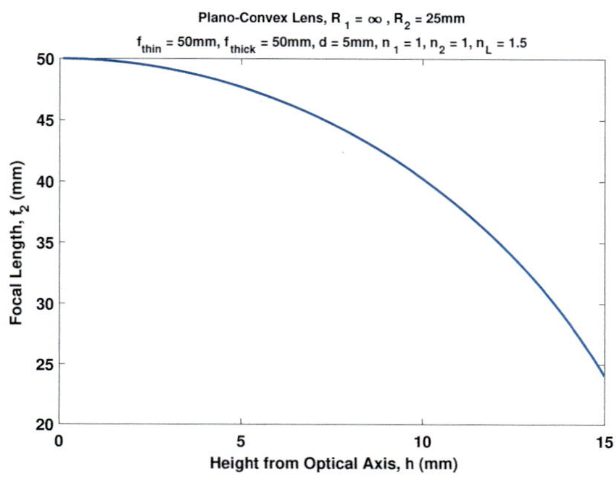

FIGURE 2.105: The focal distance f_2 of a plano-convex lens of refractive index $n_L = 1.50$, radius $R_2 = 25\,\text{mm}$, and thickness $d = 5\,\text{mm}$ surrounded by air.

getting refracted from both lens surfaces. However, when $d = 10\,\text{mm}$ then all parallel rays above $h_{max,TIR} = R_2 n_2/n_L = 16.67\,\text{mm}$ are getting totally internally reflected at the right lens surface and are not shown. In the later case $h_{max} = 20\,\text{mm}$.

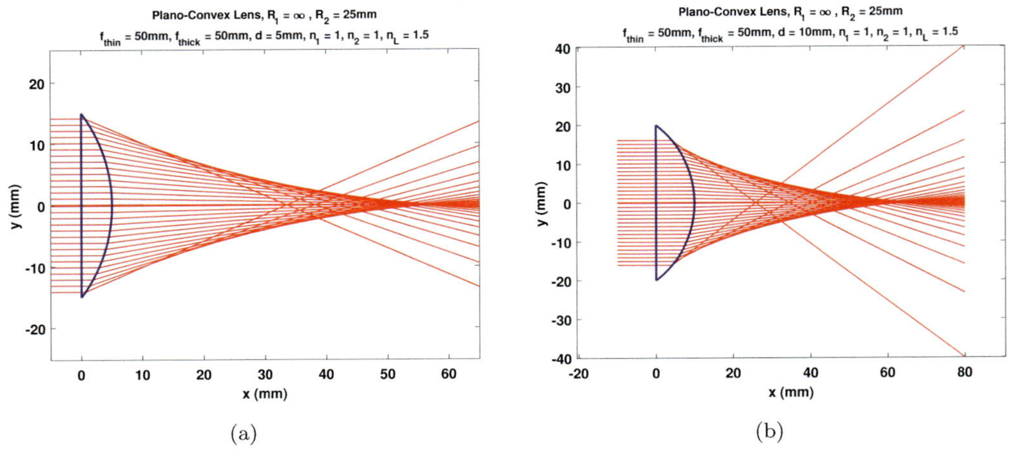

FIGURE 2.106: (a) The ray diagram of a plano-convex lens of refractive index $n_L = 1.50$, radius $R_2 = 25\,\text{mm}$, and thickness $d = 5\,\text{mm}$ surrounded by air. (b) Same as (a) but for thickness $d = 10\,\text{mm}$. Observe that for $h > h_{max}(d) = 16.67\,\text{mm}$ there are no refracted rays (due to total internal reflection at the right lens surface).

2.9.3.2 Convex-Plano Lens

Consider a convex-plano thick lens with radius R_1 (where $R_2 = \infty$) situated between two homogeneous media with refractive indices n_1 and n_2, respectively. The lens has a refractive index n_L ($n_L > n_1, n_2$) and a thickness d. A monochromatic ray is incident from the spherical surface parallel to the optical axis at a height h as shown in Fig. 2.107. The focal distance f_2 is defined as the point where the refracted ray crosses the optical axis (also shown in Fig. 2.107). The distance f_2 can be precisely determined without approximations and compared to the paraxial approximation prediction. In this section, the distance f_2 is derived as a function of the ray height h. Let ϕ_1 be the angle of incidence of a parallel

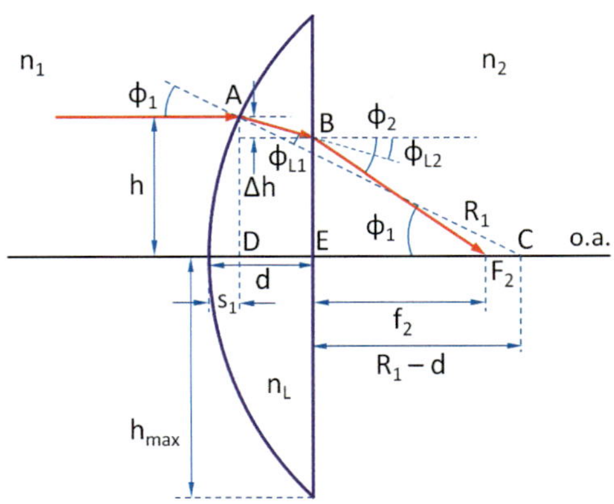

FIGURE 2.107: A convex-plano lens of refractive index n_L, radius R_1, and thickness d between two homogeneous media of refractive indices n_1 and n_2, respectively. A ray parallel to the optical axis is incident on the spherical refractive surface of the lens at height h, and after refracting though the lens, intersects the optical axis at distance f_2.

ray that strikes the spherical left surface of the convex-plano lens. Ignoring the reflected ray, the refracted ray inside the lens forms an angle ϕ_{L1} with respect to the normal at the surface (indicated by the radius CA). This refracted ray is then further refracted upon exiting the lens through the flat right surface. The angle of incidence on the flat surface is ϕ_{L2}, and the final emergent ray forms an angle ϕ_2 relative to the optical axis. These angles are interrelated through Snell's law, yielding the following relationships:

$$\begin{aligned}
\sin\phi_1 &= \frac{h}{R_1} \qquad \text{(from orthogonal triangle } CAD\text{)} \\
\sin\phi_{L1} &= \frac{n_1}{n_L}\sin\phi_1 = \frac{n_1}{n_L}\frac{h}{R_1} \\
\phi_{L2} &= \phi_1 - \phi_{L1} \\
\sin\phi_2 &= \frac{n_L}{n_2}\sin\phi_{L2} = \frac{n_L}{n_2}\sin(\phi_1 - \phi_{L1}).
\end{aligned} \tag{2.163}$$

From Fig. 2.107 the maximum allowable height h_{max} is given by:

$$h_{max} = \left[R_1^2 - (R_1 - d)^2\right]^{1/2} = \left[R_1(2R_1 - d)\right]^{1/2}. \tag{2.164}$$

The focal distance f_2 can now be determined from the following steps:

$$\begin{aligned}
s_1 &= R_1 - \sqrt{R_1^2 - h^2}, \\
\Delta h &= (d - s_1)\tan\phi_{L2} = (d - s_1)\tan(\phi_1 - \phi_{L1}), \\
BE &= h - \Delta h, \\
f_2 &= \frac{BE}{\tan\phi_2} = \frac{h - \Delta h}{\tan\phi_2} \implies
\end{aligned}$$

$$f_2 = \frac{h - \left(d - R_1 + \sqrt{R_1^2 - h^2}\right)\tan\left[\sin^{-1}\left(\frac{h}{R_1}\right) - \sin^{-1}\left(\frac{n_1 h}{n_L R_1}\right)\right]}{\tan\left\{\sin^{-1}\left[\frac{n_L}{n_2}\sin\left[\sin^{-1}\left(\frac{h}{R_1}\right) - \sin^{-1}\left(\frac{n_1 h}{n_L R_1}\right)\right]\right]\right\}}. \tag{2.165}$$

For the equation determining f_2 to remain valid, the angle ϕ_{L2} (the angle within the lens at the flat right surface) must be less than the critical angle for total internal reflection at that surface, which depends on the refractive indices of the lens material (n_L) and the surrounding medium on the right side (n_2). This condition imposes a restriction on the maximum allowable height, $h_{max}(d)$, and can be expressed by the following inequality:

$$\sin^{-1}\left(\frac{h_{max}(d)}{R_1}\right) - \sin^{-1}\left(\frac{n_1}{n_L}\frac{h_{max}(d)}{R_1}\right) < \sin^{-1}\left(\frac{n_2}{n_L}\right). \tag{2.166}$$

Example 2.9.4:

Consider a convex-plano thick lens with the following numerical parameters: $R_1 = 25\,\text{mm}$, $n_1 = n_2 = 1.0$ (air), $n_L = 1.5$ (glass), and $d = 5\,\text{mm}$. The spherical aberration $f_2(h)$ is sought along with ray diagrams for $d = 5\,\text{mm}$ and $d = 20\,\text{mm}$.

Solution: The corresponding spherical aberration $f_2 = f_2(h)$ is shown in Fig. 2.108. For comparison, the results for the plano-convex lens are also depicted in the same figure, represented by a dashed line. The plano-convex lens is essentially the flipped (mirror image) configuration of the convex-plano lens. The reduced spherical aberration in the convex-plano lens is clearly apparent. In this setup, the maximum height $h_{max} = 15\,\text{mm}$ [Eq. 2.164)] coincides with the maximum allowable height $h_{max}(d)$ [Eq. (2.166)]. The paraxial

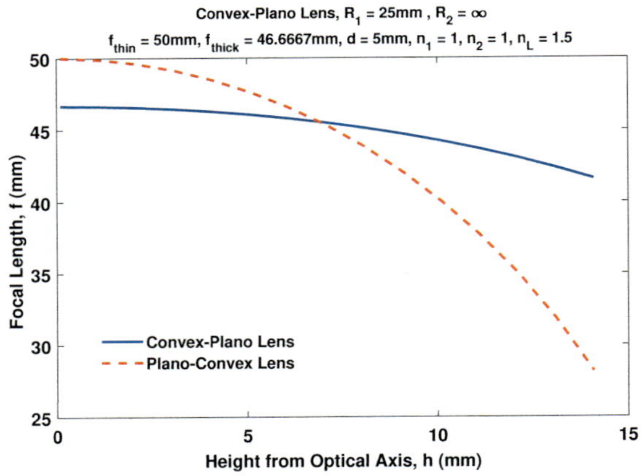

FIGURE 2.108: The focal distance f_2 of a convex-plano lens of refractive index $n_L = 1.50$, radius $R_1 = 25\,\text{mm}$, and thickness $d = 5\,\text{mm}$ surrounded by air is shown as a function of the ray's height h. In the same diagram the focal distance of a plano-convex lens with the same characteristics (just the flipped convex-plano lens) is also shown with the dashed line. The advantage of the convex-plane geometry is evident since it has much smaller spherical aberration.

approximation values can be calculated from the $ABCD$ matrix theory for a thick lens. Using the matrix approach for a thick lens of left radius R_1 and right radius $R_2 \to \infty$ the corresponding matrix elements for the convex-plano lens are:

$$A = 1 + \left(\frac{n_1}{n_L} - 1\right)\frac{d}{R_1},$$

$$C = \left(\frac{n_L}{n_2} - 1\right)\frac{1}{R_2} + \frac{n_L}{n_2}\left(\frac{n_1}{n_L} - 1\right)\frac{1}{R_1} + \frac{d}{R_1 R_2}\frac{(n_L - n_2)(n_1 - n_L)}{n_2 n_L}$$

$$\xrightarrow{R_2 \to \infty} \quad C = \frac{n_1 - n_L}{n_2}\frac{1}{R_1}.$$

The resulting $f_{2,par}$ for the paraxial approximation (measured from the right vertex of the lens) is given by:

$$f_{2,par} = -\frac{A}{C} = \left[1 + \left(\frac{n_1}{n_L} - 1\right)\frac{d}{R_1}\right]\frac{n_2}{n_L - n_1} R_1. \tag{2.167}$$

With these values, the paraxial focal length $f_{2,par}$ can be calculated, yielding $f_{2,par} = 46.67\,\text{mm}$ for the previously specified parameters (noting that the appropriate sign convention is applied, with $R_1 = +25\,\text{mm}$). As illustrated in Fig. 2.108, for small values of h, the focal distance $f_2 \simeq f_{2,par}$, aligning with the expectations based on the paraxial approximation. In Fig. 2.109(a), the ray diagram for a convex-plano lens with $d = 5\,\text{mm}$ is displayed, showing all refracted rays up to the maximum allowable height h without any total internal reflection. For comparison, Fig. 2.109(b) illustrates a similar ray diagram for a thicker lens

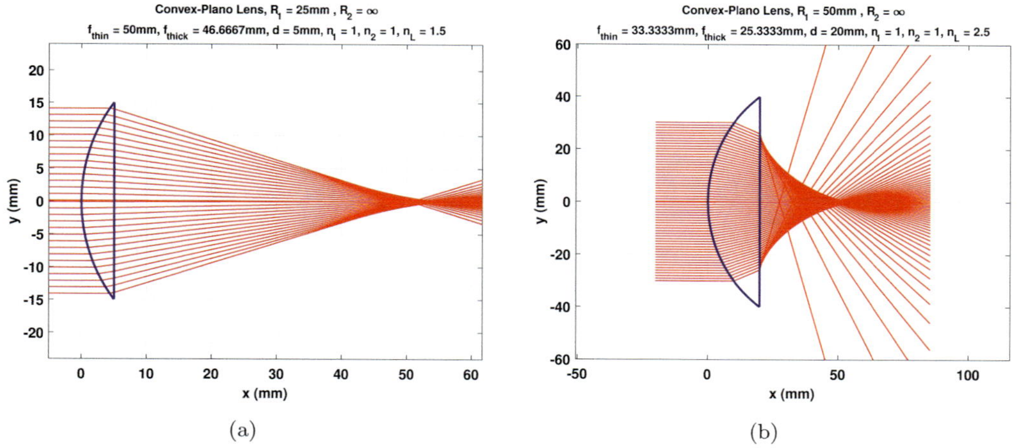

FIGURE 2.109: (a) The ray diagram of a convex-plano lens of refractive index $n_L = 1.50$, radius $R_2 = 25\,\text{mm}$, and thickness $d = 5\,\text{mm}$ surrounded by air. (b) Same as (a) but for thickness $d = 20\,\text{mm}$, $n_L = 2.5$, and $R_1 = 50\,\text{mm}$. In this case some rays can not exit from the right flat surface of the lens due to total internal reflection.

with $d = 20\,\text{mm}$, $n_L = 2.5$ and $R_1 = 50\,\text{mm}$. In this second case, some rays corresponding to higher values of h experience total internal reflection at the flat surface of the lens, preventing them from refracting out of the lens. As a result, only the permissible range of ray heights is shown in the diagram. While the theoretical maximum h_{max} is 40 mm, the practical allowable height due to total internal reflection constraints for $d = 20\,\text{mm}$ is $h_{max}(d = 20\,mm) \simeq 30.6\,\text{mm}$.

2.9.3.3 Biconvex Lens

The spherical aberration for a biconvex lens, where a ray parallel to the optical axis at height h_1 is incident from the left, can be calculated by finding the focal distance f_2 as a function of h_1. The lens has a left radius R_1, right radius R_2, thickness d, and refractive index n_L, and is surrounded by homogeneous media of refractive indices n_1 and n_2 as shown in Fig. 2.110(a). To determine f_2, the angle of incidence ϕ_1 on the left spherical surface must first be obtained from triangle ADC_1. Then, applying Snell's law allows for calculating the refraction angle ϕ_2 inside the lens. These steps are repeated for the second surface, ultimately determining the exit angle and the location at which the ray crosses the optical axis. The procedure is specified next:

$$\phi_1 = \sin^{-1}\left(\frac{h_1}{R_1}\right), \tag{2.168}$$

$$\phi_2 = \sin^{-1}\left(\frac{n_1 \sin\phi_1}{n_L}\right) = \sin^{-1}\left(\frac{n_1}{n_L}\frac{h_1}{R_1}\right). \tag{2.169}$$

The distance $s_1 = R_1 - (R_1^2 - h_1^2)^{1/2}$, and the distance CC_2 can be determined from the equation

$$CC_2 = \frac{h_1}{\tan(\phi_1 - \phi_2)} + R_2 - d + s_1. \tag{2.170}$$

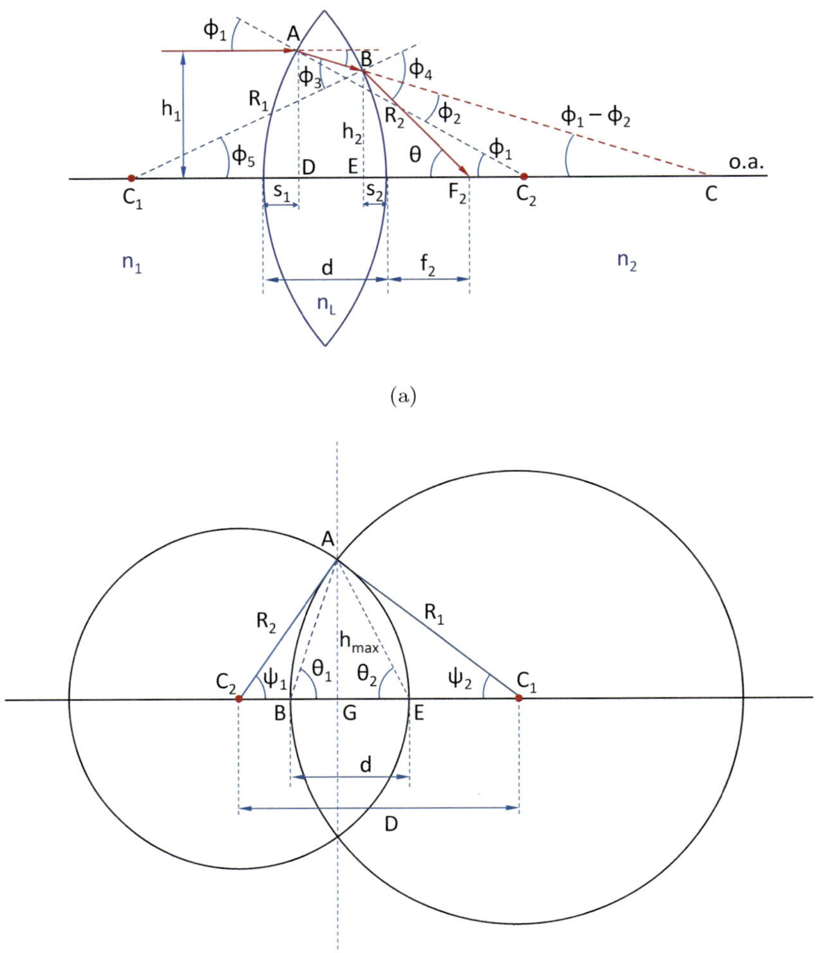

(b)

FIGURE 2.110: (a) A biconvex lens of refractive index n_L, left radius R_1, right radius R_2, and thickness d between two homogeneous media of refractive indices n_1 and n_2, respectively. A ray parallel to the optical axis is incident on the left spherical surface of the lens at height h_1 and after refracting though the lens intersects the optical axis at distance f_2. (b) Determination of critical geometric characteristics of a biconvex lens of left radius R_1, right radius R_2 and thickness d.

Using the law of sines in triangle C_2BC, the angle ϕ_3 (angle of incidence of the ray at the second spherical refractive surface) and its refracted counterpart ϕ_4 can be determined:

$$\phi_3 = \sin^{-1}\left(\frac{CC_2}{R_2}\sin(\phi_1 - \phi_2)\right), \tag{2.171}$$

$$\phi_4 = \sin^{-1}\left(\frac{n_L \sin\phi_3}{n_2}\right). \tag{2.172}$$

The angle ϕ_5 can be determined from triangle C_2BC as $\phi_5 = \phi_3 - (\phi_1 - \phi_2)$. Then the height h_2 that the ray intersects the right refractive surface is $h_2 = R_2 \sin\phi_5 = R_2 \sin[\phi_3 - (\phi_1 - \phi_2)]$. The angle θ can be determined from triangle CBF_2 as $\theta = (\phi_4 - \phi_3) + (\phi_1 - \phi_2)$. The required focal distance f_2 for the biconvex lens can be expressed in terms of the intermediate distances $EF_2 = h_2/\tan\theta$, and $s_2 = R_2 - (R_2^2 - h_2^2)^{1/2}$. Thus, the focal distance f_2 is given by:

$$f_2 = EF_2 - s_2 = \frac{h_2}{\tan(\phi_4 - \phi_3 + \phi_1 - \phi_2)} - R_2 + \left(R_2^2 - h_2^2\right)^{1/2}. \tag{2.173}$$

This equation provides f_2 as a function of the ray height h_1, accounting for the lens geometry and the refraction angles. In the special case of $R_1 \to \infty$, the lens effectively becomes a plano-convex lens. Here, the following conditions apply: $\phi_1 = 0$, $\phi_2 = 0$, and $s_1 = 0$. The angle of refraction at the second surface, ϕ_3, is given by $\phi_3 = \sin^{-1}(h_1/R_2)$, matching the setup described for the plano-convex lens and corresponding to the spherical aberration analysis in Section 2.9.3.1. In the special case that $R_2 \to \infty$, the lens becomes a convex-plano lens. Here, the center of curvature C_2 effectively approaches infinity, resulting in $\phi_5 = 0$, $\phi_3 = \phi_1 - \phi_2$, and $s_2 = 0$. The distance ED simplifies to $ED = d - s_1$, and the ray height h_2 on the flat surface is given by $h_2 = h_1 - \tan(\phi_1 - \phi_2)(ED)$, aligning with the spherical aberration analysis for the convex-plano lens (of Sec. 2.9.3.2).

The maximum height h_{max} of a biconvex lens can be derived based on the geometry illustrated in Fig. 2.110(b). Let $D = R_1 + R_2 - d$ denote the separation between the centers of curvature, C_1 and C_2, of the two spherical refractive surfaces. Using this setup, the following relationships can be established:

$$\begin{aligned}
\psi_1 &= \cos^{-1}\left(\frac{R_2^2 + D^2 - R_1^2}{2R_2D}\right), \\
\psi_2 &= \cos^{-1}\left(\frac{R_1^2 + D^2 - R_2^2}{2R_1D}\right), \\
h_{max} &= R_1 \sin\psi_2 = R_2 \sin\psi_1.
\end{aligned} \tag{2.174}$$

In order to avoid total internal reflection at point B (following similar steps with the ones presented in 2.9.3.1) the following inequality should hold:

$$\begin{aligned}
&\frac{CC_2}{R_2}\sin(\phi_1 - \phi_2) \le \sin\phi_{3,cr} = \frac{n_2}{n_L} \Longrightarrow \\
&\quad d_{min}(h_1) \le d \le 2\min(R_1, R_2) \quad \text{where,} \\
&\quad d_{min}(h_1) = \frac{h_1}{\tan(\phi_1 - \phi_2)} + R_2 + R_1 - \left(R_1^2 - h_1^2\right)^{1/2} - R_2\frac{n_2}{n_L}\frac{1}{\sin(\phi_1 - \phi_2)}.
\end{aligned} \tag{2.175}$$

If $d < d_{min}(h_1 = h_{max})$ then total internal reflection occurs at the right spherical refractive surface of the biconvex lens.

This book's analysis of thick lenses assumes that the lens thickness corresponds to the natural thickness formed by the intersection of two spheres with radii R_1 and R_2. However, in practical applications, any lens (including a biconvex lens) can be fabricated with a chosen thickness by shaping a piece of material with spherical surfaces of radii R_1 and R_2 on either side. This design flexibility allows lenses to be made thicker or thinner as needed while maintaining the desired radii of curvature.

Example 2.9.5:

Consider a biconvex lens with the following parameters: $R_1 = R_2 = 50\,\text{mm}$ (no sign convention is used in this analysis), $n_L = 1.50$, $n_1 = n_2 = 1.0$, and a thickness $d = 5\,\text{mm}$ or $d = 20\,\text{mm}$. Determine the spherical aberration $f_2(h)$ along with the corresponding ray diagrams.

Solution: The resulting spherical aberration expressed by the focal distance f_2 is shown in Fig. 2.111. In the case of $d = 5$ mm the resulting $h_{max} = 15.6125$ mm and $d_{min}(h_{max}) = -116.02$ mm and therefore for any allowable height h_1 of the incident ray, no total internal reflection will occur on the second refractive surface of the lens. In the case of $d = 20$ mm the resulting $h_{max} = 30$ mm and $d_{min}(h_{max}) = 40.42$ mm and therefore total internal reflection could occur on the second refractive surface of the lens. The maximum height in this case to avoid the total internal reflection can be specified by the equation $d_{min}(h_1) = d$ and results in $h_1 = 26.48$ mm. These results can be verified by the ray diagrams shown in Figs. 2.112(a) and 2.112(b). The paraxial approximation focal distance can be found from $f_{2,par} = -A/C$ where the parameters A and C are given in Eqs. (2.76) and (2.78) for a thick lens, respectively. The paraxial focal distance $f_{2,par} = 49.1525$ mm for $d = 5$ mm and $f_{2,par} = 46.4286$ mm for $d = 20$ mm.

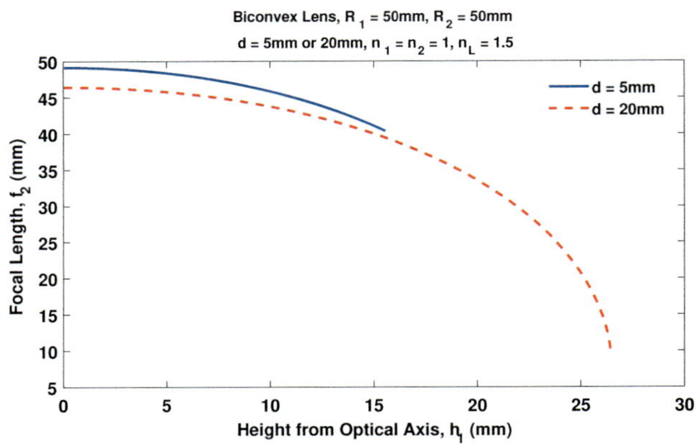

FIGURE 2.111: The focal distance f_2 (as function of incident ray's height h_1) of a biconvex lens with parameters $n_L = 1.50$, radii $|R_1| = |R_2| = 50$ mm, thickness $d = 5$ mm, or $d = 20$ mm.

2.9.3.4 Positive-Meniscus Lens

A positive meniscus lens consists of two spherical surfaces, with both surfaces either convex ($R_1, R_2 > 0$ and $R_1 < R_2$) or both concave ($R_1, R_2 < 0$ and $|R_2| < |R_1|$) assuming light propagation from left to right (using the sign convention). The lens has thickness d, refractive index n_L is surrounded by homogeneous media of refractive indices n_1 and n_2 as shown in Fig. 2.113. For this lens, a ray parallel to the optical axis at height h_1 is incident from the left. The goal is to determine the spherical aberration by calculating the focal distance f_2, as measured from the right vertex V_2 of the positive meniscus lens, as a function of the ray height h_1. The procedure for determining the spherical aberration in a positive meniscus lens is similar to that of the biconvex lens, with slight modifications arising from the different geometrical configuration of the lens. To begin, the angle of incidence ϕ_1 on the left spherical refractive surface and its corresponding refracted angle ϕ_2 as the ray enters the lens can be determined using triangle ADC_1 and Snell's law. The relationships between the angles can be derived as follows:

$$\phi_1 = \sin^{-1}\left(\frac{h_1}{R_1}\right),$$

$$\phi_2 = \sin^{-1}\left(\frac{n_1 \sin\phi_1}{n_L}\right) = \sin^{-1}\left(\frac{n_1}{n_L}\frac{h_1}{R_1}\right).$$

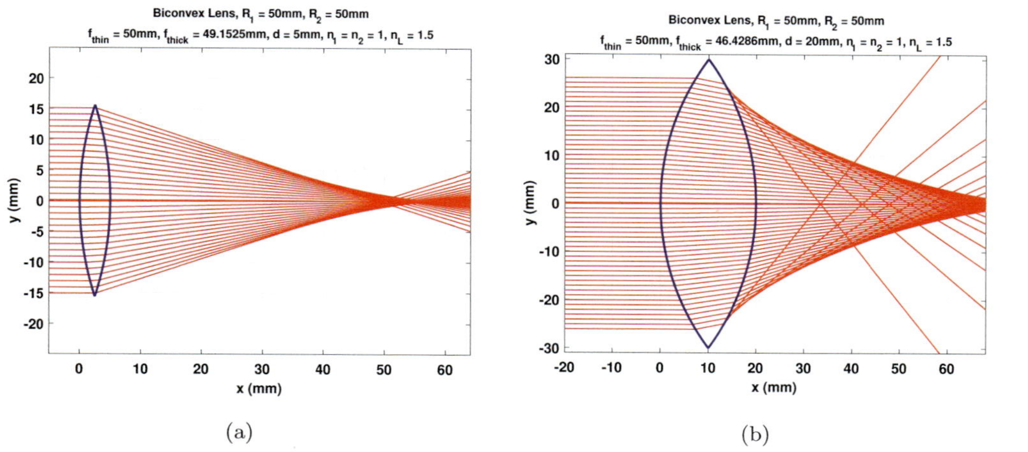

FIGURE 2.112: (a) The ray diagram of a biconvex lens of refractive index $n_L = 1.50$, radii $|R_1| = |R_2| = 50\,\text{mm}$, thickness $d = 5\,\text{mm}$, and $h_{max} = 15.6125\,\text{mm}$ surrounded by air. In this case no total reflection occurs at the right lens boundary. (b) The ray diagram of the biconvex lens with parameters same as in (a) but of thickness $d = 20\,\text{mm}$. In the latter case $h_{max} = 30\,\text{mm}$. In order to avoid total internal reflection at the right lens boundary $h_{max}(d) = h_1 = 26.48\,\text{mm}$. Therefore rays with $h > h_1$ are not drawn.

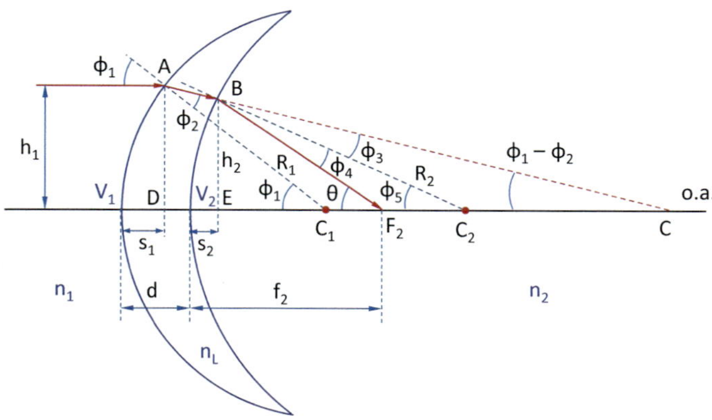

FIGURE 2.113: A positive meniscus lens of refractive index n_L, left radius R_1, right radius R_2 ($R_1 < R_2$) and thickness d between two homogeneous media of refractive indices n_1 and n_2, respectively. A ray parallel to the optical axis is incident on the left surface of the lens at height h_1 and after refracting though the lens intersects the optical axis at distance f_2 (second focal point F_2).

The distance $s_1 = R_1 - (R_1^2 - h_1^2)^{1/2}$, and the distance CC_1 can be determined from triangle ACC_1. It is straightforward using simple trigonometry to write the following equations:

$$CC_1 = R_1 \frac{\sin\phi_2}{\sin(\phi_1 - \phi_2)},$$

$$CC_2 = CC_1 - R_2 + (R_1 - d), \qquad \text{for } R_1 < R_2$$

$$CC_2 = CC_1 + R_1 - (R_2 - d), \qquad \text{for } |R_1| > |R_2|$$

$$\sin\phi_3 = \frac{CC_2}{R_2}\sin(\phi_1-\phi_2), \qquad \phi_4 = \sin^{-1}\left(\frac{n_L\sin\phi_3}{n_2}\right),$$

$$h_2 = R_2\sin(\phi_1-\phi_2+\phi_3), \qquad s_2 = R_2 - \left(R_2^2 - h_2^2\right)^{1/2},$$

$$\theta = (\phi_1-\phi_2)+(\phi_3+\phi_4),$$

$$f_2 = s_2 + \frac{h_2}{\tan\theta}. \tag{2.176}$$

The maximum height $h_{1,max}$ for a positive meniscus lens can be determined using the same

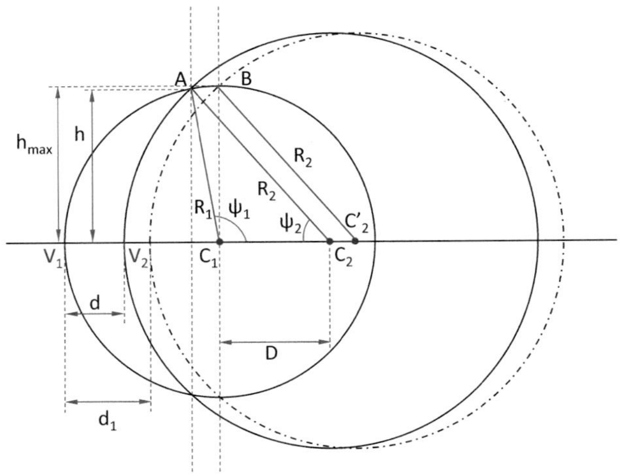

FIGURE 2.114: Determination of critical geometrical characteristics of a positive meniscus lens of left radius R_1, right radius R_2 ($R_1 < R_2$), and thickness d.

procedure as for the biconvex lens. By following similar steps as outlined in Sec. 2.9.3.3 and referring to Fig. 2.114, the maximum allowable height can be found. An interesting special case arises when $h_{1,max} = R_1$ and $d = d_1$, as shown in Fig. 2.114. In this case, the sphere of radius R_2 is represented by a dashed-dotted line. The corresponding value for d_1 is given by $d_1 = R_1 - R_2 + \left(R_2^2 - R_1^2\right)^{1/2}$. To avoid total internal reflection at point B (where the ray exits the lens), the following inequality must hold:

$$\frac{CC_2}{R_2}\sin(\phi_1-\phi_2) \le \sin\phi_{3,cr} = \frac{n_2}{n_L} \Longrightarrow$$

$$d_{min}(h_1) \le d \le 2\min(R_1,R_2) \quad \text{for } R_1 < R_2 \text{ where,} \tag{2.177}$$

$$d_{min}(h_1) = R_1\frac{\sin\phi_2}{\sin(\phi_1-\phi_2)} - R_2 + R_1 - R_2\frac{n_2}{n_L}\frac{1}{\sin(\phi_1-\phi_2)}.$$

$$0 < d \le d_{max}(h_1) \quad \text{for } |R_1| > |R_2| \text{ where,} \tag{2.178}$$

$$d_{max}(h_1) = -R_1\frac{\sin\phi_2}{\sin(\phi_1-\phi_2)} + R_2 - R_1 + R_2\frac{n_2}{n_L}\frac{1}{\sin(\phi_1-\phi_2)}.$$

If $d < d_{min}(h_1 = h_{1,max})$ for $R_1 < R_2$, then total internal reflection occurs at the right spherical refractive surface of the positive meniscus lens. Similarly, if $d > d_{max}(h_1 = h_{1,max})$ for $|R_1| > |R_2|$ then total internal reflection occurs at the right spherical refractive surface of the positive meniscus lens.

Example 2.9.6:

Consider a positive meniscus lens with the following parameters: $R_1 = 50\,\text{mm}$ and $R_2 = 100\,\text{mm}$ (no sign convention is used in this analysis), $n_L = 1.50$, $n_1 = n_2 = 1.0$, and a thickness $d = 10\,\text{mm}$. The spherical aberration $f_2(h)$ is sought along with the corresponding ray diagram. For comparison, analyze the lens with flipped its surfaces (in this case $|R_1| = 100\,\text{mm}$ and $|R_2| = 50\,\text{mm}$). According to the sign convention, the radii should be negative in the flipped lens.

Solution: The resulting spherical aberrations expressed by the focal distance f_2 (as a function of the incident ray's height h) are shown in Fig. 2.115 for the original meniscus and its flipped surfaces counterpart. In the case of $d = 10\,\text{mm}$ the resulting $h_{max} = 37.99\,\text{mm}$ and $d_{min}(h_{max}) = -176.84\,\text{mm}$ and therefore for any allowable height h_1 of the incident ray no total internal reflection will occur on the second refractive surface of the lens when $R_1 < R_2$. For the case that $R_1 > R_2$ (flipped lens) then $d_{max} = 10.05\,\text{mm}$ and since $d = 10\,\text{mm}$ again no total internal reflection in the right-hand surface of the lens will occur. These results can be verified by the ray diagrams shown in Figs. 2.116(a) and 2.116(b). The paraxial approximation focal distance can be found from $f_{2,par} = -A/C$ where the parameters A and C are given in Eqs. (2.76) and (2.78) for a thick lens, respectively. The paraxial focal distance $f_{2,par} = 175\,\text{mm}$ for $R_1 < R_2$ and $f_{2,par} = 193.75\,\text{mm}$ for $R_1 > R_2$ (flipped lens). From Fig. 2.115 it is evident that the flipped lens ($|R_1| = 100\,\text{mm}$ and $|R_2| = 50\,\text{mm}$) experiences larger spherical aberration than the $R_1 = 50\,\text{mm}$ and $R_2 = 100\,\text{mm}$ case.

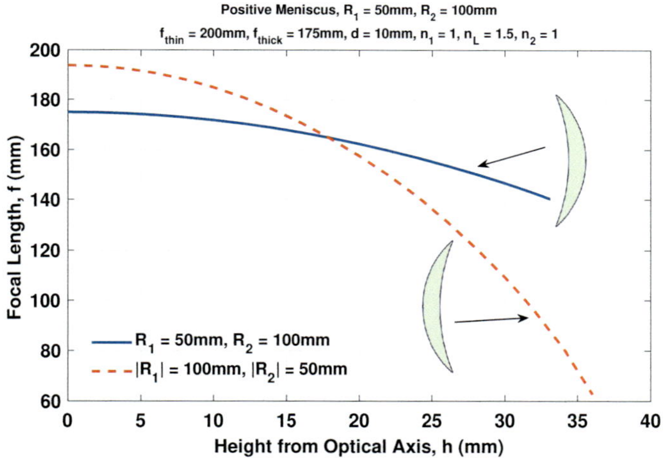

FIGURE 2.115: The spherical aberration, of a positive meniscus lens of refractive index $n_L = 1.50$, left radius $R_1 = 50\,\text{mm}$, right radius $R_2 = 100\,\text{mm}$ ($R_1 < R_2$), and thickness $d = 10\,\text{mm}$ in air $n_1 = n_2 = 1.0$. The second focal distance, as measured from right vertex V_2, is shown for $R_1 = 50\,\text{mm}$, and $R_2 = 100\,\text{mm}$ (solid line) and for $|R_1| = 100\,\text{mm}$, and $|R_2| = 50\,\text{mm}$ (dashed line). The shapes of the lenses are also inserted for better visualization.

2.9.3.5 Plano-Concave Lens

Consider a plano-concave lens of radius R_2 ($R_1 = \infty$) between two homogeneous materials of refractive indices n_1 and n_2, respectively. The refractive index of the lens is again n_L ($n_L > n_1, n_2$). The lens thickness is d (at the center of the lens). The maximum thickness occurs at the edges of the lens and is d_{max}. A monochromatic ray is incident on the flat surface of the lens parallel to the optical axis at a height h as it is shown in Fig. 2.117.

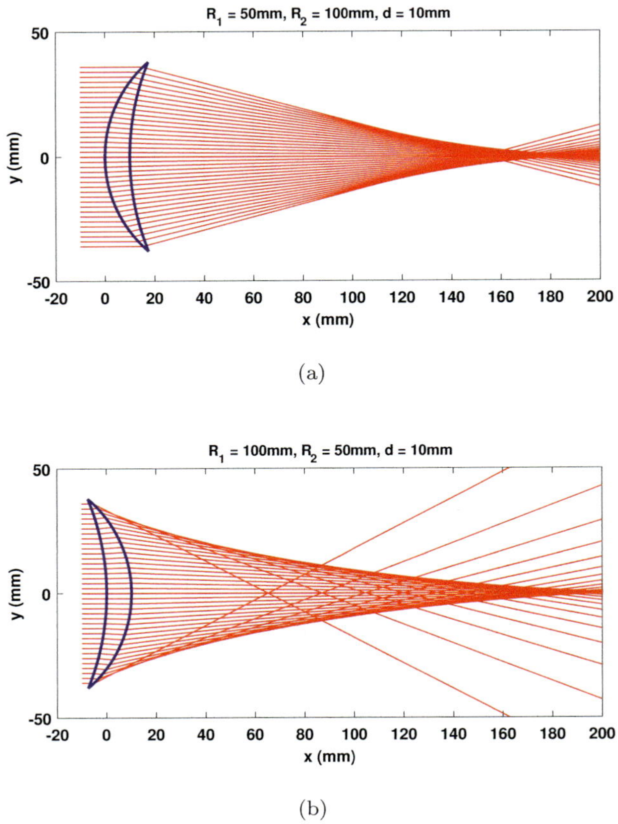

FIGURE 2.116: (a) The ray diagram of a positive meniscus lens with $n_L = 1.50$, left radius $R_1 = 50\,\text{mm}$, right radius $R_2 = 100\,\text{mm}$ ($R_1 < R_2$), and thickness $d = 10\,\text{mm}$ in air $n_1 = n_2 = 1.0$. (b) The ray diagram of a positive meniscus lens with $n_L = 1.50$, left radius $R_1 = -100\,\text{mm}$, right radius $R_2 = -50\,\text{mm}$ (with sign convention, $|R_1| > |R_2|$), and thickness $d = 10\,\text{mm}$ in air $n_1 = n_2 = 1.0$.

The goal is to determine (without paraxial approximations) the distance f_2 at which the extension of the refracted ray intersects the optical axis. In this case the finally refracted ray diverges from the optical axis and only its extension (virtual ray) intersects the optical axis at F_2. This analysis allows the precise calculation of f_2 as a function of h, which can then be compared with the prediction obtained using the paraxial approximation, where the focal distance remains independent of h. Following the exact approach presented in subsection 2.9.3.1, and the same equations Eq. (2.154)–(2.156) the focal length f_2 is given by:

$$f_2 = -\frac{h}{\tan\left[\sin^{-1}\left(\dfrac{n_L}{n_2}\dfrac{h}{R_2}\right) - \sin^{-1}\left(\dfrac{h}{R_2}\right)\right]} + R_2 - \left(R_2^2 - h^2\right)^{1/2}. \tag{2.179}$$

The only difference from Eq. (2.156) is the reversal of the sign to correspond to a diverging lens focal length. Equation (2.179) is valid for angles ϕ_1 where ϕ_2 exists, i.e., when no total internal reflection takes place at point B. From Fig. 2.117 the maximum allowable height

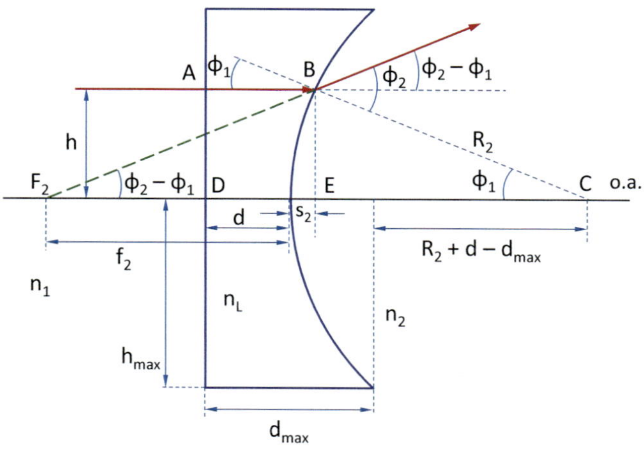

FIGURE 2.117: A plano-concave lens of refractive index n_L, radius R_2 and thickness d between two homogeneous media of refractive indices n_1 and n_2, respectively. A ray parallel to the optical axis is incident on the flat surface of the lens at height h and after refracting though the lens its extension (virtual ray) intersects the optical axis at distance f_2. At the flat surface there is no angular change in refraction due to normal incidence. However, there is an angular change of the ray in refraction that occurs at the spherical surface and the ray's new direction is denoted by the angle ϕ_2 and diverges from the optical axis.

$h_{max}(d, d_{max})$ can be expressed as:

$$h_{max}(d, d_{max}) = \left[R_2^2 - (R_2 + d - d_{max})^2\right]^{1/2} = \left[(d_{max} - d)\,(2R_2 + d - d_{max})\right]^{1/2}. \tag{2.180}$$

To avoid total internal reflection at point B (following similar steps as those presented in 2.9.3.1), the following inequality must hold:

$$(d_{max} - d)^2 - 2R_2(d_{max} - d) + R_2^2\,\frac{n_2^2}{n_L^2} \geq 0 \Longrightarrow$$

$$d \geq d_{max} - d_1, \quad \text{where,} \quad d_1 = \frac{R_2}{n_L}\left[n_L - \left(n_L^2 - n_2^2\right)^{1/2}\right]. \tag{2.181}$$

Example 2.9.7:

Consider a plano-concave lens with the parameters similar to the values used in 2.9.3.1: $R_2 = 25\,\text{mm}$, $n_1 = n_2 = 1.0$ (air), $n_L = 1.5$ (glass), $d = 5\,\text{mm}$, and $d_{max} = 12\,\text{mm}$. Determine the focal distance $f_2(h)$ as function of the height h, and show the corresponding ray diagram.

Solution: The corresponding spherical aberration $f_2 = f_2(h)$ is presented in Fig. 2.118(a). For this case, the critical thickness is determined as $d_1 = 6.3661\,\text{mm}$. Given that $d > d_{max} - d_1 = 3.6339\,\text{mm}$, total internal reflection occurs for any permissible value of h, that is, for $-h_{max} \leq h \leq h_{max}$ where $h_{max} = h_{max}(d, d_{max}) = 17.3494\,\text{mm}$ [from Eq. (2.180)] and $h_{max,TIR} = 16.67\,\text{mm}$. The paraxial approximation values can be determined from the $ABCD$ matrix theory for a thick lens. Following the same procedure as in Sec. 2.9.3.1, the paraxial focal distance is found to be $f_{2,par} = -50\,\text{mm}$ or the specified numerical values (where the sign convention is applied, i.e., $R_2 = +25\,\text{mm}$). As expected, Fig. 2.118(a) shows that for small values of h, $f_2 \simeq f_{2,par}$. For comparison, the results of the flipped lens (concave-plano) are also depicted in the same figure, highlighting the flipped lens's reduced spherical aberration. The ray diagram for this configuration is shown in Fig. 2.118(b). Since $d > d_{max} - d_1$, all allowable values of h are refracting. Additionally, the ray diagram for the flipped lens is presented in Fig. 2.118(c).

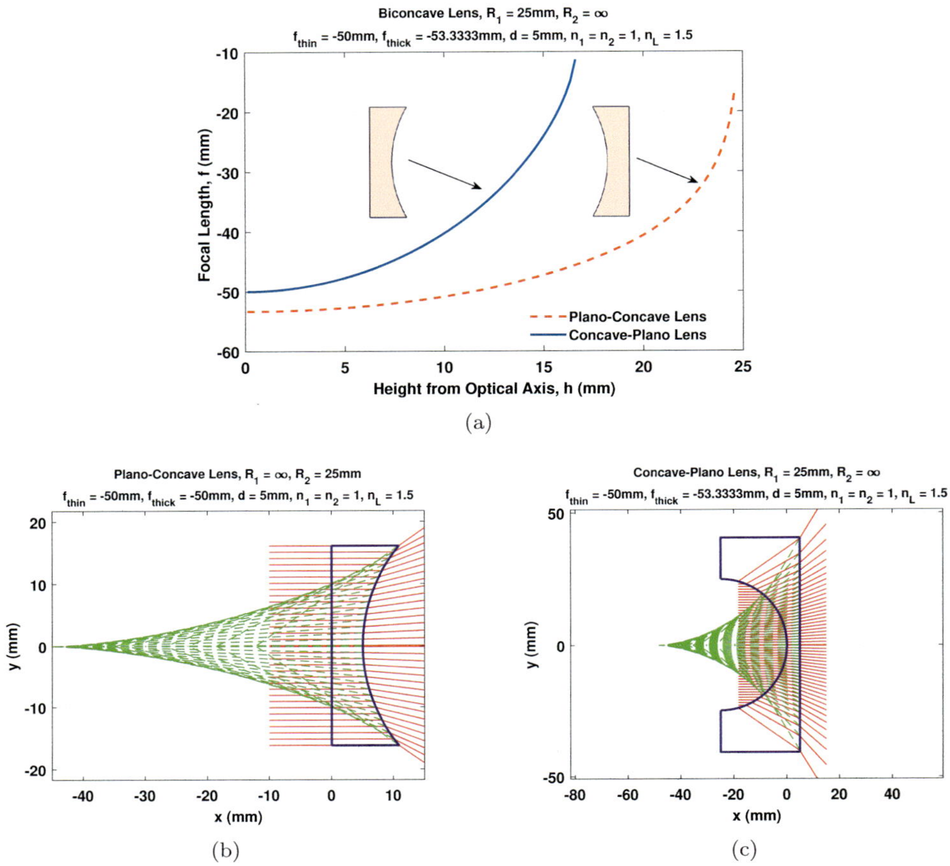

FIGURE 2.118: (a) The focal distance f_2 of a plano-concave lens of refractive index $n_L = 1.50$, radius $R_2 = 25$ mm, thickness $d = 5$ mm, and $d_{max} = 12$ mm surrounded by air. In the same figure the resulting spherical aberration for the flipped lens (concave-plano) is also shown. (b) The ray diagram of the plano-concave lens with parameters of (a). The extensions of the refracted rays that pass through the various F_2 points (virtual rays) are shown with the green dashed lines. (c) The ray diagram of the concave-plano lens with parameters of (a). The extensions of the refracted rays that pass through the various F_2 points (virtual rays) are also shown with the green dashed lines.

2.9.3.6 Biconcave Lens

A biconcave lens of left radius R_1, right radius R_2, thickness d, and refractive index n_L is situated in homogeneous media with refractive indices as shown in Fig. 2.119. A ray, parallel to the optical axis of the lens and at height h_1, is incident from the left, as shown in the figure. To quantify the spherical aberration, the focal distance f_2 (measured from the right vertex V_2 of the biconcave lens) is determined as a function of the height h_1. Since f_2 s measured to the left of V_2, it will be assigned a negative value to align with the conventions established for thin lenses.

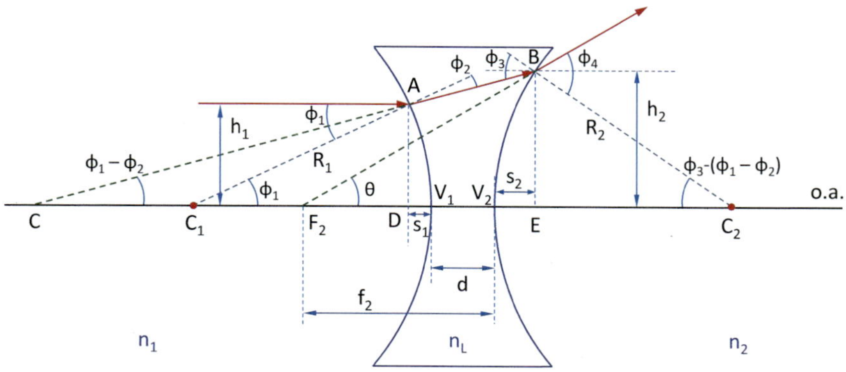

FIGURE 2.119: A biconcave lens of refractive index n_L, left radius R_1, right radius R_2, and thickness d between two homogeneous media of refractive indices n_1 and n_2, respectively. A ray parallel to the optical axis is incident on the left surface of the lens at height h_1 and after refracting through the lens, its extension (virtual ray) intersects the optical axis at distance f_2.

The approach used here closely follows the one outlined for the biconvex lens in Sec. 2.9.3.3). The angle of incidence ϕ_1 at the left spherical refractive surface, along with the corresponding refracted angle ϕ_2 within the lens, can be derived using triangle ADC_1 and Snell's law as follows:

$$\phi_1 = \sin^{-1}\left(\frac{h_1}{R_1}\right), \qquad \phi_2 = \sin^{-1}\left(\frac{n_1 \sin\phi_1}{n_L}\right) = \sin^{-1}\left(\frac{n_1}{n_L}\frac{h_1}{R_1}\right). \tag{2.182}$$

The distance s_1 is given by $s_1 = R_1 - (R_1^2 - h_1^2)^{1/2}$, while the distance CC_2 can be determined from the equation:

$$CC_2 = \frac{h_1}{\tan(\phi_1 - \phi_2)} + R_2 + d + s_1.$$

Using the law of sines in triangle C_2BC the angle ϕ_3 (angle of incidence of the ray at the second spherical refractive surface) and its refracted counterpart ϕ_4 can also be determined:

$$\phi_3 = \sin^{-1}\left(\frac{CC_2}{R_2}\sin(\phi_1 - \phi_2)\right), \qquad \phi_4 = \sin^{-1}\left(\frac{n_L \sin\phi_3}{n_2}\right). \tag{2.183}$$

The height h_2 that the ray intersects the right refractive surface is $h_2 = R_2 \sin[\phi_3 - (\phi_1 - \phi_2)]$. The angle θ can be determined from triangle C_2BF_2 as $\theta = (\phi_4 - \phi_3) + (\phi_1 - \phi_2)$. The distance $EF_2 = h_2/\tan\theta$, the distance $s_2 = R_2 - (R_2^2 - h_2^2)^{1/2}$, and finally the desired distance f_2 is given by:

$$f_2 = EF_2 - s_2 = \frac{h_2}{\tan(\phi_4 - \phi_3 + \phi_1 - \phi_2)} - R_2 + \left(R_2^2 - h_2^2\right)^{1/2}. \tag{2.184}$$

Since the point F_2 is to the left of V_2, the f_2 must be negative to maintain the sign convention applied for thin lenses. Thus, the distance V_2F_2 is designated as $-f_2$ (negative). In the special case where $R_1 \to \infty$, $\phi_1 = 0$, $\phi_2 = 0$, $s_1 = 0$, and $\phi_3 = \sin^{-1} h_1/R_2$. This situation corresponds to the plano-concave lens spherical aberration presented in Sec. 2.9.3.5. In the other special case where $R_2 \to \infty$, $C_2 \to \infty$, $\phi_3 = \phi_1 - \phi_2$, $s_2 = 0$, $ED = d + s_1$, $h_2 = h_1 + \tan(\phi_1 - \phi_2)(ED)$. This configuration corresponds to the concave-plano lens spherical aberration.

The maximum height $h_1 = h_{max}$ of a biconvex lens can be determined from the $\min(R_1, R_2)$. However, the practical upper limit for h_1 is set by avoiding total internal reflection at the second (right) spherical surface of the lens. This condition can be established in a similar manner as for the biconvex lens. To prevent total internal reflection at point B, the following inequality must hold:

$$\frac{CC_2}{R_2}\sin(\phi_1 - \phi_2) \leq \sin\phi_{3,cr} = \frac{n_2}{n_L} \Longrightarrow$$

$$0 \leq d \leq d_{max}(h_1) \quad \text{where,} \tag{2.185}$$

$$d_{max}(h_1) = -\frac{h_1}{\tan(\phi_1 - \phi_2)} - R_2 - R_1 + \left(R_1^2 - h_1^2\right)^{1/2} + R_2\frac{n_2}{n_L}\frac{1}{\sin(\phi_1 - \phi_2)}.$$

If $d > d_{max}(h_1 = h_{max})$ then total internal reflection occurs at the right spherical refractive surface of the biconcave lens.

Example 2.9.8:

Consider a biconcave thick lens with the following numerical parameters: $R_1 = R_2 = 50\,\text{mm}$ (no sign convention is used in this analysis), $n_L = 1.50$, $n_1 = n_2 = 1.0$, and a thickness $d = 5\,\text{mm}$. The spherical aberration $f_2(h)$ is sought as a function of the ray height h. In addition, a ray diagram showing rays entering at different heights h is needed. This diagram will illustrate the diverging effect of the lens and highlight variations in the focal distance f_2 with increasing ray height, characteristic of spherical aberration in a biconcave lens.
Solution: The resulting spherical aberration, as represented by the focal distance f_2, is depicted in Fig. 2.120(a). For a lens thickness of $d = 5\,\text{mm}$, the maximum allowable ray height without encountering total internal reflection is $h_{max} = 23.395\,\text{mm}$ since $d_{max}(h_{max}) \simeq 5\,\text{mm}$. Therefore, any ray height $h_1 > 23.395\,\text{mm}$ will result in total internal reflection at the second (right) refractive surface of the lens. This observation is corroborated by the ray diagram shown in Fig. 2.120(b), where rays beyond $h = h_1 > 23.395\,\text{mm}$ fail to refract outwards. Additionally, Fig. 2.120(a) confirms this, as no values of f_2 are provided for $h = h_1 > 23.395\,\text{mm}$, indicating the occurrence of total internal reflection. The paraxial focal distance $f_{2,par}$ can be calculated using $f_{2,par} = -A/C$, where the parameters A and C are derived from the $ABCD$ matrix elements for a thick lens, as given in Eqs. (2.76) and (2.78). For $d = 5\,\text{mm}$, the paraxial focal distance is $f_{2,par} = -50.8197\,\text{mm}$, which aligns closely with the values of f_2 in Fig. 2.120(a) for small ray heights h.

2.9.3.7 Negative Meniscus Lens

In the case of a negative meniscus lens, the configuration consists of a left radius R_1, right radius R_2, thickness d, and refractive index n_L, surrounded by media with refractive indices n_1 and n_2, as shown in Fig. 2.121. The radii R_1 and R_2 are considered positive when following the standard sign convention for the lens, while for the reversed lens, both radii R_1 and R_2 will be negative. However, similarly to previous cases, the magnitudes of the radii are utilized. A monochromatic ray parallel to the optical axis is incident from the left at a height h_1, as shown in the diagram. To determine the spherical aberration, the focal distance $f_2 = F_2V_2$ as measured from the right vertex (V_2) of the biconcave lens is needed as a function of the height h_1. Since the focal distance f_2 is measured to the left of V_2, it is considered negative to adhere to the conventions for thin lenses. The procedure to calculate $f_2(h_1)$ follows similarly to other lens configurations, with the main steps involving the use of Snell's law at both surfaces of the lens and the geometry of the rays passing through the lens. This allows the determination of the focal point of the ray as a function of its incident height h_1. The angle of incidence ϕ_1 on the left spherical refractive surface and its refractive

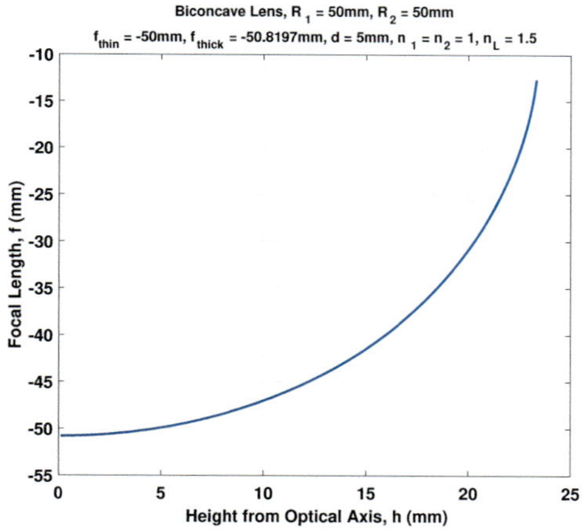

(a)

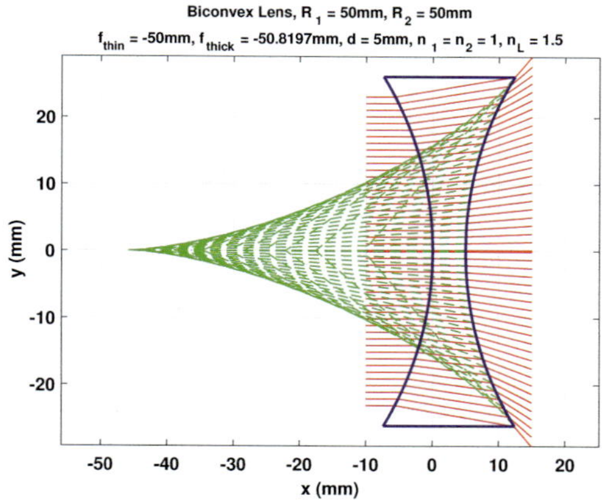

(b)

FIGURE 2.120: (a) The focal distance f_2 of a biconcave lens of refractive index $n_L = 1.50$, radius $R_1 = R_2 = 50\,\text{mm}$, thickness $d = 5\,\text{mm}$, surrounded by air. (b) The ray diagram of the biconcave lens with parameters of (a). The extensions of the refracted rays that pass through the various F_2 points (virtual rays) are shown with the green dashed lines.

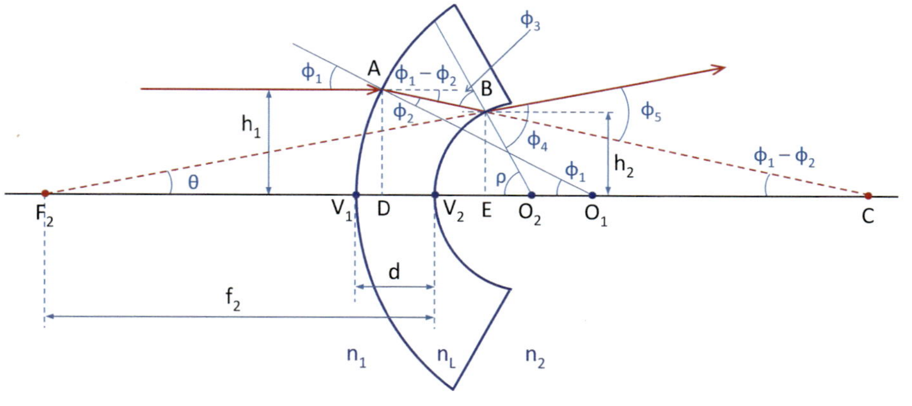

FIGURE 2.121: A negative meniscus lens of refractive index n_L, left radius R_1, right radius R_2 ($R_1, R_2 > 0$ according to the sign convention), and thickness d between two homogeneous media of refractive indices n_1 and n_2, respectively. A ray parallel to the optical axis is incident on the left surface of the lens at height h_1, and after refracting through the lens, its extension (virtual ray) intersects the optical axis at distance $f_2 = F_2V_2$ measured from the vertex V_2. Since F_2 is to the left of V_2, the focal distance $f_2 < 0$ as expected for a negative lens.

counterpart ϕ_2 inside the lens can be determined from triangle ADO_1 and Snell's law as follows:

$$\phi_1 = \sin^{-1}\left(\frac{h_1}{R_1}\right), \qquad \phi_2 = \sin^{-1}\left(\frac{n_1 \sin\phi_1}{n_L}\right) = \sin^{-1}\left(\frac{n_1}{n_L}\frac{h_1}{R_1}\right). \tag{2.186}$$

The distance s_1 is given by $s_1 = R_1 - (R_1^2 - h_1^2)^{1/2}$, while the distance O_1O_2 is $O_1O_2 = R_1 - R_2 - d$. The distance $CO_2 = CD - R_2 - d + s_1$ and $CD = h_1/\tan(\phi_1 - \phi_2)$ as it can be easily verified from the figure. The angle ϕ_3 can be determined from the law of sines (triangle O_2BC) and is given by:

$$\sin\phi_3 = (CO_2)\,\frac{\sin(\phi_1-\phi_2)}{R_2}.$$

The refraction at point B specifies the angle $\phi_4 = \sin^{-1}[(n_L \sin\phi_2)/n_2]$. The angle $\rho = \phi_3 + \phi_1 - \phi_2$. Consequently, $\theta = \phi_5 - (\phi_1 - \phi_2)$, and $h_2 = R_2 \sin\rho = R_2 \sin(\phi_3 + \phi_1 - \phi_2)$. Finally, $F_2E = h_2/\tan\theta$, and the focal distance (in magnitude) is given by:

$$f_2 = F_2E - V_2E = \frac{h_2}{\tan\theta} - \left(R_2 - \sqrt{R_2^2 - h_2^2}\right), \qquad (2.187)$$

where f_2 is the absolute value of the focal distance since the actual $f_2 = -|V_2F2|$ is negative as it is measured to the left of vertex V_2. Furthermore, h_2 was specified from the previous equations. For a negative meniscus lens, the ray exiting from the right lens surface must obey the condition that the angle ϕ_3 between the refracted ray and the optical axis should not exceed the critical angle between the lens medium and the surrounding medium on the right. This condition ensures that the ray does not undergo total internal reflection at the right surface of the lens. Mathematically, this condition is given by: $\phi_3 \leq \sin^{-1}(n_2/n_L)$. In the case that $R_1 > R_2 > 0$ (as in Fig. 2.121) there is a maximum acceptable height $h_{1,max}$ that can be determined only numerically by solving the equation $\sin\phi_3(h_{1,max}) = n_2/n_L$. In the case that $|R_1| < |R_2|$ (where $R_1, R_2 < 0$) the maximum height is simply given by the absolute value of the left radius, $h_{1,max} = |R_1|$.

Example 2.9.9:

Consider a negative meniscus with the following numerical data: $R_1 = 250$ mm, and $R_2 = 50$ mm, $n_L = 1.5$, $n_1 = n_2 = 1.0$, and thickness $d = 10$ mm. Also consider the case of the flipped lens where $R_1 = -50$ mm, and $R_2 = -250$ mm. Find the spherical aberration $f_2(h)$ as a function of the ray height h for both lenses as well as the corresponding ray diagrams.
Solution: According to the previous analysis, the spherical aberration of a negative meniscus is shown in Fig. 2.122. In this figure, both the spherical aberration of a lens with $R_1 = 250$ and $R_2 = 50$ mm, and the corresponding one of a lens with $R_1 = -50$ and $R_2 = -250$ mm, are shown. Both lenses have a thickness $d = 10$ mm and $n_L = 1.5$, with $n_1 = n_2 = 1$. Both lenses have a paraxial thin lens focal length of $f_2 = -125$ mm/. From the figure is evident that the lens with $R_1 = -50$ and $R_2 = -250$ mm suffers less spherical aberration. Also in the case of the lens with $R_1 = 250$ and $R_2 = 50$ mm, the maximum permissible height $h_{1,max} = 36.16$ mm in order to avoid total internal reflection at the right surface of the lens. Ray diagrams of these two lenses are shown in Figs. 2.123.

2.9.3.8 Lens Shape Factor

Observing the common thin lens equation [see Eq. (2.47)], it is evident that the same focal length, f, can be achieved with various combinations of the radii of curvature R_1 and R_2 of the left and right spherical surfaces. Even if the various choices predict the same focal length under the paraxial approximation, they may have significant differences in terms of the spherical aberration that they can exhibit. A measure of the shape of the lens can be defined via the *Coddington shape factor*, σ [20, 62]. The Coddington shape factor, σ, is used to describe the shape of a lens, and it plays a crucial role in determining the spherical aberration. The shape factor σ is defined as:

$$\sigma = \frac{R_2 + R_1}{R_2 - R_1}, \qquad (2.188)$$

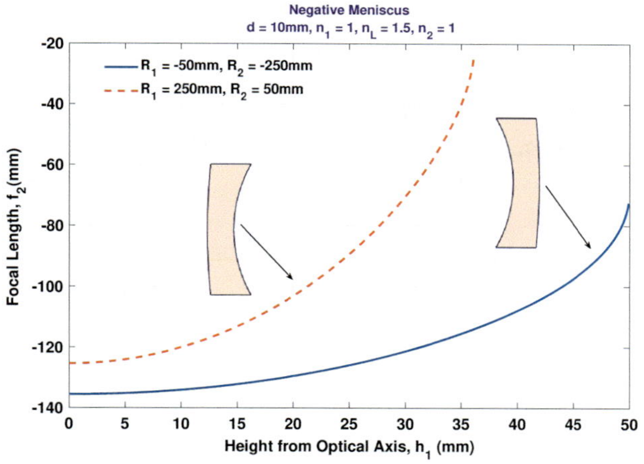

FIGURE 2.122: The spherical aberration of a negative meniscus lens of refractive index $n_L = 1.5$, left radius $R_1 = -50$ or $+250\,\text{mm}$, right radius $R_2 = -250$ or $+50\,\text{mm}$, and thickness $d = 10\,\text{mm}$ between two homogeneous media of refractive indices $n_1 = n_2 = 1$. The focal distance f_2 as a function of the height h_1 of the incident ray is shown for both cases. Obviously the lens with $R_1 = -50\,\text{mm}$ and $R_2 = -250\,\text{mm}$ suffers less spherical aberration.

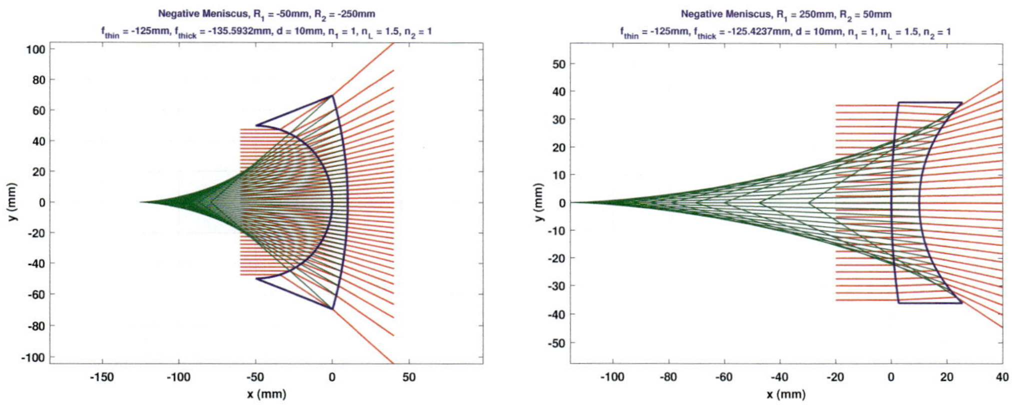

(a) (b)

FIGURE 2.123: (a) A ray diagram showing the spherical aberration of a negative meniscus of $R_1 = -50$ and $R_2 = -250\,\text{mm}$, $d = 10\,\text{mm}$, and $n_L = 1.5$, with $n_1 = n_2 = 1$. (b) A similar ray diagram for the lens with $R_1 = 250$ and $R_2 = 50\,\text{mm}$, and all other parameters the same.

where in the above equation the sign convention is used for the representation of R_1 and R_2. From Eq. (2.188), and from the thin lens equation, the following relations can be found:

$$R_2 = R_1\,\frac{\sigma+1}{\sigma-1} \Longrightarrow$$

$$R_1 = f\left(\frac{n_L - n}{n}\right)\left(\frac{2}{\sigma+1}\right), \tag{2.189}$$

$$R_2 = f\left(\frac{n_L - n}{n}\right)\left(\frac{2}{\sigma-1}\right). \tag{2.190}$$

For example, a thin lens of $f = 50\,\text{mm}$ and $n_L = 1.50$ (glass) surrounded by air ($n_1 = n_2 = n = 1.0$) can be realized by $R_1 = 50\,\text{mm}$, $R_2 = -50\,\text{mm}$ (with $\sigma = 0$, biconvex lens), and with $R_1 = 50/3 = 16.67\,\text{mm}$, $R_2 = 50\,\text{mm}$ (with $\sigma = 2.0$, positive meniscus lens). A series of $f = 50\,\text{mm}$ thin lenses of various shape factors can be seen in Fig. 2.124. In the same figure are also shown the shapes of negative lenses of $f = -50\,\text{mm}$ for various σ with the same refractive indices and thickness as the positive ones. In Fig. 2.124, the horizontal and vertical scale differences between the lenses are adjusted so that all lenses appear to have the same height, allowing for easy visual comparison of their shapes. In Fig. 2.125, the same lenses are shown with consistent scaling along both axes, providing a true geometric comparison of the lenses.

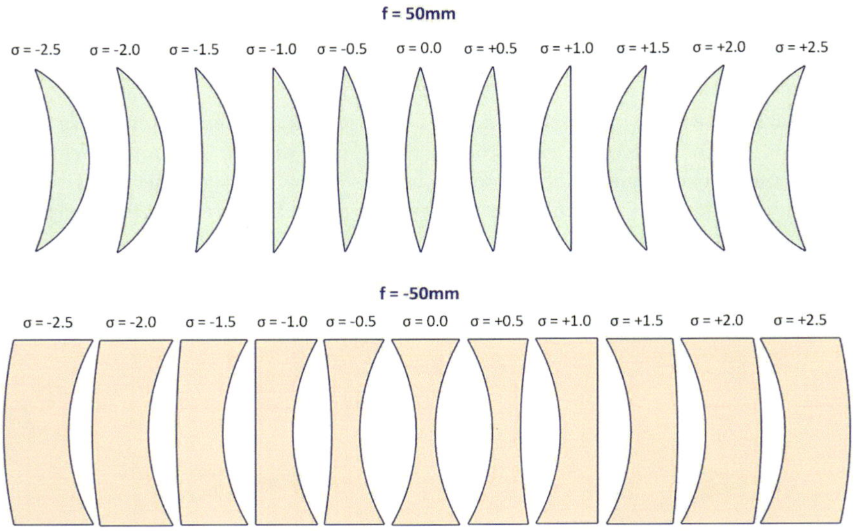

FIGURE 2.124: The shape of various lenses based on the shape factor σ. The upper lenses are positive with focal distance $f = +50\,\text{mm}$ and shape factors varying from $\sigma = -2.5$ to $+2.5$ in steps of 0.5. The lower lenses are negative with focal distance $f = -50\,\text{mm}$ and shape factors varying from $\sigma = -2.5$ to $+2.5$ in steps of 0.5. The refractive index of the lenses is $n_L = 1.5$ and of the surrounding medium is air ($n_1 = n_2 = 1.0$) while all their thicknesses are set to $d = 5\,\text{mm}$. For the negative lenses the scale of each lens differ and this is the reason that even if all have the same thickness in the middle, $d = 5\,\text{mm}$, they appear thicker as $|\sigma|$ increases. This was done to avoid showing negative lenses that look excessively thick at their edges. The positive lenses have also slightly different scales too. Therefore, in this figure only the shape of each lens is important, and no direct comparison of radii of different lenses can be done. The scales have been adjusted in order for all lenses to appear of the same height.

In the following analysis, the spherical aberration of a family of positive lenses is studied, with the goal of determining the optimum shape factor. The key parameters for these lenses are a paraxial focal length of $f = 100\,\text{mm}$, thickness $d = 20\,\text{mm}$, and refractive index $n_L = 1.517$, while the surrounding medium is air, i.e., $n_1 = n_2 = 1$. This example was presented in Jenkins and White [62] and is analyzed here too. The incident ray is parallel to the optical axis and is incident at a height of $h = 10\,\text{mm}$. Using Eqs. (2.189) and (2.190), the radii R_1 and R_2 for each lens of a given shape factor σ can be determined. The shape factor σ is varying between -2 and $+2$. Using Eq. (2.173) for biconvex lenses and Eq. (2.176) for positive meniscus lenses, the distance $f_2(h, \sigma)$ is determined (see Fig. 2.110(a) for biconvex lens and Fig. 2.113 for positive meniscus lenses). The distance $f_2(h, \sigma)$ is the point that the exiting from each lens ray, crosses the optical axis and is measured from

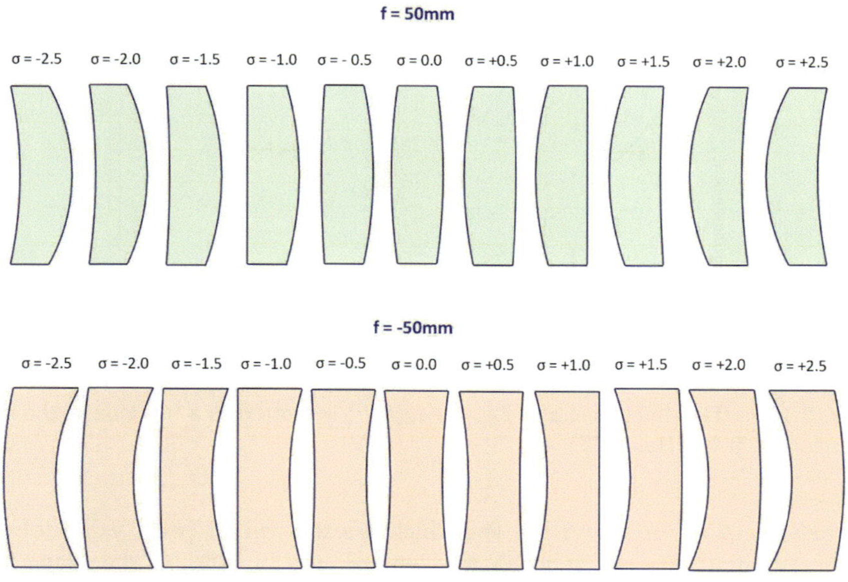

FIGURE 2.125: The same as Fig. 2.124 but with all the lenses shown on the same horizontal and vertical scale and with the same thickness and diameter (height).

the right surface of each lens. This distance is compared to the corresponding paraxial distance $q = f_{2,par} = -A/C$ [see Sec. 2.7.6, Fig. 2.47, and Eq. (2.68)]. For the thick lens $f_{2,par} = -(1/C)\,[1 + (d/R_1)(n_1 - n_L)/n_L]$. Then the difference $\Delta = f_{2,par} - f_2(h, \sigma)$ is calculated, and the results are plotted in Fig. 2.126. From this figure, the optimum shape factor can be estimated to be $\sigma_{opt} \simeq 0.655$. This result is consistent with the findings in Jenkins and White's work [62], where the optimum shape factor was approximated to be 0.7. Using the 3rd-order aberration approximation for thin lens listed in the Pedrotti textbook [20], the optimum shape factor can be approximated by:

$$\sigma_{opt} = -2\,\frac{n_L^2 - 1}{n_L + 2}\,\lim_{s\to\infty}\left\{\frac{s' - s}{s' + s}\right\} = 2\,\frac{n_L^2 - 1}{n_L + 2}. \tag{2.191}$$

Applying Eq. (2.191) for $n_L = 1.517$, the predicted optimum shape factor is $\sigma_{opt} = 0.74$. This value overestimates the exact optimum shape factor of 0.655, which was derived from the more precise analysis. The overestimation occurs because the 3rd-order approximation is only valid for thin lenses, where the lens thickness d is small compared to the radii of curvature R_1 and R_2. In thicker lenses, such as those considered in the analysis, higher-order aberrations become more significant, and the 3rd-order approximation no longer accurately captures the behavior of spherical aberration. Thus, while the approximation provides a quick estimate, it tends to overestimate the optimum shape factor for lenses with greater thicknesses.

In order to test the optimum shape for different lenses, another family of lenses was examined, all with a focal length of $f = 50$ mm, refractive index $n_L = 1.50$, and surrounded by air ($n_1 = n_2 = 1$). In this case, the lens thickness d and the height h of the incident ray were varied. Again the difference $\Delta = f_{2,par} - f_2(h, \sigma)$ was plotted as a function of the shape factor σ or various heights ($h = 2.5$ mm, 3.75 mm, and 5 mm) and lens thickness $d = 5$ mm as shown in Fig. 2.127(a). From the plot, it was found that the optimum shape factor σ_{opt} for all heights is approximately $\sigma_{opt} \simeq 0.67$. This suggests that the optimal lens shape is consistent across different incident ray heights. Furthermore, when the height of the incident

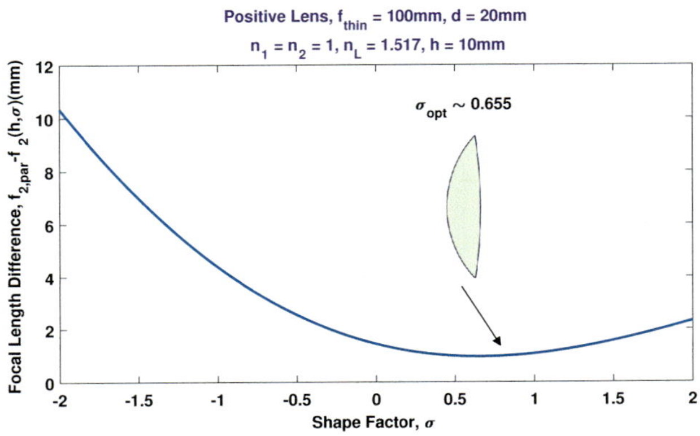

FIGURE 2.126: The difference $\Delta = f_{2,par} - f_2(h,\sigma)$ is plotted as a function of the shape factor σ. Similar to Fig. 9G of Ref. [62].

ray was fixed at $h = 5\,\text{mm}$ and the lens thickness was varied (with values of $d = 5\,\text{mm}$, $7.5\,\text{mm}$, and $10\,\text{mm}$), the difference Δ was plotted as a function of the shape factor. The results, shown in Fig. 2.127(b), reveal that the lens thickness does not significantly affect the optimum shape factor. This further indicates that the optimum shape factor is primarily determined by the refractive index and geometry of the lens, rather than its thickness.

To examine the effect of the shape factor on the spherical aberration of negative lenses, the analysis was conducted using the same parameters as in the case of positive lenses shown in Fig. 2.126 (from Ref. [62]), with the key difference being that the paraxial focal length for the negative lens was assumed to be $f_{2,par} = -100\,\text{mm}$. Equations (2.189) and (2.190) are used again for the determination of the negative lens parameters with the shape factor σ varying between -2 and $+2$. The height of the incident ray was fixed at 10 mm, and all lenses had the same thickness of $d = 20\,\text{mm}$. The refractive index of all lenses is $n_L = 1.517$, and the surrounding medium is air. The difference $\Delta = f_{2,par} - f_2(h,\sigma)$ was plotted as a function of the shape factor, and the results are shown in Fig. 2.128. From this plot, it was observed that the optimum shape factor for the negative lens is approximately $\sigma_{opt} \simeq 0.82$. Additionally, the shape of the lens corresponding to the optimum shape factor was shown as an inset in the figure, providing a visual representation of the lens shape that minimizes spherical aberration for the given parameters.

To further test the optimum shape factor for negative lenses, a family of negative lenses was examined, similar to the approach used for positive lenses. The lenses in this family had a focal distance of $f = -50\,\text{mm}$, with a refractive index of $n_L = 1.5$, and surrounded by air ($n_1 = n_2 = 1$). Both the thickness, d, of the lenses and the height, h, of the incident ray were allowed to vary. The difference $\Delta = f_{2,par} - f_2(h,\sigma)$ was plotted as a function of the shape factor σ, with the results shown in Fig. 2.129(a) for $d = 5\,\text{mm}$ and for $h = 2.5, 3.75$, and 5 mm. From the plot, it was observed that the optimum shape factor for the negative lens is approximately $\sigma \simeq 0.755$, and it was found that this value is not significantly affected by the height of the incident ray. Next, the height of the incident ray was fixed at $h = 5\,\text{mm}$ and the lens thickness d was varied with $d = 5, 7.5$, and 10 mm. The difference Δ was again plotted as a function of the shape factor, w with the results shown in Fig. 2.129(b). From this analysis, it was concluded that the optimum shape factor is slightly influenced by the lens thickness. Specifically, for $d = 5\,\text{mm}$ $\sigma_{opt} \simeq 0.755$, for $d = 7.5\,\text{mm}$ $\sigma_{opt} \simeq 0.775$, and for $d = 10\,\text{mm}$ $\sigma_{opt} \simeq 0.795$. This indicates a trend where the optimum shape factor increases slightly with increasing lens thickness for negative lenses.

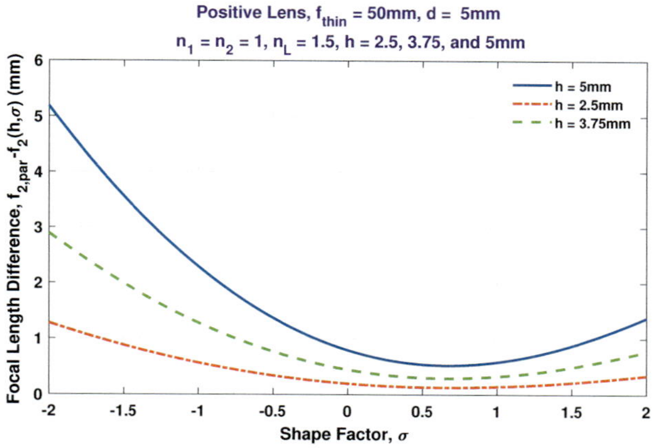

(a)

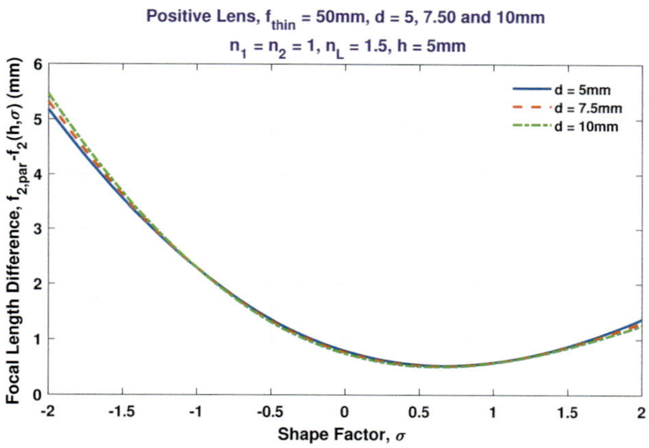

(b)

FIGURE 2.127: (a) The difference $\Delta = f_{2,par} - f_2(h, \sigma)$ is plotted as a function of the shape factor σ for $d = 5\,\text{mm}$ and three values of height $h = 2.5, 3.75$, and $5\,\text{mm}$. (b) The difference $\Delta = f_{2,par} - f_2(h, \sigma)$ is plotted as a function of the shape factor σ for $d = 5, 7.5$, and $10\,\text{mm}$ and $h = 5\,\text{mm}$.

2.9.4 Chromatic Aberrations

In the previous sections, the analysis of aberrations was conducted under monochromatic light assumptions, meaning the aberrations discussed apply to light of a single wavelength. However, many optical systems, such as cameras, microscopes, and telescopes, operate with white light, which contains a broad range of wavelengths. Additionally, materials used to fabricate optical systems, such as optical glasses, exhibit material dispersion. This means that the refractive index of the material varies with the wavelength of light. This variation in the refractive index as a function of wavelength can lead to chromatic aberrations even in systems that are paraxial or Gaussian in nature. Chromatic aberrations can be categorized into two main types: (1) Chromatic aberration dependence of paraxial imaging systems, which refers to the way that the refractive power of the optical system surfaces changes with wavelength; and (2) Wavelength dependence of all the discussed monochromatic aberrations (such as spherical aberration, coma, astigmatism, field curvature, and distortion). The amount of these aberrations changes as the wavelength of light varies due to the wavelength-dependent refractive index of the lens material.

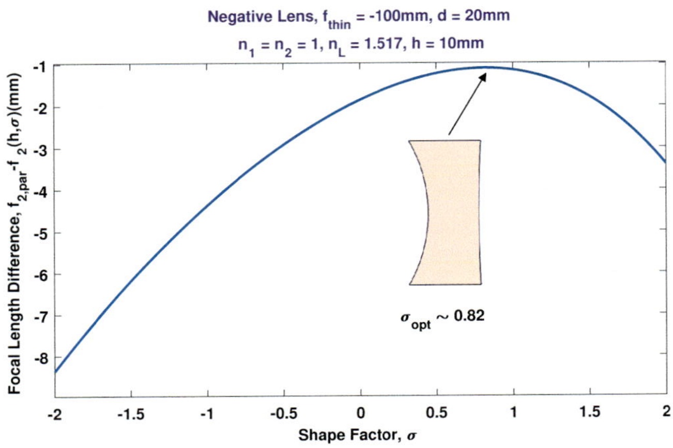

FIGURE 2.128: The difference $\Delta = f_{2,par} - f_2(h, \sigma)$ is plotted as a function of the shape factor σ. Similar to Fig. 9G of Ref. [62] but for a negative lens with $f = -100\,\text{mm}$.

The analysis of chromatic aberrations in optical systems can be performed in two ways: For the first category (chromatic dependence of paraxial systems), the primary focus is on how the system's imaging properties, such as focal length, change with the wavelength. For the second category (wavelength dependence of all aberrations), the analysis can be extended by incorporating the wavelength-dependent refractive indices of the materials involved. This requires revisiting the previous monochromatic aberration analysis while considering how the refractive indices of the media change with wavelength. The study of chromatic aberrations, especially in the context of paraxial optics, is well-documented in textbooks such as those by Klein and Furtak, and Pedrotti [20, 43], where the effects of dispersion on optical systems are explored in detail.

2.9.4.1 Chromatic Aberrations of a Spherical Refractive Interface

Assume a convex spherical interface of radius of curvature R between two media of refractive indices n_1 and n_2, as it is indicated in Fig. 2.130. Without loss of generality it is assumed that only the right medium exhibits normal dispersion, i.e., $n_2 = n_2(\lambda_0)$. According to the paraxial theory for a real object at distance s to the left, the position of the image and its magnification are given by Eqs. (2.42) and (2.45), which are written below for the case that $n_2 = n_2(\lambda_0)$:

$$\frac{n_1}{s} + \frac{n_2(\lambda_0)}{s'(\lambda_0)} = \frac{n_2(\lambda_0) - n_1}{R} \Longrightarrow s'(\lambda_0) = \frac{n_2(\lambda_0) R s}{s[n_2(\lambda_0) - n_1] - n_1 R},$$

$$m(\lambda_0) = \frac{h'(\lambda_0)}{h} = -\frac{n_1}{n_2(\lambda_0)} \frac{s'(\lambda_0)}{s} = -\frac{n_1 R}{s[n_2(\lambda_0) - n_1] - n_1 R}.$$

To demonstrate the chromatic aberrations of a spherical refractive surface, let's assume the following parameters: $n_1 = 1$ (air), $n_s = n_2(\lambda_0)$ (the refractive index of the glass material at the λ_0), $R = 100\,\text{mm}$, $s = 300\,\text{mm}$, and $h = 5\,\text{mm}$. The glass material could either be: (a) low-dispersion borosilicate crown glass (BK7) or (b) high-dispersion flint glass (Schott N-SF11). To model the dispersion of these materials, we can use the Sellmeier equation, which is often used to describe the wavelength dependence of the refractive index for optical materials. The Sellmeier coefficients for both materials are listed in Table 2A.1. Given these parameters, the exact results for the chromatic aberrations (both longitudinal and transverse) can be calculated based on simple trigonometry, using the equations derived

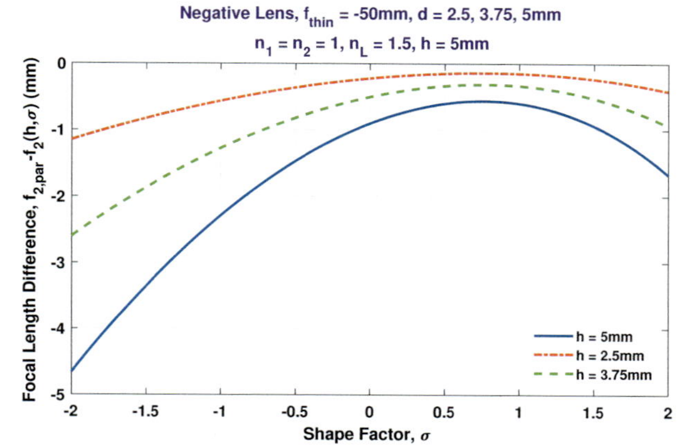

(a)

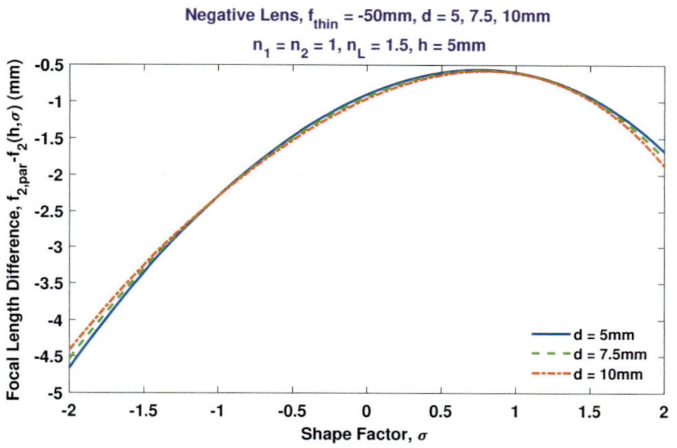

(b)

FIGURE 2.129: (a) The difference $\Delta = f_{2,par} - f_2(h, \sigma)$ is plotted as a function of the shape factor σ for $d = 5$ mm and three values of height $h = 2.5, 3.75$, and 5 mm. (b) The difference $\Delta = f_{2,par} - f_2(h, \sigma)$ is plotted as a function of the shape factor σ for $d = 5, 7.5$, and 10 mm and $h = 5$ mm.

from paraxial optics for the refractive surface (look at Problem 18). The results are summarized in Table 2.10. In this table the longitudinal, LCA, and the transverse, TCA, chromatic aberrations are listed. LCA describes the difference in the image position for different wavelengths. For a spherical refractive surface, the change in focal length due to the dispersion of the material causes different wavelengths to converge at different points along the optical axis. This leads to LCA, which is the difference in image distance for a given object height between two wavelengths. (typically, red and blue light). TCA refers to the difference in the transverse position (image height) for rays of different wavelengths. As the focal length changes for different wavelengths, the transverse magnification also varies. Typically, for both LCA and TCA, two wavelengths at the edges of the visible spectrum are used, such as red and blue light. In the case described, the paraxial theory is used to calculate these aberrations, and the results are expected to show the worst chromatic aberrations in the paraxial approximation due to the wavelength dependence of the refractive index. It's important to note that these chromatic aberrations should be symmetric for $\pm\theta$ due to the

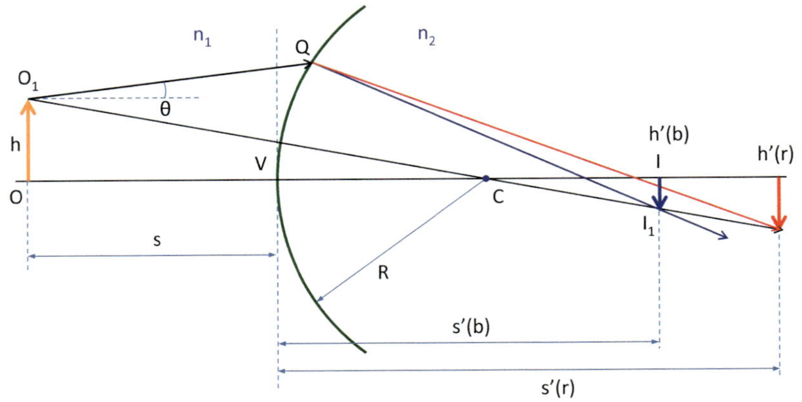

FIGURE 2.130: A spherical refractive interface between materials with refractive indices n_1 and n_2. The radius of curvature of the spherical interface is R. It is assumed that only the refractive index $n_2 = n_2(\lambda_0)$ exhibits normal dispersion. The position of the image depends on the wavelength, i.e., $s' = s'(\lambda_0)$. For normal material dispersion, the image position for the blue light, $s'(b)$ is formed closer to point V than the image of the red light $s'(r)$ as it is indicated in the figure. The height of the image $h'(\lambda_0)$ is also wavelength dependent. The image heights for the blue and the red light are denoted as $h'(b)$ and $h'(r)$, respectively. The above are true for paraxial and non-paraxial conditions.

height of the object $\pm h$. However, in Table 2.10, the height is fixed at $h = 5\,\text{mm}$ and the angle of the incident ray is changing, causing the asymmetry shown between $\pm\theta$.

In the case that the right material (of refractive index n_2) represents Crown Glass (BK7) the image position $s'(\lambda_0)$ and the magnification $m(\lambda_0)$ are shown as a function of the freespace wavelength λ_0 in Figs. 2.131(a) and 2.131(b), respectively. The paraxial results as well as the exact ones are included for $\theta = 0, \pm 5°$. Of course, the paraxial results do not depend on the angle θ. Similar results are shown in Figs. 2.132(a) and 2.132(b) in the case that the right region is comprised of flint glass (Schott N-SF11).

2.9.4.2 Chromatic Aberrations of a Thin Lens

In this section, the chromatic aberration of a thin lens is investigated, assuming the lens material is dispersive, while the surrounding medium (usually air) is homogeneous and non-dispersive. The optical power of a thin lens is given by [see Eq. (2.47)]:

$$\frac{1}{f} = \frac{n(\lambda_0) - n_0}{n_0}\left(\frac{1}{R_1} - \frac{1}{R_2}\right), \tag{2.192}$$

where n_0 is the refractive index of the surrounding medium (for air $n_0 = 1$), and $n(\lambda_0)$ is the refractive index of the dispersive medium of the lens. The R_1 and R_2 are the radii of the left and right spherical surfaces of the thin lens, respectively. The image position s' for an object at distance s from the left of the lens vertex is given by the thin lens equation and is wavelength dependent since the optical power of the thin lens is also wavelength dependent. For demonstration purposes, the image position s', the focal length f, and the magnification m, are presented as functions of the freespace wavelength for both a low-dispersion crown glass (BK7) and for a high dispersion flint glass (Schott N-SF11) in Figs. 2.133(a), 2.133(b), and 2.133(c), respectively. Of course, it is straightforward to follow the analysis presented in Sec. 2.9.3.3 to calculate the spherical aberration of a thick biconvex lens of thickness d for a

TABLE 2.10: Spherical refractive surface chromatic aberrations ($n_1 = 1$, $s = 300\,\text{mm}$, $R = 100\,\text{mm}$, $h = 5\,\text{mm}$)

Method	$LCA = \frac{s'(r) - s'(b)}{s'(b)}$ (%)	$TCA = \frac{m(r) - m(b)}{m(b)}$ (%)
Borosilicate Crown Glass (BK7)		
Paraxial	8.62	9.89
Exact $\theta = 0°$	8.59	9.87
Exact $\theta = 5°$	7.23	8.49
Exact $\theta = -5°$	7.81	9.07
Exact $\theta = 10°$	4.58	5.82
Exact $\theta = -10°$	5.41	6.65
Flint Glass (Schott N-SF11)		
Paraxial	12.14	16.80
Exact $\theta = 0°$	12.12	16.78
Exact $\theta = 5°$	10.96	15.57
Exact $\theta = -5°$	11.47	16.11
Exact $\theta = 10°$	8.06	12.56
Exact $\theta = -10°$	9.08	13.61

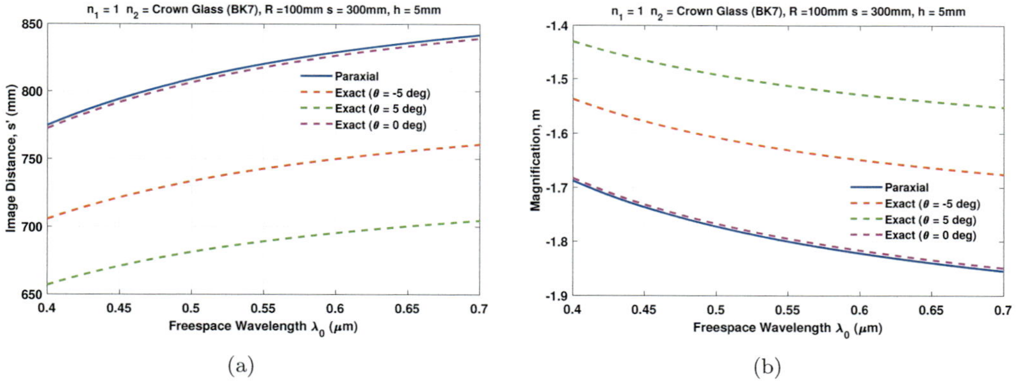

FIGURE 2.131: (a) The image position s' and the magnification m as a function of the wavelength λ_0 for low dispersion borosilicate crown glass (BK7) material for the right medium of the spherical interface of Fig. 2.130. The refractive index to the left is $n_1 = 1$, $s = 300\,\text{mm}$ and $R = 100\,\text{mm}$. The paraxial solutions are indicated by the solid lines while the exact solutions are denoted with the dashed lines. In the exact cases, the solutions depend on the angle θ (with respect to the optical axis) that the ray from point O_1 meets the spherical interface at point Q (see Fig. 2.130).

parallel to the optical axis ray incident at a height h. Using $R_1 = 50\,\text{mm}$, and $R_2 = -50\,\text{mm}$, $d = 7.5\,\text{mm}$, and $h = 5\,\text{mm}$, the exact focal length of the lens as a function of the freespace wavelength is shown in Fig. 2.134(a) for BK7 glass and in Fig. 2.134(b) for Schott N-SF11 glass.

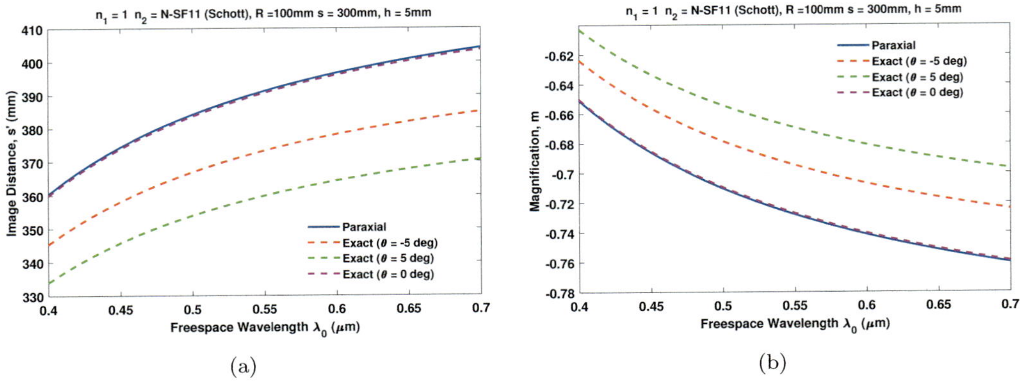

FIGURE 2.132: (a) The image position s' and the magnification m as a function of the wavelength λ_0 for high dispersion flint glass (Schott N-SF11) material for the right medium of the spherical interface of Fig. 2.130. The refractive index to the left is $n_1 = 1$, $s = 300\,\text{mm}$ and $R = 100\,\text{mm}$. The paraxial solutions are indicated by the solid lines while the exact solutions are denoted with the dashed lines. In the exact cases the solutions depend on the angle θ (with respect to the optical axis) that the ray from point O_1 meets the spherical interface at point Q (see Fig. 2.130).

2.9.4.3 Chromatic Aberrations of a Pair of Thin Lenses

Consider initially a cemented doublet composed of two thin lenses, labeled 1 and 2, respectively [20,43]. If P_1 and P_2 represent the optical powers of each lens, then the total optical power, P, of the doublet is given by:

$$P = P_1 + P_2 = [n_1(\lambda_0) - 1] \underbrace{\left(\frac{1}{R_{11}} - \frac{1}{R_{12}}\right)}_{= K_1} + [n_2(\lambda_0) - 1] \underbrace{\left(\frac{1}{R_{21}} - \frac{1}{R_{22}}\right)}_{= K_2} =$$

$$= [n_1(\lambda_0) - 1] K_1 + [n_2(\lambda_0) - 1] K_2. \tag{2.193}$$

Here, $n_1(\lambda_0)$ and $n_2(\lambda_0)$ denote the refractive indices of the materials forming lenses 1 and 2, respectively, at the freespace wavelength λ_0. The radii of curvature for the left and right surfaces of each lens are given by R_{i1} and R_{i2}, where $i = 1$ refers to lens 1 and $i = 2$ to lens 2. The constants K_1 and K_2 are wavelength-independent, as they depend solely on the radii of curvature of each thin lens. The variation in the optical power of the doublet due to changes in wavelength can be expressed as:

$$\Delta P = \Delta P_1 + \Delta P_2 = \Delta n_1(\lambda_0) K_1 + \Delta n_2(\lambda_0) K_2.$$

The differentials Δn_1 and Δn_2 can be defined in terms of the refractive indices at visible wavelengths near the edges of the spectrum. Specifically, as outlined in Sec. 2.4.1, the refractive index of each lens material at the characteristic wavelengths $\lambda_F = 486.1\,\text{nm}$, $\lambda_C = 656.3\,\text{nm}$, and $\lambda_D = 589.2\,\text{nm}$ can be used. Thus, $\Delta n_1 = n_1(\lambda_F) - n_1(\lambda_C)$ and $\Delta n_2 = n_2(\lambda_F) - n_2(\lambda_C)$. The differential ΔP can be expressed as follows:

$$\Delta P = \frac{n_1(\lambda_F) - n_1(\lambda_C)}{n_1(\lambda_D) - 1} [n_1(\lambda_D) - 1] K_1 + \frac{n_2(\lambda_F) - n_2(\lambda_C)}{n_2(\lambda_D) - 1} [n_2(\lambda_D) - 1] K_2$$

$$= \frac{P_1}{V_1} + \frac{P_2}{V_2}, \tag{2.194}$$

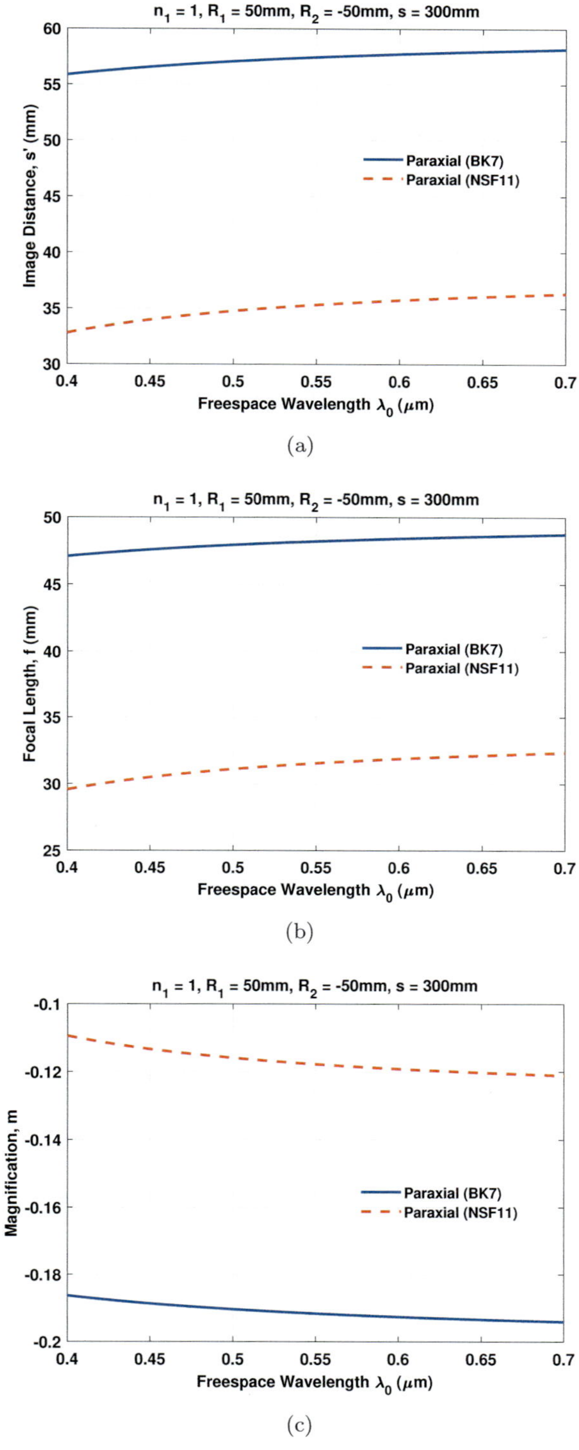

FIGURE 2.133: (a) The image position s' as a function of the wavelength λ_0 for low- and high-dispersion glass material for the thin lens medium. The surrounding medium is air ($n_0 = 1$). The radii of curvature (with the sign convention) are $R_1 = 50\,\text{mm}$ and $R_2 = -50\,\text{mm}$, while the object distance is $s = 300\,\text{mm}$. The paraxial solutions are only shown. (b) Same as (a) but for the focal length f. (c) Same as (a) but or magnification m.

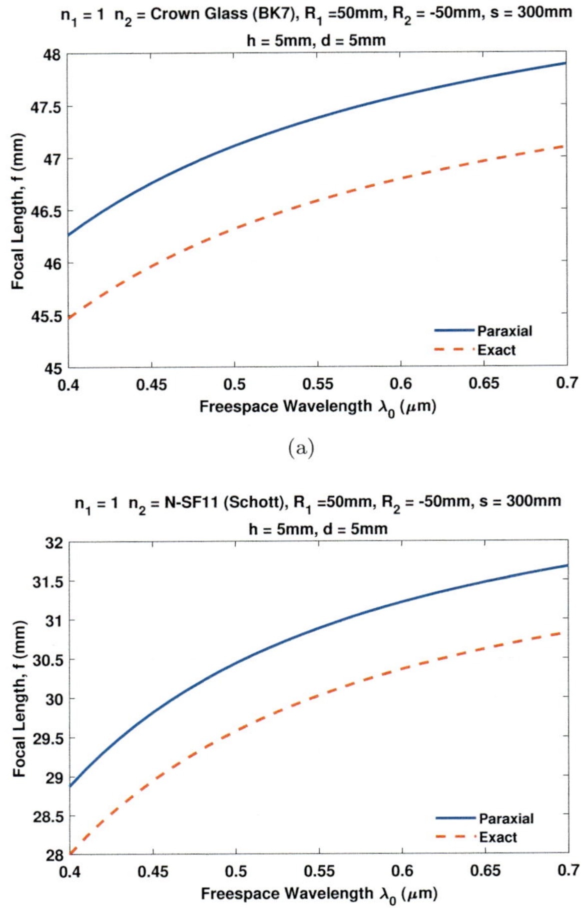

(a)

(b)

FIGURE 2.134: (a) The focal distance (measured from the right vertex of the lens) of a thick biconvex lens with $R_1 = 50\,\text{mm}$ and $R_2 = -50\,\text{mm}$, $d = 7.5\,\text{mm}$, and $h = 5\,\text{mm}$ in air with lens material low dispersion borosilicate glass (BK7). The paraxial focal length is also shown as measured from the right vertex of the lens. (b) Same as (a) but for lens material from high dispersion flint glass (Schott N-SF11).

where the V_1 and V_2 are the Abbe numbers of the lens material [see Eq. (2.15)]. For the doublet to maintain the same optical power at the two edge wavelengths, λ_F and λ_C, it is necessary that $\Delta P = 0$. This condition implies:

$$\Delta P = \frac{P_1}{V_1} + \frac{P_2}{V_2} = 0 \Longrightarrow \frac{P_2}{P_1} = -\frac{V_2}{V_1}. \tag{2.195}$$

Solving Eqs. (2.193) and (2.195) it is easily found:

$$P_1 = \frac{V_1}{V_1 - V_2} P, \tag{2.196}$$

$$P_2 = -\frac{V_2}{V_1 - V_2} P, \tag{2.197}$$

$$K_1 = \frac{P_1}{n_1(\lambda_D) - 1}, \qquad K_2 = \frac{P_2}{n_2(\lambda_D) - 1}. \tag{2.198}$$

From the last equation, the radii of curvature for the lenses can be determined, considering that $R_{12} = R_{21}$. If the first lens is biconvex, then $R_{11} = -R_{12}$. However, there is flexibility in selecting the lens shape to maintain the desired optical power. This flexibility allows for adjustments, such as minimizing spherical aberration (see Sec. 2.9.3.8).

Example 2.9.10:

Consider the design of an achromatic doublet lens with a focal length of $f = 100\,\text{mm}$, composed of low-dispersion borosilicate crown glass (BK7) and high-dispersion flint glass (Schott N-SF11). The crown glass lens is assumed to be biconvex. The radii of curvature for both lenses must be specified to ensure that the doublet behaves approximately achromatically between the λ_F and λ_C lines. Assuming an object positioned at a distance $s = 300\,\text{mm}$ from the left side, determine the image position s', the effective focal length f, and the magnification m, as functions of the freespace wavelength within the visible range.
Solution: The refractive indices of the two glass materials, BK7 for the left lens (lens No. 1), and N-SF11 for the right lens (lens No. 2) are determined using their Sellmeier coefficients as listed in Table 2A.1 of Appendix 2A. For lens No. 1, the values are $n_1(\lambda_F) = 1.5224$, $n_1(\lambda_D) = 1.5167$, $n_1(\lambda_C) = 1.5143$, and $V_1 = 64.1332$. The corresponding values for lens No. 2 are $n_2(\lambda_F) = 1.8065$, $n_1(\lambda_D) = 1.7845$, $n_1(\lambda_C) = 1.7760$, and $V_1 = 25.6611$. Using Eqs. (2.196) and (2.197), the optical powers of the two lenses are found to be $P_1 = 16.6701\,\text{D}$ and $P_2 = -6.6701\,\text{D}$ (D stands for Diopters). Naturally, $P_1 + P_2 = P = 10\,\text{D}$ with $P = 1/f = 10\,\text{D}$. Assuming the first lens is biconvex, $R_{11} = -R_{12}$ and thus $R_{11} = 2[n_1(\lambda_D) - 1]/P_1 = 0.062\,\text{m} = 62\,\text{mm}$. The second lens, cemented to the first, must have $R_{21} = R_{12} = -62\,\text{mm}$. Consequently, $1/R_{22} = 1/R_{21} - P_2/[n_2(\lambda_D) - 1] = -7.6277\,m^{-1}$, yielding $R_{22} = -131.1\,\text{mm}$. Using paraxial theory and treating the doublet as a thin lens, the image position s', effective focal length f, and magnification m can be calculated as functions of the freespace wavelength, given an object distance of $s = 300\,\text{mm}$ from the left lens of the doublet. The results are illustrated in Figs. 2.135(a), 2.135(b), and 2.135(c), where the nearly achromatic behavior of the doublet is evident.

Naturally, a similar approach can be applied to a separated doublet, as commonly described in textbooks such as Refs. [20,43,62]. Let the left thin lens be labeled as lens No. 1, with focal length f_1, and the right thin lens as lens No. 2, with focal length f_2. The separation between the two lenses is d. Using the matrix method (see Sec. 2.7), the optical power P of the two-lens combination is given by $P = P_1 + P_2 - dP_1P_2$ where $P_1 = 1/f_1$ and $P_2 = 1/f_2$. Following the same approach with the three wavelengths λ_F, λ_D, and λ_C as before, the optical powers are defined at the central wavelength λ_D. Thus, $P_1 = [n_1(\lambda_D) - 1]K_1$ and $P_2 = [n_2(\lambda_D) - 1]K_2$, where K_1 and K_2 incorporate the geometrical properties of the lenses, as indicated in Eq. (2.193). By taking the differential of P with respect to wavelength, the following equations can be derived:

$$P = P_1 + P_2 - dP_1P_2 \tag{2.199}$$

$$\Delta P = \Delta P_1 + \Delta P_2 - d(P_2\Delta P_1 + P_1\Delta P_2) = 0 \Longrightarrow$$

$$= \frac{P_1}{V_1} + \frac{P_2}{V_2} - dP_1P_2\left(\frac{1}{V_1} + \frac{1}{V_2}\right) = 0, \tag{2.200}$$

where $\Delta P_i = P_i/V_i$ ($i = 1, 2$) and V_i are the Abbe numbers of the lenses material. For a given total power of the system, P, and separation distance d, the above equations can be easily solved and their solution is given by:

$$P_2 = \frac{-b \pm \sqrt{b^2 - 4ac}}{2a} \tag{2.201}$$

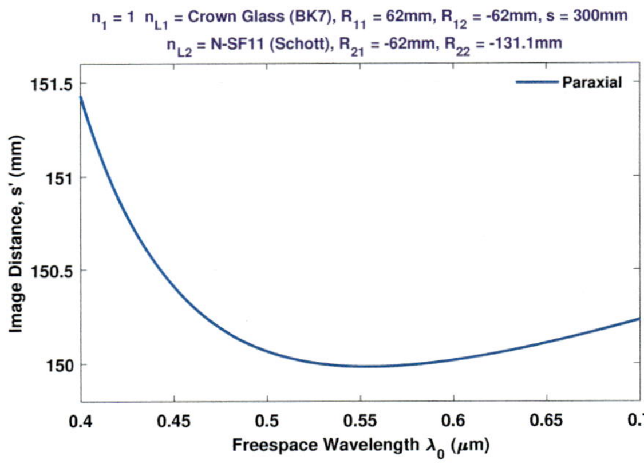

(a)

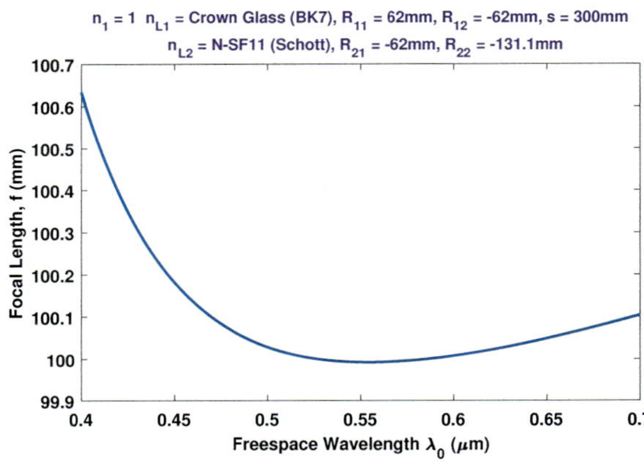

(b)

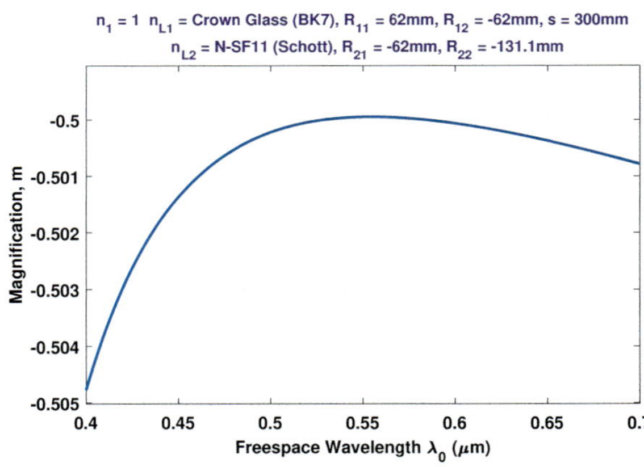

(c)

FIGURE 2.135: (a) The image position s' as a function of the wavelength λ_0 for a cemented doublet made from two lenses of glass material BK7 (left lens) and Schott N-SF11 (right lens). The surrounding medium is air ($n_0 = 1$). The radii of curvature (with the sign convention) are $R_{11} = 62\,\text{mm}$ and $R_{12} = -62\,\text{mm}$, $R_{21} = -62\,\text{mm}$, and $R_{22} = -131.1\,\text{mm}$, while the object distance is $s = 300\,\text{mm}$. The paraxial solutions are only shown assuming thin lenses. (b) Same as (a) but for the focal length f. (c) Same as (a) but or magnification m.

$$a = \frac{d}{V_1}, \qquad b = \frac{1}{V_2} - \frac{1}{V_1} - dP\left(\frac{1}{V_1} + \frac{1}{V_2}\right), \qquad c = \frac{P}{V_1}$$

$$P_1 = \frac{P - P_2}{1 - dP_2}. \tag{2.202}$$

A valid solution requires the discriminant $b^2 - 4ac > 0$. When the discriminant is positive, two real solutions are obtained: one corresponding to a real image and the other to a virtual image. If the lens parameters are chosen to make the lens combination wavelength-insensitive (from a paraxial perspective), the distance d can be selected to satisfy Eq. (2.200). The values of d and the resulting optical power P are then given by:

$$d = \frac{f_1 V_1 + f_2 V_2}{V_1 + V_2}, \tag{2.203}$$

$$P = \frac{P_1 V_1 + P_2 V_2}{V_1 + V_2}. \tag{2.204}$$

In the special case that the two lenses are made from the same material, i.e., when $V_1 = V_2 = V$, the separation distance d is given by $d = (f_1 + f_2)/2$. The combined optical power P is then $P = (f_1 + f_2)/2f_1 f_2$, the focal length of the combination is $f = 1/P = f_1 f_2/d$.

Example 2.9.11:

Consider the design of an achromatic separated doublet with an effective focal length of $f = 100\,\text{mm}$, comprised of two thin lenses made from the same low-dispersion borosilicate crown glass (BK7). The separation between the lenses is chosen to be $d = 200\,\text{mm}$. The geometric parameters K_i $(i = 1, 2)$ for both lenses must be specified in order for the doublet to behave approximately achromatically between the λ_F and λ_C lines. Assuming an object placed at a distance $s = 300\,\text{mm}$ from the left lens, the task is to determine the image position s', the effective focal length f, and the magnification m, as functions of the freespace wavelength in the visible range. Repeat the procedure in the case where the second lens is made from high-dispersion flint glass (Schott N-SF11) instead of BK7.

Solution: The refractive indices of the two glass materials are determined similarly to example 2.9.10. Using Eqs. (2.201) and (2.202), the optical powers for the two lenses are found to be: $P_2 = 0.0171\,mm^{-1}$ (or $0.0029\,mm^{-1}$) and $P_1 = 0.0029\,mm^{-1}$ (or $0.0171\,mm^{-1}$). The first solution produces a real inverted image and this is chosen as an example in Figs. 2.136(a), 2.136(b), and 2.136(c). These figures clearly demonstrate the achromatic behavior of the separated doublet. The geometric parameters of the two lenses are $K_1 = P_1/[n_1(\lambda_D) - 1] = 0.0056682\,mm^{-1}$ and $K_2 = P_2/[n_2(\lambda_D) - 1] = 0.033037\,mm^{-1}$. The radii of the two lenses can be specified once the shape factor is given (see Sec. 2.9.3.8). The image position s' was determined using the matrix method and is given by the formula: $s' = -(As + B)/(Cs + D)$ where $A = 1 - dP_1$, $B = d$, $C = -P_1 - P_2 + dP_1P_2$, and $D = 1 - dP_2$. Repeating the same procedure in case the material of the right lens is from high dispersion flint glass Schott N-SF11, the following solution is obtained: $P_2 = 0.0255\,mm^{-1}$ (or $0.0020\,mm^{-1}$) and $P_1 = 0.0038\,mm^{-1}$ (or $0.0132\,mm^{-1}$). The first solution produces a real inverted image, which is used as an example in Figs. 2.137(a), 2.137(b), and 2.137(c). As before, the achromatic behavior of the separated doublet is clearly demonstrated in these figures. The geometric parameters for this case are $K_1 = 0.0073206\,mm^{-1}$ and $K_2 = 0.032555\,mm^{-1}$.

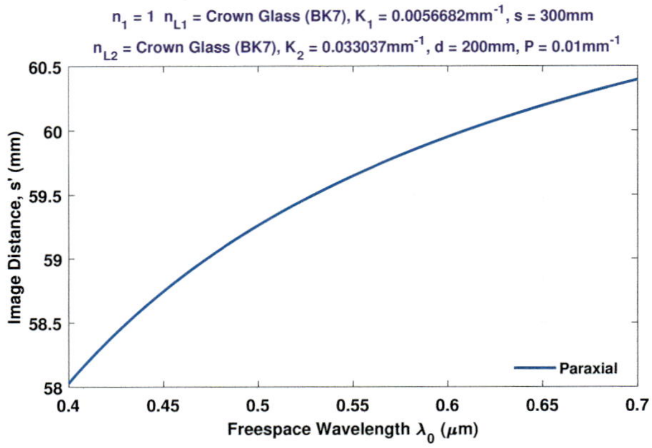

(a)

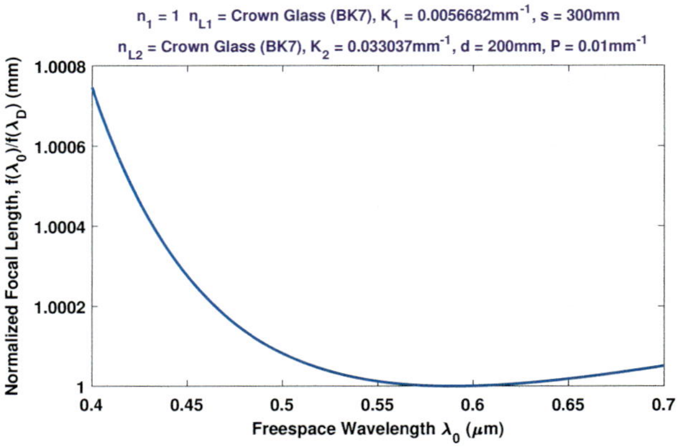

(b)

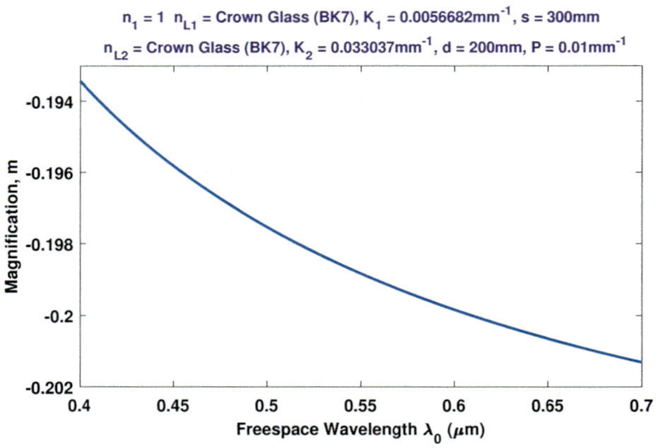

(c)

FIGURE 2.136: (a) The image position s' as a function of the wavelength λ_0 for a separated doublet made from two lenses of glass material BK7. The surrounding medium is air ($n_0 = 1$). The geometric parameters are $K_1 = 0.0056682\,mm^{-1}$ and $K_2 = 0.033037\,mm^{-1}$. The object distance is $s = 300\,\text{mm}$. The paraxial solutions are shown assuming thin lenses. (b) Same as (a) but for the normalized focal length $f/f(\lambda_D)$. (c) Same as (a) but or magnification m.

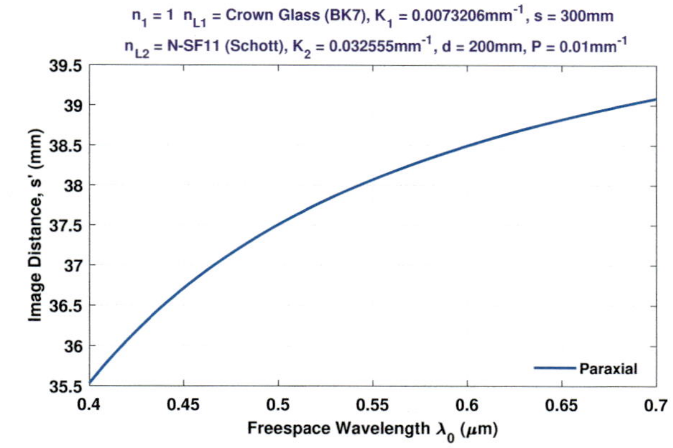

(a)

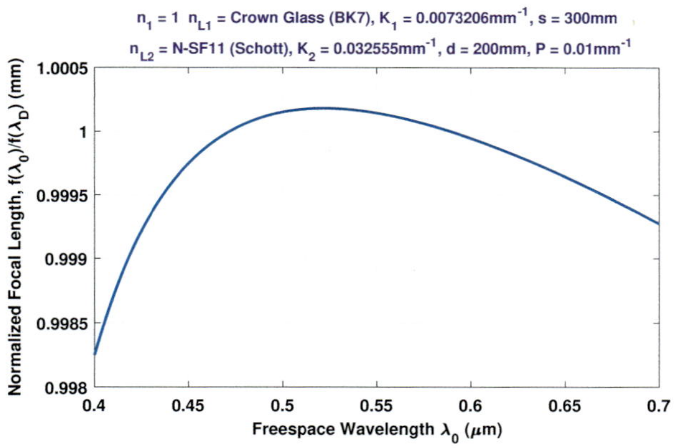

(b)

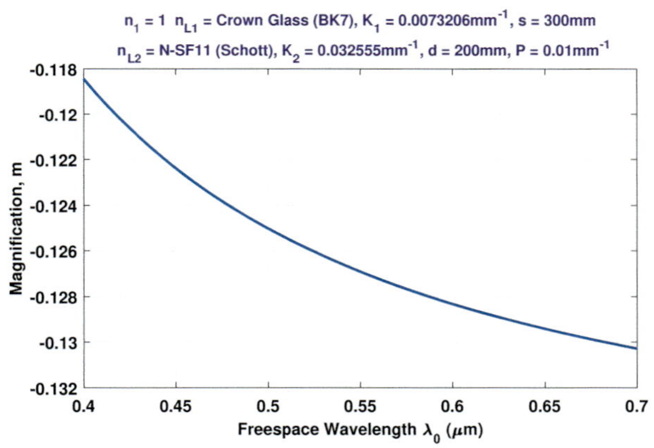

(c)

FIGURE 2.137: (a) The image position s' as a function of the wavelength λ_0 for a separated doublet made from two lenses of glass material BK7 (left lens) and Schott N-SF11 (right lens). The surrounding medium is air ($n_0 = 1$). The geometric parameters are $K_1 = 0.0073206\, mm^{-1}$ and $K_2 = 0.032555\, mm^{-1}$. The object distance is $s = 300\,\text{mm}$. The paraxial solutions are shown assuming thin lenses. (b) Same as (a) but for the normalized focal length $f/f(\lambda_D)$. (c) Same as (a) but or magnification m.

2.10 Exact Ray Tracing

In the design of practical optical systems, it is often necessary to trace rays that are not paraxial. An example of this is the study of various ray aberrations, as discussed in Sec. 2.9, where higher-order approximations, such as the third-order Seidel aberrations, may not be sufficient, and "exact" ray tracing becomes necessary. An example of exact ray paths was provided in Sec. 2.9.2.2, using Fermat's principle. However, a more systematic approach is often required, especially when dealing with systems that include multiple spherical refractive surfaces and where the rays are skewed. The term "exact" means that no approximations are made within the framework of geometrical optics theory. There are numerous references in the literature concerning exact ray tracing, also known as skew ray tracing (see for example Refs. [64–68]), as well as in various textbooks [36, 69–72]. In this section a simple approach presented by W. B. Zimmerman [73] is followed.

2.10.1 Single Spherical Refractive Interface

In this approach, a vector ray representation is utilized. The basic geometry is illustrated in Figs. 2.138(a) and 2.138(b). In these figures, it is assumed that the ray originates from point $P(x, y, z)$, where its coordinates are referenced to the coordinate system xyz shown. A spherical refractive surface of radius R has its vertex V positioned at a distance $(OV) = t$ from the origin. The ray intersects the spherical surface at point $Q(x_1, y_1, z_1)$. The coordinates (x_1, y_1, z_1) and the length $(PQ) = T$ are to be determined, along with the angle of incidence θ_1 (see Fig. 2.138(b)). The refractive indices before and after the spherical interface are n_1 and n_2, respectively. Additionally, the direction of the ray $\overrightarrow{PQ}$ is known and can be described via the unit vector $\hat{u}_1 = u_{1x}\hat{\imath}_x + u_{1y}\hat{\imath}_y + u_{1z}\hat{\imath}_z$. The following vector relation can then be written:

$$\overrightarrow{OQ} + \overrightarrow{QC} = \overrightarrow{OC} \Longrightarrow \overrightarrow{QC} = \overrightarrow{OC} - \overrightarrow{OQ} \Longrightarrow$$

$$\overrightarrow{QC} = -x_1\hat{\imath}_x - y_1\hat{\imath}_y + (R - z_1)\hat{\imath}_z \tag{2.205}$$

$$\overrightarrow{OP} + \overrightarrow{PC} + \overrightarrow{QC} = \overrightarrow{OC} \Longrightarrow$$

$$x_1 = x + Tu_{1x}, \tag{2.206}$$

$$y_1 = y + Tu_{1y}, \tag{2.207}$$

$$z_1 = z + Tu_{1z} - t. \tag{2.208}$$

It is important to note that $\overrightarrow{OC} = (t + R)\hat{\imath}_z$, and that the coordinate z_1 is measured from vertex V. To evaluate the coordinates (x_1, y_1, z_1), it is necessary to first find T. Let $\overrightarrow{VM}$ be the vector perpendicular to the ray $\overrightarrow{PQ}$ from point V, with the distance $(PM) = e$. The following vector equation can now be written:

$$\overrightarrow{OP} + e\hat{u}_1 = t\hat{\imath}_z + \overrightarrow{VM} \Longrightarrow e\hat{u}_1 = t\hat{\imath}_z + \overrightarrow{VM} - \overrightarrow{OP} \Longrightarrow$$

$$e\,(\hat{u}_1 \cdot \hat{u}_1) = t\,(\hat{\imath}_z \cdot \hat{u}_1) + \overrightarrow{VM} \cdot \hat{u}_1 - \overrightarrow{OP} \cdot \hat{u}_1 \Longrightarrow$$

$$e = tu_{1z} - (xu_{1x} + yu_{1y} + zu_{1z})\,. \tag{2.209}$$

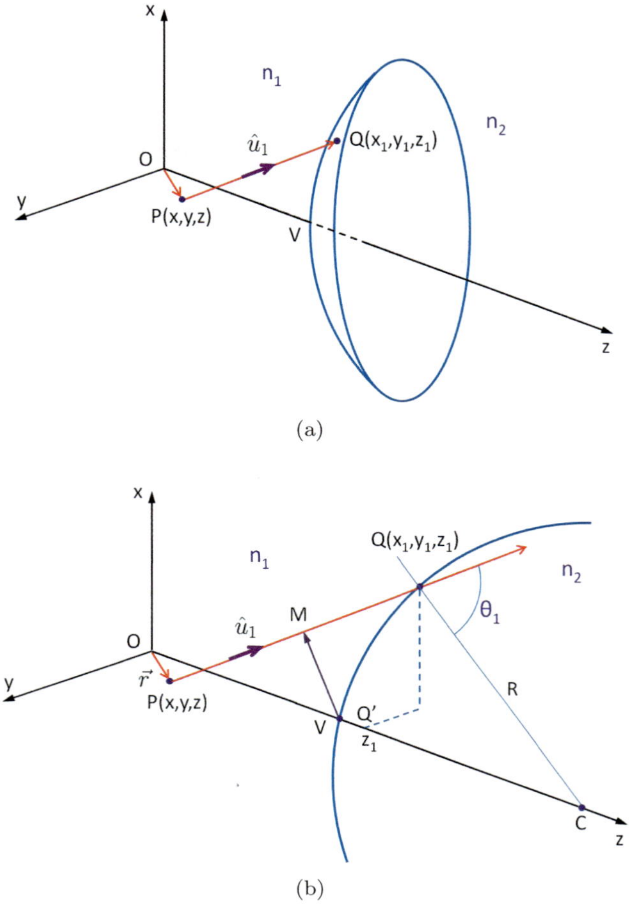

FIGURE 2.138: (a) Diagram for the exact ray tracing of a ray emanating from point $P(x, y, z)$ and incident at the spherical refractive surface of radius R between two media of refractive indices n_1 and n_2 at point $Q(x_1, y_1, z_1)$. The direction of the incident ray is specified by the unit vector $\hat{u}_1$. (b) The same diagram but simplified in order to define various vectors and distances. The vertex V of the spherical surface is located at distance t from the origin, i.e., $(OV) = t$. The vector $\overrightarrow{VM}$ is perpendicular to the incident ray $\overrightarrow{PQ}$. The center of the spherical surface is at C and the angle of incidence on the spherical surface of the incident ray is θ_1. The vector $\overrightarrow{OP} = \vec{r} = x\hat{\imath}_x + y\hat{\imath}_y + z\hat{\imath}_z$ specifies point P.

Of course, the orthogonality condition $\overrightarrow{VM} \cdot \hat{u}_1 = 0$ has been applied. Similarly, from the previous equation, the z-component of vector $\overrightarrow{VM}$, M_z ($M_z = \hat{\imath}_z \cdot \overrightarrow{VM}$) can be determined:

$$\overrightarrow{VM} = \overrightarrow{OP} + e\hat{u}_1 - t\hat{\imath}_z \Longrightarrow$$

$$M_z = \overrightarrow{VM} \cdot \hat{\imath}_z = z + eu_{1z} - t. \tag{2.210}$$

From the last equation, the following condition can be derived:

$$\overrightarrow{VM} - e\hat{u}_1 = \overrightarrow{OP} - t\hat{\imath}_z \Longrightarrow$$

$$-e = \left(\overrightarrow{OP} - t\hat{\imath}_z\right) \cdot \hat{u}_1, \quad \text{and} \tag{2.211}$$

$$(VM)^2 = \overrightarrow{VM} \cdot \overrightarrow{VM} = \left(\overrightarrow{OP} - t\hat{\imath}_z\right) \cdot \left(\overrightarrow{OP} - t\hat{\imath}_z\right) + 2\left(\overrightarrow{OP} - t\hat{\imath}_z\right) \cdot e\hat{u}_1 + e^2$$

$$(VM)^2 = x^2 + y^2 + z^2 + t^2 - e^2 - 2te, \tag{2.212}$$

where in the last expression Eq. (2.211) was used. At this point the angle θ_1 which is needed for applying the Snell's law is introduced in the calculation procedure. The following vector equation can be written:

$$\overrightarrow{VM} + (T-e)\hat{u}_1 + \overrightarrow{QC} = \overrightarrow{VC} \Longrightarrow (T-e)\hat{u}_1 \;=\; \overrightarrow{VC} - \overrightarrow{VM} - \overrightarrow{QC} \Longrightarrow$$

$$(T-e)\left(\hat{u}_1 \cdot \hat{u}_1\right) = R\left(\hat{\imath}_z \cdot \hat{u}_1\right) - \overrightarrow{QC} \cdot \hat{u}_1 \;=\; Ru_{1z} - R\cos\theta_1 \Longrightarrow$$

$$T = e + \frac{u_{1z} - \cos\theta_1}{C}, \tag{2.213}$$

where $C = 1/R$ is the curvature of the spherical surface. Manipulating the last equation the $\cos\theta_1$ can be determined as follows:

$$\overrightarrow{VM} - \overrightarrow{VC} = -\left[(T-e)\hat{u}_1 + \overrightarrow{QC}\right] \Longrightarrow$$

$$\left(\overrightarrow{VM} - R\hat{\imath}_z\right) \cdot \left(\overrightarrow{VM} - R\hat{\imath}_z\right) = \left[(T-e)\hat{u}_1 + \overrightarrow{QC}\right] \cdot \left[(T-e)\hat{u}_1 + \overrightarrow{QC}\right] \Longrightarrow$$

$$VM^2 - 2M_zR = (T-e)^2 + 2(T-e)R\cos\theta_1 \;=$$

$$= \left(\frac{u_{1z} - \cos\theta_1}{C}\right)^2 + 2\left(\frac{u_{1z} - \cos\theta_1}{C}\right)\frac{1}{C}\cos\theta_1 \Longrightarrow$$

$$\cos\theta_1 = \left\{u_{1z}^2 - C\left[C(VM)^2 - 2M_z\right]\right\}^{1/2}. \tag{2.214}$$

In Eq. (2.214) the positive sign was chosen since the angle $\theta_1 < \pi/2$. Since for flat surfaces $R = \infty$ and $C = 0$, to avoid the division by zero in Eq. (2.213), it can be re-written as follows [73]:

$$T = e + \frac{u_{1z} - \cos\theta_1}{C} = e + \frac{u_{1z} - \cos\theta_1}{C}\frac{u_{1z} + \cos\theta_1}{u_{1z} + \cos\theta_1} =$$

$$= e + \frac{u_{1z}^2 - \cos^2\theta_1}{C(u_{1z} + \cos\theta_1)} \Longrightarrow$$

$$T = e + \frac{C(VM)^2 - 2M_z}{u_{1z} + \cos\theta_1}, \tag{2.215}$$

where Eq. (2.214) was used. The last equation does not diverges when $C = 0$. Since T has been determined, the coordinates (x_1, y_1, z_1) of point Q can be found from Eqs. (2.206)-(2.208). At this point it is helpful to examine the equation for T in terms of the sign convention used for the spherical interface (for convex surfaces $R > 0$ and for concave surfaces $R < 0$). Consider Fig. 2.139 which includes either the convex case for point Q, or the concave case for point Q'. Of course, only one of these should exist for the spherical interface. Revisiting Eq. (2.213) it can be written in the form:

$$T = e + Ru_{1z} - \begin{bmatrix} +R\cos\theta_1 \\ -R\cos\theta_1' \end{bmatrix},$$

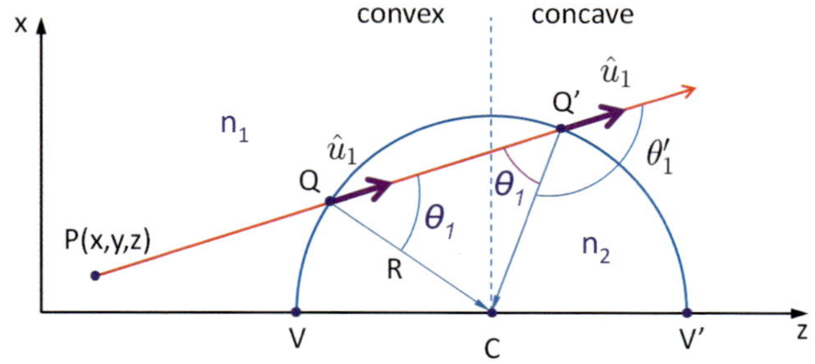

FIGURE 2.139: A side-view diagram of the spherical interface in the case of convex (point Q) or concave (point Q') surface in order to examine the sign convention [73].

where the upper part of the equation holds for the convex surface and the lower part for the concave one. Since $\cos\theta_1' = \cos(\pi - \theta_1) = -\cos\theta_1$ both cases can be treated using equation:

$$T = e + Ru_{1z} - R\cos\theta_1,$$

where the sign convention is included in R, and θ_1 is always the angle of incidence at the spherical interface. Now, the only remaining step is to apply Snell's law at the spherical interface and determine the direction of the ray after its refraction. To do this, consider Fig. 2.140, where the unit vector $\hat{u}_2 = u_{2x}\hat{\imath}_x + u_{2y}\hat{\imath}_y + u_{2z}\hat{\imath}_z$ represents the direction of the refracted ray at point Q, which needs to be determined. At this stage, the vector reflection and refraction equations presented in Appendix 2B should be used. Specifically, the unit

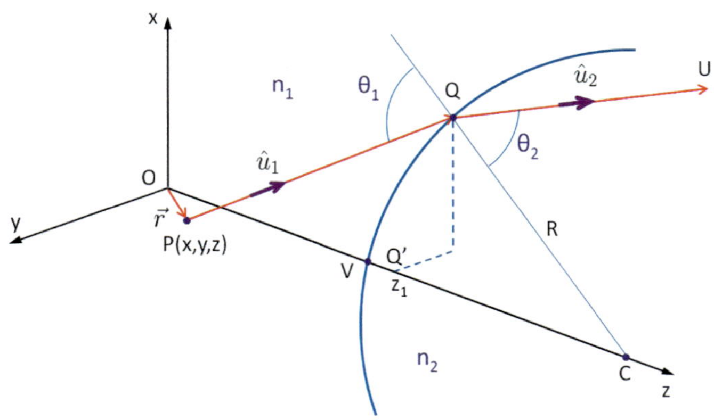

FIGURE 2.140: A diagram of the spherical interface refraction of the incident ray PQ.

vector $\hat{u}_2 = \hat{t}$, as it is presented in Eq. (2B.2), is to be found. To align with the variables used in this section, it is important to note the following equivalence: $\hat{n} = -\overrightarrow{QC}/R = -C\overrightarrow{QC}$. Additionally, from the same Appendix, we have $\cos\theta_2 = [1 - q^2(1 - \cos^2\theta_1)]^{1/2}$ and $\cos^2\theta_1 = C^2(\overrightarrow{QC}\cdot\hat{u}_1)^2$, where $q = n_1/n_2$. Applying Eq. (2B.2) results in the following:

$$\hat{u}_2 = q\hat{u}_1 + \underbrace{[-q\cos\theta_1 + \cos\theta_2]}_{=G} C\overrightarrow{QC} \Longrightarrow$$

$$u_{2x} = qu_{1x} - CGx_1, \tag{2.216}$$

$$u_{2y} = qu_{1y} - CGy_1, \tag{2.217}$$

$$u_{2z} = qu_{1z} - CGz_1 + G, \qquad \text{where} \tag{2.218}$$

$$G = \cos\theta_2 - q\cos\theta_1. \tag{2.219}$$

It is reminded that $C = 1/R$ (with the sign convention) and $\overrightarrow{QC}$ is given in Eq. (2.205). At this point the procedure can be generalized for any arbitrary number of spherical refractive interfaces.

2.10.2 Multiple Spherical Refractive Interfaces

The procedure described in Sec. 2.10.1 can be extended to handle multiple sequential spherical refractive interfaces. In this case, the output point Q and the refracted ray direction given by $\hat{u}_2$ replace the initial point P and direction $\hat{u}_1$ for the next refractive spherical surface. The situation is depicted in Fig. 2.141, where N spherical interfaces are involved. Each refractive surface has a radius of curvature R_i (with sign convention, and $i = 1, 2, \cdots, N$). The distances between vertices V_i of each spherical surface are denoted by t_i, and the refractive indices for each surface are labeled n_i. The initial ray is emanating from point $P(x, y, z)$ and has a direction specified by the unit vector $\hat{u}_1$. The following vectors, $\tilde{C}$, $\tilde{t}$, $\tilde{n}$, can be defined as they contain the necessary information for the exact ray tracing of initial ray PQ_1:

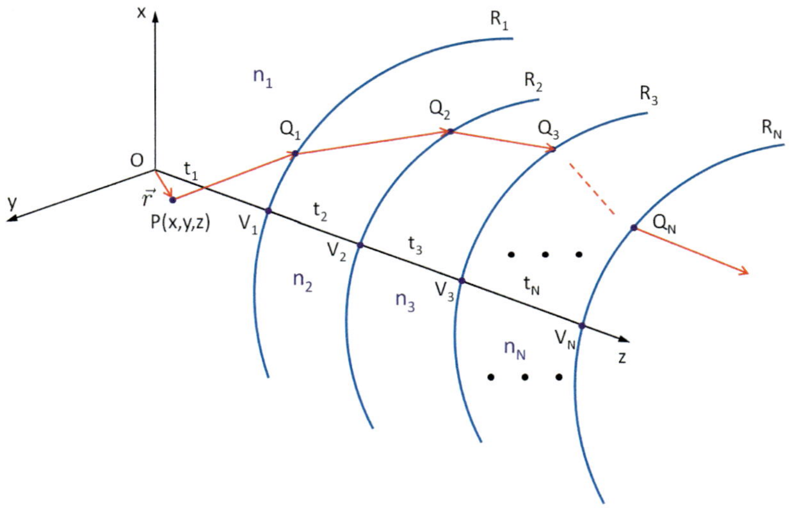

FIGURE 2.141: A diagram of multiple sequential spherical interfaces that can be used to trace the incident ray PQ_1. Each refractive surface has a radius of R_i $(i = 1, 2, \cdots, N)$ and a vertex V_i. The distances between vertexes are t_i. The final point is Q_N.

$$\tilde{C} = [C_1, C_2, \cdots, C_N], \tag{2.220}$$

$$\tilde{t} = [t_1, t_2, \cdots, t_N], \tag{2.221}$$

$$\tilde{n} = [n_1, n_2, \cdots, n_N], \tag{2.222}$$

where $C_i = 1/R_i$ and $i = 1, 2, \cdots, N$. The summarized flow chart of the calculation procedure can be presented in the following steps:

(a) Initial conditions: $P(x, y, z)$ and $\hat{u}_1 = u_{1x}\hat{\imath}_x + u_{1y}\hat{\imath}_y + u_{1z}\hat{\imath}_z$.

(b) Form vector $\tilde{C}$, $\tilde{t}$, $\tilde{n}$, for $i = 1, 2, \cdots, N$, spherical interfaces using Eqs. (2.220)–(2.222).

(c) Start the iterative procedure from the first interface (let $m = 1$ represent the first spherical interface). Set $t = t_m$.
(d) Calculate e from Eq. (2.209).
(e) Calculate M_z from Eq. (2.210).
(f) Calculate $(VM)^2$ from Eq. (2.212).
(g) Calculate $\cos\theta_1$ from Eq. (2.214).
(h) Calculate T from Eq. (2.215).
(i) Calculate x_1, y_1, and z_1 from Eqs. (2.206), (2.207), and (2.208), respectively.
(j) If $m = N$, Stop (finish iterative procedure). Otherwise calculate $q = n_m/n_{m+1}$, and $\cos\theta_2 = [1 - q^2(1 - \cos^2\theta_1)]^{1/2}$.
(k) Calculate u_{2x}, u_{2y} and u_{2z} from Eqs. (2.216), (2.217), and (2.218), respectively.
(l) Set $m = m+1$ (for the next interface) and $x = x_1$, $y = y_1$, $z = z_1$, $u_{1x} = u_{2x}$, $u_{1y} = u_{2y}$, and $u_{1z} = u_{2z}$, and $t = t_m$. Go to (d) and continue to the next interface.

It should be mentioned that the z_1 coordinates are measured from the vertex of the corresponding refractive surface. In order to illustrate the procedure some examples are presented next.

Example 2.10.1:

In this example the exact tracing of a single ray incident on a thick biconvex lens is presented. This example is chosen from Ref. [73]. The initial conditions are $x = 1\,\text{mm}$, $y = 2\,\text{mm}$, and $z = 2\,mm$ and $u_{1x} = 0.2$, $u_{1y} = 0.2$ and $u_{1z} = [1 - u_{1x}^2 - u_{1y}^2]^{1/2} = 0.9591663$. The radii of curvature are $R_1 = 20\,mm$ and $R_2 = -10\,\text{mm}$. The output is selected at a planar surface after the right side of the lens and at a distance of 12 mm. The refractive index of the lens is $n = 1.5$ and the surrounding medium is air.

Solution: Initially the vectors $\tilde{C}$, $\tilde{t}$, $\tilde{n}$ are formed ($N = 3$, the two spherical interfaces and the planar output surface to the right of the lens). All distances are in mm:

$$\tilde{C} = [1/20, \ -1/10, \ 0]\,,$$

$$\tilde{t} = [10, \ 5, \ 12]\,,$$

$$\tilde{n} = [1, \ 1.5, \ 1]\,.$$

Applying the procedure listed previously the points Q_1, Q_2, and Q_3 are determined (coordinates x_1, y_1, and z_1) along with the ray direction after refraction via u_{2x}, u_{2y}, and u_{2z}. The results are summarized in Table 2.11. It is mentioned that in the last planar boundary, no refraction occurs.

TABLE 2.11: Exact ray tracing results of Example 2.10.1

Interface	$x_1(mm)$	$y_1(mm)$	$z_1(mm)$	u_{2x}	u_{2y}	u_{2z}
$m = 1$	2.784852	3.784852	0.559848	0.082467	0.064202	0.994524
$m = 2$	3.041337	3.984530	−1.347036	−0.052866	−0.135217	0.989431
$m = 3$	2.328194	2.163140	0	−0.052866	−0.135217	0.989431

Example 2.10.2:

In this example, the behavior of rays through a spherical refractive interface is examined using the exact ray tracing approach. The parameters provided are based on those from Example 2.9.2, which was treated using Fermat's principle and the 3rd-order Seidel approximation. It is based on the geometry of Fig. 2.100. The radius of the spherical refractive interface is $R = 100\,\text{mm}$, and the refractive indices are $n_1 = 1$ and $n_2 = 1.5$. The object height is $h = 5\,\text{mm}$, and its distance is $s = 300\,\text{mm}$ to the left of the interface. The paraxial image distance is $s'_p = 900\,\text{mm}$ [as it can be easily determined from Eq. (2.42)]. The results shown in Figs. 2.101, 2.102(a), 2.102(b), 2.103(a), and 2.103(b) (the exact curves and not the 3rd-order Seidel approximations) should be verified using the exact ray tracing procedure.

Solution: Initially the vectors $\tilde{C}$, $\tilde{t}$, $\tilde{n}$ are formed ($N = 2$, the one spherical interface and the planar output surface to the right of the spherical interface at distance $t_2 = s'_p = 900\,\text{mm}$:

$$\tilde{C} = [1/100,\ 0]\,,$$

$$\tilde{t} = [300, 900]\,,$$

$$\tilde{n} = [1,\ 1.5]\,.$$

Point P is the point O_1 in Fig. 2.100. Therefore, $x = -h = -5\,\text{mm}$, $y = 0$,and $z = 0$. In the first case meridional rays are considered that are varying in x_p (shown in Fig. 2.100). In order to find the $\hat{u}_1$ unit vector, the following relations should be used:

$$z_p = R - \sqrt{R^2 - x_p^2},$$

$$u_{1x} = \frac{x_p + h}{\left[(x_p + h)^2 + (z_p + s)^2\right]^{1/2}},$$

$$u_{1y} = 0,$$

$$u_{1z} = \frac{s + z_p}{\left[(x_p + h)^2 + (z_p + s)^2\right]^{1/2}},$$

where x_p varies in $[-5\,mm, 5\,mm]$. The paraxial image height is $h' = h(s'_p/s)(n_1/n_2) = 10\,\text{mm}$. In order to relate the parameters x_1, y_1, z_1 and u_{2x}, u_{2y}, u_{2z} of the ray tracing

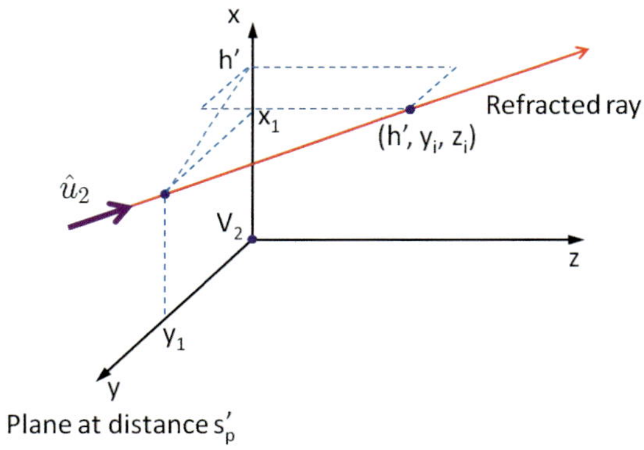

FIGURE 2.142: The ray passing from the plane at $s'_p = 900\,\text{mm}$ after it gets refracted at spherical interface. It is used to relate the exact ray tracing results to the s' and y_i used in example 2.9.2.

procedure with the results found in example 2.9.2 using Fermat's principle, the use of Fig. 2.142 is necessary. The ray shown in this figure is the one that arrives at the plane which is at the paraxial image distance $s'_p = 900\,\text{mm}$. The refracted ray can be represented by the equation:

$$\vec{r} = \lambda\hat{u}_2 + x_1\hat{\imath}_x + y_1\hat{\imath}_y + z_1\hat{\imath}_z \;=\; (x_1 + \lambda u_{2x})\,\hat{\imath}_x + (y_1 + \lambda u_{2y})\,\hat{\imath}_y + (z_1 + \lambda u_{2z})\,\hat{\imath}_z.$$

The coordinates of the refracted ray when its x-component is equal to h' are needed. These coordinates can be determined as follows:

$$x_1 + \lambda u_{2x} = h' \Longrightarrow \lambda = \frac{h' - x_1}{u_{2x}},$$

$$y_i = y_1 + \lambda u_{2y} = y_1 + \frac{h' - x_1}{u_{2x}} u_{2y},$$

$$z_i = s' \;=\; s'_p + \lambda u_{2z} = s'_p + \frac{h' - x_1}{u_{2x}} u_{2z},$$

where the parameter λ of the refracted ray is determined by requiring its x coordinate to be $x_i = h'$ as it was used in Sec. 2.9.2.2. This is necessary in order to compare the exact ray tracing results with the Fermat's procedure that was utilized in Sec. 2.9.2.2 and the example 2.9.2. Using the equations presented before for z_p and u_{1x}, u_{1y}, and u_{1z} and relating the results of the exact ray tracing algorithm, s' is calculated when $h' = 10\,\text{mm}$. The results are shown in Fig. 2.143, and as expected, are identical to the exact results of Fig. 2.101. In this case all the rays are meridional, and $y_i = 0$ for all refracted rays. The next case is when

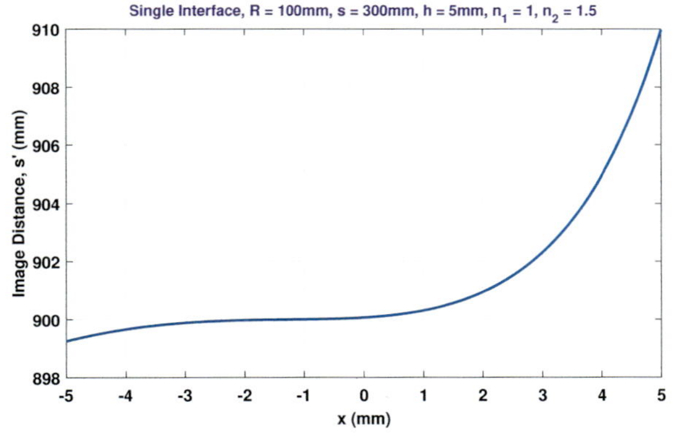

FIGURE 2.143: The image distance s' when the image height becomes h' for incident rays emanating from point P (equivalent to point O_1 in Fig. 2.100) as a function of distance x_p (presented with x). The results are identical to the ones shown in Fig. 2.101.

the rays emanating from point P (equivalent to O_1) have directions specified by varying y_p for $x_p = 0$. In this case the initial point is as before and the ray directions are given by:

$$z_p = R - \sqrt{R^2 - y_p^2},$$

$$u_{1x} = \frac{h}{\left[h^2 + y_p^2 + (s + z_p)^2\right]^{1/2}},$$

$$u_{1y} = \frac{y_p}{\left[h^2 + y_p^2 + (s + z_p)^2\right]^{1/2}},$$

$$u_{1z} = \frac{s + z_p}{\left[h^2 + y_p^2 + (s + z_p)^2\right]^{1/2}},$$

where y_p varies in $[-5\,mm, 5\,mm]$. The paraxial image height is again $h' = h(s'_p/s)(n_1/n_2) = 10\,\text{mm}$. The the parameters x_1, y_1, z_1 and u_{2x}, u_{2y}, u_{2z} of the ray tracing procedure are

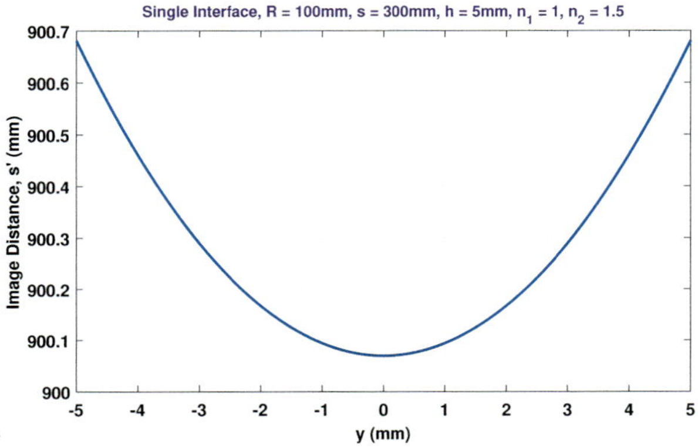

(a)

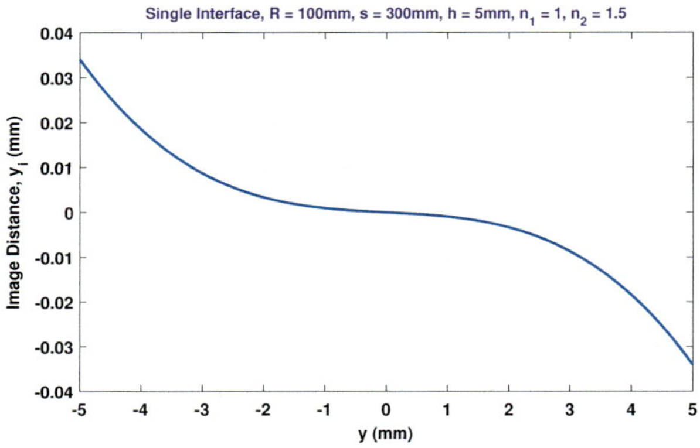

(b)

FIGURE 2.144: (a) The image distance s' when the image height becomes h' for incident rays emanating from point P (equivalent to point O_1 in Fig. 2.100) as a function of distance y_p (presented with y). The results are identical to the ones shown in Fig. 2.102(a). (b) The resulting y_i aberration for y_p varying. The results are identical to the ones shown in Fig. 2.102(b).

related similarly to s', y_i. The results are shown in Figs. 2.144(a) and 2.144(b) for s' and y_i, respectively. The last case in this example is when again all rays are emanating from point P (equivalent to O_1) and have directions specified by a constant $r_p = (x_p^2 + y_p^2)^{1/2} = 5\,\text{mm}$ and varying angle ϕ_p (see Fig. 2.99(b)). In this case the initial point is as before and the ray directions are given by:

$$z_p = R - \sqrt{R^2 - r_p^2},$$

$$u_{1x} = \frac{r_p \cos\phi_p + h}{\left[h^2 + r_p^2 + (s + z_p)^2\right]^{1/2}},$$

$$u_{1y} = \frac{r_p \sin\phi_p}{\left[h^2 + r_p^2 + (s + z_p)^2\right]^{1/2}},$$

$$u_{1z} = \frac{s + z_p}{\left[h^2 + r_p^2 + (s + z_p)^2\right]^{1/2}},$$

where ϕ_p varies in $[0, 2\pi]$. The paraxial image height is again $h' = h(s'_p/s)(n_1/n_2) = 10\,\text{mm}$. The parameters x_1, y_1, z_1 and u_{2x}, u_{2y}, u_{2z} of the ray tracing procedure are related similarly to s' and y_i. The results are illustrated in Figs. 2.145(a) and 2.145(b) for s' and y_i, respectively.

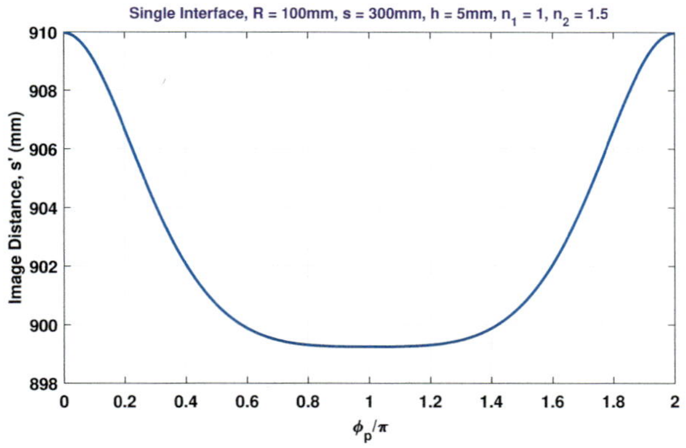

(a)

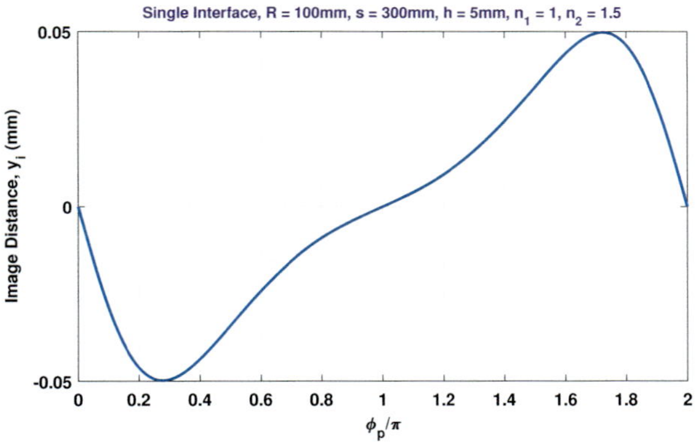

(b)

FIGURE 2.145: (a) The image distance s' when the image height becomes h' for incident rays emanating from point P (equivalent to point O_1 in Fig. 2.100) as a function of angle ϕ_p for $r_p = 5\,\text{mm}$. The results are identical to the ones shown in Fig. 2.103(a). (b) The resulting y_i aberration for ϕ_p varying. The results are identical to the ones shown in Fig. 2.103(b).

Example 2.10.3:

In this example the spherical aberration of a biconvex lens is presented. The lens has radii $R_1 = 50\,\text{mm}$ and $R_2 = -50\,\text{mm}$, a thickness of $d = 5\,\text{mm}$, and a refractive index of $n_L = 1.50$. The surrounding medium is air. Rays parallel to the optical axis are incident from the left at a height specified by the variable x. The goal is to determine the focal distance f_2 (shown in Fig. 2.146) as a function of the height x. Furthermore, the transverse spherical aberration, TSA, should also be determined.

Solution: Initially the vectors $\tilde{C}$, $\tilde{t}$, $\tilde{n}$ are formed. In this configuration, $N = 3$ (the two spherical interfaces and the planar output surface to the right of the lens at distance $t_3 = f_{2,par}$, where $f_{2,par}$ is the paraxial focal distance, which can be easily determined using

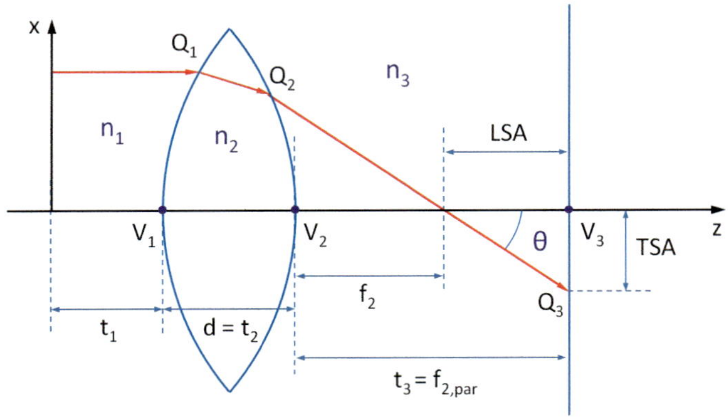

FIGURE 2.146: The geometry of a biconvex lens for applying the exact ray tracing method. The goal is the exact determination of the longitudinal spherical aberration (LSA) expressed via the effective distance f_2 and the transverse spherical aberration (TSA). Both are to be determined as functions of the parallel ray height x.

paraxial optics and is equal to 49.1525 mm):

$$\tilde{C} = [1/50, \ -1/50, \ 0],$$
$$\tilde{t} = [10, \ 5, \ 49.1525],$$
$$\tilde{n} = [1, \ 1.5, \ 1].$$

The distance $t_1 = 10$ mm is arbitrarily selected to be 10 mm (it does not play any role since the rays are parallel to the optical axis). All incident rays are parallel to the optical axis, and therefore the coordinates of initial point P are $(x, 0, 0)$. Since the rays are parallel to the optical axis (z-axis) $u_{1x} = u_{1y} = 0$ and $u_{1z} = 1$. From Fig. 2.146 the $TSA = x_1$ for the Q_3 point. The $LSA = TSA/\tan\theta = TSA(\cos\theta/\sin\theta) = TSA\ u_{2z}/(1 - u_{2z}^2)^{1/2}$. Then, the focal distance $f_2 = f_{2,par} + LSA$ (it is reminded that TSA has negative sign in the figure since it is below V_3). The resulting f_2 and TSA as functions of the height x of the incident rays are shown in Figs. 2.147(a) and 2.147(b), respectively. The results of Fig. 2.147(a) are

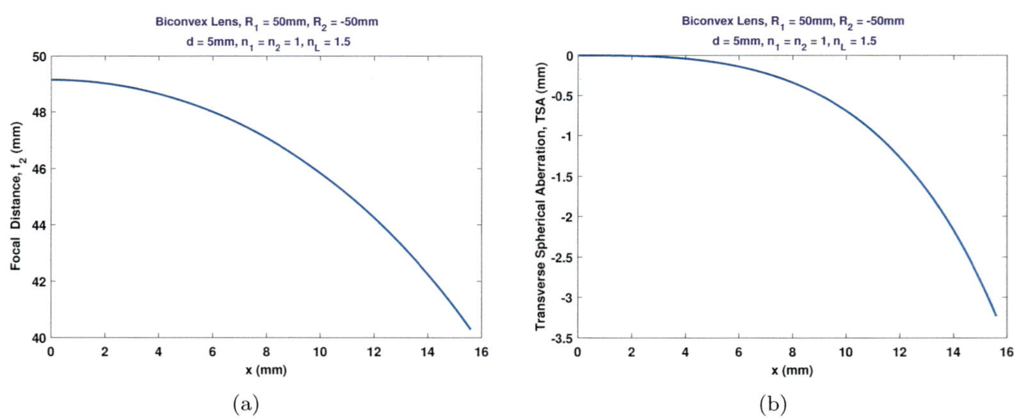

FIGURE 2.147: (a) The spherical aberration of a biconvex thick lens with paramaters: $R_1 = 50$ mm, $R_2 = -50$ mm, $d = 5$ mm, and $n_L = 1.50$ with surrounding medium air. The focal distance measured from the right vertex of the lens is shown. (b) The transverse spherical aberration of the same lens as a function of the parallel ray height again.

identical with the ones presented for the same lens in Fig. 2.111 using the trigonometric exact approach.

2.11 Problems

1 **Parallel Plane Plate:** A ray of light strikes a plane parallel plate made of a material with a refractive index n_2 at an angle of incidence α. The plate has a thickness d, and it is surrounded by a medium with a refractive index n_1. (a) Determine the lateral displacement D of the ray after it passes through the plate as a function of d, n_1, n_2, and α. (b) For a plane parallel plate of fused quartz ($n_2 = 1.46$) in the air ($n_1 = 1.0$), calculate D or an angle of incidence $\alpha = 67.5$ deg and plate thickness $d = 0.525$ inches. (c) Make a plot of the displacement D (in cm) as a function of the angle of incidence α (in deg) for d as in part (b). (Note: 1 inch = 2.54cm).

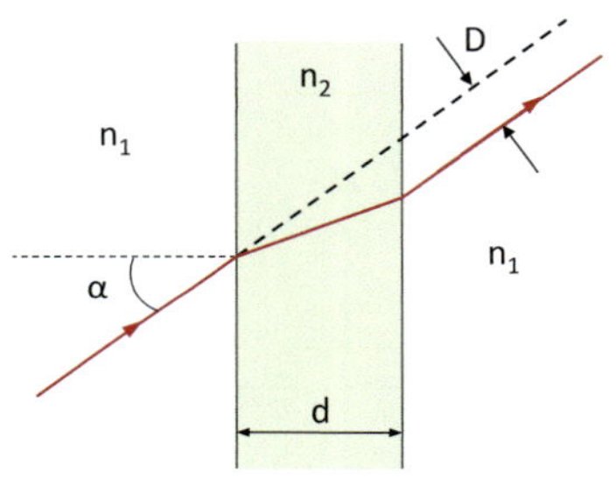

FIGURE 2P.1: A parallel plane plate of thickness d and refractive index n_2.

2 **Cartesian Ovoid:** (a) Determine the Cartesian ovoid for perfect imaging at a refractive surface when the object point O is positioned at a distance $s_o = 5$ cm to the left of vertex V_1 in air (where $n_o = 1.0$). The real image point I is located at a distance $s_i = 7.5$ cm to the right of V_1 within a refractive material of $n_i = 1.35$. Determine the equation describing the intersection of the ovoid with the xy-plane, assuming the point $(x = 0, y = 0)$ coincides with the object's position. Find the position of V_2 and create a graphical representation of the ovoid in the xy-plane. (b) Repeat part (a) for a virtual image at I by setting $s_i = -7.5$ cm (positioned to the left of V_1). Use Figs. 2.29(a) and 2.29(b) or reference in interpreting the geometric points mentioned.

3 **Dispersing Prism:** Sellmeier's formula provides a reliable approximation of the refractive index, $n(\lambda_0)$, as a function of the freespace wavelength λ_0. This formula is given as follows:

$$n^2(\lambda_0) = 1 + \sum_i \frac{B_i \lambda_0^2}{\lambda_0^2 - C_i}.$$

An equilateral prism made from *Schott multiple purpose glass* is used in a spectrometer, and the Sellmeier coefficients B_i and C_i are provided in Table 2A.1. (a) Determine the minimum angle of deviation of the prism as a function of the freespace wavelength in the visible spectrum (400 nm to 700 nm). Present your findings in a graphical form. (b) Calculate the dispersive power of the prism, both approximate and exact. (c) Evaluate the minimum angular dispersion $d\delta_{min}/d\lambda_0$ (in deg/μm) as a function of the freespace wavelength, and present a graphical representation of this dispersion.

4 **Thin Lenses System:** (a) A 10 mm tall object is placed 150 mm to the left of a positive thin lens with a focal length of 100 mm. Another negative thin lens with a focal length of −75 mm is

positioned 250 mm to the left of the positive lens. Calculate the location of the image, measuring its distance from the negative lens. Determine the transverse magnification and specify whether the image is real or virtual. Illustrate the imaging by the two-lens system with a ray diagram. (b) Now, consider that the distance L between the two thin lenses varies, while only the right lens is moved, keeping the left lens stationary. Let $0 < L < 400$ mm. Plot the image position as a function of L, measured from the center of the right lens. Identify the ranges of L in which the final image is real or virtual.

5 Focal Length Measurement for a Positive Thin Lens: To measure the focal length of a positive thin lens, one can utilize the fact that if a conjugate object-image pair O and I are separated by a distance L (where $L > 4f$), then there exist two distinct lens positions along the axis, separated by a distance d, that will produce the same conjugate points. Derive an expression for the focal length f of the lens as a function of L and d.

6 Focal Length Measurement for a Negative Thin Lens: The following procedure describes a method to measure the focal distance of a negative lens. In this process, a positive lens with focal distance f_p is positioned at a distance s_p from an object of height h. An initial real image of height h' is then formed at a distance s'_p on a screen, as illustrated in Fig. 2.2(a). With the

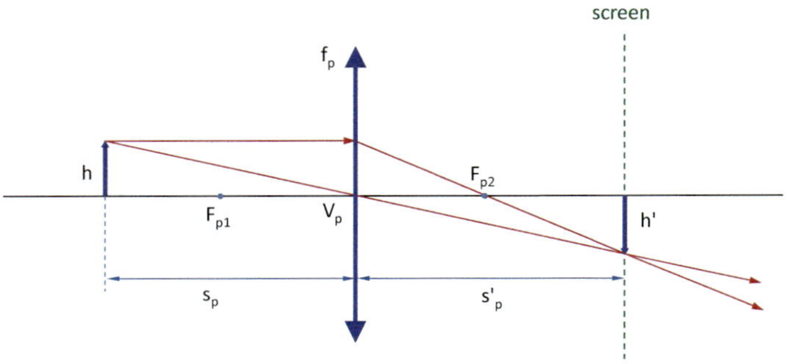

(a)

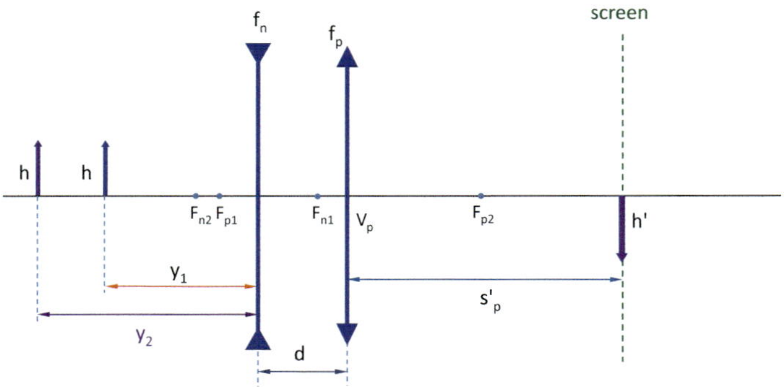

(b)

FIGURE 2P.2: Configurations for measurement of the focal length of a negative lens. (a) Initial configuration with the help of a positive lens of focal length f_p. (b) Configuration using both the negative and the positive lens.

positions of the positive lens and the screen fixed, a thin negative lens (with unknown focal length f_n) is introduced between the positive lens and the object. The negative lens is placed to the left of the positive lens, at a distance d, positioned closer to the positive lens than to the object, as illustrated in Fig. 2.2(b). Initially, the object is located at a distance y_1 from the negative lens. However, when the negative lens is inserted, the image no longer forms on the screen at the original distance s'_p from the positive lens. To restore the image on the screen, the object is repositioned to a new distance y_2 from the negative lens. Determine the focal distance f_n of the negative lens based on the measured distances y_1 and y_2. Additionally, create a ray diagram to illustrate the image formation by the two-lens system.

7 Basics of Rainbow Formation: This problem seeks to develop an understanding of the basic mechanisms behind rainbow formation in the atmosphere. A sample rainbow is displayed in Fig. 2P.3, where both the primary rainbow (lower and brighter) and the secondary rainbow (higher and fainter) can be observed. Note the reversal of color order in the two rainbows. The primary rainbow results from incident light rays undergoing two refractions and a single internal reflection within each water droplet, as illustrated schematically in Fig. 2P.4a. Meanwhile, the secondary rainbow is generated by incident light rays that undergo two refractions and two internal reflections within each droplet, as shown in Fig. 2P.4b. To justify these effects, it is

FIGURE 2P.3: The photograph of a rainbow. Both the primary and the secondary rainbows are shown. The picture can be found at the following url site: http://atgbcentral.com/data/out/98/4799387-rainbow-pics.jpg

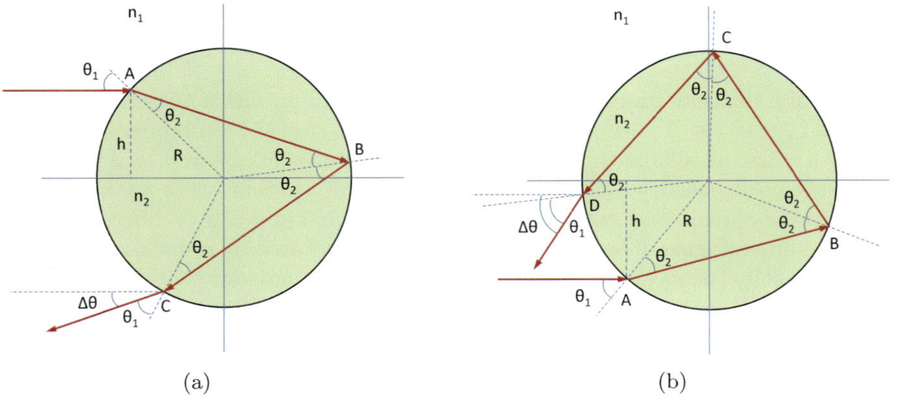

FIGURE 2P.4: (a) The primary rainbow forms from incident beams that have two refractions and one reflection inside a water droplet. (b) The secondary rainbow forms from incident beams that have two refractions and two reflections inside the water droplet. It is assumed that the water droplet has a refractive index of n_2 while the surrounding medium (air) has a refractive index of n_1. The droplet is approximated as a sphere of radius R and the incident light beam forms an incident angle of θ_1.

necessary to consider the frequency (or wavelength) dependence of the refractive index of water. The refractive index of water as a function of the free-space wavelength λ_0 (expressed in microns) is given by the following Sellmeier's equation. A plot of the refractive index is shown in Fig. 2P.5:

$$n^2(\lambda_0) = 1 + \frac{5.672526103 \times 10^{-1}\lambda_0^2}{\lambda_0^2 - 5.085550461 \times 10^{-3}} + \frac{1.736581125 \times 10^{-1}\lambda_0^2}{\lambda_0^2 - 1.814938654 \times 10^{-2}} + \frac{2.121531502 \times 10^{-2}\lambda_0^2}{\lambda_0^2 - 2.617260739 \times 10^{-2}} + \frac{1.138493213 \times 10^{-1}\lambda_0^2}{\lambda_0^2 - 1.073888649 \times 10^{+1}}$$

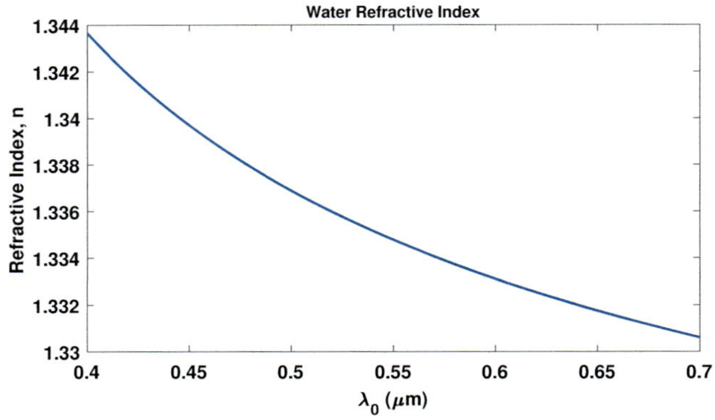

FIGURE 2P.5: The refractive index of water in the visible range according to M. Daimon and A. Masumura, *Applied Optics*, 46, pp. 3811–3820 (2007).

(a) Evaluate the angular deviation $\Delta\theta$ shown in Figs. 2P.4a and 2P.4b as a function of the ratio h/R (which is related to the incident angle θ_1) for a water droplet of radius R. For simplicity, assume that all light rays lie within the same meridian of the spherical droplet, thereby avoiding the need for 3D incident rays. Vary h/R in the range between 0 and 1 for both cases. Plot $\Delta\theta$ as a function of the ratio h/R for various wavelengths in the visible range ($0.380\,\mu$m to $0.720\,\mu$m). What do you observe? Assume that the refractive index of the air $n_1 = 1.0$. (b) Find the equation that defines the extremum of $\Delta\theta$ and the corresponding angle of incidence θ_1 for both the primary and the secondary rainbow. Make a plot of both $\Delta\theta_{ext}$ and $\theta_{1,ext}$ as a function of wavelength for the given range, using the provided Sellmeier equation for water.

8 **Focusing of a lenslike medium:** Assume a lenslike medium with a refractive index distribution of the form $n(r) = n_0[1 - (k_2/2k_0)r^2]$ (where $k_0 = 2\pi/\lambda_0$ with λ_0 representing the free space wavelength). The lenslike medium extends along the z-axis from $z = 0$ to $z = \ell$. Beyond $z = \ell$, there is a homogeneous medium of refractive index n_{out}. (a) Show that a family of parallel rays (parallel to the optical axis) entering the lens-like medium at $z = 0$ at different heights (r_0) will converge upon emerging at $z = \ell$ to a common focus located at a distance F from the $z = \ell$ boundary. Specify the inequalities between the problem parameters that determine whether F is positive, negative, or zero. (b) For the following values: $n_0 = 1.50$, $n_{out} = 1.00$, $\lambda_0 = 0.5\,\mu$m, plot the focal distance F as a function of the lenslike medium length ℓ for two cases: $k_2 = 0.01\,\mu\text{m}^{-1}$ and $k_2 = 0.001\,\mu\text{m}^{-1}$. Since the focal distance is periodic with respect to ℓ vary ℓ only within the first period. (c) For the same parameters $n_0 = 1.50$, $n_{out} = 1.00$, $\lambda_0 = 0.5\,\mu$m, $k_2 = 0.01\,\mu\text{m}^{-1}$, assuming an initial ray height $r_0 = 10\,\mu$m, plot the ray path through the system for lens-like medium lengths $\ell = 150\,\mu$m, $250\,\mu$m, and $300\,\mu$m.

9 **Mirage Simulation:** Assume that on very hot days, the air near the ground becomes thinner due to higher temperatures, and it thickens with increasing distance from the ground as the temperature decreases. Under these conditions, the refractive index of the air (neglecting dispersion) can be approximated by the equation: $n(r) = n(x, y) = n_0[1 + kr]$, where $k > 0$,

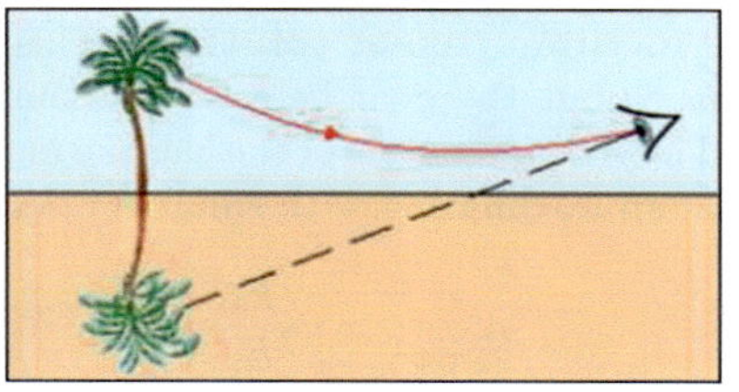

FIGURE 2P.6: A mirage inferior schematic. (from url-link: `http://images.math.cnrs.fr/Le-chant-de-la-Terre.html?lang=es`).

and $r = (x^2 + y^2)^{1/2}$ represents the radial distance from the ground. The z-axis is positioned along the ground, as depicted in the accompanying diagram. For the numerical implementation, assume the following values: $n_0 = 1.00022$ and $k = 1.3333 \times 10^{-5}\,m^{-1}$. These parameters are derived from the reference: B. Edlen, "The refractive index of air," *Metrologia*, vol. 2, pp. 71–80, 1966. (a) Assume that a ray emanates at height h with an initial angle θ, as shown in the diagram. The angle θ is positive when the ray points to negative x (as in the figure) and negative when the ray points in positive x. Let θ vary between $-3/60$ deg-$30/60$ deg in steps of $3/60$ deg. Furthermore, assume that $h = 2$ m. For each ray, determine the path it follows due to the varying refractive index, for a maximum distance of 1 km along the z direction. (b) Now, repeat the above calculations, but for a different initial height $h = 3$ m.

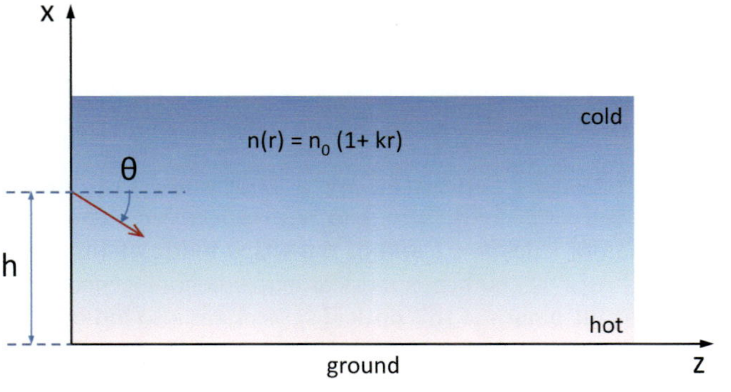

FIGURE 2P.7: Basic geometry for studying the mirage effect.

Hints: If a ray's route hits the ground it is interrupted, i.e., the ray is absorbed by the ground. For the solution of the eikonal equation the boundary conditions $(dr/dt)(t = 0) = n(r)(dr/ds)(t = 0)$ are needed. The eikonal equation can be solved numerically using Runge-Kutta techniques. For example, Matlab's "ode45" function can easily solve nonlinear systems of differential equations (other similar routines in various computational packages can be used). Check references:
A. Sharam et al. "Tracing rays through graded-index media: a new method," *Appl. Opt.*, vol. 21, No. 6, pp. 984–987, Mar. 15, 1982.
A. Sharam, "Computing optical path length in gradient-index media: a fast and accurate method," *Appl. Opt.*, vol. 24, No. 24, pp. 4367–4370, Dec. 15, 1985.
If the refractive index is described by the relationship $n(r) = n(x) = n_0(1 + kx)$ which is more appropriate for the given problem, then an analytical solution to the eikonal equation exists. In this case, you are asked to specify (using a graphical representation) the difference between the analytical and numerical solutions (obtained via the Runge-Kutta method) as a function of the distance z. Consider various values of the initial height h and angle θ to observe how the discrepancy between the two solutions changes with these parameters. (c) Repeat question (a) but making usage of Snell's law only this time instead of the usage of the eikonal equation.

10 Propagation in a lenslike medium (exact solution): This problem is essentially the same physical scenario as Problem 8, but the goal here is to demonstrate the differences between various ray approximations. The *ABCD* matrix of the lenslike medium was based on the following differential equation (paraxial approximation with small $k_2/2k_0$):

$$n_0 \frac{d^2 r}{dz^2} \simeq n_0 \frac{k_2}{k_0} r. \tag{2.11.1}$$

A slightly better paraxial approximation is the following:

$$n_0 \left(1 - \frac{k_2}{2k_0} r^2\right) \frac{d^2 r}{dz^2} \simeq n_0 \frac{k_2}{k_0} r. \tag{2.11.2}$$

Finally, the exact eikonal equation (without the paraxial approximation) is:

$$\frac{d}{ds}\left(n \frac{d\vec{r}}{ds}\right) = \vec{\nabla} n. \tag{2.11.3}$$

Equation (2.11.1) can be solved analytically as it was shown in the textbook using the *ABCD* matrix. However, equations (2.11.2) and (2.11.3) can be solved only numerically using Runge-Kutta methods. For example, Matlab's "ode45" function is well-suited to solve nonlinear systems of differential equations, though other similar routines in various computational packages can also be used. Solve all three equations numerically and plot the ray position inside the lenslike medium as a function of the distance z, where z varies from 0 to approximately $\sim 800\,\mu\text{m}$. The values of k_2, n_0, and λ_0 are as defined in Problem 8. (a) For initial conditions $r_0 = 10\,\mu\text{m}$ and $r_0' = 0$, plot the ray paths corresponding to the three solutions in the same diagram. (b) Repeat (a) for the initial conditions $r_0 = 10\,\mu\text{m}$ and $r_0' = 1$.
Hint: For the solution of the eikonal equation (2.11.3) refer to the journal paper: A. Sharma *et al.*, "Tracing rays through graded-index media: a new method," *Applied Optics*, vol. 21, No. 6, pp. 984–987, Mar. 15, 1982.

11 Two Hemi-Spherical Lenses separated by a Distance: Two hemispherical lenses, with radii R_1 and R_2, have refractive indices n_1 and n_2, respectively. The two lenses are separated by a homogeneous region of thickness t and of refractive index n_t (as shown in Fig. 2P.8). The material to the left and right of the hemispherical lenses is homogeneous with a refractive index of n_0. The input and output planes of this optical system are also indicated in Fig. 2P.8. (a) Find the matrix *ABCD* of this optical system. (b) Given the following parameters: $|R_1| = 50\,\text{mm}$, $|R_2| = 50\,\text{mm}$, $n_1 = 1.50$, $n_2 = 1.50$, $n_t = 1.00$, $n_0 = 1.00$, and $0 \leq t \leq 100\,\text{mm}$, plot the focal distance f_2 of the optical system as a function of t. Additionally, create a graphical representation of the distances of the principal points H_1 and H_2 as measured from the input plane as a function of t. (c) Repeat (b) when $n_t = 1.50$ while all other parameters remain unchanged.

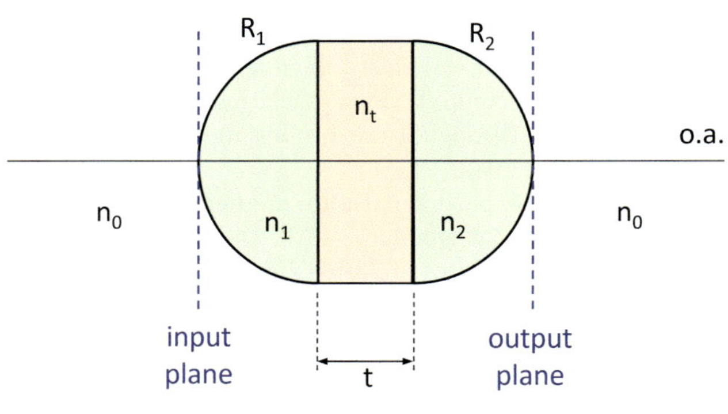

FIGURE 2P.8: The geometry of two hemispherical lenses separated by a distance t.

12 Nikon Photographic Lens: The Nikon photographic lens, Nikkor 100 mm, is described in US patent 1984/4448497. This patent outlines various implementations of the system, which are shown in the first two figures. The implementation used in this problem corresponds to the configuration described in the first embodiment. The system consists of 7 thick lenses. The relevant data from the patent, including figures and a table, are shown in Fig. 2P.9. In Fig.

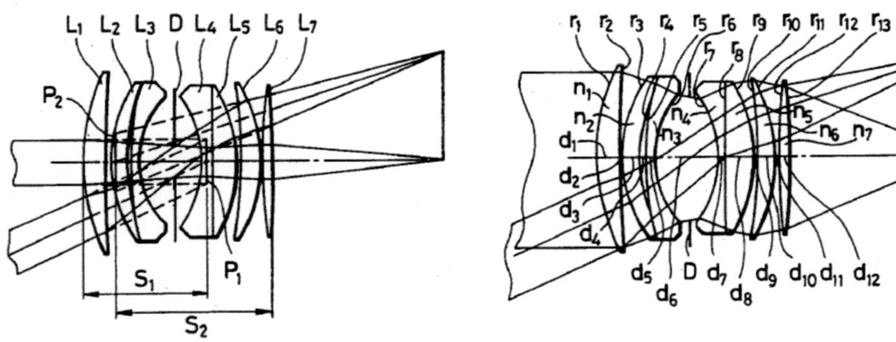

First Embodiment

Focal length f = 100 F-number 1.4 Angle of view 2ω = 46°

		d-line	
$r_1 = +78.360$	$d_1 = 9.8837$	$n_1 = 1.79668$	$\nu_1 = 45.5$
$r_2 = +469.477$	$d_2 = 0.1938$		
$r_3 = +50.297$	$d_3 = 9.1085$	$n_2 = 1.77279$	$\nu_2 = 49.4$
$r_4 = +74.376$	$d_4 = 2.9457$		
$r_5 = +138.143$	$d_5 = 2.3256$	$n_3 = 1.67270$	$\nu_3 = 32.2$
$r_6 = +34.326$	$d_6 = 29.0698$		
$r_7 = -34.407$	$d_7 = 1.9380$	$n_4 = 1.7400$	$\nu_4 = 28.3$
$r_8 = -2906.977$	$d_8 = 12.4031$	$n_5 = 1.77279$	$\nu_5 = 49.4$
$r_9 = -59.047$	$d_9 = 0.3876$		
$r_{10} = -150.021$	$d_{10} = 8.3333$	$n_6 = 1.78797$	$\nu_6 = 47.5$
$r_{11} = -57.890$	$d_{11} = 0.1938$		
$r_{12} = +284.630$	$d_{12} = 5.0388$	$n_7 = 1.78797$	$\nu_7 = 47.5$
$r_{13} = -253.217$			

Back focal length 74.1
Full length of lens l = 81.8217
$f_1 = 116.8$
$f_2 = 172.6$
$f_6 = 115.0$

FIGURE 2P.9: The geometry of the Nikon Photographic lens and its components data from US Patent 1984/4448497.

2P.9 the system of the 7 thick lenses is shown with lenses labeled L1 to L7 from left to right. The table contains the details of the lenses and their separations: The first column lists the spherical surface radii (in mm), with positive or negative values according to the convention for convex and concave surfaces. These radii correspond to the left and right surfaces of each lens, listed from left to right. The second column lists the separations (in mm) between the spherical surfaces of adjacent lenses. It is noted that lenses L4 and L5 are cemented, so the right surface of L4 coincides with the left surface of L5. The third column gives the refractive index of each lens material at the wavelength of the $d-line$ ($\lambda_0 = 587.6$ nm). The last column lists the Abbe number of the lens material, though material dispersion will not be used in this problem. All the calculations of the problem are according to the paraxial optics approximation. (a) At the bottom of the table, the focal lengths of lenses L1, L2, and L6, (f_1, f_2, and f_6), are given in mm. Verify these focal lengths of the table. (b) Find the cardinal points ($H_1, F_1, N_1, H_2, N_2, F_2$) of the lens system with input plane on the left surface of the first lens (L1), and output plane on the right surface of the last lens (L7). Determine the distance from the right surface of the L7 lens at which an image of an object at infinity will be formed. Verify that this value corresponds to the "back focal length" listed in the table. (c) Repeat part (b) for an object placed at a distance of 1 meter from the left surface of L1. Determine how much the entire lens system needs to be shifted (to the left or to the right) such that the image is formed correctly at the "back focal length" found in part (b), where the film or digital detector is positioned. Now assume that a "full frame" format (for either film or a digital detector) is used, with a Circle of Confusion $CoC = 0.029\,\text{mm}$. If an *f-number* of 1.4 is used, calculate the hyperfocal distance H and the depth-of-field (DoF) for the object distance of one meter.

13 Keplerian and Galilean Telescope: Consider two thin lenses that can be utilized to form a basic Keplerian or a Galilean telescope. The focal distance of the objective lenses is $f_o = 300\,\text{mm}$ (for either the Keplerian or the Galilean telescope). The eyepiece lens has a focal length $f_e = 25\,\text{mm}$ (for the Keplerian telescope) or $f_e = -25\,\text{mm}$ (for the Galilean telescope). Both telescopes are used to image an object at a long distance (at infinity), where they will have the same angular magnification, given by $m = f_o/f_e$. However, the image formed by the Keplerian telescope will be an inverted virtual image, whereas the image formed by the Galilean telescope will be an upright virtual image. (a) For both telescopes, calculate the minimum object distance required to form a virtual image. (b) Let O_{min} be the larger of the minimum object distances for the two telescopes to form virtual images. Choose an object distance of $2O_{min}$ for both telescopes. Using the $ABCD$ matrix method, draw the ray diagram showing the image formation for each telescope. Also, find the position and transverse magnification of the final virtual image in both cases.

14 Human Eye Accommodation: The simplified diagram of the human eye is shown in Fig. 2P.10. The parameters given are the following:
Refractive Indices: $n_0 = 1$, $n_1 = 1.376$, $n_2 = 1.336$, $n_3 = 1.417$, and $n_4 = 1.336$.
Distances: d_0 = depends on the object distance from the cornea, $d_1 = 0.55\,\text{mm}$ (distance from cornea to anterior chamber), $d_2 = 3.26\,\text{mm}$ (distance from anterior chamber to lens surface), d_3 = depends on the accommodation level of the human lens (it is the thickness of the human lens), $d_4 = 16.60\,\text{mm}$ (distance from lens to retina).
Radii of Curvature: $R_1 = 7.7\,\text{mm}$ (radius of curvature of the cornea's anterior surface), $R_2 = 6.8\,\text{mm}$ (radius of curvature of the cornea's posterior surface), R_3 and R_4 depend on the accommodation level (i.e., curvature of the human lens changes with focus). From the book of Atchison and Smith [74], the following simplified approximate formulas are given:

$$x(A) = 1.052A - 0.00531A^2 + 0.000048564A^3,$$

$$d_3 = 3.6 + 0.4\,\frac{x(A)}{10.87} \qquad \text{(in mm)},$$

$$\frac{1}{R_3} = \frac{1}{10} - \left(\frac{1}{10} - \frac{1}{5.333}\right)\frac{x(A)}{10.87}, \qquad (\text{in mm}^{-1}),$$

$$\frac{1}{R_4} = -\frac{1}{6} - \left(-\frac{1}{6} + \frac{1}{5.333}\right)\frac{x(A)}{10.87}, \qquad (\text{in mm}^{-1}),$$

(a) Determine the $ABCD$ matrix with the input plane the plane of the object (in distance d_0 from the cornea) and the output plane the plane of the image inside the vitreous humor (area of refractive index n_4). What is the image distance as measured from the right surface of the human lens? What is the refractive power n_4/f_2 of the human eye? (b) Determine the parameter A (of the approximate formulas provided) such that the image always forms on the retina surface, i.e., for $d_4 = 16.60\,\text{mm}$, for object distance (i) $d_0 = 25\,\text{m}$, (ii) $d_0 = 1.5\,\text{m}$, and (iii) $d_0 = 0.25\,\text{m}$. In these three cases find the radii of curvature R_3 and R_4, the thickness d_3, as well as the refractive power n_4/f_2 of the human eye.

15 Optical System with Stops: An optical system is comprised of three thin lenses as shown in Fig. 2P.11. The focal lengths of the thin lenses are $f_1 = f_3 = 50\,\text{mm}$, and $f_2 = -50\,\text{mm}$. The distances $d_1 = d_2 = d_3 = 10\,\text{mm}$. There is a stop in the middle between lenses $L1$ and

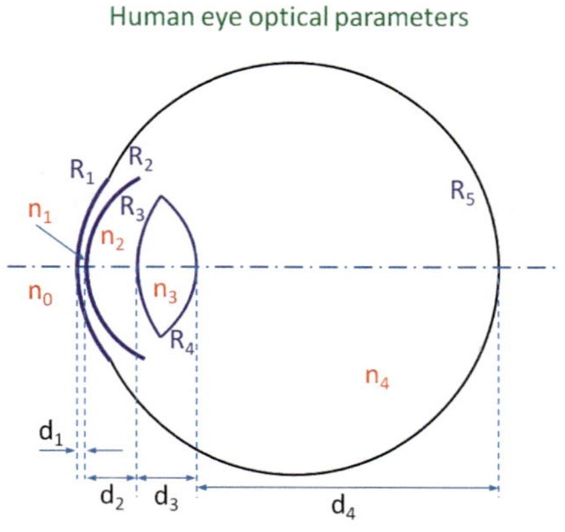

FIGURE 2P.10: The basic diagram of the human eye with the most important optical parameters shown.

$L2$ of diameter $D = 10\,\text{mm}$. The stop that is in the image plane (second focal plane of the three-thin-lens optical system enclosed by the dashed box) has a diameter of $D_s = 20\,\text{mm}$. The thin lenses have all equal diameters $D_1 = D_2 = D_3 = 49\,\text{mm}$. The object is infinity (i.e., parallel rays enter the system from the left). (a) Determine the principal points of the system, assuming the input plane is at the left surface of $L1$ and the output plane is at the right surface of $L3$. Use the $ABCD$ matrix method to calculate the position of the principal points. (b) Determine the distance q, which is the image distance from the final lens $L3$. (c) Identify the aperture stop (AS) of the system. Given the system configuration, describe where it is located and its diameter. (d) Find the entrance and exit pupils and their diameters. (e) Determine the field stop and the corresponding entrance and exit windows. (f) Find the field of view (in object space) and the f-number of the system.

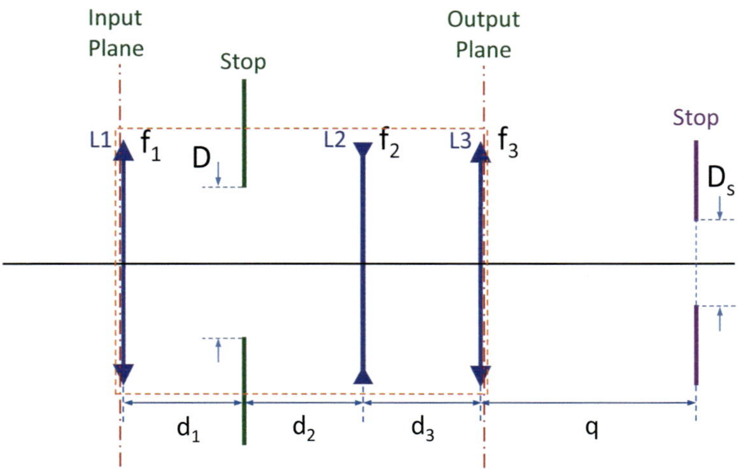

FIGURE 2P.11: A system of three thin lenses with stops. The right most stop represents the size of the detector which is placed at the second focal plane of the system.

16 Refractive Surface without Spherical Aberration: The goal of this problem is to design a refractive surface between two media that eliminates spherical aberration (refer to Fig. 2P.12). This means that parallel rays incident on the surface will all focus on the same point, at distance

f from the vertex of the surface, regardless of their incident height. (a) Determine the equation of the refractive surface $y = y(x)$ between two media with refractive indices n_1 and n_2 (where $n_2 > n_1$) that does not suffer from spherical aberration. Show that the surface is an ellipse and define its equation in the xy plane. In addition, plot the surface in the xy plane. Assume that $n_1 = 1.0$, $n_2 = 1.5$, and $f = 50\,\text{mm}$. (b) Now in order to validate the design assume a ray incident on the designed surface from the left at a height h. The ray refracts at the surface and intersects the optical axis at a point F which is a distance f from the vertex of the surface. Verify that all rays, regardless of their incident height, focus at the same point F. Create ray diagrams for several rays at different heights h, illustrating that all rays converge at the focal point F.

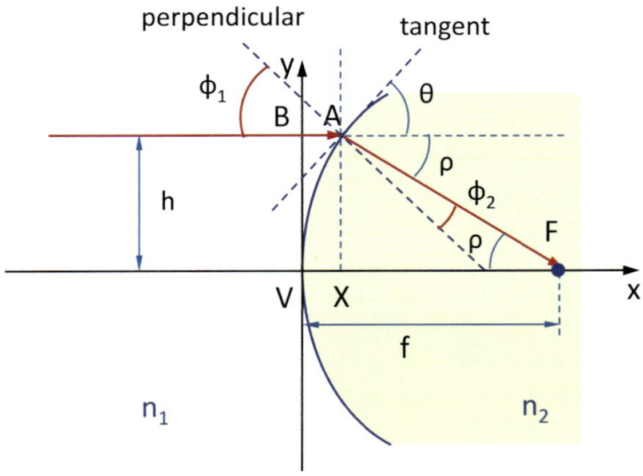

FIGURE 2P.12: The refractive surface $y = y(x)$ that is sought without spherical aberration.

17 Spherical Aberration: Consider a family of thin lenses of focal length f (for all the lenses of the family). The shape factor $[\sigma = (R_2 + R_1)/(R_2 - R_1)]$ of these lenses varies in the interval $[-2, +2]$. Calculate the focal distance difference between the paraxial focal distance and the focal distance including the 3rd-order spherical aberration as a function of the shape factor (it should be a curve similar to the one shown in Fig. 2P.13). The curve should be calculated using Equation A and suitably using Equation B (for both refractive surfaces of the lens). Assume that $n_1 = 1$, $n_2 = 1.517$, $f = 10\,\text{cm}$. Furthermore, assume that the object is at great ("infinite") distance and examine the cases that $h = 1\,\text{cm}$, $h = 5$ cm, and $h = 10\,\text{cm}$. Determine and plot the curves that correspond to Equation A (from Ref. [20]) and Equation B (from Ref. [62]) for all three cases of h as a function of σ. Determine the optimum shape factor for the minimization of the spherical aberration and compare it with the curves that you have determined in the previous questions.
Equation A: Pedrotti equation for spherical aberration (3rd-order correction) of a thin lens of refractive index n in air:

$$\frac{1}{s'(h)} - \frac{1}{s'(0)} = \frac{h^2}{8f_p^3}\frac{1}{n(n-1)}\left[\frac{n-2}{n+1}\sigma^2 + 4(n+1)p\sigma + (3n+2)(n-1)p^2 + \frac{n^3}{n-1}\right],$$

where $p = (s' - s)/(s' + s)$ and s' the paraxial image distance and f_p the paraxial focal distance.
Equation B: Jenkins and White equation for spherical refractive surface of radius R (and refractive indices n_1 and n_2) [see Eq. (2.139)]

$$\frac{n_1}{s} + \frac{n_2}{s'(h)} = \frac{n_2 - n_1}{R} + \frac{h^2}{2}\left[\frac{R}{f_p}\frac{n_1^2}{n_2}\left(\frac{1}{R} + \frac{1}{s}\right)^2\left(\frac{1}{R} + \frac{n_1 + n_2}{n_1}\frac{1}{s}\right)\right],$$

where $n_2/f_p = (n_2 - n_1)/R$ and f_p is the paraxial f_2 distance.

18 Spherical aberration of spherical refractive surface for an object of a given height: Consider a spherical refractive interface between two media with refractive indices n_1 and n_2,

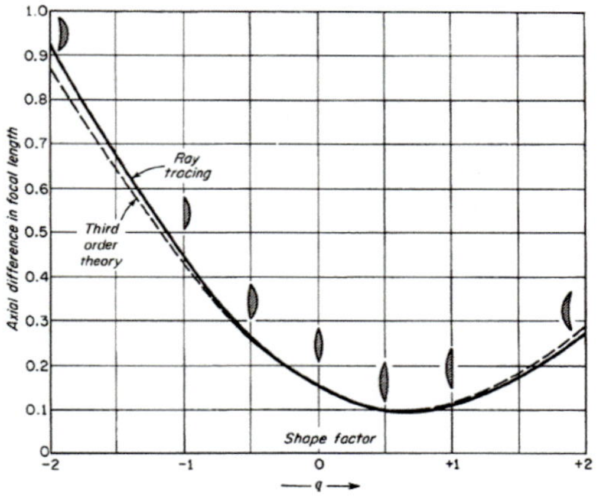

FIGURE 2P.13: The spherical aberration of lenses of different shape but with the same focal length f. This is figure 9G from Ref. [62].

respectively. The radius of curvature of the refractive surface is R. An object of height h is located at a distance s from the left of the surface as is shown in Fig. 2P.14. A meridional ray O_1Q, originating from the object, intersects the refractive surface at point Q (all points are in the same plane). The image of O_1 is denoted as I_1, and the angle of the ray O_1Q with respect to the optical axis is θ. Assume the following numerical parameters: $s = 300\,\text{mm}$, $R = 100\,\text{mm}$, $n_1 = 1$, $n_2 = 1.5$, and $h = 5\,\text{mm}$. (a) Find the range of values of the angle θ that are possible for the ray to pass through the spherical refractive surface. Assume that if the ray emanates from the object and points downward, the angle θ is considered negative. (b) Derive the image location $s'(\theta)$ as function of the angle θ for all possible angles. (c) Calculate the magnification $m(\theta)$ as function of the angle θ for all possible angles.

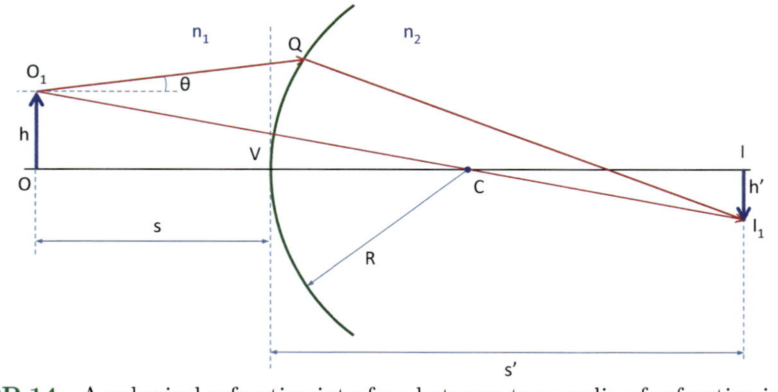

FIGURE 2P.14: A spherical refractive interface between two media of refractive indices n_1 and n_2. The radius of curvature of the surface is R. An object of height h is located at a distance s from the left of the surface's vertex V.

19 Spherical aberration of a plano-convex lens: Consider a plano-convex lens of radius $R_2 = 25\,\text{mm}$ ($R_1 = \infty$) in air ($n_1 = 1$ and $n_2 = 1$). The refractive index of the lens is $n_L = 1.5$ and its thickness is $d = 5\,\text{mm}$. A monochromatic ray is incident on the flat surface parallel to the optical axis at a height h as it is shown in Fig. 2P.15 (see Sec. 2.9.3.1). (a) Using the parameters given, calculate the focal distance f_2 of the plano-convex lens for all allowable values of h. What is the maximum allowable value of h? Furthermore, make a graphical representation of $f_2 = f_2(h)$

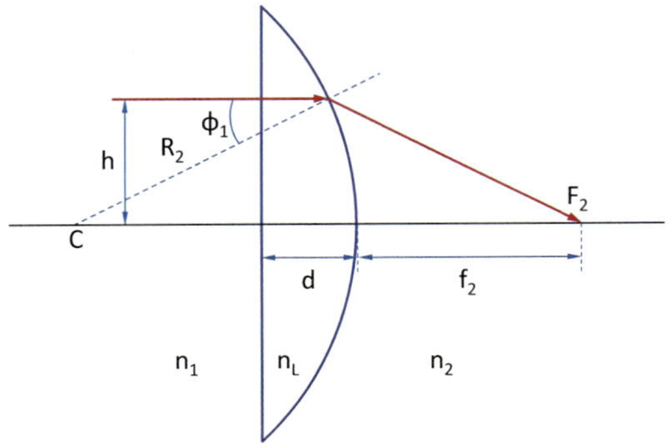

FIGURE 2P.15: Spherical aberration of a plano-convex lens (see Sec. 2.9.3.1).

for the permitted range of h values. Show the paraxial approximation also in the same plot. (b) Create a ray diagram illustrating the paths of rays incident on the lens at various allowable heights h and show the resulting image formation. Repeat the ray diagram for lens thickness of $d = 10\,\text{mm}$.

20 **Spherical aberration of a cemented doublet (exact ray tracing):** A cemented doublet lens system, depicted in Fig. 2P.16 (adapted from Ref. [73]), has the following specifications: Left Lens: Radius of curvature $R_{11} = 27.11\,\text{mm}$, $R_{12} = -15.25\,\text{mm}$, thickness $t_2 = 5\,\text{mm}$, and refractive index $n_2 = 1.50$. Right Lens: Radius of curvature $R_{21} = -15.25\,\text{mm}$, $R_{12} = -45.75\,\text{mm}$, thickness $t_3 = 5\,\text{mm}$, and refractive index $n_3 = 1.60$. Rays parallel to the optical axis originate from point P at varying heights x. The distance from the object to the left lens vertex is $t_1 = 10\,\text{mm}$, while the paraxial focal distance of the system, $t_4 = f_{2,par}$, is measured from point V_3. The surrounding medium is air, with refractive indices $n_1 = n_4 = 1.0$. Using the exact ray tracing method outlined in Sec. 2.10, determine the true focal distance f_2, of the doublet lens. Additionally, calculate the longitudinal spherical aberration (LSA) and the transverse spherical aberration (TSA) as functions of the height x, for x varying in the interval $[0, 6\,mm]$. Present the results in a table for integer values of x (i.e., $x = 0, 1, \cdots, 6\,\text{mm}$) and provide graphical representations for a more detailed set of x values within the same range.

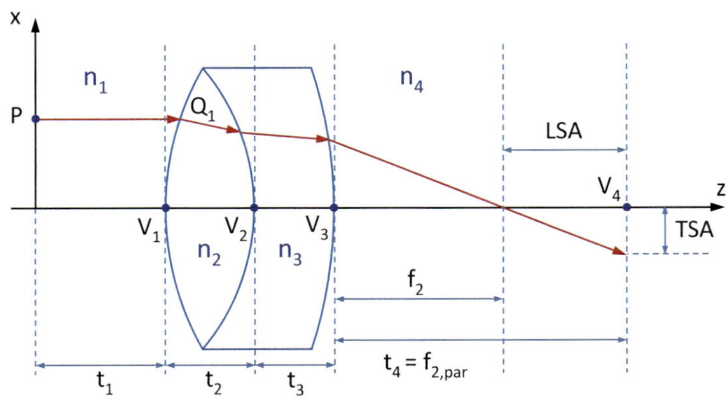

FIGURE 2P.16: A cemented doublet lens geometry for the exact calculation of its spherical aberration.

21 Five Spherical Refractive Surfaces (paraxial and exact ray tracing): This problem is an example of exact ray tracing presented in Ref. [67] and its configuration is shown in Fig. 2P.17. The object point, P, is located on the optical axis at a distance $t_1 = 12\,\text{cm}$ from the first surface. The system parameters shown in Fig. 2P.17 are: $R_1 = 10\,\text{cm}$, $R_2 = -8\,\text{cm}$, $R_3 = 12\,\text{cm}$, $R_4 = -10\,\text{cm}$, $R_5 = -8\,\text{cm}$, $t_2 = 5\,\text{cm}$, $t_3 = 5\,\text{cm}$, $t_4 = 8\,\text{cm}$, and $t_5 = 1\,\text{cm}$. The refractive indices are: $n_1 = 1.0$, $n_2 = 1.2$, $n_3 = 1.0$, $n_4 = 1.5$, $n_5 = 1.0$, and $n_6 = 1.2$. A ray originating from point P an angle θ with the optical axis. (a) First, determine the paraxial image position s'_p, measured from the vertex V_5, along with the transverse magnification. (b) Then, consider a set of rays with angles θ varying within the interval $[0, 20°]$. Using exact ray tracing, compute the image position s' as a function of the angle θ. Subsequently, calculate the longitudinal spherical aberration (LSA) and transverse spherical aberration (TSA) for these rays as functions of θ. Provide graphical representations of the results for s', LSA, and TSA as functions of θ.

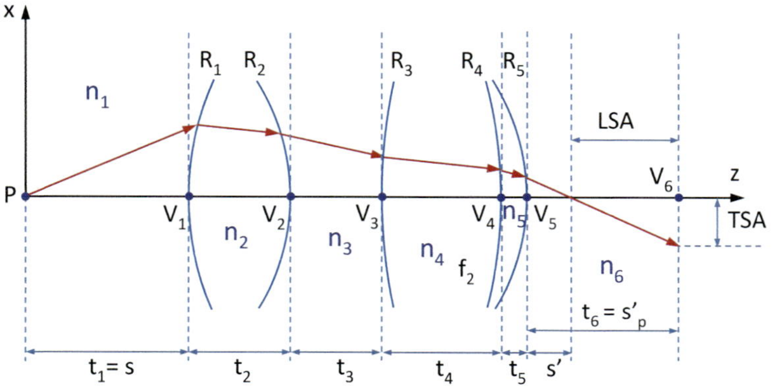

FIGURE 2P.17: The 5-spherical surfaces example adapted from Ref. [67] for the Problem 21.

Appendices

2A Sellmeier Equation

The macroscopic response of a material to electromagnetic radiation is governed by the averaged behavior of its microscopic electric and magnetic dipoles. These volume-averaged effects are represented by the material's polarization and magnetization. However, for most optical materials, which are typically nonmagnetic, the material's interaction with electromagnetic waves is primarily due to polarization, i.e., its electric dipoles. This polarization is related to the material's refractive index n (see *Physical and Electromagnetic Optics* Chapter, Chapter 3). In this appendix, we assume materials that are isotropic, linear, nonmagnetic, and homogeneous, so the refractive index can be treated as a scalar quantity. The material's response also varies with the frequency (or wavelength) of the electromagnetic radiation, a property known as *material dispersion*, expressed as $n(\omega)$ [or equivalently $n(\lambda_0)$]. In addition, most materials absorb some fraction of the electromagnetic power propagating through them, characterized by the extinction coefficient κ. For optical materials, absorption losses are generally low within the frequency ranges of interest.

Dielectric materials generally exhibit frequency-dependent regions where absorption becomes significant, resulting from molecular, ionic, or electronic resonances in the material's electric dipoles. Due to the interdependence between the refractive index n and the extinction coefficient κ (as described by the Kramers-Kronig relations), both n and κ are functions of the frequency (or wavelength) of the electromagnetic radiation. While there are no exact analytical expressions describing this dependence, a widely used empirical relationship has been established between the refractive index n of a material and the freespace wavelength λ_0. This relationship, known as the Sellmeier equation, was proposed in 1871 by Wolfgang Sellmeier [75], who observed that the wavelength dependence of the refractive index is strikingly similar for many transparent and non-transparent media in the visible and near-infrared ranges. The Sellmeier equation represents the refractive index as a series of single-dipole oscillator terms of the form:

$$n^2(\lambda_0) = 1 + \sum_{i=1}^{m} \frac{A_i \lambda_0^2}{\lambda_0^2 - B_i}. \tag{2A.1}$$

The coefficients A_i and B_i ($i = 1, \cdots, m$) in the Sellmeier equation are known as the *Sellmeier coefficients* and are determined experimentally for each material. These coefficients reflect the material's unique dispersion characteristics. Table 2A.1 provides representative Sellmeier coefficients for common optical materials with $m = 3$.

2B Vector Formulation of Reflection and Refraction

The vector formulation for reflection and refraction at a smooth boundary will be presented in this appendix, based on the work of Kloos (2007) [38]. In the derivation, Fig. 2B.1 is utilized. The smooth boundary, denoted as S, separates two homogeneous isotropic media with refractive indices n_1 and n_2. An incident ray, characterized by the unit vector $\hat{\imath}$, strikes the boundary S. The unit normal vector at the point of incidence on S is given by $\hat{n}$. The reflected and refracted rays are described by their unit vectors $\hat{r}$ and $\hat{t}$, respectively.

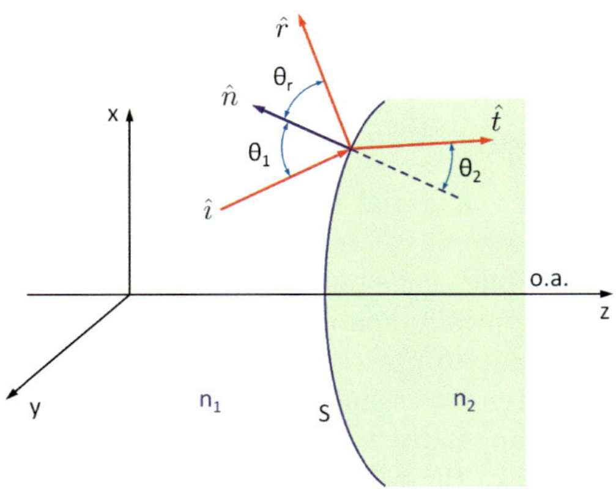

FIGURE 2B.1: Geometry to specify the vector formulation of reflection and transmission at a smooth interface between two media with refractive indices n_1 and n_2. The directions of the incident, reflected, and transmitted rays are specified by the unit vectors $\hat{\imath}$, $\hat{r}$, and $\hat{t}$, respectively.

TABLE 2A.1: Example Sellmeier Coefficients (from Ref. [76]) for common optical materials (the wavelength is expressed in microns).

Material	A_1	B_1 (μm^2)	A_2	B_2 (μm^2)	A_3	B_3 (μm^2)
Borosilicate Crown Glass (BK7)	1.03961212	$6.00069867 \times 10^{-3}$	0.231792344	$2.00179144 \times 10^{-2}$	1.01046945	103.560653
Fused Silica (quartz)	0.696166300	$4.67914826 \times 10^{-3}$	0.407942600	$1.35120631 \times 10^{-2}$	0.897479400	97.9340025
Schott Glass (multiple purpose)	1.3182408	8.79×10^{-3}	0.0244	6.09×10^{-2}	1.08915181	110
Schott Glass (SF11) (High Dispersion)	1.73759695	13.188707×10^{-3}	0.313747346	$6.23068142 \times 10^{-2}$	1.89878101	155.23629
Magnesium Fluoride (MgF_2)	0.48755108	1.8821784×10^{-3}	0.39875031	$0.89518885 \times 10^{-2}$	2.3120353	566.13559
Silicon Nitride (Si_3N_4)	3.0249	1.8317078×10^{-2}	40314	1537208.18	-	-

According to Fermat's principle, the law of reflection states:

$$\theta_1 = \theta_r \Longrightarrow \sin\theta_1 = \sin\theta_r \Longrightarrow$$
$$\sin(\pi - \theta_1) = \sin\theta_r \Longrightarrow \hat{\imath} \times \hat{n} = \hat{r} \times \hat{n} \Longrightarrow$$
$$\hat{n} \times (\hat{\imath} \times \hat{n}) = \hat{n} \times (\hat{r} \times \hat{n}) \Longrightarrow$$
$$(\hat{n}\cdot\hat{n})\hat{\imath} - (\hat{n}\cdot\hat{\imath})\hat{n} = (\hat{n}\cdot\hat{n})\hat{r} - (\hat{n}\cdot\hat{r})\hat{n} \Longrightarrow$$
$$\hat{\imath} - (\hat{n}\cdot\hat{\imath})\hat{n} = \hat{r} - (\hat{n}\cdot\hat{r})\hat{n} = \hat{r} + (\hat{n}\cdot\hat{\imath})\hat{n} \Longrightarrow$$
$$\hat{r} = \hat{\imath} - 2(\hat{n}\cdot\hat{\imath})\hat{n}, \tag{2B.1}$$

where $\hat{n}\cdot\hat{\imath} = \cos(\pi - \theta_1) = -\cos\theta_1 = -\cos\theta_r = -\hat{n}\cdot\hat{r}$ was used. Similarly the vector form of Snell's law can be derived. Specifically the following steps can be followed:

$$n_1 \sin\theta_1 = n_2 \sin\theta_2 \Longrightarrow n_1 \sin(\pi - \theta_1) = n_2 \sin\theta_2 \Longrightarrow$$
$$n_2\,(\hat{\imath} \times \hat{n}) = n_2\,(\hat{t} \times \hat{n}) \Longrightarrow$$
$$\frac{n_1}{n_2}\,[\hat{n} \times (\hat{\imath} \times \hat{n})] = \hat{n} \times (\hat{t} \times \hat{n}) \Longrightarrow$$
$$\frac{n_1}{n_2}\,[(\hat{n}\cdot\hat{n})\hat{\imath} - (\hat{n}\cdot\hat{\imath})\hat{n}] = (\hat{n}\cdot\hat{n})\hat{t} - (\hat{n}\cdot\hat{t})\hat{n} \Longrightarrow$$
$$\frac{n_1}{n_2}\,[\hat{\imath} - (\hat{n}\cdot\hat{\imath})\hat{n}] = \hat{t} - (\hat{n}\cdot\hat{t})\hat{n} = \hat{t} - \cos(\pi - \theta_2)\,\hat{n} \Longrightarrow$$
$$\frac{n_1}{n_2}\,[\hat{\imath} - (\hat{n}\cdot\hat{\imath})\hat{n}] = \hat{t} + \cos\theta_2\,\hat{n} = \hat{t} + \left[1 - \frac{n_1^2}{n_2^2}\sin^2\theta_1\right]^{1/2} \hat{n} \Longrightarrow$$
$$\frac{n_1}{n_2}\,[\hat{\imath} - (\hat{n}\cdot\hat{\imath})\hat{n}] = \hat{t} + \left[1 - \frac{n_1^2}{n_2^2}\left(1 - \cos^2\theta_1\right)\right]^{1/2} \hat{n} \Longrightarrow$$
$$\frac{n_1}{n_2}\,[\hat{\imath} - (\hat{n}\cdot\hat{\imath})\hat{n}] = \hat{t} + \left[1 - \frac{n_1^2}{n_2^2}\left(1 - (\hat{n}\cdot\hat{\imath})^2\right)\right]^{1/2} \hat{n} \Longrightarrow$$
$$\hat{t} = \frac{n_1}{n_2}\hat{\imath} - \left[\frac{n_1}{n_2}(\hat{n}\cdot\hat{\imath}) + \left[1 - \frac{n_1^2}{n_2^2}\left(1 - (\hat{n}\cdot\hat{\imath})^2\right)\right]^{1/2}\right]\hat{n}. \tag{2B.2}$$

Summarizing, given $\hat{\imath}$ and $\hat{n}$ the unit vector of the reflected and of the refracted ray are given by Eqs. (2B.1) and (2B.2), respectively. It can be easily shown that $n_2\hat{t} - n_1\hat{\imath}$ is parallel or anti-parallel to $\hat{n}$. The proof proceeds as follows:

$$(n_2\hat{t} - n_1\hat{\imath}) \times (-\hat{n}) = -n_2\,(\hat{t} \times \hat{n}) + n_1\,(\hat{\imath} \times \hat{n}) =$$
$$= -n_2 \sin\theta_2 \hat{u}_p + n_1 \sin\theta_1 \hat{u}_p = 0,$$

where $\hat{u}_p$ is the unit vector perpendicular to the plane of incidence specified by the incident ray, the refracted ray, and the normal to the interface specified by the unit vector $\hat{n}$. Thus, $n_2\hat{t} - n_1\hat{\imath} = F\hat{n}$ where F is a constant (positive for parallel to $\hat{n}$ and negative for anti-parallel to $\hat{n}$). The constant F can be determined as follows:

$$(n_2\hat{t} - n_1\hat{\imath}) \cdot (\hat{n}) = n_2\,(\hat{t}\cdot\hat{n}) - n_1\,(\hat{\imath}\cdot\hat{n}) =$$
$$= n_2 \cos\theta_2 - n_1 \cos\theta_1 = n_2\,(\cos\theta_2 - q\cos\theta_1) = F,$$

where $q = n_1/n_2$.

To illustrate the application of these equations, consider the refraction and reflection of a ray parallel to the optical axis incident on a convex spherical refractive interface. In this case, the incident ray approaches the interface at a specific height relative to the optical axis. When it meets the spherical surface, it will partially reflect back and partially refract into the second medium, following Snell's Law. By applying the vector formulations developed earlier, the directions of both the reflected and refracted rays can be calculated. Consider an incident ray parallel to the optical axis at a height h from it, as shown in

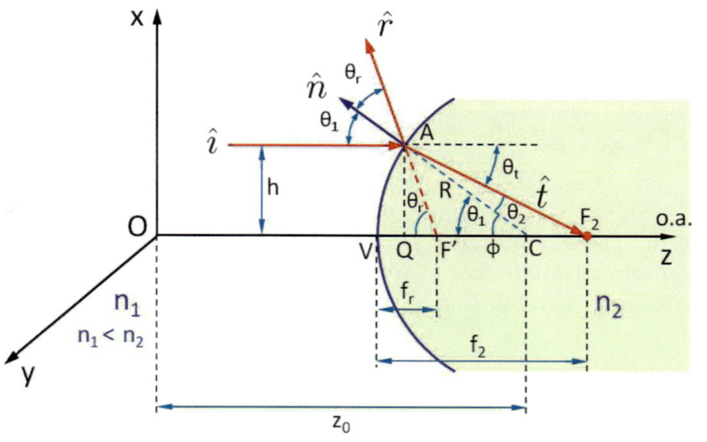

FIGURE 2B.2: Example of vector formulation of reflection and refraction laws. A spherical interface of radius $R = 100\,\text{mm}$, between two dielectric media with refractive indices $n_1 = 1$ (air) and $n_2 = 1.5$ (glass) is considered. The center C of the spherical interface is at $z_0 > R$. The directions of the incident, reflected, and transmitted rays are specified by the unit vectors $\hat{\imath}$, $\hat{r}$, and $\hat{t}$, respectively.

Fig. 2B.2. The unit vector of this incident ray is $\hat{\imath} = [0, 0, 1] = \hat{\imath}_z$. In the coordinate system of Fig. 2B.2, the convex refractive spherical surface is defined by the equation: $F(x, y, z) = [x^2 + y^2 + (z - z_0)^2]^{1/2} - R = 0$, where R is the radius of curvature, and the surface center is located at $(0, 0, z_0)$. For simplicity, assume that the ray lies within the xz plane ($y = 0$). To apply Eqs. (2B.1) and (2B.2), it is necessary to determine the unit normal vector $\hat{n}$ at the point A where the ray intersects the spherical surface. This unit vector $\hat{n}$ is derived as follows:

$$\hat{n} = \vec{\nabla} F|_A = \frac{\partial F}{\partial x}\,\hat{\imath}_x + \frac{\partial F}{\partial y}\,\hat{\imath}_y + \frac{\partial F}{\partial z}\,\hat{\imath}_z = \frac{x}{R}\,\hat{\imath}_x + \frac{y}{R}\,\hat{\imath}_y + \frac{z - z_0}{R}\,\hat{\imath}_z$$

$$= \frac{h}{R}\,\hat{\imath}_x + 0\,\hat{\imath}_y + \frac{z_0 - R\cos\phi - z_0}{R}\,\hat{\imath}_z = \frac{h}{R}\,\hat{\imath}_x - \cos\phi\,\hat{\imath}_z = \frac{h}{R}\,\hat{\imath}_x - \sqrt{1 - \frac{h^2}{R^2}}\,\hat{\imath}_z, \quad (2B.3)$$

where $\sin\phi = h/R$ and $\cos\phi = (1 - \sin^2\phi)^{1/2}$ were used.

Using the expressions for the incident ray vector $\hat{\imath}$ and the unit normal vector $\hat{n}$ in Eqs. (2B.1) and (2B.2), the unit vectors for the reflected ray $\hat{r}$ and the refracted ray $\hat{t}$ can be determined. The inner product $\hat{t} \cdot \hat{\imath}_z = \cos\theta_t$, where θ_t is the angle of refraction, allows the distance f_2 for the refracted ray to be expressed as: $f_2 = QF_2 + QV = h/\tan\theta_t + [R - (R^2 - h^2)^{1/2}]$. Similarly, the distance f_r for the reflected ray can be determined. The angle $\theta_1 + \theta_r = 2\theta_1 = 2\theta_r$ can be derived from the inner product $\hat{r} \cdot (-\hat{\imath}_z) = \cos 2\theta_r$. The distance for the reflected ray is $f_r = QF' + QV = h/\tan 2\theta_r + [R - (R^2 - h^2)^{1/2}]$. The resulting distances f_2 and f_r as functions of the incident ray height h are shown in Figs. 2.3(a) and

2.3(b), respectively. These results can also be verified geometrically-trigonometrically, as demonstrated in Sec. 2.9.

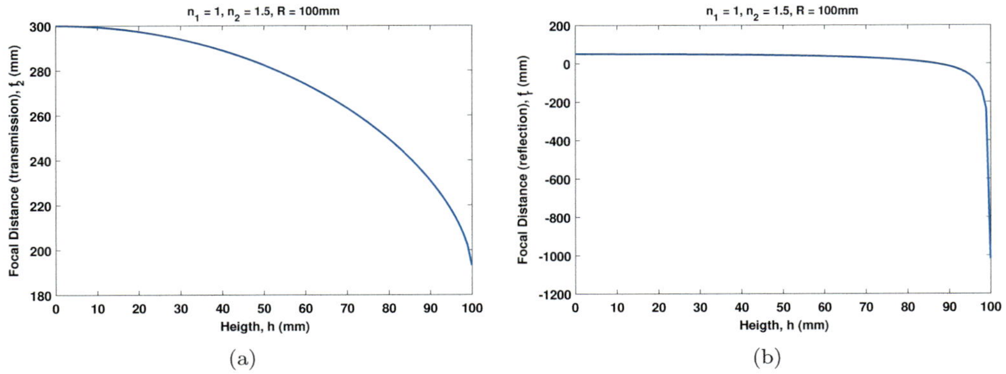

FIGURE 2B.3: (a) The focal distance f_2 (refracted ray) as a function of the incident ray height h for the spherical refractive interface of $R = 100\,\text{mm}$, with $n_1 = 1$ and $n_2 = 1.5$. (b) The focal distance f_r (reflected ray) as a function of the incident ray height h.

2C Notes on F-Number

The f-number is a critical parameter in optical systems that influences several key aspects of system performance. In general, the f-number affects or controls the following [77–79]:

(a) The amount of light incident on the image plane: The f-number determines the diameter of the entrance pupil relative to the focal length of the system. A lower f-number allows more light to pass through, increasing the brightness of the image at the image plane.

(b) The depth of focus and the depth of field: A smaller f-number results in a shallower depth of field, meaning only objects at a particular distance from the optical system will be in sharp focus. Conversely, a larger f-number increases the depth of field, bringing more of the scene into focus.

(c) The optical system resolution and optical transfer function: The f-number impacts the resolution of the optical system, with a smaller f-number potentially leading to a higher resolution and better performance in terms of spatial frequency response but potentially introducing more aberrations.

(d) The level of aperture-related aberrations: The aperture size, controlled by the f-number, affects the extent of optical aberrations such as spherical aberration, coma, and astigmatism. Larger apertures (smaller f-numbers) tend to increase the magnitude of these aberrations.

(e) The level of vignetting: The f-number influences the amount of vignetting, which occurs when the edges of the optical system's aperture are obstructed, resulting in a decrease in brightness toward the edges of the image. Larger f-numbers can reduce vignetting by ensuring a smaller entrance pupil and more even illumination.

In summary, the f-number plays a fundamental role in balancing light transmission, image sharpness, resolution, aberration control, and overall image quality.

The f-number for a camera optical system is traditionally defined as the ratio of the lens's focal length (more precisely effective focal length) f to the diameter of the entrance aperture D [see Eq. (2.81)]. To generalize the definition of the f-number, let's consider the optical system of a thin lens in air, as shown in Fig. 2C.1a [77], which forms an image of an infinitely distant object at the image plane located at point F. In this simple system, the focal length f is indicated, and the thickness of the lens is neglected, assuming it's a thin lens. The diameter of the entrance aperture is denoted as D. In the setup shown in

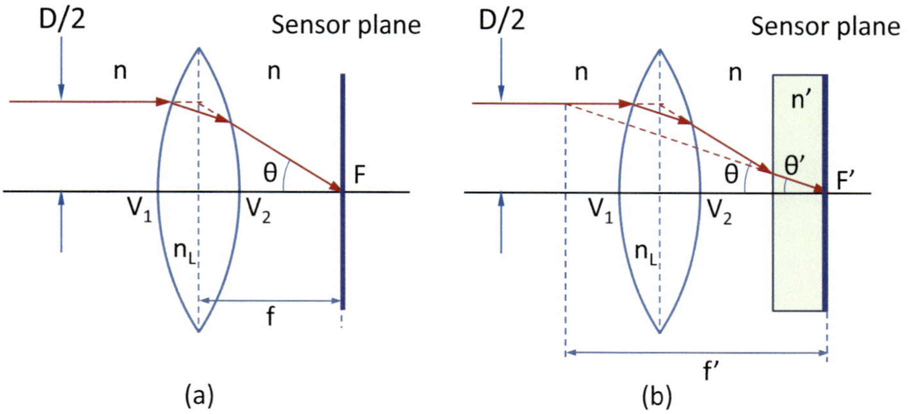

FIGURE 2C.1: (a) The f-number for an image in air. (b) The f-number for an image inside thin plate. D is the diameter of the entrance aperture of the system.

Fig. 2C.1b, a thin dielectric plate is placed over the sensor plane, with refractive indices n (for air) and n' (for the plate). The plate causes a change in the effective focal length of the system, denoted as f', due to refraction at the boundary between air and the plate. To derive the relationship between the original focal length f (without the plate) and the effective focal length f' (with the plate), we can consider the effect of the plate on the light rays passing through the system. The relation between f and f' can be derived using the refraction at the air-plate interface and applying the concept of effective focal length in the presence of a material of different refractive index. Specifically, the following relations can be easily verified:

$$\begin{aligned}
\tan\theta &= \frac{D}{2f} \quad \Longrightarrow \quad \sin\theta = \frac{D}{2f}, \\
\tan\theta' &= \frac{D}{2f'} \quad \Longrightarrow \quad \sin\theta' = \frac{D}{2f'}, \\
\frac{f'}{f} &= \frac{n'}{n},
\end{aligned} \tag{2C.1}$$

where the paraxial approximation ($\sin\theta \simeq \tan\theta$) in conjunction to the Snell's law ($n\sin\theta = n'\sin\theta'$) were utilized. In order to take into account the possible varying refractive indices between the object and image spaces, the following definition for the f-number could be used:

$$\text{f-number} = \frac{n}{2n'\sin\theta'} = \frac{n}{2NA}, \tag{2C.2}$$

where $NA = n'\sin\theta' = n\sin\theta$ is the numerical aperture of the optical system. Based on Eq. (2C.2), it is straightforward to show that under the paraxial approximation and with equal indices of refraction in the object and image space, the f-number simplifies to the form in Eq. (2.81), that is used for the basic camera.

The f-number was defined for an object at infinity. In the case of a finite object distance, the effective f-number, denoted as f-number$_e$, can be defined as follows [77–79]:

$$\text{f-number}_e = \text{f-number}\,(1-m)\,, \tag{2C.3}$$

where m is the optical system's transverse magnification. A more general definition of the effective f-number takes into account the potential pupil magnification:

$$\text{f-number}_e = \text{f-number}\left(1-\frac{m}{m_p}\right), \tag{2C.4}$$

where m_p is the optical system's pupil magnification, defined as the ratio of the exit pupil diameter (aperture stop as seen from the image) to the entrance pupil diameter (aperture stop as seen by the object). However, since m_p is usually unknown, it is often assumed to be unity, and in this case, Eq. (2C.4) simplifies to Eq. (2C.3).

2D Intricacies of 3-rd Order Imaging Equation

Equation (2.139) was derived by applying several approximations to its original form, which is expressed by Eq. (2.138). It is valuable to explore the solution for the image distance s' directly from Eq. (2.138) without any intermediate approximations except the initial one, which involves retaining only third-order terms in the Taylor expansions. For convenience, Eq. (2.138) is shown below:

$$\left(\frac{n_1}{s}+\frac{n_2}{s'}-\frac{n_2-n_1}{R}\right)\left(1+\frac{h^2}{2R^2}\right)=\frac{h^2}{2}\left[\frac{n_1}{s}\left(\frac{1}{s}+\frac{1}{R}\right)^2+\frac{n_2}{s'}\left(\frac{1}{s'}-\frac{1}{R}\right)^2\right]. \tag{2D.1}$$

The above equation is a third-order polynomial with respect to the variable $1/s'$. It can be easily transformed into the following form:

$$a_0\psi^3+a_1\psi^2+a_2\psi+a_3=0, \qquad \text{where,} \quad \psi=\frac{1}{s'}, \tag{2D.2}$$

$$a_0=\frac{h^2}{2}n_2,$$

$$a_1=-\frac{h^2 n_2}{R},$$

$$a_2=\frac{h^2 n_2}{2R^2}-\left(1+\frac{h^2}{2R^2}\right)n_2,$$

$$a_3=\frac{h^2}{2}\frac{n_1}{s}\left(\frac{1}{s}+\frac{1}{R}\right)^2-\left(1+\frac{h^2}{2R^2}\right)\left(\frac{n_1}{s}-\frac{n_2-n_1}{R}\right).$$

The above equation has either three real roots or one real root, depending on the discriminant. The discriminant of the above polynomial is given by:

$$\Delta=18a_0a_1a_2a_3-27a_0^2a_3^2-4a_0a_2^3-4a_1^3a_3+a_1^2a_2^2. \tag{2D.3}$$

If $\Delta<0$, then Eq. (2D.2) has a single real root. Let this root be z_1, in which case the image position is given by $s'=1/z_1$. If $\Delta>0$, then Eq. (2D.2) has three real solutions, which we denote as z_1, z_2, and z_3. The possible image positions are $s' = 1/z_i$ for $i = 1, 2, 3$. However, only one of these solutions is physically correct. The correct solution is selected as the image position that is closest to the approximate solution given by Eq. (2.139). Similarly to Eq. (2.140), the exact third-order solution predicts an angle $\theta_{\|}$ when $s < n_1 R/(n_2 - n_1)$. This angle is found by determining the value of $h = h_{\|}$ that makes $a_3 = 0$, and this value is given by:

$$h_{\|} = \left\{ \frac{2\left(\frac{n_1}{s} - \frac{n_2 - n_1}{R}\right)}{\frac{n_1}{s}\left(\frac{1}{R} + \frac{1}{s}\right)^2 - \frac{1}{R^2}\left(\frac{n_1}{s} - \frac{n_2 - n_1}{R}\right)} \right\}^{1/2}. \tag{2D.4}$$

Then, the corresponding angle $\theta_{\|}$ is given by Eq. (2.141).

3

Physical and Electromagnetic Optics

3.1 Introduction

Historically, light was initially perceived as consisting of particles propagating along straight lines, embodying the concept of rectilinear light propagation. This perspective prevailed due to the consistent alignment of experimental observations with this theory. The difficulty in discerning wave effects stemmed from the limited visibility of such effects, given the small wavelength of light in the visible part of the electromagnetic spectrum where experiments were conducted. The theory of rectilinear light propagation and the particle nature of light endured for centuries, adapting to accommodate minor adjustments while remaining in accord with experimental observations. This persistence continued until the early 19th century when new experimental findings introduced effects that defied interpretation within the framework of rectilinear light propagation. Notably, wave effects, previously observed by Newton (such as Newton's interference rings) and Grimaldi (initial diffraction effects), had been explained through modifications to the rectilinear theory. However, the turning point came with Young's two-slit experiment conducted between 1801 and 1802, which proved instrumental in revealing the wave nature of light. Subsequent experimental evidence indicated that light behaves as a transverse wave, a departure from the prevailing understanding of longitudinal waves like sound waves. In the case of light, the waves vibrate in a plane perpendicular to the direction of propagation in free space, such as air. Further experiments unveiled the vector property of light in its transverse vibrations, highlighting the vector nature of the electromagnetic field. This revelation paved the way for the development of the electromagnetic theory of light, encapsulated in Maxwell's equations, marking a profound shift in our understanding of the nature of light.

In the optical classification illustrated in Fig. 3.1 [19], the field of *Wave or Physical Optics* represents the wave theory of light that does not account for light polarization, treating the optical field as a scalar wavefunction.

Wave/Physical optics is a superset of Geometrical Optics, which is the rectilinear light theory discussed earlier. Notably, as the wavelength tends to zero ($\lambda \to 0$), wave optics converges to geometrical optics. However, practical applications of geometrical optics do not require the wavelength to approach zero; it is sufficient for the wavelength to be much smaller than the dimensions of the encountered objects. This explains why rectilinear light theory persisted from ancient times until the early nineteenth century. Wave/Physical Optics, while more versatile than Geometrical Optics, still fall short in explaining phenomena where light polarization is crucial. For instance, in situations like the reflection and transmission of light on a planar boundary between two dielectrics, the vectorial nature of the electromagnetic wave must be considered. The theory that encompasses these vectorial aspects is *Electromagnetic Optics*, which is a superset of both Wave/Physical Optics and Geometrical Optics, as depicted in Fig. 3.1. Wave/Physical Optics can be considered as an approximation of Electromagnetic Optics when the polarization of the light field can be neglected.

DOI: 10.1201/9781003490333-3

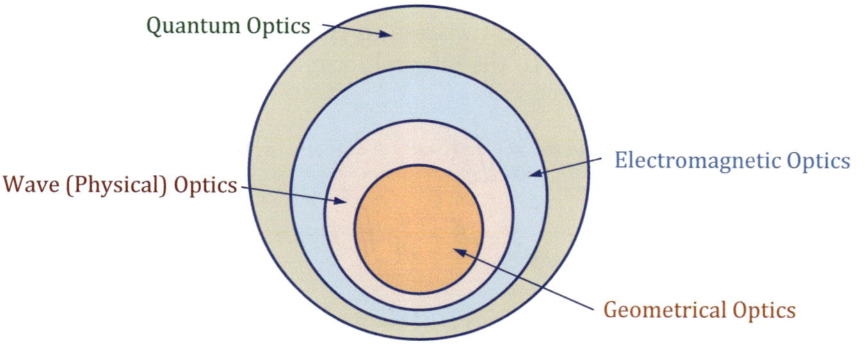

FIGURE 3.1: A diagram of the various theories of optics starting with quantum optics, which is the most rigorous, up to Geometrical (ray) optics which is the simplest. Every circle inside a larger circle is a subset of the larger circle (from Ref. [19]).

This chapter introduces fundamental concepts in Wave/Physical Optics and Electromagnetic Optics. It begins with a concise review of Maxwell's equations and their straightforward solutions, focusing on plane waves in different materials. The discussion then shifts to the quantification of light polarization. Next, the phenomenon of wave interference is explored, including basic geometries such as Young's two-slit experimental setup. Following this, the chapter examines fundamental interferometers like the Fabry-Perot, Michelson, and Mach-Zehnder interferometers. A brief overview of optical interference filters and their design is provided, with particular attention to antireflection (AR) coatings. Anisotropic interfaces and multilayers are also studied. Additionally, an introduction to the basics of light coherence is included, along with a special section on incoherent or mixed (coherent/incoherent) multilayer structures. Finally, the chapter concludes with an analysis of Gaussian beams.

3.2 Short Review of Electromagnetism

The foundation of electromagnetic field theory lies in Maxwell's equations. These equations articulate the interconnections of electromagnetic fields and their relationship to electric charges and currents, which serve as the sources of these fields. The most prevalent representation of Maxwell's equations is as follows:

$$\vec{\nabla} \times \vec{\mathcal{E}} = -\frac{\partial \vec{\mathcal{B}}}{\partial t}, \tag{3.1}$$

$$\vec{\nabla} \times \vec{\mathcal{H}} = \vec{\mathcal{J}} + \frac{\partial \vec{\mathcal{D}}}{\partial t}, \tag{3.2}$$

$$\vec{\nabla} \cdot \vec{\mathcal{D}} = \varrho, \tag{3.3}$$

$$\vec{\nabla} \cdot \vec{\mathcal{B}} = 0, \tag{3.4}$$

where $\vec{\mathcal{E}}(\vec{r},t)$ represents the electric field (in V/m), $\vec{\mathcal{D}}(\vec{r},t)$ represents the displacement vector (or the electric flux density vector in C/m), $\vec{\mathcal{H}}(\vec{r},t)$ represents the magnetic field (in A/m), $\vec{\mathcal{B}}(\vec{r},t)$ represents the magnetic flux density (in Tesla = Wb/m^2), $\vec{\mathcal{J}}(\vec{r},t)$ represents the electric current density (in A/m^2), and $\rho(\vec{r},t)$ represents the electric charge density (in C/m^3). The above equations are written in the form of differential equations while an integral form can also be used. In the integral form the Maxwell's equations are written as follows:

$$\oint_C \vec{\mathcal{E}} \cdot d\vec{\ell} = -\frac{d}{dt} \iint_S \vec{\mathcal{B}} \cdot d\vec{S}, \quad \text{(Faraday Law)}, \tag{3.5}$$

$$\oint_C \vec{\mathcal{H}} \cdot d\vec{\ell} = \int_S \vec{\mathcal{J}} \cdot d\vec{S} + \frac{d}{dt} \iint_S \vec{\mathcal{D}} \cdot d\vec{S}, \quad \text{(Ampere Law)}, \tag{3.6}$$

$$\oiint_S \vec{\mathcal{D}} \cdot d\vec{S} = \iiint_V \varrho dV, \quad \text{(Gauss Law)}, \tag{3.7}$$

$$\oiint_S \vec{\mathcal{B}} \cdot d\vec{S} = 0, \quad \text{(Absence of Magnetic Monopoles)}, \tag{3.8}$$

where in the first two equations C is a closed contour and S is a surface ending in that contour, while the orientation of its unit vector is compatible with the right-hand rule when the direction that the contour is traced. In the last two of the above integral equations, S is a closed surface that defines a volume V and its unit vector points away from the surface.

In finite or infinite spaces that encompass various materials, the utilization of the differential form of Maxwell's equations necessitates the incorporation of boundary conditions when applied at interfaces between different media. These boundary conditions in electromagnetics can be readily derived from the integral form of Maxwell's equations by reducing the contours, surfaces, and volumes to points on the boundaries between distinct regions. The resultant equations, commonly referred to as boundary conditions, are as follows:

$$\hat{\imath}_n \times (\vec{\mathcal{E}}_+ - \vec{\mathcal{E}}_-)\Big|_S = 0, \tag{3.9}$$

$$\hat{\imath}_n \times (\vec{\mathcal{H}}_+ - \vec{\mathcal{H}}_-)\Big|_S = \vec{\mathcal{K}}, \tag{3.10}$$

$$\hat{\imath}_n \cdot (\vec{\mathcal{D}}_+ - \vec{\mathcal{D}}_-)\Big|_S = \sigma, \tag{3.11}$$

$$\hat{\imath}_n \cdot (\vec{\mathcal{B}}_+ - \vec{\mathcal{B}}_-)\Big|_S = 0, \tag{3.12}$$

where $\vec{\mathcal{K}}$ is the possible surface current density (in A/m) and σ is the possible surface charge density (in C/m^2) on the boundary, respectively. The $\hat{\imath}_n$ is the unit vector normal to the boundary and pointing to the "+" region (refer to Fig. 3.2).

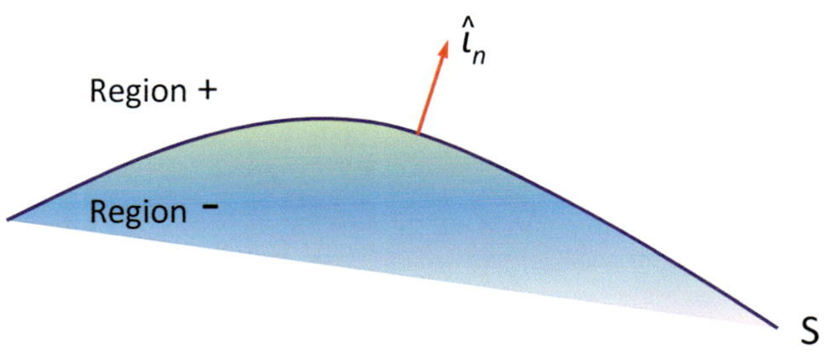

FIGURE 3.2: A boundary surface, S, between two media with different electromagnetic properties, denoted by regions "+" and "−," respectively.

Another equation derived from Maxwell's equations is the continuity equation, representing the conservation of electric charge. This equation can be expressed in both differential and integral forms as follows:

$$\vec{\nabla}\cdot\vec{\mathcal{J}} + \frac{\partial\rho}{\partial t} = 0 \Longleftrightarrow \oiint_S \vec{\mathcal{J}}\cdot d\vec{S} + \frac{d}{dt}\iiint_V \rho dV = 0, \tag{3.13}$$

while its corresponding boundary condition is:

$$\hat{\imath}_n\cdot(\vec{\mathcal{J}}_+ - \vec{\mathcal{J}}_-)\Big|_S = -\vec{\nabla}_2\cdot\vec{\mathcal{K}} - \frac{\partial\sigma}{\partial t}, \tag{3.14}$$

where $\vec{\nabla}_2\cdot\vec{\mathcal{K}}$ denotes the two-dimensional divergence of the current density.

Maxwell's equations can also be expressed in the time-harmonic form, employing phasors. For instance, the real electric field, $\vec{\mathcal{E}}(\vec{r},t)$, and its phasor representation, $\vec{E}(\vec{r},\omega)$, are connected through the relationship $\vec{\mathcal{E}}(\vec{r},t) = \mathrm{Re}\{\vec{E}(\vec{r},\omega)\exp(j\omega t)\}$, where ω denotes the radial (angular) frequency of the electromagnetic field. This same phasor representation can be applied to all electromagnetic fields and sources. The phasor representation can be considered a special case of the Fourier transform for a sinusoidal time-varying electromagnetic field. Utilizing either the Fourier transform of Eqs. (3.1)–(3.4) or their phasor representation, these equations can be expressed in the following time-harmonic form:

$$\vec{\nabla}\times\vec{E}(\vec{r},\omega) = -j\omega\vec{B}(\vec{r},\omega), \tag{3.15}$$

$$\vec{\nabla}\times\vec{H}(\vec{r},\omega) = \vec{J}(\vec{r},\omega) + j\omega\vec{D}(\vec{r},\omega), \tag{3.16}$$

$$\vec{\nabla}\cdot\vec{D}(\vec{r},\omega) = \rho(\vec{r},\omega), \tag{3.17}$$

$$\vec{\nabla}\cdot\vec{B}(\vec{r},\omega) = 0. \tag{3.18}$$

The electromagnetic fields are also connected through the constitutive equations. For instance, in freespace, these constitutive equations are expressed as:

$$\vec{\mathcal{D}}(\vec{r},t) = \epsilon_0\vec{\mathcal{E}}(\vec{r},t) \Longleftrightarrow \vec{D}(\vec{r},\omega) = \epsilon_0\vec{E}(\vec{r},\omega), \tag{3.19}$$

$$\vec{\mathcal{B}}(\vec{r},t) = \mu_0\vec{\mathcal{H}}(\vec{r},t) \Longleftrightarrow \vec{B}(\vec{r},\omega) = \mu_0\vec{H}(\vec{r},\omega), \tag{3.20}$$

where $\epsilon_0 = 8.854187817\times10^{-12}\,\mathrm{F/m}$ is the permittivity and $\mu_0 = 4\pi\times10^{-7}\,\mathrm{H/m}$ is the permeability of freespace, respectively. In a linear, homogeneous, and isotropic medium, the material response, represented through polarization, magnetization, and conductivity (or resistivity in conductive materials), is governed by relationships with the electromagnetic fields. These relationships can be expressed as follows for such media:

$$\vec{\mathcal{P}}(\vec{r},t) = \epsilon_0\int_0^\infty G_e(\tau)\vec{\mathcal{E}}(\vec{r},t-\tau)d\tau \Longleftrightarrow \vec{P}(\vec{r},\omega) = \epsilon_0\chi_e(\omega)\vec{E}(\vec{r},\omega), \tag{3.21}$$

$$\vec{\mathcal{M}}(\vec{r},t) = \int_0^\infty G_m(\tau)\vec{\mathcal{H}}(\vec{r},t-\tau)d\tau \Longleftrightarrow \vec{M}(\vec{r},\omega) = \chi_m(\omega)\vec{H}(\vec{r},\omega), \tag{3.22}$$

$$\vec{\mathcal{J}}(\vec{r},t) = \int_0^\infty G_c(\tau)\vec{\mathcal{E}}(\vec{r},t-\tau)d\tau \Longleftrightarrow \vec{J}(\vec{r},\omega) = \sigma(\omega)\vec{E}(\vec{r},\omega), \tag{3.23}$$

where G_e, G_m, and G_c are kernels that represent the material's memory effect with respect to the electromagnetic fields. The integration intervals from 0 to ∞ ensure causality, i.e., polarization cannot depend on future values of the electric field. The parameters χ_e, χ_m, and σ denote the material's electric susceptibility, magnetic susceptibility, and conductivity, respectively, which are the Fourier transforms of the kernels G_e, G_m, and G_c. By applying these relations in the frequency domain, the constitutive equations can be expressed as follows:

$$\vec{D}(\vec{r},\omega) = \epsilon_0 \vec{E}(\vec{r},\omega) + \vec{P}(\vec{r},\omega) = \epsilon_0 \left[1 + \chi_e(\omega)\right] \vec{E}(\vec{r},\omega) = \epsilon(\omega)\vec{E}(\vec{r},\omega), \tag{3.24}$$

$$\vec{B}(\vec{r},\omega) = \mu_0[\vec{H}(\vec{r},\omega) + \vec{M}(\vec{r},\omega)] = \mu_0 \left[1 + \chi_m(\omega)\right] \vec{H}(\vec{r},\omega) = \mu(\omega)\vec{H}(\vec{r},\omega), \tag{3.25}$$

$$\vec{J}(\vec{r},\omega) = \sigma(\omega)\vec{E}(\vec{r},\omega), \tag{3.26}$$

where $\epsilon(\omega) = \epsilon_0(1 + \chi_e) = \epsilon_0\varepsilon_r(\omega)$, is the material permittivity and ε_r is the material relative permittivity. The material's permeability is given by $\mu(\omega) = \mu_0(1+\chi_m) = \mu_0\mu_r(\omega)$ with μ_r representing the relative permeability. The dependence of macroscopic parameters like permittivity, permeability, and conductivity on frequency is known as *dispersion.* In practice, dispersion is present in all real materials. However, in an idealized, dispersion-free material, these parameters are independent of frequency, and Eqs. (3.24)–(3.26) remain applicable in the time domain. For materials that are linear, homogeneous, and anisotropic, the constitutive equations can be expressed as follows:

$$\vec{D}(\vec{r},\omega) = \tilde{\epsilon}(\omega)\vec{E}(\vec{r},\omega), \tag{3.27}$$

$$\vec{B}(\vec{r},\omega) = \tilde{\mu}(\omega)\vec{H}(\vec{r},\omega), \tag{3.28}$$

$$\vec{J}(\vec{r},\omega) = \tilde{\sigma}(\omega)\vec{E}(\vec{r},\omega), \tag{3.29}$$

where $\tilde{\epsilon}$, $\tilde{\mu}$, and $\tilde{\sigma}$ are the permittivity, permeability, and conductivity tensors (3×3 matrices in this case). More complex constitutive equations can describe real materials, including bianisotropic materials, where $\vec{D}$ depends on both $\vec{E}$ and $\vec{H}$, and similarly, $\vec{B}$ depends on both $\vec{H}$ and $\vec{E}$. Furthermore, real materials may exhibit additional complexities, such as anisotropy, dispersion or non-dispersion, inhomogeneity, and nonlinearity. In anisotropic or bianisotropic materials, properties vary with direction, while dispersive materials have frequency-dependent macroscopic parameters. In inhomogeneous materials, parameters vary spatially, as seen in regions like holographic gratings. For nonlinear materials, the polarization, magnetization, and/or conductivity have nonlinear dependencies on the electromagnetic fields.

3.2.1 Wave Equation

Assuming wave propagation in an isotropic, homogeneous, dispersionless, lossless, linear medium with permittivity ϵ and permeability μ, it is straightforward to eliminate either the electric or magnetic field from Maxwell's equations, Eqs. (3.1)-(3.4), to obtain the three-dimensional wave equation. For example, by eliminating the magnetic field from Maxwell's equations, the following expression is obtained:

$$\nabla^2 \vec{\mathcal{E}} - \frac{1}{u^2}\frac{\partial^2 \vec{\mathcal{E}}}{\partial t^2} = 0, \tag{3.30}$$

where $u = 1/\sqrt{\epsilon\mu}$ is the phase velocity of the wave. A similar equation can be derived with respect to the magnetic field in case that the electric field is eliminated. Equation (3.30) is a three-dimensional wave equation which can be decomposed into three scalar equations of the form:

$$\nabla^2 f - \frac{1}{u^2}\frac{\partial^2 f}{\partial t^2} = 0, \qquad \text{where} \quad f = \mathcal{E}_x, \mathcal{E}_y, \mathcal{E}_z. \tag{3.31}$$

For simplicity, assume that $f = f(z,t)$, representing a wave traveling in the z direction. Then the previous equation simplifies to the following form:

$$\frac{\partial^2 f}{\partial z^2} - \frac{1}{u^2}\frac{\partial^2 f}{\partial t^2} = 0. \tag{3.32}$$

It is straightforward to show that this equation accepts solutions of the form $f(z,t) = F_+ g_+(z - ut) + F_- g_-(z + ut)$, where $g_+(*)$ and $g_-(*)$ are any functions that are at least twice differentiable. It follows that the g_+ solution represents a wave traveling along the $+z$-axis with phase velocity u, while the g_- represents a wave traveling along the $-z$-axis with the same phase velocity u. For instance, for a point A shown in Fig. 3.3, the $g_+(z - ut)$ solution satisfies the equation $dz - udt = 0 \Longrightarrow dz/dt = u$, indicating propagation along the $+z$ direction. Similarly, if A is a point of $g_-(z + ut)$ solution, it satisfies the equation $dz + udt = 0 \Longrightarrow dz/dt = -u$, indicating propagation along the $-z$ direction with velocity u. In the special case where $f(z,t) = \mathcal{E}x(z,t) = E{+}\cos[k(z - ut)] + E_-\cos[k(z + ut)]$

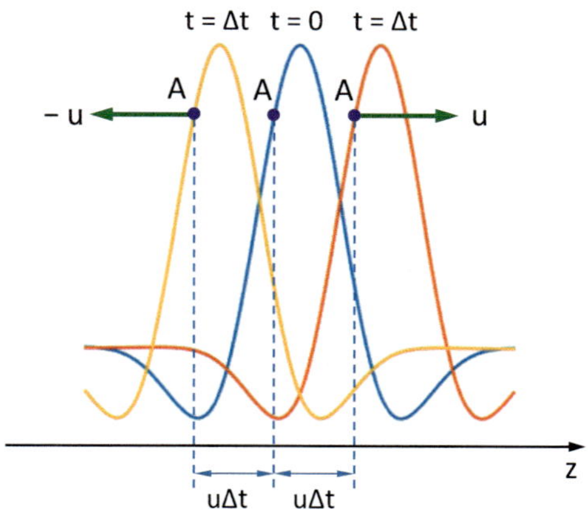

FIGURE 3.3: A waveform of the form $g_\pm(kz \mp \omega t) = \exp[-0.2(kz \mp \omega t)^2]\cos(kz \mp \omega t)$ that is comprised of a forward z propagating term, $g_+(kz - \omega t)$, and a backward $-z$ propagating term, $g_-(kz + \omega t)$.

and $g_\pm = \cos[k(z \pm ut)] = \cos(kz \pm \omega t)$ (where the spatial period of this wave defines the wavelength λ such that $k\lambda = 2\pi$), the following equations can be written:

$$\lambda = \frac{2\pi}{k}, \qquad \text{and} \qquad k = \frac{\omega}{u},$$

$$u = \frac{1}{\sqrt{\epsilon\mu}} = \frac{c}{n} \qquad \text{and} \qquad n = \sqrt{\epsilon_r \mu_r}.$$

In the provided equations, ω represents the radial frequency ($\omega = 2\pi\nu$, where ν is the frequency) of the wave. Additionally, ϵ_r and μ_r denote the relative permittivity and relative permeability of the medium, respectively. The term k represents the wavenumber, n is the refractive index, and $c = 2.99792458 \times 10^8\, m/s$ is the velocity of light in freespace. The magnetic field can be determined from Eq. (3.1), assuming no static field components, and is given by:

$$\mathcal{H}_y(z,t) = E_+ \frac{k}{\omega\mu}\cos[k(z - ut)] - E_- \frac{k}{\omega\mu}\cos[k(z + ut)],$$

$$= \frac{E_+}{Z}\cos[k(z - ut)] - \frac{E_-}{Z}\cos[k(z + ut)], \qquad \text{where} \qquad Z = \sqrt{\frac{\mu}{\epsilon}}, \tag{3.33}$$

with Z representing the wave impedance (in Ω). This simple solution is a special case of a *plane wave*. It earns this designation because the phasefront (or wavefront) of the wave remains constant in planes perpendicular to the direction of propagation. Equation (3.30) can be expressed in time-harmonic form, which is more practical for most problems in the optical frequency regime. By utilizing the phasor (or Fourier transform) of the electric (or magnetic) field, the wave equation can be rewritten as:

$$\nabla^2 \vec{E} + k_0^2 \epsilon_r \mu_r \vec{E} = 0. \tag{3.34}$$

The last equation is also recognized as the *Helmholtz equation*. In the next section, the plane wave solution is generalized and defined for both isotropic and anisotropic media.

3.2.2 General Plane-Wave Solutions

3.2.2.1 Isotropic Materials

In this section, we assume that the materials of interest are lossless dielectrics (with zero conductivity, i.e., $\sigma = 0$). Additionally, it is assumed that there are no sources (no electric charges, $\rho = 0$, and no electric currents, $\vec{J} = 0$). This implies that the electromagnetic fields were generated at an infinite distance from the areas of interest. Furthermore, all media are considered linear, homogeneous, isotropic, and nonmagnetic, which is typical for optical materials. Solutions of the form (time-harmonic) $\vec{E}(\vec{r}, \omega) = \vec{E}_0 \exp(-j\vec{k}\cdot\vec{r})$ are sought (phasors). The real field can be determined from $\vec{\mathcal{E}}(\vec{r}, t) = \text{Re}\{\vec{E}_0 \exp(-j\vec{k}\cdot\vec{r}) \exp(j\omega t)\}$. This form of solution for the electric field represents a plane wave because the locus of constant phase (wavefront) is an infinite plane perpendicular to the direction of propagation, defined by the wavevector $\vec{k}$. Utilizing the constitutive equations [Eqs. (3.24) and (3.25)], Maxwell's equations can be expressed in the following form:

$$\vec{k} \times \vec{E} = \omega \mu_0 \vec{H}, \tag{3.35}$$

$$\vec{k} \times \vec{H} = -\omega \epsilon_0 \varepsilon_r \vec{E}, \tag{3.36}$$

$$\vec{k} \cdot \vec{E} = 0, \tag{3.37}$$

$$\vec{k} \cdot \vec{H} = 0. \tag{3.38}$$

By eliminating the magnetic field from the above equations, it is straightforward to show that the following equation must be satisfied:

$$\left[\vec{k} \cdot \vec{k} - \omega^2 \epsilon_0 \mu_0 n^2\right] \vec{E} = 0, \tag{3.39}$$

with $n^2 = \varepsilon_r$ and n defined as the index of refraction of the material (since $\mu_r = 1$ for nonmagnetic media). Equation (3.39) represents the wave equation (or Helmholtz equation) for plane wave solutions in the time-harmonic form. For nontrivial solutions to exist, it is necessary for the following dispersion equation to be satisfied:

$$\vec{k} \cdot \vec{k} - \omega^2 \epsilon_0 \mu_0 n^2 = \vec{k} \cdot \vec{k} - k_0^2 n^2 = 0, \tag{3.40}$$

where $k_0 = \omega\sqrt{\epsilon_0 \mu_0} = \omega/c = 2\pi/\lambda_0$, with k_0, c, λ_0 being the freespace wavenumber, the freespace light velocity, and the freespace wavelength, respectively. As implied by Eqs. (3.35)-(3.38), the electric field, the magnetic field, and the wavevector form a right-handed orthogonal system of vectors. Furthermore, Eq. (3.40) represents a sphere in

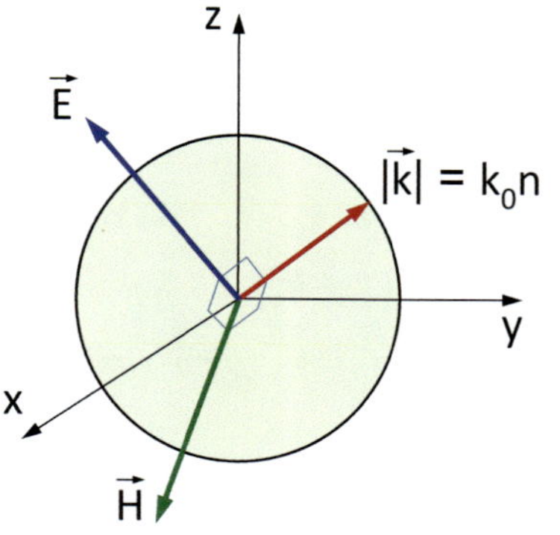

FIGURE 3.4: A plane wave dispersion sphere of radius $|\vec{k}| = k_0 n$. The electric field, the magnetic field, and the wavevector form a right-handed orthogonal system.

wavevector space of radius $k_0 n$. In other words, for any direction of propagation of an electromagnetic wave in an isotropic, homogeneous, and linear medium, the index of refraction is constant, while the polarization remains perpendicular to the direction of propagation. The wavevector surface (sphere) is illustrated in Fig. 3.4. Two orthogonal polarizations, referred to as eigen-polarizations, are typically recognized, as they propagate inside the medium without being altered. For isotropic media, the selection of the two orthogonal polarizations is arbitrary, as there are infinite pairs of orthogonal polarizations for each direction of propagation. However, this situation changes significantly in anisotropic materials.

3.2.2.2 Anisotropic Materials

Let us now assume that the materials under consideration are lossless, homogeneous, linear, and anisotropic [80]. In such materials, the permittivity is represented as a tensor (a 3×3 matrix), denoted as $\tilde{\epsilon} = \epsilon_0 \tilde{\varepsilon}_r$. The elements of this tensor are given by $\epsilon_{ij} = \epsilon_0 \varepsilon_{r,ij}$, where $i, j = x, y, z$, and $\varepsilon_{r,ij}$ are the corresponding components of the relative permittivity tensor. It should be noted that when the permittivity tensor is expressed in its principal axes system, it assumes a diagonal form: $\tilde{\epsilon} = \epsilon_0 \text{diag}[\varepsilon_{r,xx}, \varepsilon_{r,yy}, \varepsilon_{r,zz}]$ where "diag" denotes a diagonal matrix. If two of the diagonal elements of the relative permittivity are equal while the third differs, the material is classified as *uniaxial*, possessing a single optic axis. Conversely, if all three diagonal elements are distinct, the material is termed *biaxial*, possessing two optic axes. To determine the plane wave solutions for a general anisotropic (biaxial) material, it is necessary to solve the time-harmonic form of Maxwell's equations:

$$\vec{k} \times \vec{E} = \omega \mu_0 \vec{H}, \tag{3.41}$$

$$\vec{k} \times \vec{H} = -\omega \epsilon_0 \tilde{\varepsilon}_r \vec{E}, \tag{3.42}$$

$$\vec{k} \cdot \vec{D} = \epsilon_0 \vec{k} \cdot \left(\tilde{\varepsilon}_r \vec{E} \right) = 0, \tag{3.43}$$

$$\vec{k} \cdot \vec{H} = 0. \tag{3.44}$$

By eliminating the magnetic field, it can be readily demonstrated that the electric field must satisfy the following equation:

$$\vec{k}(\vec{k} \cdot \vec{E}) - (\vec{k} \cdot \vec{k})\vec{E} = -k_0^2 \tilde{\varepsilon}_r \vec{E}. \tag{3.45}$$

The last equation represents the wave equation for a general anisotropic material in the context of a plane wave solution. By expressing the wavevector, the electric field, and the relative permittivity tensor in the Cartesian x, y, z coordinate system, the following equation must be satisfied:

$$\begin{bmatrix} k_0^2\varepsilon_{r,xx} - (k_y^2+k_z^2) & k_xk_y + k_0^2\varepsilon_{r,xy} & k_xk_z + k_0^2\varepsilon_{r,xz} \\ k_yk_x + k_0^2\varepsilon_{r,yx} & k_0^2\varepsilon_{r,yy} - (k_x^2+k_z^2) & k_yk_z + k_0^2\varepsilon_{r,yz} \\ k_zk_x + k_0^2\varepsilon_{r,zx} & k_zk_y + k_0^2\varepsilon_{r,zy} & k_0^2\varepsilon_{r,zz} - (k_x^2+k_y^2) \end{bmatrix} \begin{bmatrix} E_x \\ E_y \\ E_z \end{bmatrix} = 0 \Longrightarrow$$

$$\Longrightarrow \left[\tilde{\mathcal{A}}(k_x,k_y,k_z)\right] \begin{bmatrix} E_x \\ E_y \\ E_z \end{bmatrix} = 0. \quad (3.46)$$

For the last equation to allow nontrivial solutions, it is essential that the determinant of the

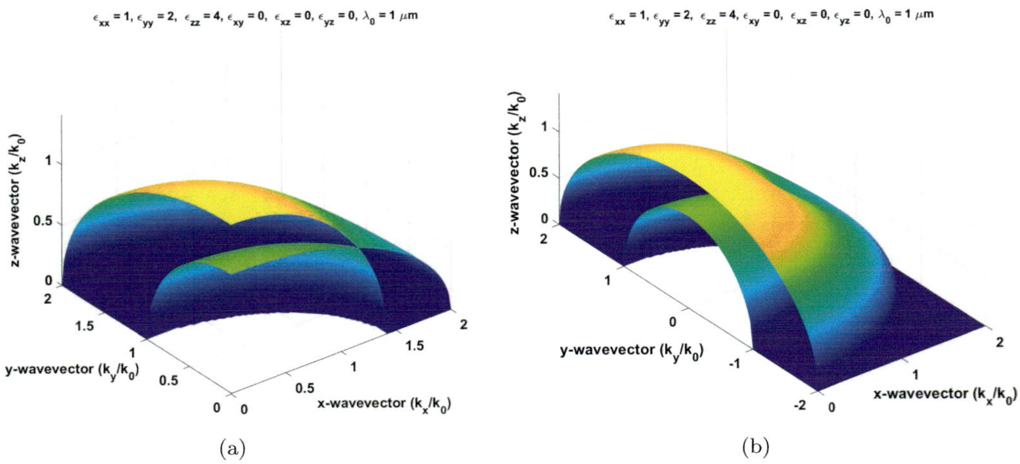

FIGURE 3.5: The wavevector surface of a biaxial material with $\varepsilon_{xx} = 1$, $\varepsilon_{yy} = 2$, $\varepsilon_{zz} = 4$, $\varepsilon_{xy} = \varepsilon_{xz} = \varepsilon_{yz} = 0$, and freespace wavelength $\lambda_0 = 1.0\,\mu$m. (a) One-eighth of the wavevector surface is shown. Observe how the two wavevector sheets intersect. (b) One-quarter of the wavevector surface is shown. The intersection of the two-sheet surface defines one of the optic axes. There is a similar intersection point in the symmetrical quarter which defines the second optic axes.

matrix $\tilde{\mathcal{A}}$ equals zero. In mathematical terms, the following condition must be satisfied [81]:

$$det\left[\tilde{\mathcal{A}}(k_x,k_y,k_z)\right] = det\left[k_0^2\tilde{\varepsilon}_r - k^2\tilde{I} + \vec{k}\vec{k}\right] = 0. \quad (3.47)$$

In the given equation, "*det*" refers to the determinant of the associated matrix. This equation defines the wavevector surface for an anisotropic material and serves as an analog to Eq. (3.40), which applies to the isotropic case. The term $\tilde{I}$ denotes a 3×3 identity matrix, while $\vec{k}\vec{k}$ represents a dyadic product, resulting in a 3×3 matrix. The wavevector surface described by Eq. (3.47) is considerably more complex than the spherical surface characteristic of isotropic materials. Specifically, it forms a two-sheeted surface, with the two sheets intersecting at certain points. These intersection points correspond to the material's two optic axes. To facilitate understanding, a representative case is illustrated in Fig. 3.5. For each wavevector (k_x, k_y, k_z) that satisfies Eq. (3.47), the corresponding refractive index can be calculated as $n = (k_x^2+k_y^2+k_z^2)^{1/2}/k_0$, whereas the polarization of the electric field vector is determined from the null space of the matrix $\tilde{A}(k_x, k_y, k_z)$. If the wavevector is expressed in a rectangular coordinate system, with its direction specified by the azimuthal, ϕ, and polar, θ, angles, then:

$$\vec{k} = k_0 n(a_x \hat{\imath}_x + a_y \hat{\imath}_y + a_z \hat{\imath}_z) = k_0 n(\sin\theta\cos\phi\ \hat{\imath}_x + \sin\theta\sin\phi\ \hat{\imath}_y + \cos\theta\ \hat{\imath}_z), \tag{3.48}$$

where n is the refractive index experienced by the plane wave as it propagates through the medium in the specified direction. By substituting the wavevector components from Eq. (3.48) into Eq. (3.47), the following bi-quadratic equation for n can be derived:

$$An^4 + Bn^2 + C = 0, \qquad \text{where,} \tag{3.49}$$

$$\begin{aligned}
A = {} & a_x^4\varepsilon_{r,xx} + a_y^4\varepsilon_{r,yy} + a_z^4\varepsilon_{r,zz} + a_x^2a_y^2\varepsilon_{r,xx} + a_x^2a_z^2\varepsilon_{r,xx} + a_x^2a_y^2\varepsilon_{r,yy} + a_y^2a_z^2\varepsilon_{r,yy} \\
& + a_x^2a_z^2\varepsilon_{r,zz} + a_y^2a_z^2\varepsilon_{r,zz} + 2a_xa_y^3\varepsilon_{r,xy} + 2a_x^3a_y\varepsilon_{r,xy} + 2a_xa_z^3\varepsilon_{r,xz} + 2a_x^3a_z\varepsilon_{r,xz} \\
& + 2a_ya_z^3\varepsilon_{r,yz} + 2a_y^3a_z\varepsilon_{r,yz} + 2a_xa_ya_z^2\varepsilon_{r,xy} + 2a_xa_y^2a_z\varepsilon_{r,xz} + 2a_x^2a_ya_z\varepsilon_{r,yz}, \\
B = {} & -a_x^2\varepsilon_{r,xx}\varepsilon_{r,yy} + a_x^2\varepsilon_{r,xz}^2 + a_y^2\varepsilon_{r,xy}^2 + a_y^2\varepsilon_{r,yz}^2 + a_z^2\varepsilon_{r,xz}^2 + a_z^2\varepsilon_{r,yz}^2 + a_x^2\varepsilon_{r,xy}^2 - a_y^2\varepsilon_{r,xx}\varepsilon_{r,yy} \\
& - a_x^2\varepsilon_{r,xx}\varepsilon_{r,zz} - a_z^2\varepsilon_{r,xx}\varepsilon_{r,zz} - a_y^2\varepsilon_{r,yy}\varepsilon_{r,zz} - a_z^2\varepsilon_{r,yy}\varepsilon_{r,zz} + 2a_xa_y\varepsilon_{r,xz}\varepsilon_{r,yz} \\
& - 2a_xa_y\varepsilon_{r,xy}\varepsilon_{r,zz} + 2a_xa_z\varepsilon_{r,xy}\varepsilon_{r,yz} - 2a_xa_z\varepsilon_{r,xz}\varepsilon_{r,yy} \\
& + 2a_ya_z\varepsilon_{r,xy}\varepsilon_{r,xz} - 2a_ya_z\varepsilon_{r,xx}\varepsilon_{r,yz}, \\
C = {} & \varepsilon_{r,xx}\varepsilon_{r,yy}\varepsilon_{r,zz} - \varepsilon_{r,zz}\varepsilon_{r,xy}^2 + 2\varepsilon_{r,xy}\varepsilon_{r,xz}\varepsilon_{r,yz} - \varepsilon_{r,yy}\varepsilon_{r,xz}^2 - \varepsilon_{r,xx}\varepsilon_{r,yz}^2.
\end{aligned}$$

Equation (3.49) always yields two real, positive solutions, corresponding to the two extraordinary waves that can propagate in each specified direction. The polarization of the associated electric field can be determined from Eq. (3.46). It should be noted that the eigenvectors of $\vec{D}$ are orthogonal to each other. The previous equation simplifies when the relative permittivity is expressed in its principal axis system, making all off-diagonal elements zero ($\varepsilon_{r,xy} = \varepsilon_{r,xz} = \varepsilon_{r,yz} = 0$). In this case, Eq. (3.49) can be rewritten as:

$$\begin{aligned}
a_x^2n^2(n^2 - \varepsilon_{r,yy})(n^2 - \varepsilon_{r,zz}) + a_y^2n^2(n^2 - \varepsilon_{r,xx})(n^2 - \varepsilon_{r,zz}) + a_z^2n^2(n^2 - \varepsilon_{r,xx})(n^2 - \varepsilon_{r,yy}) \\
= (n^2 - \varepsilon_{r,xx})(n^2 - \varepsilon_{r,yy})(n^2 - \varepsilon_{r,zz}).
\end{aligned} \tag{3.50}$$

The last equation can be expressed in the most common compact form [80]:

$$\frac{a_x^2}{n^2 - \varepsilon_{r,xx}} + \frac{a_y^2}{n^2 - \varepsilon_{r,yy}} + \frac{a_z^2}{n^2 - \varepsilon_{r,zz}} = \frac{1}{n^2}, \tag{3.51}$$

though the latter expression cannot be used along the principal axes of the coordinate system.

For a uniaxial material, the permittivity tensor can be represented (in its principal axis system) as $\tilde{\epsilon} = \epsilon_0 \mathrm{diag}[\varepsilon_O, \varepsilon_O, \varepsilon_E] = \epsilon_0 \mathrm{diag}[n_O^2, n_O^2, n_E^2]$, where n_O and n_E denote the ordinary and the principal extraordinary refractive indices, respectively. The direction of the unique optic axis is indicated by the unit vector $\hat{c} = c_x\hat{\imath}_x + c_y\hat{\imath}_y + c_z\hat{\imath}_z$. By decomposing all vectors in Eqs. (3.41)-(3.44) into components along the optic axis and those transverse to the optic axis (for instance, for the electric field, these are E_c and $\vec{E}_t$, respectively), they can be expressed in the following form [82]:

$$(\vec{k}_t \times \hat{c})E_c + k_c(\hat{c} \times \vec{E}_t) = \omega\mu_0\vec{H}_t, \tag{3.52}$$

$$\vec{k}_t \times \vec{E}_t = \omega\mu_0 H_c \hat{c}, \tag{3.53}$$

$$(\vec{k}_t \times \hat{c})H_c + k_c(\hat{c} \times \vec{H}_t) = -\omega\epsilon_0 n_O^2 \vec{E}_t, \tag{3.54}$$

$$\vec{k}_t \times \vec{H}_t = -\omega\epsilon_0 n_E^2 E_c \hat{c}, \tag{3.55}$$

$$n_O^2 \vec{k}_t \cdot \vec{E}_t + n_E^2 k_c E_c = 0, \tag{3.56}$$

$$\vec{k}_t \cdot \vec{H}_t + k_c H_c = 0. \tag{3.57}$$

Manipulating the above equations, it is straightforward to derive the wavevector surface equation equivalent to Eq. (3.47) for the uniaxial material. This is written as follows:

$$\left[\vec{k} \cdot \vec{k} - k_0^2 n_O^2\right] \left[n_O^2 \vec{k} \cdot \vec{k} + (n_E^2 - n_O^2)(\vec{k} \cdot \hat{c})^2 - k_0^2 n_O^2 n_E^2\right] = 0. \tag{3.58}$$

From Eq. (3.58) the two possible solutions become apparent. The corresponding polarizations are determined as follows:

$$\vec{k} \cdot \vec{k} - k_0^2 n_O^2 = 0 \Longrightarrow \vec{E} \cdot \hat{c} = E_c = 0, \qquad \text{ordinary wave}, \tag{3.59}$$

$$n_O^2 \vec{k} \cdot \vec{k} + (n_E^2 - n_O^2)(\vec{k} \cdot \hat{c})^2 - k_0^2 n_O^2 n_E^2 = 0, \Longrightarrow \vec{H} \cdot \hat{c} = H_c = 0. \qquad \text{extraordinary wave.} \tag{3.60}$$

To facilitate a better understanding, a sample case is illustrated in Fig. 3.6, which is similar to Fig. 3.5 but for a uniaxial material. Half of the wavevector surfaces along with their optic axes are displayed in Figs. 3.7a and 3.7b for the biaxial and uniaxial cases presented previously. For comparison purposes, a figure similar to Figs. 3.5 and 3.6 is provided in Fig. 3.8 for an isotropic material with unity relative permittivity.

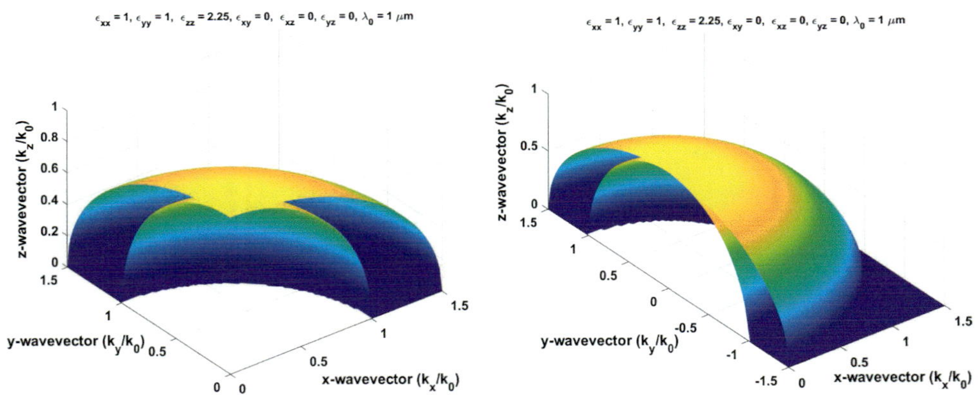

(a) (b)

FIGURE 3.6: The wavevector surface of a uniaxial material with $\varepsilon_{xx} = n_O^2 = 1$, $\varepsilon_{yy} = n_O^2 = 1$, $\varepsilon_{zz} = n_E^2 = 2.25$, $\varepsilon_{xy} = \varepsilon_{xz} = \varepsilon_{yz} = 0$, and freespace wavelength $\lambda_0 = 1.0\,\mu\text{m}$ are illustrated. (a) One-eighth of the wavevector surface is shown. The one surface is a sphere [Eq. (3.59)] while the other is an ellipsoid [Eq. (3.60)]. Observe how the two wavevector sheets become tangent in this case. (b) One-quarter of the wavevector surface is shown. The two sheet surfaces become tangent in the direction of the optic axis which in this case is the z-axis.

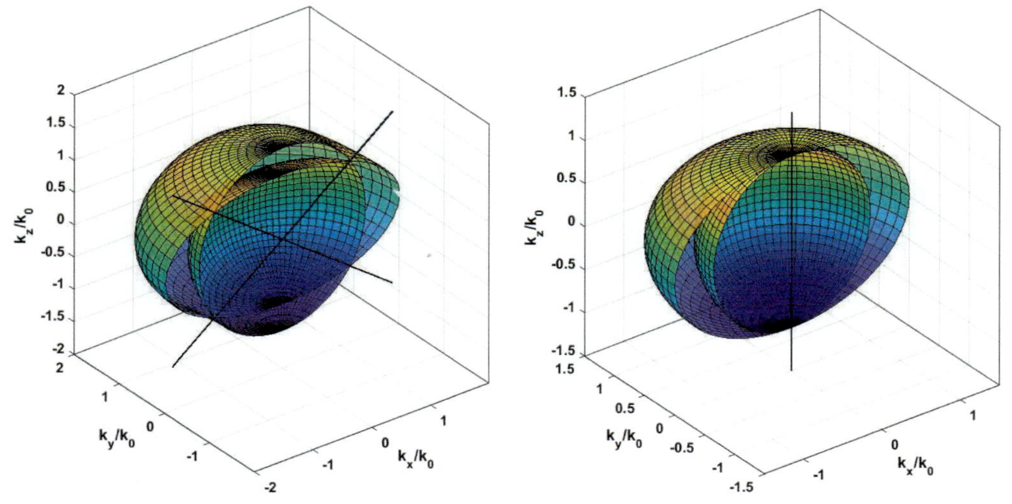

FIGURE 3.7: (a) The wavevector surface of a biaxial material with $\varepsilon_{xx} = 1$, $\varepsilon_{yy} = 2$, $\varepsilon_{zz} = 4$, $\varepsilon_{xy} = \varepsilon_{xz} = \varepsilon_{yz} = 0$, and freespace wavelength $\lambda_0 = 1.0\,\mu\text{m}$ is presented. Half of the wavevector surface is shown. Observe the two wavevector sheets that intersect in four points that define the two optic axes of the biaxial material. (b) The wavevector surface of a uniaxial material with $\varepsilon_{xx} = 1$, $\varepsilon_{yy} = 1$, $\varepsilon_{zz} = 2.25$, $\varepsilon_{xy} = \varepsilon_{xz} = \varepsilon_{yz} = 0$, and freespace wavelength $\lambda_0 = 1.0\,\mu\text{m}$. Half of the wavevector surface is shown. Observe the ellipsoidal and spherical wavevector sheets that touch only at two points where the single optic axis is defined.

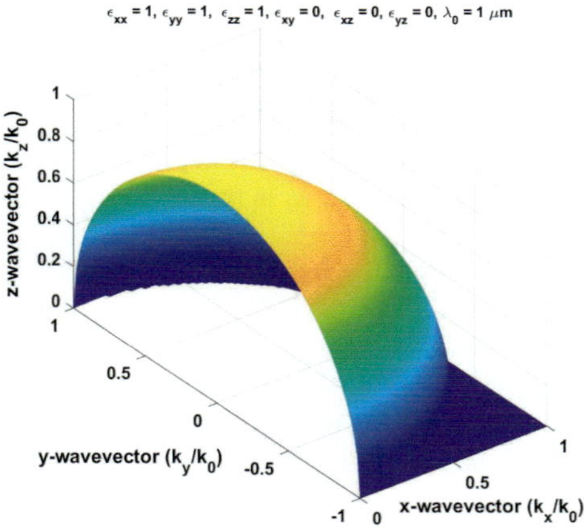

FIGURE 3.8: The wavevector surface of an isotropic material with $\varepsilon = n^2 = 1$, and freespace wavelength $\lambda_0 = 1.0\,\mu\text{m}$ is illustrated. One-quarter of the wavevector surface is shown. The two-sheeted surface collapses to a single spherical surface.

3.2.2.3 The Index Ellipsoid

The index ellipsoid is used to determine the eigen-polarizations and corresponding refractive indices for a given direction of propagation in an anisotropic crystal. The index ellipsoid, also known as the optical indicatrix, represents a normalized surface of constant electromagnetic energy density. It is well-established (and reviewed in Sec. 3.2.4) that in a homogeneous, lossless, and linear medium (such as a dielectric), the electromagnetic energy density is given by $\langle w_{em}\rangle = (1/4)\mathrm{Re}\{\vec{E}\cdot\vec{D}^*\}$. By substituting $\vec{E} = \epsilon_0^{-1}[\varepsilon_r]^{-1}\vec{D}$, the electromagnetic energy density can be expressed in the principal axis system as:

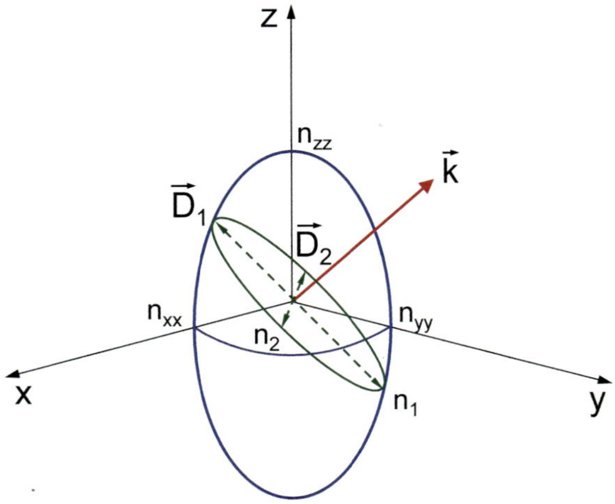

FIGURE 3.9: The index ellipsoid in the general biaxial case where $n_{xx} \neq n_{yy} \neq n_{zz} \neq n_{xx}$, shown in the principal axes system. A random direction of a plane wave wavevector is also shown. The intersection of the plane perpendicular to the wavevector forms an ellipse with its axes specifying the two eigen-polarizations $\vec{D}_1$ and $\vec{D}_2$ with n_1 and n_2 (semi-axes of the ellipse) the corresponding refractive indices.

$$4w_{em}\epsilon_0 = \frac{D_x^2}{\varepsilon_{r,xx}} + \frac{D_x^2}{\varepsilon_{r,yy}} + \frac{D_z^2}{\varepsilon_{r,zz}} = \frac{D_x^2}{n_{xx}^2} + \frac{D_y^2}{n_{yy}^2} + \frac{D_z^2}{n_{zz}^2}. \tag{3.61}$$

If in the above equation the term $\vec{D}/\sqrt{4w_{em}\epsilon_0}$ is replaced by $\vec{r} = x\hat{\imath}_x + y\hat{\imath}_y + z\hat{\imath}_z$, the resulting equation becomes:

$$\frac{x^2}{n_{xx}^2} + \frac{y^2}{n_{yy}^2} + \frac{z^2}{n_{zz}^2} = 1, \tag{3.62}$$

and represents the index ellipsoid in the xyz coordinate system which is shown in Fig. 3.9. The index ellipsoid is primarily used to determine the two refractive indices and the two corresponding directions of $\vec{D}$ associated with the two independent plane waves that can propagate along an arbitrary direction $\vec{k}$ in a crystal (as illustrated in Fig. 3.9). This process involves determining the intersection ellipse between a plane through the origin that is normal to the direction of propagation $\vec{k}$ and the index ellipsoid. The two axes of the intersection ellipse are equal in length to $2n_1$ and $2n_2$, where n_1 and n_2 are the two corresponding indices of refraction. These axes are parallel to the directions of the eigen-polarization vectors $\vec{D}_1$ and $\vec{D}_2$, which represent the two allowed solutions of the Maxwell equations [Eqs. (3.41)-(3.44)]. These indices (n_1, n_2) and the corresponding eigen-polarization directions $(\vec{D}_1, \vec{D}_2)$ are also referred to as the "slow wave" (for $max(n_1, n_2)$) and the "fast wave" (for $min(n_1, n_2)$). The index ellipsoid with an intersection ellipse, is depicted in Fig. 3.9, and a detailed proof can be found in Ref. [80].

If the index ellipsoid is expressed in a generalized coordinate system (not the principal axes system) then it will have the form of a generalized ellipsoid in the xyz coordinate system and its equation is given by:

$$\begin{bmatrix} x & y & z \end{bmatrix}^T [\mathcal{A}] \begin{bmatrix} x \\ y \\ z \end{bmatrix} = \begin{bmatrix} x & y & z \end{bmatrix}^T \begin{bmatrix} \frac{1}{n_{xx}^2} & \frac{1}{n_{xy}^2} & \frac{1}{n_{xz}^2} \\ \frac{1}{n_{yx}^2} & \frac{1}{n_{yy}^2} & \frac{1}{n_{yz}^2} \\ \frac{1}{n_{zx}^2} & \frac{1}{n_{zy}^2} & \frac{1}{n_{zz}^2} \end{bmatrix} \begin{bmatrix} x \\ y \\ z \end{bmatrix} = 1, \tag{3.63}$$

where $\mathcal{A} = [\varepsilon_r]^{-1}$ is known as the *impermeability matrix*. It is reminded that the impermeability matrix is also symmetric since it is the inverse of the symmetric relative permittivity matrix ($n_{uv} = n_{vu}$, where $u \neq v$ and $u, v = x, y, z$).

In the case of uniaxial material, where $n_{xx} = n_{yy} = n_O$ (ordinary index), and $n_{zz} = n_E$ (extraordinary index), the index ellipsoid is an ellipsoid of revolution around z-axis (optic axis). In the latter case, if the direction of the electromagnetic wavevector forms an angle θ with the optic axis, the extraordinary refractive index, $n_e(\theta)$ is given by:

$$\frac{\cos^2\theta}{n_O^2} + \frac{\sin^2\theta}{n_E^2} = \frac{1}{n_e^2(\theta)}, \tag{3.64}$$

while of course the ordinary index remains always equal to n_O.

3.2.3 Reflection and Transmission at a Planar Boundary

This section reviews the reflection and transmission of a plane wave at a planar boundary between two isotropic dielectrics. In Fig. 3.10, the boundary (corresponding to xy plane) between two dielectrics with permittivities $\epsilon_1 = \epsilon_0\varepsilon_{r1} = \epsilon_0 n_1^2$ and $\epsilon_2 = \epsilon_0\varepsilon_{r2} = \epsilon_0 n_2^2$ is shown. A plane wave is incident at angle θ_1 on the boundary, inducing both a reflected and a transmitted wave. The electric and magnetic fields in region 1 (to the left of the boundary) and region 2 (to the right of the boundary) can be expressed as follows:

$$\vec{E}_1 = \vec{E}_i \exp(-j\vec{k}_i \cdot \vec{r}) + \vec{E}_r \exp(-j\vec{k}_r \cdot \vec{r}), \tag{3.65}$$

$$\vec{H}_1 = \frac{1}{\omega\mu_0}\left(\vec{k}_i \times \vec{E}_i\right) \exp(-j\vec{k}_i \cdot \vec{r}) + \frac{1}{\omega\mu_0}\left(\vec{k}_r \times \vec{E}_r\right) \exp(-j\vec{k}_r \cdot \vec{r}), \tag{3.66}$$

$$\vec{E}_2 = \vec{E}_t \exp(-j\vec{k}_t \cdot \vec{r}), \tag{3.67}$$

$$\vec{H}_2 = \frac{1}{\omega\mu_0}\left(\vec{k}_t \times \vec{E}_t\right) \exp(-j\vec{k}_t \cdot \vec{r}), \qquad \text{where} \tag{3.68}$$

$$\vec{k}_i = k_0 n_1 \left[\sin\theta_1\ \hat{\imath}_x + \cos\theta_1\ \hat{\imath}_z\right], \qquad \vec{k}_r = k_0 n_1 \left[\sin\theta_r\ \hat{\imath}_x - \cos\theta_r\ \hat{\imath}_z\right],$$

$$\vec{k}_t = k_0 n_2 \left[\sin\theta_2\ \hat{\imath}_x + \cos\theta_2\ \hat{\imath}_z\right],$$

where $\vec{E}_i$, $\vec{E}_r$, and $\vec{E}_t$ correspond to the incident, reflected, and transmitted electric field amplitudes (vector in general), while $\vec{k}_i$, $\vec{k}_r$, and $\vec{k}_t$ correspond to the incident, reflected, and transmitted wavevectors. Naturally, the wavevectors in each region satisfy Eq. (3.40). By applying the continuity of the tangential components of the electric field at the boundary,

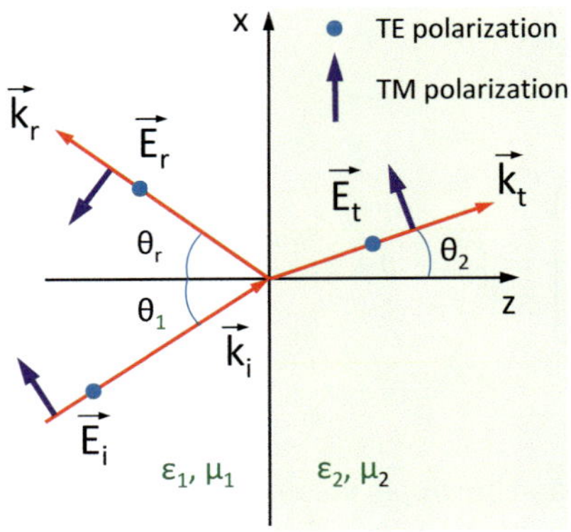

FIGURE 3.10: A planar boundary between two isotropic, nonmagnetic, dielectrics with permittivities ϵ_1 and ϵ_2 (or equivalently refractive indices n_1 and n_2). The plane of incidence is the xz plane in this simple case. The green electric field direction (along the y-axis) corresponds to the *TE* polarization, while the blue electric field direction (in the xz-plane) corresponds to the *TM* polarization. The angle of incidence is θ_1, the angle of reflection is $\theta_r = \theta_1$, and the angle of refraction is θ_2.

the following necessary condition must be fulfilled:

$$(\vec{k}_i \cdot \vec{r})_{z=0} = (\vec{k}_r \cdot \vec{r})_{z=0} = (\vec{k}_t \cdot \vec{r})_{z=0} \Longrightarrow k_{ix} = k_{rx} = k_{tx} \Longrightarrow$$
$$k_0 n_1 \sin\theta_1 = k_0 n_1 \sin\theta_r = k_0 n_2 \sin\theta_2. \qquad (3.69)$$

The last equation represents the well-known *phase-matching condition.* From this condition it follows that $\theta_r = \theta_1$ (reflection angle equals incident angle) and $n_1 \sin\theta_1 = n_2 \sin\theta_2$ (Snell's law). To determine the unknown amplitude coefficients of the reflected and transmitted waves, the complete boundary condition for the electric field must be applied, alongside the analogous boundary condition for the magnetic field. In general, any arbitrary polarization of the incident field can be decomposed into two components: one where the electric field is perpendicular to the plane of incidence (xz plane in this case), referred to as $\perp$ or *TE* polarization, and one where the electric field is parallel to the plane of incidence, referred to as $\parallel$ or *TM* polarization. The plane of incidence is defined by the direction of the incident wavevector and the direction of the normal to the interface (i.e., the plane of incidence is specified by these two intersecting lines). A generalized polarization formulation is presented in a later section.

TE Polarization Case:

In the case of *TE* polarization Eqs. (3.65)–(3.68) become:

$$\vec{E}_1 = \hat{\imath}_y \left[E_i \exp(-j\vec{k}_i \cdot \vec{r}) + E_r \exp(-j\vec{k}_r \cdot \vec{r})\right], \qquad (3.70)$$

$$\vec{H}_1 = \frac{1}{Z_1}\left\{E_i\left[-\hat{\imath}_x \cos\theta_1 + \hat{\imath}_z \sin\theta_1\right]\exp(-j\vec{k}_i \cdot \vec{r}) + E_r\left[\hat{\imath}_x \cos\theta_r + \hat{\imath}_z \sin\theta_r\right]\exp(-j\vec{k}_r \cdot \vec{r})\right\}, \qquad (3.71)$$

$$\vec{E}_2 = \hat{\imath}_y E_t \exp(-j\vec{k}_t \cdot \vec{r}), \qquad (3.72)$$

$$\vec{H}_2 = \frac{1}{Z_1} E_t\left[-\hat{\imath}_x \cos\theta_2 + \hat{\imath}_z \sin\theta_2\right]\exp(-j\vec{k}_t \cdot \vec{r}), \qquad (3.73)$$

where $Z_i = Z_0/n_i$ $(i = 1, 2)$ and $Z_0 = \sqrt{\mu_0/\epsilon_0}$ is the freespace wave impedance. The continuity of the tangential electric and magnetic field components at $z = 0$ gives the following equations:

$$E_i + E_r = E_t \Longrightarrow 1 + \frac{E_r}{E_i} = \frac{E_t}{E_i},$$

$$n_1 \left[-E_i \cos\theta_1 + E_r \cos\theta_r\right] = -n_2 E_t \cos\theta_2 \Longrightarrow -1 + \frac{E_r}{E_i}\frac{\cos\theta_r}{\cos\theta_1} = -\frac{n_2}{n_1}\frac{E_t}{E_i}\frac{\cos\theta_2}{\cos\theta_1}.$$

Defining as $r_{TE} = E_r/E_i$, the reflection coefficient for *TE* polarization, and $t_{TE} = E_t/E_i$, the transmission coefficient for *TE* polarization, and using from the phase matching conditions that $\theta_r = \theta_1$, the Fresnel equations for *TE* polarization can be easily derived and are shown in Eqs. (3.78) and (3.79) respectively.

TM Polarization Case:

In the case of *TM* polarization Eqs. (3.65)–(3.68) become:

$$\vec{E}_1 = E_i \left[\hat{\imath}_x \cos\theta_1 - \hat{\imath}_z \sin\theta_1\right] \exp(-j\vec{k}_i \cdot \vec{r}) - E_r \left[\hat{\imath}_x \cos\theta_r + \hat{\imath}_z \sin\theta_r\right] \exp(-j\vec{k}_i \cdot \vec{r}) \tag{3.74}$$

$$\vec{H}_1 = \hat{\imath}_y \left[\frac{E_i}{Z_1} \exp(-j\vec{k}_i \cdot \vec{r}) + \frac{E_r}{Z_1} \exp(-j\vec{k}_r \cdot \vec{r})\right], \tag{3.75}$$

$$\vec{E}_2 = E_t \left[\hat{\imath}_x \cos\theta_2 - \hat{\imath}_z \sin\theta_2\right] \exp(-j\vec{k}_t \cdot \vec{r}), \tag{3.76}$$

$$\vec{H}_2 = \hat{\imath}_y \frac{E_t}{Z_2} \exp(-j\vec{k}_t \cdot \vec{r}), \tag{3.77}$$

The continuity of the tangential electric and magnetic field components at $z = 0$ gives the following equations:

$$E_i \cos\theta_1 + E_r \cos\theta_r = E_t \cos\theta_2 \Longrightarrow 1 + \frac{E_r}{E_i}\frac{\cos\theta_r}{\cos\theta_1} = \frac{E_t}{E_i}\frac{\cos\theta_2}{\cos\theta_1},$$

$$n_1 \left(E_i + E_r\right) = n_2 E_t \Longrightarrow 1 + \frac{E_r}{E_i} = \frac{n_2}{n_1}\frac{E_t}{E_i}.$$

Defining as $r_{TM} = E_r/E_i$, the reflection coefficient for *TM* polarization, and $t_{TM} = E_t/E_i$, the transmission coefficient for *TM* polarization, and using again that $\theta_r = \theta_1$, the Fresnel equations for *TM* polarization can be easily derived and are shown in Eqs. (3.80) and (3.81) respectively.

3.2.3.1 Fresnel Equations

The reflection and transmission coefficients r and t for both *TE* and *TM* polarizations are generally expressed in the form of the Fresnel equations which are summarized below:

$$r_{TE} = r_\perp = \frac{E_r}{E_i} = \frac{n_1 \cos\theta_1 - n_2 \cos\theta_2}{n_1 \cos\theta_1 + n_2 \cos\theta_2}, \tag{3.78}$$

$$t_{TE} = t_\perp = \frac{E_t}{E_i} = \frac{2n_1 \cos\theta_1}{n_1 \cos\theta_1 + n_2 \cos\theta_2}, \tag{3.79}$$

$$r_{TM} = r_\parallel = \frac{E_r}{E_i} = \frac{n_2 \cos\theta_1 - n_1 \cos\theta_2}{n_2 \cos\theta_1 + n_1 \cos\theta_2}, \tag{3.80}$$

$$t_{TM} = t_\parallel = \frac{E_t}{E_i} = \frac{2n_1 \cos\theta_1}{n_2 \cos\theta_1 + n_1 \cos\theta_2}. \tag{3.81}$$

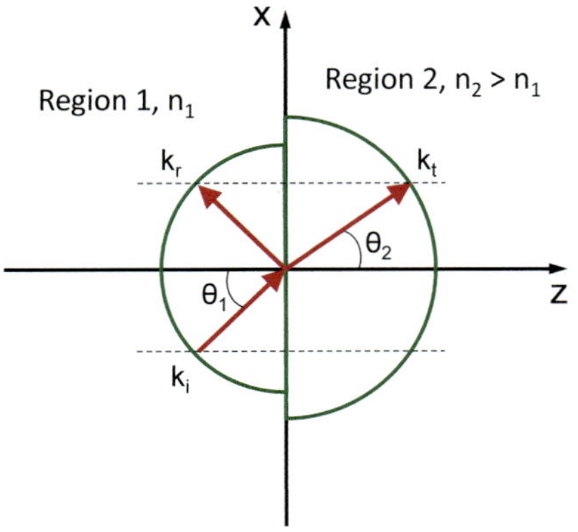

(a)

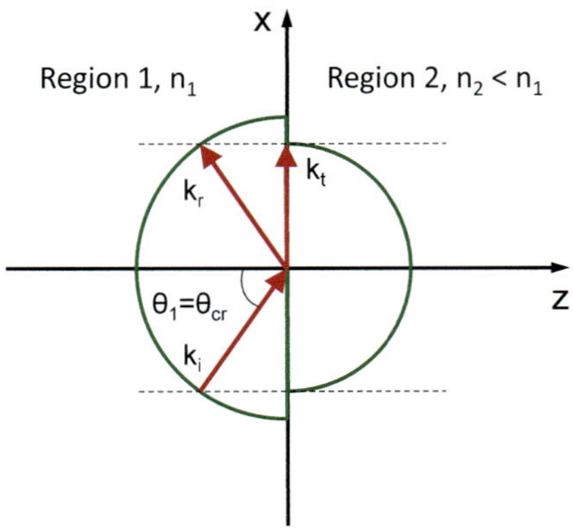

(b)

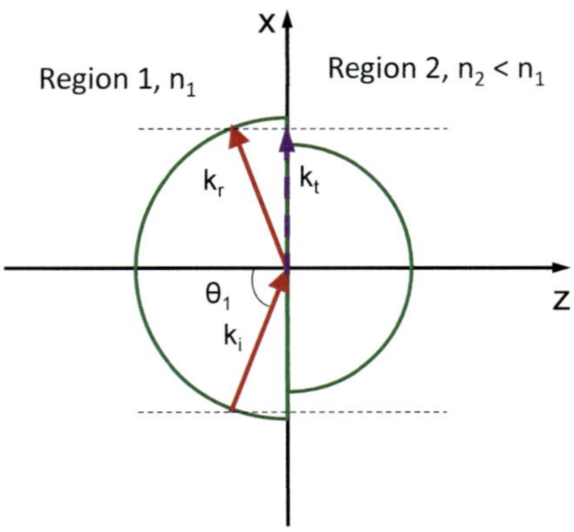

(c)

FIGURE 3.11: The wavevector surface of an interface between two dielectrics. (a) Case of incidence from a low to higher refractive index ($n_1 < n_2$). (b) Case of incidence from a high to a lower refractive index ($n_1 > n_2$) and at $\theta_1 = \theta_{cr}$. (c) Case of incidence from a high to a lower refractive index ($n_1 > n_2$) and at $\theta_1 > \theta_{cr}$.

There are several notable points regarding the Fresnel equations when $n_1 < n_2$ (for example, propagation from air to glass). In this case, the wavevector diagram illustrated in Fig. 3.11a applies, showing the phase-matching condition for the boundary between the two media. For this case the Brewster angle can also be defined as the angle for which the reflected wave vanishes. For nonmagnetic materials this can occur only in the case of *TM* ($\|$) polarization, and it is defined as $r_{_{TM}}(\theta_1 = \theta_B) = 0 \Longrightarrow \theta_B = \tan^{-1}(n_2/n_1)$. As an example, the reflection and transmission coefficients are shown in Figs. 3.12 and 3.13 as functions of the angle of incidence θ_1 for the case of $n_1 = 1.0$ and $n_2 = 1.5$.

The situation becomes more intriguing when $n_1 > n_2$ (for example, propagation from glass to air). In this case, the Brewster angle is defined in a similar manner as before and exists only for *TM* polarization. However, for both polarizations, a critical angle can also be defined. From Snell's law, the critical angle θ_{cr}, is the angle of incidence at which the refraction angle becomes 90 degrees. Therefore, the critical angle is given by: $\theta_{cr} = \sin^{-1}(n_2/n_1)$, which exists only under the condition $n_1 > n_2$. This situation is depicted in

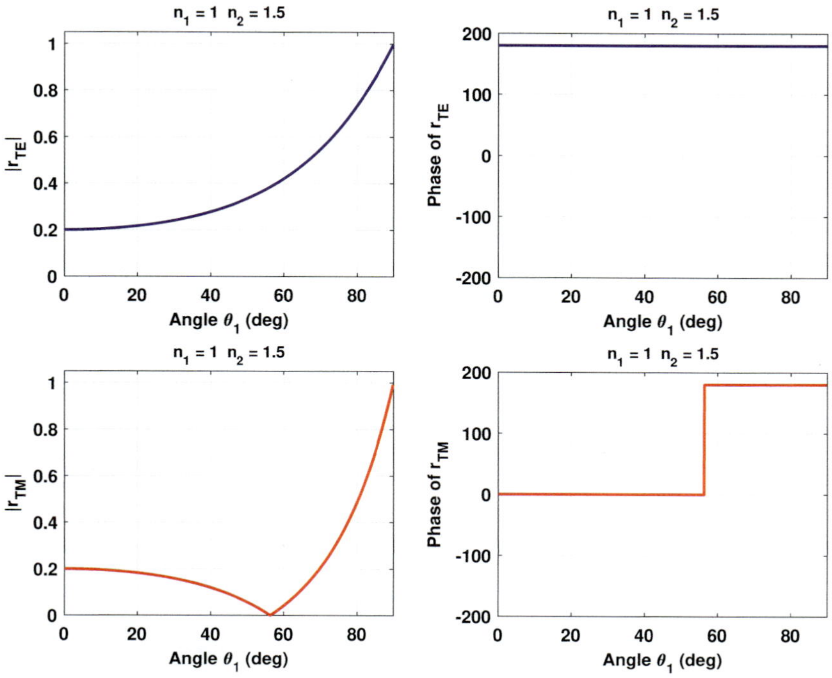

FIGURE 3.12: Reflection coefficient (amplitude and phase) as a function of the angle of incidence θ_1 for *TE* and *TM* polarization in the case of $n_1 < n_2$. The Brewster angle $\theta_B = \tan^{-1}(1.5/1) = 56.31^\circ$ is evident in the *TM* polarization case.

Fig. 3.11b, where, due to the phase-matching condition [Eq. (3.69)], the x-component of the transmitted wavevector satisfies $k_{tx} = k_0 n_2$, while the z-component, k_{tz}, becomes zero. Consequently, the transmitted electric field described by Eq. (3.67), becomes independent of z. It is worth noting that this situation can occur only in theory, as at the critical angle, the transmitted field in region 2 would be constant, requiring infinite energy from the electromagnetic wave. In practice, the more relevant case is when the angle of incidence exceeds the critical angle ($\theta_1 > \theta_{cr}$). For this condition, the phase-matching condition leads to the conclusion that:

$$k_{tx} = k_0 n_1 \sin\theta_1 > k_0 n_1 \sin\theta_{cr} = k_0 n_2. \tag{3.82}$$

From the last equation and Eq. (3.40) for region 2, it is deduced that (for $\theta_1 > \theta_{cr}$):

$$k_{tz}^2 = k_0^2 n_2^2 - k_{tx}^2 < 0 \Longrightarrow k_{tz} = \pm j\sqrt{k_{tx}^2 - k_0^2 n_2^2} = \pm j k_0 \sqrt{n_1^2 \sin^2\theta_1 - n_2^2} = \pm j\gamma_t. \tag{3.83}$$

This implies that the z-component of the transmitted wavevector becomes purely imaginary. In the wavevector diagram, this situation is depicted in Fig. 3.11c, where the dashed purple arrow representing the transmitted wavevector is complex. There is some ambiguity in selecting the sign on the right-hand side of Eq. (3.83). To ensure that the transmitted field described by Eq. (3.67) represents a physical field, the negative sign must be selected. Consequently, Eq. (3.67) becomes:

$$\vec{E}_2 = \vec{E}_t \exp[-k_0\sqrt{n_1^2 \sin^2\theta_1 - n_2^2}z] \exp[-jxk_0 n_1 \sin\theta_1] = \vec{E}_t e^{-\gamma_t z} \exp[-jxk_0 n_1 \sin\theta_1]. \tag{3.84}$$

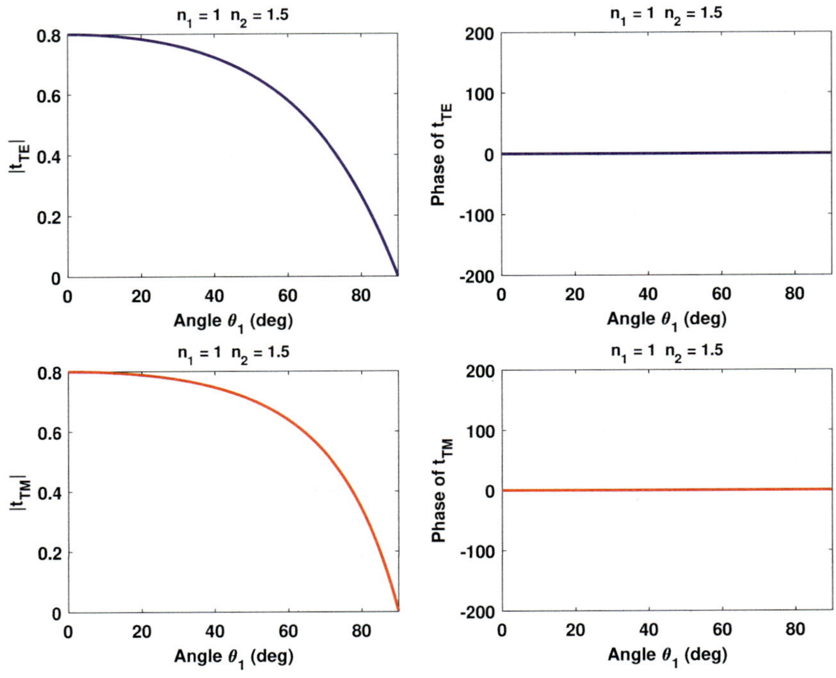

FIGURE 3.13: Transmission coefficient (amplitude and phase) as a function of the angle of incidence θ_1 for *TE* and *TM* polarization in the case of $n_1 < n_2$.

The corresponding real field can be determined from the above phasor as (assuming for simplicity that $\vec{E}_i$ is real):

$$\vec{\mathcal{E}}_2(x, z, t) = |t| E_i \hat{u}_t e^{-\gamma_t z} \cos\left[\omega t - x k_0 n_1 \sin\theta_1 + \phi(\theta_1)\right], \tag{3.85}$$

where $t = |t|e^{j\phi}$ is the complex transmission coefficient and $\hat{u}_t$ is the unit vector of the transmitted electric field. For example, $\hat{u}_t = \hat{u}_{TE} = \hat{\imath}_y$ for *TE* polarization, and $\hat{u}_t = \hat{u}^t_{TM} = \cos\theta_2 \hat{\imath}_x - \sin\theta_2 \hat{\imath}_z$ for *TM* polarization. However, in the case of *TM* polarization, $\hat{u}^t_{TM}$ has an imaginary component that should suitably (with the appropriate phase shift) be incorporated in the latter equation of the real transmitted electric field. However, for simplicity, the form of Eq. (3.85) does not take into account the complex component of $\hat{u}_t$. The resulting transmitted field in this case is referred to as the "evanescent field" or "evanescent wave," which decays exponentially into region 2. It is easy to show that the evanescent wave does not transfer real power in the z direction. This is why this phenomenon is known as *total internal reflection*, as all the power is reflected back into the incident region. An example evanescent field (its real part for $t = 0$ for *TE* polarization) is shown in Fig. 3.14 for $n_1 = 1.5$, $n_2 = 1.0$, and $\theta_1 = 65°$.

In the case of total internal reflection, the reflection and transmission coefficients derived from the Fresnel equations become complex. It is useful to define the reflection coefficients in this context, which can be described by the following equations:

$$r_{TE} = \frac{E_r}{E_i} = 1 e^{j2\phi_{TE}(\theta_1)} = 1 \exp\left[j2\tan^{-1}\left\{ \frac{\sqrt{n_1^2 \sin^2\theta_1 - n_2^2}}{n_1 \cos\theta_1} \right\} \right], \tag{3.86}$$

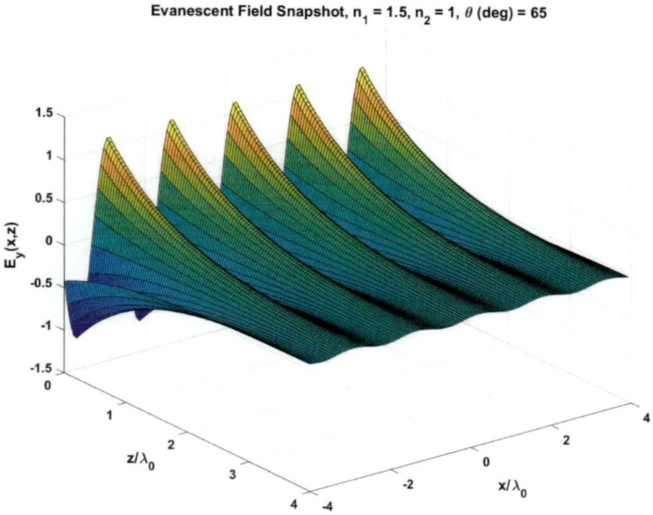

FIGURE 3.14: The real part of a *TE* polarized evanescent field that is generated in region 2 ($n_2 = 1.0$) from an incident *TE* polarized wave from region 1 ($n_1 = 1.5$) for an angle of incidence $\theta_1 = 65^\circ$ ($> \theta_{cr} = 41.8^\circ$).

$$r_{_{TM}} = \frac{E_r}{E_i} = 1e^{j2\phi_{TM}(\theta_1)} = 1\exp\left[j2\tan^{-1}\left\{\frac{n_1^2}{n_2^2}\frac{\sqrt{n_1^2\sin^2\theta_1 - n_2^2}}{n_1\cos\theta_1}\right\}\right]. \tag{3.87}$$

From the previous equations it is evident that when $\theta_1 > \theta_{cr}$, the reflection coefficients have unit magnitude (thus all incident power is reflected back) and phase that depends on θ_1 and varies from 0 to π. The transmission coefficients become also complex in this case and are given by:

$$t_{_{TE}} = \frac{E_t}{E_i} = |t_{_{TE}}|e^{j\phi_{TE}(\theta_1)} = \frac{2n_1\cos\theta_1}{\left(n_1^2 - n_2^2\right)^{1/2}}\exp\left[j\tan^{-1}\left\{\frac{\left(n_1^2\sin^2\theta_1 n_2^2\right)^{1/2}}{n_1\cos\theta_1}\right\}\right], \tag{3.88}$$

$$t_{_{TM}} = \frac{E_t}{E_i} = |t_{_{TM}}|e^{j\phi_{TM}(\theta_1)} =$$

$$\frac{2n_1 n_2\cos\theta_1}{\left(n_2^4\cos^2\theta_1 + n_1^4\sin^2\theta_1 - n_1^2 n_2^2\right)^{1/2}}\exp\left[j\tan^{-1}\left\{\frac{n_1^2}{n_2^2}\frac{\left(n_1^2\sin^2\theta_1 - n_2^2\right)^{1/2}}{n_1\cos\theta_1}\right\}\right]. \tag{3.89}$$

As an example, the reflection and transmission coefficients are shown in Figs. 3.15 and 3.16 as functions of the angle of incidence θ_1, for the case of $n_1 = 1.5$ and $n_2 = 1.0$.

3.2.4 Electromagnetic Field Energy and Power - Poynting's Theorem

The Poynting theorem is a fundamental concept in electromagnetism. It describes the energy conservation in the electromagnetic field. It is named after English physicist John Henry Poynting who first formulated it in 1884. The Poynting theorem states that the rate of energy transfer per unit volume from a region in space equals the rate of work done on the charges of the region and the energy flux leaving that region. From Maxwell's equations

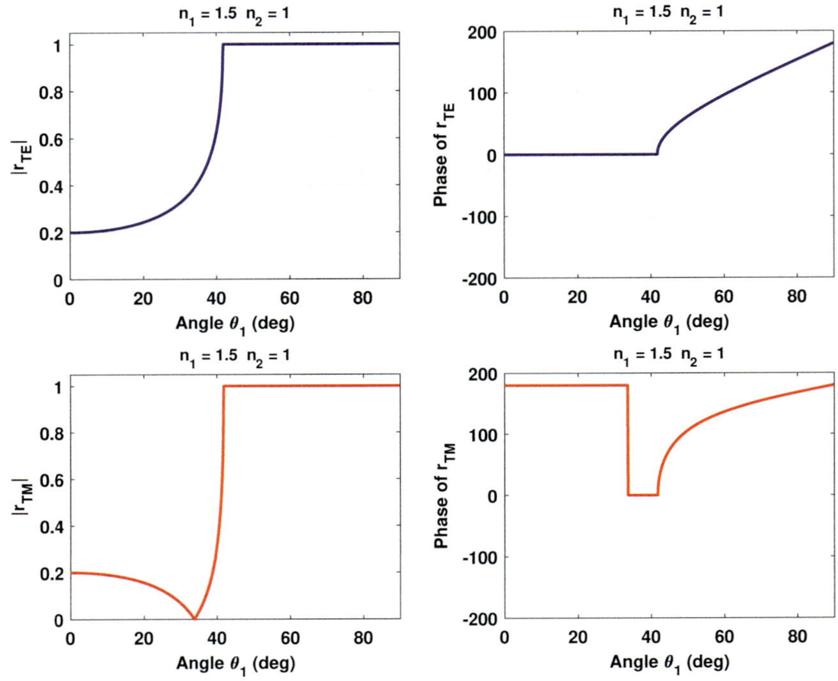

FIGURE 3.15: Reflection coefficient as a function of the angle of incidence θ_1 for *TE* and *TM* polarization in the case of $n_1 > n_2$. The Brewster angle $\theta_B = \tan^{-1}(1/1.5) = 33.69°$ is evident in the *TM* polarization case. For both *TE* and *TM* polarization cases, the critical angle $\theta_{cr} = \sin^{-1}(1/1.5) = 41.81°$ is present. For $\theta_1 > \theta_{cr}$ the reflection coefficients become complex, of unity magnitude and phase of $2\phi_{TE}(\theta_1)$ or $2\phi_{TM}(\theta_1)$.

[Eqs. (3.1)–(3.4)] the following relations can be written:

$$\vec{\mathcal{H}} \cdot \left(\vec{\nabla} \times \vec{\mathcal{E}}\right) = -\vec{\mathcal{H}} \cdot \frac{\partial \vec{\mathcal{B}}}{\partial t},$$

$$\vec{\mathcal{E}} \cdot \left(\vec{\nabla} \times \vec{\mathcal{H}}\right) = \vec{\mathcal{E}} \cdot \vec{\mathcal{J}} + \vec{\mathcal{E}} \cdot \frac{\partial \vec{\mathcal{D}}}{\partial t}.$$

Then using the vector identity $\vec{\nabla} \cdot (\vec{A} \times \vec{B}) = \vec{B} \cdot (\vec{\nabla} \times \vec{A}) - \vec{A} \cdot (\vec{\nabla} \times \vec{B})$ it is straightforward to show:

$$\begin{aligned} -\vec{\nabla} \cdot (\vec{\mathcal{E}} \times \vec{\mathcal{H}}) &= \vec{\mathcal{E}} \cdot \vec{\mathcal{J}} + \vec{\mathcal{E}} \cdot \frac{\partial \vec{\mathcal{D}}}{\partial t} + \vec{\mathcal{H}} \cdot \frac{\partial \vec{\mathcal{B}}}{\partial t} = \\ &= \vec{\mathcal{E}} \cdot \vec{\mathcal{J}} + \frac{\partial}{\partial t}\left(\frac{\epsilon_0}{2} \vec{\mathcal{E}} \cdot \vec{\mathcal{E}}\right) + \frac{\partial}{\partial t}\left(\frac{\mu_0}{2} \vec{\mathcal{H}} \cdot \vec{\mathcal{H}}\right) + \vec{\mathcal{E}} \cdot \frac{\partial \vec{\mathcal{P}}}{\partial t} + \mu_0 \vec{\mathcal{H}} \cdot \frac{\partial \vec{\mathcal{M}}}{\partial t}. \end{aligned} \tag{3.90}$$

The last expression represents the Poynting's theorem in point (differential) form. In case that a closed surface S is chosen that encloses a volume V, the Poynting's theorem can also

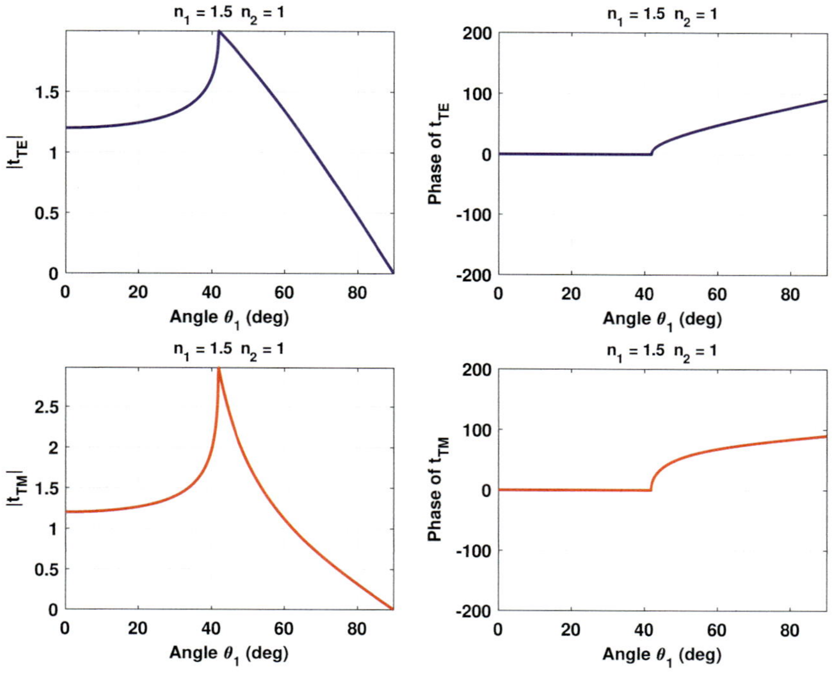

FIGURE 3.16: Transmission coefficient as a function of the angle of incidence θ_1 for *TE* and *TM* polarization in the case of $n_1 > n_2$. For $\theta_1 > \theta_{cr}$ the transmission coefficient becomes complex.

be written in its integral form as follows:

$$-\oiint_S \left(\vec{\mathcal{E}} \times \vec{\mathcal{H}}\right) \cdot d\vec{S} = \iiint_V \vec{\mathcal{E}} \cdot \vec{\mathcal{J}} dV + \iiint_V \frac{\partial}{\partial t}\left(\frac{\epsilon_0}{2}\vec{\mathcal{E}} \cdot \vec{\mathcal{E}} + \frac{\mu_0}{2}\vec{\mathcal{H}} \cdot \vec{\mathcal{H}}\right) dV + \iiint_V \vec{\mathcal{E}} \cdot \frac{\partial \vec{\mathcal{P}}}{\partial t} dV + \iiint_V \mu_0 \vec{\mathcal{H}} \cdot \frac{\partial \vec{\mathcal{M}}}{\partial t} dV, \tag{3.91}$$

where the vector $\vec{\mathcal{N}} = \vec{\mathcal{E}} \times \vec{\mathcal{H}}$ is defined as the Poynting vector (W/m^2). The left-hand side of Eq. (3.91) represents the total power entering the volume V via the closed surface S. The right-hand side terms represent the ohmic losses expended in volume V, the rate of increase of the vacuum electromagnetic energy in volume V, the power expended in electric dipoles in volume V, and the power expended in magnetic dipoles in volume V, respectively.

If the medium is linear with respect to its electric and magnetic properties and experiences negligible dispersion, the electromagnetic energy density can be defined as follows:

$$w_{em} = w_e + w_m = \frac{1}{2}\left(\vec{\mathcal{E}} \cdot \vec{\mathcal{D}} + \vec{\mathcal{H}} \cdot \vec{\mathcal{B}}\right) = \frac{1}{2}\left(\epsilon|\vec{\mathcal{E}}|^2 + \mu|\vec{\mathcal{H}}|^2\right). \tag{3.92}$$

Then Eqs. (3.90) and (3.91) can be written in the following forms:

$$-\vec{\nabla} \cdot \left(\vec{\mathcal{E}} \times \vec{\mathcal{H}}\right) = \vec{\mathcal{E}} \cdot \vec{\mathcal{J}} + \frac{\partial w_{em}}{\partial t}, \tag{3.93}$$

$$-\oiint_S \left(\vec{\mathcal{E}} \times \vec{\mathcal{H}}\right) \cdot d\vec{S} = \iiint_V \left[\vec{\mathcal{E}} \cdot \vec{\mathcal{J}} + \frac{\partial w_{em}}{\partial t}\right] dV. \tag{3.94}$$

The differential or integral form of Eqs. (3.93) or (3.94) embodies the physical concept that the time rate of change of electromagnetic energy within a certain volume, in addition to the total work done by the fields on the sources within the volume, is equated to the energy flowing in through the boundary surfaces of the volume per unit time. It is also customary to express $\vec{\mathcal{J}} = \vec{\mathcal{J}}_\sigma + \vec{\mathcal{J}}_s$ (where $\vec{\mathcal{J}}_\sigma$ arises from conductivity current, and $\vec{\mathcal{J}}_s$ originates from free current sources of electromagnetic radiation). Equations (3.93) and (3.94) are formulations of the conservation of electromagnetic energy.

It is noteworthy that the definition of the Poynting vector (and the electromagnetic energy density) is not unique, as discussed in references [83–87]. For example, if $\vec{\mathcal{N}}' = \vec{\mathcal{N}} + \vec{\mathcal{T}}$ is used, where $\vec{\nabla} \cdot \vec{\mathcal{T}} = 0$, then $\vec{\mathcal{N}}'$ still satisfies Poynting's theorem, i.e., Eqs. (3.90) and (3.91). Consequently, the point-by-point interpretation of the Poynting vector may lead to peculiar results (as illustrated in, for example, Refs. [83, 86]).

Certainly, the scenario of a dispersionless medium is considered ideal. However, in reality, all materials exhibit dispersion, characterized by frequency-dependent parameters, and may experience losses. To articulate Poynting's theorem in the dispersive case, it becomes necessary to represent all fields in the frequency domain. This is achieved by applying the Fourier transform to all field quantities. For example, the relationship between the real electric field in the time domain and its corresponding complex electric field in the frequency domain is established through the following Fourier transform pair:

$$\vec{E}(\vec{r}, \omega) = \int_{-\infty}^{+\infty} \vec{\mathcal{E}}(\vec{r}, t) e^{-j\omega t} dt, \tag{3.95}$$

$$\vec{\mathcal{E}}(\vec{r}, t) = \frac{1}{2\pi} \int_{-\infty}^{+\infty} \vec{E}(\vec{r}, \omega) e^{+j\omega t} d\omega, \tag{3.96}$$

where the same holds for all field quantities. Since the time-domain fields are real, it is straightforward to show that $\vec{E}(\vec{r}, -\omega) = \vec{E}^*(\vec{r}, \omega)$ (where the "*" denotes complex conjugate) and $\epsilon^*(\omega) = \epsilon(-\omega)$, with similar arguments holding for the magnetic field counterparts. The permittivity and the permeability are, in general, complex functions of ω and can be written in the form:

$$\epsilon(\omega) = \epsilon'(\omega) - j\epsilon''(\omega) = \epsilon_0 \left[\varepsilon_r'(\omega) - j\varepsilon_r''(\omega)\right], \tag{3.97}$$

$$\mu(\omega) = \mu'(\omega) - j\mu''(\omega) = \mu_0 \left[\mu_r'(\omega) - j\mu_r''(\omega)\right], \tag{3.98}$$

where the real and the imaginary parts of the permittivity and the permeability are denoted as primed or double-primed terms, respectively. The "−" sign in the imaginary part is compatible with the Fourier transform definition as well as with the phasors definitions (in many physics textbooks the "+" sign is selected and opposite signs in the exponents of the Fourier transforms and phasors). The doubled-primed terms denote losses that the electromagnetic field suffers as it passes through a medium at frequencies near resonances where the $\varepsilon_r''(\omega) > 0$ and $\mu_r''(\omega) > 0$ terms can represent absorption. When the frequency of the electromagnetic wave is far from the resonances then $\varepsilon_r''(\omega) \simeq \mu_r''(\omega) \simeq 0$, then this frequency range is generally refer to as "transparency range."

In most realistic situations, even monochromatic radiation has a spectrum differing from a delta function (even lasers exhibit broadening of their emitted spectrum). Therefore [88–90], it is common to consider the electric and magnetic fields of the form:

$$\vec{\mathcal{E}}(t) = \vec{\mathcal{E}}_0(t) \cos(\omega_0 t + \phi_e), \tag{3.99}$$

$$\vec{\mathcal{H}}(t) = \vec{\mathcal{H}}_0(t) \cos(\omega_0 t + \phi_h), \tag{3.100}$$

where for simplicity the spatial dependence is implied. The $\vec{\mathcal{E}}_0(t)$ and $\vec{\mathcal{H}}_0(t)$ are slowly varying amplitudes (their Fourier transforms do not possess high-frequency terms), ω_0 is the main frequency (high frequency) of oscillation of the electromagnetic field, and ϕ_e, ϕ_h are phase constants which can be space dependent. Using the approach of Jackson [88] it can be shown that:

$$\left\langle \vec{\mathcal{E}}\cdot\frac{\partial\vec{\mathcal{D}}}{\partial t}+\vec{\mathcal{H}}\cdot\frac{\partial\vec{\mathcal{B}}}{\partial t}\right\rangle_{HF}=\omega_0\epsilon''(\omega_0)\langle\vec{\mathcal{E}}\cdot\vec{\mathcal{E}}\rangle_{HF}+\omega_0\mu''(\omega_0)\langle\vec{\mathcal{H}}\cdot\vec{\mathcal{H}}\rangle_{HF}+\frac{\partial u_{eff}}{\partial t}, \tag{3.101}$$

$$u_{eff}=\frac{1}{2}\frac{d}{d\omega}\Big(\omega\epsilon'(\omega)\Big)|_{\omega_0}\left\langle\vec{\mathcal{E}}\cdot\vec{\mathcal{E}}\right\rangle_{HF}+\frac{1}{2}\frac{d}{d\omega}\Big(\omega\mu'(\omega)\Big)|_{\omega_0}\left\langle\vec{\mathcal{H}}\cdot\vec{\mathcal{H}}\right\rangle_{HF}, \tag{3.102}$$

$$\left\langle\vec{\mathcal{E}}\cdot\vec{\mathcal{E}}\right\rangle_{HF}=\frac{1}{2}\vec{\mathcal{E}}_0\cdot\vec{\mathcal{E}}_0, \tag{3.103}$$

$$\left\langle\vec{\mathcal{H}}\cdot\vec{\mathcal{H}}\right\rangle_{HF}=\frac{1}{2}\vec{\mathcal{H}}_0\cdot\vec{\mathcal{H}}_0, \tag{3.104}$$

where $\langle *\rangle_{HF}$ denotes a high-frequency averaging (over a time period $2\pi/\omega_0$). The term u_{eff} denotes the effective electromagnetic energy density, while the first two terms of the right-hand side of Eq. (3.101) account for dissipation losses. Consequently, the differential form of Poynting's theorem in the presence of dispersion and losses can be expressed as:

$$-\vec{\nabla}\cdot\left(\left\langle\vec{\mathcal{E}}\times\vec{\mathcal{H}}\right\rangle_{HF}\right)=\left\langle\vec{\mathcal{E}}\cdot\vec{\mathcal{J}}\right\rangle_{HF}+\omega_0\epsilon''(\omega_0)\left\langle\vec{\mathcal{E}}\cdot\vec{\mathcal{E}}\right\rangle_{HF}+\omega_0\mu''(\omega_0)\left\langle\vec{\mathcal{H}}\cdot\vec{\mathcal{H}}\right\rangle_{HF}+\frac{\partial u_{eff}}{\partial t}. \tag{3.105}$$

The time-averaged energy densities in the case of dispersive media can be written as [88–90]

$$\langle w_e\rangle=\frac{1}{4}\frac{d}{d\omega}\Big(\omega\epsilon'(\omega)\Big)\left\langle\vec{\mathcal{E}}_0\cdot\vec{\mathcal{E}}_0\right\rangle=\frac{1}{4}\frac{d}{d\omega}\Big(\omega\epsilon'(\omega)\Big)\mathrm{Re}\{\vec{E}\cdot\vec{E}^*\}, \tag{3.106}$$

$$\langle w_m\rangle=\frac{1}{4}\frac{d}{d\omega}\Big(\omega\mu'(\omega)\Big)\left\langle\vec{\mathcal{H}}_0\cdot\vec{\mathcal{H}}_0\right\rangle=\frac{1}{4}\frac{d}{d\omega}\Big(\omega\mu'(\omega)\Big)\mathrm{Re}\{\vec{H}\cdot\vec{H}^*\}, \tag{3.107}$$

3.2.4.1 Time-Harmonic Fields

In the time-harmonic form of Poynting's theorem, where all field components are expressed as phasors, the electric field $\vec{\mathcal{E}}$ s related to its phasor $\vec{E}$ through the following expression: $\vec{\mathcal{E}}(\vec{r},t)=\mathrm{Re}\{\vec{E}(\vec{r},\omega)\exp(j\omega t)\}$. In this case, the time average Poynting vector, which represents the average power per unit area transported by the electromagnetic wave, is given by:

$$\left\langle\vec{\mathcal{N}}\right\rangle=\frac{1}{T}\int_0^T\vec{\mathcal{N}}dt=\frac{1}{2}\mathrm{Re}\left\{\vec{E}\times\vec{H}^*\right\}=\mathrm{Re}\{\vec{S}\}, \tag{3.108}$$

where $\vec{S}=(1/2)\vec{E}\times\vec{H}^*$ is the complex Poynting vector. By manipulating Maxwell's equations in a manner akin to the derivation of Eq. (3.90), it can be demonstrated with ease that:

$$-\vec{\nabla}\cdot\vec{S}=\frac{1}{2}\vec{E}\cdot\vec{J}^*-\frac{1}{2}j\omega\vec{E}\cdot\vec{D}^*+\frac{1}{2}j\omega\vec{H}^*\cdot\vec{B}. \tag{3.109}$$

Using Eqs. (3.24), (3.25), (3.26), (3.97), and (3.98), and $\vec{J}=\vec{J}_s+\sigma(\omega)\vec{E}$ (where $\vec{J}_s$ represents source currents), the previous equation can be written as follows [where the conductivity $\sigma(\omega)$ is assumed real]:

$$-\vec{\nabla}\cdot\vec{S}=\frac{1}{2}\vec{E}\cdot\vec{J}_s^*+\frac{1}{2}\sigma(\omega)|\vec{E}|^2+\frac{1}{2}\omega\epsilon''(\omega)|\vec{E}|^2+\frac{1}{2}\omega\mu''(\omega)|\vec{H}|^2+\frac{j}{2}\omega[-\epsilon'(\omega)|\vec{E}|^2+\mu'(\omega)|\vec{H}|^2]. \tag{3.110}$$

The last equation is the Poynting theorem in its differential form for time-harmonic fields. The Poynting theorem can be expressed in its integral form by integrating Eq. (3.110) over a closed surface $\mathcal{S}$ surrounding a volume V. The resulting expression is:

$$-\oiint_{S} \vec{S}\cdot d\vec{\mathcal{S}} = \frac{1}{2}\iiint_{V} \vec{E}\cdot\vec{J}_s^{*} dV + \frac{1}{2}\iiint_{V} \sigma(\omega)|\vec{E}|^2 dV + \frac{1}{2}\iiint_{V} \omega\Big[\epsilon''(\omega)|\vec{E}|^2 + \mu''(\omega)|\vec{H}|^2\Big] dV + j\frac{\omega}{2}\iiint_{V} \Big[-\epsilon'(\omega)|\vec{E}|^2 + \mu'(\omega)|\vec{H}|^2\Big] dV. \tag{3.111}$$

By separating the real and imaginary parts of the equation for the time-harmonic Poynting theorem, the power equilibrium can be expressed as follows:

$$P + P_\ell + P_{dis} = P_s, \tag{3.112}$$

$$Q + Q_{em} = Q_s, \quad \text{where,} \tag{3.113}$$

$$P = \text{Re}\Big\{\oiint_{S} \vec{S}\cdot d\vec{\mathcal{S}}\Big\}, \qquad P_\ell = \frac{1}{2}\iiint_{V} \sigma(\omega)|\vec{E}|^2 dV,$$

$$P_{dis} = \frac{1}{2}\iiint_{V} \omega\Big[\epsilon''(\omega)|\vec{E}|^2 + \mu''(\omega)|\vec{H}|^2\Big] dV,$$

$$P_s = \frac{1}{2}\iiint_{V} \text{Re}\Big\{-\vec{E}\cdot\vec{J}_s^{*}\Big\} dV, \qquad Q = \text{Im}\Big\{\oiint_{S} \vec{S}\cdot d\vec{\mathcal{S}}\Big\},$$

$$Q_{em} = \frac{\omega}{2}\iiint_{V} \Big[-\epsilon'(\omega)|\vec{E}|^2 + \mu'(\omega)|\vec{H}|^2\Big] dV, \quad \text{and}$$

$$Q_s = \frac{1}{2}\iiint_{V} \text{Im}\Big\{-\vec{E}\cdot\vec{J}_s^{*}\Big\} dV.$$

Equation (3.112) signifies the equilibrium of real power, where P denotes the power exiting volume V, P_ℓ represents the power consumed by ohmic losses (conduction currents) within V, P_{dis} accounts for the power dissipated into volume V due to material dielectric and magnetic losses, and P_s stands for the power delivered by the sources in volume V. Equation (3.113) discloses the equilibrium of reactive power, with Q denoting the reactive power exiting volume V, Q_{em} representing the reactive power stored in volume V (comprising a capacitive part due to the electric field and an inductive part due to the magnetic field), and Q_s indicating the reactive power delivered by the sources in volume V.

In the time-harmonic fields case, the time-average electric and magnetic energy densities become:

$$\langle w_e \rangle = \frac{1}{4}\text{Re}\{\vec{E}\cdot\vec{D}^{*}\}, \tag{3.114}$$

$$\langle w_m \rangle = \frac{1}{4}\text{Re}\{\vec{H}\cdot\vec{B}^{*}\}. \tag{3.115}$$

In the case of a plane electromagnetic wave, the time-averaged electric energy density is equal to the time-averaged magnetic energy density, and the total electromagnetic energy density is given by:

$$\langle w_{em} \rangle = 2\langle w_e \rangle = 2\langle w_m \rangle = \frac{1}{2}\text{Re}\{\vec{E}\cdot\vec{D}^{*}\} = \frac{1}{2}\text{Re}\{\vec{H}\cdot\vec{B}^{*}\}. \tag{3.116}$$

Furthermore, for a plane electromagnetic wave the time-averaged Poynting vector is expressed as:

$$\left\langle \vec{\mathcal{N}} \right\rangle = \mathrm{Re}\{\vec{S}\} = \vec{P} = \frac{1}{2Z}|\vec{E}|^2\hat{k}, \tag{3.117}$$

where $Z = \sqrt{\mu/\epsilon}$ is the wave impedance of the medium, and $\hat{k}$ is the unit vector in the direction of propagation. The above equation is valid for dispersionless, lossless, homogeneous, linear, and isotropic media. Furthermore, when a plane wave is incident at a planar boundary between two linear, isotropic, lossless, and non-magnetic regions application of Poynting's theorem gives the following power conservation equations for the direction normal to the boundary:

$$\frac{P_r}{P_i} + \frac{P_t}{P_i} = |r_{_{TE}}|^2 + |t_{_{TE}}|^2 \frac{\mathrm{Re}\{n_2 \cos\theta_2\}}{n_1 \cos\theta_1} = 1, \qquad \text{TE Polarization} \tag{3.118}$$

$$\frac{P_r}{P_i} + \frac{P_t}{P_i} = |r_{_{TM}}|^2 + |t_{_{TM}}|^2 \frac{\mathrm{Re}\{n_2^* \cos\theta_2\}}{n_1 \cos\theta_1} = 1, \qquad \text{TM Polarization,} \tag{3.119}$$

where P_i, P_r, P_t are the incident, reflected, and transmitted powers (normal to the boundary). The term n_2^* represents the complex conjugate of the refractive index. These values are generally complex, especially when region 2 comprises a lossy material such as a metal. The final equations straightforwardly assert that the sum of the percentage of reflected power and the percentage of transmitted power equals unity for the reflection/refraction at a planar boundary. In the scenario where the polarization of the plane wave comprises a mixture of *TE* and *TM* polarizations, the power balance needs modification. However, for isotropic media, the two polarizations can be treated independently. A brief analysis of a generalized incident wave polarization is addressed in the following section.

3.2.5 Generalized Plane-Wave Reflection/Transmission at a Planar Boundary

The geometry of the planar interface between two isotropic media is once again illustrated in Fig. 3.10. In this case, the polarization of the incident wave includes both *TE* (perpendicular) and *TM* (parallel) components. Consequently, with this formulation, any elliptical and generally polarized incident wave can be addressed. The corresponding incident wave can be described by its electric field phasor:

$$\vec{E}_i = E_{_{TE}}\hat{u}_{_{TE}} \exp(-j\vec{k}_i \cdot \vec{r}) + E_{_{TM}}\hat{u}_{_{TM}} \exp(-j\vec{k}_i \cdot \vec{r}), \quad \text{where} \tag{3.120}$$

$$\hat{u}_{_{TE}} = \hat{\imath}_y,$$

$$\hat{u}_{_{TM}} = \cos\theta_1 \; \hat{\imath}_x - \sin\theta_1 \; \hat{\imath}_z,$$

$$\vec{k}_i = k_0\sqrt{\epsilon_{r1}\mu_{r1}} \left(\sin\theta_1 \; \hat{\imath}_x + \cos\theta_1 \; \hat{\imath}_z\right),$$

and ϵ_{r1}, μ_{r1} are the relative permittivity and relative permeability of region 1 (in case of non-magnetic medium $\mu_{r1} \simeq 1$). The $E_{_{TE}}$ and $E_{_{TM}}$ complex amplitude terms can represent any, elliptical in general, polarization. The reflected electric field phasor is given by:

$$\vec{E}_r = E_{_{TE}} r_{_{TE}} \hat{u}^r_{_{TE}} \exp(-j\vec{k}_r \cdot \vec{r}) + E_{_{TM}} r_{_{TM}} \hat{u}^r_{_{TM}} \exp(-j\vec{k}_r \cdot \vec{r}), \quad \text{where} \tag{3.121}$$

$$\hat{u}^r_{_{TE}} = \hat{u}_{_{TE}} = \hat{\imath}_y,$$

$$\hat{u}^r_{_{TM}} = -\cos\theta_1 \; \hat{\imath}_x - \sin\theta_1 \; \hat{\imath}_z,$$

$$\vec{k}_r = k_0\sqrt{\epsilon_{r1}\mu_{r1}} \left(\sin\theta_1 \; \hat{\imath}_x - \cos\theta_1 \; \hat{\imath}_z\right),$$

where r_{TE} and r_{TM} are the amplitude reflection coefficients (complex in general). Therefore, the total electric field phasor in region 1 can be expressed as:

$$\vec{E}_1 = E_{TE}\hat{u}_{TE}\exp(-j\vec{k}_i\cdot\vec{r}) + E_{TM}\hat{u}_{TM}\exp(-j\vec{k}_i\cdot\vec{r}) +$$
$$E_{TE}r_{TE}\hat{u}_{TE}\exp(-j\vec{k}_r\cdot\vec{r}) + E_{TM}r_{TM}\hat{u}^r_{TM}\exp(-j\vec{k}_r\cdot\vec{r}). \tag{3.122}$$

The corresponding magnetic field phasor in region 1 can straightforwardly be determined and is given by:

$$\vec{H}_1 = -\frac{E_{TE}}{Z_1}\hat{u}_{TM}\exp(-j\vec{k}_i\cdot\vec{r}) + \frac{E_{TM}}{Z_1}\hat{u}_{TE}\exp(-j\vec{k}_i\cdot\vec{r}) -$$
$$\frac{E_{TE}r_{TE}}{Z_1}\hat{u}^r_{TM}\exp(-j\vec{k}_r\cdot\vec{r}) + \frac{E_{TM}r_{TM}}{Z_1}\hat{u}_{TE}\exp(-j\vec{k}_r\cdot\vec{r}), \tag{3.123}$$

where $Z_1 = (\mu_1/\epsilon_1)^{1/2}$ is the wave impedance of region 1. The electric and magnetic field phasors in region 2 are similarly expressed as:

$$\vec{E}_2 = t_{TE}E_{TE}\hat{u}_{TE}\exp(-j\vec{k}_t\cdot\vec{r}) + t_{TM}E_{TM}\hat{u}^t_{TM}\exp(-j\vec{k}_t\cdot\vec{r}), \tag{3.124}$$

$$\vec{H}_2 = -\frac{t_{TE}E_{TE}}{Z_2}\hat{u}^t_{TM}\exp(-j\vec{k}_t\cdot\vec{r}) + \frac{t_{TM}E_{TM}}{Z_2}\hat{u}_{TE}\exp(-j\vec{k}_t\cdot\vec{r}), \tag{3.125}$$

$$\vec{k}_t = k_0\sqrt{\epsilon_{r2}\mu_{r2}}\left(\sin\theta_2\;\hat{\imath}_x + \cos\theta_2\;\hat{\imath}_z\right),$$

$$\hat{u}^t_{TM} = \cos\theta_2\;\hat{\imath}_x - \sin\theta_2\;\hat{\imath}_z,$$

where ϵ_{r2}, μ_{r2} are the relative permittivity and relative permeability of region 2 (in case of non-magnetic medium $\mu_{r2} \simeq 1$), and t_{TE} and t_{TM} are the amplitude transmission coefficients (complex in general).

Applying the boundary conditions at the $z = 0$ planar interface for the tangential electric and magnetic field components, the following equations can be straightforwardly derived:

$$\sqrt{\epsilon_{r1}\mu_{r1}}\sin\theta_1 = \sqrt{\epsilon_{r2}\mu_{r2}}\sin\theta_2, \tag{3.126}$$

$$1 + r_{TE} = t_{TE}, \tag{3.127}$$

$$\cos\theta_1 - r_{TM}\cos\theta_1 = t_{TM}\cos\theta_2, \tag{3.128}$$

$$-\frac{\cos\theta_1}{Z_1} + \frac{\cos\theta_1 r_{TE}}{Z_1} = -\frac{\cos\theta_2 t_{TE}}{Z_2}, \tag{3.129}$$

$$\frac{1}{Z_1} + \frac{r_{TM}}{Z_1} = \frac{t_{TM}}{Z_2}. \tag{3.130}$$

Solving the aforementioned equations, the amplitude reflection and transmission coefficients can be readily calculated in this general formulation and are expressed as:

$$r_{TE} = \frac{Z_2\cos\theta_1 - Z_1\cos\theta_2}{Z_2\cos\theta_1 + Z_1\cos\theta_2}, \tag{3.131}$$

$$t_{TE} = \frac{2Z_2\cos\theta_1}{Z_2\cos\theta_1 + Z_1\cos\theta_2}, \tag{3.132}$$

$$r_{TM} = \frac{Z_1\cos\theta_1 - Z_2\cos\theta_2}{Z_1\cos\theta_1 + Z_2\cos\theta_2}, \tag{3.133}$$

$$t_{TM} = \frac{2Z_2\cos\theta_1}{Z_1\cos\theta_1 + Z_2\cos\theta_2}. \tag{3.134}$$

The last equations are a generalized equivalent to Eqs. (3.78), (3.79), (3.80), and (3.81), which are applicable for non-magnetic media. It is worth noting that region 2 can be lossy, and in such cases, ϵ_2 and μ_2 may become complex.

To assess the percentage of power reflected and the percentage of power transmitted, the z-components of the complex Poynting vectors must be evaluated. Utilizing the previously defined phasor fields and after some manipulations, the following z-components of the complex Poynting vectors for regions 1 and 2 can be determined:

$$S_{1z} = \frac{\cos\theta_1}{Z_1}\left\{|E_{TE}|^2 + |E_{TM}|^2 - |E_{TE}|^2|r_{TE}|^2 - |E_{TM}|^2|r_{TM}|^2 - \right.$$
$$|E_{TE}|^2 r_{TE}^* \exp\left(+j(\vec{k}_r - \vec{k}_i)\cdot\vec{r}\right) - |E_{TE}|^2 r_{TE} \exp\left(-j(\vec{k}_r - \vec{k}_i)\cdot\vec{r}\right) +$$
$$\left.|E_{TM}|^2 r_{TM}^* \exp\left(+j(\vec{k}_r - \vec{k}_i)\cdot\vec{r}\right) - |E_{TM}|^2 r_{TM} \exp\left(-j(\vec{k}_r - \vec{k}_i)\cdot\vec{r}\right)\right\} \tag{3.135}$$

$$S_{2z} = \frac{1}{Z_2^*}\left\{|t_{TE}|^2|E_{TE}|^2(\cos\theta_2)^* + |t_{TM}|^2|E_{TM}|^2\cos\theta_2\right\}. \tag{3.136}$$

By equating the real parts of the z-components of the Poynting vectors, the normalized reflected and transmitted powers can be expressed as:

$$\frac{P_r}{P_i} = \frac{|r_{TE}|^2|E_{TE}|^2 + |r_{TM}|^2|E_{TM}|^2}{|E_{TE}|^2 + |E_{TM}|^2}, \tag{3.137}$$

$$\frac{P_t}{P_i} = \frac{Z_1}{\cos\theta_1}\,\frac{|t_{TE}|^2|E_{TE}|^2\mathrm{Re}\{(\cos\theta_2)^*/Z_2^*)\} + |t_{TM}|^2|E_{TM}|^2\mathrm{Re}\{\cos\theta_2/Z_2^*)\}}{|E_{TE}|^2 + |E_{TM}|^2}, \tag{3.138}$$

where $\mathrm{Re}\{(\cos\theta_2)^*/Z_2^*\} = \mathrm{Re}\{\cos\theta_2/Z_2\}$. The last two equations take the simple form of Eqs. (3.118) and (3.119) in the case of the *TE* only ($E_{TM} = 0$) or *TM* only ($E_{TE} = 0$) polarization, respectively.

3.3 Plane-Wave Polarization

Electromagnetic fields, such as the electric and magnetic fields of a light wave, are vectorial. The simplest possible solution for the electromagnetic field is the plane wave solution. In this case, the electric and magnetic fields are perpendicular to the direction of propagation, representing what is known as a transverse (TEM) electromagnetic wave. Assuming, for simplicity, a plane wave propagating along the z-axis, as shown in Fig. 3.17(a), its electric field phasor $\vec{E}$ can be expressed in the following form:

$$\vec{E} = \vec{E}_0\exp(-jkz) = [E_x\hat{\imath}_x + E_y\hat{\imath}_y]\exp(-jkz), \tag{3.139}$$

where $k = \omega\sqrt{\epsilon\mu} = \omega/u$ is the wavenumber (u the phase velocity of the wave), and $E_x = E_{0x}\exp(j\phi_x)$, $E_y = E_{0y}\exp(j\phi_y)$ are the complex amplitudes of the two possible components of the electric field vector. A similar equation can also be written for the magnetic field. Since the phasor of the electric field is, in general, complex the real electric field can be expressed in the form:

$$\vec{\mathcal{E}} = E_{0x} \cos(\omega t - kz + \phi_x)\,\hat{\imath}_x + E_{0y} \cos(\omega t - kz + \phi_y)\,\hat{\imath}_y = \mathcal{E}_x(z,t)\hat{\imath}_x + \mathcal{E}_y(z,t)\hat{\imath}_y. \quad (3.140)$$

The polarization of this plane wave is, by definition, the trajectory traced by the electric field vector at a specific z-plane as a function of time. From a physical standpoint, the polarization of the electromagnetic wave determines the direction of the Lorentz force that the wave exerts on a charged particle [20]. Assuming that the phase angles ϕ_x and ϕ_y do not vary with time, the trajectory that the electric field traverses at a constant spatial position as a function of time can be calculated as follows:

$$\mathcal{E}_x(z,t) = E_{0x} \cos\underbrace{(\omega t - kz + \phi_x)}_{a(t,z)} = E_{0x}\cos[a(t,z)],$$

$$\mathcal{E}_y(z,t) = E_{0y} \cos(\omega t - kz + \phi_y) = E_{0y}\cos[a(t,z) + \underbrace{(\phi_y - \phi_x)}_{\phi}] \Longrightarrow$$

$$\begin{aligned}\frac{\mathcal{E}_x(z,t)}{E_{0x}} &= \cos[a(t,z)],\\ \frac{\mathcal{E}_y(z,t)}{E_{0y}} &= \cos[a(t,z)+\phi] = \cos[a(t,z)]\cos\phi - \sin[a(t,z)]\sin\phi \Longrightarrow \end{aligned} \quad (3.141)$$

$$\begin{aligned}\left(\frac{\mathcal{E}_y(z,t)}{E_{0y}} - \frac{\mathcal{E}_x(z,t)}{E_{0x}}\cos\phi\right)^2 &= \sin^2[a(t,z)]\sin^2\phi = \\ &= \left(1 - \frac{\mathcal{E}_x^2(z,t)}{E_{0x}^2}\right)\sin^2\phi \Longrightarrow\end{aligned}$$

$$\frac{\mathcal{E}_x^2(z,t)}{E_{0x}^2} + \frac{\mathcal{E}_y^2(z,t)}{E_{0y}^2} - 2\,\frac{\mathcal{E}_x(z,t)}{E_{0x}}\,\frac{\mathcal{E}_y(z,t)}{E_{0y}}\cos\phi = \sin^2\phi, \quad (3.142)$$

where the last equation represents an ellipse in the xy plane as it is shown in Fig. 3.17(b). The ellipse described by Eq. (3.142) can be transformed into a diagonal form through a linear transformation from the xyz to the $x_1y_1z_1$ coordinate system. The relationship of the unit vectors between the two systems is described by the following relations:

$$\begin{aligned}\hat{\imath}_x &= \hat{\imath}_{x_1}\cos\alpha - \hat{\imath}_{y_1}\sin\alpha,\\ \hat{\imath}_y &= \hat{\imath}_{x_1}\sin\alpha + \hat{\imath}_{y_1}\cos\alpha,\\ \hat{\imath}_z &= \hat{\imath}_{z_1},\\ \mathcal{E}_x &= \cos\alpha\ \mathcal{E}_{x_1} - \sin\alpha\ \mathcal{E}_{y_1},\\ \mathcal{E}_y &= \sin\alpha\ \mathcal{E}_{x_1} + \cos\alpha\ \mathcal{E}_{y_1}.\end{aligned}$$

Utilizing the last two equations in Eq. (3.142), it can be expressed in the $x_1y_1z_1$ coordinate system. In this system, the angle α can be chosen in such a manner as to eliminate the coefficient of the $\mathcal{E}_{x_1}\mathcal{E}_{y_1}$ term, resulting in:

$$\alpha = \frac{1}{2}\tan^{-1}\left[\frac{2E_{0x}E_{0y}}{E_{0x}^2 - E_{0y}^2}\cos\phi\right], \tag{3.143}$$

$$\frac{1}{E_{0x_1}^2} = \left(\frac{\cos^2\alpha}{E_{0x}^2} + \frac{\sin^2\alpha}{E_{0y}^2} - \frac{\sin^2\alpha}{E_{0x}E_{0y}}\cos\phi\right)\frac{1}{\sin^2\phi},$$

$$\frac{1}{E_{0y_1}^2} = \left(\frac{\sin^2\alpha}{E_{0x}^2} + \frac{\cos^2\alpha}{E_{0y}^2} + \frac{\sin^2\alpha}{E_{0x}E_{0y}}\cos\phi\right)\frac{1}{\sin^2\phi},$$

$$\frac{\mathcal{E}_{x_1}^2(z,t)}{E_{0x_1}^2} + \frac{\mathcal{E}_{y_1}^2(z,t)}{E_{0y_1}^2} = 1. \tag{3.144}$$

Since the new semi-axes are the E_{0x_1} and E_{0y_1} the ellipticity can be defined as $e = E_{min}/E_{max}$, where $E_{min} = \min\{E_{0x_1}, E_{0y_1}\}$ and $E_{max} = \max\{E_{0x_1}, E_{0y_1}\}$.

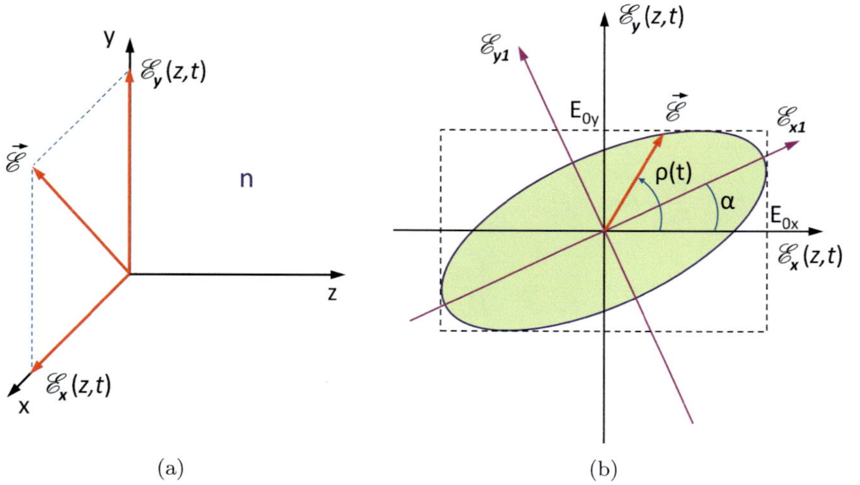

FIGURE 3.17: (a) A plane wave propagating along the axis z with its electric field components $\mathcal{E}_x$ and $\mathcal{E}_y$. (b) A general polarization ellipse describing the polarization evolution of a plane wave propagating along the z-axis.

Understanding the direction of the electric field motion along the ellipse is crucial. The rotation of the electric field around the ellipse can be either counter-clockwise or clockwise for an observer facing the wave. In the case of counter-clockwise rotation, the polarization is termed *right-handed elliptical polarization* (RHEP). This designation comes from the right-hand rule, where the thumb points in the direction of wave propagation, and the curling of fingers indicates the spatial rotation of the field. Conversely, if the rotation is clockwise, the polarization is called *left-handed elliptical polarization* (LHEP), following the left-hand rule. It's important to note a convention discrepancy where some texts consider the observer behind the wave, looking at it as it moves away. In this situation, clockwise rotation would be right-handed polarization, and counter-clockwise rotation would be left-handed polarization. The choice of convention depends on the observer's position. In this book, the assumption is made that the observer is in front of the wave, observing it propagate towards them. This conforms to the *Institute of Electrical and Electronics Engineers (IEEE)* standard, commonly used in the engineering community. The alternative convention is more prevalent in the physics community.

The clockwise and counter-clockwise rotation is contingent on the phase difference angle $\phi = \phi_y - \phi_x$. If $\phi > 0$ and $\phi \neq m\pi$ (where m is a positive integer), then the rotation is clockwise, resulting in left-handed polarization. Conversely, if $\phi < 0$ and $\phi \neq -m\pi$ (where m is a positive integer), then the rotation is clockwise, resulting in right-handed polarization. This distinction can be easily grasped by determining the angle $\rho(t)$ (see Fig. 3.17(b)):

$$\rho = \rho(t) = \left.\frac{\mathcal{E}_y}{\mathcal{E}_x}\right|_z = \frac{E_{0y}}{E_{0x}} \frac{\cos[a(t,z)+\phi]}{\cos[a(t,z)]}. \tag{3.145}$$

An example of an RHEP polarization is shown in Fig. 3.18(a) where the parameters are defined in the figure caption.

3.3.1 Special Cases of Polarization

Some special cases can be derived from the general form of Eq. (3.142). These are summarized next.

3.3.1.1 Linear Polarization

Assume that $\phi = m\pi$ where m an integer and E_{0x} and E_{0y} arbitrary. In this case Eq. (3.142) becomes:

$$\frac{\mathcal{E}_x}{E_{0x}} = +\frac{\mathcal{E}_y}{E_{0y}} \Longrightarrow \frac{\mathcal{E}_y}{\mathcal{E}_x} = \frac{E_{0y}}{E_{0x}} = \tan\rho(t), \qquad \text{for } m = 2q$$

$$\frac{\mathcal{E}_x}{E_{0x}} = -\frac{\mathcal{E}_y}{E_{0y}} \Longrightarrow \frac{\mathcal{E}_y}{\mathcal{E}_x} = -\frac{E_{0y}}{E_{0x}} = \tan[\pi-\rho(t)], \qquad \text{for } m = 2q+1$$

If $\phi = 2q\pi$ (where q is an integer), then the linear polarization is defined by the constant angle ρ with respect to the x-axis, as illustrated in Fig. 3.19(a). In the case where $\phi = (2q+1)\pi$, the linear polarization is determined by the constant angle $\pi - \rho$. The electric field at a constant z oscillates along the directions specified by the angle ρ (or $\pi - \rho$). Given that the electric field oscillates along a constant inclination line, this polarization is termed *linear polarization* (LP). To fully specify a linear polarization, the angle ρ must be provided, as indicated in the equations above.

3.3.1.2 Circular Polarization

Another special case is when $E_{0x} = E_{0y} = E_0$ and $\phi = \pm\pi/2$. Under these assumptions, Eq. (3.142) becomes:

$$\mathcal{E}_x^2 + \mathcal{E}_y^2 = E_0^2,$$

$$\tan\rho(t) = \frac{\mathcal{E}_y}{\mathcal{E}_x} = +\tan(\omega t - kz + \phi_x) \qquad \text{for } \phi = -\frac{\pi}{2},$$

$$\tan\rho(t) = \frac{\mathcal{E}_y}{\mathcal{E}_x} = -\tan(\omega t - kz + \phi_x) \qquad \text{for } \phi = +\frac{\pi}{2}.$$

In this case the ellipse becomes a circle and is called *circular polarization.* An example is illustrated in Fig. 3.19(b). When the angle ρ increases over time (for the case of $\phi = -\pi/2$), the rotation is counter-clockwise, defining the polarization as *right-handed circular*

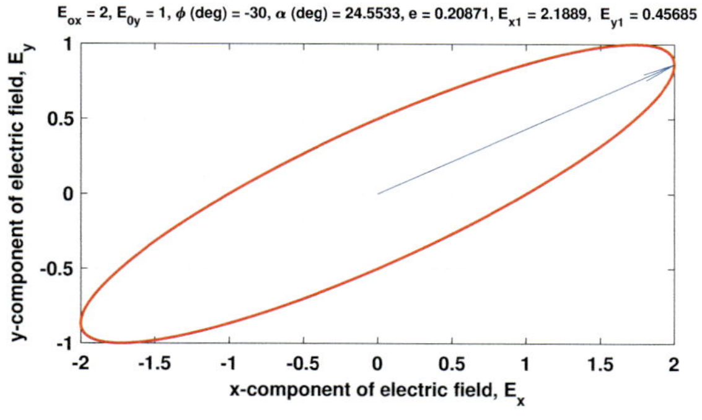

(a)

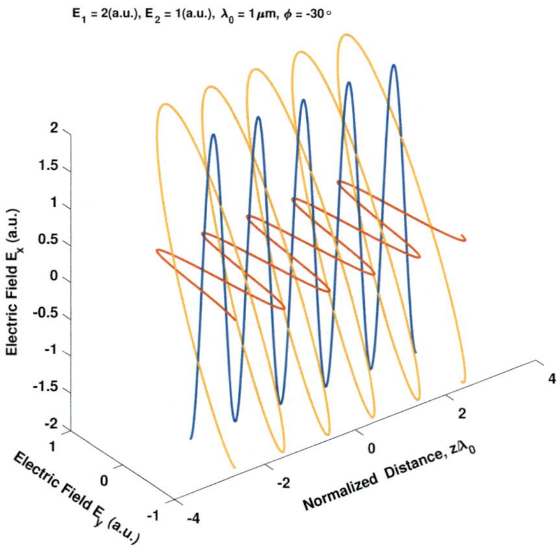

(b)

FIGURE 3.18: (a) An example of an elliptical polarization, for $E_{0x} = 2$, $E_{0y} = 1$ (could be Volts/meter or any electric field units), $\phi = -30°$ (corresponding to RHEP wave). The ellipticity is $e = 0.20871$ while the angle $\alpha = 24.5533°$. (b) A snapshot of the electric field in space showing the x and y electric field components and the total electric field that traverses the polarization ellipse in space.

polarization (RHCP). Conversely, when ρ decreases over time (for the case of $\phi = +\pi/2$), the rotation is clockwise, defining the polarization as *left-handed circular polarization* (LHCP). The phase angle can be generalized so that when $\phi = (2m - \frac{1}{2})\pi$, it corresponds to RHCP, and when $\phi = (2m + \frac{1}{2})\pi$, it corresponds to LHCP (where m is an integer).

3.3.1.3 Special Elliptical Polarization

Another special case is when E_{0x}, E_{0y} are arbitrary and $\phi = \pm\pi/2$. In this case Eq. (3.142) becomes:

$$\frac{\mathcal{E}_x^2}{E_{0x}^2} + \frac{\mathcal{E}_y^2}{E_{0y}^2} = 1,$$

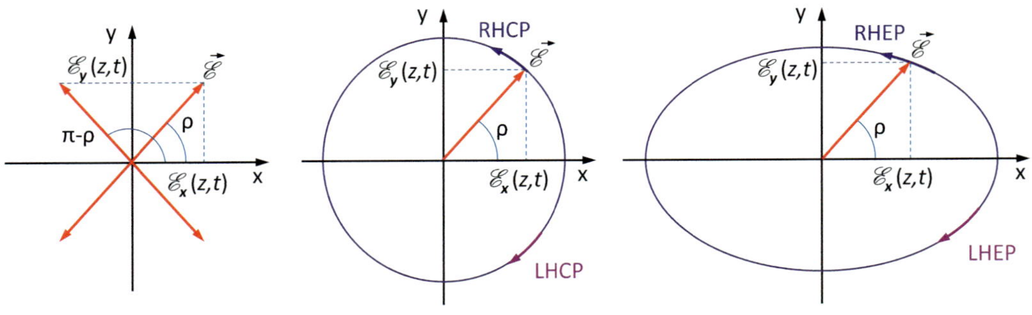

FIGURE 3.19: (a) A plane wave propagating along the axis z with linear polarization. In this case its electric field components are $\mathcal{E}_x$ and $\mathcal{E}_y$ with arbitrary amplitudes E_{0x}, E_{0y}, and $\phi = m\pi$. (b) A plane wave propagating along the axis z with circular polarization. In this case its electric field components are $\mathcal{E}_x$ and $\mathcal{E}_y$ with equal amplitudes $E_{0x} = E_{0y} = E_0$ and $\phi = -\pi/2$ for RHCP or $\phi = +\pi/2$ for LHCP. (c) A plane wave propagating along the axis z with special elliptical polarization. In this case its electric field components are $\mathcal{E}_x$ and $\mathcal{E}_y$ with unequal amplitudes $E_{0x} > E_{0y}$ and $\phi = -\pi/2$ for RHEP or $\phi = +\pi/2$ for LHEP.

$$\tan\rho(t) = \frac{\mathcal{E}_y}{\mathcal{E}_x} = +\frac{E_{0y}}{E_{0x}}\tan(\omega t - kz + \phi_x) \qquad \text{for} \quad \phi = -\frac{\pi}{2},$$

$$\tan\rho(t) = \frac{\mathcal{E}_y}{\mathcal{E}_x} = -\frac{E_{0y}}{E_{0x}}\tan(\omega t - kz + \phi_x) \qquad \text{for} \quad \phi = +\frac{\pi}{2}.$$

Under these conditions, the ellipse is diagonalized in the xyz coordinate system. This represents a special case of general elliptical polarization. An example is depicted in Fig. 3.19(c). Once again, as the angle ρ increases over time (for the case of $\phi = -\pi/2$), the rotation is counter-clockwise, defining the polarization as *right-handed elliptical polarization* (RHEP). Conversely, when ρ decreases over time (for the case of $\phi = +\pi/2$), the rotation is clockwise, defining the polarization as *left-handed elliptical polarization* (LHEP). Similar to circular polarization, the phase angle can be generalized so that $\phi = (2m - \frac{1}{2})\pi$ corresponds to RHEP and $\phi = (2m + \frac{1}{2})\pi$ corresponds to LHEP (where m is an integer).

Frequently, light emitted by various natural sources, such as the sun, is characterized by random polarization. Mathematically, this randomness can be described by considering the phase angles $\phi_x(t)$ and $\phi_y(t)$ as random variables. This inherent randomness is a result of photons being emitted through spontaneous processes, leading to general polarization randomness. This type of light is commonly referred to as *randomly polarized* or simply *unpolarized* [20]. When a portion of the light exhibits polarization, it is termed *partially polarized*.

Polarized light can be generated from unpolarized light using various methods. The most common approaches involve (a) leveraging the reflection properties of light, (b) exploiting the transmission properties of light, (c) utilizing dichroism, (d) taking advantage of double refraction and the properties of anisotropic crystals, and (e) employing scattering mechanisms [62].

3.4 Propagation in Lossy Media

In the preceding sections, the propagation of plane waves in isotropic media without losses was predominantly discussed. However, real media inherently possess losses, the extent of which depends on the free-space wavelength employed. Even dielectric materials can exhibit significant losses in specific wavelength ranges. These losses manifest as the attenuation of the optical wave as it traverses the lossy dielectric medium. In contrast, metals, with their free electrons capable of absorbing electromagnetic energy from incident optical fields, incur losses resulting in Joule heating and subsequent attenuation of the optical field. This leads to shallow penetration depths into the metal. The high conductivity of metals contributes to substantial attenuation within them, rendering most metals opaque to optical waves [91]. Despite this, metals play a crucial role in optics, serving as highly effective mirrors. Apart from dielectrics and metals, semiconductors find widespread use in optoelectronic applications (such as lasers, diodes, detectors, etc.) and can also exhibit losses due to absorption in interband or intraband transitions.

From an electromagnetic perspective, material losses are often effectively modeled using a wavelength-dependent complex relative permittivity or an equivalent complex refractive index. Subsequent subsections introduce simple theories used for the study of dielectrics (e.g., the classical electron oscillator model) and metals (e.g., the Drude model) to define their corresponding complex relative permittivity. More accurate models necessitate concepts from quantum mechanics, which are beyond the scope of this book. For semiconductor materials, more precise models based on quantum mechanics are essential to determine their relative permittivity. Interested readers are encouraged to consult relevant textbooks [92–96].

3.4.1 Classical Electron Oscillator Model for Dielectrics

For materials lacking free electric charge carriers, such as dielectrics, the simple classical electron oscillator model proves useful [97, 98]. In this model, the atom is conceptualized with an electronic cloud surrounding the nucleus, as illustrated schematically in Fig. 3.20. When subjected to an external electric field $\vec{E}(t)$, forces act on both the nucleus and the electronic cloud. In Fig. 3.20, the nucleus is assumed to be in a constant position, while the center of charge of the electronic cloud is displaced due to the electric force, thereby creating an electric dipole.

Given that the nucleus is positively charged, it exerts a restoring force, denoted as $-K\vec{s}(t)$, acting on the electron cloud. This model assumes that the electron cloud can mimic the behavior of a single electron, particularly in situations involving a single electric dipole transition. The classical equation of motion for the single electron, described by the distance $\vec{s}(t)$, can be defined as follows:

$$\begin{aligned} m_e\frac{d^2\vec{s}}{dt^2} &= -e\vec{E}(t) - b\frac{d\vec{s}}{dt} - K\vec{s}(t), \\ \frac{d^2\vec{s}}{dt^2} &= -\frac{e}{m_e}\vec{E}(t) - \frac{b}{m_e}\frac{d\vec{s}}{dt} - \frac{K}{m_e}\vec{s}(t), \end{aligned} \tag{3.146}$$

where m_e denotes the electron mass, e represents the electron charge, K stands for the spring constant of the restoring force, $\vec{E}(t)$ is the externally applied time-dependent electric field, and the term proportional to the velocity, $d\vec{s}/dt$, introduces a damping term to account for the inevitable losses in the oscillator. This damping factor encompasses various sources of motion loss, including random collisions with other atoms, coupling to vibrations in a crystal (electron-phonon coupling), emission of light (radiative relaxation), and energy transfer to other electrons (electron-electron interactions). Equation (3.146) can be written

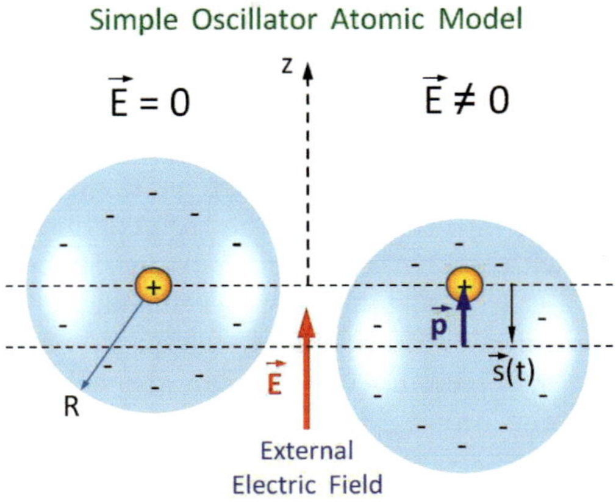

FIGURE 3.20: A simple classical oscillator model of an atom under the influence of an external electric field $\vec{E}$. On the left part no external electric field is present. The atom is represented with an electron cloud and the nucleus.

in the form:

$$\frac{d^2\vec{s}}{dt^2} + q\frac{d\vec{s}}{dt} + \omega_0^2\vec{s}(t) = -\frac{e}{m_e}\vec{E}(t), \tag{3.147}$$

where $\omega_0 = (K/m_e)^{1/2}$ is the resonance frequency, and q is the damping coefficient. The homogeneous part of Eq. (3.147) has a solution of the form [97] (assuming that $\omega_0 > q/2$):

$$s(t) = \exp\left[(-q/2)(t-t_0)\right]\left\{A\exp[-j\omega_e(t-t_0)] + B\exp[+j\omega_e(t-t_0)]\right\},$$

where $\omega_e = [\omega_0^2 - (q/2)^2]^{1/2}$ and A, B are constants that depend on the initial conditions for $s(t_0)$ and $ds(t_0)/dt$. In this case the scalar form of the homogeneous part of Eq. (3.147) is used. The term $\tau = 1/q$ is the energy decay time of the classical oscillator [97]. However, usually the time-harmonic solution of Eq. (3.147) is needed due to the very high optical frequencies. In the time-harmonic case Eq. (3.147) becomes for $\vec{E}(t) = \text{Re}\{\vec{E}(\omega)e^{j\omega t}\}$ and $\vec{s}(t) = \text{Re}\{\vec{S}(\omega)e^{j\omega t}\}$:

$$-\omega^2\vec{S}(\omega) + j\omega q\vec{S}(\omega) + \omega_0^2\vec{S}(\omega) = -\frac{e}{m_e}\vec{E}(\omega) \implies \vec{S}(\omega) = \frac{-e/m_e}{(\omega_0^2-\omega^2)+j\omega q}\vec{E}(\omega). \tag{3.148}$$

The displacement $\vec{S}(\omega)$ creates a microscopic electric dipole moment $\vec{p}(\omega)$ which is given by:

$$\vec{p}(\omega) = e\left[-\vec{S}(\omega)\right] = \frac{e^2/m_e}{(\omega_0^2-\omega^2)+j\omega q}\vec{E}(\omega). \tag{3.149}$$

The medium polarization P [see Eq. (3.21)] is defined as the electric dipoles per unit volume. If it is assumed that there are N electric dipoles per unit volume, then the macroscopic medium time-harmonic polarization is given by:

$$\vec{P}(\omega) = \frac{Ne^2/m_e}{(\omega_0^2 - \omega^2) + j\omega q}\vec{E}(\omega) = \underbrace{\epsilon_0 \left\{ \frac{Ne^2/m_e\epsilon_0}{(\omega_0^2 - \omega^2) + j\omega q} \right\}}_{= \chi_e(\omega)} \vec{E}(\omega) = \epsilon_0 \chi_e(\omega)\vec{E}(\omega) \Longrightarrow$$

$$\chi_e(\omega) = \chi_e'(\omega) - j\chi_e''(\omega) = \frac{\omega_p^2(\omega_0^2 - \omega^2)}{(\omega_0^2 - \omega^2)^2 + \omega^2 q^2} - j\frac{q\omega\omega_p^2}{(\omega_0^2 - \omega^2)^2 + \omega^2 q^2}, \tag{3.150}$$

where $\omega_p = (Ne^2/m_e\epsilon_0)^{1/2}$ is known as the *plasma frequency.* Using Eq. (3.24) the complex relative permittivity $\varepsilon'(\omega) - j\varepsilon''(\omega)$ can be determined as follows:

$$\varepsilon'(\omega) = 1 + \frac{\omega_p^2(\omega_0^2 - \omega^2)}{(\omega_0^2 - \omega^2)^2 + \omega^2 q^2} \tag{3.151}$$

$$\varepsilon''(\omega) = \frac{q\omega\omega_p^2}{(\omega_0^2 - \omega^2)^2 + \omega^2 q^2}. \tag{3.152}$$

The complex relative permittivity is related to the complex refractive index by the expression $(n - j\kappa)^2 = \varepsilon' - j\varepsilon''$. Using Eqs. (3.151) and (3.152), the real part of the refractive index, n, and the extinction index, κ, can be derived and are expressed as:

$$n = \left\{ \frac{1}{2} \left\{ \left[(\varepsilon')^2 + (\varepsilon'')^2 \right]^{1/2} + \varepsilon' \right\} \right\}^{1/2}, \tag{3.153}$$

$$\kappa = \left\{ \frac{1}{2} \left\{ \left[(\varepsilon')^2 + (\varepsilon'')^2 \right]^{1/2} - \varepsilon' \right\} \right\}^{1/2}. \tag{3.154}$$

An illustrative calculation is presented in Figs. 3.21(a) and 3.21(b) for the electric susceptibilities and the refractive indices (n and κ) corresponding to the transition of a He-Ne laser with a wavelength of $\lambda_0 = 0.6328\,\mu$m. The density of the electric dipoles is $N = 1.22 \times 10^{19}$ atoms/m^3, and $q = 2\pi \times 1.5 \times 10^9$ rad/s. In cases where the transition is exceptionally sharp, $\omega \simeq \omega_0$, and the electric susceptibilities can be approximated by:

$$\chi_e''(\omega) \simeq \frac{\omega_p^2}{4\omega_0} \frac{q}{(\omega_0 - \omega)^2 + (q/2)^2},$$

$$\chi_e'(\omega) \simeq \frac{\omega_p^2}{4\omega_0} \frac{2(\omega_0 - \omega)}{(\omega_0 - \omega)^2 + (q/2)^2} = \frac{2(\omega_0 - \omega)}{q}\chi_e''(\omega).$$

Here, the dependence of χ_e'' is evidently of a Lorentzian type. It is important to note that the external field $\vec{E}(\omega)$ may not be equivalent to the local field, $\vec{E}_{loc}(\omega)$, in the proximity of the atomic oscillator. The approximation $\vec{E}_{loc}(\omega) \simeq \vec{E}(\omega)$ is reasonable when the density of atomic oscillators is low, as in a gas medium (small N). However, when the density of atomic oscillators is high, a local field correction should be applied [99].

In the realm of real materials, more realistic models are necessary to accurately predict their macroscopic response in terms of the relative complex permittivity. Real materials often exhibit multiple resonance frequencies corresponding to various possible oscillators. Therefore, a more comprehensive expression for the electric susceptibility can be formulated as [42, 98]:

$$\chi_e(\omega) = \sum_i \frac{\omega_p^2 f_i}{(\omega_{0i}^2 - \omega^2) + j\omega q_i}, \tag{3.155}$$

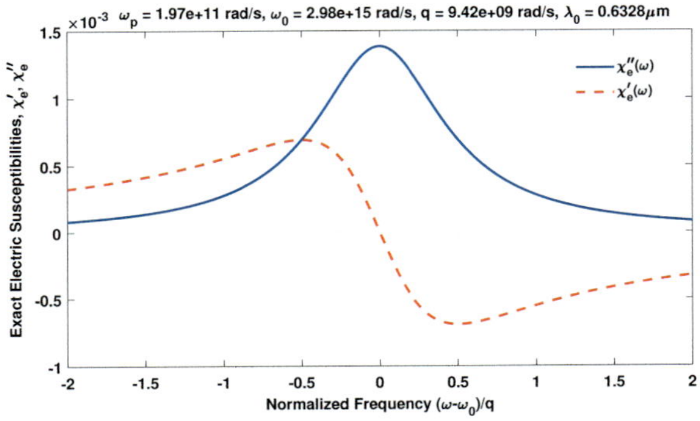

(a)

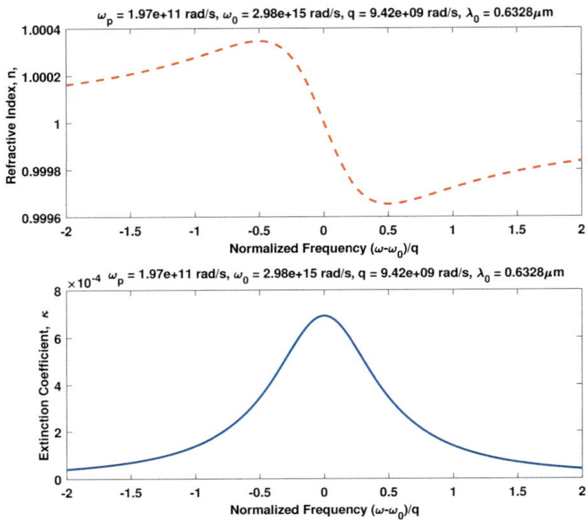

(b)

FIGURE 3.21: (a) The electric susceptibilities, $\chi_e'(\omega)$ and $\chi_e''(\omega)$ as a function of the normalized frequency $(\omega - \omega_0)/q$ based on the classical electron oscillator model. The density of electric dipoles is $N = 1.22 \times 10^{19}$ atoms$/m^3$ while $q = 2\pi \times 1.5 \times 10^9$ rad/s. The resonance frequency corresponds to the He-Ne laser freespace wavelength of 0.6328 μm. (b) The refractive index $n(\omega)$ and the extinction coefficient $\kappa(\omega)$ based on the classical electron oscillator model of the He-Ne laser for the transition in (a).

where ω_{0i} $(i = 1, 2, \cdots)$ represents the diverse resonance frequencies of the oscillators in the medium, f_i denotes the oscillator strengths (satisfying the equality $\sum_i f_i = 1$), and q_i corresponds to the various damping constants of the oscillators. Equation (3.155) holds for cases with small oscillator densities, such as in gases. For denser materials, where a local field correction is necessary, the resulting relative permittivity for isotropic materials with a spherical dipole shape can be expressed using the equation [42, 99]:

$$\frac{\varepsilon - 1}{\varepsilon + 2} = \frac{1}{3} \sum_i \frac{\omega_p^2 f_i}{(\omega_{0i}^2 - \omega^2) + j\omega q_i}. \tag{3.156}$$

The aforementioned equations find application in describing various dielectric materials through the Sellmeier formula, which expresses the refractive index and extinction coefficient as functions of the free-space wavelength (refer to Appendix 2A in the *Geometrical Optics* chapter, Chapter 2). Additionally, it is important to highlight that the real and imaginary components of the electric susceptibility are intrinsically linked through the Kramers-Kronig relations, which are discussed in detail in Appendix 3A.

3.4.2 Drude Model for Metals

In metals, electrons are not bound to specific ions as in dielectrics; instead, they are free to move under the influence of an externally applied electric field. As a result, the applied electric field causes the free electrons to move, generating microscopic currents rather than oscillations, as there is no restoring force. This behavior is well described by the Drude model [42,91,94,98,99], which assumes that free electrons, influenced by the Lorentz force, experience collisions with other electrons or metal ions. In the Drude model, these collisions occur at a rate of $1/\tau$, where τ represents the relaxation time of the free electrons in the metal [94]. The equation of motion for a free electron under this model is given by [94]:

$$m_e \frac{d^2\vec{s}}{dt^2} + \frac{m_e}{\tau}\frac{d\vec{s}}{dt} = -e\vec{E}(t), \tag{3.157}$$

where $\vec{s}$ describes the path of the motion of the free electron in this case. The term $m_e(d\vec{s}/dt) = m_e\vec{v} = \vec{p}$ is the free electron momentum. Expressing the Eq. (3.157) in terms of the velocity $\vec{v}$ the equation is written as:

$$\frac{d\vec{v}}{dt} + \frac{1}{\tau}\vec{v}(t) = -\frac{e}{m_e}\vec{E}(t). \tag{3.158}$$

For the case of a time-harmonic applied electric field, the previous equation can be rewritten as:

$$\vec{V}(\omega) = -\tau\frac{e/m_e}{1+j\omega\tau}\,\vec{E}(\omega), \tag{3.159}$$

where as before $\vec{E}(t) = \mathrm{Re}\{\vec{E}(\omega)e^{j\omega t}\}$ and $\vec{v}(t) = \mathrm{Re}\{\vec{V}(\omega)e^{j\omega t}\}$. At this point it is necessary to introduce from the electromagnetic theory the definition of the electric current density $\vec{\mathcal{J}}$ [100,101]:

$$\vec{\mathcal{J}} = \lim_{\Delta V\to 0}\sum_j q_j\vec{v}_j \simeq \vec{v}\,\underbrace{\left\{\lim_{\Delta V\to 0}\sum_j q_j\right\}}_{=\,\rho} = \rho\vec{v}, \tag{3.160}$$

where ΔV tends to zero from a macroscopic point of view and ρ is the volume charge density. In metals the volume charge density of free electrons is $\rho = N_e(-e)$ (since all $q_j = -e$) where N_e the number of free electrons per unit volume. This free electron density is given by $N_e = N_{Avo}\rho_m Z/A$ where $N_{Avo} = 6.02214076 \times 10^{23}$ units/mole is Avogadro's number, ρ_m is the mass density of the metal (in gr/cm^3), Z is the number of free electrons that each atom contributes, and A is the atomic mass of the metal [94]. If Eq. (3.160) is transformed into the time-harmonic form and the velocity is replaced by Eq. (3.159), the time-harmonic current density becomes:

$$\vec{J}(\omega) = N_e(-e)\vec{V}(\omega) = \underbrace{\left\{\tau\frac{N_e e^2/m_e}{1+j\omega\tau}\right\}}_{=\,\sigma(\omega)}\vec{E}(\omega) = \sigma(\omega)\vec{E}(\omega). \tag{3.161}$$

The term $\sigma(\omega)$ is the metal conductivity (in $1/\Omega$ m) [see also Eq. (3.23)], and Eq. (3.161) represents the microscopic Ohm's law [100,101]. Using Eq. (3.16) from the Maxwell's equations, the following relation is implied:

$$\vec{\nabla} \times \vec{H}(\omega) = \vec{J}(\omega) + j\omega\vec{D}(\omega) = \vec{J}(\omega) + j\omega\epsilon_0\vec{E}(\omega)$$

$$= j\omega\epsilon_0 \underbrace{\left[1 + \frac{\sigma(\omega)}{j\epsilon_0\omega}\right]}_{=\,\varepsilon_c(\omega)} \vec{E}(\omega) = j\omega\epsilon_0\varepsilon_c(\omega)\vec{E}(\omega), \qquad \text{where}$$

$$\varepsilon_c(\omega) = \varepsilon'(\omega) - j\varepsilon''(\omega) = 1 + \frac{\sigma(\omega)}{j\epsilon_0\omega} = 1 + \frac{1}{j\omega\epsilon_0}\,\tau\frac{N_e e^2/m_e}{1 + j\omega\tau} \Longrightarrow \tag{3.162}$$

$$\varepsilon'(\omega) = 1 - \frac{\omega_p^2\tau^2}{1 + \omega^2\tau^2}, \tag{3.163}$$

$$\varepsilon''(\omega) = \frac{\omega_p^2\tau}{1 + \omega^2\tau^2}, \tag{3.164}$$

where $\omega_p^2 = N_e e^2/m_e\epsilon_0$ is the plasma frequency. The corresponding refractive index $n(\omega)$ and extinction coefficient $\kappa(\omega)$ can be easily determined using Eqs. (3.163) and (3.164) in conjunction with Eqs. (3.153) and (3.154).

Example 3.4.1:

As an illustrative example, the case of gold as a metal is examined within the Drude model formulation. The refractive index $n(\omega)$ and the extinction coefficient $\kappa(\omega)$ are investigated as functions of the frequency of the incident light for two frequency ranges: (a) when the center wavelength is $\lambda_0 = 0.6328\,\mu$m, and (b) when the center wavelength is $\lambda_0 = 0.01\,\mu$m. The relaxation time of the free electrons in gold is specified as $\tau = 3 \times 10^{-14}$ sec [94].
Solution: Initially, consider an incident light wavelength of $\lambda_0 = 0.6328\,\mu$m. For this wavelength, the angular frequency is $\omega_0 = 2\pi(c/\lambda_0) \simeq 2.98 \times 10^{15}$ rad/s, where c denotes the speed of light in freespace. For gold, the following parameters are used: $Z = 1$, $A = 196.96657$, and $\rho_m = 19.32\,\text{gr/cm}^3$. The corresponding plasma frequency is calculated as $\omega_p = (N_e e^2/m_e\epsilon_0)^{1/2} \simeq 1.37 \times 10^{16}$ rad/s. These values yield $N_e = 5.90 \times 10^{28}$ electrons/m^3. Utilizing Eqs. (3.163) and (3.164), along with Eqs. (3.153) and (3.154), the refractive index $n(\omega)$ and the extinction coefficient $\kappa(\omega)$ can be determined. The results are presented in Fig. 3.22(a) as functions of the normalized frequency $(\omega - \omega_0)/\omega_p$. In this frequency range, which lies below the plasma frequency, gold exhibits significant losses, as indicated by the high values of the extinction coefficient. In case (b) the results are shown in Fig. 3.22(b) as functions of the normalized frequency $(\omega - \omega_0)/\omega_p$. However, in this case $\omega_0 \simeq 1.88\times^{17}$ rad/s $> \omega_p$ and gold appears transparent since κ is negligible and $n \simeq 1$.

The fundamental assumptions underlying the Drude model are as follows [94]: (a) The disregard of electron-electron interactions, known as the *independent electron approximation*; (b) The omission of electron-ion interactions, termed the *free electron approximation*; (c) Consideration of collisions as instantaneous events with a probability rate $1/\tau$; (d) The assumption that electrons achieve thermal equilibrium solely through collisions. Although the Drude model provides a foundational framework, more sophisticated approaches offer greater accuracy in describing the optical properties of metals. One such refinement is the Drude-Lorentz model, which accounts for interband transitions [102–104]. In this model, the complex relative permittivity of a metal is expressed as:

$$\varepsilon_c(\omega) = 1 + \frac{\tau\omega_p^2}{j\omega\,(1 + j\omega\tau)} + \sum_{i=1}^{k} \frac{f_i\omega_p^2}{\omega_i^2 - \omega^2 + j(\omega/\tau_i)}. \tag{3.165}$$

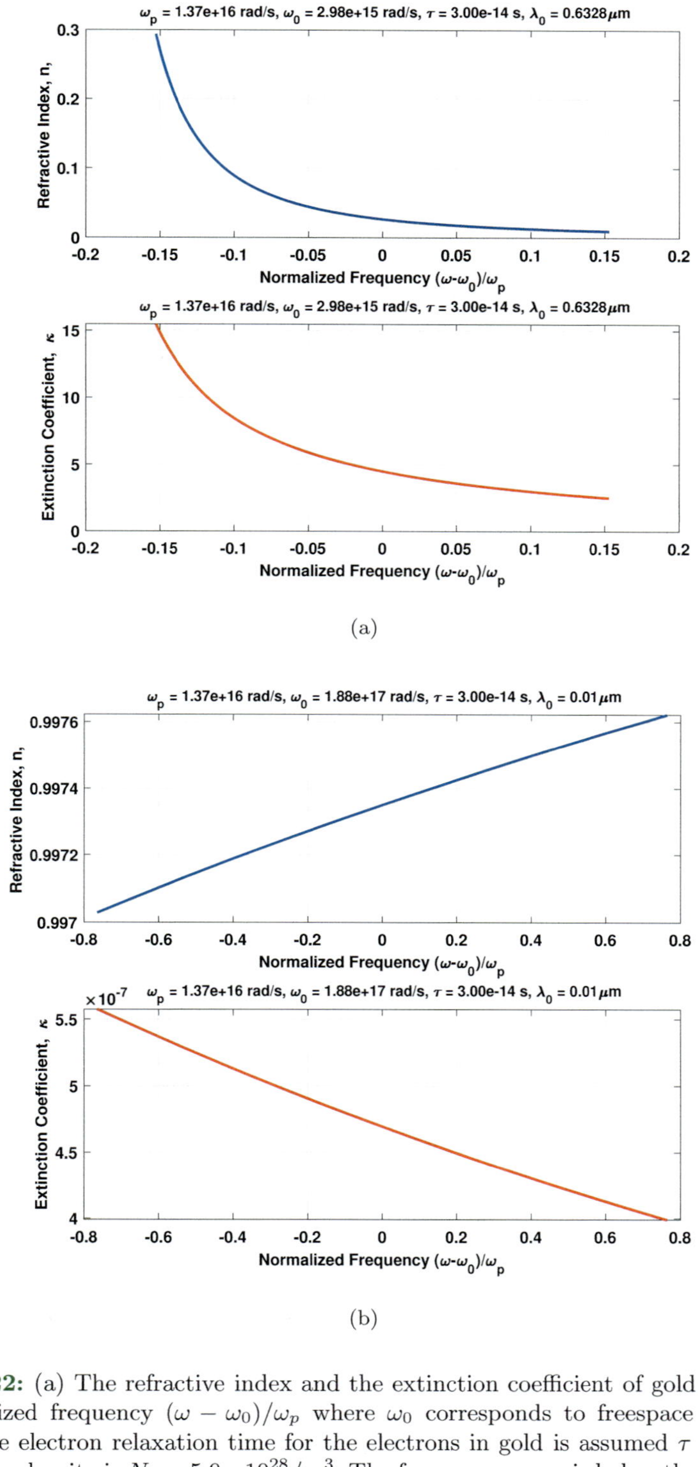

FIGURE 3.22: (a) The refractive index and the extinction coefficient of gold (Au) as functions of the normalized frequency $(\omega - \omega_0)/\omega_p$ where ω_0 corresponds to freespace wavelength $\lambda_0 = 0.6328\,\mu$m. The electron relaxation time for the electrons in gold is assumed $\tau = 3 \times 10^{-14}$ s and the free electron density is $N_e = 5.9 \times 10^{28}/m^3$. The frequency range is below the plasma frequency. (b) Similar results but for ω_0 corresponding to freespace wavelength $\lambda_0 = 0.01\,\mu$m and frequency range above the plasma frequency.

The summation term represents the interband correction contributions, where k typically takes values of either 5 or 6 [102]. An example calculation of the refractive index n and the extinction coefficient κ based on Eq. (3.165) and the parameters provided in Example 3.4.1 is shown in Fig. 3.23 for gold. Evidently, the Drude-Lorentz model yields results that differ significantly from those of the simpler Drude model, demonstrating its enhanced accuracy. The numerical parameters required for applying Eq. (3.165) for gold can be found in Ref. [102].

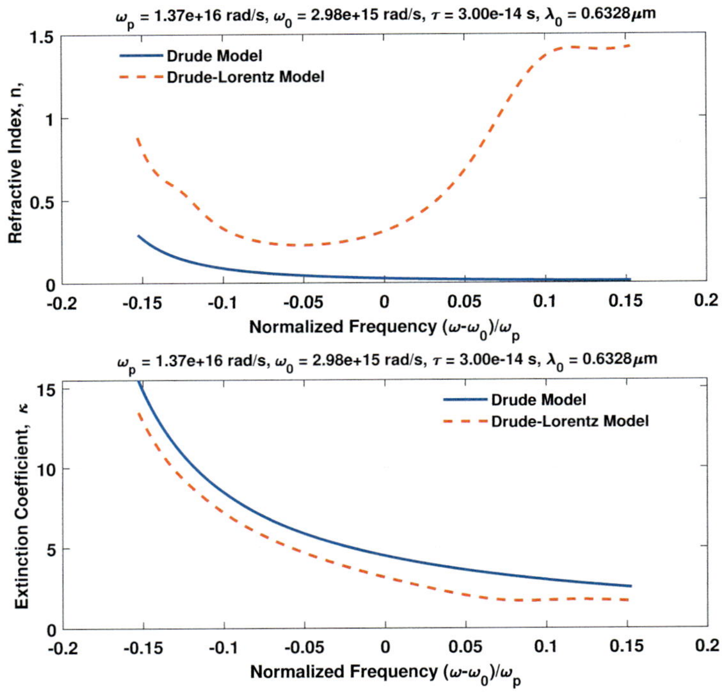

FIGURE 3.23: The refractive index and the extinction coefficient of gold (Au) as functions of the normalized frequency $(\omega - \omega_0)/\omega_p$ where ω_0 corresponds to freespace wavelength $\lambda_0 = 0.6328\,\mu$m. The electron relaxation time in gold is assumed $\tau = 3 \times 10^{-14}$ s and the free electron density is $N_e = 5.9 \times 10^{28}/m^3$. The frequency range is below the plasma frequency. The results of the Drude model (solid lines) and the Drude-Lorentz model (dashed lines) are shown.

3.4.3 Propagation of Plane Waves in Lossy Media

As demonstrated in the previous section, the presence of losses in a medium can be accounted for in the time-harmonic Maxwell equations through the introduction of a complex permittivity $\epsilon_c = \epsilon_0(\varepsilon' - j\varepsilon'')$. Consequently, in the absence of free charges, Maxwell's equations can be expressed as:

$$\vec{\nabla} \times \vec{E}(\vec{r}, \omega) = -j\omega\mu\vec{H}(\vec{r}, \omega), \tag{3.166}$$

$$\vec{\nabla} \times \vec{H}(\vec{r}, \omega) = j\omega\epsilon_c\vec{E}(\vec{r}, \omega), \tag{3.167}$$

$$\vec{\nabla} \cdot \vec{E}(\vec{r}, \omega) = 0, \tag{3.168}$$

$$\vec{\nabla} \cdot \vec{H}(\vec{r}, \omega) = 0. \tag{3.169}$$

resulting in the following Helmholtz's equation:

$$\nabla^2 \vec{E} + \underbrace{k_0^2 (\varepsilon' - j\varepsilon'') \mu_r}_{k_c^2} \vec{E} = \nabla^2 \vec{E} + \underbrace{k_0^2 (\varepsilon' - j\varepsilon'') \mu_r}_{-\gamma^2} \vec{E} = 0. \tag{3.170}$$

This is analogous to Eq. (3.34), differing primarily in the use of the complex relative permittivity. The solution of Eq. (3.170) resembles that of Eq. (3.34), with the key distinction being the incorporation of a complex wavenumber k_c. Furthermore, a complex propagation constant γ is defined, such that:

$$\gamma^2 = (\alpha + j\beta)^2 = -k_c^2 = -k_0^2 (\varepsilon' - j\varepsilon'') \mu_r. \tag{3.171}$$

Specifically, the resulting time-harmonic plane wave solution takes the form of a phasor:

$$\vec{E} = \vec{E}_0 \exp(-j\vec{k}_c \cdot \vec{r}) = \vec{E}_0 \exp[-jk_c(\hat{k} \cdot \vec{r})], \qquad \text{where} \quad \vec{k}_c = k_c\hat{k}, \tag{3.172}$$

$$k_c^2 = \omega^2 \epsilon_0 \mu_0 \mu_r (\varepsilon' - j\varepsilon'') = k_0^2 (n - j\kappa)^2, \tag{3.173}$$

$$(n - j\kappa)^2 = \mu_r (\varepsilon' - j\varepsilon''), \tag{3.174}$$

$$\gamma^2 = -k_c^2 = -k_0^2 (n - j\kappa)^2 \Longrightarrow \gamma = (\alpha + j\beta) = jk_c = jk_0(n - j\kappa) \Longrightarrow \tag{3.175}$$

$$\alpha = k_0 \kappa, \tag{3.176}$$

$$\beta = k_0 n. \tag{3.177}$$

Typically, for optical materials, $\mu_r \simeq 1$, and $\hat{k}$ is the unit vector (real) indicating the direction of propagation of the single plane wave. The term α denotes the attenuation constant, while β represents the propagation constant. In the case of a lossless material ($\kappa = 0$), the conditions $\alpha = 0$ and $\beta = k = k_0 n$ hold, as expected from Eq. (3.40). For a lossy material, the relative complex permittivity or complex refractive index is typically provided, as discussed in Sections 3.4.1 and 3.4.2). Thus, the electric field phasor representing a plane wave propagating through a lossy medium can be expressed as:

$$\vec{E} = \vec{E}_0 \exp[-jk_c(\hat{k} \cdot \vec{r})] = \vec{E}_0 \exp[-(\alpha + j\beta)(\hat{k} \cdot \vec{r})] \Longrightarrow$$

$$\vec{\mathcal{E}} = \text{Re}\{\vec{E} e^{j\omega t}\} = \vec{E}_0 \exp[-\alpha(\hat{k} \cdot \vec{r})] \cos[\omega t - \beta(\hat{k} \cdot \vec{r})].$$

Here, for simplicity, the amplitude term $\vec{E}_0$ is assumed to be real. Clearly, during propagation inside a lossy medium, the amplitude of the electric field experiences decay with distance. In this context, the penetration depth δ can be defined as the distance at which the intensity of the field, $I = |\vec{\mathcal{E}}|^2$, decays to $1/e$ of its initial value. Using this definition, $\delta = |\hat{k} \cdot \Delta \vec{r}| = 1/2\alpha$. The wavelength inside the lossy medium is defined as the spatial period of the cosine term in the electric field expression and is given by: $\lambda = |\hat{k} \cdot \Delta \vec{r}| = 2\pi/\beta$. The phase velocity of the wave is given by $u_p = |\hat{k} \cdot \vec{u}_p| = \omega/\beta$. It is essential to note that in a medium of finite width, both solutions of Helmholtz's equation are typically needed, not just the one selected here as an example. For a single plane wave inside a lossy medium, the time-harmonic Maxwell's equations can be expressed as:

$$\vec{k}_c \times \vec{E} = \omega\mu\vec{H}, \tag{3.178}$$

$$\vec{k}_c \times \vec{H} = -\omega\epsilon_c\vec{E}, \tag{3.179}$$

$$\vec{k}_c \cdot \vec{E} = 0, \tag{3.180}$$

$$\vec{k}_c \cdot \vec{H} = 0. \tag{3.181}$$

Then the resulting magnetic field phasor can be expressed in the form:

$$\vec{H} = \frac{1}{\omega\mu}\vec{k}_c \times \vec{E} = \frac{k_c}{\omega\mu}\hat{k} \times \vec{E} = \frac{1}{Z_c}\hat{k} \times \vec{E}, \tag{3.182}$$

$$Z_c = \frac{\omega\mu}{k_c} = \frac{j\omega\mu}{\alpha + j\beta} = \left(\frac{\mu}{\epsilon_c}\right)^{1/2} = |Z_c|e^{j\Phi}, \tag{3.183}$$

Here, Z_c represents the complex wave impedance of the lossy medium, which introduces a phase shift between the electric and magnetic fields within the medium.

3.4.4 Reflection and Refraction of a Plane Wave in Lossy Media

In this section, the reflection and refraction at a planar interface between a lossless dielectric and a lossy medium are analyzed. The geometry is shown in Fig. 3.24, where the right medium is characterized as lossy (isotropic) and represented by the complex refractive index $n_2 - j\kappa_2$. The analysis in Sec. 3.2.3 applies to a complex refractive index, with the only modification being the substitution of n_2 with $n_2 - j\kappa_2$. The resulting Fresnel equations (refer to Sec. 3.2.3.1) can then be expressed as:

$$r_{TE} = r_{\perp} = \frac{E_r}{E_i} = \frac{n_1\cos\theta_1 - (n_2 - j\kappa_2)\cos\theta_2}{n_1\cos\theta_1 + (n_2 - j\kappa_2)\cos\theta_2}, \tag{3.184}$$

$$t_{TE} = t_{\perp} = \frac{E_t}{E_i} = \frac{2n_1\cos\theta_1}{n_1\cos\theta_1 + (n_2 - j\kappa_2)\cos\theta_2}, \tag{3.185}$$

$$r_{TM} = r_{\parallel} = \frac{E_r}{E_i} = \frac{(n_2 - j\kappa_2)\cos\theta_1 - n_1\cos\theta_2}{(n_2 - j\kappa_2)\cos\theta_1 + n_1\cos\theta_2}, \tag{3.186}$$

$$t_{TM} = t_{\parallel} = \frac{E_t}{E_i} = \frac{2n_1\cos\theta_1}{(n_2 - j\kappa_2)\cos\theta_1 + n_1\cos\theta_2}, \tag{3.187}$$

where $\cos\theta_2$ is complex and θ_2 satisfies the Snell's law (phase matching condition) $n_1\sin\theta_1 = (n_2 - j\kappa_2)\sin\theta_2$. The phasor of the transmitted field can be written as:

$$\begin{aligned}\vec{E}_2 &= \vec{E}_t\exp(-j\vec{k}_{2c}\cdot\vec{r}) = \vec{E}_t\exp[-j(xk_{2c}\sin\theta_2 + zk_{2c}\cos\theta_2)] \\ &= \vec{E}_t\exp\{-j\left[xk_0(n_2 - j\kappa_2)\sin\theta_2 + zk_0(n_2 - j\kappa_2)\cos\theta_2\right]\} \\ &= \vec{E}_t\exp\{-j\left[xk_0n_1\sin\theta_1 + zk_0(n_2 - j\kappa_2)\cos\theta_2\right]\}.\end{aligned} \tag{3.188}$$

Determining the term $(n_2 - j\kappa_2)\cos\theta_2$ requires careful consideration. Specifically, it can be expressed as follows:

$$\begin{aligned}(n_2 - j\kappa_2)\cos\theta_2 &= \pm(n_2 - j\kappa_2)\left(1 - \sin^2\theta_2\right)^{1/2} = \pm\left[(n_2 - j\kappa_2)^2 - (n_2 - j\kappa_2)^2\sin^2\theta_2\right]^{1/2} \\ &= \pm\left[(n_2 - j\kappa_2)^2 - n_1^2\sin^2\theta_1\right]^{1/2}.\end{aligned} \tag{3.189}$$

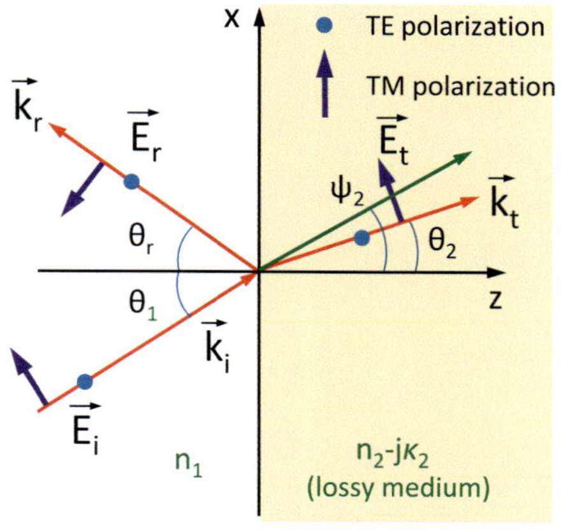

FIGURE 3.24: A planar interface between a dielectric (left region with refractive index n_1) and a lossy medium (right medium with refractive index $n_2 - j\kappa_2$). The incident plane wave can have any polarization. The directions for *TE* and *TM* polarization of the electric field are shown. In this context, the angle θ_2 is complex, reflecting the attenuation in the lossy medium, while ψ_2 is the real angle at which the phase fronts propagate in the lossy region.

The sign in Eq. (3.189) should be selected to represent a decaying field as z increases in region 2 of Fig. 3.10. From the expression for the electric field phasor in region 2, given by Eq. (3.188), the decaying field condition is satisfied if:

$$\operatorname{Im}\{(n_2 - j\kappa_2)\cos\theta_2\} < 0. \tag{3.190}$$

Therefore, the sign in Eq. (3.189) is selected in order to satisfy inequality (3.190). The right term of Eq. (3.189) is complex and can be analyzed as follows:

$$\begin{aligned}
\pm\left[(n_2 - j\kappa_2)^2 - n_1^2\sin^2\theta_1\right]^{1/2} &= \pm\Big[\underbrace{(n_2^2 - \kappa_2^2 - n_1^2\sin^2\theta_1}_{=B} - j\underbrace{2n_2\kappa_2}_{=C}\Big]^{1/2} \\
&= \pm(B - jC)^{1/2} = \pm\left(B^2 + C^2\right)^{1/4}\exp\left[-\frac{1}{2}\tan^{-1}(C/B)\right] \\
&= \pm(p_2 - jq_2) \qquad \text{where}
\end{aligned}$$

$$p_2 = \left(B^2 + C^2\right)^{1/4}\cos\left[\frac{1}{2}\tan^{-1}\left(\frac{C}{B}\right)\right], \tag{3.191}$$

$$q_2 = \left(B^2 + C^2\right)^{1/4}\sin\left[\frac{1}{2}\tan^{-1}\left(\frac{C}{B}\right)\right], \tag{3.192}$$

$$(n_2 - j\kappa_2)\cos\theta_2 = P_2 - jQ_2 = \pm(p_2 - jq_2), \tag{3.193}$$

where according to inequality (3.190) if $q_2 > 0$ the "+" sign is selected and alternatively if $q_2 < 0$ the "−" sign is selected in Eq. (3.193). The terms P_2 and Q_2 are the p_2 and q_2 that result in the correct sign selection such that $Q_2 > 0$. Then the term $(n_2 - j\kappa_2)\cos\theta_2$ is appropriately used in the Fresnel equations for the calculation of the reflection or transmission coefficients. For the *TM* polarization case the term $n_1\cos\theta_2 = [n_1/(n_2 - j\kappa_2)](n_2 - j\kappa_2)\cos\theta_2$ is used. At this point, let us reconsider the transmitted electric field phasor:

$$\begin{aligned}
\vec{E}_2 &= \vec{E}_t\exp\{-j\left[xk_0(n_2 - j\kappa_2)\sin\theta_2 + zk_0(n_2 - j\kappa_2)\cos\theta_2\right]\} \\
&= \vec{E}_t\exp\{-j\left[xk_0n_1\sin\theta_1 + zk_0(\pm 1)(p_2 - jq_2)\right]\}
\end{aligned}$$

$$= \vec{E}_t \exp\{-j\,[xk_0n_1\sin\theta_1 + zk_0(P_2 - jQ_2)]\}$$

$$= \vec{E}_t \exp\{-j\,[xk_0n_1\sin\theta_1 + zk_0P_2]\}\exp\{-k_0Q_2z\}$$

$$= \vec{E}_t \exp\{-j\,[S(x\sin\psi_2 + z\cos\psi_2]\}\exp\{-k_0Q_2z\} \qquad \text{where,}$$

$$S = k_0\left[n_1^2\sin^2\theta_1 + P_2^2\right]^{1/2},$$

$$\sin\psi_2 = \frac{k_0n_1\sin\theta_1}{S}, \tag{3.194}$$

$$\cos\psi_2 = \frac{k_0P_2}{S}. \tag{3.195}$$

The terms P_2 and Q_2 were defined previously, and the angle ψ_2 is a real angle that indicates the direction of propagation of the decaying field wavefronts (refer to Fig. 3.24).

The reflectivities r_{TE} and r_{TM} given in Eqs. (3.184) and (3.186) are always complex in the case of a lossy material. In other words, the reflectivities can be expressed as $r_{TE} = |r_{TE}|\exp(j\phi_{TE})$ and $r_{TM} = |r_{TM}|\exp(j\phi_{TM})$, where ϕ_{TE} and ϕ_{TM} are the phase shifts experienced by the reflected light in the *TE* and *TM* polarizations, respectively. For incident light with general polarization, the phase difference between the *TE* and *TM* components is $\phi_{TM} - \phi_{TE}$. When linearly polarized light is incident upon the lossy medium, the reflected light remains linearly polarized only if $\phi_{TM} - \phi_{TE} = 0°$ or $180°$ (refer to Sec. 3.3). For all other values of $\phi_{TM} - \phi_{TE}$, the reflected light becomes elliptically polarized except for normal incidence ($\theta_1 = 0°$) and grazing incidence ($\theta_1 = 90°$). Specifically, at normal incidence, $r_{TE}(\theta_1 = 0°) = A\exp(j\phi_{TE})$ and $r_{TM}(\theta_1 = 0°) = -A\exp(j\phi_{TE}) = A\exp[j(\phi_{TE} + \pi)]) = A\exp(j\phi_{TM})$, resulting in $\phi_{TM} - \phi_{TE} = \pi$. In the latter expressions $A = [(n_1-n_2)^2+\kappa_2^2]^{1/2}/[(n_1+n_2)^2+\kappa_2^2]^{1/2}$ and the "-1" factor in the $r_{TM}(\theta_1 = 0°)$ is due to the opposite direction of E_i and E_r at normal incidence in the *TM* polarization (refer to Fig. 3.24). Furthermore, $r_{TE}(\theta_1 = 90°) = -1 = 1\exp(j\pi)$ and $r_{TM}(\theta_1 = 90°) = -1 = 1\exp(j\pi)$, resulting in $\phi_{TM} - \phi_{TE} = 0$. Between these two extreme cases of incident angle, there exists an angle $\theta_1 = \theta_p$, referred to as the *principal angle of incidence* [91], such that $\phi_{TM} - \phi_{TE} = \pi/2$. At this angle of incidence, θ_p, linearly polarized light is, in general, elliptically polarized after reflection, with the axis of polarization parallel and perpendicular to the plane of incidence. For certain special linear polarizations, it is possible for the reflected light to be circularly polarized if $|r_{TE}|E_{TE} = |r_{TM}|E_{TM}$, where E_{TE} and E_{TM} are the *TE* and *TM* components of the incident electric field, respectively. For any angle of incidence, the *azimuthal angle* ψ is defined as $\psi = \tan^{-1}(|r_{TM}|/|r_{TE}|)$ [91]. When $\theta_1 = \theta_p$, the corresponding azimuthal angle becomes the *principal azimuthal angle*, ψ_p.

From a practical standpoint, determining θ_p and ψ_p and $\Delta\phi = (\phi_{TM} - \phi_{TE})$ is typically easier experimentally. These values can then be used to determine n_2 and κ_2, rather than the reverse process. In fact, measurement of r_{TE} and r_{TM} (or various combinations of their terms) can lead to the determination of the complex refractive index, $n_2 - j\kappa_2$ of the lossy material. This determination can be achieved as follows: By using Eqs. (3.184) and (3.186), it is straightforward to derive the following equation:

$$\frac{1 - \tan\psi E^{-j\Delta\phi}}{1 + \tan\psi E^{-j\Delta\phi}} = \frac{r_{TM} - r_{TE}}{r_{TM} + r_{TE}} = \frac{\left[(n_2 - jk_2)^2 - n_1^2\right]\cos\theta_1\cos\theta_2}{n_1(n_2 - j\kappa_2)(\cos^2\theta_1 - \cos^2\theta_2)}$$

$$= -\frac{\cos\theta_1\cos\theta_2}{\sin\theta_1\sin\theta_2} = -\frac{\left[(n_2 - j\kappa_2)^2 - n_1^2\sin^2\theta_1\right]^{1/2}}{n_1\sin\theta_1\tan\theta_1},$$

where Snell's law was also used in the above derivation. The left-hand term of the previous equation can be expressed in the following form by applying some basic trigonometric identities:

$$\frac{1-\tan\psi E^{-j\Delta\phi}}{1+\tan\psi E^{-j\Delta\phi}}=\frac{\cos 2\psi-\sin^2 2\psi\sin^2\Delta\phi+j\sin 4\psi\sin\Delta\phi}{1+\sin 2\psi\cos\Delta\phi}.$$

From the last two equations, the complex refractive index can be related to the angle ψ_0 and $\Delta\phi$ as follows:

$$n_2^2-\kappa_2^2=n_1^2\sin^2\theta_1\left[1+\tan^2\theta_1\frac{\cos^2 2\psi-\sin^2 2\psi\sin^2\Delta\phi}{(1+\sin 2\psi\cos\Delta\phi)^2}\right],$$

$$2n_2\kappa_2=\frac{n_1^2\sin^2\theta_1\tan^2\theta_1\sin 4\psi\sin\Delta\phi}{(1+\sin 2\psi\cos\Delta\phi)^2}.$$

Applying the previous equations for for $\theta_1=\theta_p$, $\psi=\psi_p$, and $\Delta\phi=\pi/2$, we obtain:

$$n_2^2-\kappa_2^2=n_1^2\sin^2\theta_p\left[1+\tan^2\theta_p\cos 4\psi_p\right], \tag{3.196}$$

$$2n_2\kappa_2=n_1^2\sin^2\theta_p\tan^2\theta_p\sin 4\psi_p \tag{3.197}$$

Using the measured values of ψ_p and θ_p and Eqs. (3.196) and (3.197), the n_2 and κ_2 can be determined in a manner similar to the procedure outlined for Eqs. (3.153) and (3.154). Typically, the left region consists of air, and consequently, $n_1\simeq 1$ [91].

One question that arises is how the angles θ_p and ψ_p can be measured. Recalling the key fact that $\Delta\phi=\phi_{TM}-\phi_{TE}=\pi/2$ at the principal angle of incidence, means that this elliptically polarized light can be converted to linearly polarized light by a quarter-wave plate with its fast and slow axes oriented perpendicular and parallel to the plane of incidence. Therefore, in the experimental procedure a polarizer can be utilized to produce incident light linearly polarized at 45° with respect to the plane of incidence. The reflected light is then analyzed with the quarter-wave plate followed by a second polarizer (analyzer). Rotating the analyzer allows a minimum in the transmission to be found. Then the angle of incidence is changed and the process is repeated. When the angle of incidence that produces complete extinction of the reflected light is found, that angle corresponds to θ_p and the orientation of the analyzer to ψ_p. An alternative method for measuring these quantities is to illuminate the lossy sample with circularly polarized light. Then the reflected beam is extinguished, the angle of incidence corresponds to the principal angle of incidence θ_p and the orientation of the polarizer gives the principal azimuthal angle ψ_p.

Example 3.4.2:

Consider the reflection at a planar interface between air and gold at the optical freespace wavelength of $\lambda_0=0.5893\,\mu$m. Using the refractive index information from Ref. [76] which is based on Ref. [105], the complex refractive index of gold is given by $n_2-j\kappa_2=0.18488-j2.8672$. (a) Calculate the normalized reflected power as a function of the angle of incidence θ_1 for both *TE* and *TM* polarizations. (b) Determine the phasefront angle ψ_2 inside gold as a function of the angle of incidence θ_1. (c) Plot the phase angles ϕ_{TE} and ϕ_{TM} and their difference $\phi_{TM}-\phi_{TE}$ as a function of the angle of incidence θ_1. (d) Determine the principal angle of incidence θ_p and the principal azimuthal angle ψ_p for gold.

Solution: (a) Using Eqs. (3.184) and (3.186) the reflection coefficients r_{TE} and r_{TM} are determined and the power reflectivities $|r_{TE}|^2$ and $|r_{TM}|^2$ can be determined. The results are plotted in Fig. 3.25(a). It is observed that the case of *TM* polarization exhibits a minimum for $\theta_1\simeq 69.05°$ instead of having a Brewster angle in the dielectric case. (b) Using Eqs. (3.194) and (3.195) the angle ψ_2 is presented in Fig. 3.25(b) as a function of the angle of incidence θ_1.

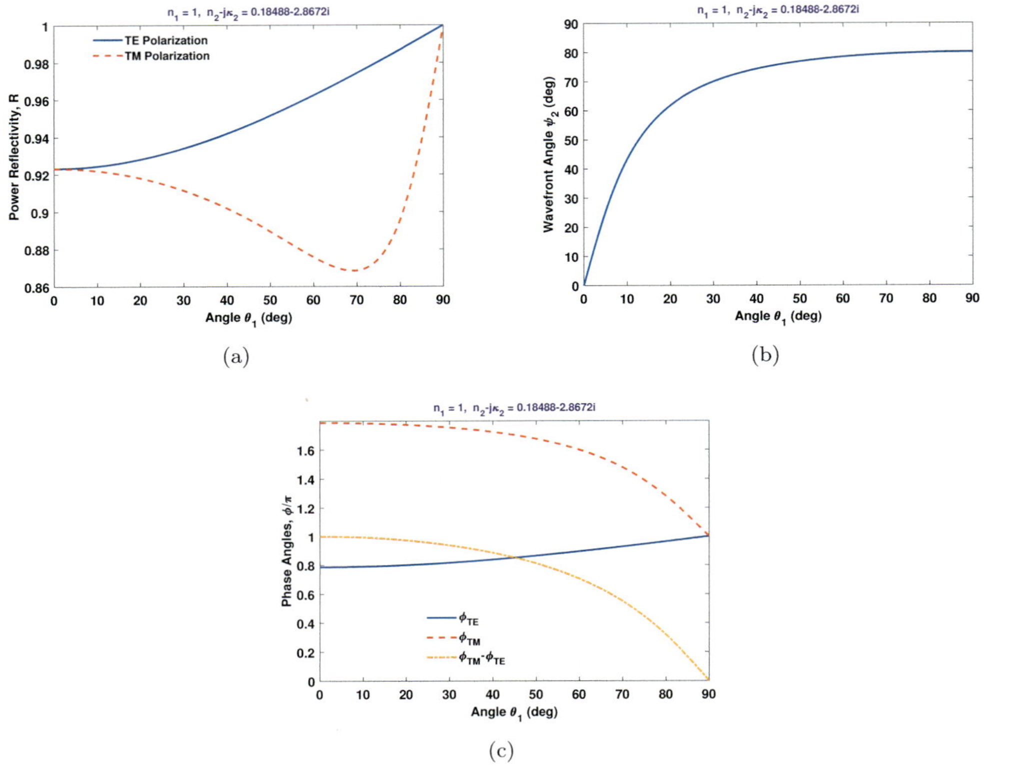

FIGURE 3.25: (a) The power reflectivity, $R = |r|^2$, at a planar interface between air and gold at $\lambda_0 = 0.5893\,\mu$m as a function of the angle of incidence. (b) The wavefront propagation direction angle ψ_2 inside gold region. (c) The reflection coefficient phases ϕ_{TE} and ϕ_{TM} for TE and TM polarizations, respectively. The difference $\phi_{TM} - \phi_{TE}$ is also shown as function of the angle of incidence.

(c) Using the coefficients r_{TE} and r_{TM} determined in part (a), the phase angles ϕ_{TE} and ϕ_{TM} and their difference $\phi_{TM} - \phi_{TE}$ can be determined as a function of the angle of incidence θ_1. The angles ϕ_{TE} and ϕ_{TM} as well as their difference $\phi_{TM} - \phi_{TE}$, are shown in Fig. 3.25(c).
(d) From the data presented in part (c) it is straightforward to estimate $\theta_p \simeq 72.51°$ and $\psi_p = \tan^{-1}(|r_{TM}(\theta_p)|/|r_{TE}(\theta_p)|) = 43.338°$. Using these estimates for the angles and Eqs. (3.196) and (3.197) the refractive index of the gold can be estimated as $n_2 \simeq 0.1849$ and $\kappa_2 \simeq 2.8678$. More accuracy can be achieved if the angles θ_p and ψ_p are estimated with better accuracy.

3.5 Light Interference

When two or more plane waves share the same temporal frequency $\nu = \omega/2\pi$ (as a result of a common monochromatic source, such as a laser), they can interact and interfere where they overlap. This phenomenon is referred to as *interference*. The interaction of waves leads to localized modifications in the overall intensity of light. *Destructive interference* occurs when the resulting intensity is smaller than the sum of the individual intensities of the plane waves (or even zero). On the contrary, *constructive interference* takes place when the resulting intensity is greater than the sum of the individual intensities of the plane waves. It is important to note that this explanation applies to general monochromatic waves; however, for simplicity, plane waves are primarily used in this section to illustrate various interference phenomena.

Observing interference of light at optical frequencies is challenging due to the small wavelength of visible light. This difficulty contributed to the corpuscular theory of light, considering light as particles, prevailing in the 1800s. However, Thomas Young's groundbreaking experiment in 1801-1802 demonstrated light interference, solidifying the wave theory of light [10, 20, 62]. To comprehend the phenomenon of interference, some simple cases, including the experiment conducted by Thomas Young, are presented next.

3.5.1 Two Collinear Plane-Waves Interference

3.5.1.1 Plane-Waves of the Same Frequency

Consider two plane waves of the same temporal frequency, characterized by electric field phasors $\vec{E}_1$ and $\vec{E}_2$, both originating from the same monochromatic light source, such as a laser. Let us assume that one of the waves, $\vec{E}_2$, acquires a constant phase difference relative to $\vec{E}_1$. One possible implementation of this interference is depicted in Fig. 3.26. In this setup, two beam splitters divide the laser beam into $\vec{E}_{10}$ and $\vec{E}_{02}$ (via beam splitter $BS1$) and into $\vec{E}_1$ and $\vec{E}_2$ (via beam splitter $BS2$). A more detailed analysis of the operation of a beam splitter (BS) is provided in Appendix 3B. Due to the difference in optical path length between the two waves, a phase difference ϕ is introduced. The resulting interference between $\vec{E}_1$ and $\vec{E}_2$ is then examined.

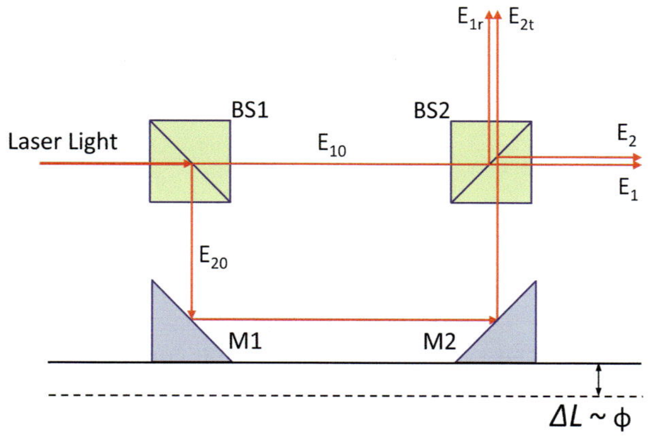

FIGURE 3.26: A possible implementation of the interference between two plane waves propagating in the same direction (along the z-axis) with a phase difference of ϕ is illustrated. The glass cubes BS_i $(i = 1, 2)$ are beam splitters of equal power and the M_i $(i = 1, 2)$ represent perfect mirrors. The fields E_1 and E_2 are the ones of interest, as they are the ones that interfere. Additionally, the waves E_{1r} and E_{2t}, also undergo interference.

The two plane waves $\vec{E}_1$ and $\vec{E}_2$, propagate along the same direction (assumed to be along the z-axis) and in the same space. As a result, they overlap and interfere. These two plane waves can be represented by the following phasors:

$$\vec{E}_1 = \hat{e}_1 \; E_1 \exp(-jk_0 z),$$

$$\vec{E}_2 = \hat{e}_2 \; E_2 \exp(-jk_0 z - j\phi),$$

where $\hat{e}_i$ and E_i $(i = 1, 2)$ represent the unit polarization vectors and the amplitudes of the two plane waves, respectively. Without loss of generality, the amplitudes are assumed to be real, as any phase difference can be incorporated either into the phase ϕ or into the polarization vectors $\hat{e}_i$'s $i = 1, 2$. When the two waves are linearly polarized the polarization vectors $\hat{e}_i$ can be considered real. However, for general elliptical polarizations, the polarization vectors $\hat{e}_i$ are complex. To quantify the interference of the waves, we can start by considering the power density expressed by the Poynting vector. In this case, the corresponding time-averaged Poynting vector can be written as:

$$\begin{aligned}
\vec{P} &= \frac{1}{2}\text{Re}\left\{\left(\vec{E}_1 + \vec{E}_2\right) \times \left(\vec{H}_1^* + \vec{H}_2^*\right)\right\} = \\
&= \frac{1}{2}\text{Re}\left\{\vec{E}_1 \times \vec{H}_1^*\right\} + \frac{1}{2}\text{Re}\left\{\vec{E}_2 \times \vec{H}_2^*\right\} + \frac{1}{2}\text{Re}\left\{\vec{E}_1 \times \vec{H}_2^*\right\} + \frac{1}{2}\text{Re}\left\{\vec{E}_2 \times \vec{H}_1^*\right\} = \\
&= \frac{1}{2Z}\left|\vec{E}_1\right|^2 \vec{k}_1 + \frac{1}{2Z}\left|\vec{E}_2\right|^2 \vec{k}_2 + \frac{1}{2\omega\mu}\text{Re}\left\{\left(\vec{E}_1 \cdot \vec{E}_2^*\right)\vec{k}_2 - \left(\vec{E}_1 \cdot \vec{k}_2\right)\vec{E}_2^*\right\} + \\
&\quad \frac{1}{2\omega\mu}\text{Re}\left\{\left(\vec{E}_2 \cdot \vec{E}_1^*\right)\vec{k}_1 - \left(\vec{E}_2 \cdot \vec{k}_1\right)\vec{E}_1^*\right\}.
\end{aligned}$$

Since the two waves are propagating in the same direction and in the same medium, we have $\vec{k}_1 = \vec{k}_2 = \vec{k} = k\hat{k}$. Considering that for each plane wave, $\vec{E}_i \cdot \vec{k}_i = 0$ [as per Eq. (3.37)], the average Poynting vector of the superposition of the two plane waves becomes:

$$\begin{aligned}
\vec{P} &= \frac{1}{2Z}\left|\vec{E}_1\right|^2 \hat{k} + \frac{1}{2Z}\left|\vec{E}_2\right|^2 \hat{k} + \frac{1}{2Z}\text{Re}\left\{\left(\vec{E}_1 \cdot \vec{E}_2^*\right)\hat{k} + \left(\vec{E}_2 \cdot \vec{E}_1^*\right)\hat{k}\right\} = \\
&= \frac{1}{2Z}\left|\vec{E}_1\right|^2 \hat{k} + \frac{1}{2Z}\left|\vec{E}_2\right|^2 \hat{k} + \frac{1}{2Z}2\text{Re}\left\{\left(\vec{E}_1 \cdot \vec{E}_2^*\right)\right\}\hat{k} = \\
&= \frac{1}{2Z}\left[\left|\vec{E}_1\right|^2 + \left|\vec{E}_2\right|^2 + 2\text{Re}\left\{\left(\vec{E}_1 \cdot \vec{E}_2^*\right)\right\}\right]\hat{k} = \\
&= \frac{1}{2Z}\left[E_1^2 + E_2^2 + 2E_1E_2|\hat{e}_1 \cdot \hat{e}_2^*|\cos(\phi + \phi_{12})\right]\hat{k} = \\
&= \frac{1}{2Z}\underbrace{\left[\left[E_1^2 + E_2^2 - 2E_1E_2|\hat{e}_1 \cdot \hat{e}_2^*|\right] + 4E_1E_2|\hat{e}_1 \cdot \hat{e}_2^*|\cos^2\left(\frac{\phi + \phi_{12}}{2}\right)\right]}_{=\,I(\phi)}\hat{k}.
\end{aligned} \tag{3.198}$$

It is essential to note that the symbol "*" denotes the complex conjugate in all preceding equations. The term ϕ_{12} is derived from $(\hat{e}_1 \cdot \hat{e}_2^*) = |\hat{e}_1 \cdot \hat{e}_2^*| \exp(j\phi_{12})$. Without loss of generality, it can be assumed that $\phi_{12} = 0$, as this phase can be incorporated into ϕ. Therefore, the intensity takes the form $I = I(\phi)$. The minima in intensity occur for $\phi = (2m + 1)\pi$, while the maxima occur for $\phi = 2m\pi$ (where m is an integer). If $\hat{e}_1 \cdot \hat{e}_2^* = 0$, there is no variation in $I(\phi)$, and the two waves do not interfere due to their orthogonal polarizations. A straightforward example of $I(\phi)$ as a function of ϕ is presented in Fig. 3.27(a) (when $E_1 = E_2 = 1$ and $\hat{e}_1 \cdot \hat{e}_2^* = 1$ - equal amplitudes, same polarization state) and in Fig. 3.27(b) (when $E_1 = 1$, $E_2 = 2$, and $\hat{e}_1 \cdot \hat{e}_2^* = 1$ - unequal amplitudes, same polarization state). The quality of the interference fringes can be described by the *fringe visibility*, V, which is defined as:

$$V = \frac{I_{max} - I_{min}}{I_{max} + I_{min}}, \tag{3.199}$$

where I_{max} and I_{min} are the maximum and the minimum of $I(\phi)$ as a function of ϕ. It is evident that the highest visibility is unity and occurs when the two amplitudes E_1 and E_2

are equal and the polarizations of the two plane waves are the same. This is the case shown in Fig. 3.27(a). Conversely, in the case depicted in Fig. 3.27(b), the visibility is $V = 0.8 < 1$ due to the unequal amplitudes. In situations where $I_{\max} = I_{\min}$, indicating no intensity variation, the visibility is $V = 0$, and no fringes are observed.

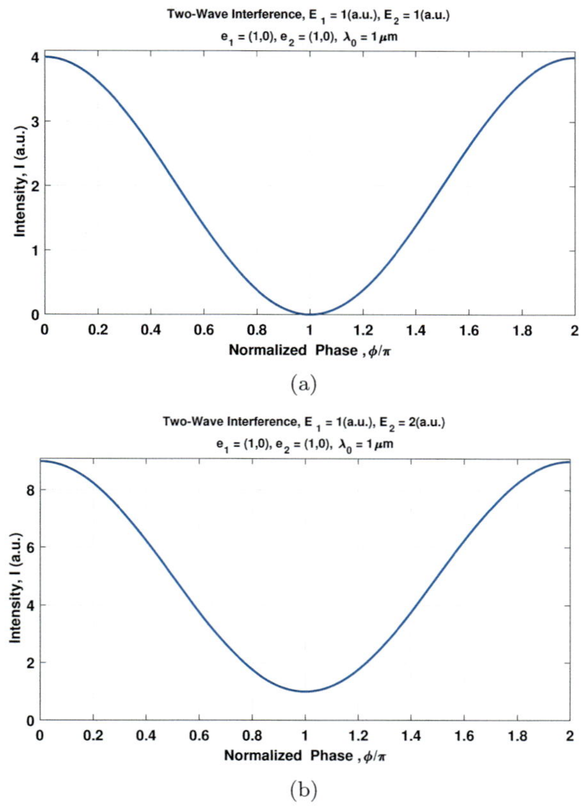

FIGURE 3.27: The interference of two plane waves propagating along the z-axis. (a) $E_1 = E_2 = 1$, $\hat{e}_1 = [1, 0] = \hat{e}_2$. The fringe visibility in this case is $V = 1$. (b) $E_1 = 1$, $E_2 = 2$, $\hat{e}_1 = \hat{\imath}_x = \hat{e}_2$. The fringe visibility in this case is $V = 0.8$.

The choice of the Poynting vector to describe interference is not entirely successful. This limitation becomes apparent when considering the case of two counter-propagating waves with equal amplitudes and polarizations. Specifically, when $E_1 = E_2 = E_0$, $\vec{k}_1 = k\hat{z} = -\vec{k}_2$, and $\hat{e}_1 \cdot \hat{e}_2^* = 1$ (for instance, $\hat{e}_1 = \hat{\imath}_y = \hat{e}_2$), Eq. (3.198) becomes for any value of ϕ:

$$\vec{P} = \frac{1}{2Z}|\vec{E}_0|^2\hat{\imath}_z - \frac{1}{2Z}|\vec{E}_0|^2\hat{\imath}_z + \frac{1}{2\omega\mu}\mathrm{Re}\left\{\left(\vec{E}_1 \cdot \vec{E}_2^*\right)(-\hat{\imath}_z)\right\} + \frac{1}{2\omega\mu}\mathrm{Re}\left\{\left(\vec{E}_2 \cdot \vec{E}_1^*\right)\hat{\imath}_z\right\} =$$

$$= \frac{1}{2\omega\mu}\mathrm{Re}\left\{\left(\vec{E}_2 \cdot \vec{E}_1^*\right) - \left(\vec{E}_1 \cdot \vec{E}_2^*\right)\right\}\hat{\imath}_z = \frac{1}{2\omega\mu}\mathrm{Re}\left\{2j\mathrm{Im}\left\{\vec{E}_2 \cdot \vec{E}_1^*\right\}\right\}\hat{\imath}_z = 0.$$

Therefore, even though the two waves form a standing wave pattern and generate fringes, the average Poynting vector is zero, as expected. This is because there is no net power transfer in any direction. For this reason, interference is typically presented in terms of electromagnetic wave intensity (or irradiance), which can be expressed as:

$$I = \frac{c}{n}\langle w_{em}\rangle, \tag{3.200}$$

where c is the freespace light velocity, n is the refractive index of the medium (assumed homogeneous), and w_{em} is the electromagnetic energy density [refer to Eq. (3.92)], and the $\langle * \rangle$ represents the time average of the electromagnetic energy density. In the case of time-harmonic fields, Eq. (3.200) becomes:

$$I = \frac{c}{n}\left(\frac{1}{4}\epsilon|\vec{E}|^2 + \frac{1}{4}\mu|\vec{H}|^2\right). \tag{3.201}$$

Now applying Eq. (3.201) in the case of the two interfering plane waves, it is straightforward to show:

$$I = \frac{c}{n}\frac{1}{4}\left(\epsilon|\vec{E}_1 + \vec{E}_2|^2 + \mu|\vec{H}_1 + \vec{H}_2|^2\right) \qquad \text{where}$$

$$|\vec{E}_1 + \vec{E}_2|^2 = |\vec{E}_1|^2 + |\vec{E}_2|^2 + 2\mathrm{Re}\left\{\vec{E}_1 \cdot \vec{E}_2^*\right\},$$

$$|\vec{H}_1 + \vec{H}_2|^2 = |\vec{H}_1|^2 + |\vec{H}_2|^2 + 2\mathrm{Re}\left\{\vec{H}_1 \cdot \vec{H}_2^*\right\},$$

$$\epsilon|\vec{E}_1|^2 = \mu|\vec{H}_1|^2, \qquad \epsilon|\vec{E}_2|^2 = \mu|\vec{H}_2|^2,$$

$$\vec{H}_1 \cdot \vec{H}_2^* = \frac{1}{Z^2}\left[(\hat{k}_1 \cdot \hat{k}_2)(\vec{E}_1 \cdot \vec{E}_2^*) - (\hat{k}_2 \cdot \vec{E}_1)(\hat{k}_1 \cdot \vec{E}_2^*)\right].$$

The last equation reveals an intriguing possibility. The interference term of the electric field component is expressed by the term $\vec{E}_1 \cdot \vec{E}_2^*$, while the interference term of the magnetic field component is expressed by the term $\vec{H}_1 \cdot \vec{H}_2^*$. However, it is noteworthy that in general, it is possible to have $\vec{E}_1 \cdot \vec{E}_2^* = 0$ (indicating no interference for the electric field) and $\vec{H}_1 \cdot \vec{H}_2^* \neq 0$ (indicating interference for the magnetic field), or vice versa. Moreover, in some cases, both terms can be equal to zero (indicating no interference for either the electric or magnetic field) or both terms non-zero (indicating interference for both the electric and magnetic fields). Remarkably, there are situations where there is no interference for either the electric or magnetic field, yet interference in the electromagnetic field helicity is still possible, and this effect can be detected by chiral molecules [106, 107].

In summary, for the purposes of classical light interference, the energy density of the electric field will be used henceforth. This choice is made because most detectors, including the human eye, predominantly respond to electric dipole interactions [97, 107] within the visible range of the electromagnetic spectrum. Therefore, the light intensity resulting from the interference of an arbitrary number N of plane waves in a homogeneous medium with permittivity ϵ should be described by the following equation:

$$I = \frac{c}{n}\frac{1}{4}\epsilon\left|\sum_{i=1}^{N}\vec{E}_i\right|^2 = \frac{c}{n}\frac{1}{4}\epsilon\sum_{i=1}^{N}\sum_{j=1}^{N}\left(\vec{E}_i \cdot \vec{E}_j^*\right) = \frac{1}{4Z}\sum_{i=1}^{N}\sum_{j=1}^{N}\left(\vec{E}_i \cdot \vec{E}_j^*\right). \tag{3.202}$$

It is straightforward to apply Eq. (3.202) in the case of two collinearly propagating waves with a phase difference ϕ and find the interference, which is also shown in Eq. (3.198) (apart from a factor of 2 due to the use of only the electric field energy density). Also, if Eq. (3.202) is applied in the case of two counter-propagating waves of equal amplitudes ($E_1 = E_2 = E_0$, $\vec{k}_1 = k\hat{\imath}_z = -\vec{k}_2$, and $\hat{e}_1 = \hat{\imath}_y = \hat{e}_2$), then the resulting intensity is

$$I = \frac{1}{4Z}\,4E_0^2\cos^2\left(\frac{2kz + \phi}{2}\right),$$

which obviously reveals the interference pattern of the resulting standing wave.

3.5.1.2 Plane-Waves of Different Frequencies

Now, let's consider two collinear interfering plane waves with different frequencies, ω_1 and ω_2, respectively. The electric fields of these two waves can be represented as follows:

$$\vec{\mathcal{E}}_1 = E_1 \mathrm{Re}\{\hat{e}_1 e^{-jk_1 z} e^{+j\omega_1 t}\},$$

$$\vec{\mathcal{E}}_2 = E_2 \mathrm{Re}\{\hat{e}_2 e^{-jk_2 z} e^{-j\phi} e^{+j\omega_2 t}\},$$

where k_1 and k_2 are the wavenumbers, and $\hat{e}_1$, $\hat{e}_2$, the polarization vectors (could be complex for general elliptical polarization) of the two plane waves. The angle ϕ is a phase difference between them. Actually, the angle ϕ is the phase of the complex in general term $(\hat{e}_1 \cdot \hat{e}_2^*)$. The time-domain intensity, $\mathcal{I}(z,t)$, is then given by:

$$\mathcal{I}(z,t) = \frac{1}{2Z}|\vec{\mathcal{E}}_1 + \vec{\mathcal{E}}_2|^2 = \frac{1}{2Z}\left[\vec{\mathcal{E}}_1 \cdot \vec{\mathcal{E}}_1 + \vec{\mathcal{E}}_2 \cdot \vec{\mathcal{E}}_2 + 2\vec{\mathcal{E}}_1 \cdot \vec{\mathcal{E}}_2\right] \qquad \text{where,}$$

$$\vec{\mathcal{E}}_1 \cdot \vec{\mathcal{E}}_1 = \frac{E_1^2}{4}\left[(\hat{e}_1 \cdot \hat{e}_1)e^{-j2k_1 z}e^{j2\omega_1 t} + (\hat{e}_1^* \cdot \hat{e}_1^*)e^{+j2k_1 z}e^{-j2\omega_1 t} + (\hat{e}_1 \cdot \hat{e}_1^*) + (\hat{e}_1^* \cdot \hat{e}_1)\right],$$

$$\vec{\mathcal{E}}_2 \cdot \vec{\mathcal{E}}_2 = \frac{E_2^2}{4}\left[(\hat{e}_2 \cdot \hat{e}_2)e^{-j2k_2 z}e^{j2\omega_2 t}e^{-2j\phi} + (\hat{e}_2^* \cdot \hat{e}_2^*)e^{+j2k_2 z}e^{-j2\omega_2 t}e^{+2j\phi} + (\hat{e}_2 \cdot \hat{e}_2^*) + (\hat{e}_2^* \cdot \hat{e}_2)\right],$$

$$\vec{\mathcal{E}}_1 \cdot \vec{\mathcal{E}}_2 = \frac{E_1 E_2}{4}\left[(\hat{e}_1 \cdot \hat{e}_2)e^{-j(k_1+k_2)z}e^{j(\omega_1+\omega_2)t}e^{-j\phi} + (\hat{e}_1^* \cdot \hat{e}_2^*)e^{+j(k_1+k_2)z}e^{-j(\omega_1+\omega_2)t}e^{+j\phi} + \right.$$
$$\left.(\hat{e}_1 \cdot \hat{e}_2^*)e^{-j(k_1-k_2)z}e^{j(\omega_1-\omega_2)t}e^{+j\phi} + (\hat{e}_1^* \cdot \hat{e}_2)e^{+j(k_1-k_2)z}e^{-j(\omega_1-\omega_2)t}e^{-j\phi}\right].$$

The time average intensity $I(z) = \langle \mathcal{I}(z,t) \rangle$ becomes:

$$I(z) = \langle \mathcal{I}(z,t) \rangle = \frac{1}{2Z}\left[\langle \vec{\mathcal{E}}_1 \cdot \vec{\mathcal{E}}_1 \rangle + \langle \vec{\mathcal{E}}_2 \cdot \vec{\mathcal{E}}_2 \rangle + 2\langle \vec{\mathcal{E}}_1 \cdot \vec{\mathcal{E}}_2 \rangle\right] =$$
$$= \frac{1}{4Z}\left[E_1^2 + E_2^2\right].$$

From the last equation, it's evident that the interference term is missing regardless of the polarization vectors, due to the terms oscillating in time with frequencies $\omega_1 \pm \omega_2$. Therefore, no interference is observed when the frequencies of the two plane waves are different. A special case arises when $|\omega_1 - \omega_2| \ll \omega_1, \omega_2$, where a slow time-varying interference pattern could be observed. However, for optical frequencies, this time-varying interference pattern is challenging to observe due to the very large values of the optical frequencies and their corresponding difference.

In conclusion, an interference pattern between two collinear interfering plane waves is observable if the waves are monochromatic and their polarizations are not orthogonal. However, strictly monochromatic sources do not exist in practice. Even the best lasers are not purely monochromatic, as they have a relatively narrow spectrum. Therefore, to observe an interference pattern, the optical path length difference (as shown in Fig. 3.26) between the two waves should be smaller than the coherence length of the source from which the two waves originate. If the optical path difference between the two waves is larger than the coherence length, no interference pattern can be observed. More information about the coherence length and coherence in general is presented in Sec. 3.9.

3.5.2 Two Non-Collinear Plane-Waves Interference

3.5.2.1 Plane-Waves of the Same Frequency

In this case, it is assumed that two plane waves interfere in a homogeneous material, but their wavevectors lie in the xz plane, as shown in Fig. 3.28(a). Therefore, the two-wave interference is analyzed in the plane defined by their wavevectors. A possible implementation of this interference type is shown in Fig. 3.28(b). The interference of the two plane waves can be described by Eq. (3.202), which in this case results in

$$I = \frac{1}{4Z}\left[|\vec{E}_1|^2 + |\vec{E}_2|^2 + 2\text{Re}\left\{\vec{E}_1\cdot\vec{E}_2^*\right\}\right] =$$
$$= \frac{1}{4Z}\left\{E_1^2 + E_2^2 + 2E_1E_2|(\hat{e}_1\cdot\hat{e}_2^*)|\cos\left[(\vec{k}_1-\vec{k}_2)\cdot\vec{r}+\phi_{12}\right]\right\} =$$
$$\frac{1}{4Z}\left[E_1^2 + E_2^2 - 2E_1E_2|(\hat{e}_1\cdot\hat{e}_2^*)| + 4E_1E_2|(\hat{e}_1\cdot\hat{e}_2^*)|\cos^2\left(\frac{(\vec{k}_1-\vec{k}_2)\cdot\vec{r}+\phi_{12}}{2}\right)\right], \quad (3.203)$$

where $\vec{r} = x\hat{\imath}_x + y\hat{\imath}_y + z\hat{\imath}_z$ is the position vector, and ϕ_{12} is defined as previously (refer to Sec. 3.5.1.1).

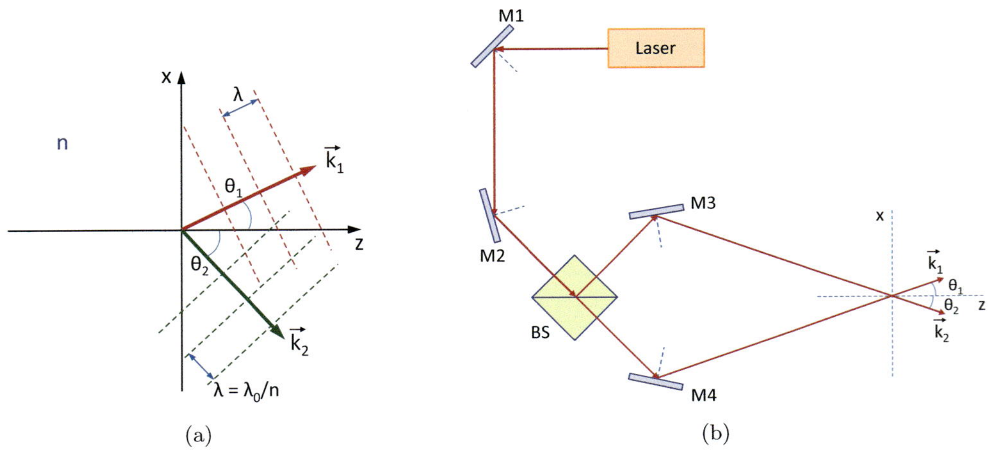

FIGURE 3.28: (a) Two plane waves propagating in an isotropic homogeneous medium of refractive index n. For simplicity, the wavevectors $\vec{k}_1$ and $\vec{k}_2$ lie in the xz plane. The wavevectors form angles θ_1 and θ_2 with respect to the z-axis. The counter-clockwise angles are considered positive and the clockwise angles are considered negative. In this figure $\theta_1 > 0$ and $\theta_2 < 0$. Their polarization unit vectors are $\hat{e}_1$ and $\hat{e}_2$, respectively, and they are perpendicular to their corresponding wavevectors. The polarization vectors are not shown in this figure. (b) Possible implementation of the interference of two non-collinear plane waves shown in (a). The angles θ_1 and θ_2 can be adjusted by the mirrors M_3 and M_4. BS denotes a beam splitter.

Without loss of generality, the phase shift ϕ_{12} can be considered zero (this will always be the case for linear polarizations). From Eq. (3.203), the maxima and minima of the interference pattern can be easily extracted. These are summarized below:

$$(\vec{k}_1-\vec{k}_2)\cdot\vec{r} = k_0 n\left[(\cos\theta_1-\cos\theta_2)z + (\sin\theta_1-\sin\theta_2)x\right] = \begin{cases} 2m\pi, & \text{(max)}, \\ (2m+1)\pi & \text{(min)}. \end{cases} \quad (3.204)$$

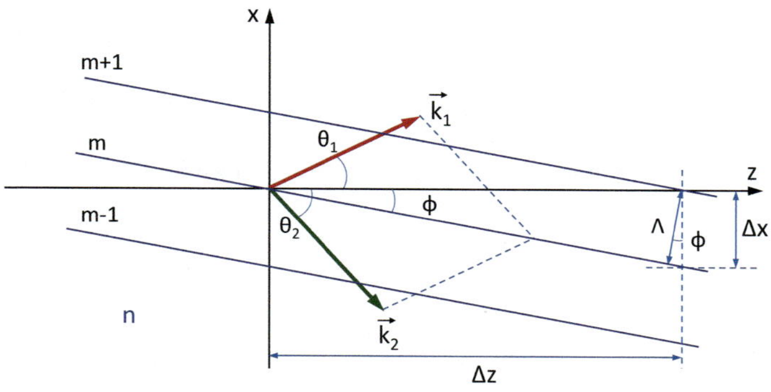

FIGURE 3.29: Two plane waves propagating in a isotropic homogeneous medium of refractive index n. For simplicity the wavevectors $\vec{k}_1$ and $\vec{k}_2$ lie again in the xz plane. The wavevectors form angles θ_1 and θ_2 with respect to the z-axis. The direction of the fringes are in the direction of the resultant of $\vec{k}_1$ and $\vec{k}_2$. The maximum intensity fringes for $m-1$, m and $m+1$ are shown. The period Λ is the spacing between two maximum (or minimum) intensity fringes.

The interference fringes align with the resultant of the two wavevectors $\vec{k}_{tot} = \vec{k}_1 + \vec{k}_2$, as depicted in Fig. 3.29. The periodicity of the fringes is the distance between either two consecutive maximum fringes of intensity or two successive minimum fringes. This periodicity is perpendicular to $\vec{k}_{tot}$. The period, Λ, can be easily determined with the help of Fig. 3.29 as follows:

$$Cz_m + Sx_m - m\frac{\lambda_0}{n} = 0, \qquad \text{for } m\text{-th maximum,}$$

$$Cz_{m+1} + Sx_{m+1} - (m+1)\frac{\lambda_0}{n} = 0, \qquad \text{for } (m+1)\text{-th maximum,}$$

$$C = \cos\theta_1 - \cos\theta_2,$$

$$S = \sin\theta_1 - \sin\theta_2,$$

$$\tan\phi = -\frac{C}{S},$$

$$C\underbrace{(z_{m+1} - z_m)}_{=\Delta z} + S\underbrace{(x_{m+1} - x_m)}_{=\Delta x} = \frac{\lambda_0}{n},$$

$$\text{for } \Delta x = 0 \ \rightarrow \ C\Delta z = \frac{\lambda_0}{n}, \ \text{ and for } \Delta z = 0 \ \rightarrow \ S\Delta x = \frac{\lambda_0}{n},$$

$$\Lambda = \Delta x\cos\phi = \Delta z|\sin\phi| \Longrightarrow \Lambda = \frac{\lambda_0/n}{\sqrt{[(\cos\theta_1 - \cos\theta_2)^2 + (\sin\theta_1 - \sin\theta_2)^2]}}. \tag{3.205}$$

In the case where the two wavevectors are symmetric with respect to the z-axis (i.e., when $\theta_1 = \theta = -\theta_2$), then $\Lambda = \lambda_0/(2n\sin\theta)$. As a result, by varying the angle θ, the fringe periodicity Λ can be adjusted. This is the basic idea behind the interferometric grating fabrication methodology, where two interfering plane waves produce a periodic intensity modulation. In a photosensitive medium, this modulation results in a periodic variation of the refractive index, forming a volume grating. Alternatively, using suitable etching techniques, removal of the photosensitized material can result in surface-relief grating.

Example 3.5.1:

Consider two interfering plane waves (in the xz plane) with the following parameters: $\theta_1 = -\theta_2 = 30°$, $E_1 = E_2 = E_0$, $\hat{e}_1 = \hat{\imath}_y = \hat{e}_2$, $n = 1$ (air), and $\lambda_0 = 0.6328\mu$m (He-Ne red wavelength). Find the interference pattern, the fringe spacing, and the locus of bright and dark fringes in the xz plane.

Solution: The resulting interference pattern can be determined using Eq. (3.203), and is presented in Fig. 3.30(a). The period of the bright and dark fringes is $\Lambda = 0.6328\,\mu$m and is calculated via Eq. (3.205). In addition, the root loci of the bright $[(\vec{k}_1 - \vec{k}_2)\cdot\vec{r} = 2m\pi]$ and dark $[\vec{k}_1 - \vec{k}_2)\cdot\vec{r} = (2m+1)\pi]$ fringes, are shown in Fig. 3.30(b) for $m = -10, -9, \cdots, 9, 10$. The colored representation of Fig. 3.30(a) was based on a simple algorithm, mapping a visible wavelength to an $[R, G, B]$ equivalent color. The procedure is defined in Appendix 3F.

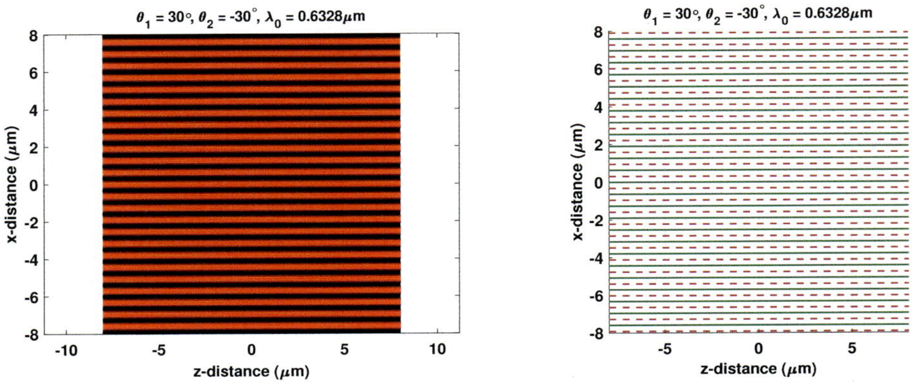

(a) (b)

FIGURE 3.30: The interference of two plane waves propagating in air in the xz plane. (a) $E_1 = E_2 = E_0$, $\hat{e}_1 = \hat{\imath}_y = \hat{e}_2$, $\theta_1 = 30°$, $\theta_2 = -30°$, and $\lambda_0 = 0.6328\mu$m. (b) The locus of points of maximum interference (solid lines), $r_2 - r_1 = m\lambda_0$, and of minimum interference (dashed lines), $r_2 - r_1 = (2m+1)(\lambda_0/2)$ for the same parameters as in (a). The color of the interference fringes is based on the algorithm defined in Appendix 3F.

3.5.2.2 Plane-Waves of Different Frequencies

When two non-collinear plane waves with different frequencies interfere, their electric fields can be represented as follows:

$$\vec{\mathcal{E}}_1 = E_1 \text{Re}\{\hat{e}_1 e^{-j\vec{k}_1\cdot\vec{r}} e^{+j\omega_1 t}\},$$

$$\vec{\mathcal{E}}_2 = E_2 \text{Re}\{\hat{e}_2 e^{-j\vec{k}_2\cdot\vec{r}} e^{+j\omega_2 t}\},$$

where $\vec{k}_1$ and $\vec{k}_2$ are the wavevectors, and $\hat{e}_1$, $\hat{e}_2$ the polarization vectors of the two plane waves, respectively. Following the same procedure as in the previous section, the following equations can be written:

$$\mathcal{I}(z,t) = \frac{1}{2Z}|\vec{\mathcal{E}}_1 + \vec{\mathcal{E}}_2|^2 = \frac{1}{2Z}\left[\vec{\mathcal{E}}_1\cdot\vec{\mathcal{E}}_1 + \vec{\mathcal{E}}_2\cdot\vec{\mathcal{E}}_2 + 2\vec{\mathcal{E}}_1\cdot\vec{\mathcal{E}}_2\right] \qquad \text{where,}$$

$$\vec{\mathcal{E}}_1\cdot\vec{\mathcal{E}}_1 = \frac{E_1^2}{4}\left[(\hat{e}_1\cdot\hat{e}_1)e^{-j2\vec{k}_1\cdot\vec{r}}e^{j2\omega_1 t} + (\hat{e}_1^*\cdot\hat{e}_1^*)e^{+j2\vec{k}_1\cdot\vec{r}}e^{-j2\omega_1 t} + (\hat{e}_1\cdot\hat{e}_1^*) + (\hat{e}_1^*\cdot\hat{e}_1)\right],$$

$$\vec{\mathcal{E}}_2 \cdot \vec{\mathcal{E}}_2 = \frac{E_2^2}{4}\left[(\hat{e}_2 \cdot \hat{e}_2)e^{-j2\vec{k}_2\cdot\vec{r}}e^{j2\omega_2 t} + (\hat{e}_2^* \cdot \hat{e}_2^*)e^{+j2\vec{k}_2\cdot\vec{r}}e^{-j2\omega_2 t} + (\hat{e}_2 \cdot \hat{e}_2^*) + (\hat{e}_2^* \cdot \hat{e}_2)\right],$$

$$\vec{\mathcal{E}}_1 \cdot \vec{\mathcal{E}}_2 = \frac{E_1 E_2}{4}\left[(\hat{e}_1 \cdot \hat{e}_2)e^{-j(\vec{k}_1+\vec{k}_2)\cdot\vec{r}}e^{j(\omega_1+\omega_2)t} + (\hat{e}_1^* \cdot \hat{e}_2^*)e^{+j(\vec{k}_1+\vec{k}_2)\cdot\vec{r}}e^{-j(\omega_1+\omega_2)t} + \right.$$

$$\left.(\hat{e}_1 \cdot \hat{e}_2^*)e^{-j(\vec{k}_1-\vec{k}_2)\cdot\vec{r}}e^{j(\omega_1-\omega_2)t} + (\hat{e}_1^* \cdot \hat{e}_2)e^{+j(\vec{k}_1-\vec{k}_2)\cdot\vec{r}}e^{-j(\omega_1-\omega_2)t}\right] \Longrightarrow$$

$$\langle\vec{\mathcal{E}}_1 \cdot \vec{\mathcal{E}}_1\rangle = \frac{E_1^2}{2}, \qquad \langle\vec{\mathcal{E}}_2 \cdot \vec{\mathcal{E}}_2\rangle = \frac{E_2^2}{2}, \qquad \langle\vec{\mathcal{E}}_1 \cdot \vec{\mathcal{E}}_2\rangle = 0, \qquad \Longrightarrow$$

$$I(z) = \langle\mathcal{I}(z,t)\rangle = \frac{1}{4Z}\left[E_1^2 + E_2^2\right].$$

This implies again that no interference pattern is observed. Even when the source of the two plane waves is a laser, which typically has the narrowest spectral width of all known sources, interference patterns will only be observed under specific conditions: (a) For interference to occur, the polarization vectors of the two plane waves must not be orthogonal. If the polarizations are orthogonal, the waves cannot interfere constructively or destructively, leading to the absence of an observable interference pattern; (b) the optical path length difference between the two waves must be smaller than the coherence length of the laser. The coherence length determines the maximum distance over which the waves remain in phase and can produce interference. If the optical path difference exceeds the coherence length, the phase relationship between the waves will be lost, and no interference pattern will form.

3.5.2.3 Interfering Pairs of Plane-Waves of Different Frequencies

Consider now the interference of two pairs of non-collinear plane waves. Each pair is similar to the setup depicted in Fig. 3.28(a), where both pairs share the same geometric configuration with respect to their angles in the xyz coordinate system. However, each pair has its own distinct frequency: ω_1 for the first pair and ω_2 for the second pair. The electric fields of the two pairs of interfering waves can be represented as follows:

$$\vec{\mathcal{E}}_1 = E_1(\omega_1)\mathrm{Re}\left\{\hat{e}_1(\omega_1)e^{-j\vec{k}_1(\omega_1)\cdot\vec{r}}e^{+j\omega_1 t}\right\} + E_1(\omega_2)\mathrm{Re}\left\{\hat{e}_1(\omega_2)e^{-j\vec{k}_1(\omega_2)\cdot\vec{r}}e^{+j\omega_2 t}\right\},$$

$$\vec{\mathcal{E}}_2 = E_2(\omega_1)\mathrm{Re}\left\{\hat{e}_2(\omega_1)e^{-j\vec{k}_2(\omega_1)\cdot\vec{r}}e^{+j\omega_1 t}\right\} + E_2(\omega_2)\mathrm{Re}\left\{\hat{e}_2(\omega_2)e^{-j\vec{k}_2(\omega_2)\cdot\vec{r}}e^{+j\omega_2 t}\right\},$$

where $\vec{k}_1(\omega_i)$ and $\vec{k}_2(\omega_i)$ $(i = 1, 2)$ are the wavevectors of the two pairs at their respective frequencies, and $\hat{e}_1(\omega_i)$, $\hat{e}_2(\omega_i)$ are polarization vectors associated with each pair. The wavevectors $\vec{k}_1(\omega_i)$ and $\vec{k}_2(\omega_i)$ form the same angles θ_1 and θ_2, respectively, for both frequencies. For simplicity, and without loss of generality, the two amplitudes E_1 and E_2 are considered real. Following the same procedure as outlined in the previous sections, the following equations can be written:

$$\mathcal{I}(\vec{r},t) = \frac{1}{2Z}|\vec{\mathcal{E}}_1 + \vec{\mathcal{E}}_2|^2 = \frac{1}{2Z}\left[\vec{\mathcal{E}}_1 \cdot \vec{\mathcal{E}}_1 + \vec{\mathcal{E}}_2 \cdot \vec{\mathcal{E}}_2 + 2\vec{\mathcal{E}}_1 \cdot \vec{\mathcal{E}}_2\right] \qquad \text{where,}$$

$$\vec{\mathcal{E}}_1 \cdot \vec{\mathcal{E}}_1 = \frac{E_1^2(\omega_1)}{4}\left[\hat{e}_1(\omega_1) \cdot \hat{e}_1(\omega_1)\right]e^{-j2\vec{k}_1(\omega_1)\cdot\vec{r}}e^{+j2\omega_1 t} +$$

$$\frac{E_1^2(\omega_1)}{4}\left[\hat{e}_1^*(\omega_1) \cdot \hat{e}_1^*(\omega_1)\right]e^{+j2\vec{k}_1(\omega_1)\cdot\vec{r}}e^{-j2\omega_1 t} +$$

$$\frac{E_1^2(\omega_2)}{4}\left[\hat{e}_1(\omega_2)\cdot\hat{e}_1(\omega_2)\right]e^{-j2\vec{k}_1(\omega_2)\cdot\vec{r}}e^{+j2\omega_2 t}+$$

$$\frac{E_1^2(\omega_2)}{4}\left[\hat{e}_1^*(\omega_2)\cdot\hat{e}_1^*(\omega_2)\right]e^{+j2\vec{k}_1(\omega_2)\cdot\vec{r}}e^{-j2\omega_2 t}+$$

$$\frac{E_1^2(\omega_1)}{4}\left[\hat{e}_1(\omega_1)\cdot\hat{e}_1^*(\omega_1)\right]+$$

$$\frac{E_1(\omega_1)E_1(\omega_2)}{4}\left[\hat{e}_1(\omega_1)\cdot\hat{e}_1(\omega_2)\right]e^{-j[\vec{k}_1(\omega_1)+\vec{k}_1(\omega_2)]\cdot\vec{r}}e^{+j(\omega_1+\omega_2)t}+$$

$$\frac{E_1(\omega_1)E_1(\omega_2)}{4}\left[\hat{e}_1(\omega_1)\cdot\hat{e}_1^*(\omega_2)\right]e^{-j[\vec{k}_1(\omega_1)-\vec{k}_1(\omega_2)]\cdot\vec{r}}e^{+j(\omega_1-\omega_2)t}+$$

$$\frac{E_1^2(\omega_1)}{4}\left[\hat{e}_1^*(\omega_1)\cdot\hat{e}_1(\omega_1)\right]+$$

$$\frac{E_1(\omega_1)E_1(\omega_2)}{4}\left[\hat{e}_1^*(\omega_1)\cdot\hat{e}_1(\omega_2)\right]e^{+j[\vec{k}_1(\omega_1)-\vec{k}_1(\omega_2)]\cdot\vec{r}}e^{-j(\omega_1-\omega_2)t}+$$

$$\frac{E_1(\omega_1)E_1(\omega_2)}{4}\left[\hat{e}_1^*(\omega_1)\cdot\hat{e}_1^*(\omega_2)\right]e^{+j[\vec{k}_1(\omega_1)+\vec{k}_1(\omega_2)]\cdot\vec{r}}e^{-j(\omega_1+\omega_2)t}+$$

$$\frac{E_1(\omega_2)E_1(\omega_1)}{4}\left[\hat{e}_1(\omega_2)\cdot\hat{e}_1(\omega_1)\right]e^{-j[\vec{k}_1(\omega_1)+\vec{k}_1(\omega_2)]\cdot\vec{r}}e^{+j(\omega_1+\omega_2)t}+$$

$$\frac{E_1(\omega_2)E_1(\omega_1)}{4}\left[\hat{e}_1(\omega_2)\cdot\hat{e}_1^*(\omega_1)\right]e^{+j[\vec{k}_1(\omega_1)-\vec{k}_1(\omega_2)]\cdot\vec{r}}e^{-j(\omega_1-\omega_2)t}+$$

$$\frac{E_1^2(\omega_2)}{4}\left[\hat{e}_1^*(\omega_2)\cdot\hat{e}_1(\omega_2)\right]+$$

$$\frac{E_1(\omega_2)E_1(\omega_1)}{4}\left[\hat{e}_1^*(\omega_2)\cdot\hat{e}_1(\omega_1)\right]e^{-j[\vec{k}_1(\omega_1)-\vec{k}_1(\omega_2)]\cdot\vec{r}}e^{+j(\omega_1-\omega_2)t}+$$

$$\frac{E_1(\omega_2)E_1(\omega_1)}{4}\left[\hat{e}_1^*(\omega_2)\cdot\hat{e}_1^*(\omega_1)\right]e^{+j[\vec{k}_1(\omega_1)+\vec{k}_1(\omega_2)]\cdot\vec{r}}e^{-j(\omega_1+\omega_2)t}+$$

$$\frac{E_1^2(\omega_2)}{4}\left[\hat{e}_1^*(\omega_2)\cdot\hat{e}_1(\omega_2)\right]\Longrightarrow$$

$$\langle\vec{\mathcal{E}}_1\cdot\vec{\mathcal{E}}_1\rangle=\frac{E_1^2(\omega_1)}{2}+\frac{E_1^2(\omega_2)}{2},$$

where the $\langle * \rangle$ denotes the time average value. In a similar fashion, the time average values of $\langle\vec{\mathcal{E}}_2\cdot\vec{\mathcal{E}}_2\rangle$ and $\langle\vec{\mathcal{E}}_1\cdot\vec{\mathcal{E}}_2\rangle$ are given by:

$$\langle\vec{\mathcal{E}}_2\cdot\vec{\mathcal{E}}_2\rangle=\frac{E_2^2(\omega_1)}{2}+\frac{E_2^2(\omega_2)}{2},$$

$$\begin{aligned}\langle\vec{\mathcal{E}}_1\cdot\vec{\mathcal{E}}_2\rangle=&\frac{E_1(\omega_1)E_2(\omega_2)}{2}\left|\hat{e}_1(\omega_1)\cdot\hat{e}_2^*(\omega_1)\right|\cos\left[(\vec{k}_1(\omega_1)-\vec{k}_2(\omega_1)]\cdot\vec{r}+\phi_{12}(\omega_1)\right]+\\&\frac{E_1(\omega_2)E_2(\omega_2)}{2}\left|\hat{e}_1(\omega_2)\cdot\hat{e}_2^*(\omega_2)\right|\cos\left[(\vec{k}_1(\omega_2)-\vec{k}_2(\omega_2)]\cdot\vec{r}+\phi_{12}(\omega_2)\right],\end{aligned}$$

where $\phi_{12}(\omega_i)$ is the possible phase angle of the $\hat{e}_1(\omega_i)\cdot\hat{e}_2^*(\omega_i)$ polarization terms $(i = 1, 2)$. Then the resulting time-average intensity pattern is given by:

$$I(\vec{r}) = \langle \mathcal{I}(\vec{r},t)\rangle = \frac{1}{4Z}\Big\{E_1^2(\omega_1) + E_2^2(\omega_1)$$

$$+ 2E_1(\omega_1)E_2(\omega_1)\cos\Big[(\vec{k}_1(\omega_1) - \vec{k}_2(\omega_1)]\cdot\vec{r} + \phi_{12}(\omega_1)\Big] +$$

$$E_1^2(\omega_2) + E_2^2(\omega_2) + 2E_1(\omega_2)E_2(\omega_2)\cos\Big[(\vec{k}_1(\omega_2) - \vec{k}_2(\omega_2)]\cdot\vec{r} + \phi_{12}(\omega_2)\Big]\Big\}, \quad (3.206)$$

where the overall intensity is the sum of the two interference patterns for the two separate frequencies ω_1 and ω_2. This is equivalent to the incoherent addition of the two individual interference patterns.

3.5.2.4 Two Interfering Plane-Waves from a Narrowband Source

Consider the situation where the two plane waves depicted in Figs. 3.28(a) and 3.28(b) originate from a narrowband source, such as a laser. While the laser spectrum is narrow, it is not perfectly monochromatic. In this section, the effect of the frequency spread of the two interfering waves is investigated. For simplicity, let us assume that each pair of plane waves, characterized by wavevectors $\vec{k}_1(\omega)$ and $\vec{k}_2(\omega)$, shares the same linear polarization, i.e., $\hat{e}_1(\omega) = \hat{e}_2(\omega) = \hat{e}$, and the same amplitude, $E_1(\omega) = E_2(\omega) = E_0(\omega)$. These assumptions are reasonable when the two interfering waves originate from the same laser source. Under these conditions, the resulting electric field can be expressed as:

$$\vec{\mathcal{E}}(\vec{r},t) = \hat{e}\int_\omega E_0(\omega)\left[\mathrm{Re}\left\{\left(e^{-j\vec{k}_1(\omega)\cdot\vec{r}} + e^{-j\vec{k}_2(\omega)\cdot\vec{r}}\right)e^{+j\omega t}\right\}\right]d\omega$$

$$= \hat{e}\int_\omega E_0(\omega)\left[\cos\left(\vec{k}_1(\omega)\cdot\vec{r} - \omega t\right) + \cos\left(\vec{k}_2(\omega)\cdot\vec{r} - \omega t\right)\right]d\omega$$

$$= \hat{e}\int_\omega 2E_0(\omega)\left\{\cos\left[\frac{\left(\vec{k}_1(\omega) - \vec{k}_2(\omega)\right)\cdot\vec{r}}{2}\right] + \cos\left[\frac{\left(\vec{k}_1(\omega) + \vec{k}_2(\omega)\right)\cdot\vec{r}}{2} + \omega t\right]\right\}d\omega.$$

In the last equation, the trigonometric identity $\cos\alpha + \cos\beta = 2\cos[(\alpha-\beta)/2]\cos[(\alpha+\beta)/2]$ was employed. At this point, some additional assumptions are introduced. Given that the source's spectrum is narrow, the integration is performed over the range $\omega_c - \Delta\omega/2 \leq \omega \leq \omega_c + \Delta\omega/2$, where ω_c denotes the center frequency, and $\Delta\omega$ represents the spectrum width (typically, for practical laser sources, $\Delta\omega \ll \omega_c$). Since optical frequencies are on the order of hundreds of terahertz ($1\,THz = 10^{12}$ Hz), it can be assumed that the wavevectors of the two waves, $\vec{k}_1(\omega)$ and $\vec{k}_2(\omega)$, are nearly the same for all ω in the integration range. Specifically, $\vec{k}_1(\omega) + \vec{k}_2(\omega) \simeq \vec{k}_1(\omega_c) + \vec{k}_2(\omega_c)$, and $\omega \simeq \omega_c$ in the time-varying term of the last equation. As a result, the electric field of the interference of the two waves can be expressed as:

$$\vec{\mathcal{E}}(\vec{r},t) = 2\hat{e}\cos\left[\frac{\left(\vec{k}_1(\omega_c) + \vec{k}_2(\omega_c)\right)\cdot\vec{r}}{2} + \omega_c t\right]\int_{\omega_c-\Delta\omega/2}^{\omega_c+\Delta\omega/2} E_0(\omega)\cos\left[\frac{\left(\vec{k}_1(\omega) - \vec{k}_2(\omega)\right)\cdot\vec{r}}{2}\right]d\omega$$

where,

$$E_0(\omega) = \frac{E_0}{\Delta\omega}rect\left(\frac{\omega - \omega_c}{\Delta\omega}\right) \qquad \text{for} \qquad \omega_c - \frac{\Delta\omega}{2} \leq \omega \leq \omega_c + \frac{\Delta\omega}{2} \qquad \text{and,}$$

$$[\vec{k}_1(\omega) - \vec{k}_2(\omega)]\cdot\vec{r} = \frac{\omega}{c}n\left[(\cos\theta_1 - \cos\theta_2)z + (\sin\theta_1 - \sin\theta_2)x\right] = A\omega. \quad (3.207)$$

In the last equation it was assumed that the spectral density of the narrowband source is uniform, and it is represented by the $rect(u)$ function (rectangle function), which is unity for $|u| \leq 1/2$ and zero otherwise. It is essential to note that the product $E_0(\omega)d\omega$ has dimensions of electric field. More realistic spectral densities, such as those of Lorentzian or Gaussian shape, could be substituted for $E_0(\omega)$. With the aforementioned assumptions and definitions, it is straightforward to perform the integration over frequency:

$$\vec{\mathcal{E}}(\vec{r},t) = 2\hat{e}E_0 \cos\left[\frac{\left(\vec{k}_1(\omega_c)+\vec{k}_2(\omega_c)\right)\cdot\vec{r}}{2} + \omega_c t\right] \underbrace{\frac{1}{\Delta\omega}\int_{\omega_c-\Delta\omega/2}^{\omega_c+\Delta\omega/2} \cos\left(\frac{A\omega}{2}\right) d\omega}_{=B}$$

where,

$$B = \frac{2}{A}\left\{\sin\left[\frac{A}{2}\left(\omega_c + \frac{\Delta\omega}{2}\right)\right] - \sin\left[\frac{A}{2}\left(\omega_c - \frac{\Delta\omega}{2}\right)\right]\right\} =$$

$$\frac{2}{A}2\sin\left(\frac{A}{4}\Delta\omega\right)\cos\left(\frac{A}{4}2\omega_c\right) = \Delta\omega\frac{\sin\left(\frac{A}{4}\Delta\omega\right)}{\frac{A}{4}\Delta\omega}\cos\left(\frac{A}{2}\omega_c\right).$$

In the previous derivation, the trigonometric identity $\sin\alpha - \sin\beta = 2\sin[(\alpha-\beta)/2]\cos[(\alpha+\beta)/2]$ was utilized. Additionally, it was assumed that the medium is dispersionless, meaning that the refractive index $n(\omega)$ is constant and equal to n. In the case of a strictly monochromatic source, where $E_0(\omega) = E_0\delta(\omega-\omega_c)$, the total electric field simplifies to the same result obtained for two interfering waves of the same frequency. This is realized by taking the limit $lim_{x\to 0}\{\sin x/x\} = 1$, where $x = A\Delta\omega/4$. Thus, the intensity of the interference of the waves is given by:

$$I(\vec{r}) = \frac{1}{2Z}\langle\vec{\mathcal{E}}\cdot\vec{\mathcal{E}}\rangle = \frac{1}{2Z}4E_0^2\frac{\sin^2\left(\frac{A}{4}\Delta\omega\right)}{(\frac{A}{4})^2\Delta\omega^2}\cos^2\left(\frac{A}{2}\omega_c\right)\left\langle\cos\left[\frac{\left(\vec{k}_1(\omega_c)+\vec{k}_2(\omega_c)\right)\cdot\vec{r}}{2} + \omega_c t\right]\right\rangle =$$

$$= \frac{1}{4Z}\,4E_0^2\frac{\sin^2\left(\frac{A}{4}\Delta\omega\right)}{(\frac{A}{4})^2\Delta\omega^2}\cos^2\left(\frac{A}{2}\omega_c\right). \tag{3.208}$$

The last equation in the case of symmetric interfering waves, i.e., when $\theta_1 = -\theta_2 = \theta$, takes the form [108]:

$$I(x,z) = \frac{1}{4Z}\,4E_0^2\frac{\sin^2\left(\frac{A}{4}\Delta\omega\right)}{(\frac{A}{4})^2\Delta\omega^2}\cos^2\left(\frac{A}{2}\omega_c\right) =$$

$$= \frac{1}{4Z}\,4E_0^2\,\frac{\sin^2\left(\pi\frac{n}{c}\sin\theta\Delta\nu x\right)}{(\pi\frac{n}{c}\sin\theta\Delta\nu x)^2}\cos^2\left(k_c n\sin\theta x\right) =$$

$$= \frac{1}{4Z}\,4E_0^2\,\mathrm{sinc}^2\left(\frac{n}{c}\sin\theta\Delta\nu x\right)\,\cos^2\left(k_c n\sin\theta x\right), \tag{3.209}$$

where $\mathrm{sinc}(u) = \sin(\pi u)/\pi u$.

Example 3.5.2:

Consider a pair of plane waves originating from a narrowband He-Ne laser. The angles of the plane waves are $\theta_1 = -\theta_2 = 10\,\text{deg}$. The He-Ne center wavelength is $\lambda_0 = 0.6328\,\mu\text{m}$. The laser bandwidth is $\Delta\nu \simeq 10^{10}\,\text{Hz}$. This bandwidth corresponds to a coherence length of $\ell_c \simeq c/\Delta\nu = 30\,\text{cm}$, which is typical for commercial multimode He-Ne lasers (for comparison, single-mode He-Ne lasers can have a coherence length of the order of tenths or hundred of meters). Determine the interference pattern that is created by the two plane waves.
Solution: The observable range of interference fringes is effectively determined by the width of the main lobe of the *sinc* function [where $sinc(x) = \sin(\pi x)/(\pi x)$] in Eq. (3.209). The normalized width $(\Delta x/\lambda_0)$ is calculated as: $\Delta x/\lambda_0 = 2c/(\lambda_0 \Delta\nu)(1/n\sin\theta) = 2 \times 2.7282 \times 10^5$. The resulting interference pattern within this range of $\Delta x/\lambda_0$ is illustrated in Fig. 3.31(a). The lower plot in this figure highlights the *sinc* function alone, where the predicted width of the main lobe is evident. For comparison, a hypothetical case of a He-Ne laser with a significantly broader bandwidth ($\Delta\nu = 10^{12}\,\text{Hz}$ corresponding to a coherence length of $\ell_c = c/\Delta\nu \simeq 0.3\,\text{cm}$) is shown in Fig. 3.31(b). In this case, the normalized width of the *sinc* function reduces to $\Delta x/\lambda_0 = 2 \times 2.7282 \times 10^3\lambda_0$. In the top plot of Fig. 3.31(b), the intensity described by Eq. (3.209) is shown, where the rapid intensity variations appear as a filled-color curve beneath the *sinc* envelope. The fringe period for both cases is determined by the cosine term in Eq. (3.209) and is calculated as $\Lambda = \lambda_0/(2n\sin\theta) = 1.822\,\mu\text{m}$. This value is confirmed by examining the interference patterns in Figs. 3.31(a) and 3.31(b).

3.5.3 Two-Slit Young's Experiment

The two-slit experiment performed by Thomas Young [10] in 1801-1802 was the cornerstone of the wave theory of light. Up to that time, the corpuscular theory of light was generally accepted, with Isaac Newton being its main proponent. Young's two-slit interference experiment was conducted using white light and a small pinhole allowing sunlight to pass through. Using a mirror, the thin beam of light from the pinhole was adjusted to propagate horizontally over a table. Instead of using two slits, Young employed a thin card in the middle of the beam to divide it into two interfering beams. He observed the interference pattern on a viewing screen, resembling the interference pattern of water waves in a pond. To simplify the theory, most textbooks describe the two-slit experiment using the setup shown in Fig. 3.32. Monochromatic light falls from the left onto a pinhole. The resulting light from the pinhole is in the form of a spherical wave (see Appendix 3C) and possesses both temporal and spatial coherence.

Assuming that the distance between the pinhole plane and the plane containing the two slits S_1 and S_2 is significantly larger than the wavelength of the light, the wavefront arriving at the slits' plane can be approximated as a plane wave. The two slits, separated by a distance d, are assumed to have widths much smaller than their separation and act as sources of initially cylindrical waves (refer to Appendix 3D). However, when the observation screen is placed at a distance L (where L is much larger than the wavelength of light), the cylindrical waves emitted from the slits can be approximated as plane waves. Since both waves originate from the same pinhole S, the two interfering waves, represented in phasor form as $\vec{E}_1$ and $\vec{E}_2$, share the same polarization. Even with unpolarized light, interference patterns can still be observed. Additionally, the amplitudes of $\vec{E}_1$ and $\vec{E}_2$ are approximately equal, as the distances r_1 and r_2 from the slits S_1 and S_2 to the observation point $P(x)$ are much larger than x, the lateral displacement of point P from the z-axis. In this analysis, diffraction effects at each slit are neglected, with these effects deferred to the chapter on *Scalar Diffraction*, Chapter 5. Based on the above assumptions, the following simplified

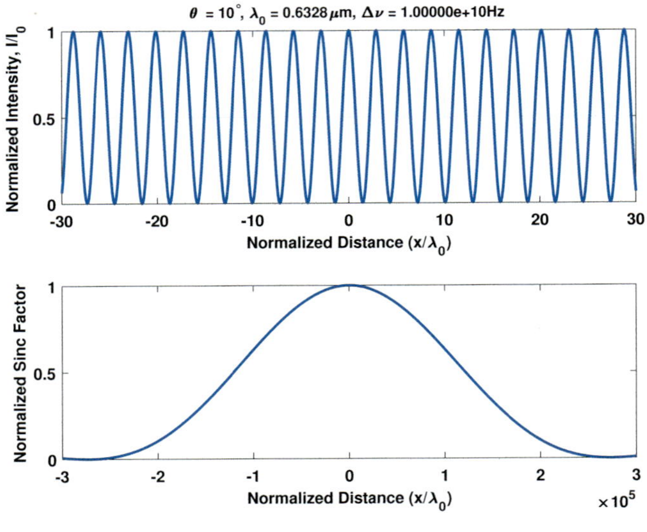

(a)

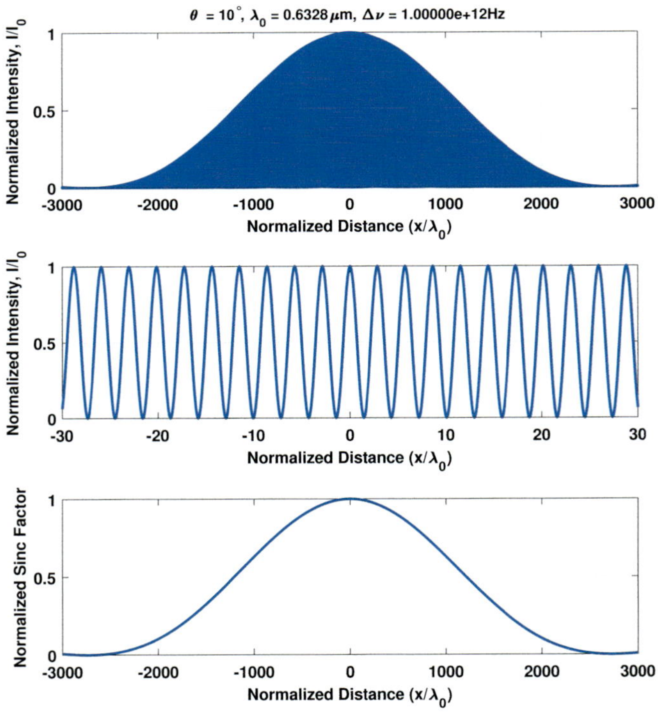

(b)

FIGURE 3.31: (a) Two narrowband He-Ne plane waves interfering in air. The wavevectors form an angle $\theta_1 = -\theta_2 = \theta = 10^\circ$ with respect to the z-axis. The center wavelength is $\lambda_0 = 0.6328\,\mu$m and the bandwidth is $\Delta\nu = 10^{10}$ Hz. The fringes are observed for about $\pm 2.7282 \times 10^5 \lambda_0$, which corresponds to the width of the main lobe of the *sinc* function. (b) Same as (a) but with a laser bandwidth of $\Delta\nu = 10^{12}$ Hz (which is extreme for a He-Ne laser). The fringes are observed for about $\pm 2728.2\lambda_0$.

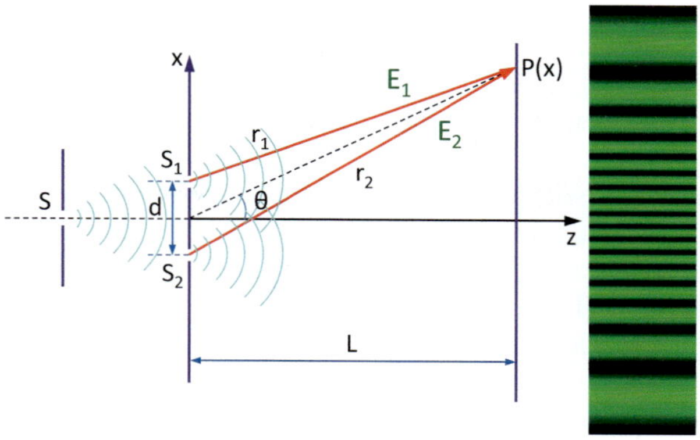

FIGURE 3.32: The configuration of Young's double-slit experiment involves monochromatic light passing through a small slit S, which acts as a spatially coherent source. This light illuminates two closely spaced small slits, S_1 and S_2, separated by a distance d. The light emanating from S_1 and S_2 interferes to produce an interference pattern on a screen placed at a distance L from the plane containing the two slits. A point $P(x)$ on the screen lies at a lateral distance x from the optical axis, which is aligned with the z-axis.

equations can be written:

$$\vec{E}_1 = \hat{e}E_0 e^{-jkr_1},$$

$$\vec{E}_2 = \hat{e}E_0 e^{-jkr_2},$$

$$I(x) = \frac{1}{4Z}|\vec{E}_1 + \vec{E}_2|^2 = \frac{1}{4Z}\left\{2E_0^2 + 2E_0^2 \cos\left[k(r_2 - r_1)\right]\right\} =$$

$$= \frac{1}{4Z}\, 4E_0^2 \cos^2\left[\frac{k(r_2 - r_1)}{2}\right], \qquad \text{where,} \tag{3.210}$$

$$r_1 = \left[\left(x - \frac{d}{2}\right)^2 + L^2\right]^{1/2}, \qquad r_2 = \left[\left(x + \frac{d}{2}\right)^2 + L^2\right]^{1/2}.$$

The unit vector $\hat{e}$ represents the common polarization vector of the two interfering waves, while E_0 denotes their shared amplitude. The wavenumber is expressed as $k = k_0 n = (2\pi/\lambda_0)n$, where λ_0 is the freespace wavelength, and n the refractive index of the medium in which the waves interfere. It is evident that even if $\hat{e} = \hat{e}(t)$ changes randomly with time, as occurs with unpolarized light, the interference pattern will remain unaffected. This is because the polarization of both interfering waves is always identical. The positions of the maxima and minima of the interference pattern are determined by the following conditions:

$$k(r_2 - r_1) = k_0 n(r_2 - r_1) = \begin{cases} 2m\pi, & \text{for maxima,} \\ (2m+1)\pi, & \text{for minima.} \end{cases} \tag{3.211}$$

In general, the geometric loci of the maxima and minima of the interference pattern are hyperbolas in the xz plane. These hyperbolas arise due to the constant phase difference between the two interfering waves, which defines the condition for constructive or destructive interference.

Example 3.5.3:

An example of the root locus for maxima and minima is illustrated in Fig. 3.33, using the parameters $\lambda_0 = 0.5145\,\mu$m, slit separation $d = 5\,\mu$m and distance to the screen $L = 20\,\mu$m. These values are not realistic but are intentionally chosen to highlight a portion of the root loci for clarity. If realistic values for d and L were used, the resulting hyperbolas would be much more numerous and closely spaced. Solid hyperbolas correspond to locations of maximum interference (bright fringes), while dashed hyperbolas represent locations of minimum interference (dark fringes).

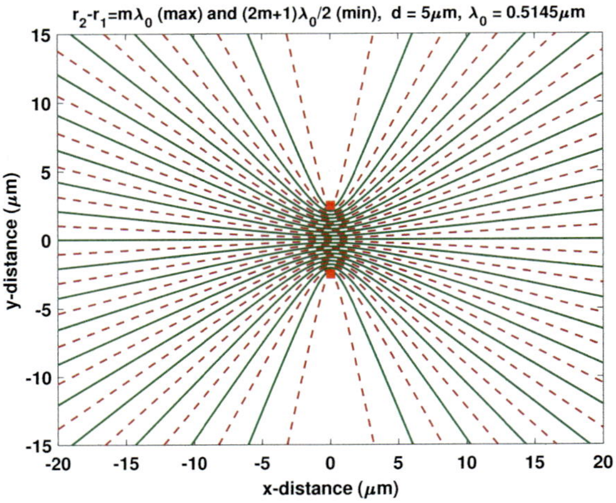

FIGURE 3.33: The root loci of the path differences for intensity maxima ($r_2 - r_1 = m\lambda_0$, shown as solid lines), and intensity minima ($r_2 - r_1 = (2m+1)\lambda_0/2$, shown as dashed lines) in Young's double-slit experiment in the air are depicted. Monochromatic light with a freespace wavelength of $\lambda_0 = 0.5145\,\mu$m is assumed. The slit separation is $d = 5, \mu$m, and the distance to the observation screen is $L = 20, \mu$m. The two squares indicate the positions of the two slits.

The points (lines in the xz plane) of maximum and minimum interference on the viewing screen can be easily determined from Eq. (3.211). Specifically, the positions of the bright fringes, denoted as x_m^B, and dark fringes denoted as x_m^D (m is an integer and $\lambda = \lambda_0/n$), on the screen are given by:

$$x_m^B = \frac{m\lambda}{2}\left[\frac{4L^2 + d^2 - m^2\lambda^2}{d^2 - m^2\lambda^2}\right]^{1/2}, \tag{3.212}$$

$$x_m^D = \frac{(2m+1)\lambda}{4}\left[\frac{16L^2 + 4d^2 - 4m^2\lambda^2 - 4m\lambda^2 - \lambda^2}{4d^2 - 4m^2\lambda^2 - 4m\lambda^2 - \lambda^2}\right]^{1/2}. \tag{3.213}$$

For the practical case where $L \gg d \gg \lambda$, It is better to repace λ, with $\lambda = \lambda_0/n$, the following simplifications can be made $[(1+x)^{1/2} \simeq 1 + x/2,\ \ |x| \ll 1]$:

$$r_1 = \left[\left(x - \frac{d}{2}\right)^2 + L^2\right]^{1/2} \simeq L + \frac{1}{2L}\left(x - \frac{d}{2}\right)^2,$$

$$r_2 = \left[\left(x + \frac{d}{2}\right)^2 + L^2\right]^{1/2} \simeq L + \frac{1}{2L}\left(x + \frac{d}{2}\right)^2,$$

$$r_2 - r_1 \simeq x\frac{d}{L} \qquad \text{resulting in}$$

$$(r_2 - r_1) \simeq x\frac{d}{L} = \begin{cases} m\dfrac{\lambda_0}{n}, & \text{for maxima,} \\ \\ (2m+1)\dfrac{\lambda_0}{2n}, & \text{for minima.} \end{cases}$$

$$x_m^B \simeq m\frac{\lambda_0}{n}\frac{L}{d}, \tag{3.214}$$

$$x_m^D \simeq (2m+1)\frac{\lambda_0}{2n}\frac{L}{d}. \tag{3.215}$$

Indeed, when the distance from the slits to the screen (L) is much larger than both the slit separation (d) and the wavelength (λ), i.e., when $L \gg d \gg \lambda$, it is straightforward to simplify the exact equations (3.212) and (3.213) for the positions of the maxima and minima to their approximate forms of Eqs. (3.214) and (3.215).

Example 3.5.4:

An example interference pattern is shown in Figs. 3.34(a) and 3.34(b) for monochromatic light with wavelength $\lambda_0 = 0.5145\,\mu$m (green light from Argon-Ion laser), the refractive index $n = 1$ (air), slit separation $d = 5\,\mu$m, and screen distance $L = 20\,\mu$m. While these dimensions are unrealistic for practical experiments, they illustrate the variation in fringe periodicity due to the hyperbolic root loci. A more realistic example is shown in Figs. 3.35(a) and 3.35(b), where the same wavelength is used, but with slit separation $d = 50\,\mu$m, and screen distance $L = 1000\,\mu$m. In this case, the fringes on the screen are nearly periodic, with a fringe spacing of $\lambda_0 L/d = 10.29\,\mu$m, as clearly shown in the figures.

3.5.3.1 Two-Slit Young's Experiment with White Light

As discussed earlier, interference fringes can be observed in the two-slit experiment even with multichromatic, i.e., white light. Let's examine what happens when the plane waves arriving at the two slits S_1 and S_2 are spatially coherent (due to the pinhole S) but are summations of plane waves of various wavelengths in the visible spectrum that comprise white light (as sunlight). As shown in Sec. 3.5.2.3 for two pairs of plane waves of differing frequencies, the intensities for each wavelength are added incoherently. If $I(x, \omega)$ is the interference of the subset of pairs of plane waves of frequency ω (in the visible range) at position x on the viewing screen, then it is expressed by [based on Eq. (3.210)]:

$$I(x, \omega) = I_{0,\omega} \cos^2\left[\frac{\omega}{c} n \frac{(r_2 - r_1)}{2}\right], \tag{3.216}$$

where $I_{0,\omega}$ the maximum intensity at frequency ω. The overall intensity at position x is the incoherent addition of all $I(x, \omega)$. Therefore, the overall intensity $I(x)$ is given by:

$$I(x) = \int_{\omega_c - \frac{\Delta\omega}{2}}^{\omega_c + \frac{\Delta\omega}{2}} I_0(\omega) \cos^2\left[\frac{\omega}{c} n \frac{(r_2 - r_1)}{2}\right] d\omega = \int_{\omega_c - \frac{\Delta\omega}{2}}^{\omega_c + \frac{\Delta\omega}{2}} I_0(\omega) \cos^2\left[A(x, L)\omega\right] d\omega,$$

$$\text{where, } A(x, L) = \frac{n}{c}\frac{(r_2 - r_1)}{2}, \qquad \text{and} \qquad I_0(\omega) d\omega = I_{0,\omega},$$

where ω_c is the center frequency and $\Delta\omega$ is the bandwidth of the white light. For simplicity it is assumed that the white light is of equal intensity all over the bandwidth, i.e., $I_0(\omega)\Delta\omega = I_0$

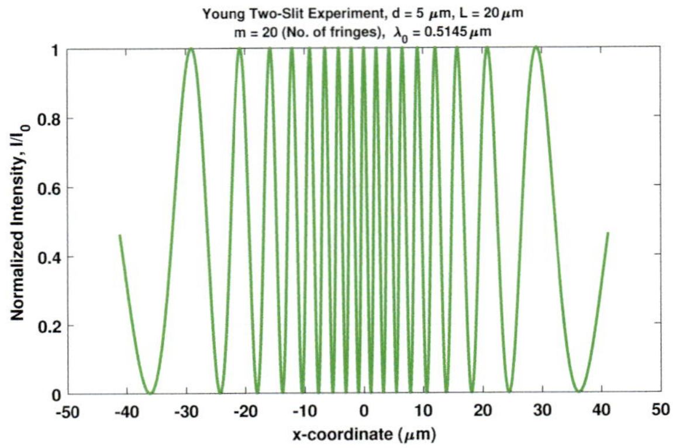

(a)

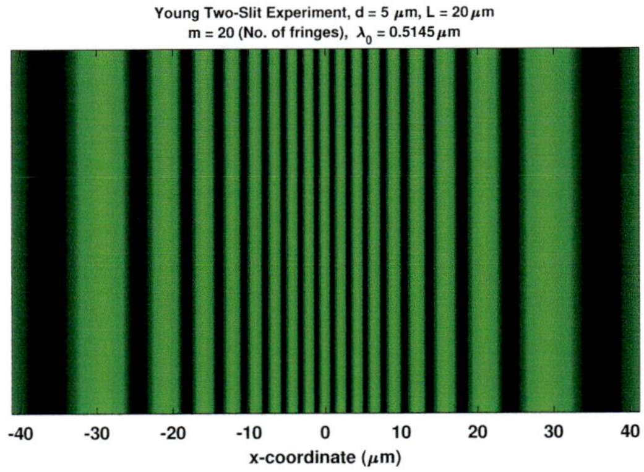

(b)

FIGURE 3.34: The interference pattern produced by two slits for monochromatic light of $\lambda_0 = 0.5145\mu m$ with $d = 5\,\mu m$ and $L = 20\,\mu m$. (a) The normalized intensity pattern on the screen as a function of x. (b) Same as in (b) with the colored representation of the intensity fringes that should be observed on the screen. The color of the interference fringes is based on the algorithm defined in Appendix 3F.

(where $I_0(\omega)$ has dimensions of $W/m^2/Hz$ and I_0 has dimensions of W/m^2). Then, the resulting $I(x)$ can be determined as:

$$I(x) = \frac{I_0}{\Delta\omega}\int_{\omega_c-\frac{\Delta\omega}{2}}^{\omega_c+\frac{\Delta\omega}{2}} \cos^2\left[A(x,L)\omega\right] d\omega =$$

$$= \frac{I_0}{2}\left\{1 + \frac{\sin(A(x,L)\Delta\omega)}{A\Delta\omega}\cos\left[A(x,L)2\omega_c\right]\right\}. \tag{3.217}$$

When $r_2 - r_1 \simeq xd/L$ and $\Delta\omega = 2\pi\Delta\nu$ then $I(x)$ becomes:

$$I(x) = \frac{I_0}{2}\left[1 + sinc\left(\frac{n}{c}x\frac{d}{L}\Delta\nu\right)\cos\left(\frac{n}{c}x\frac{d}{L}2\pi\nu_c\right)\right], \tag{3.218}$$

$$\Delta\nu = c\left(\frac{1}{\lambda_{0,min}} - \frac{1}{\lambda_{0,max}}\right), \qquad \nu_c = \frac{c}{\lambda_{0,c}},$$

where $\lambda_{0,c}$, $\lambda_{0,min}$, $\lambda_{0,max}$, are the center, the minimum, and the maximum wavelength of the white light, respectively. For example, if the white light source has a flat spectrum over

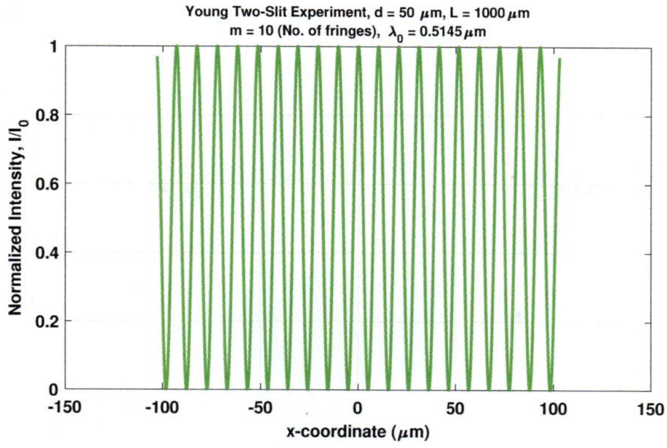

(a)

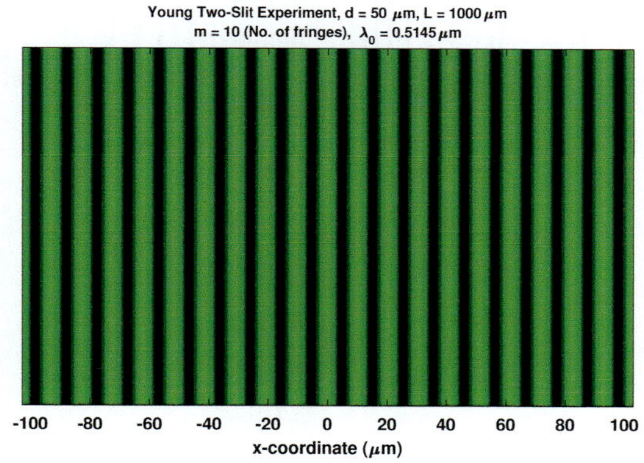

(b)

FIGURE 3.35: The interference pattern produced by two slits for monochromatic light of $\lambda_0 = 0.5145\mu$m with $d = 50\,\mu$m and $L = 1000\,\mu$m. (a) The normalized intensity pattern on the screen as a function of x. (b) Same as in (b) with the colored representation of the intensity fringes that should be observed on the screen. Again, the color of the interference fringes is based on the algorithm defined in Appendix 3F.

the range between $0.4\,\mu$m and $0.7\,\mu$m, then the resulting intensity $I(x)$ is shown in Fig. 3.36(a). his intensity is comprised of an infinite number of wavelengths and corresponding colors. In order to approximate the color fringes of the white light interference pattern, a $[R(\lambda_0), G(\lambda_0), B(\lambda_0)]$ set (tristimulus values) is assigned to each wavelength λ_0, according to the algorithm described in Appendix 3F. If $[R(x), G(x), B(x)]$ are the tristimulus values at point x on the viewing screen, they can be estimated by the equation:

$$[R(x), G(x), B(x)] \simeq \sum_{\lambda_{0,i}} \frac{I(x, \lambda_{0,i})}{I(x)} [R(\lambda_{0,i}), G(\lambda_{0,i}), B(\lambda_{0,i})], \tag{3.219}$$

where $I(x, \lambda_{0,i}) \to I(x, \omega_i)$. This approach assigns a color to each point on the screen based on the spectral distribution of the light and the interference pattern. The resulting colored fringe pattern based on the approximation described is shown in Fig. 3.36(b) (without diffraction effects) and in Fig. 3.36(c) (with diffraction effects, which are presented in the *Scalar Diffraction* chapter, Chapter 5). It is evident that at the center of the pattern, the color is white because all colors have a maximum intensity at that point, resulting in a white color. This effect was achieved by adjusting the contrast of the colors. However, since the minima for the different colors occur at different positions, this leads to the pseudo-color pattern shown. The interference fringes for each color are not perfectly aligned, causing the observed shift in color across the screen, with different wavelengths (colors) having maxima and minima at different spatial locations. This effect results in the characteristic rainbow-like pattern (here is like a pseudo-color pattern) often seen in white light interference experiments.

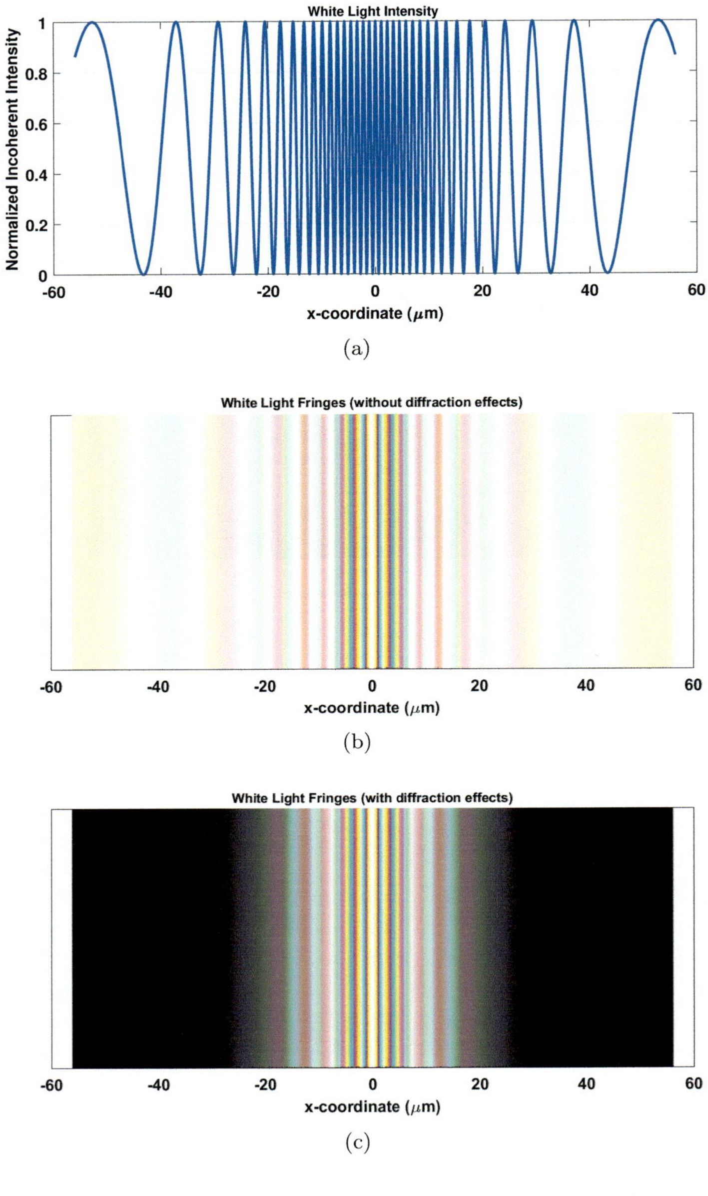

FIGURE 3.36: The interference of two slits for flat spectrum white light ($0.4\,\mu m \leq \lambda_0 \leq 0.7\,\mu m$), with slit separation $d = 5\,\mu\text{m}$ and screen distance $L = 20\,\mu\text{m}$. (a) The normalized incoherent intensity pattern on the screen as a function of x. (b) Pseudo-colored interference fringes should be observed on the screen based on the intensity of (a) (without taking diffraction effects into account). (c) Same as (b) but with diffraction effects.

3.5.4 Lloyd's Mirror

Interference fringes, similar to those produced by Young's two-slit experiment, can also appear in other optical configurations involving a single light source. This can occur via reflection, refraction, or wavefront division, where virtual sources produce interference patterns [20, 42]. Common examples of such configurations include Fresnel's double mirror, Fresnel's double prism, and Lloyd's mirror [20, 42]. In this section we focus on the Lloyd's mirror, which is shown in Fig. 3.37(a). In this setup, a point source S (or a line source for two-dimensional problems) is placed in front of a flat reflective surface (the mirror). The mirror creates a virtual image S' of the source, which acts as the second source of waves. These waves interfere with the waves emanating from the real source S. The key difference from the Young's double-slit experiment is that in Lloyd's mirror configuration, the optical path difference must be modified by adding a phase difference of π due to the external reflection experienced by the waves from the virtual source S'. This phase difference leads to a shift in the interference pattern, but the overall nature of the interference is similar to that of the two-slit experiment.

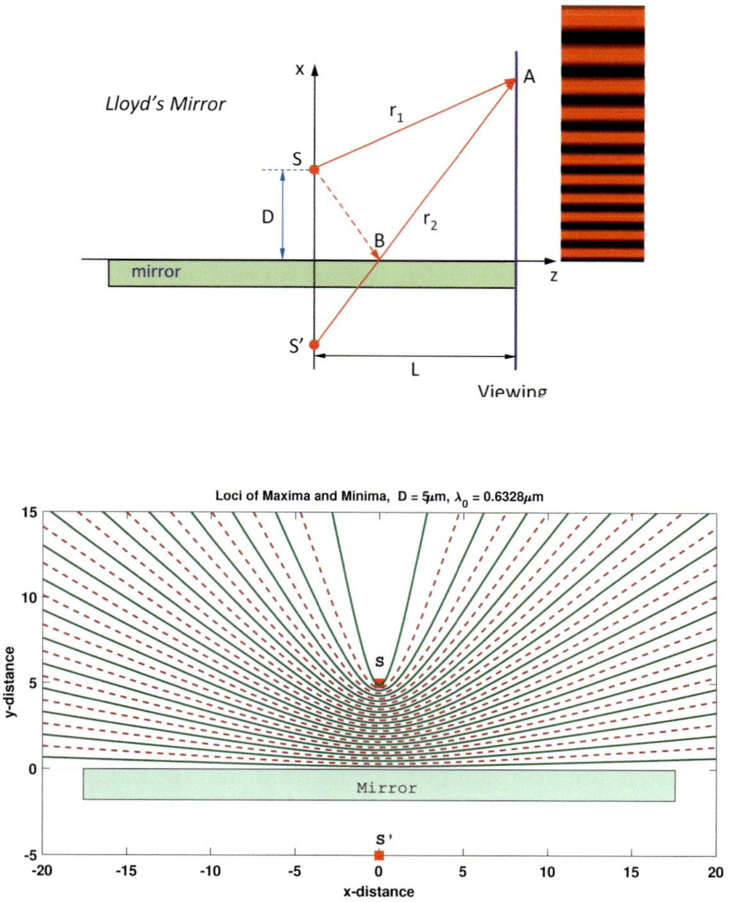

(b)

FIGURE 3.37: The Lloyd's mirror interference configuration. A point source S is shown above a mirror. (b) The loci for maximum interference (solid lines) and minimum interference (dashed lines). The red squares denote the positions of the source S and virtual source S'.

To find the maxima of the interference pattern in the Lloyd's mirror configuration, the condition for constructive interference must be satisfied. For *TE* polarization (*TE* is assumed since for *TM* polarization the reflection coefficient changes sign around the Brewster's angle), the condition for constructive interference is based on the optical path difference between the two waves—one coming directly from the point source S and the other from the virtual image S'. The interference maxima occur when:

$$k_0\left(\Delta_{path} + \Delta_{refl}\right) = 2\pi m \Longrightarrow k_0 n(r_2 - r_1) + \pi = 2\pi m,$$

where $k_0 = 2\pi/\lambda_0$, Δ_{path} is the optical path difference, and Δ_{refl} is the equivalent optical path difference due to reflection on the mirror. Then, the resulting equation for the calculation of the maxima of the interference pattern becomes:

$$k_0 n(r_2 - r_1) = (2m+1)\pi \Longrightarrow r_2 - r_1 = (2m+1)\frac{\lambda_0}{2n}.$$

Expressing the distances in the xyz coordinate system results in the following equation for the maxima:

$$\sqrt{(x+D)^2 + y^2 + z^2} - \sqrt{(x-D)^2 + y^2 + z^2} = (2m+1)\frac{\lambda_0}{2n}, \qquad m = 0, 1, \cdots .$$

Similarly, the minima can be determined by the following equation:

$$\sqrt{(x+D)^2 + y^2 + z^2} - \sqrt{(x-D)^2 + y^2 + z^2} = m\frac{\lambda_0}{n}, \qquad m = 0, 1, \cdots .$$

For example, let's assume that $D = 5\,\mu\text{m}$, $L = 20\,\mu\text{m}$, $\lambda_0 = 0.6328\,\mu\text{m}$ (He-Ne laser), and $n = 1$. The root loci of the maxima (green lines) and the minima (red lines) of this configuration are shown in Fig. 3.37(b). The corresponding intensity pattern observed on the viewing screen is given by:

$$I(x) = \frac{1}{4Z}\,4E_0^2\cos^2\left[\frac{k_0(r_2 - r_1) + \pi}{2}\right].$$

Using the parameters defined earlier the normalized intensity pattern and its fringes are shown in Fig. 3.38(a) and 3.38(b), respectively.

3.6 Interferometers

Optical interferometers are precision instruments that harness the principle of wave interference to measure a wide range of physical quantities, including distances, wavelengths, refractive indices, and surface irregularities. By analyzing the coherent superposition of light waves to produce interference patterns, these devices allow for the extraction of highly accurate information about the quantities being measured.

Interferometers are indispensable tools across various fields. In metrology, they are used for high-precision measurements of lengths, angles, surface profiles, and refractive indices. In astronomy, interferometers facilitate the measurement of stellar diameters, the study of binary star systems and exoplanets, and the detection of gravitational waves. Optical communications benefit from interferometers in applications such as optical modulation, filtering, and fiber characterization. In spectroscopy, interferometers are central to techniques

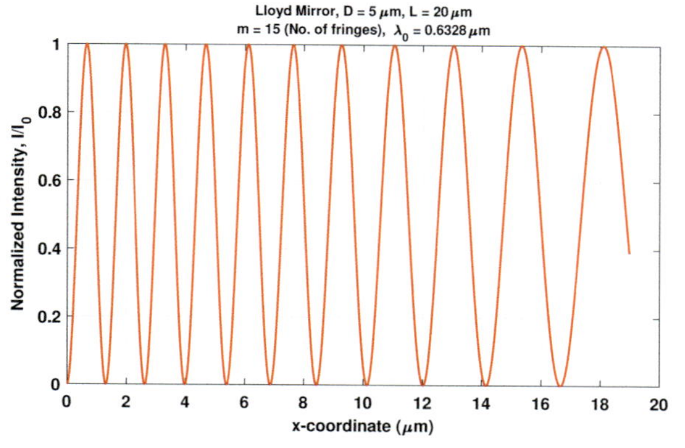

(a)

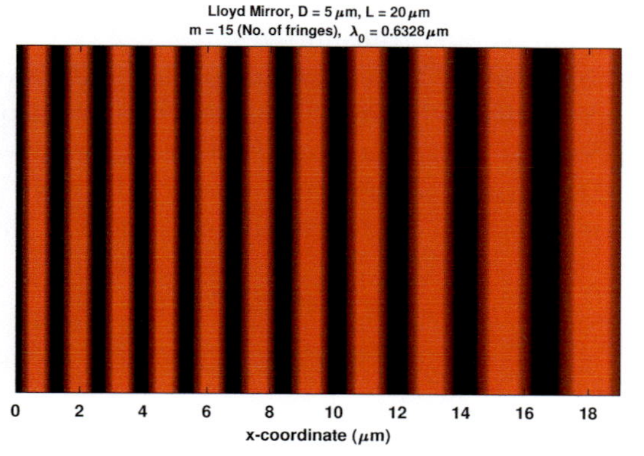

(b)

FIGURE 3.38: The Lloyd's mirror interference pattern on the viewing screen for $D = 5\,\mu\text{m}$, $L = 20\,\mu\text{m}$, $\lambda_0 = 0.6328\,\mu\text{m}$ (He-Ne laser), and $n = 1$. (a) The normalized intensity interference pattern on the screen plane. (b) The Lloyd's mirror interference pattern on the viewing screen. The color of the interference fringes is based on the algorithm defined in Appendix 3F.

like Fourier-Transform Infrared (FTIR) spectroscopy, providing high-resolution analysis of light's spectral properties for chemical and molecular studies. Material science relies on interferometers to measure strain and stress, detect cracks, and analyze structural integrity.

Biomedical applications of interferometers are significant, particularly in optical coherence tomography (OCT), which offers non-invasive imaging of tissues, and in interferometry-based microscopes that provide high-resolution analysis of biological samples. In the semiconductor and electronics industries, interferometers enable wafer inspection, lithography, and testing of microelectromechanical systems (MEMS). The aerospace and defense sectors use interferometers in gyroscopic navigation systems and for testing precision optical components. In environmental and atmospheric sciences, interferometers allow for high-resolution spectroscopy to monitor atmospheric gases, pollutants, and weather phenomena. Industrial applications include non-contact surface inspection, product quality control, and hologram analysis.

Interferometers also play a crucial role in research and development, advancing quantum optics, exploring fundamental physical principles, and enabling precision experiments. Additionally, they are used in art and cultural heritage preservation, providing non-invasive imaging techniques to study and analyze the structure and composition of historical artifacts and paintings.

This section introduces the fundamentals of the most common optical interferometers, including the Fabry-Perot etalon and Fabry-Perot interferometer, the Michelson interferometer, and the Mach-Zehnder interferometer. An introduction to Fourier-Transform Infrared (FTIR) spectroscopy is also provided.

3.6.1 Fabry-Perot Etalon

The general asymmetric Fabry-Perot etalon consists of a homogeneous dielectric layer, typically lossless, with a refractive index n_2, positioned between two additional lossless dielectric materials with refractive indices n_1 and n_3, as is shown in Fig. 3.39. While the central layer is often assumed to be lossless, it can exhibit either loss or gain, especially in the context of laser cavities. Two primary methodologies are commonly employed to analyze the Fabry-Perot etalon's behavior. The first approach is based on physical optics arguments. It involves examining each individual reflected and transmitted wave. By summing these components, the overall reflected and transmitted waves can be determined. This approach provides a straightforward, intuitive understanding of the etalon's behavior. The other approach is purely electromagnetic. It employs the fundamental principles of electromagnetics, such as boundary conditions for electric and magnetic fields, to derive the etalon's behavior. It offers a rigorous, mathematically precise description of the wave interactions within and across the dielectric interfaces. Both approaches are presented in the next sections.

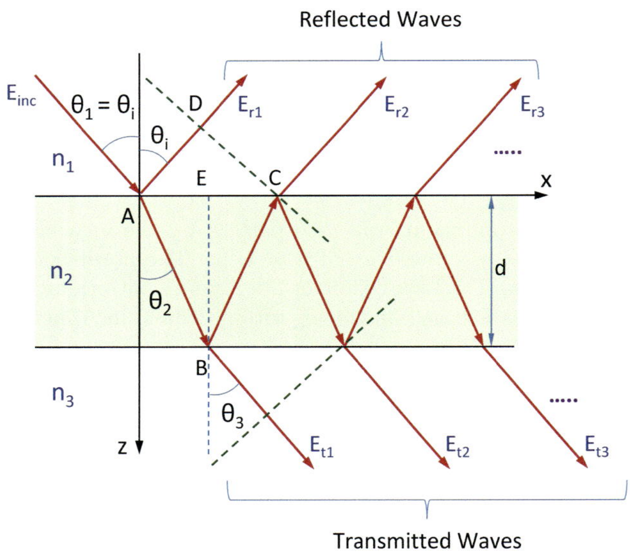

FIGURE 3.39: A general asymmetric Fabry-Perot interferometer comprises a thin dielectric plate with refractive index n_2 and thickness d, positioned between two homogeneous dielectric media with refractive indices n_1 and n_3. A plane wave incident on the structure from the top strikes the upper interface at an angle $\theta_1 = \theta_i$. The incident wave is assumed to be either *TE* or *TM* polarized. The diagram shows examples of the reflected and transmitted waves resulting from multiple reflections within the thin plate. These waves undergo constructive or destructive interference depending on their relative phase differences. The overall reflected field is the result of the infinite series of partial reflections at the two interfaces. The overall transmitted field is similarly composed of an infinite series of partial transmissions.

3.6.1.1 Physical Optics Analysis

Based on Fig. 3.39, the phase difference between E_{rm} and $E_{r(m+1)}$ $(m = 1, 2, \cdots)$ can be determined under the assumption that all layers are lossless. For example, the phase difference between E_{r1} and E_{r2} can be calculated by evaluating the difference in their optical path lengths. Specifically, the E_{r1} wave propagates from point A to point D, accumulating a phase of $k_0 n_1(AD)$. Simultaneously, the E_{r2} wave propagates from point A to point C, accumulating a phase of $k_0 n_2[(AB) + (BC)]$. Denoting δ as the phase difference between E_{r1} and E_{r2}, the following equations can be derived:

$$\delta = k_0 n_2[(AB) + (BC)] - k_0 n_1(AD), \qquad \text{where,}$$

$$(AB) = (BC) = \frac{d}{\cos\theta_2},$$

$$(AD) = (AC)\sin\theta_1 = 2(AE)\sin\theta_1 = 2d\tan\theta_2\sin\theta_1, \qquad \text{resulting in,}$$

$$\delta = 2k_0 n_2 d\cos\theta_2. \tag{3.220}$$

Snell's law was applied in the derivation of the last equation. The phase difference between E_{r1} and E_{r2} is consistent for any two consecutive reflected waves, E_{rm} and $E_{r(m+1)}$. Similarly, this same phase shift applies to any two consecutive transmitted waves, E_{tm} and $E_{t(m+1)}$. Building on this principle, the following relations can be readily expressed for the reflected waves:

$$\begin{aligned}
E_{r1} &= r_{12}E_i, \\
E_{r2} &= t_{12}t_{21}r_{23}e^{-j\delta}E_i, \\
E_{r3} &= t_{12}t_{21}r_{21}r_{23}^2 e^{-j2\delta}E_i, \\
\vdots &= \quad \vdots \\
E_{rN} &= t_{12}t_{21}r_{21}^{N-2}r_{23}^{N-1}e^{-j(N-1)\delta}E_i, \\
\vdots &= \quad \vdots
\end{aligned}$$

By summing up all the reflected waves, the resulting overall reflected wave is formed:

$$E_r = \sum_{m=1}^{\infty} E_{rm} = r_{12}E_i + t_{12}t_{21}r_{23}e^{-j\delta}E_i\left[1 + r_{21}r_{23}e^{-j\delta} + r_{21}^2 r_{23}^2 e^{-j2\delta} + \cdots\right] =$$

$$= r_{12}E_i + t_{12}t_{21}r_{23}e^{-j\delta}E_i\frac{1}{1 - r_{21}r_{23}e^{-j\delta}} \Longrightarrow$$

$$\frac{E_r}{E_i} = \frac{r_{12}(1 - r_{21}r_{23}e^{-j\delta}) + t_{12}t_{21}r_{32}e^{-j\delta}}{1 - r_{21}r_{23}e^{-j\delta}}. \tag{3.221}$$

Using a similar approach, the equivalent equation for the overall transmitted wave can be readily derived:

$$E_t = \sum_{m=1}^{\infty} E_{tm} = t_{12}t_{23}E_i e^{-jk_0 n_2 d/\cos\theta_2}\left[1 + r_{21}r_{23}e^{-j\delta} + r_{21}^2 r_{23}^2 e^{-j2\delta} + \cdots\right] \Longrightarrow$$

$$\frac{E_t}{E_i} = e^{-jk_0 n_2 d/\cos\theta_2}\frac{t_{12}t_{23}}{1 - r_{21}r_{23}e^{-j\delta}}. \tag{3.222}$$

In the above derivation, r_{ij} and t_{ij} represent the reflection and transmission coefficients for light transitioning from layer i to layer j. At this stage, it is customary to compute the intensities. To ensure that the sum of the reflected and transmitted intensities equals the incident intensity, the Poynting theorem must be applied. Accordingly, the intensities are defined based on the Poynting vector. The following expressions provide the relevant intensities, where $r_{pq} = |r_{pq}|e^{j\phi_{pq}}$ with $p, q = 1, 2, 3$:

$$\frac{I_t}{I_i} = \frac{P_t}{P_i} = \frac{n_3 \cos\theta_3}{n_1 \cos\theta_1} \frac{|E_t|^2}{|E_i|^2} = \frac{n_3 \cos\theta_3}{n_1 \cos\theta_1} \frac{|t_{12}|^2 |t_{23}|^2}{1 + |r_{21}|^2 |r_{23}|^2 - 2|r_{21}||r_{23}| \cos\delta'}, \quad \text{where}$$
$$\delta' = \delta - \phi_{12} - \phi_{32}. \tag{3.223}$$

Then, using Eqs. (3.118) or (3.119), the following relations for the transmission coefficients hold:

$$|t_{12}|^2 = \frac{n_1 \cos\theta_1}{n_2 \cos\theta_2} \left(1 - |r_{12}|^2\right),$$

$$|t_{23}|^2 = \frac{n_2 \cos\theta_2}{n_3 \cos\theta_3} \left(1 - |r_{23}|^2\right),$$

Furthermore, the Stokes relations [109] can be used to simplify the expression. The Stokes relations are presented in Appendix 3E. Specifically, the relations $|r_{ij}|^2 = |r_{ji}|^2$ hold. Incorporating this information, the expression for I_t/I_i can be written as:

$$\frac{I_t}{I_i} = \frac{(1 - |r_{21}|^2)(1 - |r_{23}|^2)}{1 + |r_{21}|^2 |r_{23}|^2 - 2|r_{21}||r_{23}| \cos\delta'} =$$
$$\frac{(1 - |r_{21}|^2)(1 - |r_{23}|^2)}{(1 - |r_{21}||r_{23}|)^2 + 4|r_{21}||r_{23}| \sin^2(\delta'/2)} = \frac{(1 - R_1)(1 - R_2)}{(1 - \sqrt{R_1 R_2})^2 + 4\sqrt{R_1 R_2} \sin^2(\delta'/2)}. \tag{3.224}$$

In the latter equation, the expressions $R_1 = |r_{21}|^2 = |r_{12}|^2$ and $R_2 = |r_{23}|^2 = |r_{32}|^2$ represent the power reflectivities.

The expression of normalized reflection intensity, I_r/I_i, can also be obtained from Eq. (3.221). Specifically, it is given by:

$$\frac{I_r}{I_i} = \frac{P_r}{P_i} = \frac{|E_r|^2}{|E_i|^2} = \left| \frac{r_{12}(1 - r_{21} r_{23} e^{-j\delta}) + t_{12} t_{21} r_{32} e^{-j\delta}}{1 - r_{21} r_{23} e^{-j\delta}} \right|^2. \tag{3.225}$$

The above expression has not been expanded, as it becomes rather cumbersome. However, in lossless Fabry-Perot etalons, the sum of the normalized intensities, $I_r/I_i + I_t/I_i = 1$, ensures power conservation. Therefore, for the lossless case, the expression for the reflected normalized intensity can be more easily found as follows:

$$\frac{I_r}{I_i} = \frac{P_r}{P_i} = 1 - \frac{I_t}{I_i} = \frac{(|r_{21}| - |r_{23}|)^2 + 4|r_{21}||r_{23}| \sin^2(\delta'/2)}{(1 - |r_{21}||r_{23}|)^2 + 4|r_{21}||r_{23}| \sin^2(\delta'/2)}$$
$$= \frac{(\sqrt{R_1} - \sqrt{R_2})^2 + 4\sqrt{R_1 R_2} \sin^2(\delta'/2)}{(1 - \sqrt{R_1 R_2})^2 + 4\sqrt{R_1 R_2} \sin^2(\delta'/2)}. \tag{3.226}$$

In the previous derivation the reflection coefficients r_{pq} and transmission coefficients t_{pq} were assumed to be complex, reflecting the potential use of multilayered stacks to achieve the enhanced reflective properties of a Fabry-Perot etalon. In such cases, particular care must be taken when applying the Stokes relations to simplify the expressions for E_r/E_i and

E_t/E_i. The maxima and minima of the reflected and transmitted intensities can be readily determined from Eqs. (3.226) and (3.224), and they are summarized as follows:

$$\left(\frac{I_t}{I_i}\right)_{max} = \frac{(1-R_1)(1-R_2)}{(1-\sqrt{R_1R_2})^2}, \qquad \text{for } \delta' = \delta - \phi_{21} - \phi_{23} = 2m\pi,$$

$$\left(\frac{I_t}{I_i}\right)_{min} = \frac{(1-R_1)(1-R_2)}{(1+\sqrt{R_1R_2})^2}, \qquad \text{for } \delta' = \delta - \phi_{21} - \phi_{23} = (2m+1)\pi,$$

$$\left(\frac{I_r}{I_i}\right)_{max} = \frac{(\sqrt{R_1}+\sqrt{R_2})^2}{(1+\sqrt{R_1R_2})^2}, \qquad \text{for } \delta' = \delta - \phi_{21} - \phi_{23} = (2m+1)\pi,$$

$$\left(\frac{I_r}{I_i}\right)_{min} = \frac{(\sqrt{R_1}-\sqrt{R_2})^2}{(1-\sqrt{R_1R_2})^2}, \qquad \text{for } \delta' = \delta - \phi_{21} - \phi_{23} = 2m\pi.$$

Typically, the symmetric Fabry-Perot etalon is employed. In the symmetric configuration, the refractive indices of the regions surrounding the thin layer are identical, i.e., $n_1 = n_3 = n_0$. Consequently, $r_{21} = r_{23} = r$ (with $\phi_{12} = \phi_{32} = \phi$), and $R_1 = R_2 = R = r^2$. Under these conditions, the normalized reflected and transmitted intensities are expressed as:

$$\frac{I_r}{I_i} = \frac{4r^2\sin^2(\delta'/2)}{(1-r^2)^2 + 4r^2\sin^2(\delta'/2)} = \frac{4R\sin^2(\delta'/2)}{(1-R)^2 + 4R\sin^2(\delta'/2)}, \tag{3.227}$$

$$\frac{I_t}{I_i} = \frac{(1-r^2)^2}{(1-r^2)^2 + 4r^2\sin^2(\delta'/2)} = \frac{(1-R)^2}{(1-R)^2 + 4R\sin^2(\delta'/2)}. \tag{3.228}$$

In many textbooks, for simplicity, the reflection coefficient r is assumed to be real, leading to $\phi = 0$ and $\delta' = \delta$. However, in this section, the reflection coefficient, even for the symmetric Fabry-Perot etalon, is retained as complex, with the phase ϕ explicitly included. The corresponding maxima and minima of the reflected and transmitted intensities for the symmetric Fabry-Perot etalon are given by:

$$\left(\frac{I_t}{I_i}\right)_{max} = 1, \qquad \text{for } \delta' = \delta - 2\phi = 2m\pi,$$

$$\left(\frac{I_t}{I_i}\right)_{min} = \frac{(1-R)^2}{(1+R)^2}, \qquad \text{for } \delta' = \delta - 2\phi = (2m+1)\pi,$$

$$\left(\frac{I_r}{I_i}\right)_{max} = \frac{(4R^2}{(1+R)^2}, \qquad \text{for } \delta' = \delta - 2\phi = (2m+1)\pi,$$

$$\left(\frac{I_r}{I_i}\right)_{min} = 0, \qquad \text{for } \delta' = \delta - 2\phi = 2m\pi.$$

Examples of Fabry-Perot etalons are given in Figs. 3.40(a) (lossless asymmetric) and 3.40(b) (lossless symmetric). The reflected and transmitted intensities are plotted as functions of the phase angle δ'. It can be observed that, for a lossless symmetric Fabry-Perot etalon, the transmission peaks become sharper and narrower as the power reflectivities $R_1 = R_2 = R$ increase. In practice, this is typically achieved by using either thin metallic layers with high reflectivity or multilayer dielectric stacks designed to exhibit very high reflectivity at a specific wavelength or over a defined wavelength range. This is the primary reason why the complex nature of the reflection and transmission coefficients was preserved in the analysis.

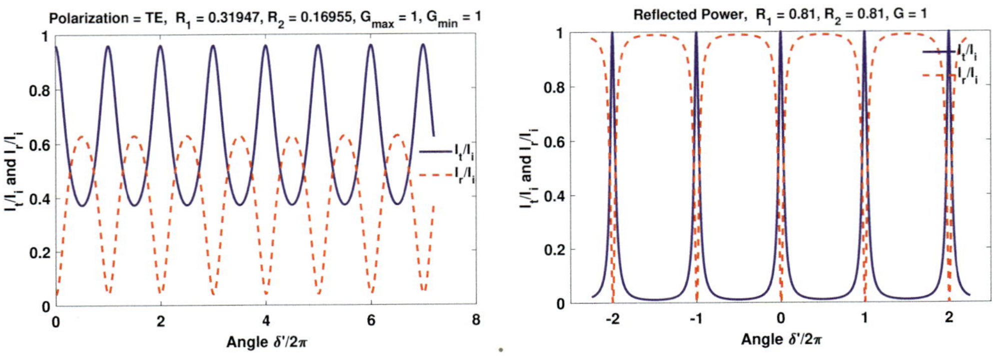

(a) (b)

FIGURE 3.40: (a) The normalized reflected and transmitted intensities for an asymmetric Fabry-Perot etalon with $n_1 = 1.0$, $n_2 = 3.6$ (silicon), $n_3 = 1.5$ (glass) $\theta_1 = 0°$, $\lambda_0 = 1\,\mu$m, and $d = 0$ to $1\,\mu$m, as a function of the phase angle δ'. In this case, $\delta' = \delta$ since the reflection and transmission coefficients are real. (b) The normalized reflected and transmitted intensities for a symmetric Fabry-Perot etalon with $r_{21} = r_{23} = r = 0.9$ ($R_1 = R_2 = R = 0.81$) as a function of the phase angle δ'.

To generalize the above approach for a middle layer with a complex refractive index, a refractive index of the form $\tilde{n}_2 = n_2 - j\kappa_2$ is considered. This representation accounts for a layer with losses when $\kappa_2 > 0$, or a layer with gain when $\kappa_2 < 0$ [consistent with the convention that a wave propagating in the positive z direction is described by $e^{j(\omega t - kz)}$]. In this case, the phase factor of waves propagating through the middle layer includes an amplitude term. This term is less than unity for losses and greater than unity for gain. Specifically, the phase factor becomes:

$$\begin{aligned}\exp[-jk_0\tilde{n}_2(z\cos\theta_2 + x\sin\theta_2)] &= \exp(-jk_0\tilde{n}_2 z\cos\theta_2)\exp(-jk_0\tilde{n}_2 x\sin\theta_2)\\ &= \underbrace{\exp(k_0\mathrm{Im}\{\tilde{n}_2\cos\theta_2\}\,z)}_{=\,g(z)}\exp(-jk_0\mathrm{Re}\{\tilde{n}_2\cos\theta_2\}\,z)\\ &\quad\exp(-jk_0 n_1 x\sin\theta_1).\end{aligned}$$

Following the same procedure as in the case of a lossless middle layer, the total reflected and transmitted fields can be determined by summing the individual contributions from E_{rN} and E_{tN} ($N = 1, 2, \cdots$). For a complex refractive index $\tilde{n}_2$, it is straightforward to verify that these terms are given by:

$$\begin{aligned}E_{rN} &= E_i t_{12} t_{21} r_{21}^{N-2} r_{23}^{N-1} g^{2(N-1)} e^{-j(N-1)\delta},\\ E_{tN} &= E_i e^{-jk_0 d\mathrm{Re}\{\tilde{n}_2/\cos\theta_2\}} t_{12} t_{23} (r_{21} r_{23})^{N-1} g^{2N-1} e^{-j(N-1)\delta}, \quad \text{where,}\\ \delta &= 2k_0 d\mathrm{Re}\{\tilde{n}_2\cos\theta_2\},\\ g &= g(z=d) = \exp\left(k_0\mathrm{Im}\{\tilde{n}_2\cos\theta_2\}\,d\right).\end{aligned}$$

By summing the contributions of E_{rN} and E_{tN}, the total reflected and transmitted fields for the case of a Fabry-Perot etalon with a middle layer of complex refractive index are obtained as follows:

$$\frac{E_r}{E_i} = \frac{r_{12}(1 - g^2 r_{21} r_{23} e^{-j\delta}) + g^2 t_{12} t_{21} r_{32} e^{-j\delta}}{1 - g^2 r_{21} r_{23} e^{-j\delta}}, \tag{3.229}$$

$$\frac{E_t}{E_i} = e^{-jk_0 d \mathrm{Re}\{\tilde{n}_2/\cos\theta_2\}} \frac{g t_{12} t_{23}}{1 - g^2 r_{21} r_{23} e^{-j\delta}}. \tag{3.230}$$

At this point, it is essential to note that all reflection and transmission coefficients are generally complex due to the gain or loss in the middle layer of thickness d. Nevertheless, the Stokes relations, as outlined in Appendix 3E, remain valid. Care must be exercised in computing the normalized reflected and transmitted efficiencies to account for the effects of gain or loss. The normalized reflected and transmitted intensities are given by:

$$\frac{I_r}{I_i} = \frac{\left|r_{12}(1 - g^2 r_{21} r_{23} e^{-j\delta}) + g^2 t_{12} t_{21} r_{32} e^{-j\delta}\right|^2}{\left(1 - G|r_{21}|\,|r_{23}|\right)^2 + 4G|r_{21}|\,|r_{23}|\sin^2(\delta'/2)}, \tag{3.231}$$

$$\begin{aligned}\frac{I_t}{I_i} &= \frac{n_3\cos\theta_3}{n_1\cos\theta_1}\frac{g^2|t_{12}|^2|t_{23}|^2}{\left(1 - G|r_{21}|\,|r_{23}|\right)^2 + 4G|r_{21}|\,|r_{23}|\sin^2(\delta'/2)} = \\ &= \frac{g^2\left(1 - |r_{21}|^2\right)\left(1 - |r_{23}|^2\right)}{\left(1 - G|r_{21}|\,|r_{23}|\right)^2 + 4G|r_{21}|\,|r_{23}|\sin^2(\delta'/2)} = \\ &= \frac{G(1 - R_1)(1 - R_2)}{\left(1 - G\sqrt{R_1 R_2}\right)^2 + 4G\sqrt{R_1 R_2}\sin^2(\delta'/2)}.\end{aligned} \tag{3.232}$$

The angle $\delta' = \delta - \phi_{21} - \phi_{23}$ as before. The reflection and transmission coefficients r_{21}, r_{23}, t_{12}, t_{21}, and t_{23} can be determined using the Fresnel equations [Eqs. (3.78)–(3.81)]. When there is no loss or gain (i.e., $g = 1$ and $\mathrm{Re}\{\tilde{n}_2\cos\theta_2\} = n_2\cos\theta_2$), the resulting expressions for the normalized reflected and transmitted intensities reduce to those found in Eq. (3.226) and Eq. (3.224), respectively. These equations describe the behavior of lossless middle layers. However, in the case of gain (i.e., when $\kappa_2 < 0$), the situation changes. There is a maximum allowable gain beyond which the system becomes unstable. Specifically, the denominator of the normalized reflected and transmitted intensities in Eqs. (3.231) and (3.232) can become zero when $g^2_{max} = 1/(|r_{21}||r_{23}|) \Longrightarrow G_{max} = 1/\sqrt{R_1 R_2}$, where $G = g^2$. This condition indicates a threshold gain, $G_{th} = G_{max}$, at which self-oscillation can occur, leading to the possibility of a Fabry-Perot etalon becoming a laser cavity. This threshold gain is crucial in laser physics and defines the onset of lasing in Fabry-Perot laser cavities [39].

Examples of asymmetric Fabry-Perot etalon responses with loss or gain are shown in Figs. 3.41(a) (medium with loss) and 3.41(b) (medium with gain as a laser/optical amplifier cavity). These figures illustrate the impact of a complex refractive index (due to loss or gain) in the middle layer of the etalon on the reflection and transmission properties. In the case of a medium with loss (Fig. 3.41(a)), the transmitted intensity typically decreases, and the reflection is affected by the specific properties of the loss (i.e., the imaginary part of the refractive index). Conversely, in a medium with gain (Fig. 3.41(b)), the transmission peaks become sharper, and there is an increase in transmitted intensity, which is characteristic of the amplification of the optical signal within the etalon.

3.6.1.2 Electromagnetic Optics Analysis

In this section, the electromagnetic analysis of the Fabry-Perot etalon is presented. The appropriate configuration of the Fabry-Perot etalon is shown in Fig. 3.42. The incident

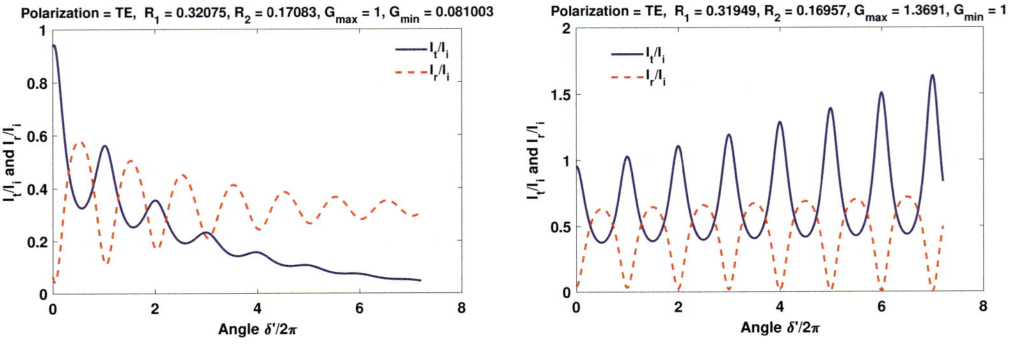

(a) (b)

FIGURE 3.41: (a) Normalized reflected and transmitted intensities for an asymmetric Fabry-Perot etalon with loss and parameters: $n_1 = 1.0$, $n_2 = 3.6 - j0.2$, $n_3 = 1.5$ $\theta_1 = 0°$, $\lambda_0 = 1\,\mu$m, and d varying from 0 to 1 μm. The normalized reflected and transmitted intensities are calculated as functions of the phase angle δ'. It is evident that the transmitted intensity decreases as the thickness of the lossy layer increases while the reflected intensity converges to a constant value. (b) Normalized reflected and transmitted intensities for an asymmetric Fabry-Perot etalon with gain and parameters: $n_1 = 1.0$, $n_2 = 3.6 + j0.025$, $n_3 = 1.5$ $\theta_1 = 0°$, $\lambda_0 = 1\,\mu$m, and $d = 0$ to $1\,\mu$m. The normalized reflected and transmitted intensities are calculated as functions of the phase angle δ'. It is evident that the transmitted and the reflected intensities increase as the thickness of the active layer increases.

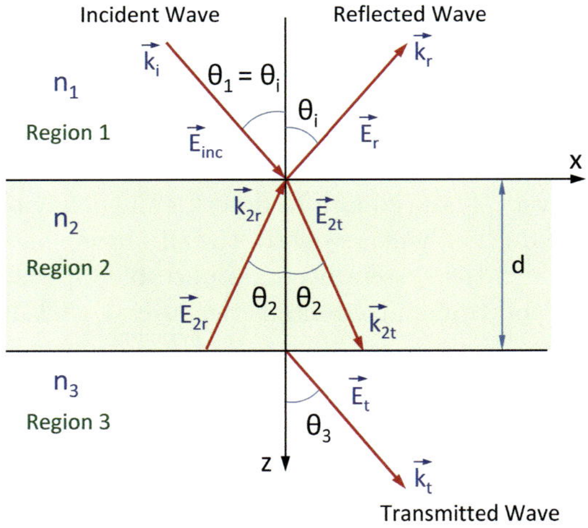

FIGURE 3.42: A general asymmetric Fabry-Perot interferometer consists of a thin plate with refractive index n_2 and thickness d, sandwiched between two homogeneous materials with refractive indices n_1 and n_3. The incident plane wave, coming from the top, strikes at an angle $\theta_1 = \theta_i$ and is assumed to be either *TE* or *TM* polarized. The resulting reflected and transmitted waves are shown, with these waves being the result of an infinite summation due to multiple interference effects, as illustrated in Fig. 3.39. This configuration is suitable for the electromagnetic analysis of the Fabry-Perot etalon.

plane wave is described by its electric or magnetic field component (the electric field is shown, for example, in Fig. 3.42). Similarly, the reflected, transmitted, as well as forward (positive z direction propagating) and backward (negative z direction propagating) waves in region 2, are all represented by their corresponding electric field vectors.

Since all the materials are homogeneous, isotropic, and linear, the polarization of the incident wave determines the polarization of all the subsequent waves. Therefore, if the incident wave is either *TE* or *TM* polarized, the same polarization will apply to all other excited waves. For simplicity, let's assume initially that the incident wave is *TE*-polarized, with the polarization unit vector $\hat{u}_{TE} = \hat{\imath}_y$. Then, the electric fields in each of the regions 1, 2, and 3 (see Fig. 3.42), $\vec{E}_1$, $\vec{E}_2$, and $\vec{E}_3$, respectively, can be expressed as follows:

$$\vec{E}_1 = \vec{E}_i + \vec{E}_r = \hat{\imath}_y \left[E_i e^{-j\vec{k}_i \cdot \vec{r}} + E_r e^{-j\vec{k}_r \cdot \vec{r}} \right], \qquad \text{where}$$

$$\vec{k}_i = k_0 n_1 (\hat{\imath}_x \sin\theta_1 + \hat{\imath}_z \cos\theta_1), \qquad \vec{k}_r = k_0 n_1 (\hat{\imath}_x \sin\theta_1 - \hat{\imath}_z \cos\theta_1),$$

$$\vec{E}_2 = \vec{E}_{2t} + \vec{E}_{2r} = \hat{\imath}_y \left[E_{2t} e^{-j\vec{k}_{2t} \cdot \vec{r}} + E_{2r} e^{-j\vec{k}_{2r} \cdot \vec{r}} \right], \qquad \text{where}$$

$$\vec{k}_{2t} = k_0 n_2 (\hat{\imath}_x \sin\theta_2 + \hat{\imath}_z \cos\theta_2), \qquad \vec{k}_{2r} = k_0 n_2 (\hat{\imath}_x \sin\theta_2 - \hat{\imath}_z \cos\theta_2),$$

$$\vec{E}_3 = \vec{E}_t = \hat{\imath}_y E_t e^{-j\vec{k}_t \cdot (\vec{r} - d\hat{\imath}_z)}, \qquad \text{where}$$

$$\vec{k}_t = k_0 n_3 (\hat{\imath}_x \sin\theta_3 + \hat{\imath}_z \cos\theta_3).$$

The corresponding magnetic fields can be easily determined using the Maxwell's equations ($\vec{\nabla} \times \vec{E} = -j\omega\mu_0 \vec{H}$, where non-magnetic media are assumed). The corresponding tangential to the dielectric boundaries magnetic field components are then given by:

$$H_{1x} = \frac{1}{Z_1} \left[-E_i \cos\theta_1 e^{-jk_0 n_1 (x \sin\theta_1 + z \cos\theta_1)} + E_r \cos\theta_1 e^{-jk_0 n_1 (x \sin\theta_1 - z \cos\theta_1)} \right],$$

$$H_{2x} = \frac{1}{Z_2} \left[-E_{2t} \cos\theta_2 e^{-jk_0 n_2 (x \sin\theta_2 + z \cos\theta_2)} + E_{2r} \cos\theta_2 e^{-jk_0 n_1 (x \sin\theta_2 - z \cos\theta_2)} \right],$$

$$H_{3x} = -\frac{1}{Z_3} E_t \cos\theta_3 e^{-jk_0 n_3 (x \sin\theta_3 + (z-d) \cos\theta_3)}.$$

In the magnetic field expressions, $Z_i = Z_0 / n_i$ $(i = 1, 2, 3)$, and $Z_0 = \sqrt{\mu_0 / \epsilon_0} = 376.73\,\Omega$ is the wave impedance of freespace. By applying the boundary conditions at $z = 0$ and $z = d$, which require the continuity of the tangential electric and magnetic field components, the following equations can be derived:

$$\begin{bmatrix} 1 & 1 \\ -n_1 \cos\theta_1 & n_1 \cos\theta_1 \end{bmatrix} \begin{bmatrix} E_i \\ E_r \end{bmatrix} = \begin{bmatrix} 1 & 1 \\ -n_2 \cos\theta_2 & n_2 \cos\theta_2 \end{bmatrix} \begin{bmatrix} E_{2t} \\ E_{2r} \end{bmatrix},$$

$$\begin{bmatrix} e^{-j\beta_2 d} & e^{+j\beta_2 d} \\ -n_2 \cos\theta_2 e^{-j\beta_2 d} & n_2 \cos\theta_2 e^{+j\beta_2 d} \end{bmatrix} \begin{bmatrix} E_{2t} \\ E_{2r} \end{bmatrix} = \begin{bmatrix} 1 & 1 \\ -n_3 \cos\theta_3 & n_3 \cos\theta_3 \end{bmatrix} \begin{bmatrix} E_t \\ 0 \end{bmatrix},$$

where, $\beta_2 = k_0 n_2 \cos\theta_2$. The last expressions can be rewritten as:

$$\begin{bmatrix} E_{2t} \\ E_{2r} \end{bmatrix} = \underbrace{\begin{bmatrix} \frac{1}{2}\left(1 + \frac{n_1 \cos\theta_1}{n_2 \cos\theta_2}\right) & \frac{1}{2}\left(1 - \frac{n_1 \cos\theta_1}{n_2 \cos\theta_2}\right) \\ \frac{1}{2}\left(1 - \frac{n_1 \cos\theta_1}{n_2 \cos\theta_2}\right) & \frac{1}{2}\left(1 + \frac{n_1 \cos\theta_1}{n_2 \cos\theta_2}\right) \end{bmatrix}}_{= \tilde{M}_{2,1}} \begin{bmatrix} E_i \\ E_r \end{bmatrix},$$

$$\begin{bmatrix} E_t \\ 0 \end{bmatrix} = \underbrace{\begin{bmatrix} \frac{1}{2}\left(1+\frac{n_2\cos\theta_2}{n_3\cos\theta_3}\right)e^{-j\beta_2 d} & \frac{1}{2}\left(1-\frac{n_2\cos\theta_2}{n_3\cos\theta_3}\right)e^{+j\beta_2 d} \\ \frac{1}{2}\left(1-\frac{n_2\cos\theta_2}{n_3\cos\theta_3}\right)e^{-j\beta_2 d} & \frac{1}{2}\left(1+\frac{n_2\cos\theta_2}{n_3\cos\theta_3}\right)e^{+j\beta_2 d} \end{bmatrix}}_{=\tilde{M}_{3,2}} \begin{bmatrix} E_{2t} \\ E_{2r} \end{bmatrix},$$

$$\begin{bmatrix} E_t \\ 0 \end{bmatrix} = \tilde{M}_{3,2}\tilde{M}_{2,1}\begin{bmatrix} E_i \\ E_r \end{bmatrix} = \begin{bmatrix} m_{11} & m_{12} \\ m_{21} & m_{22} \end{bmatrix}\begin{bmatrix} E_i \\ E_r \end{bmatrix} \Longrightarrow$$

$$\frac{E_r}{E_i} = r = -\frac{m_{21}}{m_{22}}, \tag{3.233}$$

$$\frac{E_t}{E_i} = t = m_{11} - \frac{m_{12}m_{21}}{m_{22}}, \qquad \text{where,} \tag{3.234}$$

$$K_{12} = \frac{n_1\cos\theta_1}{n_2\cos\theta_2}, \qquad K_{23} = \frac{n_2\cos\theta_2}{n_3\cos\theta_3},$$

$$m_{11} = \frac{1}{4}\left[(1+K_{12})(1+K_{23})e^{-j\beta_2 d} + (1-K_{12})(1-K_{23})e^{+j\beta_2 d}\right],$$

$$m_{12} = \frac{1}{4}\left[(1-K_{12})(1+K_{23})e^{-j\beta_2 d} + (1+K_{12})(1-K_{23})e^{+j\beta_2 d}\right],$$

$$m_{21} = \frac{1}{4}\left[(1+K_{12})(1-K_{23})e^{-j\beta_2 d} + (1-K_{12})(1+K_{23})e^{+j\beta_2 d}\right],$$

$$m_{22} = \frac{1}{4}\left[(1-K_{12})(1-K_{23})e^{-j\beta_2 d} + (1+K_{12})(1+K_{23})e^{+j\beta_2 d}\right].$$

Then, the normalized reflected and transmitted intensities are given by:

$$\frac{I_r}{I_i} = |r|^2 = \frac{|m_{21}|^2}{|m_{22}|^2}, \tag{3.235}$$

$$\frac{I_t}{I_i} = \frac{n_3\cos\theta_3}{n_1\cos\theta_1}|t|^2 = \frac{n_3\cos\theta_3}{n_1\cos\theta_1}\left[|m_{11}|^2 + \frac{|m_{12}|^2|m_{21}|^2}{|m_{22}|^2} - 2\mathrm{Re}\left\{\frac{m_{11}^* m_{12}m_{21}}{m_{22}}\right\}\right]. \tag{3.236}$$

The above results are valid for the *TE* polarization case. The *TM* polarization case can be treated in a similar manner. In this case, the roles of the electric and magnetic fields are reversed. Specifically, the electric fields are replaced by the magnetic fields, and the magnetic fields are replaced by the electric fields. This leads to the following equations:

$$\frac{H_r}{H_i} = r_H = -\frac{m'_{21}}{m'_{22}},$$

$$\frac{H_t}{H_i} = t_H = m'_{11} - \frac{m'_{12}m'_{21}}{m'_{22}}, \qquad \text{where,}$$

$$K'_{12} = \frac{n_2\cos\theta_1}{n_1\cos\theta_2}, \qquad K'_{23} = \frac{n_3\cos\theta_2}{n_2\cos\theta_3},$$

$$m'_{11} = \frac{1}{4}\left[(1+K'_{12})(1+K'_{23})e^{-j\beta_2 d} + (1-K'_{12})(1-K'_{23})e^{+j\beta_2 d}\right],$$

$$m'_{12} = \frac{1}{4}\left[(1-K'_{12})(1+K'_{23})e^{-j\beta_2 d} + (1+K'_{12})(1-K'_{23})e^{+j\beta_2 d}\right],$$

$$m'_{21} = \frac{1}{4}\left[(1 + K'_{12})(1 - K'_{23})e^{-j\beta_2 d} + (1 - K'_{12})(1 + K'_{23})e^{+j\beta_2 d}\right],$$

$$m'_{22} = \frac{1}{4}\left[(1 - K'_{12})(1 - K'_{23})e^{-j\beta_2 d} + (1 + K'_{12})(1 + K'_{23})e^{+j\beta_2 d}\right].$$

Note that in the above equations, the reflection and transmission coefficients are defined in terms of the magnetic field amplitudes. It is straightforward to redefine the reflection and transmission coefficients in terms of the electric fields as follows:

$$r' = \frac{E_r}{E_i} = \frac{-Z_1 H_r}{Z_1 H_i} = -r_H = \frac{m'_{21}}{m'_{22}}, \tag{3.237}$$

$$t' = \frac{E_t}{E_i} = \frac{Z_3 H_t}{Z_1 H_i} = \frac{n_1}{n_3} t_H = \frac{n_1}{n_3}\left[m'_{11} - \frac{m'_{12} m'_{21}}{m'_{22}}\right]. \tag{3.238}$$

Then, the resulting normalized reflected and transmitted intensities for *TM* polarization are given by:

$$\frac{I_r}{I_i} = |r'|^2 = \frac{|m'_{21}|^2}{|m'_{22}|^2}, \tag{3.239}$$

$$\frac{I_t}{I_i} = \frac{n_3 \cos\theta_3}{n_1 \cos\theta_1} |t'|^2 = \frac{n_1 \cos\theta_3}{n_3 \cos\theta_1}\left[|m'_{11}|^2 + \frac{|m'_{12}|^2 |m'_{21}|^2}{|m'_{22}|^2} - 2\mathrm{Re}\left\{\frac{m'^*_{11} m'_{12} m'_{21}}{m'_{22}}\right\}\right]. \tag{3.240}$$

The electromagnetic approach can also be applied to situations involving gain or loss in the middle layer. In such cases, the refractive index becomes complex, as discussed in Sec. 3.6.1.1, along with $\cos\theta_2$. All preceding equations remain valid even for complex n_2 and $\cos\theta_2$. The examples shown in Figs. 3.40(a) and 3.40(b), as well as in Figs. 3.41(a) and 3.41(b), can be easily verified using the complete electromagnetic approach. This approach can also be readily extended to handle an arbitrary number of intermediate layers by appropriately multiplying the $\tilde{M}_{i,j}$ matrices. A more generalized treatment of this procedure is presented in the section on optical interference filters (refer to Sec. 3.7).

3.6.2 Fabry-Perot Interferometer

A typical implementation of a Fabry-Perot interferometer is shown in Fig. 3.43. The setup consists of two glass prisms, P_1 and P_2, with their inner surfaces precisely parallel, as illustrated. To achieve high reflectivity, these parallel surfaces are coated with either a thin layer of a high-reflectance metal (such as silver or gold) or a multilayer stack designed for high reflectance. This coating effectively transforms the inner surfaces into mirrors, M_1 and M_2, with intensity reflectivities of R_1 and R_2, respectively, typically approaching unity. The mirrors generally exhibit complex amplitude reflectivities expressed as $r_i = \sqrt{R_i} e^{j\phi_i}$ $(i = 1, 2)$, even though the intermediate layer, having a thickness d and refractive index n_2, is lossless. The material between the prisms is typically air (i.e., $n_2 = 1$). Light entering the interferometer from the left undergoes multiple reflections and interference within the cavity, with the transmitted intensity being the primary quantity of interest. The non-parallel sides of the prisms, P_1 and P_2, are typically coated with anti-reflective layers to minimize undesired reflections. Additionally, any reflections from these non-parallel sides do not affect the interferometer's response, as they cannot produce multiple-wave interference due to their non-parallel orientation. For tuning purposes, one of the mirrors (prism P_2 in the figure) is often designed to be horizontally movable. This mobility allows for precise adjustment of the effective thickness d of the interferometer cavity, enabling fine control over its optical properties.

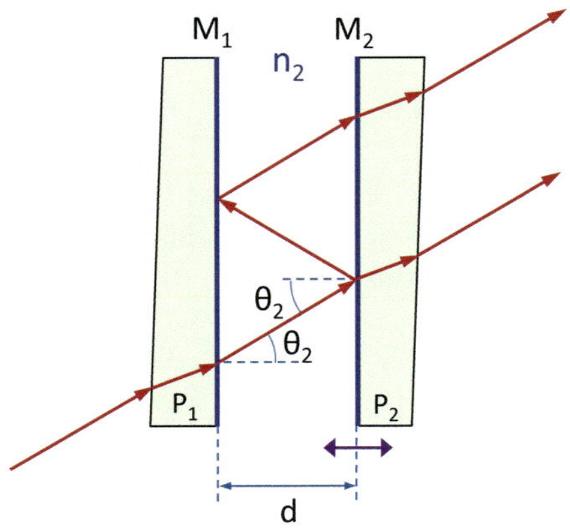

FIGURE 3.43: A typical implementation of a Fabry-Perot interferometer. Two prisms P_1 and P_2 have their inner surfaces parallel as shown. In order to achieve high reflectivity the parallel surfaces of the two prisms have a thin deposit of high reflectance metal (such as silver), or a high reflectance thin multilayer stack. This practically makes the two mirrors M_1 and M_2 to have effective intensity reflectivities of R_1 and R_2. The material between the prisms is usually air (i.e., $n_2 = 1$). Some representative rays are shown to denote the multiple interference effects for transmission only.

The normalized reflected intensity of a Fabry-Perot etalon is expressed by Eq. (3.224) in the asymmetric case and Eq. (3.228) in the symmetric case. The same equation can be applied to the Fabry-Perot interferometer, where the intensity reflectivities R_1 and R_2 correspond to the left and right mirror, respectively. The normalized transmitted intensity is then given by:

$$\frac{I_t}{I_i} = \frac{(1-R_1)(1-R_2)}{(1-\sqrt{R_1R_2})^2 + 4\sqrt{R_1R_2}\sin^2\left(\dfrac{\delta-\phi_1-\phi_2}{2}\right)} = \frac{\dfrac{(1-R_1)(1-R_2)}{(1-\sqrt{R_1R_2})^2}}{1+F\sin^2\left(\dfrac{\delta-\phi_1-\phi_2}{2}\right)} \quad \text{where,}$$

$$F = \frac{4\sqrt{R_1R_2}}{(1-\sqrt{R_1R_2})^2}, \tag{3.241}$$

for the asymmetric Fabry-Perot case. However, for the symmetric case ($r_1 = r_2 = \sqrt{R}e^{j\phi}$), the latter equations simplify to:

$$\frac{I_t}{I_i} = \frac{(1-R)^2}{(1-R)^2 + 4R\sin^2\left(\dfrac{\delta}{2}-\phi\right)} = \frac{1}{1+F\sin^2\left(\dfrac{\delta}{2}-\phi\right)}, \quad \text{where}$$

$$F = \frac{4R}{(1-R)^2}. \tag{3.242}$$

The quantity F, as defined in Eqs. (3.241) or (3.242), represents the *coefficient of finesse* for the Fabry-Perot interferometer. The coefficient of finesse, F, is highly sensitive to the power reflectivity of the interferometer mirrors and can range from zero to infinity. The dependence of the coefficient of finesse on the intensity reflectivity, $R = R_1 = R_2$ (for the symmetric case), is illustrated in Fig. 3.44 for values of R up to 0.90.

The coefficient of finesse F can also be interpreted as a measure of the fringe contrast between the maximum and minimum normalized transmitted intensities, as discussed in

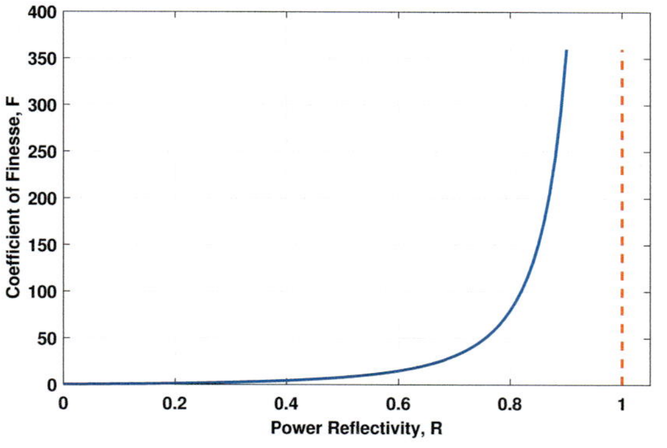

FIGURE 3.44: The coefficient of finesse, F, of a symmetric Fabry-Perot interferometer as a function of the power reflectivity, R, of its mirrors. The coefficient of finesse tends to infinity as the power reflectivity tends to unity.

Ref. [20]. Another measure of fringe contrast, analogous to the fringe visibility defined in Eq. (3.199), is given by:

$$\frac{(I_t/I_i)_{max} - (I_t/I_i)_{min}}{(I_t/I_i)_{min}} = F. \tag{3.243}$$

This measure applies to the Fabry-Perot interferometer, where $(I_t/I_i)_{min} = (I_t/I_i)_{max}/(1+F)$. Unlike the two-slit Young's experiment, the minimum transmitted intensity can never be exactly zero. A typical normalized transmitted intensity as a function of light frequency is depicted in Fig. 3.45. As shown in Sec. 3.6.1.1, the maxima in the transmission intensity are governed by the condition $\delta - (\phi_1 + \phi_2) = 2m\pi$. Using the definition of δ in Eq. (3.220), the resonant frequencies and corresponding wavelengths can be readily derived. These are expressed as:

$$\nu_m = \overbrace{\left(m + \frac{\phi_1 + \phi_2}{2\pi}\right)}^{= m_{eff}} \frac{c}{2dn_2 \cos\theta_2} = m_{eff}\, \frac{c}{2dn_2 \cos\theta_2}, \tag{3.244}$$

$$\lambda_{0m} = \frac{c}{\nu_m} = \frac{1}{\left(m + \dfrac{\phi_1 + \phi_2}{2\pi}\right)}\, 2dn_2 \cos\theta_2 = \frac{1}{m_{eff}}\, 2dn_2 \cos\theta_2. \tag{3.245}$$

The effective number m_{eff} becomes an integer m when the phase factors $\phi_1 + \phi_2 = 2\ell\pi$ (ℓ an integer), or, more simply, when r_i represents real and positive reflectivities. The *free spectral range* (FSR) can be defined as the spacing between two successive transmission peaks, either in the frequency or wavelength domain. The FSR of a Fabry-Perot interferometer is given by:

$$FSR_\nu = \Delta\nu = \nu_{m+1} - \nu_m = \frac{c}{2dn_2 \cos\theta_2}, \tag{3.246}$$

$$FSR_\lambda = \Delta\lambda = \lambda_{0m} - \lambda_{0,m+1} = \frac{2dn_2 \cos\theta_2}{m_{eff}(m_{eff}+1)} = \frac{\lambda_{0m}}{m_{eff}+1} = \frac{\lambda_{0,m+1}}{m_{eff}}, \tag{3.247}$$

or around λ_{0m},

$$FSR_\lambda = \Delta\lambda = \frac{\lambda_{0m}^2}{c}|\Delta\nu| = \frac{\lambda_{0m}^2}{2dn_2 \cos\theta_2} = \frac{\lambda_{0m}}{m_{eff}}. \tag{3.248}$$

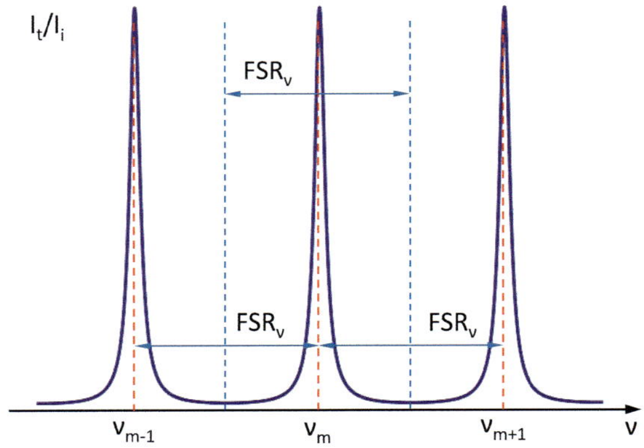

FIGURE 3.45: The transmission response of a symmetric Fabry-Perot interferometer as a function of the light frequency for relatively high mirror reflectivity R. Three resonance frequencies, ν_{m-1}, $\nu_,$, and ν_{m+1} are identified. The free spectral range in the frequency domain is $FSR_\nu = \nu_m - \nu_{m-1} = \nu_{m+1} - \nu_m$.

3.6.2.1 Minimum Angular Separation

The *full width at half maximum*, FWHM, of the normalized transmission around a peak value can be defined as $\Delta\delta_{1/2} = 2\Delta\delta$, as illustrated in Fig. 3.46. The phase difference $\Delta\delta$ can be determined as follows:

$$\frac{I_t}{I_i} = \frac{(I_t/I_i)_{max}}{1 + F\sin^2\left(\dfrac{2\pi m + \Delta\delta}{2}\right)} = \frac{1}{2}\left(\frac{I_t}{I_i}\right)_{max} \Longrightarrow$$

$$\sin\left(\frac{\Delta\delta}{2}\right) = \frac{1}{\sqrt{F}} \Longrightarrow \Delta\delta = 2\sin^{-1}\left(\frac{1}{\sqrt{F}}\right) \Longrightarrow \Delta\delta_{1/2} = 4\sin^{-1}\left(\frac{1}{\sqrt{F}}\right) \simeq \frac{4}{\sqrt{F}}.$$

The last equation is valid for large values of F, which are typically encountered in practical Fabry-Perot interferometers, where the coefficient of finesse is high. This condition often holds in interferometers with mirrors exhibiting high reflectivity, leading to sharp transmission peaks and small linewidths (narrow FWHM).

Fabry-Perot interferometers can function as spectrum analyzers through the tuning of the angular phase angle δ. The typical tuning involves a slight adjustment of one of the two mirrors, for example, mirror M_2 in Fig. 3.43. The minimum wavelength separation that a Fabry-Perot interferometer can detect is shown in Fig. 3.46. It is assumed that the interferometer is calibrated to operate around the m-th peak, which corresponds to a freespace wavelength of λ_{1m}. The limit of detectability for a slightly different wavelength is represented by the right curve. The minimum detectable wavelength separation, $\Delta\lambda_{\min}$, corresponds to the phase separation $\Delta\delta_{1/2}$. Quantitatively, this wavelength separation can be determined as follows:

$$m_{eff}\lambda_{1m} + \frac{\Delta\delta_{1/2}}{k_1} = m_{eff}\lambda_{2m} \Longrightarrow$$

$$\Delta\lambda_{min} = \lambda_{2m} - \lambda_{1m} = \frac{\lambda_{1m}}{m_{eff}}\frac{2}{\pi\sqrt{F}}. \tag{3.249}$$

At this point, the *finesse*, $\mathcal{F}$, of the Fabry-Perot interferometer can be defined as the ratio of the free spectral range to the full width at half maximum of the transmission peaks.

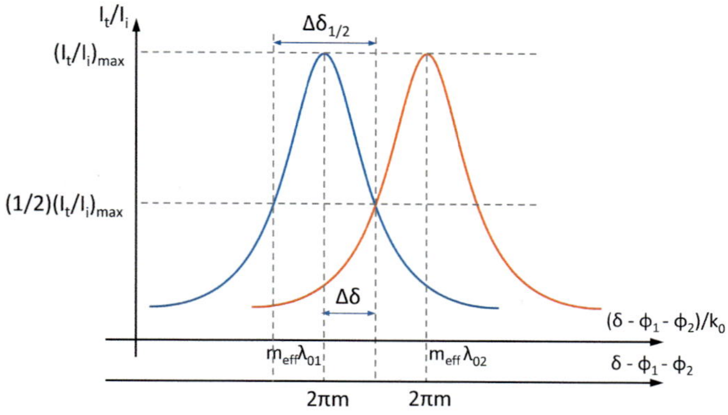

FIGURE 3.46: The definition of the minimum phase angle separation $\Delta\delta_{min}$ of a Fabry-Perot interferometer. There are two horizontal scales, one for the phase angle $\delta - \phi_1 - \phi_2$, and one for the optical path length $(\delta - \phi_1 - \phi_2)/k_0$. For real reflection coefficients $\phi_1 = \phi_2 = 0$.

Mathematically, it is given as a function of the coefficient of finesse as:

$$\mathcal{F} = \frac{\pi\sqrt{F}}{2}. \tag{3.250}$$

A typical value of the finesse for most ordinary Fabry-Perot interferometers is approximately 30 [42]. Utilizing the definition of finesse, the minimum wavelength separation, $\Delta\lambda_{min}$, and the corresponding minimum frequency separation, $\Delta\nu_{min}$, can be expressed as:

$$\Delta\lambda_{min} = \frac{\lambda_{0m}}{m_{eff}}\frac{1}{\mathcal{F}} = \frac{FSR_\lambda}{\mathcal{F}}, \tag{3.251}$$

$$\Delta\nu_{min} = \frac{\Delta\nu}{\mathcal{F}} = \frac{FSR_\nu}{\mathcal{F}}. \tag{3.252}$$

The minimum wavelength and frequency separations are constrained by the Fabry-Perot finesse, as evident from the above equations. Based on these results, the resolution criterion in wavelength is $\Delta\lambda \geq \Delta\lambda_{min}$, and in frequency, is $\Delta\nu \geq \Delta\nu_{min}$. The *resolving power*, denoted as $\mathcal{R}$, of a Fabry-Perot interferometer (operating around λ_{0m} or equivalently around ν_m) can be defined as the ratio of the central wavelength (or frequency) to the minimum detectable wavelength (or frequency) separation:

$$\mathcal{R} = \frac{\lambda_{0m}}{\Delta\lambda_{min}} = m_{eff}\mathcal{F} = \frac{\nu_m}{\Delta\nu_{min}}. \tag{3.253}$$

Indeed, for the spectrum analysis of optical sources, particularly lasers, achieving large values of the resolving power $\mathcal{R}$ is highly desirable. Large values of $\mathcal{R}$ can be obtained by increasing the number of the resonant mode m (and consequently of the effective m_{eff}) and by improving the finesse $\mathcal{F}$. The value of m increases with the thickness d of the interferometer, as a larger cavity length allows for more resonant modes. Additionally, the finesse $\mathcal{F}$ improves as the mirror reflectivities approach unity, meaning that mirrors with higher reflectivity result in sharper resonance peaks. Thus, optimizing both the thickness of the interferometer cavity and the reflectivities of the mirrors can lead to a significant improvement in the resolving power of Fabry-Perot interferometers, making them more effective for high-resolution spectral analysis.

3.6.3 Michelson's Interferometer

The Michelson interferometer, invented by Albert Michelson around 1881, holds profound historical and scientific significance. In 1887, it was pivotal in the attempt to detect the Earth's motion through the hypothesized luminiferous ether, a medium that many physicists of the time believed to be the carrier of light waves. The Michelson-Morley experiment, conducted with this interferometer, is celebrated as one of history's most famous "unsuccessful" experiments, as it failed to confirm the existence of the ether. Despite this outcome, the Michelson interferometer became a cornerstone of modern physics. It provided critical experimental evidence supporting the special theory of relativity, contributed to the discovery and measurement of hyperfine structures in spectral lines, and enabled the detection of the Moon's tidal effects on Earth, among numerous other breakthroughs [20]. In recent years, a variant of the Michelson interferometer was instrumental in confirming the existence of gravitational waves, as predicted by the general theory of relativity. This milestone was achieved through the *Laser Interferometer Gravitational-Wave Observatory* (LIGO) [13,110]. The Michelson interferometer remains a versatile and powerful instrument with wide-ranging applications in fundamental physics research.

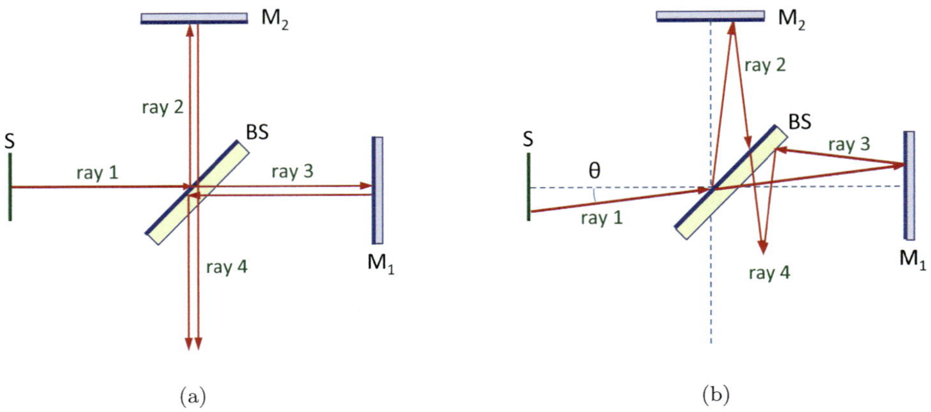

FIGURE 3.47: (a) The basic configuration of Michelson's interferometer. The mirrors M_1 and M_2 are assumed exactly perpendicular to each other. The left segment S denotes an extended light source. The beam splitter (BS) equally divides the incident ray 1 into ray 2 and ray 3. These rays after reflection from the mirrors, interfere in the direction of ray 4. The rays 2 and 3 are shown slightly displaced for better visualization of the interferometer's operation. (b) The same configuration with the initial ray at an angle θ with respect to the horizontal direction.

The fundamental configuration of the Michelson interferometer is shown in Fig. 3.47(a). A light ray (ray 1) from the extended source S is split by the beam splitter (BS) into rays 2 and 3. The beam splitter is assumed to equally distribute the incident power into the directions of rays 2 and 3. Each of these rays travels toward mirrors M_2 and M_1, respectively, get reflected, and then passes through the beam splitter again, where reversed ray 2 is transmitted, and reversed ray 3 is reflected. As a result, both rays propagate in the direction specified by ray 4. The differing path lengths of rays 2 and 3 enable interference. A micro-positioner allows parallel movement of at least one mirror to its normal, while tilt adjustments align one mirror relative to the other. These adjustments control the optical path difference between the two rays. As shown in Fig. 3.47(a), ray 3 traverses the beam splitter three times, whereas ray 2 passes through it only once. For sources emitting multiple wavelengths, this difference introduces wavelength dependence in the optical path, making it challenging to equalize the paths for all wavelengths using mirror adjustments alone.

Compensation techniques are employed to address this issue, as illustrated in Figs. 3.48a and 3.48b. In the left configuration, a compensator plate C, made of the same material and dimensions as the beam splitter (excluding the reflective layer), is inserted into the path of ray 2. In the right configuration, a cube beam splitter is used, with the reflective layer positioned along the cube's diagonal plane. This ensures that both rays pass through the same amount of material, eliminating the need for additional compensation.

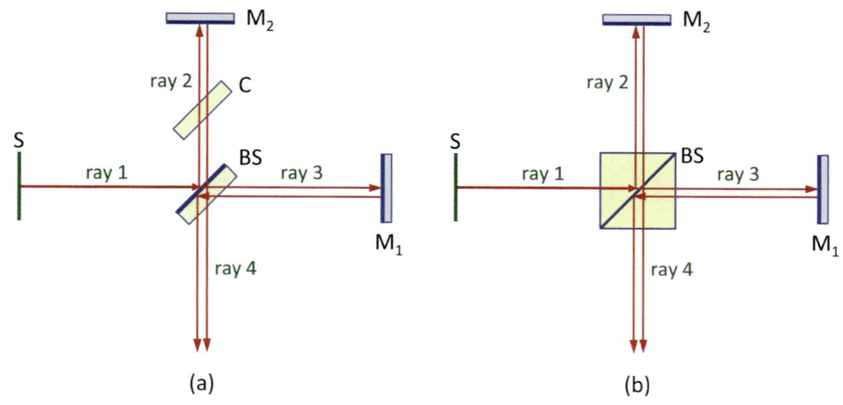

FIGURE 3.48: The basic configuration of Michelson's interferometer with dispersion compensation. (a) A compensator C is used in the path of ray 2 to adjust the optical path length between rays 2 and 3. The compensator is from the same material as the beam splitter but with a stripped reflection layer. (b) Another configuration for dispersion compensation using a cubic prism beam splitter.

The Michelson interferometer undergoes quantitative analysis under the assumption that mirrors M_1 and M_2 have been precisely adjusted to be perpendicular to each other. Folding the configuration along the optical axis of mirror M_2 results in the virtual image of the source S and mirror M_1, denoted as $M_{1'}$, as shown in Fig. 3.49. The optical path difference d between the mirrors is indicated, and a ray emanating from the source point S at an angle θ to the normal of M_2 and $M_{1'}$ is illustrated. Reflected rays from the virtual mirror $M_{1'}$ and mirror M_2 interfere with the direction of ray 4 (see Fig. 3.47(a)). These reflected rays are associated with the virtual sources S_1 and S_2, corresponding to reflections from $M_{1'}$ and M_2, respectively. The optical path difference between these reflected rays can be computed analogously to the approach for the Fabry-Perot etalon [see Eq. (3.220)]. Furthermore, phase shifts resulting from reflections must be considered. Ray 2 undergoes two external reflections, while ray 3 undergoes only one. Thus, an additional phase shift of π arises from this imbalance in reflection phases. The resulting phase shift δ is then expressed as:

$$\delta = k_0 2d \cos\theta + \pi, \tag{3.254}$$

where it was assumed that the surrounding medium is air and has a unity refractive index. The π phase shift is indeed present in the beam splitter of Fig. 3.48a. However, in the case of the cubic beam splitter of Fig. 3.48b, the π phase shift must not be present. Detailed explanations regarding these phase shifts and the modeling of beam splitters are provided in Appendix 3B. In both cases, the beam splitters are assumed to operate symmetrically.

The resulting interference intensity pattern is given by [see Eq. (3.202)]:

$$I = I_0 \cos^2\left(\frac{2\pi}{\lambda_0} d\cos\theta + \frac{\pi}{2}\right) = \frac{I_0}{2}\left[1 + \cos\left(\frac{2\pi}{\lambda_0} 2d\cos\theta + \pi\right)\right], \tag{3.255}$$

where $I_0 = |E_0|^2/4Z$ and E_0 is the amplitude of the incident wave of ray 1 (see Fig. 3.48) and Z is the wave impedance of the surrounding medium which is usually air. In the case of asymmetric beam-splitter operation, the two interfering waves may have different intensities (and corresponding amplitudes). Assuming the same polarization for the two waves, which is typical for a Michelson interferometer, the resulting interference intensity is given by [see Eq. (3.202) and Eq. (3.203)]:

$$\begin{aligned} I &= \frac{1}{4Z}\left\{E_1^2 + E_2^2 + 2E_1E_2\cos\left[\frac{2\pi}{\lambda_0} 2d\cos\theta + \pi\right]\right\} = \\ &= I_1 + I_2 + 2\sqrt{I_1 I_2}\cos\left[\frac{2\pi}{\lambda_0} 2d\cos\theta + \pi\right], \end{aligned} \tag{3.256}$$

where $I_i = E_i^2/4Z$ $(i = 1, 2)$ is the intensity of each of the two interfering waves in the direction of ray 4. The corresponding maxima and minima in the interference pattern are given by:

$$2d\cos\theta = \begin{cases} (2m+1)\dfrac{\lambda_0}{2}, & \text{for maxima,} \\ m\lambda_0, & \text{for minima.} \end{cases} \tag{3.257}$$

The maximum value of m occurs when $\theta = 0$. When a lens with focal length f is used, the interference fringes manifest as concentric bright and dark rings in the lens's focal plane. Based on the geometry shown in Fig. 3.49, the radii of the bright or dark fringes can be calculated as $r_m = f\tan\theta_m$, where $\theta_m = \theta$ satisfies Eq. (3.257) for bright (maxima) or dark (minima) fringes. Example interference patterns are displayed at the bottom of Fig. 3.49 for He-Ne light ($\lambda_0 = 0.6328\,\mu$m) and $d = 1$ cm for the bright center pattern (corresponding to $m_{max} \simeq 31605$), and $d \simeq 2$ cm for the dark center pattern (corresponding to $m_{max} \simeq 63211$).

Now, let us consider the case where mirrors M_1 and M_2 have been adjusted to be tilted relative to each other, making an angle 2ϕ. When the configuration of Fig. 3.47(a) is folded in the direction of the optical axis of mirror M_2, a virtual image of the source S and mirror M_1 (denoted again as $M_{1'}$) is shown in the configuration depicted in Fig. 3.50(a). In this tilted configuration, it is assumed that the optical path difference between the centers of the two mirrors is zero. This implies that the center of $M_{1'}$ coincides with the center of M_2, and the tilting of the mirrors introduces a spatially varying optical path difference across their surfaces. This variation leads to the formation of linear interference fringes in the output plane. Specifically, the virtual sources S_1 and S_2 are parallel to the interference observation plane. The interference geometry now resembles the configuration of Young's two-slit experiment, with the role of the slits being played by the two virtual sources S_1 and S_2. As the rays emanate from the source point S and are reflected off mirrors $M_{1'}$ (M_1) and M_2, they travel toward the observation plane and interfere in the direction of ray 4 (as shown in Fig. 3.47(a)). Since S_1 and S_2 are the virtual sources from which the rays appear to emanate, we can apply the two-slit interference formula [Eq. (3.210)] to describe the resulting intensity pattern. The intensity at any point in the observation plane will be

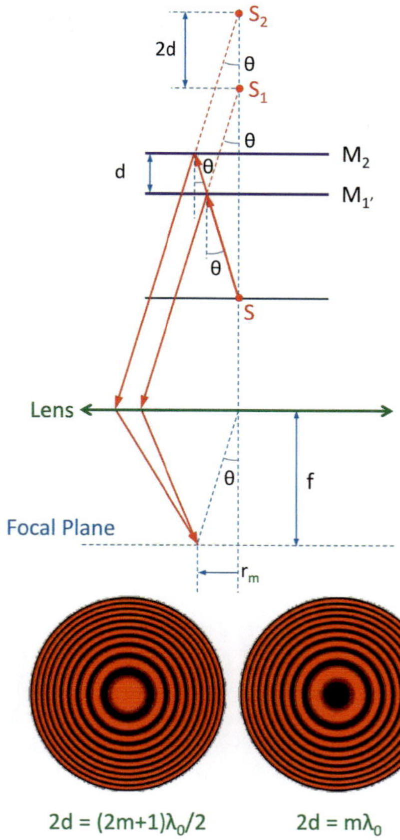

FIGURE 3.49: Michelson's interferometer configuration folded along the axis of mirror M_2. The virtual image of mirror M_1, denoted as $M_{1'}$, is formed by reflection through the beam splitter. Mirrors M_1 and M_2 are perpendicular to each other, and their geometrical path length difference is d. The source point S is folded along the axis of mirror M_2, with virtual source points S_1 and S_2 corresponding to reflections from mirrors $M_{1'}$ and M_2, respectively. A lens with focal length f focuses rays onto its focal plane, forming ring-shaped interference patterns. Two sample patterns are depicted: one with a bright central fringe ($2d = (2m+1)\lambda_0/2$) and one with a dark central fringe ($2d = (2m+1)\lambda_0/2$). The inclination of the rays is θ, and the radius of the m-th bright (or dark) ring is r_m.

given by:

$$I = I_0 \cos^2\left[\frac{2\pi}{\lambda_0}\left(\frac{r_{S_1} - r_{S_2}}{2}\right) + \frac{\pi}{2}\right], \qquad \text{where,} \tag{3.258}$$

$$r_{S_1} - r_{S_2} \simeq \underbrace{(4b\sin\phi)}_{= d_{eff}} \sin\theta = d_{eff}\sin\theta,$$

where $b = b_1 = b_2$ in this symmetric case. When the point S is at different distances from the two mirrors, i.e., when $b_1 \neq b_2$, then $d_{\text{eff}} = 2[(b_1 - b_2)^2 + 4b_1 b_2 \sin^2\phi]^{1/2}$, and the fringes observed in the screen plane are a tilted version of the hyperbolic fringes observed in general in the Young's configuration.

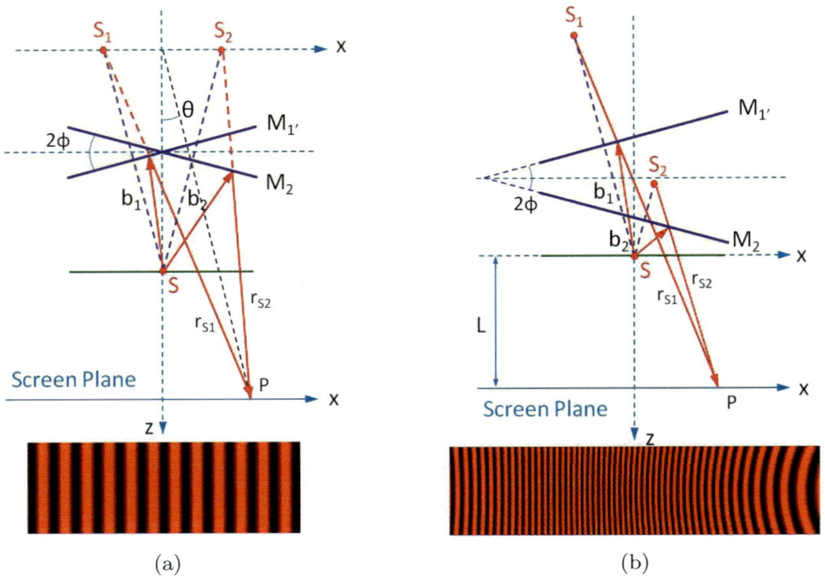

FIGURE 3.50: (a) A folded along the axis of mirror M_2 Michelson's interferometer configuration. The mirror $M_{1'}$ is the virtual image of mirror M_1 with respect to the beam splitter. The centers of the mirrors are exactly the same distance from the beam splitter but they are tilted with respect to each other forming an angle 2ϕ. The point S is a point of the source which has been folded on the axis of mirror M_2. The points S_1 and S_2 are the virtual source points of point S with respect to the mirrors $M_{1'}$ and M_2. It is assumed that the zero optical path length (the middle in the screen plane has a maximum intensity (compensated for the phase angle of π). The distances of point S from the two mirrors are equal in this case, i.e., $b_1 = b_2$. The angle θ is similar to the one shown in Fig. 3.32. (b) The case that there is a mirror separation and a tilt at the same time.

In the case of unequal beam-splitter operation (assuming E_1 and E_2 amplitudes and the same polarization of the two interfering waves), Eq. (3.258) can be generalized into:

$$I = \frac{1}{4Z}\left\{E_1^2 + E_2^2 + 2E_1E_2\cos\left[\frac{2\pi}{\lambda_0}(r_{S_1} - r_{S_2}) + \pi\right]\right\} =$$
$$= I_1 + I_2 + 2\sqrt{I_1 I_2}\cos\left[\frac{2\pi}{\lambda_0}(r_{S_1} - r_{S_2}) + \pi\right]. \quad (3.259)$$

Example 3.6.1:

The purpose of this example is to demonstrate the interference patterns resulting from the equation given in Eq. (3.258), specifically illustrating how the tilt (and or separation) of the mirrors affects the observed fringe patterns. In this setup, the distances r_{S_1} and r_{S_2}, which correspond to the paths from the observation point to each of the virtual sources, are crucial for determining the phase difference between the two interfering beams. The geometry depicted in Fig. 3.50(b) provides the necessary framework to quantify these distances and the resulting phase shift. Specifically, r_{S_1} and r_{S_2}, can be expressed as:

$$r_{S_1} = \left[(x + 2b_1\sin\phi)^2 + y^2 + (L + 2b_1\cos\phi)^2\right]^{1/2},$$

$$r_{S_2} = \left[(x - 2b_2\sin\phi)^2 + y^2 + (L + 2b_2\cos\phi)^2\right]^{1/2},$$

where (x, y, z) is any point in the $z = L$ plane. The origin of the coordinate system used is centered at point S (source point) of Fig. 3.50(b). From Eq. (3.258) the maxima in the $z = L$ plane occur when $r_{S_1} - r_{S_2} = (2m+1)\lambda_0/2$ and the minima occur when $r_{S_1} - r_{S_2} = m\lambda_0$ where m is an integer. The positions of the maxima and minima of the interference pattern along the x-axis (in the $z = L$ plane and for a given y) can be determined analytically by solving the interference condition, and are given by:

$$x_m = \frac{B \pm A_m\sqrt{D}}{C}, \qquad \text{where,}$$

$$B = \left(A_m^2 + x_1x_2\right)(x_1 + x_2) + (x_1 - x_2)\left(z_1^2 - z_2^2\right) - x_1^3 - x_2^3,$$

$$D = A_m^4 - 2A_m^2\left(z_1^2 + z_2^2\right) - 2A_m^2(x_1 - x_2)^2 + \left(z_1^2 - z_2^2\right)^2 + 2\left(z_1^2 + z_2^2\right)^2(x_1 - x_2)^2 + (x_1 - x_2)^4 - 4y^2\left[A_m^2 - (x_1 - x_2)^2\right],$$

$$C = 2\left[A_m^2 - (x_1 - x_2)^2\right],$$

$$x_1 = -2b_1\sin\phi, \quad x_2 = 2b_2\sin\phi, \quad z_1 = L + 2b_1\sin\phi, \quad z_2 = L + 2b_2\sin\phi,$$

$$A_m = \begin{cases} (2m+1)\dfrac{\lambda_0}{2}, & \text{for maxima,} \\ m\lambda_0, & \text{for minima.} \end{cases}$$

where x_m is the position of the m-th maximum (or minimum). Certainly, the above equation is valid only when $D \geq 0$. For negative values of D, there is neither a minimum nor a maximum for the given parameters. An example interference pattern is illustrated in

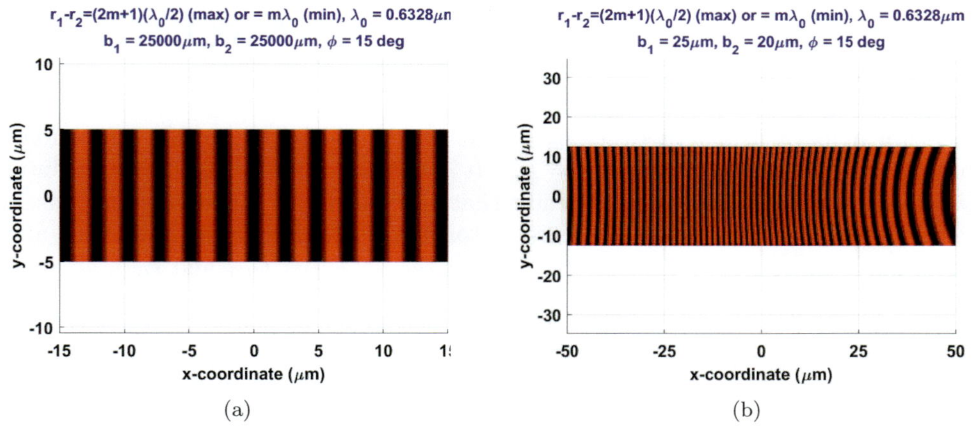

FIGURE 3.51: (a) The interference pattern observed at the plane $z = L = 5\,\text{cm}$, with $b_1 = b_2 = 2.5\,\text{cm}$ and $\phi = 15\,\text{deg}$. In this case only a tilt of $2\phi = 30\,\text{deg}$ between the two mirrors exist. (b) The interference pattern observed at the plane $z = L = 10\,\mu m$, with $b_1 = 25\,\mu m$, $b_2 = 20\,\mu m$, and $\phi = 15\,\text{deg}$. In this case there are a separation of $2(b_1 - b_2)\sin\phi = 2.588\,\mu m$, and a tilt of $2\phi = 30\,\text{deg}$ between the two mirrors. The distances are selected unrealistically small in order to better capture the shape of the interference fringes.

Fig. 3.51(a) for the symmetric case where $b_1 = b_2 = 2.5\,\text{cm}$, $\phi = 15\,\text{deg}$, $L = 5\,\text{cm}$, and $\lambda_0 = 0.6328\,\mu\text{m}$. It is straightforward to verify, using the previous equations for the location of the maxima and minima of the interference pattern, that the separation between the bright or dark fringes is $2.424\,\mu\text{m}$.

Another example pattern is presented in Fig. 3.51(b). In this case, the various distances were deliberately selected as unrealistically small to clearly illustrate the variation in the interference pattern over a compact area. Under more realistic conditions, the same pattern would be observed over a much larger spatial extent. In this specific case, the interference pattern is shown in the observation plane $z = L = 10\,\mu m$, with the parameters $b_1 = 25\,\mu m$, $b_2 = 20\,\mu m$, and a tilt angle of $\phi = 15\,\text{deg}$. The combined effect of mirror separation and tilt results in both a separation of $2(b_1 - b_2)\sin\phi = 2.588\,\mu m$ and an angular displacement of $2\phi = 30\,\text{deg}$ between the two mirrors. Using the equations for the positions of maxima and minima, it is straightforward to compute the periodicity of the resulting bright and dark fringes. The fringe spacing varies between $1.45\,\mu\text{m}$ and $4.60\,\mu\text{m}$, reflecting the influence of both the tilt and separation effects on the interference pattern.

3.6.4 Mach-Zehnder Interferometer

The Mach-Zehnder Interferometer (MZI) operates based on the principle of amplitude division, allowing the measurement of phase shifts between two beams due to variations in the optical path length or the introduction of a sample. Named after Ludwig Mach [111] and Ludwig Zehnder [112], it remains a vital tool in precision interferometry. A schematic of the MZI is presented in Fig. 3.52(a), which utilizes asymmetric lossless beam splitters, and Fig. 3.52(b), incorporating symmetric lossless cubic beam splitters. Assume that the incident beam enters from the lower left in both configurations, with its polarization being *TE*, i.e., with the electric field perpendicular to the plane of the figures shown, and with an amplitude of E_0. Focusing initially on the configuration in Fig. 3.52(a), the beam splitter BS_1 divides the incoming electric field amplitude. Using the formalism of the beam splitter matrix (detailed in Appendix 3B), the resulting electric field amplitudes after the division are given as:

$$\begin{bmatrix} E_{01} \\ E_{03} \end{bmatrix} = \frac{1}{\sqrt{2}} \begin{bmatrix} 1 & 1 \\ -1 & 1 \end{bmatrix} \begin{bmatrix} E_0 \\ 0 \end{bmatrix} = \begin{bmatrix} E_0/\sqrt{2} \\ -E_0/\sqrt{2} \end{bmatrix}.$$

The beam in path 3 undergoes a phase shift introduced by the transfer function $e^{-j\phi}$, resulting in $E_{04} = e^{-j\phi}E_{03}$ while $E_{02} = E_{01}$. It is assumed that the mirrors do not introduce any additional phase shifts (or equivalently, that both mirrors produce identical phase shifts, which will manifest as a common phase factor across all electric fields without affecting relative phase differences). The electric fields at the detectors E_{D1} and E_{D2}, after passing through the second beam splitter BS_2, are expressed as:

$$\begin{bmatrix} E_{D2} \\ E_{D1} \end{bmatrix} = \frac{1}{\sqrt{2}} \begin{bmatrix} 1 & 1 \\ -1 & 1 \end{bmatrix} \begin{bmatrix} E_{02} \\ E_{04} \end{bmatrix} = \frac{E_0}{2} \begin{bmatrix} 1 - e^{-j\phi} \\ -1 - e^{-j\phi} \end{bmatrix} = E_0 e^{-j\phi/2} \begin{bmatrix} j\sin\phi/2 \\ -\cos\phi/2 \end{bmatrix}.$$

From the last equation the resulting intensities incident on the two detectors I_{D1} and I_{D2} are given by:

$$I_{D1} = |E_{D1}|^2 = |E_0|^2 \cos^2\frac{\phi}{2}, \tag{3.260}$$

$$I_{D2} = |E_{D1}|^2 = |E_0|^2 \sin^2\frac{\phi}{2}. \tag{3.261}$$

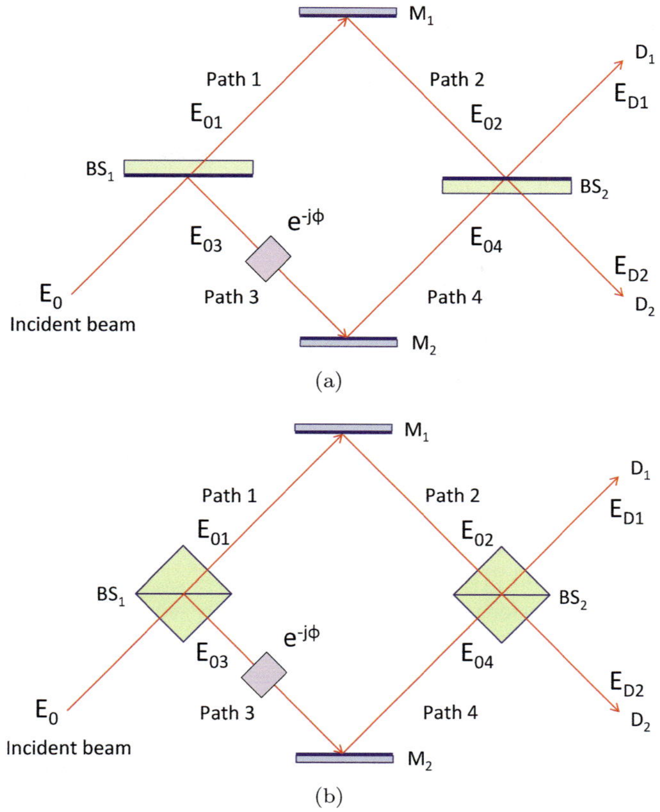

FIGURE 3.52: (a) A simple geometry of a Mach-Zehnder interferometer using asymmetric dielectric beam splitters. The M_1 and M_2 represent ideal mirrors, the BS_1 and BS_2 represent lossless asymmetric beam splitters, and D_1 and D_2 are detectors. The box in the lower left path introduces a phase shift ϕ represented by the $e^{-j\phi}$ transfer function to the beam field. The E_0 denotes the incident electric field, while E_{01}, E_{02}, E_{03}, E_{04}, denote the electric field amplitude in the same numbered path. The E_{D1} and E_{D2} denote the electric field amplitudes arriving at detectors D_1 and D_2, respectively. (b) A simple geometry of a Mach-Zehnder interferometer using symmetric dielectric cubic beam splitters. In this setup, BS_1 and BS_2 are lossless symmetric cubic beam splitters.

When the phase shift $\phi = 0$, light constructively interferes in the direction of detector D_1 and destructively interferes in the direction of detector D_2. This outcome is straightforward to understand based on the phase changes at the beam splitters. The electric field E_{03} undergoes a π phase shift to the external reflection at BS_1. Similarly, E_{02} (where $E_{01} = E_{02}$) experiences an additional π phase shift due to the external reflection at BS_2 in the direction of D_1 detector. Therefore, there is constructive interference in the direction of D_1 and destructive interference in the direction of D_2. Since the interferometer is lossless, the total power is conserved, satisfying $I_{D1} + I_{D2} = |E_0|^2 = I_0$, as expected.

For the symmetric cubic beam splitters in Fig. 3.52(b), similar principles apply. The phase shift behavior and resulting interference can be analyzed using the beam splitter matrix provided in Appendix 3B. Specifically, the resulting electric fields at detectors at detectors D_1 and D_2 are given by:

$$\begin{bmatrix} E_{01} \\ E_{03} \end{bmatrix} = \frac{1}{\sqrt{2}} \begin{bmatrix} 1 & -j \\ -j & 1 \end{bmatrix} \begin{bmatrix} E_0 \\ 0 \end{bmatrix} = \begin{bmatrix} E_0/\sqrt{2} \\ -jE_0/\sqrt{2} \end{bmatrix},$$

$$\begin{bmatrix} E_{D2} \\ E_{D1} \end{bmatrix} = \frac{1}{\sqrt{2}} \begin{bmatrix} 1 & -j \\ -j & 1 \end{bmatrix} \begin{bmatrix} E_{02} \\ E_{04} \end{bmatrix} = \frac{E_0}{2} \begin{bmatrix} 1 - e^{-j\phi} \\ -j(1 + e^{-j\phi}) \end{bmatrix} = E_0 e^{-j\phi/2} \begin{bmatrix} j\sin\phi/2 \\ -\cos\phi/2 \end{bmatrix},$$

where again $E_{04} = e^{-j\phi}E_{03}$ and $E_{02} = E_{01}$. The resulting detector intensities, even for the symmetric cubic beam splitter configuration, are still given by Eqs. (3.260) and (3.261), as in the case of asymmetric beam splitters. In all these cases, it has been assumed that the beam splitters are ideal 50/50 lossless devices. However, if the beam splitter deviates from a 50/50 split, perfect constructive or destructive interference might not be achieved, leading to an imbalance in the observed intensities. The Mach-Zehnder interferometer is versatile and can also be used with extended sources or configurations involving tilts of mirrors or beam splitters, as discussed in [91]. Its flexibility in fringe localization offers a unique advantage over other interferometers, enabling applications where precise control of fringe positioning is essential [91].

The Mach-Zehnder interferometer (MZI) is a highly versatile optical instrument with applications spanning numerous fields. It plays a critical role in fundamental research areas such as quantum mechanics, quantum computation, quantum entanglement, quantum logic, and quantum cryptography. Beyond fundamental physics, it is extensively used for studying density fluctuations in fluids, such as in wind tunnel gas dynamics, and for precise refractive index measurements. In recent years, the MZI has been increasingly adapted for multi-purpose sensing applications, particularly in its fiber optics and integrated optics implementations [113, 114]. Additionally, it forms the backbone of practical electro-optic modulators, which are widely employed for amplitude modulation of laser sources in optical communication systems [115, 116].

3.6.5 Fourier Transform Spectroscopy

Fourier transform spectroscopy is an advanced technique used for acquiring emission spectra of optical sources or absorption spectra of materials subjected to illumination. Unlike traditional dispersion-based methods, this approach leverages light interference to derive spectral information. The foundation of this technique lies in the Fourier-transform-pair relationship between the interferogram of a source or substance and its spectrum [117]. This versatile technique is widely applied in the visible and infrared regions of the spectrum. In the infrared domain, it is commonly referred to as Fourier Transform Infrared (FTIR) spectroscopy. Nonetheless, the methodology is not limited to these regions and can, in principle, be extended to other parts of the electromagnetic spectrum.

A typical Fourier transform spectrometer is often realized using a Michelson interferometer configuration, as depicted in Fig. 3.48. In this setup, one mirror remains stationary while the other is movable. As the movable mirror shifts, the detector captures the interference intensity, producing an interferogram. By applying a Fourier transform to this interferogram, one can extract either the emission spectrum of the source or the absorption spectrum of the sample, provided the spectral characteristics of the illumination source are known.

The analysis of the Fourier transform spectrometer relies on the foundational principles established for the Michelson interferometer. The primary distinction lies in the assumption of a multichromatic source instead of a monochromatic one. According to Eq. (3.256), the interference intensity for a multichromatic source can be expressed as:

$$I(x(t)) = \int_0^\infty \left[I_1(k) + I_2(k) + 2\sqrt{I_1(k)I_2(k)} \cos[kx(t) + \pi] \right] dk, \tag{3.262}$$

where I_1 and I_2 are the intensities of the interfering beams, $k = 2\pi/\lambda_0$, and $x(t) = 2d(t)$ represents the optical path difference, with $d(t)$ being the displacement of the movable mirror as a function of time during the scan. In spectroscopy, it is common to describe the wavelength in terms of the "wavenumber" $\bar{\nu} = 1/\lambda_0$ (typically expressed in cm^{-1}). For simplicity, an ideal (wavelength-independent) beam splitter/beam combiner is assumed, and the phase shift π is compensated. Defining $I_1(k) = I_2(k) = I_0(k)/4$, the last equation can be expressed as:

$$I(x) = \frac{1}{2}\int_0^\infty I_0(k)\left[1 + \cos(kx)\right] dk = \frac{1}{2}\int_0^\infty I_0(k)dk + \frac{1}{2}\int_0^\infty I_0(k)\cos(kx)dk, \tag{3.263}$$

where $I_0(k) = S(k)$ corresponds to the spectrum of the source, and $I_0(k)dk = S(k)dk$ represents the intensity of the source within the wavenumber (or frequency) range between k and $k + dk$. The first term of Eq. (3.263) is constant and does not affect the spectrum calculation. By applying the transformation $\bar{I}(x) = 2I(x) - I(0)$ [117, 118], the following equation is obtained:

$$\begin{aligned}\bar{I}(x) &= \int_0^\infty S(k)\cos(kx)dk = 2\int_{-\infty}^{+\infty} [S(k)\cos(kx) + jS(k)\sin(kx)]dk \\ &= 2\int_{-\infty}^{+\infty} S(k)e^{jkx}dk = 2\,\mathcal{F}^{-1}\left\{S(k)\right\}\Big|_x \Longrightarrow \end{aligned} \tag{3.264}$$

$$S(k) = \frac{1}{2}\mathcal{F}\left\{\bar{I}(x)\right\}\Big|_k = \frac{1}{2}\int_{-\infty}^{+\infty} \bar{I}(x)e^{-jkx}dx. \tag{3.265}$$

The even symmetry of $S(k)$ and $\cos(kx)$, along with the odd symmetry of $\sin(kx)$, was utilized in the derivation. The symbols $\mathcal{F}$ and $\mathcal{F}^{-1}$ represent the Fourier transform and its inverse, respectively. The relationship between $\bar{I}(x)$ and $S(k)$ through the Fourier transform pair forms the core of Fourier transform spectroscopy. By performing the Fourier transform of the interferogram, the spectrum of the source can be obtained. This spectrum can also be expressed as $S(\bar{\nu})$ using the wavenumber $\bar{\nu} = 1/\lambda_0$. It is important to note that the factor of 1/2 is theoretically retained for consistency. However, in practical situations, any constant factor is typically irrelevant and can be disregarded.

The resolution of a Fourier transform spectrometer can be assessed by using a monochromatic source and examining the width of the measured spectrum. Ideally, the measured spectrum would consist of a single spectral line. However, in practice, the measured spectrum is centered at the wavenumber $\bar{\nu}_0 = 1/\lambda_0$ and exhibits a full width at half maximum $\Delta\bar{\nu}$. This broadening arises due to the finite length of the interferogram. In theory, an infinite interferogram would be required to achieve perfect resolution. However, from a practical perspective, the interferogram can only be measured over a finite optical path difference, limited by the maximum displacement of the moving mirror. This finite measurement introduces a windowing effect in the optical path domain, which can be mathematically expressed as follows:

$$\bar{S}(k) = \frac{1}{2}\int_{-\infty}^{+\infty} rect\left(\frac{x - d_{max}}{2d_{max}}\right)\bar{I}(x)e^{-jkx}dx, \tag{3.266}$$

where $\bar{S}(k)$ is the source spectrum due to the effect of windowing and $rect(x)$ is the "rectangle" function [119]. From Fourier transform theory [119], it is straightforward to show that:

$$\bar{S}(k) = \frac{1}{2}\, S(k) * \mathcal{F}\left\{rect\left(\frac{x - d_{max}}{2d_{max}}\right)\right\} = \frac{1}{2}\, S(k) * \left[e^{-jkd_{max}}\,(2d_{max})\frac{\sin(kd_{max})}{kd_{max}}\right], \tag{3.267}$$

where "*" denotes convolution. Consequently, even for a monochromatic source where $S(k) = \delta(k - k_0)$ (where $\delta(u)$ is the delta function), the measured spectrum, influenced by windowing, is given by $|\bar{S}(k)| = d_{max}|\sin[(k - k_0)d_{max}]/(k - k_0)d_{max}|$. The FWHM of the resulting $\sin(x)/x$ function can be numerically determined to be approximately 2×1.89549. Thus, the spectral width of the single spectral line, as determined by the windowing effect, is expected to be:

$$\Delta\bar{\nu} = \frac{2 \times 1.89549}{2\pi d_{max}} = \frac{0.60335}{d_{max}}, \tag{3.268}$$

where d_{max} is the maximum displacement of the moving mirror. In practice, the Fourier transform of Eq. (3.265) is typically computed using techniques like the Discrete Fourier Transform (DFT) or its computationally efficient counterpart, the Fast Fourier Transform (FFT) [120]. The discrete step size, Δd, of the moving mirror directly influences the maximum frequency of the spectrum that can be resolved. This relationship is governed by the sampling theorem [119, 120], which is expressed as:

$$\frac{1}{\Delta d} > 2\bar{\nu}_{max} = \frac{2}{\lambda_{min}} \Longrightarrow \frac{N}{2d_{max}} > \frac{2}{\lambda_{min}} \Longrightarrow N > d_{max}\frac{4}{\lambda_{min}}. \tag{3.269}$$

Here N represents the total number of samples (typically 2^m when utilizing the FFT), and λ_{min} denotes the shortest wavelength within the spectrum. This wavelength corresponds to the maximum frequency, which is critical for estimating the unknown spectrum. It is important to note that d varies symmetrically from $-d_{max}$ to $+d_{max}$, resulting in the factor of "4" in Eq. (3.269). To further clarify the procedure, three illustrative examples are presented below.

Example 3.6.2:

In this example, a monochromatic source is assumed with a wavelength of $\lambda_0 = 1\,\mu m$ and unit amplitude, i.e., $S(\lambda) = \delta(\lambda - \lambda_0)$. For an interferogram with a maximum optical path difference of $d_{max} = 5\lambda_0 = 0.5 \times 10^{-5}$ m, both the interferogram and its calculated spectrum are shown in Figs. 3.53(a) and 3.53(b), respectively. The minimum number of points required, as calculated from Eq. (3.269), is 20. Therefore, for the FFT, the minimum power-of-two number of points is $2^5 = 32$. However, to achieve better spectral accuracy, $2^{12} = 4096$ points were used in Figs. 3.53(a) and 3.53(b). The expected spectral resolution, as per Eq. (3.268), is approximately $\sim 0.12\,\mu m^{-1}$ for $d_{max} = 5\lambda_0$. This resolution is represented by the dashed line in Fig. 3.53(b). When the interferogram length is increased to $d_{max} = 100\lambda_0 = 10 \times 10^{-5}$ m, the calculated spectrum is depicted in Fig. 3.53(b) as a solid, narrower line. The minimum number of points required, according to Eq. (3.269), is 400, and the nearest power of two for the FFT is $2^9 = 512$. Once again, $2^{12} = 4096$ points were used for the computation to ensure better accuracy. The expected resolution in this case, as per Eq. (3.268), is approximately $\sim 0.006\,\mu m^{-1}$, which is 20 times finer than the resolution for $d_{max} = 5\lambda_0$. This improvement in resolution is clearly evident when comparing the dashed line to the solid narrow line in Fig. 3.53(b).

Example 3.6.3:

This example illustrates the interferogram of a source with two discrete wavelengths. The source spectrum is modeled as $S(\lambda) = 2\delta(\lambda - \lambda_1) + \delta(\lambda - \lambda_2)$, with $\lambda_1 = 1\,\mu m$ and $\lambda_2 =$

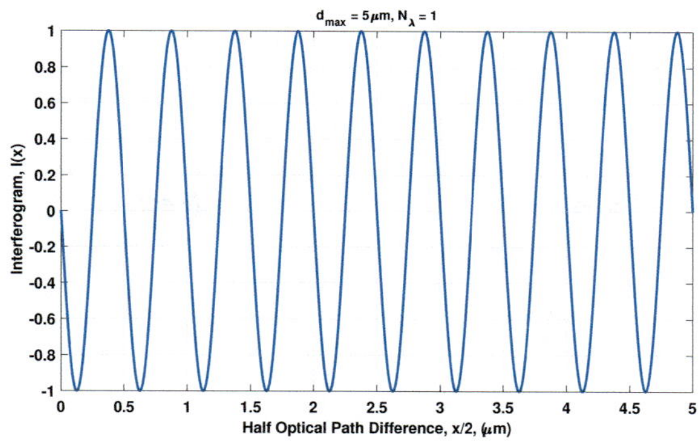

(a)

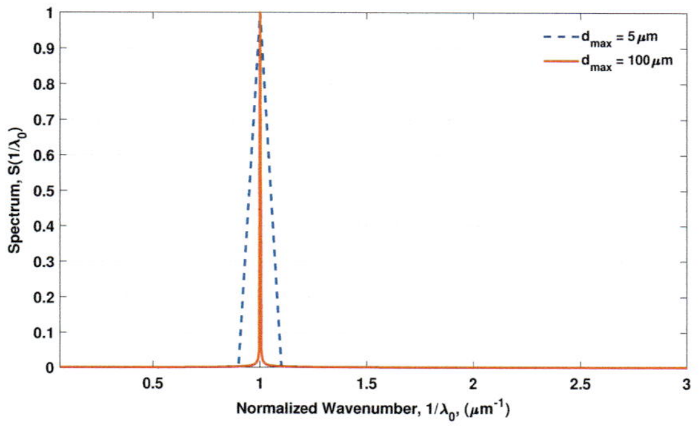

(b)

FIGURE 3.53: (a) The interferogram of a monochromatic source of wavelength $\lambda_0 = 1\,\mu$m. The maximum mirror displacement is $d_{max} = 5\,\mu$m. The intensity is assumed unity (in arbitrary units). (b) The corresponding spectrum obtain via FFT with $N = 2^{12}$ sampling points. The dashed curve represents the spectrum for $d_{max} = 5\,\mu$m, while the solid curve corresponds to $d_{max} = 100\,\mu$m. It is evident that the spectral resolution improves significantly as d_{max} increases, resulting in narrower spectral features.

$1.25\,\mu$m. For an interferogram of $d_{max} = 5\lambda_0 = 0.5 \times 10^{-5}$ m, both the interferogram and its corresponding spectrum are shown in Figs. 3.54(a) and 3.54(b), respectively, using $N = 2^{12} = 4096$ points, as in the previous example. The expected resolution remains the same as in Example 3.6.2, as the same windowing function is applied (dashed line). For $d_{max} = 100\lambda_0 = 10 \times 10^{-5}$ m, the calculated spectrum is presented in Fig. 3.54(b) (solid line). As expected, the spectral resolution improves by a factor of 20 due to the larger window size. Additionally, the relative amplitudes of the two wavelengths in the spectrum correctly reflect their intended ratio.

Example 3.6.4:

This example involves a source with a Lorentzian spectrum. Specifically, the spectrum $S(\nu)$ is assumed to be:

$$S(\nu) = \frac{(\Delta\nu/2)^2}{(\nu - \nu_0)^2 + (\Delta\nu/2)^2},$$

where $\nu_0 = c/\lambda_0 = 2.99792458 \times 10^{14}$ Hz (corresponding to $\lambda_0 = 1\,\mu$m) and $\Delta\nu = 1.5 \times 10^9$ Hz. This continuous spectrum was approximated by 1201 monochromatic sources equally spaced in the interval $\nu_0 - 20\Delta\nu < \nu < \nu_0 + 20\Delta\nu$. To achieve a high-resolution

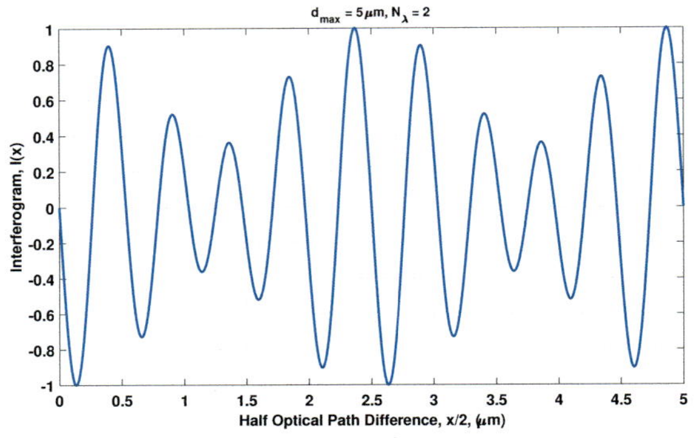

(a)

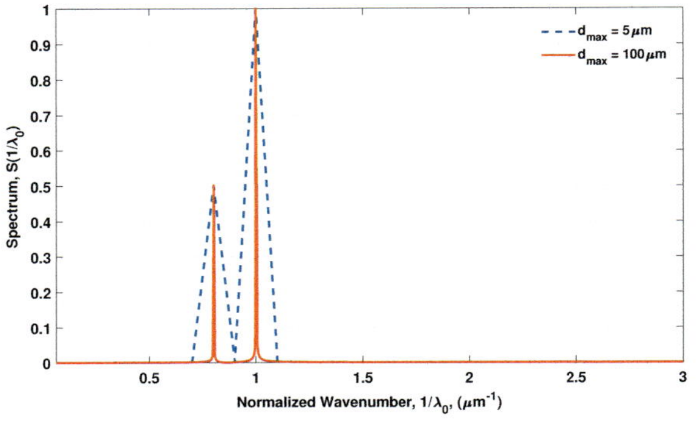

(b)

FIGURE 3.54: (a) The interferogram of a source emitting at two discrete wavelengths, $\lambda_1 = 1\,\mu\text{m}$ and $\lambda_2 = 1.25\,\mu\text{m}$. The maximum mirror displacement is $d_{max} = 5\,\mu\text{m}$. The corresponding intensities are $I(\lambda_1) = 2$ and $I(\lambda_2) = 1$ (in arbitrary units). (b) The spectrum obtained via FFT with $N = 2^{12}$ sampling points. The dashed curve represents the case of $d_{max} = 5\,\mu\text{m}$, while the solid curve represents $d_{max} = 100\,\mu\text{m}$. The resolution of the spectrum increases significantly with the larger d_{max} value, as expected.

spectrum, a large window of $d_{max} = 200000\lambda_0 = 200000\,\mu\text{m} = 20\,\text{cm}$ was employed. The interferogram and the corresponding spectrum are shown in Figs. 3.55(a) and 3.55(b), respectively. The large optical path difference was necessary to produce the exponentially decaying interferogram observed in Fig. 3.55(a). Based on Eq. (3.269), the minimum number of sampling points required was $N > 800000$, corresponding to $2^{20} = 1048576$ points. For enhanced precision, $2^{23} = 8388608$ points were used in the calculations. The resulting spectrum, $S(\bar{\nu})$, is displayed in Fig. 3.55(b) (dashed line), alongside the theoretical Lorentzian distribution of the input spectrum (dotted line). Despite the large window size, the calculated spectrum remains slightly broader than the theoretical distribution due to the finite interferogram length and the non-zero initial and final values. This discrepancy can be mitigated using non-rectangular windowing techniques. For instance, applying a Hanning window [120] to the interferogram significantly reduces spectral broadening, as shown by the solid line in Fig. 3.55(b), which closely matches the theoretical curve (dotted line). For more detailed information about Fourier Transform Spectroscopy and related windowing techniques, the reader is referred to Refs. [118, 121].

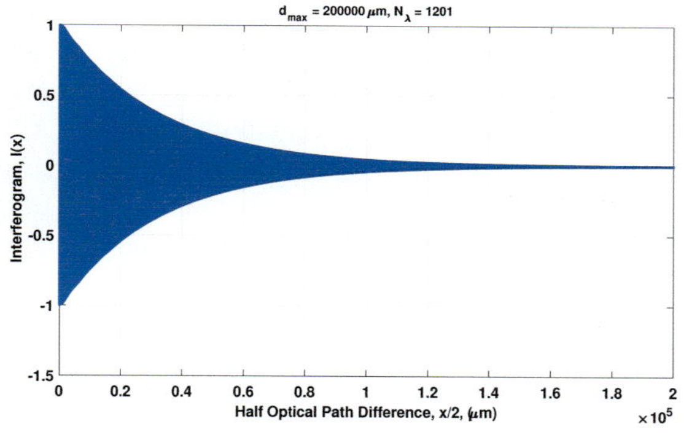

(a)

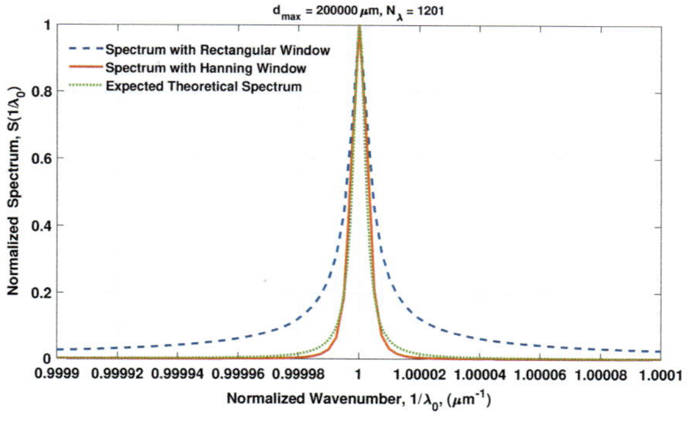

(b)

FIGURE 3.55: (a) The interferogram of a source with a Lorentzian spectrum. The intensity profile, $I(\lambda)$ (in arbitrary units), follows a Lorentzian line-shape with center frequency $\lambda_0 = 1\,\mu$m and a full-width at half-maximum (FWHM) corresponding to $\Delta\nu = 1.5 \times 10^9$ Hz. (b) The corresponding spectrum obtain via FFT with $N = 2^{23}$ sampling points for $d_{max} = 200000\,\mu$m. The dashed curve corresponds to a rectangular window, while the solid curve corresponds to a Hanning window. The dotted curve represents the theoretically expected spectrum. It is evident that the spectrum with the Hanning window better approximates the theoretically expected spectrum compared to the rectangular window.

3.7 Optical Interference Filters

The history of optical thin films dates back to 1817 when Fraunhofer first discovered the antireflection effect of a single layer on glass. Over time, thin-film optical filters have evolved significantly, progressing from simple metallic coatings to intricate multilayer stacks of dielectric materials that precisely control the reflection, transmission, and absorption of light [122, 123]. The functionality of these filters is based on the principles of multiple interference, making them indispensable in modern optical systems to enhance performance. Thin-film interference filters provide critical properties such as high reflectance, antireflection, and wavelength-band control, playing a pivotal role in defining the performance of optical components used in high-power laser systems, premium binoculars, telescopes, cameras, displays, and numerous other applications.

Optical thin films and coatings have found widespread applications across diverse fields, including telecommunications, photovoltaics, photonic devices, sensors, lasers, displays and lighting, security devices, ophthalmic technology, automotive optics, architecture, and biomedical technologies. With increasing industrial demands, the range of applications for thin-film coatings is expected to continue expanding in the future.

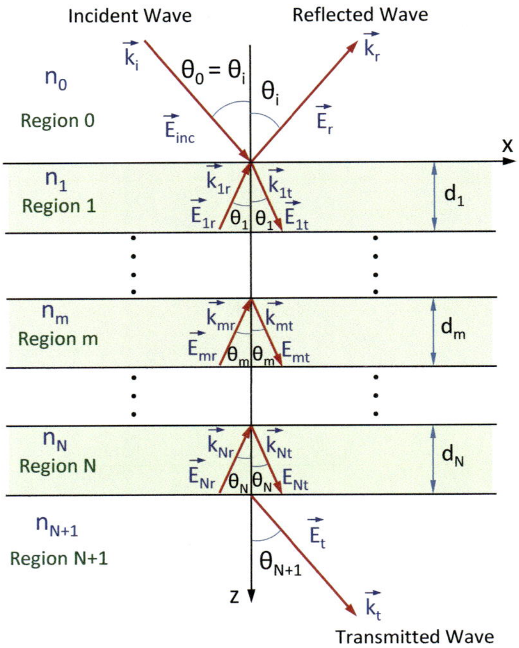

FIGURE 3.56: A general multilayer thin-film filter consists of N layers sandwiched between two regions: the input region (region 0) and the output region (region $N+1$). Each of the intermediate regions, labeled $1, 2, \cdots, N$, corresponds to the filter layers. The refractive index and thickness of the m-th layer are denoted by n_m and d_m, respectively. Within each layer, light can propagate in the forward direction ($+z$ propagating) or the backward direction ($-z$ propagating), and the corresponding electric fields describe the polarization state. All layers are assumed to be homogeneous, linear, and isotropic.

3.7.1 Transfer Matrix Method for Optical Multilayers

As previously mentioned, the operation and design of thin-film optical filters and coatings rely on the multiple interference effects of light. To extend the electromagnetic principles derived for a single-layer Fabry-Perot etalon (see Sec. 3.6.1.2), the procedure is generalized to encompass any number of layers situated between the top and bottom regions of a multilayer system. The geometry of such a generalized multilayer thin film optical filter is shown in Fig. 3.56. For simplicity, the incident light is initially assumed to exhibit transverse electric *TE* polarization. Building upon the methodology outlined in Sec. 3.6.1.2, the forward, $E_{(m+1),t}$, and backward, $E_{(m+1),r}$, electric field amplitude coefficients in the $(m+1)$-th region can be directly related to their respective counterparts $E_{m,t}$ and $E_{m,r}$, in the m-th region. This can be expressed as follows:

$$\begin{bmatrix} E_{(m+1),t} \\ E_{(m+1),r} \end{bmatrix} = \underbrace{\begin{bmatrix} \frac{1}{2}(1+K_{m,m+1})e^{-j\beta_m d_m} & \frac{1}{2}(1-K_{m,m+1})e^{+j\beta_m d_m} \\ \frac{1}{2}(1-K_{m,m+1})e^{-j\beta_m d_m} & \frac{1}{2}(1+K_{m,m+1})e^{+j\beta_m d_m} \end{bmatrix}}_{=\tilde{M}_{m+1,m}} \begin{bmatrix} E_{m,t} \\ E_{m,r} \end{bmatrix},$$

$$\text{where } K_{m,m+1} = \frac{n_m \cos\theta_m}{n_{m+1}\cos\theta_{m+1}},$$

and $\beta_m = k_0 n_m \cos\theta_m$ with $m = 0, 1, 2, \cdots, N$ (for $m = 0$ it is assumed that $d_0 = 0$). By systematically combining these tabulated electric field components across each layer, it becomes possible to establish a direct relationship between the electric field components in the output region ($m = N+1$) and those in the input region ($m = 0$). This connection is achieved through a cumulative process, where the contributions from all intermediate layers are accounted for in sequence. The resulting relationship is:

$$\begin{bmatrix} E_t \\ 0 \end{bmatrix} = \tilde{M}_{N+1,N}\tilde{M}_{N,N-1}\cdots\tilde{M}_{2,1}\tilde{M}_{1,0}\begin{bmatrix} E_i \\ E_r \end{bmatrix} = \begin{bmatrix} m_{11} & m_{12} \\ m_{21} & m_{22} \end{bmatrix}\begin{bmatrix} E_i \\ E_r \end{bmatrix} \Longrightarrow$$

$$\frac{E_r}{E_i} = r = -\frac{m_{21}}{m_{22}}, \tag{3.270}$$

$$\frac{E_t}{E_i} = t = m_{11} - \frac{m_{12}m_{21}}{m_{22}}. \tag{3.271}$$

The last two equations are in the same form as Eqs. (3.233) and (3.234) of the Fabry-Perot etalon but in this case the m_{ij} matrix elements result from the multiplication of $(N+1)$ matrices $\tilde{M}_{m+1,m}$ (for $m = 0, 1, \cdots, N$). Then the normalized reflected and transmitted intensities are given by:

$$\frac{I_r}{I_i} = |r|^2 = \frac{|m_{21}|^2}{|m_{22}|^2}, \tag{3.272}$$

$$\frac{I_t}{I_i} = \frac{n_{N+1}\cos\theta_{N+1}}{n_0\cos\theta_0}|t|^2 =$$

$$= \frac{n_{N+1}\cos\theta_{N+1}}{n_0\cos\theta_0}\left[|m_{11}|^2 + \frac{|m_{12}|^2|m_{21}|^2}{|m_{22}|^2} - 2\mathrm{Re}\left\{\frac{m_{11}^* m_{12} m_{21}}{m_{22}}\right\}\right]. \tag{3.273}$$

For the case of *TM* polarization, the analysis follows the same principles as for *TE* polarization, but with the electric field components replaced by their magnetic field counterparts. In this situation, the magnetic field amplitude coefficients of the forward, $H_{(m+1),t}$, and backward $H_{(m+1),r}$, waves in the $(m+1)$-th region, are related with their counterparts $H_{m,t}$ and $H_{m,r}$, in the m-th region, via the equation:

$$\begin{bmatrix} H_{(m+1),t} \\ H_{(m+1),r} \end{bmatrix} = \underbrace{\begin{bmatrix} \frac{1}{2}\left(1+K'_{m,m+1}\right)e^{-j\beta_m d_m} & \frac{1}{2}\left(1-K'_{m,m+1}\right)e^{+j\beta_m d_m} \\ \frac{1}{2}\left(1-K'_{m,m+1}\right)e^{-j\beta_m d_m} & \frac{1}{2}\left(1+K'_{m,m+1}\right)e^{+j\beta_m d_m} \end{bmatrix}}_{=\tilde{M}'_{m+1,m}} \begin{bmatrix} H_{m,t} \\ H_{m,r} \end{bmatrix},$$

$$\text{where} \quad K'_{m,m+1} = \frac{n_{m+1}\cos\theta_m}{n_m\cos\theta_{m+1}}, \tag{3.274}$$

and β_m is defined as before with $m = 0, 1, 2, \cdots, N$ (again for $m = 0$ it is assumed $d_0 = 0$). By multiplying these tabulated magnetic field components sequentially through the multilayer structure, the magnetic field components in the output region ($m = N + 1$) can be directly related to those in the input region ($m = 0$). The resulting equation is given by:

$$\begin{bmatrix} H_t \\ 0 \end{bmatrix} = \tilde{M}'_{N+1,N}\tilde{M}'_{N,N-1}\cdots\tilde{M}'_{2,1}\tilde{M}'_{1,0}\begin{bmatrix} H_i \\ H_r \end{bmatrix} = \begin{bmatrix} m'_{11} & m'_{12} \\ m'_{21} & m'_{22} \end{bmatrix}\begin{bmatrix} H_i \\ H_r \end{bmatrix} \Longrightarrow$$

$$r' = \frac{E_r}{E_i} = \frac{-Z_0 H_r}{Z_0 H_i} = -r_H = \frac{m'_{21}}{m'_{22}}, \tag{3.275}$$

$$t' = \frac{E_t}{E_i} = \frac{Z_{N+1}H_t}{Z_0 H_i} = \frac{n_0}{n_{N+1}}t_H = \frac{n_0}{n_{N+1}}\left[m'_{11} - \frac{m'_{12}m'_{21}}{m'_{22}}\right]. \tag{3.276}$$

Then, the normalized reflected and transmitted intensities are expressed as:

$$\frac{I_r}{I_i} = |r'|^2 = \frac{|m'_{21}|^2}{|m'_{22}|^2}, \tag{3.277}$$

$$\begin{aligned}\frac{I_t}{I_i} &= \frac{n_{N+1}\cos\theta_{N+1}}{n_0\cos\theta_0}|t'|^2 = \\ &= \frac{n_0\cos\theta_{N+1}}{n_{N+1}\cos\theta_0}\left[|m'_{11}|^2 + \frac{|m'_{12}|^2|m'_{21}|^2}{|m'_{22}|^2} - 2\mathrm{Re}\left\{\frac{m'^{*}_{11}m'_{12}m'_{21}}{m'_{22}}\right\}\right].\end{aligned} \tag{3.278}$$

3.7.2 Scattering Matrix Method for Optical Multilayers

The transfer matrix method, as outlined in Sec. 3.7.1, is a versatile and widely used approach for analyzing electromagnetic scattering and diffraction in various problems. It is particularly favored for its straightforward implementation and efficiency. However, a notable disadvantage of this method is its potential for numerical instability, especially when dealing with finite regions that involve large real exponential terms. These terms can arise in situations where a layer exhibits significant losses (or even gains) or when evanescent waves are excited in a layer with thicknesses on the order of several wavelengths. The source of this instability lies in how the transfer matrix method treats both forward and backward traveling waves identically, leading to large real exponential factors in the matrix elements. This can result in numerical overflows, making the method unsuitable for some cases. To address these issues, the *Scattering Matrix Method* has been proposed as an alternative approach. Initially introduced for quantum mechanical tunneling problems [124, 125], it has also found applications in electromagnetic analysis [126, 127]. It is important to note that in scientific literature, the term "scattering matrix method" is sometimes used interchangeably with the transfer matrix method, leading to potential confusion. In the following, the distinctions between these two methods are clarified.

To illustrate the Scattering Matrix Method, consider a multilayer optical structure, as shown in Fig. 3.57. The structure consists of N layers between the input region (layer 0) and the output region (layer $N + 1$). All layers are assumed to be homogeneous and isotropic. A plane wave with either *TE* or *TM* polarization is incident from the left at an angle θ_0. The refractive index of each layer is denoted by n_i ($i = 0, 1, \cdots, N + 1$), where all layers except the input region may have a complex refractive index. The thickness of each layer is denoted by d_i ($i = 1, \cdots, N$). The electric field amplitudes in each region are represented as $E^+_{i+1,L}$, $E^-_{i+1,L}$, $E^+_{i+1,R}$, and $E^-_{i+1,R}$ where the subscript L or R indicates the left or right edge of the $i+1$ layer, respectively. The superscripts "+" and "−" denote the electric fields propagating in the forward (positive z) or backward (negative z) directions. This setup forms the foundation for applying the Scattering Matrix Method to analyze the system.

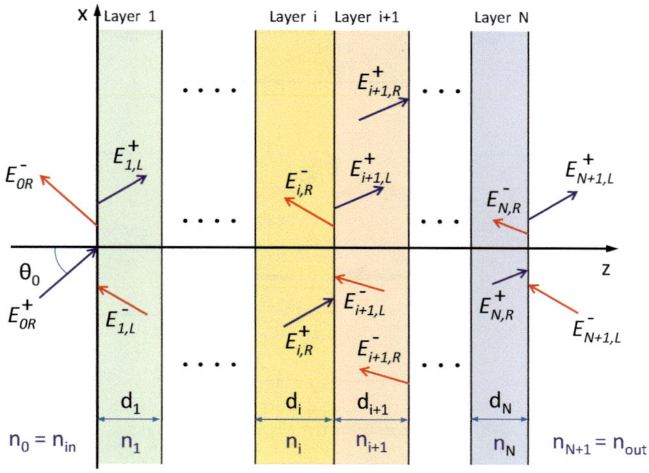

FIGURE 3.57: The geometry of a multilayer optical structure consisting of N layers is used in the scattering matrix method analysis.

At this stage, let us focus on the interface between layer i and layer $i+1$. The fundamental concept behind the scattering matrix method is to establish a relationship between the outgoing fields from the interface (E^-_{iR} and $E^+_{i+1,L}$) and the incoming fields towards the interface (E^+_{iR} and $E^-_{i+1,L}$), which is achieved through a scattering matrix, denoted as $\tilde{S}^{i/i+1}$, as shown schematically in Fig. 3.58(a). Therefore, the elements of the matrix $\tilde{S}^{i/i+1}$, $S^{i/i+1}_{11}$, $S^{i/i+1}_{12}$, $S^{i/i+1}_{21}$, and $S^{i/i+1}_{22}$, need to be determined. This can be done by using the Fresnel equations (refer to Sec. 3.2.3.1), which describe the reflection and transmission of light at an interface between two media. The polarization of the incident field must also be considered to account for the proper behavior of the waves at the interface. Specifically, the following relations can be written:

$$\begin{bmatrix} E^-_{iR} \\ E^+_{i+1,L} \end{bmatrix} = \begin{bmatrix} r_{i,i+1} & t_{i+1,i} \\ t_{i,i+1} & r_{i+1,i} \end{bmatrix} \begin{bmatrix} E^+_{iR} \\ E^-_{i+1,L} \end{bmatrix} = \underbrace{\begin{bmatrix} S^{i/i+1}_{11} & S^{i/i+1}_{12} \\ S^{i/i+1}_{21} & S^{i/i+1}_{22} \end{bmatrix}}_{=\tilde{S}^{i/i+1}} \begin{bmatrix} E^+_{iR} \\ E^-_{i+1,L} \end{bmatrix}$$

$$= \tilde{S}^{i/i+1} \begin{bmatrix} E^+_{iR} \\ E^-_{i+1,L} \end{bmatrix}, \tag{3.279}$$

where $r_{i,i+1}$ and $t_{i,i+1}$ are the reflection and transmission coefficients going from i towards $i+1$ regions and $r_{i+1,i}$ and $t_{i+1,i}$ are their counterparts going from $i+1$ toward i regions. The boundary conditions for the electric and magnetic fields are satisfied using the Fresnel coefficients, depending on the polarization of the incident field. One of the main challenges in the scattering matrix method is the cascading of two consecutive scattering matrices, which is not as straightforward as the simple multiplication used in the transfer matrix method. To illustrate this, consider the setup where the electric fields of two cascaded scattering matrices are involved (see Fig. 3.58(b)). The electric fields in the first region are related to the outgoing and incoming fields through a scattering matrix that describes the

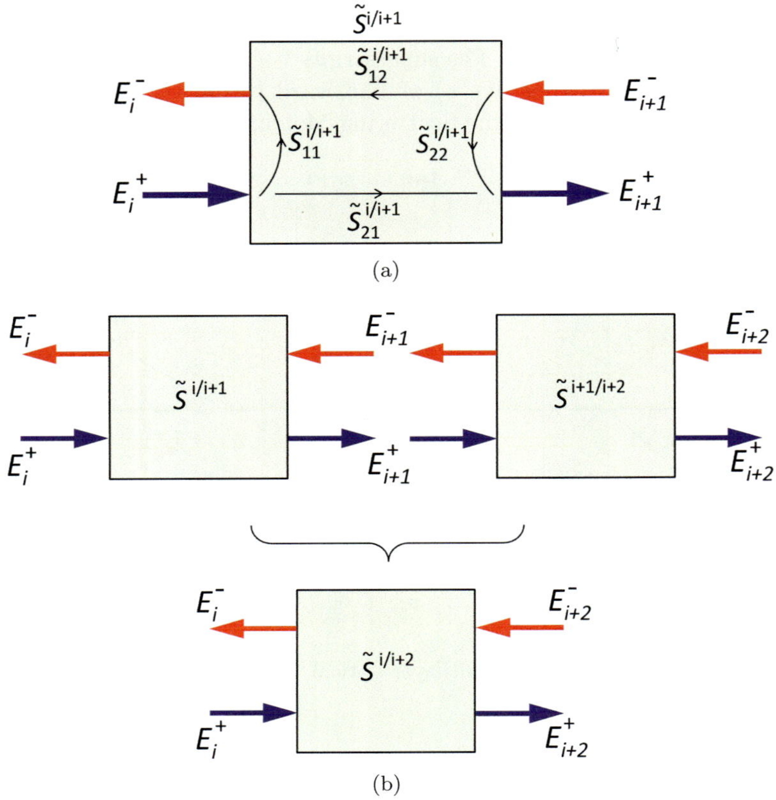

FIGURE 3.58: (a) A single block described by the scattering matrix $\tilde{S}^{i/i+1}$ between the i and $i+1$ layer. (b) Combination of two sequential scattering matrices.

interaction at the interface between the layers. Similarly, another scattering matrix governs the relationship between the electric fields in the second region. Let the electric fields of two cascaded scattering matrices be represented by the following equations (where the subscripts L and R are omitted for simplicity):

$$\begin{bmatrix} E_i^- \\ E_{i+1}^+ \end{bmatrix} = \underbrace{\begin{bmatrix} S_{11}^{i/i+1} & S_{12}^{i/i+1} \\ S_{21}^{i/i+1} & S_{22}^{i/i+1} \end{bmatrix}}_{=\tilde{S}^{i/i+1}} \begin{bmatrix} E_i^+ \\ E_{i+1}^- \end{bmatrix}$$

$$\begin{bmatrix} E_{i+1}^- \\ E_{i+2}^+ \end{bmatrix} = \underbrace{\begin{bmatrix} S_{11}^{i+1/i+2} & S_{12}^{i+1/i+2} \\ S_{21}^{i+1/i+2} & S_{22}^{i+1/i+2} \end{bmatrix}}_{=\tilde{S}^{i+1/i+2}} \begin{bmatrix} E_{i+1}^+ \\ E_{i+2}^- \end{bmatrix}.$$

The main goal is to determine the scattering matrix $\tilde{S}_{i/i+2}$ between i and $i+2$, which is defined by:

$$\begin{bmatrix} E_i^- \\ E_{i+2}^+ \end{bmatrix} = \underbrace{\begin{bmatrix} S_{11}^{i/i+2} & S_{12}^{i/i+2} \\ S_{21}^{i/i+2} & S_{22}^{i/i+2} \end{bmatrix}}_{=\tilde{S}^{i/i+2}} \begin{bmatrix} E_i^+ \\ E_{i+2}^- \end{bmatrix}. \tag{3.280}$$

To find the elements of the combined scattering matrix $\tilde{S}^{i/i+2}$ the E^+_{i+1} and E^-_{i+1} can be written in terms of E^+_i and E^-_{i+2} (from the previous equations specifying $\tilde{S}^{i/i+1}$ and $\tilde{S}^{i+1/i+2}$) and then the resulting elements of the combined scattering matrix are given by:

$$S_{11}^{i/i+2} = S_{11}^{i/i+1} + S_{12}^{i/i+1}\left[1 - S_{11}^{i+1/i+2}S_{22}^{i/i+1}\right]^{-1} S_{11}^{i+1/i+2}S_{21}^{i/i+1}, \tag{3.281}$$

$$S_{12}^{i/i+2} = S_{12}^{i/i+1}\left[1 - S_{11}^{i+1/i+2}S_{22}^{i/i+1}\right]^{-1} S_{12}^{i+1/i+2}, \tag{3.282}$$

$$S_{21}^{i/i+2} = S_{21}^{i/i+1}\left[1 - S_{22}^{i/i+1}S_{11}^{i+1/i+2}\right]^{-1} S_{21}^{i+1/i+2}, \tag{3.283}$$

$$S_{22}^{i/i+2} = S_{22}^{i/i+1} + S_{21}^{i/i+1}\left[1 - S_{22}^{i/i+1}S_{11}^{i+1/i+2}\right]^{-1} S_{22}^{i+1/i+2}S_{12}^{i/i+1}. \tag{3.284}$$

As an example let's consider the scattering matrix between E^-_{iR} and $E^+_{i+1,R}$ (consider the boundaries of the right side of layer i and of the right side of layer $i+1$, implying that the two electric fields are outgoing) and E^+_{iR} and $E^-_{i+1,R}$ (implying that these two are the incoming fields). Rewriting equation (3.279) and relating the electric fields at the left and right sides of layer $i+1$, the following expressions hold:

$$\begin{bmatrix} E^-_{iR} \\ E^+_{i+1,L} \end{bmatrix} = \begin{bmatrix} S_{11}^{i/i+1} & S_{12}^{i/i+1} \\ S_{21}^{i/i+1} & S_{22}^{i/i+1} \end{bmatrix} \begin{bmatrix} E^+_{iR} \\ E^-_{i+1,L} \end{bmatrix},$$

$$\begin{bmatrix} E^+_{i+1,R} \\ E^-_{i+1,L} \end{bmatrix} = \begin{bmatrix} 0 & \exp(-j\beta_{i+1}d_{i+1}) \\ \exp(-j\beta_{i+1}d_{i+1}) & 0 \end{bmatrix} \begin{bmatrix} E^-_{i+1,R} \\ E^+_{i+1,L} \end{bmatrix}. \tag{3.285}$$

The last equation can be written from the straightforward relations $E^+_{i+1,R} = E^+_{i+1,L}e^{-j\beta_{i+1}d_{i+1}}$ (positive z traveling wave), and $E^-_{i+1,R} = E^-_{i+1,L}e^{+j\beta_{i+1}d_{i+1}}$ (negative z traveling wave). Then, the scattering matrix $\tilde{S}^{i+1/i+2}$ has the following elements: $S_{11}^{i+1/i+2} = 0$, $S_{12}^{i+1/i+2} = e^{-j\beta_{i+1}d_{i+1}}$, $S_{21}^{i+1/i+2} = e^{-j\beta_{i+1}d_{i+1}}$, and $S_{22}^{i+1/i+2} = 0$. Applying Eqs. (3.281)–(3.284), the combined scattering matrix that takes into account the refraction at the i and $i+1$ interface as well as the propagation of length d_{i+1}, can be easily determined and is given by:

$$\begin{bmatrix} E^-_{iR} \\ E^+_{i+1,R} \end{bmatrix} = \begin{bmatrix} r_{i,i+1} & t_{i+1,i}\exp(-j\beta_{i+1}d_{i+1}) \\ t_{i,i+1}\exp(-j\beta_{i+1}d_{i+1}) & r_{i+1,i}\exp(-j2\beta_{i+1}d_{i+1}) \end{bmatrix} \begin{bmatrix} E^+_{iR} \\ E^-_{i+1,R} \end{bmatrix}. \tag{3.286}$$

Following the same recursive procedure described by Eqs. (3.281)–(3.284), it is straightforward to find the overall scattering matrix relating the outgoing fields from the optical structure E^-_{0R}, $E^+_{N+1,L}$, to the incoming fields E^+_{0R}, $E^-_{N+1,L}$, as follows:

$$\begin{bmatrix} E^-_{0R} \\ E^+_{N+1,L} \end{bmatrix} = \underbrace{\begin{bmatrix} S_{11} & S_{12} \\ S_{21} & S_{22} \end{bmatrix}}_{=\tilde{S}} \begin{bmatrix} E^+_{0R} \\ E^-_{N+1,L} \end{bmatrix}, \tag{3.287}$$

where $\tilde{S}$ is the overall scattering matrix and S_{ij} $(i,j=1,2)$ are its elements. In the case where there is only a left-incident field (i.e., for $E^-_{N+1,L}=0$), the resulting reflection and transmission coefficients for the whole multilayer structure are defined by:

$$r = \frac{E_{0R}^-}{E_{0R}^+} = S_{11}, \tag{3.288}$$

$$t = \frac{E_{N+1,L}^+}{E_{0R}^+} = S_{21}. \tag{3.289}$$

It should be noted that since the Fresnel coefficients can describe either *TE* or *TM* polarizations, the resulting procedure is applicable to both polarization types, provided the appropriate Fresnel equations are used. The normalized reflected and transmitted intensities can be determined in a manner similar to the equations for the *TE* polarization, Eqs. (3.272) and (3.273), or the corresponding equations for the *TM* polarization.

Example 3.7.1:

To demonstrate the scattering matrix method, let's consider a single air layer ($n_1 = 1.0$) between two semi-infinite homogeneous regions of silicon ($n_0 = n_2 = 3.6$). In this case, we have $N = 1$. The air layer thickness is assumed to be $d_1 = 20\,\mu$m (chosen to be large for demonstration purposes). The incident polarization is *TE* and the freespace wavelength is $\lambda_0 = 0.5\,\mu$m. The reflection and transmission coefficients of the structure are to be determined as functions of the angle of incidence θ_0.
Solution: In this case Eqs. (3.288) and (3.289) are used for the determination of the reflection, r and transmission, t, coefficients of the structure under the scattering matrix method. Similarly, Eqs. (3.270) and (3.271) are used under the transfer matrix method. Since the single slab is made of air, there is a critical angle $\theta_{cr} = \sin^{-1}(1/3.6) = 16.1276$ deg. Beyond the critical angle, evanescent waves are excited in the air slab, which, in the case of the transfer matrix method, can cause numerical instabilities. The real and imaginary parts of the reflection and transmission coefficients are shown in Fig. 3.59(a) when the scattering matrix method is utilized. The results for angles of incidence in the range $0 \leq \theta_0 < \theta_{cr}$ are identical for both the scattering and the transfer matrix methods, as no evanescent waves are excited. In the range $10 \leq \theta_0 \leq \theta_{cr} = 16.1276$ deg no evanescent waves are excited in the air slab and significant interference oscillations are observed. However, for $\theta_0 > \theta_{cr}$ only evanescent waves are excited in the air slab and tunnel to the output region. Due to the large thickness of the air slab, the transmission coefficient approaches zero. The corresponding results using the transfer matrix method are shown in Fig. 3.59(b). It is evident that for angles of incidence $\theta_0 >\sim 56$ deg, the transfer matrix method fails to compute the reflection and transmission coefficients due to numerical overflows.

3.7.3 Single-Layer Antireflection Coating

The multilayer thin-film analysis approach is initially applied to the case of a single-layer antireflection (AR) coating. By considering a single intermediate layer between the input and output regions, the fundamental principles of thin-film interference can be explored in a straightforward manner.

Antireflection coatings have a significant impact on optical technology, often surpassing all other types of coatings in terms of production volume [122]. In some applications, the primary objective of antireflection coatings is to reduce surface reflection. In other cases, however, not only must surface reflection be minimized but the transmittance must also be enhanced. A key example of this is the various lenses used in optical imaging systems. The overall transmittance of even a modest number of uncoated optical elements in series can be quite low, despite individual elements achieving transmittances of around 90%.

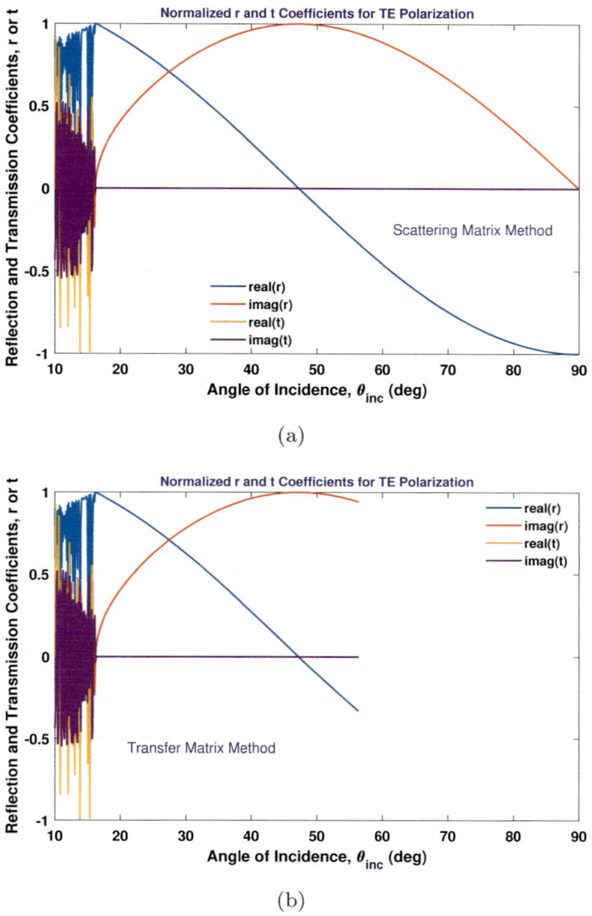

FIGURE 3.59: (a) The reflection, r, and transmission, t, coefficients of single air slab ($n_1 = 1.0$) between silicon homogeneous regions $n_0 = n_2 = 3.6$ as functions of the angle of incidence θ_0. The air slab thickness is $d_1 = 20\,\mu$m and the freespace wavelength is $\lambda_0 = 0.5\,\mu$m. The polarization is assumed to be *TE*. (b) The same results calculated using the transfer matrix method. It is observed that for angles $\theta_0 > 56$ deg the results cannot be calculated due to numerical overflows.

Additionally, some of the reflected light from the various surfaces may eventually reach the focal plane, thereby degrading image quality. This is particularly critical in zoom lenses used in television and photography, which may contain 20 or more elements and would be largely ineffective without antireflection coatings [122].

The case of a single-layer antireflecting coating is presented first. The goal of applying a single-layer coating to a substrate (such as the surface of a lens) is to completely eliminate reflected light. To analyze and design this layer, Fig. 3.42 is employed. In this figure, region 3 represents the substrate, region 1 corresponds to the surrounding medium (typically air), and region 2 is the desired single-layer coating. For a given wavelength, polarization, and angle of incidence, the characteristics of the coating (i.e., its refractive index, n_2, and thickness, d) must be determined. Initially, assume the incident polarization is *TE*. Using Eq. (3.233), the reflection coefficient can be expressed in the following form:

$$r = -\frac{(1 - K_{12}K_{23})\cos(\beta_2 d) + j(K_{23} - K_{12})\sin(\beta_2 d)}{(1 + K_{12}K_{23})\cos(\beta_2 d) + j(K_{23} + K_{12})\sin(\beta_2 d)}, \tag{3.290}$$

where $\beta_2 = k_0 n_2 \cos\theta_2$. To eliminate the reflection coefficient, the following equations must be satisfied, using the definitions of K_{12} and K_{23} as presented after Eq. (3.233):

$$(1 - K_{12}K_{23})\cos(\beta_2 d) = 0 \Longrightarrow \left(1 - \frac{n_2\cos\theta_2}{n_1\cos\theta_1}\right)\cos(\beta_2 d) = 0,$$

$$(K_{23} - K_{12})\sin(\beta_2 d) = 0 \Longrightarrow \left(\frac{n_3\cos\theta_3}{n_2\cos\theta_2} - \frac{n_2\cos\theta_2}{n_1\cos\theta_1}\right)\sin(\beta_2 d) = 0.$$

Since $n_1\sin\theta_1 = n_3\sin\theta_3 = n_2\sin\theta_2$ from Snell's law, the solution of the above equations results in:

$$\beta_2 d = (2m+1)\frac{\pi}{2},$$

$$n_2^2\cos^2\theta_2 = n_1 n_3\cos\theta_1\cos\theta_3 \Longrightarrow N = n_2\cos\theta_2 = \sqrt{n_1 n_3\cos\theta_1\cos\theta_3},$$

and with the help of Snell's law,

$$n_2 = \sqrt{n_1^2\sin^2\theta_1 + N^2}, \tag{3.291}$$

$$d = (2m+1)\frac{\lambda_0}{4\sqrt{n_1 n_3\cos\theta_1\cos\theta_3}}, \qquad \text{where } m = 0, 1, \cdots \tag{3.292}$$

Following similar steps, the solution for the *TM* polarization results in:

$$\beta_2 d = (2m+1)\frac{\pi}{2},$$

$$\frac{\cos^2\theta_2}{n_2^2} = \frac{\cos\theta_1}{n_1}\frac{\cos\theta_3}{n_3} \Longrightarrow N = \frac{n_2}{\cos\theta_2} = \sqrt{\frac{n_1}{\cos\theta_1}\frac{n_3}{\cos\theta_3}},$$

and via Snell's law,

$$n_2 = \left\{\frac{N^2}{2}\left[1 \pm \left(1 - \frac{4n_1^2\sin^2\theta_1}{N^2}\right)^{1/2}\right]\right\}^{1/2}, \tag{3.293}$$

$$d = (2m+1)\frac{\lambda_0}{4n_2^2/N}, \qquad \text{where } m = 0, 1, \cdots. \tag{3.294}$$

From the two solutions provided in Eq. (3.293), typically only one will yield a refractive index greater than unity, and this should be the selected solution. In the case of normal incidence, where $\theta_1 = \theta_2 = \theta_3 = 0$, the resulting solution for the antireflecting coating is (independent of the polarization) and is given by:

$$n_2 = \sqrt{n_1 n_3}, \tag{3.295}$$

$$d = (2m+1)\frac{\lambda_0}{4\sqrt{n_1 n_3}}, \qquad \text{where } m = 0, 1, \cdots. \tag{3.296}$$

Example 3.7.2:

As an example, the design of a single-layer antireflecting (AR) coating on a flat amorphous silicon substrate is presented. The refractive index data for amorphous silicon were obtained from Refs. [76,128,129] (corresponding to Data Set 1) or from Ref. [130] (corresponding to

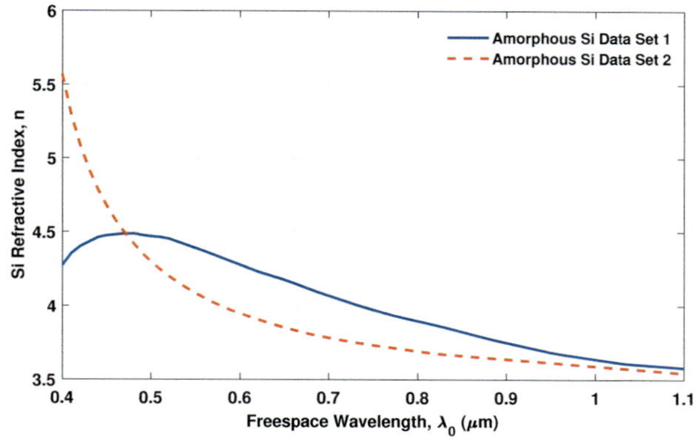

FIGURE 3.60: Refractive index data of amorphous silicon in the range of freespace wavelengths between $0.4\,\mu$m and $1.1\,\mu$m. The data Set 1 is from Refs. [76, 128, 129] while the data Set 2 is from Ref. [130].

Data Set 2). The variation of the refractive index as a function of freespace wavelength is shown in Fig. 3.60 for both Data Set 1 and Data Set 2, respectively. There is noticeable variation between the data sets, particularly in the lower wavelength region. It is assumed that the AR layer is designed for a design wavelength of $\lambda_d = 0.6\,\mu$m and for normal incidence of the incident light. Using Data Sets 1 and 2, the refractive index of amorphous silicon at the design wavelength is $n_3(\lambda_d) = 4.2760$, for data Set 1, and $n_3(\lambda_d) = 3.9473$, for data Set 2. Applying Eqs. (3.295) and (3.296), the AR coating characteristics are $n_2 = 2.0678$ and $d = 0.0725\,\mu$m for to data Set 1, and $n_2 = 2.4333$ and $d = 0.0616\,\mu$m for data Set 2. The minimum AR thickness was selected (i.e., $m = 0$).

The normalized intensity reflectivity of the amorphous silicon in air, with and without the AR coating, as a function of freespace wavelength, is shown in Figs. 3.61(a) and 3.61(b) for Data Sets 1 and 2, respectively. It is evident that the reflectivity is zero at the design wavelength of $\lambda_d = 0.6\,\mu$m. However, as the wavelength deviates from the design wavelength, the reflectivity increases.

Another practical issue with this approach is that a material with the refractive index n_2 predicted by the theory may not exist or may not be suitable for growth on the substrate. This is why multilayer coatings are predominantly used for practical AR coating designs [122]. However, before delving into the numerical design of multilayer antireflection coatings, the double-layer antireflection coating is presented, as an analytical solution is also available in this case.

3.7.4 Double-Layer Antireflection Coating

This section presents the theory of the double-layer antireflection (AR) coating, building upon the foundational work of Schuster in 1949 [131]. The double-layer AR coating offers enhanced flexibility in selecting materials to achieve antireflection. The general configuration of the double-layer AR coating is depicted in Fig. 3.62(a), which represents a specific case of the multilayer filter shown in Fig. 3.56 for $N = 2$. While the analysis assumes normal incidence, the results can be easily generalized to oblique incidence. For the analysis, it is

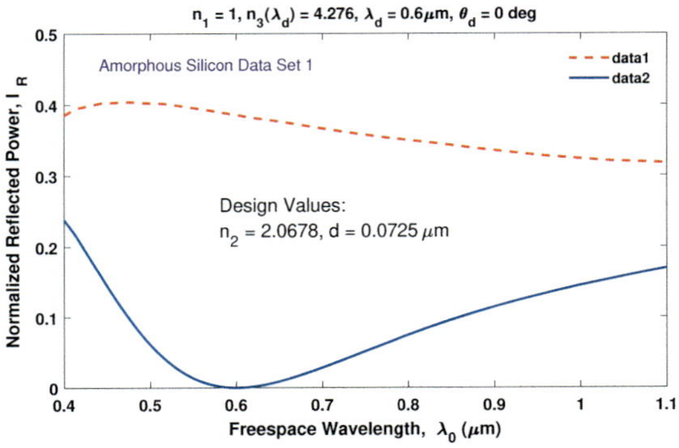

(a)

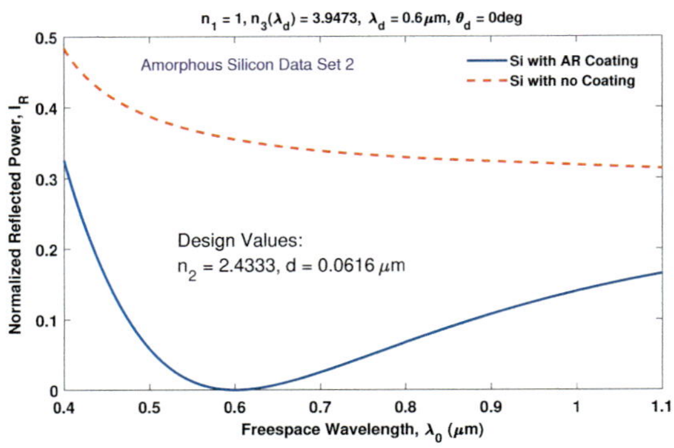

(b)

FIGURE 3.61: (a) The performance of an antireflecting (AR) layer on an amorphous silicon substrate in air. The normalized reflected intensity is shown as a function of the freespace wavelength. The refractive index of amorphous silicon was selected according to Refs. [76, 128, 129]. The design wavelength is $\lambda_d = 0.6\,\mu$m at normal incidence. The parameters of the AR layer are given by $n_2 = 2.0678$ and $d = 0.0725\,\mu$m (for minimum thickness). (b) Same as (a) but with the refractive index of amorphous silicon selected according to Ref. [130] with the same design wavelength and incidence angle as in (a). The parameters of the AR layer are given by $n_2 = 2.4333$ and $d = 0.0616\,\mu$m (again for minimum thickness).

more efficient to use the equivalence between homogeneous isotropic layers and transmission line segments, as illustrated at the top of Fig. 3.62(a) and summarized in Appendix 3G. Employing standard transmission line theory (see, for instance, Ref. [132]), the following equations can be directly derived for the impedances $Z_{2,eff}$ and $Z_{1,eff}$ as observed from the points indicated in the diagram in Fig. 3.62(a). The resulting equations are:

$$Z_{2,eff} = Z_2 \frac{Z_3\cos(\beta_2 d_2) + jZ_2\sin(\beta_2 d_2)}{Z_2\cos(\beta_2 d_2) + jZ_3\sin(\beta_2 d_2)} \Longrightarrow$$

$$\hat{Z}_{2,eff} = \frac{Z_{2,eff}}{Z_v} = \frac{\dfrac{1}{n_3}\cos(\beta_2 d_2) + j\dfrac{1}{n_2}\sin(\beta_2 d_2)}{\cos(\beta_2 d_2) + j\dfrac{n_2}{n_3}\sin(\beta_2 d_2)}, \tag{3.297}$$

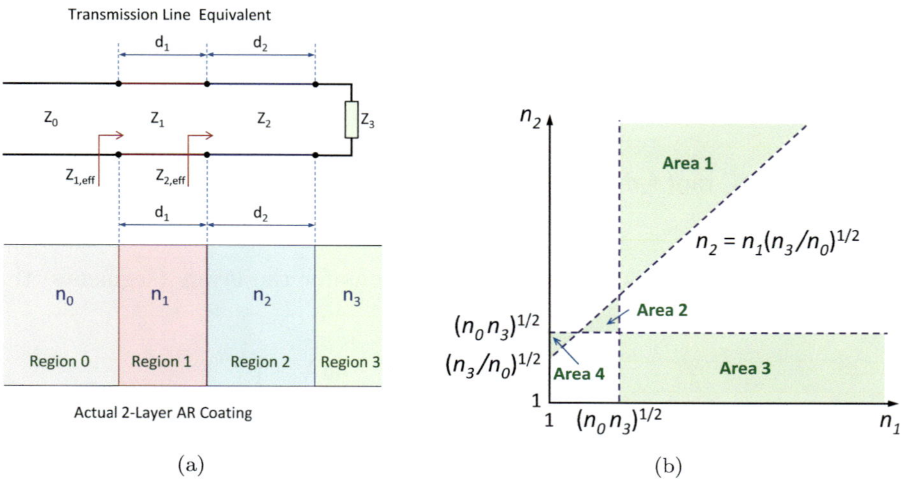

FIGURE 3.62: (a) A general double-layer antireflection coating on a substrate of refractive index n_3. The input region has a refractive index n_0. The refractive indices n_i and the thicknesses d_i (where $i = 1, 2$) are sought to obtain zero reflectance at a design wavelength. On the top, the transmission line equivalent circuit is shown along with the effective impedances $Z_{1,eff}$ and $Z_{2,eff}$ from the points observed. The characteristic impedances of the transmission line segments are equal to the wave impedances, $Z_m = Z_v/n_m$ of the corresponding homogeneous regions (where $m = 0, 1, 2, 3$ and $Z_v = \sqrt{\mu_0/\epsilon_0}$ the wave impedance of freespace). (b) The feasible regions for the refractive indices n_1 and n_2 are depicted. These regions are shaded and labeled 1 through 4, as identified in Ref. [131].

$$Z_{1,eff} = Z_1 \frac{Z_{2,eff}\cos(\beta_1 d_1) + jZ_1 \sin(\beta_1 d_1)}{Z_1 \cos(\beta_1 d_1) + jZ_{2,eff}\sin(\beta_1 d_1)} \Longrightarrow$$

$$\hat{Z}_{1,eff} = \frac{Z_{1,eff}}{Z_v} = \frac{\dfrac{Z_{2,eff}}{Z_v}\cos(\beta_1 d_1) + j\dfrac{1}{n_1}\sin(\beta_1 d_1)}{\cos(\beta_1 d_1) + j\dfrac{Z_{2,eff}}{Z_v} n_1 \sin(\beta_2 d_2)}, \tag{3.298}$$

where $Z_v \simeq 376.73\,\Omega$ is the freespace wave impedance, $\beta_1 = k_0 n_1$, and $\beta_2 = k_0 n_2$ (it is reminded that normal incidence is assumed). To achieve zero reflection at the interface between region 0 and region 1, the transmission line equivalent must ensure zero reflection between the input section of the transmission line, with characteristic impedance Z_0, and the remainder of the transmission line, which consists of the segments d_1 (with characteristic impedance Z_1), d_2 (with characteristic impedance Z_2), and the load of impedance Z_3. Thus, it is sufficient to satisfy the condition $Z_{1,eff} = Z_0$. From this requirement, the following equations can be derived:

$$Z_0 = Z_{1,eff} \Longrightarrow \frac{1}{n_0} = \frac{\hat{Z}_{2,eff}\cos(\beta_1 d_1) + j\dfrac{1}{n_1}\sin(\beta_1 d_1)}{\cos(\beta_1 d_1) + j\hat{Z}_{2,eff} n_1 \sin(\beta_2 d_2)} \Longrightarrow$$

$$\hat{Z}_{2,eff} = \frac{\cos(\beta_1 d_1) - j\dfrac{n_0}{n_1}\sin(\beta_1 d_1)}{n_0 \cos(\beta_1 d_1) - jn_1 \sin(\beta_1 d_1)}, \qquad \text{and using Eq. (3.297),}$$

$$\frac{\frac{1}{n_3}\cos(\beta_2 d_2) + j\frac{1}{n_2}\sin(\beta_2 d_2)}{\cos(\beta_2 d_2) + j\frac{n_2}{n_3}\sin(\beta_2 d_2)} = \frac{\cos(\beta_1 d_1) - j\frac{n_0}{n_1}\sin(\beta_1 d_1)}{n_0\cos(\beta_1 d_1) - jn_1\sin(\beta_1 d_1)} \Longrightarrow$$

$$\tan(\beta_1 d_1)\tan(\beta_2 d_2) = \frac{1 - \frac{n_0}{n_3}}{\frac{n_1}{n_2} - \frac{n_0 n_2}{n_1 n_3}}, \qquad \text{and} \qquad \frac{\tan(\beta_1 d_1)}{\tan(\beta_2 d_2)} = \frac{\frac{n_2}{n_3} - \frac{n_0}{n_2}}{\frac{n_0}{n_1} - \frac{n_1}{n_3}}.$$

From the last two equations, the following solutions for the layers (segments) thicknesses can be found:

$$\tan(\beta_1 d_1) = \pm\left\{\frac{\left(1 - \frac{n_0}{n_3}\right)\left(\frac{n_2}{n_3} - \frac{n_0}{n_2}\right)}{\left(\frac{n_1}{n_2} - \frac{n_0 n_2}{n_1 n_3}\right)\left(\frac{n_0}{n_1} - \frac{n_1}{n_3}\right)}\right\}^{1/2} = \pm\sqrt{\frac{AC}{BD}}, \tag{3.299}$$

$$\tan(\beta_2 d_2) = \pm\left\{\frac{\left(1 - \frac{n_0}{n_3}\right)\left(\frac{n_0}{n_1} - \frac{n_1}{n_3}\right)}{\left(\frac{n_1}{n_2} - \frac{n_0 n_2}{n_1 n_3}\right)\left(\frac{n_2}{n_3} - \frac{n_0}{n_2}\right)}\right\}^{1/2} = \pm\sqrt{\frac{AD}{BC}}, \tag{3.300}$$

where the parameters A, B, C, and D are given by:

$$A = 1 - \frac{n_0}{n_3},$$

$$B = \frac{n_1}{n_2} - \frac{n_0 n_2}{n_1 n_3},$$

$$C = \frac{n_2}{n_3} - \frac{n_0}{n_2},$$

$$D = \frac{n_0}{n_1} - \frac{n_1}{n_3}.$$

From a practical perspective, the input layer is typically air, while the substrate layer has a higher refractive index. Consequently, $n_3 > n_0$, which implies $A > 0$. For Eqs. (3.299) and (3.300) to yield real solutions, the product $BCD > 0$ must hold [131,133]. The specific values of n_1 and n_2 that nullify the parameters B, C, and D are as follows:

$$B = 0 \Longrightarrow n_2 = n_1\sqrt{\frac{n_3}{n_0}},$$

$$C = 0 \Longrightarrow n_2 = \sqrt{n_0 n_3},$$

$$D = 0 \Longrightarrow n_1 = \sqrt{n_0 n_3}.$$

The lines representing the conditions where the parameters B, C, and D are zero are shown in Fig. 3.62(b). The requirement that $BCD > 0$ is summarized in Table 3.1. The shaded regions in Fig. 3.62(b) illustrate the parameter areas where solutions for the refractive indices of the double-layer coating are possible. These regions demonstrate the increased flexibility offered by the double-layer design in selecting material refractive indices. When $n_0 = 1$, region 4 vanishes because $\sqrt{n_3 n_0}$ and $\sqrt{n_3/n_0}$ coincide along the n_2 axis. Along the slanted line, where $B = 0$, the tangents in Eqs. (3.299) and (3.300) become infinite, leading to the condition $\beta_i d_i = (2m_i + 1)(\pi/2)$ (for $i = 1, 2$ and m_i a non-negative integer). For solutions along the horizontal line $C = 0$, Eq. (3.299) becomes zero while Eq. (3.300) becomes infinite, resulting in $\beta_1 d_1 = m_1\pi$ and $\beta_2 d_2 = (2m_2 + 1)(\pi/2)$. Alternatively, along the vertical line $D = 0$, Eq. (3.299) becomes infinite and Eq. (3.300) becomes zero, leading to $\beta_1 d_1 = (2m_1 + 1)(\pi/2)$ and $\beta_2 d_2 = m_2\pi$. For solutions within the shaded regions, the layer thicknesses can be calculated using Eqs. (3.299) and (3.300).

TABLE 3.1: Signs of B, C, and D that make $BCD > 0$ (Refs. [131] and [133]).

B	C	D	Solution Area (see Fig. 3.62(b))
+	+	+	**Area 1**
+	−	−	**Area 4**
−	+	−	**Area 2**
−	−	+	**Area 3**

Example 3.7.3:

An example design of a two-layer antireflection (AR) coating on amorphous silicon in air is presented. In Sec. 3.7.3, the use of a single-layer coating was demonstrated. In this case, the refractive index of the amorphous silicon is represented by data Set 2 [130], and the design wavelength for zero reflection is again chosen as $\lambda_d = 0.6\,\mu$m. At this wavelength, the refractive index of amorphous silicon is $n_3 = 3.9473$. A solution along the $B = 0$ line, where both layers are designed to be quarter-wavelength thick, is selected. Assuming $n_1 = 2.0$, the second layer's refractive index is calculated as $n_2 = n_1\sqrt{n_3/n_0} = 3.9736$, where $n_0 = 1.0$. The layer thicknesses are determined by the quarter-wavelength condition, $d_1 n_1 = \lambda_d/4 = n_2 d_2$ (with $m_1 = m_2 = 0$ for minimum thicknesses). This results in $d_1 = 0.0750\,\mu$m and $d_2 = 0.0377\,\mu$m. The normalized reflected intensity as a function of the freespace wavelength is shown in Fig. 3.63, alongside results for the single-layer coating and the uncoated silicon substrate for comparison. It is evident that, for amorphous silicon, the two-layer AR coating offers a slight improvement in reducing reflectance compared to the single-layer design. However, the primary advantage lies in the increased flexibility in selecting practical and available materials for the coating layers.

3.7.5 Optimization of General Antireflection Coatings

While single-layer and double-layer antireflection coatings provide rigorous analytical solutions, their practicality is often limited. This is because the required refractive indices derived from theoretical calculations may correspond to materials that do not exist or are unsuitable for deposition. As a result, the design of multilayer AR coatings typically begins with a pre-selection of available materials that meet fabrication and application constraints. The task then becomes determining the optimal arrangement of these materials, including their sequence, number of layers, and respective thicknesses. Since analytical solutions are generally unavailable for multilayer systems with numerous unknowns, optimization techniques are used to achieve the desired performance. Global optimization methods are particularly effective in addressing this complex design problem. Several approaches for thin-film filter optimization are summarized in Ref. [134]. This section focuses on three prominent methods for designing multilayer AR coatings: (a) the simulated annealing method [135,136]; (b) the genetic algorithm method [137,138]; and (c) the particle

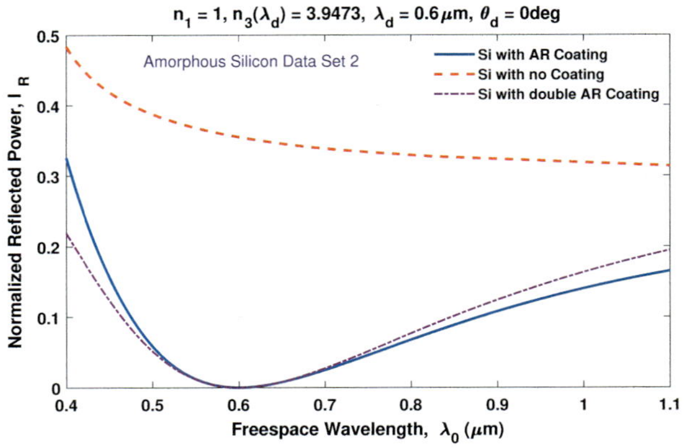

FIGURE 3.63: Performance of a double-layer antireflecting coating on an amorphous silicon substrate in air. The normalized reflected intensity is plotted as a function of the freespace wavelength. Results for the double-layer AR coating are compared against those for a single-layer AR coating and an uncoated silicon substrate. The double layer has been designed for $\lambda_d = 0.6\,\mu$m, and $n_0 = 1.0$ (air) and $n_3 = 3.9473$ (according to data Set 2 Ref. [130]). The solution was selected for quarter-wavelength thicknesses $d_1 n_1 = \lambda_d/4 = n_2 d_2$ and $n_1 = 2.0$ and $n_2 = n_1\sqrt{n_3/n_0} = 3.9736$ (along the slanted line of quarter-wave thicknesses of area 1 of Fig. 3.62(b)).

swarm optimization method [139, 140]. Although the details of these methods are not discussed here, interested readers can refer to Ref. [135] for simulated annealing, Ref. [137] for genetic algorithms, and Ref. [139] for particle swarm optimization. In the design examples that follow, these optimization methods were implemented using the MATLAB Optimization Toolbox [141], which provides a robust framework for solving the complex optimization problems associated with multilayer AR coatings.

Consider initially the examples 3.7.2 (used in Sec. 3.7.3) and example 3.7.3 (used in Sec. 3.7.4) as coatings on a silicon substrate. For simplicity, the dispersion of silicon will not be considered. The silicon's refractive index is assumed to be $n_s = 3.9473$ (which was the value considered at $\lambda_0 = 0.6\,\mu$m) and is assumed constant over the range of $0.4\,\mu$m to $1.1\,\mu$m. To proceed with the design of the antireflection coatings, a merit function is required to evaluate the performance of the coating in minimizing reflection over a specified wavelength range. In this case, the target range is from $\lambda_{01} = 0.45\,\mu\text{m} \leq \lambda_0 \leq 0.75\,\mu\text{m} = \lambda_{02}$, around the design wavelength of $\lambda_0 = 0.6\,\mu$m. The merit function, F, is defined as:

$$F = F(d_1, n_1, \cdots d_N, n_N) = \frac{100}{\lambda_{02} - \lambda_{01}} \int_{\lambda_{01}}^{\lambda_{02}} R(\lambda; d_1, n_1, \cdots, d_N, n_N) d\lambda$$

$$\simeq 100 \left\{ \frac{1}{N_w} \sum_{\lambda_i} R(\lambda_i; d_1, n_1, \cdots, d_N, n_N) \right\}. \tag{3.301}$$

In this context, d_i, n_i represent the thickness and refractive index of layer i, respectively, while N denotes the total number of layers in the antireflection coating. The merit function F expresses the average reflected intensity in percent. The summation in the merit function corresponds to the approximation of the integral over the wavelength range, evaluated at N_w wavelength points, λ_i. The number of λ_i is arbitrary though a common choice is to select N_w equally spaced wavelengths, with a separation of $\Delta\lambda = (\lambda_{02} - \lambda_{01})/N_w$, where λ_{01} and λ_{02} define the boundaries of the wavelength range of interest. The goal of the function F is to minimize the average reflectance of the coating over the specified wavelength range. By optimizing the design parameters (such as the thicknesses and refractive indices of the layers), the objective is to achieve the lowest possible reflection and, thus, maximize the antireflective performance of the coating.

From the definition of the objective function F, it is evident that an ideal design would achieve a merit function of zero, which would represent its absolute minimum, since $F \geq 0$ at all times (represents average intensity). However, the best achievable outcome is to minimize the merit function, thereby resulting in very small reflectance within the specified wavelength range. Given that the materials to be used must be predefined, the optimization will be performed over the layer thicknesses d_i, with the number of layers remaining arbitrary.

In the initial attempt, only two layers are considered, with refractive indices $[n_1, n_2] = [n_L, n_H] = [1.46, 2.40]$. The optimization parameters are the layer thicknesses d_1 and d_2, with practical bounds enforced: $0.01\,\mu m \leq d_i \leq 1.0\,\mu m$ for $i = 1, 2$. For this design, $N_w = 100$ wavelength points are sampled within the range of interest. All three global optimization methods used yielded the same optimal result: $d_1 = 0.0701\,\mu m$ and $d_2 = 0.0476\,\mu m$ with a minimum merit function of $F_{min} = 1.2764\%$, which corresponds to the average normalized reflected intensity in the design interval $0.45\,\mu m \leq \lambda_0 \leq 0.75\,\mu m$. The spectral response of the antireflection (AR) coating on silicon is shown in Fig. 3.64, where it is evident that this design performs better than previous ones that achieved zero reflectivity only for the design wavelength $\lambda_d = 0.6\,\mu m$. Since there are only two design parameters in this case, it is easy to visualize the merit function F as a function of d_1 and d_2, as shown in Figs. 3.65(a) and 3.65(b). It is worth noting that using the reverse combination of refractive indices, $[n_1, n_2] = [n_H, n_L] = [2.40, 1.46]$, results in a significantly worse design, with $d_1 = 0.0282\,\mu m$ and $d_2 = 0.0306\,\mu m$ yielding $F_{min} = 3.281\%$, as it is shown in Fig. 3.64.

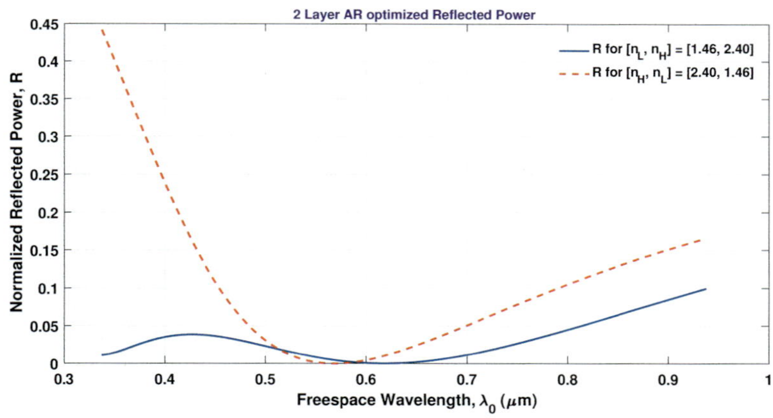

FIGURE 3.64: The performance (normalized reflected power) of an optimized antireflective 2-layer filter in the range $0.45\,\mu m \leq \lambda_0 \leq 0.75\,\mu m$, on silicon based on alternating layers of silicon oxide ($n_L = 1.46$) and titanium oxide ($n_H = 2.40$). Normal incidence is assumed and consequently, the results for *TE* and *TM* incident polarization coincide. All three optimization methods gave the same results, i.e., $d_1 = 0.0701\,\mu m$ and $d_2 = 0.0476\,\mu m$ which yield an average reflected intensity of 1.2764% in the interval of interest. In the same figure the normalized reflected intensity for the reverse layer combination [2.40, 1.46] with optimized thicknesses of $d_1 = 0.0282\,\mu m$ and $d_2 = 0.0306\,\mu m$ yielding $F_{min} = 3.281\%$, which is obviously worse than the initial layer combination.

A three-layer antireflection (AR) coating design was next attempted using three materials: silicon oxide, $n_L = 1.46$, hafnium oxide, $n_M = 1.90$, and titanium oxide, $n_H = 2.40$. The objective was to determine the thicknesses d_1, d_2, d_3 that minimize the merit function F. All optimizations algorithms used converged to the solution $[d_1, d_2, d_3] =$

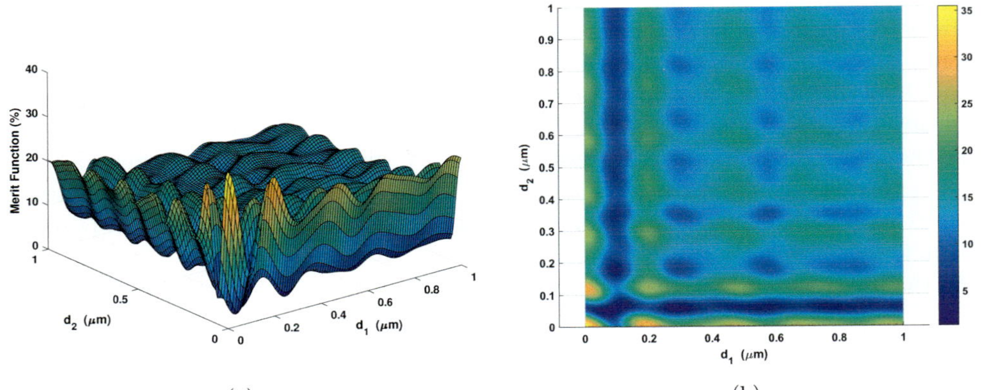

(a) (b)

FIGURE 3.65: (a) The merit function F (average percentage of reflected intensity in the interval $0.45\,\mu m \leq \lambda_0 \leq 0.75\,\mu m$) as a function of the two design parameters, layer thicknesses d_1 and d_2 (corresponding to $n_L = 1.46$ and $n_H = 2.40$). (b) Shaded version of the merit function where the region of global minimum is more evident.

$[0.0632, 0.0100, 0.0432]\,\mu$m, with a minimum merit function of $F_{min} = 1.3246\%$. As before, bounds were set for the layer thicknesses. The resulting performance of the three-layer AR coating is summarized in Fig. 3.66. It is apparent that the performance of the three-layer coating is very close to that of the two-layer coating, although slightly worse. This is primarily due to the nonzero lower limit imposed on the layer thicknesses, $d_{min} = 0.01\,\mu m$. As a result, the second layer thickness is set to the minimum value. Consequently, the three-layer design, given the pre-specified materials and order, does not outperform the two-layer design.

From the previous designs, it is evident that the selection and order of materials used in the filter can significantly impact performance. Therefore, the optimization algorithm can be adapted to handle both continuous (layer thicknesses) and discrete (materials) variables. To achieve this, the merit function was modified to account for discrete material choices, following the method proposed by Kitayama [142]. The modified merit function incorporates the material selection along with the layer thicknesses, allowing the optimization process to explore both continuous and discrete design variables. This approach ensures that the optimization algorithm can evaluate a wider range of material combinations while minimizing the reflected intensity across the desired wavelength range. The resulting modified merit function becomes:

$$F = F(d_1, n_1, \cdots d_N, n_N) = 100 \left\{ \frac{1}{N_w} \sum_{\lambda_i} R(\lambda_i; d_1, n_1, \cdots d_N, n_N) \right\} + A\Phi, \qquad (3.302)$$

where,

$$\Phi = \Phi(n_1, n_2, \cdots, n_N) = \frac{1}{2} \sum_{i=1}^{N} \left\{ \sin\left[\frac{2\pi\,(n_i - 0.25(n_{D,i,j+1} + 3n_{D,i,j})}{n_{D,i,j+1} - n_{D,i,j}} \right] + 1 \right\}.$$

The set of available refractive indices is denoted by $\{n_D\} = \{n_{D,1}, \cdots, n_{D,M}\}$, which represents the refractive indices of M materials available for the coating layers. The refractive index of the i-th layer, n_i, must lie between two consecutive values in the discrete set, specifically $n_{D,i,j} \leq n_i \leq n_{D,i,j+1}$ where $n_{D,i,j}$ and $n_{D,i,j+1}$ are two consecutive refractive

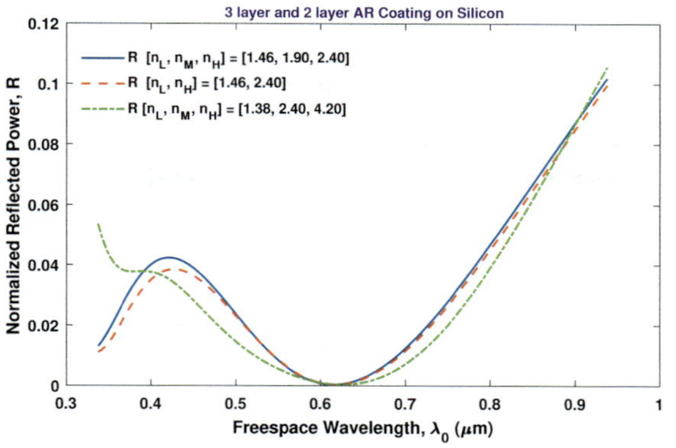

FIGURE 3.66: The performance (normalized reflected power) of an optimized three-layer AR filter in the wavelength range $0.45\,\mu m \leq \lambda_0 \leq 0.75\,\mu m$ on silicon, based on alternating layers of silicon oxide ($n_L = 1.46$), hafnium oxide ($n_M = 1.90$), and titanium oxide ($n_H = 2.40$). Normal incidence is assumed, and consequently, the results for *TE* and *TM* incident polarization coincide. All three optimization methods yielded the same results for the layer thicknesses: $d_1 = 0.0632\,\mu m$, $d_2 = 0.0100\,\mu m$, and $d_3 = 0.0432\,\mu m$, resulting in an average reflected intensity of 1.3246% in the specified wavelength range. The figure also shows the normalized reflected intensity for the $[1.46, 2.40]$ layer combination (dashed line). The three-layer filter performance with design material set, $\{n_D\} = \{1.38, 1.46, 1.90, 2.40, 4.20\}$ is also shown (dashed-dotted line). The optimized coating in this case has $[n_1, n_2, n_3] = [1.38, 2.40, 4.20]$ and $[d_1, d_2, d_3] = [0.0759, 0.0506, 0.0648]\,\mu m$ with an average reflected intensity of 0.855%.

indices from the design set. The function Φ becomes zero when all refractive indices are chosen from the set $\{n_D\}$, ensuring that the optimization algorithms only select materials from the available set. The factor A is typically a large number used to enforce that the refractive indices must belong to the discrete set $\{n_D\}$. For this example, the discrete material refractive indices set is defined as $\{n_D\} = \{1.38, 1.46, 1.90, 2.40, 4.20\}$, where the refractive index $n_D = 1.38$ corresponds to magnesium fluoride, and $n_D = 4.20$ corresponds to germanium. The extinction coefficients of the materials are neglected for this demonstration. After optimization, the resulting 3-layer AR filter has the following parameters: $[n_1, n_2, n_3] = [1.38, 2.40, 4.20]$ and $[d_1, d_2, d_3] = [0.0759, 0.0506, 0.0648]\,\mu$m with an average reflected intensity of 0.855% in the range of interest. The performance of this optimized 3-layer AR coating is shown in Fig. 3.66, alongside the previous 3-layer design and a 2-layer design. From the figure, it is clear that incorporating the material set into the optimization significantly improved the AR coating performance.

It is important to note that as the number of free variables increases, different optimization methods may yield different results. The merit function in a multi-variable space may contain many local minima, and because the optimization algorithms are stochastic in nature, they may converge to slightly different solutions each time they are run. Therefore, multiple trials may be necessary to ensure the best optimization result.

The final design example demonstrates the application of the optimization procedure to a multilayer AR coating for solar cells. This design, originally presented in Ref. [138], utilizes a genetic algorithm for the coating design process. The objective is to create a multilayer AR coating consisting of alternating layers of zinc sulfide ($n_L = 2.20$) and germanium ($n_H = 4.20$) that achieves minimal reflectivity (ideally zero) in the wavelength range $\lambda_{min} = 7.7\,\mu m \leq \lambda_0 \leq 12.3\,\mu m = \lambda_{max}$. For practical reasons, the layer thicknesses d_i $(i = 1, \cdots, N)$ are constrained between $0.01\,\mu m$ and $1.00\,\mu m$. Two cases were analyzed: (1) a design with 17 layers ($N = 17$); and (2) a design with 18 layers ($N = 18$).

The three optimization methods used, with the merit function defined in Eq. (3.301) and $N_w = 101$ sampled wavelengths, produced different results with similar performances. The results for the designs are summarized in Tables 3.2 (for $N = 17$) and 3.3 (for $N = 18$), and their spectral performance is shown in Fig. 3.67. If the layer thicknesses are allowed to extend up to 1.40 μm, a slightly better design can be achieved with the following thicknesses for $N = 18$: $d_i = [1.088, 0.607, 0.513, 0.111, 1.245, 1.228, 0.293, 0.213, 0.171, 1.355, 0.134, 1.4, 0.033, 1.344, 0.178, 0.413, 0.16]$ μm and average reflectivity of 0.6999%, which is very close to the results published in Ref. [138].

Even though only AR coatings have been designed in this section, the same optimization approach can be applied to any type of optical filter, such as low-pass, high-pass, band-pass, or band-reject filters. By appropriately modifying the merit functions in Eq. (3.301) or Eq. (3.302), this procedure can be adapted for both transmission and reflection performance optimization, making it a versatile tool for a wide range of optical filter designs.

TABLE 3.2: Antireflection coating on silicon comprised of two alternating materials (17 layers, $\{n_c 16[n_L, n_H] n_L n_s\}$, $n_c = 1.0$ (air), $n_s = 4.0$ (silicon), $n_L = 2.20$ (zinc sulfide), and $n_H = 4.20$ (germanium). Optimized using Particle Swarm, Genetic Algorithm, and Simulated Annealing.

	Particle Swarm		**Genetic Algorithm**		**Simulated Annealing**	
Layer	**d(μm)**	**n**	**d(μm)**	**n**	**d(μm)**	**n**
Air	-	1	-	1		1
1	1.0000	2.20	1.0000	2.20	0.9780	2.20
2	0.7534	4.20	0.7605	4.20	0.7473	4.20
3	0.5023	2.20	0.6089	2.20	0.5008	2.20
4	0.0794	4.20	0.0955	4.20	0.1226	4.20
5	1.0000	2.20	0.9999	2.20	0.8769	2.20
6	0.0100	4.20	0.8149	4.20	1.0000	4.20
7	0.2550	2.20	0.2624	2.20	0.7634	2.20
8	0.7020	4.20	0.1965	4.20	0.0157	4.20
9	0.1008	2.20	0.9994	2.20	0.9876	2.20
10	0.0104	4.20	0.1785	4.20	1.0000	4.20
11	0.0104	2.20	0.0777	2.20	0.9525	2.20
12	0.2768	4.20	0.9993	4.20	0.0100	4.20
13	0.6161	2.20	0.8098	2.20	0.4105	2.20
14	0.0100	4.20	0.1389	4.20	0.0100	4.20
15	0.9106	2.20	0.6611	2.20	0.6083	2.20
16	0.1293	4.20	0.2747	4.20	0.2274	4.20
17	0.3566	2.20	0.1176	2.20	0.1551	2.20
Si	-	4.00	-	4.00		4.00
AR Filter Total Thickness (μm)	**6.7232**	-	**8.9957**	-	**9.3661**	-
Average R (%)	**0.9224**		**0.7857**		**1.2527**	

TABLE 3.3: Antireflection coating on silicon comprised of two alternating materials (18 layers, $\{n_c 18[n_L, n_H] n_s\}$, $n_c = 1.0$ (air), $n_s = 4.0$ (silicon), $n_L = 2.20$ (zinc sulfide), and $n_H = 4.20$ (germanium). Optimized using Particle Swarm, Genetic Algorithm, and Simulated Annealing.

	Particle Swarm		**Genetic Algorithm**		**Simulated Annealing**	
Layer	**d(μm)**	**n**	**d(μm)**	**n**	**d(μm)**	**n**
Air	-	1	-	1		1
1	1.0000	2.20	1.0000	2.20	0.0332	2.20
2	0.0100	4.20	0.0100	4.20	0.0100	4.20
3	0.0768	2.20	0.0881	2.20	1.0000	2.20
4	0.0100	4.20	0.6240	4.20	0.6481	4.20
5	0.0100	2.20	0.0100	2.20	0.0100	2.20
6	0.6194	4.20	0.0100	4.20	0.0100	4.20
7	0.5157	2.20	0.3862	2.20	0.4980	2.20
8	0.1544	4.20	0.1614	4.20	0.1525	4.20
9	1.0000	2.20	1.0000	2.20	1.0000	2.20
10	0.7567	4.20	0.0100	4.20	0.7630	4.20
11	0.0814	2.20	0.3198	2.20	0.0810	2.20
12	0.3644	4.20	0.2402	4.20	0.3584	4.20
13	0.9999	2.20	0.2187	2.20	1.0000	2.20
14	0.1632	4.20	0.0100	4.20	0.1635	4.20
15	0.4908	2.20	0.0100	2.20	0.4866	2.20
16	0.4507	4.20	0.8424	4.20	0.4431	4.20
17	0.0131	2.20	0.0638	2.20	0.0111	2.20
18	0.7137	4.20	0.1007	4.20	0.7228	4.20
Si	-	4.00	-	4.00		4.00
AR Filter Total Thickness (μm)	**7.4303**	-	**5.1051**	-	**7.3914**	-
Average R (%)	**0.8099**		**1.1344**		**0.8197**	

3.8 Anisotropic Interfaces and Multilayers

Anisotropic materials are indispensable in many optical applications because they offer unique capabilities to manipulate light in ways that isotropic materials cannot. These materials exhibit direction-dependent optical properties, enabling precise control over light propagation, polarization, and phase. This directional dependence makes them crucial for devices and systems that require specialized light management.

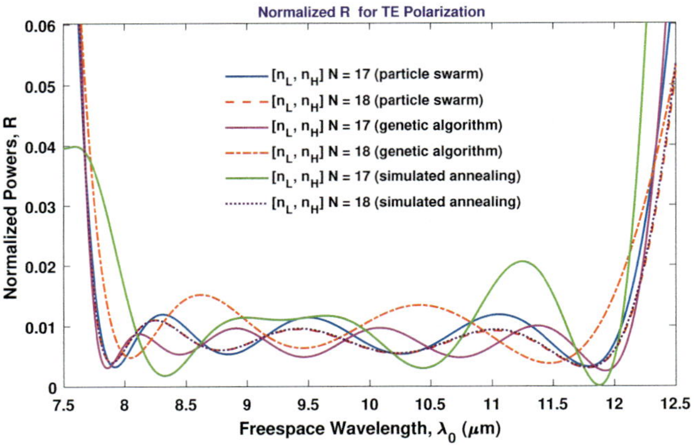

FIGURE 3.67: The performance (normalized reflected power) of an optimized antireflective filter in the wavelength range $7.7\,\mu m \leq \lambda_0 \leq 12.3\,\mu m$, on silicon, is based on alternating layers of zinc sulfide ($n_L = 2.2$) and germanium ($n_H = 4.2$) [138]. Normal incidence is assumed, and consequently, the results for *TE* and *TM* incident polarization coincide. Two cases are considered: (a) A filter with 17 layers and (b) a filter with 18 layers. The global optimization methods used are: Particle swarm, Genetic algorithm, and Simulated annealing.

The applications of anisotropic materials in optics are vast and varied. Examples include polarizers, waveplates, liquid crystal displays, optical modulators, polarization-maintaining fibers, anisotropic waveguides, and nonlinear applications such as second-harmonic generation and optical parametric oscillators. Additionally, anisotropic materials are used in photonic crystals, optical metamaterials, polarization-sensitive detectors for imaging, ellipsometry, chirped mirrors for dispersion control and pulse shaping, diffractive optics, anisotropic thin films, smart windows, and adaptive optics. These materials play a fundamental role in providing advanced control over light's polarization, phase, direction, and intensity, enabling the development of high-performance optical devices essential in telecommunications, imaging, displays, and laser systems.

Given their extensive applications, the study of anisotropic materials is critical. Accurate electromagnetic analysis of interfaces and multilayers composed of anisotropic materials is essential for the design and optimization of these devices. The general plane wave solutions in anisotropic media, discussed in Sec. 3.2.2.2, provide the foundation for understanding light behavior in such media. This section offers a brief summary of the methodologies required to study anisotropic structures, including interfaces and multilayers, emphasizing the importance of precise analysis in the development of advanced optical devices.

3.8.1 Berreman's 4×4 Matrix Formulation

The 4×4 matrix method is a powerful technique used in the study of wave propagation through anisotropic materials, especially in the fields of optics and electromagnetics [143–145]. This method is particularly useful for solving Maxwell's equations in layered media where each layer may have different anisotropic properties. Let $\tilde{\varepsilon}$ (ε_{uv} for $u, v = x, y, z$) be the relative permittivity tensor of a homogeneous anisotropic region, and $\vec{E} = E_x\,\hat{\imath}_x + E_y\,\hat{\imath}_y + E_z\,\hat{\imath}_z$, $\vec{H} = H_x\,\hat{\imath}_x + H_y\,\hat{\imath}_y + H_z\,\hat{\imath}_z$ the electric and magnetic fields of a plane wave of wavevector $\vec{k} = k_x\,\hat{\imath}_x + k_y\,\hat{\imath}_y + k_z\,\hat{\imath}_z$, propagating in the region. The corresponding Maxwell's equations (3.41) and (3.42) are:

$$\vec{k} \times \vec{E} = \omega\mu_0 \vec{H},$$

$$\vec{k} \times \vec{H} = -\omega\epsilon_0 \tilde{\varepsilon} \vec{E}.$$

The equations provided earlier form a system of six equations that relate the components of the electric and magnetic fields. Given that the plane wave propagates obliquely along the z-axis (either in the z or $-z$ direction) and considering that any planar interfaces are parallel to the xy plane (see Fig. 3.68), the z-components of the electric and magnetic fields can be expressed in terms of their x-, y-components as follows:

$$H_z = \frac{1}{\omega\mu_0}\left(k_x E_y - k_y E_x\right), \tag{3.303}$$

$$E_z = \frac{1}{\omega\epsilon_0\varepsilon_{zz}}\left(k_y H_x - k_x H_y\right) - \frac{\varepsilon_{zx}}{\varepsilon_{zz}} E_x - \frac{\varepsilon_{zy}}{\varepsilon_{zz}} E_y. \tag{3.304}$$

Thus, eliminating the E_z and H_z field components, the rest of the Maxwell's equations can be easily written in the following matrix form:

$$\tilde{\boldsymbol{M}}\tilde{\boldsymbol{V}} = \frac{k_z}{k_0}\tilde{\boldsymbol{V}}, \qquad \text{where,} \tag{3.305}$$

$$\tilde{\boldsymbol{V}} = \begin{bmatrix} E_x & H_y & E_y & H_x \end{bmatrix}^T, \tag{3.306}$$

where "T" denotes transpose, and the 4×4 matrix $\tilde{\boldsymbol{M}}$ is given by:

$$\tilde{\boldsymbol{M}} =$$

$$\begin{bmatrix} -\dfrac{\varepsilon_{zx}}{\varepsilon_{zz}}\dfrac{k_x}{k_0} & Z_0\left[1 - \dfrac{k_x^2}{k_0^2}\dfrac{1}{\varepsilon_{zz}}\right] & -\dfrac{\varepsilon_{zy}}{\varepsilon_{zz}}\dfrac{k_x}{k_0} & Z_0\dfrac{k_x k_y}{k_0^2}\dfrac{1}{\varepsilon_{zz}} \\ \dfrac{1}{Z_0}\left[\varepsilon_{xx} - \dfrac{\varepsilon_{xz}\varepsilon_{zx}}{\varepsilon_{zz}} - \dfrac{k_y^2}{k_0^2}\right] & -\dfrac{\varepsilon_{xz}}{\varepsilon_{zz}}\dfrac{k_x}{k_0} & \dfrac{1}{Z_0}\left[\varepsilon_{xy} - \dfrac{\varepsilon_{xz}\varepsilon_{zy}}{\varepsilon_{zz}}\dfrac{k_x k_y}{k_0^2}\right] & -\dfrac{\varepsilon_{yz}}{\varepsilon_{zz}}\dfrac{k_y}{k_0} \\ -\dfrac{\varepsilon_{zx}}{\varepsilon_{zz}}\dfrac{k_y}{k_0} & -Z_0\dfrac{k_x k_y}{k_0^2}\dfrac{1}{\varepsilon_{zz}} & -\dfrac{\varepsilon_{zy}}{\varepsilon_{zz}}\dfrac{k_y}{k_0} & Z_0\left[\dfrac{k_y^2}{k_0^2}\dfrac{1}{\varepsilon_{zz}} - 1\right] \\ \dfrac{1}{Z_0}\left[\dfrac{\varepsilon_{yz}\varepsilon_{zx}}{\varepsilon_{zz}} - \varepsilon_{yx} - \dfrac{k_x k_y}{k_0^2}\right] & \dfrac{\varepsilon_{yz}}{\varepsilon_{zz}}\dfrac{k_x}{k_0} & \dfrac{1}{Z_0}\left[\dfrac{\varepsilon_{yz}\varepsilon_{zy}}{\varepsilon_{zz}} - \varepsilon_{yy} + \dfrac{k_x^2}{k_0^2}\right] & -\dfrac{\varepsilon_{yz}}{\varepsilon_{zz}}\dfrac{k_y}{k_0} \end{bmatrix}$$

with $Z_0 = (\mu_0/\epsilon_0)^{1/2}$ is the freespace characteristic impedance. Equation (3.305) is an eigenvalue/eigenvector equation that generally yields four eigenvalues with their corresponding eigenvectors. These solutions are categorized into two groups: forward-propagating (or evanescent) waves, which move in the $+z$ direction, and backward-propagating (or evanescent) waves, which move in the $-z$ direction. The classification is determined by the z-component of the Poynting vector, $P_z = \text{Re}\{S_z\} = (1/2)\text{Re}\{(E_x H_y^* - E_y H_x^*)\}$, as detailed in Sec. 3.2.4. For propagating waves $P_z \neq 0$. If $P_z \neq 0$, the eigenvector $\tilde{\boldsymbol{V}}_+ = [E_x\, H_y\, E_y\, H_x]^T$ represents a forward-propagating wave, with its eigenvalue satisfying $\text{Re}\{k_{z,+}\} > 0$. In lossless media, the eigenvalue $k_{z,+}$ is purely real. On the other hand, if $P_z < 0$, the eigenvector $\tilde{\boldsymbol{V}}_- = [E_x\, H_y\, E_y\, H_x]^T$ represents a backward-propagating wave, with its eigenvalue satisfying $\text{Re}\{k_{z,-}\} < 0$. Similarly, the eigenvalue $k_{z,-}$ is real in lossless media. If $P_z = 0$ then the wave is evanescent. In this case, the wave is non-propagating, and the eigenvalue

$k_z = k_{zr} + jk_{zi}$ is complex. For lossless media, $k_{zr} = 0$ and k_z is purely imaginary. A forward evanescent wave corresponds to $k_{zi} < 0$, while a backward evanescent wave corresponds to $k_{zi} > 0$ (requirement of the exponential factor $\exp(-jk_z\, z)$ to decay with z). Based on these characteristics, the eigensolutions of Eq. (3.305) can be systematically sorted into forward and backward, propagating or evanescent waves, providing a clear framework for analyzing the wave propagation in the medium. This is represented in the following expressions:

$$\tilde{\boldsymbol{V}} = \begin{bmatrix} \tilde{\boldsymbol{V}}_{1+} & \tilde{\boldsymbol{V}}_{2+} & \tilde{\boldsymbol{V}}_{1-} & \tilde{\boldsymbol{V}}_{2-} \end{bmatrix} \begin{bmatrix} e^{-jk_{z1+}z} & 0 & 0 & 0 \\ 0 & e^{-jk_{z2+}z} & 0 & 0 \\ 0 & 0 & e^{-jk_{z1-}z} & 0 \\ 0 & 0 & 0 & e^{-jk_{z2-}z} \end{bmatrix} \begin{bmatrix} A_{1+} \\ A_{2+} \\ A_{1-} \\ A_{2-} \end{bmatrix} e^{\vec{k}_t \cdot \vec{r}_t}. \tag{3.307}$$

The eigenvectors $\tilde{\boldsymbol{V}}_{1+}$ and $\tilde{\boldsymbol{V}}_{2+}$ correspond to forward-propagating or evanescent waves, characterized by their associated eigenvalues k_{z1+} and k_{z2+}, which represent the z-components of the wavevector for these forward waves. Similarly, $\tilde{\boldsymbol{V}}_{1-}$ and $\tilde{\boldsymbol{V}}_{2-}$ correspond to backward-propagating or evanescent waves with eigenvalues k_{z1-} and k_{z2-}, describing the z-direction wavevector components for the backward waves. The amplitudes of these eigensolutions in the anisotropic region are denoted by the coefficients A_{1+}, A_{2+}, A_{1-}, and A_{2-}. These coefficients depend on the specific excitation conditions, including the incident field characteristics and the boundary conditions at the interfaces of the anisotropic region. The term $\vec{k}_t \cdot \vec{r}_t = k_x x + k_y y$ represents the transverse wavevector component, which remains constant across interfaces due to the continuity of the tangential components of the wavevector. The z-components of the electric and magnetic fields associated with these eigensolutions can be determined from Eqs. (3.304) and (3.303).

3.8.2 Anisotropic Interface

A planar interface between two general anisotropic media with relative permittivity tensors $\tilde{\varepsilon}_1$ and $\tilde{\varepsilon}_2$ is illustrated in Fig. 3.68. In this setup, a plane wave characterized by the wavevector $\vec{k}_{inc}$ is incident on the interface, traveling from the top region (associated with $\tilde{\varepsilon}_1$) towards the bottom region (associated with $\tilde{\varepsilon}_2$). This case is critical for understanding the behavior of light as it traverses boundaries between anisotropic materials, where the directionality of the material properties significantly influences the reflection and transmission of the wave.

It is assumed that the polarization and the wavevector of the incident plane wave are known. For the coordinate system used, the electric field of the incident plane wave can be written as:

$$\vec{E}_{inc} = E_0\, \hat{e}_{ie}\, \exp[-j(k_{x,inc}x + k_{y,inc}y + k_{z,inc}z)], \qquad \text{where,} \tag{3.308}$$

$$k_{x,inc} = -k_0 n_{ie} \cos\phi_{inc} \sin\theta_{inc},$$

$$k_{y,inc} = -k_0 n_{ie} \sin\phi_{inc} \sin\theta_{inc},$$

$$k_{z,inc} = k_0 n_{ie} \cos\theta_{inc}.$$

In this context, E_0 represents the amplitude of the incident electric field, which is typically normalized to unity for simplicity. The vector $\hat{e}_{ie}$ denotes the polarization unit vector, also known as the eigen-polarization, which describes the orientation of the electric field. Meanwhile, n_{ie} refers to the refractive index corresponding to the specific direction of wave

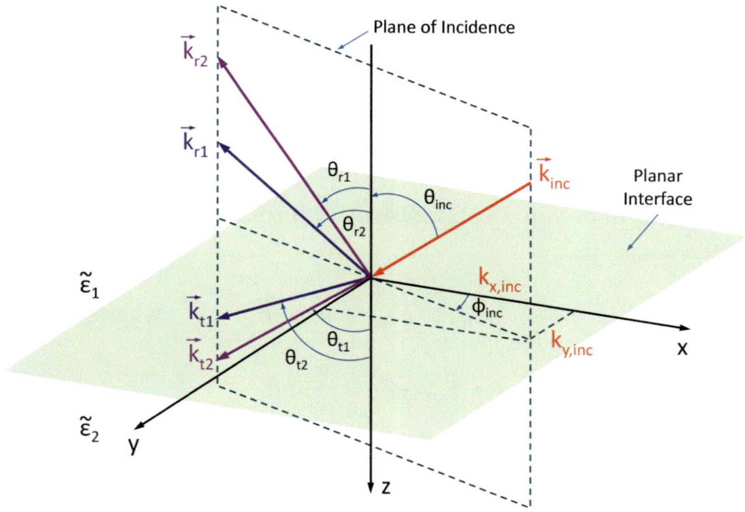

FIGURE 3.68: A planar anisotropic interface between two general anisotropic media with relative permittivity tensors $\tilde{\varepsilon}_1$ and $\tilde{\varepsilon}_2$. A plane wave is incident from the top region at a polar angle θ_{inc} and azimuthal angle ϕ_{inc}. The wavevectors of the possible reflected waves $\vec{k}_{r1}$ and $\vec{k}_{r2}$ and the possible transmitted waves $\vec{k}_{t1}$ and $\vec{k}_{t2}$ are shown.

propagation within the anisotropic medium. The calculation of $\hat{e}_{ie}$ and n_{ie} is performed using the analysis of Sec. 3.8.1. Therefore, $\hat{e}_{ie} = \tilde{\boldsymbol{V}}_{1,1+}$ or $\hat{e}_{ie} = \tilde{\boldsymbol{V}}_{1,2+}$, i.e., one of the two permissible eigen-polarizations in region 1 (top region) that propagate along the $+z$-axis for the given angles θ_{inc} and ϕ_{inc}.

3.8.2.1 Isotropic-Biaxial Interface

In the simpler case of an interface between an isotropic medium and a biaxial medium, the analysis can be streamlined by leveraging the symmetry and uniformity of the isotropic region. In this situation, the top region (region 1 in Fig. 3.68) is isotropic, meaning that its relative permittivity tensor $\tilde{\varepsilon}_1$ simplifies to $\tilde{\varepsilon}_1 = \varepsilon_1 \tilde{I} = n_1^2 \tilde{I}$, where $\tilde{I}$ is a 3×3 identity matrix. Consequently, the refractive index n_{ie} in the isotropic region is simply $n_{ie} = n_1 = \sqrt{\varepsilon_1}$. Given the isotropic nature of region 1, the reflected wavevectors $\vec{k}_{r1}$ and $\vec{k}_{r2}$ are identical and denoted as $\vec{k}_r$. This results in a single reflected wavevector, simplifying the boundary conditions and the expression of the fields in the isotropic region. The electric field in region 1, which includes both the incident and reflected components, can be written as:

$$\vec{E}_1 = \hat{u}\exp(-j\vec{k}_{inc}\cdot\vec{r}) + \left(r_{TE}\hat{u}^r_{TE} + r_{TM}\hat{u}^r_{TM}\right)\exp(-j\vec{k}_r\cdot\vec{r}), \quad \text{where} \tag{3.309}$$

$$\hat{u} = E_{TE}\hat{u}_{TE}\,\sin\psi + E_{TM}\hat{u}_{TM}\,\cos\psi \;=\; u_x\,\hat{\imath}_x + u_y\,\hat{\imath}_y + u_z\,\hat{\imath}_z,$$

$$\hat{u}_{TE} = \hat{u}^r_{TE} \;=\; -\sin\phi_{inc}\,\hat{\imath}_x + \cos\phi_{inc}\,\hat{\imath}_y,$$

$$\hat{u}_{TM} = \cos\theta_{\text{inc}}\cos\phi_{inc}\,\hat{\imath}_x + \cos\theta_{\text{inc}}\sin\phi_{inc}\,\hat{\imath}_y + \sin\theta_{inc}\,\hat{\imath}_z,$$

$$\hat{u}^r_{TM} = -\cos\theta_{\text{inc}}\cos\phi_{inc}\,\hat{\imath}_x - \cos\theta_{\text{inc}}\sin\phi_{inc}\,\hat{\imath}_y + \sin\theta_{inc}\,\hat{\imath}_z,$$

$$r_{TE}\hat{u}^r_{TE} + r_{TM}\hat{u}^r_{TM} \;=\; u_{rx}\,\hat{\imath}_x + u_{ry}\,\hat{\imath}_y + u_{rz}\,\hat{\imath}_z.$$

The polarization angle ψ is a key parameter that defines the orientation of the electric field vector of the incident wave relative to the plane of incidence. Specifically, $\psi = 90°$ corresponds to *TE* polarization and $\psi = 0°$ corresponds to *TM* polarization. The constants E_{TE}

and E_{TM}, which are generally complex, ensure that the unit vector $\hat{u}$ (defining the direction of the electric field) maintains its unit magnitude. These constants allow representation of any type of elliptical polarization for the incident wave. The reflection coefficients r_{TE} and r_{TM}, correspond to the *TE* and *TM* polarizations, respectively. These coefficients are generally complex, reflecting both the amplitude and phase changes that occur upon reflection at the interface.

The magnetic field in region 1 can be easily determined from the Maxwell's equations, and is expressed as:

$$\vec{H}_1 = \frac{1}{\omega\mu_0}\left[\vec{k}_{inc} \times \vec{E}_{inc} + \vec{k}_r \times \vec{E}_r\right], \qquad \text{where,} \tag{3.310}$$

$$\vec{E}_{inc} = \hat{u}\exp(-j\vec{k}_{inc}\cdot\vec{r}), \qquad \text{and,}$$

$$\vec{E}_r = (u_{rx}\,\hat{\imath}_x + u_{ry}\,\hat{\imath}_y + u_{rz}\,\hat{\imath}_z)\exp(-j\vec{k}_r\cdot\vec{r}).$$

The bottom region (region 2) is a general biaxial medium. Following the analysis presented in Sec. 3.8.1, the electric and magnetic fields in region 2 can be expressed as:

$$\vec{E}_2 = t_1\hat{e}_{t1}\exp(-j\vec{k}_{t1}\cdot\vec{r}) + t_2\hat{e}_{t2}\exp(-j\vec{k}_{t2}\cdot\vec{r}), \tag{3.311}$$

$$\vec{H}_2 = \frac{1}{\omega\mu_0}\left[\vec{k}_{t1} \times \vec{E}_{t1} + \vec{k}_{t2} \times \vec{E}_{t2}\right]. \tag{3.312}$$

In Eqs. (3.311) and (3.312), $\vec{k}_{t1}$ and $\vec{k}_{t2}$ represent the two permissible wavevectors in the $+z$ direction, while $\hat{e}_{t1}$ and $\hat{e}_{t2}$ are their corresponding eigen-polarizations. The unknown coefficients t_1 and t_2 are the transmission coefficients for the two eigen-polarizations. To find the wavevectors $\vec{k}_{t1}$ and $\vec{k}_{t2}$, and their corresponding eigen-polarizations, $\hat{e}_{t1}$ and $\hat{e}_{t2}$, the eigenvalue-eigenvector Eq. (3.305) must be applied. The following phase-matching conditions, which are part of the boundary conditions of the problem, should be utilized:

$$k_{t1,x} = k_{t2,x} = k_{rx} = k_{x,inc} = -k_0 n_1 \cos\phi_{inc}\sin\theta_{inc},$$

$$k_{t1,y} = k_{t2,y} = k_{ry} = k_{y,inc} = -k_0 n_{ie} \sin\phi_{inc}\sin\theta_{inc},$$

where $\vec{k}_{tw} = k_{tw,x}\,\hat{\imath}_x + k_{tw,y}\,\hat{\imath}_y + k_{tw,z}\,\hat{\imath}_z$ $(w = 1,2)$ and $\vec{k}_r = k_{rx}\,\hat{\imath}_x + k_{ry}\,\hat{\imath}_y + k_{rz}\,\hat{\imath}_z$. Knowledge of the x and y components of the wavevector ensures the determination of the $k_{t1,z}$, and $k_{t2,z}$ components using the eigenvalue/eigenvector equation [Eq. (3.305)]. Along with the $k_{t1,z}$, and $k_{t2,z}$ (eigenvalues) the corresponding polarization vectors $\hat{e}_{t1}$, and $\hat{e}_{t2}$ (eigenvectors) can also be determined. An alternative method for calculating $k_{t1,z}$, and $k_{t2,z}$ and the corresponding eigenvectors $\hat{e}_{t1}$ and $\hat{e}_{t2}$ is presented in Appendix 3H [81]. The k_{rz} component is determined from the isotropic dispersion relation, Eq. (3.40). The two possible reflected polarizations in the isotropic region are defined by the unit vectors $\hat{u}^r_{TE}$ and $\hat{u}^r_{TM}$. Given the eigenvectors of the electric field $\hat{e}_{t1}$ and $\hat{e}_{t2}$, the eigenvectors of the magnetic field, $\vec{h}_{t1}$ and $\vec{h}_{t2}$, can be easily determined from Maxwell's equations. I.e., $\vec{h}_{ti} = (1/Z_0)[(\vec{k}_i/k_0) \times \hat{e}_{ti}]$ for $i = 1,2$. To determine the unknown coefficients r_{TE}, r_{TM}, t_1 and t_2 the continuity of the tangential components of the electric and magnetic fields at the interface must be applied. Specifically, this means that the conditions $E_{1x} = E_{2x}$, $E_{1y} = E_{2y}$, $H_{1x} = H_{2x}$, and $H_{1y} = H_{2y}$ must hold at the boundary $(z = 0)$. These four equations correspond to a linear system that determines the coefficients r_{TE}, r_{TM}, t_1, and t_2 [refer to Eq. (3H.3) of Appendix 3H].

At this point, power conservation at the interface is analyzed to determine the percentage of the incident power that is reflected and transmitted into the two possible polarizations in regions 1 and 2. This conservation is a direct consequence of Poynting's theorem. The time-averaged reflected Poynting vector in region 1, $\langle\vec{\mathcal{N}}_{r1}\rangle$, is calculated as:

$$\langle\vec{\mathcal{N}}_{r1}\rangle = \frac{1}{2}\text{Re}\left\{\vec{E}_r \times \vec{H}_r^*\right\} = \frac{1}{2\omega\mu_0}\left[|r_{TE}|^2 + |r_{TM}|^2\right]\vec{k}_r.$$

Then, the percentage of the reflected power into the *TE* and *TM* reflected polarizations can be defined as:

$$\frac{P_{rz}}{P_{z,inc}} = P_{r_{TE}} + P_{r_{TM}} = \frac{-\langle\vec{\mathcal{N}}_1\rangle \cdot \hat{\imath}_z}{(1/2\omega\mu_0)|E_0|^2\vec{k}_{inc}\cdot\hat{\imath}_z} = |r_{TE}|^2 + |r_{TM}|^2 \tag{3.313}$$

The incident electric field is normalized such that $E_0 = 1$. The percentages of the reflected power into the *TE* and *TM* polarizations, denoted as $P_{r_{TE}}$ and $P_{r_{TM}}$, respectively, can be calculated directly from the reflection coefficients. However, the calculation of the transmitted power percentage is more complex due to the biaxial nature of region 2. The time-averaged Poynting vector in region 2, $\langle\vec{\mathcal{N}}_2\rangle$, is determined using the following expression:

$$\begin{aligned}
\langle\vec{\mathcal{N}}_2\rangle &= \frac{1}{2}\text{Re}\left\{\vec{E}_2 \times \vec{H}_2^*\right\} \\
&= \frac{1}{2\omega\mu_0}\Bigg\{ \left[|t_1|^2\text{Re}\{\vec{k}_{t1}^*\} - |t_1|^2\text{Re}\left\{(\hat{e}_{t1}\cdot\vec{k}_{t1}^*)\hat{e}_{t1}^*\right\}\right] + \\
&\qquad \left[|t_2|^2\text{Re}\{\vec{k}_{t2}^*\} - |t_2|^2\text{Re}\left\{(\hat{e}_{t2}\cdot\vec{k}_{t2}^*)\hat{e}_{t2}^*\right\}\right] + \\
&\quad \text{Re}\left\{t_1t_2^*(\hat{e}_{t1}\cdot\hat{e}_{t2}^*)\vec{k}_{t2} - t_1t_2^*(\hat{e}_{t1}\cdot\vec{k}_{t2}^*)\hat{e}_{t2}^*\right\} + \text{Re}\left\{t_2t_1^*(\hat{e}_{t2}\cdot\hat{e}_{t1}^*)\vec{k}_{t1} - t_2t_1^*(\hat{e}_{t2}\cdot\vec{k}_{t1}^*)\hat{e}_{t1}^*\right\}\Bigg\}.
\end{aligned}$$

The last two terms in the above expression are the cross terms between the two transmitted waves, $t_1\hat{e}_{t1}\exp[-j\vec{k}_{t1}\cdot\vec{r}]$ and $t_2\hat{e}_{t2}\exp[-j\vec{k}_{t2}\cdot\vec{r}]$. Then, the percentage of the transmitted power can be defined as:

$$\frac{P_{tz}}{P_{z,inc}} = \frac{\langle\vec{\mathcal{N}}_2\rangle\cdot\hat{\imath}_z}{(1/2\omega\mu_0)|E_0|^2\vec{k}_{inc}\cdot\hat{\imath}_z} = P_{t1} + P_{t2} + P_{t,cross}, \qquad \text{where,} \tag{3.314}$$

$$P_{t1} = \frac{1}{k_0 n_1\cos\theta_{inc}}\left[|t_1|^2\text{Re}\{\vec{k}_{t1}^*\}\cdot\hat{\imath}_z - |t_1|^2\text{Re}\left\{(\hat{e}_{t1}\cdot\vec{k}_{t1}^*)\hat{e}_{t1}^*\right\}\cdot\hat{\imath}_z\right],$$

$$P_{t2} = \frac{1}{k_0 n_1\cos\theta_{inc}}\left[|t_2|^2\text{Re}\{\vec{k}_{t2}^*\}\cdot\hat{\imath}_z - |t_2|^2\text{Re}\left\{(\hat{e}_{t2}\cdot\vec{k}_{t2}^*)\hat{e}_{t2}^*\right\}\cdot\hat{\imath}_z\right],$$

$$\begin{aligned}
P_{t,cross} = \frac{1}{k_0 n_1\cos\theta_{inc}}\Bigg[&\text{Re}\left\{t_1t_2^*(\hat{e}_{t1}\cdot\hat{e}_{t2}^*)\vec{k}_{t2} - t_1t_2^*(\hat{e}_{t1}\cdot\vec{k}_{t2}^*)\hat{e}_{t2}^*\right\}\cdot\hat{\imath}_z + \\
&\text{Re}\left\{t_2t_1^*(\hat{e}_{t2}\cdot\hat{e}_{t1}^*)\vec{k}_{t1} - t_2t_1^*(\hat{e}_{t2}\cdot\vec{k}_{t1}^*)\hat{e}_{t1}^*\right\}\cdot\hat{\imath}_z\Bigg].
\end{aligned}$$

Power conservation, as stated by Poynting's theorem, dictates that the total power of the incident wave must equal the sum of the powers carried by the reflected and transmitted waves. This is expressed as:

$$P_{r_{TE}} + P_{r_{TM}} + P_{t1} + P_{t2} + P_{t,cross} = 1. \tag{3.315}$$

Typically, the cross-term $P_{t,cross}$, which accounts for interference effects between the two transmitted polarizations, is negligible or zero.

Example 3.8.1:

Consider an interface between an isotropic medium (air) and a biaxial medium. The isotropic medium is characterized by a permittivity tensor $\tilde{\varepsilon}_1 = \tilde{I}$ (identity matrix), while the biaxial medium is described by $\tilde{\varepsilon}_2 = diag(2, 5, 8)$ and rotated with respect to the xyz coordinate system using the angles $\psi_1 = 45°$ and $\psi_2 = 45°$ (see Appendix 3H) [146,147]. The magnetic permeability of both media is μ_0. A plane wave of freespace wavelength $\lambda_0 = 1\,\mu$m is incident from the air onto the biaxial medium at an angle of incidence θ_{inc} (with an azimuthal angle $\phi_{inc} = 0°$). The polarization of the incident wave is either *TE* or *TM*. The objective is to: (a) Calculate the absolute values of the coefficients $r_{_{TE}}, r_{_{TM}}, t_1$, and t_2 as functions of the angle of incidence θ_{inc} (with $\phi_{inc} = 0°$) for both polarizations. (b) Determine the normalized reflected and transmitted powers for each polarization state. (c) Verify power conservation at the interface.

Solution: Using Eq. (3H.5) the relative permittivity of the biaxial region can be found as:

$$\tilde{\varepsilon}_2 = \begin{bmatrix} 4.25 & 2.25 & 1.0607 \\ 2.25 & 4.25 & 1.0607 \\ 1.0607 & 1.0607 & 6.50 \end{bmatrix},$$

For *TE* incident polarization, the angle $\psi = 90°$, while for *TM* polarization, $\psi = 0°$. Using the previously outlined methodology, the complex coefficients $r_{_{TE}}, r_{_{TM}}, t_1$, and t_2 are calculated as functions of the angle of incidence θ_{inc}. The results of these computations are shown in Fig. 3.69(a) and 3.69(c) which correspond to *TE* and *TM* polarizations, respectively. A distinctive feature of the isotropic-biaxial interface is the coupling of polarization states during reflection. Specifically: When a *TE* wave is incident, the reflected wave includes contributions from both *TE* and *TM* polarizations. Similarly, for a *TM* incident wave, the reflected wave also contains both *TE* and *TM* components. This cross-polarization phenomenon is a direct consequence of the anisotropic nature of the biaxial medium, which induces coupling between orthogonal polarization components upon reflection.

Another notable effect occurs for *TM* incident polarization, where a Brewster angle is observed around $\theta_{inc} \simeq 58.3°$, as evident from the plot of $|r_{_{TM}}|$ in Fig. 3.69(c). Using Eqs. (3.313) and (3.314) the normalized reflected powers $P_{r_{TE}}$, $P_{r_{TM}}$, as well as the normalized transmitted powers P_{t1}, and P_{t2}, can be calculated. In this case, $P_{cross} = 0$ for all angles of incidence. The results for these normalized reflected and transmitted powers are presented in Figs. 3.72(b) and 3.72(d) for *TE* and *TM* incident polarizations, respectively. These figures also include the sum of the normalized efficiencies, confirming that power conservation is upheld, as the biaxial layer is lossless.

3.8.2.2 Biaxial-Biaxial Interface

The analysis of the general biaxial-biaxial planar interface closely follows the procedure outlined for the isotropic-biaxial interface. The main distinction lies in the fact that the incident wave in region 1 must now satisfy Maxwell's equations for a biaxial medium instead of an isotropic one. Consequently, the electric and magnetic fields in region 1 can be

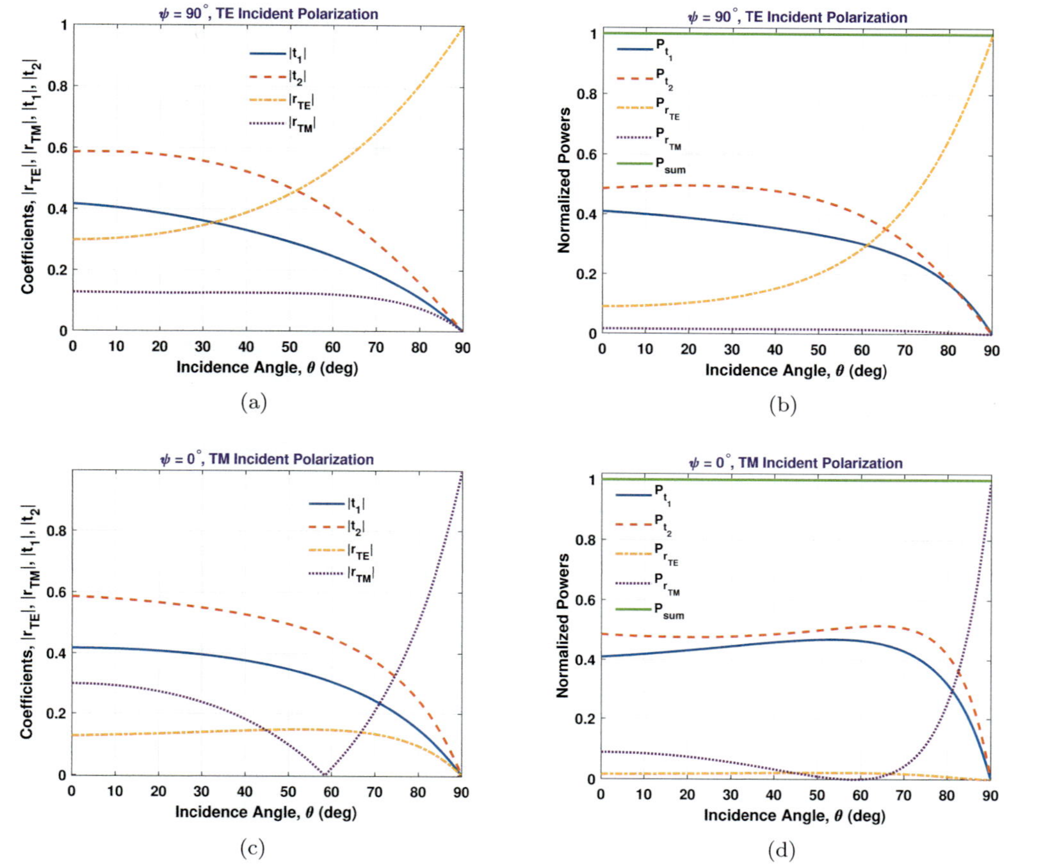

FIGURE 3.69: An example isotropic-biaxial interface [146, 147] (a) Absolute values of the coefficients r_{TE}, r_{TM}, t_1 and t_2 for *TE* incident polarization. (b) The normalized powers for *TE* incident polarization. (c) Same as (a) but for *TM* incident polarization. (d) Same as (b) but for *TM* incident polarization.

expressed as:

$$\vec{E}_1 = \underbrace{E_0 \hat{e}_{ie} \exp(-j\vec{k}_{inc} \cdot \vec{r})}_{\vec{E}_0} + \underbrace{r_1 \hat{e}_{r1} \exp(-j\vec{k}_{r1} \cdot \vec{r})}_{\vec{E}_{r1}} + \underbrace{r_2 \hat{e}_{r2} \exp(-j\vec{k}_{r2} \cdot \vec{r})}_{\vec{E}_{r2}}, \tag{3.316}$$

$$\vec{H}_1 = \frac{1}{\omega\mu_0}\left[\vec{k}_{inc} \times \vec{E}_0 + \vec{k}_{r1} \times \vec{E}_{r1} + \vec{k}_{r2} \times \vec{E}_{r2}\right]. \tag{3.317}$$

In the above expressions E_0 is typically set to unity as in the case of the incident wave in the isotropic region. The unit vector $\hat{e}_{ie}$ represents an eigen-polarization corresponding to the direction of the incident wave in region 1, which is now a biaxial medium. The associated refractive index n_{ie} is an eigenvalue that depends on the angles of incidence θ_{inc} and ϕ_{inc} [refer to Eq. (3.308)]. Generally, there are two possible solutions for n_{ie} and $\hat{e}_{ie}$. These solutions are distinguished as the "inner" and "outer" solutions, where the "inner" solution corresponds to the smaller refractive index and the "outer" solution corresponds to the larger refractive index, based on the wavevector surface described in the discussion of plane waves in anisotropic media in Sec. 3.2.2.2. The reflection coefficients r_1, r_2 relate to the two possible reflected eigen-polarizations in region 1, each associated with different wavevectors $\vec{k}_{r1}$ and $\vec{k}_{r2}$, as illustrated in Fig. 3.68. The expressions for the electric and magnetic fields in region 2 are given by Eqs. (3.311) and (3.312).

The corresponding phase-matching conditions in this case become:

$$k_{t1,x} = k_{t2,x} = k_{r1,x} = k_{r2,x} = k_{x,inc} = -k_0 n_{ie} \cos\phi_{inc} \sin\theta_{inc},$$

$$k_{t1,y} = k_{t2,y} = k_{r1,y} = k_{r2,y} = k_{y,inc} = -k_0 n_{ie} \sin\phi_{inc} \sin\theta_{inc},$$

Knowledge of the x and y components of the wavevectors enables the determination of the $k_{r1,z}$, $k_{r2,z}$, $k_{t1,z}$ and $k_{t2,z}$ components through the eigenvalue/eigenvector equation, i.e., Eq. (3.305). With these components, all the wavevectors $\vec{k}_{r1}$, $\vec{k}_{r2}$, $\vec{k}_{t1}$, and $\vec{k}_{t2}$, and their corresponding eigen-polarizations, $\hat{e}_{r1}$, $\hat{e}_{r2}$, $\hat{e}_{t1}$, and $\hat{e}_{t2}$, can be fully determined. To determine the unknown coefficients r_1, r_2, t_1, and t_2 the continuity of the tangential components of the electric and magnetic fields at the interface must be applied, similarly to the isotropic-biaxial interface [refer to Eq. (3H.4) of Appendix 3H]. Solving this system will yield the values of the unknown coefficients, r_1, r_2, t_1, and t_2.

At this point, the power conservation at the interface is examined to determine the percentage of incident power that is reflected and transmitted into the two possible eigen-polarizations in the corresponding regions 1 and 2. Since region 1 is now also biaxial, the incident z-component of the Poynting vector is given by:

$$P_{z,inc} = \frac{1}{2}\mathrm{Re}\left\{\vec{E}_{inc} \times \vec{H}^*_{inc}\right\} \cdot \hat{\imath}_z = \frac{1}{2\omega\mu_0}\left[\vec{k}_{inc} - \mathrm{Re}\left\{(\hat{e}_{ie} \cdot \vec{k}_{inc})\hat{e}^*_{ie}\right\}\right] \cdot \hat{\imath}_z. \tag{3.318}$$

It is important to note that the wavevector $\vec{k}_{inc}$ is real since region 1, the input region, is always assumed to be lossless. Furthermore, due to the biaxial nature of region 1, $\vec{k}_{inc} \cdot \hat{e}_{ie} \neq 0$, as the wavevector is generally orthogonal to the displacement field vector rather than the electric field vector. The reflected time-average Poynting vector in region 1, $\langle\vec{\mathcal{N}}_{r1}\rangle$, is calculated as:

$$\begin{aligned}\langle\vec{\mathcal{N}}_{r1}\rangle = \frac{1}{2}\mathrm{Re}\left\{\vec{E}_r \times \vec{H}^*_r\right\} = \frac{1}{2\omega\mu_0}\Bigg\{ &\left[|r_1|^2\mathrm{Re}\{\vec{k}^*_{r1}\} - |r_1|^2\mathrm{Re}\left\{(\hat{e}_{r1} \cdot \vec{k}^*_{r1})\hat{e}^*_{r1}\right\}\right] + \\ &\left[|r_2|^2\mathrm{Re}\{\vec{k}^*_{r2}\} - |r_2|^2\mathrm{Re}\left\{(\hat{e}_{r2} \cdot \vec{k}^*_{r2})\hat{e}^*_{r2}\right\}\right] + \\ &\mathrm{Re}\left\{r_1 r_2^*(\hat{e}_{r1} \cdot \hat{e}^*_{r2})\vec{k}_{r2} - r_1 r_2^*(\hat{e}_{r1} \cdot \vec{k}^*_{r2})\hat{e}^*_{r2}\right\} + \\ &\mathrm{Re}\left\{r_2 r_1^*(\hat{e}_{r2} \cdot \hat{e}^*_{r1})\vec{k}_{r1} - r_2 r_1^*(\hat{e}_{r2} \cdot \vec{k}^*_{r1})\hat{e}^*_{r1}\right\}\Bigg\}.\end{aligned}$$

The last two terms in the expression for $\langle\vec{\mathcal{N}}_{r1}\rangle$, are cross terms arising from the interaction between the two reflected waves, $r_1\hat{e}_{r1}\exp[-j\vec{k}_{r1} \cdot \vec{r}]$ and $r_2\hat{e}_{r2}\exp[-j\vec{k}_{r2} \cdot \vec{r}]$. In the biaxial case, correctly applying Poynting's theorem along the z-direction necessitates the computation of the total Poynting vector, $\langle\vec{\mathcal{N}}_1\rangle$, where $\vec{E}_1$ and $\vec{H}_1$ are defined in Eqs. (3.316) and (3.317). In this case, additional cross-product terms between the incident wave and each of the reflected waves may also emerge and need to be accounted for. This total cross-product term, normalized by the incident power $P_{inc,z}$, s referred to as $P_{r,cross}$. Consequently, the percentage of reflected power can be expressed as:

$$\frac{P_{rz}}{P_{z,inc}} = \frac{-\langle\vec{\mathcal{N}}_1\rangle \cdot \hat{\imath}_z}{P_{z,inc}} = P_{r1} + P_{r2} + P_{r,cross}, \tag{3.319}$$

where P_{r1} and P_{r2} are the powers associated with the two reflected waves, and $P_{r,cross}$ accounts for any cross-product contributions. The percentage of the transmitted power is given again by Eq. (3.314). Summarizing, power conservation can be expressed as:

$$P_{r1} + P_{r2} + P_{r,cross} + P_{t1} + P_{t2} + P_{t,cross} = 1. \qquad (3.320)$$

Usually, the terms $P_{r,cross}$ and $P_{t,cross}$ which represent cross-product contributions between different reflected and transmitted waves, respectively, are zero.

Example 3.8.2:

Consider a biaxial-biaxial interface between two homogeneous regions, where the biaxial material is identical in both regions but oriented differently [148]. The relative permittivity matrix of the material in the principal axes system is given by: $\tilde{\varepsilon}_0 = diag(1.2^2, 1.7^2, 2.2^2)$. In region 1, the principal axes are oriented with respect to the (x, y, z) coordinate system by the Euler angles $\theta = 70°$, $\phi = 90°$, and $\psi = -90°$ (refer to Appendix 3H, Sec. 3H.3). In region 2, the Euler angles are $\theta = \phi = \psi = 30°$. Given a plane wave incident at an angle θ_{inc}, the following tasks must be performed: (a) Calculate the absolute values of the reflection and transmission coefficients r_1, r_2, t_1, and t_2 as functions of the incidence angle, ranging from $-30°$ to $+30°$. (b) Calculate the normalized powers of the reflected and transmitted waves as functions of the angle of incidence. (c) Verify power conservation at the interface. (d) Investigate the cases where the incident plane wave is the "outer" wave (with the largest refractive index for the direction of propagation) or the "inner" wave (with the smallest refractive index for the direction of propagation).

Solution: Using Eq. (3H.6) the relative permittivities of regions 1 and 2 can be determined as:

$$\tilde{\varepsilon}_1 = \begin{bmatrix} 4.4423 & 0 & 1.0927 \\ 0 & 2.89 & 0 \\ 1.0927 & 0 & 1.8377 \end{bmatrix}, \qquad \tilde{\varepsilon}_2 = \begin{bmatrix} 2.5992 & -0.8361 & 0.2288 \\ -0.8361 & 2.3089 & -1.0242 \\ 0.2288 & -1.0242 & 4.2619 \end{bmatrix},$$

According to Fig. 3.68 the range $0° \leq \theta_{inc} \leq 30°$ corresponds to an azimuthal angle of $\phi_{inc} = 180°$. Conversely, the range $-30° \leq \theta_{inc} \leq 0°$ is achieved with $0° \leq \theta_{inc} \leq 30°$ and $\phi = 0°$. Applying the analysis of Sec. 3.8.2.2 the reflection and transmission coefficients r_1, r_2, t_1, and t_2 are calculated as functions of the incidence angle. The results are summarized in Figs. 3.70(a), 3.70(b), 3.70(c), and 3.70(d). In these figures, solid lines represent the case of the "outer" incident wave, while dashed lines correspond to the "inner" incident wave. The jumps observed in the coefficients are attributed to transitions on the wavevector surface between the inner and outer sheets, as described in Ref. [148]. The normalized powers are depicted in Fig. 3.70(e) for the case of an "outer" incident plane wave. The case for the "inner" incident wave is illustrated in Fig. 3.70(f). In all situations, power conservation is confirmed, ensuring that the total reflected and transmitted power equals the incident power.

3.8.3 Anisotropic Mutlilayers

The extension of the biaxial-biaxial single interface to a multilayer planar stack of general anisotropic media follows a similar approach to that presented in Sec. 3.7, albeit with additional complexity due to the presence of anisotropy. The incident plane wave is characterized in the same manner as for the single interface scenario (as depicted in Fig. 3.68) with the angles θ_{inc} and ϕ_{inc}. The polarization of the incident wave corresponds to an allowed eigen-polarization $\hat{e}_{ie}$ of the input region (region 0). The expression of the electric and magnetic

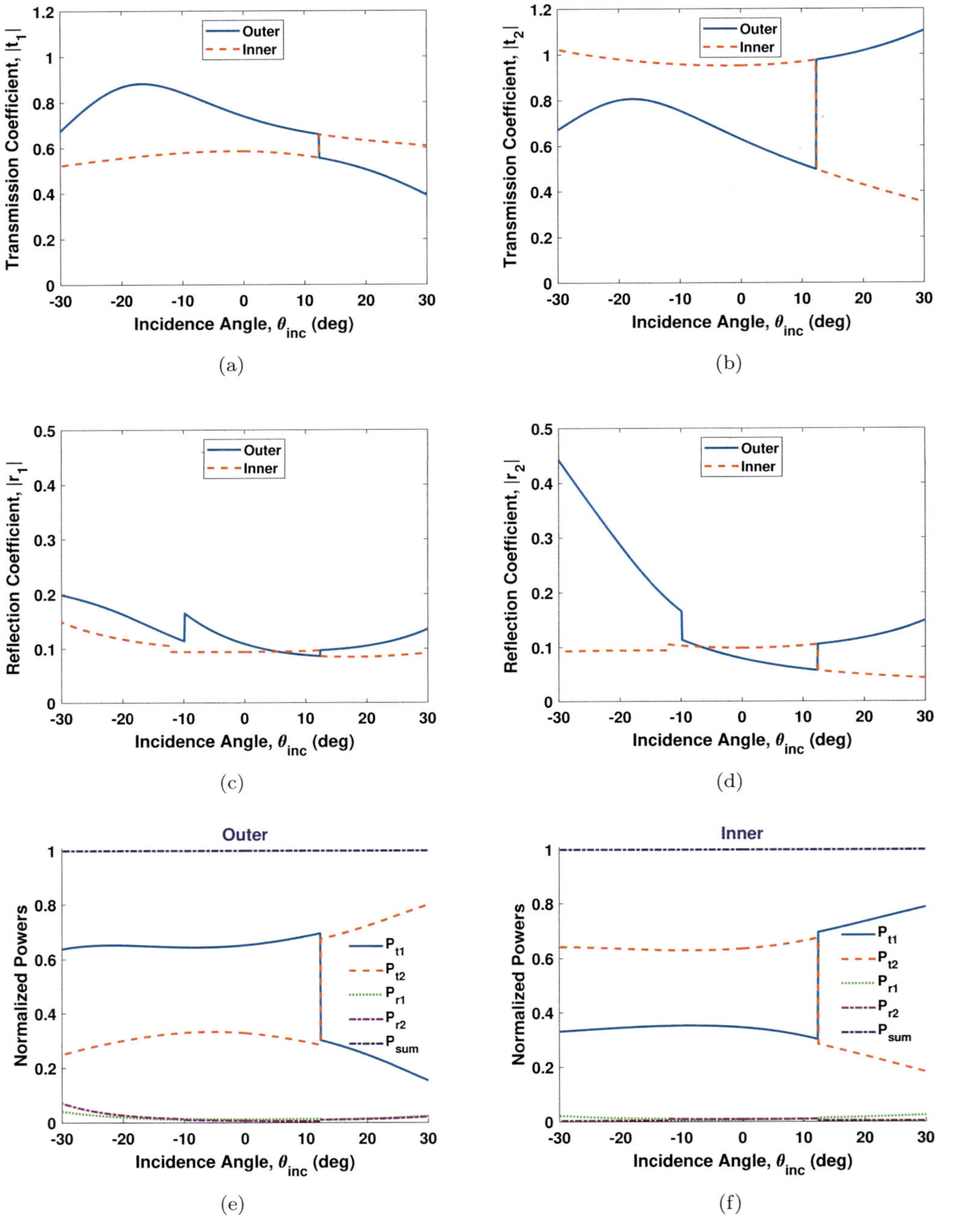

FIGURE 3.70: An example of a biaxial-biaxial interface [148]. The parameters are listed in Example 3.8.2. (a) Absolute values of the coefficient t_1 as a function of the angle of incidence θ_{inc} for both "outer" and "inner" incidence wave. (b) Same as (a) but for t_2. (c) Same as (a) but for r_1. (d) Same as (a) but for r_2. (e) Normalized powers as functions of the angle of incidence for "outer" incident wave. (f) Same as (e) but for "inner" incident wave.

field phasors in each of the m regions (where $m = 0, 1, 2, \cdots, N+1$) are expressed as:

$$\vec{E}_0 = \underbrace{E_0 \hat{e}_{ie} \exp(-j\vec{k}_{inc} \cdot \vec{r})}_{= \vec{E}_0} + \underbrace{r_{01} \hat{e}_{0r1} \exp(-j\vec{k}_{0r1} \cdot \vec{r})}_{= \vec{E}_{0r1}} + \underbrace{r_{02} \hat{e}_{0r2} \exp(-j\vec{k}_{0r2} \cdot \vec{r})}_{= \vec{E}_{0r2}}, \quad (3.321)$$

$$\vec{H}_0 = \frac{1}{\omega\mu_0} \left[\vec{k}_{inc} \times \vec{E}_0 + \vec{k}_{0r1} \times \vec{E}_{0r1} + \vec{k}_{0r2} \times \vec{E}_{0r2} \right], \qquad \text{for } m = 0, \quad (3.322)$$

$$\vec{E}_m = \underbrace{r_{m1} \hat{e}_{mr1} \exp[-j\vec{k}_{mr1} \cdot (\vec{r} - s_m \hat{\imath}_z)]}_{= \vec{E}_{mr1}} + \underbrace{r_{m2} \hat{e}_{mr2} \exp[-j\vec{k}_{mr2} \cdot (\vec{r} - s_m \hat{\imath}_z)]}_{= \vec{E}_{mr2}} +$$

$$\underbrace{t_{m1} \hat{e}_{mt1} \exp[-j\vec{k}_{mt1} \cdot (\vec{r} - s_m \hat{\imath}_z)]}_{= \vec{E}_{mt1}} + \underbrace{t_{m2} \hat{e}_{mt2} \exp[-j\vec{k}_{mt2} \cdot (\vec{r} - s_m \hat{\imath}_z)]}_{= \vec{E}_{mt2}}, \quad (3.323)$$

$$\vec{H}_m = \frac{1}{\omega\mu_0} \left[\vec{k}_{mr1} \times \vec{E}_{mr1} + \vec{k}_{mr2} \times \vec{E}_{mr2} + \vec{k}_{mt1} \times \vec{E}_{mt1} + \vec{k}_{mt2} \times \vec{E}_{mt2} \right], \quad (3.324)$$

$$\text{for } m = 1, 2, \cdots, N,$$

$$\vec{E}_{N+1} = \underbrace{t_{N+1,1} \hat{e}_{N+1,t1} \exp[-j\vec{k}_{N+1,t1} \cdot (\vec{r} - s_{N+1} \hat{\imath}_z)]}_{= \vec{E}_{N+1,t1}} +$$

$$\underbrace{t_{N+1,2} \hat{e}_{N+1,t2} \exp[-j\vec{k}_{N+1,t2} \cdot (\vec{r} - s_{N+1} \hat{\imath}_z)]}_{= \vec{E}_{N+1,t2}}, \quad (3.325)$$

$$\vec{H}_{N+1} = \frac{1}{\omega\mu_0} \left[\vec{k}_{N+1,t1} \times \vec{E}_{N+1,t1} + \vec{k}_{N+1,t2} \times \vec{E}_{N+1,t2} \right], \qquad \text{for } m = N+1. \quad (3.326)$$

In the above equations, $\hat{e}_{mu1}$ and $\hat{e}_{mu2}$ ($u = r, t$) represent the two allowable eigen-polarizations for the forward ($u = t$) and backward ($u = r$) plane waves in each region m. The corresponding wavevectors, $\vec{k}_{mu1}$ and $\vec{k}_{mu2}$, are the two permissible forward ($u = t$) and backward ($u = r$) wavevectors in each region m. These wavevectors are determined by applying the phase-matching conditions along with the dispersion equation, similar to the approach used in the simpler single-interface problem. The reflection coefficients r_{m1} and r_{m2}, as well as the transmission coefficients t_{m1} and t_{m2} need to be determined to fully characterize the wave interactions within each region. The parameter $s_m = \sum_{i=1}^{m-1} d_m$ represents the cumulative thickness of all preceding layers up to region m, with $s_0 = s_1 = 0$ indicating the starting reference point.

The corresponding phase-matching conditions in this case become:

$$k_{mt1,x} = k_{mt2,x} = k_{mr1,x} = k_{mr2,x} = k_{x,inc} = -k_0 n_{ie} \cos\phi_{inc} \sin\theta_{inc},$$

$$k_{mt1,y} = k_{mt2,y} = k_{mr1,y} = k_{mr2,y} = k_{y,inc} = -k_0 n_{ie} \sin\phi_{inc} \sin\theta_{inc},$$

where $m = 0, 1, 2, \cdots, N, N+1$, and n_{ie} is the corresponding refractive index for the incident wave in region 0. Knowledge of the x and y components of the wavevectors enables the determination of the $k_{mr1,z}$, $k_{mr2,z}$, $k_{mt1,z}$, and $k_{mt2,z}$ components through the eigenvalue/eigenvector equation, Eq. (3.305), as it was done in all previously examined cases. With these components, all the wavevectors $\vec{k}_{mr1}$, $\vec{k}_{mr2}$, $\vec{k}_{mt1}$, and $\vec{k}_{mt2}$, and their corresponding eigen-polarizations, $\hat{e}_{mr1}$, $\hat{e}_{mr2}$, $\hat{e}_{mt1}$, and $\hat{e}_{mt2}$, can be fully determined.

To determine the unknown reflection and transmission coefficients, the continuity of the tangential components of the electric and magnetic fields at all interfaces must be applied, similar to the procedure used for the single interface problem. It is convenient to use Eq. (3.307) to express the tangential fields. If $\tilde{\boldsymbol{V}}_m(z)$ represents the electric and magnetic tangential field components at a position z of region m, then they can be represented by the following expression:

$$\tilde{\boldsymbol{V}}_{\boldsymbol{m}}(z) = \tilde{\boldsymbol{W}}_{\boldsymbol{m}} \underbrace{\begin{bmatrix} e^{-jk_{mr1,z}(z-s_m)} & 0 & 0 & 0 \\ 0 & e^{-jk_{mr2,z}(z-s_m)} & 0 & 0 \\ 0 & 0 & e^{-jk_{mt1,z}(z-s_m)} & 0 \\ 0 & 0 & 0 & e^{-jk_{mt2,z}(z-s_m)} \end{bmatrix}}_{=\tilde{\boldsymbol{\Lambda}}_{\boldsymbol{m}}(z)} \overbrace{\begin{bmatrix} r_{m1} \\ r_{m2} \\ t_{m1} \\ t_{m2} \end{bmatrix}}^{=\tilde{\boldsymbol{C}}_{\boldsymbol{m}}}$$

$$\Longrightarrow \quad \tilde{\boldsymbol{V}}_{\boldsymbol{m}}(z) = \tilde{\boldsymbol{W}}_{\boldsymbol{m}} \tilde{\boldsymbol{\Lambda}}_{\boldsymbol{m}}(z) \tilde{\boldsymbol{C}}_{\boldsymbol{m}}. \tag{3.327}$$

The $e^{\vec{k}_t \cdot \vec{r}_t}$ dependence of the tangential field components has been omitted since it is the same for every region m due to the phase-matching conditions. The matrix $\tilde{\boldsymbol{W}}_{\boldsymbol{m}} = [\tilde{\boldsymbol{V}}_{\boldsymbol{mr1}}\ \tilde{\boldsymbol{V}}_{\boldsymbol{mr2}}\ \tilde{\boldsymbol{V}}_{\boldsymbol{mt1}}\ \tilde{\boldsymbol{V}}_{\boldsymbol{mt2}}]$, contains the appropriate eigenvector components from the solution of Eq. (3.305). The indices "1" and "2" in $k_{mr1,z}$, $k_{mr2,z}$, $k_{mt1,z}$, and $k_{mt2,z}$, could correspond to the "outer" and the "inner" wavevectors, respectively. In the case of an isotropic region, these indices could represent the *TE* and *TM* polarization waves.

Using Eq. (3.327) the boundary conditions of the multilayer anisotropic stack of Fig. 3.71 can be written as follows:

$$\tilde{\boldsymbol{V}}_{\boldsymbol{0}}(z = s_1^-) = \tilde{\boldsymbol{V}}_{\boldsymbol{1}}(z = s_1^+), \qquad \text{boundary at } z = 0,$$

$$\tilde{\boldsymbol{V}}_{\boldsymbol{m}}(z = s_{m+1}^-) = \tilde{\boldsymbol{V}}_{\boldsymbol{m+1}}(z = s_{m+1}^+), \qquad \text{boundary at } z = s_{m+1}, \quad (m = 1, 2, \cdots, N-1),$$

$$\tilde{\boldsymbol{V}}_{\boldsymbol{N}}(z = s_{N+1}^-) = \tilde{\boldsymbol{V}}_{\boldsymbol{N+1}}(z = s_{N+1}^+), \qquad \text{boundary at } z = s_{N+1}.$$

Alternately, the set of boundary conditions can be written as:

$$\tilde{\boldsymbol{W}}_{\boldsymbol{0}} \tilde{\boldsymbol{\Lambda}}_{\boldsymbol{0}}(z = s_1^-) \tilde{\boldsymbol{C}}_{\boldsymbol{0}} = \tilde{\boldsymbol{W}}_{\boldsymbol{1}} \tilde{\boldsymbol{\Lambda}}_{\boldsymbol{1}}(z = s_1^+) \tilde{\boldsymbol{C}}_{\boldsymbol{1}},$$

$$\tilde{\boldsymbol{W}}_{\boldsymbol{m}} \tilde{\boldsymbol{\Lambda}}_{\boldsymbol{m}}(z = s_{m+1}^-) \tilde{\boldsymbol{C}}_{\boldsymbol{m}} = \tilde{\boldsymbol{W}}_{\boldsymbol{m+1}} \tilde{\boldsymbol{\Lambda}}_{\boldsymbol{m+1}}(z = s_{m+1}^+) \tilde{\boldsymbol{C}}_{\boldsymbol{m+1}}, \quad (m = 1, 2, \cdots, N-1),$$

$$\tilde{\boldsymbol{W}}_{\boldsymbol{N}} \tilde{\boldsymbol{\Lambda}}_{\boldsymbol{N}}(z = s_{N+1}^-) \tilde{\boldsymbol{C}}_{\boldsymbol{N}} = \tilde{\boldsymbol{W}}_{\boldsymbol{N+1}} \tilde{\boldsymbol{\Lambda}}_{\boldsymbol{N+1}}(z = s_{N+1}^+) \tilde{\boldsymbol{C}}_{\boldsymbol{N+1}}.$$

The resulting set of equations forms a linear system consisting of $4+4(N-1)+4 = 4(N+1)$ equations with $2 + 4N + 2 = 4(N + 1)$ unknowns, encompassing all the reflection and transmission coefficients. This system can be cast into the standard form $\tilde{\boldsymbol{A}}\tilde{\boldsymbol{x}} = \tilde{\boldsymbol{b}}$ and solved directly. However, employing the transfer matrix approach can significantly simplify the system of equations. By eliminating all the intermediate coefficients $\tilde{\boldsymbol{C}}_{\boldsymbol{m}}$ (for $m = 1, 2, \cdots, N$), it becomes possible to directly relate the unknown coefficients in the input region (region 0), $\tilde{\boldsymbol{C}}_{\boldsymbol{0}}$, with those in the output region (region $N+1$), $\tilde{\boldsymbol{C}}_{\boldsymbol{N+1}}$, as follows:

$$\tilde{\boldsymbol{C}}_{\boldsymbol{N+1}} = \underbrace{\left[\tilde{\boldsymbol{W}}_{\boldsymbol{N+1}}^{\boldsymbol{-1}} \tilde{\boldsymbol{W}}_{\boldsymbol{N}} \tilde{\boldsymbol{\Lambda}}_{\boldsymbol{N}}(z = s_{N+1}^-)\right] \left[\tilde{\boldsymbol{W}}_{\boldsymbol{N}}^{\boldsymbol{-1}} \tilde{\boldsymbol{W}}_{\boldsymbol{N-1}} \tilde{\boldsymbol{\Lambda}}_{\boldsymbol{N-1}}(z = s_N^-)\right] \cdots \left[\tilde{\boldsymbol{W}}_{\boldsymbol{1}}^{\boldsymbol{-1}} \tilde{\boldsymbol{W}}_{\boldsymbol{0}}\right]}_{=\tilde{\boldsymbol{T}}} \tilde{\boldsymbol{C}}_{\boldsymbol{0}}$$

$$\tilde{\boldsymbol{C}}_{\boldsymbol{N+1}} = \tilde{\boldsymbol{T}} \tilde{\boldsymbol{C}}_{\boldsymbol{0}}. \tag{3.328}$$

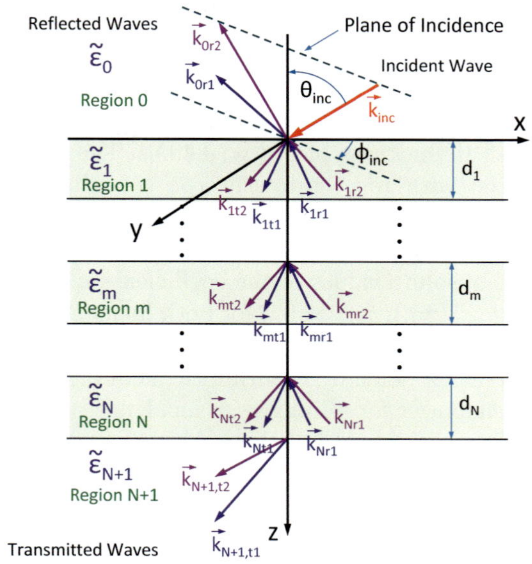

FIGURE 3.71: A planar anisotropic N-layer stack of general anisotropic media with relative permittivity tensors $\tilde{\varepsilon}_m$ $(m = 0, 1, 2, \cdots, N+1)$. A plane wave is incident from the top region at a polar angle θ_{inc} and azimuthal angle ϕ_{inc}. The wavevectors of the possible reflected waves $\vec{k}_{mr1}$ and $\vec{k}_{mr2}$ and the possible transmitted waves $\vec{k}_{mt1}$ and $\vec{k}_{mt2}$ are shown $(m = 0, 1, 2, \cdots, N+1)$.

where $\tilde{\boldsymbol{T}}$ is the 4×4 transfer matrix. In the previous derivation the relations $\tilde{\boldsymbol{\Lambda}}_{\boldsymbol{0}}(z = s_1^-) = \tilde{\boldsymbol{\Lambda}}_{\boldsymbol{N+1}}(z = s_{N+1}^+) = \tilde{\boldsymbol{I}}$ were also utilized. Equation (3.328) can be re-written in the form:

$$\begin{bmatrix} 0 \\ 0 \\ t_{N+1,1} \\ t_{N+1,2} \end{bmatrix} = \begin{bmatrix} \tilde{\boldsymbol{T}}_{\boldsymbol{11}} & \tilde{\boldsymbol{T}}_{\boldsymbol{12}} \\ \tilde{\boldsymbol{T}}_{\boldsymbol{21}} & \tilde{\boldsymbol{T}}_{\boldsymbol{22}} \end{bmatrix} \begin{bmatrix} r_{01} \\ r_{02} \\ E_1 \\ E_2 \end{bmatrix}, \tag{3.329}$$

where $E_1 = E_0$ and $E_2 = 0$ if the "outer" incident wave is selected, and $E_1 = 0$, $E_2 = E_0$ if the "inner" incident wave is selected. The 2×2 matrices $\tilde{\boldsymbol{T}}_{\boldsymbol{ij}}$ $(i, j = 1, 2)$ are submatrices of the overall transfer matrix $\tilde{\boldsymbol{T}}$. The solution for the r_{01}, r_{02}, $t_{N+1,1}$, and $t_{N+1,2}$ is then given by:

$$\begin{bmatrix} r_{01} \\ r_{02} \end{bmatrix} = -\tilde{\boldsymbol{T}}_{\boldsymbol{11}}^{\boldsymbol{-1}} \tilde{\boldsymbol{T}}_{\boldsymbol{12}} \begin{bmatrix} E_1 \\ E_2 \end{bmatrix}, \tag{3.330}$$

$$\begin{bmatrix} t_{N+1,1} \\ t_{N+1,2} \end{bmatrix} = \left[\tilde{\boldsymbol{T}}_{\boldsymbol{22}} - \tilde{\boldsymbol{T}}_{\boldsymbol{21}} \tilde{\boldsymbol{T}}_{\boldsymbol{11}}^{\boldsymbol{-1}} \tilde{\boldsymbol{T}}_{\boldsymbol{12}} \right] \begin{bmatrix} E_1 \\ E_2 \end{bmatrix}. \tag{3.331}$$

By utilizing $\tilde{\boldsymbol{C}}_{\boldsymbol{0}}$ and $\tilde{\boldsymbol{C}}_{\boldsymbol{N+1}}$, all coefficients $\tilde{\boldsymbol{C}}_{\boldsymbol{m}}$ $(m = 1, 2, \cdots N)$ can be determined using the boundary conditions established previously. This approach enables the determination of the electromagnetic fields in every region of the multilayer structure. Once the fields are known, the percentage of reflected and transmitted powers can be calculated in the same manner as for single interfaces.

Example 3.8.3:

Consider an biaxial layer surrounded by air. In this case, $\tilde{\varepsilon}_0 = \tilde{\varepsilon}_2 = \tilde{I}$ (identity matrix). The biaxial layer has a relative permittivity tensor in its principal axes system of $\tilde{\varepsilon}_1 = diag(3, 4, 5)$, and a physical thickness of s $d = 0.40\lambda_0$. The principal axes system of the biaxial medium is rotated with respect to the xyz coordinate system using the angles $\psi_1 = 30°$ and $\psi_2 = 75°$ (see Appendix 3H). A plane wave of freespace wavelength $\lambda_0 = 1\,\mu$m is incident from the air onto the biaxial layer. The polarization of the incident wave is either *TE* or *TM*. Compute the absolute values of the coefficients $r_{_{TE}}, r_{_{TM}}, t_1$, and t_2 as functions of the angle of incidence θ_{inc} (with $\phi_{inc} = 0°$) for both incident polarizations. Calculate and plot the normalized powers for the reflected and the transmitted waves as functions of θ_{inc}. Verify that power conservation holds by ensuring the sum of the reflected and transmitted powers equals the incident power for all angles of incidence.
Solution: This example is a special case of the analysis presented in Sec. 3.8.3 for a single layer, i.e., for $N = 1$. Using Eq. (3H.5) the relative permittivity of the biaxial layer can be found as:

$$\tilde{\varepsilon}_1 = \begin{bmatrix} 4.1663 & 0.3125 & 0.4183 \\ 0.3125 & 3.0837 & 0.1121 \\ 0.4183 & 0.1121 & 4.75 \end{bmatrix},$$

For *TE* incident polarization, the angle $\psi = 90°$, while for *TM* polarization, $\psi = 0°$. Using the analysis presented in Sec. 3.8.3, the complex coefficients $r_{_{TE}}, r_{_{TM}}, t_1$, and t_2 can be computed as functions of the angle of incidence, θ_{inc}. The results of these calculations are shown in Fig. 3.72(a) and 3.72(c), corresponding to *TE* and *TM* incident polarizations, respectively. An interesting observation for the biaxial layer sandwiched between air is that when a *TE* wave is incident, the reflected wave includes components of both *TE* and *TM* polarizations. Similarly, for a *TM* incident wave, the reflected waves contain both *TE* and *TM* polarizations. This cross-polarization effect arises from the anisotropic properties of the biaxial layer and is influenced by the layer's thickness.

Another notable effect is the presence of a Brewster-like angle for *TM* incident polarization, observed at approximately $\theta_{inc} \simeq 62.7°$, as evident from the plot of $|r_{_{TM}}|$ in Fig. 3.72(c). Using Eqs. (3.313) and (3.314), the normalized reflected powers $P_{r_{TE}}$, $P_{r_{TM}}$, as well as the normalized transmitted powers P_{t1}, and P_{t2}, can be calculated. In this case, $P_{cross} = 0$ for all angles of incidence. The results for these normalized reflected and transmitted powers are presented in Figs. 3.69(b) and 3.69(d) for *TE* and *TM* incident polarizations, respectively. These figures also include the sum of the normalized efficiencies, illustrating that power conservation is upheld across the interface. Specifically, the total reflected and transmitted powers equal the incident power, confirming that no energy is lost during the reflection and transmission processes, as expected from the Poynting theorem [see Eq. (3.315)].

3.8.4 Approximate Jones Calculus Method

Jones calculus offers a straightforward, approximate, and systematic framework for analyzing the behavior of light as it interacts with a sequence of anisotropic plates and polarizers. Introduced by R. Clark Jones in 1941 [149], this method represents the polarization state of light using a two-component vector, while each optical element is modeled as a 2×2 matrix. The overall response of the optical system is derived by sequentially multiplying the 2×2 matrices corresponding to each optical element in the order in which light encounters them. The resulting polarization state of the output is then calculated by applying the overall

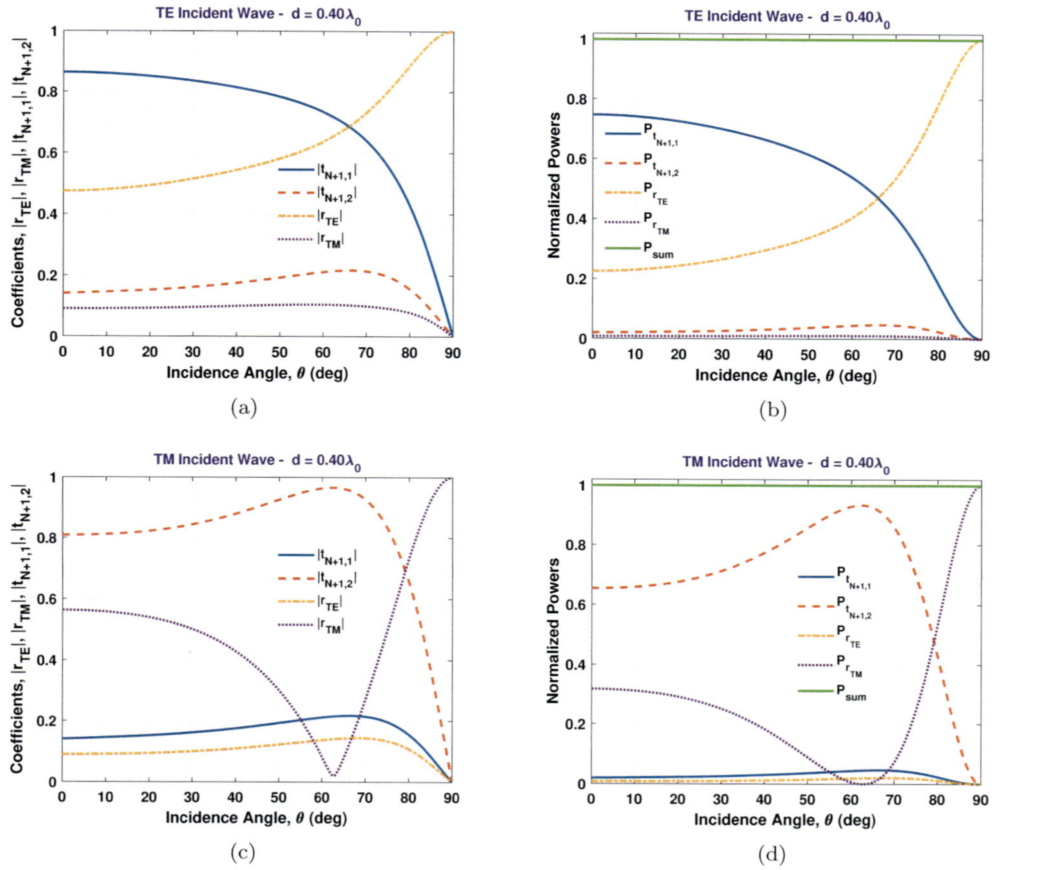

FIGURE 3.72: An example of a biaxial layer surrounded by air [146, 147]. (a) Absolute values of the coefficients r_{TE}, r_{TM}, t_1, and t_2 for *TE* incident polarization. (b) The normalized powers for *TE* incident polarization. (c) Same as (a) but for *TM* incident polarization. (d) Same as (b) but for *TM* incident polarization.

system matrix to the input polarization state vector. This method offers a straightforward approach for determining the evolution of the polarization state of light as it interacts with various optical components. Jones calculus proves particularly valuable in applications involving polarized light, including the design of polarizing filters, the characterization of birefringent materials, and optical communications, where precise control and manipulation of polarization states are essential.

When light propagates through anisotropic media, it consists of a linear superposition of two orthogonally polarized waves, each associated with its eigen-polarization and corresponding refractive index. These waves are well-defined for each direction of propagation, as discussed earlier. In the context of common retardation plates, one of the principal axes is typically aligned along the z direction, as illustrated in Fig. 3.73. The other two principal axes (named as "slow" axis and "fast" axis) lie in the plane of the plate (xy plane in Fig. 3.73). In Jones calculus, all reflections and multiple interference effects are neglected, which is generally a valid approximation. This is because the main surfaces of these plates are usually coated with antireflective layers, reducing reflections. Additionally, the main surfaces of the plates are often not optically parallel, further justifying the neglect of multiple interference effects. The polarization state of the incident electromagnetic wave can be

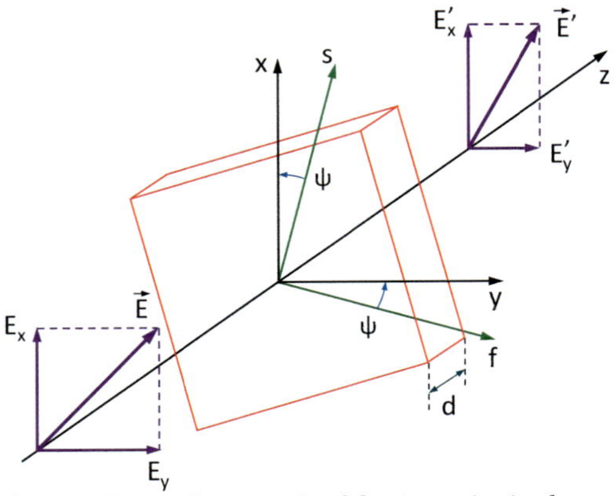

FIGURE 3.73: A retardation plate, characterized by its principal axes s (slow), f (fast), and z, is depicted within the laboratory coordinate system xyz. The slow and fast axes are oriented at an angle ψ with respect to x and y axes, respectively. The thickness of the plate is d, measured along the z direction. It is assumed that the angle psi is measured in the positive sense following the right-handed coordinate system convention.

represented as follows:

$$\vec{E} = \begin{bmatrix} E_x \\ E_y \end{bmatrix},$$

which is compatible with the decomposition of the incident field shown in Fig. 3.73. This incident wave should be decomposed into the eigen-polarizations (along f and s axes). This can be accomplished as follows:

$$\begin{bmatrix} E_s \\ E_f \end{bmatrix} = \underbrace{\begin{bmatrix} \cos\psi & \sin\psi \\ -\sin\psi & \cos\psi \end{bmatrix}}_{=\tilde{\boldsymbol{R}}(\psi)} \begin{bmatrix} E_x \\ E_y \end{bmatrix} = \tilde{\boldsymbol{R}}(\psi) \begin{bmatrix} E_x \\ E_y \end{bmatrix}.$$

The eigen-polarizations E_s and E_f propagate into the plate of thickness d, independent of each other according to the equation:

$$\begin{bmatrix} E_s' \\ E_f' \end{bmatrix} = \begin{bmatrix} \exp(-jk_0 n_s d) & 0 \\ 0 & \exp(-jk_0 n_f d) \end{bmatrix} \begin{bmatrix} E_s \\ E_f \end{bmatrix},$$

where the primed values E_s' and E_f' denote the eigen-polarizations at the end of the propagation distance d, n_s and n_f are the slow and fast axis refractive indices, and k_0 the freespace wavenumber. Now, by introducing the retardation $\Gamma = k_0(n_s - n_f)d$ and the phase angle $\Phi = (1/2)k_0(n_s + n_f)d$, the previous equation can be rewritten as:

$$\begin{bmatrix} E_s' \\ E_f' \end{bmatrix} = \exp(-j\Phi) \begin{bmatrix} \exp(-j\Gamma/2) & 0 \\ 0 & \exp(+j\Gamma/2) \end{bmatrix} \begin{bmatrix} E_s \\ E_f \end{bmatrix}. \tag{3.332}$$

Converting the last equation into the xy coordinate system yields:

$$\begin{bmatrix} E_x' \\ E_y' \end{bmatrix} = \begin{bmatrix} \cos\psi & -\sin\psi \\ \sin\psi & \cos\psi \end{bmatrix} \begin{bmatrix} E_s' \\ E_f' \end{bmatrix} = \tilde{\boldsymbol{R}}(-\psi) \begin{bmatrix} E_s' \\ E_f' \end{bmatrix}.$$

Here, the primed fields represent the electric field components at the exit of the plate, expressed in the xy coordinate system. By combining all the steps, it becomes possible to relate the output electric field components E_x' and E_y' to the input components E_x and E_y. This results in the following expression:

$$\begin{bmatrix} E_x' \\ E_y' \end{bmatrix} = \tilde{\boldsymbol{R}}(-\psi)\exp(-j\Phi)\begin{bmatrix} \exp(-j\Gamma/2) & 0 \\ 0 & \exp(+j\Gamma/2) \end{bmatrix}\tilde{\boldsymbol{R}}(\psi)\begin{bmatrix} E_x \\ E_y \end{bmatrix}$$
$$= \tilde{\boldsymbol{W}}(\psi,\Gamma)\begin{bmatrix} E_x \\ E_y \end{bmatrix}, \qquad (3.333)$$

where the overall 2×2 matrix $\tilde{\boldsymbol{W}}(\psi,\Gamma)$ is given by:

$$\tilde{\boldsymbol{W}}(\psi,\Gamma) = e^{-j\Phi}\begin{bmatrix} e^{-j\frac{\Gamma}{2}}\cos^2\psi + e^{+j\frac{\Gamma}{2}}\sin^2\psi & -j\sin\frac{\Gamma}{2}\sin 2\psi \\ -j\sin\frac{\Gamma}{2}\sin 2\psi & e^{-j\frac{\Gamma}{2}}\sin^2\psi + e^{+j\frac{\Gamma}{2}}\cos^2\psi \end{bmatrix}. \qquad (3.334)$$

It is important to note that, since the plate is lossless (and reflections are neglected), the electromagnetic power density, as described by the Poynting vector, remains constant after the light passes through the plate. This implies that $|E_x'|^2+|E_y'|^2 = |E_x|^2+|E_y|^2$, a property that can be readily verified using Eq. (3.333).

An ideal polarizer, with its transmission axis aligned parallel to the x-axis, can be represented by the following Jones matrix:

$$\tilde{\boldsymbol{P}}_{\boldsymbol{x}} = e^{-j\Theta}\begin{bmatrix} 1 & 0 \\ 0 & 0 \end{bmatrix}, \qquad (3.335)$$

while an ideal polarizer along the y-axis is represented by the Jones matrix

$$\tilde{\boldsymbol{P}}_{\boldsymbol{y}} = e^{-j\Theta}\begin{bmatrix} 0 & 0 \\ 0 & 1 \end{bmatrix}. \qquad (3.336)$$

The angle Θ represents a phase shift accumulated due to the finite size of the plate along the z-axis. Finally, an ideal polarizer with its transmission axis oriented at an angle Ψ with respect to the x-axis, can be described by the following Jones matrix:

$$\tilde{\boldsymbol{P}}(\Psi) = e^{-j\Theta}\begin{bmatrix} \cos^2\Psi & -\sin\Psi\cos\Psi \\ -\sin\Psi\cos\Psi & \sin^2\Psi \end{bmatrix}, \qquad (3.337)$$

where $\tilde{\boldsymbol{P}}(\Psi = 0) = \tilde{\boldsymbol{P}}_{\boldsymbol{x}}$ and $\tilde{\boldsymbol{P}}(\Psi = \pi/2) = \tilde{\boldsymbol{P}}_{\boldsymbol{y}}$, correspond to special cases of polarizers oriented along the x and the y-axis, respectively.

Example 3.8.4:

Consider a half-wave plate with a retardation $\Gamma = \pi = k_0(n_s - n_f)d$, where the slow axis is oriented at an angle $\psi = \pi/4$ with respect to the x-axis. The incident plane wave is polarized along the y-axis, such that $E_x = 0$ and $E_y = E_0 \neq 0$. Using Jones calculus, determine the polarization of the output wave after passing through the half-wave plate.

Solution: The corresponding Jones matrix for the half-wave plate, with $\Gamma = \pi$ and $\psi = \pi/4$, is given by Eq. (3.334), and is equal to:

$$\tilde{\boldsymbol{W}}(\psi = \frac{\pi}{4}, \Gamma = \pi) = e^{-j\Phi}\begin{bmatrix} 0 & -j \\ -j & 0 \end{bmatrix}.$$

The resulting field at the output of the half-wave plate becomes:

$$\begin{bmatrix} E'_x \\ E'_y \end{bmatrix} = e^{-j\Phi} \begin{bmatrix} 0 & -j \\ -j & 0 \end{bmatrix} \begin{bmatrix} 0 \\ E_0 \end{bmatrix} = e^{-j\Phi} \begin{bmatrix} -jE_0 \\ 0 \end{bmatrix} = -je^{-j\Phi} \begin{bmatrix} E_0 \\ 0 \end{bmatrix}.$$

The last equation reveals that the y-polarized plane wave was transformed into an x-polarized plane wave by the half-wave plate. If the incident polarization is right-handed circularly polarized, the input electric field is given by: $[E_x, E_y]^T = E_0[1, -j]^T$. Now, applying the Jones matrix for the half-wave plate [Eq. (3.334)] it is straightforward to show that $[E'_x, E'_y]^T = E_0[1, +j]^T$. This corresponds to a left-handed circularly polarized plane wave, as the polarization state has been flipped by the half-wave plate.

Example 3.8.5:

In this example, a retardation plate is sandwiched between two x polarizers, as shown in Fig. 3.74. The angle $\psi = \pi/4$ represents the orientation of the slow axis of the retardation plate with respect to the x-axis. Assume that the incident wave is polarized along the x-axis. Using Jones calculus, determine how the polarization state evolves as the light passes through the system of polarizers and the retardation plate. Additionally, calculate the intensity modulation caused by the retardation plate.

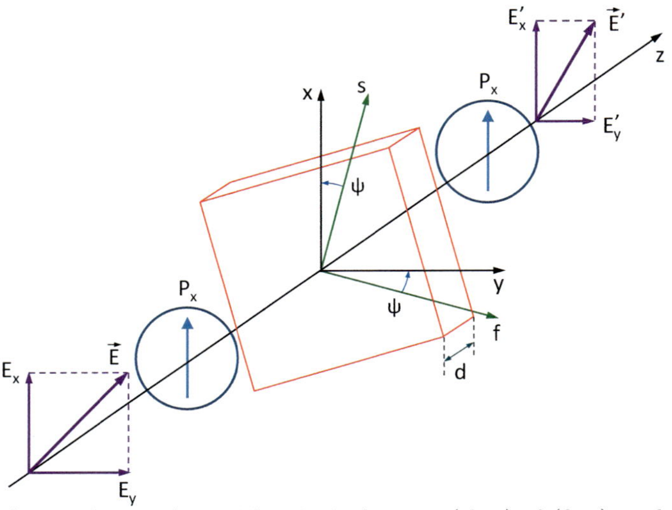

FIGURE 3.74: A retardation plate with principal axes s (slow), f (fast), and z, is shown in the laboratory coordinate system xyz. The slow and fast axes are at an angle ψ with respect to x and y axes, respectively. The thickness of the plate is d along the z direction. Two parallel x-polarizers are inserted in front of the plate and after the plate.

Solution: The resulting polarization vector after the output polarizer is given by:

$$\begin{bmatrix} E'_x \\ E'_y \end{bmatrix} = e^{-j(\Phi+\Theta)} \begin{bmatrix} 1 & 0 \\ 0 & 0 \end{bmatrix} \begin{bmatrix} \cos\frac{\Gamma}{2} & -j\sin\frac{\Gamma}{2} \\ -j\sin\frac{\Gamma}{2} & \cos\frac{\Gamma}{2} \end{bmatrix} \begin{bmatrix} 1 \\ 0 \end{bmatrix} = e^{-j(\Phi+\Theta)} \begin{bmatrix} \cos\frac{\Gamma}{2} \\ 0 \end{bmatrix}.$$

From the last equation the output intensity can be related to the input intensity as follows:

$$\frac{I_{out}}{I_{in}} = \frac{|E'_x|^2 + |E'_y|^2}{|E_x|^2 + |E_y|^2} = \cos^2\frac{\Gamma}{2}. \tag{3.338}$$

The last equation represents the modulation of the output intensity as a function of the retardation angle Γ. This modulation can be controlled by varying Γ through the electro-optic effect, which allows for the manipulation of the index of refraction of the material in response to an applied electric field. This variation of Γ directly influences the phase shift between the slow and fast axes of the retardation plate, thereby modulating the polarization state and the intensity of the transmitted light.

3.9 Coherence Theory Basics

Up to this point, most discussions on interference assumed strictly monochromatic and perfectly coherent light sources. Coherence refers to the predictable repeatability of a wave in both time and space. However, perfect coherence is unattainable in practical light sources for several reasons. One reason is that any light source, including the most advanced lasers, has a finite bandwidth, making it impossible to achieve strict monochromaticity. This finite bandwidth leads to random temporal variations in the wave's amplitude and phase, as it results from the superposition of numerous monochromatic components with random amplitudes and phases. These variations yield partial temporal coherence (or temporal incoherence). Furthermore, in the case of an extended light source, even if it is strictly monochromatic, the Huygens wavelets emitted from different parts of the source are not in perfect phase agreement. This discrepancy results in partial spatial coherence (or spatial incoherence). Since practical light sources exhibit both finite bandwidth and finite spatial extent, temporal and spatial coherence are inherently interrelated. Nevertheless, coherence effects are traditionally classified into temporal and spatial categories for analytical convenience. Qualitative examples of temporal and spatial coherence are shown in Figs. 3.75(a) and 3.75(b), respectively. In Fig. 3.75(a) a Gaussian distribution of the source's frequency spectrum illustrates the effect of temporal coherence. In Fig. 3.75(b) a Gaussian spatial phase distribution demonstrates spatial coherence effects. This section introduces the basics of optical coherence theory. For a more rigorous analysis, readers are encouraged to consult specialized references, such as those by Born and Wolf, Goodman, and Marathay [91, 150, 151].

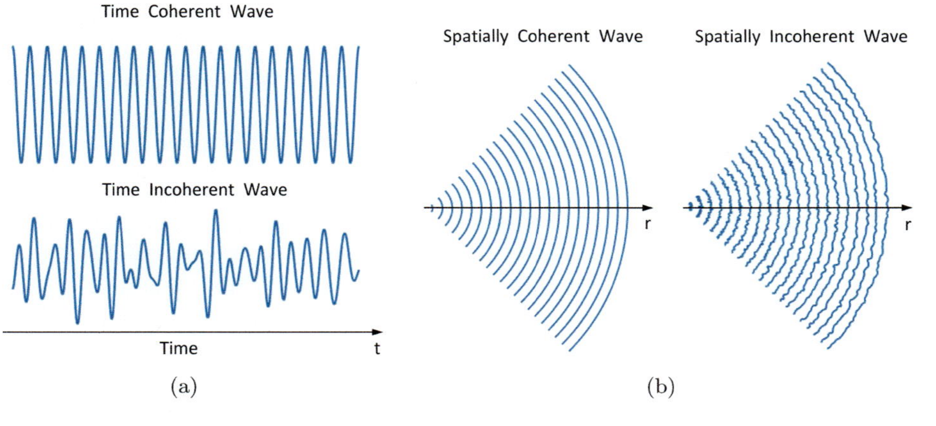

(a) (b)

FIGURE 3.75: (a) Qualitative comparison of a temporally coherent wave and a temporally incoherent wave. (b) Qualitative comparison of a spatially coherent wave and a spatially incoherent one.

3.9.1 Interference Fringes of Multichromatic Light Sources

The fundamental cause of temporal coherence is the finite bandwidth of the light source. To illustrate this effect, consider the Michelson interferometer (refer to Sec. 3.6.3) under two situations: (a) A light source consisting of two nearly equal wavelengths. (b) A light source with a narrow spectral bandwidth.

3.9.1.1 Interference Fringes of a Doublet Light Source

In this case, a doublet light source is considered, emitting two nearly equal wavelengths λ_1 and λ_2 where $\Delta\lambda = |\lambda_1 - \lambda_2| \ll \lambda_1, \lambda_2$. By varying the distance d (the displacement between the two mirrors M_1 and M_2 of Fig. 3.49), the visibility of the interference circular fringes is observed. For small values of d, the fringes are relatively sharp and well-defined. However, as d increases, the fringes gradually fade, and their visibility decreases. Assuming that the beam splitter (dielectric with compensator) of the Michelson interferometer operates symmetrically, the circular fringes for the doublet light source can be modeled using Eq. (3.255) as:

$$I = I(\lambda_1) + I(\lambda_2) = I_0 \cos^2\left(\frac{2\pi}{\lambda_1} d\cos\theta + \frac{\pi}{2}\right) + I_0 \cos^2\left(\frac{2\pi}{\lambda_2} d\cos\theta + \frac{\pi}{2}\right), \tag{3.339}$$

where I_1 and I_2 are the intensities corresponding to the two wavelengths λ_1 and λ_2. For simplicity, it is assumed that the amplitudes of the two wavelength components are equal, i.e., $I_{01} = I_{02} = I_0$. The term $\cos\theta = \cos\theta_m = f/(f^2 + r_m^2)^{1/2}$ corresponds to the m-th circular fringe, where f is the focal length of the lens used (see Fig. 3.49). To numerically evaluate Eq. (3.339), the sodium light source is considered, which has a doublet in the visible region with $\lambda_1 = 588.995\,\text{nm}$ and $\lambda_2 = 589.592\,\text{nm}$ [152]. The resulting normalized intensity I/I_0 is shown in Fig. 3.76(a) for $d = 146\,\mu\text{m}$ (dotted line), $d = 200\,\mu\text{m}$ (dashed line), and $d = 300\,\mu\text{m}$ (solid line) (adapted Ref. [152]). In the case of $d = 146\,\mu\text{m}$, the two interference fringes set $I(\lambda_1)/I_0$ and $I(\lambda_2)/I_0$ are almost out of phase leading to very low fringe visibility. Applying Eq. (3.199) the visibility is only $V = 0.0248$. In contrast for $d = 200\,\mu\text{m}$, the fringe visibility increases to $V = 0.5549$ while in the case of $d = 300\,\mu\text{m}$ the fringes are almost perfect with $V = 0.9974$. Chromatic representations of the circular fringes for these three cases are shown in Fig. 3.76(b). The variation of I/I_0 from almost flat ($d = 146\,\mu\text{m}$) to almost perfectly visible ($d = 300\,\mu\text{m}$) can repeat as the distance d varies. However, for real sources (even those that are nearly monochromatic), the wavelength content of the source causes the fringes in the Michelson interferometer to vanish when the mirror displacement d exceeds a value commonly referred to as the *coherence length* of the light source.

3.9.1.2 Interference Fringes of a Narrow-band Light Source

In this section, the fringe visibility in Michelson's interferometer using a narrow-band light source is analyzed. A light source with a spectral intensity $I(\nu)$ (in $W/m^2/Hz$) is considered, where $I(\nu)$ is centered at a frequency ν_0 and has a narrow spectral width $\Delta\nu$ such that $\Delta\nu \ll \nu_0$ [91, 152]. Using Eq. (3.255) for the narrow band light source, the interference pattern formed by the two beams in the Michelson interferometer can be expressed as:

$$\begin{aligned} I(\tau) &= \frac{1}{2}\int_{-\infty}^{+\infty} I(\nu)\left[1 + \cos\left(2\pi\nu\frac{2d\cos\theta}{c}\right)\right] d\nu, \\ &= \int_{-\infty}^{+\infty} I_0(\nu)\left[1 + \cos\left(2\pi\nu\tau\right)\right] d\nu, \end{aligned}$$

where $I_0(\nu) = (1/2)I(\nu)$, and $\tau = 2d\cos\theta/c$ represents the time delay between the two beams introduced by the mirror displacement d. The phase shift of $\pi/2$, which arises from

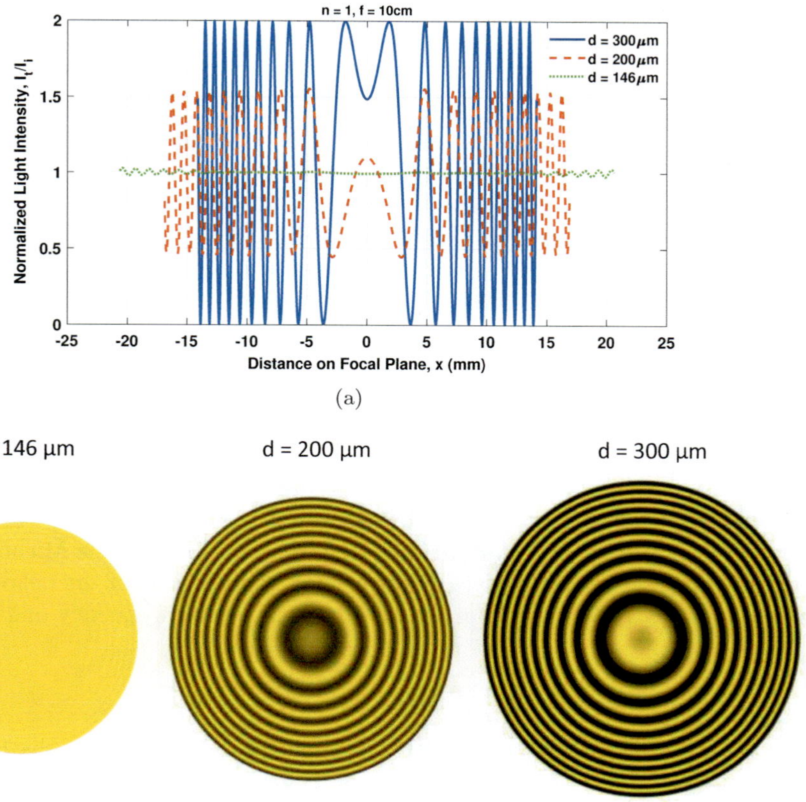

(a)

(b)

FIGURE 3.76: (a) Circular fringes of the Michelson interferometer with a sodium doublet source. The cases of mirror length difference $d = 146\,\mu\text{m}$ (dotted line), $d = 200\,\mu\text{m}$ (dashed line), and $d = 300\,\mu\text{m}$ (solid line) are shown. (b) Qualitative circular fringes for the $d = 146\,\mu\text{m}$ ($V = 0.0248$), $d = 200\,\mu\text{m}$ ($V = 0.5549$), and $d = 300\,\mu\text{m}$ ($V = 0.9974$) cases. In the case of $d = 146\,\mu\text{m}$, there are practically no visible fringes while in the case of $d = 300\,\mu\text{m}$ are the most visible fringes. For all cases approximately 10 circular fringes are shown.

the specifics of the beam splitters used in the interferometer, is omitted for simplicity in this analysis. Furthermore, the omitted phase shift only introduces a shift in the fringes and does not affect their visibility. Now let's define $\nu = \nu_0 + \nu'$, where ν' represents the frequency deviation from the center frequency ν_0. Substituting this into the previous equation, the intensity can be rewritten as:

$$I(\tau) = \int_{-\infty}^{+\infty} I_0(\nu')\,[1 + \cos(2\pi(\nu_0 + \nu')\tau)]\,d\nu' =$$

$$= \underbrace{\int_{-\infty}^{+\infty} I_0(\nu')d\nu'}_{=P} + \cos(2\pi\nu_0\tau)\underbrace{\int_{-\infty}^{+\infty} I_0(\nu')\cos(2\pi\nu'\tau)d\nu'}_{=C(\tau)} -$$

$$\sin(2\pi\nu_0\tau)\underbrace{\int_{-\infty}^{+\infty} I_0(\nu')\sin(2\pi\nu'\tau)d\nu'}_{=S(\tau)} \Longrightarrow$$

$$I(\tau) = P + C(\tau)\cos(2\pi\nu_0\tau) - S(\tau)\sin(2\pi\nu_0\tau). \tag{3.340}$$

In the above expressions, $I_0(\nu')$ is defined as the spectral intensity of the light source, with the frequency ν' measured from the center frequency ν_0.

Example 3.9.1:

A narrow-band light source is considered, with its spectrum resembling the gain curve of a He-Ne laser. When operating below threshold, the emitted light is due to spontaneous emission, characterized by a broader spectrum. However, when lasing (above threshold), the spectrum narrows significantly. For the below-threshold condition, the light source spectrum is centered at a free-space wavelength of $\lambda_0 = 0.6328\,\mu$m, corresponding to a center frequency ν_0. The spectral density of this He-Ne source is approximated by a Gaussian function of the form (normalized):

$$\frac{I_0(\nu)}{I_0} = \frac{1}{\sqrt{2\pi}\Delta\nu} e^{-(\nu-\nu_0)^2/2(\Delta\nu)^2},$$

where $\nu_0 = c/\lambda_0$ (c is the freespace light velocity) and $\Delta\nu = 1.5 \times 10^9$ Hz (usually this broadening is due to the Doppler effect in the He-Ne lasers). The spectrum is normalized such that $I_0 = \int_{-\infty}^{+\infty} I_0(\nu')d\nu'$. Assume that this source is used in the Michelson interferometer, and the interference intensity, $I(\tau)$, is sought as a function of the delay time τ. Repeat the procedure for a Lorentzian source with the same center frequency and the same $\Delta\nu$ of the form:

$$\frac{I_0(\nu)}{I_0} = \frac{2}{\pi\Delta\nu} \frac{1}{(\nu-\nu_0)^2 + (\Delta\nu/2)^2},$$

where again the area under the Lorentzian is unity. Compare the results for the Gaussian and Lorentzian sources, highlighting how the coherence time τ_c and the visibility of the interference fringes depend on the spectral shape.

Solution: To apply Eq. (3.340) for the Gaussian spectrum, the terms $C(\tau)$ and $S(\tau)$ must be calculated. These terms represent the real and imaginary components of the Fourier transform of the spectral density $I_0(\nu)$. Specifically:

$$C(\tau) = \int_{-\infty}^{+\infty} I_0(\nu')\cos(2\pi\nu'\tau)d\nu' = \frac{1}{2}\left[\mathcal{F}\{I_0(\nu')\} + \mathcal{F}\{I_0(\nu')\}^*\right], \quad \text{and}$$

$$S(\tau) = \int_{-\infty}^{+\infty} I_0(\nu')\sin(2\pi\nu'\tau)d\nu' = \frac{-j}{2}\left[\mathcal{F}\{I_0(\nu')\} - \mathcal{F}\{I_0(\nu')\}^*\right],$$

where $\mathcal{F}$ denotes a Fourier transform and "*" denotes complex conjugate. For the Gaussian spectrum, the Fourier transform of the spectral intensity is: $\mathcal{F}\{I_0(\nu')\} = \exp(-2\pi^2(\Delta\nu)^2\tau^2)$. The spectrum is shown in Fig. 3.77(a) while the interference intensity $I(\tau)$ is shown in Fig. 3.77(b). In the latter, the oscillation frequency has been artificially reduced by a factor of 7500 to make the oscillations visible alongside the reduction in fringe visibility as τ increases. The relation between $\Delta\nu$ and Δk is given by $\Delta\nu = c\Delta(1/\lambda) = c\Delta k$. For the Lorentzian spectrum, the Fourier transform is: $\mathcal{F}\{I_0(\nu')\} = \exp(-\pi\Delta\nu|\tau|)$. The Lorentzian spectrum and its resulting interference intensity $I(\tau)$ are shown in Figs. 3.78(a) and 3.78(b), respectively. In both cases, the spectra are symmetric about the center frequency, which results in $S(\tau) = 0$. The envelope of $I(\tau)$ corresponds to the Fourier transform of $C(\tau)$, and this is evident in Figs. 3.77(b) (Gaussian) and 3.78(b) (Lorentzian).

From the interference intensities shown in Figs. 3.77(b) (Gaussian spectrum) and 3.78(b) (Lorentzian spectrum), it is evident that the visibility fringes reduce as the displacement d (or equivalently the delay time τ) increases. The visibility as a function of τ, $V(\tau)$, can be determined in a straightforward manner since the functions $C(\tau)$ and $S(\tau)$ are slowly

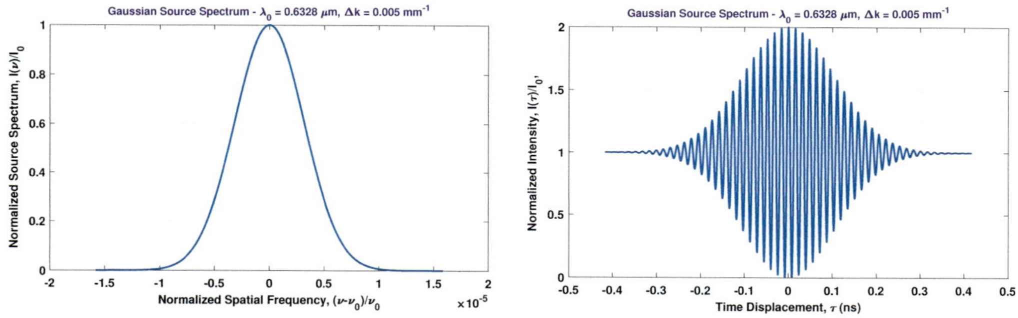

(a) (b)

FIGURE 3.77: (a) A He-Ne laser source with a normalized Gaussian spectral intensity is shown as a function of the normalized frequency $(\nu - \nu_0)/\nu_0$. The center frequency corresponds to the freespace wavelength of $\lambda_0 = 0.6328\,\mu$m, while the Gaussian spectrum is characterized by a bandwidth of $\Delta\nu = 1.5 \times 10^9$ Hz equivalent to $\Delta k = 0.005\,\text{mm}^{-1}$. (b) The normalized interference intensity $I(\tau)/I_0$ of the Michelson interferometer for the Gaussian spectrum He-Ne source is plotted as a function of the delay time τ. The oscillation frequency is artificially reduced by a factor of 7500 to make the periodic intensity oscillations and the reduction in fringe visibility with increasing τ simultaneously visible.

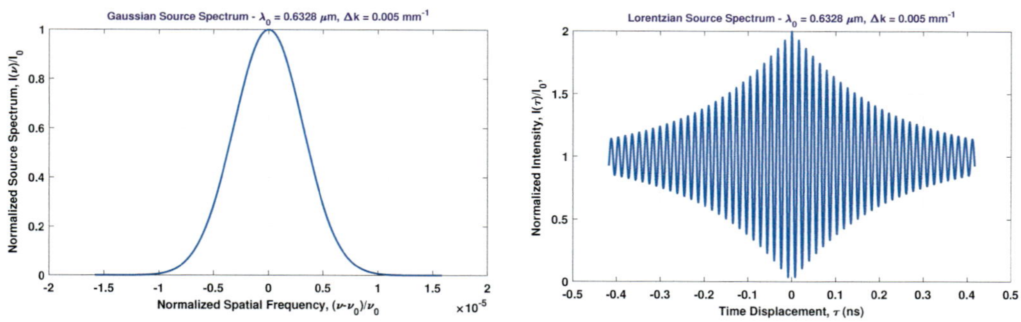

(a) (b)

FIGURE 3.78: (a) A He-Ne laser source with a normalized Lorentzian spectral intensity is shown as a function of the normalized frequency $(\nu - \nu_0)/\nu_0$. The center frequency corresponds to the freespace wavelength of $\lambda_0 = 0.6328\,\mu$m, while the Lorentzian spectrum is characterized by a bandwidth of $\Delta\nu = 1.5 \times 10^9$ Hz, equivalent to $\Delta k = 0.005\,\text{mm}^{-1}$. (b) The normalized interference intensity $I(\tau)/I_0$ of the Michelson interferometer for the Lorentzian spectrum He-Ne source is plotted as a function of the delay time τ. The oscillation frequency is artificially reduced by a factor of 7500 to make the periodic intensity oscillations and the reduction in fringe visibility with increasing τ simultaneously visible.

varying, i.e., $dC/d\tau \simeq dS/d\tau \simeq 0$ [91, 152]. The maxima and the minima of $I(\tau)$ [of Eq. (3.340)] can be calculated from the solution of the following equation:

$$\frac{dI(\tau)}{d\tau} = 0 \Longrightarrow -C(\tau)\sin(2\pi\nu_0\tau)2\pi\nu_0 - S(\tau)\cos(2\pi\nu_0\tau)2\pi\nu_0 = 0 \Longrightarrow$$

$$\frac{\sin(2\pi\nu_0\tau)}{\cos(2\pi\nu_0\tau)} = -\frac{S(\tau)}{C(\tau)} \Longrightarrow \quad \tan(2\pi\nu_0\tau) = \frac{-S(\tau)}{C(\tau)} = \frac{S(\tau)}{-C(\tau)} \Longrightarrow$$

$$\cos(2\pi\nu_0\tau) = \pm\frac{C}{\sqrt{C^2+S^2}}, \qquad \sin(2\pi\nu_0\tau) = \pm\frac{S}{\sqrt{C^2+S^2}},$$

where the maxima are located for values of $2\pi\nu_0\tau$ in the fourth quadrant, and the minima are located for values in the second quadrant, respectively. It is reminded that in the previous expressions the functions $S(\tau)$ and $C(\tau)$ are almost constant as the τ varies near a maximum or a minimum. Then, the resulting maxima and minima of $I(\tau)$ are given by [from Eq. (3.340)]:

$$I_{max}(\tau) \simeq P + \sqrt{C^2 + S^2}, \tag{3.341}$$

$$I_{min}(\tau) \simeq P - \sqrt{C^2 + S^2}. \tag{3.342}$$

Then, the visibility, $V(\tau)$, can be expressed as:

$$V(\tau) = \frac{I_{max}(\tau) - I_{min}(\tau)}{I_{max}(\tau) + I_{min}(\tau)} \simeq \frac{\sqrt{C^2 + S^2}}{P}. \tag{3.343}$$

In the case of a symmetrical (around a center frequency) intensity spectrum, $S(\tau) = 0$, and the visibility is simply $V(\tau) = |C(\tau)|/P$. If the intensity spectrum is normalized to the unity area then $P = 1$. In general, though this is not a necessary requirement. For band-limited light sources the spectrum has practically a finite bandwidth and correspondingly, its Fourier transform has also a finite bandwidth. Therefore, $C(\tau \to \infty) \to 0$ and similarly $S(\tau \to \infty) \to 0$. As a result $V(\tau \to \infty) = 0$ and the fringes disappear completely. In this example, where a Gaussian and a Lorentzian (both symmetrical) spectra were utilized, the visibility becomes:

$$V(\tau) = \begin{cases} \exp\left(-2\pi^2\Delta\nu^2\tau^2\right) & \text{for Gaussian spectrum,} \\ \exp\left(-\pi\Delta\nu|\tau|\right) & \text{for Lorentzian spectrum.} \end{cases} \tag{3.344}$$

The visibility $V(\tau)$ is shown in Fig. 3.79(a) for the Gaussian intensity spectrum of this example. Similarly, the $V(\tau)$ is shown in Fig. 3.79(b) for a Lorentzian intensity spectrum. From these diagrams, it is apparent that the visibility reduces rapidly with increasing τ. When the visibility drops to $1/e$ of its maximum value of $V(\tau = 0) = 1$, a time τ_c (called "coherence time") can be defined as follows for the Gaussian and the Lorentzian spectrum profiles, respectively:

$$\tau_c = \begin{cases} \dfrac{1}{\sqrt{2}\pi\Delta\nu} & \text{for Gaussian spectrum,} \\ \dfrac{1}{\pi\Delta\nu} & \text{for Lorentzian spectrum.} \end{cases} \tag{3.345}$$

The selection of the visibility decreasing to the $1/e$ value for defining the coherence time τ_c is indeed somewhat arbitrary, but it is commonly used because it provides a clear, practical threshold for observing the decay of temporal coherence. This definition captures the essence of the time over which the light maintains its ability to interfere. In general, the coherence time τ_c is a statistical measure of the temporal coherence of the light source and is directly related to the bandwidth of the source's spectral intensity. The relationship can be expressed as:

$$\tau_c \simeq \frac{1}{\Delta\nu}, \tag{3.346}$$

where $\Delta\nu$ is the spectral width (or bandwidth) of the light source. This indicates that a broader spectral width (i.e., a larger $\Delta\nu$) corresponds to a shorter coherence time, and a narrower spectral width leads to a longer coherence time. This connection between the coherence time and the bandwidth is a fundamental result of Fourier theory, which relates the temporal and spectral properties of a light source.

The coherence length ℓ_c is the spatial counterpart to the coherence time τ_c, and it is defined as: $\ell_c = c\tau_c$. This defines the characteristic distance over which the light remains coherent, analogous to how τ_c defines the temporal coherence. In the context of the Michelson interferometer, the coherence length corresponds to the path difference $2d\cos\theta$ between the two arms of the interferometer for the interference fringes to practically disappear. When the path difference exceeds the coherence length, the light in both arms is no longer temporally coherent with respect to each other, leading to a reduction or complete disappearance of the interference fringes. This results in a drop in the visibility of the fringes as the displacement d (or equivalently the time delay τ_c) increases beyond a certain point, which corresponds to the coherence length.

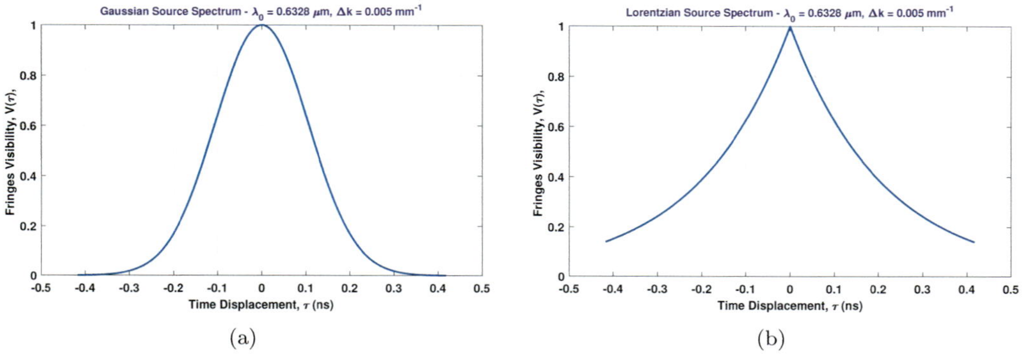

FIGURE 3.79: (a) The visibility $V(\tau)$ as a function of time delay τ for the Gaussian spectral intensity of example 3.9.1. The corresponding coherence time according to Eq. (3.345) is $\tau_c \simeq 0.15$ ns and the coherence length is $\ell_c \simeq 4.5$ cm. (b) The visibility $V(\tau)$ as a function of time delay τ for the Lorentzian spectral intensity of example 3.9.1. The corresponding coherence time according to Eq. (3.345) is $\tau_c \simeq 0.21$ ns and the coherence length is $\ell_c \simeq 6.4$ cm.

3.9.2 Time-Domain Representation of Temporal Coherence

In the previous sections, the interference fringes produced by a light source (a doublet or a narrowband spectrum source) in the Michelson interferometer were examined, with the analysis performed in the frequency domain. For the study of light coherence, the vectorial nature of the electromagnetic field can be ignored since most optical instruments (including the human eye) cannot distinguish the various field components, and they respond only to the average light intensity [151] (see also Sec. 3.5.1.1). Therefore, a multichromatic scalar electric field (real) of a light source can be represented in the time domain as [91, 153]:

$$\mathcal{E}_r(t) = \int_{-\infty}^{+\infty} E(\nu)e^{j2\pi\nu t}d\nu, \tag{3.347}$$

where $E(\nu)$ is the Fourier transform of $\mathcal{E}_r(t)$. The term $E(\nu)e^{j2\pi\nu t}d\nu$ can be considered as the plane wave contribution at frequencies between ν and $\nu+d\nu$, with the spatial coordinates suppressed for now. Since the electric field $\mathcal{E}_r(t)$ is real, it is straightforward to show that $E(-\nu) = E^*(\nu)$ (the "*" denotes complex conjugate). This implies that all the spectral information is contained in the positive frequencies only [91, 153]. Using these properties, the multichromatic electric field can be written as:

$$\begin{aligned}
\mathcal{E}_r(t) &= \int_0^{+\infty} E(\nu)e^{j2\pi\nu t}d\nu + \int_{-\infty}^{0} E(\nu)e^{j2\pi\nu t}d\nu \\
&= \int_0^{+\infty} E(\nu)e^{j2\pi\nu t}d\nu + \int_0^{+\infty} E(-\nu)e^{-j2\pi\nu t}d\nu \\
&= \int_0^{+\infty} \left[E(\nu)e^{j2\pi\nu t} + E^*(\nu)e^{-j2\pi\nu t}\right] d\nu \\
&= 2\int_0^{+\infty} |E(\nu)| \cos\left[2\pi\nu t + \phi(\nu)\right] d\nu,
\end{aligned} \tag{3.348}$$

where $E(\nu) = |E(\nu)| \exp[j\phi(\nu)]$. At this point, another real multichromatic light field can be defined as follows:

$$\mathcal{E}_i(t) = 2\int_0^{+\infty} |E(\nu)| \sin\left[2\pi\nu t + \phi(\nu)\right] d\nu. \tag{3.349}$$

Then, a complex multichromatic electric light field, $\tilde{E}(t)$, can be defined as:

$$\tilde{E}(t) = \mathcal{E}_r(t) + j\mathcal{E}_i(t) = 2\int_0^{+\infty} E(\nu)e^{+2\pi\nu t}d\nu, \tag{3.350}$$

where $E(\nu) = \mathcal{F}\{\mathcal{E}_r(t)\}$ with $\mathcal{F}$ denoting Fourier transform. This resembles the concept of the phasor. It can be shown [91, 150] that $\mathcal{E}_r(t)$ and $\mathcal{E}_i(t)$ are related to each other via the following relations:

$$\mathcal{E}_i(t) = +\frac{1}{\pi}\mathcal{PV}\int_{-\infty}^{+\infty} \frac{\mathcal{E}_r(t')}{t-t'}dt' = +\mathcal{H}\{\mathcal{E}_r\} = +\mathcal{E}_r(t) * \frac{1}{\pi t}, \tag{3.351}$$

$$\mathcal{E}_r(t) = -\frac{1}{\pi}\mathcal{PV}\int_{-\infty}^{+\infty} \frac{\mathcal{E}_i(t')}{t-t'}dt' = -\mathcal{H}\{\mathcal{E}_i\} = -\mathcal{E}_i(t) * \frac{1}{\pi t}, \tag{3.352}$$

where $\mathcal{PV}$ denotes the principal value of the corresponding integral and $\mathcal{H}$ the Hilbert transform of the corresponding function. It is noted that Eqs. (3.351) and (3.352) differ slightly from the ones presented in Ref. [91] due to the selection of the sign of the exponential term of the Fourier transform. Using Eqs. (3.350), (3.351), and (3.352), it can be shown (utilizing also Parceval's theorem) [91] that:

$$\int_{-\infty}^{+\infty} \mathcal{E}_r(t)^2 dt = \int_{-\infty}^{+\infty} \mathcal{E}_i(t)^2 dt = \frac{1}{2}\int_{-\infty}^{+\infty} |\tilde{E}(t)|^2 dt = 2\int_0^{+\infty} |E(\nu)|^2 d\nu. \tag{3.353}$$

In most practical applications, the light sources are considered to have appreciable spectral components around a center frequency ν_0 with a spectral width $\Delta\nu \ll \nu_0$. In this case, the complex multichromatic electric field can be written as:

$$\tilde{E}(t) \simeq A(t)e^{j[2\pi\nu_0 t + \Phi(t)]}. \tag{3.354}$$

Using Eqs. (3.354) and (3.350) the real field of a quasi-monochromatic light source can be represented as [43, 91, 153]:

$$\mathcal{E}_r(t) = A(t)\cos\left[2\pi\nu_0 t + \Phi(t)\right]. \tag{3.355}$$

The electric field described in this way is a more general representation of a light source, where both the amplitude $A(t)$ and phase $\Phi(t)$ vary randomly over time, but with rates much slower compared to the $\cos(2\pi\nu_0 t)$ term. This randomness arises due to the inherent nature of the light emission process, which involves the superposition of many individual photon emissions from a large number of atoms or molecules in the light source. Even highly coherent sources like lasers exhibit some degree of randomness in their phase and amplitude, leading to small fluctuations over time. Because the functions $A(t)$ (an envelope of the electric field of the light source) and the phase $\Phi(t)$ are random variables, the light source electric field can be described in terms of its statistical properties. On the other hand, the frequency representation of the electric field of the light source of Eq. (3.348) can be written as:

$$\mathcal{E}_r(t) = \int_0^{+\infty} \mathcal{A}(\nu)\cos\left[2\pi\nu t + \varphi(\nu)\right] d\nu \simeq \sum_i \mathcal{A}(\nu_i)\cos\left[2\pi\nu_i t + \varphi(\nu_i)\right], \tag{3.356}$$

where the factor "2" was incorporated into the amplitude term. For example, in the case of Fig. 3.75(a), $\mathcal{A}(\nu)$ was assumed to follow a Gaussian distribution (with standard deviation of 10% of its mean value) and $\varphi(\nu)$ to follow a uniform distribution (in the $-\pi$ to $+\pi$ interval). The number of sample frequencies, ν_i, (randomly selected using also a Gaussian distribution around a center mean frequency, ν_0, with a standard deviation of 10% of the mean frequency) was of the order of 100.

Before proceeding, it is useful to illustrate the representation of the electric field of the light source using either Eq. (3.355) or Eq. (3.356). To demonstrate this, let us assume the following parameters for the random variables: the amplitude $\mathcal{A}(\nu)$ is modeled as a Gaussian random variable with a mean value of unity (arbitrary unit for electric field amplitude) and a standard deviation of $\sigma_{\mathcal{A}} = 0.1$. The center frequency is selected to be $\nu_0 = 100\,Hz$, chosen arbitrarily for demonstration purposes. The number of sampled frequencies in the summation of Eq. (3.356) is on the order of 500 or more. These frequencies are distributed according to a Gaussian distribution around the center frequency ν_0, as done in Fig. 3.75(a).

Furthermore, the phase spectrum $\varphi(\nu)$ follows a uniform distribution in the interval $[-\pi, \pi]$. The sampled values, $A(t_i)$ of the random variable $A(t)$ and $\Phi(t_i)$ of the random variable $\Phi(t)$, follow Gaussian and uniform distributions, respectively. The sampled times t_i are generated according to a Poisson process with an event rate of $1/\lambda = 2$ events per period, chosen arbitrarily. The results of these simulations are shown in Fig. 3.80(a), which corresponds to the usage of Eq. (3.356), and in Fig. 3.80(b), corresponding to the usage of Eq. (3.355). Clearly, the two spectral representations exhibit significant differences. To narrow the spectral width of the sources, the standard deviation of the frequency distribution was reduced to $1\,Hz$ in the representation based on Eq. (3.356), and the Poisson process rate was reduced to $1/\lambda = 0.1$ events per period in the representation based on Eq. (3.355). The results of these adjustments are shown in Fig. 3.80(c) for the frequency sampling representation and in Fig. 3.80(d) for the time random variable representation. In this case, the spectra of the two representations show much better agreement. However, it is important to note that the frequency and time domain representations are not strictly equivalent. Despite the improved agreement in the narrowband case, the two approaches remain distinct in how they model the light source's electric field.

As discussed in Sec. 3.5, all light detectors (including the human eye) respond to light intensity (irradiance), which is based on the energy density of the electric field. In the case of a statistically varying light field, the intensity $I(t_0)$ at time t_0 in the time domain can be defined as:

$$I(t_0) = \langle \mathcal{E}_r^2(t) \rangle = \lim_{T\to\infty} \frac{1}{T} \int_{t_0-T/2}^{t_0+T/2} \mathcal{E}_r^2(t)dt. \tag{3.357}$$

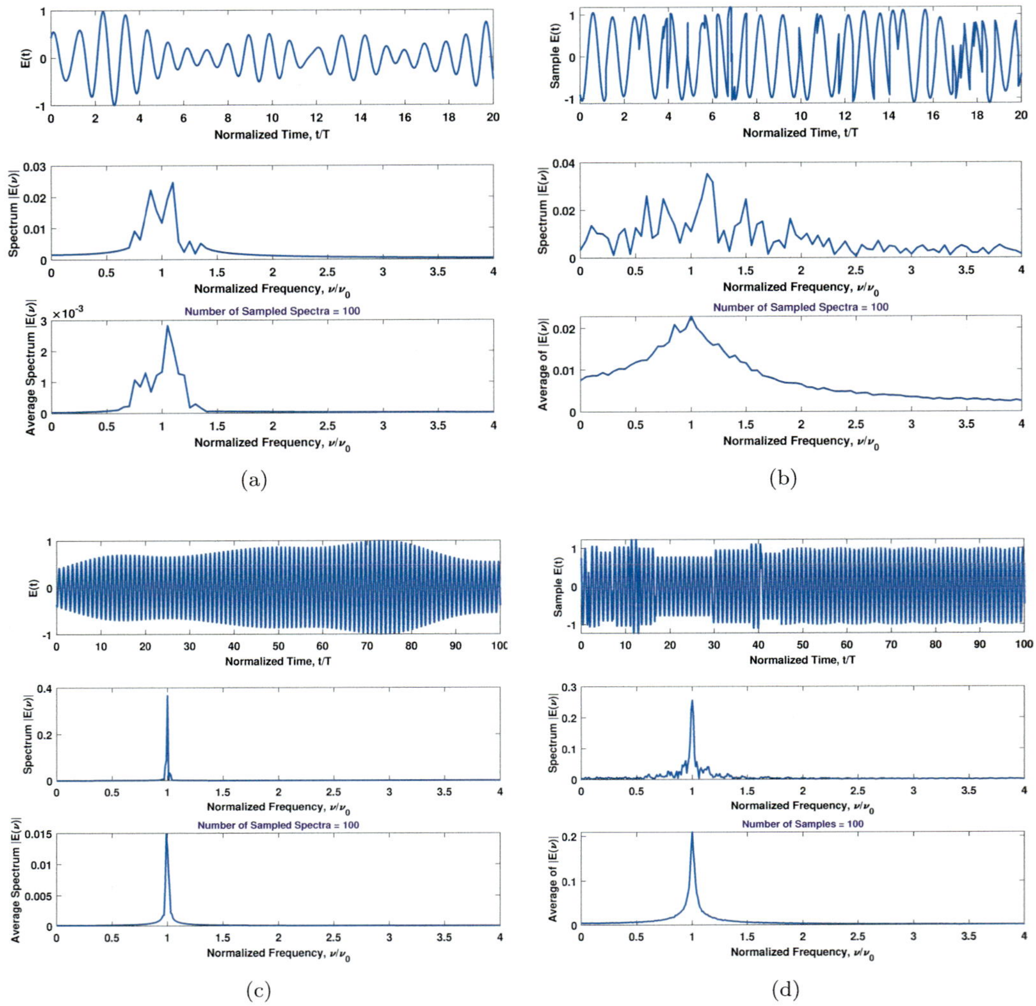

FIGURE 3.80: (a) The electric field $\mathcal{E}(t)$ calculated using Eq. (3.356) with mean frequency $\nu_0 = 100\,Hz$, and standard deviation $\sigma_\nu = 10\,Hz$, for a time of 20 periods $T = 1/\nu_0$. The amplitude has a mean value of unity and a standard deviation of 0.1. At the top is the sampled electric field in the time domain, in the middle is its spectrum, and at the bottom is the average spectrum of 100 such samples. (b) The electric field $\mathcal{E}(t)$ calculated using Eq. (3.355) with mean frequency $\nu_0 = 100\,Hz$, for a time of 20 periods $T = 1/\nu_0$. The amplitude has a mean value of unity and a standard deviation of 0.1. The sampled times t_i follow a Poisson process with $1/\lambda = 2$ events/period. At the top is the sampled electric field in the time domain, in the middle is its spectrum, and at the bottom is the average spectrum of 100 such samples. (c) Same as (a) but for $\sigma_\nu = 1\,Hz$, and for a time of 100 periods $T = 1/\nu_0$. (d) Same as (b) but with $1/\lambda = 0.1$ events/period, and for a time of 100 periods $T = 1/\nu_0$.

The brackets in the previous equation denote time averaging as defined. In optics [91], the randomness of the light field is usually considered stationary, meaning that the average intensity, $I(t_0) = I$, is independent of the randomly selected time t_0. Furthermore, the light field, viewed as a random process, is *ergodic*, which means that the ensemble average of the light intensity, $\bar{I}$, is equal to the time average [91, 150]. This relationship can be expressed mathematically as:

$$\bar{I} = \iint \mathcal{E}_r^2 p_A(A) p_\Phi(\Phi) dA d\Phi \simeq \frac{1}{N}\sum_{i=1}^{N} \mathcal{E}_r^2(t; A_i, \Phi_i) = I, \tag{3.358}$$

where p_A and p_Φ are the probability densities of the random variables A and Φ of Eq. (3.355). As shown in Eq. (3.358), the ensemble average of $\mathcal{E}_r^2(t)$ corresponds to the average of a finite number of samples of $\mathcal{E}_r^2(t)$, where A and Φ take random values according to their respective probability densities. For instance, in Fig. 3.80, the ensemble average of the spectrum $E(\nu)$ was computed using $N = 100$ samples.

The average intensity of the stationary optical field can be written as:

$$I = \langle \mathcal{E}_r^2(t)\rangle = \lim_{T\to\infty}\frac{1}{T}\int_{-T/2}^{T/2} \mathcal{E}_r^2(t)dt = \lim_{T\to\infty}\frac{1}{T}\left\{\int_{-T/2}^{T/2} \mathcal{E}_{rT}^2(t)dt\right\}, \tag{3.359}$$

where $\mathcal{E}_{rT}(t)$ is a truncated version of $\mathcal{E}_r(t)$ that obtains nonzero values only in the interval $[-T/2, T/2]$ [91]. Using Eq. (3.353) the following relations can be written:

$$\begin{aligned}\lim_{T\to\infty}\frac{1}{T}\left\{\int_{-T/2}^{T/2} \mathcal{E}_{rT}^2(t)dt\right\} &= \lim_{T\to\infty}\frac{1}{T}\left\{\int_{-\infty}^{+\infty} \mathcal{E}_{rT}^2(t)dt\right\} = \lim_{T\to\infty}\frac{1}{T}\left\{2\int_{0}^{+\infty} |E_T(\nu)|^2 d\nu\right\}\\ &= \lim_{T\to\infty}\left\{2\int_0^{+\infty}\frac{|E_T(\nu)|^2}{T}d\nu\right\} = \lim_{T\to\infty}\left\{2\int_0^{+\infty} S_T(\nu)d\nu\right\},\end{aligned}$$

where $E_T(\nu)$ is the Fourier transform of $\mathcal{E}_{rT}(t)$. The function $S_T(\nu)$ in many practical cases may not converge to a limit [91]. The factor of "2" in the denominator of $|E_T(\nu)|^2/2T$ in Ref. [91] arises from the time interval $[-T, T]$ used there, as opposed to the $[-T/2, T/2]$ interval used here [see Eq. (3.359)]. However, assuming the ergodicity and stationarity of the optical field, it is implied that the ensemble average of many spectral samples of the random field can approximate the spectrum of the random field intensity. For instance, in Fig. 3.80, the ensemble average of the spectrum $E(\nu)$ was computed using $N = 100$ samples. Based on the concept of the ensemble average, the *power spectrum* $S(\nu)$ of the random process describing the light field $\mathcal{E}_r(t)$ can be defined as:

$$S(\nu) = \lim_{T\to\infty} S_T(\nu) = \lim_{T\to\infty}\left\{\frac{|E_T(\nu)|^2}{T}\right\}, \tag{3.360}$$

Summarizing, $\mathcal{E}_r(t)$ represents the time-representation of the random light field, and $S(\nu)d\nu = |E(\nu)|^2 d\nu$ is the contribution to its intensity from the frequency range $(\nu, \nu + d\nu)$. Furthermore, the time averages of $\mathcal{E}_r^2(t)$ and $\mathcal{E}_i^2(t)$ can be expressed in the form:

$$\langle \mathcal{E}_r^2(t)\rangle = \langle \mathcal{E}_i^2(t)\rangle = \frac{1}{2}\langle |\tilde{E}(t)|^2\rangle = 2\int_0^\infty S(\nu)d\nu. \tag{3.361}$$

In the last equation the following relations between the Fourier transforms of $\mathcal{E}_r(t)$ and $\mathcal{E}_i(t)$ were used [based on Eqs. (3.351) and (3.352)]:

$$\mathcal{F}\{\mathcal{E}_i(t)\} = +\mathcal{F}\{\mathcal{E}_r(t)\}(-j\text{sgn}(\nu)) = -j\text{sgn}(\nu)E(\nu), \tag{3.362}$$

$$\mathcal{F}\{\mathcal{E}_r(t)\} = j\text{sgn}(\nu)\mathcal{F}\{\mathcal{E}_i(t)\}, \tag{3.363}$$

where sgn(ν) is the sign function [i.e., sgn$(\nu) = 1$ when $\nu > 0$ and sgn$(\nu) = -1$ when $\nu < 0$], and the term $-j\text{sgn}(\nu)$ is the Fourier transform of $1/\pi t$.

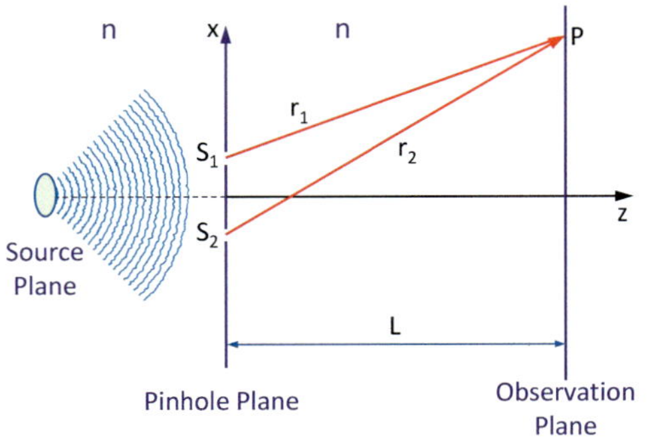

FIGURE 3.81: The Young's two-slit setup for investigating optical coherence.

3.9.3 Correlation Functions of Optical Fields

In this section, the experimental setup of Young's two-slit experiment is considered (see Sec. 3.5.3), utilizing an extended multichromatic light source, as illustrated in Fig. 3.81 [91,151]. Let $\mathcal{E}_r(S_i, t)$ represent the optical field (assumed to be scalar) at pinhole S_i $(i = 1, 2)$ at time t. The observable (measurable) intensity at pinhole S_i, denoted as $I(S_i)$, is determined using Eq. (3.353) as follows:

$$I(S_i) = \langle \mathcal{E}_r^2(S_i, t)\rangle = \frac{1}{2}\left\langle \left|\tilde{E}(S_i, t)\right|^2 \right\rangle, \qquad (i = 1, 2). \tag{3.364}$$

The complex light field at point P on the observation plane may be expressed as [91,150,151]:

$$\tilde{E}(P, t) = K_1 \tilde{E}(S_1, t - t_1) + K_2 \tilde{E}(S_2, t - t_2), \tag{3.365}$$

where $t_i = r_i/u$ and $u = c/n$ (light velocity in homogeneous medium of refractive index n, $i = 1, 2$). The t_i terms represent the time delays for the optical fields from the pinholes to reach point P. The factors K_1 and K_2 are inversely proportional to the distances r_1 and r_2, respectively. These factors also depend on the size of the pinholes, the geometry of the setup, and other parameters but remain independent of time t. According to Huygens' principle and scalar diffraction theory, it can be shown that K_1 and K_2 are purely imaginary [91]. The observable light intensity at point P, denoted as $I(P)$, is then expressed as:

$$\begin{aligned} I(P) &= \langle \mathcal{E}_r^2(P, t)\rangle = \frac{1}{2}\left\langle \left|\tilde{E}(P, t)\right|^2 \right\rangle \\ &= \frac{1}{2}|K_1|^2 \left\langle \left|\tilde{E}(S_1, t - t_1)\right|^2 \right\rangle + \frac{1}{2}|K_2|^2 \left\langle \left|\tilde{E}(S_2, t - t_2)\right|^2 \right\rangle + \\ &\quad \frac{1}{2}K_1 K_2^* \left\langle \tilde{E}(S_1, t - t_1)\tilde{E}^*(S_2, t - t_2)\right\rangle + \frac{1}{2}K_1^* K_2 \left\langle \tilde{E}^*(S_1, t - t_1)\tilde{E}(S_2, t - t_2)\right\rangle . \end{aligned}$$

Since the optical field is assumed to be stationary, the following relations can be expressed:

$$\begin{aligned} I(S_i) &= I_i = \frac{1}{2}\left\langle \left|\tilde{E}(S_i, t - t_i)\right|^2 \right\rangle, \qquad (i = 1, 2) \\ \Gamma(\vec{r}_1, \vec{r}_2, \tau) &= \Gamma_{12}(\tau) = \frac{1}{2}\left\langle \tilde{E}(S_1, t - t_1)\tilde{E}^*(S_2, t - t_2)\right\rangle = \frac{1}{2}\left\langle \tilde{E}(\vec{r}_1, t + \tau)\tilde{E}^*(\vec{r}_2, t)\right\rangle \\ \Gamma^*(\vec{r}_1, \vec{r}_2, \tau) &= \Gamma_{12}^*(\tau) = \frac{1}{2}\left\langle \tilde{E}^*(S_1, t - t_1)\tilde{E}(S_2, t - t_2)\right\rangle = \frac{1}{2}\left\langle \tilde{E}^*(\vec{r}_1, t + \tau)\tilde{E}(\vec{r}_2, t)\right\rangle . \end{aligned} \tag{3.366}$$

The function $\Gamma(\vec{r}_1, \vec{r}_2, \tau) = \Gamma_{12}(\tau)$ is defined as the *mutual coherence function* with $\tau = t_2 - t_1$, while $\vec{r}_i$ denotes the position of pinhole S_i $(i = 1, 2)$. The mutual coherence function is a fundamental concept in the theory of partial coherence. It represents the cross-correlation of the light field at point $\vec{r}_1(S_1)$ at time $t + \tau$ with the light field at point $\vec{r}_2(S_2)$ at time t. Due to the stationarity of the light field, this cross-correlation depends only on the time difference τ. When $\tau = 0$, i.e., when the correlation between the two points S_1 and S_2 is evaluated at the same time, the *mutual intensity* (or *equal time coherence function*), J_{12}, is defined as:

$$J_{12} = J(\vec{r}_1, \vec{r}_2) = \Gamma(\vec{r}_1, \vec{r}_2, \tau = 0) = \frac{1}{2}\left\langle \tilde{E}(\vec{r}_1, t)\tilde{E}^*(\vec{r}_2, t)\right\rangle. \tag{3.367}$$

When the two points coincide, i.e., when $\vec{r}_1 = \vec{r}_2$ then the function Γ becomes:

$$\Gamma_{ii}(\tau) = \frac{1}{2}\left\langle \tilde{E}(\vec{r}_i, t + \tau)\tilde{E}^*(\vec{r}_i, t)\right\rangle, \qquad (i = 1, 2), \tag{3.368}$$

and is defined as the *self-coherence function* of the light at point $\vec{r}_i$, representing its autocorrelation at that point. When $\tau = 0$, it is evident that $\Gamma_{ii}(0) = I_i$, where I_i s the intensity at point $\vec{r}_i$. Using the above definitions, the observable intensity at point P can be expressed as:

$$I(P) = \frac{1}{2}|K_1|^2 I_1 + \frac{1}{2}|K_2|^2 I_2 + |K_1 K_2| \text{Re}\left\{\Gamma(\vec{r}_1, \vec{r}_2, \tau)\right\}. \tag{3.369}$$

The mutual coherence function can be normalized [91, 150] as follows:

$$\gamma(\vec{r}_1, \vec{r}_2, \tau) = \gamma_{12}(\tau) = \frac{\Gamma(\vec{r}_1, \vec{r}_2, \tau)}{\sqrt{\Gamma_{11}(0)}\sqrt{\Gamma_{22}(0)}} = \frac{\Gamma_{12}(\tau)}{\sqrt{I_1 I_2}} = |\gamma_{12}(\tau)|e^{j\phi_{12}(\tau)}. \tag{3.370}$$

The function $\gamma_{12}(\tau)$ is defined as the *complex degree of coherence* of the optical field. It can be shown [91] that $0 \leq |\gamma_{12}(\tau)| \leq 1$. Furthermore, analogous to $\gamma_{12}(\tau)$, a normalized version of J_{12} can be defined as:

$$j_{12} = \gamma_{12}(\tau = 0) = \frac{J_{12}}{\sqrt{J_{11}}\sqrt{J_{22}}} = \frac{J_{12}}{\sqrt{I_1}\sqrt{I_2}}. \tag{3.371}$$

The function j_{12} is called the *equal time degree of coherence.* It can be shown that $0 \leq |j_{12}| \leq 1$ which is a special case of the inequality satisfied by by $\gamma_{12}(\tau)$. When $\vec{r}_1 = \vec{r}_2 = \vec{r}$, the function $\gamma(\vec{r}, \vec{r}, \tau) = \gamma(\tau)$ is related to the degree of self-correlation (in time) of the light source at a point in space. Summarizing,

$$\gamma(\tau) = \frac{\Gamma_{ii}(\tau)}{\Gamma_{ii}(0)} = \frac{\Gamma_{ii}(\tau)}{I_i}. \tag{3.372}$$

The first two terms of Eq. (3.369) represent the intensity at point P when one of the pinholes is open and the other is closed. For example, the first term of Eq. (3.369) represents the observable intensity at point P when the S_2 pinhole is closed. Conversely, the second term represents the observable intensity at point P when the S_1 pinhole is closed. Therefore, it is customary to define the observable intensities at point P with a single pinhole opened as follows:

$$I_1(P) = \frac{1}{2}|K_1|^2 I_1, \tag{3.373}$$

$$I_2(P) = \frac{1}{2}|K_2|^2 I_2. \tag{3.374}$$

It is important to note that the factor 1/2 in the above expressions arises from the definition of the complex field $\tilde{E}$ in Eq. (3.350). This factor does not appear in the standard definition of intensity using phasors, as presented in Sec. 3.5.3. Using the values of the intensities from Eqs. (3.373) and (3.374), the observable intensity at point P can also be expressed as:

$$I(P) = I_1(P) + I_2(P) + 2\sqrt{I_1(P)I_2(P)}\ \mathrm{Re}\{\gamma_{12}(\tau)\},$$
$$= I_1(P) + I_2(P) + 2\sqrt{I_1(P)I_2(P)}\ |\gamma_{12}(\tau)|\cos[\phi_{12}(\tau)]. \tag{3.375}$$

This final equation expresses the general interference law for stationary optical fields [91]. It shows that interference depends on the intensities of the two interfering beams and the real part of their complex degree of coherence. Equation (3.375) is also valid in cases where the two interfering beams are produced not by wavefront division (as in Young's setup) but by amplitude division (as in Michelson's interferometer). In the latter case, the term $\gamma_{12}(\tau)$ should be replaced by $\gamma_{11}(\tau)$. The factor $\gamma_{12}(\tau)$ [or $\gamma_{11}(\tau)$] of Eq. (3.375) controls the quality of the interference fringes. When $|\gamma_{12}(\tau)| = 1$, the intensity at point P is the same as that obtained for purely coherent light [see Eq. (3.210)]. In contrast, when $|\gamma_{12}(\tau)| = 0$, there is no interference term, and no fringes are observed, indicating that the light is completely incoherent. When $0 < |\gamma_{12}(\tau)| < 1$, the light field is considered *partially coherent*. In the case of Young's setup, the phase term of Eq. (3.375) can be expressed as $\phi_{12}(\tau) = k(r_2 - r_1) + \alpha_{12}(\tau)$, where $k = 2\pi/\bar{\lambda}$. The term $\bar{\lambda}$ denotes the average wavelength, and the term $\alpha_{12}(\tau)$ represents the statistical randomness of the light source [91]. Thus, Eq. (3.375) can be rewritten as:

$$I(P) = |\gamma_{12}(\tau)| \underbrace{\left[I_1(P) + I_2(P) + 2\sqrt{I_1(P)I_2(P)}\cos\left(k(r_2 - r_1) + \alpha_{12}(\tau)\right)\right]}_{= I_{coh}} +$$
$$[1 - |\gamma_{12}(\tau)|] \underbrace{[I_1(P) + I_2(P)]}_{= I_{incoh}}.$$

Therefore, the intensity at point P can be expressed as a mixture of coherent I_{coh} and incoherent I_{incoh} light [91]. The ratio of coherent to incoherent intensity is given by $I_{coh}/I_{incoh} = |\gamma_{12}(\tau)|/[1 - |\gamma_{12}(\tau)|]$. Furthermore, it is clear that the ratio of coherent to total intensity is $I_{coh}/I_{tot} = |\gamma_{12}(\tau)|$, where $I_{tot} = I_{coh} + I_{incoh}$. For practical experimental arrangements, it is often assumed that $I_1 \simeq I_2 = I_0$. In this case, the maximum and the minimum observable intensities at point P are:

$$I_{max} = 2I_0\left[1 + |\gamma_{12}(\tau_+)|\right],$$
$$I_{min} = 2I_0\left[1 - |\gamma_{12}(\tau_-)|\right],$$

where τ_+ and τ_- are the consecutive times that I_{max} and I_{min} occurs respectively. Typically, γ_{12} varies slowly compared to $I(P)$, so it can be assumed that $|\gamma_{12}(\tau_+)| \simeq |\gamma_{12}(\tau_-)| = |\gamma_{12}(\tau)|$. This is similar to the effect discussed in Eqs. (3.341) and (3.342). In this case, the fringe visibility $V(\tau)$ can be expressed as:

$$V(\tau) = \frac{I_{max}(\tau) - I_{min}(\tau)}{I_{max}(\tau) + I_{min}(\tau)} \simeq |\gamma_{12}(\tau)|. \tag{3.376}$$

Therefore, when $|\gamma_{12}(\tau)| \simeq 1$, the fringes are of high quality, and the light field at the pinholes S_1 and S_2 is coherent. On the other hand, when $|\gamma_{12}(\tau)| \simeq 0$ the fringes are of poor quality (or non-existent), and the light at the pinholes S_1 and S_2 is incoherent.

3.9.4 Correlation Functions in the Frequency Domain

In this section, the mutual coherence function is presented in the frequency domain. From the definition of $\Gamma_{12}(\tau)$ in Eq. (3.366), its Fourier transform can be expressed as follows:

$$\begin{aligned}\mathcal{F}\{\Gamma(\vec{r}_1,\vec{r}_2,\tau)\} &= \frac{1}{2}\mathcal{F}\left\{\left\langle \tilde{E}(\vec{r}_1,t+\tau)\tilde{E}^*(\vec{r}_2,t)\right\rangle\right\}\\ &= \frac{1}{2}\mathcal{F}\left\{\left\langle\left[\mathcal{E}_r(\vec{r}_1,t+\tau)+j\mathcal{E}_i(\vec{r}_1,t+\tau)\right]\left[\mathcal{E}_r(\vec{r}_2,t)-j\mathcal{E}_i(\vec{r}_2,t)\right]\right\rangle\right\}\\ &= \frac{1}{2}\mathcal{F}\left\{\left\langle\left[\mathcal{E}_r(\vec{r}_1,t+\tau)\mathcal{E}_r(\vec{r}_2,t)\right]\right\rangle\right\}+\frac{1}{2}\mathcal{F}\left\{\left\langle\left[\mathcal{E}_i(\vec{r}_1,t+\tau)\mathcal{E}_i(\vec{r}_2,t)\right]\right\rangle\right\}+\\ &\quad \frac{j}{2}\mathcal{F}\left\{\left\langle\left[\mathcal{E}_i(\vec{r}_1,t+\tau)\mathcal{E}_r(\vec{r}_2,t)\right]\right\rangle\right\}-\frac{j}{2}\mathcal{F}\left\{\left\langle\left[\mathcal{E}_r(\vec{r}_1,t+\tau)\mathcal{E}_i(\vec{r}_2,t)\right]\right\rangle\right\}.\end{aligned}$$

The Fourier transforms of the four-time averages can be calculated using the concept of the ensemble averages in the spectral domain, as presented in Sec. 3.9.2. The resulting Fourier transforms are as follows [where the Fourier transform of a correlation was utilized i.e., $\mathcal{F}\{\int f(t+\tau)g(t)dt\} = \mathcal{F}\{f\}\mathcal{F}^*\{g\}$]:

$$\begin{aligned}\frac{1}{2}\mathcal{F}\left\{\left\langle\left[\mathcal{E}_r(\vec{r}_1,t+\tau)\mathcal{E}_r(\vec{r}_2,t)\right]\right\rangle\right\} &= \lim_{T\to+\infty}\left(\frac{E(\vec{r}_1,\nu)E^*(\vec{r}_2,\nu)}{T}\right),\\ \frac{1}{2}\mathcal{F}\left\{\left\langle\left[\mathcal{E}_i(\vec{r}_1,t+\tau)\mathcal{E}_i(\vec{r}_2,t)\right]\right\rangle\right\} &= \mathrm{sgn}^2(\nu)\lim_{T\to+\infty}\left(\frac{E(\vec{r}_1,\nu)E^*(\vec{r}_2,\nu)}{T}\right),\\ \frac{1}{2}\mathcal{F}\left\{\left\langle\left[\mathcal{E}_i(\vec{r}_1,t+\tau)\mathcal{E}_r(\vec{r}_2,t)\right]\right\rangle\right\} &= -j\mathrm{sgn}(\nu)\lim_{T\to+\infty}\left(\frac{E(\vec{r}_1,\nu)E^*(\vec{r}_2,\nu)}{T}\right),\\ \frac{1}{2}\mathcal{F}\left\{\left\langle\left[\mathcal{E}_r(\vec{r}_1,t+\tau)\mathcal{E}_i(\vec{r}_2,t)\right]\right\rangle\right\} &= +j\mathrm{sgn}(\nu)\lim_{T\to+\infty}\left(\frac{E(\vec{r}_1,\nu)E^*(\vec{r}_2,\nu)}{T}\right).\end{aligned}$$

Therefore, the Fourier transform of $\Gamma(\vec{r}_1,\vec{r}_2,\tau)$ can be expressed in the form:

$$\mathcal{F}\{\Gamma(\vec{r}_1,\vec{r}_2,\tau)\} = 2G_{12}(\nu)+2\mathrm{sgn}(\nu)G_{12}(\nu), \tag{3.377}$$

where

$$G(\vec{r}_1,\vec{r}_2,\nu) = G_{12}(\nu) = \lim_{T\to+\infty}\left(\frac{E(\vec{r}_1,\nu)E^*(\vec{r}_2,\nu)}{T}\right), \tag{3.378}$$

and the ensemble average is implied in the above definition. The function $G_{12}(\nu)$ is known as the *mutual spectral density*. Obviously, Eq. (3.379) reveals that $\mathcal{F}\{\Gamma(\vec{r}_1,\vec{r}_2,\tau)\}$ is nonzero for positive frequencies, meaning it is an analytic signal. A special case arises when the spectra correspond to the same point in space, i.e., when $\vec{r}_1 = \vec{r}_2$. In this case, the Fourier transform of the self-coherence function can be defined as:

$$G_{ii}(\nu) = \mathcal{F}\{\Gamma_{ii}(\tau)\} = 2S(\vec{r}_i,\nu)+2\mathrm{sgn}(\nu)S(\vec{r}_i,\nu), \qquad (i=1,2), \tag{3.379}$$

where $S(\vec{r}_i,\nu)$ is the power spectrum [defined in Eq. (3.360)] of the light field at position $\vec{r}_i$. Obviously, $G_{ii}(\nu)$ is one sided Fourier transform, since it is nonzero only for positive frequencies. Since both $\mathcal{F}\{\Gamma(\vec{r}_1,\vec{r}_2,\tau)\}$ and $\mathcal{F}\{\Gamma_{ii}(\tau)\}$ are non-zero only for positive frequencies, the relationships between τ-domain and ν-domain are:

$$G_{12}(\nu) = \frac{1}{4}\mathcal{F}\{\Gamma(\vec{r}_1, \vec{r}_2, \tau)\}, \qquad (\nu \geq 0) \tag{3.380}$$

$$S(\vec{r}_i, \nu) = S_i(\nu) = \frac{1}{4}\mathcal{F}\{\Gamma_{ii}(\tau)\}, \qquad (\nu \geq 0). \tag{3.381}$$

In addition, the *spectral degree of coherence* can be defined as:

$$g(\vec{r}_1, \vec{r}_2, \nu) = g_{12}(\nu) = \frac{G_{12}(\nu)}{\sqrt{S(\vec{r}_1, \nu)}\sqrt{S(\vec{r}_2, \nu))}}. \tag{3.382}$$

It can be shown [91] that g_{12} satisfies the inequality $0 \leq |g_{12}| \leq 1$, analogous to the inequality observed by γ_{12}. The spectral degree of coherence is a measure of the correlations between the spectral components of a light source at positions $\vec{r}_1$ and $\vec{r}_2$. Since $|\gamma(\vec{r}_i, \vec{r}_i, 0)| = 1$ and $|\gamma(\vec{r}_1, \vec{r}_2, 0)| \leq 1$, the magnitude of the $\gamma(\vec{r}_1, \vec{r}_2, 0)$ can be regarded as a measure of the degree of spatial coherence of the light source. In real light sources, a larger distance $|\vec{r}_1 - \vec{r}_2|$ results in a smaller degree of spatial coherence. This is analogous to the behavior of $\gamma(\vec{r}_i, \vec{r}_i, \tau) = \gamma(\tau)$ when longer τ values result in a smaller degree of temporal coherence. There is a relationship between the complex degree of coherence (which characterizes correlations in the space-time domain) and the spectral degree of coherence (which characterizes correlations in the space-frequency domain) [154]. Specifically, the following relationships hold:

$$\gamma_{12}(\tau) = \int_{-\infty}^{+\infty} \sqrt{s_1(\nu)}\sqrt{s_2(\nu)}\; g_{12}(\nu) e^{j2\pi\nu\tau} d\nu, \tag{3.383}$$

$$g_{12}(\nu) = \frac{1}{\sqrt{s_1(\nu)}\sqrt{s_2(\nu)}} \int_{-\infty}^{+\infty} \gamma_{12}(\tau) e^{-j2\pi\nu\tau} d\tau, \tag{3.384}$$

$$\text{where} \qquad s_i(\nu) = \frac{S_i(\nu)}{\int S_i(\nu) d\nu}, \qquad (i = 1, 2).$$

3.9.4.1 Propagation of Mutual Coherence Function

The mutual coherence function $\Gamma_{12}(\tau) = \Gamma(\vec{r}_1, \vec{r}_2, \tau)$ can be shown to propagate according to the wave equation [91,150,151]. To present a simple proof of this property, the operator ∇_i^2 (where $i = 1, 2$ referring to either $\vec{r}_1$ or $\vec{r}_2$, respectively) is applied to $\Gamma(\vec{r}_1, \vec{r}_2, \tau)$. From here, the following expressions can be easily written:

$$\nabla_1^2 \Gamma(\vec{r}_1, \vec{r}_2, \tau) = \frac{1}{2}\left\langle \left[\nabla_1^2 \tilde{E}(\vec{r}_1, t_1 = t + \tau) \tilde{E}^*(\vec{r}_2, t_2 = t)\right]\right\rangle, \tag{3.385}$$

$$\nabla_2^2 \Gamma(\vec{r}_1, \vec{r}_2, \tau) = \frac{1}{2}\left\langle \left[\tilde{E}(\vec{r}_1, t_1 = t + \tau) \nabla_2^2 \tilde{E}^*(\vec{r}_2, t_2 = t)\right]\right\rangle. \tag{3.386}$$

Of course, the real field $\mathcal{E}_r(\vec{r}_i, t)$ satisfies the following wave equation:

$$\nabla_i^2 \mathcal{E}_r - \frac{1}{u^2}\frac{\partial^2 \mathcal{E}_r}{\partial t_i^2} = 0 \Longrightarrow \nabla_i^2 E + k^2 E = 0, \qquad i = 1, 2.$$

where u is the phase velocity of the electromagnetic field in the homogeneous medium where it propagates and $k = 2\pi\nu/u$ the corresponding wavenumber. The above wave equation is also written in the frequency domain (Helmholtz equation), where E is the Fourier transform of $\mathcal{E}_r$ [see Eq. (3.347)]. Since $\mathcal{E}_r$ and $\mathcal{E}_i$ are related via a Hilbert transform [Eqs. (3.351),

(3.352)], their Fourier transforms are also related by $\mathcal{F}\{\mathcal{E}_i\} = -j\text{sgn}(\nu)E(\nu)$ [see Eq. (3.362)]. Therefore, the Fourier transform of $\mathcal{E}_i$ satisfies the same Helmholtz equation as $\mathcal{E}_r$. This implies that both $\mathcal{E}_r$ and $\mathcal{E}_i$ satisfy the same wave equation. Since $\tilde{E}$ is related via Eq. (3.350) to $\mathcal{E}_r$ and $\mathcal{E}_i$, it also satisfies the same wave equation. Therefore,

$$\nabla_i^2 \tilde{E} - \frac{1}{u^2}\frac{\partial^2 \tilde{E}}{\partial t_i^2} = 0 \qquad i = 1, 2.$$

Replacing the Laplacian operators in Eqs. (3.385) and (3.386), the following expressions can be derived:

$$\nabla_1^2 \Gamma(\vec{r}_1, \vec{r}_2, \tau) = \frac{1}{2u^2}\left\langle \left[\frac{\partial^2 \tilde{E}(\vec{r}_1, t_1)}{\partial t_1^2}\tilde{E}^*(\vec{r}_2, t_2)\right]\right\rangle = \frac{1}{2u^2}\frac{\partial^2}{\partial t_1^2}\left\langle \left[\tilde{E}(\vec{r}_1, t_1)\tilde{E}^*(\vec{r}_2, t_2)\right]\right\rangle,$$

$$\nabla_2^2 \Gamma(\vec{r}_1, \vec{r}_2, \tau) = \frac{1}{2u^2}\left\langle \left[\tilde{E}(\vec{r}_1, t_1)\frac{\partial^2 \tilde{E}^*(\vec{r}_2, t_2)}{\partial t_2^2}\right]\right\rangle = \frac{1}{2u^2}\frac{\partial^2}{\partial t_2^2}\left\langle \left[\tilde{E}(\vec{r}_1, t_1)\tilde{E}^*(\vec{r}_2, t_2)\right]\right\rangle.$$

If a function $f(t_1, t_2) = f(\tau = t_1 - t_2)$, i.e., is stationary, as is the function $\Gamma(\vec{r}_1, \vec{r}_2, \tau)$, then it is straightforward to show that $\partial f/\partial t_1 = \partial f/\partial \tau$ and $\partial f/\partial t_2 = -\partial f/\partial \tau$ and for the second partial derivatives $\partial^2 f/\partial^2 t_1 = \partial^2 f/\partial^2 \tau = \partial^2 f/\partial^2 t_2$. Therefore, the mutual coherence function satisfies the following wave equation:

$$\nabla_i^2 \Gamma(\vec{r}_1, \vec{r}_2, \tau) - \frac{1}{u^2}\frac{\partial^2 \Gamma(\vec{r}_1, \vec{r}_2, \tau)}{\partial \tau^2}, \qquad i = 1, 2, \tag{3.387}$$

which is similar to the wave equation satisfied by the real electric field $\mathcal{E}_r$. It can be shown [150] that the mutual spectral density $G_{12}(\nu) = G(\vec{r}_1, \vec{r}_2, \nu)$ satisfies the following Helmholtz equation:

$$\nabla_i^2 G(\vec{r}_1, \vec{r}_2, \nu) + k^2 G(\vec{r}_1, \vec{r}_2, \nu) = 0, \qquad i = 1, 2. \tag{3.388}$$

A special case of Eq. (3.388) is the Helmholtz equation satisfied by the mutual intensity J_{12},

$$\nabla_i^2 J_{12} + \bar{k}^2 J_{12} = 0, \qquad i = 1, 2, \tag{3.389}$$

where $\bar{k} = 2\pi\bar{\nu}/u = 2\pi/\bar{\lambda}$ with $\bar{\nu}$ the center frequency of the light source (and $\bar{\lambda}$ the center wavelength).

3.9.4.2 Van Cittert-Zernike Theorem

Since the theory of scalar diffraction is covered in a later chapter (Chapter 5), the Van Cittert-Zernike theorem is introduced in this section without a proof. The interested reader can consult more specialized texts, such as Refs. [91, 150, 151]. The Van Cittert-Zernike theorem pertains to extended incoherent sources, such as the one shown schematically in Fig. 3.82. Consider a point Σ on the plane of the extended light source with coordinates $(x', y', z = 0)$. At a distance z away from the source, there is an observation plane. Let S_1 and S_2 be two points on the observation plane with coordinates (x_1, y_1, z) and (x_2, y_2, z), respectively. The Van Cittert-Zernike theorem relates the mutual intensity J_{12} between points S_1 and S_2 with the intensity I of the completely incoherent light source as follows:

$$J_(x_1, y_1; x_2, y_2) = J_{12} \simeq A\int_S I(x', y')\frac{e^{-j\bar{k}(R_1 - R_2)}}{R_1 R_2}dx'dy', \tag{3.390}$$

where S is the surface of the extended source, A is a constant term, and $\bar{k}$ is the average wavenumber as it was defined in Eq. (3.389). The proof [150,151] is based on the propagation

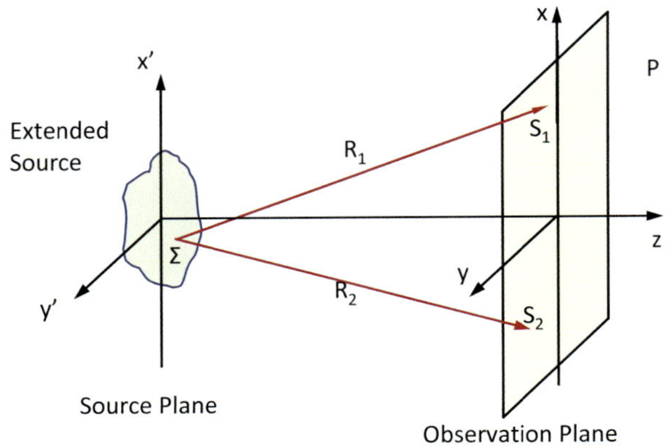

FIGURE 3.82: An extended incoherent light source emits radiation, and the coherence of the light is examined at the observation plane.

properties of J_{12} (satisfies Helmholtz equation), the Green's function of Helmholtz equation in homogeneous media, and the Rayleigh-Sommerfeld's formulations of scalar diffraction for $z \gg d$ (refer to the *Scalar Diffraction Theory* chapter, Chapter 5). It can be shown that the resulting complex degree of coherence j_{12}, is given by:

$$j_{12} = \frac{e^{-j\psi} \iint I(x',y') \exp[+j(\bar{k}/z)(x'\Delta x + y'\Delta y)] dx' dy'}{\iint I(x',y') dx' dy'} = \frac{e^{-j\psi} \mathcal{F}\{I(x',y')\}\Big|_{f_x = \Delta x/\bar{\lambda} z,\, f_y = \Delta y/\bar{\lambda} z}}{\iint I(x',y') dx' dy'}, \tag{3.391}$$

where $\psi = \dfrac{\bar{k}}{2z}\left[(x_2^2 + y_2^2) - (x_1^2 + y_1^2)\right]$, $\Delta x = x_2 - x_1$, and $\Delta y = y_2 - y_1$.

3.9.4.3 Example Application of Van Cittert-Zernike Theorem

The application of the Van Cittert-Zernike theorem, as described, involves a scenario where the spatial coherence of light from an extended incoherent source is examined at an observation plane. The example setup typically involves a geometrical arrangement where the light source is an extended object, complex degree of coherence j_{12} between two points on the observation plane is related to the intensity distribution of the source via the Van Cittert-Zernike theorem [see Eq. 3.391)]. This is an example of the application of the Van Cittert-Zernike theorem that was adapted from Refs. [91, 150] and was originally presented in Ref. [155]. The example geometry is illustrated in Fig. 3.83. An extended completely incoherent light source has a circular shape of radius a_s. Light from this source arrives at the pinhole plane, which is at a distance L_s from the source plane. Assume that the pinholes S_1 and S_2, have radius s are symmetrically located with respect to the z-axis, and are separated by distance d. The light emanating from the two pinholes interferes at an observation plane located at distance $L_p \gg d$. The light intensity pattern at point P, denoted as $I(x)$ is sought. Two distinct cases are discussed. In the first case the diffraction effects from the two pinholes are neglected, while in the second case the diffraction effects (in the far field

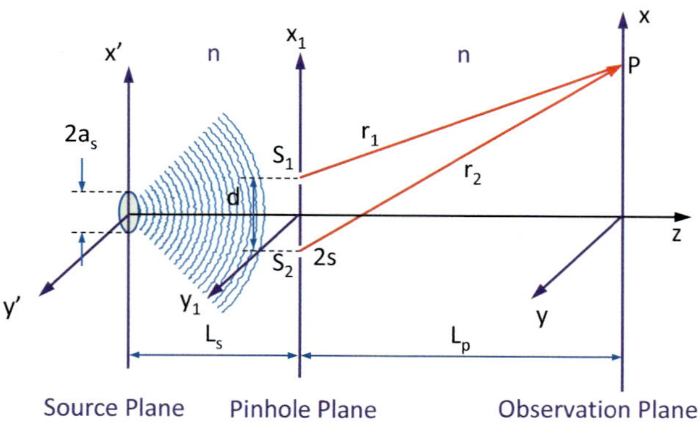

FIGURE 3.83: An incoherent extended source of circular shape in front of two pinholes. This is an example adapted from Ref. [91].

regime) are included. The intensity distribution of the light source is assumed uniform and can be expressed by:

$$I(x', y') = I_0 \text{circ}\left(\frac{\sqrt{x'^2 + y'^2}}{a_s}\right). \tag{3.392}$$

The function "circ" is a simple indicator function commonly used in signal processing and diffraction theory. It is defined as: $\text{circ}(w) = 1$ for $w < 1$ and $\text{circ}(w) = 0$ for $w > 1$. In addition, $\text{circ}(1) = 1/2$. In order to apply Eq. (3.375), it is necessary to know $\gamma_{12}(0) = j_{12}$ which can be determined from the Van Cittert-Zernike theorem, Eq. (3.391). The Fourier transform of the $\text{circ}(\rho/a_s)$ function is $\mathcal{F}\{\text{circ}(\rho/a_s)\} = a_s J_1[2\pi a_s(f_x^2+f_y^2)^{1/2}]/(f_x^2+f_y^2)^{1/2}$, where $\rho = (x'^2 + y'^2)^{1/2}$, and f_x, f_y are the spatial frequencies in the x' and y' directions, respectively (2D Fourier transform), while J_1 is the Bessel function of the first order. Since the pinholes S_1 and S_2 are positioned symmetrically with respect to the z-axis $x_1 = x_2 = d/2$ and $y_1 = y_2 = 0$, resulting in $\Delta x = d$ and $\Delta y = 0$. Then, the factor ψ of Eq. (3.391) becomes zero and the complex degree of coherence j_{12} becomes:

$$j_{12} = \frac{2J_1\left(\frac{2\pi a_s}{\bar{\lambda} L_s} d\right)}{\frac{2\pi a_s}{\bar{\lambda} L_s} d}. \tag{3.393}$$

Applying Eq. (3.375) with j_{12} from Eq. (3.393) and neglecting the diffraction from each pinhole the time-averaged intensity $I(x)$ can be expressed by:

$$I(x) = 2I_0 \left\{ 1 + \left| \frac{2J_1\left(\frac{2\pi a_s}{\bar{\lambda} L_s} d\right)}{\frac{2\pi a_s}{\bar{\lambda} L_s} d} \right| \cos\left(\bar{k}(r_2 - r_1) + \phi\right) \right\}, \tag{3.394}$$

where $\phi = 0$ if $j_{12} \geq 0$ and $\phi = \pi$ if $j_{12} < 0$. Since the distance $L_p \gg d$ it can be easily shown that the factor $\bar{k}(r_2 - r_1)$ is approximately given by $(2\pi/\bar{\lambda})(xd/L_p)$. Due to symmetry, it was assumed that $I_1(P) = I_2(P) = I_0$. For the numerical implementation, the following parameters are used: $\bar{\lambda} = 0.579\,\mu\text{m}$, $L_s = 1.52\,\text{m}$, $a_s = 0.045\,\text{mm}$, $L_p = 1.5\,\text{m}$. It is worth mentioning that the first zero of $J_1(w)$ (excluding $w = 0$) occurs at $w_1 = 3.83170597$. Therefore, when the argument of J_1 reaches w_1, no interference fringes will be observed. For the given numerical values, this happens when $d = 11.9267\,\text{mm}$. The results are shown in Fig. 3.84.

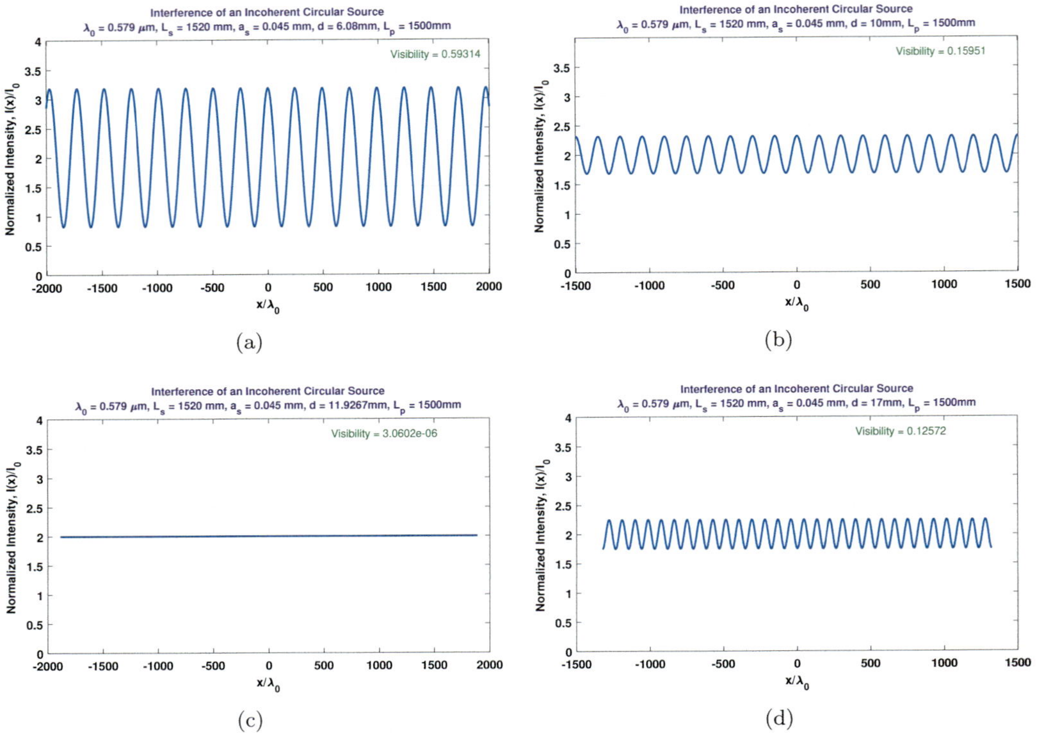

FIGURE 3.84: For all cases the fringe visibility is shown on the plots. The diffraction from the pinholes is neglected [150]. (a) The time-averaged intensity $I(x)$ for $d = 6.08$ mm. (b) The time-averaged intensity $I(x)$ for $d = 10$ mm. (c) The time-averaged intensity $I(x)$ for $d = 11.9267$ mm (no interference fringes expected). (d) The time-averaged intensity $I(x)$ for $d = 17$ mm.

When the diffraction from each pinhole (in the far-field) is taken into account, the only change from the previous analysis is that the intensity at point P becomes: $I_1(P) = I_2(P) = 4J_1^2(2\pi sx/\bar{\lambda}L_p)/[2\pi sx/\bar{\lambda}L_p]^2$. This result is derived in the chapter on *Scalar Diffraction*, Chapter 5, specifically for circular aperture far-field Fraunhofer diffraction. In this case, $I(x)$ can be expressed as:

$$I(x) = 2\left[\frac{2J_1\left(\dfrac{2\pi s}{\bar{\lambda}L_p}x\right)}{\dfrac{2\pi s}{\bar{\lambda}L_p}x}\right]^2\left\{1+\left|\frac{2J_1\left(\dfrac{2\pi a_s}{\bar{\lambda}L_s}d\right)}{\dfrac{2\pi a_s}{\bar{\lambda}L_s}d}\right|\cos\left(\bar{k}(r_2-r_1)+\phi\right)\right\}, \tag{3.395}$$

Using the same numerical parameters, the time-averaged intensity $I(x)$ can be calculated for various values of the pinhole separation d. The results are shown in Fig. 3.85.

3.9.4.4 Mathematical Definitions of Coherence Time and Coherence Area

From the analysis up to this point, it is clear that the coherence time τ_c (and the corresponding coherence length ℓ_c) are statistical variables, and a single value holds only statistical significance. The power-equivalent coherence length can be defined as [19, 150]:

$$\tau_c = \int_{-\infty}^{+\infty}|\gamma(\tau)|^2 d\tau = \frac{1}{|\Gamma_{ii}(0)|^2}\int_{-\infty}^{+\infty}|\Gamma_{ii}(\tau)|^2\, d\tau. \tag{3.396}$$

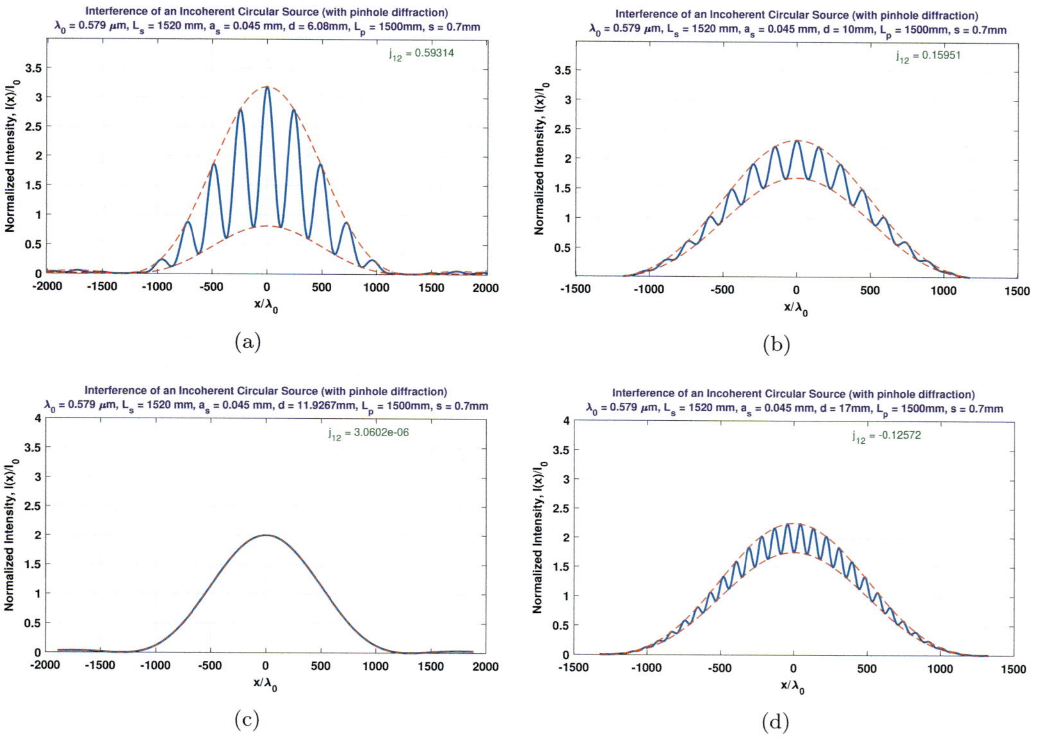

FIGURE 3.85: For all cases the complex degree of coherence j_{12} is shown on the plots. The diffraction from the pinholes is now taken care (Fraunhofer diffraction) [91]. (a) The time-averaged intensity $I(x)$ for $d = 6.08\,\text{mm}$. (b) The time-averaged intensity $I(x)$ for $d = 10\,\text{mm}$. (c) The time-averaged intensity $I(x)$ for $d = 11.9267\,\text{mm}$ (no interference fringes expected). (d) The time-averaged intensity $I(x)$ for $d = 17\,\text{mm}$.

Equation (3.396) can be expressed also in the frequency domain using Parseval's theorem and Eq. (3.381). Furthermore, $\Gamma_{ii}(0) = \int_0^{+\infty} 4S_i(\nu)d\nu$ [from the Fourier transform relation of Eq. (3.381)]. Thus, τ_c can be written as:

$$\tau_c = \frac{1}{|\Gamma_{ii}(0)|^2} \int_0^{+\infty} |S_i(\nu)|^2\, d\nu = \frac{\displaystyle\int_0^{+\infty} |S_i(\nu)|^2\, d\nu}{\left|\displaystyle\int_0^{+\infty} S_i(\nu) d\nu\right|^2}. \tag{3.397}$$

As an example, consider the normalized Gaussian power spectrum of a light source [150]:

$$S_i(\nu) = \frac{2\sqrt{\ln 2}}{\sqrt{\pi}\Delta\nu} \exp\left[-\left(2\sqrt{\ln 2}\,\frac{(\nu - \bar{\nu})}{\Delta\nu}\right)^2\right],$$

where $\Delta\nu$ is the half power (Full Width at Half Maximum, FWHM) bandwidth. Using the Fourier transform properties and that $\mathcal{F}\{\exp(-ax^2)\} = (\sqrt{\pi/a})\exp(-\pi^2 f^2/a)$ the following expressions can be easily obtained:

$$\int_0^{+\infty} |S_i(\nu)|^2\,d\nu = \frac{2\sqrt{\ln 2}}{\pi}\frac{1}{\Delta\nu},$$

$$\left|\int_0^{+\infty} S_i(\nu)d\nu\right|^2 = 1,$$

$$\tau_c = \frac{2\sqrt{\ln 2}}{\pi}\frac{1}{\Delta\nu}. \tag{3.398}$$

Of course, the same expression for the coherence time can be derived using the definition in Eq. (3.396) if the $\gamma(\tau)$ is determined. Using the properties of the Fourier transform, it is straightforward to show that:

$$\gamma(\tau) = \exp\left[-\frac{\pi^2\tau^2\Delta\nu^2}{4\ln 2}\right]e^{-j2\pi\bar{\nu}\tau},$$

which, when inserted into Eq. (3.396), yields Eq. (3.398).

The coherence area, like coherence time, also has a statistical meaning. A similar definition of the coherence area A_c can be given using $|j_{12}|$, which depends only on the coordinate differences Δx and Δy of two points $\vec{r}_1$ and $\vec{r}_2$ in the xy-plane of Fig. 3.82. Therefore, the coherence area A_c is defined as:

$$A_c = \int_{-\infty}^{+\infty}\int_{-\infty}^{+\infty} |j_{12}|^2 d(\Delta x)d(\Delta y). \tag{3.399}$$

For an extended completely incoherent source, the expression of A_c can be simplified [150]. Specifically, using Eq. (3.391) the following expressions can be written:

$$A_c = \iint \left| \frac{\iint I(x',y')\exp[+j(\bar{k}/z)(x'\Delta x + y'\Delta y)]dx'dy'}{\iint I(x',y')dx'dy'} \right|^2 d(\Delta x)d(\Delta y)$$

$$= \frac{\iint dx'dy' \iint I(x',y')I(x'',y'')dx''dy'' \iint F(x,x',y,y',\Delta x,\Delta y)d(\Delta x)d(\Delta y)}{\left|\iint I(x',y')dx'dy'\right|^2},$$

where $F(x,x',y,y',\Delta x,\Delta y) = \exp\left\{+j\left[\frac{\bar{k}}{z}\Big((x'-x'')\Delta x + (y'-y'')\Delta y\Big)\right]\right\}.$

In the previous expression, all double integrals are assumed to range from $-\infty$ to $+\infty$. Using the property of the delta function, $\int_{-\infty}^{+\infty}\exp[j2\pi(x_1-x_2)u]du = \delta(x_1-x_2)$, the following relationship can be derived (where $\bar{k} = 2\pi/\bar{\lambda}$, and $\bar{\lambda}$ is the center wavelength of the source):

$$\iint \exp[+j(\bar{k}/z)((x'-x'')\Delta x + (y'-y'')\Delta y)]d(\Delta x)d(\Delta y) = (\bar{\lambda}z)^2\delta(x'-x'')\delta(y'-y'').$$

Using the last expression, the coherence area A_c can be written as:

$$A_c = (\bar{\lambda}z)^2\frac{\iint I^2(x',y')dx'dy'}{\left|\iint I(x',y')dx'dy'\right|^2}. \tag{3.400}$$

For a uniform distribution incoherent extended source of area A_s, Eq. (3.400) becomes:

$$A_c = \frac{(\bar{\lambda}z)^2}{A_s} = \frac{\bar{\lambda}^2}{\Omega_s}, \tag{3.401}$$

where Ω_s is the solid angle subtended by the source from the center of the observation region.

As an example, the coherence area of the Sun on earth can be calculated. The Sun's blackbody temperature is about 5800° K which results in a peak freespace wavelength $\lambda_0 \simeq 0.5\,\mu$m. This can be considered the center wavelength of the Sun's radiation arriving on Earth, i.e., $\bar{\lambda} = 0.5\,\mu$m. The distance of the Sun from Earth (on the average) is about 150×10^6 km. This means that $z = 1.5 \times 10^{11}$ m. Finally, the Sun's radius is $r = 696000$ km. Therefore, Sun's area is $A_s = \pi r^2$. Using these values in Eq. (3.401) the resulting coherence area of the Sun as an extended incoherent source is $A_c \simeq 3.69 \times 10^{-3}\,\text{mm}^2$, which is quite small.

3.10 Incoherent Reflection and Transmission

Multilayer thin-film optical structures have a wide range of applications, including anti-reflection coatings, high-reflection coatings, and interference optical filters, among others. The electromagnetic analysis of such structures, as discussed in Sec. 3.7, employs the transfer matrix method. This technique allows for efficient computation of the optical response and is well-suited for integration into the design process. In this analysis, all layers are assumed to have flat boundaries, with thicknesses on the order of the wavelength of the incident light and smaller than the light's coherence length. As a result, the treatment is coherent, accounting for all interference effects between the multiple parallel and flat interfaces.

However, in real-world simulations, it is sometimes necessary to model certain layers as coherent (with interference effects considered) and others as incoherent (ignoring interference effects). The incoherent treatment is typically applied to thick layers, where the thickness is much larger than the wavelength or the coherence length of the light source. In this case, only the light intensity is considered, without including phase information. In practice, many thin-film multilayer structures combine coherent and incoherent layers. Various methodologies have been proposed in the literature to handle these mixed structures effectively [156–168]. This section addresses some straightforward examples involving single or multiple incoherent layers. Additionally, the treatment of mixed coherent/incoherent structures is presented.

3.10.1 Single Incoherent Layer

Assume a single incoherent layer with refractive index n_2 and thickness d, positioned between two homogeneous isotropic media of refractive indices n_1 and n_3, respectively, as shown in Fig. 3.86. In this configuration, interference effects within the layer are ignored due to its incoherent nature. Consequently, the phase information of the light is not considered, and only the intensities of the reflected and transmitted beams are analyzed. Light with intensity I_i is incident from the left region onto the single incoherent layer at an angle $\theta_i = \theta_1$, as is shown in Fig. 3.86. A portion of the incident light intensity is sequentially refracted and reflected at the two interfaces, as illustrated. Let R_{12}, R_{21}, R_{23} denote the intensity reflectances at the interfaces between regions 1-2, 2-1, and 2-3, respectively. From

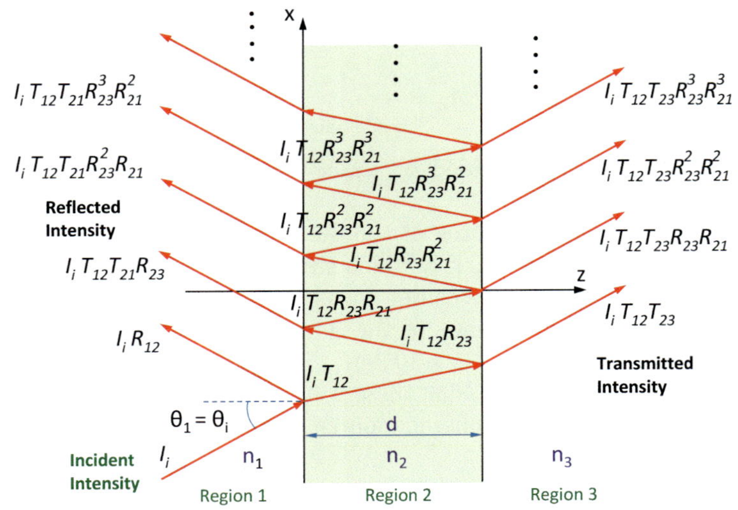

FIGURE 3.86: A single incoherent layer of thickness d and refractive index n_2 between two homogeneous isotropic media of refractive indices n_1 and n_3, respectively. In this figure the layer is assumed lossless.

Fresnel equations and Stokes relations, $R_{12} = R_{21}$. Additionally, the intensity transmittances at the interfaces are denoted as T_{12}, T_{21}, T_{23} between regions 1-2, 2-1, and 2-3, respectively. At each interface, power conservation ensures that $R_{ij} + T_{ij} = 1$. In Fig. 3.86 the various intensities of partially reflected and transmitted rays are depicted. Based on these definitions, the following equations can be straightforwardly written:

$$\begin{aligned}
I_r &= I_i R_{12} + I_i T_{12} T_{21} R_{23} + I_i T_{12} T_{21} R_{23}^2 R_{12} + I_i T_{12} T_{21} R_{23}^3 R_{12}^2 + \cdots \Longrightarrow \\
\frac{I_r}{I_i} &= R_{12} + T_{12} T_{21} R_{23} \Big[1 + R_{12} R_{23} + (R_{12} R_{23})^2 + (R_{12} R_{23})^3 + \cdots \Big] \\
&= R_{12} + \frac{T_{12} T_{21} R_{23}}{1 - R_{12} R_{23}} = \frac{R_{12}(1 - R_{12} R_{23}) + T_{12} T_{21} R_{23}}{1 - R_{12} R_{23}} = \frac{R_{12} + R_{23} - 2R_{12} R_{23}}{1 - R_{12} R_{23}},
\end{aligned} \tag{3.402}$$

$$\begin{aligned}
I_t &= I_i T_{12} T_{23} + I_i T_{12} T_{23} R_{23} R_{21} + I_i T_{12} T_{23} R_{23}^2 R_{21}^2 + I_i T_{12} T_{23} R_{23}^3 R_{21}^3 \cdots \Longrightarrow \\
\frac{I_t}{I_i} &= T_{12} T_{23} \Big[1 + R_{21} R_{23} + (R_{21} R_{23})^2 + (R_{21} R_{23})^3 + \cdots \Big] \\
&= \frac{T_{12} T_{23}}{1 - R_{12} R_{23}} = \frac{(1 - R_{12})(1 - R_{23})}{1 - R_{12} R_{23}}.
\end{aligned} \tag{3.403}$$

Naturally, the normalized intensity reflectivities and transmissivities depend on the polarization of the incident light. Alternatively, the coherent analysis presented in Sec. 3.6.1 can be applied as follows: The phase angle δ, as defined in Eq. (3.220) can be expressed as:

$$\delta(m) = 2k_0 n_2 d \cos\theta_2 + \phi_r(m), \tag{3.404}$$

where $\phi_r(m)$ is a uniform random variable taking values in the interval $[0, 2\pi]$, and m is an index representing the random samples. Using Eqs. (3.226) and (3.224), the coherent calculations of $(I_r(m)/I_i)_c$ and $(I_t(m)/I_i)_c$ (where the subscript "c" denotes coherent intensities) can be performed for each sample corresponding to a specific random phase $\phi_r(m)$. By averaging the coherent results over a large number of samples $\phi_r(m)$, the incoherent values of I_r/I_i and I_t/I_i, as determined by Eqs. (3.402) and (3.403), should be obtained. I.e., the following relations can be written:

$$\frac{I_r}{I_i} = \lim_{N\to\infty} \left\{ \frac{1}{N} \sum_{m=1}^{N} \left(\frac{I_r(m)}{I_i} \right)_c \right\}, \tag{3.405}$$

$$\frac{I_t}{I_i} = \lim_{N\to\infty} \left\{ \frac{1}{N} \sum_{m=1}^{N} \left(\frac{I_t(m)}{I_i} \right)_c \right\}. \tag{3.406}$$

Alternatively, the incoherent intensities can be also determined from coherent calculations averaged over the phase angle ϕ_r of Eq. (3.404). Specifically, the incoherent values of I_r/I_i and I_t/I_i are given by [160]:

$$\frac{I_r}{I_i} = \frac{1}{2\pi} \left\{ \int_{\phi_r=0}^{2\pi} \left(\frac{I_r(\phi_r)}{I_i} \right)_c d\phi_r \right\}, \tag{3.407}$$

$$\frac{I_t}{I_i} = \frac{1}{2\pi} \left\{ \int_{\phi_r=0}^{2\pi} \left(\frac{I_t(\phi_r)}{I_i} \right)_c d\phi_r \right\}. \tag{3.408}$$

Example 3.10.1:

A single incoherent layer of silicon (Si) with refractive index $n_2 = 3.6$ is considered in air ($n_1 = n_3 = 1$). The layer has a thickness of $d = 1\,\mu$m, while the wavelength of the incident light is $\lambda_0 = 1\,\mu$m. Note that in the lossless incoherent analysis, the thickness of the layer does not influence the results. The goal is to: (a) Calculate the coherent and incoherent intensity reflectivities (I_r/I_i) and transmissivities (I_t/I_i) as functions of the angle of incidence. (b) Verify the usage of Eqs. (3.405) and (3.406) by computing the incoherent reflectivities and transmissivities using coherent calculations averaged over $N = 100$ and $N = 2000$ samples. (c) Compare the results with those obtained using Eqs. (3.407) and (3.408).

Solution: The coherent normalized intensities are first determined using Eqs. (3.226) and (3.224) as functions of the angle of incidence $\theta_i = \theta_1$, with the random phase set to $\phi_r(m) = 0$. The resulting curves are displayed in Fig. 3.87(a). Subsequently, the incoherent normalized intensities (I_r/I_i and I_t/I_i) are calculated using Eqs. (3.402) and (3.403). The results are shown in Figs. 3.87(b) and 3.87(c), which also include the outcomes obtained by applying Eqs. (3.405) and (3.406) for $N = 100$ or $N = 2000$ samples. It is evident that as N increases, the results converge to their incoherent counterparts. Moreover, the incoherent results differ significantly from the coherent ones, highlighting the sensitivity of coherent results to layer thickness, while the incoherent results remain independent of thickness (for the lossless layer considered here). The application of Eqs. (3.407) and (3.408) yield results practically identical to those obtained in Figs. 3.87(b) and 3.87(c) using Eqs. (3.402) and (3.403).

Equations (3.402) and (3.403) can be generalized in the case of a lossy medium in region 2 of Fig. 3.86. The lossy medium can be represented with a complex refractive index $n_2 - j\kappa_2$ in this case (with $\kappa_2 > 0$ for loss or $\kappa_2 < 0$ for gain). Following the same procedure as the one presented for the lossless case, the resulting incoherent reflected and transmitted intensities

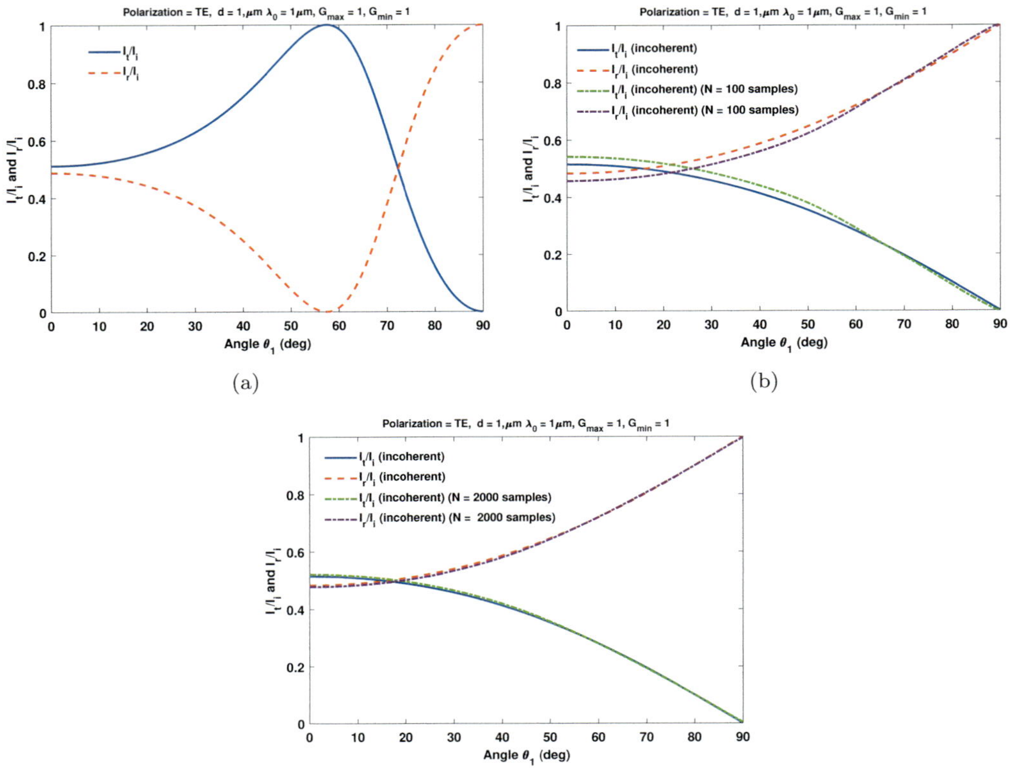

(a) (b) (c)

FIGURE 3.87: (a) The reflected and the transmitted normalized intensities of a single coherent slab of $n_2 = 3.6$ in air $n_1 = n_3 = 1.0$ of thickness $d = 1\,\mu$m and $\lambda_0 = 1\,\mu$m for *TE* polarization, as functions of the angle of incidence θ_1. (b) The reflected and the transmitted normalized intensities of the same single slab treated incoherently and with the average of $N = 100$ random phase samples of the coherent approach. (c) The reflected and the transmitted normalized intensities of the same single slab treated incoherently and with the average of $N = 2000$ random phase samples. It is evident that the average of the coherent approach with the random phase converges to the incoherent results as the number of samples increases.

are given by:

$$\frac{I_r}{I_i} = \frac{R_{12} + R_{23}e^{-2\alpha_I d} - 2R_{12}R_{23}e^{-2\alpha_I d}}{1 - R_{12}R_{23}e^{-2\alpha_I d}}, \tag{3.409}$$

$$\frac{I_t}{I_i} = \frac{(1 - R_{12})(1 - R_{23})e^{-\alpha_I d}}{1 - R_{12}R_{23}e^{-2\alpha_I d}}, \tag{3.410}$$

where $\alpha_i = 2\alpha = 2k_0 \mathrm{Im}\{(n_2 - j\kappa_2)\cos\theta_2\}d$ represents the attenuation constant (or gain constant in the case of an active incoherent layer) for the intensity.

For demonstration, the refractive index of region 2 in Example 3.10.1 is modified to $n_2 = 3.6 - j0.1$ (significant loss) and $n_2 = 3.6 - j0.01$ (smaller loss). The corresponding coherent and incoherent results are illustrated in Figs. 3.88(a) and 3.88(b), respectively. The results clearly show that as the loss in the medium increases, the difference between the coherent and incoherent results diminishes. This effect is due to the reduction in the amplitude of the multiple reflected and transmitted rays, which decreases interference effects. Furthermore, the validity of Eqs. (3.405), (3.406), (3.407), and (3.408) extends to the case of a lossy

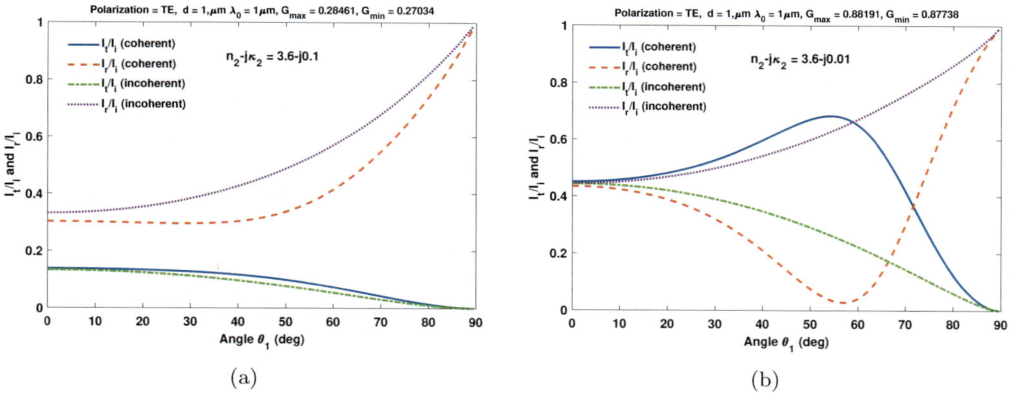

FIGURE 3.88: (a) The reflected and transmitted normalized intensities for a single coherent lossy slab with refractive index $n_2 = 3.6 - j0.1$, thickness $d = 1\,\mu$m, surrounded by air ($n_1 = n_3 = 1.0$), and illuminated with light of freespace wavelength $\lambda_0 = 1\,\mu$m, are calculated for *TE* polarization as functions of the angle of incidence θ_1. The corresponding incoherent normalized intensities are also included for comparison. (b) The same analysis is performed for $n_2 = 3.6 - j0.01$, where the material exhibits lower loss.

slab as well, ensuring consistency across the coherent and incoherent frameworks even with significant absorption.

3.10.2 Multiple Cascaded Incoherent Layers

In this section, the incoherent analysis from Sec. 3.10.1 is extended to address the case of multiple cascaded incoherent layers. The geometry of the stratified structure is illustrated in Fig. 3.89. The structure varies only along a single direction, assumed to be the z-axis. It consists of N cascaded incoherent layers, each with a refractive index n_i (potentially complex, $n_i - j\kappa_i$, for $i = 1, 2, \cdots, N$) and a thickness d_i. At each interface, the electric fields propagating in the $+z$ and $-z$ are shown. For instance, at the interface between the i-th and the $i+1$-th layers, the electric fields E^+_{iR} and $E^+_{i+1,L}$ propagate in the $+z$ direction, while the fields E^-_{iR} and $E^-_{i+1,L}$ propagate in the $-z$ direction. Here the subscripts R and L denote the fields on the "right" and "left" sides of the i-th or $i+1$-th layers, respectively. Since the analysis is incoherent, the corresponding field intensities are used instead of the fields themselves. These intensities are defined as $I^+_{iR} = |E^+_{iR}|^2$, $I^+_{i+1,L} = |E^+_{i+1,L}|^2$, $I^-_{iR} = |E^-_{iR}|^2$, and $I^-_{i+1,L} = |E^-_{i+1,L}|^2$, for $i = 0, 1, 2, \cdots N, N+1$. Notably, the intensities for the i-th layer are evaluated at its right boundary, while those for the $i+1$-th layer are evaluated at its left boundary. The $0R$-terms represent the intensities at the input region (on the right side of the input region), while the $(N+1, L)$-terms correspond to the intensities at the output region (on the left side of the output region). Next, we define the Fresnel reflection and transmission coefficients for both *TE* and *TM* polarizations. Specifically, $r_{i,i+1}$, and $t_{i,i+1}$ denote the reflection and transmission coefficients for light incident from layer i towards layer $i+1$, while $r_{i+1,i}$, and $t_{i+1,i}$ denote the reflection and transmission coefficients for light incident from layer $i+1$ toward layer i.

The following equations describe the relationships between the light intensities at the interface between the i-th and i_1-th layer:

$$I^-_{iR} = |r_{i,i+1}|^2 I^+_{iR} + |t_{i+1,i}|^2 I^-_{i+1,L},$$

$$I^+_{i+1,L} = |t_{i,i+1}|^2 I^+_{iR} + |r_{i+1,i}|^2 I^-_{i+1,L},$$

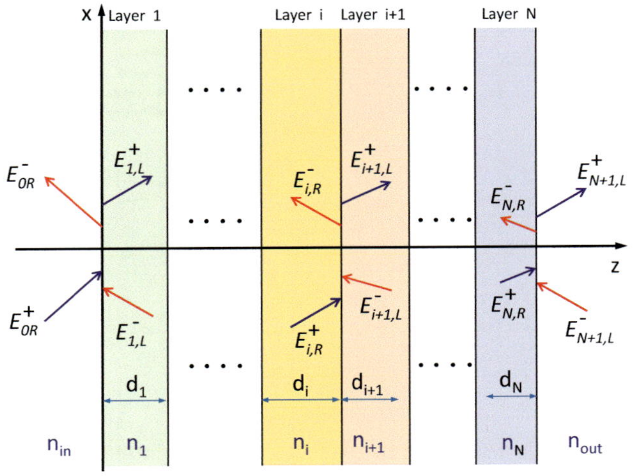

FIGURE 3.89: An optical multilayer structure comprised of N cascaded incoherent layers of thickness d_i and refractive index n_i $(i = 1, 2, \cdots, N)$, positioned between two homogeneous media with refractive indices n_{in} and n_{out}, respectively.

assuming that E^+_{iR} and $E^-_{i+1,L}$ are incident at the interface, producing $E^+_{i+1,L}$ and E^-_{iR}. The last equations can be rearranged to relate the intensities of the i-th layer with the intensities of the $i+1$-th layer. These relationships can be compactly represented in the following matrix form:

$$\begin{bmatrix} I^+_{iR} \\ I^-_{iR} \end{bmatrix} = \underbrace{\frac{1}{|t_{i,i+1}|^2}\begin{bmatrix} 1 & -|r_{i+1,i}|^2 \\ |r_{i,i+1}|^2 & |t_{i,i+1}|^2|t_{i+1,i}|^2 - |r_{i,i+1}|^2|r_{i+1,i}|^2 \end{bmatrix}}_{=\tilde{S}_{i,i+1}} \begin{bmatrix} I^+_{i+1,L} \\ I^-_{i+1,L} \end{bmatrix}$$

$$\begin{bmatrix} I^+_{iR} \\ I^-_{iR} \end{bmatrix} = \tilde{S}_{i,i+1} \begin{bmatrix} I^+_{i+1,L} \\ I^-_{i+1,L} \end{bmatrix}. \tag{3.411}$$

For the $(i+1)$-th layer the intensities at the right and left boundaries are related by:

$$\begin{bmatrix} I^+_{i+1,L} \\ I^-_{i+1,L} \end{bmatrix} = \underbrace{\begin{bmatrix} \exp[+2\alpha_{i+1}d_{i+1}] & 0 \\ 0 & \exp[-2\alpha_{i+1}d_{i+1}] \end{bmatrix}}_{=\tilde{D}_{i+1}} \begin{bmatrix} I^+_{i+1,R} \\ I^-_{i+1,R} \end{bmatrix} = \tilde{D}_{i+1} \begin{bmatrix} I^+_{i+1,R} \\ I^-_{i+1,R} \end{bmatrix}. \tag{3.412}$$

The parameter $\alpha_{i+1} = \text{Im}\{(n_{i+1} - j\kappa_{i+1})\cos\theta_{i+1}\}$ represents the intensity attenuation coefficient (or gain, in the case of an active medium). The term $\cos\theta_{i+1}$ is determined from the application of Snell's law at the interfaces, with θ_0 corresponding to the angle of incidence of the light from the input region. For lossless layers, the matrix $\tilde{D}_{i+1}$ becomes a 2×2 unity matrix. Using these definitions, the following relations can be derived [161,162]:

$$\begin{bmatrix} I^+_{0R} \\ I^-_{0R} \end{bmatrix} = \underbrace{\tilde{S}_{0,1}\tilde{D}_1\tilde{S}_{1,2}\tilde{D}_2\cdots\tilde{S}_{N-1,N}\tilde{D}_N\tilde{S}_{N,N+1}}_{=\tilde{S}} \begin{bmatrix} I^+_{N+1,L} \\ I^-_{N+1,L} \end{bmatrix} = \begin{bmatrix} S_{11} & S_{12} \\ S_{21} & S_{22} \end{bmatrix} \begin{bmatrix} I^+_{N+1,L} \\ I^-_{N+1,L} \end{bmatrix}, \tag{3.413}$$

where S_{ij} $(i, j = 1, 2)$ are the four elements of the overall matrix $\tilde{S}$. Using the last equation, it is straightforward to define the following intensity reflection and transmission coefficients:

$$R = \left. \frac{I_{0R}^-}{I_{0R}^+} \right|_{I_{N+1,L}^-=0} = \frac{S_{21}}{S_{11}}, \tag{3.414}$$

$$T = \left. \frac{I_{N+1,L}^+}{I_{0R}^+} \right|_{I_{N+1,L}^-=0} = \frac{1}{S_{11}}, \tag{3.415}$$

$$R' = \left. \frac{I_{N+1,L}^+}{I_{N+1,L}^-} \right|_{I_{0R}^+=0} = -\frac{S_{12}}{S_{11}}, \tag{3.416}$$

$$T' = \left. \frac{I_{0R}^-}{I_{N+1,L}^-} \right|_{I_{0R}^+=0} = \frac{S_{11}S_{22} - S_{12}S_{21}}{S_{11}} = \frac{det[\tilde{S}]}{S_{11}}, \tag{3.417}$$

where "$det[\tilde{S}]$" denotes the determinant of $\tilde{S}$. For incident light from the left at an angle of incidence θ_0 the percentages of the reflected and transmitted powers (in the incoherent case) are given by:

$$\frac{I_r}{I_i} = R = \frac{S_{21}}{S_{11}}, \tag{3.418}$$

$$\frac{I_t}{I_i} = T\,\frac{\mathrm{Re}\{n_{out}\cos\theta_{N+1}\}}{n_{in}\cos\theta_0} = \frac{\mathrm{Re}\{n_{out}\cos\theta_{N+1}\}}{n_{in}\cos\theta_0}\frac{1}{S_{11}}, \tag{3.419}$$

where θ_{N+1} is the angle of the transmitted field in the output region.

The incoherent normalized intensities can alternatively be determined from their coherent counterparts, $(I_r/I_i)_c$ and $(I_t/I_i)_c$, as calculated using Eqs. (3.272) and (3.273) (for *TE* polarization) or Eqs. (3.277) and (3.278) (for *TM* polarization), based on the transfer matrix method presented in Sec. 3.7. These expressions are given by:

$$\frac{I_r}{I_i} = \lim_{M\to\infty}\left\{\frac{1}{M}\sum_{m=1}^{M}\left(\frac{I_r(\beta_i'(m))}{I_i}\right)_c\right\}, \tag{3.420}$$

$$\frac{I_t}{I_i} = \lim_{M\to\infty}\left\{\frac{1}{M}\sum_{m=1}^{M}\left(\frac{I_t(\beta_i'(m))}{I_i}\right)_c\right\}. \tag{3.421}$$

where $\beta_i'(m) = \mathrm{Re}\{k_0(n_i - j\kappa_i)\cos\theta_i\} + \phi_{ri}(m)$, and $\phi_{ri}(m)$ represents the m-th of the M samples of the random phase ϕ_{ir} associated with the i-th layer, uniformly distributed within the interval $[0, 2\pi]$ (for $m = 1, 2, \cdots, M$). Alternatively, the incoherent normalized intensities can be determined by integrating over the phase angles ϕ_i, which are incorporated into $\beta_i'(\phi_i) = \mathrm{Re}\{k_0(n_i - j\kappa_i)\cos\theta_i\} + \phi_i$, as follows:

$$\frac{I_r}{I_i} = \frac{1}{(2\pi)^N}\left\{\int_{\phi_{r1}=0}^{2\pi}\cdots\int_{\phi_{rN}=0}^{2\pi}\left(\frac{I_r(\phi_{r1},\phi_{r2},\cdots,\phi_{rN})}{I_i}\right)_c d\phi_{r1}d\phi_{r2}\cdots d\phi_{rN}\right\}, \tag{3.422}$$

$$\frac{I_t}{I_i} = \frac{1}{(2\pi)^N}\left\{\int_{\phi_{r1}=0}^{2\pi}\cdots\int_{\phi_{rN}=0}^{2\pi}\left(\frac{I_t(\phi_{r1},\phi_{r2},\cdots,\phi_{rN})}{I_i}\right)_c d\phi_{r1}d\phi_{r2}\cdots d\phi_{rN}\right\}. \tag{3.423}$$

Unfortunately, independently treating N phase angles becomes computationally intensive, particularly when the number of layers in the optical structure is large. To address this challenge, the following approximation is proposed:

$$\frac{I_r}{I_i} \simeq \frac{1}{2\pi}\left\{\int_{\phi_r=0}^{2\pi}\left(\frac{I_r(\phi_{r1}=\phi_{r2}=\cdots=\phi_{rN}=\phi)}{I_i}\right)_c d\phi_r\right\}, \tag{3.424}$$

$$\frac{I_t}{I_i} \simeq \frac{1}{2\pi}\left\{\int_{\phi_r=0}^{2\pi}\left(\frac{I_t(\phi_{r1}=\phi_{r2}=\cdots=\phi_{rN}=\phi_r)}{I_i}\right)_c d\phi_r\right\}. \tag{3.425}$$

Example 3.10.2:

To demonstrate the application of the preceding analysis, a two-layer ($N = 2$) incoherent multilayer structure is analyzed in air ($n_0 = 1$ and $n_3 = 1$). The example parameters are: refractive index $n_1 = 1.5$ (SiO_2) and thickness $d_1 = 1.0\,\mu$m, for layer 1, and refractive index $n_2 = 3.6$ (Si) and thickness $d_2 = 4.0\mu$m, for layer 2. Assume that all layers are lossless, rendering their thicknesses irrelevant in the incoherent analysis. The freespace wavelength of the incident light is $\lambda_0 = 1\,\mu$m. The tasks are as follows: (a) Calculate the coherent and incoherent normalized intensities as functions of the angle of incidence. (b) Verify the incoherent results using Eqs. (3.420) and (3.421) for $M = 100$ and $M = 1000$ random phase samples. (c) Test the validity of the results with Eqs. (3.424) and (3.425). Assume both *TE* and *TM* polarizations.

Solution: Initially, the coherent normalized intensities, are calculated as functions of the angle of incidence θ_0 using Eqs. (3.272) and (3.273) for *TE* polarization, or Eqs. (3.277) and (3.278) for *TM* polarization, following the transfer matrix procedure. The resulting coherent intensity curves are presented in Fig. 3.90(a). Subsequently, the incoherent normalized intensities, I_r/I_i and I_t/I_i, are determined using Eqs. (3.418) and (3.419). These incoherent results for *TE* polarization are displayed in Figs. 3.90(b) and 3.90(c), calculated for $M = 100$ or $M = 1000$ random phase samples. Additionally, incoherent normalized intensities derived using Eqs. (3.420) and (3.421) are included for comparison. The results demonstrate that as the number of random samples M increases, the computed values converge toward the incoherent intensity limits. However, larger angles of incidence require higher M values to achieve this convergence.

The results obtained via the ϕ_r integration method, using Eqs. (3.424) and (3.425), are also plotted. These approximate values show some deviation from the incoherent intensities, unlike the single-layer case where a closer match is observed. For all cases, the incoherent results differ significantly from their coherent counterparts. Finally, for *TM* polarization, similar trends are observed, with results shown in Fig. 3.90(d). Notably, the incoherent intensities for *TM* polarization deviate significantly from those of *TE* polarization due to the distinct Fresnel coefficients associated with the two polarizations.

It should be noted that the use of Eqs. (3.418) and (3.419) is independent of the layer thicknesses, assuming the layers are lossless. However, when averaging the coherent intensities to compute the incoherent ones, special attention is required when the layer thicknesses become large relative to the wavelength. In such cases, a larger number of random samples M may be necessary to achieve accurate results. Additionally, the approximate Eqs. (3.424) and (3.425) may not provide accurate results in these situations.

3.10.3 Multiple Mixed Coherent-Incoherent Layers

In real applications, there are optical structures where some layers behave incoherently (particularly when their thickness exceeds the coherence length of the incident light, or

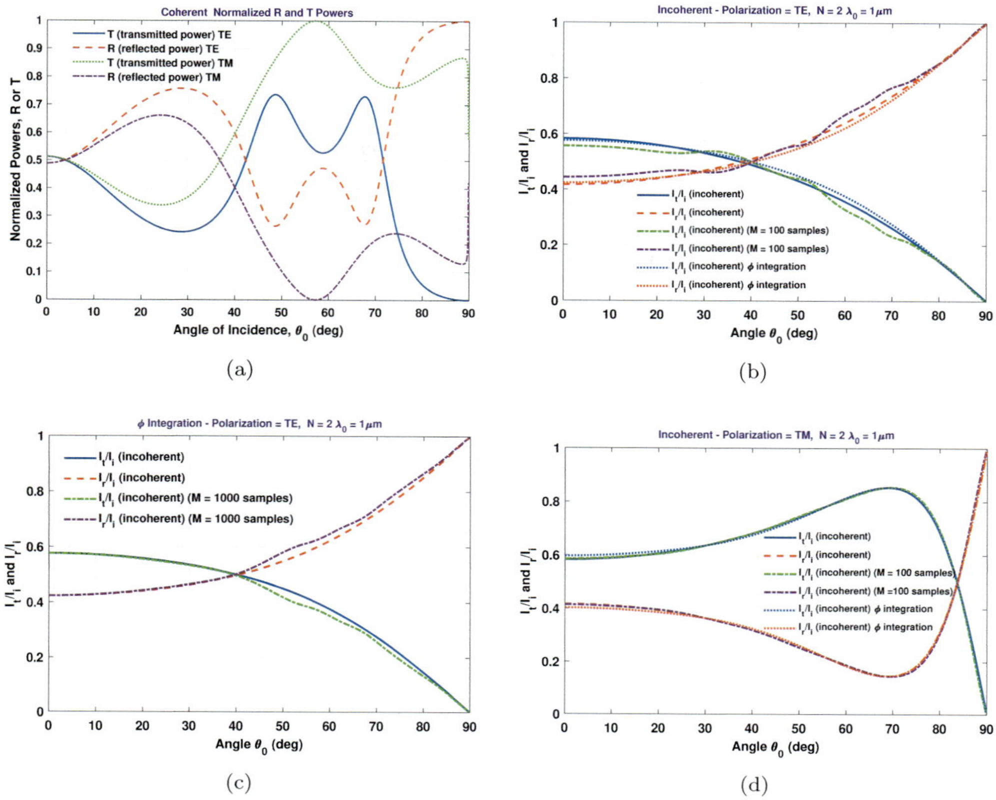

FIGURE 3.90: (a) The coherent reflected and the transmitted normalized intensities for a 2-layer ($N = 2$) multilayer structure with refractive indices of $n_1 = 1.5$, $n_2 = 3.6$, and surrounding media refractive indices $n_0 = n_3 = 1.0$ (air). The layer thicknesses are $d_1 = 1\,\mu\text{m}$, $d_2 = 4\,\mu\text{m}$, and the freespace wavelength is $\lambda_0 = 1\,\mu\text{m}$. Results are presented for both *TE* and *TM* polarizations, as functions of the angle of incidence θ_0. (b) The incoherent reflected and transmitted normalized intensities of the same structure for *TE* polarization. Calculations are performed using direct methods, random sampling ($M = 100$), and averaging based on the coherent counterparts. (c) The same analysis as in (b), but with $M = 1000$ number of random samples. Results using the ϕ_r integration are not shown, as they are identical to those in (b). (d) The same analysis as in (b) for *TM* polarization, with the number of random samples set to $M = 100$.

when there are thickness variations that surpass this coherence length), while others (usually thinner layers) behave coherently. Therefore, instead of treating all layers as either coherent or incoherent, a mixed coherent-incoherent modeling approach is necessary to accurately predict the performance of these structures. For example, an N-layer mixed coherent-incoherent structure is shown in Fig. 3.91. In this case, a group of coherent layers is embedded within the other layers that behave incoherently. Specifically, layers i through $i + M - 1$ form an M-layer coherent stack, surrounded by the remaining incoherent layers. The analysis can easily be extended to accommodate any number of coherent groups (with sizes $M = 1, 2, \cdots$) placed at various positions among the incoherent layers. Additionally, the analysis presented next can also handle the special cases where all layers are treated as coherent or all as incoherent. The case of coherent layers can be handled using the transfer matrix approach outlined in Sec. 3.7. To maintain consistency with the approach presented in Sec. 3.10.2 for incoherent layers, an equivalent alternate method is introduced in the following section.

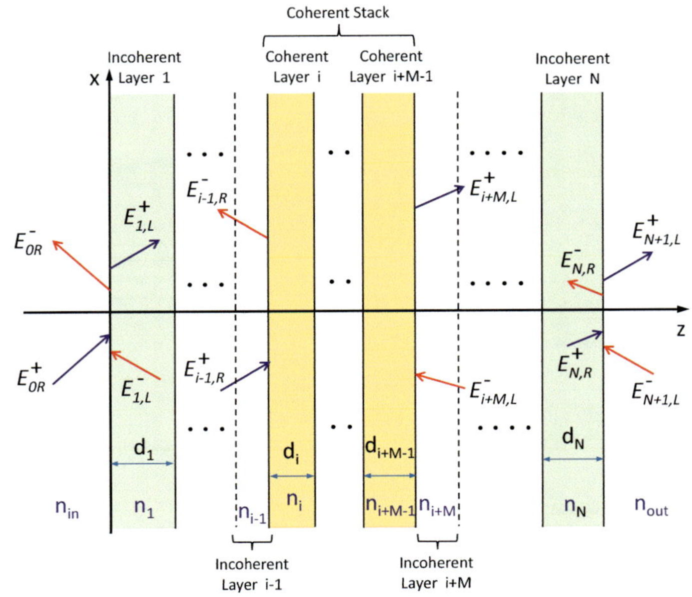

FIGURE 3.91: An optical structure composed of N mixed coherent-incoherent layers, each with thickness d_i and refractive index n_i ($i = 1, 2, \cdots, N$), is placed between two homogeneous media with refractive n_{in} and n_{out}, respectively. In this figure, the stack of layers between indices i and $i - M + 1$ forms an M-layer coherent combination, representing a coherent group. In general, an arbitrary number of coherent groups can be present within the structure, with the remaining layers assumed to behave incoherently.

3.10.3.1 Multiple Coherent Layers: Alternate Transfer Matrix Approach

For this alternate analysis, Fig. 3.89 is used, assuming that all layers are coherent. Let's focus on the interface between the i-th and $(i + 1)$-th layers. In a similar manner to the intensity equations presented in Sec. 3.10.2, the following equations can be written for the electric field amplitudes at the $i/i + 1$ interface:

$$E_{iR}^{-} = r_{i,i+1}E_{iR}^{+} + t_{i+1,i}E_{i+1,L}^{-},$$

$$E_{i+1,L}^{+} = t_{i,i+1}E_{iR}^{+} + r_{i+1,i}E_{i+1,L}^{-}.$$

It is reminded that r_{ij} and t_{ij} are the Fresnel reflection and transmission coefficients from layer i to layer j. Assuming that the electric fields E_{iR}^{+} and $E_{i+1,L}^{-}$ are incident at the interface, producing the $E_{i+1,L}^{+}$ and E_{iR}^{-}, the corresponding equations can be rearranged to relate the electric field amplitudes of layer i with those of layer $i + 1$. This relationship can be represented in the following matrix form:

$$\begin{bmatrix} E_{iR}^{+} \\ E_{iR}^{-} \end{bmatrix} = \underbrace{\frac{1}{t_{i,i+1}} \begin{bmatrix} 1 & -r_{i+1,i} \\ r_{i,i+1} & t_{i,i+1}t_{i+1,i} - r_{i,i+1}r_{i+1,i} \end{bmatrix}}_{=\tilde{S}_{i,i+1}^{c}} \begin{bmatrix} E_{i+1,L}^{+} \\ E_{i+1,L}^{-} \end{bmatrix}$$

$$\begin{bmatrix} E_{iR}^{+} \\ E_{iR}^{-} \end{bmatrix} = \tilde{S}_{i,i+1}^{c} \begin{bmatrix} E_{i+1,L}^{+} \\ E_{i+1,L}^{-} \end{bmatrix}, \tag{3.426}$$

where $\tilde{S}_{i,i+1}^{c}$ is the analogous to $\tilde{S}_{i,i+1}$ matrix used for intensities in the incoherent cascaded layers. For the $(i+1)$-layer the electric field amplitudes at the right and left boundaries are related by:

$$\begin{bmatrix} E_{i+1,L}^{+} \\ E_{i+1,L}^{-} \end{bmatrix} = \underbrace{\begin{bmatrix} \exp[+j\beta_{i+1}d_{i+1}] & 0 \\ 0 & \exp[-j\beta_{i+1}d_{i+1}] \end{bmatrix}}_{=\tilde{D}_{i+1}^{c}} \begin{bmatrix} E_{i+1,R}^{+} \\ E_{i+1,R}^{-} \end{bmatrix} = \tilde{D}_{i+1}^{c} \begin{bmatrix} E_{i+1,R}^{+} \\ E_{i+1,R}^{-} \end{bmatrix}. \tag{3.427}$$

The parameter $\beta_{i+1} = (n_{i+1} - j\kappa_{i+1})\cos\theta_{i+1}$ is the complex in general propagation constant in layer $i+1$, where $\cos\theta_{i+1}$ is determined from Snell's law at the interfaces, and θ_0 represents the angle of incidence of the light from the input region. To relate the electric fields in the input region to those in the output region, the following equations can be written:

$$\begin{bmatrix} E_{0R}^{+} \\ E_{0R}^{-} \end{bmatrix} = \underbrace{\tilde{S}_{0,1}^{c}\tilde{D}_{1}^{c}\tilde{S}_{1,2}^{c}\tilde{D}_{2}^{c}\cdots\tilde{S}_{N-1,N}^{c}\tilde{D}_{N}^{c}\tilde{S}_{N,N+1}^{c}}_{=\tilde{S}^{c}} \begin{bmatrix} E_{N+1,L}^{+} \\ E_{N+1,L}^{-} \end{bmatrix} = \begin{bmatrix} S_{11}^{c} & S_{12}^{c} \\ S_{21}^{c} & S_{22}^{c} \end{bmatrix} \begin{bmatrix} E_{N+1,L}^{+} \\ E_{N+1,L}^{-} \end{bmatrix}. \tag{3.428}$$

The terms S_{ij}^{c} $(i, j = 1, 2)$ are the four elements of the overall matrix $\tilde{S}^{c}$. Equation (3.428) can now be expressed in its intensity-analogous relation as follows:

$$\begin{aligned} \begin{bmatrix} I_{0R}^{+} \\ I_{0R}^{-} \end{bmatrix} &= \frac{1}{|t_{0,N+1}|^2} \begin{bmatrix} 1 & -|r_{N+1,0}|^2 \\ |r_{0,N+1}|^2 & |t_{0,N+1}t_{N+1,0}|^2 - |r_{0,N+1}r_{N+1,0}|^2 \end{bmatrix} \begin{bmatrix} I_{N+1,L}^{+} \\ I_{N+1,L}^{-} \end{bmatrix} \\ &= \underbrace{\begin{bmatrix} |S_{11}^{c}|^2 & -|S_{21}^{c}|^2 \\ |S_{12}^{c}|^2 & \dfrac{|det[S^{c}]|^2 - |S_{12}^{c}|^2|S_{21}^{c}|^2}{|S_{11}^{c}|^2} \end{bmatrix}}_{=\tilde{C}_{0,N+1}} \begin{bmatrix} I_{N+1,L}^{+} \\ I_{N+1,L}^{-} \end{bmatrix} = \tilde{C}_{0,N+1} \begin{bmatrix} I_{N+1,L}^{+} \\ I_{N+1,L}^{-} \end{bmatrix}, \end{aligned} \tag{3.429}$$

where $r_{0,N+1} = E_{0R}^{-}/E_{0R}^{+} = S_{21}^{c}/S_{11}^{c}$, $t_{0,N+1} = E_{N+1,L}^{+}/E_{0R}^{+} = 1/S_{11}^{c}$, $r_{N+1,0} = E_{N+1,L}^{+}/E_{N+1,L}^{-} = -S_{12}^{c}/S_{11}^{c}$ and $t_{N+1,0} = E_{0R}^{-}/E_{N+1,L}^{-} = det[S^{c}]/S_{11}^{c}$. At this point, the formulation of a coherent cascaded stack of layers is compatible with the approach used in the case of the incoherent stack. Therefore, the procedure for analyzing a mixed coherent and incoherent stack can now be presented. This involves treating coherent layers using the transfer matrix method while applying the incoherent analysis for the incoherent layers and combining the two approaches to model a structure with mixed coherent and incoherent behavior.

3.10.3.2 Multiple Coherent/Incoherent Layers: Transfer Matrix Approach

Now, the special structure in Fig. 3.91 is used to treat both incoherent and coherent layers. In this specific case, it is assumed that there is a single stack of M coherent layers, extending from layer i up to the layer $i + M - 1$. The light intensities at the input and output of the coherent stack can be related using Eq. (3.429), as follows:

$$\begin{bmatrix} I_{i-1,R}^{+} \\ I_{i-1,R}^{-} \end{bmatrix} = \tilde{C}_{i-1,i+M} \begin{bmatrix} I_{i+M,L}^{+} \\ I_{i+M,L}^{-} \end{bmatrix}. \tag{3.430}$$

Now, using the approach outlined in Sec. 3.10.2, the following equations can be easily written to relate the input and output light intensities of the mixed structure:

$$\begin{bmatrix} I_{0R}^{+} \\ I_{0R}^{-} \end{bmatrix} = \underbrace{\tilde{S}_{0,1}\tilde{D}_1\tilde{S}_{1,2}\tilde{D}_2\cdots\tilde{D}_{i-1}\underbrace{\tilde{C}_{i-1,i+M}}_{\text{coherent stack}}\tilde{D}_{i+M}\cdots\tilde{S}_{N-1,N}\tilde{D}_N\tilde{S}_{N,N+1}}_{=\tilde{S}} \begin{bmatrix} I_{N+1,L}^{+} \\ I_{N+1,L}^{-} \end{bmatrix}$$

$$= \tilde{S}\begin{bmatrix} I_{N+1,L}^{+} \\ I_{N+1,L}^{-} \end{bmatrix} = \begin{bmatrix} \mathcal{S}_{11} & \mathcal{S}_{12} \\ \mathcal{S}_{21} & \mathcal{S}_{22} \end{bmatrix}\begin{bmatrix} I_{N+1,L}^{+} \\ I_{N+1,L}^{-} \end{bmatrix}. \tag{3.431}$$

Using the elements of the final matrix $\tilde{S}$ in Eqs. (3.418) and (3.419), the normalized reflection and transmission intensities can be determined. The matrix $\tilde{S}$ can be easily calculated for an arbitrary number of coherent groups, N_g where each group consists of M_g layers, and these groups are interspersed among $N - \sum_{g=1}^{N_g} M_g$ incoherent layers. Consequently, the mixed structure can be analyzed. The index g denotes the number of coherent groups, each of which has M_g layers ($M_g \geq 1$).

As stated, Eqs. (3.420) and (3.421) can be used for any incoherent layer and averaged over a number of random samples, M. However, an interesting approach proposed in Ref. [166] suggests that instead of averaging a large number of random samples, equidistant phases can be added and averaged over a smaller number of samples. Specifically, instead of using random phases ϕ_{ri} uniformly distributed in the interval $[0, 2\pi]$, the phases ϕ_{ri} take specific values from the discrete set $[0, 2\pi/N_s, \cdots, (N_s - 1)2\pi/N_s]$ consisting of N_s values. In the case of N_{inc} incoherent layers, the phases for each layer should take all possible combinations from the set, resulting in $N_s^{N_{inc}}$ combinations of samples to average. For example, if $N_s = 3$ and $N_{inc} = 2$, then only $3^2 = 9$ combinations of samples would need to be averaged. As the values of N_s and N_{inc} increase, the number of equidistant samples, $N_s^{N_{inc}}$, becomes computationally equivalent to averaging M random samples. Furthermore, an alternative approach involves using Eqs. (3.422) and (3.423) for the incoherent-only layers of the mixed structure.

Example 3.10.3:

Consider a multilayer optical structure consisting of $N = 5$, layers with refractive indices $n_1 = 1.5$, $n_2 = 2.1$, $n_3 = 3.6$, $n_4 = 2.1$, and $n_5 = 1.5$, surrounded by air ($n_0 = n_6 = 1.0$). The layer thicknesses are $d_1 = 1000\,\mu$m, $d_2 = 0.5\,\mu$m, $d_3 = 2\,\mu$m, $d_3 = 0.5\,\mu$m, and $d_5 = 1000\,\mu$m with an incident freespace wavelength of $\lambda_0 = 1\,\mu$m. The first and fifth layers are considered incoherent ($N_{inc} = 2$), while the second, third, and fourth layers are considered coherent ($N_c = 3$). The system is subjected to *TE*-polarized light. Determine the normalized reflected and transmitted intensities as a function of the angle of incidence θ_0 using the following methods: (a) Mixed Coherent-Incoherent Modeling with the application of Eq. (3.431). (b) Use the coherent layer analysis to model the multilayer structure, incorporating the averaging of $M = 500$ random samples to account for phase variations of the first and fifth layers. (c) Instead of random phase sampling, apply the equidistant phase averaging methodology to the incoherent layers (1 and 5), where the phase angles are chosen from a discrete set of equidistant values (as presented in Sec. 3.10.3.2).

Solution: To calculate the normalized reflected and transmitted intensities for the given multilayer structure, the methodology presented in Sec. 3.10.3.2 is applied. This method takes into account the coherent and incoherent layers within the structure, providing the reflection and transmission intensities as a function of the angle of incidence. The results of the calculations are shown in Fig. 3.92(a), where the normalized reflection and transmission intensities are plotted for *TE*-polarized light as a function of θ_0.

By applying Eqs. (3.420) and (3.421) for the two incoherent layers (1 and 5) and averaging over a number of random samples $M = 500$, the resulting normalized intensities are shown in Fig. 3.92(b). The plot reveals many small oscillations, which are caused by the relatively large thickness of the incoherent layers. These oscillations indicate the need for a larger number of random samples to adequately suppress them and obtain a smoother result.

Finally, the equidistant sampling method proposed in Ref. [166] is applied. Specifically, $N_s = 7$ is selected, and the phases ϕ_{r1} (for incoherent layer 1) and ϕ_{r5} (for incoherent layer 5) are chosen from the set $[0, 2\pi/7, 4\pi/7, 6\pi/7, 8\pi/7, 10\pi/7, 12\pi/7]$. All possible permutations $N_s^{N_{inc}} = 7^2 = 49$ combinations of these phases are calculated and averaged. The results are presented in Fig. 3.92(c), where it is evident that the equidistant sampling significantly improves the estimation of the normalized intensities compared to the results shown in Fig. 3.92(a).

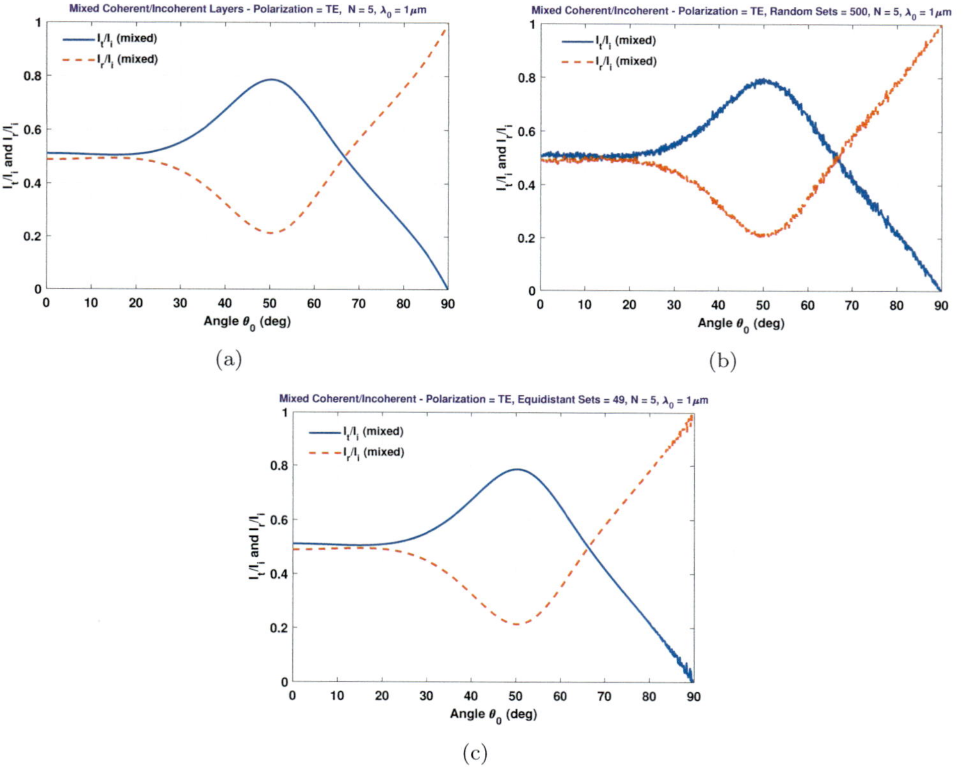

FIGURE 3.92: (a) Reflected and transmitted normalized intensities of a $N = 5$ multilayer structure as a function of the angle of incidence θ_0 for *TE* polarization. The layers have refractive indices of $n_1 = 1.5$, $n_2 = 2.1$, $n_3 = 3.6$, $n_4 = 2.1$, and $n_5 = 1.5$, in air $n_0 = n_6 = 1.0$. The layer thicknesses are $d_1 = 1000\,\mu$m, $d_2 = 0.5\,\mu$m, $d_3 = 2\,\mu$m, $d_4 = 0.5\,\mu$m, $d_5 = 1000\,\mu$m and the wavelength is $\lambda_0 = 1\,\mu$m. The layers 1 and 5 are considered incoherent ($N_{inc} = 2$), while layers 2, 3, and 4 are considered coherent. (b) Reflected and transmitted normalized intensities of the same structure, calculated with random sampling of their coherent counterparts. The number of random samples is $M = 500$. (c) Reflected and transmitted normalized intensities calculated using the equidistant sampling methodology with $N_s = 7$ samples per incoherent layer, resulting in an average of $N_s^{N_{inc}} = 49$ samples.

Example 3.10.4:

A multilayer structure with both coherent and incoherent layers, as well as some loss, is analyzed in this problem. The structure is a simplified example of a solar cell, based on the design adapted from Ref. [162]. The structure consists of $N = 4$ layers with the following properties: A thick incoherent glass layer of refractive index $n_1 = 1.5$ and thickness $d_1 = 1000\,\mu\text{m}$. A thin indium tin oxide (ITO) layer of refractive index $n_2 = 2$ and thickness $d_2 = 0.2\,\mu\text{m}$, an amorphous Si layer of refractive index $n_3 = 4.4 - j0.14$ and thickness $d_3 = 0.655\,\mu\text{m}$, and an aluminum layer of refractive index $n_4 = 1.5 - j6.8$ and thickness $d_4 = 0.5\,\mu\text{m}$. The layers 2, 3, and 4 are considered coherent, while layer 1 is incoherent. The input and output regions are assumed to be air ($n_0 = n_5 = 1$). The task is to calculate the normalized absorbance of the structure in the wavelength range $\lambda_0 = 0.4 - 0.8\,\mu\text{m}$. The light is incident normally at the structure ($\theta_0 = 0$), meaning both polarizations behave identically. For simplicity, the dispersion of the materials is neglected, and refractive indices are taken as average values over the specified wavelength range.

Solution: Using the methodology presented in Sec. 3.10.3.2, the normalized reflected and transmitted intensities are calculated for the given multilayer structure as a function of the free-space wavelength, λ_0. The normalized absorbance is then determined using the relationship: $A = 1-(I_r/I_i)-(I_t/I_i)$. Due to the aluminum layer, the transmitted intensity is negligible. The results are shown in Fig. 3.93(a) as a function of λ_0. To account for the incoherent glass layer, Eqs. (3.420) and (3.421) are applied, averaging over $M = 500$ random samples. The normalized absorbance resulting from this approach is shown in Fig. 3.93(b). While the random sampling produces good agreement with the mixed coherent/incoherent theory, minor oscillations persist due to the thickness of the glass layer. These oscillations could be further reduced by increasing the number of random samples.

Finally, the equidistant sampling methodology proposed in Ref. [166] is employed. With $N_s = 5$, the phase ϕ_{r1} (for the incoherent glass layer) takes values from the discrete set $[0, 2\pi/5, 4\pi/5, 6\pi/5, 8\pi/5]$. The resulting $N_s^{N_{inc}} = 5^1 = 5$ samples are averaged. The normalized absorbance obtained from this approach is displayed in Fig. 3.93(c). Remarkably, this method achieves a more accurate estimation of the normalized absorbance compared to random sampling, using only five samples. The results of Fig. 3.93(c) are practically identical to those of Fig. 3.93(a).

3.10.3.3 Partially Coherent Layers in Multilayer Optical Structures

This section examines the effect of partial coherence in certain layers of multilayer optical structures. The methodology follows the approaches outlined in Refs. [158, 161, 169]. Partial coherence (or partial incoherence) is introduced through the macroscopic roughness of layer interfaces, which causes scattering of electromagnetic radiation incident on the optical structure. Typically, the crystallization of a deposited film leads to rough interfaces, as discussed in [158]. The analysis involves a modification of the Fresnel coefficients, based on the assumption of Gaussian variations in interface irregularities with zero mean and a standard deviation Z [158, 161, 169]. For the interface between layer i and layer $i+1$, the modified reflection and transmission coefficients are given by:

$$r_{i,i+1} = r_{i,i+1}^{(0)} \exp\left[-2\left(\frac{2\pi}{\lambda_0} Z_i n_i\right)^2\right] = a\, r_{i,i+1}^{(0)} \tag{3.432}$$

$$r_{i+1,i} = r_{i+1,i}^{(0)} \exp\left[-2\left(\frac{2\pi}{\lambda_0} Z_i n_{i+1}\right)^2\right] = b\, r_{i+1,i}^{(0)} \tag{3.433}$$

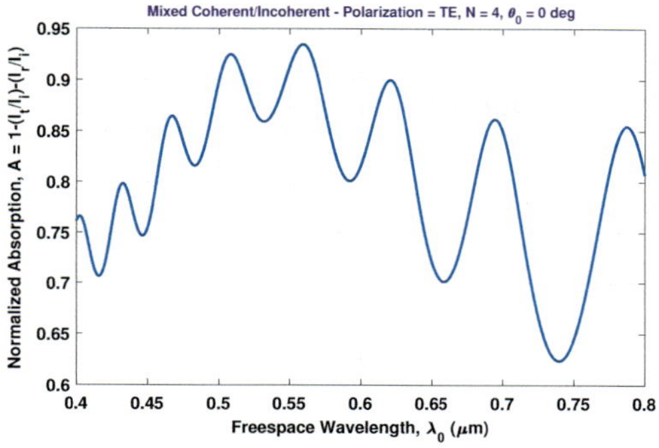

(a)

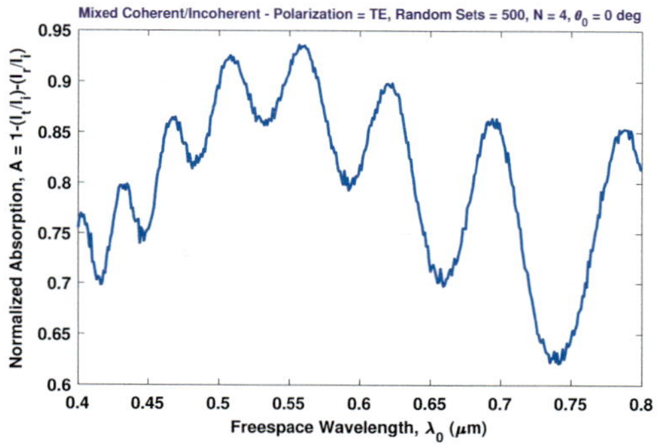

(b)

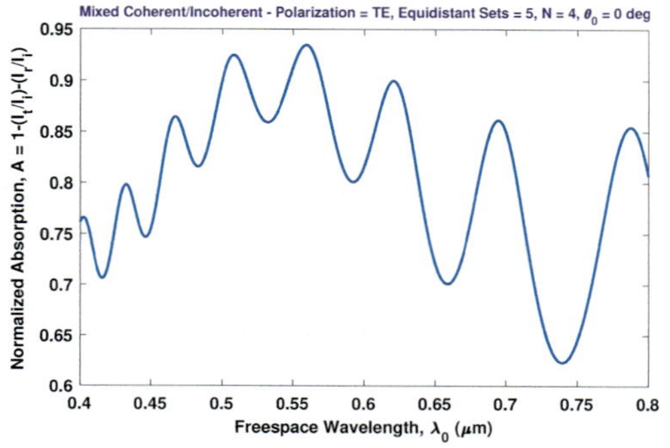

(c)

FIGURE 3.93: (a) The normalized reflected and transmitted intensities of a multilayer structure with $N = 4$ layers are presented as functions of the freespace wavelength λ_0 under normal incidence ($\theta_0 = 0$). The refractive indices of the layers are $n_1 = 1.5$, $n_2 = 2$, $n_3 = 4.4 - j0.14$, $n_4 = 1.5 - j6.8$, with input and output regions being air ($n_0 = n_5 = 1.0$). The corresponding layer thicknesses are $d_1 = 1000\,\mu\text{m}$, $d_2 = 0.2\,\mu\text{m}$, $d_3 = 0.655\,\mu\text{m}$, and $d_4 = 0.5\,\mu\text{m}$. Layer 1, with a thickness of $1000\,\mu\text{m}$, is treated as incoherent ($N_{inc} = 1$), whereas layers 2, 3, and 4 are treated as coherent. (b) The normalized incoherent reflected and transmitted intensities of the same multilayer structure are calculated using random sampling of their coherent counterparts. A total of $M = 500$ random samples is used. (c) As in (b), but the incoherent intensities are computed using $N_s = 5$ equidistant samples per incoherent layer, resulting in an average of $N_s^{N_{inc}} = 5$ samples.

$$t_{i,i+1} = t_{i,i+1}^{(0)} \exp\left[-\frac{1}{2}\left(\frac{2\pi}{\lambda_0} Z_i (n_{i+1} - n_i)\right)^2\right] = c\, t_{i,i+1}^{(0)} \tag{3.434}$$

$$t_{i+1,i} = c\, t_{i+1,i}^{(0)}. \tag{3.435}$$

Here, $r_{i,i+1}^{(0)}$, $r_{i+1,i}^{(0)}$, $t_{i,i+1}^{(0)}$, and $t_{i+1,i}^{(0)}$ represent the Fresnel coefficients for perfectly smooth interfaces, while Z_i denotes the standard deviation of the irregularities at the i-th layer. Consequently, the matrix $\tilde{S}_{i,i+1}^c$ in Eq. (3.426) becomes:

$$\tilde{S}_{i,i+1}^c = \frac{1}{t_{i,i+1}^{(0)}} \begin{bmatrix} \Delta_{11} & -\Delta_{12} r_{i+1,i}^{(0)} \\ \Delta_{21} r_{i,i+1}^{(0)} & \Delta_{22} t_{i,i+1}^{(0)} t_{i+1,i}^{(0)} - \Delta_{22}' r_{i,i+1}^{(0)} r_{i+1,i}^{(0)} \end{bmatrix}, \tag{3.436}$$

$$\Delta_{11} = \frac{1}{c}, \qquad \Delta_{12} = \frac{b}{c},$$

$$\Delta_{21} = \frac{a}{c}, \qquad \Delta_{22} = c, \qquad \Delta_{22}' = \frac{ab}{c}.$$

The coefficients a, b, and c are defined in Eqs. (3.432)-(3.435). Using these modified matrices, the procedure remains identical to the one described in Sec. 3.10.3. It should be noted that, in this case, even if all layers are lossless, the presence of rough interfaces causes the sum of the normalized reflected and transmitted intensities to deviate from unity. The resulting normalized absorption, given by $A = 1-(I_r/I_i)-(I_t/I_i)$, represents the scattering loss due to the rough interfaces. If all layers have perfectly smooth interfaces, the modified Fresnel coefficients revert to their original forms, and no scattering loss occurs in the absence of material absorption.

Alternatively, the method of sampling and averaging multiple coherent results can be applied to simulate partial coherence. However, in this approach, the sum of the normalized reflected and transmitted intensities remains unity for lossless layers. For instance, the phase of the i-th layer can be modified as follows: as follows:

$$\delta_i(m) = \frac{2\pi}{\lambda_0} n_i \cos\theta_i d_i + \phi_i(m), \tag{3.437}$$

$$\phi_i(m) = \left(\frac{(Z_i^2 + Z_{i+1}^2)^{1/2}}{\lambda_0} n_i \cos\theta_i \right) \frac{m 2\pi}{N_s}, \qquad m = 0, 1, \cdots N_s - 1.$$

Here, N_s represents the number of equidistant samples, as described in Sec. 3.10.3.2, and $(Z_i^2 + Z_{i+1}^2)^{1/2}$ is the standard deviation of the i-th layer's effective width, accounting for the roughness of its two interfaces. An alternative approach, as suggested in [160], involves varying the thickness of the i-th layer according to a Gaussian distribution of the form $d_i(m) = d_i + \Delta d_i(m) = d_i + (1/\sigma\sqrt{2\pi}) \exp(-\Delta d_i^2/2\sigma^2)$, where $\sigma = (Z_i^2 + Z_{i+1}^2)^{1/2}$. The normalized reflected and transmitted intensities are then calculated by averaging over M samples. In this method, the sum of the normalized reflected and transmitted intensities remains unity for lossless media.

Example 3.10.5:

Consider the example structure described in Ref. [158], consisting of a silicon layer ($N = 1$) on a semi-infinite SiO_2 (glass) substrate. The refractive index of the silicon layer is $n_1 = 3.42$, the input region (air) has $n_0 = 1$, and the refractive index of the SiO_2 substrate is $n_2 = 1.52$. The thickness of the silicon layer is $d = 2\,\mu$m. (a) Using the method outlined in Sec. 3.10.3.3, calculate the normalized reflected intensity as a function of the freespace wavenumber in the range of $1000-5000\,cm^{-1}$ for the following interface roughness scenarios: $Z_1 = Z_2 = 0$ (smooth interfaces), $Z_1 = 0.1\,\mu$m, $Z_2 = 0$ (air-Si interface rough and Si/SiO_2 smooth), $Z_1 = 0$, $Z_2 = 0.1\,\mu$m (air-Si interface smooth and Si/SiO_2 rough), $Z_1 = Z_2 = 0.1\,\mu$m, (both interfaces rough). (b) For the roughness scenarios specified above, calculate the normalized absorption due to scattering from the rough interfaces as a function of the freespace wavenumber in the range of $1000-5000\,cm^{-1}$. (c) Apply the phase averaging sampling approach to compute the normalized reflected intensity as a function the freespace wavenumber in the range of $1000-5000\,cm^{-1}$ for the case $Z_1 = Z_2 = 0.1\,\mu$m, (both interfaces rough).

Solution: (a) Apply Eqs. (3.432)-(3.435) and the modified matrices $\tilde{S}_{01}^c$ and $\tilde{S}_{12}^c$ to calculate the normalized reflected intensity as a function of $1/\lambda_0$. The results for the specified cases are shown in Fig. 3.94(a), which match the findings presented in Ref. [158]. It is observed that the worst-case scenario occurs when both interfaces are rough. Conversely, the case of smooth interfaces corresponds to treating the entire structure as fully coherent.

(b) In this case the normalized absorbance $A = 1-(I_r/I_i)-(I_t/I_i)$ is calculated based on the analysis from part (a), and the results are shown in Fig. 3.94(b). As expected, in the case of smooth interfaces, the scattering loss is zero.

(c) In this case, assume $N_s = 25$ samples are used to simulate the partial coherence of the silicon layer, as described in Eq. (3.437). Alternatively, $M = 2000$ samples are averaged, assuming a Gaussian variation in the silicon layer thickness. The results are shown in Fig. 3.94(c) as a function of the freespace wavenumber. The sample averaging preserves power in the case of lossless media, as is the case in this example. The Gaussian width variation appears to better simulate the original problem compared to the equidistant phase variation, which is more suitable for modeling a uniform distribution of width variation, rather than a Gaussian distribution.

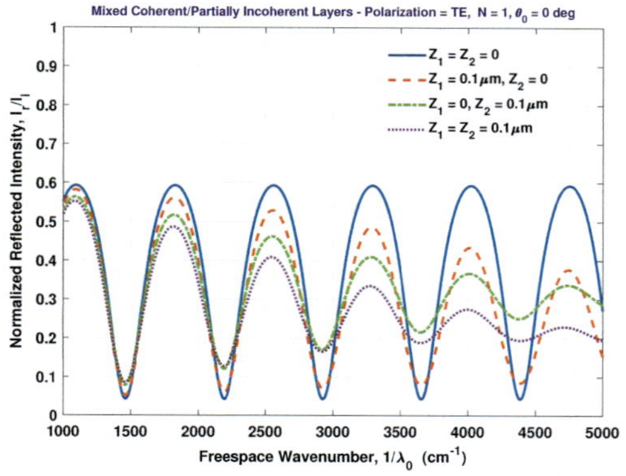

(a)

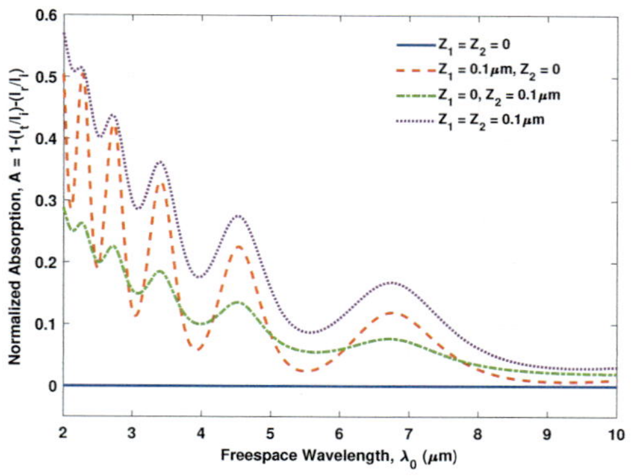

(b)

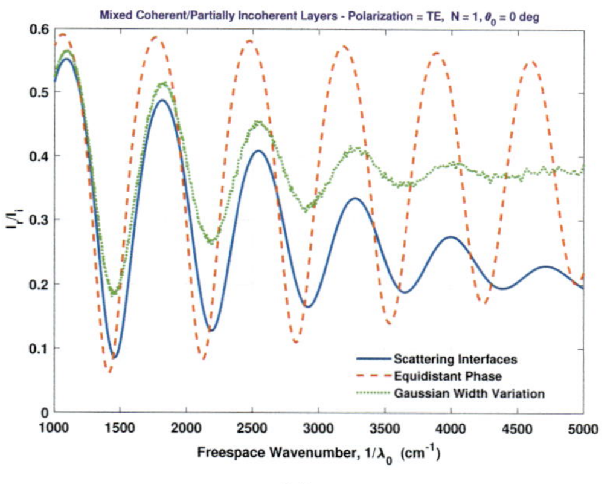

(c)

FIGURE 3.94: (a) The normalized reflected intensity of a silicon Si slab ($N = 1$) structure as a function of the freespace wavenumber $1/\lambda_0$ for normal incidence ($\theta_0 = 0$). The silicon layer has a refractive index $n_1 = 3.42$, with air in the input region $n_0 = 1$, and SiO_2 in the output region with $n_2 = 1.52$. The silicon layer thickness is $d_1 = 2\,\mu\text{m}$, with interface variations Z_1 and Z_2 as shown in the legend. This example is presented in Ref. [158]. (b) The normalized absorbance due to interface scattering for the cases described in (a). (c) The normalized reflected intensity in the case of equidistant phase samples and with a Gaussian distribution of layer thickness averaged over $M = 2000$ samples. The use of coherent layer analysis, averaging either with phase or layer thickness samples, retains power, in contrast to the case in (a) (which is also included for $Z_1 = Z_2 = 0.1\mu\text{m}$).

3.11 Gaussian Beams

A ray does not convey any information about the electromagnetic field's amplitude, phase, or spatial extent. Additionally, while plane waves are simple solutions to Maxwell's equations, they are of infinite extent and cannot accurately model the beam emitted by a laser source. Practical laser beams are usually well-collimated and of finite size. Therefore, a more precise representation of optical beams and laser resonators is required. This can be achieved by using an approximate solution to the wave equation, known as the "paraxial approximation," leading to the corresponding "paraxial wave equation." This approximate yet analytical solution of the paraxial wave equation is referred to as a "Gaussian beam" [39,97]. Gaussian beam solutions align well with the characteristics of laser cavities and provide an effective representation of laser beams in most practical situations.

Most optical beams that propagate through free space can be characterized as TEM (Transverse Electromagnetic) waves, meaning their electric and magnetic fields are transverse to the direction of propagation. From Maxwell's equations, it is well established that in a source-free region, Gauss's law can be expressed in the following form for isotropic homogeneous media [see Eq. (3.3)]:

$$\vec{\nabla} \cdot \vec{D} = 0 \Longrightarrow \vec{\nabla} \cdot \vec{E} = 0, \tag{3.438}$$

where, $\vec{D}$ and $\vec{E}$ are the displacement vector phasor (electric flux density) and the electric field vector phasor, respectively. Splitting the divergence operator into a transverse and a longitudinal component (assuming that z is the propagation direction), the previous equation can be written as [39]:

$$\vec{\nabla} \cdot \vec{E} = 0 \Longrightarrow \vec{\nabla}_t \cdot \vec{E}_t + \frac{\partial E_z}{\partial z} = 0, \tag{3.439}$$

where $\vec{E} = \vec{E}_t + E_z \hat{z}$, with $\vec{E}_t$ and E_z representing the transverse and longitudinal electric field components, respectively. Using the approach described by Verdeyen [39] and assuming that the most significant phase variation along the z direction is of the form $\exp(-jkz)$ [where the wavenumber $k = k_0 n = (\omega/c)n = (2\pi/\lambda_0)$, with n the medium refractive index, ω the radial frequency of the electromagnetic wave, λ_0 is the freespace wavelength, and c the speed of light in freespace], the following approximations can be made at optical frequencies:

$$\frac{\partial E_z}{\partial z} \simeq -j\frac{2\pi}{\lambda_0} n E_z, \tag{3.440}$$

$$\vec{\nabla}_t \cdot \vec{E}_t = \lim_{\Delta S \to 0} \left\{ \frac{1}{\Delta S} \oint_{\Delta \ell} \vec{E}_t \cdot \hat{\imath}_{\perp \ell} d\ell \right\} \simeq \lim_{D \to 0} \left\{ \frac{4}{\pi D^2} |\vec{E}_t| \pi D \right\} \simeq \frac{|\vec{E}_t|}{D}. \tag{3.441}$$

In the latter expressions, D represents the electromagnetic beam's finite diameter. Using the last two approximations, the approximate relation between the longitudinal and the transverse electric field components becomes:

$$|E_z| \simeq \frac{\lambda_0}{2\pi n D} |\vec{E}_t|. \tag{3.442}$$

The ratio $\lambda_0/(2\pi n D)$ is indeed small ($< 10^{-3}$) for visible wavelengths ($0.4\,\mu m \lesssim \lambda_0 \lesssim 0.7\,\mu m$) and beam diameters of the order of a centimeter. Therefore, the longitudinal electric field component E_z can be neglected as compared to the transverse electric field component, resulting in a TEM solution of the electromagnetic field of a Gaussian beam. Consequently, a solution for the electromagnetic beam of the form:

$$\vec{E}(x,y,z) = \hat{t}E_0\psi(x,y,z)\exp(-jkz), \tag{3.443}$$

is sought, where $\hat{t}$ is a unit vector in the transverse direction (in the xy plane) that is independent of the x, y, z variables. Then, the electric field of Eq. (3.443) should satisfy the scalar Helmholtz's equation:

$$\nabla^2 E + k_0^2 n^2 E = 0. \tag{3.444}$$

Inserting the form of the electric field of Eq. (3.443) into Eq. (3.444) results in the following equation:

$$\nabla_t^2\psi - j2k\frac{\partial\psi}{\partial z} + \frac{\partial^2\psi}{\partial z^2} = 0, \tag{3.445}$$

where ∇_t is the transverse gradient operator. At this point it is assumed that the function ψ is slowly varying along the z direction since the "fast" z-variation is included in the $\exp(-jkz)$ term. If $|\partial^2\psi/\partial z^2| \ll 2k|\partial\psi/\partial z|$ (slow variation of ψ along z as compared to the wavelength), then the second derivative of ψ with respect to z in Eq. (3.445) can be neglected, resulting in the following equation, which is known as the "paraxial wave equation":

$$\nabla_t^2\psi - j2k\frac{\partial\psi}{\partial z} \simeq 0, \qquad \text{if} \quad \left|\frac{\partial^2\psi}{\partial z^2}\right| \ll 2k\left|\frac{\partial\psi}{\partial z}\right|. \tag{3.446}$$

3.11.1 Fundamental Gaussian Beam Solution

In order to find the simplest possible solution, Eq. (3.446) is expressed in the cylindrical coordinates, assuming azimuthal independence. This means that $\psi = \psi(r_T, z)$ where $r_T = (x^2+y^2)^{1/2}$ is the transverse radial distance. Then, Eq. (3.446) can be rewritten as:

$$\frac{1}{r_T}\frac{\partial}{\partial r_T}\left(r_T\frac{\partial\psi}{\partial r_T}\right) - j2k\frac{\partial\psi}{\partial z} = 0. \tag{3.447}$$

Now, a solution of the following form is sought:

$$\psi(r_T, z) = \exp\left\{-j\left[P(z) + \frac{kr_T^2}{2q(z)}\right]\right\}, \tag{3.448}$$

where $P(z)$ and $q(z)$ are unknown functions that must be determined in order to satisfy Eq. (3.447). Substituting the last equation in Eq. (3.447) results in:

$$\left\{\left[\frac{k^2}{q^2(z)}\left(q'(z)-1\right)\right]r_T^2 - 2k\left[P'(z) + \frac{j}{q(z)}\right]r_T^0\right\}\psi = 0, \qquad \forall r_T \Longrightarrow \tag{3.449}$$

$$\frac{dq}{dz} = q'(z) = 1, \tag{3.450}$$

$$\frac{dP}{dz} = P'(z) = -\frac{j}{q(z)}. \tag{3.451}$$

It is straightforward to show that the solution of Eq. (3.450) is of the form $q(z) = z + jz_0$, where z_0 is real. This form is necessary for the electromagnetic beam's intensity, $I \sim |\psi|^2$, to exhibit transverse r_T-dependence [39]. This can be easily understood by considering $|\psi(r_T, z=0)|^2 = \exp(-kr_T^2/z_0)$. An imaginary value of z_0 would result in no radial dependence of $|\psi(r_T, z=0)|^2$, which is not an acceptable solution for a beam with transverse confinement. The variable z_0 is known as "Rayleigh distance" or "Rayleigh range" and can be related to the "minimum spot size" or "minimum beam waist," w_0, of the Gaussian beam solution. If $|\psi(r_T, z=0)| = \exp(-r_T^2/w_0^2)$, which represents a Gaussian profile of the beam in the transverse direction, w_0 and z_0 can be related by:

$$w_0^2 = \frac{2z_0}{k} = \frac{\lambda_0 z_0}{n\pi} \Longrightarrow z_0 = \frac{\pi n w_0^2}{\lambda_0}, \tag{3.452}$$

where λ_0 is the wavelength of the beam in freespace. This relationship connects the minimum beam waist w_0 to the Rayleigh range z_0, defining the beam's spreading behavior. Using the general solution for $q(z)$ the $1/q$ factor can be written in the following form:

$$\frac{1}{q(z)} = \frac{z}{z^2+z_0^2} - j\frac{z_0}{z^2+z_0^2} = \frac{1}{R(z)} - j\frac{\lambda_0}{\pi n w^2(z)}, \tag{3.453}$$

$$w^2(z) = w_0^2\left[1+\left(\frac{z}{z_0}\right)^2\right], \tag{3.454}$$

$$R(z) = z\left[1+\left(\frac{z_0}{z}\right)^2\right]. \tag{3.455}$$

The parameter $q(z)$ is also referred as a complex radius of curvature of the Gaussian beam. Using $q(z)$ the solution for $P(z)$ can be easily obtained from Eq. (3.451) as:

$$jP(z) = \ln\left[1 - j\left(\frac{z}{z_0}\right)\right] = \ln\left[1+\left(\frac{z}{z_0}\right)^2\right]^{1/2} - j\tan^{-1}\left(\frac{z}{z_0}\right). \tag{3.456}$$

Applying the solutions for $q(z)$ and $P(z)$ into Eq. (3.448), the complete expression of the fundamental lowest-order (named as TEM_{00}) Gaussian beam is summarized as follows:

$$E_{00}(r_T, z) = \mathcal{E}_0 \underbrace{\frac{w_0}{w(z)}\exp\left[-\frac{r_T^2}{w^2(z)}\right]}_{\text{amplitude factor}} \underbrace{\exp\left\{-j\left[kz - \tan^{-1}\left(\frac{z}{z_0}\right)\right]\right\}}_{\text{longitudinal phase factor}} \underbrace{\exp\left[-j\frac{kr_T^2}{2R(z)}\right]}_{\text{radial phase factor}}. \tag{3.457}$$

A simple Gaussian beam solution is depicted in Fig. 3.95(a). Additionally, sample radial profiles of the beam at various distances from its center ($z = 0$) are illustrated in Fig. 3.95(b). A snapshot of the Gaussian beam's electric field at a specific time is shown in Fig. 3.96(a), while its intensity $|E_{00}(r_T, z)|^2$ is represented in Fig. 3.96(b).

3.11.2 Gaussian Beam Divergence

The Gaussian beam, like all electromagnetic wave solutions, exhibits diffraction effects, which cause the beam to expand as it propagates away from the plane $z = 0$, where it has its smallest waist size w_0. This expansion is described by the beam waist $w(z)$, which increases with z according to Eq. (3.454). For large distances, specifically when $z \gg z_0$, the beam waist from Eq. (3.454) approaches:

$$w(z) \simeq \frac{w_0 z}{z_0} = \frac{\lambda_0 z}{\pi n w_0},$$

and the divergence angle θ (shown in Fig. 3.95(a), can be easily determined by the following equation:

$$\tan\theta = \frac{dw(z)}{dz} = \frac{\lambda_0}{\pi n w_0} \Longrightarrow \Delta\theta = 2\theta \simeq \frac{2\lambda_0}{\pi n w_0} = \frac{4}{\pi}\frac{\lambda_0/n}{D_0}. \tag{3.458}$$

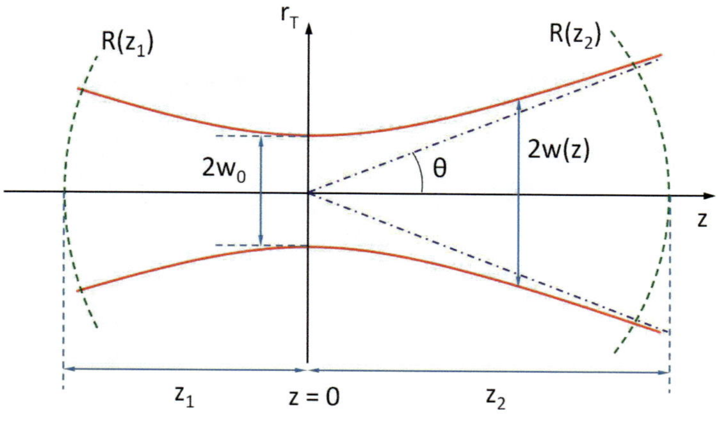

(a)

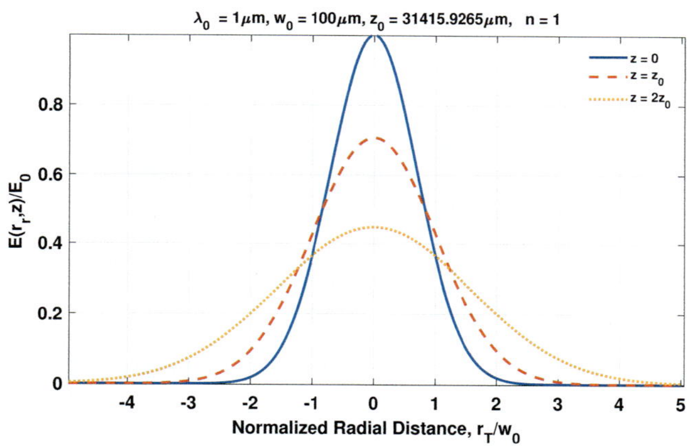

(b)

FIGURE 3.95: (a) A graphical representation of a Gaussian beam. The minimum beam waist, w_0, as well the beam waist at distance z, $w(z)$, and the radii of curvature, $R(z_1)$ and $R(z_2)$ at two, at distances z_1 and z_2 (opposite to $z = 0$) are shown. (b) Example profiles of a Gaussian beam along the radial direction for various z's and for $\lambda_0 = 1\,\mu$m, $w_0 = 100\,\mu$m, and $n = 1$.

The angle $\Delta\theta$ represents the asymptotic angular expansion of the Gaussian beam, and $D_0 = 2w_0$ is its minimum diameter, defined by the $1/e$ field points. In the final equation, the approximation $\tan\theta \simeq \theta$ was used, which is valid for relatively small θ's.

3.11.3 Gaussian Beam Power

Another important property of the Gaussian beam is that the solution described by Eq. (3.457) conserves power. This can be demonstrated by calculating the total power crossing any arbitrary transverse plane at a distance z. The total power carried by the Gaussian beam is given by:

$$P(z) = \frac{1}{2Z}\int_{-\infty}^{+\infty}\int_{-\infty}^{+\infty}|E_{00}|^2 dxdy = \frac{1}{2Z}\int_{-\infty}^{+\infty}\int_{0}^{2\pi}|E_{00}(r_T, z)|^2 r_T dr_T d\phi = \frac{1}{2Z}|\mathcal{E}_0|^2\left[\frac{\pi w_0^2}{2}\right],$$

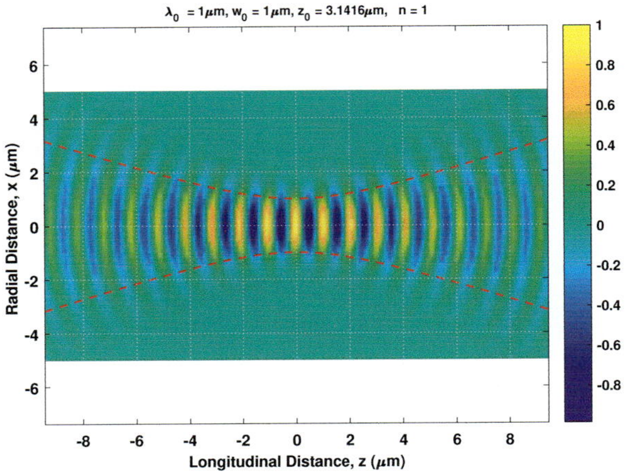

(a)

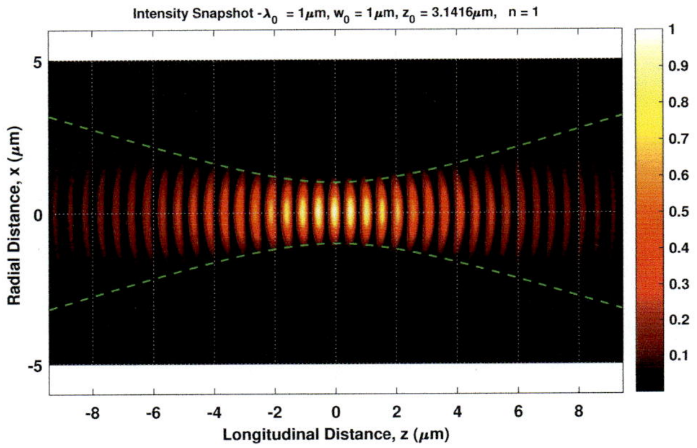

(b)

FIGURE 3.96: (a) A graphical representation of a Gaussian beam snapshot in time. The real electric field is shown in space for an instant of time. (b) The intensity $|E_{00}(r_T, z)|^2$ snapshot. The example parameters are for $\lambda_0 = 1\,\mu\text{m}$, $w_0 = 1\,\mu\text{m}$, and $n = 1$.

where Z is the wave impedance of the homogeneous medium in which the beam propagates. Indeed, the power carried by the Gaussian beam remains constant as it propagates, which is expected for any viable electromagnetic solution in a lossless medium. This invariance with respect to z confirms the conservation of power in the beam's propagation.

3.11.4 Gaussian Beam Radius of Curvature

At this point let's comment on the radial phase shift factor observed in Eq. (3.457) of the fundamental Gaussian beam solution. The scalar representation of a spherical wave, in general, is given by:

$$E_{sph} = \frac{1}{r}\exp(-jkr), \quad \text{where}$$

$$r = \sqrt{x^2 + y^2 + z^2} = \sqrt{r_T^2 + z^2} = z\sqrt{1 + \frac{r_T^2}{z^2}}.$$

For $z \gg r_T$ the following approximations can be made in the expression of the scalar spherical wave:

$$r \simeq z + \frac{1}{2}\frac{r_T^2}{z} \simeq z + \frac{1}{2}\frac{r_T^2}{r}, \quad \text{and}$$

$$E_{sph} \simeq \frac{1}{r}\exp(-jkz)\exp\left(-j\frac{kr_T^2}{2r}\right). \tag{3.459}$$

In the context of the Gaussian beam, a similar radial phase shift factor appears, but the beam's profile is Gaussian rather than spherical. The phase term in the Gaussian beam solution introduces a radial phase shift that contributes to the beam's overall shape and propagation characteristics. By comparing the radial phase of the spherical wave with the term in Eq. (3.457), it becomes evident that the wavefront of the fundamental Gaussian beam, with its radius of curvature $R(z)$ evolves in a manner similar to the spherical wavefront. However, unlike a spherical wave, the Gaussian beam's wavefront curvature varies with the distance z from the beam's center. Depending on the distance z from the beam's center, the following approximations for $R(z)$ can be made (where z_c represents the position of the center of the spherical wavefront):

$$R(z) \simeq \begin{cases} +\infty, & \text{centered at} \quad z_c = \mp\infty, \quad \text{for} \quad |z| \ll z_0, \\ \mp 2z_0, & \text{centered at} \quad z_c = \mp z_0 \quad \text{for} \quad |z| = z_0, \\ z, & \text{centered at} \quad z_c = 0, \quad \text{for} \quad |z| \gg z_0. \end{cases} \tag{3.460}$$

These approximations reflect how the Gaussian beam's wavefront curvature evolves with distance from the beam's center, showing the transition from a spherical to a nearly planar wavefront as the beam propagates.

3.11.5 Gaussian Beam Longitudinal Phase Shift—Gouy's Shift

The longitudinal phase shift of the fundamental Gaussian beam, given by $\Phi(z) = kz - \tan^{-1}(z/z_0)$, differs from the phase shift of a plane wave (or equivalently, a spherical wave) due to the additional term "$\tan^{-1}(z/z_0)$." This term is known as the "Gouy's phase shift," named after M. Gouy who first discovered this phenomenon [170, 171]. For a fundamental Gaussian beam, there is a π phase shift as it passes through its focal plane when propagating from $z = -\infty$ to $z = +\infty$, relative to the phase shift experienced by an ideal plane wave. The Gouy phase shift is intrinsic to all spatially confined electromagnetic fields in the transverse direction and can be derived from the transverse momentum of the monochromatic wave solution [172]. The phase velocity of the Gaussian beam, which can exceed the speed of light in free space, is connected to this phase shift. According to the definition provided by Born and Wolf [91], the phase velocity, u_p, of the fundamental Gaussian beam is given by:

$$u_p = \frac{\omega}{d\Phi/dz} = \frac{c/n}{1 - \dfrac{\lambda_0 z_0}{2\pi n}\dfrac{1}{z^2 + z_0^2}}. \tag{3.461}$$

In Verdeyen [39], an approximate expression of the phase velocity is given by:

$$u_p = \frac{\omega}{\Phi/z} = \frac{c/n}{1 - \dfrac{\lambda_0}{2\pi n z} \tan^{-1}\left(\dfrac{z}{z_0}\right)}. \tag{3.462}$$

For comparison, the two-phase velocities defined in Eqs. (3.461) and (3.462) are illustrated in Fig. 3.97. It is observed that both phase velocities exceed the speed of light in free space (superluminal phase velocities) near the focal plane of the Gaussian beam. This phenomenon is not surprising [173], as it can also occur in other electromagnetic contexts, such as when X-rays undergo total internal reflection on crystals. This does not contravene the fundamental laws of physics, which state that the speed of light in a vacuum is the ultimate speed limit. The superluminal phase velocities observed do not correspond to the transmission speed of information or signals. Instead, any information or signal propagates at a speed that remains below the speed of light in free space.

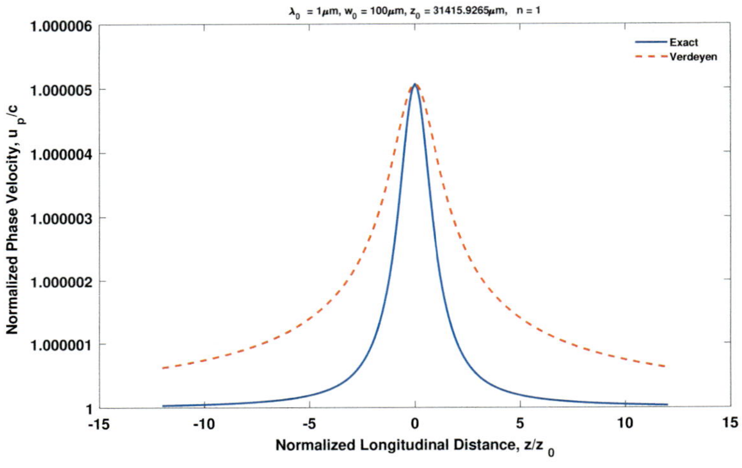

FIGURE 3.97: Phase velocity of the fundamental Gaussian beam along the longitudinal direction z for $\lambda_0 = 1\,\mu$m, $w_0 = 100\,\mu$m, and $n = 1$. Both the exact Eq. (3.461) and the approximate Eq. (3.462) solutions are shown.

3.11.6 Higher-Order Hermite-Gaussian Beams

The fundamental Gaussian beam solution presented in Sec. 3.11.1 assumes no azimuthal dependence, which means it has cylindrical symmetry around the propagation axis. However, real laser resonators can produce more complex beam profiles. In Cartesian coordinates, higher-order Gaussian beam solutions can be obtained by solving the paraxial wave equation, Eq. (3.446). The family of these solutions, known as TEM_{mp} Hermite-Gaussian beams is detailed in Refs. [97, 174]. The resulting electric field is given by:

$$E_{mp}(x,y,z) = \mathcal{E}_{mp}^{HG} H_m\left(\frac{\sqrt{2}x}{w(z)}\right) H_p\left(\frac{\sqrt{2}y}{w(z)}\right) \frac{w_0}{w(z)} \exp\left[-\frac{x^2+y^2}{w^2(z)}\right]$$
$$\exp\left\{-j\left[kz - (1+m+p)\tan^{-1}\left(\frac{z}{z_0}\right)\right]\right\} \exp\left[-j\frac{k(x^2+y^2)}{2R(z)}\right], \tag{3.463}$$

where,

$$w^2(z) = w_0^2\left(1+\frac{z^2}{z_0^2}\right), \qquad R(z) = z\left(1+\frac{z_0^2}{z^2}\right), \qquad H_m(u) = (-1)^m e^{u^2}\frac{d^m}{du^m}\left(e^{-u^2}\right),$$

$$H_{m+1}(u) = 2uH_m(u) - 2mH_{m-1}(u), \quad \text{with} \quad H_0(u) = 1 \quad \text{and} \quad H_1(u) = 2u,$$

where the functions H_m, H_p are Hermite polynomials of order m and p, respectively ($m, p = 0, 1, 2, \cdots$). It is interesting to note that the beam waist, $w(z)$, the radius of curvature $R(z)$, the Rayleigh distance z_0, and the minimum beam waist w_0, are identical to those in the fundamental Gaussian beam solution. Consequently, these higher-order beams exhibit the same divergence and power conservation properties as the fundamental Gaussian beam. However, their Gouy phase shift, Φ_{Gouy}, is given by:

$$\Phi_{Gouy}(z) = (1+m+p)\tan^{-1}\left(\frac{z}{z_0}\right). \tag{3.464}$$

Similarly, their phase velocity is also affected and is expressed as:

$$u_p = \frac{\omega}{d\Phi/dz} = \frac{c/n}{1-(1+m+p)\dfrac{\lambda_0 z_0}{2\pi n}\dfrac{1}{z^2+z_0^2}}. \tag{3.465}$$

The intensity patterns, $|E_{mp}(x, y, z = 0)|^2$, of various Hermite-Gaussian beams are shown in their focal plane in Fig. 3.98.

3.11.7 Higher-Order Laguerre-Gaussian Beams

An alternative and equivalent family of solutions to the paraxial wave equation in cylindrical coordinates, which includes azimuthal dependence, is known as TEM$_{pm}$ Laguerre-Gaussian beams. The electric field components for these paraxial solutions are provided below [97,174]:

$$E_{pm}(r_T, \phi, z) = \mathcal{E}_{pm}^{LG}\frac{w_0}{w(z)}\left(\frac{\sqrt{2}r_T}{w(z)}\right)^{|m|} L_p^{|m|}\left(\frac{2r_T^2}{w^2(z)}\right)\exp\left[-\frac{r_T^2}{w^2(z)}\right]$$

$$\exp\left\{-j\left[kz-(1+|m|+2p)\tan^{-1}\left(\frac{z}{z_0}\right)\right]\right\}\exp(jm\phi)\exp\left[-j\frac{kr_T^2}{2R(z)}\right], \tag{3.466}$$

where,

$$w^2(z) = w_0^2\left(1+\frac{z^2}{z_0^2}\right),$$

$$R(z) = z\left(1+\frac{z_0^2}{z^2}\right), \qquad L_p^m(u) = \frac{e^u}{p!}u^{-m}\frac{d^p}{du^p}\left(e^{-u}u^{p+m}\right),$$

$$(k+1)L_{k+1}^p(u) = (2k+1+p-u)L_k^p(u) - (k+p)L_{k-1}^p(u), \quad \text{with}$$

$$L_0^m(u) = 1 \quad \text{and} \quad L_1^m(u) = -u+(m+1),$$

and the functions $L_p^m(u)$ are the Laguerre polynomials, with the indices p and m being integers taking values $p = 0, 1, 2, \cdots$, and $m = 0, \pm 1, \pm 2, \cdots$. The integer m in the Laguerre-Gaussian beam solution is associated with the azimuthal dependence through the term $\exp(jm\phi)$ in Eq. (3.466). Linear combinations of Laguerre-Gaussian beams with $\pm m$ terms

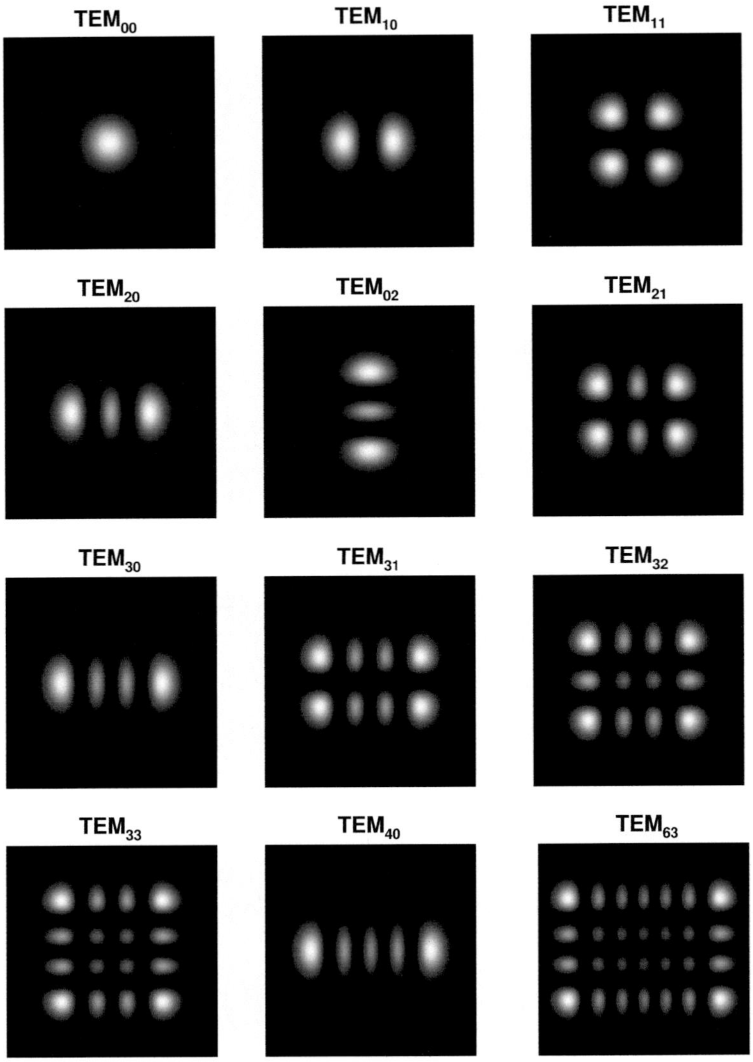

FIGURE 3.98: Intensity diagrams, $|E_{mp}(x, y, z = 0)|^2$, of some example Hermite-Gaussian beams for various values of m and p.

yield dependencies of $\cos(m\phi)$ and $\sin(m\phi)$. Unlike Hermite-Gaussian modes, Laguerre-Gaussian beams with $m \neq 0$ possess a unique property: they carry non-zero orbital angular momentum [174]. Specifically, a Laguerre-Gaussian beam with $m \neq 0$ has an orbital angular momentum of $m\hbar$ per photon [174]. Consequently, such beams can induce rotation in an absorbing dielectric particle due to the conservation of angular momentum. Laguerre-Gaussian beams are not typically generated by standard laser resonators, though they can be produced in lasers with extremely well-aligned mirrors [174]. Any Laguerre-Gaussian beam can be expressed as a linear combination of Hermite-Gaussian beams and vice versa, as both types form complete sets [97, 174]. The Gouy phase shift, Φ_{Gouy}, and phase velocity for Laguerre-Gaussian beams follow similar expressions to those for Hermite-Gaussian beams. The intensity patterns, $|E_{pm}(r_T, \phi, z = 0)|^2$, of various Laguerre-Gaussian beams are shown in their focal plane in Fig. 3.99, illustrating different values of m and p, as well as $\cos(m\phi)$ or $\sin(m\phi)$ dependencies.

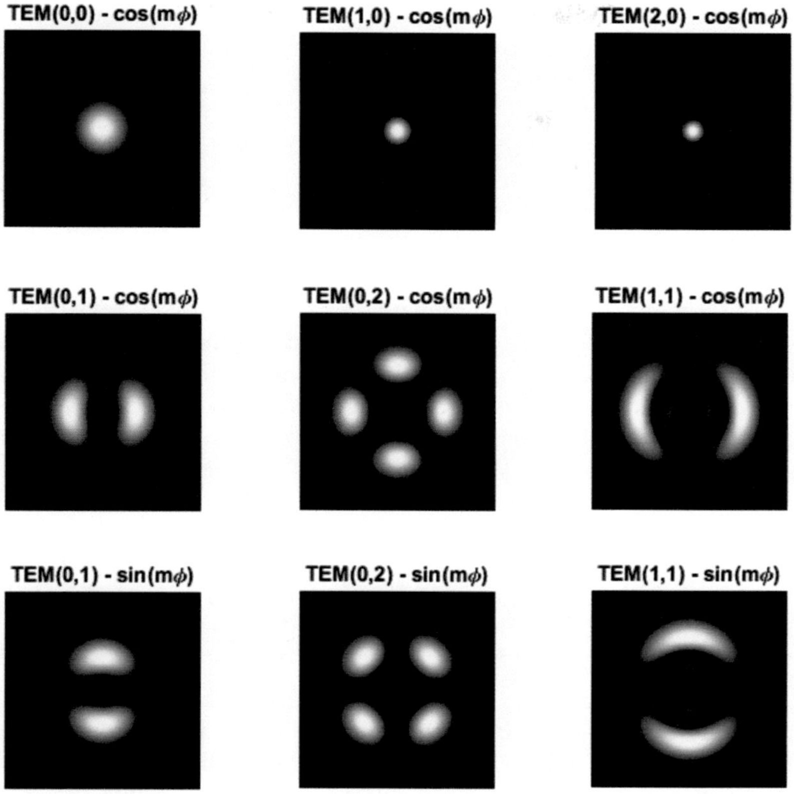

FIGURE 3.99: Intensity diagrams, $|E_{pm}(r_T, \phi, z = 0)|^2$, of some example Laguerre-Gaussian beams for various values of m and p, and $\cos(m\phi)$ or $\sin(m\phi)$ dependencies.

3.11.8 Validity of the Paraxial Approximation

An interesting question regarding Gaussian beams, whether Hermite-Gaussian or Laguerre-Gaussian, is the range of validity of the solutions derived from the paraxial wave equation [Eq. (3.446)]. Reference [175] employs an integral method to assess the validity of paraxial beams by comparing the vector Helmholtz equation with the paraxial equation. In this study, a paraxiality parameter $\mathcal{P}$ is defined as:

$$\mathcal{P} = 1 - \frac{N+1}{k_0^2 n^2 w_0^2}, \qquad \text{where} \tag{3.467}$$

$$N = m + p, \qquad \text{for Hermite-Gaussian beams,}$$
$$N = 2p + |m|, \qquad \text{for Laguerre-Gaussian beams.}$$

The closer the paraxiality parameter $\mathcal{P}$ is to unity, the more accurate the paraxial approximation is. This means that the solutions of the Helmholtz and paraxial equations are very similar, and the paraxial solutions consistently provide meaningful and physical results.

3.11.9 The ABCD Law for Gaussian Beams

If q is the complex radius of curvature of a Gaussian beam [see Eq. (3.453)] then the $ABCD$ law as used in geometrical optics (refer to Chapter of *Geometrical Optics*, Chapter 2), can be applied to Gaussian beams [39]. For example, if q_{in} is the input complex radius of curvature of a Gaussian beam entering an optical system (see Fig. 3.100(a)), which is described by the paraxial $ABCD$ matrix, then the complex radius of curvature q_{out} of the emerging Gaussian beam from the system is given by:

$$q_{out} = \frac{Aq_{in} + B}{Cq_{in} + D}, \quad \text{or} \tag{3.468}$$

$$\frac{1}{q_{out}} = \frac{C + D(1/q_{in})}{A + B(1/q_{in})}. \tag{3.469}$$

Thus, the $ABCD$ matrix approach, originally derived for paraxial rays (i.e., rays at small angles relative to the propagation axis) in geometrical optics, is also applicable to paraxial Gaussian beams. This is due to the fact that geometrical optics rays are perpendicular to spherical wavefronts, and the spherical wavefronts are associated with Gaussian beam solutions. The transformation described in Eqs. (3.468) and (3.469) can be readily demonstrated in specific cases, such as the propagation of a Gaussian beam in a homogeneous medium or when a Gaussian beam passes through a thin lens. However, to the best of the author's knowledge, there is no formal proof for the general case involving an optical system described by an $ABCD$ matrix. Nevertheless, these equations are extremely useful for analyzing Gaussian beams passing through optical systems, provided the paraxial approximation is valid.

Example 3.11.1:

As an example of applying the $ABCD$ transformation rule for a Gaussian beam, consider the system shown in Fig. 3.100(b). A Gaussian beam is incident on a thin lens with a focal length f from the left (assuming light propagation from left to right). The Gaussian beam is at its minimum spot size, w_{01}, at the left surface of the thin lens. The goal is to determine the focal position L and the minimum spot size w_{03} of the Gaussian beam emerging from the thin lens. For numerical implementation, assume the following parameters: The Gaussian beam is produced by a He-Ne laser with a free-space wavelength of $\lambda_0 = 0.6328\,\mu$m and output power of $P = 10\,$mW (which is a typical value for a He-Ne laser). The minimum beam waist of the incident Gaussian beam is $w_{01} = 1\,$cm (indicating it is a well-collimated beam incident on the thin lens). The thin lens has a focal length of $f = 10\,$cm.

Solution: Initially the $ABCD$ elements of the thin lens are $A = 1$, $B = 0$, $C = -1/f$, and $D = 1$. Since the incident Gaussian beam is focused on the left side of the thin lens, the complex radius of curvature q_1 is given by $1/q_1 = -j\lambda_0/\pi n w_{01}^2$ (where $R_1 = R(z = 0) = \infty$). Applying Eq. (3.469) the complex radius of curvature q_2 on the right surface of the thin lens is given by:

$$\frac{1}{q_2} = \frac{C + D(1/q_1)}{A + B(1/q_1)} = -\frac{1}{f} + \frac{1}{q_1} = -\frac{1}{f} - j\frac{\lambda_0}{\pi n w_{01}^2} = a + jb,$$

where $a = -1/f$ and $b = -\lambda_0/\pi n w_{01}^2 = -1/z_{01}$, with z_{01} the Rayleigh distance of the Gaussian beam incident from the left. From the right surface of the thin lens, the emerging Gaussian beam propagates through a homogeneous medium for a distance L (which is still unknown), where it is expected to focus. In a homogeneous medium, the corresponding $ABCD$ matrix elements are $A = 1$, $B = L$, $C = 0$, and $D = 1$. Therefore, applying Eq.

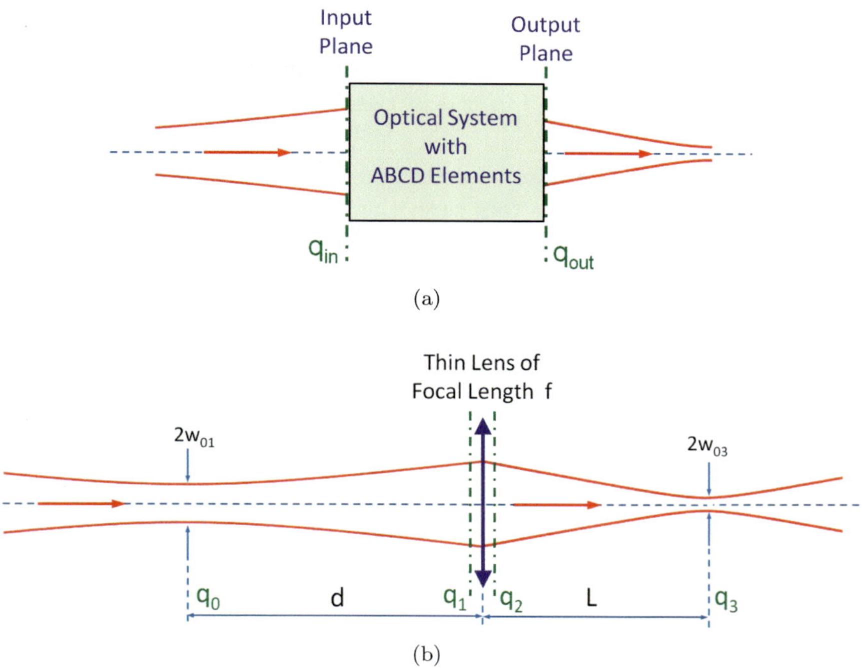

FIGURE 3.100: (a) A Gaussian beam with a complex radius of curvature of q_{in} at the input plane of an optical system that can be described with the paraxial $ABCD$ matrix. The Gaussian beam exiting the optical system has an output complex radius of curvature of q_{out}. (b) An example optical system comprised of a thin lens of focal length f. The Gaussian beam incident from the left is at its minimum waist, w_{01}, on the left surface of the thin lens.

(3.469) once more, it can be shown that:

$$\frac{1}{q_3} = \frac{C + D(1/q_2)}{A + B(1/q_2)} = \frac{1/q_2}{1 + L/q_2} = \frac{a + jb}{(1 + aL) + jbL} =$$
$$= \frac{a(1 + aL) + Lb^2}{(1 + aL)^2 + L^2b^2} - j\frac{-b}{(1 + aL)^2 + L^2b^2} =$$
$$= \frac{1}{R_3} - j\frac{\lambda_0}{\pi n w_{03}^2}. \tag{3.470}$$

However, at $z = L$ the emerging Gaussian beam is expected to be focused. Therefore, the radius of curvature at this point, R_3 is infinite, meaning $1/R_3 = 0$. As a result, the real part of the equation for $1/q_3$ should be zero. Solving this equation for L gives:

$$a(1 + aL) + Lb^2 = 0 \quad \Longrightarrow L = \frac{-a}{a^2 + b^2} \quad \Longrightarrow$$
$$L = \frac{f}{1 + \left(\dfrac{f}{z_{01}}\right)^2}. \tag{3.471}$$

Once L is determined, the minimum spot size w_{03} of the Gaussian beam at the focal point can be found from the imaginary part of Eq. (3.470). After performing the necessary algebra,

w_{03} is given by:

$$w_{03} = w_{01} \frac{f/z_{01}}{\left[1 + \left(\frac{f}{z_{01}}\right)^2\right]^{1/2}}. \tag{3.472}$$

The complex curvature $1/q_3$ can also be determined directly from $1/q_1$ by using the properties of the $ABCD$ matrices. For this case, the combined $ABCD$ matrix elements for the thin lens and the homogeneous propagation that should be used in Eq. (3.469) are $A = 1 - L/f$, $B = L$, $C = -1/f$, and $D = 1$. At this point it is informative to apply the numerical values given. The Rayleigh distance of the incident Gaussian beam is $z_{01} = \pi n w_{01}^2/\lambda_0 = 496.459\,m \simeq 0.5\,\text{km}$! (it is also assumed that $n = 1$, i.e., the beam propagates in air). The focal distance L is easily found from Eq. (3.471) and is equal to $L \simeq 10\,cm = f$. Therefore, the result is the one expected for a thin lens under the paraxial geometrical optics approximation. This happens due to the well-collimated and relatively wide incident Gaussian beam. The minimum beam waist of the emerging from the thin lens Gaussian beam is given by Eq. (3.472) and is $w_{03} = 2.0143\,\mu\text{m}$! Despite the small focused beam waist, the Gaussian beam solution remains valid and reliable since the paraxiality parameter $\mathcal{P} = 1 - 1/k_0^2 n w_{03}^2 \simeq 0.9975$, is still very close to unity [175]. Finally, the power density at the focal plane is $I_3 \simeq P/(\pi w_{03}^2) = 7.8454 \times 10^8\,W/m^2 = 78.454\,KW/cm^2$. This significant power density, despite the moderate power of the laser, demonstrates how an optical system (in this case, a simple thin lens) can focus a laser beam to a very small area, which can be used to achieve nonlinear effects such as cutting a target.

3.11.10 Two-Dimensional Gaussian Beams

A two-dimensional Gaussian beam is often used to model beams in optical systems with cylindrical symmetry, such as those passing through cylindrical lenses or beams propagating in the plane of a single axis (e.g., in laser beam shaping systems). A two-dimensional Gaussian beam is a simplification of the full three-dimensional Gaussian beam, focusing only on the variation of the beam intensity in one transverse direction (say, the x-direction), while assuming the beam does not vary in the other transverse direction (the y-direction) [176].

Two-dimensional Gaussian beams are crucial in situations where the optical system exhibits cylindrical symmetry or when the beam's behavior in one transverse direction is of primary interest. While three-dimensional Gaussian beams are ideal for modeling full 3D beam propagation, the 2D Gaussian beam simplifies the analysis, making it more computationally efficient and physically relevant for many real-world applications in optics, especially in systems where variations in the transverse direction are minimal or constrained.

The analysis is very similar to that of the three-dimensional case. Assume the beam intensity varies only along the x-direction and is uniform along the y-direction. The 2D paraxial wave equation can be derived from the 3D counterpart of Eq. (3.446) and expressed as:

$$\frac{\partial^2 \psi}{\partial x^2} - j2k\frac{\partial \psi}{\partial z} \simeq 0, \quad \text{if} \quad \left|\frac{\partial^2 \psi}{\partial z^2}\right| \ll 2k \left|\frac{\partial \psi}{\partial z}\right|, \tag{3.473}$$

where $y(x, z)$ is the slowly varying amplitude of the beam as in Eq. (3.443). Looking again for the fundamental solution of the form of Eq. (3.448), it is straightforward to find that the following equations should be satisfied:

$$\frac{dq}{dz} = q'(z) = 1, \tag{3.474}$$

$$\frac{dP}{dz} = P'(z) = -\frac{j}{2q(z)}. \tag{3.475}$$

The solutions of the last equations are indeed very similar to those in the 3D case, with the primary difference being the dimensionality of the transverse variation. The resulting fundamental solution for the 2D Gaussian beam can be written as:

$$E_0(x,z) = \mathcal{E}_0 \sqrt{\frac{w_0}{w(z)}} \exp\left[-\frac{x^2}{w^2(z)}\right] \exp\left\{-j\left[kz - \frac{1}{2}\tan^{-1}\left(\frac{z}{z_0}\right)\right]\right\} \exp\left[-j\frac{kx^2}{2R(z)}\right], \tag{3.476}$$

where the complex radius of curvature $q(z)$, the beam waist $w(z)$, the minimum beam waist w_0, and the wavefront radius of curvature $R(z)$ are given by the same equations as in the 3D case. This 2D Gaussian beam solution represents the spatial and phase evolution of the electric field in the paraxial regime, where the intensity distribution is confined to one transverse dimension (x) and is uniform along the other transverse dimension (y). Higher-order solutions, analogous to the 3D Hermite-Gaussian modes discussed in Sec. 3.11.6, can also be derived following similar principles [176].

Summarizing, the 2D Gaussian beam model simplifies optical analysis by focusing on beam behavior in one transverse direction, making it ideal for systems with cylindrical symmetry or applications like laser fiber coupling, slit illumination, and line scanning. Neglecting variations in the second transverse direction allows for efficient analysis and design of systems where one-dimensional beam shaping, such as in cylindrical lenses or micromachining, is the primary concern. This simplification retains essential propagation characteristics while streamlining the optimization of optical systems for specialized tasks.

3.12 Problems

1 **Reflected and Transmitted Powers:** Sellmeier's formula describes the refractive index of a material $n(\lambda_0)$ as a function of the freespace wavelength λ_0. It is expressed as:

$$n^2(\lambda_0) = 1 + \sum_i \frac{B_i \lambda_0^2}{\lambda_0^2 - C_i}$$

where for fused silica (quartz), the Sellmeier's coefficients are: $B_1 = 0.696166300$, $B_2 = 0.407942600$, $B_3 = 0.897479400$, $C_1 = 4.67914826 \times 10^{-3}\,\mu m^2$, $C_2 = 1.35120631 \times 10^{-2}\,\mu m^2$, and $C_3 = 97.9340025\,\mu m^2$. Consider light from a monochromator incident at an angle θ (from air) on a planar and smooth surface of fused silica. The polarization of the incident wave may be either *TE* or *TM*. (a) For an incident wave with a freespace wavelength $\lambda_0 = 633\,\text{nm}$, generate graphical representations of the reflection and transmission coefficients, as well as the percentages of reflected and transmitted powers, as functions of the angle of incidence ($0 \le \theta \le 90\,\text{deg}$) for both polarizations. (b) For an incident angle fixed at $\theta = 30\,deg$, produce graphical representations of the reflection and transmission coefficients, as well as the percentages of reflected and transmitted powers, as functions of the freespace wavelength ($0.1\,\mu m \le \lambda_0 \le 2.0\,\mu m$) for both polarizations.

2 **Plane Waves in Anisotropic Media:** Consider a plane wave propagating in a linear, homogeneous, non-magnetic, and anisotropic medium characterized by a permittivity tensor $\tilde{\epsilon}$. (a) Show that the wavevector $\vec{k}$ always points in the direction of $\vec{D} \times \vec{B}$. (b) Prove that the magnitude of the wavevector is given by: $\vec{k} \cdot \vec{k} = \omega^2 \mu_0 [(\vec{D} \cdot \vec{D})/(\vec{D} \cdot \vec{E})]$. (c) Demonstrate that the complex Poynting vector, $\vec{S} = (1/2)\vec{E} \times \vec{H}^*$, can point in a direction different from that of the wavevector $\vec{k}$.

3 **Index Ellipsoid:** Assume a plane-wave with wavevector $\vec{k}$ propagating in a linear, homogeneous, non-magnetic, and anisotropic medium with permittivity $\tilde{\epsilon}$. The wavevector can be expressed as:

$\vec{k} = k_0 n\hat{k} = k_0 n(a_x\hat{\imath}_x + a_y\hat{\imath}_y + a_z\hat{\imath}_z)$, where k_0 is the freespace wavenumber, n is the refractive index along the direction of propagation of $\vec{k}$, and $\hat{k}$ is the unit vector in the propagation direction (with a_x, a_y, and a_z as directional cosines). The permittivity is $\tilde{\epsilon} = \epsilon_0\tilde{\varepsilon}$ where $\tilde{\varepsilon}$ is the relative permittivity tensor. Let $[\mathcal{A}] = \tilde{\varepsilon}^{-1}$ denote the inverse relative permittivity tensor (impermeability tensor). For numerical implementation, use the following parameters: $n_{xx} = 1.552$, $n_{yy} = 1.582$, $n_{zz} = 1.588$, and $a_x = a_y = a_z = 1/\sqrt{3}$. (a) Show that the displacement vector $\vec{D}$ satisfies the eigenvalue/eigenvector equation:

$$\hat{k} \times \left[\hat{k} \times \left([\mathcal{A}]\vec{D}\right)\right] + \frac{1}{n^2}\vec{D} = 0.$$

Solve this equation in the principal axes system xyz shown in Fig. 3.9. (b) Since $\vec{D}$ is always perpendicular to *veck* it is convenient to use a new coordinate system with one axis coinciding with the direction of propagation of the wave (assume the unit vector in this direction as $\hat{\imath}_3 = \hat{k}$). Then, denote the two new transverse axes by 1 and 2, with unit vectors $\hat{\imath}_1$ and $\hat{\imath}_2$, respectively. In this new coordinate system $\hat{k} = [0\ 0\ 1]^T = \hat{\imath}_3$. Furthermore, assume that $\vec{D}$ and $[\mathcal{A}]$ are now expressed in this new coordinate system. Since $\vec{k} \cdot \vec{D} = 0$, the displacement vector can be expressed as $\vec{D} = [D_1\ D_2\ 0]^T = D_1\hat{\imath}_1 + D_2\hat{\imath}_2$. Show that the eigenvalue/eigenvector equation can be written as:

$$\begin{bmatrix} \alpha_{11} & \alpha_{12} \\ \alpha_{21} & \alpha_{22} \end{bmatrix} \vec{D}_t = [\mathcal{A}_t]\vec{D}_t = \frac{1}{n^2}\vec{D}_t,$$

where $[\mathcal{A}_t]$ is the transverse impermeability tensor (symmetric) and $\vec{D}_t = [D_1\ D_2]^T$. The above equation is again an eigenvalue/eigenvector equation in two dimensions. The two eigenvalues and the two eigenvectors correspond to the two refractive indices for the direction of propagation and their corresponding eigen-polarizations. Find the two eigen-indices and the two eigen-polarizations. Show that the two eigen-polarizations are orthogonal to each other. (c) Now show that the ellipse which is defined from the index ellipsoid as an intersection of the index ellipsoid with a plane perpendicular to $\hat{k}$ and passing through the origin has semi-axes with lengths equal to the squares of the eigen-indices. In addition, show that the directions of the two semi-axes correspond to the two eigen-polarizations.

4 **Newton Rings:** Newton's rings are formed due to interference between the partially reflected light beams (beams 1 and 2) between a plano-convex lens and a flat planar optical surface, as shown in Fig. 3P.1. Given the following parameters: Diameter of the 10th bright ring is 7.89 mm freespace wavelength of the incident light is 546.1 nm, and refractive index of the lens is $n = 1.50$. Determine: (a) The radius of curvature R of the spherical surface of the lens. (b) The focal distance of the lens.

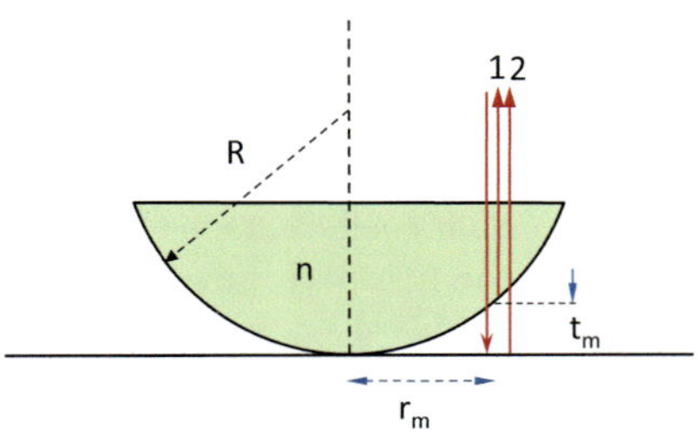

FIGURE 3P.1: A spherical plano-convex lens of radius R and refractive index n producing Newton interference rings.

5 **Lloyd's Mirror**: Lloyd's mirror experiment creates interference patterns similar to those observed in Young's double-slit experiment (refer to Sec. 3.5.4 and Fig. 3.37(a)). The setup includes a monochromatic point source S emitting rays, some of which reach point A directly, while others reflect off a mirror before interfering at A. The surrounding medium is air.

(a) Determine the equations for the positions of interference maxima (constructive interference) and minima (destructive interference) in 3D space. Express these as equations $f(x, y, z) = 0$ in the xyz coordinate system. (b) In the xz plane ($y = 0$), derive the loci of points where interference maxima and minima occur. For numerical implementation, use the parameters $D = 5\,\mu m$ (the vertical separation between the source point S and the mirror) and $\lambda_0 = 1\,\mu m$. Graphically represent the loci for the region $-20\,\mu m < z < 20\,\mu m$ and $0\,\mu m < x < 15\,\mu m$. (c) At the fixed plane $z = L = 20\,\mu m$, determine the exact and approximate x-coordinates of the first 5 maxima (constructive interference) and the first 5 minima (destructive interference) in the xz-plane. Compare the numerical values obtained for the two cases.

6 **Spherical and Plane Wave Interference**: The interference of a spherical wave and a plane wave in free space results in a pattern of maxima and minima. The spherical wave originates at the origin $(0, 0, 0)$ (see Fig. 3P.2) and has the form $E_{sw} = (E_0/r)\exp(-jk_0 r)$, where k_0 is the wavenumber of freespace and $r = (x^2+y^2+z^2)^{1/2}$. The plane wave propagates with a wavevector $\vec{k} = k_0(\cos\theta\hat{i}_x + \sin\theta\hat{i}_y)$, and is expressed as $E_{pw} = E_0\exp[-jk_0(x\cos\theta + y\sin\theta)]$. Both waves are assumed to have the same wavelength λ_0 and collinear polarization.

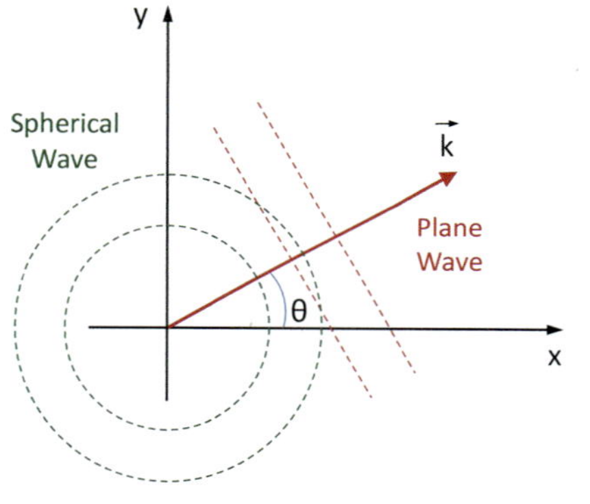

FIGURE 3P.2: The interference of a spherical wave and a plane wave at an angle θ in the xyz coordinate system.

(a) Determine equations that define the positions of interference maxima (constructive interference) and minima (destructive interference) in the 3D xyz coordinate system. Express these as equations of the form $f(x, y, z) = 0$. (b) For the xy plane ($z = 0$), derive the loci of points where interference maxima and minima occur. Assume the freespace wavelength $\lambda_0 = 1\,\mu m$ and the angle $\theta = 30\,deg$. Graphically represent the loci of maxima and minima for the region $-20\,\mu m \leq x \leq 20\,\mu m$ and $-20\,\mu m \leq y \leq 20\,\mu m$. Include loci for several values of the optical path difference. (c) For the line $x = 20\,\mu m$ in the xy-plane ($z = 0$), determine the points along the y direction where maxima (constructive interference) and minima (destructive interference) occur. Use the parameters from part (b).

7 **Young's Double-Slit Experiment**: In a Young's Double-Slit Experiment setup (see Sec. 3.5.3 and Fig. 3.32), white spatially coherent light (wavelength range $380\,nm - 780\,nm$) is directed through two slits with a separation of $d = 0.5\,mm$. The observation screen is located $L = 50\,cm$ away from the slits. A hole in the screen is positioned $1\,mm$ away from the central line.
(a) Which wavelengths will be absent in the light coming from the hole? (b) Which wavelengths will have strong intensity in the light coming from the hole?

8 **Three-Plane-Wave Interference**: Three coherent plane waves of the same frequency interfere inside a homogeneous, isotropic, and linear medium. This problem is based on information included in the journal paper: J. L. Stay and T. K. Gaylord, "Three-beam-interference lithography: contrast and crystallography," Appl. Opt. vol. 47, no. 18, pp. 3221–3230, Jun. 20, 2008. The wavevectors of the three plane waves are $\vec{k}_i$ (where $i = 1, 2, 3$). For example, one plane wave is

shown in Fig. 3P.3. Every plane wave can be written in the following phasor form:

$$\vec{E}_i = E_i \hat{e}_i \exp\left[-j\left(\vec{k}_i \cdot \vec{r} + \phi_{0i}\right)\right]$$

where $\vec{r}$ is the position vector, E_i is the amplitude of plane wave i (assume that these are real numbers), ϕ_{0i} is a phase, and $\hat{e}_i$ is the polarization vector of a plane wave i (assume that these vectors are real so that all plane waves are linearly polarized).

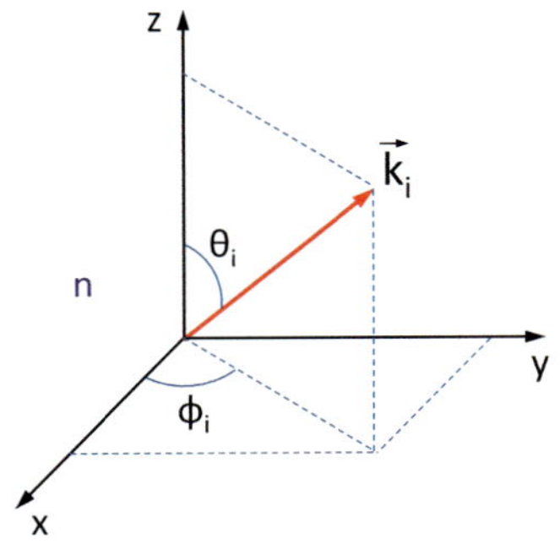

FIGURE 3P.3: A single plane wave with wavevector $\vec{k}_i$ expressed in the xyz coordinate system.

(a) Find the intensity, $I(x, y, z) = |\vec{E}|^2$, resulting from the interference of the three plane waves in space (x, y, z). (b) The interference is characterized by three terms that correspond to the interference between each pair of waves. Derive the equations for the maximum interference of each pair (waves 1 and 2, waves 1 and 3, and waves 2 and 3). Show that the equation for the third term could result from the equations of the first two. Therefore, the maximum total interference of the three plane waves is comprised of lines that are the intersections of the equations (which represent planes) of the maximum interference of the two terms discussed previously. If the lines of maximum interference are required to be parallel to the z-axis, find the condition that the three wavevectors should satisfy. (c) Assume the three plane waves are interfering in the air with the refractive index $n = 1$ and freespace wavelength $\lambda_0 = 1\,\mu m$. The following wavevectors and polarization vectors are given ($k_0 = 2\pi/\lambda_0$):

$$\vec{k}_1 = k_0\left(-\frac{1}{3}\hat{\imath}_x - \frac{\sqrt{3}}{3}\hat{\imath}_y - \frac{\sqrt{5}}{3}\hat{\imath}_z\right), \qquad \vec{k}_2 = k_0\left(-\frac{2}{3}\hat{\imath}_x - 0\hat{\imath}_y - \frac{\sqrt{5}}{3}\hat{\imath}_z\right),$$

$$\vec{k}_3 = k_0\left(-\frac{1}{3}\hat{\imath}_x + \frac{\sqrt{3}}{3}\hat{\imath}_y - \frac{\sqrt{5}}{3}\hat{\imath}_z\right),$$

$$\hat{e}_1 = -\frac{\sqrt{3}}{2}\hat{\imath}_x - \frac{1}{2}\hat{\imath}_y - 0\hat{\imath}_z, \qquad \hat{e}_2 = +0.2821\hat{\imath}_x - 0.9256\hat{\imath}_y + 0.2523\hat{\imath}_z,$$

$$\hat{e}_3 = +0.9426\hat{\imath}_x + 0.2185\hat{\imath}_y - 0.2523\hat{\imath}_z,$$

$$E_1 = 1, \qquad E_2 = E_3 = \frac{\sqrt{2}}{2}, \qquad \phi_{01} = \phi_{02} = \phi_{02} = 0.$$

Make a graphical representation of the intensity $I = |\vec{E}|^2$ in the xy plane. Then, calculate numerically the visibility of the fringes produced, $V = (I_{max} - I_{min})/(I_{max} + I_{min})$, as well as the angles θ_i and ϕ_i of the three plane waves and verify that their polarizations are compatible with plane waves.

9 **Interference Filter**: In some application, it is desired to use a one-half wavelength thick interference filter [as the etalon shown in Fig. 3P.4 with $d = \lambda/2$ or $d = (2m+1)\lambda/2$ where m is an integer]. The task is to calculate the angular tuning rate of the one-half wavelength thick interference filter, in a setup where the filter is designed to scan the wavelength range around a central wavelength λ_c. The interference filter has a specified resonant wavelength λ_s where $\lambda_s > \lambda_c$. The half wavelength layer in the interference filter has an index of refraction of n. If

the interference filter is designed for $\lambda_s = 520\,nm$ and has $n = 1.4$, calculate the angular tuning rate for the argon laser wavelength $\lambda_c = 514.5\,nm$.

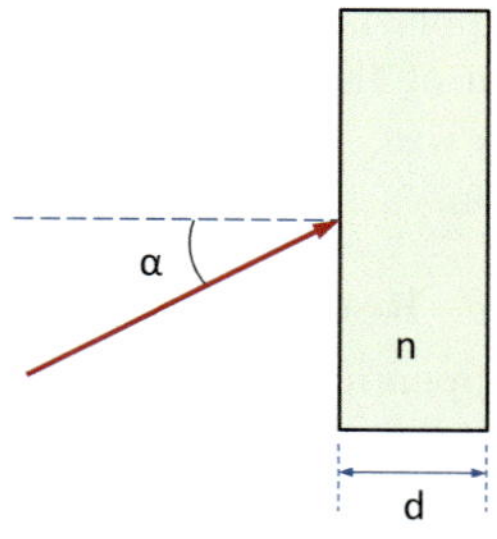

FIGURE 3P.4: A simple etalon-based interference filter of refractive index n and thickness d.

10 Fabry-Perot Interference Filter: A beam of white light (which has a continuous spectrum between 400 nm to 700 nm) is incident at an angle of 45 deg onto a set of parallel glass plates separated by a thin air layer with a thickness of 0.001 cm. The reflected light passes through a prism spectrometer. How many and which dark lines will be seen in the spectrometer?

11 Fabry-Perot Interference Filter: A Fabry-Perot interferometer is comprised of a transparent plate of refractive index $n = 4.50$ and thickness of 2 cm. The parallel surfaces of the plate have a power reflectivity of ($R = r^2$) of 90%. If this interferometer is used in the range of freespace wavelength of 546 nm determine: (a) the highest order bright fringe (largest value of integer m) that can be observed, (b) the ratio of maximum to minimum transmission T_{max}/T_{min}, and (c) the resolving power of the interferometer.

12 Single Antireflection Layer: In this problem the objective is to design an anti-reflective (AR) layer between two materials with refractive indices n_1 and n_2, respectively (see Fig. 3P.5. The incident plane wave has a freespace wavelength of λ_d and the light is incident at an angle θ_1. From the electromagnetic analysis of this problem, it is recommended to use the transmission line theory as it was presented in Appendix 3G. For the numerical implementation, assume that $\lambda_d = 500\,nm$ (design wavelength), $n_1 = 1.0$, and $n_2 = 2.0$.

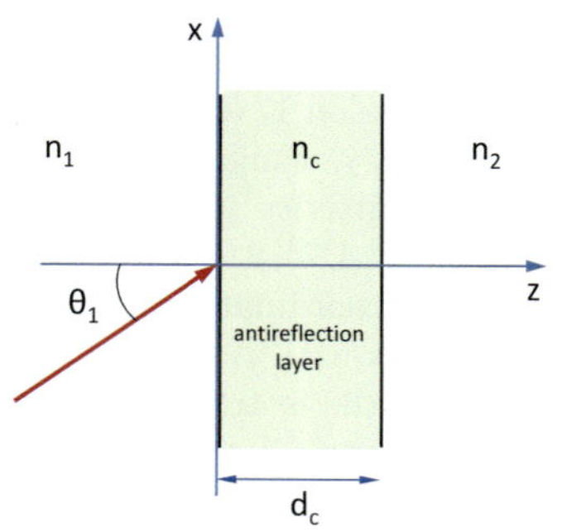

FIGURE 3P.5: A single layer antireflection coating.

(a) Initially address the design of the antireflection layer for normal incidence, i.e., for $\theta_1 = 0\,\text{deg}$. Determine the minimum thickness d_c as well as the refractive index n_c of the single antireflection layer. Make a graphical representation of the reflected power as a function of the freespace wavelength in the range $300\,nm \leq \lambda_0 \leq 700\,nm$, when the antireflection layer is designed for λ_d. Furthermore, make a graphical representation of the reflected power as a function of the angle of incidence θ_1 ($-90\,\text{deg} \leq \theta_1 \leq +90\,\text{deg}$) when the designed wavelength is used. (b) Then, tackle the design of the antireflection layer for an angle of incidence $\theta_1 = 45\,deg$. Determine the minimum thickness d_c as well as the refractive index n_c of the antireflection layer for either *TE* or *TM* polarization. Make a graphical representation of the reflected power as a function of the freespace wavelength in the range $300\,nm \le \lambda_0 \le 700\,nm$ when the antireflection layer was designed for λ_d (separate for each polarization). Furthermore, make a graphical representation of the reflected power as a function of the angle of incidence θ_1 ($-90\,\text{deg} \le \theta_1 \le +90\,\text{deg}$) for both polarizations, when the design was done for *TE* polarization and correspondingly when the design was done for *TM* polarization.

13 Relative Permittivity Tensor Rotations: Prove that the matrices $\tilde{\boldsymbol{R}}(\psi_1, \psi_2)$ and $\tilde{\boldsymbol{M}}^T(\theta, \phi, \psi)$ are as presented in Appendix 3H, Sec. 3H.3. Furthermore, prove that the relative permittivity $\tilde{\varepsilon}$ in the (x, y, z) coordinate system is related to the diagonal relative permittivity $\tilde{\varepsilon}_0$ in the principal axes system, by Eqs. (3H.5) and (3H.6).

14 Isotropic-Biaxial Planar Interface: Consider an isotropic-biaxial interface between a dielectric $\tilde{\varepsilon}_1 = 6\tilde{I}$, and an anisotropic medium with $\tilde{\varepsilon}_2 = diag(4, 4, 3)$. The principal axes system of the anisotropic medium is rotated with respect to the xyz coordinate system using the angles $\psi_1 = 30°$ and $\psi_2 = 0°$ (see Appendix 3H). A plane wave of freespace wavelength $\lambda_0 = 1\,\mu\text{m}$ is incident from the isotropic to the biaxial medium. The polarization of the incident wave is either *TE* or *TM*. Find the absolute values of the coefficients r_{TE}, r_{TM}, t_1, and t_2 as functions of the angle of incidence θ_{inc} (with $\phi_{inc} = 60°$) for both polarizations. Furthermore, calculate and plot the normalized powers related to the reflected and the transmitted waves.

15 Biaxial-Isotropic Planar Interface: Consider a biaxial-isotropic interface between an anisotropic medium (region 1) with relative permittivity $\tilde{\varepsilon}_1$ and an isotropic region (region 2) with relative permittivity $\tilde{\varepsilon}_2 = n_2^2\tilde{I}$. The analysis of this problem follows the approach outlined in Sec. 3.8.2.1. In this case the electric field vector in the regions 1 and 2 can be expressed by:

$$\vec{E}_1 = E_0\hat{e}_{ie}\exp(-j\vec{k}_{inc}\cdot\vec{r}) + r_1\hat{e}_{r1}\exp(-j\vec{k}_{r1}\cdot\vec{r}) + r_2\hat{e}_{r2}\exp(-j\vec{k}_{r2}\cdot\vec{r}),$$

$$\vec{E}_2 = \left(t_{TE}\hat{u}^t_{TE} + t_{TM}\hat{u}^t_{TM}\right)\exp(-j\vec{k}_t\cdot\vec{r}),$$

where r_1 and r_2 are the reflection coefficients for the two possible eigen-polarizations $\hat{e}_{r1}$ and $\hat{e}_{r2}$, respectively. The coefficients t_{TE} and t_{TM} are the transmission coefficients into the isotropic region for *TE* and *TM* polarizations, respectively. The amplitude E_0 can be set equal to unity. Formulate the problem to address the solution for the reflection and transmission coefficients, r_1, r_2, t_{TE}, and t_{TM}, in a manner analogous to the one used in Sec. 3.8.2.1.

16 Biaxial-Isotropic Planar Interface (Numerical Case): Consider an interface between a biaxial-isotropic anisotropic medium (region 1) and an isotropic medium (region 2). The relative permittivity of region 1 is denoted by $\tilde{\varepsilon}_1$, while that of region 2 is given by $\tilde{\varepsilon}_2 = n_2^2\tilde{I}$, where $n_2 = 1$. In region 1, the relative permittivity in the principal axes system of the anisotropic medium is represented by $\tilde{\varepsilon}_0 = diag(1.2^2, 1.7^2, 2.2^2)$. The principal axes system of the anisotropic medium is rotated relative to the xyz coordinate system using Euler angles $\theta = 75°$, $\phi = 75°$ and $\psi_2 = -75°$ [148] (refer to Appendix 3H). A plane wave of freespace wavelength $\lambda_0 = 1\,\mu\text{m}$ is incident from the biaxial medium to the isotropic medium. The incident wave is classified as either "outer" or "inner." The goal is to find the absolute values of the reflection coefficients r_1, and r_2, as well as the total transmission coefficient $t = (|t_{TE}|^2 + |t_{TM}|^2)^{1/2}$, as functions of the angle of incidence $-25° \le \theta_{inc} \le 25°$ (with $\phi_{inc} = 0°$ or $180°$ as in Example 3.8.2), for both incident waves. Additionally, calculate and plot the normalized powers associated with the reflected and transmitted waves.

17 Magneto-optic Layer: This problem investigates the reflection properties of a magneto-optic film sandwiched between glass layers. The study is based on the work presented in Refs. [177] and [178]. Consider a $TbFe$ magneto-optic layer of thickness d, with glass superstrate and substrate, each having a refractive index of $n = 1.5$. The relative permittivity tensor of the magneto-optic layer is specified by the following elements: $\varepsilon_{xx} = \varepsilon_{yy} = \varepsilon_{zz} = -2.6 - j20$, $\varepsilon_{xy} = \varepsilon_{yx}^* = -0.04 - j0.33$, and $\varepsilon_{xz} = \varepsilon_{zx} = \varepsilon_{yz} = \varepsilon_{zy} = 0$. The freespace wavelength of the incident plane wave is given by $\lambda_0 = 0.83\,\mu$m. The task is to find the magnitudes of the reflection coefficients as a function of the angle of incidence θ_{inc} (with $\phi_{inc} = 0$) for both *TE* and *TM* incident polarization. Analyze the cases for $d = 0.01\,\mu$m, and $d = 0.10\,\mu$m.

18 Quarter-Wave Plate (Jones Calculus): A linearly polarized electromagnetic wave is incident normally at at $z = 0$ on the xy-face of a crystal, with propagation along its z-axis. The xyz coordinate system is aligned with the principal axes of the crystal. The corresponding permittivity tensor is diagonal, with elements ϵ_{xx}, ϵ_{yy}, and ϵ_{zz}. The wave is initially polarized such that it has equal components along the x and y axes. The task is to determine the state of polarization of the wave at the plane $z = z_0$, where

$$k_0(n_{xx} - n_{yy})z_0 = \frac{\pi}{2},$$

with $\epsilon_{ww} = \epsilon_0 n_{ww}^2$ $(w = x, y, z)$. Assume that $n_{xx} > n_{yy}$, with $n_{xx} = n_s$ (the slow axis along x direction), and $n_{yy} = n_f$ (the fast axis along the y direction).

19 Polarization Transformation by a Wave Plate (Jones Calculus): A wave plate is characterized by its phase retardation Γ and the azimuth angle ψ. (a) Find the polarization state of the emerging beam, assuming the incident beam is polarized along the x direction. Represent the resulting polarization state using a complex number. Specifically, define the ratio $p = E_y/E_x = re^{j\varphi}$, where E_x and E_y are the components of the electric field in the x- and y-directions. (b) The polarization state of the incident x-polarized beam is represented by a point at the origin of the complex plane. Show that the transformed polarization state can be anywhere in the complex plane, provided Γ can be varied from 0 to 2π and ψ can be varied from 0 to $\pi/2$. Physically, this implies that any polarization state can be produced from linearly polarized light, provided a suitable wave plate is available. To demonstrate this, find the wave plate parameters Γ and ψ for any given r and φ. For a numerical example, calculate the values of Γ and ψ for the (r, φ) pairs $(1, \pi/2)$, $(1, -\pi/2)$, $(\infty, 0)$, and $(2, -\pi/2)$. Additionally, plot r and φ as a function of Γ and ψ. (c) Show that the Jones matrix W of a wave plate is unitary, i.e., $W^\dagger W = 1$, where $W^\dagger = (W^T)^*$, with T denoting transposition and $*$ denoting complex conjugate. (d) Let $\vec{V}_1'$ and $\vec{V}_2'$ be the transformed Jones vectors of $\vec{V}_1$ and $\vec{V}_2$, respectively. Show that if $\vec{V}_1$ and $\vec{V}_2$ are orthogonal, so are $\vec{V}_1'$ and $\vec{V}_2'$ (vectors $\vec{A}$ and $\vec{B}$ are orthogonal if $\vec{A}\cdot\vec{B}^* = 0$).

20 Coherence Time: Find the coherence time τ_c of a Lorentzian power spectrum source with normalized spectrum is given by:

$$S(\nu) = \frac{\Delta\nu}{2\pi} \frac{1}{(\nu - \bar{\nu})^2 + (\Delta\nu/2)^2},$$

where $\bar{\nu}$ is the central frequency, and $\Delta\nu$ is the bandwidth of the source. Determine the self-correlation function $\gamma(\tau)$ of this source. Next, repeat the same questions for a source with a rectangular power spectrum, normalized as follows:

$$S(\nu) = \frac{1}{\Delta\nu}\ \mathrm{rect}\left(\frac{\nu - \bar{\nu}}{\Delta\nu}\right),$$

where $\mathrm{rect}(x) = 1$ for $|x| \leq 1$ and zero otherwise.

21 Coherence Area: Find the coherence area of the light from the red giant Betelgeuse that arrives on Earth. Assume that Betelgeuse has an approximate blackbody temperature of 3500° K. The radius of Betelgeuse is approximately $764R_{sun}$ where the Sun's radius is $R_{sun} = 696000$ km. The distance of Betelguese from Earth is approximately 548 light years.

22 Mutual Intensity: The mutual intensity of a light source at an observation xy-plane is given by:

$$J_{12}(x_1, y_1; x_2, y_2) = I_0 \exp\left[-\frac{(x_1^2 + y_1^2)}{W^2}\right] \exp\left[-\frac{(x_2^2 + y_2^2)}{W^2}\right] \exp\left[-\frac{(x_1 - x_2)^2}{r_c^2}\right] \exp\left[-\frac{(y_1 - y_2)^2}{r_c^2}\right],$$

where W and r_c are known constants. (a) Assume that we are looking at points either on the x-axis or on the y-axis. In this case, make a 2D plot of $J_{12}(x_1, x_2)$ (if points on the x-axis are of interest) or $J_{12}(y_1, y_2)$ (if points on the y-axis are of interest). Plot the function $J_{12}(x_1, x_2)$ or $J_{12}(y_1, y_2)$ on the x_1x_2 or y_1y_2 plane and comment on the parameters W and r_c. For the numerical implementation of the plots, use the cases $(W, r_c) = (2, 0.5)$, $(W, r_c) = (2, 1)$, $(W, r_c) = (2, 2)$, $(W, r_c) = (2, 4)$, and $(W, r_c) = (4, 0.5)$. (b) Find the complex degree of coherence j_{12}. (c) Calculate the coherence area A_c of this source on the observation plane.

23 Multiple Incoherent Layers: Consider a three-layer ($N = 3$) incoherent multilayer structure surrounded by air ($n_0 = 1$ and $n_4 = 1$). The refractive indices and thicknesses of the layers are as follows: $n_1 = 1.5$ (SiO_2) with thickness $d_1 = 0.5\,\mu$m, for layer 1, $n_2 = 3.6$ (Si) and thickness $d_2 = 1.0\mu$m, for layer 2, Assume that all layers are lossless, rendering their thicknesses irrelevant in the incoherent analysis. The freespace wavelength of the incident light is $\lambda_0 = 1\,\mu$m. The tasks are as follows: (a) Calculate the coherent and incoherent normalized intensities as functions of the angle of incidence. (b) Verify the incoherent results using Eqs. (3.420) and (3.421) for $M = 100$ and $M = 1000$ random phase samples. (c) Test the validity of the results with Eqs. (3.424) and (3.425). Assume both *TE* and *TM* polarizations. (d) Repeat parts (a) to (c) when the layer thicknesses are multiplied by a factor of 100. Discuss and interpret the observed results.

24 Mixed Coherent-Incoherent Layers: This problem focuses on the analysis of organic solar cells (OSC), inspired by Ref. [164]. The OSC structure comprises a multilayer optical stack with $N = 5$ layers, each characterized by the following refractive indices: $n_1 = 1.51$m (glass layer), $n_2 = 2 - j0.01$ indium tin oxide (ITO) layer), $n_3 = 1.60 - j0.02$ (organic PEDOT:PSS layer), $n_4 = 1.99 - j0.28$ (butyric acid methyl ester P3HT:PCBM layer), and $n_5 = 2.5$ (spacer layer of TiO_x/Aluminum layer). The input region has a refractive index $n_0 = 1.0$ (air), and the output region consists of silver (Ag) with $n6 = 0.71 - j5.37$. The layer thicknesses are: $d_1 = 2000\,\mu$m, $d_2 = 0.05\,\mu$m, $d_3 = 0.12\,\mu$m, $d_3 = 0.1\,\mu$m, and $d_5 = 0.08\,\mu$m. The incident free-space wavelength is $\lambda_0 = 0.55\,\mu$m which corresponds to the average wavelength of white light. The first layer is treated as incoherent ($N_{inc} = 1$), while the remaining four layers are considered coherent ($N_c = 4$). The system is illuminated by either *TE* or *TM*-polarized light. Determine the normalized reflected and transmitted intensities as a function of the angle of incidence θ_0 using the following methods: (a) Mixed Coherent-Incoherent Modeling with the application of Eq. (3.431). (b) Use the coherent layer analysis to model the multilayer structure, incorporating the averaging of $M = 300$ random samples to account for the phase variations of the first incoherent layer. (c) Instead of random phase sampling, apply the equidistant phase averaging method to the incoherent layer 1, selecting phase angles from a discrete set of $N_s = 5$ as detailed in Sec. 3.10.3.2.

25 Gaussian Beam Focusing: A Gaussian beam is incident on a thin lens of focal length f where the lens is placed in an arbitrary position relative to the input beam. Specifically, denote with d the distance of the minimum waist, w_{01}, of the incident Gaussian beam from the lens. The freespace wavelength is λ_0, and the refractive index of the surrounding medium is n. (a) Calculate the location of the minimum waist of the output beam, L, and its minimum spot size, w_{03}. Assume that the minimum waist of the incident beam is to the left of the lens. (b) Make a graph of the location of the minimum waist, L, and of the minimum waist w_{03}, of the output beam with respect to d, for the following parameters: $d \in [10,\ 250]$mm, $f = 100\,$mm, $\lambda_0 = 1.0\,\mu$m, $w_{01} = 1\,$mm, and $n = 1$. (c) Draw a picture of the input and output Gaussian beam envelope [i.e., $\pm w(z)$] for the following parameters: $d = 15\,$mm, $f = 100\,$mm, $\lambda_0 = 1.0\,\mu$m, $w_{01} = 1\,$mm, and $n = 1$.

26 Focusing of a Gaussian beam on a target through a lens: A Gaussian beam with wavelength λ_0 is incident on a lens placed at distance $z = \ell$, as shown in Fig. 3P.6. Calculate the lens focal length, f, so that the output beam has its minimum waist at the front surface of the sample crystal, which is placed at distance L from the lens. Show that given ℓ and L, there are two solutions. Show graphs of the two solutions as a function of the input Gaussian beam minimum waist, w_0. For the numerical implementation, assume the following parameters: $\lambda_0 = 1\,\mu\text{m}$, $n = 1.0$, $\ell = 15\,\text{mm}$, $L = 100\,\text{mm}$, $w_0 \in [0.02,\ 10]\text{mm}$.

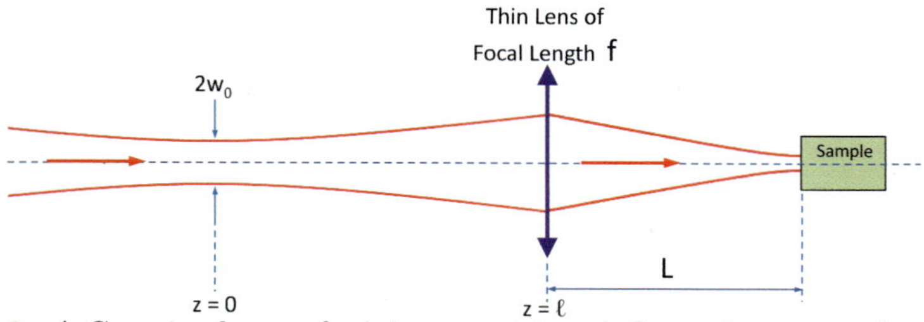

FIGURE 3P.6: A Gaussian beam of minimum waist w_0 is focused on a sample with the help of a thin lens of focal length f.

Appendices

3A Kramers-Kronig Relations

In this Appendix, a short proof of the Kramers-Kronig relations is presented. The proof is based on the causality principle and is adapted from Ref. [179]. From Eq. (3.21), the medium polarization should be nonzero only after the external field is applied. This implies that the medium is causal, i.e., it does not respond before the external field is applied. This means that $G_e(t) = 0$ for $t < 0$. Since the Fourier transform of $G_e(t)$ is the electric susceptibility $\chi_e(\omega)$, the following relation can be written:

$$\chi_e(\omega) = \mathcal{F}\left\{G_e(t)\right\} = \int_{-\infty}^{+\infty} G_e(t) e^{-j\omega t} dt,$$

where, of course, the previous integral can be restricted only to non-negative times due to the principle of causality. Due to causality, the previous Fourier transform can be rewritten with the introduction of the unit step function, $u(t)$ [$u(t) = 1$ for $t > 0$ and $u(t) = 0$ for $t < 0$]:

$$\chi_e(\omega) = \mathcal{F}\left\{G_e(t)u(t)\right\} = \int_{-\infty}^{+\infty} G_e(t)u(t) e^{-j\omega t} dt, \tag{3A.1}$$

Using the properties of the Fourier transform, the following relations can be easily written:

$$\begin{aligned}
\chi_e(\omega) &= \mathcal{F}\left\{G_e(t)u(t)\right\} = \frac{1}{2\pi}\mathcal{F}\{G_e(t)\} * \mathcal{F}\{u(t)\} = \\
&= \frac{1}{2\pi}\chi_e(\omega) * \mathcal{F}\{u(t)\} = \frac{1}{2\pi}\chi_e(\omega) * \left[\pi\delta(\omega) + \frac{1}{j\omega}\right] \\
&= \frac{1}{2\pi}\left\{\int_{-\infty}^{+\infty} \frac{\chi_e(\omega')}{j(\omega - \omega')} d\omega' + \pi \int_{-\infty}^{+\infty} \chi_e(\omega')\delta(\omega - \omega') d\omega'\right\} \\
&= \frac{1}{2\pi}\left\{\int_{-\infty}^{+\infty} \frac{\chi_e(\omega')}{j(\omega - \omega')} d\omega' + \pi\chi_e(\omega)\right\} \implies
\end{aligned}$$

$$\chi_e(\omega) = \frac{1}{\pi}\int_{-\infty}^{+\infty} \frac{\chi_e(\omega')}{j(\omega-\omega')}d\omega' = \frac{1}{\pi}\int_{-\infty}^{0} \frac{\chi_e(\omega')}{j(\omega-\omega')}d\omega' + \frac{1}{\pi}\int_{0}^{+\infty} \frac{\chi_e(\omega')}{j(\omega-\omega')}d\omega'$$
$$= \frac{1}{\pi}\int_{-\infty}^{0} \frac{\chi_e(-\omega'')}{j(\omega+\omega'')}(-d\omega'') + \frac{1}{\pi}\int_{0}^{+\infty} \frac{\chi_e(\omega')}{j(\omega-\omega')}d\omega'$$
$$= \frac{1}{\pi}\int_{0}^{+\infty} \frac{\chi_e^*(\omega'')}{j(\omega+\omega'')}d\omega'' + \frac{1}{\pi}\int_{0}^{+\infty} \frac{\chi_e(\omega')}{j(\omega-\omega')}d\omega',$$

where in the last equation, due to causality, $\chi_e(-\omega) = \chi_e^*(\omega)$ (with a^* denoting the complex conjugate of a), and $f * g$ denotes the convolution of functions f and g. Now, it is assumed that $\chi_e(\omega) = \chi_e'(\omega) - j\chi_e''(\omega)$, and the last equation can be rewritten as:

$$\chi_e(\omega) = \chi_e'(\omega) - j\chi_e''(\omega) = \frac{1}{\pi}\int_{0}^{+\infty} \frac{\chi_e^*(\omega')}{j(\omega+\omega')}d\omega' + \frac{1}{\pi}\int_{0}^{+\infty} \frac{\chi_e(\omega')}{j(\omega-\omega')}d\omega',$$
$$= \frac{1}{j\pi}\int_{0}^{\infty} \frac{[\chi_e(\omega') - j\chi_e''(\omega'')](\omega+\omega') + [\chi_e(\omega') + j\chi_e''(\omega'')](\omega-\omega')}{\omega^2-\omega'^2}.$$

Then, solving for the $\chi_e'(\omega)$ and $\chi_e''(\omega)$, the following expressions are derived:

$$\chi_e'(\omega) = -\frac{2}{\pi}\int_{0}^{\infty} \frac{\chi_e''(\omega')\omega'}{\omega^2-\omega'^2}d\omega', \tag{3A.2}$$

$$\chi_e''(\omega) = -\frac{2}{\pi}\int_{0}^{\infty} \frac{\chi_e'(\omega')\omega}{\omega^2-\omega'^2}d\omega'. \tag{3A.3}$$

The last two equations express the Kramers-Kronig relations and are valid not only for χ_e but for any causal function which can represent the impulse response of a linear system. Some caution is needed in the usage of Eqs. (3A.2) and (3A.3) since the principal value of the integrals is assumed. The principal value of an improper integral is defined as:

$$\mathcal{PV}\int_{-\infty}^{+\infty} \frac{f(x)}{x-x_0}dx = \lim_{\epsilon\to 0}\left\{\int_{-\infty}^{+x_0-\epsilon} \frac{f(x)}{x-x_0}dx + \int_{x_0+\epsilon}^{+\infty} \frac{f(x)}{x-x_0}dx\right\},$$

where $f(x_0) \neq 0$ and $\mathcal{PV}$ indicate the principal value. An interesting feature of the Kramers-Kronig relations is that there is no possibility of having a real, purely lossless material since, in this case, both $\chi_e'(\omega) = \chi_e''(\omega) = 0$. Therefore, any real medium will experience some loss. However, the losses are concentrated near the resonance frequencies, and away from them, the losses can be negligible.

3B Simple Classical Beam Splitter Analysis

Assume a lossless beam splitter [180–184] with two incoming waves E_1 and E_2, as shown in Fig. 3B.1. The incident fields are assumed to be scalar for simplicity, implying the same constant polarization. Each incoming wave is partially reflected and partially transmitted by the beam splitter, creating the fields E_{1r}, E_{1t}, E_{2r}, and E_{2t} that produce the outgoing fields $E_1' = E_{1t} + E_{2r}$, and $E_2' = E_{2t} + E_{1r}$. The incoming and outgoing waves can be related by the following equation:

$$\begin{bmatrix} E_1' \\ E_2' \end{bmatrix} = \begin{bmatrix} E_{1t} + E_{2r} \\ E_{2t} + E_{1r} \end{bmatrix} = \begin{bmatrix} t_{11} & r_{21} \\ r_{12} & t_{22} \end{bmatrix}\begin{bmatrix} E_1 \\ E_2 \end{bmatrix} = \tilde{R}\begin{bmatrix} E_1 \\ E_2 \end{bmatrix} \tag{3B.1}$$

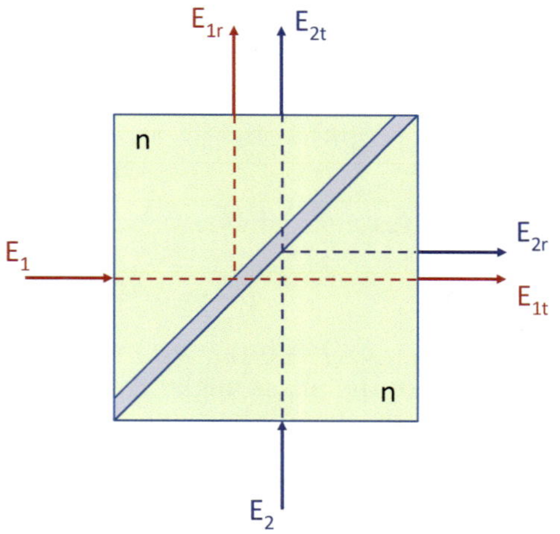

FIGURE 3B.1: The basic beam splitter geometry.

where the r_{12}, t_{11}, r_{21}, and t_{22} are complex reflection and transmission coefficients relating the outgoing with the incoming waves. As a special case, if $E_1 \neq 0$ and $E_2 = 0$, the resulting fields are $E_1' = t_{11}E_1$ and $E_2' = r_{12}E_1$. Since the beam splitter is considered to be lossless, the power should be conserved, and therefore $|E_1'|^2 + |E_2'|^2 = (|t_{11}|^2 + |r_{12}|^2)|E_1|^2$, which implies that $|t_{11}|^2 + |r_{12}|^2 = 1$. Similarly, if $E_1 = 0$ and $E_2 \neq 0$, then $E_1' = r_{21}E_2$ and $E_2' = t_{22}E_2$, and consequently $|t_{22}|^2 + |r_{21}|^2 = 1$. In the general case that $E_1 \neq 0$ and $E_2 \neq 0$ then:

$$\begin{aligned}
|E_1'|^2 + |E_2'|^2 &= (t_{11}E_1 + r_{21}E_2)\,(t_{11}^*E_1^* + r_{21}^*E_2^*) + (r_{12}E_1 + t_{22}E_2)\,(r_{12}^*E_1^* + t_{22}^*E_2^*) \\
&= \left(|t_{11}|^2 + |r_{12}|^2\right)|E_1|^2 + \left(|t_{22}|^2 + |r_{21}|^2\right)|E_2|^2 + \\
&\qquad 2\mathrm{Re}\left\{E_1^*E_2(r_{21}t_{11}^* + r_{12}^*t_{22}\right\} = |E_1|^2 + |E_2|^2 \Longrightarrow \\
&\qquad r_{21}t_{11}^* + r_{12}^*t_{22} = 0,
\end{aligned}$$

where the E^* denotes the complex conjugate of E, and the lossless property of the beam splitter was again utilized. Summarizing, for the lossless beam splitter, the following conditions should be satisfied:

$$|t_{11}|^2 + |r_{12}|^2 = 1, \tag{3B.2}$$

$$|t_{22}|^2 + |r_{21}|^2 = 1, \tag{3B.3}$$

$$r_{21}t_{11}^* + r_{12}^*t_{22} = 0. \tag{3B.4}$$

Now, assume that $t_{ii} = |t|e^{j\delta_{ii}}$ $(i = 1, 2)$ and $r_{i\ell} = |r|e^{j\delta_{i\ell}}$ [where $(i, \ell) = (1, 2)$ or $(i, \ell) = (2, 1)$]. Furthermore, in order to satisfy Eqs. (3B.2) and (3B.3), it is necessary that $|r|^2 + |t|^2 = 1$. Also, the phase angles can be re-defined as $\delta_{11} = \delta_0 - \delta_t$, $\delta_{22} = \delta_0 + \delta_t$, and $\delta_{21} = \delta_0 + \delta_r$. The last phase angle δ_{12} must be chosen in such a way as to satisfy Eq.

(3B.4). Therefore,

$$r_{21}t_{11}^* + r_{12}^* t_{22} = 0 \Longrightarrow,$$

$$|t||r|e^{j(\delta_{21}-\delta_{11})}\left[1 + e^{j(\delta_{22}-\delta_{21}+\delta_{11}-\delta_{12})}\right] = 0 \Longrightarrow$$

$$(\delta_{22} - \delta_{21}) + (\delta_{11} - \delta_{12}) = \pi \Longrightarrow$$

$$\delta_{12} = \delta_0 - \delta_r - \pi.$$

For symmetric beam splitters $(\delta_{22} - \delta_{21}) = (\delta_{11} - \delta_{12}) = \pi/2$. Then, the angles δ_0, δ_t and δ_r can be expressed as a function of the phase angles $\delta_{i\ell}$ as follows:

$$\delta_0 = \frac{1}{2}(\delta_{22} + \delta_{11}), \tag{3B.5}$$

$$\delta_t = \frac{1}{2}(\delta_{22} - \delta_{11}), \tag{3B.6}$$

$$\delta_r = \frac{1}{2}(\delta_{21} - \delta_{12} - \pi). \tag{3B.7}$$

Of course, the term π can be replaced with $(2m+1)\pi$ (where m an integer). Now, defining $|t| = \cos\theta$ and $|r| = \sin\theta$ (with $0 \leq \theta \leq \pi/2$, so both terms remain positive), the lossless beam splitter matrix can be written as:

$$\tilde{R} = e^{j\delta_0}\begin{bmatrix} \cos\theta e^{-j\delta_t} & \sin\theta e^{j\delta_r} \\ -\sin\theta e^{-j\delta_r} & \cos\theta e^{+j\delta_t} \end{bmatrix}. \tag{3B.8}$$

For a 50/50 symmetric beam splitter, assume $\theta = \pi/4$ (which means the reflection and transmission coefficients are equal), and $\delta_0 = \delta_t = \delta_r = 0$ (which means the reflection and transmission coefficients are real). Then, the resulting matrix $\tilde{R}$ defined in Eq. (3B.8) becomes:

$$\tilde{R} = \frac{1}{\sqrt{2}}\begin{bmatrix} 1 & 1 \\ -1 & 1 \end{bmatrix}. \tag{3B.9}$$

In this case, the phase angles become $\delta_{11} = \delta_{22} = \delta_{21} = 0$ and $\delta_{12} = -\pi$ (or π, representing one external reflection) which characterizes a dielectric symmetric beam splitter [184]. An alternate symmetric beam splitter matrix can be defined assuming $\theta = \pi/4$ and $\delta_0 = \pi/2$, $\delta_t = 0$, $\delta_r = -\pi/2$. Then, the resulting matrix $\tilde{R}$ defined in Eq. (3B.8) becomes:

$$\tilde{R} = \frac{1}{\sqrt{2}}\begin{bmatrix} 1 & -j \\ -j & 1 \end{bmatrix}. \tag{3B.10}$$

In this situation, the phase angles become $\delta_{11} = \delta_{22} = \pi/2$, $\delta_{21} = \delta_{12} = 0$. This configuration, could correspond to a symmetric cube beam splitter (with anti-reflecting coatings as shown in Fig. 3B.1) where the thin middle layer is region 2 and the surrounding media are regions 1 and 3 (made from the same material). Then, using Eqs. (3.221) and (3.222), it is straightforward to show that $\arg\{(E_r/E_i)(E_t/E_i)^*\} = \delta_r - \delta_t = \pm\pi/2$, which implies that there is a $\pm\pi/2$ phase shift between the reflected and the transmitted beams as it is evident from the matrix of Eq. (3B.10) [184].

For collinear wave interference, assume that $E_1 = E_0$ and $E_2 = E_0 e^{-j\phi}$. Then, using Eq. (3B.9), the following outgoing field intensities can be calculated after some simple manipulations:

$$|E_1'|^2 = 2|E_0|^2 \cos^2 \frac{\phi}{2}, \tag{3B.11}$$

$$|E_2'|^2 = 2|E_0|^2 \sin^2 \frac{\phi}{2}, \tag{3B.12}$$

It is apparent that the power conservation holds since $|E_1'|^2+|E_2'|^2 = 2|E_0|^2 = |E_1|^2+|E_2|^2$. When $\phi = 2m\pi$ then $|E_1'|^2 = 2|E_0|^2$, while $|E_2'|^2 = 0$, and all the power propagates in the direction of E_1' with no power into E_2' direction. Of course, the opposite happens in the case that $\phi = (2m+1)\pi$. Similarly, if the matrix of Eq. (3B.10) is utilized, then the following outgoing field intensities can be found:

$$|E_1'|^2 = |E_0|^2 (1 - \sin\phi), \tag{3B.13}$$

$$|E_2'|^2 = |E_0|^2 (1 + \sin\phi), \tag{3B.14}$$

Again power conservation is retained as expected due to the lossless beam splitter.

3C Vectorial Spherical Waves

An electric dipole with sinusoidal time-varying charge is positioned that the origin of the coordinate system as shown in Fig. 3C.1. The sinusoidally time-varying current density $\vec{\mathcal{J}}$ can be expressed as:

$$\vec{\mathcal{J}}(\vec{r},t) = \hat{\imath}_z I_0 h \cos(\omega t)\delta(x,y,z) \Longrightarrow \vec{J}(\vec{r},\omega) = \hat{\imath}_z I_0 h \delta(x)\delta(y)\delta(z), \tag{3C.1}$$

where I_0 is the amplitude of the oscillating current, h is the dipole length (which is considered much smaller than the magnitude of the position vector $\vec{r}$ where the radiating field

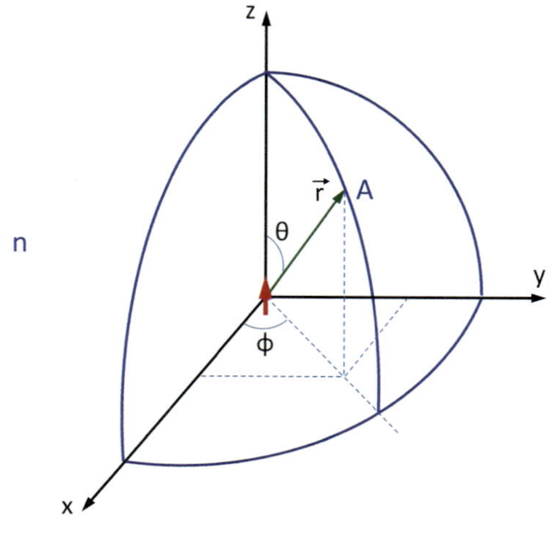

FIGURE 3C.1: Geometry of a radiating electric dipole oriented along the z-axis and placed at the origin of the coordinate system. The dipole radiates inside a homogeneous, isotropic, linear, dispersionless medium of refractive index n.

is sought), ω is the oscillation frequency, $\vec{J}$ is the phasor of $\vec{\mathcal{J}}$, and δ the impulse delta function. The retarded vector magnetic potential, $\vec{\mathcal{A}}(\vec{r},t)$, can be written as [86, 185]:

$$\vec{\mathcal{A}}(\vec{r},t) = \mu \int_V \vec{\mathcal{J}}\left(\vec{r}\,',t-\frac{R}{u}\right)\frac{1}{4\pi R}dV' \Longrightarrow \vec{A}(\vec{r}) = \mu \int_V \vec{J}(\vec{r}\,')\frac{e^{-jkR}}{4\pi R}dV', \tag{3C.2}$$

with,

$$R = |\vec{r}-\vec{r}\,'| = \sqrt{(x-x')^2+(y-y')^2+(z-z')^2},$$

where $\vec{A}(\vec{r})$ is the phasor of the vector potential in the time-harmonic form, and $\vec{r}\,'$ represents the position of the sources (currents). By inserting the current density of Eq. (3C.1) into Eq. (3C.2), the phasor of the vector potential in the case of the radiating electric dipole becomes:

$$\vec{A}(\vec{r}) = \mu \int_V \hat{\imath}_z I_0 h \delta(x',y',z')\frac{e^{-jkR}}{4\pi R}dV' = \hat{\imath}_z \mu I_0 h \frac{e^{-jkr}}{4\pi r} = \hat{\imath}_z A_z(r), \tag{3C.3}$$

where $r = (x^2+y^2+z^2)^{1/2}$ is the distance of the point (x,y,z) (where the radiation field from the dipole is sought) from the origin, where the electric dipole is positioned. The magnetic field $\vec{H}(\vec{r})$ can be easily determined from the following equation:

$$\vec{H} = \frac{1}{\mu}\vec{\nabla}\times\vec{A} = \frac{1}{r^2\sin\theta}\begin{vmatrix} \hat{\imath}_r & r\hat{\imath}_\theta & r\sin\theta\hat{\imath}_\phi \\ \dfrac{\partial}{\partial r} & \dfrac{\partial}{\partial\theta} & \dfrac{\partial}{\partial\phi} \\ A_r & rA_\theta & 0 \end{vmatrix}, \tag{3C.4}$$

where $A_r = A_z\cos\theta$ and $A_\theta = -A_z\sin\theta$. Therefore, the solution for the magnetic field is given by:

$$H_r = 0,$$

$$H_\theta = 0,$$

$$H_\phi = I_0 h \frac{e^{-jkr}}{4\pi r}\left(jk+\frac{1}{r}\right)\sin\theta. \tag{3C.5}$$

The electric field can be easily determined from the magnetic field from Ampère-Maxwell law, $\vec{\nabla}\times\vec{H} = j\omega\epsilon\vec{E}$. Then, the resulting electric field is expressed by:

$$E_r = I_0 h \frac{e^{-jkr}}{4\pi r}\left(\frac{2Z}{r}+\frac{2}{j\omega\epsilon r^2}\right)\cos\theta \tag{3C.6}$$

$$E_\theta = I_0 h \frac{e^{-jkr}}{4\pi r}\left(j\omega\mu+\frac{Z}{r}+\frac{1}{j\omega\epsilon r^2}\right)\sin\theta \tag{3C.7}$$

$$E_\phi = 0.$$

The corresponding time-domain electric and magnetic fields can be found from their phasors as follows:

$$\mathcal{H}_\phi = \frac{E_0}{Z}\frac{\sin\theta}{kr}\left[\sin(kr-\omega t)+\frac{\cos(kr-\omega t)}{kr}\right], \tag{3C.8}$$

$$\mathcal{E}_r = E_0\frac{2\cos\theta}{(kr)^2}\left[\cos(kr-\omega t)-\frac{\sin(kr-\omega t)}{kr}\right], \tag{3C.9}$$

$$\mathcal{E}_\theta = E_0 \frac{\sin\theta}{kr}\left[\sin(kr-\omega t) + \frac{\cos(kr-\omega t)}{kr} - \frac{\sin(kr-\omega t}{(kr)^2}\right], \tag{3C.10}$$

$$\text{where,} \quad E_0 = \frac{I_0 h}{4\pi}\frac{k^3}{\omega\epsilon}, \quad \text{and} \quad Z = \sqrt{\frac{\mu}{\epsilon}}.$$

The electric field lines obey the following differential equation:

$$d\vec{r}\times\vec{\mathcal{E}} = 0 \Longrightarrow \frac{dr}{d\theta} = \frac{r\mathcal{E}_r}{\mathcal{E}_\theta} \Longrightarrow \frac{d(kr)}{d\theta} = \frac{(kr)\mathcal{E}_r}{\mathcal{E}_\theta} = -2\cot\theta\frac{Q(kr-\omega t)}{Q'(kr-\omega t)},$$

$$\text{where,} \quad Q(kr-\omega t) = \cos(kr-\omega t) - \frac{\sin(kr-\omega t)}{kr},$$

$$Q'(kr-\omega t) = \frac{dQ}{d(kr)} = -\left[\sin(kr-\omega t) + \frac{\cos(kr-\omega t)}{kr} - \frac{\sin(kr-\omega t}{(kr)^2}\right],$$

$$d(kr)\frac{Q'}{Q} = -2\cot\theta d\theta \Longrightarrow \frac{dQ}{Q} = -2\frac{d(\sin\theta)}{\sin\theta} \Longrightarrow$$

$$d\left[\ln(Q\sin^2\theta)\right] = 0 \Longrightarrow Q(kr-\omega t)\sin^2\theta = C, \tag{3C.11}$$

where C is a constant. The last equation represents the contours of the electric field lines as a function of r for a given time t. These field lines provide a visual representation of the dipole radiation pattern. For the magnetic field, which has only a ϕ-component, its contours can be directly obtained from Eq. (3C.5). A simple snapshot in time of a radiating electric dipole is shown in Fig. 3C.2.

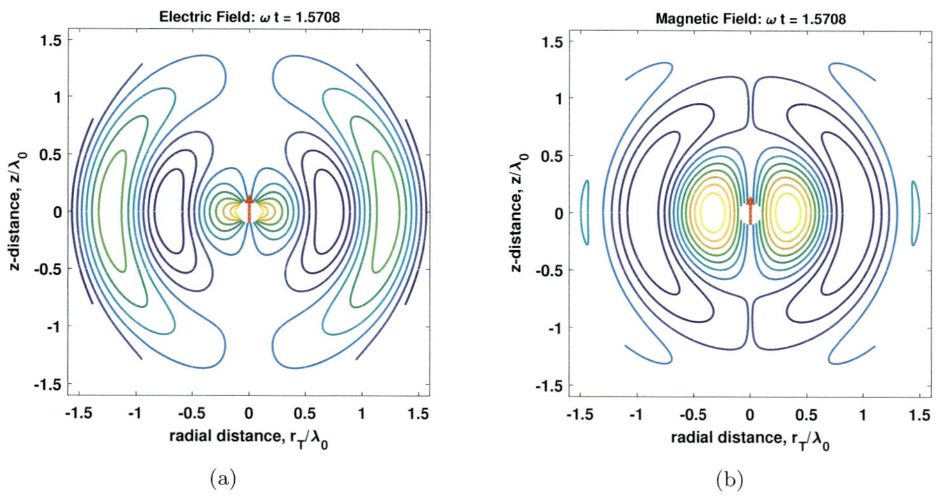

FIGURE 3C.2: Snapshots in space of a radiating electric dipole at $\omega t = \pi/2$. (a) The normalized electric field contours. (b) The normalized magnetic field contours (it has only ϕ component).

In the far-field region where $kr \gg 1$, the electromagnetic wave radiated by the electric dipole can be simplified by neglecting terms proportional to $1/(kr)^2$ and smaller. The fields are dominated by the radiative components, which vary as $1/r$. The electric and magnetic fields in spherical coordinates are given by:

$$E_r \simeq 0,$$

$$E_\theta \simeq \underbrace{j\frac{I_0 h}{4\pi}k^2 Z}_{E_0}\frac{e^{-jkr}}{kr}\sin\theta, \tag{3C.12}$$

$$H_\phi \simeq \frac{E_0}{Z}\frac{e^{-jkr}}{kr}\sin\theta = \frac{E_0}{Z}E_\theta. \tag{3C.13}$$

In the far-field region, where $kr \gg 1$, the electromagnetic wave radiated by an electric dipole becomes a transverse electromagnetic (TEM) wave with spherical wavefronts. In this regime, the electric field and magnetic field are perpendicular to each other and to the direction of wave propagation (r), confirming the transverse nature of the wave. The azimuthal symmetry of the dipole ensures that the resulting wavefronts are spherical and depend only on the radial distance r and the polar angle θ, with no variation in the azimuthal angle ϕ. These waves diminish in amplitude with distance, following an inverse proportionality to r, reflecting the spherical spreading of energy. This far-field behavior, characterized by the symmetry and transverse nature of the fields, makes the radiation pattern consistent with the properties of spherical waves. The power radiated exhibits a doughnut-shaped angular distribution, with maximum intensity perpendicular to the dipole axis and no radiation along its axis ($\theta = 0$ or π).

In the far-field region, specifically along the $\theta = \pi/2$ direction (perpendicular to the dipole axis), the scalar counterpart of the radiation field takes the form $\exp(-jkr)/(kr)$. This expression, which captures the amplitude and phase variation of the radiated wave, is commonly referred to as a spherical wave. It reflects the fundamental nature of wave propagation from a point-like source, where the wavefronts are concentric spheres centered on the origin of the radiation.

3D Vectorial Cylindrical Waves

An infinite line current with sinusoidal time variation is positioned along the z-axis of a Cartesian coordinate system, as shown in Fig. 3D.1. The time-varying current density $\vec{\mathcal{J}}$ of this current can be expressed as:

$$\vec{\mathcal{J}}(\vec{r},t) = \hat{\imath}_z I_0 \cos(\omega t)\delta(x,y) \Longrightarrow \vec{J}(\vec{r}_T,\omega) = \hat{\imath}_z I_0 \delta(x)\delta(y), \tag{3D.1}$$

where I_0 is the amplitude of the oscillating current, $\vec{r}_T = r_T(\cos\phi\hat{\imath}_x + \sin\phi\hat{\imath}_y)$, r_T is the radial distance on the xy plane, ω is the oscillation frequency, $\vec{J}$ is the phasor of $\vec{\mathcal{J}}$, and δ the impulse delta function, as previously.

Inserting the current density from Eq. (3D.1) into Eq. (3C.2), the phasor of the vector potential in the case of the radiating infinite line current becomes:

$$\vec{A}(\vec{r}) = \mu\int_V \hat{\imath}_z I_0 \delta(x',y')\frac{e^{-jkR}}{4\pi R}dV' = \hat{\imath}_z \mu I_0 \int_{-\infty}^{+\infty}\frac{e^{-jk[r_T^2+(z-z')^2]^{1/2}}}{4\pi[r_T^2+(z-z')^2]^{1/2}}dz' = \hat{\imath}_z A_z(r_T), \tag{3D.2}$$

where $r_T = (x^2+y^2)^{1/2}$ is the radial distance of the point (x,y,z) (point A where the radiation from the dipole is sought) from the z-axis, where the line current is positioned.

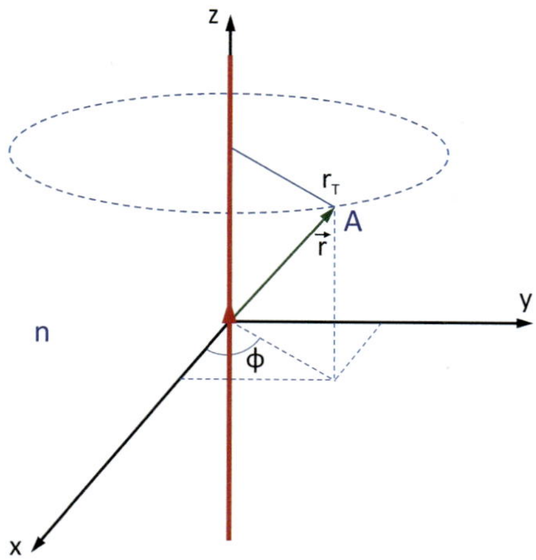

FIGURE 3D.1: Geometry of a radiating infinite line current oriented and placed along the z-axis of the coordinate system. The line current radiates inside a homogeneous, isotropic, linear, dispersionless medium of refractive index n.

The integral in Eq. (3D.2) can be determined using Hankel functions [186, 187], which are commonly employed to solve problems involving cylindrical symmetry in electromagnetic radiation. Therefore, the integral is equal to:

$$\int_{-\infty}^{+\infty} \frac{e^{-jk[r_T^2+(z-z')^2]^{1/2}}}{4\pi[r_T^2+(z-z')^2]^{1/2}} dz' = -\frac{j}{4} H_0^{(2)}(kr_T), \tag{3D.3}$$

where $H_0^{(2)}$ is the Hankel function of zero order of the second kind (radiating away from the current line). The magnetic field $\vec{H}(\vec{r})$ can be determined from the following equation:

$$\vec{H} = \frac{1}{\mu}\vec{\nabla} \times \vec{A} = \frac{1}{r_T} \begin{vmatrix} \hat{\imath}_{r_T} & r_T \hat{\imath}_\phi & \hat{\imath}_z \\ \dfrac{\partial}{\partial r_T} & \dfrac{\partial}{\partial \phi} & \dfrac{\partial}{\partial z} \\ 0 & 0 & A_z \end{vmatrix}, \tag{3D.4}$$

resulting in the following components:

$$\begin{aligned} H_{r_T} &= 0, \\ H_\phi &= \frac{j}{4} I_0 k \frac{dH_0^{(2)}}{d(kr_T)}, \\ H_z &= 0. \end{aligned} \tag{3D.5}$$

The electric field can also be calculated from the Ampère-Maxwell equation and is given by:

$$\begin{aligned} E_{r_T} &= 0, \\ E_\phi &= 0, \\ E_z &= Z \frac{I_0}{4} \frac{1}{r_T} \left[\frac{dH_0^{(2)}}{d(kr_T)} + kr_T \frac{d^2 H_0^{(2)}}{d(kr_T)^2} \right]. \end{aligned} \tag{3D.6}$$

Using properties of the Hankel functions [specifically, $dH_0^{(2)}/dz = -H_1^{(2)}$ and $z(dH_1^{(2)}/dz) + H_1^{(2)} = zH_0^{(2)}$], the expressions for the electric and magnetic field phasors and their real-time harmonic-counterparts become:

$$\vec{E} = -\hat{\imath}_z \frac{I_0 k}{4} Z H_0^{(2)}(kr_T) \implies \vec{\mathcal{E}} = -\hat{\imath}_z \frac{I_0 k}{4} Z\text{Re}\left\{H_0^{(2)}(kr_T)e^{j\omega t}\right\}, \tag{3D.7}$$

$$\vec{H} = -\hat{\imath}_\phi \frac{I_0 k}{4} H_1^{(2)}(kr_T) \implies \vec{\mathcal{H}} = -\hat{\imath}_\phi \frac{I_0 k}{4} \text{Re}\left\{H_1^{(2)}(kr_T)e^{j\omega t}\right\}, \tag{3D.8}$$

where $H_1^{(2)}$ is the Hankel function of order "1" of the second kind (radiating away from the line). When $kr_T \gg 1$ the Hankel functions can be approximated by their asymptotic expressions for large arguments, yielding the following electric and magnetic field phasors:

$$\vec{E} = -\hat{\imath}_z \frac{I_0 k}{4} Z \sqrt{\frac{2}{\pi k r_T}} e^{-jkr_T} e^{j\pi/4}, \tag{3D.9}$$

$$\vec{H} = +\hat{\imath}_\phi \frac{I_0 k}{4} \sqrt{\frac{2}{\pi k r_T}} e^{-jkr_T} e^{j\pi/4}. \tag{3D.10}$$

It is evident that the radiating electromagnetic wave is a transverse electromagnetic wave (TEM), and when $kr_T \gg 1$, the wave forms cylindrical wavefronts. Due to the azimuthal and z-axis symmetry of the infinite line current radiation problem, the resulting waves at far distances are referred to as cylindrical waves, with a phasor form of $\exp(-jkr_T)/\sqrt{kr_T}$. This form highlights the characteristic radial decay of the wave amplitude with distance.

In the far-field, the radiation exhibits cylindrical symmetry, meaning that the wavefronts propagate outward from the current along circular paths around the z-axis. As the distance increases, the wave exhibits the typical $1/\sqrt{r_T}$ decay in amplitude, as opposed to the spherical decay seen for dipole radiation.

A simple snapshot in time of the radiating infinite line current is shown in Fig. 3D.2. These snapshots correspond to the near-field region, where the fields are more complex due

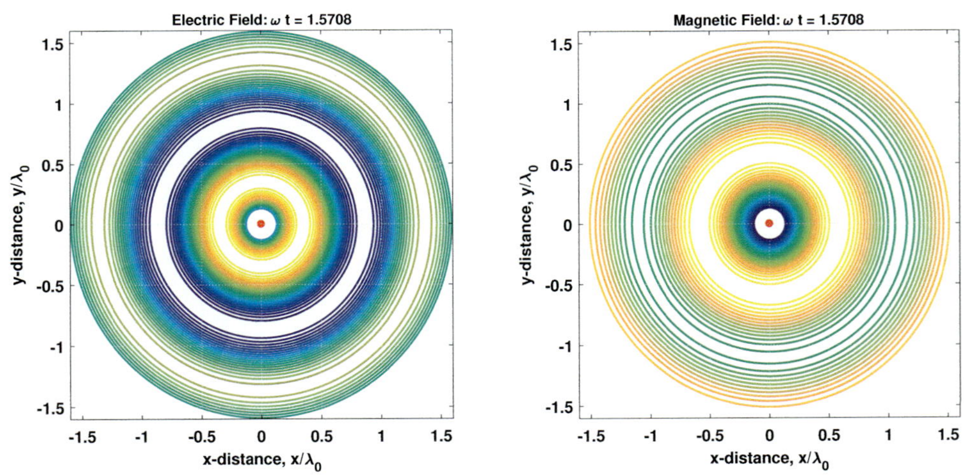

(a) (b)

FIGURE 3D.2: Snapshots in space of a radiating infinite line current along the z-axis at $\omega t = \pi/2$. (a) The normalized electric field contours (there is only a z component). (b) The normalized magnetic field contours (there is only a ϕ component).

to the proximity to the source. In the far-field, the radiation becomes more uniform as the cylindrical wavefronts evolve.

3E Stokes Relations

The Stokes relations between the reflection and transmission coefficients, as described by Stokes in 1849 [109] and further explored in textbooks like Ref. [42], offer an alternative approach to the familiar Fresnel equations. While the Stokes relations are implied by the Fresnel equations, they provide a more elegant formulation that relates the reflection and transmission coefficients for wave propagation from one medium to another and vice versa. Consider a plane wave of amplitude E_i incident on the planar interface between two dielectric, isotropic, homogeneous media with refractive indices n_1 and n_2, as shown in Fig. 3E.1(a). Here, it is assumed without loss of generality that $n_2 > n_1$. Let r and t represent the reflection and transmission coefficients, respectively, for a wave incident from the left medium (with refractive index n_1) onto the right medium (with refractive index n_2). Similarly, let r' and t' denote the reflection and transmission coefficients for a wave incident from the right medium (with refractive index n_2) onto the left medium (with refractive index n_1).

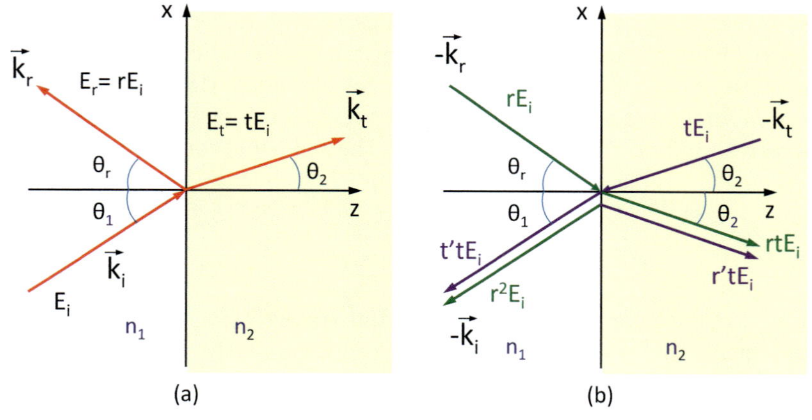

FIGURE 3E.1: A planar interface between two homogeneous, isotropic, linear media of refractive indices n_1 and n_2. (a) An incident plane wave of amplitude E_i and wavevector $\vec{k}_i$ generates a reflected wave of amplitude rE_i and wavevector $\vec{k}_r$ and a transmitted plane wave of amplitude tE_i and wavevector $\vec{k}_t$. (b) Time-reversed fields. An incident from the right plane wave of amplitude tE_i and wavevector $-\vec{k}_t$ generates a transmitted wave of amplitude $tt'E_i$ and a reflected wave of amplitude $tr'E_i$ as shown. Furthermore, an incident from the left plane wave of amplitude rE_i and wavevector $-\vec{k}_r$ generates a transmitted wave of amplitude rtE_i and a reflected wave of amplitude r^2E_i as shown.

The Stokes relations are based on the time-reversal invariance of classical electromagnetic fields [42,109,188–190]. This principle suggests that when the directions of the reflected and transmitted waves are reversed (i.e., when the incident beams are directed in the opposite direction), the resulting fields should correspond to the original incident wave traveling in the reverse direction, as depicted in Fig. 3E.1(b). Initially consider that both media are lossless, and all the reflection and transmission coefficients (r, r', t, and t') are real.

From Fig. 3E.1(b), the incident wave from the left has an amplitude rE_i, and the incident wave from the right has an amplitude tE_i. These waves interact with the interface and are reflected and transmitted as shown in the diagram. The reflected and transmitted waves from the rE_i wave are r^2E_i and rtE_i, respectively. Similarly, the reflected and transmitted waves from the tE_i wave are $r'tE_i$ and $tt'E_i$. Due to the time-reversal symmetry of the classical electromagnetic field, these reflected and transmitted waves should result in the backward-traveling incident wave E_i. This means that, under time-reversal, the amplitudes and coefficients must satisfy the relationships:

$$rt + r't = 0 \Longrightarrow r' = -r, \tag{3E.1}$$

$$tt' + r^2 = 1. \tag{3E.2}$$

The Stokes relations are indeed independent of the polarization of light. As mentioned, these relations can be derived by directly applying the Fresnel equations for both *TE* and *TM* polarizations for both left-to-right (r, t) and right-to-left (r', t') incidences. Thus, the Stokes relations in Eqs. (3E.1) and (3E.2) should hold in any situation, even if there are gain, loss, or evanescent waves in one of the regions [191, 192].

In cases where there is loss or gain, the time-reversal symmetry approach discussed earlier may not hold, particularly when dealing with evanescent waves. However, the Stokes relations are still valid because they can be derived directly from the Fresnel equations, regardless of whether the media involved are lossless, lossy, or have gain. These relations are particularly useful in the study of Fabry-Perot etalons, where one or both media may exhibit loss or gain. Moreover, similar relations can be generalized for stratified media, where different layers of materials with varying refractive indices are involved. For lossless stratified structures, the Stokes relations can be extended [188, 191, 193–196], offering a powerful tool in analyzing complex multilayered optical systems. For such lossless stratified media, the corresponding Stokes relations take the following form:

$$rt^* + (r')^*t = 0, \qquad \text{or} \qquad r'(t')^* + r^*t' = 0, \tag{3E.3}$$

$$t^*t' + r^*r = 1, \qquad \text{or} \qquad (t')^*t + (r')^*r' = 1. \tag{3E.4}$$

In the case of stratified media, the reflection and transmission coefficients r, t correspond to light incident on the structure from left to right. Similarly, r' and t' are reflection and transmission coefficients for light incident on the stratified structure from right to left. In the right pairs of Eqs. (3E.3) and (3E.4), there is a simple interchange between r, t with r' and t'. The notation w^* denotes the complex conjugate of w. It is evident from Eq. (3E.4) that $t^*t' = 1 - |r|^2 = (t^*t')^* = (t')^*t = |r'|^2$.

3F RGB Values for a Visible Wavelength

The color to represent a monochromatic light of wavelength λ_0 (where $380\,nm \leq \lambda_0 \leq 780\,nm$) in a graph is selected by using its approximate R, G, B, values (see the R, G, and B definitions in Chapter 4). To generate color representations for monochromatic light, an algorithm takes the wavelength λ_0 as input and calculates the corresponding RGB values which are normalized to a range between 0 and 1. These RGB values can be used in graphical representations of intensity or interference patterns involving visible light, making the plots visually meaningful by showing the "true" color of the light. The algorithm was originally implemented in FORTRAN by Prof. Dan Bruton, and is found at url cite: `http://www.physics.sfasu.edu/astro/color/spectra.html`. The listed FORTRAN code was converted into MatLab code [141] and is listed below. The resulting normalized R, G, B, values corresponding to any visible wavelength based on this algorithm are shown in Fig. 3F.1.

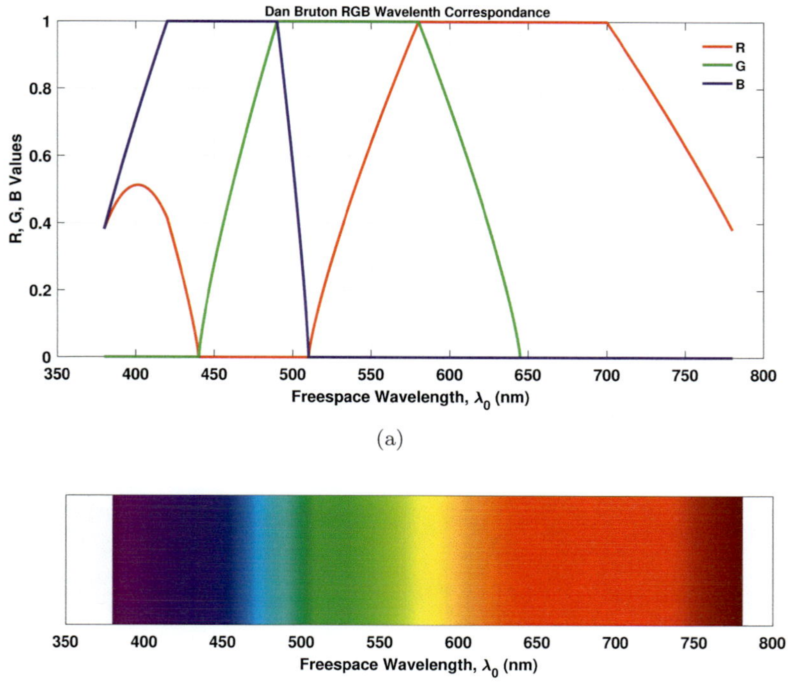

(a)

(b)

FIGURE 3F.1: (a) The normalized R, G, B values corresponding to a freespace wavelength λ_0 according to Prof. Dan Bruton's algorithm. (b) The approximate colors of the wavelengths in the visible spectrum.

Matlab Conversion Code Utilized:

```
function [R, G, B] = find_RGB3(lambda0)

% ------------------------------------------------------------
%     Convert freespace wavelength from microns to nanometers
%     Based on code by Dan Bruton:
%     http://www.physics.sfasu.edu/astro/color/spectra.html
% ------------------------------------------------------------
%     Inputs:
%     lambda0 = freespace wavelength in microns
%     Outputs:
%     R, G, B = in 0 to 1 values
% ------------------------------------------------------------

gamma  = 0.80;
lambda = 1000*lambda0;

if lambda >= 380 & lambda < 440
    R = -(lambda - 440.) / (440. - 380.);
    G = 0.0;
    B = 1.0;
elseif lambda >= 440 & lambda < 490
    R = 0.0;
    G = (lambda - 440.) / (490. - 440.);
    B = 1.0;
elseif lambda >= 490 & lambda < 510
```

```
    R = 0.0;
    G = 1.0;
    B = -(lambda - 510.) / (510. - 490.);
elseif lambda >= 510 & lambda < 580
    R = (lambda - 510.) / (580. - 510.);
    G = 1.0;
    B = 0.0;
elseif lambda >= 580 & lambda < 645
    R = 1.0;
    G = -(lambda - 645.) / (645. - 580.);
    B = 0.0;
elseif lambda >= 645 & lambda <= 780
    R = 1.0;
    G = 0.0;
    B = 0.0;
else
    R = 0.0;
    G = 0.0;
    B = 0.0;
end

%
%     Adjust intensity near visible spectrum limits
%     and add gamma correction
%

if lambda > 700
    factor = 0.3 + 0.7*(780-lambda)/(780-700);
elseif    lambda < 420
    factor = 0.3 + 0.7*(lambda-380)/(420-380);
else
    factor = 1;
end


R = (factor*R)^gamma;
G = (factor*G)^gamma;
B = (factor*B)^gamma;
```

3G Single-Phase Transmission Line and Analogy to Thin-Film Reflection/Refraction

Consider a single-phase (two-conductor) transmission line that extends infinitely along the z-direction, as shown in Fig. 3G.1. For a differential length Δz, the transmission line is characterized by several key parameters [132]: (a) resistance per unit length R (in Ω/m), (b) the inductance per unit length, L (in H/m), (c) the conductance per unit length, G (in $S/m = 1/\Omega m$), and (d) the capacitance per unit length, C (in F/m). The equivalent circuit of the transmission line between z and $z+\Delta z$ is shown at the bottom part of Fig. 3G.1. From this equivalent circuit, it is straightforward to derive the well-known telegrapher's equations. These equations describe the propagation of voltage and current along the transmission line, accounting for the resistive, inductive, capacitive, and conductive properties of the transmission line. These equations are:

$$-\frac{\partial v(z,t)}{\partial z} = Ri(z,t) + L\,\frac{\partial i(z,t)}{\partial t}, \tag{3G.1}$$

$$-\frac{\partial i(z,t)}{\partial z} = Gv(z,t) + C\,\frac{\partial v(z,t)}{\partial t}. \tag{3G.2}$$

Using the time-harmonic form of the equations and corresponding phasors, where $V(z)$ represents the voltage and $I(z)$ represents the current, it is straightforward to rewrite the

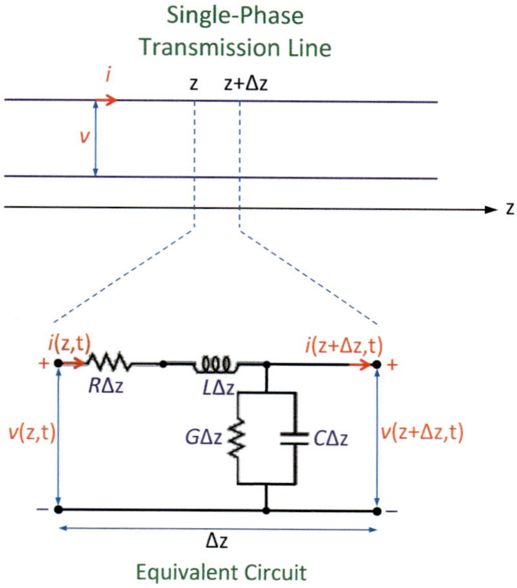

FIGURE 3G.1: An infinitely long single-phase transmission line, extending along the direction of z-axis, and its distributed-parameter equivalent circuit between z and $z + \Delta z$.

telegrapher's equations in the following form:

$$-\frac{dV(z)}{dz} = RI(z) + j\omega LI(z) = (R + j\omega L)I(z),$$

$$-\frac{dI(z)}{dz} = GV(z) + j\omega CV(z) = (G + j\omega C)V(z).$$

Eliminating either the voltage or the current yields:

$$\frac{d^2V(z)}{dz^2} = \gamma^2 V(z), \tag{3G.3}$$

$$\frac{d^2I(z)}{dz^2} = \gamma^2 I(z), \qquad \text{where} \tag{3G.4}$$

$$\gamma = \alpha + j\beta = \sqrt{(R + j\omega L)(G + j\omega C)},$$

where γ is a complex propagation constant with α denoting the attenuation constant and β the phase propagation constant. For a lossless transmission line, where $R = G = 0$, the propagation constant simplifies to $\gamma = j\beta = j\omega\sqrt{LC}$. It is worth noting that the equations for voltage and current, Eqs. (3G.3) and (3G.4), are analogous to the Helmholtz equation governing propagation in a homogeneous, isotropic medium. The solutions to these equations can be expressed as:

$$V(z) = V_+ e^{-\gamma z} + V_- e^{+\gamma z}, \tag{3G.5}$$

$$I(z) = I_+ e^{-\gamma z} + I_- e^{+\gamma z} = \frac{V_+}{Z} e^{-\gamma z} - \frac{V_-}{Z} e^{+\gamma z} \tag{3G.6}$$

$$\text{where,} \qquad Z = \sqrt{\frac{R + j\omega L}{G + j\omega C}}. \tag{3G.7}$$

The parameter Z represents the characteristic impedance of the transmission line measured in ohms (Ω). The solutions for voltage and current on the transmission line are analogous to the solutions for the electric and magnetic fields inside a homogeneous isotropic medium. At this point, let's focus on a segment of the transmission line with length d and characteristic impedance Z, as shown in Fig. 3G.2. This segment is connected to a load, and we will examine its behavior and properties in more detail. The voltage and the current at $z = d$

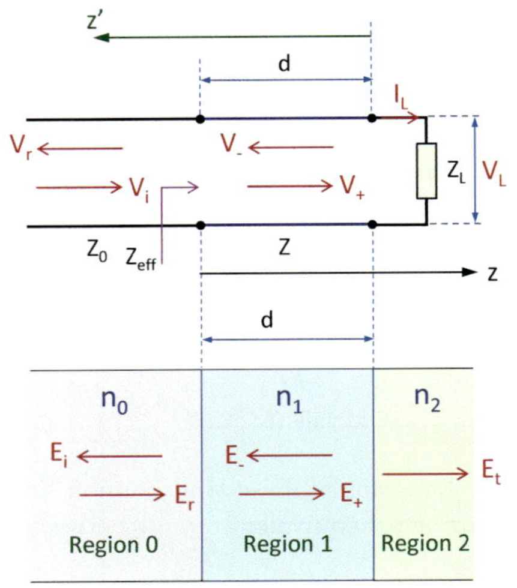

FIGURE 3G.2: A single-phase transmission line, with a load Z_L. The transmission line segment of length d and characteristic impedance Z is used to match the load with the characteristic impedance Z_0 of the left transmission line. On the bottom the equivalent structure containing the region 0 which corresponds to the transmission line of characteristic impedance Z_0, the region 1 which corresponds to the segment of characteristic impedance Z, and the output region 2 which corresponds to the load Z_L.

can be written as:

$$V(z = d) = V_+ e^{-\gamma d} + V_- e^{+\gamma d} = V_L,$$

$$I(z = d) = \frac{V_+}{Z} e^{-\gamma d} - \frac{V_-}{Z} e^{+\gamma d} = I_L.$$

Solving for the V_+ and V_- in terms of V_L and I_L, results in the following equations:

$$V(z) = \frac{I_L(Z_L + Z)}{2} \left[e^{-\gamma(z-d)} + \frac{Z_L - Z}{Z_L + Z} e^{+\gamma(z-d)} \right] =$$

$$\frac{I_L(Z_L + Z)}{2} \left[e^{-\gamma(z-d)} + \Gamma e^{+\gamma(z-d)} \right]$$

$$V(z') = \frac{I_L(Z_L + Z)}{2} \left[e^{+\gamma z'} + \Gamma e^{-\gamma z'} \right],$$

$$I(z) = \frac{I_L(Z_L + Z)}{2Z} \left[e^{-\gamma(z-d)} - \frac{Z_L - Z}{Z_L + Z} e^{+\gamma(z-d)} \right] =$$

$$\frac{I_L(Z_L + Z)}{2Z} \left[e^{-\gamma(z-d)} - \Gamma e^{+\gamma(z-d)} \right]$$

$$I(z') = \frac{I_L(Z_L + Z)}{2Z}\left[e^{-\gamma(z-d)} - \Gamma e^{+\gamma(z-d)}\right], \qquad \text{where,}$$

$$z' = d - z, \qquad \text{and}$$

$$\Gamma = \frac{Z_L - Z}{Z_L + Z}, \tag{3G.8}$$

$$Z(z') = \frac{V(z')}{I(z')} = Z\,\frac{Z_L\cosh(\gamma z') + Z\sinh(\gamma z')}{Z\cosh(\gamma z') + Z_L\sinh(\gamma z')}. \tag{3G.9}$$

The variable Γ represents the reflection coefficient at the load position in Fig. 3G.2. To prevent reflections at the load, the reflection coefficient must be zero, i.e., $\Gamma = 0$. For a lossless transmission line, where $R = G = 0$ and $\gamma = j\beta$ (with $\alpha = 0$), the impedance $Z(z')$ can be written as:

$$Z(z') = Z\,\frac{Z_L\cos(\beta z') + jZ\sin(\beta z')}{Z\cos(\beta z') + jZ_L\sin(\beta z')}. \tag{3G.10}$$

The reflection coefficient at the beginning of the segment of length d of Fig. 3G.2 can be defined in a similar manner as:

$$\Gamma = \frac{Z_{eff} - Z_0}{Z_{eff} + Z_0}, \qquad \text{where,}$$

$$Z_{eff} = Z(z' = d) = Z\,\frac{Z_L\cos(\beta d) + jZ\sin(\beta d)}{Z\cos(\beta d) + jZ_L\sin(\beta d)}, \qquad \text{resulting in,}$$

$$\Gamma = \frac{Z(Z_L - Z_0)\cos(\beta d) - j(Z^2 - Z_0 Z_L)\sin(\beta d)}{Z(Z_L + Z_0)\cos(\beta d) + j(Z^2 + Z_0 Z_L)\sin(\beta d)}. \tag{3G.11}$$

To demonstrate the equivalence between the transmission line model (top of Fig. 3G.2) and the thin-film structure (bottom of Fig. 3G.2), the results from Sections 3.6.1.2 and 3.7.3 will be utilized. The equation defining the reflection coefficient r between region 0 and region 1 is given by Eq. (3.290) for *TE* polarization (as an example, though the polarization is irrelevant in the case of normal incidence). Using $\tilde{Z}_i = Z_i/\cos\theta_i = Z_v/(n_i\cos\theta_i)$ (for *TE* polarization, where θ_i the angle with respect to the z-axis of waves inside the region i, for $i = 0, 1, 2$, and Z_v is the wave impedance of freespace) the reflection coefficient of the bottom structure of Fig. 3G.2, the reflection coefficient for the structure shown at the bottom of Fig. 3G.2, can easily be written as:

$$r = \frac{\tilde{Z}_1(\tilde{Z}_2 - \tilde{Z}_0)\cos(\beta_1 d) - j(\tilde{Z}_1^2 - \tilde{Z}_0\tilde{Z}_2)\sin(\beta_1 d)}{\tilde{Z}_1(\tilde{Z}_2 + \tilde{Z}_0)\cos(\beta_1 d) + j(\tilde{Z}_1^2 + \tilde{Z}_0\tilde{Z}_2)\sin(\beta_1 d)}. \tag{3G.12}$$

Comparing Eqs. (3G.11) and (3G.12), they are identical provided that the following analogies are used:

$$Z_0 \longrightarrow \tilde{Z}_0 = \frac{Z_0}{\cos\theta_0},$$

$$Z \longrightarrow \tilde{Z}_1 = \frac{Z_1}{\cos\theta_1},$$

$$Z_L \longrightarrow \tilde{Z}_2 = \frac{Z_2}{\cos\theta_1},$$

$$\beta \longrightarrow \beta_1 = k_0 n_1\cos\theta_1.$$

Thus, as anticipated, the transmission line model can be effectively used to analyze reflection from a stack of thin films. For off-normal TM incident polarization, the only modification is that $\tilde{Z}_i = Z_i \cos\theta_i = (Z_v/n_i)\cos\theta_i$ $(i = 0, 1, 2)$. This approach can be extended to any number of thin layers by appropriately adding transmission line segments, allowing for a more comprehensive analysis of multilayer structures.

3H Anisotropic Interfaces: Additional Information

For a biaxial region of relative permittivity $\tilde{\varepsilon}$, the dispersion equation for a plane wave solution is given by Eq. (3.47). The determinant of the dispersion equation can be written in the form of polynomial with respect to the longitudinal component k_z as follows:

$$C_1 k_z^4 + C_2 k_z^3 + C_3 k_z^2 + C_4 k_z + C_5 = 0, \tag{3H.1}$$

where the coefficients C_i $(1, 2, \cdots, 5)$ are given by:

$$C_1 = -\varepsilon_{zz} k_0^2,$$

$$C_2 = -k_0^2 \left[(\varepsilon_{xz} + \varepsilon_{zx}) k_x + (\varepsilon_{yz} + \varepsilon_{zy} k_y) \right],$$

$$C_3 = (\varepsilon_{xx}\varepsilon_{zz} - \varepsilon_{xz}\varepsilon_{zx} + \varepsilon_{yy}\varepsilon_{zz} - \varepsilon_{yz}\varepsilon_{zy}) k_0^4 - \\ \left[(\varepsilon_{xx} + \varepsilon_{zz}) k_x^2 + (\varepsilon_{yy} + \varepsilon_{zz}) k_y^2 + (\varepsilon_{xy} + \varepsilon_{yx}) k_x k_y \right] k_0^2,$$

$$C_4 = \left[(\varepsilon_{xz}\varepsilon_{yy} - \varepsilon_{xy}\varepsilon_{yz} - \varepsilon_{yx}\varepsilon_{zy} + \varepsilon_{yy}\varepsilon_{zx}) k_x + (\varepsilon_{xx}\varepsilon_{yz} - \varepsilon_{xz}\varepsilon_{yx} + \varepsilon_{xx}\varepsilon_{zy} - \varepsilon_{xy}\varepsilon_{zx}) \right] k_0^4 + \\ - \left[(\varepsilon_{xz} + \varepsilon_{zx}) k_x^3 + (\varepsilon_{yz} + \varepsilon_{zy}) k_y^3 + (\varepsilon_{xz} k_y + \varepsilon_{yz} k_x + \varepsilon_{zx} k_y + \varepsilon_{zy} k_x) k_x k_y \right] k_0^2,$$

$$C_5 = (\varepsilon_{xx}\varepsilon_{yz}\varepsilon_{zy} - \varepsilon_{xx}\varepsilon_{yy}\varepsilon_{zz} + \varepsilon_{xy}\varepsilon_{yx}\varepsilon_{zz} - \varepsilon_{xy}\varepsilon_{yz}\varepsilon_{zx} - \varepsilon_{xz}\varepsilon_{yx}\varepsilon_{zy} + \varepsilon_{xz}\varepsilon_{yy}\varepsilon_{zx}) k_0^6 + \\ \big(\varepsilon_{xx}\varepsilon_{yy} k_x^2 - \varepsilon_{xy}\varepsilon_{yx} k_x^2 + \varepsilon_{xx}\varepsilon_{yy} k_y^2 - \varepsilon_{xy}\varepsilon_{yx} k_y^2 + \varepsilon_{xx}\varepsilon_{zz} k_x^2 - \varepsilon_{xz}\varepsilon_{zx} k_x^2 + \varepsilon_{yy}\varepsilon_{zz} k_y^2 - \\ \varepsilon_{yz}\varepsilon_{zy} k_y^2 + \varepsilon_{xy}\varepsilon_{zz} k_x k_y - \varepsilon_{xz}\varepsilon_{zy} k_x k_y + \varepsilon_{yx}\varepsilon_{zz} k_x k_y - \varepsilon_{yz}\varepsilon_{zx} k_x k_y \big) k_0^4 + \\ \big(- \varepsilon_{xx} k_x^4 - \varepsilon_{yy} k_y^4 - \varepsilon_{xx} k_x^2 k_y^2 - \varepsilon_{yy} k_x^2 k_y^2 - \varepsilon_{xy} k_x k_y^3 - \varepsilon_{xy} k_x^3 k_y - \varepsilon_{yx} k_x k_y^3 - \varepsilon_{yx} k_x^3 k_y \big) k_0^2.$$

Solving this polynomial equation gives the possible values for k_z, which correspond to the different propagation waves within the biaxial medium. Two of the solutions correspond to forward waves (along the $+z$ direction) and two to backward waves (along the $-z$ direction). The eigenvectors can be calculated based on the information presented in Ref. [81]. Specifically, the electric field eigen-polarization vector can be determined from the following expression:

$$\vec{e} = \text{adj} \left[k_0^2 \tilde{\varepsilon} - k^2 \tilde{I} \right] \vec{k}, \tag{3H.2}$$

where "adj" specifies the adjoint matrix and $\tilde{I}$ is the 3×3 identity matrix. Equation (3H.2) is valid when $det[k_0^2 \tilde{\varepsilon} - k^2 \tilde{I}] \neq 0$. In the case that $det[k_0^2 \tilde{\varepsilon} - k^2 \tilde{I}] = 0$, the eigenvector must be found differently. Specifically, $\vec{e}$ becomes the eigenvector of $k_0^2 \tilde{\varepsilon}\, \vec{e}_i = k_i^2 \vec{e}_i$, under the assumption that the magnetic field is perpendicular to $\vec{e}_i$ (i.e., when $\vec{H} \cdot \vec{e}_i = 0$). If $\vec{H} \cdot \vec{e}_i \neq 0$ the wavevector is aligned along an optic axis, and the directions of the electric and magnetic fields must satisfy Maxwell's equations. The calculated eigenvector should be normalized to become a unit vector, $\hat{e} = \vec{e}/|\vec{e}|$. Using Eq. (3.305) simplifies this process considerably, allowing for a straightforward determination of the wave properties in anisotropic media. It is reminded that the waves could be either propagating or evanescent depending on the excitation conditions. The selection of the forward and backward waves is accomplished with the utilization of the z-component of the Poynting vector, as it was discussed in Sec. 3.8.1.

3H.1 Isotropic-Biaxial Interfaces Boundary Conditions

The continuity of the tangential components of the electric and magnetic fields at the interface between the isotropic and biaxial media must be applied. Specifically, this means that the conditions $E_{1x} = E_{2x}$, $E_{1y} = E_{2y}$, $H_{1x} = H_{2x}$, and $H_{1y} = H_{2y}$ must hold. Using Eqs. (3.309), (3.310), (3.311), and (3.312), the following equations represent the boundary conditions of the problem:

$$u_x - r_{TE}\sin\phi_{inc} - r_{TM}\cos\theta_{inc}\cos\phi_{inc} = t_1 e_{t1x} + t_2 e_{t2x},$$

$$u_y + r_{TE}\cos\phi_{inc} - r_{TM}\cos\theta_{inc}\sin\phi_{inc} = t_1 e_{t1y} + t_2 e_{t2y},$$

$$(k_{y,inc}u_z - k_{z,inc}u_y) + k_{y,inc}r_{TM}\sin\theta_{inc} - k_{rz}(r_{TE}\cos\phi_{inc} - r_{TM}\sin\phi_{inc}\cos\theta_{inc}) =$$
$$t_1(k_{y,inc}e_{t1z} - k_{t1,z}e_{t1y}) + t_2(k_{y,inc}e_{t2z} - k_{t2,z}e_{t2y}),$$

$$(k_{z,inc}u_x - k_{x,inc}u_z) - k_{x,inc}r_{TM}\sin\theta_{inc} - k_{rz}(r_{TE}\sin\phi_{inc} + r_{TM}\cos\phi_{inc}\cos\theta_{inc}) =$$
$$t_1(k_{t1,z}e_{t1x} - k_{x,inc}e_{t1z}) + t_2(k_{t2,z}e_{t2x} - k_{x,inc}e_{t2z}),$$

where $\hat{e}_{ti} = e_{tix}\,\hat{\imath}_x + e_{tiy}\,\hat{\imath}_y + e_{tiz}\,\hat{\imath}_z$ $(i = 1, 2)$. In the above relations, it was assumed that $E_0 = 1$. The latter four equations can be formed into the following linear system of equations:

$$\tilde{\boldsymbol{A}}_{IB}\tilde{\boldsymbol{x}}_{IB} = \boldsymbol{b}_{IB}, \qquad \text{where,} \tag{3H.3}$$

$$\tilde{\boldsymbol{x}}_{IB} = [r_{TE}\quad r_{TE}\quad t_1\quad t_2]^T,$$

$$\boldsymbol{b}_{IB} = [u_x\quad u_y\quad (k_{y,inc}u_z - k_{z,inc}u_y)\quad (k_{z,inc}u_x - k_{x,inc}u_z)]^T.$$

The 4×4 matrix $\tilde{\boldsymbol{A}}_{IB}$ is given by:

$$\tilde{\boldsymbol{A}}_{IB} =$$
$$\begin{bmatrix} \sin\phi_{inc} & \cos\theta_{inc}\cos\phi_{inc} & e_{t1x} & e_{t2x} \\ -\cos\phi_{inc} & \cos\theta_{inc}\sin\phi_{inc} & e_{t1y} & e_{t2y} \\ k_{rz}\cos\phi_{inc} & k_{yz} & k_{y,inc}e_{t1z} - k_{t1,z}e_{t1y} & k_{y,inc}e_{t2z} - k_{t2,z}e_{t2y} \\ k_{rz}\sin\phi_{inc} & k_{xz} & k_{t1,z}e_{t1x} - k_{x,inc}e_{t1z} & k_{t2,z}e_{t2x} - k_{x,inc}e_{t2z} \end{bmatrix},$$

where $k_{yz} = -k_{y,inc}\sin\theta_{inc} - k_{rz}\cos\theta_{inc}\sin\phi_{inc}$,

$$k_{xz} = k_{x,inc}\sin\theta_{inc} + k_{rz}\cos\theta_{inc}\cos\phi_{inc}.$$

3H.2 Biaxial-Biaxial Interfaces Boundary Conditions

The continuity of the tangential components of the electric and magnetic fields at the interface between two biaxial media must be applied. Specifically, this means that the conditions

$E_{1x} = E_{2x}$, $E_{1y} = E_{2y}$, $H_{1x} = H_{2x}$, and $H_{1y} = H_{2y}$ must hold. Using Eqs. (3.316), (3.317), (3.311), and (3.312), the following equations represent the boundary conditions of the problem:

$$
\begin{aligned}
e_{iex} + r_1 e_{r1x} + r_2 e_{r2x} &= t_1 e_{t1x} + t_2 e_{t2x}, \\
e_{iey} + r_1 e_{r1y} + r_2 e_{r2y} &= t_1 e_{t1y} + t_2 e_{t2y}, \\
(k_{y,inc} e_{iez} - k_{z,inc} e_{iey}) + r_1 (k_{y,inc} e_{r1z} - k_{r1,z} e_{r1y}) + r_2 (k_{y,inc} e_{r2z} - k_{r2,z} e_{r2y}) &= \\
t_1 (k_{y,inc} e_{t1z} - k_{t1,z} e_{t1y}) &+ \\
t_2 (k_{y,inc} e_{t2z} - k_{t2,z} e_{t2y}) &, \\
(k_{z,inc} e_{iex} - k_{x,inc} e_{iez}) + r_1 (k_{t1,z} e_{r1x} - k_{x,inc} e_{r1z}) + r_2 (k_{r2,z} e_{r2x} - k_{x,inc} e_{r2z}) &= \\
t_1 (k_{t1,z} e_{t1x} - k_{x,inc} e_{t1z}) &+ \\
t_2 (k_{t2,z} e_{t2x} - k_{x,inc} e_{t2z}) &,
\end{aligned}
$$

where $\hat{e}_{ie} = e_{iex}\hat{\imath}_x + e_{iey}\hat{\imath}_y + e_{iez}\hat{\imath}_z$, $\hat{e}_{uw} = e_{uwx}\hat{\imath}_x + e_{uwy}\hat{\imath}_y + e_{uwz}\hat{\imath}_z$ (with $u = r, t$ and $w = 1, 2$). The latter four equations can be formed into the following linear system of equations:

$$
\begin{aligned}
&\tilde{\boldsymbol{A}}_{BB}\tilde{\boldsymbol{x}}_{BB} = \boldsymbol{b}_{BB}, \qquad \text{where,} \\
&\tilde{\boldsymbol{x}}_{BB} = \begin{bmatrix} r_1 & r_2 & t_1 & t_2 \end{bmatrix}^T, \\
&\boldsymbol{b}_{BB} = \begin{bmatrix} e_{iex} & e_{iey} & (k_{y,inc} e_{iez} - k_{z,inc} e_{iey}) & (k_{z,inc} e_{iex} - k_{x,inc} e_{iez}) \end{bmatrix}^T.
\end{aligned} \tag{3H.4}
$$

The 4×4 matrix $\tilde{\boldsymbol{A}}_{BB}$ is given by:

$$
\tilde{\boldsymbol{A}}_{BB} =
$$

$$
\begin{bmatrix}
-e_{r1x} & -e_{r2x} & e_{t1x} & e_{t2x} \\
-e_{r1y} & -e_{r2y} & e_{t1y} & e_{t2y} \\
k_{r1,z} e_{r1y} - k_{y,inc} e_{r1z} & k_{r2,z} e_{r2y} - k_{y,inc} e_{r2z} & k_{y,inc} e_{t1z} - k_{t1,z} e_{t1y} & k_{y,inc} e_{t2z} - k_{t2,z} e_{t2y} \\
k_{x,inc} e_{r1z} - k_{r1,z} e_{r1x} & k_{x,inc} e_{r2z} - k_{r2,z} e_{r2x} & k_{t1,z} e_{t1x} - k_{x,inc} e_{t1z} & k_{t2,z} e_{t2x} - k_{x,inc} e_{t2z}
\end{bmatrix}
$$

3H.3 Relative Permittivity Tensor Rotations

In many practical problems, it is essential to express the relative permittivity tensor of a biaxial medium in the coordinate system used to analyze the interaction of light, particularly in scenarios involving reflection and transmission at a planar interface. Consider the following coordinate systems: (a) The principal axes system (x_0, y_0, z_0). In this coordinate system, the relative permittivity tensor is diagonal, simplifying the analysis of wave propagation within the medium. (b) The (x, y, z) coordinate system used to describe the physical problem of reflection and transmission at a planar interface involving the biaxial medium. The orientation of the biaxial medium relative to this system is defined by the angles ψ_1 and ψ_2 that describe the rotations needed to align the principal axes with the main coordinate system, as illustrated in Fig. 3.1(a). The relative permittivity tensor, initially diagonal in the (x_0, y_0, z_0) system, must be transformed into the (x, y, z) coordinate system to accurately describe how the biaxial medium interacts with light in this new orientation. The angle ψ_1 defines a counterclockwise rotation of the (x_0, y_0, z_0) system around the x_0

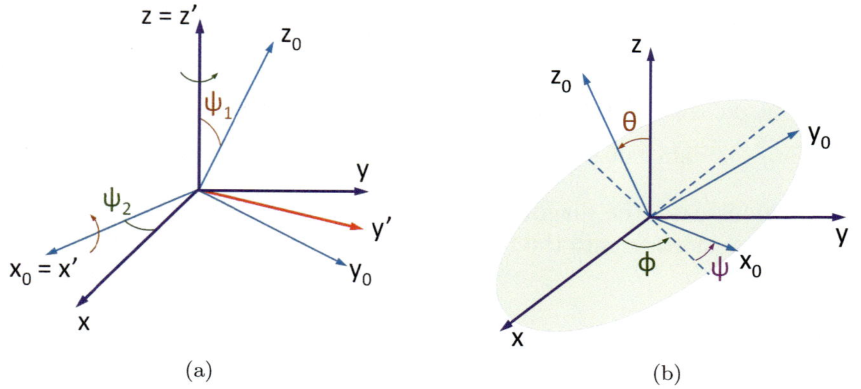

FIGURE 3H.1: (a) The coordinate system (x_0, y_0, z_0) represents the principal axes system of a biaxial medium. The (x, y, z) is the coordinate system used to study reflection and transmission via a planar interface involving the biaxial medium. The (x', y', z') is an auxiliary coordinate system. The angles ψ_1 and ψ_2 relate the (x, y, z) and (x_0, y_0, z_0) coordinate systems. (b) The relation between the (x, y, z) coordinate system with the principal axes system (x_0, y_0, z_0) using the Euler's angles, θ, ϕ, and ψ.

axis resulting in the $(x' = x_0, y', z')$ coordinate system. Furthermore, the angle ψ_2 defines an additional counterclockwise rotation around the $z' = z$, resulting in the final (x, y, z) coordinate system which is used the reflection/transmission analysis. The matrix $\tilde{\boldsymbol{R}}$ relating the unit vectors of the (x, y, z) coordinate system with the unit vectors of the (x_0, y_0, z_0) coordinate system can be expressed as follows:

$$\begin{bmatrix} \hat{\imath}_x \\ \hat{\imath}_y \\ \hat{\imath}_z \end{bmatrix} = \underbrace{\begin{bmatrix} \cos\psi_2 & \cos\psi_1 \sin\psi_2 & \sin\psi_1 \sin\psi_2 \\ -\sin\psi_2 & \cos\psi_1 \cos\psi_2 & \sin\psi_1 \cos\psi_2 \\ 0 & -\sin\psi_1 & \cos\psi_1 \end{bmatrix}}_{= \tilde{\boldsymbol{R}}(\psi_1, \psi_2)} \begin{bmatrix} \hat{\imath}_{x0} \\ \hat{\imath}_{y0} \\ \hat{\imath}_{z0} \end{bmatrix}.$$

The $\hat{\imath}_{x_0}, \hat{\imath}_{y_0}$, and $\hat{\imath}_{z_0}$ are the unit vectors of the (x_0, y_0, z_0) coordinate system. Then, the diagonal relative permittivity tensor $\tilde{\varepsilon}_0$ in the principal axes system, can be transformed into $\tilde{\varepsilon}$ in the (x, y, z) coordinate system, as follows:

$$\tilde{\varepsilon} = \tilde{\boldsymbol{R}}(\psi_1, \psi_2)\, \tilde{\varepsilon}_0\, \tilde{\boldsymbol{R}}^T(\psi_1, \psi_2), \tag{3H.5}$$

where "T" denotes the transpose matrix.

An alternative approach to relate the principal axes system (x_0, y_0, z_0) with the (x, y, z) coordinate system is through the Euler's angles θ, ϕ, and ψ as shown in Fig. 3.1(b). The unit vectors of the principal axes system (x_0, y_0, z_0) are connected to the unit vectors of the (x, y, z) coordinate system by the following relationships [197]:

$$\begin{bmatrix} \hat{\imath}_{x0} \\ \hat{\imath}_{y0} \\ \hat{\imath}_{z0} \end{bmatrix} = \tilde{\boldsymbol{M}}(\theta, \phi, \psi) \begin{bmatrix} \hat{\imath}_x \\ \hat{\imath}_y \\ \hat{\imath}_z \end{bmatrix}.$$

where the matrix $\tilde{\boldsymbol{M}}(\theta, \phi, \psi)$ is given by:

$$\tilde{\boldsymbol{M}} = \begin{bmatrix} \cos\psi\cos\phi - \cos\theta\sin\phi\sin\psi & \cos\psi\sin\phi + \cos\theta\cos\phi\sin\psi & \sin\psi\sin\theta \\ -\sin\psi\cos\phi - \cos\theta\sin\phi\cos\psi & -\sin\psi\sin\phi + \cos\theta\cos\phi\cos\psi & \cos\psi\sin\theta \\ \sin\theta\sin\phi & -\sin\theta\cos\phi & \cos\theta \end{bmatrix}.$$

As in the previous case, the diagonal relative permittivity tensor $\tilde{\varepsilon}_0$ in the principal axes system can be transformed into the relative permittivity tensor $\tilde{\varepsilon}$ in the (x, y, z) coordinate system, as follows:

$$\tilde{\varepsilon} = \tilde{\boldsymbol{M}}^T(\theta, \phi, \psi)\,\tilde{\varepsilon}_0\,\tilde{\boldsymbol{M}}(\theta, \phi, \psi). \tag{3H.6}$$

4

Light Measurement and Emission: Radiometry, Photometry, and Blackbody Radiation

4.1 Introduction

Light measurement and emission are fundamental concepts across various fields, including science, engineering, and daily life. Light measurement involves quantifying the properties of light, such as intensity, brightness, color, and spectrum. Accurate light measurement is essential in applications such as photography, lighting design, optical engineering, and environmental monitoring. Understanding how light behaves, propagates, and interacts with materials enhances the precision of methods used to measure these properties.

Light emission refers to the process by which light is produced from sources like the sun, light bulbs, lasers, and other luminous objects. Light can be described both classically, as electromagnetic radiation involving oscillating electric and magnetic fields, and quantum mechanically, in terms of photons and energy transitions. In the classical sense, light is generated by accelerating charged particles (such as electrons), which leads to the release of electromagnetic radiation. Quantum mechanically, light emission occurs when photons are emitted as particles transition between energy states. There are three primary radiative processes, often referred to as Einstein's processes [39, 97, 198]:

- **Spontaneous Emission:** This process occurs when an excited particle spontaneously transitions to a lower energy state, releasing energy in the form of an emitted photon. Examples include fluorescent lamps and light-emitting diodes (LEDs).
- **Absorption (or Stimulated Absorption):** In this process, a particle absorbs a photon, causing its energy level to rise to a higher state. Light must be present for the particle to absorb a photon.
- **Stimulated Emission:** When an excited particle is perturbed by an external photon with matching energy, it emits an identical photon. This results in two photons with the same energy, phase, and polarization. Stimulated emission is the fundamental principle behind lasers (Light Amplification by Stimulated Emission of Radiation).

Another form of light emission is thermal radiation, where objects at high temperatures emit light. The spectrum of this emission depends on the object's temperature, as seen in incandescent bulbs or sunlight. This phenomenon was crucial in the development of quantum theory by Max Planck, particularly in explaining blackbody radiation.

Radiometry and photometry are scientific disciplines dedicated to the measurement of different aspects of light. Radiometry focuses on quantifying the power of electromagnetic radiation across all wavelengths, whereas photometry is concerned with measuring light as perceived by human vision.

DOI: 10.1201/9781003490333-4

The connection between light emission and its measurement is fundamental, as tools and techniques for quantification allow for the analysis of light's properties. Whether light originates from natural sources like the sun or artificial sources such as bulbs and lasers, its characteristics are critical to a wide range of applications, from everyday illumination to advanced scientific research. Moreover, measuring light to optimize human perception is essential for achieving comfortable and effective lighting in buildings, public spaces, displays, and other environments.

A thorough understanding of the blackbody radiator theory is also crucial, as it underpins many concepts related to light emission and measurement. This theory, which links classical thermodynamics with quantum physics, describes how objects emit radiation as a function of their temperature. It provides key insights into light sources such as stars and incandescent bulbs.

This chapter will focus on the foundational principles of radiometry and photometry, which are essential for the study and measurement of light. It will also introduce the basics of color quantification and measurement, which are vital in applications like display calibration, lighting design, and studies of visual perception.

Additionally, the theory of the blackbody radiator will be explored in detail. This concept played a pivotal role in the development of quantum mechanics by offering a deeper understanding of thermal radiation and the mechanisms of light emission based on temperature. It bridged the gap between classical and modern physics, providing a framework for interpreting light in both theoretical and practical contexts. Together, these topics will offer a comprehensive overview of the methods used to measure and describe light, serving both practical applications and theoretical studies.

4.2 Radiometry and Photometry

4.2.1 Introduction

Radiometry is the science of measuring electromagnetic radiation, specifically in terms of the energy transferred from a source, through a medium, to a receiver. Radiometry treats electromagnetic radiation across the entire spectrum, without giving preferential treatment to any specific wavelength (see Fig. 4.1). Traditionally, radiometry is based on the principles of geometrical optics to describe how energy propagates from a source into the surrounding space. This approach assumes that radiation behaves as incoherent electromagnetic fields, meaning that effects such as coherence, interference, and diffraction are not taken into account. For most radiometric applications, this simplification is sufficient, as these additional complexities are typically unnecessary for practical energy measurement tasks [199].

Radiometry is divided into various regions of the electromagnetic spectrum, with each region using similar measurement techniques. Examples include ultraviolet radiometry, intermediate-infrared radiometry, far-infrared radiometry, and microwave radiometry, all of which are considered distinct subfields. However, these subfields differ from radiometry focused on the visible and near-visible regions of the spectrum.

The subdivision of radiometry that focuses on the measurement of electromagnetic radiation in the visible and near-visible ranges is called *Photometry*. Photometry is unique in that it is based on human perception of light and colors, making it distinct from other forms of radiometry that focus purely on the physical properties of radiation.

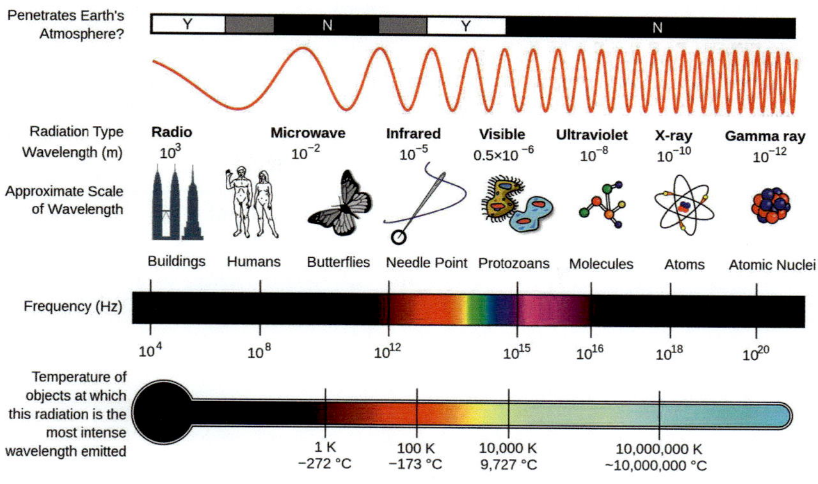

FIGURE 4.1: The electromagnetic spectrum from the radio-wave regime up to the gamma-ray regime. (Based on File: EM Spectrum3-new.jpg by NASA, taken from url-link: https://upload.wikimedia.org/wikipedia/commons/thumb/c/cf/EM_Spectrum_Properties_edit.svg/675px-EM_Spectrum_Properties_edit.svg.png?20160523015848).

4.2.2 Basic Radiometric and Photometric Quantities

An essential aspect of optical system design is its efficiency in transferring light. Achieving this requires the ability to accurately quantify the energy emitted by a source or received by a detector. Various quantities are employed to describe the amount of light, or electromagnetic radiation more broadly, leaving a source or reaching a receiver. Additionally, multiple systems of units are used to represent these measurements, depending on the context and application.

In this section, quantities related to radiometry are prefixed with the term *radiant* and carry the subscript "e" (for electromagnetic). Conversely, quantities related to photometry are prefixed with the term, *luminous* and carry the subscript "v" (for visible). The primary radiometric and photometric quantities include: the radiant/luminous energy (Q_e and Q_v, respectively), the radiant/luminous power (Φ_e and Φ_v, respectively), the radiant/luminous intensity (I_e and I_v, respectively), the radiance/luminance (L_e and L_v, respectively), the irradiance/illuminance (E_e and E_v, respectively), and the radiant/luminous emittance (M_e and M_v respectively). Radiant and luminous intensity, radiance/luminance, and radiant/luminous emittance are typically associated with sources of radiation. In contrast, irradiance and illuminance are usually associated with receivers of radiation. Throughout this chapter, the most widely accepted symbols are used to represent these quantities. However, it should be noted that some of these symbols may represent different physical quantities in other chapters of this book.

All these quantities, along with their definitions and corresponding SI (System International) units, are summarized in Table 4.1. In the definitions of Table 4.1, $d\Omega$ represents the differential solid angle, dA_s refers to the differential area of an extended source, $dA_{s\perp}$ is the projected source area in a direction of observation ($dA_{s\perp} = dA_s \cos\theta$ where θ is the observation angle relative to the normal of the source surface). In the case of photometric quantities, especially illuminance and luminance, several other non-SI units are also in use. The most important of these non-SI units are summarized in Table 4.2 for reference purposes.

TABLE 4.1: Radiometric and Photometric Quantities

Radiometric Quantity	**Symbol**	**Unit**	**Definition**	**Photometric Quantity**	**Symbol**	**Unit**
Radiant Energy	Q_e	Joule		Luminous Energy	Q_v	Talbot
Radiant Power	Φ_e	Watt	$\Phi = \frac{dQ}{dt}$	Luminous Power	Φ_v	Lumen (lm)
Radiant Intensity	I_e	$\frac{\text{Watt}}{\text{sr}}$	$I = \frac{d\Phi}{d\Omega}$	Luminous Intensity	I_v	candela $\text{cd} = \frac{\text{lm}}{\text{sr}}$
Radiance	L_e	$\frac{\text{Watt}}{\text{m}^2\text{sr}}$	$L = \frac{d^2\Phi}{d\Omega dA_{s\perp}}$	Luminance	L_v	$\frac{\text{lm}}{\text{m}^2\text{sr}}$
Irradiance	E_e	$\frac{\text{Watt}}{\text{m}^2}$	$E = \frac{d\Phi}{dA}$	Illuminance	E_v	$\text{Lux} = \frac{\text{lm}}{\text{m}^2}$
Radiant Exitance	M_e	$\frac{\text{Watt}}{\text{m}^2}$	$M = \frac{d\Phi}{dA_s}$	Luminous Exitance	M_v	$\frac{\text{lm}}{\text{m}^2}$

TABLE 4.2: Common Non-SI units of Illuminance and Luminance

Quantity	**SI unit**	**Non-SI Unit**	**Conversion to SI (K)** (non-SI unit) = K×(SI unit)
Illuminance, E_v	Lux	Phot	10000
		Footcandle	10.764
Luminance, L_v	cd/m^2	Stilb	10000
		Apostilb	$1/\pi = 0.3183$
		Lambert	$10^4/\pi = 3183$
		MilliLambert	$10/\pi = 3.183$
		FootLambert	3.426
		Nit	1
		Skot	$10^{-3}/\pi$
		cd/ft^2	10.764
		cd/in^2	1550

4.2.3 Point Source

A point light source (within an isotropic homogeneous medium) typically radiates energy uniformly in all directions. As a result, its intensity $I = d\Phi/d\Omega$, where Φ is the radiant power and Ω is the solid angle, can be considered constant over the surrounding space. A simple representation of a point light source is depicted in Fig. 4.2, and the intensity, I, is given by:

$$I = \frac{d\Phi}{d\Omega} = \frac{\Phi}{\Omega}. \tag{4.1}$$

This equation assumes that the light source radiates energy equally in all directions, distributing the radiant power Φ over the surface area of a sphere with a radius determined by the distance from the source. Now consider a planar surface positioned at a distance

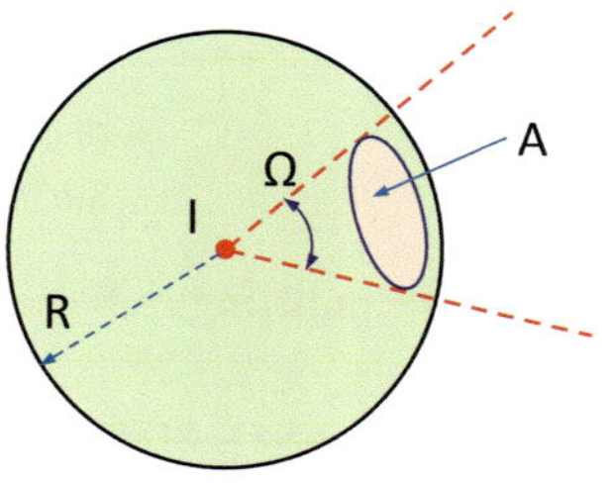

FIGURE 4.2: A point light source emitting radiation isotropically into surrounding space. The intensity of the point source is I.

R from an isotropic point source of intensity I, as shown in Fig. 4.3. The objective is to determine the irradiance E (or illuminance, in the case of visible light) at various points on the planar surface. For a differential area dA that directly faces the point source (i.e., at an angle $\theta = 0$), the irradiance received by this area is given by:

$$E(\theta = 0) = \frac{d\Phi}{dA} = \frac{d\Phi}{R^2 d\Omega} = \frac{d\Phi}{d\Omega}\frac{1}{R^2} = \frac{I}{R^2} = E_0. \tag{4.2}$$

To determine the irradiance at a differential area dA' located at an angle $\theta \neq 0$ with respect to the point source (as shown in Fig. 4.3), both the distance and the orientation of the area relative to the source must be accounted for. The irradiance $E(\theta, \phi)$ due to the point source at this position can be determined as follows (where ϕ is the azimuthal angle):

$$E(\theta, \phi) = \frac{d\Phi'}{dA'} = \frac{d\Phi'}{dA/\cos\theta} = \frac{d\Phi'}{d\Omega' R'^2}\cos\theta = \frac{I}{R^2/\cos^2\theta}\cos\theta = \frac{I}{R^2}\cos^3\theta = E_0\cos^3\theta. \tag{4.3}$$

The last equation reveals that the irradiance received by the differential area on the plane decreases with the cube of the cosine of the angle θ, i.e., the irradiance fall-off is governed by the $\cos^3\theta$ term. This fall-off is illustrated in Fig. 4.4, where the more gradual $\cos\theta$ fall-off is also shown as a reference. The $\cos^3\theta$ dependence indicates that the irradiance drops off more sharply as the angle θ increases, meaning the surface receives significantly less light as it tilts further away from being perpendicular to the point source. This effect becomes more pronounced at larger angles. Additionally, since the point source is isotropic, the irradiance is independent of the azimuthal angle ϕ. This means that the irradiance at a given distance and angle θ from the source is the same in all directions around the source in the horizontal plane. The fall-off behavior is solely a function of the zenith (polar) angle θ.

4.2.4 Extended Source

For an extended source, the radiance (or luminance), L, is the key quantity used to describe the amount of light emitted in a specific direction from the source. Referring to the geometric configuration illustrated in Fig. 4.5, the radiance L is defined as:

$$L = \frac{d^2\Phi}{d\Omega dA_{s\perp}} = \frac{d^2\Phi}{d\Omega dA_s\cos\theta} = \frac{dI}{dA_s\cos\theta}. \tag{4.4}$$

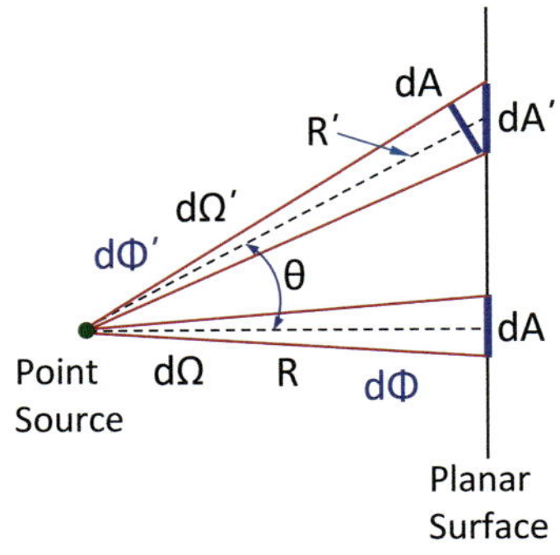

FIGURE 4.3: A point light source emitting radiation in front of a planar surface.

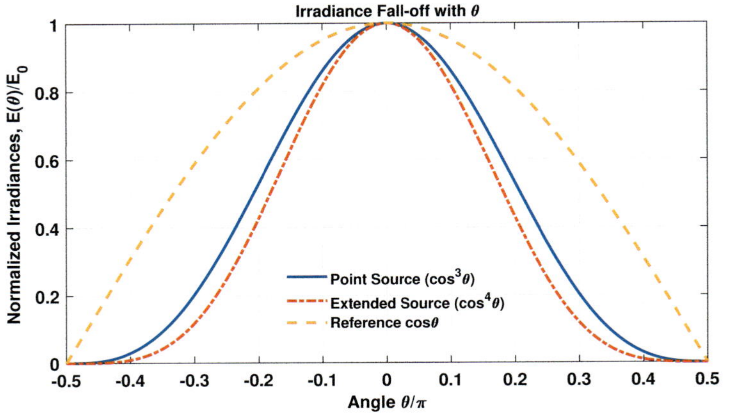

FIGURE 4.4: The fall-off of irradiance as a function of the angle θ for a point source and an extended source. The cosine curve is shown for reference purposes.

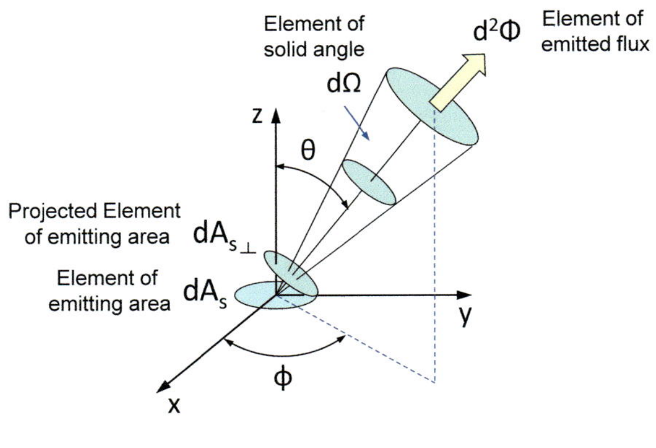

FIGURE 4.5: An extended source of differential element dA_s, emitting radiation flux of $d^2\Phi$ inside a solid angle $d\Omega$. The differential element $dA_{s\perp}$ corresponds to the projected emitting area in the direction specified by the azimuthal angle ϕ and the polar angle θ.

It is important to note that differential elements in radiance calculations can indeed depend on more than one variable. Therefore, the radiance in Eq. (4.4) can be expressed using partial derivatives [200]. For simplicity, the notation used here retains a straightforward form, but it should be understood that the actual expressions account for all relevant variables.

For instance, the solid angle Ω is generally a function of both polar and azimuthal angles, $\Omega = \Omega(\theta, \phi)$, and can be represented by $d\Omega = \sin\theta d\theta d\phi$ in a spherical coordinate system. Additionally, the surface area A_s might also vary with spatial coordinates, $A_s = A_s(x, y, z)$, where the variables x, y, z define the geometry of the radiating source surface. This approach ensures that all dependencies are considered, providing accurate and comprehensive measurements of radiance across different geometries and observation angles.

From the previous definition, it is evident that when an extended source is observed at a grazing angle (i.e., when $\theta \to \pi/2$), the radiance or luminance tends to infinity. This implies that the source would appear brighter as the observation angle θ increases. However, this contradicts everyday experience, where extended sources typically appear dimmer when observed at angles closer to the grazing angle. A special case arises when the radiance or luminance remains constant regardless of the angle θ. Such a source is known as a *Lambertian* source. For a Lambertian source, the intensity and radiance are given by:

$$I(\theta) = I_0 \cos\theta, \tag{4.5}$$

$$L(\theta) = L_0, \tag{4.6}$$

where I_0 and L_0 are constants. In this case, the radiance (or luminance) remains constant across all observation angles, providing a uniform appearance regardless of the angle of view. This is characteristic of ideal diffusive or matte surfaces, where the light is scattered evenly in all directions, ensuring a consistent perceived brightness from any angle. A polar plot of the intensity and radiance of a Lambertian source is shown in Fig. 4.6.

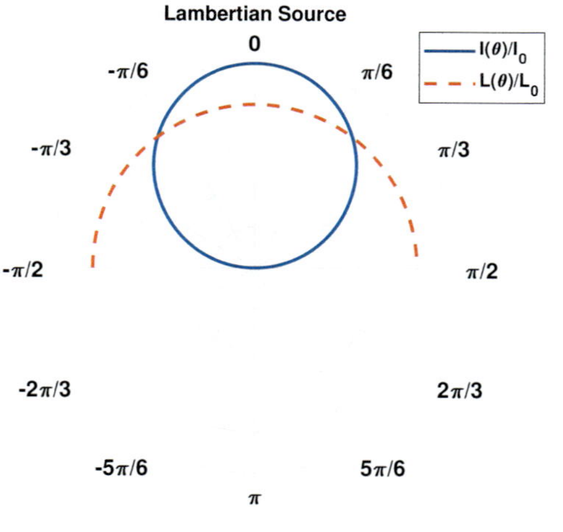

FIGURE 4.6: The normalized intensity $I(\theta)/I_0$ and the normalized radiance $L(\theta)/L_0$ of a Lambertian source in a polar diagram.

Now, consider a planar surface at a distance R from an extended source of radiance L as shown in Fig. 4.7. The goal is to determine the irradiance (or illuminance) at various points on the planar surface. Consider a differential area dA on the planar surface, located at an angle $\theta = 0$ (directly facing the extended source). Additionally, consider an elementary area dA_s on the extended source. The irradiance dE received at the plane on the area dA is given by:

$$dE(\theta = 0) = \frac{d^2\Phi}{dA} = \frac{L d\Omega dA_s}{dA} = L\,\frac{dA\,dA_s}{R^2\,dA} = L\,\frac{dA_s}{R^2}.$$

If the extended source is Lambertian and has a size with dimensions much smaller than the distance from the planar surface R, the irradiance can be calculated by integrating over the

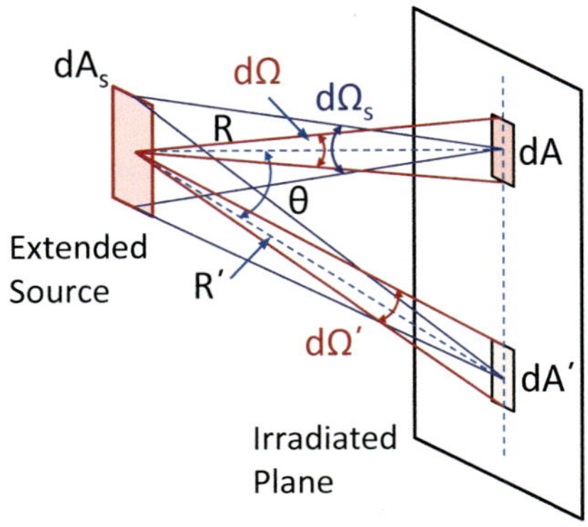

FIGURE 4.7: An extended light source emitting radiation in front of a planar surface.

surface area of the source. Then, $E(\theta = 0)$ is expressed as follows:

$$E(\theta = 0) = E_0 = \int_{A_s} L \frac{dA_s}{R^2} = L \int_{A_s} \frac{dA_s}{R^2} \simeq L\Omega_s, \tag{4.7}$$

where Ω_s is the solid angle subtended by the extended source area as seen from the center of the differential area dA. The irradiance at the infinitesimal area dA' can be determined as follows:

$$dE(\theta) = \frac{d^2\Phi'}{dA'} = \frac{Ld\Omega' dA_s \cos\theta}{dA'} = L \frac{dA' \cos\theta}{R'^2} \frac{dA_s \cos\theta}{dA'} = L \frac{dA_s}{R^2} \cos^4\theta,$$

where in the last equation the relation $R' = R/\cos\theta$ was utilized. Assuming again that the extended source is Lambertian and of dimensions much smaller than the distance R the irradiance at the plane in the direction of θ is given by:

$$E(\theta) = \int_{A_s} L \frac{dA_s}{R^2} \cos^4\theta = L \int_{A_s} \frac{dA_s}{R^2} \cos^4\theta \simeq L\Omega_s \cos^4\theta = E_0 \cos^4\theta. \tag{4.8}$$

The last equation is referred to as the irradiance cosine-fourth-power fall-off. Thus, the irradiance from an extended Lambertian source decreases more rapidly than that from a point source, which decreases according to a cosine-cube fall-off, as demonstrated in Eq. (4.3). This difference in fall-off rates is illustrated in Fig. 4.4, where both the simple cosine fall-off and the point source fall-off are shown for comparison. The irradiance of the emitted light is isotropic in the azimuthal direction, meaning that no preferential direction exists in the plane of the surface. Therefore, any azimuthal angle ϕ variation is absent in the Lambertian case, and it only depends on the polar angle θ.

4.2.5 Conservation Laws for Radiance/Luminance

Let's consider an infinitesimal beam propagating through a lossless, isotropic, and homogeneous medium, as depicted in Fig. 4.8. The beam intersects two infinitesimal fictitious surfaces, dA_1 and dA_2 as shown. The normals of these surfaces form angles θ_1 and θ_2 with respect to the ray direction, respectively. Since the areas are infinitesimal, there is no significant difference in the radiance L_1 among rays diverging from any point on dA_1 and the radiance L_2 among rays converging to any point on dA_2 [201, 202]. The radiant (or luminous) infinitesimal power (flux) $d^2\Phi_1$ of the beam leaving surface dA_1 is given by:

$$d^2\Phi_1 = L_1 dA_1 \cos\theta_1 d\Omega_1 = L_1 dA_1 \cos\theta_1 \left(\frac{dA_2 \cos\theta_2}{R^2}\right) = L_1 \frac{dA_1 dA_2 \cos\theta_1 \cos\theta_2}{R^2}.$$

Similarly, the radiant (or luminous) infinitesimal power $d^2\Phi_2$ of the beam leaving surface dA_2 is given by:

$$d^2\Phi_2 = L_2 dA_2 \cos\theta_2 d\Omega_2 = L_2 dA_2 \cos\theta_2 \left(\frac{dA_1 \cos\theta_1}{R^2}\right) = L_2 \frac{dA_2 dA_1 \cos\theta_2 \cos\theta_1}{R^2}.$$

Since the medium is lossless, the total power exiting surface dA_1 must equal the total power entering surface dA_2, which implies that $d^2\Phi_1 = d^2\Phi_2$. Equating the two expressions for $d^2\Phi_1$ and $d^2\Phi_2$, we get:

$$L_1 = L_2. \tag{4.9}$$

Because no restrictions were imposed on the areas dA_1 and dA_2 of the beam or their centers C_1 and C_2, Eq. (4.9) must apply to any pair of points along the beam, assuming a lossless, isotropic, and homogeneous medium. This property is known as radiance (or luminance) invariance or radiance (or luminance) conservation [201, 202]. The radiance L is invariant along the propagation of a beam in a lossless, isotropic, and homogeneous medium. This principle is crucial in various optical systems, particularly when dealing with light propagation through clear materials like air, glass, or water, where the radiance remains unchanged during the propagation unless external factors (like absorption or scattering) are involved.

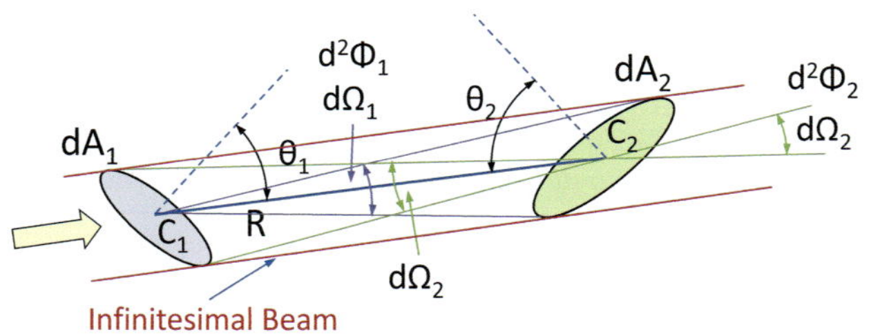

FIGURE 4.8: An infinitesimal beam carrying radiant/luminous flux inside a lossless isotropic and homogeneous medium. The beam is crossing two infinitesimal surfaces dA_1 and dA_2 separated by a distance R. The planes of the infinitesimal surfaces are at angles θ_1 and θ_2 with respect to their normals. The points C_1 and C_2 are the centers of the infinitesimal areas.

When considering a beam refracting at the interface between two different lossless, isotropic, and homogeneous media, radiance conservation must be taken into account. This principle ensures that the radiance remains constant in the direction of propagation across the interface, but the distribution of radiance may change due to the difference in refractive indices of the two media. The setup is illustrated in Fig. 4.9. An infinitesimal area dA on the smooth boundary surface S between the two media is examined [201]. Using similar arguments as before, the infinitesimal power incident on the area dA of the boundary can be expressed as:

$$d^2\Phi_1 = L_1 dA \cos\theta_1 d\Omega_1 = L_1 dA \cos\theta_1 \sin\theta_1 d\theta_1 d\phi,$$

where L_1 is the radiance (or luminance) from the side of dA facing the medium of refractive index n_1 and ϕ is the azimuthal angle on the tangential plane to the boundary at the

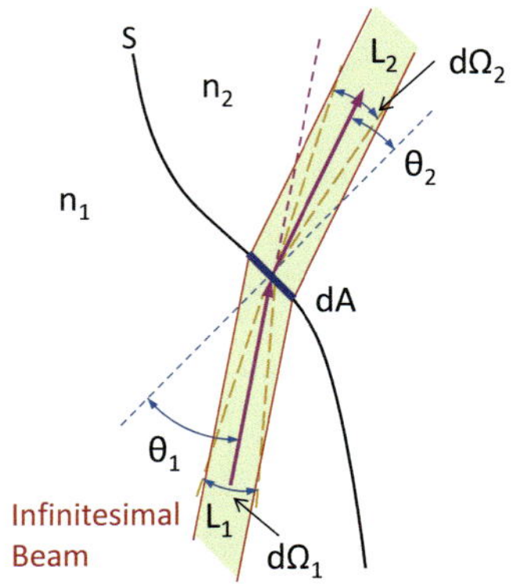

FIGURE 4.9: An infinitesimal beam carrying radiant/luminous flux refracting at the interface between two lossless isotropic and homogeneous media of refractive indices n_1 and n_2 respectively. The beam is crossing an infinitesimal surface dA at the boundary between the two media.

position of dA. Similarly, the power that is refracting into the medium of refractive index n_2 is given by:

$$d^2\Phi_2 = L_2 dA \cos\theta_2 d\Omega_2 = L_2 dA \cos\theta_2 \sin\theta_2 d\theta_2 d\phi,$$

where L_2 is the radiance (or luminance) from the side of dA facing medium of refractive index n_2. If it is assumed that the power reflection coefficient at the boundary is ρ (for the angle of incidence θ_1, neglecting polarization or averaging polarization), then the relation between $d^2\Phi_1$ and $d^2\Phi_2$ should be:

$$d^2\Phi_2 = (1-\rho)d^2\Phi_1 \Longrightarrow$$

$$L_2 dA \cos\theta_2 \sin\theta_2 d\theta_2 d\phi = (1-\rho)L_1 dA \cos\theta_1 \sin\theta_1 d\theta_1 d\phi \Longrightarrow$$

$$L_2 \cos\theta_2 \sin\theta_2 d\theta_2 = (1-\rho)L_1 \cos\theta_1 \sin\theta_1 d\theta_1.$$

From Snell's law it is straightforward to show that $n_1 \cos\theta_1 d\theta_1 = n_2 \cos\theta_2 d\theta_2$. Then, applying Snell's law the last equation results in:

$$\frac{L_2}{n_2^2} = (1-\rho)\frac{L_1}{n_1^2}, \tag{4.10}$$

where Eq. (4.10) represents the *radiance (or luminance) conservation via an interface* between two media. If reflections can be neglected ($\rho \simeq 0$), then the relationship $L_1/n_1^2 = L_2/n_2^2$ holds. Of course, if $n_1 = n_2$, Eq. (4.10) simplifies to Eq. (4.9) (where $\rho = 0$, since there is no refractive index change).

The exitance, M, can be expressed in terms of the radiance as follows:

$$M = \frac{d\Phi}{dA_s} = \int L \cos\theta \, d\Omega. \tag{4.11}$$

For Lambertian sources the radiance L is constant and Eq. (4.11) becomes:

$$M = \frac{d\Phi}{dA_s} = L \int_{\theta=0}^{\pi/2} \int_{\phi=0}^{2\pi} \cos\theta \sin\theta d\theta d\phi = L\pi. \tag{4.12}$$

In the previous derivation, it was assumed that the source emits in the upper half-space. Using the exitances, the conservation law of Eq. (4.10) can also be expressed as:

$$\frac{M_2}{n_2^2} = (1-\rho)\frac{M_1}{n_1^2}. \tag{4.13}$$

4.3 Basics of Photometry

Photometry deals with the measurement of light in terms of its perceived brightness to the human eye, focusing on the visible and near-visible parts of the electromagnetic spectrum. To understand photometry, it's essential to first explore how the human eye responds to light. A simplified diagram of the human eye is shown in Fig. 4.10. Light enters through the cornea, passes through the aqueous humor, iris, and eye lens, and then moves through the vitreous humor before reaching the retina, which acts as the photodetector of the eye. The retina contains two types of photoreceptors: *cone* cells and *rod* cells. These photoreceptors convert absorbed photon energy into electrochemical signals that are transmitted via the optic nerve to the back area of the brain, where they are processed to create visual perception. The cone cells are primarily responsible for the day vision (which means good illumination conditions) while the rod cells are mainly responsible for the night vision (which means poor illumination conditions). The cone-based vision is called *photopic vision* while the rod-based vision is called *scotopic vision*. There is also an intermediate vision regime called *mesopic vision*, where both cone and rod cells are active, and the distinctions between the roles of cones and rods are less clear. A visual representation of these vision regimes is depicted in Fig. 4.11 (adapted from Ref. [203]).

In practical terms, photopic and scotopic vision are distinguished by the range of luminance in which they function. Photopic vision generally dominates in luminance levels of approximately $3\,cd/m^2$ or higher, while scotopic vision is prevalent at luminance levels of about $0.003\,cd/m^2$ or lower [204]. The mesopic vision regime occurs in between, covering luminance levels from $0.003\,cd/m^2$ to $3\,cd/m^2$. However, as indicated in Fig. 4.11 these boundaries are not strictly defined and can vary among individuals. Rod cells, which are more sensitive, require at least 5–14 photons to respond at very low luminance levels and become insensitive when luminance drops to around $10^{-6}\,cd/m^2$. In contrast, cone cells need approximately 100–1000 photons to elicit a response [204]. For comparison, high-speed photographic film needs four or more photons to induce a reaction in its fine silver halide grains. Thus, the detectivity of human eye rod cells is comparable to that of photographic film. On the other hand, modern CCD arrays can achieve a sensitivity of about 1 photon per pixel per second across a sensor with 1 million active pixels [205]. Rod cells are sensitive

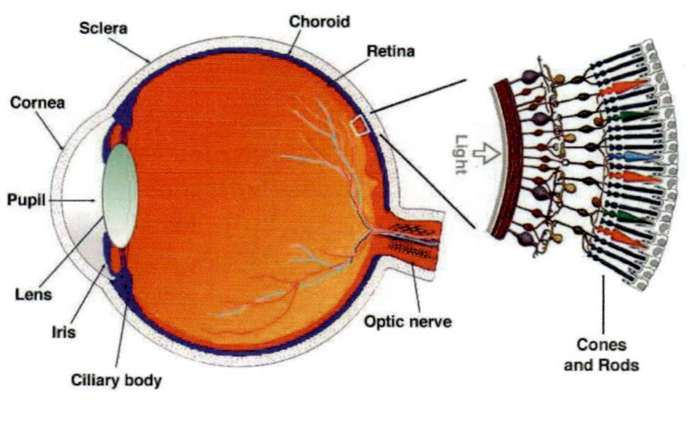

FIGURE 4.10: A diagram of the human eye with its photoreceptors (three types of cones and rods) (from url-link: `http://www.blueconemonochromacy.org/how-the-eye-functions/`).

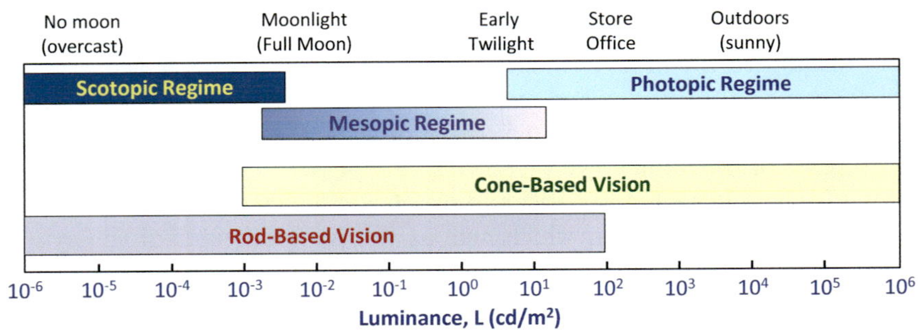

FIGURE 4.11: Approximate ranges of vision regimes (slightly modified from Ref. [203]).

across a broad range of the visible spectrum, whereas cone cells are specialized for specific parts of it. Cone cells are categorized into three main types: (a) The red-type (R-type) cone cells (or equivalently long wavelength L-cone cells), which are more sensitive in the red-green part of the visible spectrum; (b) the green-type (G-type) cone cells (or equivalently medium wavelength M-cone cells) which are more sensitive in the green-blue part of the visible spectrum; and finally (c) the blue-type (B-type) cone cells (or equivalently short wavelength S-cone cells), which are more sensitive in the blue-violet part of the visible spectrum. The spectral sensitivity of rod and cone cells is illustrated by their absorption characteristics, as shown in Fig. 4.12(a). Rod cells are more prevalent in the retina, with peak density located away from the central vision region (fovea centralis). In contrast, cone cells are fewer in number but are concentrated in the central part of the retina. The fovea centralis, a small area about 1.5 mm in diameter, has the highest density of cone cells. The distribution of cone and rod cells is depicted in Fig. 4.12(b). Notably, there is a region in the retina where neither cone nor rod cells are present, known as the blind spot, where the optic nerve exits the eye. On average, the human retina contains approximately 100 million rod cells and 7 million cone cells [204]. The relative distribution of the cone types varies, with the typical ratio of red-type, green-type, and blue-type cone cells being approximately 32:16:1 [204]. This indicates a significantly higher number of red-type and green-type cone cells compared to blue-type cone cells.

4.3.1 Human Eye Response and Its Luminous Efficiency

The human eye functions as a sophisticated photodetector, converting radiant energy (photons) into electrochemical signals that are then processed by the brain. As discussed previously, the human photodetector system comprises both cone cells and rod cells. The spectral sensitivity of these cells is illustrated in Fig. 4.12(a). These cells ultimately produce outputs in terms of brightness and color responses. However, the response of the human eye is not based on physical measurements but on the subjective sensations of brightness and color perceived under various lighting conditions. To quantify the spectral sensation of the human eye, a matching method is employed [204]. This approach involves comparing the luminous power Φ_v of a test light with an arbitrary wavelength to the sensation produced by a reference light of a specific wavelength. By measuring the radiant power Φ_e of the test light under these matching conditions, the luminous power and radiant power can be related through the following equation:

$$\Phi_v = K\Phi_e, \tag{4.14}$$

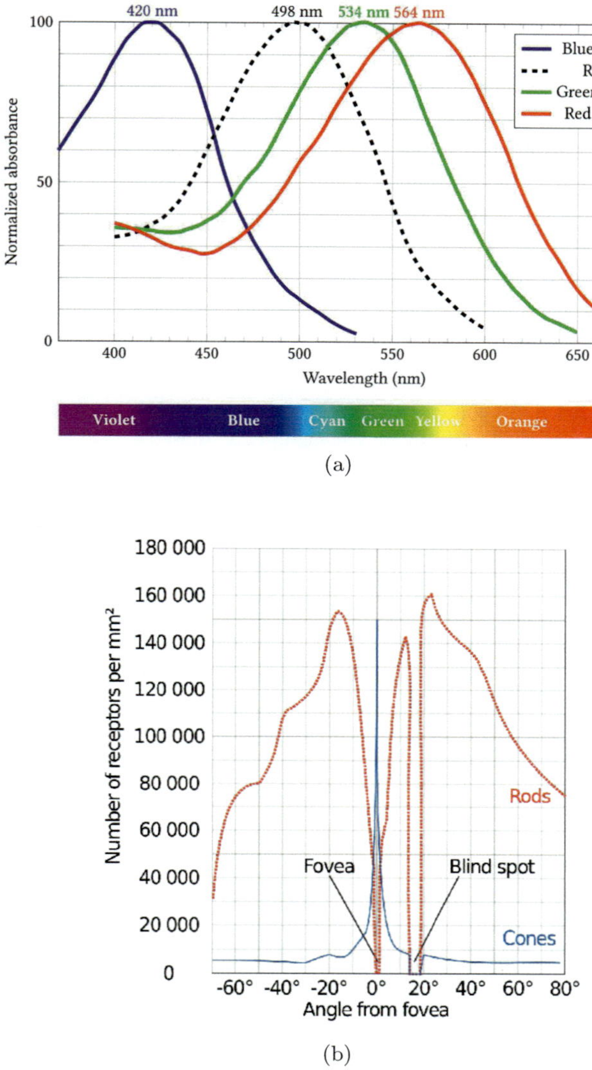

FIGURE 4.12: (a) Normalized human photoreceptor absorbances for different wavelengths of light (from url-link: `https://en.wikipedia.org/wiki/Photoreceptor_cell`). (b) The density of photoreceptors (cones and rods) of the human eye (from url-link: `https://en.wikipedia.org/wiki/Photoreceptor_cell`).

where K is a measure of the brightness sensation per unit radiant power and could be defined as the responsivity of the human eye.

Several methods exist for matching the test light to the reference light [204]. Among these, the flicker method is one of the most common and practical. In this method, a reference light of wavelength λ_{ref} and a test light of wavelength λ_{test} are alternated in the visual field of an individual. The perceived color can exhibit flickering due to the alternating lights. A specific frequency of alternation can be found where flickering is noticeable when there is a difference in brightness between the two lights. In this regime, the brightness of the test light is adjusted until flickering disappears, indicating that the two lights are perceived as matching. The condition at which flickering ceases is considered to be the point where the two lights are perceptually equivalent. Based on this technique and similar

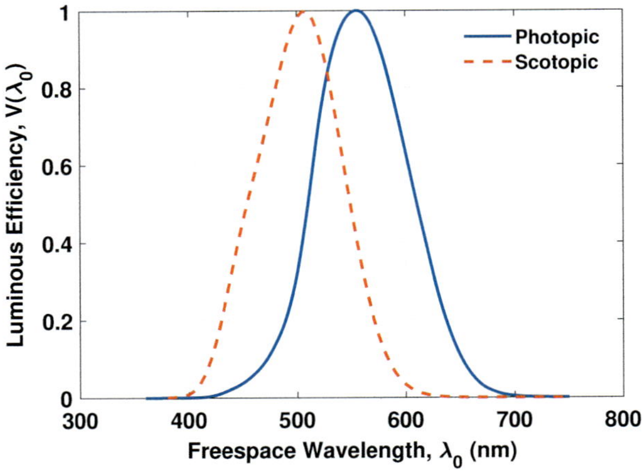

FIGURE 4.13: Luminous efficiency curve under good illumination conditions (photopic) and poor illumination conditions (scotopic). Numerical values per 5nm wavelength intervals are listed in Appendix 4A and Table 4A.1.

methods, the value of K in Eq. (4.14) can be determined. The quantity K is known as the luminous efficacy of radiation and is dependent on the wavelength $K(\lambda_0)$. If K_m represents the maximum value of $K(\lambda_0)$, then we have the relationship $K(\lambda_0) = K_m V(\lambda_0)$, where $V(\lambda_0)$ is referred to as the *luminous efficiency*. Consequently, the following equations can be written:

$$K(\lambda_0) = K_m V(\lambda_0), \tag{4.15}$$

$$K'(\lambda_0) = K'_m V'(\lambda_0), \tag{4.16}$$

where the non-primed variables correspond to photopic vision conditions, and the primed variables to scotopic vision conditions, respectively.

The standard photopic (or scotopic) luminous efficiency function $V(\lambda_0)$ [206, 208] (or $V'(\lambda_0)$ [209]) is derived from a complex combination of measurement data obtained through various methods and sources. These luminous efficiency functions are shown as functions of free-space wavelength in Fig. 4.13. The uncertainty surrounding these functions, especially for $V(\lambda_0)$, is highlighted by the significant variations observed among different studies (some diverging by as much as a factor of ten in the violet range [206, 210]). It is noted that the standard luminous efficiency function $V(\lambda_0)$ tends to underestimate sensitivity at short wavelengths. The constants K_m and K'_m of Eqs. (4.15) and (4.16) have been set equal to $683\,lm/W$ and $1700\,lm/W$, respectively. The peak of the $V(\lambda_0)$ luminous efficiency occurs at $\lambda_0 = 555\mu$m, while $V'(\lambda_0)$ peaks at $\lambda_0 = 507\,\mu$m. Consequently, $683\,lm$ is defined as equivalent to one Watt of radiant power at the free-space wavelength of $\lambda_0 = 555\,\mu$m. Since the initial publication of luminous efficiency data by the CIE in 1924, several updates and improvements have been made. However, for the purposes of this introductory material, no additional curves are discussed. Interested readers are encouraged to refer to Refs. [204] and [211] for more detailed information.

Equations (4.15) and (4.16) can be generalized for the conversion of the radiant power of a light source into luminous power. If $\Phi_e(\lambda_0)$ is the spectral radiant power of a light source, then its luminous power is given by:

$$\Phi_v = 683(lm/W) \int_{\lambda_0} \Phi_e(\lambda_0) V(\lambda_0) d\lambda_0, \qquad \text{Photopic Vision,} \tag{4.17}$$

$$\Phi_v = 1700(lm/W) \int_{\lambda_0} \Phi_e(\lambda_0) V'(\lambda_0) d\lambda_0, \qquad \text{Scotopic Vision,} \tag{4.18}$$

where the integrals span the entire spectrum of the light source. For monochromatic sources, such as most lasers, the conversion between radiant power and luminous power can be simplified due to the Dirac delta function representation of spectral radiant power. If the spectral radiant power is given by $\Phi_e(\lambda_0) = P_0\delta(\lambda_0 - \lambda_{0L})$, where λ_{0L} is the wavelength of the monochromatic source and P_0 its corresponding spectral power, then the luminous power Φ_v can be expressed as $\Phi_v = \kappa P_0 V(\lambda_{0L})\delta(\lambda_0 - \lambda_{0L})$, where $\kappa = K_m = 683\, lm/W$ for the photopic regime or $\kappa = K'_m = 1700\, lm/W$ for the scotopic regime. Here, $V(\lambda_{0L})$ is the luminous efficiency function at the wavelength λ_{0L}. In this case, the delta function simplifies the integration process, resulting in the direct multiplication of the spectral power by the luminous efficacy and the luminous efficiency at the specific wavelength.

The situation for the mesopic vision regime is more complex, as there is no universally accepted standard luminous efficiency curve like those established for the photopic and scotopic regimes. Recent proposals have attempted to address this gap [212–214]. For example, Ref. [213] introduces a definition for mesopic luminous efficiency, $V_{mes,m}(\lambda_0)$, which can be expressed as:

$$M(m)V_{mes,m}(\lambda_0) = mV(\lambda_0) + (1 - m)V'(\lambda_0). \tag{4.19}$$

Here, m is the adaptation coefficient, which according to Ref. [213] can be determined

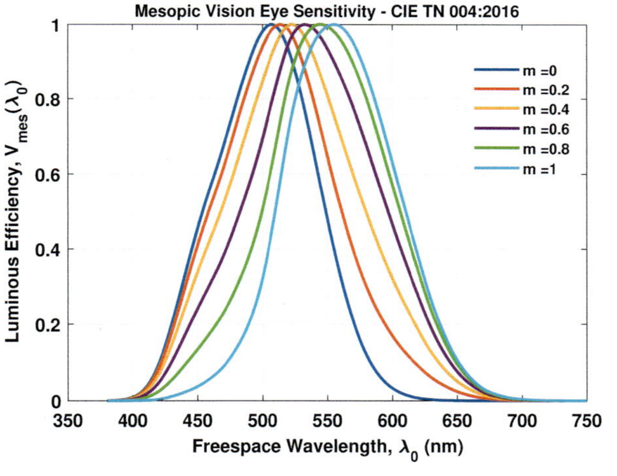

FIGURE 4.14: Mesopic luminous efficiency curves for various values of the adaptation coefficient. The $m = 1$ corresponds to photopic luminous efficiency and the $m = 0$ corresponds to the scotopic luminous efficiency.

from the photopic adaptation luminance and spectral characteristics of the visual adaptation field, and $M(m)$ is a normalization function such as the mesopic luminous efficiency $V_{mes,m}(\lambda_0)$ can have a maximum of unity. This mesopic luminous efficiency should be used for the mesopic vision regime where the luminance is approximately between $0.003\, cd/m^2$ and $3\, cd/m^2$. This function reflects a combination of photopic and scotopic sensitivities, adapted to the intermediate light levels characteristic of mesopic vision. Various methods have been used to derive this function, considering factors such as the relative contributions of rod and cone cells under mesopic conditions. Unlike the well-defined photopic and scotopic luminous efficiency curves, the mesopic luminous efficiency $V_{mes,m}(\lambda_0)$ is not standardized and can vary based on recent research. For $m = 1$ the mesopic luminous efficiency becomes the photopic one, $V(\lambda_0)$, while for $m = 0$ it becomes the scotopic one, $V'(\lambda_0)$. The spectral mesopic luminous efficiencies for various values of the adaptation coefficient m are shown in Fig. 4.14. The maximum mesopic luminous efficacy $K_{mes,m}$ can be determined from the following equation [213]:

$$K_{mes,m} = \frac{683}{V_{mes,m}(\lambda_0 = \lambda_{0,p})}, \qquad \text{lm/W}, \tag{4.20}$$

where $\lambda_{0,p} = 555\,\text{nm}$, i.e., the peak wavelength of the photopic luminous efficiency. The values of the normalization function $M(m)$ and of the maximum mesopic luminous efficacy $K_{mes,m}$ are shown as functions of the adaptation coefficient m in Figs. 4.15(a) and 4.15(b), respectively. It can be observed that the maximum mesopic efficacy for $m = 1$ is $683\,lm/W$ and for $m = 0$ is $1700\,lm/W$ respectively, as expected for the photopic and scotopic vision regimes. Finally, the resulting luminous power of a light source in the mesopic regime can be determined by:

$$\Phi_v = K_{mes,m}(lm/W) \int_{\lambda_0} \Phi_e(\lambda_0) V_{mes,m}(\lambda_0) d\lambda_0, \qquad \text{Mesopic Vision.} \tag{4.21}$$

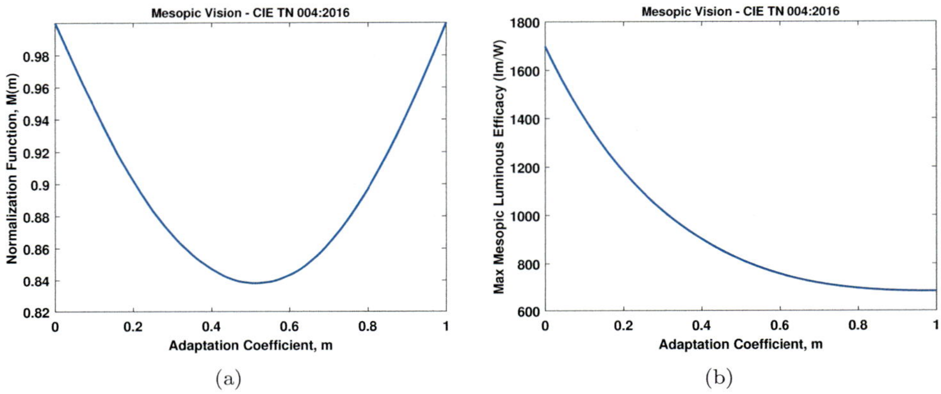

FIGURE 4.15: (a) The normalization function $M(m)$ as a function of the adaptation coefficient m. (b) The maximum mesopic efficacy $K_{mes}(m)$ (in lm/W) as a function of the adaptation coefficient m.

Example 4.3.1:

Find the luminous power of a 50 mW He-Ne laser operating at a freespace wavelength of $\lambda_0 = 632.8\,\text{nm}$.

Solution: For this simple example, Eq. (4.17) should be utilized. The laser radiant spectral power is $\Phi_e(\lambda_0) = P_0\delta(\lambda_0 - \lambda_L)$, where $P_0 = 50 \times 10^{-3}\,\text{W}$, and $\lambda_L = 632.8\,\text{nm}$. To apply Eq. (4.17), the luminous (photopic) efficiency $V(\lambda_L)$ needs to be determined. The value of $V(\lambda_L)$ is estimated via linear interpolation from the values listed in Table 4A.1. From the table, $V(630nm) = 0.298086$ and $V(635nm) = 0.241690$. Therefore,

$$V(632.8) = V(630) + \frac{V(635) - V(630)}{635 - 630}(632.8 - 630) = 0.2665.$$

Consequently, the luminous power is:

$$\Phi_v = 683\,(lm/W) \times 50 \times 10^{-3}\,W \times 0.2665 = 9.1\,lm.$$

Example 4.3.2:

Equal energy (actually power) white light of 1 W is distributed in the wavelength interval $360\,nm \leq \lambda_0 \leq 750\,nm$. Find the luminous power of the white light.
Solution: Using again Eq. (4.17) the luminous power, Φ_v, can be determined as follows:

$$\Phi_v = 683\left(\frac{lm}{W}\right)\int_{\lambda_0} V(\lambda_0)\Phi_e(\lambda_0)d\lambda_0 \quad \Longrightarrow$$
$$\Phi_v \simeq 683\left(\frac{lm}{W}\right)\sum_i V(\lambda_{0,i})\Phi_e(\lambda_{0,i})\Delta\lambda_0$$
$$= 683\left(\frac{lm}{W}\right)\sum_{i=1}^{N}\frac{V(\lambda_{0,i})+V(\lambda_{0,i+1})}{2}\,\frac{1\,W}{\lambda_{max}-\lambda_{min}}\left(\frac{1}{nm}\right)\Delta\lambda_0(nm)$$
$$= 187.1336\,lm.$$

Using a wavelength interval of $\Delta\lambda_0 = 1\,\text{nm}$, the values of $\lambda_{0,i}$ are given by $\lambda_{0,i} = 360 + (i-1)\Delta\lambda_0$ for $i = 1, 2, \cdots, 391$. The $V(\lambda_{0i})$ values corresponding to this interval were obtained from cite url:`http://www.cvrl.org`. Alternatively, a coarser interval of $\Delta\lambda_0 = 5\,\text{nm}$ was used, with values from Table 4A.1, resulting in $\Phi_v = 197.2485\,lm$, for the wavelength range $380\,nm \leq \lambda_0 \leq 750\,nm$. The lower wavelength bound was adjusted to match the data in Table 4A.1. The observed difference is attributed to the redistribution of the 1 W power over a shorter wavelength range. For comparison, using $\Delta\lambda_0 = 1\,\text{nm}$ in the range $380\,nm \leq \lambda_0 \leq 750\,nm$ yielded $\Phi_v = 197.2483\,lm$, demonstrating a negligible difference from the coarser interval.

If the spectral power of the source is evenly distributed over the wavelength range $400\,nm \leq \lambda_0 \leq 700\,nm$, the resulting luminous power is $\Phi_v = 243.1340\,lm$. This value is higher than the previous results, as the 1 W power is confined to a narrower wavelength range where the luminous efficiency function $V(\lambda_0)$ exhibits its primary distribution.

4.4 Colorimetry Basics

Colorimetry is the science and technology dedicated to assessing, quantifying, and measuring color as perceived by the human eye. In this field, subjective color descriptions are replaced by objective numerical values. The foundations of colorimetry were established in the early 19th century through the work of Young, Helmholtz, and Maxwell, who identified the principles of additive and subtractive color mixing and proposed the trichromatic theory of human color vision. The trichromatic theory is based on the three types of cone photoreceptors in the human retina (L-type, M-type, and S-type) and their responses to various stimuli. According to empirical laws, light incident on the retina is absorbed by these three cone types, and color matching between two stimuli occurs when the cone responses to both stimuli are identical. This approach is known as the *receptor-level theory of color matching* [215]. Given the three cone types, it is reasonable to expect that three values could specify the visual effects produced by any spectral distribution of stimuli. Consequently, the trichromatic theory is based on the experimental observation that nearly any color can be reproduced by combining three primary colors, typically red, green, and blue. This principle serves as the foundation for numerous modern technologies, including color televisions, computer monitors, and color photography [204].

Another competing theory is the opponent-color theory, proposed by E. Hering in 1878 [204]. This theory is based on three types of photoreceptors that respond to red-green, yellow-blue, and white-black opponencies (these pairs constitute the opponent colors). According to this theory, the perception of any color by the human eye depends on the responses of these opponent-color photoreceptors.

For the purposes of this introductory material, the focus is on the trichromatic theory, which remains the prevailing model. The field of colorimetry was formally established in 1931 when the International Commission for Illumination (*Commission Internationale de l' Éclairage, CIE*) standardized color measurement through the use of color-matching functions and the corresponding chromaticity diagram [216].

4.4.1 RGB Color Space

To quantify the color of a stimulus (optical source), it is necessary to understand the spectral response of human cone photoreceptors. However, in the 1920s, when the science of colorimetry began, direct measurements of these sensitivities were not feasible. As a result, they were inferred solely from psychophysical color-matching experiments. Since the 1980s, advancements in experimental techniques have allowed for direct measurement of cone sensitivities, and these measurements have shown strong agreement with the earlier psychophysically derived results [217–221]. The cone spectral sensitivities, as shown in Fig. 4.16, align well with these findings. To characterize a color stimulus, the CIE has developed standardized functions known as CIE color matching functions. To establish the color

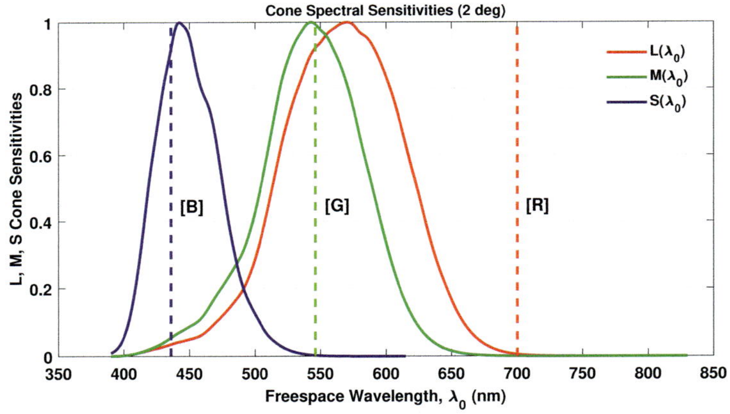

FIGURE 4.16: The spectral sensitivities of the human cone photoreceptors. The vertical lines show the three primaries used at the CIE 1931 measurements, [***R***] at 700 nm, [***G***] at 546.1 nm, and [***B***] at 435.8 nm (the data for this figure were obtained from url-link: `http://www.cvrl.org/`).

matching functions through psychophysical experiments, a measurement system was developed based on the setup shown in Fig. 4.17. A monochromatic test light illuminates the test field area, where a human observer can adjust the relative radiant powers of the red [***R***], green [***G***], and blue [***B***] primary lights (the three primaries used) to match the color of the test light in the comparison field area. This process, known as color matching, allows the observer to make the two lights appear identical. The unit for the radiant power of the primaries is chosen so that one unit from each primary matches the color of equal-energy white (more accurately, equal-power white), which refers to a light source with constant radiant power per wavelength interval. According to the CIE [216,222], the color matching functions were determined using three specific monochromatic primary colors: $\lambda_R = 700.0\,\text{nm}$ for R (red), $\lambda_G = 546.1\,\text{nm}$ for G (green), and $\lambda_B = 435.8\,\text{nm}$ for B (blue). The green and blue wavelengths correspond to prominent lines in the mercury discharge spectrum, aiding in wavelength calibration, while the red wavelength was selected for being in a region of the spectrum where hue changes slowly with wavelength [223].

The three color-matching functions are derived from a series of matching experiments in which a human observer adjusts the radiant powers of the three primary lights to match a series of monochromatic lights across the visible spectrum [203]. In cases where a monochromatic color cannot be matched using only the three primaries in the comparison field, a portion of one of the primary colors is added to the test field by activating a primary in that area. In such cases, the relative intensity of the corresponding primary is considered negative.

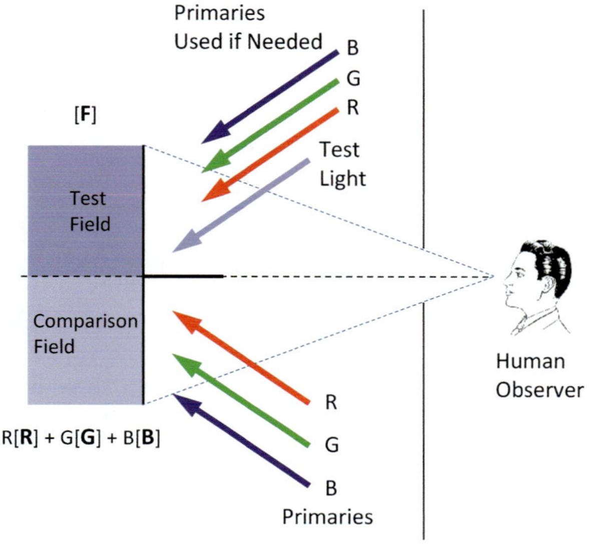

FIGURE 4.17: The color matching measurement procedure. A monochromatic test light $[\boldsymbol{F}]$ illuminates the test field. The test light is matched using a combination of the three primaries [summation of red (R), green (G), and blue (B), $[\boldsymbol{F}] = R[\boldsymbol{R}] + G[\boldsymbol{G}] + B[\boldsymbol{B}]$] in the comparison field according to the human observer that adjusts the weights of the three primaries. In some cases usage of primaries in the test field is required for matching the test color. This is the case of a negative weighting factor of a primary. The angle θ represents the observer's viewing angle. In the initial psychophysical experiments, this angle was set to 2 degrees.

Let $[\boldsymbol{F}]$ represent the stimulus of a given color [204,215]. The stimuli of the three primary sources are the $[\boldsymbol{R}]$, $[\boldsymbol{G}]$, and $[\boldsymbol{B}]$, corresponding to the red, green, and blue primaries, respectively. Assume the $[\boldsymbol{F}(\lambda_0)]$ is the stimulus of the test monochromatic color shown in Fig. 4.17. Then, according to Refs. [204] and [215], the following equation can be written:

$$[\boldsymbol{F}(\lambda_0)] = \bar{r}(\lambda_0)[\boldsymbol{R}] + \bar{g}(\lambda_0)[\boldsymbol{G}] + \bar{b}(\lambda_0)[\boldsymbol{B}], \tag{4.22}$$

where $\bar{r}(\lambda_0), \bar{g}(\lambda_0), \bar{b}(\lambda_0)$ are defined as the *color matching functions* at wavelength λ_0.

In the trichromatic color mixing system, it is essential to determine the color matching functions, which represent the relative weights of the three primary colors. The equality sign in Eq. (4.22) indicates that a visual match is achieved for the human eye between the test light at λ_0, and the linear mixture of the primaries $[\boldsymbol{R}]$, $[\boldsymbol{G}]$, and $[\boldsymbol{B}]$. Additionally, it is assumed that Grassmann's empirical rules of analogy and addition are followed [204,215]. To determine the color matching functions, the following assumptions are made [215]: (a) The cone signals are independent of photon wavelength, which aligns with experimental data. (b) The stimulus intensity and size are large enough to avoid small neural signals. (c) The human visual system can compare signals from spatially separated cones. (d) The perceived color at a given spatial location depends on the number of photons absorbed by the cones at that location. Moreover, color matching experiments should be conducted under consistent background conditions, visual angle, illumination, and adaptation state of the eyes.

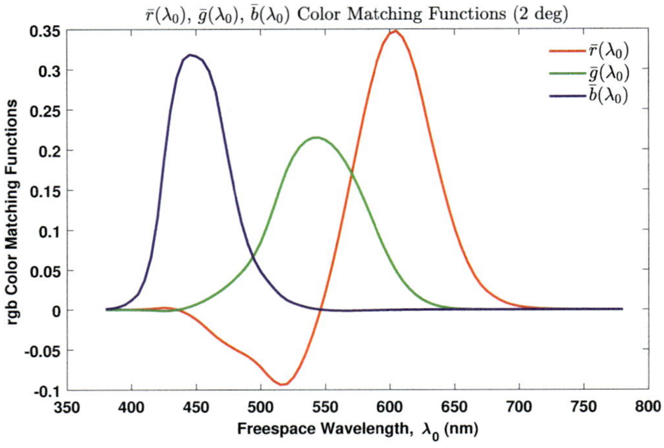

FIGURE 4.18: The $\bar{r}(\lambda_0)$, $\bar{g}(\lambda_0)$, $\bar{b}(\lambda_0)$ color matching functions (CIE 1931 - 2 degrees).

Equation (4.22) can be viewed as a vector equation in a three-dimensional vector space spanned by the primaries $[\boldsymbol{R}]$, $[\boldsymbol{G}]$, and $[\boldsymbol{B}]$ [204]. This three-dimensional space is used for the geometrical representation of colors and is referred to as a color space. Any color $[\boldsymbol{F}]$ (not necessarily monochromatic) can be located within this color space at a point defined by the matching amounts of $[\boldsymbol{R}]$, $[\boldsymbol{G}]$, and $[\boldsymbol{B}]$. Specifically, $[\boldsymbol{F}] = R[\boldsymbol{R}] + G[\boldsymbol{G}] + B[\boldsymbol{B}]$, where R, G, and B are the matching amounts, known as the *tristimulus values* of the color stimulus $[\boldsymbol{F}]$. For a monochromatic stimulus $[\boldsymbol{F}]$, the tristimulus values correspond directly to the color matching functions. Since the spectral power distribution of any test light can be decomposed into a sum of monochromatic components, the tristimulus values for any light source can be determined. For a light source with a power spectrum $P(\lambda_0)$ (power per wavelength), the R, G, and B tristimulus values can be obtained using the following equations:

$$
\begin{aligned}
R &= k\int_{\lambda_0} \bar{r}(\lambda_0)P(\lambda_0)d\lambda_0, \\
G &= k\int_{\lambda_0} \bar{g}(\lambda_0)P(\lambda_0)d\lambda_0, \qquad (4.23) \\
B &= k\int_{\lambda_0} \bar{b}(\lambda_0)P(\lambda_0)d\lambda_0,
\end{aligned}
$$

where k is a constant used to make R, G, and B dimensionless. The intersection (r, g, b) of the vector $[\boldsymbol{F}]$ with the unit plane $R + G + B = 1$ is commonly used to express color $[\boldsymbol{F}]$. This normalized representation is given by the following equations:

$$
\begin{aligned}
r &= \frac{R}{R+G+B}, \\
g &= \frac{G}{R+G+B}, \qquad (4.24)
\end{aligned}
$$

$$b = \frac{B}{R+G+B},$$

where the r, g, b are defined as the *rgb chromaticity coordinates* and specify the attributes of a color stimulus independently of its radiant power. Since $r+g+b=1$, only two of the three values are needed to define the color $[\boldsymbol{F}]$ (for example r and g can be used). A diagram that shows only two of the values, such as r and g (or any two of the r, g, b) is called *chromaticity diagram.* An example of such a diagram for r and g is shown in Fig. 4.20. The chromaticity coordinates of color $[\boldsymbol{F}]$ correspond to a point within chromaticity diagram. The chromaticity coordinates of monochromatic light are known as spectral chromaticity coordinates and correspond to the boundary of the chromaticity diagram, which is also referred to as the spectrum locus. The straight line connecting the two ends of the spectrum locus forms the line of purples. The point where $r = g = 1/3 = b$ is the equal energy white point. The chromaticity coordinates of any real color are enclosed within the area formed by the spectrum locus and the line of purples. During the tests for determining the rgb chromaticity

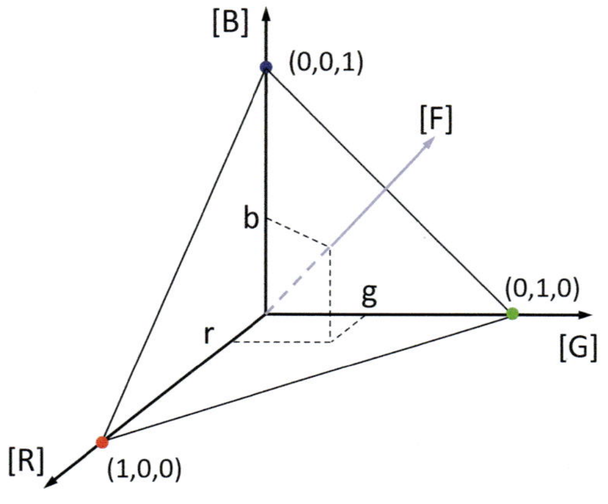

FIGURE 4.19: The r, g, b chromaticity coordinates of color stimulus $[F]$.

coordinates for various monochromatic colors, the three primaries $[\boldsymbol{R}]$, $[\boldsymbol{G}]$, and $[\boldsymbol{B}]$ were scaled so that the mixture of unit quantities of these primaries matched the equal energy (power) spectrum. The radiant powers of the three primaries were found to be in the ratios of 72.0962:1.3791:1 (for $[\boldsymbol{R}]$, $[\boldsymbol{G}]$, and $[\boldsymbol{B}]$, respectively). Using the photopic luminous efficiency function $V(\lambda_0)$ with linear interpolation, the luminous efficiencies for the primary wavelengths were determined to be: $V(\lambda_0 = 700\,nm) = 0.004102$, $V(\lambda_0 = 546.1\,nm) = 0.984425$, and $V(\lambda_0 = 435.8\,nm) = 0.017774$. The corresponding ratios of the three primaries in lumens for the equal energy spectrum were 1.0000:4.5907:0.0601 (for $[\boldsymbol{R}]$, $[\boldsymbol{G}]$, and $[\boldsymbol{B}]$, respectively). This means that $1.0000+4.5907+0.0601 = 5.6508$ lumens of white light (equal energy spectrum) can be matched by 1.0000 lumens of $[\boldsymbol{R}]$, 4.5907 lumens of $[\boldsymbol{G}]$, and 0.0601 lumens of $[\boldsymbol{B}]$. To determine the $r(\lambda_0), g(\lambda_0), b(\lambda_0)$ color matching functions, it was assumed that the linear combination matches the photopic luminous efficiency function $V(\lambda_0)$ (shown in Fig. 4.13). Specifically, $V(\lambda_0) = L_R\bar{r}(\lambda_0)+L_G\bar{g}(\lambda_0)+L_B\bar{b}(\lambda_0)$ where the ratio of luminous powers of the three primaries was set to $L_R : L_G : L_B = 1.0000 : 4.5907 : 0.0601$. These values are related to the relative intensities of the primaries as depicted in the measurements (Fig. 4.17). The resulting color matching functions were then plotted as functions of the free space wavelength λ_0 of the test color, as shown in Fig. 4.18. These functions are crucial in matching colors and quantifying the perceived light in the CIE 1931 color space [216].

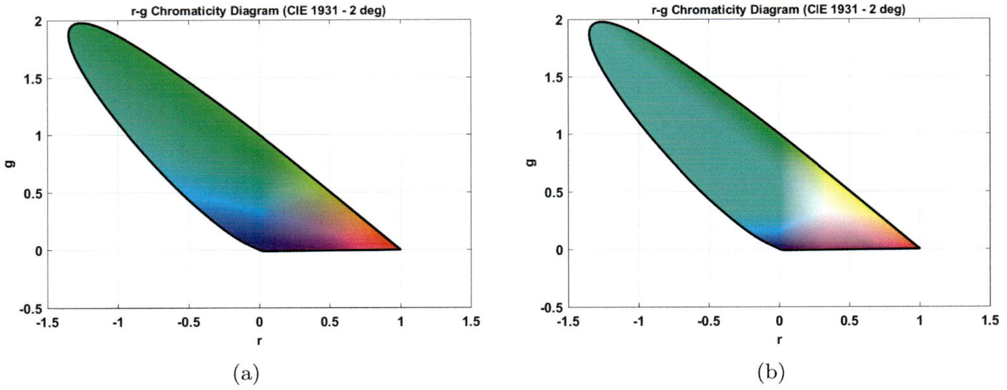

FIGURE 4.20: (a) The CIE r-g chromaticity diagram (from 1931, 2-deg data) where the white point has $r = g = b = 1/3$, and its representative color in the sRGB standard is determined by $R = G = B = 1/3$. In this case the white point looks grey due to its low illumination, and accordingly all other colors in the diagram. (b) Same as (a) but with $R = G = B = 1$ (in sRGB standard), which corresponds to the correctly illuminated white.

4.4.2 XYZ Color Space

In order to avoid the negative values that the $\bar{r}(\lambda_0)$, $\bar{g}(\lambda_0)$, and $\bar{b}(\lambda_0)$ color matching functions could sometimes attain, the CIE published the $\bar{x}(\lambda_0)$, $\bar{y}(\lambda_0)$, $\bar{z}(\lambda_0)$ color matching functions, which are all greater than or equal to zero for any wavelength in the visible spectrum. These $\bar{x}(\lambda_0)$, $\bar{y}(\lambda_0)$, $\bar{z}(\lambda_0)$, new color matching functions were adopted in CIE 1931 for the standard observer with a 2-degree viewing angle. They were derived from a mathematical transformation of the $\bar{r}(\lambda_0)$, $\bar{g}(\lambda_0)$, and $\bar{b}(\lambda_0)$ ones. More details about this transformation process can be found in references such as Refs. [215] and [222]. The basic steps for transforming between two color spaces (for example, from XYZ and RGB) are described in Appendix 4B. The three color-matching functions $\bar{x}(\lambda_0)$, $\bar{y}(\lambda_0)$, $\bar{z}(\lambda_0)$ reflect the fact that human color vision is trichromatic, meaning that any color can be described by just three variables, similar to the $\bar{r}(\lambda_0)$, $\bar{g}(\lambda_0)$, $\bar{b}(\lambda_0)$ color matching functions. These functions are dimensionless. Importantly, the $\bar{y}(\lambda_0)$ color matching function was chosen to match the photopic luminous efficiency function of the human eye, so $\bar{y}(\lambda_0) = V(\lambda_0)$. The CIE 1931 $\bar{x}(\lambda_0)$, $\bar{y}(\lambda_0)$, $\bar{z}(\lambda_0)$ color matching functions are shown in Fig. 4.21 [216]. Numerical values for the more recent 2-degree observer functions (2006), defined at 5 nm intervals, can be found in Appendix 4A and Table 4A.4 Similarly to the definition of the RGB tristimulus values in Eq. (4.23), the tristimulus values X, Y, Z for a light source with power spectrum $P(\lambda_0)$ can be defined as:

$$
\begin{aligned}
X &= k \int_{\lambda_0} \bar{x}(\lambda_0) P(\lambda_0) d\lambda_0, \\
Y &= k \int_{\lambda_0} \bar{y}(\lambda_0) P(\lambda_0) d\lambda_0, \\
Z &= k \int_{\lambda_0} \bar{z}(\lambda_0) P(\lambda_0) d\lambda_0,
\end{aligned}
\tag{4.25}
$$

where k is a constant to make X, Y, and Z dimensionless. Frequently, k may be selected to make Y of a specific value or maximum in order to arrange the luminance level. Furthermore, the corresponding xyz chromaticity coordinates can be defined similarly to rgb by the following expressions:

$$x = \frac{X}{X+Y+Z},$$
$$y = \frac{Y}{X+Y+Z}, \quad (4.26)$$
$$z = \frac{Z}{X+Y+Z},$$

Again since $x + y + z = 1$ only two of the chromaticity coordinates (for example x and y) are needed (by convention x and y are selected). It can be shown that for the equal energy white color the tristimulus X, Y, and Z values are related to the tristimulus R, G, and B values by the following linear transformation [204, 222]:

$$\begin{bmatrix} X \\ Y \\ Z \end{bmatrix} = \begin{bmatrix} 2.768892 & 1.751748 & 1.130160 \\ 1.000000 & 4.590700 & 0.060100 \\ 0 & 0.056508 & 5.594292 \end{bmatrix} \begin{bmatrix} R \\ G \\ B \end{bmatrix}. \quad (4.27)$$

Some details on determining the matrix for this transformation are presented in Appendix 4B. Because the color matching functions are the tristimulus values of monochromatic light, the above transformation holds between $[\bar{x}(\lambda_0), \bar{y}(\lambda_0), \bar{z}(\lambda_0)]$ and $[\bar{r}(\lambda_0), \bar{g}(\lambda_0), \bar{b}(\lambda_0)]$ also. The CIE chromaticity diagram is depicted in Fig. 4.22. The curved boundary of this "shoe-

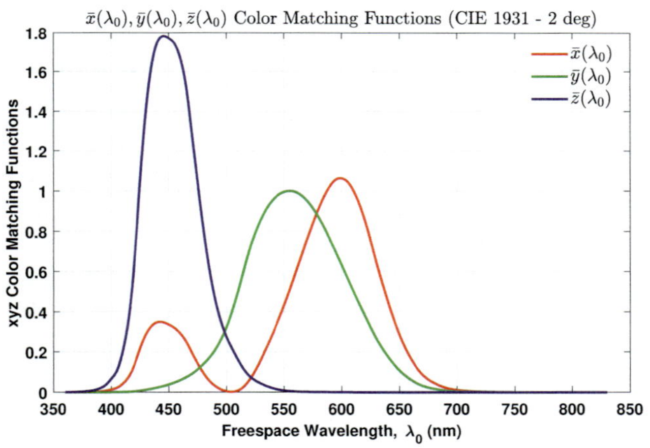

FIGURE 4.21: The $\bar{x}(\lambda_0)$, $\bar{y}(\lambda_0)$, $\bar{z}(\lambda_0)$ color matching functions (CIE 1931 - 2 degrees).

horse" shaped diagram corresponds to monochromatic colors, while the bottom edge, known as the "line of purples," represents colors that cannot be produced by monochromatic light. Points within the diagram, specified by the (x, y) chromaticity coordinates, represent colors visible to the human eye. White light is located near the center of the chromaticity diagram. When white light is represented by the point $x_w = y_w = z_w = 1/3$, it is referred to as the equal-energy point. However, other white point locations can be used depending on the definition of white light and the spectral distribution of the light source. For example, the D65 illuminant, corresponding to a blackbody temperature of $T = 6504°$ K, is commonly used in the sRGB color system. For this illuminant, the chromaticity coordinates of white light are $x_w = 0.3127$, $y_w = 0.3290$ and $z_w = 0.3583$. In Fig. 4.22(a) the (x, y) chromaticity coordinates are transformed into R, G, B coordinates, assuming that $y = Y$ (where Y

represents the luminance of the color). Figure 4.22(b) shows an adjustment of the luminance from x, y, z values point by point, such that $X = x/w_{max}$, $Y = y/w_{max}$, and $Z = z/w_{max}$, where $w_{max} = \max\{x, y, z\}$. This adjustment results in a brighter version of Fig. 4.22(a), with white color centrally located.

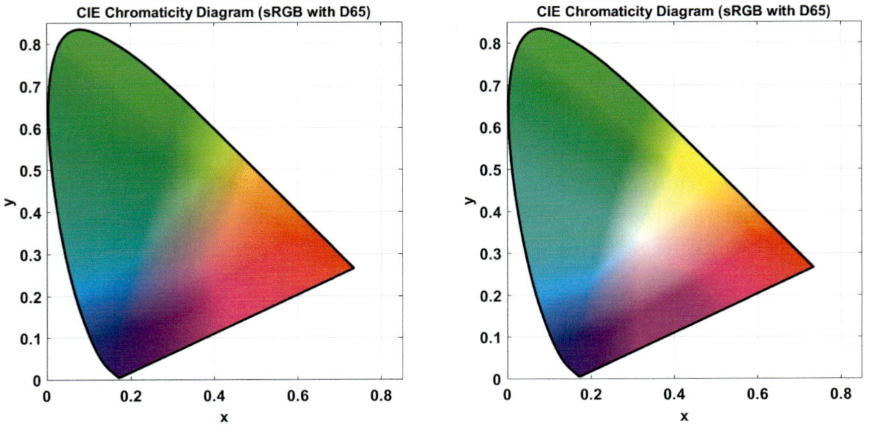

(a) (b)

FIGURE 4.22: (a) The CIE chromaticity diagram for $X = x$, $Y = y$, and $Z = z$. (b) The CIE chromaticity diagram for $X = x/w_{max}$, $Y = y/w_{max}$, and $Z = z/w_{max}$ where $w_{max} = \max\{x, y, z\}$.

4.4.3 Mixture of Colors, Dominant and Complimentary Wavelengths

Assume that $[\boldsymbol{A}]$ and $[\boldsymbol{B}]$ are two colors with chromaticity coordinates (x_a, y_a) and (x_b, y_b), respectively, which are represented by points A and B in the chromaticity diagram of Fig. 4.23(a). The chromaticity coordinates of the additive mixture of the two colors with weighting factors a and b can be represented by $a[\boldsymbol{A}] + b[\boldsymbol{B}]$ which for simplicity is shown as $aA + bB$. It can be proven that the chromaticity coordinates of this mixture lie on the line segment connecting points A and B in the chromaticity diagram, as shown in Fig. 4.23(a) [215]. If X_a, Y_a, Z_a and X_b, Y_b, Z_b are the tristimulus values of colors $[\boldsymbol{A}]$ and $[\boldsymbol{B}]$, respectively, then the tristimulus values of the mixture of the colors are $aX_a + bX_b$, $aY_a + bY_b$, and $aZ_a + bZ_b$. The resulting chromaticity coordinates of the mixture are:

$$\begin{aligned} x &= \frac{aX_a + bX_b}{a(X_a + Y_a + Z_a) + b(X_b + Y_b + Z_b)} \\ &= \underbrace{\frac{a(X_a + Y_a + Z_a)}{a(X_a + Y_a + Z_a) + b(X_b + Y_b + Z_b)}}_{c_a} x_a + \underbrace{\frac{b(X_b + Y_b + Z_b)}{a(X_a + Y_a + Z_a) + b(X_b + Y_b + Z_b)}}_{c_b} x_b \\ &= c_a x_a + c_b x_b, \qquad \text{and similarly,} \end{aligned} \tag{4.28}$$

$$y = c_a y_b + c_b y_b. \tag{4.29}$$

It is reminded that $x_a = X_a/(X_a + Y_a + Z_a)$ and similarly for x_b, y_a, and y_b. Since by definition $c_a + c_b = 1$, it is straightforward to show that the point (x, y) is on the AB line.

This observation indicates that the chromaticities of all physically realizable colors are confined within the convex spectral locus in the chromaticity diagram. Ideally, the horseshoe-shaped spectral locus should exhibit convexity. However, detailed numerical studies have identified slight deviations from perfect convexity [224].

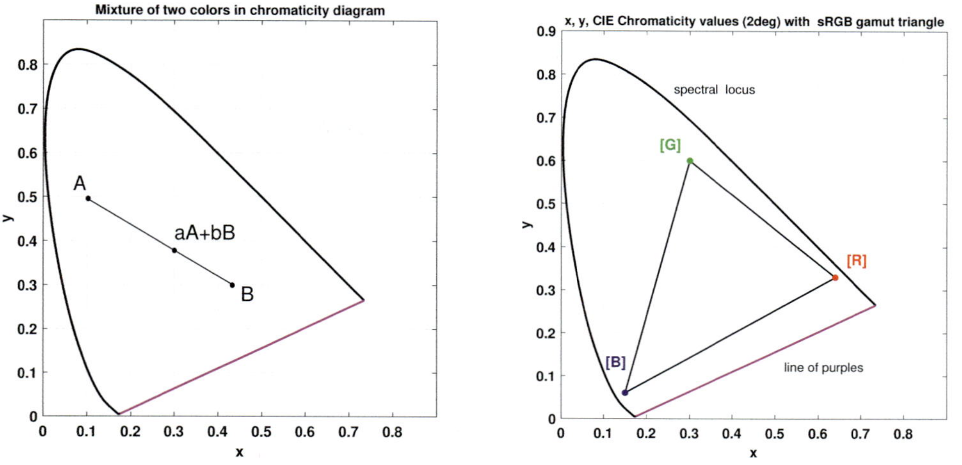

(a) (b)

FIGURE 4.23: (a) The locus of the chromaticity coordinates of the additive mixture of colors A and B represented by $aA + bB$ is on the line connecting the chromaticity coordinates of A and B. (b) The gamut triangle for the *sRGB* color system placed in the CIE 1931 chromaticity diagram. The vertices of the triangle correspond to the three primaries. The spectral locus and the line of purples are also shown. Only colors with color coordinates within the triangle can be represented by the *sRGB* color system.

In the context of color reproduction, the term *gamut* (or *color gamut*) defines the subset of colors that a specific device, such as a computer monitor, is capable of displaying. For example, the widely used *sRGB* color system, designed for standard monitors, represents its gamut as a triangular region in the chromaticity diagram, as illustrated in Fig. 4.23(b). The corners of this triangle correspond to the chromaticity coordinates of the three primary colors, $[\boldsymbol{R}]$, $[\boldsymbol{G}]$, and $[\boldsymbol{B}]$. Any linear combination or additive mixture of these primary colors produces colors with chromaticities that lie inside the triangular region. This triangle defines the limit of the colors reproducible by the *sRGB* color system.

The CIE recommends some characteristic points in the chromaticity diagram that are defined relative to an illuminant W which is usually a white color light [215, 223]. Consider two stimuli with chromaticity coordinates represented by points A and B in the chromaticity diagram, as shown in Fig. 4.24. The *dominant wavelength* of color A is the point that is specified by the pure color of wavelength λ_d defined by point $D(\lambda_d)$ on the spectral locus. This point is specified by the intersection of the line WA with the spectral locus if such an intersection exists, as is the case for color A shown in the diagram. This dominant wavelength (pure color), when mixed with the illuminant W matches color A. In the case that the line WA does not intersects the spectral locus, as is the case of the line WB for color B, then the *complimentary wavelength* for the color B, λ_c is defined and is represented by point $C(\lambda_c)$. This is the intersection of the line BW with the spectral locus. The complimentary color when additively is mixed with color B (with appropriate proportions) produces the white color W. The *excitation purity*, p_e, is defined as the ratio of $WA/WD(\lambda_d)$. When p_e is near unity the color represented by point A is near the spectral locus. In the case that only the complimentary wavelength is defined the excitation purity is defined as the ratio WB/WB'. In this case excitation purity near unity means that the color B is near the line of purples. When spectral purity is near zero then the color A or B is near the white color W.

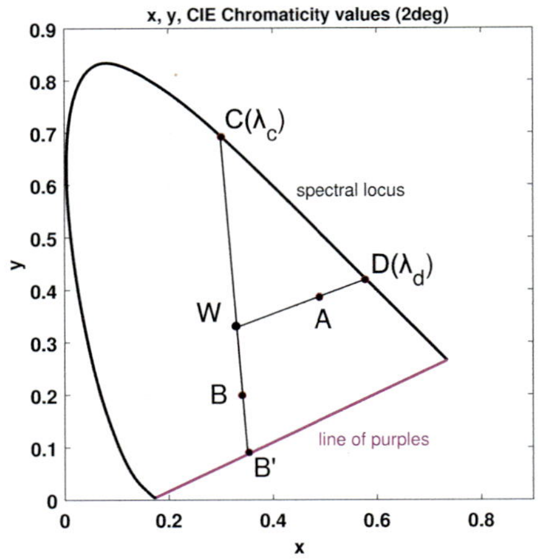

FIGURE 4.24: CIE chromaticity diagram showing dominant, λ_d, and complimentary wavelengths λ_c, relative to the illuminant W (usually white color).

Example 4.4.1:

Assume a color stimulus $[\boldsymbol{A}]$ represented by the chromaticity coordinates $(x_A = 0.40, y_A = 0.45)$. The white point W is appropriate for the *sRGB* color system, and it is represented by the chromaticity coordinates $(x_w = 0.312713, y_w = 0.329016)$. (a) Determine the dominant wavelength color stimulus $[\boldsymbol{A}]$. (b) Calculate the purity of excitation of $[\boldsymbol{A}]$. (c) Find the mixture of the dominant wavelength stimulus and the white light stimulus that reproduces the color stimulus $[\boldsymbol{A}]$, represented by point A in the chromaticity diagram.

Solution: According to the definition of the dominant wavelength, as illustrated in Fig. 4.24, the intersection of the line WA with the spectral locus is required. The color stimulus $[\boldsymbol{A}]$ is depicted in Fig. 4.25(a), where point A, along with points W (the white point) and D (the dominant wavelength point), are shown. In this context, the spectral locus is represented by the coordinates $(x_b(\lambda_0), y_b(\lambda_0))$ on the chromaticity diagram. These points correspond to different wavelengths along the boundary of the chromaticity diagram. The coordinates of the spectral locus $(x_b(\lambda_0), y_b(\lambda_0))$ can be easily calculated from the equations, $x_b(\lambda_0) = \bar{x}(\lambda_0)/[\bar{x}(\lambda_0) + \bar{y}(\lambda_0) + \bar{z}(\lambda_0)]$, and $y_b(\lambda_0) = \bar{y}(\lambda_0)/[\bar{x}(\lambda_0) + \bar{y}(\lambda_0) + \bar{z}(\lambda_0)]$. The line connecting points W and A is described by the equation, $y = y_w + [(y_A - y_w)/(x_A - x_w)](x - x_w)$. The intersection point (if it exists) must satisfy the equation: $y_b(\lambda_d) = y_w + [(y_A - y_w)/(x_A - x_w)](x_b(\lambda_d) - x_w)$. Since there is no known analytic expression for the spectral locus, this intersection must be determined numerically. In this example, global optimization (minimization) of the function:

$$F(\lambda_d) = [y_b(\lambda_d) - y_w] - \frac{(y_A - y_w)}{(x_A - x_w)} [x_b(\lambda_d) - x_w],$$

was performed using Matlab [141] routines. Special care must be taken if the line *WA* intersects the spectral locus at two points. In such cases, the point closer to A should be selected. Using this procedure, the intersection point D was determined with chromaticity coordinates $x_D = 0.46241$ and $y_D = 0.5365$, corresponding to the dominant wavelength $\lambda_d = 572.6286\,\text{nm}$. The excitation purity is given by:

$$p_e = \frac{WA}{WD} = \frac{\left[(x_A - x_w)^2 + (y_A - y_w)^2\right]^{1/2}}{\left[(x_D - x_w)^2 + (y_D - y_w)^2\right]^{1/2}} = 0.58311.$$

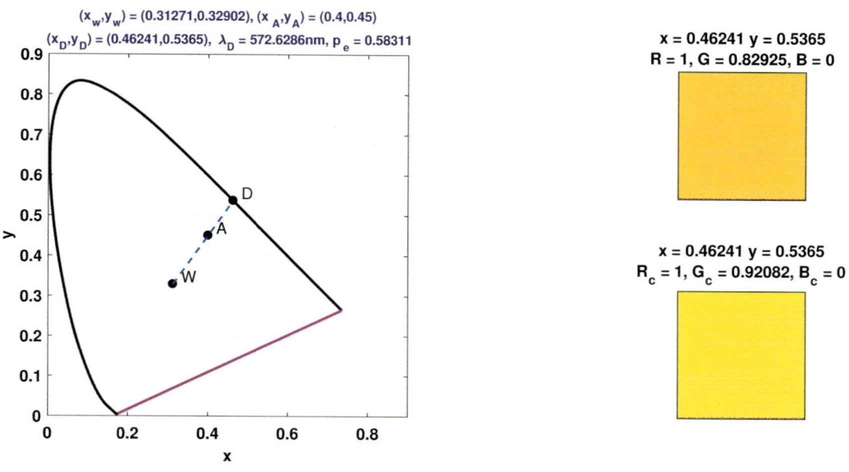

(a) (b)

FIGURE 4.25: (a) CIE chromaticity diagram showing dominant, $\lambda_d = 572.6286\,\text{nm}$, on the spectral locus with $x_D = 0.46241$ and $y_D = 0.5365$, for color A with chromaticity coordinates $(x_A = 0.40, y_A = 0.45)$ relative to the illuminant W (D65 standard illuminant) with chromaticity coordinates $(x_w = 0.312713, y_w = 0.329016)$. (b) The corresponding color of the dominant wavelength in the *sRGB* color system without (upper) and with (lower) gamma correction.

To determine the appropriate mixture of the dominant wavelength and white light to produce color $[\boldsymbol{A}]$, the following equations should be satisfied [see Eqs. (4.28) and (4.29)],

$$x_A = c_w x_w + c_D x_D = c_w x_w + (1 - c_w)x_D \Longrightarrow c_w = \frac{x_D - x_A}{x_D - x_w} \simeq 0.4169$$

$$y_A = c_w y_w + c_D y_D = c_w y_w + (1 - c_w)y_D \Longrightarrow c_w = \frac{y_D - y_A}{y_D - y_w} \simeq 0.4169.$$

In this case, $c_D = 0.5831$ represents the proportion of the dominant wavelength in the mixture, while $c_W = 0.4169$ corresponds to the proportion of the white light.

Finally, in Fig. 4.25(b) the *sRGB* color representation of the dominant wavelength is shown. The upper color corresponds to the color without gamma correction, while the lower one includes gamma correction [225] (refer to Problem 14 of this chapter). The color R, G, B, coordinates were determined using the transformation described in Appendix 4B. The tristimulus values X_D, Y_D, and Z_D of the dominant wavelength were calculated as: $X_D = x_D/\max\{x_D, y_D, 1 - x_D - y_D\}$, $Y_D = y_D/\max\{x_D, y_D, 1 - x_D - y_D\}$, and $Z_D = (1 - x_D - y_D)/\max\{x_D, y_D, 1 - x_D - y_D\}$.

4.4.4 Colors Differences and Uniform Color Spaces

The CIE 1931 xyz color coordinates were not specifically designed with the goal of uniform color differences. As a result, two colors with small differences in their tristimulus values could appear either significantly different or nearly indistinguishable, depending on their location in the xyz coordinates. MacAdam conducted experiments to measure color differences for various chromaticities in the (x, y) color coordinate plane and found significant variation depending on the region of the chromaticity diagram. He quantified these differences using elliptic contours, now known as *MacAdam ellipses* [226], which represent regions where colors appear nearly identical. Larger ellipses indicate areas in the chromaticity diagram where colors are practically indistinguishable. Due to these variations, researchers have attempted

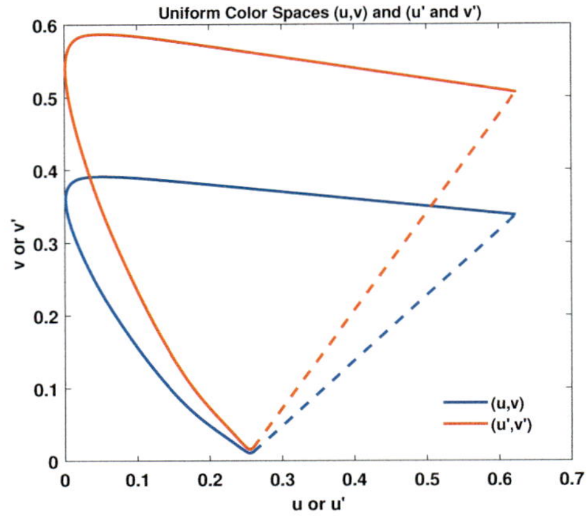

FIGURE 4.26: The (u, v) and the (u', v') uniform color spaces.

to develop color spaces where color differences are more uniform, aiming to make MacAdam ellipses circular with uniform radii. One of the early attempts to achieve this was made by MacAdam [227] himself, who introduced a transformation to define the (u, v) color space. After a series of studies, the CIE recommended the CIE 1976 (u', v') uniform color space to address these discrepancies. The transformations between the (x, y) and (u, v) or (u', v') color systems are given as follows [215, 222, 227]:

$$u = \frac{4x}{-2x + 12y + 3}, \quad v = \frac{6x}{-2x + 12y + 3}, \tag{4.30}$$

$$u' = \frac{4x}{-2x + 12y + 3}, \quad v' = \frac{9x}{-2x + 12y + 3}. \tag{4.31}$$

The resulting plots are shown in Fig. 4.26. It is important to note that none of the transformations fully succeeded in transforming the MacAdam ellipses into perfectly uniform circles. However, the transformations discussed above result in color spaces that are much closer to uniformity, providing an improved representation of color differences across different regions of the chromaticity diagram.

Other color spaces proposed by the CIE in 1976 are the *CIELUV* and *CIELAB* systems, which also take luminance into account, along with metrics for color differences [215, 222]. For a deeper understanding of these color systems and newer proposals, interested readers are encouraged to consult the aforementioned references.

4.5 Blackbody Radiation

4.5.1 Introduction and Electromagnetic Modes Density

A blackbody is an idealized object that absorbs all incident electromagnetic radiation, regardless of frequency or angle of incidence. In thermal equilibrium, a blackbody re-emits all absorbed radiation, with the emitted energy depending solely on its temperature. The intensity of this emitted radiation is directly related to the body's thermal energy at equilibrium. The development of blackbody radiation theory holds historical significance, as it ultimately led to the discovery of quantum theory [228]. Key figures in this development include Josef Stefan, Ludwig Boltzmann, Wilhelm Wien, and Max Planck. Their contributions were pivotal in shaping our understanding of radiation and quantum mechanics. A detailed account of their biographies and the methods they employed is provided by Crepeau [229].

Early experimental studies revealed that the emissivity of a blackbody is dependent on both frequency and temperature. A measure of this emissivity is represented by the term $\rho(\nu, T)$, which denotes the density of radiation energy, i.e., energy per unit volume per unit frequency (J/m^3Hz) at a given absolute temperature T and frequency ν. Initial theoretical attempts to describe this relationship relied on Maxwell's equations, which were highly successful at the time, to determine the density of electromagnetic modes and, consequently, $\rho(\nu, T)$. For instance, in 1896, Wilhelm Wien used a simple model to derive an expression for $\rho(\nu, T)$, which contributed to the foundational understanding of blackbody radiation. Wien's expression is given below:

$$\rho(\nu, T) = \alpha\nu^3 \exp(-\beta\nu/T), \tag{4.32}$$

where α, β were constants. Wien hypothesized that radiation was emitted by molecules following a Maxwellian velocity distribution and that the wavelength of the emitted radiation depended solely on the molecules' velocities [229]. While this approach successfully described radiation at higher frequencies, the resulting equation failed to match experimental data at lower frequencies, indicating that a more comprehensive theory was needed to fully explain blackbody radiation across all frequency ranges.

In June 1900, Lord Rayleigh proposed a model based on the behavior of electromagnetic wave modes within a cavity, where each mode corresponded to a specific frequency and could exchange energy in a continuous manner. He applied the standard electromagnetic theory for a cuboid cavity resonator (see Fig. 4.27) with perfectly conducting walls. Using this framework, he derived the following dispersion equation, which describes the relationship between the frequency of the electromagnetic waves and the physical properties of the cavity [101, 132, 185] (see Appendix 4C for a detailed proof):

$$\left(\frac{m\pi}{a}\right)^2 + \left(\frac{p\pi}{b}\right)^2 + \left(\frac{q\pi}{d}\right)^2 = \left(\frac{2\pi\nu}{c}\right)^2 n^2, \tag{4.33}$$

where n is the index of refraction of the medium, and a, b, and d are the dimensions of the cuboid cavity resonator in the x, y, and z directions, and m, p, q are positive integers. If it is assumed, for simplicity, that the cavity is a cube with side lengths $a = b = d$, the previous equation can be simplified to:

$$m^2 + p^2 + q^2 = \left(\frac{2\nu}{c}\right)^2 a^2 n^2 = \left(\frac{2\nu n a}{c}\right)^2. \tag{4.34}$$

To count the electromagnetic modes up to frequency ν, it is necessary to evaluate the number of modes that fit within one-eighth of the sphere in (m, p, q)-space as illustrated in Fig. 4.28. This is due to the fact that negative values of m, p, and q do not represent distinct modes (refer to Appendix 4C). Thus, the total number of electromagnetic modes, $N(\nu)$, can be determined as follows:

$$N(\nu) = \frac{(1/8)\text{ cavity volume}}{\text{volume of a mode}} = \frac{(1/8)(4/3)\pi(2\nu a n/c)^3}{1 \times 1 \times 1} = \frac{4}{3}\pi\frac{\nu^3 n^3 a^3}{c^3}. \tag{4.35}$$

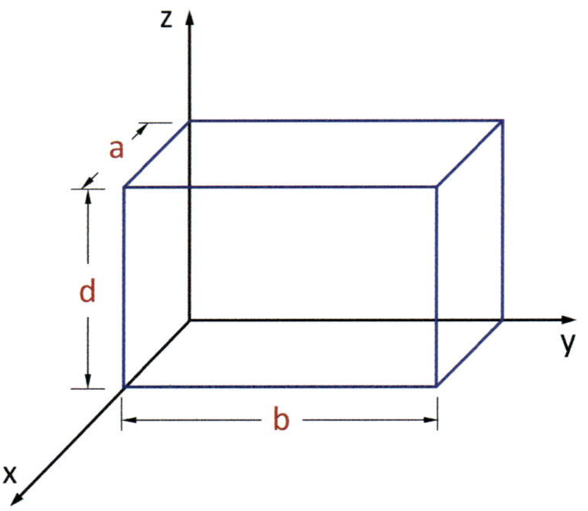

FIGURE 4.27: Cavity box (a cuboid) for the determination of the density of electromagnetic modes.

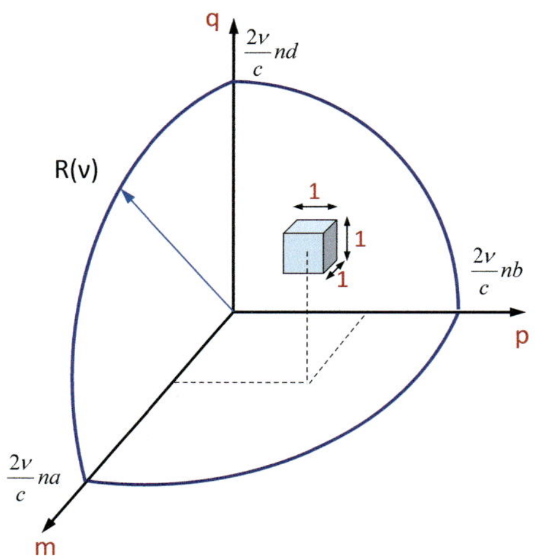

FIGURE 4.28: The one-eighth of the sphere in the *mpq* space for the determination of the number of electromagnetic modes up to frequency ν.

Due to the degeneracy of the transverse electric (TE) and transverse magnetic (TM) modes, the number of electromagnetic modes must be multiplied by a factor of 2. Therefore, the total number of electromagnetic modes per unit volume, $\mathcal{N}(\nu)$, is given by:

$$\mathcal{N}(\nu) = \frac{N(\nu)}{Volume = a^3} = \frac{8}{3}\pi\frac{\nu^3 n^3}{c^3}. \tag{4.36}$$

Then the density of electromagnetic modes per frequency becomes:

$$\frac{d\mathcal{N}(\nu)}{d\nu} = \frac{8\pi\nu^2 n^3}{c^3}. \tag{4.37}$$

In the case of dispersion, where the refractive index n depends on the frequency ν of the electromagnetic radiation, the previous equation must be modified to account for this dependence. Specifically, with $n = n(\nu)$, the total number of electromagnetic modes per unit volume per unit frequency is influenced by this relationship. To incorporate this, the mode density equation is adjusted. When the frequency dependence of the refractive index is considered, the total number of electromagnetic modes per unit volume can be expressed as [39]:

$$\frac{d\mathcal{N}(\nu)}{d\nu} = \frac{8\pi\nu^2 n^2 \left(n + \nu \dfrac{dn}{d\nu}\right)}{c^3} = \frac{8\pi\nu^2 n^2 n_g}{c^3}. \tag{4.38}$$

The group refractive index, denoted by n_g, is defined as $n_g = n + \nu(dn/d\nu) = n - \lambda_0(dn/d\lambda_0)$, where λ_0 is the free-space wavelength of the electromagnetic radiation. The group refractive index is significant in materials like semiconductors and optical fibers, where the refractive index varies notably with frequency (or wavelength). The refractive index is assumed to be independent of frequency (or wavelength) for simplicity in the remainder of this section. This assumption facilitates the analysis and calculations, allowing focus on the primary effects without the added complexity of dispersion.

4.5.2 Energy Associated with Electromagnetic Modes

Rayleigh assigned an energy $k_BT/2$ to each electromagnetic mode ($k_BT/2$ for the electric field oscillation and $k_BT/2$ for the magnetic field oscillation, where $k_B = 1.380649 \times 10^{-23} J/^\circ K$ is Boltzmann's constant). More rigorously, the average energy of each electromagnetic mode can be determined using Boltzmann's statistics [198]. According to these statistics, the probability $p(E)$ that the energy of each electromagnetic mode is between E and $E + dE$ is given by:

$$p(E)\,dE = A \exp\left(-\frac{E}{k_BT}\right) dE, \tag{4.39}$$

where A is a normalization constant that can be easily found from the normalization of $p(E)$ to ensure it represents a probability density function. Therefore, the constant A is given by:

$$\int_0^\infty p(E)\,dE = 1 \Longrightarrow A = \frac{1}{\displaystyle\int_0^\infty \exp(-E/k_BT)dE} = \frac{1}{k_BT}. \tag{4.40}$$

Then, the average energy of each electromagnetic mode can be determined from the expression:

$$\langle E \rangle = \int_0^\infty E\,p(E)\,dE = \int_0^\infty E \frac{1}{k_BT} \exp\left(-\frac{E}{k_BT}\right) dE = k_BT. \tag{4.41}$$

Using Eq. (4.37) and the average energy of Eq. (4.41), the electromagnetic energy density per unit frequency $\rho(\nu, T)$ becomes:

$$\rho(\nu, T) = \frac{d\mathcal{N}(\nu)}{d\nu}\langle E \rangle = \frac{8\pi\nu^2 n^3}{c^3} k_BT. \tag{4.42}$$

The last equation is known as the Rayleigh-Jeans distribution of blackbody radiation and fails dramatically in the ultraviolet part of the spectrum, historically referred to as the "ultraviolet catastrophe." This inadequacy is evident in the Rayleigh-Jeans curve presented in Fig. 4.29.

Planck employed purely thermodynamic entropy arguments to derive an improved equation for Wien's distribution law, as shown in Eq. (4.32). The derived equation took the following form [229]:

$$\rho(\nu, T) = \frac{C\nu^3}{\exp(\beta\nu/T) - 1}. \tag{4.43}$$

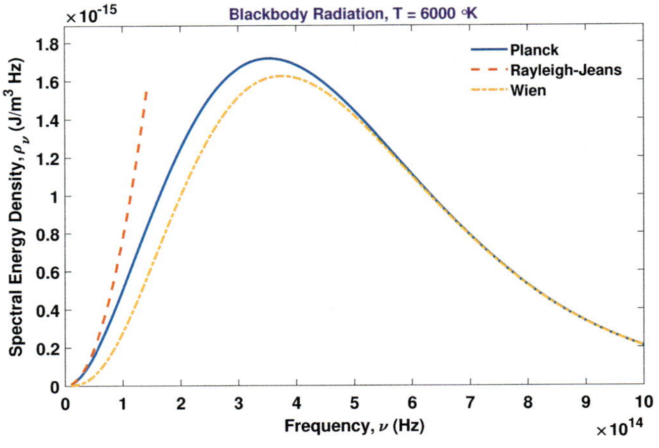

FIGURE 4.29: Blackbody radiation for $T = 6000°$ K. The initial theories by Rayleigh-Jeans and Wien are also shown for comparison.

It has been suggested that Planck discovered his famous constant (h) on the evening of October 7, 1900 [228]. In his analysis, Planck incorporated additional experimental data provided by Heinrich Reubens and Ferdinand Kurlbaum, s well as Wien's formula. He deduced an expression in Eq. (4.43) that effectively "fitted" all the available experimental data. This formula is now known as the blackbody radiation formula, given by:

$$\rho(\nu, T) = \frac{8\pi\nu^2 n^3}{c^3} \frac{h\nu}{\exp(h\nu/k_B T) - 1}. \tag{4.44}$$

Here, $h = 6.62607015 \times 10^{-34} Joule \cdot sec$ is known as Planck's constant. The above expression reduces to Wien's formula for high frequencies (i.e., when $h\nu/k_B T \gg 1$) and to the Rayleigh-Jeans formula for low frequencies (i.e., when $h\nu/k_B T \ll 1$). An example of Planck's radiation formula is presented in Fig. 4.29, alongside the Rayleigh-Jeans and Wien's approximations for a blackbody at an absolute temperature of $T = 6000°K$.

After deriving his formula, Planck sought to understand its physical basis. The determination of the density of electromagnetic modes proved challenging to argue. As a result, he shifted his focus to the average energy per electromagnetic mode. Following discussions with Boltzmann about how discrete equal energy values could be distributed among molecules, Planck proposed that electromagnetic energy at frequency ν could only occur in multiples of the step size $h\nu$, a quantum of energy (later known as a photon). Consequently, the energy of the electromagnetic modes was expressed as $E_i = ih\nu$ where $i = 0, 1, 2, \cdots$. Energies between $ih\nu$ and $(i+1)h\nu$ were not allowed. Planck then applied Boltzmann's statistics to calculate the average energy of an electromagnetic mode. Given that E_0, E_1, E_2, $\ldots$, represent the allowed energies, the probability of an electromagnetic mode having energy E_i is determined by:

$$p(E_i) = A \exp\left(-\frac{E_i}{k_B T}\right). \tag{4.45}$$

The normalization constant A can be determined as:

$$\sum_{i=0}^{\infty} p(E_i) = 1 \Longrightarrow A = \frac{1}{\sum\limits_{i=0}^{\infty} \exp(-E_i/k_B T)} = 1 - \exp\left(-\frac{h\nu}{k_B T}\right). \tag{4.46}$$

Then, the average energy $\langle E \rangle$ of an electromagnetic mode can be calculated as follows:

$$\langle E \rangle = A \sum_{i=0}^{\infty} E_i \exp\left(-\frac{E_i}{k_B T}\right) = A\left[1h\nu e^{-h\nu/k_B T} + 2h\nu e^{-2h\nu/k_B T} + \cdots\right] =$$

$$= A\,\frac{h\nu \exp(-h\nu/k_B T)}{[1-\exp(-h\nu/k_B T)]^2} = \frac{h\nu}{\exp(h\nu/k_B T)-1}. \tag{4.47}$$

Using the calculation of the average energy of an electromagnetic mode, Planck's formula can be rewritten to provide the physical meaning of each of its terms. The formula can be expressed as:

$$\rho(\nu,T) = \underbrace{\frac{8\pi\nu^2 n^3}{c^3}}_{\substack{\text{Number of em modes}\\ \text{per volume per frequency}}} \overbrace{h\nu}^{\text{photon energy}} \underbrace{\frac{1}{\exp(h\nu/k_B T)-1}}_{\text{Number of photons/mode}}. \tag{4.48}$$

Later, Planck employed the energy discretization of the quantum oscillator, expressed as $E_i = [(1/2)+i]h\nu$ (for $i = 0, 1, 2, \cdots$). This formulation introduces what is now known as zero-point energy, which represents the lowest energy state of a quantum oscillator. According to the Heisenberg uncertainty principle, this lowest energy cannot be zero. In this case, the average energy of an electromagnetic mode can be calculated similarly to the previous approach [as in Eq. (4.47)], leading to:

$$\langle E \rangle = \frac{h\nu}{2} + \frac{h\nu}{\exp(h\nu/k_B T)-1}, \tag{4.49}$$

However, in the above equation, the zero-point energy term $h\nu/2$ leads to an infinite radiation density $\rho(\nu, T)$ and should not be included in the expression for blackbody radiation energy density [230–236]. One simplistic explanation for the absence of the zero-point energy term is that the photons emitted or absorbed by the blackbody radiator are associated with transitions between energy states, described by $E_i - E_{i'} = (i - i')h\nu = \ell h\nu$ (where $\ell = 1, 2, \cdots$). Therefore, the initial assumption of Planck, as stated in Eq. (4.47), should be employed instead. As a general comment, zero-point energy, or vacuum energy, remains a controversial topic in modern physics.

The density of electromagnetic modes can also be expressed per wavelength (in free space) and is given by:

$$\frac{d\mathcal{N}(\lambda_0)}{d\lambda_0} = -\frac{8\pi n^3}{\lambda_0^4}, \tag{4.50}$$

and the corresponding density of electromagnetic radiation of a blackbody per wavelength (freespace) becomes:

$$\rho(\lambda_0, T) = \frac{8\pi n^3}{\lambda_0^4}\frac{hc/\lambda_0}{\exp(hc/\lambda_0 k_B T)-1}. \tag{4.51}$$

4.5.3 Blackbody Radiated Power, Radiance, and Exitance

In the literature, the blackbody radiation formula is often expressed in terms of the radiant exitance (or radiant emittance) of the blackbody, which is measured in units of power per unit area (i.e., W/m^2). Radiant exitance quantifies the total power emitted by a source in a hemisphere (toward the direction of emission) per unit area of the source [see Eq. (4.11)]. The Poynting vector represents the power per unit area of the electromagnetic radiation and is given by: $P_{avg} = (1/2Z)|E|^2$ where E is the electric field amplitude of the electromagnetic wave, and $Z = \sqrt{\mu_0/n^2\epsilon_0} = Z_0/n$ is the intrinsic impedance of the non-magnetic homogeneous isotropic medium in which the electromagnetic radiation propagates (with Z_0 is the freespace wave impedance). The energy density of the electromagnetic radiation can be expressed as $w_{em} = (1/2)n^2\epsilon_0|E|^2$. Thus, the average power per unit area can be written as $P_{avg} = (c/n)w_{em}$. The energy density between frequencies ν and $\nu + d\nu$ (or equivalently between λ_0 and $\lambda_0 + d\lambda_0$) is given by $dw_{em} = \rho(\nu, T)d\nu = \rho(\lambda_0, T)d\lambda_0$. Consequently, the power per unit area (between ν and $\nu + d\nu$ or equivalently between λ_0 and $\lambda_0 + d\lambda_0$), dP_{avg}, can be determined as follows:

$$dP_{avg} = \frac{8\pi n^2\nu^2}{c^2}\frac{h\nu}{\exp(h\nu/k_BT) - 1}d\nu = P_{avg,\nu}d\nu, \tag{4.52}$$

$$dP_{avg} = \frac{8\pi n^2 c}{\lambda_0^4}\frac{hc/\lambda_0}{\exp(hc/\lambda_0 k_BT) - 1}d\lambda_0 = P_{avg,\lambda_0}d\lambda_0. \tag{4.53}$$

The radiance L [measured in W/m^2sr, where sr represents steradian; refer to Eq. (4.4)] is defined as $L = d^2P/dA_\perp/d\Omega$ where d^2P represents the differential electromagnetic power emitted by the source (blackbody radiator) in a specified direction, $dA_\perp$ is the differential area element of the source perpendicular to the specified direction of propagation, and $d\Omega$ is the differential solid angle within which the differential power is propagated [202]. A blackbody emits radiation equally in all directions, making it appear equally bright from any observation angle. Consequently, the radiance L is constant and independent of the observation angle. Therefore, a blackbody is always considered a Lambertian source [202]. By integrating the blackbody radiance over all solid angles, the following relationship can be derived: $\int_\Omega Ld\Omega = L\int_\Omega d\Omega = 4\pi L = \int(d^2P/dA_\perp) = P_{avg}$. The spectral radiance, dL_s of a blackbody radiator between ν and $\nu + d\nu$ or λ_0 and $\lambda_0 + d\lambda_0$ can be expressed as:

$$dL_s = \frac{dP_{avg}}{4\pi} = \frac{2n^2\nu^2}{c^2}\frac{h\nu}{\exp(h\nu/k_BT) - 1}d\nu = L_{s,\nu}d\nu, \qquad \text{and} \tag{4.54}$$

$$dL_s = \frac{dP_{avg}}{4\pi} = \frac{2n^2 c}{\lambda_0^4}\frac{hc/\lambda_0}{\exp(hc/\lambda_0 k_BT) - 1}d\lambda_0 = L_{s,\lambda_0}d\lambda_0. \tag{4.55}$$

This expression provides a relationship between the spectral radiance of a blackbody and the parameters involved in its emission characteristics, particularly at different frequencies and wavelengths. From radiometry it can be established that the radiance L and the radiant exitance (emittance) M (in W/m^2) of a blackbody (or a Lambertian source in general) are related by the equation $M = L\pi$ [see Eq. (4.12)]. Consequently, $dM = \pi dL = M_\nu(\nu)d\nu = M_{\lambda_0}(\lambda_0)d\lambda_0 = M_\lambda(\lambda)d\lambda$, where M_ν, M_{λ_0}, and M_λ are the spectral exitances in $W/m^2/Hz$, $W/m^2/m$ (in freespace wavelength) and $W/m^2/m$ (inside medium wavelength), respectively. In addition, $c = \lambda_0\nu$ and $c/n = \lambda\nu = (\lambda_0/n)\nu$. For example, the radiant spectral exitance (power/unit area/frequency $= W/m^2/Hz$) of a blackbody radiator is given by:

$$M_\nu(\nu) = \frac{2\pi n^2\nu^2}{c^2}\frac{h\nu}{\exp(h\nu/k_BT) - 1}. \tag{4.56}$$

The same spectral exitance expressed per wavelength (in freespace or in medium of propagation) interval (power/unit area/wavelength $= W/m^2/m$) is given by:

$$M_{\lambda_0} = \frac{2\pi n^2 c}{\lambda_0^4}\frac{hc/\lambda_0}{\exp(hc/\lambda_0 k_BT) - 1}, \tag{4.57}$$

$$M_\lambda = \frac{2\pi c}{n^2\lambda^4}\frac{hc/\lambda}{\exp(hc/\lambda n k_BT) - 1}. \tag{4.58}$$

Integrating the above equations over all frequencies (or wavelengths) allows for the determination of the radiant exitance M of a blackbody radiator at temperature T. This relationship is known as Stefan's law and is expressed by the following equation:

$$M = \int_0^\infty M_{\lambda_0} d\lambda_0 = \left(\frac{2\pi^5 k_B^4}{15h^3c^2}\right) n^2 T^4 = \sigma n^2 T^4, \tag{4.59}$$

where $\sigma = 5.670374419 \times 10^{-8}\, W/m^2\, {}^\circ K^{-4}$ is the Stefan-Boltzmann constant (usually the refractive index is considered that of vacuum or air, i.e., $n \simeq 1$). This law indicates that the radiant exitance of a blackbody is proportional to the fourth power of its absolute temperature, highlighting the significant increase in emitted radiation with increasing temperature. The maxima of the blackbody radiator curve can be found by solving the equation derived from Planck's radiation law, which is typically expressed as:

$$\frac{dM_{\lambda_0}(\lambda_{0,max})}{d\lambda_0} = 0 \Rightarrow \frac{hc}{\lambda_{0,max} k_B T} = 4.965114231 \Rightarrow \lambda_{0,max} T = 2897.772\, \mu m {}^\circ K. \tag{4.60}$$

The last part of the above equation represents how the peak of the blackbody radiation shifts with the temperature, and it is known as *Wien's displacement law*. An example of M_{λ_0} for $T = 6000\,{}^\circ$K along with the corresponding Wien's displacement law is illustrated in Fig. 4.30. A similar Wien's displacement law can be defined for M_ν. In this case, the maxima for M_ν can be found by:

$$\frac{dM_\nu(\nu_{max})}{d\nu} = 0 \Rightarrow \frac{h\nu_{max}}{k_B T} = 2.82143937 \Rightarrow \frac{\nu_{max}}{T} = 5.878926 \times 10^{10}\, Hz/{}^\circ K. \tag{4.61}$$

The derivation of Wien's displacement law is presented in Appendix 4D with the utilization

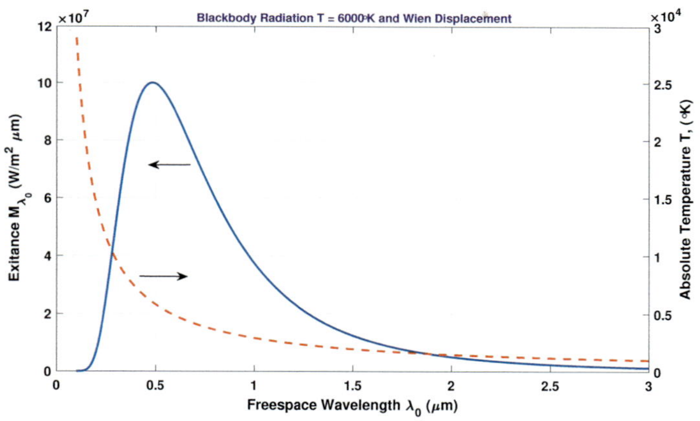

FIGURE 4.30: Blackbody radiation spectral exitance (emittance), $M_{\lambda_0}(\lambda_0)$, for $T = 6000^\circ$ K as a function of freespace wavelength. The Wien's displacement law is also shown for the same wavelength range. The maximum of M_{λ_0} occurs for $\lambda_{0,max} = 0.483\,\mu m$.

of Lambert's function. An example of M_ν for $T = 6000\,{}^\circ$K along with the corresponding Wien's displacement law is shown in Fig. 4.31. The peaks of $M_{\lambda_0}(\lambda_{0,max})$ (the spectral radiant exitance per unit wavelength) and $M_\nu(\nu_{max})$ (the spectral radiant exitance per unit frequency) are indeed related, but they are not directly connected through the equation $\lambda_{0,max}\nu_{max} = c$ because the spectral distributions are defined over different measures (wavelength versus frequency). An important point that warrants discussion is the inclusion of the refractive index in Eq. (4.59). It is noted that n refers to the refractive index (assuming no dispersion) of the medium within the blackbody cavity, as seen in the derivation of

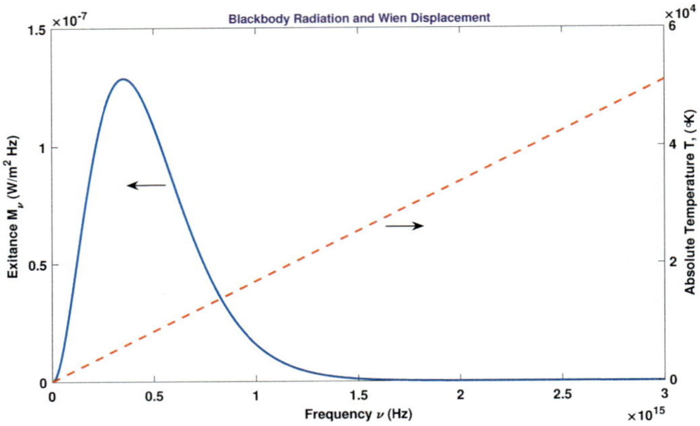

FIGURE 4.31: Blackbody radiation spectral exitance (emittance), $M_\nu(\nu)$, for $T = 6000°$ K as a function of frequency. The Wien's displacement law is also shown for the same frequency range. The maximum of M_ν occurs for $\nu_{max} = 3.52 \times 10^{14}\ Hz$.

the density of electromagnetic modes inside the cuboid cavity. However, the electromagnetic energy radiated by the blackbody propagates outward. In many textbooks, the refractive index is omitted from Eq. (4.59) under the assumption that the blackbody emits radiation into a vacuum or air (where $n_{air} \simeq 1$). This omission is justified by the conservation of radiance across two homogeneous media with different refractive indices, as discussed in [237, 238] and elaborated in Sec. 4.2.5. From Eq. (4.13) (assuming the power reflection coefficient $\rho = 0$), it follows that $M_1/n_1^2 = M_2/n_2^2 = M_0$, where M_0 represents the exitance in vacuum. Returning to Eq. (4.59), it is clear that $M_i = n_i^2 M_0$ where $M_0 = \sigma T^4$, implying that for $n_i > 1$ the exitance radiated from a dielectric medium into air could exceed M_0. However, this is not the case because part of the energy emitted within the medium with refractive index n_i is reflected back at the medium-air interface due to total internal reflection. According to Snell's law, the maximum angle refracted into the air is $\theta_{max} = \sin^{-1}(1/n_i)$, which represents the critical angle. Thus, only radiation within a cone of apex angle θ_{max} is refracted into the air (neglecting reflection losses). The radiated power, dP, for an area dA within the accepted cone can then be calculated, as shown in Fig. 4.9, and is given by:

$$dP = \int_0^{2\pi} \int_0^{\theta_{max}} L_i dA \cos\theta \sin\theta d\phi d\theta = 2\pi L_i dA \frac{\sin^2\theta_{max}}{2}$$

$$= \pi L_i dA \frac{1}{n_i^2} = \frac{M_i}{n_i^2} dA = M_0 dA.$$

Consequently, the total power (per unit area) emitted by a blackbody radiator is independent of the refractive index of the medium. This conclusion holds under the assumption that all reflections (below critical angle) are neglected in the analysis. A more comprehensive treatment, including the effects of reflections, can be found in Refs. [238] and [239], where additional details on radiative transfer and reflection phenomena are discussed.

4.5.4 Thermal Radiation Basics

The blackbody is an ideal radiator, and its basic properties are as follows: (a) The radiation emitted by a blackbody is isotropic, homogeneous, and unpolarized, meaning the radiation is completely diffuse. (b) The blackbody radiation at a given wavelength (or frequency) depends solely on the temperature. (c) A blackbody emits more radiation than any other object at the same temperature, making it the upper limit to the amount of radiation that a real body may emit at a given temperature. Because a blackbody radiator is both a perfect absorber and a perfect emitter, it serves as a reference to compare the radiative properties of real surfaces. At any given freespace wavelength λ_0, the emissivity $\varepsilon(\lambda_0, T)$ is defined as the ratio of the actual emitted radiant exitance (or radiance) $M^a_{\lambda_0}$ (or $L^a_{\lambda_0}$) to the emitted radiant exitance of a blackbody M_{λ_0} (or L_{λ_0}), as follows:

$$\varepsilon(\lambda_0, T) = \frac{M^a_{\lambda_0}(\lambda_0, T)}{M_{\lambda_0}(\lambda_0, T)} = \frac{L^a_{\lambda_0}(\lambda_0, T)}{L_{\lambda_0}(\lambda_0, T)}. \tag{4.62}$$

Equation (4.62) is valid only for Lambertian emitters, where $M = L\pi$ and the radiance L is direction-independent, as mentioned earlier. Typically, the emissivity of a real object is defined using the ratio of radiances rather than exitances because, in real objects, radiation depends not only on wavelength and temperature but also on the direction of emission. The most fundamental emissivity of a real object is the spectral directional emissivity, $\varepsilon(\lambda_0, T, \theta, \phi)$, where θ and ϕ are the polar and azimuthal angles specifying the direction of emission [240]. When the radiance is integrated over the hemisphere of emission, the spectral hemispherical emissivity, $\varepsilon(\lambda_0, T)$, is defined using the ratio of exitances as in Eq. (4.62) [240].

Emissivity measures how effectively a body radiates energy at a given wavelength, temperature, and direction. For simplicity, throughout this chapter, the term emissivity will refer to the spectral hemispherical emissivity. Emissivity values for real substances range from zero to one ($0 \leq \varepsilon(\lambda_0, T) \leq 1$). A gray body (and diffused body) is a substance whose emissivity is independent of wavelength, i.e., $\varepsilon(\lambda_0, T) = \varepsilon(T)$. In contrast, in the atmosphere, clouds and gases exhibit emissivities that can vary significantly with wavelength due to different absorption lines. On the other hand, the ocean surface, with near-unit emissivity in the visible spectrum, closely resembles a blackbody.

For a body in local thermodynamic equilibrium, the amount of thermal energy emitted must equal the energy absorbed. Otherwise, the body would experience a temperature change over time, contradicting the assumption of equilibrium. Consequently, materials that are strong absorbers at a given wavelength are also strong emitters at that wavelength. Likewise, weak absorbers correspond to weak emitters.

All bodies are continuously exposed to electromagnetic radiation. When radiation is incident on a slab of semi-transparent material, as illustrated in Fig. 4.32, the fraction of the radiation that is reflected is termed *reflectivity*, denoted by R. The fraction transmitted through the material is referred to as *transmissivity*, T, while the portion absorbed is known as *absorptivity*, A. The nature of reflectivity may vary, depending on the angle of incidence, and can be categorized as either specular (mirror-like reflection) or diffuse (uniform reflection in all backward directions). Here, R, T, and T are assumed to represent average values over all directions and wavelengths of the incident radiation [240]. The conservation of energy requires that R, T, and A obey the equation:

$$R + T + A = 1. \tag{4.63}$$

For ideal blackbodies, the reflected and transmitted powers are zero, i.e., $R = T = 0$, which leads to $A = 1$, meaning all incident radiation is absorbed. For opaque materials, where $T = 0$, the relationship simplifies to $R + A = 1$. In the case of semi-transparent materials, such as gases, the reflectivity is negligible ($R = 0$), and thus $T + A = 1$.

4.5.4.1 Kirchoff's Law

Consider a small body with surface area S, temperature T, emissivity ε, and absorptivity A, fully contained within a large isothermal volume of the same temperature T, as illustrated

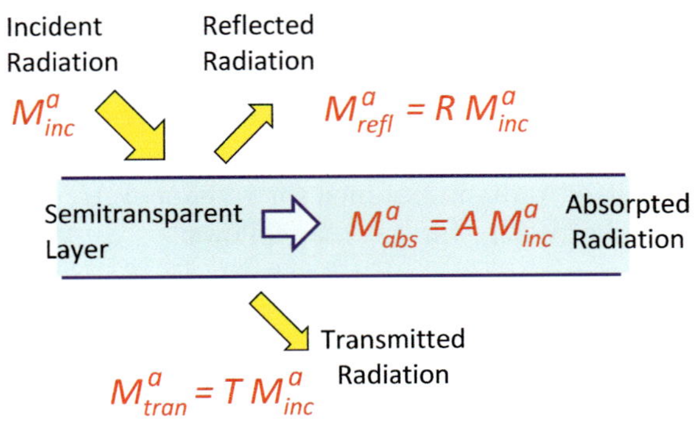

FIGURE 4.32: The incident, M^a_{inc}, reflected, $M^a_{refl} = RM^a_{inc}$, transmitted, $M^a_{tran} = TM^a_{inc}$, and absorbed, $M^a_{abs} = AM^a_{inc}$, powers, respectively, by a semi-transparent homogeneous material slab. The R, T, and A, represent the reflectivity, transmissivity, and absorptivity of the material, respectively.

in Fig. 4.33 [240]. The large isothermal volume can be modeled as a blackbody cavity, and the small body, being of negligible size relative to the cavity, does not influence the thermal state of the blackbody enclosure. The radiation incident on the small body is

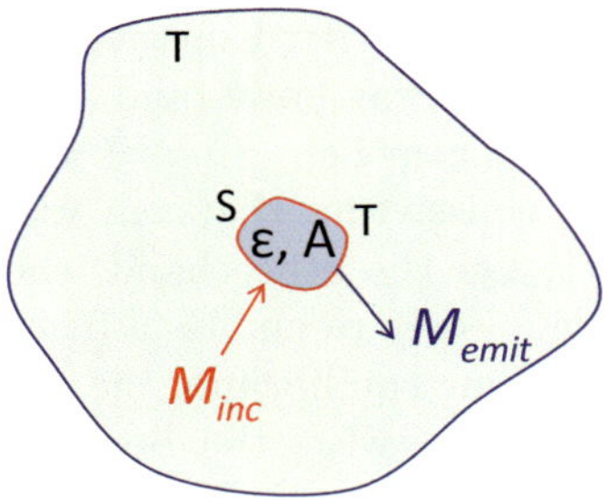

FIGURE 4.33: A small body of temperature T, surface area S, emissivity ε, and absorptivity A is inside a large isothermal enclosure of temperature T also.

$M_{inc} = \sigma T^4$ according to Stefan-Boltzmann law. The radiation absorbed by the small body is $M_{abs} = AM_{inc} = A\sigma T^4$. On the other hand, the radiation emitted by the small body is $M_{emit} = \varepsilon M_{inc} = \varepsilon\sigma T^4$. Considering thermodynamic equilibrium, the absorbed and emitted radiation must be equal, leading to $A\sigma T^4 = \varepsilon\sigma T^4$, which simplifies to the well-known Kirchhoff's law:

$$\varepsilon(T) = A(T), \tag{4.64}$$

where both the emissivity as well as the absorptivity are average values over all wavelengths and directions. In general, the Kirchhoff's law can be generalized as $\varepsilon(\lambda_0, T, \theta, \phi) = A(\lambda_0, T, \theta, \phi)$.

4.5.4.2 Color Temperature

Blackbody radiation is also used to establish a color scale as a function of absolute temperature. The *color temperature* of a light source is defined as the temperature of a blackbody that has the closest spectral distribution to that of the source. However, color temperature is strictly applicable to blackbody radiators since their spectral response is completely determined by temperature. Real sources, on the other hand, do not emit perfect blackbody spectra. To address this, the concept of *Correlated Color Temperature* (CCT) has been proposed [241, 242]. The CCT measures a light source's color appearance by considering the proximity of its chromaticity coordinates to the blackbody locus. Consequently, the CCT serves as a single numerical value to characterize the color of a light source, rather than requiring two chromaticity coordinates for quantitative specification. The CCT is measured in the u-v chromaticity coordinates (see Appendix 4E). For example, the chromaticity coordinates x, y of a light source can be determined from its spectrum and the color matching functions $\bar{x}(\lambda_0)$, $\bar{y}(\lambda_0)$, and $\bar{z}(\lambda_0)$. The resulting x, y chromaticity coordinates are then transformed into the u, v coordinates [222] (Appendix 4E). If $u(T)$ and $v(T)$ represent the u, v chromaticity coordinates of a blackbody radiator, the CCT of the light source is defined as the temperature CCT $= T_{CCT}$ that minimizes the distance of point (u, v) from the Planckian locus (the locus of blackbody radiators in the $u - v$ diagram). Mathematically, T_{CCT} is defined as:

$$\text{CCT} = T_{CCT} = \min_T \left\{ d = \left[(u - u(T))^2 + (v - v(T))^2 \right]^{1/2} \right\}. \tag{4.65}$$

The latter equation finds the intersection of the normal to the Planckian locus from the

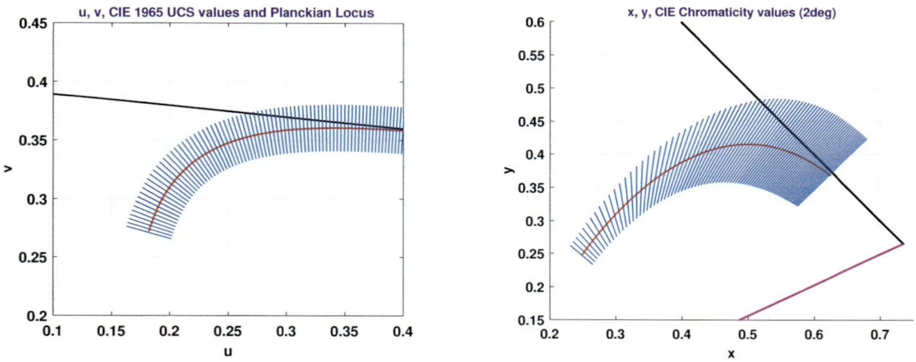

(a) (b)

FIGURE 4.34: (a) Isotemperature lines in uv chromaticity diagram for $\Delta uv = 0.02$. The solid black line is part of the boundary of the chromaticity space. The red line is the Planckian locus. The temperatures (b) Isotemperature lines in xy chromaticity diagram for $\Delta uv = 0.02$. The solid black line is part of the boundary of the chromaticity space. The red line is the Planckian locus and the purple line is part of the *line of purples*.

point (u, v). This intersection defines the correlated color temperature (CCT). Typically, isotemperature lines are drawn on the chromaticity diagram by specifying the maximum acceptable distance, Δuv, from the Planckian locus in the u-v diagram. These lines can be easily transformed into the standard x-y chromaticity diagram. For example, isotemperature lines with $\Delta uv = \pm 0.02$ are shown in Figs. 4.34(a) and 4.34(b) in the u-v and in the x-y diagrams, respectively. The isotemperature lines represent temperatures ranging from 1515 °K (first isotemperature line from the right) to $\sim$ 24959 °K (first isotemperature line from the left). The temperatures were selected to differ by $10\,mired = 10 MK^{-1}$ (micro reciprocal degrees, $MK^{-1} = 10^6\,\text{K}/T$, with T the color temperature in degrees K) according to Ref. [243]. The details of the calculation of the isotemperature lines is presented in Appendix 4E.

Color temperature and correlated color temperature are characteristics of visible light and are utilized in various fields, including lighting, photography, publishing, manufacturing, and astrophysics. Correlated color temperatures above 5000° K are referred to as "cool colors" (appearing bluish), while color temperatures in the range of 2700–3000° K are termed "warm colors" (appearing yellowish). For example, the sun has a typical correlated color temperature of approximately 5780° K.

4.5.5 Cosmic Microwave Background Radiation (CMB)

The discovery and measurement of the Cosmic Microwave Background (CMB) radiation hold significant importance in modern cosmology, as they provided critical information about the early universe, including the amounts of regular matter, dark matter, and dark energy, as well as the universe's age, flatness, and the Hubble constant. Observations of the CMB radiation currently confirmed the "Big Bang" theory and offer a glimpse into the young universe (however, there is ongoing research for understanding the early universe). In the early moments after the Big Bang, the temperature of the universe was extraordinarily high. For instance, about three minutes post-Big Bang, the temperature reached approximately $10^9\,{}^\circ K$! At these extreme temperatures, protons, electrons, and neutrons coexisted with photons, forming an opaque plasma fog. No atoms had yet formed, and photons were continuously absorbed and re-emitted, unable to escape the plasma. It took roughly 380,000 years for the universe to cool to about $3000\,{}^\circ K$. At this temperature, the first hydrogen and helium atoms formed, allowing the universe to become transparent to photons. This epoch is known in cosmology as photon decoupling. The early photons that escaped from this nascent universe are still traveling today, constituting the faint CMB radiation that permeates all of space. Due to the expansion of the universe, these early photons, which would have adhered to a blackbody radiation pattern at $T = 3000\,{}^\circ\mathrm{K}$, have cooled to a present temperature of approximately $T = 2.72548 \pm 0.00057\,{}^\circ K$ [244]! This cooling is a result of the cosmological redshift caused by the universe's expansion from the age of 380,000 years to its current age of about 13.8 billion years. This observed redshift corresponds to an expansion factor of roughly 1100 since the universe first became transparent [245]. Thus, the CMB photons provide a snapshot of the universe before the formation of stars. The Cosmic Microwave Background (CMB) radiation was first predicted in 1948 by Ralph Alpher and Robert Herman [246]. However, the initial observation of the CMB radiation was made in 1964 by Arno Penzias and Robert Woodrow Wilson at the Crawford Hill location of Bell Telephone Laboratories. Penzias and Wilson employed a horn antenna for their radio astronomy and satellite communication experiments. Despite their diligent measurements, they consistently detected a background microwave radiation that they could not explain. After discussing their results with researchers at Princeton University, including R. Dicke and his team, they realized that they had measured the CMB radiation. Their findings were published in 1965 [247], alongside a theoretical paper from the Princeton group [248]. For their groundbreaking discovery, Penzias and Wilson were awarded the Nobel Prize in Physics in 1978.

Since Penzias and Wilson's discovery of the Cosmic Microwave Background (CMB), increasingly precise measurements have been conducted. NASA's (National Aeronautics and Space Administration) COBE satellite (1989), WMAP (2001), and ESA's Planck satellite (2009) provided progressively accurate data on CMB radiation. Planck's measurements, shown in Fig. 4.35(a), reveal temperature fluctuations, known as CMB radiation anisotropies, of $1/10^6 - 1/10^5$ across the sky, indicating early universe inhomogeneities [245]. Additionally, Far Infrared Absolute Spectrometer (FIRAS) data from COBE, shown in Fig. 4.35(b), matches the blackbody Planck formula for $T = 2.725,{}^\circ\mathrm{K}$, confirming CMB as the most perfect blackbody radiator measured.

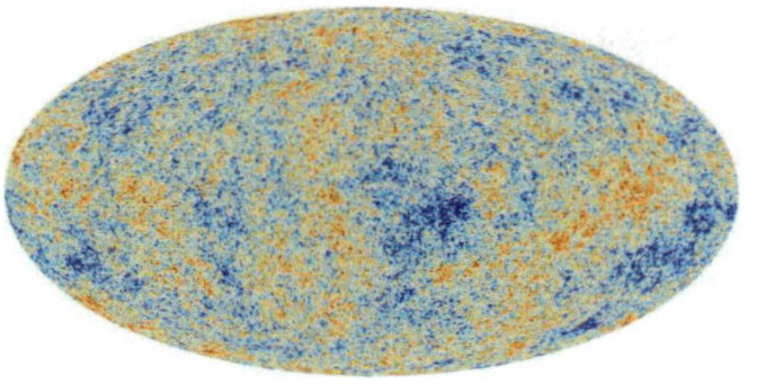

(a)

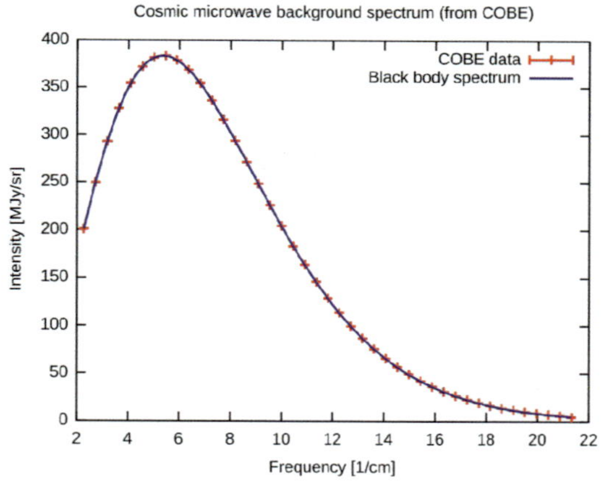

(b)

FIGURE 4.35: (a) The Cosmic Microwave Background as seen from the Planck satellite. Credit to European Space Agency (ESA). The picture was taken from url-link: `https://www.esa.int/ESA_Multimedia/Images/2013/03/Planck_CMB`). (b) The cosmic microwave background radiation spectrum measured by the FIRAS instrument on the COBE satellite. This is the most precisely measured blackbody spectrum in nature. The crosses are the measurement points, and the continuous line is the blackbody radiation theoretical curve for $T = 2.725\,^\circ K$. The intensity is measured in MJy/sr where $1\,MJy = 10^6\,Jy$ and $1\,Jy = 1\,Jansky = 10^{-26}W/m^2/Hz$. (the picture, "By Quantum Doughnut—Own work, Public Domain" was taken from url-link: `https://commons.wikimedia.org/w/index.php?curid=12958270`).

4.6 Problems

1 Photometry-Lighting: A space is illuminated by two identical lamps that are placed at a height of $h = 8$ m, as shown in Fig. 4P.1. The horizontal distance between the lamps is $L = 12$ m. Assume that the lamps emit light uniformly in all directions. (a) Determine the ratio of the illuminance at a point directly below each lamp on the ground (such as point S_1) to the illuminance at the midpoint between the two lamps on the ground (point S_2). (b) If the illuminance at point S_2 is required to be 100 lux, find the luminous intensity of each lamp in candelas (lumen/sr). Additionally, specify the luminous power of each lamp in lumens.

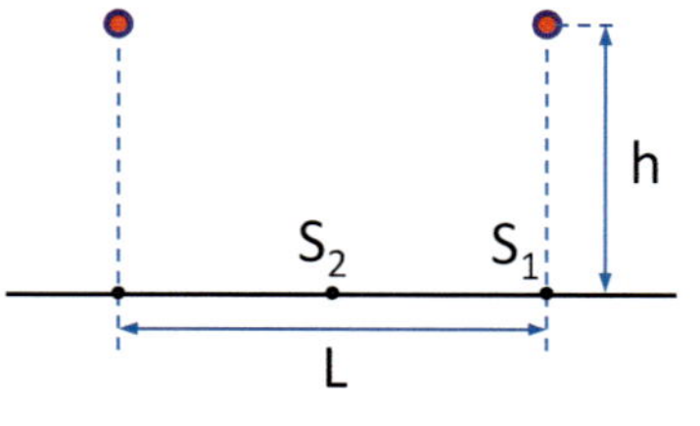

FIGURE 4P.1: Two light lamps illuminating a space.

2 Photometry-LED: Three light-emitting diodes (LEDs) are experimentally tested in a laboratory using a photodetector. The first SiC LED emits light at 480 nm (blue color), and its radiation is measured at an angle of $\theta = 10^\circ$ as shown in Fig. 4P.2. The second GaP LED emits at 565 nm (yellow-green color), and its radiation is measured at an angle of $\theta = 75^\circ$. The third AlGaP LED emits at 620 nm (orange color), and its radiation is measured at an angle of $\theta = 20^\circ$. All three

LEDs have the same total optical power of 5 mW and are considered to emit as Lambertian sources. The surface area of each LED is $A_s = 1\,\text{mm}^2$. The photodetector used in the measurements has an effective surface area of $A_d = 4\,\text{mm}^2$ and is located at a distance of $L_d = 100\,\text{mm}$ from each LED. (a) Determine the luminous intensity (in lm/sr) measured by the photodetector for each LED under the specified measurement angle θ. Which LED appears brighter under its measurement angle? (b) Calculate the luminance of each LED in lumens/sr/mm^2.

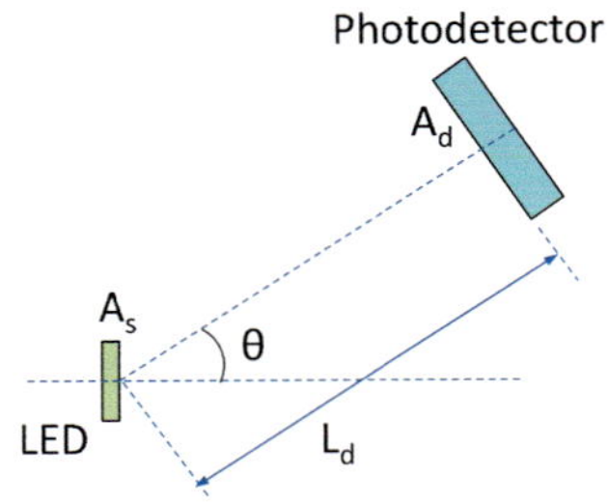

FIGURE 4P.2: LED measurement setup.

3 Photometry-LED: Three light-emitting diodes (LEDs) are positioned in space as depicted in Fig. 4P.3. These LEDs are observed by a human located at the origin, as shown in the figure. The distance from each LED to the observer's eye is 1000 mm. LED 1 positioned at an angle of $\theta_1 = 30°$, is a SiC LED that emits light at 480 nm (blue color). LED 2, placed at an angle of $\theta_1 = 60°$, is a GaP:N LED that emits light at 565 nm (yellow-green color). LED 3, located at an angle of $\theta_1 = 90°$, is a AlGaP LED that emits light at 620 nm (orange color). Each LED emits an optical power of 50 mW and is assumed to radiate as a Lambertian source. The axis of each LED is aligned vertically, as indicated in Fig. 4P.3). Determine the luminous intensity (in lm/sr) of each LED as perceived by the observer. Additionally, identify which LED appears brightest and which appears dimmest.

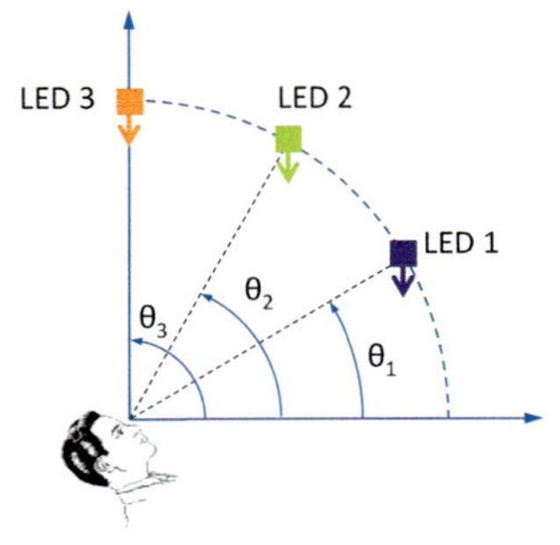

FIGURE 4P.3: Three LEDs in space.

4 Photometry-Radiometry: An illuminated panel switch has a square, flat surface with dimensions of 5 mm by 5 mm. By filtering the light from an incandescent bulb, it emits red light with a wavelength of 630 nm. The emitted light is assumed to follow a Lambertian distribution. The radiant optical power emitted is 10^{-6} watts. Calculate (a) the luminous power, (b) the luminous intensity normal to the surface of the switch, and (c) the luminance.

5 Photometry-Brightness Comparison: In an optical laboratory, the outputs of two lasers are observed with the eye under identical conditions (for example, the laser beams could be directed onto a diffusing screen). The first laser is a 10 mW Helium-Cadmium laser with a wavelength of 441.6 nm. The second laser is a 1.5 mW Helium-Neon laser with a wavelength of 632.8 nm. All conditions are the same except for the wavelengths and radiant powers. Which laser would appear brighter? Provide calculations to support your conclusion.

6 Photometric Quantities: An argon ion laser emits 2.8 watts of power at a wavelength of 514.5 nm. The beam has a diameter of 1.4 mm at the laser's output. The beam diverges and strikes a screen located 10 m away, with the beam axis perpendicular to the screen. (a) Calculate the luminous power on the screen. (b) Calculate the illuminance on the screen, in footcandles, within the illuminated region. For simplicity, assume the beam has a uniform intensity across its cross-section (rather than a Gaussian profile).

7 Photometric Quantities: A Liquid-Crystal (LC) computer projector has a rated output of 1800 lumens. The projector emits equal luminous power in each of the three primary colors. The freespace wavelengths corresponding to the primary colors are $\lambda_R = 670$ nm (for red), $\lambda_G = 550$ nm (for green), and $\lambda_B = 440$ nm (for blue). Calculate the radiant power (in watts) for each of the three primary colors, as well as the total radiant power emitted by the projector. Now, assume the projector illuminates a fully reflective screen with dimensions $1.80m \times 1.20m$. The screen can be modeled as a Lambertian source. Determine the luminance of the reflective screen in cd/m^2.

8 Liquid Crystal Projector: A Liquid-Crystal (LC) computer projector exhibits a luminous efficacy of 6 lumens per watt, with an electrical power consumption of 300 W. (a) Determine the luminous efficiency of the projector. (b) Calculate the projector's luminous power in lumens. (c) Assuming the optical power of the projector is evenly distributed among the three primary colors R, G, and B (with wavelengths $\lambda_R = 630$ nm, $\lambda_G = 515$ nm, and $\lambda_B = 470$ nm), determine the electromagnetic power in watts corresponding to each of these basic colors.

9 Argon-Ion Laser Beam: An Argon-Ion laser beam can be modeled as a fundamental Gaussian beam with a free-space wavelength of $\lambda_0 = 514.5$ nm. The optical output power of the laser is 1 W, and the minimum waist of the Gaussian beam at the laser output is $w_0 = 2.5$ mm. (a) Determine the divergence angle, $\Delta\theta$, of the Gaussian beam, both in radians and degrees. Additionally, calculate the Rayleigh distance of the beam. What is the beam width at a distance of 1 km? (b) The luminous intensity (in lm/sr) of the laser beam is modeled as $I(\theta) = I_0 \cos^n(\theta)$ where n is a positive integer and θ is the polar angle from the laser axis. Determine the integer n and the constant I_0 for this laser beam, making all necessary assumptions. (c) Assuming that the entire optical power of the laser is confined within the solid angle of a cone with apex angle $\Delta\theta$, calculate the luminance of the laser in lm/sr/m^2.

10 Blackbody Radiation: The radiation emitted by the Sun can be approximated by the radiation of a blackbody, with its maximum energy density per unit wavelength occurring at $\lambda_{0m} = 501.4$ nm. (a) Calculate the electromagnetic power per unit area (in W/m^2) that reaches Earth from the Sun. (b) Assuming the Earth is in thermodynamic equilibrium and radiates as a blackbody, determine the Earth's equivalent temperature in degrees Kelvin. (c) For a more accurate estimate in the previous part, account for the reflection of solar radiation by the Earth. The average reflection coefficient of the Sun's radiation by Earth (known as "albedo") is about 30%. Considering Earth's albedo, what would its equivalent temperature be in degrees Kelvin? (Given parameters: Sun's mean radius is $R_s = 696000$ km, Earth's mean radius is $R_e = 6371$ km, and the mean distance between Sun and Earth is $D = 149600000$ km.)

11 Blackbody Radiation: The radiation emitted by the Sun can be approximated by that of a blackbody with an average surface temperature of 5505° C. Assuming that each planet (celestial body in general) in our solar system is in thermodynamic equilibrium and radiates as a blackbody, calculate the mean temperature of each planet in degrees Kelvin. Consider the average power reflection coefficient (albedo) of each planet's surface with respect to the Sun's radiation. For simplicity, neglect the greenhouse effect, planetary atmospheric pressure, and any other phenomena that could affect the average temperature of each planet. All the required data are provided in Table 4.3.

12 Blackbody Radiation-Moon: Unlike Earth, the Moon has no atmosphere, so there is no greenhouse effect. Additionally, the dark side of the Moon is very cold, while the bright side is

TABLE 4.3: Data for Problem 11

Celestial Body	Average Radius (km)	Average Distance from the Sun ($\times 10^6$ km)	Average Albedo
Sun	695700	-	-
Mercury	2439.5	57.9	0.12
Venus	6052	108.2	0.75
Earth	6378	149.6	0.30
Mars	3396	227.9	0.16
Jupiter	71492	778.6	0.34
Saturn	60268	1433.5	0.34
Uranus	25559	2872.5	0.30
Neptune	24764	4495.1	0.29
Pluto	1185	5906.4	0.40

much warmer. Consider two models for estimating surface temperatures on the Moon's bright side. The Moon's radius is 1737 km, and its distance from the Sun is 149.6×10^6 km (approximately the mean distance between Earth and the Sun). The radiation emitted by the Sun can be modeled as blackbody radiation with an average surface temperature of 5505° C. The Sun's radius is 695700 km. (a) Assume that all the solar power incident on the Moon is radiated as thermal radiation from the half of the Moon's surface that is sunlit (the bright side). Estimate the temperature of the Moon's bright surface, considering the Moon's albedo, which is assumed to be constant at 0.12. (b) The model in part (a) is an oversimplification, as it is unrealistic for the entire sunlit side of the Moon to maintain a uniform temperature. A more accurate model assumes that each specific point on the Moon's surface absorbs solar energy and radiates it as thermal energy. Consider a $1m^2$ area of the Moon's surface directly facing the Sun, absorbing sunlight and re-emitting it as thermal radiation. Estimate the surface temperature of this area, again considering the Moon's albedo.
Note: This problem was adapted from an online source and modified.

13 **Blackbody Radiation-Greenhouse Effect:** A single-layer atmosphere surrounding a planet, such as Earth, can be used to illustrate the basics of the greenhouse effect. In this simplified model, the atmospheric layer acts like an invisible blanket wrapped around the planet. Visible sunlight is assumed to pass through the atmosphere without scattering or absorption (a crude approximation), and the planetary albedo remains constant at 0.30 (as noted for Earth in the previous table, 4.3). The surface net shortwave flux is the same as at the top of the atmosphere. The atmosphere only affects the infrared (IR) radiation flux emitted by the Earth's surface, absorbing part of it to heat the atmosphere, while the rest is transmitted and escapes into space. The atmosphere is modeled as a "gray body" with constant emissivity (and absorptivity) ε which is independent of the free-space wavelength, with $0 < \varepsilon < 1$. In equilibrium, the surface temperature of the Earth is T_s, and the atmospheric temperature is T_a, both assumed to be constant in this simple model. A diagram showing the single-layer atmosphere and the associated radiation fluxes is provided below (Fig. 4P.4). It is also given that the solar radiation arriving at Earth, as measured by satellites outside Earth, is approximately $1360\,W/m^2$. (a) Derive expressions for the Earth's surface temperature T_s and atmospheric temperature T_a as functions of the emissivity ε. (b) If the average surface temperature of the Earth is known to be 15° C, calculate the emissivity ε of the atmosphere. (c) Recalculate parts (a) and (b), accounting for the finite thickness of the atmosphere, h. For numerical implementation, assume $h = 100$ km and the Earth's radius $R_e = 6378$ km.

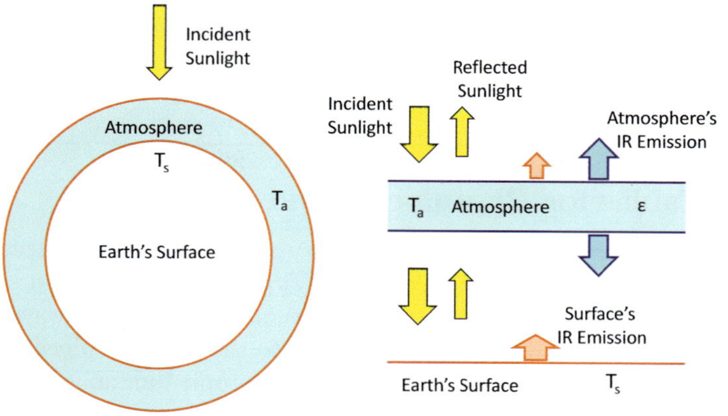

FIGURE 4P.4: A simple model of the Earth and its atmosphere for the explanation of the greenhouse effect.

14 Blackbody Color: Calculate the chromaticity coordinates of an ideal blackbody radiator as a function of its temperature. Assume that the temperature range is between 1000° K to 12000° K in steps of 100° K. For this problem the color matching functions, $\bar{x}(\lambda_0)$, $\bar{y}(\lambda_0)$, $\bar{z}(\lambda_0)$, are required. These data can be found at the following URL: `http://www.cvrl.org` and in Table 4A.4 of Appendix 4A. a) Create a plot of the chromaticity coordinates embedded in the CIE chromaticity diagram as a function of the temperature of the blackbody radiator. (b) Tabulate the chromaticity coordinates of a blackbody for temperatures ranging from $T = 1000° - 10000°$ K in steps of 1000° K. (c) Now, represent the perceived color of the blackbody, as seen by the human eye, as a function of its temperature. For this representation, the standard RGB color space will be utilized. The conversion matrix $\tilde{M}$ that transforms the X, Y, and Z, tristimulus values to R, G, and B values is given by:

$$\tilde{M} = \begin{bmatrix} 3.2404542 & -1.5371385 & -0.4985314 \\ -0.9692660 & 1.8760108 & 0.04215560 \\ 0.0556434 & -0.2040259 & 1.0572252 \end{bmatrix}.$$

Since R, G, B should be in the range from 0 to 1, normalize the calculated R, G, and B by dividing them by their maximum value. To enhance color perception, a nonlinear correction is then applied to the normalized R, G, and B values, which is given by:

$$W' = \begin{cases} 12.92W & \text{for} \quad W < 0.0031308 \\ 1.055W^{1/2.4} - 0.055 & \text{for} \quad W \geq 0.0031308 \end{cases}$$

where W is any of R, G, and B. The final R, G, and B values should be clipped to zero if they are negative and clipped to one if they exceed unity. Based on the information provided, attempt to represent the perceived color of a blackbody as a function of its temperature, both with and without the gamma correction.

15 Blackbody Radiation—Human Body: Assume that the average human body can be modeled as a blackbody radiator with a temperature of 36.6° C and an approximate surface area of one square meter. (a) Create a graphical representation of the spectral radiant emittance $M_{\lambda_0}(\lambda_0)$ in $W/m^2/\mu m$ (with the wavelength axis expressed in microns). Identify the wavelength at which $M_{\lambda_0}(\lambda_0)$ is maximized. b) Calculate the total power emitted by the human body between wavelengths of 10 μm-10.1 μm. Determine what percentage of the total emitted power this represents. Additionally, calculate the average number of photons emitted per second by the human body in this wavelength range. (c) Calculate the total power emitted by the human body in the visible spectrum, i.e., between wavelengths of 0.38 μm –0.78 μm. Determine what percentage of the total emitted power this represents. Also, calculate the average number of photons emitted per second by the human body in the visible spectrum.

Appendices

4A Useful Tables for Photometry and Colorimetry

In this Appendix, some useful tables for photometry and colorimetry are listed. These tables were obtained from the following url-link: `http://www.cvrl.org/`. The first table provides luminous efficiency values for photopic and scotopic conditions at 5 nm wavelength intervals. The photopic luminous efficiency values correspond to the 2-degree physiologically relevant values consistent with the 1924 standard [206]. The scotopic luminous efficiency values are based on the 1951 CIE results.

TABLE 4A.1: Luminous Efficiency Values (per 5nm)

Freespace Wavelength (nm)	Photopic Values $V(\lambda_0)$	Scotopic Values $V'(\lambda_0)$
380	3.9000×10^{-5}	5.890×10^{-4}
385	6.4000×10^{-5}	1.108×10^{-3}
390	1.2000×10^{-4}	2.209×10^{-3}
395	2.1700×10^{-4}	4.530×10^{-3}
400	3.9600×10^{-4}	9.290×10^{-3}
405	6.4000×10^{-4}	1.852×10^{-2}
410	1.2100×10^{-3}	3.484×10^{-2}
415	2.1800×10^{-3}	6.040×10^{-2}
420	4.0000×10^{-3}	9.660×10^{-2}
425	7.3000×10^{-3}	1.436×10^{-1}
430	1.1600×10^{-2}	1.998×10^{-1}
435	1.6840×10^{-2}	2.625×10^{-1}
440	2.3000×10^{-2}	3.281×10^{-1}
445	2.9800×10^{-2}	3.931×10^{-1}
450	3.8000×10^{-2}	4.550×10^{-1}
455	4.8000×10^{-2}	5.130×10^{-1}
460	6.0000×10^{-2}	5.670×10^{-1}
465	7.3900×10^{-2}	6.200×10^{-1}
470	9.0980×10^{-2}	6.760×10^{-1}
475	1.1260×10^{-1}	7.340×10^{-1}
480	1.3902×10^{-1}	7.930×10^{-1}
485	1.6930×10^{-1}	8.510×10^{-1}
490	2.0802×10^{-1}	9.040×10^{-1}
495	2.5860×10^{-1}	9.490×10^{-1}
500	3.2300×10^{-1}	9.820×10^{-1}
505	4.0730×10^{-1}	9.980×10^{-1}
510	5.0300×10^{-1}	9.970×10^{-1}
515	6.0820×10^{-1}	9.750×10^{-1}
520	7.1000×10^{-1}	9.350×10^{-1}
525	7.9320×10^{-1}	8.800×10^{-1}
530	8.6200×10^{-1}	8.110×10^{-1}
535	9.1485×10^{-1}	7.330×10^{-1}
540	9.5400×10^{-1}	6.500×10^{-1}

Luminous Efficiency Values (per 5nm) (Continued)

Freespace Wavelength (nm)	Photopic Values $V(\lambda_0)$	Scotopic Values $V'(\lambda_0)$
545	9.8030×10^{-1}	5.640×10^{-1}
550	9.9495×10^{-1}	4.810×10^{-1}
555	1.0000	4.020×10^{-1}
560	9.9500×10^{-1}	3.288×10^{-1}
565	9.7860×10^{-1}	2.639×10^{-1}
570	9.5200×10^{-1}	2.076×10^{-1}
575	9.1540×10^{-1}	1.602×10^{-1}
580	8.7000×10^{-1}	1.212×10^{-1}
585	8.1630×10^{-1}	8.990×10^{-2}
590	7.5700×10^{-1}	6.550×10^{-2}
595	6.9490×10^{-1}	4.690×10^{-2}
600	6.3100×10^{-1}	3.315×10^{-2}
605	5.6680×10^{-1}	2.312×10^{-2}
610	5.0300×10^{-1}	1.593×10^{-2}
615	4.4120×10^{-1}	1.088×10^{-2}
620	3.8100×10^{-1}	7.370×10^{-3}
625	3.2100×10^{-1}	4.970×10^{-3}
630	2.6500×10^{-1}	3.335×10^{-3}
635	2.1700×10^{-1}	2.235×10^{-3}
640	1.7500×10^{-1}	1.497×10^{-3}
645	1.3820×10^{-1}	1.005×10^{-3}
650	1.0700×10^{-1}	6.770×10^{-4}
655	8.1600×10^{-2}	4.590×10^{-4}
660	6.1000×10^{-2}	3.129×10^{-4}
665	4.4580×10^{-2}	2.146×10^{-4}
670	3.2000×10^{-2}	1.480×10^{-4}
675	2.3200×10^{-2}	1.026×10^{-4}
680	1.7000×10^{-2}	7.150×10^{-5}
685	1.1920×10^{-2}	5.010×10^{-5}
690	8.2100×10^{-3}	3.533×10^{-5}
695	5.7230×10^{-3}	2.501×10^{-5}
700	4.1020×10^{-3}	1.780×10^{-5}
705	2.9290×10^{-3}	1.273×10^{-5}
710	2.0910×10^{-3}	9.140×10^{-6}
715	1.4840×10^{-3}	6.600×10^{-6}
720	1.0470×10^{-3}	4.780×10^{-6}
725	7.4000×10^{-4}	3.482×10^{-6}
730	5.2000×10^{-4}	2.546×10^{-6}
735		1.870×10^{-6}
740		1.379×10^{-6}
745		1.022×10^{-6}
750		7.600×10^{-7}
755		5.670×10^{-7}
760		4.250×10^{-7}
765		3.196×10^{-7}
770		2.413×10^{-7}
775		1.829×10^{-7}

Luminous Efficiency Values (per 5nm) (Continued)

Freespace Wavelength (nm)	Photopic Values $V(\lambda_0)$	Scotopic Values $V'(\lambda_0)$
780		1.390×10^{-7}

The photopic luminous efficiency values corresponding to the 2-degree physiologically relevant luminous efficiency values and consistent with the Stockman and Sharpe cone fundamentals [249, 250], are listed below.

TABLE 4A.2: Luminous Efficiency Values (Stockman & Sharpe) (per 5nm)

Freespace Wavelength (nm)	Photopic Values (1951) $V(\lambda_0)$
390	4.14616×10^{-4}
395	1.05965×10^{-3}
400	2.45219×10^{-3}
405	4.97172×10^{-3}
410	9.07986×10^{-3}
415	1.42938×10^{-2}
420	2.02737×10^{-2}
425	2.61211×10^{-2}
430	3.31904×10^{-2}
435	4.15794×10^{-2}
440	5.03366×10^{-2}
445	5.74339×10^{-2}
450	6.47235×10^{-2}
455	7.23834×10^{-2}
460	8.51482×10^{-2}
465	1.06014×10^{-1}
470	1.29896×10^{-1}
475	1.53507×10^{-1}
480	1.78805×10^{-1}
485	2.06483×10^{-1}
490	2.37916×10^{-1}
495	2.85068×10^{-1}
500	3.48354×10^{-1}
505	4.27760×10^{-1}
510	5.20497×10^{-1}
515	6.20626×10^{-1}
520	7.18089×10^{-1}
525	7.94645×10^{-1}
530	8.57580×10^{-1}
535	9.07135×10^{-1}
540	9.54468×10^{-1}
545	9.81411×10^{-1}
550	9.89023×10^{-1}
555	9.99461×10^{-1}
560	9.96774×10^{-1}
565	9.90255×10^{-1}

Luminous Efficiency Values (Stockman & Sharpe) (per 5nm) (Continued)

Freespace Wavelength (nm)	Photopic Values (1951) $V(\lambda_0)$
570	9.73261×10^{-1}
575	9.42457×10^{-1}
580	8.96361×10^{-1}
585	8.58720×10^{-1}
590	8.11587×10^{-1}
595	7.54479×10^{-1}
600	6.91855×10^{-1}
605	6.27007×10^{-1}
610	5.58375×10^{-1}
615	4.89595×10^{-1}
620	4.22990×10^{-1}
625	3.60924×10^{-1}
630	2.98086×10^{-1}
635	2.41690×10^{-1}
640	1.94312×10^{-1}
645	1.54740×10^{-1}
650	1.19312×10^{-1}
655	8.97959×10^{-2}
660	6.67104×10^{-2}
665	4.89970×10^{-2}
670	3.55998×10^{-2}
675	2.55422×10^{-2}
680	1.80794×10^{-2}
685	1.26157×10^{-2}
690	8.66128×10^{-3}
695	6.02768×10^{-3}
700	4.19594×10^{-3}
705	2.91086×10^{-3}
710	1.99556×10^{-3}
715	1.36702×10^{-3}
720	9.44727×10^{-4}
725	6.53705×10^{-4}
730	4.55597×10^{-4}
735	3.17974×10^{-4}
740	2.21745×10^{-4}
745	1.56557×10^{-4}
750	1.10393×10^{-4}
755	7.82744×10^{-5}
760	5.57886×10^{-5}
765	3.98188×10^{-5}
770	2.86018×10^{-5}
775	2.05126×10^{-5}
780	1.48724×10^{-5}
785	1.08000×10^{-5}
790	7.86392×10^{-6}
795	5.73694×10^{-6}
800	4.21160×10^{-6}

Luminous Efficiency Values (Stockman & Sharpe) (per 5nm) (Continued)

Freespace Wavelength (nm)	Photopic Values (1951) $V(\lambda_0)$
805	3.10656×10^{-6}
810	2.28679×10^{-6}
815	1.69315×10^{-6}
820	1.26256×10^{-6}
825	9.42251×10^{-7}
830	7.05386×10^{-7}

In the next table, the values of the *rgb* color matching functions, $\bar{r}(\lambda_0)$, $\bar{g}(\lambda_0)$, $\bar{b}(\lambda_0)$, are listed per 5 nm intervals. These values are based on the work of Stiles and Burns [251].

TABLE 4A.3: Color Matching Functions (per 5nm)

λ_0(nm)	$\bar{r}(\lambda_0)$	$\bar{g}(\lambda_0)$	$\bar{b}(\lambda_0)$
390	1.83970×10^{-3}	-4.53930×10^{-4}	1.21520×10^{-2}
395	4.61530×10^{-3}	-1.04640×10^{-3}	3.11100×10^{-2}
400	9.62640×10^{-3}	-2.16890×10^{-3}	6.23710×10^{-2}
405	1.89790×10^{-2}	-4.43040×10^{-3}	1.31610×10^{-1}
410	3.08030×10^{-2}	-7.20480×10^{-3}	2.27500×10^{-1}
415	4.24590×10^{-2}	-1.25790×10^{-2}	3.58970×10^{-1}
420	5.16620×10^{-2}	-1.66510×10^{-2}	5.23960×10^{-1}
425	5.28370×10^{-2}	-2.12400×10^{-2}	6.85860×10^{-1}
430	4.42870×10^{-2}	-1.99360×10^{-2}	7.96040×10^{-1}
435	3.22200×10^{-2}	-1.60970×10^{-2}	8.94590×10^{-1}
440	1.47630×10^{-2}	-7.34570×10^{-3}	9.63950×10^{-1}
445	-2.33920×10^{-3}	1.36900×10^{-3}	9.98140×10^{-1}
450	-2.91300×10^{-2}	1.96100×10^{-2}	9.18750×10^{-1}
455	-6.06770×10^{-2}	4.34640×10^{-2}	8.24870×10^{-1}
460	-9.62240×10^{-2}	7.09540×10^{-2}	7.85540×10^{-1}
465	-1.37590×10^{-1}	1.10220×10^{-1}	6.67230×10^{-1}
470	-1.74860×10^{-1}	1.50880×10^{-1}	6.10980×10^{-1}
475	-2.12600×10^{-1}	1.97940×10^{-1}	4.88290×10^{-1}
480	-2.37800×10^{-1}	2.40420×10^{-1}	3.61950×10^{-1}
485	-2.56740×10^{-1}	2.79930×10^{-1}	2.66340×10^{-1}
490	-2.77270×10^{-1}	3.33530×10^{-1}	1.95930×10^{-1}
495	-2.91250×10^{-1}	4.05210×10^{-1}	1.47300×10^{-1}
500	-2.95000×10^{-1}	4.90600×10^{-1}	1.07490×10^{-1}
505	-2.97060×10^{-1}	5.96730×10^{-1}	7.67140×10^{-2}
510	-2.67590×10^{-1}	7.01840×10^{-1}	5.02480×10^{-2}
515	-2.17250×10^{-1}	8.08520×10^{-1}	2.87810×10^{-2}
520	-1.47680×10^{-1}	9.10760×10^{-1}	1.33090×10^{-2}
525	-3.51840×10^{-2}	9.84820×10^{-1}	2.11700×10^{-3}
530	1.06140×10^{-1}	1.03390	-4.15740×10^{-3}
535	2.59810×10^{-1}	1.05380	-8.30320×10^{-3}
540	4.19760×10^{-1}	1.05120	-1.21910×10^{-2}
545	5.92590×10^{-1}	1.04980	-1.40390×10^{-2}
550	7.90040×10^{-1}	1.03680	-1.46810×10^{-2}

Color Matching Functions (per 5nm) (Continued)

λ_0(nm)	$\bar{r}(\lambda_0)$	$\bar{g}(\lambda_0)$	$\bar{b}(\lambda_0)$
555	1.00780	9.98260×10^{-1}	-1.49470×10^{-2}
560	1.22830	9.37830×10^{-1}	-1.46130×10^{-2}
565	1.47270	8.80390×10^{-1}	-1.37820×10^{-2}
570	1.74760	8.28350×10^{-1}	-1.26500×10^{-2}
575	2.02140	7.46860×10^{-1}	-1.13560×10^{-2}
580	2.27240	6.49300×10^{-1}	-9.93170×10^{-3}
585	2.48960	5.63170×10^{-1}	-8.41480×10^{-3}
590	2.67250	4.76750×10^{-1}	-7.02100×10^{-3}
595	2.80930	3.84840×10^{-1}	-5.74370×10^{-3}
600	2.87170	3.00690×10^{-1}	-4.27430×10^{-3}
605	2.85250	2.28530×10^{-1}	-2.91320×10^{-3}
610	2.76010	1.65750×10^{-1}	-2.26930×10^{-3}
615	2.59890	1.13730×10^{-1}	-1.99660×10^{-3}
620	2.37430	7.46820×10^{-2}	-1.50690×10^{-3}
625	2.10540	4.65040×10^{-2}	-9.38220×10^{-4}
630	1.81450	2.63330×10^{-2}	-5.53160×10^{-4}
635	1.52470	1.27240×10^{-2}	-3.16680×10^{-4}
640	1.25430	4.50330×10^{-3}	-1.43190×10^{-4}
645	1.00760	9.66110×10^{-5}	-4.08310×10^{-6}
650	7.86420×10^{-1}	-1.96450×10^{-3}	1.10810×10^{-4}
655	5.96590×10^{-1}	-2.63270×10^{-3}	1.91750×10^{-4}
660	4.43200×10^{-1}	-2.62620×10^{-3}	2.26560×10^{-4}
665	3.24100×10^{-1}	-2.30270×10^{-3}	2.15200×10^{-4}
670	2.34550×10^{-1}	-1.87000×10^{-3}	1.63610×10^{-4}
675	1.68840×10^{-1}	-1.44240×10^{-3}	9.71640×10^{-5}
680	1.20860×10^{-1}	-1.07550×10^{-3}	5.10330×10^{-5}
685	8.58110×10^{-2}	-7.90040×10^{-4}	3.52710×10^{-5}
690	6.02600×10^{-2}	-5.67650×10^{-4}	3.12110×10^{-5}
695	4.14800×10^{-2}	-3.92740×10^{-4}	2.45080×10^{-5}
700	2.81140×10^{-2}	-2.62310×10^{-4}	1.65210×10^{-5}
705	1.91170×10^{-2}	-1.75120×10^{-4}	1.11240×10^{-5}
710	1.33050×10^{-2}	-1.21400×10^{-4}	8.69650×10^{-6}
715	9.40920×10^{-3}	-8.57600×10^{-5}	7.43510×10^{-6}
720	6.51770×10^{-3}	-5.76770×10^{-5}	6.10570×10^{-6}
725	4.53770×10^{-3}	-3.90030×10^{-5}	5.02770×10^{-6}
730	3.17420×10^{-3}	-2.65110×10^{-5}	4.12510×10^{-6}

The next table contains the values of the xyz color matching functions, $\bar{x}(\lambda_0)$, $\bar{y}(\lambda_0)$, $\bar{z}(\lambda_0)$ per 5 nm intervals. These values correspond to the 2-deg CIE functions transformed from the CIE 2006 2-deg LMS cone fundamentals.

TABLE 4A.4: Color Matching Functions (per 5nm)

λ_0(nm)	$\bar{x}(\lambda_0)$	$\bar{y}(\lambda_0)$	$\bar{z}(\lambda_0)$
390	3.769647×10^{-3}	4.146161×10^{-4}	1.847260×10^{-2}
395	9.382967×10^{-3}	1.059646×10^{-3}	4.609784×10^{-2}

Color Matching Functions (per 5nm) (Continued)

λ_0(nm)	$\bar{x}(\lambda_0)$	$\bar{y}(\lambda_0)$	$\bar{z}(\lambda_0)$
400	2.214302×10^{-2}	2.452194×10^{-3}	1.096090×10^{-1}
405	4.742986×10^{-2}	4.971717×10^{-3}	2.369246×10^{-1}
410	8.953803×10^{-2}	9.079860×10^{-3}	4.508369×10^{-1}
415	1.446214×10^{-1}	1.429377×10^{-2}	7.378822×10^{-1}
420	2.035729×10^{-1}	2.027369×10^{-2}	1.051821
425	2.488523×10^{-1}	2.612106×10^{-2}	1.305008
430	2.918246×10^{-1}	3.319038×10^{-2}	1.552826
435	3.227087×10^{-1}	4.157940×10^{-2}	1.748280
440	3.482554×10^{-1}	5.033657×10^{-2}	1.917479
445	3.418483×10^{-1}	5.743393×10^{-2}	1.918437
450	3.224637×10^{-1}	6.472352×10^{-2}	1.848545
455	2.826646×10^{-1}	7.238339×10^{-2}	1.664439
460	2.485254×10^{-1}	8.514816×10^{-2}	1.522157
465	2.219781×10^{-1}	1.060145×10^{-1}	1.428440
470	1.806905×10^{-1}	1.298957×10^{-1}	1.250610
475	1.291920×10^{-1}	1.535066×10^{-1}	9.991789×10^{-1}
480	8.182895×10^{-2}	1.788048×10^{-1}	7.552379×10^{-1}
485	4.600865×10^{-2}	2.064828×10^{-1}	5.617313×10^{-1}
490	2.083981×10^{-2}	2.379160×10^{-1}	4.099313×10^{-1}
495	7.097731×10^{-3}	2.850680×10^{-1}	3.105939×10^{-1}
500	2.461588×10^{-3}	3.483536×10^{-1}	2.376753×10^{-1}
505	3.649178×10^{-3}	4.277595×10^{-1}	1.720018×10^{-1}
510	1.556989×10^{-2}	5.204972×10^{-1}	1.176796×10^{-1}
515	4.315171×10^{-2}	6.206256×10^{-1}	8.283548×10^{-2}
520	7.962917×10^{-2}	7.180890×10^{-1}	5.650407×10^{-2}
525	1.268468×10^{-1}	7.946448×10^{-1}	3.751912×10^{-2}
530	1.818026×10^{-1}	8.575799×10^{-1}	2.438164×10^{-2}
535	2.405015×10^{-1}	9.071347×10^{-1}	1.566174×10^{-2}
540	3.098117×10^{-1}	9.544675×10^{-1}	9.846470×10^{-3}
545	3.804244×10^{-1}	9.814106×10^{-1}	6.131421×10^{-3}
550	4.494206×10^{-1}	9.890228×10^{-1}	3.790291×10^{-3}
555	5.280233×10^{-1}	9.994608×10^{-1}	2.327186×10^{-3}
560	6.133784×10^{-1}	9.967737×10^{-1}	1.432128×10^{-3}
565	7.016774×10^{-1}	9.902549×10^{-1}	8.822531×10^{-4}
570	7.967750×10^{-1}	9.732611×10^{-1}	5.452416×10^{-4}
575	8.853376×10^{-1}	9.424569×10^{-1}	3.386739×10^{-4}
580	9.638388×10^{-1}	8.963613×10^{-1}	2.117772×10^{-4}
585	1.051011	8.587203×10^{-1}	1.335031×10^{-4}
590	1.109767	8.115868×10^{-1}	8.494468×10^{-5}
595	1.143620	7.544785×10^{-1}	5.460706×10^{-5}
600	1.151033	6.918553×10^{-1}	3.549661×10^{-5}
605	1.134757	6.270066×10^{-1}	2.334738×10^{-5}
610	1.083928	5.583746×10^{-1}	1.554631×10^{-5}
615	1.007344	4.895950×10^{-1}	1.048387×10^{-5}
620	9.142877×10^{-1}	4.229897×10^{-1}	0.000000
625	8.135565×10^{-1}	3.609245×10^{-1}	0.000000
630	6.924717×10^{-1}	2.980865×10^{-1}	0.000000

Color Matching Functions (per 5nm) (Continued)

λ_0(nm)	$\bar{x}(\lambda_0)$	$\bar{y}(\lambda_0)$	$\bar{z}(\lambda_0)$
635	5.755410×10^{-1}	2.416902×10^{-1}	0.000000
640	4.731224×10^{-1}	1.943124×10^{-1}	0.000000
645	3.844986×10^{-1}	1.547397×10^{-1}	0.000000
650	2.997374×10^{-1}	1.193120×10^{-1}	0.000000
655	2.277792×10^{-1}	8.979594×10^{-2}	0.000000
660	1.707914×10^{-1}	6.671045×10^{-2}	0.000000
665	1.263808×10^{-1}	4.899699×10^{-2}	0.000000
670	9.224597×10^{-2}	3.559982×10^{-2}	0.000000
675	6.639960×10^{-2}	2.554223×10^{-2}	0.000000
680	4.710606×10^{-2}	1.807939×10^{-2}	0.000000
685	3.292138×10^{-2}	1.261573×10^{-2}	0.000000
690	2.262306×10^{-2}	8.661284×10^{-3}	0.000000
695	1.575417×10^{-2}	6.027677×10^{-3}	0.000000
700	1.096778×10^{-2}	4.195941×10^{-3}	0.000000
705	7.608750×10^{-3}	2.910864×10^{-3}	0.000000
710	5.214608×10^{-3}	1.995557×10^{-3}	0.000000
715	3.569452×10^{-3}	1.367022×10^{-3}	0.000000
720	2.464821×10^{-3}	9.447269×10^{-4}	0.000000
725	1.703876×10^{-3}	6.537050×10^{-4}	0.000000
730	1.186238×10^{-3}	4.555970×10^{-4}	0.000000
735	8.269535×10^{-4}	3.179738×10^{-4}	0.000000
740	5.758303×10^{-4}	2.217445×10^{-4}	0.000000
745	4.058303×10^{-4}	1.565566×10^{-4}	0.000000
750	2.856577×10^{-4}	1.103928×10^{-4}	0.000000
755	2.021853×10^{-4}	7.827442×10^{-5}	0.000000
760	1.438270×10^{-4}	5.578862×10^{-5}	0.000000
765	1.024685×10^{-4}	3.981884×10^{-5}	0.000000
770	7.347551×10^{-5}	2.860175×10^{-5}	0.000000
775	5.259870×10^{-5}	2.051259×10^{-5}	0.000000
780	3.806114×10^{-5}	1.487243×10^{-5}	0.000000
785	2.758222×10^{-5}	1.080001×10^{-5}	0.000000
790	2.004122×10^{-5}	7.863920×10^{-6}	0.000000
795	1.458792×10^{-5}	5.736935×10^{-6}	0.000000
800	1.068141×10^{-5}	4.211597×10^{-6}	0.000000
805	7.857521×10^{-6}	3.106561×10^{-6}	0.000000
810	5.768284×10^{-6}	2.286786×10^{-6}	0.000000
815	4.259166×10^{-6}	1.693147×10^{-6}	0.000000
820	3.167765×10^{-6}	1.262556×10^{-6}	0.000000
825	2.358723×10^{-6}	9.422514×10^{-7}	0.000000
830	1.762465×10^{-6}	7.053860×10^{-7}	0.000000

4B Color Space Transformation

Let a color $[\boldsymbol{F}]$ represented in both RGB and XYZ color coordinate systems. Then the following equation can be written:

$$[\boldsymbol{F}] = R_F[\boldsymbol{R}] + G_F[\boldsymbol{G}] + B_F[\boldsymbol{B}] = X_F[\mathbf{X}] + Y_F[\mathbf{Y}] + Z_F[\boldsymbol{Z}], \tag{4B.1}$$

Now the primaries $[\boldsymbol{R}]$, $[\boldsymbol{G}]$, and $[\boldsymbol{B}]$ can be expressed in the XYZ color system as follows:

$$[\boldsymbol{R}] = X_R[\mathbf{X}] + Y_R[\mathbf{Y}] + Z_R[\boldsymbol{Z}],$$
$$[\boldsymbol{G}] = X_G[\mathbf{X}] + Y_G[\mathbf{Y}] + Z_G[\boldsymbol{Z}],$$
$$[\boldsymbol{B}] = X_B[\mathbf{X}] + Y_B[\mathbf{Y}] + Z_B[\boldsymbol{Z}].$$

The last equations can be written in the matrix form:

$$\begin{bmatrix} [\boldsymbol{R}] \\ [\boldsymbol{G}] \\ [\boldsymbol{B}] \end{bmatrix} = \begin{bmatrix} X_R & Y_R & Z_R \\ X_G & Y_G & Z_G \\ X_B & Y_B & Z_B \end{bmatrix} \begin{bmatrix} [\mathbf{X}] \\ [\mathbf{Y}] \\ [\mathbf{Y}] \end{bmatrix} = \tilde{M} \begin{bmatrix} [\mathbf{X}] \\ [\mathbf{Y}] \\ [\boldsymbol{Z}] \end{bmatrix}. \tag{4B.2}$$

In the above equation the primaries resemble the unit vectors of the corresponding color system. Equation (4B.1) can also be written as:

$$\begin{bmatrix} R_F & G_F & B_F \end{bmatrix} \begin{bmatrix} [\boldsymbol{R}] \\ [\boldsymbol{G}] \\ [\boldsymbol{B}] \end{bmatrix} = \begin{bmatrix} X_F & Y_F & Z_F \end{bmatrix} \begin{bmatrix} [\mathbf{X}] \\ [\mathbf{Y}] \\ [\boldsymbol{Z}] \end{bmatrix} \Longrightarrow$$

$$\begin{bmatrix} R_F & G_F & B_F \end{bmatrix} \tilde{M} \begin{bmatrix} [\mathbf{X}] \\ [\mathbf{Y}] \\ [\boldsymbol{Z}] \end{bmatrix} = \begin{bmatrix} X_F & Y_F & Z_F \end{bmatrix} \begin{bmatrix} [\mathbf{X}] \\ [\mathbf{Y}] \\ [\boldsymbol{Z}] \end{bmatrix} \Longrightarrow$$

$$\begin{bmatrix} X_F \\ Y_F \\ Z_F \end{bmatrix} = \tilde{M}^T \begin{bmatrix} R_F \\ G_F \\ B_F \end{bmatrix} = \begin{bmatrix} X_R & X_G & X_B \\ Y_R & Y_G & Y_B \\ Z_R & Z_G & Z_B \end{bmatrix} \begin{bmatrix} R_F \\ G_F \\ B_F \end{bmatrix} \tag{4B.3}$$

The transformation matrix of Eq. (4B.3) can be written as follows using the color space coordinates. Specifically if (x, y, z) the color coordinates the tristimuls values can be expressed as $X = Y(x/y)$, $Z = Y(z/y) = Y(1 - x - y)/y$:

$$\tilde{M}_t = \begin{bmatrix} X_R & X_G & X_B \\ Y_R & Y_G & Y_B \\ Z_R & Z_G & Z_B \end{bmatrix} = \begin{bmatrix} Y_R\dfrac{x_r}{y_r} & Y_G\dfrac{x_g}{y_g} & Y_B\dfrac{x_b}{y_b} \\ Y_R & Y_G & Y_B \\ Y_R\dfrac{z_r}{y_r} & Y_G\dfrac{z_g}{y_g} & Y_B\dfrac{z_b}{y_b} \end{bmatrix} \Longrightarrow$$

$$\begin{bmatrix} X_R & X_G & X_B \\ Y_R & Y_G & Y_B \\ Z_R & Z_G & Z_B \end{bmatrix} = \begin{bmatrix} \dfrac{x_r}{y_r} & \dfrac{x_g}{y_g} & \dfrac{x_b}{y_b} \\ 1 & 1 & 1 \\ \dfrac{z_r}{y_r} & \dfrac{z_g}{y_g} & \dfrac{z_b}{y_b} \end{bmatrix} \begin{bmatrix} Y_R & 0 & 0 \\ 0 & Y_G & 0 \\ 0 & 0 & Y_B \end{bmatrix} = \tilde{M}' \begin{bmatrix} Y_R & 0 & 0 \\ 0 & Y_G & 0 \\ 0 & 0 & Y_B \end{bmatrix}$$

Now assume that the white color in the XYZ system has tristimulus values (X_w, Y_w, Z_w). The white color in the RGB system has tristimulus values $(1, 1, 1)$. Applying Eq. (4B.3) for the white color the luminances Y_R, Y_G, and Y_B can be determined as follows:

$$\begin{bmatrix} X_w \\ Y_w \\ Z_w \end{bmatrix} = \tilde{M}' \begin{bmatrix} Y_R & 0 & 0 \\ 0 & Y_G & 0 \\ 0 & 0 & Y_B \end{bmatrix} \begin{bmatrix} 1 \\ 1 \\ 1 \end{bmatrix} = \tilde{M}' \begin{bmatrix} Y_R \\ Y_G \\ Y_B \end{bmatrix} \Longrightarrow$$

$$\begin{bmatrix} Y_R \\ Y_G \\ Y_B \end{bmatrix} = \left(\tilde{M}'\right)^{-1} \begin{bmatrix} X_x \\ Y_w \\ Z_w \end{bmatrix}. \tag{4B.4}$$

Then the transformation matrix between the RGB and XYZ color spaces is given by:

$$\tilde{M}_t = \begin{bmatrix} X_R & X_G & X_B \\ Y_R & Y_G & Y_B \\ Z_R & Z_G & Z_B \end{bmatrix} = \tilde{M}' \begin{bmatrix} Y_R & 0 & 0 \\ 0 & Y_G & 0 \\ 0 & 0 & Y_B \end{bmatrix}. \tag{4B.5}$$

For an example let's consider the initial CIE 1931 chromaticity diagram. In this case the primaries that were used in the psychophysical experiments where $\lambda_0 = 700\,\text{nm}$ for $[\boldsymbol{R}]$, $\lambda_0 = 546.1\,\text{nm}$ for $[\boldsymbol{G}]$, and $\lambda_0 = 435.8\,\text{nm}$ for $[\boldsymbol{B}]$, while the white point was selected to be the equal energy point. Therefore, $x_w = y_w = z_w = 1/3$. Using the values for the color matching functions $\bar{x}(\lambda_0)$, $\bar{y}(\lambda_0)$, and $\bar{z}(\lambda_0)$, for the three primaries used (using linear interpolation between the values listed in the tables) the following color coordinates can be calculated: $(x_R, y_R) = (0.7347, 0.2653)$ (for red primary), $(x_G, y_G) = (0.2737, 0.7174)$ (for green primary), and $(x_B, y_B) = (0.1665, 0.0089)$ (for blue primary). Using $X_w = Y_w(x_w/y_w)$, $Z_w = Y_w(1 - x_w - y_w)/y_w$ and $Y_w = 1$, the following matrix $\tilde{M}_t$ can be determined via Eqs. (4B.4) and (4B.5):

$$\tilde{M}_t = \begin{bmatrix} 0.4902 & 0.3049 & 0.1999 \\ 0.1770 & 0.8123 & 0.0107 \\ 0.0000 & 0.0101 & 0.9900 \end{bmatrix}. \tag{4B.6}$$

In order to make the luminances of the test stimuli the same in both RGB and XYZ color systems, the elements of the matrix $\tilde{M}_t$ of Eq. (4B.6) should be multiplied by the $1.0000 + 4.5907 + 0.0601 = 5.6508$ lumens. Then the resulting matrix becomes:

$$\tilde{M}_t = \begin{bmatrix} 2.7700 & 1.7511 & 1.1297 \\ 1.0003 & 4.5903 & 0.0602 \\ 0.0000 & 0.0569 & 5.5940 \end{bmatrix}. \tag{4B.7}$$

The last matrix is very close to the transformation matrix defined Eq. (4.27) which is the standard matrix provided by CIE [222]. The differences could be due to the differences in the values of the chromaticity coordinates of the primaries.

Another example is to determine the transformation matrix between the XYZ and the $sRGB$ (standard RGB) color spaces. For the $sRGB$ the chromaticity coordinates of the primaries in the XYZ are given as: $(x_R, y_R) = (0.64, 0.33)$, $(x_G, y_G) = (0.30, 0.60)$, and $(x_B, y_B) = (0.15, 0.06)$. The white point was selected to be the D65 standard illuminant [222] with chromaticity coordinates $x_w = 0.312713$ and $y_w = 0.329016$. Using again $Y_w = 1$ the following transformation matrix can be determined with the help of Eqs. (4B.4) and (4B.5):

$$\tilde{M}_t = \begin{bmatrix} 0.412411 & 0.357585 & 0.180454 \\ 0.212649 & 0.715169 & 0.072182 \\ 0.019332 & 0.119195 & 0.950390 \end{bmatrix}. \tag{4B.8}$$

The above matrix is very close to various $sRGB$ to XYZ matrix transformations appearing in the literature and the internet.

4C Determination of Electromagnetic Modes in Rectangular Metallic Cavities

The purpose of this Appendix is to review the determination of electromagnetic modes within a rectangular cavity, which is assumed to have perfectly conducting walls. The material filling the cavity is considered homogeneous, linear, and isotropic [101, 132, 185]. The approach presented here is largely independent of the traditional solutions for rectangular metallic waveguides, which are commonly used to determine cavity modes. The rectangular cavity and the corresponding coordinate system are illustrated in Fig. 4.27. It is assumed that the TE_{mpq} modes are sought, meaning that $E_z = 0$ while all other field components E_x, E_y, H_x, H_y, and H_z can be nonzero. Each field component satisfies the Helmholtz equation:

$$\nabla^2 S + k_0^2 n^2 S = \left(\frac{\partial^2}{\partial x^2} + \frac{\partial^2}{\partial y^2} + \frac{\partial^2}{\partial z^2}\right) S + k_0^2 n^2 S = 0, \tag{4C.1}$$

where $S = E_x, E_y, H_x, H_y, H_z$, $k_0 = \omega/c = 2\pi/\lambda_0$, is the freespace wavenumber, and n is the refractive index of the material inside the cavity. Due to the rectangular geometry of the cavity, we can use the method of separation of variables to solve for the fields. In this method, we assume that the field components can be written as the product of three functions, each depending only on one of the spatial coordinates: $S(x, y, z) = X(x)Y(y)Z(z)$, where $X(x) = A\cos(k_x x) + B\sin(k_x x)$, $Y(y) = C\cos(k_y y) + D\sin(k_y y)$, and $Z(z) = E\cos(k_z z) + F\sin(k_z z)$, with $k_x^2 + k_y^2 + k_z^2 = k_0^2 n^2$. From the two curl Maxwell's equations $\nabla \times \vec{E} = -j\omega\mu_0\vec{H}$, and $\nabla \times \vec{H} = +j\omega\epsilon_0 n^2 \vec{E}$, for the TE_{mpq} modes the following equations are derived:

$$E_x = \frac{1}{j\omega\epsilon_0 n^2}\left(\frac{\partial H_z}{\partial y} - \frac{\partial H_y}{\partial z}\right), \tag{4C.2}$$

$$E_y = \frac{1}{j\omega\epsilon_0 n^2}\left(\frac{\partial H_x}{\partial z} - \frac{\partial H_z}{\partial x}\right), \tag{4C.3}$$

$$E_z = \frac{1}{j\omega\epsilon_0 n^2}\left(\frac{\partial H_y}{\partial x} - \frac{\partial H_x}{\partial y}\right) = 0, \tag{4C.4}$$

$$H_x = +\frac{1}{j\omega\mu_0}\frac{\partial E_y}{\partial z}, \tag{4C.5}$$

$$H_y = -\frac{1}{j\omega\mu_0}\frac{\partial E_x}{\partial z}, \tag{4C.6}$$

$$H_z = -\frac{1}{j\omega\mu_0}\left(\frac{\partial E_y}{\partial x} - \frac{\partial E_x}{\partial y}\right). \tag{4C.7}$$

Since the cavity is surrounded by perfectly conducting walls, the boundary conditions for the various field components dictate that the normal components of the magnetic field at the walls are zero, while the tangential components of the electric field at the boundaries are also zero. These conditions can be expressed by the following equations:

$$H_x(x=0,y,z) = H_x(x=a,y,z) = 0,$$

$$H_y(x,y=0,z) = H_y(x,y=b,z) = 0,$$

$$H_z(x,y,z=0) = H_z(x,y,z=d) = 0,$$

$$E_x(x,y=0,z) = E_x(x,y=b,z) = E_x(x,y,z=0) = E_x(x,y,z=d) = 0,$$

$$E_y(x=0,y,z) = E_y(x=a,y,z) = E_y(x,y,z=0) = E_y(x,y,z=d) = 0,$$

where it is reminded that for the TE_{mpq} modes $E_z = 0, \forall\ x,y,z$. To satisfy the boundary conditions for the H_x component, the solution $X(x) = \sin(k_{xm}x)$ where $k_{xm} = (m\pi/a)$ and $m = 1,2,\cdots$, ($m = 0$ is excluded since it gives a trivial solution for H_x, i.e., it makes $H_x = 0$) is used. Similarly, for H_y component to satisfy the boundary conditions, the $Y(y) = \sin(k_{yp}y)$ where $k_{yp} = (p\pi/b)$ and $p = 1,2,\cdots$, ($p = 0$ is again excluded since it gives a trivial solution for H_y, i.e., it makes $H_y = 0$) solution is used. Summarizing, the solutions for H_x and H_y take the following form:

$$H_x(x,y,z) = \sin\left(\frac{m\pi}{a}x\right)Y_1(y)Z_1(z), \qquad \text{with } \left(\frac{m\pi}{a}\right)^2 + k_y^2 + k_z^2 = k_0^2 n^2, \tag{4C.8}$$

$$H_y(x,y,z) = X_2(x)\sin\left(\frac{p\pi}{b}y\right)Z_2(z), \qquad \text{with } k_x^2 + \left(\frac{p\pi}{b}\right)^2 + k_z^2 = k_0^2 n^2. \tag{4C.9}$$

In order to force the E_z field component to be zero from Eq. (4C.4) the following should hold $\forall\ x,y,z$,

$$\frac{1}{j\omega\epsilon_0 n^2}\left\{\frac{dX_2}{dx}\sin\left(\frac{p\pi}{b}y\right)Z_2(z)\right\} = \frac{1}{j\omega\epsilon_0 n^2}\left\{\sin\left(\frac{m\pi}{a}x\right)\frac{dY_1}{dy}Z_1(z)\right\} \quad \forall x,y,z. \tag{4C.10}$$

Using $X_2(x) = A_2\cos(k_x x) + B_2\sin(k_x x)$ and $Y_1(y) = C_1\cos(k_y y) + D_1\sin(k_y y)$ it is straightforward to show that $B_2 = 0 = D_1$, and $k_x = (m\pi/a)$, $k_y = (p\pi/b)$, and the coefficients of the $Z_1(z)$ and $Z_2(z)$ are related in such a way that the field components H_x and H_y are expressed by the following equations:

$$H_x(x,y,z) = \sin\left(\frac{m\pi}{a}x\right)\cos\left(\frac{p\pi}{b}y\right)[E_1\cos(k_z z) + F_1\sin(k_z z)], \tag{4C.11}$$

$$H_y(x,y,z) = \cos\left(\frac{m\pi}{a}x\right)\sin\left(\frac{p\pi}{b}y\right)\frac{p\pi/b}{m\pi/a}[E_1\cos(k_z z) + F_1\sin(k_z z)], \tag{4C.12}$$

where, of course $(m\pi/a)^2 + (p\pi/b)^2 + k_z^2 = k_0^2 n^2$. Now in order to satisfy the boundary condition for the H_z field component the following solution is valid:

$$H_z(x,y,z) = H_{0z}X_3(x)Y_3(y)\sin\left(\frac{q\pi}{d}z\right). \tag{4C.13}$$

In the latter expression $q = 1,2,\cdots$, (where the $q = 0$ value is again excluded in order to avoid the trivial zero solution for H_z). From the z-dependence of H_z it is implied that the

E_x and E_y field components have the following form due to Eq. (4C.7):

$$E_x(x, y, z) = E_{0x} X_1(x) Y_1(y) \sin\left(\frac{q\pi}{d} z\right), \tag{4C.14}$$

$$E_y(x, y, z) = E_{0y} X_2(x) Y_2(y) \sin\left(\frac{q\pi}{d} z\right), \tag{4C.15}$$

where E_{0x}, E_{0y} are amplitude constants. Applying Eqs. (4C.5) and (4C.6) for the H_x and H_y components, respectively, in conjunction with Eqs. (4C.11), (4C.12), (4C.14), and (4C.15), the following conditions must be satisfied: $\forall\ x, y, z$:

$$\sin\left(\frac{m\pi}{a} x\right) \cos\left(\frac{p\pi}{b} y\right) [E_1 \cos(k_z z) + F_1 \sin(k_z z)] = + \frac{1}{j\omega\mu_0} \left[E_{0y} \frac{q\pi}{d} \cos\left(\frac{q\pi}{d} z\right) X_2(x) Y_2(y)\right],$$

$$\cos\left(\frac{m\pi}{a} x\right) \sin\left(\frac{p\pi}{b} y\right) \frac{p\pi/b}{m\pi/a} [E_1 \cos(k_z z) + F_1 \sin(k_z z)] = - \frac{1}{j\omega\mu_0} \left[E_{0x} \frac{q\pi}{d} \cos\left(\frac{q\pi}{d} z\right) X_1(x) Y_1(y)\right].$$

From the last two equations the following solutions for the E_x, E_y, H_x, H_y fields can be obtained:

$$H_x = \frac{E_{0y}}{j\omega\mu_0} \left(\frac{q\pi}{d}\right) \sin\left(\frac{m\pi}{a} x\right) \cos\left(\frac{p\pi}{b} y\right) \cos\left(\frac{q\pi}{d} z\right),$$

$$H_y = \frac{E_{0y}}{j\omega\mu_0} \frac{p\pi/b}{m\pi/a} \left(\frac{q\pi}{d}\right) \cos\left(\frac{m\pi}{a} x\right) \sin\left(\frac{p\pi}{b} y\right) \cos\left(\frac{q\pi}{d} z\right),$$

$$E_x = -E_{0y} \frac{p\pi/b}{m\pi/a} \cos\left(\frac{m\pi}{a} x\right) \sin\left(\frac{p\pi}{b} y\right) \sin\left(\frac{q\pi}{d} z\right),$$

$$E_y = E_{0y} \sin\left(\frac{m\pi}{a} x\right) \cos\left(\frac{p\pi}{b} y\right) \sin\left(\frac{q\pi}{d} z\right).$$

The last component to be determined is the H_z. Using the solutions for E_x and E_y as well as Eq. (4C.7) and Eq. (4C.13) the following solution for H_z is obtained:

$$H_z = -\frac{E_{0y}}{j\omega\mu_0} \frac{a}{m\pi} \left[\left(\frac{m\pi}{a}\right)^2 + \left(\frac{p\pi}{b}\right)^2\right] \cos\left(\frac{m\pi}{a} x\right) \cos\left(\frac{p\pi}{b} y\right) \sin\left(\frac{q\pi}{d} z\right).$$

In order to write the equations in the usual format found in the literature [132, 185] the coefficient of the H_z component can be defined as $C = -(E_{0y}/j\omega\mu_0)(a/m\pi)k_c^2$ where $k_c^2 = (m\pi/a)^2 + (p\pi/b)^2$. Using C as the free parameter in the expressions of the fields, the TE_{mpq} mode field components are summarized in the following form:

<u>TE_{mpq} Modes :</u>

$$E_x = C \frac{j\omega\mu_0}{k_c^2} \left(\frac{p\pi}{b}\right) \cos\left(\frac{m\pi}{a} x\right) \sin\left(\frac{p\pi}{b} y\right) \sin\left(\frac{q\pi}{d} z\right), \tag{4C.16}$$

$$E_y = -C \frac{j\omega\mu_0}{k_c^2} \left(\frac{m\pi}{a}\right) \sin\left(\frac{m\pi}{a} x\right) \cos\left(\frac{p\pi}{b} y\right) \sin\left(\frac{q\pi}{d} z\right), \tag{4C.17}$$

$$E_z = 0, \tag{4C.18}$$

$$H_x = -C\frac{1}{k_c^2}\left(\frac{m\pi}{a}\right)\left(\frac{q\pi}{d}\right)\sin\left(\frac{m\pi}{a}x\right)\cos\left(\frac{p\pi}{b}y\right)\cos\left(\frac{q\pi}{d}z\right), \tag{4C.19}$$

$$H_y = -C\frac{1}{k_c^2}\left(\frac{p\pi}{b}\right)\left(\frac{q\pi}{d}\right)\cos\left(\frac{m\pi}{a}x\right)\sin\left(\frac{p\pi}{b}y\right)\cos\left(\frac{q\pi}{d}z\right), \tag{4C.20}$$

$$H_z = C\cos\left(\frac{m\pi}{a}x\right)\cos\left(\frac{p\pi}{b}y\right)\sin\left(\frac{q\pi}{d}z\right). \tag{4C.21}$$

In a similar manner, the solutions for the TM_{mpq} modes modes can be determined, where $H_z = 0$. These solutions are summarized as follows for completeness:

$\underline{TM_{mpq}\ \text{Modes}:}$

$$E_x = -D\frac{1}{k_c^2}\left(\frac{m\pi}{a}\right)\left(\frac{q\pi}{d}\right)\cos\left(\frac{m\pi}{a}x\right)\sin\left(\frac{p\pi}{b}y\right)\sin\left(\frac{q\pi}{d}z\right), \tag{4C.22}$$

$$E_y = -D\frac{1}{k_c^2}\left(\frac{p\pi}{b}\right)\left(\frac{q\pi}{d}\right)\sin\left(\frac{m\pi}{a}x\right)\cos\left(\frac{p\pi}{b}y\right)\sin\left(\frac{q\pi}{d}z\right), \tag{4C.23}$$

$$E_z = D\sin\left(\frac{m\pi}{a}x\right)\sin\left(\frac{p\pi}{b}y\right)\cos\left(\frac{q\pi}{d}z\right), \tag{4C.24}$$

$$H_x = D\frac{j\omega\epsilon_0 n^2}{k_c^2}\left(\frac{p\pi}{b}\right)\sin\left(\frac{m\pi}{a}x\right)\cos\left(\frac{p\pi}{b}y\right)\cos\left(\frac{q\pi}{d}z\right), \tag{4C.25}$$

$$H_y = -D\frac{j\omega\epsilon_0 n^2}{k_c^2}\left(\frac{m\pi}{a}\right)\cos\left(\frac{m\pi}{a}x\right)\sin\left(\frac{p\pi}{b}y\right)\cos\left(\frac{q\pi}{d}z\right), \tag{4C.26}$$

$$H_z = 0. \tag{4C.27}$$

where now D has been selected as the free parameter coefficient. For both TE_{mpq} and TM_{mpq} modes the dispersion relation and the corresponding resonance frequencies are given by the following equations:

$$k_0^2 n^2 = \left(\frac{m\pi}{a}\right)^2 + \left(\frac{p\pi}{b}\right)^2 + \left(\frac{q\pi}{d}\right)^2, \tag{4C.28}$$

$$\omega_{mnq} = \frac{c}{n}\sqrt{\left(\frac{m\pi}{a}\right)^2 + \left(\frac{p\pi}{b}\right)^2 + \left(\frac{q\pi}{d}\right)^2}. \tag{4C.29}$$

4D Lambert Function

Let the function $y = f(x) = xe^x$ and its inverse $x = f^{-1}(y) = W(xe^x)$. The inverse function $W(xe^x) = x$ is known as the Lambert function [252,253]. A graphical representation of $W(x)$ is shown in Fig. 4D.1. It is straightforward to show that the function W is defined in the interval $-1/e \leq x \leq \infty$. In the interval $-1/e \leq x \leq 0$ there are two possible solutions that are characterized as the branches W_0 and W_{-1}. The Lambert function can also be extended into the complex domain, resulting in an infinite number of branches W_k with k an integer. However, the focus here is on the real solutions represented by the branches W_0 and W_{-1}. The Lambert function is now available in several computing platforms as *Maple, Mathematica, Matlab* and others. Using the Lambert function, solutions to certain transcendental equations can be obtained. A general transcendental equation of the form

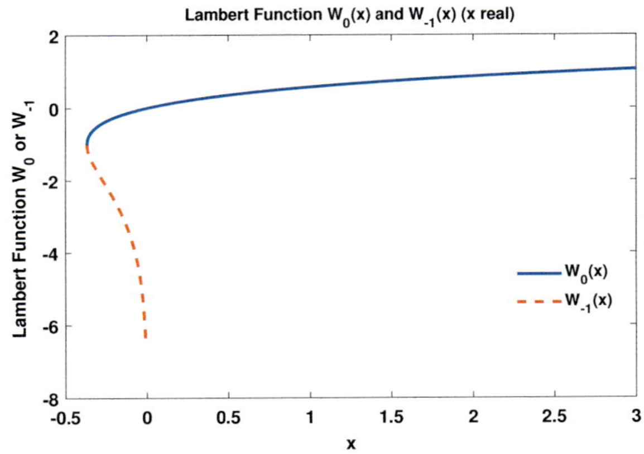

FIGURE 4D.1: Lambert function $W(x)$ showing only the real branches. The solid line corresponds to the $W_0(x)$-branch while the dash line corresponds to $W_{-1}(x)$-branch.

$a^{bx} = cx + d$ (where $a > 0$) can be solved [253]. To solve this equation, the transformation $-t = bx + bd/c$, can be utilized. In this case the equation becomes:

$$a^{bx} = cx + d \implies a^{-t-bd/c} = \frac{c}{b}\left(-t - \frac{bd}{c}\right) + d = -\frac{tc}{b} \implies$$

$$a^{-t-(bd/c)} = -\frac{tc}{b} \implies ta^t = -\frac{b}{c}a^{-(bd/c)} \implies$$

$$te^{t\ln a} = -\frac{b}{c}a^{-(bd/c)} \implies t(\ln a)e^{t\ln a} = -(\ln a)\frac{b}{c}a^{-(bd/c)} \implies$$

$$t\ln a = W\left(-(\ln a)\frac{b}{c}a^{-(bd/c)}\right) \implies$$

$$-t = -\frac{1}{\ln a}W\left(-(\ln a)\frac{b}{c}a^{-(bd/c)}\right) \implies$$

$$bx + \frac{bd}{c} = -\frac{1}{\ln a}W\left(-(\ln a)\frac{b}{c}a^{-(bd/c)}\right) \implies$$

$$x = -\frac{d}{c} - \frac{1}{b\ln a}W\left(-(\ln a)\frac{b}{c}a^{-(bd/c)}\right). \tag{4D.1}$$

If x is replaced by $x' = -x$ the equation becomes $a^{-bx'} = (-c)x' + d$ and Eq. (4D.1) becomes:

$$x' = \frac{d}{c} + \frac{1}{b\ln a}W\left[-(\ln a)\frac{b}{c}a^{-(bd/c)}\right] \tag{4D.2}$$

With the above information the Wien's displacement law can be easily verified [Eq. (4.60)]. The maximum exitance can be determined from Eq. (4.57) taking the derivative $dM_{\lambda_0}/d\lambda_0 = 0$. Setting $x' = hc/\lambda_0 k_B T$ the derivative can be computed as $dM_{\lambda_0}/d\lambda_0 = (dM_{\lambda_0}/dx)(dx/d\lambda_0) = 0 \implies Ax'^6(5e^{x'} - 5 - x'e^{x'}) = 0$, where A a constant. Therefore, the maximum can be found from the solution of the transcendental equation $5e^{x'} - 5 - x'e^{x'} = 0 \implies e^{-x'} = (-1/5)x' + 1$. Applying Eq. (4D.2) for $a = e = \exp(1)$, $b = 1$, $c = 1/5$, $d = 1$, the following solution is found:

$$x'_m = 5 + W_0(-5e^{-5}) = 4.965114231744276 \implies \lambda_{0,max}T = \frac{k_B x'_m}{hc} = 2897.721\,\mu m\,^\circ K.$$

Similarly, starting from Eq. (4.56) the maximum exitance in terms of frequency can be found from the zero of the derivative $dM_\nu/d\nu = 0$. Setting $x' = h\nu/k_BT$ the derivative zeroes for $A(3e^{x'} - 3 - x'e^{x'}) = 0 \implies e^{-x'} = (-1/3)x' + 1$. Applying Eq. (4D.2) for $a = e = \exp(1)$, $b = 1$, $c = 1/3$, $d = 1$, the following solution is found:

$$x'_m = 3 + W_0(-3e^{-3}) = 2.821439372122079 \implies \frac{\nu_{max}}{T} = \frac{k_B x'_m}{h} = 5.878926\times10^{10}\, Hz/°K.$$

4E Isotemperature Lines Calculation

The isotemperature lines were initially proposed by Judd [241] and were defined in the u-v chromaticity diagram proposed by MacAdam [227]. In this Appendix, a simple method for computing the isotemperature lines is presented. The u-v chromaticity coordinates are related to the x-y chromaticity coordinates (or the X, Y, and Z tristimulus values) through the following relations [227]:

$$u = \frac{4x}{-2x + 12y + 3} = \frac{4X}{X + 15Y + 3Z}, \tag{4E.1}$$

$$v = \frac{6y}{-2x + 12y + 3} = \frac{6Y}{X + 15Y + 3Z}. \tag{4E.2}$$

Let a point (u, v) be located in the u-v chromaticity coordinates diagram, as shown in Fig. 4E.1. This point can be represented by the position vector $\vec{r} = u\hat{\imath}_u + v\hat{\imath}_v$. The locus of points that correspond to a blackbody radiator of temperature T can easily be calculated in the x-y chromaticity diagram by first determining the X, Y, and Z tristimulus values:

$$X(T) = k\int M_{\lambda_0}(\lambda_0, T)\bar{x}(\lambda_0)d\lambda_0,$$

$$Y(T) = k\int M_{\lambda_0}(\lambda_0, T)\bar{y}(\lambda_0)d\lambda_0,$$

$$Z(T) = k\int M_{\lambda_0}(\lambda_0, T)\bar{x}(\lambda_0)d\lambda_0,$$

for each temperature T. The term $M_{\lambda_0}(\lambda_0, T)$ is the blackbody radiant exitance [Eq, (4.57)], the $\bar{x}, \bar{y}, \bar{z}$ are the color matching functions, and λ_0 is the freespace wavelength. Then the chromaticities coordinates $x(T)$ and $y(T)$ can be determined from $x(T) = X(T)/[X(T) + Y(T) + Z(T)]$ and $y(T) = X(T)/[X(T) + Y(T) + Z(T)]$, respectively. Using Eqs. (4E.1) and (4E.2) the values of $(u(T), v(T))$ corresponding to the points of the Planckian locus can be found. A part of this Planckian locus is shown in Fig. 4E.1 by the red solid line.

The Planckian locus can be represented by the vector $\vec{f}(T) = u(T)\hat{\imath}_u + v(T)\hat{\imath}_v$. This provides a parametric representation of the Planckian locus as a function of temperature T. The normal line from the point (u, v) to the Planckian locus, known as the isotemperature line at temperature T, is represented by the vector $\vec{r}_{uv}$. At the intersection of the isotemperature line with the Planckian locus, the tangential unit vector is defined as $\hat{\imath}_{f'} = \vec{f}'/|\vec{f}'|$ with $\vec{f}' = d\vec{f}/dT$. Therefore, the tangential and normal unit vectors at this intersection are

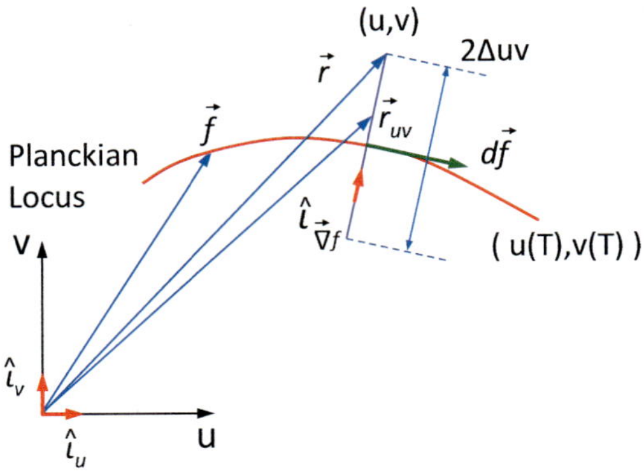

FIGURE 4E.1: Diagram used for the isotemperature line determination. The solid red line is the Planckian locus in the u-v chromaticity diagram. The line specified by $\vec{r}_{uv}$ is one isotemperature line which is perpendicular to the Planckian locus at the CCT of the light source represented by (u, v) point.

expressed as follows:

$$\hat{\imath}_{f'} = \frac{\vec{f}'}{|\vec{f}'|} = a_u\hat{\imath}_u + a_v\hat{\imath}_v, \qquad \text{with,}$$

$$a_u = \frac{du(T)}{dT}, \qquad \text{and} \qquad a_v = \frac{dv(T)}{dT},$$

$$\hat{\imath}_{\vec{\nabla} f} = -a_v\hat{\imath}_u + a_u\hat{\imath}_v = -\frac{dv(T)}{dT}\hat{\imath}_u + \frac{du(T)}{dT}\hat{\imath}_v.$$

Then, the isotemperature line can be described by the following expression:

$$\begin{aligned}\vec{r}_{uv}(T,t) &= u_{_{ISO}}(T,t)\hat{\imath}_u + v_{_{ISO}}(T,t)\hat{\imath}_v \\ &= u(T)\hat{\imath}_u + v(T)\hat{\imath}_v + \hat{\imath}_{\vec{\nabla} f}(\Delta uv)t, \end{aligned} \tag{4E.3}$$

where t a parameter taking values in the interval $t \in [-1, 1]$ and Δuv the maximum acceptable distance of the chromaticity point (u, v) from the Planckian locus at temperature T (which is the point's correlated color temperature, CCT). The $u_{_{ISO}}(T, t)$ and $v_{_{ISO}}(T, t)$ are the chromaticity coordinates of the isotemperature line of temperature T. The required derivatives du/dT and dv/dT can be determined from the following expressions [by differentiating Eqs. (4E.1) and (4E.2)]:

$$u'(T) = \frac{du}{dT} = \frac{4X'(T)\left[X(T) + 15Y(T) + 3Z(T)\right] - 4X(T)\left[X'(T) + 15Y'(T) + 3Z'(T)\right]}{\left[X(T) + 15Y(T) + 3Z(T)\right]^2},$$

$$v'(T) = \frac{dv}{dT} = \frac{6Y'(T)\left[X(T) + 15Y(T) + 3Z(T)\right] - 6Y(T)\left[X'(T) + 15Y'(T) + 3Z'(T)\right]}{\left[X(T) + 15Y(T) + 3Z(T)\right]^2},$$

where the $W'(T)$ $(W = X, Y, Z)$ are calculated from the following equations:

$$W'(T) = k\int \frac{M_{\lambda_0}(\lambda_0, T)}{dT}\bar{w}(\lambda_0)d\lambda_0,$$

$$\frac{M_{\lambda_0}(\lambda_0, T)}{dT} = \frac{2\pi h^2 c^3}{\lambda_0^6 k_B}\frac{1}{\left(e^{hc/\lambda_0 k_B T} - 1\right)^2}\, e^{hc/\lambda_0 k_B T}\frac{1}{T^2}.$$

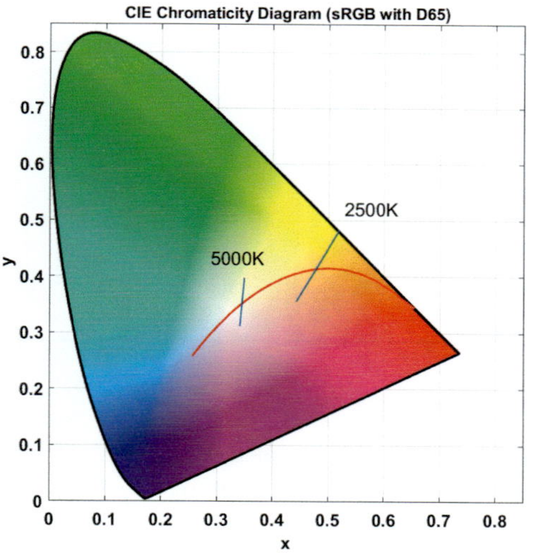

FIGURE 4E.2: The CIE 1931 chromaticity diagram with the Planckian locus. Two sample isotemperature lines are shown for temperatures 2500 °K and 5000 °K and $\Delta uv = \pm 0.02$.

Here, $\bar{w}(\lambda_0)$ ($\bar{w} = \bar{x}, \bar{y}, \bar{z}$) are the color matching functions. Then, the points of the isotemperature line at T can be transformed into the x-y chromaticity diagram via the Eqs. (4E.1) and (4E.2) which are converted to express x and y as functions of u and v:

$$x = \frac{3u}{2(u - 4v + 2)}, \tag{4E.4}$$

$$y = \frac{v}{u - 4v + 2}. \tag{4E.5}$$

Two isotemperature lines for $T = 2500$ °K and $T = 5000$ °K, with $\Delta uv = 0.02$ are shown in Fig. 4E.2. The isotemperature lines shown in Figs. 4.34(a) and 4.34(b) are shown both in the u-v and in the x-y coordinates. It is apparent that in the u-v diagram the isotemperature lines are normal to the Planckian locus as expected from their definition. The temperatures selected in Figs. 4.34(a) and 4.34(b) are according to the work of Kelly in Ref. [243] where the difference between the isotemperature lines was selected according to the inverse temperature difference as it was proposed initially by Priest in Ref. [254]. Actually, these isotemperature lines were selected to differ successively by $10\, MK^{-1}$ (microreciprocal degrees). I.e., the initial temperature was selected to be $T_0 = 1515°$ K and the rest of the temperatures were selected by the rule:

$$10^6 \left(\frac{1}{T_i} - \frac{1}{T_{i+1}} \right) = 10\, MK^{-1},$$

where $i = 0, 1, \cdots, 62$. According to this scheme the highest isotemperature line (first from the left) is $T \simeq 24959\,°$ K. A more practical problem is that given a source's chromaticity coordinates to determines its CCT. The interested reader should consult Refs. [242,255–257] among many available, for more details on this subject.

5

Scalar Diffraction

5.1 Introduction

Diffraction is yet another manifestation of the wave nature of light. In essence, diffraction and interference are closely related concepts, both arising from the wave properties of electromagnetic radiation. However, diffraction typically refers to phenomena that occur when a wave (not limited to optical or electromagnetic waves) encounters an obstacle or an aperture (such as a slit). In the optical domain, diffraction is characterized by the bending of light around the edges of obstacles or apertures and into the geometrical shadow regions. Observations of shadow edges reveal that some electromagnetic radiation penetrates into the otherwise dark areas of the geometrical shadow. This deviation from the predictions of geometrical optics is termed diffraction. The term "diffraction" was introduced by the Italian scientist Francesco Maria Grimaldi, who was also the first to record precise observations of the phenomenon in 1660. His findings were published posthumously in 1665. Information regarding the first observation of diffraction by Grimaldi can be found in Ref. [258].

A rigorous (vectorial) diffraction theory relies on solving Maxwell's equations under the appropriate boundary conditions associated with the obstacle or aperture. The field scattered by the obstacle or aperture interferes with the incident field to produce the total resulting diffracted field. In general, the application of this rigorous theory to most diffraction problems does not yield analytical solutions. Consequently, vectorial numerical methods, such as finite-difference or finite-element techniques, are required to analyze light diffraction. Nevertheless, for many diffraction problems, a simpler scalar diffraction theory can provide meaningful results that align well with experimental observations.

The scalar diffraction theory is founded on Christian Huygens' principle [9]. Huygens intuitively proposed that if each point on the wavefront of a light beam is treated as a source of a "secondary" spherical wavelet, the wavefront at any later moment can be determined by constructing the "envelope" of these secondary wavelets. Huygens' intuitive concepts were significantly refined by Augustin-Jean Fresnel [24, 259], who enhanced Huygens' envelope construction by incorporating Young's principle of interference. By making certain assumptions, albeit somewhat arbitrary, regarding the effective amplitudes and phases of Huygens' secondary sources, and allowing these secondary wavelets to interfere with one another, Fresnel succeeded in calculating the light distribution in diffraction patterns with remarkable accuracy.

Rather than simply using wavefronts to construct new wavefronts, the scalar theory of diffraction assumes that Huygens' wavelets interfere to form the resultant wavefront [91,260]. The ideas proposed by Huygens and Fresnel were placed on a firmer mathematical foundation in 1882 by Gustav Kirchhoff [25, 261], who demonstrated that the amplitudes and phases attributed to the secondary sources by Fresnel were logical consequences of the wave nature of light. Kirchhoff's mathematical formulation relied on two assumptions regarding the boundary values of light incident on the surface of an aperture or obstacle placed in its path. However, these assumptions were later shown to be inconsistent by

DOI: 10.1201/9781003490333-5

Poincaré [262] and Sommerfeld [263]. Despite this inconsistency, Kirchhoff's formulation, also referred to as the Huygens-Fresnel principle, remains a valuable first approximation in scalar diffraction theory. Remarkably, it often provides results that agree closely with experimental observations under most practical conditions.

In this chapter, Kirchhoff's scalar theory is presented using Green's theorem and Green's functions for light propagation in homogeneous media. The theory was subsequently refined by Sommerfeld, who addressed one of Kirchhoff's problematic assumptions about the light field at the aperture or obstacle boundary by leveraging the properties of Green's functions. These refined formulations, known as the Rayleigh-Sommerfeld formulations, are presented as the primary framework for the scalar theory of diffraction in this chapter.

It is important to emphasize that both the Kirchhoff and Rayleigh-Sommerfeld theories incorporate significant simplifications and approximations. Most notably, light is treated as a scalar field phenomenon rather than a vector field. This approach entirely neglects the coupling of the electric and magnetic field components, as dictated by Maxwell's equations and boundary conditions, which generally cannot be treated independently. Fortunately, experiments have shown that the scalar theory provides relatively accurate results, particularly under two key conditions: (a) the diffracting aperture or obstacle is large compared to the wavelength of light, and (b) the diffracted fields are not observed in close proximity to the aperture or obstacle.

In this chapter, the scalar diffraction theory is developed systematically. It begins with the Helmholtz-Kirchhoff integral theorem, which provides the foundational theoretical framework. The two Rayleigh-Sommerfeld formulations are then derived, followed by an examination of the Fresnel and Fraunhofer approximations. A concise description of the plane wave spectrum technique is included, along with a derivation of Babinet's principle for scalar diffraction.

The theory is subsequently applied to various geometries, including circular and elliptical apertures, obstacles, and annuli, with numerous numerical examples provided. Circular geometries are explored first due to the availability of analytical solutions along the propagation axis. The study then extends to two-dimensional slit apertures and their array counterparts, followed by an analysis of rectangular, triangular, and other polygonal apertures and obstacles, supported by appropriate numerical examples.

The chapter concludes with an exploration of Fresnel zones and Fresnel zone plates, as well as the application of scalar theory to periodic structures, including both one-dimensional and two-dimensional gratings. To enhance completeness and facilitate a deeper understanding of the concepts introduced, several appendices are provided at the end of the chapter. This systematic presentation aims to ensure a thorough comprehension of scalar diffraction theory and its practical applications.

5.2 Helmholtz' and Kirchhoff's Integral Theorem

The Kirchhoff integral theorem for scalar diffraction appropriately applies Green's theorem [91,264,265]. It enables the expression of the scalar field at an arbitrary point P_0 (as shown in Fig. 5.1) within a volume V, based on the scalar field and its first derivative on an arbitrary closed surface enclosing P_0. For the volume V enclosed by the surface S, the complex functions $U(\vec{r})$ and $G(\vec{r})$. which possess continuous, single-valued first and second partial derivatives within V and on S, satisfy the following relation, known as the second

Green's identity or Green's theorem [88]:

$$\iiint_V \left(U\nabla^2 G - G\nabla^2 U\right) dv = \iint_S \left(U\frac{\partial G}{\partial n} - G\frac{\partial U}{\partial n}\right) ds, \tag{5.1}$$

where $\partial/\partial n$ denotes the partial derivative in the outward normal direction of surface S (refer to Fig. 5.1).

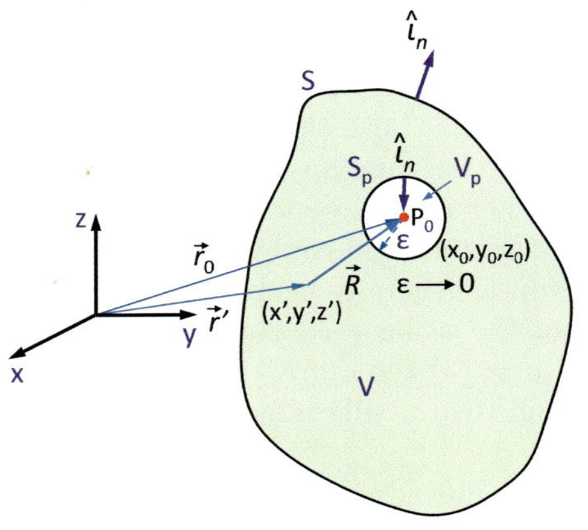

FIGURE 5.1: Geometrical configuration used for the derivation of Helmholtz-Kirchhoff's integral theorem [91, 264, 265].

Let the function $U(\vec{r})$ satisfy the three-dimensional scalar Helmholtz equation inside an isotropic, linear and homogeneous volume V:

$$\nabla^2 U + k^2 U = 0, \tag{5.2}$$

where k is the wavenumber inside the volume V. The function $G(\vec{r})$ is chosen to be the three-dimensional Green's function (as detailed in Appendix 5A). Consequently, the function $G(\vec{r})$ also satisfies the scalar Helmholtz equation:

$$\nabla^2 G + k^2 G = -\delta(\vec{r}_0 - \vec{r}\,'). \tag{5.3}$$

Here, $\vec{r}_0$ denotes the position vector of the point of interest P_0, while $\vec{r}\,'$ represents the position vector of a point source. Since the Green's function becomes infinite when the source point coincides with P_0, it is necessary to exclude the singularity at P_0 to apply Green's theorem effectively. To achieve this, a small sphere of radius ϵ, with surface S_p, centered at P_0 is excluded from the volume V, creating a new volume $V' = V - V_p$ where V_p is the volume of the excluded sphere. In this configuration, Green's theorem is applied to the volume V', which lies between the outer surface S and the surface S_p of the excluded sphere. Within the volume V', the Green's function $G(\vec{r})$ satisfies the homogeneous scalar Helmholtz equation, as the right-hand side of Eq. (5.3) becomes zero. Applying the Green's theorem for U and G in V', Eq. (5.1) transforms into:

$$\iiint_{V'} \left(U\nabla^2 G - G\nabla^2 U\right) dv = \iint_{S+S_p} \left(U\frac{\partial G}{\partial n} - G\frac{\partial U}{\partial n}\right) ds, \qquad \text{where,}$$

$$\iiint_{V'} \left(U\nabla^2 G - G\nabla^2 U\right) dv = \iiint_{V'} \left(U(-k^2 G) - G(-k^2 U)\right) dv = 0, \qquad \text{and}$$

$$\iint_{S+S_p}\left(U\frac{\partial G}{\partial n}-G\frac{\partial U}{\partial n}\right)ds=0\Longrightarrow$$

$$\iint_S\left(U\frac{\partial G}{\partial n}-G\frac{\partial U}{\partial n}\right)ds=-\iint_{S_p}\left(U\frac{\partial G}{\partial n}-G\frac{\partial U}{\partial n}\right)ds. \tag{5.4}$$

Using $G=\exp(-jk|\vec{r}_0-\vec{r}\,'|)/4\pi|\vec{r}_0-\vec{r}\,'|$, and defining $\vec{R}=\vec{r}_0-\vec{r}\,'$, (refer to Appendix 5A and Fig. 5.1), the factors required for computing the integral over the shrinking surface S_p are provided below. It is understood that the integration is performed with respect to the primed variables:

$$\frac{\partial G}{\partial n}=\vec{\nabla}'G\cdot\hat{\imath}_n,\qquad\text{where,}$$

$$\frac{\partial G}{\partial w'}=\left(jk+\frac{1}{R}\right)\frac{\exp(-jkR)}{4\pi R}\,\frac{w_0-w'}{R},\quad (w=x,y,z),$$

$$R=\left[(x_0-x')^2+(y_0-y')^2+(z_0-z')^2\right]^{1/2},$$

$$\hat{\imath}_R=\frac{x_0-x'}{R}\hat{\imath}_x+\frac{y_0-y'}{R}\hat{\imath}_y+\frac{z_0-z'}{R}\hat{\imath}_z,\qquad\text{and}\quad \hat{\imath}_n=+\hat{\imath}_R,$$

$$\vec{\nabla}'G=\frac{\partial G}{\partial x'}\hat{\imath}_x+\frac{\partial G}{\partial y'}\hat{\imath}_y+\frac{\partial G}{\partial z'}\hat{\imath}_z=+\left(jk+\frac{1}{R}\right)\frac{\exp(-jkR)}{4\pi R}\hat{\imath}_R,$$

$$\lim_{\epsilon\to 0}\left\{-\iint_{S_p}\left(U\frac{\partial G}{\partial n}-G\frac{\partial U}{\partial n}\right)ds\right\}=$$

$$-\lim_{\epsilon\to 0}\left\{\left[U(\vec{r}_0)\left(jk+\frac{1}{\epsilon}\right)\frac{e^{-jk\epsilon}}{4\pi\epsilon}\underbrace{(\hat{\imath}_R\cdot\hat{\imath}_n)}_{+1}-\frac{e^{-jk\epsilon}}{4\pi\epsilon}\frac{\partial U}{\partial n}\right]4\pi\epsilon^2\right\}=$$

$$=-U(\vec{r}_0). \tag{5.5}$$

Then, using the last equation in Eq. (5.4), the field at P_0 can be expressed as:

$$U(\vec{r}_0)=\iint_S\left(G\frac{\partial U}{\partial n}-U\frac{\partial G}{\partial n}\right)ds'.$$

Setting P_0 as any point (x_0,y_0,z_0) inside volume V, the field at P_0 can be expressed in terms of the field and its normal derivative on the boundary surface S as:

$$U(\vec{r}_0)=U(x_0,y_0,z_0)=\iint_S\left(G\frac{\partial U}{\partial n}-U\frac{\partial G}{\partial n}\right)ds',$$

$$=\iint_S\left[\frac{\exp(-jkR)}{4\pi R}\frac{\partial U}{\partial n}-U\left(jk+\frac{1}{R}\right)\frac{\exp(-jkR)}{4\pi R}(\hat{\imath}_R\cdot\hat{\imath}_n)\right]ds', \tag{5.6}$$

$$\text{where}\quad R=\left[(x_0-x')^2+(y_0-y')^2+(z_0-z')^2\right]^{1/2}.$$

It should be noted that the integration is performed over the primed variables, which represent points on the boundary surface S. The last equation is known as the integral theorem of Helmholtz and Kirchhoff [265]. To evaluate this equation, complete knowledge of the boundary field and its normal derivative on S is required. However, in most practical problems, this information is not fully available. Nevertheless, by using this result, along with the Sommerfeld radiation condition and certain approximations, the fundamental outcomes of scalar diffraction analysis can be derived.

5.2.1 Application of Helmholtz' and Kirchhoff's Integral Theorem

At this point, let us consider the diffraction of a scalar field (representing a light wave) by an aperture in an infinite opaque screen, as shown in Fig. 5.2(a). It is assumed that the field is incident on the aperture from the left. The coordinate system is defined with the z-axis (for $z > 0$) as the symmetry axis, perpendicular to the aperture, as illustrated. A random point P_0 is selected along the z-axis, where the diffracted field from the aperture is to be evaluated. The Helmholtz and Kirchhoff integral theorem, presented previously, should be applied. Consider the closed surface $S = S_1 \cup S_2$, which consists of an infinite plane, denoted by surface S_1, positioned at an infinitesimal distance to the right of the aperture, and a segment of a spherical surface, denoted by surface S_2, centered at P_0 with radius $R \to \infty$. As the radius R approaches infinity, the spherical part of the surface becomes a hemisphere. The surfaces S_1 and S_2 are joined at $\pm\infty$ (on the screen plane) to form the closed surface S, as depicted in Fig. 5.2(a).

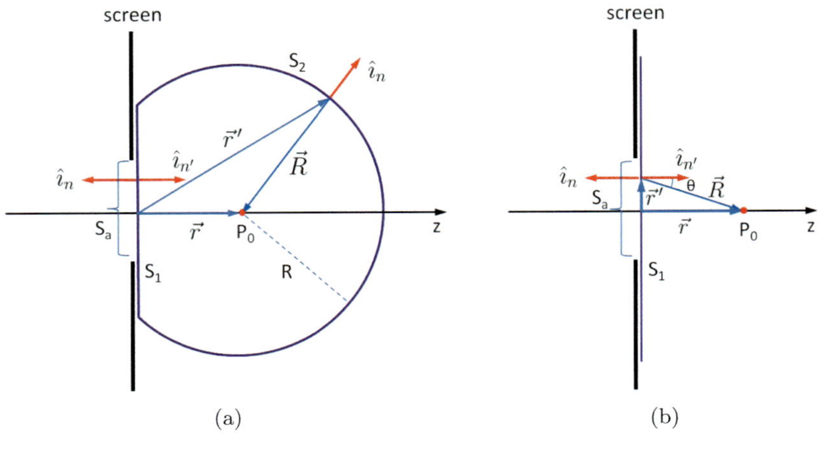

FIGURE 5.2: (a) Application of Helmholtz-Kirchhoff's integral theorem on a screen aperture. The point P_0 is assumed to be on the z-axis. The radius R of the closed boundary surface tends to infinity. The area of the aperture is S_a. (b) Detailed consideration using only the remaining surface S_1 and definition of the angle θ.

By applying the Helmholtz-Kirchhoff theorem to the closed surface S, the scalar field at P_0 can be calculated as:

$$\begin{aligned} U(P_0) &= \iint_{S_1 \cup S_2} \left(G\frac{\partial U}{\partial n} - U\frac{\partial G}{\partial n} \right) ds', \\ &= \iint_{S_1} \left(G\frac{\partial U}{\partial n} - U\frac{\partial G}{\partial n} \right) ds' + \iint_{S_2} \left(G\frac{\partial U}{\partial n} - U\frac{\partial G}{\partial n} \right) ds', \end{aligned}$$

where,

$$\iint_{S_2}\left(G\frac{\partial U}{\partial n}-U\frac{\partial G}{\partial n}\right)ds'=\lim_{R\to\infty}\left\{\iint_{S_2}\left[\frac{e^{-jkR}}{4\pi R}\frac{\partial U}{\partial R}-U\left(jk+\frac{1}{R}\right)\frac{e^{-jkR}}{4\pi R}\underbrace{(\hat{\imath}_R\cdot\hat{\imath}_n)}_{=-1}\right]ds'\right\}=$$

$$=\lim_{R\to\infty}\left\{\left[\frac{e^{-jkR}}{4\pi R^2}\underbrace{R\left(\frac{\partial U}{\partial R}+jkU\right)}_{\to 0}+\frac{e^{-jkR}}{4\pi R^2}\underbrace{U}_{\to 0}\right]4\pi R^2\right\}=0. \tag{5.7}$$

In the last equation, the Sommerfeld's radiation conditions were utilized [see Eqs. (5B.2) and (5B.4) of Appendix 5B]. Therefore, the diffracted field at point P_0 is given by the integral over the surface S_1 only. Consequently,

$$U(P_0)=\iint_{S_1}\left(G\frac{\partial U}{\partial n}-U\frac{\partial G}{\partial n}\right)ds'=$$
$$=\iint_{S_1}\left[\frac{e^{-jkR}}{4\pi R}\frac{\partial U}{\partial n}-U\left(jk+\frac{1}{R}\right)\frac{e^{-jkR}}{4\pi R}(\hat{\imath}_R\cdot\hat{\imath}_n)\right]ds'. \tag{5.8}$$

In the last equation, the values of the field, U, and its normal derivative $\partial U/\partial n$, on the surface S_1 are required. In general, these values are not known, as the total field on S_1 is the sum of the incident and scattered fields. However, for most cases, it is a reasonable approximation to assume that, on the aperture, the field is approximately equal to the incident field, and the same holds for its normal derivative. Additionally, since the screen is opaque, the field values and their normal derivatives are set to zero on the surface $S_1 - S_a$ (where S_a is the aperture area), i.e., for all points not on the aperture. These boundary conditions are known as the Kirchhoff boundary conditions [265]. Therefore, Eq. (5.8) can be written as:

$$U_K(P_0)=U(P_0)\simeq\iint_{S_a}\left[\frac{e^{-jkR}}{4\pi R}\frac{\partial U_{inc}}{\partial n}-U_{inc}\left(jk+\frac{1}{R}\right)\frac{e^{-jkR}}{4\pi R}(\hat{\imath}_R\cdot\hat{\imath}_n)\right]ds'=$$
$$=\iint_{S_a}\frac{e^{-jkR}}{4\pi R}\left[\frac{\partial U_{inc}(x',y',0)}{\partial n}+U_{inc}(x',y',0)\left(jk+\frac{1}{R}\right)\cos\theta\right]dx'dy', \tag{5.9}$$

where $\cos\theta=\dfrac{z_0}{R}$.

Here, U_{inc} and $\partial U_{inc}/\partial n$ represent the incident field and its normal derivative on the aperture ($z'=0$). Of course, the same approach can be applied even for a point P_0 not on the z-axis. Therefore, Eq. (5.9) is valid for any point P_0 with $z_0>0$ (and $z_0<\infty$).

The integral appearing in Eq. (5.9) is a two-dimensional integral, as the integration is performed over the aperture area S_a. Several publications have attempted to transform this two-dimensional integral into a one-dimensional line integral over the boundary of the aperture area. Initially, Rabinowicz [266], Miyamoto and Wolf [267,268], and Marchand and Wolf [269,270] proposed this approach. Later, Asvestas [271,272] presented a more general formulation that was valid even in the geometric-shadow boundary region. These alternative methods can be used to express scalar diffraction by apertures in terms of line integrals.

However, as they are quite complex, they are not presented in this chapter. Interested readers can refer to the cited references for more details on this approach.

5.2.1.1 Case of Incident Spherical Wave: Fresnel-Kirchhoff Formulation

This analysis is a special case of the previous expression of Eq. (5.9), where the incident wave is a spherical wave emanating from point S, as shown in Fig. 5.3. The Helmholtz-Kirchhoff approximation allows the aperture surface S_a to be non-planar, as depicted in this figure. In contrast, the Rayleigh-Sommerfeld formulations presented in the next section (Sec. 5.3) require a planar aperture surface S_a. The various position vectors are illustrated in Fig. 5.3 with respect to the coordinate system xyz.

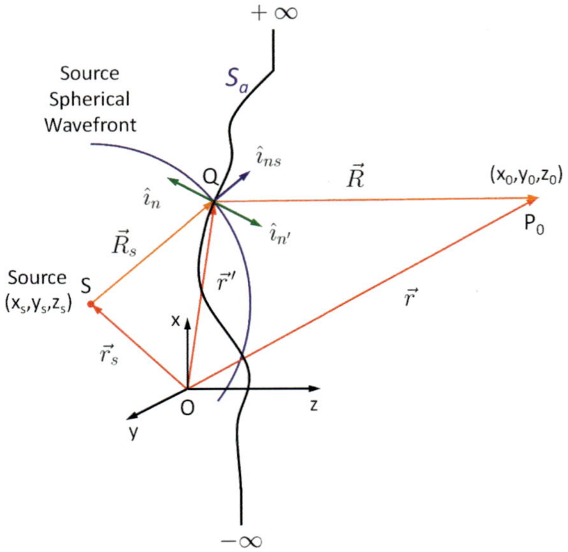

FIGURE 5.3: Application of Helmholtz-Kirchhoff's integral theorem for the case that the incident wave is a spherical wave emanating from point S (from Refs. [91, 264, 265]). The surface S_a could be a non-planar surface (extended in the direction of the aperture from $-\infty$ to $+\infty$) as opposed to the planar surface S_a shown in Fig. 5.2(a).

The incident spherical wave and its normal derivative, according to the Kirchhoff approximation, can be expressed as:

$$U_{inc} = U_0 \frac{e^{-jkR_s}}{R_s}, \qquad \text{and,}$$

$$\frac{\partial U_{inc}}{\partial n} = \vec{\nabla} U_{inc} \cdot \hat{\imath}_n = -\vec{\nabla} U_{inc} \cdot \hat{\imath}_{n'} = \left(jk + \frac{1}{R_s} \right) \frac{e^{-jkR_s}}{R_s} \left(\hat{\imath}_{ns} \cdot \hat{\imath}_{n'} \right), \qquad \text{where}$$

$$R_s = \left[(x' - x_s)^2 + (y' - y_s)^2 + (z' - z_s)^2 \right]^{1/2}, \quad \text{and}$$

$$\hat{\imath}_{ns} = \frac{\vec{R}_s}{R_s} = \frac{\vec{r}' - \vec{r}_s}{R_s} = \frac{x' - x_s}{R_s} \hat{\imath}_x + \frac{y' - y_s}{R_s} \hat{\imath}_y + \frac{z' - z_s}{R_s} \hat{\imath}_z.$$

The coordinates (x_s, y_s, z_s) represent the location of the origin of the incident spherical wave (point S in Fig. 5.3). Substituting U_{inc} and $\partial U_{inc}/\partial n$ into Eq. (5.9) yields the following expression for the diffracted field $U_{FK}(P_0)$:

$$U_{FK}(P_0) = U_K(x_0, y_0, z_0) \simeq jkU_0 \iint_{S_a} \frac{e^{-jk(R+R_s)}}{4\pi R R_s} [(\hat{\imath}_{ns} \cdot \hat{\imath}_{n'}) + (\hat{\imath}_R \cdot \hat{\imath}_{n'})]\, dS'$$

$$= j\frac{U_0}{\lambda} \iint_{S_a} \frac{e^{-jk(R+R_s)}}{R R_s} \left[\frac{(\hat{\imath}_{ns} \cdot \hat{\imath}_{n'}) + (\hat{\imath}_R \cdot \hat{\imath}_{n'})}{2}\right] dS', \qquad \text{where} \qquad (5.10)$$

$$R = \left[(x_0 - x')^2 + (y_0 - y')^2 + (z_0 - z')^2\right]^{1/2}, \quad \text{and}$$

$$\hat{\imath}_R = \frac{\vec{R}}{R} = \frac{\vec{r} - \vec{r}\,'}{R} = \frac{x_0 - x'}{R}\hat{\imath}_x + \frac{y_0 - y'}{R}\hat{\imath}_y + \frac{z_0 - z'}{R}\hat{\imath}_z.$$

It should be emphasized that the terms $1/R$ and $1/R_s$ were neglected, assuming that these terms are much smaller than k in the integral expression. This approximation is almost always valid when the point P_0 is several wavelengths away from the aperture, which is typically the case of interest. A special case of Eq. (5.10) arises when the surface S represents the incident spherical wavefront. In this case, $U_{FK}(P_0)$ can be written as:

$$U_{FK}(P_0) = U_K(x_0, y_0, z_0) = j\frac{U_0}{\lambda}\frac{e^{-jkR_s}}{R_s} \iint_{S_a} \frac{e^{-jkR}}{R} \underbrace{\left[\frac{1 + (\hat{\imath}_R \cdot \hat{\imath}_{n'})}{2}\right]}_{\substack{\text{Inclination} \\ \text{Factor}}} dS', \qquad (5.11)$$

since $\hat{\imath}_{ns} = \hat{\imath}_{n'}$ in this case. This form of the Helmholtz-Kirchhoff formulation is also known in the literature as the *Fresnel-Kirchhoff* formulation.

5.3 The Rayleigh-Sommerfeld Formulations

The previously defined Helmholtz-Kirchhoff diffraction formula for the diffraction of light by an aperture has provided remarkably accurate results in experimental measurements, particularly in situations where the dimensions of the aperture are large compared to the light wavelength. However, Kirchhoff's boundary conditions introduce some inconsistencies. These inconsistencies arise from the specification of both the field and its normal derivative in Eq. (5.9), a topic discussed in many textbooks [273–276].

In fact, these conditions together over-determine the diffraction problem and have sparked significant scientific interest over the past century. Sommerfeld [273] demonstrated that imposing both the field and its normal derivative to be zero on the opaque part of the screen results in the field vanishing over the entire screen, including the aperture area. Another well-known inconsistency is that the Helmholtz-Kirchhoff equations with Kirchhoff's boundary conditions fail to satisfy the assumed conditions on the plane of the aperture.

The inconsistencies of Kirchhoff's boundary conditions were addressed by Sommerfeld, who eliminated the need to impose both conditions simultaneously by carefully selecting the Green's functions. The resulting formulations are generally known as the Rayleigh-Sommerfeld theory of scalar diffraction. Both Rayleigh-Sommerfeld formulations are consistent with their assumed boundary conditions, as demonstrated in Sec. 5.5.1.

At this point, let us define the Green's functions used by Sommerfeld based on the geometry shown in Fig. 5.4. The source points, denoted by $\vec{r}\,'$ and $\vec{r}\,''$, are symmetric with

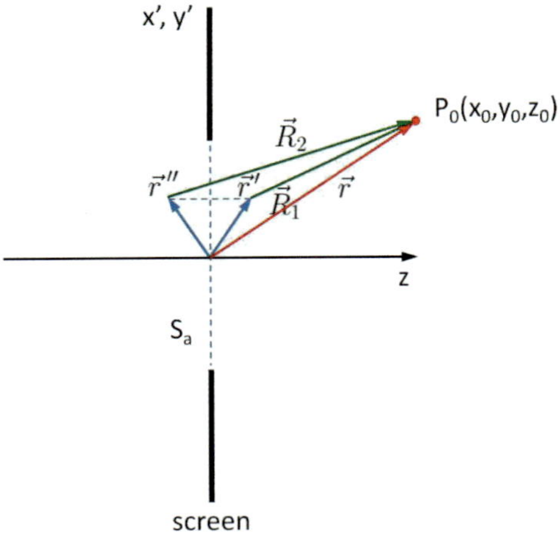

FIGURE 5.4: Definitions of Dirichlet and Neumann Green's functions for the aperture diffraction problem [264, 265].

respect to the $x'y'$ plane of the aperture. In this context, the following Dirichlet, G_D, and Neumann, G_N, Green's functions can be defined:

$$G_D(\vec{r};\vec{r}',\vec{r}'') = \frac{\exp(-jk|\vec{r}-\vec{r}'|)}{4\pi|\vec{r}-\vec{r}'|} - \frac{\exp(-jk|\vec{r}-\vec{r}''|)}{4\pi|\vec{r}-\vec{r}''|} =$$

$$= \frac{\exp(-jkR_1)}{4\pi R_1} - \frac{\exp(-jkR_2)}{4\pi R_2}, \tag{5.12}$$

$$G_N(\vec{r};\vec{r}',\vec{r}'') = \frac{\exp(-jk|\vec{r}-\vec{r}'|)}{4\pi|\vec{r}-\vec{r}'|} + \frac{\exp(-jk|\vec{r}-\vec{r}''|)}{4\pi|\vec{r}-\vec{r}''|} =$$

$$= \frac{\exp(-jkR_1)}{4\pi R_1} + \frac{\exp(-jkR_2)}{4\pi R_2}, \qquad \text{where,} \tag{5.13}$$

$$\vec{R}_1 = (x_0 - x')\hat{\imath}_x + (y_0 - y')\hat{\imath}_y + (z_0 - z')\hat{\imath}_z, \qquad R_1 = |\vec{R}_1|,$$

$$\vec{R}_2 = (x_0 - x')\hat{\imath}_x + (y_0 - y')\hat{\imath}_y + (z_0 + z')\hat{\imath}_z, \qquad R_2 = |\vec{R}_2|.$$

The corresponding normal derivatives of the above Green's functions (see Figs. 5.2) can be easily determined and are given by:

$$\frac{\partial G_D}{\partial n} = \vec{\nabla}' G_D \cdot \hat{\imath}_n, \qquad \hat{\imath}_n = -\hat{\imath}_z,$$

$$\vec{\nabla}' G_D = \left(jk + \frac{1}{R_1}\right)\frac{\exp(-jkR_1)}{4\pi R_1}\left(\frac{x_0 - x'}{R_1}\hat{\imath}_x + \frac{y_0 - y'}{R_1}\hat{\imath}_y + \frac{z_0 - z'}{R_1}\hat{\imath}_z\right) -$$

$$\left(jk + \frac{1}{R_2}\right)\frac{\exp(-jkR_2)}{4\pi R_2}\left(\frac{x_0 - x'}{R_2}\hat{\imath}_x + \frac{y_0 - y'}{R_2}\hat{\imath}_y - \frac{z_0 + z'}{R_2}\hat{\imath}_z\right),$$

$$\left.\frac{\partial G_D}{\partial n}\right|_{z'=0} = -\left(jk + \frac{1}{R}\right)\frac{\exp(-jkR)}{4\pi R}\frac{2z_0}{R}, \qquad \text{where,} \tag{5.14}$$

$$R = [(x_0 - x')^2 + (y_0 - y')^2 + z_0^2]^{1/2}, \qquad \text{while,}$$

$$G_D|_{z'=0} = 0. \tag{5.15}$$

Similarly, for the G_N function the following equations can be written:

$$\frac{\partial G_N}{\partial n} = \vec{\nabla}' G_N \cdot \hat{\imath}_n, \qquad \hat{\imath}_n = -\hat{\imath}_z,$$

$$\vec{\nabla}' G_N = \left(jk + \frac{1}{R_1}\right)\frac{\exp(-jkR_1)}{4\pi R_1}\left(\frac{x_0 - x'}{R_1}\hat{\imath}_x + \frac{y_0 - y'}{R_1}\hat{\imath}_y + \frac{z_0 - z'}{R_1}\hat{\imath}_z\right) +$$
$$\left(jk + \frac{1}{R_2}\right)\frac{\exp(-jkR_2)}{4\pi R_2}\left(\frac{x_0 - x'}{R_2}\hat{\imath}_x + \frac{y_0 - y'}{R_2}\hat{\imath}_y - \frac{z_0 + z'}{R_2}\hat{\imath}_z\right),$$

$$\left.\frac{\partial G_N}{\partial n}\right|_{z'=0} = 0, \qquad \text{while,} \tag{5.16}$$

$$G_N|_{z'=0} = \frac{\exp(-jkR)}{2\pi R}. \tag{5.17}$$

Based on the above selection of the Green's functions, the two Rayleigh-Sommerfeld formulations (RS-1 and RS-2) can be defined as:

$$U_{RS1}(P_0) = \iint_{S_a}\left[G_D\frac{\partial U_{inc}}{\partial n} - U_{inc}\frac{\partial G_D}{\partial n}\right]_{z'=0} dx'dy' =$$
$$= \iint_{S_a}\left[-U_{inc}\frac{\partial G_D}{\partial n}\right]_{z'=0} dx'dy' =$$
$$= \iint_{S_a}\left[U_{inc}\left(jk + \frac{1}{R}\right)\frac{\exp(-jkR)}{2\pi R}\frac{z_0}{R}\right]_{z'=0} dx'dy', \qquad \text{and} \tag{5.18}$$

$$U_{RS2}(P_0) = \iint_{S_a}\left[G_N\frac{\partial U_{inc}}{\partial n} - U_{inc}\frac{\partial G_N}{\partial n}\right]_{z'=0} dx'dy' =$$
$$= \iint_{S_a}\left[G_N\frac{\partial U_{inc}}{\partial n}\right]_{z'=0} dx'dy' = \iint_{S_a}\left[\frac{\exp(-jkR)}{2\pi R}\frac{\partial U_{inc}}{\partial n}\right]_{z'=0} dx'dy'. \tag{5.19}$$

RS-1 uses the Dirichlet Green's function for calculating the diffracted field, while RS-2 uses the Neumann Green's function. Both formulations involve integrating over the aperture surface and require the evaluation of either the field or its normal derivative at the aperture. These formulations provide a more accurate description of diffraction, addressing the inconsistencies found in the Kirchhoff approach. It is straightforward to show that the Helmholtz-Kirchhoff formulation (with the Kirchhoff's boundary conditions) and the Rayleigh-Sommerfeld formulations are interrelated by:

$$U_K(\vec{r}) = \frac{1}{2}\left\{U_{RS1}(\vec{r}) + U_{RS2}(\vec{r})\right\}. \tag{5.20}$$

5.3.1 The Rayleigh-Sommerfeld Formulations in Two Dimensions

In the case of two-dimensional diffraction problems, where the aperture is assumed to be infinite along the y (or y') direction, the diffraction problem simplifies to dependence on the x and z coordinates. The source points specified by $\vec{r}'$ and $\vec{r}''$ are symmetric with respect to the x'-axis (instead of the $x'y'$ plane) of the aperture, and they are assumed to lie in the xz plane due to the two-dimensional nature of the problem.

For this geometry, the Dirichlet Green's function G_D^{2d}, and the Neumann Green's function G_N^{2d}, are defined based on specific boundary conditions. The Dirichlet Green's function is characterized by being zero on the boundary, while the Neumann Green's function is defined such that its normal derivative is zero on the boundary. These Green's functions can be derived using the results from Appendix 5A and are given by:

$$G_D^{2d}(\vec{r};\vec{r}',\vec{r}'') = -\frac{j}{4}H_0^{(2)}(kR_{T_1}) + \frac{j}{4}H_0^{(2)}(kR_{T_2}), \tag{5.21}$$

$$G_N^{2d}(\vec{r};\vec{r}',\vec{r}'') = -\frac{j}{4}H_0^{(2)}(kR_{T_1}) - \frac{j}{4}H_0^{(2)}(kR_{T_2}), \tag{5.22}$$

$$R_{T_1} = \left[(x_0 - x')^2 + (z_0 - z')^2\right]^{1/2},$$

$$R_{T_2} = \left[(x_0 - x')^2 + (z_0 + z')^2\right]^{1/2},$$

The corresponding normal derivatives of the above Green's functions (see Fig. 5.2) can be easily determined and are given by:

$$\frac{\partial G_D^{2d}}{\partial n} = \vec{\nabla}' G_D \cdot \hat{\imath}_n = -\frac{\partial G_D^{2d}}{\partial z'} =$$

$$\left.\frac{\partial G_D^{2d}}{\partial n}\right|_{z'=0} = \frac{k}{2j}H_1^{(2)}(kR_T)\frac{z_0}{R_T},$$

$$G_N^{2d}|_{z'=0} = -\frac{j}{2}H_0^{(2)}(kR_T), \qquad \text{where}$$

$$R_T = [(x_0 - x')^2 + z_0^2]^{1/2}, \qquad \text{while,}$$

$$G_D^{2d}|_{z'=0} = 0, \qquad \left.\frac{\partial G_N^{2d}}{\partial n}\right|_{z'=0} = 0.$$

Based on the selection of the Green's functions, the two Rayleigh-Sommerfeld formulations in their two-dimensional version can be defined as follows:

$$U_{RS1}(P_0) = \int_{x'} \left[G_D^{2d}\frac{\partial U_{inc}}{\partial n} - U_{inc}\frac{\partial G_D^{2d}}{\partial n}\right]_{z'=0} dx' =$$

$$= \int_{x'} \left[U_{inc}\frac{k}{2j}H_1^{(2)}(kR_T)\frac{z_0}{R_T}\right]_{z'=0} dx', \qquad \text{and} \tag{5.23}$$

$$U_{RS2}(P_0) = \int_{x'} \left[G_N^{2d}\frac{\partial U_{inc}}{\partial n} - U_{inc}\frac{\partial G_N^{2d}}{\partial n}\right]_{z'=0} dx' =$$

$$= \int_{x'} \left[-\frac{j}{2}H_0^{(2)}(kR_T)\frac{\partial U_{inc}}{\partial n}\right]_{z'=0} dx', \tag{5.24}$$

The Helmholtz-Kirchhoff formulation, with the Kirchhoff boundary conditions, can also be determined by applying the relation given in Eq. (5.20), which holds true in the two-dimensional case as well.

5.4 Fresnel and Fraunhofer Approximations

In the preceding sections, the Helmholtz-Kirchhoff and Rayleigh-Sommerfeld formulations for scalar diffraction were discussed. These formulations provide reasonably accurate results despite neglecting the vector nature of the electromagnetic field and approximating the field and/or its normal derivative on the surface S using the values of the incident field. The focus now shifts to commonly used approximations that simplify the mathematical complexity of the diffraction integrals. These approximations, known as the *Fresnel* and *Fraunhofer* approximations, are extensively applied in various domains involving wave propagation. Both approximations are examined in the context of aperture diffraction, starting from the Rayleigh-Sommerfeld 1 formulation (RS-1), as expressed in Eq. (5.18).

5.4.1 Fresnel Approximation

Using the RS-1 formulation the diffracted field, $U(x_0, y_0, z_0)$ at point $P_0(x_0, y_0, z_0)$ (with $z_0 > 0$), is given by:

$$U(x_0, y_0, z_0) = \iint_{S_a} \left[U_{inc} \left(jk + \frac{1}{R} \right) \frac{\exp(-jkR)}{2\pi R} \frac{z_0}{R} \right]_{z'=0} dx'dy', \qquad \text{where,}$$

$$R = \left[(x_0 - x')^2 + (y_0 - y')^2 + z_0^2 \right]^{1/2} = z_0 \left[1 + \frac{(x_0 - x')^2}{z_0^2} + \frac{(y_0 - y')^2}{z_0^2} \right]^{1/2} \simeq$$

$$\simeq z_0 \left[1 + \frac{(x_0 - x')^2 + (y_0 - y')^2}{2z_0^2} \right] = z_0 + \frac{(x_0 - x')^2 + (y_0 - y')^2}{2z_0}.$$

In the previous equation, the Taylor series approximation $(1+X)^{1/2} \simeq 1 + X/2$ was employed under the condition $|X| \ll 1$. For distances z_0 satisfying $kR \gg 1$, substituting the approximate expression for R into the diffraction integral allows the diffracted field $U(x_0, y_0, z_0)$ to be expressed as follows:

$$U(x_0, y_0, z_0) \simeq jk \frac{\exp(-jkz_0)}{2\pi z_0} \iint_{S_a} \left[\exp\left\{ -jk \left[\frac{(x_0 - x')^2}{2z_0} + \frac{(y_0 - y')^2}{2z_0} \right] \right\} U_{inc}(x', y') \right] dx'dy', \tag{5.25}$$

where $z_0 \simeq R$ was also used for relatively large distances from the aperture. The above equation defines the Fresnel diffraction approximation. The range of z_0 values for which Eq. (5.25) remains valid can be determined from the neglected terms in the Taylor series expansion of the exponential term in the diffraction integral. Specifically, the first neglected term in the series $(1 + X)^{1/2} = 1 + X/2 - X^2/8 + \cdots$ is the $X^2/8$. In the context of the diffraction integral, the expression for X is given by:

$$X = \frac{(x_0 - x')^2 + (y_0 - y')^2}{z_0^2},$$

and taking into account the constant term kz_0 multiplying the exponent, the following inequality should be satisfied:

$$kz_0 \frac{X^2}{8} = kz_0 \frac{1}{8} \left[\frac{(x_0 - x')^2 + (y_0 - y')^2}{z_0^2} \right]^2 \ll 1.$$

If $D = \max\{[(x_0-x')^2+(y_0-y')^2]^{1/2}\}$ represents the maximum transverse distance between a source point and the observation point, the Fresnel approximation remains valid when the following inequality is satisfied:

$$z_0 \gg \left(\frac{\pi}{4\lambda_0}D^4\right)^{1/3}. \tag{5.26}$$

This condition ensures that the higher-order terms neglected in the Taylor expansion of the distance R are sufficiently small, thereby maintaining the accuracy of the Fresnel approximation. This inequality is actually stricter than necessary [265, 277], as rapid phase variations may not significantly affect the integral in certain cases. More detailed discussions on the validity of the Fresnel approximation can be found in various references, such as [278–281]. It is common to introduce the Fresnel number, N_F, defined as follows [19]:

$$N_F = \frac{D^2}{\lambda_0 z_0}. \tag{5.27}$$

Based on the definition of the Fresnel number, the condition for applying the Fresnel approximation can be expressed as:

$$\frac{\pi}{4}N_F\theta_{max}^2 \ll 1, \tag{5.28}$$

where $\theta_{max} = D/z_0$ and corresponds to the maximum observation angle in radians. corresponds to the maximum observation angle in radians. Typically, for Fresnel diffraction, $N_F \gtrsim 1$, while for Fraunhofer diffraction, $N_F \ll 1$.

5.4.1.1 Fresnel Approximation in the Two-Dimensional Case

The Fresnel approximation in the two-dimensional case can be readily derived from the corresponding RS-1 formulation given in Eq. (5.23). Starting with the two-dimensional RS-1 integral, the Fresnel approximation is expressed as:

$$\begin{aligned} U(x_0, z_0) &= \int_{x'} \left[U_{inc}\frac{k}{2j}H_1^{(2)}(kR_T)\frac{z_0}{R_T}\right]_{z'=0} dx' \\ &\simeq \frac{k}{2j}e^{j3\pi/4}\sqrt{\frac{2}{\pi k z_0}}\int_{x'} \exp(-jkR_T)U_{inc}(x')dx' \\ &\simeq j\sqrt{\frac{k}{2\pi z_0}}e^{-jkz_0}\int_{x'} \exp\left[-j\frac{k}{2z_0}(x'-x_0)^2\right]U_{inc}(x')dx' \end{aligned} \tag{5.29}$$

$$\text{with the condition} \qquad z_0 \gg \left[\left(\frac{\pi}{4\lambda}\right)(x_0-x')^4\right]^{1/3}. \tag{5.30}$$

In the derivation of Eq. (5.29), the asymptotic expansion of the Hankel function for large arguments was applied. Specifically, the expression $H_m^{(2)}(z) \simeq (2/\pi z)^{1/2}\exp[-j(z-m(\pi/2)-\pi/4)]$ was used for large $|z|$. Additionally, a similar to 3D case Taylor expansion was utilized for the distance R_T to simplify its expression under the Fresnel approximation conditions.

5.4.2 Fraunhofer Approximation

For the Fraunhofer approximation, further simplification of the phase factor in the Fresnel integral, Eq. (5.25), is required. Specifically,

$$k\left[\frac{(x_0-x')^2}{2z_0}+\frac{(y_0-y')^2}{2z_0}\right]=k\left[\frac{x_0^2+y_0^2}{2z_0}+\frac{x'^2+y'^2}{2z_0}-\frac{x_0x'+y_0y'}{z_0}\right]=$$

$$\simeq k\left[\frac{x_0^2+y_0^2}{2z_0}-\frac{x_0x'+y_0y'}{z_0}\right],$$

$$\text{if}\quad k\left(\frac{x'^2+y'^2}{2z_0}\right)\ll 1\Longrightarrow z\gg\frac{kD_{max}^2}{2}=\frac{\pi}{\lambda}D_{max}^2\simeq\frac{Area}{\lambda},\tag{5.31}$$

where D_{max} denotes the maximum dimension of the aperture and "*Area*" corresponds to the aperture area. Using the last approximation in Eq. (5.25), the Fraunhofer diffraction integral, U_{ff}, can be derived, and is given by:

$$\begin{aligned}U_{ff}(x_0,y_0,z_0)&=jk\frac{\exp(-jkz_0)}{2\pi z_0}\exp\left(-jk\frac{x_0^2+y_0^2}{2z_0}\right)\\&\iint_{S_a}\left[\exp\left\{+jk\left(\frac{x_0x'+y_0y'}{z_0}\right)\right\}U_{inc}(x',y')\right]dx'dy'\\&=jk\frac{\exp(-jkz_0)}{2\pi z_0}\exp\left(-jk\frac{x_0^2+y_0^2}{2z_0}\right)\\&\iint_{-\infty}^{\infty}\exp\left\{+jk\left(\frac{x_0x'+y_0y'}{z_0}\right)\right\}[t_a(x',y')U_{inc}(x',y')]\,dx'dy'\\&=jk\frac{\exp(-jkz_0)}{2\pi z_0}\exp\left(-jk\frac{x_0^2+y_0^2}{2z_0}\right)\mathcal{F}\Big\{t_a(x',y')U_{inc}(x',y')\Big\}\Bigg|_{\substack{k_x=kx_0/z_0\\k_y=ky_0/z_0}}.\end{aligned}\tag{5.32}$$

Here, the symbol $\mathcal{F}$ denotes the Fourier transform. The spatial radial frequencies k_x and k_y correspond to the wavevector components in the x and y directions, respectively. The function $t_a(x',y')$ is the amplitude transmittance function of the aperture (defined in the next section) that permits the integration limits to vary from $-\infty$ to $+\infty$. This approximation, also known as the *far-field approximation*, asserts that the diffracted field in the far-field region is a scaled version of the Fourier transform of the incident field at the aperture.

A key distinction of the Fraunhofer approximation is that it breaks the spatial invariance seen in the Fresnel and Rayleigh-Sommerfeld formulations [265]. In the Fraunhofer case, shifting the aperture's position does not affect the diffracted pattern. In contrast, both the Fresnel and Rayleigh-Sommerfeld formulations preserve this spatial invariance, meaning that the diffraction pattern shifts along with the aperture. Thus, while the Fraunhofer approximation simplifies the diffraction calculation, it sacrifices the spatial invariance of the diffraction pattern that is maintained in the other two formulations.

5.4.2.1 Fraunhofer Approximation Using the Boundary Waves Approach

In the last paragraph of Sec. 5.2.1, it was mentioned that there are approaches where the 2D integral of the Helmholtz-Kirchhoff formulation can be transformed into a 1D line integral over the boundary of the aperture. In the case of the Fraunhofer approximation with a normally incident plane wave, the transformation is simpler, and it is presented here to illustrate the approach. The method followed is the one introduced by Komrska [282]. In this analysis Fig. 5.5 is used. Using Eq. (5.32), the diffracted field (for a normally incident

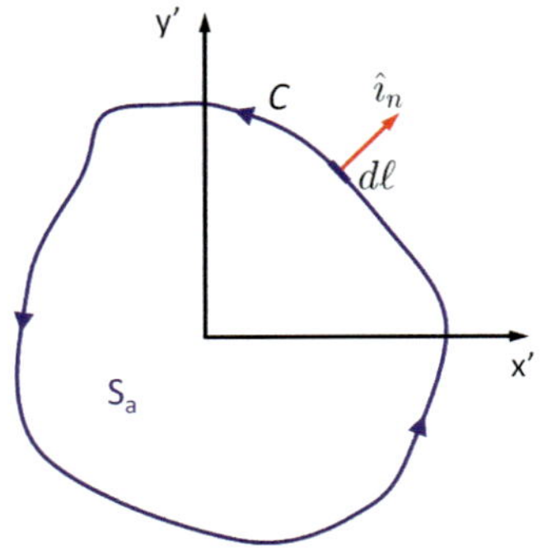

FIGURE 5.5: An aperture of area S_a is located on an opaque screen at $z = 0$, with the aperture enclosed by the closed contour C that defines its boundary. A normally incident plane wave is assumed, and the coordinate system on the aperture plane is defined as (x', y'). The differential length along the contour C is represented by $d\ell$. The unit vector $\hat{\imath}_n$ is the normal unit vector perpendicular to the contour C at the position of $d\ell$.

plane wave) can be written as:

$$U_{ff}(x_0, y_0, z_0) = jk\frac{\exp(-jkz_0)}{2\pi z_0}\exp\left(-jk\frac{x_0^2 + y_0^2}{2z_0}\right)U_0\underbrace{\iint_{S_a}\exp\left\{+jk\left(\frac{x_0x' + y_0y'}{z_0}\right)\right\}dx'dy'}_{=\,S(k_x,k_y)}$$

$$= jkU_0\frac{\exp(-jkz_0)}{2\pi z_0}\exp\left(-jk\frac{x_0^2 + y_0^2}{2z_0}\right)S(k_x, k_y)\Big|_{\substack{k_x = kx_0/z_0 \\ k_y = ky_0/z_0}}.$$

At this point, the two-dimensional integral $S(k_x, k_y)$ is converted into a line integral. To proceed, let's define the following variables:

$$\vec{p} = k_x\hat{\imath}_x + k_y\hat{\imath}_y,$$

$$\vec{\xi} = x'\hat{\imath}_x + y'\hat{\imath}_y,$$

$$S(k_x, k_y) = \iint_{S_a}\exp(+j\vec{p}\cdot\vec{\xi})dx'dy',$$

$$\Phi(\vec{\xi}) = \exp(+j\vec{p}\cdot\vec{\xi}).$$

It is evident that the function $\Phi(\vec{\xi})$ satisfies the Helmholtz's wave equation ($\nabla^2\Phi + p^2\Phi = 0$, with $p^2 = \vec{p}\cdot\vec{p}$). Therefore, the function $S(k_x, k_y)$ can be written as:

$$S(k_x, k_y) = \iint_{S_a}\left(-\frac{1}{p^2}\nabla^2\Phi\right)dx'dy' = -\frac{1}{p^2}\iint_{S_a}\nabla^2\Phi dx'dy'.$$

It should be mentioned that the previous substitution is valid only when $p \neq 0$. In the case of $p = 0$ (which corresponds to $k_x = k_y = 0$), $S(0, 0) = S_a$, meaning that the Fourier transform becomes simply the area of the aperture, as is evident from the definition of

$S(k_x, k_y)$. At this point, the two-dimensional form of the divergence theorem is utilized. For a vector function $\vec{A}$ (depending on two variables, x' and y') the following relation holds:

$$\iint_{S_a} \left(\vec{\nabla}\cdot\vec{A}\right) dx'dy' = \oint_C \vec{A}\cdot\hat{\imath}_n d\ell. \tag{5.33}$$

Here, S_a is the area enclosed by the contour C in the $x'y'$ plane, as shown in Fig. 5.5. At this stage, the function $\vec{A}$ is defined as $\vec{A} = \vec{\nabla}\Phi = j\vec{p}\Phi(\vec{\xi})$. By applying the two-dimensional divergence theorem, $S(k_x, k_y)$ can be expressed as follows:

$$S(k_x, k_y) = -\frac{j}{p^2}\oint_C \Phi(\vec{\xi})\vec{p}\cdot\hat{\imath}_n d\ell = -\frac{j}{p^2}\oint_C \exp(+j\vec{p}\cdot\vec{\xi})\,(\vec{p}\cdot\hat{\imath}_n)\,d\ell. \tag{5.34}$$

Based on the last equation, the Fraunhofer approximation can be written as:

$$U_{ff}(x_0, y_0, z_0) = \frac{k}{p^2}U_0\frac{\exp(-jkz_0)}{2\pi z_0}\exp\left(-jk\frac{x_0^2+y_0^2}{2z_0}\right)\oint_C \exp(+j\vec{p}\cdot\vec{\xi})\,(\vec{p}\cdot\hat{\imath}_n)\,d\ell, \tag{5.35}$$

where $\vec{p} = k\frac{x_0}{z_0}\hat{\imath}_x + k\frac{y_0}{z_0}\hat{\imath}_y$.

Equivalent representations for the $S(k_x, k_y)$ term, as provided by Saga [283, 284], are as follows:

$$S(k_x, k_y) = \frac{+j}{k_y}\oint_C \exp(+jk_x x' + k_y y')dx', \tag{5.36}$$

$$S(k_x, k_y) = \frac{-j}{k_x}\oint_C \exp(+jk_x x' + k_y y')dy'. \tag{5.37}$$

In the last equations, $y' = y'(x')$ or $x' = x'(y')$ represent the equation of the contour C. These equations describe the parametric relationship between the coordinates along the contour. The choice of parametric form depends on how the contour is described and the nature of the aperture.

5.4.2.2 Fraunhofer Approximation in the Two-Dimensional Case

The Fraunhofer approximation in the two-dimensional case can be easily obtained from the corresponding Fresnel formulation of Eq. (5.29) as follows:

$$\begin{aligned}
U_{ff}(x_0, z_0) &= j\sqrt{\frac{k}{2\pi z_0}}e^{-jkz_0}\int_{x'} \exp\left[-j\frac{k}{2z_0}\left(x'-x_0\right)^2\right]U_{inc}(x')dx' \\
&\simeq j\sqrt{\frac{k}{2\pi z_0}}e^{-jkz_0}\exp\left(-j\frac{k}{2z_0}x_0^2\right)\int_{x'} \exp\left[+j\frac{kx_0}{z_0}x'\right]U_{inc}(x')dx' \\
&\simeq j\sqrt{\frac{k}{2\pi z_0}}e^{-jkz_0}\exp\left(-j\frac{k}{2z_0}x_0^2\right)\int_{-\infty}^{+\infty} \exp\left[+j\frac{kx_0}{z_0}x'\right]\left[t_a(x')U_{inc}(x')\right]dx' \\
&= j\sqrt{\frac{k}{2\pi z_0}}e^{-jkz_0}\exp\left(-j\frac{k}{2z_0}x_0^2\right)\mathcal{F}\left\{t_a(x')U_{inc}(x')\right\}\Big|_{k_x=kx_0/z_0}.
\end{aligned} \tag{5.38}$$

$$\text{with the condition} \qquad z_0 \gg \frac{\pi}{\lambda}x'^2 \simeq \frac{Length^2}{\lambda}. \tag{5.39}$$

In the last inequality the "Length" corresponds to the maximum width of the aperture.

5.5 Plane Wave Spectrum

The concept of the plane wave spectrum is valuable for understanding the decomposition of a scalar field into its constituent plane waves, especially in the context of diffraction. The plane wave spectrum provides a way to represent a complex field as a sum of simpler plane waves using the two-dimensional Fourier transform. This approach not only facilitates the analysis of diffraction but also places scalar diffraction theory within the framework of linear systems theory. In this context, the material through which the field propagates is assumed to be linear, isotropic, homogeneous, and non-dispersive [265]. To define the plane wave spectrum, assume that the incident field is a scalar field $U(x', y', z' = 0)$, as shown in Fig. 5.2(a), where the field is propagating in the positive z-direction. The field at the aperture $z = 0$ can be expressed using the two-dimensional Fourier transform (refer to Appendix 5D). According to this transform, the incident field $U(x', y', z' = 0)$, can be decomposed into the following summation of plane wave components:

$$U(x', y', z' = 0) = \frac{1}{(2\pi)^2} \int_{-\infty}^{+\infty} \int_{-\infty}^{+\infty} \tilde{U}(k_x, k_y; 0) e^{-j(k_x x' + k_y y')} dk_x dk_y,$$

where,

$$\tilde{U}(k_x, k_y; 0) = \int_{-\infty}^{+\infty} \int_{-\infty}^{+\infty} U(x', y', z' = 0) e^{+j(k_x x' + k_y y')} dx' dy' = \mathcal{F}\left\{U(x', y', z' = 0)\right\}.$$

Here, $\tilde{U}(k_x, k_y; 0)$ represents the two-dimensional Fourier transform of the incident field $U_{inc}(x', y', 0)$. The variables k_x and k_y denote the transverse wavevector components of the plane wave spectrum, which characterize the direction of propagation of each plane wave in the x- and y-directions, respectively. To determine the field at a point $P(x_0, y_0, z_0)$ it is necessary to find its plane wave spectrum at this point. Denote this plane wave spectrum at the point P as $\tilde{U}(k_x, k_y; z_0)$. Then, the following expressions can be written:

$$U(x_0, y_0, z_0) = \frac{1}{(2\pi)^2} \int_{-\infty}^{+\infty} \int_{-\infty}^{+\infty} \tilde{U}(k_x, k_y; z_0) e^{-j(k_x x_0 + k_y y_0)} dk_x dk_y,$$

where,

$$\tilde{U}(k_x, k_y; z_0) = \int_{-\infty}^{+\infty} \int_{-\infty}^{+\infty} U(x', y', z_0) e^{+j(k_x x' + k_y y')} dx' dy'.$$

The propagating field should satisfy the Helmholtz equation. Therefore, the following steps can be written [265]:

$$\nabla^2 U(x, y, z) + k^2 U(x, y, z) = 0 \Longrightarrow$$

$$\frac{1}{(2\pi)^2} \int_{-\infty}^{+\infty} \int_{-\infty}^{+\infty} \left[\nabla^2 \left(\tilde{U}(k_x, k_y; z) e^{+j(k_x x + k_y y)}\right) + k^2 \tilde{U}(k_x, k_y; z)\right] dk_x dk_y = 0 \Longrightarrow$$

$$\frac{1}{(2\pi)^2} \int_{-\infty}^{+\infty} \int_{-\infty}^{+\infty} \left[\frac{d^2 \tilde{U}}{dz^2} + (k^2 - k_x^2 - k_y^2)\tilde{U}\right] dk_x dk_y = 0 \Longrightarrow$$

$$\frac{d^2\tilde{U}}{dz^2} + \underbrace{(k^2 - k_x^2 - k_y^2)}_{= k_z^2}\tilde{U} = 0 \Longrightarrow \tilde{U}(k_x, k_y; z) = \tilde{U}(k_x, k_y; 0)e^{-jk_z z},$$

$$\text{where} \qquad k_z = \begin{cases} \left(k^2 - k_x^2 - k_y^2\right)^{1/2}, & \text{when } k_x^2 + k_y^2 \le k^2, \\ -j\left(k_x^2 + k_y^2 - k^2\right)^{1/2}, & \text{when } k_x^2 + k_y^2 > k^2, \end{cases} \tag{5.40}$$

where the signs have been selected for propagation along the positive z-direction (and the $e^{i\omega t}$ time dependence is used throughout this chapter). In the case that k_z becomes imaginary, the exponentially decreasing solution has been chosen to correspond to an evanescent wave (with decreasing amplitude). Using the plane wave spectrum at $z = z_0$, the propagated field from $z = 0$ to $z = z_0$ is expressed as:

$$U(x_0, y_0, z_0) = \frac{1}{(2\pi)^2}\int_{-\infty}^{+\infty}\int_{-\infty}^{+\infty} \tilde{U}(k_x, k_y; 0)e^{-j(k_x x_0 + k_y y_0 + k_z z_0)}dk_x dk_y. \tag{5.41}$$

In order to apply the previous equation to the planar aperture of Fig. 5.2(a), the definition of the aperture amplitude transmittance function, $t_a = t_a(x', y')$ must be introduced. This function is defined as:

$$t_a(x', y') = \frac{U_t(x', y', 0^+)}{U_{inc}(x', y', 0^-)}, \tag{5.42}$$

where $U_t(x', y', 0^+)$ is the transmitted field from the planar aperture at $z = 0^+$, and $U_{inc}(x', y', 0^-)$ is the incident field at $z = 0^-$. In practice, the calculation of the amplitude transmittance function is challenging, and typically, approximations similar to the Kirchhoff approximations are used. Alternatively, vectorial electromagnetic methods (primarily numerical) can be employed for the evaluation of t_a. Since $U_t(x', y', 0^+) = t_a(x', y')U_{inc}(x', y', 0^-)$, the corresponding plane wave spectrum for an infinitesimal distance after the aperture is given by the Fourier transform of $U_t(x', y', 0^+)$. Therefore, the plane wave spectrum at $z = 0^+$ is given by:

$$\tilde{U}_t(k_x, k_y) = \mathcal{F}\left\{t_a(x', y')U_{inc}(x', y', 0^-)\right\} = \frac{1}{(2\pi)^2}\,\tilde{t}_a(k_x, k_y) ** \tilde{U}_{inc}(k_x, k_y; 0), \quad \text{where,}$$

$$\tilde{t}_a(k_x, k_y) = \mathcal{F}\left\{t_a(x', y')\right\}, \quad \text{and,}$$

$$\tilde{U}_{inc}(k_x, k_y; 0) = \mathcal{F}\left\{U_{inc}(x', y', 0^-)\right\},$$

where "**" denotes a two-dimensional convolution. It is reminded the Fourier transform property of a product, $\mathcal{F}\{f(x,y)g(x,y)\} = (1/(2\pi)^2)\tilde{f}(k_x, k_y) ** \tilde{g}(k_x, k_y)$, where $\tilde{f}$ and $\tilde{g}$ are the Fourier transforms of f and g, respectively. Then, Eq. (5.41) can be applied to calculate the diffracted field at any point $P(x_0, y_0, z_0)$ with $z_0 > 0$. In summary, the diffracted field can be expressed in the plane wave spectrum approach as:

$$U(x_0, y_0, z_0) = \frac{1}{(2\pi)^2}\int_{-\infty}^{+\infty}\int_{-\infty}^{+\infty}\left[\frac{\tilde{U}_{inc}(k_x, k_y; 0) ** \tilde{t}_a(k_x, k_y)}{(2\pi)^2}\right]e^{-j(k_x x_0 + k_y y_0 + k_z z_0)}dk_x dk_y. \tag{5.43}$$

It is simple to show that the plane wave spectrum technique for studying scalar diffraction is identical to the Rayleigh-Sommerfeld formulations under the same Kirchhoff assumptions. This was initially proven by Sherman [285]. To demonstrate the approach and familiarize with the use of the Fourier transform, it will be shown that the Rayleigh-Sommerfeld

2 formulation is identical with the plane wave spectrum approach. Starting from the RS-2 integral, the following steps can be written:

$$
\begin{aligned}
U_{RS_2}(x_0, y_0, z_0) &= \iint_{S_a} \left[\frac{\exp(-jkR)}{2\pi R} \frac{\partial U_{inc}}{\partial n} \right]_{z'=0} dx'dy' = \\
&= \iint_{-\infty}^{+\infty} \underbrace{\left[\frac{\exp(-jkR)}{2\pi R} \right]}_{= G(x_0-x', y_0-y'; z_0)} \underbrace{\left[-\frac{\partial U_{inc}(x', y'; z')}{\partial z'} \Big|_{z'=0} t_a(x', y') \right]}_{= f_2(x', y')} dx'dy' = \\
&= \iint_{-\infty}^{+\infty} G(x_0 - x', y_0 - y'; z_0) f_2(x', y') dx'dy' = G(x, y, z_0) ** f_2(x, y),
\end{aligned}
$$

$$
\text{with } R = \left[(x_0 - x')^2 + (y_0 - y')^2 + z_0^2 \right]^{1/2}, \tag{5.44}
$$

where in the last expression the diffracted field was written in the form of a two-dimensional convolution of the G function with the f_2 function. The amplitude transmittance function, $t_a(x', y')$, was used to convert the integration over the entire $x'y'$ plane. According to Kirchhoff boundary conditions, $t_a(x', y') = 1$, for $(x', y') \in S_a$ and $t_a(x', y') = 0$, for $(x', y') \notin S_a$. Taking the Fourier transform of the above expression and applying the convolution theorem for the Fourier transform, it is straightforward to write the following expressions:

$$
\begin{aligned}
\tilde{U}_{RS_2}(k_x, k_y; z_0) &= \tilde{G}(k_x, k_y; z_0) \tilde{f}_2(k_x, k_y), \qquad \text{where,} \\
\tilde{U}_{RS_2}(k_x, k_y; z_0) &= \mathcal{F} \left\{ U_{RS_2}(x, y, z_0) \right\}, \\
\tilde{G}(k_x, k_y; z_0) &= \mathcal{F} \left\{ G(x, y, z_0) \right\}, \\
\tilde{f}_2(k_x, k_y) &= \mathcal{F} \left\{ f_2(x, y) \right\} = \mathcal{F} \left\{ -\frac{\partial U_{inc}(x', y'; 0)}{\partial z'} t_a(x', y') \right\} = \\
&= \frac{1}{(2\pi)^2} \left\{ \left[jk_z \tilde{U}_{inc}(k_x, k_y; 0) \right] ** \tilde{t}_a(k_x, k_y) \right\}.
\end{aligned}
$$

The Fourier transform of the $G(x, y; z)$ can be found if the Weyl's [286] (refer to Appendix 5C) plane wave expansion is used. Then,

$$
\begin{aligned}
\mathcal{F} \left\{ G(x, y, z_0) \right\} &= \iint_{-\infty}^{+\infty} \left(\frac{e^{-jkr}}{2\pi r} \right) e^{+j(k_x x + k_y y)} dxdy = \\
&= \iint_{-\infty}^{+\infty} \left[\frac{-j}{(2\pi)^2} \iint_{-\infty}^{+\infty} \frac{e^{-j(k'_x x + k'_y y + k'_z z_0)}}{k'_z} dk'_x dk'_y \right] e^{+j(k_x x + k_y y)} dxdy = \\
&= \iint_{-\infty}^{+\infty} \left[\frac{-j}{(2\pi)^2} \frac{e^{-jk'_z z_0}}{k'_z} dk'_x dk'_y \right] \underbrace{\iint_{-\infty}^{+\infty} e^{-j(k'_x - k_x)x} e^{-j(k'_y - k_y)y} dxdy}_{= (2\pi)^2 \delta(k'_x - k_x) \delta(k'_y - k_y)} = \\
&= -j \iint_{-\infty}^{+\infty} \frac{e^{-jk'_z z_0}}{k'_z} \delta(k'_x - k_x) \delta(k'_y - k_y) dk'_x dk'_y = -j \frac{e^{-jk_z z_0}}{k_z}.
\end{aligned} \tag{5.45}
$$

Using the last result the plane wave spectrum $\tilde{U}_{RS_2}(k_x, k_y; z)$ is given by ($z_0 > 0$):

$$\tilde{U}_{RS_2}(k_x, k_y; z_0) = \frac{e^{-jk_z z_0}}{k_z}\left[\frac{k_z \tilde{U}_{inc}(k_x, k_y; 0) ** \tilde{t}_a(k_x, k_y)}{(2\pi)^2}\right] \Longrightarrow$$

$$U_{RS_2}(x_0, y_0, z_0) = \frac{1}{(2\pi)^2}\iint_{-\infty}^{+\infty} \underbrace{\frac{1}{k_z}\left[\frac{k_z \tilde{U}_{inc}(k_x, k_y; 0) ** \tilde{t}_a(k_x, k_y)}{(2\pi)^2}\right]}_{=\tilde{U}_{RS_2}(k_x, k_y; 0)} e^{-j(k_x x_0 + k_y y_0 + k_z z_0)} dk_x dk_y. \tag{5.46}$$

The last integral is similar to the plane wave spectrum propagation as shown in Eq. (5.43), where the corresponding plane wave spectrum, under the Kirchhoff approximation, is $\tilde{U}_{RS_2}(k_x, k_y; 0)$. It is important to note that the terms k_z, in general, do not cancel each other, because the one inside the two-dimensional convolution depends on the convolution integration variables.

It is interesting to prove also the equivalence of the Rayleigh-Sommerfeld 1 formulation with the plane wave spectrum approach. Following a similar approach, as for RS-2 case, the following steps can be written:

$$U_{RS_1}(x_0, y_0, z_0) = \iint_{S_a}\left[\left(jk + \frac{1}{R}\right)\frac{\exp(-jkR)}{2\pi R}\frac{z_0}{R}U_{inc}\right]_{z'=0} dx'dy' =$$

$$= \iint_{-\infty}^{\infty} \underbrace{\left[-\frac{d}{dz_0}\left(\frac{\exp(-jkR)}{2\pi R}\right)\right]}_{=G_1(x_0 - x', y_0 - y'; z_0)} \underbrace{\left[U_{inc}\Big|_{z'=0} t_a(x', y')\right]}_{=f_1(x', y')} dx'dy' =$$

$$= \iint_{-\infty}^{+\infty} G_1(x_0 - x', y_0 - y'; z_0) f_1(x', y') dx'dy' = G_1(x, y, z_0) ** f_1(x, y).$$

Taking again the Fourier transform of the last expression and using the convolution theorem for the Fourier transform, it is straightforward to show:

$$\tilde{U}_{RS_1}(k_x, k_y; z_0) = \tilde{G}_1(k_x, k_y; z_0)\tilde{f}_1(k_x, k_y), \qquad \text{where,}$$

$$\tilde{U}_{RS_1}(k_x, k_y; z_0) = \mathcal{F}\{U_{RS_1}(x, y, z_0)\},$$

$$\tilde{G}_1(k_x, k_y; z_0) = +jk_z \mathcal{F}\{G(x, y, z_0)\} = e^{-jk_z z_0},$$

$$\tilde{f}_1(k_x, k_y) = \mathcal{F}\{f_1(x, y)\} = \mathcal{F}\{U_{inc}(x', y'; 0) t_a(x', y')\} =$$

$$= \frac{1}{(2\pi)^2}\left\{\tilde{U}_{inc}(k_x, k_y; 0) ** \tilde{t}_a(k_x, k_y)\right\}.$$

where for the Fourier transform of G_1 the result of the Fourier transform of G was used. Using the last result the plane wave spectrum $\tilde{U}_{RS_1}(k_x, k_y; z_0)$ is given by ($z_0 > 0$):

$$\tilde{U}_{RS_1}(k_x, k_y; z_0) = e^{-jk_z z_0}\left[\tilde{U}_{inc}(k_x, k_y, ; 0) ** \tilde{t}_a(k_x, k_y)\right] \Longrightarrow$$

$$U_{RS_1}(x_0, y_0, z_0) = \frac{1}{(2\pi)^2}\iint_{-\infty}^{+\infty} \underbrace{\left[\frac{\tilde{U}_{inc}(k_x, k_y, ; 0) ** \tilde{t}_a(k_x, k_y)}{(2\pi)^2}\right]}_{=\tilde{U}_{RS_1}(k_x, k_y; 0)} e^{-j(k_x x_0 + k_y y_0 + k_z z_0)} dk_x dk_y. \tag{5.47}$$

Here, the last integral is similar to the plane wave spectrum propagation of Eq. (5.43), with $\tilde{U}_{RS_1}(k_x, k_y; 0)$ representing the plane wave spectrum under the Kirchhoff approximation. It is evident that Eqs. (5.46) and (5.47) are not identical due to the difference in the kernel of the corresponding integrals, which represent the plane wave spectrum of the Kirchhoff approximations for the two Rayleigh-Sommerfeld formulations.

5.5.1 Limiting Cases

In this section some limiting cases of the plane-wave spectrum approach are presented. Initially, let's assume that there is no aperture, i.e., the incident wave U_{inc} propagates from $z' = 0$ to any $z_0 > 0$ unobstructed. In this case, the aperture amplitude transmittance function is $t_a(x, y) = 1, \forall x, y$. Then, the Fourier transform of $t_a(x, y)$ is $\tilde{t}_a(k_x, k_y) = (2\pi)^2\delta(k_x)\delta(k_y)$. Consequently, Eq. (5.47) for the U_{RS_1} diffracted field, and Eq. (5.46) for the U_{RS_2} diffracted field, reduce to:

$$U_{RS_1}(x_0, y_0, z_0) = \frac{1}{(2\pi)^2}\iint_{-\infty}^{+\infty} \tilde{U}_{inc}(k_x, k_y, ; 0)e^{-j(k_x x_0 + k_y y_0 + k_z z_0)}dk_x dk_y = U_{inc}(x_0, y_0, z_0),$$

$$U_{RS_2}(x_0, y_0, z_0) = \frac{1}{(2\pi)^2}\iint_{-\infty}^{+\infty} \tilde{U}_{inc}(k_x, k_y, ; 0)e^{-j(k_x x_0 + k_y y_0 + k_z z_0)}dk_x dk_y = U_{inc}(x_0, y_0, z_0).$$

The above equations reveal that the diffracted field from the RS-1 and RS-2 formulations is predicted correctly as the propagating incident field in the case where there is no aperture. This result is consistent with the expectation that in the absence of any obstruction (aperture), the field should simply propagate without diffraction.

Another special case occurs when $z_0 \to 0$. In this situation, both RS-1 and RS-2 formulations should yield the Kirchhoff boundary conditions, as these formulations are based on them. For the case of RS-1 formulation as $z_0 \to 0$, the function $\tilde{G}_1(k_x, k_y; z_0 \to 0)$ tends to unity. This results in $\tilde{U}_{RS_1}(k_x, k_y, z_0 \to 0) \to \tilde{f}_1(k_x, k_y)$. This implies that $U_{RS_1}(x_0, y_0, z_0 \to 0) \to U_{inc}(x, y; 0)t_a(x, y)$ which is the expected Kirchhoff boundary condition for the RS-1 formulation.

Similarly, for the RS-2 formulation, the following relations can be written:

$$\tilde{G}(k_x, k_y; z_0 \to 0) \longrightarrow \frac{-j}{k_z},$$

$$-\frac{\partial U_{RS_2}(x_0, y_0, z_0 \to 0)}{\partial z} \longrightarrow \frac{1}{(2\pi)^2}\iint_{-\infty}^{+\infty} \frac{-j}{k_z}\left[\frac{k_z\tilde{U}_{inc}(k_x, k_y; 0) ** \tilde{t}_a(k_x, k_y)}{(2\pi)^2}\right]$$
$$(-jk_z)e^{-j(k_x x_0 + k_y y_0)}dk_x dk_y =$$
$$= -\frac{\partial U_{inc}(x, y; 0)}{\partial z}t_a(x, y).$$

The last expression is indeed the Kirchhoff boundary condition used in the RS-2 formulation. Therefore, it has been demonstrated that both the RS-1 and RS-2 formulations are consistent with their boundary assumptions, confirming the claim made in Sec. 5.3, where this was stated without proof.

5.6 Babinet's Principle for Scalar Diffraction

Consider two diffractive planes at $z = 0$, as shown in Fig. 5.6. The apertures on these planes are referred to as "complementary" when the transparent regions of one plane (see surface S_a of Fig. 5.6a) precisely correspond to the opaque regions of the other plane (see surface S_a of Fig. 5.6b), and vice versa [42]. Assume that an incident wave, $U_{inc}(x', y', z' = 0)$, propagates from the left toward each of the apertures. Let U_1 denote the resulting diffracted field at the point $P(x_0, y_0, z_0)$ when the aperture in Fig. 5.6a, with an open area S_a, is present. Similarly, let U_2 represent the diffracted field at $P(x_0, y_0, z_0)$ when the complementary aperture defined by the opaque region S_a of Fig. 5.6b is present. If both apertures are simultaneously present, resulting in no opaque regions, the diffracted field at $P(x_0, y_0, z_0)$ must equal the incident wave $U_{inc}(x_0, y_0, z_0)$. This concept, known as the *Babinet's principle*, was first proposed in 1837 [287]. Babinet's principle can be mathematically expressed as [42, 91, 273]:

$$U_1(x_0, y_0, z_0) + U_2(x_0, y_0, z_0) = U_{inc}(x_0, y_0, z_0). \tag{5.48}$$

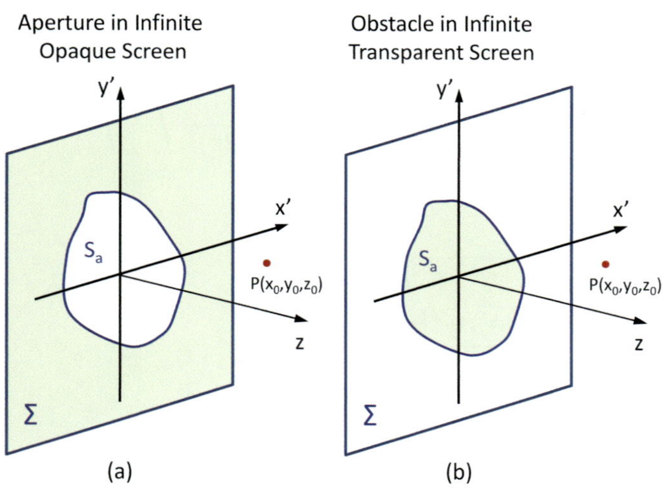

FIGURE 5.6: (a) An aperture (opening) in an infinite opaque screen. (b) The complementary aperture is comprised of an obstacle, exactly the same with the aperture of (a), in an infinite transparent screen. In both geometries a random point $P(x_0, y_0, z_0)$ (with $z_0 > 0$) is shown.

Babinet's principle can be readily proven using the Rayleigh-Sommerfeld second formulation (RS-2), as follows:

$$U_1(x_0, y_0, z_0) = \iint_{S_a} \left[\frac{\exp(-jkR)}{2\pi R} \frac{\partial U_{inc}}{\partial n} \right]_{z'=0} dx'dy',$$

$$U_2(x_0, y_0, z_0) = \iint_{\Sigma - S_a} \left[\frac{\exp(-jkR)}{2\pi R} \frac{\partial U_{inc}}{\partial n} \right]_{z'=0} dx'dy',$$

$$U_1(x_0, y_0, z_0) + U_2(x_0, y_0, z_0) = \underbrace{\iint_{\Sigma} \left[\frac{\exp(-jkR)}{2\pi R} \frac{\partial U_{inc}}{\partial n} \right]_{z'=0} dx'dy'}_{= I_\Sigma}. \tag{5.49}$$

Here, Σ represents the infinite surface of the $x'y'$ plane at $z' = 0$. At this stage, it must be demonstrated that the final integral equals $U_{inc}(x_0, y_0, z_0)$. To establish this result, the Weyl expansion of a spherical wave into a plane wave spectrum is employed [286]. For completeness, a concise proof of Weyl's formula is provided in Appendix 5C. To begin, the incident field and its normal derivative are expressed in terms of their respective plane wave spectra as follows:

$$U_{inc}(x', y', z' = 0) = \frac{1}{(2\pi)^2} \int_{-\infty}^{+\infty} \int_{-\infty}^{+\infty} \tilde{U}(k_x, k_y; 0) e^{-j(k_x x' + k_y y')} dk_x dk_y \Longrightarrow$$

$$\left.\frac{\partial U_{inc}}{\partial n}\right|_{z'=0} = -\left.\frac{\partial U_{inc}}{\partial z'}\right|_{z'=0} = \frac{1}{(2\pi)^2} \int_{-\infty}^{+\infty} \int_{-\infty}^{+\infty} jk_z \tilde{U}(k_x, k_y; 0) e^{-j(k_x x' + k_y y')} dk_x dk_y.$$

Here, $\tilde{U}(k_x, k_y; 0)$ represents the plane wave spectrum of U_{inc} at $z' = 0$. Substituting the above expansion, along with the Weyl expansion, into Eq. (5.49), the following expressions are obtained:

$$I_\Sigma = \iint_\Sigma \left[\frac{-j}{(2\pi)^2} \int_{-\infty}^{+\infty} \int_{-\infty}^{+\infty} \frac{e^{-j[k_x(x_0 - x') + k_y(y_0 - y') + k_z |z_0|]}}{k_z} dk_x dk_y \right]$$

$$\left[\frac{1}{(2\pi)^2} \int_{-\infty}^{+\infty} \int_{-\infty}^{+\infty} jk'_z \tilde{U}(k'_x, k'_y; 0) e^{-j(k'_x x' + k'_y y')} dk'_x dk'_y \right] dx' dy' =$$

$$= \frac{-j}{(2\pi)^2} \int_{-\infty}^{+\infty} \int_{-\infty}^{+\infty} \frac{e^{-j[k_x x_0 + k_y y_0 + k_z |z_0|]}}{k_z} dk_x dk_y \left[\frac{1}{(2\pi)^2} \int_{-\infty}^{+\infty} \int_{-\infty}^{+\infty} jk'_z \tilde{U}(k'_x, k'_y; 0) dk'_x dk'_y \right]$$

$$\underbrace{\left[\iint_\Sigma e^{-j[(k'_x - k_x)x' + (k'_y - k_y)y']} dx' dy' \right]}_{= (2\pi)^2 \delta(k'_x - k_x) \delta(k'_y - k_y)} =$$

$$= \frac{1}{(2\pi)^2} \int_{-\infty}^{+\infty} \int_{-\infty}^{+\infty} \frac{e^{-j[k_x x_0 + k_y y_0 + k_z |z_0|]}}{k_z} dk_x dk_y$$

$$\int_{-\infty}^{+\infty} \int_{-\infty}^{+\infty} k'_z \tilde{U}(k'_x, k'_y; 0) \delta(k'_x - k_x) \delta(k'_y - k_y) dk'_x dk'_y$$

$$= \frac{1}{(2\pi)^2} \int_{-\infty}^{+\infty} \int_{-\infty}^{+\infty} \frac{e^{-j[k_x x_0 + k_y y_0 + k_z |z_0|]}}{k_z} k_z \tilde{U}(k_x, k_y; 0) dk_x dk_y = U_{inc}(x_0, y_0, z_0).$$

The proof presented above can be generalized to curved apertures as well [273], rather than being restricted to planar ones as demonstrated here. Since the proof was derived using the Rayleigh-Sommerfeld second formulation, it is equally valid for the Rayleigh-Sommerfeld first formulation (RS-1), as well as the Helmholtz-Kirchhoff, Fresnel, and Fraunhofer diffraction approximations [288]. However, some caution is required when the incident field has an infinite extent, such as a plane wave, particularly in the context of the Fraunhofer approximation. If the initial aperture is finite, i.e., $S_a < \infty$, the complementary aperture has an infinite area, and the Fraunhofer approximation generally fails to satisfy the conditions necessary for its strict application (as illustrated later in the case of a circular obstacle). Nonetheless, in practice, Fraunhofer results often remain meaningful even under such circumstances.

Although Eq. (5.48) can be expressed in terms of the amplitudes of the diffracted fields, it does not hold for their intensities (irradiances) due to the presence of an interference term. Specifically, from Eq. (5.48), it is straightforward to derive the following:

$$I_1 + I_2 + \frac{1}{2Z}\text{Re}\{U_1 U_2^*\} = I_{inc},$$

where $I_i = |U_i|^2/4Z$ ($i = 1, 2$, or *"inc"*), and Z is the wave impedance of the homogeneous space. However, there are cases, particularly within the Fraunhofer diffraction regime, where U_{inc} may be nearly zero across the far field, except within very small regions (e.g., a narrow laser beam or light focused by a lens). In such cases, $U_1 + U_2 = 0$ implies $U_1 = -U_2$, leading to $I_1 = I_2$, except in those small regions where the incident field is nonzero. Finally, it is worth noting that Babinet's principle can be extended to the vectorial representation of the electromagnetic field [86, 91]. Examples of applications of Babinet's principle are presented in the following sections of this chapter.

It is noted that in all the numerical examples presented in the various sections of this chapter, relatively small-sized apertures or obstacles are employed, typically on the order of 10–20 wavelengths. This choice is made to simplify the numerical integrations (1D and 2D) required for the calculations. Larger aperture or obstacle sizes would significantly increase computation times, which is not the primary objective of these examples. Instead, the focus is on illustrating the different formulations and their application to a variety of aperture and obstacle configurations.

5.7 Scalar Diffraction by a Circular/Elliptical Apertures, Obstacles, and Annuli

This section presents simple applications of scalar diffraction involving circular or elliptical apertures, obstacles, and annuli. These diffracting structures are discussed first because, in certain cases (such as the circular aperture or obstacle), analytical solutions to the Rayleigh-Sommerfeld formulations exist along the z-axis (the propagation axis). In contrast, rectangular apertures lack analytical solutions for the Rayleigh-Sommerfeld formulations and are therefore addressed in a later section. The incident wave is assumed to be either a plane wave (typically normally incident) or a spherical wave (either diverging or converging).

5.7.1 Circular Aperture with Incident Plane Wave

Initially, the diffraction of an incident plane wave on a circular aperture in an opaque screen is examined. The geometry of this setup is illustrated in Fig. 5.7. In this figure, the incident wave may either be a spherical wave originating from point S (a diverging wave) or a plane wave, which is obtained as a limiting case when $s_0 \to -\infty$. If point S lies to the right of the aperture, the spherical wave becomes converging.

5.7.1.1 On-Axis Points

Assume that a plane wave $U = U_0 \exp(-jkz)$ is incident from the left onto a planar circular aperture in an opaque screen (see Fig. 5.7). In this case, the Rayleigh-Sommerfeld integrals can be determined analytically, as described in several references in the literature [275, 289–291]. By applying Eq. (5.18) for points P located on the positive z-axis, the following

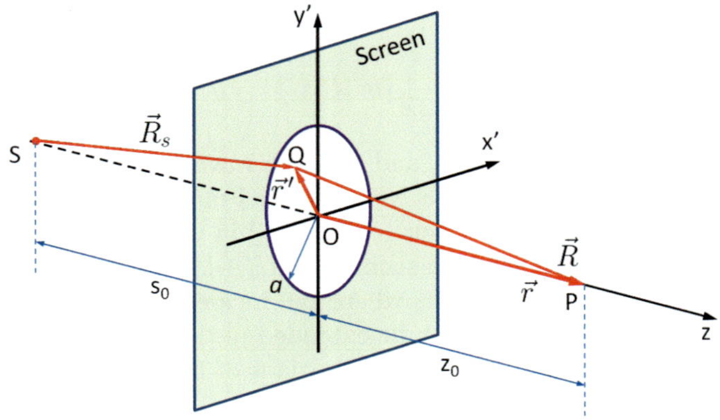

FIGURE 5.7: A circular aperture of radius a on an opaque screen. The diffracted field for the on-axis point P is sought. The point S specifies the origin of a diverging spherical wave that is incident on the aperture. In the case of an incident plane wave the point S moves to $-\infty$ ($s_0 \to +\infty$).

expression can be derived:

$$U_{RS_1}(P) = U_{RS_1}(z_0) = \iint_{S_a} \left[U_0 \exp(-jkz')|_{z'=0} \left(jk + \frac{1}{R} \right) \frac{\exp(-jkR)}{2\pi R} \frac{z}{R} \right]_{z'=0} dx'dy' =$$

$$= \int_0^a \int_0^{2\pi} U_0 \left(jk + \frac{1}{R} \right) \frac{\exp(-jkR)}{2\pi R} \frac{z_0}{R} r'_T dr'_T d\phi' =$$

$$= - \int_0^a \int_0^{2\pi} U_0 \frac{d}{dR} \left(\frac{\exp(-jkR)}{2\pi R} \right) \frac{z_0}{R} r'_T dr'_T d\phi',$$

$$\text{with} \quad R = \sqrt{z_0^2 + r_T'^2}, \quad \text{resulting in}$$

$$U_{RS_1}(P) = U_{RS_1}(z_0) = -2\pi U_0 \int_0^a \frac{d}{dR} \left(\frac{\exp(-jkR)}{2\pi R} \right) \frac{z_0}{R} r'_T dr'_T, \quad \text{where} \quad 2RdR = 2r'_T dr'_T,$$

$$U_{RS_1}(P) = U_{RS_1}(z_0) = -2\pi U_0 z_0 \int_0^a \frac{d}{dR} \left(\frac{\exp(-jkR)}{2\pi R} \right) \frac{1}{R} RdR$$

$$= -2\pi U_0 z_0 \left[\frac{\exp(-jkR)}{2\pi R} \right]_{r'_T=0}^{r'_T=a} =$$

$$= -2\pi U_0 z_0 \left[\frac{\exp(-jk\sqrt{a^2+z_0^2})}{2\pi\sqrt{a^2+z_0^2}} - \frac{\exp(-jkz_0)}{2\pi z_0} \right] =$$

$$= U_0 \left[\exp(-jkz_0) - \frac{z_0}{\sqrt{a^2+z_0^2}} \exp\left(-jk\sqrt{a^2+z_0^2} \right) \right]. \tag{5.50}$$

The corresponding intensity of the diffracted field is proportional to $|U|^2$ and is expressed as:

$$I_{RS_1}(z_0) = \frac{|U_0|^2}{4Z}\left[1+\frac{z_0^2}{a^2+z_0^2} - \frac{2z_0}{\sqrt{a^2+z_0^2}}\cos\left(k\sqrt{a^2+z_0^2}-kz_0\right)\right] =$$

$$= I_0\left[1+\frac{z_0^2}{a^2+z_0^2} - \frac{2z_0}{\sqrt{a^2+z_0^2}}\cos\left(k\sqrt{a^2+z_0^2}-kz_0\right)\right]. \qquad (5.51)$$

Here, $I_0 = |U_0|^2/4Z$ and Z the wave impedance of the surrounding medium. It is noteworthy that the Rayleigh-Sommerfeld 1 formulation is self-consistent, as $U_{RS_1}(z_0 = 0) = U_0 = U_{inc}(z = 0)$. Now, let us derive the corresponding result for the Rayleigh-Sommerfeld 2 formulation. By applying Eq. (5.19) for points P Now let's derive the corresponding result for the Rayleigh-Sommerfeld 2 formulation. Using Eq. (5.19) for points P on the positive z-axis, the following expression can be obtained:

$$U_{RS_2}(P) = U_{RS_2}(z_0) = \iint_{S_a}\left[\frac{\exp(-jkR)}{2\pi R}\frac{\partial U_{inc}}{\partial n}\right]_{z=0} dx'dy' =$$

$$= \int_0^a\int_0^{2\pi}\frac{\exp(-jkR)}{2\pi R}(jkU_0)r'_T\,dr'_T\,d\phi' =$$

$$= jkU_0 2\pi\int_0^a\frac{\exp(-jkR)}{2\pi R}R\,dR = jkU_0\left[\frac{\exp(-jkR)}{-jk}\right]_{r'_T=0}^{r'_T=a} =$$

$$= U_0\left[\exp(-jkz_0) - \exp\left(-jk\sqrt{a^2+z_0^2}\right)\right]. \qquad (5.52)$$

The corresponding intensity along the positive z-axis is given by:

$$I_{RS_2}(z_0) = \frac{|U_0|^2}{4Z}\left[2-2\cos\left(k\sqrt{a^2+z_0^2}-kz_0\right)\right] = I_0\left[2-2\cos\left(k\sqrt{a^2+z_0^2}-kz_0\right)\right]. \qquad (5.53)$$

It is also worth noting that the Rayleigh-Sommerfeld 2 formulation is self-consistent, as $\partial U_{RS_2}/\partial n|_{z_0=0} = -\partial U_{RS_2}/\partial z|_{z_0=0} = jkU_0 = \partial U_{inc}/\partial n|_{z=0}$. The Helmholtz-Kirchhoff formulation, with Kirchhoff boundary conditions, is given by Eq. (5.20) and can be written as:

$$U_K(P) = U_K(z_0) = \frac{1}{2}\left\{U_{RS1}(z_0) + U_{RS2}(z_0)\right\} =$$

$$= U_0\left[\exp(-jkz_0) - \frac{1}{2}\left(1+\frac{z_0}{\sqrt{a^2+z_0^2}}\right)\exp\left(-jk\sqrt{a^2+z_0^2}\right)\right]. \qquad (5.54)$$

Then, the corresponding intensity of the Helmholtz-Kirchhoff solution, U_K is given by:

$$I_K(z_0) = \frac{|U_0|^2}{4Z}\left[1+\frac{1}{4}\left(1+\frac{z_0}{\sqrt{a^2+z_0^2}}\right)^2 - \left(1+\frac{z_0}{\sqrt{a^2+z_0^2}}\right)\cos\left(k\sqrt{a^2+z_0^2}-kz_0\right)\right] =$$

$$= I_0\left[1+\frac{1}{4}\left(1+\frac{z_0}{\sqrt{a^2+z_0^2}}\right)^2 - \left(1+\frac{z_0}{\sqrt{a^2+z_0^2}}\right)\cos\left(k\sqrt{a^2+z_0^2}-kz_0\right)\right]. \qquad (5.55)$$

It is important to note that the Helmholtz-Kirchhoff formulation, $U_K(z_0)$ is not self-consistent since at the aperture, $U_K(z_0 = 0) = U_0[1-(1/2)\exp(-jka)] \neq U_0 = U_{inc}(z' = 0)$. A similar inconsistency arises for the normal derivative, $\partial U_K/\partial n = -\partial U_K/\partial z|_{z_0=0} = U_0[jk + (1/2a)\exp(-jka)] \neq jkU_0 = \partial U_{inc}/\partial n|_{z=0}$.

It is interesting to examine the intensity envelopes for the various formulations. From Eq. (5.51), the envelope of the Rayleigh-Sommerfeld first formulation, denoted as $\mathcal{I}^{\pm}_{RS_1}$, is given by:

$$\mathcal{I}^{\pm}_{RS_1} = I_0 \left(1 \pm \frac{z_0}{\sqrt{a^2 + z_0^2}}\right)^2.$$

Similarly, the intensity envelopes for the Rayleigh-Sommerfeld 2 formulation, $\mathcal{I}^{\pm}_{RS_2}$, are:

$$\mathcal{I}^{+}_{RS_2} = 4I_0, \quad \text{and} \mathcal{I}^{-}_{RS_2} = 0.$$

Finally, the intensity envelopes of the Helmholtz-Kirchhoff formulation, $\mathcal{I}^{\pm}_{K}$, are:

$$\mathcal{I}^{\pm}_{K} = I_0 \left[1 \pm \frac{1}{2}\left(1 + \frac{z_0}{\sqrt{a^2 + z_0^2}}\right)\right]^2.$$

Example 5.7.1:

Determine and plot the intensity distribution along the z-axis of a circular aperture of radius $a = 10\,\mu$m, given a normally incident plane wave of freespace wavelength of $\lambda_0 = 1\,\mu$m,
Solution: The corresponding intensity curves along the z-axis are presented in Figs. 5.8(a) and 5.8(b) for $\lambda_0 = 1\,\mu$m. In this case, $ka = 20\pi$ representing a special case where the radius is an integer multiple of the wavelength. However, similar results (not shown) can be obtained even if the radius is not an integer multiple of wavelength, for example, when $\lambda = 0.6328\,\mu$m (then, $ka = 31.6056\pi$). The intensity curves are derived using the Rayleigh-Sommerfeld 1 (RS-1) and 2 (RS-2) formulations, as well as the Helmholtz-Kirchhoff (H-K) formulation. Notable differences between these formulations are observed at distances near the aperture. For relatively large distances from the aperture, specifically when $z_0 > 5a$, all formulations converge to nearly identical results. In Fig. 5.8(b), the same results are presented on a logarithmic scale to highlight the differences in intensity for distances up to $1000\,\mu m = 1000\lambda_0 = 100a$. It is also of interest to identify the points along the z-axis where the minima and maxima of the diffracted intensity occur. These extrema points are the same for all three formulations (RS-1, RS-2, and H-K) and are given by:

$$z_m^+ = \frac{a^2 - \left(m + \frac{1}{2}\right)^2 \lambda^2}{(2m+1)\lambda}, \qquad \text{for maxima, with } m < \frac{1}{2}\left(\frac{2a}{\lambda} - 1\right),$$

$$z_m^- = \frac{a^2 - m^2\lambda^2}{2m\lambda}, \qquad \text{for minima, with } m < \frac{a}{\lambda}.$$

Here, λ represents the wavelength within the medium surrounding the space for $z_0 > 0$ (which, in the case of air, is approximately equal to λ_0). Using the above equations, it is straightforward to determine the number and positions of the maxima and minima of the diffracted intensity along the z-axis. These are illustrated in Fig. 5.8.

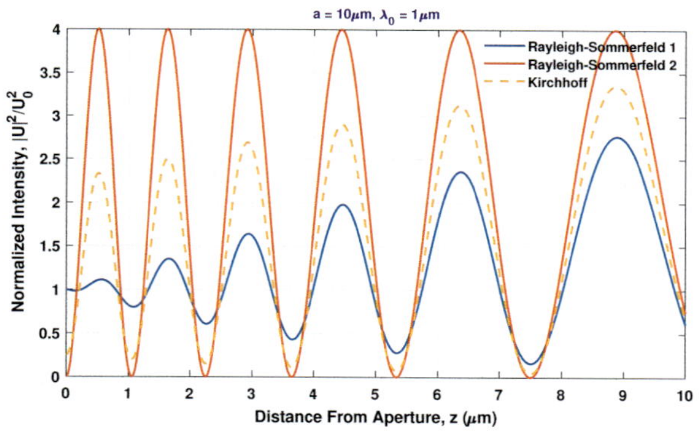

(a)

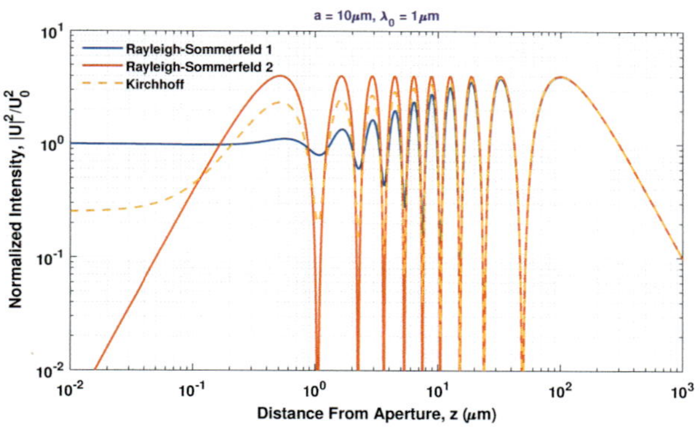

(b)

FIGURE 5.8: (a) The diffracted field normalized intensity, $|U|^2/|U_0|^2$, along the z-axis for a plane wave incident on a circular aperture of radius $a = 10\,\mu m$. The freespace wavelength is $\lambda_0 = 1.0\,\mu m$. The results were obtained using the Rayleigh-Sommerfeld formulations 1 and 2 and the Kirchhoff with the Kirchhoff's boundary conditions. (b) Same as (a) but in logarithmic scale and for larger values of z.

It is of interest to calculate the corresponding diffracted field along the positive z-axis using both the Fresnel and the Fraunhofer approximations. Starting from Eq. (5.25), it is straightforward to write the following expressions (with $x_0 = y_0 = 0$):

$$U(z_0) \simeq jk\frac{\exp(-jkz_0)}{2\pi z_0}\iint_{S_a}\left[\exp\left\{-jk\left[\frac{(x_0-x')^2}{2z_0}+\frac{(y_0-y')^2}{2z_0}\right]\right\}U_{inc}(x',y')\right]dx'dy' =$$

$$= jk\frac{\exp(-jkz_0)}{2\pi z_0}U_0\int_0^a\int_0^{2\pi}\exp\left[-jk\frac{r_T'^2}{2z_0}\right]r_T'dr_T'd\phi' =$$

$$= jk\frac{\exp(-jkz_0)}{2\pi z_0}U_0 2\pi\int_0^a\exp\left[-jk\frac{r_T'^2}{2z_0}\right]\frac{1}{2}dr_T'^2$$

$$= U_0\exp(-jkz_0)\left[1-\exp\left(-jk\frac{a^2}{2z_0}\right)\right], \qquad \text{for } z_0 \gg \left(\frac{\pi}{4\lambda_0}a^4\right)^{1/3}. \tag{5.56}$$

The Fraunhofer approximation for points on the positive z-axis can also be derived starting from Eq. (5.32). Specifically, for $x_0 = y_0 = 0$, the expression becomes:

$$U(z_0) = jk\frac{\exp(-jkz_0)}{2\pi z_0}\exp\left(-jk\frac{x_0^2+y_0^2}{2z_0}\right)\iint_{S_a}\left[\exp\left\{+jk\left(\frac{x_0x'+y_0y'}{z_0}\right)\right\}U_{inc}(x',y')\right]dx'dy' =$$

$$= jk\frac{\exp(-jkz_0)}{2\pi z_0}\int_0^a\int_0^{2\pi} U_0 r_T' dr_T' d\phi' =$$

$$= jkU_0\frac{\exp(-jkz_0)}{z_0}\frac{a^2}{2}, \qquad \text{for } z_0 \gg \frac{\pi}{\lambda}a^2. \tag{5.57}$$

For example, the Rayleigh-Sommerfeld 1 (RS-1) formulation, along with the Fresnel and

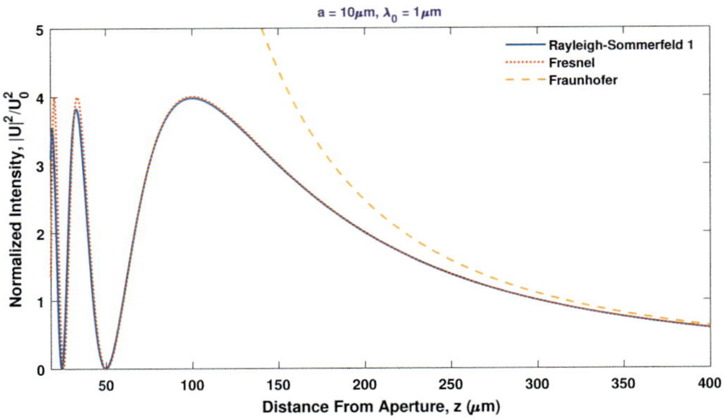

FIGURE 5.9: The diffracted field normalized intensity, $|U|^2/|U_0|^2$, along the z-axis for a plane wave normally incident on a circular aperture of radius $a = 10\,\mu m$. The freespace wavelength is $\lambda_0 = 1.0\,\mu m$. The results were obtained using the Rayleigh-Sommerfeld formulations 1, the Fresnel approximation, and the Fraunhofer approximation are shown. It is mentioned that for the range of z shown the Rayleigh-Sommerfeld 1 and 2 formulations as well as the Helmholtz-Kirchhoff formulation practically coincide.

Fraunhofer approximations, are shown for the diffracted field along the z-axis when a plane wave is normally incident on a circular aperture with a radius of $a = 10\,\mu m$, and $\lambda_0 = \lambda = 1\,\mu m$ (parameters from Example 5.7.1). The limits at which the Fresnel and Fraunhofer approximations hold are $19.8776\,\mu m$ and $314.1593\,\mu m$, respectively. The corresponding diffracted field is shown in Fig. 5.9. The results for the RS-2 and H-K formulations coincide with those from the RS-1 formulation within the range of z values shown in the figure.

5.7.1.2 Off-Axis Points

In this section, the diffraction of a plane wave by a circular aperture is extended to points in any off-axis direction. The geometric configuration is illustrated in Fig. 5.10. The point $P(x_0, y_0, z_0)$ represents an arbitrary point of interest for $z_0 \geq 0$. In this case, the diffraction integrals for the RS-1 and RS-2 formulations cannot be determined analytically as in the on-axis situation. Therefore, either approximations are needed or numerical evaluation of the integrals is required [291, 292]. For example, the diffraction integrals for the RS-1, RS-2

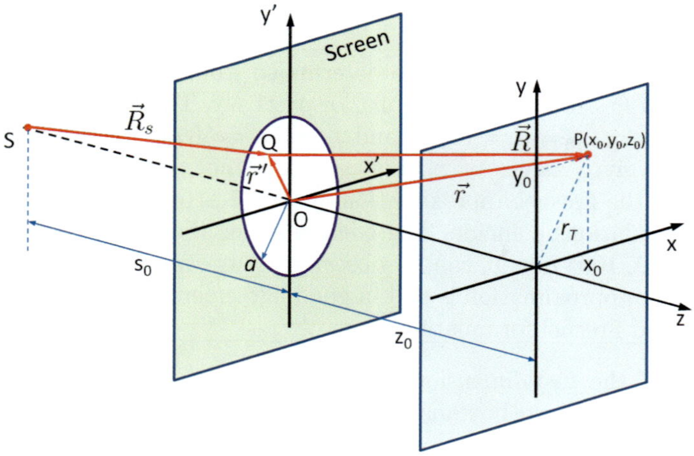

FIGURE 5.10: A circular aperture of radius a on an opaque screen. The diffracted field for the off-axis point $P(x_0, y_0, z_0)$ is sought. The point S specifies the origin of a diverging spherical wave that is incident on the aperture. In the case of an incident plane wave the point S moves to $-\infty$ ($s_0 \to +\infty$).

formulations, and the Fresnel approximation, for the point P are given by:

$$U_{RS_1}(x_0, y_0, z_0) = \int_0^a \int_0^{2\pi} U_0 \left(jk + \frac{1}{R} \right) \frac{\exp(-jkR)}{2\pi R} \frac{z_0}{R} r'_T dr'_T d\phi', \tag{5.58}$$

$$U_{RS_2}(x_0, y_0, z_0) = \int_0^a \int_0^{2\pi} \frac{\exp(-jkR)}{2\pi R} (jkU_0) r'_T dr'_T d\phi', \tag{5.59}$$

$$U(x_0, y_0, z_0) = jk \frac{\exp(-jkz_0)}{2\pi z_0} \int_0^a \int_0^{2\pi} \exp \left\{ -jk \frac{r_T^2 + r_T'^2 - 2r_T r'_T \cos(\phi - \phi')}{2z_0} \right\} U_0 r'_T dr'_T d\phi' =$$

$$= jk \frac{\exp(-jkz_0)}{z_0} \int_0^a \exp \left(-jk \frac{r_T^2 + r_T'^2}{2z_0} \right) U_0 J_0 \left(\frac{kr_T r'_T}{z_0} \right) r'_T dr'_T, \tag{5.60}$$

where

$$R = \left[r_T^2 + r_T'^2 - 2r_T r'_T \cos(\phi - \phi') + z_0^2 \right]^{1/2}, \quad \text{and,}$$

$$r_T = (x_0^2 + y_0^2)^{1/2}, \quad \phi = \tan^{-1}\left(\frac{y_0}{x_0} \right), \quad \phi' = \tan^{-1}\left(\frac{y'}{x'} \right).$$

All of the integrals can be determined through two-dimensional numerical integration, except for the one in the Fresnel approximation, which, through the use of the Bessel function, has been transformed into a one-dimensional integral. The two-dimensional integration can be reduced to one-dimensional integration using the procedure defined by Dubra and Ferrari [291]. However, modern numerical packages, such as MATLAB [141], can easily perform these integrations. Additionally, the H-K formulation integral can be computed from the RS-1 and RS-2 formulations using Eq. (5.20).

Example 5.7.2:

In this example, the diffraction of an incident plane wave with a freespace wavelength of $\lambda_0 = 1\,\mu m$, by a circular aperture of radius $a = 10\,\mu m$ is considered. The goal is to determine the normalized intensity in the radial direction at various distances, specifically at $z_0 = 10, 50, 300\,\mu$m.

Solution: The resulting intensities can be determined from Eqs. (5.58), (5.59), and (5.60), for RS-1, RS-2, and Fresnel approximations, respectively. These results are shown in Figs. 5.11(a) and 5.11(b) for $z_0 = a = 10\,\mu m$ and $z_0 = 5a = 50\,\mu m$, as a function of the radial distance r_T for the given values of z_0. For $z_0 = 300\,\mu$m, the Fraunhofer results are also included, alongside the Fresnel approximation, in Fig. 5.11(c). In this plot, the RS-1, RS-2, and Kirchhoff formulations coincide. The constraint for the Fraunhofer approximation to hold is $z \gg (\pi a^2)/\lambda_0 100\pi\,\mu$m. It can be observed that, even if this condition is not fully met, the Fraunhofer approximation is still in good agreement with the more accurate RS-1, RS-2, Kirchhoff, and Fresnel formulations.

When $kR \gg 1$, the two-dimensional integrals for the Rayleigh-Sommerfeld 1 and 2 formulations can be conveniently transformed into one-dimensional integrals, as follows:

$$U_{RS_1}(x_0, y_0, z_0) = \int_0^a \int_0^{2\pi} U_0 \left(jk + \frac{1}{R}\right) \frac{\exp(-jkR)}{2\pi R} \frac{z_0}{R} r'_T dr'_T d\phi' \Longrightarrow$$

$$U^a_{RS_1}(x_0, y_0, z_0) \simeq jkz_0 \int_0^a \frac{\exp(-jkR_0)}{R_0^2} U_0 J_0 \left(\frac{kr_T r'_T}{R_0}\right) r'_T dr'_T, \qquad \text{and,} \qquad (5.61)$$

$$U_{RS_2}(x_0, y_0, z_0) = \int_0^a \int_0^{2\pi} \frac{\exp(-jkR)}{2\pi R} (jkU_0) r'_T dr'_T d\phi' \Longrightarrow$$

$$U^a_{RS_2}(x_0, y_0, z_0) \simeq jk \int_0^a \frac{\exp(-jkR_0)}{R_0} U_0 J_0 \left(\frac{kr_T r'_T}{R_0}\right) r'_T dr'_T, \qquad \text{where,} \qquad (5.62)$$

$$R_0 = \left(r_T^2 + r_T'^2 + z_0^2\right)^{1/2}, \qquad \text{and} \qquad \frac{k}{8}\left(\frac{2r_T r'_T}{R_0^2}\right)^2 \ll 1,$$

where $U^a_{RS_1}$ and $U^a_{RS_2}$ represent the approximate one-dimensional integrals for the Rayleigh-Sommerfeld 1 and 2 formulations, respectively.

Example 5.7.3:

Use the approximate expressions of RS-1 and RS-2 formulations to determine the normalized intensity in the radial direction. The parameters are the same as those in Example 5.7.2. Calculate the approximate intensities for $z_0 = 10, 50\,\mu$m and compare them with their exact expressions.

Solution: The calculated intensities are shown in Figs. 5.12(a) and 5.12(b) for $z_0 = 10\,\mu m$ and $z_0 = 50\,\mu m$, respectively. It is evident that, for the $z_0 = 50\,\mu$m distance, the approximate formulations are in very close agreement with the exact two-dimensional integral formulations.

The Fraunhofer approximation, $U(x_0, y_0, z_0) = U_{ff}$, even in the off-axis case, can be determined analytically as follows:

$$U_{ff} = jk\frac{\exp(-jkz_0)}{2\pi z_0} \exp\left(-jk\frac{r_T^2}{2z_0}\right) \int_0^a \int_0^{2\pi} \left[\exp\left\{+jk\left(\frac{r_T r'_T \cos(\phi - \phi')}{z_0}\right)\right\} U_0\right] r'_T dr'_T d\phi' =$$

$$= jk\frac{\exp(-jkz_0)}{2\pi z_0} \exp\left(-jk\frac{r_T^2}{2z_0}\right) U_0 \int_0^a 2\pi J_0\left(\frac{kr_T r'_T}{z_0}\right) r'_T dr'_T =$$

$$= jk\frac{\exp(-jkz_0)}{2\pi z_0} \exp\left(-jk\frac{r_T^2}{2z_0}\right) U_0 2\pi \frac{aJ_1\left(kar_T/z_0\right)}{kr_T/z_0} \Longrightarrow$$

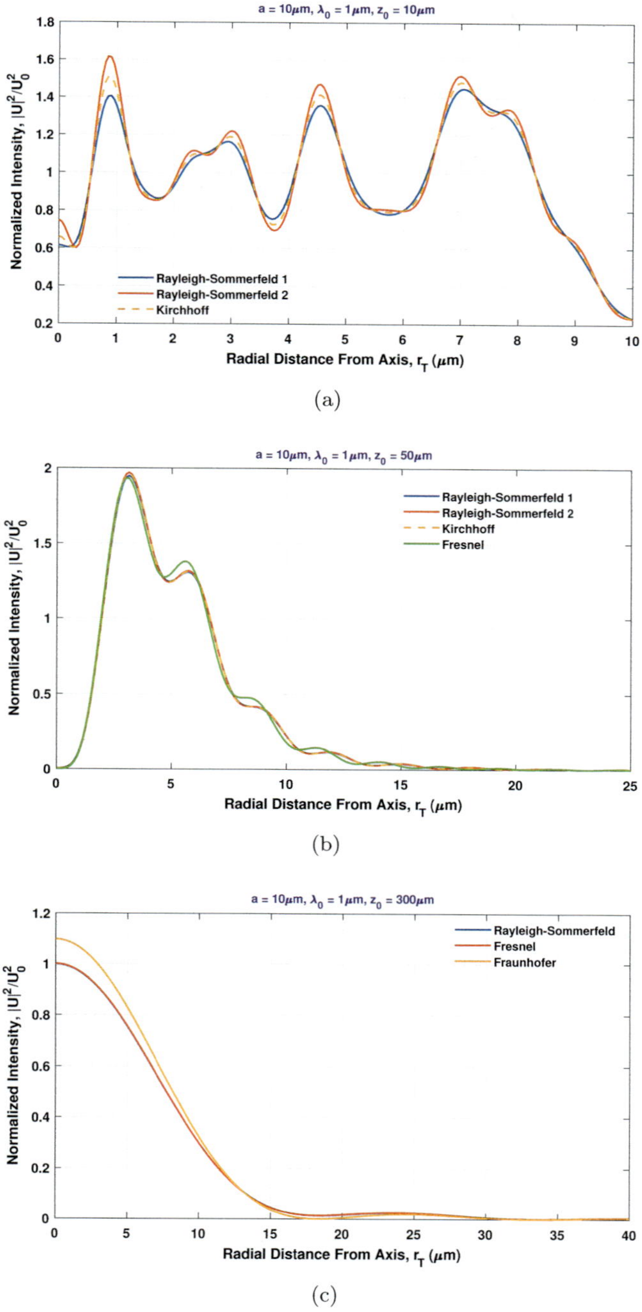

FIGURE 5.11: The diffracted field normalized intensity, $|U|^2/|U_0|^2$, along the radial direction for a plane wave incident on a circular aperture of radius $a = 10\,\mu m$ and $\lambda_0 = 1.0\,\mu m$. The results were obtained using the Rayleigh-Sommerfeld formulations 1 and 2 and the Helmholtz-Kirchhoff with the Kirchhoff's boundary conditions. Fresnel and Fraunhofer results are also shown for larger distances from the aperture. (a) $z_0 = 10\,\mu m$. (b) $z_0 = 50\,\mu m$. (c) $z_0 = 300\,\mu m$.

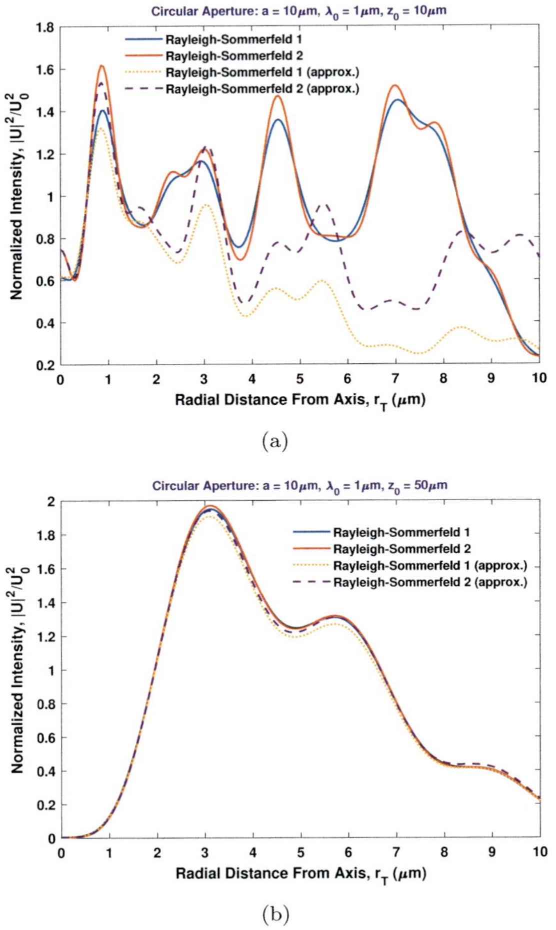

FIGURE 5.12: The diffracted field normalized intensity, $|U|^2/|U_0|^2$, along the radial direction for a plane wave normally incident on a circular aperture of radius $a = 10\,\mu m$ and $\lambda_0 = 1.0\,\mu m$. The results were obtained using the Rayleigh-Sommerfeld formulations 1 and 2 and their approximate one-dimensional integral formulations $U^a_{RS_1}$ and $U^a_{RS_2}$, respectively. (a) $z_0 = 10\,\mu m$. (b) $z_0 = 50\,\mu m$.

$$U_{ff} = jU_0 \exp(-jkz_0) \exp\left[-jk\frac{r_T^2}{2z_0}\right] \frac{aJ_1\left(kar_T/z_0\right)}{r_T}, \quad \text{and the intensity is} \tag{5.63}$$

$$I_{ff} = \frac{|U_0|^2}{4Z}\frac{k^2a^4}{z_0^2}\left(\frac{J_1\left(kar_T/z_0\right)}{kar_T/z_0}\right)^2 = I_0\left(\frac{2J_1\left(kar_T/z_0\right)}{kar_T/z_0}\right)^2. \tag{5.64}$$

Here, $I_0 = (|U_0|^2/4Z)(k^2a^4/4z_0^2)$ represents the maximum value of the diffracted intensity. In the above derivation, several properties of the Bessel functions were utilized [293]. Specifically, the following properties were used: $\int_0^{2\pi} \exp[jz\cos(\phi - \phi')]d\phi' = 2\pi J_0(z)$ and $\int J_0(az)zdz = zJ_1(az)/a$. Furthermore, I_0 corresponds to the maximum value of the diffracted intensity, which occurs along the z-axis ($r_T = 0$). The intensity pattern of $J_1^2(z)/z^2$, commonly referred to as the *Airy pattern*, resembles the $\text{sinc}^2(z) = \sin^2(\pi z)/(\pi z)^2$, although the Bessel function does not have regularly spaced zeros like the sine function. The first zero (excluding the origin) of the J_1 function occurs at 3.831706.

Therefore, the first zero of the diffracted intensity occurs when:

$$\frac{kar_T}{z_0} = 3.831706 \Longrightarrow \frac{r_T}{z_0} = 1.21967\frac{\lambda}{2a} \simeq 1.22\frac{\lambda}{2a}. \tag{5.65}$$

Example 5.7.4:

Use the Fraunhofer approximation to calculate the Airy pattern for a circular aperture with a radius of $a = 10\,\mu$m at distances of $z_0 = 300\,\mu$m, and $2000\,\mu$m for an incident plane wave with a freespace wavelength of $\lambda_0 = 1\,\mu$m.

Solution: The Fraunhofer approximation (Airy pattern) can be determined from Eq. (5.63) and is depicted in Fig. 5.11(c) for a distance of $z_0 = 300\,\mu m$ from the aperture. The results obtained using the Rayleigh-Sommerfeld formulations, along with the Fresnel approximation, are also included for comparison. For the Fraunhofer approximation to be valid, $z_0 \gg (\pi/\lambda)a^2 = 314.16\,\mu m$ (corresponding to a Fresnel number $N_F = 1/3$), a condition that is clearly not satisfied for the selected z_0. The resulting pattern is radially symmetric, as expected due to the azimuthal symmetry of the aperture. For $z_0 = 2000\,\mu m \gg 314.16\,\mu m$ ($N_F = 1/20$), the Airy disk pattern forms and is shown in Fig. 5.13.

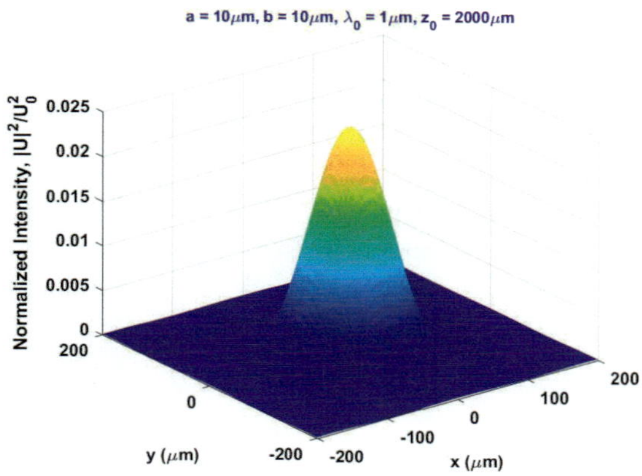

FIGURE 5.13: The far-field Airy disk pattern of the diffracted field normalized intensity from a circular aperture of radius $a = 10\,\mu m$ at a distance $z_0 = 2000\,\mu m$ (again $\lambda_0 = 1\,\mu m$).

Example 5.7.5:

Apply the boundary wave approach presented in Sec. 5.4.2.1 for a circular aperture with a radius of a. Perform this analysis to demonstrate the line integration method for the Fraunhofer approximation.

Solution: Equation (5.35) will be applied for the circular aperture of radius a. By applying this equation, the following steps can be derived, starting with the $S(k_x, k_y)$ factor and transitioning to a polar coordinate system, which is well-suited for the circular aperture geometry:

$$S(k_x, k_y) = -\frac{j}{p^2}\oint_C \Phi(\vec{\xi})\,\vec{p}\cdot\hat{\imath}_n d\ell,$$

$$\vec{\xi} = a\cos\phi'\,\hat{\imath}_x + a\sin\phi'\,\hat{\imath}_y,$$

$$\vec{p} = p\cos\psi\,\hat{\imath}_x + p\sin\psi\,\hat{\imath}_y,$$

$$\hat{\imath}_n = \cos\phi'\,\hat{\imath}_x + \sin\phi'\,\hat{\imath}_y,$$

$$S(k_x, k_y) = -\frac{j}{p^2}\int_0^{2\pi} \exp\left[+jpa\cos(\phi' - \psi)\right] p\cos(\phi' - \psi) a d\phi' =$$

$$= -\frac{j}{p} aj2\pi J_1(pa) = \frac{1}{p} a2\pi J_1(pa),$$

$$\text{where} \quad \int_0^{2\pi} e^{jz\cos\theta}\cos\theta d\theta = j2\pi J_1(z), \quad \text{was used.}$$

By setting $p = kr_T/z_0$ and substituting $S(k_x, k_y)$ into Eq. (5.35), it can be straightforwardly demonstrated that Eq. (5.63) is obtained. For demonstration purposes, let us now utilize Eq. (5.37) for the case of a circular aperture. In this case, the following substitutions are made: $k_y = p\sin\psi$, $k_x = p\cos\psi$, $x' = a\cos\phi'$, $y' = a\sin\phi'$, $dy' = a\cos\phi' d\phi'$. Substituting these into Eq. (5.37), the resulting expression becomes:

$$S(k_x, k_y) = \frac{-j}{k_x}\oint_C \exp(+jk_x x' + k_y y') dy' =$$

$$= \frac{-j}{p\cos\psi}\int_0^{2\pi} \exp\left[+jpa\cos(\phi' - \psi)\right] a\cos\phi' d\phi' =$$

$$= \frac{-j}{p\cos\psi} aj2\pi J_1(pa)\cos\psi = \frac{1}{p} a2\pi J_1(pa).$$

Again, by setting $p = kr_T/z_0$ and substituting $S(k_x, k_y)$ into Eq. (5.35), it is straightforward to demonstrate that Eq. (5.63) is obtained. In this derivation, the following integrals were also employed:

$$\int_0^{2\pi} \exp\left[+jz\cos(\phi - \psi)\right]\cos\phi \, d\phi = j2\pi J_1(z)\cos\psi,$$

$$\int_0^{2\pi} \exp\left[+jz\cos(\phi - \psi)\right]\sin\phi \, d\phi = j2\pi J_1(z)\sin\psi.$$

5.7.1.3 Plane-Wave-Spectrum Approach

At this stage, it is insightful to approach the diffraction of a plane wave by a circular aperture using the plane wave spectrum method introduced in Sec. 5.5. Let us begin with Eq. (5.47). This equation requires the computation of the two-dimensional Fourier transforms of the incident plane wave, $\tilde{U}_{inc}(k_x, k_y; 0)$, and the aperture amplitude transmittance function $\tilde{t}_a(k_x, k_y)$. Since the incident plane wave propagates normally to the circular aperture, it can be expressed as $U(x', y'; z') = U_0 e^{-jkz'}$. The circular aperture can be modeled by the $circ(r_T/a)$ function, which equals unity when $r_T \leq a$ and zero otherwise. Using these definitions, the following Fourier transform pairs can be established:

$$U(x', y'; z') = U_0 e^{-jkz'} \longrightarrow \tilde{U}(k_x, k_y; 0) = U_0(2\pi)^2\delta(k_x)\delta(k_y),$$

$$t_a(x', y') = circ\left(\frac{r_T}{a}\right) \longrightarrow \tilde{t}_a(k_x, k_y) = 2\pi\frac{aJ_1(k_t a)}{k_t}, \quad \text{where,}$$

$$r_T = (x'^2 + y'^2)^{1/2} \quad \text{and} \quad k_t = (k_x^2 + k_y^2)^{1/2}.$$

Here, J_1 is the Bessel function of the first kind of order one. Consequently, the plane wave spectrum $\tilde{U}_{RS_1}(k_x, k_y)$ can be expressed as:

$$\tilde{U}_{RS_1}(k_x, k_y) = \frac{1}{(2\pi)^2}\left[U_0(2\pi)^2\delta(k_x)\delta(k_y)\right] ** \left[2\pi\frac{aJ_1\left(\sqrt{k_x^2+k_y^2}\;a\right)}{\sqrt{k_x^2+k_y^2}}\right] =$$

$$= U_0 2\pi a\frac{J_1\left(\sqrt{k_x^2+k_y^2}\;a\right)}{\sqrt{k_x^2+k_y^2}}.$$

By inserting the above results into Eq. (5.47), the following steps can be derived:

$$U_{RS_1}(x_0, y_0, z_0) = \frac{1}{(2\pi)^2}\int_{-\infty}^{+\infty}\int_{-\infty}^{+\infty} U_0 2\pi a\frac{J_1\left(\sqrt{k_x^2+k_y^2}\;a\right)}{\sqrt{k_x^2+k_y^2}} e^{-j(k_x x_0 + k_y y_0 + k_z z_0)} dk_x dk_y =$$

$$= \frac{1}{(2\pi)^2}\int_0^{+\infty}\int_0^{2\pi} U_0 2\pi a\frac{J_1(k_t a)}{k_t} e^{-j[k_t r_T \cos(\psi-\phi)]} e^{-jk_z z_0} k_t dk_t d\psi =$$

$$= \frac{1}{(2\pi)^2}\int_0^{+\infty} U_0 2\pi a J_1(k_t a)\, e^{-jk_z z_0} dk_t\left[2\pi J_0(k_t r_T)\right] =$$

$$= U_0 a\int_0^{+\infty} J_1(k_t a)\, J_0(k_t r_T) e^{-jk_z z_0} dk_t, \tag{5.66}$$

where k_z is given by Eq. (5.40). Clearly, for $k_t > k$ the $e^{-jk_z z_0}$ becomes a decaying exponential. Therefore, for $z_0 \gg \lambda$ the evanescent components of the spectrum do not contribute to the diffracted field. In this case, $r_T = (x_0^2 + y_0^2)^{1/2}$.

For completeness, let us present the plane wave expansion method for the case of the Rayleigh-Sommerfeld 2 spectrum, which arises due to the corresponding Kirchhoff boundary condition (see Eq. (5.46)). The plane wave spectrum $\tilde{U}_{RS_2}(k_x, k_y)$ can be determined as:

$$\tilde{U}_{RS_2}(k_x, k_y) = \frac{1}{(2\pi)^2}\frac{1}{k_z}\left[k_z(2\pi)^2 U_0\delta(k_x)\delta(k_y)\right] ** \left[2\pi\frac{aJ_1\left(\sqrt{k_x^2+k_y^2}\;a\right)}{\sqrt{k_x^2+k_y^2}}\right] =$$

$$= \frac{k}{k_z}U_0 2\pi a\frac{J_1\left(\sqrt{k_x^2+k_y^2}\;a\right)}{\sqrt{k_x^2+k_y^2}}.$$

Inserting the above results into Eq. (5.46), the following steps can be written:

$$U_{RS_2}(x_0, y_0, z_0) = \frac{1}{(2\pi)^2}\int_{-\infty}^{+\infty}\int_{-\infty}^{+\infty} U_0 2\pi a\frac{k}{k_z}\frac{J_1\left(\sqrt{k_x^2+k_y^2}\;a\right)}{\sqrt{k_x^2+k_y^2}} e^{-j(k_x x_0 + k_y y_0 + k_z z_0)} dk_x dk_y =$$

$$= \frac{1}{(2\pi)^2}\int_0^{+\infty}\int_0^{2\pi} U_0 2\pi a\frac{k}{k_z}\frac{J_1(k_t a)}{k_t} e^{-j[k_t r_T \cos(\psi-\phi)]} e^{-jk_z z_0} k_t dk_t d\psi =$$

$$= \frac{1}{(2\pi)^2}\int_0^{+\infty} U_0 2\pi a\frac{k}{k_z} J_1(k_t a)\, e^{-jk_z z_0} dk_t\left[2\pi J_0(k_t r_T)\right] =$$

$$= U_0 k a\int_0^{+\infty} \frac{1}{k_z} J_1(k_t a)\, J_0(k_t r_T) e^{-jk_z z_0} dk_t. \tag{5.67}$$

Example 5.7.6:

Compare the intensity of the plane wave spectrum method with the RS-1, RS-2, and Kirchhoff formulations for a circular aperture of radius $a = 10\,\mu$m and a normally incident plane wave of freespace wavelength of $\lambda_0 = 1\,\mu$m, at $z_0 = 10\,\mu$m.

Solution: The plane wave spectrum results are obtained through numerical integration of Eqs. (5.66) and (5.67). The resulting intensities are shown in Fig. 5.14, alongside the intensities from the RS-1, RS-2, and Kirchhoff formulations in the radial direction. As expected, the plane-wave spectrum results are virtually identical to those of the RS-1 and RS-2 formulations.

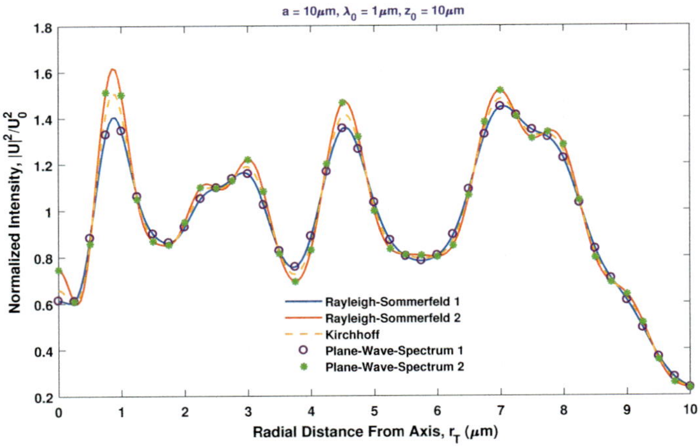

FIGURE 5.14: The diffracted field normalized intensity, $|U|^2/|U_0|^2$, along the radial direction for a plane wave normally incident on a circular aperture of radius $a = 10\,\mu m$ and $\lambda_0 = 1.0\,\mu m$. The results were obtained using the Rayleigh-Sommerfeld formulations 1 and 2, and the Helmholtz-Kirchhoff formulation with Kirchhoff's boundary conditions. The plane-wave spectrum results, as given by Eqs. (5.66) and (5.67), are also shown with symbols and coincide with the RS formulations, as proven in Sec. 5.5.

5.7.2 Circular Aperture with Incident Diverging Spherical Wave

5.7.2.1 On-Axis Points

In this section, the diffraction of an incident diverging spherical wave by a circular aperture is examined. The geometric configuration is illustrated in Fig. 5.7. In this scenario, the incident wave is a diverging spherical wave emanating from a point S, located at a distance $z = -s_0$ (to the left of the circular aperture). The distance s_0 is always considered positive and specifies the center of the diverging spherical wave. The diverging spherical wave can be represented as a scalar field: $U_{inc} = U_0 \exp(-jkR_s)/R_s$, where $R_s = [x'^2 + y'^2 + (z + s_0)^2]^{1/2} = [r_T'^2 + (z + s_0)^2]^{1/2}$. As a reminder, in Fig. 5.7, the coordinate system is denoted by $x'y'z$, with $z = 0$ corresponding to the center of the circular aperture. In this case, the Rayleigh-Sommerfeld integrals can be determined analytically, as described in various references in the literature [289, 291, 294]. Applying Eq. (5.18) to points P located on the positive z-axis (see Fig. 5.7), the following expression can be derived:

$$U_{RS_1}(P) = U_{RS_1}(z_0) = \iint_{S_a} \left[U_0 \frac{\exp(-jkR_s)}{R_s}\bigg|_{z=0} \left(jk + \frac{1}{R}\right) \frac{\exp(-jkR)}{2\pi R}\frac{z}{R}\right] dx'dy' =$$

$$= -\int_0^a \int_0^{2\pi} U_0 \frac{\exp(-jkR_s)}{R_s} \frac{d}{dR}\left(\frac{\exp(-jkR)}{2\pi R}\right) \frac{z_0}{R} r'_T dr'_T d\phi',$$

$$= -2\pi \int_0^a U_0 \frac{\exp(-jkR_s)}{R_s} \frac{d}{dR}\left(\frac{\exp(-jkR)}{2\pi R}\right) \frac{z_0}{R} r'_T dr'_T,$$

$$\text{with } R = \sqrt{z_0^2 + r_T'^2}, \qquad \text{and } R_s = \sqrt{s_0^2 + r_T'^2}.$$

However, through direct differentiation, it can be demonstrated [294] that:

$$\int \frac{\exp(-jkR_s)}{R_s} \frac{1}{R}\frac{d}{dR}\left(\frac{\exp(-jkR)}{R}\right) r'_T dr'_T = \frac{1}{R}\frac{\exp\left[-jk(R+R_s)\right]}{R+R_s}. \tag{5.68}$$

By utilizing the preceding remarkable result, U_{RS_1} can be analytically determined as follows:

$$U_{RS_1}(z_0) = -U_0 z_0 \left[\frac{1}{R}\frac{\exp\left[-jk(R+R_s)\right]}{R+R_s}\right]_0^a =$$

$$= U_0 \left[\frac{\exp\left[-jk(z_0+s_0)\right]}{z_0+s_0} - \frac{z_0}{\sqrt{a^2+z_0^2}} \frac{\exp\left[-jk\left(\sqrt{a^2+z_0^2}+\sqrt{a^2+s_0^2}\right)\right]}{\sqrt{a^2+z_0^2}+\sqrt{a^2+s_0^2}}\right]. \tag{5.69}$$

The intensity of the diffracted field, which is proportional to $|U_{RS_1}|^2$, is expressed as:

$$I_{RS_1}(z_0) = \frac{|U_0|^2}{4Z}\left[\frac{1}{S_0^2} + \frac{F_1^2}{S_1^2} - \frac{2F_1}{S_0 S_1}\cos\left[k(S_1 - S_0)\right]\right] \qquad \text{where,} \tag{5.70}$$

$$S_0 = z_0 + s_0, \qquad S_1 = \sqrt{a^2+z_0^2} + \sqrt{a^2+s_0^2}, \qquad F_1 = \frac{z_0}{\sqrt{z_0^2+a^2}}.$$

It is customary to plot the relative intensity, defined as the ratio of the intensity at the diffracted point to the intensity of the incident spherical wave at the origin (the center of the circular aperture). The intensity of the incident spherical wave at the origin is given by $I_{sw} = |U_0|^2/4Zs_0^2$ (where Z represents the wave impedance of the medium). Accordingly, the relative intensity can be expressed as:

$$\frac{I_{RS_1}(z_0)}{I_{sw}} = s_0^2\left[\frac{1}{S_0^2} + \frac{F_1^2}{S_1^2} - \frac{2F_1}{S_0 S_1}\cos\left[k(S_1 - S_0)\right]\right]. \tag{5.71}$$

The case of the Rayleigh-Sommerfeld 2 formulation can also be determined analytically, as follows:

$$U_{RS_2}(z_0) = \iint_{S_a}\left[\frac{\exp(-jkR)}{2\pi R}\frac{\partial U_{inc}}{\partial n}\right]_{z'=0} dx'dy' =$$

$$= \int_0^a \int_0^{2\pi} \frac{\exp(-jkR)}{2\pi R}\left(-\frac{\partial U_{inc}}{\partial z}\right) r'_T dr'_T d\phi' =$$

$$= -\int_0^a \int_0^{2\pi} \frac{\exp(-jkR)}{2\pi R} U_0 \left[\frac{d}{dR_s}\left(\frac{\exp(-jkR_s)}{R_s}\right)\frac{s_0}{R_s}\right] r'_T dr'_T d\phi' =$$

$$= -U_0 s_0 \int_0^a \frac{\exp(-jkR)}{R} \frac{1}{R_s}\frac{d}{dR_s}\left(\frac{\exp(-jkR_s)}{R_s}\right) r'_T dr'_T =$$

$$= U_0 \left[\frac{\exp\left[-jk(z_0+s_0)\right]}{z_0+s_0} - \frac{s_0}{\sqrt{a^2+s_0^2}} \frac{\exp\left[-jk\left(\sqrt{a^2+z_0^2}+\sqrt{a^2+s_0^2}\right)\right]}{\sqrt{a^2+z_0^2}+\sqrt{a^2+s_0^2}}\right]. \tag{5.72}$$

Then, the corresponding intensity of the diffracted field is proportional to $|U_{RS_2}|^2$ and is given by:

$$I_{RS_2}(z_0) = \frac{|U_0|^2}{4Z}\left[\frac{1}{S_0^2} + \frac{F_2^2}{S_1^2} - \frac{2F_2}{S_0 S_1}\cos\left[k(S_1 - S_0)\right]\right] \quad \text{or,} \tag{5.73}$$

$$\frac{I_{RS_2}(z_0)}{I_{sw}} = s_0^2\left[\frac{1}{S_0^2} + \frac{F_2^2}{S_1^2} - \frac{2F_2}{S_0 S_1}\cos\left[k(S_1 - S_0)\right]\right], \quad \text{where,} \tag{5.74}$$

$$F_2 = \frac{s_0}{\sqrt{s_0^2 + a^2}}.$$

The Helmholtz-Kirchhoff result can be readily derived once more from Eq. (5.20) and is expressed as:

$$U_K(z_0) = U_0 \left[\frac{\exp\left[-jk(z_0+s_0)\right]}{z_0+s_0} - F\frac{\exp\left[-jk\left(\sqrt{a^2+z_0^2}+\sqrt{a^2+s_0^2}\right)\right]}{\sqrt{a^2+z_0^2}+\sqrt{a^2+s_0^2}}\right], \tag{5.75}$$

$$\text{where} \quad F = \frac{1}{2}\left[\frac{z_0}{\sqrt{a^2+z_0^2}} + \frac{s_0}{\sqrt{a^2+s_0^2}}\right].$$

The relative intensity of the diffracted field, which is proportional to $|U_K|^2$, is expressed as:

$$\frac{I_K(z_0)}{I_{sw}} = s_0^2\left[\frac{1}{S_0^2} + \frac{F^2}{S_1^2} - \frac{2F}{S_0 S_1}\cos\left[k(S_1 - S_0)\right]\right], \tag{5.76}$$

where S_0, S_1, and F were defined previously.

Example 5.7.7:

Apply the previous equations to study the on-axis diffraction of a diverging spherical wave by a circular aperture with a radius $a = 10\,\mu m$. The incident diverging spherical wave, with a freespace wavelength $\lambda_0 = 1\,\mu m$, emanates from $z = -s_0$, where s_0 takes the values $50\,\mu$m or $100\,\mu$m. Calculate and plot the relative intensities I/I_{sw}, as functions of the distance along the z-axis.

Solution: The resulting normalized intensities, I/I_{sw}, plotted on a logarithmic scale, are shown in Figs. 5.15(a) and 5.15(b), respectively. Note that the intensities are presented as functions of $z = z_0$. It is worth mentioning that the results for $s_0 = 50\,\mu m$ are consistent with those in Heurtley's work [289], while the results for $s_0 = 100\,\mu m$ do not align in the same way. However, in Heurtley's paper, s_0 is listed as $100\,\mu m$ instead of $50\,\mu m$, which could be a typographical error.

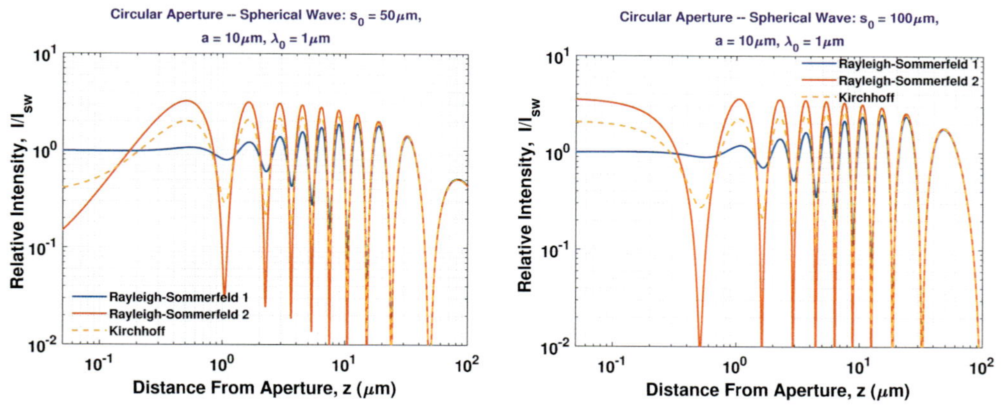

(a) (b)

FIGURE 5.15: The normalized diffracted intensity, I/I_{sw}, along the z-axis for a spherical wave incident on a circular aperture of radius $a = 10\,\mu m$. The freespace wavelength is $\lambda_0 = 1.0\,\mu m$. The results were obtained using the Rayleigh-Sommerfeld formulations 1 and 2 and the Helmholtz-Kirchhoff with the Kirchhoff's boundary conditions. (a) The spherical wave is emanating from $z = -50\,\mu m$. (b) The spherical wave is emanating from $z = -100\,\mu m$.

From the expressions for the relative intensities, the positions of the extrema can be approximated. It is clear that in all the relative intensity expressions, the oscillating term is represented by $\cos[k(S_1 - S_0)]$. Therefore, the maxima and minima can be approximated by $k(S_1 - S_0) = (2m+1)\pi$ and $k(S_1 - S_0) = 2m\pi$, respectively. These expressions are approximate, as the exact positions of the maxima and minima depend on the coefficients of the cosine term as well as the non-cosine terms, all of which vary with z_0. The approximate expressions for the maxima, z_m^+, and the minima, z_m^-, are:

$$z_m^+ = \frac{a^2 - A^2}{2A}, \qquad \text{where,} \quad A = \left(m + \frac{1}{2}\right)\lambda + s_0 - \sqrt{s_0^2 + a^2},$$

$$m > \left(\sqrt{s_0^2 + a^2} - s_0\right)\frac{1}{\lambda} - \frac{1}{2}, \qquad \text{and,}$$

$$\left(-a - s_0 + \sqrt{s_0^2 + a^2}\right)\frac{1}{\lambda} - \frac{1}{2} < m < \left(a - s_0 + \sqrt{s_0^2 + a^2}\right)\frac{1}{\lambda} - \frac{1}{2},$$

$$z_m^- = \frac{a^2 - B^2}{2B}, \qquad \text{where,} \quad B = m\lambda + s_0 - \sqrt{s_0^2 + a^2},$$

$$m > \left(\sqrt{s_0^2 + a^2} - s_0\right)\frac{1}{\lambda}, \qquad \text{and,}$$

$$\left(-a - s_0 + \sqrt{s_0^2 + a^2}\right)\frac{1}{\lambda} < m < \left(a - s_0 + \sqrt{s_0^2 + a^2}\right)\frac{1}{\lambda}.$$

By applying the above expressions, the maxima and minima of the example case shown in Fig. 5.15(a) can be verified. From the inequalities for m, it is straightforward to determine that $1 \leq m \leq 10$ for both maxima and minima. However, the minimum for $m = 1$ occurs at $z \simeq 5100\,\mu m$ which is not shown in the diagram (it can be verified). It is worth noting that in Ref. [289], it is claimed that the minima of the RS-2 formulation are zeros. However, upon examining Eq. (5.74), it is evident that for a zero, the following condition must be satisfied:

$$\cos\left[k\left(S_1 - S_0\right)\right] = \frac{\dfrac{1}{S_0^2} + \dfrac{F_2^2}{S_1^2}}{\dfrac{2F_2}{S_0 S_1}} \geq 1 \qquad \text{implying,}$$

$$k(S_1 - S_0) = 2\pi m \qquad \text{and} \quad \frac{1}{S_0} = \frac{F_2}{S_1}.$$

However, the last equality cannot be satisfied for any positive z_0, as it would result in $z_0 + s_0 = 0$, which is impossible for a positive z_0. Therefore, the minima cannot be zeros. Consequently, for all formulations, the minima are not generally zeros.

The Fresnel and Fraunhofer approximations in this case can also be determined in a manner similar to that in Sec. 5.7.1.1. Specifically, for the Fresnel approximation, the following expressions can be written:

$$U(z_0) = jk\frac{\exp(-jkz_0)}{2\pi z_0}\int_0^a\int_0^{2\pi}\exp\left(-jk\frac{r_T'^2}{2z_0}\right)\left[U_0\frac{\exp\left(-jk\sqrt{r_T'^2+s_0^2}\right)}{\sqrt{r_T'^2+s_0^2}}\right]r_T'\,dr_T'\,d\phi' \simeq$$

$$= jkU_0\frac{\exp\left[-jk\left(z_0+s_0\right)\right]}{z_0 s_0}\int_0^a \exp\left[-jk\frac{r_T'^2}{2}\left(\frac{1}{z_0}+\frac{1}{s_0}\right)\right]r_T'\,dr_T' =$$

$$= U_0\frac{\exp\left[-jk\left(z_0+s_0\right)\right]}{z_0+s_0}\left\{1-\exp\left[-jk\frac{a^2}{2}\left(\frac{1}{z_0}+\frac{1}{s_0}\right)\right]\right\}, \tag{5.77}$$

$$\text{with} \quad z_0 \gg \left(\frac{\pi}{4\lambda_0}a^4\right)^{1/3}, \qquad \text{and} \quad s_0 \gg a.$$

The corresponding relative intensity for points along the z-axis, under the Fresnel approximation, is given by:

$$\frac{I(z_0)}{I_{sw}} = s_0^2\frac{2}{(z_0+s_0)^2}\left\{1-\cos\left[k\frac{a^2}{2}\left(\frac{1}{z_0}+\frac{1}{s_0}\right)\right]\right\}. \tag{5.78}$$

From the Fresnel approximation, it is straightforward to predict the positions of the minima and maxima based on the cosine argument term. It is important to note that these will be good approximations only if the conditions for the Fresnel formulation are satisfied. Finally, the Fraunhofer approximation can also be determined as follows:

$$U_{ff_1} = jk\frac{\exp(-jkz_0)}{2\pi z_0}\int_0^a\int_0^{2\pi}\left[U_0\frac{\exp\left(-jk\sqrt{r_T'^2+s_0^2}\right)}{\sqrt{r_T'^2+s_0^2}}\right]r_T'\,dr_T'\,d\phi' \simeq$$

$$= jU_0 k\frac{\exp\left[-jk\left(z_0+s_0\right)\right]}{z_0 s_0}\frac{a^2}{2}, \qquad \text{for} \quad z_0, s_0 \gg \frac{\pi}{\lambda}a^2. \tag{5.79}$$

In the previous expression, an approximation of the incident spherical wave was used, assuming $s_0 \gg r_T'$. However, if this approximation is not applied, the resulting Fraunhofer field, U_{ff_2} is given by:

$$U_{ff_2} = jk\frac{\exp(-jkz_0)}{2\pi z_0}\int_0^a\int_0^{2\pi}\left[U_0\frac{\exp(-jkR_s)}{R_s}\right]r'_T dr'_T d\phi' =$$

$$= jk\frac{\exp(-jkz_0)}{z_0}\int_0^a U_0\frac{\exp(-jkR_s)}{R_s}R_s dR_s =$$

$$= jk\frac{\exp(-jkz_0)}{z_0}U_0\left[\frac{\exp(-jkR_s)}{-jk}\right]_{r'_T=0}^{r'_T=a} =$$

$$= U_0\frac{\exp(-jkz_0)}{z_0}\left[\exp(-jks_0) - \exp\left(-jk\sqrt{a^2+s_0^2}\right)\right]. \tag{5.80}$$

The corresponding Fraunhofer relative intensities are expressed as:

$$\frac{I_{ff_1}(z_0)}{I_{sw}} = \frac{k^2}{z_0^2}\frac{a^4}{4}, \tag{5.81}$$

$$\frac{I_{ff_2}(z_0)}{I_{sw}} = 2\frac{s_0^2}{z_0^2}\left\{1-\cos\left[k\left(\sqrt{s_0^2+a^2}-s_0\right)\right]\right\}. \tag{5.82}$$

It is straightforward to show that when $s_0 \gg a$ and $s_0 \gg \pi a^2/\lambda$, the two previous expressions coincide, i.e., $I_{ff_2} \to I_{ff_1}$. In general, the I_{ff_1} expression overestimates the relative intensity. Example plots comparing the RS-1 (which in this range coincides with RS-2 and H-K), Fresnel, and Fraunhofer formulations (1 and 2) are shown in Figs. 5.16(a) and 5.16(b). When $s_0 = 100\,\mu$m, the inequality $s_0 \gg \pi a^2/\lambda \simeq 314$ is not satisfied, leading to a significant difference between I_{ff_1} and I_{ff_2} (see Fig. 5.16(a)). On the other hand, when $s_0 = 500\,\mu$m, the inequality is weakly satisfied, and the two Fraunhofer intensities are nearly identical (see Fig. 5.16(b)).

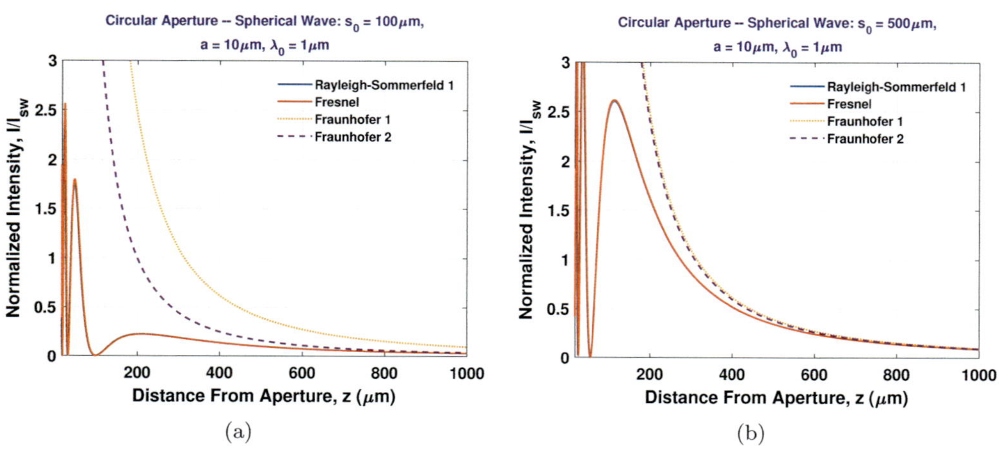

FIGURE 5.16: The diffracted field relative intensity, I/I_{sw}, along the axis z for a diverging spherical wave incident on a circular aperture of radius $a = 10\,\mu m$. The freespace wavelength is $\lambda_0 = 1.0\,\mu m$. The results were obtained using the Rayleigh-Sommerfeld formulation 1, the Fresnel approximation, and the Fraunhofer (both 1 and 2) approximations. It is mentioned that for the range of z shown, the Rayleigh-Sommerfeld 1 and 2 formulations as well as the Helmholtz-Kirchhoff formulation practically coincide. (a) The spherical wave is emanating from $z = -100\,\mu m = -s_0$. (b) The spherical wave is emanating from $z = -500\,\mu m = -s_0$.

5.7.2.2 Off-Axis Points

In this section, the diffraction of a diverging spherical wave by a circular aperture is further explored, but now for points in any off-axis direction. The geometric configuration is shown in Fig. 5.10. The point $P(x_0, y_0, z_0)$ represents any arbitrary point (for $z_0 \geq 0$) of interest. In this case, the diffraction integrals for all formulations cannot be determined analytically as was done in the previous on-axis cases. Instead, either approximations are required or numerical evaluation of the integrals is necessary [291, 292]. For example, the diffraction integrals for the RS-1, RS-2 formulations, and the Fresnel approximation, for the arbitrary point P are given by:

$$U_{RS_1}(x_0, y_0, z_0) = \int_0^a \int_0^{2\pi} \left(jk + \frac{1}{R}\right) \frac{\exp(-jkR)}{2\pi R} \frac{z_0}{R} \left[U_0 \frac{\exp(-jkR_s)}{R_s}\right] r'_T dr'_T d\phi', \quad (5.83)$$

$$U^a_{RS_1}(x_0, y_0, z_0) \simeq jkz_0 \int_0^a \frac{\exp(-jkR_0)}{R_0^2} J_0\left(\frac{kr_T r'_T}{R_0}\right) \left[U_0 \frac{\exp(-jkR_s)}{R_s}\right] r'_T dr'_T, \quad (5.84)$$

$$U_{RS_2}(x_0, y_0, z_0) = \int_0^a \int_0^{2\pi} \frac{\exp(-jkR)}{2\pi R} \left[U_0 \left(jk + \frac{1}{R_s}\right) \frac{\exp(-jkR_s)}{R_s} \frac{s_0}{R_s}\right] r'_T dr'_T d\phi', (5.85)$$

$$U^a_{RS_2}(x_0, y_0, z_0) \simeq \int_0^a \frac{\exp(-jkR)}{2\pi R} J_0\left(\frac{kr_T r'_T}{R_0}\right) \left[U_0 \left(jk + \frac{1}{R_s}\right) \frac{\exp(-jkR_s)}{R_s} \frac{s_0}{R_s}\right] r'_T dr'_T, \quad (5.86)$$

$$U(x_0, y_0, z_0) = jkU_0 \frac{\exp(-jkz_0)}{z_0} \int_0^a \exp\left[-jk \frac{r_T^2 + r'^2_T}{2z_0}\right] J_0\left(\frac{kr_T r'_T}{R_0}\right) \frac{\exp(-jkR_s)}{R_s} r'_T dr'_T, \quad (5.87)$$

where,

$$R = \left[r_T^2 + r'^2_T - 2r_T r'_T \cos(\phi - \phi') + z_0^2\right]^{1/2}, \qquad R_s = \left(r'^2_T + s_0^2\right)^{1/2}, \quad \text{and,}$$

$$r_T = (x_0^2 + y_0^2)^{1/2}, \quad r'_T = (x'^2 + y'^2)^{1/2}, \quad \phi = \tan^{-1}\left(\frac{y_0}{x_0}\right), \quad \phi' = \tan^{-1}\left(\frac{y'}{x'}\right).$$

All integrals can be determined through two-dimensional numerical integration (for the exact RS-1 and RS-2 formulations) or via one-dimensional integration (for the approximate RS-1 and RS-2 formulations, respectively, and the Fresnel approximation). The H-K formulation integral can be computed from RS-1 and RS-2 using Eq. (5.20). It is important to note that the incident spherical wave or its normal derivative is required at $z = z' = 0$ in the previous integrals.

Example 5.7.8:

Evaluate the normalized intensities, I_{RS_1}, I_{RS_2} (and their 1D-integral approximations), as well as the Kirchhoff, Fresnel, and Fraunhofer counterparts, of a diverging spherical wave diffracted by a circular aperture with a radius of $a = 10\,\mu$m. The incident diverging spherical wave emanates from $z = -s_0 = -50\,\mu m$, and has a freespace wavelength $\lambda_0 = 1\,\mu m$. Calculate the radial distribution at $z_0 = 10$ and $50\,\mu$m.

Solution: The normalized I_{RS_1}, I_{RS_2}, $I^a_{RS_1}$, $I^a_{RS_2}$ are shown in Figs. 5.17(a) and 5.17(b) for $z_0 = 10\,\mu$m and $z_0 = 50\,\mu m$, respectively, as functions of the radial direction. The approximate expressions show significant variation for smaller z_0, while for larger z_0, the differences are much smaller. The normalized I_{RS_1}, I_{RS_2}, along with Kirchhoff and Fresnel formulations, are shown in Figs. 5.18(a) (for $z_0 = 10\,\mu$m) and 5.18(b) (for $z_0 = 50\,\mu$m) as

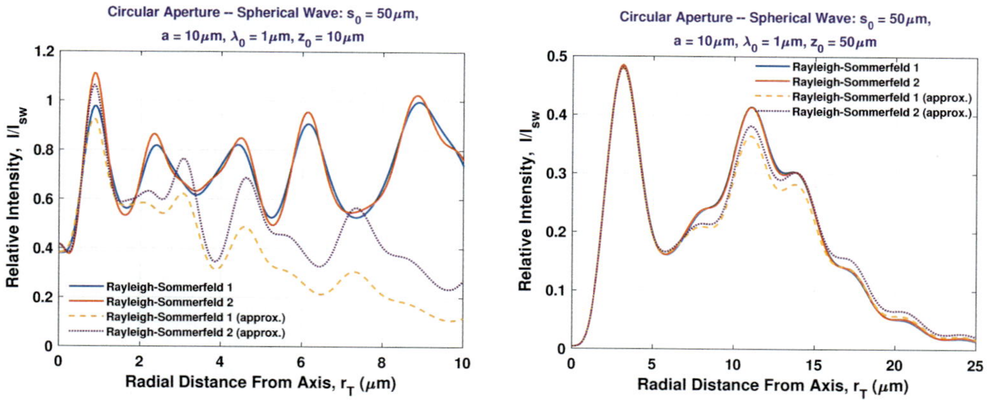

(a) (b)

FIGURE 5.17: The diffracted field relative intensity, I/I_{sw}, along the radial direction for a spherical wave emanating from $z = -s_0$, ($s_0 = 50\,\mu m$), incident on a circular aperture of radius $a = 10\,\mu m$, for $\lambda_0 = 1.0\,\mu m$. The results were obtained using the Rayleigh-Sommerfeld formulations 1 and 2 and the approximate one-dimensional integral formulations $U^a_{RS_1}$ and $U^a_{RS_2}$, respectively. (a) $z_0 = 10\,\mu m$. (b) $z_0 = 50\,\mu m$.

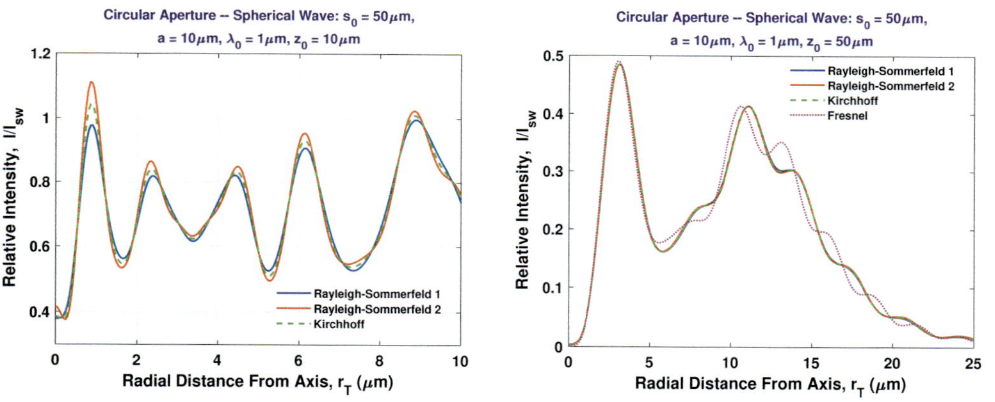

(a) (b)

FIGURE 5.18: The diffracted field relative intensity along the radial direction for a spherical wave emanating from $z = -s_0$, ($s_0 = 50\,\mu m$) incident on a circular aperture of radius $a = 10\,\mu m$ for $\lambda_0 = 1.0\,\mu m$. The results obtained using the Rayleigh-Sommerfeld formulations 1 and 2 and the Helmholtz-Kirchhoff with the Kirchhoff's boundary conditions. Fresnel results are also shown for the larger distance from the aperture. (a) $z_0 = 10\,\mu m$. (b) $z_0 = 50\,\mu m$.

functions of the radial distance r_T. The Fraunhofer approximation, $U_{ff} = U(x_0, y_0, z_0)$, cannot be determined analytically as was done in the plane wave case, where it resulted in the well-known Airy disk pattern. However, the two-dimensional integration can be reduced to a one-dimensional form. Specifically,

$$U_{ff} = jk\frac{\exp(-jkz_0)}{z_0}\exp\left(-jk\frac{r_T^2}{2z_0}\right)U_0\int_0^a J_0\left(\frac{kr_Tr'_T}{z_0}\right)\left[U_0\frac{\exp(-jkR_s)}{R_s}\right]r'_Tdr'_T. \quad (5.88)$$

The numerical calculation of the Fraunhofer approximation is shown in Fig. 5.19(a) for a circular aperture of radius $a = 10\,\mu m$. The distance from the aperture is $z_0 = 300\,\mu m$.

The Rayleigh-Sommerfeld results, along with the Fresnel approximation, are also shown for comparison. For the Fraunhofer approximation to be valid, $z_0 \gg (\pi/\lambda)a^2 = 314.16\,\mu m$ ($N_F = 1/3$), which for the selected z_0 is obviously not satisfied. It is interesting to note that the diffracted pattern in the radial direction for the diverging spherical wave differs substantially from the case of the incident plane wave. Furthermore, the difference between the more exact approaches (RS and Fresnel) and the Fraunhofer approximation is significant, even near the axis (not shown in Fig.5.19(a)). To demonstrate this, another example is shown in Fig. 5.19(b) for $z_0 = 1000\,\mu m$, where the Fraunhofer approximation is much closer to the RS and Fresnel formulations. Again, for the diverging spherical wave with $s_0 = 50\,\mu m$, the comparison with the Airy disk pattern is dramatically different. Of course, it is straightforward to show that when s_0 becomes larger (for example, $s_0 = 1000\,\mu m$) the Fraunhofer pattern converges to the Airy disk pattern, since for all practical purposes, the incident spherical wave resembles a plane wave.

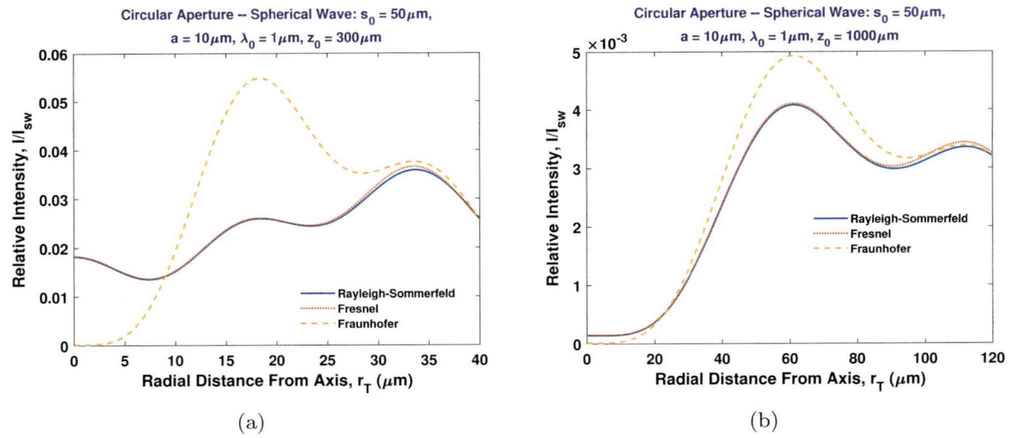

FIGURE 5.19: The diffracted field relative intensity, I/I_{sw}, along the radial direction for a spherical wave emanating from $z = -s_0$, ($s_0 = 50\,\mu m$) incident on a circular aperture of radius $a = 10\,\mu m$ for$\lambda_0 = 1.0\,\mu m$. The results were obtained using the Rayleigh-Sommerfeld formulation 1 (RS-1, RS-2, and H-K coincide in these cases). Fresnel and Fraunhofer results are also shown. (a) $z_0 = 300\,\mu m$. (b) $z_0 = 1000\,\mu m$.

5.7.2.3 Plane-Wave-Spectrum Approach

At this point, it is interesting to approach the diffraction of a spherical wave by a circular aperture using the plane wave spectrum method, as presented in Sec. 5.5. We begin by using Eq. (5.47). In this equation, it is necessary to determine the two-dimensional Fourier transforms of the incident plane wave, $\tilde{U}_{inc}(k_x, k_y; 0)$, and of the aperture amplitude transmittance function, $\tilde{t}_a(k_x, k_y)$. Since the incident spherical wave is normal to the circular aperture (its center lies on the z-axis), it has the form $U(x', y', z') = U_0 e^{-jkR_s}/R_s$. Then, the following Fourier transform pairs can be written (refer to Sec. 5.5 for the Fourier transform of the spherical wave):

$$U(x', y'; z') = U_0 \frac{e^{-jkR_s}}{R_s} \longrightarrow \tilde{U}(k_x, k_y; 0) = -jU_0 2\pi \frac{e^{-jk_z s_0}}{k_z},$$

$$t_a(x', y') = circ\left(\frac{r_T}{a}\right) \longrightarrow \tilde{t}_a(k_x, k_y) = 2\pi \frac{aJ_1(k_t a)}{k_t}, \quad \text{where,}$$

$$r_T = (x'^2 + y'^2)^{1/2} \quad \text{and} \quad k_t = (k_x^2 + k_y^2)^{1/2},$$

$$k_z^2 = k^2 - k_x^2 - k_y^2 \quad \text{and} \quad R_0 = \left[x'^2 + y'^2 + (z' + s_0)^2\right]^{1/2}. \tag{5.89}$$

Inserting the above Fourier transforms into Eq. (5.47) the following integral expression can be written:

$$U_{RS_1}(x_0, y_0, z_0) = \frac{1}{(2\pi)^2} \int_{-\infty}^{+\infty}\int_{-\infty}^{+\infty} \left[\frac{1}{(2\pi)^2}\left(-jU_0 2\pi \frac{e^{-jk_z s_0}}{k_z}\right) ** \left(2\pi \frac{aJ_1(k_t a)}{k_t}\right)\right] e^{-j(k_x x_0 + k_y y_0 + k_z z_0)} dk_x dk_y.$$

The two-dimensional convolution (with respect to k_x and k_y) inside the last integral expression cannot be determined analytically. Therefore, it is more appropriate to perform the propagation using the Discrete Fourier Transform (DFT) or its faster counterpart, the Fast Fourier Transform (FFT) [119,120]. The diffracted field can then be determined using the following procedure:

$$U_{RS_1}(x_m, y_n; z_0) = FFT^{-1}\left\{\tilde{U}(k_{xp}, k_{yq})\exp(-jk_{zpq}z_0)\right\}, \qquad \text{where,}$$

$$\tilde{U}(k_{xp}, k_{yq}) = FFT\left\{U_{inc}(x', y'; z = z' = 0)t_a(x', y')\right\}$$

$$k_{zpq} = \begin{cases} \left(k^2 - k_{xp}^2 - k_{yq}^2\right)^{1/2}, & \text{when } k_{xp}^2 + k_{yq}^2 \leq k^2, \\ -j\left(k_{xp}^2 + k_{yq}^2 - k^2\right)^{1/2}, & \text{when } k_{xp}^2 + k_{yq}^2 > k^2, \end{cases}$$

$$\text{with } k_{xp} = p\frac{2\pi}{L_x}, \quad k_{yq} = q\frac{2\pi}{L_y}, \quad x_m = m\frac{L_x}{M}, \quad y_n = n\frac{L_y}{N}.$$

Here, $m, p = -M/2, \cdots (M/2) - 1$ and $n, q = -N/2, \cdots (N/2) - 1$, with x_m, y_n representing the discretization points in the spatial domain, and k_{xp}, k_{yq} representing the spatial radial frequencies in the spatial frequency domain. The variables L_x, L_y correspond to the window lengths in the x and y directions, respectively. For further details, the reader is referred to Appendix 5D.

Example 5.7.9:

In this example, apply the plane-wave-spectrum technique to study the diffraction of a diverging spherical wave from a circular aperture of radius $a = 10\,\mu$m. The incident diverging spherical wave has a freespace wavelength of $\lambda_0 = 1\,\mu m$ and is emanating from $z = -50\,\mu m = -s_0$ (similarly to previous examples). Calculate the normalized radial intensities using the plane-wave spectrum, the Rayleigh-Sommerfeld (RS) formulation, and the Fresnel approximation for $z_0 = 300\,\mu$m.

Solution: The results from the plane-wave spectrum approach using the FFT show reasonable agreement with the Rayleigh-Sommerfeld (RS-1) and Fresnel formulations for the diffraction of a diverging spherical wave through a circular aperture. As seen in Fig. 5.20, the normalized radial intensity for the plane-wave spectrum (PWS) method aligns closely with the RS-1 and Fresnel approximations when using $M = N = 512 = 2^9$ discretization points and windowing lengths $L_x = L_y = 500\,\mu m$. However, increasing the number of discretization points to $M = N = 2^{12} = 4096$ and expanding the windowing lengths to $L_x = L_y = 1200\,\mu$m improves the agreement further, making the PWS FFT approach even closer to the RS-1 results. The small oscillations that remain in the intensity could be minimized by optimizing the FFT parameters. This demonstrates that the plane-wave spectrum method can provide a highly accurate approximation for diffraction problems, especially with sufficient resolution in the numerical discretization.

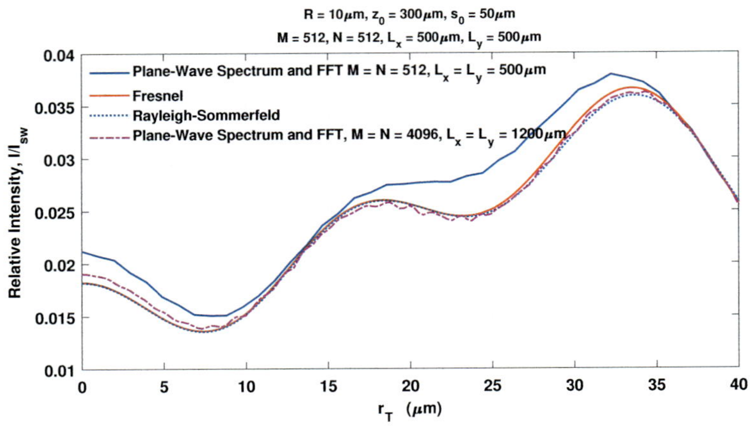

FIGURE 5.20: The diffracted field relative intensity along the radial direction for a spherical wave incident on a circular aperture of radius $a = 10\,\mu m$ and $\lambda_0 = 1.0\,\mu m$. The spherical wave is emanating from $z = -s_0 = -50\,\mu m$. The diffracted field is shown in the radial direction for $z_0 = 300\,\mu m$. The results obtained using the Rayleigh-Sommerfeld formulation (for this z_0 the formulations 1 and 2 coincide), the Fresnel approximation, and the plane-wave-spectrum propagation method.

5.7.3 Circular Obstacle with Incident Plane Wave

Consider a plane wave $U = U_0 \exp(-jkz)$ incident from the left onto a planar circular obstacle in a transparent screen (see Fig. 5.21). In this situation, the Rayleigh-Sommerfeld integrals can be determined analytically, for on-axis points, as described in various references in the literature [275, 289, 291, 294]. By applying Eqs. (5.18), (5.19), and (5.25) for a point $P(x_0, y_0, z_0)$ with $z_0 > 0$, the following expressions (on- or off-axis points) can be written for the RS-1 and RS-2 formulations, as well as for the Fresnel approximation:

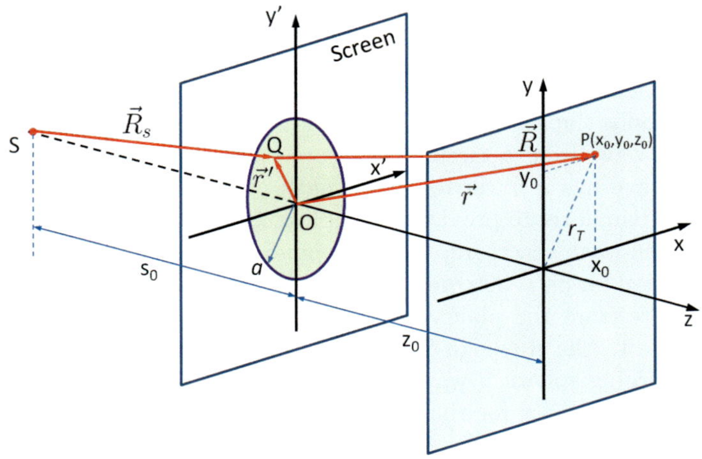

FIGURE 5.21: A circular obstacle of radius a on an transparent screen. The diffracted field at an arbitrary point $P(x_0, y_0, z_0)$ is sought ($z_0 > 0$). The point S specifies the origin of a diverging spherical wave that could be incident on the aperture. In the case of an incident plane wave the point S moves to $-\infty$ ($s_0 \to +\infty$).

$$U_{RS_1}(x_0, y_0, z_0) = \int_a^{+\infty}\int_0^{2\pi} U_0\left(jk + \frac{1}{R}\right)\frac{\exp(-jkR)}{2\pi R}\frac{z_0}{R}\, r'_T\, dr'_T\, d\phi',$$

$$U_{RS_2}(x_0, y_0, z_0) = \int_a^{+\infty}\int_0^{2\pi} \frac{\exp(-jkR)}{2\pi R}(jkU_0)\, r'_T\, dr'_T\, d\phi',$$

$$U(x_0, y_0, z_0) = jk\frac{\exp(-jkz_0)}{z_0}\int_a^{+\infty}\exp\left(-jk\frac{r_T^2 + r_T'^2}{2z_0}\right)U_0 J_0\left(\frac{kr_T r'_T}{z_0}\right) r'_T\, dr'_T, \quad (5.90)$$

where,

$$R = \left[r_T^2 + r_T'^2 - 2r_T r'_T\cos(\phi - \phi') + z_0^2\right]^{1/2}, \qquad \text{and,}$$

$$r_T = (x_0^2 + y_0^2)^{1/2}, \quad r'_T = (x'^2 + y'^2)^{1/2}, \quad \phi = \tan^{-1}\left(\frac{y_0}{x_0}\right), \quad \phi' = \tan^{-1}\left(\frac{y'}{x'}\right),$$

The definite integrals above do not need to be evaluated if Babinet's principle [Eq. (5.48)] is applied. Therefore, by combining the results from Sec. 5.7.1.1 with Babinet's principle, the following analytical expressions for the on-axis diffracted field can be derived:

$$U_{RS_1}(z_0) = U_0\frac{z_0}{\sqrt{a^2 + z_0^2}}\exp\left(-jk\sqrt{a^2 + z_0^2}\right), \quad (5.91)$$

$$I_{RS_1}(z_0) = \frac{|U_0|^2}{4Z}\frac{z_0^2}{a^2 + z_0^2}, \quad (5.92)$$

$$U_{RS_2}(z_0) = U_0\exp\left(-jk\sqrt{a^2 + z_0^2}\right), \quad (5.93)$$

$$I_{RS_2}(z_0) = \frac{|U_0|^2}{4Z}, \quad (5.94)$$

$$U_K(z_0) = \frac{U_0}{2}\exp\left(-jk\sqrt{a^2 + z_0^2}\right)\left(1 + \frac{z_0}{\sqrt{a^2 + z_0^2}}\right), \quad (5.95)$$

$$I_K(z_0) = \frac{|U_0|^2}{16Z}\left(1 + \frac{z_0}{\sqrt{a^2 + z_0^2}}\right)^2, \quad (5.96)$$

while for the Fresnel approximation,

$$U(z_0) = U_0\exp(-jkz_0)\exp\left(-jk\frac{a^2}{2z_0}\right), \quad (5.97)$$

$$I(z_0) = \frac{|U_0|^2}{4Z}. \quad (5.98)$$

Example 5.7.10:

Calculate the normalized intensity along the z-axis of a plane wave normally incident on a circular obstacle with a radius of $a = 10\,\mu$m at a freespace wavelength of $\lambda_0 = 0.5145\,\mu m$. Repeat the procedure to determine the normalized intensity along the radial direction for $z_0 = 50\,\mu$m.

Solution: By applying the previous equations, the normalized intensity [using RS-1, RS-2, and Kirchhoff (equivalently Huygens-Kirchhoff or H-K] formulations) along the z-axis is shown in Fig. 5.22. Notably, along the z-axis there are no oscillations in the normalized intensity, in contrast to the case of a circular aperture. Additionally, there is no complete darkness along the axis behind the circular obstacle (except for at $z_0 = 0^+$, where the intensity is zero) [295]. The presence of light in the traditional shadow region behind a circular obstacle is referred to as the "spot of Arago." In 1818, Fresnel presented his work on the wave theory of light to the French Academy. Poisson, a proponent of the corpuscular theory of light, predicted, based on Fresnel's theory, that a bright spot should appear at the center of the shadow (along the z-axis) on an illuminated disk (circular obstacle). Poisson found this prediction absurd and intended to discredit Fresnel's wave theory of light. However, the prediction was experimentally confirmed by Arago, who thus validated Fresnel's wave theory. Due to Arago's pivotal confirmation of Fresnel's theory and its historical significance, this bright spot is referred to in the literature as the "spot of Arago."

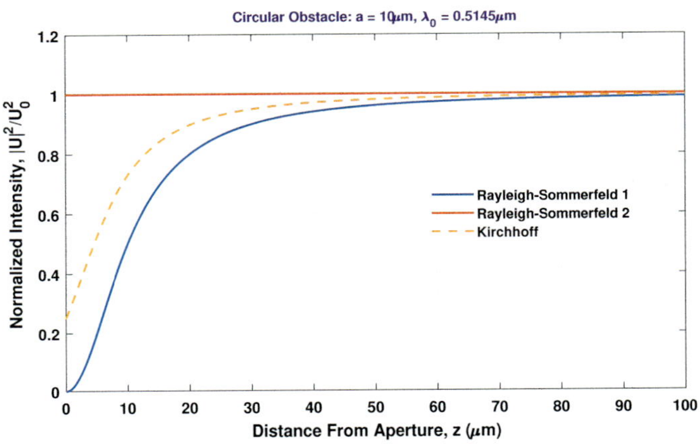

FIGURE 5.22: The diffracted field normalized intensity, $|U|^2/|U_0|^2$, along the z-axis for an incident plane wave on a circular obstacle of radius a. The freespace wavelength is $\lambda_0 = 0.5145\,\mu m$. The RS-1, RS-2, and H-K results are shown.

By again applying Babinet's principle in conjunction with the results from Sec. 5.7.1.2, the diffracted field and its normalized intensity in the radial direction can be readily determined numerically. The normalized diffracted intensity for the current circular obstacle is depicted in Figs. 5.23(a) and 5.23(b) for $z_0 = 50\,\mu m$. The presence of the Arago spot is clearly evident in Fig. 5.23(b).

It is worth noting that the conditions for the Fraunhofer approximation are not satisfied in the case of a circular obstacle due to the infinite extent of the integration region in Eq. (5.32). Consequently, if one attempts to calculate the integral using the Fraunhofer approximation, as follows:

$$U_{ff} = jk\frac{\exp(-jkz_0)}{2\pi z_0}\exp\left(-jk\frac{r_T^2}{2z_0}\right)U_0\int_a^\infty 2\pi J_0\left(\frac{kr_T r'_T}{z_0}\right)r'_T\,dr'_T \longrightarrow \infty,$$

it diverges. However, using the Babinet's principle [Eq. (5.48)] the diffracted field in the far-field can be determined as:

$$U_{ff} = U_0\exp(-jkz_0) - U_{ff}^a = U_0\exp(-jkz_0)\left[1 - j\exp\left(-jk\frac{r_T^2}{2z_0}\right)\frac{aJ_1(kar_T/z_0)}{r_T}\right], \tag{5.99}$$

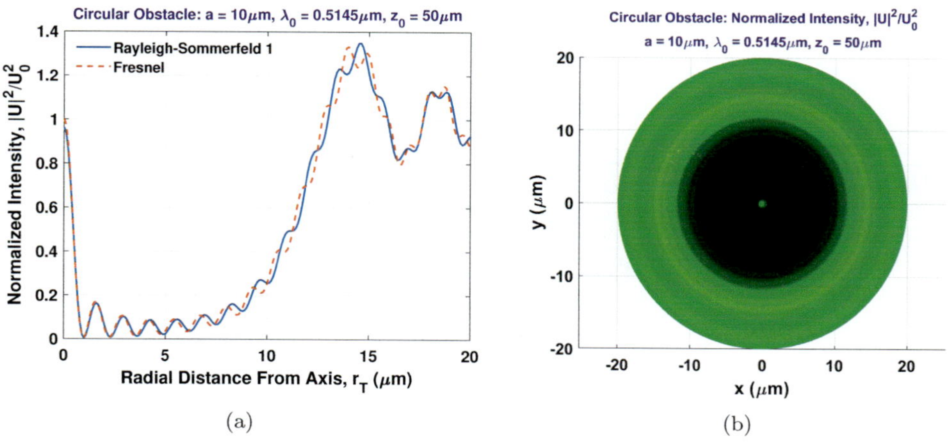

FIGURE 5.23: (a) The diffracted field normalized intensity, $|U|^2/|U_0|^2$, along the radial direction (at $z_0 = 50\,\mu m$) for a plane wave incident on a circular obstacle of radius $a = 10\,\mu m$. The freespace wavelength is $\lambda_0 = 0.5145\,\mu m$. The results obtained using the RS-1 formulation and the Fresnel approximation, are shown. The RS-2 and the H-K formulations practically coincide. (b) Same as (a) in the xy plane at $z_0 = 50\,\mu m$ where the Arago spot is evident along the z-axis.

where U^a_{ff} is the Fraunhofer approximation for the circular aperture [see Eq. (5.63)]. Then, the corresponding far-field intensity is given by:

$$I_{ff} = \frac{|U_0|^2}{4Z}\left[1 + 2\frac{ka^2}{z_0}\left(\frac{J_1(kar_T/z_0)}{kar_T/z_0}\right)\sin\left(k\frac{r_T^2}{2z_0}\right) + \frac{k^2a^4}{z_0^2}\left(\frac{J_1(kar_T/z_0)}{kar_T/z_0}\right)^2\right]. \quad (5.100)$$

5.7.4 Circular Obstacle with Incident Diverging Spherical Wave

Assume that a diverging spherical wave $U = U_0\exp(-jkR_s)/R_s$, where $R_s = (x'^2 + y'^2 + s_0^2)^{1/2}$, is incident from the left on a planar circular obstacle in a transparent screen (see Fig. 5.21). For on-axis points, the Rayleigh-Sommerfeld integrals can be determined analytically, as described in various references in the literature [275,289,291,294]. By applying Babinet's principle in conjunction with the results for the on-axis diffracted field from a circular aperture described in Sec. 5.7.2.1, the following equations can be derived for a point $P(x_0 = 0, y_0 = 0, z_0)$ with $z_0 > 0$, using the RS-1 and RS-2 formulations as well as the Fresnel approximation:

$$U_{RS_1}(z_0) = U_0\frac{z_0}{\sqrt{a^2+z_0^2}}\frac{\exp\left[-jk\left(\sqrt{a^2+z_0^2}+\sqrt{a^2+s_0^2}\right)\right]}{\sqrt{a^2+z_0^2}+\sqrt{a^2+s_0^2}}, \quad (5.101)$$

$$\frac{I_{RS_1}(z_0)}{I_{sw}} = s_0^2\frac{z_0^2}{a^2+z_0^2}\frac{1}{\left[\sqrt{a^2+z_0^2}+\sqrt{a^2+s_0^2}\right]^2}, \quad (5.102)$$

$$U_{RS_2}(z_0) = U_0\frac{s_0}{\sqrt{a^2+s_0^2}}\frac{\exp\left[-jk\left(\sqrt{a^2+z_0^2}+\sqrt{a^2+s_0^2}\right)\right]}{\sqrt{a^2+z_0^2}+\sqrt{a^2+s_0^2}}, \quad (5.103)$$

$$\frac{I_{RS_2}(z_0)}{I_{sw}} = \frac{s_0^4}{a^2+s_0^2}\frac{1}{\left[\sqrt{a^2+z_0^2}+\sqrt{a^2+s_0^2}\right]^2}, \quad (5.104)$$

$$U_K(z_0) = \frac{U_0}{2}\left[\frac{z_0}{\sqrt{a^2+z_0^2}} + \frac{s_0}{\sqrt{a^2+s_0^2}}\right]\frac{\exp\left[-jk\left(\sqrt{a^2+z_0^2}+\sqrt{a^2+s_0^2}\right)\right]}{\sqrt{a^2+z_0^2}+\sqrt{a^2+s_0^2}}, \tag{5.105}$$

$$\frac{I_K(z_0)}{I_{sw}} = \frac{s_0^2}{4}\left[\frac{z_0}{\sqrt{a^2+z_0^2}} + \frac{s_0}{\sqrt{a^2+s_0^2}}\right]^2 \frac{1}{\left[\sqrt{a^2+z_0^2}+\sqrt{a^2+s_0^2}\right]^2}, \tag{5.106}$$

while for the Fresnel approximation,

$$U(z_0) = U_0\frac{\exp\left[-jk\left(z_0+s_0\right)\right]}{z_0+s_0}\exp\left[-jk\frac{a^2}{2}\left(\frac{1}{z_0}+\frac{1}{s_0}\right)\right], \tag{5.107}$$

$$\frac{I(z_0)}{I_{sw}} = s_0^2\frac{1}{(z_0+s_0)^2}. \tag{5.108}$$

From the previous results, it is evident that the diffracted intensity along the z-axis does not oscillate as it was observed in the case of the circular aperture. Moreover, the Arago spot reappears, similar to the scenario with an incident plane wave.

Example 5.7.11:

Calculate the normalized intensities I_{RS_1}/I_{sw}, I_{RS_2}/I_{sw}, and I_K/I_{sw} resulting from the diffraction of a diverging spherical wave incident normally on a circular obstacle with a radius of $a = 10\,\mu$m. The diverging spherical wave originates from $z = -s_0 = -50\,\mu$m and has a freespace wavelength of $\lambda_0 = 0.5145\,\mu$m. Calculate these intensities along the z-axis. Repeat the calculations for the intensities along the radial direction at $z_0 = 50\,\mu$m.
Solution: The normalized (relative) intensities are determined analytically using Eqs. (5.102), (5.104), and (5.106) for the RS-1, RS-2, and Kirchhoff formulations, respectively. The results are presented in Fig. 5.24. The relative intensities are defined as I/I_{sw} where $I_{sw} = |U_0|^2/(4Zs_0^2)$, and I represents the intensities for RS-1, RS-2, and Kirchhoff formulations. By applying Babinet's principle in conjunction with the results from Sec. 5.7.2.2,

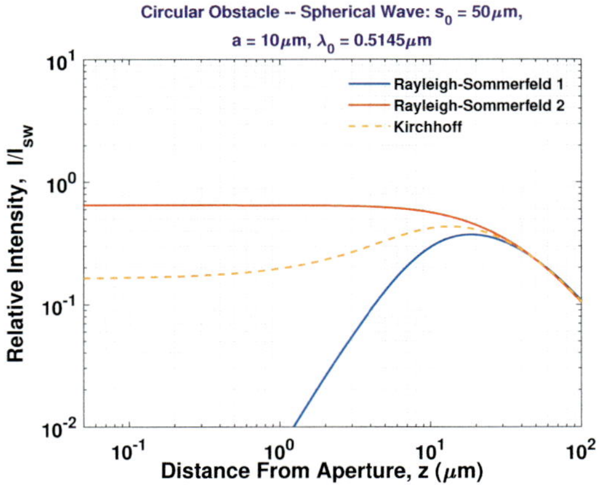

FIGURE 5.24: The diffracted field relative intensity, I/I_{sw}, along the z-axis of an incident diverging spherical wave (with $s_0 = 50\,\mu m$) on a circular obstacle of radius $a = 10\,\mu$m. The freespace wavelength is $\lambda_0 = 0.5145\,\mu m$. The RS-1, RS-2, and Kirchhoff results are shown.

the diffracted field and its normalized intensity in the radial direction can be determined numerically. As an example, the relative diffracted intensity for a diverging spherical wave

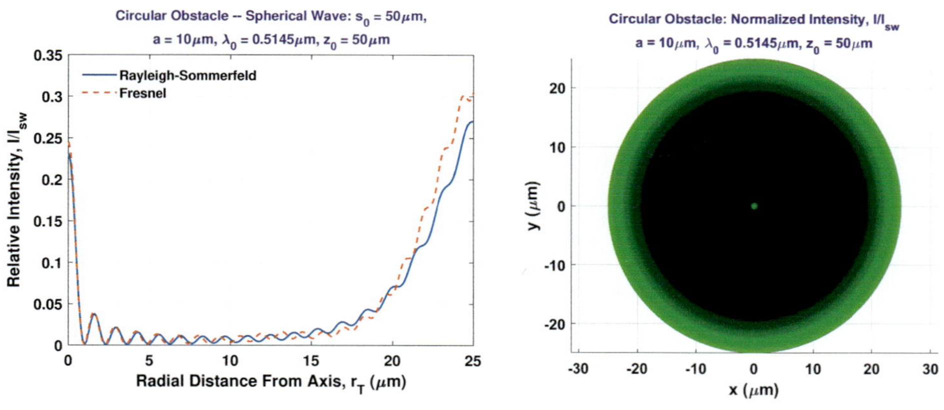

(a) (b)

FIGURE 5.25: (a) The diffracted field relative intensity, I/I_{sw}, along the radial direction (at $z_0 = 50\,\mu m$) for a diverging spherical wave (with $s_0 = 50\,\mu m$) incident on a circular obstacle of radius $a = 10\,\mu m$. The freespace wavelength is $\lambda_0 = 0.5145\,\mu m$. The results obtained using the RS-1 formulation and Fresnel approximation, are shown. The RS-2 and the H-K formulations practically coincide with RS-1. (b) Same as (a) in the xy plane at $z_0 = 50\,\mu m$ where the Arago spot is evident along the z-axis.

($s_0 = 50\,\mu m$) incident on a circular obstacle is depicted in Figs. 5.25(a) and 5.25(b) for $z_0 = 50\,\mu m$. The presence of the Arago spot is clearly visible in Fig. 5.25(b).

It is worth noting that the conditions for the Fraunhofer approximation are, once again, not satisfied in the case of the circular obstacle due to the infinite extent of the integration region in Eq. (5.32). However, by applying Babinet's principle [Eq. (5.48)], the diffracted field in the far field can still be determined as:

$$U_{ff} = U_0 \, \frac{\exp\left(-jk\sqrt{r_T^2 + z_0^2}\right)}{\sqrt{r_T^2 + z_0^2}} -$$

$$jk\frac{\exp(-jkz_0)}{z_0} \exp\left(-jk\frac{r_T^2}{2z_0}\right) U_0 \int_0^a J_0\left(\frac{kr_T r_T'}{z_0}\right)\left[U_0 \frac{\exp(-jkR_s)}{R_s}\right] r_T' dr_T', \quad (5.109)$$

where the Fraunhofer approximation for the circular aperture [Eq. (5.88)] has been applied. In the final expression, the integral must once again be evaluated numerically.

5.7.5 Circular Annulus with Incident Plane Wave

Assume that a plane wave $U = U_0 \exp(-jkz)$ is incident from the left on a planar circular annulus (either as an aperture or an obstacle) in a transparent screen (see Fig. 5.21). The inner radius of the annulus is a, and the outer radius is b. As in previous cases, the Rayleigh-Sommerfeld integrals can be determined analytically [275, 289, 291, 294] only for the on-axis scenario where $P(x_0 = 0, y_0 = 0, z_0 \geq 0)$. Using Eqs. (5.18), (5.19), and (5.25), the following analytical expressions can be derived for a point along the z-axis: for RS-1 and RS-2 formulations, the Kirchhoff formulation [Huygens-Kirchhoff or H-K being the average of RS-1 and RS-2, as described in Eq. (5.20)], as well as for the Fresnel approximation.

5.7.5.1 Case of the Annular Aperture

For the on-axis points exact analytical expressions can be found as follows:

$$U_{RS_1}(z_0) = U_0 \left[\frac{z_0}{\sqrt{a^2+z_0^2}} \exp\left(-jk\sqrt{a^2+z_0^2}\right) - \frac{z_0}{\sqrt{b^2+z_0^2}} \exp\left(-jk\sqrt{b^2+z_0^2}\right) \right], \tag{5.110}$$

$$I_{RS_1}(z_0) = \frac{|U_0|^2}{4Z} \left[\frac{z_0^2}{a^2+z_0^2} + \frac{z_0^2}{b^2+z_0^2} - \frac{2z_0^2}{\sqrt{(a^2+z_0^2)(b^2+z_0^2)}} \cos\left(k\sqrt{b^2+z_0^2} - k\sqrt{a^2+z_0^2}\right) \right], \tag{5.111}$$

$$U_{RS_2}(z_0) = U_0 \left[\exp\left(-jk\sqrt{a^2+z_0^2}\right) - \exp\left(-jk\sqrt{b^2+z_0^2}\right) \right], \tag{5.112}$$

$$I_{RS_2}(z_0) = \frac{|U_0|^2}{4Z} \left[2 - 2\cos\left(k\sqrt{b^2+z_0^2} - k\sqrt{a^2+z_0^2}\right) \right], \tag{5.113}$$

$$U(z_0) = U_0 \exp(-jkz_0) \left[\exp\left(-jk\frac{a^2}{2z_0}\right) - \exp\left(-jk\frac{b^2}{2z_0}\right) \right], \tag{5.114}$$

$$I(z_0) = \frac{|U_0|^2}{4Z} \left[2 - 2\cos\left(\frac{k}{2z_0}(b^2-a^2)\right) \right]. \tag{5.115}$$

The maxima and the minima in the diffracted intensity along the z-axis can be readily approximated using the following equations:

$$\sqrt{b^2+z_0^2} - \sqrt{a^2+z_0^2} \simeq \begin{cases} \left(m+\frac{1}{2}\right)\lambda = K^+ & \text{for maxima,} \\ m\lambda = K^- & \text{for minima,} \end{cases}$$

$$z_m^{\pm} \simeq \frac{1}{2K^{\pm}} \left[(b^2-a^2)^2 + (K^{\pm})^4 - 2(a^2+b^2)(K^{\pm})^2 \right]^{1/2}. \tag{5.116}$$

Here, z_m^+ and z_m^- represent the approximate positions of the intensity maxima and minima, respectively. The exact positions of the maxima and minima cannot be determined analytically due to the z-dependence of the amplitude terms in Eqs. (5.111) and (5.113). The number of minima and maxima can be estimated by ensuring that the square root in the $z_m^{\pm}$ solutions remains real. In the radial direction, no analytical solutions exist for RS-1, RS-2, H-K, or Fresnel approximations. These values can be computed numerically using integrals similar to those in Eqs. (5.58)–(5.62), where the radial integration is performed between the inner radius a and the outer radius b. Approximate expressions for RS-1 and RS-2 can also be derived analogously to Eqs. (5.61) and (5.62). The Fraunhofer approximation, however, can be determined analytically in a manner similar to Eq. (5.63) and is expressed as:

$$U_{ff} = jU_0 \exp(-jkz_0) \exp\left(-jk\frac{r_T^2}{2z_0}\right) \left[\frac{bJ_1(kbr_T/z_0) - aJ_1(kar_T/z_0)}{r_T} \right], \tag{5.117}$$

$$I_{ff} = \frac{|U_0|^2}{4Z} \left[\frac{bJ_1(kbr_T/z_0) - aJ_1(kar_T/z_0)}{r_T} \right]^2. \tag{5.118}$$

Example 5.7.12:

Find the on-axis and off-axis (at $z_0 = 50\,\mu$m) diffracted normalized intensities of a plane wave incident on an annulus aperture of inner radius $a = 5\,\mu$m and outer radius $b = 15\,\mu$m. The freespace wavelength of the plane wave is $\lambda_0 = 1\,\mu m$.

Solution: The normalized intensities for RS-1, RS-2, and Kirchhoff formulations along the z-axis are illustrated in Fig. 5.26(a). Similarly, the normalized intensities for RS-1 and Fresnel approximations along the radial direction at $z_0 = 50\,\mu$m are shown in Fig. 5.26(b). Using Eq. (5.116), the positions of 10 minima are predicted as $z_m^- = (99.37, 48.74, 31.44, 22.45, 16.77, 12.72, 9.56, 6.87, 4.32, 0)\,\mu$m, and 9 maxima are predicted as $z_m^+ = (65.73, 38.43, 26.35, 19.33, 14.60, 11.06, 8.18, 5.61, 2.89)\,\mu$m. These predictions align well with the results presented in Fig. 5.26(a). It is noteworthy that the largest values of z_m^- and z_m^+ values correspond to the lowest order, $m = 1$.

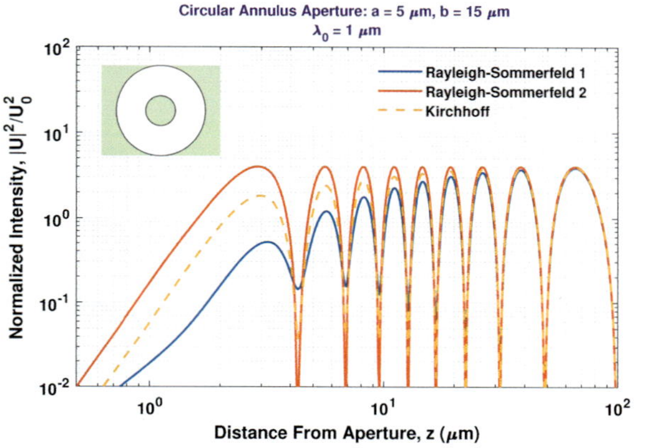

(a)

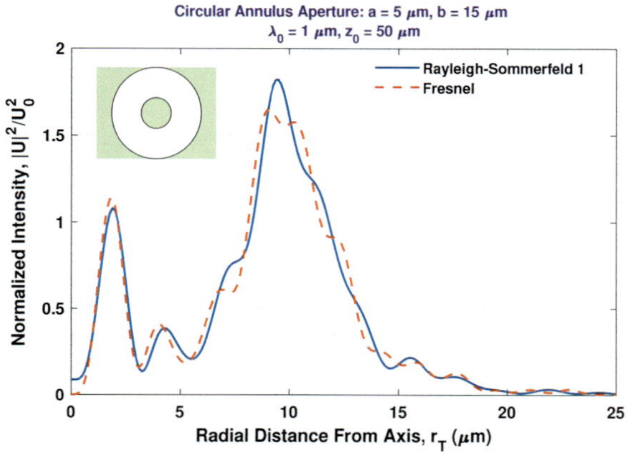

(b)

FIGURE 5.26: (a) The diffracted field normalized intensity, $|U|^2/|U_0|^2$, along the z-axis of an incident plane wave on a circular annular aperture of inner radius $a = 5\,\mu m$ and outer radius $b = 15\,\mu m$. The freespace wavelength is $\lambda_0 = 1\,\mu m$. The RS-1, RS-2, and Kirchhoff results are shown. (b) The diffracted field normalized intensity along the radial direction at $z_0 = 50\,\mu m$ for the aperture of (a). The RS-1 and Fresnel results are shown.

5.7.5.2 Case of the Annular Obstacle

In the case of the annular obstacle, which is complementary to the annular aperture, the RS-1, RS-2, and Fresnel approximations can be analytically calculated using Babinet's principle for points along the z-axis:

$$U_{RS_1}(z_0) = U_0\left[\exp(-jkz_0) - \frac{z_0}{\sqrt{a^2+z_0^2}}\exp\left(-jk\sqrt{a^2+z_0^2}\right) + \frac{z_0}{\sqrt{b^2+z_0^2}}\exp\left(-jk\sqrt{b^2+z_0^2}\right)\right], \tag{5.119}$$

$$\begin{aligned} I_{RS_1}(z_0) = \frac{|U_0|^2}{4Z}\Bigg[&1 + \frac{z_0^2}{a^2+z_0^2} + \frac{z_0^2}{b^2+z_0^2} - 2\frac{z_0}{\sqrt{a^2+z_0^2}}\cos\left(k\sqrt{a^2+z_0^2} - kz_0\right) \\ &+ 2\frac{z_0}{\sqrt{b^2+z_0^2}}\cos\left(k\sqrt{b^2+z_0^2} - kz_0\right) \\ &- \frac{2z_0^2}{\sqrt{(a^2+z_0^2)(b^2+z_0^2)}}\cos\left(k\sqrt{b^2+z_0^2} - k\sqrt{a^2+z_0^2}\right)\Bigg], \end{aligned} \tag{5.120}$$

$$U_{RS_2}(z_0) = U_0\left[\exp(-jkz_0) - \exp\left(-jk\sqrt{a^2+z_0^2}\right) + \exp\left(-jk\sqrt{b^2+z_0^2}\right)\right], \tag{5.121}$$

$$\begin{aligned} I_{RS_2}(z_0) = \frac{|U_0|^2}{4Z}\Bigg[&3 - 2\cos\left(k\sqrt{a^2+z_0^2} - kz_0\right) + 2\cos\left(k\sqrt{b^2+z_0^2} - kz_0\right) \\ &- 2\cos\left(k\sqrt{b^2+z_0^2} - k\sqrt{a^2+z_0^2}\right)\Bigg], \end{aligned} \tag{5.122}$$

$$U(z_0) = U_0\exp(-jkz_0)\left[1 - \exp\left(-jk\frac{a^2}{2z_0}\right) + \exp\left(-jk\frac{b^2}{2z_0}\right)\right], \tag{5.123}$$

$$I(z_0) = \frac{|U_0|^2}{4Z}\left[3 - 2\cos k\left(z_0 - \frac{a^2}{2z_0}\right) + 2\cos k\left(z_0 - \frac{b^2}{2z_0}\right) - 2\cos\left(\frac{k}{2z_0}(b^2-a^2)\right)\right]. \tag{5.124}$$

From the above equations, it is evident that the intensity along the z-axis for an annular obstacle is more intricate than its annular aperture counterpart due to the presence of three oscillating terms with distinct z-dependent arguments. Consequently, predicting the positions of the intensity minima and maxima is not as straightforward as in the case of the annular aperture. For points along the radial direction (off-axis), the analysis can be conducted by employing Babinet's principle in conjunction with the off-axis results of the annular aperture. This approach involves determining the complementary field distribution numerically or analytically, depending on the chosen approximation. The Fraunhofer approximation for the annular obstacle can similarly be derived using Babinet's principle and the Fraunhofer results for the annular aperture. The resulting expression is:

$$U_{ff} = U_0 \exp(-jkz_0) \left\{ 1 - j \exp\left(-jk\frac{r_T^2}{2z_0}\right) \left[\frac{bJ_1(kbr_T/z_0) - aJ_1(kar_T/z_0)}{r_T} \right] \right\}, \quad (5.125)$$

$$I_{ff} = \frac{|U_0|^2}{4Z} \left\{ 1 + \left[\frac{bJ_1(kbr_T/z_0) - aJ_1(kar_T/z_0)}{r_T} \right]^2 - 2\frac{bJ_1(kbr_T/z_0) - aJ_1(kar_T/z_0)}{r_T} \sin\left(k\frac{r_T^2}{2z_0}\right) \right\}. \quad (5.126)$$

Example 5.7.13:

Find the on-axis and off-axis (at $z_0 = 50\,\mu$m) diffracted normalized intensities of a plane wave incident on an annulus obstacle of inner radius $a = 5\,\mu$m and outer radius $b = 15\,\mu$m. The freespace wavelength of the plane wave is $\lambda_0 = 1\,\mu m$.

Solution: The normalized intensities calculated using the RS-1, RS-2, and Kirchhoff formulations are presented in Fig. 5.27(a) for points along the z-axis. These results illustrate the intricate variations in intensity due to the multiple oscillatory terms characteristic of the annular obstacle's diffraction pattern. Additionally, the normalized intensities determined using the RS-1 and Fresnel approximations for the radial direction at $z_0 = 50\,\mu$m are depicted in Fig. 5.27(b). These results highlight the radial intensity distribution and further emphasize the distinct diffraction characteristics of the annular obstacle.

5.7.6 Circular Annulus with Incident Diverging Spherical Wave

In this section, a diverging spherical wave of the form $U = U_0 \exp(-jkR_s)/R_s$, where $R_s = (x'^2 + y'^2 + s_0^2)^{1/2}$, is incident from the left on a planar circular annulus (which may either be an aperture or an obstacle). As in the previous cases, the Rayleigh-Sommerfeld integrals can be derived analytically for points along the z-axis [275, 289, 291, 294].

5.7.6.1 Case of the Annular Aperture

For the on-axis points exact analytical expressions can be found as follows:

$$U_{RS_1}(z_0) = U_0 \left[\frac{z_0}{\sqrt{a^2+z_0^2}} \frac{\exp\left[-jk\left(\sqrt{a^2+z_0^2}+\sqrt{a^2+s_0^2}\right)\right]}{\sqrt{a^2+z_0^2}+\sqrt{a^2+s_0^2}} - \frac{z_0}{\sqrt{b^2+z_0^2}} \frac{\exp\left[-jk\left(\sqrt{b^2+z_0^2}+\sqrt{b^2+s_0^2}\right)\right]}{\sqrt{b^2+z_0^2}+\sqrt{b^2+s_0^2}} \right], \quad (5.127)$$

$$\frac{I_{RS_1}(z_0)}{I_{sw}} = s_0^2 \left[\frac{F_1^2}{S_1^2} + \frac{F_2^2}{S_2^2} - \frac{2F_1F_2}{S_1S_2} \cos\left[k(S_2 - S_1)\right] \right], \quad \text{where,} \quad (5.128)$$

$$F_1 = \frac{z_0}{\sqrt{a^2+z_0^2}}, \qquad F_2 = \frac{z_0}{\sqrt{b^2+z_0^2}},$$

$$S_1 = \sqrt{a^2+z_0^2} + \sqrt{a^2+s_0^2}, \qquad S_2 = \sqrt{b^2+z_0^2} + \sqrt{b^2+s_0^2},$$

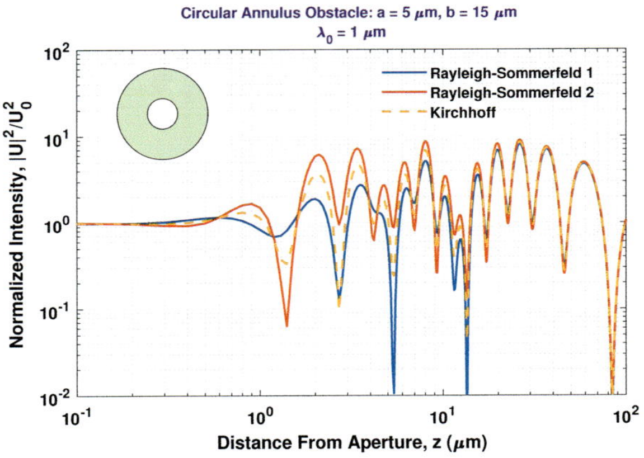

(a)

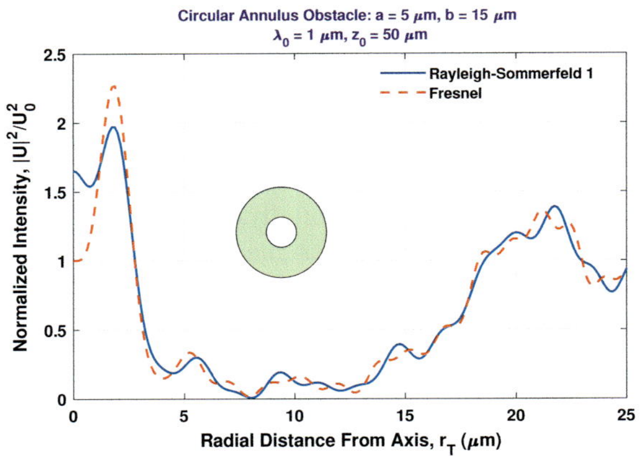

(b)

FIGURE 5.27: (a) The diffracted field normalized intensity, $|U|^2/|U_0|^2$, along the z-axis of an incident plane wave on a circular annular obstacle of inner radius $a = 5\,\mu m$ and outer radius $b = 15\,\mu m$. The freespace wavelength is $\lambda_0 = 1\,\mu m$. The RS-1, RS-2, and Kirchhoff results are shown. (b) The diffracted field normalized intensity along the radial direction at $z_0 = 50\,\mu m$ for the obstacle of (a).

$$U_{RS_2}(z_0) = U_0\left[\frac{s_0}{\sqrt{a^2+s_0^2}}\frac{\exp\left[-jk\left(\sqrt{a^2+z_0^2}+\sqrt{a^2+s_0^2}\right)\right]}{\sqrt{a^2+z_0^2}+\sqrt{a^2+s_0^2}} - \frac{s_0}{\sqrt{b^2+s_0^2}}\frac{\exp\left[-jk\left(\sqrt{b^2+z_0^2}+\sqrt{b^2+s_0^2}\right)\right]}{\sqrt{b^2+z_0^2}+\sqrt{b^2+s_0^2}}\right], \tag{5.129}$$

$$\frac{I_{RS_2}(z_0)}{I_{sw}} = s_0^2\left[\frac{G_1^2}{S_1^2}+\frac{G_2^2}{S_2^2}-\frac{2G_1G_2}{S_1S_2}\cos\left[k(S_2-S_1)\right]\right], \qquad \text{where,} \tag{5.130}$$

$$G_1 = \frac{s_0}{\sqrt{a^2+s_0^2}}, \qquad G_2 = \frac{s_0}{\sqrt{b^2+s_0^2}},$$

$$U(z_0) = U_0\frac{\exp\left[-jk(z_0+s_0)\right]}{z_0+s_0}\left[\exp\left(-jk\frac{a^2}{2}\mathcal{A}\right)-\exp\left(-jk\frac{b^2}{2}\mathcal{A}\right)\right], \tag{5.131}$$

$$\frac{I(z_0)}{I_{sw}} = s_0^2\left[2 - 2\cos\left(\frac{k}{2}(b^2 - a^2)\mathcal{A}\right)\right], \qquad \text{with} \tag{5.132}$$

$$I_{sw} = \frac{|U_0|^2}{4Z}\frac{1}{s_0^2}, \qquad \text{and} \quad \mathcal{A} = \frac{1}{z_0} + \frac{1}{s_0}.$$

The maxima and the minima in the diffracted intensity along the z-axis can also be easily approximated from the following equation:

$$\sqrt{b^2 + z_0^2} - \sqrt{a^2 + z_0^2} \simeq \begin{cases} \left(m + \frac{1}{2}\right)\lambda - \sqrt{s_0^2 + b^2} + \sqrt{s_0^2 + a^2} = K^+ & \text{for maxima,} \\ m\lambda - \sqrt{s_0^2 + b^2} + \sqrt{s_0^2 + a^2} = K^- & \text{for minima,} \end{cases}$$

$$z_m^{\pm} \simeq \frac{1}{2K^{\pm}}\left[(b^2 - a^2)^2 + (K^{\pm})^4 - 2(a^2 + b^2)(K^{\pm})^2\right]^{1/2}. \tag{5.133}$$

Here, z_m^+ and z_m^- represent the approximate positions of the intensity maxima and minima, respectively. The integer m should be chosen such that the solutions are real and positive. For off-axis points along the radial direction, there are no analytical solutions for the Rayleigh-Sommerfeld (RS-1 and RS-2), Kirchhoff, and Fresnel approximations. These can be readily determined from integrals such as those in Eqs. (5.83)–(5.87), where the radial integration is performed between the inner radius a and the outer radius b. Similarly, the Fraunhofer approximation can be obtained by adjusting the integration limits in Eq. (5.88).

Example 5.7.14:

Calculate the on-axis and off-axis (at $z_0 = 50\,\mu$m) relative intensities I/I_{sw} of a spherical wave incident on an annular aperture of inner radius $a = 5\,\mu$m and outer radius $b = 15\,\mu$m. The freespace wavelength of the incident spherical wave is $\lambda_0 = 1\,\mu m$ and the wave emanates from the point $z = -s_0 = -50\,\mu m$.

Solution: The relative intensities for the Rayleigh-Sommerfeld RS-1, RS-2, and Kirchhoff formulations are shown in Fig. 5.28(a) along the z-axis. Additionally, the relative intensities for RS-1 and Fresnel along the radial direction at $z_0 = 50\,\mu m$ are displayed in Fig. 5.28(b). Using Eq. (5.133), 9 minima are predicted at $z_m^- = (94.78, 47.54, 30.88, 22.12, 16.55, 12.55, 9.42, 6.75, 4.20)\,\mu$m, and 9 maxima at $z_m^+ = (63.64, 37.64, 25.93, 19.07, 14.41, 10.91, 8.05, 5.49, 2.74)\,\mu$m (up to the range shown). These predictions agree very well with the results shown in Fig. 5.28(a). It is noted that the largest values of z_m^- and z_m^+ correspond to the lowest m, $m = 1$.

5.7.6.2 Case of the Annular Obstacle

In the case of the annular obstacle, which is complementary to the annular aperture, the Rayleigh-Sommerfeld RS-1, RS-2, and Fresnel approximations can be easily calculated for on-axis points using Babinet's principle. The resulting expressions are:

$$U_{RS_1}(z_0) = U_0\left[\frac{e^{-jk(z_0+s_0)}}{z_0 + s_0} - \frac{z_0}{\sqrt{a^2 + z_0^2}}\frac{\exp\left[-jk\left(\sqrt{a^2 + z_0^2} + \sqrt{a^2 + s_0^2}\right)\right]}{\sqrt{a^2 + z_0^2} + \sqrt{a^2 + s_0^2}} + \right.$$

$$\left. \frac{z_0}{\sqrt{b^2 + z_0^2}}\frac{\exp\left[-jk\left(\sqrt{b^2 + z_0^2} + \sqrt{b^2 + s_0^2}\right)\right]}{\sqrt{b^2 + z_0^2} + \sqrt{b^2 + s_0^2}}\right], \tag{5.134}$$

$$\frac{I_{RS_1}(z_0)}{I_{sw}} = s_0^2\left[\frac{1}{(z_0 + s_0)^2} + \frac{F_1^2}{S_1^2} + \frac{F_2^2}{S_2^2} - \frac{2F_1}{S_1(z_0 + s_0)}\cos\left[k(S_1 - z_0 - s_0)\right] + \right.$$

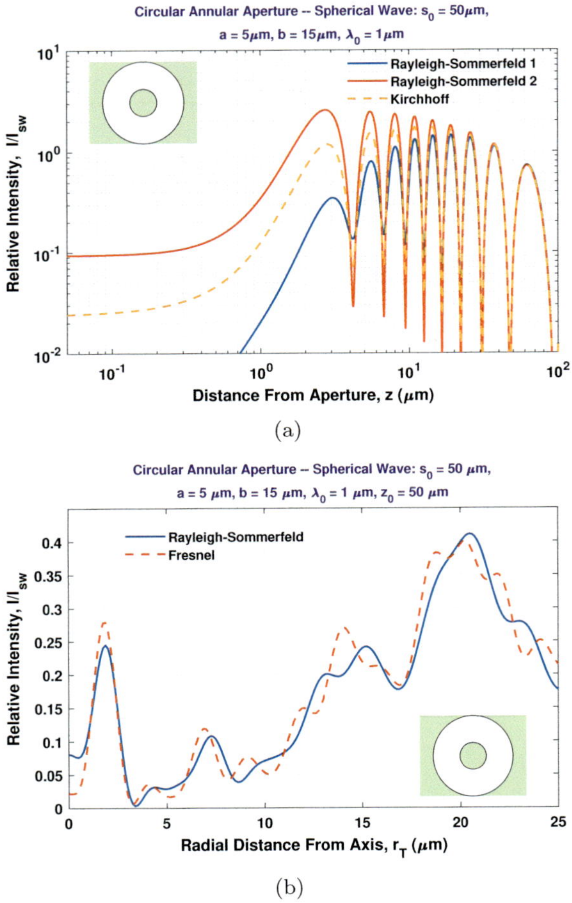

FIGURE 5.28: (a) The diffracted field relative intensity, I/I_{sw}, along the z-axis of an incident spherical wave (emanating from $z = -s_0 = -50\,\mu m$) on a circular annular aperture of inner radius $a = 5\,\mu m$ and outer radius $b = 15\,\mu m$. The freespace wavelength is $\lambda_0 = 1\,\mu m$. The RS-1, RS-2, and Kirchhoff results are shown. (b) The diffracted field normalized intensity along the radial direction at $z_0 = 50\,\mu m$ for the aperture of (a). The RS-1 and Fresnel results are shown.

$$\left. \frac{2F_2}{S_2(z_0+s_0)}\cos\left[k(S_2 - z_0 - s_0)\right] - \frac{2F_1F_2}{S_1S_2}\cos\left[k(S_2 - S_1)\right]\right], \tag{5.135}$$

$$U_{RS_2}(z_0) = U_0\left[\frac{e^{-jk(z_0+s_0)}}{z_0+s_0} - \frac{s_0}{\sqrt{a^2+s_0^2}}\frac{\exp\left[-jk\left(\sqrt{a^2+z_0^2}+\sqrt{a^2+s_0^2}\right)\right]}{\sqrt{a^2+z_0^2}+\sqrt{a^2+s_0^2}}+\right.$$
$$\left.\frac{s_0}{\sqrt{b^2+s_0^2}}\frac{\exp\left[-jk\left(\sqrt{b^2+z_0^2}+\sqrt{b^2+s_0^2}\right)\right]}{\sqrt{b^2+z_0^2}+\sqrt{b^2+s_0^2}}\right], \tag{5.136}$$

$$\frac{I_{RS_2}(z_0)}{I_{sw}} = s_0^2\left[\frac{1}{(z_0+s_0)^2} + \frac{G_1^2}{S_1^2} + \frac{G_2^2}{S_2^2} - \frac{2G_1}{S_1(z_0+s_0)}\cos\left[k(S_1 - z_0 - s_0)\right]+\right.$$
$$\left.\frac{2G_2}{S_2(z_0+s_0)}\cos\left[k(S_2 - z_0 - s_0)\right] - \frac{2G_1G_2}{S_1S_2}\cos\left[k(S_2 - S_1)\right]\right], \tag{5.137}$$

$$U(z_0) = U_0 \frac{e^{-jk(z_0+s_0)}}{z_0 + s_0} \left[1 - \exp\left(-jk\frac{a^2}{2}\mathcal{A} \right) + \exp\left(-jk\frac{b^2}{2}\mathcal{A} \right) \right], \tag{5.138}$$

$$\frac{I(z_0)}{I_{sw}} = \frac{s_0^2}{(z_0 + s_0)^2} \left[3 - 2\cos\left(k\frac{a^2}{2}\mathcal{A} \right) + 2\cos\left(k\frac{b^2}{2}\mathcal{A} \right) - 2\cos\left(\frac{k}{2}(b^2 - a^2)\mathcal{A} \right) \right]. \tag{5.139}$$

Again, $S1$, S_2, F_1, F_2, G_1, G_2, and $\mathcal{A}$, are as defined previously. Along the radial direction, Babinet's principle is applied in conjunction with the numerical evaluation of the integrals of the annular aperture.

Example 5.7.15:

Calculate the on-axis and off-axis (at $z_0 = 50\,\mu\text{m}$) relative intensities I/I_{sw} of a spherical wave incident on an annular obstacle of inner radius $a = 5\,\mu\text{m}$ and outer radius $b = 15\,\mu\text{m}$. The freespace wavelength of the incident spherical wave is $\lambda_0 = 1\,\mu m$, and the wave emanates from point $z = -s_0 = -50\,\mu m$.

Solution: The relative intensities for the Rayleigh-Sommerfeld RS-1, RS-2, and Kirchhoff formulations are shown in Fig. 5.29(a) along the z-axis. The relative intensities for RS-1 and Fresnel along the radial direction at $z_0 = 50\,\mu\text{m}$ are displayed in Fig. 5.29(b).

5.7.7 Elliptical Apertures, Obstacles, and Annuli

Assume that a plane wave $U = U_0 \exp(-jkz)$ is incident from the left on a planar elliptical aperture with x-semi-axis a and y-semi-axis b in an opaque screen (see Fig. 5.30). The approach for analyzing the scalar diffraction by the elliptical aperture follows the methodology presented in Refs. [296,297]. Using Eqs. (5.18), (5.19), and (5.25) for a random point $P(x_0, y_0, z_0)$ with $z_0 > 0$, the following equations can be derived for the Rayleigh-Sommerfeld RS-1, RS-2 formulations, and the Fresnel approximation:

$$U_{RS_1}(x_0, y_0, z_0) = \int_0^a \int_0^{2\pi} U_0 \left(jk + \frac{1}{R} \right) \frac{\exp(-jkR)}{2\pi R} \frac{z_0}{R} r'_{_T} dr'_{_T} d\phi', \tag{5.140}$$

$$U_{RS_2}(x_0, y_0, z_0) = \int_0^a \int_0^{2\pi} \frac{\exp(-jkR)}{2\pi R} (jkU_0) r'_{_T} dr'_{_T} d\phi', \tag{5.141}$$

$$U(x_0, y_0, z_0) = jk\frac{\exp(-jkz_0)}{2\pi z_0} \int_0^a \int_0^{2\pi} \exp\left\{ -jk\frac{R_t^2}{2z_0} \right\} U_0 r'_{_T} dr'_{_T} d\phi', \tag{5.142}$$

where,

$$\epsilon_e = \frac{b}{a}, \quad \text{(ellipse eccentricity)},$$

$$R = \left[r_{_T}^2 + r_{_T}'^2 \cos^2\phi' + \epsilon_e^2 r_{_T}'^2 \sin^2\phi' - 2r_{_T} r'_{_T} \cos(\phi - \phi') - 2r_{_T} r'_{_T} \epsilon_e \sin(\phi - \phi') + z_0^2 \right]^{1/2},$$

$$R_t = \left[r_{_T}^2 + r_{_T}'^2 \cos^2\phi' + \epsilon_e^2 r_{_T}'^2 \sin^2\phi' - 2r_{_T} r'_{_T} \cos(\phi - \phi') - 2r_{_T} r'_{_T} \epsilon_e \sin(\phi - \phi') \right]^{1/2},$$

and,

$$r_{_T} = \left(x_0^2 + y_0^2 \right)^{1/2}, \quad \phi = \tan^{-1}\left(\frac{y_0}{x_0} \right),$$

$$x' = r'_{_T} \cos\phi', \quad y' = \epsilon_e r'_{_T} \sin\phi', \quad r'_{_T} = \left(x'^2 + \frac{y'^2}{\epsilon_e^2} \right)^{1/2}, \quad \phi' = \tan^{-1}\left(\frac{y'}{\epsilon_e x'} \right).$$

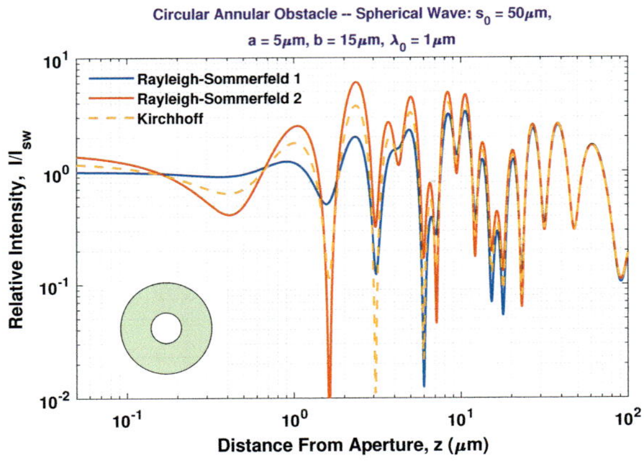

(a)

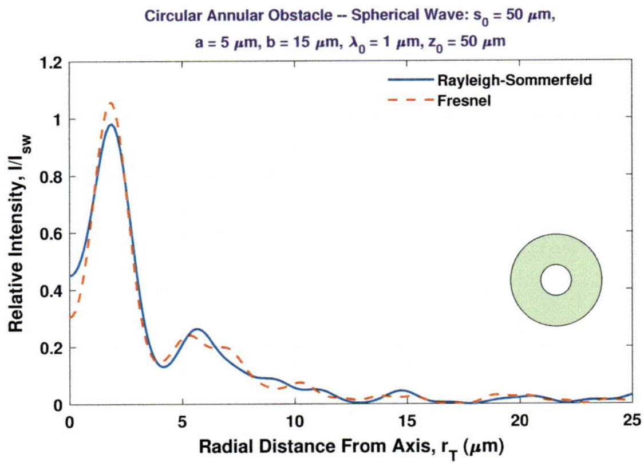

(b)

FIGURE 5.29: (a) The diffracted field relative intensity, I/I_{sw}, along the z-axis of an incident spherical wave (emanating from $z = -s_0 = -50\,\mu m$) on a circular annular obstacle of inner radius $a = 5\,\mu m$ and outer radius $b = 15\,\mu m$. The freespace wavelength is $\lambda_0 = 1\,\mu m$. The RS-1, RS-2, and Kirchhoff results are shown. (b) The diffracted field normalized intensity along the radial direction at $z_0 = 50\,\mu m$ for the obstacle of (a). The RS-1 and Fresnel results are shown.

In this case, all integrals can be evaluated using two-dimensional numerical integration. Unlike the circular aperture case, the resulting field does not exhibit azimuthal symmetry. Consequently, the fields depend on both (x_0, y_0) or equivalently (r_T, ϕ). However, the Fraunhofer approximation, $U(x_0, y_0, z_0) = U_{ff}$, can still be determined analytically as follows [296]:

$$x_0 = r_T \epsilon_e \cos\phi, \quad y_0 = r_T \sin\phi \quad \text{and} \quad x' = r'_T \cos\phi', \quad y' = r'_T \epsilon_e \sin\phi'.$$

Using the above transformation the Fraunhofer approximation can be written as:

$$U_{ff}(x_0, y_0, z_0) = jU_0 \exp(-jkz_0) \exp\left(-jk\frac{x_0^2 + y_0^2}{2z_0}\right) ab \frac{J_1\left(\dfrac{k}{z_0}\sqrt{a^2x_0^2 + b^2y_0^2}\right)}{\sqrt{a^2x_0^2 + b^2y_0^2}}, \quad (5.143)$$

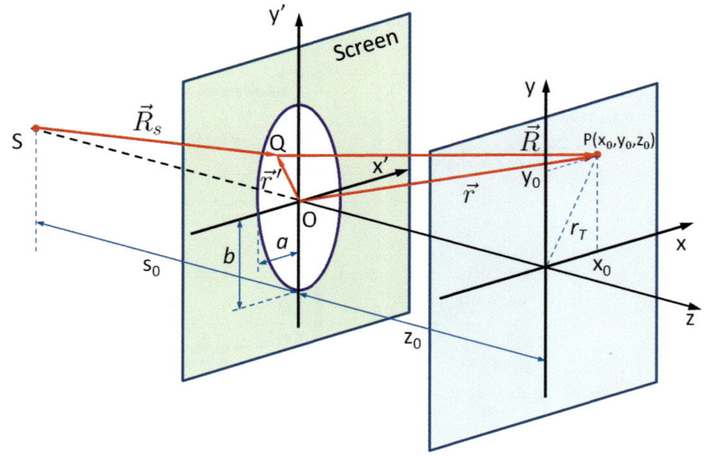

FIGURE 5.30: An elliptical aperture of semi-axes a and b in an opaque screen. The diffracted field for a general $P(x_0, y_0, z_0)$ is sought ($z_0 > 0$). The point S specifies the origin of a diverging spherical wave that could be incident onto the aperture. In the case of an incident plane wave, the point S moves to $-\infty$ ($s_0 \to +\infty$).

and the corresponding intensity is given by:

$$I_{ff}(x_0, y_0, z_0) = \frac{|U_0|^2}{4Z}\frac{k^2a^2b^2}{z_0^2}\left(\frac{2J_1\left(\dfrac{k}{z_0}\sqrt{a^2x_0^2 + b^2y_0^2}\right)}{\dfrac{k}{z_0}\sqrt{a^2x_0^2 + b^2y_0^2}}\right)^2. \tag{5.144}$$

Example 5.7.16:

Determine the normalized intensities resulting from the diffraction of a normally incident plane wave, with a free-space wavelength of $\lambda_0 = 1.0\,\mu m$, on an elliptical aperture with x-semi-axis $a = 10\,\mu m$ and y-semi-axis $a = 20\,\mu m$. Calculate the Rayleigh-Sommerfeld RS-1, RS-2, and Kirchhoff formulations at $z_0 = 10\,\mu$m for $\phi = 0°$ (along the x-axis) and $\phi = 90°$ (along the y-axis). Next, calculate the RS-1 and Fresnel approximation normalized intensities at $z_0 = 50\,\mu$m for $\phi = 0°$ (along the x-axis) and $\phi = 90°$ (along the y-axis). Finally, calculate the RS, Fresnel, and Fraunhofer normalized intensities at $z_0 = 1000\,\mu$m for $\phi = 0°$ (along the x-axis) and $\phi = 90°$ (along the y-axis).

Solution: The resulting normalized intensities for the RS-1, RS-2, Kirchhoff, Fresnel, and Fraunhofer formulations are presented in Figs. 5.31(a) to 5.31(f). For all cases, the intensities are shown along the radial direction for two orthogonal orientations ($\phi = 0°$, corresponding to the x-axis, and $\phi = 90°$, corresponding to the y-axis), and for various distances from the aperture plane. For short distances ($z_0 = 10\,\mu m$), only the accurate scalar results (RS-1, RS-2, and Kirchhoff) are displayed. The Fraunhofer results in the first four figures (Figs. 5.31(a) to 5.31(d)) are omitted since the Fraunhofer approximation does not hold. It is observed that for the largest distance from the aperture, the diffracted field extends more in the x-direction ($\phi = 0°$) than in the y-direction ($\phi = 90°$), as the smallest aperture dimension lies along the x-direction. The diffraction in the Fraunhofer regime is more clearly shown in the intensity patterns of Fig. 5.32(a) (for an elliptical aperture with x-semi-axis $a = 10\,\mu m$ and y-semi-axis $a = 20\,\mu m$), and in Fig. 5.32(b) (for an elliptical aperture of x-semi-axis $a = 20\,\mu m$ and y-semi-axis $a = 10\,\mu m$) at a distance of $z_0 = 2000\,\mu m$ (where the Fraunhofer

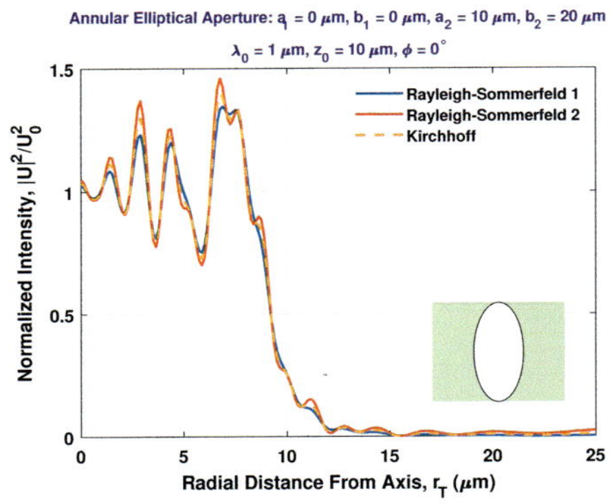

(a)

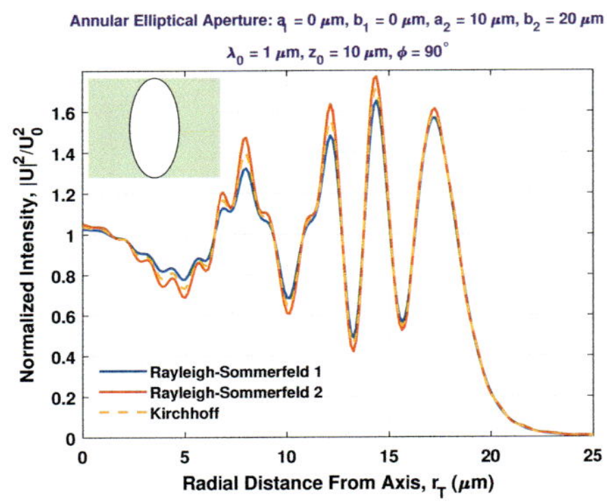

(b)

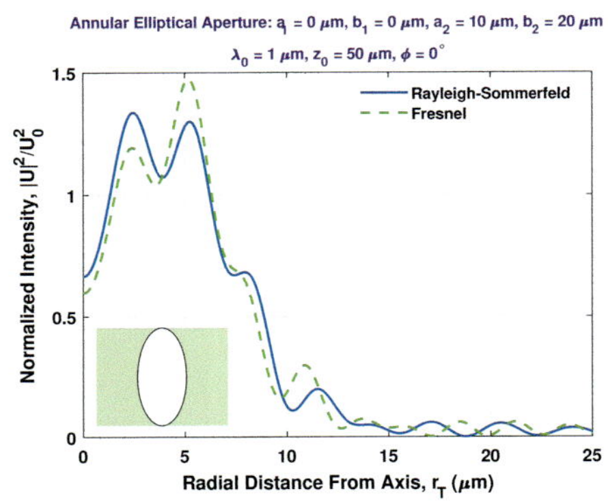

(c)

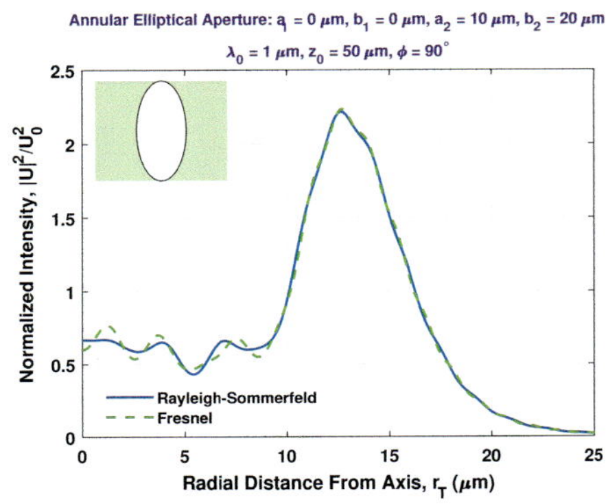

(d)

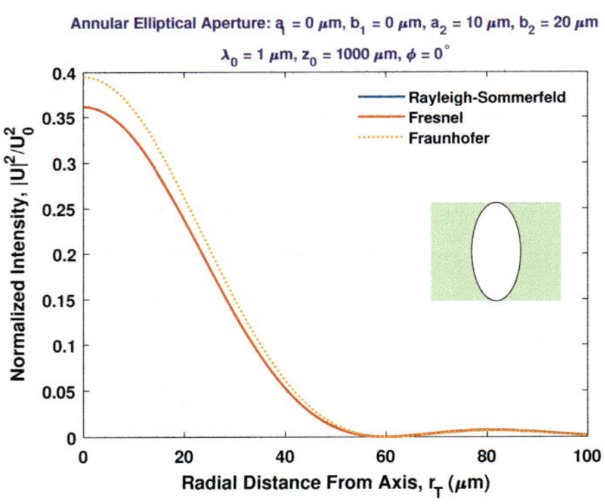

(e)

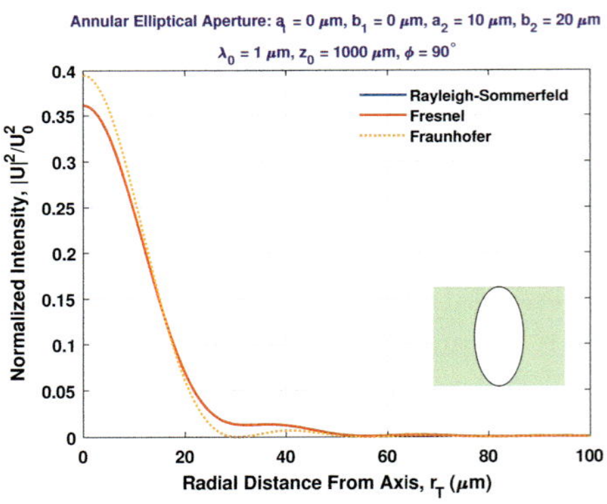

(f)

FIGURE 5.31: (a) The diffracted field normalized intensity, $|U|^2/|U_0|^2$, in the radial direction for $z_0 = 10\,\mu m$ and $\phi = 0°$. This pattern is produced by a plane wave incident on an elliptical aperture of x-semi-axis $a = 10\,\mu m$ and y-semi-axis $b = 20\,\mu m$. The freespace wavelength is $\lambda_0 = 1.0\,\mu m$. (b) Same as (a) but for $\phi = 90°$. (c) Same as (a) but for $z_0 = 50\,\mu m$. (d) Same as (c) but for $\phi = 90°$. (e) Same as (a) but for $z_0 = 1000\,\mu m$. (f) Same as (e) but for $\phi = 90°$.

approximation clearly holds, since $z_0 \gg (\pi ab/\lambda_0 = 628\,\mu m)$. It is evident that the diffracted light expands more in the direction corresponding to the smaller aperture dimension.

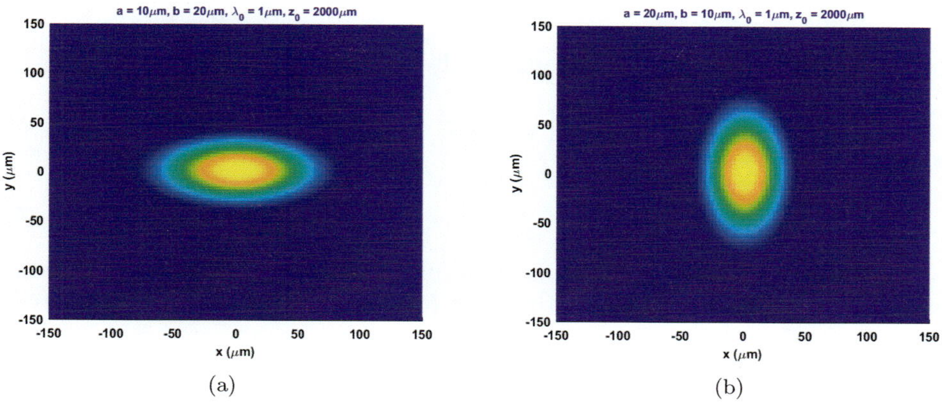

FIGURE 5.32: (a) The diffracted field normalized intensity, $|U|^2/|U_0|^2$, in the xy plane in the far-field (Fraunhofer regime) for $z_0 = 2000\,\mu m$. This pattern is produced by a plane wave incident on an elliptical aperture of x-semi-axis $a = 10\,\mu m$ and y-semi-axis $a = 20\,\mu m$. The freespace wavelength is $\lambda_0 = 1.0\,\mu m$. (b) Same as (a) but for an elliptical aperture of x-semi-axis $a = 20\,\mu m$ and y-semi-axis $a = 10\,\mu m$.

The case of an elliptical obstacle can be easily analyzed using the results for the elliptical aperture, along with Babinet's principle, as discussed in Sec. 5.7.3. Similarly, the cases of elliptical annular apertures/obstacles can be treated in an analogous manner to the method outlined in Sec. 5.7.5, assuming that both ellipses comprising the annulus have the same eccentricities, i.e., $\epsilon_{e1} = b_1/a_1 = b_2/a_2 = \epsilon_{e2}$, where a_1, a_2 and b_1, b_2 are the x-semi-axes and the y-semi-axes of the two ellipses, respectively.

5.7.8 Converging Spherical Wave Diffraction

This section presents the diffraction of an incident converging spherical wave by a circular aperture. The geometric configuration is illustrated in Fig. 5.10, where the incident wave is a spherical converging wave focused at the point S, located at a distance $z = +s_0 > 0$ (to the right of the circular aperture). The converging spherical wave can be represented as a scalar field by $U_{inc} = U_0 \exp(+jkR_s)/R_s$, where $R_s = [x'^2 + y'^2 + (z' - s_0)^2]^{1/2} = [r_T'^2 + (z' - s_0)^2]^{1/2}$. In the case of a point along the z-axis, the Rayleigh-Sommerfeld integrals can be determined analytically, as discussed in Refs. [289, 291, 294], similar to the case of a diverging incident spherical wave. Using Eq. (5.18) for points P on the positive z-axis, the following expressions can be derived:

$$
\begin{aligned}
U_{RS_1}(P) = U_{RS_1}(z_0) &= \iint_{S_a} \left[U_0 \frac{\exp(+jkR_s)}{R_s}\Big|_{z'=0} \left(jk + \frac{1}{R} \right) \frac{\exp(-jkR)}{2\pi R} \frac{z}{R} \right]_{z'=0} dx'dy' = \\
&= -\int_0^a \int_0^{2\pi} U_0 \frac{\exp(+jkR_s)}{R_s} \frac{d}{dR} \left(\frac{\exp(-jkR)}{2\pi R} \right) \frac{z_0}{R} r_T' dr_T' d\phi', \\
&= -2\pi \int_0^a U_0 \frac{\exp(+jkR_s)}{R_s} \frac{d}{dR} \left(\frac{\exp(-jkR)}{2\pi R} \right) \frac{z_0}{R} r_T' dr_T', \\
&\text{with } R = \sqrt{z_0^2 + r_T'^2}, \qquad \text{and } R_s = \sqrt{s_0^2 + r_T'^2}.
\end{aligned}
$$

However, it can be shown by direct differentiation [294] that:

$$\int \frac{\exp(+jkR_s)}{R_s}\frac{1}{R}\frac{d}{dR}\left(\frac{\exp(-jkR)}{R}\right)r'_{T}dr'_{T} = -\frac{1}{R}\frac{\exp\left[-jk(R-R_s)\right]}{R-R_s}. \tag{5.145}$$

Using the last result, the U_{RS_1} can be analytically determined as:

$$U_{RS_1}(z_0) = U_0\left[\frac{z_0}{\sqrt{a^2+z_0^2}}\frac{\exp\left[-jk\left(\sqrt{a^2+z_0^2}-\sqrt{a^2+s_0^2}\right)\right]}{\sqrt{a^2+z_0^2}-\sqrt{a^2+s_0^2}} - \frac{\exp\left[-jk(z_0-s_0)\right]}{z_0-s_0}\right]. \tag{5.146}$$

At $z_0 = s_0$, special care must be taken, as the results from Eq. (5.146) diverge. However, by starting with the RS-1 integral and noting that $R = R_s$ when $z_0 = s_0$, the diffracted field at $z_0 = s_0$ can be determined with ease. The expression for the diffracted field at this point is given by:

$$U_{RS_1}(z_0 = s_0) = -U_0\left[jk\left(\frac{s_0}{\sqrt{a^2+s_0^2}}-1\right)+\frac{1}{2s_0}\left(\frac{s_0^2}{a^2+s_0^2}-1\right)\right]. \tag{5.147}$$

The corresponding intensity of the diffracted field is proportional to $|U_{RS_1}|^2$ and is given by

$$\frac{I_{RS_1}(z_0)}{I_{sw}} = s_0^2\left[\frac{1}{S_0^2}+\frac{F_1^2}{S_1^2}-\frac{2F_1}{S_0S_1}\cos\left[k(S_1-S_0)\right]\right] \quad \text{and,} \tag{5.148}$$

$$\frac{I_{RS_1}(z_0 = s_0)}{I_{sw}} = s_0^2\left[k^2\left(\frac{s_0}{\sqrt{a^2+s_0^2}}-1\right)^2+\frac{1}{4s_0^2}\left(\frac{a^2}{a^2+s_0^2}\right)^2\right], \quad \text{where,} \tag{5.149}$$

$$S_0 = z_0 - s_0, \quad S_1 = \sqrt{a^2+z_0^2}-\sqrt{a^2+s_0^2}, \quad F_1 = \frac{z_0}{\sqrt{z_0^2+a^2}}.$$

The normalization intensity, I_{sw}, is defined similarly to the diverging wave case as $I_{sw} = |U_0|^2/4Zs_0^2$ (Z is the wave impedance of the medium). The case of the Rayleigh-Sommerfeld 2 formulation can also be determined analytically and is given by:

$$U_{RS_2}(z_0) = U_0\left[\frac{s_0}{\sqrt{a^2+s_0^2}}\frac{\exp\left[-jk\left(\sqrt{a^2+z_0^2}-\sqrt{a^2+s_0^2}\right)\right]}{\sqrt{a^2+z_0^2}-\sqrt{a^2+s_0^2}} - \frac{\exp\left[-jk(z_0-s_0)\right]}{z_0-s_0}\right], \tag{5.150}$$

$$U_{RS_2}(z_0 = s_0) = -U_0\left[jk\left(\frac{s_0}{\sqrt{a^2+s_0^2}}-1\right)+\frac{1}{2s_0}\left(\frac{s_0^2}{a^2+s_0^2}-1\right)\right]. \tag{5.151}$$

The corresponding intensity of the diffracted field is proportional to $|U_{RS_2}|^2$ and is expressed as:

$$\frac{I_{RS_2}(z_0)}{I_{sw}} = s_0^2\left[\frac{1}{S_0^2}+\frac{F_2^2}{S_1^2}-\frac{2F_2}{S_0S_1}\cos\left[k(S_1-S_0)\right]\right], \quad \text{and,} \tag{5.152}$$

$$\frac{I_{RS_2}(z_0 = s_0)}{I_{sw}} = s_0^2\left[k^2\left(\frac{s_0}{\sqrt{a^2+s_0^2}}-1\right)^2+\frac{1}{4s_0^2}\left(\frac{a^2}{a^2+s_0^2}\right)^2\right], \quad \text{where,} \tag{5.153}$$

$$F_2 = \frac{s_0}{\sqrt{s_0^2+a^2}}.$$

The Kirchhoff (also referred to as the Helmholtz-Kirchhoff) approximation can be easily determined by averaging the RS-1 and RS-2 formulations [see Eq. (5.20)]. The Fresnel and Fraunhofer approximations in this case can also be evaluated in a manner similar to the approach described in Sec. 5.7.2. Specifically, for the Fresnel approximation at axial points, the expression can be written as:

$$U(z_0) = -jU_0 \frac{\exp\left[-jk\left(z_0 - s_0\right)\right]}{z_0 - s_0} \left\{1 - \exp\left[-jk\frac{a^2}{2}\left(\frac{1}{z_0} - \frac{1}{s_0}\right)\right]\right\}, \tag{5.154}$$

$$U(z_0 = s_0) = -jkU_0 \frac{a^2}{2s_0^2}, \tag{5.155}$$

$$\text{with } z_0 \gg \left(\frac{\pi}{4\lambda_0} a^4\right)^{1/3}, \quad \text{and } s_0 \gg a.$$

The corresponding relative intensity for points along the z-axis for the Fresnel approximation is given by:

$$\frac{I(z_0)}{I_{sw}} = s_0^2 \frac{2}{(z_0 - s_0)^2} \left\{1 - \cos\left[k\frac{a^2}{2}\left(\frac{1}{z_0} - \frac{1}{s_0}\right)\right]\right\}, \tag{5.156}$$

$$\frac{I(z_0)}{I_{sw}} = \frac{1}{2}k^2 a^2.$$

For the Fraunhofer approximation, the following results along the z-axis can be easily obtained:

$$U_{ff}(z_0) = U_0 \frac{\exp(-jkz_0)}{z_0} \left[\exp\left(+jk\sqrt{a^2 + s_0^2}\right) - \exp(+jks_0)\right], \tag{5.157}$$

$$\frac{I_{ff}(z_0)}{I_{sw}} = \frac{s_0^2}{z_0^2} \left[2 - 2\cos\left[k\left(\sqrt{a^2 + s_0^2} - s_0\right)\right]\right]. \tag{5.158}$$

Example 5.7.17:

Calculate the relative intensities resulting from the diffraction of a converging spherical wave incident on a circular aperture of radius $a = 10\,\mu$m. The incident spherical wave converges to the point $z = s_0 = 50\,\mu m$ and has a freespace wavelength is $\lambda_0 = 1\,\mu m$. The intensities should be evaluated along the z-axis.

Solution: The calculated relative intensities, I/I_{sw}, are shown in Figs. 5.33(a) (linear scale) and 5.33(b) (logarithmic scale). It is evident that the diffracted intensity peaks at $z \simeq 40\,\mu m \neq s_0$. The positions of the diffracted minima and maxima can be approximated using the procedure outlined in previous sections. The peak of the diffracted intensity is challenging to determine analytically. However, as shown in Ref. [294], for a given s_0, the required aperture radius, a, can be determined such that the actual peak diffracted intensity occurs at the geometrical focus, $z = s_0$. The radius can be obtained by solving the following polynomial in terms of F_2, which is close to 1/3,

$$F_2^3 + F_2^2 + (24\pi^2 s_0^2 + 1)F_2/3 - (8\pi^2 s_0^2 - 1)/3 = 0,$$

where F_2 is defined previously. For $s_0 = 50\,\mu m$ the resulting nearest to 1/3 root is $F_{2,\rho} = 0.33333$ which leads to an aperture radius $a = [(s_0/F_{2,\rho})^2 - s_0^2]^{1/2} = 141.42\,\mu m$.

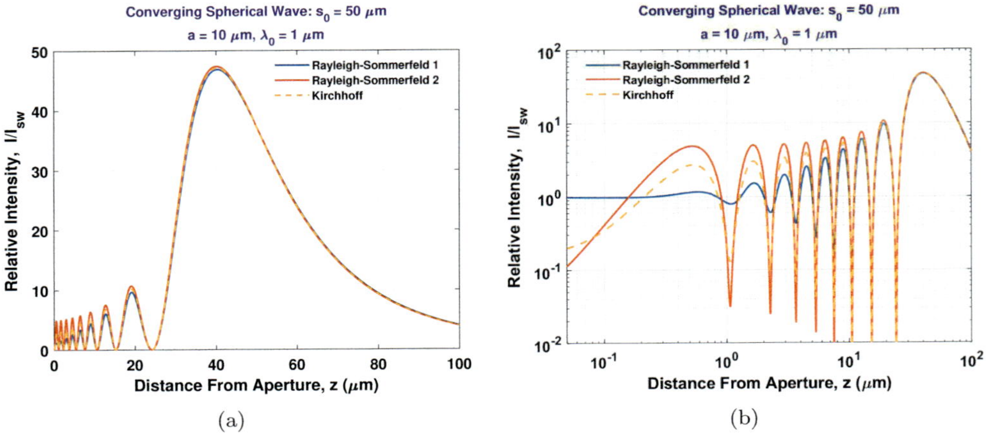

FIGURE 5.33: (a) The diffracted field relative intensity, I/I_{sw}, in the axial z direction for an incident converging (to $z = s_0 = 50\,\mu m$) spherical wave to a circular aperture of radius $a = 10\,\mu m$. The freespace wavelength is $\lambda_0 = 1.0\,\mu m$. Observe that the focal distance of the converging diffracted wave is not at $z = 50\,\mu m$ but at $z \simeq 40\,\mu m$. (b) Same as (a) but in logarithmic scale.

In the radial direction, the diffraction integrals for the RS-1 and RS-2 formulations, as well as for the Fresnel approximation, for an off-axis random point P, are given by:

$$U_{RS_1}(x_0, y_0, z_0) = \int_0^a \int_0^{2\pi} \left(jk + \frac{1}{R}\right) \frac{\exp(-jkR)}{2\pi R} \frac{z_0}{R} \left[U_0 \frac{\exp(+jkR_s)}{R_s}\right] r'_{\scriptscriptstyle T} dr'_{\scriptscriptstyle T} d\phi', \quad (5.159)$$

$$U^a_{RS_1}(x_0, y_0, z_0) \simeq jkz_0 \int_0^a \frac{\exp(-jkR_0)}{R_0^2} J_0\left(\frac{kr_{\scriptscriptstyle T} r'_{\scriptscriptstyle T}}{R_0}\right) \left[U_0 \frac{\exp(+jkR_s)}{R_s}\right] r'_{\scriptscriptstyle T} dr'_{\scriptscriptstyle T}, \quad (5.160)$$

$$U_{RS_2}(x_0, y_0, z_0) = \int_0^a \int_0^{2\pi} \frac{\exp(-jkR)}{2\pi R} \left[U_0 \left(jk - \frac{1}{R_s}\right) \frac{\exp(+jkR_s)}{R_s} \frac{s_0}{R_s}\right] r'_{\scriptscriptstyle T} dr'_{\scriptscriptstyle T} d\phi', \quad (5.161)$$

$$U^a_{RS_2}(x_0, y_0, z_0) \simeq \int_0^a \frac{\exp(-jkR)}{2\pi R} J_0\left(\frac{kr_{\scriptscriptstyle T} r'_{\scriptscriptstyle T}}{R_0}\right) \left[U_0 \left(jk - \frac{1}{R_s}\right) \frac{\exp(+jkR_s)}{R_s} \frac{s_0}{R_s}\right] r'_{\scriptscriptstyle T} dr'_{\scriptscriptstyle T}, \quad (5.162)$$

$$U(x_0, y_0, z_0) = jk\frac{e^{-jkz_0}}{z_0} U_0 \int_0^a \exp\left(-jk\frac{r^2_{\scriptscriptstyle T} + r'^2_{\scriptscriptstyle T}}{2z_0}\right) J_0\left(\frac{kr_{\scriptscriptstyle T} r'_{\scriptscriptstyle T}}{R_0}\right) \frac{\exp(+jkR_s)}{R_s} r'_{\scriptscriptstyle T} dr'_{\scriptscriptstyle T}, \quad (5.163)$$

where

$$R = \left[r^2_{\scriptscriptstyle T} + r'^2_{\scriptscriptstyle T} - 2r_{\scriptscriptstyle T} r'_{\scriptscriptstyle T} \cos(\phi - \phi') + z_0^2\right]^{1/2}, \quad R_s = \left(r'^2_{\scriptscriptstyle T} + s_0^2\right)^{1/2}, \quad \text{and}$$

$$r_{\scriptscriptstyle T} = (x_0^2 + y_0^2)^{1/2}, \quad r'_{\scriptscriptstyle T} = (x'^2 + y'^2)^{1/2}, \quad \phi = \tan^{-1}\left(\frac{y_0}{x_0}\right), \quad \phi' = \tan^{-1}\left(\frac{y'}{x'}\right).$$

Once again, all the integrals can be determined through two-dimensional numerical integration (for the exact RS-1 and RS-2 formulations) or via one-dimensional integration (for the approximate RS-1 and RS-2 formulations, $U^a_{RS_1}$ and $U^a_{RS_2}$, respectively, and the Fresnel approximation). The Kirchhoff formulation integral can be computed from the RS-1 and RS-2 results using Eq. (5.20).

Example 5.7.18:

This example uses parameters similar to those in Example 5.7.17, but this time, the relative intensities are calculated in the radial direction for $z_0 = 10\,\mu$m. Only the rigorous relative intensities are considered, as the z_0 distance is small. Repeat the procedure for $z_0 = 40\,\mu$m, incorporating the approximate RS-1 and RS-2 formulations as well.
Solution: The corresponding relative intensities, based on the previously derived off-axis equations, are shown in Fig. 5.34. For $z_0 = 40\,\mu m$, where the incident wave is focused, as evident from Fig. 5.33(a), the radial relative intensity distribution is displayed in Figs. 5.35(a) and 5.35(b). The focusing of the diffracted wave is visualized in Fig. 5.35(b). It is clear that the approximate RS-1 and RS-2 formulations (1D integrals) closely match their RS-1 and RS-2 counterparts (2D integrals).

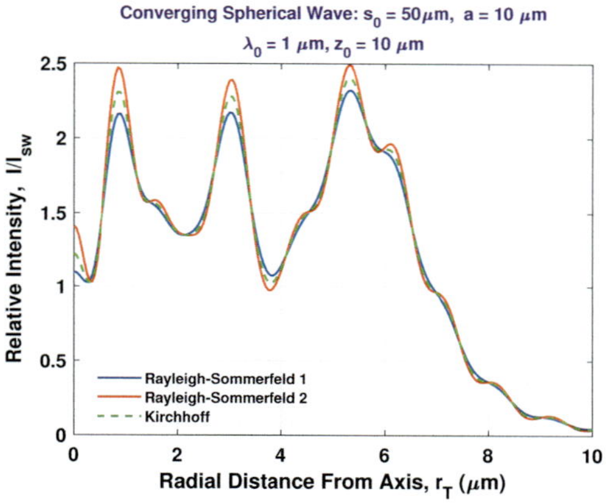

FIGURE 5.34: The diffracted field relative intensity, I/I_{sw}, in the radial r_T direction for an incident converging (to $z = s_0 = 50\,\mu m$) spherical wave to a circular aperture of radius $a = 10\,\mu m$. The freespace wavelength is $\lambda_0 = 1.0\,\mu m$, at $z_0 = 10\,\mu m$.

The results for an incident converging spherical wave can be easily derived for cases involving circular, elliptical, or annular obstacles or apertures by applying the procedures outlined in Sections 5.7.4, 5.7.6, and 5.7.7.

5.8 Scalar Diffraction by a Rectangular Apertures, Obstacles, and Annuli

In this section, several straightforward applications of scalar diffraction involving rectangular apertures, obstacles, or annuli are presented. The incident wave is typically either a plane wave (usually normal incidence) or a spherical wave (or cylindrical wave in two-dimensional problems).

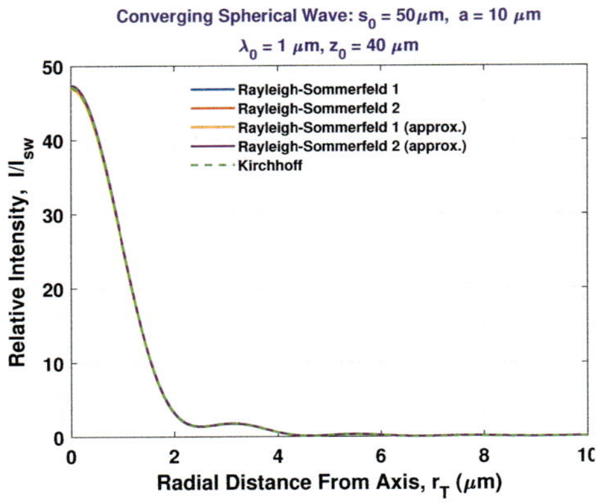

(a)

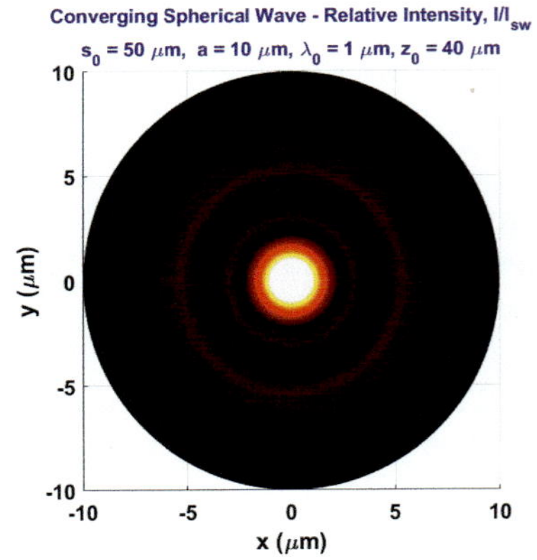

(b)

FIGURE 5.35: (a) The diffracted field relative intensity, I/I_{sw}, in the radial r_T direction for an incident converging (to $z = s_0 = 50\,\mu m$) spherical wave to a circular aperture of radius $a = 40\,\mu m$ where the peak (focus) of the converging wave occurs. The freespace wavelength is $\lambda_0 = 1.0\,\mu m$ at $z_0 = 10\,\mu m$. (b) Same as (a) in the xy plane.

5.8.1 Slit Aperture

The slit aperture problem is analyzed first. The configuration is illustrated in Fig. 5.36. This problem can be simplified to a two-dimensional case by assuming the slit extends infinitely along the y, y' directions. It is a classic example often introduced in optics textbooks (see, for example, [20, 42]), typically starting with the Fraunhofer regime. As demonstrated in Sec. 5.4.2, this regime yields a diffracted field proportional to the scaled Fourier transform of the aperture.

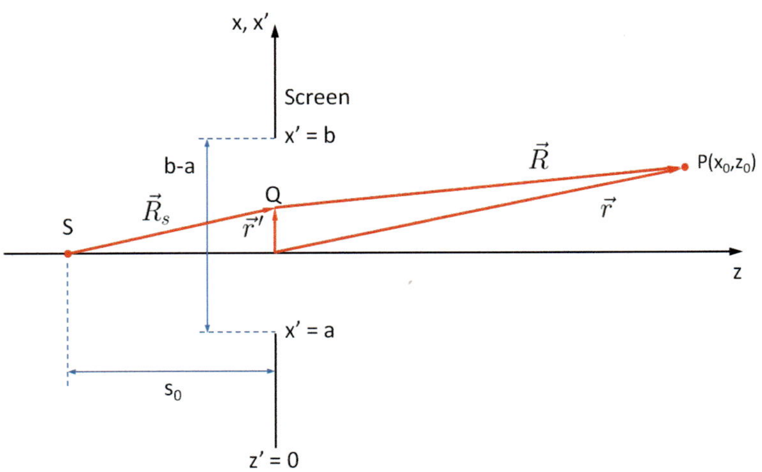

FIGURE 5.36: A slit aperture of width $b-a$ (extending from $x' = a$ to $x' = b$) is considered on an opaque screen. It is assumed that the slit extends infinitely in the y-direction. The diffracted field for the general off-axis point $P(x_0, z_0)$ is of interest. For an incident plane wave $S \to -\infty$ ($s_0 \to \infty$). For an incident diverging cylindrical wave $z = -s_0$ (as shown). For an incident converging cylindrical wave $z = s_0$ (not shown).

In this section, the slit is initially examined using the more rigorous RS-1 and RS-2 formulations. The Fresnel approximation is then introduced, along with the associated Fresnel integrals and Cornu spiral. Finally, the Fraunhofer approximation is also discussed. The incident field is assumed to be either a plane wave of the form $U_{inc} = U_0 \exp(-jkz)$ or a cylindrical wave.

5.8.1.1 Rayleigh-Sommerfeld Formulations

Since the slit aperture problem is analyzed as a two-dimensional scalar diffraction problem, the two-dimensional Green's functions, as introduced in Sec. 5.3.1, are employed. Using Eqs. (5.23) and (5.24), the diffracted fields based on the RS-1 and RS-2 formulations are expressed by the following equations:

$$U_{RS1}(x_0, z_0) = \int_a^b \left[U_{inc} \frac{k}{2j} H_1^{(2)}(kR_T) \frac{z_0}{R_T} \right]_{z'=0} dx', \qquad \text{and} \tag{5.164}$$

$$U_{RS2}(x_0, z_0) = \int_a^b \left[-\frac{j}{2} H_0^{(2)}(kR_T) \frac{\partial U_{inc}}{\partial n} \right]_{z'=0} dx', \qquad \text{where,} \tag{5.165}$$

$$R_T = \left[(x_0 - x')^2 + z_0^2 \right]^{1/2}.$$

It is noted that the $x'y'z'$ and xyz coordinate systems have a common z-axis, with $z' = 0$ in the integrals affecting only the terms related to the incident field. Both integrals must be evaluated numerically by performing the one-dimensional integration outlined in Eqs. (5.164) and (5.165). Even along the z-axis, numerical evaluation is required since no analytical solutions exist, unlike the circular aperture case, where analytical on-axis solutions are available. The Helmholtz-Kirchhoff formulation can be readily determined using Eq. (5.20). Approximate expressions for Eqs. (5.164) and (5.165) can be derived by employing the asymptotic forms of the Hankel functions. According to Ref. [293], the asymptotic expressions for Hankel functions with large arguments are given by:

$$H_\nu^{(2)}(w) \simeq \sqrt{\frac{2}{\pi w}} \exp\left[-j\left(w - \frac{\nu\pi}{2} - \frac{\pi}{4} \right) \right], \qquad \text{for} \quad w \gg 1.$$

By applying the asymptotic expressions to Eqs. (5.164) and (5.165), the following approximations are derived:

$$U_{RS1}^a(x_0, z_0) = e^{j3\pi/4} \int_a^b \left[U_{inc} \frac{k}{2j} \sqrt{\frac{2}{\pi k R_T}} \exp(-jkR_T) \frac{z_0}{R_T} \right]_{z'=0} dx', \qquad \text{and} \tag{5.166}$$

$$U_{RS2}^a(x_0, z_0) = e^{j\pi/4} \int_a^b \left[-\frac{j}{2} \sqrt{\frac{2}{\pi k R_T}} \exp(-jkR_T) \frac{\partial U_{inc}}{\partial n} \right]_{z'=0} dx'. \tag{5.167}$$

The incident field U_{inc} can represent any scalar electromagnetic field. In this section, the cases of a normally incident plane wave, as well as diverging and converging cylindrical waves, are considered. The cylindrical wave is chosen over the spherical wave to align with the two-dimensional nature of the problem under analysis. Cylindrical waves have their focal points along the z-axis. Consequently, the incident field can be expressed as:

$$U_{inc}(x', z') = \begin{cases} U_0 \exp(-jkz'), & \text{(plane wave)} \\ U_0 \dfrac{\exp(-jkR_s)}{\sqrt{R_s}}, & R_s = [x'^2 + (z' + s_0)^2]^{1/2}, \quad \text{(diverging)}, \\ U_0 \dfrac{\exp(+jkR_s)}{\sqrt{R_s}}, & R_s = [x'^2 + (z' - s_0)^2]^{1/2}, \quad \text{(converging)}, \end{cases} \tag{5.168}$$

More general incident waves can also be considered without difficulty. For example, plane waves incident at an angle or cylindrical waves with arbitrarily positioned focal points can be readily analyzed. Additionally, cylindrical Gaussian beams can be treated within the same formulation.

5.8.1.2 Fresnel and Fraunhofer Formulations

Case of Normally Incident Plane Wave:

For the Fresnel approximation, it is necessary to apply the asymptotic expression of the Hankel functions appearing in Eq. (5.166). Assuming a normally incident plane wave, as described in Eq. (5.168), the following steps are performed to derive the Fresnel approximation:

$$U(x_0, z_0) \simeq \int_a^b \left[U_{inc} \frac{k}{2j} \sqrt{\frac{2}{\pi k R_T}} \exp\left[-j\left(kR_T - \frac{3\pi}{4} \right) \right] \frac{z_0}{R_T} \right]_{z'=0} dx'$$

$$\simeq U_0 \frac{k}{2j} \sqrt{\frac{2}{\pi k z_0}} e^{j3\pi/4} e^{-jkz_0} \int_a^b \exp\left[-j\frac{k}{2z_0}(x_0 - x')^2 \right] dx'.$$

The last integral can be converted into a standard Fresnel integral by applying the following transformation:

$$\frac{\pi}{2}u^2 = \frac{k}{2z_0}(x_0 - x')^2, \quad \text{resulting in}$$

$$\int_a^b \exp\left[-j\frac{k}{2z_0}(x_0 - x')^2 \right] dx' = \sqrt{\frac{\lambda z_0}{2}} \int_{u_1}^{u_2} \exp\left(-j\frac{\pi}{2}u^2 \right) du, \quad \text{where,}$$

$$u_1 = \sqrt{\frac{2}{\lambda z_0}}(x_0 - b),\ u_2 = \sqrt{\frac{2}{\lambda z_0}}(x_0 - a). \tag{5.169}$$

The integral over u can be written as:

$$\int_0^w \exp\left(-j\frac{\pi}{2}u^2 \right) du = C(w) - jS(w), \quad \text{where}$$

$$C(w) = \int_0^w \cos\left(\frac{\pi}{2}u^2 \right) du,\ S(w) = \int_0^w \sin\left(\frac{\pi}{2}u^2 \right) du.$$

The functions $C(w)$ and $S(w)$ are known as the Fresnel integrals [20,42,91]. These functions are readily available in many modern computational software packages such as MATLAB and Mathematica. The locus of points $[C(w), \pm S(w)]$ as a function of the parameter w forms a spiral in the CS-plane, known as the *Cornu spiral.* An example of the Cornu spiral is shown in Fig. 5.37(a) for the convention of the forward-propagating wave $e^{j(kr-\omega t)}$, and in Fig. 5.37(b) for the convention of the forward-propagating wave $e^{-j(kr-\omega t)}$. Since this chapter adopts the second convention, the corresponding Cornu spiral is shown in Fig. 5.37(b). Using the Fresnel integrals, C and S, the diffracted field under the Fresnel approximation for the point $P_0(x_0, z_0)$ is given by:

$$U(x_0, z_0) = U_0 e^{j\pi/4} \sqrt{\frac{k}{2\pi z_0}} \sqrt{\frac{\lambda z_0}{2}} e^{-jkz_0} \Big\{ [C(u_2) - C(u_1)] - j\,[S(u_2) - S(u_1)] \Big\} =$$

$$= \frac{U_0}{\sqrt{2}} e^{j\pi/4} e^{-jkz_0} \Big\{ [C(u_2) - C(u_1)] - j\,[S(u_2) - S(u_1)] \Big\}. \tag{5.170}$$

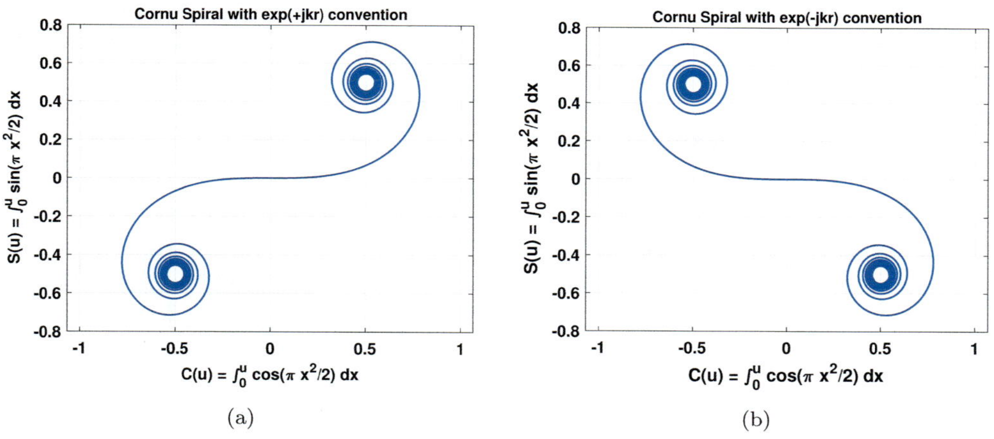

FIGURE 5.37: (a) The Cornu spiral, $[C(u), S(u)]$, corresponding to the convention of the positively propagating wave with a phasor of the form $\exp(+jkr)$. As $u \to \pm\infty$, the Cornu spiral asymptotically approaches $(\pm 0.5, \pm 0.5)$. (b) The Cornu spiral, $[C(u), -S(u)]$, corresponding to the convention of the positively propagating wave with a phasor of the form $\exp(-jkr)$. As $u \to \pm\infty$, the Cornu spiral asymptotically approaches $(\mp 0.5, \pm 0.5)$. The value of u is varying from -7 to $+7$ for both plots.

From the previous equation, it is evident that for an infinite slit, $u_2 \to +\infty$ and $u_1 \to -\infty$, yielding $U(x_0, z_0) = U_0 \exp(-jkz_0)$, as expected. Therefore, U_0 represents the amplitude of the unobstructed field. The corresponding intensity in the Fresnel approximation is given by:

$$
\begin{aligned}
I(x_0, z_0) &= \frac{|U_0|^2}{4Z}\frac{1}{2}\left\{[C(u_2) - C(u_1)]^2 + [S(u_2) - S(u_1)]^2\right\} = \\
&= \frac{I_u}{2}\left\{[C(u_2) - C(u_1)]^2 + [S(u_2) - S(u_1)]^2\right\},
\end{aligned}
\tag{5.171}
$$

where $I_u = |U_0|^2/4Z$ is the intensity of the unobstructed light. This is evident because, for an infinite slit, $u_2 \to +\infty$ and $u_1 \to -\infty$, which leads to $I(x_0, z_0) = I_u$, as expected for a propagating plane wave without any obstacle.

The Fraunhofer approximation, U_{ff}, can also be determined starting from the Fresnel approximation as follows:

$$
\begin{aligned}
U_{ff}(x_0, z_0) &\simeq \int_a^b \left[U_{inc}\frac{k}{2j}\sqrt{\frac{2}{\pi k R_T}}\exp\left[-j\left(kR_T - \frac{3\pi}{4}\right)\right]\frac{z_0}{R_T}\right]_{z'=0} dx' \\
&\simeq U_0\frac{k}{2j}\sqrt{\frac{2}{\pi k z_0}}e^{j3\pi/4}e^{-jkz_0}\int_a^b \exp\left[-j\frac{k}{2z_0}(x_0 - x')^2\right]dx' \\
&\simeq U_0 e^{j\pi/4}\sqrt{\frac{k}{2\pi z_0}}e^{-jkz_0}\exp\left(-j\frac{k}{2z_0}x_0^2\right)\int_a^b \exp\left[j\frac{k}{z_0}x_0 x'\right]dx' = \\
&= U_0 e^{j\pi/4}\sqrt{\frac{k}{2\pi z_0}}e^{-jkz_0}\exp\left(-j\frac{k}{2z_0}x_0^2\right)\int_{-\infty}^{+\infty} \exp\left[j\frac{k}{z_0}x_0 x'\right]t_a(x')dx' = \\
&= U_0 e^{j\pi/4}\sqrt{\frac{k}{2\pi z_0}}e^{-jkz_0}\exp\left(-j\frac{k}{2z_0}x_0^2\right)\mathcal{F}\left\{t_a(x')\right\}\Big|_{k_x = kx_0/z_0}.
\end{aligned}
$$

In the previous expression $t_a(x')$ is the aperture amplitude transmittance function, which can be written as $t_a(x') = \text{rect}\{[x-(a+b)/2]/(b-a)\}$. Then taking the Fourier transform of t_a, the resulting Fraunhofer result is:

$$U_{ff}(x_0, z_0) = U_0 e^{j\pi/4}\sqrt{\frac{k}{2\pi z_0}}e^{-jkz_0}\exp\left(-j\frac{k}{2z_0}x_0^2\right)$$
$$\left\{(b-a)\exp\left(j\frac{a+b}{2}\frac{kx_0}{z_0}\right)\frac{\sin\left[\frac{kx_0}{z_0}\left(\frac{b-a}{2}\right)\right]}{\frac{kx_0}{z_0}\left(\frac{b-a}{2}\right)}\right\}. \tag{5.172}$$

The corresponding far-field intensity is given by:

$$I_{ff} = \underbrace{\frac{|U_0|^2}{4Z}\frac{k}{2\pi z_0}(b-a)^2}_{=I_0(z_0)}\left\{\frac{\sin\left[\frac{kx_0}{z_0}\left(\frac{b-a}{2}\right)\right]}{\frac{kx_0}{z_0}\left(\frac{b-a}{2}\right)}\right\}^2 = I_0(z_0)\,\text{sinc}^2\left[\frac{x_0}{\lambda z_0}(b-a)\right], \tag{5.173}$$

Here, $\text{sinc}(u) = \sin(\pi u)/(\pi u)$ is the standard "sinc" function. The maximum of the far-field diffracted intensity is $I_0(z_0)$ and occurs along the z-axis, i.e., when $x_0 = 0$. The minima of the diffracted intensity can be easily determined from the zeros of the sine term (excluding the origin, which corresponds to the maximum) and are given by:

$$\frac{kx_0}{z_0}\left(\frac{b-a}{2}\right) = m\pi \Longrightarrow x_{0m}^{min} = m\lambda\frac{z_0}{b-a}, \qquad m = \pm 1, \pm 2, \cdots. \tag{5.174}$$

Between each pair of minima, there are smaller secondary maxima, the exact locations of which can only be determined numerically. However, these secondary maxima are much smaller than the main maximum and diminish as the point of interest moves away from the z-axis. These are commonly referred to as the sidelobes of the $\text{sinc}^2(*)$ function.

It should be noted that the Fraunhofer approximation destroys the space invariance property of the RS-1, RS-2, and Fresnel approximations. Consequently, shifting the aperture (i.e., shifting the "rect" function) along the x'-axis will produce the same intensity pattern as the unshifted aperture, which is physically incorrect. This limitation arises from the Fraunhofer approximation.

Example 5.8.1:

In this example, the diffraction of a plane wave with freespace wavelength $\lambda_0 = 1\,\mu$m, normally incident on a slit, is demonstrated numerically. Consider a slit with a width of $10\,\mu$m (where $a = -5\,\mu$m and $b = +5\,\mu$). The task is to calculate the normalized intensities for the RS-1, RS-2, Kirchhoff, as well as Fresnel and Fraunhofer formulations.

Solution: Applying the previous equations, the diffracted field can be determined numerically at any arbitrary point for positive z_0 values. For example, the normalized diffracted intensities $|U|^2/|U_0|^2$ are shown in Fig. 5.38(a) for the RS-1, RS-2, and Kirchhoff formulations along the x-direction for $z_0 = 1\,\mu m = \lambda_0$ (which is extremely close to the aperture). The approximate expressions of the RS-1 and RS-2 formulations are also presented alongside the exact expressions in Fig. 5.38(b). It is evident that the approximate expressions for RS-1 and RS-2 are very close to the exact results, implying that the use of the Hankel Green's functions is unnecessary for most practical cases.

Furthermore, the normalized diffracted intensities for the RS-1, Fresnel, and Fraunhofer approximations (for z_0 distances where they are meaningful) are shown for various distances

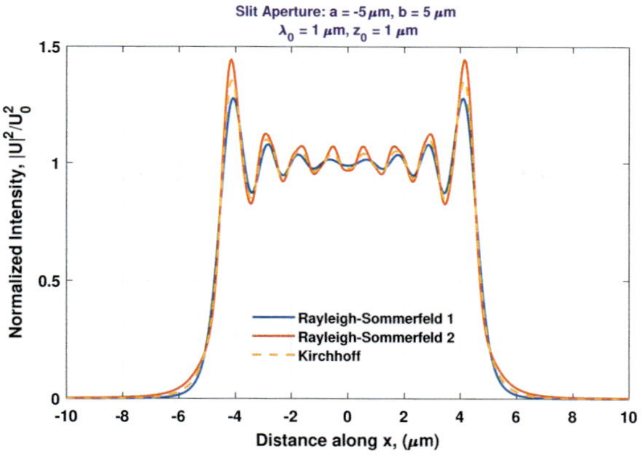

(a)

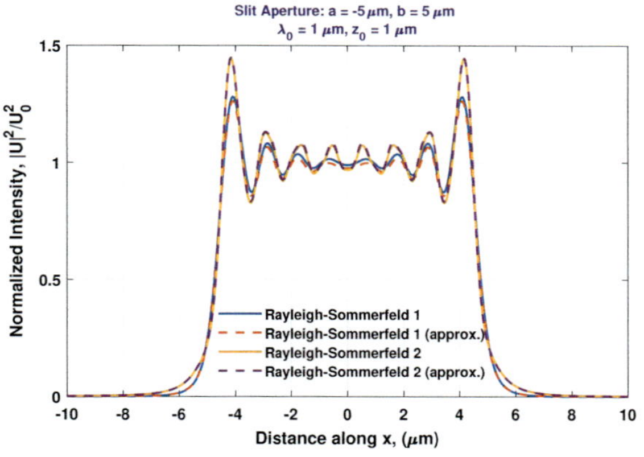

(b)

FIGURE 5.38: (a) The normalized intensity of the diffracted field, $|U|^2/|U_0|^2$, in the x-direction for $z_0 = 1\,\mu m$. This diffraction pattern is produced by a plane wave incident on a slit aperture with $a = -5\,\mu m$ and $b = 5\,\mu m$. The freespace wavelength is $\lambda_0 = 1.0\,\mu m$. The RS-1, RS-2, and Kirchhoff formulations results are shown. (b) Same as (a) but showing only the RS-1 and RS-2 formulations and their approximate counterparts.

from the slit in Fig. 5.38. If $\Delta x = x - x' = 10\,\mu\text{m}$, then $z_0 \gg [(\pi/4\lambda_0)\Delta x^4]^{1/3} = 34.13\,\mu\text{m}$. Clearly, for Figs. 5.39(a) through 5.39(d), the Fresnel approximation is not valid. Particularly at short distances from the aperture, the Fresnel approximation exhibits nonphysical ripples. For $z_0 = 50\,\mu\text{m}$, in Fig. 5.39(e), the Fresnel results coincide with the RS-1 results. From the Fraunhofer approximation, $z_0 \gg (b-a)^2/\lambda = 100\,\mu\text{m}$. Therefore, when $z_0 = 500\,\mu\text{m}$, the Fraunhofer normalized intensities are very close to the RS-1 and Fresnel counterparts.

Case of Diverging/Converging Incident Cylindrical Wave:

In this section, a diverging or converging cylindrical wave is assumed as the incident field, as described in Eq. (5.168). For simplicity, let's assume a diverging cylindrical wave with its focal point at $z = -s_0$ as shown in Fig. 5.36. The following steps can be written

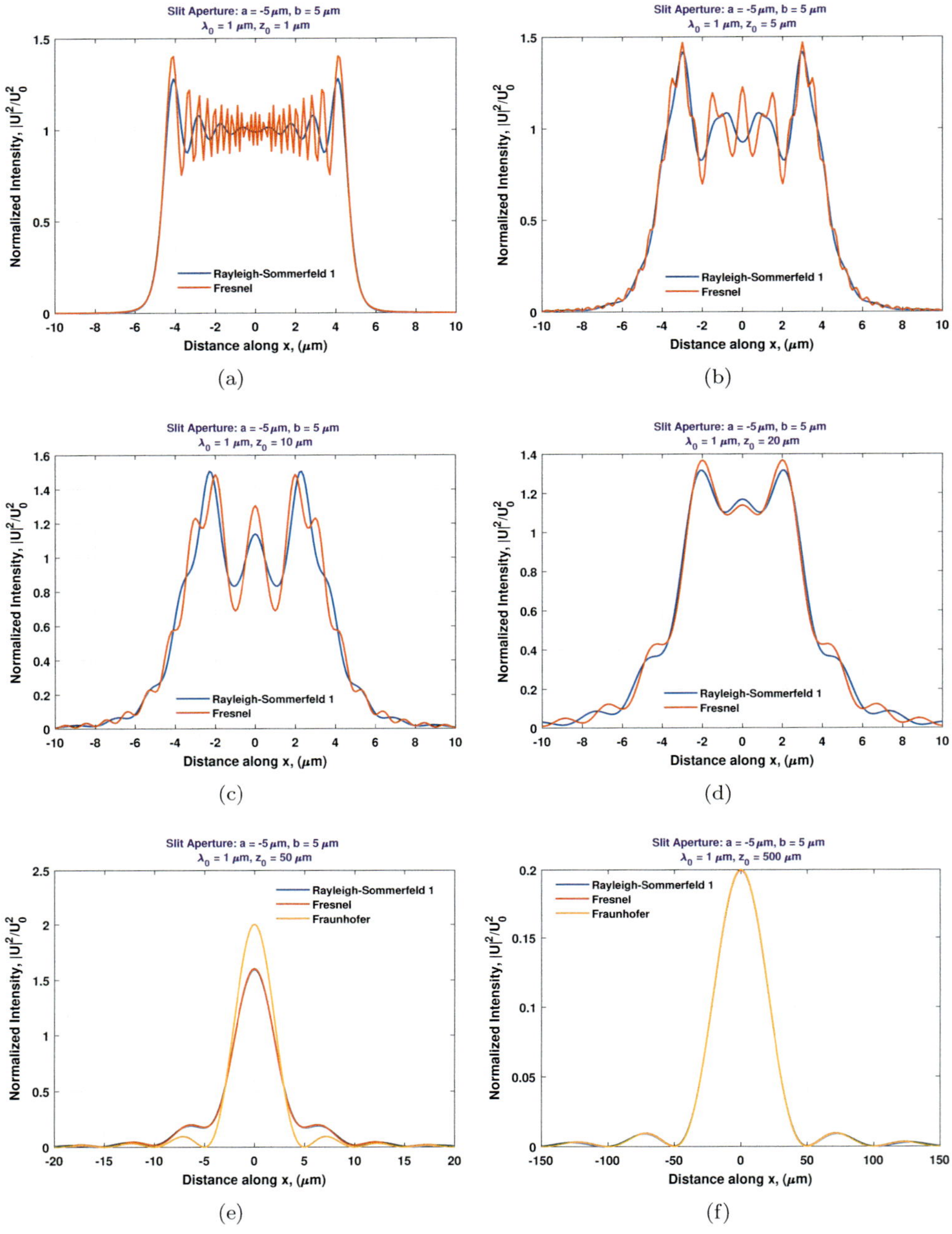

FIGURE 5.39: (a) The normalized intensity of the diffracted field, $|U|^2/|U_0|^2$, in the x-direction for $z_0 = 1\,\mu m$. This pattern is produced by a plane wave incident on a slit aperture of $a = -5\,\mu m$ and $b = 5\,\mu m$. The freespace wavelength is $\lambda_0 = 1.0\,\mu m$. (b) Same as (a) but for $z_0 = 5\,\mu m$. (c) Same as (a) but for $z_0 = 10\,\mu m$. (d) Same as (c) but for $z_0 = 20\,\mu m$. (e) Same as (a) but for $z_0 = 50\,\mu m$. (f) Same as (e) but for $z_0 = 500\,\mu m$.

for the Fresnel approximation:

$$U(x_0, z_0) \simeq U_0 \frac{k}{2j}\sqrt{\frac{2}{\pi k z_0}} e^{j3\pi/4} e^{-jkz_0} \int_a^b \exp\left[-j\frac{k}{2z_0}(x_0 - x')^2\right] \frac{\exp[-jk(x'^2 + s_0^2)^{1/2}]}{\sqrt{(x'^2 + s_0^2)^{1/2}}} dx'$$

$$\simeq U_0 \sqrt{\frac{k}{2kz_0 s_0}} e^{j\pi/4} e^{-jk(z_0+s_0)} \int_a^b \exp\left\{-j\frac{k}{2}\left[\frac{(x_0 - x')^2}{z_0} + \frac{x'^2}{s_0}\right]\right\} dx'. \quad (5.175)$$

The last integral can be decomposed into integrals of the form $I_c = \int \cos(\alpha u^2 + \beta u + \gamma) du$ and $I_s = \int \sin(\alpha u^2 + \beta u + \gamma) du$. These integrals can be expressed in terms of the Fresnel integrals $C(u)$ and $S(u)$ as follows:

$$I_c(u) = \underbrace{\sqrt{\frac{\pi}{2a}}\left[\cos\left(\frac{4\alpha\gamma - \beta^2}{4\alpha}\right) C\left(\frac{\beta + 2\alpha u}{\sqrt{2\pi\alpha}}\right) - \sin\left(\frac{4\alpha\gamma - \beta^2}{4\alpha}\right) S\left(\frac{\beta + 2\alpha u}{\sqrt{2\pi\alpha}}\right)\right]}_{= J_c(u)}$$

$$I_s(u) = \underbrace{\sqrt{\frac{\pi}{2a}}\left[\cos\left(\frac{4\alpha\gamma - \beta^2}{4\alpha}\right) S\left(\frac{\beta + 2\alpha u}{\sqrt{2\pi\alpha}}\right) + \sin\left(\frac{4\alpha\gamma - \beta^2}{4\alpha}\right) C\left(\frac{\beta + 2\alpha u}{\sqrt{2\pi\alpha}}\right)\right]}_{= J_s(u)},$$

where for this case the α, β, and γ coefficients are given by:

$$\alpha = \frac{k}{2}\left(\frac{1}{z_0} + \frac{1}{s_0}\right), \quad \beta = -k\frac{x_0}{z_0}, \quad \gamma = \frac{k}{2}\frac{x_0^2}{z_0}, \quad \text{and} \quad \frac{4\alpha\gamma - \beta^2}{4\alpha} = \frac{kx_0^2}{2(z_0 + s_0)}.$$

Then, the Fresnel approximation for the diverging cylindrical wave becomes:

$$U(x_0, z_0) \simeq U_0 \sqrt{\frac{1}{2(z_0 + s_0)}} e^{j\pi/4} e^{-jk(z_0+s_0)} \Big\{ [J_c(b) - J_c(a)] - j\,[J_s(b) - J_s(a)] \Big\}. \quad (5.176)$$

In a similar manner, it is straightforward to derive the Fresnel approximation for a converging cylindrical wave as:

$$U(x_0, z_0) \simeq U_0 \sqrt{\frac{1}{2(z_0 - s_0)}} e^{j\pi/4} e^{-jk(z_0-s_0)} \Big\{ [J_c(b) - J_c(a)] - j\,[J_s(b) - J_s(a)] \Big\}, \quad (5.177)$$

$$\alpha = \frac{k}{2}\left(\frac{1}{z_0} - \frac{1}{s_0}\right), \quad \beta = -k\frac{x_0}{z_0}, \quad \gamma = \frac{k}{2}\frac{x_0^2}{z_0}, \quad \text{and} \quad \frac{4\alpha\gamma - \beta^2}{4\alpha} = \frac{kx_0^2}{2(z_0 - s_0)}.$$

<u>A More Elegant Manner of Evaluating the Fresnel Approximation Integral:</u>
There is a more elegant way to evaluate the Fresnel integral shown previously for the case of an incident diverging or converging cylindrical wave. The approach is described in Ref. [43] for an arbitrary focal point of the cylindrical wave, and it is adapted here for the specified conditions. From the similar triangles SAC and SBP (see Fig. 5.40) it is straightforward to show that $x'_a = s_0 x_0/(z_0 + s_0) = x_0 - z_0 x_0/(z_0 + s_0)$. Based on these expressions the following equations can be written:

$$\frac{(x' - x'_a)^2}{s_0} = \frac{x'^2}{s_0} + \frac{s_0}{(z_0 + s_0)^2} x_0^2 - \frac{2x' x_0}{z_0 + s_0},$$

$$\frac{(x' - x'_a)^2}{z_0} = \frac{(x' - x_0)^2}{z_0} + \frac{z_0}{(z_0 + s_0)^2} x_0^2 + \frac{2(x' - x_0) x_0}{z_0 + s_0}.$$

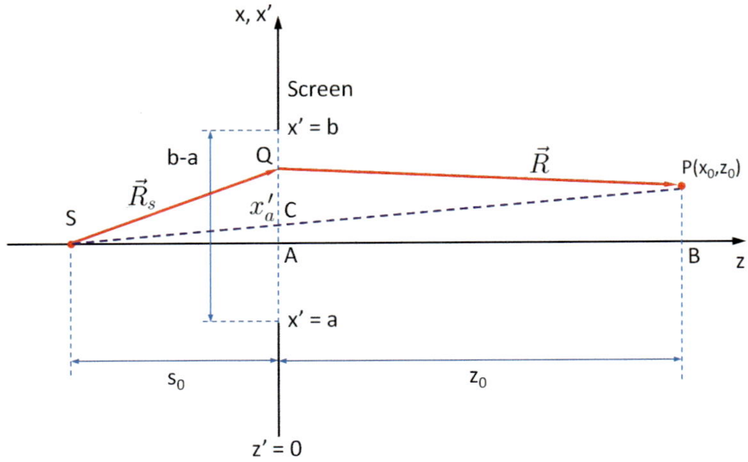

FIGURE 5.40: A slit aperture of width $b-a$ (extending from $x'=a$ to $x'=b$) on an opaque screen. It is assumed that the slit is infinite in the y-direction The diffracted field at the general off-axis point $P(x_0, z_0)$ is sought for the case of a diverging cylindrical wave $z=-s_0$ (as shown). The point x'_a is the intersection of the line between the focal point S and the point of interest P.

By adding the previous equations, it is straightforward to show that:

$$(x'-x'_a)^2\left(\frac{1}{z_0}+\frac{1}{s_0}\right)+\frac{x_0^2}{z_0+s_0}=\frac{(x_0-x')^2}{z_0}+\frac{x'^2}{s_0}.$$

By replacing the exponent inside the integral of Eq. (5.175) with the above result, the expression for the Fresnel approximation becomes:

$$U(x_0,z_0)\simeq U_0\sqrt{\frac{k}{2kz_0s_0}}e^{j\pi/4}e^{-jk(z_0+s_0)}\exp\left(-j\frac{k}{2}\frac{x_0^2}{z_0+s_0}\right)$$
$$\int_a^b \exp\left\{-j\frac{k}{2}\left[(x'-x'_a)^2\left(\frac{1}{z_0}+\frac{1}{s_0}\right)\right]\right\}dx'.$$

The last integral can be transformed into a standard Fresnel integral using the following transformation:

$$\frac{\pi}{2}u^2=\frac{k}{2d}(x'-x'_a)^2,\quad \text{resulting in}$$

$$\int_a^b \exp\left[-j\frac{k}{2d}(x'-x'_a)^2\right]dx'=\sqrt{\frac{\lambda d}{2}}\int_{u_1}^{u_2}\exp\left(-j\frac{\pi}{2}u^2\right)du,\quad \text{where,}$$

$$u_1=\sqrt{\frac{2}{\lambda d}}(a-x'_a),\quad u_2=\sqrt{\frac{2}{\lambda d}}(b-x'_a),\quad \text{and,}$$

$$\frac{1}{d}=\frac{1}{z_0}+\frac{1}{s_0},\ x'_a=\frac{s_0}{z_0+s_0}x_0.$$

Then, the resulting Fresnel approximation for an incident diverging cylindrical wave can be written as:

$$U(x_0,z_0)\simeq U_0\sqrt{\frac{1}{2(z_0+s_0)}}e^{j\pi/4}e^{-jk(z_0+s_0)}\exp\left(-j\frac{k}{2}\frac{x_0^2}{z_0+s_0}\right)F(u_1,u_2) \tag{5.178}$$

where $F(u_1,u_2)=\Big\{[C(u_2)-C(u_1)]-j\,[S(u_2)-S(u_1)]\Big\}.$

Following a similar procedure, the resulting Fresnel approximation for an incident converging cylindrical wave can be written as:

$$U(x_0,z_0) \simeq U_0\sqrt{\frac{1}{2(z_0-s_0)}}e^{j\pi/4}e^{-jk(z_0-s_0)}\exp\left(-j\frac{k}{2}\frac{x_0^2}{z_0-s_0}\right)F(u_1,u_2) \tag{5.179}$$

$$\text{with} \quad \frac{1}{d}=\frac{1}{z_0}-\frac{1}{s_0}, \qquad x_a'=\frac{-s_0}{z_0-s_0}x_0.$$

Observe that the Fresnel results for the converging cylindrical wave can be obtained from those of the diverging cylindrical wave by replacing s_0 with $-s_0$.

The Fraunhofer approximation for an incident diverging or converging cylindrical wave becomes:

$$\begin{aligned}
U_{ff}(x_0,z_0) &\simeq U_0e^{j\pi/4}\sqrt{\frac{k}{2\pi z_0}}e^{-jkz_0}\exp\left(-j\frac{k}{2z_0}x_0^2\right)\int_a^b\exp\left[j\frac{k}{z_0}x_0x'\right]\frac{\exp(\mp jkR_s)}{\sqrt{R_s}}dx' \\
&\simeq U_0e^{j\pi/4}\sqrt{\frac{k}{2\pi z_0s_0}}e^{-jk(z_0\pm s_0)}\exp\left(-j\frac{k}{2z_0}x_0^2\right)\int_a^b\exp\left[j\frac{k}{z_0}x_0x'\right]\exp\left(\mp j\frac{k}{2s_0}x'^2\right)dx' \\
&\simeq U_0e^{j\pi/4}\sqrt{\frac{k}{2\pi z_0s_0}}e^{-jk(z_0\pm s_0)}\exp\left(-j\frac{k}{2z_0}x_0^2\right) \\
&\quad\int_{-\infty}^{+\infty}\exp\left[j\frac{k}{z_0}x_0x'\right]\left\{\exp\left(\mp j\frac{k}{2s_0}x'^2\right)t_a(x')\right\}dx'.
\end{aligned} \tag{5.180}$$

The upper sign in the double-sign terms corresponds to the diverging cylindrical wave, while the lower sign corresponds to its converging counterpart. In general, the last integral is the convolution of the Fourier transforms of $t_a(x')$ and of $\exp(\mp jkx'^2/2s_0)$, which is difficult to solve analytically. Therefore, it is much easier to perform the one-dimensional integration numerically.

All the diffracted intensities can be determined from the previously specified diffracted fields and are normalized based on the incident cylindrical wave. Therefore, all the relative intensities in this case are defined as:

$$\frac{I}{I_{cw}}=\frac{|U|^2}{|U_0|^2}|s_0|, \tag{5.181}$$

where $I_{cw}=|U_0|^2/4Z|s_0|$ and U corresponds to any of the formulations presented, such as RS-1, RS-2, Kirchhoff, RS-1 approximate, RS-2 approximate, Fresnel, and Fraunhofer.

Example 5.8.2:

In this example, the diffraction of a diverging cylindrical wave by a slit aperture is demonstrated numerically. Assume that the slit has a width of $10\,\mu$m (with $a=-5\,\mu$m and $b=+5\,\mu$) and that the cylindrical wave originates from $z=-50\,\mu m=-s_0$, and has a freespace wavelength of $\lambda_0=1\,\mu m$. The relative intensities I/I_{cw} along the z-direction or x-direction are to be calculated for the RS-1, RS-2, Kirchhoff, Fresnel, and Fraunhofer formulations. Find the same relative intensities for the case of a converging cylindrical to $z=+50\,\mu m=+s_0$.

Solution: Using the previous equations, the resulting relative intensities are shown in Fig. 5.41 for various distances. Specifically, in Fig. 5.41(a), the relative diffracted intensities based on the RS-1, RS-2, and Kirchhoff formulations are presented along the z-axis. It is evident that for $z_0>10\,\mu m=10\lambda_0$, the three formulations practically coincide.

In Fig. 5.41(b), it is interesting to note that the approximate expressions for the RS-1 and RS-2 formulations [see Eqs. (5.166) and (5.167)] are practically identical to the exact RS-1 and RS-2 formulations, except at distances very close to the aperture. Therefore, the use of the asymptotic expansions of the Hankel functions does not significantly affect the resulting diffracted field for all distances of interest.

In Figs. 5.41(c), 5.41(d), 5.41(e), and 5.41(f), the relative diffracted intensities are shown in the transverse direction for various values of the distance from the aperture. The Fresnel approximation is expected to be valid for $z_0 \gg 3.413\,\mu$m. For larger distances, it is evident that even the Fraunhofer approximation gives excellent results (for $z_0 \gg 100\,\mu$m).

At this point, let's consider the case of an incident converging cylindrical wave, focusing at $z = 50\,\mu m = -s_0$, which strikes the same slit aperture. The resulting relative intensities are shown in Fig. 5.42 for various distances. Specifically, in Fig. 5.42(a), the relative diffracted intensities based on the RS-1, RS-2, and Kirchhoff formulations are presented along the z-axis. Again, it is evident that for $z_0 > 10\,\mu m = 10\lambda_0$, the three formulations practically coincide. A significant difference from the diverging cylindrical wave case is that there is a peak in the diffracted intensity (focusing) at a distance from the aperture of approximately $z_0 \simeq 24\,\mu m$. In Fig. 5.42(b), it is again clear that the approximate expressions for RS-1 and RS-2 [see Eqs. (5.166) and (5.167)] are almost identical to the exact RS-1 and RS-2 formulations, except at distances very close to the aperture. Therefore, the same conclusions drawn from the diverging cylindrical wave case also apply here. In Figs. 5.42(c), 5.42(d), 5.42(e), and 5.42(f), the relative diffracted intensities are shown in the transverse direction for various values of the distance from the aperture. For larger distances, it is evident that the Fraunhofer approximation provides excellent results. The special case is shown in Fig. 5.42(e), where the focusing effect of the converging cylindrical wave occurs.

5.8.2 Slit Obstacle

The scalar diffraction analysis for a slit obstacle is relatively straightforward, as it relies on the result derived for a slit aperture in combination with Babinet's principle. For any of the formulations presented, the diffracted field corresponding to the slit obstacle, which is complementary to the slit aperture, is given by [see Eq. (5.48)]:

$$U_{ob}(x_0, z_0) = U_{inc} - U_{ap}(x_0, z_0).$$

Here, $U_{ob}(x_0, z_0)$ represents the diffracted field produced by the slit obstacle, while $U_{ap}(x_0, z_0)$ denotes the diffracted field resulting from a slit aperture (as determined in Sec. 5.8.1). The term U_{inc} corresponds to the incident field, which, for the cases considered, is defined in Eq. (5.168). The diffracted fields may be described using any of the following formulations: RS-1, RS-2, Kirchhoff, RS-1 approximate, RS-2 approximate, Fresnel, or Fraunhofer.

Example 5.8.3:

Numerically demonstrate the diffraction of a plane wave or a cylindrical wave by a slit obstacle. The slit obstacle is assumed to have a width of $10\,\mu$m, with the boundaries at $a = -5\,\mu$m and $b = +5\,\mu$m. The incident wave can either be a plane wave, a diverging cylindrical wave emanating from $z = -50\,\mu m = -s_0$, or a converging cylindrical wave focusing at $z = +50\,\mu m = -s_0$. The freespace wavelength is specified as $\lambda_0 = 1\,\mu m$. The goal is to determine the normalized intensity $|U|^2/|U_0|^2$ in the case of a plane wave, or the relative intensity I/I_{cw} for the cylindrical wave cases. The analysis focuses on the intensity distribution along both the z-direction and the x-direction. These calculations are performed using the RS-1, RS-2, Kirchhoff, Fresnel, and Fraunhofer formulations.

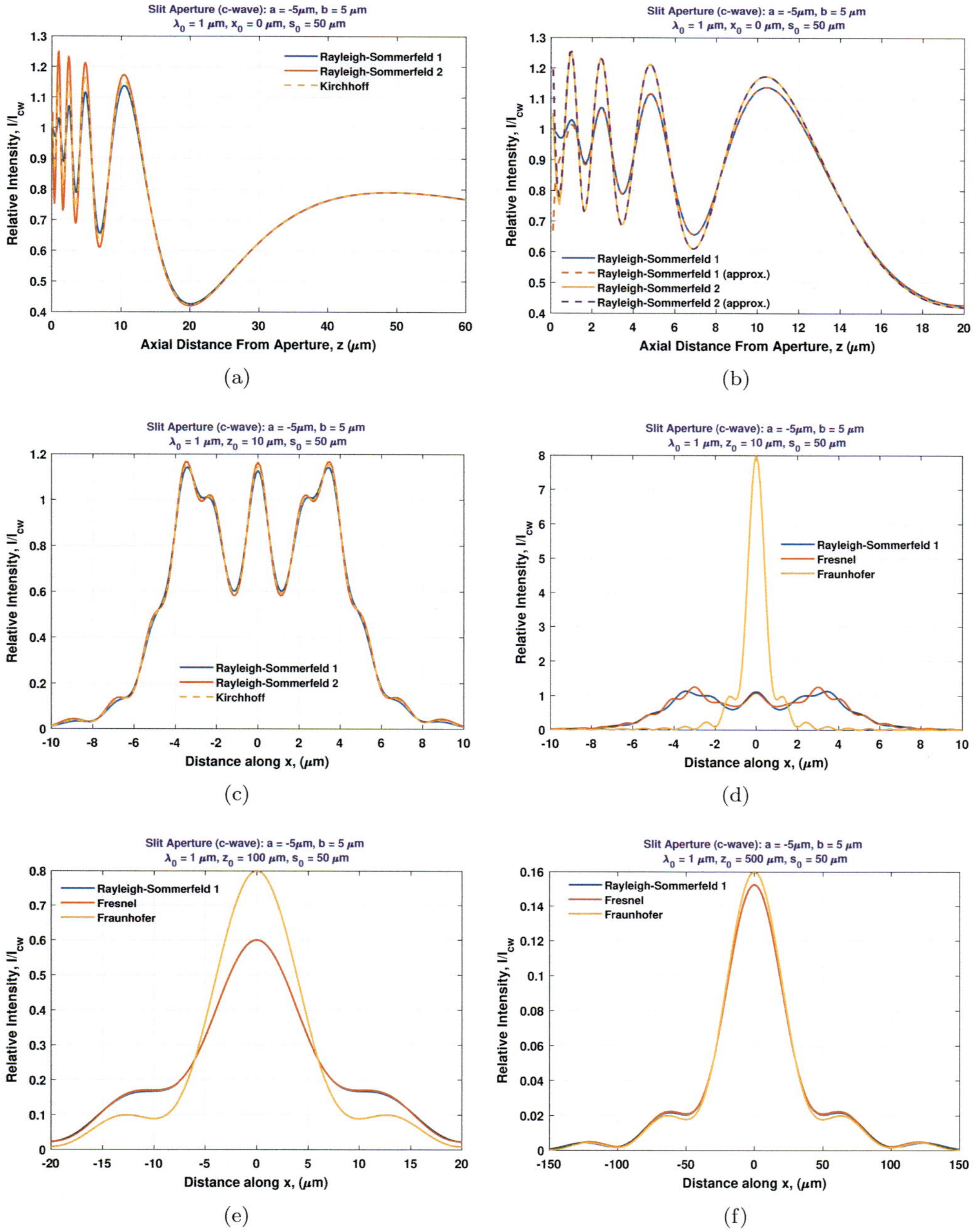

FIGURE 5.41: (a) The diffracted field relative intensity, I/I_{cw}, in the z-direction for an incident diverging cylindrical wave (originating from $z = -s_0 = -50\,\mu m$), incident on a slit aperture of $a = -5\,\mu m$ and $b = 5\,\mu m$. The freespace wavelength is $\lambda_0 = 1.0\,\mu m$. The RS-1, RS-2, and Kirchhoff formulations are shown. (b) Same as (a) but for the comparison of the RS-1 and RS-2 with their approximate formulations. (c) Same as (a) but as a function of the x direction for $z_0 = 10\,\mu m$. (d) Same as (c) but for comparison of RS-1, Fresnel, and Fraunhofer results. (e) Same as (d) but for $z_0 = 100\,\mu m$. (f) Same as (e) but for $z_0 = 500\,\mu m$.

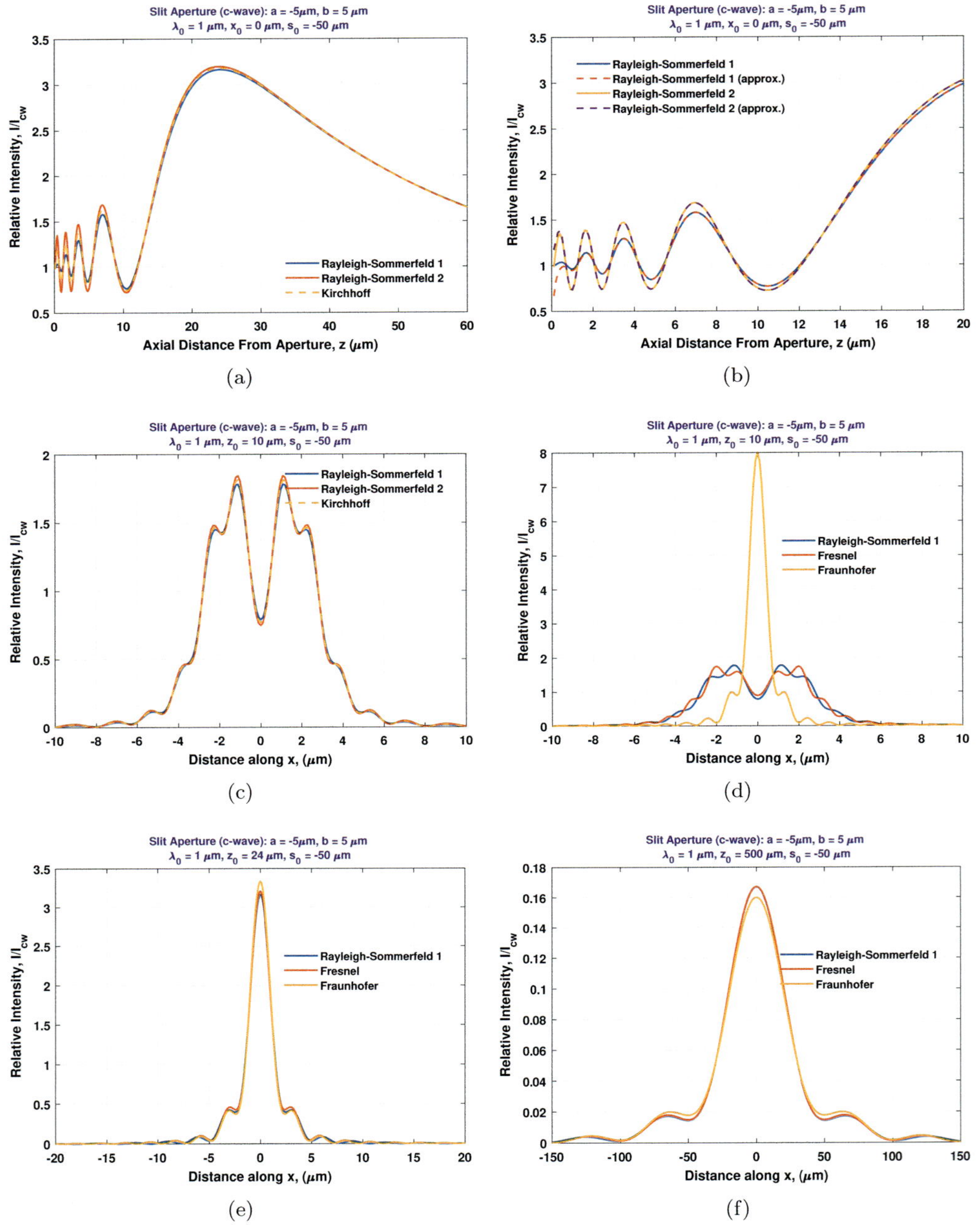

FIGURE 5.42: (a) The diffracted field relative intensity, I/I_{cw}, in the z-direction for an incident converging cylindrical wave (with focal point at $z = -s_0 = 50\,\mu m$) incident on a slit aperture of $a = -5\,\mu m$ and $b = 5\,\mu m$. The freespace wavelength is $\lambda_0 = 1.0\,\mu m$. The RS-1, RS-2, and Kirchhoff formulations are shown. (b) Same as (a) but for the comparison of the RS-1 and RS-2 with their approximate formulations. (c) Same as (a) but as a function of the x direction for $z_0 = 10\,\mu m$. (d) Same as (c) but for comparison of RS-1, Fresnel, and Fraunhofer results. (e) Same as (d) but for $z_0 = 24\,\mu m$ (focal point) of diffracted field. (f) Same as (e) but for $z_0 = 500\,\mu m$.

Solution: The intensities are summarized in Fig. 5.43. Figures 5.43(a) and 5.43(b) present the normalized diffracted intensities for an incident plane wave. Similarly, Figs. 5.43(c) and 5.43(d) display the relative diffracted intensities for an incident diverging cylindrical wave emanating from $z = -50\,\mu m = -s_0$. Finally, Figs. 5.43(e) and 5.43(f) illustrate the relative diffracted intensities for an incident converging cylindrical wave focusing at $z = 50\,\mu m = -s_0$. In the latter case, the diffracted field appears to diverge at $z_0 = 50\,\mu m$, as expected due to the converging cylindrical wave with its focal point at $z_0 = 50\,\mu m$. Beyond the focal point, rapid oscillations in the diffracted intensity become evident to the right of $z_0 = 50\,\mu$m. A more suitable incident field for such scenarios might be a cylindrical Gaussian beam, which does not exhibit divergence at any point in space.

5.8.3 Special Case: Edge Scalar Diffraction

Following the discussion of the slit aperture and obstacle, the scalar diffraction by a semi-infinite edge can be treated as a special case of either configuration. For a slit aperture, the semi-infinite edge corresponds to $a = 0$ and $b = +\infty$, while for a slit obstacle, it corresponds to $a = -\infty$ and $b = 0$. Consequently, no specialized analysis is required. The Fresnel formulation, which utilizes Fresnel integrals, does not exhibit numerical issues, as these integrals are well-defined and calculable at $\pm\infty$. In contrast, the RS-1, RS-2, and Kirchhoff formulations involve one-dimensional integration over a semi-infinite domain (for example, from 0 to ∞). This semi-infinite limit can typically be approximated by a sufficiently large finite limit, chosen to exceed the distances of interest from the edge, ensuring the accuracy of the results.

Example 5.8.4:

To illustrate the diffraction from an edge numerically, consider an edge modeled as a slit aperture with $a = 0$ and $b = +\infty$. The incident wave can be a plane wave, a diverging cylindrical wave emanating from $z = -50\,\mu m = -s_0$, or a converging cylindrical wave focusing at $z = +50\,\mu m = -s_0$. The freespace wavelength is $\lambda_0 = 1.0\,\mu m$. The objective is to determine the normalized intensity $|U|^2/|U_0|^2$ for the plane wave case or the relative intensity I/I_{cw} for cylindrical waves, along both the z-direction and the x-direction. The analysis will use the RS-1, RS-2, Kirchhoff, and Fresnel formulations. In the case of edge diffraction, the Fraunhofer regime is never reached, so Fraunhofer results are not required.
Solution: To approximate the semi-infinite integrals for the RS-1, RS-2, and Kirchhoff formulations, the semi-infinite slit is modeled with a width of $10000\,\mu m = 10000\lambda_0 = b$. The distances of practical interest near the edge are much smaller than the $10000\lambda_0$ width approximation. The resulting intensities are shown in Fig. 5.44. Specifically, in Figs. 5.44(a) and 5.44(b), the normalized diffracted intensities, $|U|^2/|U_0|^2$, are presented as a function of the x-distance at a distance from the edge $z_0 = 1\,\mu m = \lambda_0$. The RS-1, RS-2, Kirchhoff, and Fresnel results are compared. It is observed that the Fresnel results exhibit significant differences from the more accurate RS-based results at this small distance from the edge. For $z_0 = 10\,\mu m = 10\lambda_0$, the same results are shown in Figs. 5.44(c) and 5.44(d). The more rigorous results (RS-1, RS-2, and Kirchhoff) practically coincide, while the Fresnel results still exhibit some differences, especially when compared to the RS-1 results. As the distance from the edge increases, the Fresnel results approach the RS results, as expected.

Finally, in Figs. 5.44(e) and 5.44(f), the relative diffracted intensities, I/I_{cw}, are presented for an incident diverging cylindrical wave ($z = -s_0 = -50\,\mu m$) and a converging cylindrical wave ($z = s_0 = -50\,\mu m$), respectively. In general, the same conclusions as those for the incident plane wave case hold.

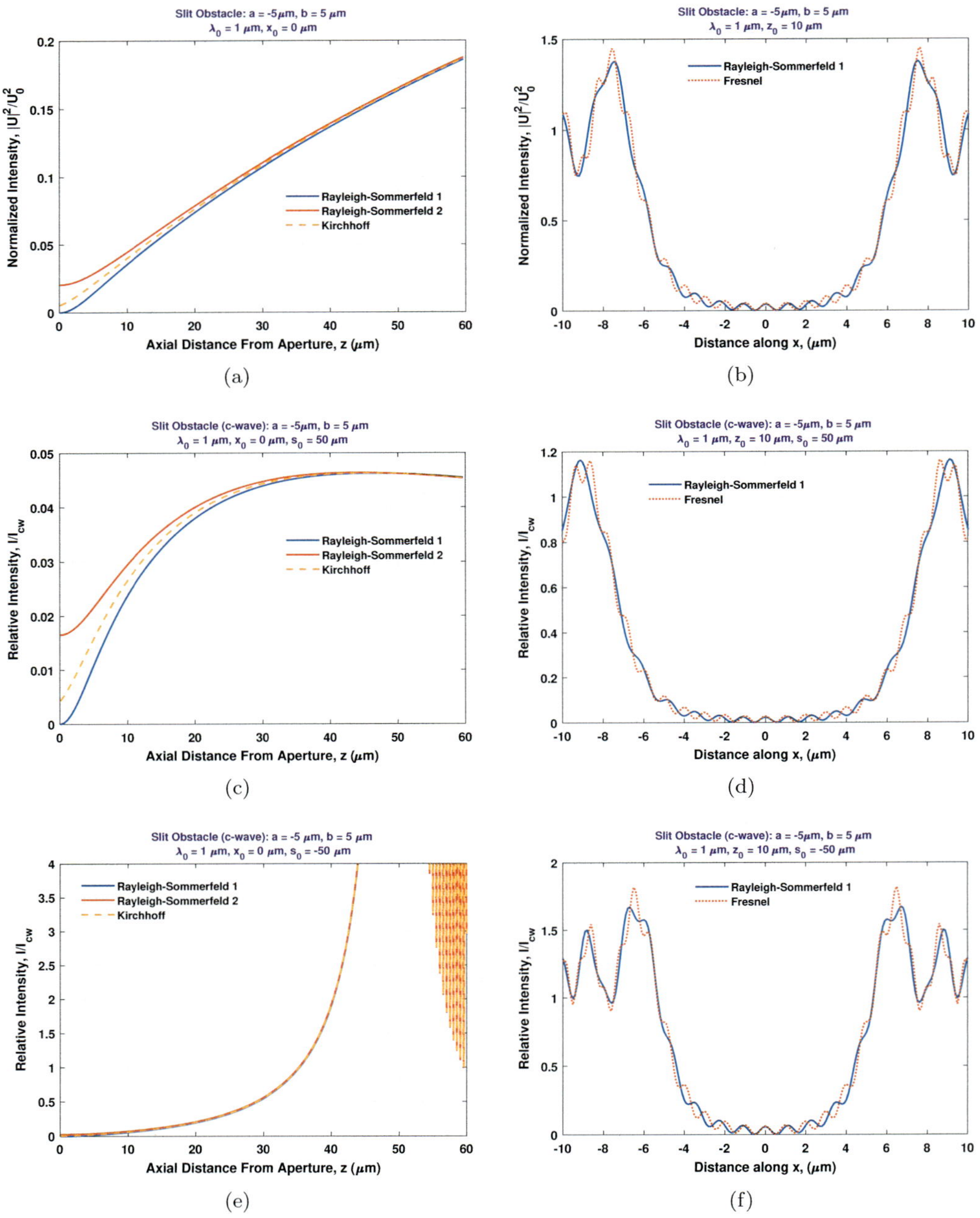

FIGURE 5.43: (a) The relative intensity of the diffracted field, $|U|^2/|U_0|^2$, is analyzed along the z-direction for an incident plane wave interacting with a slit obstacle defined by $a = -5\,\mu m$ and $b = 5\,\mu m$. The freespace wavelength is $\lambda_0 = 1.0\,\mu m$, and the results are presented using the RS-1, RS-2, and Kirchhoff formulations. (b) The same analysis as in (a) is conducted along the x-direction for $z_0 = 10\,\mu m$. In this case, the RS-1 and Fresnel formulations are used, as the Fraunhofer approximation is not valid at this distance. (c) The analysis in (a) is extended to a diverging cylindrical wave emanating from $z = -s_0 = -50\,\mu m$. The relative intensity I/I_{cw} is calculated along the z-direction. (d) Same as (c) but along the x-direction for $z_0 = 10\,\mu m$. (e) Same as (c) but for an incident converging cylindrical wave (with focal point at $z = -s_0 = 50\,\mu m$). (f) Same as (e) but along the x-direction for $z_0 = 10\,\mu m$.

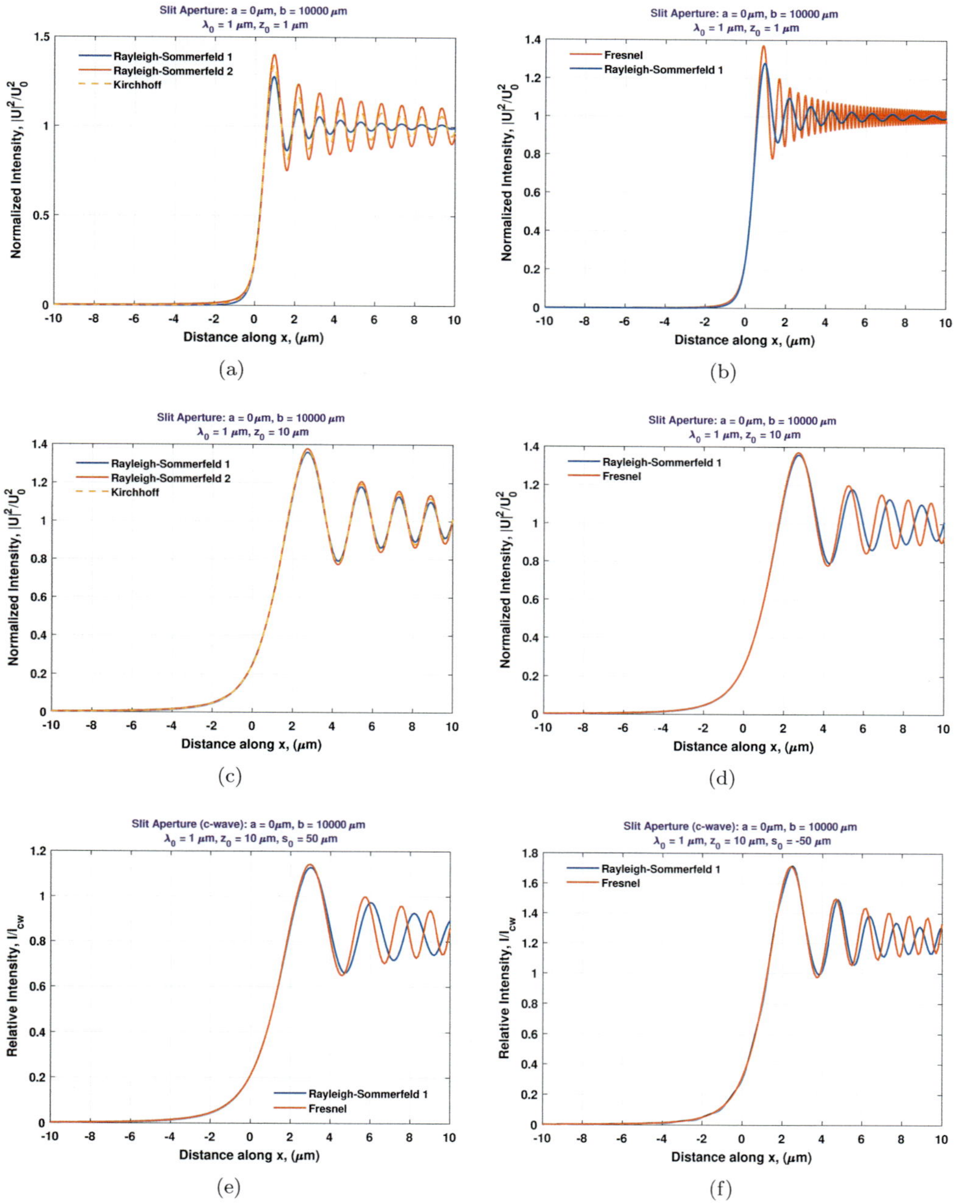

FIGURE 5.44: (a) The relative intensity of the diffracted field, $|U|^2/|U_0|^2$ in the x-direction (at $z_0 = 1\,\mu m$) for an incident plane wave on an edge ($a = 0\,\mu m$ and $b = \infty$). The freespace wavelength is $\lambda_0 = 1.0\,\mu m$. The RS-1, RS-2, and Kirchhoff formulations are shown. (b) Same as (a) but with the RS-1 and Fresnel formulations shown. The Fraunhofer is not valid at this distance and aperture size. (c) Same as (a) but for $z_0 = 10\,\mu m$. (d) Same as (b) but for $z_0 = 10\,\mu m$. (e) Same as (b) but for I/I_{cw} and an incident diverging cylindrical wave (emanating $z = -s_0 = -50\,\mu m$) and $z_0 = 10\,\mu m$. (f) Same as (e) but an incident converging cylindrical wave (focusing at $z = -s_0 = 50\,\mu m$) and $z_0 = 10\,\mu m$.

5.8.4 Multiple Slit Scalar Diffraction

The scalar diffraction by a single slit, as illustrated in Fig. 5.36, can be easily extended to accommodate any finite number N of slits on the same plane (at $z = 0$). These slits can have either equal or unequal widths, and they can be arranged at equal or unequal distances on the opaque screen. A simple diagram of the multiple-slit aperture problem is shown in Fig. 5.45. The positions of the slits are defined by their starting point $x' = a_i$ and their ending point $x' = b_i$ for each slit, where for $i = 1, 2, \cdots, N$.

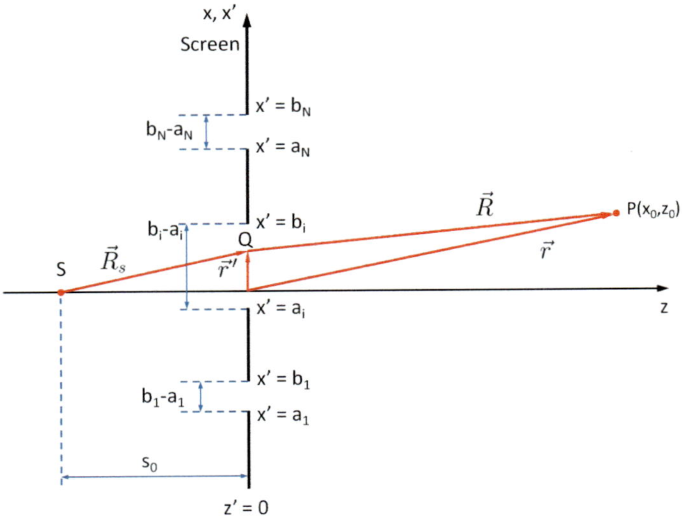

FIGURE 5.45: Multiple-slit aperture consisting of N slits, each with width $b_i - a_i$, extending from $x' = a_i$ to $x' = b_i$, for $i = 1, 2, \cdots, N$, positioned on an opaque screen. The aperture is assumed to be infinite in the y-direction. The diffracted field is calculated for a general off-axis point $P(x_0, z_0)$ with three types of incident waves: (a) a plane wave, where the source is at $s_0 \to -\infty$, (b) a diverging cylindrical wave, originating from $z = -s_0$ (as shown), and (c) a converging cylindrical wave, focused at $z = s_0$.

Equations (5.164)–(5.167) can be readily extended to the multiple slits problem as follows:

$$U_{RS1}(x_0, z_0) = \sum_{i=1}^{N} \int_{a_i}^{b_i} \left[U_{inc} \frac{k}{2j} H_1^{(2)}(kR_T) \frac{z_0}{R_T} \right]_{z'=0} dx', \qquad \text{and} \tag{5.182}$$

$$U_{RS2}(x_0, z_0) = \sum_{i=1}^{N} \int_{a_i}^{b_i} \left[-\frac{j}{2} H_0^{(2)}(kR_T) \frac{\partial U_{inc}}{\partial n} \right]_{z'=0} dx', \tag{5.183}$$

$$U_{RS1}^{a}(x_0, z_0) = e^{j3\pi/4} \sum_{i=1}^{N} \int_{a_i}^{b_i} \left[U_{inc} \frac{k}{2j} \sqrt{\frac{2}{\pi k R_T}} \exp(-jkR_T) \frac{z_0}{R_T} \right]_{z'=0} dx', \tag{5.184}$$

$$U_{RS2}^{a}(x_0, z_0) = e^{j\pi/4} \sum_{i=1}^{N} \int_{a_i}^{b_i} \left[-\frac{j}{2} \sqrt{\frac{2}{\pi k R_T}} \exp(-jkR_T) \frac{\partial U_{inc}}{\partial n} \right]_{z'=0} dx', \qquad \text{where,} \tag{5.185}$$

$$R_T = \left[(x_0 - x')^2 + z_0^2 \right]^{1/2}.$$

The incident field can either be a plane wave or a cylindrical wave, as described in Eq. (5.168). The Fresnel approximation can also be easily generalized from Eq. (5.170) and is given by:

$$U(x_0, z_0) = \frac{U_0}{\sqrt{2}} e^{j\pi/4} e^{-jkz_0} \sum_{i=1}^{N} \Big\{ [C(u_{2i}) - C(u_{1i})] - j\,[S(u_{2i}) - S(u_{1i})] \Big\}, \tag{5.186}$$

$$\text{where} \quad u_{1i} = \sqrt{\frac{2}{\lambda z_0}}(x_0 - b_i), \quad u_{2i} = \sqrt{\frac{2}{\lambda z_0}}(x_0 - a_i). \tag{5.187}$$

Finally, the Fraunhofer approximation is obtained by generalizing Eq. (5.172) and is expressed as:

$$U_{ff}(x_0, z_0) = U_0 e^{j\pi/4} \sqrt{\frac{k}{2\pi z_0}} e^{-jkz_0} \exp\left(-j\frac{k}{2z_0}x_0^2\right)$$
$$\sum_{i=1}^{N} \left\{ (b_i - a_i) \exp\left(j\frac{a_i + b_i}{2}\frac{kx_0}{z_0}\right) \frac{\sin\left[\frac{kx_0}{z_0}\left(\frac{b_i - a_i}{2}\right)\right]}{\frac{kx_0}{z_0}\left(\frac{b_i - a_i}{2}\right)} \right\}. \tag{5.188}$$

As a special case, consider the multiple-slit aperture where all N slits have equal width $s = b_i - a_i$, and are equally spaced by a distance $\Lambda = (1/2)[(a_{i+1} + b_{i+1}) - (a_i + b_i)]$. For slits positioned symmetrically with respect to the z-axis, the values of a_i and b_i are given by:

$$a_i = -\frac{N-1}{2}\Lambda + (i-1)\Lambda - \frac{s}{2},$$
$$b_i = -\frac{N-1}{2}\Lambda + (i-1)\Lambda + \frac{s}{2}, \quad \text{for} \quad i = 1, 2, \cdots, N.$$

The RS-1, RS-2, Kirchhoff, and Fresnel results are not simplified analytically but are still described by Eqs. (5.182) to (5.186). However, the Fraunhofer results can be further simplified as follows:

$$U_{ff}(x_0, z_0) = U_0 e^{j\pi/4} \sqrt{\frac{k}{2\pi z_0}} e^{-jkz_0} \exp\left(-j\frac{k}{2z_0}x_0^2\right)$$
$$\sum_{i=1}^{N} \left\{ s \exp\left[j\frac{kx_0}{z_0}\left(-\frac{N-1}{2}\Lambda + i\Lambda\right)\right] \frac{\sin\left[\frac{kx_0}{z_0}\left(\frac{s}{2}\right)\right]}{\frac{kx_0}{z_0}\left(\frac{s}{2}\right)} \right\} =$$
$$= U_0 e^{j\pi/4} \sqrt{\frac{k}{2\pi z_0}} e^{-jkz_0} \exp\left(-j\frac{k}{2z_0}x_0^2\right) \frac{\sin\left[\frac{kx_0}{z_0}\left(\frac{s}{2}\right)\right]}{\frac{kx_0}{z_0}\left(\frac{s}{2}\right)}$$
$$s \exp\left[j\frac{kx_0}{z_0}\left(-\frac{N-1}{2}\Lambda\right)\right] \sum_{i=1}^{N} \left\{ \exp\left(j\frac{kx_0}{z_0} i\Lambda\right) \right\} =$$

$$= U_0 e^{j\pi/4}\sqrt{\frac{k}{2\pi z_0}}e^{-jkz_0}\exp\left(-j\frac{k}{2z_0}x_0^2\right)s$$

$$\underbrace{\left\{\frac{\sin\left[\frac{kx_0}{z_0}\left(\frac{s}{2}\right)\right]}{\frac{kx_0}{z_0}\left(\frac{s}{2}\right)}\right\}}_{\text{slit diffraction}} \quad \underbrace{\exp\left(j\frac{kx_0}{z_0}\Lambda\right)\left\{\frac{\sin\left(N\frac{kx_0}{z_0}\frac{\Lambda}{2}\right)}{\sin\left(\frac{kx_0}{z_0}\frac{\Lambda}{2}\right)}\right\}}_{\text{array factor}}. \tag{5.189}$$

Here, for the array factor, the identity $1 + x + x^2 + \cdots + x^{N-1} = (x^N - 1)/(x - 1)$ has been utilized, and it is assumed that the slits are symmetrically positioned with respect to the z-axis. It is evident that the far-field diffraction consists of two factors: the single-slit diffraction factor and the array factor, which results from the interference of the N slits. The corresponding far-field normalized intensity is given by:

$$I_{ff} = \underbrace{\frac{|U_0|^2}{4Z}\frac{k}{2\pi z_0}s^2}_{=I_0(z_0)}\underbrace{\left\{\frac{\sin\left(\frac{kx_0}{z_0}\frac{s}{2}\right)}{\frac{kx_0}{z_0}\frac{s}{2}}\right\}^2}_{\text{slit diffraction}}\underbrace{\left\{\frac{\sin\left(N\frac{kx_0}{z_0}\frac{\Lambda}{2}\right)}{\sin\left(\frac{kx_0}{z_0}\frac{\Lambda}{2}\right)}\right\}^2}_{\text{array factor}}. \tag{5.190}$$

Of course, it is always true that $s < \Lambda$ (and usually $s \ll \Lambda$), meaning the slit diffraction "sinc" factor is wider than the array factor. The maximum of the slit factor, as well as its minima, are described by Eq. (5.174). The maxima and minima of the array factor are determined as follows:

$$N\frac{kx_0}{z_0}\frac{\Lambda}{2} = m\pi, \quad \text{and} \quad \frac{kx_0}{z_0}\frac{\Lambda}{2} = \ell\pi \Longrightarrow$$

$$m = \ell N, \quad \text{resulting} \quad x_{0m}^{max} = \frac{m}{N}\lambda\frac{z_0}{\Lambda}, \quad m = 0, \pm N, \pm 2N, \cdots \tag{5.191}$$

$$\text{and} \quad I_{ff}^{max} = I_0(z_0)N^2, \quad \text{while}$$

$$I_{ff}^{min} = 0, \quad \text{for} \quad N\frac{kx_0}{z_0}\frac{\Lambda}{2} = m\pi \Longrightarrow x_{0m}^{min} = \frac{m}{N}\lambda\frac{z_0}{\Lambda}, \quad m \in Z \text{ and } m \neq 0, \pm N, \pm 2N, \cdots.$$

This means that there are $N - 1$ minima (between two successive main maxima, for $m = 1, 2, \cdots, N - 1$). In between the $(N - 1)$ minima, there are $(N - 2)$ secondary maxima, which are generally much smaller than the main maxima and diminish as the number of slits N increases.

Example 5.8.5:

In this example, the diffraction by multiple slits is demonstrated numerically. Consider cases where the slit width is $s = 3\,\mu m$ and the slit separation is $\Lambda = 12\,\mu m$. The cases of a single slit ($N = 1$) (for comparison purposes) are considered along with the cases of $N = 2$ and $N = 3$. Assume a normally incident plane wave with a freespace wavelength of $\lambda_0 = 1\,\mu$m. The task is to determine the normalized intensities for the RS-1, RS-2, Kirchhoff, Fresnel, and Fraunhofer formulations in the x-direction for various values of z_0.
Solution: The resulting normalized diffracted intensities are shown in Fig. 5.46. Specifically, results are shown in the transverse direction for two distances from the aperture plane: (a) for $z_0 = 10\,\mu m = 10\lambda_0$ and (b) for $z_0 = 1000\,\mu m = 1000\lambda_0$. In Figs. 5.46(a) and 5.46(b), the results of the single slit are shown. For the Fresnel approximation to hold, $z_0 \gg$

$[(\pi/4\lambda_0)(x_0 + s/2)^4]^{1/3} = 1089.9\,\mu m$ (for $x_0 = 200\,\mu m$) and $436.84\,\mu m$ (for $x_0 = 100\,\mu m$). The Fraunhofer approximation is valid for $z_0 \gg (\pi/\lambda_0)s^2 = 28.27\,\mu m$. Obviously, it is necessary to satisfy the Fresnel limit first for the Fraunhofer limit to be meaningful. The corresponding Fresnel number is $N_F = s^2/\lambda z_0 = 0.009$, which implies that both Fresnel and Fraunhofer approximations are valid in the case of $z_0 = 1000\,\mu m$ for the single slit. This is evident in Fig. 5.46(b).

In the case of $z_0 = 10\,\mu m$, neither the Fraunhofer nor the Fresnel condition is satisfied, and only the RS-1, RS-2, and Kirchhoff results are shown, which are very similar. For $z_0 = 1000\,\mu m$, the Fresnel condition is still not well satisfied for distances $|x_0| \geq 200\,\mu m$, but it appears to be reasonably accurate for $|x_0| \leq 200\,\mu m$. From Fig. 5.46(b), it is evident that the Fraunhofer and Fresnel results are nearly identical, but they start to deviate from the RS-1 results as $|x_0| >\sim 300\,\mu m$, since the Fresnel condition is no longer satisfied. Also, observe that the minima (zeros) of the intensity pattern occur at $x_{0m}^{min} = \pm m\lambda(z_0/s) = m333.33\,\mu m$, as expected.

Next, let's observe the case of $N = 2$ (two slits separated by Λ and centered at the origin). The results are shown in Figs. 5.46(c) and 5.46(d) for $z_0 = 10\,\mu m = 10\lambda_0$ and for $z_0 = 1000\,\mu m = 1000\lambda_0$, respectively. For the Fresnel approximation to hold, $z_0 \gg [(\pi/4\lambda_0)(x_0 + \Lambda/2 + s/2)^4]^{1/3} = 1133.4\,\mu m$ (for $x_0 = 200\,\mu m$) and $471.60\,\mu m$ (for $x_0 = 100\,\mu m$). The Fresnel number is $N_F = (s + \Lambda)^2/\lambda z_0 = 0.225$ in the case of $z_0 = 1000\,\mu m$. The Fraunhofer approximation is valid for $z_0 \gg (\pi/\lambda_0)(\Lambda + s)^2 = 706.86\,\mu m$. In the short distance from the aperture, again only the RS-1, RS-2, and Kirchhoff results make sense and they nearly coincide with each other. For the longer distance, the same conclusions stated for the single slit hold. It is evident that as $|x_0| >\sim 200 - 250\,\mu m$, the Fresnel/Fraunhofer results start showing some deviation from the RS-1 results. Something else that is expected is that the main maxima occur at $x_{0m}^{max} = m\lambda(z_0/\Lambda) = m83.33\,\mu m$. The actual main maxima positions are slightly affected by the slit diffraction factor, and therefore, the previous positions are good approximations. These approximations become better as the ratio s/Λ becomes smaller. Also the minima occur at $x_{0m}^{min} = (m/2)\lambda(z_0/\Lambda) = m41.67\,\mu m$ (for $m \in Z$ and $m \neq 0, \pm 2, \pm 4, \cdots$). The positions of the minima are exact since they make zero the array factor and therefore the diffracted intensity. Another interesting point is that the centered maximum (which coincides with the single-slit diffraction maximum) is as expected: $I_{max} = 2^2 I_0(z_0)$, where in this case $I_0(z_0) = s^2/\lambda_0 z_0 = 9 \times 10^{-3}$.

The case of $N = 3$ (three slits separated by Λ and centered at the origin) is shown in Figs. 5.46(e) and 5.46(f) for $z_0 = 10\,\mu m = 10\lambda_0$ and for $z_0 = 1000\,\mu m = 1000\lambda_0$, respectively. For the Fresnel approximation to hold $z_0 \gg [(\pi/4\lambda_0)(x_0 + \Lambda + s/2)^4]^{1/3} = 1177.3\,\mu m$ (for $x_0 = 200\,\mu m$) and $507.02\,\mu m$ (for $x_0 = 100\,\mu m$). The Fraunhofer approximation is valid for $z_0 \gg (\pi/\lambda_0)(2\Lambda + s)^2 = 2290.2\,\mu m$. The Fresnel number is $N_F = (s + 2\Lambda)^2/\lambda z_0 = 0.729$ in the case of $z_0 = 1000\,\mu m$. It is evident from Fig. 5.46(f) that there is some difference between the Fresnel and Fraunhofer results since the condition for the Fraunhofer approximation is not satisfied. Furthermore, for $|x_0| >\sim 200\,\mu m$, even the Fresnel results start showing some deviation from the RS-1 results because the Fresnel condition is not satisfied (the Fresnel number is near 1, where the RS formulations should be used).

In this case, the main maxima occur approximately at $x_{0m}^{max} = m\lambda(z_0/\Lambda) = m83.33\,\mu m$. Also the minima occur at $x_{0m}^{min} = (m/3)\lambda(z_0/\Lambda) = m27.78\,\mu m$ (for $m \in Z$ and $m \neq 0, \pm 3, \pm 6, \cdots$). Another interesting point is that the centered maximum (which coincides with the single-slit diffraction maximum) is as expected: $I_{max} = 3^2 I_0(z_0)$, where in this case $I_0(z_0) = s^2/\lambda_0 z_0 = 9 \times 10^{-3}$.

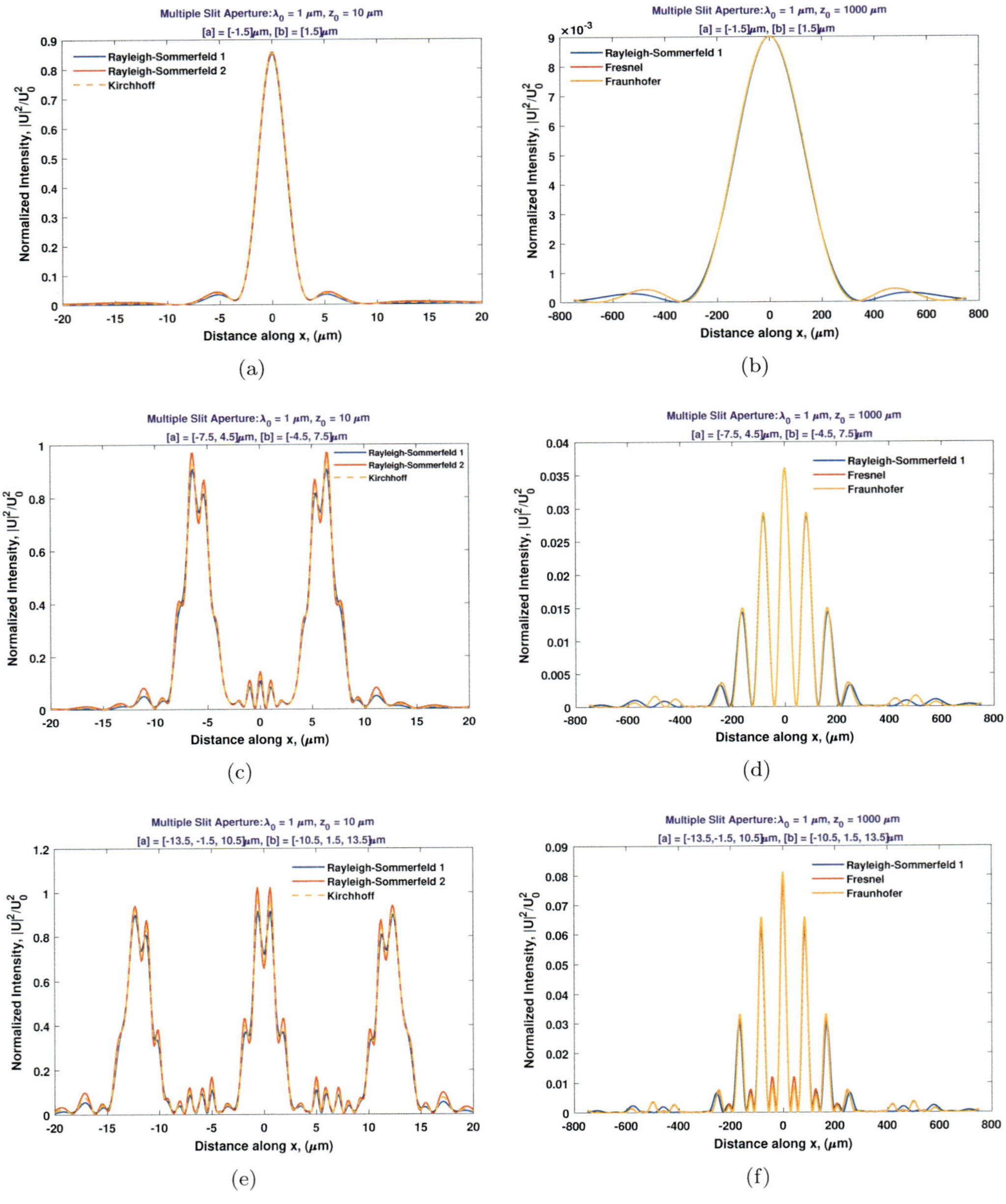

FIGURE 5.46: (a) The relative intensity of the diffracted field, $|U|^2/|U_0|^2$, in the x-direction (at $z_0 = 10\,\mu m$), for an incident plane wave on a single slit of width $s = 3\,\mu m$. The freespace wavelength is $\lambda_0 = 1.0\,\mu m$. The RS-1, RS-2, and Kirchhoff formulations are shown. (b) Same as (a) (at $z_0 = 1000\,\mu m$) but for the RS-1, Fresnel, and Fraunhofer formulations shown. (c) Same as (a) but for two slits of width $s = 3\,\mu m$ and separation $\Lambda = 12\,\mu m$. (d) Same as (b) but for $z_0 = 1000\,\mu m$. (e) Same as (a) but for three slits of width $s = 3\,\mu m$ and separation $\Lambda = 12\,\mu m$. (f) Same as (e) but $z_0 = 1000\,\mu m$.

The case of an incident cylindrical diverging or converging wave can be treated similarly to the approach used for the single slit and thus does not need to be explicitly presented. Additionally, the case of multiple-slit obstacles can be analyzed using the multiple-slit aperture results in conjunction with Babinet's principle. However, this will not be presented here, as it appears redundant at this stage.

5.8.5 Rectangular Aperture

In this section, let us consider a rectangular aperture with dimensions $2a \times 2b$, centered around the origin, as shown in Fig. 5.47. Similar to previous cases, a monochromatic wave is incident from the left on the opaque plane where the aperture is located. According to the RS-1 and RS-2 formulations, the diffracted field at point $P_0(x_0, y_0, z_0)$ (with $z_0 > 0$) is given by:

$$U_{RS1}(P_0) = \int_{-a}^{+a}\int_{-b}^{+b}\left[U_{inc}\left(jk+\frac{1}{R}\right)\frac{\exp(-jkR)}{2\pi R}\frac{z_0}{R}\right]_{z'=0} dx'dy', \qquad \text{and} \tag{5.192}$$

$$U_{RS2}(P_0) = \int_{-a}^{+a}\int_{-b}^{+b}\left[\frac{\exp(-jkR)}{2\pi R}\frac{\partial U_{inc}}{\partial n}\right]_{z'=0} dx'dy', \tag{5.193}$$

$$\text{with } R = \left[(x_0-x')^2+(y_0-y')^2+z_0^2\right]^{1/2}.$$

Both of these integrals must be evaluated numerically, performing the two-dimensional

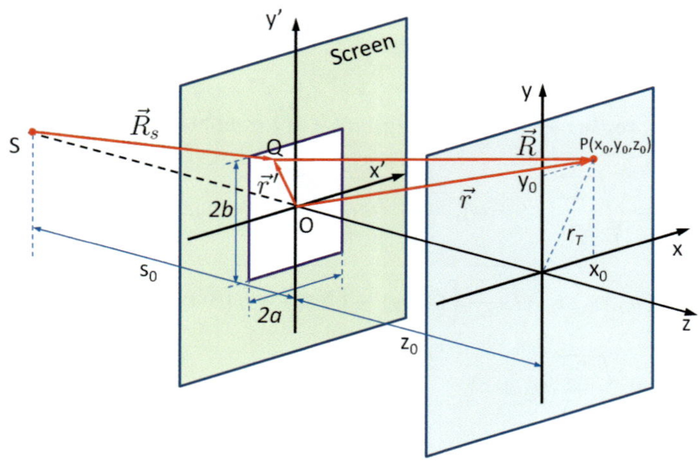

FIGURE 5.47: A rectangular aperture of dimensions $2a \times 2b$ on an opaque screen. The diffracted field for the off-axis point $P(x_0, y_0, z_0)$ is sought. The point S specifies the origin of a diverging spherical wave that is incident on the aperture. In the case of an incident plane wave, the point S moves to $-\infty$ ($s_0 = \infty$).

integration appearing in Eqs. (5.192) and (5.193). Even along the z-axis, the integrals must be determined numerically since no analytical solution exists. The Kirchhoff (or Helmholtz-Kirchhoff) formulation can again be obtained by utilizing Eq. (5.20). The incident field can be either a plane wave or a spherical wave. In this section, the cases of a normally incident plane wave, as well as a diverging or converging spherical wave, are analyzed. The spherical waves have their focal points along the z-axis. Therefore, the incident field can be described by:

$$U_{inc}(x',y',z') = \begin{cases} U_0 \exp(-jkz'), & \text{(plane wave)} \\ U_0 \dfrac{\exp(-jkR_s)}{R_s}, & R_s = [x'^2 + y'^2 + (z'+s_0)^2]^{1/2}, \quad \text{(diverging)}, \\ U_0 \dfrac{\exp(+jkR_s)}{R_s}, & R_s = [x'^2 + y'^2 + (z'-s_0)^2]^{1/2}, \quad \text{(converging)}. \end{cases} \tag{5.194}$$

Of course, as in the case of the slit, more general incident waves, $U_{inc}(x',y')$, can be considered without any difficulty. Plane waves incident at an angle or spherical waves with arbitrary focal point positions in space can easily be analyzed. Furthermore, Gaussian beams can also be treated using the same formulation.

For the Fresnel approximation, let us first consider the case of a normally incident plane wave. In this scenario, the incident field is given by $U_{inc} = U_0 \exp(-jkz')$. Substituting this into the Fresnel approximation [see Eq. (5.25)], the resulting expression can be written as:

$$U(x_0,y_0,z_0) = jk\frac{\exp(-jkz_0)}{2\pi z_0}\int_{-a}^{+a}\int_{-b}^{+b}\left[\exp\left\{-jk\left[\frac{(x_0-x')^2}{2z_0} + \frac{(y_0-y')^2}{2z_0}\right]\right\}U_0\right]dx'dy' =$$

$$= jk\frac{\exp(-jkz_0)}{2\pi z_0}U_0 \underbrace{\int_{-a}^{+a}\exp\left[-jk\frac{(x_0-x')^2}{2z_0}\right]dx'}_{=U_x}\underbrace{\int_{-b}^{+b}\exp\left[-jk\frac{(y_0-y')^2}{2z_0}\right]dy'}_{=V_y}.$$

The factors U_x and V_y can be expressed in terms of the standard Fresnel integrals using the method described in Sec. 5.8.1.2. Specifically, it can be shown that:

$$U_x = U_x(x_0,z_0) = \sqrt{\frac{\lambda z_0}{2}}\Big\{C(u_2) - C(u_1) - j\left[S(u_2) - S(u_2)\right]\Big\}, \tag{5.195}$$

$$\text{with} \quad u_2 = \sqrt{\frac{2}{\lambda z_0}}(x_0 + a), \quad u_1 = \sqrt{\frac{2}{\lambda z_0}}(x_0 - a),$$

$$V_y = V_y(y_0,z_0) = \sqrt{\frac{\lambda z_0}{2}}\Big\{C(v_2) - C(v_1) - j\left[S(v_2) - S(v_2)\right]\Big\}, \tag{5.196}$$

$$\text{with} \quad v_2 = \sqrt{\frac{2}{\lambda z_0}}(y_0 + b), \quad v_1 = \sqrt{\frac{2}{\lambda z_0}}(y_0 - b).$$

Then, the resulting Fresnel approximation becomes:

$$U(x_0,y_0,z_0) = j\frac{U_0}{2}\exp\left(-jkz_0\right)F(u_2,u_1)F(v_2,v_1), \tag{5.197}$$

$$\text{where} \quad F(u_2,u_1) = \Big\{C(u_2) - C(u_1) - j\left[S(u_2) - S(u_2)\right]\Big\},$$

$$\text{and} \quad F(v_2,v_1) = \Big\{C(v_2) - C(v_1) - j\left[S(v_2) - S(v_2)\right]\Big\}. \tag{5.198}$$

When there is no aperture, i.e., when $a, b \to +\infty$, the entire wavefront is unobstructed, and the resulting field is simply the incident field propagating along the z-axis, $U_0\exp(-jkz_0)$. The corresponding diffracted intensity is then given by the squared magnitude of the field:

$$I(x_0,y_0,z_0) = \frac{|U_0|^2}{4Z}\frac{1}{4}|F(u_2,u_1)|^2|F(v_2,v_1)|^2 = \frac{I_u}{4}|F(u_2,u_1)|^2|F(v_2,v_1)|^2. \tag{5.199}$$

Under the Fraunhofer approximation, the diffracted field from a rectangular aperture of dimensions $2a \times 2b$ can be expressed in terms of the results derived in Sec. 5.8.1.2. This approximation assumes that the observation point is in the far field, where the phase variations due to aperture geometry simplify significantly. Thus, the Fraunhofer approximation diffracted field is expressed as:

$$U_{ff}(x_0, y_0, z_0) = jkU_0 \frac{e^{-jkz_0}}{2\pi z_0} \exp\left[-j\frac{k}{2z_0}\left(x_0^2 + y_0^2\right)\right] \left\{ 4ab \frac{\sin\left(a\frac{kx_0}{z_0}\right)}{a\frac{kx_0}{z_0}} \frac{\sin\left(b\frac{ky_0}{z_0}\right)}{b\frac{ky_0}{z_0}} \right\}. \tag{5.200}$$

The corresponding Fraunhofer diffracted intensity is given by:

$$I_{ff}(x_0, y_0, z_0) = \underbrace{\frac{|U_0|^2}{4Z}\left(\frac{4ab}{\lambda z_0}\right)^2}_{= I_0(z_0)} \left\{ \frac{\sin\left(a\frac{kx_0}{z_0}\right)}{a\frac{kx_0}{z_0}} \right\}^2 \left\{ \frac{\sin\left(b\frac{ky_0}{z_0}\right)}{b\frac{ky_0}{z_0}} \right\}^2 =$$

$$= I_0(z_0)\, \mathrm{sinc}^2\left(2a\frac{x_0}{\lambda z_0}\right) \mathrm{sinc}^2\left(2b\frac{y_0}{\lambda z_0}\right). \tag{5.201}$$

Here, $\mathrm{sinc}(u) = \sin(\pi u)/(\pi u)$ is the standard "sinc" function as it was used earlier too. The maximum of the far-field diffracted intensity is $I_0(z_0)$ and occurs along the z-axis, i.e., for $x_0 = y_0 = 0$. The minima of the far-field diffracted intensity occur at positions where the sine terms in the "sinc" functions become zero, excluding the origin, which corresponds to the maximum intensity. Specifically, the positions of the minima are given by:

$$\frac{kx_0}{z_0}2a = m\pi \Longrightarrow x_{0p}^{min} = p\lambda\frac{z_0}{2a}, \qquad p = \pm 1, \pm 2, \cdots, \tag{5.202}$$

$$\frac{ky_0}{z_0}2b = q\pi \Longrightarrow y_{0q}^{min} = q\lambda\frac{z_0}{2b}, \qquad q = \pm 1, \pm 2, \cdots. \tag{5.203}$$

In the diffraction pattern of a rectangular aperture, secondary maxima appear between each pair of minima. These secondary maxima are smaller than the main maxima, and their precise locations can only be determined numerically, similar to the case of a slit aperture. These secondary peaks are often referred to as sidelobes, and their intensity diminishes as the observation point moves further from the z-axis. The sidelobes result from the $\mathrm{sinc}^2(*)$ functions that characterize the diffraction pattern.

It is important to note that the Fraunhofer approximation loses the space invariance property, unlike the RS-1, RS-2, and Fresnel approximations, which retain this characteristic. In the Fraunhofer approximation, shifting the rectangular aperture in the $x'y'$-plane does not cause a shift in the diffracted intensity. However, in the RS-1, RS-2, and Fresnel formulations, such a shift results in a corresponding shift in the diffracted intensity.

To better understand these effects and illustrate the diffraction behavior, it is now useful to apply the derived expressions in a practical example. This example will demonstrate the various diffraction patterns and highlight the differences between the various approximations in terms of intensity distribution and the behavior of sidelobes.

Example 5.8.6:

Consider a rectangular aperture with dimensions $10 \times 20\,\mu m^2$ (i.e., $a = 5\,\mu m$ and $b = 10\,\mu m$), centered at the origin. A plane wave of freespace wavelength of $\lambda_0 = 1\,\mu m$ is incident normally from the left of the aperture. The objective is to determine the normalized intensities $|U|^2/|U_0|^2$ along the x or y directions for two different distances from the aperture, $z_0 = 10\,\mu$m (short distance) and $z_0 = 1000\,\mu$m (long distance). The normalized intensities should be computed using the following diffraction formulations where applicable: RS-1, RS-2, Kirchhoff, Fresnel, and Fraunhofer.

Solution: The diffracted field at any arbitrary point $P_0(x_0, y_0, z_0)$ can be determined numerically by applying the previously discussed equations. For instance, the normalized diffracted intensities $|U|^2/|U_0|^2$ are presented along the x- and y-directions for $z_0 = 10\,\mu m = 10\lambda_0$ in Figs. 5.48(a) and 5.48(b), respectively, using the RS-1, RS-2, and Kirchhoff formulations.

For a longer distance of $z_0 = 1000\,\mu m = 1000\lambda_0$, similar results are shown in Figs. 5.48(c) and 5.48(d) along the x- and y-directions. At this distance, the Fraunhofer approximation condition is not well satisfied because $(\pi/\lambda)[(2a)^2 + (2b)^2] = 1570.8\,\mu m > 1000\,\mu m$. The corresponding Fresnel number is $N_F = [(2a)^2 + (2b)^2]/\lambda z_0 = 0.5$, which suggests the applicability of the Fresnel approximation but not the Fraunhofer approximation. However, the RS-1, Fresnel, and Fraunhofer results are very close. Additionally, the minima of the diffracted intensity along both the x- and y-directions closely match the Fraunhofer predictions from Eqs. (5.202) and (5.203). Specifically, the minima occur at $x_{0p}^{min} = p100\,\mu m$ and $y_{0q}^{min} = q50\,\mu m$ $(p, q = \pm 1, \pm 2, \cdots)$. The figures demonstrate that the minima are well approximated even when the Fraunhofer condition is not fully satisfied. The larger divergence of the diffracted beam along the x-direction, due to the smaller width of the aperture along the x-direction, is also evident.

Finally, xy-plane diffracted intensity is presented in Figs. 5.48(e) and 5.48(f) for $z_0 = 1000\,\mu m = 1000\lambda_0$, where the two-dimensional "sinc^2" diffraction pattern is visible. Figure 5.48(f) is similar to Fig. 5.48(e), except that the logarithm of the diffracted intensity is plotted to highlight the features of the pattern in the xy-plane.

The case of an incident spherical wave, whether diverging or converging, can be addressed using the corresponding expression from Eq. (5.194) inside the two-dimensional integrals of Eqs. (5.192) and (5.193). Of particular interest is the case of the Fresnel approximation, as it can be expressed in terms of Fresnel integrals, similar to the treatment of the one-dimensional case for the slit aperture in Sec. 5.8.1.2. The approach outlined in Ref. [43], which was used in Sec. 5.8.1.2 for the slit aperture, can be readily adapted for the rectangular aperture as well. This adaptation involves considering the spherical wave's curvature and focal properties, and it allows for a consistent treatment of the diffracted field using the Fresnel approximation, as the rectangular aperture's shape does not complicate the integral forms significantly compared to the slit case. This methodology enables precise numerical calculations of the diffracted field for both diverging and converging incident spherical waves.

From the similar triangles in the xz-plane (as depicted in Fig. 5.49), it is straightforward to show that $x'_a = s_0 x_0/(z_0 + s_0) = x_0 - z_0 x_0/(z_0 + s_0)$. In an analogous manner, from the similar triangles in the yz-plane, it can be shown that $y'_a = s_0 y_0/(z_0 + s_0) = y_0 - z_0 y_0/(z_0 + s_0)$. Using these transformations and following the same procedure as outlined in Sec. 5.8.1.2, the resulting equations for the diffracted field can be written in terms of these adjusted coordinates as:

$$(x' - x'_a)^2 \left(\frac{1}{z_0} + \frac{1}{s_0}\right) + \frac{x_0^2}{z_0 + s_0} = \frac{(x_0 - x')^2}{z_0} + \frac{x'^2}{s_0}.$$

$$(y' - y'_a)^2 \left(\frac{1}{z_0} + \frac{1}{s_0}\right) + \frac{y_0^2}{z_0 + s_0} = \frac{(y_0 - y')^2}{z_0} + \frac{y'^2}{s_0}.$$

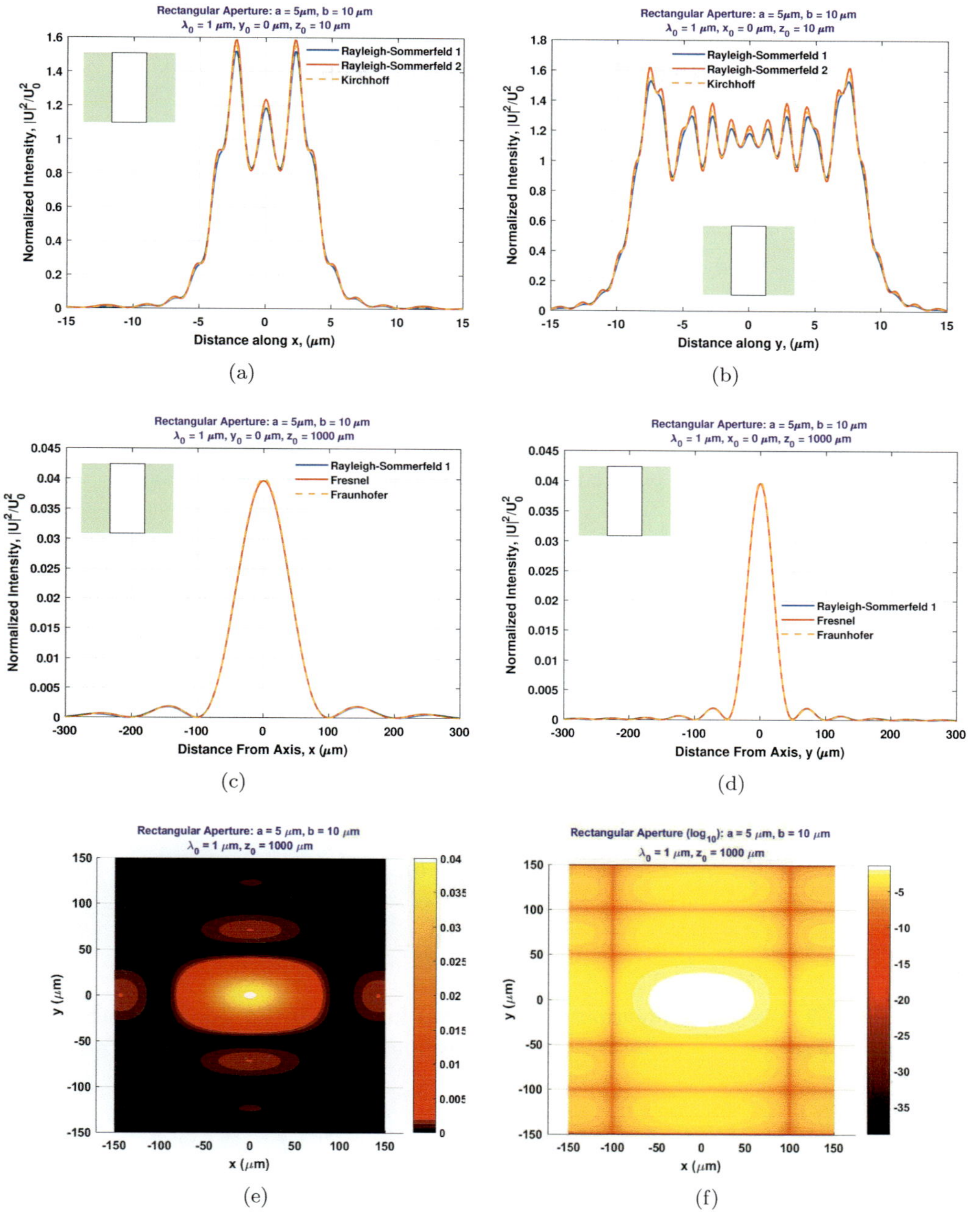

FIGURE 5.48: (a) The diffracted field normalized intensity $|U|^2/|U_0|^2$ in the x-direction (at $z_0 = 10\,\mu m$) for an incident plane wave on a rectangular aperture of dimensions $a = 5\,\mu m$ and $b = 10\,\mu m$. The freespace wavelength is $\lambda_0 = 1.0\,\mu m$. The RS-1, RS-2, and Kirchhoff formulations are shown. (b) Same as (a) but in the y-direction (c) Same as (a) but for $z_0 = 1000\,\mu m$. (d) Same as (b) but for $z_0 = 1000\,\mu m$. (e) Far field intensity in the xy-plane for $z_0 = 1000\,\mu m$. (f) Same as (e) but in logarithmic scale of intensity.

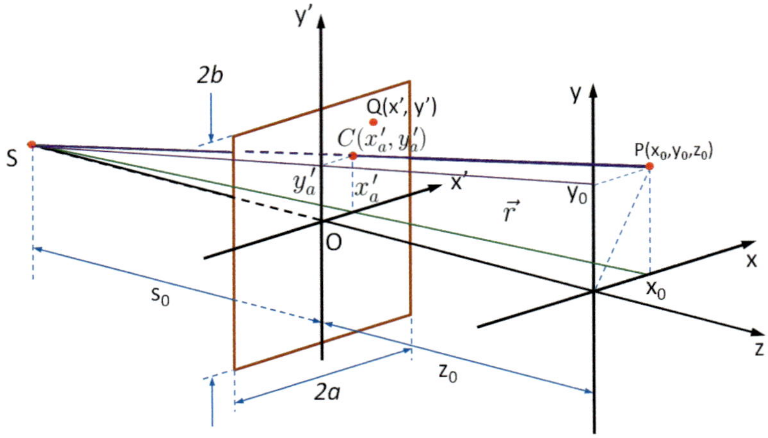

FIGURE 5.49: A rectangular aperture dimensions $2a \times 2b$ on an opaque screen. The diffracted field for the general off-axis point $P(x_0, y_0, z_0)$ is sought for the case of a diverging spherical wave emanating from $z = -s_0$ (as shown). The point $C(x'_a, y'_a)$ is the intersection of the $z = 0$ plane with the line between the focal point S and the point of interest P.

Following the procedure used in Sec. 5.8.1.2 for the slit aperture and the diverging cylindrical wave, the Fresnel approximation for the rectangular aperture can be expressed as:

$$U(x_0, y_0, z_0) = j\frac{U_0}{2}\frac{\exp\left[-jk\left(z_0 + s_0\right)\right]}{z_0 + s_0}\exp\left(-j\frac{k}{2}\frac{x_0^2 + y_0^2}{z_0 + s_0}\right)F(u_2, u_1)F(v_2, v_1), \quad \text{where} \tag{5.204}$$

$$u_2 = \sqrt{\frac{2}{\lambda d}}(x'_a + a), \quad u_1 = \sqrt{\frac{2}{\lambda d}}(x'_a - a),$$

$$v_2 = \sqrt{\frac{2}{\lambda d}}(y'_a + b), \quad v_1 = \sqrt{\frac{2}{\lambda d}}(y'_a - b),$$

$$\text{and} \ \ x'_a = \frac{s_0}{z_0 + s_0}x_0, \quad y'_a = \frac{s_0}{z_0 + s_0}y_0, \quad \frac{1}{d} = \frac{1}{z_0} + \frac{1}{s_0},$$

and $F(w_1, w_2)$ is given as a function of the Fresnel integrals in Eq. (5.198). The case of a converging spherical wave can be treated similarly, and the resulting Fresnel approximation is given by:

$$U(x_0, y_0, z_0) = -j\frac{U_0}{2}\frac{\exp\left[-jk\left(z_0 - s_0\right)\right]}{z_0 - s_0}\exp\left(-j\frac{k}{2}\frac{x_0^2 + y_0^2}{z_0 - s_0}\right)F(u_2, u_1)F(v_2, v_1), \tag{5.205}$$

where

$$u_2 = \sqrt{\frac{2}{\lambda d}}(x'_a + a), \quad u_1 = \sqrt{\frac{2}{\lambda d}}(x'_a - a),$$

$$v_2 = \sqrt{\frac{2}{\lambda d}}(y'_a + b), \quad v_1 = \sqrt{\frac{2}{\lambda d}}(y'_a - b),$$

$$\text{and} \ \ x'_a = \frac{-s_0}{z_0 - s_0}x_0, \quad y'_a = \frac{-s_0}{z_0 - s_0}y_0, \quad \frac{1}{d} = \frac{1}{z_0} - \frac{1}{s_0}.$$

Example 5.8.7:

In this example consider a square aperture with dimensions $20\mu m \times 20\mu m$ (i.e., with $a = b = 10\,\mu m$). The incident wave can be a plane wave (normally incident), a diverging spherical wave originating from $z_0 = -s_0 = -50\,\mu m$ or a converging spherical wave focused at $z_0 = -s_0 = 50\,\mu m$. The freespace wavelength for incident waves is $\lambda_0 = 1.0\,\mu m$. The goal is to calculate the normalized intensities $|U|^2/|U_0|^2$ (for plane wave) or the relative intensities I/I_{sw} (for spherical wave) along the z- or x-axis (for $z_0 = 20\,\mu\text{m}$). The RS-1 and Fresnel formulations should be applied.
Solution: Using the previously presented theory and numerical two-dimensional integration, the results of the RS-1 formulation have been initially calculated. The RS-2 and Kirchhoff results are very similar and are not shown in this example. The normalized diffracted intensity $|U|^2/|U_0|^2$ for the incident plane wave and the relative diffracted intensity I/I_{sw} for the diverging spherical wave are presented along the z-axis in Fig. 5.50(a). The case of the incident converging spherical wave is shown separately in Fig. 5.50(b), where the focusing of the diffracted wave is evident near $z_0 \simeq 43\,\mu m$. Additionally, the relative intensity is much higher for the converging spherical wave than for the diverging spherical wave or plane wave, due to the focusing of the incident wave. It is important to note that the diffracted wave is focused at $z_0 \neq s_0$ because of the diffraction effects caused by the square aperture. In Figs. 5.50(c) and 5.50(d), similar results are presented along the x-direction (and by symmetry, these results are identical along the y-direction as well). The RS-1 and Fresnel results are shown in these figures. The condition for the Fresnel diffraction approximation is not well satisfied, as $[(\pi/4\lambda)D^4]^{1/3} \simeq [(\pi/4\lambda)(20^2 + 20^2)^2]^{1/3} = 79.5\,\mu m > z_0 = 20\,\mu m$, which is greater than $z_0 = 20\,\mu\text{m}$. The Fresnel number in this case is $N_F = [(2a)^2 + (2a)^2]/\lambda z_0 = 25$ which is large, further indicating that the Fresnel condition is not satisfied for the angular deviation shown in the plots. However, even though the Fresnel condition is not strictly met, the Fresnel results show relatively good agreement with the more precise RS-1 results.

The case of a rectangular obstacle can be easily analyzed using the results from Sec. 5.8.5 in conjunction with Babinet's principle (see Sec. 5.6 and Eq. (5.48)). Furthermore, more complex apertures or obstacles with rectangular geometry can be analyzed as linear combinations of rectangular apertures or obstacles.

5.9 Scalar Diffraction by a Triangular and Polygonal Apertures and Obstacles

In this section, some straightforward applications of scalar diffraction involving triangular-shaped or polygonal-shaped apertures and obstacles are presented. The incident wave is either a plane wave (typically normally incident) or a spherical wave, as described in Eq. (5.194).

5.9.1 Scalar Diffraction by Triangular Apertures and Obstacles

As in all previous cases, a monochromatic wave is incident from the left onto the opaque plane containing the triangular aperture (see Fig. 5.51). Using the RS-1 and RS-2 formulations, the diffracted field at a point $P_0(x_0, y_0, z_0)$ (with $z_0 > 0$) can be determined through appropriate integration. Following the procedure outlined in Appendix 5D (see Sec. 5D.5

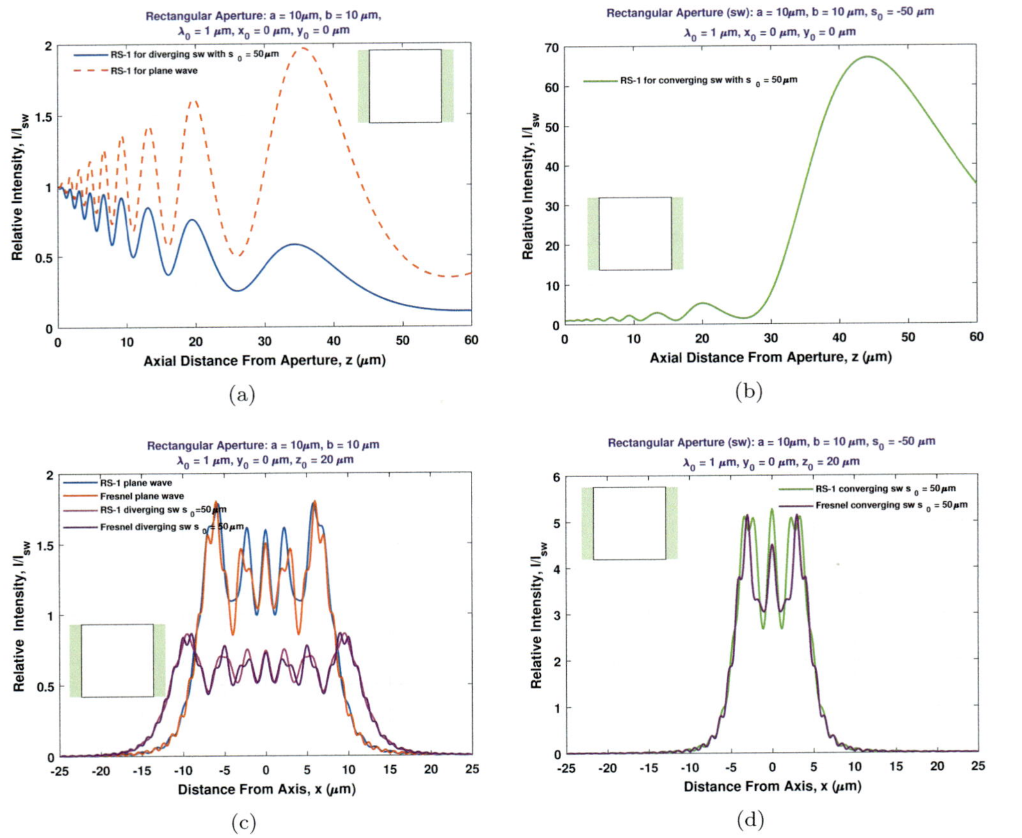

FIGURE 5.50: (a) The diffracted field normalized intensity $|U|^2/|U_0|^2$ for the incident plane wave, or I/I_{sw} for the incident diverging spherical wave (with $z_0 = -s_0 = -50\,\mu m$) along the z direction for a square aperture of size $20\mu m \times 20\mu m$ (i.e., with $a = b = 10\,\mu m$). The freespace wavelength is $\lambda_0 = 1.0\,\mu m$. The RS-1 formulations are shown. (b) Same as (a) but for an incident converging spherical wave with $z_0 = -s_0 = 50\,\mu m$. The focusing of the diffracted field is apparent. (c) Same as (a) but along the x direction for $z_0 = 20\,\mu m = 20\lambda_0$. (d) Same as (b) but along the x direction for $z_0 = 20\,\mu m = 20\lambda_0$.

and Fig. 5.5(a)), the diffraction integrals can be expressed as:

$$U_{RS1}(P_0) = \int_{x_1}^{x_2}\int_{y_{13}(x)}^{y_{12}(x)} \left[U_{inc}\left(jk + \frac{1}{R}\right)\frac{\exp(-jkR)}{2\pi R}\frac{z_0}{R}\right]_{z'=0} dx'dy' +$$

$$\int_{x_2}^{x_3}\int_{y_{13}(x)}^{y_{23}(x)} \left[U_{inc}\left(jk + \frac{1}{R}\right)\frac{\exp(-jkR)}{2\pi R}\frac{z_0}{R}\right]_{z'=0} dx'dy', \quad \text{and} \tag{5.206}$$

$$U_{RS2}(P_0) = \int_{x_1}^{x_2}\int_{y_{13}(x)}^{y_{12}(x)} \left[\frac{\exp(-jkR)}{2\pi R}\frac{\partial U_{inc}}{\partial n}\right]_{z'=0} dx'dy' +$$

$$\int_{x_2}^{x_3}\int_{y_{13}(x)}^{y_{23}(x)} \left[\frac{\exp(-jkR)}{2\pi R}\frac{\partial U_{inc}}{\partial n}\right]_{z'=0} dx'dy', \tag{5.207}$$

$$\text{with} \quad R = \left[(x_0 - x')^2 + (y_0 - y')^2 + z_0^2\right]^{1/2}.$$

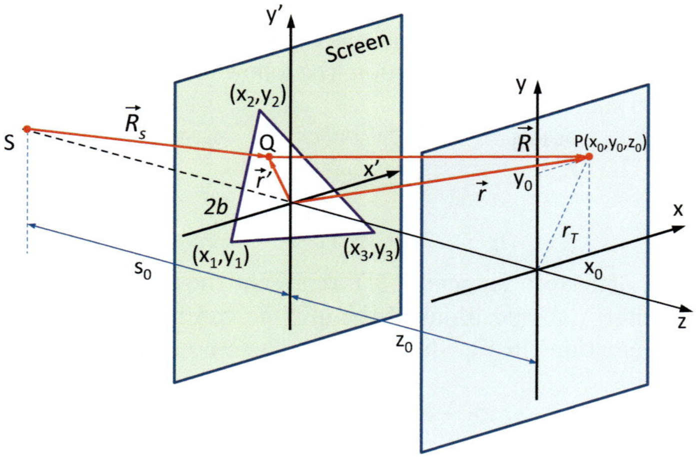

FIGURE 5.51: A triangular aperture (defined by the coordinates (x_i, y_i) of its apexes with $i = 1, 2, 3$) on an opaque screen. The diffracted field for the off-axis point $P(x_0, y_0, z_0)$ is sought. The point S specifies the origin of a diverging spherical wave that is incident on the aperture. In the case of an incident plane wave, the point S moves to $-\infty$ ($s_0 = \infty$).

The linear boundaries $y_{ij}(x)$ $(i, j = 1, 2, 3)$ are defined in Appendix 5D (Sec. 5D.5) and correspond to the three sides of the triangle. All these integrals must be evaluated numerically by performing the two-dimensional integrations given in Eqs. (5.206) and (5.207). The Kirchhoff formulation can similarly be derived using Eq. (5.20).

For the Fresnel approximation (see Eq. (5.25)), the corresponding integrals can be expressed as:

$$U(x_0, y_0, z_0) = jk\frac{\exp(-jkz_0)}{2\pi z_0}$$
$$\left\{ \int_{x_1}^{x_2}\int_{y_{13}(x)}^{y_{12}(x)} \left[\exp\left\{ -jk\left[\frac{(x_0 - x')^2}{2z_0} + \frac{(y_0 - y')^2}{2z_0} \right] \right\} U_{inc} \right] dx'dy' \right.$$
$$\left. + \int_{x_2}^{x_3}\int_{y_{13}(x)}^{y_{23}(x)} \left[\exp\left\{ -jk\left[\frac{(x_0 - x')^2}{2z_0} + \frac{(y_0 - y')^2}{2z_0} \right] \right\} U_{inc} \right] dx'dy' \right\}. \tag{5.208}$$

Even for a plane wave incidence, the above integrals cannot be expressed in terms of Fresnel integrals, as was done for the rectangular aperture. Therefore, numerical integration must again be employed.

Finally, the Fraunhofer approximation can be determined analytically (only in the case of a plane wave incidence) since it is directly related to the 2D Fourier transform of the aperture. The Fourier transform of the triangular aperture is calculated in Sec. 5D.5 of Appendix 5D [see Eqs. (5D.34), (5D.35), (5D.36), and (5D.37)]. The Fraunhofer approximation is then given by:

$$U_{ff}(x_0, y_0, z_0) = jkU_0\frac{e^{-jkz_0}}{2\pi z_0}\exp\left[-j\frac{k}{2z_0}\left(x_0^2 + y_0^2\right) \right] \mathcal{F}\left\{ t_a(x', y')U_{inc}(x', y') \right\}\Bigg|_{\substack{k_x = kx_0/z_0 \\ k_y = ky_0/z_0}}. \tag{5.209}$$

Here, $t_a(x', y')$ corresponds to the triangular amplitude transmittance function defined in Appendix 5D. Using Eq. (5D.37), the positions of the maxima in the Fraunhofer diffraction pattern can be estimated. From this equation, the following lines corresponding to the main maxima can be determined:

$$\begin{aligned} x_0 + a_{12}y_0 &= 0, \\ x_0 + a_{13}y_0 &= 0, \\ x_0 + a_{23}y_0 &= 0. \end{aligned} \tag{5.210}$$

The coefficients a_{ij}, defined in Appendix 5D, represent the slopes of the sides of the triangular aperture. Similarly, the positions of the minima can be estimated by identifying the zeros of the "sinc" functions in Eq. (5D.37). The corresponding equations for the minima are:

$$\begin{aligned} x_0 + a_{12}y_0 &= m\lambda\frac{z_0}{x_2 - x_1}, \\ x_0 + a_{13}y_0 &= p\lambda\frac{z_0}{x_2 - x_1}, \\ x_0 + a_{23}y_0 &= q\lambda\frac{z_0}{x_3 - x_2}, \\ x_0 + a_{13}y_0 &= \ell\lambda\frac{z_0}{x_3 - x_2}, \end{aligned} \tag{5.211}$$

where m, p, q, ℓ are non-zero integers. Each line represents a secondary minimum, as all terms in Eq. (5D.37) must simultaneously equal zero to produce an actual zero in the intensity pattern.

The case of the triangular obstacle can be readily analyzed using the results for the triangular aperture in conjunction with Babinet's principle, as done in previously presented cases.

5.9.2 Analytical Fraunhofer Approximation of a Triangular Aperture based on Boundary Waves

In this section, the boundary wave expression of the Fraunhofer approximation is employed to represent the Fraunhofer diffracted field from a triangular aperture in an alternative, compact, and infinity-free form. The result from Eq. (5.34) is first applied to one side of the triangle and then extended to the entire triangular aperture. This approach can similarly be applied to polygonal apertures.

For this purpose, consider a side of the triangle between apexes i and $i+1$ ($i = 1, 2, 3$ with apex 4 coinciding with apex 1). The position of each apex is specified by a position vector $\vec{r}_i = x_i\hat{\imath}_x + y_i\hat{\imath}_y$. The triangle side between the apexes i and $i+1$ is represented by the vector $\vec{a}_i = \vec{r}_{i+1} - \vec{r}_i = \hat{t}_i L_i$, where $\hat{t}_i = \vec{a}_i/|\vec{a}_i|$ is the unit vector along the i-th side, and $L_i = |\vec{a}_i|$ represents the length of the i-th side.

At this point, let's apply Eq. (5.34) for the side L_i of the triangle. The position vector along the i-th side can be expressed as $\vec{r} = \vec{r}_i + \hat{t}_i\ell$ where ℓ a parameter that takes values from zero ($\vec{r} = \vec{r}_i$) to L_i ($\vec{r} = \vec{r}_{i+1}$). The vector $\hat{\imath}_{ni}$ is the unit normal vector to the side i pointing away from the triangle. If $S_i(k_x, k_y)$ is the Fourier transform part that corresponds

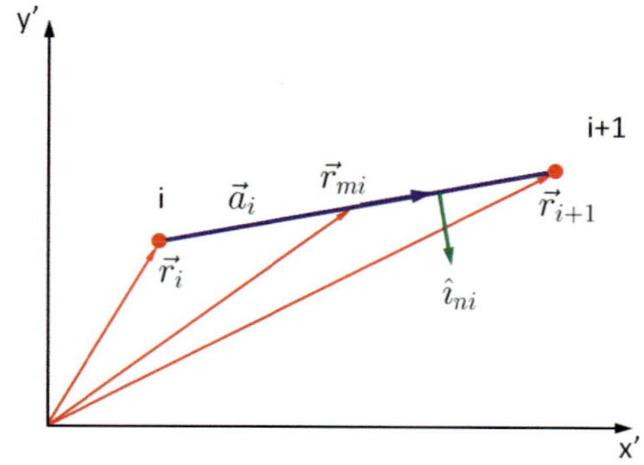

FIGURE 5.52: A side of a triangle (or a polygon, in general) is defined between two apexes, i and $i+1$. The position vectors of the apexes are denoted as $\vec{r}_i$, and the side between the apexes i and $i+1$ is specified by the vector $\vec{a}_i = \vec{r}_{i+1} - \vec{r}_i = \hat{t}_i L_i$, where $\hat{t}_i = \vec{a}_i/|\vec{a}_i|$ is the unit vector along the i-th side, and $L_i = |\vec{a}_i|$ represents the length of the i-th side. The vector $\hat{\imath}_{ni}$ is the unit normal vector to the side i pointing away from the triangle.

to the L_i side of the triangle, then it can be expressed as:

$$
\begin{aligned}
S_i(k_x, k_y) &= -\frac{j}{k_x^2+k_y^2}\int_{\ell=0}^{\ell=L_i} \exp\left(+j\vec{p}\cdot\vec{r}\right)\left(\vec{p}\cdot\hat{\imath}_{ni}\right) d\ell = \\
&= -\frac{j}{k_x^2+k_y^2}\exp\left(+j\vec{p}\cdot\vec{r}_i\right)\left(\vec{p}\cdot\hat{\imath}_{ni}\right)\int_{\ell=0}^{\ell=L_i} \exp\left(+j\vec{p}\cdot\hat{t}_i\ell\right) d\ell \\
&= -\frac{j}{k_x^2+k_y^2}\exp\left(+j\vec{p}\cdot\vec{r}_i\right)\left(\vec{p}\cdot\hat{\imath}_{ni}\right)\frac{\exp\left(+j\vec{p}\cdot\hat{t}_i L_i\right)-1}{j\vec{p}\cdot\hat{t}_i} = \\
&= -\frac{j}{k_x^2+k_y^2} L_i \exp\left(+j\vec{p}\cdot\vec{r}_{mi}\right)\left(\vec{p}\cdot\hat{\imath}_{ni}\right)\operatorname{sinc}\left(\vec{p}\cdot\hat{t}_i\frac{L_i}{2\pi}\right) \\
&= -\frac{j}{k_x^2+k_y^2} L_i \exp\left(+j\vec{p}\cdot\vec{r}_{mi}\right)\left(\vec{p}\cdot\hat{\imath}_{ni}\right)\operatorname{sinc}\left(\frac{\vec{p}\cdot\vec{a}_i}{2\pi}\right), \quad \text{where} \qquad (5.212)
\end{aligned}
$$

$$
\vec{p} = k_x\hat{\imath}_x + k_y\hat{\imath}_y = \frac{k}{z_0}\left(x_0\hat{\imath}_x + y_0\hat{\imath}_y\right), \quad \vec{r}_{mi} = \vec{r}_i + \frac{L_i}{2}\hat{t}_i, \quad \hat{\imath}_{ni} = \hat{t}_i \times \hat{\imath}_z.
$$

In the above derivation the term $\vec{p}\cdot\hat{\imath}_{ni}$ was constant along the integration path. Summing up the three sides of the triangle the Fraunhofer approximation can be written in the following form:

$$
\begin{aligned}
U_{ff}(x_0, y_0, z_0) = U_0\frac{e^{-jkz_0}}{2\pi}\frac{z_0}{k(x_0^2+y_0^2)}\exp\left[-j\frac{k}{2z_0}\left(x_0^2+y_0^2\right)\right] \\
\sum_{i=1}^{3}\left\{|\vec{a}_i|\exp\left(+j\vec{p}\cdot\vec{r}_{mi}\right)\left(\vec{p}\cdot\hat{\imath}_{ni}\right)\operatorname{sinc}\left(\frac{\vec{p}\cdot\vec{a}_i}{2\pi}\right)\right\}, \qquad (5.213)
\end{aligned}
$$

$$
\text{with} \quad U_{ff}(x_0 = 0, y_0 = 0, z_0) = jkU_0\frac{e^{-jkz_0}}{2\pi z_0}S_a,
$$

where S_a is the area of the triangle. It should be noted that for $x_0 = y_0 = 0$ (or equivalently $k_x = k_y = 0$) Eq. (5.34) is not valid. However, this special case is very simple and it is treated separately using the area of the triangle. From Eq. (5.213), it is clear that along the spatial frequency directions perpendicular to the sides of the triangle, the intensity of the diffraction pattern decreases more slowly. These directions correspond to the main maxima, as expressed in Eqs(5.210). That is, the equations (5.210) are equivalent to $\vec{p}\cdot\vec{a}_i = 0$ (for $i = 1, 2, 3$).

Example 5.9.1:

In this example, the application of Eq. (5.212) for the case of an equilateral triangle is demonstrated. Consider an equilateral triangle with side length a as shown in Fig. 5.53. The task is to calculate the $S(k_x, k_y)$ term for this configuration.

Solution: The following vectors and their corresponding inner products, which are needed for the application of Eq. (5.212) to each side of the triangle, can be easily determined as:

$$\vec{r}_1 = -\frac{a}{2}\hat{\imath}_x - \frac{a}{2\sqrt{3}}\hat{\imath}_y, \qquad \vec{r}_2 = \frac{a}{2}\hat{\imath}_x - \frac{a}{2\sqrt{3}}\hat{\imath}_y, \qquad \vec{r}_3 = \frac{a}{\sqrt{3}}\hat{\imath}_y,$$

$$\vec{a}_1 = a\hat{\imath}_x, \qquad \vec{a}_2 = -\frac{a}{2}\hat{\imath}_x + \frac{a}{2\sqrt{3}}\hat{\imath}_y, \qquad \vec{a}_3 = -\frac{a}{2}\hat{\imath}_x - \frac{a}{2\sqrt{3}}\hat{\imath}_y,$$

$$\vec{r}_{m1} = -\frac{a}{2\sqrt{3}}\hat{\imath}_y, \qquad \vec{r}_{m2} = \frac{a}{4}\hat{\imath}_x + \frac{a}{4\sqrt{3}}\hat{\imath}_y, \qquad \vec{r}_{m3} = -\frac{a}{4}\hat{\imath}_x + \frac{a}{4\sqrt{3}}\hat{\imath}_y,$$

$$\hat{\imath}_{n1} = -\hat{\imath}_y, \qquad \hat{\imath}_{n2} = \frac{\sqrt{3}}{2}\hat{\imath}_x + \frac{1}{2}\hat{\imath}_y, \qquad \hat{\imath}_{n3} = -\frac{\sqrt{3}}{2}\hat{\imath}_x + \frac{1}{2}\hat{\imath}_y,$$

$$\vec{p}\cdot\hat{\imath}_{n1} = -k_y, \qquad \vec{p}\cdot\hat{\imath}_{n2} = \frac{\sqrt{3}}{2}k_x + \frac{1}{2}k_y, \qquad \vec{p}\cdot\hat{\imath}_{n3} = -\frac{\sqrt{3}}{2}k_x + \frac{1}{2}k_y,$$

$$\vec{p}\cdot\vec{a}_1 = ak_x, \qquad \vec{p}\cdot\vec{a}_2 = -\frac{a}{2}k_x + \frac{a\sqrt{3}}{2}k_y, \qquad \vec{p}\cdot\vec{a}_3 = -\frac{a}{2}k_x - \frac{a\sqrt{3}}{2}k_y,$$

$$\vec{p}\cdot\vec{r}_{m1} = -\frac{a}{2\sqrt{3}}k_y, \qquad \vec{p}\cdot\vec{r}_{m2} = \frac{a}{4}k_x + \frac{a}{4\sqrt{3}}k_y, \qquad \vec{p}\cdot\vec{r}_{m3} = -\frac{a}{4}k_x + \frac{a}{4\sqrt{3}}k_y.$$

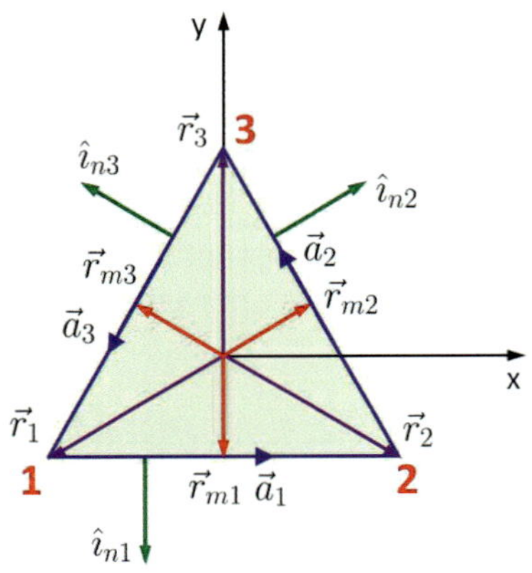

FIGURE 5.53: An equilateral triangle of side $L_i = a$ ($i = 1, 2, 3$) with its centroid coinciding with the origin of the xy coordinate system. The triangle apexes are denoted as 1, 2, and 3. The position vectors of each apex are $\vec{r}_i$, and the side between the apexes i and $i+1$ is specified by the vector $\vec{a}_i = \vec{r}_{i+1} - \vec{r}_i$, while $\hat{\imath}_{ni}$ are the unit vectors perpendicular to the i-th side. The midpoint vectors $\vec{r}_{mi}$ are also shown ($i = 1, 2, 3$).

Then, the corresponding Fourier transform, $S(k_x, k_y)$, is given by:

$$S(k_x, k_y) = -\frac{j}{k_x^2 + k_y^2} \sum_{i=1}^{3} \left\{ |\vec{a}_i| \exp\left(+j\vec{p}\cdot\vec{r}_{mi}\right)\left(\vec{p}\cdot\hat{\imath}_{ni}\right) \operatorname{sinc}\left(\frac{\vec{p}\cdot\vec{a}_i}{2\pi}\right)\right\} =$$

$$= -\frac{ja}{k_x^2 + k_y^2}\left\{ - k_y \exp\left(-jk_y\frac{a}{2\sqrt{3}}\right)\operatorname{sinc}\left(\frac{k_x a}{2\pi}\right) + \right.$$

$$\left(\frac{\sqrt{3}}{2}k_x + \frac{1}{2}k_y\right)\exp\left[j\left(\frac{a}{4}k_x + \frac{a}{4\sqrt{3}}k_y\right)\right]\operatorname{sinc}\left[\frac{1}{2\pi}\left(-\frac{a}{2}k_x + \frac{a\sqrt{3}}{2}k_y\right)\right] +$$

$$\left.\left(-\frac{\sqrt{3}}{2}k_x + \frac{1}{2}k_y\right)\exp\left[j\left(-\frac{a}{4}k_x + \frac{a}{4\sqrt{3}}k_y\right)\right]\operatorname{sinc}\left[\frac{1}{2\pi}\left(\frac{a}{2}k_x + \frac{a\sqrt{3}}{2}k_y\right)\right]\right\}$$

Of course, the special case of $k_x = k_y = 0$, is equal to the area of the equilateral triangle, i.e., $S(0, 0) = a^2\sqrt{3}/4$.

Example 5.9.2:

Consider an equilateral triangle aperture with its centroid coinciding with the origin in the $x'y'$ plane, as described in Example 5.9.1. The coordinates of the equilateral triangle's apexes are given by $(x_1 = -a/2, y_1 = -h/3)$, $(x_2 = 0, y_2 = 2h/3)$, and $(x_3 = +a/2, y_3 = -h/3)$, where $a = 5\,\mu m$ is each side of the equilateral triangle, and $h = a\sqrt{3}/2$ is its height. Assume an incident plane wave as described in Eq. (5.194) with freespace wavelength of $\lambda_0 = 1\,\mu$m. The normalized intensities $|U|^2/|U_0|^2$ are sought along x-, y-directions for $z_0 = 10\,\mu$m, and and along the z-direction for the RS-1, RS-2, Kirchhoff, Fresnel, and Fraunhofer formulations (where applicable).

<u>Solution:</u> The normalized diffracted intensities $|U|^2/|U_0|^2$ are shown in Figs. 5.54(a) and 5.54(b) as functions of the distance z_0 from the aperture for $x_0 = y_0 = 0$. The RS-1, RS-2, and Kirchhoff formulations are included. All three formulations practically coincide a few wavelengths away from the triangular aperture. In Figs. 5.54(c) and 5.54(d), the normalized diffracted intensity is shown at $z_0 = 10\,\mu m = 10\lambda_0$ along either the x-direction or the y-direction. Some asymmetry is observed in the two cases, as the x- and y-directions are not equivalent due to the positioning of the triangle in the $x'y'$-plane. The RS-1, Fresnel, and Fraunhofer formulations are compared in Fig. 5.55. These figures specifically show the diffracted intensities along the x- and y-directions for $z_0 = 500\,\mu m = 500\lambda_0$ and $y_0 = 0$ or $x_0 = 0$, respectively. As expected, some asymmetry between the two directions is observed. The Fresnel condition is satisfied for $z_0 \gg [(\pi/4\lambda)D_{max}^4]^{1/3} \simeq 1414\,\mu m$ (for $D_{max} \simeq 250 - a = 245\,\mu m$) for the maximum range of the $250\,\mu m$ shown along the x- or y-directions ($N_F = 120.05$). The corresponding Fraunhofer condition is $z_0 \gg (\pi/\lambda)a^2 = 78.5\,\mu m$ ($N_F = a^2/\lambda z_0 = 0.05$). Naturally, the Fraunhofer condition for z_0 should be larger than the Fresnel condition for consistency. Even though the Fresnel condition for the furthest points from the axis is not perfectly satisfied, very good agreement between the Fresnel and Fraunhofer approximations and the RS-1 formulation is observed.

Finally, the far-field normalized diffracted intensity pattern is shown in Figs. 5.55(c) and 5.55(d) (in logarithmic scale) for the same $z_0 = 500\,\mu m$. The lines observed, especially in the logarithmic diagram, correspond to the lines specified by Eqs. (5.210). Using these equations, it is straightforward to show that the lines with slopes of $30\,deg$, $-30\,deg$, and $90\,deg$ will contain the maxima of the diffraction pattern. The intersection of the three lines represents the major maximum, while smaller maxima appear along each of the line directions, as evident in the figures.

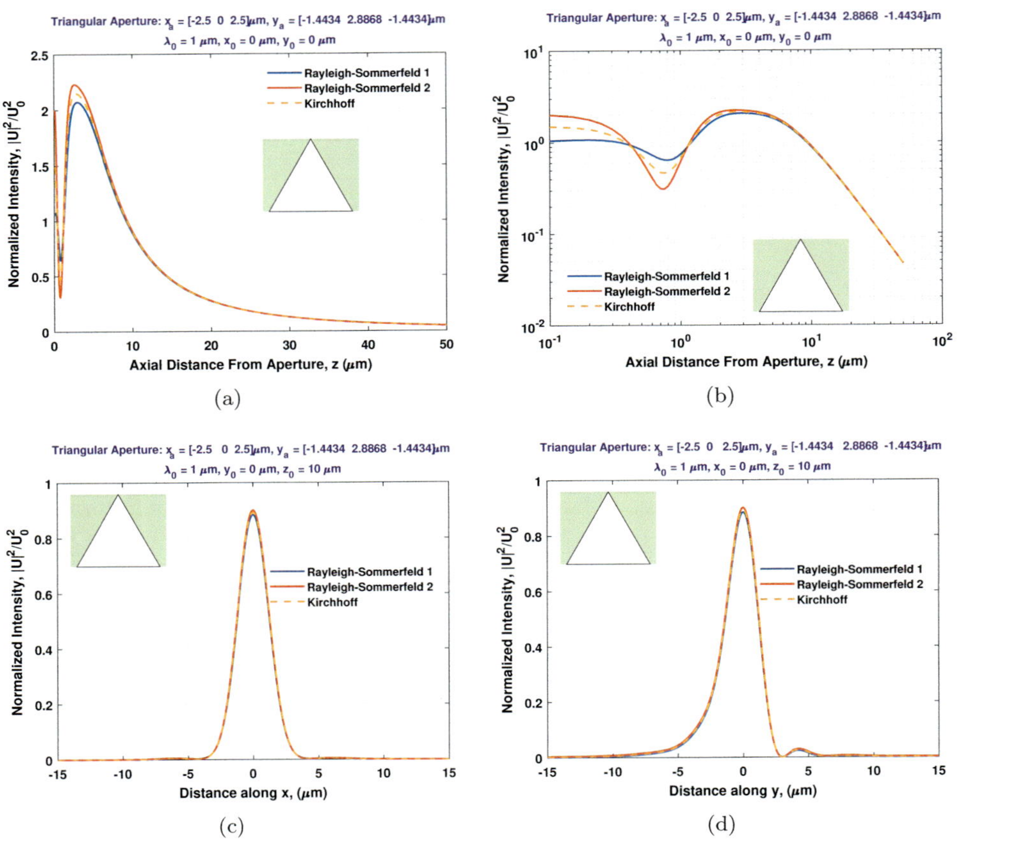

FIGURE 5.54: (a) The diffracted field normalized intensity $|U|^2/|U_0|^2$ in the z direction for an incident plane wave on an equilateral triangular aperture of side $a = 5\,\mu m$ ($x_0 = y_0 = 0$). The freespace wavelength is $\lambda_0 = 1.0\,\mu m$. The RS-1, RS-2, and Kirchhoff formulations are shown. (b) Same as (a) but in a logarithmic scale. (c) Same as (a) but along the x direction for $z_0 = 10\,\mu m$ and $y_0 = 0$. (d) Same as (c) but along the y direction for $z_0 = 10\,\mu m$ and $x_0 = 0$.

Example 5.9.3:

This example is similar to Example 5.9.2, but in this case, the equilateral triangle serves as an obstacle rather than an aperture. The dimensions of the triangle and the incident plane wave remain the same. The goal is to calculate the normalized intensities $|U|^2/|U_0|^2$ along x-, y-, and z-directions for the RS-1, RS-2, Kirchhoff, Fresnel, and Fraunhofer formulations (where applicable). For the intensities along the x- and y-directions assume again that $z_0 = 10\,\mu$m.

Solution: In the given example, the normalized diffracted intensities $|U|^2/|U_0|^2$ are presented as functions of the distance z_0 from the equilateral triangular obstacle for $x_0 = y_0 = 0$. Figures showing the RS-1, RS-2, and Kirchhoff formulations reveal that all three formulations converge at distances just a few wavelengths away from the obstacle. Additionally, the normalized diffracted intensity is shown at a distance of $z_0 = 10, \mu m = 10\lambda_0$ along the x- and y-directions in separate figures. Some asymmetry is observed between the two directions, which results from the unequal positioning of the triangle within the $x'y'$-plane.

The RS-1, Fresnel, and Fraunhofer formulations are compared in Fig. 5.57. Specifically, these figures show the diffracted intensity along the x- and y-directions for $z_0 = 500\,\mu m =$

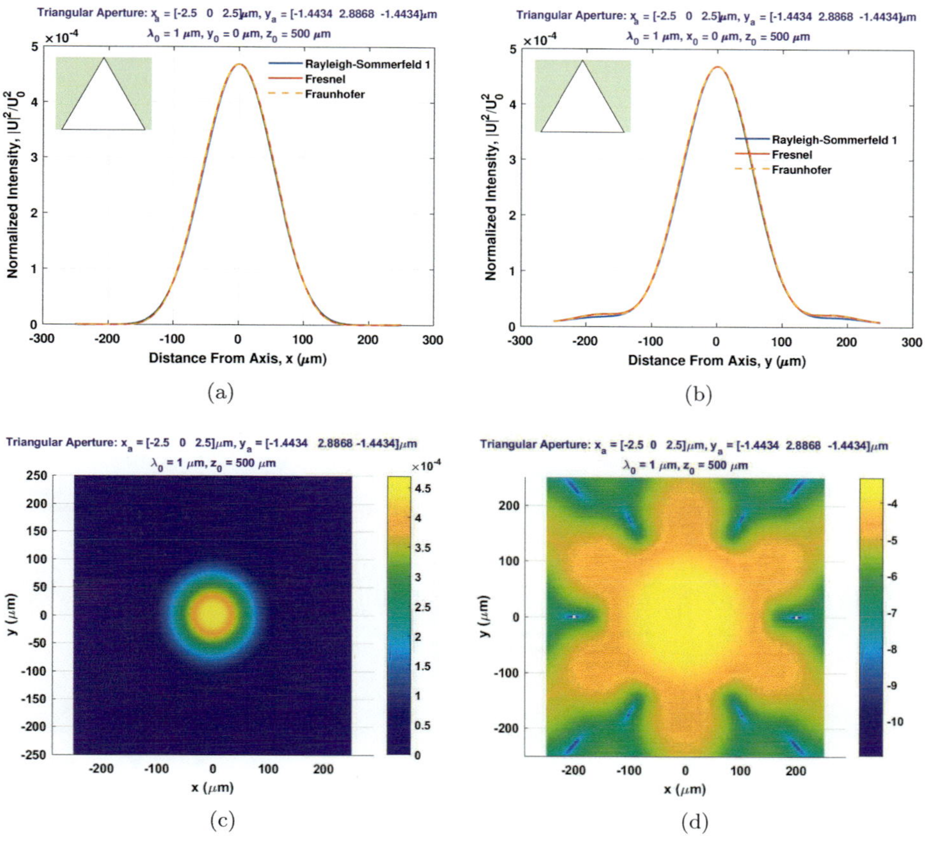

FIGURE 5.55: (a) The diffracted field normalized intensity $|U|^2/|U_0|^2$ in the x-direction is shown for an incident plane wave on an equilateral triangular aperture with a side length $a = 5\,\mu m$ (with $x_0 = y_0 = 0$) for $z_0 = 500\,\mu m = 500\lambda_0$ and $y_0 = 0$. The freespace wavelength is $\lambda_0 = 1.0\,\mu m$. The RS-1, Fresnel, and Fraunhofer formulations are shown. (b) Same as (a) but along the y-direction for $z_0 = 500\,\mu m = 500\lambda_0$ and $x_0 = 0$. (c) Fraunhofer normalized intensity pattern at $z_0 = 500\,\mu m = 500\lambda_0$. (d) Same as (c) but in logarithmic scale.

$500\lambda_0$ and $y_0 = 0$ or $x_0 = 0$, respectively. Some asymmetry is evident between the two directions due to the positioning of the triangle. The intensity oscillates around unity, as expected for a normalized intensity, but the contrast of these oscillations is small (less than 5%). Despite the mild satisfaction of the Fresnel and Fraunhofer conditions, the Fresnel and Fraunhofer approximations yield results that closely match the RS-1 formulation. For the far-field diffraction pattern at $z_0 = 500\,\mu m$, the pattern reveals fine details, though with a low contrast. Therefore, a logarithmic plot is not included, as it does not offer additional resolution.

5.9.3 Scalar Diffraction by Polygonal Apertures and Obstacles

In this section, polygonal apertures and obstacles are discussed. Any polygonal aperture can be subdivided into a summation of triangular apertures, where the area S_a of a polygonal aperture can be expressed as the sum of the areas of individual triangles: $S_a = \sum_{i=1}^{N} S_{ai}$, where N is the number of triangles the polygon is decomposed into, and S_{ai} represents the area of each triangle. It is important to note that this decomposition is not unique.

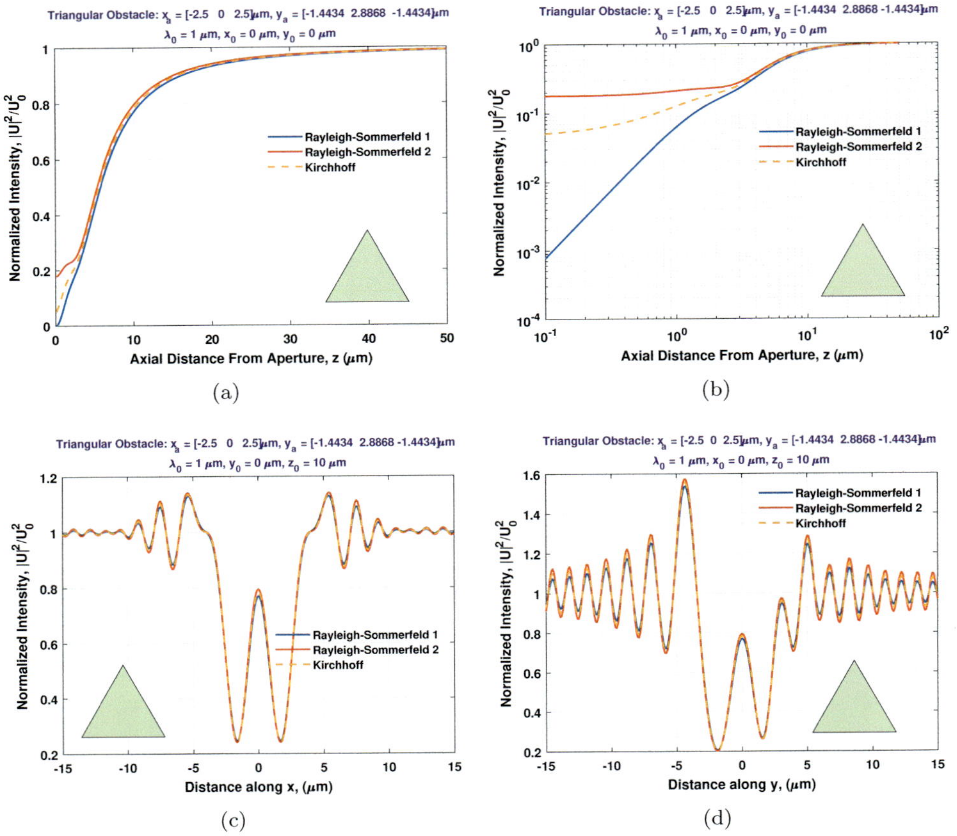

FIGURE 5.56: (a) The diffracted field normalized intensity $|U|^2/|U_0|^2$ in the z-direction for an incident plane wave on an equilateral triangular obstacle of side $a = 5\,\mu m$ ($x_0 = y_0 = 0$). The freespace wavelength is $\lambda_0 = 1.0\,\mu m$. The RS-1, RS-2, and Kirchhoff formulations are shown. (b) Same as (a) but in a logarithmic scale. (c) Same as (a) but along the x-direction for $z_0 = 10\,\mu m$ and $y_0 = 0$. (d) Same as (c) but along the y-direction for $z_0 = 10\,\mu m$ and $x_0 = 0$.

For example, a square can be divided into two equal triangles by its diagonal, or it can be decomposed into four equal triangles, with the center of the square serving as a common apex for all four triangles. The remaining apexes of these triangles are sequential apexes of the square. Regardless of how the polygon is decomposed, the results should be identical, as long as the union of all triangles results in the same polygonal area and the triangles do not overlap. With this decomposition, the RS-1 and RS-2 diffraction integrals can be written accordingly as:

$$U_{RS1}(P_0) = \sum_{i=1}^{N} \iint_{S_{ai}} \left[U_{inc} \left(jk + \frac{1}{R} \right) \frac{\exp(-jkR)}{2\pi R} \frac{z_0}{R} \right]_{z'=0} dx'dy' \tag{5.214}$$

$$U_{RS2}(P_0) = \sum_{i=1}^{N} \iint_{S_{ai}} \left[\frac{\exp(-jkR)}{2\pi R} \frac{\partial U_{inc}}{\partial n} \right]_{z'=0} dx'dy', \tag{5.215}$$

$$\text{with } R = \left[(x_0 - x')^2 + (y_0 - y')^2 + z_0^2 \right]^{1/2},$$

All of these integrals must be determined numerically, as they involve two-dimensional integrations as described in Eqs. (5.214) and (5.215). These integrals are decomposed into

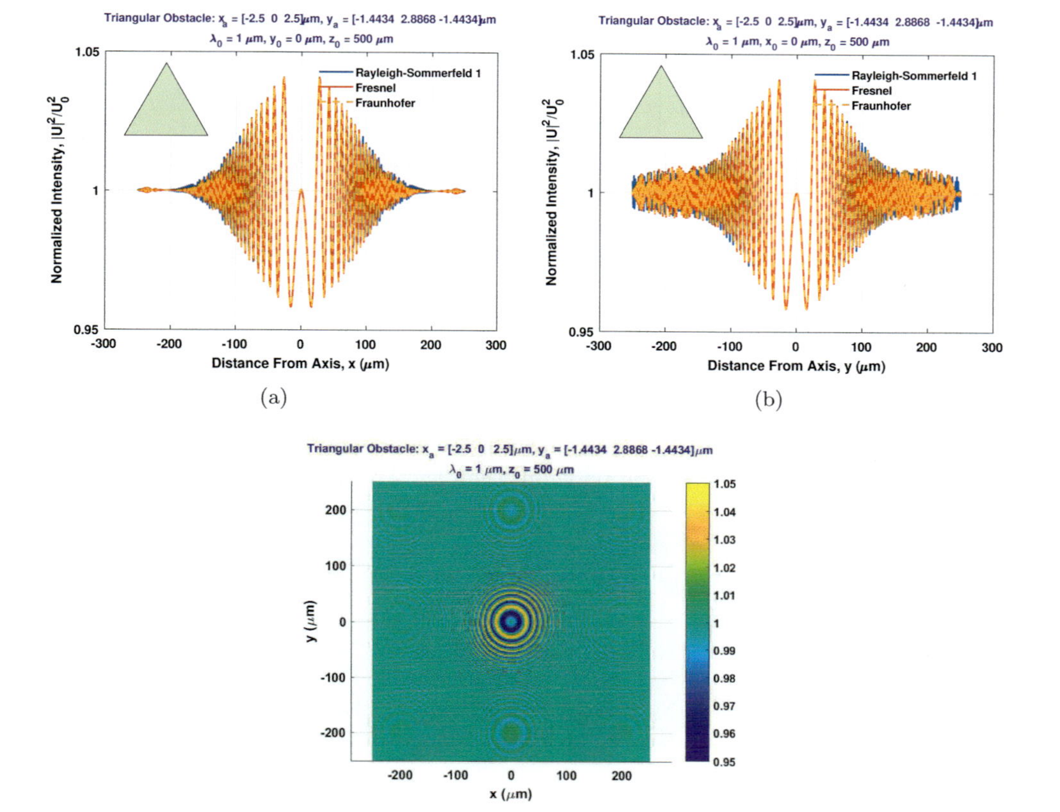

(c)

FIGURE 5.57: (a) The diffracted field normalized intensity $|U|^2/|U_0|^2$ in the x-direction for an incident plane wave on an equilateral triangular obstacle of side $a = 5\,\mu m$ for $z_0 = 500\,\mu m = 500\lambda_0$ and $y_0 = 0$. The freespace wavelength is $\lambda_0 = 1.0\,\mu m$. The RS-1, Fresnel, and Fraunhofer formulations are shown. (b) Same as (a) but along the y-direction for $z_0 = 500\,\mu m = 500\lambda_0$ and $x_0 = 0$. (c) Fraunhofer normalized intensity pattern at $z_0 = 500\,\mu m = 500\lambda_0$.

N triangular regions, and the integration process follows the methodology outlined in Sec. 5.9.1. The Kirchhoff formulation can again be obtained by applying Eq. (5.20).

As previously mentioned in Sec. 5.2.1, the surface integral can be converted into a line integral over the boundary of the aperture. However, in this chapter, the line integration approach is applied only within the Fraunhofer approximation (as discussed in Sec. 5.4.2.1), similar to the approach used for the triangular aperture.

For the Fresnel approximation [see Eq. (5.25)], the corresponding integrals can be written as:

$$U(P_0) = jk\frac{\exp(-jkz_0)}{2\pi z_0}\sum_{i=1}^{N}\iint_{S_{ai}}\left[\exp\left\{-jk\left[\frac{(x_0-x')^2}{2z_0}+\frac{(y_0-y')^2}{2z_0}\right]\right\}U_{inc}\right]dx'dy'. \tag{5.216}$$

Finally, the Fraunhofer approximation can be determined analytically (only in the case of an incident plane wave as it was shown in the case of the triangular aperture), since it is related to the 2D Fourier transform of the aperture. The Fourier transform of the polygonal aperture can be calculated as a summation of Fourier transforms of triangles (see Sec. 5D.5

of Appendix 5D). Then, the Fraunhofer approximation can be expressed as:

$$U_{ff}(x_0,y_0,z_0) = jkU_0\frac{e^{-jkz_0}}{2\pi z_0}\exp\left[-j\frac{k}{2z_0}\left(x_0^2+y_0^2\right)\right]\underbrace{\sum_{i=1}^{N}\mathcal{F}\left\{t_{ai}x',y')U_{inc}(x',y')\right\}}_{=\,S(k_x,k_y)}\Bigg|_{\substack{k_x=kx_0/z_0\\ k_y=ky_0/z_0}}. \tag{5.217}$$

In this expression, $t_{ai}(x',y')$ is the amplitude transmittance of the i-th triangle. The result is the total diffracted field from the polygonal aperture, which is the sum of the contributions from all the triangular sections that make up the aperture.

In the Fraunhofer approximation, a more compact form for the Fourier transform of a polygonal aperture can be derived by applying the boundary wave approach. This approach, as previously demonstrated for the triangular aperture (see Sec. 5.9.2), expresses the Fourier transform as a line integral over the boundary of the polygon [based on Eq. (5.212]. If N_p represents the number of sides of the polygon, then the Fourier transform $S(k_x,k_y)$ of Eq. (5.217) can be written as:

$$S(k_x,k_y) = \sum_{i=1}^{N}\mathcal{F}\left\{t_{ai}x',y')U_{inc}(x',y')\right\}\Bigg|_{\substack{k_x=kx_0/z_0\\ k_y=ky_0/z_0}}$$

$$= -\frac{j}{k_x^2+k_y^2}\sum_{i=1}^{N_p}L_i\exp\left(+j\vec{p}\cdot\vec{r}_{mi}\right)\left(\vec{p}\cdot\hat{\imath}_{ni}\right)\operatorname{sinc}\left(\frac{\vec{p}\cdot\vec{a}_i}{2\pi}\right). \tag{5.218}$$

Here, all the vectors have been defined in Sec. 5.9.2. Replacing $S(k_x,k_y)$ in Eq. (5.217), an alternate Fraunhofer approximation for a polygonal aperture is developed.

Example 5.9.4:

In this example, the Fourier transform of a regular N_p-side polygon is calculated based on the application of Eq. (5.218).

Solution: A generic regular polygon is shown in Fig. 5.58, where it is assumed that the polygon has N_p equal sides and a radius R. The N_p apexes of the polygon can be described by their position vectors $\vec{r}_i$. The angle between two successive apexes is $\theta = 2\pi/N_p$. The various vectors needed for the application of Eq. (5.218) are listed next. The first apex $i=1$, is assumed to be on the x-axis (in the figure this coincides with the i-th apex). The approach of Komrska [282] is followed using polar coordinate systems:

$$\vec{p} = k_x\hat{\imath}_x + k_y\hat{\imath}_y = k_t\left(\cos\psi\hat{\imath}_x + \sin\psi\hat{\imath}_y\right),$$

$$\vec{r}_i = R\left\{\cos\left[\frac{2\pi}{N_p}(i-1)\right]\hat{\imath}_x + \sin\left[\frac{2\pi}{N_p}(i-1)\right]\hat{\imath}_y\right\}, \quad i=1,2,\cdots,N_p,$$

$$\vec{r}_{mi} = R\cos\left(\frac{\pi}{N_p}\right)\left\{\cos\left[\frac{\pi}{N_p}(2i-1)\right]\hat{\imath}_x + \sin\left[\frac{\pi}{N_p}(2i-1)\right]\hat{\imath}_y\right\},$$

$$\hat{\imath}_{ni} = \cos\left[\frac{\pi}{N_p}(2i-1)\right]\hat{\imath}_x + \sin\left[\frac{\pi}{N_p}(2i-1)\right]\hat{\imath}_y,$$

$$\vec{a}_i = 2R\sin\left(\frac{\pi}{N_p}\right)\left\{-\sin\left[\frac{\pi}{N_p}(2i-1)\right]\hat{\imath}_x + \cos\left[\frac{\pi}{N_p}(2i-1)\right]\hat{\imath}_y\right\},$$

$$|\vec{a}_i| = L_i = 2R\sin\left(\frac{\pi}{N_p}\right).$$

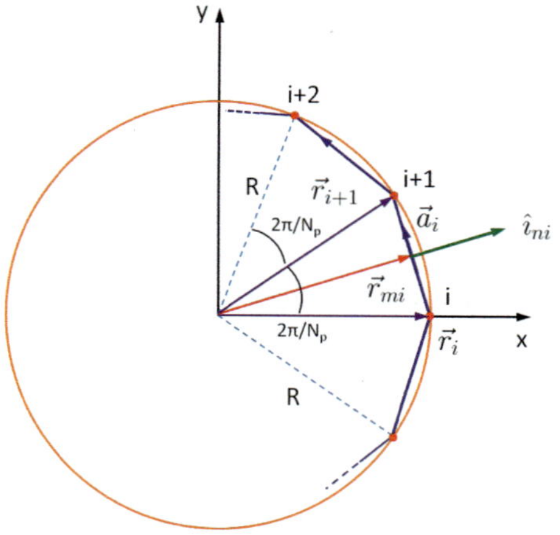

FIGURE 5.58: Part of a regular polygon of N_p sides is shown. The $i = 1$ apex is on the x-axis. The $i+N_p$ apex coincides with the i-th apex. The angular separation between two successive apexes is $2\pi/N_p$. The position vectors of each apex are $\vec{r}_i$, and the side between the apexes i and $i+1$ is specified by the vector $\vec{a}_i = \vec{r}_{i+1} - \vec{r}_i$ ($|\vec{a}_i| = a$), while $\hat{\imath}_{ni}$ are the unit vectors perpendicular to the i-th side. The midpoint vectors $\vec{r}_{mi}$ are also shown ($i = 1, 2, \cdots, N_p$). The regular polygon radius is R.

The inner products necessary for the application of Eq. (5.218) are the following:

$$\vec{p}\cdot\vec{a}_i = 2k_t R \sin\left(\frac{\pi}{N_p}\right) \sin\left[\psi - \frac{\pi}{N_p}(2i-1)\right],$$

$$\vec{p}\cdot\hat{\imath}_{ni} = k_t \cos\left[\psi - \frac{\pi}{N_p}(2i-1)\right],$$

$$\vec{p}\cdot\vec{r}_{mi} = k_t R \cos\left(\frac{\pi}{N_p}\right) \cos\left[\psi - \frac{\pi}{N_p}(2i-1)\right].$$

Inserting the previous expressions into Eq. (5.218), the Fourier transform of a regular polygon with N_p sides is given, with $S(k_x, k_y)$ is expressed in polar coordinates as $S(k_t, \psi)$. This formulation accounts for the geometry of the polygon, where k_t is the radial spatial frequency and ψ is the azimuthal angle in the spatial frequency space.

$$S(k_t, \psi) = \frac{2R}{jk_t} \sin\left(\frac{\pi}{N_p}\right) \sum_{i=1}^{N_p} \left\{ \cos\left[\psi - \frac{\pi}{N_p}(2i-1)\right] \right.$$
$$\exp\left\{ jk_t R \cos\left(\frac{\pi}{N_p}\right) \cos\left[\psi - \frac{\pi}{N_p}(2i-1)\right]\right\}$$
$$\left. \operatorname{sinc}\left\{ 2k_t R \sin\left(\frac{\pi}{N_p}\right) \sin\left[\psi - \frac{\pi}{N_p}(2i-1)\right]\right\} \right\}. \quad (5.219)$$

In Ref. [282], it is demonstrated that the diffracted Fraunhofer field diminishes as $1/k_t$ along the directions perpendicular to the sides of a regular polygon and as $1/k_t^2$ in all other directions. The directions of the major peaks in the diffracted pattern align with the lines where the "sinc" functions have a zero argument. This occurs when $\psi - (\pi/N_p)(2i-1) = \nu\pi$ (where $\nu = 0, \pm1, \pm2, \cdots$). These directions correspond to those that are perpendicular to the sides of the polygon. Equation (5.219) is not valid at $k_t = 0$. In this case, the Fourier transform reduces to the area of the regular polygon, which is given by $S(k_x = 0, k_y = 0) = S(k_t = 0, \psi) = N_p R^2 \cos(\pi/N_p)\sin(\pi/N_p)$. The equilateral triangle is a special case of the regular polygon for $N_p = 3$.

Example 5.9.5:

In this example, several sample polygonal apertures (or obstacles) are analyzed using the RS-1, RS-2, Kirchhoff, Fresnel, and Fraunhofer formulations. The selected structures include a regular hexagon with a radius $R = 5\,\mu m$, a starred-like aperture derived from a regular heptagon with $R = a = 5\,\mu m$ (where a represents the two equal sides of the isosceles triangles placed on top of each polygon side to create the starred shape), and a daisy-pattern consisting of 5 leaves formed from a regular decagon with $R = a = 5\,\mu m$. In each case, a normally incident plane wave with a freespace wavelength of $\lambda_0 = 1\,\mu m$ is considered.
Solution: To enhance the visualization of the apertures, insets are included in most plots to display the polygonal aperture. The various results are presented in Fig. 5.59. Initially, the near-field patterns are shown either along the x-direction for $z_0 = 20\,\mu m$, or along the z-axis. The RS-1, RS-2, and Kirchhoff formulations are included in each plot. Figures 5.59(a) and 5.59(b) display the results for the regular hexagon. Figures 5.59(c) and 5.59(d) show the results for the starred-like aperture from a regular heptagon, while Figs. 5.59(e) and 5.59(f) present the results for the 5-leaf daisy-pattern derived from a regular decagon.

In Fig. 5.60, similar diagrams are included for the same apertures but for $z_0 = 200\,\mu m$. The figures on the left show the normalized intensity (RS-1, Fresnel, and Fraunhofer) along the x-direction for $z_0 = 200\,\mu m$. The areas of the three apertures are $64.95\,\mu m^2$ for the regular hexagon, $136.82\,\mu m^2$ for the starred-regular heptagon, and $73.47\,\mu m^2$ for the 5-leave daisy-pattern from a regular decagon. The corresponding conditions for the Fresnel approximation are: $z_0 \gg [(\pi/\lambda)D_{max}^4]^{1/3} \simeq 36.6\,\mu m$ for the regular hexagon and $43.6\,\mu m$ for the other two apertures. Additionally, the conditions for the Fraunhofer approximation are: $z_0 \gg S_a/\lambda \simeq 64.95\,\mu m$ for the regular hexagon, $136.8\,\mu m$ for the starred-regular heptagon, and $73.5\,\mu m$ for the 5-leave daisy-pattern from a regular decagon. Therefore, in all cases, the approximations should provide good agreement with the RS-1 results, as shown in the figures. The figures on the right in Fig. 5.60 depict the pattern of the normalized amplitude of the diffracted field in the xy-plane for $z_0 = 200\,\mu m$. The directions where the diffracted field decreases as $1/k_t$ (in the xy-plane) can be predicted from Eq. (5.218) and are normal to the sides of each polygon.

Similar diagrams can be easily calculated when these polygons serve as obstacles (occulters) instead of apertures by applying the equations for the aperture and utilizing Babinet's principle, as was done for other obstacles under scalar diffraction theory. Some examples of the same geometric parameters used for the apertures but now for obstacles are shown in Fig. 5.61 to illustrate these cases. A notable feature is displayed in Fig. 5.61(c) for the starred-heptagon obstacle. The plot is asymmetric along the x-axis and is not symmetric around the origin, which is expected due to the positioning of the aperture and its symmetries. Another interesting feature is the suppression of the middle peak (equivalent to the Arago spot) as a result of the approximate apodization of its starred pattern. Similar obstacles, often referred to as occulters, have been proposed for "starshade" telescopes, which aim to photograph planets around distant star systems [298–301].

5.9.3.1 Serrated Circular Apertures-Obstacles

Another type of aperture that can be treated using the approach presented in Sec. 5.9.3 is the serrated aperture. The method is based on the decomposition of the aperture (or obstacle) into non-overlapping triangles. Serrated apertures have been proposed as a means

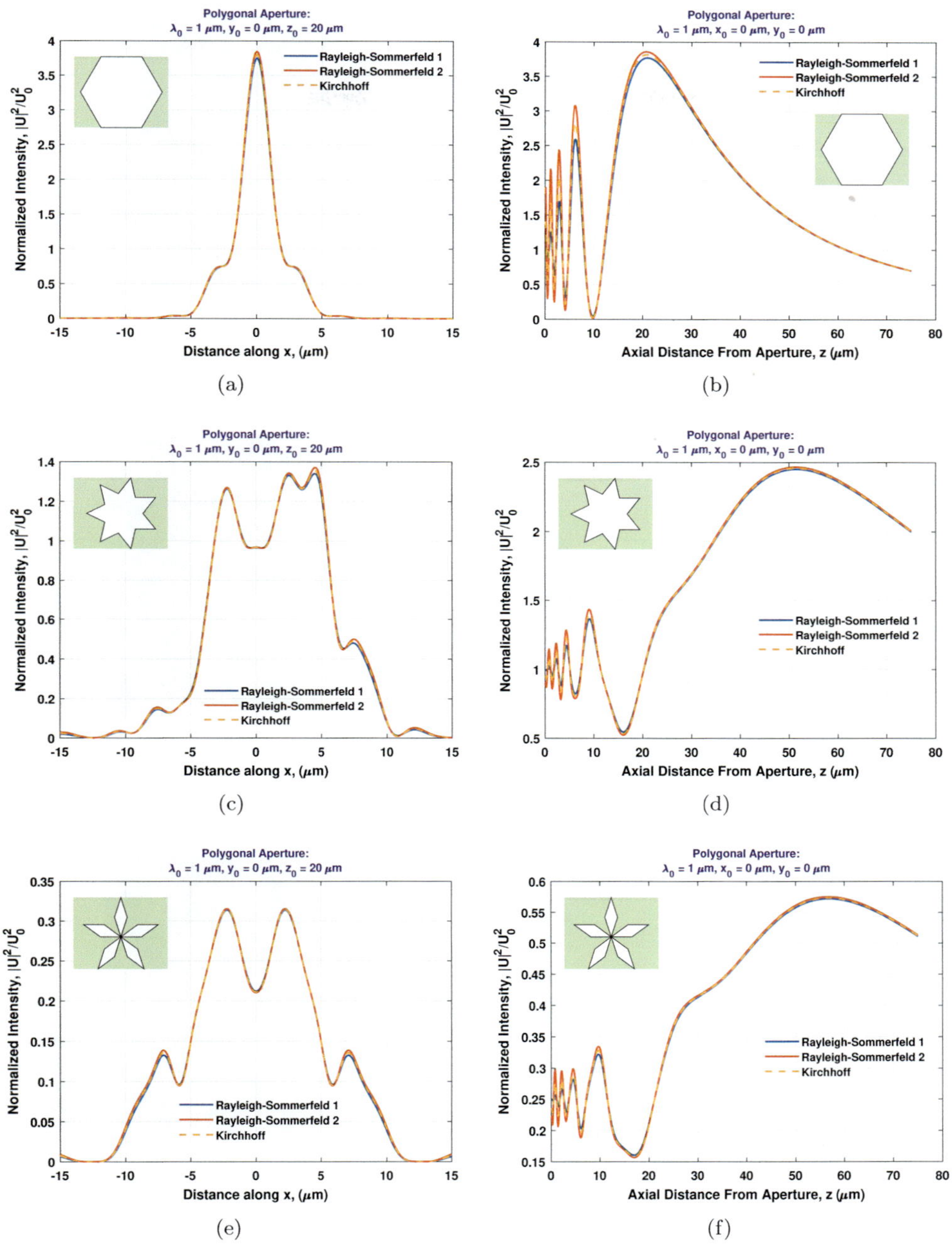

FIGURE 5.59: (a) The diffracted field normalized intensity $|U|^2/|U_0|^2$ in the x direction for an incident plane wave on a regular-hexagon aperture (shown in inset) of side $R = 5\,\mu m$ for $z_0 = 20\,\mu m = 20\lambda_0$ and $y_0 = 0$. The freespace wavelength is $\lambda_0 = 1.0\,\mu m$. (b) Same as (a) but along the z-axis for $x_0 = y_0 = 0$. (c) Same as (a) but for a starred-regular heptagon aperture (shown in inset) with side $R = a = 5\,\mu m$. (d) Same as (c) but along the z-axis for $x_0 = y_0 = 0$. (e) Same as (a) but for a daisy-pattern aperture (from regular decagon) (shown in inset) with side $R = a = 5\,\mu m$. (f) Same as (e) but along the z-axis for $x_0 = y_0 = 0$.

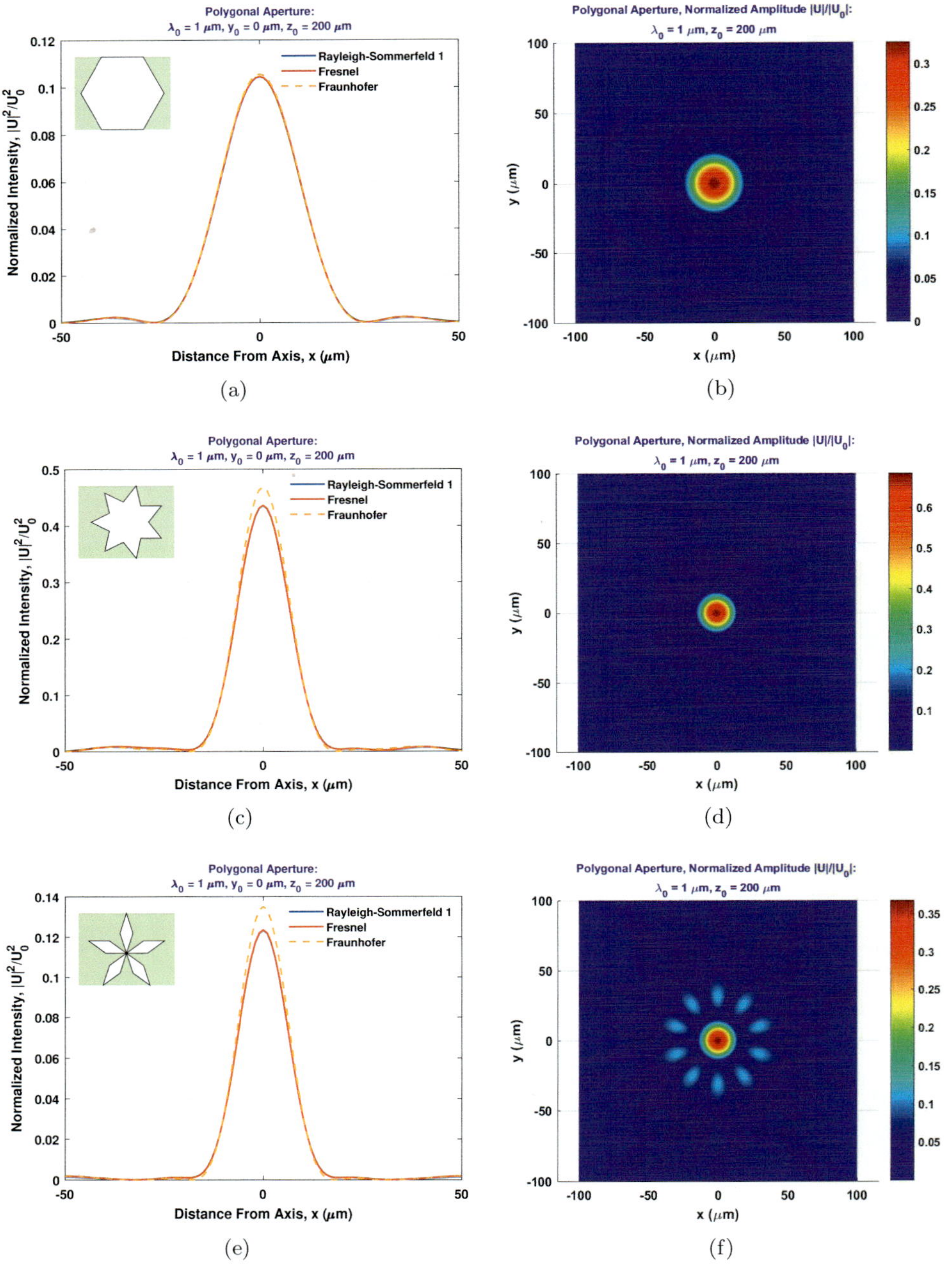

FIGURE 5.60: (a) The diffracted field normalized intensity $|U|^2/|U_0|^2$ in the x-direction for an incident plane wave on a regular-hexagon aperture (shown in inset) of side $R = 5\,\mu m$ for $z_0 = 200\,\mu m = 200\lambda_0$ and $y_0 = 0$. The freespace wavelength is $\lambda_0 = 1.0\,\mu m$. (b) Same as (a) but the normalized amplitude pattern, $|U|/|U_0|$, is shown in the xy-plane for $z_0 = 200\,\mu m = 200\lambda_0$. (c) Same as (a) but for a starred-regular-heptagon (shown in inset) with side $R = a = 5\,\mu m$. (d) Same as (c) but the normalized amplitude pattern, $|U|/|U_0|$, is shown in the xy-plane for $z_0 = 200\,\mu m = 200\lambda_0$. (e) Same as (a) but for a daisy-pattern (from regular decagon) (shown in inset) with side $R = a = 5\,\mu m$. (f) Same as (e) but the normalized amplitude pattern, $|U|/|U_0|$, is shown in the xy plane for $z_0 = 200\,\mu m = 200\lambda_0$.

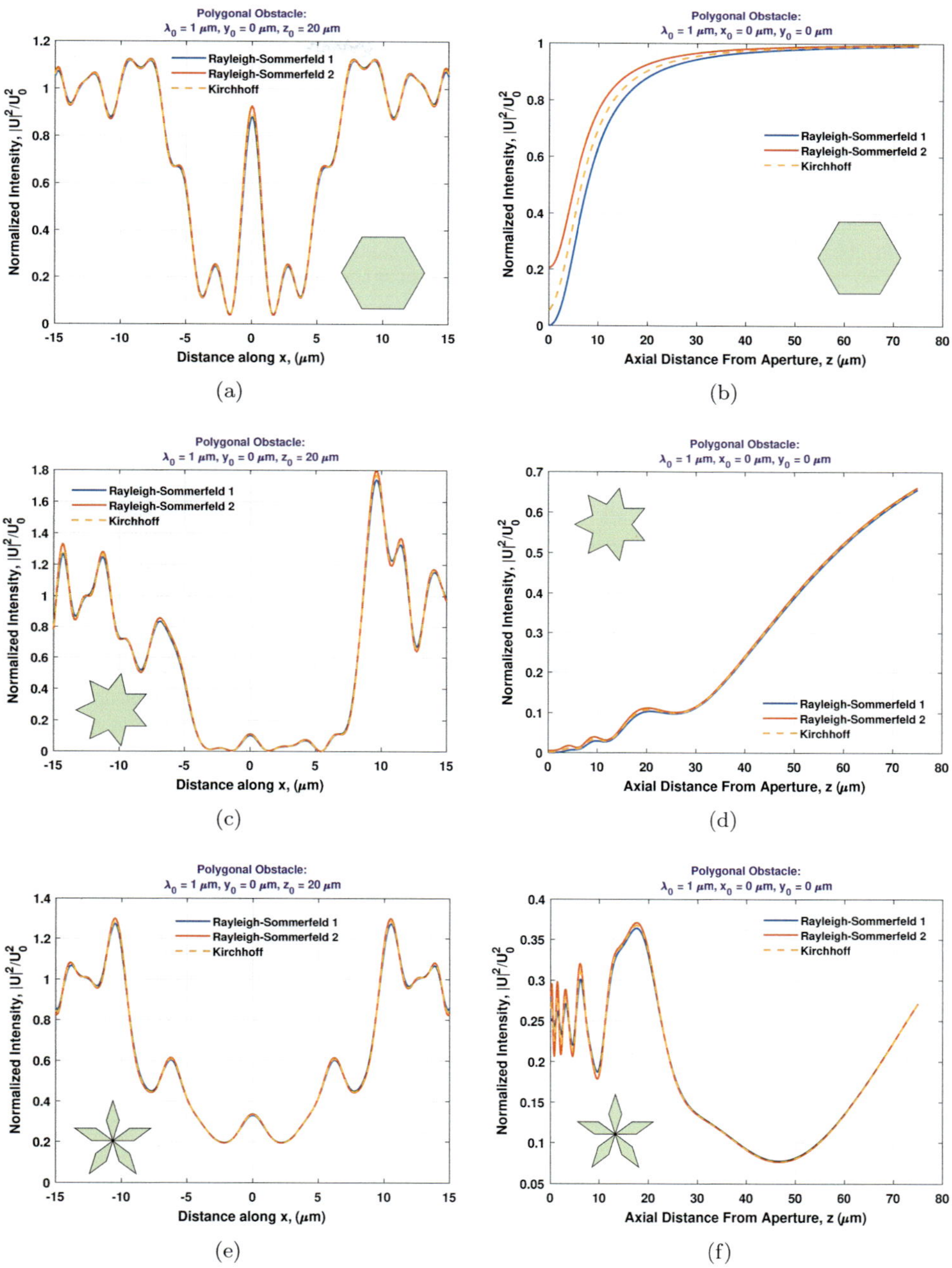

FIGURE 5.61: (a) The diffracted field normalized intensity $|U|^2/|U_0|^2$ in the x direction for an incident plane wave on a regular-hexagon obstacle (shown in inset) of side $R = 5\,\mu m$ for $z_0 = 20\,\mu m = 20\lambda_0$ and $y_0 = 0$. The freespace wavelength is $\lambda_0 = 1.0\,\mu m$. (b) Same as (a) but along the z-axis for $x_0 = y_0 = 0$. (c) Same as (a) but for a starred-regular heptagon obstacle (shown in inset) with side $R = a = 5\,\mu m$. (d) Same as (c) but along the z-axis for $x_0 = y_0 = 0$. (e) Same as (a) but for a daisy-pattern obstacle (from regular decagon) (shown in inset) with side $R = a = 5\,\mu m$. (f) Same as (e) but along the z-axis for $x_0 = y_0 = 0$.

of reducing ripples caused by diffraction [302–307]. Consider a serrated circular aperture with sinusoidal undulation. The radius of the circular aperture is denoted by a, and the amplitude of the sinusoidal undulation is r_u. The overall radius of the aperture can be described by the following equation:

$$R(\phi) = a + r_u \cos(M\phi), \qquad (0 \leq \phi \leq 2\pi), \tag{5.220}$$

where M is the number of periods of the sinusoid on the periphery of the circle with radius a. The free parameters of the serration are the height r_u and the number of periods M. By sampling the resulting periphery with an arbitrary number of points $[x_i = R(\phi_i)\cos\phi_i, y_i = R(\phi_i)\sin\phi_i]$ $(i = 1, 2, \cdots, N_p)$, an equivalent N_p-polygon is formed, which can be treated exactly as in Sec. 5.9.3. Additionally, randomly serrated apertures can also be analyzed [305, 306]. In this case, the radius R is given by $R(\phi) = a + r_u rand(\phi)$ where $rand$ is a random variable that causes the periphery of the aperture to vary randomly. Using Babinet's principle, it is straightforward to extend the analysis to serrated obstacles. The following example demonstrates the simple analysis presented.

Example 5.9.6:

Assume a basic circular aperture of radius $a = 10\,\mu$m. A plane wave of freespace wavelength of $\lambda_0 = 1.0\,\mu$m is incident normally from the left of the aperture. The aperture can be serrated using Eq. (5.220). Use various cases and calculate the normalized intensities at $z_0 = 40\,\mu$m using the RS-1 and Fresnel formulations.

Solution: Initially, using the results from Sec. 5.7, the normalized intensities for the perfectly circular aperture are shown in Fig. 5.62(a) along the x-axis at $z_0 = 40\,\mu$m. Both the RS-1 and Fresnel intensities are plotted. For the range of x values shown, the distance from the aperture should be much greater than $z_0 = [(\pi/4\lambda_0)(30 + a/2)^4]^{1/3} = 105.6\,\mu$m, which is not satisfied. Consequently, differences between the RS-1 and Fresnel results are evident.

Next, a serrated aperture with $M = 16$ and $r_u = 1\,\mu$m is considered. The resulting normalized intensities are shown in Fig. 5.62(b). It is observed that the ripples of the pattern are reduced. When the number of periods of the undulation is increased to $M = 32$ while keeping $r_u = 1\,\mu$m, the resulting intensities are shown in Fig. 5.62(c). Doubling the number of undulations further smoothes the diffracted intensities.

In both Figs. 5.62(b) and 5.62(c), the serrated aperture is symmetric with respect to the x-axis, so the normalized intensities are symmetrical along the x-axis. When the number of periods is increased to $M = 33$ with the same r_u, a small asymmetry in the intensities with respect to the x-axis is observed, as shown in Fig. 5.62(d).

Next, the number of periods is set to $M = 32$ and r_u is increased to $r_u = 1.5\,\mu$m. The resulting intensities are shown in Fig. 5.62(e), where further smoothing of the diffraction pattern is observed.

Finally, a random variation of the aperture radius is introduced using a uniform distribution function $rand(*)$ and 96 values of ϕ. The resulting normalized intensities are shown in Fig. 5.62(f), with the random variation of the serrated aperture shown in the inset of the figure. The random variation further smooths the ripples of the diffraction pattern, though the result is more asymmetrical.

5.10 Fresnel Zones and Fresnel Plates

Fresnel proposed an ingenious technique for calculating the diffracted field through circular apertures. In his method, the field could be approximately determined by summing the

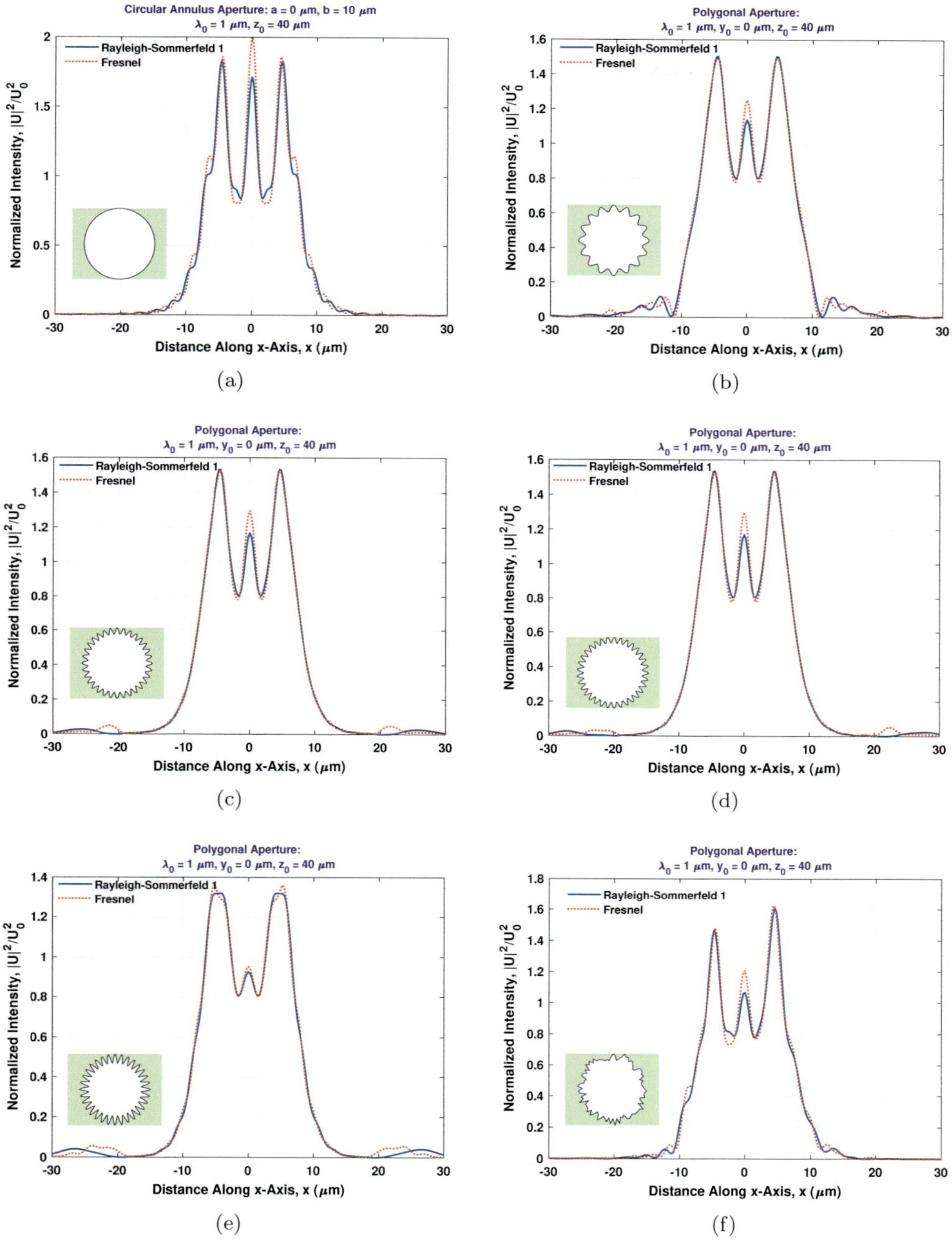

FIGURE 5.62: The diffracted field normalized intensity $|U|^2/|U_0|^2$ in the x-direction for an incident plane wave on serrated circular apertures (shown in inset) of radius $a = 10\,\mu m$ for $z_0 = 40\,\mu m = 40\lambda_0$ and $y_0 = 0$. The freespace wavelength is $\lambda_0 = 1.0\,\mu$m. (a) Completely circular aperture without serration. (b) Serrated aperture with $M = 16$ and $r_u = 1.0\,\mu$m. (c) Serrated aperture with $M = 32$ and $r_u = 1.0\,\mu$m. (d) Serrated aperture with $M = 33$ and $r_u = 1.0\,\mu$m. (e) Serrated aperture with $M = 32$ and $r_u = 1.5\,\mu$m. (f) Randomly serrated aperture with $r_u = 1.5\,\mu$m.

contributions of successive zones, without the need to evaluate a specific integral such as the one presented in Eq. (5.11). In Fresnel's approach, the circular aperture was divided into concentric circular zones, with the inner and outer radii of each zone chosen to correspond to a phase shift of π. Initially, Fresnel considered an incident diverging spherical wave to define these zones, which are now known as *Fresnel zones.*

This section examines two distinct cases: (a) the simpler case of a normally incident plane wave, and (b) the case originally considered by Fresnel, involving an incident spherical wave with its center along the z-axis.

5.10.1 Fresnel Zones Definition for a Normally Incident Plane Wave

The definition of Fresnel zones for the case of a normally incident plane wave can be established using the geometry illustrated in Fig. 5.63. Let ρ_m and ρ_{m-1} denote the outer and the inner radius, respectively, of the m-th Fresnel zone, with the assumption that $\rho_0 = 0$. By definition, the radii of the m-th Fresnel zone are determined by the following relationship:

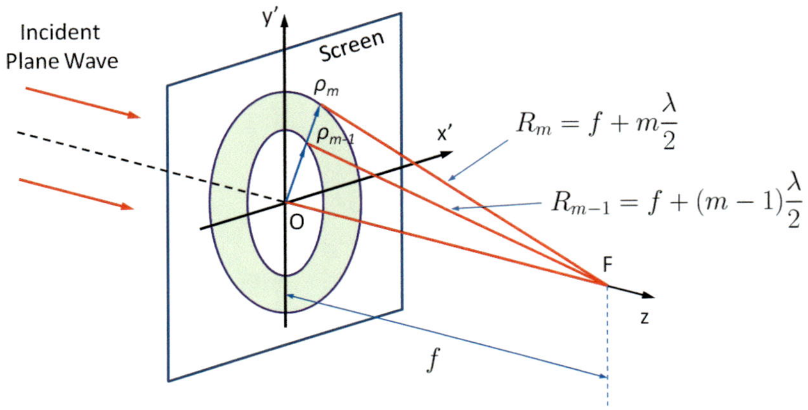

FIGURE 5.63: An infinite planar aperture is divided into circular annuli based on the definitions of the radii ρ_m, where $m = 1, 2, \cdots N$. The integer m, along with the radii ρ_m and ρ_{m-1}, specifies the m-th zone and its outer and inner radial boundaries, respectively. The radii ρ_m $(m = 0, 1, 2, \cdots)$ are defined such that the distance from any point on the circumference of a zone boundary to the focal point F equals to $f + m(\lambda/2)$. The normally incident plane wave, traveling from the left of the aperture, is described by $U_{inc} = U_0 e^{-jkz}$.

$$\rho_m^2 + f^2 = R_m^2 = \left(f + m\frac{\lambda}{2}\right)^2 \Longrightarrow$$

$$\rho_m = \left(m\lambda f + m^2\frac{\lambda^2}{4}\right)^{1/2}, \qquad m = 0, 1, 2, \cdots. \tag{5.221}$$

The m-th Fresnel zone is defined as the circular region bounded by an inner radius ρ_{m-1} and an outer radius ρ_m (where $m = 1, 2, \cdots$). The area of the m-th Fresnel zone, S_m, is straightforwardly calculated as:

$$S_m = \pi(\rho_m^2 - \rho_{m-1}^2) = \pi\left[\lambda f + \frac{\lambda^2}{4}(2m-1)\right] \simeq \pi\lambda f, \tag{5.222}$$

since usually $\lambda \ll f$. The final result implies that the circular aperture is divided into nearly equal-area circular zones. At this stage, the Fresnel-Kirchhoff equation, Eq. (5.11), can be applied for the specific case of a normally incident plane wave described as $U_{inc} = U_0 \exp(-jkz)$. Consequently, the diffracted field along the z-axis is expressed as follows (noting that $\hat{\imath}_R \cdot \hat{\imath}_{n'} = f/R$):

$$U_{FK}(f) = U_0 jk \iint_{S_a} \frac{e^{-jkR}}{4\pi R}\left(1+\frac{f}{R}\right) dS$$

$$= \frac{U_0 jk}{2} \sum_{m=1}^{N} \int_{\rho_{m-1}}^{\rho_m} \frac{e^{-jkR}}{R}\left(1+\frac{f}{R}\right) r_T\, dr_T = \frac{U_0 jk}{2} \sum_{m=1}^{N} \int_{\rho_{m-1}}^{\rho_m} \frac{e^{-jkR}}{R}\left(1+\frac{f}{R}\right) R dR$$

$$= \frac{U_0 jk}{2} \sum_{m=1}^{N} \left\{ \frac{\exp(-jkR_{m-1}) - \exp(-jkR_m)}{jk} + f\left[E_1(jkR_{m-1}) - E_1(jkR_m)\right] \right\},$$

$$\text{where} \quad E_1(w) = \int_w^{+\infty} \frac{e^{-t}}{t} dt.$$

In the last derivation, the relation $r_T dr_T = RdR$ was employed, and N denotes the total number of zones considered. Additionally, the function $E_1(w)$ represents the *exponential integral function* [293]. The arguments of the exponential functions and the exponential integral functions are given as $kR_{m-1} = kf + (m-1)\pi$ and $kR_m = kf + m\pi$, respectively. For large arguments, the exponential integral function can be approximated by $E_1(w) \simeq e^{-w}/w$ [293]. In most practical situations, $kf = 2\pi(f/\lambda) \gg m\pi$ (for a reasonable number of Fresnel zones). Under these conditions, $U_{FK}(f)$ can be approximated as:

$$U_{FK}(f) \simeq \frac{U_0 e^{-jkf}}{2} \sum_{m=1}^{N} \Big\{ [\exp(-j(m-1)\pi) - \exp(-jm\pi)] +$$

$$[\exp(-j(m-1)\pi) - \exp(-jm\pi)] \Big\}$$

$$= U_0 e^{-jkf} \sum_{m=1}^{N} \underbrace{[\exp(-j(m-1)\pi) - \exp(-jm\pi)]}_{= V_{m-1}^m}$$

$$= U_0 e^{-jkf} \Big\{ \underbrace{[1-(-1)]}_{= V_0^1} + \underbrace{[(-1)-1]}_{= V_1^2} + \underbrace{[1-(-1)]}_{= V_2^3} + \cdots \Big\} = \sum_{m=1}^{N} [U_{FK}]_{m-1}^m \Longrightarrow$$

$$U_{FK}(f) \simeq \begin{cases} 2U_0 e^{-jkf}, & \text{for odd } N, \\ 0, & \text{for even } N. \end{cases} \tag{5.223}$$

Here, $[U_{FK}]_{m-1}^m$ represents the contribution of the m-th Fresnel zone. The approximation made for the exponential integral function essentially assumes $f/R \simeq 1$, meaning that the inclination factor is neglected. If this approximation is not applied, the diffracted field of the first zone, $[U_{FK}]_0^1 = U_0 e^{-jkf} V_0^1$, is not entirely canceled by the diffracted field of the second zone, $[U_{FK}]_1^2 = U_0 e^{-jkf} V_1^2$. A similar partial cancellation occurs for subsequent zones. As a result, the total field exhibits distinct behavior based on whether the number of zones is odd or even. For an odd number of zones, the resulting field satisfies $|U_{FK}(f)| < 2U_0$, whereas for an even number of zones, the resulting field satisfies $|U_{FK}(f)| > 0$.

In the preceding Fresnel-Kirchhoff result for $U_{FK}(f)$, the value of the diffracted field along the z-axis was determined. However, analytical solutions for a circular annulus in this context have already been provided in Sec. 5.7.5. Consequently, the results obtained using the RS-1, RS-2, and Fresnel formulations are also presented here. It should be noted that the Kirchhoff results closely approximate the Fresnel-Kirchhoff results. These two formulations would yield identical outcomes if the $1/R$ term were not neglected in the Fresnel-Kirchhoff approximation. The diffracted fields derived from the RS-1 and RS-2 formulations, as well as the Fresnel approximation, are summarized as follows:

$$\begin{aligned} U_{RS_1}(f) &= U_0 f \sum_{m=1}^{N} \left[\frac{e^{-jkR_{m-1}}}{R_{m-1}} - \frac{e^{-jkR_m}}{R_m}\right] \\ &= \sum_{m=1}^{N} \underbrace{U_0 f e^{-jkf}\left[\frac{1}{R_{m-1}} e^{-j(m-1)\pi} - \frac{1}{R_m} e^{-jm\pi}\right]}_{=\,[U_{RS_1}]_{m-1}^{m}} = \sum_{m=1}^{N} [U_{RS_1}]_{m-1}^{m}, \end{aligned} \tag{5.224}$$

$$\begin{aligned} U_{RS_2}(f) &= U_0 \sum_{m=1}^{N} \left[e^{-jkR_{m-1}} - e^{-jkR_m}\right] \\ &= \sum_{m=1}^{N} \underbrace{U_0 \left[e^{-j(m-1)\pi} - e^{-jm\pi}\right]}_{=\,[U_{RS_2}]_{m-1}^{m}} = \sum_{m=1}^{N} [U_{RS_2}]_{m-1}^{m}, \end{aligned} \tag{5.225}$$

$$\begin{aligned} U_{Fre}(f) &= U_0 e^{-jkf} \sum_{m=1}^{N} \left[\exp\left(-j\frac{k}{2f}\rho_{m-1}^2\right) - \exp\left(-j\frac{k}{2f}\rho_m^2\right)\right] \\ &= \sum_{m=1}^{N} \underbrace{U_0 e^{-jkf}\left[\exp\left(-j\frac{k}{2f}\rho_{m-1}^2\right) - \exp\left(-j\frac{k}{2f}\rho_m^2\right)\right]}_{=\,[U_{Fre}]_{m-1}^{m}} = \sum_{m=1}^{N} [U_{Fre}]_{m-1}^{m}. \end{aligned} \tag{5.226}$$

All the previous expressions converge to the form given in Eq. (5.223) when $R_m \simeq f$ in the amplitude terms and $\rho_m \simeq m\lambda f$. From these results, an intriguing possibility arises. Since the diffracted fields from two successive zones approximately cancel each other, one could consider blocking (i.e., setting the transmittance to zero) either all even or all odd zones. For instance, if all even zones are blocked, the resulting diffracted field $U_{FK}^{odd}(f)$ would be given by:

$$U_{FK}^{odd}(f) \simeq \begin{cases} 2U_0 e^{-jkf}\dfrac{N}{2} = NU_0 e^{-jkf}, & \text{for even } N, \\ 2U_0 e^{-jkf}\dfrac{N+1}{2} = (N+1)U_0 e^{-jkf}, & \text{for odd } N. \end{cases} \tag{5.227}$$

On the other hand, if all odd zones are blocked, the resulting diffracted field $U_{FK}^{even}(f)$ is given by:

$$U_{FK}^{even}(f) \simeq \begin{cases} 2U_0 e^{-jkf}\dfrac{N}{2} = NU_0 e^{-jkf}, & \text{for even } N, \\ 2U_0 e^{-jkf}\dfrac{N-1}{2} = (N-1)U_0 e^{-jkf}, & \text{for odd } N. \end{cases} \tag{5.228}$$

The corresponding intensities can be expressed as:

$$I_{FK}^{odd}(f) \simeq \begin{cases} \dfrac{|U_0|^2}{4Z}N^2 = N^2 I_0, & \text{for even } N, \\ \dfrac{|U_0|^2}{4Z}(N+1)^2 = (N+1)^2 I_0, & \text{for odd } N, \end{cases} \tag{5.229}$$

$$I_{FK}^{even}(f) \simeq \begin{cases} \dfrac{|U_0|^2}{4Z}N^2 = N^2 I_0, & \text{for even } N, \\ \dfrac{|U_0|^2}{4Z}(N-1)^2 = (N-1)^2 I_0, & \text{for odd } N. \end{cases} \tag{5.230}$$

The previous results suggest that by blocking half of the Fresnel zones, the circular aperture can function as a converging lens, with the focused intensity being approximately N^2 times greater than the intensity of the incident plane wave. However, due to the neglected inclination factor, the actual intensities will be slightly smaller than the approximate values predicted by Eqs. (5.229) and (5.230).

The aperture plates, formed by blocking either the odd or the even Fresnel zones, produce a focusing effect, and these plates are known in the literature as *Fresnel Zone Plates*. These plates behave similarly to converging lenses. The key difference compared to ordinary refractive converging lenses is that they can exhibit secondary focusing points at distances $f/(2s+1)$, where $s = 1, 2, \cdots$. To understand why this occurs, it is useful to examine Fig. 5.64. In this figure, the focus is on the initial m-th zone, defined based on the primary focal

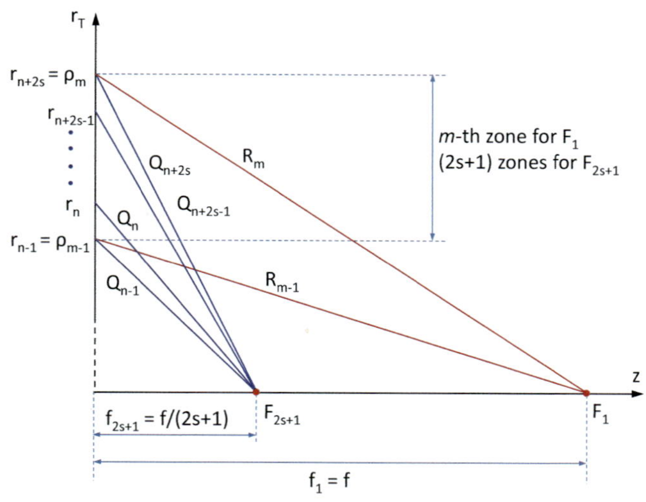

FIGURE 5.64: The annulus with inner radius ρ_{m-1} and outer radius ρ_m represent the m-th Fresnel zone for the initial focal point $F = F_1$. This zone can be subdivided into $2s+1$ zones for the secondary focal point F_{2s+1}, for which $f_{2s+1} = f/(2s+1)$. The $r_{n-1}, r_n, \cdots, r_{n+2s}$ represent the radii of the new $(2s+1)$ Fresnel zones defined within the initial m-th zone.

distance $f_1 = f$, according to Eq. (5.221). Consider a secondary focal point F_{2s+1}, for which $f_{2s+1} = f/(2s+1)$. Due to the smaller focal distance, the radii of the newly defined Fresnel zones become smaller, and the number of zones increases. Specifically, for point F_{2s+1}, the initial m-th zone is now divided into $(2s+1)$ new zones. Since the number of these new zones is odd, their effect on the new focal point F_{2s+1} is positive, and it is approximately equal to U_{n-1}^{n}, which represents the effect of the first zone of the newly divided initial m-th zone (the integer n is defined below).

If the number of zones into which the initial m-th zone is divided is even, its effect on the secondary focal point will be approximately zero because the zones cancel each other out. Thus, only secondary focal points at F_{2s+1} (i.e., the odd secondary points) can serve as valid secondary focal points. This qualitative explanation can also be quantified by using the approximate expression for the zone boundaries, specifically when $\rho_m^2 \simeq m\lambda f$. The following relations can then be derived:

$$\rho_{m-1}^2 \simeq (m-1)\lambda f = (m-1)(2s+1)\lambda \, \frac{f}{2s+1} = (n-1)\lambda \, \frac{f}{2s+1} = r_{n-1}^2,$$

$$\text{where} \quad n-1 = (m-1)(2s+1),$$

$$\rho_m^2 \simeq m\lambda f = m(2s+1)\lambda \, \frac{f}{2s+1} = r_{n+2s}^2, \quad \text{and}$$

$$r_{n-1+i}^2 \simeq (n-1+i)\lambda \, \frac{f}{2s+1}, \qquad i = 0, 1, 2, \cdots, (2s+1).$$

The new Fresnel zone boundaries, defined by r_{n-1+i}, specify $(2s+1)$ zones within the initial m-th zone. Since the number of these new zones is odd, they contribute positively to the diffracted field at the point F_{2s+1}. This same effect occurs for any of the initial m zones, where $m = 1, 2, \cdots N$, with N representing the total number of initial zones for the primary focal point $F = F_1$. For the secondary point F_{2s+1}, there are a total of $N(2s+1)$ contributing zones, but only one out of every $(2s+1)$ zone contributes constructively at F_{2s+1}. As a result, the expected light intensity at F_{2s+1} can be approximately expressed as $I_{F_{2s+1}} \simeq N^2 I_0$, where I_0 is the intensity of the incident plane wave. However, as the points F_{2s+1} approach the plate, the inclination factors (which were neglected in this simplified analysis) increase, leading to a decrease in the expected intensity at the secondary focal points. Some examples of this behavior are presented next.

Example 5.10.1:

Consider a circular aperture of radius a. For a focal length $f = 25\,\mu m$ the radius of the aperture, a, is such that includes the first $N = 11$ Fresnel zones for an incident plane wave of freespace wavelength of $\lambda_0 = 0.6328\,\mu m$. Determine the zone boundaries as well as the $[U_{RS_1}]_{m-1}^{m}$, $[U_{RS_2}]_{m-1}^{m}$, and $[U_{FK}]_{m-1}^{m}$ terms and their successive sums.

Solution: The zone radii ρ_m for $m = 1, 2, \cdots, 11$ are calculated using the formula from Eq. (5.221), resulting in the following values: $\rho_1 = 3.99\,\mu m$, $\rho_2 = 5.66\,\mu m$, $\rho_3 = 6.95\,\mu m$, $\rho_4 = 8.05\,\mu m$, $\rho_5 = 9.03\,\mu m$, $\rho_6 = 9.93\,\mu m$, $\rho_7 = 10.75\,\mu m$, $\rho_8 = 11.53\,\mu m$, $\rho_9 = 12.27\,\mu m$, $\rho_{10} = 12.97\,\mu m$, and $\rho_{11} = 13.64\,\mu m$. The aperture radius is therefore $a = 13.64\,\mu m$. Although the aperture size in this example is impractical, it is chosen for demonstration purposes to illustrate the behavior of each zone and its subdivision into subzones.

Each zone is subdivided into 10 subzones to better visualize the contribution of each zone as well as its decomposition into smaller subzones. The resulting diffracted field at the focal point is shown in Fig. 5.65(a), using the RS-1 formulation. The contribution of each zone to the total field is represented by the vectorial addition of its subzones. Since the number of zones is odd ($N = 11$), there is a non-zero total field at the focal point, which is shown in the figure as a red vector. Further, the RS-2 and Fresnel-Kirchhoff (FK) results are shown in Figs. 5.65(b) and 5.65(c), respectively. The contribution of each zone, as well as the sum of sequential zones, is quantified and normalized ($U_0 = 1$) in Table 5.1.

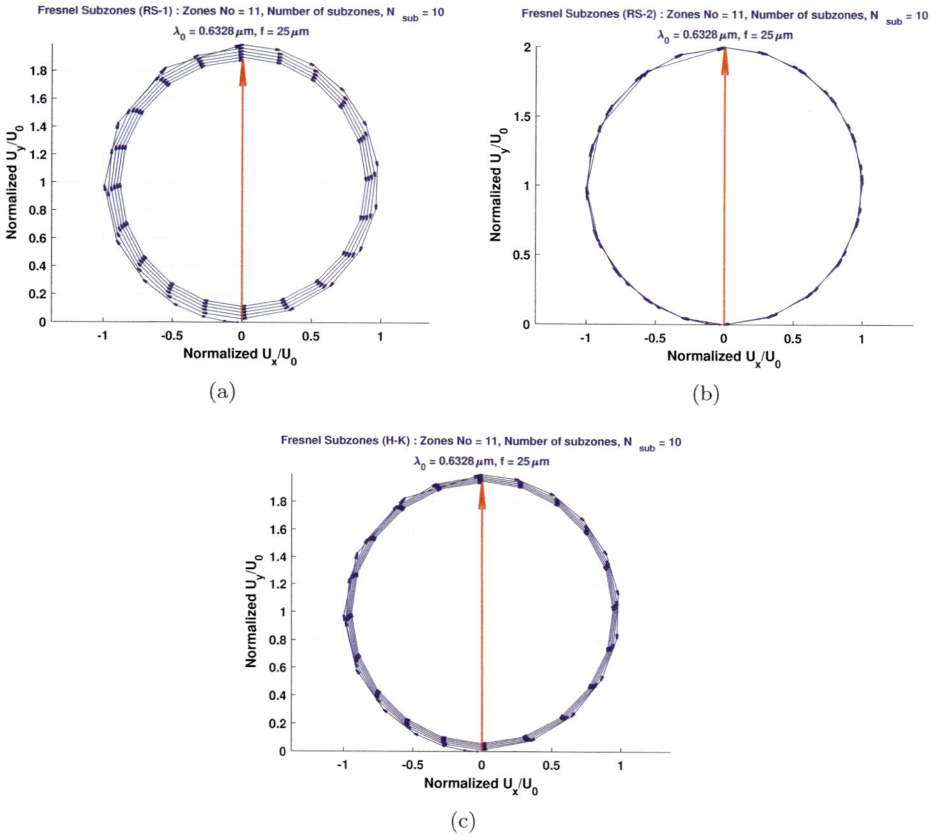

FIGURE 5.65: (a) A vectorial representation of the first 11 Fresnel zones ($N = 11$), where each zone is subdivided into 10 subzones to visualize the buildup of the diffracted field from each zone. The freespace wavelength is $\lambda_0 = \lambda = 0.6328\,\mu m$, and the primary focal distance is $f = 25\,\mu m$. The results are based on the RS-1 formulation. (b) Same as (a), but with the results from the RS-2 formulation shown. (c) Same as (a), but with the Kirchhoff (or equivalently Fresnel-Kirchhoff) formulation results shown. For all the case, the red vector represents the total field, which is the vector summation of all contributing zones.

From the table, it is evident that when an even number of zones is included, the resulting field is very small (it is exactly zero for the RS-2 formulation). On the other hand, if all even-numbered zones are blocked, the total field at the focal point becomes $11.24U_0$ (for RS-1), $12U_0$ (for RS-2), and $11.62U_0$ (for Fresnel-Kirchhoff). Alternatively, if the odd-numbered zones are blocked, the diffracted field at the focal point is reduced to $9.36U_0$ (for RS-1), $10U_0$ (for RS-2), and $9.68U_0$ (for Fresnel-Kirchhoff). These results can be easily verified in Fig. 5.66. Specifically, in Figs. 5.66(a) and 5.66(c), the effective Fresnel zones are shown in green, and the blocked zones are shown in white. In Figs. 5.66(b) and 5.66(d), the intensity of the diffracted light along the z-axis is displayed for the RS-1, RS-2, and Helmholtz-Kirchhoff formulations. As expected, the focusing effect at $z = f = 25\,\mu m$ is evident.

Furthermore, the normalized intensities align closely with the predictions provided in Table 5.1. It should be noted that as the focal distance f and the aperture radius a increase,

TABLE 5.1: Fresnel zone contributions for the first 11 zones. The freespace wavelength is $\lambda_0 = 0.6328\,\mu m$ and the primary focal point is at $z_0 = f = 25\,\mu m$. Assume $U_0 = 1$. The summation symbol represents the sum $\sum = \sum_{n=1}^{m}$. (Example 5.10.1).

Zone	$[U_{RS_1}]_{m-1}^{m}$	$\sum[U_{RS_1}]_{n-1}^{n}$	$[U_{RS_2}]_{m-1}^{m}$	$\sum[U_{RS_2}]_{n-1}^{n}$	$[U_{FK}]_{m-1}^{m}$	$\sum[U_{FK}]_{n-1}^{n}$
1	1.9857	1.9857	2	2	1.9937	1.9937
2	−1.9628	0.0247	−2	0	−1.9814	0.0123
3	1.9387	1.9634	2	2	1.9693	1.9817
4	−1.9152	0.0482	−2	0	−1.9576	0.0241
5	1.8923	1.9405	2	2	1.9461	1.9702
6	−1.8699	0.0706	−2	0	−1.9349	0.0353
7	1.8480	1.9186	2	2	1.9240	1.9593
8	−1.8267	0.0919	−2	0	−1.9133	0.0460
9	1.8058	1.8977	2	2	1.9029	1.9488
10	−1.7854	0.1123	−2	0	−1.8927	0.0562
11	1.7655	1.8778	2	2	1.8827	1.9389

the results for RS-1, RS-2, and H-K formulations practically coincide, as demonstrated in the next example.

Example 5.10.2:

In this example, the focal distance of the zone plate is initially selected to be $f = 3000\,\mu m$ and the freespace wavelength is again $\lambda_0 = 0.6328\,\mu m$. The aperture has a radius ($174.36\,\mu m$) which includes only the first 16 Fresnel zones ($N = 16$). Block either the odd or the even zones and calculate the normalized intensities along the z-axis.
Solution: By blocking either the even or the odd zones, the focusing effects shown in Figs. 5.67 are obtained. In the intensity diagrams, it is evident that all three formulations (RS-1, RS-2, and Kirchhoff) are practically identical, as expected from the previous discussion in example 5.10.1. Furthermore, even the Fresnel approximation is essentially the same as the others in the vicinity of the primary focus (though not shown in the figures).

Some differences, however, are evident near the aperture. Another observation that can be made from the intensity diagrams is the secondary focal points, which were predicted previously at distances from the aperture of $f/(2s+1)$. It is clear from Figs. 5.67(b) and 5.67(d) that secondary focusing occurs at $3000/3 = 1000\,\mu$m, $3000/5 = 600\,\mu$m, and $3000/7 = 428.6\,\mu$m. At shorter distances, the diffracted pattern along the z-axis becomes less predictable due to the varying number of new zones that are formed and the much larger inclination factors.

A similar example is the case where $f = 300000, \mu m = 30, cm$ for the same wavelength and an aperture with a radius of $a = 3080.9, \mu m = 3.0809, mm$, which includes, for the specified focal distance, the first 50 ($N = 50$) Fresnel zones. The results are shown in Fig. 5.68. It is evident that the intensity is significantly higher due to the larger number of odd or even zones contributing. It should be noted that in this example, either the even or the odd

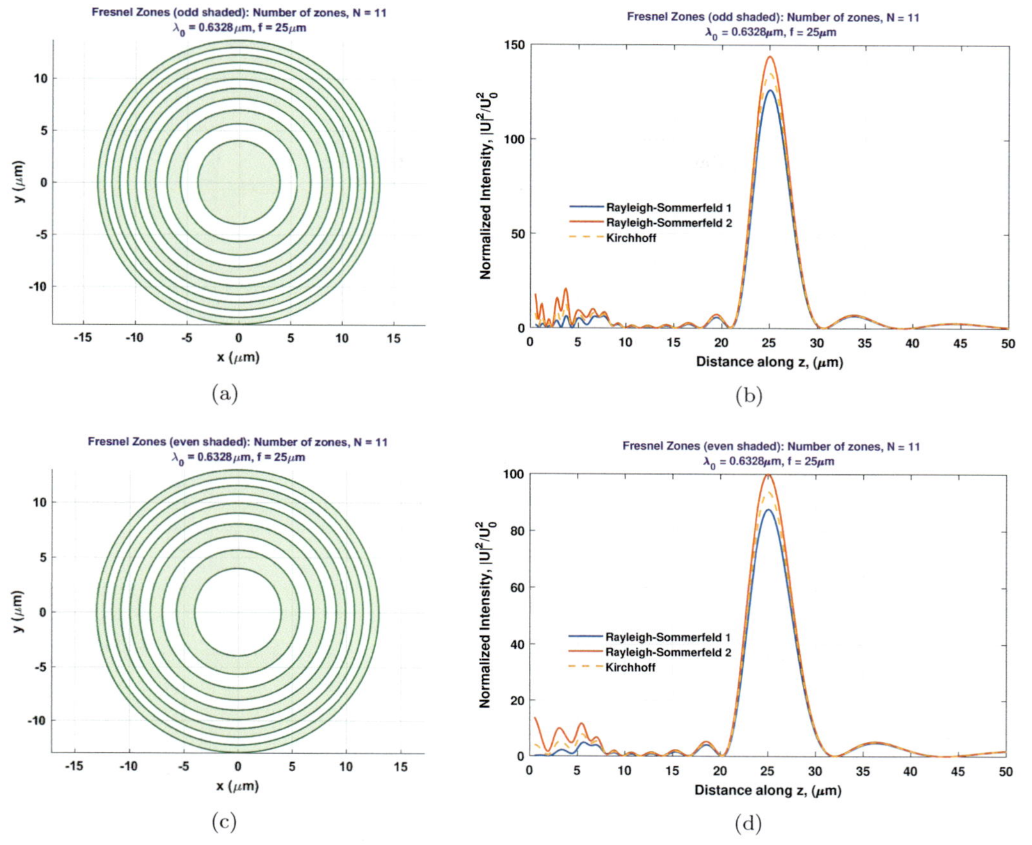

FIGURE 5.66: (a) A Fresnel zone plate with $N = 11$ zones, where the even zones are blocked (shown in white). The free-space wavelength is $\lambda_0 = 0.6328, \mu m$, and the focal distance is $f = 25, \mu m$. (b) The intensity pattern of the diffracted light along the z-axis with the even zones blocked. (c) Same as (a), but with the odd zones blocked (shown in white). (d) Same as (b), but for the odd zones blocked. In (b) and (d), the RS-1, RS-2, and Kirchhoff results are shown.

zones are blocked. As expected, the secondary focal points are observable. In this example, a more realistic focal distance was used, although the number of zones is still relatively small for the selected f. If more zones are used (by increasing the radius of the aperture), the results will be similar, but with higher intensity at the focal points due to more light passing through the aperture.

Example 5.10.3:

This case is similar to example 5.10.2 but with a fundamental difference. In the previous Fresnel zone plate examples, either the odd or the even zones were blocked to achieve focusing. In those cases, an amplitude mask could be utilized immediately after the aperture to selectively block or permit the various zones from contributing to the diffracted field. A more advanced implementation involves using a phase mask, which introduces a phase shift of π in selected zones. Apply the appropriate phase masks and study the diffraction by the resulting zone plates.

Solution: For example, instead of blocking the even zones and permitting only the odd ones, a phase shift of π can be introduced only for the even zones, while their amplitude remains unaffected. In this case, all the zones contribute to the diffracted field, but the even zones

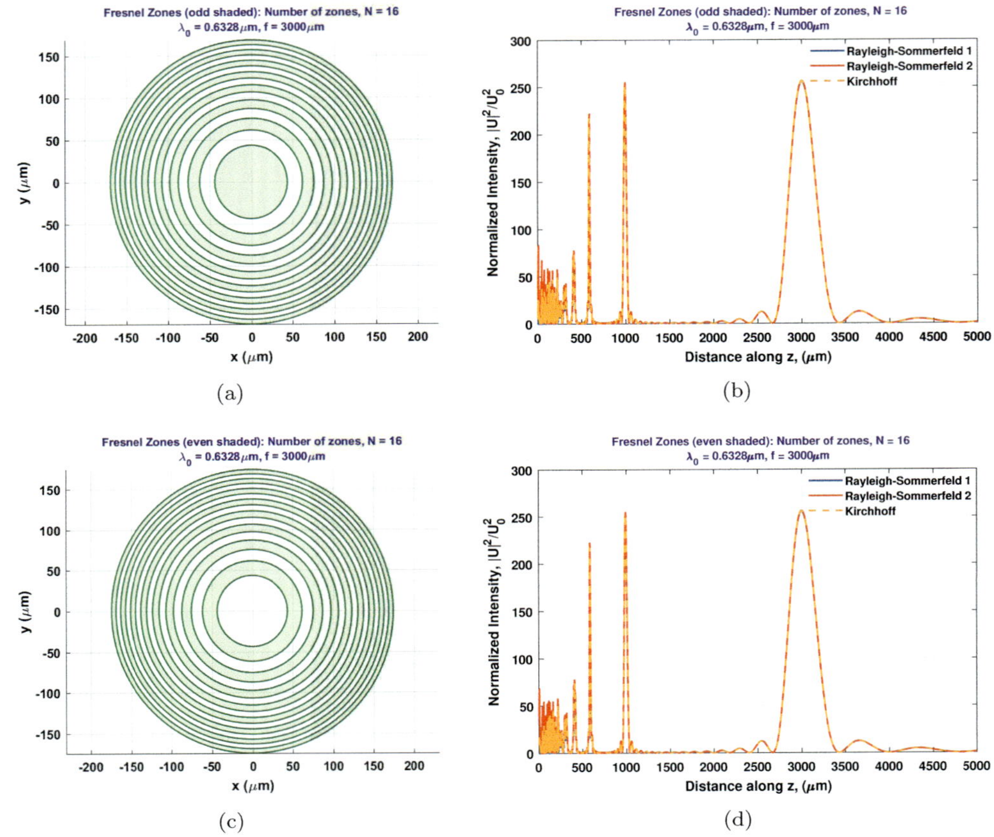

FIGURE 5.67: (a) A Fresnel zone plate with $N = 16$ zones and the even zones blocked (shown in white). The free space wavelength is $\lambda_0 = 0.6328, \mu m$, and the focal distance is $f = 3000, \mu m$. (b) The intensity pattern of the diffracted light along the z-axis with the even zones blocked. (c) Same as (a), but with the odd zones blocked (shown in white). (d) Same as (b), but for the odd zones blocked. In (b) and (d), the RS-1, RS-2, and Kirchhoff results are shown.

add a phase shift of π. This phase shift of π causes the contributions of the even zones to become in phase (constructive) with those of the odd zones, leading to higher intensities at the focal point. This can be easily demonstrated by using Eq. (5.223) in the context of the Fresnel-Kirchhoff approximation:

$$\begin{aligned} U_{FK}(f) &\simeq \sum_{m=1}^{N} e^{j\phi_m} [U_{FK}]_{m-1}^{m} \\ &= [U_{FK}]_0^1 - [U_{FK}]_1^2 + [U_{FK}]_2^3 - [U_{FK}]_3^4 + \cdots \\ &= U_0(2N)e^{-jkf} \qquad \text{for any } N. \end{aligned}$$

In the above derivation, $\phi_m = 0$ for odd zones ($m = 1, 3, 5, \cdots$) and $\phi_m = \pi$ for even zones ($m = 2, 4, 6, \cdots$). If the phase shifts are selected in the opposite manner (i.e., $\phi_m = \pi$ for odd zones and $\phi_m = 0$ for even zones), the final diffracted field at the focal point will be $U_{FK}(f) \simeq -U_0(2N)e^{-jkf}$. In both cases, the resulting intensity at the focal point is identical and equal to $I_{FK} = 4N^2(|U_0|^2/4Z) = 4N^2 I_0$, which is nearly four times larger than when half of the zones are blocked. Moreover, the phase mask results in identical

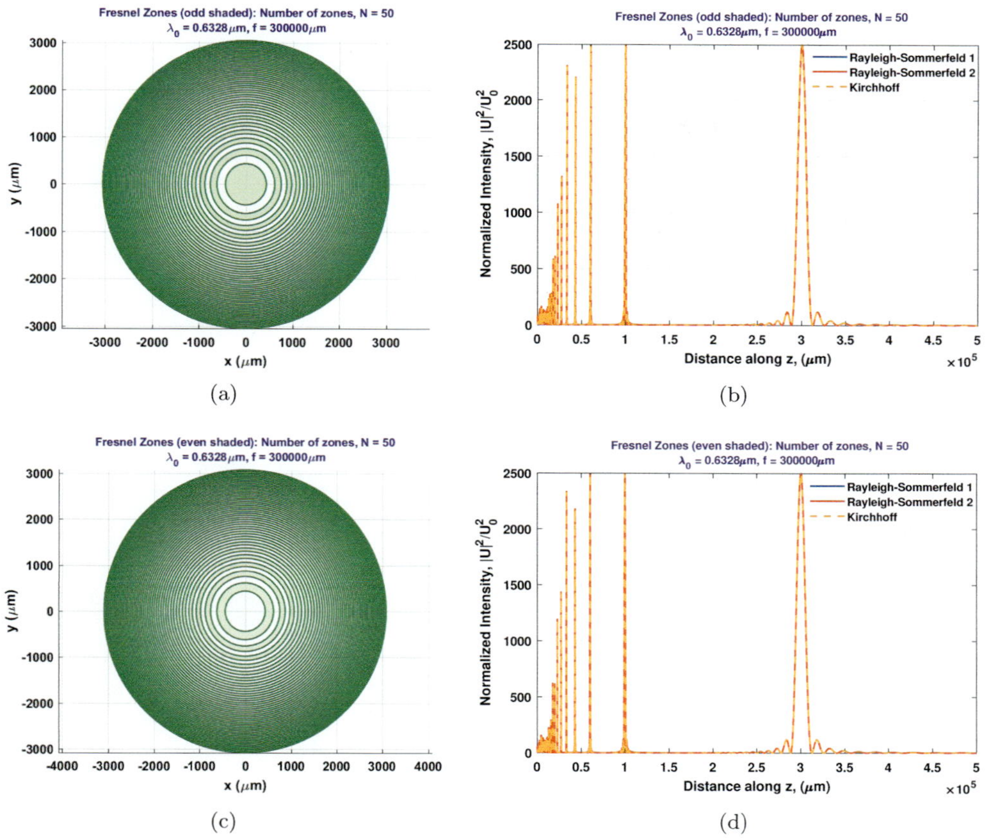

FIGURE 5.68: (a) A Fresnel zone plate with $N = 50$ zones, where the even zones are blocked (shown in white). The freespace wavelength is $\lambda_0 = 0.6328, \mu m$, and the focal distance is $f = 300000, \mu m = 30, cm$. (b) The intensity pattern of the diffracted light along the z-axis with the even zones blocked. (c) Same as (a), but with the odd zones blocked (shown in white). (d) Same as (b), but for the odd zones blocked. In (b) and (d), the RS-1, RS-2, and Kirchhoff results are shown.

diffraction intensity at all points, regardless of whether the odd/even zones or the $0/\pi$ phase shifts are selected. These conclusions hold true for all formulations, including RS-1, RS-2, Kirchhoff, and Fresnel.

A sample case is shown in Fig. 5.69, which is similar to the previous example but with the phase mask applied. In Fig. 5.69(a), the odd zones are assigned a zero phase shift ($\phi_m = 0$) and are shown in light green, while the even zones have an additional phase shift of π ($\phi_m = \pi$) and are shown in light purple. The opposite phase assignment is shown in Fig. 5.69(b), where the roles of odd and even zones are reversed. The resulting normalized intensity along the z-axis is shown in Fig. 5.69(c), where the peak normalized intensity is about four times larger than the intensity predicted for the blocked zones. It is important to note that the diffracted intensity is independent of the specific selection of odd/even zones and the $0/\pi$ phase shifts.

5.10.2 Fresnel Zones Definition for an Incident Spherical Wave

The definition of the Fresnel zones for a diverging incident spherical wave follows a similar concept to that used for a plane wave. Since the Fresnel zones are a method of evaluating diffraction at a point on the axis by a circular aperture, there are various ways to define the

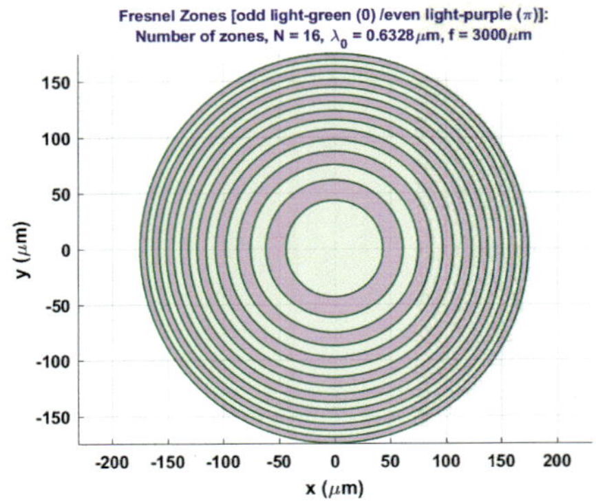

(a)

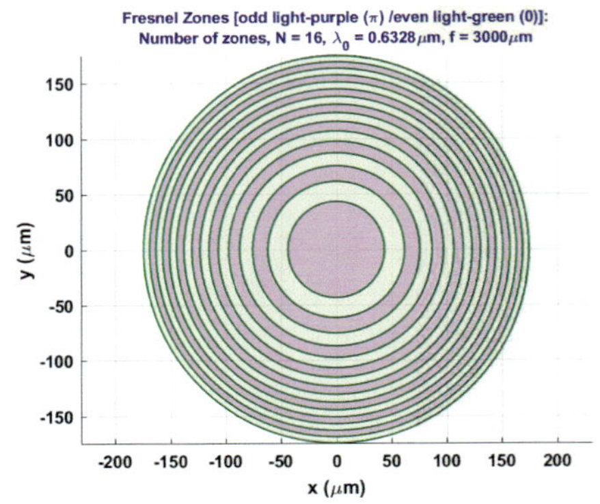

(b)

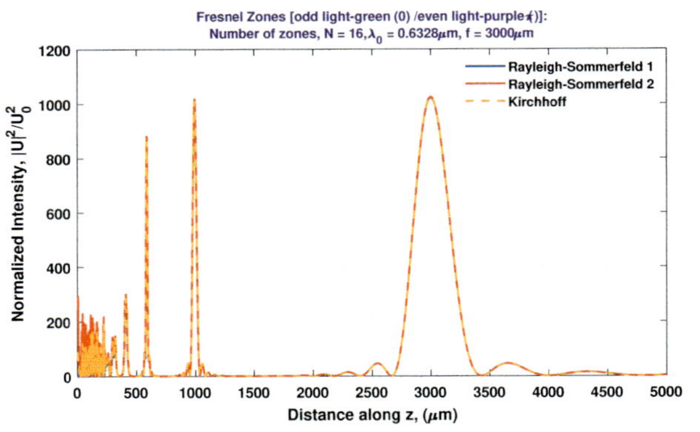

(c)

FIGURE 5.69: (a) A Fresnel zone plate with $N = 16$ zones, where the odd zones have a phase shift of 0 (light green) and the even zones have a phase shift of π (light purple). The freespace wavelength is $\lambda_0 = 0.6328, \mu m$, and the focal distance is $f = 3000\, \mu m = 0.3\, cm$. (b) Same as (a) but with the odd zones having a phase shift of π and the even zones having a phase shift of 0. (c) The intensity pattern of the diffracted light along the z-axis with either (a) or (b) Fresnel zone plate. The RS-1, RS-2, and Kirchhoff results are shown.

zones. For example, the Fresnel zones can be defined on a spherical wavefront of the incident wave, as shown in Figs. 5.70(a) and 5.70(b), an approach followed by several textbooks, including Hecht [42]. This definition is convenient when using the Kirchhoff formulation [see Eq. (5.6)] or the slightly simplified Fresnel-Kirchhoff formulation [see Eq. (5.11)], since in these cases, the integration surface S_a can be non-planar. However, when the Rayleigh-Sommerfeld formulations are used, the Fresnel zones should be defined on a planar surface, as shown in Fig. 5.71. Both cases will be treated here. Let's start with the zones defined on a spherical surface as shown in Figs. 5.70(a) and 5.70(b). The zone radius ρ_m can be determined as follows:

$$\rho_m^2 + (f + s_0 - \sqrt{s_0^2 - \rho_m^2})^2 = R_m^2 = \left(f + m\frac{\lambda}{2}\right)^2 \Longrightarrow$$

$$\rho_m = \left[\frac{s_0}{s_0 + f}\left(fm\lambda + \frac{m^2\lambda^2}{4}\right) - \left(\frac{fm\lambda + m^2\lambda^2/4}{2(s_0 + f)}\right)^2\right]^{1/2}, \tag{5.231}$$

$$\rho_m \simeq \left[\frac{s_0}{s_0 + f}\, fm\lambda\right]^{1/2}, \qquad \text{for } f, s_0 \gg m\lambda. \tag{5.232}$$

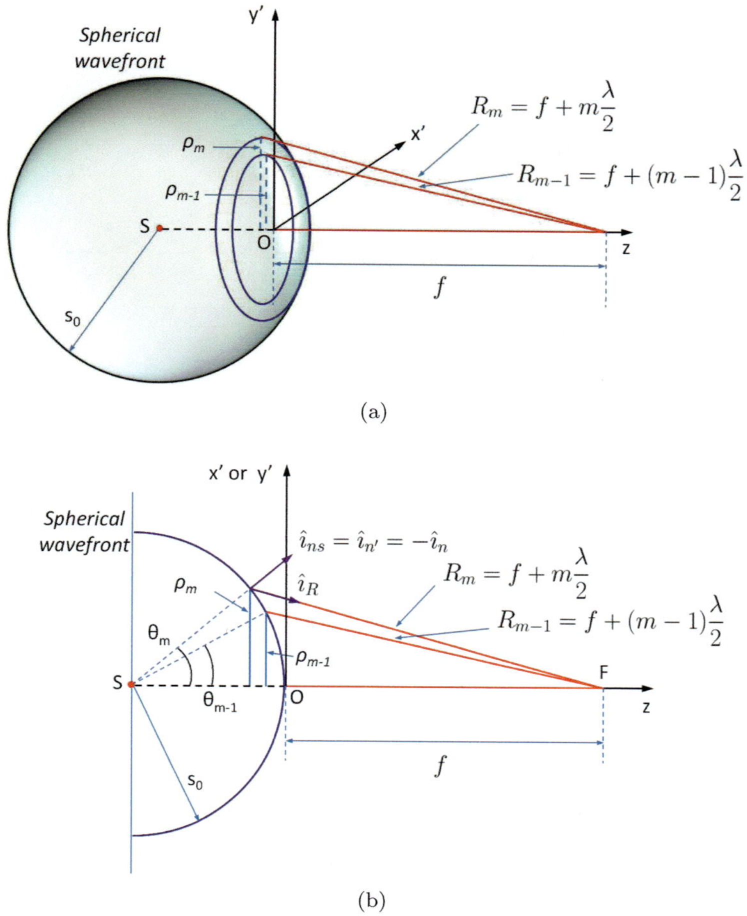

FIGURE 5.70: (a) A spherical wavefront of radius s_0 (emanating from point S on the z-axis) is divided into circular annuli (on the surface of the spherical wavefront) based on the definitions of the radii ρ_m, where $m = 1, 2, \cdots N$. The integer m and the radii ρ_m and ρ_{m-1} specify the m-th zone and its outer and inner radial boundaries, respectively. The radii ρ_m are defined such that the distance from a point on the circumference of the zone boundary to the defined focal point F equals to $f + m(\lambda/2) = R_m$. (b) Same as (a) but as a cross-section through the $x'z$ or $y'z$ plane.

Similarly, the zone radii ρ'_m for the case of the planar surface S_a are given, according to Fig. 5.71, as follows:

$$\left(f^2 + \rho'^2_m\right)^{1/2} + \left(s_0^2 + \rho'^2_m\right)^{1/2} = R_m + R_{s,m} = s_0 + f + m\frac{\lambda}{2} \Longrightarrow$$

$$\rho'_m = \frac{\left[\lambda m(4s_0 + \lambda m)(4f + \lambda m)(4f + 4s_0 + \lambda m)\right]^{1/2}}{4(2f + 2s_0 + \lambda m)}, \quad (5.233)$$

$$\rho'_m \simeq \left[\frac{s_0}{s_0 + f}\, fm\lambda\right]^{1/2}, \qquad \text{for } f, s_0 \gg m\lambda. \quad (5.234)$$

It is obvious that for practical cases with aperture sizes much larger than the wavelength, the approximate expressions for the zone radii ρ_m and ρ'_m, coincide.

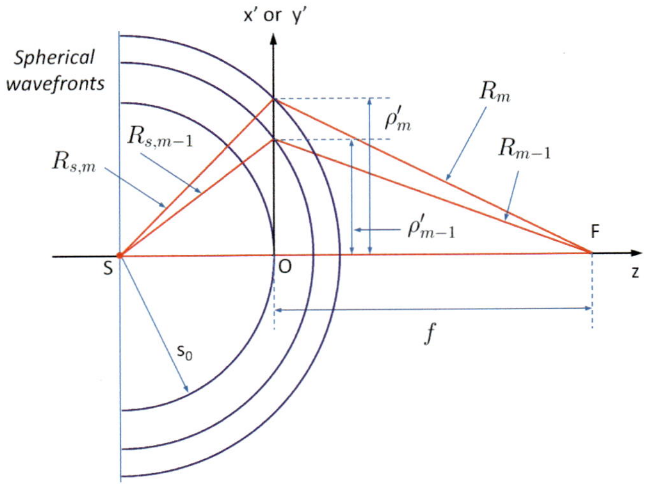

FIGURE 5.71: A spherical diverging wave (emanating from point S on the z-axis) is divided into circular annuli (on the $x'y'$-plane) based on the definitions of the radii ρ'_m, where $m = 0, 1, 2, \cdots N$ (with $\rho'_0 = 0$). The radii ρ'_m are defined in such a way as to satisfy the condition $R_{s,m} + R_m = s_0 + f + m(\lambda/2)$.

The area of the m-th Fresnel zone is defined by the spherical zone area between ρ_m and ρ_{m-1} and is given by:

$$S_m = 2\pi s_0 \left[\sqrt{s_0^2 - \rho_{m-1}^2} - \sqrt{s_0^2 - \rho_m^2} \right], \qquad m = 1, 2, \cdots \tag{5.235}$$
$$\simeq \pi \frac{s_0}{s_0 + f} f\lambda, \qquad \text{for } f, s_0 \gg m\lambda.$$

Similarly, for the case of the ρ'_m Fresnel zones on the $z = 0$ plane, the corresponding zone areas are circular annuli and are given by:

$$S'_m = \pi \left(\rho'^2_m - \rho'^2_{m-1} \right) \simeq \pi \frac{s_0}{s_0 + f} f\lambda, \qquad \text{for } f, s_0 \gg m\lambda. \tag{5.236}$$

The Fresnel-Kirchhoff equation [see Eq. (5.11)] for the case of the incident diverging spherical wave, $U_{inc} = U_0 \exp(-jks_0)/s_0$, as shown in Fig. 5.70(b), is applied to determine the diffracted field $U_{FK}(z = f)$ at the focal point. The integration is performed over the spherical wavefront of radius s_0, where the contribution from each zone is taken into account. The expression for the diffracted field takes into consideration the wavefront's curvature and the relative distances from the source to the aperture and from the aperture to the observation point on the z-axis. The resulting diffracted field is then calculated by summing the contributions from each zone, capturing the diffraction effects of the incident spherical wave as it propagates toward the focal point, and can be expressed as:

$$U_{FK}(f) = jkU_0 \frac{e^{-jks_0}}{4\pi s_0} \iint_{S_a} \frac{e^{-jkR}}{R} \Big(1 + \underbrace{\hat{\imath}_{n'} \cdot \hat{\imath}_R}_{= \cos\Theta} \Big) dS$$
$$= jkU_0 \frac{e^{-jks_0}}{4\pi s_0} s_0^2 \sum_{m=1}^{N} \int_{\theta_{m-1}}^{\theta_m} \int_0^{2\pi} \frac{e^{-jkR}}{R} (1 + \cos\Theta) \sin\theta' d\theta' d\phi'$$

$$= jkU_0 \frac{e^{-jks_0}}{2} s_0 \sum_{m=1}^{N} \int_{\theta_{m-1}}^{\theta_m} \frac{e^{-jkR}}{R} (1+\cos\Theta) \sin\theta' d\theta'$$

$$\text{where} \quad \cos\Theta = \frac{1}{R} \left[(f+s_0)\cos\theta' - s_0 \right].$$

The integral can be calculated analytically by transforming it into an integral with respect to the R variable. In this case, R varies between R_m and R_{m+1} for the m-th zone, while the angle θ' varies between θ_m and θ_{m+1}. This transformation simplifies the integration process by expressing the contributions of each zone in terms of R and the corresponding angular range. The following relations, based on the geometry of the problem, can be easily written to define the limits and the appropriate integrals:

$$R^2 = (s_0+f)^2 + s_0^2 - 2s_0(s_0+f)\cos\theta' \Longrightarrow$$

$$2RdR = 2s_0(s_0+f)\sin\theta' d\theta', \quad \text{and,}$$

$$1+\cos\Theta = 1 + \frac{1}{2s_0 R} \left[f(f+2s_0) - R^2 \right].$$

Then, the integral for $U_{FK}(f)$ can be written in the form:

$$U_{FK}(f) = jkU_0 \frac{e^{-jks_0}}{2} s_0 \sum_{m=1}^{N} \underbrace{\int_{R_{m-1}}^{R_m} \frac{e^{-jkR}}{R} \left[1 + \frac{f(f+2s_0)}{2s_0 R} - \frac{R}{2s_0} \right] \frac{RdR}{s_0(f+s_0)}}_{=\,[V_{FK}]_{m-1}^{m}}$$

$$= \sum_{m=1}^{N} \underbrace{jkU_0 \frac{e^{-jks_0}}{2} s_0 [V_{FK}]_{m-1}^{m}}_{=\,[U_{FK}]_{m-1}^{m}} = \sum_m [U_{FK}]_{m-1}^{m}. \tag{5.237}$$

The diffracted field due to the m-th only zone, $[U_{FK}]_{m-1}^{m}$, can be evaluated analytically for points along the z-axis and is given by:

$$[U_{FK}]_{m-1}^{m} = jkU_0 \frac{e^{-jks_0}}{2} s_0 \left\{ \frac{1}{s_0(f+s_0)} \int_{R_{m-1}}^{R_m} e^{-jkR}\, dR + \frac{f(f+2s_0)}{2s_0^2(f+s_0)} \int_{R_{m-1}}^{R_m} \frac{e^{-jkR}}{R}\, dR - \right.$$

$$\left. \frac{1}{s_0^2(f+s_0)} \int_{R_{m-1}}^{R_m} e^{-jkR} R\, dR \right\} =$$

$$= jkU_0 \frac{e^{-jks_0}}{2} s_0 \left\{ \frac{1}{jks_0(f+s_0)} \left(e^{-jkR_{m-1}} - e^{-jkR_m} \right) + \right.$$

$$\frac{f(f+2s_0)}{2s_0^2(f+s_0)} \left[E_1(kR_{m-1}) - E_1(kR_m) \right] -$$

$$\left. \frac{1}{k^2 s_0^2(f+s_0)} \left[e^{-jkR_m}(1+jkR_m) - e^{-jkR_{m-1}}(1+jkR_{m-1}) \right] \right\}.$$

It should be noted that in the previous equation, the $E_1(*)$ functions represent the exponential integral functions that were defined earlier in Section 5.10.1. To simplify the evaluation, it is common to approximate $[U_{FK}]_{m-1}^{m}$, especially when $R_m = f + m\lambda/2$. Using this approximation, the resulting expression for the diffracted field from the m-th zone becomes more tractable, typically involving the approximation of exponential terms and using suitable expansions for the exponential integrals. The resulting expression is:

$$[U_{FK}]_{m-1}^{m} \simeq jkU_0 \frac{s_0 e^{-jk(s_0+f)}}{2(s_0+f)} \left[\frac{1}{2s_0^2k^2} - j\frac{1}{ks_0} + j\frac{f}{2s_0^2k} - j\frac{f+2s_0}{2s_0^2k}\right] \left[e^{-j(m-1)\pi} - e^{-jm\pi}\right]$$

$$= \underbrace{jkU_0 \frac{s_0 e^{-jk(s_0+f)}}{2(s_0+f)} \left[\frac{1}{2s_0^2k^2} - j\frac{2}{ks_0}\right]}_{=U_0'} = U_0' \left[e^{-j(m-1)\pi} - e^{-jm\pi}\right], \tag{5.238}$$

$$\text{and} \quad U_0' \simeq jkU_0 \frac{s_0 e^{-jk(s_0+f)}}{2(s_0+f)} \left(-j\frac{2}{ks_0}\right) = U_0 \frac{e^{-jk(s_0+f)}}{s_0+f}, \qquad \text{since } ks_0 \gg 1.$$

Similar to the case of the incident plane wave, the diffracted fields from two successive zones approximately cancel each other. Therefore, when all even or all odd zones are blocked, the resulting diffracted field at $z = f$ becomes significantly larger. For example, if all even zones are blocked, the resulting diffracted field is substantially enhanced compared to the case where no zones are blocked. For example, if all even zones are blocked, then the resulting $U_{FK}^{odd}(f)$ is given by:

$$U_{FK}^{odd}(f) \simeq \begin{cases} 2U_0' \dfrac{N}{2} = NU_0', & \text{for even } N, \\ 2U_0' \dfrac{N+1}{2} = (N+1)U_0', & \text{for odd } N. \end{cases} \tag{5.239}$$

On the other hand, if all odd zones are blocked, the resulting diffracted field at $z = f$, denoted $U_{FK}^{even}(f)$, is given by:

$$U_{FK}^{even}(f) \simeq \begin{cases} 2U_0' \dfrac{N}{2} = NU_0', & \text{for even } N, \\ 2U_0' \dfrac{N-1}{2} = (N-1)U_0', & \text{for odd } N. \end{cases} \tag{5.240}$$

The diffracted field along the axis can be calculated analytically using the RS-1 and RS-2 formulations, as both have analytical solutions for an annular aperture, as detailed in Sec. 5.7.6. For example, the diffracted field U_{RS_1} can be determined using Eq. (5.127), while U_{RS_2} can be calculated using Eq. (5.129) at the focal point $z_0 = f$, for the RS-1 and RS-2 formulations, respectively. These are given by:

$$U_{RS_1}(f) = U_0 \sum_{m=1}^{N} \underbrace{\left[\frac{f}{R_{m-1}} \frac{e^{-jk(R_{s,m-1}+R_{m-1})}}{R_{s,m-1}+R_{m-1}} - \frac{f}{R_m} \frac{e^{-jk(R_{s,m}+R_m)}}{R_{s,m}+R_m}\right]}_{=[U_{RS_1}]_{m-1}^{m}/U_0} =$$

$$= \sum_{m=1}^{N} [U_{RS_1}]_{m-1}^{m}, \qquad \text{and} \tag{5.241}$$

$$[U_{RS_1}]_{m-1}^{m} \simeq U_0 \frac{e^{-jk(s_0+f)}}{s_0+f} \left[\frac{f}{R_{m-1}} e^{-j(m-1)\pi} - \frac{f}{R_m} e^{-jm\pi}\right]$$

$$\simeq U_0 \frac{e^{-jk(s_0+f)}}{s_0+f} \left[e^{-j(m-1)\pi} - e^{-jm\pi}\right] = U_0' \left[e^{-j(m-1)\pi} - e^{-jm\pi}\right],$$

$$U_{RS_2}(f) = U_0 \sum_{m=1}^{N} \underbrace{\left[\frac{s_0}{R_{m-1}} \frac{e^{-jk(R_{s,m-1}+R_{m-1})}}{R_{s,m-1}+R_{m-1}} - \frac{s_0}{R_m} \frac{e^{-jk(R_{s,m}+R_m)}}{R_{s,m}+R_m} \right]}_{=[U_{RS_2}]^m_{m-1}/U_0} =$$

$$= \sum_{m=1}^{N} [U_{RS_2}]^m_{m-1}, \qquad \text{and} \tag{5.242}$$

$$[U_{RS_2}]^m_{m-1} \simeq U_0 \frac{e^{-jk(s_0+f)}}{s_0+f} \left[\frac{s_0}{R_{m-1}} e^{-j(m-1)\pi} - \frac{s_0}{R_m} e^{-jm\pi} \right]$$

$$\simeq U_0 \frac{e^{-jk(s_0+f)}}{s_0+f} \left[e^{-j(m-1)\pi} - e^{-jm\pi} \right] = U_0' \left[e^{-j(m-1)\pi} - e^{-jm\pi} \right].$$

In these expressions, $[U_{RS_1}]^m_{m-1}$ and $[U_{RS_2}]^m_{m-1}$ represent the contributions from the m-th Fresnel zone to the diffracted field at $z_0 = z = f$, for the RS-1 and RS-2 formulations, respectively. The approximations $f/R_i \simeq 1$ and $s_0/R_i \simeq 1$ (where $i = m-1, m$) have been used. These approximations are valid when $\rho'_m \ll R_m, s_0$, and they become more accurate for lower-order Fresnel zones.

As a result, it is clear that the contributions from the m-th zone for the RS-1, $[U_{RS_1}]^m_{m-1}$, and RS-2, $[U_{RS_2}]^m_{m-1}$, formulations are approximately the same as those from the Fresnel-Kirchhoff (FK), $[U_{FK}]^m_{m-1}$, formulation. Therefore, the relative intensities $I_{RS_1}/I_{sw} \simeq I_{RS_2}/I_{sw} \simeq I_{FK}/I_{sw}$, with $I_{sw} = |U_0|^2/(4Zs_0^2)$ (where Z is the wave impedance of the surrounding medium). For example, the I_{FK}/I_{sw} relative intensity at the design focal point f, is given by:

$$\frac{I_{FK}^{odd}(f)}{I_{sw}} \simeq \left(\frac{s_0}{s_0+f} \right)^2 \begin{cases} N^2, & \text{for even } N, \\ (N+1)^2, & \text{for odd } N, \end{cases} \tag{5.243}$$

$$\frac{I_{FK}^{even}(f)}{I_{sw}} \simeq \left(\frac{s_0}{s_0+f} \right)^2 \begin{cases} N^2, & \text{for even } N, \\ (N-1)^2, & \text{for odd } N. \end{cases} \tag{5.244}$$

The relative intensities increase with the square of the number of odd or even zones included while blocking the even or odd zones, respectively. This is a similar result to the one obtained in the case of the incident plane wave. However, the inclination factors (the $\cos\Theta$ term) in the case of the spherical wave exhibit more variation than the corresponding factor in the plane wave case. As a result, there is a slightly larger difference between the magnitudes of successive zones in the spherical wave case compared to the plane wave case. This variation in the spherical wave formulation causes the intensity distribution to deviate more from the plane wave case, highlighting the effect of the diverging nature of the spherical wave.

It is noteworthy that the diffracted field along any axial point can be determined analytically without approximations using the formulas for the F-K, RS-1, and RS-2 formulations, as presented in Eqs. (5.237), (5.241), and (5.242). The Kirchhoff results are very close to the F-K results and can always be derived from RS-1 and RS-2 if necessary. Additionally, by following the procedure outlined in Sec. 5.7.6, the diffracted field can be calculated numerically as a sum of Fresnel zone contributions for any point (x, y, z) with $z > 0$.

When Fresnel zone plates are used with an incident plane wave, they typically act as converging lenses with secondary focal points located at distances $f_{2s+1} = f/(2s+1)$ where $s = 1, 2, \cdots$. The situation is more complex for an incident spherical (diverging) wave, but the same fundamental physical principles apply, and secondary focal points do indeed exist. Using the approximate expressions for the Fresnel zone radii, ρ_m [as provided in Eq.

(5.232)], we can derive expressions for the secondary focal points in the case of a diverging wave. Then, the following expressions can be written:

$$\rho_{m-1}^2 \simeq \frac{s_0}{s_0+f}(m-1)\lambda f = (m-1)\lambda f_e,$$

$$\text{where} \;\; f_e = \frac{s_0 f}{s_0+f} \Longrightarrow \frac{1}{f_e} = \frac{1}{s_0} + \frac{1}{f}.$$

Using the term f_e, an approach similar to the one used for the plane wave can be followed:

$$\rho_{m-1}^2 \simeq (m-1)\lambda f_e = (m-1)(2s+1)\lambda\,\frac{f_e}{2s+1} = (n-1)\lambda\,\frac{f_e}{2s+1} = r_{n-1}^2,$$

$$\text{where} \quad n-1 = (m-1)(2s+1),$$

$$\rho_m^2 \simeq m\lambda f_e = m(2s+1)\lambda\,\frac{f_e}{2s+1} = r_{n+2s}^2, \quad \text{and}$$

$$r_{n-1+i}^2 \simeq (n-1+i)\lambda\,\frac{f_e}{2s+1}, \qquad i = 0,1,2,\cdots,(2s+1).$$

Then, the resulting secondary focal points appear at distances f_{2s+1} which are given by:

$$f_{2s+1} = \frac{s_0 f_{e_{2s+1}}}{s_0 - f_{e_{2s+1}}}, \qquad \text{with} \quad f_{e_{2s+1}} = \frac{f_e}{2s+1}. \tag{5.245}$$

Example 5.10.4:

Consider a circular aperture with a design focal length of $f = 25\,\mu m$. An incident spherical wave emanates from a point source located at a distance $s_0 = 50\,\mu m$, with a freespace wavelength of $\lambda_0 = 0.6328\,\mu m$. Assume that the first $N = 11$ Fresnel zones are included in the analysis. (a) Calculate the Fresnel zone radii ρ_m and ρ'_m and determine the boundaries of each zone. (b) Compute the diffracted field contributions $[U_{RS_1}]_{m-1}^m$, $[U_{RS_2}]_{m-1}^m$, and $[U_{FK}]_{m-1}^m$, for each zone. Then, calculate the successive sums of these contributions for the first 11 zones.

Solution: The resulting Fresnel zone radii ρ_m and ρ'_m are provided in Table 5.2. The approximate values are included in the last column of the table for comparison. It is evident that, even for relatively small sizes relevant to scalar diffraction, the differences between the exact values ρ_m and ρ'_m and their approximations are minor.

To illustrate the contributions of each zone and their decomposition into subzones, each zone was divided into 10 subzones. The resulting diffracted field at the focal point is depicted in Fig. 5.72 for the RS-1, RS-2, and Fresnel-Kirchhoff (F-K) formulations. These diagrams present the vectorial addition of the subzone contributions, showing the incremental build-up of the total field from individual zone contributions. The results resemble those obtained for plane wave incidence, though the field amplitudes are reduced, consistent with the presence of the U'_0 factor. Additionally, the varying inclination factor for the spherical wavefront causes the contributions from successive zones to exhibit greater variation compared to the plane wave case. The diagrams depict the exact calculated diffracted fields.

The contributions of each Fresnel zone, normalized to U/U_{sw} where $U_{sw} = 1/s_0$, to the diffracted field at the design focal point, are quantified in Table 5.3. The table includes the contributions of each individual zone and also presents the results of blocking either the even or the odd zones. These results demonstrate the enhanced converging properties that arise when one set of zones is blocked.

TABLE 5.2: Fresnel zone radii for the first 11 zones for an incident diverging spherical wave. The freespace wavelength is $\lambda_0 = 0.6328\,\mu m$, and the primary focal point is at $z_0 = f = 25\,\mu m$. The spherical wave originates from $s_0 = 50\,\mu m$. The radii ρ_m correspond to the ones for the F-K formulation (spherical zones), while the ρ'_m correspond to the ones for RS-1 and RS-2 formulations (planar annuli).

Zone m	ρ_m (μm)	ρ'_m (μm)	Approximate ρ_m, ρ'_m (μm)
1	3.2561	3.2527	3.2476
2	4.6168	4.6072	4.5927
3	5.6689	5.6515	5.6249
4	6.5626	6.5359	6.4951
5	7.3557	7.3186	7.2618
6	8.0778	8.0294	7.9549
7	8.7466	8.6860	8.5922
8	9.3733	9.2998	9.1855
9	9.9659	9.8788	9.7427
10	10.5301	10.4288	10.2697
11	11.0702	10.9541	10.7709

Example 5.10.5:

This example builds upon Example 5.10.4, with the distinction that either the even or the odd Fresnel zones are blocked, enabling the zones to function as a converging lens. The analysis involves $N = 11$ zones, and the objective is to determine the relative intensities I/I_{sw} as functions of the distance along the z-axis. Assume initially that $f = 25\,\mu$m and then examine the case of $f = 3000\,\mu$m.

Solution: The resulting relative intensities, I/I_{sw}, along the axial direction are depicted in Fig. 5.73(a) when the even zones are blocked, and in Fig. 5.73(b) when the odd zones are blocked. These intensities were calculated using the RS-1, RS-2, Kirchhoff, and F-K formulations. The differences between the formulations are minimal. It is evident that the peak values of the relative intensities near the design focal length of $f = 25\,\mu m$ are well predicted by Eqs. (5.243) and (5.244) for the odd and even zone plates, respectively. For example, $I(f)/I_{sw} \simeq 64$ for the odd zone plate and $I(f)/I_{sw} \simeq 44.4$ for the even plate. The secondary focal points are located very close to the aperture, where the approximations no longer hold accurately.

In this case, we consider a similar setup but with a design focal length of $f = 3000\,\mu m$ and a spherical wave originating from $s_0 = 4000\,\mu m$. Again, $N = 11$ zones are included in the analysis, with either the even (odd zone plate) or the odd (even zone plate) zones blocked. The freespace wavelength is again $\lambda_0 = 0.6328\,\mu m$. The resulting relative intensities, I/I_{sw}, along the axial direction are shown in Fig. 5.74(a) when the even zones are blocked, and in Fig. 5.74(b) when the odd zones are blocked. The intensities were calculated using the RS-1, RS-2, Kirchhoff, and F-K formulations. The differences between the formulations are very small, and the curves visually coincide.

It is straightforward to verify that the peak values of the relative intensities near the design focal length of $f = 3000\,\mu m$ are predicted well by Eqs. (5.243) and (5.244) for odd

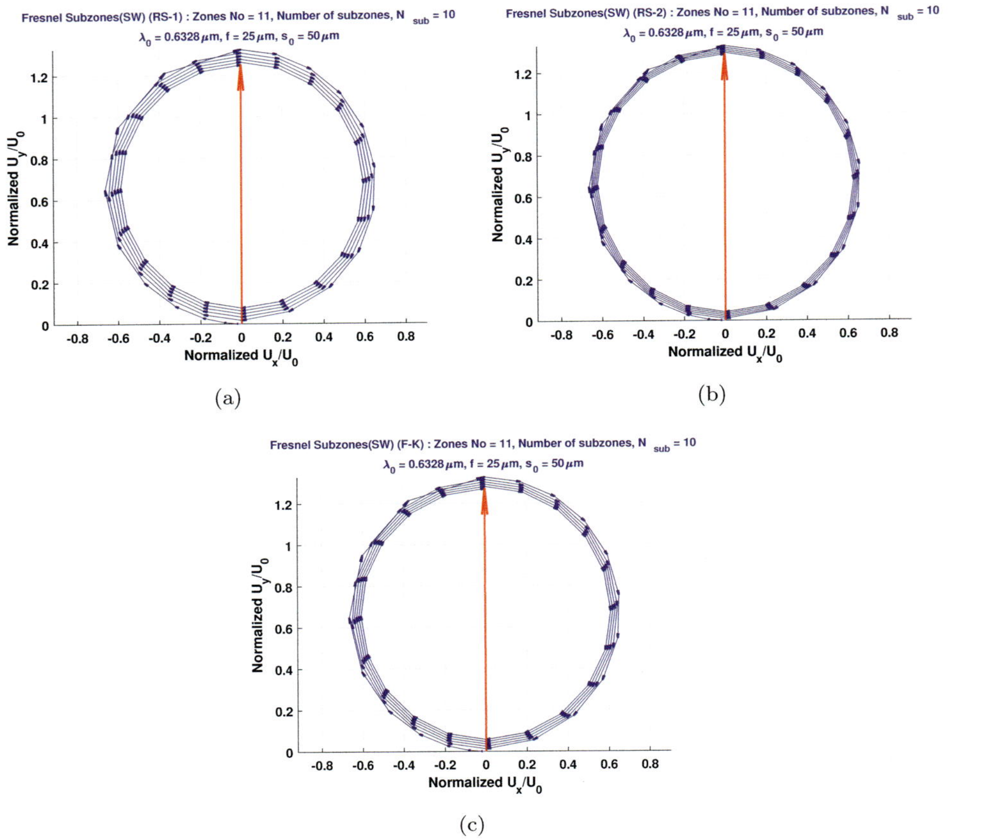

FIGURE 5.72: (a) A vectorial representation of the first 11 ($N = 11$) Fresnel zones, each subdivided into 10 subzones, illustrating the buildup of the diffracted field at the focal point. The incident wave is a diverging spherical wave originating from $s_0 = 50\,\mu m$. The freespace wavelength is $\lambda_0 = \lambda = 0.6328\,\mu m$, and the primary focal distance is $f = 25\,\mu m$. Results using the RS-1 formulation are displayed. (b) Same as (a) but using the RS-2 formulation results. (c) Same as (a) but using the F-K (Fresnel-Kirchhoff) formulation results. The red vector in each case represents the total field, calculated as the vector summation of all zone contributions.

and even zone plates, respectively. For instance, $I(f)/I_{sw} \simeq 47$ for the odd zone plate and $I(f)/I_{sw} \simeq 32.6$ for the even zone plate.

In this case, secondary focal points begin to emerge. Using Eq. (5.245), the following secondary focal points and their corresponding relative intensities are predicted for the odd plate: $f_3 = 666.67\,\mu m$ with $I/I_{sw} \simeq 105.8$, $f_5 = 375\,\mu m$ with $I/I_{sw} \simeq 120.37$, and $f_7 = 260.9\,\mu m$ with $I/I_{sw} \simeq 126.9$, and $f_9 = 200\,\mu m$ with $I/I_{sw} \simeq 130.61$.

From Fig. 5.74(a), the observed values are $f_3 \simeq 662\,\mu m$ with $I/I_{sw} \simeq 105$, $f_5 \simeq 367\,\mu m$ with $I/I_{sw} \simeq 110$, and $f_7 \simeq 250\,\mu m$ with $I/I_{sw} \simeq 57$. It is worth noting that the approximations used to derive Eqs. (5.243) and (5.245) become less accurate as the focal points move closer to the aperture. A similar analysis applies to the even zone plate as well. Specifically, the predicted secondary focal points and relative intensities for the even plate are: $f_3 = 666.67\,\mu m$ with $I/I_{sw} \simeq 73.5$, $f_5 = 375\,\mu m$ with $I/I_{sw} \simeq 83.6$, and $f_7 = 260.9\,\mu m$ with $I/I_{sw} \simeq 88.1$, and $f_9 = 200\,\mu m$ with $I/I_{sw} \simeq 90.7$. From Fig. 5.74(b) the observed values are $f_3 \simeq 660\,\mu m$ with $I/I_{sw} \simeq 73.5$, $f_5 \simeq 367\,\mu m$ with $I/I_{sw} \simeq 80$, and $f_7 \simeq 250\,\mu m$ with $I/I_{sw} \simeq 60$.

TABLE 5.3: Fresnel zone contributions for the first 11 zones for an incident diverging spherical wave originating from $s_0 = 50\,\mu m$. The freespace wavelength is $\lambda_0 = 0.6328\,\mu m$, and the primary focal point is at $z_0 = f = 25\,\mu m$. The summation symbol represents $\sum = \sum_{n=1}^{m}$. Assume $U_0 = s_0$.

Zone	$[U_{RS_1}]_{m-1}^{m}$	$\sum[U_{RS_1}]_{n-1}^{n}$	$[U_{RS_2}]_{m-1}^{m}$	$\sum[U_{RS_2}]_{n-1}^{n}$	$[U_{FK}]_{m-1}^{m}$	$\sum[U_{FK}]_{n-1}^{n}$
1	1.3250	1.3250	1.3291	1.3291	1.3270	1.3270
2	−1.3085	0.0165	−1.3208	0.0084	−1.3147	0.0124
3	1.2923	1.3088	1.3125	1.3208	1.3025	1.3149
4	−1.2764	0.0324	−1.3042	0.0166	−1.2906	0.0243
5	1.2609	1.2933	1.2961	1.3127	1.2789	1.3032
6	−1.2457	0.0476	−1.2880	0.0247	−1.2675	0.0357
7	1.2307	1.2783	1.2800	1.3047	1.2563	1.2920
8	−1.2161	0.0622	−1.2720	0.0326	−1.2453	0.0467
9	1.2018	1.2640	1.2641	1.2968	1.2345	1.2812
10	−1.1877	0.0763	−1.2563	0.0405	−1.2239	0.0574
11	1.1739	1.2502	1.2485	1.2890	1.2134	1.2708
	odd	$\sum\lvert[U]_{n-1}^{n}\rvert$		$\sum\lvert[U]_{n-1}^{n}\rvert$		$\sum\lvert[U]_{n-1}^{n}\rvert$
		7.4846		7.7303		7.6127
	even	$\sum\lvert[U]_{n-1}^{n}\rvert$		$\sum\lvert[U]_{n-1}^{n}\rvert$		$\sum\lvert[U]_{n-1}^{n}\rvert$
		6.2344		6.4413		6.3419

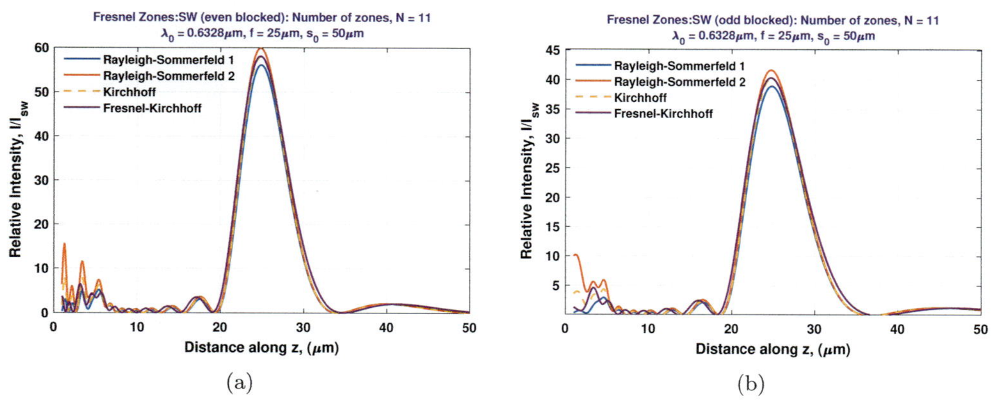

FIGURE 5.73: (a) A Fresnel zone plate with $N = 11$ zones and the even blocked. The freespace wavelength is $\lambda_0 = 0.6328\,\mu m$, and the focal distance is $f = 25\,\mu m$. The incident spherical wave originates from $s_0 = 50\,\mu m$. The relative intensity I/I_{sw} of the diffracted light along the z-axis with the even zones blocked. (b) Same as (a), but for the odd zones blocked. In all plots, the RS-1, RS-2, Kirchhoff, and Fresnel-Kirchhoff results are shown.

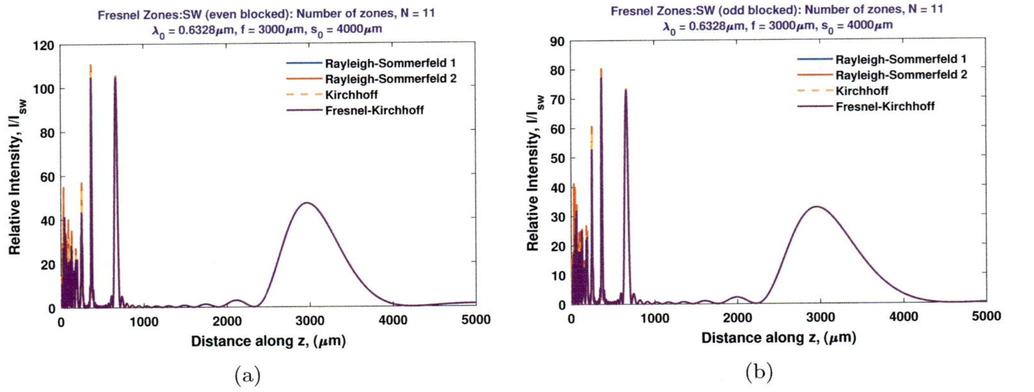

FIGURE 5.74: (a) A Fresnel zone plate with $N = 11$ zones and the even blocked. The freespace wavelength is $\lambda_0 = 0.6328\,\mu m$, and the focal distance is $f = 3000\,\mu m$. The incident spherical wave originates from $s_0 = 4000\,\mu m$. The relative intensity I/I_{sw} of the diffracted light along the z-axis with the even zones blocked. (b) Same as (a), but for the odd zones blocked. In all plots, the RS-1, RS-2, Kirchhoff, and Fresnel-Kirchhoff results are shown.

5.11 Scalar Diffraction by Gratings

A diffraction grating [308–310] is an optical element characterized by a periodic variation in its electromagnetic properties, either along one direction (one-dimensional grating) or along two directions (two-dimensional or crossed gratings). Consequently, the diffraction of light by such a grating generates, through constructive interference, a discrete set of waves referred to as diffracted orders, which propagate in both the forward and backward directions.

Optical gratings may be planar, such as slab or volume gratings, or surface-relief, such as corrugated gratings. The periodic modulation can manifest as a variation in permittivity, resulting in refractive index gratings, or in conductivity, leading to absorption gratings, or as a combination of both effects. Diffraction gratings can be fabricated using a variety of materials, including dielectrics, semiconductors, metals, and even plasmas. They may be classified based on the material properties as isotropic, uniaxial, biaxial-anisotropic, or gyrotropic. Furthermore, gratings can exhibit constant or varying modulation profiles and can be cascaded or multiplexed for more complex applications.

The diffraction of electromagnetic waves by periodic structures has long been a topic of significant interest and practical importance due to its numerous applications across a wide range of fields, including acousto-optics, electro-optics, integrated optics, spectroscopy, optical computing, optical interconnects, and quantum electronics. Examples of devices based on gratings include distributed feedback lasers, beam deflectors, beam modulators, waveguide couplers, spectral filters, wavelength multiplexers and demultiplexers, holographic beam combiners, head-up displays, photonic-crystal devices, and many others.

The scalar theory of grating diffraction simplifies the electromagnetic field by treating it as a scalar field rather than a vector field, thereby neglecting the coupling between the electric and magnetic components. The general assumptions underpinning the validity of scalar diffraction theory also apply to the analysis of gratings. Specifically, the grating's feature size must be significantly larger than the wavelength of light. Additional assumptions, as outlined in [265, 311], include the following: the medium through which light propagates should be lossless, isotropic, homogeneous, linear, and devoid of electric charges. Moreover, the diffracted field is typically evaluated at distances considerably larger than the wavelength of light from the diffraction grating. If these assumptions are not met, it becomes necessary to consider the vector nature of the electromagnetic field and to employ the vectorial form of Maxwell's equations for a more accurate analysis of grating diffraction. The treatment of vector diffraction will be presented in a subsequent chapter (Chapter 7).

This section presents the scalar diffraction theory as applied to grating diffraction, encompassing both one-dimensional and two-dimensional gratings.

5.11.1 One-Dimensional Gratings

The geometry of a simple one-dimensional (1D) grating is depicted in Fig. 5.75. The grating is assumed to be periodic along the x'-direction, with feature sizes that are significantly larger than the freespace wavelength of the incident light. The grating profile illustrated corresponds to a surface-relief type, formed by periodic corrugation at the interface between two regions with distinct electromagnetic properties. These regions are characterized by refractive indices n_i for the top region and n_o for the bottom region. The period of the grating is denoted as Λ.

A plane wave is incident from the top region at an angle θ_{inc}, as shown in the upper part of Fig. 5.75. Since the analysis adheres to scalar diffraction theory, the polarization of the incident plane wave is not a critical factor. Under the scalar diffraction framework, the grating can be represented by its transmittance function $t(x')$, which encapsulates the phase and amplitude modulation introduced by the grating while neglecting its thickness along the direction normal to its periodicity (z-direction). The grating is uniform along the y'-direction, confirming its classification as a one-dimensional grating. The transmittance function $t(x')$ is applicable when the grating primarily behaves as a transmission grating. For gratings that function primarily as reflection gratings, the reflectance function $r(x')$

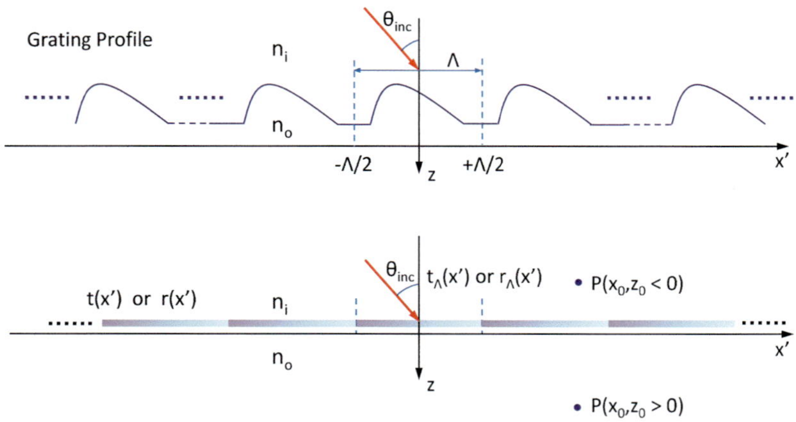

FIGURE 5.75: General geometry of a one-dimensional grating. The top figure illustrates the grating's profile, which is uniform along the y'-direction (not shown) and periodic along the x'-direction with period Λ. The transmittance and reflectance functions, $t(x')$, and $r(x')$, respectively, are periodic with period Λ. The $x'y'z'$ and xyz systems coordinate systems share a common z-axis (shown). For a single period, centered at $x' = 0$, the transmittance and reflectance functions are $t_\Lambda(x')$ and $r_\Lambda(x')$, expressed as $t_\Lambda(x') = A_t(x')\exp[-j\Phi_t(x')]$ and $r_\Lambda(x') = A_r(x')\exp[-j\Phi_r(x')]$, where $A_t(x'), A_r(x')$ represent the amplitude terms, and $\Phi_t(x'), \Phi_r(x')$ represent the phase factors. A plane wave is incident at an angle θ_{inc} relative to the z-axis. The points $P(x_0, z_0 > 0)$ and $P(x_0, z_0 < 0)$ denote arbitrary locations in the regions below and above the grating, with refractive indices n_o and n_i, respectively.

is considered. This reflectance function incorporates the phase and amplitude information corresponding to the backward propagation direction, similarly neglecting thickness variations in the z-direction. The equivalent transmittance or reflectance representation of the one-dimensional grating is illustrated in the lower part of Fig. 5.75.

The transmittance or reflectance functions of a grating over a single period, specifically within the interval $-\Lambda/2$ to $+\Lambda/2$, are defined as $t_\Lambda(x') = A_t(x')\exp[-j\Phi_t(x')]$ or $r_\Lambda(x') = A_r(x')\exp[-j\Phi_r(x')]$, where $A_t(x'), A_r(x')$ denote the amplitude modulation, and $\Phi_t(x'), \Phi_r(x')$ represent the phase modulation, respectively. These functions encapsulate the complete amplitude and phase modifications that light undergoes when it is either transmitted (transmittance function) or reflected (reflectance function) by the grating.

All the structural characteristics of the grating along the z-direction are implicitly accounted for within these functions. As such, the transmittance function $t(x')$ is analogous to the aperture transmittance function used in aperture diffraction problems. Similarly, the reflectance function $r(x')$ corresponds to an aperture function but is adjusted for backward propagation.

Given the periodic nature of both the transmittance and reflectance functions along the x'-direction, they satisfy the relations $t(x') = t(x' + m\Lambda)$ and $r(x') = r(x' + m\Lambda)$, where $m = 0, \pm1, \pm2, \cdots$. Consequently, these functions can be represented using a Fourier series expansion, as outlined in Appendix 5E:

$$t(x') = t(\vec{k}_{inc}) \sum_{m=-\infty}^{+\infty} t_m \exp[-jmKx'], \tag{5.246}$$

$$t_m = \frac{1}{\Lambda}\mathcal{F}\{t_\Lambda(x')\}\Big|_{k_x=mK} = \frac{1}{\Lambda}T_\Lambda(mK), \tag{5.247}$$

$$r(x') = r(\vec{k}_{inc}) \sum_{m=-\infty}^{+\infty} r_m \exp[-jmKx'], \tag{5.248}$$

$$r_m = \frac{1}{\Lambda}\mathcal{F}\{r_\Lambda(x')\}\Big|_{k_x=mK} = \frac{1}{\Lambda}R_\Lambda(mK). \tag{5.249}$$

Here, $K = 2\pi/\Lambda$ represents the grating vector, while $T_\Lambda(k_x)$ and $R_\Lambda(k_x)$ denote the Fourier transforms of the transmittance $t_\Lambda(x')$ and reflectance $r_\Lambda(x')$, functions, respectively. The symbol $\mathcal{F}$ indicates the Fourier transform operator. The coefficients $t(\vec{k}_{inc})$ and $r(\vec{k}_{inc})$ are the Fresnel transmission and reflection coefficients, respectively, for a planar interface between two media with refractive indices n_i and n_o. These coefficients provide an approximate means of accounting for the partial transmission and reflection effects inherent to the interface, which the scalar diffraction approach cannot directly address.

Examples of transmittance functions and their corresponding Fourier series representations are provided in Appendix 5F. These examples include amplitude-only, phase-only, blazed, multilevel, and sinusoidal grating profiles. Similar functions can also be defined for reflection gratings, provided the optical paths and phase shifts of the reflected light are taken into account. Several examples for reflection gratings are also presented in Appendix 5F. The incident plane wave, treated as a scalar, can be expressed as follows:

$$U_{inc}(x', z) = U_0 \exp[-jk\sin\theta_{inc}x'] \exp[-jk\cos\theta_{inc}z]. \tag{5.250}$$

As demonstrated in Sec. 5.5, the plane wave spectrum approach is equivalent to the Rayleigh-Sommerfeld formulations. In the case of a two-dimensional setup (since the one-dimensional transmission grating is being considered), the plane wave spectrum approach can be expressed as follows:

$$U(x_0, z_0 > 0) = \frac{1}{2\pi}\int_{-\infty}^{\infty} \tilde{U}_t(k_x; z=0)e^{-j(k_x x_0 + k_z z_0)}dk_x, \quad \text{where,}$$

$$\begin{aligned}
\tilde{U}_t(k_x; z=0) &= \mathcal{F}\left\{U_{inc}(x', z=0)t(x')\right\}\Big|_{k_x} \\
&= \mathcal{F}\left\{U_0 e^{-jk\sin\theta_{inc}x'} \sum_{m=-\infty}^{+\infty} t_m e^{-jmKx'}\right\} t(\vec{k}_{inc}) \\
&= U_0 t(\vec{k}_{inc}) \sum_{m=-\infty}^{+\infty} t_m 2\pi\delta\left(k_x - mK - k\sin\theta_{inc}\right) \\
&= 2\pi U_0 t(\vec{k}_{inc}) \sum_{m=-\infty}^{+\infty} t_m \delta\left(k_x - k_{xm}\right),
\end{aligned}$$

$$\text{where,} \quad k_{xm} = mK + k\sin\theta_{inc}.$$

Using the previous equations, the diffracted field $U(x_0, z_0 > 0)$ can be easily determined and is given by:

$$U(x_0, z_0 > 0) = U_0 t(\vec{k}_{inc}) \sum_{m=-\infty}^{+\infty} t_m \exp[-jk_{xm}x_0]\exp[-jk_{zm}z_0], \quad \text{where,} \tag{5.251}$$

$$k_{zm} = \begin{cases} \sqrt{k^2 - k_{xm}^2}, & \text{when } k \geq k_{xm}, \\ -j\sqrt{k_{xm}^2 - k^2}, & \text{when } k < k_{xm}. \end{cases}$$

Equation (5.251) shows that the diffraction of the incident plane wave by the one-dimensional grating results in a discrete set of plane waves, labeled by the integer index m, known as diffracted orders. It is clear that there exists a finite number of propagating diffracted orders, corresponding to real k_{zm} wavevector components, and an infinite number of evanescent orders, corresponding to imaginary k_{zm} wavevector components. However, for z_0 sufficiently larger than a few wavelengths, the evanescent orders do not significantly affect the diffracted field, as they rapidly decay away from the grating plane ($x'z$ plane).

In the case of the backward diffracted field (i.e., for reflection gratings), the previous equations can be modified as follows:

$$U(x_0, z_0 < 0) = \frac{1}{2\pi}\int_{-\infty}^{\infty} \tilde{U}_r(k_x; z=0)e^{-j(k_x x_0 + k_z z_0)}dk_x, \quad \text{where,}$$

$$\begin{aligned}
\tilde{U}_r(k_x; z=0) &= \mathcal{F}\left\{U_{inc}(x', z=0)r(x')\right\}\Big|_{k_x} \\
&= \mathcal{F}\left\{U_0 e^{-jk\sin\theta_{inc}x'} \sum_{m=-\infty}^{+\infty} r_m e^{-jmKx'}\right\} r(\vec{k}_{inc}) \\
&= U_0 r(\vec{k}_{inc}) \sum_{m=-\infty}^{+\infty} r_m 2\pi\delta\left(k_x - mK - k\sin\theta_{inc}\right) \\
&= 2\pi U_0 r(\vec{k}_{inc}) \sum_{m=-\infty}^{+\infty} r_m \delta\left(k_x - k_{xm}\right),
\end{aligned}$$

$$\text{where,} \quad k_{xm} = mK + k\sin\theta_{inc}.$$

Using the previous equations, the diffracted field $U(x_0, z_0 < 0)$ can be easily determined and is given by:

$$U(x_0, z_0 < 0) = U_0 r(\vec{k}_{inc}) \sum_{m=-\infty}^{+\infty} r_m \exp[-jk_{xm}x_0] \exp[+jk_{zm}z_0], \quad \text{where,} \tag{5.252}$$

$$k_{zm} = \begin{cases} \sqrt{k^2 - k_{xm}^2}, & \text{when } k \geq k_{xm}, \\ -j\sqrt{k_{xm}^2 - k^2}, & \text{when } k < k_{xm}. \end{cases}$$

It is interesting to examine the case where only a finite number of periods is retained for the one-dimensional grating. This problem, when considering only amplitude modulation, is analogous to the multiple-slit scalar diffraction presented in Sec. 5.8.4, where the RS-1 and RS-2 formulations were applied. In this case, the plane-wave spectrum method is used, similar to the approach for an infinite-periodic grating. Let us define the "*comb*" function (used in Appendix 5E) as: $\text{comb}(x; \Lambda) = \sum_{m=-\infty}^{+\infty} \delta(x - m\Lambda)$. With this, the transmittance function $t(x')$ or reflectance function $r(x')$ of a finite-period grating can be expressed as:

$$t(x') = t(\vec{k}_{inc}) \Big[t_\Lambda(x') * \text{comb}(x'; \Lambda) \Big] \text{rect}\left(\frac{x'}{L}\right), \tag{5.253}$$

$$r(x') = r(\vec{k}_{inc}) \Big[r_\Lambda(x') * \text{comb}(x'; \Lambda) \Big] \text{rect}\left(\frac{x'}{L}\right), \tag{5.254}$$

where L is the grating size, which corresponds to a finite number of periods. The "rect' function is the rectangular function, defined in Appendix 5E, and "*" denotes the one-dimensional convolution operation. With this, the Fourier transform of $t(x')$ is given by:

$$\mathcal{F}\{t(x')\}\Big|_{k_x} = \frac{1}{2\pi} \left[T_\Lambda(k_x) \frac{2\pi}{\Lambda} \sum_{m=-\infty}^{+\infty} \delta(k_x - mK) \right] * \left[L \text{ sinc}\left(\frac{k_x L}{2\pi}\right) \right] =$$

$$= L \sum_{m=-\infty}^{+\infty} \frac{1}{\Lambda} T_\Lambda(mK) \text{ sinc}\left(\frac{(k_x - mK)L}{2\pi}\right). \tag{5.255}$$

The resulting diffracted field at the general point $P(x_0, z_0 > 0)$ is given by (using the plane wave spectrum method):

$$U(x_0, z_0 > 0) = \frac{1}{2\pi} \int_{-\infty}^{\infty} \tilde{U}_t(k_x; z = 0) e^{-j(k_x x_0 + k_z z_0)} dk_x, \quad \text{where,} \tag{5.256}$$

$$\tilde{U}_t(k_x; z = 0) = \mathcal{F}\{U_{inc}(x', z = 0)t(x')\}\Big|_{k_x} = U_0 L \sum_{m=-\infty}^{+\infty} \frac{1}{\Lambda} T_\Lambda(k_{xm}) \text{ sinc}\left(\frac{(k_x - k_{xm})L}{2\pi}\right) =$$

$$= U_0 L \sum_{m=-\infty}^{+\infty} t_m \text{ sinc}\left(\frac{(k_x - k_{xm})L}{2\pi}\right), \tag{5.257}$$

where $k_{xm} = mK + k \sin\theta_{inc}$.

Of course, the wavevector components k_{zm} are again specified as in Eq. (5.251). It is interesting to show that in the limit of $L \to \infty$ Eq. (5.256) converges to Eq. (5.251). This can be accomplished in a straightforward manner by using the definition of the Dirac delta function:

$\delta(x) = \lim_{\epsilon\to 0}\{(1/\pi x)\sin(x/\epsilon)\}$ [312] or equivalently $\delta(x) = \lim_{L\to+\infty}\{(1/\pi x)\sin(Lx)\}$. The last definition of $\delta(x)$ can also be written as (using $x = \pi y$):

$$\delta(\pi y) = \lim_{L\to+\infty}\left\{\frac{1}{\pi^2 y}\sin(L\pi y)\right\} \Longrightarrow$$

$$\frac{1}{\pi}\delta(y) = \lim_{L\to+\infty}\left\{\frac{L}{\pi}\,\frac{\sin(\pi L y)}{\pi L y}\right\} \Longrightarrow$$

$$\delta(y) = \lim_{L\to+\infty}\{L\,\mathrm{sinc}(Ly)\},$$

where the property of the dleta function $\delta(ay) = (1/|a|)\delta(y)$ was employed. By applying the last equation to $\tilde{U}_t$ [Eq. (5.257)], it is straightforward to show that:

$$\lim_{L\to\infty}\left\{\tilde{U}_t(k_x; z=0)\right\} = U_0 \sum_{m=-\infty}^{+\infty} t_m \lim_{L\to\infty}\left\{L\,\mathrm{sinc}\left(\frac{(k_x - k_{xm})L}{2\pi}\right)\right\} =$$

$$= U_0 \sum_{m=-\infty}^{+\infty} t_m \delta\left(\frac{k_x - k_{xm}}{2\pi}\right)$$

$$= U_0 \sum_{m=-\infty}^{+\infty} t_m 2\pi\delta(k_x - k_{xm}).$$

Using the last equation in Eq. (5.256) results in Eq. (5.251), as expected, when the number of periods tends to infinity. Naturally, similar equations can be derived for reflection gratings, and these are left as an exercise for the reader.

It is worth noting that the Fresnel and Fraunhofer approximations are not valid for infinite one-dimensional gratings due to the infinite extent of the aperture. However, these approximations can be applied to gratings with a finite number of periods, analogous to the case of multiple-slit diffraction discussed in Sec. 5.8.4. For instance, as described in Sec. 5.4.2.2, the Fraunhofer approximation for a grating with a finite number of periods is expressed as:

$$U_{ff}(x_0, z_0) = j\sqrt{\frac{k}{2\pi z_0}}e^{-jkz_0}\exp\left(-j\frac{k}{2z_0}x_0^2\right)\mathcal{F}\left\{t(x')U_{inc}(x')\right\}\Big|_{k_x = kx_0/z_0} =$$

$$j\sqrt{\frac{k}{2\pi z_0}}e^{-jkz_0}\exp\left(-j\frac{k}{2z_0}x_0^2\right)U_0 L\sum_{m=-\infty}^{+\infty} t_m\,\mathrm{sinc}\left(\frac{(k(x_0/z_0) - k_{xm})L}{2\pi}\right).$$

The Fraunhofer approximation is valid for $z_0 \gg (\pi/\lambda_0)N^2\Lambda^2$. Naturally, this expression applies exclusively to a grating with a finite number of periods and is contingent on the stated validity constraint. Consequently, only the results provided in Eq. (5.251), derived using the plane wave spectrum method, remain applicable in the case of scalar diffraction by an infinite-period one-dimensional grating.

The diffraction efficiency of a grating is typically defined as the fraction of the incident optical power that emerges in a specific diffracted order. While the power in a given direction is commonly related to the Poynting vector, in the context of this chapter, where only scalar diffraction is considered, it is also practical to use the intensity of the diffracted field. The intensity is determined from $|U|^2/4Z$, in conjunction with Eq. (5.251). This yields the following expression for the intensity:

$$I(x_0, z_0 > 0) = \frac{|U_0|^2}{4Z} \sum_{m=-\infty}^{+\infty} \sum_{p=-\infty}^{+\infty} t_m t_p^* \exp\left[-j(k_{xm} - k_{xp})x_0\right] \exp\left[-j(k_{zm} - k_{zp}^*)z_0\right]. \tag{5.258}$$

The interference pattern described by the last equation underscores the combined contributions of all the forward diffracted orders generated by the grating. From a practical perspective, diffraction efficiency is typically evaluated in the far-field, where the individual diffracted orders are spatially separated. Under such conditions, the cross-interference terms among the various diffracted orders can be ignored. Consequently, the diffracted intensity simplifies to the summation of the intensities of the individual diffracted orders, expressed as follows (note that this expression is independent of x_0):

$$I(z_0 > 0) \simeq \frac{|U_0|^2}{4Z} \sum_{m=-\infty}^{+\infty} |t_m|^2 \exp\left[2\mathrm{Im}\{k_{zm}\}z_0\right] \simeq \frac{|U_0|^2}{4Z} \sum_{m=-m_1}^{+m_2} |t_m|^2. \tag{5.259}$$

Here, $-m_1 \leq m \leq m_2$, where $-m_1$ and m_2 denote the minimum and the maximum indices of the forward propagating diffracted orders. These represent the range of orders contributing to the intensity in the far-field region, as evanescent orders decay rapidly and do not affect the intensity for z_0 sufficiently larger than a few wavelengths. The symbol $\mathrm{Im}\{*\}$ denotes the imaginary part of the expression within the brackets. The integers m_1 and m_2 can be determined from the "grating equation" for the forward diffracted orders. The grating equation can be derived from the definition of the k_{xm} wavevector components of Eq. (5.251). Specifically:

$$k_{xm} = k \sin\theta_m = mK + k\sin\theta_{inc} \Longrightarrow$$

$$\sin\theta_m - \sin\theta_{inc} = m\frac{\lambda_0}{n\Lambda}, \quad m = 0, \pm 1, \pm 2, \cdots. \tag{5.260}$$

Here, θ_m represents the angle of the m-th forward diffracted order with respect to the z-axis. For a reflection grating, such as one coated with a thin layer of a perfect conductor, a similar equation can be derived to describe the backward diffracted orders. To determine m_1 and m_2, it is necessary to ensure that the values of $|sin\theta_m|$ remain less than or equal to unity. Consequently:

$$-m_1 = -\left(1 + \sin\theta_{inc}\right)\frac{n\Lambda}{\lambda_0} \leq m \leq \left(1 - \sin\theta_{inc}\right)\frac{n\Lambda}{\lambda_0} = m_2. \tag{5.261}$$

Here, n denotes the refractive index through which the diffracted orders propagate. In most practical applications of scalar diffraction, air is assumed to be the medium. Specifically, $n = n_o$ for transmission grating and $n = n_i$ for reflection grating.

At this point, it is useful to apply Parseval's theorem for the Fourier series [313]. Given the periodic nature of the grating's transmittance function, the theorem asserts the following:

$$\frac{1}{\Lambda}\int_{-\Lambda/2}^{+\Lambda/2} \left|t(x')\right|^2 dx' = \sum_{m=-\infty}^{+\infty} |t_m|^2. \tag{5.262}$$

The following definitions of the diffracted efficiencies are applicable to gratings with a theoretically infinite number of periods. The intensity of the transmitted field at $z_0 = 0^+$ (for a transmission grating) is given by:

$$I(z_0 = 0^+) = \frac{1}{4Z}|U(x_0, z_0 = 0^+)|^2 = \frac{1}{4Z}\frac{1}{\Lambda}\int_{-\Lambda/2}^{+\Lambda/2} |U_0 e^{-jk\sin\theta_{inc}x_0} t(x')|^2 dx' =$$

$$= |t(\vec{k}_{inc})|^2 \frac{|U_0|^2}{4Z} \sum_{m=-\infty}^{+\infty} |t_m|^2.$$

Then, the transmitted (forward) diffraction efficiency based on the intensity can be defined as follows:

$$\eta_m^t = |t(\vec{k}_{inc})|^2 |t_m|^2.$$

From Parseval's theorem for phase-only gratings, the sum $\sum_{-\infty}^{+\infty} |t_m|^2 = 1$. However, since only a finite number of the infinitely many diffracted orders can propagate, the aforementioned efficiency can be normalized in the following manner:

$$\eta_m^{t,n} = |t(\vec{k}_{inc})|^2 \frac{|t_m|^2}{\sum\limits_{m=-m_1}^{m_2} |t_m|^2} \implies \sum_{m=-m_1}^{m_2} \eta_m^{t,n} = |t(\vec{k}_{inc})|^2. \tag{5.263}$$

For a lossless medium of propagation, the power at $z_0 = 0^+$ should remain constant, as described by Poynting's theorem. However, due to evanescent orders, the intensity depends on z_0 [see Eq. (5.259)]. A more accurate definition of the diffraction efficiencies is based on Poynting's theorem. This definition is employed in the vector rigorous analysis of gratings, as discussed in the chapter on *Vector Diffraction*, Chapter 7. In this context, the definition is adapted, and the forward diffraction efficiencies can be defined as:

$$DE_m^{t,n} = |t(\vec{k}_{inc})|^2 \frac{|t_m|^2 \,(n_o \text{Re}\{\cos\theta_m\}/n_i \cos\theta_{inc})}{\sum\limits_{m=-m_1}^{m_2} |t_m|^2 (n_o \text{Re}\{\cos\theta_m\}/n_i \cos\theta_{inc})} \implies \sum_{m=-m_1}^{m_2} DE_m^{t,n} = |t(\vec{k}_{inc})|^2. \tag{5.264}$$

It is noted that the definitions of $DE_m^{t,n}$ are not affected by the evanescent orders, as these do not contribute to the power balance (observe the real part in the $\cos\theta_m$ term). Therefore, the summation could include all orders, since the evanescent orders do not contribute, and their corresponding diffraction efficiencies are zero.

It is straightforward to extend the definitions of diffraction efficiencies to reflection gratings as well. In this case, the corresponding intensity-based $\eta_m^{r,n}$ and Poynting-theorem-based $DE_m^{r,n}$ diffraction efficiencies are given by:

$$\eta_m^{r,n} = |r(\vec{k}_{inc})|^2 \frac{|r_m|^2}{\sum\limits_{m=-m_1}^{m_2} |r_m|^2} \implies \sum_{m=-m_1}^{m_2} \eta_m^{r,n} = |r(\vec{k}_{inc})|^2, \tag{5.265}$$

$$DE_m^{r,n} = |r(\vec{k}_{inc})|^2 \frac{|r_m|^2 (\text{Re}\{\cos\theta_m\}/\cos\theta_{inc})}{\sum\limits_{m=-m_1}^{m_2} |r_m|^2 (\text{Re}\{\cos\theta_m\}/\cos\theta_{inc})} \implies \sum_{m=-m_1}^{m_2} DE_m^{r,n} = |r(\vec{k}_{inc})|^2. \tag{5.266}$$

Example 5.11.1:

Consider a one-dimensional amplitude-only grating of a rectangular-groove with a filling factor of 50% (refer to Sec. 5F.1.1). The grating is illuminated by an incident plane wave with a freespace wavelength of $\lambda_0 = 0.65\,\mu$m and an angle of incidence $\theta_{inc} = 0\,deg$. The grating period is $\Lambda = 15\,\mu$m, and the propagation medium is air ($n_i = n_o = 1$). Employ the plane wave spectrum technique to compute the diffracted field along x_0-direction at a propagation distance of $z_0 = 200\,\mu$m. Perform the calculation for two cases: (a) an infinite number of grating periods and (b) a finite number of periods, where $L = N\Lambda$ with $N = 10$ or $N = 100$.

Solution: The above information clearly indicates that the propagating diffracted orders must satisfy the inequality $-(\Lambda/\lambda_0) \leq m \leq +(\Lambda/\lambda_0)$. Substituting the given values, this yields $m = 0, \pm1, \cdots \pm 23$. The analysis reveals that there are 47 propagating diffracted

TABLE 5.4: Example case of an amplitude-only one-dimensional rectangular-groove grating with a filling factor $F = 0.50$. The incident plane wave has a freespace wavelength of $\lambda_0 = 0.65\,\mu$m and the angle of incidence is $\theta_{inc} = 0\,deg$. The grating period is $\Lambda = 15\,\mu$m. The amplitude of the incident plane wave is $U_0 = 1$.

Transmittance Fourier Coefficients						
t_{-5}	t_{-3}	t_{-1}	t_0	t_{+1}	t_{+3}	t_{+5}
0.0637	−0.1061	0.3183	0.5000	0.3183	−0.1061	0.0637

orders, consistent with the inequality $-(\Lambda/\lambda_0) \leq m \leq +(\Lambda/\lambda_0)$. Regardless of the number of orders retained in Eq. (5.251), only these 47 orders contribute to the diffracted field at distances at least a few wavelengths away from the grating. This relatively large number of orders is attributed to the selected grating period, which is significantly larger than the wavelength. The amplitudes t_m of the diffracted orders are determined using the method outlined in Appendix 5F, specifically in Sec. 5F.1.1. Since $t_m = F\text{sinc}(m\pi F)$ and $F = 1/2$, only odd orders (and the zero order) have nonzero amplitudes, as all even orders are null. The first seven nonzero t_m coefficients are presented in Table 5.4.

The normalized intensity $I(x_0, z_0) = |U(x_0, z_0)|^2/|U_0|^2$, was computed at $z_0 = 200\,\mu$m along the x-axis (i.e., for varying x_0). The analysis considered both the infinite-period grating and a finite-period grating with $L = 10\Lambda$ and $L = 100\Lambda$. The results are illustrated in Fig. 5.76(a) for $L = 10\Lambda$ and Fig. 5.76(b) for $L = 100\Lambda$. For the infinite-period grating, the intensity pattern exhibits strong oscillations due to interference among the 25 nonzero propagating orders (47 total orders minus the 22 even orders with zero amplitude).

Conversely, the finite-period grating produces a distinct interference pattern. As the number of grating periods increases to $N = 100$, the normalized intensities approach those of the infinite-period case, as evidenced by Fig. 5.76(b), where $L = 100\Lambda$.

Example 5.11.2:

This example involves a phase-only grating with the geometry illustrated in Fig. 5.2(a). The grating parameters are as follows: grating period $\Lambda = 15\,\mu$, ridge height $h = 0.5\,\mu$m, filling factor $F = 0.50$, refractive index of the cover medium $n_c = 1$, and refractive index of the substrate $n_s = 1.5$. The incident plane wave has a freespace wavelength of $\lambda_0 = 0.65\,\mu$m,

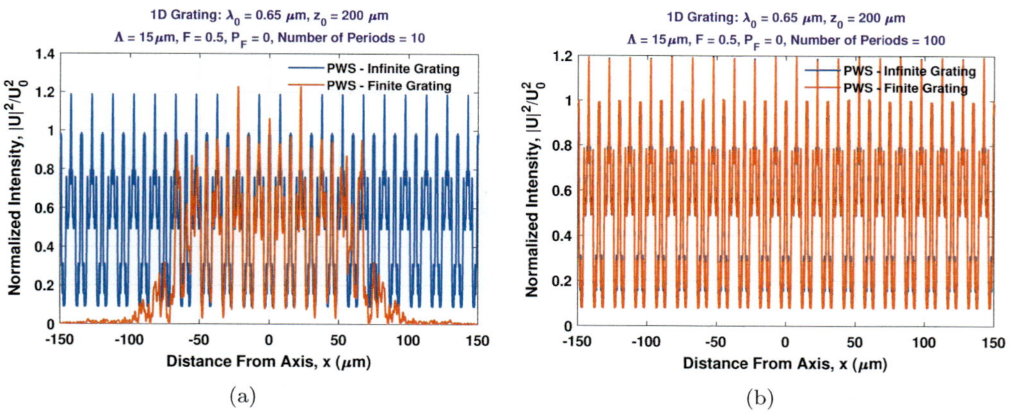

FIGURE 5.76: (a) Comparison of the normalized intensity of the diffracted field, $|U(x_0, z_0)|^2/|U_0|^2$, along the x_0-direction for a one-dimensional grating with an infinite number of periods and a finite number of periods. The grating parameters are: incident light with a freespace wavelength $\lambda_0 = 0.65\,\mu m$, grating period $\Lambda = 15\,\mu m$, incidence angle $\theta_{inc} = 0\,deg$, and grating length $L = 10\Lambda$. The observation plane is located at $z_0 = 200\,\mu$m. (b) Same as (a), but for a grating length $L = 100\Lambda$.

and the angle of incidence θ_{inc} is variable. Determine the forward diffraction efficiencies, $\eta_m^{t,n}$ and $DE_m^{t,n}$, for the diffraction orders $m = 0, 1, -1$.

Solution: The number of propagating diffracted orders can be determined using Eq. (5.261). For normal incidence ($\theta_{inc} = 0$ deg), the propagating orders correspond to the range $-34 \leq m \leq 34$. In contrast, for an incidence angle of $\theta_{inc} = 30$ deg, the propagating orders span the range $-51 \leq m \leq 17$. For intermediate angles of incidence, the range of propagating orders varies between these intervals; however, the total number of diffracted orders remains constant at 60.

For the phase-only grating parameters in this example, the diffraction efficiencies $\eta_m^{t,n}$ and $DE_m^{t,n}$ were calculated for $m = 0, 1, -1$. These results are displayed in Figs. 5.77(a) and 5.77(b), respectively. It is observed that $\eta_{+1}^{t,n}$ and $\eta_{-1}^{t,n}$ always coincide, as the coefficients t_{+1} and t_{-1} are identical [see Eq. (5F.6)].

On the other hand, the diffraction efficiencies $DE_{+1}^{t,n}$ and $DE_{-1}^{t,n}$ vary as the angle of incidence changes, since they depend on the diffraction angle corresponding to each order. Selected values of the diffraction efficiencies are presented in Table 5.5, which includes the $m = 0, 1, -1$ diffraction efficiencies for $\theta_{inc} = 0, 15$, and 30 deg. Additionally, the table provides diffraction efficiencies calculated using a well-known vectorial method (presented in the vectorial diffraction chapter, Chapter 7), the rigorous coupled-wave analysis (RCWA) [314–317]. The vectorial diffraction efficiencies are theoretically normalized and account for both forward and backward diffracted orders. Since the vectorial method depends on the incident polarization, both *TE* and *TM* diffraction efficiencies are included. It can be observed that the scalar results are in relatively good agreement with the rigorous results.

Example 5.11.3:

This example involves the diffraction of a plane wave by a reflection grating. The incident plane wave has a freespace wavelength of $\lambda_0 = 0.65\,\mu$m. The grating profile is sinusoidal, with a period $\Lambda = 15\,\mu$, and a height $h = 0.1\,\mu$m. The refractive index of the cover medium is $n_i = n_c = 1$, while the refractive index of the substrate is $n_o = n_s = 0.052225 - j4.4094$ (corresponding to silver) [76,318]. The grating profile is shown in Fig. 5F.5 in Appendix 5F.

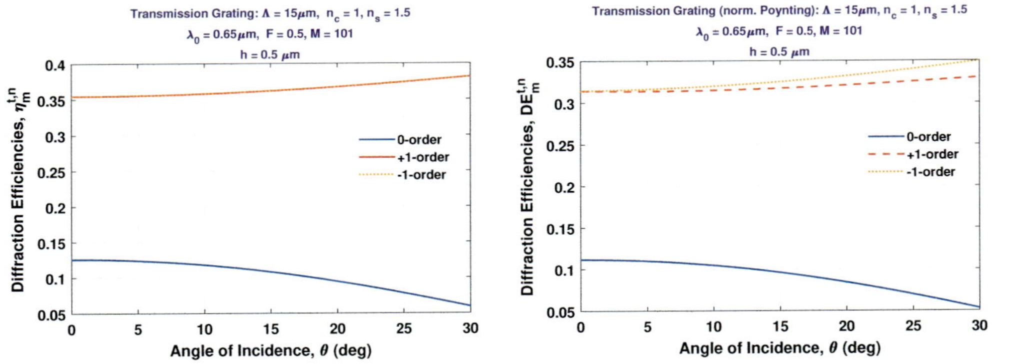

(a) (b)

FIGURE 5.77: (a) The normalized diffraction efficiencies, $\eta_m^{t,n}$ for the forward diffracted orders ($m = 0, 1, -1$) as functions of the angle of incidence θ_{inc}. The grating parameters are as follows: filling factor $F = 0.50$, grating period $\Lambda = 15\,\mu$m, height $h = 0.5\,\mu$m, cover refractive index $n_c = 1$, and substrate refractive index $n_s = 1.5$. (b) The normalized diffraction efficiencies $DE_m^{t,n}$, for the forward diffracted orders ($m = 0, 1, -1$).

TABLE 5.5: Example case of an phase-only one-dimensional grating with a filling factor $F = 0.50$. The incident plane wave has a freespace wavelength of $\lambda_0 = 0.65\,\mu$m and the angle of incidence varies. The grating period is $\Lambda = 15\,\mu$m. The amplitude of the incident plane wave is $U_0 = 1$.

		Scalar Theory		Vector Theory	
θ_{inc} (deg)	m	$\eta_m^{t,n}$	$DE_m^{t,n}$	DE_m (TE)	DE_m (TM)
	0	0.1105	0.1115	0.1257	0.1312
0	1	0.3112	0.3139	0.3402	0.3400
	−1	0.3112	0.3139	0.3402	0.3400
	0	0.0948	0.0957	0.1088	0.1150
15	1	0.3181	0.3169	0.3455	0.3467
	−1	0.3181	0.3247	0.3560	0.3497
	0	0.0525	0.0529	0.0630	0.0695
30	1	0.3372	0.3297	0.3574	0.3683
	−1	0.3372	0.3494	0.3603	0.3739

Calculate the diffraction efficiencies, $DE_m^{r,n}$ for the backward diffracted orders $m = 0, +1$, and -1 as functions of the angle of incidence θ_{inc}, over a range of incidence angles from 0 to 60 degrees.

Solution: The number of propagating backward (reflected) diffracted orders can be determined using Eq. (5.261) for $n = 1$. For normal incidence ($\theta_{inc} = 0$ deg), the propagating orders range from $m = -23$ to $m = 23$. Conversely, for an incidence angle of $\theta_{inc} = 60$ deg, the propagating orders span from $m = -43$ to $m = 3$. For intermediate incidence angles, the range of propagating reflected orders will vary between these intervals, but the total number of diffracted orders remains constant at 47.

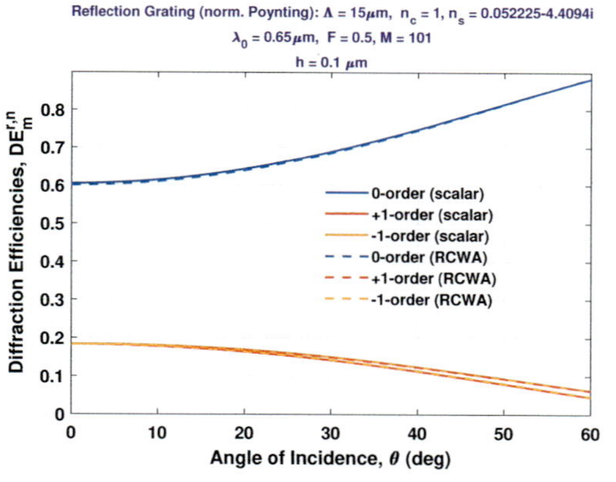

FIGURE 5.78: The normalized diffraction efficiencies $DE_m^{r,n}$, for the backward diffracted orders ($m = 0, 1, -1$), as functions of the angle of incidence θ_{inc}. The grating parameters are as follows: grating period $\Lambda = 15\,\mu$m, height $h = 0.1\,\mu$m, cover refractive index $n_c = 1$, and substrate refractive index $n_s = 0.052225 - j4.4094$ (corresponding to silver). The dashed lines correspond to the vectorial (TE polarization) results obtained using the RCWA.

Using the grating parameters and the equations provided in Sec. 5F.2.2 of Appendix 5F, the coefficients r_m can be determined, and the normalized diffraction efficiencies, $DE_m^{r,n}$ can be calculated using Eq. (5.266). These diffraction efficiencies are shown in Fig. 5.78. In the same plot, the vectorial diffraction efficiencies for TE polarization, calculated using the RCWA, are also shown with dashed lines. It is evident that the scalar and vectorial diffraction efficiencies are nearly identical.

5.11.2 Two-Dimensional Gratings

The geometry of a specialized two-dimensional grating is illustrated in Fig. 5.79. The grating consists of a periodic array of rectangular pillars, each with dimensions $F_x\Lambda_x \times F_y\Lambda_y$. The two-dimensional grating has periodicities Λ_x and Λ_y along the x'- and the y'-directions, respectively. The height of each rectangular pillar is h. An incident plane wave, characterized by its azimuthal angle ϕ_{inc} and polar angle θ_{inc}, interacts with the grating. The incident wave has a freespace wavelength of λ_0.

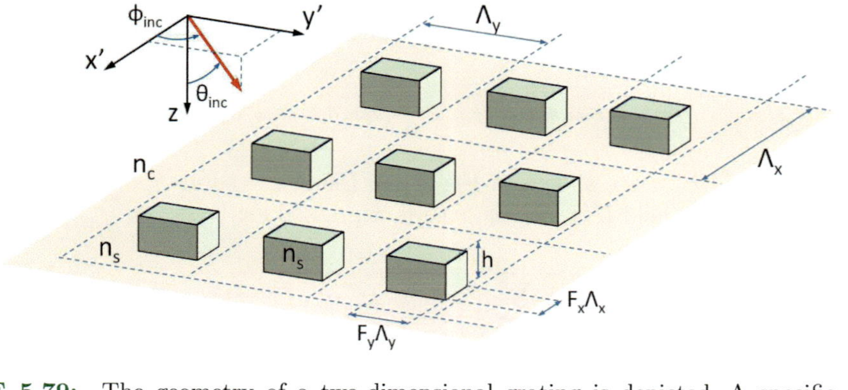

FIGURE 5.79: The geometry of a two-dimensional grating is depicted. A specific case of a rectangular-pillar grating is shown. The unit cell has dimensions $\Lambda_x \times \Lambda_y$, corresponding to periodicities Λ_x and Λ_y along the x- and the y-directions, respectively. The rectangular pillar features xy-plane dimensions of $F_x\Lambda_x$ and $F_y\Lambda_y$, with a height h.

For scalar diffraction theory to be applicable, the feature sizes are assumed to be significantly larger than the freespace wavelength. While the grating described here is a specific case, the same principles can be used to analyze any two-dimensional grating. The refractive index of the grating's cover region is denoted as n_c, while the rectangular pillars and the substrate share the refractive index n_s. According to scalar theory, the grating can be characterized either by its transmittance function $t(x', y')$ for transmission gratings or its reflectance function $r(x', y')$ for reflection gratings. These functions are periodic along the x'- and y'-directions.

To facilitate analysis, the transmittance or reflectance function for a single unit cell of the grating is defined as $t_{uc}(x', y')$ and $r_{uc}(x', y')$, respectively, analogous to the $t_\Lambda(x')$ or $r_\Lambda(x')$ definitions used for one-dimensional gratings. These functions within a unit cell can be expressed as $t_{uc}(x', y') = A_t(x', y')\exp[-j\Phi_t(x', y')]$ or $r_{uc}(x', y') = A_r(x', y')\exp[-j\Phi_r(x', y')]$, where $A_t(x', y'), A_r(x', y')$ represent the amplitude modulation, and $\Phi_t(x', y'), \Phi_r(x', y')$ represent the phase modulation, respectively.

The assumptions and implications outlined for one-dimensional gratings are equally applicable to their two-dimensional counterparts. Given the periodicity of the transmittance and reflectance functions in the x'- and y'-directions, they satisfy the relations $t(x', y') = t(x' + m_1\Lambda_x, y' + m_2\Lambda_y)$ and $r(x', y') = r(x' + m_1\Lambda_x, y' + m_2\Lambda_y)$, where $m_1, m_2 = 0, \pm 1, \pm 2, \cdots$. Consequently, these functions can be represented using a Fourier series, as described in Appendix 5E and shown below:

$$t(x', y') = t(\vec{k}_{inc}) \sum_{m_1=-\infty}^{+\infty} \sum_{m_2=-\infty}^{+\infty} t_{m_1 m_2} \exp[-jm_1 K_x x'] \exp[-jm_2 K_y y'], \tag{5.267}$$

$$t_{m_1 m_2} = \frac{1}{\Lambda_x \Lambda_y} \mathcal{F}\{t_{uc}(x', y')\}\Big|_{\substack{k_x = m_1 K_x \\ k_y = m_2 K_y}} = \frac{1}{\Lambda_x \Lambda_y} T_{uc}(m_1 K_x, m_2 K_y), \tag{5.268}$$

$$r(x', y') = r(\vec{k}_{inc}) \sum_{m_1=-\infty}^{+\infty} \sum_{m_2=-\infty}^{+\infty} r_{m_1 m_2} \exp[-jm_1 K_x x'] \exp[-jm_2 K_y y'], \tag{5.269}$$

$$r_{m_1 m_2} = \frac{1}{\Lambda_x \Lambda_y} \mathcal{F}\{r_{uc}(x', y')\}\Big|_{\substack{k_x = m_1 K_x \\ k_y = m_2 K_y}} = \frac{1}{\Lambda_x \Lambda_y} R_{uc}(m_1 K_x, m_2 K_y). \tag{5.270}$$

Here, $K_x = 2\pi/\Lambda_x$, $K_y = 2\pi/\Lambda_y$ denote the grating vector along the x- and y-direction. The terms $T_{uc}(k_x, k_y)$ and $R_{uc}(k_x, k_y)$ represent the Fourier transforms of $t_{uc}(x', y')$ and $r_{uc}(x', y')$, respectively. Some example transmittance functions and their Fourier series representation are included in Appendix 5F. The incident plane wave (scalar) can be represented by:

$$U_{inc}(x', y', z) = U_0 e^{-jk\sin\theta_{inc}\cos\phi_{inc} x'} e^{-jk\sin\theta_{inc}\sin\phi_{inc} y'} e^{-jk\cos\theta_{inc} z}. \tag{5.271}$$

The angles θ_{inc} and ϕ_{inc} are shown in Fig. 5.79. Using the same approach as in the 1D-gratings, the diffracted field $U(x_0, y_0, z_0 > 0)$ is expressed as:

$$U(x_0, y_0, z_0 > 0) = U_0\, t(\vec{k}_{inc}) \sum_{m_1=-\infty}^{+\infty} \sum_{m_2=-\infty}^{+\infty} t_{m_1 m_2} e^{-jk_{xm_1} x_0} e^{-jk_{ym_2} y_0} e^{-jk_{zm_1 m_2} z_0}, \tag{5.272}$$

$$\text{where,} \quad k_{xm_1} = m_1 K_x + k\sin\theta_{inc}\cos\phi_{inc},$$

$$k_{ym_2} = m_2 K_y + k \sin\theta_{inc} \sin\phi_{inc},$$

$$k_{zm_1m_2} = \begin{cases} \sqrt{k^2 - k_{xm_1}^2 - k_{ym_2}^2}, & \text{when } k \geq (k_{xm_1}^2 + k_{ym_2}^2)^{1/2}, \\ -j\sqrt{k_{xm_1}^2 + k_{ym_2}^2 - k^2}, & \text{when } k < (k_{xm_1}^2 + k_{ym_2}^2)^{1/2}. \end{cases}$$

Similarly, in the case of backward diffraction (reflection grating), the corresponding diffracted field $U(x_0, y_0, z_0 < 0)$ is expressed as:

$$U(x_0, y_0, z_0 < 0) = U_0\, r(\vec{k}_{inc}) \sum_{m_1=-\infty}^{+\infty} \sum_{m_2=-\infty}^{+\infty} r_{m_1m_2} e^{-jk_{xm_1}x_0} e^{-jk_{ym_2}y_0} e^{+jk_{zm_1m_2}z_0}, \tag{5.273}$$

$$k_{zm_1m_2} = \begin{cases} \sqrt{k^2 - k_{xm_1}^2 - k_{ym_2}^2}, & \text{when } k \geq (k_{xm_1}^2 + k_{ym_2}^2)^{1/2}, \\ -j\sqrt{k_{xm_1}^2 + k_{ym_2}^2 - k^2}, & \text{when } k < (k_{xm_1}^2 + k_{ym_2}^2)^{1/2}. \end{cases}$$

It is possible to extend the theory to a finite number of unit cells, as was done in the case of 1D gratings, by employing the 2D "rect" function. However, this approach is straightforward and is left as an exercise for the interested reader. The propagating diffracted orders can be determined from the grating equation, which, in the case of 2D gratings, becomes slightly more intricate. Specifically, the following expressions are applicable:

$$k_{xm_1} = k \sin\theta_{m_1m_2} \cos\phi_{m_1m_2} = m_1 K_x + k \sin\theta_{inc} \cos\phi_{inc},$$

$$k_{ym_2} = k \sin\theta_{m_1m_2} \sin\phi_{m_1m_2} = m_2 K_y + k \sin\theta_{inc} \sin\phi_{inc},$$

$$k_{xm_1}^2 + k_{ym_2}^2 \leq k^2 \implies$$

$$\left(\frac{m_1\lambda_0}{n\Lambda_x} + \sin\theta_{inc}\cos\phi_{inc}\right)^2 + \left(\frac{m_2\lambda_0}{n\Lambda_y} + \sin\theta_{inc}\sin\phi_{inc}\right)^2 \leq 1. \tag{5.274}$$

The first two equations specify the directions ($\theta_{m_1m_2}$ and $\phi_{m_1m_2}$) of the diffracted waves, assuming that they are capable of propagation. Equation (5.274) gives the pairs of (m_1, m_2) that ensure that the (m_1, m_2)-th diffracted order can propagate. The number of propagating diffracted orders, M_{prop}, in either the forward or backward direction can be estimated by the area of the ellipse in Eq. (5.274) and is approximately given by: $M_{prop} \simeq \pi n^2 \Lambda_x \Lambda_y / \lambda_0^2$.

The normalized diffraction efficiencies for a 2D grating are calculated in a manner similar to those of 1D gratings and are given by:

$$\eta_{m_1m_2}^{t,n} = |t(\vec{k}_{inc})|^2 \frac{|t_{m_1m_2}|^2}{\sum_{m_1}\sum_{m_2} |t_{m_1m_2}|^2} \implies \sum_{m_1}\sum_{m_2} \eta_{m_1m_2}^{t,n} = |t(\vec{k}_{inc})|^2, \tag{5.275}$$

$$DE_{m_1m_2}^{t,n} = |t(\vec{k}_{inc})|^2 \frac{|t_{m_1m_2}|^2 (\mathrm{Re}\{\cos\theta_{m_1m_2}\}/\cos\theta_{inc})}{\sum_{m_1}\sum_{m_2} |t_{m_1m_2}|^2 (\mathrm{Re}\{\cos\theta_{m_1m_2}\}/\cos\theta_{inc})} \implies \tag{5.276}$$

$$\sum_{m_1}^{m_2} DE_{m_1m_2}^{t,n} = |t(\vec{k}_{inc})|^2.$$

$$\eta^{r,n}_{m_1 m_2} = |r(\vec{k}_{inc})|^2 \frac{|r_{m_1 m_2}|^2}{\sum_{m_1}\sum_{m_2} |r_{m_1 m_2}|^2} \quad \Longrightarrow \quad \sum_{m_1}\sum_{m_2} \eta^{r,n}_{m_1 m_2} = |r(\vec{k}_{inc})|^2, \tag{5.277}$$

$$DE^{r,n}_{m_1 m_2} = |r(\vec{k}_{inc})|^2 \frac{|r_{m_1 m_2}|^2 (\mathrm{Re}\{\cos\theta_{m_1 m_2}\}/\cos\theta_{inc})}{\sum_{m_1}\sum_{m_2} |r_{m_1 m_2}|^2 (\mathrm{Re}\{\cos\theta_{m_1 m_2}\}/\cos\theta_{inc})} \quad \Longrightarrow \tag{5.278}$$

$$\sum_{m_1}^{m_2} DE^{r,n}_{m_1 m_2} = |r(\vec{k}_{inc})|^2.$$

It should be noted that the summations in the definitions of the normalized diffraction efficiencies are over the propagating orders only. Specifically, the summations are taken for all combinations of m_1 and m_2 that satisfy the condition given in Eq. (5.274). This ensures that only the diffracted orders capable of propagation are considered in the calculation of the diffraction efficiencies.

Example 5.11.4:

Assume a 2D grating with a rectangular pillar unit cell as shown in Fig. 5.7(a) and in Fig. 5.79. The grating parameters are $F_x = F_y = 0.50$, $\Lambda_x = \Lambda_y = 10\,\mu\text{m}$, $n_c = 1$, and $n_s = 1.5$. The pillar height h is varying in the range from 0 to $2\,\mu$m. The normally incident ($\theta_{inc} = \phi_{inc} = 0$) plane wave has a freespace wavelength of $\lambda_0 = 1\,\mu$m. Calculate the normalized diffraction efficiencies $DE^{t,n}_{00}$, $DE^{t,n}_{10}$, $DE^{t,n}_{01}$, and $DE^{r,n}_{00}$ as a function of the pillar height.

Solution: In this case, the number of propagating diffracted orders in the forward (transmitted) and backward (reflected) directions can be estimated using the equations for 2D gratings. For the forward direction, for $n = n_s = 1.5$, [using Eq. (5.274)] the number of propagating modes is $M_{prop} = 709$, and the estimate from the area of the ellipse is approximately 707. In the backward direction, for $n = n_c = 1$, the number of propagating diffracted orders is $M_{prop} = 317$, with the ellipse area estimate being approximately 314. Therefore, there is a relatively large number of diffracted orders in both directions, which will influence the diffraction efficiency calculations.

The diffraction efficiencies are calculated using the results from Appendix 5F (Sec. 5F.3.1) and the relevant equations for scalar diffraction efficiencies, Eqs. (5.276) and (5.278), both for transmission and reflection. The normalized diffraction efficiency $DE^{t,n}_{00}$ is shown as a function of the normalized pillar height h/λ_0 in Fig. 5.80(a). In addition to the scalar results, vectorial results obtained via the RCWA method are also presented for comparison. These vectorial results consider two orthogonal polarizations, with the electric field of the incident wave oriented either along the x- or the y-direction. For the $(0,0)$ diffracted orders under normal incidence, the vectorial results are identical for both polarizations due to the symmetry of the grating along the x- or the y-directions.

Additional figures display normalized diffraction efficiencies for other diffracted orders. Specifically, Fig. 5.80(b) shows the results for the forward $(1,0)$ order, Fig. 5.80(c) for the forward $(0,1)$ order, and Fig. 5.80(d) for the backward $(0,0)$ order. The vectorial results for the backward $(0,0)$ order coincide with those for the forward $(0,0)$ order for the same reasons. The vectorial results for the forward $(1,0)$ and $(0,1)$ orders, however, are not identical and depend on the polarization of the incident wave. Due to the symmetry of the grating, the results for the $(1,0)$ order with $E = E_x$ coincide with those for the $(0,1)$ order with $E = E_y$, and vice versa [the $(0,1)$ results for $E = E_x$ coincide with the results of $(1,0)$ for $E = E_y$]. The comparison between the scalar and vectorial results is generally good, with the agreement deteriorating as the pillar height increases.

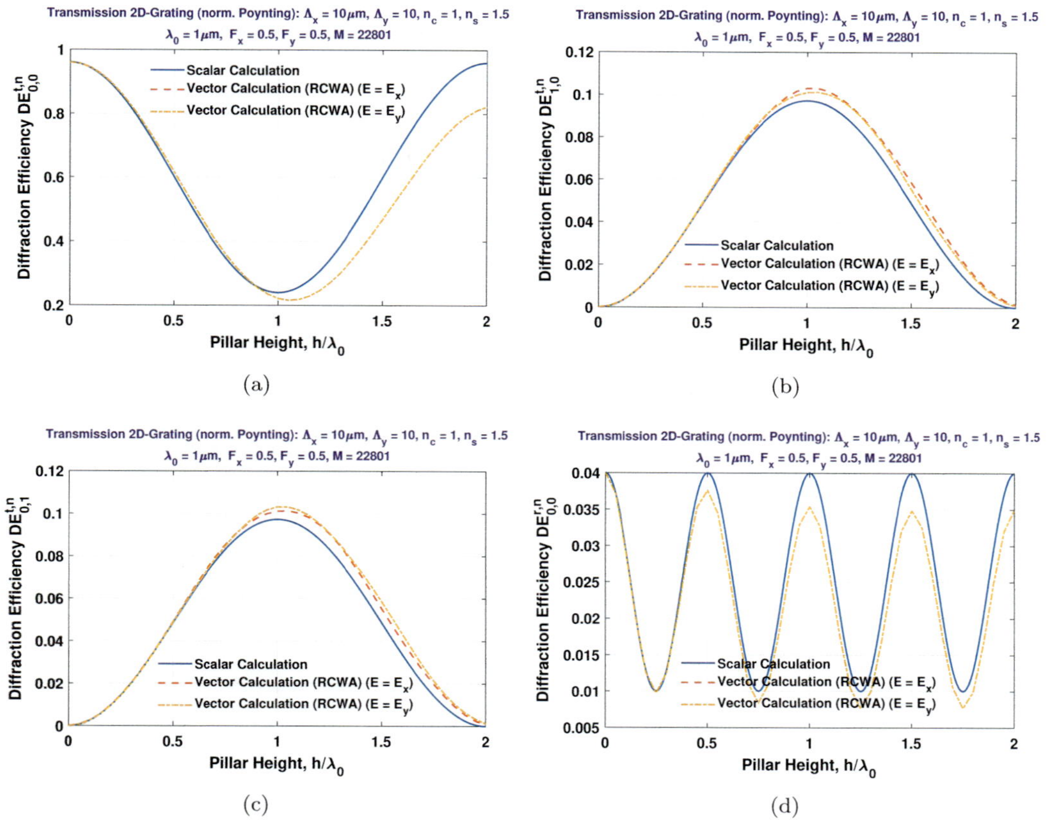

FIGURE 5.80: (a) The normalized diffraction efficiency $DE_{00}^{t,n}$ (for a 2D rectangular pillar grating) as function of the pillar height h/λ_0. The grating filling factors are $F_x = F_y = 0.50$, and the periods $\Lambda_x = \Lambda_y = 10\,\mu\text{m}$. The input refractive index is $n_c = 1$, and the substrate refractive index is $n_s = 1.5$. The freespace wavelength of the normally incident plane wave is $\lambda_0 = 1\,\mu\text{m}$. The vectorial results are calculated via the RCWA for electric field polarization $E = E_x$ (electric field along the x-direction) and $E = E_y$ (electric field along the y-direction) (b) Same as (a) but for $DE_{10}^{t,n}$. (c) Same as (a) but for $DE_{01}^{t,n}$. (d) Same as (a) but for $DE_{00}^{r,n}$ (reflected).

Example 5.11.5:

Assume a two-dimensional grating with an elliptical pillar unit cell, as shown in Fig. 5.7(b). The grating parameters are $\Lambda_x = \Lambda_y = 10\,\mu\text{m}$, with the semi-axes of the elliptical pillar given by $2a/\Lambda_x = 1/3$, $2b/\Lambda_x = 2/3$. he refractive indices of the cladding and substrate are $n_c = 1$, and $n_s = 1.5$, respectively. The pillar height h varies within the range from 0 to $2\,\mu\text{m}$. A normally incident plane wave ($\theta_{inc} = \phi_{inc} = 0$) with a freespace wavelength of $\lambda_0 = 1\,\mu\text{m}$ is considered. The task is to calculate the normalized diffraction efficiencies $DE_{00}^{t,n}$, $DE_{10}^{t,n}$, $DE_{11}^{t,n}$, and $DE_{00}^{r,n}$ as function of the pillar height.

Solution: The number of propagating modes is exactly the same as in Example 5.11.4 since the values of Λ_x, Λ_y, and λ_0 are identical.

Using the results from Appendix 5F (Sec. 5F.3.2) in conjunction with Eqs. (5.276) and (5.278), the desired diffraction efficiencies are calculated and presented in Fig. 5.81.

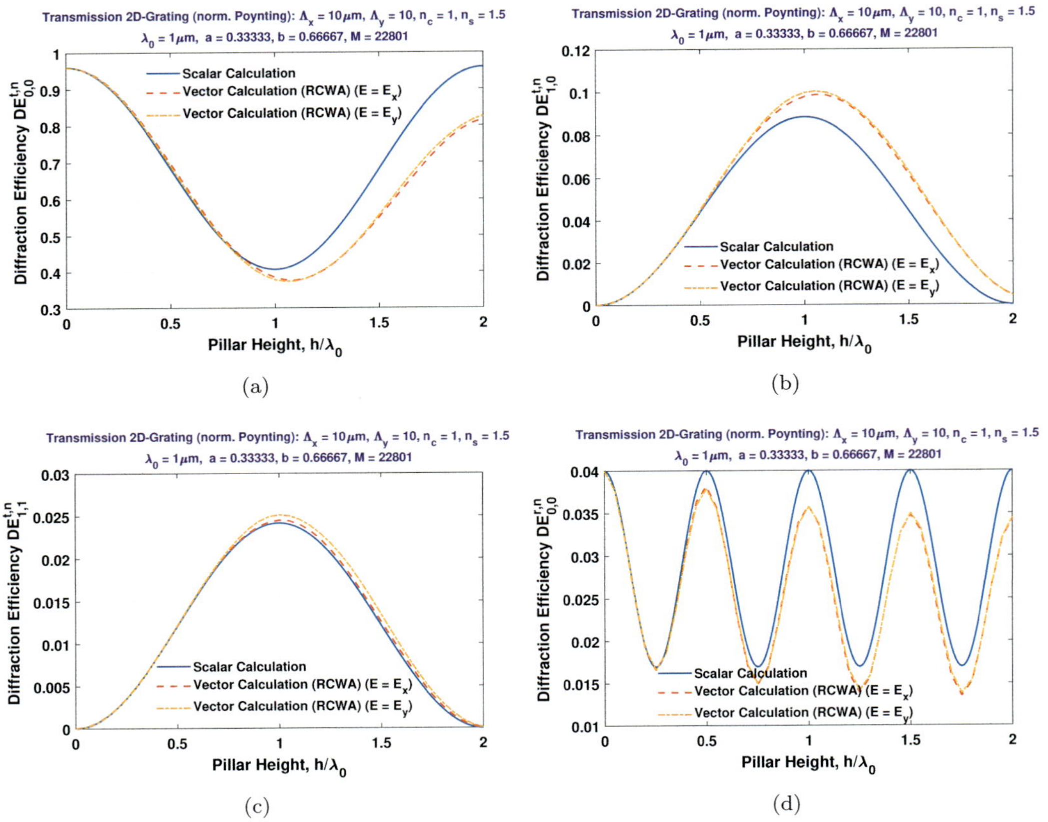

FIGURE 5.81: (a) The normalized diffraction efficiency $DE_{00}^{t,n}$ (for a 2D elliptical pillar grating) as function of the pillar height h/λ_0. the periods $\Lambda_x = \Lambda_y = 10\,\mu$m. The elliptical pillar semi-axes are $2a/\Lambda_x = 1/3$ and $2b/\Lambda_x = 2/3$. The input refractive index $n_c = 1$, and the substrate refractive index is $n_s = 1.5$. The freespace wavelength of the normally incident plane wave is $\lambda_0 = 1\,\mu$m. The vectorial results are calculated via the RCWA for electric field polarization $E = E_x$ (electric field along the x-direction) and $E = E_y$ (electric field along the y-direction) (b) Same as (a) but for $DE_{10}^{t,n}$. (c) Same as (a) but for $DE_{11}^{t,n}$. (d) Same as (a) but for $DE_{00}^{r,n}$ (reflected).

Specifically, the normalized efficiency $DE_{00}^{t,n}$ s shown in Fig. 5.81(a) as a function of the normalized pillar height h/λ_0. In addition to the scalar results, the vectorial (RCWA) results are also displayed for comparison. For the vectorial case, two orthogonal polarizations are considered as before: in the first case, the electric field of the incident wave is along the x-direction, and in the second case, it is along the y-direction.

For the $(0,0)$ diffracted orders and normal incidence, the vectorial results for the two different polarizations are not identical due to the asymmetry of the grating along the x- and y-directions. Similar normalized diffraction efficiencies for the other sought diffracted orders are shown in Figs. 5.81(b) [for the forward $(1,0)$ order], 5.81(c) [for the forward $(1,1)$ order], and 5.81(d) [for the backward $(0,0)$ order]. The vectorial results for the $(0,0)$ backward order are also slightly different due to the asymmetry of the grating.

The comparison between the vectorial and scalar results is generally good, with the agreement deteriorating as the pillar height increases, similar to the behavior observed for the rectangular pillar 2D grating.

5.12 Diffractive Lenses

Diffractive lenses are a particular subset of a broader category of devices known as diffractive optical elements (DOEs). DOEs exploit the phenomenon of diffraction to effectively control light. The literature contains numerous examples of DOEs, often referred to by various names such as binary optical elements, kinoforms, computer-generated holograms, holographic optical elements, and others. It should be noted, however, that not all of these terms refer to identical diffractive structures.

Diffractive lenses may also be referred to as Fresnel zone plates, as discussed in Section 5.10. Nevertheless, diffractive lenses have a distinct design methodology. Unlike refractive lenses, which rely on refraction, diffractive lenses are thin structures that focus light through diffraction.

The material presented in this section provides a basic introduction suitable for scalar diffraction analysis. Readers seeking a more comprehensive treatment of the topic are encouraged to consult specialized texts, such as those referenced in Refs. [311] and [319].

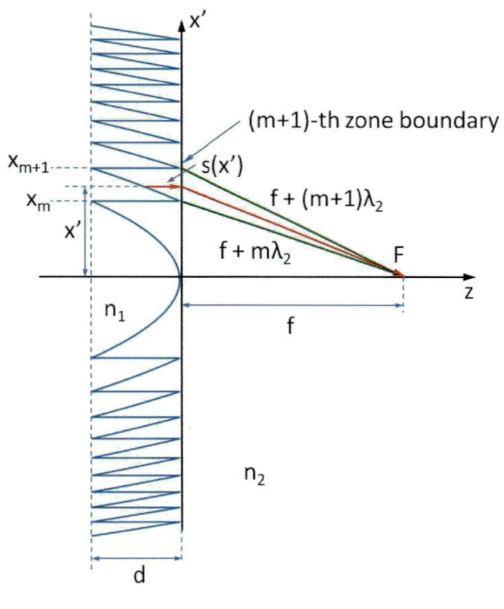

FIGURE 5.82: General geometry for illustrating the design process of a diffractive lens.

The design process employed here is based on the work presented by Buralli and Morris [320]. Referring to the geometry of a general diffractive lens illustrated in Fig. 5.82, the boundary of the m-th zone is defined such that:

$$\sqrt{f^2 + x_m^2} = m\lambda_2 + f \Longrightarrow x_m = \left(2mf\lambda_2 + m^2\lambda_2^2\right)^{1/2}. \tag{5.279}$$

Here, $\lambda_2 = \lambda_0/n_2$ represents the wavelength of light in the region to the right of the lens, $m = 0, 1, 2, \cdots$, with $x_0 = 0$, and f denotes the design focal length of the diffractive lens. It is important to note that, for a diffractive lens, the zone boundaries are defined by a phase shift of 2π, corresponding to a change of a λ_2. This contrasts with Fresnel zone plates, where the zone boundaries [as described by Eq. (5.221)] involve a phase shift of π, corresponding to a change of $\lambda_2/2$. The maximum allowed value of $\max\{s(x')\} = s_{max}$, which corresponds to a phase shift of 2π, can be determined easily by:

$$k_0 n_1 d - \{k_0 n_1 [d - s_{max}] + k_0 n_2 s_{max}\} = 2\pi \Longrightarrow s_{max} = \frac{\lambda_0}{n_1 - n_2}. \tag{5.280}$$

Let us now focus on the region where x' lies between two successive zone boundaries, $x_m \leq x' \leq x_{m+1}$, corresponding to the $(m+1)$-th zone, as shown in Fig. 5.82. For simplicity, the figure assumes $m = 1$. To ensure that light within the region $x_m \leq x' \leq x_{m+1}$ arrives at the focal point F with the same phase, the optical path for any x within this interval must be identical. This condition can be expressed as:

$$n_1 d + n_2\sqrt{f^2 + x_m^2} - \left\{n_1 [d - s_{np}(x')] + n_2 s_{np}(x') + n_2\sqrt{f^2 + x'^2}\right\} = 0 \Longrightarrow$$

$$s_{np}(x') = \frac{n_2}{n_1 - n_2}\left[\sqrt{f^2 + x'^2} - f - m\lambda_2\right], \qquad x_m \leq x' \leq x_{m+1}, \tag{5.281}$$

where Eq. (5.279) was utilized. The function $s_{np}(x')$ describes the profile of the diffractive lens and represents a non-parabolic shape, as defined in Eq. (5.281). In the literature, a simplification is often employed by approximating $(f^2 + x'^2)^{1/2} \simeq f + x'^2/2f$ for $f \gg x'$. Under this approximation, the resulting profile becomes parabolic, denoted as $s_p(x')$, and is expressed as:

$$s_p(x') = \frac{n_2}{n_1 - n_2}\left(\frac{x'^2}{2f} - m\lambda_2\right). \tag{5.282}$$

The design procedure outlined above is, of course, an approximation and is recognized in the literature [321] as neglecting the thickness of the lens. When the finite thickness of the lens is considered, the design process requires slight modifications. Specifically, to ensure that light within the region $x_m \leq x' \leq x_{m+1}$ arrives at the focal point F with the same phase, the optical path condition must be reformulated as:

$$n_1 d + n_2\sqrt{f^2 + x_m^2} - \left\{n_1 \left[d - s_{np}^t(x')\right] + n_2\sqrt{[f + s_{np}^t(x')]^2 + x'^2}\right\} = 0 \Longrightarrow$$

$$s_{np}^t(x') = \sqrt{B(x')} - A, \qquad x_m \leq x' \leq x_{m+1}, \tag{5.283}$$

$$A = \frac{n_2[n_1(f + m\lambda_2) - n_2 f]}{n_1^2 - n_2^2},$$

$$B(x') = \frac{n_2^2 x'^2 + n_2^2 f^2 - n_2^2 (f + m\lambda_2)^2}{n_1^2 - n_2^2} + A^2.$$

The function $s_{np}^t(x')$ defines the profile of the diffractive lens while accounting for the lens's thickness [320]. The preceding equation can also be expressed in the following more compact form [320]:

$$s_{np}^t(x') = \frac{c x'^2}{1 + \sqrt{1 + (n_1^2 - n_2^2)c^2 x'^2}} - \frac{m\lambda_2}{n_1 - n_2}, \qquad x_m \leq x' \leq x_{m+1}, \tag{5.284}$$

$$\text{where} \quad c = \frac{1}{(n_1 - n_2)f - m\lambda_2}.$$

The performance of diffractive lenses can be evaluated through scalar diffraction analysis, employing any of the approaches such as RS-1, RS-2, Kirchhoff, or Fresnel. In this section, the RS-1 formulation is used to analyze the performance of a designed diffractive lens.

For simplicity, the design process has been presented for cylindrical diffractive lenses. However, the design of three-dimensional diffractive lenses follows the same methodology, with the variable x' replaced by the radial coordinate $r_r = (x'^2 + y'^2)^{1/2}$, assuming azimuthal symmetry. In this context, the profile illustrated in Fig. 5.82 could be rotated about the z-axis to generate the 3D profile of the lens.

For the performance analysis of a cylindrical diffractive lens, the two-dimensional RS-1 formulation, as given by Eq. (5.23), becomes:

$$U_{RS1}(x_0, z_0) = \int_{-D/2}^{D/2} \left[\frac{k}{2j} H_1^{(2)}(kR_T) \frac{z_0}{R_T} \right] U_{inc}(x')dx',$$

$$R_T = [(x_0 - x')^2 + z_0^2]^{1/2}, \qquad U_{inc}(x') = U_0 e^{-j\Phi(x')},$$

$$\Phi(x') = k_0 s(x')(n_2 - n_1). \tag{5.285}$$

Here, D represents the diameter of the diffractive lens, which depends on the number of zones used. The phase $\Phi(x')$ approximately accounts for the profile of the lens. The profile $s(x')$ can take various forms: the parabolic profile $s_p(x')$, the non-parabolic profile $s_{np}(x')$, or the non-parabolic profile that includes thickness considerations, $s_{np}^t(x')$.

In many cases, the continuous profile within each zone can be approximated using M discrete levels. For instance, if s_{max} is the maximum height, the levels can be defined as is_{max}/M (with $i = 1, 2, \cdots M-1$). The corresponding transition points between these levels within the m-th zone, i.e., in the range $x_m \leq x' \leq x_{m+1}$, need to be determined by solving the equation $s(x_{m,i}) = iS_{max}/M$ (with $x_{m,0} = x_m$).

In general, the number of levels M can be arbitrary. However, in practice, M is often chosen as a power of two, $M = 2^\ell$, since it depends on the number of masks ℓ used during the fabrication process.

The performance analysis can then be easily conducted using the RS-1 formulation by appropriately adjusting the phase function $\Phi(x')$. For example, within the m-th zone, where $x_m \leq x' \leq x_{m+1}$, the phase function becomes $\Phi(x') = k_0[is_{max}/M](n_2 - n_1)$ for $x_{m,i-1} < x' < x_{m,i}$. At this stage, some example diffractive lens designs are presented.

Example 5.12.1:

The problem is to design a diffractive lens with a focal length of $f = 100\,\mu$m for an incident plane wave with a wavelength of $\lambda_0 = 1\,\mu$m, coming from the substrate region. The refractive index of the substrate and lens features is $n_1 = 1.5$, while the surrounding region is air, with $n_2 = 1$. The objective is to determine the profile of the diffractive lens for three different cases: non-parabolic, parabolic, and non-parabolic with thickness consideration. The width of the lens should correspond to 10 zones, and for each of these designs, the multilevel binary version should be shown with 4, 8, and 16 levels.

Solution: In this design process, a refractive lens is first designed based on Eq. (5.283) with $m = 0$. To account for 10 zones, the maximum radius of the lens is calculated using Eq. (5.279), yielding $x_{10} \simeq 45.83\,\mu$m. The resulting refractive lens profile, based on the given parameters, is shown in Fig. 5.83(a). It is noted that the thickness of this refractive lens is $20\,\mu$m.

In contrast, the maximum thickness of the diffractive lens is calculated to be $s_{max} = \lambda_0/(n_1 - n_2) = 2\,\mu$m. The profiles for the various diffractive lens designs are derived from the following equations: Eq. (5.281) for the non-parabolic design, Eq. (5.282) for the parabolic design, and Eq. (5.283) for the non-parabolic design with thickness consideration.

The resulting lens profiles for the three designs are displayed in Fig. 5.83(b), Fig. 5.83(c), and Fig. 5.83(d), corresponding to the non-parabolic, parabolic, and non-parabolic with thickness consideration designs, respectively. Additionally, the 4-level ($M = 4$) equivalent profiles are shown in the same figures.

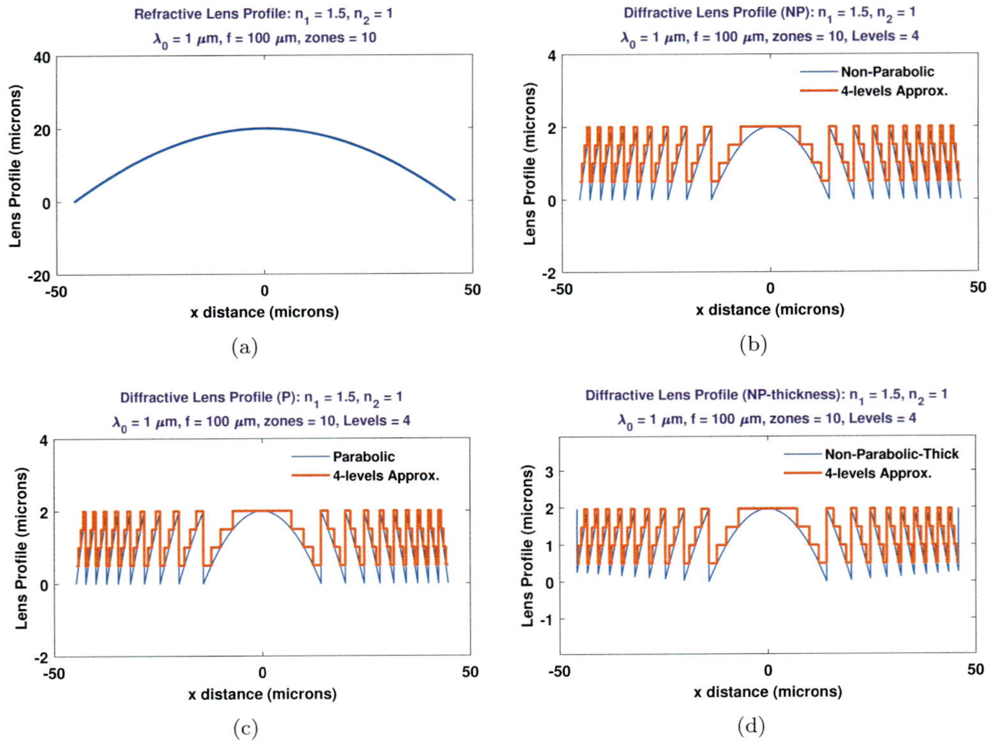

FIGURE 5.83: Profiles of a diffractive lens for $f = 100\,\mu\text{m}$, $\lambda_0 = 1\,\mu\text{m}$, $n_1 = 1.5$, and $n_2 = 1$ with 10 zones. (a) The equivalent refractive lens profile (for comparison). (b) The profile based on the non-parabolic design and its 4-level counterpart. (c) The profile based on the parabolic design and its 4-level counterpart. (d) The profile based on the non-parabolic with thickness consideration design and its 4-level counterpart.

It is observed that the non-parabolic design with thickness consideration leads to varying depths across the different zones, which may present fabrication challenges.

When the number of levels is increased to $M = 8$ or $M = 16$, the non-parabolic profiles along with the M-level equivalent discretized profiles are shown in Figs. 5.84(a) and 5.84(b), respectively.

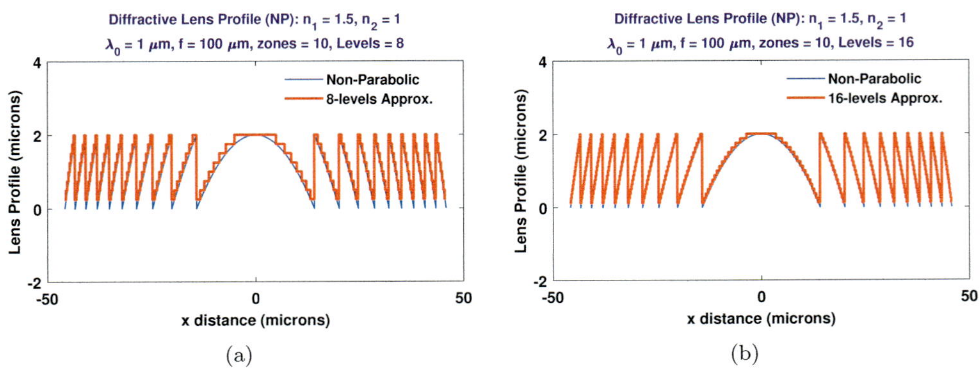

FIGURE 5.84: Profiles of a diffractive lens for $f = 100\,\mu\text{m}$, $\lambda_0 = 1\,\mu\text{m}$, $n_1 = 1.5$, and $n_2 = 1$ with 10 zones. (a) The profile based on the non-parabolic design and its 8-level counterpart. (b) The profile based on the non-parabolic design and its 16-level counterpart.

Example 5.12.2:

This example uses the same parameters for the diffractive lens as in Example 5.12.1. In this case, the normalized intensity $|U|^2/|U_0|^2$ is sought along the z-axis. The objective is to compare the various designs, as well as the discretized profiles.

Solution: Using the RS-1 formulation as described in Eq. (5.23) and with the phase function $\Phi(x')$ given by Eq. (5.285), the normalized intensity $|U|^2/|U_0|^2$ can be calculated for the various profiles. The normalized intensity for 10 zones for the non-parabolic continuous profile, as well as with the 2-, 4-, and 8-level discretized counterparts, is shown in Fig. 5.85(a). It can be observed that as the number of levels increases, the discretized profile lenses approach the performance of the continuous profile. The case of 16 levels is not shown since it is practically identical to the continuous profile. The effect of focusing near $f =$

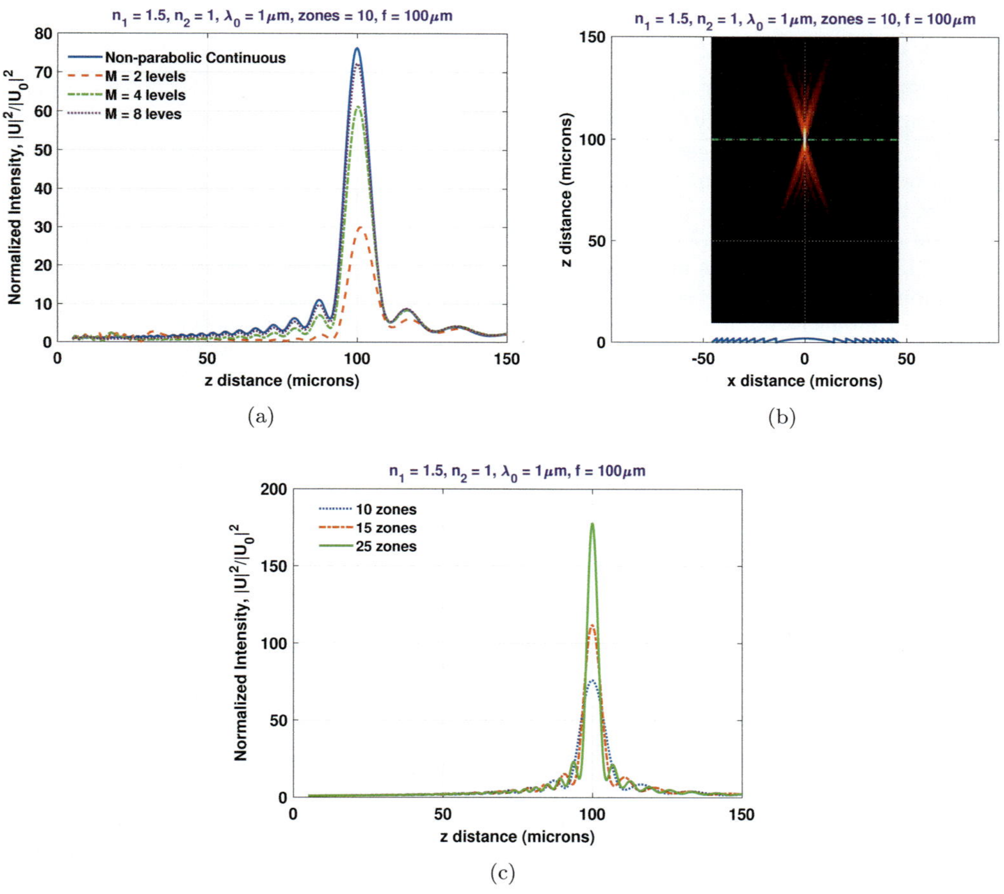

FIGURE 5.85: (a) The normalized intensity $|U|^2/|U_0|^2$ of a diffractive lens with focal length $f = 100\,\mu$m, with 10 zones, $n_1 = 1.5$ and $n_2 = 1$. The freespace wavelength is $\lambda_0 = 1\,\mu$m. The solid line represents the continuous non-parabolic design, while the dashed, dot-dashed, and dotted lines represent the discretized profiles with $M = 2$, $M = 4$, and $M = 8$ levels, respectively. (b) The image plot in the xz plane of the normalized intensity reveals the focusing effect. On the bottom of the plot, the non-parabolic continuous profile of the lens is shown in the same scale. (c) The normalized intensity $|U|^2/|U_0|^2$ of the diffractive lens (non-parabolic continuous profile) for 10, 15, and 25 zones, respectively.

100 μm is evident. Observe that the peak of the normalized intensity is about $|U|^2/|U_0|^2 = 76$.

In Fig. 5.85(b), the effect of focusing is shown in the xz plane. The lens profile is also included at the bottom of this figure in the same scale for comparison. Finally, the effect of increasing the number of zones is investigated. The maximum radius of the diffractive lens is 48.83 μm, 56.79 μm, and 75 μm, for 10, 15, and 25 zones, respectively. The resulting normalized intensities for non-parabolic profiles with 10, 15, and 25 zones are shown in Fig. 5.85(c). At the peak, the normalized intensities increase as the number of zones increases, since more light is available for focusing. Specifically, the maximum intensities at the focus are $|U|^2/|U_0|^2 = 76$, $|U|^2/|U_0|^2 = 112$, and $|U|^2/|U_0|^2 = 177$, for 10, 15, and 25 zones, respectively.

5.13 Problems

1 **Single Slit Far-Field Diffraction:** A monochromatic plane wave with a freespace wavelength of $\lambda_0 = 632.8\,\text{nm}$ is normally incident on a single-slit aperture, as illustrated in Fig. 5P.1. The slit has a width d, and a screen is placed at a distance $L = 1\,\text{m}$ along the optical axis to observe the diffraction pattern. Assume that $L \gg d$, so the Fraunhofer diffraction approximation is valid. Initially, the setup is in the air ($n = 1$). (a) If the tenth dark fringe of the diffraction pattern is observed at an angle $\theta = 6.2°$ (measured from the central axis), determine the width d of the slit. (b) If the entire setup is immersed in water ($n = 1.33$), at what angle θ will the tenth dark fringe appear? (c) For the case where the setup is immersed in water, find the positions of the normalized intensity minima on the screen. Derive the equation that specifies the positions of the normalized intensity maxima along the x-axis.

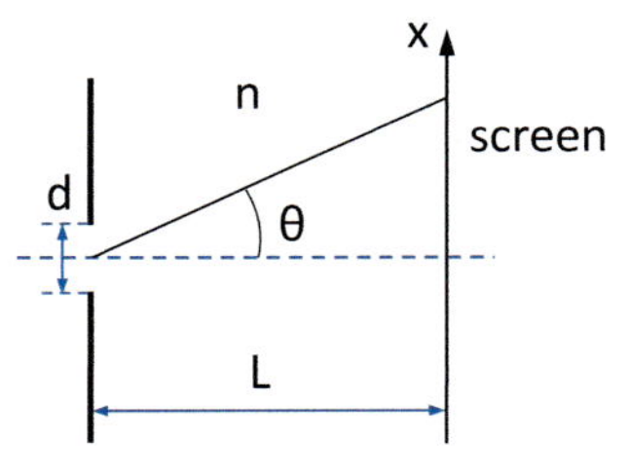

FIGURE 5P.1: The configuration for the single-slit diffraction.

2 **Young's Double Slit Experiment:** In Young's double-slit experiment with monochromatic light, the fringe pattern shown in Fig. 5P.2 is observed. The screen, which is parallel to the plane of the slits, is positioned at a distance of 3 meters from the slits. The two identical slits are separated by a distance of 0.235 mm. (a) Determine the freespace wavelength λ_0 of the monochromatic light used in the experiment. (b) Calculate the width of each slit.

3 **Three-Slit Far-Field Diffraction:** A plane wave of freespace wavelength λ_0 is normally incident on a three-slit aperture, as illustrated in Fig. 5P.3. The slits have a width s and are separated by a distance d. A screen is positioned at a distance L along the optical axis, where the diffraction pattern is observed. Assume that $L \gg d \gg s$, and the Fraunhofer diffraction approximation applies. Given $\lambda_0 = 1\,\mu\text{m}$, $s = 5\lambda_0$, $d = 50\lambda_0$, and $L = 50\,\text{cm}$: (a) Calculate the normalized intensity $|U|^2/|U_0|^2$, as a function of x on the screen. (b) Determine the points (or lines) on the screen where the normalized intensity becomes zero.

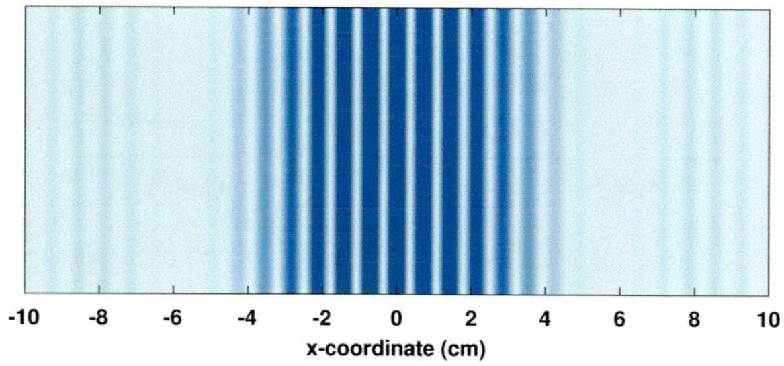

FIGURE 5P.2: The observed fringes on the screen in the double-slit Young's experiment.

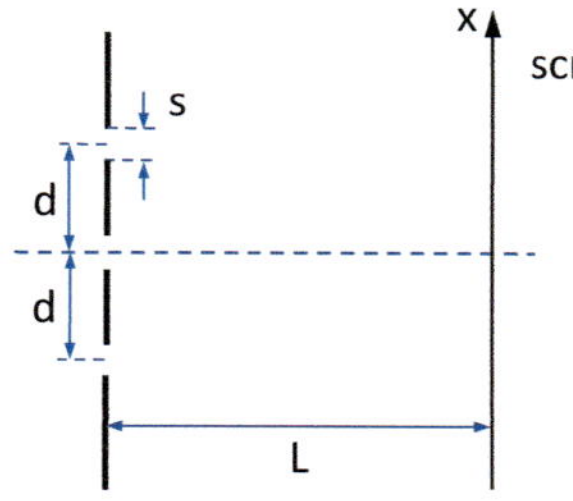

FIGURE 5P.3: The configuration for the three-slit diffraction.

4 **Far Field of a Rectangular Aperture:** A rectangular aperture of dimensions $a \times b$ ($-a/2 < x < a/2$, $-b/2 < y < b/2$) is positioned at the first focal plane of a spherical thin lens. If $a/\lambda_0 = 1$ and $b/\lambda_0 = 2$, calculate and plot the light normalized intensity, $I = |U|^2/|U_0|^2$ in the far-field (Fraunhofer regime), which appears at the second focal plane of the lens. Assume the aperture is illuminated by a plane wave propagating parallel to the optical axis of the lens.

5 **Far Field of a Circular Aperture:** A circular aperture of radius $a = 10\,\mu$m is illuminated by a plane wave of freespace wavelength $\lambda_0 = 1\,\mu$m, incident at an angle θ relative to the z-axis. Calculate the diffracted intensity at a distance $z = 500\lambda_0$ in the radial direction for $\theta = 0°, 15°, 30°, 45°$, and $60°$. *Hint:* The following integral might be helpful, $\int_0^{2\pi} \exp[\pm j(A\cos x + B\sin x)]dx = 2\pi J_0(\sqrt{A^2+B^2})$, where J_0 is the zeroth-order Bessel function of the first kind.

6 **Edge Diffraction:** A plane wave of freespace wavelength $\lambda_0 = 589$ nm is incident on an opaque edge, as illustrated in Fig. 5P.4. The wave diffracts, producing visible diffraction fringes in the Fresnel diffraction region, on a screen located at a distance $q = 120$ cm. (a) Calculate the normalized light intensity, $I = |E|^2/|E_u|^2$, at points $x_1 = 1$ mm and $x_2 = -2$ mm, where E_u is the unobstructed electric field amplitude. Assume a 2D geometry for this problem. (b) Now consider that the plane wave is replaced by a diverging cylindrical wave originating from point A on the y-axis, as shown in the figure. The cylindrical wave is monochromatic with a freespace wavelength $\lambda_0 = 589$ nm and travels a distance $p = 60$ cm before being diffracted by the edge. The diffraction fringes are again observed in the Fresnel region on the screen at a distance $q = 120$ cm. Determine the relative light intensity, $I = I/I_{cw}$, at points $x_1 = 1$ mm and $x_2 = -2$ mm, where I_{cw} represents the intensity of the unobstructed cylindrical wave.

7 **Thin-wire Diffraction:** Fresnel diffraction fringes are observed behind a thin wire of width $w = 0.37$ mm, as illustrated in Fig. 5P.5. The light source is located at a distance $s_0 = 2$ m from the wire, and an observation screen is placed at a distance $z_0 = 3$ m from the wire. The freespace wavelength of the source is $\lambda_0 = 630$ nm. (a) Determine the Fresnel diffraction normalized intensity I/I_u at point P of the screen, where I_u is the intensity in the absence of the wire. (b) Extend the calculation to any point on the screen in the same plane as shown in the figure, and generate a graphical representation of I/I_u.

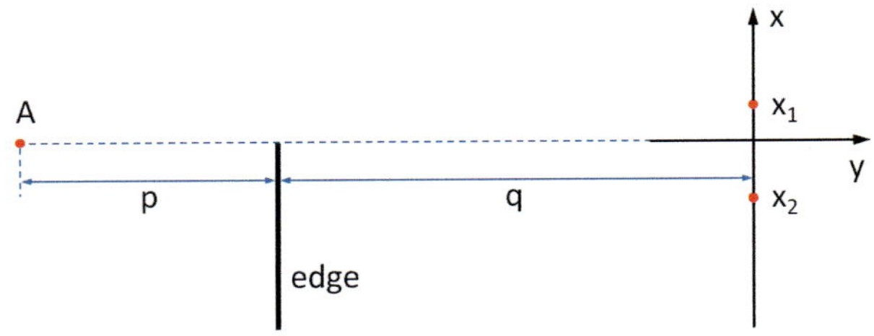

FIGURE 5P.4: The configuration for the edge diffraction Problem 6.

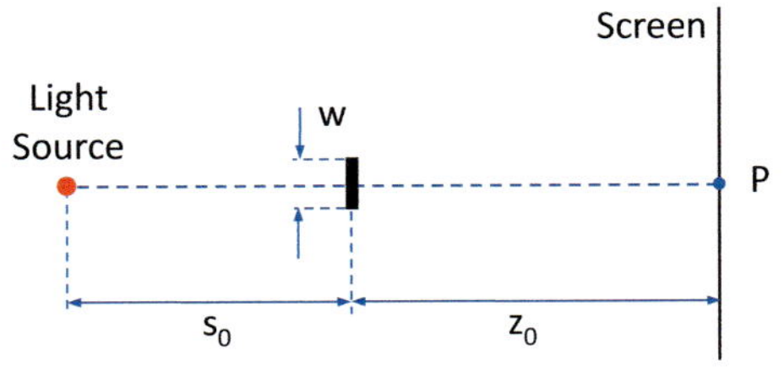

FIGURE 5P.5: The configuration for the diffraction by a thin wire.

8 **Single Slit Diffraction with Oblique Incidence:** A plane wave of freespace wavelength λ_0 is incident at an angle θ relative to the x'-axis on a single-slit aperture of width d. A screen is positioned at a distance L along the axis of the slit, where the diffraction pattern is observed. The refractive index of the material on the right side of the slit is n. (a) Using the Fraunhofer approximation, derive the expression for the normalized intensity, $|U|^2/|U_0|^2$, as a function of the distance x on the screen. Determine the positions of the intensity minima. (b) For the specific case where $\lambda_0 = 1\,\mu\text{m}$, $L = 1000\,\mu\text{m}$, $d = 10\,\mu\text{m}$, $\theta = 30°$, and $n = 1$, calculate and plot the normalized intensity as a function of x on the screen using the RS-1, Fresnel, and Fraunhofer formulations. Discuss and compare the resulting diffraction peaks from each formulation.

9 **Circular Aperture Far-Field Diffraction - Secondary Peaks:** Assume a circular aperture of radius a. The Fraunhofer diffraction pattern (far-field) is characterized by an Airy function, with the main peak occurring along the axis of the system. (a) Determine the positions of the next three secondary peaks in the diffraction pattern. (b) Calculate the normalized intensity of these secondary peaks in comparison to the main peak.

10 **Circular Aperture Diffraction with Oblique Incidence:** A plane wave of freespace wavelength λ_0 is incident at an angle θ relative to the x'-axis on a circular aperture of radius a. A screen is positioned at a distance L along the z-axis from the aperture, where the diffraction pattern is observed. The refractive index of the material to the right of the aperture is n. (a) Using the Fraunhofer approximation, derive the expression for the normalized intensity $|U|^2/|U_0|^2$ as a function of the distance x on the screen. Determine the positions of the intensity minima. (b) For the specific case where $\lambda_0 = 1\,\mu\text{m}$, $L = 1000\,\mu\text{m}$, $a = 10\,\mu\text{m}$, $\theta = 30°$, and $n = 1$, calculate and plot the normalized intensity as a function of x on the screen using the RS-1, Fresnel, and Fraunhofer formulations. Discuss and compare the resulting diffraction peaks from each formulation.

11 **Circular Obstacle Diffraction with Oblique Incidence:** A plane wave of freespace wavelength λ_0 is incident at an angle θ relative to the x'-axis on a circular obstacle with radius a. A screen is positioned at a distance L along the z-axis from the obstacle, where the diffraction pattern is observed. The refractive index of the material to the right of the obstacle is n. (a) Using the Fraunhofer approximation, derive the expression for the normalized intensity $|U|^2/|U_0|^2$ as a function of the distance x on the screen. (b) For the specific case where $\lambda_0 = 1\,\mu\text{m}$, $L = 1000\,\mu\text{m}$,

$a = 10\,\mu\text{m}$, $\theta = 30°$, and $n = 1$, calculate and plot the normalized intensity as a function of x on the screen using the RS-1, Fresnel, and Fraunhofer formulations.

12 Fresnel Zone Plate: Light of freespace wavelength of $\lambda_0 = 0.5145\,\mu\text{m}$ is normally incident on a circular aperture of radius $a = 1\,\text{cm}$. (a) How many Fresnel zones are contained within the aperture for a screen placed at a distance of 1 m? Calculate the number of zones using both the approximate and the exact expressions. (b) If the desired normalized intensity on the screen along the axis is $|U|^2/|U_0|^2 = 4$, what adjustment is needed to the aperture radius a? (c) For the same screen, find the number of Fresnel zones included when a diverging spherical wave of the same freespace wavelength is incident on the aperture. The spherical wave originates from a distance $s_0 = 2\,\text{cm}$ to the left of the aperture. (d) In the case of the diverging spherical wave, determine the number of Fresnel zones as a function of the spherical wave origin distance s_0.

13 Fresnel Zone Plate - Intensity Patterns: A circular aperture is illuminated by a plane wave of freespace wavelength $\lambda_0 = 0.5\,\mu\text{m}$. The aperture is to be considered as a Fresnel zone plate with 15 zones, and the focal distance is 200 mm. (a) Determine the aperture radius required for the circular aperture. (b) What is the corresponding normalized intensity, $|U|^2/|U_0|^2$ at the focal point? (c) If only the odd-numbered zones are transparent, what would the normalized intensity be at the focal point? (d) If only the even-numbered zones are transparent, what would the normalized intensity be at the focal point? (e) Calculate and plot the normalized intensity, $|U|^2/|U_0|^2$, along the z-axis to the right of the aperture as a function of z. Using the appropriate plots verify the results of parts (b), (c), and (d) in this case. (f) Estimate the spot size at the focal point in the focal plane. Also, plot the normalized intensity at the focal plane as a function of the radial distance to verify your prediction.

14 Grating Diffraction: An unslanted grating has a period $\Lambda = 5\,\mu\text{m}$, and an average refractive index of $n_0 = 2.0$. Its thickness is $250\,\mu\text{m}$. The grating is illuminated in air by a 2 mW He-Ne laser beam (of freespace wavelength $\lambda_0 = 0.6328\,\mu\text{m}$) at an angle of incidence of 20 deg. (a) Determine the Bragg angle in air and inside the grating region. (b) Determine the diffraction angles for the $-2, -1, 0, +1, +2$ forward diffracted orders. (c) Determine the largest positive forward diffracted order that could carry power into the forward direction.

15 Czerny-Turner Monochromator: A monochromator selects a single wavelength from a light beam by dispersing the light into its component colors. The dispersive element could be a prism or a grating. In a Czerny-Turner monochromator, a planar metallic reflection grating serves this role. A schematic of the Czerny-Turner monochromator is shown in Fig. 5P.6. To select a particular wavelength for diffraction through the exit slit, the grating is rotated about a vertical axis. The rotation angle of the grating, ρ, is the angle between the normal to the grating (indicated by a long dashed line) and the axis of the instrument (indicated by a short dashed line). The bisector of the mirror-grating-mirror angle, (2ϕ), also known as the Ebert angle, is used to define the rotation. In a particular commercial Czerny-Turner monochromator, the grating is specified as having 1200 grooves/mm, corresponding to 1200 periods/mm, or $1/\Lambda$. The monochromator has an Ebert angle of $2\phi = 9.2°$ and is filled with air. It operates in the first diffracted order. For a grating rotation of $\rho = 17.5158°$, calculate: (a) The wavelength (in nanometers) of the light passing through the exit slit of the monochromator. (b) The angular dispersion $d\lambda_0/d\rho$ (in nanometers per degree) of the light passing through the exit slit. (This problem was originally created by Prof. T. K. Gaylord, School of ECE, Georgia Tech, USA.)

16 Surface-Relief Grating Diffraction: A surface-relief rectangular-groove grating is fabricated on a glass substrate, as shown in Fig. 5P.7. The period of the grating is $\Lambda = 1\,\mu\text{m}$. An Ar-Ion laser beam of freespace wavelength $\lambda_0 = 514.5\,\text{nm}$ is incident from the air on the grating at an angle $\theta = 30°$. (a) Find the directions of all propagating diffracted orders in both the forward and backward directions. (a) Find the directions of all propagating diffracted orders in the forward and in the backward direction. (b) Determine the angular sensitivity of the first and second forward propagating orders for small changes in the incident wavelength around $\lambda_0 = 514.5\,\text{nm}$. Express the sensitivity in degrees per nanometer.

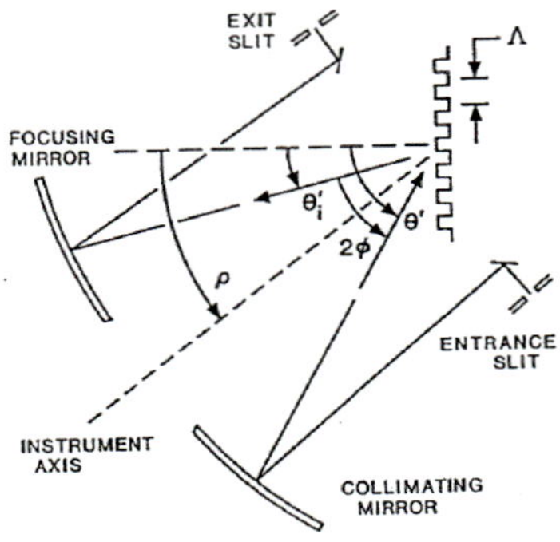

FIGURE 5P.6: The Czerny-Turner monochromator configuration.

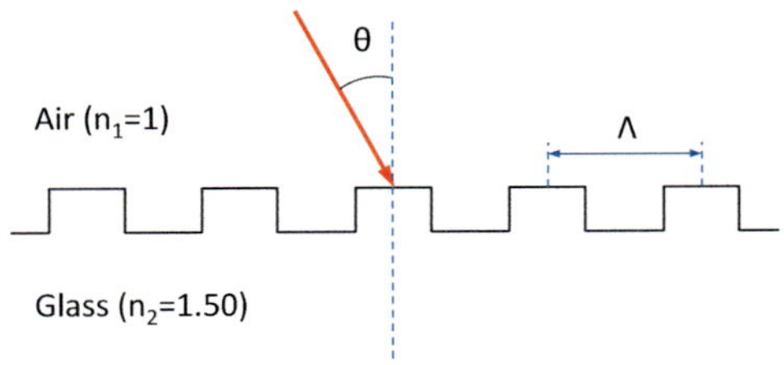

FIGURE 5P.7: The configuration of a surface-relief rectangular-groove grating of period Λ.

17 Blazed Grating Diffraction: Consider a blazed (sawtooth) grating, as illustrated in Fig. 5F.3. Determine the grating height h that maximizes the diffraction efficiency for the $m = -1$ forward diffracted order. Assume that a plane wave with a freespace wavelength of $\lambda_0 = 0.5\,\mu\text{m}$ is incident normally on the grating. The grating period is $\Lambda = 5\,\mu\text{m}$, and the refractive indices of the cover and substrate are $n_c = 1$ and $n_s = 1.5$, respectively. Additionally, calculate the angle of the tilted side of each grating period relative to the x'-axis in this configuration. Finally, determine the corresponding diffraction efficiency of the $m = 1$ forward diffracted order for the selected h.

18 Sinusoidal Grating Diffraction: Consider a sinusoidal grating, as illustrated in Fig. 5F.5. Determine the grating height h that maximizes the diffraction efficiency for the $m = \pm 1$ forward diffracted orders. Assume that a plane wave with a freespace wavelength of $\lambda_0 = 0.5\,\mu\text{m}$ is incident normally on the grating. The grating period is $\Lambda = 5\,\mu\text{m}$, and the refractive indices of the cover and substrate are $n_c = 1$ and $n_s = 1.5$, respectively. Additionally, calculate the grating height h required to maximize the diffraction efficiency for the $m = \pm 2$ forward diffracted orders.

19 Amplitude-Only Grating Diffraction: Consider an amplitude-only diffraction grating, as depicted in Fig. 5F.1, with a filling factor F and a period Λ. Assume that a plane wave with a freespace wavelength of λ_0 is incident normally on the grating. (a) Determine the filling factor F that maximizes the diffraction efficiency for the $m = \pm 1$ forward diffracted orders. (b) Now assume that the grating is no longer infinitely periodic but consists of a finite number N of periods. Derive its transmittance function $t(x')$. (c) Calculate the normalized far-field diffraction intensity on a screen located at a distance L from the grating plane, assuming that L satisfies the Fraunhofer approximation. (d) If a second light component with a freespace wavelength $\lambda_0' = \lambda_0 + \Delta\lambda_0$ is diffracted through the same grating, determine the smallest $\Delta\lambda_0$ that can be resolved on the screen using the Rayleigh criterion. (e) For the sodium doublet with $\lambda_0 = 588.995\,\text{nm}$, and $\lambda_0' = 589.592\,\text{nm}$, compute the minimum number of periods N required to resolve the doublet in the 3rd or 5th diffraction orders. (f) Determine the filling factor F of the grating that causes the 5th diffraction order to disappear.

20 Diffractive Lens Design: Design a diffractive micro-lens with a focal length of $f = 200\,\mu$m for an incident plane wave with a wavelength of $\lambda_0 = 0.85\,\mu$m, coming from the substrate region. The refractive index of the substrate and lens features is $n_1 = 1.45$, while the surrounding region is air, with $n_2 = 1$. The objective is to determine the profile of the diffractive lens for three different cases: non-parabolic, parabolic, and non-parabolic with thickness consideration. The diameter of the lens should $D = 50\,\mu$m. For each of these designs, the multilevel binary version should be shown with 4, 8, and 16 levels. In addition, calculate the normalized intensity $|U|^2/|U_0|^2$ along the z-axis for each of the designs to verify their performance.

Appendices

5A Green's Functions of Scalar Helmholtz's Equation

This appendix presents the Green's functions of the scalar Helmholtz equation within an isotropic, linear, dispersionless, and homogeneous medium in one, two, and three dimensions. The simplified sources and corresponding geometric configurations are illustrated in Figs. 5.1(a), 5.1(b), and 5.1(c) for one-, two-, and three-dimensional cases, respectively.

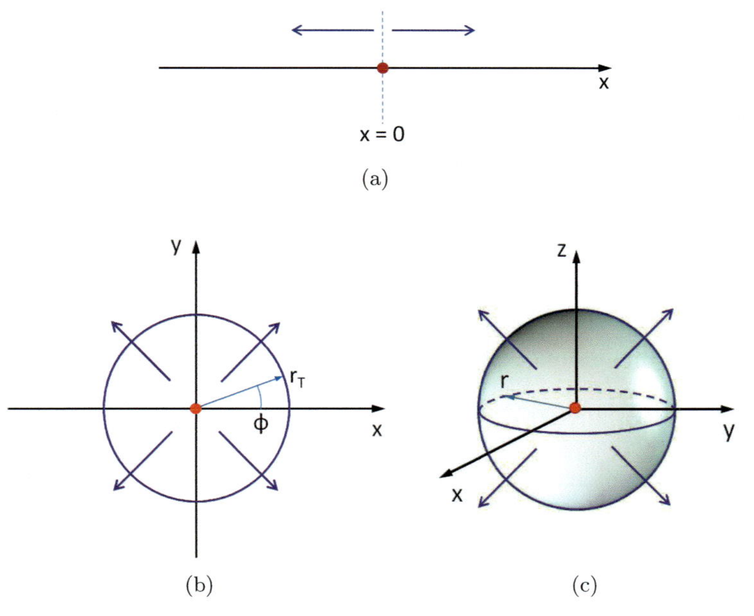

FIGURE 5A.1: (a) Geometry of a one-dimensional source located at the origin, emitting waves outward. (b) Geometry of a two-dimensional source positioned at the origin, radiating waves outward. (c) Geometry of a three-dimensional source centered at the origin, propagating waves outward.

5A.1 Green's Function in One-Dimension

The one-dimensional scalar Helmholtz equation is expressed as:

$$\frac{d^2U}{dx^2} + k^2U = 0, \tag{5A.1}$$

where $U(x)$ denotes the scalar field (representing either the electric or magnetic field) as a function of a single spatial variable, assumed to be x, and k represents the wavenumber. This configuration is schematically illustrated in Fig. 5.1(a). The corresponding Green's function satisfies the equation:

$$\frac{d^2G}{dx^2} + k^2G = -\delta(x). \tag{5A.2}$$

Here, $G(x)$ represents the Green's function for a source described by the $\delta(x)$ function, located at the origin ($x = 0$). The solutions to Eq. (5A.2) for $x > 0$ and $x < 0$, corresponding to waves propagating outward from the source at the origin, are:

$$G(x) = \begin{cases} C_1 \exp(-jkx), & \text{for } x > 0, \\ C_2 \exp(+jkx), & \text{for } x < 0. \end{cases} \tag{5A.3}$$

Assuming continuity at $x = 0$, such that $G(x \to 0^-) = G(x \to 0^+)$, it follows that $C_1 = C_2 = C$. At $x = 0$, the delta function on the right-hand side of Eq. (5A.2) must be accounted for. To determine the unknown constant C, Eq. (5A.2) is integrated over the interval $-\epsilon$ to $+\epsilon$, with $\epsilon \to 0$. Specifically,

$$\int_{-\epsilon}^{+\epsilon} \left(\frac{d^2G}{dx^2} + k^2G \right) dx = -\int_{-\epsilon}^{+\epsilon} \delta(x)dx = -1,$$

$$\text{where,} \quad \lim_{\epsilon \to 0} \left\{ \int_{-\epsilon}^{+\epsilon} k^2 G dx \right\} = \lim_{\epsilon \to 0} \left\{ k^2 C \, \frac{2 - 2e^{-jk\epsilon}}{jk} \right\} = 0,$$

$$\text{and,} \quad \lim_{\epsilon \to 0} \left\{ \int_{-\epsilon}^{+\epsilon} \frac{d}{dx} \left(\frac{dG}{dx} \right) dx \right\} = \lim_{\epsilon \to 0} \left\{ \left. \frac{dG}{dx} \right|_{+\epsilon} - \left. \frac{dG}{dx} \right|_{-\epsilon} \right\} = -2Cjk,$$

$$\text{resulting in} \quad -2Cjk = -1 \Longrightarrow C = \frac{1}{2jk}. \tag{5A.4}$$

In summary, the one-dimensional Green's function can be expressed as:

$$G(x) = G(x, 0) = -\frac{j}{2k} \exp(-jk|x|). \tag{5A.5}$$

When the source point is located at an arbitrary position, x' [i.e., when the delta function in Eq. (5A.2) is replaced by $\delta(x - x')$], the one-dimensional Green's function for the scalar Helmholtz equation becomes:

$$G(x, x') = -\frac{j}{2k} \exp(-jk|x - x'|). \tag{5A.6}$$

It is now straightforward, as shown in [88,264], to express the solution of the inhomogeneous one-dimensional Helmholtz equation as:

$$\frac{d^2U}{dx^2} + k^2U = -f(x) \Longrightarrow \tag{5A.7}$$

$$U(x) = \int_{-\infty}^{+\infty} G(x, x') f(x') dx'. \tag{5A.8}$$

5A.2 Green's Function in Two-Dimensions

The two-dimensional scalar Helmholtz equation (with azimuthal symmetry) in polar coordinates is expressed as:

$$\nabla^2 U + k^2 U = 0 \Longrightarrow \frac{1}{r_T}\frac{d}{dr_T}\left(r_T \frac{dU}{dr_T}\right) + k^2 U = 0, \tag{5A.9}$$

where $U(r_T)$ represents the scalar field (either the electric or magnetic field) as a function of the radial distance r_T in the polar coordinate system (r_T, ϕ). This configuration is schematically illustrated in Fig. 5.1(b). The corresponding Green's function satisfies the equation:

$$\nabla^2 G + k^2 G = -\delta(r_T, \phi) \Longrightarrow \frac{1}{r_T}\frac{d}{dr_T}\left(r_T \frac{dG}{dr_T}\right) + k^2 G = -\delta(r_T, \phi) = -\frac{1}{r_T}\delta(r_T)\delta(\phi). \tag{5A.10}$$

Here, $G(r_T)$ represents the Green's function for a source described by the two-dimensional $\delta(r_T, \phi)$ function located at the origin $r_T = 0$ $(\phi = 0)$. Equation (5A.10), for $r_T \neq 0$, can be written as:

$$r_T^2 \frac{d^2 G}{dr_T^2} + r_T \frac{dG}{dr_T} + k^2 r_T^2 G = 0.$$

This is a special case of the classic Bessel differential equation (with $\nu = 0$) and has a solution of the form [101]:

$$G(r_T) = C_1 H_0^{(1)}(kr_T) + C_2 H_0^{(2)}(kr_T), \tag{5A.11}$$

where $H_0^{(1)}(kr_T)$ and $H_0^{(2)}(kr_T)$ are the Hankel functions of the first and second kinds, respectively. The function $H_0^{(1)}(kr_T)$ represents a wave propagating inward toward the origin, while $H_0^{(2)}(kr_T)$ represents a wave traveling outward from the origin. Since the origin is assumed to be a source, the waves should be outgoing, and thus only the $H_0^{(2)}(kr_T)$ solution should be retained, forcing $C_1 = 0$.

At the origin, $r_T = 0$, special care is required. By integrating Eq. (5A.10) over a circular area of radius ϵ, centered at the origin of the coordinate system (and then taking the limit as $\epsilon \to 0$), the following equation is obtained:

$$\int_0^{\epsilon}\int_0^{2\pi}\left[\vec{\nabla}\cdot\left(\vec{\nabla}G\right) + k^2 G\right] r_T\, d\phi\, dr_T = -\int_0^{\epsilon}\int_0^{2\pi}\frac{1}{r_T}\delta(r_T)\delta(\phi) r_T\, d\phi\, dr_T = -1, \quad \text{where,}$$

$$\lim_{\epsilon\to 0}\left\{2\pi\int_0^{\epsilon} k^2 G r_T\, dr_T\right\} = \lim_{\epsilon\to 0}\left\{2\pi k^2 C_2 \int_0^{\epsilon} H_0^{(2)}(kr_T) r_T\, dr_T\right\} =$$

$$= 2\pi k^2 C_2 \lim_{\epsilon\to 0}\left\{\int_0^{\epsilon}\left[\frac{1}{\Gamma(1)} - j\left(\frac{2}{\pi}\ln\left(\frac{kr_T}{2}\right) + \gamma\right)\right] r_T\, dr_T\right\} =$$

$$= 2\pi k^2 C_2 \lim_{\epsilon\to 0}\left\{\frac{1}{\Gamma(1)}\frac{\epsilon^2}{2} - j\left[\frac{2}{k^2\pi}\left(-\frac{\epsilon^2}{2} + \frac{\epsilon^2}{2}\ln\left(\frac{\epsilon}{2}\right)\right) + \gamma\frac{\epsilon^2}{2}\right]\right\} = 0,$$

$$\text{and,}\quad \int_0^{\epsilon}\int_0^{2\pi}\vec{\nabla}\cdot\left(\vec{\nabla}G\right) r_T\, d\phi\, dr_T = \int_0^{\epsilon}\vec{\nabla}G\Big|_{r_T=\epsilon}\cdot\hat{\imath}_{r_T} r_T\, d\phi = \int_0^{\epsilon}\frac{dG}{dr_T}\Big|_{r_T=\epsilon}\epsilon\, d\phi =$$

$$= -2\pi C_2 k \lim_{\epsilon\to 0}\left\{H_1^{(2)}(k\epsilon)\epsilon\right\} =$$

$$= -2\pi C_2 k \lim_{\epsilon\to 0}\left\{j\frac{1}{\pi}\Gamma(1)\frac{2}{k\epsilon}\epsilon\right\} = -4jC_2,$$

$$\text{resulting in,}\quad C_2 = \frac{1}{4j} = -\frac{j}{4}. \tag{5A.12}$$

In the previous derivation, limiting expressions of the Hankel functions were used [293]. It is noted that $\Gamma(1) = 1$ (where Γ is the known gamma function [293]) and $\gamma = 0.57721566$ represents Euler's constant. Furthermore, the two-dimensional Green's theorem was applied, which states: $\iint_S \vec{\nabla}\cdot\vec{F}dS = \oint_C \vec{F}\cdot\hat{\imath}_n d\ell$, where S is a surface in the 2D space and C a contour enclosing the surface. The vector $\hat{\imath}_n$ represents the unit normal vector to the contour C. In summary, the two-dimensional Green's function can be expressed as:

$$G(r_{_T}) = G(r_{_T}, 0) = -\frac{j}{4}H_0^{(2)}(kr_{_T}). \tag{5A.13}$$

When the source line is located at an arbitrary position $(r'_{_T}, \phi')$ [i.e., when the delta function in Eq. (5A.10) is replaced by $\delta(r_{_T} - r'_{_T}, \phi - \phi')$], the two-dimensional Green's function for the Helmholtz scalar equation can be written as (for both polar and rectangular coordinate systems):

$$G(r_{_T}, r'_{_T}; \phi, \phi') = -\frac{j}{4}H_0^{(2)}(kR_T) = -\frac{j}{4}H_0^{(2)}\left(k\left[r_{_T}^2 + {r'_{_T}}^2 - 2r_{_T}r'_{_T}\cos(\phi - \phi')\right]^{1/2}\right), \tag{5A.14}$$

$$G(x, x'; y, y') = -\frac{j}{4}H_0^{(2)}(kR_T) = -\frac{j}{4}H_0^{(2)}\left(k\left[(x - x')^2 + (y - y')^2\right]^{1/2}\right). \tag{5A.15}$$

As was done in the one-dimensional case, it is now straightforward, as shown in [88, 264], to express the solution of the inhomogeneous scalar two-dimensional Helmholtz equation as:

$$\nabla^2 U + k^2 U = -f_c(r_{_T}, \phi) = -f_r(x, y) \Longrightarrow \tag{5A.16}$$

$$U(r_{_T}, \phi) = \int_0^{\infty}\int_0^{2\pi} G(r_{_T}, r'_{_T}; \phi, \phi')f_c(r'_{_T}, \phi')r'_{_T}d\phi' dr'_{_T}, \tag{5A.17}$$

$$U(x, y) = \int_{-\infty}^{+\infty}\int_{-\infty}^{+\infty} G(x, x'; y, y')f_r(x', y')dx'dy'. \tag{5A.18}$$

5A.3 Green's Function in Three-Dimensions

The three-dimensional scalar Helmholtz equation in spherical coordinates, with radial dependence only (spherical symmetry), is expressed as:

$$\nabla^2 U + k^2 U = 0 \Longrightarrow \frac{1}{r^2}\frac{d}{dr}\left(r^2\frac{dU}{dr}\right) + k^2 U = 0, \tag{5A.19}$$

where $U(r)$ represents the scalar field (electric or magnetic field) as a function of the radial distance r in the spherical coordinate system (r, θ, ϕ), as shown in Fig. 5.1(c). The corresponding Green's function is the solution to the equation:

$$\nabla^2 G + k^2 G = -\delta(\vec{r}) \Longrightarrow \frac{1}{r^2}\frac{d}{dr}\left(r^2\frac{dG}{dr}\right) + k^2 G = -\frac{1}{r^2\sin\theta}\delta(r)\delta(\theta)\delta(\phi). \tag{5A.20}$$

By substituting $G = u/r$ into Eq. (5A.20) for $r \neq 0$, the equation is transformed into the following form:

$$\frac{d^2u}{dr^2} + k^2 u = 0 \Longrightarrow u = C_1\exp(-jkr) + C_2\exp(+jkr).$$

By retaining only the outgoing wave (i.e., setting $C_2 = 0$), the solution for G for $r \neq 0$ becomes:

$$G(r) = C_1\frac{\exp(-jkr)}{r}. \tag{5A.21}$$

To determine the constant C_1, the condition at the origin must be considered. By integrating Eq. (5A.20) over the volume of a sphere with radius ϵ, centered at the origin of the coordinate system, the following conditions must be satisfied:

$$\int_0^{\epsilon}\int_0^{\pi}\int_0^{2\pi}\left[\vec{\nabla}\cdot\left(\vec{\nabla}G\right)+k^2G\right]r^2\sin\theta drd\theta d\phi$$

$$=-\int_0^{\epsilon}\int_0^{\pi}\int_0^{2\pi}\frac{1}{r^2\sin\theta}\delta(r)\delta(\theta)\delta(\phi)r^2\sin\theta drd\theta d\phi=-1, \qquad \text{where,}$$

$$\lim_{\epsilon\to 0}\left\{\int_0^{\epsilon}\int_0^{\pi}\int_0^{2\pi}k^2Gr^2\sin\theta drd\theta d\phi\right\}=4\pi C_1k^2\lim_{\epsilon\to 0}\left\{\left[\frac{e^{-jk\epsilon}(1+jk\epsilon)}{k^2}-\frac{1}{k^2}\right]\right\}=0,$$

$$\lim_{\epsilon\to 0}\left\{\int_0^{\epsilon}\int_0^{\pi}\int_0^{2\pi}\vec{\nabla}\cdot\left(\vec{\nabla}G\right)r^2\sin\theta drd\theta d\phi\right\}=\lim_{\epsilon\to 0}\left\{\int_0^{\pi}\int_0^{2\pi}\vec{\nabla}G\cdot\hat{\imath}_r\epsilon^2\sin\theta d\theta d\phi\right\}=$$

$$=\lim_{\epsilon\to 0}\left\{\int_0^{\pi}\int_0^{2\pi}\frac{dG}{dr}\epsilon^2\sin\theta d\theta d\phi\right\}=$$

$$=\lim_{\epsilon\to 0}\left\{-4\pi C_1e^{-jk\epsilon}\left[\frac{1}{\epsilon^2}+j\frac{k}{\epsilon}\right]\epsilon^2\right\}=$$

$$=-4\pi C_1,$$

$$\text{resulting in,}\quad C_1=\frac{1}{4\pi}.$$

In the previous derivation, the divergence theorem was applied. In summary, the three-dimensional Green's function can be expressed as:

$$G(r)=G(r,0)=\frac{\exp(-jkr)}{4\pi r}. \tag{5A.22}$$

When the point source is located at an arbitrary position (r',θ',ϕ') [i.e., when the delta function of Eq. (5A.20) is replaced by $\delta(r-r',\theta-\theta',\phi-\phi')$], the three-dimensional Green's function for the Helmholtz scalar equation can be written as (for both the spherical and rectangular coordinate systems):

$$G(r,r';\theta,\theta';\phi,\phi')=\frac{\exp(-jkR)}{4\pi R}, \tag{5A.23}$$

$$\text{where}\quad R=\left\{r^2+r'^2-2rr'\left[\cos\theta\cos\theta'+\sin\theta\sin\theta'\cos(\phi-\phi')\right]\right\}^{1/2},$$

$$G(x,x';y,y';z,z')=\frac{\exp(-jkR)}{4\pi R}, \tag{5A.24}$$

$$\text{where}\quad R=\left[(x-x')^2+(y-y')^2+(z-z')^2\right]^{1/2}.$$

As was done in the one-dimensional and two-dimensional cases, it is now straightforward, as shown in [88,264], to express the solution of the inhomogeneous scalar three-dimensional

Helmholtz equation as:

$$\nabla^2 U + k^2 U = -f_s(r,\theta,\phi) = -f_r(x,y,z) \Longrightarrow \tag{5A.25}$$

$$U(r,\theta,\phi) = \int_0^{\infty}\int_0^{\pi}\int_0^{2\pi} G(r,r';\theta,\theta';\phi,\phi') f_s(r',\theta',\phi') r'^2 \sin\theta' dr' d\theta' d\phi', \tag{5A.26}$$

$$U(x,y,z) = \int_{-\infty}^{+\infty}\int_{-\infty}^{+\infty}\int_{-\infty}^{+\infty} G(x,x';y,y';z,z') f_r(x',y',z') dx' dy' dz'. \tag{5A.27}$$

5B Sommerfeld's Radiation Conditions

In this Appendix, Sommerfeld's radiation condition is reviewed. In 1912, Arnold Sommerfeld introduced this condition to ensure the uniqueness of solutions to the Helmholtz equation for exterior boundary value problems involving radiation [322]. The Sommerfeld radiation condition ensures that the solutions consist solely of outgoing waves, excluding waves incoming from infinity [323].

This condition has been extensively applied in radiation problems and has inspired significant research in mathematical physics and uniqueness theorems [323–327]. In this Appendix, a simple (though not mathematically rigorous) proof of the Sommerfeld radiation condition is provided.

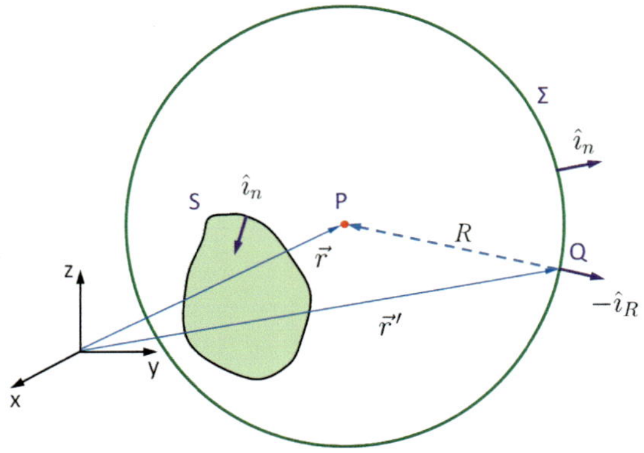

FIGURE 5B.1: The exterior of a closed surface S which encloses sources generating outgoing waves. The point P is an arbitrary point in the exterior of S. A sphere of radius R, centered at P, is considered with $R \to \infty$. The surface of the expanding sphere is denoted byΣ. The unit normal vector to the surface S is $\hat{\imath}_n$, directed outward from P (pointing to the interior of the surface S), while $\hat{\imath}_R$ represents the unit normal vector to the surface Σ.

Consider a closed surface S that encloses sources generating electromagnetic waves, which are incident on potential objects within the volume defined by S. Let P be an arbitrary point outside the closed surface S. With P as the center, imagine a sphere of radius R, sufficiently large (in the limit $R \to \infty$), such that it encompasses the volume defined by S, as illustrated in Fig. 5B.1. The surface of this sphere is denoted by Σ. Since the point P lies within the region defined by the exterior of S and the interior of Σ, the Helmholtz-Kirchhoff theorem can be applied [refer to Eq. (5.6)]. Consequently, the scalar field at P

can be expressed as:

$$U(\vec{r}) = U(P) = \iint_S \left(G\frac{\partial U}{\partial n} - U\frac{\partial G}{\partial n} \right) ds - \iint_\Sigma \left(G\frac{\partial U}{\partial R} - U\frac{\partial G}{\partial R} \right) d\sigma. \tag{5B.1}$$

Here, $G = \exp(-jk|\vec{r} - \vec{r}'|)/4\pi|\vec{r} - \vec{r}'|$ is the three-dimensional Green's function [see Eq. (5A.24)]. On the spherical surface Σ, the partial derivatives are expressed as $\partial U/\partial n = \vec{\nabla} U \cdot \hat{\imath}_n = -\vec{\nabla} U \cdot \hat{\imath}_R = -\partial U/\partial R$ and similarly $\partial G/\partial n = -\partial G/\partial R$. To ensure that waves arriving at P from infinity (incoming waves) are excluded, the contribution of the second integral in Eq. (5B.1) must vanish. In the limit $R \to \infty$, the following can be written:

$$\lim_{R\to\infty} \left\{ \iint_\Sigma \left(G\frac{\partial U}{\partial R} - U\frac{\partial G}{\partial R} \right) d\sigma \right\} = \lim_{R\to\infty} \left\{ \iint_\Sigma \left[\frac{e^{-jkR}}{4\pi R} \left[\frac{\partial U}{\partial R} - U\left(jk + \frac{1}{R} \right) \underbrace{(\hat{\imath}_R \cdot \hat{\imath}_n)}_{=\,-1} \right] \right] d\sigma \right\} =$$

$$= \lim_{R\to\infty} \left\{ \left[\frac{e^{-jkR}}{4\pi R^2} R \left(\frac{\partial U}{\partial R} + jkU \right) + \frac{e^{-jkR}}{4\pi R^2} U \right] 4\pi R^2 \right\} = 0 \Longrightarrow$$

$$\lim_{R\to\infty} \left\{ R \left(\frac{\partial U}{\partial R} + jkU \right) \right\} = 0, \qquad \text{and} \tag{5B.2}$$

$$\lim_{R\to\infty} \{U\} = 0. \tag{5B.3}$$

The last equation, $\lim_{R\to\infty} \{U\} = 0$, is often expressed by Sommerfeld as $\lim_{R\to\infty} \{RU\} < \infty$. This condition is commonly referred to as the *Sommerfeld finiteness condition*. Rigorous mathematical proofs of these conditions can be found in the literature [324–327].

For the purposes of this Appendix, it suffices to consider that the power associated with the scalar field must remain finite on the infinitely large sphere Σ. This requirement can be expressed as:

$$\lim_{R\to\infty} \left\{ \frac{1}{4Z} \iint_\Sigma |U|^2 d\sigma \right\} = \lim_{R\to\infty} \left\{ \frac{1}{4Z} |U|^2 4\pi R^2 \right\} < \infty \Longrightarrow$$

$$\lim_{R\to\infty} \{R|U|\} < \infty. \tag{5B.4}$$

The last equation represents Sommerfeld's finiteness condition. The Sommerfeld radiation conditions, as presented, apply to the three-dimensional case. For one- and two-dimensional scenarios, the conditions are expressed as:

$$\lim_{R\to\infty} \left\{ R^{(n-1)/2} \left(\frac{\partial U}{\partial R} + jkU \right) \right\} = 0, \qquad \text{and} \tag{5B.5}$$

$$\lim_{R\to\infty} \left\{ R^{(n-1)/2} |U| \right\} < \infty, \tag{5B.6}$$

where $n = 1, 2, 3$ denotes the space dimensionality. Similar radiation conditions have been formulated for vector electromagnetic fields and are known as the Silver-Müller radiation conditions [328].

5C Weyl's Plane Wave Representation of a Spherical Wave

The Weyl representation of a spherical wave through a plane wave spectrum is discussed in this Appendix [286]. The approach adopted follows the formulation by Mandel and

Wolf [329], with slight modifications to align with the $\exp[-j(kz - \omega t)]$ representation of a traveling wave propagating along the positive z-axis, as used herein.

Initially, consider a spherical wave emanating from the origin of a (x, y, z) coordinate system. The diverging spherical wave can be represented as:

$$U(\vec{r}) = U_0 \frac{\exp(-jk|\vec{r}|)}{|\vec{r}|} = U_0 \frac{\exp(-jkr)}{r}, \tag{5C.1}$$

where $r = |\vec{r}| = (x^2 + y^2 + z^2)^{1/2}$ is the distance of the field point (x, y, z) to the origin, and U_0 a constant that, without loss of generality, can be set to unity. At any point other than the origin, U satisfies the scalar Helmholtz equation. A representation of U in the form of a plane wave spectrum for the regions $z > 0$ and $z < 0$ is sought (in the z-direction, it is arbitrarily chosen to divide the space into two semi-infinite regions). Let us now consider that the desired representation for the $z > 0$ half space can be written as:

$$\frac{\exp(-jkr)}{r} = \frac{1}{(2\pi)^2} \int_{-\infty}^{+\infty} \int_{-\infty}^{+\infty} A(k_x, k_y) e^{-j(k_x x + k_y y + k_z z)} dk_x dk_y, \quad \text{where,} \tag{5C.2}$$

$$k_z = \begin{cases} \left(k^2 - k_x^2 - k_y^2\right)^{1/2}, & \text{when } k_x^2 + k_y^2 \le k^2, \\ -j\left(k_x^2 + k_y^2 - k^2\right)^{1/2}, & \text{when } k_x^2 + k_y^2 > k^2. \end{cases}$$

If the function $A(k_x, k_y)$ exists, then the previous equation will also hold in the limit as $z \to 0$. Therefore,

$$\frac{\exp\left[-jk(x^2 + y^2)\right]}{\sqrt{x^2 + y^2}} = \frac{1}{(2\pi)^2} \int_{-\infty}^{+\infty} \int_{-\infty}^{+\infty} A(k_x, k_y) e^{-j(k_x x + k_y y)} dk_x dk_y \Longrightarrow$$

$$A(k_x, k_y) = \int_{-\infty}^{+\infty} \int_{-\infty}^{+\infty} \frac{\exp\left(-jk\sqrt{x^2 + y^2}\right)}{\sqrt{x^2 + y^2}} e^{+j(k_x x + k_y y)} dx dy. \tag{5C.3}$$

In the last equation, the definition of the inverse 2D Fourier transform was applied. The integral in Eq. (5C.3) can be computed analytically [329]. First, we transition from the rectangular coordinate system to a polar coordinate system for the pairs (x, y) and (k_x, k_y). Consequently, the following substitutions can be made:

$$x = \rho \cos\phi, \quad y = \rho \sin\phi,$$

$$k_x = k_t \cos\psi, \quad k_y = k_t \sin\psi,$$

where both angles ϕ and ψ vary from 0 to 2π. Then, Eq. (5C.3) can be written in the form:

$$A(k_x, k_y) = \int_{\rho=0}^{+\infty} \int_{\phi=0}^{2\pi} \frac{e^{-jk\rho}}{\rho} e^{+jk_t\rho\cos(\phi-\psi)} \rho d\rho d\phi =$$

$$= \int_{\rho=0}^{+\infty} \frac{e^{-jk\rho}}{\rho} \rho d\rho \underbrace{\int_{\phi=0}^{2\pi} e^{+jk_t\rho\cos(\phi-\psi)} d\phi}_{=\,2\pi J_0(k_t\rho)} = 2\pi \int_{\rho=0}^{+\infty} \left(\frac{e^{-jk\rho}}{\rho}\right) J_0(k_t\rho) \rho d\rho. \tag{5C.4}$$

The function J_0 is the Bessel function of the first kind of zero order [293]. The last integral represents the Hankel transform of the function $e^{-jk\rho}/\rho$, and it is given by [329, 330]:

$$\int_{\rho=0}^{+\infty} \left(\frac{e^{-jk\rho}}{\rho}\right) J_0(k_t\rho)\rho d\rho = \begin{cases} \dfrac{-j}{\sqrt{k^2-k_t^2}} & \text{when } 0<k_t<k, \\ \dfrac{1}{\sqrt{k_t^2-k^2}}, & \text{when } 0<k<k_t. \end{cases} \tag{5C.5}$$

By inserting the expression for $A(k_x, k_y)$ from Eq. (5C.4) into Eq. (5C.2), the following plane wave spectrum of the spherical wave is obtained:

$$\frac{\exp(-jkr)}{r} = \frac{-j}{2\pi}\int_{-\infty}^{+\infty}\int_{-\infty}^{+\infty} \frac{e^{-j(k_x x+k_y y+k_z z)}}{k_z} dk_x dk_y, \qquad \text{for } z>0. \tag{5C.6}$$

The above result can be generalized to be valid for the negative z semi-infinite space ($z<0$) as well. The final representation of Weyl's plane wave expansion can then be written as:

$$\frac{\exp(-jkr)}{r} = \frac{-j}{2\pi}\int_{-\infty}^{+\infty}\int_{-\infty}^{+\infty} \frac{e^{-j(k_x x+k_y y+k_z |z|)}}{k_z} dk_x dk_y. \tag{5C.7}$$

In the case where the source of the spherical wave is at the point $\vec{r}\,'$, the last expression can be generalized as follows:

$$\frac{\exp(-jk|\vec{r}-\vec{r}\,'|)}{|\vec{r}-\vec{r}\,'|} = \frac{-j}{2\pi}\int_{-\infty}^{+\infty}\int_{-\infty}^{+\infty} \frac{e^{-j\left[k_x(x-x')+k_y(y-y')+k_z|z-z'|\right]}}{k_z} dk_x dk_y. \tag{5C.8}$$

5D Two-Dimensional Fourier Transform

5D.1 Fourier Transform Definition

The Fourier transform can be extended to multiple dimensions in a manner analogous to the one-dimensional case [119]. For scalar diffraction from two-dimensional apertures, two-dimensional Fourier transforms are particularly useful for describing far-field diffraction (Fraunhofer regime). Let a point in $\mathbb{R}^n$ be defined as the n-tuple $\mathbf{r} = (x_1, x_2, \cdots, x_n)$. The n-dimensional Fourier transform of a function $f(\mathbf{r}) = f(x_1, x_2, \cdots, x_n)$ and its inverse can be defined as follows:

$$F(\mathbf{k}) = \mathcal{F}\{f(\mathbf{r})\} = \int_{-\infty}^{+\infty}\int_{-\infty}^{+\infty}\cdots\int_{-\infty}^{+\infty} f(\mathbf{r})e^{+j\mathbf{k}\cdot\mathbf{r}} d\mathbf{r}, \tag{5D.1}$$

$$f(\mathbf{r}) = \mathcal{F}^{-1}\{F(\mathbf{k})\} = \frac{1}{(2\pi)^n}\int_{-\infty}^{+\infty}\int_{-\infty}^{+\infty}\cdots\int_{-\infty}^{+\infty} F(\mathbf{k})e^{-j\mathbf{k}\cdot\mathbf{r}} d\mathbf{k}, \tag{5D.2}$$

$$\text{with} \quad \mathbf{k}\cdot\mathbf{r} = x_1k_1 + x_2k_2 + \cdots + x_nk_n,$$

Here, $\mathbf{k} = (k_1, k_2, \cdots, k_n)$ represents the radial spatial frequency n-tuple and $\mathbf{k} \in \mathbb{R}^n$. The application of the n-dimensional Fourier transform and its inverse for two dimensions, using the common notation for two-dimensional space [$\mathbf{r} = (x, y)$, $\mathbf{k} = (k_x, k_y)$], are expressed as follows:

$$F(k_x, k_y) = \mathcal{F}\{f(x, y)\} = \int_{-\infty}^{+\infty}\int_{-\infty}^{+\infty} f(x, y)e^{+j(xk_x+yk_y)} dx dy, \tag{5D.3}$$

$$f(x, y) = \mathcal{F}^{-1}\{F(k_x, k_y)\} = \frac{1}{(2\pi)^2}\int_{-\infty}^{+\infty}\int_{-\infty}^{+\infty} F(k_x, k_y)e^{-j(xk_x+yk_y)} dk_x dk_y. \tag{5D.4}$$

The signs in the exponents of the Fourier transform and its inverse can also be interchanged. In this Appendix, the signs have been chosen such that the plane-wave spectrum corresponds to waves of the form $\exp(-j\mathbf{k}\cdot\mathbf{r})$, which propagate in the positive $\mathbf{r}$ direction. Linear transformations can be employed to extend the straightforward application of 2D Fourier transforms to more complex cases.

5D.2 Linear Transformation in Two Dimensions

The scaling property of the Fourier transform in one dimension can be extended to two or higher dimensions. If $\mathbf{x}$, $\mathbf{u} \in \mathbb{R}^2$, a linear transformation of the form $\mathbf{u} = [T]\mathbf{r}$ can be considered, where $[T]$ is a 2×2 matrix with elements independent of x, y). This transformation can be expressed as:

$$\mathbf{u} = [T]\mathbf{r} \Longrightarrow \begin{bmatrix} u_1 \\ u_2 \end{bmatrix} = \begin{bmatrix} t_{11} & t_{12} \\ t_{21} & t_{22} \end{bmatrix} \begin{bmatrix} x \\ y \end{bmatrix}. \tag{5D.5}$$

Under the transformation $[T]$, a vector $\mathbf{r}$ and a function $f(\mathbf{r})$ are mapped as follows:

$$\mathbf{r} \xrightarrow{[T]} \mathbf{u}, \tag{5D.6}$$

$$f(\mathbf{r}) \xrightarrow{[T]} f\left([T]^{-1}\mathbf{u}\right). \tag{5D.7}$$

The function $f\left([T]^{-1}\mathbf{u}\right)$ can be redefined as a function of the variable $\mathbf{r}$ (by renaming $\mathbf{u}$ to $\mathbf{r}$) in the form $f\left([T]^{-1}\mathbf{r}\right) = f([A]\mathbf{r})$, where $[A] = [T]^{-1}$. The Fourier transform $\mathcal{F}\{f([A]\mathbf{r})\}$ is then given by [331]:

$$\mathcal{F}\{f([A]\mathbf{r})\} = G(\mathbf{k}) = \frac{1}{|det([A])|} F\left([A]^{-T}\mathbf{k}\right). \tag{5D.8}$$

Here, $F = \mathcal{F}\{f\}$ represents the Fourier transform of of f, $det([A])$ denotes the determinant of the matrix $[A]$, and $[A]^{-T}$ represents the inverse transpose of $[A]$, defined as, $[A]^{-T} = \left([A]^{-1}\right)^T$.

5D.3 Rectangular and Slanted Parallelogram Shaped Apertures

In this section, the Fourier transform of a slanted parallelogram pillar function, which could represent the amplitude transmittance function of a slanted parallelogram aperture, is calculated using the scaling property shown in Eq. (5D.8). To begin, recall that the $\text{rect}(x/\text{a}, y/\text{b})$ function, representing a rectangular aperture amplitude function, can be defined as follows:

$$\text{rect}\left(\frac{x}{\text{a}}, \frac{y}{\text{b}}\right) = \begin{cases} 1, & |x| \le \text{a}/2, \text{and } |y| \le \text{b}/2 \\ 0, & |x| > \text{a}/2, \text{and } |y| > \text{b}/2 \end{cases} \tag{5D.9}$$

The Fourier transform of the $\text{rect}(x/\text{a}, y/\text{b})$ function can be easily calculated and is given by the expression:

$$\mathcal{F}\left\{\text{rect}\left(\frac{x}{\text{a}}, \frac{y}{\text{b}}\right)\right\} = \text{ab}\,\frac{\sin(k_x\text{a}/2)}{k_x\text{a}/2}\,\frac{\sin(k_y\text{b}/2)}{k_y\text{b}/2}, \tag{5D.10}$$

where $k_x = 2\pi f_x$, $k_y = 2\pi f_y$, and f_x, f_y are the spatial frequencies along the x- and y-directions, respectively.

A rectangle represented by the function $\mathrm{rect}(x/\mathrm{a}, y/\mathrm{b})$ is shown in Fig. 5D.1. The linear transformation $[T]$ that maps the rectangle into the slanted parallelogram $ABCD$ needs to be determined. The parallel sides of the slanted parallelogram are spaced by a and b, as shown in the figure. The acute angle $C\hat{D}A$ of the slanted parallelogram is $90° - \phi_1 - \phi_2$, where ϕ_1 and ϕ_2 represent the angles that vectors $\vec{v}_1$ and $\vec{v}_2$ make with respect to the x-axis (measured positive counterclockwise) and the y-axis (measured positive clockwise), respectively.

The transformation should be such that vectors $\vec{w}_1$ and $\vec{w}_2$ are mapped into vectors $\vec{v}_1$ and $\vec{v}_2$, respectively. From Fig. 5D.1, it is straightforward to define these vectors as follows (note that $s_1 = \mathrm{a}/[\cos(\phi_1 + \phi_2)]$ and $s_2 = \mathrm{b}/[\cos(\phi_1 + \phi_2)]$):

$$\vec{w}_1 = w_{1x}\,\hat{\imath}_x + w_{1y}\,\hat{\imath}_y = \frac{\mathrm{a}}{2}\,\hat{\imath}_x, \tag{5D.11}$$

$$\vec{w}_2 = w_{2x}\,\hat{\imath}_x + w_{2y}\,\hat{\imath}_y = \frac{\mathrm{b}}{2}\,\hat{\imath}_y, \tag{5D.12}$$

$$\vec{v}_1 = v_{1x}\,\hat{\imath}_x + v_{1y}\,\hat{\imath}_y = \frac{\mathrm{a}}{2\cos(\phi_1 + \phi_2)}\left[\cos\phi_1\,\hat{\imath}_x + \sin\phi_1\,\hat{\imath}_y\right], \tag{5D.13}$$

$$\vec{v}_2 = v_{2x}\,\hat{\imath}_x + v_{2y}\,\hat{\imath}_y = \frac{\mathrm{b}}{2\cos(\phi_1 + \phi_2)}\left[\cos\phi_2\,\hat{\imath}_x + \sin\phi_2\,\hat{\imath}_y\right]. \tag{5D.14}$$

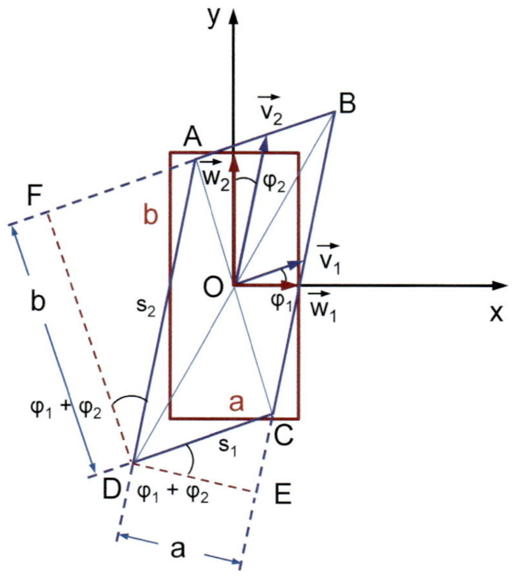

FIGURE 5D.1: Transformation of a rectangular of size a × b into a slanted parallelogram with parallel-side distances equal to a and b, respectively. The angles ϕ_1 and ϕ_2 measured from the x-axis (positive counterclockwise) and y-axis (positive clockwise), respectively, specify the shape of the slanted parallelogram. The vertices A, B, C, D correspond to the corners of the slanted parallelogram. The vectors $\vec{w}_1$ and $\vec{w}_2$ of the rectangle are transformed into $\vec{v}_1$ and $\vec{v}_2$ of the slanted parallelogram.

Since the vectors $\vec{w}_1$ and $\vec{w}_2$ are mapped into vectors $\vec{v}_1$ and $\vec{v}_2$, respectively, the following equations can be written:

$$\vec{v}_1 = [T]\,\vec{w}_1 \Longrightarrow \begin{bmatrix} v_{1x} \\ v_{1y} \end{bmatrix} = \begin{bmatrix} t_{11} & t_{12} \\ t_{21} & t_{22} \end{bmatrix} \begin{bmatrix} w_{1x} \\ w_{1y} \end{bmatrix}, \tag{5D.15}$$

$$\vec{v}_2 = [T]\,\vec{w}_2 \Longrightarrow \begin{bmatrix} v_{2x} \\ v_{2y} \end{bmatrix} = \begin{bmatrix} t_{11} & t_{12} \\ t_{21} & t_{22} \end{bmatrix} \begin{bmatrix} w_{2x} \\ w_{2y} \end{bmatrix}. \tag{5D.16}$$

In order to find the elements of the transformation matrix, the last equations can be written, after rearranging, in the form:

$$\begin{bmatrix} v_{1x} \\ v_{2x} \end{bmatrix} = \begin{bmatrix} w_{1x} & w_{1y} \\ w_{2x} & w_{2y} \end{bmatrix} \begin{bmatrix} t_{11} \\ t_{12} \end{bmatrix} \Longrightarrow \begin{bmatrix} t_{11} \\ t_{12} \end{bmatrix} = \begin{bmatrix} w_{1x} & w_{1y} \\ w_{2x} & w_{2y} \end{bmatrix}^{-1} \begin{bmatrix} v_{1x} \\ v_{2x} \end{bmatrix}, \tag{5D.17}$$

$$\begin{bmatrix} v_{1y} \\ v_{2y} \end{bmatrix} = \begin{bmatrix} w_{1x} & w_{1y} \\ w_{2x} & w_{2y} \end{bmatrix} \begin{bmatrix} t_{21} \\ t_{22} \end{bmatrix} \Longrightarrow \begin{bmatrix} t_{21} \\ t_{22} \end{bmatrix} = \begin{bmatrix} w_{1x} & w_{1y} \\ w_{2x} & w_{2y} \end{bmatrix}^{-1} \begin{bmatrix} v_{1y} \\ v_{2y} \end{bmatrix}. \tag{5D.18}$$

Applying Eqs. (5D.17)-(5D.18) using the parameters of the vectors defined in Eqs. (5D.11)-(5D.14), the transformation matrix $[T]$ and its inverse can be determined as follows:

$$[T] = \begin{bmatrix} t_{11} & t_{12} \\ t_{21} & t_{22} \end{bmatrix} = \begin{bmatrix} \dfrac{\cos\phi_1}{\cos(\phi_1+\phi_2)} & \dfrac{\sin\phi_2}{\cos(\phi_1+\phi_2)} \\[2ex] \dfrac{\sin\phi_1}{\cos(\phi_1+\phi_2)} & \dfrac{\cos\phi_2}{\cos(\phi_1+\phi_2)} \end{bmatrix}, \tag{5D.19}$$

$$[A] = [T]^{-1} = \begin{bmatrix} a_{11} & a_{12} \\ a_{21} & a_{22} \end{bmatrix} = \begin{bmatrix} \cos\phi_2 & -\sin\phi_2 \\ -\sin\phi_1 & \cos\phi_1 \end{bmatrix}. \tag{5D.20}$$

Of course, the above equations are valid if $\phi_1+\phi_2 \neq 90\,deg$. In the case of $\phi_1+\phi_2 = 90\,deg$, the slanted parallelogram collapses since the angle $C\hat{D}A$ becomes zero. According to Eqs. (5D.7) and (5D.8) the slanted parallelogram function $f(x, y)$ and its Fourier transform are given by the following equations:

$$\begin{aligned} f(x,y) &= \mathrm{rect}\left(\frac{a_{11}x + a_{12}y}{\mathrm{a}}, \frac{a_{21}x + a_{22}y}{\mathrm{b}}\right) = \\ &\mathrm{rect}\left(\frac{\cos\phi_2 x - \sin\phi_2 y}{\mathrm{a}}, \frac{-\sin\phi_1 x + \cos\phi_1 y}{\mathrm{b}}\right), \end{aligned} \tag{5D.21}$$

$$\begin{aligned} F(k_x, k_y) &= \frac{\mathrm{ab}}{|det\left([A]\right)|} \frac{\sin\left[(t_{11}k_x + t_{21}k_y)\,\mathrm{a}/2\right]}{(t_{11}k_x + t_{21}k_y)\,\mathrm{a}/2} \frac{\sin\left[(t_{12}k_x + t_{22}k_y)\,\mathrm{b}/2\right]}{(t_{12}k_x + t_{22}k_y)\,\mathrm{b}/2} = \\ &\frac{\mathrm{ab}}{|\cos(\phi_1+\phi_2)|} \frac{\sin\left[\dfrac{\cos\phi_1 k_x + \sin\phi_1 k_y}{\cos(\phi_1+\phi_2)}\dfrac{\mathrm{a}}{2}\right]}{\dfrac{\cos\phi_1 k_x + \sin\phi_1 k_y}{\cos(\phi_1+\phi_2)}\dfrac{\mathrm{a}}{2}} \frac{\sin\left[\dfrac{\sin\phi_2 k_x + \cos\phi_2 k_y}{\cos(\phi_1+\phi_2)}\dfrac{\mathrm{b}}{2}\right]}{\dfrac{\sin\phi_2 k_x + \cos\phi_2 k_y}{\cos(\phi_1+\phi_2)}\dfrac{\mathrm{b}}{2}}. \end{aligned} \tag{5D.22}$$

The last equation can also be written in the following compact form [332]:

$$F(\mathbf{k}) = F(k_x, k_y) = \frac{\mathrm{ab}}{|\sin\theta|}\,\mathrm{sinc}\left(\frac{\mathrm{a}}{2\pi}\frac{\hat{v}_1\cdot\mathbf{k}}{\sin\theta}\right)\mathrm{sinc}\left(\frac{\mathrm{b}}{2\pi}\frac{\hat{v}_2\cdot\mathbf{k}}{\sin\theta}\right), \tag{5D.23}$$

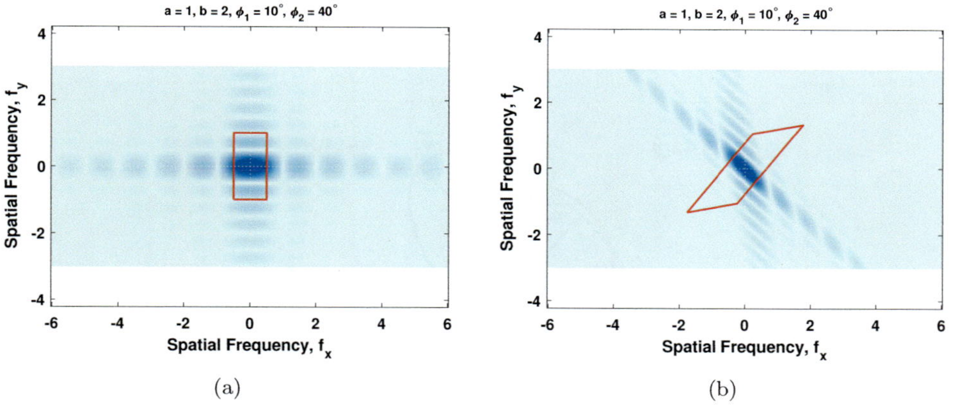

FIGURE 5D.2: (a) The Fourier transform, $|\mathcal{F}\{\mathrm{rect}(x/\mathrm{a}, y/\mathrm{b})\}|$, of a rectangular pillar function (of size a × b, with a = 1, and b = 2 in arbitrary units) in two dimensions. The rectangle is shown as an inset inside the Fourier transform for better understanding. (b) The Fourier transform of a slanted parallelogram, produced from the initial rectangle using a linear transformation with $\phi_1 = 10\,deg$ and $\phi_2 = 40\,deg$. The slanted parallelogram is shown as an inset in the Fourier transform figure. It can be observed that the main "arms" of the Fourier pattern are perpendicular to the parallelogram sides.

where $\mathrm{sinc}(x) = \sin(\pi x)/(\pi x)$, and $\hat{v}_1$, $\hat{v}_2$ are the unit vectors of $\vec{v}_1$, $\vec{v}_2$, respectively, while $\theta = 90° - (\phi_1 + \phi_2)$. An example function of a rectangle and its slanted parallelogram mapping through the linear transformation is shown in Fig. 5D.2.

5D.4 Circular and Elliptical Shaped Apertures

In this section, the Fourier transform of a rotated ellipsoidal pillar function (which could represent the amplitude transmittance function of an elliptical aperture) is determined based on the scaling property shown in Eq. (5D.8). Initially, it is reminded that the $circ(r/r_0)$ (circular aperture amplitude transmittance function) can be defined as follows:

$$circ\left(\frac{r}{r_0}\right) = \begin{cases} 1, & \dfrac{r}{r_0} \leq 1, \quad \text{or} \quad \dfrac{\sqrt{x^2+y^2}}{r_0} \leq 1, \\ 0, & \dfrac{r}{r_0} > 1, \quad \text{or} \quad \dfrac{\sqrt{x^2+y^2}}{r_0} > 1. \end{cases} \tag{5D.24}$$

The Fourier transform of the $circ(r/r_0)$ function can be easily calculated and is given by the expression:

$$\mathcal{F}\left\{circ\left(\frac{r}{r_0}\right)\right\} = F(k_x, k_y) = \frac{2\pi r_0}{\sqrt{k_x^2 + k_y^2}}\, J_1\left(r_0\sqrt{k_x^2 + k_y^2}\right), \tag{5D.25}$$

where $k_x = 2\pi f_x$, $k_y = 2\pi f_y$, and f_x, f_y are the spatial frequencies along the x- and y-directions, respectively, and J_1 is the Bessel function of the first kind of order one.

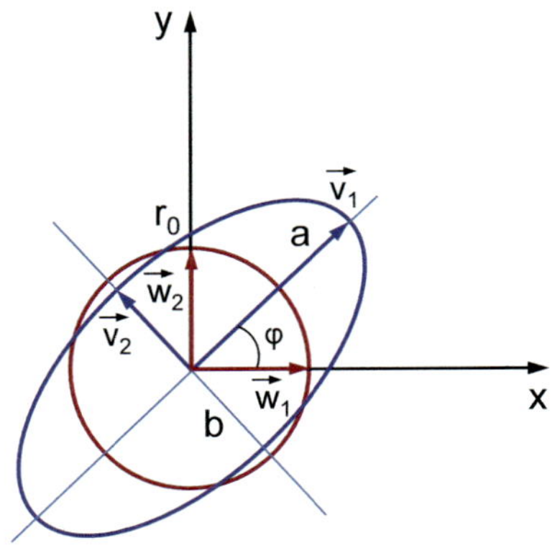

FIGURE 5D.3: Transformation of a circle of radius r_0 into an ellipse of semi-axes equal to a and b, respectively. The angle ϕ, as measured from the x-axis (positive counterclockwise) specifies the orientation of the ellipse. The vectors $\vec{w}_1$ and $\vec{w}_2$ of the circle are transformed into $\vec{v}_1$ and $\vec{v}_2$ of the ellipse.

A circle of radius r_0 represented by the function $circ(r/r_0)$ is shown in Fig. 5D.3. The linear transformation $[T]$ that transforms the circle into the rotated ellipse shown at an angle ϕ (see Fig. 5D.3) needs to be determined. The semi-axes of the ellipse are denoted by a and b, respectively, as shown in the figure. Following the same approach as before, the transformation should be such that vectors $\vec{w}_1$ and $\vec{w}_2$ are mapped into vectors $\vec{v}_1$ and $\vec{v}_2$, respectively. From Fig. 5D.3 it is straightforward to define these vectors as follows:

$$\vec{w}_1 = w_{1x}\ \hat{\imath}_x + w_{1y}\ \hat{\imath}_y = r_0\ \hat{\imath}_x, \tag{5D.26}$$

$$\vec{w}_2 = w_{2x}\ \hat{\imath}_x + w_{2y}\ \hat{\imath}_y = r_0\ \hat{\imath}_y, \tag{5D.27}$$

$$\vec{v}_1 = v_{1x}\ \hat{\imath}_x + v_{1y}\ \hat{\imath}_y = \mathrm{a}\left(\cos\phi\ \hat{\imath}_x + \sin\phi\ \hat{\imath}_y\right), \tag{5D.28}$$

$$\vec{v}_2 = v_{2x}\ \hat{\imath}_x + v_{2y}\ \hat{\imath}_y = \mathrm{b}\left(-\sin\phi\ \hat{\imath}_x + \cos\phi\ \hat{\imath}_y\right). \tag{5D.29}$$

Using the above equations in a manner analogous to the procedure described previously for the rectangle and the slanted parallelogram, the transformation matrix and its inverse can be determined as follows:

$$[T] = \begin{bmatrix} t_{11} & t_{12} \\ t_{21} & t_{22} \end{bmatrix} = \begin{bmatrix} \dfrac{\mathrm{a}}{r_0}\cos\phi & -\dfrac{\mathrm{b}}{r_0}\sin\phi \\ \dfrac{\mathrm{a}}{r_0}\sin\phi & \dfrac{\mathrm{b}}{r_0}\cos\phi \end{bmatrix}, \tag{5D.30}$$

$$[A] = [T]^{-1} = \begin{bmatrix} a_{11} & a_{12} \\ a_{21} & a_{22} \end{bmatrix} = \begin{bmatrix} \dfrac{r_0}{\mathrm{a}}\cos\phi & \dfrac{r_0}{\mathrm{a}}\sin\phi \\ -\dfrac{r_0}{\mathrm{b}}\sin\phi & \dfrac{r_0}{\mathrm{b}}\cos\phi \end{bmatrix}. \tag{5D.31}$$

According to Eqs. (5D.7) and (5D.8), the rotated ellipse function $f(x,y)$ and its Fourier transform are given by the following equations:

$$f(x,y) = circ\left[\frac{[(a_{11}x + a_{12}y)^2 + (a_{21}x + a_{22}y)^2]^{1/2}}{r_0}\right] =$$

$$circ\left[\left(\frac{\cos^2\phi}{\mathrm{a}^2} + \frac{\sin^2\phi}{\mathrm{b}^2}\right)x^2 + \left(\frac{\sin^2\phi}{\mathrm{a}^2} + \frac{\cos^2\phi}{\mathrm{b}^2}\right)y^2 + \right.$$

$$\left. 2\left(\frac{1}{\mathrm{a}^2} - \frac{1}{\mathrm{b}^2}\right)xy\cos\phi\sin\phi\right], \tag{5D.32}$$

$$F(k_x,k_y) = \frac{1}{|det([A])|}\frac{2\pi r_0 J_1\left(r_0\left[(t_{11}k_x + t_{21}k_y)^2 + (t_{12}k_x + t_{22}k_y)^2\right]^{1/2}\right)}{\left[(t_{11}k_x + t_{21}k_y)^2 + (t_{12}k_x + t_{22}k_y)^2\right]^{1/2}} =$$

$$= \frac{2\pi\mathrm{ab}J_1\left(\sqrt{g(k_x,k_y)}\right)}{\sqrt{g(k_x,k_y)}}, \tag{5D.33}$$

where, $g(k_x,k_y) = \left[\left(\mathrm{a}^2\cos^2\phi + \mathrm{b}^2\sin^2\phi\right)k_x^2 + \left(\mathrm{a}^2\sin^2\phi + \mathrm{b}^2\cos^2\phi\right)k_y^2 + \right.$

$$\left.\left(\mathrm{a}^2 - \mathrm{b}^2\right)k_x k_y\cos\phi\sin\phi\right].$$

The Fourier transform patterns of an example circle and the corresponding rotated ellipse are shown in Fig. 5D.4.

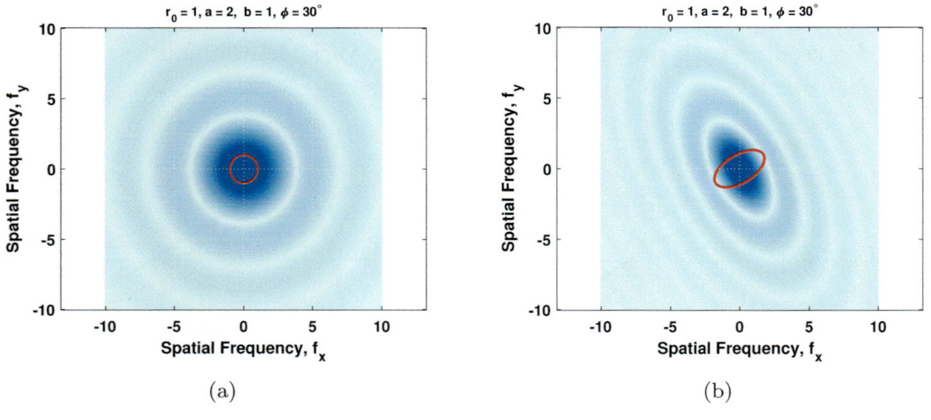

FIGURE 5D.4: (a) The Fourier transform, $|\mathcal{F}\{circ(r/r_0)\}|$, of a circular pillar function with a radius $r_0 = 1$, (in arbitrary units) in two dimensions. The circle is shown as an inset inside the Fourier transform for better understanding. (b) The Fourier transform of an elliptical pillar, which is produced from the initial circular pillar using a linear transformation with $\phi = 30\,deg$, $\mathrm{a} = 2$, and $\mathrm{b} = 1$. The elliptical cross-section of the elliptical pillar is shown as an inset in the Fourier transform figure. It can be observed that the main "arms" of the pattern are perpendicular to the main axes of the elliptical pillar.

5D.5 Triangular Shaped Apertures

In this section, the special case of the Fourier transform of a triangular pillar (which could represent the amplitude transmittance function of a triangular aperture) is developed. Assume a triangle in the xy-plane as shown in Fig. 5.5(a). The triangle is specified by the coordinates of its apexes (x_i, y_i) where $i = 1, 2, 3$. The apexes have been named in such a manner as to satisfy the condition $x_1 \leq x_2 \leq x_3$. Then the Fourier transform of the triangular pillar is given by appropriate application of Eq. (5D.3):

$$F(k_x, k_y) = \int_{x_1}^{x_2} \int_{y_{13}(x)}^{y_{12}(x)} e^{+j(k_x x + k_y y)} dx dy + \int_{x_2}^{x_3} \int_{y_{13}(x)}^{y_{23}(x)} e^{+j(k_x x + k_y y)} dx dy =$$

$$\int_{x_1}^{x_2} e^{jk_x x} \frac{e^{jk_y y_{12}(x)} - e^{jk_y y_{13}(x)}}{jk_y} dx + \int_{x_2}^{x_3} e^{jk_x x} \frac{e^{jk_y y_{23}(x)} - e^{jk_y y_{13}(x)}}{jk_y} dx =$$

$$= \frac{1}{jk_y} \int_{x_1}^{x_2} \left[e^{jk_x x + jk_y (a_{12} x + b_{12})} - e^{jk_x x + jk_y (a_{13} x + b_{13})} \right] dx +$$

$$\frac{1}{jk_y} \int_{x_2}^{x_3} \left[e^{jk_x x + jk_y (a_{23} x + b_{23})} - e^{jk_x x + jk_y (a_{13} x + b_{13})} \right] dx$$

where, $y_{12}(x) = a_{12} x + b_{12}$, $y_{13}(x) = a_{13} x + b_{13}$. $y_{23}(x) = a_{23} x + b_{23}$,

$$a_{12} = \frac{y_2 - y_1}{x_2 - x_1}, \qquad b_{12} = y_1 - a_{12} x_1,$$

$$a_{13} = \frac{y_3 - y_1}{x_3 - x_1}, \qquad b_{13} = y_1 - a_{13} x_1,$$

$$a_{23} = \frac{y_3 - y_2}{x_3 - x_2}, \qquad b_{23} = y_2 - a_{23} x_2.$$

The last equation is valid only when $k_y \neq 0$. The specific case where $k_y = 0$ will be analyzed separately. Proceeding with the integration over the x-variable, the Fourier transform can

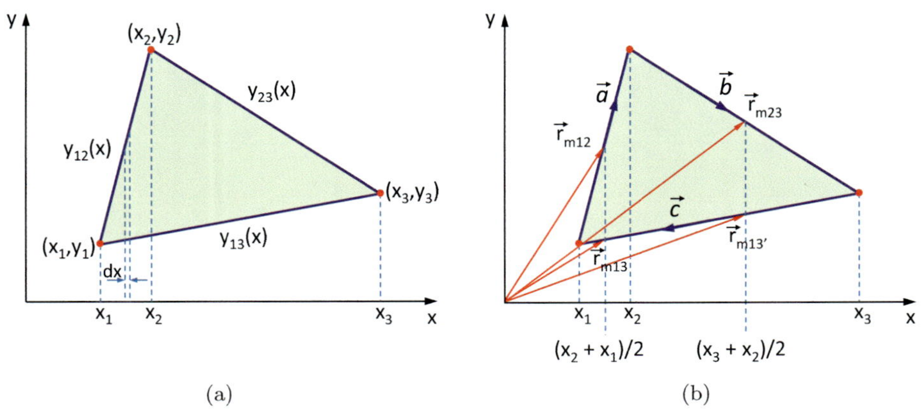

FIGURE 5D.5: (a) The integration process for calculating the 2D Fourier transform of a triangular aperture is described. The triangle is defined by the coordinates of its vertices (x_i, y_i) where $i = 1, 2, 3$. The vertices are ordered such that $x_1 \leq x_2 \leq x_3$. (b) The various vectors involved in the coordinate-free formulation of the 2D Fourier transform are defined and described for the transformation.

be readily derived as follows:

$$F(k_x, k_y \neq 0) = \frac{x_2 - x_1}{jk_y} \left\{ e^{j[k_x x_{m12} + k_y y_{12}(x_{m12})]} \mathrm{sinc}\left[(k_x + k_y a_{12}) \frac{x_2 - x_1}{2\pi} \right] \right.$$

$$\left. - e^{j[k_x x_{m12} + k_y y_{13}(x_{m12})]} \mathrm{sinc}\left[(k_x + k_y a_{13}) \frac{x_2 - x_1}{2\pi} \right] \right\} +$$

$$\frac{x_3 - x_2}{jk_y} \left\{ e^{j[k_x x_{m23} + k_y y_{23}(x_{m23})]} \mathrm{sinc}\left[(k_x + k_y a_{23}) \frac{x_3 - x_2}{2\pi} \right] \right.$$

$$\left. - e^{j[k_x x_{m23} + k_y y_{13}(x_{m23})]} \mathrm{sinc}\left[(k_x + k_y a_{13}) \frac{x_3 - x_2}{2\pi} \right] \right\}, \tag{5D.34}$$

$$\text{where, } x_{m12} = \frac{x_2 + x_1}{2}, \qquad x_{m23} = \frac{x_3 + x_2}{2}, \qquad \mathrm{sinc}(u) = \frac{\sin(\pi u)}{\pi u}.$$

The case where $k_y = 0$ can be straightforwardly calculated as follows:

$$F(k_x, 0) = \int_{x_1}^{x_2} \int_{y_{13}(x)}^{y_{12}(x)} e^{+jk_x x} dx dy + \int_{x_2}^{x_3} \int_{y_{13}(x)}^{y_{23}(x)} e^{+jk_x x} dx dy =$$

$$= \frac{e^{jk_x x_2}}{k_x^2} \left[(a_{12} - a_{13}) - jk_x(b_{12} - b_{13}) - jk_x(a_{12} - a_{13})x_2 \right] -$$

$$\frac{e^{jk_x x_1}}{k_x^2} \left[(a_{12} - a_{13}) - jk_x(b_{12} - b_{13}) - jk_x(a_{12} - a_{13})x_1 \right] +$$

$$\frac{e^{jk_x x_3}}{k_x^2} \left[(a_{23} - a_{13}) - jk_x(b_{23} - b_{13}) - jk_x(a_{23} - a_{13})x_3 \right] -$$

$$\frac{e^{jk_x x_2}}{k_x^2} \left[(a_{23} - a_{13}) - jk_x(b_{23} - b_{13}) - jk_x(a_{23} - a_{13})x_2 \right]. \tag{5D.35}$$

Finally, the special case where $k_x = k_y = 0$ is expressed as:

$$F(0, 0) = \frac{1}{2} \Big| x_1(y_2 - y_3) + x_2(y_3 - y_1) + x_3(y_1 - y_2) \Big|. \tag{5D.36}$$

The final expression, as anticipated, represents the area of the triangle. At this stage, let us attempt to rewrite Eq. (5D.34) in a more concise form. To achieve this, the following vectors are introduced:

$$\vec{r}_i = x_i \hat{\imath}_x + y_i \hat{\imath}_y, \qquad i = 1, 2, 3,$$

$$\vec{a} = \vec{r}_2 - \vec{r}_1 = (x_2 - x_1)\hat{\imath}_x + (y_2 - y_1)\hat{\imath}_y,$$

$$\vec{b} = \vec{r}_3 - \vec{r}_2 = (x_3 - x_2)\hat{\imath}_x + (y_3 - y_2)\hat{\imath}_y,$$

$$\vec{c} = \vec{r}_1 - \vec{r}_3 = (x_1 - x_3)\hat{\imath}_x + (y_1 - y_3)\hat{\imath}_y,$$

$$\vec{r}_{m12} = \frac{1}{2} [\vec{r}_2 + \vec{r}_1] = \frac{1}{2} [(x_2 + x_1)\hat{\imath}_x + (y_2 + y_1)\hat{\imath}_y] = \vec{r}_1 + \frac{\vec{a}}{2},$$

$$\vec{r}_{m23} = \frac{1}{2} [\vec{r}_3 + \vec{r}_2] = \frac{1}{2} [(x_3 + x_2)\hat{\imath}_x + (y_3 + y_2)\hat{\imath}_y] = \vec{r}_2 + \frac{\vec{b}}{2},$$

$$\vec{r}_{m13} = \vec{r}_1 - \frac{s}{2}\vec{c}, \qquad \vec{r}_{m13'} = \vec{r}_1 + \frac{1-s}{2}\vec{c},$$

$$\vec{k} = k_x\hat{\imath}_x + k_y\hat{\imath}_y, \qquad s = \frac{x_2 - x_1}{x_3 - x_1}.$$

Using the definitions provided above, an alternative form of Eq. (5D.34) can be expressed as follows:

$$\begin{aligned} F(k_x, k_y \neq 0) = \frac{(\vec{r}_3 - \vec{r}_1)\cdot\hat{\imath}_x}{j\vec{k}\cdot\hat{\imath}_y} \Bigg\{ & s\exp\left[j\vec{k}\cdot\left(\vec{r}_1 + \frac{\vec{a}}{2}\right)\right] \operatorname{sinc}\left(\frac{\vec{k}\cdot\vec{a}}{2\pi}\right) - \\ & s\exp\left[j\vec{k}\cdot\left(\vec{r}_1 + s\frac{\vec{a}+\vec{b}}{2}\right)\right] \operatorname{sinc}\left(s\frac{\vec{k}\cdot(\vec{a}+\vec{b})}{2\pi}\right) + \\ & (1-s)\exp\left[j\vec{k}\cdot\left(\vec{r}_2 + \frac{\vec{b}}{2}\right)\right] \operatorname{sinc}\left(\frac{\vec{k}\cdot\vec{b}}{2\pi}\right) - \\ & (1-s)\exp\left[j\vec{k}\cdot\left(\vec{r}_3 - (1-s)\frac{\vec{a}+\vec{b}}{2}\right)\right] \operatorname{sinc}\left[(1-s)\frac{\vec{k}\cdot(\vec{a}+\vec{b})}{2\pi}\right] \Bigg\}. \end{aligned} \tag{5D.37}$$

At this point, it is useful to provide an example calculation. Figures 5.6(a) and 5.6(c) illustrate the magnitude of the 2D Fourier transforms for an equilateral triangle with a side length of $a = 5\,m$, and an orthogonal isosceles triangle with equal side lengths of $a = 5\,m$. The units are arbitrary and can be adapted as needed. In Figures 5.6(b) and 5.6(d), the base-10 logarithm of the magnitude of the Fourier transform is displayed to enhance the visibility of finer details. The insets of Figures 5.6(a) and 5.6(c) illustrate the actual triangles, positioned with their centroids at the origin of the coordinate system. It is worth noting that shifting the triangles in the xy-plane results only in a phase shift in the Fourier transform (as dictated by Fourier transform properties), leaving the magnitude of the Fourier transform unchanged.

More compact forms of the Fourier transform for a triangular-shaped pillar function, with applications to Fraunhofer diffraction, have been presented by R. M. Sillitto and W. Sillitto [333, 334]. Extensions of the triangular-shaped pillar function to polygonal shape counterparts have also been published by Komrska [282,335,336] and Lee and Mittra [337]. However, many of these compact expressions exhibit singularities, where the expressions tend to infinity [such as the expression in Eq. (5D.37) when $k_y = 0$].

A notable exception is the formula provided by Komrska [282], which is based on the line integral expression derived from Fraunhofer's approximation (see Sec. 5.4.2.1). This approach is detailed in Sec. 5.9.

The previously presented expressions can be generalized for polygonal-shaped functions, as these can always be decomposed into a sum of triangular shapes. The procedure outlined in this appendix avoids the discontinuities encountered in the Fourier domain by the compact forms. Consequently, the method presented is general and applicable to a broad range of polygonal-shaped apertures.

5D.6 Summary of 1D and 2D Fourier Transforms

This section provides a summary of the definitions for 1D and 2D Fourier transforms, along with some commonly used Fourier transform pairs for the reader's convenience. Additionally,

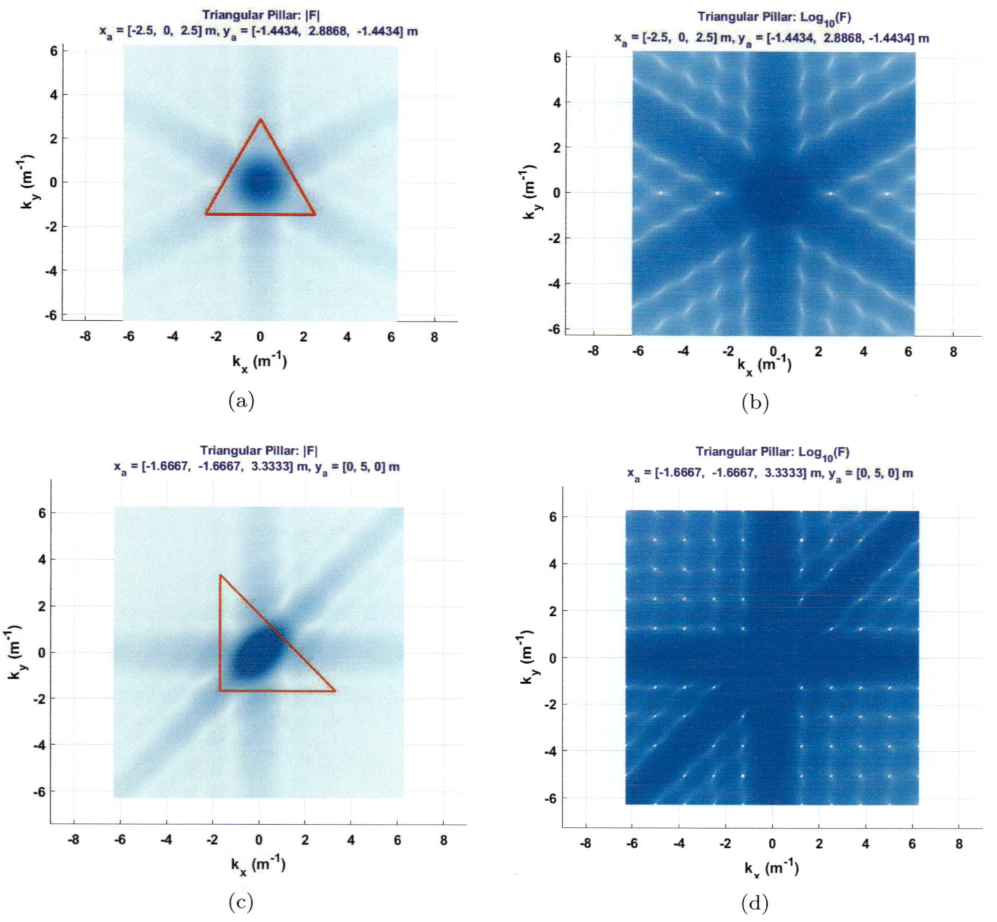

FIGURE 5D.6: In all plots, the square magnitude of the Fourier transform is displayed. (a) The 2D Fourier transform of an equilateral triangle with a side length of $a = 5m$. (b) The same as (a), but presented in a logarithmic scale (base 10). (c) The 2D Fourier transform of an orthogonal isosceles triangle with equal side lengths of $a = 5m$. (d)The same as (c), but presented in a logarithmic scale (base 10).

the section specifies the sign convention employed in the exponential terms of the Fourier transforms. It is noted that the 1D and 2D variable pairs are as follows: $(x \longleftrightarrow k_x = 2\pi u)$ for the 1D case, and $(x \longleftrightarrow k_x = 2\pi u, y \longleftrightarrow k_y = 2\pi v)$ for the 2D case.

It is helpful to define the Fourier transform, its inverse, and the convolution and correlation functions. These definitions are provided below.

$$f(x) = \int_{-\infty}^{+\infty} F(u)e^{-j2\pi ux}du \;=\; \frac{1}{2\pi}\int_{-\infty}^{+\infty} F(k_x)e^{-jk_x x}dk_x$$

$$F(u) = \int_{-\infty}^{+\infty} f(x)e^{+j2\pi ux}dx \;=\; \int_{-\infty}^{+\infty} f(x)e^{+jk_x x}dx$$

$$f(x) * g(x) = \int_{-\infty}^{+\infty} f(x')g(x-x')dx' \quad \text{(convolution)}$$

$$f(x) \star g(x) = \int_{-\infty}^{+\infty} f(x')g(x+x')dx' \quad \text{(correlation)}$$

Another important property is Parseval's theorem, which can be expressed as follows:

$$\int_{-\infty}^{+\infty} |f(x)|^2 dx = \int_{-\infty}^{+\infty} |F(u)|^2 du = \frac{1}{2\pi}\int_{-\infty}^{+\infty} |F(k_x)|^2 dk_x.$$

TABLE 5D.1: One-Dimensional Fourier Transform Properties

$f(x)$	$F(u)$	$F(k_x)$
$f(x)+g(x)$	$F(u)+G(u)$	$F(k_x)+G(k_x)$
$f(ax)$	$(1/\|a\|)F(u/a)$	$\frac{1}{\|a\|}F(k_x/a)$
$\frac{1}{\|a\|}f(x/a)$	$F(au)$	$F(ak_x)$
$f(x-a)$	$e^{+j2\pi au}F(u)$	$e^{+jk_x a}F(k_x)$
$e^{-j2\pi ax}f(x)$	$F(u-a)$	$F(k_x-2\pi a)$
$f\left(\frac{x-a}{b}\right)$	$\|b\|F(bu)e^{+j2\pi au}$	$\|b\|F(bk_x)e^{+jk_x a}$
$\frac{df}{dx}$	$-j2\pi uF(u)$	$-jk_xF(k_x)$
$j2\pi xf(x)$	$\frac{dF}{du}$	$2\pi\frac{dF}{dk_x}$
$f(x)*g(x)$	$F(u)G(u)$	$F(k_x)G(k_x)$
$f(x)g(x)$	$F(u)*G(u)$	$\frac{1}{2\pi}F(k_x)*G(k_x)$
$f(x)\star g(x)$	$F^*(u)G(u)$	$F^*(k_x)G(k_x)$
$f^*(x)$	$F^*(-u)$	$F^*(-k_x)$
$f^*(-x)$	$F^*(u)$	$F^*(k_x)$
$f(-x)$	$F(-u)$	$F(-k_x)$
$F(x)$	$f(-u)$	$f(-k_x)$
$F(-x)$	$f(u)$	$f(k_x)$

TABLE 5D.2: Common One-Dimensional Fourier Transform Pairs

$f(x)$	$F(u)$	$F(k_x)$
$\text{rect}(x) = \begin{cases} 1 & (\|x\| < 1/2) \\ 0 & (\|x\| > 1/2) \end{cases}$	$\text{sinc}(u) = \dfrac{\sin(\pi u)}{\pi u}$	$\text{sinc}(k_x) = \dfrac{\sin(k_x/2)}{k_x/2}$
$\text{tri}(x) = \begin{cases} 1 - \|x\| & (\|x\| < 1) \\ 0 & (\|x\| > 1) \end{cases}$	$\text{sinc}^2(u) = \dfrac{\sin^2(\pi u)}{(\pi u)^2}$	$\text{sinc}^2(k_x) = \dfrac{\sin^2(k_x/2)}{(k_x/2)^2}$
$e^{-ax^2} \quad (a > 0)$	$\sqrt{\dfrac{\pi}{a}}\, e^{-\pi^2 u^2/a}$	$\sqrt{\dfrac{\pi}{a}}\, e^{-k_x^2/4a}$
$e^{\pm jax^2} \quad (a > 0)$	$\sqrt{\dfrac{\pi}{a}}\, e^{\mp j\pi^2 u^2/a} e^{\pm j\pi/4}$	$\sqrt{\dfrac{\pi}{a}}\, e^{\mp jk_x^2/4a} e^{\pm j\pi/4}$
$\delta(x)$	1	1
$\delta(x - a)$	$e^{+j2\pi au}$	$e^{+jk_x a}$
$e^{-j2\pi ax}$	$\delta(u - a)$	$2\pi\delta(k_x - 2\pi a)$
$\cos(2\pi ax)$	$\dfrac{1}{2}[\delta(u + a) + \delta(u - a)]$	$\pi[\delta(k_x + 2\pi a) + \delta(k_x - 2\pi a)]$
$\sin(2\pi ax)$	$\dfrac{1}{2j}[\delta(u + a) - \delta(u - a)]$	$\dfrac{\pi}{j}[\delta(k_x + 2\pi a) - \delta(k_x - 2\pi a)]$
$\text{comb}(x) = \sum_{n=-\infty}^{\infty} \delta(x - n)$	$\text{comb}(u) = \sum_{n=-\infty}^{\infty} \delta(u - n)$	$\text{comb}(k_x) = 2\pi \sum_{n=-\infty}^{\infty} \delta(k_x - 2\pi n)$

TABLE 5D.3: Two-Dimensional Fourier Transform: Definition and Properties

$$f(x, y) = \int_{-\infty}^{+\infty} \int_{-\infty}^{+\infty} F(u, v) e^{-j2\pi(ux+vy)} du dv$$

$$F(u, v) = \int_{-\infty}^{+\infty} \int_{-\infty}^{+\infty} f(x, y) e^{+j2\pi(ux+vy)} dx dy$$

$$f(x, y) * *g(x, y) = = \int_{-\infty}^{+\infty} \int_{-\infty}^{+\infty} f(x', y') g(x - x', y - y') dx' dy'$$

$$f(x, y) \star\star g(x, y) = \int_{-\infty}^{+\infty} \int_{-\infty}^{+\infty} f(x', y') g(x + x', y + y') dx' dy'$$

$f(x, y)$	$F(u, v)$	$F(k_x, k_y)$
$f(x, y) + g(x, y)$	$F(u, v) + G(u, v)$	$F(k_x, k_y) + G(k_x, k_y)$
$f(ax, by)$	$\dfrac{1}{\|a\|\|b\|} F(u/a, v/b)$	$\dfrac{1}{\|a\|\|b\|} F(k_x/a, k_y/b)$
$f(x - a, y - b)$	$e^{+j2\pi(au+bv)} F(u, v)$	$e^{+j(k_x a + k_y b)} F(k_x, k_y)$
$e^{-j2\pi(ax+by)} f(x, y)$	$F(u - a, v - b)$	$F(k_x - 2\pi a, k_y - 2\pi b)$
$f(x, y) g(x, y)$	$F(u, v) * *G(u, v)$	$\dfrac{1}{(2\pi)^2} F(k_x, k_y) * *G(k_x, k_y)$
$f(x, y) * *g(x, y)$	$F(u, v) G(u, v)$	$F(k_x, k_y) G(k_x, k_y)$
$f(x, y) \star\star g(x, y)$	$F^*(u, v) G(u, v)$	$F^*(k_x, k_y) G(k_x, k_y)$
$f^*(x, y)$	$F^*(-u, -v)$	$F^*(-k_x, -k_y)$
$f^*(-x, -y)$	$F^*(u, v)$	$F^*(k_x, k_y)$
$f(-x, -y)$	$F(-u, -v)$	$F(-k_x, -k_y)$

TABLE 5D.4: Common Two-Dimensional Fourier Transform Pairs

$f(x,y)$	$F(u,v)$
$\mathrm{rect}(x,y)=\mathrm{rect}(x)\mathrm{rect}(y)$	$\mathrm{sinc}(u,v)=\mathrm{sinc}(u)\mathrm{sinc}(v)=\dfrac{\sin(\pi u)}{\pi u}\dfrac{\sin(\pi v)}{\pi v}$
$\mathrm{tri}(x,y)=\mathrm{tri}(x)\mathrm{tri}(y)$	$\mathrm{sinc}^2(u,v)=\mathrm{sinc}^2(u)\mathrm{sinc}^2(v)$
$e^{-a(x^2+y^2)}\quad (a>0)$	$\dfrac{\pi}{a}e^{-\pi^2(u^2+v^2)/a}$
$e^{\pm ja(x^2+y^2)}$	$\pm j\dfrac{\pi}{a}e^{\mp j\pi^2(u^2+v^2)/a}$
$\delta(x,y)$	1
$\delta(x-a,y-b)$	$e^{+j2\pi au}e^{+j2\pi bv}$
$e^{-j2\pi(ax+by)}$	$\delta(u-a,v-b)$
$\cos(2\pi(ax+by))$	$\dfrac{1}{2}[\delta(u+a,v+b)+\delta(u-a,v-b)]$
$\sin(2\pi(ax+by))$	$\dfrac{1}{2j}[\delta(u+a,v+b)-\delta(u-a,v-b)]$
$\mathrm{comb}(x,y)=\mathrm{comb}(x)\mathrm{comb}(y)$	$\mathrm{comb}(u,v)=\mathrm{comb}(u)\mathrm{comb}(v)$
$\mathrm{circ}(r/a)=\begin{cases}1 & (r=\sqrt{x^2+y^2}<a)\\ 0 & (r=\sqrt{x^2+y^2}>a)\end{cases}$	$a\dfrac{J_1(2\pi a\sqrt{u^2+v^2})}{\sqrt{u^2+v^2}}$
$f(x,y)$	**$F(k_x,k_y)$**
$\mathrm{rect}(x,y)=\mathrm{rect}(x)\mathrm{rect}(y)$	$\dfrac{\sin(k_x/2)}{k_x/2}\dfrac{\sin(k_y/2)}{k_y/2}$
$\mathrm{tri}(x,y)=\mathrm{tri}(x)\mathrm{tri}(y)$	$\dfrac{\sin^2(k_x/2)}{(k_x/2)^2}\dfrac{\sin^2(k_y/2)}{(k_y/2)^2}$
$e^{-a(x^2+y^2)}\quad (a>0)$	$\dfrac{\pi}{a}e^{-(k_x^2+k_y^2)/4a}$
$e^{\pm ja(x^2+y^2)}$	$\pm j\dfrac{\pi}{a}e^{\mp j(k_x^2+k_y^2)/4a}$
$\delta(x,y)$	1
$\delta(x-a,y-b)$	$e^{+jk_xa}e^{+jk_yb}$
$e^{-j2\pi(ax+by)}$	$(2\pi)^2\delta(k_x-2\pi a,k_y-2\pi b)$
$\cos(2\pi(ax+by))$	$\dfrac{(2\pi)^2}{2}[\delta(k_x+2\pi a,k_y+2\pi b)+\delta(k_x-2\pi a,k_y-2\pi b)]$
$\sin(2\pi(ax+by))$	$\dfrac{(2\pi)^2}{2j}[\delta(k_x+2\pi a,k_y+2\pi b)-\delta(k_x-2\pi a,k_y-2\pi b)]$
$\mathrm{comb}(x,y)=\mathrm{comb}(x)\mathrm{comb}(y)$	$\mathrm{comb}(k_x,k_y)=\mathrm{comb}(k_x)\mathrm{comb}(k_y)$
$\mathrm{circ}(r/a)=\begin{cases}1 & (r=\sqrt{x^2+y^2}<a)\\ 0 & (r=\sqrt{x^2+y^2}>a)\end{cases}$	$2\pi a\dfrac{J_1\left(a\sqrt{k_x^2+k_y^2}\right)}{\sqrt{k_x^2+k_y^2}}$

5E Fourier Series

5E.1 Fourier Series in One-Dimension

Assume a periodic function $f(x)$ defined on $\mathbb{R}$, with Λ being its period. Due to the periodicity, the following can be written:

$$f(x + m\Lambda) = f(x), \qquad \forall\, m \in \mathbb{Z} \tag{5E.1}$$

The Fourier series expansion of $f(x)$ can be written as [119, 120]:

$$f(x) = \sum_m c_m e^{-jmKx}, \quad \text{with,} \tag{5E.2}$$

$$c_m = \frac{1}{\Lambda}\int_{x_0}^{x_0+\Lambda} f(x) e^{+jmKx} dx, \tag{5E.3}$$

where $K = 2\pi/\Lambda$ and x_0 is an arbitrary point. It is noted that the signs in the exponential terms for Eqs. (5E.2) and (5E.3) could be reversed, as they are commonly presented in various textbooks. The same sign reversal is applied in the Fourier transform as well, with the goal of ensuring that the $\exp(-jkz)$ wave corresponds to the wave propagating in the positive z-direction. This sign convention is consistently followed in this Appendix.

A convenient function to describe periodic structures, such as 1D gratings, is the comb(x) function, also known as the sampling function. In one dimension, it can be defined as:

$$\text{comb}(x) = S_R(x) = \sum_m \delta(x - m\Lambda), \tag{5E.4}$$

where δ is the Dirac delta function in one dimension. Now, assume that the "unit cell" (the repeated part of the periodic function) of the periodic function $f(x)$ is denoted as $h(x)$, where $x \in (-\Lambda/2, +\Lambda/2)$. Then, the periodic function $f(x)$ can be represented as:

$$f(x) = h(x) * \text{comb}(x) = \sum_m h(x - m\Lambda), \tag{5E.5}$$

where "*" denotes the one-dimensional convolution. Since the function comb(x) is also periodic, it can be represented as a Fourier series and can be written as:

$$\text{comb}(x) = S_R(x) = \frac{1}{\Lambda}\sum_m e^{-jmKx}. \tag{5E.6}$$

The corresponding Fourier transform of the comb(x) function can be easily written as:

$$\mathcal{F}\left\{\text{comb}(x)\right\}\Big|_{k_x} = \frac{2\pi}{\Lambda}\sum_m \delta(k_x - mK). \tag{5E.7}$$

By taking the Fourier transform of Eq. (5E.5), the following steps can be written:

$$\mathcal{F}\left\{f(x)\right\} = F(k_x) = \mathcal{F}\left\{h(x)\right\}\mathcal{F}\left\{\text{comb}(x)\right\} =$$

$$= H(k_x)\frac{2\pi}{\Lambda}\sum_m \delta(k_x - mK) = \frac{2\pi}{\Lambda}\sum_m H(mK)\delta(k_x - mK), \tag{5E.8}$$

where $H(k_x) = \mathcal{F}\{h(x)\}$. By taking the inverse Fourier transform of Eq. (5E.8), the following expression can be written:

$$f(x) = \frac{1}{\Lambda}\sum_m H(mK)e^{-jmKx}. \tag{5E.9}$$

Comparing Eqs. (5E.2) and (5E.9), it is evident that the Fourier series coefficients can be written as functions of the Fourier transform of the unit cell. Therefore, they can be expressed as:

$$c_m = \frac{1}{\Lambda}H(mK). \tag{5E.10}$$

Of course, the last result is evident when considering Eq. (5E.3) and recognizing that for $x_0 = -\Lambda/2$, the integral corresponds to the Fourier transform of $h(x)$ evaluated at $k_x = mK$.

5E.2 Fourier Series in Two-Dimensions

Similar to the Fourier transform, the Fourier series expansion of a periodic function can be generalized from one dimension to the n dimensions. Let the function $f(\mathbf{r}) = f(x_1, x_2, \cdots, x_n)$ be periodic in $\mathbb{R}^n$. This can be written as:

$$f(x_1 + m_1\Lambda_1, x_2 + m_2\Lambda_2, \cdots, x_n + m_n\Lambda_n) = f(x_1, x_2, \cdots, x_n), \quad \forall\, m_1, m_2, \cdots, m_n \in \mathbb{Z},$$

$$\text{or} \quad f\left(\mathbf{r} + \boldsymbol{\Lambda}\,[m]\right) = f(\mathbf{r}), \tag{5E.11}$$

$$\text{where} \quad [m] = diag(m_1, m_2, \cdots, m_n), \ \text{and} \quad \boldsymbol{\Lambda} = \left[\Lambda_1, \Lambda_2, \cdots, \Lambda_n\right],$$

where Λ_i is the period in the x_i-direction ($i = 1, 2, \cdots, n$). The Fourier series expansion of the periodic function $f(\mathbf{r})$ is given by:

$$f(\mathbf{r}) = \sum_{m_1}\sum_{m_2}\cdots\sum_{m_n} c_{m_1,m_2,\cdots,m_n} e^{-j[m]\mathbf{K}\cdot\mathbf{r}}, \tag{5E.12}$$

$$\text{where} \quad [m]\mathbf{K}\cdot\mathbf{r} = m_1K_1x_1 + m_2K_2x_2 \cdots m_nK_nx_n,$$

$$c_{m_1,m_2,\cdots,m_n} = \frac{1}{V}\int\!\!\int_V\!\cdots\!\int f(\mathbf{r})e^{+j[m]\mathbf{K}\cdot\mathbf{r}}d\mathbf{r}, \tag{5E.13}$$

and K_i are elements of the reciprocal lattice vector, V corresponds to the volume of the n-dimensional unit cell, and the integration is performed over the volume of the unit cell.

Of course, for two-dimensional periodic apertures, the two-dimensional Fourier series is of interest. Therefore, applying Eqs. (5E.11) and (5E.12) for $n = 2$ and using the standard terminology, the Fourier series of a two-dimensionally periodic function is:

$$f(x, y) = \sum_{m_1}\sum_{m_2} c_{m_1,m_2}e^{-j(m_1K_1x + m_2K_2y)}, \quad \text{with} \tag{5E.14}$$

$$c_{m_1,m_2} = \frac{1}{\Lambda_x\Lambda_y}\int_{-\Lambda_x/2}^{\Lambda_x/2}\int_{-\Lambda_y/2}^{\Lambda_y/2} f(x, y)e^{+j(m_1K_1x + m_2K_2y)}dxdy. \tag{5E.15}$$

Now, similar steps to those presented in the one-dimensional case are followed. Consider again the sampling function comb(x, y) in two dimensions, which can be written as follows:

$$\text{comb}(x, y) = \sum_{m_1}\sum_{m_2}\delta(x - m_1\Lambda_x, y - m_2\Lambda_x) = \sum_{m_1}\sum_{m_2}\delta(x - m_1\Lambda_x)\delta(y - m_2\Lambda_x), \tag{5E.16}$$

where $\delta(x, y)$ is the impulse function in two dimensions. Now, if it is assumed that the unit cell of a periodic function $f(x, y)$ is denoted by the function $h(x, y)$ (where $x \in [-\Lambda_x/2, \Lambda_x/2]$

and $y \in [-\Lambda_y/2, \Lambda_y/2]$), then $f(x,y)$ can be produced by the two-dimensional convolution of $S_R(x,y) = \text{comb}(x,y)$ with $h(x,y)$, expressed as:

$$f(x,y) = h(x,y) ** \text{comb}(x,y) = h(x,y) ** S_R(x,y) = \sum_{m_1}\sum_{m_2} h(x - m_1\Lambda_x, y - m_2\Lambda_y). \tag{5E.17}$$

Since the function $\text{comb}(x,y)$ is periodic, it can be written in the form of the following Fourier series:

$$\text{comb}(x,y) = \frac{1}{\Lambda_x\Lambda_y}\sum_{m_1}\sum_{m_2} e^{-j(m_1K_1x + m_2K_2y)}, \tag{5E.18}$$

where $K_1 = 2\pi/\Lambda_x$ and $K_2 = 2\pi/\Lambda_y$ (for a rectangular lattice). Then the Fourier transform of comb can be easily determined as:

$$\mathcal{F}\{\text{comb}(x,y)\}(k_x,k_y) = \frac{(2\pi)^2}{\Lambda_x\Lambda_y}\sum_{m_1}\sum_{m_2}\delta(k_x - m_1K_1)\delta(k_y - m_2K_2). \tag{5E.19}$$

Now let's take the Fourier transform of Eq. (5E.17):

$$\begin{aligned}\mathcal{F}\{f(x,y)\} = F(k_x,k_y) = \mathcal{F}\{h(x,y)\}\,\mathcal{F}\{\text{comb}(x,y)\}(k_x,k_y) = \\ H(k_x,k_y)\,\frac{(2\pi)^2}{\Lambda_x\Lambda_y}\sum_{m_1}\sum_{m_2}\delta(k_x - m_1K_1)\delta(k_y - m_2K_2) = \\ \frac{(2\pi)^2}{\Lambda_x\Lambda_y}\sum_{m_1}\sum_{m_2} H(m_1K_1, m_2K_2)\delta(k_x - m_1K_1)\delta(k_y - m_2K_2).\end{aligned} \tag{5E.20}$$

Taking the inverse Fourier transform of the last equation, the following expression for $f(x,y)$ can be obtained:

$$f(x,y) = \frac{1}{\Lambda_x\Lambda_y}\sum_{m_1}\sum_{m_2} H(m_1K_1, m_2K_2)e^{-j(m_1K_1x + m_2K_2y)}. \tag{5E.21}$$

Comparing Eqs. (5E.14) and (5E.21), the Fourier series coefficients can be found in terms of the Fourier transform of the unit cell instead of the integration in the unit cell shown in Eq. (5E.15), i.e.,

$$c_{m_1,m_2} = \frac{1}{\Lambda_x\Lambda_y}H(m_1K_1, m_2K_2), \tag{5E.22}$$

where the last expression can be easier to use, especially when the unit cell is derived after a linear transformation of a simpler unit cell. The result of Eq. (5E.22) is evident if one looks at the definition of c_{m_1,m_2} in Eq. (5E.15) and realizes that $f(x,y) = h(x,y)$ in the region of integration of the central unit cell.

Up to this point, the periodic lattice was considered to be rectangular. However, there could be applications (such as in photonic crystals) where oblique lattices are needed. Therefore, in the following section, the generalization of the Fourier series expansion in oblique lattices is presented.

5E.2.1 Oblique Lattice

An oblique two-dimensional periodic grid is shown in Fig. 5E.1. The basis vectors $\vec{a}_1$ and $\vec{a}_2$ in the two grid directions can be specified by their magnitudes Λ_1 and Λ_2 and their angles with respect to the x-axis, θ_1 and θ_2, respectively, as shown in the figure. The reciprocal lattice vectors $\vec{K}_1$ and $\vec{K}_2$ can be defined such that they satisfy the relations $\vec{a}_i \cdot \vec{K}_j = 2\pi\delta_{ij}$ where δ_{ij} is the Kronecker's symbol ($\delta_{ii} = 1$ and $\delta_{ij} = 0$ for $i \neq j$, and $i,j = 1,2$). Now, the sampling function $S_O(\mathbf{r})$ has to be used to reflect the oblique grid. Therefore, $S_O(\mathbf{r})$ and its

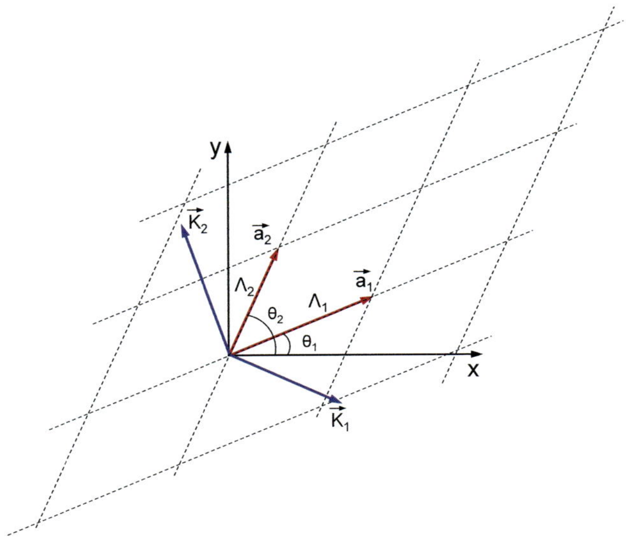

FIGURE 5E.1: An oblique two-dimensional periodic grid with basis vectors $\vec{a}_1$ and $\vec{a}_2$. The basis vectors are specified by their magnitudes Λ_1 and Λ_2, respectively, and their angles with respect to the x-axis, θ_1 and θ_2, respectively (measured positive in the counterclockwise direction). The reciprocal lattice basis vectors $\vec{K}_1$ and $\vec{K}_2$ are also shown.

Fourier transform are given by:

$$S_O(\mathbf{r}) = S_O(x, y) = \sum_{m_1}\sum_{m_2} \delta\left(\mathbf{r} - m_1\vec{a}_1 - m_2\vec{a}_2\right), \tag{5E.23}$$

$$\mathcal{F}\left\{S_O(\mathbf{r})\right\}(\mathbf{k}) = \frac{(2\pi)^2}{|\vec{a}_1 \times \vec{a}_2|} \sum_{m_1}\sum_{m_2} \delta\left(\mathbf{k} - m_1\vec{K}_1 - m_2\vec{K}_2\right). \tag{5E.24}$$

Then, the periodic function $f(\mathbf{r})$ in the periodic oblique grid and its Fourier series expansion can be written as follows:

$$f(\mathbf{r}) = h(\mathbf{r}) **S_O(\mathbf{r}), \tag{5E.25}$$

$$f(\mathbf{r}) = f(x, y) = \sum_{m_1}\sum_{m_2} c_{m_1,m_2} e^{-j(m_1\vec{K}_1\cdot\mathbf{r} + m_2\vec{K}_2\cdot\mathbf{r})}, \tag{5E.26}$$

$$\text{where,}\quad c_{m_1,m_2} = \frac{1}{|\vec{a}_1 \times \vec{a}_2|}\mathcal{F}\left\{h(\mathbf{r})\right\}\left(\mathbf{k} = m_1\vec{K}_1 + m_2\vec{K}_2\right)$$

$$= \frac{1}{|\vec{a}_1 \times \vec{a}_2|} H\left(k_x = m_1K_{1x} + m_2K_{2x}, k_y = m_1K_{1y} + m_2K_{2y}\right), \tag{5E.27}$$

where $H(k_x, k_y)$ is the Fourier transform of the unit cell function $h(x, y)$.

5F Diffraction Grating Profiles

In this appendix, some common grating profiles are analyzed in terms of the Fourier series representation of their transmittance or reflectance function (for transmission or reflection gratings, respectively). Both one-dimensional and two-dimensional common grating profiles are presented.

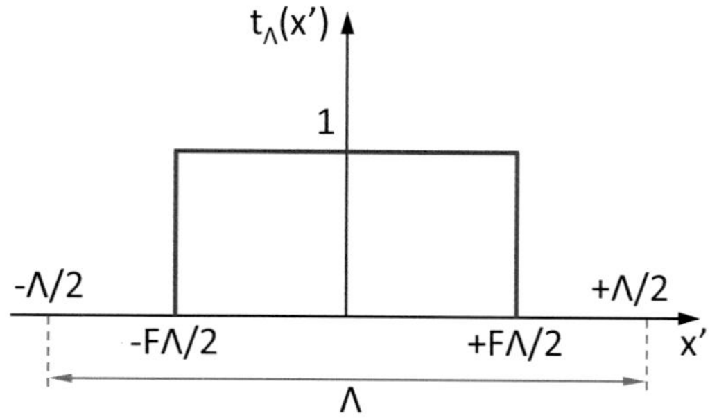

FIGURE 5F.1: The transmittance function of a single period, $t_\Lambda(x')$, of a rectangular amplitude-only grating is shown. The filling factor F represents the percentage of a period that has unit amplitude. The grating period is denoted as Λ.

5F.1 One-Dimensional Transmission Gratings

Five distinct transmission grating profiles are presented in this section. These are: (a) the profile of a rectangular amplitude-only grating, (b) the profile of a rectangular surface-relief phase-only grating, (c) the profile of a blazed surface-relief (sawtooth) phase-only grating, (d) the profile of a binary multilevel grating with both amplitude and phase dependence, and (e) the profile of a sinusoidal surface-relief grating. In the following cases, the variable x' (instead of the usual x) is used to match the coordinate system defined in the figures of the scalar diffraction sections (aperture coordinate system).

5F.1.1 Rectangular Amplitude-Only Grating

The transmittance function for a single period, $t_\Lambda(x')$, of a rectangular grating with amplitude-only modulation, is shown in Fig. 5F.1. The amplitude of the single-period transmittance function is unity in F percentage of the grating period and zero elsewhere. The factor F is usually referred to as the "filling factor." The transmittance function of the grating $t(x')$ can be expressed as a convolution of $t_\Lambda(x')$ with a train of delta functions (similar to the standard "comb" function), as follows:

$$t(x') = t_\Lambda(x') * \sum_m \delta(x' - m\Lambda) = \sum_m t_m e^{-jmKx'},$$

where $K = 2\pi/\Lambda$, and therefore, $t(x' + m\Lambda) = t(x')$, with $m = 0, \pm1, \pm2, \cdots$. o determine the Fourier series coefficients t_m of the periodic transmittance function $t(x')$, the Fourier transform of $t_\Lambda(x')$ is utilized, following Eq. (5E.10). This results in the following expressions for the Fourier coefficients:

$$T_\Lambda(k_x) = \mathcal{F}\{t_\Lambda(x')\} = F\Lambda \frac{\sin(k_x F\Lambda/2)}{k_x F\Lambda/2} = F\Lambda \operatorname{sinc}\left(\frac{k_x F\Lambda}{2\pi}\right), \qquad (5F.1)$$

$$t_m = \frac{1}{\Lambda} T_\Lambda(mK) = F \frac{\sin(mF\pi)}{mF\pi} = F \operatorname{sinc}(mF), \qquad (5F.2)$$

where $\operatorname{sinc}(x) = \sin(\pi x)/(\pi x)$.

5F.1.2 Rectangular Surface-Relief Phase-Only Grating

A rectangular surface-relief grating is depicted in Fig. 5.2(a). For generality, and to allow for later application to more complex profiles, the rectangular ridge of a single period is offset relative to the origin. The refractive index of the grating, encompassing both the substrate and ridges, is denoted by n_s, while the refractive index of the region above the grating and within its grooves is n_c. The grating period is Λ, and the ridge occupies a fraction of the period represented by w_1. The fractions of the period corresponding to the grooves on the left and right sides of the ridge are denoted by w_3 and w_2, respectively. The parameter $w_1 = F$ is also referred to as the filling factor, with the sum of the groove fractions given by $w_2 + w_3 = 1 - F$. The fraction s_1 of the period represents the offset of the ridge relative to the origin, while s_3 and s_2 denote the center offsets of the grooves to the left and right of the ridge, respectively.

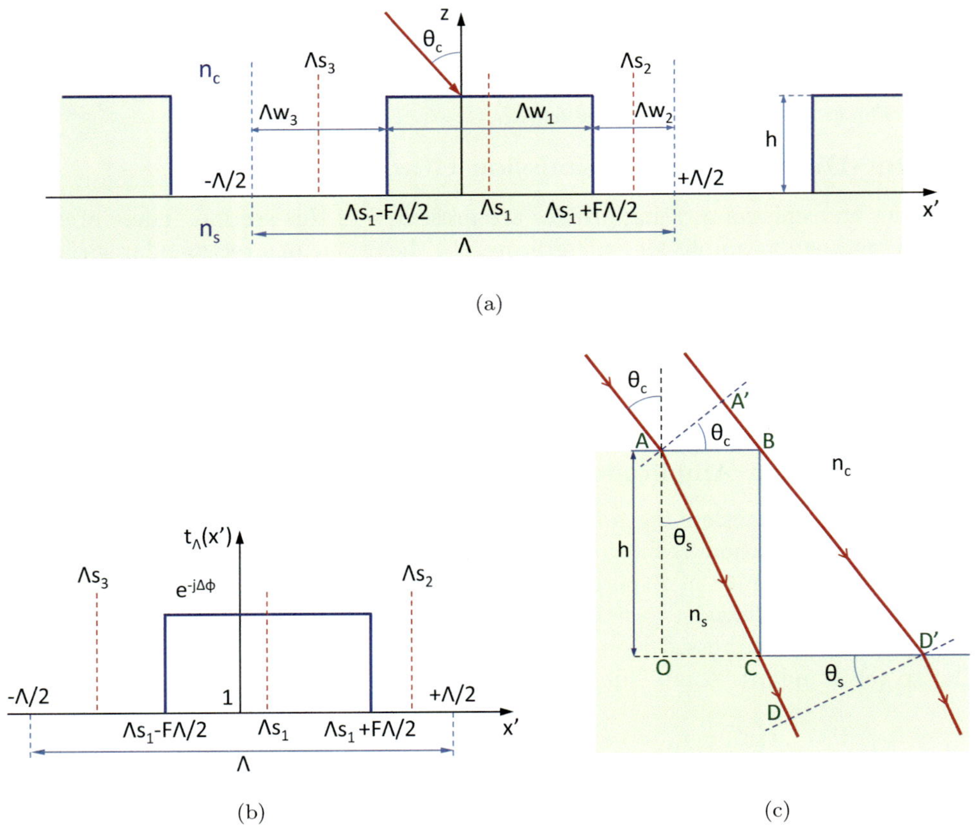

FIGURE 5F.2: (a) The geometry of a rectangular surface-relief, phase-only one-dimensional grating is illustrated. The grating period is Λ, the angle of incidence is θ_c, and the refractive indices of the groove and ridge regions are n_c and n_s, respectively. The ridge height is h. (b) The transmittance function of a single period, $t_\Lambda(x')$, of the rectangular surface-relief, phase-only grating is shown. The filling factor $F = w_1$ represents the fraction of the period occupied by the ridge material. Within the ridge region, the transmittance function takes the value $e^{-j\Delta\Phi}$, while in the remainder of the grating period, it assumes a value of $+1$ (indicating zero phase). (c) The basic configuration used to calculate the phase angle $\Delta\Phi$ is also presented.

The transmittance function for a single period, $t_\Lambda(x')$, of the rectangular grating with phase-only modulation is illustrated in Fig. 5.2(b). The amplitude of the single-period transmittance function, $t_\Lambda(x')$ is unity, while its phase is $\Delta\phi$ over a fraction $F = w_1$ of the grating period and zero over the remainder of the period. In other words, this corresponds to a phase shift of $\Delta\phi$ over a segment of length $F\Lambda$ within the grating period and a phase of zero over the remaining segment of length $\Lambda(1-F)$. The phase difference $\Delta\phi$ can be determined with the aid of Fig. 5.2(c) and the analysis provided by Pommet *et al.* [338], which has been referenced in numerous works, including [339–341]. Specifically, $\Delta\Phi$ can be expressed as:

$$\Delta\Phi = k_0\left[n_s(AC + DC) - n_c(A'B + BD')\right], \quad \text{where,}$$
$$AB = h\tan\theta_s, \qquad A'B = AB\sin\theta_c,$$
$$BD' = \frac{h}{\cos\theta_c}, \qquad AC = \frac{h}{\cos\theta_s},$$
$$CD = CD'\sin\theta_s, \qquad CD' = h\tan\theta_c, \quad \text{resulting in}$$
$$\Delta\Phi = k_0 h\left(n_s\cos\theta_s - n_c\cos\theta_c\right). \tag{5F.3}$$

The angles θ_c and θ_s represent the angle of incidence in the cover region and the angle within the grating ridge material, respectively. These angles are related by Snell's law, expressed as $n_c\sin\theta_c = n_s\sin\theta_s$.

Using the value of $\Delta\Phi$ provided in Eq. (5F.3), the transmittance function of a single-period grating, $t_\Lambda(x')$, can be expressed as:

$$t_\Lambda(x') = e^{-j\Delta\Phi}\text{rect}\left(\frac{x' - s_1\Lambda}{w_1\Lambda}\right) + \text{rect}\left(\frac{x' - s_2\Lambda}{w_2\Lambda}\right) + \text{rect}\left(\frac{x' - s_3\Lambda}{w_3\Lambda}\right). \tag{5F.4}$$

Using the Fourier Transform pairs of Table 5D.2 it is straightforward to write the following equations:

$$T_\Lambda(k_x) = \mathcal{F}\left\{t_\Lambda(x)\right\}$$
$$= e^{-j\Delta\Phi}w_1\Lambda\,\frac{\sin(k_x w_1\Lambda/2)}{k_x w_1\Lambda/2}e^{jk_x s_1\Lambda} + w_2\Lambda\,\frac{\sin(k_x w_2\Lambda/2)}{k_x w_2\Lambda/2}e^{jk_x s_2\Lambda} +$$
$$w_3\Lambda\,\frac{\sin(k_x w_3\Lambda/2)}{k_x w_2\Lambda/2}e^{jk_x s_3\Lambda}$$
$$s_2 = \frac{s_1}{2} + \frac{1+F}{4}, \qquad s_3 = \frac{s_1}{2} - \frac{1+F}{4},$$
$$w_1 = F, \quad w_2 = \frac{1-F}{2} - s_1, \quad w_3 = \frac{1-F}{2} + s_1,$$
$$t_m = \frac{1}{\Lambda}T_\Lambda(mK)$$
$$= e^{-j\Delta\Phi}w_1\frac{\sin(m\pi w_1)}{m\pi w_1}e^{jm2\pi s_1} + w_2\frac{\sin(m\pi w_2)}{m\pi w_2}e^{jm2\pi s_2} +$$
$$w_3\frac{\sin(m\pi w_3)}{m\pi w_3}e^{jm2\pi s_3}, \quad \text{for } m \neq 0 \tag{5F.5}$$
$$t_0 = w_1 e^{-j\Delta\Phi} + w_2 + w_3 = Fe^{-j\Delta\Phi} + 1 - F, \quad \text{for } m = 0.$$

In the special case where the rectangular grating ridge is symmetric with respect to the origin, the parameters take the values $s_1 = 0$, $s_2 = (1+F)/4$, $s_3 = -(1+F)/4$, $w_1 = F$, $w_2 = w_3 = (1-F)/2$. The resulting Fourier coefficients are then given by:

$$t_m = \frac{\sin(m\pi F)}{m\pi}\left(e^{-j\Delta\Phi} - 1\right), \qquad \text{for } m \neq 0 \tag{5F.6}$$
$$t_0 = Fe^{-j\Delta\Phi} + 1 - F, \qquad \text{for } m = 0.$$

5F.1.3 Blazed (Sawtooth) Grating

A blazed (sawtooth profile) surface-relief grating is depicted in Fig. 5F.3. The refractive index of the sawtooth profile, as well as the substrate, is n_s, while the material surrounding the top of the grating has a refractive index of n_c. The maximum height of the sawtooth within one period is represented by h. The transmittance function for a single period, $t_\Lambda(x')$,

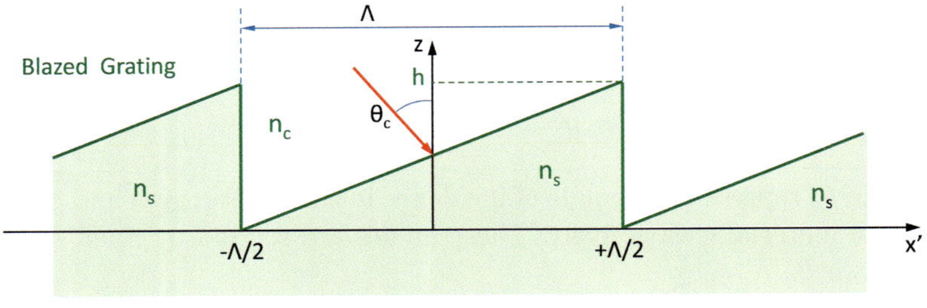

FIGURE 5F.3: The geometry of a blazed surface-relief, phase-only grating is illustrated. The grating period is denoted by Λ, and the maximum height of the blazed grating is h. The refractive index of the grating material is n_s, while the material surrounding the grating has a refractive index of n_c. The angle of incidence is represented by θ_c.

of this blazed grating exhibits linear phase-only modulation, which can be expressed as:

$$t_\Lambda(x') = e^{-j\phi_\Lambda(x')}\text{rect}\left(\frac{x'}{\Lambda}\right), \quad \text{where,} \tag{5F.7}$$

$$\phi_\Lambda(x') = 2\pi\beta\left(\frac{x' + \Lambda/2}{\Lambda}\right),$$

$$\beta = \frac{d}{\lambda_0}(n_s \cos\theta_s - n_c \cos\theta_c).$$

Here, Λ represents the grating period, λ_0 is the freespace wavelength, and θ_c and θ_s are the angles of incidence in media with refractive indices n_c and n_s, respectively. The dependence of β on the angle of incidence is approximately incorporated following the approach used for the rectangular groove surface-relief grating. Using the same methodology as in the previous sections, it is straightforward to derive the following equations for the Fourier coefficients of the transmittance function:

$$T_\Lambda(k_x) = \mathcal{F}\{t_\Lambda(x')\} = e^{-j\pi\beta}\Lambda\,\frac{\sin[(k_x + \frac{2\pi\beta}{\Lambda})\frac{\Lambda}{2}]}{(k_x + \frac{2\pi\beta}{\Lambda})\frac{\Lambda}{2}} = e^{-j\pi\beta}\Lambda\,\text{sinc}\left(\frac{k_x}{2\pi}\Lambda + \beta\right) \tag{5F.8}$$

$$t_m = \frac{1}{\Lambda}T_\Lambda(mK) = e^{-j\pi\beta}\,\frac{\sin[\pi(m+\beta)]}{\pi(m+\beta)} = e^{-j\pi\beta}\,\text{sinc}(m+\beta). \tag{5F.9}$$

The maximum Fourier coefficient occurs when $m + \beta = 0$ which leads to the condition $(h/\lambda_0)(n_s \cos\theta_s - n_c \cos\theta_c) + m = 0$, where m can only take integer values.

5F.1.4 Binary Multilevel Grating

Surface-relief gratings can be fabricated using phase masks and etching techniques in multiple steps. When M phase masks (where M is a positive integer) are used sequentially, the resulting grating profile features $2^M = N$ distinct phase levels. These gratings are commonly referred to as binary multilevel gratings [342]. Using this technique, various continuous-profile surface-relief gratings can be fabricated by approximating the continuous profile with a piecewise constant approximation, employing $2^M = N$ phase levels. An

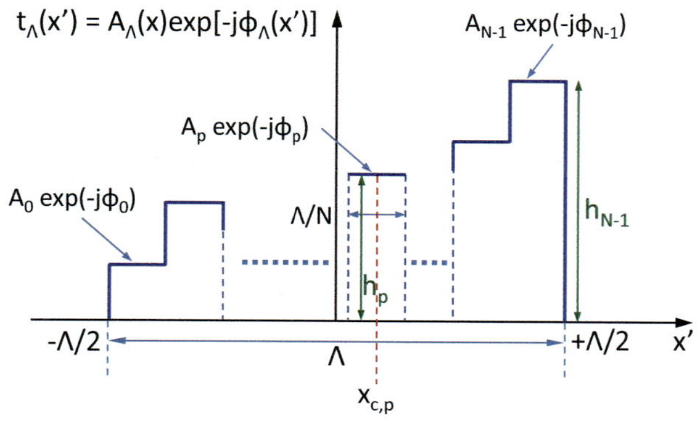

FIGURE 5F.4: The transmittance function of a single period, $t_\Lambda(x') = A_\Lambda(x')e^{\phi_\Lambda(x')}$, of a binary multilevel grating. Both the amplitude $A_\Lambda(x)$ and the phase $\phi_\Lambda(x')$ can vary. The grating period is divided into N equal segments of width Λ/N. The middle point of the p-th segment is denoted as $x_{c,p}$. The grating period is given by Λ.

example of such a multiphase profile is shown in Fig. 5F.4 (only one period is depicted). The transmittance function $t_\Lambda(x') = A_\Lambda(x')\exp[-j\phi_\Lambda(x')]$ for a single period of a binary multilevel grating (with N levels) [342], featuring both phase and amplitude modulation, is shown in Fig. 5F.4. Typically, when $A_\Lambda(x') = 1$, the grating exhibits phase-only variation (in the context of the scalar theory of diffraction based on the transmittance approach). However, if $A_\Lambda(x') < 1$, some amplitude variation due to material losses may need to be considered. The transmittance function $t_\Lambda(x')$, within the period, can be expressed in the following form:

$$t_\Lambda(x') = \sum_{p=0}^{N-1} A_p e^{-j\phi_p} \text{rect}\left(\frac{x' - x_{c,p}}{\Delta x}\right), \qquad \text{where,} \tag{5F.10}$$

$$x_{c,p} = \frac{\Lambda}{N}\left(p + \frac{1}{2}\right),$$

$$\Delta x = \frac{\Lambda}{N},$$

$$\phi_p = \frac{2\pi}{\lambda_0} h_p (n_s \cos\theta_s - n_c \cos\theta_c),$$

$$\text{rect}\left(\frac{x}{\text{a}}\right) = \begin{cases} 1, & |x| \le \text{a}/2, \\ 0, & |x| > \text{a}/2. \end{cases}$$

In the above equation, the phase factor ϕ_p can be determined from the height of the segment h_p, while the angle dependence follows a similar approach to that used for the rectangular grating in Sec. 5F.1.2. Using the same methodology as in the previous sections, it is straightforward to derive the following equations for the Fourier coefficients of $t(x')$:

$$T_\Lambda(k_x) = \mathcal{F}\left\{t_\Lambda(x)\right\} = \frac{\Lambda}{N} \sum_{p=0}^{N-1} A_p e^{-j\phi_p} \,\text{sinc}\left(\frac{k_x}{2\pi}\frac{\Lambda}{N}\right) e^{jk_x x_{c,p}} \tag{5F.11}$$

$$t_m = \frac{1}{\Lambda} T_\Lambda(mK) = \frac{1}{N}\,\text{sinc}\left(\frac{m}{N}\right) \sum_{p=0}^{N-1} A_p e^{-j\phi_p} \exp\left[j2\pi\frac{m}{N}\left(p + \frac{1}{2}\right)\right]. \tag{5F.12}$$

5F.1.5 Sinusoidal-Profile Grating

Another example of a 1D surface-relief grating is the sinusoidal profile, as shown in Fig. 5F.5. The grating height is given by $h(x) = (h/2)[1 + \sin(Kx)]$, with $K = 2\pi/\Lambda$ and Λ is the grating period. The refractive index of the grating material is n_s, while the refractive index of the surrounding material is n_c. The transmittance function $t_\Lambda(x') = \exp[-j\phi_\Lambda(x')]$ for a single period, is expressed as:

$$t_\Lambda(x') = e^{-j\phi_\Lambda(x')} \operatorname{rect}\left(\frac{x'}{\Lambda}\right), \tag{5F.13}$$

$$\phi_\Lambda(x') = \frac{2\pi}{\lambda_0} h(x')\left(n_s \cos\theta_s - n_c \cos\theta_c\right),$$

$$h(x') = \frac{h}{2}\left[1 + \sin\left(Kx'\right)\right].$$

The phase factor in Eq. (5F.13) can be expressed as follows:

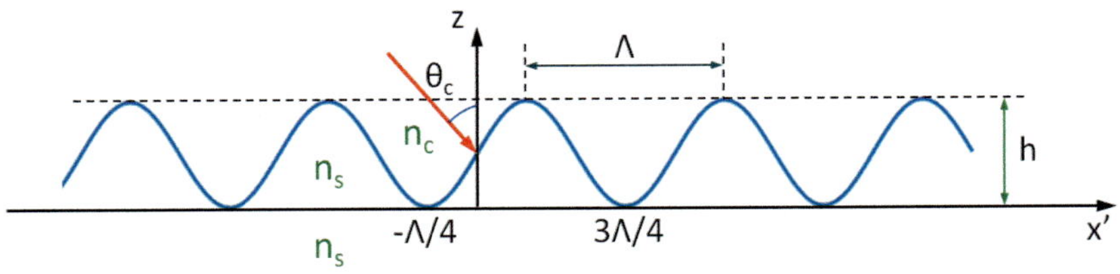

FIGURE 5F.5: The geometry of a sinusoidal-profile surface-relief grating is considered. The grating period is Λ, and the maximum height of the sinusoidal grating is h. The refractive index of the grating material is n_s, while the material surrounding the grating has a refractive index of n_c. The angle of incidence is denoted by θ_c.

$$e^{-j\phi_\Lambda(x')} = \exp\Big[\underbrace{-j\,k_0\frac{h}{2}\left(n_s\cos\theta_s - n_c\cos\theta_c\right)}_{=\alpha}\Big]\exp\left[-jk_0\frac{h}{2}\left(n_s\cos\theta_s - n_c\cos\theta_c\right)\sin(Kx')\right]$$

$$= \exp(-j\alpha)\exp\left[-j\alpha\sin(Kx')\right] = \exp(-j\alpha)\sum_{q=-\infty}^{+\infty} J_q(\alpha)\exp\left(-jqKx'\right), \tag{5F.14}$$

where J_q are the Bessel functions of the first kind of order q. Using the same approach as in the previous sections, it is straightforward to derive the following equations for the Fourier coefficients coefficients of $t(x')$:

$$T_\Lambda(k_x) = \mathcal{F}\left\{t_\Lambda(x')\right\} = e^{-j\alpha}\sum_{q=-\infty}^{+\infty} J_q(\alpha)\left[\frac{2\sin[(k_x - qK)\Lambda/2]}{k_x - qK}\right],$$

$$t_m = \frac{1}{\Lambda}T_\Lambda(mK) = e^{-j\alpha}J_m(\alpha). \tag{5F.15}$$

5F.2 One-Dimensional Reflection Gratings

The procedure for determining the Fourier expansions of reflection gratings is similar to that followed for transmission gratings. Therefore, only two profiles are summarized for demonstration purposes. These are: (a) the profile of a rectangular-shaped reflection grating, and (b) the profile of a sinusoidal-shaped reflection grating.

5F.2.1 Rectangular Surface-Relief Reflection Grating

The geometry of a rectangular-shaped (surface-relief) grating is shown in Fig. 5F.6. The grating parameters are depicted in the figure. Two parallel rays, labeled 1 and 2, are also shown. Additionally, the incident wavefront AB and the reflected wavefront ED are also shown. These can be used to determine the phase difference between the two rays due to their different optical paths. The phase difference is given by $\phi = k_0 n_c(BC + CD - AE)$. Using simple trigonometry, it can be shown that $\phi = k_0 n_c 2h \cos\theta_c$, where θ_c is the angle of incidence. Then, the reflectance function of the grating $r_\Lambda(x')$ is given by:

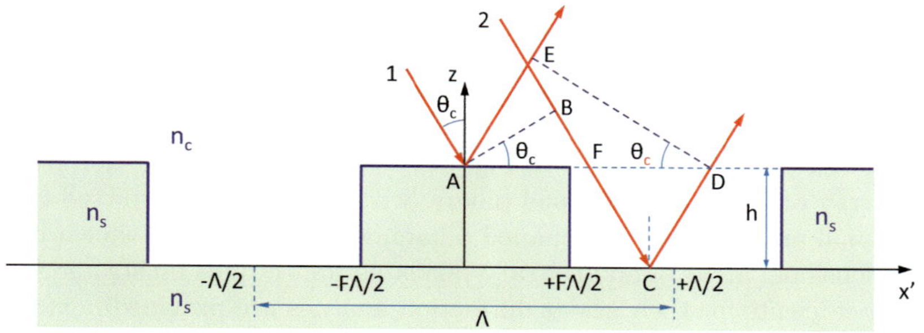

FIGURE 5F.6: The rectangular profile of a 1D reflection grating is shown. The grating period is Λ and the ridge height is h. The refractive index of the medium above the grating is n_c while the refractive index of the grating material and the substrate is n_s (typically a highly reflective metal). The filling factor is denoted by F. Two parallel rays, labeled 1 and 2, are incident and are used to calculate the phase difference due to the difference in optical path lengths.

$$r_\Lambda(x') = \begin{cases} 1 & -\dfrac{F\Lambda}{2} < x' < -\dfrac{F\Lambda}{2}, \\ e^{-j\phi} & \text{elsewhere}, \end{cases}$$

Using the same approach as in the previous sections, it is straightforward to derive the following equations for the Fourier coefficients of $r_\Lambda(x')$:

$$R_\Lambda(k_x) = \mathcal{F}\{r_\Lambda(x')\} = \frac{2}{k_x}\sin\left(k_x\frac{F\Lambda}{2}\right)(1 - e^{-j\phi}) + \frac{2}{k_x}\sin\left(k_x\frac{\Lambda}{2}\right)e^{-j\phi},$$

$$r_m = \frac{1}{\Lambda}R_\Lambda(mK) = \frac{1}{m\pi}\sin(m\pi F)(1 - e^{-j\phi}), \quad m \neq 0, \tag{5F.16}$$

$$r_0 = F + (1 - F)e^{-j\phi}, \quad m = 0. \tag{5F.17}$$

5F.2.2 Sinusoidal-Profile Reflection Grating

The geometry of a sinusoidal-profile reflection grating is the same as that shown in Fig. 5F.5, with the difference that the refractive index n_s of the grating and its substrate is highly reflective. The profile of the grating can again be described by the equation $h(x') = (h/2)[1 + \sin(Kx')]$. The resulting phase shift between two parallel rays reflected at different points on the surface is found in a similar manner to the rectangular reflection grating of Sec. 5F.2.1 and is given by $\phi(x') = k_0 n_c 2h(x')\cos\theta_c$. Then, the single-period reflectance function $r_\Lambda(x')$ of the grating is given by:

$$r_\Lambda(x') = e^{-j\phi(x')} = \exp\Big(-j\underbrace{k_0 h n_c\cos\theta_c}_{=\,\alpha}\Big)\exp\left[-jk_0 h n_c\cos\theta_c\sin(Kx')\right]$$

$$= \exp(-j\alpha)\exp\left[-j\alpha\sin(Kx')\right] = \exp(-j\alpha)\sum_{q=-\infty}^{+\infty} J_q(\alpha)\exp\left(-jqKx'\right). \tag{5F.18}$$

Using the same approach as in the previous sections, it is straightforward to derive the following equations for the Fourier coefficients of $r_\Lambda(x')$:

$$R_\Lambda(k_x) = \mathcal{F}\{r_\Lambda(x')\} = e^{-j\alpha} \sum_{q=-\infty}^{+\infty} J_q(\alpha) \left[\frac{2\sin[(k_x - qK)\Lambda/2]}{k_x - qK} \right],$$

$$r_m = \frac{1}{\Lambda} R_\Lambda(mK) = e^{-j\alpha} J_m(\alpha). \tag{5F.19}$$

5F.3 Two-Dimensional Transmission/Reflection Gratings

In this section, three different unit cells of 2D gratings are analyzed in terms of their Fourier expansion coefficients. Both transmission and reflection gratings are considered. The three unit cell profiles are shown in Fig. 5F.7. Specifically, the section includes a two-dimensional array of rectangular-shaped pillars, with the unit cell depicted in Fig. 5.7(a), a two-dimensional array of elliptical-shaped pillars, with the unit cell shown in Fig. 5.7(b), and a two-dimensional array of an arbitrary-pixelated unit cell, as illustrated in Fig. 5.7(c). The Fourier coefficients for their scalar diffraction analysis are provided next.

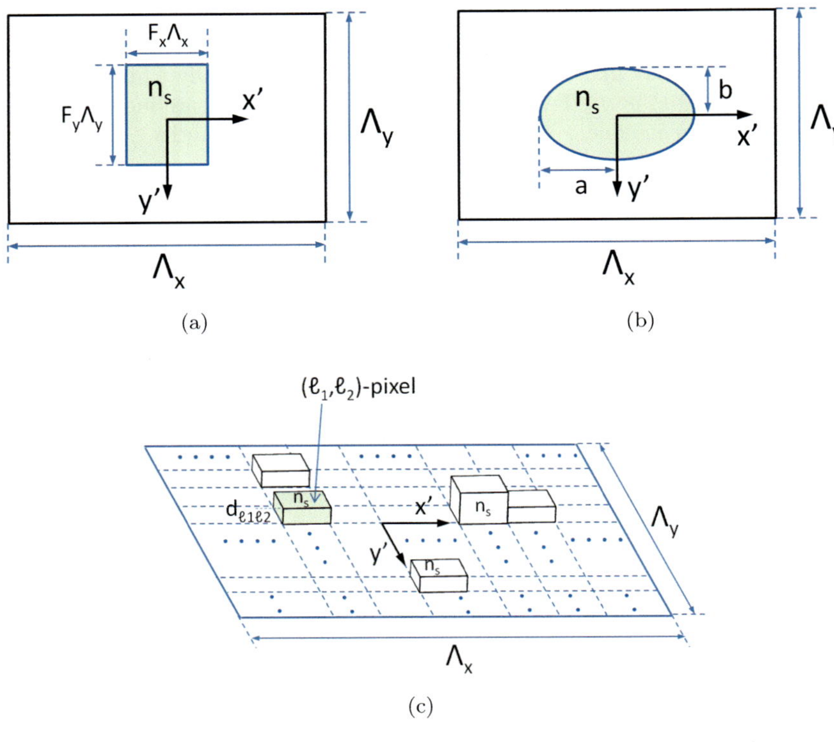

FIGURE 5F.7: (a) Top view of the unit cell of rectangular pillar 2D grating. The filling factors along the x'- and the y'-directions are denoted by F_x and F_y, respectively. (b) Top view of the unit cell of an elliptical pillar 2D grating. The semi-axes of the ellipse are a and b along the x'- and y'-directions, respectively. (c) A perspective view of the unit cell of an arbitrarily pixelated 2D grating. The periods along the x'- and the y'-directions are Λ_x and Λ_y, respectively. The cover refractive index is n_c and the substrate and pixel refractive index is n_s. Variable-height pixels are allowed to represent multilevel 2D gratings.

5F.3.1 Rectangular-Pillar 2D-Grating

In the case of the rectangular pillar 2D unit cell, as shown in Fig. 5.7(a), the transmittance $t_{uc}(x, y)$ and reflectance $r_{uc}(x, y)$ functions of a unit cell as given by:

$$t_{uc}(x', y') = \text{rect}(\frac{x'}{F_x\Lambda_x}, \frac{y'}{F_y\Lambda_y})(e^{-j\phi_t} - 1) + \text{rect}(\frac{x'}{\Lambda_x}, \frac{y'}{\Lambda_y}) \quad (5F.20)$$

$$r_{uc}(x', y') = \text{rect}(\frac{x'}{F_x\Lambda_x}, \frac{y'}{F_y\Lambda_y})(1 - e^{-j\phi_r}) + e^{-j\phi_r}\text{rect}(\frac{x'}{\Lambda_x}, \frac{y'}{\Lambda_y}), \quad (5F.21)$$

where $\text{rect}(x, y) = 1$ if $|x|, |y| \leq 1/2$ and $\text{rect}(x, y) = 0$ if $|x|, |y| > 1/2$, $\phi_t = (2\pi/\lambda_0)h(n_s \cos\theta_s - n_c \cos\theta_c)$ and h is the height of the rectangular pillar shown in Fig. 5.7(a), and $\phi_r = (2\pi/\lambda_0)2hn_c \cos\theta_c$. From the above equations the harmonic coefficients $t_{m_1m_2}$ and $r_{m_1m_2}$ (where m_1, m_2 are integers) can be easily determined and are given by:

$$t_{m_1m_2} = (e^{-j\phi_t} - 1)\,\frac{\sin(\pi m_1 F_x)}{\pi m_1}\,\frac{\sin(\pi m_2 F_y)}{\pi m_2} + \frac{\sin(\pi m_1)}{\pi m_1}\frac{\sin(\pi m_2)}{\pi m_2}, \quad (5F.22)$$

$$r_{m_1m_2} = (1 - e^{-j\phi_r})\,\frac{\sin(\pi m_1 F_x)}{\pi m_1}\,\frac{\sin(\pi m_2 F_y)}{\pi m_2} + e^{-j\phi_r}\frac{\sin(\pi m_1)}{\pi m_1}\frac{\sin(\pi m_2)}{\pi m_2}. \quad (5F.23)$$

In the last equations, the terms $sinc(m_1)sinc(m_2) = \delta(m_1)\delta(m_2)$ (where δ is the discrete delta function). This implies that these terms are zero for any $m_1 \neq 0$ or $m_2 \neq 0$ and unity when $m_1 = m_2 = 0$. Therefore, these terms could be replaced by the corresponding delta function.

5F.3.2 Elliptical-Pillar 2D-Grating

In the case of the elliptical pillar 2-D unit cell shown in Fig. 5.7(b), the corresponding unit cell transmittance $t_{uc}(x', y')$ and reflectance $r_{uc}(x', y')$ functions, are given by equations similar to Eqs. (5F.20) and (5F.21) respectively, where the $\text{rect}(x', y')$ function is replaced with a function that describes the ellipse $x'^2/a^2 + y'^2/b_2 = 1$. This function can be represented by $\text{ellip}(x', y')$ where $\text{ellip}(x', y') = 1$ for (x', y') inside the ellipse and zero otherwise. Then, the corresponding $t_{uc}(x', y')$ and $r_{uc}(x', y')$ are given by:

$$t_{uc}(x', y') = \text{ellip}(\frac{x'}{a}, \frac{y'}{b})(e^{-j\phi_t} - 1) + \text{rect}(\frac{x'}{\Lambda_x}, \frac{y'}{\Lambda_y}) \quad (5F.24)$$

$$r_{uc}(x', y') = \text{ellip}(\frac{x'}{a}, \frac{y'}{b})(1 - e^{-j\phi_r}) + e^{-j\phi_r}\text{rect}(\frac{x'}{\Lambda_x}, \frac{y'}{\Lambda_y}), \quad (5F.25)$$

where $\phi_t = (2\pi/\lambda_0)h(n_s \cos\theta_s - n_c \cos\theta_c)$, h is the height of the elliptical pillar shown in Fig. 5.7(b), and $\phi_r = (2\pi/\lambda_0)2hn_c \cos\theta_c$. At this point the Fourier transform of the ellipse function is needed. From the definition of the 2D Fourier transform,

$$\mathcal{F}\{\text{ellip}(x', y')\} = \int_{-\infty}^{+\infty}\int_{-\infty}^{+\infty} \text{ellip}(x', y')e^{+j(k_x x' + k_y y')}dx'dy'.$$

The above integral can be changed using the transformation $x' = r'_T \cos\phi'$ and $y' = r'_T \epsilon_e \sin\phi'$, where $\epsilon_e = b/a$ (similar transformation was utilized in Sec. 5.7.7). Then the Fourier transform integral can be rewritten as:

$$\mathcal{F}\{\text{ellip}(x'y')\} = \int_0^a \int_0^{2\pi} \exp\left[j(k_x r'_T \cos\phi' + k_y r'_T \epsilon_e \sin\phi')\right] |J| dr'_T d\phi'.$$

The term $|J| = \epsilon_e r'_T$ represents the Jacobian for the transformation. Next, the transformation in the spatial frequency domain is defined as follows: $k_x = k_t \epsilon_e \cos\psi$ and $k_y = k_t \sin\psi$ which implies that $k_t = [(k_x/\epsilon_e)^2 + k_y^2]^{1/2}$. Now, the Fourier transform integral becomes:

$$\mathcal{F}\{\text{ellip}(x', y')\} = \int_0^a \int_0^{2\pi} \exp\left[jk_t \epsilon_e r'_T \cos(\psi - \phi')\right] \epsilon_e r'_T dr'_T d\phi' = \int_0^a 2\pi J_0(k_t \epsilon_e r'_T)\epsilon_e dr'_T$$

$$= \epsilon_e 2\pi \left(\frac{r'_T J_1(k_t \epsilon_e r'_T)}{k_t \epsilon_e} \right) \Big|_0^a = 2\pi \frac{a J_1(k_r \epsilon_e a)}{k_t}$$

$$= 2\pi ab \frac{J_1\left(\sqrt{a^2 k_x^2 + b^2 k_y^2}\right)}{\sqrt{a^2 k_x^2 + b^2 k_y^2}}. \tag{5F.26}$$

In the above derivation the properties $\int_0^{2\pi} \exp[jz\cos(\phi - \phi')]d\phi' = 2\pi J_0(z)$ and $\int J_0(az)zdz = zJ_1(az)/a$ were employed [293]. Now, it is straightforward to show that the harmonic coefficients $t_{m_1 m_2}$ and $r_{m_1 m_2}$ can be expressed as:

$$t_{m_1 m_2} = \frac{(e^{-j\phi_t} - 1)}{\Lambda_x \Lambda_y} \frac{ab}{\left[\left(\frac{am_1}{\Lambda_x}\right)^2 + \left(\frac{bm_2}{\Lambda_y}\right)^2\right]^{1/2}} J_1 \left[2\pi \left[\left(\frac{am_1}{\Lambda_x}\right)^2 + \left(\frac{bm_2}{\Lambda_y}\right)^2\right]^{1/2}\right]$$
$$+ \delta(m_1)\delta(m_2) \tag{5F.27}$$

$$r_{m_1 m_2} = \frac{(1 - e^{-j\phi_r})}{\Lambda_x \Lambda_y} \frac{ab}{\left[\left(\frac{am_1}{\Lambda_x}\right)^2 + \left(\frac{bm_2}{\Lambda_y}\right)^2\right]^{1/2}} J_1 \left[2\pi \left[\left(\frac{am_1}{\Lambda_x}\right)^2 + \left(\frac{bm_2}{\Lambda_y}\right)^2\right]^{1/2}\right]$$
$$+ e^{-j\phi_r}\delta(m_1)\delta(m_2). \tag{5F.28}$$

5F.3.3 Arbitrarily Pixelated 2D-Grating

For the arbitrarily-pixelated 2D grating, the unit cell consists of a grid of pixels with variable heights, as shown in Fig. 5.7(c). The corresponding unit cell transmittance and reflectance functions, $t_{uc}(x', y')$ and $r_{uc}(x', y')$ functions are given by:

$$t_{uc}(x', y') = \sum_{\ell_1} \sum_{\ell_2} \text{rect}\left(\frac{x' - a_x^{\ell_1,\ell_2}}{F_x^{\ell_1,\ell_2}\Lambda_x}, \frac{y' - a_y^{\ell_1,\ell_2}}{F_y^{\ell_1,\ell_2}\Lambda_y}\right)(e^{-j\phi_t^{\ell_1,\ell_2}} - 1) + \text{rect}\left(\frac{x'}{\Lambda_x}, \frac{y'}{\Lambda_y}\right), \tag{5F.29}$$

$$r_{uc}(x', y') = \sum_{\ell_1} \sum_{\ell_2} \text{rect}\left(\frac{x' - a_x^{\ell_1,\ell_2}}{F_x^{\ell_1,\ell_2}\Lambda_x}, \frac{y' - a_y^{\ell_1,\ell_2}}{F_y^{\ell_1,\ell_2}\Lambda_y}\right)(1 - e^{-j\phi_r^{\ell_1,\ell_2}}) + e^{-j\phi_r^{\ell_1,\ell_2}}\text{rect}\left(\frac{x'}{\Lambda_x}, \frac{y'}{\Lambda_y}\right). \tag{5F.30}$$

The phase functions are defined (approximately) as previously and are given by: $\phi_t^{\ell_1,\ell_2} = (2\pi/\lambda_0)\, d_{\ell_1,\ell_2}(n_s \cos\theta_s - n_c \cos\theta_s)$, with d_{ℓ_1,ℓ_2} being the height of the (ℓ_1, ℓ_2)-th pixel shown in Fig. 5.7(c), and $\phi_r^{\ell_1,\ell_2} = (2\pi/\lambda_0)2d_{\ell_1,\ell_2}n_c \cos\theta_c$. The coefficients $a_w^{\ell_1,\ell_2}$ $(w = x, y)$ are defined as $a_x^{\ell_1,\ell_2} = [-\Lambda_x/2 + (2\ell_1 - 1)L_x/2]$ and $a_y^{\ell_1,\ell_2} = [-\Lambda_y/2 + (2\ell_2 - 1)L_y/2]$. From the above equations, the harmonic coefficients $t_{m_1 m_2}$ and $r_{m_1 m_2}$ can be easily determined and are given by:

$$t_{m_1 m_2} = \sum_{\ell_1} \sum_{\ell_2} t_{m_1 m_2}^{\ell_1,\ell_2} \exp[-j2\pi(m_1 a_x^{\ell_1,\ell_2} + m_2 a_y^{\ell_1,\ell_2})], \tag{5F.31}$$

$$r_{m_1 m_2} = \sum_{\ell_1} \sum_{\ell_2} r_{m_1 m_2}^{\ell_1,\ell_2} \exp[-j2\pi(m_1 a_x^{\ell_1,\ell_2} + m_2 a_y^{\ell_1,\ell_2})], \tag{5F.32}$$

where $t_{m_1 m_2}^{\ell_1,\ell_2}$ and $t_{m_1 m_2}^{\ell_1,\ell_2}$ are given by equations similar to Eqs. (5F.22) and (5F.23), respectively.

6

Basics of Fourier Optics

6.1 Introduction

Geometric optics is a useful tool for analyzing the imaging properties of optical systems, including image location, magnification, and aberrations. However, a comprehensive description of an optical system requires the inclusion of the wave nature of light and associated phenomena, such as diffraction. These wave phenomena fundamentally determine the resolving power of optical devices, image contrast, and the effects of spatial filters.

One approach within the domain of wave optics considers the Fourier spectrum of the object and examines how these spectral components propagate through the optical system. This method, known as *Fourier Optics*, applies Fourier transforms and spatial frequency analysis to explore light propagation and manipulation within optical systems. It provides a robust mathematical framework for understanding the behavior of light as it interacts with optical components such as lenses, mirrors, diffraction gratings, and other elements. The principles of Fourier optics are also grounded in linear systems theory [19, 265].

Fourier optics finds numerous practical applications across a range of fields. It plays a crucial role in the design and optimization of imaging systems such as microscopes, telescopes, and cameras by predicting image quality, resolution, and aberrations to enhance system performance. In holography, it underpins the creation of three-dimensional images through interference patterns, with uses in security, data storage, and medical imaging.

In spectroscopy, Fourier-transform techniques analyze light's frequency components, aiding in chemical analysis and astronomy. Optical signal processing benefits from Fourier optics in tasks like image filtering, encryption, and pattern recognition, enabling advanced data manipulation and communication. The behavior of diffraction gratings, essential in spectroscopy and wavelength selection, is also explained through Fourier principles.

Remote sensing utilizes synthetic aperture radar (SAR) techniques based on Fourier analysis to generate high-resolution images of Earth's surface. X-ray crystallography relies on Fourier methods to interpret diffraction patterns, revealing atomic structures critical to drug development, materials science, and biology. Interferometry exploits Fourier optics for precise measurements in applications like gravitational wave detection and non-destructive testing.

Medical imaging techniques such as computed tomography (CT) and magnetic resonance imaging (MRI) use Fourier-based algorithms to reconstruct detailed images. In laser beam shaping, Fourier optics is applied to customize beam profiles for specific purposes, including material processing. Light propagation in optical fibers and waveguides is analyzed with Fourier methods to improve data transmission and photonic circuit efficiency.

In adaptive optics, Fourier-based approaches correct for atmospheric distortions or aberrations, enhancing image sharpness in astronomy and microscopy. Digital image processing employs Fourier transforms for tasks like compression, noise reduction, and image enhancement. Nonlinear optics also leverages Fourier techniques to study light-matter interactions, supporting developments in laser sources, frequency conversion, and parametric amplification.

DOI: 10.1201/9781003490333-6

The purpose of this chapter is to provide a concise introduction to the fundamental theory of Fourier optics. For a more detailed and advanced treatment, readers are encouraged to consult specialized texts, such as Refs. [265, 343–346].

The fundamental concept in Fourier optics is the application of the Fourier transform to describe how a complex optical field, such as a light wave, can be decomposed into simpler components, each corresponding to a specific spatial frequency. The two-dimensional Fourier transform was previously introduced in an appendix to the *Scalar Diffraction* chapter, Chapter 5. For clarity and convenience, it is redefined here along with its inverse as follows:

$$F(f_x, f_y) = \mathcal{F}\{f(x,y)\} = \int_{-\infty}^{+\infty}\int_{-\infty}^{+\infty} f(x,y)e^{+j2\pi(f_x x + f_y y)}dxdy, \tag{6.1}$$

$$f(x,y) = \mathcal{F}^{-1}\{F(f_x, f_y)\} = \int_{-\infty}^{+\infty}\int_{-\infty}^{+\infty} F(f_x, f_y)e^{-j2\pi(f_x x + f_y y)}df_x df_y. \tag{6.2}$$

Here, x, y represent the two-dimensional spatial coordinates, while f_x, f_y denote their corresponding spatial frequencies. It is often convenient to express these frequencies in terms of radial spatial frequencies, defined as $k_x = 2\pi f_x$, $k_y = 2\pi f_y$, drawing an analogy with time signals and the radial frequency $\omega = 2\pi f$, where f is the temporal frequency.

6.2 Spatial Frequency Domain and Its Implications

In the definition of the 2D Fourier transform and its inverse, as expressed in Eqs. (6.1) and (6.2), the terms of interest are the $\exp[\pm j2\pi(f_x x + f_y y)]$ functions, incorporating spatial frequencies f_x and f_y. These spatial frequencies play a crucial role in various scientific disciplines, including optics, vision, optical signal processing, material science, astrophysics, and more. The concept of spatial frequency serves as a universal measure across these fields, describing the rate of change of a pattern or signal in space. It quantifies how frequently a repeating pattern occurs within a given unit of space. This fundamental idea is instrumental in analyzing and understanding diverse phenomena, making spatial frequency a key parameter in the study of patterns and signals in different scientific domains.

From Eq. (6.2), the term $F(f_x, f_y)\exp[-j2\pi(f_x x + f_y y)]$ represents the contribution of the spatial frequencies f_x and f_y at point (x, y) in the signal. For example, if $f(x, y)$ corresponds to an image, this term is part of the decomposition of the image into its constituent spatial frequencies. It is useful to examine the properties of the elementary function $\exp[-j2\pi(f_x x + f_y y)]$. The phase of this function becomes multiple of 2π along straight lines in the xy-plane, as shown in Fig. 6.1(a). For any constant phase, the locus of of $\exp[-j2\pi(f_x x + f_y y)]$ forms an infinite set of parallel lines. These lines of constant phase, corresponding to a multiple of 2π, can be described by the following equations:

$$y = -\frac{f_x}{f_y}x + \frac{m}{f_y}, \qquad \text{where} \quad m \in \mathbb{Z}.$$

These lines of equal phase are oriented perpendicular to the vector $\vec{k}_t = k_x\hat{\imath}_x + k_y\hat{\imath}_y = 2\pi(f_x\hat{\imath}_x + f_y\hat{\imath}_y)$, as shown in Fig. 6.1(a). The angle θ of the equal phase lines is given by $\tan\theta = k_y/k_x = f_y/f_x$. The periodicity of $\exp[-j2\pi(f_x x + f_y y)]$ in the xy-plane can be determined as follows: Let $\vec{k}_t \cdot \vec{r} = \phi$, where $\vec{r} = x\hat{\imath}_x + y\hat{\imath}_y$ is the position vector in the xy-plane. For the case where $\phi = 2\pi\ell$ ($\ell \in \mathbb{Z}$), the set of parallel lines has a phase which is

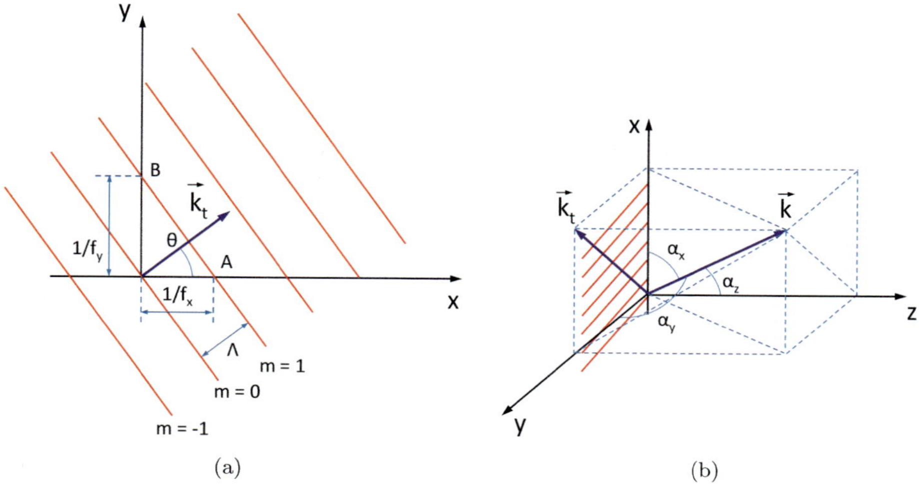

FIGURE 6.1: (a) Lines of equivalent phase (multiple of 2π) in the xy-plane of the function $e^{-j2\pi(f_x x+f_y y)}$. (b) The plane wave wavevector components that emanates from the xy-plane due to the $e^{-j2\pi(f_x x+f_y y)}$ term. The angles α_x, α_y, and α_z are related to the directional cosines of the propagating plane wave along the z-axis.

multiple of 2π. If $\hat{k}_t = \vec{k}_t/|\vec{k}_t|$ then the period Λ can be determined from the relation:

$$\vec{k}_t \cdot \left(\vec{r} + \Lambda\hat{k}_t\right) = 2\pi + \phi \Longrightarrow \Lambda = \frac{2\pi}{|\vec{k}_t|} = \frac{1}{\sqrt{f_x^2 + f_y^2}}. \tag{6.3}$$

It is evident that as spatial frequencies increase, the periodicity Λ decreases. Geometrically, this periodicity can be determined using the equal phase lines corresponding to $m = 0$ and $m = 1$ in Fig. 6.1(a). Considering the x-intercept (point A) and y-intercept (point B) of the $m = 1$ equal phase line, which are $1/f_x$ and $1/f_y$ respectively, the distance between the $m = 0$ and $m = 1$ lines represents the periodicity Λ. Using the area of the orthogonal triangle formed by the origin and the x- and y-intercepts of the $m = 1$ equal phase line, the following is valid: $(1/2)\Lambda[(1/f_x)^2+(1/f_y)^2]^{1/2} = (1/2)(1/f_x)(1/f_y) \Longrightarrow \Lambda = 1/(f_x^2+f_y^2)^{1/2}$. This expression aligns with the previously established result [refer to Eq. (6.3)], providing a geometric interpretation of the periodicity in terms of the spatial frequencies f_x and f_y. Now it is interesting to relate the spatial frequencies f_x and f_y to the propagation of the corresponding plane wave that passes through (x, y) point. The z-component of the wavevector for this plane wave is expressed as:

$$\begin{aligned} k_z &= \left(k^2 - k_x^2 - k_y^2\right)^{1/2} = \left[k^2 - (2\pi f_x)^2 - (2\pi f_y)^2\right]^{1/2} \\ &= k\left[1 - \left(\frac{2\pi f_x}{k}\right)^2 - \left(\frac{2\pi f_y}{k}\right)^2\right]^{1/2} = k\left[1 - (\lambda f_x)^2 - (\lambda f_y)^2\right]^{1/2}, \end{aligned}$$

since $k = 2\pi/\lambda$. It is assumed that the wave propagates in the positive z-direction. The directional cosine angles α_x, α_y and α_z (see Fig. 6.1(b)) of the propagating wavevector $\vec{k}$ can be expressed as:

$$\alpha_x = \cos^{-1}\{\lambda f_x\}, \tag{6.4}$$

$$\alpha_y = \cos^{-1}\{\lambda f_y\}, \tag{6.5}$$

$$\alpha_z = \cos^{-1}\left\{\left[1 - (\lambda f_x)^2 - (\lambda f_y)^2\right]^{1/2}\right\}. \tag{6.6}$$

It is crucial to note that a plane wave can effectively propagate along the z-direction only if $(\lambda f_x)^2 + (\lambda f_y)^2 \leq 1$. If this condition is exceeded, the wave becomes evanescent. As spatial frequencies increase, the angle α_z increases approaching $\pi/2$. This indicates that the spatial frequency content in the xy-plane plays a decisive role in determining the propagation angles of plane waves along z-axis. Higher spatial frequencies correspond to larger propagation angles, approaching the maximum limit where $\alpha_z \longrightarrow \pi/2$. Conversely, lower spatial frequencies result in smaller angles, tending toward zero as spatial frequencies approach zero ($\alpha_z \to 0$). This relationship highlights the direct link between the spatial frequency distribution and the propagation angles of plane waves in the z-direction.

Example 6.2.1:

This example is adopted with some modifications from Ref. [43]. Assume a 1D-grating which is represented in the xy-plane by the function $f(x, y) = s_0 + s_1 \cos(2\pi f_0 x)$, where s_0 and s_1 are known amplitude constants and f_0 is the fundamental spatial frequency. Determine all possible directions for the propagation of corresponding plane waves along the positive z-direction. Assume the free-space wavelength is $\lambda = 1\,\mu$m, and evaluate the cases where $f_0 = 500\,\text{mm}^{-1}$ and $f_0 = 750\,\text{mm}^{-1}$.

Solution: The function $f(x, y)$ can be expressed as a sum of exponential terms using Euler's formula as:

$$f(x, y) = s_0 + \frac{s_1}{2} e^{j2\pi f_0 x} + \frac{s_1}{2} e^{-j2\pi f_0 x}.$$

This implies that $f_x = 0$, or $\pm f_0$ and $f_y = 0$. These values represent the spatial frequency content of the grating function. Using Eqs. (6.4)-(6.6), which relate the spatial frequencies f_x and f_y to the directional cosines of the corresponding propagating plane waves, the angles of propagation for all possible directions can be determined. These results are summarized in Table 6.1. Figure 6.2 schematically illustrates the diffracted wavevectors corresponding to the spatial frequency components. Each wavevector represents a plane wave propagating in the z-direction with angles determined by the given spatial frequencies.

TABLE 6.1: Results of Example 6.2.1

f_x (1/mm)	f_y (1/mm)	λf_x	λf_y	α_x (deg)	α_y (deg)	α_z (deg)
0	0	0	0	90	90	0
500	0	0.50	0	60	90	30
−500	0	−0.50	0	120	90	30
0	0	0	0	90	90	0
750	0	0.75	0	41.41	90	48.59
−750	0	−0.75	0	138.59	90	48.59

6.3 Linear System Approach and Plane Wave Spectrum

Plane waves are essential for describing complex optical fields, and defining the propagation of such fields in a homogeneous isotropic medium is crucial. This can be achieved using the plane wave spectrum approach, which was discussed in the "Plane Wave Spectrum" section

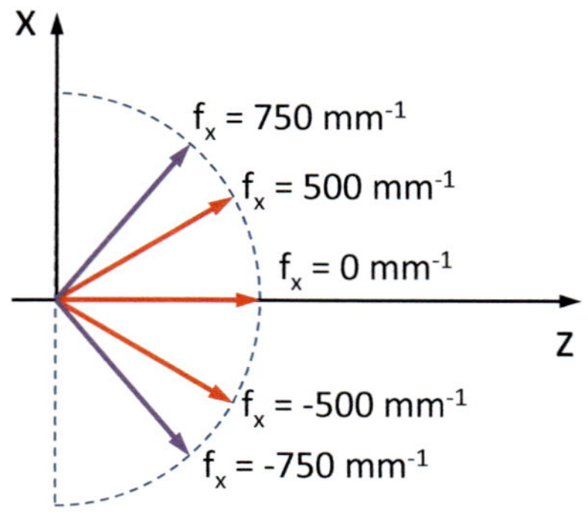

FIGURE 6.2: The diffracted directions of the plane waves emanating from the xy-plane for an image of the form $f(x,y) = s_0 + s_1\cos(2\pi f_0 x)$. The cases of $f_0 = 500\,\text{mm}^{-1}$ and $f_0 = 750\,\text{mm}^{-1}$ are shown. Obviously higher spatial frequencies result in larger α_z angles. The values of all angles of this example are listed in Table 6.1.

of the *Scalar Diffraction* chapter (see Sec. 5.5). The approach relies on concepts from linear systems theory.

Key elements in this framework include the impulse response function, which represents the system's reaction to an impulse or point source, and the transfer function, which characterizes the system's response to spatial frequencies, as outlined by Saleh and Teich [19]. To address this general problem, a block diagram (shown in Fig. 6.3) is used. This diagram captures the critical components involved in describing the linear system aspects of optical propagation, offering a structured approach to analyzing optical field behavior in a homogeneous isotropic medium.

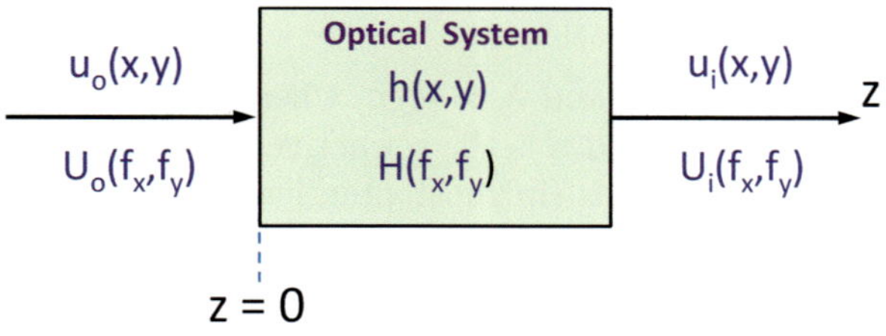

FIGURE 6.3: An optical field described by $u_o(x,y)$ (object or input field) in the spatial domain, and $U_o(f_x,f_y)$ in the spatial frequency domain, passes through an optical system. The optical system is assumed to be linear and shift-invariant and is represented by its impulse response $h(x,y)$ or by its transfer function $H(f_x,f_y)$. The image (or output) optical field is described as $u_i(x,y)$ or $U_i(f_x,f_y)$ in the spatial and in the spatial frequency domains, respectively. The origin of the xyz-coordinate system is at the left edge of the optical system.

The functions $u_o(x,y)$ and $u_i(x,y)$ represent the object (input) and image (output) optical fields in the spatial domain. The z-axis is the axis of propagation, and the xy-plane is transverse to the propagation direction. The optical fields can also be expressed in the spatial frequency domain using the Fourier transform [see Eq. (6.1)], with $U_o(f_x,f_y) = \mathcal{F}\{u_o(x,y)\}$ for the input field, and $U_i(f_x,f_y) = \mathcal{F}\{u_i(x,y)\}$ for the output field. The linear shift-invariant optical system is described by its impulse response $h(x,y)$ or its transfer function $H(f_x,f_y)$. Assume that the optical system corresponds to the propagation over a distance z_0 in a homogeneous isotropic medium of refractive index n. According to linear systems theory, the following relationships hold:

$$u_i(x, y; z_0) = u_o(x, y) ** h(x, y) \tag{6.7}$$

$$U_i(f_x, f_y; z_0) = U_o(f_x, f_y)H(f_x, f_y), \tag{6.8}$$

where "**" denotes the 2D convolution operation. Using the results of the plane wave spectrum section, the output optical field $u_i(x, y; z_0)$ at z_0 is given by:

$$u_i(x, y; z_0) = \frac{1}{(2\pi)^2} \int_{-\infty}^{+\infty} \int_{-\infty}^{+\infty} U_o(k_x, k_y; z_0 = 0) e^{-jk_x x} e^{-jk_y y}$$
$$\exp\left[-jz_0\sqrt{k^2 - k_x^2 - k_y^2}\,\right] dk_x dk_y.$$

The input optical field in the spatial frequency domain is given by $U_o(k_x, k_y; z_0 = 0) = \mathcal{F}\{u_o(x, y; z_0 = 0)\}$, where $k_w = 2\pi f_w$ represents the wavevector components of the incident (object) optical field, and f_w corresponds to its spatial frequencies along the x-, y-, and z-directions. Rewriting the last equation using the spatial frequencies f_x, f_y the output (image) optical field can be expressed as:

$$u_i(x, y; z_0) = \int_{-\infty}^{+\infty}\int_{-\infty}^{+\infty} \underbrace{U_o(k_x, k_y; 0) \exp\left[-j\frac{2\pi z_0}{\lambda}\sqrt{1 - (\lambda f_x)^2 - (\lambda f_y)^2}\,\right]}_{U_i(f_x, f_y; z_0) \;=\; U_o(f_x, f_y; 0) H(f_x, f_y)}$$
$$e^{-j2\pi f_x x} e^{-j2\pi f_y y} df_x df_y$$
$$= \int_{-\infty}^{+\infty}\int_{-\infty}^{+\infty} U_i(f_x, f_y; z_0) e^{-j2\pi f_x x} e^{-j2\pi f_y y} df_x df_y = \mathcal{F}^{-1}\{U_i(f_x, f_y; z_0)\}.$$

In the previous equation, $k = 2\pi/\lambda$ and $\lambda = \lambda_0/n$, where λ_0 is the freespace wavelength of the incident monochromatic field and n is the refractive index of the homogeneous medium in which the optical field propagates (in most practical cases $n \simeq 1$ - air medium). The transfer function $H(f_x, f_y)$ of the optical system, which describes the propagation over a distance z_0 in a homogeneous, isotropic medium, is given by:

$$H(f_x, f_y) = \begin{cases} \exp\left[-j\frac{2\pi z_0}{\lambda}\sqrt{1 - (\lambda f_x)^2 - (\lambda f_y)^2}\,\right], & \sqrt{(\lambda f_x)^2 + (\lambda f_y)^2} \leq 1, \\ 0, & \sqrt{(\lambda f_x)^2 + (\lambda f_y)^2} > 1. \end{cases} \tag{6.9}$$

In the case where $[(\lambda f_x)^2 + (\lambda f_y)^2]^{1/2} > 1$, the corresponding waves are evanescent, which means that they decay exponentially and do not contribute to the output optical field over practical propagation distances z. The impulse response $h(x, y)$ of the system can be determined by taking the inverse Fourier transform of the transfer function $H(f_x, f_y)$, expressed as: $h(x, y) = \mathcal{F}^{-1}\{H(f_x, f_y)\}$. However, the computation of the inverse Fourier transform can often be bypassed by utilizing the Rayleigh-Sommerfeld formulation. According to the plane wave spectrum section of Scalar Diffraction, the impulse response $h(x, y)$ can be expressed as:

$$h(x - x', y - y', z_0) = -2\frac{dG}{dz},$$

where $G = e^{-jkR}/4\pi R$ is the 3D Green's function of homogeneous isotropic space and $R = [(x - x')^2 + (y - y')^2 + z_0^2]^{1/2}$. By performing the necessary differentiation, the impulse response can be determined as:

$$h(x-x',y-y';z_0) = -\frac{e^{-jkR}}{j\lambda R}\left(1+\frac{1}{jkR}\right)\frac{z}{R} \Longrightarrow$$

$$h(x,y;z_0) = -\frac{e^{-jk(x^2+y^2+z_0^2)^{1/2}}}{j\lambda(x^2+y^2+z_0^2)^{1/2}}\left[1+\frac{1}{jk(x^2+y^2+z_0^2)^{1/2}}\right]\frac{z}{(x^2+y^2+z_0^2)^{1/2}}. \tag{6.10}$$

The impulse response can be alternatively determined by direct evaluation of the inverse Fourier transform as described by Stark [347]. It is worth mentioning that the impulse response function $h(x,y;z_0)$ has dimensions of inverse square meters (m^{-2}). On the other hand, the transfer function $H(f_x,f_y)$ is dimensionless! A usual approximation of Eqs. (6.9) and (6.10) is in the case that $[(\lambda f_x)^2+(\lambda f_y)^2]^{1/2} \ll 1$. Approximating the square root, the following expressions can be written:

$$H(f_x,f_y) = \exp\left[-j\frac{2\pi z_0}{\lambda}\right]\exp\left[+j\pi\lambda z_0(f_x^2+f_y^2)\right], \tag{6.11}$$

$$h(x,y) = j\frac{e^{-jkz_0}}{\lambda z_0}\exp\left[-j\frac{k}{2z_0}(x^2+y^2)\right] = jk\frac{e^{-jkz_0}}{2\pi z_0}\exp\left[-j\frac{k}{2z_0}(x^2+y^2)\right]. \tag{6.12}$$

The last approximation is the Fresnel approximation which result in the following expression for the output (image) field $u_i(x,y;z_0)$ as a function of the input (object) field:

$$u_i(x,y;z_0) = h(x,y;z_0) ** u_o(x,y;0) =$$

$$\int_{-\infty}^{+\infty}\int_{-\infty}^{+\infty} h(x-x',y-y';z_0)u_o(x',y';0)dx'dy' =$$

$$jk\frac{e^{-jkz_0}}{2\pi z_0}\int_{-\infty}^{+\infty}\int_{-\infty}^{+\infty}\exp\left\{-j\frac{k}{2z_0}\left[(x-x')^2+(y-y')^2\right]\right\}u_o(x',y';0)dx'dy'. \tag{6.13}$$

Equation (6.13) is identical to the Fresnel approximation integral presented in the *Scalar Diffraction* chapter, Chapter 5. However, in this section, the derivation was based on linear system theory. This approach emphasizes the relationship between the input and output fields through the system's transfer function and impulse response, offering a different perspective on the propagation process compared to the traditional diffraction-based derivation. The linear system framework provides a more general and systematic way to analyze the transformation of optical fields.

6.4 Fourier Transform Property of a Thin Lens

The objective of this section is to examine the transformative properties of thin lenses. To achieve this, the transmittance function of a thin lens is determined. Like many other thin optical elements, including phase plates, prisms, and gratings, the effect of a thin lens on wave propagation can be described through its transmittance function, expressed as $t(x,y) = e^{-j\phi(x,y)}$.

The transmittance function establishes a relationship between the optical field incident on the lens from the left, denoted as $u_o(x, y)$, and the field emerging from the right, denoted as $u_i(x, y)$, through the equation $u_i(x, y) = t(x, y)u_o(x, y)$. It is important to emphasize that the transmittance function differs from the impulse response function of the lens. Although the thin lens, described by its transmittance function, constitutes a linear system, it is inherently shift-variant.

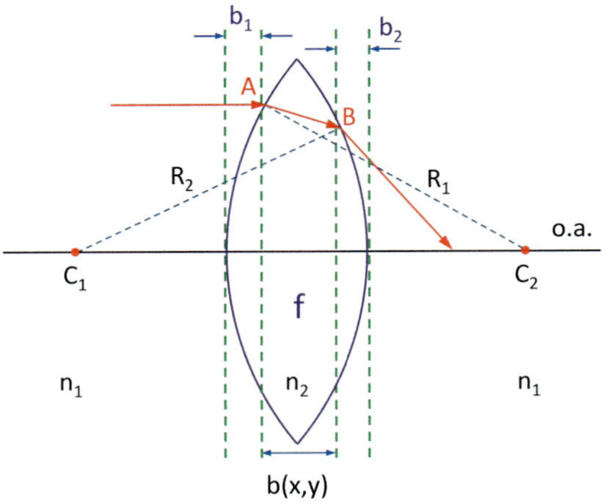

FIGURE 6.4: A thin lens (with exaggerated thickness) of refractive index n_2 surrounded by an isotropic medium of refractive index n_1. The radii of the spherical left and right surfaces are R_1 and R_2, respectively, and they obey the sign convention introduced in the *Geometrical Optics* chapter, Chapter 2. In this case, $R_1 > 0$ and $R_2 < 0$.

The phase function $\phi(x, y)$ is derived using the geometric configuration depicted in Fig. 6.4. Consider a ray entering the thin lens at point $A(x, y)$ and exiting at point $B(x'y')$. The phase shift $\phi(x, y)$ from the input plane to the output plane of the lens is given by:

$$\phi(x, y) = k_0 n_2 b(x, y) + k_0 n_1[b_0 - b(x, y)] = k_0 n_1 b_0 + k_0(n_2 - n_1)b(x, y),$$

where $b_0 = b_1 + b_2 + b(x, y)$. The distances b_1 and b_2 can be calculated as follows:

$$b_1 = |R_1| - \sqrt{R_1^2 - (x^2 + y^2)} \simeq \frac{x^2 + y^2}{2|R_1|},$$

$$b_2 = |R_2| - \sqrt{R_2^2 - (x'^2 + y'^2)} \simeq \frac{x'^2 + y'^2}{2|R_1|} \simeq \frac{x^2 + y^2}{2|R_2|}.$$

Since the lens is assumed to have negligible thickness, it is reasonable to approximate $x^2 + y^2 \simeq x'^2 + y'^2$. Points C_1 and C_2 correspond to the centers of the right and left spherical surfaces of the lens, respectively. The distance $b(x, y)$, which represents the optical path contribution through the lens, can then be expressed as:

$$b(x, y) = b_0 - \frac{x^2 + y^2}{2}\left(\frac{1}{|R_1|} + \frac{1}{|R_2|}\right) = b_0 - \frac{x^2 + y^2}{2}\left(\frac{1}{R_1} - \frac{1}{R_2}\right).$$

In the last equation, the sign convention for the radii R_1 and R_2 was applied, ensuring consistency with the geometric optics framework. Using the expression for $b(x, y)$, the phase function $\phi(x, y)$ can be written as:

$$\phi(x, y) = k_0 n_2 b_0 - k_0 n_1 \frac{x^2 + y^2}{2} \underbrace{\frac{(n_2 - n_1)}{n_1}\left(\frac{1}{R_1} - \frac{1}{R_2}\right)}_{= 1/f} = k_0 n_2 b_0 - k\frac{x^2 + y^2}{2f}.$$

In the last expression, the definition of the thin lens focal distance, f, was also utilized, along with the relation $k_0 n_1 = k$. Summarizing, the transmittance function of the thin lens can be written as:

$$t(x,y) = \exp(-jk_0 n_2 b_0) \exp\left[jk\frac{x^2+y^2}{2f}\right]. \tag{6.14}$$

The factor $\exp(-jk_0 n_2 b_0)$, being independent of the x, y coordinates, is typically neglected as it does not influence the spatial variation of the optical field. It is worth noting that Eq. (6.14) applies to both converging ($f > 0$) and diverging ($f < 0$) thin lenses.

However, Eq. (6.14) does not account for the finite lateral extent of the lens aperture. To incorporate this limitation, the pupil function $P(x,y)$ is introduced, where $P(x,y) = 1$ within the aperture area and $P(x,y) = 0$ outside it. A commonly used pupil is the "circ" function, which represents a circular aperture of radius a as $P(x,y) = circ(r/a)$ [where $r = (x^2+y^2)^{1/2}$]. Thus, the overall transmittance function of the thin lens can be expressed as:

$$t(x,y) = P(x,y) \exp\left[jk\frac{x^2+y^2}{2f}\right]. \tag{6.15}$$

6.5 Simple Systems with a Single Converging Thin Lens

This section examines three straightforward optical configurations involving a single converging thin lens. These configurations are as follows: (a) When the input optical field propagates towards the thin lens, and the objective is to compute the field distribution at the focal plane. (b) When the input optical field is positioned at a finite distance from the thin lens, and the goal is to determine the field distribution at the focal plane. (c) When the input optical field is located behind the lens, illuminated by a plane wave, with the aim of calculating the field distribution at the focal plane. These optical setups are analyzed to gain insight into and characterize the behavior of the optical system in each case, highlighting the role of a single converging thin lens.

6.5.1 Input Optical Field Against a Thin Lens

The geometry illustrated in Fig. 6.5 is now considered. An object optical field, denoted by $u_o(x_o, y_o)$, is incident on a converging thin lens with a focal length f. The aperture located to the left of the lens represents the pupil function $P(x_o, y_o)$. Both the thin lens and the aperture are assumed to lie in the $x_o y_o$-plane. The objective is to determine the optical field at a focal distance $z = f$ to the right of the lens. Based on the Fresnel diffraction integral, the optical field $u_f(x_f, y_f)$ at the focal plane is given by:

$$\begin{aligned} u_f(x_f,y_f) &= jk\frac{e^{-jkf}}{2\pi f}\int_{-\infty}^{+\infty}\int_{-\infty}^{+\infty} \exp\left\{-j\frac{k}{2f}\left[(x_f-x_o)^2+(y_f-y_o)^2\right]\right\} \times \\ &\quad P(x_o,y_o)t(x_o,y_o)u_o(x_o,y_o)dx_o dy_o \\ &= jk\frac{e^{-jkf}}{2\pi f}\exp\left[-j\frac{k}{2f}(x_f^2+y_f^2)\right]\int_{-\infty}^{+\infty}\int_{-\infty}^{+\infty}\exp\left[j\frac{2\pi}{\lambda f}(x_f x_o + y_f y_o)\right] \times \\ &\quad P(x_o,y_o)u_i(x_o,y_o)dx_o dy_o \\ &= jk\frac{e^{-jkf}}{2\pi f}\exp\left[-j\frac{k}{2f}(x_f^2+y_f^2)\right]\mathcal{F}\{P(x_o,y_o)u_o(x_o,y_o)\}\Big|_{\substack{f_x=x_f/\lambda f \\ f_y=y_f/\lambda f}}. \end{aligned} \tag{6.16}$$

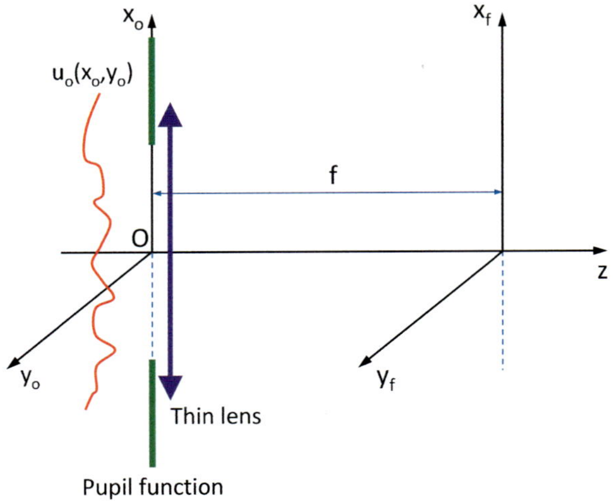

FIGURE 6.5: An input (object) optical field $u_o(x_o, y_o)$ is incident from the left on a converging thin lens. The distance of the lens from the origin is practically zero (in the figure is exaggerated). The aperture to the left of the lens comprise the pupil function $P(x_o, y_o)$. The focal distance of the thin lens is f. The coordinate system at the focal plane of the lens is denoted by $x_f y_f z$.

Therefore, the resulting field is related to the Fourier transform of $P(x_o, y_o)u_o(x_o, y_o)$. If the incident optical field has a smaller extent than the pupil function, the factor $P(x_o, y_o)$ in the Fourier transform can be neglected. In this case, the optical field at the focal plane of the lens closely approximates the Fourier transform of the incident optical field. This equivalence resembles the Fraunhofer diffraction pattern of the incident field at a distance f, as opposed to much larger distances that meet the criteria for the Fraunhofer diffraction regime (see the *Scalar Diffraction* chapter, Chapter 5).

6.5.2 Input Optical Field at Distance in Front of a Thin Lens

Now, let's examine the case illustrated in Fig. 6.6 [265]. The incident (object) field, denoted as $u_o(x_o, y_o)$, is positioned just to the right of the origin O. It may encompass the effect of the aperture shown and could contain information related to a 2D object or transparency. The field propagates over a distance d before reaching the thin lens. The objective is to determine the optical field at the focal plane of the lens, denoted as $u_f(x_f, y_f)$.

Using Eq. (6.16) (where the pupil function is neglected), the optical field at the focal plane can be expressed as (assuming the Fresnel-parabolic approximation holds):

$$u_f(x_f, y_f) = jk\frac{e^{-jkf}}{2\pi f} \exp\left[-j\frac{k}{2f}(x_f^2 + y_f^2)\right] \mathcal{F}\{u_\ell(x_\ell, y_\ell)\}\Big|_{\substack{f_x = x_f/\lambda f, \\ f_y = y_f/\lambda f}} \tag{6.17}$$

where $u_\ell(x_\ell, y_\ell)$ is the optical field that is incident to the left of the lens. This field can be easily determined using Eq. (6.8) in conjunction with Eq. (6.11). Then, the Fourier transform $U_\ell(f_x, f_y) = \mathcal{F}\{u_\ell(x_\ell, y_\ell)\}$ can be determined by:

$$U_\ell(f_x, f_y) = U_o(f_x, f_y)H(f_x, f_y) = U_o(f_x, f_y) \exp\left[-j\frac{2\pi d}{\lambda}\right] \exp\left[+j\pi\lambda d(f_x^2 + f_y^2)\right].$$

Substituting, the last Fourier transform in Eq. (6.17), the resulting field at the focal plane becomes:

$$u_f(x_f, y_f) = jk\frac{e^{-jk(f+d)}}{2\pi f} \exp\left[-j\frac{k}{2f}\left(1 - \frac{d}{f}\right)(x_f^2 + y_f^2)\right] U_o\left(\frac{x_f}{\lambda f}, \frac{y_f}{\lambda f}\right). \tag{6.18}$$

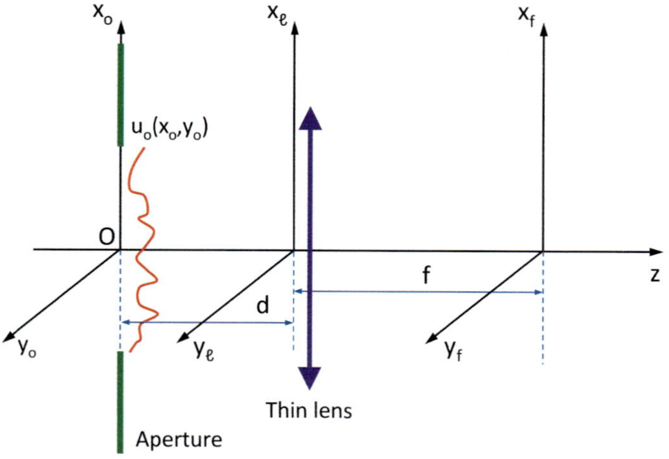

FIGURE 6.6: An input optical field $u_o(x_o, y_o)$ is incident from the left on a converging thin lens. The effect of any aperture is included in $u_o(x_o, y_o)$. The distance of the lens from the origin is d, and the focal distance of the thin lens is f. The coordinate system on the thin lens plane is denoted by $x_\ell y_\ell z$, while the coordinate system at the focal plane of the lens is denoted by $x_f y_f z$, as previously defined.

It is noticeable that the quadratic phase factor can be eliminated if $d = f$. In this case, the field at the focal plane of the lens simplifies to:

$$u_f(x_f, y_f) = j\frac{e^{-jk2f}}{\lambda f}U_o\left(\frac{x_f}{\lambda f}, \frac{y_f}{\lambda f}\right). \tag{6.19}$$

Hence, when placing the incident field on the first focal plane of the lens ($f > 0$), the field at the second focal plane directly corresponds to the Fourier transform of the incident field. It is important to note that in this analysis, the effect of the finite extent of the lens via its pupil function was neglected. In cases where this effect is considered, Eq. (6.17) should be modified as follows:

$$u_f(x_f, y_f) = jk\frac{e^{-jkf}}{2\pi f}\exp\left[-j\frac{k}{2f}(x_f^2 + y_f^2)\right]\mathcal{F}\{u_\ell(x_\ell, y_\ell)P(x_\ell, y_\ell)\}\Big|_{\substack{f_x = x_f/\lambda f \\ f_y = y_f/\lambda f}}. \tag{6.20}$$

Then, the Fourier transform of the product can be calculated using the convolution in the frequency domain, and is given by:

$$\begin{aligned}\mathcal{F}\{u_\ell(x_\ell, y_\ell)P(x_\ell, y_\ell)\} &= U_\ell(f_x, f_y) ** \tilde{P}(f_x, f_y) = [U_o(f_x, f_y)H(f_x, f_y)] ** \tilde{P} \\ &= \left\{U_o(f_x, f_y)\exp\left[-j\frac{2\pi d}{\lambda}\right]\exp\left[+j\pi\lambda d(f_x^2 + f_y^2)\right]\right\} ** \tilde{P},\end{aligned}$$

where $\tilde{P} = \mathcal{F}\{P(x_o, y_o)\}$. In general, the resulting field, $u_f(x_f, y_f)$, at the focal plane becomes:

$$u_f(x_f, y_f) = jk\frac{e^{-jk(f+d)}}{2\pi f}\exp\left[-j\frac{k}{2f}\left(1 - \frac{d}{f}\right)(x_f^2 + y_f^2)\right]$$

$$\left[\int_{-\infty}^{+\infty}\int_{-\infty}^{+\infty} U_o(f_x - f'_x, f_y - f'_y)e^{j\pi\lambda d(f_x'^2 + f_y'^2)}e^{j2\pi\lambda d(f_x f'_x + f_y f'_y)}\tilde{P}(f'_x, f'_y)df'_x df'_y\right]_{\substack{f_x = x_f/\lambda f \\ f_y = y_f/\lambda f}}. \tag{6.21}$$

In the latter case, even if $d = f$, the resulting field at the focal plane is not simply the Fourier transform of $u_o(x, y)$. However, if the pupil function $P(x_o, y_o) = 1$ across the entire lens plane (and assuming the lens is sufficiently large to encompass the entire field u_ℓ), then the Fourier transform of the pupil function becomes $\tilde{P}(f_x, f_y) = \delta(f_x, f_y)$, and Eq. (6.21) collapses into Eq. (6.18) as expected.

In Goodman [265] an approximate expression based on geometrical optics approximation is used. In this case a point (x_f, y_f) of the focal plane is projected via the center of the pupil aperture to the $(-(d/f)x_f, -(d/f)y_f)$ point. Then the pupil is projected into the input plane and modifies the incident field like $u_i(x, y)P(x + (d/f)x_f, y + (d/f)y_f)$. Then, Eq. (6.18) is modified to the form:

$$u_f(x_f, y_f) = jk\frac{e^{-jk(f+d)}}{2\pi f} \exp\left[-j\frac{k}{2f}\left(1 - \frac{d}{f}\right)(x_f^2 + y_f^2)\right] U_o'\left(\frac{x_f}{\lambda f}, \frac{y_f}{\lambda f}\right), \tag{6.22}$$

where,

$$U_o'\left(\frac{x_f}{\lambda f}, \frac{y_f}{\lambda f}\right) = \int_{-\infty}^{+\infty}\int_{-\infty}^{+\infty} \left\{u_o(x_o, y_o)P\left(x_o + \frac{d}{f}x_f, y_o + \frac{d}{f}y_f\right)\right\}$$
$$\exp\left[+j\frac{2\pi}{\lambda f}(x_f x_o + y_f y_o)\right] dx_o dy_o.$$

6.5.3 Input Optical Field Placed Behind a Thin Lens

Another configuration to be considered involves placing the object field (implicitly assumed to be produced by transparency) after (to the right of) a converging thin lens [265]. The geometric configuration is illustrated in Fig. 6.7. The lens aperture corresponds to the pupil function, which is initially neglected in the analysis. The incident field from the left of the thin lens is a normally incident plane wave represented as $u = u_0 e^{-jkz}$. The optical field generated by the plane wave impinging on the lens is calculated under the assumption that the Fresnel-parabolic approximation holds and can be expressed as:

$$u_\ell(x_\ell, y_\ell) = u_0 t(x_\ell, y_\ell) = u_0 \exp\left[j\frac{k}{2f}(x_\ell^2 + y_\ell^2)\right].$$

The optical field in front of the object transparency, denoted as $u_-(x_o, y_o)$, is obtained through straightforward propagation over a distance of $(f - d)$. Initially expressed in the spatial frequency domain, it is then transformed into the spatial domain as:

$$\begin{aligned} U_-(f_x, f_y) &= U_\ell(f_x, f_y)H(f_x, f_y; f - d) \\ &= u_0\left[j\lambda f e^{-j\pi\lambda f(f_x^2+f_y^2)}\right]\left[e^{-jk(f-d)}\ e^{j\pi\lambda(f-d)(f_x^2+f_y^2)}\right] = \\ &= ju_0\lambda f e^{-jk(f-d)} e^{-j\pi\lambda d(f_x^2+f_y^2)} \Longrightarrow \end{aligned}$$

$$u_-(x_o, y_o) = u_0\frac{f}{d}e^{-jk(f-d)} \exp\left[j\frac{k}{2d}(x_o^2 + y_o^2)\right].$$

The field to the right of the object transparency is generated by multiplying $u_-(x_o, y_o)$ with the object transparency, $u_o(x_o, y_o)$, as $u_+(x_o, y_o) = u_-(x_o, y_o)u_o(x_o, y_o)$. Subsequently, the

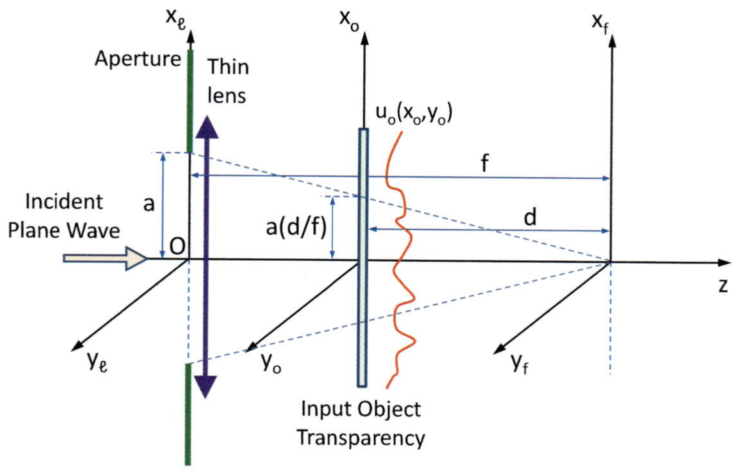

FIGURE 6.7: In this configuration [265] the input (object) optical field $u_o(x_o, y_o)$ is in the form of a transparency placed to the right of a converging thin lens of focal distance f. The lens aperture is also shown. The distance of the lens from the origin is negligible. The coordinate system on the thin lens plane is denoted by $x_\ell y_\ell z$, while the coordinate system at the focal plane of the lens is denoted by $x_f y_f z$, as previously. The coordinate system of the object transparency is denoted by $x_o y_o z$. The distance of the object transparency from the focal plane is d.

field at the focal plane, denoted as $u_f(x_f, y_f)$, can be calculated as:

$$u_f(x_f, y_f) = u_+ ** h(d) = ju_0 \frac{f}{d}\left(\frac{1}{\lambda d}\right) \exp\left[-j\frac{k}{2d}(x_f^2 + y_f^2)\right]$$

$$\int_{-\infty}^{+\infty}\int_{-\infty}^{+\infty} u_o(x_o, y_o) \exp\left[j\frac{2\pi}{\lambda d}(x_f x_o + y_f y_o)\right] dx_o dy_o \Longrightarrow$$

$$u_f(x_f, y_f) = ju_0 \frac{f}{d}\left(\frac{1}{\lambda d}\right) \exp\left[-j\frac{k}{2d}(x_f^2 + y_f^2)\right] \mathcal{F}\left\{u_o(x_o, y_o)\right\}\Big|_{\substack{f_x = x_f/\lambda d \\ f_y = y_f/\lambda d}}. \tag{6.23}$$

It is noteworthy that the resulting field at the focal plane bears a resemblance to a scaled Fourier transform of the transparency (object) field. The scaling factor can be adjusted by varying the distance d. Notably, when $d = f$, Eq. (6.23) becomes identical to Eq. (6.16) (without the effect of the pupil function and for $u_0 = 1$).

When accounting for the aperture in front of the lens, the procedure remains the same, but the final result becomes more intricate. Specifically, the lens field is defined as $u_\ell(x_\ell, y_\ell) = u_0 P(x_\ell, y_\ell) t(x_\ell, y_\ell)$. Subsequently, the corresponding $u_-(x, y)$, $u_+(x, y)$, and $u_f(x_f, y_f)$ are given by:

$$u_-(x_o, y_o) = [u_0 P(x_\ell, y_\ell) t(x_\ell, y_\ell)] ** h(x_o, y_o; f - d),$$

$$u_+(x_o, y_o) = u_-(x_o, y_o) u_o(x_o, y_o),$$

$$u_f(x_f, y_f) = u_+(x_o, y_o) ** h(x_f, y_f; d).$$

In this case, the final result cannot be cast into a simple form as in the case of the infinite aperture. If the calculations are performed, the ultimate expression in terms of integrals is given by:

$$u_f(x_f, y_f) = -u_0 \frac{e^{-jkf}}{\lambda^2 d(f-d)} \exp\left[-j\frac{k}{2d}(x_f^2 + y_f^2)\right] \int_{-\infty}^{+\infty}\int_{-\infty}^{+\infty} L(x_o, y_o) u_o(x_o, y_o) \times$$
$$\exp\left[-j\frac{kf}{2d(f-d)}(x_o^2 + y_o^2)\right] \exp\left[j\frac{2\pi}{\lambda d}(x_o x_f + y_o y_f)\right] dx_o dy_o, \tag{6.24}$$

$$\text{where,} \quad L(x_o, y_o) = \int_{-\infty}^{+\infty}\int_{-\infty}^{+\infty} P(x', y') \exp\left[-j\frac{kd}{2f(f-d)}(x'^2 + y'^2)\right]$$
$$\exp\left[j\frac{2\pi}{\lambda(f-d)}(x_o x' + y_o y')\right] dx' dy'.$$

It can be demonstrated straightforwardly that the intricate expression in Eq. (6.24) simplifies to Eq. (6.23) when $P(x_\ell, y_\ell) = 1$. Once again, Goodman [265] has provided a much simpler expression based on a geometrical optics approximation by projecting the pupil function into the object plane $x_o y_o x$. From Fig. 6.7, it can be shown that $x_\ell = (f/d)x_o$ and similarly $y_\ell = (f/d)y_o$. This approximation leads to the following result:

$$u_f(x_f, y_f) = j u_0 \frac{f}{d}\left(\frac{1}{\lambda d}\right) \exp\left[-j\frac{k}{2d}(x_f^2 + y_f^2)\right] \mathcal{F}\left\{u_o(x_o, y_o) P\left(\frac{f}{d}x_o, \frac{f}{d}y_o\right)\right\}\Bigg|_{\substack{f_x = x_f/\lambda d \\ f_y = y_f/\lambda d}}. \tag{6.25}$$

6.6 Impulse Response and Coherent Point Spread Function of a Converging Thin Lens

The calculation of the impulse response function for a converging thin lens is based on the imaging geometry illustrated in Fig. 6.8. For simplicity, it is assumed that the point object source is located at the origin of the object coordinate system, labeled as $x_o y_o z$. In this setup, the object function is expressed as $u_o(x_o, y_o) = A\delta(x_o, y_o)$, where A serves as a scaling factor that adjusts for dimensions, given that the delta function has dimensions of m^{-2}. For example, if $u_o(x_o, y_o)$ represents an electric field (measured in V/m), the units of A are in Vm. This methodology, consistent with previous approaches, employs the Fresnel-parabolic approximation. As a result, the optical field immediately before the lens aperture plane, denoted as $u_{\ell-}(x_\ell, y_\ell)$, is given by [see Eq. (6.12)]:

$$u_{\ell-}(x_\ell, y_\ell) = A\delta(x_o, y_o) ** h(x_\ell, y_\ell; s_o) = jA\frac{e^{-jks_o}}{\lambda s_o} \exp\left[-j\frac{k}{2s_o}(x_\ell^2 + y_\ell^2)\right].$$

Infinitesimally after the lens, the optical field is modified by the lens's transmission function. It is represented as:

$$u_{\ell+}(x_\ell, y_\ell) = u_{\ell-}(x_\ell, y_\ell) P(x_\ell, y_\ell) t(x_\ell, y_\ell)$$
$$= jA\frac{e^{-jks_o}}{\lambda s_o} P(x_\ell, y_\ell) \exp\left[-j\frac{k}{2}\left(\frac{1}{s_o} - \frac{1}{f}\right)(x_\ell^2 + y_\ell^2)\right],$$

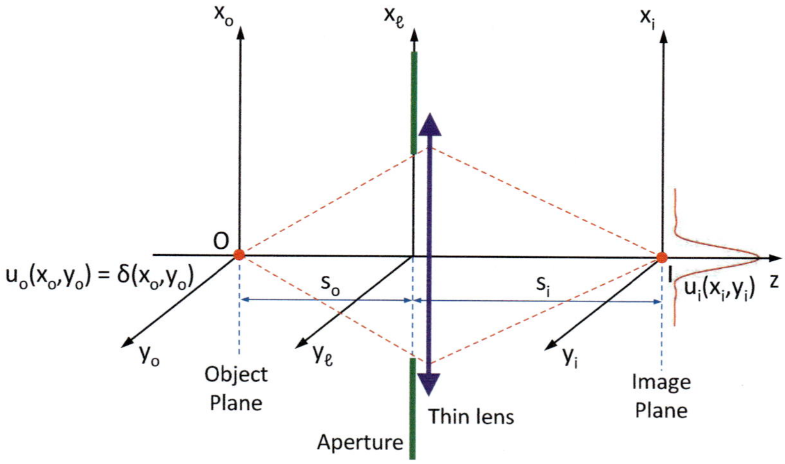

FIGURE 6.8: The geometric configuration for the computation of the impulse response of a converging thin lens for imaging. The object plane is the $x_o y_o$, the lens plane is the $x_\ell y_\ell$, and the image plane is the $x_i y_i$. All coordinate systems have a common z-axis. The object distance from the lens is s_o while the image distance is s_i. The lens and its aperture are located at the origin of the $x_\ell y_\ell z$ coordinate system. The input is a point source at the origin of the object coordinate system.

where $t(x_\ell, y_\ell)$ and $P(x_\ell, y_\ell)$ are the transmittance and the pupil function of the thin lens, respectively. Then the optical field at the image plane, which is the final impulse response of the system, $h(x_i, y_i; 0, 0)$, becomes:

$$u_i(x_i, y_i) = h(x_i, y_i; 0, 0) = u_{\ell+}(x_\ell, y_\ell) ** h(x_i, y_i; s_i)$$

$$= -A\frac{e^{-jk(s_o+s_i)}}{\lambda^2 s_o s_i} \exp\left[-j\frac{k}{2s_i}(x_i^2 + y_i^2)\right] \times$$

$$\int_{-\infty}^{+\infty}\int_{-\infty}^{+\infty} P(x_\ell, y_\ell) \exp\left[-j\frac{k}{2}\left(\frac{1}{s_o} + \frac{1}{s_i} - \frac{1}{f}\right)(x_\ell^2 + y_\ell^2)\right] \exp\left[j\frac{2\pi}{\lambda s_i}(x_i x_\ell + y_i y_\ell)\right] dx_\ell dy_\ell \tag{6.26}$$

Equation (6.26) represents the impulse response of the system for a point source located at the origin of the object plane. To extend this to the case of a point source situated at a general position in the object plane, the object function is expressed as $u_o(x'_o, y'_o) = A\delta(x'_o - x_o, y'_o - y_o)$. Here, the prime notation $x'_o y'_o$ is used for coordinates in the object plane, while (x_o, y_o) specifies the location of the input point source. Following the same procedure as before, the optical field at the image plane, which represents the generalized impulse response of the system $h(x_i, y_i; x_o, y_o)$, is formulated according to [265]:

$$h(x_i, y_i; x_o, y_o) = -\frac{e^{-jk(s_o+s_i)}}{\lambda^2 s_o s_i} \exp\left[-j\frac{k}{2s_i}(x_i^2 + y_i^2)\right] \exp\left[-j\frac{k}{2s_o}(x_o^2 + y_o^2)\right] \times$$

$$\int_{-\infty}^{+\infty}\int_{-\infty}^{+\infty} P(x_\ell, y_\ell) \exp\left[-j\frac{k}{2}\left(\frac{1}{s_o} + \frac{1}{s_i} - \frac{1}{f}\right)(x_\ell^2 + y_\ell^2)\right] \times$$

$$\exp\left\{j\frac{2\pi}{\lambda}\left[\left(\frac{x_i}{s_i} + \frac{x_o}{s_o}\right)x_\ell + \left(\frac{y_i}{s_i} + \frac{y_o}{s_o}\right)y_\ell\right]\right\} dx_\ell dy_\ell. \tag{6.27}$$

From Eq. (6.27), the most problematic term is the quadratic phase term within the integral. To address this, the following defocus parameter can be defined:

$$D = \frac{1}{s_o} + \frac{1}{s_i} - \frac{1}{f}. \tag{6.28}$$

When the defocus parameter is zero ($D = 0$), the distances s_o, s_i, and f satisfy the geometric imaging condition of a thin lens under the paraxial approximation. In this case, the impulse response function of the thin lens is also referred to as the *coherent Point Spread Function* (coherent PSF). Typically, the quadratic phase factors outside the integral in Eq. (6.27) do not vary significantly in imaging systems with acceptable levels of diffraction-induced blurring [265]. Furthermore, since most light detectors are sensitive to intensity rather than phase, these phase factors are often neglected. Consequently, Eq. (6.27) can be simplified to:

$$h(x_i, y_i; x_o, y_o) \simeq -\frac{1}{\lambda^2 s_o s_i} \int_{-\infty}^{+\infty}\int_{-\infty}^{+\infty} P(x_\ell, y_\ell) \exp\left[-j\frac{k}{2}D\left(x_\ell^2 + y_\ell^2\right)\right] \times$$
$$\exp\left\{j\frac{2\pi}{\lambda}\left[\left(\frac{x_i}{s_i} + \frac{x_o}{s_o}\right)x_\ell + \left(\frac{y_i}{s_i} + \frac{y_o}{s_o}\right)y_\ell\right]\right\} dx_\ell dy_\ell. \tag{6.29}$$

If the geometrical optics imaging condition holds ($D = 0$), the quadratic phase factor inside the integral of Eq. (6.29) is eliminated. As a result, the coherent PSF becomes proportional to a scaled version of the Fourier transform of the pupil function. The expression for $h(x_i, y_i; x_o, y_o)$ corresponds to a linear space-variant system, signifying that the relationship between the image and the object is described by a superposition integral rather than a convolution integral [265]. This spatial variation stems from the inversion and magnification inherent in the imaging process. To convert the image-object relationship into a convolution, a change of variables is required, as demonstrated below [265]. By employing the magnification factor from geometrical optics, $m = -s_i/s_o$, the equation can be reformulated as:

$$h(x_i, y_i; x_o, y_o) \simeq -\frac{1}{\lambda^2 s_o s_i} \int_{-\infty}^{+\infty}\int_{-\infty}^{+\infty} P(x_\ell, y_\ell) \times$$
$$\exp\left\{j\frac{2\pi}{\lambda s_i}\left[(x_i - m x_o)x_\ell + (y_i - m y_o)y_\ell\right]\right\} dx_\ell dy_\ell$$
$$= -\frac{1}{\lambda^2 s_o s_i} \mathcal{F}\left\{P(x_\ell, y_\ell)\right\}\Big|_{\substack{f_x = (x_i - m x_o)/\lambda s_i \\ f_y = (y_i - m y_o)/\lambda s_i}}$$
$$= -\frac{1}{\lambda^2 s_o s_i} \tilde{P}\left(\frac{x_i - m x_o}{\lambda s_i}, \frac{y_i - m y_o}{\lambda s_i}\right), \tag{6.30}$$

where it is reminded that $\tilde{P}(f_x, f_y) = \mathcal{F}\{P(x, y)\}$. In the case that geometrical optics approximation is valid, and for $\lambda \to 0$, it can be shown straightforwardly [265] that the coherent PSF can be expressed as:

$$\lim_{\lambda \to 0}\{h(x_i, y_i; x_o, y_o)\} \simeq \frac{1}{|m|}\,\delta\left(x_o - \frac{x_i}{m}, y_o - \frac{y_i}{m}\right).$$

In this case, the optical field at the image plane is $u_i(x_i, y_i) = u_o(x_o, y_o) ** h(x_i, y_i; x_o, y_o) = (1/|m|)u_o(x_i/m, y_i/m)$, which is the prediction of the geometrical optics. Equation (6.30) can be rewritten using the following transformations:

$$\tilde{x}_o = m x_o, \quad \tilde{y}_o = m y_o \quad \Longrightarrow$$

$$h(x_i, y_i; \tilde{x}_o, \tilde{y}_o) \simeq -\frac{1}{\lambda^2 s_o s_i} \int_{-\infty}^{+\infty}\int_{-\infty}^{+\infty} P(x_\ell, y_\ell) \exp\left\{ j\frac{2\pi}{\lambda s_i}\left[(x_i - \tilde{x}_o)\, x_\ell + (y_i - \tilde{y}_o)\, y_\ell\right] \right\} dx_\ell dy_\ell.$$

For $\quad f_{x\ell} = \dfrac{x_\ell}{\lambda s_i}, \quad f_{y\ell} = \dfrac{y_\ell}{\lambda s_i} \quad \Longrightarrow$

$$h(x_i, y_i; \tilde{x}_o, \tilde{y}_o) \simeq m \int_{-\infty}^{+\infty}\int_{-\infty}^{+\infty} P(\lambda s_i f_{x\ell}, \lambda s_i f_{y\ell}) \exp\left\{ j2\pi\left[(x_i - \tilde{x}_o)\, f_{x\ell} + (y_i - \tilde{y}_o)\, f_{y\ell}\right] \right\} df_{x\ell} df_{y\ell},$$

$$h(x_i, y_i; \tilde{x}_o, \tilde{y}_o) \simeq |m|\, \tilde{h}(x_i - \tilde{x}_o, y_i - \tilde{y}_o) \Longrightarrow \tilde{h}(x_i - \tilde{x}_o, y_i - \tilde{y}_o) = \frac{1}{|m|}\, h(x_i, y_i; \tilde{x}_o, \tilde{y}_o). \tag{6.31}$$

Here, m is replaced with $|m|$ because the sign of h, representing a constant phase factor, is not of interest. Using the definition of the simplified (at focus) coherent PSF, the image and the object fields are related as follows due to the linearity of the optical wave propagation (superposition integral [265]):

$$\begin{aligned} u_i(x_i, y_i) &= \int_{-\infty}^{+\infty}\int_{-\infty}^{+\infty} h(x_i, y_i; x_o, y_o) u_o(x_o, y_o) dx_o dy_o \\ &= \int_{-\infty}^{+\infty}\int_{-\infty}^{+\infty} |m|\, \tilde{h}(x_i - \tilde{x}_o, y_i - \tilde{y}_o) u_o(x_o, y_o) dx_o dy_o \\ &= \int_{-\infty}^{+\infty}\int_{-\infty}^{+\infty} |m|\, \tilde{h}(x_i - \tilde{x}_o, y_i - \tilde{y}_o) u_o\left(\frac{\tilde{x}_o}{m}, \frac{\tilde{y}_o}{m}\right) \frac{1}{m^2} d\tilde{x}_o d\tilde{y}_o \\ &= \tilde{h}(x_i, y_i) ** \left[\frac{1}{|m|}\, u_o\left(\frac{x_i}{m}, \frac{y_i}{m}\right)\right]. \end{aligned} \tag{6.32}$$

$$\text{where} \quad \tilde{h}(x_i, y_i) = \frac{1}{|m|}\frac{1}{(\lambda s_i)^2} \mathcal{F}\left\{P(x_\ell, y_\ell)\right\}\Big|_{f_x = x_i/\lambda s_i,\, f_y = y_i/\lambda s_i} = \frac{1}{|m|}\frac{1}{(\lambda s_i)^2} \tilde{P}\left(\frac{x_i}{\lambda s_i}, \frac{y_i}{\lambda s_i}\right).$$

The result of Eq. (6.32) reveals that the effect of diffraction is to convolve the ideal inverted and scaled image predicted by geometrical optics with the Fourier transform of the scaled pupil function of the thin lens [see Eq. (6.30)]. This convolution serves as a smoothing operation, which can lead to the attenuation of fine details of the object in the image plane. The diffraction introduces a blurring effect, which limits the resolution of the optical system and causes the finer features of the object to become less distinguishable in the image.

Example 6.6.1:

Consider the coherent point spread function of Eq. (6.26) which corresponds to a point object at the origin of the object plane. Also assume that the pupil function of the thin lens is circular of radius a. The purpose of this example is to quantify the effect of defocusing parameter to the coherent PSF function. Therefore, plot the coherent PSF with $D = 0$ and with $D \neq 0$. Assume that the thin lens has a focal length of $f = 10\,\text{cm}$, $a = 10\,\text{cm}$ and that the object distance is $s_o = 50\,\text{cm}$. The freespace wavelength of the monochromatic source is $\lambda = 0.55\,\mu\text{m}$.

Solution: First, let us determine the image position using geometrical optics (aberration-free). By applying the lens equation $1/s_o + 1/s_i = 1/f$, the image distance s_i can be derived as $s_i = s_o f/(s_o - f) = 12.5\,\text{cm}$. The defocus parameter, as given by Eq. (6.28), is non-zero ($D \neq 0$) if the image plane distance s_i deviates from 12.5 cm. To calculate the coherent point spread function (PSF), $h(x_i, y_i; 0, 0)$ from Eq. (6.26), it is useful to rewrite it in cylindrical coordinates and exploit the azimuthal symmetry of the system. Let us introduce $x_i = r_i \cos\phi_i$, $y_i = r_i \sin\phi_i$, for the image coordinate system, and $x_\ell = r_\ell \cos\phi_\ell$, $y_\ell = r_\ell \sin\phi_\ell$ for the lens coordinate system. Additionally, the pupil function is given by $P(r_\ell, \phi_\ell) = \text{circ}(r_\ell/a)$.

$$h(x_i, y_i; 0, 0) = \frac{e^{-jk(s_o+s_i)}}{\lambda^2 s_o s_i} \exp\left[-j\frac{k}{2s_i} r_i^2\right] \times$$

$$\int_0^{+\infty}\int_0^{2\pi} \text{circ}\left(\frac{r_\ell}{a}\right) \exp\left[-j\frac{k}{2} D r_\ell^2\right] \exp\left[j\frac{2\pi}{\lambda s_i} r_i r_\ell \cos(\phi_i - \phi_\ell)\right] r_\ell dr_\ell d\phi_\ell$$

$$= \frac{e^{-jk(s_o+s_i)}}{\lambda^2 s_o s_i} \exp\left[-j\frac{k}{2s_i} r_i^2\right] \int_0^{a} \exp\left[-j\frac{k}{2} D r_\ell^2\right] 2\pi J_0\left(\frac{2\pi r_i}{\lambda s_i} r_\ell\right) r_\ell dr_\ell.$$

The integral in the last expression can only be computed numerically when $D \neq 0$. In the case of perfect focusing, where $D = 0$ the integral simplifies to the well-known $2J_1(x)/x$ form, whose square magnitude is recognized as the Airy function. The numerically calculated $|h(r_i)|$ is normalized and visualized in Figs. 6.9. Specifically, the magnitude $|h|$ is shown in Fig. 6.9(a) for $s_i = 12.49995\,\text{cm}$ and $D \simeq -3.2 \times 10^{-7}\,\text{cm}^{-1}$. The same result would be obtained if $s_i = 12.50005\,\text{cm}$ and $D \simeq +3.2 \times 10^{-7}\,\text{cm}^{-1}$. In Fig. 6.9(b) the normalized magnitude of the coherent PSF is shown for $s_i = 12.49994\,\text{cm}$ and $D \simeq -3.84 \times 10^{-7}\,\text{cm}^{-1}$. It is observed that a slight increase in defocus contributes to the deterioration of the PSF. Finally, Figs. 6.9(c) (for $s_i = 12.49990\,\text{cm}$ and $D \simeq -6.4 \times 10^{-7}\,\text{cm}^{-1}$) and 6.9(d) (for $s_i = 12.49987\,\text{cm}$ and $D \simeq -8.3 \times 10^{-7}\,\text{cm}^{-1}$) depict the widening and smoothing of the PSF, resulting in increased image blurring.

6.7 Imaging Systems Comprised of Four or Two Converging Thin Lenses

In this section, the analysis begins with an optical imaging system consisting of four converging thin lenses. The geometric configuration is depicted in Fig. 6.10. The first two lenses (L1a and L1b) have a focal length f_1, while the other two lenses (L2a and L2b) have a focal length f_2. The middle lenses (L1b and L2a), along with the system aperture, are effectively

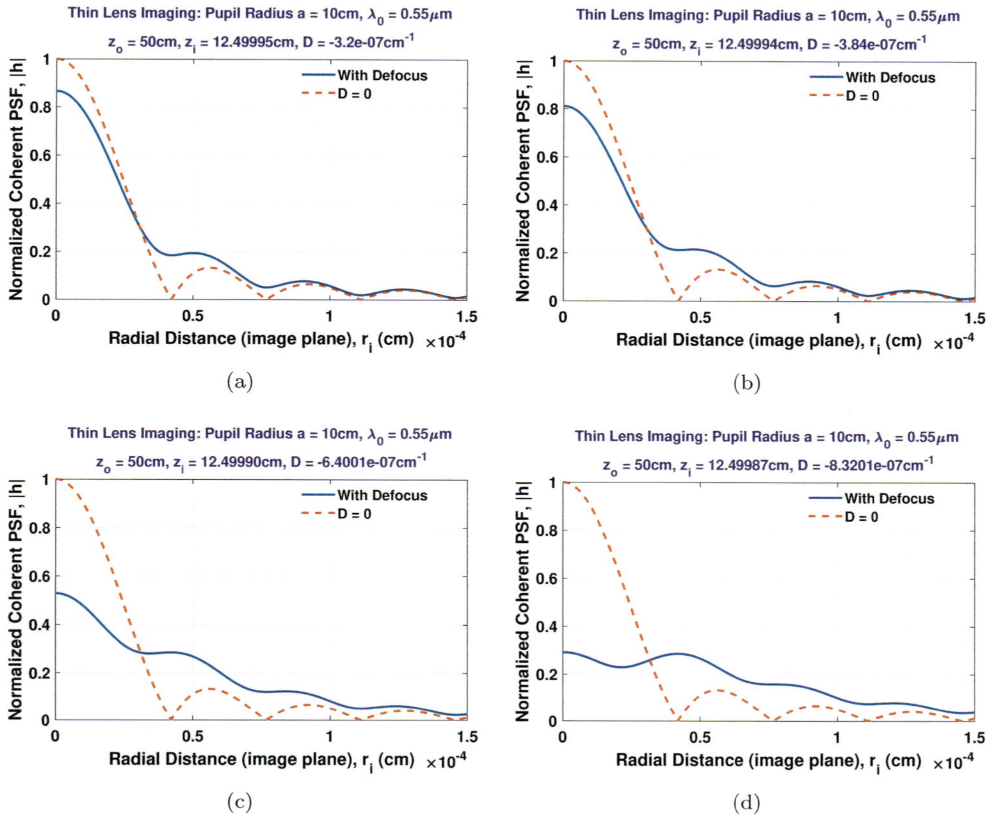

FIGURE 6.9: Normalized coherent PSF function for a converging thin lens of $f = 10\,\text{cm}$, aperture radius $a = 10\,\text{cm}$, object distance $s_o = 50\,\text{cm}$, and image distance $s_i = 12.5\,\text{cm}$. The freespace wavelength of the monochromatic source is $\lambda = 0.55\,\mu\text{m}$. (a) A comparison is made between the focused PSF and the defocused PSF with $s_i = 12.49995\,\text{cm}$. (b) A comparison is made between the focused PSF and the defocused PSF with $s_i = 12.49994\,\text{cm}$. (c)A comparison is made between the focused PSF and the defocused PSF with $s_i = 12.49990\,\text{cm}$. (d) A comparison is made between the focused PSF and the defocused PSF with $s_i = 12.49987\,\text{cm}$.

collocated at the $x_\ell y_\ell$ plane. The object field incident from the left is denoted as $u_o(x_o, y_o)$. The optical field immediately before lens L1b, denoted $u_{\ell-}(x_\ell, y_\ell)$, is determined from Eq. (6.16) as:

$$u_{\ell-}(x_\ell, y_\ell) = j\frac{e^{-jkf_1}}{\lambda f_1} \exp\left[-j\frac{k}{2f_1}(x_\ell^2 + y_\ell^2)\right] \mathcal{F}\{u_o(x_o, y_o)\}\Big|_{\substack{f_x = x_\ell/\lambda f_1 \\ f_y = y_\ell/\lambda f_1}}.$$

The optical field infinitesimally to the right of the aperture, $u_{\ell+}(x_\ell, y_\ell)$, is expressed as:

$$u_{\ell+}(x_\ell, y_\ell) = j\frac{e^{-jkf_1}}{\lambda f_1} P(x_\ell, y_\ell)\mathcal{F}\{u_o(x_o, y_o)\}\Big|_{\substack{f_x = x_\ell/\lambda f_1 \\ f_y = y_\ell/\lambda f_1}} = j\frac{e^{-jkf_1}}{\lambda f_1} P(x_\ell, y_\ell) U_o\left(\frac{x_\ell}{\lambda f_1}, \frac{y_\ell}{\lambda f_1}\right).$$

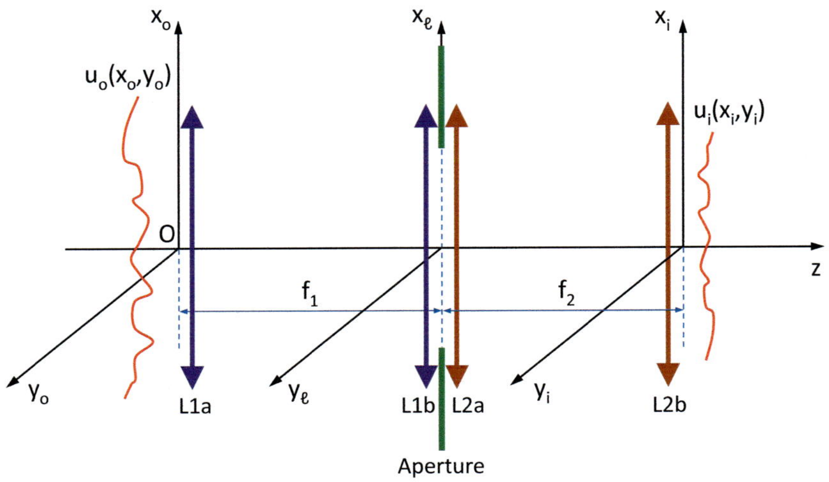

FIGURE 6.10: The geometric configuration of an imaging system comprised of two converging thin lenses (L1a and L1b) of focal length f_1, and two converging thin lenses (L2a and L2b) of focal length f_2, arranged in tandem. The object plane is the $x_o y_o$-plane, the two middle lenses plane is the $x_\ell y_\ell$-plane, and the image plane is the $x_i y_i$-plane. All coordinate systems have a common z-axis. The L1a and L1b lenses are separated by a distance f_1, while the L2a and L2b lenses are separated by a distance f_2. The lenses L1b and L2a, along with the aperture, are essentially located in the $x_\ell y_\ell$-plane.

This field can be propagated to the right in a similar manner by applying Eq. (6.16) once again. The resulting field to the right of the L2b lens is given by:

$$
\begin{aligned}
u_i(x_i, y_i) &= j\frac{e^{-jkf_2}}{\lambda f_2}\mathcal{F}\{u_{\ell+}(x_\ell, y_\ell)\}\Big|_{\substack{f_x = x_i/\lambda f_2 \\ f_y = y_i/\lambda f_2}} \\
&= -\frac{e^{-jk(f_1+f_2)}}{\lambda^2 f_1 f_2}\mathcal{F}\left\{U_o\left(\frac{x_\ell}{\lambda f_1}, \frac{y_\ell}{\lambda f_1}\right) P(x_\ell, y_\ell)\right\}\Bigg|_{\substack{f_x = x_i/\lambda f_2 \\ f_y = y_i/\lambda f_2}} \\
&= -\frac{e^{-jk(f_1+f_2)}}{\lambda^2 f_1 f_2}\left[\mathcal{F}\left\{U_o\left(\frac{x_\ell}{\lambda f_1}, \frac{y_\ell}{\lambda f_1}\right)\right\} ** \tilde{P}(f_x, f_y)\right]\Bigg|_{\substack{f_x = x_i/\lambda f_2 \\ f_y = y_i/\lambda f_2}} \\
&= -\frac{e^{-jk(f_1+f_2)}}{\lambda^2 f_1 f_2}\left[(\lambda f_1)^2 u_o(-\lambda f_1 f_x, -\lambda f_1 f_y) ** \tilde{P}(f_x, f_y)\right]\Bigg|_{\substack{f_x = x_i/\lambda f_2 \\ f_y = y_i/\lambda f_2}} \\
&= -e^{-jk(f_1+f_2)}\frac{f_1}{f_2}\int_{-\infty}^{+\infty}\int_{-\infty}^{+\infty} u_o\left[-\lambda f_1\left(\frac{x_i}{\lambda f_2} - f'_x\right), -\lambda f_1\left(\frac{y_i}{\lambda f_2} - f'_y\right)\right] \times \\
&\qquad \tilde{P}(f'_x, f'_y)\,df'_x df'_y.
\end{aligned}
$$

At this point, the following transformations are applied: $\lambda f_2 f'_x = x'_i$ and $\lambda f_2 f'_y = y'_i$. Furthermore, let $M = f_2/f_1$. Then, the last expression can then be rewritten as:

$$u_i(x_i, y_i) = -e^{-jk(f_1+f_2)} \int_{-\infty}^{+\infty}\int_{-\infty}^{+\infty} \underbrace{\left[\frac{1}{M} u_o\left(-\frac{x_i - x'_i}{M}, -\frac{y_i - y'_i}{M}\right)\right]}_{= u_g(x_i - x'_i, y_i - y'_i)} \times$$

$$\underbrace{\left[\frac{1}{(\lambda f_2)^2} \tilde{P}\left(\frac{x'_i}{\lambda f_2}, \frac{y'_i}{\lambda f_2}\right)\right]}_{= h(x'_i, y'_i)} dx'_i dy'_i$$

$$= -e^{-jk(f_1+f_2)} \left[\frac{1}{M} u_o\left(-\frac{x_i}{M}, -\frac{y_i}{M}\right)\right] ** \left[\frac{1}{(\lambda f_2)^2} \tilde{P}\left(\frac{x_i}{\lambda f_2}, \frac{y_i}{\lambda f_2}\right)\right] \tag{6.33}$$

$$= \left[-e^{-jk(f_1+f_2)}\right] u_g(x_i, y_i) ** h(x_i, y_i).$$

Typically, the constant phase term $-e^{-jk(f_1+f_2)}$ can be neglected.

A similar configuration of an imaging system using two thin lenses instead of four is shown in Fig. 6.11. This configuration is commonly referred to as a $4f$-system. The system in Fig. 6.11 is equivalent to the system in Fig. 6.10 if vignetting effects are neglected. Using Eq. (6.19), the optical fields immediately before the aperture plane, $u_{p-}(x_p, y_p)$, and immediately after the aperture plane, $u_{p+}(x_p, y_p)$, are given by:

$$u_{p-}(x_p, y_p) = j\frac{e^{-jk2f_1}}{\lambda f_1} U_o\left(\frac{x_p}{\lambda f_1}, \frac{y_p}{\lambda f_1}\right).$$

$$u_{p+}(x_p, y_p) = j\frac{e^{-jk2f_1}}{\lambda f_1} U_o\left(\frac{x_p}{\lambda f_1}, \frac{y_p}{\lambda f_1}\right) P(x_p, y_p).$$

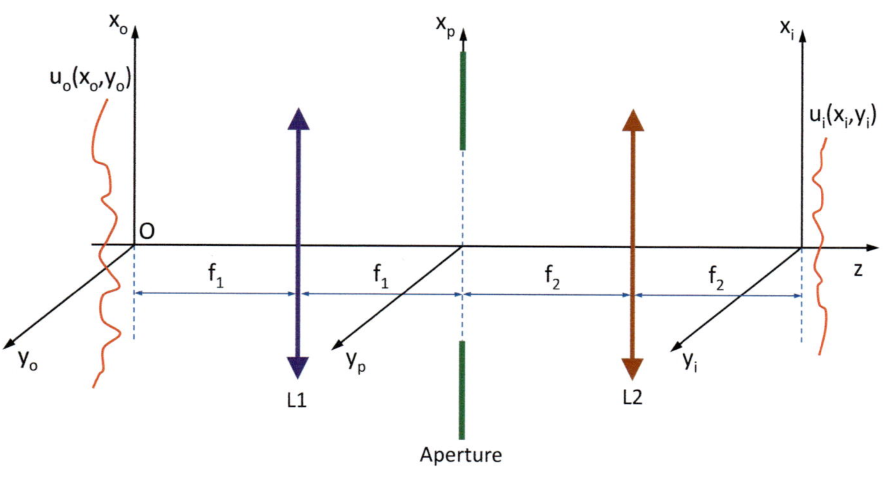

FIGURE 6.11: The geometric configuration of an imaging system (also known as a $4f$-system) comprised of two converging thin lenses (L1 and L2) of focal lengths f_1 and f_2, respectively. The object plane is the $x_o y_o$-plane, the aperture plane is the $x_p y_p$-plane, and the image plane is the $x_i y_i$-plane. All coordinate systems have a common z-axis.

Then using exactly the same approach that was followed in the derivation of Eq. (6.33), the resulting image field is given by:

$$u_i(x_i, y_i) = -e^{-jk2(f_1+f_2)} \left[\frac{1}{M} u_o\left(-\frac{x_i}{M}, -\frac{y_i}{M}\right)\right] ** \left[\frac{1}{(\lambda f_2)^2} \tilde{P}\left(\frac{x_i}{\lambda f_2}, \frac{y_i}{\lambda f_2}\right)\right]$$
$$= \left[-e^{-jk2(f_1+f_2)}\right] u_g(x_i, y_i) ** h(x_i, y_i). \tag{6.34}$$

The only difference between Eqs. (6.33) and (6.34) lies in the constant phase factor, which is typically neglected. Of course, in the case of vignetting, the two systems are only approximately equivalent. Again the factor $u_g(x_i, y_i) = (1/M)u_o(-x_i/M, -y_i/M)$ represents the geometrical image, and $h(x_i, y_i)$ accounts for the diffraction effects introduced by the pupil aperture of the lens. It is instructive to use the thin lens equation from geometrical optics to verify u_g. Applying the thin lens equation for the first lens, $1/(s_1 = f_1) + 1/s_1' = 1/f_1 \longrightarrow s_1' = \infty$. Then, applying the equation for the second lens, $1/[s_2 = -\infty - (f_1 + f_2)] + 1/s_2' = 1/f_2 \longrightarrow s_2' = f_2$. Therefore, the final image is located at the $x_i y_i$-plane. The magnification is given by $m = (-s_1'/s_1)(-s_2'/s_2) = (-\infty/f_1))(-f_2/(-\infty)) = -f_2/f_1 = -M$. Thus, $u_g = (1/M)u_o(-x_i/M, -y_i/M)$, which corresponds to an inverted image. The sign of u_g is retained positive, as the constant phase shifts are not of interest.

Example 6.7.1:

This example, originating from the lecture notes of Prof. W. T. Rhodes from his 1984 Fourier Optics class [348], illustrates a two-point object for the introduction of incoherent imaging. Assume that the object field for the configuration of Fig. 6.11 is $u_o(x_o, y_o) = \delta(x_o, y_o) + \delta(x_o - a, y_o)e^{j\theta(t)}$ where $a > 0$ and $\theta(t)$ a time-varying phase function. The objective of this example is to determine the image field $u_i(x_i, y_i)$, and its intensity $I_i(x_i, y_i)$ for various cases of $\theta(t)$. For simplicity assume that $f_1 = f_2 = f = 10\,\text{cm}$. The pupil function is a square of side $L = 1\,\text{cm}$. The freespace wavelength is $\lambda_0 = 1\,\mu\text{m}$ and the shift $a = \lambda f/L = 0.001\,\text{mm}$.

Solution: According to Eq. (6.34), since $M = f_2/f_1 = 1$, the geometrical optics field is given by $u_g(x_i, y_i) = u_o(-x_i, -y_i) = \delta(-x_i, -y_i) + \delta(-x_i - a, -y_i)e^{j\theta(t)} = \delta(x_i, y_i) + \delta(x_i + a, y_i)e^{j\theta(t)}$. On the other hand, the pupil function is $P(x_\ell, y_\ell) = \text{rect}(x_\ell/L, y_\ell/L)$, and the coherent PSF is expressed as:

$$h(x_i, y_i) = \left(\frac{L}{\lambda_0 f}\right)^2 \text{sinc}\left(L\frac{x_i}{\lambda_0 f}, L\frac{y_i}{\lambda_0 f}\right).$$

The resulting image field is $u_i = u_g ** h$ (neglecting the constant phase factor) and is given by:

$$u_i(x_i, y_i) = \left(\frac{L}{\lambda_0 f}\right)^2 \Bigg\{\underbrace{\text{sinc}\left(L\frac{x_i}{\lambda_0 f}, L\frac{y_i}{\lambda_0 f}\right)}_{=u_1} + \underbrace{\text{sinc}\left(L\frac{x_i + a}{\lambda_0 f}, L\frac{y_i}{\lambda_0 f}\right)e^{j\theta(t)}}_{=u_2}\Bigg\}.$$

Since, in general, $\theta(t)$ could be a random function (as in the case of an incoherent object field), the intensity in the image plane can be determined using time averaging and is equal to:

$$I_i(x_i, y_i) = \langle |u_i(x_i, y_i)|^2 \rangle$$
$$= \left(\frac{L}{\lambda_0 f}\right)^4 \left\{ |u_1|^2 + |u_2|^2 + 2\text{Re}\{u_1 u_2^*\} \langle \cos\theta(t) \rangle \right\},$$

where $\langle * \rangle$ denotes time averaging. At this point, some special cases are considered. The first case is when $\theta(t) = 0$ (for any t), which implies $\langle \cos\theta(t) \rangle = 1$. The resulting intensity in the image plane along the x_i-axis is shown in Fig. 6.12(a). This outcome illustrates the coherent addition of the u_1 and u_2 fields that are in phase.

The second special case is when $\theta(t) = \pi/2$ (for any t), leading to $\langle \cos\theta(t) \rangle = 0$. The resulting intensity in the image plane along the x_i-axis is shown in Fig. 6.12(b). This scenario illustrates the coherent addition of the u_1 and u_2 fields that are out of phase by $\pi/2$.

The third special case occurs when $\theta(t) = \pi$ (for any t), resulting in $\langle \cos\theta(t) \rangle = -1$. The corresponding intensity in the image plane along the x_i-axis is presented in Fig. 6.12(c). In this case, the outcome reflects the coherent addition of the u_1 and u_2 fields that are out of phase by π.

If $\theta = \omega t$ (where ω is a known radial frequency), then the $\langle \cos\theta(t) \rangle = \langle \cos(\omega t) \rangle = 0$, resulting in the same outcome as the case of $\theta = \pi/2$. A similar result is obtained when $\theta(t)$ is a random variable with zero average. In such cases, the resulting intensity can be expressed as:

$$I_i(x_i, y_i) = \left(\frac{L}{\lambda_0 f}\right)^4 \left\{|u_1|^2 + |u_2|^2\right\} = I_1 + I_2.$$

Here, $I_i = (L/\lambda_0 f)^4 |u_i|^2$ (with $i = 1, 2$) represent the light intensities due to each point of the object. The scenario of $\langle \cos\theta(t) \rangle = 0$ corresponds to the case of incoherent imaging where the image intensity is the sum of the individual intensities of each point of the object. Conversely, when $\theta(t)$ is constant, there is a constant phase shift between the two object point sources, and the resulting intensity includes an interference term, $2\mathrm{Re}\{u_1 u_2^*\} \langle \cos\theta(t) \rangle$.

Coherent imaging generally involves the use of light sources that maintain a fixed phase relationship across their wavefronts. This coherence enables the formation of interference patterns when the light waves interact, providing fine details about the observed object. In contrast, incoherent imaging employs light sources where the phase relationships between different parts of the wavefront are either random or time-varying, such as in the case of $\theta(t) = \omega t$. In this regime, interference effects are less pronounced, leading to differences in the imaging characteristics.

The choice between coherent and incoherent imaging depends on the specific goals of the application or investigation. Coherent methods are well-suited for situations requiring high-resolution and detailed information, leveraging the interference effects to extract fine details. On the other hand, incoherent methods are typically more practical for general imaging tasks, offering simplicity and ease of implementation. This balance between resolution and practicality underscores the importance of selecting the appropriate imaging approach for each scientific or applied context.

6.8 Monochromatic Diffraction-Limited Imaging of an Optical System

In geometrical optics the entrance pupil (EnP, image of aperture stop to all elements preceding it) and the exit pupil (ExP, image of aperture stop to all elements following it), provide the cone of light entering and exiting the system. In the absence of aberrations, the entering and exiting wavefronts are spherical, as illustrated in Fig. 6.13 under the assumptions of paraxial or Gaussian optics.

The effects of diffraction are particularly significant in the propagation of light from the object plane to the entrance pupil, or equivalently from the exit pupil to the image

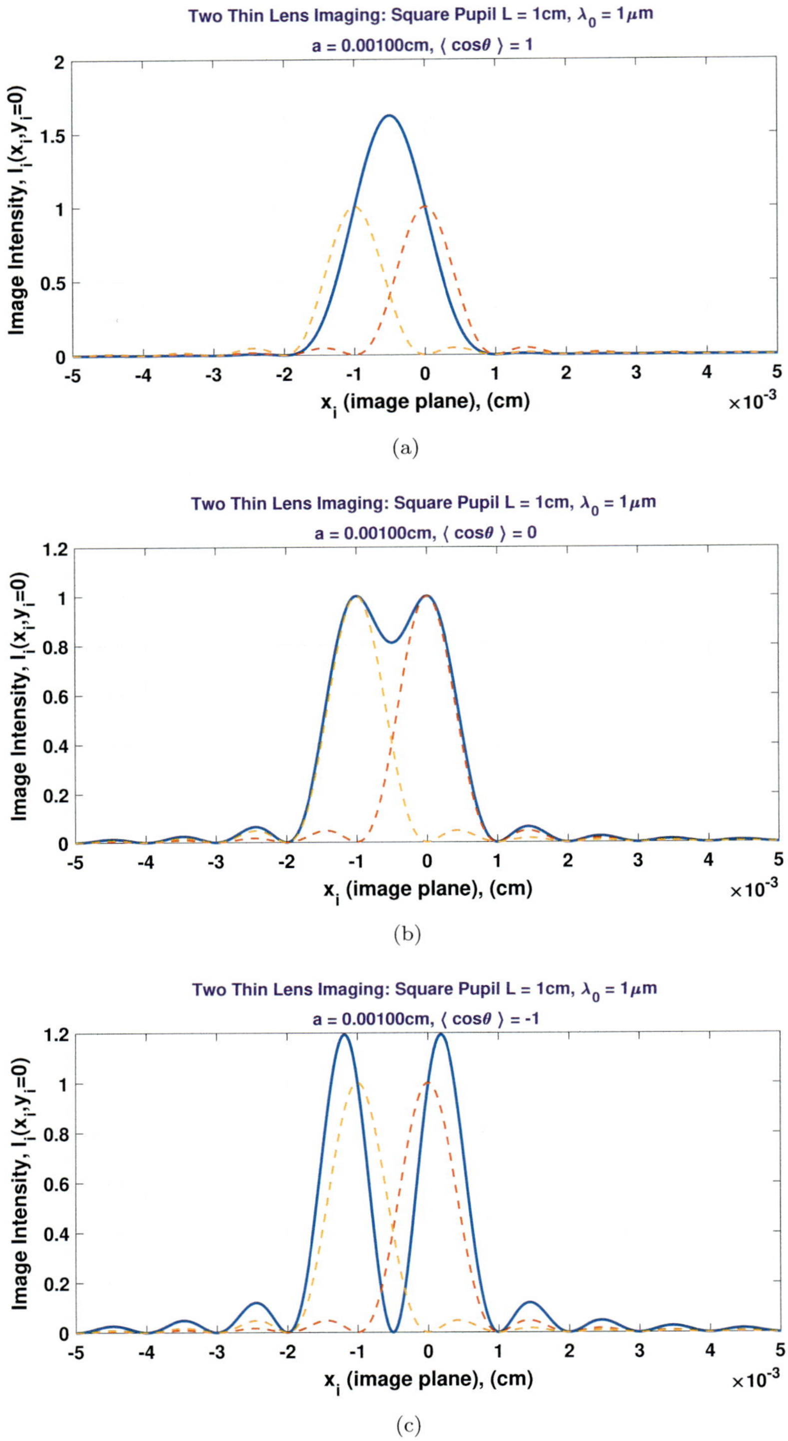

FIGURE 6.12: Normalized image intensity $I_1(x_i, y_i = 0)$ for a two-thin lens imaging system with $f_1 = f_2 = f = 10\,\text{cm}$. The pupil function is a square aperture of side length of $L = 1\,\text{cm}$. The shift of the second point of the object is selected as $a = \lambda f/L = 0.001\,\text{mm}$, where the freespace wavelength of the monochromatic source is $\lambda = 1\,\mu\text{m}$. The dashed lines correspond to the intensities due to each single point of the object. (a) Intensity when $\langle\cos\theta\rangle = 1$. (b) Intensity when $\langle\cos\theta\rangle = 0$. (c) Intensity when $\langle\cos\theta\rangle = -1$.

plane [265]. Figure 6.13 illustrates the case of a point source located at (x_{oa}, y_{oa}) in the object plane, emitting a diverging spherical wave. A corresponding converging spherical wave exits the exit pupil and converges to (x_{ia}, y_{ia}) in the image plane. Under the assumptions of geometrical optics and without considering aberrations, these points are conjugate, ensuring an undistorted image. However, when aberrations are introduced, the wavefront converging from the exit pupil becomes distorted, deviating from the ideal spherical shape. For a more comprehensive understanding of these phenomena, particularly the role of aberrations, detailed discussions can be found in advanced optics textbooks such as Born and Wolf [91]. A concise overview of the effects of aberrations on the frequency response of optical systems is provided in Sec. 6.11. The spherical wavefronts at the EnP and ExP can be expressed as

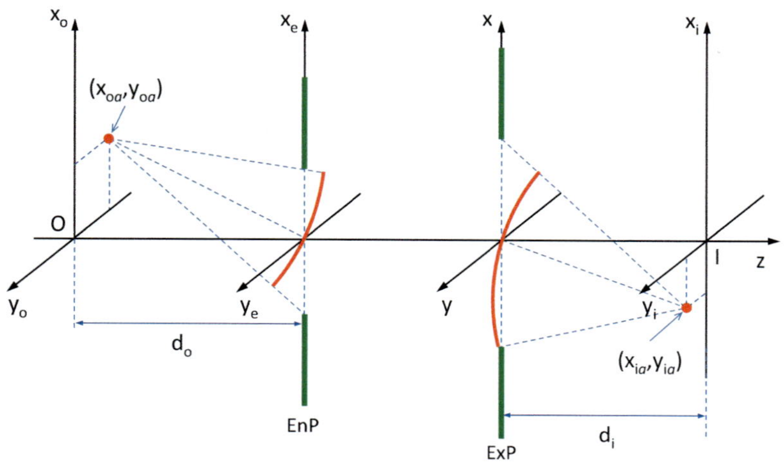

FIGURE 6.13: The entrance (EnP) and exit (ExP) pupils of an optical system. The diverging spherical wave entering the system at the EnP from the (x_{oa}, y_{oa}) object point is shown on the left. The converging spherical wave exiting from the ExP towards the image point (x_{ia}, y_{ia}) is shown on the right. The d_o and d_i are the distances between the object plane and the EnP plane and the ExP plane and image plane, respectively.

(using the parabolic - Fresnel approximation):

$$A\frac{e^{-jkR_o}}{R_o} \simeq A\frac{e^{-jkd_o}}{d_o}\exp\left\{-j\frac{k}{2d_o}\left[(x_e - x_{oa})^2 + (y_e - y_{oa})^2\right]\right\},$$

$$A\frac{e^{+jkR_i}}{R_i} \simeq A\frac{e^{+jkd_i}}{d_i}\exp\left\{+j\frac{k}{2d_i}\left[(x - x_{ia})^2 + (y - y_{ia})^2\right]\right\},$$

where (x_e, y_e) is a point at the EnP plane and (x, y) is a point at the ExP plane. The term A is the amplitude of the optical field wave. For example, in case that it represents an electric field, A has dimensions of volts and the term Ae^{-jkR_o}/R_o has dimensions of electric field (V/m). At this point, the interest is on the exiting field infinitesimally after the ExP plane, u^+_{ExP}, which can be expressed as:

$$u^+_{ExP}(x, y) = AP(x, y)\frac{e^{+jkd_i}}{d_i}\exp\left\{+j\frac{k}{2d_i}\left[(x - x_{ia})^2 + (y - y_{ia})^2\right]\right\},$$

where $P(x,y)$ is the ExP aperture. Then, the field at the image plane is easily determined by:

$$
\begin{aligned}
u_i(x_i, y_i) &= u^+_{ExP}(x, y) ** h(x, y; d_i) \\
&= jA\frac{e^{jkd_i}}{\lambda d_i^2} \exp\left[+j\frac{k}{2d_i}\left(x_{ia}^2 + y_{ia}^2\right)\right] \exp\left[-j\frac{k}{2d_i}\left(x_i^2 + y_i^2\right)\right] \times \\
&\int_{-\infty}^{+\infty}\int_{-\infty}^{+\infty} P(x', y') \exp\left\{j\frac{k}{d_i}\left[(x_i - x_{ia})x' + (y_i - y_{ia})y'\right]\right\} dx'dy'.
\end{aligned}
$$

When the ExP aperture $P(x,y)$ is considered of infinite extent, the previous expression simplifies as follows [neglecting the e^{jkd_i} phase factor, and using the delta function properties, such as $\delta(x-a) = \int_{-\infty}^{+\infty} e^{j2\pi f(x-a)} df$, and $\delta(ax) = (1/|a|)\delta(x)$]:

$$
\begin{aligned}
u_i(x_i, y_i) &= jA\lambda\delta(x_i - x_{ia}, y_i - y_{ia}) \\
&= jA\lambda\delta(mx_o - mx_{oa}, my_o - my_{oa}) \\
&= jA\lambda\frac{1}{|m|^2}\delta(x_o - x_{oa}, y_o - y_{oa}) \\
&= j\,\underbrace{\lambda\frac{A}{|m|^2}\delta\left(\frac{\tilde{x}_o - \tilde{x}_{oa}}{m}, \frac{\tilde{y}_o - \tilde{y}_{oa}}{m}\right)}_{= u_g} = j\,u_g\left(\frac{\tilde{x}_o}{m}, \frac{\tilde{y}_o}{m}\right) = ju_g\left(\frac{x_i}{m}, \frac{y_i}{m}\right),
\end{aligned}
$$

where m is the transverse magnification of geometrical optics under the paraxial approximation, $m = x_i/x_o = y_i/y_o$, and $\tilde{x}_o = mx_o$ and $\tilde{y}_o = my_o$, as it was used before. Notice that $m < 0$ and the geometrical image is inverted as expected. The delta function $\delta(x_o - x_{oa}, y_o - y_{oa})$ is the point input source used. For finite extent of ExP aperture $P(x,y)$, and neglecting all phase factors, the image field $u_i(x_i, y_i)$ can be expressed via a convolution integral as:

$$
u_i(x_i, y_i) = h(x_i, y_i) ** u_g(\tilde{x}_o, \tilde{y}_o) = \int_{-\infty}^{+\infty}\int_{-\infty}^{+\infty} h(x_i - \tilde{x}'_o, y_i - \tilde{y}'_o) u_g(\tilde{x}'_o, \tilde{y}'_o) d\tilde{x}'_o d\tilde{y}'_o, \tag{6.35}
$$

$$
\begin{aligned}
\text{where,}\quad h(x_i, y_i) &= \frac{1}{(\lambda d_i)^2}\int_{-\infty}^{+\infty}\int_{-\infty}^{+\infty} P(x', y') \exp\left\{j\frac{k}{d_i}\left(x_i x' + y_i y'\right)\right\} dx'dy' \\
&= \frac{1}{(\lambda d_i)^2}\mathcal{F}\left\{P(x, y)\right\}\Big|_{\substack{f_x = x_i/\lambda d_i \\ f_y = y_i/\lambda d_i}}.
\end{aligned} \tag{6.36}
$$

The dimensions of $h(x_i, y_i)$ in Eq. (6.36) are indeed consistent with m^{-2}. In the widely used textbook on Fourier Optics by J. W. Goodman [265], the dimensional consistency is addressed through the introduction of factor A. There, the prefactor in front of the Fourier transform is expressed as $A/\lambda d_i$, instead of $1/(\lambda d_i)^2$ as presented here. This difference highlights variations in normalization conventions used across different treatments of the subject.

In a diffraction-limited optical system, the image formation process can be conceptualized as the convolution of the geometrical optics image (obtained under the paraxial approximation) with the Fourier transform of the exit pupil. The function $h(x_i, y_i)$ serves as the impulse response or coherent Point Spread Function (PSF) of the optical system.

This function encapsulates the system's response to a point source in the object plane and plays a crucial role in determining the system's imaging characteristics, particularly in the presence of diffraction.

6.9 Polychromatic Diffraction-Limited Imaging of an Optical System

In all preceding sections, the assumption of monochromatic light was employed. This assumption implied that the optical field was spatially and temporally coherent. Spatial coherence implies that the phase between any two points of the object field is constant and does not randomly vary with time (see Coherence section, Sec. 3.9 in *Physical and Electromagnetics Optics* chapter, Chapter 3). Furthermore, it implies that the frequency bandwidth of the optical field, denoted as $\Delta\nu$, is practically zero, resulting in infinite coherence time and coherence length.

In practical optical systems, however, the object field exhibits statistical variations due to the use of polychromatic light and the finite spatial extent of objects. Such variations influence the imaging system's performance. For a simplified analysis of these effects, the object field illuminated by polychromatic light can be represented as a complex phasor: $\tilde{U}_g(t) = A(t)e^{j2\pi\nu_0 t + j\Phi(t)}$ where ν_0 is the central frequency of the object field, while $A(t)$ and $\Phi(t)$ are generally considered random variables. In this representation, the real field can be expressed in the form $u_g = A(t)\cos[2\pi\nu_0 t + \Phi(t)]$ (suppressing at this point the spatial dependence). This representation is valid when the spectral width $\Delta\nu$ of the field is rather narrow relative to the central frequency, i.e., $\Delta\nu \ll \nu_0$. Under such conditions, the analysis focuses on statistical time-averaged fields, as instantaneous field values are less meaningful in a fluctuating system.

The time-varying image field, considering these statistical variations, extends the convolution relationship from Eq. (6.35). It can be expressed as:

$$u_i(x_i, y_i; t) = \int_{-\infty}^{+\infty}\int_{-\infty}^{+\infty} h(x_i - \tilde{x}_o, y_i - \tilde{y}_o) u_g(\tilde{x}_o, \tilde{y}_o; t - \tau) d\tilde{x}_o d\tilde{y}_o,$$

where τ is the time delay associated with propagation of the light field from object point $(\tilde{x}_o, \tilde{y}_o)$ to image point (x_i, y_i). Then, the time-average image intensity, $I_i(x_i, y_i)$, is given by:

$$\begin{aligned}
I_i(x_i, y_i) &= \left\langle |u_i(x_i, y_i; t)|^2 \right\rangle = \left\langle u_i(x_i, y_i; t) u_i^*(x_i, y_i; t) \right\rangle \\
&= \int_{-\infty}^{+\infty}\int_{-\infty}^{+\infty} d\tilde{x}_o d\tilde{y}_0 \int_{-\infty}^{+\infty}\int_{-\infty}^{+\infty} d\tilde{x}'_o d\tilde{y}'_o h(x_i - \tilde{x}_o, y_i - \tilde{y}_o) h^*(x_i - \tilde{x}'_o, y_i - \tilde{y}'_o) \times \\
&\quad \left\langle u_g(\tilde{x}_o, \tilde{y}_o; t - \tau_1) u_g^*(\tilde{x}'_o, \tilde{y}'_o; t - \tau_2) \right\rangle.
\end{aligned}$$

From coherence section (refer to *Physical and Electromagnetics Optics* chapter, Chapter 3) the time-average term can be expressed in the following form:

$$\begin{aligned}
\left\langle u_g(\tilde{x}_o, \tilde{y}_o; t - \tau_1) u_g^*(\tilde{x}'_o, \tilde{y}'_o; t - \tau_2) \right\rangle &= \frac{1}{2} \left\langle \tilde{U}_g(\tilde{x}_o, \tilde{y}_o; t - \tau_1) \tilde{U}_g^*(\tilde{x}'_o, \tilde{y}'_o; t - \tau_2) \right\rangle \\
&= \Gamma(\tilde{x}_o, \tilde{y}_o; \tilde{x}'_o, \tilde{y}'_o; \tau).
\end{aligned}$$

Here, Γ is the mutual coherence function and the optical field is assumed to be stationary and $\tau = \tau_1 - \tau_2$. It is important to note that further assumptions are required at this stage. For a given image point (x_i, y_i) the coherent PSF, h, is non-zero only over a small region around it. Consequently, the integrand is non-zero only when $(\tilde{x}_o, \tilde{y}_o)$ and $(\tilde{x}'_o, \tilde{y}'_o)$ are very close to each other [265]. In this case, the time delays τ_1 and τ_2 can be considered almost identical (implying $\tau \simeq 0$), and the time-averaged intensity $I_i(x_i, y_i)$ can be rewritten as:

$$\begin{aligned}
I_i(x_i, y_i) &= \int_{-\infty}^{+\infty}\int_{-\infty}^{+\infty} d\tilde{x}_o d\tilde{y}_o \int_{-\infty}^{+\infty}\int_{-\infty}^{+\infty} d\tilde{x}'_o d\tilde{y}'_o h(x_i - \tilde{x}_o, y_i - \tilde{y}_o) h^*(x_i - \tilde{x}'_o, y_i - \tilde{y}'_o) \times \\
&\quad J_{12}(\tilde{x}_o, \tilde{y}_o; \tilde{x}'_o, \tilde{y}'_o),
\end{aligned}$$

where $J_{12} = \Gamma(\tilde{x}_o, \tilde{y}_o; \tilde{x}'_o, \tilde{y}'_o, \tau = 0)$ is the *mutual intensity function* (or equal time coherence function), which is a measure of the spatial coherence of the object field. For perfect spatial coherence of the object fields of two point sources of the object, $u_g(\tilde{x}_o, \tilde{y}_o)$ and $u_g(\tilde{x}'_o, \tilde{y}'_o)$, are related to each other by a constant term and $J_{12} = u_g(\tilde{x}_o, \tilde{y}_o) u_g^*(\tilde{x}'_o, \tilde{y}'_o)$. Then, the time-averaged intensity becomes:

$$\begin{aligned}
I_i(x_i, y_i) &= \int_{-\infty}^{+\infty}\int_{-\infty}^{+\infty} h(x_i - \tilde{x}_o, y_i - \tilde{y}_o) u_g(\tilde{x}_o, \tilde{y}_o) d\tilde{x}_o d\tilde{y}_o \times \\
&\quad \int_{-\infty}^{+\infty}\int_{-\infty}^{+\infty} h^*(x_i - \tilde{x}'_o, y_i - \tilde{y}'_o) u_g^*(\tilde{x}'_o, \tilde{y}'_o) d\tilde{x}'_o d\tilde{y}'_o \\
&= \left| \int_{-\infty}^{+\infty}\int_{-\infty}^{+\infty} h(x_i - \tilde{x}_o, y_i - \tilde{y}_o) u_g(\tilde{x}_o, \tilde{y}_o) d\tilde{x}_o d\tilde{y}_o \right|^2 = |u_i(x_i, y_i)|^2, \qquad (6.37)
\end{aligned}$$

implying that a spatially coherent system is linear with respect to the optical field (amplitude). On the other hand, when the object field is spatially incoherent the mutual intensity is given by $J_{12} = K u_g(\tilde{x}_o, \tilde{y}_o) u_g^*(\tilde{x}'_o, \tilde{y}'_o) \delta(\tilde{x}_o - \tilde{x}'_o, \tilde{y}_o - \tilde{y}'_o)$ where K is a real constant (with dimensions of m^2 since the delta function has units and makes sense only in an integral form). In the latter case, the time-averaged intensity at the image point is written as:

$$\begin{aligned}
I_i(x_i, y_i) &= \int_{-\infty}^{+\infty}\int_{-\infty}^{+\infty} d\tilde{x}_o d\tilde{y}_o \int_{-\infty}^{+\infty}\int_{-\infty}^{+\infty} h(x_i - \tilde{x}_o, y_i - \tilde{y}_o) h^*(x_i - \tilde{x}'_o, y_i - \tilde{y}'_o) \times \\
&\quad K u_g(\tilde{x}_o, \tilde{y}_o) u_g^*(\tilde{x}'_o, \tilde{y}'_o) \delta(\tilde{x}_o - \tilde{x}'_o, \tilde{y}_o - \tilde{y}'_o) d\tilde{x}'_o d\tilde{y}'_o \\
&= K \int_{-\infty}^{+\infty}\int_{-\infty}^{+\infty} |h(x_i - \tilde{x}_o, y_i - \tilde{y}_o)|^2 |u_g(\tilde{x}_o, \tilde{y}_o)|^2 d\tilde{x}_o d\tilde{y}_o \\
&= K|h(x_i, y_i)|^2 ** \underbrace{|u_g(\tilde{x}_o, \tilde{y}_o)|^2}_{= I_g} = K|h(x_i, y_i)|^2 ** I_g(\tilde{x}_o, \tilde{y}_o). \qquad (6.38)
\end{aligned}$$

Equation (6.38) indicates that, for a spatially incoherent object field, the image intensity is obtained by convolving the intensity of the impulse response, $|h|^2$, with the intensity predicted by geometrical optics under the paraxial approximation, I_g. This result highlights a fundamental distinction between coherent and incoherent imaging systems. In the case of spatially incoherent imaging, the system exhibits linearity with respect to the intensity of the object field, rather than its amplitude. This behavior stems from the statistical nature of spatial incoherence, where the phases of the object field at different points are uncorrelated. As a result, interference effects average out, and the image intensity is directly influenced by the spatial distribution of intensity in the object field and the system's point spread function. This property is particularly useful in applications where phase information is either unavailable or irrelevant, simplifying the analysis and design of optical systems used for general-purpose imaging.

6.10 Spatial Frequency Response of Diffraction-Limited Coherent or Incoherent Optical Systems

In this section, the spatial frequency response of both coherent and incoherent optical systems is investigated under the assumption of aberration-free optical systems. This analysis is essential for understanding how different optical systems handle spatial frequencies and how they can be used to capture various details of the object being imaged.

6.10.1 Frequency Response of Coherent Optical Systems

For a coherent optical system, the image field is given by Eq. (6.35) as:

$$u_i(x_i, y_i) = h(x_i, y_i) ** u_g(\tilde{x}_o, \tilde{y}_o) \Longrightarrow$$

$$U_i(f_x, f_y) = H(f_x, f_y) U_g(f_x, f_y) \qquad \text{where,}$$

$$H(f_x, f_y) = \mathcal{F}\left\{h(x_i, y_i)\right\} = \mathcal{F}\left\{\frac{1}{(\lambda d_i)^2}\mathcal{F}\left\{P\left(\frac{x_i}{\lambda d_i}, \frac{y_i}{\lambda d_i}\right)\right\}\right\}$$

$$= P\left(-\lambda d_i f_x, -\lambda d_i f_y\right). \tag{6.39}$$

In the above derivation, Eq. (6.36) was employed. The function $H(f_x, f_y)$ is the *amplitude transfer function* of the coherent optical imaging system. It is reminded that the function $P(x, y)$ is the ExP function, and for the majority of optical systems, is symmetrical in x and y directions. When this symmetry is centered at the origin, the amplitude transfer function can be expressed as $H(f_x, f_y) \sim P\left(\lambda d_i f_x, \lambda d_i f_y\right)$. This relationship reveals that the spatial frequency response of the coherent system is determined by the geometry of the pupil, and the system's ability to pass certain spatial frequencies depends on the size and shape of the exit pupil.

Generally, the ExP function $P(x, y)$ is unity within a specific region and zero outside of it. This implies a finite passband in the spatial frequency domain [265]. The sharp decrease in frequency response at the edge of the pupil reflects the system's inability to pass spatial frequencies beyond a certain limit, which is a key factor in determining the system's resolution and image quality. This concept is critical when evaluating the performance of optical systems, as the amplitude transfer function governs how different spatial frequencies are transmitted, impacting the overall image formed by the system.

Example 6.10.1:

Some common symmetrical functions that can serve as ExP functions include rectangular and the circular pupil functions:

$$P(x, y) = \text{rect}\left(\frac{x}{2L_x}, \frac{y}{2L_y}\right) = \text{rect}\left(\frac{x}{2L_x}\right)\text{rect}\left(\frac{y}{2L_x}\right),$$

$$P(x, y) = \text{circ}\left(\frac{r_T}{a}\right) = \text{circ}\left(\frac{\sqrt{x^2 + y^2}}{a}\right).$$

Determine for these simple pupil functions the aberrations-free amplitude transfer functions and the corresponding cutoff spatial frequencies. Assume that the ExP is at a distance d_i from the image plane. In addition, determine the maximum possible directional angle α_z for plane waves propagating along the z direction. For the numerical implementation assume

$2L_x = 2L_y = 50\,\text{mm}$ (for a square ExP), and $a = 50\,\text{mm}$ (for a circular ExP). For both cases assume $d_i = 100\,\text{mm}$.

Solution: Using the definition of the amplitude transfer function $H(f_x, f_y)$ from Eq. (6.39), and neglecting any constant amplitude factors as well as exploiting the symmetry of the pupil functions, the following amplitude transfer functions H's are found:

$$H(f_x, f_y) = \text{rect}\left(\frac{\lambda d_i f_x}{2L_x}\right)\text{rect}\left(\frac{\lambda d_i f_y}{2L_x}\right),$$

$$H(f_x, f_y) = \text{circ}\left(\frac{\lambda d_i}{a}\sqrt{f_x^2 + f_y^2}\right).$$

It is straightforward to find the cutoff spatial frequencies. For example in the rectangular pupil case $\lambda d_i f_w/2L_w < 1/2 \Longrightarrow f_w < L_w/\lambda d_i$ with $w = x, y$. The maximum cutoff frequency is $f_{cutoff} = \max\{L_x/\lambda d_i, L_y/\lambda d_i\}$. Similarly, in the case of the circular pupil function $f_{curoff} = a/\lambda d_i$. Using the numerical values given $f_{cutoff} = 250\,\text{mm}^{-1}$ for the square ExP, and $f_{cutoff} = 500\,\text{mm}^{-1}$ for the circular ExP. On the other hand, using Eq. (6.6) maximum α_z angles are $\alpha_z \simeq 20.7\,\text{deg}$ and $\alpha_z = 30\,\text{deg}$, for the square and the circular ExP, respectively.

6.10.2 Frequency Response of Incoherent Optical Systems

In the case of an incoherent imaging system, the time-averaged intensity is given by Eq. (6.38) as:

$$I_i(x_i, y_i) = K|h(x_i, y_i)|^2 ** I_g(\tilde{x}_o, \tilde{y}_o) \Longrightarrow$$
$$\mathcal{F}\{I_i\} = K\mathcal{F}\{|h|^2\}\mathcal{F}\{I_g\}.$$

At this point, it is customary to normalize the intensity spectra by their zero-spatial-frequency value. Since the intensity functions are always non-negative, the Fourier transform of the intensity, function reaches its maximum at the origin. This can be shown as follows: $\iint I(x,y)\exp[+j2\pi(xf_x + yf_y)]dxdy \leq \iint |I(x,y)||\exp[+j2\pi(xf_x + yf_y)]|dxdy = \iint I(x,y)dxdy = \mathcal{F}\{I\}(f_x = f_y = 0)$. Next, the following Fourier transform normalizations are performed:

$$\mathcal{I}_i(f_x, f_y) = \frac{\displaystyle\int_{-\infty}^{+\infty}\int_{-\infty}^{+\infty} I_i(x,y)\exp\left[+j2\pi(xf_x + yf_y)\right]dxdy}{\displaystyle\int_{-\infty}^{+\infty}\int_{-\infty}^{+\infty} I_i(x,y)dxdy},$$

$$\mathcal{I}_g(f_x, f_y) = \frac{\displaystyle\int_{-\infty}^{+\infty}\int_{-\infty}^{+\infty} I_g(x,y)\exp\left[+j2\pi(xf_x + yf_y)\right]dxdy}{\displaystyle\int_{-\infty}^{+\infty}\int_{-\infty}^{+\infty} I_g(x,y)dxdy},$$

$$\mathcal{H}(f_x, f_y) = \frac{\displaystyle\int_{-\infty}^{+\infty}\int_{-\infty}^{+\infty} |h(x,y)|^2\exp\left[+j2\pi(xf_x + yf_y)\right]dxdy}{\displaystyle\int_{-\infty}^{+\infty}\int_{-\infty}^{+\infty} |h(x,y)|^2 dxdy}. \tag{6.40}$$

Neglecting the constant K (the units are not affected by the normalizations) the following equation can be written in the spatial frequency domain:

$$\mathcal{I}_i(f_x, f_y) = \mathcal{H}(f_x, f_y)\mathcal{I}_g(f_x, f_y), \tag{6.41}$$

where the function $\mathcal{H}(f_x, f_y)$ is known as the *Optical Transfer Function* (OTF) of the imaging system. The modulus $|\mathcal{H}(f_x, f_y)|$ is known as the *Modulation Transfer Function* (MTF). Observe that $\mathcal{H}(f_x, f_y)$ specifies the complex weighting factor applied by the system to the (f_x, f_y) spatial frequency relative to the weighting factor applied to the zero-frequency component. Using the property of the Fourier transform of an autocorrelation the OTF can be written as:

$$\mathcal{H}(f_x, f_y) = \frac{\int_{-\infty}^{+\infty}\int_{-\infty}^{+\infty} |h(x,y)|^2 \exp\left[+j2\pi(xf_x + yf_y)\right] dxdy}{\int_{-\infty}^{+\infty}\int_{-\infty}^{+\infty} |h(x,y)|^2 dxdy}$$

$$= \frac{H(f_x, f_y) \star\star\, H(f_x, f_y)}{\int_{-\infty}^{+\infty}\int_{-\infty}^{+\infty} |h(x,y)|^2 dxdy} = \frac{\int_{-\infty}^{+\infty}\int_{-\infty}^{+\infty} H^*(f'_x, f'_y)H(f_x + f'_x, f_y + f'_y)df'_x df'_y}{\int_{-\infty}^{+\infty}\int_{-\infty}^{+\infty} |h(x,y)|^2 dxdy} \tag{6.42}$$

$$= \frac{\int_{-\infty}^{+\infty}\int_{-\infty}^{+\infty} H^*(w_x - f_x/2, w_y - f_y/2)H(w_x + f_x/2, w_y + f_y/2)dw_x dw_y}{\int_{-\infty}^{+\infty}\int_{-\infty}^{+\infty} |h(x,y)|^2 dxdy}. \tag{6.43}$$

Therefore, the OTF is the autocorrelation function of the amplitude transfer function. It is straightforward to show that the OTF satisfies the following properties: (a) $\mathcal{H}(0,0) = 1$, (b) $\mathcal{H}(-f_x, -f_y) = \mathcal{H}^*(f_x, f_y)$, using Eq. (6.43). (c) A third interesting property is $|\mathcal{H}(f_x, f_y)| \leq |\mathcal{H}(0,0)| = 1$. A proof of this property is given in Ref. [265] and it is based on Cauchy-Schwarz inequality and on the 2D Parseval's theorem. From the Cauchy-Schwarz inequality for any two 2D functions $f(x,y)$ and $g(x,y)$ the following holds: $|\iint fg dxdy|^2 \leq \iint |f|^2 dxdy \iint |g|^2 dxdy$. Setting $f = H^*(w_x - f_x/2, w_y - f_y/2)$ and $g = H(w_x + f_x/2, w_y + f_y/2)$ the following inequality holds:

$$\left|\int_{-\infty}^{+\infty}\int_{-\infty}^{+\infty} H^*\left(w_x - \frac{f_x}{2}, w_y - \frac{f_y}{2}\right) H\left(w_x + \frac{f_x}{2}, w_y + \frac{f_y}{2}\right) dw_x dw_y\right|^2 \leq$$

$$\int_{-\infty}^{+\infty}\int_{-\infty}^{+\infty} \left|H^*\left(w_x - \frac{f_x}{2}, w_y - \frac{f_y}{2}\right)\right|^2 dw_x dw_y \times$$

$$\int_{-\infty}^{+\infty}\int_{-\infty}^{+\infty} \left|H\left(w_x + \frac{f_x}{2}, w_y + \frac{f_y}{2}\right)\right|^2 dw_x dw_y =$$

$$\int_{-\infty}^{+\infty}\int_{-\infty}^{+\infty} \left|H\left(w'_x, w'_y\right)\right|^2 dw'_x dw'_y \int_{-\infty}^{+\infty}\int_{-\infty}^{+\infty} \left|H\left(w''_x, w''_y\right)\right|^2 dw''_x dw''_y =$$

$$\left[\int_{-\infty}^{+\infty}\int_{-\infty}^{+\infty} \left|H\left(w'_x, w'_y\right)\right|^2 dw'_x dw'_y\right]^2 = \left[\int_{-\infty}^{+\infty}\int_{-\infty}^{+\infty} \left|h\left(x', y'\right)\right|^2 dx'dy'\right]^2,$$

where for the last part the 2D Parseval's theorem was utilized. From the last expression the property (c) $|\mathcal{H}(f_x, f_y)| \leq |\mathcal{H}(0,0)| = 1$ is easily implied if it is divided by the $\iint |h(x', y')|^2 dx'dy'$ term.

It is worth noting that the analytical evaluation of the autocorrelation in Eq. (6.42) can often be challenging, especially for complex transfer functions. To address this, it is practical to compute the autocorrelation in a discrete form, enabling numerical evaluation and analysis. The transfer function $H(f_{xk}, f_{y\ell})$ is sampled at equally spaced discrete spatial frequencies f_{xk} and $f_{y\ell}$, defined as $f_{xk} = f_{x,min} + (k-1)\Delta f_x$ and $f_{y\ell} = f_{y,min} + (\ell - 1)\Delta f_y$ with $\Delta f_x = (f_{x,max} - f_{x,min})/(N_x - 1)$ and $\Delta f_y = (f_{y,max} - f_{y,min})/(N_y - 1)$. The parameters $f_{x,max}$, $f_{y,max}$, $f_{x,min}$, $f_{y,min}$ define the range of spatial frequencies sampled along the x- and y-directions. The variables N_x and N_y indicate the number of samples along these directions. By discretizing the spatial frequencies, the optical transfer function (OTF) can be expressed in a form suitable for numerical computation, ensuring that the analysis remains practical even for complex or irregular pupil functions. The resulting discrete OTF can be expressed in the following form:

$$\mathcal{H}(f_{xm}, f_{yn}) = \frac{\sum_k \sum_\ell H^*(f_{xk}, f_{y\ell}) H(f_{xm} + f_{xk}, f_{yn} + f_{y\ell})}{\sum_k \sum_\ell |H(f_{xk}, f_{y\ell})|^2}, \tag{6.44}$$

where $-(N_x - 1) \leq m \leq (N_x - 1)$ and $-(N_y - 1) \leq n \leq (N_y - 1)$, and the discrete spatial frequencies are $(f_{xm}, f_{yn}) = (m\Delta f_x, n\Delta f_y)$. The size of the discrete OTF is $(2N_x - 1) \times (2N_y - 1)$ while the size of the discrete transfer function is $N_x \times N_y$.

Example 6.10.2:

The objective of this example is to compute the OTF of an optical system with an exit pupil of a rectangular shape, characterized by dimensions $2L_x \times 2L_y$. The freespace wavelength is denoted as λ, and the distance of the Exit Pupil (ExP) from the image plane is d_i. For the numerical implementation, specific values are selected: $2L_x = 50\,\text{mm}$, $2L_y = 80\,\text{mm}$, $d_i = 100\,\text{mm}$ and $\lambda = 1\,\mu\text{m}$.

Solution: To theoretically evaluate the OTF function, it is helpful to reference Fig. 6.14(a). According to Eq. (6.42), the functions $H(w_x, w_y)$ (centered at the origin) and its shifted counterpart $H(w_x + f_x, w_y + f_y)$ [centered at $(-f_x, -f_y)$)] are shown. Since the ExP is assumed rectangular it can be represented using the "rect" function, as $\text{rect}(x/2L_x, y/2L_y)$, and the corresponding $H(w_x, w_y)$ and $H(w_x + f_x, w_y + f_y)$ are then expressed as:

$$H(w_x, w_y) = P(\lambda d_i w_x, \lambda d_i w_y) = \text{rect}\left(\frac{\lambda d_i}{2L_x} w_x, \frac{\lambda d_i}{2L_y} w_y\right),$$

$$H(w_x + f_x, w_y + f_y) = \text{rect}\left(\frac{\lambda d_i}{2L_x}(w_x + f_x), \frac{\lambda d_i}{2L_y}(w_y + f_y)\right).$$

The autocorrelation of Eq. (6.42) is non-zero when the functions $H(w_x, w_y)$ and $H(w_x + f_x, w_y + f_y)$ overlap as in the shaded area of Fig. 6.14(a). Since the 'rect' function is nonzero when $|\lambda d_i w_x / 2L_x| \leq 1/2$ and $|\lambda d_i w_y / 2L_y| \leq 1/2$ there is an overlap when the following conditions hold:

$$\max\left\{-\frac{L_x}{\lambda d_i}, -\frac{L_x}{\lambda d_i} - f_x\right\} \leq w_x \leq \min\left\{\frac{L_x}{\lambda d_i}, \frac{L_x}{\lambda d_i} - f_x\right\}$$

$$\max\left\{-\frac{L_y}{\lambda d_i}, -\frac{L_y}{\lambda d_i} - f_y\right\} \leq w_y \leq \min\left\{\frac{L_y}{\lambda d_i}, \frac{L_y}{\lambda d_i} - f_y\right\}$$

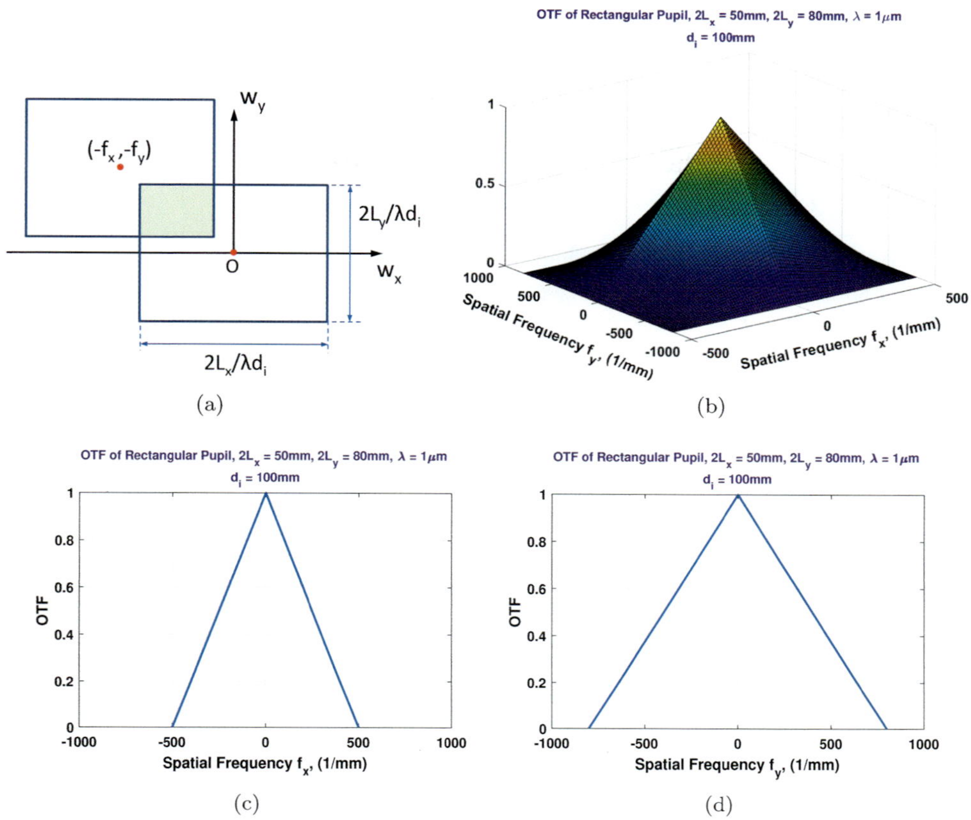

FIGURE 6.14: (a) The diagram illustrating the calculation of the OTF of a rectangular-shaped exit pupil. The lengths of the rectangular along the w_x- and w_y-directions are $2L_x/\lambda d_i$ and $2L_y/\lambda d_i$, respectively. The distance d_i is the distance of the ExP from the image plane and λ is the freespace wavelength. (b) The calculated OTF (it is the same as MTF since OTF is real) for $2L_x = 50\,\text{mm}$ $2L_y = 80\,\text{mm}$, $\lambda = 1\,\mu\text{m}$, and $d_i = 100\,\text{mm}$ in the $f_x f_y$-plane. (c) Same as (b) but with OTF shown along the f_x spatial frequency direction. (d) Same as (c) but along the f_y spatial frequency direction.

For example, when $f_x \geq 0$ and $f_y \geq 0$ then $-(L_x/\lambda d_i) \leq w_x \leq (L_x/\lambda d_i) - f_x$ and similarly for w_y resulting in a 'Shaded Area' $= [(2L_x/\lambda d_i) - f_x][(2L_y/\lambda d_i) - f_y]$. Summarizing, using the overlap conditions and the normalization, it is straightforward to show that the sought OTF is given by:

$$\mathcal{H}(f_x, f_y) = \begin{cases} \left(1 - \dfrac{\lambda d_i}{2L_x}|f_x|\right)\left(1 - \dfrac{\lambda d_i}{2L_y}|f_y|\right) & |f_x| \leq \dfrac{2L_x}{\lambda d_i}, \quad |f_y| \leq \dfrac{2L_y}{\lambda d_i} \\ 0 & \text{otherwise.} \end{cases} \tag{6.45}$$

It is also common to express Eq. (6.45) using the "tri" (triangle function) which is defined as $\text{tri}(x) = 1 - |x|/L$ when $|x| \leq 1$ and is zero otherwise. In this case the OTF can be expressed as $\mathcal{H}(f_x, f_y) = \text{tri}(f_x/a_x)\text{tri}(f_y/a_y)$ where $a_w = 2L_w/\lambda d_i$ $(w = x, y)$. Now let's consider a numerical example with $2L_x = 50\,\text{mm}$, $2L_y = 80\,\text{mm}$, $d_i = 100\,\text{mm}$, and $\lambda = 1\,\mu\text{m}$. The results are shown in Fig. 6.14(b) in the $f_x f_y$-plane. The limits that the OTF is nonzero are $\pm L_x/\lambda d_i = \pm 500/\text{mm}$ and $\pm L_y/\lambda d_i = \pm 800/\text{mm}$. The OTF along the f_x-direction for $f_y = 0$ is presented in Fig. 6.14(c) where the triangular shape of the OTF is evident. Similarly, the OTF along the f_y-direction for $f_x = 0$ is presented in Fig. 6.14(d). Since the dimensions of the pupil in the x- and y-directions are not equal the OTF differs along the f_x and f_y spatial frequencies directions. The maximum directional cosine angle α_z is given by Eq. (6.6) and is $\alpha_z = \cos^{-1}\{[1 - (L_x/d_i)^2 - (L_y/d_i)^2]^{1/2}\} = 28.14\,\text{deg}$.

Example 6.10.3:

The objective of this example is to calculate the OTF of an optical system with exit pupil of a circular shape with radius a. The freespace wavelength is λ and the distance of the ExP from the image plane is d_i. For the numerical implementation, specific values are selected: $a = 10\,\text{mm}$, $d_i = 100\,\text{mm}$ and $\lambda = 1\,\mu\text{m}$.

Solution: For the theoretical evaluation of the Optical Transfer Function (OTF) corresponding to a circular-shaped exit pupil, reference to Fig. 6.15(a) is essential. The calculation is guided by Eq. (6.42), which involves the autocorrelation of the amplitude transfer function $H(w_x, w_y)$. The function $H(w_x, w_y)$ is centered at the origin, while its shifted counterpart $H(w_x + f_x, w_y)$ is centered at $(-f_x, 0)$. This simplification, which assumes alignment along the w_x-axis, is justified due to the cylindrical symmetry of the OTF when derived from a

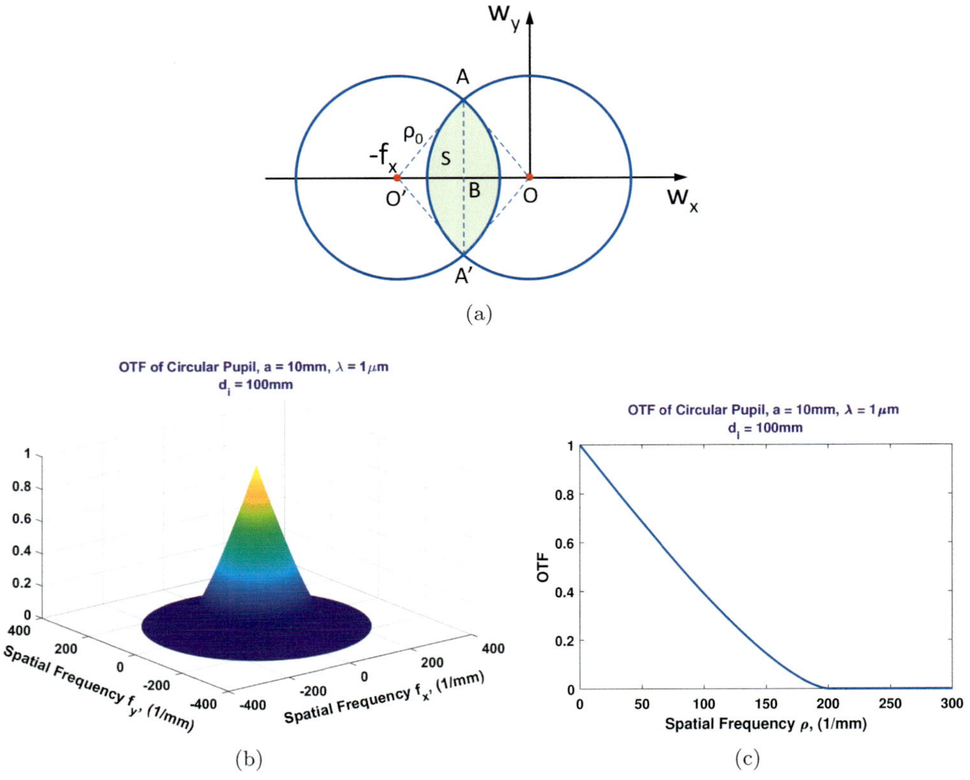

FIGURE 6.15: (a) Diagram illustrating the calculation of the OTF for a circular-shaped exit pupil. The radius of the circle in the $w_x w_y$-plane is $a/\lambda d_i$, where a is the physical radius of the pupil, λ is the freespace wavelength, and d_i is the distance of the ExP from the image plane. (b) The calculated OTF (identical to the MTF since the OTF is real) for a circular pupil with $a = 10\,\text{mm}$, $\lambda = 1\,\mu\text{m}$, and $d_i = 100\,\text{mm}$ in the $f_x f_y$-plane. (c) Same as (b), but the OTF is shown along the radial spatial frequency direction, $\rho = (f_x^2 + f_y^2)^{1/2}$, reflecting the cylindrical symmetry of the circular pupil.

circular pupil function. The following expressions can be written:

$$H(w_x, w_y) = P(\lambda d_i w_x, \lambda d_i w_y) = \text{circ}\left(\frac{\lambda d_i}{a}\sqrt{w_x^2 + w_y^2}\right),$$

$$H(w_x + f_x, w_y) = \text{circ}\left(\frac{\lambda d_i}{a}\sqrt{(w_x + f_x)^2 + w_y^2}\right).$$

The autocorrelation described in Eq. (6.42) remains non-zero only when the functions $H(w_x, w_y)$ and $H(w_x + f_x, w_y)$ intersect, as represented by the shaded region in Fig. 6.15(a). The 'circ' function is nonzero within the boundary defined by $(\lambda d_i/a)(w_x^2 + w_y^2)^{1/2} = (\lambda d_i/a)\rho = \rho/\rho_0 \leq 1$ with $\rho = (w_x^2 + w_y^2)^{1/2}$, and $\rho_0 = a/\lambda d_i$. The intersection or overlap of these functions occurs for $\rho \leq 2\rho_0$.

The shaded area S is determined by summing the areas of two circular sectors and subtracting the areas of two triangles. For instance, one-half of S corresponds to the area of the sector $O'AA'$ minus the area of the triangle $O'AA'$. This calculation yields the expression $S = 2\rho_0^2 \cos^{-1}(f_x/2\rho_0) - f_x[\rho_0^2 - (f_x/2)^2]^{1/2}$. In summary, based on the overlap conditions and appropriate normalization, it is straightforward to derive the expression for the OTF, which accounts for the geometrical and spatial constraints of the system. The resulting OTF is given by:

$$\mathcal{H}(\rho) = \begin{cases} \dfrac{2}{\pi}\left\{\cos^{-1}\left(\dfrac{\rho}{2\rho_0}\right) - \dfrac{\rho}{2\rho_0}\left[1 - \left(\dfrac{\rho}{2\rho_0}\right)^2\right]^{1/2}\right\} & \rho \leq \dfrac{2a}{\lambda d_i} = 2\rho_0, \\ 0 & \text{otherwise.} \end{cases} \tag{6.46}$$

In the last expression, f_x was replaced by $\rho = (f_x^2 + f_y^2)^{1/2}$ due to the cylindrical symmetry. Let us now consider a numerical example with parameters $a = 10\,\text{mm}$, $d_i = 100\,\text{mm}$, and $\lambda = 1\,\mu\text{m}$. The results of this example are presented in Fig. 6.15(b), which illustrates the OTF in the $f_x f_y$-plane.

The maximum spatial frequency, beyond which the OTF becomes zero, is determined as $\rho = 2\rho_0 = 200\,/\text{mm}$. Additionally, the OTF along the radial direction ρ is displayed in Fig. 6.15(c), capturing the decay behavior due to spatial frequency limitations. The maximum angle α_z, is calculated as $\alpha_z = \cos^{-1}\{[1 - (2a/d_i)^2]^{1/2}\} = 11.54\,\text{deg}$ [refer to Eq. (6.6)].

Example 6.10.4:

The exit pupil function of an incoherent optical system is represented by two square regions, as shown in Fig. 6.16(a). Each square has a side length of ℓ, and the separation between their centers is $2s$. Given that the distance of the exit pupil from the image plane is d_i and the freespace wavelength (center wavelength) is λ, the objective is to compute and plot the resulting optical transfer function (OTF). For the numerical implementation, use the following parameters: $\ell = 50\,\text{mm}$, $2s = 120\,\text{mm}$, $d_i = 100\,\text{mm}$, and $\lambda = 1\,\mu\text{m}$. This example is adapted from Ref. [348].

Solution: To simplify the analysis, it is convenient to express the pupil function using more fundamental functions. This can be achieved by employing again the "rect" function in combination with convolutions involving the "delta" function, as represented in the following form:

$$P(x, y) = \text{rect}\left(\frac{x}{\ell}, \frac{y}{\ell}\right) ** \left[\delta(x + s, y) + \delta(x - s, y)\right].$$

The amplitude transfer function, as described by Eq. (6.39), is expressed as follows:

$$H(f_x, f_y) = P(-\lambda d_i f_x, -\lambda d_i f_y) = P(\lambda d_i f_x, \lambda d_i f_y)$$
$$= \text{rect}\left(\frac{\lambda d_i}{\ell} f_x, \frac{\lambda d_i}{\ell} f_y\right) ** \left[\delta\left(\lambda d_i f_x + s, y\right) + \delta\left(\lambda d_i f_x - s, y\right)\right].$$

At this point, the autocorrelation of H is required for the calculation of the OTF. Since

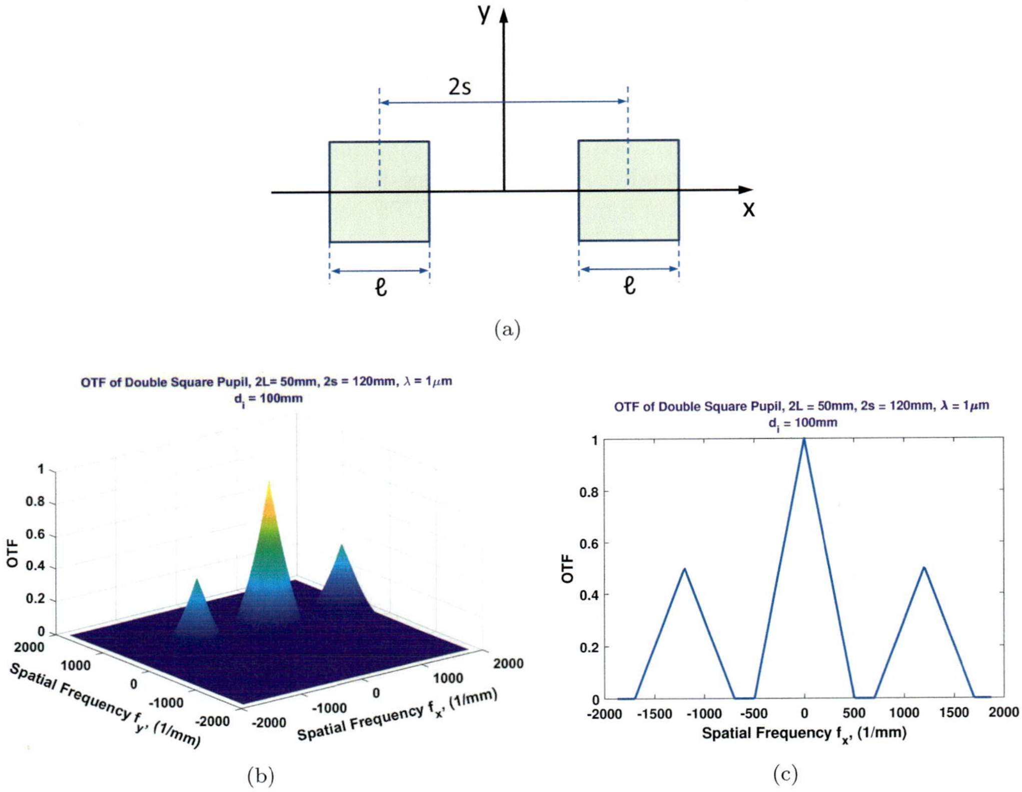

FIGURE 6.16: (a) Diagram of the double-square exit pupil function, where the side length of each square is ℓ and the separation between the square centers is $2s$. (b) The calculated OTF for $\ell = 50$ mm, $2s = 120$ mm, $\lambda = 1\,\mu$m, and $d_i = 100$ mm in the $f_x f_y$-plane. (c) The same as (b), but with the OTF shown along the f_x spatial frequency direction.

the autocorrelation of the "rect" function is already known from Example 6.10.2, it is advantageous to utilize the following identity: $(f ** g) \star\star (f ** g) = (f \star\star f) ** (g \star\star g)$. This identity is straightforward to prove in the spatial frequency domain. Specifically, $\mathcal{F}\{(f ** g) \star\star (f ** g)\} = \mathcal{F}^*\{(f ** g)\}\mathcal{F}\{(f ** g)\} = (FG)^*(FG) = |F|^2|G|^2 = \mathcal{F}\{(f \star\star f)\}\mathcal{F}\{(g \star\star g)\} = \mathcal{F}\{(f \star\star f) ** (g \star\star g)\}$. The following choice for the functions f and g is then selected:

$$f(f_x, f_y) = \text{rect}\left(\frac{\lambda d_i}{\ell} f_x, \frac{\lambda d_i}{\ell} f_y\right)$$

$$g(f_x, f_y) = \left[\delta\left(\lambda d_i f_x + s, y\right) + \delta\left(\lambda d_i f_x - s, y\right)\right].$$

The autocorrelation of H can then be expressed as:

$$H(f_x, f_y) \star\star H(f_x, f_y) = (f \star\star f) ** (g \star\star g) \quad \text{where,}$$

$$f \star\star f = \left(\frac{\ell}{\lambda d_i}\right)^2 \text{tri}\left(\frac{\lambda d_i}{\ell} f_x, \frac{\lambda d_i}{\ell} f_y\right), \quad \text{and,}$$

$$g \star\star g = \delta\left(\lambda d_i f_x + 2s, y\right) + 2\delta(\lambda d_i f_x, \lambda d_i f_y) + \delta\left(\lambda d_i f_x - 2s, y\right).$$

The convolution $\text{tri}(af_x/\ell, af_y/\ell) ** \delta(af_x + b, af_y) = \text{tri}[(af_x + b)/\ell]\text{tri}(af_y/\ell)$ with $a = \lambda d_i$, and $b = 0, \pm 2s$. The normalization of the autocorrelation of H leads to the following OTF:

$$\mathcal{H}(f_x, f_y) = \frac{1}{2}\text{tri}\left(\frac{\lambda d_i f_x + 2s}{\ell}, \frac{\lambda d_i f_y}{\ell}\right) + \text{tri}\left(\frac{\lambda d_i f_x}{\ell}, \frac{\lambda d_i f_y}{\ell}\right) + \frac{1}{2}\text{tri}\left(\frac{\lambda d_i f_x + 2s}{\ell}, \frac{\lambda d_i f_y}{\ell}\right). \tag{6.47}$$

From the preceding equation, it becomes clear that the OTF has nonzero values within the following intervals, determined by the limits of the "tri" functions:

$$-\frac{\ell + 2s}{\lambda d_i} \leq f_x \leq \frac{\ell - 2s}{\lambda d_i},$$

$$-\frac{\ell}{\lambda d_i} \leq f_x \leq \frac{\ell}{\lambda d_i},$$

$$-\frac{2s - \ell}{\lambda d_i} \leq f_x \leq \frac{2s + \ell}{\lambda d_i},$$

$$-\frac{\ell}{\lambda d_i} \leq f_y \leq \frac{\ell}{\lambda d_i}.$$

It is evident that, for the three "tri" functions to remain non-overlapping, the condition $s > \ell$ must be satisfied. For the numerical implementation, the following parameters are given: $\ell = 50\,\text{mm}$, $2s = 120\,\text{mm}$ (thus ensuring $s > \ell$ and that the three "tri" functions do not have overlap), $d_i = 100\,\text{mm}$, and $\lambda = 1\,\mu\text{m}$. The OTF in the $f_x f_y$-plane is shown in Fig. 6.16(b). The three "tri" functions are clearly visible, with the middle one exhibiting twice the peak value of the two smaller ones. The OTF along the f_x-direction is presented in Fig. 6.16(c). The limiting points (cutoff spatial frequencies) of the OTF along the f_x-direction are: $\pm(\ell + 2s)/\lambda d_i = \pm 1700\,/\text{mm}$, $\pm\ell/\lambda d_i = \pm 500\,/\text{mm}$, and $\pm(2s - \ell)/\lambda d_i = \pm 700\,/\text{mm}$, as can be observed in the figure. The corresponding maximum α_z angles for the various cutoff spatial frequencies are:

$$\left(f_x = \frac{\ell}{\lambda d_i}, \; f_y = \frac{\ell}{\lambda d_i}\right) \longrightarrow \alpha_z = \cos^{-1}\left\{\left[1 - 2\left(\frac{\ell}{d_i}\right)^2\right]^{1/2}\right\} = 45°,$$

$$\left(f_x = \frac{\ell}{\lambda d_i}, \; f_y = \frac{\ell}{\lambda d_i}\right) \longrightarrow \alpha_z = \cos^{-1}\left\{\left[1 - \left(\frac{2s - \ell}{d_i}\right)^2 - \left(\frac{\ell}{d_i}\right)^2\right]\right\} = 59.34°,$$

$$\left(f_x = \frac{2s + \ell}{\lambda d_i}, \; f_y = \frac{\ell}{\lambda d_i}\right) \longrightarrow \alpha_z = \cos^{-1}\left\{\left[1 - \left(\frac{2s + \ell}{d_i}\right)^2 - \left(\frac{\ell}{d_i}\right)^2\right]\right\} \rightarrow$$

$$\text{evanescent wave.}$$

In the last case, no α_z angle exists, as the wave corresponding to $f_x = (2s + \ell)/\lambda d_i$ becomes evanescent. For $f_y = 500\,\text{mm}^{-1}$, the maximum f_x must satisfy the condition $f_x^2 + f_y^2 \leq 1/\lambda^2$. This yields, $f_{x,max} = 866.03\,\text{mm}^{-1}$.

6.11 Aberration Effects in Frequency Response of Optical Systems

In Section 6.8, it was initially assumed that the wavefront at ExP of an optical system was purely spherical (refer to Fig. 6.13). However, for this assumption to hold, the optical system must be entirely free of aberrations. Aberrations, which are discussed in the *Geometrical Optics* chapter (Chapter 2) represent deviations from ideal optical behavior and often manifest as image blurring or distortion.

In practical optical systems, aberrations are nearly always present, leading to a degradation in image quality. This degradation influences the spatial frequency properties, introducing phase errors, variations in focal length, and other imperfections. In addition to reducing image contrast and distorting the shape of the image, these aberrations result in non-uniform spatial frequency responses. As a consequence, the system's ability to accurately reproduce high-frequency patterns and resolve fine details is compromised. Engineers and optical designers must address and correct these aberrations to optimize the spatial frequency response of optical systems for specific applications.

The impact of aberrations can be accounted for by introducing a deforming phase plate at the ExP, as discussed by Goodman [265]. The deviation of the wavefront phase in ExP can be quantified using the function $W(x, y)$, which represents an effective path length error, as illustrated in Fig. 6.17. The difference between the phase Φ_{sp} of the spherical reference

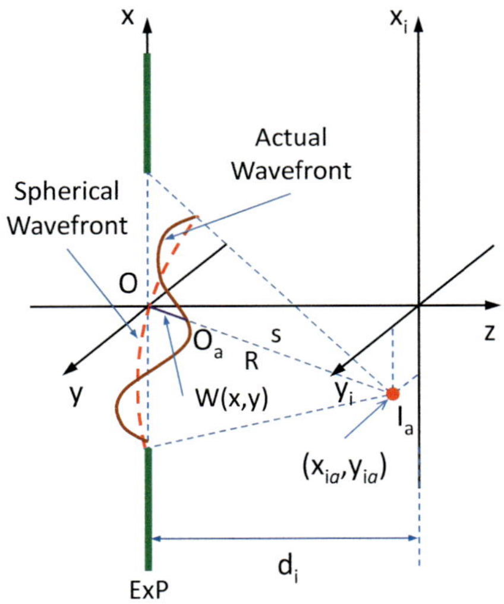

FIGURE 6.17: The actual (due to aberrations) wavefront at the exit pupil of an optical system. The spherical wavefront corresponds to an aberration-free equivalent system. The distance $W(x, y)$ specifies the aberration function used. The point I_a, (x_{ia}, y_{ia}) is the ideal image of an input object point. The distance $OI_a = R$ and the distance $O_aI_a = s$.

wavefront, and the phase Φ_{ac} of the actual wavefront is given by $\Phi_{sp} - \Phi_{ac} = -kR - (-ks) = -k(R - s) = -kW(x, y)$ (in the figure the difference for the point $x = 0, y = 0$ is shown). Of course, the term $nW(x, y)$ is the optical path error. Using this phase factor the generalized pupil function, $\mathcal{P}(x, y)$, can be expressed as:

$$\mathcal{P}(x, y) = P(x, y)e^{jkW(x,y)}, \tag{6.48}$$

where $P(x, y)$ describes the ExP function without the aberrations effect. For coherent optical systems, as described in Sec. 6.10.1, the optical transfer function, $H(f_x, f_y)$ from Eq. (6.39), under the influence of aberrations, is expressed in the following form:

$$H(f_x, f_y) = \mathcal{P}(-\lambda d_i f_x, -\lambda d_i f_y) \sim \; P(\lambda d_i f_x, \lambda d_i f_y) e^{jkW(\lambda d_i f_x, \lambda d_i f_y)}. \tag{6.49}$$

It was again assumed that the pupil function is symmetric. From the previous equation, the bandwidth limitation, which refers to the spatial cutoff frequencies of the amplitude transfer function, is entirely determined by the ExP function $P(x, y)$ and remains unaffected by aberrations. The impact of aberrations is solely reflected in the introduction of phase variations within the passband. These phase distortions can degrade the performance of the optical system, even though they do not directly affect the cutoff spatial frequencies of the amplitude transfer function.

In the case of incoherent optical systems, as discussed in Sec. 6.10.2, the effect of aberrations is more complex. In this context, the influence of aberrations on the optical transfer function is of primary concern. The optical transfer function (OTF) is still given by Eqs. (6.42) or (6.43). At this point, the normalization factor, $\mathcal{N}$, used for the OTF can be derived (using Parseval's identity). It is given as:

$$\begin{aligned}
\mathcal{N} &= \int_{-\infty}^{+\infty}\int_{-\infty}^{+\infty} |h(x, y)|^2 dx dy = \int_{-\infty}^{+\infty}\int_{-\infty}^{+\infty} |H(f_x, f_y)|^2 df_x df_y \\
&= \int_{-\infty}^{+\infty}\int_{-\infty}^{+\infty} |P(\lambda d_i f_x, \lambda d_i f_y) e^{jkW(\lambda d_i f_x, \lambda d_i f_y)}|^2 df_x df_y \\
&= \int_{-\infty}^{+\infty}\int_{-\infty}^{+\infty} |P(\lambda d_i f_x, \lambda d_i f_y)|^2 df_x df_y.
\end{aligned}$$

Since $P(x, y)$ is unity in the aperture region and zero elsewhere, $\mathcal{N}$ represents the area in the spatial frequency domain corresponding to the aperture of the ExP function. Consequently, the OTF can be written as [using Eq. (6.42)]:

$$\begin{aligned}
\mathcal{H}(f_x, f_y) &= \frac{1}{\mathcal{N}} \int_{-\infty}^{+\infty}\int_{-\infty}^{+\infty} H^*(f'_x, f'_y) H(f_x + f'_x, f_y + f'_y) df'_x df'_y \\
&= \frac{1}{\mathcal{N}} \int_{-\infty}^{+\infty}\int_{-\infty}^{+\infty} P\left(\lambda d_i f'_x, \lambda d_i f'_y\right) P\left(\lambda d_i (f_x + f'_x), \lambda d_i (f_y + f'_y)\right) \times \\
&\qquad \exp\left[-jkW\left(\lambda d_i f'_x, \lambda d_i f'_y\right)\right] \times \\
&\qquad \exp\left[+jkW\left(\lambda d_i (f_x + f'_x), \lambda d_i (f_y + f'_y)\right)\right] df'_x df'_y.
\end{aligned} \tag{6.50}$$

It can be demonstrated straightforwardly using Schwarz's inequality that $|\mathcal{H}_a(f_x, f_y)|^2 \leq |\mathcal{H}(f_x, f_y)|^2$ [265], where $\mathcal{H}_a$ and $\mathcal{H}$ represent the OTF with aberrations and without aberrations, respectively. Consequently, aberrations cannot increase the contrast of any spatial component of the image; rather, they typically lead to a reduction in contrast.

In general, the absolute spatial cutoff frequency remains unaffected by aberrations. However, significant aberrations can diminish the high spatial frequency response of the optical system. Additionally, aberrations can cause the OTF to assume negative values in certain frequency bands, a phenomenon that is absent in aberration-free optical systems. Negative OTF values indicate that image components at these spatial frequencies undergo a contrast reversal. Contrast reversal is characterized by the transformation of intensity peaks into intensity nulls, and vice versa [265].

At this point, it is useful to define the *Strehl ratio* of an optical system. The Strehl ratio is a parameter commonly used in optics to quantify the quality of an optical system, particularly with respect to imaging or wavefront aberrations. It is a dimensionless number that compares the peak intensity of an aberrated optical system to the peak intensity of an ideal, diffraction-limited optical system. A Strehl ratio of 1 indicates a perfect, diffraction-limited system, while lower values indicate a degradation in performance due to optical imperfections or aberrations. The Strehl ratio, denoted as S, is defined as the ratio of the peak point spread function (PSF) intensity of the actual optical system to the peak PSF intensity of the ideal diffraction-limited system:

$$S = \frac{I_a^{max}}{I^{max}} = \frac{|h_a(0,0)|^2}{|h(0,0)|^2}, \tag{6.51}$$

where, I_a^{max} is the actual (aberrated) maximum intensity of the PSF of the optical system, and I^{max} is the corresponding maximum PSF intensity of the diffraction-limited optical system without aberrations. It is important to note that the maximum PSF intensity occurs at the origin $(0,0)$. This can be easily understood from Eq. (6.36) and the properties of the Fourier transform.

By utilizing Eq. (6.40), an alternate expression for the Strehl ratio can be derived, as shown in the following:

$$S = \frac{\iint \mathcal{H}_a(f_x, f_y) df_x df_y}{\iint \mathcal{H}(f_x, f_y) df_x df_y}, \tag{6.52}$$

where $\mathcal{H}_a$ is the aberrated OTF, and $\mathcal{H}$ is the diffraction-limited OTF, which is free from aberrations. More advanced discussions on the Strehl ratio can be found in Ref. [58].

In practical terms, a Strehl ratio close to 1 indicates a high-quality optical system with minimal aberrations, whereas a lower Strehl ratio suggests a system with more significant aberrations and reduced imaging performance. The Strehl ratio serves as a useful metric in fields such as astronomy, microscopy, and adaptive optics, where the quality of the optical system is crucial for obtaining clear and detailed images.

Example 6.11.1:

An example of an incoherent optical system exhibiting a focusing error, specifically defocus, as an aberration, is considered. In this case, a straightforward mathematical treatment is employed to derive analytical results. The system is assumed to have an image point on the z-axis at a distance d, where $d > d_i$ (with $z = d_i$ representing the position of the image plane). The spherical reference wavefront and the actual wavefronts are depicted in Fig. 6.18(a). The ExP is assumed to have a rectangular shape with dimensions $2L_x \times 2L_y$. The objective is to determine the corresponding optical transfer function and the Strehl ratio. For the numerical implementation, specific values are chosen, such as $2L_x = 2L_y = 50\,\text{mm}$, and the term $(1/2)(1/d_i - 1/d)L_x^2/\lambda$ is varied across different values, including 0, 0.25, 0.50, 0.75, and 1.

Solution: Initially, the function $W(x,y)$ needs to be determined. According to the previous definitions, the phase difference is given by $-kW(x,y) = \Phi_{sp} - \Phi_{ac} = -k(x^2+y^2+d_i^2)^{1/2} + k(x^2+y^2+d^2)^{1/2}$. For the practical case where $d, d_i \gg x, y$, a parabolic approximation is employed, and the function $kW(x,y)$ is expressed as follows:

$$kW(x,y) = \frac{k}{2}\left(\frac{1}{d_i} - \frac{1}{d}\right)(x^2+y^2), \tag{6.53}$$

where the constant phase term of $k(d_i - d)$ was neglected. Now, define the terms,

$$L_{mx} = \frac{1}{2}\left(\frac{1}{d_i} - \frac{1}{d}\right) L_x^2,$$

$$L_{my} = \frac{1}{2}\left(\frac{1}{d_i} - \frac{1}{d}\right) L_y^2.$$

Using the defined terms, $kW(x, y)$ can be expressed in the following form:

$$kW(x, y) = k\left(L_{mx}\frac{x^2}{L_x^2} + L_{my}\frac{y^2}{L_y^2}\right). \tag{6.54}$$

The ExP function without the aberration defocus can be expressed by the "rect" function as $P(x, y) = \text{rect}(x/2L_x, y/2L_y)$. Substituting this $P(x, y)$ function along with the corresponding $kW(x, y)$ function into Eq. (6.50), the calculation is restricted to regions where there is overlap between $P(\lambda d_i f_x', \lambda d_i f_y')$ and $P(\lambda d_i(f_x + f_x'), \lambda d_i(f_y' + f_y))$. As discussed in Example 6.10.2, the overlap condition dictates that, for instance, when $f_x \geq 0$ and $f_y \geq 0$, the limits for f_x' are $-(L_x/\lambda d_i) \leq f_x' \leq (L_x/\lambda d_i) - f_x$, and similar limits apply for f_y'. Consequently, the OTF can be expressed as:

$$\mathcal{H}(f_x, f_y) = \frac{1}{\mathcal{N}} \int_{-L_x/\lambda d_i}^{L_x/\lambda d_i - f_x} \int_{-L_y/\lambda d_i}^{L_y/\lambda d_i - f_y} \exp\left[-jkW\left(\lambda d_i f_x', \lambda d_i f_y'\right)\right] \times$$
$$\exp\left[+jkW\left(\lambda d_i(f_x + f_x'), \lambda d_i(f_y + f_y')\right)\right] df_x' df_y'.$$

Since the integrals with respect to f_x' and f_y' are identical and separable, it is possible to work only with the f_x' integral and the result can be extrapolated to f_y' integrals by replacing the x subscripts with the y subscripts. Therefore, the f_x' integral, I_x, with the substitution of the kW function becomes:

$$\begin{aligned} I_x &= \int_{-L_x/\lambda d_i}^{L_x/\lambda d_i - f_x} \exp\left[-jk\frac{L_{mx}}{L_x^2}(\lambda d_i)^2 f_x'^2\right] \exp\left[+jk\frac{L_{mx}}{L_x^2}(\lambda d_i)^2\left(f_x' + f_x\right)^2\right] df_x' \\ &= \exp\left[+jk\frac{L_{mx}}{L_x^2}(\lambda d_i)^2 f_x^2\right] \int_{-L_x/\lambda d_i}^{L_x/\lambda d_i - f_x} \exp\left[+jk\frac{L_{mx}}{L_x^2}(\lambda d_i)^2\left(2f_x f_x'\right)\right] df_x' \\ &= a_x\left(1 - \frac{f_x}{a_x}\right) \text{sinc}\left[\frac{8L_{mx}}{\lambda}\frac{f_x}{a_x}\left(1 - \frac{f_x}{a_x}\right)\right], \end{aligned}$$

where $a_x = 2L_x/\lambda d_i$. A similar result, for the f_y' integral I_y, is obtained for $f_y > 0$ with $a_y = 2L_y/\lambda d_i$. The normalization factor can be easily shown to be $\mathcal{N} = a_x a_y$. Combining the results for I_x and I_y, the OTF for any pair of spatial frequencies (f_x, f_y) can be expressed as:

$$\mathcal{H}(f_x, f_y) = \text{tri}\left(\frac{f_x}{a_x}\right) \text{tri}\left(\frac{f_y}{a_y}\right) \text{sinc}\left[\frac{8L_{mx}}{\lambda}\frac{f_x}{a_x}\left(1 - \frac{|f_x|}{a_x}\right)\right] \text{sinc}\left[\frac{8L_{my}}{\lambda}\frac{f_y}{a_y}\left(1 - \frac{|f_y|}{a_y}\right)\right]. \tag{6.55}$$

In the last equation, "tri" refers to the triangle function (previously defined), and "sinc" is the sinc function, denoted as $\text{sinc}(x) = \sin(\pi x)/(\pi x)$. For the numerical implementation using the specified parameters for a square ExP ($2L_x = 2L_y$), the OTF along the f_x spatial frequency direction is depicted in Fig. 6.18(b). The spatial frequency axis has been normalized as f_x/a_x. A noteworthy observation is that as the defocus error increases, the OTF may exhibit negative values, leading to contrast reversal.

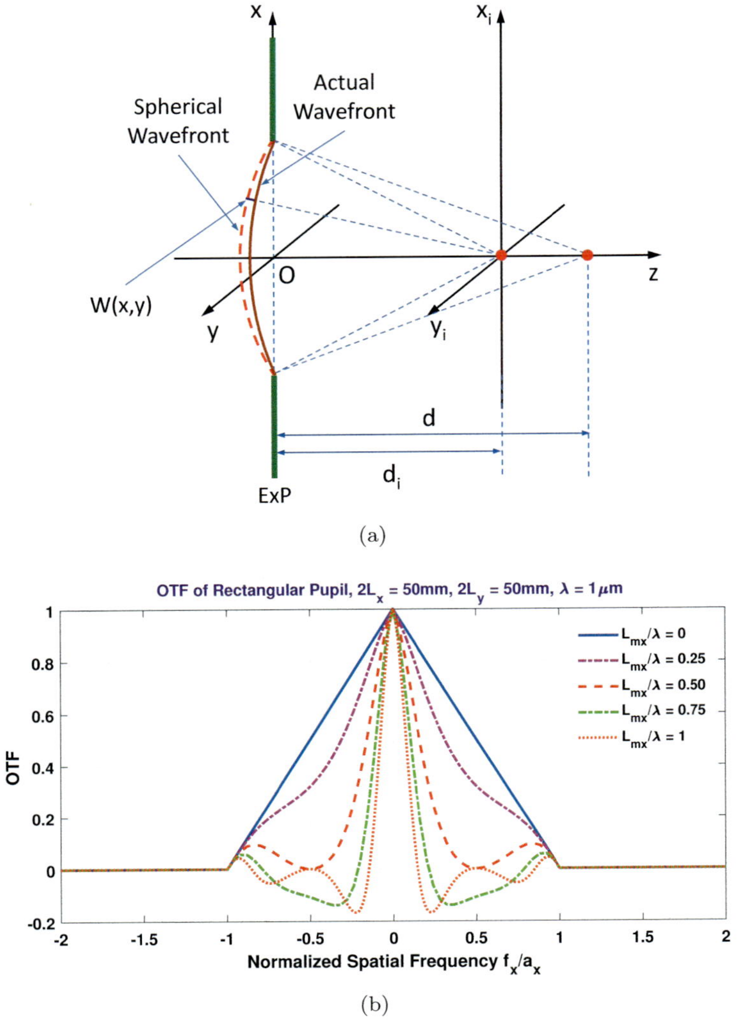

FIGURE 6.18: (a) Geometry in order to define the aberration defocus of an optical system. (b) The OTF of an incoherent optical system with square ExP (of dimensions $2L \times 2L$) is plotted along the normalized spatial frequency f_x/a_x, where $a_x = 2L/\lambda d_i$. The OTF is shown for various defocus parameter values L_{mx}/λ, illustrating the effect of defocus aberration on spatial frequency response [265].

To better comprehend the level of defocus corresponding to $L_{mx}/\lambda = 1$, the following parameters are considered: $\lambda = 0.55\,\mu\text{m}$, $d_i = 100\,\text{mm}$, and $L_x = L_y = 25\,\text{mm}$. Assuming $d = d_i + \Delta$, where Δ specifies the defocus, it can be straightforwardly determined that $\Delta = 0.0176\,\text{mm}$ when $L_{mx}/\lambda = 1$.

The Strehl ratio, S, can be readily calculated using Eq. (6.52). For values of $(1/2)(1/d_i - 1/d)L_x^2/\lambda$ equal to 0.25, 0.50, 0.75, and 1, the corresponding S values are 0.80, 0.39, 0.12, and 0.09, respectively. These results illustrate the pronounced effect of defocus aberration on the Strehl ratio and, consequently, the optical system's imaging performance.

6.11.1 Pupil Apodization Effects on the Frequency Response of Optical Systems

The point spread function (PSF) of a diffraction-limited optical system may exhibit side-lobes in the image of any single object point. While these effects are typically negligible, there are specific situation, such as in astronomy [265], where they become significant. For instance, resolving a faint star situated near a bright star requires the side-lobes to be addressed to achieve adequate separation of the two objects. To mitigate these effects, a technique known as *apodization* is employed.

Pupil apodization is a method used in optical systems to improve image quality and correct certain optical imperfections. It involves modifying the light distribution across the aperture of the optical system, often through the application of specialized attenuation masks or filters. One widely adopted apodization technique is Gaussian apodization, which utilizes a Gaussian intensity profile to reduce the light intensity near the edges of the pupil.

The influence of apodization on the frequency response of optical systems is relevant for both coherent and incoherent imaging. In coherent optical systems, the effect is relatively straightforward, as the amplitude transfer function $H(f_x, f_y)$ is directly related to the pupil function [see Eq. (6.39)]. Specifically, $H(f_x, f_y) = P(\lambda d_i f_x, \lambda d_i f_y)G(\lambda d_i f_x, \lambda d_i f_y)$, where $G(x, y) = e^{-A(x^2+y^2)}$ represents the Gaussian apodization function. Consequently, the coherent PSF can be expressed as $h(x, y) = \mathcal{F}^{-1}\{PG\} = \mathcal{F}^{-1}\{P\} ** \mathcal{F}^{-1}\{G\}$, where the convolution of the inverse Fourier transform of the pupil function P with the inverse Fourier transform of G (which is itself Gaussian) effectively reduces the PSF side-lobes.

For incoherent optical systems, the relationship between the optical transfer function (OTF) and the pupil function modulated by the Gaussian apodization is more complex, owing to the normalized autocorrelation involved [see Eq. (6.42)]. In this case, $\mathcal{H} = (PG) \star \star\, (PG)/\mathcal{N}$, where $\mathcal{N}$ represents the normalization constant.

The effects on spatial frequency can be better understood through illustrative examples, which are provided in the following discussion.

Example 6.11.2:

Consider a coherent optical system with a rectangular pupil function, $P(x, y)$, defined as $P(x, y) = \text{rect}(x/2L_x, y/2L_y)$. The system incorporates a Gaussian apodization function given by $G(x, y) = e^{-A(x^2+y^2)}$. The task is to determine the coherent point spread function (PSF) resulting from this configuration. For the numerical implementation, the following parameters should be used: $2L_x = 2L_y = 25\,\text{mm}$, $d_i = 100\,\text{mm}$, and $\lambda = 1\,\mu\text{m}$. In addition, assume $AL_x^2 = \kappa$ and $\kappa = 2$.

Solution: As it was presented in Sec. 6.11.1, the coherent PSF is given by:

$$h(x, y) = \mathcal{F}^{-1}\left\{\text{rect}\left(\frac{\lambda d_i}{2L_x} f_x, \frac{\lambda d_i}{2L_y} f_y\right) \exp\left[-A(\lambda d_i)^2 \left(f_x^2 + f_y^2\right)\right]\right\}$$

$$= \mathcal{F}^{-1}\left\{\text{rect}\left(\frac{\lambda d_i}{2L_x} f_x, \frac{\lambda d_i}{2L_y} f_y\right)\right\} ** \mathcal{F}^{-1}\left\{\exp\left[-A(\lambda d_i)^2 \left(f_x^2 + f_y^2\right)\right]\right\}$$

$$= \left[\frac{2L_x}{\lambda d_i}\frac{2L_y}{\lambda d_i}\text{sinc}\left(\frac{2L_x}{\lambda d_i}x\right)\text{sinc}\left(\frac{2L_y}{\lambda d_i}y\right)\right] ** \left[\frac{\pi}{A(\lambda d_i)^2}\exp\left[-\frac{\pi^2}{A(\lambda d_i)^2}(x^2 + y^2)\right]\right].$$

It is noteworthy that the dimensions of $h(x, y)$ remain consistent as m^{-2}, as expected. To visualize the effects of Gaussian apodization, the PSF is examined along either the x- or the y-direction. For simplicity, let us consider $y = 0$ and observe the PSF along the x-direction.

The normalized PSF without apodization is presented in Fig. 6.19(a), alongside the normalized Gaussian function (scaled to a unit maximum). The effect of apodization is evaluated by performing a convolution between these two functions. The resulting normalized coherent PSF, incorporating Gaussian apodization, is shown in Fig. 6.19(b). It is evident that the apodized PSF exhibits reduced side-lobes compared to the un-apodized PSF, demonstrating the effectiveness of the Gaussian apodization in improving the optical system's performance.

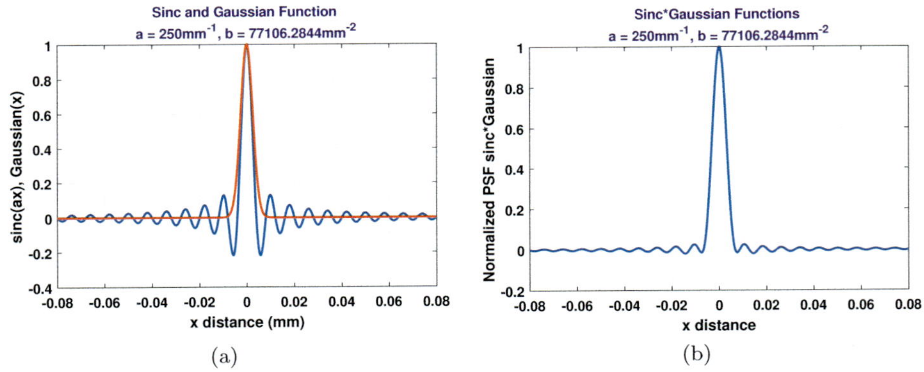

FIGURE 6.19: (a) The coherent PSF of a square pupil function with parameters $2L_x = 2L_y = 25\,\text{mm}$, $d_i = 100\,\text{mm}$, and $\lambda = 1\,\mu\text{m}$, is shown without apodization. In the same plot, the inverse Fourier transform of the Gaussian apodization function, $\mathcal{F}^{-1}\{G\}$, is also presented. The variables are defined as $a = 2L_x/(\lambda d_i)$ and $b = \pi^2/A(\lambda d_i)^2$ with $A = \kappa/L_x^2$. In this example $\kappa = 2$. (b) The coherent PSF after Gaussian apodization is shown, illustrating the suppression of side-lobes compared to the un-apodized case.

Example 6.11.3:

Consider again the rectangular pupil function $P(x, y) = \text{rect}(x/2L_x, y/2L_y)$ of example 6.11.2, with the same Gaussian apodization function $G(x, y) = e^{-A(x^2+y^2)}$. For this optical system, assume the system is incoherent, and the objective is to determine the optical transfer function (OTF). For the numerical implementation, use the following specific values: $2L_x = 2L_y = 50\,\text{mm}$, $d_i = 100\,\text{mm}$, and $\lambda = 1\,\mu\text{m}$. In addition, assume $AL_x^2 = \kappa$ and $\kappa = 0, 1, 2, 3$.

Solution: To determine the OTF, it is necessary to compute the normalized autocorrelation of the H function. In this case, the autocorrelation integral is separable in the x- and y-directions, which simplifies the calculation. By solving for one direction, the results can then be applied to the other direction. For this example, the x-direction is chosen for the calculation. First, the normalized factor, required for the application of Eq. (6.42) must be determined. If $\mathcal{N}_x$ represents the normalization factor in the x-direction, it can be expressed as follows:

$$\mathcal{N}_x = \int_{-L_x/\lambda d_i}^{+L_x/\lambda d_i} \exp\left[-A(\lambda d_i f_x')^2\right] \exp\left[-A(\lambda d_i f_x')^2\right] df_x'$$

$$= \frac{\sqrt{\pi}}{2\sqrt{2A}(\lambda d_i)} \left[\text{erf}\left(\sqrt{2A}L_x\right) - \text{erf}\left(-\sqrt{2A}L_x\right)\right],$$

where "erf" is the standard error function defined as $\text{erf}(x) = (2/\sqrt{\pi}) \int_0^x e^{-t^2} dt$. The auto-correlation integral along the x-direction I_x (for $f_x \geq 0$) is given by:

$$I_x = \int_{-L_x/\lambda d_i}^{+L_x/\lambda d_i - f_x} \exp\left[-A\left(\lambda d_i f_x'\right)^2\right] \exp\left[-A\left(\lambda d_i\left(f_x + f_x'\right)\right)^2\right] df_x'$$

$$= \frac{\sqrt{\pi}}{2\sqrt{2A}(\lambda d_i)} \exp\left[-2AL_x^2\left(\frac{f_x}{a_x}\right)^2\right] \times$$

$$\left\{\text{erf}\left[\sqrt{2A}L_x\left(1-\frac{f_x}{a_x}\right)\right] - \text{erf}\left[-\sqrt{2A}L_x\left(1-\frac{f_x}{a_x}\right)\right]\right\},$$

where $a_x = 2L_x/(\lambda d_i)$ as before. By normalizing I_x with $\mathcal{N}_x$, the following OTF is derived (for $f_x \geq 0$):

$$\mathcal{H}(f_x \geq 0, f_y = 0) = \exp\left[-2AL_x^2\left(\frac{f_x}{a_x}\right)^2\right] \frac{\text{erf}\left[\sqrt{2A}L_x\left(1-\frac{f_x}{a_x}\right)\right] - \text{erf}\left[-\sqrt{2A}L_x\left(1-\frac{f_x}{a_x}\right)\right]}{\text{erf}\left(\sqrt{2A}L_x\right) - \text{erf}\left(-\sqrt{2A}L_x\right)}.$$

It is straightforward to calculate $\mathcal{H}$ for negative spatial frequencies, as it exhibits symmetry

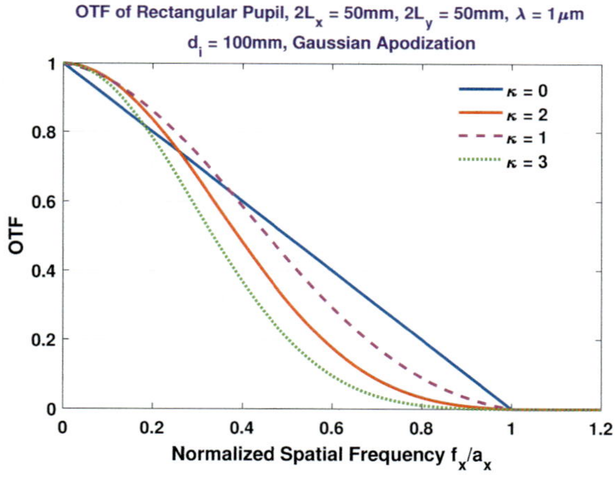

FIGURE 6.20: The effect of Gaussian pupil apodization on the OTF of the optical system with a square pupil function of dimensions $2L_x \times 2L_y$ ($L_x = L_y$). The variable κ is related to the Gaussian function as $e^{-AL_x^2} = e^{-\kappa}$.

in the (f_x, f_y)-plane. For the numerical implementation, $2L_x = 2L_y = 50$ mm, $d_i = 100$ mm, and $\lambda = 1\,\mu$m. The resulting OTF as a function of the normalized spatial frequency f_x/a_x is presented in Fig. 6.20.

The impact of Gaussian apodization is to enhance the performance at low spatial frequencies while reducing the influence of higher ones. A gradual change in the Gaussian apodization function ($\kappa = 1$) results a milder effect, whereas a more abrupt change ($\kappa = 3$) produces a more pronounced impact. When $\kappa = 0$, no apodization is applied. For $\kappa = 2$, the Gaussian drops to $1/e^2$ at the edge ($\pm L_x$) of the pupil [265].

The discrete form of the Optical Transfer Function (OTF) [refer to Eq. (6.44)] proves useful for comparison in this context. The number of discretization points in the f_x-direction is denoted as N_x, and can range from 25 to 400 points. For this case, the parameter κ is set to 2, while the other parameters remain the same as in the previous example. In Fig. 6.21(a), the OTF is presented for both the analytical solution and the discrete autocorrelation approach with $N_x = 25$. The blue line corresponds to the discrete results, and even with 25 points, it closely aligns with the analytical curve. Figure 6.21(b) displays the relative percent error. As expected, the error diminishes with an increase in the number of discretization points ($N_x = 25, 50, 100, 200,$ and 400). It is noticeable that the error only increases when the OTF tends towards zero (or very small values). Importantly, the OTF calculated via the discrete autocorrelation can be generalized in cases where the analytical result is not feasible.

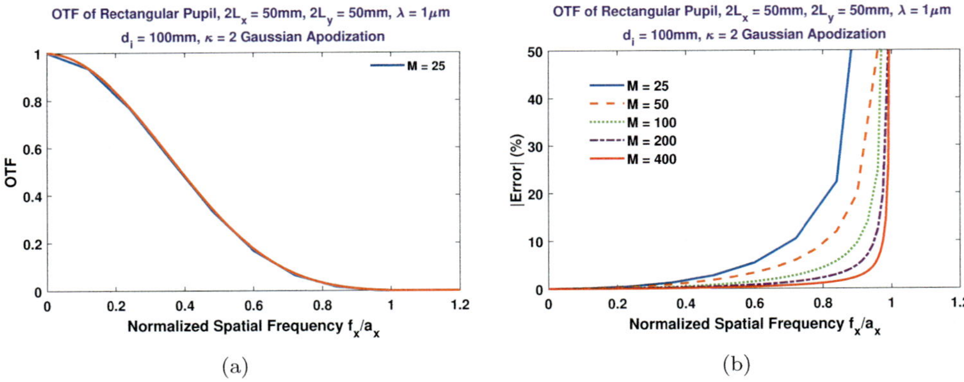

FIGURE 6.21: (a) The analytical OTF [Eq. (6.42)] and the discrete OTF [Eq. (6.44), with $M = N_x = 25$] for a Gaussian apodization with $\kappa = 2$. (b) The percent error between the analytical and the discrete OTF as a function of the normalized spatial frequency with the number of the discretization points, $M = N_x$ as a parameter.

Example 6.11.4:

In this example, consider a circular pupil function with a radius of a, represented as $P(r_T) = \text{circ}(r_T/a)$, combined with a Gaussian apodization function of the form $G(r_T) = e^{-Ar_T^2}$. The optical system is assumed to be incoherent, and the investigation focuses on the optical transfer function (OTF). For numerical implementation, let $a = 10$,mm, $d_i = 100$,mm, and $\lambda = 1, \mu$m. Furthermore, assume $Aa^2 = \kappa$ with $\kappa = 0, 0.5, 1, 2, 3$.

Solution: As in the previous example, the normalization factor $\mathcal{N}$ is first determined as follows, where $\rho = (f_x^2 + f_y^2)^{1/2}$.

$$\mathcal{N} = \int_0^{a/\lambda d_i}\int_0^{2\pi} \exp\left[-A(\lambda d_i \rho')^2\right] \exp\left[-A(\lambda d_i \rho')^2\right] \rho' d\rho' d\phi = \frac{2\pi}{4A(\lambda d_i)^2}\left[1 - \exp\left(-2Aa^2\right)\right].$$

Next, the autocorrelation of H is required. Since the pupil is circular, the autocorrelation will exhibit azimuthal symmetry. The autocorrelation integral can be written as follows:

$$I_\rho(f_x, f_y) = \int_{-a/\lambda d_i}^{a/\lambda d_i}\int_{-a/\lambda d_i}^{a/\lambda d_i} \text{circ}\left(\frac{\lambda d_i}{a}\sqrt{f_x'^2 + f_y'^2}\right) \text{circ}\left(\frac{\lambda d_i}{a}\sqrt{(f_x + f_x')^2 + (f_y + f_y')^2}\right) \times$$
$$\exp\left[-A(\lambda d_i)^2\left(f_x'^2 + f_y'^2\right)\right] \exp\left[-A(\lambda d_i)^2\left((f_x + f_x')^2 + (f_y + f_y')^2\right)\right] df_x' df_y'.$$

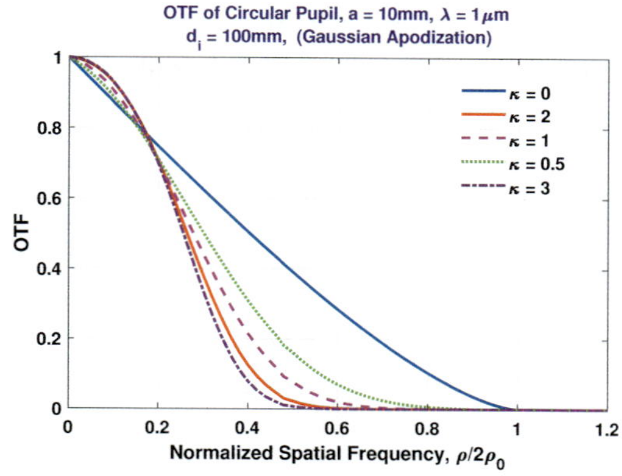

FIGURE 6.22: The effect of Gaussian pupil apodization on the OTF of an incoherent optical system with a circular pupil function of radius a. The variable κ is related with Gaussian function as $e^{-Aa^2} = e^{-\kappa}$.

The last integral may not be analytically solvable. However, it can be readily evaluated using numerical integration. Due to the azimuthal symmetry of the autocorrelation, it suffices to compute the optical transfer function (OTF) along either the f_x- or f_y-direction, i.e., $I_\rho(f_x, 0) = I_\rho(0, f_y) = I_\rho(\rho)$.

For the numerical integration, the parameters specified in this example were used. The resulting OTF, $I_\rho(\rho)/\mathcal{N}$, is shown in Fig. 6.22 as a function of the normalized spatial frequency $\rho/2\rho_0$, where $\rho_0 = a/\lambda d_i$.

In the case of a circular pupil function, the effect of Gaussian apodization is similar to that observed with a square pupil. However, in the circular case, the enhancement of lower frequencies is less pronounced, while the attenuation of higher frequencies is more significant.

6.12 Comparison of Coherent and Incoherent Optical Imaging Systems

Coherent and incoherent optical systems represent two distinct types of imaging systems that differ fundamentally in how they manage the phase of light waves and the resulting interference patterns.

Coherent systems preserve the phase information of light waves. In these systems, the relative phase of different parts of the optical wavefront is maintained and can be precisely controlled. This capability allows the observation and manipulation of interference phenomena, such as diffraction and interference patterns. Coherent systems are commonly employed in applications where precise control of phase and interference is essential, including holography, interferometry, and certain types of microscopy.

In contrast, incoherent systems do not retain the phase information of light waves. As light propagates through the system, phase information is either lost or randomized. These systems are typically used in applications where phase information is not critical, and only the intensity or brightness of light is of interest. Examples include most conventional photography, everyday lighting systems, and simple imaging systems like the human eye.

In summary, the key distinction between coherent and incoherent optical systems lies in their treatment of the phase of light waves. Coherent systems maintain phase information, enabling precise control and manipulation of interference patterns, whereas incoherent systems disregard phase information and primarily focus on the intensity or brightness of light. The choice between these systems depends on the specific requirements of the optical application or experiment.

It is generally believed that the characteristics and distinctions between coherent and incoherent imaging systems are well understood [349]. For coherent imaging:

(a) The imaging operation is linear with respect to the complex amplitude of the optical field.
(b) The object is typically illuminated with coherent light, such as that from a laser.
(c) If the entrance pupil of the system is located in the far field of the object and the paraxial approximation holds, the imaging process becomes shift-invariant. In this case, the transfer function for the complex amplitude is represented by a scaled version of the system's pupil function.
(d) The corresponding coherent point spread function is generally complex-valued.

On the other hand, for incoherent imaging:

(a) The imaging operation is linear with respect to the optical wave intensity.
(b) The object is either self-luminous or illuminated by a spatially incoherent light source.
(c) No specific condition is imposed on the distance between the entrance pupil and the object.
(d) The transfer function for optical intensity is represented by a scaled version of the autocorrelation of the exit pupil function.
(e) The corresponding incoherent point spread function (PSF) is real and non-negative.

Another crucial assumption underlying the presented analysis warrants attention. Although an imaging system typically consists of multiple optical elements, each with its own distinct optical aperture, the approach adopted here simplifies the system into a single "black box." In this representation, only the projected image of the system's limiting aperture, whether of the entrance pupil (EnP) or the exit pupil (ExP), is used to account for diffraction effects [350]. This approach, based on the paraxial approximation, provides a linear and shift-invariant description of the imaging process, applicable to both spatially coherent and incoherent optical systems. The implications of these approximations are examined in the context of a 4f imaging system in the study by Kelly [350].

When superficially comparing coherent and incoherent optical imaging systems, one might consider their cutoff spatial frequencies. However, such a comparison is not entirely valid. In a coherent system, the cutoff frequency of the amplitude transfer function, $H(f_x, f_y)$, defines the maximum spatial frequency for the image amplitude. In contrast, for an incoherent system, the cutoff frequency of the optical transfer function (OTF), $\mathcal{H}(f_x, f_y)$, determines the maximum spatial frequency for the image intensity. Consequently, any comparison between the two systems must be conducted in terms of a common metric, which is the image intensity [265].

Even when comparing intensities, ambiguity arises because image quality is often subjective and influenced by the human observer's perception. Nonetheless, it is both interesting and instructive to investigate intensity comparisons between coherent and incoherent optical systems. In the case of a coherent system, the image intensity, I_i^{coh}, is determined using Eq. (6.35) as follows:

$$I_i^{coh}(x_i, y_i) = |h(x_i, y_i) ** u_g(\tilde{x}_o, \tilde{y}_o)|^2 \Longrightarrow \mathcal{F}\left\{I_i^{coh}\right\} = (HU_g) \star\star (HU_g). \tag{6.56}$$

It is important to recall that $H = \mathcal{F}\{h\}$ and $U_g = \mathcal{F}\{u_g\}$. In an incoherent optical system, the image intensity, I_i^{inc}, is given by Eq. (6.38) as follows:

$$I_i^{inc}(x_i, y_i) = K\,|h(x_i, y_i)|^2 ** \,|u_g(\tilde{x}_o, \tilde{y}_o)|^2 \Longrightarrow \mathcal{F}\left\{I_i^{inc}\right\} \sim (H \star\star H)\,(U_g \star\star U_g)\,. \tag{6.57}$$

In this context, the factor K was omitted, as it does not influence the cutoff frequencies. While the last two equations do not provide a definitive conclusion regarding which system is superior, they underscore significant differences in the frequency response between coherent and incoherent systems. To illustrate these distinctions, consider the following instructive example adapted from Ref. [265].

Example 6.12.1:

In this example, consider a square exit pupil function of side length $2L$, represented as $P(x, y) = \text{rect}(x/2L, y/2L)$. Two distinct objects, A and B, are analyzed with transmittances given by:

$$\text{Object } A: \quad t_A(x_o, y_o) = \cos(2\pi f_s x_o),$$

$$\text{Object } B: \quad t_B(x_o, y_o) = |\cos(2\pi f_s x_o)|\,.$$

For simplicity, unity magnification is assumed. Additionally, let $f_s < f_0 = 2L/\lambda d_i$ (with $2f_s > f_0$), where λ represents the central freespace wavelength (for the incoherent case), and d_i denotes the distance between the exit pupil and the image plane. Compare the frequency response of a coherent and an incoherent system for each of the two objects.

Solution: Let us first study the two systems for an object with transmittance $t_A(x_o, y_o) = \cos(2\pi f_s x_o)$. The amplitude transfer function H and the Fourier transform of the object field U_g are given by:

$$H(f_x, f_y) = \text{rect}\left(\lambda d_i \frac{f_x}{2L}, \lambda d_i \frac{f_y}{2L}\right),$$

$$U_g(f_x, f_y) = \mathcal{F}\{t_A\} = \frac{1}{2}\left[\delta(f_x + f_s) + \delta(f_x - f_s)\right]\delta(f_y).$$

Then, using Eq. (6.56), the Fourier transform of the image intensity, $\mathcal{F} I_i^{coh}$, and the corresponding image intensity, I_i^{coh}, are readily determined as:

$$\mathcal{F}\left\{I_i^{coh}\right\} = \frac{1}{2}\left[\frac{1}{2}\delta(f_x + 2f_s) + \delta(f_x) + \frac{1}{2}\delta(f_x - 2f_s)\right]\delta(f_y),$$

$$I_i^{coh}(x_i, y_i) = \frac{1}{2}\cos\left(2\pi 2 f_s x_i\right) + \frac{1}{2} = \cos^2\left(2\pi f_s x_i\right).$$

The plot of the image intensity of the coherent system is shown in Fig. 6.23 (solid line) and its visibility is unity. For the incoherent system case, the following can be easily calculated:

$$H \star\star H = 4f_0^2 \text{tri}\left(\lambda d_i \frac{f_x}{2L}\right)\text{tri}\left(\lambda d_i \frac{f_y}{2L}\right),$$

$$U_g \star\star U_g = \frac{1}{2}\left[\frac{1}{2}\delta(f_x + 2f_s) + \delta(f_x) + \frac{1}{2}\delta(f_x - 2f_s)\right]\delta(f_y).$$

Then, using Eq. (6.57), the Fourier transform of the image intensity, $\mathcal{F} I_i^{inc}$, and the corresponding image intensity, I_i^{inc}, are readily determined as:

$$\mathcal{F}\left\{I_i^{inc}\right\} \sim 4f_0^2 \frac{1}{2}\left[\frac{1}{2}\left(1 - \frac{f_s}{f_0}\right)\delta(f_x + 2f_s) + \delta(f_x) + \frac{1}{2}\left(1 - \frac{f_s}{f_0}\right)\delta(f_x - 2f_s)\right]\delta(f_y),$$

$$I_i^{inc}(x_i, y_i) \sim 4f_0^2\left[\left(1 - \frac{f_s}{f_0}\right)\cos^2\left(2\pi f_s x_i\right) + \frac{1}{2}\frac{f_s}{f_0}\right].$$

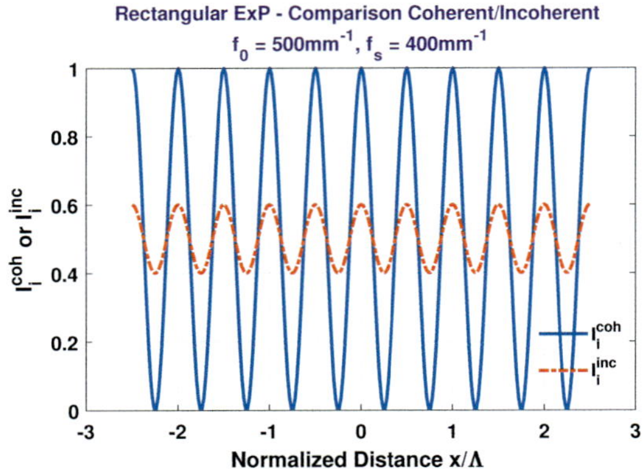

FIGURE 6.23: The image intensity I_i^{coh} and I_i^{inc} for a coherent and incoherent optical system with a square ExP and object transmittance $t_A(x_o, y_o) = \cos(2\pi f_s x_o)$. The spatial frequencies selected are $f_0 = 2L/\lambda d_i = 500\,\text{mm}^{-1}$ and $f_s = 400\,\text{mm}^{-1}$. The visibility of the coherent system is unity while the visibility of the incoherent is 0.2.

The plot of the image intensity for the incoherent system is also presented in Fig. 6.23 (dash-dot line). The visibility of the image is given by $V = 1 - f_s/f_0 = 0.2$, as clearly demonstrated in the figure. It is evident that, for this object, the performance of the coherent system is superior.

Now, the object B with transmittance $t_B(x_o, y_o) = |\cos(2\pi f_s x_o)|$ is analyzed. It is convenient to represent t_B using the following Fourier series:

$$|\cos(2\pi f_s x_o)| = \frac{2}{\pi} \sum_{m=-\infty}^{\infty} \frac{(-1)^m}{1-(2m)^2} \exp\left[j2\pi(2mf_s)x_o\right].$$

The amplitude transfer function H and the Fourier transform of the object field U_g are expressed as:

$$H(f_x, f_y) = \text{rect}\left(\lambda d_i \frac{f_x}{2L}, \lambda d_i \frac{f_y}{2L}\right),$$

$$U_g(f_x, f_y) = \mathcal{F}\{t_B\} = \left\{\frac{2}{\pi} \sum_{m=-\infty}^{\infty} \frac{(-1)^m}{1-(2m)^2} \delta\left(f_x - 2mf_s\right)\right\} \delta(f_y).$$

Then using Eq. (6.56) the Fourier transform of the image intensity, $\mathcal{F}\{I_i^{coh}\}$ and the corresponding image intensity I_i^{coh} are easily found to be:

$$HU_g = \frac{2}{\pi}\delta(f_x)\delta(f_y),$$

$$\mathcal{F}\left\{I_i^{coh}\right\} = \left(\frac{2}{\pi}\right)^2 \delta(f_x)\delta(f_y),$$

$$I_i^{coh}(x_i, y_i) = \left(\frac{2}{\pi}\right)^2.$$

Since $2f_s > f_0$, the image in the coherent system does not exhibit any variations!

For the incoherent system case, the following can be easily calculated:

$$H \star\star H = 4f_0^2 \text{tri}\left(\lambda d_i \frac{f_x}{2L}\right) \text{tri}\left(\lambda d_i \frac{f_y}{2L}\right),$$

$$U_g \star\star U_g = \left\{\left(\frac{2}{\pi}\right)^2 \sum_{m=-\infty}^{\infty} \sum_{n=-\infty}^{\infty} \frac{(-1)^m}{1-(2m)^2} \frac{(-1)^n}{1-(2n)^2} \delta\left[f_x + 2(m-n)f_s\right]\right\} \delta(f_y).$$

Then, using Eq. (6.57), the Fourier transform of the image intensity, $\mathcal{F}I_i^{inc}$, and the corresponding image intensity, I_i^{inc}, are easily found to be (it is known that $f_0 < 2f_s < 2f_0$):

$$\mathcal{F}\left\{I_i^{inc}\right\} \sim 4f_0^2 \left(\frac{2}{\pi}\right)^2 \left[\sum_{m=-\infty}^{\infty} \frac{1}{(1-4m^2)^2} \delta(f_x) + \right.$$

$$\sum_{n=-\infty}^{\infty} \frac{-1}{(1-4n^2)(1-4(n+1)^2)} \delta(f_x - 2f_s) +$$

$$\left. \sum_{n=-\infty}^{\infty} \frac{-1}{(1-4n^2)(1-4(n+1)^2)} \delta(f_x + 2f_s)\right] \delta(f_y)$$

$$\sim 4f_0^2 \left(\frac{2}{\pi}\right)^2 \left[\frac{\pi^2}{8} \delta(f_x) + \frac{\pi^2}{16} \delta(f_x - 2f_s) + \frac{\pi^2}{16} \delta(f_x + 2f_s)\right] \delta(f_y),$$

$$I_i^{inc}(x_i, y_i) \sim 4f_0^2 \left[\left(1 - \frac{f_s}{f_0}\right) \cos^2(2\pi f_s x_i) + \frac{1}{2} \frac{f_s}{f_0}\right].$$

In the above derivation the summations $\sum_{m=-\infty}^{\infty} 1/(1-4m^2)^2 = \pi^2/8$ and $\sum_{m=-\infty}^{\infty} (-1)/[(1-4m^2)(1-4(m+1)^2)] = \pi^2/16$ were used. It is evident that the response of the incoherent system is the same as in the case of object A, and consequently, in this case, the performance of the incoherent system surpasses that of the coherent system.

6.13 Selective Applications of Fourier Optics

In this section, a concise overview of selected applications in Fourier optics is provided. The applications chosen are: (a) the Zernike phase contrast microscopy, (b) the basics of holography, (c) spatial filtering, and (d) the VanderLugt correlator.

6.13.1 Zernike Phase Contrast Microscopy

In optical microscopy, certain objects of interest are nearly transparent, and when light passes through them, primarily the phase of the light is affected. Such objects are not easily observable with conventional microscopes. Frits Zernike [351] introduced a phase-contrast technique to address this challenge. Zernike phase-contrast microscopy enhances the imaging of transparent objects by converting phase variations within the specimen into intensity differences in the resulting image.

Zernike phase-contrast microscopy involves inserting a specialized phase plate into the optical pathway. Typically, this phase plate is a ring-shaped optical element that introduces a phase shift into the transmitted light. The phase plate imparts a phase shift to the light passing through the outer areas of the specimen (the annular region), while leaving the light passing through the central portion of the specimen unchanged. As the phase-shifted light and the unaltered light (from the central region) combine, they interfere with each other. The phase shift introduced by the phase plate results in either constructive or destructive interference, leading to variations in brightness within the final image. The interference pattern generated by the phase plate improves the contrast within the image. Previously imperceptible structures within the specimen, which would have been challenging to observe in a conventional brightfield image, become more visible due to the differences in brightness.

Zernike phase-contrast microscopy is especially valuable for imaging biological specimens, such as living cells, which are frequently transparent and would be challenging to study using conventional brightfield microscopy. It allows for the visualization of internal structures and details without the necessity of staining or other sample preparation methods that could modify or harm the specimen. In the following elementary analysis, for simplic-

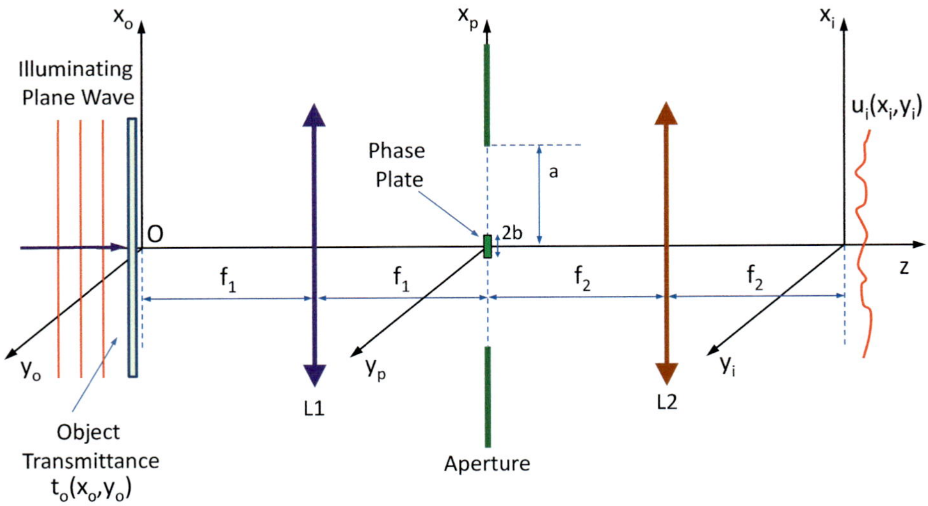

FIGURE 6.24: A 4f simple setup to explain the basics of Zernike phase contrast microscopy method. The phase plate has a diameter of $2b$, while the circular aperture at the right focal plane of L1 has a diameter of $2a$. It is assumed the $a \gg b$.

ity, the phase plate is modeled as a small disk with a deposited dielectric, rather than as a ring [265]. The fundamental concept of Zernike's phase contrast is illustrated in a straightforward setup, shown in the 4f system of Fig. 6.24. In this system, the small phase plate is positioned at the center of the focal plane of the left lens (objective).

Consider a transparent object with a transmittance function $t_o(x_o, y_o) = e^{-j\phi(x_o,y_o)}$. When the object is illuminated by a coherent plane wave, the object field can be expressed as $u_o(x_o, y_o) = Ae^{-j\phi(x_o,y_o)}$, where A the amplitude of the plane wave. If, however, incoherent light is used, the intensity becomes $I_g = |u_g|^2 = |Ae^{-j\phi(\tilde{x}_o/m,\tilde{y}_o/m)}|^2 = |A|^2$, and no image of the object can be formed in the image plane.

Thus, coherent illumination is essential (or at least partially coherent), as the phase shift introduced by the plate causes interference between the phase-shifted portion and the non-phase-shifted portion of the light. It is assumed at this stage that the object exhibits a weak phase, meaning the object field can be expressed as [265]:

$$u_o(x_o, y_o) = Ae^{-j\phi(x_o,y_o)}Ae^{-j\phi_0}e^{-j\Delta\phi(x_o,y_o)} \simeq Ae^{-j\phi_0}\left[1 - j\Delta\phi(x_o, y_o)\right], \tag{6.58}$$

where ϕ_0 represents the average phase shift through the transparent object and $\Delta\phi(x_o, y_o)$ denotes the small variations in the phase. According to Eq. (6.34), the image field can be expressed as $u_i(x_i, y_i) \sim u_g(x_i, y_i) * *h(x_i, y_i)$ (neglecting the phase factor), where $u_g = (1/M)u_o(-x_i/M, -y_i/M)$. The amplitude transfer function, $H(f_x, f_y)$, of the system depicted in Fig. 6.24 is given by:

$$H(f_x, f_y) = P(-\lambda f_2 f_x, -\lambda f_2 f_y) = P(\lambda f_2 f_x, \lambda f_2 f_y),$$
$$P(x_\ell, y_\ell) = \text{circ}\left(\frac{r_T}{a}\right) + (e^{-j\theta} - 1)\text{circ}\left(\frac{r_T}{b}\right),$$

where θ is the phase shift introduced by the phase plate of diameter $2b$. Therefore, the resulting amplitude transfer function and coherent PSF (cPSF) are given by:

$$H(f_x, f_y) = \text{circ}\left(\frac{\lambda f_2}{a}\sqrt{f_x^2 + f_y^2}\right) + (e^{-j\theta} - 1)\text{circ}\left(\frac{\lambda f_2}{b}\sqrt{f_x^2 + f_y^2}\right)$$
$$h(x_i, y_i) = \pi\left(\frac{a}{\lambda f_2}\right)^2 \text{jinc}\left(\frac{2a}{\lambda f_2}r_T\right) + (e^{-j\theta} - 1)\pi\left(\frac{b}{\lambda f_2}\right)^2 \text{jinc}\left(\frac{2b}{\lambda f_2}r_T\right),$$

where,

$$\text{jinc}(\rho) = \frac{2J_1(\pi\rho)}{\pi\rho}, \qquad \text{and} \quad r_T = \sqrt{x_i^2 + y_i^2}.$$

At this stage, and to simplify the calculations, it is assumed that $a \gg b$, and the first "jinc" function can be approximated by the delta function, thereby simplifying the cPSF as follows [352]:

$$h(x_i, y_i) = \delta(x_i, y_i) + (e^{-j\theta} - 1)\pi\left(\frac{b}{\lambda f_2}\right)^2 \text{jinc}\left(\frac{2b}{\lambda f_2}r_T\right).$$

It is important to note that the dimensions of the coherent PSF remain consistent, scaling as m^{-2}, as anticipated. For simplicity, let us assume a magnification factor of $M = 1$ and that $u_g(x_i, y_i) \simeq u_o(x_i, y_i)$. This assumption is justified, as any inversion of the sample is unlikely to influence the contrast enhancement provided by the phase plate. Under these conditions, the image field can be expressed by the following equation:

$$\begin{aligned}
u_i(x_i, y_i) &= u_o(x_i, y_i) ** h(x_i, y_i) \\
&= Ae^{\phi(x_i, y_i)} ** \left[\delta(x_i, y_i) + (e^{-j\theta} - 1)\pi\left(\frac{b}{\lambda f_2}\right)^2 \text{jinc}\left(\frac{2b}{\lambda f_2}r_T\right)\right]. \\
&= Ae^{\phi(x_i, y_i)} + (e^{-j\theta} - 1)\pi\left(\frac{b}{\lambda f_2}\right)^2 A\left[e^{\phi(x_i, y_i)} ** \text{jinc}\left(\frac{2b}{\lambda f_2}r_T\right)\right] \\
&\simeq Ae^{\phi(x_i, y_i)} + (e^{-j\theta} - 1)\pi\left(\frac{b}{\lambda f_2}\right)^2 Ae^{-j\phi_0}\left[[1 - j\Delta\phi(x_i, y_i)] ** \text{jinc}\left(\frac{2b}{\lambda f_2}r_T\right)\right].
\end{aligned}$$

At this point, the intensity at the image plane is determined. This intensity is expressed as $I_i = |u_i|^2 = |u_o(x_i, y_i) ** h(x_i, y_i)|^2$. In the subsequent calculations, terms of second order with respect to $\Delta\phi$ are neglected. Following some straightforward steps, the image intensity can be formulated as:

$$I_i(x_i, y_i) = |A|^2 + |A|^2 \left[\pi\left(\frac{b}{\lambda f_2}\right)^2\right]^2 [2(1-\cos\theta)] \left[1 ** \text{jinc}\left(\frac{2b}{\lambda f_2} r_T\right)\right]^2 +$$

$$(e^{j\theta} - 1)|A|^2 \left[\pi\left(\frac{b}{\lambda f_2}\right)^2\right] (1 - j\Delta\phi) \left[1 ** \text{jinc}\left(\frac{2b}{\lambda f_2} r_T\right) + j\Delta\phi ** \text{jinc}\left(\frac{2b}{\lambda f_2} r_T\right)\right] +$$

$$(e^{-j\theta} - 1)|A|^2 \left[\pi\left(\frac{b}{\lambda f_2}\right)^2\right] (1 + j\Delta\phi) \left[1 ** \text{jinc}\left(\frac{2b}{\lambda f_2} r_T\right) - j\Delta\phi ** \text{jinc}\left(\frac{2b}{\lambda f_2} r_T\right)\right].$$

By further manipulating the last expression and using the relation $1 * *\text{jinc}(2br_T/(\lambda f_2)) = (1/\pi)(\lambda f_2/b)^2$ [where the integral $\int_0^\infty J_1(x)dx = 1$ was utilized] the resulting image intensity I_i can be expressed as:

$$I_i(x_i, y_i) = |A|^2 [1 + 2\sin\theta\Delta\phi(x_i, y_i)] +$$

$$|A|^2 \left[\pi\left(\frac{b}{\lambda f_2}\right)^2\right] [2\sin\theta - 2\cos\theta + 2] \left[\Delta\phi ** \text{jinc}\left(\frac{2b}{\lambda f_2} r_T\right)\right]. \quad (6.59)$$

Therefore, the intensity at the image plane includes a linear term with respect to $\Delta\phi(x_i, y_i)$, which reveals the phase information of the transparent object. The final term in Eq. (6.59) corresponds to the halo that may arise due to the leakage of diffracted light from the edges of the phase plate [352]. In most textbooks, this term is omitted, and the resulting image intensity is expressed as:

$$I_i(x_i, y_i) \simeq |A|^2 [1 + 2\sin\theta\Delta\phi(x_i, y_i)]. \quad (6.60)$$

If $\theta = \pm\pi/2$ the image intensity becomes $I_i = |A|^2[1 \pm 2\Delta\phi(x_i, y_i)]$, where the positive sign corresponds to *positive phase contrast*, and the negative sign corresponds to *negative phase contrast* [91, 265]. It is also possible to introduce partial absorption into the phase plate, which can improve the contrast [265]. In this case, the term $e^{j\theta}$ should be replaced by $\alpha e^{j\theta}$, where $\alpha \leq 1$ accounts for the loss due to the phase plate. If $\alpha = 0$, the resulting configuration corresponds to the *dark ground method* [91].

6.13.2 Holography Basics

The concept of holography was first introduced by Dennis Gabor in 1948 [353]. This process is also referred to as wavefront reconstruction [265,354]. Gabor demonstrated that, by using a suitable reference wave to interact with the light scattered from an object, it is possible to record both the amplitude and phase information of the object, despite the limitation of recording media that respond only to intensity variations. Consequently, holography is a technique that enables the recording and reconstruction of three-dimensional images.

Holography finds a wide range of applications across various fields, owing to its unique ability to capture and reproduce three-dimensional images with remarkable realism and depth. Notable applications of holography include its use in arts and entertainment, security and authentication, medical imaging, non-destructive testing through holographic interferometry, microscopy, virtual reality (VR) and augmented reality (AR) technologies, head-up displays, holographic data storage, and more, as the field continues to evolve. This section provides a basic introduction to the fundamental principles of holography. For a more in-depth exploration, specialized textbooks such as Refs. [355–357] can be consulted.

The basic recording and reconstruction geometry of an on-axis hologram, as originally proposed by Dennis Gabor, is illustrated in Figs. 6.25(a) and 6.25(b), respectively. At this stage, it is assumed that the object is nearly transparent and thin, and the wave scattered by it due to an incident reference wave is depicted on the left side of Fig. 6.25(a).

The transmitted field of the object can be expressed as $u_o(x, y, z) = t_o(x, y)u_r(x, y, z)$, where $t_o(x, y)$ represents the transmittance of the thin object, and $u_r(x, y, z)$ denotes the reference wave. The reference wave is typically either a plane wave or a spherical wave. The recording plate is positioned at the xy-plane, centered at the origin of the xyz coordinate system as shown. This plate captures the interference pattern created by the object and reference waves within its volume. Certainly, the response of the recording plate (composed

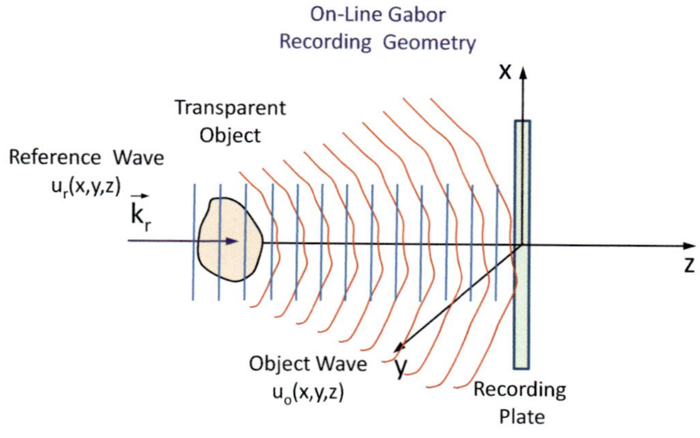

(a)

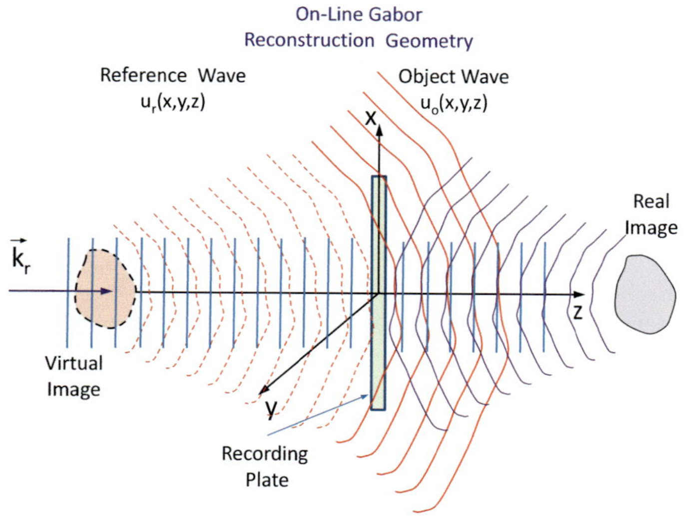

(b)

FIGURE 6.25: (a) The basic recording process of an on-axis hologram (Gabor's initial approach). The object wave is $u_o(x, y, z)$ while the reference wave is $u_r(x, y, z)$. The recording plate is positioned at the origin of the xyz coordinate system. (b) The reconstruction process (playback) of a recorded on-axis hologram. The virtual image of the object is formed along with an in-line real image.

of materials such as photopolymers, photorefractive media, dichromated gelatin, or similar) [265] is generally nonlinear. Nevertheless, it is commonly assumed that the developed recording plate exhibits an amplitude transmittance of the following form:

$$t(x, y) = t_b + \beta I(x, y) = t_b + \beta |u(x, y)|^2. \tag{6.61}$$

Here, t_b represents a uniform bias of the plate, and β is a material-dependent constant that can be positive (indicating positive transparency) or negative (indicating negative transparency). The term $I(x, y) = |u(x, y)|^2$ denotes the intensity of the interference pattern formed by the object and reference wave within the recording plate, which is assumed to be thin and located at $z = 0$.

The object and reference wave interact, and their interference pattern is captured within the thin recording plate situated at the $z = 0$ plane. The resulting fields and intensity at this plane can be expressed as follows:

$$\begin{aligned}
u_r(x, y; z = 0) &= |u_r(x, y; z = 0)| e^{-j\phi_r(x,y;z=0)}, \\
u_o(x, y; z = 0) &= |u_o(x, y; z = 0)| e^{-j\phi_o(x,y;z=0)}, \\
I(x, y; z = 0) &= |u_r(x, y; z = 0) + u_o(x, y; z = 0)|^2 \\
&= |u_r(x, y; z = 0)|^2 + |u_o(x, y; z = 0)|^2 + \\
&\quad u_r(x, y; z = 0) u_o^*(x, y; z = 0) + u_r^*(x, y; z = 0) u_o(x, y; z = 0), \\
I(x, y; z = 0) &= |u_r|^2 + |u_o|^2 + u_r u_o^* + u_r^* u_o.
\end{aligned} \tag{6.62}$$

In the last equation for the recorded intensity, the spatial dependence is implied for the sake of simplifying the notation. Assuming that the recording plate has been developed such that its transmittance function can be expressed via Eq. (6.61), illuminating the plate with the same reference wave during the reconstruction process, as shown in Fig. 6.25(b), generates the following diffracted field:

$$\begin{aligned}
u_d(x, y; z = 0) &= t(x, y) u_r(x, y; z = 0) \\
&= [t_b + \beta I(x, y; z = 0)]\, u_r(x, y; z = 0) \\
&= \left(t_b + \beta |u_r|^2 + \beta |u_o|^2\right) u_r + \beta |u_r|^2 u_o + \beta u_o^* u_r^2.
\end{aligned} \tag{6.63}$$

Clearly, the first term in the expression for $u_d(x, y; z = 0)$ corresponds to the transmitted version of the reference wave passing through the recording plate. The second term, which is proportional to the object wave u_o, is associated with the virtual image of the object, as illustrated in Fig. 6.25(b). Finally, the last term, proportional to the conjugate of the object wave u_o^*, forms a real image of the object alongside the virtual image.

In the case where the reference wave is a plane wave (see Fig. 6.25(a)), we have $u_r(x, y, z) = E_r e^{-j2\pi f_z z} = E_r e^{-j2\pi z/\lambda}$. The diffracted field produced by the recording plate is then given by:

$$u_d(x, y; z = 0) = \left(t_b + \beta |E_r|^2 + \beta |u_o|^2\right) E_r + \beta |E_r|^2 u_o(x, y, z = 0) + \beta E_r^2 u_o^*(x, y; z = 0).$$

The on-axis (Gabor) hologram generates both a virtual and a real image of the transparent object, which are aligned along the z-axis as shown in Fig. 6.25(b). These images are referred to as *twin images* [265]. The term $\beta E_r |u_o|^2$ in Eq. (6.63) affects both the virtual and real images of the object. This term can be safely neglected when $|u_o| \ll E_r$, which can be ensured by using a reference wave with higher intensity than the object wave. Typically, a reference/object wave intensity ratio between 3 and 10 yields satisfactory results.

However, the on-axis hologram has several limitations, including the requirement for a thin, transparent object. Moreover, a significant drawback is the generation of twin in-line images, making their separation impossible. For example, an observer viewing the virtual image simultaneously perceives the defocused real image as well [265]. Many of these issues associated with on-axis holograms are addressed by off-axis holograms, as initially proposed by Leith and Upatnieks [358, 359]. The next section provides a brief overview of off-axis holography.

6.13.2.1 Off-Axis Holography

The key distinction of the off-axis hologram, compared to the on-axis hologram, is that the object field is independent of the reference wave. Additionally, instead of being collinear with the object-recording plate axis, an oblique reference wave is utilized in off-axis holography.

The fundamental recording and reconstruction geometry of an off-axis hologram is depicted in Figs. 6.26(a) and 6.26(b), respectively. In this context, the scattered wave from the object (shown on the left side of Fig. 6.26(a)) can be expressed as $u_o(x, y, z) = |u_o(x, y, z)|e^{-j\phi_o(x,y,z)}$. At the same time, the separate oblique reference wave is denoted as $u_r(x, y, z) = |u_r(x, y, z)|e^{-j\phi_r(x,y,z)}$. Once again, the reference wave is typically either a plane wave or a spherical wave.

Assuming that the recording plate is centered in the xy-plane at the origin of the xyz coordinate system, it captures the interference pattern generated by the object and reference waves within its volume. Similar to the on-axis hologram, the recording plate is considered thin in off-axis holography. The intensity on the recording plate is given by Eq. (6.62) and can be expressed as:

$$\begin{aligned} I(x, y; z = 0) &= |u_r|^2 + |u_o|^2 + u_r u_o^* + u_r^* u_o, \\ &= |u_r|^2 + |u_o|^2 + 2|u_r||u_o| \cos\left[\phi_r(x, y; z = 0) - \phi_o(x, y; z = 0)\right]. \end{aligned} \tag{6.64}$$

The intensity expression reveals that both the amplitude and the phase of the object are recorded. The term ϕ_r indicates the direction of the reference wave. Under certain conditions, particularly when a plane wave is used as the reference, the amplitude and phase distributions of the object can be unambiguously separated.

Assuming that the recording plate has been developed, its transmittance function can again be expressed using Eq. (6.61). When illuminated with the same reference wave during the reconstruction process, the following field is generated [see Eq. (6.63)]:

$$u_d(x, y; z = 0) = \left(t_b + \beta|u_r|^2 + \beta|u_o|^2\right) u_r + \beta|u_r|^2 u_o + \beta u_o^* u_r^2.$$

As in the case of the on-axis hologram, the first term in the expression for $u_d(x, y; z = 0)$ represents a transmitted version of the oblique reference wave through the recording plate, with its amplitude modulated by the object wave. The second term is proportional to the object wave u_o and corresponds to the virtual image of the object, as illustrated in Fig. 6.26(b). Finally, the last term is proportional to the conjugate of the object wave u_o^*, directed according to u_r^2, and forms a real image of the object in a direction distinct from that of the virtual image.

In the case where the reference wave is a plane wave with a wavevector in the xz-plane at an angle ϕ (see Fig. 6.26(a)), $u_r(x, y, z) = E_r e^{-j2\pi(f_x x + f_z z)} = E_r e^{-j2(\pi/\lambda)(-x \sin\phi + z\cos\phi)}$,

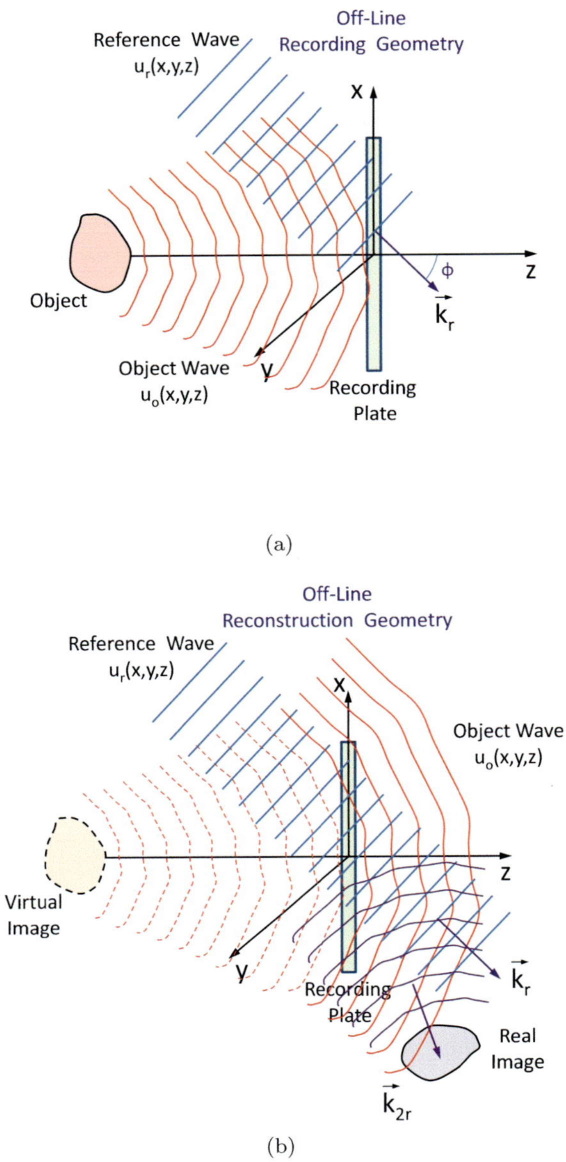

FIGURE 6.26: (a) The fundamental recording process of an off-axis hologram involves the object wave $u_o(x, y, z)$ and the reference oblique wave $u_r(x, y, z)$. The recording plate is situated at the origin of the xyz coordinate system. (b) During the reconstruction process (playback) of a recorded off-axis hologram, a virtual image of the object is formed, along with a real image, but in different directions. It's important to note that the position of the real image is only qualitatively correct.

the field diffracted by the recording plate is given by:

$$u_d(x, y; z = 0) = \underbrace{\left(t_b + \beta|E_r|^2\right) E_r e^{-j2(\pi/\lambda)(-x\sin\phi)}}_{= u_1(x,y)} +$$

$$\underbrace{\beta|u_o|^2 E_r e^{-j2(\pi/\lambda)(-x\sin\phi)}}_{= u_2(x,y)} +$$

$$\underbrace{\beta|E_r|^2 u_o(x,y,z=0)}_{=u_3(x,y)} +$$

$$\underbrace{\beta E_r^2 u_o^*(x,y;z=0) e^{-j2(\pi/\lambda)(-2x\sin\phi)}}_{=u_4(x,y)} \Longrightarrow$$

$$u_d(x,y;z=0) = u_1(x,y) + u_2(x,y) + u_3(x,y) + u_4(x,y).$$

The fields u_1, u_2, u_3, and u_4 diffracted by the recorded plate at the $z = 0$ plane will propagate in the homogeneous region to the right of the recording plate, following the plane wave spectrum approach. The u_1 and u_2 fields propagate in the direction of the reference plane wave, while the u_3 field is proportional to the object wave. Finally, the u_4 field propagates in the direction of the double spatial frequency (f_x) of the reference wave. To spatially separate these fields, it is convenient to use their Fourier transforms. The Fourier transforms of the u_i fields ($i = 1, 2, 3, 4$) are given by:

$$U_1(f_x,f_y) = \mathcal{F}\{u_1(x,y)\} = \left(t_b + \beta|E_r|^2\right) E_r\ \delta(f_x + f_0, f_y),$$

$$U_2(f_x,f_y) = \mathcal{F}\{u_2(x,y)\} = \beta E_r\left[U_o(f_x,f_y) \star\star\ U_o(f_x,f_y)\right]\ \delta(f_x + f_0, f_y),$$

$$U_3(f_x,f_y) = \mathcal{F}\{u_3(x,y)\} = \beta|E_r|^2 U_o(f_x,f_y),$$

$$U_4(f_x,f_y) = \mathcal{F}\{u_4(x,y)\} = \beta E_r^2\left[U_o^*(-f_x,-f_y)\right] ** \ \delta(f_x + 2f_0, f_y).$$

$$\text{where,}\quad f_0 = \frac{\sin\phi}{\lambda}, \qquad \text{and}\quad U_o(f_x,f_y) = \mathcal{F}\{u_o(x,y)\}.$$

At this point, let's assume that the spatial-frequency extent of the object wave is limited,

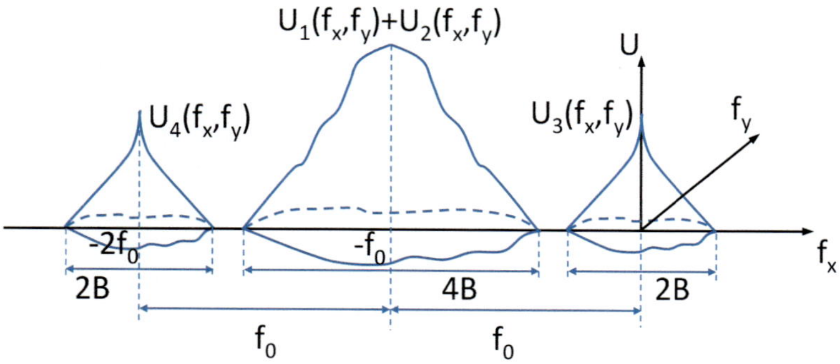

FIGURE 6.27: The output of the off-axis hologram in the reconstruction process shown in the spatial frequency domain.

with a maximum width in the f_x spatial frequency denoted as $2B$. This constraint is qualitatively depicted in Fig. 6.27 with the $U_3(f_x,f_y)$ surface. The fields $U_1(f_x,f_y) + U_2(f_x,f_y)$ are centered at $f_x = -f_0$ and have a spatial extent in the f_x direction of $4B$ due to the correlation operation included in U_2. Additionally, $U_4(f_x,f_y)$ shares the same characteristics as $U_3(f_x,f_y)$ but is centered at $f_x = -2f_0$.

A qualitative representation of all the hologram output fields during the reconstruction process is shown in the spatial frequency domain in Fig. 6.27. From this figure, it is clear that in order to successfully separate $U_3(f_x,f_y)$ from $U_1(f_x,f_y) + U_2(f_x,f_y)$, the following condition must be met:

$$-f_0 + 2B \leq -B \Longrightarrow f_0 \geq 3B \Longrightarrow \phi \geq \sin^{-1}(3B\lambda). \tag{6.65}$$

Under some special conditions the angle ϕ condition can be relaxed [265].

6.13.2.2 Image Location and Magnification

In this section, the primary objective is to determine the position and magnification of the point image resulting from an object point recorded via a spherical diverging wave on a holographic recording plate. It is important to note that since any object can be decomposed into point sources, the findings for a point object can be extended to more complex, realistic objects.

The recording geometry is depicted in Fig. 6.28(a) [265, 356, 360]. The object point is described by a diverging spherical wave centered at (x_o, y_o, z_o), with $z_o < 0$. Similarly, the reference wave is also a diverging spherical wave centered at (x_r, y_r, z_r), with $z_r < 0$. It is essential to reiterate that, according to the convention in this context, a diverging (scalar) spherical wave u_{sph} is represented by: $u_{sph} = E_0 e^{-jkR}/R$ with $R = [(x - x_0)^2 + (y - y_0)^2 + (z - z_0)^2]^{1/2}$. Utilizing the Fresnel approximation (paraxial/parabolic), the diverging spherical wave is expressed as (with $z = 0$ at the recording plate):

$$u_{sph} \simeq \frac{E_0}{|z_0|} \exp[-jk|z_0|] \exp\left\{-j\frac{k}{2|z_0|}\left[(x - x_0)^2 + (y - y_0)^2\right]\right\},$$

implying that in case that of negative z_0 the absolute value can be replaced by $|z_0| = -z_0$. Summarizing the Fresnel approximation of a spherical plane wave at the $z = 0$ plane is given by:

$$u_{sph} \simeq \frac{E_0}{|z_0|} \begin{cases} \exp[-jkz_0] \exp\left\{-j\dfrac{k}{2z_0}\left[(x - x_0)^2 + (y - y_0)^2\right]\right\}, & z_0 > 0, \\ \exp[+jkz_0] \exp\left\{+j\dfrac{k}{2z_0}\left[(x - x_0)^2 + (y - y_0)^2\right]\right\}, & z_0 < 0. \end{cases}$$

Analogous expressions apply to a spherical wave converging to a point. In general, when the phase factor $k/2z_0$ is positive, the term $\exp[-j(k/2z_0)[(x - x_0)^2 + (y - y_0)^2]]$ represents a diverging spherical wave. Conversely, when this factor is negative, the term $\exp[-j(k/2z_0)[(x - x_0)^2 + (y - y_0)^2]]$ represents a converging spherical wave. Using this convention, the reference and object waves, as well as the intensity at the recording plate, can be expressed as:

$$u_r(x, y) = \frac{E_r}{|z_r|} \exp[+jk_1 z_r] \exp\left\{+j\frac{k_1}{2z_r}\left[(x - x_r)^2 + (y - y_r)^2\right]\right\},$$

$$u_o(x, y) = \frac{E_o}{|z_o|} \exp[+jk_1 z_o] \exp\left\{+j\frac{k_1}{2z_o}\left[(x - x_o)^2 + (y - y_o)^2\right]\right\},$$

$$I(x, y) = |u_r|^2 + |u_o|^2 + u_o u_r^* + u_o^* u_r.$$

where $z_r < 0$ and $z_o < 0$ as it is implied by Fig. 6.28(a) and $k_1 = 2\pi/\lambda_1$ where λ_1 the wavelength of the light source used. The transmittance of the recording plate is given by Eq. (6.61). Now, assume that the reconstruction process is carried out using another diverging spherical wave originating from the point (x_p, y_p, z_p) with wavelength λ_2. Again, applying the Fresnel approximation for the spherical wave, it can be represented as:

$$u_p(x, y) = \frac{B}{|z_p|} \exp[+jk_2 z_p] \exp\left\{+j\frac{k_2}{2z_p}\left[(x - x_p)^2 + (y - y_p)^2\right]\right\},$$

assuming again that $z_p < 0$ and $k_2 = 2\pi/\lambda_2$ (see Fig. 6.28(b)). The hologram output can be obtained by multiplying $t(x, y)$ and $u_p(x, y)$. The significant components resulting from

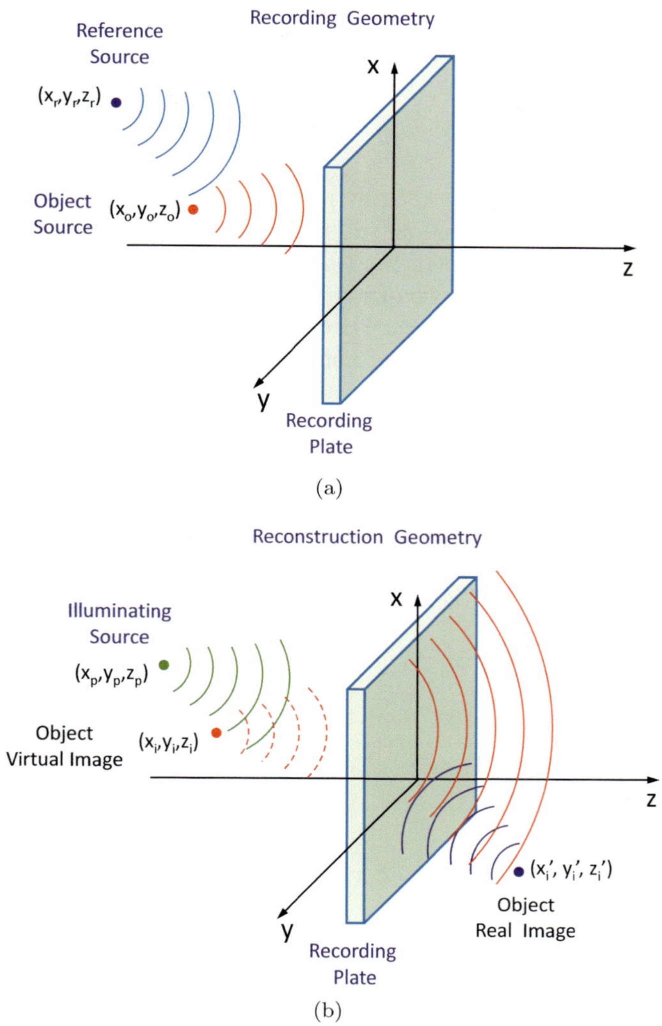

FIGURE 6.28: (a) Illustration of a generalized recording process for an off-axis hologram employing a spherical diverging reference wave centered at (x_r, y_r, z_r). The object wave is assumed to represent a point on the object, acting as a diverging spherical wave with center (x_o, y_o, z_o). The recording plate is situated at the origin of the xyz coordinate system. (b) Depiction of the reconstruction process for the generalized hologram. The illuminating spherical wave may differ in wavelength and position from the reference wave used in the recording process. For instance, a diverging spherical wave with center (x_p, y_p, z_p) is shown. In the reconstruction, a virtual image of the object is formed, alongside a real image. The positions of these images are represented qualitatively by (x_i, y_i, z_i) for the virtual image and (x_i', y_i', z_i') for the real image. It's important to note that these positions are only presented in a qualitative manner.

this multiplication are as follows:

$$(tu_p)\Big|_{u_r^* u_o} \sim \exp\left\{-j\frac{k_1}{2z_r}\left[(x-x_r)^2+(y-y_r)^2\right]\right\}\exp\left\{+j\frac{k_1}{2z_o}\left[(x-x_o)^2+(y-y_o)^2\right]\right\}$$
$$\exp\left\{+j\frac{k_2}{2z_p}\left[(x-x_p)^2+(y-y_p)^2\right]\right\}, \tag{6.66}$$

$$(tu_p)\Big|_{u_r u_o^*} \sim \exp\left\{+j\frac{k_1}{2z_r}\left[(x-x_r)^2+(y-y_r)^2\right]\right\}\exp\left\{-j\frac{k_1}{2z_o}\left[(x-x_o)^2+(y-y_o)^2\right]\right\}$$
$$\exp\left\{+j\frac{k_2}{2z_p}\left[(x-x_p)^2+(y-y_p)^2\right]\right\}. \qquad (6.67)$$

Only the varying phase factors have been included in the final expressions. The term $(tu_p)|_{u_r^* u_o}$ represents the component of tu_p that involves the term $u_r^* u_o$, and similarly, $(tu_p)|_{u_r u_o^*}$ represents the component of tu_p that involves the term $u_r u_o^*$.

Of course, the multiplication of two or more spherical factors generally results in a non-spherical wave [360], i.e., aberrations occur. However, by applying the Fresnel approximation to the spherical waves, these transmitted components contain only linear and quadratic terms, indicating that the transmitted factors resemble the Fresnel approximations of a spherical wave emanating from the hologram. The spherical wave produced by the hologram, corresponding to the term $(tu_p)\big|_{u_r^* u_o}$, can be written as:

$$u_i(x,y) = \frac{B}{|z_i|}\exp[+jk_2 z_i]\exp\left\{+j\frac{k_2}{2z_i}\left[(x-x_i)^2+(y-y_i)^2\right]\right\},$$

with $z_i < 0$. Comparing this term with Eq. (6.66) the following values for z_i, x_i, and y_i can easily found:

$$\frac{1}{z_i} = \frac{1}{z_p} + \frac{\lambda_2}{\lambda_1}\frac{1}{z_o} - \frac{\lambda_2}{\lambda_1}\frac{1}{z_r}, \qquad (6.68)$$

$$x_i = \frac{z_i}{z_p}x_p + \frac{\lambda_2}{\lambda_1}\frac{z_i}{z_o}x_o - \frac{\lambda_2}{\lambda_1}\frac{z_i}{z_r}x_r, \qquad (6.69)$$

$$y_i = \frac{z_i}{z_p}y_p + \frac{\lambda_2}{\lambda_1}\frac{z_i}{z_o}y_o - \frac{\lambda_2}{\lambda_1}\frac{z_i}{z_r}y_r. \qquad (6.70)$$

In the special case where $\lambda_1 = \lambda_2$ (i.e., the recording and playback wavelengths are the same) and $(x_p, y_p, z_p) = (x_r, y_r, z_r)$ (i.e., the recording and playback spherical waves are the same), it follows from Eqs. (6.68)-(6.70) that $z_i = z_o < 0$, $x_i = x_o$, and $y_i = y_o$. Therefore, the term $(tu_p)|_{u_r^* u_o}$ in this case produces a virtual image of the point object, as shown in Fig. 6.28(b). In general, when $z_i < 0$, a virtual image of the point object is produced, whereas if $z_i > 0$, a real image is produced by this hologram term.

Using the same approach the term $(tu_p)|_{u_r u_o^*}$ produces a spherical wave with z_i', x_i', and y_i' given by the following equations:

$$\frac{1}{z_i'} = \frac{1}{z_p} - \frac{\lambda_2}{\lambda_1}\frac{1}{z_o} + \frac{\lambda_2}{\lambda_1}\frac{1}{z_r}, \qquad (6.71)$$

$$x_i' = \frac{z_i'}{z_p}x_p - \frac{\lambda_2}{\lambda_1}\frac{z_i'}{z_o}x_o + \frac{\lambda_2}{\lambda_1}\frac{z_i'}{z_r}x_r, \qquad (6.72)$$

$$y_i' = \frac{z_i'}{z_p}y_p - \frac{\lambda_2}{\lambda_1}\frac{z_i'}{z_o}y_o + \frac{\lambda_2}{\lambda_1}\frac{z_i'}{z_r}y_r. \qquad (6.73)$$

In the special case that $\lambda_1 = \lambda_2$ and $(x_p, y_p, z_p) = (x_r, y_r, z_r)$ then from Eqs. (6.71)-(6.73) it is obvious that $z_i' = z_o z_r/(2z_o - z_r)$ which is positive if $2|z_o| < |z_r|$. Furthermore, $x_i' = (2x_r z_o - x_o z_r)/(2z_o - z_r)$ and $y_i' = (2y_r z_o - y_o z_r)/(2z_o - z_r)$. Therefore, the term $(tu_p)|_{u_r u_o^*}$ in this case produces a real image of the point object as is shown in Fig. 6.28(b). In general, when $z_i' > 0$ a real image of the point object is produced while if $z_i' < 0$ a virtual image is produced by this hologram term.

The axial and transverse magnifications of the hologram can be easily derived from the previous image location equations [265, 356, 360]. Specifically, the transverse magnification, M_t, and the axial magnification, M_a, corresponding to the $(tu_p)|_{u_r^* u_o}$ term are given by:

$$M_t = \left|\frac{\partial x_i}{\partial x_o}\right| = \left|\frac{\partial y_i}{\partial y_o}\right| = \left|\frac{\lambda_2}{\lambda_1}\frac{z_i}{z_o}\right| = \left|1 - \frac{z_o}{z_r} + \frac{\lambda_1}{\lambda_2}\frac{z_o}{z_p}\right|^{-1} \tag{6.74}$$

$$M_a = \left|\frac{\partial z_i}{\partial z_o}\right| = \frac{\lambda_1}{\lambda_2}\frac{1}{\left(1 - \dfrac{z_o}{z_r} + \dfrac{\lambda_1}{\lambda_2}\dfrac{z_o}{z_p}\right)^2} = \frac{\lambda_1}{\lambda_2}M_t^2. \tag{6.75}$$

The corresponding magnifications, M_t' and M_a' for the $(tu_p)|_{u_r u_o^*}$ term are similarly given by

$$M_t' = \left|\frac{\partial x_i'}{\partial x_o}\right| = \left|\frac{\partial y_i'}{\partial y_o}\right| = \left|\frac{\lambda_2}{\lambda_1}\frac{z_i'}{z_o}\right| = \left|-1 + \frac{z_o}{z_r} + \frac{\lambda_1}{\lambda_2}\frac{z_o}{z_p}\right|^{-1} \tag{6.76}$$

$$M_a' = \left|\frac{\partial z_i'}{\partial z_o}\right| = \frac{\lambda_1}{\lambda_2}\frac{1}{\left(-1 + \dfrac{z_o}{z_r} + \dfrac{\lambda_1}{\lambda_2}\dfrac{z_o}{z_p}\right)^2} = \frac{\lambda_1}{\lambda_2}M_t'^2. \tag{6.77}$$

It is evident that, in general, the axial and transverse magnifications are not equal. This is particularly important when imaging three-dimensional objects. The difference in magnifications can result in distorted images, both real and virtual. One potential solution to this issue is to scale the hologram itself, as discussed in Ref. [265].

Example 6.13.1:

In order to demonstrate the previous equations about the location of point images and their corresponding magnifications, the following numerical example is presented. Assume that during the recording process, the freespace wavelength is $\lambda_1 = 0.5\,\mu$m. The position of the object point is $(x_o, y_o, z_o) = (-1, 0, -10)$ (cm) while the origin position of the emanating reference spherical wave is $(x_r, y_r, z_r) = (1, 0, -25)$ (cm). Make a table of the position (x_i, y_i, z_i) of the virtual and real $(xI-', y_i', z_i')$ images along with their transverse and longitudinal magnifications for the following reconstruction cases (all coordinates in cm): (a) $\lambda_2 = 0.50\,\mu$m, $(x_p, y_p, z_p) = (1, 0, -25)$, (b) $\lambda_2 = 0.50\,\mu$m, $(x_p, y_p, z_p) = (1, 0, -30)$, (c) $\lambda_2 = 0.50\,\mu$m, $(x_p, y_p, z_p) = (1, 0, -20)$, (d) $\lambda_2 = 0.50\,\mu$m, $(x_p, y_p, z_p) = (1, 0, -10)$, and (e) $\lambda_2 = 0.65\,\mu$m, $(x_p, y_p, z_p) = (1, 0, -25)$.
Solution: Using Eqs. (6.68)-(6.70) for the coordinates of the virtual image, and Eqs. (6.74)-(6.75) for the corresponding transverse and axial magnifications, the entries in Table 6.2 are calculated. To determine the corresponding parameters of the real image, Eqs. (6.71)-(6.73) were used for the coordinates, and Eqs. (6.76)-(6.77) for the magnifications. It is observed that only in case (a) do the virtual images of the point hologram remain undistorted and in the initial position of the object point. Additionally, in case (d), even the real image becomes virtual, as $z_i' = -25 < 0$.

6.13.3 Spatial Filtering

Spatial filtering in Fourier optics represents a sophisticated technique used to selectively modify or enhance specific spatial frequencies within an optical image. It plays a pivotal role in tasks such as noise reduction, image enhancement, and the extraction of particular features from an optical image. The spatial filter acts as a versatile tool for tailoring the final image's characteristics based on desired spatial frequency components.

TABLE 6.2: Point Hologram Images: $\lambda_1 = 0.5\,\mu$m, $x_o = -1$ cm, $y_o = 0$, $z_o = -10$ cm, $x_r = 1$ cm, $y_r = 0$, $z_r = -25$ cm.

λ_2 (μm)	(x_p, y_p, z_p) (cm)	(x_i, y_i, z_i) (cm)	M_t	M_a	(x_i', y_i', z_i') (cm)	M_t'	M_a'
0.50	$(1, 0, -25)$	$(-1, 0, -10)$	1	1	$(-9, 0, 50)$	5	25
0.50	$(2, 0, -30)$	$(-0.78, 0, -10.71)$	1.07	1.15	$(-7.75, 0, 37.50)$	3.75	14.06
0.50	$(2, 0, -20)$	$(-0.36, 0, -9.09)$	0.90	0.83	$(-24, 0, 100)$	10	100
0.50	$(1, 0, -10)$	$(-0.25, 0, -6.25)$	0.63	0.39	$(6, 0, -25)$	2.5	6.25
0.65	$(1, 0, -25)$	$(-1.20, 0, -8.47)$	1.10	0.93	$(-5.84, 0, 26.32)$	3.42	9.00

For example, the implementation of spatial filtering can be achieved by positioning a spatial filter in the Fourier plane of a 4f-imaging system, as shown in Fig. 6.29. In this configuration, the object, represented as a two-dimensional transparency, is illuminated by a collimated wave (typically a plane wave). The object's characteristics are described by $u_o(x_o, y_o)$. Through Fourier transformation, facilitated by the first lens, the amplitude and phase information of the object is made accessible for manipulation.

To accomplish this, a spatial filter, denoted as $F(x_p, y_p)$, is strategically placed in the spatial frequency plane (x_p, y_p). Subsequently, a second lens, positioned beyond the spatial frequency plane, is employed to produce the Fourier transform of $U_o(f_x, f_y)F(f_x, f_y)$. This process generates the filtered image $u_i(x_i, y_i)$ of the object, where $U_o(f_x, f_y)$ represents the Fourier transform of the object transparency. This advanced setup allows for precise control over spatial frequency components, enabling the fine-tuning of optical information and significantly improving image quality.

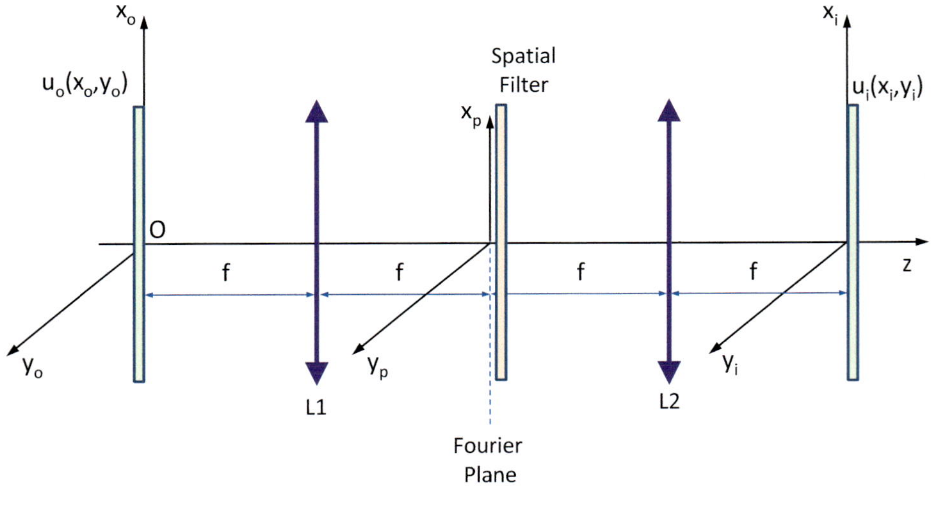

FIGURE 6.29: A 4f-imaging system comprised of two thin lenses of focal length f. The object field $u_o(x_o, y_o)$ can be spatially filtered in the Fourier plane and the filtered object is imaged at the image plane.

The process of spatial filtering entails a refined adjustment of image content through the deliberate placement of masks, or spatial filters, in the Fourier plane. These masks are instrumental in altering the amplitude and/or phase of specific spatial frequency components.

Among the most notable are several widely used types of optical filters:

- Low pass Filtering: This filter facilitates the passage of low spatial frequencies while attenuating higher frequencies. It is particularly effective for suppressing noise or smoothing out fine details in an image. A low-pass filter is usually characterized by a small circular aperture positioned in the center of the Fourier plane.
- High pass Filtering: High-pass filters allow high spatial frequencies to pass while attenuating lower frequencies. This type of filtering is instrumental in enhancing fine details and accentuating edges, earning it the moniker of an edge enhancement filter. A high-pass filter is manifested by a dark spot strategically placed in the center of the Fourier plane.
- Image correction: Spatial filtering extends beyond the conventional to address less trivial tasks, such as image correction. For instance, periodic noise, arising from repetitive patterns superimposed on the original image, can be mitigated. Such patterns often result from electrical or electromechanical interference during the image-capturing process.

To illustrate the optical filtering process, let's consider the following simple example. The form of the object transparency is assumed as follows (adapted and expanded from Ref. [361]):

$$u_o(x_o, y_o) = c + a\cos(2\pi f_1 x_o) + b\cos(2\pi f_2 y_o),$$

where f_1 and f_2 are two spatial frequencies (assuming that $f_2 > f_1$). The purpose of the filter is to eliminate the spatial frequency along the y_o direction, i.e., the f_2. According to Eq. (6.19) the optical field at the Fourier plane of the lens is given by:

$$u_p(x_p, y_p) = j\frac{e^{-j2kf}}{\lambda f}U_o\left(\frac{x_p}{\lambda f}, \frac{y_p}{\lambda f}\right),$$

$$U_o(f_x, f_y) = \mathcal{F}\{u_o\} = c\delta(f_x, f_y) + \frac{a}{2}\left[\delta(f_x + f_1, f_y) + \delta(f_x - f_1, f_y)\right] +$$

$$\frac{b}{2}\left[\delta(f_x, f_y + f_2) + \delta(f_x, f_y - f_2)\right].$$

In order to suppress the unwanted f_2 spatial frequency the following filter can be used in the Fourier plane:

$$F(f_x, f_y) = 1 - \delta(f_x, f_y + f_2) - \delta(f_x, f_y - f_2).$$

Multiplication of U_o by F easily results in:

$$U_o(f_x, f_y)F(f_x, f_y) = c\delta(f_x, f_y) + \frac{a}{2}\left[\delta(f_x + f_1, f_y) + \delta(f_x - f_1, f_y)\right].$$

Then, the second lens produces at its image plane the following field:

$$u_i(x_i, y_i) \sim c + a\cos(2\pi f_1 x_i).$$

The f_2 spatial frequency component is evidently suppressed. In the final step of the derivation, products involving delta functions appear. It is assumed that the condition $\delta(f_x, f_y)\delta(f_x + f_a, f_y + f_b) = 0$ holds when f_a and f_b are not equal to zero. While this assumption lacks mathematical rigor, it significantly simplifies the calculations.

In practical situations, the physical object typically has finite dimensions. This finite extent can be modeled by multiplying the object function with a rectangular function, denoted as $\text{rect}(x_o/L, y_o/L)$, where L represents the size of the object transparency in the

x_o and y_o directions. Within this framework, the filter function is correspondingly adjusted to:

$$F(f_x, f_y) = 1 - \mathrm{rect}\left(\frac{f_x}{w}, \frac{f_y + f_2}{w}\right) - \mathrm{rect}\left(\frac{f_x}{w}, \frac{f_y - f_2}{w}\right).$$

Here, w is sufficiently narrow in the Fourier plane to eliminate the "sinc" functions introduced by $\mathrm{rect}(x_o/L, y_o/L)$. To utilize the preceding results, the parameters were assumed to be set as follows: $a = b = 1$, $c = 2$, $f_1 = 16$, and $f_2 = 25$. These values were applied in a scenario where the input transparency was represented as a 512×512 digital image generated by:

$$u_o(x_o, y_o) = c + a\cos\left(2\pi f_1 \frac{x_o}{512}\right) + b\cos\left(2\pi f_2 \frac{y_o}{512}\right),$$

with $x_o, y_o = 1, 2, \cdots, 512$. The top row of plots in Fig. 6.30 illustrates the initial transparency, its corresponding spectrum, and the applied spatial filter. The parameter w was set to 6 pixels. The resulting image and its spectrum are shown in the bottom row of plots.

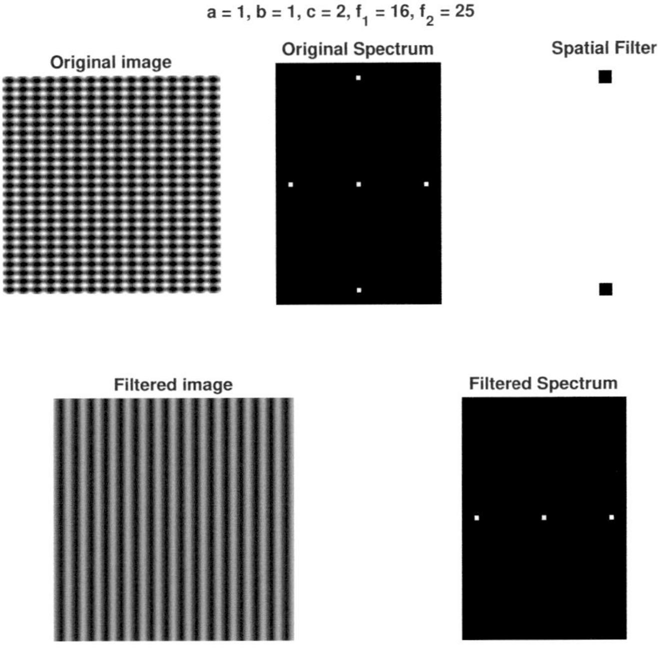

FIGURE 6.30: Application of spatial filtering at the Fourier plane of a periodic in x_o and y_o image transparency. The parameters are listed in the figure title.

Example 6.13.2:

In this example, an image, represented as a transparency, is considered. The chosen image is a grayscale photograph with dimensions of 4048×4048 pixels, depicting the author. The top-left plot of Fig. 6.31(a) displays this photograph, denoted as $u_o(x_o, y_o)$. In the Fourier plane, the Fourier transform of the object is formed according to Eq. (6.19). The spatial filter is assumed to be either (a) a circular aperture with radius a or (b) a circular stop with radius a. In both cases, the corresponding filtered images are to be determined.

Solution: For the 4f imaging system employed, the magnification is unity ($m = 1$) and $u_o = u_g$ (see Sec. 6.8). Using Eqs. (6.35), (6.36), and (6.39), the resulting image field is expressed as:

$$u_i(x_i, y_i) = h(x_i, y_i) ** u_o(x_i, y_i), \quad \text{where}$$

$$U_i(f_x, f_y) = H(f_x, f_y)U_o(f_x, f_y),$$

$$H(f_x, f_y) = \mathcal{F}\{h_i(x_i, y_i)\} = P(-\lambda f f_x, -\lambda f f_y).$$

For case (a), $P(x_p, y_p) = \text{circ}(r_T/a)$. The resulting image field can then be expressed as:

$$u_i(x_i, y_i) = u_o(-x_i, -y_i) ** \left[\pi\left(\frac{a}{\lambda f}\right)^2 \text{jinc}\left(\frac{2a}{\lambda f}\sqrt{x_i^2 + y_i^2}\right)\right].$$

The "jinc" function, known as the Airy pattern function, was defined in Sec. 6.13.1. Analogous to the "sinc" function in polar coordinates, the "jinc" function produces a similar effect, and its convolution with the object function introduces a blurring of the object's details. As a result, the circular aperture of the spatial filter acts as a low-pass filter.

In this implementation, a radius a of 20 pixels is assumed. The results are summarized in Fig. 6.31(a). The top row shows the original image, its spectrum in the spatial frequency domain, and the spatial filter. The bottom row presents the filtered image (non-inverted) and its corresponding spectrum. As expected, the image demonstrates blurring due to the low-pass filtering, which suppresses high spatial frequencies. The spectra are computed using the two-dimensional Fast Fourier Transform (FFT), with only the regions containing significant values displayed.

For case (b), the only difference lies in the spatial filter, which is now configured as a circular stop and modeled by $P(r_T) = 1 - \text{circ}(r_T/a)$. By following the same procedure, the resulting image field is expressed as:

$$u_i(x_i, y_i) \sim u_o(-x_i, -y_i) - u_o(-x_i, -y_i) ** \left[\pi\left(\frac{a}{\lambda f}\right)^2 \text{jinc}\left(\frac{2a}{\lambda f}\sqrt{x_i^2 + y_i^2}\right)\right].$$

The operation described above functions as an edge detector, highlighting the edges of the object image by truncating the low spatial frequencies using the circular stop. The implementation of this procedure for the same original image as in case (a) is shown in Fig. 6.31(b). Similar information to that presented in case (a) is provided, with the only difference being that the stop radius is now set to 10 pixels. The resulting high-pass filtered image accentuates the edges, as expected.

Example 6.13.3:

In this example, the image used as an object in Example 6.13.2 has been distorted by periodic noise, described by $u_{od}(x_o, y_o) = u_o(x_o, y_o) + a\cos(2\pi f_0 y)$. With $a = 0.25$, $\cos(2\pi f_0 y/4048)$, $f_0 = 16$, and $y_o = 1, 2, \ldots, 4048$, the top-left image of Fig. 6.32 is generated. Notably, the periodic noise appears as horizontal lines superimposed on the original image. To correct this distortion, an appropriate spatial filter is applied in the Fourier plane.
Solution: The Fourier transform of the object field is now expressed as:

$$U_{od}(f_x, f_y) = U_o(f_x, f_y) + \frac{a}{2}\left[\delta(f_x, f_y + f_0) + \delta(f_x, f_y - f_0)\right].$$

The distorted spectrum is clearly visible, exhibiting two delta spikes, as shown in the middle plot of the top row in Fig. 6.32. To restore the image, it is essential to truncate these two

(a)

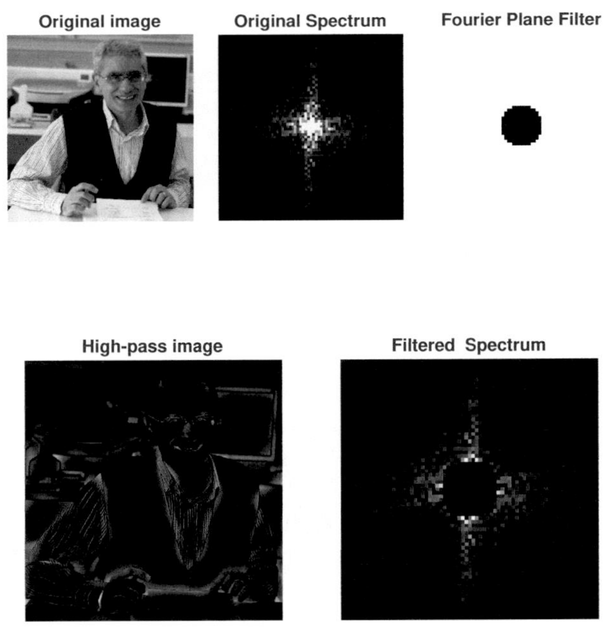

(b)

FIGURE 6.31: (a) Low-pass filtering of the original image with a circular aperture of radius of 20 pixels in the Fourier plane. (b) High-pass filtering of the original image with a circular stop of radius of 10 pixels in the Fourier plane.

delta spikes in the Fourier plane. This can be accomplished by selecting a spatial filter with the following form:

$$P(x_p, y_p) = 1 - \text{rect}\left(\frac{x_p}{w}, \frac{y_p + y_0}{w}\right) - \text{rect}\left(\frac{x_p}{w}, \frac{y_p - y_0}{w}\right),$$

$$H(f_x, f_y) = P(-\lambda f f_x, -\lambda f f_y) =$$

$$1 - \text{rect}\left(\lambda f \frac{f_x}{w}, \lambda f \frac{f_y + f_0}{w}\right) - \text{rect}\left(\lambda f \frac{f_x}{w}, \lambda f \frac{f_y - f_0}{w}\right),$$

$$\text{where} \quad f_0 = \frac{y_0}{\lambda f}.$$

The purpose of incorporating the two rectangular-shaped stops is to block the delta spikes in the spectrum of the distorted image caused by periodic noise. In this case, w is set to 4 pixels. The spatial filter is shown in the top-right plot of Fig. 6.32. The resulting filtered (restored) image is presented in the bottom-left plot of the same figure, with its spectrum displayed on the right. Clearly, the image has been significantly improved, although some minor distortion remains. This residual distortion occurs because the optical implementation with the stops not only truncates the delta spikes but also eliminates certain spatial frequencies from the original image. It is important to note that more advanced filters are available in digital image processing for better restoration. However, this example focuses on a simple optical solution using a Fourier plane filter.

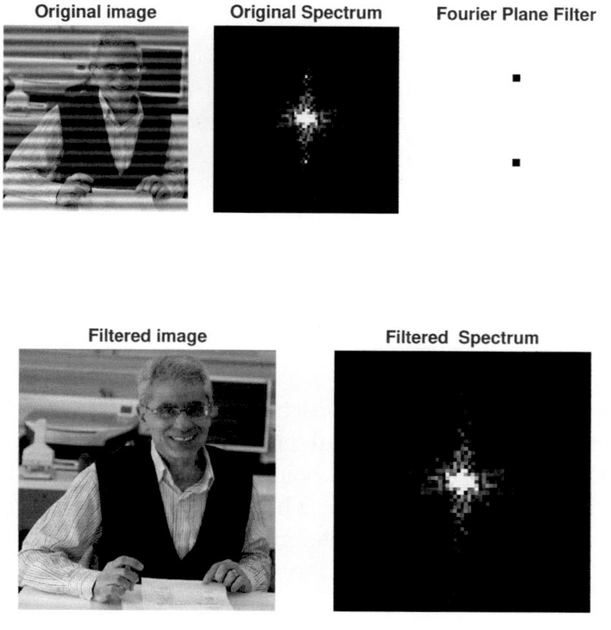

FIGURE 6.32: The restoration of a periodically distorted image via filtering in the Fourier plane.

6.13.4 VanderLugt Correlator

An optical correlator is a device that uses optical techniques to perform correlation operations. Correlation is a mathematical operation that measures the similarity between two signals. In the context of optical correlators, this often involves comparing an input image with a reference image.

The VanderLugt correlator, developed by Anthony VanderLugt in 1964 [362], is a well-known optical correlator. It employs a spatial light modulator to implement matched filtering, a technique for pattern recognition. Matched filtering is particularly useful for detecting specific patterns or objects in an image.

The basic idea behind the VanderLugt correlator is to leverage the Fourier transform properties of lenses to perform correlation in the frequency domain. The optical system utilizes a lens and a holographic filter implemented on a spatial light modulator. The Fourier transform of the input image is multiplied by the complex conjugate of the Fourier transform of the reference image, yielding the correlation in the spatial domain. These correlators are widely used in image processing, pattern recognition, and other applications where rapid and efficient image correlation is required.

The VanderLugt correlator operates in a coherent two-step process. In the first step, a hologram is recorded in the Fourier plane of a positive thin lens through the interference of a reference plane wave at an angle θ. The recording process is illustrated in Fig. 6.33(a). Suppose the object transparency, illuminated coherently with a collimated beam of light, generates the object field $f(x_o, y_o)$. The reference plane wave is produced with the help of the prism shown in the figure. The reference plane wave, deflected by the prism, takes the form $r\exp(-j2\pi\alpha x_p)$ in the Fourier $x_p y_p$-plane. Therefore, the total field at the Fourier plane, denoted as $u_p(x_p, y_p)$, can be written as:

$$u_p(x_p, y_p) = \frac{j}{\lambda f} e^{-j2kf} F\left(\frac{x_p}{\lambda f}, \frac{y_p}{\lambda f}\right) + re^{-j2\pi\alpha x_p},$$

where $F = \mathcal{F}\{f\}$. The holographic plate placed in the Fourier plane records the intensity of the incident coherent light, resulting in a transmittance described by the following expression:

$$\begin{aligned} t(x_p, y_p) &= t_0 + \beta\left|\frac{j}{\lambda f} e^{-j2kf} F\left(\frac{x_p}{\lambda f}, \frac{y_p}{\lambda f}\right) + re^{-j2\pi\alpha x_p}\right|^2 \\ &= t_0 + \beta\left[\frac{1}{(\lambda f)^2}|F|^2 + |r|^2 + \frac{jr^*}{\lambda f} e^{-j2kf} F e^{+j2\pi\alpha x_p} + \frac{-jr}{\lambda f} e^{+j2kf} F^* e^{-j2\pi\alpha x_p}\right]. \end{aligned} \tag{6.78}$$

In the second phase of the process, reconstruction is performed, as shown in Fig. 6.33(b). A 4f-imaging system is used, with an input field corresponding to an object transparency $g(x_o, y_o)$. During this reconstruction, the holographic plate recorded in the first phase is placed at the Fourier plane of the system. The resulting field immediately after the holographic plate, denoted as $u_p^+(x_p, y_p)$, is then given by:

$$\begin{aligned} u_p^+(x_p, y_p) &= \left[\frac{j}{\lambda f} e^{-j2kf} G\left(\frac{x_p}{\lambda f}, \frac{y_p}{\lambda f}\right)\right] t(x_p, y_p) \\ &= \frac{j}{\lambda f} e^{-j2kf}\left(t_0 + \beta|r|^2\right) G + \frac{j\beta}{\lambda f} e^{-j2kf} \frac{G|F|^2}{(\lambda f)^2} + \\ &\quad \frac{j\beta}{\lambda f} e^{-j2kf} \frac{r^* GF}{\lambda f} e^{+j2\pi\alpha x_p} + \frac{-j\beta}{\lambda f} e^{+j2kf} \frac{rGF^*}{\lambda f} e^{-j2\pi\alpha x_p} \Longrightarrow \end{aligned}$$

$$u_p^+(x_p, y_p) \simeq G + G|F|^2 + GFe^{+j2\pi\alpha x_p} + GF^* e^{-j2\pi\alpha x_p}. \tag{6.79}$$

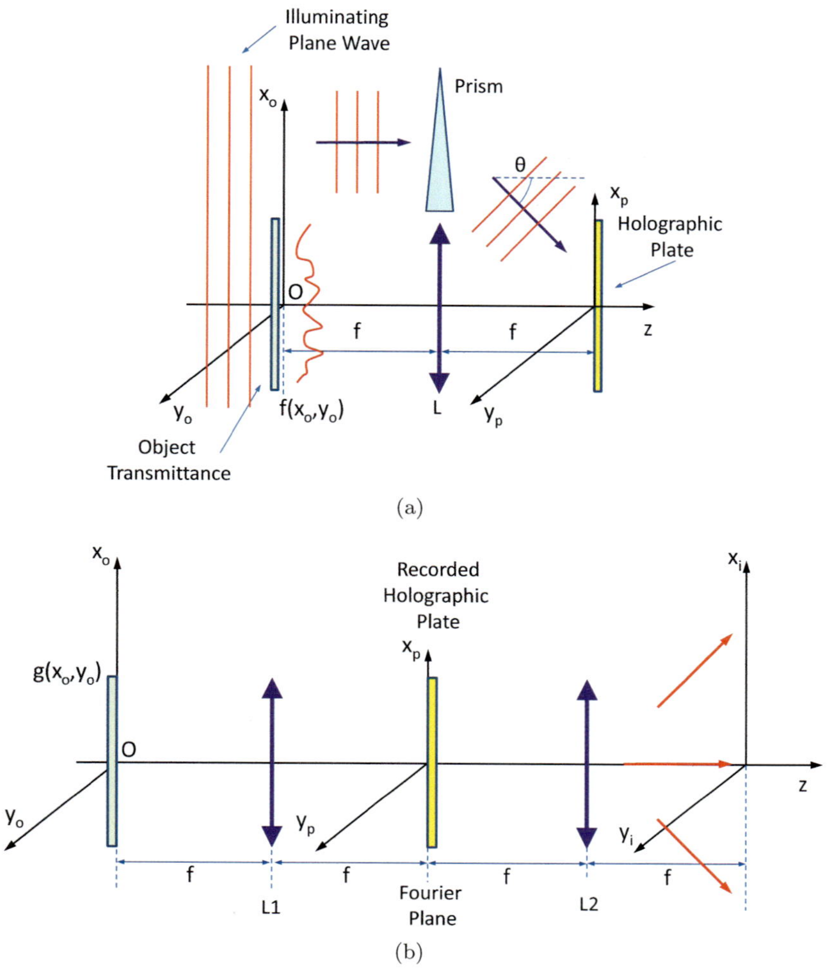

FIGURE 6.33: (a) The recording process of the VanderLugt correlator. (b) The reconstruction process of the VanderLugt correlator. The three red arrows indicate the directions in which the reconstructed images appear in the $x_i y_i$ plane.

In the last expression $u_p^+(x_p, y_p)$ has been simplified by omitting the constant amplitude and phase terms. The function G represents the Fourier transform of g. The optical field at the image plane (neglecting constant amplitude and phase terms) can be expressed in the form:

$$\begin{aligned} u_i(x_i, y_i) &\simeq \mathcal{F}\left\{u_p^+(x_p, y_p)\right\}\Big|_{\substack{f_x = x_i/\lambda f,\\ f_y = y_i/\lambda f}} \\ &= g(-x_i, -y_i) + g(-x_i, -y_i) ** \left[f(-x_i, -y_i) \star\star f(-x_i, -y_i)\right] + \\ &\quad g(-x_i - \lambda f\alpha, -y_i) ** f(-x_i - \lambda f\alpha, -y_i) + \\ &\quad g(-x_i + \lambda f\alpha, -y_i) \star\star f(-x_i + \lambda f\alpha, -y_i). \end{aligned} \tag{6.80}$$

The first two terms of Eq. (6.80) represent the zero-order diffraction terms, which appear at the origin of the image plane $x_i y_i$. The third term corresponds to a displaced convolution between the g and f functions, located near $-\lambda f\alpha$ from the origin. Finally, the fourth term denotes the cross-correlation between g and f, positioned near $+\lambda f\alpha$.

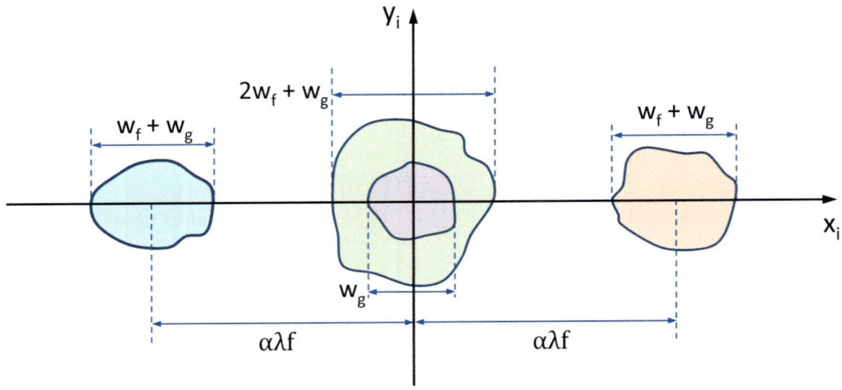

FIGURE 6.34: The produced images areas in the image plane x_iy_i during the reconstruction process of the VaderLugt correlator.

It is reasonable to assume that the functions $f(x_o, y_o)$ and $g(x_o, y_o)$ are localized along the x_o- and y_o-directions. Let w_f and w_g represent the widths along the x_o-direction where f and g have non-zero values. The qualitative representation of the spatial extent of each of the four terms in Eq. (6.80) is shown in Fig. 6.34. The width in the x_i-direction of the convolution or cross-correlation of f and g is given by $w_f + w_g$. The autocorrelation of f has a width of $2w_f$. To ensure the images remain separable in the image plane, the following condition must be met:

$$\alpha\lambda f \geq \frac{2w_f + w_g}{2} + \frac{w_f + w_g}{2} \Longrightarrow \alpha = \frac{\sin\theta}{\lambda} \geq \frac{1}{\lambda f}\left(w_g + \frac{3}{2}w_f\right). \tag{6.81}$$

6.14 Problems

1 **2D-Fourier Transforms:** The 2D functions $f(x, y)$ displayed in Fig. 6P.1 are characterized by a unity value within the shaded area and zero values elsewhere. The 2D Fourier transforms $F(f_x, f_y)$ of these functions are to be determined.

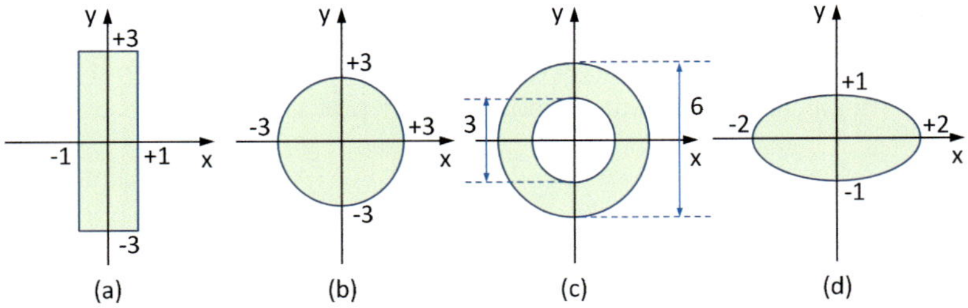

FIGURE 6P.1: The footprints of various 2D functions $f(x, y)$ are characterized by unity values in the shaded areas and zero values elsewhere.

2 **2D-Fourier Transforms:** Find the Fourier transforms of the following 2D functions $f(x, y)$: (a) $f(x, y) = \text{rect}(x/4, y/4)[1 + \cos(2\pi 5x)]$; (b) The function shown in Fig. 6P.2 which is unity in the shaded area and zero elsewhere.

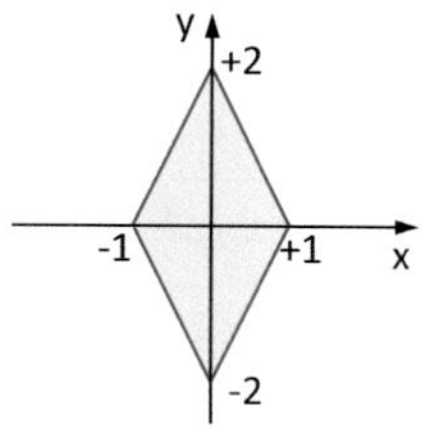

FIGURE 6P.2: The rhombus footprint of a 2D function $f(x, y)$ is characterized by unity values in the shaded area and zero values elsewhere.

3 Directional Cosines and Spatial Frequencies: Assume that the wavevector of a plane wave makes an angle $\alpha_x = 20°$ with the x-axis and an angle $\alpha_y = 85°$ with the y-axis. Find the three directional cosines of this plane wave (propagating towards the positive z-axis). Furthermore, find the spatial frequencies of this plane wave along the x- and y-directions if the freespace wavelength is $\lambda = 1\,\mu$m.

4 Grating and Plane-Wave Spectrum: Upon illuminating a large grating with a monochromatic plane wave, a complex object wave is generated, expressed as $u_o(x_o, y_o) = 1 + \cos(2\pi f_0 x_o)$. (a) Show that this field corresponds to three distinct plane waves and specify them. Assume that the freespace wavelength of the incident monochromatic plane wave is $\lambda = 0.5, \mu$m and $f_0 = 12,$mm^{-1}. (b) When the object field, produced by the initial illumination, encounters a square aperture with a width of 2 mm, the resulting plane wave spectrum undergoes modification due to diffraction effects. What is the modified by the aperture plane wave spectrum?

5 Optical Field Upon a Lens with Two-Rectangular Slits Aperture: Consider the configuration in which a monochromatic plane wave of unit amplitude illuminates a double-slit aperture consisting of two rectangular openings. The widths of these openings are denoted by L_x in the x-direction and L_y in the y-direction. The openings are separated by a distance D along the y-axis, measured between their centers. The aperture is positioned adjacent to a thin lens with a focal length f, similar to the aperture illustrated in Fig. 6.5. The objective of this problem is to determine the intensity distribution at the focal plane of the lens.

6 Optical Field Upon a Lens with an Annular Aperture: Consider a scenario where a unit amplitude monochromatic plane wave illuminates an annular aperture, of outer diameter D and inner diameter d. The aperture is positioned adjacent to a positive thin lens with a focal length of f, similarly to the one depicted in Fig. 6.5. (a) Determine the intensity pattern at the focal plane of the lens. (b) Show that in the limit as D approaches d the intensity pattern is proportional to $(d/\lambda f)^2 J_0^2(\pi d r_T/\lambda f)$. (*Hint:* Utilize properties of Bessel functions)

7 Optical Field Upon a Lens and a Ronchi Ruling: A Ronchi ruling, sometimes referred to as a "square-wave grating," is a glass plate ruled with parallel, equally spaced, opaque lines. The transmittance of this ruling is illustrated in Fig. 6P.3. When positioned against a square aperture of width L (where it is assumed that $L \gg \Lambda$), the combination of the Ronchi ruling and the aperture is placed adjacent to a thin lens with a focal length of f (similarly to Fig. 6.5 for example). (a) Determine the intensity pattern, $I(x_f, y_f)$, at the focal plane of the lens. (b) Sketch the intensity pattern $I(x_f, y_f = 0)$ along the x_f-axis in the focal plane.

8 A Four-Lens System with a Three-Rectangles Aperture: The aperture function of the four-lens imaging system consists of three square openings of size L at a separation of $2L$, as depicted in Fig. 6P.4. (a) Determine the coherent amplitude transfer function, accompanied by a sketch illustrating the various cutoff frequencies and other relevant distances. (b) Determine the incoherent impulse response function as well as its graphical representation. (c) Calculate the optical transfer function and provide a graphical representation of the modulation transfer function.

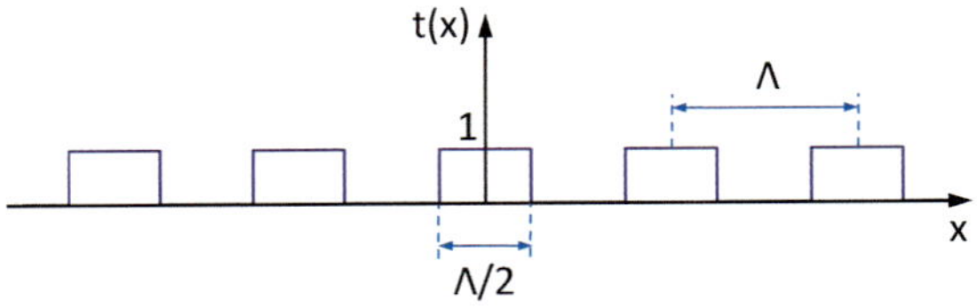

FIGURE 6P.3: The transmittance $t(x)$ of a Ronchi ruling of period Λ.

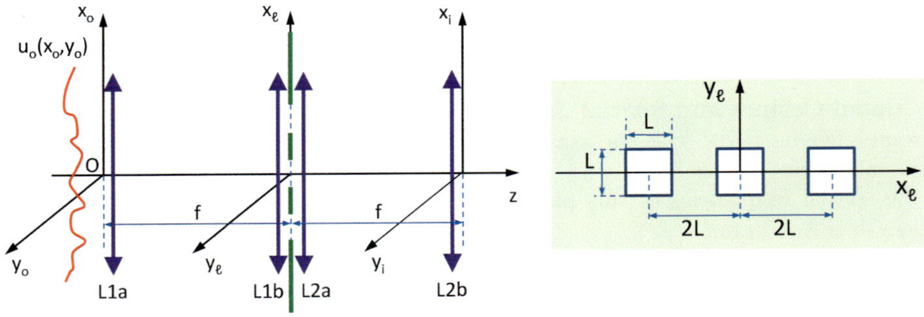

FIGURE 6P.4: A configuration of an imaging system with four lenses (4F system) with focal distance f, and separation distance f and an aperture function implemented by encompassing three equal square openings of size L spaced at intervals of $2L$.

9 A Two-Lens System with a Two-Circular Openings Aperture: Consider the imaging system depicted in Fig. 6P.4, excluding lenses L1a and L2b. The aperture function now comprises two small, pinhole-like circular openings separated by a distance S in the x_ℓ-direction. The object is quasi-monochromatic with a mean wavelength λ and spatially incoherent. The object has a rectangular shape with an intensity distribution of $I_o(x_o, y_o) = \text{rect}(x_o/W, y_o/H)$ [348]. (a) Sketch the cross-section of the optical transfer function in the f_x-direction, labeling the location of spatial frequency passbands. (b) Find the normalized spatial frequency spectrum $\mathcal{I}_g(f_x, f_y)$ of the geometric optics image distribution and sketch its f_x-axis cross-section. (c) Assuming that the pinholes are sufficiently small to be represented by delta functions, if the pinholes are close together, the image pattern consists of a high-contrast sinusoidal fringe pattern. As the pinholes move apart, the fringe contrast decreases until the fringes disappear completely. Determine the spacing S_0 at which they first disappear. (d) Once again assuming that the pinholes can be represented by delta functions, what is the image intensity, up to a proportionality constant, as a function of the pinhole separation S?

10 A Two-Lens System with a Three-Circular Openings Aperture: Consider the imaging system illustrated in Fig. 6P.4, excluding lenses L1a and L2b. The aperture function now features three small, circular openings separated by a distance S in the x_ℓ-direction, each with a diameter D. The object is quasi-monochromatic with a mean wavelength λ and spatially incoherent. (a) Provide a perspective view sketch of the optical transfer function (OTF) of the system, labeling the location of spatial frequency passbands along with their relative heights. (b) Now, assume that light passing through the top-most opening undergoes a half-wavelength retardation, introducing a 180° phase shift across that portion of the aperture function. Sketch a suitable cross-section of the corresponding OTF.

11 Incoherent Imaging System with a Square Annular Pupil: In an incoherent imaging system, the pupil function is formed by a square annular aperture, as illustrated in Fig. 6.5(a). The aperture has a side length of D, and within it, there is an inner square stop with a width of d (modified from Ref. [265]). Assuming that $D/2 \leq d \leq D$ and that $\lambda d_i = 0.1\,\text{mm}^2$: (a) Determine the impulse response function $|h|^2$. (b) Determine and sketch the cross-section of the optical transfer function (OTF) along the f_x spatial frequency direction. (c) Employ the numerical correlation method to sketch the OTF in the $f_x f_y$-plane. (d) Determine the OTF in the special case where $d \to D$.

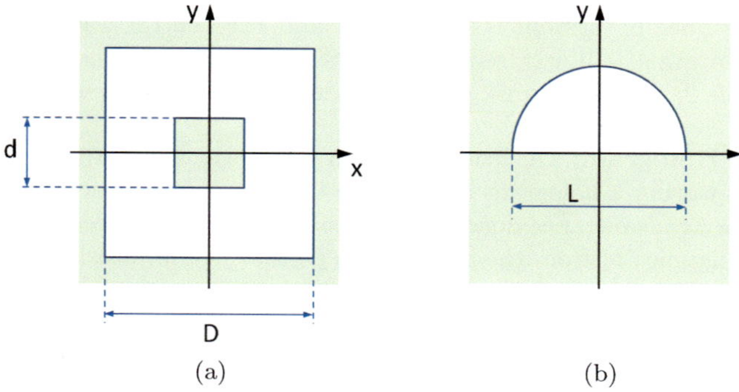

FIGURE 6P.5: (a) An annular square-shaped aperture. The large side is D while the inner stop side is d. (b) The semi-circular aperture of diameter L.

12 Incoherent Imaging System with a Semi-Circular Pupil: In an incoherent imaging system, the pupil function is formed by a semi-circular aperture, as illustrated in Fig. 6.5(b). The diameter of the aperture has a length of L (modified from Ref. [265]). Assuming that $\lambda d_i = 0.1\,\text{mm}^2$: (a) Determine and sketch the cross-sections of the optical transfer function (OTF) along the f_x- and f_y spatial frequency directions. (b) Specify and sketch the cross-section of the OTF along a general direction in the $f_x f_y$-plane. (c) Employ the numerical correlation method to sketch the OTF in the $f_x f_y$-plane.

13 Incoherent Imaging with an Extended Source: Assuming that all thin lenses in the configuration of Fig. 6P.6 have the same focal length f, find the expression for the minimum source diameter D required for fully incoherent imaging of the object transparency $t_o(x_o)$. The answer should be expressed in terms of the focal length f, the aperture diameter W, and the spatial cut-off frequency f_c of the object transmittance $t_o(x_o, y_o) = t_o(x_o)$ (i.e., it depends only on x_o) [348].

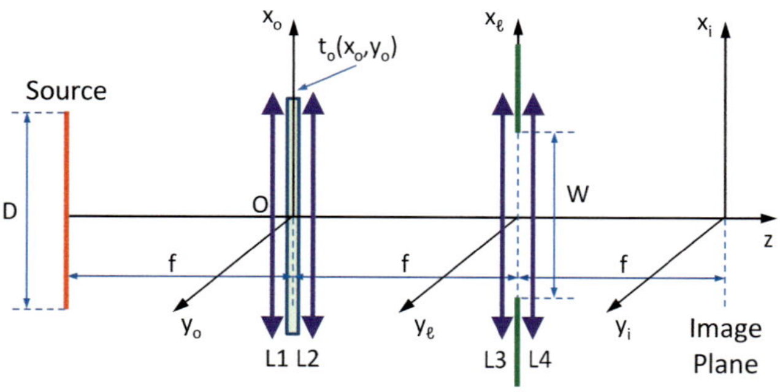

FIGURE 6P.6: An imaging configuration for an object transparency $t_o(x_o, y_o)$ via an extended source of diameter D. The aperture at the Fourier plane (between lenses L3 and L4) has a diameter of W.

14 Zernike Phase-Contrast Configuration with Absorption Plate: For this problem, assume the Zernike phase-contrast configuration depicted in Fig. 6.24. In this case, the phase-shifting dot-shaped plate is also partially absorbing, with intensity transmittance α in the range $0 < \alpha < 1$. Determine the image intensity distribution in this case. Also, consider methods for specifying the value of α that could optimize the image in terms of its contrast [348].

15 Image Contrast Calculation: A thin lens with a focal length of $f = 100\,\text{mm}$ and an f-number of 4 (implying a circular aperture in front or in the back of the lens) is employed to image a cosine grating of the form $t_o(x_o, y_o) = 1 + a\cos(2\pi f_0 x_o)$, where $f_0 = 100$ lines/mm and $a = 0.5$. The imaging system is incoherent, and the central wavelength is $\lambda = 500\,\text{nm}$. The system configuration is illustrated in Fig. 6.8. For magnifications with real images $|m| = |s_i/s_o| = 1, 0.25, 0.5, 2$, and 4, the task is to calculate the image contrast.

16 Lensless Fourier Hologram: The lensless Fourier transform configuration depicted in Fig. 6P.7 is employed for capturing a hologram of a square transparency object with a side length of L, featuring amplitude transmittance denoted by $t_o(x_o, y_o)$. During the recording phase, the object is positioned at a distance z_1 from the holographic plate. A small positive lens (L1) is utilized to implement the point source located at (x_r, y_r). In the subsequent reconstruction phase, images are acquired using a positive lens (L2) with a focal length of f [348]. (a) What is the magnification M of the reconstructed images? (b) Show that the amplitude of the zero-order image term (i.e., the light lying between the two first-order images) is proportional to the autocorrelation of the function $t_o(x/M, y/M)\exp[j(\pi/(\lambda z_1 M^2))(x^2 + y^2)]$ plus a central diffraction-limited spot. (c) To ensure no overlap of the zero-order light with first-order images, the reference point source (x_r, y_r) should be positioned away from the center of the object transparency. Determine the specific distance by ensuring sufficient separation between the first-order images and avoiding interference.

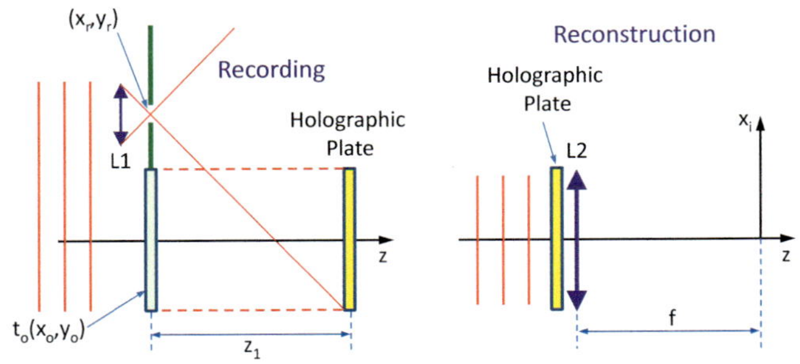

FIGURE 6P.7: On the left: The recording configuration of a lensless Fourier hologram. On the right: The reconstruction process.

17 Two-Lens Imaging System: Consider the 2-lens imaging system illustrated in Fig. 6.11. The object diameter is $w = 2\,\text{cm}$, and the aperture (pupil) diameter is $\ell = 1\,\text{cm}$. The focal lengths are $f_1 = 3\,\text{cm}$ and $f_2 = 4\,\text{cm}$. The object is spatially coherent and quasi-monochromatic [348]. (a) Assuming a mean wavelength of $\lambda = 500\,\text{nm}$, what is the highest spatial frequency present in the image amplitude distribution? (b) In connection with (a), what is the corresponding object spatial frequency? (c) What should be the size of the L1 lens to prevent vignetting?

18 Two-Lens Imaging System: Assume the coherent optical system depicted in Fig. 6.8(a). Both lenses have the same focal length f. The objective is to show that the image amplitude distribution is given by:

$$u_i(x_i, y_i) = A\exp[-jka(x_i^2 + y_i^2)][t_o(-x_i, -y_i) ** h(x_i, y_i)],$$

where the factor A depends on the point source parameters. In doing so, find the constant a in the quadratic phase term as well as the impulse response h.

19 Gabor Zone Plate: The geometry often referred to as a Gabor zone plate is depicted in Fig. 6.8(b). A holographic plate is illuminated by a normally incident plane wave with amplitude A. Additionally, a weak expanding spherical wave, generated by the forward scattering of the incident plane wave by a point scatterer, is incident on the plate. (a) Assuming the scattering point is on the z-axis at a distance z_1 from the holographic plate, use Fresnel approximations to

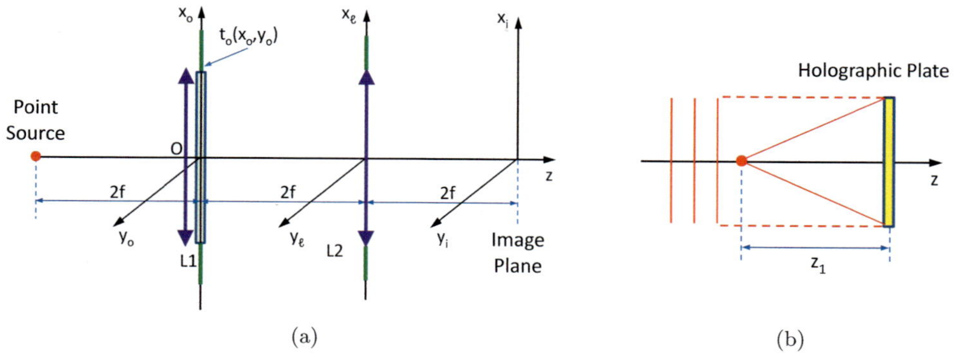

FIGURE 6P.8: (a) An imaging system of two lenses of focal length f separated by $2f$ distances. (b) Recording of a Gabor zone plate.

model the expanding spherical wave and derive an expression for the intensity of the total wave distribution $I_o(x, y)$ incident on the holographic plate. (b) Based on the result from (a), derive an expression for the amplitude $u_t(x, y)$ transmitted by the developed holographic plate when illuminated by a normally incident plane wave. (c) Clearly describe, in physical terms, each of the terms in the transmitted wave field from (b).

20 Defocus Aberration: Consider an incoherent optical system with a square aperture of size L experiencing defocus aberration. The optical transfer function (OTF) exhibits a zero under certain conditions. The criteria for the OTF to have a zero are crucial to understanding the system's behavior. To determine the minimum defocus, denoted as L_m (refer to Example 6.11.1), at which the zero of the OTF occurs, and the corresponding spatial frequency, perform the following steps: (a) Identify the conditions leading to the zero in the OTF. (b) Calculate the minimum defocus, L_m, for which the zero of the OTF occurs. (c) Determine the spatial frequency at which the zero in the OTF takes place. (d) Additionally, create a cross-sectional plot of the OTF along the f_x-direction for specific values of defocus, such as $L_m = 0, \lambda/4, \lambda/2, 3\lambda/4, \lambda$, and $5\lambda/4$. This will provide a visual representation of how the OTF changes with defocus. (e) For medium defocus, approximately locate the point of maximum contrast reversal. This information is valuable in understanding the system's performance and behavior under different defocus conditions.

7

Vector Diffraction

7.1 Introduction

Vector diffraction in optics refers to the extension of classical scalar diffraction theory to account for the vector nature of light. In traditional scalar diffraction theory, light is modeled as a scalar wave, disregarding its polarization and the vector characteristics of electromagnetic fields. Electromagnetic vector diffraction theory incorporates the vectorial nature of the electric and magnetic fields, providing a more comprehensive framework. This approach is particularly significant when analyzing the behavior of light or other electromagnetic waves interacting with objects or structures whose dimensions are comparable to the wavelength of the light (or radiation in general).

Vector diffraction theory is indispensable for understanding how light behaves when it interacts with diffractive optical elements, such as gratings, microlens arrays, and other microstructures. These elements can alter the phase, amplitude, and polarization of light, and vector diffraction offers a more precise prediction of these effects. In optical systems with high numerical apertures, where the angles of light rays become significant, vector diffraction becomes particularly critical. It provides a more accurate description of light behavior in such cases, enabling a deeper understanding of phenomena such as optical aberrations.

When the dimensions of optical elements approach or fall below the wavelength of light, vector diffraction becomes essential. Scalar theory may fail to accurately describe phenomena such as optical scattering, focusing, and polarization conversion at the subwavelength scale.

In summary, vector diffraction in optics extends the traditional scalar diffraction theory to provide a more comprehensive understanding of how light propagates, interacts with optical elements, and behaves in complex optical systems. This is particularly important in situations involving polarization-sensitive materials, microstructures, and subwavelength optics. The theory finds applications in various fields, including optics, telecommunications, and electromagnetic imaging, and is essential for designing and analyzing optical components such as lenses, gratings, and other devices that manipulate electromagnetic waves.

A thorough understanding of electromagnetic vector diffraction theory is indispensable for engineers, physicists, and researchers involved in the design and analysis of optical systems and devices. It offers a more accurate and detailed description of electromagnetic wave behavior than scalar theories, especially in scenarios where the vector nature of the fields cannot be neglected.

Analytically solving Maxwell's equations for complex structures is often highly challenging or even impossible. Consequently, numerical methods, such as finite-difference time-domain (FDTD), finite-difference frequency-domain (FDFD), and finite element method (FEM), are widely employed to simulate and analyze electromagnetic vector diffraction.

DOI: 10.1201/9781003490333-7

Analyzing vector diffraction in optics involves various methods and techniques to solve the vectorial wave equation and understand the behavior of light waves with polarization effects. Here are some common analysis methods used for vector diffraction:

- **Vectorial Diffraction Integrals:** These are extensions of the integrals used in scalar diffraction theory, such as the Kirchhoff integrals, Stratton-Chu formulations, Kottler and Franz diffraction integrals, among others. These integrals account for both the incident and diffracted waves and are applicable to a wide range of optical configurations [104, 363–369]. Moreover, some of these methods serve as the foundation for volume integral numerical techniques [370, 371]. Common approximations of these integrals involve replacing the total field with the incident field at apertures, similar to the approximations made in scalar diffraction theory. Additionally, approximations of the Green's function give rise to Fresnel and Fraunhofer formulations.
- **Finite-Difference Time-Domain/Frequency-Domain (FDTD/FDFD):** FDTD is a numerical method that discretizes both space and time to solve Maxwell's equations. It is a versatile approach for studying vector diffraction in complex geometries, including three-dimensional structures [372]. On the other hand, FDFD is a numerical method that discretizes space and solves Maxwell's equations in the frequency domain [373,374].
- **Finite Element Method (FEM) and Finite Volume Method (FVM):** FEM is a numerical technique where the spatial domain of the problem is discretized into small elements. It is used to solve partial differential equations, including Maxwell's equations, by employing variational principles to minimize an error function. FEM is highly flexible and can be applied to analyze vector diffraction in various optical configurations [373, 375]. In the finite volume method (FVM), the computational space is divided into control volumes. Maxwell's equations are then integrated over these control volumes and, with the aid of the divergence theorem, are transformed into surface integrals. These surface integrals are subsequently converted into a set of linear equations. Another numerical method is the Discontinuous Galerkin (DG) method, which combines features of both FEM and FVM. The FEM itself can be categorized as either a differential or an integral method.
- **Method of Moments (MoM):** The method of moments converts the integral form of Maxwell's equations into a matrix equation by approximating the unknown fields using basis functions [373].
- **Transmission Line Matrix (TLM) Method:** The Transmission Line Matrix (TLM) method models electromagnetic fields using equivalent transmission lines. It is a space- and time-discretizing method in which the computational domain is divided into a grid of nodes, with each node representing a small section of a transmission line [376].
- **Boundary Element Method (BEM):** The Boundary Element Method (BEM) is an integral equation-based numerical method that uses the Green's function to discretize the boundaries of objects and solve for the vectorial electromagnetic field on these boundaries. It is particularly well-suited for problems with open boundaries [377–379]. BEM can also be combined with the Finite Element Method (FEM) to create hybrid formulations [380]. There are several variations of these methods, including the Electric Field Integral Equation (EFIE) method, the Magnetic Field Integral Equation (MFIE) method, and the Combined Field Integral Equation (CFIE) method [370, 371, 381].
- **Fourier Modal Method (FMM) and Rigorous-Coupled Wave Analysis (RCWA):** The Fourier Modal Method (FMM) decomposes electromagnetic fields into Floquet-Fourier series. It is particularly effective for analyzing periodic structures and is capable of handling both scalar and vector diffraction problems [382]. The Rigorous

Coupled-Wave Analysis (RCWA) is a semi-analytical method in computational electromagnetics that is most commonly applied to solve scattering problems involving periodic dielectric (or metallic) structures. As a Fourier-space method, it represents devices and fields as a sum of spatial harmonics [314, 316, 317].

- **Plane Wave Spectrum Method:** The plane wave spectrum approach is a mathematical technique used to analyze and simulate wave propagation, particularly in the context of electromagnetic waves. In this method, the electromagnetic field, especially for apertures, is decomposed into an infinite sum of plane waves. These plane waves propagate through homogeneous media and, when combined with the fast Fourier transform, allow for the prediction of diffracted fields. This approach provides an efficient and convenient means to model the behavior of electromagnetic waves in complex optical systems and structures [104, 383].

Certainly, vectorial diffraction of electromagnetic waves encompasses a variety of methods beyond those discussed in this chapter. The methods presented here represent only a subset, and for a more comprehensive exploration of specific techniques, readers are encouraged to consult specialized texts in the fields of electromagnetics and optics literature. A diverse array of methodologies exists, each tailored to different configurations and applications within vector diffraction, contributing to the richness of this field of study.

This chapter begins with a brief overview of vector diffraction integrals and their application to planar apertures, which are of particular significance in optical applications. It also covers common approximations of these integral formulations. As a specific example, the chapter examines vector diffraction by a circular aperture on a perfectly conducting screen within the far-field approximation. Additionally, Babinet's principle is extended to vectorial electromagnetic fields and applied to the diffraction by a perfectly conducting disk in the far-field. The vectorial plane-wave spectrum method is also reviewed.

Building upon this theoretical foundation, the chapter explores exact numerical methods, including the two-dimensional boundary element method (BEM) and the two-dimensional finite-difference frequency-domain (FDFD) method. The chapter concludes with a review of rigorous coupled-wave analysis (RCWA) for the vectorial diffraction of light by grating structures, encompassing both one-dimensional and two-dimensional periodic gratings. It also discusses the effective medium theory (EMT) for subwavelength gratings and illustrates Kogelnik's two-wave coupled-wave approximate theory. Finally, a brief introduction to the theory of photonic crystals is also included.

7.2 Vector Diffraction Integrals

Assuming a closed surface S with no charge or current sources within the enclosed volume V (as depicted in Fig. 7.1), the Kirchhoff diffraction integral, originally derived for scalar diffraction and the scalar field U, is expressed as follows:

$$U(\vec{r}_0) = U(x_0, y_0, z_0) = \iint_S \left(G\frac{\partial U}{\partial n} - U\frac{\partial G}{\partial n} \right) ds'. \tag{7.1}$$

The function G is the Green's function for the three-dimensional homogeneous space $G(\vec{r}_0, \vec{r}') = e^{-jk|\vec{r}_0 - \vec{r}'|}/4\pi|\vec{r}_0 - \vec{r}'|$ (see Scalar Diffraction Appedices). At this point, let's consider a vectorial electric field given by $\vec{E} = E_x\hat{\imath}_x + E_y\hat{\imath}_y + E_z\hat{\imath}_z$, which satisfies the Helmholtz equation $\nabla^2\vec{E} + k^2\vec{E} = 0$ inside the volume V. Each component E_w $(w = x, y, z)$

of the electric field also satisfies the Helmholtz equation. Therefore, Eq. (7.1) is fulfilled for each component of the electric field, where $U = E_w$ $(w = x, y, z)$. Combining these three scalar equations into a vector expression, the Kirchhoff diffraction integral is derived:

$$\vec{E}(\vec{r}_0) = \vec{E}(x_0, y_0, z_0) = \iint_S \left(G\frac{\partial \vec{E}}{\partial n} - \vec{E}\frac{\partial G}{\partial n} \right) ds'. \tag{7.2}$$

Certainly, a similar equation can be derived for the vectorial magnetic field, as Eq. (7.1) is

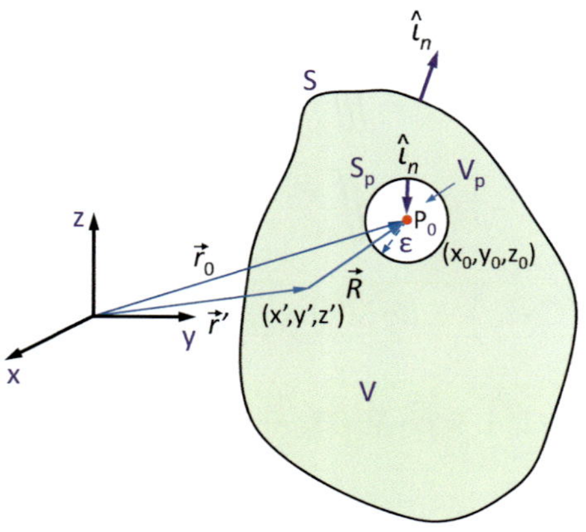

FIGURE 7.1: A closed surface S that does not include any charge or current sources. A point $P_0(x_0, y_0, z_0)$ is in the interior of the volume enclosed by S.

also satisfied for $U = H_w$ $(w = x, y, z)$. Thus, the Kirchhoff's vector diffraction formula for the magnetic field is given by:

$$\vec{H}(\vec{r}_0) = \vec{H}(x_0, y_0, z_0) = \iint_S \left(G\frac{\partial \vec{H}}{\partial n} - \vec{H}\frac{\partial G}{\partial n} \right) ds'. \tag{7.3}$$

A more rigorous derivation is provided in Chapter 18 of Ref. [104], where charge and current sources are also considered. At this point, Kirchhoff's vector diffraction formulas are transformed into the well-known Stratton-Chu formulas [104, 363–369].

Let's manipulate Kirchhoff's diffraction integral for the E_w electric field component $(w = x, y, z)$. It is important to note that $\partial G/\partial n = \vec{\nabla}'G \cdot \hat{\imath}_n$ (where $\vec{\nabla}'$ operates on the primed variables $\vec{r}'$) and $\vec{\nabla}(fg) = f\vec{\nabla}g + g\vec{\nabla}f$. Then, the integral for E_w can be expressed as follows (the divergence theorem is also utilized):

$$\begin{aligned} E_w(\vec{r}_0) &= \iint_S \left(G\frac{\partial E_w}{\partial n} - E_w\frac{\partial G}{\partial n} \right) ds' = \iint_S \left[G\left(\vec{\nabla}'E_w \cdot \hat{\imath}_n\right) - E_w\left(\vec{\nabla}'G \cdot \hat{\imath}_n\right)\right] ds' \\ &= \iint_S \left[-2E_w\left(\vec{\nabla}'G \cdot \hat{\imath}_n\right) + \vec{\nabla}'\left(GE_w\right) \cdot \hat{\imath}_n \right] ds' \\ &= -2\iint_S E_w\left(\vec{\nabla}'G \cdot \hat{\imath}_n\right) ds' + \iiint_V \vec{\nabla}' \cdot \left(\vec{\nabla}'\left(GE_w\right)\right) dv' \\ &= -2\iint_S E_w\left(\vec{\nabla}'G \cdot \hat{\imath}_n\right) ds' + \iiint_V \nabla'^2\left(GE_w\right) dv'. \end{aligned}$$

Using the last equation for all three electric field components, it can be expressed in the following form:

$$\vec{E}(\vec{r}_0) = -2\iint_S \vec{E}\left(\vec{\nabla}'G\cdot\hat{\imath}_n\right)ds' + \iiint_V \nabla'^2\left(G\vec{E}\right)dv'.$$

Then, utilizing the vector identity $\vec{\nabla}\times\vec{\nabla}\times\vec{A} = \vec{\nabla}\left(\vec{\nabla}\cdot\vec{A}\right) - \nabla^2\vec{A}$ as well as Gauss' law ($\vec{\nabla}\cdot\vec{E} = 0$), $\iiint_V \vec{\nabla}\times\vec{A}dv = \iint_S \hat{\imath}_n\times\vec{A}ds$, and $\iiint_V \vec{\nabla}f dv = \iint_S f\hat{\imath}_n ds$, the expression for the electric field becomes [384]:

$$\begin{aligned}
\vec{E}(\vec{r}_0) &= -2\iint_S \vec{E}\left(\vec{\nabla}'G\cdot\hat{\imath}_n\right)ds' + \iiint_V\left[\vec{\nabla}'\left(\vec{\nabla}'\cdot(G\vec{E})\right) - \vec{\nabla}'\times\vec{\nabla}'\times(G\vec{E})\right]dv' \\
&= -2\iint_S \vec{E}\left(\vec{\nabla}'G\cdot\hat{\imath}_n\right)ds' + \iint_S \vec{\nabla}'\cdot(G\vec{E})\hat{\imath}_n ds' - \iint_S \hat{\imath}_n\times\left(\vec{\nabla}'\times(G\vec{E})\right)ds' \\
&= \iint_S\left[-2\vec{E}\left(\vec{\nabla}'G\cdot\hat{\imath}_n\right) + \vec{E}\cdot\vec{\nabla}'G\hat{\imath}_n - \hat{\imath}_n\times\vec{\nabla}'G\times\vec{E} + j\omega(\hat{\imath}_n\times\vec{B})G\right]ds' \\
&= \iint_S\left[-\vec{E}\left(\vec{\nabla}'G\cdot\hat{\imath}_n\right) + \vec{E}\cdot\vec{\nabla}'G\hat{\imath}_n - (\vec{E}\cdot\hat{\imath}_n)\vec{\nabla}'G + j\omega(\hat{\imath}_n\times\vec{B})G\right]ds' \\
&= -\iint_S\left[\left(\hat{\imath}_n\times\vec{E}\right)\times\vec{\nabla}'G + \left(\vec{E}\cdot\hat{\imath}_n\right)\vec{\nabla}'G - j\omega\mu(\hat{\imath}_n\times\vec{H})G\right]ds'.
\end{aligned} \tag{7.4}$$

In the above derivation, the Faraday's law ($\vec{\nabla}\times\vec{E} = -j\omega\vec{B} = -j\omega\mu\vec{H}$), and the vector identity $\vec{a}\times\vec{b}\times\vec{c} = (\vec{a}\cdot\vec{c})\vec{b} - (\vec{a}\cdot\vec{b})\vec{c}$, were also utilized. Using the same procedure, a similar expression for the magnetic field can also be derived:

$$\vec{H}(\vec{r}_0) = -\iint_S\left[\left(\hat{\imath}_n\times\vec{H}\right)\times\vec{\nabla}'G + \left(\vec{H}\cdot\hat{\imath}_n\right)\vec{\nabla}'G + j\omega\epsilon(\hat{\imath}_n\times\vec{E})G\right]ds'. \tag{7.5}$$

Equations (7.4) and (7.5) represent the Stratton-Chu diffraction integrals. In some references the unit vector $\hat{\imath}_n$ is defined in the opposite direction [104], or the Green's function is defined as $G(\vec{r}_0,\vec{r}') = e^{+jk|\vec{r}_0-\vec{r}'|}/|\vec{r}_0-\vec{r}'|$ [365, 384].

The Kirchhoff and Stratton-Chu vector expressions for the electric and magnetic fields can be transformed into vector expressions that rely solely on the tangential fields at the boundary of the surface S. These transformed expressions are commonly referred to as Kottler expressions [385] and are given by:

$$\vec{E}(\vec{r}_0) = -\frac{1}{j\omega\epsilon}\iint_S\left[k^2G\left(\hat{\imath}_n\times\vec{H}\right) + \left[\left(\hat{\imath}_n\times\vec{H}\right)\cdot\vec{\nabla}'\right]\vec{\nabla}'G + j\omega\epsilon\left(\hat{\imath}_n\times\vec{E}\right)\times\vec{\nabla}'G\right]ds', \tag{7.6}$$

$$\vec{H}(\vec{r}_0) = \frac{1}{j\omega\mu}\iint_S\left[k^2G\left(\hat{\imath}_n\times\vec{E}\right) + \left[\left(\hat{\imath}_n\times\vec{E}\right)\cdot\vec{\nabla}'\right]\vec{\nabla}'G - j\omega\epsilon\left(\hat{\imath}_n\times\vec{H}\right)\times\vec{\nabla}'G\right]ds'. \tag{7.7}$$

The proof of the Kottler diffraction integrals can be found in Refs. [104, 365, 385]. For simplicity, the detailed proof is omitted here.

Using the Stratton-Chu equations (7.4) and (7.5), the Franz [386] diffraction formulas can be derived and are given by [104, 365]:

$$\vec{E}(\vec{r}_0) = -\frac{1}{j\omega\epsilon}\vec{\nabla}\times\left[\vec{\nabla}\times\left(\iint_S G\left(\hat{\imath}_n\times\vec{H}\right)ds'\right)\right] - \vec{\nabla}\times\left[\iint_S G\left(\hat{\imath}_n\times\vec{E}\right)ds'\right], \tag{7.8}$$

$$\vec{H}(\vec{r}_0) = \frac{1}{j\omega\mu}\vec{\nabla}\times\left[\vec{\nabla}\times\left(\iint_S G\left(\hat{\imath}_n\times\vec{E}\right)ds'\right)\right] - \vec{\nabla}\times\left[\iint_S G\left(\hat{\imath}_n\times\vec{H}\right)ds'\right]. \tag{7.9}$$

It is worth noting that the operator $\vec{\nabla}$ acts on the non-primed variables $\vec{r}_0$. For simplicity, the detailed proof is omitted here. In Eqs. (7.8) and (7.9), the electric and magnetic fields at point $\vec{r}_0$ are expressed solely in terms of the tangential fields on the surface S. More general Kirchhoff, Stratton-Chu, Kottler, and Franz vector diffraction integrals, including current and charge sources are thoroughly presented in Ch. 18 of Ref. [104].

7.2.1 Vector Diffraction Integrals for Apertures

Similar to the analysis in scalar diffraction theory, we focus on the case of an aperture in an infinite opaque screen, as depicted in Fig. 7.2. In this case, the surface $S = S_1 \cup S_2$, where S_1 is the xy-plane (containing the aperture) and S_2 is a spherical surface with radius $R \to \infty$. It is important to note that the integral expressions of Kirchhoff, Stratton-Chu, Kottler, and Franz are equivalent only for the case of a closed surface [104]. In the aperture case, the integrals over the S_2 surface do not contribute to the diffracted field when the Silver-Müller radiation conditions [387, 388] are applied. A summary of the Silver-Müller radiation conditions is presented in Appendix 7A. These conditions are analogous to the Sommerfeld radiation conditions used in scalar diffraction analysis.

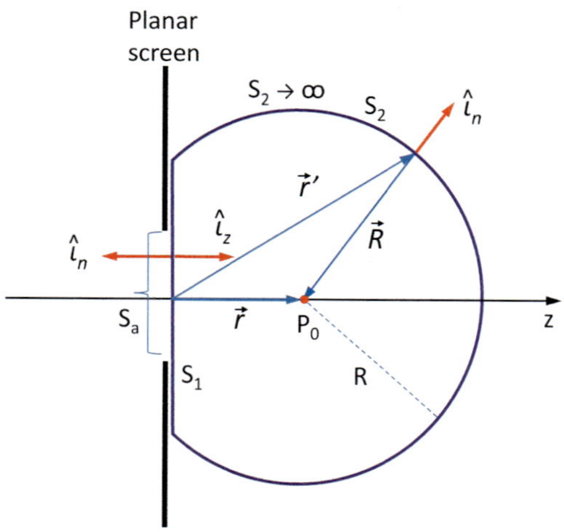

FIGURE 7.2: A closed surface $S = S_1 \cup S_2$ that does not include any charge or current sources. The S_1 part involves the xy plane where the aperture is and the S_2 part is part of a sphere of radius that tends to infinity. The area S_a represents the aperture area.

Using the geometry of Fig. 7.2, the Kirchhoff [Eqs. (7.2) and (7.3)], Stratton-Chu [Eqs. (7.4) and (7.5)], Kottler [Eqs. (7.6) and (7.7)], and Franz [Eqs. (7.8) and (7.9)] diffraction formulas have contributions only from the S_1 surface, as the integral over the S_2 surface (with $R \to \infty$) is zero according to the Silver-Müller radiation conditions. For an open surface S_a (Fig. 7.2) bounded by a contour C (assuming that the screen is opaque in the

$S_1 - S_a$ area, and therefore the integral contributions over the $S_1 - S_a$ area are zero), the Kirchhoff, Stratton-Chu, Kottler, and Franz formulations are not equivalent (see Refs. [104, 369]) since contour integrals over C must be added to the Kirchhoff and Stratton-Chu formulations. However, these contour integrals usually have a minor effect and can be ignored, allowing the diffracted field to be alternatively calculated using any of the following formulations:

$$\vec{E}(\vec{r}) = \iint_{S_a} \left(-G\frac{\partial \vec{E}}{\partial n'} + \vec{E}\frac{\partial G}{\partial n'} \right) ds', \quad \text{(Kirchhoff)}, \tag{7.10}$$

$$\vec{E}(\vec{r}) = \iint_{S_a} \left[\left(\hat{\imath}_{n'} \times \vec{E}\right) \times \vec{\nabla}' G + \left(\vec{E} \cdot \hat{\imath}_{n'}\right) \vec{\nabla}' G - j\omega\mu(\hat{\imath}_{n'} \times \vec{H}) G \right] ds', \quad \text{(Stratton-Chu)}, \tag{7.11}$$

$$\vec{E}(\vec{r}) = \frac{1}{j\omega\epsilon} \iint_{S_a} \left[k^2 G \left(\hat{\imath}_{n'} \times \vec{H}\right) + \left[\left(\hat{\imath}_{n'} \times \vec{H}\right) \cdot \vec{\nabla}' \right] \vec{\nabla}' G + j\omega\epsilon \left(\hat{\imath}_{n'} \times \vec{E}\right) \times \vec{\nabla}' G \right] ds', \tag{7.12}$$

(Kottler),

$$\vec{E}(\vec{r}) = \frac{1}{j\omega\epsilon} \vec{\nabla} \times \left[\vec{\nabla} \times \left(\iint_{S_a} G \left(\hat{\imath}_{n'} \times \vec{H}\right) ds' \right) \right] + \vec{\nabla} \times \left[\iint_{S_a} G \left(\hat{\imath}_{n'} \times \vec{E}\right) ds' \right], \quad \text{(Franz)}, \tag{7.13}$$

where the equations for the electric field were presented. Similar expressions hold for the magnetic field as well. In the above expressions, the $\hat{\imath}_{n'} = \hat{\imath}_z = -\hat{\imath}_n$ was used. In these integral expressions, the electric and magnetic fields are the total fields, which are not known at the aperture, as was the case in scalar diffraction theory. For instance, the total electric field can be expressed as $\vec{E} = \vec{E}_{inc} + \vec{E}_d$, where $\vec{E}_{inc}$ represents the incident field in the absence of the aperture screen, and $\vec{E}_d$ represents the diffracted field by the aperture in both $z > 0$ and $z < 0$ regions. However, the diffracted field $\vec{E}_d$ is unknown. Therefore, similar approximations to those used in scalar diffraction theory can be employed, which unfortunately lead to comparable inconsistencies. For example, the total field on the aperture can be approximated by the incident field, as was done in the scalar diffraction case. Another important consideration is that the integrals in Eqs. (7.10)-(7.13) apply to the entire surface S_1 and not just to the aperture area S_a. The assumption regarding the opaque screen does not directly involve any electromagnetic boundary conditions. However, with perfectly conducting screens, there may be induced surface charges and currents from the boundary conditions, which cannot be ignored, as nonzero magnetic fields can be present on the conductive screen [88].

Now, assuming that the surface S_1 is planar, Sommerfeld Green's functions can be employed. These functions satisfy either Dirichlet conditions, $G_D(\vec{r}; \vec{r}' = \vec{r}'') = 0$, with $G_D(\vec{r}; \vec{r}', \vec{r}'') = G(\vec{r}; \vec{r}') - G(\vec{r}; \vec{r}'')$, or Neumann conditions, $\partial G_N/\partial n(\vec{r}; \vec{r}' = \vec{r}'') = 0$, with $G_N(\vec{r}; \vec{r}', \vec{r}'') = G(\vec{r}; \vec{r}') + G(\vec{r}; \vec{r}'')$ (refer to the *Scalar Diffraction* chapter (Chapter 5) for details). Applying Kirchhoff's vector formula, Eq. (7.10) for the S_1 planar surface for $\vec{U} = \vec{E}$ or $\vec{H}$ (for each component), the following equations can be written:

$$\vec{U}(x, y, z) = \iint_{S_1} \left[-\vec{U}\,\frac{\partial G_D}{\partial n} \right]_{z'=0} ds' = -2 \iint_{S_1} \left[\vec{U}\,\frac{\partial G}{\partial n} \right]_{z'=0} ds', \tag{7.14}$$

$$\vec{U}(x, y, z) = \iint_{S_1} \left[G_N\,\frac{\partial \vec{U}}{\partial n} \right]_{z'=0} ds' = 2 \iint_{S_1} \left[G\,\frac{\partial \vec{U}}{\partial n} \right]_{z'=0} ds', \tag{7.15}$$

since $\partial G_D/\partial n|_{z'=0} = 2\partial G/\partial n|_{z'=0}$ and $G_N|_{z'=0} = 2G|_{z'=0}$. The above expressions can be simplified as follows. Assume that $\vec{U} = \vec{E}$. Then, the following expressions can be written:

$$\begin{aligned}
\vec{E} &= 2\iint_{S_1}\left[-\vec{E}\frac{\partial G}{\partial n}\right]_{z'=0} ds' = -2\iint_{S_1}\left[\vec{E}\left(\hat{\imath}_n\cdot\vec{\nabla}'G\right)\right]_{z'=0} ds' \\
&= -2\iint_{S_1}\left[\left(\hat{\imath}_n\times\vec{E}\right)\times\vec{\nabla}'G+\left(\vec{E}\cdot\vec{\nabla}'G\right)\hat{\imath}_n\right]ds' \\
&= -2\iint_{S_1}\vec{\nabla}G\times\left(\hat{\imath}_n\times\vec{E}\right)ds' - 2\iint_{S_1}\left[\vec{\nabla}'\cdot\left(G\vec{E}\right)-G\underbrace{\vec{\nabla}'\cdot\vec{E}}_{=0}\right]\hat{\imath}_n ds' \\
&= -2\iint_{S_1}\left[\vec{\nabla}\times\left(\hat{\imath}_n\times\vec{E}\right)G-\underbrace{G\vec{\nabla}\times\left(\hat{\imath}_n\times\vec{E}\right)}_{=0}\right]ds' - 2\iint_{S_1}\vec{\nabla}'\cdot\left(G\vec{E}\right)ds' \\
&= -2\vec{\nabla}\times\iint_{S_1}\left(\hat{\imath}_n\times\vec{E}\right)Gds' - 2\iint_{S_1}\vec{\nabla}'\cdot\left(G\vec{E}\right)ds'.
\end{aligned}$$

It is reminded that the operator ∇ acts only on non-primed variables (point of interest) and the operator ∇' acts only on primed variables (points on the S_1). Also the following identities were utilized [384]: $(\vec{a}\times\vec{b})\times\vec{c} = (\vec{c}\cdot\vec{a})\vec{b} - (\vec{c}\cdot\vec{b})\vec{a}$, $\vec{\nabla}\cdot(f\vec{F}) = f\vec{\nabla}\cdot\vec{F} + \vec{F}\cdot\vec{\nabla}f$, and $\vec{\nabla}\times(f\vec{F}) = f\vec{\nabla}\times\vec{F} + \vec{\nabla}f\times\vec{F}$, with f a scalar and $\vec{F}$ a vector function. The last integral in the latter equation can also be shown that has zero contribution. Initially, the term $\vec{\nabla}'\cdot(G\vec{E}) = \partial(GE_x)/\partial x' + \partial(GE_y)/\partial y' + \partial(GE_z)/\partial z'$. By subtracting Eqs. (7.14) and (7.15) for the z component of the electric field, it is straightforward to show that $\int\int_{S_1}\partial(GE_z)/\partial z' ds' = 0$ (it is reminded that $\hat{\imath}_n = -\hat{\imath}_z$). The integral over the other two electric field components can be written as:

$$I_{xy} = \iint_{S_1}\left[\frac{\partial(GE_x)}{\partial x'}+\frac{\partial(GE_y)}{\partial y'}\right]ds' = \iint_{S_1}\vec{\nabla}'_2\cdot(G\vec{E})ds' = \oint_C G\vec{E}\cdot\hat{\imath}_{\perp n}d\ell = 0, \quad (7.16)$$

since in the contour C, which is at an infinite distance in the S_1 plane, both G and $\vec{E}$ tend to zero. The unit vector $\hat{\imath}_{\perp n}$ is perpendicular to the contour boundary and lies in the $x'y'$ plane. The term $\vec{\nabla}'_2\cdot\vec{F}$ is the two-dimensional divergence of the vector function $\vec{F}$. Summarizing, the electric field can be written in the following form:

$$\vec{E}(\vec{r}) = 2\vec{\nabla}\times\left\{\iint_{S_1}\left(\hat{\imath}_z\times\vec{E}(\vec{r}')\right)G(\vec{r};\vec{r}')ds'\right\}. \quad (7.17)$$

The procedure can be repeated for the magnetic field, which can also be expressed in a similar form as:

$$\vec{H}(\vec{r}) = 2\vec{\nabla}\times\left\{\iint_{S_1}\left(\hat{\imath}_z\times\vec{H}(\vec{r}')\right)G(\vec{r};\vec{r}')ds'\right\}. \quad (7.18)$$

These equations which are similar to the scalar Rayleigh-Sommerfeld expressions are known as the m-theory and e-theory integral expressions [369].

7.2.2 Diffraction by a Circular Aperture

In Fig. 7.3, a circular aperture with radius a is positioned on a perfectly conducting screen at the $z = 0$ plane [88, 384]. A plane wave $\vec{E}_i = \vec{E}_0\exp(-j\vec{k}\cdot\vec{r})$ is incident from the left on the aperture. The angles θ_i and ϕ_i specify the direction of the incident wavevector $\vec{k}$, and the amplitude $\vec{E}_0$ characterizes the polarization of the incident plane wave. The observation point $P(x, y, z) = P(r, \theta, \phi)$ is on the right of the aperture. Since the screen is assumed to be a perfect conductor, the tangential component of the total electric field should be zero, i.e., $\hat{z} \times \vec{E} = 0$. As a result, the integral of Eq. (7.17) is restricted only to the area of the circular aperture of radius a. Furthermore, the electric field on the aperture is approximated by the incident field (Kirchhoff approximation). Then the resulting field on the right of the aperture is given by (S_a is the area of the circular aperture):

$$\vec{E}(\vec{r}) \simeq 2\vec{\nabla} \times \left\{ \iint_{S_a} \left(\hat{\imath}_z \times \vec{E}_i(\vec{r}\,') \right) G(\vec{r}; \vec{r}\,') ds' \right\}. \tag{7.19}$$

Following the approach of Refs. [88] and [384], the Green's function is approximated assum-

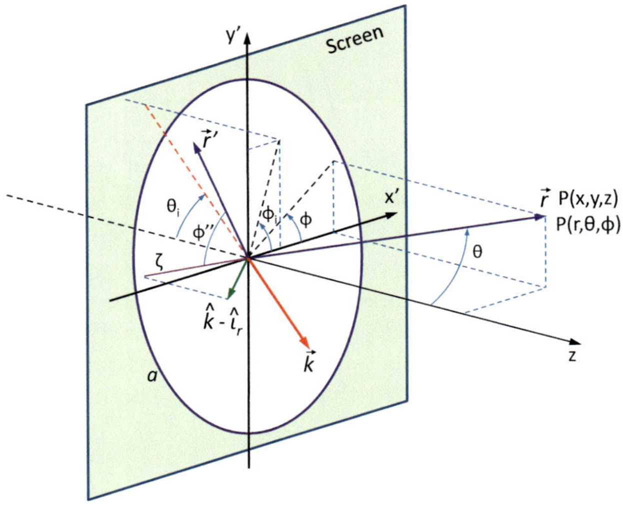

FIGURE 7.3: A circular aperture of radius a on an opaque perfectly conducting screen at $z = 0$ plane. The aperture is illuminated by a plane wave $\vec{E}_i = \vec{E}_0 \exp(-j\vec{k}\cdot\vec{r})$. The angles θ_i and ϕ_i specify the direction of the incident wavevector $\vec{k}$, and the amplitude $\vec{E}_0$ characterizes the polarization of the incident plane wave. For the wavevector $\vec{k}$ shown the corresponding azimuthal angle is $\phi_i + \pi$. The observation point $P(x, y, z) = P(r, \theta, \phi)$ is denoted by the position vector $\vec{r}$. The position vector $\vec{r}\,'$ denotes the integration points in the aperture area. The ζ denotes the projection of the $\hat{k} - \hat{\imath}_r$ vector on the $x'y'$-plane [384]. The $\hat{k}$ and $\hat{\imath}_r$ are the unit vectors of $\vec{k}$ and $\vec{r}$, respectively.

ing that the observation point P is sufficiently far from the aperture plane, $R = |\vec{r} - \vec{r}\,'| \gg a$. Then, the distance R can be approximated as:

$$R = |\vec{r} - \vec{r}\,'| = \left[r^2 + r'^2 - 2\vec{r}\cdot\vec{r}\,' \right]^{1/2} = r\left[1 + \left(\frac{r}{r'} \right)^2 - 2\hat{\imath}_r \cdot \frac{\vec{r}\,'}{r} \right]^{1/2}$$

$$\simeq r\left[1 - \frac{1}{2} 2\hat{\imath}_r \cdot \frac{\vec{r}\,'}{r} \right] = r - \hat{\imath}_r \cdot \vec{r}\,'.$$

Then, the Green's function can be approximated as:

$$G(\vec{r}, \vec{r}\,') = \frac{e^{-jk|\vec{r} - \vec{r}\,'|}}{4\pi|\vec{r} - \vec{r}\,'|} \simeq \frac{\exp[-jk(r - \hat{\imath}_r \cdot \vec{r}\,')]}{4\pi r}.$$

Inserting the approximate expression of the Green's function in Eq. (7.19), the electric field expression becomes:

$$\vec{E}(\vec{r}) \simeq 2\vec{\nabla} \times \left\{ (\hat{\imath}_z \times \vec{E}_0) \frac{e^{-jkr}}{4\pi r} \iint_{S_a} \exp\left[-jk(\hat{k} - \hat{\imath}_r) \cdot \vec{r}' \right] dx' dy' \right\}. \tag{7.20}$$

The unit vectors $\hat{k}$ and $\hat{\imath}_r$ are given by (refer to Fig. 7.3):

$$\hat{k} = \sin\theta_i \cos\phi_i \hat{\imath}_x + \sin\theta_i \sin\phi_i \hat{\imath}_y + \cos\theta_i \hat{\imath}_z,$$

$$\hat{\imath}_r = \sin\theta \cos\phi \hat{\imath}_x + \sin\theta \sin\phi \hat{\imath}_y + \cos\theta \hat{\imath}_z.$$

On the other hand, the vector (on the aperture) $\vec{r}'$ is given as $\vec{r}' = r'(\cos\phi' \hat{\imath}_x + \sin\phi' \hat{\imath}_y)$. The inner product $(\hat{k} - \hat{\imath}_r) \cdot \vec{r}' = r' \zeta \cos\phi''$ [384], where the variable ζ is given by $\zeta = [\sin^2\theta_i + \sin^2\theta - 2\sin\theta_i \sin\theta \cos(\phi_i - \phi)]^{1/2}$. Then, the integral in Eq. (7.20) can be calculated as follows:

$$\iint_{S_a} \exp\left[-jk(\hat{k} - \hat{\imath}_r) \cdot \vec{r}' \right] dx' dy' = \int_{r'=0}^{a} \int_{\phi''=0}^{2\pi} \exp[-jk\zeta \cos\phi'' r'] r' dr' d\phi'' =$$

$$= \int_{r'=0}^{a} 2\pi J_0(k\zeta r') r' dr' = 2\pi \frac{a J_1(k\zeta a)}{k\zeta},$$

where J_0 and J_1 are the Bessel functions of the first kind. Additionally, the following properties properties were utilized: $\int_0^{2\pi} e^{-jx\cos\phi} d\phi = 2\pi J_0(x)$ and $\int_0^a x J_0(\beta x) dx = a J_1(\beta a)/\beta$. Thus, Eq. (7.20) can be rewritten as follows:

$$\vec{E}(\vec{r}) \simeq 2\vec{\nabla} \times \left\{ (\hat{\imath}_z \times \vec{E}_0) \frac{e^{-jkr}}{4\pi r} 2\pi \frac{a J_1(k\zeta a)}{k\zeta} \right\}.$$

Using the identity $\vec{\nabla} \times (\psi \vec{A}) = \psi \vec{\nabla} \times \vec{A} + \vec{\nabla}\psi \times \vec{A}$ with $\vec{A} = (\hat{\imath}_z \times \vec{E}_0)$ (independent of r) and $\psi = e^{-jkr}/2r$, the last integral after a few manipulations becomes (Fraunhofer regime):

$$\vec{E}(\vec{r}) \simeq -\frac{a J_1(k\zeta a)}{k\zeta} \left(jk + \frac{1}{r} \right) \frac{e^{-jkr}}{r} \left[(\vec{E}_0 \cdot \hat{\imath}_r)\hat{\imath}_z - (\hat{\imath}_r \cdot \hat{\imath}_z)\vec{E}_0 \right]$$

$$\simeq -j \frac{a J_1(k\zeta a)}{\zeta} \frac{e^{-jkr}}{r} \left[(\vec{E}_0 \cdot \hat{\imath}_r)\hat{\imath}_z - (\hat{\imath}_r \cdot \hat{\imath}_z)\vec{E}_0 \right]. \tag{7.21}$$

At this point, it is instructive to apply the last equation to the case of a normally incident plane wave from the left ($z < 0$) of a circular aperture (Fig. 7.3). In this situation, $\phi_i = \theta_i = 0$ and $\zeta = \sin\theta$. The incident plane wave is assumed to be linearly polarized with $\vec{E}_0 = \hat{\imath}_x E_0$. The vector component of Eq. (7.21) becomes $[(\vec{E}_0 \cdot \hat{\imath}_r)\hat{\imath}_z - (\hat{\imath}_r \cdot \hat{\imath}_z)\vec{E}_0] = E_0[\sin\theta\cos\phi\,\hat{\imath}_z - \cos\theta\,\hat{\imath}_x]$. In order to find the vector diffracted field along the z-axis, $\theta = 0$, which implies that $\zeta = 0$. Using the known limit $\lim_{x\to 0}[J_1(ax)/x] = a/2$, the resulting diffracted field is given by:

$$\vec{E}(\vec{r}) \simeq jk \frac{a^2}{2} \frac{e^{-jkz}}{z} E_0\, \hat{\imath}_x,$$

which is identical to the scalar diffraction field result in the Fraunhofer regime. The vector diffracted field along the y'-direction can be determined if $\phi = \pi/2$ and the θ angle is allowed to vary. In this case the vector diffracted electric field becomes:

$$\vec{E}(\vec{r}) \simeq -j\frac{aJ_1(ka\sin\theta)}{\sin\theta}\frac{e^{-jkr}}{r}E_0\left[\sin\theta\cos\phi\,\hat{\imath}_z - \cos\theta\,\hat{\imath}_x\right]$$

$$\simeq -j\frac{aJ_1(ka\sin\theta)}{\sin\theta}\frac{e^{-jkz}}{z}e^{-jkr_T^2/2z}E_0\left[\sin\theta\cos\phi\,\hat{\imath}_z - \cos\theta\,\hat{\imath}_x\right].$$

With the substitution of $\sin\theta = r_T/\sqrt{r_T^2+z^2} \simeq r_T/z$, the last equation is equivalent to the corresponding expression of the scalar theory if the vectorial part is neglected. Therefore, the polarization of the vector diffracted field varies along the y'-axis (since θ varies). This result, of course, cannot be predicted by scalar theory.

It is interesting to examine the case of an incident plane wave at an angle. Assume $\phi_i = 0$ and $\theta_i \neq 0$. The incident wavevector is $\vec{k} = k_0[\sin\theta_i\,\hat{\imath}_x + \cos\theta_i\,\hat{\imath}_z]$. The polarization of the incident plane wave can be either *TE* ($\vec{E}_0 = E_0\,\hat{\imath}_y$) or *TM* [$\vec{E}_0 = E_0(\cos\theta_i\,\hat{\imath}_x - \sin\theta_i\,\hat{\imath}_z)$]. The value of the parameter ζ becomes $\zeta = (\sin^2\theta_i + \sin^2\theta - 2\sin\theta_i\sin\theta\cos\phi)^{1/2}$. The vectorial part for the *TE* and *TM* polarizations becomes:

$$\left[(\vec{E}_0\cdot\hat{\imath}_r)\hat{\imath}_z - (\hat{\imath}_r\cdot\hat{\imath}_z)\vec{E}_0\right] = \begin{cases} E_0\left[\sin\theta\cos\phi\,\hat{\imath}_z - \cos\theta\,\hat{\imath}_x\right], & (TE), \\ E_0\left[\cos\theta_i\sin\theta\cos\phi\,\hat{\imath}_z - \cos\theta_i\cos\theta\,\hat{\imath}_x\right], & (TM). \end{cases}$$

The corresponding intensity $I = (1/4Z)|\vec{E}|^2$ is given as (Z is the characteristic impedance of the surrounding medium):

$$I = \begin{cases} a^2\dfrac{|E_0|^2}{4Z}\left(\dfrac{J_1(ka\zeta)}{\zeta}\right)^2\dfrac{1}{r^2}\left(\sin^2\theta\sin^2\phi + \cos^2\theta\right), & (TE), \\ a^2\dfrac{|E_0|^2}{4Z}\left(\dfrac{J_1(ka\zeta)}{\zeta}\right)^2\dfrac{1}{r^2}\cos^2\theta_i\left(\sin^2\theta\cos^2\phi + \cos^2\theta\right), & (TM). \end{cases} \tag{7.22}$$

Example 7.2.1:

At this point a numerical calculation is helpful. Consider a circular aperture of radius $a = 10\,\mu$m on an opaque perfectly conducting screen at the $z = 0$ plane. The aperture is illuminated by a *TE* or *TM* polarized plane wave with $\theta_i = 45°$, $\phi_i = 0°$, and $\lambda_0 = 1\,\mu$m. Determined the vectorial diffracted intensity at $z = 500\,\mu$m along the radial direction.
Solution: The intensity of the vectorial diffracted electric field is determined using Eq. (7.22) and is depicted in Fig. 7.4 for $\phi = 0$. The results obtained from the Fraunhofer (scalar) theory are also included. The Fraunhofer scalar theory, introduced in the *Scalar Diffraction* chapter (Chapter 5), can be readily adapted to account for a plane wave incident at an angle (refer to the Problem Section of the *Scalar Diffraction* chapter, Chapter 5).

As indicated by Eq. (7.22), the intensity of the *TM* polarization is scaled by a factor of $\cos^2\theta_i$ relative to the *TE* polarization intensity. For smaller radial distances, the *TE* results closely align with those predicted by the Fraunhofer (scalar) theory.

If the range of radial distance values is extended (not shown in Fig. 7.4), the peak intensity predicted by the vectorial analysis appears at $r_T \simeq 500\,\mu$m for both *TE* and *TM* polarizations. In contrast, the peak intensity predicted by the scalar Fraunhofer theory occurs at $r_T \simeq 353\,\mu$m. The location of the scalar theory peak can be determined using the equation $k(r_T/z) - k\sin\theta_i = 0 \Longrightarrow r_T = z\sin\theta_i = 353.55\,\mu$m.

Similarly, the peak of the vectorial intensity can be estimated from Eq. (7.22), which occurs approximately when $\zeta = 0 \Longrightarrow r_T = z\tan\theta_i = 500\,\mu$m (assuming $\phi_i = 0$).

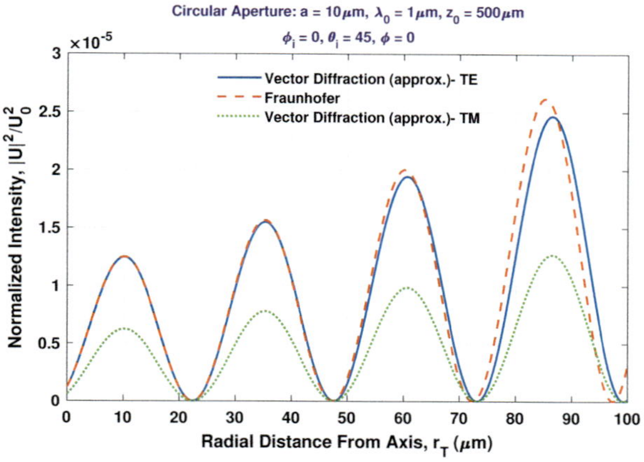

FIGURE 7.4: A circular aperture of radius $a = 10\,\mu$m on an opaque perfectly conducting screen at the $z = 0$ plane. The aperture is illuminated by a plane wave with $\theta_i = 45°$ and $\phi_i = 0°$. The polarization can be either *TE* or *TM*. The freespace wavelength is $\lambda_0 = 1\,\mu$m and the diffracted intensity is shown in the radial direction (for $\phi = 0$) at $z = 500\,\mu$m. Both *TE* and *TM* intensities are shown along with the scalar Fraunhofer result.

7.3 Babinet's Principle for Vector Fields

Babinet's principle was originally introduced in the *Scalar Diffraction* chapter (Chapter 5, Sec. 5.6) for scalar fields. For scalar electromagnetic fields, the principle states that the sum of the field produced by the aperture and the field produced by its complementary screen (obstacle) equals the scalar incident field. However, Babinet's principle for the vectorial electromagnetic field is significantly more complex [88, 104, 389, 390].

Consider a thin, perfectly conducting screen with an aperture, as illustrated in Fig. 7.5a. The area of the aperture is denoted by S_A, while the area of the metallic portion is denoted by S_M. The screen is positioned in the $z = 0$ plane. Conversely, the complementary screen, depicted in Fig. 7.5b, has an aperture area $S_A^c = S_M$ and a metallic area $S_M^c = S_A$. In both situations, the incident field propagates from left to right.

Let $(\vec{E}_i, \vec{H}_i)$, $(\vec{E}_s, \vec{H}_s)$, and $(\vec{E}_t, \vec{H}_t)$ represent the incident, scattered, and total fields, respectively, for the screen shown in Fig. 7.5a. Similarly, let $(\vec{E}_i^c, \vec{H}_i^c)$, $(\vec{E}_s^c, \vec{H}_s^c)$, and $(\vec{E}_t^c, \vec{H}_t^c)$ denote the incident, scattered, and total fields, respectively, for the complementary screen shown in Fig. 7.5b.

By definition, the total field is the sum of the incident and scattered fields in each region of either the original or the complementary screen. As outlined in Appendix 7B, the incident fields for the complementary screen are chosen to be the dual fields [for $\xi = -\pi/2$ in Eq. (7B.1)]. Consequently, the incident fields for the complementary screen are given as follows (where Z represents the characteristic impedance of the surrounding homogeneous and isotropic medium):

$$\vec{E}_i^c = -Z\vec{H}_i, \qquad \vec{H}_i^c = Z^{-1}\vec{E}_i. \tag{7.23}$$

Then, using the boundary conditions specified by Eqs. (7C.3) and (7C.4) of Appendix 7C, the following equations can be obtained:

$$\hat{\imath}_z \times \vec{E}_t = 0, \quad (x, y) \in S_M, \tag{7.24}$$

$$\hat{\imath}_z \times \vec{H}_t = \hat{\imath}_z \times \vec{H}_i, \quad (x, y) \in S_A, \tag{7.25}$$

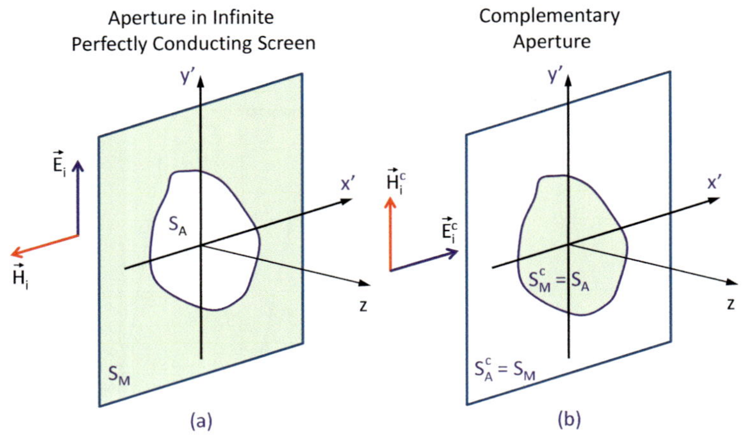

FIGURE 7.5: (a) A thin perfectly conducting planar screen with an aperture of area S_A. The area of the metallic part is S_M. The screen is located in the $z = 0$ plane. The incident electromagnetic field from the left is $(\vec{E}_i, \vec{H}_i)$. (b) The complementary screen where the metallic part area is $S_M^c = S_A$ and the.aperture area is $S_A^c = S_M$. The incident electromagnetic field from the left is $(\vec{E}_i^c, \vec{H}_i^c)$ where these fields are the dual of the ones in (a).

$$\hat{\imath}_z \times \vec{E}_t^c = 0, \quad (x, y) \in (S_M^c = S_A), \tag{7.26}$$

$$\hat{\imath}_z \times \vec{H}_t^c = \hat{\imath}_z \times \vec{H}_i^c, \quad (x, y) \in (S_A^c = S_M). \tag{7.27}$$

At this point, with the help of Eqs. (7.23) and Eq. (7.27), the following condition can be written:

$$\hat{\imath}_z \times Z\vec{H}_t^c = \hat{\imath}_z \times Z\vec{H}_i^c = \hat{\imath}_z \times \vec{E}_i, \quad (x, y) \in (S_A^c = S_M).$$

By adding Eq. (7.24) to the preceding equation, the following condition is obtained:

$$\hat{\imath}_z \times \left[Z\vec{H}_t^c + \vec{E}_t\right] = \hat{\imath}_z \times \vec{E}_i, \quad (x, y) \in S_M.$$

By similarly manipulating Eqs. (7.25) and (7.26), the following conditions are derived:

$$\hat{\imath}_z \times \left[\vec{H}_t - Z^{-1}\vec{E}_t^c\right] = \hat{\imath}_z \times \vec{H}_i. \quad (x, y) \in S_A.$$

From the last two equations, Babinet's principle can be expressed through the following two conditions, which relate the fields of the original and complementary screens:

$$Z\vec{H}_t^c + \vec{E}_t = \vec{E}_i, \quad (x, y) \in S_M \tag{7.28}$$

$$\vec{H}_t - Z^{-1}\vec{E}_t^c = \vec{H}_i, \quad (x, y) \in S_A. \tag{7.29}$$

At this point, the uniqueness theorem [391] must be applied to determine the fields at every point on the right-hand side ($z \geq 0$) of the screen. This theorem states that if the tangential electric fields are known on one part of a closed boundary (assuming the boundary is completed to the right by an infinite sphere) and the tangential magnetic fields are known on the remaining part, the electromagnetic field is uniquely determined. Additionally, it is necessary to assume slight losses in the region of interest and impose the Silver-Müller radiation conditions. Based on these assumptions, it can be demonstrated [104] that the following conditions hold for any point $(x, y, z \geq 0)$:

$$Z\vec{H}_t^c + \vec{E}_t = \vec{E}_i, \qquad (x, y, z \geq 0) \tag{7.30}$$

$$\vec{H}_t - Z^{-1}\vec{E}_t^c = \vec{H}_i, \qquad (x, y, z \geq 0). \tag{7.31}$$

Considering that the total fields are the sum of the incident and scattered fields, the following alternative expressions of Babinet's principle can be easily derived from Eqs. (7.30) and (7.31):

$$\vec{E}_s = -Z\vec{H}_t^c, \qquad \vec{H}_s = Z^{-1}\vec{E}_t^c, \qquad (x, y, z \geq 0) \tag{7.32}$$

$$\vec{E}_s^c = Z\vec{H}_t, \qquad \vec{H}_s^c = -Z^{-1}\vec{E}_t, \qquad (x, y, z \geq 0). \tag{7.33}$$

Example 7.3.1:

Consider a circular perfectly conducting disk of radius $a = 10\,\mu$m at $z = 0$ plane. The disk is illuminated by a *TE* or *TM* polarized plane wave with $\theta_i = 0°$, $\phi_i = 0°$, and $\lambda_0 = 1\,\mu$m. Determined the vectorial diffracted intensity along the longitudinal direction for $z > 500\,\mu$m (far field approximation). Furthermore, calculate the vectorial diffracted intensity along the radial directions for $z = 500\,\mu$m.

Solution: For this problem, the Babinet's principle is used. The initial configuration considers a circular aperture with radius $a = 10, \mu$m on a perfectly conducting screen. The complementary problem is that of a conducting disk. Since the incident plane wave is normally incident, both *TE* and *TM* polarizations are diffracted in the same manner by either the aperture or the disk. Therefore, without loss of generality, it is assumed that the incident electric field is polarized along the x-direction, $\vec{E}_0 = E_0\hat{\imath}_x$, and is given as $\vec{E}_i = E_0\hat{\imath}_x\, e^{-jkz}$. Then, the corresponding incident magnetic field is given as $\vec{H}_i = (E_0/Z)\hat{\imath}_y\, e^{-jkz}$. The scattered electric and magnetic fields along the z-direction are expressed as:

$$\vec{E}_s = j\frac{ka^2}{2}\frac{e^{-jkz}}{z}\, E_0\,\hat{\imath}_x,$$

$$\vec{H}_s = -\frac{1}{j\omega\mu}\vec{\nabla} \times \vec{E}_s = j\frac{ka^2}{2}\frac{e^{-jkz}}{z}\frac{E_0}{Z}\,\hat{\imath}_y. \tag{7.34}$$

The complementary problem corresponds to a perfectly conducting disk located in the $z = 0$ plane. However, the disk is illuminated by the dual incident fields: $\vec{E}_i^c = -Z\vec{H}_i = E_0\hat{\imath}_y\, e^{-jkz}$ and $\vec{H}_i^c = Z^{-1}\vec{E}_i = (E_0/Z)\hat{\imath}_x\, e^{-jkz}$ (with polarization rotated compared to the initial problem). The scattered electric field, $\vec{E}_s^c$, and magnetic field, $\vec{H}_s^c$, by the disk are related by Babinet's principle as:

$$\vec{E}_s^c = Z\left(\vec{H}_i + \vec{H}_s\right) = E_0 e^{-jkz}\hat{\imath}_y\left[1 + j\frac{ka^2}{2}\frac{1}{z}\right],$$

$$\vec{H}_s^c = -Z^{-1}\left(\vec{E}_i + \vec{E}_s\right) = -\frac{E_0}{Z}\,\hat{\imath}_x e^{-jkz}\left[1 + j\frac{ka^2}{2}\frac{1}{z}\right].$$

The intensity of the diffraction by the disk along the z-axis in the far-field can then be determined as:

$$I_{ff}^c = \frac{|\vec{E}_s^c|^2}{4Z} = \frac{|E_0|^2}{4Z}\left[1 + \left(\frac{ka^2}{2}\frac{1}{z}\right)^2\right].$$

The far-field intensity along the z-axis is identical to the intensity corresponding to scalar diffraction. Next, the far-field vectorial intensity along the radial direction at $z = 500\,\mu$m is sought. In this case, the scattered magnetic field by a circular aperture is again required. Instead of using Maxwell's equations to derive the magnetic field from the electric field, it is much simpler to use Eq. (7.18) and proceed in a manner analogous to the derivation of Eq. (7.21) for the electric field. The resulting scattered magnetic field by the circular aperture is then given by:

$$\vec{H}_s(\vec{r}) \simeq -j\frac{aJ_1(k\zeta a)}{\zeta}\frac{e^{-jkr}}{r}\left[(\vec{H}_0\cdot\hat{\imath}_r)\hat{\imath}_z - (\hat{\imath}_r\cdot\hat{\imath}_z)\vec{H}_0\right], \tag{7.35}$$

where $\zeta = \sin\theta$. Since for the circular aperture $\vec{E}_0 = E_0\,\hat{\imath}_x$, it follows that $\vec{H}_0 = (E_0/Z)\,\hat{\imath}_y$. Therefore, the last equation can be written as:

$$\begin{aligned}\vec{H}_s(\vec{r}) &\simeq -j\frac{aJ_1(k\zeta a)}{\zeta}\frac{e^{-jkr}}{r}\frac{E_0}{Z}\left[\sin\theta\sin\phi\,\hat{\imath}_z - \cos\theta\,\hat{\imath}_y\right] \\ &\simeq -j\frac{aJ_1(k\zeta a)}{\zeta}\frac{e^{-jkz}}{z}\exp\left(-jk\frac{r_T^2}{2z}\right)\frac{E_0}{Z}\left[\sin\theta\sin\phi\,\hat{\imath}_z - \cos\theta\,\hat{\imath}_y\right].\end{aligned}$$

According to Babinet's principle, the scattered electric field by the perfectly conducting disk (using the dual incident fields) is given by:

$$\vec{E}_s^c = Z\left(\vec{H}_i + \vec{H}_s\right) = E_{sy}^c\,\hat{\imath}_y + E_{sz}^c\,\hat{\imath}_z, \qquad \text{where,}$$

$$E_{sy}^c = E_0 e^{-jkz}\left[1 + j\frac{aJ_1(k\zeta a)}{\zeta}\exp\left(-jk\frac{r_T^2}{2z}\right)\frac{\cos\theta}{z}\right]$$

$$E_{sz}^c = -jE_0 e^{-jkz}\frac{aJ_1(k\zeta a)}{\zeta}\exp\left(-jk\frac{r_T^2}{2z}\right)\frac{\sin\theta\sin\phi}{z}.$$

Then the vector diffracted intensity is given by $I_{ff}^c = (|E_0|^2/4Z)(|E_{sy}^c|^2 + |E_{sz}^c|^2)$. This intensity is shown in Fig. 7.6, along with the result from scalar diffraction theory for the Fraunhofer regime. It can be observed that the vectorial intensity is calculated along the y-axis ($\phi = \pi/2$) and exhibits a similar dependence along the radial direction when compared to its scalar counterpart. However, the vectorial and scalar intensities are not closely matched, except along the z-axis.

7.4 Vectorial Plane-Wave Spectrum

The method of the plane-wave spectrum (or angular spectrum) was initially introduced for scalar fields in the *Scalar Diffraction* chapter (Chapter 5). In this section, the method is extended to the vectorial electromagnetic case [86, 104, 383]. Any electromagnetic field can be decomposed into a superposition of plane waves using Fourier transform techniques. Thus, complex waveforms can be represented as sums (typically infinite) of plane waves.

Let $\vec{E}_{xy}(x, y; z = 0)$ be a known, transverse to the z-axis (tangential to the $z = 0$ plane), electric field vector. Therefore, the $\vec{E}_{xy}$ field can be expressed as $\vec{E}_{xy} = E_x\,\hat{\imath}_x + E_y\,\hat{\imath}_y$ at the $z = 0$ plane. This electric field can be easily decomposed into the following summation of

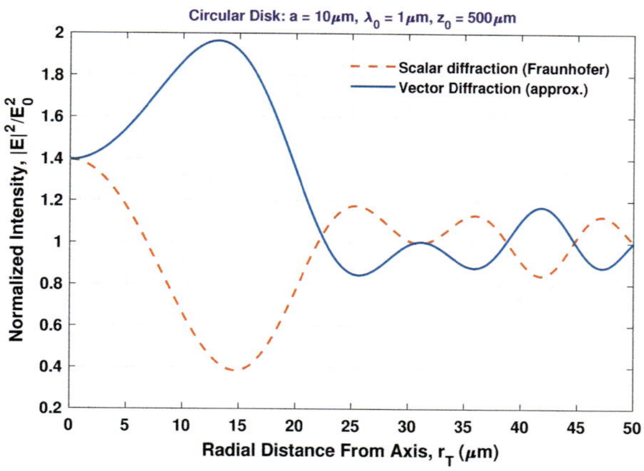

FIGURE 7.6: The far-field intensity for a circular disk illuminated by a normally incident plane wave using the dual fields to the circular aperture. The intensity is shown along the radial direction (along the y-axis, $\phi = \pi/2$) at $z = 500\,\mu$m. The solid line is the vectorial intensity while the dashed line is the scalar theory Fraunhofer intensity.

plane waves:

$$\vec{E}_{xy}(x, y; z = 0) = E_x(x, y; z = 0)\,\hat{\imath}_x + E_y(x, y; z = 0)\,\hat{\imath}_y$$

$$= \frac{1}{(2\pi)^2} \int_{-\infty}^{+\infty}\int_{-\infty}^{+\infty} \vec{\boldsymbol{E}}_{xy}(k_x, k_y) e^{-j(k_x x + k_y y)} dk_x dk_y, \quad \text{where,} \tag{7.36}$$

$$\vec{\boldsymbol{E}}_{xy}(k_x, k_y) = \mathcal{F}\left\{\vec{E}_{xy}(x, y; z = 0)\right\} = \int_{-\infty}^{+\infty}\int_{-\infty}^{+\infty} \vec{E}_{xy}(x, y; z = 0) e^{j(k_x x + k_y y)} dx dy. \tag{7.37}$$

Here, $\vec{\boldsymbol{E}}_{xy}(k_x, k_y) = \boldsymbol{E}_x(k_x, k_y)\,\hat{\imath}_x + \boldsymbol{E}_y(k_x, k_y)\,\hat{\imath}_y$, represents the Fourier transform of $\vec{E}_{xy}(x, y; z = 0)$. In this chapter, boldface vector fields generally denote the Fourier transforms of vector fields , unless otherwise specified. By applying the vectorial Helmholtz equation in a manner analogous to the scalar case, the electric field propagating along the z-direction can be written as:

$$\vec{E}(x, y; z) = \frac{1}{(2\pi)^2} \int_{-\infty}^{+\infty}\int_{-\infty}^{+\infty} \vec{\boldsymbol{E}}(k_x, k_y) e^{-j(k_x x + k_y y + k_z z)} dk_x dk_y, \tag{7.38}$$

where k_z is given by (for propagation to the positive z-direction):

$$k_z = \begin{cases} \left(k^2 - k_x^2 - k_y^2\right)^{1/2}, & \text{when } k_x^2 + k_y^2 \le k^2, \\ -j\left(k_x^2 + k_y^2 - k^2\right)^{1/2}, & \text{when } k_x^2 + k_y^2 > k^2. \end{cases} \tag{7.39}$$

Of course, the case of real k_z corresponds to propagating plane waves, while the case of the imaginary k_z corresponds to evanescent waves. However, the difference from the scalar case is that for vectorial plane waves, Gauss's law (source-free) must be satisfied. As a result, the electric field components of the plane waves and their wavevector components must satisfy the equation $\vec{\nabla} \cdot \vec{\boldsymbol{E}} = 0 \longrightarrow k_x \boldsymbol{E}_x + k_y \boldsymbol{E}_y + k_z \boldsymbol{E}_z = 0$, where $\vec{k} = k_x\,\hat{\imath}_x + k_y\,\hat{\imath}_y + k_z\,\hat{\imath}_z$. The corresponding propagated vector electric field for a distance $z \geq 0$ can then be written as:

$$\vec{E}(x,y;z) = \frac{1}{(2\pi)^2}\int_{-\infty}^{+\infty}\int_{-\infty}^{+\infty}\left[\boldsymbol{E}_x\,\hat{\imath}_x + \boldsymbol{E}_y\,\hat{\imath}_y - \frac{k_x\boldsymbol{E}_x + k_y\boldsymbol{E}_y}{k_z}\,\hat{\imath}_z\right]e^{-j(k_x x + k_y y + k_z z)}dk_x dk_y. \tag{7.40}$$

Since the plane wave components, $\vec{\boldsymbol{E}}(k_x, k_y)$, satisfy Maxwell's equations, the propagated vector electric field, $\vec{E}(x, y; z)$, does as well (due to linear superposition). Clearly, for $z = 0$ the tangential electric field continuity is satisfied. The vector magnetic field can be easily determined from Maxwell's equation $\vec{\nabla} \times \vec{E} = -j\omega\mu\vec{H}$, resulting in the following expression:

$$\vec{H}(x,y;z) = \frac{1}{(2\pi)^2}\frac{1}{Zk}\int_{-\infty}^{+\infty}\int_{-\infty}^{+\infty}\vec{k}\times\left[\boldsymbol{E}_x\,\hat{\imath}_x + \boldsymbol{E}_y\,\hat{\imath}_y - \frac{k_x\boldsymbol{E}_x + k_y\boldsymbol{E}_y}{k_z}\,\hat{\imath}_z\right]$$

$$e^{-j(k_x x + k_y y + k_z z)}dk_x dk_y, \tag{7.41}$$

where Z is the characteristic impedance of the medium where the wave is propagating. Since the magnetic field is also comprised of a collection of plane waves, it too satisfies Maxwell's equations. Naturally, a similar procedure could be followed by using the tangential magnetic fields at the boundary instead of the electric fields. The resulting expressions for the propagated magnetic and electric fields [104] are:

$$\vec{H}(x,y;z) = \frac{1}{(2\pi)^2}\int_{-\infty}^{+\infty}\int_{-\infty}^{+\infty}\left[\boldsymbol{H}_x\,\hat{\imath}_x + \boldsymbol{H}_y\,\hat{\imath}_y - \frac{k_x\boldsymbol{H}_x + k_y\boldsymbol{H}_y}{k_z}\,\hat{\imath}_z\right]e^{-j(k_x x + k_y y + k_z z)}dk_x dk_y, \tag{7.42}$$

$$\vec{E}(x,y;z) = -\frac{1}{(2\pi)^2}\frac{Z}{k}\int_{-\infty}^{+\infty}\int_{-\infty}^{+\infty}\vec{k}\times\left[\boldsymbol{H}_x\,\hat{\imath}_x + \boldsymbol{H}_y\,\hat{\imath}_y - \frac{k_x\boldsymbol{H}_x + k_y\boldsymbol{H}_y}{k_z}\,\hat{\imath}_z\right]$$

$$e^{-j(k_x x + k_y y + k_z z)}dk_x dk_y. \tag{7.43}$$

In the last equations $\boldsymbol{H}_w$ $(z = x, y)$ are the Fourier transforms of the transverse magnetic components at the $z = 0$ plane. It is important to note that in order to apply the vectorial plane wave spectrum (VPWS) method, the Kirchhoff approximation is again necessary for the fields on the aperture.

Example 7.4.1:

Example 7.2.1 is used again, but the analysis should now be performed using the vectorial plane wave spectrum (VPWS) method. All parameters are identical to those in example 7.2.1. The new task is to determine the vectorial diffracted intensity at $z = 500\,\mu$m along the radial direction.

Solution: For $\phi_i = 0$, the components of the incident plane wave wavevector are $k_x = k\sin\theta_i\cos\phi_i = k\sin\theta_i$, $k_y = k\sin\theta_i\sin\phi_i = 0$, and $k_z = k\cos\theta_i$. Additionally, the required polarization vectors are $\hat{u}_{TE} = \cos\phi_y\,\hat{\imath}_y - \sin\phi_i\,\hat{\imath}_x = \hat{\imath}_y$, and $\hat{u}_{TM} = \cos\phi_i\cos\theta_i\,\hat{\imath}_x + \sin\phi_i\cos\theta_i\,\hat{\imath}_y - \sin\theta_i\,\hat{\imath}_z = \cos\theta_i\,\hat{\imath}_x - \sin\theta_i\,\hat{\imath}_z$, for *TE* and *TM* polarization, respectively. Therefore, the tangential components of the incident electric fields on the $z = 0$ plane are: $E_x(x, y; z = 0) = 0$, and $E_y(x, y; z = 0) = E_0 e^{-jk\sin\theta_i x}w(r_T)$ in the case of *TE* polarization, and $E_x(x, y; z = 0) = E_0\cos\theta_i e^{-jk\sin\theta_i x}w(r_T)$ and $E_y(x, y; z = 0) = 0$ in the case of *TM* polarization, where $w(r_T) = \text{circ}(r_T/a)$ and "circ" represents the circle function of the aperture. Using Eq. (7.37), the 2D Fourier transforms $\vec{\boldsymbol{E}}_x(k_x, k_y)$ and $\vec{\boldsymbol{E}}_y(k_x, k_y)$ can be determined. Finally, by utilizing Eq. (7.40), the electric field components at distance z away from the circular aperture can be computed. The vectorial intensity $I(x, y; z) = (1/4Z)|\vec{E}|^2$ can then be easily determined.

For the numerical values provided in this example, the number of points used in the 2D FFT (Fast Fourier transforms) was $2^{12} = 4096$ in each direction, while the sampled area in the xy-plane was a square with dimensions $L_x \times L_y$ where $L_x = L_y = 1300\,\mu\text{m}$. The resulting vectorial intensities are shown for both polarizations in Fig. 7.7 as a function of the distance x. The results for the vectorial intensities of Eq. (7.22) practically coincide with those of the VPWS method and are not shown. Furthermore, the scalar Fraunhofer intensity is also presented. The peak of the scalar theory differs substantially from the vector theory, as discussed in Example 7.2.1.

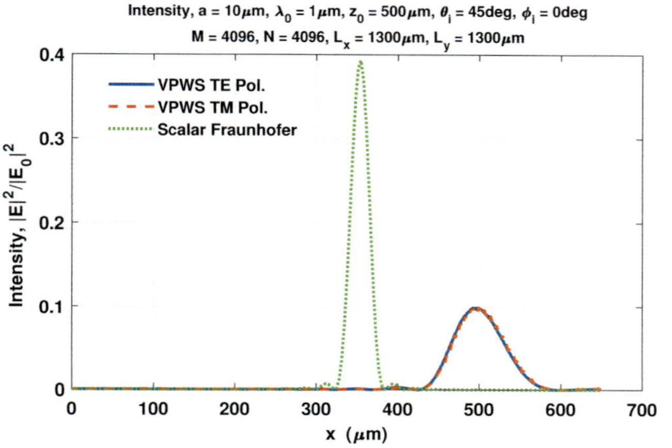

FIGURE 7.7: A circular aperture of radius $a = 10\,\mu\text{m}$ on an opaque perfectly conducting screen at $z = 0$ plane. The aperture is illuminated by a plane wave with $\theta_i = 45°$ and $\phi_i = 0°$. The polarization can be either *TE* or *TM*. The freespace wavelength is $\lambda_0 = 1\,\mu\text{m}$ and the diffracted intensity is shown in the radial direction (for $\phi = 0$) at $z = 500\,\mu\text{m}$. Both *TE* and *TM* intensities are calculated by the vectorial plane wave spectrum (VPWS) method and are shown along with the scalar Fraunhofer result.

7.5 Integral Methods Hierarchy

There is a plethora of integral methods available for studying the vector diffraction of electromagnetic waves. The purpose of this section is not to classify all existing methods exhaustively but to provide an overview of the most common approaches. These methods are divided into two main categories:

- **Exact Methods:** These methods solve Maxwell's equations in integral form using numerical procedures without making any theoretical approximations.
- **Approximate Methods:** These methods involve some common approximations, such as Kirchhoff's approximation for boundary fields (usually at an aperture) or approximations of the corresponding Green's functions.

In Fig. 7.8, a simple hierarchy diagram is shown. Some of the integral methods have been discussed briefly in previous sections, such as the vector integral formulations and the VPWS method. Others, such as the finite-element method, are not covered here and can be subjects of specialized books. In this chapter, the 2D BEM method is briefly presented along with a few numerical examples.

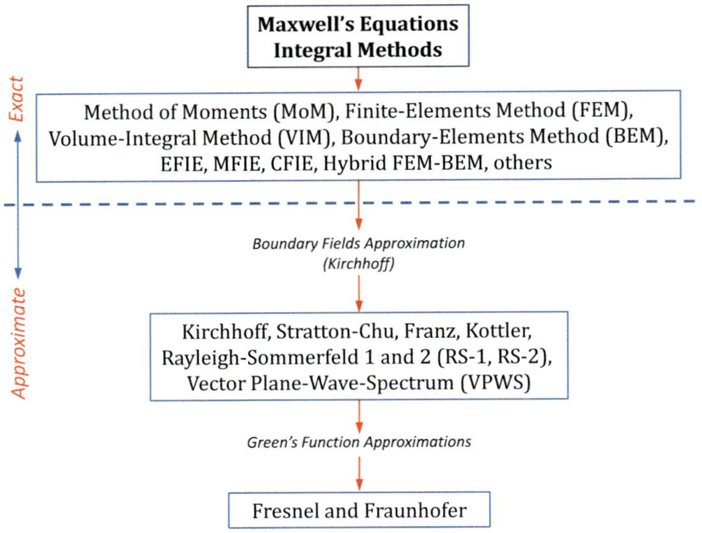

FIGURE 7.8: A simple hierarchical diagram of some common integral diffraction methods. The dashed horizontal line separates the rigorous (exact) from the approximate methods.

7.5.1 2D Boundary Element Method (BEM)

The Boundary Element Method (BEM) is a numerical technique for solving linear partial differential equations that have been reformulated as integral equations. This method is particularly advantageous for addressing problems involving complex geometries and boundary conditions. Unlike domain-based methods, such as the Finite Element Method (FEM), BEM requires the discretization of only the boundary of the domain, rather than the entire domain itself [378, 392–395].

Consider initially a closed two-dimensional homogeneous object (region S_1) with permittivity ϵ_1, surrounded by an infinite homogeneous region S_2 with permittivity ϵ_2, as

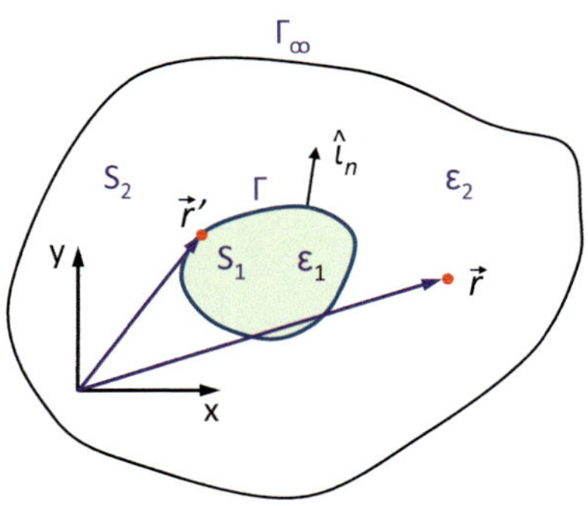

FIGURE 7.9: A closed 2D region S_1 with permittivity ϵ_1 surrounded by region S_2 (which closes at infinity with boundary Γ_∞) and permittivity ϵ_2. Both regions are homogeneous, isotropic, and non-magnetic.

depicted in Fig. 7.9. An electromagnetic wave is incident from the surrounding region and is diffracted by the two-dimensional object. Both regions are assumed to be isotropic, allowing two distinct polarizations to be analyzed: (a) *TE* polarization, where the electric field is given by $\vec{E} = E_z(x,y)\hat{\imath}_z$, and (b) *TM* polarization, where the magnetic field is given by $\vec{H} = H_z(x,y)\hat{\imath}_z$. For either polarization, the electric or magnetic field will be denoted as $\phi = E_z$ or $\phi = H_z$ in the following analysis [395]. Using the Green's identity (as in the case of scalar diffraction) it is straightforward to show [393]:

$$-\phi_1(\vec{r}) = \int_\Gamma \left[\phi_1(\vec{r}')\,\vec{\nabla}' G_1(\vec{r}_1,\vec{r}') - p_1 G_1(\vec{r}_1,\vec{r}')\,\vec{\nabla}'\phi_1(\vec{r}')\right]\cdot\hat{\imath}_n\,d\ell, \quad \vec{r}_1 \in S_1 \tag{7.44}$$

$$\phi_2(\vec{r}_2) = \phi^{\text{inc}}(\vec{r}_2) + \int_\Gamma \left[\phi_\Gamma(\vec{r}')\,\vec{\nabla}' G_2(\vec{r}_2,\vec{r}') - p_2 G_2(\vec{r}_2,\vec{r}')\,\vec{\nabla}'\phi_2(\vec{r}')\right]\cdot\hat{\imath}_n\,d\ell = 0, \quad \vec{r}_2 \in S_2, \tag{7.45}$$

$$G_i(\vec{r}_i,\vec{r}') = -\frac{j}{4}\,H_0^{(2)}\left(k_i\,|\vec{r}_i - \vec{r}'|\right), \quad (i = 1,2). \tag{7.46}$$

Here, ϕ_i represents the total field (electric for *TE* polarization and magnetic for *TM* polarization) in region i, while ϕ_{inc} denotes the incident field (electric or magnetic) arriving from Γ_∞. The parameter p_i is defined as $p_i = 1$ for *TE* polarization and $p_i = \epsilon_i/\epsilon_0$ for *TM* polarization, where $i = 1,2$. The vectors $\vec{r}_1$, $\vec{r}_2$, and $\vec{r}'$ correspond to the position vectors of points located in regions 1, 2, and on the boundary Γ, respectively. The symbol $H_0^{(2)}$ denotes the zeroth-order Hankel function of the second kind, and $k_i = \omega\sqrt{\epsilon_i\mu_0}$ represents the wavenumber in region i.

For the case of an open region S_1, as depicted in Fig. 7.10(a), the half-space regions S_1 and S_2 are also assumed to be isotropic and homogeneous. The interface Γ between the two dielectrics is infinite, with a normal unit vector $\hat{\imath}_n$. It is further assumed that the incident field, with angular frequency ω, originates from region S_1 and is diffracted by the boundary Γ.

Using Maxwell's equations and applying Green's formula to regions S_1, S_2, and contours C_1, C_2 (Fig. 7.10(a)), while considering the radiation condition (in 2D) for the scattered field and incorporating the asymptotic approximations for the Hankel function, the following integral equations are derived:

$$-\phi_1(\vec{r}_1) + \int_\Gamma \left[\phi_1(\vec{r}')\,\vec{\nabla}' G_1(\vec{r}_1,\vec{r}') - p_1 G_1(\vec{r}_1,\vec{r}')\,\vec{\nabla}'\phi_1(\vec{r}')\right]\hat{\imath}_n\cdot d\ell = -\phi^{\text{inc}}(\vec{r}_1), \quad \vec{r}_1 \in S_1 \tag{7.47}$$

$$\phi_2(\vec{r}_2) + \int_\Gamma \left[\phi_2(\vec{r}')\,\vec{\nabla}' G_2(\vec{r}_2,\vec{r}') - p_2 G_2(\vec{r}_2,\vec{r}')\,\vec{\nabla}'\phi_2(\vec{r}')\right]\hat{\imath}_n\cdot d\ell = 0, \quad \vec{r}_2 \in S_2. \tag{7.48}$$

In the above equations, it is observed that the contributions of the integrals over the $C_i - \Gamma$ portion of the contours are zero. Equations (7.44)-(7.45) and (7.47)-(7.48) can also be applied to points on the boundary Γ, i.e., when $\vec{r}_1 = \vec{r}_2 = \vec{r}_\Gamma$. The boundary conditions at Γ are given as [395]:

$$\phi_1(\vec{r}_\Gamma) = \phi_2(\vec{r}_\Gamma) \equiv \phi_\Gamma, \tag{7.49}$$

$$\frac{1}{p_1}\vec{\nabla}\phi_1\cdot\hat{\imath}_n = \frac{1}{p_2}\vec{\nabla}\phi_2\cdot\hat{\imath}_n \equiv \psi_\Gamma. \tag{7.50}$$

For the computation of the integrals over Γ at the boundary points $\vec{r}_\Gamma$, special care must be taken due to the singularities of the Green's function when $\vec{r} = \vec{r}_\Gamma = \vec{r}'$. In such

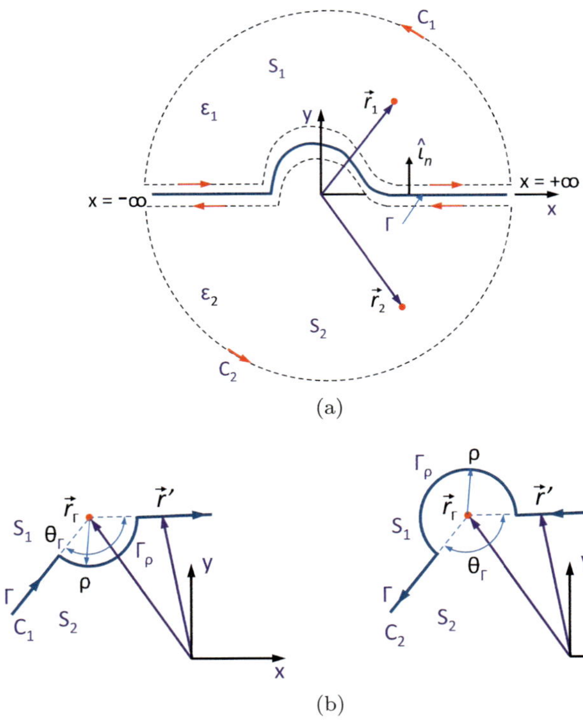

FIGURE 7.10: (a) A boundary line Γ between two 2D regions S_1 and S_2. Both regions are homogeneous, isotropic, and non-magnetic with permittivities of ϵ_1 and ϵ_2, respectively. (b) Treatment of a boundary node for the principal value integrals calculations.

cases, principal value integrals must be employed for the corresponding integrals. Example boundary points are illustrated in Fig. 7.10(b), where arcs of radius $\rho \to 0$ are shown around the integration point $\vec{r}_\Gamma$. Using Appendix 7D, Eqs. (7.44)-(7.45) and (7.47)-(7.48) for points on the boundary Γ can be expressed as:

$$\left(\frac{\theta_\Gamma}{2\pi} - 1\right) \phi_\Gamma(\vec{r}_\Gamma) + \mathcal{PV}\!\int_\Gamma \left[\phi_\Gamma(\vec{r}^{\,\prime}_\Gamma)\, \vec{\nabla}' G_1(\vec{r}_\Gamma, \vec{r}^{\,\prime}_\Gamma) - p_1 G_1(\vec{r}_\Gamma, \vec{r}^{\,\prime}_\Gamma)\, \vec{\nabla}' \phi_\Gamma(\vec{r}^{\,\prime}_\Gamma)\right] \hat{\imath}_n \cdot d\ell$$
$$= -\phi^{\text{inc}}(\vec{r}_\Gamma), \tag{7.51}$$

$$\frac{\theta_\Gamma}{2\pi}\, \phi_\Gamma(\vec{r}_\Gamma) + \mathcal{PV}\!\int_\Gamma \left[\phi_\Gamma(\vec{r}^{\,\prime}_\Gamma)\, \vec{\nabla}' G_2(\vec{r}_\Gamma, \vec{r}^{\,\prime}_\Gamma) - p_2 G_2(\vec{r}_\Gamma, \vec{r}^{\,\prime}_\Gamma)\, \vec{\nabla}' \phi_1(\vec{r}^{\,\prime}_\Gamma)\right] \hat{\imath}_n \cdot d\ell = 0. \tag{7.52}$$

Here, the angle θ_Γ is shown in Fig. 7.10(b), and $\mathcal{PV}$ denotes the Cauchy principal value of the corresponding integrals. By dividing the boundary Γ into a series of elements (either linear or quadratic), the field ϕ_Γ and its derivative ψ_Γ on each element can be expanded within the element as (refer to Appendix 7D for details):

$$\phi_\Gamma = \boldsymbol{N}^T \boldsymbol{\phi}_\Gamma, \tag{7.53}$$

$$\psi_\Gamma = \boldsymbol{N}^T \boldsymbol{\psi}_\Gamma. \tag{7.54}$$

Here, $\boldsymbol{\phi}_\Gamma$ and $\boldsymbol{\psi}_\Gamma$ are matrices containing the field values and field-derivative values at the element nodes, respectively. The next step in the BEM process is to discretize the integral equations presented in Eqs. (7.51)-(7.52). This discretization is achieved by employing the expansions in Eqs. (7.53) and (7.54), which result in the following set of linear equations (refer to Appendix 7D for details on the case of quadratic elements):

$$\left(\frac{\theta_j}{2\pi}-1\right)\phi_{\Gamma j}+\sum_{i=1}^{M}\left\{\left[\boldsymbol{h}^{\boldsymbol{I}}\right]_i^T[\boldsymbol{\phi}]_i-\left[\boldsymbol{g}^{\boldsymbol{I}}\right]_i^T[\boldsymbol{\psi}]_i\right\}=-\phi^{\mathrm{inc}}(\vec{r}_{\Gamma j}),\quad j=1,2,\cdots M \tag{7.55}$$

$$\frac{\theta_j}{2\pi}\phi_{\Gamma j}+\sum_{i=1}^{M}\left\{\left[\boldsymbol{h}^{\boldsymbol{II}}\right]_i^T[\boldsymbol{\phi}]_i-\left[\boldsymbol{g}^{\boldsymbol{II}}\right]^T[\boldsymbol{\psi}]_i\right\}=0,\quad j=1,2,\cdots M. \tag{7.56}$$

In the preceding equations, M represents the number of nodes on the boundary Γ, while $[\boldsymbol{\phi}]$ and $[\boldsymbol{\psi}]$ are column vectors of dimension $M\times 1$ that contain the unknown field and field-derivative values at the nodes. Equations (7.55) and (7.56) form a system of $2M$ linear equations with $2M$ unknowns (the node field and field-derivative values), which can be written as:

$$[\boldsymbol{A}]\begin{bmatrix}[\boldsymbol{\phi}]\\[\boldsymbol{\psi}]\end{bmatrix}=[\boldsymbol{b}], \tag{7.57}$$

that can be solved. The matrix $[\boldsymbol{A}]$ is of dimension $2M\times 2M$. The solution to the system of equations specifies the field and field-derivative values at all boundary nodes. Subsequently, the field and field-derivatives at any point on the boundary Γ can be determined using the expansions in Eqs. (7.53) and (7.54). The field values can then be computed using the discretized versions of Eqs. (7.44)-(7.45) and Eqs. (7.47)-(7.48) for any point in regions 1 or 2. Other field components (magnetic for *TE* polarization and electric for *TM* polarization) can be calculated via Maxwell's equations and numerical differentiation.

Example 7.5.1:

A dielectric cylinder with its axis along the z-direction is illuminated by a *TE*-polarized plane wave, $\vec{E}_i=\hat{\imath}_z\,E_0e^{+jk_1y}$ (with $E_0=1\,V/m$), where the freespace wavelength is $\lambda_0=1.0\,\mu\mathrm{m}$ and the wavenumber is $k_1=(2\pi/\lambda_0)n_1$. The plane wave is incident from the top of the cylinder and propagates in the $-y$ direction. The structure corresponds to a closed region 2, similar to Fig. 7.9. The refractive index of the cylinder is $n_2=2$, and the surrounding medium is air, with $n_1=1$. The objective is to find the magnitude of the total electric field E_z along the y-axis using the BEM approach and compare it with the analytical solution presented in Ref. [101], as summarized in Appendix 7D, Sec. 7D.4.

Solution: The cylinder structure is a special case of the closed-format geometry shown in Fig. 7.9. The boundary of the cylinder is approximated by a polygon with $N=100$ sides. By applying the BEM method, the boundary fields ϕ_Γ and their field derivatives ψ_Γ can be determined. The field at any point in space can then be computed using the integral equations (7.44) and (7.45), along with numerical integration of the required integrals (which, for points not on the boundary Γ, do not possess singularities).

For the analytical expressions of the electric field, the maximum number of Bessel functions used was 50, ensuring the convergence of the field series (see Appendix 7D). The resulting magnitude of the electric field $|E_z|$ along the y-axis is shown in Fig. 7.11, where excellent agreement between the BEM and analytical results is observed.

Example 7.5.2:

Design a 10-zone 2D-diffractive lens with a focal length of $f=100\,\mu\mathrm{m}$ for a plane wave incident from a medium of refractive index $n_1=1.5$ to air ($n_2=1$). The freespace wavelength

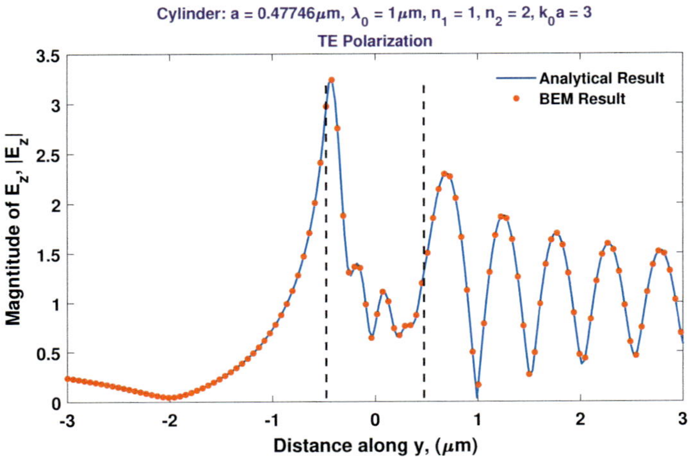

FIGURE 7.11: The electric field magnitude, $|E_z|$, along the y-direction for a dielectric cylinder illuminated by a plane wave of freespace wavelength of $\lambda_0 = 1.0\,\mu$m. The refractive index of the cylinder is $n_2 = 2$ and the surrounding medium is air, $n_1 = 1$. The cylinder has its axis along the z-direction and its radius is $a = 0.47746\,\mu$m. Both the BEM (dots) and the analytical (solid line) results are shown.

is $\lambda_0 = 1\,\mu$m. The plane wave is normally incident on the diffractive lens from the bottom (region 1) and propagates along the $+y$ direction into region 2, which is air. The goal of this example is to determine the intensity of the diffracted light in air along the y-axis for $x = 0$, using the BEM approach for both *TE* and *TM* polarization. The results obtained from the vectorial approach will then be compared with those derived from scalar theory.
Solution: Using the design process of Buralli and Morris [320] (see the *Scalar Diffraction* chapter, Chapter 5, Sec. 5.12), the continuous profile diffractive lens shown in Fig. 7.12(a) can be easily designed. This structure follows the open boundaries format, as depicted in Fig. 7.10(a). The region below the diffractive lens corresponds to region S_1 with refractive index $n_1 = 1.5$, while the region above the lens corresponds to region S_2 with refractive index $n_2 = 1$. The incident plane wave is $\vec{F} = \hat{\imath}_z\, F_0 e^{-k_1 y}$, where $\vec{F} = \vec{E}$ represents the electric field for *TE* polarization, and $\vec{F} = \vec{H}$ represents the magnetic field for *TM* polarization. Furthermore, $k_1 = k_0 n_1$, where k_0 is the freespace wavenumber.

The boundary of the diffractive lens is discretized into elements with a length smaller than $\lambda_0/5n_1$. Additionally, some extra space was added to the boundary Γ on the left and right (to minimize the edge effects of the infinite boundary truncation), resulting in a total width of the diffractive lens of $100\,\mu$m. A total of 973 elements were used. The resulting intensity, $I = |E_z/E_0|^2$ (for *TE* polarization) and $I = |H_z/H_0|^2$ (for *TM* polarization), are shown in Fig. 7.12(b).

In addition, using the approach described in the *Scalar Diffraction* chapter (Chapter 5), the scalar intensity, $I = |U/U_0|^2$ (where U represents either the electric or the magnetic field), is also shown in the same figure. The scalar results fall between the vectorial results.

The BEM method presented in this section is also known as the Electric Field Integral Equation (EFIE) method in the case of $\phi = E_z$ (*TE* polarization), or the Magnetic Field Integral Equation (MFIE) method in the case of $\phi = H_z$ (*TM* polarization). The presented BEM approach can be easily extended to multilayered closed or open 2D boundaries. In these cases, several modeling considerations need to be addressed. One issue involves the non-physical resonance phenomena associated with integral multilayered diffraction problems

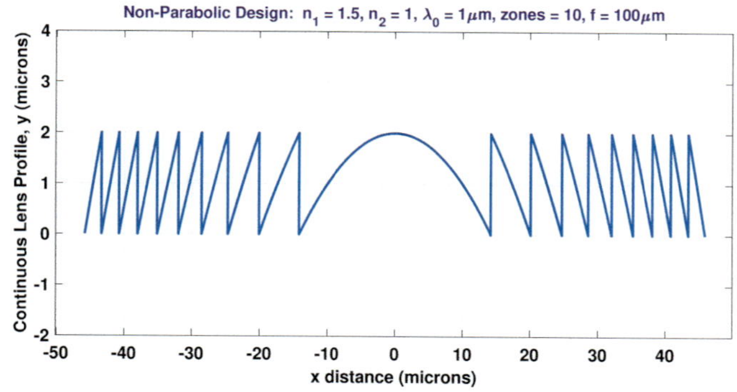

(a)

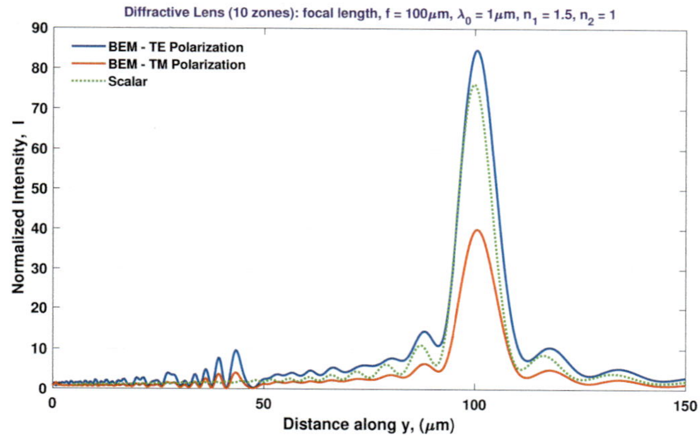

(b)

FIGURE 7.12: (a) The non-parabolic diffractive lens profile with 10 zones. (b) The normalized intensity $I = |E_z/E_0|^2$ (for *TE* polarization) or $I = |H_z/H_0|^2$ (for *TM* polarization), calculated via BEM, are shown along the y axis (for $x = 0$) for a 10-zone diffractive lens with design focal distance of $f = 100\,\mu$m. The scalar intensity $I = |U/U_0|^2$ is also shown.

[396]. Another issue is the potential breakdown of the discretized integral equations for structures with very thin features (compared to the wavelength). Most of these effects can be mitigated by using the Combined Field Integral Equation (CFIE) method, which combines the EFIE and MFIE methods. For more details, see Ref. [396].

Additionally, the BEM method can be extended to 3D scattering problems, where, instead of contour boundaries, boundary areas are used, and the corresponding 3D Green's functions are applied [379, 397].

7.6 Differential Methods Hierarchy

Similar to integral methods, numerous differential methods are available for vector diffraction. As with integral methods, the purpose of this chapter is not to provide an exhaustive list but to present some of the most commonly used techniques. Like integral methods, differential methods can also be categorized as either "Exact" and "Approximate." The

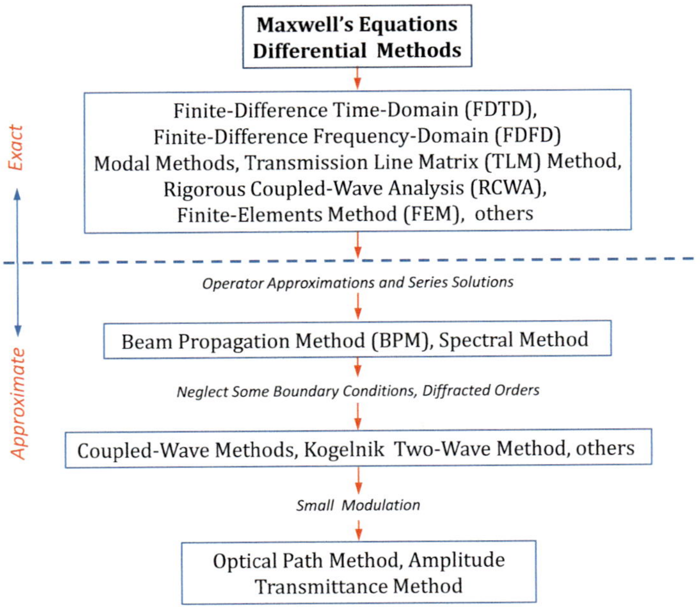

FIGURE 7.13: A simple hierarchical diagram of the most common differential diffraction methods. The dashed horizontal line separates the rigorous (exact) from the approximate methods.

exact methods do not involve any approximations in Maxwell's equations, whereas the approximate methods incorporate various approximations depending on the operators used, simplifications of certain boundary conditions, the exclusion of some diffracted orders, and other factors. Figure 7.13 presents a simple hierarchy diagram. Among the exact methods, the Finite-Difference Time-Domain (FDTD), Finite-Difference Frequency-Domain (FDFD), Modal methods and the Transmission Line Method (TLM) are included. Additionally, the Rigorous Coupled-Wave Analysis (RCWA) method is listed, which is both exact and highly versatile for diffraction by periodic structures.

In contrast, the approximate methods include the Beam Propagation Method (BPM), the Spectral method, Kogelnik's two-wave method, and others. Similar to the discussion on integral methods, only a few of these differential methods are covered in some detail. Specifically, the 2D FDFD method and the RCWA method are discussed. Several details regarding both methods are provided in the appendices of this chapter. Additionally, a few numerical examples are included to demonstrate the applicability of these methods.

7.6.1 Finite-Difference Frequency-Domain Method (FDFD)

The following analysis is based on Ref. [398]. The finite-difference frequency-domain (FDFD) method is a numerical solution to Maxwell's curl equations in their time-harmonic form, employing second-order accurate finite-difference approximations for all spatial derivatives on a Yee's grid [399]. Among the many references available, the specific FDFD approach utilized and modified in this section is described by Rumpf [374, 400].

The 2D-FDFD geometry is illustrated in Fig. 7.14. The actual computational region, of dimensions $L_x \times L_y$, is enclosed by a uniaxial anisotropic perfectly matched layer (UPML) [401], designed to absorb all waves exiting the computational domain. Within the computational region, various 2D objects, assumed to be infinite or very long in the z-direction, are represented as polygons. Smooth boundaries can also be effectively approximated by polygonal lines with an arbitrarily large number of segments.

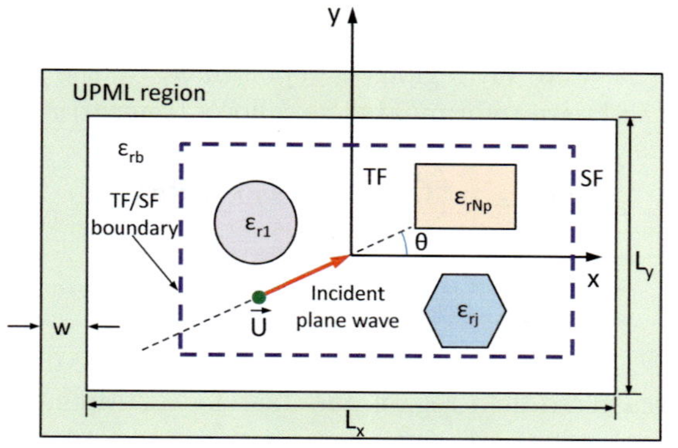

FIGURE 7.14: The basic 2D geometry for the 2D FDFD method. The computational region, of dimensions $L_x \times L_y$, is surrounded by the UPML region of width w. Inside the computational region various possible scatterers are shown along with the incident field plane wave $\vec{U}$. The dashed line parallelogram represents the boundary between Total Field (TF) and Scattered Field (SF).

The surrounding relative permittivity of the actual computational region is denoted as ε_{rb}, while the relative permittivity inside polygon j is represented by ε_{rj} ($j = 1, \cdots, N_p$, where N_p is the total number of polygons). The incident plane wave is also depicted in the same figure. By employing the normalized magnetic field defined by Rumpf [374, 400], $\vec{H}' = -jZ_0\vec{H}$ (where Z_0 is the freespace impedance), Maxwell's curl equations, which are valid even in the UPML regions, can be expressed as (see Appendix 7E):

$$\vec{\nabla} \times \vec{E} = k_0 \left[\tilde{s}(x,y)\right] \left[\tilde{\mu}_r(x,y)\right] \vec{H}' = k_0 \left[\tilde{\mu}_r'(x,y)\right] \vec{H}', \tag{7.58}$$

$$\vec{\nabla} \times \vec{H}' = k_0 \left[\tilde{s}(x,y)\right] \left[\tilde{\epsilon}_r(x,y)\right] \vec{E} = k_0 \left[\tilde{\epsilon}_r'(x,y)\right] \vec{E}. \tag{7.59}$$

Here, $\vec{E}$ and $\vec{H}'$ represent the electric field and the normalized magnetic field, respectively. The parameter $k_0 = 2\pi/\lambda_0$, is the wavenumber in freespace, with λ_0 denoting the freespace wavelength of the incident field (a plane wave in this case). The term $[\tilde{\mu}'(x,y)] = [\tilde{s}(x,y)][\tilde{\mu}_r(x,y)]$ is a 3×3 diagonal relative permeability matrix. In non-magnetic media, this matrix reduces to the identity matrix, except in the UPML region. Similarly, $[\tilde{\epsilon}_r'(x,y)] = [\tilde{s}(x,y)][\tilde{\epsilon}_r(x,y)]$ represents a 3×3 diagonal relative permittivity matrix, and $[\tilde{s}(x,y)]$ is the UPML tensor [400]. The incident field is given by $\vec{U}_{inc} = \hat{z}U_0 \exp[-jk_0\sqrt{\epsilon_{rb}}(x\cos\theta + y\sin\theta)]$, where $\vec{U} = \vec{E} = \hat{\imath}_z E_z$ for *TE* polarization and $\vec{U} = \vec{H}' = \hat{\imath}_z H_z'$ for *TM* polarization.

By utilizing the derivative operators $\tilde{\boldsymbol{D}}_{\boldsymbol{x}}^{\boldsymbol{E}}$, $\tilde{\boldsymbol{D}}_{\boldsymbol{y}}^{\boldsymbol{E}}$, $\tilde{\boldsymbol{D}}_{\boldsymbol{x}}^{\boldsymbol{H}}$, and $\tilde{\boldsymbol{D}}_{\boldsymbol{y}}^{\boldsymbol{H}}$, Eqs. (7.58) and (7.59) can be expressed in the following matrix form (refer to Appendix 7E, Sec. 7E.1):

$$\underbrace{\left\{\tilde{\boldsymbol{\epsilon}}_{\boldsymbol{rzz}} + \tilde{\boldsymbol{D}}_{\boldsymbol{x}}^{\boldsymbol{H}} \tilde{\boldsymbol{\mu}}_{\boldsymbol{ryy}}^{-1} \tilde{\boldsymbol{D}}_{\boldsymbol{x}}^{\boldsymbol{E}} + \tilde{\boldsymbol{D}}_{\boldsymbol{y}}^{\boldsymbol{H}} \tilde{\boldsymbol{\mu}}_{\boldsymbol{rxx}}^{-1} \tilde{\boldsymbol{D}}_{\boldsymbol{y}}^{\boldsymbol{E}}\right\}}_{=\tilde{\boldsymbol{A}}_{\boldsymbol{TE}}} \tilde{\boldsymbol{E}}_{\boldsymbol{z}} = 0 \Longrightarrow \tilde{\boldsymbol{A}}_{\boldsymbol{TE}} \tilde{\boldsymbol{E}}_{\boldsymbol{z}} = 0, \tag{7.60}$$

$$\underbrace{\left\{\tilde{\boldsymbol{\mu}}_{\boldsymbol{rzz}} + \tilde{\boldsymbol{D}}_{\boldsymbol{x}}^{\boldsymbol{E}} \tilde{\boldsymbol{\epsilon}}_{\boldsymbol{ryy}}^{-1} \tilde{\boldsymbol{D}}_{\boldsymbol{x}}^{\boldsymbol{H}} + \tilde{\boldsymbol{D}}_{\boldsymbol{y}}^{\boldsymbol{E}} \tilde{\boldsymbol{\epsilon}}_{\boldsymbol{rxx}}^{-1} \tilde{\boldsymbol{D}}_{\boldsymbol{y}}^{\boldsymbol{H}}\right\}}_{=\tilde{\boldsymbol{A}}_{\boldsymbol{TM}}} \tilde{\boldsymbol{H}}_{\boldsymbol{z}}' = 0 \Longrightarrow \tilde{\boldsymbol{A}}_{\boldsymbol{TM}} \tilde{\boldsymbol{H}}_{\boldsymbol{z}}' = 0. \tag{7.61}$$

All matrices appearing in Eqs. (7.60) and (7.61) are defined in Appendix 7E, Sec. 7E.1. Care must be taken with the positions of the grid used in each of the material (relative permittivity or permeability) matrices to ensure compatibility with Yee's cell [374, 400], as each field component is located at a different position within Yee's cell (see Fig. 7E.1).

By applying the total-field/scattered-field (TF/SF) formulation [374, 400], where the boundary between the SF and TF regions is represented by the dashed rectangle in Fig. 7.14, Eqs. (7.60) and (7.61) are reformulated as follows (refer to Appendix 7E, Sec. 7E.2):

$$\tilde{\boldsymbol{A}}_{\boldsymbol{TE}}\tilde{\boldsymbol{E}}_{\boldsymbol{z}} = \left(\tilde{\boldsymbol{Q}}\tilde{\boldsymbol{A}}_{\boldsymbol{TE}} - \tilde{\boldsymbol{A}}_{\boldsymbol{TE}}\tilde{\boldsymbol{Q}}\right)\tilde{\boldsymbol{E}}_{\boldsymbol{z,inc}}, \tag{7.62}$$

$$\tilde{\boldsymbol{A}}_{\boldsymbol{TM}}\tilde{\boldsymbol{H}}'_{\boldsymbol{z}} = \left(\tilde{\boldsymbol{Q}}\tilde{\boldsymbol{A}}_{\boldsymbol{TM}} - \tilde{\boldsymbol{A}}_{\boldsymbol{TM}}\tilde{\boldsymbol{Q}}\right)\tilde{\boldsymbol{H}}'_{\boldsymbol{z,inc}}. \tag{7.63}$$

Here, $\tilde{\boldsymbol{Q}}$ is a $N_xN_y \times N_xN_y$ diagonal matrix with "ones" on the diagonal corresponding to grid positions in the scattered-field region and "zeros" corresponding to grid positions in the total-field region [400]. The vectors $\tilde{\boldsymbol{E}}_{\boldsymbol{z,inc}}$ and $\tilde{\boldsymbol{H}}'_{\boldsymbol{z,inc}}$ are of size $N_xN_y \times 1$ and contain the values of the incident field at all grid points.

At this stage, the vectors $\tilde{\boldsymbol{E}}_{\boldsymbol{z}}$ and $\tilde{\boldsymbol{H}}'_{\boldsymbol{z}}$, which represent the unknowns of the scattering problem, contain either the scattered field (for grid points in the scattered-field region) or the total field (for grid points in the total-field region). The total or scattered fields at any grid point can then be determined by solving the linear system. The remaining field components (H'_x and H'_y for *TE* polarization, and E_x and E_y for *TM* polarization) can be directly calculated from the matrix form of Maxwell's equations [374, 400], as detailed in Appendix 7E, Sec. 7E.1.

It should be noted that caution is required because the different field components are defined at different positions within the Yee's cell. It is also important to highlight that all matrices in the aforementioned equations are either diagonal or banded, which makes sparse-matrix techniques an essential and efficient tool for implementing the FDFD algorithm. When grid points lie at boundaries between different materials, various averaging techniques can be employed to improve the accuracy of the FDFD solution [398].

Example 7.6.1:

A dielectric cylinder, aligned along the z-axis, is illuminated by a plane wave of the form $\vec{U}_i = \hat{\imath}_z U_0 e^{-jk_1 y}$, where $U_0 = E_0 = 1\,V/m$ for *TE* polarization or $U_0 = H_0 = 1\,A/m$ for *TM* polarization. The freespace wavelength is given as $\lambda_0 = 0.6328\,\mu$m and the surrounding refractive index is $n_1 = 1$. The incident region wavenumber is $k_1 = (2\pi/\lambda_0)n_1$. The cylinder has a refractive index $n_2 = \sqrt{2}$, and the wave propagates in the $+y$-direction, incident from the bottom of the cylinder. The task involves determining the magnitude of the total field (E_z for *TE* polarization or H_z for *TM* polarization) along the y-axis using the FDFD method for grid spacings $\Delta = \Delta_x = \Delta_y = \lambda_1/N$, where $\lambda_1 = \lambda_0/n_1$. Two values of N are to be used: $N = 10$ and $N = 20$ for *TE* polarization, and $N = 15$ and $N = 30$ for *TM* polarization. The FDFD results should be compared with the analytical solution provided in Ref. [101], as summarized in Appendix 7D, Sec. 7D.4.

Solution: The structure under consideration is a special case of the one depicted in Fig. 7.14, where the surrounding medium is air with $n_1 = \sqrt{\epsilon_{rb}} = 1$, and the cylinder is represented by a polygon with $n_2 = \sqrt{\epsilon_{N1}} = \sqrt{2}$. The *TE* or *TM* polarized plane wave is incident from the bottom of the cylinder and propagates in the $+y$-direction. To absorb the diffracted waves, a UPML region of width $w = \lambda_0$ was employed (see Fig. 7.14). Details of the UPML region can be found in Appendix 7E. The computational domain was chosen to be $L_x = L_y = 3\,\mu$m (excluding the UPML region). For the *TE* polarization, the grid size $\Delta x \times \Delta_y$ was set to $\Delta = \Delta x = \Delta y = \lambda_0/N$, with $N = 10$ or 20. In contrast, for the *TM* polarization, the grid was chosen such that $\Delta = \Delta x = \Delta y = \lambda_0/N$, with $N = 15$ or 30.

Using the FDFD method described in Sec. 7.6.1, the diffraction of the plane wave by the infinite cylinder was studied. The resulting magnitudes of E_z/H_z (for *TE*/*TM* polarization) are shown in Figs. 7.15(a) and 7.15(b), respectively. The dots and crosses indicate the FDFD results for the two discretization cases for each polarization, while the solid line represents the analytical result obtained using Bessel functions up to the 50th order. The magnitude of the electric or magnetic field is plotted along the y-direction for x near the center of the cylinder (note that x may not be exactly zero, as it depends on the computational region and grid size). The analytical results were computed at the same (x, y) points as the FDFD results.

The insets in the two plots show 2D color maps of the scattering for enhanced visualization. In both cases, the plane wave is incident from the bottom and propagates towards the top.

From the FDFD results, it is evident that the smaller grid sizes ($\Delta = \lambda_0/20$ for *TE* polarization and $\Delta = \lambda_0/30$ for *TM* polarization) yield better agreement with the analytical results. Nevertheless, even with the larger grid sizes, the FDFD results remain sufficiently accurate.

The focusing effect of the cylinder is evident from the peak of the diffracted magnitude. In the case of *TE* polarization, the focusing occurs near the top edge of the cylinder, whereas for *TM* polarization, the focusing happens slightly inside the top edge of the cylinder. It is of interest to calculate the focal distance of the cylinder using geometric optics. If the cylinder is treated as a thick cylindrical lens with $R_1 = -R_2 = a = 0.5\,\mu\text{m}$ and a thickness of $d = 2a = 1\,\mu\text{m}$, the focal distance is given by $f_2 \simeq 0.85\,\mu\text{m}$, and $q = 0.3536\,\mu\text{m}$ (refer to the *Geometrical Optics* chapter, Chapter 2). This focal line (instead of point due to the cylindrical nature of the lens) lies outside the lens, whereas the vectorial results indicate that the focal line is located either at the cylinder surface (for *TE*) or within its interior (for *TM*). This discrepancy is expected, given the size of the lens relative to the freespace wavelength. For geometric optics to be valid, the size of the lens should be much larger than the wavelength.

Example 7.6.2:

Consider a three-ridge structure (2D stripes, as the structure is infinite along the z-axis) placed on top of a substrate with a refractive index of $n_2 = \sqrt{2}$. The ridges have a square profile with dimensions $S_x \times S_y = 0.15\,\mu\text{m} \times 0.15\,\mu\text{m}$, and they share the substrate refractive index n_2. The central ridge is located at the origin, $(x, y) = (0, 0)$. The period of the ridges is $\Lambda = 0.3\,\mu\text{m}$, representing a finite-period grating. A normally incident plane wave (*TE* polarized) is described by $\vec{E}_{\text{inc}} = \hat{z}E_0 e^{+jk_1 y}$, where $k_1 = (2\pi/\lambda_0)n_1$, and the wave is incident from the top (air with $n_1 = 1.0$) and propagates along the negative y-axis. The freespace wavelength is $\lambda_0 = 0.633\,\mu\text{m}$. The objective of this example is to find the magnitude of the total electric field E_z along the y-axis using the FDFD method for two grid resolutions, $\Delta = \Delta_x = \Delta_y = \lambda_0/N$ (for example, $N = 30$ and $N = 60$), and compare the results with the solution obtained using the Boundary Element Method described in Sec. 7.5.1.

Solution: The structure in this example is a special case of the one shown in Fig. 7.14, where the surrounding region is again air with $n_1 = \sqrt{\epsilon_{rb}} = 1$, and the three ridges are represented by three polygons ($N_p = 3$) with $\epsilon_{N_i} = \sqrt{2}$ for $i = 1, 2, 3$. A *TE*-polarized plane wave is incident from the top of the three ridges and propagates in the $-y$-direction. To absorb the diffracted waves, a UPML region is utilized, similar to the one used in Example 7.6.1, with a width of $w = \lambda_0$.

The computational domain was chosen to be $L_x = L_y = 2.5\,\mu\text{m}$ (excluding the UPML region). The grid size $\Delta x \times \Delta_y$ was set to $\Delta = \Delta x = \Delta y = \lambda_0/N$, with $N = 30$ or 60. A smaller grid size is required due to the small dimensions of the ridges relative to the wavelength of light.

Special care is needed for the incident field in this case, as there are two semi-infinite regions. The simple plane wave $\vec{E}_{\text{inc}} = \hat{z}E_0 e^{+jk_1 y}$ will not provide an accurate solution.

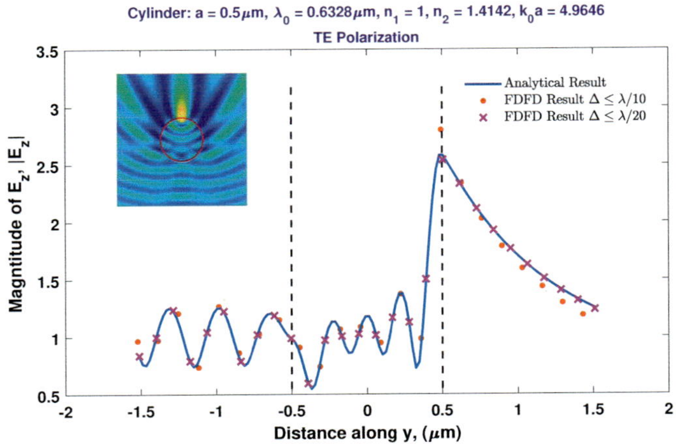

(a)

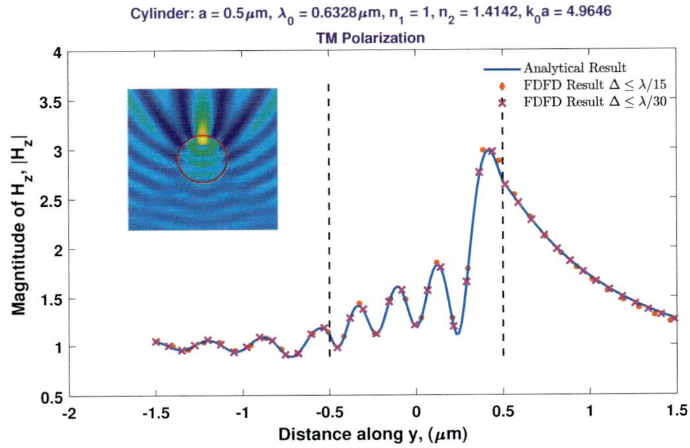

(b)

FIGURE 7.15: Scattering of a plane wave propagating along the y-direction, $\vec{U} = \hat{\imath}_z\, U_0 e^{-jk_1 y}$ with $k_1 = k_0 n_1$ ($k_0 = 2\pi/\lambda_0$) and $\lambda_0 = 0.6328\,\mu$m. $U = E$ (*TE* polarization) or $U = H$ (*TM* polarization). The refractive indices of the surrounding region and the cylinder are $n_1 = 1$ and $n_2 = \sqrt{2}$, respectively. The cylinder radius is $a = 0.5\,\mu$m. The plane wave is incident from the bottom of the cylinder and propagates along $+y$ direction. (a) *TE* polarization case. (b) *TM* polarization case.

Therefore, the following field should be used for the incident wave in the FDFD method:

$$\vec{E}_{\rm inc} = \hat{\imath}_z \begin{cases} E_0 e^{+jk_1 y} + r_{_{TE}} E_0 e^{-jk_1 y}, & y \geq 0, \\ t_{_{TE}} E_0 e^{+jk_1 y}, & y \leq 0, \end{cases} \tag{7.64}$$

where $r_{_{TE}}$ and $t_{_{TE}}$ are the *TE* reflection and transmission coefficients at the interface between regions of refractive indices n_1 and n_2. For more details, refer to Refs. [374, 400].

By applying the FDFD method described in Sec. 7.6.1, the diffraction of the plane wave by the three-ridge structure was studied. The resulting magnitude of E_z is shown in Fig. 7.16. The dots and crosses represent the FDFD results for the two discretization cases, while the solid line corresponds to the BEM result with boundary elements $\ell_e \leq \lambda/8$. The magnitude of the electric field is presented along the y-direction for $x = 0$. The inset shows

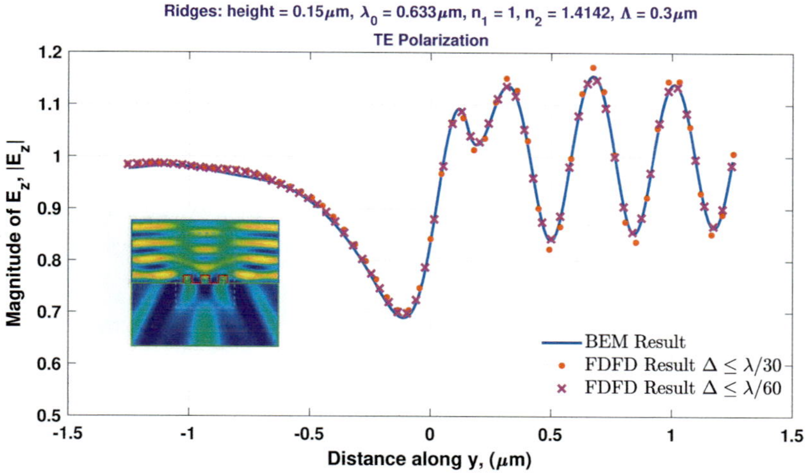

FIGURE 7.16: Scattering of a plane wave propagating along the $-y$-direction, $\vec{E} = \hat{\imath}_z\, E_0 e^{+jk_1 y}$ with $k_1 = k_0 n_1$ ($k_0 = 2\pi/\lambda_0$), and $\lambda_0 = 0.633\,\mu$m, by a three-ridge structure. The magnitude of the electric field, $|E_z|$, as function of the distance y is shown for $x = 0$. The inset shows the magnitude of the electric field in the xy-plane along with the three-ridges boundary. The dashed line in the inset shows the TF/SF boundary. The plane wave is incident from the top. The average scheme of Gaussian-blurring in conjunction to anisotropic smoothing [398] was the one that gave slightly better results compared to no averaging.

the 2D color graph of the scattering for better visualization. The three-ridge structure is also depicted in the inset, along with the TS/SF boundary used (indicated by the dashed line in the inset).

For the FDFD results, Gaussian blurring in conjunction with anisotropic smoothing was utilized [398], which yielded slightly better results than those obtained without permittivity averaging. It is evident that the case with $\Delta \leq \lambda/60$ performs better, as it captures the fine features of the structure more accurately.

7.7 Vector Diffraction by Gratings

The diffraction of electromagnetic waves by periodic structures remains a topic of great interest and practical significance due to its wide range of applications in fields such as acousto-optics, electro-optics, integrated optics, spectroscopy, optical computing, optical interconnects, and quantum electronics. Example grating devices include distributed feedback lasers, beam deflectors, beam modulators, waveguide couplers, spectral filters, wavelength multiplexers and demultiplexers, holographic beam combiners, head-up displays, photonic-crystal devices, and many others.

Optical gratings can be planar (slab or volume gratings) or surface-relief (corrugated gratings). The periodic modulation may occur in the permittivity (resulting in index of refraction gratings), in conductivity (affecting absorption), or in a combination of these properties. Diffraction gratings can be fabricated in dielectrics, semiconductors, metals, or even plasmas. They may be isotropic, uniaxial, biaxial-anisotropic, or gyrotropic, depending on the material properties. Gratings can have constant or varying modulation and may be cascaded or multiplexed.

The rigorous coupled-wave analysis (RCWA) [314,316,317] and modal analysis are exact methods for solving Maxwell's equations without any approximations. However, they are numerically intensive and do not provide much intuitive insight into the design of diffractive gratings. Over the years, several approximate methods have been developed to study diffraction by gratings, some of which offer analytical solutions under various approximations. A schematic diagram of the most popular differential methods is shown in Fig. 7.17. Some of the most widely used approximate methods include Kogelnik's two-wave coupled-wave approach [402], the Raman-Nath approach [403–407], and the transmittance method [265,408].

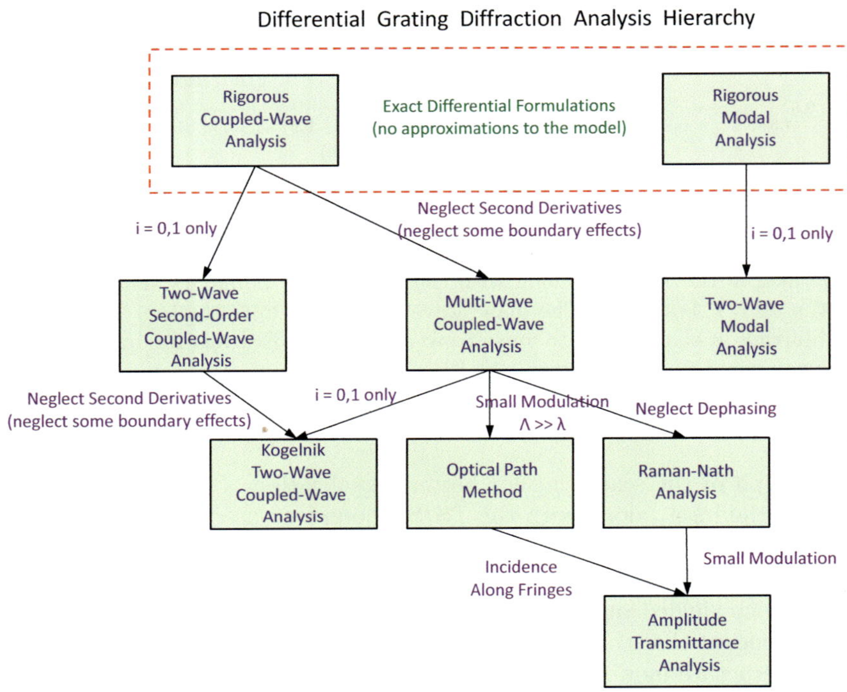

FIGURE 7.17: A diagram showing the hierarchy of the major differential theories used for grating diffraction analysis.

7.7.1 Grating Equation

When light is incident on a periodic structure (grating), it is diffracted into specific directions known as diffracted orders, both in the forward and backward directions. The grating equation describes the relationship between the angles at which light is diffracted by a grating (characterized by its periodicity Λ) and the wavelength of the light.

A simple volume diffraction grating is depicted in Figs. 7.18(a) and 7.18(b) for the 3D and 2D cases, respectively. The incident light is a plane wave. To account for anisotropic cases, the polarization is defined by the angle ψ of the $\vec{D}$ vector. Regions 1 and 3 are the boundary regions, assumed to be isotropic with refractive indices n_1 and n_3, respectively, while region 2 is the grating region (which can be anisotropic).

In Fig. 7.18(b), the 0-th and i-th forward and backward diffracted orders are shown. The integer i can take values $i = 0, \pm1, \pm2, \ldots$. The incident wavevector is $\vec{k}_{inc}$ and the angle of incidence is specified by θ_{inc} and α in the 3D case, or by θ_{inc} (with $\alpha = 0$) in the

2D case, respectively. For simplicity, the derivation of the grating equation is performed for the 2D case. Later, the grating equation will also be derived based on the phase-matching conditions of the electromagnetic field.

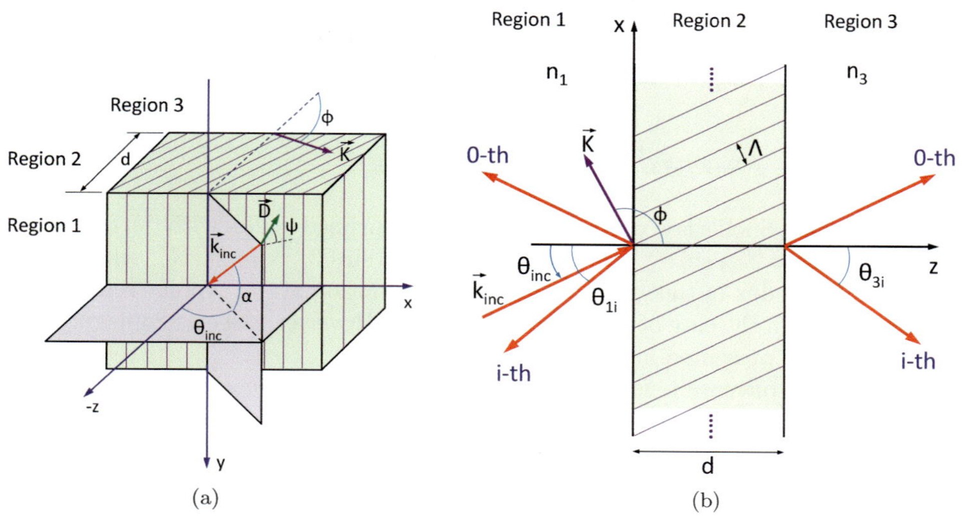

FIGURE 7.18: Diffraction of a plane wave by a volume grating. (a) 3D representation case showing the incident plane wave. The polarization of the plane wave is specified by the $\vec{D}$ vector to account for anisotropic cases. (b) 2D representation case where the 0-th and the i-th forward and backward diffracted orders are shown.

A simple ray approach is used to derive the grating equation. For this derivation, refer to Fig. 7.19a (for forward diffracted orders) and Fig. 7.19b (for backward diffracted orders). All the angles shown are assumed to be within the grating region (region 2). These internal angles are denoted with a prime, in contrast to the non-primed angles, which are defined in the boundary regions 1 and 3.

The grating vector $\vec{K}$ has a magnitude of $2\pi/\Lambda$ and a slant angle ϕ, as shown. The diagrams are presented in the xz-plane. The angle of incidence inside the grating region is θ'_{inc}. Two rays are shown in each case, along with a wavefront depicted by a red dashed

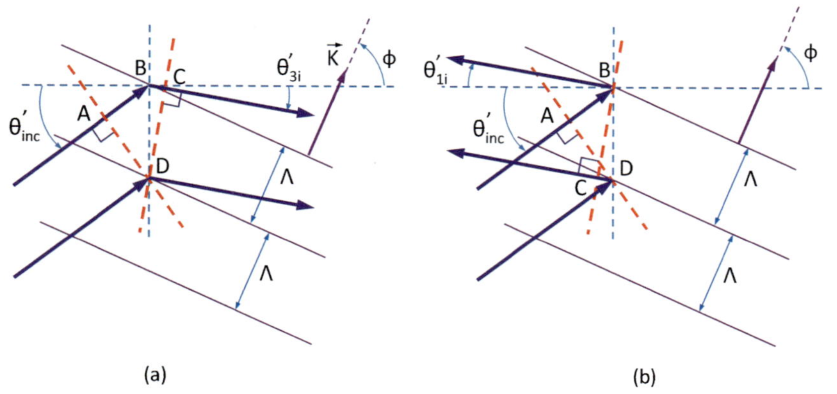

FIGURE 7.19: Path diagrams of an incident ray inside the grating region that is getting diffracted. The condition for constructive wavefront formation leads to the grating equation. (a) Case of a forward diffracted order. (b) Case of a backward diffracted order.

line. The incident wavefront AD is diffracted into the DC wavefront at an angle θ'_{3i} for the forward diffracted case. For the diffracted order to be formed, the optical path difference must be an integer multiple of the freespace wavelength to ensure constructive interference. This condition is expressed as (note that $DB = \Lambda/\sin\phi$):

$$n_0(AB + BC) = i\lambda_0 \Longrightarrow n_0\left(\frac{\Lambda}{\sin\phi}\sin\theta'_{inc} + \frac{\Lambda}{\sin\phi}\sin\theta'_{3i}\right) = i\,\lambda_0 \Longrightarrow$$

$$n_1\sin\theta_{inc} - n_3\sin\theta_{3i} = i\,\frac{\lambda_0}{\Lambda}\sin\phi. \tag{7.65}$$

In the last equation, Snell's law was used to replace the primed angles with the corresponding non-primed ones, i.e., $n_1\sin\theta_{\rm inc} = n_0\sin\theta'{\rm inc}$ and $n_3\sin\theta 3i = n_0\sin\theta'_{3i}$, where n_0 is the average refractive index in the grating region. The definition of the average refractive index may be ambiguous in the case of surface-relief gratings. However, the electromagnetic phase-matching condition presented later will address this issue.

In addition, a convention for measuring these angles was introduced. All angles are measured from the z-axis and are considered positive in the counter-clockwise direction (CCW). This accounts for the negative sign appearing in Eq. (7.65) for the θ'_{3i} angle. Similarly, using Fig. 7.19b, the corresponding grating equation for the backward diffracted orders is given by:

$$n_0(AB - DC) = i\lambda_0 \Longrightarrow n_0\left(\frac{\Lambda}{\sin\phi}\sin\theta'_{inc} - \frac{\Lambda}{\sin\phi}\sin\theta'_{1i}\right) = i\,\lambda_0 \Longrightarrow$$

$$n_1\sin\theta_{inc} + n_1\sin\theta_{1i} = i\,\frac{\lambda_0}{\Lambda}\sin\phi. \tag{7.66}$$

Since Eqs. (7.65) and (7.66) are valid for any integer value of i, there is only a finite number of diffracted orders that correspond to real diffraction angles θ_{3i} and θ_{1i}. These are the propagating diffracted orders. All other orders correspond to complex diffraction angles that characterize evanescent (non-propagating) diffracted orders.

Example 7.7.1:

Consider a lithium-niobate volume grating in the air ($n_1 = n_3 = 1.0$), where the average refractive index of the grating is $n_0 = 2.337$. The grating period is given as $\Lambda = 2\,\mu$m, and the slant angle is $\phi = 90°$ (i.e., the grating is unslanted). The angle of incidence is $\theta_{inc} = 10°$, and the freespace wavelength of the incident plane wave is $\lambda_0 = 0.5145\,\mu$m. The objective is to determine the angles of all the propagating forward and backward diffracted orders.

Solution: Using equations (7.65) and (7.66), it is straightforward to determine the directions of all propagating diffracted orders, both in the forward and backward directions. The propagating orders must satisfy the following inequalities (for real angles):

$$\frac{n_3\Lambda}{\lambda_0}\left(\frac{n_1}{n_3}\sin\theta_{inc} - 1\right) \le i \le \frac{n_3\Lambda}{\lambda_0}\left(\frac{n_1}{n_3}\sin\theta_{inc} + 1\right), \quad \text{(forward orders)},$$

$$\frac{n_1\Lambda}{\lambda_0}(\sin\theta_{inc} - 1) \le i \le \frac{n_1\Lambda}{\lambda_0}(\sin\theta_{inc} + 1), \quad \text{(backward orders)}.$$

Applying the inequalities above, the propagating orders in both the forward and backward directions satisfy $-3.21 \le i \le 4.56$. This is confirmed in Table 7.1, which also includes the corresponding diffraction angles θ_{3i} and θ_{1i} in degrees.

TABLE 7.1: Diffraction Angles of Example 7.7.1

i	Forward Angles θ_{3i} (deg)	Backward Angles θ_{1i} (deg)
-4	evanescent	evanescent
-3	70.9788	-70.9788
-2	43.4837	-43.4837
-1	25.5246	-25.5246
0	10.0000	-10.0000
$+1$	-4.7956	4.7956
$+2$	-19.9288	19.9288
$+3$	-36.7341	36.7341
$+4$	-58.7986	58.7986
$+5$	evanescent	evanescent

7.7.2 Rigorous Coupled-Wave Analysis (RCWA)

From the hierarchy diagram presented in Fig. 7.17, the Rigorous Coupled-Wave Analysis (RCWA) is regarded as an exact method, as it requires no approximations of Maxwell's equations or the boundary conditions. A brief overview of the method is provided in this section. The presentation is based on Refs. [314–317, 409]. The fundamental concept of the RCWA is to represent the electromagnetic fields in the grating region as a summation of non-homogeneous waves, each of which is not a solution to Maxwell's equations in the region. However, the summation is constructed in such a way that it satisfies Maxwell's equations within the periodic region. The electric and magnetic fields in the grating region (region 2) are expressed as:

$$\vec{E}_2 = \sum_i \vec{S}_i(z) \exp(-j\vec{\sigma}_i \cdot \vec{r}), \tag{7.67}$$

$$\vec{H}_2 = \left(\frac{\epsilon_0}{\mu_0}\right)^{1/2} \sum_i \vec{U}_i(z) \exp(-j\vec{\sigma}_i \cdot \vec{r}), \tag{7.68}$$

where $\vec{\sigma}_i = \vec{k}_{inc} - i\vec{K}$ $(i = 0, \pm 1, \pm 2, \cdots)$ are the Floquet vectors arising due to the periodic structure [315]. The Floquet vectors are shown in Fig. 7.20. The grating vector is defined as $\vec{K} = K_x \hat{\imath}_x + K_z \hat{\imath}_z = (2\pi/\Lambda)(\sin\phi\, \hat{\imath}_x + \cos\phi\, \hat{\imath}_z)$. The freespace impedance $Z_0 = (\mu_0/\epsilon_0)^{1/2}$ is used in the magnetic field expression, ensuring that both $\vec{S}_i$ and $\vec{U}_i$ have the same units (electric field units). On the other hand, the electromagnetic field expressions in regions (1 and 3) consist of a summation of plane waves and are given by:

$$\vec{E}_1 = \vec{E}_{inc} + \sum_i \vec{R}_i \exp(-j\vec{k}_{1i} \cdot \vec{r}), \tag{7.69}$$

$$\vec{E}_3 = \sum_i \vec{T}_i \exp(\left[-j\vec{k}_{3i} \cdot (\vec{r} - d\hat{\imath}_z)\right], \tag{7.70}$$

$$\vec{H}_w = -\frac{1}{j\omega\mu_0} \vec{\nabla} \times \vec{E}_w, \qquad w = 1, 3.$$

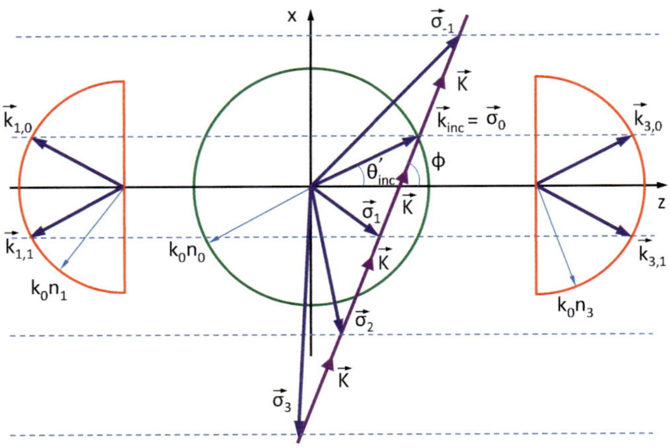

(a)

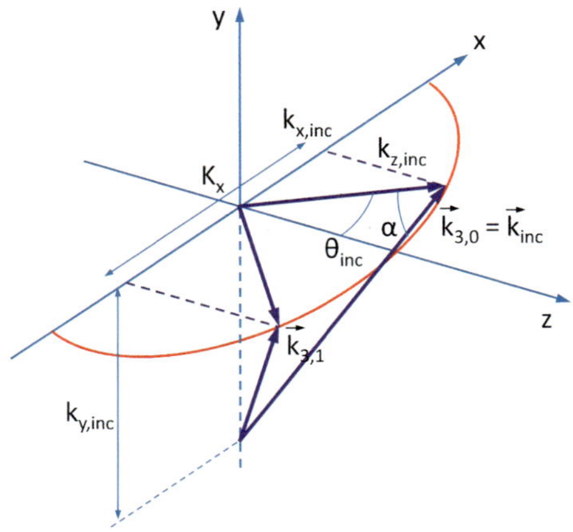

(b)

FIGURE 7.20: The Floquet condition representation. (a) This is the case of incidence with $k_{y,inc} = 0$ ($\alpha = 0$). Both the first forward and the first backward diffracted orders are shown at the boundary to the grating regions. Additionally, the various Floquet vectors $\vec{\sigma}_i$ are depicted in the grating region. The refractive index n_0 again represents the average refractive index within the grating region. (b) This case corresponds to conical incidence with $k_{y,inc} \neq 0$ ($\alpha \neq 0$). Only the first diffracted order in the forward direction is shown for region 3. The red boundary represents part of the wavevector sphere in region 3 and has a radius of $(k_0^2 n_3^2 - k_{y,inc}^2)^{1/2}$.

In the preceding equations, $\vec{R}i$ and $\vec{T}i$ represent the unknown vector amplitudes of the diffracted plane waves in the backward and forward directions, respectively. The wavevectors of the backward and forward diffracted orders are denoted by $\vec{k}_{1i}$ and $\vec{k}_{3i}$, respectively. Additionally, $\vec{E}_{inc}$ represents the electric field of the incident plane wave. To solve Maxwell's equations, it is crucial to satisfy the boundary conditions at the grating boundaries. To ensure this, the following phase-matching requirements are established:

$$\left(\vec{k}_{1i}\right)_x = (\vec{\sigma}_i)_x \Longrightarrow k_0 n_1 \sin\theta_{1i} \;=\; k_{x,inc} - iK_x, \tag{7.71}$$

$$\left(\vec{k}_{3i}\right)_x = (\vec{\sigma}_i)_x \Longrightarrow k_0 n_3 \sin\theta_{3i} \;=\; k_{x,inc} - iK_x, \tag{7.72}$$

$$\left(\vec{k}_{1i}\right)_y = (\vec{\sigma}_i)_y \Longrightarrow k_0 n_1 \sin\alpha_{1i} = k_{y,inc}, \tag{7.73}$$

$$\left(\vec{k}_{3i}\right)_y = (\vec{\sigma}_i)_y \Longrightarrow k_0 n_3 \sin\alpha_{3i} = k_{y,inc}. \tag{7.74}$$

The angles α_{wi} ($w = 1, 3$) are defined in a manner similar to α, as shown in Fig. 7.20(b). In the case of non-conical diffraction, $\alpha = \alpha_{1i} = \alpha_{3i} = 0$. The equations (7.71) and (7.72) are equivalent to equations (7.65) and (7.66) defined earlier, based on constructive interference (the grating equation). However, in this case, no average refractive index of the grating region is required. It should be noted that $K_x = (2\pi/\Lambda)\sin\phi$.

At this point, to simplify the analysis, we will focus on non-conical diffraction ($\alpha = 0$) and consider a basic surface-relief rectangular-groove grating, as illustrated in Fig. 7.21. A detailed analysis of the conical case involving anisotropic and multiple cascaded gratings is presented in Ref. [409] for readers interested in further details. Two cases are considered:

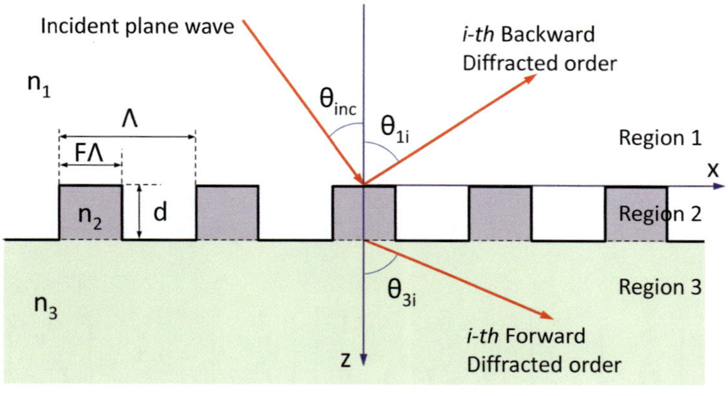

FIGURE 7.21: A diagram of a simple surface-relief rectangular-groove grating of period Λ, filling factor F, and depth d. The refractive indices of the cover, ridges, and substrate are denoted as n_1, n_2, and n_3, respectively.

(a) the case of a *TE*-polarized incident wave, and (b) the case of a *TM*-polarized incident wave.

7.7.2.1 RCWA for a Surface-Relief Grating: TE Polarization Case

In the case of *TE* polarization, the incident and diffracted fields are all linearly polarized, with their electric fields aligned along the y-direction. Consequently, the electric fields and the magnetic fields tangential to the boundaries in regions 1, 2, and 3 are given by:

$$\vec{E}_1 = \hat{\imath}_y E_0 \exp\left[-jk_0 n_1(x\sin\theta_{inc} + z\cos\theta_{inc})\right] + \hat{\imath}_y \sum_i R_i \exp\left[-j(k_{xi}x + k_{1,zi}z)\right],$$

$$k_{xi} = k_0 n_1 \sin\theta_{inc} - iK,$$

$$k_{1,zi} = \begin{cases} -(k_0^2 n_1^2 - k_{xi}^2)^{1/2} & \text{for } k_0 n_1 \geq |k_{xi}| \text{ (propagating orders)}, \\ +j(k_{xi}^2 - k_0^2 n_1^2)^{1/2} & \text{for } k_0 n_1 < |k_{xi}| \text{ (evanescent orders)}, \end{cases}$$

$$H_{1x} = -\frac{k_0 n_1 \cos\theta_{inc}}{\omega\mu_0} \exp\left[-jk_0 n_1(x\sin\theta_{inc} + z\cos\theta_{inc})\right] - \frac{1}{\omega\mu_0}\sum_i R_i k_{1,zi} \exp\left[-j(k_{xi}x + k_{1,zi}z)\right]$$

$$\vec{E}_2 = \hat{\imath}_y \sum_i S_{yi}(z) \exp\left[-jk_{xi}x\right],$$

$$H_{2x} = \left(\frac{\epsilon_0}{\mu_0}\right)^{1/2} \sum_i U_{xi}(z) \exp\left[-jk_{xi}x\right],$$

$$\vec{E}_3 = \hat{\imath}_y \sum_i T_i \exp\left[-j(k_{xi}x + k_{3,zi}(z-d))\right],$$

$$k_{3,zi} = \begin{cases} +(k_0^2 n_3^2 - k_{xi}^2)^{1/2} & \text{for } \; k_0 n_3 \geq |k_{xi}| \; \text{(propagating orders)}, \\ -j(k_{xi}^2 - k_0^2 n_3^2)^{1/2} & \text{for } \; k_0 n_3 < |k_{xi}| \; \text{(evanescent orders)}, \end{cases}$$

$$H_{3x} = -\frac{1}{\omega\mu_0} \sum_i T_i k_{3,zi} \exp\left[-j(k_{xi}x + k_{3,zi}(z-d))\right].$$

The expressions for the electromagnetic fields in the grating region are derived from Eqs. (7.67) and (7.68), with $K_z = 0$, since the surface-relief grating is unslanted. In the following derivation, it is assumed that $E_0 = 1$ V/m. Consequently, the amplitudes R_i and T_i are normalized as R_i/E_0 and T_i/E_0, respectively.

By substituting the expressions for the electric and magnetic fields in the grating region into Maxwell's equations, the following system of coupled first-order differential equations can be formulated [see Eq. (7F.3) in Appendix 7F for details and definitions of the various matrices]:

$$\frac{d}{dz'}\begin{bmatrix} \tilde{\boldsymbol{S}}_{\boldsymbol{y}} \\ \tilde{\boldsymbol{U}}_{\boldsymbol{x}} \end{bmatrix} = j\begin{bmatrix} \tilde{0} & \tilde{\boldsymbol{I}} \\ \left(\tilde{\boldsymbol{E}} - \tilde{\boldsymbol{K}}_{\boldsymbol{x}}^2\right) & \tilde{0} \end{bmatrix}\begin{bmatrix} \tilde{\boldsymbol{S}}_{\boldsymbol{y}} \\ \tilde{\boldsymbol{U}}_{\boldsymbol{x}} \end{bmatrix} = j\tilde{\boldsymbol{A}}_{\boldsymbol{TE}}\begin{bmatrix} \tilde{\boldsymbol{S}}_{\boldsymbol{y}} \\ \tilde{\boldsymbol{U}}_{\boldsymbol{x}} \end{bmatrix} \tag{7.75}$$

It is worth mentioning that for the surface-relief grating $K_z = 0$ (unslanted), which results in $\tilde{\boldsymbol{K}}_{\boldsymbol{z}} = 0$ in Eq. (7F.3). The system of the first-order coupled differential equations can be easily solved using the eigenvalues λ_i and eigenvectors $\vec{w}_i$ (with $i = 1, 2, \cdots, 2(2M+1)$) of matrix $\tilde{\boldsymbol{A}}_{\boldsymbol{TE}}$. Therefore, the solution of Eq. (7.75) is expressed as:

$$\begin{bmatrix} \tilde{\boldsymbol{S}}_{\boldsymbol{y}} \\ \tilde{\boldsymbol{U}}_{\boldsymbol{x}} \end{bmatrix} = \begin{bmatrix} \tilde{\boldsymbol{W}}_{\boldsymbol{11}} & \tilde{\boldsymbol{W}}_{\boldsymbol{12}} \\ \tilde{\boldsymbol{W}}_{\boldsymbol{21}} & \tilde{\boldsymbol{W}}_{\boldsymbol{22}} \end{bmatrix}\begin{bmatrix} e^{jk_0 z \tilde{\boldsymbol{\Lambda}}_{\boldsymbol{1}}} & \tilde{0} \\ \tilde{0} & e^{jk_0(z-d)\tilde{\boldsymbol{\Lambda}}_{\boldsymbol{2}}} \end{bmatrix}\begin{bmatrix} \tilde{\boldsymbol{C}}_{\boldsymbol{1}} \\ \tilde{\boldsymbol{C}}_{\boldsymbol{2}} \end{bmatrix}, \tag{7.76}$$

where $\tilde{\boldsymbol{C}}_{\boldsymbol{1}}$ and $\tilde{\boldsymbol{C}}_{\boldsymbol{2}}$ are $(2M+1) \times 1$ matrices that contain the unknown coefficients from the solution of the system of the coupled differential equations. The element $\tilde{\boldsymbol{0}}$ is a $(2M+1) \times (2M+1)$ zero matrix. The matrices $\tilde{\boldsymbol{W}}_{\boldsymbol{mn}}$ $(m, n = 1, 2)$ are of size $(2M+1) \times (2M+1)$ and are comprised of the elements of the eigenvectors of $\tilde{\boldsymbol{A}}_{\boldsymbol{TE}}$. Finally, $\tilde{\boldsymbol{\Lambda}}_{\boldsymbol{m}}$ $(m = 1, 2)$ are diagonal matrices of size $(2M+1) \times (2M+1)$ containing the corresponding eigenvalues of $\tilde{\boldsymbol{A}}_{\boldsymbol{TE}}$.

The diffraction problem can be solved once the coefficients $\tilde{\boldsymbol{C}}_{\boldsymbol{1}}$, $\tilde{\boldsymbol{C}}_{\boldsymbol{2}}$, and the amplitudes R_i and T_i are determined. This can be accomplished by applying the boundary conditions of the electromagnetic problem. Using Eqs. (7F.8) and (7F.11) (derived in Appendix 7F, Sec. 7F.2), the following system of $2(2M+1)$ linear equations can be formulated:

$$\tilde{\boldsymbol{B}}\tilde{\boldsymbol{X}} = \tilde{\boldsymbol{b}}, \tag{7.77}$$

where $\tilde{\boldsymbol{B}}$ is a $4(2M+1) \times 4(2M+1)$ matrix with elements derived from Eqs. (7F.8)-(7F.11) and $\tilde{\boldsymbol{X}} = [\tilde{\boldsymbol{R}}^T, \tilde{\boldsymbol{C}}_{\boldsymbol{1}}^T, \tilde{\boldsymbol{C}}_{\boldsymbol{2}}^T, \tilde{\boldsymbol{T}}^T]^T$. The matrices $\tilde{\boldsymbol{R}}$ and $\tilde{\boldsymbol{T}}$ contain the unknown R_i and

T_i amplitudes and are also defined in Appendix 7F. The matrix $\tilde{\boldsymbol{b}} = [\tilde{\boldsymbol{I}}_{\boldsymbol{1}}^T, \tilde{\boldsymbol{I}}_{\boldsymbol{2}}^T, \tilde{\boldsymbol{0}}^T, \tilde{\boldsymbol{0}}^T]^T$ contains the known parameters due to the incident plane wave. In the last matrix the elements $\tilde{\boldsymbol{0}}$ represent $(2M+1) \times 1$ zero matrices. The superscript "T" denotes matrix transposition. Solving the system of equations determines all the unknowns R_i, T_i, $\tilde{\boldsymbol{C}}_{\boldsymbol{1}}$, and $\tilde{\boldsymbol{C}}_{\boldsymbol{2}}$ which specify completely the electromagnetic field everywhere in the structure of Fig. 7.21.

It is very important to mention that the eigenvalues of matrix $\tilde{\boldsymbol{A}}_{\boldsymbol{TE}}$ form always pairs $\pm\lambda_i$, as they correspond to z-propagation constants in the positive and negative directions. This can be formally demonstrated. If λ' is an eigenvalue of $\tilde{\boldsymbol{A}}_{\boldsymbol{TE}}$, then it satisfies the equation $det(\lambda'\tilde{\boldsymbol{I}} - \tilde{\boldsymbol{A}}_{\boldsymbol{TE}}) = 0$, where det denotes the matrix determinant. The last determinant can be written as:

$$det\left(\lambda'\tilde{\boldsymbol{I}} - \tilde{\boldsymbol{A}}_{\boldsymbol{TE}}\right) = det\begin{bmatrix} \lambda'\tilde{\boldsymbol{I}} & -\tilde{\boldsymbol{I}} \\ -\left(\tilde{\boldsymbol{E}} - \tilde{\boldsymbol{K}}_{\boldsymbol{x}}^{\boldsymbol{2}}\right) & \lambda'\tilde{\boldsymbol{I}} \end{bmatrix} = 0 \Longrightarrow$$

$$det\left(\lambda'\tilde{\boldsymbol{I}}\right) det\left[\lambda'\tilde{\boldsymbol{I}} - \frac{1}{\lambda'}\left(\tilde{\boldsymbol{E}} - \tilde{\boldsymbol{K}}_{\boldsymbol{x}}^{\boldsymbol{2}}\right)\right] = 0 \Longrightarrow det\left[\lambda'^2\tilde{\boldsymbol{I}} - \left(\tilde{\boldsymbol{E}} - \tilde{\boldsymbol{K}}_{\boldsymbol{x}}^{\boldsymbol{2}}\right)\right] = 0.$$

The determinant of a block matrix has been utilized, with the assumption that $\lambda' \neq 0$. From this determinant, it is evident that $\pm\lambda'$ are both solutions if λ' is a solution. In fact, the size of the matrix in the last determinant is half the size of the original matrix $\tilde{\boldsymbol{A}}_{\boldsymbol{TE}}$. However, since both $\pm\lambda'$ are eigenvalues, the total number of eigenvalues is equal to the size of $\tilde{\boldsymbol{A}}_{\boldsymbol{TE}}$, i.e., $2(2M+1)$.

It is straightforward to show that the actual eigenvalues are $\lambda = j\lambda'$ due to the preceding "j" factor in Eq. (7.75). The eigenvalues are organized into two diagonal matrices, $\tilde{\boldsymbol{\Lambda}}_{\boldsymbol{m}}$ $(m = 1, 2)$ as follows: The eigenvalues, λ_{1i}, that satisfy the condition $\text{Re}\{j\lambda_{1i}\} < 0$ form the diagonal matrix $\tilde{\boldsymbol{\Lambda}}_{\boldsymbol{1}}$, while the eigenvalues $\lambda_{2i} = -\lambda_{1i}$, satisfy the condition $\text{Re}\{j\lambda_{2i}\} > 0$ and form the diagonal matrix $\tilde{\boldsymbol{\Lambda}}_{\boldsymbol{2}}$. The eigenvalues λ_i that have $\text{Re}\{j\lambda_i\} = 0$, belong to $\tilde{\boldsymbol{\Lambda}}_{\boldsymbol{1}}$ and the $-\lambda_i$ belong to $\tilde{\boldsymbol{\Lambda}}_{\boldsymbol{2}}$, or vice-versa since those cannot cause any numerical overflows.

This sorting of the eigenvalues ensures that for the eigenvalues, λ_{1i} of $\tilde{\boldsymbol{\Lambda}}_{\boldsymbol{1}}$, all terms $e^{j\lambda_{1i}z}$ have real parts that are either evanescently decaying or have unity magnitude (when $\text{Re}\{j\lambda_i\} = 0$). Similarly, for the eigenvalues λ_{2i} of $\tilde{\boldsymbol{\Lambda}}_{\boldsymbol{2}}$, all terms $e^{j\lambda_{2i}(z-d)}$ have real parts that are evanescently decaying or have unity magnitude (when $\text{Re}\{j\lambda_i\} = 0$) since $z \leq d$. By sorting the eigenvalues and their corresponding eigenvectors in this manner, numerical overflows in elements of matrix $\tilde{\boldsymbol{B}}$ are avoided, ensuring that the RCWA remains numerically stable and robust [316].

In grating diffraction problems, the diffraction efficiencies for both the forward and backward orders represent the fraction of the total electromagnetic power transmitted into each diffracted order. These efficiencies are derived using Poynting's theorem, which relates the power flow to the electric and magnetic fields. These diffraction efficiencies are defined as follows:

$$DE_i^r = |R_i|^2 \frac{\text{Re}\{-k_{1,zi}\}}{k_0 n_1 \cos\theta_{inc}}, \tag{7.78}$$

$$DE_i^t = |T_i|^2 \frac{\text{Re}\{k_{3,zi}\}}{k_0 n_1 \cos\theta_{inc}}, \tag{7.79}$$

where DE_i^r and DE_i^t are the backward and forward diffraction efficiencies of the i-th order, respectively. The real part in the numerator of these expressions makes it clear that the evanescent orders have zero diffraction efficiencies, as expected.

7.7.2.2 RCWA for a Surface-Relief Grating: TM Polarization Case

In the *TM* polarization case, the incident and diffracted fields are all linearly polarized with their magnetic fields aligned along the y-direction. Consequently, the magnetic fields and the electric fields tangential to the boundaries in regions 1, 2, and 3 are given by:

$$\vec{H}_1 = \hat{\imath}_y H_0 \exp\left[-jk_0 n_1 (x \sin\theta_{inc} + z\cos\theta_{inc})\right] + \hat{\imath}_y \sum_i R_i \exp\left[-j(k_{xi}x + k_{1,zi}z)\right],$$

$$E_{1x} = -\frac{k_0 n_1 \cos\theta_{inc}}{\omega\epsilon_0 n_1^2} \exp\left[-jk_0 n_1 (x\sin\theta_{inc} + z\cos\theta_{inc})\right] - \frac{1}{\omega\epsilon_0 n_1^2}\sum_i R_i k_{1,zi} \exp\left[-j(k_{xi}x + k_{1,zi}z)\right]$$

$$\vec{H}_2 = \hat{\imath}_y \sum_i U_{yi}(z)\exp\left[-jk_{xi}x\right],$$

$$E_{2x} = \left(\frac{\mu_0}{\epsilon_0}\right)^{1/2} \sum_i S_{xi}(z)\exp\left[-jk_{xi}x\right],$$

$$\vec{H}_3 = \hat{\imath}_y \sum_i T_i \exp\left[-j(k_{xi}x + k_{3,zi}(z-d))\right],$$

$$E_{3x} = -\frac{1}{\omega\epsilon_0 n_3^2}\sum_i T_i k_{3,zi}\exp\left[-j(k_{xi}x + k_{3,zi}(z-d))\right].$$

The quantities k_{xi}, $k_{1,zi}$, and $k_{3,zi}$ are defined exactly as in Sec. 7.7.2.1. In the following derivation, it is assumed that $H_0 = 1$ V/m. As a result, the amplitudes R_i and T_i are normalized as R_i/H_0 and T_i/H_0, as was done in the *TE* case. By substituting the expressions for the electric and magnetic fields in the grating region into Maxwell's equations, the following system of coupled first-order differential equations can be derived [see Eq. (7F.18) of Appendix 7F, noting that $\tilde{\boldsymbol{K}}_{\boldsymbol{z}} = 0$ for the unslanted grating]:

$$\frac{d}{dz'}\begin{bmatrix} \tilde{\boldsymbol{U}}_{\boldsymbol{y}} \\ \tilde{\boldsymbol{S}}_{\boldsymbol{x}} \end{bmatrix} = j\begin{bmatrix} \tilde{0} & -\tilde{\boldsymbol{A}}^{-1} \\ \tilde{\boldsymbol{K}}_{\boldsymbol{x}}\tilde{\boldsymbol{E}}^{-1}\tilde{\boldsymbol{K}}_{\boldsymbol{x}} - \tilde{\boldsymbol{I}} & \tilde{0} \end{bmatrix}\begin{bmatrix} \tilde{\boldsymbol{S}}_{\boldsymbol{y}} \\ \tilde{\boldsymbol{U}}_{\boldsymbol{x}} \end{bmatrix} = j\tilde{\boldsymbol{A}}_{\boldsymbol{TM}}\begin{bmatrix} \tilde{\boldsymbol{S}}_{\boldsymbol{y}} \\ \tilde{\boldsymbol{U}}_{\boldsymbol{x}} \end{bmatrix} \tag{7.80}$$

The solution of Eq. (7.80) is similar to the one of the *TE* polarization case, and is given as:

$$\begin{bmatrix} \tilde{\boldsymbol{U}}_{\boldsymbol{y}} \\ \tilde{\boldsymbol{S}}_{\boldsymbol{x}} \end{bmatrix} = \begin{bmatrix} \tilde{\boldsymbol{W}}_{\boldsymbol{11}} & \tilde{\boldsymbol{W}}_{\boldsymbol{12}} \\ \tilde{\boldsymbol{W}}_{\boldsymbol{21}} & \tilde{\boldsymbol{W}}_{\boldsymbol{22}} \end{bmatrix}\begin{bmatrix} e^{jk_0 z\tilde{\boldsymbol{\Lambda}}_{\boldsymbol{1}}} & \tilde{0} \\ \tilde{0} & e^{jk_0(z-d)\tilde{\boldsymbol{\Lambda}}_{\boldsymbol{2}}} \end{bmatrix}\begin{bmatrix} \tilde{\boldsymbol{C}}_{\boldsymbol{1}} \\ \tilde{\boldsymbol{C}}_{\boldsymbol{2}} \end{bmatrix}, \tag{7.81}$$

where the matrices $\tilde{\boldsymbol{W}}_{\boldsymbol{mn}}$ $(m, n = 1, 2)$ are of size $(2M+1)\times(2M+1)$ and are comprised of the elements of the eigenvectors of $\tilde{\boldsymbol{A}}_{\boldsymbol{TM}}$ matrix. The eigenvalues are arranged following the procedure outlined in the *TE* polarization case. By applying the boundary conditions, a system of linear equations can be formulated, similar to the one in Eq. (7.77). Solving this system determines all the unknowns R_i, T_i, $\tilde{\boldsymbol{C}}_{\boldsymbol{1}}$, and $\tilde{\boldsymbol{C}}_{\boldsymbol{2}}$, which fully characterize the electromagnetic field throughout the structure shown in Fig. 7.21. The diffraction efficiencies are then given by Eqs. (7.78) and (7.79).

7.7.3 Bragg Condition

The Bragg condition is a fundamental concept in the diffraction of light waves by a periodic structure. It provides the criterion for constructive interference, which results in diffraction maxima. When applied to a diffraction grating, the Bragg condition relates the geometry of the grating to the wavelength of the incident waves and the angles of incidence and diffraction. It is important to note that while the Bragg condition is necessary for efficient diffraction, it is not sufficient by itself.

The Bragg condition can be derived by initially considering Fig. 7.19a for forward diffraction. The Bragg condition can be understood as the scenario where the incident beam and the forward diffracted beam inside the grating form the same angle with respect to the grating fringes. In other words, the angles of the incident and diffracted rays with respect to the grating fringes must be equal. For the m-th forward diffracted order, this condition is satisfied if the angle θ'_{3m} meets the relation $\theta'_{3m} = 2\phi - \pi - \theta'_{inc}$ (using the sign convention that counterclockwise angles are positive). By substituting this angle θ'_{3m} into the grating equation, Eq. (7.65), the following condition can be easily derived:

$$2n_0 \cos(\phi - \theta'_{inc}) = m\frac{\lambda_0}{\Lambda}. \tag{7.82}$$

The last expression is referred to as the Bragg condition. In the case of an unslanted grating ($\phi = \pi/2$), Eq. (7.82) simplifies to $2n_1 \sin\theta_{inc} = m\lambda_0/\Lambda$, where Snell's law is also applied. The same Bragg condition can also be derived when the m-th backward order is considered, using a diagram similar to Fig. 7.19b.

Another way to visualize the Bragg condition is by using simplified wavevector diagrams. In Figs. 7.22a and 7.22b, wavevector diagrams satisfying the Bragg condition are shown for forward diffraction (transmission grating) and backward diffraction (reflection grating). The key idea in these diagrams is that the diffracted Floquet wavevector $\vec{\sigma}_m$ satisfies the same dispersion equation as the incident wave, i.e., $|\vec{\sigma}_m| = k_0 n_0$. This condition can be mathematically expressed as:

$$\vec{\sigma}_m \cdot \vec{\sigma}_m = (\vec{k}_{inc} - m\vec{K}) \cdot (\vec{k}_{inc} - m\vec{K}) \;=\; k_0^2 n_0^2. \tag{7.83}$$

The last equation can be easily converted into Eq. (7.82) by substituting $\vec{K} = (2\pi/\Lambda)(\sin\phi, \hat{\imath}_x + \cos\phi, \hat{\imath}_z)$ and noting that $\vec{k}inc \cdot \vec{k}inc = k_0^2 n_0^2$. This wavevector surface interpretation of the Bragg condition is particularly useful, as it can be straightforwardly generalized to the case of anisotropic materials [409]. It should be noted, however, that these wavevector surfaces are not entirely accurate in the grating region due to the periodic variation of the grating's relative permittivity. For volume gratings, though, this approximation

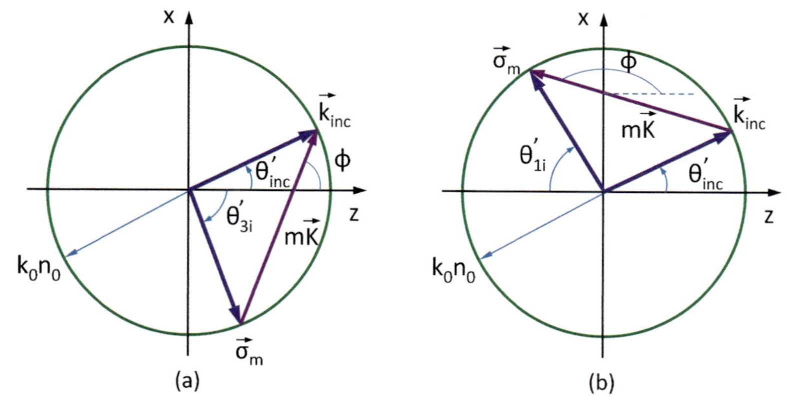

FIGURE 7.22: A simple wavevector diagram inside the grating region representing the Bragg condition. (a) The case of a transmission grating, and (b) the case of a reflection grating.

is justified, as the variation of the grating's refractive index around its average value, n_0, is small.

Example 7.7.2:

This example is adapted from Ref. [410]. Consider a surface-relief grating as shown in Fig. 7.21. The parameters of the grating are: $n_1 = 1$, $n_2 = n_3 = 1.5$, $\Lambda = 1\,\mu\text{m}$, $d = 0.75\,\mu\text{m}$, $F = 0.50$, $\theta_{inc} = 0$ (normally incident plane wave), and $\lambda_0 = 1\,\mu\text{m}$. The task is to find the forward diffraction efficiencies of the propagating orders as a function of the number of orders $2M + 1$ retained in the RCWA. Consider both TE and TM polarizations.
Solution: Using equation (7.65), it is straightforward to determine that the propagating orders satisfy $-1.5 \leq i \leq 1.5$. Therefore, the requested forward diffraction efficiencies are the DE_0^t and $DE_{\pm 1}^t$ ($DE_{-1}^t = DE_{+1}^t$ due to the grating symmetry and normal incidence). By selecting the number of diffracted orders, $2M+1$, ranging from 3 to 55, the RCWA was implemented, and the desired diffraction efficiencies were calculated and plotted in Fig. 7.23 for both TE and TM polarizations. It should be noted that, based on the RCWA formulation presented here, the number of diffracted orders must be odd. It can be observed that the diffraction efficiencies converge very quickly for $2M + 1 \geq 11$.

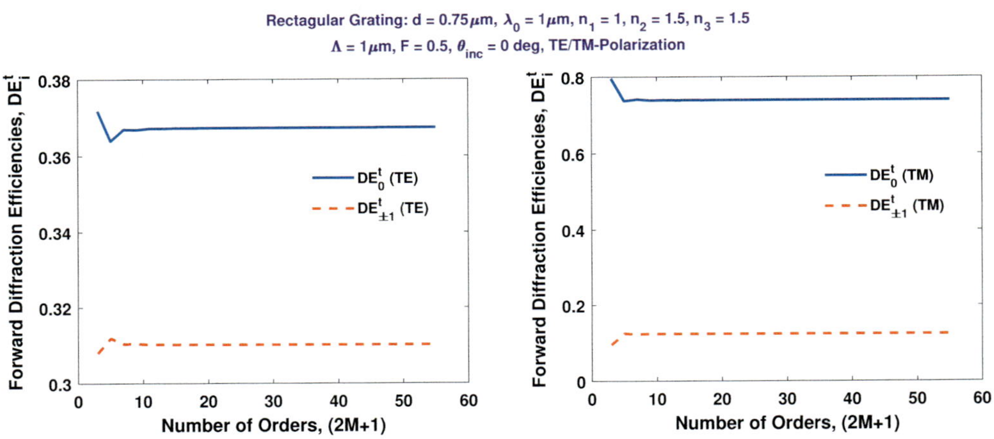

FIGURE 7.23: The forward diffraction efficiencies, DE_0^t and $DE_{\pm 1}^t$ as a function of the number of diffracted orders $2M+1$ for TE polarization (left) and TM polarization (right). The parameters of the grating are listed in the title and in Example 7.7.2.

Example 7.7.3:

This example is also adapted from Ref. [410]. Consider again a surface-relief grating as shown in Fig. 7.21. The parameters of the grating in this example are: $n_1 = 1$, $n_2 = n_3 = 0.22 - j6.71$ (which corresponds to gold), $\Lambda = 1\,\mu\text{m}$, $d = 1.0\,\mu\text{m}$, $F = 0.50$, $\theta_{inc} = 30°$, and $\lambda_0 = 1\,\mu\text{m}$. The task is to find the backward diffraction efficiencies of the propagating orders as a function of the number of orders $2M + 1$ retained in the RCWA. Both TE and TM polarizations should be considered.
Solution: All the diffracted orders in the forward direction are absorbed into the lossy gold substrate of the grating and, therefore, are not of interest. The propagating backward (reflected) diffracted orders from Eq. (7.66) should satisfy $-0.5 \leq i \leq 1.5$. Therefore, the requested backward diffraction efficiencies are DE_0^r and DE_{+1}^r.

By selecting the number of diffracted orders, $2M + 1$, varying between 3 and 55, the RCWA was implemented, and the desired diffraction efficiencies were calculated and plotted in Fig. 7.24. It can be observed that the diffraction efficiencies do not converge as quickly as in the dielectric grating case. Specifically, convergence is practically achieved for $2M+1 \geq 31$.

The reason for the higher number of diffracted orders necessary for convergence is that the electromagnetic fields exhibit more abrupt changes due to the gold-air interfaces. It should be emphasized that the convergence shown here for the *TM* polarization case is much faster than that presented in Ref. [410], where convergence is not fully obtained even for $2M+1 \simeq 125$. The presented efficiencies converge more quickly because the RCWA implements the approach of Ref. [411], which was not known at the time of Ref. [410].

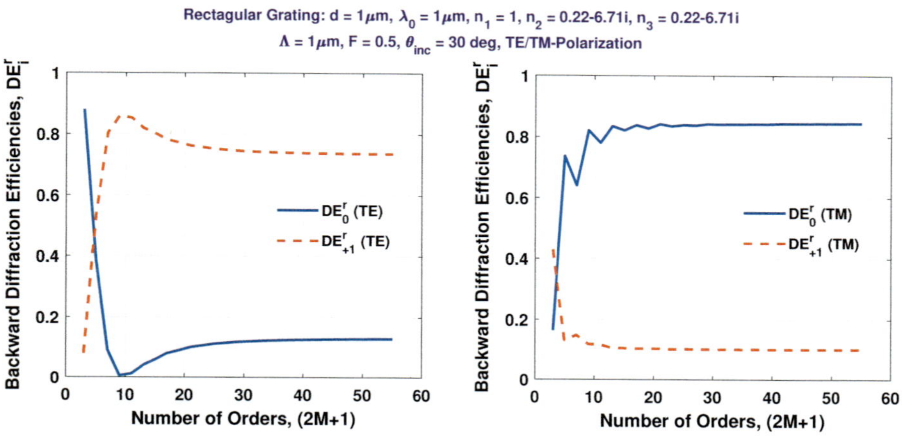

FIGURE 7.24: The backward diffraction efficiencies, DE_0^r and DE_{+1}^r as a function of the number of diffracted orders $2M+1$ for *TE* polarization (left) and *TM* polarization (right). The parameters of the grating are listed in the title and in Example 7.7.3.

Example 7.7.4:

This is an example of a slanted holographic grating that could be used for optical interconnections (see Ref. [410]). The grating geometry is represented by Fig. 7.18(b), where the angle $\alpha = 0$. The refractive index of region 1 is $n_1 = 1.4527$, and that of region 3 is $n_3 = 1.510$. The holographic (volume) grating has a sinusoidal modulation of its relative permittivity of the form $\varepsilon = \varepsilon_o + \varepsilon_1 \cos(\vec{K} \cdot \vec{r})$, where $\varepsilon_o = n_0^2 = 1.4527^2$, and $\varepsilon_1 = 0.005$ or 0.05 (two cases). The grating vector is $\vec{K} = (2\pi/\Lambda)(\cos\phi\,\hat{\imath}_z + \sin\phi\,\hat{\imath}_x)$, with $\Lambda \simeq 0.72\,\mu$m (to satisfy the first-order Bragg condition) and $\phi = 66.35°$. The grating thickness d is either $d = 200\,\mu$m (for $\varepsilon_1 = 0.005$) or $d = 20.3\,\mu$m (for $\varepsilon_1 = 0.05$). A *TE*-polarized plane wave of freespace wavelength $\lambda_0 = 0.84\,\mu$m is incident normally from region 1 into the grating ($\vec{E}_{inc} = \hat{\imath}_y E_0 e^{-jk_0 n_1 z}$). The objective is to find the diffraction efficiency of the Bragg diffracted order in the forward direction as a function of the angle of incidence.

<u>Solution:</u> For the given set of parameters, the Bragg condition expressed by Eq. (7.82) is satisfied for $m = 1$. The actual value of the corresponding grating period is $\Lambda = 0.7207225, \mu$m. Then, the RCWA can be applied with a total number of diffraction orders of $2M+1 = 9$ (since this is a holographic grating, the RCWA converges much more quickly compared to the surface-relief grating cases).

Using equations (7.65) and (7.66), it is straightforward to determine the directions of all propagating diffracted orders, both in the forward and backward directions. The propagating orders must satisfy the following inequalities (for real angles):

$$\frac{n_3\Lambda}{\lambda_0 \sin\phi}\left(\frac{n_1}{n_3}\sin\theta_{inc} - 1\right) \leq i \leq \frac{n_3\Lambda}{\lambda_0 \sin\phi}\left(\frac{n_1}{n_3}\sin\theta_{inc} + 1\right), \qquad \text{(forward)}, \tag{7.84}$$

$$\frac{n_1\Lambda}{\lambda_0 \sin\phi}\left(\sin\theta_{inc} - 1\right) \leq i \leq \frac{n_1\Lambda}{\lambda_0 \sin\phi}\left(\sin\theta_{inc} + 1\right), \qquad \text{(backward)}. \tag{7.85}$$

Using the above inequalities, for angles of incidence less than 10°, it can be easily verified

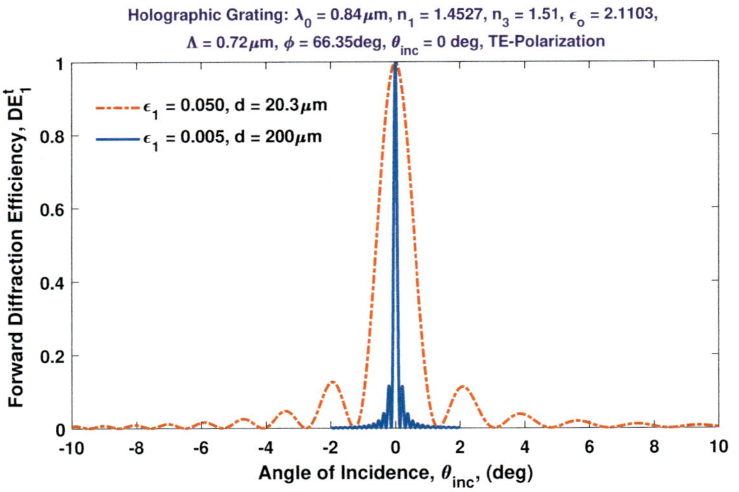

FIGURE 7.25: The forward diffraction efficiency of the first order, DE_1^t, as a function of the angle of incidence θ_{inc} for a slanted holographic grating and for *TE* polarization. The solid curve represents the small modulation $\varepsilon_1 = 0.005$ while the dash-dot line represents the larger modulation $\varepsilon_1 = 0.05$. The parameters of the grating are listed in the title and in Example 7.7.4.

that the only propagating orders, both in the forward and backward directions, are the -1, 0, and $+1$. The order that satisfies the first-order Bragg condition is the $+1$ forward diffracted order. The diffraction efficiency for this order, DE_{+1}^t, is plotted as a function of the angle of incidence in Fig. 7.25.

The case of $\varepsilon_1 = 0.005$ (solid curve) exhibits a highly sensitive response to variations in the angle of incidence around normal incidence. It should be noted that the thickness of the volume grating was chosen as $d = 200, \mu$m to ensure that the diffraction efficiency of the Bragg order approaches 100%. This behavior is characteristic of "thick" gratings, as will be discussed later using Kogelnik's approximate two-wave theory (see the hierarchy of grating theories in Fig. 7.17).

On the other hand, the case of $\varepsilon_1 = 0.05$ (dash-dot curve) shows a less sensitive response to variations in the angle of incidence around normal incidence. A similar response can be observed if the incident wave is *TM* polarized.

7.7.4 Cascaded Gratings

The RCWA can be employed to analyze surface-relief gratings with general profiles, such as the one depicted in Fig. 7.26. To apply the method, the grating is divided into a series of cascaded rectangular-groove surface-relief gratings, enabling a piece-wise approximation of the grating profile to an arbitrary degree of accuracy. A rather crude approximation of the grating profile is illustrated in Fig. 7.26. This approximation allows the RCWA to handle complex grating geometries systematically while maintaining computational efficiency and flexibility in modeling intricate grating profiles.

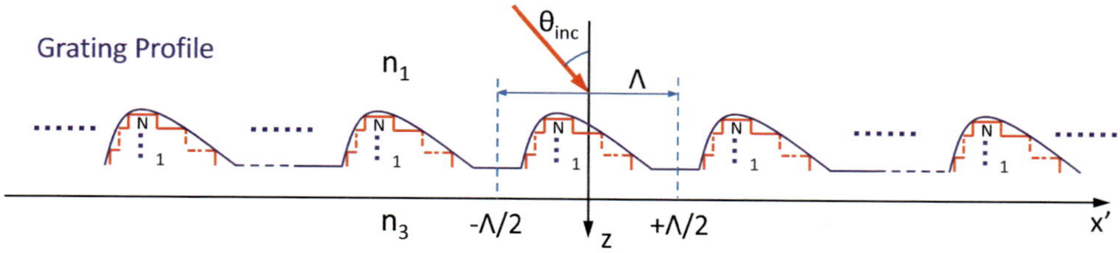

FIGURE 7.26: A surface-relief grating of the general profile and its decomposition into N rectangular-groove surface-relief gratings.

The analysis of such a structure is analogous to the one presented for the single grating. For each grating, Eq. (7.75) (for *TE* polarization) or Eq. (7.80) (for *TM* polarization) can be applied, and the electromagnetic fields in each grating region can be expressed using Eq. (7.76) or Eq. (7.81), respectively [in conjunction with Eqs. (7.67) and (7.68)]. The field expressions in the boundary regions remain consistent with those presented in Secs. 7.7.2.1 and 7.7.2.2.

By applying the boundary conditions for the continuity of the tangential field components, a system of linear equations can be formed and solved for all the unknown coefficients that define the electromagnetic fields throughout the grating structure. The definitions of the diffraction efficiencies remain the same as those provided in Eqs. (7.78) and (7.79). For additional details, the reader is referred to Ref. [317]. The same approach can also be employed for analyzing cascaded volume holographic gratings, as demonstrated in Ref. [409].

Example 7.7.5:

This example examines a dielectric surface-relief grating with a sinusoidal profile, as described in Ref. [412]. The surface relief profile is defined by the function $d(x) = (d/2) + (d/2)\cos(Kx)$, where $K = 2\pi/\Lambda$ and Λ denotes the grating period. The refractive index above the grating is $n_1 = 1$, while the refractive index of the grating and substrate is $n_3 = \sqrt{2.50} = 1.581139$. A plane wave, incident from above, strikes the grating at the first Bragg angle with *TE* polarization. The freespace wavelength is $\lambda_0 = \Lambda = 1\,\mu$m. The objective is to determine the forward diffraction efficiencies, DE_0^t, DE_{+1}^t, and DE_{-1}^t, as well as the backward diffraction efficiency, DE_0^r, as functions of the ratio d/Λ.

Solution: For the given set of parameters, the Bragg condition, as expressed by Eq. (7.82) for $m = 1$, indicates that $\sin\theta_{inc} = 1/2$, corresponding to $\theta_{inc} = 30°$. From the grating equation, it can be readily verified that only the 0 and +1 backward orders propagate. For the forward-propagating orders, the 0, +1, −1, and +2 orders propagate, while the remaining diffracted orders are evanescent.

The sinusoidal grating profile is approximated using a cascaded series of $N = 25$ rectangular-groove gratings, as illustrated for one period in Fig. 7.27(a). The diffraction efficiencies of the specified orders are plotted as a function of d/Λ in Fig. 7.27(b). While a larger number of cascaded gratings can be employed, the diffraction efficiencies are practically converged for $N = 25$. The total number of diffracted orders considered was $2M + 1 = 35$, which is more than sufficient for convergence in the case of a dielectric grating with *TE* polarization.

It is noteworthy that the size of the linear system of equations in this case is $(2N + 2)(2M + 1) \times (2N + 2)(2M + 1) = 1820 \times 1820$. Despite the size of the system, the solution is computed very efficiently.

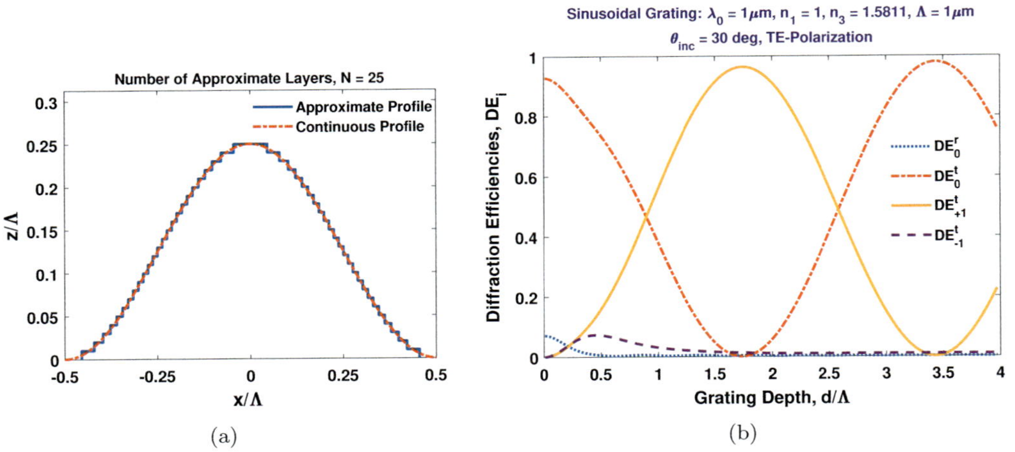

FIGURE 7.27: (a) One period of the sinusoidal dielectric grating with a continuous profile is depicted alongside its piecewise approximation using 25 cascaded rectangular-groove gratings. (b) The forward diffraction efficiencies DE_0^t, DE_{+1}^t, DE_{-1}^t, and the backward diffraction efficiency DE_0^r for a sinusoidal grating are shown as functions of the normalized grating depth, d/Λ. The incident plane wave is TE polarized and strikes the grating at an angle of $\theta_{inc} = 30^\circ$. This grating configuration is adapted from Ref. [412].

7.7.5 Three-Dimensional Incidence (Conical Diffraction)

Extending the RCWA analysis to three-dimensional incidence, often referred to as *conical diffraction*, is a straightforward process. A simple rectangular-groove surface-relief grating is illustrated in Fig. 7.28, along with the incident wavevector. The plane of incidence is defined by the dashed lines in the figure. The polarization of the incident plane wave can assume any elliptical form, with the vectors $\vec{E}_{TE}$ and $\vec{E}_{TM}$ representing the special cases of TE and TM polarizations in the context of 3D incidence.

The electromagnetic field expressions in the grating region are provided by Eqs. (7.67) and (7.68). Additionally, the electric fields in the superstrate region (region 1) and substrate region (region 3) are described by Eqs. (7.69) and (7.70), respectively. The magnetic fields in these regions can be derived using Maxwell's equations. In this 3D case, unlike the 2D TE and TM cases analyzed previously, the unknown reflection and transmission coefficients $\vec{R}_i$ and $\vec{T}_i$ are vector quantities. The incident electric field, $\vec{E}_{inc}$, can be expressed in the following form:

$$\vec{E}_{inc} = \vec{u}e^{-j\vec{k}_{inc}\cdot\vec{r}} = \left(\vec{E}_{TE} + \vec{E}_{TM}\right) e^{-j\vec{k}_{inc}\cdot\vec{r}} = \left(E_{TE}\,\hat{a}_{TE} + E_{TM}\,\hat{a}_{TM}\right) e^{-j\vec{k}_{inc}\cdot\vec{r}}, \quad \text{with} \tag{7.86}$$

$$\begin{aligned}
\hat{a}_{TE} &= -\sin\phi_{inc}\,\hat{\imath}_x - \cos\phi_{inc}\,\hat{\imath}_y, \\
\hat{a}_{TM} &= -\cos\phi_{inc}\cos\theta_{inc}\,\hat{\imath}_x + \sin\phi_{inc}\cos\theta_{inc}\,\hat{\imath}_y - \sin\theta_{inc}\,\hat{\imath}_z, \\
\vec{k}_{inc} &= k_0 n_1\left(-\cos\phi_{inc}\sin\theta_{inc}\,\hat{\imath}_x + \sin\phi_{inc}\sin\theta_{inc}\,\hat{\imath}_y + \cos\theta_{inc}\,\hat{\imath}_z\right),
\end{aligned} \tag{7.87}$$

where the unit vectors $\hat{a}_{TE}$ (perpendicular to the plane of incidence) and $\hat{a}_{TM}$ (in the plane of incidence and perpendicular to the incident wavevector) denote the directions of TE and TM polarization, respectively. The amplitudes E_{TE} and E_{TM} are in general complex, accounting for the possibility of elliptical polarization.

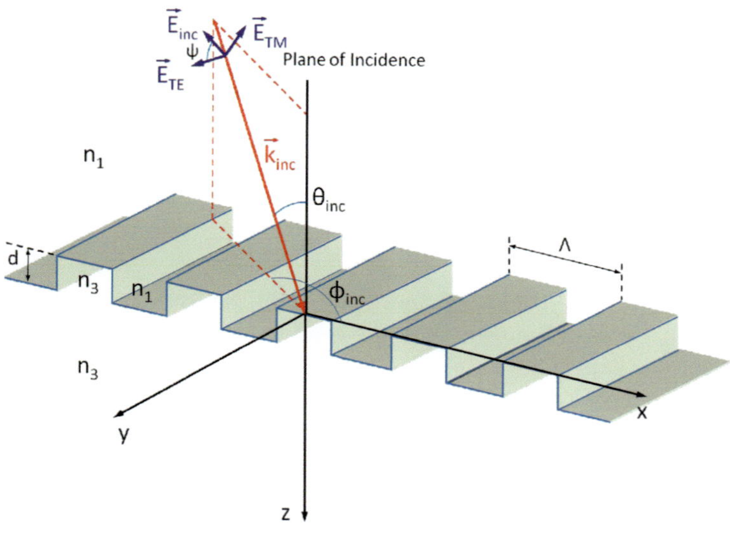

FIGURE 7.28: A rectangular-groove surface-relief grating for the case of conical diffraction. The plane of incidence with the incident plane wave are shown.

Using exactly the same approach presented in Appendix 7F for the 2D *TE* and *TM* cases, it is straightforward to express the tangential electric and magnetic field components $S_y(z)$, $S_x(z)$, $U_y(z)$, and $U_x(z)$ in the following set of first-order coupled differential equations:

$$\frac{d}{dz'}\begin{bmatrix}\tilde{\boldsymbol{S}}_y\\ \tilde{\boldsymbol{S}}_x\\ \tilde{\boldsymbol{U}}_y\\ \tilde{\boldsymbol{U}}_x\end{bmatrix} = j\begin{bmatrix} -\tilde{\boldsymbol{K}}_z & \tilde{0} & \tilde{\boldsymbol{K}}_y\tilde{\boldsymbol{E}}^{-1}\tilde{\boldsymbol{K}}_x & \tilde{\boldsymbol{I}}-\tilde{\boldsymbol{K}}_y\tilde{\boldsymbol{E}}^{-1}\tilde{\boldsymbol{K}}_x\\ \tilde{0} & -\tilde{\boldsymbol{K}}_z & \tilde{\boldsymbol{K}}_x\tilde{\boldsymbol{E}}^{-1}\tilde{\boldsymbol{K}}_x-\tilde{\boldsymbol{I}} & -\tilde{\boldsymbol{K}}_x\tilde{\boldsymbol{E}}^{-1}\tilde{\boldsymbol{K}}_y\\ -\tilde{\boldsymbol{K}}_y\tilde{\boldsymbol{K}}_x & \tilde{\boldsymbol{K}}_y\tilde{\boldsymbol{K}}_y-\tilde{\boldsymbol{A}}^{-1} & -\tilde{\boldsymbol{K}}_z & \tilde{0}\\ \tilde{\boldsymbol{E}}-\tilde{\boldsymbol{K}}_x\tilde{\boldsymbol{K}}_x & \tilde{\boldsymbol{K}}_x\tilde{\boldsymbol{K}}_y & \tilde{0} & -\tilde{\boldsymbol{K}}_z\end{bmatrix}\begin{bmatrix}\tilde{\boldsymbol{S}}_y\\ \tilde{\boldsymbol{S}}_x\\ \tilde{\boldsymbol{U}}_y\\ \tilde{\boldsymbol{U}}_x\end{bmatrix}$$

$$\frac{d[\tilde{\boldsymbol{V}}]}{dz'} = j\tilde{\boldsymbol{\mathcal{A}}}[\tilde{\boldsymbol{V}}], \tag{7.88}$$

where the matrices $\tilde{\boldsymbol{K}}_x$, $\tilde{\boldsymbol{K}}_z$, $\tilde{\boldsymbol{E}}$, and $\tilde{\boldsymbol{A}}$ are defined as in Appendix 7F, while $\tilde{\boldsymbol{K}}_y$ is a diagonal matrix of size $(2M+1)\times(2M+1)$ with diagonal elements equal to $k_{y,inc}/k_0$. In the case of unslanted grating $\tilde{\boldsymbol{K}}_z = 0$. The matrix $\tilde{\boldsymbol{V}} = [\tilde{\boldsymbol{S}}_y^T, \tilde{\boldsymbol{S}}_x^T, \tilde{\boldsymbol{U}}_y^T, \tilde{\boldsymbol{U}}_x^T]^T$, while the overall matrix $\tilde{\boldsymbol{\mathcal{A}}}$ is the matrix comprised of the block matrices shown. In the case of 2D incidence $\phi_{inc} = 0$ or π and $k_{y,inc} = 0$. Then Eq. (7.88) easily transforms into either Eq. (7.75) (for *TE* polarization) or Eq. (7.80) (for *TM* polarization). The solution of the system of coupled differential equations can be again expressed in terms of the eigenvectors and the eigenvalues of matrix $\tilde{\boldsymbol{\mathcal{A}}}$, using the same procedure to prevent numerical overflows and to ensure the stability and robustness of the RCWA method. Specifically, the following equation can be written:

$$[\tilde{\boldsymbol{V}}] = \begin{bmatrix}\tilde{\boldsymbol{W}}_{11} & \tilde{\boldsymbol{W}}_{12}\\ \tilde{\boldsymbol{W}}_{21} & \tilde{\boldsymbol{W}}_{22}\end{bmatrix}\begin{bmatrix} e^{jk_0 z\tilde{\boldsymbol{\Lambda}}_1} & \tilde{0}\\ \tilde{0} & e^{jk_0(z-d)\tilde{\boldsymbol{\Lambda}}_2}\end{bmatrix}\begin{bmatrix}\tilde{\boldsymbol{C}}_1\\ \tilde{\boldsymbol{C}}_2\end{bmatrix}, \tag{7.89}$$

with the difference now lies in the size of $[\tilde{\boldsymbol{V}}]$ which is $4(2M+1)\times 1$, while all other block matrices are twice as large, with dimensions $2(2M+1)\times 2(2M+1)$.

The backward and forward diffraction efficiencies can be defined using the Poynting's theorem as follows:

$$DE_i^r = -\frac{\mathrm{Re}\left\{|\vec{R}_i|^2 k_{1,zi}^* + (k_{1,zi} - k_{1,zi}^*)|R_{zi}|^2\right\}}{|\vec{u}|^2 k_0 n_1 \cos\theta_{inc}}, \tag{7.90}$$

$$DE_i^t = \frac{\mathrm{Re}\left\{|\vec{T}_i|^2 k_{3,zi}^* + (k_{3,zi} - k_{3,zi}^*)|T_{zi}|^2\right\}}{|\vec{u}|^2 k_0 n_1 \cos\theta_{inc}}, \tag{7.91}$$

where DE_i^r and DE_i^t are the backward and forward diffraction efficiencies of the i-th order, respectively. It can be easily shown that for evanescent diffracted orders the diffraction efficiencies are zero as expected.

The preceding analysis can, of course, be easily extended to any general-profile surface-relief grating [317], or any holographic grating with an arbitrary slant angle and varying modulation, by simply decomposing the structure into a finite number of cascaded gratings [409].

Example 7.7.6:

This example involves a dielectric surface-relief grating with rectangular grooves. The grating period is given by $\Lambda = 1\,\mu$m. The refractive index above the grating is $n_1 = 1$, while the refractive index of both the grating and the substrate is $n_3 = \sqrt{2.50} = 1.581139$. The filling factor (duty cycle) is $F = 0.50$. A plane wave is incident from above at an angle $\theta_{inc} = 30°$, with the azimuthal angle ϕ_{inc} varying from 180° to 0° in decrements of 30°. The polarization is transverse electric (*TE*), and the freespace wavelength is $\lambda_0 = \Lambda = 1\,\mu$m. Generate a table that lists the backward and forward diffraction efficiencies (of the propagating orders) for each azimuthal angle ϕ_{inc}.

Solution: To determine which backward and forward diffraction orders are propagating, it is necessary to generalize the grating equation for the case of conical diffraction. It is straightforward to demonstrate that for the diffraction orders to be propagating, the index i (which characterizes the diffraction order) should satisfy the following inequality:

$$-\frac{\Lambda}{\lambda_0 \sin\phi}\left[A_f + \cos\phi_{inc}\sin\theta_{inc}\right] \le i \le \frac{\Lambda}{\lambda_0 \sin\phi}\left[A_f - \cos\phi_{inc}\sin\theta_{inc}\right], \quad \text{(forward)} \tag{7.92}$$

$$\text{where} \quad A_f = \left(n_3^2 - n_1^2 \sin^2\theta_{inc}\sin^2\phi_{inc}\right)^{1/2},$$

$$-\frac{n_1\Lambda}{\lambda_0 \sin\phi}\left[A_b + \cos\phi_{inc}\sin\theta_{inc}\right] \le i \le \frac{n_1\Lambda}{\lambda_0 \sin\phi}\left[A_b - \cos\phi_{inc}\sin\theta_{inc}\right], \quad \text{(bakward)} \tag{7.93}$$

$$\text{where} \quad A_b = \left(1 - \sin^2\theta_{inc}\sin^2\phi_{inc}\right)^{1/2}.$$

For the specified surface-relief grating, the slant angle is $\phi = 90°$. By applying inequalities (7.92) and (7.93), the propagating orders can be determined. Depending on the azimuthal angle ϕ_{inc}, the backward orders can include at most two of the -1, 0, and $+1$ orders (always including the zero order, which corresponds to the specular reflection). Similarly, the forward orders can include at most four of the -2, -1, 0, $+1$, and $+2$ orders (with the zero order representing the refracted wave from the interface). The RCWA method was applied with $2M + 1 = 35$ orders, which is more than sufficient to ensure convergence for the dielectric grating with *TE* polarization. The calculated diffraction efficiencies are summarized in Table 7.2. As expected from symmetry considerations, the cases of $\phi_{inc} =$

TABLE 7.2: Diffraction Efficiencies Example 7.7.6, where "e" implies evanescent order.

ϕ_{inc}	Backward Orders			Forward Orders				
(*deg*)	DE^r_{-1}	DE^r_0	DE^r_{+1}	DE^t_{-2}	DE^r_{-1}	DE^r_0	DE^r_{+1}	DE^r_{+2}
180	e	0.0421	0.0097	e	0.0204	0.2045	0.6726	0.0507
150	e	0.0449	0.0065	e	0.0187	0.2199	0.7099	e
120	e	0.0245	0.0081	e	0.0317	0.3574	0.5783	e
90	e	0.0308	e	e	0.2043	0.5607	0.2043	e
60	0.0081	0.0245	e	e	0.5783	0.3574	0.0317	e
30	0.0065	0.0449	e	e	0.7099	0.2199	0.0187	e
0	0.0097	0.0421	e	0.0507	0.6726	0.2045	0.0204	e

(180° or 0°), ϕ_{inc} = (150° or 30°), and ϕ_{inc} = (120° or 60°), exhibit symmetry, with an interchange between the non-zero diffracted orders. This is anticipated due to the even symmetry of the grating profile. Orders that become evanescent under these conditions are denoted by the letter "e." As expected, evanescent diffracted orders do not carry power and therefore have zero diffraction efficiencies.

7.7.6 Two-Dimensionally Periodic (Crossed) Gratings

Two-dimensionally periodic (crossed) diffraction gratings constitute a particularly large class of diffractive optical elements (DOEs). Extensive research has been dedicated to designing these gratings using scalar diffraction analysis. An overview of various design methods for generating one- and two-dimensional arrays is provided by Mait [413]. Additionally, Vasara *et al.* [414] explored one- and two-dimensional array generators (such as Dammann and other pixelated gratings) using a simulated annealing optimization in conjunction with scalar diffraction analysis. Other notable applications and design approaches for two-dimensionally periodic gratings include coherent laser addition [415], multilevel array generators [416], and optimized kinoforms for efficient fan-out elements [417–419].

Several rigorous methods have been developed for the precise analysis of two-dimensionally periodic (crossed) gratings. Most of these methods are based on the generalization of their one-dimensional counterparts, such as rigorous coupled-wave analysis [420–422], which is summarized in this section. Other methods include the extension of Chandezon's method [423], the Fourier modal method [424, 425] (which is similar to the RCWA), Rayleigh-Fourier-based solution methods [426, 427], and regularized Waterman/Rayleigh methods [428].

The incident plane wave can be described exactly as expressed by Eq. (7.86), allowing for the treatment of any general elliptical polarization. A simple surface-relief crossed grating with rectangular pillars is illustrated in Fig. 7.29. Due to the Floquet condition, the wavevector components generated by the doubly periodic structure in the directions of periodicity are given by:

$$k_{x,i_1} = k_{x,inc} - i_1 K_x = -k_0 n_1 \cos\phi_{inc} \sin\theta_{inc} - i_1 K_x, \tag{7.94}$$

$$k_{y,i_2} = k_{y,inc} - i_2 K_y = \quad k_0 n_1 \sin\phi_{inc} \sin\theta_{inc} - i_2 K_y, \tag{7.95}$$

where $K_x = 2\pi/\Lambda_x$ and $K_y = 2\pi/\Lambda_y$ are the two grating vectors corresponding to the x- and y-periodicities, and Λ_x, Λ_y are the periodicities along the x- and y-directions, respectively.

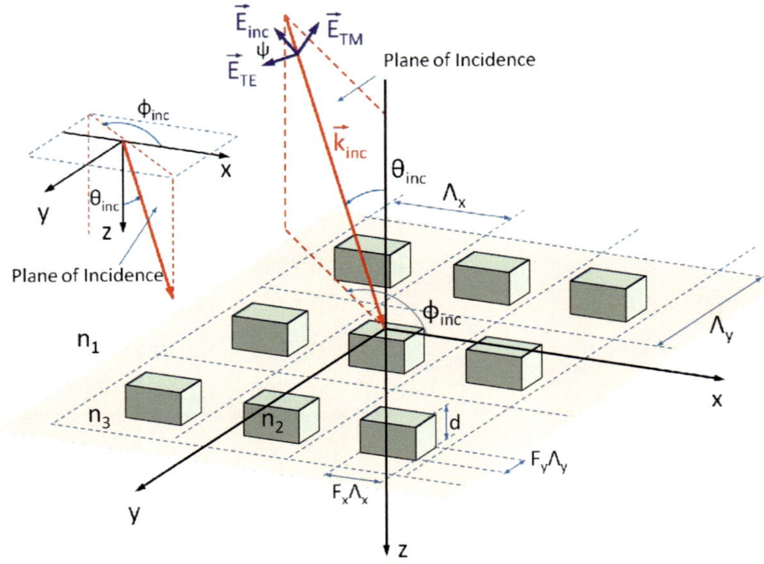

FIGURE 7.29: A rectangular-pillar surface-relief cross grating with periodicities of Λ_x and Λ_y along the x and y directions, respectively. The plane of incidence with the incident plane wave are also shown.

The indices i_1, i_2 are integers that can take any values $0, \pm 1, \pm 2, \ldots$, as in the case of one-dimensional gratings. Let an arbitrary (i_1, i_2)-th forward or backward propagating order, as shown in Fig. 7.30. For propagating orders, either in the forward or backward directions,

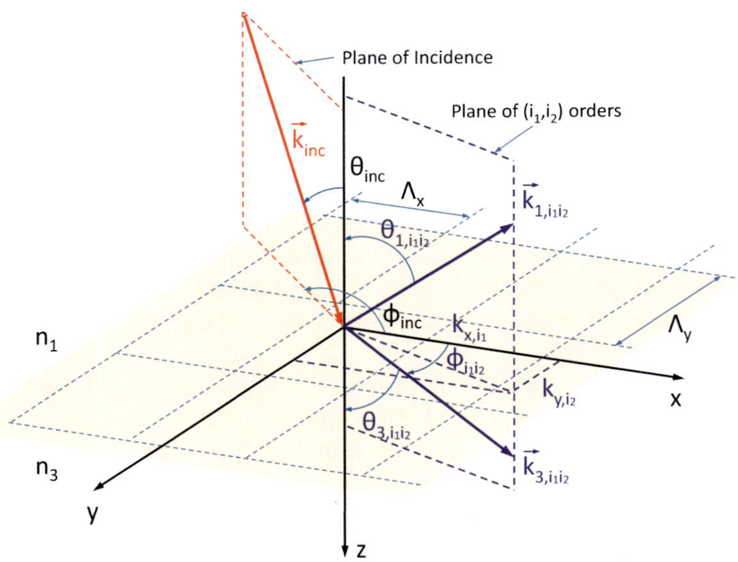

FIGURE 7.30: A diagram shown the (i_1, i_2)-th forward, $\vec{k}_{3,i_1 i_2}$, and backward, $\vec{k}_{1,i_1 i_2}$, diffracted orders. For simplicity, the grating details are not included. The various angles are also defined.

the following inequalities for i_1 and i_2 must hold:

$$k_{3z,i_1i_2}^2 = k_0^2 n_3^2 - k_{x,i_1}^2 - k_{y,i_2}^2 \geq 0, \tag{7.96}$$

$$k_{1z,i_1i_2}^2 = k_0^2 n_1^2 - k_{x,i_1}^2 - k_{y,i_2}^2 \geq 0. \tag{7.97}$$

The angles of the (i_1, i_2)-th forward and backward propagating order can be easily determined from the equations:

$$\theta_{w,i_1i_2} = \sin^{-1}\left(\frac{\sqrt{k_{x,i_1}^2 + k_{y,i_2}^2}}{k_0 n_w}\right), \qquad \phi_{i1i2} = \tan^{-1}\left(\frac{k_{y,i_2}}{k_{x,i_1}}\right),$$

where $w = 1$ for backward, and $w = 3$ for forward diffracted order.

For the crossed-grating case, the electromagnetic field expressions in the boundary regions (1, top region, and 3, bottom region) are composed of double summations of plane waves and are given by:

$$\vec{E}_1 = \vec{E}_{inc} + \sum_{i_1}\sum_{i_2} \vec{R}_{i_1i_2} \exp(-j\vec{k}_{1i_1i_2} \cdot \vec{r}), \tag{7.98}$$

$$\vec{E}_3 = \sum_{i_1}\sum_{i_2} \vec{T}_{i_1i_2} \exp(\left[-j\vec{k}_{3i_1i_2} \cdot (\vec{r} - d\hat{\imath}_z)\right], \tag{7.99}$$

$$\vec{H}_w = -\frac{1}{j\omega\mu_0}\vec{\nabla} \times \vec{E}_w, \qquad w = 1, 3,$$

$$\vec{k}_{wi_1i_2} = k_{x,i_1}\,\hat{\imath}_x + k_{y,i_2}\,\hat{\imath}_y + k_{wz,i_1i_2}\,\hat{\imath}_z, \qquad w = 1, 3.$$

Note that in this case the vector reflection and transmission coefficients $\vec{R}_{i_1i_2}$ and $\vec{T}_{i_1i_2}$ depend on the two integers i_1 and i_2. Similarly, the expressions of the electromagnetic fields in the grating region are as follows:

$$\vec{E}_2 = \sum_{i_1}\sum_{i_2} \vec{S}_{i_1i_2}(z) \exp(-j\vec{\sigma}_{i_1i_2} \cdot \vec{r}), \tag{7.100}$$

$$\vec{H}_2 = \left(\frac{\epsilon_0}{\mu_0}\right)^{1/2} \sum_{i_1}\sum_{i_2} \vec{U}_{i_1i_2}(z) \exp(-j\vec{\sigma}_{i_1i_2} \cdot \vec{r}), \tag{7.101}$$

$$\text{where} \quad \vec{\sigma}_{i_1i_2} = k_{x,i_1}\,\hat{\imath}_x + k_{y,i_2}\,\hat{\imath}_y,$$

where the $\vec{S}_{i_1i_2}(z)$ and $\vec{U}_{i_1i_2}(z)$ represent the complex varying vector field amplitudes within the grating region, analogous to those in the case of one-dimensional gratings. By utilizing the electromagnetic field expressions mentioned above, the solutions to Maxwell's equations in the grating region can be expressed in the following form (refer to Appendix 7F, Sec. 7F.5):

$$\frac{d\tilde{\boldsymbol{V}}}{dz'} = j\tilde{\boldsymbol{A}}\tilde{\boldsymbol{V}},$$

where $\tilde{\boldsymbol{V}} = [\tilde{\boldsymbol{S}}_{\boldsymbol{x}}^T, \tilde{\boldsymbol{S}}_{\boldsymbol{y}}^T, \tilde{\boldsymbol{U}}_{\boldsymbol{x}}^T, \tilde{\boldsymbol{U}}_{\boldsymbol{y}}^T]^T$ is a matrix of size $4(2M+1)^2 \times 1$ and $\tilde{\boldsymbol{X}}_{\boldsymbol{w}}$ ($X = S, U$ and $w = x, y$) are defined in Appendix 7F and Sec. 7F.5. The solution of the system of coupled differential equations is expressed in terms of the eigenvalues and eigenvectors of $\tilde{\boldsymbol{A}}$, similar to the approach used in the conical diffraction case, as described by Eq. (7.89). The key difference in the crossed grating case is that the size of $[\tilde{\boldsymbol{V}}]$ is $4(2M+1)^2 \times 1$, and all other block matrices are larger, with dimensions $2(2M+1)^2 \times 2(2M+1)^2$. Then, the boundary conditions of the problem can be cast into a system of linear equations of size $10(2M+1)^2 \times 10(2M+1)^2$ as the unknowns include the $\vec{R}_{i_1,i_2}$, corresponding to $3(2M+1)^2$ terms, the $\vec{T}_{i_1,i_2}$ with $3(2M+1)^2$ terms, and the $\tilde{\boldsymbol{C}}_{\boldsymbol{1}}$ and $\tilde{\boldsymbol{C}}_{\boldsymbol{2}}$ which consist of $4(2M+1)^2$ terms. At the $z=0$ and $z=d$ boundaries, the continuity of the tangential electric and magnetic field components (along the x- and y-directions) provides $4(2M+1)^2+4(2M+1)^2 = 8(2M+1)^2$ equations. Since the solutions in regions 1 and 3 are plane waves, there are $2(2M+1)^2$ more equations of the form $\vec{k}_{1,i_1i_2} \cdot \vec{R}_{i_1i_2} = 0$ and $\vec{k}_{3,i_1i_2} \cdot \vec{T}_{i_1i_2} = 0$. Thus, the number of unknowns and the number of equations both equal to $10(2M+1)^2$. The system's size can always be reduced by properly manipulating the equations representing the boundary conditions. Solving the system of equations provides the solution for the electromagnetic fields in all regions of the grating structure. This involves determining the reflection and transmission coefficients as well as the field distributions within the grating itself.

The backward and forward diffraction efficiencies can be defined, similarly to previous cases, using the Poynting's theorem as follows:

$$DE^r_{i_1i_2} = -\frac{\mathrm{Re}\left\{|\vec{R}_{i_1i_2}|^2 k^*_{1,zi_1i_2} + (k_{1,zi_1i_2} - k^*_{1,zi_1i_2})|R_{zi_1i_2}|^2\right\}}{|\vec{u}|^2 k_0 n_1 \cos\theta_{inc}}, \tag{7.102}$$

$$DE^t_{i_1i_2} = \frac{\mathrm{Re}\left\{|\vec{T}_{i_1i_2}|^2 k^*_{3,zi_1i_2} + (k_{3,zi_1i_2} - k^*_{3,zi_1i_2})|T_{zi_1i_2}|^2\right\}}{|\vec{u}|^2 k_0 n_1 \cos\theta_{inc}}, \tag{7.103}$$

where $DE^r_{i_1i_2}$ and $DE^t_{i_1i_2}$ are the backward and forward diffraction efficiencies of the (i_1, i_2)-th order, respectively. It can be easily shown that for evanescent diffracted orders the diffraction efficiencies are again zero as expected.

Of course, the preceding analysis can be readily extended to any general surface-relief grating profile through a simple decomposition into a finite number of cascaded gratings [429]. Furthermore, a similar analysis can be applied to superposed volume holographic gratings, which may, in general, be anisotropic [430, 431].

Example 7.7.7:

To apply the 2D RCWA for a given antireflection rectangular pillar surface-relief grating [420], consider the following setup: Region 1 is air, with a refractive index $n_1 = 1$, and region 3 is a glass substrate with a refractive index $n_3 = 1.5$. The grating filling factor (percentage of the grating surface occupied by the pillars) for achieving antireflective properties at normal incidence is $F^2 = 0.53$ (assuming $F_x = F_y = F$). The freespace wavelength is $\lambda_0 = 1.0\,\mu$m. The grating periodicities are $\Lambda_x/\lambda_0 = \Lambda_y/\lambda_0 = 0.1$ or $\Lambda_x/\lambda_0 = \Lambda_y/\lambda_0 = 0.5$. Determined the diffraction efficiencies of the $(0,0)$-th forward and backward orders as a function of the pillar height d. Assume that d is varying between 0 and $0.5\,\mu$m.

Solution: The rectangular pillar 2D surface-relief grating has filling factors $F_x = F_y = F = \sqrt{0.53} = 0.728$. Using Eqs. (7.97) and (7.96), it is straightforward to demonstrate that the only propagating orders in both the forward and backward directions are the $(0,0)$-th orders. This is valid for both $\Lambda_x/\lambda_0 = \Lambda_y/\lambda_0 = 0.1$ and $\Lambda_x/\lambda_0 = \Lambda_y/\lambda_0 = 0.5$. These types of gratings are referred to as high-spatial frequency (or subwavelength) gratings and behave as a uniform, generally anisotropic slab. For the RCWA calculation, $(2M+1)^2 = 13^2 = 169$ orders were used, implying that the indices range from -6 to $+6$ for both i_1 and i_2. This number of orders is sufficient to ensure the convergence of the diffraction efficiencies.

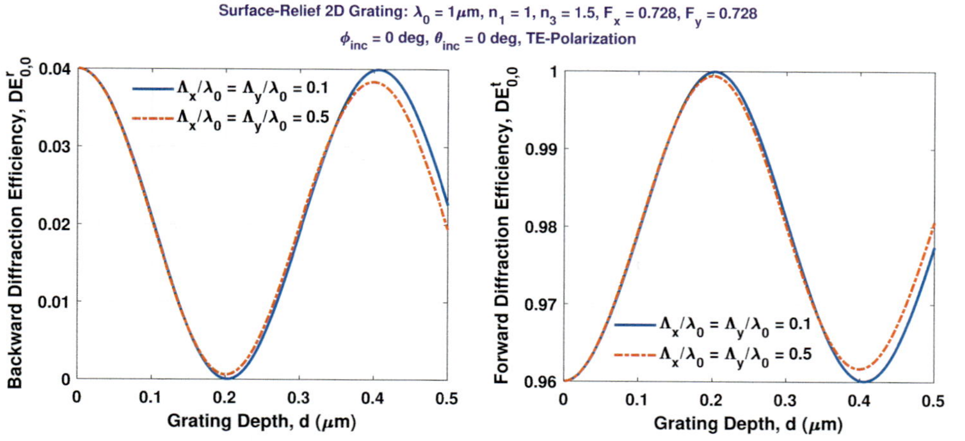

FIGURE 7.31: A rectangular-pillar surface-relief crossed-grating with periodicities of $\Lambda_x = \Lambda_y = 0.1\mu$m along the x- and y-directions. A plane wave of freespace wavelength $\lambda_0 = 1.0\mu$m is normally incident. The refractive indices are $n_1 = 1.0$ and $n_2 = n_3 = 1.5$. The filling factors are $F_x = F_y = 0.728$ [420]. The cases of $\Lambda_x/\lambda_0 = \Lambda_y/\lambda_0 = 0.1$ (solid lines) and $\Lambda_x/\lambda_0 = \Lambda_y/\lambda_0 = 0.5$ (dash-dot lines) are shown.

The resulting backward and forward diffraction efficiencies, denoted as $DE^r_{0,0}$ and $DE^t_{0,0}$, respectively, are depicted in Fig. 7.31 as a function of the pillar height d.

It is observed that the antireflection properties are slightly improved for the case of $\Lambda_x/\lambda_0 = \Lambda_y/\lambda_0 = 0.1$. The optimal pillar height for achieving the lowest reflectivity in the crossed grating is approximately $d \simeq 0.204\,\mu$m. This result can be understood by considering the theory of a single antireflection layer. In this scenario, the effective refractive index of the uniform layer should be $n_2 = \sqrt{n_1 n_3} = \sqrt{1 \times 1.5} = 1.2247$. The optimal smaller pillar height, corresponding to a quarter-wavelength layer, is therefore $d = \lambda/4n_2 = \lambda_0/(4n_2) = 0.204\,\mu$m. This theoretical prediction aligns well with the RCWA results shown in Fig. 7.31.

Example 7.7.8:

In this problem, a crossed grating composed of a two-dimensional periodic array of cylindrical pillars is analyzed [429]. The grating has identical periodicities in the x- and y-directions, given by $\Lambda_x = \Lambda_y = \Lambda$, where Λ/λ_0 can take values of 10, 5, or 2.5. The refractive index of the superstrate (region 1) is $n_1 = 1.0$, while both the grating pillars and the substrate have a refractive index of $n_2 = n_3 = 1.5$. A plane wave with a freespace wavelength of $\lambda_0 = 1.0\,\mu$m is normally incident on the grating ($\theta_{inc} = \phi_{inc} = 0°$). The incident wave is TE-polarized ($\psi = 0°$). However, due to the normal incidence and the symmetric structure of the grating, the diffraction efficiencies for TE and TM polarizations are identical. Determine the propagating diffracted orders and plot the forward diffraction efficiencies for the $(1, 0)$-th and $(1, 1)$-th orders as functions of the pillar height d, within the range $0 \leq d/\lambda_0 \leq 2$.
Solution: Using Eqs. (7.97) and (7.96), it is straightforward to determine the propagating orders in the forward and backward directions.

For example, in the case of $\Lambda/\lambda_0 = 10$, there are 317 backward propagating orders and 709 forward propagating orders. In the case of $\Lambda/\lambda_0 = 5$, there are 81 backward propagating orders and 177 forward propagating orders. Finally, in the case of $\Lambda/\lambda_0 = 2.5$, there are 21 backward and 45 forward propagating orders. In this last case, the forward propagating orders are: $(\pm3, \pm2)$, $(\pm3, \pm1)$, $(\pm3, 0)$, $(\pm2, \pm3)$, $(\pm2, \pm2)$, $(\pm2, \pm1)$, $(\pm2, 0)$, $(\pm1, \pm3)$, $(\pm1, \pm2)$, $(\pm1, \pm1)$, $(\pm1, 0)$, and $(0, \pm3)$, $(0, \pm2)$, $(0, \pm1)$, $(0, 0)$. For the $(1, 0)$-th

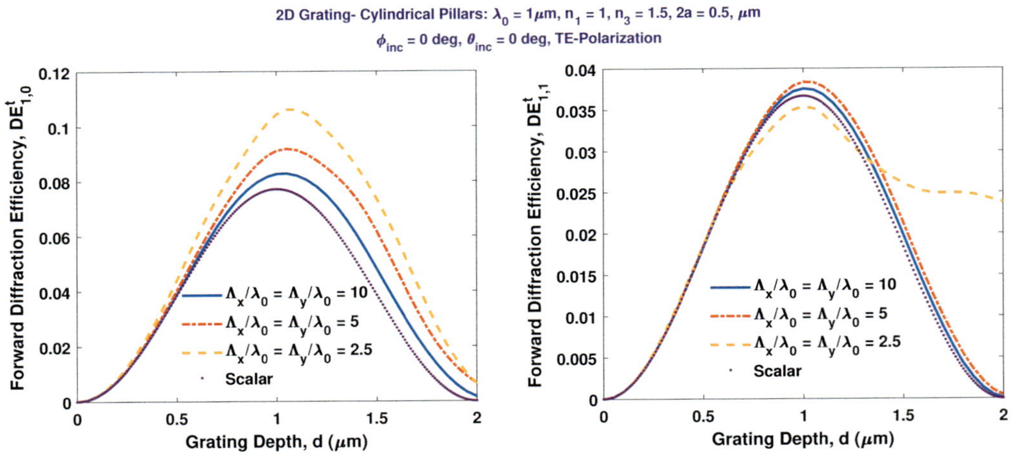

FIGURE 7.32: A cylindrical-pillar surface-relief crossed grating with periodicities $\Lambda_x = \Lambda_y = \Lambda$ along the x- and y-axes is considered. A plane wave with a freespace wavelength of $\lambda = 1.0\,\mu$m is incident normally on the grating. The refractive indices of the regions are $n_1 = 1.0$ for the superstrate and $n_2 = n_3 = 1.5$ for the grating and substrate. The radius of the cylindrical pillars is given by $a/\Lambda = 0.25$ [429]. The diffraction efficiencies of the $(1, 0)$-th and $(1, 1)$-th forward diffracted orders are plotted as functions of the pillar height d. The cases for $\Lambda/\lambda_0 = 10$, 5, and 2.5 are presented. Additionally, scalar diffraction efficiencies are included for comparison.

and $(1, 1)$-th forward diffraction orders that are of interest, the number of orders retained in the RCWA was $21^2 = 441$ $(-10 \leq i_1, i_2 \leq 10)$. These diffraction efficiencies are plotted as functions of the pillar height d in Fig. 7.32. The three cases of $\Lambda/\lambda_0 = 10$, 5, and 2.5 are presented. Furthermore, the scalar diffraction efficiencies are included in the same plots for comparison. It can be seen that when the feature size, $2a/\lambda_0 = 5$ (largest), the scalar results predict the grating diffraction efficiencies quite well.

7.7.7 High-Spatial-Frequency (Subwavelength) Gratings and Effective Medium Theory

A class of optical components requires diffraction gratings with grating periods smaller than the freespace wavelength. When the period is sufficiently small, no forward or backward diffracted orders are generated except for the specular (zero-order) ones. These gratings are referred to as high-spatial-frequency or subwavelength gratings.

Subwavelength gratings exhibit intriguing properties and serve as valuable tools for optical designers. In particular, two distinct characteristics can be exploited [432]: (a) Subwavelength gratings typically behave as slabs of artificial uniaxial material, with optic axes oriented parallel to the direction of periodicity and principal refractive indices determined by the grating geometry. (b) Subwavelength gratings can be designed to produce sharp resonances in the specular orders.

Examples of such subwavelength optical structures include waveguide couplers, subwavelength antireflection structures, resonance structures, gratings used in distributed-feedback and distributed-Bragg-reflector laser configurations, and wavelength-division multiplexing.

Let us consider the electromagnetic diffraction by a rectangular-groove subwavelength surface-relief grating, as illustrated in Fig. 7.33. Assume initially that light is normally incident with TE polarization (also referred to as $E \perp K$) and that the grating period is much smaller than the wavelength of light. In this case, the electric field is oriented in the

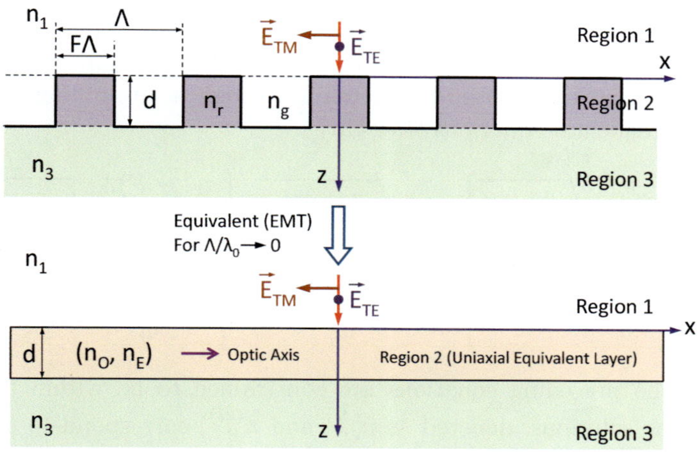

FIGURE 7.33: A rectangular-groove surface-relief 1D grating and its effective-medium-theory (EMT) equivalent homogeneous uniaxial layer model of the same depth d. The refractive indices n_O and n_E represent the ordinary and the extraordinary index of the equivalent layer.

$\hat{\imath}_y$ direction. Since the electric field is tangential to the ridge-groove interfaces, it can be approximated as a constant within one period. However, the electric displacement vector, $\vec{D} = \epsilon\vec{E}$, is discontinuous at these interfaces. A spatially averaged displacement field can be defined as $\vec{D}_{avg} = \epsilon_0[Fn_r^2 + (1-F)n_g^2)]\vec{E}$, where n_r and n_g are the refractive indices of the ridge and groove, respectively [432]. The term in brackets represents the average relative permittivity of the periodic medium.

Similarly, for *TM* polarization (also referred to as $E \parallel K$), the displacement vector is given by $\vec{D} = -\hat{\imath}_x D$ and can be considered approximately constant over one period due to its continuity at the ridge-groove interfaces. In this case, however, the electric field is discontinuous. Spatial averaging of the electric field over one period yields $\vec{E}_{avg} = [F/n_r^2 + (1-F)/n_g^2]\vec{D}/\epsilon_0$. Here, the average relative permittivity of the grating region is the inverse of the term in brackets. The two average permittivities corresponding to the two limiting polarizations can be used to define the effective refractive indices, namely the ordinary and extraordinary indices:

$$n_O^{(1)} = \left[Fn_r^2 + (1-F)n_g^2\right]^{1/2}, \tag{7.104}$$

$$n_E^{(1)} = \left[\frac{F}{n_r^2} + \frac{1-F}{n_g^2}\right]^{-1/2}. \tag{7.105}$$

From the two average refractive indices defined (commonly referred to as "first-order" effective refractive indices), it is evident that the periodic medium exhibits different optical behaviors for the two considered polarizations. This anisotropy, known as "form birefringence" [91], allows the periodic medium to be approximated as a homogeneous uniaxial slab, as illustrated in Fig. 7.33.

It should be noted that the derivation of the first-order effective refractive indices is largely heuristic and is strictly valid in the limit $\Lambda/\lambda_0 \to 0$. To relax this condition, an analysis presented by Collin [433] and Rytov [434] is based on the exact eigenvalue equation for the effective index at any angle of incidence θ_{inc} (assuming $\phi_{inc} = 0$). This analysis is summarized in Appendix 7G, where the resulting eigenvalue equations are given by Eq. (7G.5) for *TE* polarization and Eq. (7G.6) for *TM* polarization.

These eigenvalue equations were later generalized to account for conical incidence ($\phi_{inc} \neq 0$) by Li [435] and Haggans et al. [436]. When applied to normal incidence, it can be shown that the ordinary and extraordinary refractive indices are obtained as solutions of the following transcendental equations [434, 437]:

$$\sqrt{n_r^2 - n_O^2} \tan\left[\frac{\pi F \Lambda}{\lambda_0}\sqrt{n_r^2 - n_O^2}\right] = -\sqrt{n_g^2 - n_O^2} \tan\left[\frac{\pi(1-F)\Lambda}{\lambda_0}\sqrt{n_g^2 - n_O^2}\right], \quad (7.106)$$

$$\frac{\sqrt{n_r^2 - n_E^2}}{n_r^2} \tan\left[\frac{\pi F \Lambda}{\lambda_0}\sqrt{n_r^2 - n_E^2}\right] = -\frac{\sqrt{n_g^2 - n_E^2}}{n_g^2} \tan\left[\frac{\pi(1-F)\Lambda}{\lambda_0}\sqrt{n_g^2 - n_E^2}\right]. \quad (7.107)$$

The solutions of the preceding equations are constrained to lie within the range $n_g < n_O, n_E < n_r$. These solutions, denoted as $n_O^{(H)}$ and $n_E^{(H)}$, correspond to the higher-order ordinary and extraordinary refractive indices of the homogeneous uniaxial layer that is equivalent to the grating.

Rytov [434] further provided the second-order ordinary and extraordinary refractive indices by employing a Taylor series expansion of the tangents in Eqs. (7.106) and (7.107). These second-order refractive indices, denoted as $n_O^{(2)}$ and $n_E^{(2)}$, are given by:

$$n_O^{(2)} = \left[[n_O^{(1)}]^2 + \frac{1}{3}\left[\frac{\pi\Lambda}{\lambda_0}F(1-F)\right]^2 (n_r^2 - n_g^2)^2\right]^{1/2}, \quad (7.108)$$

$$n_E^{(2)} = \left[[n_E^{(1)}]^2 + \frac{1}{3}\left[\frac{\pi\Lambda}{\lambda_0}F(1-F)\right]^2 \left(\frac{1}{n_r^2} - \frac{1}{n_g^2}\right)^2 [n_E^{(1)}]^6 [n_O^{(1)}]^2\right]^{1/2}. \quad (7.109)$$

For conical incidence, the roots of the transcendental equations (7G.5) for *TE* polarization and (7G.6) for *TM* polarization are no longer directly equivalent to $n_O^{(H)}$ and $n_E^{(H)}$. In this case, ordinary and extraordinary refractive indices can still be defined, but they become dependent on the incidence angles θ_{inc} and ϕ_{inc}. Consequently, the homogeneous layer equivalent to the grating varies with these angles of incidence. The corresponding ordinary and extraordinary refractive indices in this case are given by [436]:

$$n_O^{(H)}(\theta_{inc}, \phi_{inc}) = \left[n_1^2 \sin^2\theta_{inc}\cos^2\phi_{inc} + n_1^2 \sin^2\theta_{inc}\sin^2\phi_{inc} + n_{eff,TE}^2\right]^{1/2}, \quad (7.110)$$

$$n_E^{(H)}(\theta_{inc}, \phi_{inc}) = \left[\frac{n_1^2 \sin^2\theta_{inc}\sin^2\phi_{inc} + n_{eff,TM}^2}{1 - n_1^2 \sin^2\theta_{inc}\cos^2\phi_{inc}/[n_O^{(H)}(\theta_{inc}, \phi_{inc})]^2}\right]^{1/2}, \quad (7.111)$$

where $n_{eff,TE}$ and $n_{eff,TM}$ are the lowest order eigenvalue of Eqs. (7G.5) and (7G.6), respectively.

In previously established Effective Medium Theories (EMT), each pair of (n_O, n_E) can be used to approximate the periodic grating region as a homogeneous uniaxial layer, provided that the grating is subwavelength and produces only specular orders. Instead of employing the RCWA to study the diffraction of a plane wave by the subwavelength grating, one can utilize anisotropic layer analysis to approximate the diffracted fields. Moreover, EMT provides optical engineers with a powerful tool for designing diffractive structures that exhibit either antireflection properties [438–440] or reflection/transmission resonances [441, 442].

Example 7.7.9:

Consider a rectangular-groove surface-relief grating with refractive indices $n_r = 3.5$ (silicon) and $n_g = 1.0$ (air) [432, 443]. A plane wave with a freespace wavelength of $\lambda_0 = 1.0\,\mu$m is normally incident on the grating. (a) Compute and plot the ordinary and extraordinary refractive indices (first-order, second-order, and higher-order) as functions of the grating filling factor F for $\lambda_0/\Lambda = 5$. (b) Compute and plot the ordinary and extraordinary refractive indices (first-order, second-order, and higher-order) as functions of the wavelength-to-period ratio, λ_0/Λ, within the range $2 \leq \lambda_0/\Lambda \leq 5$, while maintaining a grating filling factor of $F = 0.5$.

Solution: For normal incidence, the higher-order refractive indices are independent of the angle of incidence. By using Eqs. (7.104) and (7.105) for the first-order, Eqs. (7.108) and (7.109) for the second-order, and Eqs. (7.106) and (7.107) for the higher-order, all ordinary and extraordinary indices can be determined. The results are illustrated in Figs. 7.34(a) and 7.34(b). As expected, all indices in Fig. 7.34(a) converge smoothly to n_g for $F = 0$ and

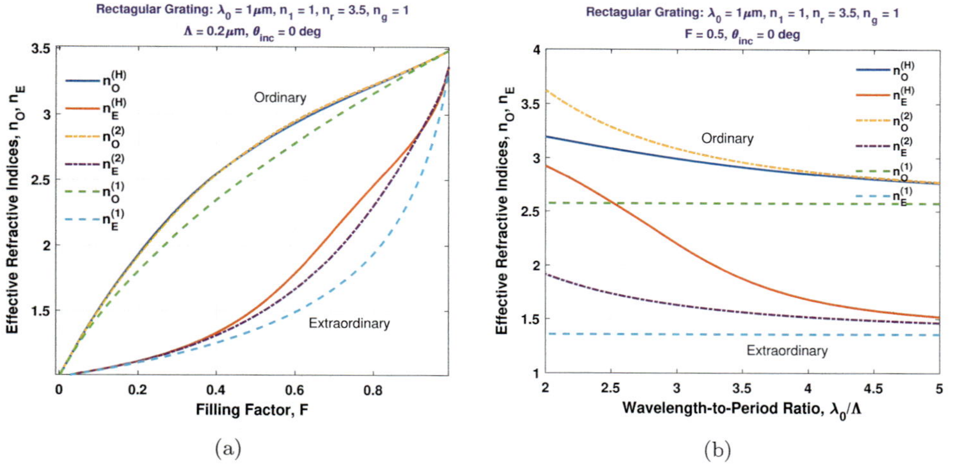

FIGURE 7.34: The effective medium refractive indices $n_O^{(H)}$, $n_E^{(H)}$, $n_O^{(2)}$, $n_E^{(2)}$, and $n_O^{(1)}$, $n_E^{(1)}$ for a rectangular-groove surface-relief grating with $n_r = 3.5$ and $n_g = 1.0$. The freespace wavelength is $\lambda_0 = 1.0\,\mu$m and the angle of incidence is $\theta_{inc}0^\circ$. (a) The refractive indices as a function of the filling factor F for $\Lambda = 0.2\,\mu$m ($\lambda_0/\Lambda = 5$). (b) The refractive indices as a function of the grating period for $F = 0.5$.

to n_r for $F = 1$, since in these limiting cases, the grating effectively disappears. For the given $\lambda_0/\Lambda = 5$ ratio, the second-order and higher-order indices are nearly identical for the ordinary index, but show some differences for the extraordinary index. As the λ_0/Λ ratio increases, all ordinary and extraordinary indices converge to common values, as suggested by the results in Fig. 7.34(b).

Additionally, the effective birefringence $n_O - n_E$ ranges from zero at the filling factor extremes to values greater than unity when $F = 0.5$. This induced birefringence is orders of magnitude larger than the natural birefringence observed in crystalline materials such as calcite or quartz.

Example 7.7.10:

To design an antireflection rectangular-groove surface-relief grating on a silicon substrate with an air superstrate, we can apply the first-order ordinary and extraordinary refractive indices derived from the Effective Medium Theory (EMT). The goal is to minimize the reflectance at normal incidence for a plane wave with a freespace wavelength of $\lambda_0 = 1.5\,\mu$m [432]. The refractive indices of the silicon substrate and the grating ridges are $n_r = n_3 = 3.5$, while the air superstrate and the grating grooves have refractive indices $n_1 = n_g = 1.0$. Consider both *TE* ($E \perp K$) and *TM* ($E \parallel K$) polarizations separately, and calculate the specular reflectance of each designed grating as a function of the angle of incidence ($0° \leq \theta_{inc} \leq 45°$, with $\phi_{inc} = 0°$). The specular reflectance should be determined using the EMT uniaxial slab model, considering first-order, second-order, and higher-order effective refractive indices. Additionally, calculate the specular reflectance using the RCWA for comparison. Consider grating periods of $\Lambda = 0.15\,\mu$m and $\Lambda = 0.375\,\mu$m.

Solution: To design the antireflection grating, the impedance matching approach presented in the section on Optical Interference Filters is utilized. The first-order refractive indices $n_O^{(1)}$ and $n_E^{(1)}$ from Eqs. (7.104) and (7.105) are employed. By adjusting the filling factor F, the effective refractive index, either $n_O^{(1)}$ or $n_E^{(1)}$, can be made equal to the impedance matching condition $\sqrt{n_1 n_3} = 1.8708$, thereby minimizing reflection.

For the two polarizations, the resulting filling factors, F_{TE} and F_{TM}, can be easily determined as:

$$F_{TE} = \frac{n_1 n_3 - n_g^2}{n_r^2 - n_g^2} = \frac{n_g}{n_r + n_g} \simeq 0.2222,$$

$$F_{TM} = \frac{n_r^2}{n_1 n_3} \frac{n_1 n_3 - n_g^2}{n_r^2 - n_g^2} = \frac{n_r}{n_r + n_g} \simeq 0.7778.$$

For the antireflection condition, the grating depth must be an odd multiple of a quarter wavelength. Hence, the depth d can be expressed as $d = (2m+1)\lambda_0/(4\sqrt{n_1 n_3}) \simeq (2m+1)0.2005\,\mu$m (where $m = 0, 1, \cdots$). To align with the analysis presented in Ref. [432], we set $m = 1$, yielding a grating depth (and equivalent homogeneous uniaxial medium thickness) of $d \simeq 0.6015\,\mu$m.

For the electromagnetic analysis of the equivalent homogeneous uniaxial medium, the 4×4 method by Berreman [143] was employed. In the RCWA, 21 diffracted orders were considered to ensure the convergence of the diffraction efficiencies for the propagating orders. For $\Lambda = 0.15\,\mu$m, only the specular orders propagate (DE_0^r and DE_0^t). However, for $\Lambda = 0.375\,\mu$m and $\theta_{inc} > 30°$, the +1-th forward diffracted order also propagates.

The resulting specular diffraction efficiencies are shown in Fig. 7.35. For *TE* polarization with $\lambda_0/\Lambda = 10$, the second-order EMT refractive indices closely match the RCWA results. The reflected power is less than 0.5%, but it does not reach zero at normal incidence, contrary to the prediction of the first-order refractive indices. Similar observations hold for *TM* polarization with $\lambda_0/\Lambda = 10$.

For *TE* polarization with $\lambda_0/\Lambda = 4$, deviations from the RCWA results increase, as expected. Nevertheless, the second-order EMT indices still approximate the RCWA results well across the range of angles considered. Since the antireflection grating was designed using first-order indices, its antireflection performance is less effective compared to the $\lambda_0/\Lambda = 10$ case. The small spike in the RCWA results arises due to the onset of the +1-th forward diffracted order for $\theta_{inc} > 30°$.

In the *TM* polarization with $\lambda_0/\Lambda = 4$, the antireflection properties degrade, but the higher-order and second-order EMT methods still track the RCWA results satisfactorily.

7.7.7.1 Effective Medium Theory for Subwavelength Crossed Gratings

When dealing with a subwavelength crossed grating, as depicted in Fig. 7.29, the application of effective medium theory (EMT) becomes more complex than in the case of a

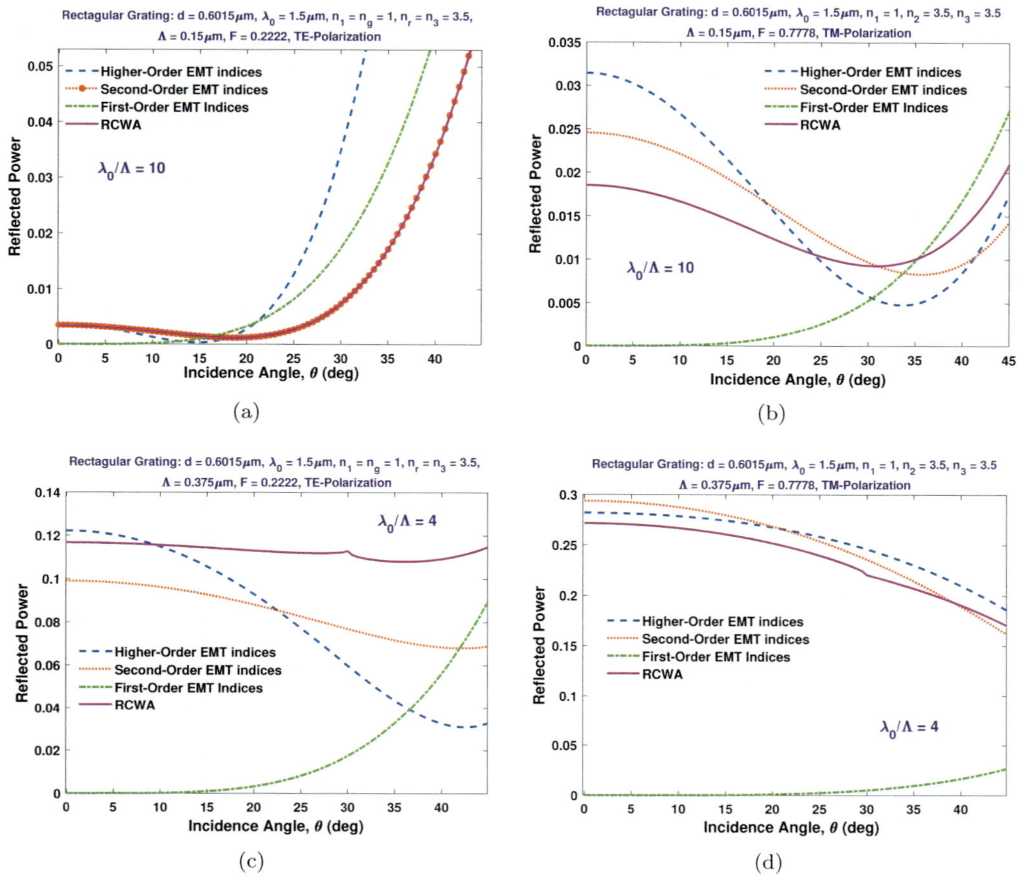

FIGURE 7.35: The normalized specular reflection of the antireflection gratings, DE_0^r, as a function of the angle of incidence, computed with the first-order, second-order, and higher-order EMT and the RCWA. (a) *TE* polarization and $\lambda_0/\Lambda = 10$. (b) *TM* polarization and $\lambda_0/\Lambda = 10$. (c) *TE* polarization and $\lambda_0/\Lambda = 4$. (d) *TM* polarization and $\lambda_0/\Lambda = 4$.

one-dimensional (1D) grating. This complexity arises from the two-dimensional (2D) periodic structure, which introduces additional challenges in modeling the grating's electromagnetic behavior. Relatively few studies have focused on developing EMT models specifically for 2D crossed gratings. However, even simplified EMT models can provide a reasonable approximation of the electromagnetic response of these structures, offering valuable insights without the need for solving the exact, often computationally intensive, problem.

One of the initial approaches was proposed by Motamedi *et al.* [444], who defined effective indices by averaging the first-order indices derived from 1D EMT. This method provided a basic approximation but lacked the accuracy required for more complex designs. Grann *et al.* [445] introduced a more sophisticated approach, utilizing the RCWA to determine the EMT refractive indices along the x- and y-directions. The key idea behind this approach is to monitor the amplitude and phase of the reflected or transmitted field, or the reflected/transmitted power, to derive the effective indices. This method generally results in an equivalent homogeneous biaxial layer with principal axes aligned along the x-, y-, and z-directions. Kikuta *et al.* [446] further advanced this concept by developing a generalized approach, also based on RCWA, that specifies a more accurate equivalent biaxial layer. In their approach, the principal axes of the layer are numerically defined through the RCWA, providing improved precision in the EMT model.

Despite these advancements, closed-form solutions for the effective indices of an equivalent biaxial layer in 2D crossed gratings are not readily available due to the inherent complexity of the geometry. This contrasts with 1D gratings, where characteristic equations can be derived and solved to obtain usable indices. In this section, a simple first-order index approach is presented, based on 1D EMT theory, specifically for the case of normal incidence. For those seeking more accurate EMT models, the references provided offer detailed methodologies and insights into more sophisticated approaches.

To derive the simple first-order refractive indices, consider the grating depicted in Fig. 7.29. Assume a normally incident plane wave with polarization aligned along the x-axis, where $\theta_{inc} = 0$, $\phi_{inc} = 0$, and $\psi = 0$. By applying the same principles used for 1D gratings, the effective relative permittivity, ε'_{xx}, along the x-direction can be expressed as follows, considering the periodicity along the x-axis:

$$\varepsilon'_{xx} = \left[\frac{F_x}{n_r^2} + \frac{1 - F_x}{n_g^2}\right]^{-1}.$$

On the other hand, when considering the periodicity along the y-axis, the effective relative permittivity along the x-direction can be expressed as:

$$\varepsilon_{xx} = F_y\varepsilon'_{xx} + (1 - F_y)n_g^2 \Longrightarrow$$

$$n_{xx} = n_g \left[\frac{(1 - F_x + F_x F_y)n_r^2 + F_x(1 - F_y)n_g^2}{F_x n_g^2 + (1 - F_x)n_r^2}\right]^{1/2}, \tag{7.112}$$

where $n_{xx} = \sqrt{\varepsilon_{xx}}$ is the effective refractive index along the x-direction. If the normally incident plane wave is now polarized along the y-direction, following the same reasoning, the effective relative permittivity along the y-direction can be derived as:

$$n_{yy} = n_g \left[\frac{(1 - F_y + F_x F_y)n_r^2 + F_y(1 - F_x)n_g^2}{F_y n_g^2 + (1 - F_y)n_r^2}\right]^{1/2}, \tag{7.113}$$

where $n_{yy} = \sqrt{\varepsilon_{yy}}$ is the effective refractive index along the y-direction. The relative permittivity along the z-direction also needs to be determined. In this case, it is assumed that the pillars are infinitely long, and the incident wave is propagating along the y-axis while being polarized along the z-direction. Then, when examining the x-periodicity, the effective permittivity, ε'_{zz}, can be defined as:

$$\varepsilon'_{zz} = F_x n_r^2 + (1 - F_x)n_g^2.$$

Then, for the same wave, when examining the y-periodicity, the final effective relative permittivity along the z-direction is given by:

$$\varepsilon_{zz} = F_y\varepsilon'_{zz} + (1 - F_y)n_g^2 \Longrightarrow n_{zz} = \left[F_x F_y n_r^2 + (1 - F_x F_y)n_g^2\right]^{1/2}, \tag{7.114}$$

where $n_{zz} = \sqrt{\varepsilon_{zz}}$ is the effective refractive index along the z-direction. Then, a biaxial (in general) homogeneous layer can be defined with its relative permittivity matrix expressed as:

$$[\tilde{\varepsilon}] = \begin{bmatrix} n_{xx}^2 & 0 & 0 \\ 0 & n_{yy}^2 & 0 \\ 0 & 0 & n_{zz}^2 \end{bmatrix}. \tag{7.115}$$

Observe that the principal axes of the equivalent biaxial layer coincide with the x, y, and z axes of the coordinate system used. For a symmetric crossed grating where $F_x = F_y = F$, the resulting effective indices simplify as:

$$n_O = n_{xx} = n_{yy} = n_g \left[\frac{(1 - F + F^2)n_r^2 + F(1 - F)n_g^2}{F n_g^2 + (1 - F)n_r^2} \right]^{1/2}, \tag{7.116}$$

$$n_E = n_{zz} = \left[F^2 n_r^2 + (1 - F^2)n_g^2 \right]^{1/2}, \tag{7.117}$$

and the relative permittivity matrix becomes $[\bar{\varepsilon}] = diag[n_O^2, n_O^2, n_E^2]$ which represents a uniaxial homogeneous equivalent layer.

Example 7.7.11:

Consider a rectangular-pillar crossed grating with $n_1 = n_g = 1.0$ and $n_3 = n_r = 3.0$ [445]. The incident plane wave has a freespace wavelength of $\lambda_0 = 1.0\,\mu$m. The grating periods are identical and equal to $\Lambda_x = \Lambda_y = \Lambda = 0.001\,\mu$m. What matters, however, is the ratio $\Lambda/\lambda_0 = 0.001$, rather than the actual values of λ_0 or Λ. For this subwavelength crossed grating, determine the specular normalized reflected power as a function of the angle of incidence θ_{inc} (with $\phi_{inc} = 0°$) using the effective medium theory and the RCWA. Examine two cases: (a) A symmetric grating with $F_x = F_y = 0.7597$; (b) An asymmetric grating with $F_x = 0.40$ and $F_y = 0.80$.
Solution: a) In this case, the analysis focuses on the symmetric 2D subwavelength grating. For the given ratio $\Lambda/\lambda_0 = 0.001$, only the specular backward and forward diffracted orders, $(0, 0)$, are propagating. Since $F_x = F_y = F = 0.7597$, the effective indices, obtained from

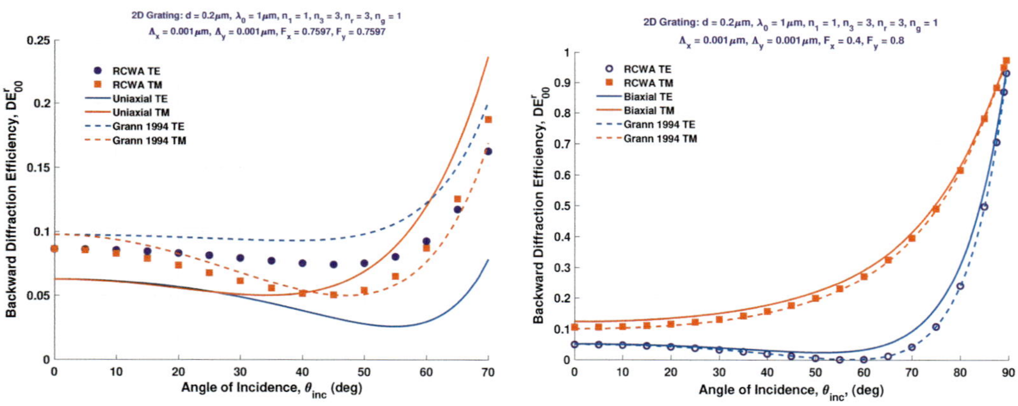

(a) (b)

FIGURE 7.36: The normalized specular reflected power, DE_{00}^r, of a crossed grating, as a function of the angle of incidence θ_{inc} with $\phi_{inc} = 0°$. The normalized power is computed with the simple 2D EMT and the RCWA [445]. (a) Symmetric 2D grating with $F_x = F_y = 0.7597$. (b) Asymmetric 2D grating with $F_x = 0.40$ and $F_y = 0.80$.

Eqs. (7.116) and (7.117), are $n_O = 1.6062$ and $n_E = 2.3701$. By applying the 4×4 matrix approach, the normalized reflected power as a function of the angle of incidence can be determined. These results are represented by solid curves in Fig. 7.36(a), corresponding to *TE* ($\psi = 90°$) and *TM* ($\psi = 0°$) polarizations. The normalized reflected power, DE_{00}^r, calculated with the RCWA, is shown by the solid circles for *TE* polarization and squares for *TM* polarization. The number of diffracted orders retained in the RCWA was $15^2 =$

225, ensuring the convergence of the propagating diffracted orders. Additionally, using the information provided in Ref. [445], with $n_O = 1.732$ and $n_E = 2.374$, the dashed lines were also calculated for both polarizations. However, there appears to be some inconsistency in the data provided in Ref. [445], as the dashed lines should nearly coincide with the RCWA results. Despite the heuristic nature of the definitions of n_O and n_E, the uniaxial equivalent layer prediction shows relatively good agreement with the RCWA results.
(b) In this case, the analysis focuses on the asymmetric 2D subwavelength grating. The grating parameters are the same as in case (a), except that the filling factors now become $F_x = 0.40$ and $F_y = 0.80$. Consequently, the effective indices derived from Eqs. (7.112), (7.113), and (7.114) are $n_{xx} = 1.2006$, $n_{yy} = 1.4088$, and $n_{zz} = 1.8868$. Therefore, the equivalent EMT layer is now biaxial. By applying the 4×4 matrix approach again, the normalized reflected power as a function of the angle of incidence can be determined. These results are represented by solid curves in Fig. 7.36(b), corresponding to *TE* ($\psi = 90°$) and *TM* ($\psi = 0°$) polarizations. The RCWA results are shown as solid circles and squares for the two polarizations, respectively. Additionally, using the information provided in Ref. [445], with $n_{xx} = 1.2464$, $n_{yy} = 1.5311$, and $n_{zz} = 1.8868$, the dashed lines were also calculated for both polarizations. It can be observed that the dashed lines practically coincide with the RCWA results, as expected from Ref. [445]. However, even the simple heuristic model presented here, based on the biaxial equivalent layer, shows very good agreement with the RCWA results.

7.7.8 Kogelnik's Two-Wave Coupled-Wave Analysis

The two-wave coupled wave analysis, commonly referred to as Kogelnik's theory [402], is an approximate differential technique illustrated in the third row of the diagram in Fig. 7.17. This method involves neglecting the second derivatives in the electric field envelope functions, which leads to some boundary conditions, such as those for the magnetic field, not being used. This approach is generally applied successfully in the case of volume diffraction gratings. Additionally, instead of considering the infinite number of diffracted orders that could exist in both the forward and backward directions, only the 0-order (forward) and the 1-order (forward for transmission gratings or backward for reflection gratings) are considered. To minimize errors due to the non-utilization of all boundary conditions, it is assumed that the refractive indices of the input (n_1) and the output (n_3) regions match the average refractive index of the grating region. This implies that $n_1 = n_3 = n_0 = \sqrt{\varepsilon_0}$ (where ε_0 represents the average relative permittivity of the grating region), as depicted in Figs. 7.37. The relative permittivity of the grating region is assumed to vary sinusoidally and is expressed as:

$$\varepsilon(\vec{r}) = \varepsilon_0 + \varepsilon_1 \cos(\vec{K} \cdot \vec{r}),$$

where ε_1 is the relative permittivity modulation amplitude, and $\vec{K}$ is the grating vector. The corresponding refractive index in the grating region is given by:

$$n(\vec{r}) = \sqrt{\varepsilon} = n_0 + n_1 \cos(\vec{K} \cdot \vec{r}) \simeq \sqrt{\varepsilon_0} + \frac{\varepsilon_1}{2\sqrt{\varepsilon_0}} \cos(\vec{K} \cdot \vec{r}),$$

which implies that $n_0 = \sqrt{\varepsilon_0}$, and $2n_0 n_1 = \varepsilon_1$. The analysis extends to both transmission and reflection gratings, examining the behavior under both *TE* and *TM* polarizations.

7.7.8.1 TE Polarization Case

In this case, the electric field is assumed to be *TE*-polarized, meaning it is polarized along the y-direction (according to Fig. 7.37a). Denoting $R(z) = S_0(z)$ and $S(z) = S_1(z)$ as the

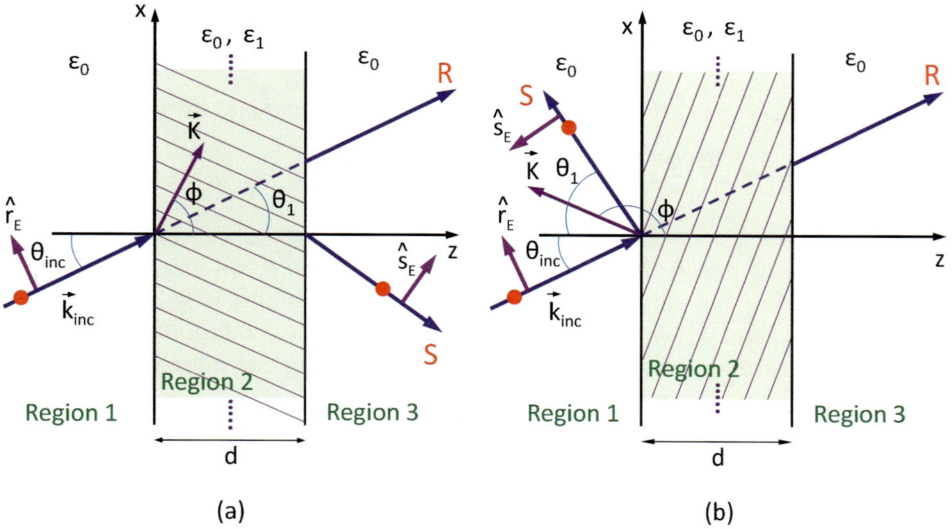

FIGURE 7.37: Kogelnik's [402] two-wave theory geometry as applied to a volume grating is illustrated. The incident, transmitted (R-wave), and diffracted first-order (S-wave) wavevectors are depicted, along with the grating vector $\vec{K}$. The red dots indicate the direction of the unit vector of the electric field for TE polarization ($\hat{\imath}_y$). The vectors $\hat{r}_E$ and $\hat{s}_E$ represent the unit vectors of the electric field for TM polarization. This is shown for both (a) the transmission grating case and (b) the reflection grating case.

amplitudes of the 0-order and 1-order forward diffracted orders (refer again to Fig. 7.37a), the field within the grating region can be expressed as:

$$\vec{E} = \hat{\imath}_y E(x,z) = \hat{\imath}_y \left[R(z)\exp(-j\vec{k}_{inc}\cdot\vec{r}) + S(z)\exp(-j\vec{\sigma}_1\cdot\vec{r}) \right], \quad \text{where,} \tag{7.118}$$

$$\vec{k}_{inc}\cdot\vec{r} = \rho_x x + \rho_z z = (k_0 n_0 \sin\theta)x + (k_0 n_0 \cos\theta)z,$$

$$\vec{\sigma}_1 = \vec{k}_{inc} - \vec{K},$$

$$\vec{\sigma}_1\cdot\vec{r} = \sigma_x x + \sigma_z z = (k_0 n_0 \sin\theta - iK_x)x + (k_0 n_0\cos\theta - iK_z)z,$$

$$\vec{K} = K_x\hat{\imath}_x + K_z\hat{\imath}_z = \frac{2\pi}{\Lambda}\left(\sin\phi\,\hat{\imath}_x + \cos\phi\,\hat{\imath}_z\right).$$

The electric field in Eq. (7.118), should satisfy the Helmholtz equation in the grating region. Therefore, the following expressions can be written:

$$\nabla^2 E + k_0^2\varepsilon(\vec{r})E = 0 \implies$$

$$\left[R'' - j2\rho_z R' - (\rho_x^2+\rho_z^2)R\right]\exp(-j\vec{k}_{inc}\cdot\vec{r}) + \left[S'' - j2\sigma_z S' - (\sigma_x^2+\sigma_z^2)S\right]\exp(-j\vec{\sigma}_1\cdot\vec{r}) +$$

$$k_0^2\left[\varepsilon_0\frac{\varepsilon_1}{2}\left(\exp(j\vec{K}\cdot\vec{r}) + \exp(-j\vec{K}\cdot\vec{r})\right)\right]\times$$

$$\left[R(z)\exp(-j\vec{k}_{inc}\cdot\vec{r}) + S(z)\exp(-j\vec{\sigma}_1\cdot\vec{r})\right] = 0.$$

Retaining only the "synchronous" terms from the last equation, specifically those with exponential factors involving $\vec{\sigma}_1 = \vec{k}_{inc} - \vec{K}$, and $\vec{k}_{inc} = \vec{\sigma}_1 + \vec{K}$ (since the expansion of the electric field includes only the 0-order and 1-order waves), the Helmholtz equation simplifies to:

$$\left[R'' - j2\rho_z R' - (\rho_x^2 + \rho_z^2)R + k_0^2\varepsilon_0 R + k_0^2\frac{\varepsilon_1}{2}S\right]\exp(-j\vec{k}_{inc}\cdot\vec{r}) +$$

$$\left[S'' - j2\sigma_z S' - (\sigma_x^2 + \sigma_z^2)S + k_0^2\varepsilon_0 S + k_0^2\frac{\varepsilon_1}{2}R\right]\exp(-j\vec{\sigma}_1\cdot\vec{r}) = 0. \tag{7.119}$$

In this context, where the double-primed terms denote second derivatives with respect to z and the primed terms denote first derivatives with respect to z, we proceed by neglecting the second derivatives, simplifying to $R'' \simeq 0 \simeq S''$. Additionally, it is important to note that $\rho_x^2 + \rho_z^2 = k_0^2\varepsilon_0$, and $\sigma_x^2 + \sigma_z^2 = \sigma_1^2 = |\vec{\sigma}_1|^2$. By utilizing these relations in the Helmholtz equation (after discarding the second derivatives) Eq. (7.119), the following set of coupled differential equations is derived:

$$-j2\rho_z R' + k_0^2\frac{\varepsilon_1}{2}S = 0, \tag{7.120}$$

$$-j2\sigma_z S' + \left[k_0^2\varepsilon_0 - |\vec{\sigma}_1|^2\right]S + k_0^2\frac{\varepsilon_1}{2}R = 0. \tag{7.121}$$

At this point, Kogelnik's [402] notation is adopted. Defining the coupling coefficient $\kappa = \pi n_1/\lambda_0$, the normalized incident z-wavevector component $c_R = k_{inc,z}/k_0 n_0 = \cos\theta_{inc}$, the normalized diffracted z-wavevector component $c_S = \sigma_z/k_0 n_0 = \cos\theta_{inc} - (K\cos\phi)/(k_0 n_0)$, and the dephasing factor $\vartheta = (k_0^2 n_0^2 - |\vec{\sigma}_1|^2)/(2k_0 n_0)$, the coupled wave equations can be written as:

$$c_R R' = -j\kappa S, \tag{7.122}$$

$$c_S S' + j\vartheta S = -j\kappa R. \tag{7.123}$$

For a transmission grating, $c_S > 0$ (since the diffracted 1-order propagates in the forward direction), and the required initial conditions are $R(z=0) = 1$ and $S(z=0) = 0$. Using these initial conditions, Eqs. (7.122) and (7.123) can be easily solved. The resulting $S(z)$ can be written as:

$$S(z) = -j\left(\frac{c_R}{c_S}\right)^{1/2}\exp\left[-j\xi\frac{z}{d}\right]\frac{\sin\left[\left(\gamma^2+\xi^2\right)^{1/2}\frac{z}{d}\right]}{\left(1+\dfrac{\xi^2}{\gamma^2}\right)^{1/2}}, \quad \text{where,} \tag{7.124}$$

$$\gamma = \frac{\pi n_1 d}{\lambda_0(c_R c_S)^{1/2}}, \qquad \xi = \frac{\vartheta d}{2c_S}.$$

Here, γ is often referred to as the grating "strength," reflecting the ability of the grating to efficiently couple the incident and diffracted waves. The parameter ξ is related to the dephasing factor ϑ and measures the degree of mismatch from the ideal Bragg condition. The diffraction efficiency of the S wave (1-order) can be defined as before using the Poynting theorem and is given by:

$$DE_1^t = \frac{c_R}{c_S}\frac{|S(z=d)|^2}{|R(z=0)|^2} = \frac{\sin^2\left[\left(\gamma^2+\xi^2\right)^{1/2}\right]}{1+\dfrac{\xi^2}{\gamma^2}}. \tag{7.125}$$

In the case where the Bragg condition is satisfied, $\xi = 0$, the diffraction efficiency simplifies to the following expression:

$$DE_1^t = \sin^2\gamma. \tag{7.126}$$

This last equation highlights that the diffraction efficiency depends primarily on the grating strength and can potentially reach 100%, indicating complete power transfer from the incident wave to the diffracted wave.

For a reflection grating, $c_S < 0$ (since the diffracted 1-order propagates in the backward direction), and the required initial conditions are $R(z=0)=1$ and $S(z=d)=0$. Using these initial conditions, Eqs. (7.122) and (7.123) can be easily solved. The resulting $S(z)$ can be written as:

$$S(z) = +j\left|\frac{c_R}{c_S}\right|^{1/2} \exp\left[+j\xi\frac{z}{d}\right] \frac{\sinh\left[\left(\gamma^2-\xi^2\right)^{1/2}\left(\frac{z}{d}-1\right)\right]}{\left(1-\frac{\xi^2}{\gamma^2}\right)^{1/2}\cosh\left[\left(\gamma^2-\xi^2\right)^{1/2}\right]+j\frac{\xi}{\gamma}\sinh\left[\left(\gamma^2-\xi^2\right)^{1/2}\right]}, \tag{7.127}$$

where, $\gamma = \dfrac{\pi n_1 d}{\lambda_0\left|c_R c_S\right|^{1/2}}, \qquad \xi = -\dfrac{\vartheta d}{2c_S}.$

Here, γ and ϑ have the same meanings as in the transmission grating case. The diffraction efficiency of the backward S wave (1-order) can be defined as before using the Poynting theorem and is given by:

$$DE_1^r = \frac{c_R}{|c_S|}\frac{|S(z=0)|^2}{|R(z=0)|^2} = \frac{1}{1+\left(1-\frac{\xi^2}{\gamma^2}\right)\Big/\sinh^2\left[\left(\gamma^2-\xi^2\right)^{1/2}\right]}. \tag{7.128}$$

In the case that the Bragg condition is satisfied, $\xi = 0$, the diffraction efficiency simplifies into the following expression:

$$DE_1^r = \tanh^2\gamma, \tag{7.129}$$

implying that the power that can be diffracted into the backward 1-order can asymptotically reach the limit of 100% as the grating strength increases.

7.7.8.2 TM Polarization Case

The *TM* polarization case is slightly more complex. As shown in Fig. 7.37, the electric field of the R-wave is polarized along the $\hat{r}_E$ direction, and the S-wave is polarized along the $\hat{s}_E$ direction. Therefore, the electric field in the grating region can be expressed as:

$$\vec{E} = \hat{r}_E R(z)\exp(-j\vec{k}_{inc}\cdot\vec{r}) + \hat{s}_E S(z)\exp(-j\vec{\sigma}_1\cdot\vec{r}), \tag{7.130}$$

where all the parameters are defined as in the *TE* polarization case. The fundamental assumption in the above equation is that the polarization vectors $\hat{r}_E$ and $\hat{s}_E$ are z-independent. In this case, the Helmholtz equation becomes:

$$\nabla^2\vec{E} - \vec{\nabla}(\vec{\nabla}\cdot\vec{E}) + k_0^2\varepsilon(\vec{r})\vec{E} = 0.$$

Inserting Eq. (7.130) into the Helmholtz equation and following the same procedure as in the *TE* polarization case, the following set of coupled differential equations can be derived. This derivation uses only the "synchronous" terms and neglects the second derivatives of $R(z)$ and $S(z)$ [402]:

$$c_R R' = -j\kappa(\hat{r}_E\cdot\hat{s}_E)S, \tag{7.131}$$

$$c_S S' + j\vartheta S = -j\kappa(\hat{r}_E\cdot\hat{s}_E)R. \tag{7.132}$$

Equations (7.131) and (7.132) are identical to Eqs. (7.122) and (7.123) if the coupling constant κ is replaced by $\kappa(\hat{r}_E \cdot \hat{s}_E)$. Therefore, all solutions presented in the *TE* polarization case for transmission and reflection gratings are valid for the *TM* polarization case by substituting γ with $\gamma_{TM} = \gamma(\hat{r}_E \cdot \hat{s}_E)$ for the grating strength. Since $(\hat{r}_E \cdot \hat{s}_E) \leq 1$, it is clear that $\gamma_{TM} \leq \gamma$, implying that the same grating is performing less efficiently in the *TM* polarization case compared to the *TE* polarization for isotropic gratings. Additionally, it can be easily shown that $(\hat{r}_E \cdot \hat{s}_E) = \cos(\theta_{inc} - \theta_1)$.

7.7.9 "Thick" and "Thin" Gratings

Before the development of rigorous vectorial electromagnetic methods, diffraction gratings were commonly categorized as "thin," "thick," or falling outside these categories. Generally, "thin" gratings, which have a small thickness, could be analyzed using the approximate transmittance method (as illustrated in Fig. 7.17). Examples of this approach include the scalar methods used by Raman and Nath for analyzing diffraction gratings [403–407]. These "thin" gratings typically do not exhibit strong wavelength and angular selectivity, meaning their performance does not depend significantly on the Bragg condition.

Conversely, "thick" gratings, which generally have a larger thickness (mostly holographic volume gratings), cannot be analyzed using simple transmittance methods. Instead, Kogelnik's two-wave coupled-wave method [402] is commonly employed. These gratings exhibit strong angular and wavelength selectivity, meaning their performance strongly depends on the Bragg condition.

Gaylord and Moharam have provided clarifications on the definitions of "thin" and "thick" gratings [447–449], based on a parameter Q'. Specifically, a volume grating can be characterized as "thin" or "thick" if the following conditions are satisfied:

$$Q'\gamma \ll 1 \quad \longrightarrow \quad \text{"thin" grating,} \tag{7.133}$$

$$\varrho = \frac{Q'}{2\gamma} \gg 1 \quad \longrightarrow \quad \text{"thick" grating,} \tag{7.134}$$

$$\text{where,} \quad Q' = \frac{2\pi\lambda_0 d}{\varepsilon_0^{1/2}\Lambda^2 \cos\theta_{inc}}.$$

The parameter γ is the grating strength as defined in Sec. 7.7.8. The parameter ε_0 is the average relative permittivity of the grating region. For "thick" gratings, where Kogelnik's two-wave coupled-wave theory is applied, there is an estimate of the error associated with the neglected diffracted orders. Specifically, the sum of the diffraction efficiencies of all neglected propagating diffracted orders is $\sum_{i \neq 0,1} DE_i < 1/\varrho^2$. If neither of the above conditions is met, more rigorous diffraction analysis methods, such as RCWA, should be employed.

Example 7.7.12:

Design a "thick" unslanted volume transmission grating on lithium niobate ($LiNbO_3$) for an incident plane wave of freespace wavelength $\lambda_0 = 1.06\,\mu$m and $\theta_{inc} = 5°$ (with $\phi_{inc} = 0°$). The polarization of the plane wave is *TE*, meaning the electric field is polarized along the y-direction. The refractive indices in the three regions are $n_1 = n_0 = n_3 = 2.155313$. For simplicity, it is assumed that the grating region behaves isotropically with an average refractive index of n_0, despite the uniaxial properties of the $LiNbO_3$ crystal. Determine the grating period required to satisfy the first-order Bragg condition for the given parameters. It is also assumed that the volume grating is sinusoidally modulated with a modulation amplitude of $n_1 = 0.588444 \times 10^{-3}$. Next, calculate the smallest grating thickness d necessary to achieve the maximum diffraction efficiency for the first Bragg order. Then, compute and plot the

first-order diffraction efficiency as a function of the angle of incidence around the Bragg angle of $\theta_B = \theta_{inc} = 5°$. Additionally, plot the diffraction efficiency of the Bragg order as a function of the grating thickness d for $\theta_{inc} = \theta_B$, $\theta_{inc} = \theta_B - 0.25°$, and $\theta_{inc} = \theta_B - 0.5°$. Finally, assess whether the characterization of the grating as "thick" is valid.
Solution: The grating periodicity, Λ, can be determined from the Bragg condition, specifically from Eq. (7.82) for $m = 1$, which gives $\Lambda = 2.82143\,\mu$m. With the given parameters and the calculated grating periodicity, the grating parameter ϱ is found to be 111.29 [refer to Eq. (7.134)]. This value indicates that the grating is classified as a "thick" grating, making it appropriate for analysis using Kogelnik's two-wave coupled-wave theory. The error introduced by neglecting all diffracted orders other than the zeroth and first is given by $\sum_{i \neq 0,1} DE_i < 1/\varrho^2 = 8 \times 10^{-5}$.

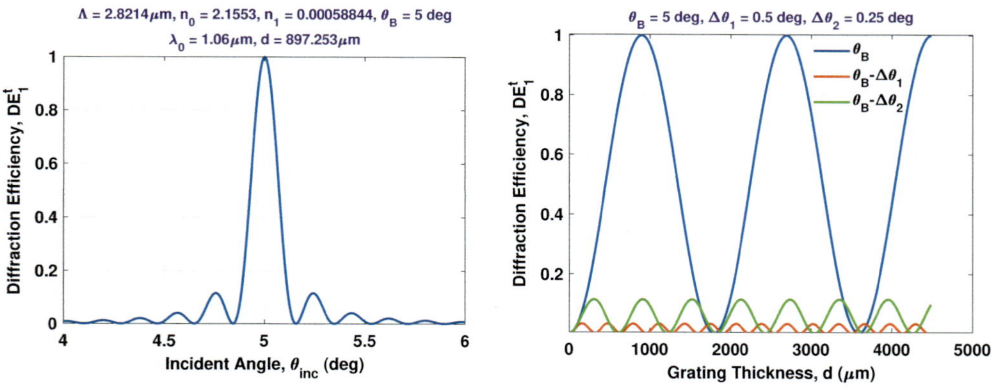

FIGURE 7.38: The first order transmitted efficiency, DE_1^t, of a a "thick" grating with $\phi = 90°$ (unslanted), $n_0 = 2.155313$, $n_1 = 0.588444 \times 10^{-3}$. The incident plane wave is TE polarized and $\theta_{inc} = 5°$. (a) DE_1^t as a function of the angle of incidence. (b) DE_1^t as a function of the grating thickness.

According to Kogelnik's analysis, the diffraction efficiency of the first Bragg order can be determined using Eq. (7.126). To achieve maximum diffraction efficiency, the minimum grating thickness is such that $\gamma = \pi/2$. Given that $\gamma = \pi n_1 d / \lambda_0 \cos\theta_{inc}$ (at the Bragg condition), the required grating thickness is found to be $d = 897.25\,\mu$m.

To analyze the diffraction efficiency of the first Bragg order as a function of the angle of incidence, Eq. (7.125) should be used. The resulting diffraction efficiency is shown in Fig. 7.38(a). The figure clearly demonstrates that the grating is highly sensitive to the angle of incidence, justifying its classification as a "thick" grating.

Furthermore, the diffraction efficiency as a function of the grating thickness is also determined from Eq. (7.125) and is depicted in Fig. 7.38(b). When $\theta_{inc} = \theta_B = 5°$, the diffraction efficiency follows a sinusoidal pattern with unit amplitude. As the angle of incidence deviates from the Bragg condition, the diffraction efficiency continues to exhibit a sinusoidal behavior, but with a significantly reduced amplitude and a shorter period.

Example 7.7.13:

This example is similar to the previous one, but the design now focuses on a "thick" slanted reflection grating. Assume the same material parameters n_1, n_0, n_3 and freespace wavelength λ_0 as in Example 7.7.12. The slant angle is given as $\phi = 30°$, and the angle of incidence is

$\theta_{inc} = 0°$. In this case, the grating thickness is specified as $d = 750\,\mu$m. Again, the polarization is assumed to be TE. Determine the grating period required to satisfy the first-order Bragg condition for the given parameters. Next, calculate and plot the first-order diffraction efficiency as a function of the angle of incidence around the Bragg angle of $\theta_B = \theta_{inc} = 0°$. Additionally, plot the diffraction efficiency of the Bragg order as a function of the grating thickness d for $\theta_{inc} = \theta_B$, $\theta_{inc} = \theta_B - 0.075°$, and $\theta_{inc} = \theta_B - 0.0125°$. Finally, assess whether the characterization of the grating as "thick" is valid.
Solution: The grating periodicity, Λ, can be determined from the Bragg condition, specifically from Eq. (7.82) for $m = 1$, which gives $\Lambda = 0.28395\,\mu$m. With the given parameters and the calculated grating periodicity, the grating parameter ϱ is found to be 7768.8 [refer to Eq. (7.134)]. This value indicates that the grating is definitively characterized as a "thick" grating, making it suitable for analysis using Kogelnik's two-wave coupled-wave theory. The error introduced by neglecting all other diffracted orders is given by $\sum_{i\neq 0,1} DE_i < 1/varrho^2 = 1.7 \times 10^{-8}$.

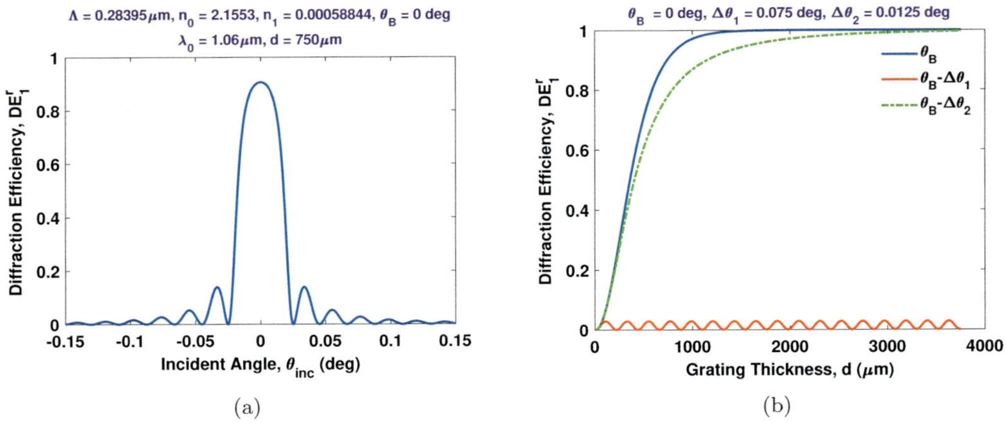

FIGURE 7.39: The first order reflected efficiency, DE_1^r, of a a "thick" grating with $\phi = 30°$ (slanted), $n_0 = 2.155313$, $n_1 = 0.588444 \times 10^{-3}$. The incident plane wave is TE polarized and $\theta_{inc} = 0°$. (a) DE_1^r as a function of the angle of incidence. (b) DE_1^r as a function of the grating thickness.

According to Kogelnik's analysis, the diffraction efficiency of the first Bragg order can be determined using Eq. (7.129). To analyze the diffraction efficiency of the first Bragg order as a function of the angle of incidence, Eq. (7.128) should be used. The resulting diffraction efficiency is shown in Fig. 7.39(a). The figure clearly demonstrates that the grating is highly sensitive to the angle of incidence, further justifying its classification as a "thick" grating. In fact, this grating is more sensitive (or angularly selective) than the transmission grating in Example 7.7.12, due to its larger ϱ value.

Furthermore, the diffraction efficiency as a function of the grating thickness is determined from Eq. (7.128) and is illustrated in Fig. 7.39(b). When $\theta_{inc} = \theta_B = 0°$, the diffraction efficiency follows a hyperbolic tangent squared response. When the angle of incidence deviates from the Bragg condition, the diffraction efficiency can either continue to follow the square of a hyperbolic tangent or exhibit sinusoidal-like behavior, but with a smaller magnitude and shorter period. This sinusoidal-like behavior is predicted from Eq. (7.128) when $|\xi| > \gamma$.

7.8 Introduction to the Theory of Photonic Crystals

Photonic crystals are structures whose properties vary periodically in one, two, or three dimensions. This periodicity is typically achieved by alternating between two dielectric materials with differing relative permittivities, as shown schematically in Fig. 7.40. The concept of photonic crystal structures was first introduced by Eli Yablonovitch [450,451] to suppress spontaneous emission in semiconductor lasers. A photonic crystal is an artificial structure that manipulates light using frequency bandgaps, analogous to how semiconductor crystals control electrons through energy bandgaps. In semiconductor crystals, the periodicity is on the order of a few angstroms, corresponding to an atomic-scale periodic array. In contrast, photonic bandgap materials involve similar structures at a much larger scale. Specifically, in photonic crystals (or photonic bandgap materials), the periodicity must be comparable to the wavelength of the electromagnetic radiation for which the crystals are designed. For example, for visible light, the periodicity should be on the submicron scale, while for microwaves, it should be in the millimeter range.

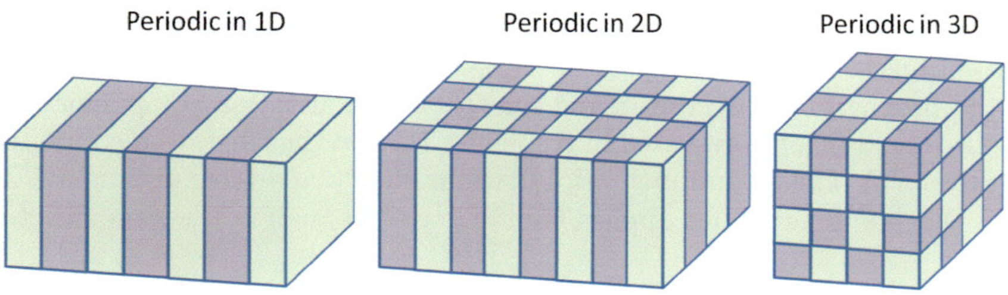

FIGURE 7.40: Example geometric configurations of one-dimensional, two-dimensional, and three-dimensional photonic crystals comprised of two alternating materials [452].

In the early 1990s, the applications of photonic crystals (PCs) were limited. However, today, they are widely utilized across various fields, with the potential for even broader applications in the future. Current examples include integrated photonics, where PCs are employed in filters, couplers, modulators, and resonators. They are also used in photonic crystal fibers, which have applications in sensing, as well as in the creation of highly efficient light-emitting diodes. Furthermore, PCs are integral to the development of artificial tunable metamaterials, radio-frequency antennas, reflectors, and more. This section will explore some of the techniques used to analyze and study photonic crystals, beginning with one-dimensional photonic crystals and expanding to cover two-dimensional structures.

7.8.1 One-Dimensional Photonic Crystals

A basic one-dimensional periodic photonic crystal (PC) is illustrated in Fig. 7.41. This structure consists of a repeated sequence of two homogeneous, isotropic layers, each with distinct relative permittivities, ε_1 and ε_2. The period of the 1D PC, denoted by Λ, represents the length of one complete unit cell. Within this unit cell, the first material has a thickness expressed as a fraction of the period, $F\Lambda$, where $0 \leq F \leq 1$. Consequently, the thickness of the second layer is $(1-F)\Lambda$. The direction of wave propagation is along the z-axis, which is aligned with the direction of periodicity.

In this section, two methods for determining the photonic bandstructure of a one-dimensional photonic crystal (PC) will be discussed. The first method uses a layered

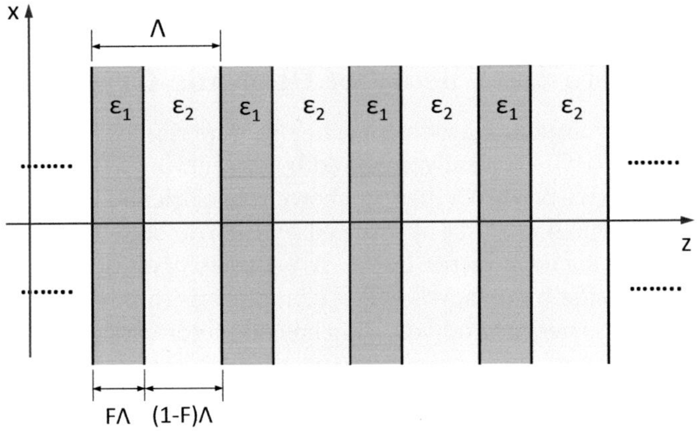

FIGURE 7.41: The geometric configuration of a one-dimensional photonic crystal comprised by two alternating homogeneous isotropic materials.

electromagnetic approach combined with Bloch's theorem, leading to a transcendental equation. The solutions to this equation provide the allowed frequency values for a given wavevector along the propagation direction z and for specific polarization [145]. This approach is similar to the modal theory presented in Appendix 7G and the effective medium theory discussed in Sec. 7.7.7. The second method, known as the "plane wave expansion" approach, also utilizes Bloch's theorem but applies Fourier series to analyze the bandstructure [452].

7.8.1.1 Electromagnetics of Periodic Layered Media—Yeh Approach

In this subsection, the approach of Yeh [145] will be presented. Figure 7.42 illustrates two consecutive unit cells, labeled as the n-th and $(n+1)$-th cells, along with a wavevector diagram. The unit cell length, Λ, corresponds to the periodicity of the 1D photonic crystal (PC). For simplicity, a two-dimensional incident wavevector, $\vec{k} = k_x\,\hat{\imath}_x + k_z\,\hat{\imath}_z$, is assumed. Extending the theory to include a three-dimensional wavevector is straightforward. The z-component of the wavevector differs in the two distinct regions with relative permittivities ε_1 and ε_2, denoted as k_{1z} and k_{2z}, respectively. The x-component, k_x, remains conserved across each region due to the phase-matching condition, which is a consequence of the electromagnetic boundary conditions. The wavevector components satisfy the plane wave dispersion equation in each region of the unit cell: $k_x^2 + k_{iz}^2 = k_0^2\varepsilon_i = k_0^2 n_i^2$ (where $i = 1, 2$), where n_i is the corresponding refractive index, $k_0 = \omega/c$ is the freespace wavenumber, and c is the velocity of light in free space. Two independent polarizations can be considered: *TE* polarization, where the electric field is oriented along the y-direction (perpendicular to the figure plane), and *TM* polarization, where the magnetic field is oriented along the y-direction.

Assuming the *TE* polarization is of interest, the electric field in each region of the n-th unit cell can be expressed as:

$$\vec{E} = \hat{\imath}_y e^{-jk_x x}\begin{cases} \mathrm{a}_n e^{-jk_{1z}[z-(n-1)\Lambda]} + b_n e^{+jk_{1z}[z-(n-1)\Lambda]}, & (n-1)\Lambda \le z \le (n-1)\Lambda + \mathrm{a}, \\ c_n e^{-jk_{1z}[z-(n-1)\Lambda-\mathrm{a}]} + d_n e^{+jk_{1z}[z-(n-1)\Lambda-\mathrm{a}]}, & (n-1)\Lambda + \mathrm{a} \le z \le n\Lambda, \end{cases} \tag{7.135}$$

where the amplitude coefficients $\mathrm{a_n}$, b_n, c_n, and d_n correspond to waves propagating in the $+z$ or $-z$ direction. The parameter $\mathrm{a} = F\Lambda$, while $b = (1-F)\Lambda$, as indicated in Fig. 7.42. Using Maxwell's equations, it is straightforward to verify that the tangential magnetic field

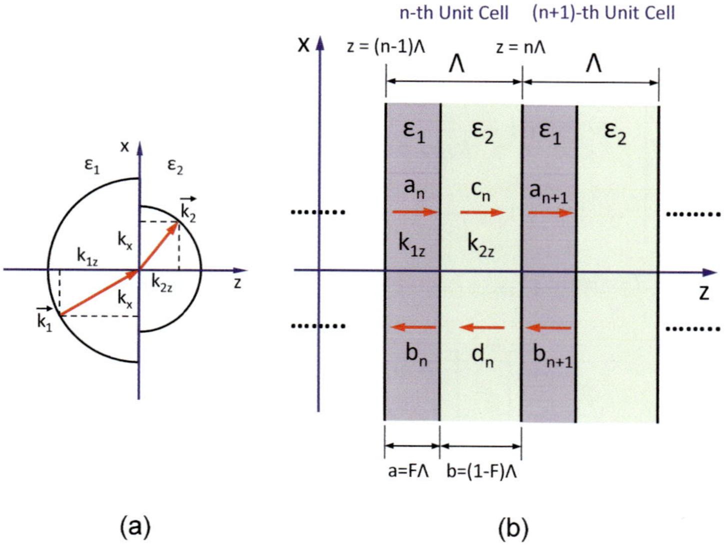

FIGURE 7.42: (a) A two-dimensional wavevector diagram that shows the x and z wavevector components in regions of relative permittivity ε_1 and ε_2, respectively. The phase matching condition guarantees that the tangential to the interfaces components are the same (k_x components). (b) A set of two consecutive unit cells in a layered one-dimensional periodic photonic crystal comprised by two alternating homogeneous isotropic materials of relative permittivities ε_1 and ε_2. The n-th unit cell starts at $z = (n-1)\Lambda$ while the $(n+1)$-th unit cell starts at $z = n\Lambda$. The thicknesses of the two homogeneous layers comprising the unit cell are $\mathrm{a} = F\Lambda$ and $b = (1-F)\Lambda$, respectively.

is given by $H_x = -\partial E_y/\partial z$, where $\vec{E} = E_y\hat{\imath}_y$. Consequently, the tangential magnetic field can be expressed as:

$$H_x = -\frac{e^{-jk_x x}}{\omega\mu_0}\begin{cases} k_{1z}\mathrm{a}_n e^{-jk_{1z}[z-(n-1)\Lambda]} + k_{1z}b_n e^{+jk_{1z}[z-(n-1)\Lambda]}, \\ \qquad\qquad \text{for } (n-1)\Lambda \leq z \leq (n-1)\Lambda + \mathrm{a}, \\ k_{2z}c_n e^{-jk_{1z}[z-(n-1)\Lambda-\mathrm{a}]} + k_{2z}d_n e^{+jk_{1z}[z-(n-1)\Lambda-\mathrm{a}]}, \\ \qquad\qquad \text{for } (n-1)\Lambda + \mathrm{a} \leq z \leq n\Lambda. \end{cases} \tag{7.136}$$

Using the continuity of the tangential electric and magnetic field components at $z = (n-1)\Lambda + \mathrm{a}$, the following matrix equation can be derived (where $\mathrm{a} = F\Lambda$ and $b = (1-F)\Lambda$):

$$\begin{bmatrix} e^{-jk_{1z}\mathrm{a}} & e^{+jk_{1z}\mathrm{a}} \\ -k_{1z}e^{-jk_{1z}\mathrm{a}} & k_{1z}e^{+jk_{1z}\mathrm{a}} \end{bmatrix}\begin{bmatrix} \mathrm{a}_n \\ b_n \end{bmatrix} = \begin{bmatrix} 1 & 1 \\ -k_{2z} & k_{2z} \end{bmatrix}\begin{bmatrix} c_n \\ d_n \end{bmatrix},$$

and solving for a_n and b_n as functions of c_n and d_n it can be written in the following form:

$$\begin{bmatrix} \mathrm{a}_n \\ b_n \end{bmatrix} = \frac{1}{2}\begin{bmatrix} e^{+jk_{1z}\mathrm{a}}\left(1+\dfrac{k_{2z}}{k_{1z}}\right) & e^{+jk_{1z}\mathrm{a}}\left(1-\dfrac{k_{2z}}{k_{1z}}\right) \\ e^{-jk_{1z}\mathrm{a}}\left(1-\dfrac{k_{2z}}{k_{1z}}\right) & e^{-jk_{1z}\mathrm{a}}\left(1+\dfrac{k_{2z}}{k_{1z}}\right) \end{bmatrix}\begin{bmatrix} c_n \\ d_n \end{bmatrix}. \tag{7.137}$$

Using exactly the same approach it is possible to relate the c_n and d_n coefficients to the a_{n+1} and b_{n+1} coefficients and derive an equation analogous to Eq. (7.137):

$$\begin{bmatrix} c_n \\ d_n \end{bmatrix} = \frac{1}{2} \begin{bmatrix} e^{+jk_{2z}b}\left(1+\dfrac{k_{1z}}{k_{2z}}\right) & e^{+jk_{2z}b}\left(1-\dfrac{k_{1z}}{k_{2z}}\right) \\ e^{-jk_{2z}b}\left(1-\dfrac{k_{1z}}{k_{2z}}\right) & e^{-jk_{2z}b}\left(1+\dfrac{k_{1z}}{k_{2z}}\right) \end{bmatrix} \begin{bmatrix} \mathrm{a}_{n+1} \\ b_{n+1} \end{bmatrix}. \tag{7.138}$$

Eliminating c_n and d_n from Eqs. (7.137) and (7.138), the following equation is readily obtained:

$$\begin{bmatrix} \mathrm{a}_n \\ b_n \end{bmatrix} = \begin{bmatrix} A_{TE} & B_{TE} \\ C_{TE} & D_{TE} \end{bmatrix} \begin{bmatrix} \mathrm{a}_{n+1} \\ b_{n+1} \end{bmatrix}, \tag{7.139}$$

where a_{n+1} and b_{n+1} are the coefficients of the electric field forward and backward wave in the first region of the $(n+1)$-th unit cell (see Fig. 7.42) and the coefficients A_{TE}, B_{TE}, C_{TE}, and D_{TE} are given by:

$$A_{TE} = e^{+jk_{1z}\mathrm{a}}\left[\cos(k_{2z}b) + j\frac{1}{2}\left(\frac{k_{2z}}{k_{1z}} + \frac{k_{1z}}{k_{2z}}\right)\sin(k_{2z}b)\right],$$

$$B_{TE} = \frac{1}{2}je^{+jk_{1z}\mathrm{a}}\left(\frac{k_{2z}}{k_{1z}} - \frac{k_{1z}}{k_{2z}}\right)\sin(k_{2z}b),$$

$$C_{TE} = -\frac{1}{2}je^{-jk_{1z}\mathrm{a}}\left(\frac{k_{2z}}{k_{1z}} - \frac{k_{1z}}{k_{2z}}\right)\sin(k_{2z}b),$$

$$D_{TE} = e^{-jk_{1z}\mathrm{a}}\left[\cos(k_{2z}b) - j\frac{1}{2}\left(\frac{k_{2z}}{k_{1z}} + \frac{k_{1z}}{k_{2z}}\right)\sin(k_{2z}b)\right].$$

It can be easily demonstrated that $A_{TE}D_{TE} - B_{TE}C_{TE} = 1$, confirming that the matrix in Eq. (7.139) is unimodular. It is important to note that the same approach applies to the case of *TM* polarization. The primary difference lies in the field used for the analysis: instead of using the electric field as in Eq. (7.135), the magnetic field is employed for the *TM* case. By following the same procedure, Eq. (7.139) can be derived for the *TM* polarization, with the corresponding coefficients A_{TM}, B_{TM}, C_{TM}, and D_{TM} given by:

$$A_{TM} = e^{+jk_{1z}\mathrm{a}}\left[\cos(k_{2z}b) + j\frac{1}{2}\left(\frac{k_{2z}/\varepsilon_2}{k_{1z}/\varepsilon_1} + \frac{k_{2z}/\varepsilon_2}{k_{1z}/\varepsilon_1}\right)\sin(k_{2z}b)\right],$$

$$B_{TM} = \frac{1}{2}je^{+jk_{1z}\mathrm{a}}\left(\frac{k_{2z}/\varepsilon_2}{k_{1z}/\varepsilon_1} - \frac{k_{2z}/\varepsilon_2}{k_{1z}/\varepsilon_1}\right)\sin(k_{2z}b),$$

$$C_{TM} = -\frac{1}{2}je^{-jk_{1z}\mathrm{a}}\left(\frac{k_{2z}/\varepsilon_2}{k_{1z}/\varepsilon_1} - \frac{k_{2z}/\varepsilon_2}{k_{1z}/\varepsilon_1}\right)\sin(k_{2z}b),$$

$$D_{TM} = e^{-jk_{1z}\mathrm{a}}\left[\cos(k_{2z}b) - j\frac{1}{2}\left(\frac{k_{2z}/\varepsilon_2}{k_{1z}/\varepsilon_1} + \frac{k_{2z}/\varepsilon_2}{k_{1z}/\varepsilon_1}\right)\sin(k_{2z}b)\right].$$

The propagation of electromagnetic waves in periodic media is analogous to the behavior of electrons in crystalline solids, which also form periodic lattices. This similarity allows the application of concepts such as Bloch waves, Brillouin zones, and forbidden bands to photonic crystals (PCs). In the case of a one-dimensional PC, the relative permittivity is periodic, i.e., $\varepsilon(z + \Lambda) = \varepsilon(z)$. Due to this periodicity, and in accordance with Floquet's theorem, the solutions to the wave equation in such a periodic medium must take the following form:

$$\vec{E} = \hat{\imath}_y\, E_{k_z}(x,z) = \hat{\imath}_y\, E_{k_z}(z)e^{-jk_x x}, \qquad \text{for } TE \text{ Polarization,} \tag{7.140}$$

$$\vec{H} = \hat{\imath}_y\, H_{k_z}(x,z) = \hat{\imath}_y\, H_{k_z}(z)e^{-jk_x x}, \qquad \text{for } TM \text{ Polarization,} \tag{7.141}$$

where $E_{k_z}(z)$ [or equivalently $H_{k_z}(z)$] are periodic functions with a period Λ. Therefore, $E_{k_z}(z+\Lambda) = E_{k_z}(z)$ for *TE* polarization, and $H_{k_z}(z+\Lambda) = H_{k_z}(z)$ for *TM* polarization. For simplicity, let us focus on the case of *TE* polarization. The results for *TM* polarization follow a similar reasoning. Assume that the coordinate z is within the range $(n-1)\Lambda \leq z \leq (n-1)\Lambda + \mathrm{a}$, meaning that the point z lies within the n-th unit cell. Then, by applying Eqs. (7.135), (7.140), and the periodicity condition, it can be shown that:

$$E_{k_z}(x, z+\Lambda) = \left[\mathrm{a}_{n+1}e^{-jk_{1z}[(z+\Lambda)-n\Lambda]} + b_{n+1}e^{+jk_{1z}[(z+\Lambda)-n\Lambda]}\right] e^{-jk_x x} =$$

$$E_{k_z}(x,z)e^{-jk_z\Lambda} = \left[\mathrm{a}_n e^{-jk_{1z}[z-(n-1)\Lambda]} + b_n e^{+jk_{1z}[z-(n-1)\Lambda]}\right] e^{-jk_x x}e^{-jk_z\Lambda}.$$

The last equation can be written in the following matrix form with the help of Eq. (7.139):

$$\begin{bmatrix} A_{TE} & B_{TE} \\ C_{TE} & D_{TE} \end{bmatrix} \begin{bmatrix} \mathrm{a}_{n+1} \\ b_{n+1} \end{bmatrix} = e^{jk_z\Lambda} \begin{bmatrix} \mathrm{a}_{n+1} \\ b_{n+1} \end{bmatrix}. \tag{7.142}$$

The last equation clearly demonstrates that the factor $e^{jk_z\Lambda}$ is an eigenvalue of the matrix formed by the coefficients A_{TE}, B_{TE}, C_{TE}, and D_{TE}. Specifically, this matrix relates the fields at the boundaries of the unit cell, and its eigenvalues correspond to the propagation characteristics of the wave through the periodic structure. The eigenvalue can be evaluated as follows:

$$e^{jk_z\Lambda} = \frac{A_{TE}+D_{TE}}{2} \pm \left\{\left[\frac{A_{TE}+D_{TE}}{2}\right]^2 - 1\right\}^{1/2}. \tag{7.143}$$

The same equation holds for the *TM* polarization with the substitution of the A_{TE}, B_{TE}, C_{TE}, and D_{TE} with A_{TM}, B_{TM}, C_{TM}, and D_{TM}, respectively.

From Eq. (7.143), it is evident that when $|(A_p + D_p)/2| < 1$ (where $p = TE$ or $p = TM$), k_z becomes real, meaning that the Bloch waves in Eqs. (7.140) and (7.141) propagate in the z-direction. Conversely, if $|(A_p + D_p)/2| > 1$, k_z becomes complex, indicating that the Bloch waves are evanescent, and propagation along the z-direction does not occur. The frequencies where k_z is complex correspond to the photonic band gaps. The edges of these bands are determined by the condition $|(A_p + D_p)/2| = 1$. For the case of real k_z, the following equation can be derived from Eq. (7.143) [145]:

$$\cos(k_z\Lambda) = \cos(k_{1z}\mathrm{a})\cos(k_{2z}b) - \frac{1}{2}\left[\frac{k_{2z}}{k_{1z}} + \frac{k_{1z}}{k_{2z}}\right]\sin(k_{1z}\mathrm{a})\sin(k_{2z}b), \quad (TE), \tag{7.144}$$

$$\cos(k_z\Lambda) = \cos(k_{1z}\mathrm{a})\cos(k_{2z}b) - \frac{1}{2}\left[\frac{k_{2z}/\varepsilon_2}{k_{1z}/\varepsilon_1} + \frac{k_{1z}/\varepsilon_1}{k_{2z}/\varepsilon_2}\right]\sin(k_{1z}\mathrm{a})\sin(k_{2z}b), \quad (TM). \tag{7.145}$$

It should be noted that $k_{iz} = \left[(\omega/c)^2\varepsilon_i - k_x^2\right]^{1/2}$ (for $i = 1, 2$, and $\varepsilon_i = n_i^2$). For simplicity, it is assumed that $k_y = 0$. In the case of on-axis propagation ($k_x = k_y = 0$), the last two equations become identical and simplify to the following form:

$$\cos[k_z(\omega)\Lambda] = \cos\left(\frac{\omega}{c}n_1\mathrm{a}\right)\cos\left(\frac{\omega}{c}n_2 b\right) - \frac{1}{2}\left[\frac{n_2}{n_1} + \frac{n_1}{n_2}\right]\sin\left(\frac{\omega}{c}n_1\mathrm{a}\right)\sin\left(\frac{\omega}{c}n_2 b\right). \tag{7.146}$$

The dispersion relation $k_z = k_z(\omega)$, or the more commonly used $\omega = \omega(k_z)$, as given by Eq. (7.146), describes how the frequency ω depends on the Bloch wavevector k_z. In the on-axis case, this relation is independent of the polarization of the Bloch wave. Typically, it is sufficient to consider k_z within the first Brillouin zone, which spans the interval $[-\pi/\Lambda, \pi/\Lambda]$, as the various bands fold back into this zone. Band diagrams are often presented in the form $\omega = \omega(k_z)$, using normalized frequencies and wavenumbers for clarity and ease of comparison across different structures. The normalized frequency $\tilde{\omega}$ and normalized wavenumber $\tilde{k}_z$ are given by the following equations:

$$\tilde{\omega} = \frac{\omega}{cK}, \tag{7.147}$$

$$\tilde{k}_z = \frac{k_z}{K}. \tag{7.148}$$

Using the normalized variables $\tilde{\omega}$ and $\tilde{k}_z$, and noting that $\vec{K} = K, \hat{z} = 2(\pi/\Lambda), \hat{z}$ is the grating vector (in optical grating terminology, or the reciprocal lattice vector in semiconductor band structure terminology), Eq. (7.146) can be rewritten in the following normalized form:

$$\cos(\tilde{k}_z 2\pi) = \cos(\tilde{\omega} n_1 2\pi F) \cos[\tilde{\omega} n_2 2\pi(1-F)] - \frac{1}{2}\left[\frac{n_2}{n_1} + \frac{n_1}{n_2}\right] \sin(\tilde{\omega} n_1 2\pi F) \sin[\tilde{\omega} n_2 2\pi(1-F)]. \tag{7.149}$$

The solutions to this equation provide the allowed and forbidden frequencies (photonic bands and band gaps) for the given structure. The photonic band structure is typically plotted as $\tilde{\omega}$ versus $\tilde{k}_z$, illustrating the relationship between the normalized frequency and the normalized wavenumber. This plot reveals the bands where light can propagate and the gaps where it cannot.

Example 7.8.1:

Consider a one-dimensional photonic crystal (as shown in Fig. 7.41) with alternating layers of GaAs and air, having relative permittivities $\varepsilon_1 = 13$ and $\varepsilon_2 = 1$, respectively, and a filling factor $F = 0.2$ (the percentage of a period occupied by GaAs) [452]. Calculate the band diagrams for both on-axis ($k_x = k_y = 0$) and off-axis ($k_x \neq 0$ and $k_y = 0$) propagation, including the first three bands. Determine the range in which the propagation bands in the off-axis case lie in the $\tilde{\omega}$-$\tilde{k}_x$ plane for various values of k_z.

Solution: For on-axis propagation in a one-dimensional photonic crystal (PC), the band structure is typically obtained by solving the dispersion relation, Eq. (7.149), numerically, as described. The resulting band structure, which is independent of polarization for the on-axis case, reveals the frequency bands where electromagnetic waves can propagate through the crystal, as well as the gaps (bandgaps) where propagation is forbidden.

In the specific case presented, the first three bands are shown in Fig. 7.43(a), with the first bandgap occurring between the normalized frequencies $\tilde{\omega} = 0.203$ and $\tilde{\omega} = 0.454$. This bandgap is a characteristic feature of one-dimensional photonic crystals, resulting from the periodic variation in the relative permittivity of the materials that make up the crystal. The existence of bandgaps is a general property of one-dimensional PCs, as long as the relative permittivity ratio $\varepsilon_1/\varepsilon_2$ is not equal to one. A smaller ratio $\varepsilon_1/\varepsilon_2$ leads to smaller bandgaps, as discussed in Ref. [452].

When considering off-axis propagation, where the wavevector component k_x is non-zero but $k_y = 0$, the degeneracy between *TE* and *TM* polarizations is lifted. This means that the dispersion relations for *TE* [Eq. (7.144)] and *TM* [Eq. (7.145)] polarized waves will differ,

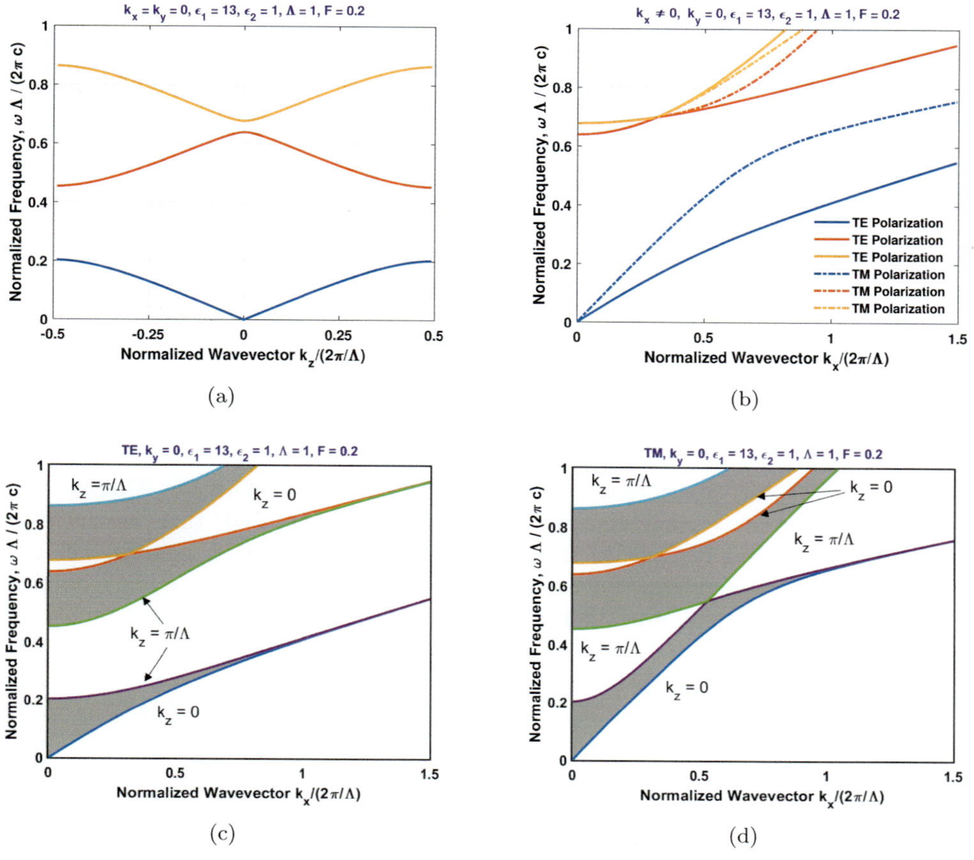

FIGURE 7.43: (a) The band diagram (ω vs. k_z) of a 1D photonic crystal consisting of alternating layers of GaAs and air, with relative permittivities $\varepsilon_1 = 13$ and $\varepsilon_2 = 1$, respectively, and a filling fraction $F = 0.2$, for the case of on-axis propagation ($k_x = k_y = 0$). The first three bands are shown. (b) The band diagram (ω vs. k_x) of the 1D photonic crystal from (a), for the case of off-axis propagation ($k_x \neq 0, k_y = 0$) for both *TE* and *TM* polarizations. Again, the first three bands are shown. (c) Regions in $\tilde{\omega}$-$\tilde{k}_x$ space where propagating bands exist for *TE* polarized waves. (d) Same as (c), but for *TM* polarization.

leading to distinct band structures for each polarization. The band diagrams for this off-axis case are illustrated in Fig. 7.43(b). It should be noted that the normalized wavevectors $\tilde{k}_{iz} = k_{iz}/K = [(\omega/cK)^2\varepsilon_i - (k_x/K)^2]^{1/2}$ (for $i = 1, 2$) are used similarly to the on-axis case. Notably, in the off-axis scenario, the solutions of the corresponding dispersion relations, Eqs. (7.144) and (7.145), indicate that no frequency bandgaps exist for either *TE* or *TM* polarization. This absence of bandgaps in the off-axis case highlights how the direction of wave propagation within the crystal significantly influences the photonic band structure. The shaded gray regions in Figs. 7.43(c) and 7.43(d) represent the regions of the $\tilde{\omega}$-$\tilde{k}_x$ plane where propagating bands can exist for the photonic crystal, for *TE* and *TM* polarization, respectively. The curves representing the first three bands for $k_z = 0$ (as shown in Fig. 7.43(b)) and for $k_z = \pi/\Lambda$ are plotted on top of the shaded regions.

The shaded gray regions in the figures effectively illustrate the range of frequencies and wave vectors where the photonic crystal allows the propagation of electromagnetic waves. The band curves for $k_z = 0$ and $k_z = \pi/\Lambda$ provide the boundaries, while the interior of the shaded regions indicates the continuous spectrum of allowed modes for intermediate k_z values.

In the case where $|(A_p + D_p)/2| > 1$ (where $p = TE$ or $p = TM$), the k_z wavevector becomes complex, as mentioned previously, i.e., $k_z = k_{zr} + jk_{zi}$. Substituting this into Eq. (7.143), which describes the eigenvalue condition for the photonic crystal, the equation becomes:

$$e^{jk_{zr}\Lambda}e^{-k_{zi}\Lambda} = \left| \frac{A_p + D_p}{2} \pm \left\{ \left[\frac{A_p + D_p}{2} \right]^2 - 1 \right\}^{1/2} \right| e^{j\pi m}, \tag{7.150}$$

where m is an integer. From the last equation, the solutions for k_{zr} and k_{zi} are:

$$k_{zr} = m\frac{\pi}{\Lambda}, \qquad m = \pm 1, \pm 2, \cdots \tag{7.151}$$

$$k_{zi} = -\frac{1}{\Lambda} \ln \left\{ \left| \frac{A_p + D_p}{2} \pm \left\{ \left[\frac{A_p + D_p}{2} \right]^2 - 1 \right\}^{1/2} \right| \right\}. \tag{7.152}$$

The imaginary part, k_{zi}, describes the exponential attenuation of the wave. This attenuation indicates that the wave is evanescent within the photonic crystal for this frequency range, meaning it decays exponentially as it propagates along the z-axis, resulting in no real propagation within the bandgap. Therefore, the equation in the complex k_z case emphasizes the condition under which photonic bandgaps occur. The frequencies corresponding to these conditions do not support propagating Bloch waves in the photonic crystal, leading to the formation of forbidden bands (bandgaps).

In this case, the two real roots of Eq. (7.143) are inverses of each other. The root that is greater than unity corresponds to the evanescent wave propagating in the $+z$ direction ($k_{zi} < 0$), while the root that is smaller than unity corresponds to the evanescent wave propagating in the $-z$ direction ($k_{zi} > 0$). These evanescent waves characterize the behavior of electromagnetic waves within the bandgap of a photonic crystal, where propagation is forbidden, and energy cannot be transmitted through the crystal in these frequency ranges.

7.8.1.2 Plane Wave Expansion Method for 1D Photonic Crystals

The Plane Wave Expansion (PWE) method is a widely used computational technique for analyzing the band structure of photonic crystals. It is particularly effective for solving Maxwell's equations within periodic structures, allowing for the determination of the crystal's band structure, or dispersion relation. The PWE method works by expanding both the fields and material properties, such as permittivity, into Fourier series. The resulting eigenvalue problem is then solved to obtain the allowed frequencies corresponding to different wave vectors.

According to Joannopoulos *et al.* [452] and Sakoda [453], using the magnetic field rather than the electric field in the formulation of the problem is more efficient. This efficiency arises from the simpler implementation of the transversality condition for the magnetic field, which reflects the absence of magnetic monopoles, or Gauss' law for magnetic fields, $\vec{\nabla} \cdot \vec{H} = 0$. As a result, by applying Maxwell's equations, the electric field can be eliminated, leading to the following Helmholtz equation with respect to the magnetic field:

$$\vec{\nabla} \times \left(\frac{1}{\varepsilon(\vec{r})} \vec{\nabla} \times \vec{H}(\vec{r}) \right) = \left(\frac{\omega}{c} \right)^2 \vec{H}(\vec{r}). \tag{7.153}$$

In the preceding Helmholtz equation, ε represents the periodically varying relative permittivity along the z-direction. According to Bloch's theorem, the magnetic field in a 1D periodic photonic crystal can be expressed in the form:

$$\vec{H} = e^{-j\vec{k}\cdot\vec{r}} \vec{H}_{k_z}(z) = e^{-jk_x x} e^{-jk_y y} e^{-jk_z z} \vec{H}_{k_z}(z), \tag{7.154}$$

where the $\exp[j(\omega t - \vec{k}\cdot\vec{r})]$ dependence is as commonly assumed. The magnetic field $\vec{H}_{k_z}(z)$ is a periodic function in the z-direction with periodicity Λ. Therefore, it can be expanded in a Fourier series, allowing for the distinction between the two orthogonal polarizations, *TE* and *TM*. In this case, the magnetic field can be expressed as:

$$\vec{H} = e^{-j\vec{k}\cdot\vec{r}}\vec{H}_{k_z}(z) = e^{-jk_x x}e^{-jk_y y}e^{-jk_z z}\sum_{m=-\infty}^{+\infty}\left[A_{1m}\hat{h}_m^{TE} + A_{2m}\hat{h}_m^{TM}\right]e^{-jmKz}. \quad (7.155)$$

In the given equation, $K = 2\pi/\Lambda$, represents the grating vector, and m is an integer denoting the different Fourier components. The unit vectors $\hat{h}_m^{TE}$ and $\hat{h}_m^{TM}$ are defined to represent the spatial components of the magnetic field for the *TE* and *TM* polarizations, respectively. To satisfy the transversality condition, $\vec{\nabla}\cdot\vec{H} = 0$, the unit vectors must be perpendicular to the wavevectors $\vec{k}_m = k_x\hat{\imath}_x + k_y\hat{\imath}_y + (k_z + mK)\hat{\imath}_z = k_x\hat{\imath}_x + k_y\hat{\imath}_y + k_{zm}\hat{\imath}_z$ (where $k_{zm} = k_z + mK$). Additionally, these unit vectors should be perpendicular to each other. This can be achieved by selecting $\hat{h}_m^{TM}$ and $\hat{h}_m^{TE}$ to satisfy the following conditions:

$$\hat{h}_m^{TM} = \frac{\vec{k}_m \times \hat{\imath}_z}{|\vec{k}_m \times \hat{\imath}_z|} = \frac{k_y\hat{\imath}_x - k_x\hat{\imath}_y}{\sqrt{k_x^2 + k_y^2}}, \quad (7.156)$$

$$\hat{h}_m^{TE} = \frac{\vec{k}_m \times \hat{h}_m^{TM}}{|\vec{k}_m \times \hat{h}_m^{TM}|} = \frac{k_x k_{zm}\hat{\imath}_x + k_y k_{zm}\hat{\imath}_y - (k_x^2 + k_y^2)\hat{\imath}_z}{\sqrt{k_x^2 + k_y^2}\sqrt{k_x^2 + k_y^2 + k_{zm}^2}}. \quad (7.157)$$

In the case of normal incidence ($k_x = k_y = 0$), the unit vectors $\hat{h}_m^{TM}$ and $\hat{h}_m^{TE}$ indeed have infinitely many possible solutions. One of these solutions will be treated separately later.

Using Eqs. (7.156) and (7.157), the magnetic field can be expressed for each polarization in the following form, with the transversality condition being automatically satisfied:

$$\vec{H}_{TE} = e^{-jk_x x}e^{-jk_y y}e^{-jk_z z}\sum_{m=-\infty}^{+\infty} C_{1m}\left[k_x k_{zm}\hat{\imath}_x + k_y k_{zm}\hat{\imath}_y - (k_x^2 + k_y^2)\hat{\imath}_z\right]e^{-jmKz}, \quad (7.158)$$

$$\vec{H}_{TM} = e^{-jk_x x}e^{-jk_y y}e^{-jk_z z}\sum_{m=-\infty}^{+\infty} C_{2m}\left[k_y\hat{\imath}_x - k_x\hat{\imath}_y\right]e^{-jmKz}. \quad (7.159)$$

In the last two equations the normalization factors from the unit vectors have been incorporated into the new coefficients C_{1m} and C_{2m}.

In the case of on-axis propagation, $k_x = k_y = 0$, the *TE* and *TM* polarizations become degenerate. This means that the distinction between the polarizations disappears, and the magnetic field can be expressed in a unified form as:

$$\vec{H} = \hat{\imath}_y\, e^{-jk_z z}\sum_{m=-\infty}^{+\infty} C_m e^{-jmKz}, \quad (7.160)$$

where instead of $\hat{\imath}_y$ any unit vector perpendicular to z-axis could be used. The relative permittivity of the 1D PC is periodic, i.e., $\varepsilon(z + \Lambda) = \varepsilon(z)$ and consequently the terms ε

and $1/\varepsilon$ can be expanded into a Fourier series as follows:

$$\varepsilon(z) = \sum_{p=-\infty}^{+\infty} \alpha_p e^{-jpKz} \quad \text{where,} \tag{7.161}$$

$$\alpha_p = F\varepsilon_1 + (1-F)\varepsilon_2, \quad p = 0,$$

$$\alpha_p = \frac{1}{j2\pi p}(\varepsilon_1 - \varepsilon_2)\left(1 - e^{-j2\pi pF}\right), \quad p \neq 0.$$

$$\frac{1}{\varepsilon(z)} = \sum_{p=-\infty}^{+\infty} \gamma_p e^{-jpKz} \quad \text{where,} \tag{7.162}$$

$$\gamma_p = \frac{F}{\varepsilon_1} + \frac{1-F}{\varepsilon_2}, \quad p = 0,$$

$$\gamma_p = \frac{1}{j2\pi p}\left(\frac{1}{\varepsilon_1} - \frac{1}{\varepsilon_2}\right)\left(1 - e^{-j2\pi pF}\right), \qquad p \neq 0.$$

Substituting Eqs. (7.160) and (7.162) into Eq. (7.153) and performing some algebraic manipulations, the resulting equation simplifies to the following eigenvalue equation:

$$\sum_{m=-\infty}^{+\infty} \gamma_{\ell-m} k_{zm} k_{z\ell} C_m = \left(\frac{\omega}{c}\right)^2 C_\ell. \tag{7.163}$$

Then for each value of k_z the corresponding frequency ω can be determined from the eigenvalues λ of Eq. (7.163) using the relation $\omega = c\sqrt{\lambda}$. This allows the construction of the $\omega = \omega(k_z)$ band diagram. To numerically solve Eq. (7.163), the number of Fourier components must be truncated to a finite value. After this truncation, Eq. (7.163) can be expressed in matrix form as follows:

$$\left[\tilde{\boldsymbol{K}}_z \tilde{\boldsymbol{\Gamma}} \tilde{\boldsymbol{K}}_z\right] \tilde{\boldsymbol{C}} = \left(\frac{\omega}{c}\right)^2 \tilde{\boldsymbol{C}} = \lambda \tilde{\boldsymbol{C}}, \quad \text{where} \tag{7.164}$$

$$\tilde{\boldsymbol{K}}_z = \operatorname{diag}\left[k_{z(-M)}, k_{z(-(M-1))}, \cdots, k_{z0}, k_{z1}, \cdots, k_{zM}\right], \quad \text{and}$$

$$\tilde{\boldsymbol{C}} = \left[C_{-M}, C_{-(M-1)}, \cdots, C_0, C_1, \cdots, C_{+M}\right]^T.$$

The matrix $\tilde{\boldsymbol{\Gamma}} = [\gamma_{i-j}]$ is a Toeplitz matrix of size $N \times N$ (as the one given in Appendix 7F, Sec. 7F.1), $N = 2M + 1$ with M being the maximum harmonic term retained in the expansion of Eq. (7.162), $\tilde{\boldsymbol{K}}_z$ is a diagonal matrix size $N \times N$, and $\tilde{\boldsymbol{C}}$ is a vector containing the coefficients C_m of size $N \times 1$.

For the case of the off-axis propagation, Eqs. (7.158) and (7.159) must be utilized. By substituting Eqs. (7.158) or (7.159) and (7.162) into Eq. (7.153) and manipulating the resulting equations, the following eigenvalue equations for the off-axis *TE* and *TM* polarization cases can be derived straightforwardly:

$$\sum_{m=-\infty}^{+\infty} \gamma_{\ell-m}\left[k_x^2 + k_y^2 + k_{zm}^2\right] C_{1m} = \left(\frac{\omega}{c}\right)^2 C_{1\ell}, \quad (TE), \tag{7.165}$$

$$\sum_{m=-\infty}^{+\infty} \gamma_{\ell-m}\left[k_x^2 + k_y^2 + k_{zm} k_{z\ell}\right] C_{2m} = \left(\frac{\omega}{c}\right)^2 C_{2\ell}, \quad (TM). \tag{7.166}$$

The last two equations can also be written in a matrix form assuming again a finite number of harmonic terms, M, in the Fourier series expansions. This leads to the following matrix equations:

$$\left\{\tilde{\mathbf{\Gamma}}\left[(k_x^2+k_y^2)\tilde{\boldsymbol{I}}+\tilde{\boldsymbol{K}}_{\boldsymbol{z}}^{\boldsymbol{2}}\right]\right\}\tilde{\boldsymbol{C}}_{\mathbf{1}}=\left(\frac{\omega}{c}\right)^2\tilde{\boldsymbol{C}}_{\mathbf{1}}=\lambda_{TE}\tilde{\boldsymbol{C}}_{\mathbf{1}},\quad (TE), \tag{7.167}$$

$$\left[(k_x^2+k_y^2)\tilde{\mathbf{\Gamma}}+\tilde{\boldsymbol{K}}_{\boldsymbol{z}}\tilde{\mathbf{\Gamma}}\tilde{\boldsymbol{K}}_{\boldsymbol{z}}\right]\tilde{\boldsymbol{C}}_{\mathbf{2}}=\left(\frac{\omega}{c}\right)^2\tilde{\boldsymbol{C}}_{\mathbf{2}}=\lambda_{TM}\tilde{\boldsymbol{C}}_{\mathbf{2}},\quad (TM), \tag{7.168}$$

where,

$$\tilde{\boldsymbol{C}}_{\mathbf{1}}=\left[C_{1(-M)},C_{1(1-M)},\cdots,C_{10},C_{11},\cdots,C_{1M}\right]^T,$$

$$\tilde{\boldsymbol{C}}_{\mathbf{2}}=\left[C_{2(-M)},C_{2(1-M)},\cdots,C_{20},C_{21},\cdots,C_{2M}\right]^T,$$

with $\tilde{\boldsymbol{I}}$ being the unity matrix of size $N\times N$, and $\tilde{\mathbf{\Gamma}}$ is the Toeplitz matrix defined previously. The λ_{TE} and λ_{TM} are the eigenvalues corresponding to the respective polarizations.

It is well known, even from grating analysis, that Fourier expansions of discontinuous functions, such as the permittivity in a 1D photonic crystal, can lead to poor convergence. Initially, Li [411] demonstrated that the traditional Laurent's direct rule used in Fourier factorization for grating theory suffers from convergence issues when the two functions in the product are simultaneously discontinuous (such as permittivity and a field component). To address this, Li [411] proposed an inverse Laurent's factorization rule, which significantly improves convergence. This approach was also employed in the rigorous coupled-wave analysis discussed in Sec. 7.7. Building on this methodology, Antos [454] applied similar principles to 1D and 2D photonic crystals, significantly enhancing the convergence of the plane wave expansion method. The eigenvalue equations, modified according to Antos, for both on-axis and off-axis propagation (and for both *TE* and *TM* polarizations), are presented below:

$$\left[\tilde{\boldsymbol{K}}_{\boldsymbol{z}}\tilde{\boldsymbol{E}}^{-\mathbf{1}}\tilde{\boldsymbol{K}}_{\boldsymbol{z}}\right]\tilde{\boldsymbol{C}}=\left(\frac{\omega}{c}\right)^2\tilde{\boldsymbol{C}},\quad \text{(on axis)}, \tag{7.169}$$

$$\left\{\tilde{\boldsymbol{E}}^{-\mathbf{1}}\left[(k_x^2+k_y^2)\tilde{\boldsymbol{I}}+\tilde{\boldsymbol{K}}_{\boldsymbol{z}}^{\ 2}\right]\right\}\tilde{\boldsymbol{C}}_{\mathbf{1}}=\left(\frac{\omega}{c}\right)^2\tilde{\boldsymbol{C}}_{\mathbf{1}},\quad (TE), \tag{7.170}$$

$$\left[(k_x^2+k_y^2)\tilde{\mathbf{\Gamma}}+\tilde{\boldsymbol{K}}_{\boldsymbol{z}}\tilde{\boldsymbol{E}}^{-\mathbf{1}}\tilde{\boldsymbol{K}}_{\boldsymbol{z}}\right]\tilde{\boldsymbol{C}}_{\mathbf{2}}=\left(\frac{\omega}{c}\right)^2\tilde{\boldsymbol{C}}_{\mathbf{2}},\quad (TM). \tag{7.171}$$

Here, $\tilde{\boldsymbol{E}}=[\alpha_{i-j}]$ is a Toeplitz matrix of size $N\times N$ analogous to $\tilde{\mathbf{\Gamma}}$. To demonstrate the plane wave expansion technique an example follows.

Example 7.8.2:

The Example 7.8.1 is treated again using the plane wave expansion method (with $M=15$ and $N=31$ harmonics) and is compared to Yeh's approach. The first three bands are included as in Example 7.8.1.

Solution: The plane wave expansion method was applied using $M=15$ harmonics (resulting in $N=31$ terms). The eigenvalue equations Eq. (7.164) for on-axis propagation, and Eqs. (7.167) and (7.168) for off-axis propagation with $k_x\neq 0$ and $k_y=0$ in the *TE* and *TM* polarizations were solved.

Additionally, the fast Fourier factorization method proposed by Antos [454] was employed via Eq. (7.169) (on-axis), Eq. (7.170) (*TE*) and Eq. (7.171) (TM), yielding improved performance. Furthermore, results were also calculated using Yeh's electromagnetic approach, as presented in Sec. 7.8.1.1. Yeh's results are essentially exact since they do not rely

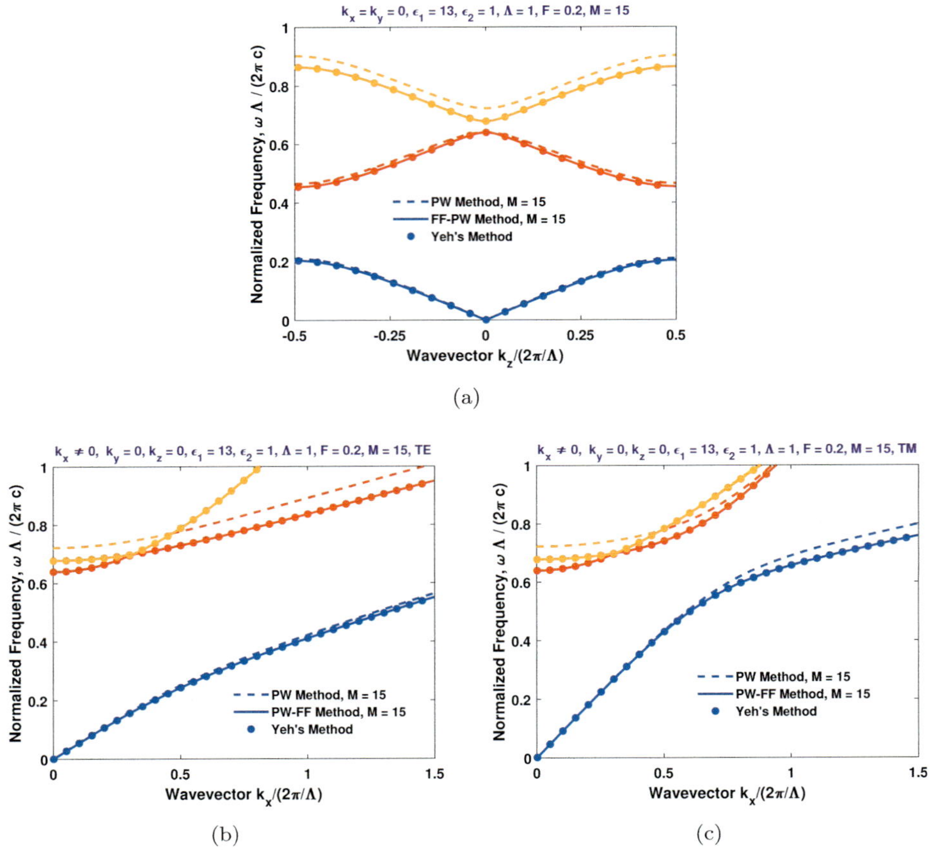

FIGURE 7.44: (a) The band diagram (ω vs. normalized k_z) of a 1D PC consisting of alternating layers of GaAs and air with relative permittivities of $\varepsilon_1 = 13$ and $\varepsilon_2 = 1$, respectively, and $F = 0.2$, in the case of on-axis propagation ($k_x = k_y = 0$). The first 3 bands are shown. The dash lines correspond to the PWE method with $N = 15$ harmonics. The solid lines correspond to the PWE method with Antos fast factorization with $M = 15$ harmonics. The solid dots lines correspond to Yeh' method. (b) The band diagram (ω vs. k_x) of the 1D PC of (a), in the case of off-axis propagation ($k_x \neq 0, k_z = k_y = 0$) for *TE* polarization. (c) The band diagram (ω vs. k_x) of the 1D PC of (a), in the case of off-axis propagation ($k_x \neq 0, k_z = k_y = 0$) for *TM* polarization.

on a finite number of harmonics and depend only on the accuracy of solving the transcendental Eqs. (7.146), (7.144), and (7.145). The calculated band diagrams are summarized in Figs. 7.44(a) (on-axis), 7.44(b) (off-axis *TE*), and 7.44(c) (off-axis *TM*). In these figures, the dashed lines represent the plane wave method results with $M = 15$ harmonics, while the solid lines correspond to the plane wave method with Antos' fast factorization with $M = 15$ harmonics. The solid dots indicate Yeh's method results. These figures clearly show that the improved convergence plane wave expansion method by Antos converges to Yeh's results with only $M = 15$ harmonics, whereas the traditional approach has not fully converged, particularly for the higher bands.

7.8.2 Two-Dimensional Photonic Crystals

Two-dimensional photonic crystals (PCs) display periodicity in two dimensions, typically along the x- and y-directions. The periodic structure of a 2D photonic crystal can be characterized using 2D periodic lattices, analogous to the lattice structures in semiconductor theory. The most general form of a Bravais lattice in two dimensions is the oblique (or triangular) lattice [455]. When the angle between the two lattice vectors is 90 degrees, the lattice assumes a more symmetric configuration, resulting in either a rectangular or square Bravais lattice, depending on whether the unit cell vectors are of unequal or equal lengths, respectively. The rectangular lattice can be further classified into primitive rect-

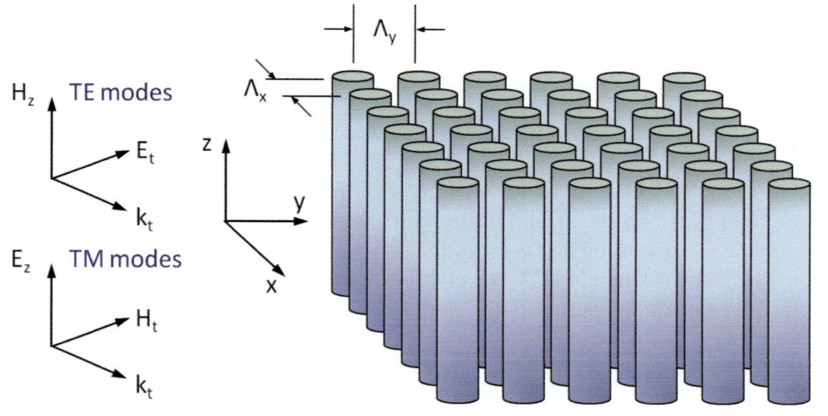

FIGURE 7.45: Example configuration of a two-dimensional photonic crystal comprised of infinite cylinders aligned along the z-axis and spaced according to a rectangular grid. The orientation of the coordinate system is shown along with the electric and magnetic fields and transverse wavevectors of the corresponding *TE* and *TM* modes.

angular lattice and a centered rectangular lattice, the latter featuring an additional lattice point at its center. The centered rectangular lattice can be viewed as a primitive triangular lattice with reduced symmetry. Furthermore, when the lattice vectors of an oblique (triangular) lattice are of equal length and the angle between them is either 60 degrees or 120 degrees, a hexagonal lattice is formed, resembling an equilateral triangular lattice. A typical two-dimensional photonic crystal (PC) based on a rectangular lattice, consisting of infinitely long dielectric cylinders aligned along the z-direction, is shown in Fig. 7.45. The terminology *TE*-mode and *TM*-mode are defined later in the text.

In this analysis, the focus is on rectangular and equilateral triangular lattices. However, the methodology can be easily extended to more general oblique lattices. These two specific lattice types are shown in Figs. 7.46 and 7.47, along with their corresponding reciprocal lattices [455]. The coordinate axis system is illustrated in Fig. 7.45, where the periodicity occurs in the xy-plane, and the structures are uniform along the z-direction.

Each grid point, $\vec{R}$, in the two-dimensional lattices, can be represented in the following form:

$$\vec{R} = m_1\vec{a}_1 + m_2\vec{a}_2, \quad \text{where} \quad m_1, m_2 \in \mathbb{Z}, \tag{7.172}$$

where $\vec{a}_1$ and $\vec{a}_2$ are the basis vectors of the two-dimensional grid. In the reciprocal lattice [455], the grid points are expressed as:

$$\vec{K} = m_1\vec{K}_1 + m_2\vec{K}_2, \quad \text{where} \quad m_1, m_2 \in \mathbb{Z}, \quad \text{and} \tag{7.173}$$

$$\vec{K}_1 = 2\pi\frac{\vec{a}_2 \times \vec{a}_3}{\vec{a}_1 \cdot (\vec{a}_2 \times \vec{a}_3)}, \quad \vec{K}_2 = 2\pi\frac{\vec{a}_3 \times \vec{a}_1}{\vec{a}_1 \cdot (\vec{a}_2 \times \vec{a}_3)}.$$

where $\vec{a}_3 = \Lambda_z\hat{z}$ (with Λ_z arbitrary). The lattice and reciprocal lattice basis vectors, and the unit cell area are summarized in Table 7.3 for the rectangular and the equilateral grids.

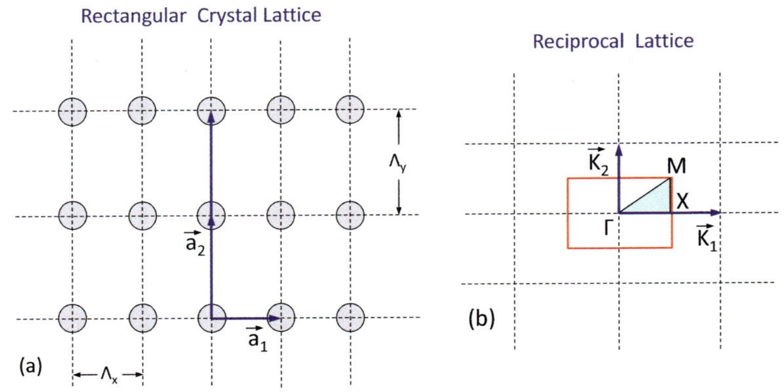

FIGURE 7.46: (a) A top view of an array of cylinders arranged in a rectangular lattice with periodicities Λ_x and Λ_y along the x- and y-directions, respectively. (b) The corresponding reciprocal lattice, showing the reciprocal lattice vectors $\vec{K}_1$ and $\vec{K}_2$, with the first Brillouin zone outlined by a red rectangle. The shaded triangle (ΓXM) within the first Brillouin zone represents the irreducible Brillouin zone. The symmetry points Γ, X, and M are also indicated.

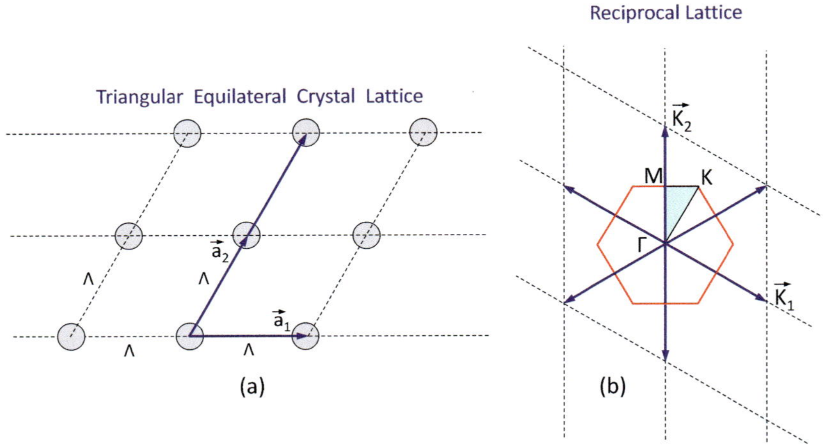

FIGURE 7.47: (a) A top view of an array of cylinders arranged in a hexagonal lattice, which is equivalent to an equilateral triangle lattice (an oblique lattice with equal lattice vectors and a 60 deg angle between them) with periodicity Λ. (b) The corresponding reciprocal lattice, showing the reciprocal lattice vectors, $\vec{K}_1$ and $\vec{K}_2$, with the first Brillouin zone outlined by a red hexagon. The shaded triangle (ΓMK) within the first Brillouin zone represents the irreducible Brillouin zone. The symmetry points Γ, M and K are also indicated.

As in the one-dimensional photonic crystal (discussed in Sec. 7.8.1.2), the Helmholtz equation for the magnetic field, given by Eq. (7.153), is employed. This approach is chosen because the operator is Hermitian, ensuring that the eigenvalues are real. Unlike the 1D case, there is no direct counterpart to Yeh's method, as introduced in Sec. 7.8.1.1. According to Bloch's theorem, the magnetic field in the two-dimensional periodic photonic crystal can be expressed in the following form:

$$\vec{H} = e^{j\vec{k}\cdot\vec{r}}\vec{H}_{\mathbf{K}}(x,y) = e^{-jk_x x}e^{-jk_y y}e^{-jk_z z}\vec{H}_{\mathbf{K}}(x,y). \tag{7.174}$$

The magnetic field $\vec{H}_{\mathbf{K}}(x,y)$ is a periodic in the xy-plane with a periodicity defined by the basis vectors $\vec{a}_1$ and $\vec{a}_2$. This means that $\vec{H}_{\mathbf{K}}(x,y) = \vec{H}_{\mathbf{K}}(\vec{r}) = \vec{H}_{\mathbf{K}}(\vec{r}+\vec{R})$, where

TABLE 7.3: Basis vectors and their reciprocal lattices basis vectors for a rectangular grid and an equilateral grid. The parameters Λ_x and Λ_y are the periodicities along the x and y directions, respectively, for the rectangular lattice. In contrast, Λ denotes the distance between nearest-neighbor lattice points in the equilateral lattice.

	Rectangular Grid	Equilateral Grid
$\vec{a}_1$	$\Lambda_x \hat{\imath}_x$	$\Lambda \hat{\imath}_x$
$\vec{a}_2$	$\Lambda_y \hat{\imath}_y$	$\Lambda\left(\frac{1}{2}\hat{\imath}_x + \frac{\sqrt{3}}{2}\hat{\imath}_y\right)$
$\vec{K}_1$	$\frac{2\pi}{\Lambda_x}\hat{\imath}_x$	$\frac{2\pi}{\Lambda}\left(\hat{\imath}_x - \frac{\sqrt{3}}{3}\hat{\imath}_y\right)$
$\vec{K}_2$	$\frac{2\pi}{\Lambda_y}\hat{\imath}_y$	$\frac{2\pi}{\Lambda}\frac{2\sqrt{3}}{3}\hat{\imath}_y$
$\lvert\vec{a}_1 \times \vec{a}_2\rvert$	$\Lambda_x \Lambda_y$	$\Lambda^2 \frac{\sqrt{3}}{2}$

$\vec{R}$ is a lattice vector. For convenience, the reciprocal lattice vector $\vec{K}$ from Eq. (7.173) is sometimes denoted as $\mathbf{K}$ when used as a subscript. Due to its periodicity, the magnetic field can be expanded into a two-dimensional Fourier series (refer to the Appendices "Two-Dimensional Fourier Transform" (Appendix 5D) and "Fourier Series" (Appendix 5E) in the *Scalar Diffraction* chapter, Chapter 5).

To satisfy the transversality condition $\vec{\nabla} \cdot \vec{H} = 0$, which implies that $(\vec{k} + \vec{K}) \cdot \vec{H}_{\mathbf{K}} = 0$, it must follow that $\vec{k} + \vec{K}$ is perpendicular to $\vec{H}_{\mathbf{K}}$. By selecting the unit vectors $\hat{h}_{\mathbf{K}1}$, $\hat{h}_{\mathbf{K}2}$, and $\hat{h}_{\mathbf{K}3} = \vec{k} + \vec{K}/|\vec{k} + \vec{K}|$, as shown in Fig. 7.48, the magnetic field can be expressed in the following form:

$$\vec{H} = \sum_{m_1}\sum_{m_2}\left[C_{1,m_1m_2}\hat{h}_{\mathbf{K}1} + C_{2,m_1m_2}\hat{h}_{\mathbf{K}2}\right]e^{-j(\vec{k}+m_1\vec{K}_1+m_2\vec{K}_2)\cdot\vec{r}} =$$

$$= \sum_{\mathbf{K}}\left[C_{1\mathbf{K}}\hat{h}_{\mathbf{K}1} + C_{2\mathbf{K}}\hat{h}_{\mathbf{K}2}\right]e^{-j(\vec{k}+\vec{K})\cdot\vec{r}} = \sum_{\mathbf{K}}\left[C_{1\mathbf{K}}\hat{h}_{\mathbf{K}1} + C_{2\mathbf{K}}\hat{h}_{\mathbf{K}2}\right]e^{-j\vec{k}_{\mathbf{K}}\cdot\vec{r}}, \quad (7.175)$$

where $\vec{k}_{\mathbf{K}} = \vec{k} + \vec{K} = \vec{k} + m_1\vec{K}_1 + m_2\vec{K}_2 = k_{\mathbf{K}x}\hat{\imath}_x + k_{\mathbf{K}y}\hat{\imath}_y + k_{\mathbf{K}z}\hat{\imath}_z$. Using the definitions shown in Fig. 7.48 the unit vectors are defined according to the following equations:

$$\hat{h}_{\mathbf{K}3} = \frac{\vec{k} + \vec{K}}{|\vec{k} + \vec{K}|}, \quad (7.176)$$

$$\hat{h}_{\mathbf{K}1} = \frac{\vec{h}_{\mathbf{K}3} \times \hat{z}}{|\vec{h}_{\mathbf{K}3} \times \hat{z}|} = \frac{\hat{x}k_{\mathbf{K}y} - \hat{y}k_{\mathbf{K}x}}{\sqrt{k_{\mathbf{K}x}^2 + k_{\mathbf{K}y}^2}}, \quad (7.177)$$

$$\hat{h}_{\mathbf{K}2} = \frac{\vec{h}_{\mathbf{K}3} \times \vec{h}_{\mathbf{K}1}}{|\vec{h}_{\mathbf{K}3} \times \vec{h}_{\mathbf{K}1}|} = \frac{k_{\mathbf{K}x}k_{\mathbf{K}z}\hat{x} + k_{\mathbf{K}y}k_{\mathbf{K}z}\hat{y} - (k_{\mathbf{K}x}^2 + k_{\mathbf{K}y}^2)\hat{z}}{\sqrt{k_{\mathbf{K}x}^2 + k_{\mathbf{K}y}^2}\sqrt{k_{\mathbf{K}x}^2 + k_{\mathbf{K}y}^2 + k_{\mathbf{K}z}^2}}. \quad (7.178)$$

The relative permittivity function (and its inverse) is also periodic in the xy-plane, sharing the same periodicity as the magnetic field. As a result, the inverse relative permittivity can

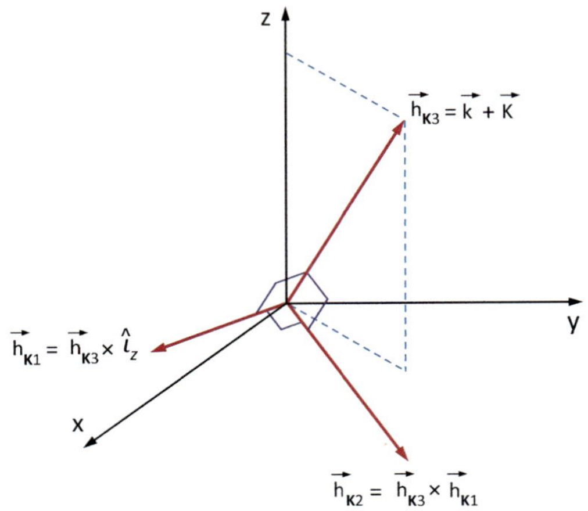

FIGURE 7.48: To satisfy the transversality condition of the magnetic field, the vectors $\vec{h}_{\mathbf{K}1}$, $\vec{h}_{\mathbf{K}2}$, $\vec{h}_{\mathbf{K}3}$ are determined. Here, $\hat{h}_{\mathbf{K}1}$, $\hat{h}_{\mathbf{K}2}$, $\hat{h}_{\mathbf{K}3}$ are the unit vectors corresponding to $\vec{h}_{\mathbf{K}1}$, $\vec{h}_{\mathbf{K}2}$, $\vec{h}_{\mathbf{K}3}$, respectively. For simplicity $\vec{K}$ is denoted as $\mathbf{K}$.

be expressed in a two-dimensional Fourier series expansion as:

$$\frac{1}{\varepsilon} = \sum_{m_1'} \sum_{m_2'} \gamma_{m_1' m_2'} e^{-j(m_1'\vec{K}_1 + m_2'\vec{K}_2)\cdot\vec{r}} = \sum_{\mathbf{K}'} \gamma_{\mathbf{K}'} e^{-j\vec{K}'\cdot\vec{r}}. \tag{7.179}$$

By substituting Eqs. (7.175) and (7.179) into Eq. (7.153), and considering that $\vec{\nabla}\cdot\vec{H} = j(\vec{k}+\vec{K})\cdot\vec{H}$ and $\vec{\nabla}\times\vec{H} = j(\vec{k}+\vec{K})\times\vec{H}$, the following vector equation can be derived:

$$\sum_{\mathbf{K}'} \gamma_{\mathbf{K}-\mathbf{K}'} \left[\left\{ \left(\vec{k}+\vec{K}\right) \times \left[\left(\vec{k}+\vec{K}'\right) \times \hat{h}_{\mathbf{K}1} \right] \right\} C_{1\mathbf{K}'} + \left\{ \left(\vec{k}+\vec{K}\right) \times \left[\left(\vec{k}+\vec{K}'\right) \times \hat{h}_{\mathbf{K}2} \right] \right\} C_{2\mathbf{K}'} \right] = \left(\frac{\omega}{c}\right)^2 \left[C_{1\mathbf{K}} \hat{h}_{\mathbf{K}1} + C_{2\mathbf{K}} \hat{h}_{\mathbf{K}2} \right]. \tag{7.180}$$

By projecting the left-hand side of the above equation onto $\hat{h}_{\mathbf{K}1}$ and $\hat{h}_{\mathbf{K}2}$, and appropriately applying the vector identity $\vec{C}\cdot(\vec{A}\times\vec{B}) = \vec{B}\cdot(\vec{C}\times\vec{A}) = \vec{A}\cdot(\vec{B}\times\vec{C})$, the following set of equations can be derived:

$$\sum_{\mathbf{K}'} \gamma_{\mathbf{K}-\mathbf{K}'} \left\{ \left[\left(\vec{k}+\vec{K}'\right) \times \hat{h}_{\mathbf{K}'1} \right] \cdot \left[\left(\vec{k}+\vec{K}\right) \times \hat{h}_{\mathbf{K}1} \right] C_{1\mathbf{K}'} + \left[\left(\vec{k}+\vec{K}'\right) \times \hat{h}_{\mathbf{K}'2} \right] \cdot \left[\left(\vec{k}+\vec{K}\right) \times \hat{h}_{\mathbf{K}1} \right] C_{2\mathbf{K}'} \right\} = \left(\frac{\omega}{c}\right)^2 C_{1\mathbf{K}},$$

$$\sum_{\mathbf{K}'} \gamma_{\mathbf{K}-\mathbf{K}'} \left\{ \left[\left(\vec{k}+\vec{K}'\right) \times \hat{h}_{\mathbf{K}'1} \right] \cdot \left[\left(\vec{k}+\vec{K}\right) \times \hat{h}_{\mathbf{K}2} \right] C_{1\mathbf{K}'} + \left[\left(\vec{k}+\vec{K}'\right) \times \hat{h}_{\mathbf{K}'2} \right] \cdot \left[\left(\vec{k}+\vec{K}\right) \times \hat{h}_{\mathbf{K}2} \right] C_{2\mathbf{K}'} \right\} = \left(\frac{\omega}{c}\right)^2 C_{2\mathbf{K}}.$$

Utilizing the definitions of the unit vectors, it is easy to realize that $\left(\vec{k}+\vec{K}'\right) \times \hat{h}_{\mathbf{K}'1} = |\vec{k}+\vec{K}'|\hat{h}_{\mathbf{K}'2}$, $\left(\vec{k}+\vec{K}'\right) \times \hat{h}_{\mathbf{K}'2} = -|\vec{k}+\vec{K}'|\hat{h}_{\mathbf{K}'1}$, $\left(\vec{k}+\vec{K}\right) \times \hat{h}_{\mathbf{K}1} = |\vec{k}+\vec{K}|\hat{h}_{\mathbf{K}2}$, $\left(\vec{k}+\vec{K}\right) \times \hat{h}_{\mathbf{K}2} = -|\vec{k}+\vec{K}|\hat{h}_{\mathbf{K}1}$. Then, the latter set of equations can be written in the form [453]:

$$\sum_{\mathbf{K}'} \gamma_{\mathbf{K}-\mathbf{K}'} |\vec{k}+\vec{K}||\vec{k}+\vec{K}'| \begin{bmatrix} \hat{h}_{\mathbf{K}'2} \cdot \hat{h}_{\mathbf{K}2} & -\hat{h}_{\mathbf{K}'1} \cdot \hat{h}_{\mathbf{K}2} \\ -\hat{h}_{\mathbf{K}'2} \cdot \hat{h}_{\mathbf{K}1} & \hat{h}_{\mathbf{K}'1} \cdot \hat{h}_{\mathbf{K}1} \end{bmatrix} \begin{bmatrix} C_{1\mathbf{K}'} \\ C_{2\mathbf{K}'} \end{bmatrix} = \left(\frac{\omega}{c}\right)^2 \begin{bmatrix} C_{1\mathbf{K}} \\ C_{2\mathbf{K}} \end{bmatrix}. \tag{7.181}$$

The last equation is also applicable to three-dimensional photonic crystals, provided that the reciprocal lattice vectors $\vec{K}$ and $\vec{K}'$ are defined to account for the three-dimensional periodicity. However, in this chapter, the focus remains on two-dimensional photonic crystals. Equation (7.181) represents an eigenvalue/eigenvector problem of the form $\tilde{\boldsymbol{A}}[C_{1\mathbf{K}}, C_{2\mathbf{K}}]^T = \lambda[C_{1\mathbf{K}}, C_{2\mathbf{K}}]^T$, where the eigenfrequency $\omega = \omega(\vec{k}) = c\sqrt{\lambda}$, similar to the case of one-dimensional photonic crystals.

Of particular interest is the case where the wavevector $\vec{k}$ of the incident wave is confined to the xy-plane. In this case, $\vec{k} = k_x\hat{\imath}_x + k_y\hat{\imath}_y$, and $\vec{K}$, as defined by Eq. (7.173), also lies within the xy-plane. Consequently, Eqs. (7.176)-(7.178) simplify to:

$$\hat{h}_{\mathbf{K}3} = \frac{\vec{k}+\vec{K}}{|\vec{k}+\vec{K}|} = \frac{\hat{\imath}_x k_{\mathbf{K}x} + \hat{\imath}_y k_{\mathbf{K}y}}{\sqrt{k_{\mathbf{K}x}^2 + k_{\mathbf{K}y}^2}},$$

$$\hat{h}_{\mathbf{K}1} = \frac{\vec{h}_{\mathbf{K}3} \times \hat{z}}{|\vec{h}_{\mathbf{K}3} \times \hat{\imath}_z|} = \frac{\hat{\imath}_x k_{\mathbf{K}y} - \hat{\imath}_y k_{\mathbf{K}x}}{\sqrt{k_{\mathbf{K}x}^2 + k_{\mathbf{K}y}^2}},$$

$$\hat{h}_{\mathbf{K}2} = \frac{\vec{h}_{\mathbf{K}3} \times \vec{h}_{\mathbf{K}1}}{|\vec{h}_{\mathbf{K}3} \times \vec{h}_{\mathbf{K}1}|} = -\hat{\imath}_z.$$

Using the latter simplified unit vectors, Eq. (7.181) can be written in the form:

$$\sum_{\mathbf{K}'} \gamma_{\mathbf{K}-\mathbf{K}'} |\vec{k}+\vec{K}||\vec{k}+\vec{K}'| \begin{bmatrix} 1 & 0 \\ 0 & \hat{h}_{\mathbf{K}'1} \cdot \hat{h}_{\mathbf{K}1} \end{bmatrix} \begin{bmatrix} C_{1\mathbf{K}'} \\ C_{2\mathbf{K}'} \end{bmatrix} = \left(\frac{\omega}{c}\right)^2 \begin{bmatrix} C_{1\mathbf{K}} \\ C_{2\mathbf{K}} \end{bmatrix}. \tag{7.182}$$

From the above equation, it is evident that the two polarizations represented by $\hat{h}_{\mathbf{K}1}$ and $\hat{h}_{\mathbf{K}2}$ in Eq. (7.175) are independent. Since $\hat{h}_{\mathbf{K}2} = -\hat{\imath}_z$ (the sign does not significantly affect the result), these two independent polarizations can be analyzed separately. The polarization corresponding to $\hat{h}_{\mathbf{K}2}$ indicates that the magnetic field is aligned along the z-direction, with the electric field lying in the xy-plane. This polarization is referred to as the "*TE*-mode," as illustrated in Fig. 7.45. Conversely, the polarization corresponding to $\hat{h}_{\mathbf{K}1}$ suggests that the magnetic field lies in the xy-plane, with the electric field aligned along the z-axis. This polarization is known as the "*TM*-mode." The independent eigenvalue/eigenvector equations for these two polarizations can be derived by separating the components of Eq. (7.182) as follows:

$$\sum_{\mathbf{K}'} \gamma_{\mathbf{K}-\mathbf{K}'} \left(\vec{k}+\vec{K}\right) \cdot \left(\vec{k}+\vec{K}'\right) C_{2\mathbf{K}'} = \left(\frac{\omega}{c}\right)^2 C_{2\mathbf{K}}, \quad \textit{TE}\text{-mode}, \tag{7.183}$$

$$\sum_{\mathbf{K}'} \gamma_{\mathbf{K}-\mathbf{K}'} |\vec{k}+\vec{K}||\vec{k}+\vec{K}'| C_{1\mathbf{K}'} = \left(\frac{\omega}{c}\right)^2 C_{1\mathbf{K}}, \quad \textit{TM}\text{-mode}, \tag{7.184}$$

where the equality $\hat{h}_{\mathbf{K}'1}\cdot\hat{h}_{\mathbf{K}1} = \hat{h}_{\mathbf{K}'3}\cdot\hat{h}_{\mathbf{K}3}$ was used for Eq. (7.183). It can be easily verified that Eq. (7.184) is identical to the eigenvalue equation derived using the E_z wave equation (see, for example, Ref. [453]). Equations (7.183) and (7.184) represent an infinite set of equations dependent on the integers m_1 and m_2. By truncating this infinite set to a finite number of plane waves, these equations can be rewritten in the form:

$$\sum_{m_1'=-M_1}^{+M_1}\sum_{m_2'=-M_2}^{+M_2}\gamma_{m_1-m_1',m_2-m_2'}\left(k_{\mathbf{K}xm_1m_2}k_{\mathbf{K}xm_1'm_2'}+k_{\mathbf{K}ym_1m_2}k_{\mathbf{K}ym_1'm_2'}\right)C_{2m_1'm_2'} = \left(\frac{\omega}{c}\right)^2 C_{2m_1m_2}, \tag{7.185}$$

$$\sum_{m_1'=-M_1}^{+M_1}\sum_{m_2'=-M_2}^{+M_2}\gamma_{m_1-m_1',m_2-m_2'}\sqrt{k^2_{\mathbf{K}xm_1m_2}+k^2_{\mathbf{K}ym_1m_2}}\sqrt{k^2_{\mathbf{K}xm_1'm_2'}+k^2_{\mathbf{K}ym_1'm_2'}}\,C_{1m_1'm_2'} = \left(\frac{\omega}{c}\right)^2 C_{1m_1m_2}. \tag{7.186}$$

It is important to note that $\vec{k}_{\mathbf{K}} = \vec{k}+\vec{K}$, and M_i ($i = 1,2$) represents the maximum number of plane waves used in the i-th direction. Specifically, m_i ranges from $-M_i$ to $+M_i$, including all integers in between, i.e., $m_i = -M_i, -(M_i-1), \cdots, -1, 0, 1, \cdots, M_i-1, M_i$.

The last two equations can also be expressed in matrix form, analogous to Eqs. (7.165) and (7.166) from the one-dimensional case:

$$\left\{[K_x]\,\tilde{\mathbf{\Gamma}}\,[K_x]+[K_y]\,\tilde{\mathbf{\Gamma}}\,[K_y]\right\}\tilde{C}_2 = \left(\frac{\omega}{c}\right)^2\tilde{C}_2, \tag{7.187}$$

$$\left\{[K]\,\tilde{\mathbf{\Gamma}}\,[K]\right\}\tilde{C}_1 = \left(\frac{\omega}{c}\right)^2\tilde{C}_1, \tag{7.188}$$

where, $\tilde{C}_1 = \left[C_{1(-M_1,-M_2)}, C_{1(-M_1,1-M_2)}, \cdots, C_{1(00)}, \cdots, C_{1(M_1,M_2)}\right]^T,$

$$\tilde{C}_2 = \left[C_{2(-M_1,-M_2)}, C_{2(-M_1,1-M_2)}, \cdots, C_{2(00)}, \cdots, C_{2(M_1,M_2)}\right]^T,$$

$$[K_w] = \text{diag}\left[k_{\mathbf{K}w(-M_1,-M_2)}, k_{\mathbf{K}w(-M_1,1-M_2)}, \cdots, k_{\mathbf{K}w(00)}, \cdots, k_{\mathbf{K}w(M_1,M_2)}\right], \quad (w = x, y),$$

$$[K] = \text{diag}\left[k_{\mathbf{K}(-M_1,-M_2)}, k_{\mathbf{K}(-M_1,1-M_2)}, \cdots, k_{\mathbf{K}(00)}, \cdots, k_{\mathbf{K}(M_1,M_2)}\right],$$

$$k_{\mathbf{K}(m_1,m_2)} = \sqrt{k^2_{\mathbf{K}xm_1m_2}+k^2_{\mathbf{K}ym_1m_2}},$$

where $\tilde{\mathbf{\Gamma}}$ is a block-Toeplitz matrix (with elements coefficients of the Fourier series expansion of $1/\varepsilon$) corresponding to the 2D case (analogous to the matrix in the 2D-periodic gratings of Sec. 7.7.6 and Appendix 7F, Sec. 7F.5). It is worth noting that in this case, the Antos factorization [454] is not presented due to its complexity and its dependence on the specific shape of the unit cell. However, the interested reader could consult Ref. [454] for more information.

Since the incident wavevector lies in the xy-plane, it is customary to vary its k_x and k_y components along the edges of the irreducible Brillouin zone, as illustrated in Figs. 7.46b and 7.47b for the rectangular and equilateral grids, respectively. The specific ranges of k_x and k_y while moving along the edges of the irreducible Brillouin zone are summarized in Table 7.4. At this point, some example 2D photonic crystals are presented.

TABLE 7.4: Edges of Irreducible Brillouin Zone for a rectangular grid and an equilateral grid.

Rectangular Grid		Equilateral Grid	
$\Gamma \to X$	$k_x : 0 \to \frac{\pi}{\Lambda_x}$	$\Gamma \to M$	$k_x = 0$
	$k_y = 0$		$k_y : 0 \to \frac{2\pi}{\Lambda}\frac{1}{\sqrt{3}}$
$X \to M$	$k_x = \frac{\pi}{\Lambda_x}$	$M \to K$	$k_x : 0 \to \frac{2\pi}{\Lambda}\frac{1}{3}$
	$k_y : 0 \to \frac{\pi}{\Lambda_y}$		$k_y = \frac{2\pi}{\Lambda}\frac{1}{\sqrt{3}}$
$M \to \Gamma$	$k_x : \frac{\pi}{\Lambda_x} \to 0$	$K \to \Gamma$	$k_x : \frac{2\pi}{\Lambda}\frac{1}{3} \to 0$
	$k_y = k_x \frac{\Lambda_x}{\Lambda_y}$		$k_y = k_x\sqrt{3}$

Example 7.8.3:

This example involves a rectangular array (with $\Lambda_x = \Lambda_y = \Lambda$) of square cross-section air prism columns embedded in a dielectric with a relative permittivity of 8.9. The air square columns have dimensions in the xy-plane of $\mathrm{a}/\Lambda_x = \mathrm{b}/\Lambda_y = 0.835$. The task is to calculate the $\tilde{\omega}$ vs. $\tilde{k}$ band diagrams following the edges $\Gamma \to X \to M \to \Gamma$ of the irreducible Brillouin zone, for both *TE*- and *TM*-modes. This example is adapted and reworked from Ref. [452].
Solution: Using the rectangular grid geometry depicted in Fig. 7.46, the Fourier series expansion of the inverse dielectric constant $1/\varepsilon$ can be straightforwardly calculated, as outlined in the Appendices "Two-Dimensional Fourier Transform" (Appendix 5D) and "Fourier Series" (Appendix 5E) in the *Scalar Diffraction* chapter (Chapter 5). In this analysis, 1681 plane waves were utilized ($M_1 = M_2 = 20$), achieving reasonable convergence for this number of plane waves. By applying Eqs. (7.187) (for TE modes) and (7.188) (for TM modes), the band structure can be calculated, as shown in Fig. 7.49. The inset illustrates a portion of the 2D photonic crystal, where the dark blue color represents the dielectric material and the colorless squares represent the air columns. It can be observed that no photonic bandgap exist for either *TE*- or *TM*-modes.

Example 7.8.4:

This example involves a rectangular array (with $\Lambda_x = \Lambda_y = \Lambda$) of circular cross-section dielectric prism columns of relative permittivity of 8.9 surrounded by air. The radius of the dielectric cylinders is $r_0/\Lambda = 0.2$. Calculate the $\tilde{\omega}$ vs. $\tilde{k}$ band diagrams following the edges $\Gamma \to X \to M \to \Gamma$ of the irreducible Brillouin zone, for both *TE*- and *TM*-modes. This example is also adapted and reworked from Ref. [452].

Solution: Using again the rectangular grid geometry depicted in Fig. 7.46, the Fourier series expansion of the inverse dielectric constant $1/\varepsilon$ can be straightforwardly calculated, as outlined in the Appendices "Two-Dimensional Fourier Transform" (Appendix 5D) and

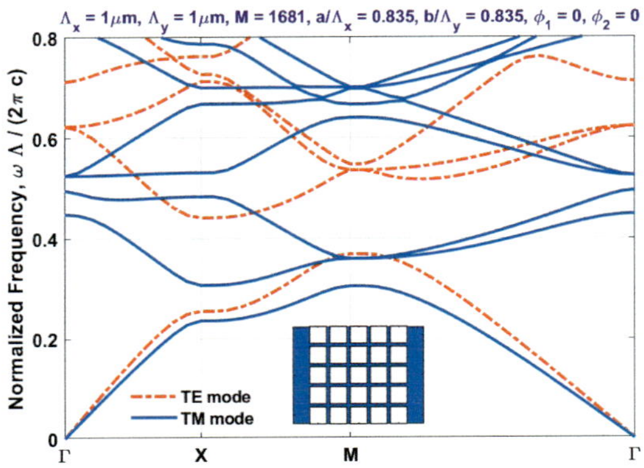

FIGURE 7.49: The band diagram ($\tilde{\omega}$ vs. $\tilde{k}$) for a 2D photonic crystal consisting of a rectangular array of air prism columns (with square cross-sections) embedded in a dielectric material with a relative permittivity of 8.9 is presented. The dimensions of the air prisms are such that $a/\Lambda_x = b/\Lambda_y = 0.835$, with equal grid periodicity $\Lambda_x = \Lambda_y = \Lambda$. Results for both *TE* and TM modes are displayed. The inset provides a visual sample of the structure. The wavevector k varies along the edges of the irreducible Brillouin zone, specifically following the path $\Gamma \to X \to M \to \Gamma$. In this analysis, 1681 plane waves were employed ($M_1 = M_2 = 20$) [452].

"Fourier Series" (Appendix 5E) in the *Scalar Diffraction* Chapter. In this case the Fourier coefficients correspond to the cylindrical pillars. Similarly to Example 7.8.3, in this analysis, 1681 plane waves were utilized ($M_1 = M_2 = 20$), achieving reasonable convergence for this number of plane waves. By applying the same equations as of Example 7.8.3, the band structure can be calculated, as shown in Fig. 7.50. The inset again illustrates a portion of the 2D photonic crystal, where the dark blue color represents the dielectric cylindrical pillars surrounded by air (colorless region). It can be observed that there is no photonic bandgap *TE*-modes. However, this 2D PC has a complete band gap for *TM*-modes between the first and the second bands.

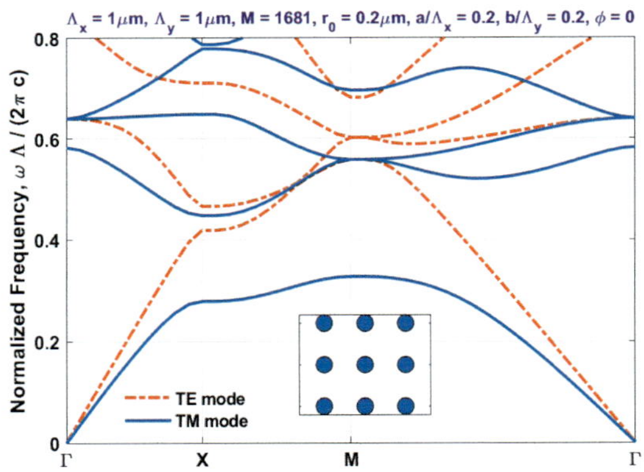

FIGURE 7.50: The band diagram ($\tilde{\omega}$ vs. $\tilde{k}$) of a 2D PC consisting of a rectangular array of dielectric rods (of relative permittivity of 8.9) and air, with rod radius $r_0/\Lambda = 0.2$ and equal periodicities in x and y directions ($\Lambda_x = \Lambda_y = \Lambda$). Both *TE* and TM mode results are shown. The inset shows a sample of the structure. The wavevector k varies along the edges of the irreducible Brillouin zone, i.e $\Gamma \to X \to M \to \Gamma$. In these results, 1681 plane waves were used ($M_1 = M_2 = 20$) [452].

Example 7.8.5:

Now consider an equilateral grid of circular cross-section air cylinders embedded in dielectric of relative permittivity of 13. Calculate the $\tilde{\omega}$ vs. $\tilde{k}$ band diagrams following the edges $\Gamma \to M \to K \to \Gamma$ of the irreducible Brillouin zone, for both *TE*- and *TM*-modes. The radius of the circular air cylinders is $r_0/\Lambda = 0.48$. This example is also adapted and reworked from Ref. [452].

Solution: In this case, the equilateral grid geometry depicted in Fig. 7.47 is utilized, along with the Fourier series expansion of the inverse dielectric constant $1/\varepsilon$. Similarly to the previous examples, the band structure can be calculated, as illustrated in Fig. 7.51 for both *TE*- and *TM*-modes. The inset provides a sample of the structure, where the darker areas correspond to regions with a higher refractive index. Notably, in this configuration, a photonic bandgap exists for both polarizations simultaneously! In this analysis, 1681 plane wave components were employed ($M_1 = M_2 = 20$), achieving practical convergence primarily for the lower bands. To illustrate the convergence effect, some bands are shown in green for the case of 961 plane waves ($M_1 = M_2 = 15$). The curves with fewer plane waves generally lie above those with a higher number of plane waves, with the discrepancies increasing for higher bands. Therefore, for greater accuracy, especially at higher frequency bands, an even larger number of plane waves may be required.

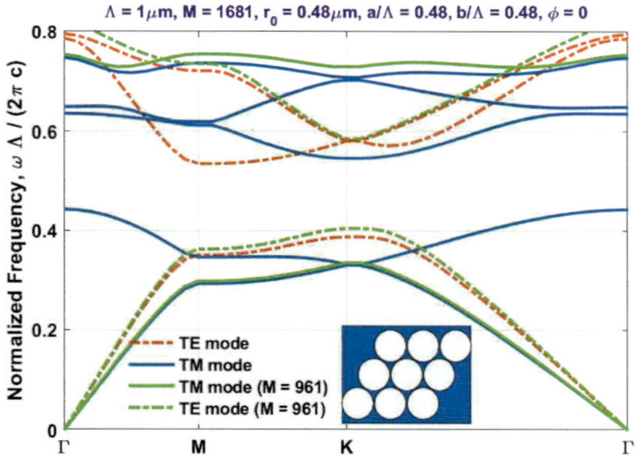

FIGURE 7.51: The band diagram ($\tilde{\omega}$ vs. $\tilde{k}$) of a 2D PC consisting of an equilateral array of air cylinders (of circular cross-section) in a dielectric of relative permittivity of 13, with air-cylinders radius $r_0/\Lambda = 0.48$ and and grid periodicity Λ. Both *TE*- and *TM*-mode results are shown. The inset shows a sample of the structure. The wavevector k varies along the edges of the irreducible Brillouin zone, i.e $\Gamma \to M \to K \to \Gamma$. In these results, plane waves were used ($M_1 = M_2 = 20$). In green color some bands are shown for 961 plane waves. It can be seen that in this case the convergence of the plane-wave method is slow.

Example 7.8.6:

To demonstrate the applicability of the method to more complicated unit cells, another example is presented. (a) The first case involves an equilateral array of air-columns with a slanted parallelogram cross-section, embedded in a dielectric with a relative permittivity of 8.9. The parameters of the air columns are a/$\Lambda = 0.75$, b/$\Lambda = 0.75$, $\phi_1 = 0°$, and $\phi_2 = 30°$. (b) The second case involves a rectangular array of elliptical cross-section dielectric pillars (with a relative permittivity of 8.9) in air. The parameters of the elliptical pillars are $r_0/\Lambda = 0.2$, a/$\Lambda = 0.4$, b/$\Lambda = 0.15$, and $\phi = 45°$, while the grid periodicities in the x- and

y-directions are equal, $\Lambda = \Lambda_x = \Lambda_y$. For both cases, the geometrical data refer to the Appendix "Two-Dimensional Fourier Transform" (Appendix 5D) of the *Scalar Diffraction* chapter. The task is to calculate the band structure $\tilde{\omega}$ vs. $\tilde{k}$ for both *TE*- and *TM*-modes.
Solution: In case (a), the equilateral grid geometry depicted in Fig. 7.47 is utilized, while in case (b), the rectangular grid geometry shown in Fig. 7.46 is employed. In both cases, consult the Appendices "Two-Dimensional Fourier Transform" (Appendix 5D) and "Fourier Series" (Appendix 5E) in the *Scalar Diffraction* chapter for the correlation between the geometrical data and the Fourier series expansion of the inverse relative permittivity $1/\varepsilon$.

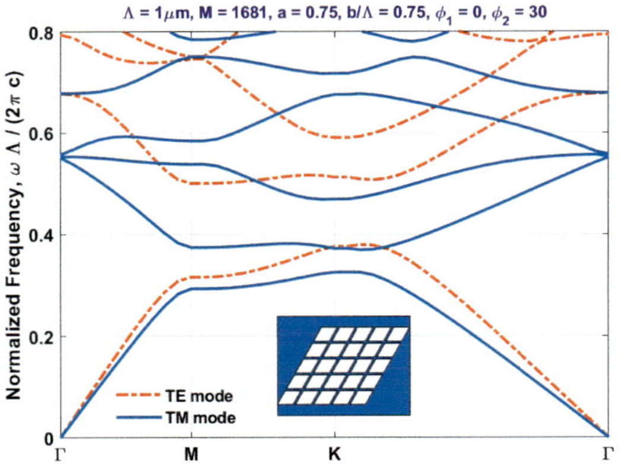

(a)

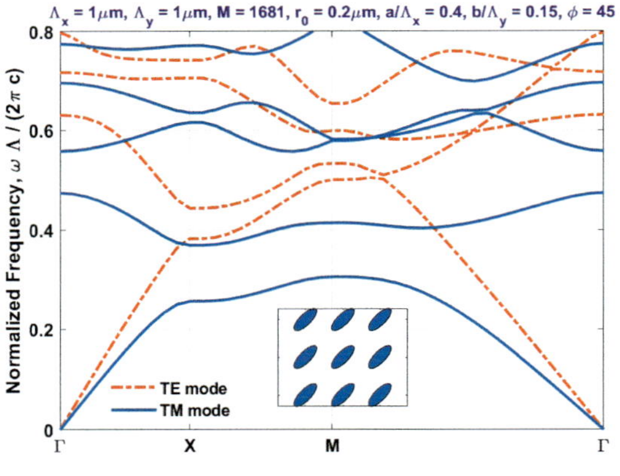

(b)

FIGURE 7.52: The band diagrams ($\tilde{\omega}$ vs. $\tilde{k}$) for *TE*- and *TM*-modes are shown for cases (a) and (b), with 1681 plane waves employed ($M_1 = M_2 = 20$). The wavevector k varies along the edges of the corresponding irreducible Brillouin zone. (a) 2D PC consisting of an equilateral array of air columns of slanted parallelogram cross-section embedded in a dielectric of relative permittivity of 8.9. The parameters of the air columns are $a/\Lambda = 0.75$, $b/\Lambda = 0.75$ and $\phi_1 = 0°$, and $\phi_2 = 30°$. (b) Similar to (a) for a 2D PC consisting of a rectangular array of elliptical pillars of relative permittivity of 8.9, in air. The parameters of the elliptical pillars are $r_0/\Lambda = 0.2$, $a/\Lambda = 0.4$, $b/\Lambda = 0.15$ and $\phi = 45°$, while the grid periodicities in x- and y-directions are equal, $\Lambda = \Lambda_x = \Lambda_y$.

Following the methodology outlined in the previous examples, the band structure can be calculated and is illustrated in Fig. 7.52(a) for case (a), and Fig. 7.52(b) for case (b). Both *TE*- and *TE*-mode results are included, with 1681 plane waves employed in the calculations

($M_1 = M_2 = 20$). The insets in both plots show a sample of the 2D photonic crystal, where the darker color represents regions with higher relative permittivity.

In case (a), a simultaneous photonic bandgap is observed for both *TE*- and *TM*-modes. However, in case (b), a bandgap is observed only for the *TM*-mode.

7.9 Problems

1 **Diffraction by a Circular Aperture:** Consider a circular aperture with a radius of $a = 10\,\mu$m. A plane wave, either *TE*- or *TM*-polarized, is incident at an angle of $\theta = 45°$ ($\phi = 0$) from the left of the aperture. The freespace wavelength is $\lambda_0 = 0.6328\,\mu$m. Using the approximate equation (7.22), calculate the diffracted intensity along the x-axis for $y = 0$ at $z_0 = 500\,\mu$m. Additionally, employ the vectorial plane wave spectrum method to compute the same intensity distribution.

2 **Diffraction by a Square Aperture:** Consider a square aperture of dimensions $a \times a$, where $a = 10, \mu$m. A *TE*-polarized plane wave, with its electric field oriented along the y-axis, is normally incident from the left of the aperture. The freespace wavelength is $\lambda_0 = 0.6328\,\mu$m. Utilize the vectorial plane wave spectrum method to compute the diffracted intensity along the x- and y-axes at a distance of $z_0 = 500\,\mu m$ to the right of the aperture. Compare the resulting vectorial intensity distribution with the intensity obtained using the scalar Fraunhofer approximation.

3 **Diffraction of a Plane Wave by a Square-Cross-Section Infinite Cylinder** Consider an infinite cylinder with a square cross-section, aligned along the z-axis and centered at the origin of the xyz coordinate system (Fig. 7P.1). The refractive index of the cylinder is n_2, while the surrounding medium has a refractive index of n_1. A plane wave $\vec{k}_{inc}$ is incident at an angle θ. Using the BEM and FDFD methods, analyze the diffraction of the plane wave by the dielectric square cylinder. Assume $n_1 = 1$ and $n_2 = 2$, a freespace wavelength of $\lambda_0 = 1\,\mu$m, an incident angle of $\theta = 0°$, and *TE* polarization with the electric field oriented along the z-axis. The side length of the square is $a = 5\lambda_0$. Compare the electric field distributions along the x- and y-axes as computed by the FDFD and BEM methods. Repeat the analysis for $\theta = 30°$.

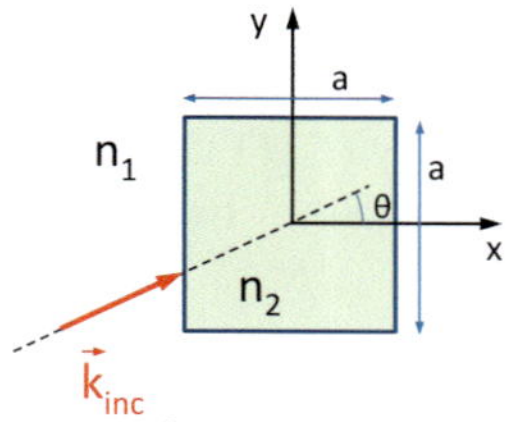

FIGURE 7P.1: A square-cross-section infinite cylinder of refractive index n_2 is surrounded by a homogeneous medium of refractive index n_1. A plane wave $\vec{k}_{inc}$ is incident at an angle θ.

4 **Grating Equation:** Consider a grating with a period of $\Lambda = 1.1765\,\mu$m. The superstrate is air ($n_1 = 1$), and the substrate is glass ($n_3 = 1.5$). The freespace wavelength of the incident plane wave is $\lambda_0 = 0.6328\,\mu$m. For the range of incident angles $0° \leq \theta_{inc} \leq 90°$ with $\phi_{inc} = 180°$, determine and plot the diffracted angles θ_{3i} and θ_{1i} for all forward- and backward-propagating orders, respectively, as functions of the angle of incidence.

5 **Diffraction Efficiencies in the Conical Diffraction Case:** Prove equations (7.90) and (7.91) for the case of conical diffraction. Additionally, demonstrate that the diffraction efficiencies of the evanescent orders are zero. Further, determine how Eqs. (7.90) and (7.91) simplify for propagating orders. Discuss the implications when region 3 is lossy.

6 Grating Equation in the Conical Diffraction Case: Generalize the grating equation for conical diffraction in the case of a one-dimensional grating with a period Λ, as illustrated in Fig. 7.28. Assume that the incident plane wave has angles of incidence θ_{inc} and ϕ_{inc} and a freespace wavelength of λ_0. Determine the propagating diffracted orders in both the backward and forward directions. Apply these results to the specific case where $n_1 = 1.0$, $n_3 = \sqrt{2.5}$, $\lambda_0 = \Lambda = 1\,\mu$m, $\theta_{inc} = 30°$, and $\phi_{inc} = 180°$ and $\phi_{inc} = 90°$.

7 Eigenvalues in the Grating Region in the Conical Diffraction Case: Prove that the eigenvalues λ of the matrix $\tilde{\boldsymbol{A}}$ in Eq. (7.88) occur in pairs as $\pm\lambda$.

8 EMT Higher-Order Refractive Indices: Prove that Eqs. (7.106) and (7.107) can be derived from Eqs. (7G.5) and (7G.6) in the case of normal incidence, i.e., $\theta_{inc} = 0$. Furthermore, demonstrate that Eqs. (7.106) and (7.107) reduce to the first-order refractive indices in the asymptotic limit $\Lambda/\lambda_0 \to 0$. Additionally, derive Eqs. (7.108) and (7.109). *Hint:* Consult Ref. [434].

9 Design of an Antireflective Rectangular-Groove Surface-Relief Grating: Design an antireflective (AR) rectangular-groove surface-relief grating on a silicon substrate ($n_r = n_3 = 3.5$) with an air superstrate ($n_1 = n_g = 1$) for normal incidence, *TE* polarization, and a freespace wavelength of $\lambda = 1.5\,\mu$m. Assume that the grating period is $\Lambda = 0.15\,\mu$m. Use effective medium theory (EMT) with second-order refractive indices to design the grating. Compare the resulting design with its counterpart based on first-order refractive indices. Compute and plot the specularly reflected power as a function of the angle of incidence ($0° \leq \theta_{inc} \leq 45°$ with $\phi_{inc} = 0$). Include in the graphs the results obtained using first-order, second-order, and higher-order EMT, as well as those based on rigorous coupled-wave analysis (RCWA). Repeat the design and performance analysis for $\Lambda = 0.375\,\mu$m. Determine the maximum ratio λ_0/Λ for which the design using second-order refractive indices remain valid.

10 Design of an Antireflective Rectangular-Groove Surface-Relief Grating: Design an anti-reflective (AR) grating, as described in Problem 9, but for *TM* polarization. Address the same tasks as in Problem 9.

11 Eigenvalues of the Grating Matrix and Comparison with the EMT First-order, Second-order, and Higher-order Refractive Indices: Assume a rectangular-groove surface-relief grating with the following parameters: $n_1 = n_g = 1.0$, $n_3 = n_r = 3.5$, and $F = 0.2222$. Consider either *TE* or *TM* polarization for the normally incident plane wave with a freespace wavelength of $\lambda_0 = 1.5\,\mu$m. Compare the eigenvalues of the matrix $\tilde{\boldsymbol{A}}_{\boldsymbol{TE}}$ in Eq. (7.75) (for *TE* incident polarization) with the refractive indices obtained using the first-order, second-order, and higher-order effective medium theory (EMT). Tabulate the results as a function of the ratio λ_0/Λ in the range of 2 to 10 with a unit step. Repeat the analysis for the case of *TM* incident polarization.

12 Design of a Guided-Mode Resonance Rectangular-Groove Surface-Relief Grating (TE polarization): The primary objective of this problem is to design a resonance-guided-mode grating filter, as described in [456]. The grating should be a rectangular-groove surface-relief grating with a reflection resonance at $\lambda_0 = 0.55\,\mu$m for a normally incident *TE*-polarized plane wave. The superstrate is air ($n_1 = 1$), and the substrate is silica ($n_3 = 1.52$). The ridges of the grating are composed of zirconium dioxide ($n_2 = 2.1$), and the grooves are made of hafnium dioxide ($n_g = 2.0$). The filling factor is assumed to be $F = 0.50$. To design this grating, the period must be chosen such that it excites the first guided mode in the grating region, assuming the region behaves like a slab waveguide. The effective index of this mode can be found by modeling the grating region as an effective medium theory (EMT) layer with the first-order ordinary index given by $n_O^{(1)} = [Fn_r^2 + (1-F)n_g^2]^{1/2}$. The equivalent waveguide propagation constant is denoted as β, which can be found numerically by solving the slab waveguide transcendental resonance equation. If the method for calculating β is not known, it can be expressed as $\beta = k_0 N_{eff}$, where $k_0 = 2\pi/\lambda_0$ (with $\lambda_0 = 0.55\,\mu$m), and $N_{eff} = 2.05060966544$ for a grating depth of $d = 0.134\,\mu$m (which is approximately $\lambda_0/2N_{eff}$) [456]. It is recommended to verify the provided propagation constant. Determine the required period of the grating, and the specular reflective power, DE_0^r, by applying the RCWA to verify the grating resonance at the design wavelength. It is important to note that the EMT approach using a uniaxial homogeneous layer does not apply to this grating (explain).

13 Design of an Antireflective Rectangular-Pillar crossed-Surface-Relief Grating: Design a symmetric antireflective rectangular-pillar crossed-grating using the first-order effective indices for normal incidence. Determine the filling factor F and the depth d. Assume the following parameters: $n_1 = n_g = 1.0$, $n_3 = n_r = 3.5$, and $\Lambda_x/\lambda_0 = \Lambda_y/\lambda_0 = 0.001$. Validate the design using RCWA by generating a table of the normalized reflected power in the specular direction, DE_{00}^r, as a function of the incident angle θ_{inc} (with $\phi_{inc} = 0°$). Consider the following values for θ_{inc}: $0°, 5°, 10°, 15°, 30°, 45°$, and $60°$. Perform the analysis for both *TE* ($\psi = 90°$) and *TM* ($\psi = 0°$) polarizations.

14 Kogelnik's Two-Wave Theory (*TE* polarization): Prove the equations (7.124) and (7.125) by solving the coupled differential equations (7.122) and (7.123) for the case of a transmission grating.

15 Kogelnik's Two-Wave Theory (*TE* polarization): Prove equations (7.127) and (7.128) by solving the coupled-differential equations (7.122) and (7.123) for the case of a reflection grating.

16 Kogelnik's Two-Wave Theory (*TM* polarization): Prove the set of the coupled-differential equations (7.131) and (7.132), for the case of *TM* polarization.

17 Design of a "Thick" Transmission Grating for *TM* incident Polarization: Repeat Example 7.7.12 for the case of *TM* incident polarization. All parameters remain the same as in Example 7.7.12, with the exception of the polarization. Compare the results obtained for the *TM* polarization with those for the *TE* polarization.

18 Design of a "Thick" Reflection Grating for *TM* incident Polarization: Repeat Example 7.7.13 for the case of *TM* incident polarization. All parameters remain the same as in Example 7.7.13, except polarization. Compare the results obtained for the *TM* polarization with those for the *TE* polarization.

19 Examine the Behavior of a "Thick" Reflection Grating: Consider the grating defined in Example 7.7.13. Calculate the minimum deviation from the Bragg angle necessary for the diffraction efficiency of the grating to transition from hyperbolic-tangent-like behavior to sinusoidal-like behavior. Perform this analysis for both *TE* and *TM* polarizations.

20 Design of a Transmission Grating In a Lithium Niobate Waveguide: Consider a volume holographic unslanted grating fabricated in a titanium-diffused lithium niobate slab waveguide for *TE* incident polarization. Assume the effective average refractive index of the grating is $n_0 = 2.234$. The effective refractive indices in regions 1 and 3 are $n_1 = n_3 = n_0$. The grating period is $\Lambda = 500\,\text{nm}$, and its thickness is $d = 0.1\,\text{mm}$. The objective is to have 50% of the incident power diffracted into the first forward diffracted order, while the remaining 50% remains in the specular transmitted (0th) order. Light of freespace wavelength $\lambda_0 = 0.6328\,\mu\text{m}$ is used. (a) Should the grating operate in the "thin" or "thick" regime? (b) What should be the angle of incidence (in degrees) in order to achieve the desired beam division? (c) To ensure that all other diffracted beams carry less than 1% of the power, what range of values is acceptable for the modulation index n_1? (d) To obtain a diffraction efficiency of 50% in the first forward order, what value should n_1 take? Is this value consistent with the result from part (c)? (e) Estimate the allowable error in the angle of incidence (in degrees) to maintain the diffraction efficiency within 1%, i.e., $DE_1^t(\theta_{inc} + \Delta\theta)/DE_1^t(\theta_{inc}) < 1\%$. Here, DE_1^t refers to the first-order forward diffraction efficiency.

21 One-Dimensional Photonic Crystals: Prove the eigenvalue equations given in Eq. (7.163) and its matrix equivalent in Eq. (7.164). Furthermore, prove the eigenvalue equations in Eq. (7.165) for off-axis *TE* polarization and in Eq. (7.166) for off-axis *TM* polarization, along with their matrix equivalent counterparts, Eqs. (7.167) and (7.168), respectively.

22 One-Dimensional Photonic Crystals: Consider a 1D photonic crystal made from two alternating dielectric layers with dielectric constants $\varepsilon_1 = 17.88$ (InSb) and $\varepsilon_2 = 2.33$ (SiO_2) [457]. Assume a filling factor of $F = 0.25$. Determine the regions in the $\tilde{\omega}$-$\tilde{k}_x$ plane where propagating waves are allowed, similar to Example 7.8.1, and include the first five zones in your analysis. Additionally, calculate the first five bands for the following cases: (a) $k_x \neq 0, k_z = k_y = 0$, and (b) $k_x \neq 0, k_z = \pi/\Lambda, k_y = 0$.

23 Two-Dimensional Photonic Crystals: Prove the general eigenvalue equation given in Eq. (7.181) for a 2D photonic crystal. Furthermore, derive Eqs. (7.182) for the case where the wavevector $\vec{k}$ lies in the xy-plane. Then, show that in this special case, the *TE*- and *TM*-mode eigenvalue equations are given by Eqs. (7.185) and (7.186), respectively. Finally, prove the matrix equivalents of these equations, i.e., Eqs. (7.187) and (7.188).

24 Band Structure Calculation: Determine the band structure for an equilateral grid of the following 2D photonic crystals. In all cases, the dielectric material has a relative permittivity of 8.9, and the grid periodicities in the x- and y-directions are equal, $\Lambda = \Lambda_x = \Lambda_y$. All geometrical characteristics are defined in Appendix "Two-Dimensional Fourier Transform" (Appendix 5D) of the *Scalar Diffraction* chapter. The wavevector k should vary along the edges of the corresponding irreducible Brillouin zone, as shown in Example 7.8.6. The geometrical parameters are: $r_0/\Lambda = 0.2$, $a/\Lambda = 0.4$, and $b/\Lambda = 0.2$. (a) Dielectric pillars with elliptical cross-sections suspended in air for $\phi = 45°$. (b) Air pillars with elliptical cross-sections embedded in dielectric material for $\phi = 45°$. (c) Dielectric pillars with elliptical cross-sections suspended in air for $\phi = 135°$. (d) Air pillars with elliptical cross-sections embedded in dielectric material for $\phi = 135°$. Comment on and compare the calculated band structures.

Appendices

7A Silver-Müller Radiation Conditions

In this Appendix, the Silver-Müller radiation conditions are reviewed [388]. These conditions are applied to the scattered electromagnetic field when evaluating diffraction integrals over the infinite sphere (corresponding to surface S_2 in Fig. 7.2). It is important to note that the Silver-Müller conditions are analogous to Sommerfeld's radiation conditions, which are used in scalar diffraction analysis. Initially, the Sommerfeld conditions, as reviewed in Appendix B on Scalar Diffraction, are generalized for the vector scattered electromagnetic field. For convenience, these conditions are summarized here and are given by:

$$\lim_{R\to\infty}\left\{R\left(\frac{\partial U}{\partial R}+jkU\right)\right\}=0, \qquad \text{and}$$

$$\lim_{R\to\infty}\{RU\}<\infty.$$

By applying the previous conditions to the electric and magnetic field components E_w and H_w (where $w = x, y, z$), and expressing them in vector form, the following radiation conditions are derived:

$$\lim_{R\to\infty}\left\{R\left(\frac{\partial\vec{E}}{\partial R}+jk\vec{E}\right)\right\}=0,\qquad\text{and}\tag{7A.1}$$

$$\lim_{R\to\infty}\left\{R|\vec{E}|\right\}<\infty,\tag{7A.2}$$

$$\lim_{R\to\infty}\left\{R\left(\frac{\partial\vec{H}}{\partial R}+jk\vec{H}\right)\right\}=0,\qquad\text{and}\tag{7A.3}$$

$$\lim_{R\to\infty}\left\{R|\vec{H}|\right\}<\infty.\tag{7A.4}$$

Similar radiation conditions can be derived using the Stratton-Chu diffraction integrals [e.g., Eq. (7.4) for the scattered electric field] on the infinite spherical surface S_2, as shown in Fig. 7.2. Since the scattered waves reaching point P should not originate from S_2, the contribution $\vec{E}_{S_2}$ of the Stratton-Chu integral for the electric field must be zero:

$$\vec{E}_{S_2}=\lim_{R\to\infty}\left\{-\iint_{S_2}\underbrace{\left[\left(\hat{\imath}_n\times\vec{E}\right)\times\vec{\nabla}'G+\left(\vec{E}\cdot\hat{\imath}_n\right)\vec{\nabla}'G-j\omega\mu(\hat{\imath}_n\times\vec{H})G\right]}_{=I}ds'\right\}=0.$$

At this point, let us focus on the integrand of the above integral. By replacing $\vec{\nabla}'G=[jk+(1/R)]G\hat{\imath}_R=-[jk+(1/R)]G\hat{\imath}_n$, the integrand can be expressed as follows:

$$\begin{aligned}I&=-\left(\hat{\imath}_n\times\vec{E}\right)\times\hat{\imath}_n\left(jk+\frac{1}{R}\right)G-\left(\vec{E}\cdot\hat{\imath}_n\right)\hat{\imath}_n\left(jk+\frac{1}{R}\right)G-j\omega\mu(\hat{\imath}_n\times\vec{H})G\\&=-\left(jk+\frac{1}{R}\right)G\left[\hat{\imath}_n\times(\hat{\imath}_n\times\vec{E})-(\vec{E}\cdot\hat{\imath}_n)\hat{\imath}_n\right]-j\omega\mu(\hat{\imath}_n\times\vec{H})G\\&=-\left(jk+\frac{1}{R}\right)G\vec{E}-j\omega\mu(\hat{\imath}_n\times\vec{H})G\\&=-\left\{jk\frac{e^{-jkR}}{4\pi R^2}\left[R(\vec{E}+Z(\hat{\imath}_n\times\vec{H})\right]+\frac{e^{-jkR}}{4\pi R^2}\vec{E}\right\}.\end{aligned}$$

By replacing I in the expression for $\vec{E}_{S_2}$ and integrating over the infinite sphere to nullify the integral, the following radiation conditions must be satisfied:

$$\lim_{R\to\infty}\left\{R\left[\vec{E}+Z\left(\hat{\imath}_n\times\vec{H}\right)\right]\right\}=0,\qquad\text{and}\tag{7A.5}$$

$$\lim_{R\to\infty}\left\{\vec{E}\right\}=0.\tag{7A.6}$$

If the Stratton-Chu equation for the magnetic field [Eq. (7.5)] is applied in a similar manner, it can be shown that the following radiation conditions must be satisfied on the infinite sphere S_2:

$$\lim_{R\to\infty}\left\{R\left[\vec{H}-\frac{1}{Z}\left(\hat{\imath}_n\times\vec{E}\right)\right]\right\}=0,\qquad\text{and}\tag{7A.7}$$

$$\lim_{R\to\infty}\left\{\vec{H}\right\}=0.\tag{7A.8}$$

The radiation conditions expressed by Eqs. (7A.5)-(7A.8) are known as the Silver-Müller radiation conditions [387, 388].

7B Electromagnetic Fields Duality

In this Appendix, the *Duality Principle* of electromagnetic fields is summarized [88, 104, 390]. To apply the duality principle, it is necessary to introduce fictitious magnetic charge density ρ_m and magnetic current density $\vec{J}_m$. Assuming isotropic and homogeneous media, Maxwell's equations in their time-harmonic form can be written as:

$$\vec{\nabla} \times \vec{E} = -\vec{J}_m - j\omega\mu\vec{H},$$
$$\vec{\nabla} \times \vec{H} = \vec{J} + j\omega\epsilon\vec{E},$$
$$\vec{\nabla} \cdot \vec{E} = \frac{\rho}{\epsilon},$$
$$\vec{\nabla} \cdot \vec{H} = \frac{\rho_m}{\mu}.$$

With the introduction of the fictitious ρ_m and $\vec{J}_m$, Maxwell's equations attain complete symmetry and the following substitution (duality transformation) leaves the four Maxwell's equations invariant:

$$\vec{E} \longrightarrow \vec{H}, \qquad \vec{H} \longrightarrow -\vec{E}, \qquad \epsilon \longrightarrow \mu, \qquad \mu \longrightarrow \epsilon,$$
$$\vec{J} \longrightarrow \vec{J}_m, \qquad \rho \longrightarrow \rho_m, \qquad \vec{J}_m \longrightarrow -\vec{J}, \qquad \rho_m \longrightarrow -\rho.$$

The last duality transformation is a special case of the generalized duality transformation presented in [88, 104]:

$$\begin{bmatrix} \vec{E}^d & Z\vec{J}^d & Z\rho^d \\ Z\vec{H}^d & \vec{J}_m^d & \rho_m^d \end{bmatrix} = \begin{bmatrix} \cos\xi & \sin\xi \\ -\sin\xi & \cos\xi \end{bmatrix} \begin{bmatrix} \vec{E} & Z\vec{J} & Z\rho \\ Z\vec{H} & \vec{J}_m & \rho_m \end{bmatrix}. \tag{7B.1}$$

In this context, any variable with the superscript X^d represents the dual variable of X, and $Z = (\mu/\epsilon)^{1/2}$ denotes the characteristic impedance of the medium in which the electromagnetic field propagates. The parameter ξ is any angle in the interval $[0, 2\pi]$. For instance, the initially presented dual quantities are obtained from Eq. (7B.1) for $\xi = \pi/2$. It is important to note that in the duality transformation of Eq. (7B.1), ϵ and μ remain unchanged, as the characteristic impedance is utilized.

7C Planar Conducting Screen Aperture: Boundary Conditions

Consider an infinitely thin, perfectly conducting planar screen with an aperture in the xy-plane of the Cartesian coordinate system xyz. Let S_A represent the aperture area and S_M the conducting area. An electromagnetic field $(\vec{E}_i, \vec{H}_i)$ is incident from the left $(z < 0)$. Let $\vec{E}_s$ and $\vec{H}_s$ denote the scattered electric and magnetic fields, which may propagate either to the left or to the right of the screen. The scattered fields are generated by the induced surface charges and currents on the metallic part of the screen due to the incident field.

Since the screen is planar, the induced charges depend only on the x and y coordinates, denoted as $\sigma(x, y)$. Similarly, the induced currents, which also have only x and y components, are expressed as $\vec{K} = K_x(x,y)\hat{x} + K_y(x,y)\hat{y}$. Using the concept of scalar potential, Φ, and vector potential, $\vec{A}$, the scattered electric and magnetic fields can be written as [88,104]:

$$\vec{E}_s = -\vec{\nabla}\Phi - j\omega\vec{A}, \qquad \vec{H}_s = \frac{1}{\mu}\vec{\nabla} \times \vec{A}.$$

The potentials at any point (x, y, z) in the homogeneous isotropic region surrounding the aperture and the screen are given by:

$$\vec{A}(x,y,z) = \frac{\mu}{4\pi}\int_{S_M} \frac{\vec{K}(x',y')e^{-jkR}}{R}dx'dy', \qquad \Phi(x,y,z) = \frac{1}{4\pi\epsilon}\int_{S_M} \frac{\sigma(x',y')e^{-jkR}}{R}dx'dy'.$$

$$\text{where} \quad R = \left[(x-x')^2 + (y-y')^2 + z^2\right]^{1/2}.$$

The vector potential $\vec{A}$ does not have a z-component since there are no z-directed currents. Additionally, both the scalar and vector potentials are symmetric functions of z, i.e., $\Phi(x,y,z) = \Phi(x,y,-z)$ and $A_w(x,y,z) = A_w(x,y,-z)$ for $w = x, y$. The scattered electric field is given by:

$$E_{sx}(x,y,z) = -\frac{\partial\Phi(x,y,z)}{\partial x} - j\omega A_x(x,y,z),$$

$$E_{sy}(x,y,z) = -\frac{\partial\Phi(x,y,z)}{\partial y} - j\omega A_y(x,y,z),$$

$$E_{sz}(x,y,z) = -\frac{\partial\Phi(x,y,z)}{\partial z}.$$

From the last equations, it is straightforward to conclude that E_{sx} and E_{sy} are even functions of z, while E_{sz} is an odd function of z. (It is worth noting that the derivative of an even function is an odd function.) That is,

$$E_{sx}(x,y,z) = E_{sx}(x,y,-z), \quad E_{sy}(x,y,z) = E_{sy}(x,y,-z), \quad E_{sz}(x,y,z) = -E_{sz}(x,y,-z). \tag{7C.1}$$

Similarly the scattered magnetic field is given by:

$$H_{sx}(x,y,z) = -\frac{1}{\mu}\frac{\partial A_y(x,y,z)}{\partial z},$$

$$H_{sy}(x,y,z) = \frac{1}{\mu}\frac{\partial A_x(x,y,z)}{\partial z},$$

$$H_{sz}(x,y,z) = \frac{1}{\mu}\left(\frac{\partial A_y(x,y,z)}{\partial x} - \frac{\partial A_x(x,y,z)}{\partial y}\right).$$

Using the same arguments as for the scattered electric field, the following symmetries are deduced:

$$H_{sx}(x,y,z) = -H_{sx}(x,y,-z), \quad H_{sy}(x,y,z) = -H_{sy}(x,y,-z), \quad H_{sz}(x,y,z) = H_{sz}(x,y,-z). \tag{7C.2}$$

The total field, $(\vec{E}_t, \vec{H}_t)$, to the left or to the right of the screen can be defined as $\vec{E}_t = \vec{E}_i + \vec{E}_s$ and $\vec{H}_t = \vec{H}_i + \vec{H}_s$ [104]. The boundary conditions on the metallic area S_M and the aperture area S_A are given by:

$$\hat{\imath}_z \times \vec{E}_t(z = 0^-) = 0, \qquad \hat{\imath}_z \cdot \vec{H}_t(z = 0^-) = 0, \quad (x, y) \in S_M$$

$$\hat{\imath}_z \times \left[\vec{E}_t(z = 0^+) - \vec{E}_t(z = 0^-)\right] = 0, \qquad \hat{\imath}_z \cdot \left[\vec{E}_t(z = 0^+) - \vec{E}_t(z = 0^-)\right] = 0, \quad (x, y) \in S_A$$

$$\hat{\imath}_z \times \left[\vec{H}_t(z = 0^+) - \vec{H}_t(z = 0^-)\right] = 0, \qquad \hat{\imath}_z \cdot \left[\vec{H}_t(z = 0^+) - \vec{H}_t(z = 0^-)\right] = 0, \quad (x, y) \in S_A.$$

The continuity conditions and the odd symmetry (with respect to z) of E_{sz}, H_{sx}, and H_{sy} imply that these components vanish on the $z = 0$ plane. Therefore, the following boundary conditions can be written as:

$$\hat{\imath}_z \times \vec{E}_t = 0, \qquad \hat{\imath}_z \cdot \vec{H}_t = 0, \quad (x, y) \in S_M \tag{7C.3}$$

$$\hat{\imath}_z \times \vec{H}_t = \hat{\imath}_z \times \vec{H}_i, \qquad \hat{\imath}_z \cdot \vec{E}_t = \hat{\imath}_z \cdot \vec{E}_i, \quad (x, y) \in S_A. \tag{7C.4}$$

Some of these conditions are utilized in the derivation of Babinet's principle in Sec. 7.3.

7D Boundary Element Method Details

7D.1 Integral Boundary Equations

The principal values of the integrals appearing in Eqs. (7.44)-(7.45) and (7.47)-(7.48) are required. For instance, the principal value of the integral in Eq. (7.47) is evaluated as follows:

$$\mathcal{PV}\int_{\Gamma} \left[\phi_1(\vec{r}_\Gamma')\,\hat{\imath}_n \cdot \vec{\nabla}' G_1(\vec{r}_\Gamma, \vec{r}\,') - p_1 G_1(\vec{r}_\Gamma, \vec{r}_\Gamma')\psi_\Gamma\right] d\ell =$$

$$\int_{\Gamma-\Gamma_\rho} \left[\phi_1(\vec{r}_\Gamma')\,\hat{\imath}_n \cdot \vec{\nabla}' G_1(\vec{r}_\Gamma, \vec{r}\,') - p_1 G_1(\vec{r}_\Gamma, \vec{r}_\Gamma')\psi_\Gamma\right] d\ell +$$

$$\lim_{\rho \to 0} \int_{\Gamma_\rho} \left[\phi_1(\vec{r}_\Gamma')\,\hat{\imath}_n \cdot \vec{\nabla}' G_1(\vec{r}_\Gamma, \vec{r}_\Gamma') - p_1 G_1(\vec{r}_\Gamma, \vec{r}_\Gamma')\psi_\Gamma\right] d\ell,$$

where "$\mathcal{PV}$" denotes the principal value, indicating that the singularity is excluded by modifying the integration contour with a circular arc of radius ρ around the point $\vec{r}\Gamma$, as illustrated in Fig. 7.10(b). It is recalled that G_1 is given by Eq. (7.46). To evaluate the principal value integral, the following asymptotic expansions of the Hankel functions are required [293]: (a) $\lim_{z\to 0} H_0^{(2)}(z) = 1 - j(2/\pi)[\gamma + \ln(z/2)]$, and (b) $\lim_{z\to 0} H_1^{(2)}(z) = (z/2) + j(2/\pi z)$ where γ is the Euler's constant. Moreover, the factor $\vec{\nabla}' G_1(\vec{r}_\Gamma, \vec{r}_\Gamma') \cdot \hat{\imath}_n$ can be readily determined as:

$$\vec{\nabla}' G_1(\vec{r}_\Gamma, \vec{r}_\Gamma') \cdot \hat{\imath}_n = \vec{\nabla}' G_1(\vec{r}_\Gamma, \vec{r}_\Gamma') \cdot \hat{\imath}_\rho = \frac{k_1}{4j} H_1^{(2)}(k_1\rho), \quad \text{where,}$$

$$\hat{\imath}_\rho = \frac{x - x'}{\rho}\hat{\imath}_x + \frac{y - y'}{\rho}\hat{\imath}_y,$$

$$\rho = [(x - x')^2 + (y - y')^2]^{1/2}. \tag{7D.1}$$

In the above equations (x, y) represents the boundary point $\vec{r}_\Gamma$ and (x', y') denotes a point $\vec{r}'_\Gamma$ on Γ_ρ arc. The two terms of the integral over Γ_ρ can then be evaluated as follows:

$$\lim_{\rho\to 0}\int_{\Gamma_\rho}\left[\phi_1(\vec{r}'_\Gamma)\,\hat{\imath}_n\cdot\vec{\nabla}'G_1(\vec{r}_\Gamma,\vec{r}'_\Gamma)\right]d\ell \simeq \lim_{\rho\to 0}\left\{\phi_1(\vec{r}_\Gamma)\frac{k_1}{4j}\left(\frac{k_1\rho}{2}+j\frac{2}{\pi k_1\rho}\right)\rho\theta_\Gamma\right\} = \phi_1(\vec{r}_\Gamma)\frac{\theta_\Gamma}{2\pi},$$

$$\lim_{\rho\to 0}\int_{\Gamma_\rho}\left[-p_1G_1(\vec{r}_\Gamma,\vec{r}'_\Gamma)\psi_\Gamma\right]d\ell \simeq -p_1\lim_{\rho\to 0}\left\{\psi_\Gamma\frac{1}{4j}\left[1-j\frac{2}{\pi}\left[\gamma+\ln\left(\frac{k_1\rho}{2}\right)\right]\right]\rho\theta_\Gamma\right\} = 0.$$

Then Eq. (7.47) can be written as:

$$\left(\frac{\theta_\Gamma}{2\pi}-1\right)\phi_\Gamma(\vec{r}_\Gamma)+ \mathcal{PV}\!\!\int_\Gamma\left[\phi_\Gamma(\vec{r}'_\Gamma)\,\vec{\nabla}'G_1(\vec{r}_\Gamma,\vec{r}'_\Gamma)-p_1G_1(\vec{r}_\Gamma,\vec{r}'_\Gamma)\,\vec{\nabla}'\phi_\Gamma(\vec{r}'_\Gamma)\right]\hat{\imath}_n\cdot d\ell = -\phi^{\text{inc}}(\vec{r}_\Gamma). \tag{7D.2}$$

Similarly, Eq. (7.48) can be expressed as follows:

$$\frac{\theta_\Gamma}{2\pi}\phi_\Gamma(\vec{r}_\Gamma)+\mathcal{PV}\!\!\int_\Gamma\left[\phi_\Gamma(\vec{r}'_\Gamma)\,\vec{\nabla}'G_2(\vec{r}_\Gamma,\vec{r}'_\Gamma)-p_2G_2(\vec{r}_\Gamma,\vec{r}'_\Gamma)\,\vec{\nabla}'\phi_1(\vec{r}'_\Gamma)\right]\hat{\imath}_n\cdot d\ell = 0. \tag{7D.3}$$

7D.2 Linear and Quadratic Boundary Elements

The boundary Γ in Figs. 7.9 and 7.10(a) is divided into small segments, referred to as elements. Within each element, both the field ϕ_Γ and its normal to the boundary derivatives ψ_Γ can be expressed in terms of the field and field-derivative values at the element nodes. Different approximations for ϕ_Γ and ψ_Γ can be employed within each element. The most

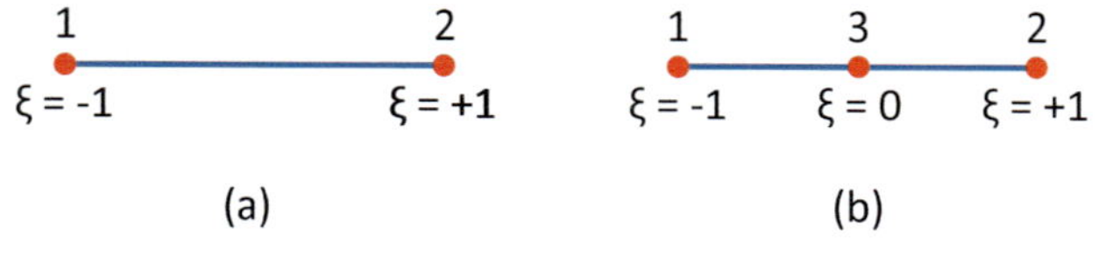

FIGURE 7D.1: Example of a boundary element. (a) Linear element. (b) Quadratic element.

common choices are linear and quadratic elements, as illustrated in Fig. 7D.1 [392]. The selection between linear and quadratic elements depends on the required accuracy and available computational resources. Linear elements are simpler and computationally less demanding, though they may necessitate a finer mesh to achieve the same accuracy as quadratic elements. Quadratic elements, while more complex, can offer higher accuracy with fewer elements.

7D.2.1 Linear Boundary Elements

In the case of linear elements, the field ϕ_Γ and its normal derivative ψ_Γ are approximated as linear functions within each element. Consider an element with nodes 1 and 2 (see Fig. 7D.1a). The field values within the element can be expressed as $\phi_\Gamma(\xi) = a\xi + b$, where ξ is a local coordinate within the element. If $\phi_{\Gamma 1}$ and $\phi_{\Gamma 2}$ denote the field values at nodes 1 and

2, respectively, then it is straightforward to show that:

$$\phi_\Gamma(\xi) = \begin{bmatrix} \dfrac{1-\xi}{2} & \dfrac{1+\xi}{2} \end{bmatrix} \begin{bmatrix} \phi_{\Gamma 1} \\ \phi_{\Gamma 2} \end{bmatrix} = \begin{bmatrix} N_1(\xi) \\ N_2(\xi) \end{bmatrix}^T \begin{bmatrix} \phi_{\Gamma 1} \\ \phi_{\Gamma 2} \end{bmatrix}, \tag{7D.4}$$

where "T" indicates transpose of a matrix and $N_1(\xi) = (1-\xi)/2$ and $N_2(\xi) = (1+\xi)/2$. A similar expression can be derived if the field normal derivative values within the element are approximated as $\psi_\Gamma(\xi) = a'\xi + b'$, where $\psi_{\Gamma 1}$ and $\psi_{\Gamma 2}$ represent the field derivative values at nodes 1 and 2, respectively.

7D.2.2 Quadratic Boundary Elements

In the case of quadratic elements, the field ϕ_Γ and its normal to the boundary derivative ψ_Γ are approximated as quadratic functions within each element. Consider an element with nodes 1, 2, and 3 (see Fig. 7D.1b). The field values within the element can be expressed as $\phi_\Gamma(\xi) = a\xi^2 + b\xi + c$, where ξ is again a local coordinate within the element. If $\phi_{\Gamma 1}$, $\phi_{\Gamma 2}$, and $\phi_{\Gamma 3}$ denote the field values at nodes 1, 2, and 3, respectively, then it is straightforward to show that:

$$\phi_\Gamma(\xi) = \begin{bmatrix} \dfrac{\xi(\xi-1)}{2} & \dfrac{\xi(\xi+1)}{2} & 1-\xi^2 \end{bmatrix} \begin{bmatrix} \phi_{\Gamma 1} \\ \phi_{\Gamma 2} \\ \phi_{\Gamma 3} \end{bmatrix} = \begin{bmatrix} N_1(\xi) \\ N_2(\xi) \\ N_3(\xi) \end{bmatrix}^T \begin{bmatrix} \phi_{\Gamma 1} \\ \phi_{\Gamma 2} \\ \phi_{\Gamma 3} \end{bmatrix}, \tag{7D.5}$$

In this case, $N_1(\xi) = \xi(\xi-1)/2$, $N_2(\xi) = \xi(\xi+1)/2$, and $N_3(\xi) = 1-\xi^2$. Again similar expression can be obtained if the field normal derivatives values within the element can be approximated as $\psi_\Gamma(\xi) = a'\xi^2 + b'\xi + c'$, with $\psi_{\Gamma 1}$, $\psi_{\Gamma 2}$ and $\psi_{\Gamma 3}$ the field derivative values at the nodes 1, 2 and 3, respectively.

Summarizing, in both linear and quadratic boundary element approximations, the field and its normal to the boundary derivative values within the element can be expressed in the form:

$$\phi_\Gamma(\xi) = \boldsymbol{N}^T \boldsymbol{\phi}_\Gamma,$$

$$\psi_\Gamma(\xi) = \boldsymbol{N}^T \boldsymbol{\psi}_\Gamma,$$

where $\boldsymbol{N}$ is the matrix appearing in Eqs. (7D.4) and (7D.5) and $\boldsymbol{\phi}_\Gamma$, $\boldsymbol{\psi}_\Gamma$, are the matrices containing the node field and field-normal-derivatives values, respectively.

7D.3 Discretization of Integral Equations

In this section the elements $[\boldsymbol{h}^I]_i$, $[\boldsymbol{g}^I]_i$, $[\boldsymbol{h}^{II}]_i$, and $[\boldsymbol{g}^{II}]_i$ in Eqs. (7.55) and (7.56) are determined based on the approach presented in Ref. [392]. Assume that the elements $[\boldsymbol{h}^I]_i$, $[\boldsymbol{g}^I]_i$, $[\boldsymbol{h}^{II}]_i$, and $[\boldsymbol{g}^{II}]_i$ are to be determined for the node i shown in Fig. 7D.2. The integrals involve integration over the boundary elements according to Eqs. (7.51) and (7.52). Assume that the contribution from the e-th element is being sought. The required terms are $\int_e \phi(\partial G_s/\partial n)d\ell$ and $\int_e \psi G_s d\ell$ ($s = 1, 2$ for regions 1 and 2, respectively). Initially, assume that the i-th node is not a node of the e-th element. Using the local element coordinate ξ,

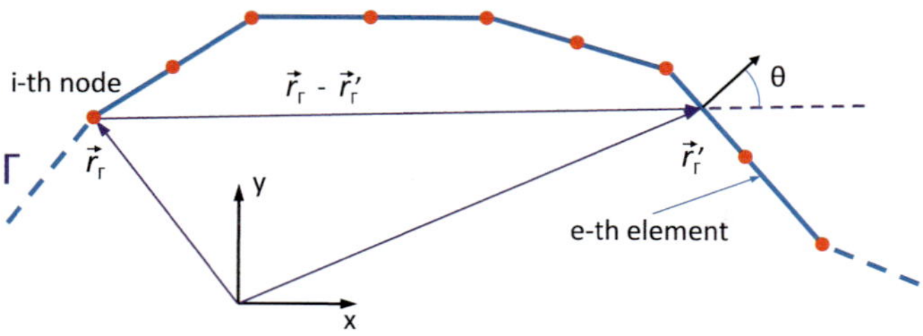

FIGURE 7D.2: A close-up of the boundary Γ, composed of boundary elements, is shown. The i-th node and the e-th boundary element are labeled.

the following expressions can be written:

$$\int_e \phi \frac{\partial G_s}{\partial n} d\ell = \int_{-1}^{+1} \left[N_1(\xi) \quad N_2(\xi) \quad N_3(\xi) \right] \begin{bmatrix} \phi_{e1} \\ \phi_{e2} \\ \phi_{e3} \end{bmatrix} \left(\frac{-k_s}{4j} \right) H_1^{(2)} \left[k_s |\vec{r}_\Gamma - \vec{r}'_\Gamma \right] \cos\theta \, \frac{\ell_e}{2} d\xi =$$

$$= \begin{bmatrix} [\boldsymbol{h}^S]_{\mathbf{e1}} \\ [\boldsymbol{h}^S]_{\mathbf{e2}} \\ [\boldsymbol{h}^S]_{\mathbf{e3}} \end{bmatrix}^T \begin{bmatrix} \phi_{e1} \\ \phi_{e2} \\ \phi_{e3} \end{bmatrix}, \tag{7D.6}$$

where $S = I$ or II (for region 1 and 2, respectively), k_s is the wavenumber of region S, ℓ_e is the length of the e-th element, and the integration is over the e-th element only. The ϕ_{ej} ($j = 1, 2, 3$) are the field values at nodes 1, 2, and 3 of the e-th element and θ is shown in Fig. 7D.2.

Similarly, the elements due to $\int_e \psi G_s d\ell$ term are given by:

$$\int_e \psi G_s d\ell = \int_{-1}^{+1} \left[N_1(\xi) \quad N_2(\xi) \quad N_3(\xi) \right] \begin{bmatrix} \psi_{e1} \\ \psi_{e2} \\ \psi_{e3} \end{bmatrix} \left(\frac{1}{4j} \right) H_0^{(2)} \left[k_s |\vec{r}_\Gamma - \vec{r}'_\Gamma \right] \frac{\ell_e}{2} d\xi =$$

$$= \begin{bmatrix} [\boldsymbol{g}^S]_{\mathbf{e1}} \\ [\boldsymbol{g}^S]_{\mathbf{e2}} \\ [\boldsymbol{g}^S]_{\mathbf{e3}} \end{bmatrix}^T \begin{bmatrix} \psi_{e1} \\ \psi_{e2} \\ \psi_{e3} \end{bmatrix}, \tag{7D.7}$$

where ψ_{ej} ($j = 1, 2, 3$) denote the field normal-derivatives at nodes 1, 2, and 3 of the e-th element. Equations (7D.6) and (7D.7) are valid if the node i does not belong to the e-th element. However, if the i-th node corresponds to the $e1$ (left node) of the e-th element, then the principal values of the corresponding integrals must be used. In this case [392]:

$$\mathcal{PV}\int_e \phi \frac{\partial G_s}{\partial n} d\ell = 0,$$

$$\mathcal{PV}\int_e \psi G_s d\ell = \lim_{\rho \to 0} \frac{\ell_e}{2} \int_\rho^2 \left[a_e(\eta - 1)^2 + b_e(\eta - 1) + c_e \right] \frac{1}{4j} H_0^{(2)} \left(\frac{\eta k_s \ell_e}{2} \right) d\eta$$

$$= \frac{\ell_e}{2} \left\{ (a_e - b_e + c_e) I_0(2) + (-2a_e + b_e) \left[I_1(2) - \frac{2}{\pi k_s^2 \ell_e^2} \right] + a_e I_2(2) \right\},$$

since $\theta = \pi/2$ in this case ($s = 1, 2$). Additionally, $\eta = \xi + 1$, and a_e, b_e, and c_e are the coefficients of the expansion $\psi_\Gamma(\xi) = a_e\xi^2 + b_e\xi + c_e$ over the e-th element, which can be expressed as functions of ψ_{e1}, ψ_{e2}, and ψ_{e3}. The remaining variables used are given by [392]:

$$I_0(\eta) = \int_0^\eta \frac{1}{4j} H_0^{(2)}\left(\frac{\eta' k_s \ell_e}{2}\right) d\eta' =$$

$$= -\frac{\eta}{4}\left\{\sum_{k=1}^{\infty} \frac{(-1)^k (k_s \ell_e \eta/4)^{2k}}{(k!)^2(2k+1)}\left[\gamma + \ln\left(\frac{k_s \ell_e \eta}{4}\right) - \frac{1}{2k+1} - \sum_{\nu=1}^{k}\frac{1}{\nu}\right] + j\right\},$$

$$I_1(\eta) = \frac{1}{4j}\frac{2\eta}{k_s \ell_e} H_1^{(2)}\left(\frac{\eta k_s \ell_e}{2}\right),$$

$$I_2(\eta) = \left(\frac{2}{k_s \ell_e}\right)^2 \left[\frac{k_s \ell_e}{2}\frac{\eta^2}{4j} H_1^{(2)}\left(\frac{\eta k_s \ell_e}{2}\right) + \frac{\eta}{4j} H_0^{(2)}\left(\frac{\eta k_s \ell_e}{2}\right) - I_0(\eta)\right].$$

In the last expression, γ is Euler's gamma constant. Similar expressions can be derived if the node i coincides with either the $e2$ or $e3$ node of the e-th element. Complete expressions can be found in Ref. [392]. Therefore, all elements of $[\boldsymbol{h}^I]_i$, $[\boldsymbol{g}^I]_i$, $[\boldsymbol{h}^{II}]_i$, and $[\boldsymbol{g}^{II}]_i$ in Eqs. (7.55) and (7.56) can be computed, and the linear system in Eq. (7.57) can be formed and solved.

7D.4 Scattering by a Cylinder

This analysis is a summary of the corresponding problem presented in Ref. [101]. An infinite homogeneous isotropic cylinder of radius a and refractive index $\tilde{n}_1$ (which may be complex to account for lossy materials) is illuminated by a *TE*-polarized plane wave $\vec{E}_i = \hat{\imath}_z E_0 \exp[-jk_0 n_2(x\sin\theta - y\cos\theta)]$, as shown in Fig. 7D.3. The freespace wavelength is λ_0, and the surrounding medium has a refractive index of n_2.

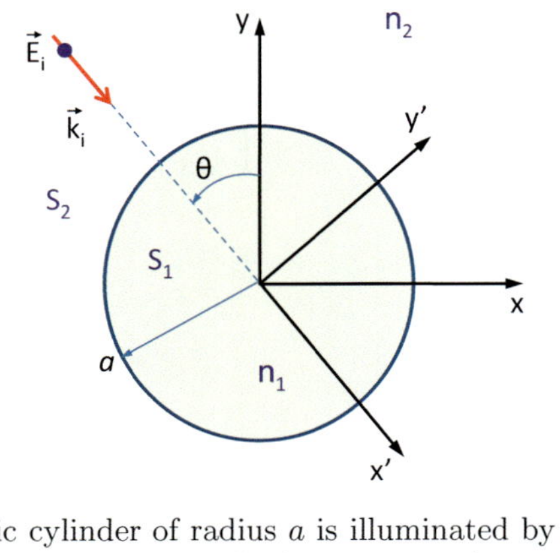

FIGURE 7D.3: A dielectric cylinder of radius a is illuminated by a plane wave. The refractive indices of the surrounding medium and the cylinder are n_1 and n_2, respectively.

It is more convenient to use the rotated coordinate system ($x', y', z' = z$), as shown in Fig. 7D.3. In the (x', y', z) coordinate system, the incident plane wave is expressed as $\vec{E}_i = \hat{\imath}_z E_0 \exp[-jk_0 n_2 x']$. For a cylinder, it is also convenient to use the cylindrical coordinate system (r_T, ϕ, z) with $x' = r_T\cos\phi$ and $y' = r_T\sin\phi$. In the cylindrical coordinate system, the incident plane wave can be written as:

$$\vec{E}_i = \hat{\imath}_z E_0 \exp[-jk_0 n_2 r_{_T} \cos\phi] = \hat{\imath}_z\, E_0 \sum_{m=-\infty}^{\infty} j^{-m} J_m(k_0 n_2 r_{_T}) e^{jm\phi}$$

$$= \hat{\imath}_z\, E_0 \sum_{m=0}^{\infty} q_m j^{-m} J_m(k_0 n_2 r_{_T}) \cos(m\phi), \tag{7D.8}$$

where J_m are the Bessel functions, $q_0 = 1$ and $q_m = 2$ for $m \geq 1$, and $J_{-m} = (-1)^m J_m$. The corresponding incident magnetic field can be determined from Maxwell's equations and can e expressed as:

$$\vec{H}_{r_T} = +\frac{E_0}{j\omega\mu_0}\frac{1}{r_T}\sum_{m=0}^{\infty} q_m m j^{-m} J_m(k_0 n_2 r_{_T}) \sin(m\phi), \tag{7D.9}$$

$$\vec{H}_\phi = +\frac{E_0 k_0 n_2}{j\omega\mu_0}\sum_{m=0}^{\infty} q_m j^{-m} J'_m(k_0 n_2 r_{_T}) \cos(m\phi). \tag{7D.10}$$

Here, J'_m denotes the derivative of the Bessel function J_m. Using the solutions of the Helmholtz equation in homogeneous isotropic regions in cylindrical coordinates [101], the scattered electric field in region S_2, $\vec{E}_2^s$, and the total electric field inside the cylinder, $\vec{E}_1$ (region S_1), can be written as:

$$\vec{E}_1 = \hat{\imath}_z\, E_0 \sum_{-\infty}^{+\infty} \left[b_m J_m(k_0 \tilde{n}_1 r_{_T}) + c_m Y_m(k_0 \tilde{n}_1 r_{_T})\right] e^{jm\phi}, \tag{7D.11}$$

$$\vec{E}_2^s = \hat{\imath}_z\, E_0 \sum_{-\infty}^{+\infty} a_m H_m^{(2)}(k_0 n_2 r_{_T}) e^{jm\phi}, \tag{7D.12}$$

where Y_m is the Bessel function of the second kind, and $H_m^{(2)}$ is the Hankel function. he ϕ components of the corresponding magnetic fields are then given by:

$$H_{1,\phi} = \frac{E_0 k_0 \tilde{n}_1}{j\omega\mu_0}\sum_{-\infty}^{+\infty} \left[b_m J'_m(k_0 \tilde{n}_1 r_{_T}) + c_m Y'_m(k_0 \tilde{n}_1 r_{_T})\right] e^{jm\phi}, \tag{7D.13}$$

$$H_{2,\phi}^s = \frac{E_0 k_0 n_2}{j\omega\mu_0}\sum_{-\infty}^{+\infty} a_m H_m^{\prime(2)}(k_0 n_2 r_{_T}) e^{jm\phi}, \tag{7D.14}$$

where again the primed functions denote the first derivatives. The coefficients $c_m = 0$ are chosen to avoid infinite fields inside the cylinder. Then, by using the continuity of the tangential electric and magnetic field components at $r_{_T} = a$, a system of equations for a_m and b_m can be defined and solved. The resulting coefficients a_m and b_m are given by:

$$a_m = j^{-m}\frac{J_m(k_0\tilde{n}_1 a) J'_m(k_0 n_2 a) - (\tilde{n}_1/n_2) J'_m(k_0\tilde{n}_1 a) J_m(k_0 n_2 a)}{(\tilde{n}_1/n_2) J'_m(k_0\tilde{n}_1 a) H_m^{(2)}(k_0 n_2 a) - J_m(k_0\tilde{n}_1 a) H_m^{\prime(2)}(k_0 n_2 a)}, \tag{7D.15}$$

$$b_m = j^{-m}\frac{H_m^{(2)}(k_0 n_2 a) J'_m(k_0 n_2 a) - J_m(k_0 n_2 a) H_m^{\prime(2)}(k_0 n_2 a)}{(\tilde{n}_1/n_2) J'_m(k_0\tilde{n}_1 a) H_m^{(2)}(k_0 n_2 a) - J_m(k_0\tilde{n}_1 a) H_m^{\prime(2)}(k_0 n_2 a)}. \tag{7D.16}$$

Therefore, the expressions for the electric and magnetic fields in both regions of the problem are given analytically. These field expressions can be easily written in the original coordinate system (x, y, z), which is the one utilized in the BEM approach. Similar expressions can also be derived for the *TM* polarization [101].

7E FDFD Method Details

7E.1 Discretization of Maxwell's Equations

For the 2D FDFD method, *TE* polarization implies that $\vec{E} = \hat{\imath}_z, E_z(x,y)$ and $\vec{H}' = H_x'(x,y)\hat{\imath}_x + H_y'(x,y)\hat{\imath}_y$. The coordinates x and y are normalized as $x' = k_0 x$ and $y' = k_0 y$, where $k_0 = 2\pi/\lambda_0$ denotes the freespace wavenumber. Considering a grid of $N_x \times N_y$ points, with N_x points along the x-direction and N_y points along the y-direction, Eqs. (7.58) can be discretized using Yee's cell, as illustrated in Fig. 7E.1a, yielding the following formulation:

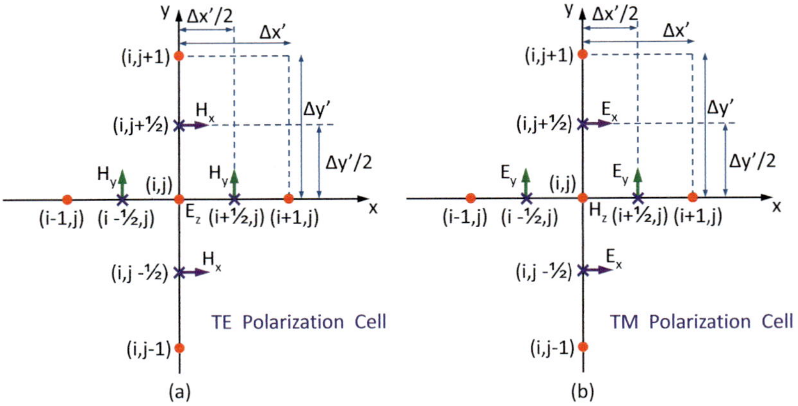

FIGURE 7E.1: The 2D Yee's cells for (a) *TE* polarization, and (b) for *TM* Polarization.

$$\frac{E_z^{i,j+1} - E_z^{i,j}}{\Delta y'} = \mu'^{\,i,j+1/2}_{rxx} H'^{\,i,j+1/2}_x$$

$$-\frac{E_z^{i+1,j} - E_z^{i,j}}{\Delta x'} = \mu'^{\,i+1/2,j}_{ryy} H'^{\,i+1/2,j}_y$$

$$\frac{H'^{\,i+1/2,j}_y - H'^{\,i-1/2,j}_y}{\Delta x'} + \frac{H'^{\,i,j+1/2}_x - H'^{\,i,j-1/2}_x}{\Delta y'} = \epsilon^{i,j}_{rzz} E_z^{i,j},$$

Similarly, for *TM* polarization, the fields are given by $\vec{H}' = \hat{\imath}_z, H_z'(x,y)$ and $\vec{E} = E_x(x,y)\hat{\imath}_x + E_y(x,y)\hat{\imath}_y$. In this case, Eq. (7.59) can be discretized using Yee's cell, as illustrated in Fig. 7E.1b, leading to the following formulation:

$$\frac{H'^{\,i,j}_z - H'^{\,i,j-1}_z}{\Delta y'} = \epsilon^{i,j-1/2}_{rxx} {E_x}^{i,j-1/2}$$

$$-\frac{H'^{\,i,j}_z - H'^{\,i-1,j}_z}{\Delta x'} = \epsilon^{i-1/2,j}_{ryy} {E_y}^{i-1/2,j}$$

$$\frac{{E_y}^{i+1/2,j} - {E_y}^{i-1/2,j}}{\Delta x'} + \frac{{E_x}^{i,j+1/2} - {E_x}^{i,j-1/2}}{\Delta y'} = \mu'^{\,i,j}_{rzz} H'^{\,i,j}_z,$$

Applying the discretized equations at each grid point and utilizing the derivative matrices defined in Refs. [374, 400], the equations for *TE* polarization can be expressed as:

$$\tilde{\boldsymbol{D}}_y^{\boldsymbol{E}} \tilde{\boldsymbol{E}}_z = \tilde{\boldsymbol{\mu}}_{\boldsymbol{rxx}} \tilde{\boldsymbol{H}}'_x,$$

$$-\tilde{\boldsymbol{D}}_x^{\boldsymbol{E}} \tilde{\boldsymbol{E}}_z = \tilde{\boldsymbol{\mu}}_{\boldsymbol{ryy}} \tilde{\boldsymbol{H}}'_y,$$

$$\tilde{\boldsymbol{D}}_x^{\boldsymbol{H}} \tilde{\boldsymbol{H}}'_y - \tilde{\boldsymbol{D}}_y^{\boldsymbol{H}} \tilde{\boldsymbol{H}}'_x = \tilde{\boldsymbol{\epsilon}}_{\boldsymbol{rzz}} \tilde{\boldsymbol{E}}_z,$$

where $\tilde{\boldsymbol{E}}_z$ is a $1 \times N_x N_y$ column vector containing the E_z field values at each grid node. The matrices $\tilde{\boldsymbol{\mu}}_{\boldsymbol{rxx}}$, $\tilde{\boldsymbol{\mu}}_{\boldsymbol{ryy}}$, and $\tilde{\boldsymbol{\epsilon}}_{\boldsymbol{rzz}}$ are $N_x N_y \times N_x N_y$ diagonal matrices that represent the material parameters at the corresponding grid points. The above equations can then be rewritten as:

$$\tilde{\boldsymbol{H}}'_x = \tilde{\boldsymbol{\mu}}_{\boldsymbol{rxx}}^{-1} \tilde{\boldsymbol{D}}_y^{\boldsymbol{E}} \tilde{\boldsymbol{E}}_z, \tag{7E.1}$$

$$\tilde{\boldsymbol{H}}'_y = -\tilde{\boldsymbol{\mu}}_{\boldsymbol{ryy}}^{-1} \tilde{\boldsymbol{D}}_x^{\boldsymbol{E}} \tilde{\boldsymbol{E}}_z, \tag{7E.2}$$

$$\left\{ \tilde{\boldsymbol{\epsilon}}_{\boldsymbol{rzz}} + \tilde{\boldsymbol{D}}_x^{\boldsymbol{H}} \tilde{\boldsymbol{\mu}}_{\boldsymbol{ryy}}^{-1} \tilde{\boldsymbol{D}}_x^{\boldsymbol{E}} + \tilde{\boldsymbol{D}}_y^{\boldsymbol{H}} \tilde{\boldsymbol{\mu}}_{\boldsymbol{rxx}}^{-1} \tilde{\boldsymbol{D}}_y^{\boldsymbol{E}} \right\} \tilde{\boldsymbol{E}}_z = 0 \Longrightarrow \tilde{\boldsymbol{A}}_{\boldsymbol{TE}} \tilde{\boldsymbol{E}}_z = 0. \tag{7E.3}$$

Similarly, the corresponding equations for the *TM* polarization case are given by:

$$\tilde{\boldsymbol{E}}_x = \tilde{\boldsymbol{\epsilon}}_{\boldsymbol{rxx}}^{-1} \tilde{\boldsymbol{D}}_y^{\boldsymbol{H}} \tilde{\boldsymbol{H}}'_z, \tag{7E.4}$$

$$\tilde{\boldsymbol{E}}_y = -\tilde{\boldsymbol{\epsilon}}_{\boldsymbol{ryy}}^{-1} \tilde{\boldsymbol{D}}_x^{\boldsymbol{H}} \tilde{\boldsymbol{H}}'_z, \tag{7E.5}$$

$$\left\{ \tilde{\boldsymbol{\mu}}_{\boldsymbol{rzz}} + \tilde{\boldsymbol{D}}_x^{\boldsymbol{E}} \tilde{\boldsymbol{\epsilon}}_{\boldsymbol{ryy}}^{-1} \tilde{\boldsymbol{D}}_x^{\boldsymbol{H}} + \tilde{\boldsymbol{D}}_y^{\boldsymbol{E}} \tilde{\boldsymbol{\epsilon}}_{\boldsymbol{rxx}}^{-1} \tilde{\boldsymbol{D}}_y^{\boldsymbol{H}} \right\} \tilde{\boldsymbol{H}}'_z = 0 \Longrightarrow \tilde{\boldsymbol{A}}_{\boldsymbol{TM}} \tilde{\boldsymbol{H}}'_z = 0. \tag{7E.6}$$

The derivative matrices have dimensions $N_x N_y \times N_x N_y$, with most of their elements being zero except for the diagonal and a subdiagonal. For example, the matrices $\tilde{\boldsymbol{D}}_x^{\boldsymbol{E}}$ and $\tilde{\boldsymbol{D}}_y^{\boldsymbol{E}}$ take the form:

$$\tilde{\boldsymbol{D}}_x^{\boldsymbol{E}} = \frac{1}{\Delta x'} \begin{bmatrix} -1 & 1 & 0 & \cdots & \cdots & \cdots & 0 \\ 0 & -1 & 1 & 0 & \cdots & \cdots & 0 \\ \vdots & \ddots & \ddots & \ddots & \ddots & \ddots & \vdots \\ 0 & 0 & 0 & -1 & 0 & \ddots & \vdots \\ \vdots & \ddots & \ddots & \ddots & \ddots & \ddots & \vdots \\ 0 & \cdots & \cdots & \cdots & \cdots & -1 & 1 \\ 0 & \cdots & \cdots & \cdots & \cdots & \cdots & -1 \end{bmatrix} \tag{7E.7}$$

$$\tilde{\boldsymbol{D}}_y^{\boldsymbol{E}} = \frac{1}{\Delta y'} \begin{bmatrix} -1 & 0 & \cdots & 1 & \cdots & \cdots & 0 \\ 0 & -1 & \cdots & 0 & 1 & \cdots & 0 \\ \vdots & \ddots & \ddots & \ddots & \ddots & \ddots & \vdots \\ 0 & 0 & 0 & -1 & \ddots & \ddots & 1 \\ \vdots & \ddots & \ddots & \ddots & \ddots & \ddots & \vdots \\ 0 & \cdots & \cdots & \cdots & \cdots & -1 & 0 \\ 0 & \cdots & \cdots & \cdots & \cdots & \cdots & -1 \end{bmatrix}. \tag{7E.8}$$

Similarly, the matrices $\tilde{\boldsymbol{D}}_{\boldsymbol{x}}^{\boldsymbol{H}}$ and $\tilde{\boldsymbol{D}}_{\boldsymbol{y}}^{\boldsymbol{H}}$ can be defined, as described in Refs. [374,400]. These matrices contain only a few nonzero elements, and sparsity techniques are employed to store them efficiently in a computer. For instance, the following excerpt provides a MATLAB implementation of the derivative matrices:

Matlab Code for Derivative Matrices:

```
%     ——    MatLab implementation of DE_x and DE_y
%     ——    sparse matrices of size NxNy x NxNy
%     ——    (Nx points along x and Ny points along y)

% ————————————————————————————————————————
%        DE_x — calculations
% ————————————————————————————————————————

DE_x = sparse(Nx*Ny,Nx*Ny);
% ——— add first diagonal ———
DE_x = spdiags(-ones(Nx*Ny,1),0,DE_x);
% ——— add second diagonal ———
DE_x = spdiags(+ones(Nx*Ny,1),+1,DE_x);

% ——— zero the iNx,iNx+1 points ———

for  ny = 1:Ny-1
     neq = ny*Nx;
     DE_x(neq,neq+1) = 0;
end

% ————————————————————————————————————————
%        DE_y — calculations
% ————————————————————————————————————————

DE_y  = sparse(Nx*Ny,Nx*Ny);
DE_y  = spdiags(-ones(Nx*Ny,1),  0, DE_y);
DE_y  = spdiags(+ones(Nx*Ny,1),+Nx, DE_y);
```

7E.2 Total-Field/Scattered-Field (TF/SF) Boundary

The total-field/scattered-field (TF/SF) boundary [374,400], illustrated in Fig. 7.14, typically encloses the scatterers. The field outside the TF/SF boundary corresponds to the scattered field, while the field inside the boundary represents the total field, which is the sum of the scattered and incident fields. Care must be taken when formulating the system of linear equations (7E.3) and (7E.6). Consider a cell located near the TF/SF boundary, as depicted in Fig. 7E.2 for *TE* polarization. The corresponding discretized Maxwell's equations for this cell are given by:

$$\frac{E_{sz}^{i,j} - E_{z}^{i,j-1}}{\Delta y'} = \mu_{rxx}^{\prime\, i,j-1/2} H_{x}^{\prime\, i,j-1/2}$$

$$-\frac{E_{sz}^{i,j} - E_{sz}^{i-1,j}}{\Delta x'} = \mu_{ryy}^{\prime\, i-1/2,j} H_{sy}^{\prime\ i-1/2,j}$$

$$\frac{H_{sy}^{\prime\ i+1/2,j} - H_{sy}^{\prime\ i-1/2,j}}{\Delta x'} + \frac{H_{sx}^{\prime\ i,j+1/2} - H_{x}^{\prime\, i,j-1/2}}{\Delta y'} = \epsilon_{rzz}^{i,j} E_{sz}^{i,j},$$

where the field components F_{sw} (with $F = E$ or $F = H'$) represent the scattered fields, while the corresponding F_w components represent the total fields. It is noted that the above equations are formulated for the bottom half of the cell to ensure a mix of total and scattered fields. Since the node (i, j) lies within the scattered-field (SF) region, the equations can be

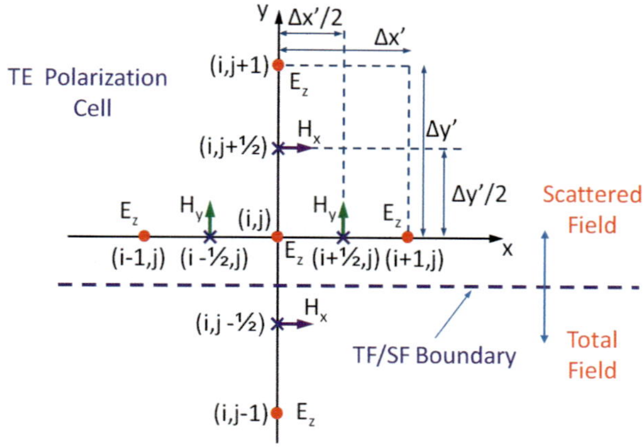

FIGURE 7E.2: The 2D Yee's cell for *TE* polarization near the TF/SF boundary.

rewritten to include only scattered fields as follows:

$$\frac{E_{sz}^{i,j} - \overbrace{\left(E_z^{i,j-1} - E_{z,\text{inc}}^{i,j-1}\right)}^{=E_{sz}^{i,j-1}}}{\Delta y'} = \mu'^{\,i,j-1/2}_{rxx} \overbrace{\left(H'^{\,i,j-1/2}_x - H'^{\,i,j-1/2}_{x,\text{inc}}\right)}^{=H'^{\,i,j-1/2}_{sx}}$$

$$-\frac{E_{sz}^{i,j} - E_{sz}^{i-1,j}}{\Delta x'} = \mu'^{\,i-1/2,j}_{ryy} H'^{\,i-1/2,j}_{sy}$$

$$\frac{H'^{\,i+1/2,j}_{sy} - H'^{\,i-1/2,j}_{sy}}{\Delta x'} + \frac{H'^{\,i,j+1/2}_{sx} - \overbrace{\left(H'^{\,i,j-1/2}_x - H'^{\,i,j-1/2}_{x,\text{inc}}\right)}^{=H'^{\,i,j-1/2}_{sx}}}{\Delta y'} = \epsilon_{rzz}^{i,j} E_{sz}^{i,j}.$$

Therefore, in the first and third equations, the incident field components are subtracted to isolate the scattered fields. To efficiently implement this, a masking matrix, $\tilde{\boldsymbol{Q}}$ [374, 400], is used. This diagonal matrix has dimensions $N_x N_y \times N_x N_y$ and contains "1" on the diagonal for points (i, j) that correspond to scattered fields, and "0" for points (i, j) that correspond to total fields. As a result, the source field $E_{z,\text{inc}}$ can be separated into the incident field in the total field region (TF), $E_{z,\text{inc}}^{TF}$, and the incident field in the scattered field region, $E_{z,\text{inc}}^{SF}$, as follows:

$$\tilde{\boldsymbol{E}}_{\boldsymbol{z},\textbf{inc}}^{\boldsymbol{TF}} = \left(\tilde{\boldsymbol{I}} - \tilde{\boldsymbol{Q}}\right) \tilde{\boldsymbol{E}}_{\boldsymbol{z},\textbf{inc}},$$

$$\tilde{\boldsymbol{E}}_{\boldsymbol{z},\textbf{inc}}^{\boldsymbol{SF}} = \tilde{\boldsymbol{Q}} \tilde{\boldsymbol{E}}_{\boldsymbol{z},\textbf{inc}},$$

where the incident field, $\tilde{\boldsymbol{E}}_{\boldsymbol{z},\textbf{inc}}$, is a matrix (or vector) of size $N_x N_y \times 1$. Some scattered field (SF) equations contain terms from the total field (TF). These terms can be corrected as follows:

$$\tilde{\boldsymbol{Q}} \tilde{\boldsymbol{A}}_{\boldsymbol{TE}} \tilde{\boldsymbol{E}}_{\boldsymbol{z}} - \tilde{\boldsymbol{Q}} \tilde{\boldsymbol{A}}_{\boldsymbol{TE}} \left(\tilde{\boldsymbol{I}} - \tilde{\boldsymbol{Q}}\right) \tilde{\boldsymbol{E}}_{\boldsymbol{z},\textbf{inc}} = 0.$$

Similarly, some total field (TF) equations contain terms from the scattered field (SF), and their correction is:

$$\left(\tilde{\boldsymbol{I}} - \tilde{\boldsymbol{Q}}\right) \tilde{\boldsymbol{A}}_{\boldsymbol{TE}} \tilde{\boldsymbol{E}}_{\boldsymbol{z}} + \left(\tilde{\boldsymbol{I}} - \tilde{\boldsymbol{Q}}\right) \tilde{\boldsymbol{A}}_{\boldsymbol{TE}} \tilde{\boldsymbol{Q}} \tilde{\boldsymbol{E}}_{\boldsymbol{z},\textbf{inc}} = 0.$$

Performing both corrections is equivalent to adding all the previous equations, resulting in:

$$\tilde{\boldsymbol{A}}_{\boldsymbol{TE}}\tilde{\boldsymbol{E}}_{\boldsymbol{z}} = \left(\tilde{\boldsymbol{Q}}\tilde{\boldsymbol{A}}_{\boldsymbol{TE}} - \tilde{\boldsymbol{A}}_{\boldsymbol{TE}}\tilde{\boldsymbol{Q}}\right)\tilde{\boldsymbol{E}}_{\boldsymbol{z,inc}}. \tag{7E.9}$$

Of course, the same procedure can be applied to the *TM* polarization case as well.

7E.3 Uniaxial Perfectly Matched Layer (UPML)

For the application of the FDFD method in open boundary structures, such as the one depicted in Fig. 7.14, it is critical to employ a procedure to eliminate unwanted reflections from the boundary of the computational domain. This can be achieved using the Uniaxial Perfectly Matched Layer (UPML). Initially, the Perfectly Matched Layer (PML) was developed by Berenger [458] for the FDTD method. Later, the UPML was introduced [401,459], again for the FDTD method. However, the UPML can also be applied to the FDFD method. In this appendix, a brief introduction to the fundamentals of the UPML is provided [374].

From the Fresnel equations, which describe the reflection and transmission at an interface (between the computational domain and the absorbing region), it becomes clear that if both media are isotropic, it is not possible to eliminate reflections for both polarizations and all angles of incidence by adjusting the parameters of the absorbing region. More flexibility is required, and this can be achieved by introducing anisotropy. Therefore, the absorbing region is made diagonally anisotropic in both permittivity and permeability.

Assuming that the computational region is initially a vacuum, and following the approach outlined in Refs. [374] and [459], the relative permittivity matrix $[\tilde{\epsilon}_{r2}]$, and the relative permeability matrix $[\tilde{\mu}_{r2}]$ of the absorbing (UPML) region are given by:

$$[\tilde{\epsilon}_{r2}] = [\tilde{\mu}_{r2}] = \begin{bmatrix} a & 0 & 0 \\ 0 & b & 0 \\ 0 & 0 & c \end{bmatrix}.$$

Initially, it is assumed that the interface between the computational domain and the UPML lies in the xy-plane, with the incident plane wave propagating along the z-direction. Following the approach of Sacks [459], it can be demonstrated that the phase-matching condition (Snell's law) takes the form $\sin\theta_1 = \sqrt{bc}\sin\theta_2$, where θ_1 and θ_2 are the propagation angles (measured relative to the normal z-direction) of the plane wave in the computational and UPML regions, respectively. Through a straightforward electromagnetic analysis, the reflection and transmission coefficients between the vacuum and the anisotropic UPML region for *TE* and *TM* polarization can be determined as:

$$r_{_{TE}} = \frac{\sqrt{a}\cos\theta_1 - \sqrt{b}\cos\theta_2}{\sqrt{a}\cos\theta_1 + \sqrt{b}\cos\theta_2}, \qquad r_{_{TM}} = \frac{\sqrt{b}\cos\theta_2 - \sqrt{a}\cos\theta_1}{\sqrt{b}\cos\theta_2 + \sqrt{a}\cos\theta_1}.$$

Selecting $bc = 1$ ensures that $\theta_1 = \theta_2$, making the reflection coefficients $r_{_{TE}}$ and $r_{_{TM}}$ independent of the angle of incidence. Further, by choosing $a = b$, both $r_{_{TE}}$ and $r_{_{TM}}$ become zero for any polarization and angle of incidence. Consequently, the elements of the relative permittivity and permeability of the UPML are given by $(a, b, c) = (a, a, 1/a)$. Setting $a = s_z$ (due to the z-propagating waves being considered), the relative permittivity and permeability matrices are then expressed as:

$$[\tilde{\epsilon}_{r2}] = [\tilde{\mu}_{r2}] = \begin{bmatrix} s_z & 0 & 0 \\ 0 & s_z & 0 \\ 0 & 0 & 1/s_z \end{bmatrix}.$$

If the same procedure is applied for propagation along the x-direction (with the yz interface) or along the y-direction (with the xz interface), similar tensors are obtained. To absorb all waves at all types of interfaces, these tensors are multiplied, resulting in the UPML relative permittivity and permeability matrices of the following form:

$$[\tilde{\epsilon}_{r2}] = [\tilde{\mu}_{r2}] = [\tilde{s}] = \begin{bmatrix} s_y s_z/s_x & 0 & 0 \\ 0 & s_x s_z/s_y & 0 \\ 0 & 0 & s_x s_y/s_z \end{bmatrix}. \tag{7E.10}$$

The matrix $[\tilde{s}]$ is set to unity outside the UPML region and must be defined within the UPML region. When incorporating $[\tilde{s}]$ into Maxwell's equations (assuming that $\vec{H}' = -jZ_0\vec{H}$, where Z_0 is the free-space impedance), the equations can be written as follows (for 2D geometry, it is implied that $s_z = 1$):

$$\vec{\nabla} \times \vec{E} = k_0 \, [\tilde{s}] \, [\tilde{\mu}_r] \, \vec{H}' = k_0 \, [\tilde{\mu}_r'] \, \vec{H}',$$

$$\vec{\nabla} \times \vec{H}' = k_0 \, [\tilde{s}] \, [\tilde{\epsilon}_r] \, \vec{E} = k_0 \, [\tilde{\epsilon}_r'] \, \vec{E}.$$

Here, the relative permittivity and relative permeability matrices are:

$$[\tilde{\epsilon}_r'] = [\tilde{s}] \, [\tilde{\epsilon}_r] = \begin{bmatrix} \frac{s_y}{s_x}\epsilon_{rxx} & 0 & 0 \\ 0 & \frac{s_x}{s_y}\epsilon_{ryy} & 0 \\ 0 & 0 & s_x s_y \epsilon_{rzz} \end{bmatrix},$$

$$[\tilde{\mu}_r'] = [\tilde{s}] \, [\tilde{\mu}_r] = \begin{bmatrix} \frac{s_y}{s_x}\mu_{rxx} & 0 & 0 \\ 0 & \frac{s_x}{s_y}\mu_{ryy} & 0 \\ 0 & 0 & s_x s_y \mu_{rzz} \end{bmatrix}.$$

Now, the parameters s_x and s_y (for the 2D case) need to be determined. These parameters are complex in order to introduce loss inside the UPML region and absorb all outgoing waves from the structure. The selection of s_x and s_y in Examples 7.6.1 and 7.6.2 was carried out using the method described in Ref. [374].

If x_ℓ and y_ℓ are defined as local coordinates that measure the distance (from the computational domain boundary) to various grid points inside the UPML region, the $s_x(x_\ell)$ and $s_y(y_\ell)$ can be selected as follows:

$$s_x(x_\ell) = a_x(x_\ell) \left[1 - j\frac{Z_0}{k_0} \, \sigma_x(x_\ell) \right], \tag{7E.11}$$

$$s_y(y_\ell) = a_y(y_\ell) \left[1 - j\frac{Z_0}{k_0} \, \sigma_y(y_\ell) \right], \tag{7E.12}$$

where σ_x and σ_y are artificial conductivities of the form $\sigma_u(u_\ell) = \sigma_{max} \sin^2(\pi u_\ell / 2w)$, with $u = x, y$ (note that w is the width of the UPML region). The maximum artificial conductivity, σ_{max}, is selected to be $1/\Omega\text{m}$. The dimensionless parameters a_x and a_y are given by $a_u(u_\ell) = 1 + a_{max}(u_\ell/w)^p$, with $u = x, y$, and $0 \leq a_{max} \leq 5$, and $3 \leq p \leq 5$. In the FDFD examples presented in this book, $p = 3$ and $a_{max} = 3$ are used. Further details can be found in Ref. [374].

7F RCWA Method Details

7F.1 Maxwell's Equations in Grating Region—TE Polarization

Using Maxwell's equations, the electric and magnetic fields in the grating region for the *TE* polarization case can be expressed as:

$$E_{2y} = \frac{1}{j\omega\epsilon_0\varepsilon(x,z)}\left(\frac{\partial H_{2x}}{\partial z} - \frac{\partial H_{2z}}{\partial x}\right),$$

$$H_{2x} = \frac{1}{j\omega\mu_0}\frac{\partial E_{2y}}{\partial z},$$

$$H_{2z} = -\frac{1}{j\omega\mu_0}\frac{\partial E_{2y}}{\partial x},$$

where $\varepsilon(x,z)$ represents the periodic relative permittivity in the grating region. By using the expansions from Eqs. (7.67) and (7.68) for the fields in terms of non-homogeneous plane waves, the equation for H_{2x} can be written as follows:

$$j\omega\mu_0\left(\frac{\epsilon_0}{\mu_0}\right)^{1/2}\sum_i U_{xi}(z)e^{-jk_{xi}x+jiK_z z} = \sum_i\left[\frac{dS_{yi}}{dz} + jiK_z S_{yi}\right]e^{-jk_{xi}x+jiK_z z},$$

where $K_z = (2\pi/\Lambda)\cos\phi$ (refer to Fig. 7.18(b)). The last equation must hold for any x, z within the grating region. Therefore, the coefficients of the exponential terms on both sides must be equal. This condition leads to the following:

$$\frac{1}{k_0}\frac{dS_{yi}}{dz} = jU_{xi} - ji\frac{K_z}{k_0}S_{yi}.$$

By setting the normalized length $z' = k_0 z$ and assuming a finite number of diffracted orders, $-M \leq i \leq M$ (for a total of $2M+1$ diffracted orders), the last equation can be rewritten in the following form:

$$\frac{d\tilde{\boldsymbol{S}}_{\boldsymbol{y}}}{dz'} = j\tilde{\boldsymbol{U}}_{\boldsymbol{x}} - j\tilde{\boldsymbol{K}}_{\boldsymbol{z}}\tilde{\boldsymbol{S}}_{\boldsymbol{y}}. \tag{7F.1}$$

Here, $\tilde{\boldsymbol{S}}_{\boldsymbol{y}}$ and $\tilde{\boldsymbol{U}}_{\boldsymbol{x}}$ are matrices of size $(2M+1)\times 1$ defined as: $\tilde{\boldsymbol{S}}_{\boldsymbol{y}} = [S_{y,-M}, S_{y,-M+1}, \cdots, S_{y,M}]^T$, and $\tilde{\boldsymbol{U}}_{\boldsymbol{x}} = [U_{x,-M}, U_{x,-M+1}, \cdots, U_{x,M}]^T$. The matrix $\tilde{\boldsymbol{K}}_{\boldsymbol{z}}$ is a diagonal matrix of size $(2M+1)\times(2M+1)$ defined as $\tilde{\boldsymbol{K}}_{\boldsymbol{z}} = (K_z/k_0)\mathrm{diag}(-M, -M+1, \cdots, M-1, M)$. Eliminating H_{2z} from the equation of the electric field, it is easy to show:

$$j\omega\epsilon_0\varepsilon(x,z)E_{2y} = \frac{\partial H_{2x}}{\partial z} + \frac{1}{j\omega\mu_0}\frac{\partial^2 E_{2y}}{\partial x^2}.$$

At this stage, it is necessary to use the Fourier series expansion of the periodic relative permittivity. It is given by:

$$\varepsilon(x,z) = \sum_h \varepsilon_h \exp[jh(K_x x + K_z z)],$$

where ε_h represents the relative permittivity h-th harmonic. By inserting the field expansions and the relative permittivity expansion into the equation derived for the electric field earlier, it can be transformed into the following form:

$$j\omega\epsilon_0 \sum_h \varepsilon_h e^{jh(K_x x+K_z z)} \sum_m S_{ym} e^{-jk_{xm}x+jmK_z z} =$$

$$= \left(\frac{\epsilon_0}{\mu_0}\right)^{1/2} \sum_i \left[\frac{dU_{xi}}{dz} + jiK_z U_{xi}\right] e^{-jk_{xi}x+jiK_z z} +$$

$$\frac{1}{j\omega\mu_0} \sum_i (-k_{xi}^2 S_{yi}) e^{-jk_{xi}x+jiK_z z} \Longrightarrow$$

$$j\omega\epsilon_0 \sum_i \left[\sum_m \varepsilon_{i-m} S_{ym}\right] e^{-jk_{xi}x+jiK_z z}$$

$$= \sum_i \left\{\left(\frac{\epsilon_0}{\mu_0}\right)^{1/2} \left[\frac{dU_{xi}}{dz} + jiK_z U_{xi}\right] - \frac{k_{xi}^2}{j\omega\mu_0} S_{yi}\right\} e^{-jk_{xi}x+jiK_z z}.$$

Again, the last equation must hold for any x, z within the grating region. This implies:

$$\frac{1}{k_0}\frac{dU_{xi}}{dz} = j\sum_m \varepsilon_{i-m} S_{ym} - ji\frac{K_z}{k_0} U_{xi} - j\frac{k_{xi}^2}{k_0^2} S_{yi}.$$

The last equation can be expressed in matrix form, similar to Eq. (7F.1), as:

$$\frac{d\tilde{\boldsymbol{U}}_{\boldsymbol{x}}}{dz'} = j\left(\tilde{\boldsymbol{E}} - \tilde{\boldsymbol{K}}_{\boldsymbol{x}}^{\boldsymbol{2}}\right)\tilde{\boldsymbol{S}}_{\boldsymbol{y}} - j\tilde{\boldsymbol{K}}_{\boldsymbol{z}}\tilde{\boldsymbol{U}}_{\boldsymbol{x}}. \tag{7F.2}$$

The matrix $\tilde{\boldsymbol{E}}$ is a Toeplitz $(2M+1) \times (2M+1)$ matrix of the following form:

$$\tilde{\boldsymbol{E}} = \begin{bmatrix} \varepsilon_0 & \varepsilon_{-1} & \varepsilon_{-2} & \cdots & \varepsilon_{-M} & 0 & \cdots & 0 \\ \varepsilon_{+1} & \varepsilon_0 & \varepsilon_{-1} & \varepsilon_{-2} & \cdots & & & \vdots \\ \varepsilon_{+2} & \varepsilon_{+1} & \varepsilon_0 & \varepsilon_{-1} & \ddots & & & \vdots \\ \vdots & \varepsilon_{+2} & \ddots & \ddots & \ddots & \ddots & & \vdots \\ \vdots & & \ddots & \ddots & \ddots & \ddots & \varepsilon_{-2} & \vdots \\ \vdots & & & \ddots & \varepsilon_{+1} & \varepsilon_0 & \varepsilon_{-1} & \varepsilon_{-2} \\ \vdots & & & & \varepsilon_{+2} & \varepsilon_{+1} & \varepsilon_0 & \varepsilon_{-1} \\ 0 & 0 & \cdots & \varepsilon_{+M} & \cdots & \varepsilon_{+2} & \varepsilon_{+1} & \varepsilon_0 \end{bmatrix}$$

Equations (7F.1) and (7F.2) can be combined into the following system of coupled first-order differential equations:

$$\frac{d}{dz'}\begin{bmatrix} \tilde{\boldsymbol{S}}_{\boldsymbol{y}} \\ \tilde{\boldsymbol{U}}_{\boldsymbol{x}} \end{bmatrix} = j\begin{bmatrix} -\tilde{\boldsymbol{K}}_{\boldsymbol{z}} & \tilde{\boldsymbol{I}} \\ \left(\tilde{\boldsymbol{E}} - \tilde{\boldsymbol{K}}_{\boldsymbol{x}}^{\boldsymbol{2}}\right) & -\tilde{\boldsymbol{K}}_{\boldsymbol{z}} \end{bmatrix}\begin{bmatrix} \tilde{\boldsymbol{S}}_{\boldsymbol{y}} \\ \tilde{\boldsymbol{U}}_{\boldsymbol{x}} \end{bmatrix} = j\tilde{\boldsymbol{A}}_{\boldsymbol{TE}}\begin{bmatrix} \tilde{\boldsymbol{S}}_{\boldsymbol{y}} \\ \tilde{\boldsymbol{U}}_{\boldsymbol{x}} \end{bmatrix}, \tag{7F.3}$$

where $\tilde{\boldsymbol{I}}$ is the identity matrix.

7F.2 System of Boundary Conditions—TE Polarization

In this section, the formulation of the linear system of equations representing the boundary conditions of the diffraction problem is presented. The tangential electric and magnetic fields must be continuous at the two boundaries, $z = 0$ and $z = d$. These conditions are:

$$E_{1y}(z=0) = E_{2y}(z=0), \tag{7F.4}$$

$$H_{1x}(z=0) = H_{2x}(z=0), \tag{7F.5}$$

$$E_{2y}(z=d) = E_{3y}(z=d), \tag{7F.6}$$

$$H_{2x}(z=d) = H_{3x}(z=d). \tag{7F.7}$$

From Eq. (7F.4), the following equation is obtained (using the field expansions from Sec. 7.7.2.1):

$$e^{-jk_{x0}x} + \sum_i R_i e^{-jk_{xi}x} = \sum_i S_{yi}(0)e^{-jk_{xi}x} \Longrightarrow \delta_{i0} + R_i \;=\; S_{yi}(0),$$

where δ_{0i} is the Kronecker delta (which equals 1 for $i = 0$ and 0 otherwise). Using the solution for the fields in the grating region from Eq. (7.76), the last equation can be expressed in the following matrix form:

$$\tilde{\boldsymbol{I}}_{\mathbf{1}} + \tilde{\boldsymbol{R}} = \tilde{\boldsymbol{W}}_{\mathbf{11}}\tilde{\boldsymbol{C}}_{\mathbf{1}} + \tilde{\boldsymbol{W}}_{\mathbf{12}} \exp\left[-j\tilde{\boldsymbol{\Lambda}}_{\mathbf{2}}k_0 d\right] \tilde{\boldsymbol{C}}_{\mathbf{2}}, \tag{7F.8}$$

where $\tilde{\boldsymbol{I}}_{\mathbf{1}}$ is a $(2M+1)\times 1$ matrix of the form $\tilde{\boldsymbol{I}}_{\mathbf{1}} = [0,\cdots,0,1,0,\cdots,0]^T$. Also $\tilde{\boldsymbol{R}}$ is a $(2M+1)\times 1$ matrix defined as $\tilde{\boldsymbol{R}} = [R_{-M}, R_{-M+1},\cdots,R_0,\cdots,R_M]^T$, that contains all the unknown reflected amplitudes. Working in a similar manner, Eqs. (7F.5)-(7F.7) can be converted into the following matrix form equations:

$$\tilde{\boldsymbol{I}}_{\mathbf{2}} - \tilde{\boldsymbol{Z}}_{\mathbf{1}}\tilde{\boldsymbol{R}} = \tilde{\boldsymbol{W}}_{\mathbf{21}}\tilde{\boldsymbol{C}}_{\mathbf{1}} + \tilde{\boldsymbol{W}}_{\mathbf{22}} \exp\left[-j\tilde{\boldsymbol{\Lambda}}_{\mathbf{2}}k_0 d\right] \tilde{\boldsymbol{C}}_{\mathbf{2}}, \tag{7F.9}$$

$$\tilde{\boldsymbol{T}} = \tilde{\boldsymbol{Y}}\left[\tilde{\boldsymbol{W}}_{\mathbf{11}} \exp\left[+j\tilde{\boldsymbol{\Lambda}}_{\mathbf{1}}k_0 d\right] \tilde{\boldsymbol{C}}_{\mathbf{1}} + \tilde{\boldsymbol{W}}_{\mathbf{12}}\tilde{\boldsymbol{C}}_{\mathbf{2}}\right], \tag{7F.10}$$

$$-\tilde{\boldsymbol{Z}}_{\mathbf{3}}\tilde{\boldsymbol{T}} = \tilde{\boldsymbol{Y}}\left[\tilde{\boldsymbol{W}}_{\mathbf{21}} \exp\left[+j\tilde{\boldsymbol{\Lambda}}_{\mathbf{1}}k_0 d\right] \tilde{\boldsymbol{C}}_{\mathbf{1}} + \tilde{\boldsymbol{W}}_{\mathbf{22}}\tilde{\boldsymbol{C}}_{\mathbf{2}}\right], \tag{7F.11}$$

where $\tilde{\boldsymbol{T}}$ is a $(2M+1)\times 1$ matrix defined as $\tilde{\boldsymbol{T}} = [T_{-M}, T_{-M+1},\cdots,T_0,\cdots,T_M]^T$ and $\tilde{\boldsymbol{Y}}$ is a diagonal $(2M+1)\times(2M+1)$ matrix of the form $\tilde{\boldsymbol{Y}} = \mathrm{diag}[e^{j(-M)K_z d},\cdots,e^{j(+M)K_z d}]$. In addition, $\tilde{\boldsymbol{Z}}_{\mathbf{1}}$ and $\tilde{\boldsymbol{Z}}_{\mathbf{3}}$ are diagonal $(2M+1)\times(2M+1)$ matrices defined as: $\tilde{\boldsymbol{Z}}_{\mathbf{1}} = (1/k_0)\mathrm{diag}[k_{1,z(-M)},\cdots,k_{1,z(+M)}]$ and $\tilde{\boldsymbol{Z}}_{\mathbf{3}} = (1/k_0)\mathrm{diag}[k_{3,z(-M)},\cdots,k_{3,z(+M)}]$. Finally, the matrix $\tilde{\boldsymbol{I}}_{\mathbf{2}}$ is a $(2M+1)\times 1$ matrix of the form $\tilde{\boldsymbol{I}}_{\mathbf{2}} = [0,\cdots,0,-n_1\cos\theta_{inc},0,\cdots,0]^T$.

7F.3 Maxwell's Equations in Grating Region—TM Polarization

Using Maxwell's equations, the electric and magnetic fields in the grating region for the *TM* polarization case can be expressed as:

$$H_{2y} = -\frac{1}{j\omega\mu_0}\left(\frac{\partial E_{2x}}{\partial z} - \frac{\partial E_{2z}}{\partial x}\right),$$

$$E_{2x} = \frac{j}{\omega\epsilon_0\varepsilon(x,z)}\frac{\partial H_{2y}}{\partial z},$$

$$E_{2z} = -\frac{j}{\omega\epsilon_0\varepsilon(x,z)}\frac{\partial H_{2y}}{\partial x},$$

where $\varepsilon(x, z)$ is again the periodic relative permittivity in the grating region. Eliminating the E_{2z} from the above equations the following equations that involved only the tangential to the boundaries fields are derived:

$$\frac{\partial H_{2y}}{\partial z} = -j\omega\epsilon_0\varepsilon(x,z)E_{2x} \tag{7F.12}$$

$$H_{2y} = -\frac{1}{j\omega\mu_0}\frac{\partial E_{2x}}{\partial z} + \frac{j}{\omega\epsilon_0}\frac{\partial}{\partial x}\left(\frac{1}{\varepsilon(x,z)}\frac{\partial H_{2y}}{\partial x}\right) \tag{7F.13}$$

For the *TM* polarization case, to improve the convergence of RCWA (particularly for metallic gratings), the Fourier series analysis for discontinuous periodic structures [411, 460, 461] was employed. The following Fourier expansions for the relative permittivity and its inverse are used:

$$\varepsilon(x,z) = \sum_h \varepsilon_h \exp[jh(K_x x + K_z z)], \tag{7F.14}$$

$$\frac{1}{\varepsilon(x,z)} = \sum_h a_h \exp[jh(K_x x + K_z z)]. \tag{7F.15}$$

By manipulating Eqs. (7F.12) and (7F.13) in the same manner as in the *TE* polarization case, the following matrix equations are obtained:

$$\frac{d\tilde{\boldsymbol{U}}_{\boldsymbol{y}}}{dz'} = -j\tilde{\boldsymbol{K}}_{\boldsymbol{z}}\tilde{\boldsymbol{U}}_{\boldsymbol{y}} - j\tilde{\boldsymbol{A}}^{-1}\tilde{\boldsymbol{S}}_{\boldsymbol{x}}, \tag{7F.16}$$

$$\frac{d\tilde{\boldsymbol{S}}_{\boldsymbol{x}}}{dz'} = j\left(\tilde{\boldsymbol{K}}_{\boldsymbol{x}}\tilde{\boldsymbol{E}}^{-1}\tilde{\boldsymbol{K}}_{\boldsymbol{x}} - \tilde{\boldsymbol{I}}\right)\tilde{\boldsymbol{U}}_{\boldsymbol{y}} - j\tilde{\boldsymbol{K}}_{\boldsymbol{z}}\tilde{\boldsymbol{S}}_{\boldsymbol{x}}, \tag{7F.17}$$

where $\tilde{\boldsymbol{U}}_{\boldsymbol{y}}$ and $\tilde{\boldsymbol{S}}_{\boldsymbol{x}}$ are matrices of size $(2M+1)\times 1$ defined as: $\tilde{\boldsymbol{U}}_{\boldsymbol{y}} = [U_{y,-M}, U_{y,-M+1}, \cdots, U_{y,M}]^T$, and $\tilde{\boldsymbol{S}}_{\boldsymbol{x}} = [S_{x,-M}, S_{x,-M+1}, \cdots, S_{x,M}]^T$, and $\tilde{\boldsymbol{I}}$ is the identity matrix of size $(2M+1)\times(2M+1)$. The matrices $\tilde{\boldsymbol{K}}_{\boldsymbol{x}}$, $\tilde{\boldsymbol{K}}_{\boldsymbol{z}}$, and $\tilde{\boldsymbol{E}}$ are defined in Sec. 7F.1. The matrix $\tilde{\boldsymbol{A}}$ is a $(2M+1)\times(2M+1)$ Toeplitz matrix that is exactly like $\tilde{\boldsymbol{E}}$ but with elements a_h instead of ε_h. Then, Eqs. (7F.16) and (7F.17) can be combined in the following set of coupled first-order differential equations:

$$\frac{d}{dz'}\begin{bmatrix}\tilde{\boldsymbol{U}}_{\boldsymbol{y}}\\ \tilde{\boldsymbol{S}}_{\boldsymbol{x}}\end{bmatrix} = j\begin{bmatrix}-\tilde{\boldsymbol{K}}_{\boldsymbol{z}} & -\tilde{\boldsymbol{A}}^{-1}\\ \tilde{\boldsymbol{K}}_{\boldsymbol{x}}\tilde{\boldsymbol{E}}^{-1}\tilde{\boldsymbol{K}}_{\boldsymbol{x}} - \tilde{\boldsymbol{I}} & -\tilde{\boldsymbol{K}}_{\boldsymbol{z}}\end{bmatrix}\begin{bmatrix}\tilde{\boldsymbol{S}}_{\boldsymbol{y}}\\ \tilde{\boldsymbol{U}}_{\boldsymbol{x}}\end{bmatrix} = j\tilde{\boldsymbol{A}}_{\boldsymbol{TM}}\begin{bmatrix}\tilde{\boldsymbol{S}}_{\boldsymbol{y}}\\ \tilde{\boldsymbol{U}}_{\boldsymbol{x}}\end{bmatrix}. \tag{7F.18}$$

7F.4 System of Boundary Conditions—TM Polarization

In this section, the formulation of the linear system of equations that represent the boundary conditions of the diffraction problem is presented. The tangential electric and magnetic fields must be continuous at the two boundaries $z = 0$ and $z = d$. These conditions are:

$$H_{1y}(z=0) = H_{2y}(z=0), \tag{7F.19}$$

$$E_{1x}(z=0) = E_{2x}(z=0), \tag{7F.20}$$

$$H_{2y}(z=d) = H_{3y}(z=d), \tag{7F.21}$$

$$E_{2x}(z=d) = E_{3x}(z=d). \tag{7F.22}$$

By manipulating the above boundary conditions in a manner analogous to the *TE* polarization case, the following set of linear equations is obtained:

$$\tilde{\boldsymbol{I}}_{\boldsymbol{1}} + \tilde{\boldsymbol{R}} = \tilde{\boldsymbol{W}}_{\boldsymbol{11}}\tilde{\boldsymbol{C}}_{\boldsymbol{1}} + \tilde{\boldsymbol{W}}_{\boldsymbol{12}} \exp\left[-j\tilde{\boldsymbol{\Lambda}}_{\boldsymbol{2}}k_0 d\right]\tilde{\boldsymbol{C}}_{\boldsymbol{2}}, \tag{7F.23}$$

$$\tilde{\boldsymbol{I}}_{\boldsymbol{2}} + \tilde{\boldsymbol{Z}}_{\boldsymbol{1}}\tilde{\boldsymbol{R}} = \tilde{\boldsymbol{W}}_{\boldsymbol{21}}\tilde{\boldsymbol{C}}_{\boldsymbol{1}} + \tilde{\boldsymbol{W}}_{\boldsymbol{22}} \exp\left[-j\tilde{\boldsymbol{\Lambda}}_{\boldsymbol{2}}k_0 d\right]\tilde{\boldsymbol{C}}_{\boldsymbol{2}}, \tag{7F.24}$$

$$\tilde{\boldsymbol{T}} = \tilde{\boldsymbol{Y}}\left[\tilde{\boldsymbol{W}}_{\boldsymbol{11}} \exp\left[+j\tilde{\boldsymbol{\Lambda}}_{\boldsymbol{1}}k_0 d\right]\tilde{\boldsymbol{C}}_{\boldsymbol{1}} + \tilde{\boldsymbol{W}}_{\boldsymbol{12}}\tilde{\boldsymbol{C}}_{\boldsymbol{2}}\right], \tag{7F.25}$$

$$-\tilde{\boldsymbol{Z}}_{\boldsymbol{3}}\tilde{\boldsymbol{T}} = \tilde{\boldsymbol{Y}}\left[\tilde{\boldsymbol{W}}_{\boldsymbol{21}} \exp\left[+j\tilde{\boldsymbol{\Lambda}}_{\boldsymbol{1}}k_0 d\right]\tilde{\boldsymbol{C}}_{\boldsymbol{1}} + \tilde{\boldsymbol{W}}_{\boldsymbol{22}}\tilde{\boldsymbol{C}}_{\boldsymbol{2}}\right]. \tag{7F.26}$$

Here, $\tilde{\boldsymbol{R}}$, $\tilde{\boldsymbol{T}}$, $\tilde{\boldsymbol{Y}}$ and $\tilde{\boldsymbol{I}}_{\boldsymbol{1}}$ are the same as the matrices defined in Sec. 7F.2. In addition, $\tilde{\boldsymbol{Z}}_{\boldsymbol{1}}$ and $\tilde{\boldsymbol{Z}}_{\boldsymbol{3}}$ are diagonal $(2M+1)\times(2M+1)$ matrices defined as: $\tilde{\boldsymbol{Z}}_{\boldsymbol{1}} = (1/k_0 n_1^2)\text{diag}[k_{1,z(-M)},\cdots,k_{1,z(+M)}]$ and $\tilde{\boldsymbol{Z}}_{\boldsymbol{3}} = (1/k_0 n_3^2)\text{diag}[k_{3,z(-M)},\cdots,k_{3,z(+M)}]$. Finally, the matrix $\tilde{\boldsymbol{I}}_{\boldsymbol{2}}$ is a $(2M+1)\times 1$ matrix of the form $\tilde{\boldsymbol{I}}_{\boldsymbol{2}} = [0,\cdots,0,\cos\theta_{inc}/n_1,0,\cdots,0]^T$.

7F.5 Crossed-Gratings—Maxwell's Equations in Grating Region

For the crossed-grating, the relative permittivity and its inverse can be expressed by the following Fourier series:

$$\varepsilon(x,y) = \sum_{h_1}\sum_{h_2} \varepsilon_{h_1h_2} \exp[jh_1K_x x + jh_2K_y y)],$$

$$\frac{1}{\varepsilon(x,y)} = \sum_{h_1}\sum_{h_2} \alpha_{h_1h_2} \exp[jh_1K_x x + jh_2K_y y)],$$

where $\varepsilon_{h_1h_2}$ and $\alpha_{h_1h_2}$ denote the (h_1,h_2)-th harmonic of the relative permittivity and its inverse, respectively. By applying the curl Maxwell equation for the electric field, it is straightforward, following the approach used for 1D gratings, to derive the following equations by substituting the expressions from Eqs. (7.100) and (7.101):

$$-jk_{y,i_2}S_{z,i_1i_2} - \frac{dS_{y,i_1i_2}}{dz} = -jk_0U_{x,i_1,i_2},$$

$$\frac{dS_{x,i_1i_2}}{dz} + jk_{x,i_1}S_{z,i_1i_2}(z) = -jk_0U_{y,i_1,i_2},$$

$$-jk_{x,i_1}S_{y,i_1i_2}(z) + jk_{y,i_2}S_{x,i_1i_2} = -jk_0U_{z,i_1,i_2}.$$

Similarly, working with the curl equation of the magnetic field and the Fourier series of the relative permittivity the following set of equations can be derived:

$$-jk_{y,i_2}U_{z,i_1i_2} - \frac{dU_{y,i_1i_2}}{dz} = jk_0\sum_{m_1}\sum_{m_2}\varepsilon_{i_1-m_1,i_2-m_2}S_{x,m_1,m_2},$$

$$\frac{dU_{x,i_1i_2}}{dz} + jk_{x,i_1}U_{z,i_1i_2}(z) = jk_0\sum_{m_1}\sum_{m_2}\varepsilon_{i_1-m_1,i_2-m_2}S_{y,m_1,m_2},$$

$$-jk_{x,i_1}U_{y,i_1i_2}(z) + jk_{y,i_2}U_{x,i_1i_2} = jk_0\sum_{m_1}\sum_{m_2}\varepsilon_{i_1-m_1,i_2-m_2}S_{z,m_1,m_2}.$$

The electromagnetic field equations in the grating region can be expressed in matrix form, following the approach used for 1D gratings. Assuming that a finite but arbitrary number of diffracted orders are retained in the analysis ($-M \leq i_1, i_2 \leq +M$), the following matrices can be defined:

$$
\begin{aligned}
\tilde{\boldsymbol{S}}_{\boldsymbol{x}} &= [S_{x,-M,-M}, S_{x,-M,-M+1}, \cdots, S_{x,-M,M}, S_{x,-M+1,-M}, \cdots, S_{x,M,M}]^T, \\
\tilde{\boldsymbol{S}}_{\boldsymbol{y}} &= [S_{y,-M,-M}, S_{y,-M,-M+1}, \cdots, S_{y,-M,M}, S_{y,-M+1,-M}, \cdots, S_{y,M,M}]^T, \\
\tilde{\boldsymbol{U}}_{\boldsymbol{x}} &= [U_{x,-M,-M}, U_{x,-M,-M+1}, \cdots, U_{x,-M,M}, U_{x,-M+1,-M}, \cdots, U_{x,M,M}]^T, \\
\tilde{\boldsymbol{U}}_{\boldsymbol{y}} &= [U_{y,-M,-M}, U_{y,-M,-M+1}, \cdots, U_{y,-M,M}, U_{y,-M+1,-M}, \cdots, U_{y,M,M}]^T,
\end{aligned}
$$

where all these matrices have size $(2M+1)^2 \times 1$. By applying the same procedure as for one-dimensional gratings, the following set of coupled differential equations can be formulated:

$$
\begin{aligned}
\frac{d}{dz'} \underbrace{\begin{bmatrix} \tilde{\boldsymbol{S}}_{\boldsymbol{x}} \\ \tilde{\boldsymbol{S}}_{\boldsymbol{y}} \\ \tilde{\boldsymbol{U}}_{\boldsymbol{x}} \\ \tilde{\boldsymbol{U}}_{\boldsymbol{y}} \end{bmatrix}}_{=\tilde{\boldsymbol{V}}} &= j \underbrace{\begin{bmatrix} \tilde{\mathbf{0}} & \tilde{\mathbf{0}} & -\tilde{\boldsymbol{K}}_{\boldsymbol{x}} \tilde{\boldsymbol{E}}^{-1} \tilde{\boldsymbol{K}}_{\boldsymbol{y}} & \tilde{\boldsymbol{K}}_{\boldsymbol{x}} \tilde{\boldsymbol{E}}^{-1} \tilde{\boldsymbol{K}}_{\boldsymbol{x}} - \tilde{\boldsymbol{I}} \\ \tilde{\mathbf{0}} & \tilde{\mathbf{0}} & \tilde{\boldsymbol{I}} - \tilde{\boldsymbol{K}}_{\boldsymbol{y}} \tilde{\boldsymbol{E}}^{-1} \tilde{\boldsymbol{K}}_{\boldsymbol{y}} & \tilde{\boldsymbol{K}}_{\boldsymbol{y}} \tilde{\boldsymbol{E}}^{-1} \tilde{\boldsymbol{K}}_{\boldsymbol{x}} \\ \tilde{\boldsymbol{K}}_{\boldsymbol{x}} \tilde{\boldsymbol{K}}_{\boldsymbol{y}} & \tilde{\boldsymbol{E}} - \tilde{\boldsymbol{K}}_{\boldsymbol{x}} \tilde{\boldsymbol{K}}_{\boldsymbol{x}} & \tilde{\mathbf{0}} & \tilde{\mathbf{0}} \\ \tilde{\boldsymbol{K}}_{\boldsymbol{y}} \tilde{\boldsymbol{K}}_{\boldsymbol{y}} - \tilde{\boldsymbol{E}} & -\tilde{\boldsymbol{K}}_{\boldsymbol{y}} \tilde{\boldsymbol{K}}_{\boldsymbol{x}} & \tilde{\mathbf{0}} & \tilde{\mathbf{0}} \end{bmatrix}}_{=\tilde{\boldsymbol{A}}} \begin{bmatrix} \tilde{\boldsymbol{S}}_{\boldsymbol{x}} \\ \tilde{\boldsymbol{S}}_{\boldsymbol{y}} \\ \tilde{\boldsymbol{U}}_{\boldsymbol{x}} \\ \tilde{\boldsymbol{U}}_{\boldsymbol{y}} \end{bmatrix} \\
&= j \tilde{\boldsymbol{A}} \tilde{\boldsymbol{V}}, \qquad \text{(7F.27)}
\end{aligned}
$$

where the normalized coordinate $z' = k_0 z$ is used once again. All block matrices now have dimensions $(2M+1)^2 \times (2M+1)^2$. The matrix $\tilde{\mathbf{0}}$ represents a zero block matrix, while the matrices $\tilde{\boldsymbol{K}}_{\boldsymbol{x}}$ and $\tilde{\boldsymbol{K}}_{\boldsymbol{y}}$ are diagonal matrices of the form:

$$
\begin{aligned}
\tilde{\boldsymbol{K}}_{\boldsymbol{x}} = diag\Big(&\overbrace{k_{x,-M}, k_{x,-M}, \cdots, k_{x,-M}}^{2M+1}, \underbrace{k_{x,-M+1}, k_{x,-M+1}, \cdots, k_{x,-M+1}}_{2M+1}, \cdots, \\
&\overbrace{k_{x,M}, k_{x,M}, \cdots, k_{x,M}}^{2M+1}\Big), \\
\tilde{\boldsymbol{K}}_{\boldsymbol{y}} = diag\Big(&\overbrace{k_{y,-M}, k_{y,-M+1}, \cdots, k_{y,M}}^{2M+1}, \underbrace{k_{y,-M}, k_{y,-M+1}, \cdots, k_{y,M}}_{2M+1}, \cdots, \\
&\overbrace{k_{y,-M}, k_{y,-M+1}, \cdots, k_{y,M}}^{2M+1}\Big),
\end{aligned}
$$

and size $(2M+1)^2 \times (2M+1)^2$. The matrix $\tilde{\boldsymbol{E}}$ is a block Toeplitz matrix of size $(2M+1)^2 \times (2M+1)^2$ comprised of Toeplitz block matrices as follows:

$$\tilde{\boldsymbol{E}} = \begin{bmatrix} \tilde{\boldsymbol{E}}_0 & \tilde{\boldsymbol{E}}_{-1} & \tilde{\boldsymbol{E}}_{-2} & \cdots & \tilde{\boldsymbol{E}}_{-M} & \tilde{0} & \cdots & \tilde{0} \\ \tilde{\boldsymbol{E}}_{+1} & \tilde{\boldsymbol{E}}_0 & \tilde{\boldsymbol{E}}_{-1} & \tilde{\boldsymbol{E}}_{-2} & \cdots & & & \vdots \\ \tilde{\boldsymbol{E}}_{+2} & \tilde{\boldsymbol{E}}_{+1} & \tilde{\boldsymbol{E}}_0 & \tilde{\boldsymbol{E}}_{-1} & \ddots & & & \vdots \\ \vdots & \tilde{\boldsymbol{E}}_{+2} & \ddots & \ddots & \ddots & \ddots & & \vdots \\ \vdots & & \ddots & \ddots & \ddots & \ddots & \tilde{\boldsymbol{E}}_{-2} & \vdots \\ \vdots & & & \ddots & \tilde{\boldsymbol{E}}_{+1} & \tilde{\boldsymbol{E}}_0 & \tilde{\boldsymbol{E}}_{-1} & \tilde{\boldsymbol{E}}_{-2} \\ \vdots & & & & \tilde{\boldsymbol{E}}_{+2} & \tilde{\boldsymbol{E}}_{+1} & \tilde{\boldsymbol{E}}_0 & \tilde{\boldsymbol{E}}_{-1} \\ \tilde{0} & \tilde{0} & \cdots & \tilde{\boldsymbol{E}}_{+M} & \cdots & \tilde{\boldsymbol{E}}_{+2} & \tilde{\boldsymbol{E}}_{+1} & \tilde{\boldsymbol{E}}_0 \end{bmatrix},$$

where the block Toeplitz matrices $\tilde{\boldsymbol{E}}_j$ of size $(2M+1) \times (2M+1)$ are of the form:

$$\tilde{\boldsymbol{E}}_j = \begin{bmatrix} \varepsilon_{j,0} & \varepsilon_{j,-1} & \varepsilon_{j,-2} & \cdots & \varepsilon_{j,-M} & 0 & \cdots & 0 \\ \varepsilon_{j,+1} & \varepsilon_{j,0} & \varepsilon_{j,-1} & \varepsilon_{j,-2} & \cdots & & & \vdots \\ \varepsilon_{j,+2} & \varepsilon_{j,+1} & \varepsilon_{j,0} & \varepsilon_{j,-1} & \ddots & & & \vdots \\ \vdots & \varepsilon_{j,+2} & \ddots & \ddots & \ddots & \ddots & & \vdots \\ \vdots & & \ddots & \ddots & \ddots & \ddots & \varepsilon_{j,-2} & \vdots \\ \vdots & & & \ddots & \varepsilon_{j,+1} & \varepsilon_0 & \varepsilon_{j,-1} & \varepsilon_{j,-2} \\ \vdots & & & & \varepsilon_{j,+2} & \varepsilon_{j,+1} & \varepsilon_{j,0} & \varepsilon_{j,-1} \\ 0 & 0 & \cdots & \varepsilon_{j,+M} & \cdots & \varepsilon_{j,+2} & \varepsilon_{j,+1} & \varepsilon_{j,0} \end{bmatrix}.$$

Using the analysis presented in Ref. [462], improved convergence can be achieved by replacing the block matrices at positions (3,2) and (4,1) of the matrix $\tilde{\boldsymbol{\mathcal{A}}}$ in Eq. (7F.27) as follows:

$$\alpha\tilde{\boldsymbol{E}} + (1-\alpha)\tilde{\boldsymbol{A}}^{-1}\tilde{\boldsymbol{K}}_{\boldsymbol{x}}\tilde{\boldsymbol{K}}_{\boldsymbol{x}} \quad \text{instead of} \quad \tilde{\boldsymbol{E}} - \tilde{\boldsymbol{K}}_{\boldsymbol{x}}\tilde{\boldsymbol{K}}_{\boldsymbol{x}},$$

$$\tilde{\boldsymbol{K}}_{\boldsymbol{y}}\tilde{\boldsymbol{K}}_{\boldsymbol{y}} - a\tilde{\boldsymbol{A}}^{-1} - (1-\alpha)\tilde{\boldsymbol{E}} \quad \text{instead of} \quad \tilde{\boldsymbol{K}}_{\boldsymbol{y}}\tilde{\boldsymbol{K}}_{\boldsymbol{y}} - \tilde{\boldsymbol{E}}. \tag{7F.28}$$

where $\tilde{\boldsymbol{A}}$ is a block Toeplitz matrix similar to $\tilde{\boldsymbol{E}}$, but with elements corresponding to the α_{h_1,h_2} terms from the Fourier series of $1/\varepsilon(x,y)$. The parameter $0 \le \alpha \le 1$ can be selected optimally, with a suboptimal value suggested in Ref. [462], given by $\alpha = F_y\Lambda_y/(F_x\Lambda_x + F_y\Lambda_y)$ for the rectangular pillar crossed grating.

7G Effective Index of a Rectangular-Groove Surface-Relief Grating

One period of a rectangular-groove surface-relief grating is shown in Fig. 7G.1. This Appendix focuses on solving for the electromagnetic field in the grating region. The analysis follows the approach presented in Ref. [463]. It is assumed that the incident plane wave is *TE* polarized, i.e., $\vec{E}inc = \hat{\imath}_y E_0 \exp[-jk_0 n_1(x\sin\theta_{inc} + z\cos\theta_{inc})]$. The electric and magnetic fields in the grating region are denoted as $\vec{E}_w = E_w(x,z)\hat{\imath}_y$ and $\vec{H}_w = H_{wx}(x,z)\hat{\imath}_x + H_{wz}(x,z)\hat{\imath}_z$, where the subscript w refers to either r or g, indicating the ridge and groove regions of the grating period, respectively.

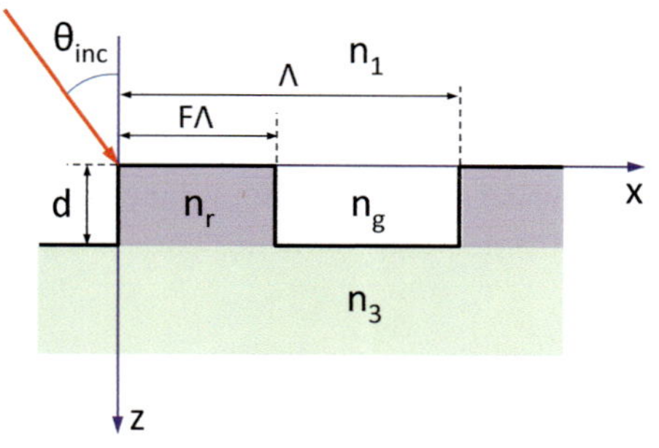

FIGURE 7G.1: One period of a rectangular-groove surface-relief one-dimensional grating of period Λ and filling factor F. The ridge and groove refractive indices are n_r and n_g, respectively. The incident plane wave can be either *TE* or *TM* polarized.

From Maxwell's equations the magnetic field components can be expressed in terms of the electric field is as follows:

$$\vec{H}_w = \frac{1}{j\omega\mu_0}\left(\frac{\partial E_w}{\partial z}\hat{\imath}_x - \frac{\partial E_w}{\partial x}\hat{\imath}_z\right), \qquad w = r, g.$$

Before solving for the electric field in the ridge and groove regions, we must first establish the boundary conditions at the interfaces between these regions, located at $x = F\Lambda$ and $x = \Lambda$. These boundary conditions are crucial for ensuring the continuity and proper transition of the electromagnetic fields across this boundary and involve the continuity of both the electric and magnetic field components. Considering that the solution in the periodic region should be Bloch periodic, the following conditions can be written:

$$E_r(F\Lambda, z) = E_g(F\Lambda, z), \tag{7G.1}$$

$$E_r(\Lambda, z) = E_g(\Lambda, z) \implies e^{jk_{x,inc}\Lambda} E_r(0, z) = E_g(\Lambda, z) \tag{7G.2}$$

$$H_{rz}(F\Lambda, z) = H_{gz}(F\Lambda, z) \implies \frac{\partial E_r(F\Lambda, z)}{\partial x} = \frac{\partial E_g(F\Lambda, z)}{\partial x}, \tag{7G.3}$$

$$H_{rz}(\Lambda, z) = H_{gz}(\Lambda, z) \implies e^{jk_{x,inc}\Lambda}\frac{\partial E_r(0, z)}{\partial x} = \frac{\partial E_g(F\Lambda, z)}{\partial x}. \tag{7G.4}$$

The electric field in either the ridge or the groove region satisfies the Helmholtz equation:

$$\frac{\partial^2 E_w}{\partial x^2} + \frac{\partial^2 E_w}{\partial z^2} + k_0^2 n_w^2 E_w = 0, \qquad w = r, g.$$

Given the rectangular geometry of the problem, the method of separation of variables can be applied to solve the Helmholtz equation in each region (ridge and groove). This approach allows us to express the electric field as $E_w(x,z) = X_w(x)Z_w(z)$ for $w = r, g$. Let us define n_{eff} as the effective refractive index along the z-direction. The solution to the Helmholtz equation then yields the following expressions:

$$X_r(x) = A_1 \cos\gamma_r x + B_1 \sin\gamma_r x,$$
$$X_g(x) = A_2 \cos(\gamma_g x - F\Lambda) + B_2 \sin(\gamma_g x - F\Lambda),$$
$$Z_r(z) = Z_g(z) = C\cos(k_0 n_{eff} z) + D\sin(k_0 n_{eff} z),$$
$$\gamma_w = k_0(n_w^2 - n_{eff}^2)^{1/2}, \qquad w = r, g.$$

It is stated that $Z_r = Z_g$ in order to satisfy the boundary conditions of Eqs. (7G.1)-(7G.4). By applying the boundary conditions, the following equations must be satisfied:

$$A_1 \cos(\gamma_r F\Lambda) + B_1 \sin(\gamma_r F\Lambda) = A_2,$$
$$-\gamma_r A_1 \sin(\gamma_r F\Lambda) + \gamma_r B_1 \cos(\gamma_r F\Lambda) = B_2\gamma_g,$$
$$e^{jk_{x,inc}\Lambda} A_1 = A_2 \cos\left[\gamma_g(1-F)\Lambda\right] + B_2 \sin\left[\gamma_g(1-F)\Lambda\right],$$
$$e^{jk_{x,inc}\Lambda}\gamma_r B_1 = -\gamma_g A_2 \sin\left[\gamma_g(1-F)\Lambda\right] + +\gamma_g B_2 \cos\left[\gamma_g(1-F)\Lambda\right].$$

The last four equations can be rewritten as the following system of equations:

$$\begin{bmatrix} \cos(\gamma_r F\Lambda) & \sin(\gamma_r F\Lambda) & -1 & 0 \\ -\gamma_r \sin(\gamma_r F\Lambda) & \gamma_r \cos(\gamma_r F\Lambda) & 0 & -\gamma_g \\ e^{jk_{x,inc}\Lambda} & 0 & -\cos\left[\gamma_g(1-F)\Lambda\right] & -\sin\left[\gamma_g(1-F)\Lambda\right] \\ 0 & \gamma_r e^{jk_{x,inc}\Lambda} & \gamma_g \sin\left[\gamma_g(1-F)\Lambda\right] & -\gamma_g \cos\left[\gamma_g(1-F)\Lambda\right] \end{bmatrix} \begin{bmatrix} A_1 \\ B_1 \\ A_2 \\ B_2 \end{bmatrix} = \begin{bmatrix} 0 \\ 0 \\ 0 \\ 0 \end{bmatrix}$$

To obtain nontrivial solutions from the last linear system, the determinant of the 4×4 matrix must be zero. This condition is readily derived as:

$$\cos(\gamma_r F\Lambda)\cos\left[\gamma_g(1-F)\Lambda\right] - \frac{1}{2}\left(\frac{\gamma_r}{\gamma_g} + \frac{\gamma_g}{\gamma_r}\right)\sin(\gamma_r F\Lambda)\sin\left[\gamma_g(1-F)\Lambda\right] = \cos(k_{x,inc}\Lambda). \tag{7G.5}$$

Using a similar procedure, an equation analogous to Eq. (7G.5) can be derived for *TM* polarization of the incident field. The resulting eigenvalue equation is given by:

$$\cos(\gamma_r F\Lambda)\cos\left[\gamma_g(1-F)\Lambda\right] - \frac{1}{2}\left(\frac{\gamma_r n_g^2}{\gamma_g n_r^2} + \frac{\gamma_g n_r^2}{\gamma_r n_g^2}\right)\sin(\gamma_r F\Lambda)\sin\left[\gamma_g(1-F)\Lambda\right] = \cos(k_{x,inc}\Lambda). \tag{7G.6}$$

Both eigenvalue equations are transcendental with respect to the unknown quantity n_{eff} for each polarization. However, they can always be solved numerically. This approach forms the core process for the modal analysis of a rectangular-groove surface-relief grating. The coefficients C, D are independent of A_1, A_2, B_1, and B_2 can be determined from the boundary conditions with the regions surrounding the grating, as detailed in Ref. [463]. The two eigenvalue equations are employed in the analysis of the Effective Medium Theory (EMT) for subwavelength gratings.

Equations (7G.5) and (7G.6) can be generalized for conical incidence (see Fig. 7.28) by replacing γ_r and γ_g with the expressions given in [435]:

$$\gamma_w = k_0\left[n_w^2 - n_1^2 \sin^2\theta_{inc}\sin^2\phi_{inc} - n_{eff}^2\right]^{1/2}, \qquad w = r, g.$$

Bibliography

[1] B. Vohnsen, "A short history of optics," *Physica Scripta*, vol. T109, pp. 75–79, 2004.

[2] O. Darrigol, *A History of Optics.* New York: Oxford University Press, 2012.

[3] I. Galili, *Scientific Knowledge as a Culture: The Pleasure of Understanding*, ch. Teaching Optics: A Historico-Philosophical Perspective. Cham, Switzerland: Springer Nature Switzerland, 2021.

[4] D. F. Vanderwerf, *The Story of Light Science.* Cham, Switzerland: Springer International Publishing, 2017.

[5] M. D. Al-Amri, M. El-Gomati, and M. S. Zubairy, *Optics in Our Time.* Cham, Switzerland: Springer International Publishing, 2016.

[6] I. Galili, *Scientific Knowledge as a Culture: The Pleasure of Understanding*, ch. Optical Image abd Vision: From Pythgoras to Kepler. Cham, Switzerland: Springer Nature Switzerland, 2021.

[7] I. Newton, *Opticks: Or a Treatise of the Reflections, Refractions, Inflections and Colours of Light.* London, England: Printed for S. Smith, and B. Walford, London, 1st ed., 1704.

[8] H. Kubbinga, "Christiaan Huygens and the foundations of optics," *Pure and Applied Optics: Journal of the European Optical Society Part A*, vol. 4, pp. 723–739, Nov. 1995.

[9] C. Huygens, *Traité de la Lumière.* Pierre van der Aa, Leyden, 1690, English Translation by S. P. Thompson, London, UK: Macmillan, 1912.

[10] T. Young, "I. The Bakerian Lecture: Experiments and calculations relative to physical optics," *Phil. Trans. R. Soc.*, vol. 94, pp. 1–16, Jan. 1804.

[11] J. C. Maxwell, *A Treatise on Electricity and Magnetism, Vol. 1.* Oxford, UK: Oxford, Clarendon Press, 1873.

[12] J. C. Maxwell, *A Treatise on Electricity and Magnetism, Vol. 2.* Oxford, UK: Oxford, Clarendon Press, 1873.

[13] B. P. Abbott *et al.*, "Observation of gravitational waves from a binary black hole merger," *Phys. Rev. Lett.*, vol. 116, p. 061102, Feb. 2016.

[14] M. Romer and I. B. Cohen, "Roemer and the first determination of the velocity of light (1676)," *Isis*, vol. 31, no. 2, pp. 327–379, 1940.

[15] K. D. Froome, "A new determination of the free-space velocity of electromagnetic waves," *Proc. R. Soc. London, Ser. A*, vol. 247, pp. 109–122, 1958.

[16] K. M. Evenson, J. S. Wells, F. R. Petersen, B. L. Danielson, G. W. Day, R. L. Barger, and J. L. Hall, "Speed of light from direct frequency and wavelength measurements of the methane-stabilized laser," *Phys. Rev. Lett.*, vol. 29, pp. 1346–1349, Nov. 1972.

[17] K. M. Evenson, "The development of direct optical frequency measurement and the speed of light," *ISA Transactions*, vol. 14, pp. 209–216, 1975.

[18] "The Nobel Prize." `https://www.nobelprize.org/prizes/lists/all-nobel-prizes/`. Accessed: 2024-09-27.

[19] B. E. A. Saleh and M. C. Teich, *Fundamentals of Photonics.* Pure and Applied Optics, Hoboken, New Jersey: J. Wiley and Sons, Inc., 2nd ed., 2007.

[20] F. L. Pedrotti, S. J., L. M. Pedrotti, and L. S. Pedrotti, *Introduction to Optics.* Upper Saddle River, NJ: Pearson Education, Inc., 3rd ed., 2007.

[21] J. Hadamard, *Lectures on Cauchy's problem in linear partial differential equations.* New Haven: Yale University Press, 1923.

[22] P. Gunther, "Huygens' principle and Hadamard's conjecture," *The Math. Intelligencer*, vol. 13, no. 2, pp. 56–63, 1991.

[23] G. P. Menzala and R. C. Charão, "Wave phenomena, Huygens' principle and 3D elastic waves," *Revista de la Unión Matemática Argentina*, vol. 37, pp. 117–126, 1991.

[24] A. Fresnel, "Sur la diffraction de la lumiére où l'on examine particulèlement le phénomène des franges colorées que présentent les ombres des corps éclairés par un point lumineux," *Ann. Chem. Phys.*, vol. 1, pp. 239–281, 1816.

[25] G. Kirchhoff, "Zur theorie del' lichtstrahlen," *Sitzltngsberichle der Kööniglich Preufßischen Akademie der Wissertschaften zu Berlin*, vol. Part II, pp. 641–669, 1882.

[26] D. A. B. Miller, "Huygens's wave propagation principle corrected," *Opt. Lett.*, vol. 16, pp. 1370–1372, Sep. 1991.

[27] O. N. Stavroudis, *The Mathematics of Geometrical and Physical Optics.* Weinheim, Germany: Wiley-VCH Verlag GmbH & Co KGaA, 2006.

[28] A. Sharma, D. V. Kumar, and A. K. Ghatak, "Tracing rays through graded-index media: a new method," *Appl. Opt.*, vol. 21, pp. 984–987, Mar. 1982.

[29] B. K. Kaushik and B. D. Sukheeja, "Intricacies of Fermat's principle," *Physics Education*, vol. 19, pp. 90–91, Mar. 1984.

[30] R. E. Hopkins, "Mirror and prisms systems," in *Applied Optics and Optical Engineering* (R. Kingslake, ed.), Vol. 3, pp. 269–308, New York, NY, USA: Academic Press, 1965.

[31] R. E. Hopkins, "Mirror and prisms systems," in *Military Standardization Handbook: Optical Design* (U. S. D. of Defense, ed.), MIL-HDBK-141, ch. 13, Defense Supply Agency, 1962.

[32] J. C. Polasek, "Matrix analysis of gimbaled mirror and prism systems," *J. Opt. Soc. Am.*, vol. 57, pp. 1193–1201, Oct. 1967.

[33] D. F. Vanderwerf, *Applied Prismatic and Reflective Optics.* Bellingham, WA, USA: SPIE Press, 2010.

[34] E. P. Mosca, "Hyperbolic mirrors, holography, and Fermat's principle," *Amer. J. Phys.*, vol. 45, pp. 402–403, Apr. 1977.

[35] G. P. Sastry, "Comment on: Hyperbolic mirrors, holography, and Fermat's principle," *Amer. J. Phys.*, vol. 47, pp. 194–195, Feb. 1979.

[36] P. D. Lin, *Advanced Geometrical Optics.* Progress in Optical Science and Photonics, Ed. J. Atai, Singapore: Springer, 2017.

[37] N. Hodgson and H. Weber, *Laser Resonators and Beam Propagation.* Optical Sciences, New York: Springer-Verlag, 3rd ed., 2005.

[38] G. Kloos, *Matrix Methods for Optical Layout.* Tutorial Texts in Optical Engineering, vol. TT77, Bellingham, WA, USA: SPIE Press, 2007.

[39] J. T. Verdeyen, *Laser Electronics.* New Jersey: Prentice Hall, 3rd ed., 1995.

[40] A. Yariv, *Optical Electronics in Modern Communications.* The Oxford Series in Electrical and Computer Engineering, New York, USA: Oxford University Press, 5th ed., 1997.

[41] "Nikon imaging products, Nikon Corporation." `http://imaging.nikon.com/history/story/0002/index.htm`. Accessed: 2019-04-23.

[42] E. Hecht, *Optics.* New Jersey: Pearson Education, Inc., 5th ed., 2017.

[43] M. V. Klein and T. E. Furtak, *Optics.* New York, NY, USA: John Wiley & Sons, Inc., 2nd ed., 1986.

[44] M. Osterman and G. B. Romer, "History and the evolution of photography," in *Focal Encyclopedia of Photography: Digital Imaging, Theory and Applications, History, and Science* (M. R. Peres, ed.), pp. 23–176, Oxford, UK: Elsevier - Focal Press, 2007.

[45] "Vermeer and the camera obscura." `http://www.essentialvermeer.com/camera_obscura/co_one.html`.

[46] N. Waltham, "CCD and CMOS sensors," in *Observing Photons in Space: A Guide to Experimental Space Astronomy* (M. C. E. Huber, A. Pauluhn, J. L. Culhane, J. G. Timothy, K. Wilhelm, and A. Zehnder, eds.), ISSI Scientific Report Series, vol. 9, pp. 391–408, New York, NY, USA: Springer, 2013.

[47] "Circle of confusion, wikipedia." `https://en.wikipedia.org/wiki/Circle_of_confusion`. Accessed: 2021-08-18.

[48] D. Falk, D. Brill, and D. Stork, *Seeing the Light: Optics in Nature, Photography, Color, Vision and Holography*, ch. 4, pp. 289–375. Brattleboro, Vermont, USA: Echo Point Books and Media, 2019.

[49] B. London, J. Stone, and J. Upton, *Photography.* Boston, MA, USA: Pearson Education, Inc., 12th ed., 2017.

[50] M. W. Davidson and M. Abramowitz, *Optical Microscopy.* American Cancer Society, 2002.

[51] R. N. Wilson, *Reflecting Telescope Optics I: Basic Design Theory and its Historical Development*, ch. 2. Astronomy and Astrophysics Library, Berlin-Heidelberg, Germany: Springer-Verlag, 2nd ed., 2004.

[52] M. D. Lallo, "Experience with the Hubble Space Telescope: 20 years of an archetype," *Opt. Engr.*, vol. 51, no. 1, pp. 1–19, 2012.

[53] P. Y. Bely, *The Design and Construction of Large Optical Telescopes.* Astronomy and Astrophysics Library, New York, USA: Springer-Verlag, 2003.

[54] J. Cheng, *The Principles of Astronomical Telescope Design.* Astrophysics and Space Science Library, vol. 360, New York, USA: Springer Science + Business Media, 2009.

[55] P. Artal, "Optics of the eye and its impact in vision: a tutorial," *Adv. Opt. Photon.*, vol. 6, pp. 340–367, Sep 2014.

[56] W. T. Welford, *Aberrations of Optical Systems.* The Adam Hilger Series on Optics and Optoelectronics, New York, NY, USA: Adam Hilger, 1986.

[57] V. N. Mahajan, *Optical Imaging and Aberrations, Part I: Ray Geometrical Optics.* Bellingham, WA, USA: SPIE Press, 1998.

[58] V. N. Mahajan, *Optical Imaging and Aberrations, Part II: Wave Diffraction Optics.* Bellingham, WA, USA: SPIE Press, 1998.

[59] H. Gross, H. Zügge, M. Peschka, and F. Blechinger, "Aberration theory and correction of optical systems," in *Handbook of Optical Systems* (H. Gross, ed.), Vol. 3, ch. 29, pp. 1–70, Weinheim, Germany: WILEY-VCH Verlag GmbH & Co. KGaA, 2007.

[60] J. Sasian, *Introduction to Aberrations in Optical Imaging Systems.* The Edinburgh Building, Cambridge, UK: Cambridge University Press, 2013.

[61] A. Lahiri, *Basic Optics: Principles and Concepts.* Amsterdam, Netherland: Elsevier, 2016.

[62] F. A. Jenkins and H. E. White, *Fundamentals of Optics.* New York: McGraw-Hill Inc., 4th ed., 1976.

[63] L. Hazra, *Foundations of Optical System Analysis and Design.* Boca Raton, FL, USA: CRC Press, Taylor & Francis Group, 2022.

[64] D. P. Feder, "Tracing of skew rays," *Journal of Research of National Bureau of Standards*, vol. 45, pp. 61–63, July 1950.

[65] P. D. Lin and T.-T. Liao, "Skew ray tracing and sensitivity analysis of geometrical optics," *Journal of Manufacturing Science and Engineering*, vol. 122, pp. 338–349, 05 1999.

[66] P. D. Lin and C.-K. Sung, "Matrix-based paraxial skew ray-tracing in 3d systems with non-coplanar optical axis," *Optik*, vol. 117, no. 7, pp. 329–340, 2006.

[67] H. A. Elagha, "Exact ray tracing formulas based on a nontrigonometric alternative to Snell's law," *J. Opt. Soc. Amer. A*, vol. 29, pp. 2679–2687, Dec. 2012.

[68] H. A. Elagha, "Generalized formulas for ray-tracing and longitudinal spherical aberration," *J. Opt. Soc. Amer. A*, vol. 34, pp. 335–343, Mar. 2017.

[69] M. J. Kidger, *Fundamental Optical Design.* Bellingham, WA, USA: SPIE Press, 2001.

[70] H. Gross, "Raytracing," in *Handbook of Optical Systems* (H. Gross, ed.), Vol. 1, ch. 5, pp. 173–228, Weinheim, Germany: WILEY-VCH Verlag GmbH & Co. KGaA, 2005.

[71] D. Malacara-Hernandez and Z. Malacara-Hernandez, *Handbook of Optical Design.* Boca Raton, FL, USA: CRC Press, Taylor & Francis Group, 3 ed., 2013.

[72] J. Sasian, *Introduction to Lens Design.* University Printing House, Cambridge, UK: Cambridge University Press, 2019.

[73] W. B. Zimmerman, "Exact ray tracing and spherical aberration." from internet, unavailable url.

[74] D. A. Atchison and G. Smith, *Optics of the Human Eye.* Multidisciplinary and Applied Optics, Boca Raton, FL, USA: CRC Press - Taylor and Francis Group, 2nd ed., 2023.

[75] W. Sellmeier, "Ueber die durch die aetherschwingungen erregten mitschwingungen der korpertheilchen und deren ruckwirkung auf die ersteren, besonders zur erklarung der dispersion und ihrer anomalien," *Annalen der Physik*, vol. 223, no. 11, pp. 386–403, 1872.

[76] "Refractive index database." `https://refractiveindex.info/`.

[77] M. R. Hatch and D. E. Stoltzmann, "The f-stops here." `https://wp.optics.arizona.edu/jpalmer/wp-content/uploads/sites/65/2018/11/The-f-stops-here.pdf`. Accessed: 2021-08-17.

[78] W. J. Smith, *Modern Optical Engineering.* New York, USA: McGraw-Hill Companies, Inc., 3rd ed., 2000.

[79] J. E. Greivenkamp, *Field Guide to Geometrical Optics.* SPIE Field Guides, vol. FG01, Bellingham, WA, USA: SPIE Press, 2004.

[80] A. Yariv and P. Yeh, *Optical Waves in Crystals.* New York, USA: J. Wiley and Sons, Inc., 1984.

[81] H. C. Chen, *Theory of Electromagnetic Waves: A Coordinate-Free Approach.* New York: McGraw-Hill, 1983.

[82] A. Knoesen, M. G. Moharam, and T. K. Gaylord, "Electromagnetic propagation at interfaces and in waveguides in uniaxial crystals," *Appl. Phys. B*, vol. 38, pp. 171–178, 1985.

[83] R. P. Feynman, R. B. Leighton, and M. L. Sands, *The Feynman Lectures on Physics.* Vol. 2, Ch. 27, Reading, MA: Addison-Wesley Publishing Company, 1989.

[84] C. S. Lai, "Alternative choice for the energy flow vector of the electromagnetic field," *Amer. J. Phys.*, vol. 49, pp. 841–843, Sept. 1981.

[85] U. Backhaus and K. Schäfer, "On the uniqueness of the vector for energy flow density in electromagnetic fields," *Amer. J. Phys.*, vol. 54, no. 3, pp. 279–280, 1986.

[86] G. S. Smith, *An Introduction to Classical Electomagnetic Radiation.* Cambridge, UK: Cambridge University Press, 1997.

[87] H. E. Puthoff, "Electromagnetic potentials basis for energy density and power flux," *Eur. J. Phys.*, vol. 37, p. 055203, Jul. 2016.

[88] J. D. Jackson, *Classical Electrodynamics.* New York, NY, USA: John Wiley & Sons, 3rd ed., 1999.

[89] L. D. Landau and E. M. Lifshitz, *Electrodynamics of Continuous Media.* Oxford, England: Pergamon Press, 2nd ed., 1960.

[90] K. J. Webb and Shivanand, "Electromagnetic field energy in dispersive materials," *J. Opt. Soc. Amer. B*, vol. 27, pp. 1215–1220, June 2010.

[91] M. Born and E. Wolf, *Principles of Optics.* New York: Cambridge University Press, 7th ed., 1999.

[92] D. L. Grennaway and G. Harbeke, *Optical Properties and Band Structure of Semiconductors.* International Series in the Science of the Solid State, Oxford, United Kingdom: Pergamon Press, 1968.

[93] J. I. Pankove, *Optical Processes in Semiconductors.* New York, NY, USA: Dover Publications, Inc., 1972.

[94] N. W. Ashcroft and A. N. Mermin, *Solid State Physics.* HRW international editions, Orlando, Florida, USA: Holt, Rinehart and Winston, 1976.

[95] H. Haug and S. W. Koch, *Quantum Theory of the Optical and Electronic Properties of Semiconductors.* 5 Toh Tuck Link, Signapore: World Scietific Publishing Co., Pte. Ltd., 5th ed., 2009.

[96] H. Kalt and C. F. Klingshirn, *Semiconductor Optics 1: Linear Optical Properties of Semiconductors.* Graduate text in Physics, Cham, Switzerland: Springer Nature Switzerland AG, 5th ed., 2019.

[97] A. E. Siegman, *Lasers.* California: University Science Books, 1986.

[98] K. D. Möller, *Optics.* Mill Valley, California, USA: University Science Books, 1988.

[99] O. Stenzel, *The Physics of Thin Film Optical Spectra: An Introduction.* Springer Series in Surfaces Sciences, vol. 44, Cham, Switzerland: Springer International Pubslishing, 2nd ed., 2015.

[100] H. A. Haus and J. R. Melcher, *Electromagnetic Fields and Waves.* Englewood Cliffs, New Jersey, USA: Prentice Hall, Inc., 1989.

[101] C. A. Balanis, *Advanced Engineering Electromagnetics.* New York, NY, USA: John Wiley & Sons, Inc., 2nd ed., 2012.

[102] A. D. Rakić, A. B. Djurišić, J. M. Elazar, and M. L. Majewski, "Optical properties of metallic films for vertical-cavity optoelectronic devices," *Appl. Opt.*, vol. 37, pp. 5271–5283, Aug. 1998.

[103] K. Prokopidis and C. Kalialakis, "Supersymmetry-inspired non-Hermitian optical couplers," *Appl. Phys. B*, vol. 117, pp. 25–32, Oct. 2014.

[104] S. J. Orfanidis, *Electromagnetic Waves and Antennas.* Sophocles J. Orfanidis, 2016.

[105] S. Babar and J. H. Weaver, "Optical constants of Cu, Ag, and Au revisited," *Appl. Opt.*, vol. 54, pp. 477–481, Jan. 2015.

[106] R. P. Cameron, S. M. Barnett, and A. M. Yao, "Discriminatory optical force for chiral molecules," *New J. Phys.*, vol. 16, p. 013020, Jan. 2014.

[107] K. C. van Kruining, R. P. Cameron, and J. B. Götte, "Superpositions of up to six plane waves without electric-field interference," *Optica*, vol. 5, pp. 1091–1098, Sept. 2018.

[108] R. L. Easton, Jr., *Fourier Methods in Imaging*, ch. 22. Wiley-IS & T Series in Imaging Science and Technology, Chichester, UK: J. Wiley and Sons, Inc., 2010.

[109] G. G. Stokes, "On the perfect blackness of the central spot in Newton's rings, and on the verification of Fresnel's formulae for the intensities of reflected and refracted rays," *Cambridge Dublin Math. J.*, vol. 4, pp. 1–14, 1849.

[110] B. P. Abbott *et al.*, "Gw170817: Observation of gravitational waves from a binary neutron star inspiral," *Phys. Rev. Lett.*, vol. 119, p. 161101, Oct. 2017.

[111] L. Mach, "Ueber einen interferenzrefraktor," *Zeitschrift für Instrumentenkunde*, vol. 12, pp. 89–93, 1892.

[112] L. Zehnder, "Ein neuer interferenzrefraktor," *Zeitschrift für Instrumentenkunde*, vol. 11, pp. 275–285, 1891.

[113] B. H. Schneider, J. G. Edwards, and N. F. Hartman, "Hartman interferometer: versatile integrated optic sensor for label-free, real-time quantification of nucleic acids, proteins, and pathogens," *Clinical Chemistry*, vol. 43, pp. 1757–1763, 09 1997.

[114] H. Mukundan, A. S. Anderson, W. K. Grace, K. M. Grace, N. F. Hartman, J. S. Martinez, and B. I. Swanson, "Waveguide-based biosensors for pathogen detection," *Sensors*, vol. 9, no. 7, pp. 5783–5809, 2009.

[115] G. Lifante, *Integrated Photonics.* New Sussex, England: John Wiley & Sons, Inc., 2003.

[116] R. G. Hunsperger, *Integrated Optics: Theory and Technology.* Optical Sciences, New York: Springer-Verlag, 2009.

[117] J. B. Bates, "Fourier transform spectroscopy," *Comp. and Maths. with Appls.*, vol. 4, no. 2, pp. 73–84, 1978.

[118] R. J. Bell, *Introductory Fourier Transform Spectroscopy.* New York, NY, USA: Academic Press, Inc., 1972.

[119] R. N. Bracewell, *The Fourier Transform and its Applications.* McGraw-Hill electrical and electronic series, New York, NY, USA: McGraw-Hill, Inc., 2 ed., 1978.

[120] O. E. Brigham, *The Fast Fourier Transform and its Applications.* Englewood Cliffs, NJ, USA: Prentice-Hall, Inc., 1988.

[121] P. R. Griffiths and J. A. de Haseth, *Fourier Transform Infrared Spectrometry.* Wiley Interscience, Hoboken, New Jersey, USA: J. Wiley & Sons Inc., 2 ed., 2007.

[122] H. A. MacCleod, *Thin-Film Optical Filters.* Series in Optics and Optoelectronics, Boca Raton, FL, USA: CRC Press, Taylor & Francis Group, 4 ed., 2010.

[123] S. W. Teare, *Optical Interference Filtes using MATLAB.* Bellingham WA, USA: SPIE Press, vol. PM299, 2019.

[124] P. A. Maksym, "Some aspects of RHEED theory," in *Thin Film Growth Techniques for Low-Dimensional Structures* (R. F. C. Farrow, S. S. P. Parkin, P. J. Dobson, J. H. Neave, and A. S. Arrott, eds.), NATO Advanced Science Institute Series, vol. 163, pp. 95–113, Springer New York, NY, USA, 1987.

[125] D. Y. K. Ko and J. C. Inkson, "Matrix method for tunneling in heterostructures: Resonant tunneling in multilayer systems," *Phys. Rev. B*, vol. 38, pp. 9945–9951, Nov. 1988.

[126] N. P. K. Cotter, T. W. Preist, and J. R. Sambles, "Scattering-matrix approach to multilayer diffraction," *J. Opt. Soc. Amer. A*, vol. 12, pp. 1097–1103, May 1995.

[127] R. C. Rumpf, "Improved formulation of scattering matrices for semi-analytical methods that is consistent with convention," *Progress in Electromagnetics Research B*, vol. 35, pp. 241–261, 2011.

[128] D. T. Pierce and W. E. Spicer, "Electronic structure of amorphous Si from photoemission and optical studies," *Phys. Rev. B*, vol. 5, pp. 3017–3029, Apr. 1972.

[129] H. Piller, "Silicon (amorphous)," in *Handbook of Optical Constants of Solids* (D. D. Palik, ed.), Volume 1, pp. 571–586, San Diego, CA, USA: Academic Press, 1998.

[130] "Filmmetrics, a KLA company." `https://www.filmetrics.com/refractive-index-database/Si/Silicon`. Accessed: 2019-06-14.

[131] K. Schuster, "Anwendung der vierpoltheorie auf die probleme der optischen reflexionsminderung, reflexionsverstarkung und der interferenzfilter. (mit 2 abbildungen)," *Annalen der Physik*, vol. 439, no. 6, pp. 352–356, 1949.

[132] D. K. Cheng, *Fields and Waves in Electromagnetics.* New York: Addison Wesley, 2nd ed., 1989.

[133] B. A. Moys, "The theory of double-layer antireflection coatings," *Thin Solid Films*, vol. 21, no. 1, pp. 145–157, 1974.

[134] J. A. Dobrowolski and R. A. Kemp, "Refinement of optical multilayer systems with different optimization procedures," *Appl. Opt.*, vol. 29, pp. 2876–2893, Jul. 1990.

[135] A. Corana, M. Marchesi, C. Martini, and S. Ridella, "Minimizing multimodal functions of continuous variables with the "simulated annealing" algorithm," *ACM Trans. Math. Softw.*, vol. 13, pp. 262–280, Sept. 1987.

[136] T. Boudet, P. Chaton, L. Herault, G. Gonon, L. Jouanet, and P. Keller, "Thin-film designs by simulated annealing," *Appl. Opt.*, vol. 35, pp. 6219–6226, Nov. 1996.

[137] D. E. Goldberg, *Genetic Algorithms in Search, Optimization and Machine Learning.* Reading, MA, USA: Addison-Wesley, Inc., 1989.

[138] S. Martin, J. Rivory, and M. Schoenauer, "Synthesis of optical multilayer systems using genetic algorithms," *Appl. Opt.*, vol. 34, pp. 2247–2254, May 1995.

[139] J. Kennedy and R. C. Eberhart, "Particle swarm optimization," in *IEEE Proceedings of ICNN'95 - International Conference on Neural Networks*, vol. 4, (Piscataway, NJ), pp. 1942–1948 vol.4, IEEE Service Center, Nov. 1995.

[140] R. I. Rabady and A. Ababneh, "Global optimal design of optical multilayer thin-film filters using particle swarm optimization," *Optik*, vol. 125, no. 1, pp. 548–553, 2014.

[141] Matlab, *version 24.1.0.2628055 (R2024a).* Natick, Massachusetts, USA: The MathWorks Inc., 2024.

[142] S. Kitayama and K. Yasuda, "A method for mixed integer programming problems by particle swarm optimization," *Electr. Eng. Jpn.*, vol. 157, no. 2, pp. 40–49, 2006.

[143] D. Berreman, "Optics in stratified and anisotropic media: 4×4 matrix formulation," *J. Opt. Soc. Amer.*, vol. 62, pp. 502–510, Apr. 1994.

[144] I. J. Hodgkinson, S. Kassam, and Q. H. Wu, "Eigenequations and compact algorithms for bulk and layered anisotropic optical media: Reflection and refraction at a crystal-crystal interface," *J. Comput. Phys.*, vol. 133, no. 1, pp. 75–83, 1997.

[145] P. Yeh, *Optical Waves in Layered Media.* Wiley Series in Pure and Applied Optics, New York, USA: J. Wiley and Sons, Inc., 2005.

[146] J. W. Graham, *Arbitrarily Oriented Boaxially Anisotropic Media: Wave Behavior and Microstrip Antennas.* PhD thesis, School of Electrical Engineering and Computer Science, Syracuse University, 2012.

[147] J. W. Graham and J. K. Lee, "Reflection and transmission from biaxially anisotropic-isotropic interfaces," *Progress In Electromagnetics Research*, vol. 136, pp. 681–702, 2013.

[148] G. D. Landry and T. A. Maldonado, "Complete method to determine transmission and reflection characteristics at a planar interface between arbitrarily oriented biaxial media," *J. Opt. Soc. Am. A*, vol. 12, pp. 2048–2063, Sep. 1995.

[149] R. C. Jones, "A new calculus for the treatment of optical systems," *J. Opt. Soc. Am.*, vol. 31, pp. 488–493, July 1941.

[150] J. W. Goodman, *Statistical Optics.* Wiley Series in Pure and Applied Optics, Hoboken, NJ, USA: John Wiley & Sons, Inc., 2nd ed., 2015.

[151] A. S. Marathay, *Elements of Optical Coherence Theory.* Pure and Applied Optics Series, New York, NY, USA: John Wiley & Sons, Inc., 1982.

[152] W. B. Zimmerman, "Interference by division of amplitude." from internet, unavailable url.

[153] K. K. Sharma, *Optics: Principles and Applications.* Burlington, MA, USA: Academic Press - Elsevier, 2006.

[154] A. T. Friberg and E. Wolf, "Relationships between the complex degrees of coherence in the space-time and in the space-frequency domains," *Opt. Lett.*, vol. 20, pp. 623–625, Mar 1995.

[155] B. J. Thompson and E. Wolf, "Two-beam interference with partially coherent light," *J. Opt. Soc. Am.*, vol. 47, pp. 895–902, Oct 1957.

[156] R. Siegel, "Net radiation method for transmission through partially transparent plates," *Solar Energy*, vol. 15, no. 3, pp. 273–276, 1973.

[157] B. Herbecke, "Coherent and incoherent reflection and transmission of multilayer structures," *Appl. Phys. B*, vol. 39, pp. 165–170, Mar. 1986.

[158] C. L. Mitsas and D. I. Siapkas, "Generalized matrix method for analysis of coherent and incoherent reflectance and transmittance of multilayer structures with rough surfaces, interfaces, and finite substrates," *Appl. Opt.*, vol. 34, pp. 1678–1683, Apr. 1995.

[159] M. C. J. Large, D. R. McKenzie, and M. I. Large, "Incoherent reflection processes: a discrete approach," *Opt. Commun.*, vol. 128, no. 4, pp. 307–314, 1996.

[160] J. S. C. Prentice, "Coherent, partially coherent and incoherent light absorption in thin-film multilayer structures," *J. Phys. D – Appl. Phys.*, vol. 33, pp. 3139–3145, Nov. 2000.

[161] C. C. Katsidis and D. I. Siapkas, "General transfer-matrix method for optical multilayer systems with coherent, partially coherent, and incoherent interference," *Appl. Opt.*, vol. 41, pp. 3978–3987, July 2002.

[162] E. Centurioni, "Generalized matrix method for calculation of internal light energy flux in mixed coherent and incoherent multilayers," *Appl. Opt.*, vol. 44, pp. 7532–7539, Dec. 2005.

[163] S. Jung, K.-Y. Kim, Y.-I. Lee, J.-H. Youn, H.-T. Moon, J. Jang, and J. Kim, "Optical modeling and analysis of organic solar cells with coherent multilayers and incoherent glass substrate using generalized transfer matrix method," *Jpn. J. Appl. Phys.*, vol. 50, p. 122301, Nov. 2011.

[164] K. Kang, S. Lee, and J. Kim, "Effect of an incoherent glass substrate on the absorption efficiency of organic solar cells at oblique incidence analyzed by the transfer matrix method with a glass factor," *Jpn. J. Appl. Phys.*, vol. 52, p. 052301, May 2013.

[165] A. Campa, J. Krc, and M. Topic, "Two approaches for incoherent propagation of light in rigorous numerical simulations," *Progress in Electromagnetics Research*, vol. 137, pp. 187–202, 2013.

[166] R. Santbergen, A. H. M. Smets, and M. Zeman, "Optical model for multilayer structures with coherent, partly coherent and incoherent layers," *Opt. Express*, vol. 21, pp. A262–A267, Mar. 2013.

[167] J. Puhan, B. Lipovsek, A. Burmen, and I. Fajfar, "An accurate representation of incoherent layers within one-dimensional thin-film multilayer structures with equivalent propagation matrices," *IEEE Photonic J.*, vol. 9, no. 5, pp. 1–12, 2017.

[168] J. Puhan, A. Burmen, T. Tuma, and I. Fajfar, "Irradiance in mixed coherent/incoherent structures: An analytical approach," *Coatings*, vol. 9, no. 9, 2019.

[169] I. Filiński, "The effects of sample imperfections on optical spectra," *Phys. Status Solidi B*, vol. 49, no. 2, pp. 577–588, 1972.

[170] M. Gouy, "Sur une propriètè nouvelle des ondes lumineuses," *C. R. Acad. Sci. Paris*, vol. 110, pp. 1251–1253, 1890.

[171] M. Gouy, "Sur la propagation anomale des ondes," *Ann. Chim. Phys. 6*, vol. XXIV, pp. 145–213, 1891.

[172] S. Feng and H. G. Winful, "Physical origin of the Gouy phase shift," *Opt. Lett.*, vol. 26, pp. 485–487, Apr. 2001.

[173] J. Weber, "Phase, group, and signal velocity," *Amer. J. Phys.*, vol. 22, pp. 618–620, 1954.

[174] P. W. Milonni and J. H. Eberly, *Laser Physics.* Hoboken, New Jersey: J. Wiley & Sons, Inc., 2010.

[175] P. Vaveliuk, B. Ruiz, and A. Lencina, "Limits of the paraxial approximation in laser beams," *Opt. Lett.*, vol. 32, pp. 927–929, Apr. 2007.

[176] P. F. Goldsmith, *Quasioptical Systems: Gaussian Beam Quasioptical Propogation and Applications.* IEEE Press Series on RF and Microwave Technology, New York, NY, USA: Wiley and IEEE Press, 1998.

[177] R. S. Weis, *Electromagnetic Transmission and Reflection Characteristics of Anisotropic Multilayered Structures.* PhD thesis, School of Electrical Engineering, Georgia Institute of Technology, 1987.

[178] M. Ruane, M. Mansuripur, and R. Rosenvold, "Measurement of reflectivities for magnetooptical media," *Appl. Opt.*, vol. 25, pp. 1946–1951, Jun. 1986.

[179] M. Schönleber, D. Klotz, and E. Ivers-Tiffée, "A method for improving the robustness of linear Kramers-Kronig validity tests," *Electrochemica Acta*, vol. 131, pp. 20–27, 2014.

[180] V. Degiorgio, "Phase shift between the transmitted and the reflected optical fields of a semireflecting lossless mirror is $\pi/2$," *Amer. J. Phys.*, vol. 48, no. 1, pp. 81–81, 1980.

[181] A. Zeilinger, "General properties of lossless beam splitters in interferometry," *Amer. J. Phys.*, vol. 49, pp. 882–883, 1981.

[182] M. W. Hamilton, "Phase shifts in multilayer dielectric beam splitters," *Amer. J. Phys.*, vol. 68, no. 2, pp. 186–191, 2000.

[183] C. H. Holbrow, E. Galvez, and M. E. Parks, "Photon quantum mechanics and beam splitters," *Amer. J. Phys.*, vol. 70, no. 3, pp. 260–265, 2002.

[184] F. Hénault, "Quantum physics and the beam splitter mystery," in *The Nature of Light: What are Photons? VI* (C. Roychoudhuri, A. F. Kracklauer, and H. D. Raedt, eds.), vol. 9570, pp. 199 – 213, International Society for Optics and Photonics, SPIE, 2015.

[185] S. Ramo, J. R. Whinnery, and T. Van Duzer, *Fields and Waves in Communications Electronics.* New York, USA: John Wiley and Sons, Inc., 3rd ed., 1993.

[186] R. E. Collin and F. J. Zucker, *Antenna Theory: Part 1.* Inter-University Electronics Series, vol. 7, New York, NY, USA: McGraw-Hill Book Company, 1969.

[187] G. Fikioris, I. Tastsoglou, and O. Bakas, *Selected Asymptotic Methods with Applications to Electromagnetics and Antennas.* Synthesis Lectures on Computational Electromagnetics, Cham, Switzerland: Springer International Publishing, 2022.

[188] R. J. Potton, "Reciprocity in optics," *Rep. Prog. Phys.*, vol. 67, pp. 717–754, Apr. 2004.

[189] D. B. Malament, "On the time reversal invariance of classical electromagnetic theory," *Studies in History and Philosophy of Science Part B: Studies in History and Philosophy of Modern Physics*, vol. 35, no. 2, pp. 295–315, 2004.

[190] F. Arntzenius and H. Greaves, "Time reversal in classical electromagnetism," *The British Journal for the Philosophy of Science*, vol. 60, no. 3, pp. 557–584, 2009.

[191] M. Nieto-Vesperinas and E. Wolf, "Generalized Stokes reciprocity relations for scattering from dielectric objects of arbitrary shape," *J. Opt. Soc. Amer. A*, vol. 3, pp. 2038–2046, Dec. 1986.

[192] R. Carminati, J. J. Sáenz, J.-J. Greffet, and M. Nieto-Vesperinas, "Reciprocity, unitarity, and time-reversal symmetry of the S matrix of fields containing evanescent components," *Phys. Rev. A*, vol. 62, p. 012712, Jun. 2000.

[193] R. Carminati, M. Nieto-Vesperinas, and J.-J. Greffet, "Reciprocity of evanescent electromagnetic waves," *J. Opt. Soc. Amer. A*, vol. 15, pp. 706–712, Mar. 1998.

[194] J. J. Monzón and L. L. Sánchez-Soto, "Lossless multilayers and Lorentz transformations: more than an analogy," *Opt. Commun.*, vol. 162, no. 1, pp. 1–6, 1999.

[195] J. J. Monzón and L. L. Sánchez-Soto, "Fully relativisticlike formulation of multilayer optics," *J. Opt. Soc. Amer. A*, vol. 16, pp. 2013–2018, Aug. 1999.

[196] J. M. Vigoureux and R. Giust, "Explicit Stoke's reciprocity relations for plane stratified media," *Opt. Commun.*, vol. 176, no. 1, pp. 1–8, 2000.

[197] H. Goldstein, *Classical Mechanics.* Addison-Wesley series in Physics, Reading, MA, USA: Addison-Wesley, Inc., 2nd ed., 1980.

[198] O. Svelto, *Principles of Lasers.* New York, NY, USA: Springer, Inc., 5th ed., 2010.

[199] E. T. Zalewski, "Radiometry and photometry," in *Devices, Measurements and Properties* (M. Bass, ed.), Handbook of Optics, vol. II, ch. 24, New York: McGraw-Hill Inc., 1995.

[200] C. A. DiMarzio, *Optics for Engineers.* Boca Raton, FL, USA: CRC Press - Taylor and Francis Group, 2012.

[201] W. R. McCluney, *Introduction to Radiometry and Photometry.* Boston: Artech House Inc., 1994.

[202] W. L. Wolfe, *Introduction to Radiometry.* Tutorial Texts in Optical Engineering, vol. TT29, Bellingham, WA, USA: SPIE Press, 1998.

[203] E. F. Schubert, *Light Emitting Diodes.* New York: Cambridge University Press, 2nd ed., 2006.

[204] N. Ohta and A. R. Robertson, *Colorimetry: Fundamentals and Applications.* Sussex, UK: J. Wiley & Sons Ltd., 2005.

[205] "Nikon Instruments, Inc., MicroscopyU, the source for microscopy education." https://www.microscopyu.com/tutorials/ccd-signal-to-noise-ratio. Accessed: 2019-03-22.

[206] CIE *Commission Internationale de l'Éclairage Proceedings (CIE), 1924*, 1926. Cambridge, MA: Cambridge University Press.

[207] CIE, *Commission Internationale de l'Éclairage Proceedings (CIE).* Cambridge, MA: Cambridge University Press, 1924.

[208] K. S. Gibson and E. P. T. Tyndall, "Visibility of radiant energy," *Scientific Papers of the Bureau of Standards*, vol. 19, pp. 131–191, 1923.

[209] CIE *Commission Internationale de l'Éclairage Proceedings (CIE)*, vol. 1, 1951. Sec. 4, vol. 3, p. 37, Bureau Central de la CIE, Paris.

[210] Y. LeGrand, *Light, Colour and Vision.* London: Chapman and Hall, 1968.

[211] C. Oleary, *Standards Colorimetry: Definitions, Algorithms and Software.* Sussex, UK: J. Wiley & Sons, Ltd., 2016.

[212] CIE, "Recommended system for mesopic photometry based on visual performance," *Commission Internationale de l'Éclairage Proceedings (CIE)*, 2010. Technical Report 191.

[213] CIE, "The use of terms and units in photometry: Implementation of the CIE system for mesopic photometry," *Commission Internationale de l'Éclairage Proceedings (CIE)*, 2016. Technical Report 004:2016.

[214] M. Shpak, P. Karha, and E. Ikonen, "Mathematical limitations of the CIE mesopic photometry system," *Lighting Research & Technology*, vol. 49, no. 1, pp. 111–121, 2017.

[215] Hsien-Che Lee, *Introduction to Color Imaging Science.* New York, NY, USA: Cambridge University Press, 2005.

[216] CIE, *Commission Internationale de l'Éclairage Proceedings (CIE).* Cambridge, MA: Cambridge University Press, 1931.

[217] J. K. Bowmaker and H. J. Dartnall, "Visual pigments of rods and cones in a human retina," *The Journal of Physiology*, vol. 298, no. 1, pp. 501–511, 1980.

[218] H. J. A. Dartnall, J. K. Bowmaker, J. D. Mollon, and H. B. Barlow, "Human visual pigments: microspectrophotometric results from the eyes of seven persons," *P. Roy. Soc. Lond. B Bio.*, vol. 220, no. 1218, pp. 115–130, 1983.

[219] D. A. Baylor, "Photoreceptor signals and vision. Proctor lecture," *Invest. Ophath. Vis. Sci.*, vol. 28, pp. 34–49, Jan. 1987.

[220] J. L. Schnapf, T. W. Kraft, and D. A. Baylor, "Spectral sensitivity of human cone photoreceptors," *Nature*, vol. 325, pp. 439–441, Jan. 1987.

[221] A. Stockman and L. T. Sharpe, "Human cone spectral sensitivities: a progress report," *Vision Research*, vol. 38, no. 21, pp. 3193–3206, 1998.

[222] CIE, "Colorimetry," *Commission Internationale de l'Éclairage Proceedings (CIE)*, 2004. Technical Report CIE:15:2004 (CIE Central Bureau, Vienna).

[223] R. W. G. Hunt and M. R. Pointer, *Measuring Colour.* IS & T Series in Imaging Science and Technology, West Sussex, UK: John Wiley and Sons, Ltd., 4th ed., 2011.

[224] S. A. Burns, "The location of optimal object colors with more than two transitions," *Color Res. Appl.*, vol. 46, no. 6, pp. 1180–1193, 2021.

[225] C. A. Poynton, "Rehabilitation of gamma," in *Human Vision and Electronic Imaging III* (B. E. Rogowitz and T. N. Pappas, eds.), vol. 3299, pp. 232–249, International Society for Optics and Photonics, SPIE, 1998.

[226] D. L. MacAdam, "Visual sensitivities to color differences in daylight," *J. Opt. Soc. Am.*, vol. 32, pp. 247–274, May 1942.

[227] D. L. MacAdam, "Projective transformations of I. C. I. color specifications," *J. Opt. Soc. Am.*, vol. 27, pp. 294–299, Aug. 1937.

[228] J. Baggott, *The Meaning of Quantum Theory.* New York: Oxford University Press, 1992.

[229] J. Crepeau, "A brief history of the T^4 radiation law," in *Proceedings of ASME: Heat Transfer in Energy Systems; Thermophysical Properties; Heat Transfer Equipment; Heat Transfer in Electronic Equipment*, vol. 1, (San Francisco, CA), pp. HT2009–88060, Heat Transfer Summer Conference, ASME, Jul. 19–23, 2009.

[230] J. L. Jiménez, L. de la Peña, and T. A. Brody, "Zero point term in cavity radiation," *Amer. J. Phys.*, vol. 48, no. 10, pp. 840–846, 1980.

[231] L. de la Peña and A. M. Cetto, "Planck's law as a consequence of the zeropoint radiation field," *Rev. Mex. Fis. (Suppl. 1)*, vol. 48, pp. 1–8, Sept. 2002.

[232] H. Vic Dannon, "Zero point energy: Planck radiation law," *Gauge Institute Journal*, vol. 1, Aug. 2005.

[233] F. Ogiba, "Planck's radiation law: Thermal excitations of vacuum induced fluctuations," *Progr. Phys.*, vol. 11, no. 2, pp. 146–148, 2015.

[234] L. Reggiani and E. Alfinito, "The puzzling of zero-point energy contribution to black-body radiation spectrum: The role of Casimir force," *Fluctuation and Noise Letters*, vol. 16, no. 04, p. 1771002, 2017.

[235] T. H. Boyer, "Blackbody radiation in classical physics: A historical perspective," *Amer. J. Phys.*, vol. 86, no. 7, pp. 495–509, 2018.

[236] T. H. Boyer, "Thermodynamics of the harmonic oscillator: derivation of the planck blackbody spectrum from pure thermodynamics," *Eur. J. Phys.*, vol. 40, p. 025101, Jan. 2019.

[237] F. Grunn and R. J. Becherer, *Optical Radiation Measurements, Volume 1, Radiometry.* New York: Academic Press, Inc., 1979.

[238] J. R. Howell, M. P. Mengüç, and R. Siegel, *Thermal Radiation Heat Transfer*, ch. 17. Boca Raton, FL, USA: CRC Press, Taylor & Francis Group, 6th ed., 2016.

[239] J. Hartmann, "Correct consideration of the index of refraction using blackbody radiation," *Opt. Express*, vol. 14, pp. 8121–8126, Sep. 2006.

[240] Y. A. Çengel and A. J. Ghajar, *Heat and Mass Transfer: Fundamentals and Applications*, ch. 12. New York: McGraw-Hill Education, 5th ed., 2015.

[241] D. B. Judd, "Estimation of chromaticity differences and nearest color temperature on the standard 1931 ici colorimetric coordinate system*," *J. Opt. Soc. Am.*, vol. 26, pp. 421–426, Nov. 1936.

[242] D. L. MacAdam, "Correlated color temperature?," *J. Opt. Soc. Am.*, vol. 67, pp. 839–840, June 1977.

[243] K. L. Kelly, "Lines of constant correlated color temperature based on MacAdam's (u,v) uniform chromaticity transformation of the CIE diagram," *J. Opt. Soc. Am.*, vol. 53, pp. 999–1002, Aug. 1963.

[244] D. J. Fixsen, "The temperature of the Cosmic Microwave Background," *Astrophys. J.*, vol. 707, pp. 916–920, nov 2009.

[245] E. Gawiser and J. Silk, "The cosmic microwave background radiation," *Phys. Rep.*, vol. 333-334, pp. 245–267, 2000.

[246] R. A. Alpher and R. Herman, "Evolution of the universe," *Nature*, vol. 162, pp. 774–775, Nov. 1948.

[247] A. A. Penzias and R. W. Wilson, "A measurement of excess antenna temperature at 4080 Mc/s," *Astrophys. J.*, vol. 142, pp. 419–421, Jul. 1965.

[248] R. H. Dicke, P. J. E. Peebles, P. G. Roll, and D. T. Wilkinson, "Cosmic black-body radiation," *Astrophys. J.*, vol. 142, pp. 414–419, Jul. 1965.

[249] A. Stockman, L. T. Sharpe, and C. Fach, "The spectral sensitivity of the human short-wavelength sensitive cones derived from thresholds and color matches," *Vision Research*, vol. 39, no. 17, pp. 2901–2927, 1999.

[250] A. Stockman and L. T. Sharpe, "The spectral sensitivities of the middle- and long-wavelength-sensitive cones derived from measurements in observers of known genotype," *Vision Research*, vol. 40, no. 13, pp. 1711–1737, 2000.

[251] W. S. Stiles and J. M. Burch, "Interim report to the Commission Internationale de l'Éclairage, Zurich, 1955, on the National Physical Laboratory's investigation of colour-matching (1955)," *Optica Acta: International Journal of Optics*, vol. 2, no. 4, pp. 168–181, 1955.

[252] R. M. Corless, G. H. Gonnet, D. E. G. Hare, D. J. Jeffrey, and D. E. Knuth, "On the Lambert W function," *Adv. Comp. Math.*, vol. 5, pp. 329–359, Dec. 1996.

[253] T. Dence, "A brief look into the Lambert W function," *Applied Mathematics*, vol. 4, no. 6, pp. 887–892, 2013.

[254] I. G. Priest, "A proposed scale for use in specifying the chromaticity of incandescent illuminants and various phases of daylight*," *J. Opt. Soc. Am.*, vol. 23, pp. 41–45, Feb. 1933.

[255] M. Krystek, "An algorithm to calculate correlated colour temperature," *Color Res. & Appl.*, vol. 10, no. 1, pp. 38–40, 1985.

[256] J. Hernández-Andrés, R. L. Lee, and J. Romero, "Calculating correlated color temperatures across the entire gamut of daylight and skylight chromaticities," *Appl. Opt.*, vol. 38, pp. 5703–5709, Sept. 1999.

[257] C. Li, G. Cui, M. Melgosa, X. Ruan, Y. Zhang, L. Ma, K. Xiao, and M. R. Luo, "Accurate method for computing correlated color temperature," *Opt. Express*, vol. 24, pp. 14066–14078, June 2016.

[258] R. Cecchini and G. Pelosi, "Diffraction: the first recorded observation," *IEEE Antennas & Prop. Mag.*, vol. 32, no. 2, pp. 27–30, 1990.

[259] A. Fresnel, "Mémoire sur la diffraction de la lumiére," *Mem. Acad. R. Sci. (France)*, vol. 5, pp. 339–475, 1826.

[260] R. D. Guenther, *Modern Optics.* New York, USA: John Wiley & Sons, 1990.

[261] G. Kirchhoff, "Zur theorie del' lichtstrahlen," *Ann. Phys.*, vol. 18, 3rd ser., pp. 663–695, 1883.

[262] H. Poincaré, *Leçons sur la thèorie mathématique de la lumière, professés pedant le premier semestre 1887-1888.* Paris: Georges Carré, 1889.

[263] A. Sommerfeld, "Mathematische theorie der diffraction," *Mathematische Annalen*, vol. 47, pp. 317–374, 1896.

[264] A. Ishimaru, *Electromagnetic Wave Propagation, Radiation, and Scattering.* Englewood Cliffs, New Jersey, USA: Prentice-Hall, Inc., 1991.

[265] J. W. Goodman, *Introduction to Fourier Optics.* New York, USA: McGraw-Hill Companies, Inc., 2nd ed., 1996.

[266] A. Rabinowicz, "The Miyamoto-Wolf diffraction wave," in *Progress in Optics, vol. IV* (E. Wolf, ed.), ch. V, pp. 198–240, Amsterdam, Netherlands: North-Holland Publishing Company, 1965.

[267] K. Miyamoto and E. Wolf, "Generalization of the Maggi-Rubinowicz theory of the boundary diffraction wave–Part I," *J. Opt. Soc. Am.*, vol. 52, pp. 615–625, Jun. 1962.

[268] K. Miyamoto and E. Wolf, "Generalization of the Maggi-Rubinowicz theory of the boundary diffraction wave–Part II," *J. Opt. Soc. Am.*, vol. 52, pp. 626–636, Jun. 1962.

[269] E. W. Marchand and E. Wolf, "Boundary diffraction wave in the domain of the Rayleigh-Kirchhoff diffraction theory," *J. Opt. Soc. Am.*, vol. 52, pp. 761–767, Jul. 1962.

[270] E. W. Marchand and E. Wolf, "Consistent formulation of Kirchhoff's diffraction theory*," *J. Opt. Soc. Am.*, vol. 56, pp. 1712–1721, Dec. 1966.

[271] J. S. Asvestas, "Line integrals and physical optics. Part I. The transformation of the solid-angle surface integral to a line integral," *J. Opt. Soc. Amer. A*, vol. 2, pp. 891–895, Jun. 1985.

[272] J. S. Asvestas, "Line integrals and physical optics. Part II. The conversion of the kirchhoff surface integral to a line integral," *J. Opt. Soc. Amer. A*, vol. 2, pp. 896–902, Jun. 1985.

[273] A. Sommerfeld, *Optics.* Lectures on Theoretical Physics, Vol. IV, New York, NY, USA: Academic Press, Inc., 1954.

[274] J. J. Stamnes, *Waves in Focal Regions.* The Adam Hilger series on Optics and Optoelectronics, Bristol, UK: CRC Press, 1986.

[275] G. Barton, *Elements of Green's Function and Propagation: Potentials, Diffusion, and Waves.* New York, NY, USA: Oxford University Press, 1989.

[276] G. Brooker, *Modern Classical Optics.* Oxford Master Series in Physics, New York, NY, USA: Oxford University Press, Inc., 2003.

[277] E. E. Kriezis, D. P. Chrissoulidis, and A. G. Papagiannakis, *Electromagnetics and Optics.* Singapore: World Scientific Publishing Co. Pte. Ltd., 1992.

[278] W. H. Southwell, "Validity of the Fresnel approximation in the near field," *J. Opt. Soc. Am.*, vol. 71, pp. 7–14, Jan. 1981.

[279] A. M. Steane and H. N. Rutt, "Diffraction calculations in the near field and the validity of the Fresnel approximation," *J. Opt. Soc. Am. A*, vol. 6, pp. 1809–1814, Dec. 1989.

[280] G. W. Forbes, "Validity of the Fresnel approximation in the diffraction of collimated beams," *J. Opt. Soc. Am. A*, vol. 13, pp. 1816–1826, Sep. 1996.

[281] S. Mezouari and A. R. Harvey, "Validity of Fresnel and Fraunhofer approximations in scalar diffraction," *J. Opt. A Pure Appl. Op.*, vol. 5, pp. S86–S91, jun 2003.

[282] J. Komrska, "Simple derivation of formulas for Fraunhofer diffraction at polygonal apertures," *J. Opt. Soc. Am.*, vol. 72, pp. 1382–1384, Oct. 1982.

[283] N. Saga and K. Tanaka, "Relationship between the Abbe transform and the Maggi-Rubinowicz transformation formulas," *J. Opt. Soc. Amer. A*, vol. 3, pp. 1450–1451, Sep. 1986.

[284] N. Saga, "New line integral expressions for fraunhofer diffraction," *Opt. Commun.*, vol. 64, no. 1, pp. 4–8, 1987.

[285] G. C. Sherman, "Application of the convolution theorem to Rayleigh's integral formulas," *J. Opt. Soc. Am.*, vol. 57, pp. 546–547, Apr. 1967.

[286] H. Weyl, "Ausbreitung elektromagnetischer wellen uber einem ebenen leiter," *Annalen der Physik*, vol. 365, no. 21, pp. 481–500, 1919.

[287] M. Babinet, "Memoires d'optique métérologique," *C. R. Acad. Sci.*, vol. 4, pp. 638–648, 1837.

[288] J. R. Jiménez and E. Hita, "Babinet's principle in scalar theory of diffraction," *Opt. Rev.*, vol. 8, pp. 495–497, Nov. 2001.

[289] J. C. Heurtley, "Scalar Rayleigh-Sommerfeld and Kirchhoff diffraction integrals: A comparison of exact evaluations for axial points∗," *J. Opt. Soc. Am.*, vol. 63, pp. 1003–1008, Aug. 1973.

[290] R. E. English and N. George, "Diffraction from a circular aperture: on-axis field strength," *Appl. Opt.*, vol. 26, pp. 2360–2363, Jun. 1987.

[291] A. Dubra and J. A. Ferrari, "Diffracted field by an arbitrary aperture," *Amer. J. Phys.*, vol. 67, no. 1, pp. 87–92, 1999.

[292] I. J. Cooper, C. J. R. Sheppard, and M. Sharma, "Numerical integration of diffraction integrals for a circular aperture," *Optik*, vol. 113, no. 7, pp. 293–298, 2002.

[293] M. Abramowitz and I. A. Stegun, *Handbook of Mathematical Functions with Formulas, Graphs, and Mathematical Tables.* Applied Mathematics Series, vol. 55, Washington, D.C., USA: National Bureau of Standards, tenth printing ed., 1972.

[294] H. Osterberg and L. W. Smith, "Closed solutions of Rayleigh's diffraction integral for axial points," *J. Opt. Soc. Am.*, vol. 51, pp. 1050–1054, Oct. 1961.

[295] J. E. Harvey and J. L. Forgham, "The spot of Arago: New relevance for an old phenomenon," *Amer. J. Phys.*, vol. 52, no. 3, pp. 243–247, 1984.

[296] Y. P. Kathuria, "Far-field radiation patterns of elliptical apertures and its annulli," *IEEE Trans. Antennas & Prop.*, vol. 31, pp. 360–364, Mar. 1983.

[297] Y. P. Kathuria, "Fresnel and far-field diffraction due to an elliptical aperture," *J. Opt. Soc. Am. A*, vol. 2, pp. 852–857, Jun. 1985.

[298] W. Cash, "Detection of earth-like planets around nearby stars using a petal-shaped occulter," *Nature*, vol. 442, pp. 51–53, July 2006.

[299] W. Cash, "Analytic modeling of starshades," *Astrophys. J.*, vol. 738, pp. 76–88, Aug. 2011.

[300] E. Cady, "Boundary diffraction wave integrals for diffraction modeling of external occulters," *Opt. Express*, vol. 20, pp. 15196–15208, July 2012.

[301] A. Harness, S. Shaklan, W. Cash, and P. Dumont, "Advances in edge diffraction algorithms," *J. Opt. Soc. Amer. A*, vol. 35, pp. 275–285, Feb. 2018.

[302] N. George and G. M. Morris, "Diffraction by serrated apertures," *J. Opt. Soc. Am.*, vol. 70, pp. 6–17, Jan. 1980.

[303] Y. Kim, H. Grebel, and D. L. Jaggard, "Diffraction by fractally serrated apertures," *J. Opt. Soc. Amer. A*, vol. 8, pp. 20–26, Jan. 1991.

[304] J. M. Auerbach and V. P. Karpenko, "Serrated-aperture apodizers for high-energy laser systems," *Appl. Opt.*, vol. 33, pp. 3179–3183, May 1994.

[305] M. M. Beal, *Serrated circular apertures: optical Fourier transforms and fractal analysis.* PhD thesis, Institute of Optics, College of Engineering and Applied Science, University of Rochester, 1990.

[306] W. Jinguo, Z. Zhanda, S. Zhe, H. Yongling, L. Hong, and L. Qiang, "Diffraction characteristics of serrated circular aperture with random radius," *High Power Laser and Particle Beams*, vol. 24, pp. 1801–1805, Aug. 2012.

[307] I. Sizova, T. Moskalev, and L. Mikheev, "Laser beam shaping with circular serrated apertures. I. Spatial filtering," *Appl. Opt.*, vol. 58, pp. 4905–4909, Jun. 2019.

[308] C. Palmer, *Diffraction Grating Handbook.* Rochester, NY, USA: Newport Corporation, 6th ed., 2005.

[309] E. G. Loewen and E. Popov, *Diffraction Gratings and Applications.* Optical Engineering, Ed. B. J. Thompson, New York, NY, USA: Marcel Dekker, Inc., 1997.

[310] J. E. Harvey and R. N. Pfisterer, "Understanding diffraction grating behavior: including conical diffraction and Rayleigh anomalies from transmission gratings," *Opt. Engr.*, vol. 58, no. 8, pp. 1–21, 2019.

[311] D. C. O'Shea, T. J. Suleski, A. D. Kathman, and D. W. Prather, *Diffractive Optics: Design, Fabrication, and Test.* Tutorial Texts in Optical Engineering, Bellingham WA, USA: SPIE Press, vol. TT62, 2004.

[312] I. M. Gel'fand and G. E. Shilov, *Generalized Functions, Vol. I, Properties and Operations.* New York and London: Academic Press, Inc., 1964.

[313] A. V. Oppenheim, A. S. Willsky, and S. H. Nawab, *Signals and Systems.* Upper Saddle River, NJ, USA: Prentice-Hall, Inc., 2nd ed., 1996.

[314] M. G. Moharam and T. K. Gaylord, "Rigorous coupled-wave analysis of planar-grating diffraction," *J. Opt. Soc. Amer.*, vol. 71, pp. 811–818, Jul 1981.

[315] T. K. Gaylord and M. G. Moharam, "Analysis and applications of optical diffraction by gratings," *Proc. IEEE*, vol. 73, no. 5, pp. 894–937, 1985.

[316] M. G. Moharam, E. B. Grann, D. A. Pommet, and T. K. Gaylord, "Formulation for stable and efficient implementation of the rigorous coupled-wave analysis of binary gratings," *J. Opt. Soc. Amer.*, vol. 12, pp. 1068–1076, May 1995.

[317] M. G. Moharam, D. A. Pommet, E. B. Grann, and T. K. Gaylord, "Stable implementation of the rigorous coupled-wave analysis for surface-relief gratings: enhanced transmittance matrix approach," *J. Opt. Soc. Amer. A*, vol. 12, pp. 1077–1086, May 1995.

[318] P. B. Johnson and R. W. Christy, "Optical constants of the noble metals," *Phys. Rev. B*, vol. 6, pp. 4370–4379, Dec. 1972.

[319] A. Wood and J. Babington, *Diffractive Lens Design.* Series in Emerging Technologies in Optics and Photonics, Bristol, UK: IOP Publishing, 2023.

[320] D. A. Buralli, G. M. Morris, and J. R. Rogers, "Optical performance of holographic kinoforms," *Appl. Opt.*, vol. 28, pp. 976–983, Mar. 1989.

[321] A. B. Buckman, *Guide-Wave Photonics.* New York: Sauders College Publishing, 1992.

[322] A. Sommerfeld, "Die greensche funktion der schwingungslgleichung," *Jahresbericht der Deutschen Mathematiker-Vereinigung*, vol. 21, pp. 309–352, 1912.

[323] S. H. Schot, "Eighty years of Sommerfeld's radiation condition," *Historia Mathematica*, vol. 19, no. 4, pp. 385–401, 1992.

[324] B. B. Baker and E. T. Copson, *The Mathematical Theory of Huygens' Principle.* London, UK: Oxford Univerity Press, 1939.

[325] F. V. Atkinson, "Lxi. on Sommerfeld's 'radiation condition'," *The London, Edinburgh, and Dublin Philosophical Magazine and Journal of Science*, vol. 40, no. 305, pp. 645–651, 1949.

[326] C. H. Wilcox, "A generalization of theorems of Rellich and Atkinson," *Proceedings of the American Mathematical Society*, vol. 7, no. 2, pp. 271–276, 1956.

[327] E. T. Copson, *Partial Differential Equations*, ch. 11. London, UK: Cambridge Univeristy Press, 1975.

[328] C. Müller, *Foundations of the Mathematical Theory of Electromagnetic Waves*, ch. VIII. Berlin, Germany: Springer-Verlag GmbH, Inc., 1969.

[329] L. Mandel and E. Wolf, *Optical Coherence and Quantum Optics.* New York, NY, USA: Cambridge University Press, 1995.

[330] L. Debnath and D. Bhatta, *Integral Transforms and Their Applications.* Boca Raton, FL, USA: Chapman and Hall/CRC - Taylor and Francis Group, 2nd ed., 2007.

[331] R. N. Bracewell, K.-Y. Chang, A. K. Jha, and Y.-H. Wang, "Affine theorem for two-dimensional fourier transform," *Electron. Lett.*, vol. 29, p. 304, Feb. 1993.

[332] B. G. Osgood, *Lectures on the Fourier transform and its applications.* Pure and Applied Undergraduate Texts, vol. 33, Providence, RI, USA: American Mathematical Society, 2019.

[333] R. M. Sillitto and W. Sillitto, "A simple Fourier approach to Fraunhofer diffraction by triangular apertures," *Opt. Acta*, vol. 22, no. 12, pp. 999–1010, 1975.

[334] W. Sillitto, "A simple Fourier approach to Fraunhofer diffraction by triangular apertures," *J. Opt. Soc. Am.*, vol. 69, pp. 765–770, May 1979.

[335] J. Komrska, "Fraunhofer diffraction at apertures in the form of regular polygons. I," *Opt. Acta*, vol. 19, no. 10, pp. 807–816, 1972.

[336] J. Komrska, "Fraunhofer diffraction at apertures in the form of regular polygons. II," *Opt. Acta*, vol. 20, no. 7, pp. 549–563, 1973.

[337] S.-W. Lee and R. Mittra, "Fourier transform of a polygonal shape function and its application in electromagnetics," *IEEE Trans. Antennas & Prop.*, vol. 31, pp. 99–103, Jan. 1983.

[338] D. A. Pommet, M. G. Moharam, and E. B. Grann, "Limits of scalar diffraction theory for diffractive phase elements," *J. Opt. Soc. Amer. A*, vol. 11, pp. 1827–1834, Jun. 1994.

[339] M. Testorf, "On the zero-thickness model of diffractive optical elements," *J. Opt. Soc. Amer. A*, vol. 17, pp. 1132–1133, Jun. 2000.

[340] X. Jing and Y. Jin, "Transmittance analysis of diffraction phase gratings," *Appl. Opt.*, vol. 50, pp. C11–C18, Mar. 2011.

[341] S. Zhou, Y. Yang, S. Hu, and X. Xu, "Scalar-based analysis of phase gratings etched in the micro/nanofabrication process," *IEEE Photonics J.*, vol. 7, pp. 1–11, Aug. 2015.

[342] G. J. Swanson, "Binary optics technology: Theoretical limits on the diffraction efficiency of multilevel diffractive optical elements," Tech. Rep. 914, Lincoln Laboratories, MIT, Boston, MA, USA, Mar. 1991.

[343] J. D. Gaskill, *Linear Systems, Fourier Transforms, and Optics.* Wiley Series in Pure and Applied Optics, New York, NY, USA: John Wiley & Sons, Inc., 1978.

[344] E. G. Steward, *Fourier Optics: An Introduction.* Mineola, NY, USA: Dover Publications, Inc., 1983.

[345] O. K. Ersoy, *Diffraction, Fourier Optics and Imaging.* Wiley Series in Pure and Applied Optics, New York, NY, USA: John Wiley & Sons, Inc., 2006.

[346] K. Khare, M. Butola, and S. Rajora, *Fourier Optics and Computational Imaging.* Cham, Switzerland: Springer International Publishing, 2nd ed., 2023.

[347] H. Stark, *Applications of Optical Fourier Transforms*, ch. Theory and Measurement of the Optical Fourier Trasnform, pp. 2–7. New York, NY, USA: Academic Press, Inc., 1982.

[348] W. T. Rhodes, "Lecture notes on Fourier Optics; School of ECE, Georgia Institute of Technology, Atlanta, USA." University Lecture, Feb. 1984.

[349] W. T. Rhodes, "When the difference between coherent and incoherent imaging begins to blur," in *Frontiers in Optics 2015*, p. FW4G.1, Optica Publishing Group, 2015.

[350] D. P. Kelly, J. T. Sheridan, and W. T. Rhodes, "Fundamental diffraction limitations in a paraxial 4-f imaging system with coherent and incoherent illumination," *J. Opt. Soc. Amer. A*, vol. 24, pp. 1911–1919, Jul. 2007.

[351] F. Zernike, "Das Phasenkontrastverfahren bei der Mikroskopischen beobachtung," *Z. Tech. Phys.*, vol. 16, pp. 454–457, 1935.

[352] Z. Yin, T. Kanade, and M. Chen, "Understanding the phase contrast optics to restore artifact-free microscopy images for segmentation," *Med. Image Anal.*, vol. 16, no. 5, pp. 1047–1062, 2012.

[353] D. Gabor, "A new microscopic principle," *Nature*, vol. 161, pp. 777–778, May 1948.

[354] F. T. S. Yu and I. C. Khoo, *Principles of Optical Engineering.* New York: J. Wiley and Sons, Inc., 1990.

[355] R. J. Collier, C. B. Burckhardt, and L. H. Lin, *Optical Holography.* San Diego, CA, USA: Academic Press, Inc., 1971.

[356] P. Hariharan, *Optical Holography: Principles, Techniques and Applications.* Cambridge Studies in Modern Optics, New York, NY, USA: Cambridge University Press, 1996.

[357] V. Toal, *Introduction to Holography.* Series in Optics and Optoelectronics, Boca Raton, FL, USA: CRC Press, Taylor & Francis Group, 2022.

[358] E. N. Leith and J. Upatnieks, "Reconstructed wavefronts and communication theory," *J. Opt. Soc. Am.*, vol. 52, pp. 1123–1130, Oct. 1962.

[359] E. N. Leith and J. Upatnieks, "Wavefront reconstruction with diffused illumination and three-dimensional objects," *J. Opt. Soc. Am.*, vol. 54, pp. 1295–1301, Nov. 1964.

[360] R. W. Meier, "Magnification and third-order aberrations in holography," *J. Opt. Soc. Am.*, vol. 55, pp. 987–992, Aug. 1965.

[361] A. K. Ghatak and K. Thyagarajan, *Contemporary Optics.* New York, NY, USA: Plenum Press, 1978.

[362] A. VanderLugt, "Signal detection by complex spatial filtering," *IEEE T. Inform. Theory*, vol. 10, pp. 139–145, Apr. 1964.

[363] J. A. Stratton and L. J. Chu, "Diffraction theory of electromagnetic waves," *Phys. Rev.*, vol. 56, pp. 99–107, Jul. 1939.

[364] J. A. Stratton, *Electromagnetic Theory.* IEEE Press Series on Electromagnetic Wave Theory, New York, NY, USA: J. Wiley and Sons, Inc., 2007.

[365] J. Kong, *Electromagnetic Wave Theory.* A Wiley Interscience publication, New York, NY, USA: John Wiley & Sons, Inc., 1990.

[366] W. Hsu and R. Barakat, "Stratton-Chu vectorial diffraction of electromagnetic fields by apertures with application to small-Fresnel-number systems," *J. Opt. Soc. Am. A*, vol. 11, pp. 623–629, Feb. 1994.

[367] A. S. Marathay and J. F. McCalmont, "Vector diffraction theory for electromagnetic waves," *J. Opt. Soc. Am. A*, vol. 18, pp. 2585–2593, Oct. 2001.

[368] J. Kim, Y. Wang, and X. Zhang, "Calculation of vectorial diffraction in optical systems," *J. Opt. Soc. Am. A*, vol. 35, pp. 526–535, Apr. 2018.

[369] J. Braat and P. Török, *Imaging Optics.* Cambridge, United Kingdom: Cambridge University Press, 2019.

[370] K. F. Warnick, *Numerical Analysis for Electromagnetic Integral Equations.* Artech House electromagnetic analysis series, Norwood, MA, USA: Artech House, 2008.

[371] J. L. Volakis and K. Sertel, *Integral Equation Methods for Electromagnetics.* Electromagnetic Waves, Raleigh, NC, USA: SciTech Publishing Inc., 2012.

[372] A. Taflove and S. C. Hagness, *Computational Electrodynamics: The Finite-Difference Time-Domain Method.* 3rd Ed., Norwood, MA: Artech House inc. , 2005.

[373] M. N. O. Sadiku, *Numerical Techniques in Electromagnetics.* Boca Raton, FL, USA: CRC Press - Taylor and Francis Group, 2nd ed., 2000.

[374] R. C. Rumpf, *Electromagnetic and Photonic Simulation for the Beginner: Finite-Difference Frequency-Domain in MATLAB®.* Applied Photonics, Norwood, MA, USA: Artech House, 2022.

[375] J. M. Jin, *The Finite Element Method in Electromagnetics.* IEEE Press, New York, NY, USA: John Wiley & Sons, Inc., 3rd ed., 2015.

[376] C. Christopoulos, *The Transmission-line Modeling Method: TLM.* IEEE/OUP series on electromagnetic wave theory, New York, NY, USA: IEEE, 1995.

[377] E. Yamashita, *Analysis Methods for Electromagnetic Wave Problems.* Analysis Methods for Electromagnetic Wave Problems, Boston, MA, USA: Artech House, 1990.

[378] P.-B. Zhou, *Numerical Analysis of Electromagnetic Fields.* Electric Energy Systems and Engineering Series, Springer Berlin Heidelberg, 2012.

[379] D. Poljak and C. A. Brebbia, *Boundary Element Methods for Electrical Engineers.* Advances in electrical engineering and electromagnetics, Southampton, UK: WIT Press, 2005.

[380] S. Salon, "The hybrid finite element-boundary element method in electromagnetics," *IEEE Trans. on Magnetics*, vol. 21, no. 5, pp. 1829–1834, 1985.

[381] A. F. Peterson, S. L. Ray, and R. Mittra, *Computational Methods for Electromagnetics.* Series on Electromagnetic Waves, New York, NY, USA: IEEE Press, 1998.

[382] L. Li, "Fourier modal method," in *Gratings: Theory and Numeric Applications* (E. Popov, ed.), pp. 13.1–13.40, Aix Marseille Université, CNRS, Centrale Marseille, Institut Fresnel, 2012.

[383] C. P. Clemmow, *The Plane Wave Spectrum Representation of Electromagnetic Fields.* IEEE-Oxford University Press Series on Electromagnetic Wave Theory, Piscataway, NJ, USA: IEEE Press, 1996.

[384] M. Mansuripur, "A tutorial on the classical theories of electromagnetic scattering and diffraction," *Nanophotonics*, vol. 10, no. 1, pp. 315–342, 2021.

[385] F. Kottler, "VIII Diffraction at a black screen. Part II: Electromagnetic Theory," *Progess in Optics*, vol. 6, pp. 331–377, Jan. 1967.

[386] W. Franz, "On the theory of diffraction," *Proceedings of the Physical Society. Section A*, vol. 63, pp. 925–939, Sept. 1950.

[387] S. Silver, *Microwave Antenna Theory and Design. Edited by S. Silver.* [Massachusetts Institute of Technology. Radiation Laboratory Series. No. 12.], New York, NY, USA: McGraw-Hill Book Company, 1949.

[388] N. A. Gumerov and R. Duraiswami, *Fast Multipole Methods for the Helmholtz Equation in Three Dimensions.* A volume in Elsevier Series in Electromagnetism, Amsterdam, Netherland: Elsevier, 2005.

[389] H. G. Booker, "Slot aerials and their relation to complementary wire aerials (Babinet's Principle)," *J. Inst. Electr. Eng. - Part IIIA: Radiolocation*, vol. 93, pp. 620–626(6), Jan. 1946.

[390] R. Janaswamy, "Duality Principle and Babinet's Principle," in *Engineering Electrodynamics*, 2053-2563, pp. 6–1 to 6–6, Bristol, UK: IOP Publishing, 2020.

[391] R. F. Harrington, *Time-harmonic Electromagnetic Fields.* Electrical and Electronic Engineering Series, New York, NY, USA: McGraw-Hill, 1961.

[392] M. Koshiba, *Optical Waveguide Theory by the Finite Element Method.* Advances in opto-electronics, Tokyo, Japan: KTK Scientific Publishers, 1993.

[393] N. Morita, "The boundary-element method," in *Analysis Methods for Electromagnetic Wave Problems* (E. Yamashita, ed.), Analysis Methods for Electromagnetic Wave Problems, Norwood, MA, USA: Artech House, 1990.

[394] D. W. Prather, M. S. Mirotznik, and J. N. Mait, "Boundary element method for vector modeling diffractive optical elements," in *Diffractive and Holographic Optics Technology II* (I. Cindrich and S. H. Lee, eds.), vol. 2404, pp. 28–39, International Society for Optics and Photonics, SPIE, 1995.

[395] J. M. Bendickson, E. N. Glytsis, and T. K. Gaylord, "Scalar integral diffraction methods: unification, accuracy, and comparison with a rigorous boundary element method with application to diffractive cylindrical lenses," *J. Opt. Soc. Amer. A*, vol. 15, pp. 1822–1837, July 1998.

[396] J. M. Bendickson, E. N. Glytsis, T. K. Gaylord, and A. F. Peterson, "Modeling considerations for rigorous boundary element method analysis of diffractive optical elements," *J. Opt. Soc. Amer. A*, vol. 18, pp. 1495–1506, July 2001.

[397] S. Rao, D. Wilton, and A. Glisson, "Electromagnetic scattering by surfaces of arbitrary shape," *IEEE Trans. on Antennas and Propagation*, vol. 30, no. 3, pp. 409–418, 1982.

[398] E. N. Glytsis, A. D. Papadopoulos, and T. T. Koutserimpas, "Review and accuracy comparison of various permittivity-averaging schemes for material discontinuities in the two-dimensional fdfd method: implementation using efficient computer graphics techniques," *Appl. Opt.*, vol. 57, pp. 7303–7313, Sep 2018.

[399] K. S. Yee, "Numerical solution of initial boundary value problems involving Maxwell's equations in isotropic media," *IEEE Trans. Antennas & Prop.*, vol. 14, pp. 302–307, May 1966.

[400] R. C. Rumpf, "Simple implementation of arbitrary shaped total-field/scattered-field regions in finite-difference frequency-domain," *Progress in Electromagnetics Research B*, vol. 36, pp. 221–248, 2012.

[401] S. D. Gedney, "An anisotropic perfectly matched layer absorbing media for the truncation of FDTD lattices," *IEEE Trans. Antennas & Prop.*, vol. 44, pp. 1630–1639, Dec. 1996.

[402] H. Kogelnik, "Coupled wave theory for thick hologram gratings," *Bell Syst. Tech. J.*, vol. 48, pp. 2909–2947, 1969.

[403] C. V. Raman and N. S. N. Nath, "The diffraction of light by high frequency sound waves: Part I," *Proc. Ind. Acad. Sci. A*, pp. 406–412, oct 1935.

[404] C. V. Raman and N. S. N. Nath, "The diffraction of light by high frequency sound waves: Part II," *Proc. Ind. Acad. Sci. A*, pp. 413–420, oct 1935.

[405] C. V. Raman and N. S. N. Nath, "The diffraction of light by high frequency sound waves: Part III," *Proc. Ind. Acad. Sci. A*, pp. 75–84, jan 1936.

[406] C. V. Raman and N. S. N. Nath, "The diffraction of light by high frequency sound waves: Part IV," *Proc. Ind. Acad. Sci. A*, pp. 119–125, feb 1936.

[407] C. V. Raman and N. S. N. Nath, "The diffraction of light by high frequency sound waves: Part V," *Proc. Ind. Acad. Sci. A*, pp. 459–465, may 1936.

[408] W. T. Cathey, *Optical Information Processing and Holography.* New York, USA: J. Wiley & Sons Inc., 1974.

[409] E. N. Glytsis and T. K. Gaylord, "Rigorous three-dimensional coupled-wave diffraction analysis of single and cascaded anisotropic gratings," *J. Opt. Soc. Am. A*, vol. 4, pp. 2061–2080, Nov. 1987.

[410] E. N. Glytsis, T. K. Gaylord, and D. L. Brundrett, "Rigorous coupled-wave analysis and applications of grating diffraction," in *Diffractive and Miniaturized Optics: A Critical Review* (S. H. Lee, ed.), vol. 10271, pp. 1027102–1–1027102–29, International Society for Optics and Photonics, SPIE, 1993.

[411] L. Li, "Use of Fourier series in the analysis of discontinuous periodic structures," *J. Opt. Soc. Am. A*, vol. 13, pp. 1870–1876, Sep 1996.

[412] M. G. Moharam and T. K. Gaylord, "Diffraction analysis of dielectric surface-relief gratings," *J. Opt. Soc. Am.*, vol. 72, pp. 1385–1392, Oct. 1982.

[413] J. N. Mait, "Design of binary-phase and multiphase fourier gratings for array generation," *J. Opt. Soc. Am. A*, vol. 7, pp. 1514–1528, Aug. 1990.

[414] A. Vasara, M. R. Taghizadeh, J. Turunen, J. Westerholm, E. Noponen, H. Ichikawa, J. M. Miller, T. Jaakkola, and S. Kuisma, "Binary surface-relief gratings for array illumination in digital optics," *Appl. Opt.*, vol. 31, pp. 3320–3336, June 1992.

[415] J. R. Leger, G. J. Swanson, and W. B. Veldkamp, "Coherent laser addition using binary phase gratings," *Appl. Opt.*, vol. 26, pp. 4391–4399, Oct. 1987.

[416] S. J. Walker and J. Jahns, "Array generation with multilevel phase gratings," *J. Opt. Soc. Am. A*, vol. 7, pp. 1509–1513, Aug. 1990.

[417] D. Prongué, H. P. Herzig, R. Dändliker, and M. T. Gale, "Optimized kinoform structures for highly efficient fan-out elements," *Appl. Opt.*, vol. 31, pp. 5706–5711, Sept. 1992.

[418] M. T. Gale, M. Rossi, H. Schütz, P. Ehbets, H. P. Herzig, and D. Prongué, "Continuous-relief diffractive optical elements for two-dimensional array generation," *Appl. Opt.*, vol. 32, pp. 2526–2533, May 1993.

[419] J. Bengtsson, "Design of fan-out kinoforms in the entire scalar diffraction regime with an optimal-rotation-angle method," *Appl. Opt.*, vol. 36, pp. 8435–8444, Nov. 1997.

[420] M. G. Moharam, "Coupled-Wave Analysis Of Two-Dimensional Dielectric Gratings," in *Holographic Optics: Design and Applications* (I. Cindrich, ed.), vol. 0883, pp. 8–11, International Society for Optics and Photonics, SPIE, 1988.

[421] S. T. Han, Y.-L. Tsao, R. M. Walser, and M. F. Becker, "Electromagnetic scattering of two-dimensional surface-relief dielectric gratings," *Appl. Opt.*, vol. 31, pp. 2343–2352, May 1992.

[422] S. Peng and G. M. Morris, "Efficient implementation of rigorous coupled-wave analysis for surface-relief gratings," *J. Opt. Soc. Am. A*, vol. 12, pp. 1087–1096, May 1995.

[423] J. B. Harris, T. W. Preist, J. R. Sambles, R. N. Thorpe, and R. A. Watts, "Optical response of bigratings," *J. Opt. Soc. Am. A*, vol. 13, pp. 2041–2049, Oct. 1996.

[424] E. Noponen and J. Turunen, "Eigenmode method for electromagnetic synthesis of diffractive elements with three-dimensional profiles," *J. Opt. Soc. Am. A*, vol. 11, pp. 2494–2502, Sept. 1994.

[425] L. Li, "New formulation of the Fourier modal method for crossed surface-relief gratings," *J. Opt. Soc. Am. A*, vol. 14, pp. 2758–2767, Oct. 1997.

[426] D. Maystre and M. Nevière, "Electromagnetic theory of crossed gratings," *J. of Opt. (Paris)*, vol. 9, pp. 301–306, Sept. 1978.

[427] V. Bagnoud and S. Mainguy, "Diffraction of electromagnetic waves by dielectric crossed gratings: a three-dimensional rayleigh–fourier solution," *J. Opt. Soc. Am. A*, vol. 16, pp. 1277–1285, June 1999.

[428] M. Bagieu and D. Maystre, "Regularized waterman and rayleigh methods: extension to two-dimensional gratings," *J. Opt. Soc. Am. A*, vol. 16, pp. 284–292, Feb. 1999.

[429] E. N. Glytsis, "Two-dimensionally-periodic diffractive optical elements: Limitations of scalar analysis," *J. Opt. Soc. Amer. A*, vol. 19, pp. 702–715, Apr. 2002.

[430] E. N. Glytsis and T. K. Gaylord, "Rigorous 3-d coupled wave diffraction analysis of multiple superposed gratings in anisotropic media," *Appl. Opt.*, vol. 28, pp. 2401–2421, June 1989.

[431] E. N. Glytsis and T. K. Gaylord, "Three-dimensional (vector) rigorous coupled-wave analysis of anisotropic grating diffraction," *J. Opt. Soc. Am. A*, vol. 7, pp. 1399–1420, Aug. 1990.

[432] D. L. Brundrett, *Analysis, Design, and Applications of Subwavelength Diffraction Gratings*. PhD thesis, School of Electrical Engineering, Georgia Institute of Technology, 1997.

[433] R. E. Collin, "Reflection and transmission at a slotted dielectric interface," *Can. J. Phys.*, vol. 34, pp. 398–411, Apr. 1956.

[434] S. M. Rytov, "Electromagnetic properties of a finely stratified medium," *Sov. Phys. JETP*, vol. 2, pp. 466–475, May 1956.

[435] L. Li, "A modal analysis of lamellar diffraction gratings in conical mountings," *J. of Modern Optics*, vol. 40, no. 4, pp. 553–573, 1993.

[436] C. W. Haggans, L. Li, and R. K. Kostuk, "Effective-medium theory of zeroth-order lamellar gratings in conical mountings," *J. Opt. Soc. Am. A*, vol. 10, pp. 2217–2225, Oct. 1993.

[437] R. E. Collin and J. Brown, "The design of quarter-wave matching layers for dielectric surfaces," *Proc. IEEE Part C*, vol. 103, pp. 153–158, Sep. 1955.

[438] T. K. Gaylord, W. E. Baird, and M. G. Moharam, "Zero-reflectivity high spatial-frequency rectangular-groove dielectric surface-relief gratings," *Appl. Opt.*, vol. 25, pp. 4562–4567, Dec. 1986.

[439] T. K. Gaylord, E. N. Glytsis, and M. G. Moharam, "Zero-reflectivity homogeneous layers and high spatial-frequency surface-relief gratings on lossy materials," *Appl. Opt.*, vol. 26, pp. 3123–3135, Aug. 1987.

[440] E. N. Glytsis and T. K. Gaylord, "Antireflection surface structure: dielectric layer(s) over a high spatial-frequency surface-relief grating on a lossy substrate," *Appl. Opt.*, vol. 27, pp. 4288–4304, Oct. 1988.

[441] R. Magnusson and S. S. Wang, "Transmission bandpass guided-mode resonance filters," *Appl. Opt.*, vol. 34, pp. 8106–8109, Dec. 1995.

[442] S. S. Wang and R. Magnusson, "Multilayer waveguide-grating filters," *Appl. Opt.*, vol. 34, pp. 2414–2420, May 1995.

[443] D. L. Brundrett, E. N. Glytsis, and T. K. Gaylord, "Homogeneous layer models for high-spatial-frequency dielectric surface-relief gratings: conical diffraction and antireflection designs," *Appl. Opt.*, vol. 33, pp. 2695–2706, May 1994.

[444] M. E. Motamedi, W. H. Southwell, and W. J. Gunning, "Antireflection surfaces in silicon using binary optics technology," *Appl. Opt.*, vol. 31, pp. 4371–4376, Aug. 1992.

[445] E. B. Grann, M. G. Moharam, and D. A. Pommet, "Artificial uniaxial and biaxial dielectrics with use of two-dimensional subwavelength binary gratings," *J. Opt. Soc. Am. A*, vol. 11, pp. 2695–2703, Oct. 1994.

[446] H. Kikuta, Y. Ohira, H. Kubo, and K. Iwata, "Effective medium theory of two-dimensional subwavelength gratings in the non-quasi-static limit," *J. Opt. Soc. Am. A*, vol. 15, pp. 1577–1585, June 1998.

[447] M. G. Moharam, T. K. Gaylord, and R. Magnusson, "Criteria for bragg regime diffraction by phase gratings," *Opt. Commun.*, vol. 32, no. 1, pp. 14–18, 1980.

[448] M. G. Moharam, T. K. Gaylord, and R. Magnusson, "Criteria for raman-nath regime diffraction by phase gratings," *Opt. Commun.*, vol. 32, no. 1, pp. 19–23, 1980.

[449] T. K. Gaylord and M. G. Moharam, "Thin and thick gratings: terminology clarification," *Appl. Opt.*, vol. 20, pp. 3271–3273, Oct. 1981.

[450] E. Yablonovitch, "Inhibited spoteneous emisssion in solid state physics and electronics," *Phys. Rev. Lett.*, vol. 58, pp. 2059–2062, May 1987.

[451] E. Yablonovitch, "Photonic crystals: Semicondictors of light," *Sci. Am.*, vol. 285, pp. 47–55, Dec. 2001.

[452] J. D. Joannopoulos, R. D. Meade, S. G. Johnson, and J. N. Winn, *Photonic Crystals: Molding the Flow of Light.* Princeton, New Jersey, USA: Princeton University Press, 1995.

[453] K. Sakoda, *Optical Properties of Photonic Crystals.* Springer Series in Optical Sciences, Heidelberg, Germany: Springer Berlin Heidelberg, 2001.

[454] R. Antos and M. Veis, "Fourier factorization in the plane wave expansion method in modeling photonic crystals," in *Photonic Crystals - Introduction, Applications and Theory* (A. Massaro, ed.), ch. 15, pp. 319–344, Intech Open Science/Open Minds, 2012.

[455] C. Kittel, *Introduction to Solid State Physics.* New York, NY, USA: John Wiley & Sons, 6th ed., 1986.

[456] S. Tibuleac and R. Magnusson, "Reflection and transmission guided-mode resonance filters," *J. Opt. Soc. Am. A*, vol. 14, pp. 1617–1626, Jul. 1997.

[457] L. Novotny and B. Hecht, *Principles of Nano-Optics.* Principles of Nano-optics, Cambridge, UK: Cambridge University Press, 2 ed., 2012.

[458] J.-P. Berenger, "A perfectly matched layer for the absorption of electromagnetic waves," *J. of Computational Physics*, vol. 114, no. 2, pp. 185–200, 1994.

[459] Z. S. Sacks, D. M. Kingsland, R. Lee, and J.-F. Lee, "A perfectly matched anisotropic absorber for use as an absorbing boundary condition," *IEEE Trans. Antennas & Prop.*, vol. 43, no. 12, pp. 1460–1463, 1995.

[460] P. Lalanne and G. M. Morris, "Highly improved convergence of the coupled-wave method for TM polarization," *J. Opt. Soc. Am. A*, vol. 13, pp. 779–784, Apr. 1996.

[461] P. Lalanne, "Convergence performance of the coupled-wave and the differential methods for thin gratings," *J. Opt. Soc. Am. A*, vol. 14, pp. 1583–1591, July 1997.

[462] P. Lalanne, "Improved formulation of the coupled-wave method for two-dimensional gratings," *J. Opt. Soc. Am. A*, vol. 14, pp. 1592–1598, July 1997.

[463] I. C. Botten, M. S. Craig, R. C. McPhedran, J. L. Adams, and J. R. Andrewartha, "The dielectric lamellar diffraction grating," *Optica Acta: International Journal of Optics*, vol. 28, no. 3, pp. 413–428, 1981.

Index

Note: **Bold** and *italic* page numbers refer to **tables** and *figures*.

C

P

Q

R

S

For Product Safety Concerns and Information please contact our EU representative GPSR@taylorandfrancis.com Taylor & Francis Verlag GmbH, Kaufingerstraße 24, 80331 München, Germany